ENCYCLOPEDIA OF PHYSICS

EDITED BY

S. FLÜGGE

VOLUME III/1

PRINCIPLES OF CLASSICAL MECHANICS AND FIELD THEORY

WITH 106 FIGURES

SPRINGER-VERLAG BERLIN HEIDELBERG GMBH
1960

HANDBUCH DER PHYSIK

HERAUSGEGEBEN VON

S. FLÜGGE

BAND III/1

PRINZIPIEN DER KLASSISCHEN MECHANIK UND FELDTHEORIE

MIT 106 FIGUREN

SPRINGER-VERLAG BERLIN HEIDELBERG GMBH
1960

ISBN 978-3-540-02547-4 ISBN 978-3-642-45943-6 (eBook)
DOI 10.1007/978-3-642-45943-6

Ursprünglich erschienen bei Springer-Verlag OHG. Berlin • Göttingen • Heidelberg 1960
Softcover reprint of the hardcover 1st edition 1960

Contents.

Classical Dynamics.

By

J. L. SYNGE.

With 57 Figures.

A. Introduction.

1. Classical dynamics defined. Its applicability. For about two centuries (1700 to 1900) physicists recognized only one dynamical theory[1]. Now three theories exist, of which the third may be subdivided:

(i) Newtonian dynamics.

(ii) Relativistic dynamics (with quantum theory excluded).

(iii) (a) Newtonian quantum dynamics, based on the absolute space and time of NEWTON.

(b) Relativistic quantum dynamics, based on the flat space-time of MINKOWSKI or the curved space-time of EINSTEIN.

The present article is confined to (i) and (ii), which are separated sharply from (iii) on the philosophical question of determinacy versus indeterminacy. But only parts of (i) and (ii) are included, statics being entirely omitted, and also the dynamics of continua (dealt with in other articles mainly in Vol. VI); in relativity only the special theory is considered, and that briefly, in view of other articles.

In fact, for present purposes classical dynamics means the dynamics[2] of particles and rigid bodies, with emphasis on general theory; the essential kinematical preliminaries are included, with finite displacements, mass geometry, force systems, and generalized coordinates.

As regards the applicability of classical dynamics, it may be said at once that Newtonian dynamics describes physical phenomena excellently under what may be called "ordinary circumstances", i.e. when applied to problems of engineering or to physical problems involving systems which are neither very large nor very small. Such discrepancies between theory and experiment as do occur may usually be traced to oversimplification in the mathematical model used (see

[1] The present article contains only incidental historical references. For the history of dynamics, see R. DUGAS, Histoire de la Mécanique (Neuchatel: Editions du Griffon 1950) and also La Mécanique au XVIIe Siècle (same publisher, 1954). Many detailed historical references will also be found in VOSS [*27*] and WHITTAKER [*28*] (see List of General References, p. 223).

[2] According to present custom, the word *mechanics* embraces *dynamics* and *statics*, dynamics dealing with systems in motion and statics with systems at rest. This usage disregards the literal meaning of *dynamics* ($\delta\tilde{\upsilon}\nu\alpha\mu\iota\varsigma$ = force), and was deprecated by Sir WILLIAM THOMSON (Lord KELVIN) and P. G. TAIT in the following words: "Keeping in view the proprieties of language, and following the example of the most logical writers, we employ the word *Dynamics* in its true sense as the science which treats of the action of *force*, whether it maintains relative rest, or produces acceleration of relative motion. The two corresponding divisions of Dynamics are thus conveniently entitled *Statics* and *Kinetics*". (Preface to Treatise on Natural Philosophy, Vol. 1, Part. 1. Cambridge: University Press 1879.) But the word *kinetics* did not catch on, perhaps on account of its too great resemblance to *kinematics*. It was, however, used in the German form (Kinetik) by GRAMMEL [*8*] p. 305 and WINKELMANN and GRAMMEL [*29*] p. 373.

Sect. 2), such as neglect of friction in the model, or the replacement of a (physically) elastic body by a (mathematically) rigid one.

Newtonian dynamics may also be used successfully in the kinetic theory of gases and in celestial mechanics (but see below). Defects in prediction appear when (i) relative speeds (u) are not small compared with the speed of light (c), or (ii) masses of atomic size are involved. Since high speeds are attainable in a laboratory only for very light particles, these two conditions are not distinct practically from one another. But we may separate them for purposes of analysis. They represent (i) the boundary where Newtonian dynamics must be replaced by relativistic dynamics, and (ii) the boundary where classical dynamics must be replaced by quantum dynamics.

Errors of order $(u/c)^2$ appear when Newtonian dynamics is applied to bodies in rapid motion. But it is not possible to assess in any such simple way the errors committed when classical dynamics is applied to problems on the atomic scale. Although quantum dynamics uses many of the old words, the mathematical concepts corresponding to them are radically different from those of classical dynamics, and one no longer attempts to formulate atomic problems in a classical way with any feeling of confidence. However, the classical concepts are not wholly abandoned even there, the conservation of momentum and energy, for example, being employed in problems of collision, annihilation or creation of particles on the atomic or subatomic scale (cf. Sect. 120 et seq.).

In celestial mechanics, Newtonian dynamics remains the standard basis for computations, and it is remarkably successful. Nevertheless certain small discrepancies between prediction and observation exist[1], the most noteworthy being an excess in the rotation of the perihelion of Mercury. This is more simply explained by EINSTEIN's general theory of relativity than by special Newtonian causes introduced to explain it. One concludes that EINSTEIN's theory is the better mathematical model[2], and that Newtonian dynamics must be used with caution in very refined calculations in celestial mechanics.

Newtonian dynamics may be used in cosmology[3], as an alternative to the general theory of relativity or MILNE's kinematical cosmology. The nature of the subject is such that it is hardly possible to say to what extent any of the proposed theories agree with observation.

However, the scientific importance of classical dynamics, Newtonian dynamics in particular, is not to be assessed solely in terms of physical predictions made directly out of it. Newtonian dynamics consists of a body of mathematical conclusions obtained by subjecting certain simple concepts to certain simple laws. In the mathematical development of the subject, general procedures are evolved (Lagrangian and Hamiltonian methods in particular) in which it is convenient to replace the original primitive concepts by more general ones (such as configuration-space and phase-space). It is found that these new mathematical concepts may be taken to represent physical concepts different from those originally envisaged, and so Newtonian dynamics gives birth to new physical conclusions by applying the mathematical ideas inherent in it outside their

---

[1] Cf. J. CHAZY: Théorie de la Relativité et la mécanique céleste, Tome 1, Chap. IV, V. Paris: Gauthier-Villars 1928. — G. C. McVITTIE: General Relativity and Cosmology, Chap. V. New York: Wiley 1956.

[2] WHITEHEAD's theory of gravitation, based on the special theory of relativity, gives the same rotation of perihelion as EINSTEIN's general theory of relativity [cf. A. S. EDDINGTON: Nature, Lond. **113**, 192 (1924); J. L. SYNGE: Proc. Roy. Soc. Lond., Ser. A **211**, 303 (1952)].

[3] For discussion and references, see H. BONDI: Cosmology, pp. 75, 172. Cambridge: University Press 1952.

original domain. Examples are the application of Lagrangian methods to electrical circuit theory, and (more striking) the application of Hamiltonian methods in the development of quantum mechanics.

To pursue this matter a little further, it may be remarked that Newtonian dynamics sets before us for solution sets of ordinary differential equations, and we might therefore classify the subject mathematically as ODE. Hamiltonian methods introduce partial differential equations of the first order, and, when so considered, dynamics might be called PDE_1. The transition to quantum theory via the SCHRÖDINGER equation involves a passage to partial differential equations of the second order, with consequent classification as PDE_2. Viewed in the light of this process of mathematical development ($\mathrm{ODE} \rightarrow \mathrm{PDE}_1 \rightarrow \mathrm{PDE}_2$), Newtonian dynamics takes on a significance much greater than that indicated by its original scope; it is the parent of new theories in which the original concepts are generalized and rarified, though never entirely lost sight of.

2. Mathematical maps or models. To adapt a famous definition of geometry, physics is what physicists do. Among physicists at large, there is comparatively little inquiry into why or how they do what they are doing, and this is not to be deprecated, because human activities are inhibited by introspection. But there are occasions when a greater danger of intellectual confusion over-shadows the danger of self-analysis. How is it, we ask, that we can tolerate the co-existence of several different dynamical theories, all purporting to describe the behaviour of one single natural world? Is one true, and the others false? Or are all false? There is no doubt that these several theories exist, because men work at them. Nature also exists. The question is: How are these theories related to nature?

A satisfying answer is not to be found in discussing whether or not a certain theory gives, or does not give, accurate predictions of the results of certain experiments. The question goes deeper, and it seems possible to approach an answer only by recognizing that, however much they may have been inspired by nature, mathematical theories are no more than maps or models of nature. A "particle" of the natural world (planet, atom or electron) is no more to be confused with the "particle" which represents it in dynamical theory than a city is to be confused with the ink-spot which represents it on a map. But even this analogy does not do justice to the immense gulf separating the natural world from mathematical treatments of it. For an ink-spot on a sheet of paper does at least exist in the natural world (like the city it represents), whereas the essence of the mathematical map or model exists only in the mind, even though the mathematical symbols are written down on paper; mathematical operations involving infinity (differentiation and integration) are purely intellectual concepts, and belong to nature only in so far as the human mind belongs to nature.

If it is admitted that mathematical models are to be sharply distinguished from nature, what, then, is their relationship to nature? The relationship seems to be based on certain *concepts*, the names of which provide a common language for all physicists, experimental and mathematical. These concepts appear as *mathematical* concepts in the model and as *physical* concepts in the direct discussion of nature. We have, as it were, a dictionary with three columns:

Name of concept	Mathematical concept	Physical concept
Mass	A positive number (m)	The quantity of matter in a body. A measure of the reluctance of a body to change its velocity. A measure of the capacity of a body to attract another gravitationally.

A sample row of entries is shown. The entries in the first two columns are complete. But the third entry is only suggestive, for the physical concept demands for its description an account of all the ways in which the idea of mass enters into our understanding of nature, and indeed no description in words can suffice since part of our appreciation of mass arises from muscular sensations and cannot be completely described.

This hypothetical dictionary is used as follows. A physical problem is first formulated in terms of physical concepts. It is then translated into mathematical concepts by using the same words, now with their mathematical meanings. Mathematical laws (usually differential equations) are found by a similar translation of physical laws, first stated in terms of physical concepts. The application of these laws to the problem in question then presents a problem in pure mathematics, and, when this problem has been solved, the conclusion is translated into terms of reality by restoring to the words their physical meanings.

Such a description of standard procedure in theoretical physics would have appeared ridiculously elaborate a century ago, at which time (even in geometry) there was no clear distinction between physical and mathematical concepts (even in the minds of mathematicians). This distinction is essential for pure mathematics today, since otherwise mathematical reasoning would be confused by contact with the confusions of nature. But physicists today may rightly and honestly dispute the distinction, since it may be their practice and wish to keep mathematical concepts inextricably mixed with physical concepts as a fertile source of new ideas. Clarity and fecundity of thought are not one thing.

If the analysis given above is accepted, it clears up the question regarding the co-existence of several dynamical theories. No one of these theories is true, any more than a map is a true representation of a country. And, as it is convenient to have a variety of maps (on different scales) in order to study the geography of a country, so it is convenient to have a number of maps or models of nature. It is easy to exaggerate the differences between the various models; under ordinary circumstances (cf. Sect. 1) they yield the same information.

It is interesting to compare the above ideas with those of BRIDGMAN[1]. According to his *operational method*, concepts are defined in terms of physical operations; *the concept is synonymous with the corresponding set of operations*. Thus, for example, the concept of absolute Newtonian time is to be abandoned because it is impossible to describe experiments by which it may be measured.

This would mean that, in the hypothetical dictionary, the first column would read "absolute time" and the second column "a number t", but the third column would be blank, there being no corresponding physical concept. But is this, in fact, the case? There are many physicists, astronomers and engineers who use the word *time* to refer to a variable t which occurs in certain equations. When they have solved the equations, they pass from their formulae to physical reality, making predictions which are sometimes of very great accuracy, as in celestial mechanics. It is clear that, in *their* dictionaries, the third column is not blank. If it were, they could not use their results for physical prediction. True, there may be no *words* in the third column; the entry may lie in the subconscious, in which a great deal of our thought takes place. But an entry of some sort there must be, since otherwise a formula assigning a numerical value to a variable t could not be translated into instructions to direct a telescope in a certain definite manner.

---

[1] P. W. BRIDGMAN: The Logic of Modern Physics, p. 5. New York: MacMillan 1951. See also P. W. BRIDGMAN: The Nature of some of our Physical Concepts. New York: Philosophical Library 1952.

Valuable as the operational method is in clarifying our ideas, it seems that physical concepts are too complex and confused for us to demand that they should always satisfy the operational test.

3. Axiomatics. The word *logic* has a wide range of meanings, according to context, from the ordinary logic of daily intercourse, through the logic of the expert diagnostician or detective, to the basic logic of twentieth century mathematics[1], and beyond that to the more recent developments of mathematical logic. Physical concepts, being by their nature vague, cannot be treated with logical rigour. On the other hand, classical dynamics, if regarded as a purely mathematical theory, admits of an axiomatic basis, as developed by HAMEL [*10*] and others[2]. Therefore it would seem right that any systematic treatment of classical dynamics should start with axioms, carefully laid down, on which the whole structure would rest as a house rests on its foundations.

The analogy to a house is, however, a false one. Theories are created in mid-air, so to speak, and develop both upward and downward. Neither process is ever completed. Upward, the ramifications can extend indefinitely; downward, the axiomatic base must be rebuilt continually as our views change as to what constitutes logical precision. Indeed, there is little promise of finality here, as we seem to be moving towards the idea that logic is a man-made thing, a game played according to rules to some extent arbitrary.

To a physicist thoroughly familiar with classical dynamics, as traditionally understood, there is an element of artificiality in the creation of a complete axiomatic base, for he knows that the axioms will be chosen to fit the theory, which he believes he understands already with reasonable clarity, and that the theory will not be changed at all as a result of the axiomatics. But when such a physicist is faced by two different theories and seeks to understand wherein they agree and wherein they differ, he is forced back towards axiomatics in order to understand the agreement and the difference. It might be said that axiomatics spring to life, and promise intellectual excitement outside a restricted circle of axiomatic specialists, only when one seriously considers the creation of new theories by changing the axioms—new theories with physical significance. Therefore, although this article does not offer a treatment of classical dynamics which can be regarded as axiomatic in the modern sense, the relationship between Newtonian dynamics and relativistic dynamics is so interesting that the next two sections are devoted to a comparison between them based on a fairly axiomatic approach.

4. Newtonian and relativistic dynamics of a particle. In this section, and the next, Newtonian dynamics will be denoted by ND and relativistic dynamics by RD.

The word *event* is common to ND and RD. The mathematical concept of an event is a set of four numbers (coordinates) or equivalently a point in a four-dimensional space-time continuum, which is in fact the totality of events. The physical concept is a sharply localized occurrence of very brief duration.

---

[1] Any logical system, if it is to avoid vicious circles, must start with undefined terms and unproved propositions. Cf. O. VEBLEN and J. W. YOUNG: Projective Geometry, Vol. 1, p. 1. Boston: Ginn 1910.

[2] For some recent work on this, and references to older work, see J. C. C. MCKINSEY, A. C. SUGAR and P. SUPPES: Axiomatic foundations of classical particle mechanics. J. Rational Mech. Anal. **2**, 253–272 (1953); J. C. C. MCKINSEY and P. SUPPES: Transformations of systems of classical particle mechanics. J. Rational Mech. Anal. **2**, 273–289 (1953); J. C. C. MCKINSEY and P. SUPPES: Philosophy and the axiomatic foundations of physics. Proc. XIth Internat. Congr. of Philosophy, Vol. VI, p. 49–54, 1953; H. RUBIN and P. SUPPES: Transformations of systems of relativistic particle mechanics. Pacif. J. Math. **4**, 563–601 (1954).

Throughout the rest of this section, and indeed throughout the whole article, concepts are considered only as mathematical concepts. As explained in Sect. 2, the corresponding physical concepts are sometimes extremely complicated; we cannot *reason* about them with a degree of precision suited to the present occasion. For these physical concepts, the reader must consult his own private three-column dictionary (cf. Sect. 2); if the required entry in the third column should happen to be non-existent, he must fill it from some other source of information.

In ND an event has an *absolute position* and an *absolute time* (t). The totality of all possible positions form an *absolute space* of three dimensions. Two absolute positions define a *distance*, and absolute space is Euclidean when this distance is used as metric. This implies the existence of coordinates (x, y, z) such that the element of distance $d\sigma$ is

$$d\sigma = (dx^2 + dy^2 + dz^2)^{\frac{1}{2}}. \tag{4.1}$$

There is a 1:1 correspondence between all possible events and the totality of tetrads (x, y, z, t), with variables in the range $-\infty$ to $+\infty$.

In RD (only the special theory of relativity is considered here) two events define a *separation*. There exist coordinates (x, y, z, t), ranging from $-\infty$ to $+\infty$, such that the separation ds between adjacent events is[1]

$$ds = |dx^2 + dy^2 + dz^2 - dt^2|^{\frac{1}{2}}. \tag{4.2}$$

There is a 1:1 correspondence between all possible events and the tetrads (x, y, z, t).

The word *particle* is common to ND and RD. The *history* of a particle is a curve in space-time (*world line*), and may be described by equations of the form

$$x = x(t), \quad y = y(t), \quad z = z(t). \tag{4.3}$$

In ND the derivatives of the functions in (4.3) (the components of *velocity*) may have any values. In RD these derivatives are bounded by

$$\left(\frac{dx}{dt}\right)^2 + \left(\frac{dy}{dt}\right)^2 + \left(\frac{dz}{dt}\right)^2 < 1, \tag{4.4}$$

so that along the world line of a particle we have

$$ds = (dt^2 - dx^2 - dy^2 - dz^2)^{\frac{1}{2}}; \tag{4.5}$$

this is called the element of *proper time*. We may use proper time as parameter on a world line, writing its equations in the form

$$x = x(s), \quad y = y(s), \quad z = z(s), \quad t = t(s), \tag{4.6}$$

instead of as in (4.3). These four functions satisfy the equation

$$\left(\frac{dt}{ds}\right)^2 - \left(\frac{dx}{ds}\right)^2 - \left(\frac{dy}{ds}\right)^2 - \left(\frac{dz}{ds}\right)^2 = 1. \tag{4.7}$$

The word *mass* is common to ND and RD (also called *proper mass* in RD, but the single word *mass* will be used here). It is a number m associated with a particle[2]; it may be constant, or it may vary along the world line.

In ND a *force* with *components* (X, Y, Z) may *act on* a particle. In that case the world line satisfies the *equations of motion*

$$\frac{d}{dt}\left(m\frac{dx}{dt}\right) = X, \quad \frac{d}{dt}\left(m\frac{dy}{dt}\right) = Y, \quad \frac{d}{dt}\left(m\frac{dz}{dt}\right) = Z. \tag{4.8}$$

[1] A factor c^2 is usually inserted before dt^2 in (4.2) (cf. Sect. 107), but we can make $c = 1$ by changing the unit in which t is measured.

[2] In RD proper mass is often denoted by m_0, the symbol m being used forrelative mass (cf. Sect. 108). Note that, throughout this article, m means proper mass.

If (X, Y, Z) are given functions of the quantities

$$m,\ x,\ y,\ z,\ t,\ \frac{dx}{dt},\ \frac{dy}{dt},\ \frac{dz}{dt}, \tag{4.9}$$

we say that the particle moves in a *given field* of force. In that case the equations of motion, together with an equation $m=m(t)$ (usually $m=\text{const}$), determine a unique world line corresponding to assigned initial values of the quantities (4.9).

In RD a *four-force* with components (X, Y, Z, T) may *act on* a particle. Then the *equations of motion* are

$$\frac{d}{ds}\left(m\frac{dx}{ds}\right)=X,\quad \frac{d}{ds}\left(m\frac{dy}{ds}\right)=Y,\quad \frac{d}{ds}\left(m\frac{dz}{ds}\right)=Z,\quad \frac{d}{ds}\left(m\frac{dt}{ds}\right)=T. \tag{4.10}$$

It follows from (4.7) that these equations of motion imply

$$\frac{dm}{ds}=T\frac{dt}{ds}-X\frac{dx}{ds}-Y\frac{dy}{ds}-Z\frac{dz}{ds}. \tag{4.11}$$

If (X, Y, Z, T) are given functions of the quantities listed in (4.9) we say that the particle moves in a *given field* of force. Then the Eqs. (4.10) determine a unique world line, and also m along it, corresponding to assigned initial values of the quantities listed in (4.9).

Let us now take $m=\text{const}$ in ND and in RD. In ND the equations of motion read

$$m\frac{d^2x}{dt^2}=X,\quad m\frac{d^2y}{dt^2}=Y,\quad m\frac{d^2z}{dt^2}=Z. \tag{4.12}$$

In RD, we have by (4.11)

$$T=X\frac{dx}{dt}+Y\frac{dy}{dt}+Z\frac{dz}{dt}, \tag{4.13}$$

so that only X, Y, Z, and not T, are to be regarded as arbitrarily assigned. The last equation in (4.10) is implied by the first three, and, if we take t for parameter, the equations of motion (4.10) may be written

$$m\frac{d^2x}{dt^2}=P,\quad m\frac{d^2y}{dt^2}=Q,\quad m\frac{d^2z}{dt^2}=R, \tag{4.14}$$

where

$$\left.\begin{aligned} P&=\frac{X}{\gamma^2}-\frac{m}{\gamma}\frac{d\gamma}{dt}\frac{dx}{dt},\\ Q&=\frac{Y}{\gamma^2}-\frac{m}{\gamma}\frac{d\gamma}{dt}\frac{dy}{dt},\\ R&=\frac{Z}{\gamma^2}-\frac{m}{\gamma}\frac{d\gamma}{dt}\frac{dz}{dt},\\ \gamma&=\frac{1}{\sqrt{1-v^2}},\\ v^2&=\left(\frac{dx}{dt}\right)^2+\left(\frac{dy}{dt}\right)^2+\left(\frac{dz}{dt}\right)^2. \end{aligned}\right\} \tag{4.15}$$

If we compare the Eqs. (4.12) for ND with (4.14) for RD, we observe only a formal change from (X, Y, Z) to (P, Q, R). However there is a difference. Suppose (X, Y, Z) independent of velocity $(dx/dt,\ dy/dt,\ dz/dt)$, as is often the case in ND. Then (P, Q, R) depend on velocity, tending to zero as v tends to unity, i.e. as γ tends to infinity. This fact, and the inequality (4.4) connected with it, distinguishes RD from ND, as far as the motion, in a *given* field of force, of a particle of constant mass is concerned.

A much more important difference between ND and RD emerges when we consider, not a single particle in a given field of force, but a *system of particles*, moving under forces which are *due to* the particles alone.

5. Newtonian and relativistic dynamics of a system. To fix the ideas, consider two physical concepts:

(i) the solar system,

(ii) a free rigid[1] body.

For the solar system, Newtonian dynamics (ND) sets up as mathematical model a system of P particles with constant masses m_i $(i = 1, 2, \ldots P)$. For the several particles we have equations of motion of the form

$$m_i \frac{d^2 x_i}{dt^2} = X_i, \quad m_i \frac{d^2 y_i}{dt^2} = Y_i, \quad m_i \frac{d^2 z_i}{dt^2} = Z_i, \qquad (i = 1, 2, \ldots P), \tag{5.1}$$

the force (X_i, Y_i, Z_i) being given, in accordance with Newton's law of gravitation, by

$$\left.\begin{aligned} X_i &= G\, m_i \sum_j \frac{m_j (x_j - x_i)}{r_{ij}^3}, \\ r_{ij}^2 &= (x_i - x_j)^2 + (y_i - y_j)^2 + (z_i - z_j)^2, \end{aligned}\right\} \tag{5.2}$$

with similar expressions for Y_i and Z_i. The summation for j is from $j = 1$ to $j = P$, with $j \neq i$; G is the gravitational constant. We have, then, in (5.1), (5.2) a set of equations adequate to determine (x_i, y_i, z_i) $(i = 1, 2, \ldots P)$ as functions of t and of the values of

$$x_i,\ y_i,\ z_i,\ \frac{dx_i}{dt},\ \frac{dy_i}{dt},\ \frac{dz_i}{dt} \qquad (i = 1, 2, \ldots P) \tag{5.3}$$

for $t = 0$.

For a free rigid body, we again take a system of P particles and the equations of motion (5.1). With these we associate conditions of rigidity:

$$(x_i - x_j)^2 + (y_i - y_j)^2 + (z_i - z_j)^2 = a_{ij}^2, \tag{5.4}$$

where a_{ji} are constants, the distances between the particles. As for the forces, they are taken to be of the form

$$X_i = \sum_j X_{ij}, \quad Y_j = \sum_j Y_{ij}, \quad Z_i = \sum_j Z_{ij}, \tag{5.5}$$

where

$$X_{ij} = -X_{ji} = A_{ij}(x_j - x_i). \tag{5.6}$$

Here $A_{ij} (= A_{ji})$ are unknown; on eliminating them, we have in (5.1), (5.4), (5.5) (5.6) a set of equations adequate to determine (x_i, y_i, z_i) $(i = 1, 2, \ldots P)$ as functions of t and the initial data (5.3), which must be chosen to satisfy (5.4) and the equations obtained by applying d/dt to (5.4).

These mathematical models of the solar system and of a rigid body are mathematically clear and physically satisfactory, in that a vast number of satisfactory physical predictions have been made by their use. But we may ask: What is the most general model of a system of particles of constant masses permissible in Newtonian dynamics?

[1] This word provides a good example of that confusion regarding physical concepts (Sect. 2) which makes it difficult to treat them logically. On one occasion, a physicist might say: "I mounted the interferometer on a rigid base" (meaning, perhaps, a block of stone). On another occasion he might say: "There are no rigid bodies" (meaning that any body is deformed by sufficiently great stress). These statements are meaningful, when taken separately; taken together, they make nonsense, and would ruin any attempt at a logical argument. In a mathematical model this sort of confusion should not be allowed to occur.

In attempting to answer this question, let us confine our attention to a system of P particles which is *closed* or *isolated* in the sense that all forces are due to these particles alone (and not to any external agency), and in which the particles are *free* in the sense that there are no rigid bonds between them. We write down $3P$ equations of motion of the form (5.1), with the understanding that the forces (X_i, Y_i, Z_i) depend only on the instantaneous state of the system. For simplicity, let us suppose that they depend only on positions and velocities, so that they are functions of the $6P$ quantities

$$x_j,\ y_j,\ z_j,\ \frac{dx_j}{dt},\ \frac{dy_j}{dt},\ \frac{dz_j}{dt} \qquad (j = 1, 2, \ldots P). \tag{5.7}$$

We ask: What functions are permissible?

Newton's Third Law[1] gives a partial answer, stating essentially that the forces exerted on one another by two particles, A and B, act on the line AB, in opposite senses and with a common magnitude. This is equivalent to saying that (X_i, Y_i, Z_i) are of the form shown in (5.5) and (5.6), but it gives no information as to the nature of the coefficients A_{ij} except for the symmetry condition $A_{ij} = A_{ji}$; they might be any functions of the quantities (5.7). A more complete answer to our question is given by the following *Axiom of homogeneity and isotropy of space: The set of equations determining the motion of the system has the same form for all coordinate systems (x, y, z) obtained from one such coordinate system by a translation and rotation of axes.*

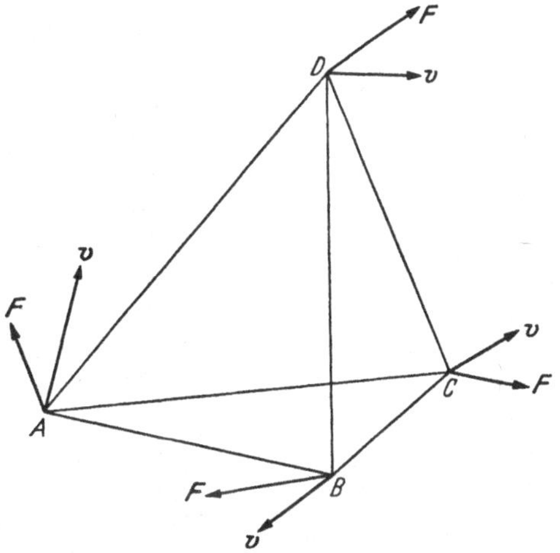

Fig. 1. Interaction in Newtonian dynamics.

To enlarge on this, we note that (4.1) determines the coordinate system (rectangular Cartesian coordinates) only to within an orthogonal transformation (cf. Sect. 9). The above axiom demands the invariance of the equations of motion under such orthogonal transformations, provided they be proper (no reflections included). Invariance under translation implies the *homogeneity* of space and invariance under rotation implies *isotropy*. Invariance under *reflection* in a plane (improper transformation) would imply equivalence of the two screw-senses.

To explore the most general type of force system satisfying the axiom, we note that the transformation considered corresponds precisely to a rigid-body displacement. *Thus the axiom is satisfied if the force system is "rigidly attached" to the instantaneous configuration of the particles.* To see what this means, consider a system of four particles as in Fig. 1. The positions at time t are A, B, C, D, and the velocities are the four vectors marked $\boldsymbol{v}$. We have to assign the four vectors marked $\boldsymbol{F}$, the forces on the particles. The axiom demands that these forces can be described in terms of the tetrahedron $ABCD$ and the four vectors $\boldsymbol{v}$, rigidly attached to the tetrahedron, in such a way that if these defining elements are moved rigidly together in space, the forces $\boldsymbol{F}$ are carried rigidly along with them. The idea is extremely simple, concepts of elementary Euclidean geometry replacing the formal equations. If we abandon Newton's Third Law, but accept

[1] Quoted in a footnote to Sect. 26.

the axiom of homogeneity and isotropy, then any force system so constructed is permissible. If we demand invariance under reflections also, then a reflection of the defining elements in a plane must reflect the forces in that plane.

NEWTON'S Third Law is consistent with the axiom of homogeneity and isotropy, but it restricts the interactions to the lines joining the particles, and thus fails to cope with electrodynamic interactions other than simple COULOMB attractions and repulsions. However, electrodynamic interactions should be treated relativistically, and, in the systematic development of Newtonian dynamics, we shall accept NEWTON'S Third Law, since otherwise we could not establish the fundamental principles of linear and angular momentum (cf. Sect. 44).

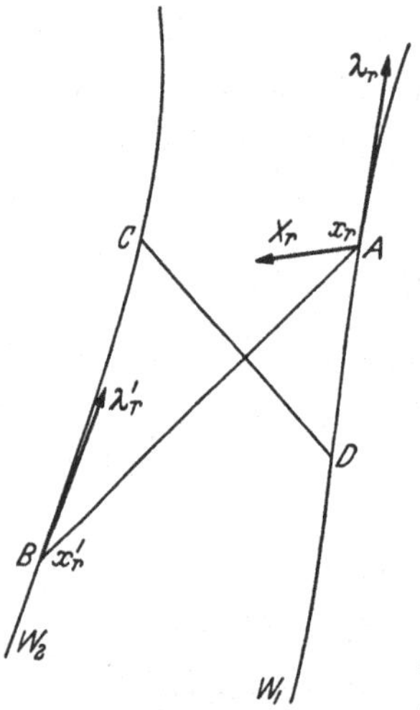

Fig. 2. Interaction in relativistic dynamics.

We now pass to the relativistic dynamics (RD) of a system. The requirement that the separation between adjacent events should have the form (4.2) limits the class of permissible coordinates (x, y, z, t) to those obtained from one such set by a LORENTZ transformation (cf. Sect. 106). Just as in ND we imposed an axiom of homogeneity and isotropy of space, so in RD we impose a similar axiom in space-time.

For a closed or isolated system of P particles, the *Axiom of homogeneity and isotropy of space-time* is as follows: *The equations determining the motion of the system must be invariant under a proper*[1] LORENTZ *transformation.*

A LORENTZ transformation may be regarded as a translation and rotation of space-time, a rigid-body transformation with "rigidity" understood in terms of separation (4.2). Therefore, just as permissible Newtonian force systems can be discussed in terms of rigid constructions in space, so permissible relativistic force systems can be discussed in terms of rigid constructions in space-time. But there the analogy between ND and RD ends. In ND we have, at any (absolute) time t, a configuration of particles with attached velocity vectors as in Fig. 1, but in RD we have only a set of world lines, and there is no obvious way to set up a unique correspondence between the events on the several world lines To correlate events with the same value of t would not be a LORENTZ-invariant procedure.

The most natural way to correlate events on the world lines is by means of null lines. Consider a system composed of two particles with constant proper masses m_1, m_2. Let W_1, W_2 be their world lines (Fig. 2). It is convenient to use Minkowskian coordinates x_r (small Latin suffixes take the values 1, 2, 3, 4 with summation understood for a repeated suffix), writing

$$x = x_1, \quad y = x_2, \quad z = x_3, \quad i t = x_4, \tag{5.8}$$

where $i = \sqrt{-1}$. Let A, with coordinates x_r, be an event on W_1. Draw the null cone[2] into the past from A as vertex; let it cut W_2 at B, with coordinates x'_r, so that BA is a null line, and we have

$$(x_r - x'_r)(x_r - x'_r) = 0. \tag{5.9}$$

---

[1] A LORENTZ transformation is a linear transformation with determinant +1 or −1. In the former case, it is *proper*; in the latter case, *improper* (cf. Sect. 106).

[2] Cf. Sect. 107.

Let

$$\lambda_r = \frac{dx_r}{ds} \text{ at } A, \quad \lambda'_r = \frac{dx'_r}{ds} \text{ at } B, \tag{5.10}$$

where ds, ds' are elements of proper time on W_1, W_2 respectively.

The pair of world lines and the events A, B on them provide us with the following vectors:

$$x_r - x'_r, \ \lambda_r, \ \lambda'_r, \ \frac{d\lambda_r}{ds}, \ \frac{d\lambda'_r}{ds'}, \ \ldots. \tag{5.11}$$

Equations of the form

$$\frac{d^2 x_r}{ds^2} = X_r, \tag{5.12}$$

for W_1, together with similar equations for W_2, will constitute a suitable formulation of the two-body problem in RD, provided that X_r is a vector constructed out of the vectors (5.11) and invariants formed out of them. Thus we might take, as a fairly simple choice, satisfying this requirement,

$$X_r = \alpha(x_r - x'_r) + \beta\lambda_r + \gamma\lambda'_r + \delta\frac{d\lambda_r}{ds} + \varepsilon\frac{d\lambda'_r}{ds'}, \tag{5.13}$$

where the coefficients are assigned functions of the invariants

$$\left.\begin{aligned} &w = (x_n - x'_n)\lambda_n, \quad w' = (x'_n - x_n)\lambda'_n, \quad \lambda_n\lambda'_n, \\ &W = (x_n - x'_n)\frac{d\lambda_n}{ds}, \quad W' = (x'_n - x_n)\frac{d\lambda'_n}{ds'} \\ &\lambda_n\frac{d\lambda'_n}{ds'}, \quad \lambda'_n\frac{d\lambda_n}{ds}, \quad \frac{d\lambda_n}{ds}\frac{d\lambda'_n}{ds'}. \end{aligned}\right\} \tag{5.14}$$

In writing down the equations for W_2, we use events C, D as in Fig. 2 instead of A, B and make appropriate changes.

The preservation of the mass m_1 imposes the condition [cf. (4.11)]

$$X_r\lambda_r = 0, \tag{5.15}$$

or

$$\alpha w - \beta + \gamma\lambda_r\lambda'_r + \varepsilon\lambda_r\frac{d\lambda'_r}{ds'} = 0; \tag{5.16}$$

the preservation of m_2 imposes a similar condition.

Although the equations as in (5.12) satisfy the condition of LORENTZ-invariance, they present a mathematical problem far more complicated than what we meet in ND. They appear to be differential equations, but they involve the two events A, B, and have the nature of difference equations on account of the "retarded" effects due to B.

The commonly accepted equations of motion for a pair of particles carrying electric charges e_1, e_2 are as in (5.12); the coefficients in (5.13) are given by[1]

$$\left.\begin{aligned} 4\pi c^2\alpha &= \frac{e_1 e_2}{w'^2}\left(\frac{W'-1}{w'}\lambda_n\lambda'_n - \lambda_n\frac{d\lambda'_n}{ds'}\right), \\ 4\pi c^2\gamma &= -\frac{e_1 e_2 w}{w'^2}\frac{W'-1}{w'}, \\ 4\pi c^2\varepsilon &= \frac{e_1 e_2 w}{w'^2}, \end{aligned}\right\} \tag{5.17}$$

the other coefficients being zero.

---

[1] The charges are measured in HEAVISIDE rational units; to change to Gaussian electrostatic units, delete the factor 4π. The factor c is the speed of light. For the derivation of these formulae, see W. PAULI, Relativitätstheorie, Encykl. d. Math. Wiss. V 19, p. 645. Leipzig und Berlin: Teubner 1920, where, however, there is a misprint in the last line—for $(X_r u^r)^3$ read $(X_r u^r)^2$. Cf. also J. L. SYNGE: Relativity: The Special Theory, pp. 394, 423. Amsterdam: North-Holland 1956.

To summarize, for a single particle in a given field of force, we have a set of three differential equations to solve, whether the dynamics be Newtonian or relativistic. But, for a system of interacting particles, the differential equations of Newtonian dynamics are replaced in relativity by differential-difference equations; these present such great mathematical difficulties that only certain limiting cases can be handled by approximate methods[1].

B. Kinematics.

I. Displacements of rigid bodies.

6. Displacements parallel to a plane. The position of a rigid body is determined by the position of any plane section of it, and so the displacements of a rigid body which are parallel to a fixed plane may be discussed in terms of the displacements of a lamina in a plane. Such a displacement may be described by stating that two points of the lamina, initially at A and B, are moved to positions A' and B'. This displacement may be broken down into a sequence of two displacements: (i) a *translation* carrying A to A', and (ii) a *rotation* about A'. Or we may start with a rotation and follow it by a translation.

It is convenient to use complex numbers ($z = x + iy$) to describe displacements in a plane. A translation, represented by a complex number t, causes a transformation $z' = z + t$. For a rotation through an angle ϑ about a point c, we have

$$z' - c = (z - c)\,e^{i\vartheta}.$$

If we use T and R as symbols for the operations of translation and rotation respectively, and denote by RT and TR (read from right to left) the combined operations, then we have the following transformations for the two orders of the operations:

$$\left.\begin{aligned} RT&: \quad z' = c + (z + t - c)\,e^{i\vartheta}, \\ TR&: \quad z' = t + c + (z - c)\,e^{i\vartheta}. \end{aligned}\right\} \tag{6.1}$$

In general these transformations are different ($RT \neq TR$), and this is perhaps the simplest example of two operations in mechanics which do not *commute*. We have $RT = TR$ only in the exceptional cases where $t = 0$ or $\vartheta = 0$.

Consider the first of (6.1). To find a fixed point for the displacement RT, we are to put $z' = z$; this gives for z the equation

$$z\,(1 - e^{i\vartheta}) = c + (t - c)\,e^{i\vartheta}. \tag{6.2}$$

Unless $\vartheta = 0$ (or a multiple of 2π) this equation has a unique solution. Therefore *every rigid plane displacement (except a pure translation) has a unique fixed point*. Equivalently, any rigid plane displacement (except a translation) can be produced by a rotation about a suitably chosen centre.

This centre can be found (and indeed the above theorem proved) by means of a simple construction. Let the displacement send A to B and B to C. Then unless the displacement is a pure translation, the right bisectors of AB and BC meet at some point D; D is the required centre.

The resultant of two plane displacements ($D_2 D_1$) is therefore the resultant of two rotations ($R_2 R_1$), and this resultant is itself a rotation (R_3); we may write

$$R_3 = R_2 R_1. \tag{6.3}$$

---

[1] Cf. C. G. DARWIN: Phil. Mag. (6) **39**, 537 (1920). — J. L. SYNGE: Proc. Roy. Soc. Lond., Ser. A **177**, 118 (1940).

Two rotations about different centres do not commute ($R_2 R_1 \neq R_1 R_2$). It is easy to construct the two resultants. Let A_1, A_2 be the centres and ϑ_1, ϑ_2 the angles of rotation. Then the angles of $R_2 R_1$ and $R_1 R_2$ are the same, viz. $2\varphi = \vartheta_1 + \vartheta_2$, and their centres are C_{21} and C_{12} as in Fig. 3; these centres are reflections of one another in the line $A_1 A_2$.

A lamina moving in a plane (or a rigid body moving parallel to a plane) has three degrees of freedom, since the position of the lamina is fixed when we know the two coordinates of any point of it and the inclination to a fixed direction of any line in it. Thus the configuration-space is of three dimensions (Sect. 62). It has the same type of connectivity as an infinite cylinder, there being one independent irreducible circuit, corresponding to a complete rotation of the lamina (Sect. 63); and it is flat with respect to the kinematical line element (Sect. 84).

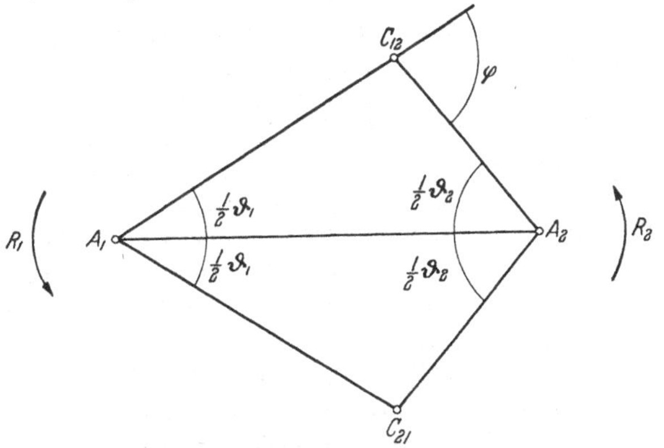

Fig. 3. Resultant of two rotations in a plane. C_{12} is the centre for $R_1 R_2$ and C_{21} the centre for $R_2 R_1$.

7. EULER's theorem. Consider a rigid body with a fixed point O. Draw a spherical surface S with unit radius and centre O. The position of the body is fixed by the positions of those points of it which lie on S, and any displacement of the body which leaves O fixed is a rigid transformation of S into itself.

EULER's theorem: *Any rigid displacement of a spherical surface into itself leaves two points of the surface fixed, these points being diametrically opposed to one another.*

This theorem can be proved (and the fixed points found) by the construction given after (6.2), the straight right bisectors in the plane being replaced by great circles on the sphere. The exceptional case (pure translation in the plane) cannot arise in the case of a sphere, for two great circles must intersect[1].

EULER's theorem may be expressed by saying that a rotation about a point is a rotation about some line through that point. This property (fixed point implies fixed line) is due to the *oddness* of the dimensionality of space.

The composition of two rotations about a point can be effected as in Fig. 3, with the straight lines replaced by great circles on the sphere. The points A_1 and A_2 are now points where the sphere is cut by the two axes of rotation with A_1 on one axis and A_2 on the other. The only essential difference is that although the angle of the resultant rotation is still 2φ, where φ is the angle so marked in Fig. 3, we now have

$$2\varphi = \vartheta_1 + \vartheta_2 - 2E, \tag{7.1}$$

E being the spherical excess of the triangle $A_1 A_2 C_{12}$.

It is clear that finite rotations about a point do not commute ($R_2 R_1 \neq R_1 R_2$), unless they are about a common line[2].

---

[1] For a proof of EULER's theorem based on the fact that a real orthogonal matrix has one eigenvalue equal to ± 1, see GOLDSTEIN [7], pp. 118–123.

[2] For a number of interesting special results on finite rotations and fuller demonstrations of the matters dealt with here, see LAMB [14], Chap. I.

A rotation is specified by three parameters, e.g. the angle of the rotation and two of the direction cosines of its axis. Therefore a rigid body with a fixed point has three degrees of freedom, the same number as for a lamina moving in a plane. But in spite of the analogies between these two systems, there is a great algebraic difference. In the plane case we can use complex numbers very simply as in (6.1), but much more complicated methods are required to deal with finite rotations about a point, as will be seen in the following sections. The configuration-space (Sect. 62) has three dimensions in both cases and its connectivity is similar (one independent irreducible circuit, Sect. 63); but, whereas the configuration-space of a lamina is flat with respect to the kinematical line element (Sect. 84), that of a body with a fixed point is curved.

8. General rigid body displacements. Let D denote any displacement of a rigid body. Suppose that D carries a point of the body from a position A to a position A'. Then D can be accomplished in two steps: (i) a translation AA', and (ii) a rotation about A'. This is the standard way of describing a general displacement, the point A of the body being called the *base-point*.

A set of orthogonal triads originally parallel to one another is changed by D into a set of orthogonal triads parallel to one another. Hence it is clear that the rotation (i.e. the direction of its axis and the angle of rotation) is independent of the choice of base-point. But the translation depends on the choice of base point.

CHASLES' theorem: *Any rigid body displacement is equivalent to a screw.*

A screw is defined as the resultant of a rotation and a translation parallel to the axis of rotation; it is easy to see that for a screw the translation and the rotation commute. To prove CHASLES' theorem, we start with any base-point and break down the translation into two translations, one (say T_1) parallel to the axis of rotation and the other (say T_2) perpendicular to that axis. Then T_2 and the rotation are plane displacements with a common plane, and may be compounded into a single rotation with a new axis parallel to the old one (cf. Sect. 6). This rotation and T_1 together form a screw. This proves CHASLES' theorem.

A general rigid body displacement can be reproduced by two half-turns, a half-turn being a rotation about a line through two right angles (equivalently, reflection in a line). The axes of the two half-turns cut perpendicularly the axis of the equivalent screw; their distance apart is half the translation of the screw; and the angle between them is half the angle of rotation of the screw[1].

A free rigid body has six degrees of freedom, three for the translation and three for the rotation. Its 6-dimensional configuration-space has one independent irreducible circuit, and is curved with respect to the kinematical line element (Sect. 84).

We have observed the analogy between a plane displacement (translation + rotation) and a displacement of a 3-dimensional rigid body with a fixed point (rotation). There is a similar analogy between the general displacement of a rigid body (translation + rotation) and a displacement of a 4-dimensional rigid body with a fixed point (rotation). We do not meet 4-dimensional rigid bodies in Newtonian physics, but, in the special theory of relativity, LORENTZ transformations with fixed origin [put $B_r = 0$ in (107.5)] may be regard as 4-dimensional rotations, modified on account of the indefinite metric of space-time.

[1] For further details and other properties of finite displacements, see LAMB [*14*], Chap. I.

9. Orthogonal matrices. The following notation will be used for matrices:

$\tilde{\boldsymbol{M}}$ is the transpose of $\boldsymbol{M}$.

$\boldsymbol{M}^\dagger$ is the complex conjugate of $\tilde{\boldsymbol{M}}$.

$\mathbf{1}$ is the unit matrix.

$\boldsymbol{M}$ is orthogonal if $\boldsymbol{M}\tilde{\boldsymbol{M}} = \mathbf{1}$ (equivalently, $\boldsymbol{M}^{-1} = \tilde{\boldsymbol{M}}$), and proper or improper according as $\det \boldsymbol{M} = 1$ or -1.

$\boldsymbol{M}$ is unitary if $\boldsymbol{M}\boldsymbol{M}^\dagger = \mathbf{1}$ (equivalently, $\boldsymbol{M}^{-1} = \boldsymbol{M}^\dagger$).

$\boldsymbol{M}$ is symmetric if $\tilde{\boldsymbol{M}} = \boldsymbol{M}$.

$\boldsymbol{M}$ is Hermitian if $\boldsymbol{M}^\dagger = \boldsymbol{M}$.

$\boldsymbol{r}$ is the column matrix (x, y, z).

Let $(\boldsymbol{I}, \boldsymbol{J}, \boldsymbol{K})$ be an orthogonal triad of unit vectors (orthonormal triad) with origin at O. Let a rigid body be given a rotation about O, and, as a result, let $(\boldsymbol{I}, \boldsymbol{J}, \boldsymbol{K})$, regarded as fixed in the body, be carried into the orthonormal triad $(\boldsymbol{i}, \boldsymbol{j}, \boldsymbol{k})$. There exists, then, a matrix of scalar products (or relative direction cosines):

$$\boldsymbol{M} = \begin{pmatrix} \boldsymbol{I}\cdot\boldsymbol{i} & \boldsymbol{I}\cdot\boldsymbol{j} & \boldsymbol{I}\cdot\boldsymbol{k} \\ \boldsymbol{J}\cdot\boldsymbol{i} & \boldsymbol{J}\cdot\boldsymbol{j} & \boldsymbol{J}\cdot\boldsymbol{k} \\ \boldsymbol{K}\cdot\boldsymbol{i} & \boldsymbol{K}\cdot\boldsymbol{j} & \boldsymbol{K}\cdot\boldsymbol{k} \end{pmatrix}. \tag{9.1}$$

Using orthogonal projections and introducing the column matrices of vectors

$$\boldsymbol{T} = \begin{pmatrix} \boldsymbol{I} \\ \boldsymbol{J} \\ \boldsymbol{K} \end{pmatrix}, \qquad \boldsymbol{t} = \begin{pmatrix} \boldsymbol{i} \\ \boldsymbol{j} \\ \boldsymbol{k} \end{pmatrix}, \tag{9.2}$$

we may exhibit the connection between the two triads in the following forms:

$$\boldsymbol{t} = \tilde{\boldsymbol{M}}\boldsymbol{T}, \qquad \boldsymbol{T} = \boldsymbol{M}\boldsymbol{t}. \tag{9.3}$$

Let $\boldsymbol{i}$ have direction cosines (l_1, m_1, n_1) relative to $(\boldsymbol{I}, \boldsymbol{J}, \boldsymbol{K})$, with a similar notation for $\boldsymbol{j}$ and $\boldsymbol{k}$. Then

$$\boldsymbol{M} = \begin{pmatrix} l_1 & l_2 & l_3 \\ m_1 & m_2 & m_3 \\ n_1 & n_2 & n_3 \end{pmatrix}, \qquad \tilde{\boldsymbol{M}} = \begin{pmatrix} l_1 & m_1 & n_1 \\ l_2 & m_2 & n_2 \\ l_3 & m_3 & n_3 \end{pmatrix}. \tag{9.4}$$

From the orthonormality of the triads we have six conditions, of which the first two are

$$l_1^2 + l_2^2 + l_3^2 = 1, \qquad l_1 m_1 + l_2 m_2 + l_3 m_3 = 0, \tag{9.5}$$

and it follows that $\boldsymbol{M}$ is an orthogonal matrix:

$$\boldsymbol{M}\tilde{\boldsymbol{M}} = \mathbf{1} = \tilde{\boldsymbol{M}}\boldsymbol{M}. \tag{9.6}$$

Hence $\det \boldsymbol{M} = \pm 1$. If the triad is not moved at all, we have $\boldsymbol{M} = \mathbf{1}$, $\det \boldsymbol{M} = 1$; therefore, by continuity, $\boldsymbol{M}$ is proper for all rotations. An improper orthogonal matrix corresponds to a rotation combined with a reflection in the origin. We shall confine the argument to rotations, although some of the formulae are applicable to improper orthogonal transformations.

So far we have avoided the use of coordinates. But now let $OXYZ$ be axes fixed in space, and let $(\boldsymbol{I}, \boldsymbol{J}, \boldsymbol{K})$ coincide with them, so that $\boldsymbol{I} = (1, 0, 0)$, $\boldsymbol{J} = (0, 1, 0)$, $\boldsymbol{K} = (0, 0, 1)$. Consider any point of the body. Let its coordinates be (x, y, z) before the rotation and (x', y', z') after it. Then the initial and final position vectors of the point are respectively

$$\left.\begin{aligned} \boldsymbol{r} &= x\boldsymbol{I} + y\boldsymbol{J} + z\boldsymbol{K}, \\ \boldsymbol{r}' &= x\boldsymbol{i} + y\boldsymbol{j} + z\boldsymbol{k}, \end{aligned}\right\} \tag{9.7}$$

so that

$$x' = \boldsymbol{r}' \cdot \boldsymbol{I} = l_1 x + l_2 y + l_3 z, \quad \text{etc.}, \tag{9.8}$$

and the coordinate transformation may be expressed in the matrix forms

$$\boldsymbol{r}' = \boldsymbol{M}\boldsymbol{r}, \quad \boldsymbol{r} = \widetilde{\boldsymbol{M}}\boldsymbol{r}', \tag{9.9}$$

the second following from the first by (9.6). Note the linearity of the transformation, a most important fact. Further, note the formal interchange of $\boldsymbol{M}$ and $\widetilde{\boldsymbol{M}}$ in the comparison of (9.3) and (9.9).

For a given rotation, the matrix $\boldsymbol{M}$ depends on the choice of the triad $(\boldsymbol{I}, \boldsymbol{J}, \boldsymbol{K})$, or, equivalently, on the choice of the axes $OXYZ$. By choosing $\boldsymbol{K}$ along the axis of rotation, we can simplify $\boldsymbol{M}$ to the form

$$\boldsymbol{M} = \begin{pmatrix} \cos\chi & -\sin\chi & 0 \\ \sin\chi & \cos\chi & 0 \\ 0 & 0 & 1 \end{pmatrix}, \tag{9.10}$$

where χ is the angle of the rotation.

When we compound a sequence of rotations, we do not, as a rule, keep the axes $OXYZ$ fixed once for all. We may follow the scheme

$$\boldsymbol{T} \xrightarrow[\boldsymbol{M}_1]{} \boldsymbol{T}_1 \xrightarrow[\boldsymbol{M}_2]{} \boldsymbol{T}_2 \xrightarrow[\boldsymbol{M}_3]{} \cdots \xrightarrow[\boldsymbol{M}_{n-1}]{} \boldsymbol{T}_{n-1} \xrightarrow[\boldsymbol{M}_n]{} \boldsymbol{t}, \tag{9.11}$$

which shows, under each passage from triad to triad, the corresponding matrix as in (9.1) or (9.4). The resultant rotation is then, as in (9.3),

$$\boldsymbol{t} = \widetilde{\boldsymbol{M}}\boldsymbol{T}, \quad \widetilde{\boldsymbol{M}} = \widetilde{\boldsymbol{M}}_n \widetilde{\boldsymbol{M}}_{n-1} \ldots \widetilde{\boldsymbol{M}}_1. \tag{9.12}$$

To get the corresponding coordinate transformation, using fixed axes coincident with $\boldsymbol{T}$, we must, as in the transition from (9.3) to (9.9), replace $\boldsymbol{M}$ by its transpose, obtaining

$$\boldsymbol{r}' = \boldsymbol{M}\boldsymbol{r}, \quad \boldsymbol{M} = \boldsymbol{M}_1 \boldsymbol{M}_2 \ldots \boldsymbol{M}_n. \tag{9.13}$$

Note that the matrices are written down here in their natural order, so that the operations are actually applied in the reverse of that order.

The difference between (9.12) and (9.13) (interchange of $\boldsymbol{M}$ and $\widetilde{\boldsymbol{M}}$) can be a source of such petty confusion. And there is yet a *third* way of looking at the rotation. We may hold a point *fixed in space*, and consider its coordinates (x, y, z) relative to the old triad $\boldsymbol{T}$ and its coordinates (x', y', z') relative to the new triad $\boldsymbol{t}$. Then the transformation is

$$\boldsymbol{r}' = \widetilde{\boldsymbol{M}}\boldsymbol{r}, \tag{9.14}$$

which is similar to (9.12).

To clarify the situation at this stage, we may sum up by saying that a proper orthogonal 3×3 matrix $\boldsymbol{M}$ can be interpreted in the following four consistent ways:

(i) A table of scalar products (9.1) of the old triad $(\boldsymbol{I}, \boldsymbol{J}, \boldsymbol{K})$ and the new triad $(\boldsymbol{i}, \boldsymbol{j}, \boldsymbol{k})$.

(ii) Rotation of an orthonormal triad, $\boldsymbol{T} \to \boldsymbol{t}$ with $\boldsymbol{t} = \widetilde{\boldsymbol{M}}\boldsymbol{T}$.

(iii) Coordinate transformation with the axes fixed in space and the point carried with the body, $\boldsymbol{r} \to \boldsymbol{r}'$ with $\boldsymbol{r}' = \boldsymbol{M}\boldsymbol{r}$.

(iv) Coordinate transformation with the point fixed in space and the axes carried with the body, $\boldsymbol{r} \to \boldsymbol{r}'$ with $\boldsymbol{r}' = \widetilde{\boldsymbol{M}}\boldsymbol{r}$.

10. Rotation in terms of its axis and angle (Eulerian parameters). An ordered orthogonal triad of vectors ($\boldsymbol{I}, \boldsymbol{J}, \boldsymbol{K}$) has two possible orientations, right-handed and left-handed. At a point on the earth's surface, we get a right-handed triad by taking $\boldsymbol{I}$ horizontal and pointing to the East, $\boldsymbol{J}$ horizontal and pointing to the North, and $\boldsymbol{K}$ directed vertically upward. In this article, all triads, including triads of coordinate axes, will be chosen right-handed, in accordance with current practice, unless specially noted otherwise.

The positive sense of rotation about $\boldsymbol{K}$ is the sense of the 90°-rotation which carries $\boldsymbol{I}$ into $\boldsymbol{J}$.

Any rotation about an axis lying on a unit vector $\boldsymbol{U}$ can be described by a symbol $[\boldsymbol{U}, \chi]$, where χ is an angle of rotation, counted positive in the positive sense just defined. The correspondence between rotations and such representations is, however, multiple valued. For, if $\boldsymbol{R}$ denotes any rotation about $\boldsymbol{U}$, we may write symbolically

$$\boldsymbol{R} = [\boldsymbol{U}, \chi + 2n\pi] = [-\boldsymbol{U}, -\chi + 2n\pi] \quad (10.1)$$

n being any integer.

We reduce the multiplicity of the correspondence by introducing a vector $\boldsymbol{V}$ and a scalar ϱ as follows:

$$\boldsymbol{V} = \boldsymbol{U} \sin \tfrac{1}{2}\chi, \quad \varrho = \cos \tfrac{1}{2}\chi. \quad (10.2)$$

Let the components of $\boldsymbol{V}$ on any axes be (λ, μ, ν); then (10.2) imply

$$\lambda^2 + \mu^2 + \nu^2 + \varrho^2 = 1. \quad (10.3)$$

It is easy to see that any set of values $(\lambda, \mu, \nu, \varrho)$ satisfying (10.3) determine a unique rotation, but that to a given rotation there correspond two sets of values of these quantities. We may write symbolically

$$\boldsymbol{R} = \{\lambda, \mu, \nu, \varrho\} = \{-\lambda, -\mu, -\nu, -\varrho\}. \quad (10.4)$$

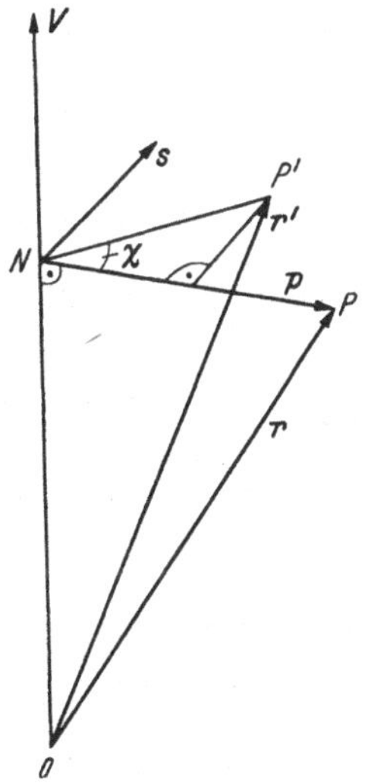

Fig. 4. Axis $\boldsymbol{V}$ and angle of rotation χ.

The quantities $(\lambda, \mu, \nu, \varrho)$ are EULER's parameters[1]. They serve to describe the configurations of a rigid body with a fixed point. For we can pass from any given initial configuration C_0 to a final configuration C by a definite rotation $\boldsymbol{R}$, and $\boldsymbol{R}$ is determined by $(\lambda, \mu, \nu, \varrho)$. In view of (10.3) and (10.4), we may make the following statements, if we regard $(\lambda, \mu, \nu, \varrho)$ as rectangular Cartesian coordinates of a point in Euclidean 4-space:

(i) Any point on the hypersphere (10.3) determines a final configuration of the body.

(ii) Any final configuration of the body determines a pair of diametrically opposed points on the hypersphere (10.3).

(iii) There is a continuous 1:1 correspondence between the final configurations of the body and the straight lines drawn through the origin in a space of four dimensions.

Since $(\lambda, \mu, \nu, \varrho)$ describe a rotation about a fixed point, the matrix $\boldsymbol{M}$ of Sect. 9 is expressible in terms of them: this is done as follows.

Let $P(\boldsymbol{r})$ and $P'(\boldsymbol{r}')$ (Fig. 4) be the initial and final positions of a point of a body which experiences the rotation (10.2). Let N be the common foot of the

[1] The notation follows F. D. MURNAGHAN: The Theory of Group Representations, p. 328. Baltimore: Johns Hopkins Press 1938. — WHITTAKER [28], p. 8, uses (ξ, η, ζ, χ).

perpendiculars dropped from P and P' on $\boldsymbol{V}$; let $\overrightarrow{NP}=\boldsymbol{p}$, and let $\boldsymbol{s}$ be a unit vector such that $(\boldsymbol{p}, \boldsymbol{s}, \boldsymbol{V})$ form a righthanded orthogonal triad. Then

$$\boldsymbol{r}' = \overrightarrow{ON} + \overrightarrow{NP'} = \overrightarrow{ON} + \boldsymbol{p}\cos\chi + \boldsymbol{s}\,p\sin\chi. \tag{10.5}$$

But

$$\boldsymbol{p} = \boldsymbol{r} - \overrightarrow{ON}, \qquad \boldsymbol{s} = \frac{\boldsymbol{V}\times\boldsymbol{p}}{Vp} = \frac{\boldsymbol{V}\times\boldsymbol{r}}{Vp},$$

and so by (10.2)

$$\left.\begin{aligned} \boldsymbol{r}' &= \boldsymbol{r}\cos\chi + \overrightarrow{ON}(1-\cos\chi) + \sin\chi\,(\boldsymbol{V}\times\boldsymbol{r})/V \\ &= \boldsymbol{r}(\varrho^2 - V^2) + 2V^2\overrightarrow{ON} + 2\varrho(\boldsymbol{V}\times\boldsymbol{r}). \end{aligned}\right\} \tag{10.6}$$

Now

$$V^2\overrightarrow{ON} = \boldsymbol{V}(\boldsymbol{V}\cdot\boldsymbol{r}), \tag{10.7}$$

and so the transformation $\boldsymbol{r}\to\boldsymbol{r}'$ is

$$\boldsymbol{r}' = \boldsymbol{r}(\varrho^2 - V^2) + 2\boldsymbol{V}(\boldsymbol{V}\cdot\boldsymbol{r}) + 2\varrho(\boldsymbol{V}\times\boldsymbol{r}), \tag{10.8}$$

or in matrix form $\boldsymbol{r}' = \boldsymbol{M}\boldsymbol{r}$, where

$$\boldsymbol{M} = \begin{pmatrix} \lambda^2-\mu^2-\nu^2+\varrho^2 & 2(\lambda\mu-\nu\varrho) & 2(\nu\lambda+\mu\varrho) \\ 2(\lambda\mu+\nu\varrho) & \mu^2-\nu^2-\lambda^2+\varrho^2 & 2(\mu\nu-\lambda\varrho) \\ 2(\nu\lambda-\mu\varrho) & 2(\mu\nu+\lambda\varrho) & \nu^2-\lambda^2-\mu^2+\varrho^2 \end{pmatrix}, \tag{10.9}$$

λ, μ, ν being the components of $\boldsymbol{V}$, so that

$$\lambda^2+\mu^2+\nu^2 = \sin^2\tfrac{1}{2}\chi, \qquad \lambda^2+\mu^2+\nu^2+\varrho^2 = 1. \tag{10.10}$$

Note that $\boldsymbol{M}$ is unchanged if we replace $(\lambda, \mu, \nu, \varrho)$ by $(-\lambda, -\mu, -\nu, -\varrho)$, as of course must be the case.

11. Eulerian angles[1]. Let a rotation about O carry the orthonormal triad $(\boldsymbol{I}, \boldsymbol{J}, \boldsymbol{K})$ into $(\boldsymbol{i}, \boldsymbol{j}, \boldsymbol{k})$. We break this rotation into three rotations (Fig. 5). First, rotate about $\boldsymbol{K}$ so as to make the new position of the plane $(\boldsymbol{I}, \boldsymbol{K})$ contain $\boldsymbol{k}$, say through an angle φ; this gives a transformation

$$(\boldsymbol{I}, \boldsymbol{J}, \boldsymbol{K}) \to (\boldsymbol{I}_1, \boldsymbol{J}_1, \boldsymbol{K}_1) \left\{\begin{aligned} \boldsymbol{I}_1 &= \quad \boldsymbol{I}\cos\varphi + \boldsymbol{J}\sin\varphi, \\ \boldsymbol{J}_1 &= -\boldsymbol{I}\sin\varphi + \boldsymbol{J}\cos\varphi, \\ \boldsymbol{K}_1 &= \boldsymbol{K}. \end{aligned}\right\} \tag{11.1}$$

Secondly, rotate about $\boldsymbol{J}_1$ to bring $\boldsymbol{K}_1$ to $\boldsymbol{k}$, say through an angle ϑ; this gives a transformation

$$(\boldsymbol{I}_1, \boldsymbol{J}_1, \boldsymbol{K}_1) \to (\boldsymbol{I}_2, \boldsymbol{J}_2, \boldsymbol{k}) \left\{\begin{aligned} \boldsymbol{I}_2 &= \boldsymbol{I}_1\cos\vartheta - \boldsymbol{K}_1\sin\vartheta, \\ \boldsymbol{J}_2 &= \boldsymbol{J}_1, \\ \boldsymbol{k} &= \boldsymbol{I}_1\sin\vartheta + \boldsymbol{K}_1\cos\vartheta. \end{aligned}\right\} \tag{11.2}$$

---

[1] The notation used here follows WHITTAKER [28], p. 9, and has the advantage that (ϑ, φ) are the usual polar angles of the vector $\boldsymbol{k}$. For alternative notations, cf. H. TIETZ, this Encyclopedia, Vol. II, p. 135; APPELL [2] II, p. 151 (he interchanges φ and ψ); GOLDSTEIN [7], p. 107 (he discusses various usages).

Finally, rotate about $\boldsymbol{k}$ to bring $\boldsymbol{I}_2$ to $\boldsymbol{i}$ and $\boldsymbol{J}_2$ to $\boldsymbol{j}$, say through an angle ψ; this gives the transformation

$$(\boldsymbol{I}_2, \boldsymbol{J}_2, \boldsymbol{k}) \rightarrow (\boldsymbol{i}, \boldsymbol{j}, \boldsymbol{k}) \left\{ \begin{aligned} \boldsymbol{i} &= \boldsymbol{I}_2 \cos\psi + \boldsymbol{J}_2 \sin\psi, \\ \boldsymbol{j} &= -\boldsymbol{I}_2 \sin\psi + \boldsymbol{J}_2 \cos\psi, \\ \boldsymbol{k} &= \boldsymbol{k}. \end{aligned} \right\} \tag{11.3}$$

The angles $(\vartheta, \varphi, \psi)$ are the Eulerian angles. Their values determine the position of the triad $(\boldsymbol{i}, \boldsymbol{j}, \boldsymbol{k})$ relative to $(\boldsymbol{I}, \boldsymbol{J}, \boldsymbol{K})$. They may be given any values whatever, but all positions of $(\boldsymbol{i}, \boldsymbol{j}, \boldsymbol{k})$ are obtained by letting them cover the following ranges:

$$\left. \begin{aligned} 0 &\leq \vartheta \leq \pi, \\ 0 &\leq \varphi < 2\pi, \\ 0 &\leq \psi < 2\pi. \end{aligned} \right\} \tag{11.4}$$

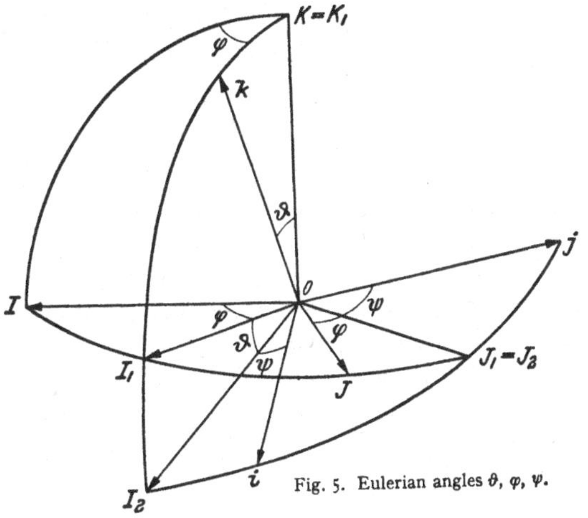

Fig. 5. Eulerian angles ϑ, φ, ψ.

From the above equations of transformation we can express $(\boldsymbol{i}, \boldsymbol{j}, \boldsymbol{k})$ as linear functions of $(\boldsymbol{I}, \boldsymbol{J}, \boldsymbol{K})$, and hence obtain a matrix $\boldsymbol{M}$ of scalar products as in (9.1) or direction cosines as in (9.4). This matrix $\boldsymbol{M}$ is exhibited compactly as follows in an abridged notation in which $c = \cos$, $s = \sin$, and the subscripts 1, 2, 3 refer to ϑ, φ, ψ respectively:

	$\boldsymbol{i}$	$\boldsymbol{j}$	$\boldsymbol{k}$
$\boldsymbol{I}$	$c_1 c_2 c_3 - s_2 s_3$	$-c_1 c_2 s_3 - s_2 c_3$	$s_1 c_2$
$\boldsymbol{J}$	$c_1 s_2 c_3 + c_2 s_3$	$-c_1 s_2 s_3 + c_2 c_3$	$s_1 s_2$
$\boldsymbol{K}$	$-s_1 c_3$	$s_1 s_3$	c_1

(11.5)

We have here an illustration of the composition of rotations according to (9.13), for we can pick out the matrices $\boldsymbol{M}_1$, $\boldsymbol{M}_2$, $\boldsymbol{M}_3$ by transposing the elements in (11.1) to (11.3), and verify directly that the matrix $\boldsymbol{M}$ of (11.5) is

$$\boldsymbol{M} = \boldsymbol{M}_1 \boldsymbol{M}_2 \boldsymbol{M}_3 = \begin{pmatrix} c_2 & -s_2 & 0 \\ s_2 & c_2 & 0 \\ 0 & 0 & 1 \end{pmatrix} \begin{pmatrix} c_1 & 0 & s_1 \\ 0 & 1 & 0 \\ -s_1 & 0 & c_1 \end{pmatrix} \begin{pmatrix} c_3 & -s_3 & 0 \\ s_3 & c_3 & 0 \\ 0 & 0 & 1 \end{pmatrix}. \tag{11.6}$$

On comparing the matrix (11.5) with (10.9), it is easy to obtain the Eulerian parameters in terms of the Eulerian angles:

$$\left. \begin{aligned} \lambda &= \varepsilon \sin\tfrac{1}{2}\vartheta \sin\tfrac{1}{2}(\psi - \varphi), \\ \mu &= \varepsilon \sin\tfrac{1}{2}\vartheta \cos\tfrac{1}{2}(\psi - \varphi), \\ \nu &= \varepsilon \cos\tfrac{1}{2}\vartheta \sin\tfrac{1}{2}(\psi + \varphi), \\ \varrho &= \varepsilon \cos\tfrac{1}{2}\vartheta \cos\tfrac{1}{2}(\psi + \varphi), \end{aligned} \right\} \tag{11.7}$$

where $\varepsilon = \pm 1$; for definiteness, we may take $\varepsilon = 1$, thus determining $(\lambda, \mu, \nu, \varrho)$ uniquely in terms of $(\vartheta, \varphi, \psi)$.

12. Quaternions. A quaternion q is of the form

$$q = a\,i + b\,j + c\,k + d, \tag{12.1}$$

where a, b, c, d are ordinary numbers (we shall take them real) and i, j, k are the quaternionic units[1], or unit vectors, satisfying the algebraic rules

$$\left.\begin{array}{c} i^2 = j^2 = k^2 = -1, \\ j\,k = -k\,j = i, \quad k\,i = -i\,k = j, \quad i\,j = -j\,i = k. \end{array}\right\} \tag{12.2}$$

The vector part Vq, the scalar part Sq, the conjugate Kq, the norm Nq, and the reciprocal q^{-1} are defined as follows[2]:

$$\left.\begin{array}{c} V\,q = a\,i + b\,j + c\,k, \quad S\,q = d, \quad q = V\,q + S\,q, \\ K\,q = -V\,q + S\,q, \quad N\,q = (q\,K\,q)^{\frac{1}{2}} = (a^2 + b^2 + c^2 + d^2)^{\frac{1}{2}}, \\ q^{-1} = \dfrac{K\,q}{(N\,q)^2}. \end{array}\right\} \tag{12.3}$$

The vector part Vq may be regarded as an ordinary vector, i, j, k being unit vectors on the coordinate axes. If $Sq = 0$, q degenerates into a vector, with $Nq = 1$ if it is a unit vector.

Any quaternion q defines a positive number h, a unit vector p, and an angle χ $(0 \leqq \chi < 2\pi)$ by the formula

$$q = h\,(\cos\tfrac{1}{2}\chi + p\sin\tfrac{1}{2}\chi). \tag{12.4}$$

Then $Nq = h$, and

$$q^{-1} = h^{-1}\,(\cos\tfrac{1}{2}\chi - p\sin\tfrac{1}{2}\chi). \tag{12.5}$$

Let r be any vector and q any quaternion. Then the formula

$$r' = q\,r\,q^{-1} \tag{12.6}$$

defines a transformation $r \to r'$ such that $Sr' = Sr = 0$ (so that r' is a vector), and this transformation is in fact a rotation about the directed axis p through an angle χ, where p and χ are defined by (12.4). This may be shown as follows[3].

It is clear that $h (= Nq)$ cancels in (12.6), and we may take $Nq = 1$, writing

$$q = \lambda\,i + \mu\,j + \nu\,k + \varrho, \quad \lambda^2 + \mu^2 + \nu^2 + \varrho^2 = 1. \tag{12.7}$$

Then

$$q^{-1} = -\lambda\,i - \mu\,j - \nu\,k + \varrho, \tag{12.8}$$

and working out (12.6) with

$$r = x\,i + y\,i + z\,k, \quad r' = x'\,i + y'\,j + z'\,k,$$

we find, in matrix notation,

$$\boldsymbol{r}' = \boldsymbol{M}\,\boldsymbol{r} \tag{12.9}$$

where $\boldsymbol{M}$ has precisely the form (10.9). It follows that (12.6) defines a rotation and that the numbers $(\lambda, \mu, \nu, \varrho)$, introduced in (12.7), are in fact the Eulerian

---

[1] Following traditional practice, ordinary type is used here for quaternions.

[2] See L. BRAND: Vector and Tensor Analysis (New York: Wiley 1947), Chap. X for a compact account of quaternions. BRAND defines the norm as qKq. C. J. JOLY, A Manual of Quaternions (London: Macmillan 1905), p. 12, writes $q\,K\,q = (T\,q)^2$, and calls Tq the *tensor* of q; but it seems advisable to avoid the historic word *tensor*, on account of its common use nowadays in a different sense in tensor calculus.

[3] For other proofs, see BRAND, *op. cit.*, p. 417, and JOLY, *op. cit.*, p. 18.

parameters (Sect. 10) of that rotation. Further, putting $h=1$ in (12.4), we have

$$\left.\begin{aligned} &Vq = p\sin\tfrac{1}{2}\chi, \quad (N\,V\,q)^2 = \lambda^2+\mu^2+\nu^2 = \sin^2\tfrac{1}{2}\chi, \\ &S\,q = \varrho = \cos\tfrac{1}{2}\chi, \end{aligned}\right\} \tag{12.10}$$

so that [cf. (10.2)] Vq is the vector $\boldsymbol{V}$ of Sect. 10 (directed axis of rotation), and the angle χ of the quaternion, as defined by (12.4), is the angle of the rotation.

13. Stereographic projection[1] and CAYLEY-KLEIN parameters. Project the sphere $x^2+y^2+z^2=1$ from the point (0, 0, 1) on the plane $z=0$ (stereographic projection); let (X, Y) be the projection of (x, y, z) (Fig. 6). Then, as is easy to see,

$$\left.\begin{aligned} &\frac{x}{X} = \frac{y}{Y} = \frac{1-z}{1}; \\ &X = \frac{x}{1-z}, \\ &Y = \frac{y}{1-z}; \\ &x = \frac{2X}{X^2+Y^2+1}, \\ &y = \frac{2Y}{X^2+Y^2+1}, \\ &z = \frac{X^2+Y^2-1}{X^2+Y^2+1}, \\ &1-z = \frac{2}{X^2+Y^2+1}. \end{aligned}\right\} \tag{13.1}$$

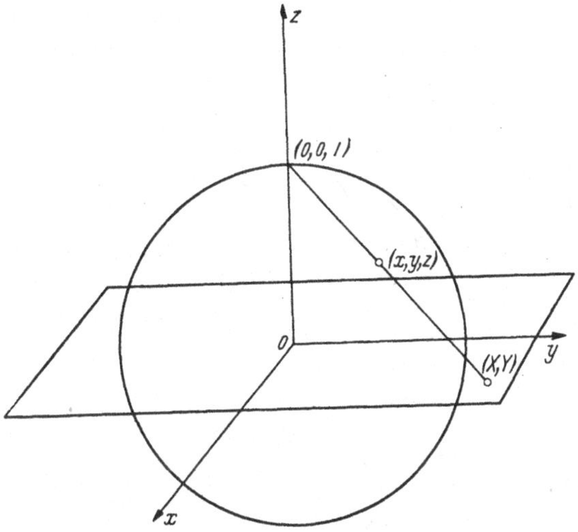

Fig. 6. Stereographic projection.

A simple calculation gives

$$dx^2+dy^2+dz^2 = \frac{4\,(dX^2+dY^2)}{(X^2+Y^2+1)^2} = \frac{4\,dZ\,d\overline{Z}}{(Z\overline{Z}+1)^2}. \tag{13.2}$$

where $Z = X+iY$ and the bar indicates the complex conjugate.

Any transformation $Z\to Z'$ induces a transformation $(x, y, z)\to(x', y', z')$ of the unit sphere into itself, and this will be a rigid transformation if it conserves $dx^2+dy^2+dz^2$, i.e. if

$$\frac{dZ'\,d\overline{Z}'}{(Z'\overline{Z}'+1)^2} = \frac{dZ\,d\overline{Z}}{(Z\overline{Z}+1)^2}. \tag{13.3}$$

It is easy to verify that this condition is satisfied by the transformation

$$Z' = \frac{pZ+q}{-\overline{q}Z+\overline{p}}, \tag{13.4}$$

where p, q are any complex numbers (stereographic parameters) satisfying

$$p\overline{p}+q\overline{q}=1, \tag{13.5}$$

so that they possess three degrees of freedom.

Calculation gives

$$\frac{1}{Z'\overline{Z}'+1} = \frac{(\overline{q}Z-\overline{p})\,(q\overline{Z}-p)}{Z\overline{Z}+1}, \tag{13.6}$$

[1] Cf. H. TIETZ: This Encyclopedia, Vol. II, p. 187.

and hence, by (13.1),

$$\left.\begin{aligned} x'+iy' &= \frac{2Z'}{Z'\bar{Z}'+1} = p^2(x+iy) - q^2(x-iy) - 2pqz, \\ x'-iy' &= \bar{p}^2(x-iy) - \bar{q}^2(x+iy) - 2\overline{pq}z, \\ z' &= \frac{Z'\bar{Z}'-1}{Z'\bar{Z}'+1} = p\bar{q}(x+iy) + \bar{p}q(x-iy) + (p\bar{p}-q\bar{q})z. \end{aligned}\right\} \tag{13.7}$$

Thus we have the transformation $\boldsymbol{r}'=\boldsymbol{M}\boldsymbol{r}$, where the matrix $\boldsymbol{M}$ is

$$\boldsymbol{M} = \begin{pmatrix} \frac{1}{2}(p^2+\bar{p}^2-q^2-\bar{q}^2) & \frac{1}{2}i(p^2-\bar{p}^2+q^2-\bar{q}^2) & -pq-\overline{pq} \\ \frac{1}{2}i(-p^2+\bar{p}^2+q^2-\bar{q}^2) & \frac{1}{2}(p^2+\bar{p}^2+q^2+\bar{q}^2) & i(pq-\overline{pq}) \\ p\bar{q}+\bar{p}q & i(p\bar{q}-\bar{p}q) & p\bar{p}-q\bar{q} \end{pmatrix}. \tag{13.8}$$

The transformation $\boldsymbol{r}\to\boldsymbol{r}'$ appears primarily as a rigid displacement of the unit sphere into itself, but, since it is linear and homogeneous, it gives a rigid rotation of all space.

Certain properties of the rotation are evident from (13.4). Thus $Z=-q/p$ gives $Z'=0$, so that the corresponding point on the unit sphere goes to $(0, 0, -1)$; also $Z=\bar{p}/\bar{q}$ gives $Z'=\infty$, so that the corresponding point goes to $(0, 0, 1)$. The axis of rotation is obtained by putting $Z'=Z$ and solving the quadratic equation so obtained.

By comparing the matrices (10.9) and (13.8), we connect the Eulerian parameters $(\lambda, \mu, \nu, \varrho)$ with the stereographic parameters (p, q)[1]; and we link up with the Eulerian angles $(\vartheta, \varphi, \psi)$ by (11.7), taking $\varepsilon=1$. There is of course an ambiguity of sign in the connection between the Eulerian parameters and the stereographic parameters, because the matrix (13.8) is not changed if we reverse the signs of both p and q. Making a choice of sign, we obtain

$$\left.\begin{aligned} p &= \varrho + i\nu = \cos\tfrac{1}{2}\vartheta\, e^{\frac{1}{2}i(\varphi+\psi)}, \\ q &= i\lambda - \mu = -\sin\tfrac{1}{2}\vartheta\, e^{\frac{1}{2}i(\varphi-\psi)}. \end{aligned}\right\} \tag{13.9}$$

The CAYLEY-KLEIN parameters $(\alpha, \beta, \gamma, \delta)$ are defined by[2]

$$\left.\begin{aligned} \alpha &= p = \varrho + i\nu = \cos\tfrac{1}{2}\vartheta\, e^{\frac{1}{2}i(\varphi+\psi)}, \\ \beta &= -iq = \lambda + i\mu = i\sin\tfrac{1}{2}\vartheta\, e^{\frac{1}{2}i(\varphi-\psi)}, \\ \gamma &= -i\bar{q} = -\lambda + i\mu = i\sin\tfrac{1}{2}\vartheta\, e^{-\frac{1}{2}i(\varphi-\psi)}, \\ \delta &= \bar{p} = \varrho - i\nu = \cos\tfrac{1}{2}\vartheta\, e^{-\frac{1}{2}i(\varphi+\psi)}. \end{aligned}\right\} \tag{13.10}$$

By virtue of (13.5) they satisfy the unimodular condition

$$\begin{vmatrix} \alpha & \beta \\ \gamma & \delta \end{vmatrix} = \alpha\delta - \beta\gamma = 1, \tag{13.11}$$

and in terms of them the matrix (10.9) or (13.8) reads

$$\boldsymbol{M} = \begin{pmatrix} \frac{1}{2}(\alpha^2+\beta^2+\gamma^2+\delta^2) & \frac{1}{2}i(\alpha^2-\beta^2+\gamma^2-\delta^2) & -i(\alpha\beta+\gamma\delta) \\ \frac{1}{2}i(-\alpha^2-\beta^2+\gamma^2+\delta^2) & \frac{1}{2}(\alpha^2-\beta^2-\gamma^2+\delta^2) & -\alpha\beta+\gamma\delta \\ i(\alpha\gamma+\beta\delta) & -\alpha\gamma+\beta\delta & \alpha\delta+\beta\gamma \end{pmatrix}. \tag{13.12}$$

---

[1] For relevant group theory, see MURNAGHAN, p. 328 of op. cit. in Sect. 10.
[2] The notation follows WHITTAKER [28], p. 12.

14. Spin matrices of PAULI. The 2×2 spin matrices of PAULI and the unit matrix are defined as follows[1]:

$$\sigma_1=\begin{pmatrix}0&1\\1&0\end{pmatrix},\quad \sigma_2=\begin{pmatrix}0&-i\\i&0\end{pmatrix},\quad \sigma_3=\begin{pmatrix}1&0\\0&-1\end{pmatrix},\quad \mathbf{1}=\begin{pmatrix}1&0\\0&1\end{pmatrix}. \tag{14.1}$$

Note that the σ's are Hermitian ($\sigma^\dagger=\sigma$) and unitary ($\sigma\sigma^\dagger=\mathbf{1}$), and that the trace[2] of each is zero.

The following multiplication table may be verified directly:

$$\left.\begin{array}{l}\sigma_1^2=\sigma_2^2=\sigma_3^2=\mathbf{1},\\ \sigma_2\sigma_3=-\sigma_3\sigma_2=i\sigma_1,\quad \sigma_3\sigma_1=-\sigma_1\sigma_3=i\sigma_2,\quad \sigma_1\sigma_2=-\sigma_2\sigma_1=i\sigma_3.\end{array}\right\} \tag{14.2}$$

Comparison with (12.2) shows that the three matrices $\tau_\varrho=-i\sigma_\varrho$ ($\varrho=1, 2, 3$) obey the quaternionic rules[3].

Any 2×2 matrix can be expressed as a linear combination of $\sigma_1, \sigma_2, \sigma_3$, and $\mathbf{1}$; for, no matter what values are given to a, b, c, d, we have

$$\begin{pmatrix}a&b\\c&d\end{pmatrix}=\tfrac{1}{2}(b+c)\,\sigma_1+\tfrac{1}{2}i(b-c)\,\sigma_2+\tfrac{1}{2}(a-d)\,\sigma_3+\tfrac{1}{2}(a+d)\,\mathbf{1}. \tag{14.3}$$

If the trace of the given matrix is zero ($a+d=0$), the last term disappears. If, further, the matrix is Hermitian (a, d real, $c=\bar{b}$), then the coefficients of the σ's are real.

The PAULI matrices are linked with three-dimensional geometry by the identity

$$\boldsymbol{P}=\begin{pmatrix}z&x-iy\\x+iy&-z\end{pmatrix}=x\sigma_1+y\sigma_2+z\sigma_3; \tag{14.4}$$

any point (x,y,z) defines such a matrix, and conversely every Hermitian 2×2 matrix with zero trace defines a point (x,y,z).

Consider now any 2×2 matrix $\boldsymbol{U}$ which is unitary ($\boldsymbol{U}\boldsymbol{U}^\dagger=\boldsymbol{U}^\dagger\boldsymbol{U}=\mathbf{1}$), but not in general Hermitian ($\boldsymbol{U}^\dagger\neq\boldsymbol{U}$). With $\boldsymbol{P}$ as in (14.4), the formula

$$\boldsymbol{P}'=\boldsymbol{U}\boldsymbol{P}\boldsymbol{U}^\dagger \tag{14.5}$$

defines a 2×2 matrix $\boldsymbol{P}'$. It is easy to show that $\boldsymbol{P}'$ is Hermitian, with zero trace, and therefore (14.5) gives a transformation $(x,y,z)\to(x',y',z')$ of space into itself. Since $\boldsymbol{U}\boldsymbol{U}^\dagger=\mathbf{1}$, we have $\det\boldsymbol{U}\det\boldsymbol{U}^\dagger=1$, and hence $\det\boldsymbol{P}'=\det\boldsymbol{P}$; therefore

$$x'^2+y'^2+z'^2=x^2+y^2+z^2, \tag{14.6}$$

so that the unit sphere is transformed into itself. Further, (14.5) gives $d\boldsymbol{P}'=\boldsymbol{U}d\boldsymbol{P}\boldsymbol{U}^\dagger$, and hence, as above,

$$dx'^2+dy'^2+dz'^2=dx^2+dy^2+dz^2; \tag{14.7}$$

the transformation is a rigid rotation about the origin[4].

---

1 The notation follows GOLDSTEIN [7], p. 116, and W. PAULI, this Encyclopedia, Vol. V, p. 109. MURNAGHAN (p. 296 of op. cit. in Sect. 10) and H. TIETZ (this Encyclopedia, Vol. II, p. 191) define the second matrix with the opposite sign.

2 The trace (or spur) of a matrix is the sum of the diagonal elements.

3 In Sect. 12 HAMILTON's notation (i, j, k) is used for the quaternionic units, but when $\sqrt{-1}$ is involved in the ordinary sense, as in (14.1), confusion may be avoided by changing to (I, J, K) or (e_1, e_2, e_3).

4 It will become clear in Sect. 15 that the transformation is proper. For a proof, see MURNAGHAN (op. cit. in Sect. 10, p. 298), where the group theory connected with the PAULI matrices is discussed.

Let χ be real; write $c=\cos\frac{1}{2}\chi$, $s=\sin\frac{1}{2}\chi$. Then the three matrices

$$\boldsymbol{U}_1(\chi)=c\mathbf{1}-is\boldsymbol{\sigma}_1,\quad \boldsymbol{U}_2(\chi)=c\mathbf{1}-is\boldsymbol{\sigma}_2,\quad \boldsymbol{U}_3(\chi)=c\mathbf{1}-is\boldsymbol{\sigma}_3 \tag{14.8}$$

are easily seen to be unitary, by (14.2). Consider the transformation

$$\boldsymbol{P}'=\boldsymbol{U}_3(\chi)\,\boldsymbol{P}\,\boldsymbol{U}_3^\dagger(\chi)=(c\mathbf{1}-is\boldsymbol{\sigma}_3)\,(x\boldsymbol{\sigma}_1+y\boldsymbol{\sigma}_2+z\boldsymbol{\sigma}_3)\,(c\mathbf{1}+is\boldsymbol{\sigma}_3). \tag{14.9}$$

On working out the algebra, using (14.2), we find that it gives

$$x'=x\cos\chi-y\sin\chi,\qquad y'=x\sin\chi+y\cos\chi,\quad z'=z, \tag{14.10}$$

which is the transformation (relative to axes fixed in space) corresponding to a rotation through angle χ about the z-axis. By the symmetry of (14.2) and (14.8), $\boldsymbol{U}_1(\chi)$ and $\boldsymbol{U}_2(\chi)$ give similar rotations about the x-axis and the y-axis respectively.

15. Connections between the PAULI matrices and the other ways of representing rotations. Let us first connect the PAULI matrices with the Eulerian angles. The three matrices in (11.6) correspond respectively to rotations as follows: φ about the z-axis, ϑ about the new y-axis, and ψ about the new z-axis. These operations correspond to $\boldsymbol{U}_3(\varphi)$, $\boldsymbol{U}_2(\vartheta)$, $\boldsymbol{U}_3(\psi)$ respectively, and so the transformation $\boldsymbol{r}'=\boldsymbol{M}\boldsymbol{r}$ of (11.6) is the same as

$$\boldsymbol{P}'=\boldsymbol{T}\boldsymbol{P}\boldsymbol{T}^\dagger, \tag{15.1}$$

where $\boldsymbol{P}$ is as in (14.4) and[1]

$$\left.\begin{aligned}\boldsymbol{T}&=\boldsymbol{U}_3(\varphi)\,\boldsymbol{U}_2(\vartheta)\,\boldsymbol{U}_3(\psi)\\&=\begin{pmatrix}e^{-\frac{1}{2}i\varphi}&0\\0&e^{\frac{1}{2}i\varphi}\end{pmatrix}\begin{pmatrix}\cos\frac{1}{2}\vartheta&-\sin\frac{1}{2}\vartheta\\\sin\frac{1}{2}\vartheta&\cos\frac{1}{2}\vartheta\end{pmatrix}\begin{pmatrix}e^{-\frac{1}{2}i\psi}&0\\0&e^{\frac{1}{2}i\psi}\end{pmatrix}\\&=\begin{pmatrix}\cos\frac{1}{2}\vartheta\,e^{-\frac{1}{2}i(\varphi+\psi)}&-\sin\frac{1}{2}\vartheta\,e^{-\frac{1}{2}i(\varphi-\psi)}\\\sin\frac{1}{2}\vartheta\,e^{\frac{1}{2}i(\varphi-\psi)}&\cos\frac{1}{2}\vartheta\,e^{\frac{1}{2}i(\varphi+\psi)}\end{pmatrix}.\end{aligned}\right\} \tag{15.2}$$

In terms of the stereographic parameters, we have, by (13.9)

$$\boldsymbol{T}=\begin{pmatrix}\bar p&\bar q\\-q&p\end{pmatrix}, \tag{15.3}$$

and so the transformation (15.1) may be written

$$\begin{pmatrix}z'&x'-iy'\\x'+iy'&z'\end{pmatrix}=\begin{pmatrix}\bar p&\bar q\\-q&p\end{pmatrix}\begin{pmatrix}z&x-iy\\x+iy&z\end{pmatrix}\begin{pmatrix}p&-\bar q\\q&\bar p\end{pmatrix}, \tag{15.4}$$

$$p\bar p+q\bar q=1.$$

The unitary character of $\boldsymbol{T}$ may be verified immediately.

In terms of the Eulerian parameters, we have, by (13.9),

$$\boldsymbol{T}=\begin{pmatrix}\varrho-i\nu&-i\lambda-\mu\\-i\lambda+\mu&\varrho+i\nu\end{pmatrix}, \tag{15.5}$$

or

$$\boldsymbol{T}=-i\lambda\boldsymbol{\sigma}_1-i\mu\boldsymbol{\sigma}_2-i\nu\boldsymbol{\sigma}_3+\varrho\mathbf{1}, \tag{15.6}$$

[1] The matrix $\boldsymbol{T}$ differs from the corresponding matrix given by GOLDSTEIN [7], p. 116, which has a factor i in two elements. This is because our rotation (11.2) is about the y-axis, whereas in GOLDSTEIN's definition of the Eulerian angles his second rotation is about the x-axis. As a result $\boldsymbol{\sigma}_1$ appears in his work where $\boldsymbol{\sigma}_2$ appears here.

so that the transformation (15.4) may be written

$$\left.\begin{aligned} x'\sigma_1 + y'\sigma_2 + z'\sigma_3 = (-i\lambda\sigma_1 - i\mu\sigma_2 - i\nu\sigma_3 + \varrho 1)\times \\ \times(x\sigma_1 + y\sigma_2 + z\sigma_3)(i\lambda\sigma_1 + i\mu\sigma_2 + i\nu\sigma_3 + \varrho 1),\ \lambda^2+\mu^2+\nu^2+\varrho^2 = 1.\end{aligned}\right\} \tag{15.7}$$

This is equivalent to the quaternionic formula (12.6):

$$\left.\begin{aligned} x'i + y'j + z'k = (\lambda i + \mu j + \nu k + \varrho)(xi + yj + zk)(-\lambda i - \mu j - \nu k + \varrho), \\ \lambda^2+\mu^2+\nu^2+\varrho^2 = 1.\end{aligned}\right\} \tag{15.8}$$

16. Infinitesimal displacements. An infinitesimal displacement of a rigid body is reducible to an infinitesimal translation and an infinitesimal rotation. On account of their infinitesimal character (all differentials of order higher than the first being neglected), the translation and the rotation commute. Indeed, any two infinitesimal displacements commute, and this makes the discussion of infinitesimal displacements comparatively simple.

To discuss an infinitesimal rotation, we take the rotation-angle χ in (10.2) to be infinitesimal, so that $V = \frac{1}{2}\chi$ and $\varrho = 1$. Let $\boldsymbol{\chi}$ be an infinitesimal vector with magnitude χ, lying on the directed axis of rotation. Then $\boldsymbol{\chi} = 2\boldsymbol{V}$, and (10.8) gives, for the infinitesimal displacement resulting from the rotation $\boldsymbol{\chi}$,

$$\boldsymbol{r}' - \boldsymbol{r} = \boldsymbol{\chi}\times\boldsymbol{r}. \tag{16.1}$$

The vector character of this equation tells us that we can compound infinitesimal rotations by adding the corresponding vectors $\boldsymbol{\chi}$ according to the parallelogram law.

In matrix form (16.1) reads

$$\boldsymbol{r}' - \boldsymbol{r} = \boldsymbol{M}\boldsymbol{r}, \qquad \boldsymbol{M} = \begin{pmatrix} 0 & -\chi_3 & \chi_2 \\ \chi_3 & 0 & -\chi_1 \\ -\chi_2 & \chi_1 & 0 \end{pmatrix}, \tag{16.2}$$

where χ_1, χ_2, χ_3 are the components of $\boldsymbol{\chi}$. The matrix is skew-symmetric; it is easy to prove directly that any orthogonal matrix which differs infinitesimally from the unit matrix, differs from it by a skew-symmetric matrix.

II. Kinematics.

17. Frames of reference. Velocity of a particle. Let S_0 be absolute fixed space (cf. Sect. 4), and let S be any rigid body, fixed or moving. S constitutes a *frame of reference*; if $Oxyz$ are any rectangular axes fixed in S, then to any event there corresponds a tetrad of numbers (x, y, z, t), t being the absolute Newtonian time (cf. Sect. 4). Kinematics deals only with relative motions, and for kinematical purposes S is as good as S_0; it is only in dynamics proper (i.e. when forces producing motion are considered) that the distinction between S and S_0 appears.

Let P be a moving point or particle. Its *velocity* relative to S is the vector

$$\boldsymbol{v} = \dot{\boldsymbol{r}} = (\dot{x}, \dot{y}, \dot{z}), \tag{17.1}$$

the dot indicating differentiation with respect to t. The velocity is tangent to the trajectory of P and of magnitude $\dot{s}$, where s is arc-length on the trajectory. The magnitude of the velocity is called *speed*.

Let x^ϱ be curvilinear coordinates in S; then the line-element of S is of the form $ds^2 = g_{\varrho\sigma}\,dx^\varrho\,dx^\sigma$, suffixes taking the values 1, 2, 3, with summation understood for a repeated suffix. The contravariant and covariant velocity vectors are respectively

$$v^\varrho = \dot{x}^\varrho, \quad v_\varrho = g_{\varrho\sigma}\dot{x}^\sigma. \tag{17.2}$$

The *physical component* of velocity in the direction of a unit vector with contravariant components λ^ϱ (and covariant components λ_ϱ) is the orthogonal projection of (17.1) on that direction, i.e. the invariant[1]

$$v(\lambda) = v_\varrho \lambda^\varrho = v^\varrho \lambda_\varrho. \tag{17.3}$$

In cylindrical coordinates (r, φ, z), the physical components of velocity along the coordinate lines are

$$(\dot{r}, r\dot{\varphi}, \dot{z}). \tag{17.4}$$

In spherical polar coordinates (r, ϑ, φ), the physical components of velocity are

$$(\dot{r}, r\dot{\vartheta}, r\sin\vartheta\,\dot{\varphi}). \tag{17.5}$$

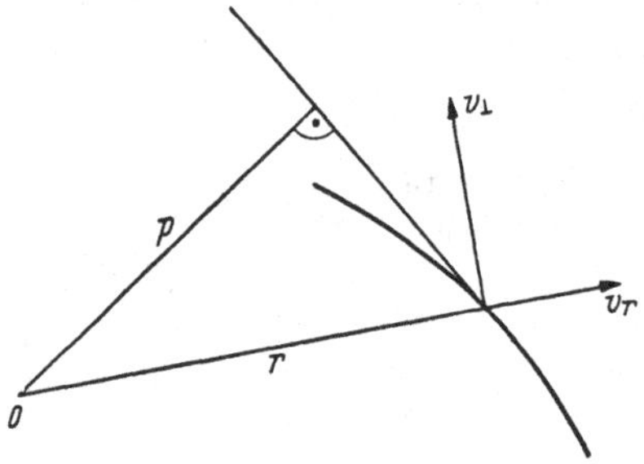

Fig. 7. (p, r) coordinates.

In (p, r) coordinates in a plane (Fig. 7), the components v_r along the radius vector and $v_\perp$ perpendicular to it are

$$v_r = \dot{r}, \quad v_\perp = -\frac{p\dot{r}}{\sqrt{r^2 - p^2}}. \tag{17.6}$$

18. Acceleration of a particle. Hodograph. The *acceleration* $\boldsymbol{a}$ of a point or particle is the rate of change of velocity:

$$\boldsymbol{a} = \dot{\boldsymbol{v}} = \ddot{\boldsymbol{r}} = (\ddot{x}, \ddot{y}, \ddot{z}). \tag{18.1}$$

Resolution along $\boldsymbol{i}$, the unit tangent vector to the trajectory, and $\boldsymbol{j}$, the unit vector along its principal normal, gives

$$\boldsymbol{a} = \dot{v}\boldsymbol{i} + \frac{v^2}{\varrho}\boldsymbol{j} = v\frac{dv}{ds}\boldsymbol{i} + \frac{v^2}{\varrho}\boldsymbol{j}, \tag{18.2}$$

where v is the speed and ϱ is the radius of curvature of the trajectory.

For curvilinear coordinates x^ϱ with $ds^2 = g_{\varrho\sigma}dx^\varrho dx^\sigma$ as in Sect. 17, the contravariant acceleration vector is[2]

$$a^\varrho = \ddot{x}^\varrho + \left\{{\varrho \atop \mu\nu}\right\}\dot{x}^\mu \dot{x}^\nu, \tag{18.3}$$

where the CHRISTOFFEL symbol is defined by

$$\left\{{\varrho \atop \mu\nu}\right\} = g^{\varrho\sigma}[\mu\nu, \sigma], \quad 2[\mu\nu, \sigma] = \frac{\partial g_{\mu\sigma}}{\partial x^\nu} + \frac{\partial g_{\nu\sigma}}{\partial x^\mu} - \frac{\partial g_{\mu\nu}}{\partial x^\sigma}, \quad g^{\varrho\mu}g_{\sigma\mu} = \delta^\varrho_\sigma. \tag{18.4}$$

Here δ^ϱ_σ is the KRONECKER delta ($=1$ if $\varrho=\sigma$, $=0$ if $\varrho\neq\sigma$). It is usually easier to calculate the covariant acceleration vector:

$$a_\varrho = \frac{d}{dt}\frac{\partial T}{\partial \dot{x}^\varrho} - \frac{\partial T}{\partial x^\varrho}, \quad T = \tfrac{1}{2}g_{\varrho\sigma}\dot{x}^\varrho\dot{x}^\sigma. \tag{18.5}$$

The physical component[3] of acceleration in the direction of a unit vector λ^ϱ is

$$a(\lambda) = a_\varrho \lambda^\varrho = a^\varrho \lambda_\varrho. \tag{18.6}$$

---

[1] If the coordinates are orthogonal ($g_{\varrho\sigma} = 0$ for $\varrho \neq \sigma$), this definition is satisfactory. But when the coordinates are oblique, it is sometimes convenient to define physical components by oblique resolution, and care is needed to avoid confusion of ideas; cf. C. TRUESDELL: Z. angew. Math. Mech. **33**, 345 (1953); **34**, 69 (1954).

[2] Cf. A. J. McCONNELL: Applications of the Absolute Differential Calculus, Chap. 17. London and Glasgow: Blackie 1931; I. S. SOKOLNIKOFF: Tensor Analysis, Chap. 4. New York: Wiley 1951; J. L. SYNGE and A. SCHILD: Tensor Calculus, Chap. 5. Toronto: University of Toronto Press 1952.

[3] See footnote in Sect. 17 with reference to oblique coordinates.

For cylindrical coordinates (r, φ, z) we have

$$ds^2 = dr^2 + r^2 d\varphi^2 + dz^2, \tag{18.7}$$

and the physical components of acceleration along the coordinate lines are

$$a_r = \ddot{r} - r\dot{\varphi}^2, \quad a_\varphi = \frac{1}{r}\frac{d}{dt}(r^2\dot{\varphi}) \quad a_z = \ddot{z}. \tag{18.8}$$

For spherical polar coordinates (r, ϑ, φ) we have

$$ds^2 = dr^2 + r^2 d\vartheta^2 + r^2 \sin^2\vartheta\, d\varphi^2, \tag{18.9}$$

and the physical components of acceleration along the coordinate lines are[1]

$$\left.\begin{aligned} a_r &= \ddot{r} - r\dot{\vartheta}^2 - r\sin^2\vartheta\,\dot{\varphi}^2, \\ a_\vartheta &= \frac{1}{r}\frac{d}{dt}(r^2\dot{\vartheta}) - r\sin\vartheta\cos\vartheta\,\dot{\varphi}^2, \\ a_\varphi &= \frac{1}{r\sin\vartheta}\frac{d}{dt}(r^2\sin^2\vartheta\,\dot{\varphi}). \end{aligned}\right\} \tag{18.10}$$

The curve with equation $\boldsymbol{r}' = \boldsymbol{v}(t)$, where $\boldsymbol{v}(t)$ is the velocity of a moving point, is called the *hodograph* of the motion. For constant velocity, the hodograph is a single point; for constant acceleration, it is a straight line; for motion with $\boldsymbol{a} = k\boldsymbol{r}/r^3$ (inverse square law), it is a circle[2].

19. Angular velocity of a rigid body. Let S_0 be a rigid body relative to which a second rigid body S moves. For simplicity of description we may think of S_0 as fixed in absolute space. Having selected a particle P_0 of S as base-point, we can describe any infinitesimal displacement of S by giving the displacement of P_0 and the rotation about P_0. The latter is an infinitesimal vector χ, as in (16.1). We define the *angular velocity* ω of S by the equation

$$\chi = \omega\, dt, \tag{19.1}$$

dt being the infinitesimal time-interval during which the displacement takes place. The velocity $\boldsymbol{v}$ of any particle P of S is then

$$\boldsymbol{v} = \boldsymbol{v}_0 + \omega \times \boldsymbol{r}, \tag{19.2}$$

where $\boldsymbol{v}_0$ is the velocity of P_0 and $\boldsymbol{r}$ the position vector of P relative to P_0.

The vectors $\boldsymbol{v}_0$, ω and $\boldsymbol{r}$ may be described by giving their components at the instant t along any orthonormal triad $\boldsymbol{T}$, which may be fixed in S_0, or in S, or may be moving relative to both S and S_0. It is usually most convenient to fix $\boldsymbol{T}$ in S, but in cases of symmetry it may be better to fix one vector of $\boldsymbol{T}$ in S and confine another to a plane fixed in S_0, as we shall see later (Sect. 56).

Since $\boldsymbol{v}_0$ is separated from ω in (19.2), we can discuss the angular velocity of S as if P_0 were fixed. To express ω in terms of the Eulerian angles (Sect. 11), we may take $\boldsymbol{T}$ to be the triad $(\boldsymbol{i}, \boldsymbol{j}, \boldsymbol{k})$ of Fig. 5, fixed in S. The displacement in time dt can be produced by infinitesimal rotations $d\vartheta$, $d\varphi$, $d\psi$ about $\boldsymbol{J}_1$, $\boldsymbol{K}$, $\boldsymbol{k}$, respectively, and so

$$\omega\, dt = \boldsymbol{J}_1 d\vartheta + \boldsymbol{K} d\varphi + \boldsymbol{k}\, d\psi. \tag{19.3}$$

---

[1] These and other accelerations may also be calculated by means of moving axes; cf. WHITTAKER [28], p. 18, where a number of special coordinate systems are discussed.

[2] Cf. MACMILLAN [17] I, p. 285.

On resolving $\boldsymbol{J}_1, \boldsymbol{K}, \boldsymbol{k}$ along $(\boldsymbol{i}, \boldsymbol{j}, \boldsymbol{k})$, we obtain

$$\left.\begin{aligned} \boldsymbol{\omega} &= \omega_1 \boldsymbol{i} + \omega_2 \boldsymbol{j} + \omega_3 \boldsymbol{k}, \\ \omega_1 &= \dot{\vartheta} \sin\psi - \dot{\varphi} \sin\vartheta \cos\psi, \\ \omega_2 &= \dot{\vartheta} \cos\psi + \dot{\varphi} \sin\vartheta \sin\psi, \\ \omega_3 &= \dot{\psi} + \dot{\varphi} \cos\vartheta. \end{aligned}\right\} \tag{19.4}$$

Similarly we can resolve on $(\boldsymbol{I}, \boldsymbol{J}, \boldsymbol{K})$, fixed in S_0 (Fig. 5), obtaining

$$\left.\begin{aligned} \boldsymbol{\omega} &= \Omega_1 \boldsymbol{I} + \Omega_2 \boldsymbol{J} + \Omega_3 \boldsymbol{K}, \\ \Omega_1 &= -\dot{\vartheta} \sin\varphi + \dot{\psi} \sin\vartheta \cos\varphi, \\ \Omega_2 &= \dot{\vartheta} \cos\varphi + \dot{\psi} \sin\vartheta \sin\varphi, \\ \Omega_3 &= \dot{\varphi} + \dot{\psi} \cos\vartheta. \end{aligned}\right\} \tag{19.5}$$

To discuss angular velocity in terms of quaternions and the Eulerian parameters, we let the quaternionic units (i, j, k) correspond to axes fixed in S_0. The position r of a particle at time t is given, as in (12.6), by

$$r = q r_0 q^{-1}, \tag{19.6}$$

where $r_0 = x_0 i + y_0 j + z_0 k$, the initial position of the particle, and

$$\left.\begin{aligned} q &= \lambda i + \mu j + \nu k + \varrho, \\ q^{-1} &= -\lambda i - \mu j - \nu k + \varrho, \\ \lambda^2 &+ \mu^2 + \nu^2 + \varrho^2 = 1. \end{aligned}\right\} \tag{19.7}$$

From (19.6) we have

$$\left.\begin{aligned} \dot{r} &= \dot{q} r_0 q^{-1} + q r_0 \frac{d}{dt}(q^{-1}) \\ &= \dot{q} q^{-1} r + r q \frac{d}{dt}(q^{-1}) \\ &= \dot{q} q^{-1} r - r \dot{q} q^{-1}. \end{aligned}\right\} \tag{19.8}$$

We note that

$$\dot{q} q^{-1} = (\dot{\lambda} i + \dot{\mu} j + \dot{\nu} k + \dot{\varrho})(-\lambda i - \mu j - \nu k + \varrho), \tag{19.9}$$

and that

$$S(\dot{q} q^{-1}) = S(q^{-1} \dot{q}) = \lambda \dot{\lambda} + \mu \dot{\mu} + \nu \dot{\nu} + \varrho \dot{\varrho} = 0, \tag{19.10}$$

so that these products are vectors.

If

$$\Omega = \Omega_1 i + \Omega_2 j + \Omega_3 k \tag{19.11}$$

is the angular velocity, resolved along the fixed axes, then, by (19.2) with $v_0 = 0$,

$$\dot{r} = \tfrac{1}{2}(\Omega r - r \Omega). \tag{19.12}$$

Comparing this with (19.8), we have

$$a r - r a = 0, \tag{19.13}$$

where

$$a = \Omega - 2 \dot{q} q^{-1}. \tag{19.14}$$

Since a is a vector and r an arbitrary vector, (19.13) implies $a = 0$, and therefore the angular velocity is

$$\Omega = 2 \dot{q} q^{-1}; \tag{19.15}$$

evaluating this product by (19.9), we find the components of angular velocity, relative to the fixed axes (i, j, k):

$$\left.\begin{aligned}\Omega_1 &= 2(\dot{\lambda}\varrho - \lambda\dot{\varrho} - \dot{\mu}\nu + \mu\dot{\nu}),\\ \Omega_2 &= 2(\dot{\mu}\varrho - \mu\dot{\varrho} - \dot{\nu}\lambda + \nu\dot{\lambda}),\\ \Omega_3 &= 2(\dot{\nu}\varrho - \nu\dot{\varrho} - \dot{\lambda}\mu + \lambda\dot{\mu}).\end{aligned}\right\} \tag{19.16}$$

By (19.15) we have

$$q^{-1}\Omega\, q = 2q^{-1}\dot{q} = \omega_1 i + \omega_2 j + \omega_3 k, \tag{19.17}$$

$\omega_1, \omega_2, \omega_3$ being defined by this equation; hence

$$\Omega = \omega_1 q i q^{-1} + \omega_2 q j q^{-1} + \omega_3 q k q^{-1}, \tag{19.18}$$

and we recognize $\omega_1, \omega_2, \omega_3$ as the components of angular velocity on that triad which initially coincided with (i, j, k). Carrying out the calculation in (19.17), we have the components of angular velocity on the moving axes[1]:

$$\left.\begin{aligned}\omega_1 &= 2(\dot{\lambda}\varrho - \lambda\dot{\varrho} + \dot{\mu}\nu - \mu\dot{\nu}),\\ \omega_2 &= 2(\dot{\mu}\varrho - \mu\dot{\varrho} + \dot{\nu}\lambda - \nu\dot{\lambda}),\\ \omega_3 &= 2(\dot{\nu}\varrho - \nu\dot{\varrho} + \dot{\lambda}\mu - \lambda\dot{\mu}).\end{aligned}\right\} \tag{19.19}$$

20. Moving axes. Absolute and relative rates of change of a vector. The theory of moving axes is often found difficult and confusing on account of the demands it makes on one's power to visualize bodies in motion. If one gets confused, the best plan is to think in terms of infinitesimal displacements, resolving the actual displacement which occurs in time dt into a set of elementary displacements, each due to a different cause. There is no question of the order in which these causes are applied, because infinitesimal displacements commute with one another. For brevity, such an analysis into infinitesimal causes is omitted in the derivation of the formulae which follow.

Let $(\boldsymbol{i}, \boldsymbol{j}, \boldsymbol{k})$ be an orthonormal triad, rotating with angular velocity $\boldsymbol{\omega}$. Let $\boldsymbol{V}$ be a vector, with components (V_1, V_2, V_3) on the rotating triad, so that

$$\boldsymbol{V} = V_1\boldsymbol{i} + V_2\boldsymbol{j} + V_3\boldsymbol{k}. \tag{20.1}$$

The rates of change of the vectors $(\boldsymbol{i}, \boldsymbol{j}, \boldsymbol{k})$ are the same whether the origin of the triad is fixed or moving; they are determined solely by the angular velocity $\boldsymbol{\omega}$. If the origin is fixed, these rates of change are the velocities of points with position vectors $(\boldsymbol{i}, \boldsymbol{j}, \boldsymbol{k})$, and so, by (19.2) we have

$$\frac{d\boldsymbol{i}}{dt} = \boldsymbol{\omega}\times\boldsymbol{i}, \quad \frac{d\boldsymbol{j}}{dt} = \boldsymbol{\omega}\times\boldsymbol{j}, \quad \frac{d\boldsymbol{k}}{dt} = \boldsymbol{\omega}\times\boldsymbol{k}. \tag{20.2}$$

Differentiating (20.1), we see that the *absolute rate of change*[2] of $\boldsymbol{V}$ is

$$\frac{d\boldsymbol{V}}{dt} = \frac{\delta\boldsymbol{V}}{\delta t} + \boldsymbol{\omega}\times\boldsymbol{V}, \tag{20.3}$$

where

$$\frac{\delta\boldsymbol{V}}{\delta t} = \frac{dV_1}{dt}\boldsymbol{i} + \frac{dV_2}{dt}\boldsymbol{j} + \frac{dV_3}{dt}\boldsymbol{k}; \tag{20.4}$$

this is the *relative* rate of change of $\boldsymbol{V}$.

[1] See WHITTAKER [28], p. 16, for an alternative derivation

[2] WHITTAKER [28], p. 17, calls it the *time-flux*.

If, in particular, $\boldsymbol{V}=\boldsymbol{\omega}$, we have

$$\frac{d\boldsymbol{\omega}}{dt}=\frac{\delta\boldsymbol{\omega}}{\delta t}, \tag{20.5}$$

the absolute and relative rates agreeing.

For the absolute second derivative we have

$$\frac{d^2\boldsymbol{V}}{dt^2}=\frac{\delta^2\boldsymbol{V}}{\delta t^2}+2\boldsymbol{\omega}\times\frac{\delta\boldsymbol{V}}{\delta t}+\frac{\delta\boldsymbol{\omega}}{\delta t}\times\boldsymbol{V}+\boldsymbol{\omega}\times(\boldsymbol{\omega}\times\boldsymbol{V}). \tag{20.6}$$

We shall apply these formulae to the calculation of velocity and acceleration. Let S_0 be fixed absolutely. Let S be a rigid body in general motion, carrying an orthonormal triad $(\boldsymbol{i},\boldsymbol{j},\boldsymbol{k})$, the origin O of this triad having a position vector $\boldsymbol{r}_0(t)$ relative to some origin in S_0. Let P be any moving point or particle, not necessarily attached to S; its absolute position vector $\boldsymbol{r}$, relative to the origin in S_0, may be written

$$\boldsymbol{r}=\boldsymbol{r}_0+\boldsymbol{r}', \tag{20.7}$$

where

$$\boldsymbol{r}'=x\boldsymbol{i}+y\boldsymbol{j}+z\boldsymbol{k}, \tag{20.8}$$

this last being in fact the vector $\overrightarrow{OP}$; (x, y, z) are the coordinates of P as judged by an observer attached to S, which is, in fact, a *moving frame of reference*. By (20.3) the absolute velocity of P is

$$\boldsymbol{v}=\frac{d\boldsymbol{r}}{dt}=\frac{d\boldsymbol{r}_0}{dt}+\frac{d\boldsymbol{r}'}{dt}=\boldsymbol{v}_0+\boldsymbol{v}'+\boldsymbol{\omega}\times\boldsymbol{r}', \tag{20.9}$$

where

$\boldsymbol{v}_0=$ absolute velocity of O,

$\boldsymbol{v}'=\dfrac{\delta\boldsymbol{r}'}{\delta t}=$ relative velocity of P, as observed by an observer carried by S; its components are $(\dot{x},\dot{y},\dot{z})$,

$\boldsymbol{\omega}=$ angular velocity of S.

If P is attached to S, then $\boldsymbol{v}'=0$; for this reason the remaining part $(\boldsymbol{v}_0+\boldsymbol{\omega}\times\boldsymbol{r}')$ is called the *velocity of transport*.

Differentiation of (20.9) gives the absolute acceleration of P in the form

$$\boldsymbol{a}=\boldsymbol{a}_0+\boldsymbol{a}'+\boldsymbol{a}_c+\boldsymbol{a}_t, \tag{20.10}$$

where

$\boldsymbol{a}_0=\dfrac{d\boldsymbol{v}_0}{dt}=$ absolute acceleration of O,

$\boldsymbol{a}'=\dfrac{\delta^2\boldsymbol{r}'}{\delta t^2}=\ddot{x}\boldsymbol{i}+\ddot{y}\boldsymbol{j}+\ddot{z}\boldsymbol{k}=$ relative acceleration,

$\boldsymbol{a}_c=2\boldsymbol{\omega}\times\boldsymbol{v}'=$ CORIOLIS (or complementary) acceleration, (20.11)

$\boldsymbol{a}_t=\dot{\boldsymbol{\omega}}\times\boldsymbol{r}'+\boldsymbol{\omega}\times(\boldsymbol{\omega}\times\boldsymbol{r}')=\dot{\boldsymbol{\omega}}\times\boldsymbol{r}'+\boldsymbol{\omega}(\boldsymbol{\omega}\cdot\boldsymbol{r}')-\boldsymbol{r}'\omega^2.$ (20.12)

Here $\dot{\boldsymbol{\omega}}=d\boldsymbol{\omega}/dt=\delta\boldsymbol{\omega}/\delta t$.

If P is attached to S, then $\boldsymbol{a}'=\boldsymbol{a}_c=0$ and

$$\boldsymbol{a}=\boldsymbol{a}_0+\boldsymbol{a}_t; \tag{20.13}$$

this part of (20.10) is called the *acceleration of transport*.

If ω is constant, then
$$\boldsymbol{a}_t = -\boldsymbol{R}\,\omega^2, \tag{20.14}$$
where $\boldsymbol{R}$ is the vector drawn perpendicularly from the axis of ω to P; this may be called *centripetal* acceleration.

III. Mass distributions and force systems.

21. Mass-centres. Moments and products of inertia. The *mass-centre* of a system of particles with masses m_i and position vectors $\boldsymbol{r}_i$ is the point
$$\boldsymbol{r} = \frac{\sum_i m_i \boldsymbol{r}_i}{\sum_i m_i}. \tag{21.1}$$
If the system is rigidly displaced, its mass-centre is carried rigidly with it.

In a uniform gravitational field, the gravitational forces acting on a system of particles are statically equivalent (or equipollent) to a single force acting at the mass-centre. For that reason, the mass-centre is commonly called the *centre of gravity*. The word *barycentre* is also used. In this article the term *mass-centre* will be used throughout.

For a continuum of density ϱ the mass-centre is defined as in (21.1), with summations replaced by integrations:
$$\boldsymbol{r} = \frac{\int \varrho\, \boldsymbol{r}\, d\tau}{\int \varrho\, d\tau} \tag{21.2}$$
This formula applies to distributions of mass over volumes, surfaces, or curves, $d\tau$ being respectively an element of volume, surface, or length, and ϱ being the appropriate density.

If a system S consists of n parts S_i $(i = 1, 2, \ldots, n)$ with masses m_i and mass-centres $\boldsymbol{r}_i$, then the mass-centre of S may be found by replacing each part S_i by a particle of mass m_i at $\boldsymbol{r}_i$, and using the formula (21.1).

The *linear moment* of a particle with respect to the origin is the vector $m\boldsymbol{r}$, and its *quadratic moment* with respect to the axes of coordinates is the matrix or tensor
$$\begin{pmatrix} m x^2 & m x y & m x z \\ m y x & m y^2 & m y z \\ m z x & m z y & m z^2 \end{pmatrix}. \tag{21.3}$$
The linear and quadratic moments of a system are given by summation or integration.

Moments and products of inertia are closely connected with the quadratic moment. The *moment of inertia* of a particle P of mass m about a line L is $m p^2$, where p is the perpendicular distance of P from L. Its *product of inertia* with respect to a pair of perpendicular planes is $m p q$, where p, q are the distances of P from the planes, taken with appropriate signs. The moments and products of inertia of systems are found by summation or integration[1]. Thus, for a discrete system of particles, the moments of inertia with respect to the coordinate axes $Oxyz$ are
$$\left.\begin{aligned} A = \sum_i m_i (y_i^2 + z_i^2), \quad B = \sum_i m_i (z_i^2 + x_i^2), \\ C = \sum_i m_i (x_i^2 + y_i^2), \end{aligned}\right\} \tag{21.4}$$

[1] If a system of total mass m has a moment of inertia I about a line L, then the *radius of gyration* k about L is defined by $m k^2 = I$.

and the products of inertia with respect to the coordinate planes are

$$F = \sum_i m_i y_i z_i, \qquad G = \sum_i m_i z_i x_i, \qquad H = \sum_i m_i x_i y_i. \tag{21.5}$$

If $\boldsymbol{V} = (l, m, n)$ is a unit vector through the origin, then the moment of inertia about $\boldsymbol{V}$ is, by the definition,

$$\left.\begin{aligned} I &= \sum_i m_i p_i^2 = \sum_i m_i |\boldsymbol{r}_i \times \boldsymbol{V}|^2 \\ &= A\,l^2 + B\,m^2 + C\,n^2 - 2F\,m\,n - 2G\,n\,l - 2H\,l\,m. \end{aligned}\right\} \tag{21.6}$$

Since I is a quantity independent of the choice of directions of coordinate axes, the elements of the symmetric inertia matrix

$$\boldsymbol{I} = \begin{pmatrix} A & -H & -G \\ -H & B & -F \\ -G & -F & C \end{pmatrix} \tag{21.7}$$

are the components of a tensor of the second order.

For a continuous distribution

$$\left.\begin{aligned} A &= \int \varrho (y^2 + z^2)\, d\tau, & B &= \int \varrho (z^2 + x^2)\, d\tau, & C &= \int \varrho (x^2 + y^2)\, d\tau, \\ F &= \int \varrho\, y\, z\, d\tau, & G &= \int \varrho\, z\, x\, d\tau, & H &= \int \varrho\, x\, y\, d\tau. \end{aligned}\right\} \tag{21.8}$$

22. Theorem of parallel axes. Principal axes of inertia. Let L be any line and L_0 a parallel line through the mass-centre of a system. The *theorem of parallel axes* states that

$$I = I_0 + m\,p^2, \tag{22.1}$$

where I, I_0 are the moments of inertia about L, L_0, respectively, m is the total mass of the system, and p the perpendicular distance between L and L_0. This is easy to prove, as is also a similar theorem for products of inertia.

From (22.1) it follows that the moment of inertia about a line L_0 passing through the mass-centre is less than that about any parallel line L.

Principal axes are defined in terms of stationary values of the moment of inertia. To each unit vector $\boldsymbol{V}$ drawn through the origin (which may be any point) there corresponds a moment of inertia I, as in (21.6). The direction cosines (l, m, n) of a vector $\boldsymbol{V}$ for which I has a stationary value satisfy

$$\left.\begin{aligned} A\,l - H\,m - G\,n &= I\,l, \\ -H\,l + B\,m - F\,n &= I\,m, \\ -G\,l - F\,m + C\,n &= I\,n, \end{aligned}\right\} \tag{22.2}$$

where I is the stationary value. The three *principal* moments of inertia at the origin are these stationary values, which are the roots of the cubic determinantal equation

$$\begin{vmatrix} A - I & -H & -G \\ -H & B - I & -F \\ -G & -F & C - I \end{vmatrix} = 0. \tag{22.3}$$

The three roots are real and positive: real because the matrix (21.7) is symmetric, and positive because the form (21.6) is positive-definite[1].

[1] Except in the case when all the mass lies on a single line through the origin; then it is semi-positive-definite, and one root is zero, while the other two are equal and positive.

The directions defined by (22.2), in which I has any one of the three values determined by (22.3), are the *principal axes of inertia* at the origin; these axes form an orthogonal triad[1], and the three planes determined by it are the *principal planes*. Products of inertia with respect to principal planes vanish, and for that reason (since it greatly simplifies the work) one uses, almost always, axes which coincide with the principal axes of inertia. Then the inertial properties are described by the three *principal moments of inertia*, A, B, C, and the moment of inertia about any line through the origin with direction cosines (l, m, n) is given by

$$I = A\,l^2 + B\,m^2 + C\,n^2. \tag{22.4}$$

For the geometrical representation of inertial properties one uses the momental ellipsoid[2] with equation

$$A\,x^2 + B\,y^2 + C\,z^2 - 2F\,y\,z - 2G\,z\,x - 2H\,x\,y = 1. \tag{22.5}$$

The moment of inertia about any line L drawn through the origin is $1/r^2$, where r is the radius vector of this ellipsoid, drawn in the direction of L.

Two mass distributions are *equimomental* if they have the same moment of inertia about any arbitrary line. It follows that two equimomental systems have the same mass-centre, the same total mass, the same principal axes of inertia at the mass-centre, and the same principal moments of inertia there. The converse is also true.

A uniform triangular plate of mass m is equimomental to a set of three particles, each of mass $m/3$, placed at the middle points of the sides. A uniform solid tetrahedron of mass m is equimomental to a set of five particles, one of mass $4m/5$ placed at the mass-centre, and the other four, each of mass $m/20$, placed at the vertices[3].

23. Linear momentum. The *linear momentum* of a particle is $m\boldsymbol{v}$, where m is the mass and $\boldsymbol{v}$ the velocity. The linear momentum $\boldsymbol{M}$ of a system is the sum of the linear momenta of its parts:

$$\boldsymbol{M} = \sum_i m_i \boldsymbol{v}_i, \qquad \boldsymbol{M} = \int \varrho\, \boldsymbol{v}\, d\tau. \tag{23.1}$$

All the theory with which we are here concerned is applicable both to discrete systems of particles and to continuous systems, so that we have two types of formulae, one involving summation and the other integration. As the passage from the one to the other is obvious, only the summation will be shown in general.

The velocity $\boldsymbol{v}_i$ of any particle of a system may be written

$$\boldsymbol{v}_i = \boldsymbol{v} + \boldsymbol{v}_i', \tag{23.2}$$

where $\boldsymbol{v}$ is the velocity of the mass-centre and $\boldsymbol{v}_i'$ the velocity relative to the mass-centre. By (21.1) we have

$$\sum_i m_i \boldsymbol{r}_i' = 0, \tag{23.3}$$

---

[1] If two of the principal moments of inertia are equal, only one member of the triad is determined, and the triad may be completed by taking any orthogonal pair orthogonal to the determined member; this is the case of *axial symmetry*. If all three principal moments of inertia are equal, any orthogonal triad is principal; this is the case of *spherical symmetry*.

[2] For further details about the momental ellipsoid and the ellipsoid of gyration (the polar reciprocal of the momental ellipsoid with respect to its centre), and for the theory of moments of inertia generally, see Routh [22] I, pp. 16–22; see also Ames and Murnaghan [1], pp. 191–196, Appell [2] II, pp. 1–17, and Jung [12].

[3] Cf. Routh [22] I, pp. 22–27.

where r'_i is the position vector relative to the mass-centre; hence

$$\sum_i m_i v'_i = 0, \tag{23.4}$$

and so, by (23.1) and (23.2), the linear momentum of the system is

$$M = m v, \tag{23.5}$$

where m is the total mass of the system. The linear momentum of any system is the same as that of a single fictitious particle, with mass equal to the total mass of the system, moving with its mass-centre.

24. Angular momentum. The *angular momentum*[1] of a particle about a point O is

$$h = r \times M = r \times m v = m (r \times v), \tag{24.1}$$

where m is the mass, v the absolute velocity[2], and r the position vector relative to O. It is, in fact, the moment of the linear momentum. The three components read

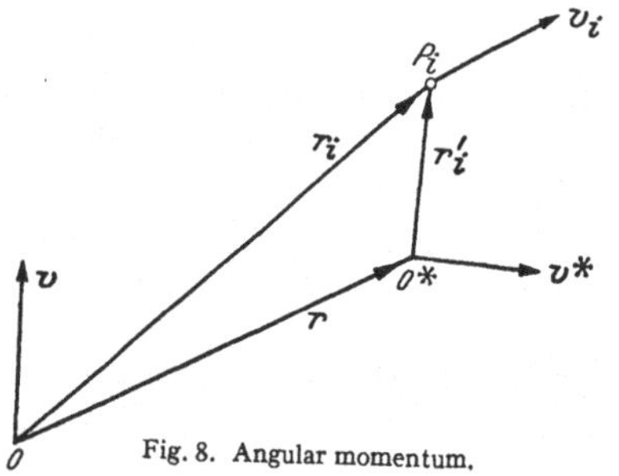

Fig. 8. Angular momentum.

$$\left.\begin{aligned} h_x &= m(y v_z - z v_y), \\ h_y &= m(z v_x - x v_z), \\ h_z &= m(x v_y - y v_x). \end{aligned}\right\} \tag{24.2}$$

The (scalar) angular momentum about a directed line through O is the orthogonal projection of h on that line. Thus, if V is a unit vector on the line, the scalar moment is

$$V \cdot (r \times M) = r \cdot (M \times V) = M \cdot (V \times r). \tag{24.3}$$

The angular momentum of a system is the sum of the angular momenta of its parts:

$$h = \sum_i r_i \times m_i v_i = \sum_i m_i (r_i \times v_i). \tag{24.4}$$

The point O with respect to which angular momentum is calculated may be fixed or moving. To investigate the effect of changing from one such point to another, consider two points O, O^* with absolute velocities v, v^*, and let $\overrightarrow{OO^*} = r$ (Fig. 8). Let there be any system of particles, a typical particle P_i having position vectors r_i, r'_i relative to O, O^*, respectively, so that

$$r_i = r + r'_i. \tag{24.5}$$

Then the angular momenta about O, O^* respectively, are

$$h = \sum_i m_i r_i \times v_i, \quad h^* = \sum_i m_i r'_i \times v_i, \tag{24.6}$$

---

[1] The old term *moment of momentum* is obsolete. Following APPELL [2] II, p. 157, one might call it *kinetic moment* (moment cinétique). However, the term *angular momentum* is now standard in English practice. But it is hard to find a suitable symbol for it, avoiding all possible confusion with symbols having accepted meanings. Following WHITTAKER's notation [28], p. 60, the symbol h is used in this article; since quantum mechanics is excluded, there can be no confusion with PLANCK's constant h. But in other contexts, a different symbol is advisable. GOLDSTEIN [7], p. 379, uses L, in spite of the standard use of L for the Lagrangian function. APPELL (loc. cit.) and PÉRÈS [20], p. 14, use σ.

[2] We are still dealing with kinematics and the absolute space of NEWTON is not involved. Absolute velocity here means velocity relative to some frame S_0 which is used throughout the argument, and is spoken of as fixed.

where $\boldsymbol{v}_i$ is the absolute velocity of P_i. Let $\boldsymbol{v}_i'$ be the velocity of P_i relative to O^*, so that

$$\boldsymbol{v}_i = \boldsymbol{v}^* + \boldsymbol{v}_i'; \tag{24.7}$$

then

$$\boldsymbol{h} = \sum_i m_i (\boldsymbol{r} + \boldsymbol{r}_i') \times (\boldsymbol{v}^* + \boldsymbol{v}_i'), \quad \boldsymbol{h}^* = \sum_i m_i \boldsymbol{r}_i' \times (\boldsymbol{v}^* + \boldsymbol{v}_i'), \tag{24.8}$$

and hence

$$\boldsymbol{h} = m \boldsymbol{r} \times \boldsymbol{v}^* + \boldsymbol{r} \times \sum_i m_i \boldsymbol{v}_i' + \boldsymbol{h}^*, \tag{24.9}$$

where m is the total mass of the system.

This formula is particularly useful when O^* is the mass-centre. For then the middle term disappears, and we have

$$\boldsymbol{h} = \boldsymbol{h}_0 + \boldsymbol{h}^*, \tag{24.10}$$

where

$$\boldsymbol{h}_0 = \boldsymbol{r} \times m \boldsymbol{v}^*, \quad \boldsymbol{h}^* = \sum_i \boldsymbol{r}_i' \times (m_i \boldsymbol{v}_i'); \tag{24.11}$$

$\boldsymbol{h}_0$ may be called the *orbital* angular momentum and $\boldsymbol{h}^*$ the *spin* angular momentum, to borrow the terms of quantum mechanics. We note that $\boldsymbol{h}_0$ is the angular momentum about O of a fictitious particle of mass m, moving with the mass-centre, and $\boldsymbol{h}^*$ is the angular momentum of the system about the mass-centre; in computing $\boldsymbol{h}^*$ it is actually a matter of indifference whether we use the velocities relative to the mass-centre, as in (24.11), or the absolute velocities.

It is well to emphasise that, in any computation of angular momentum, we need to specify (i) a frame of reference relative to which velocities are measured, and (ii) a point about which moments are taken.

For a rigid body turning about a fixed point O with angular velocity $\boldsymbol{\omega}$, the angular momentum about O is

$$\boldsymbol{h} = \sum_i m_i \boldsymbol{r}_i \times \boldsymbol{v}_i = \sum_i m_i \boldsymbol{r}_i \times (\boldsymbol{\omega} \times \boldsymbol{r}_i) = \boldsymbol{\omega} \sum_i m_i r_i^2 - \sum_i m_i \boldsymbol{r}_i (\boldsymbol{\omega} \cdot \boldsymbol{r}_i). \tag{24.12}$$

Resolution of this vector equation on any orthogonal triad gives the components

$$\left.\begin{aligned} h_1 &= A\,\omega_1 - H\,\omega_2 - G\,\omega_3, \\ h_2 &= -H\,\omega_1 + B\,\omega_2 - F\,\omega_3, \\ h_3 &= -\,G\,\omega_1 - F\,\omega_2 + C\,\omega_3, \end{aligned}\right\} \tag{24.13}$$

where $(\omega_1, \omega_2, \omega_3)$ are the components of $\boldsymbol{\omega}$ and A, B, C, F, G, H are the moments and products of inertia relative to the triad.

If the triad is principal, then these equations simplify to

$$h_1 = A\,\omega_1, \quad h_2 = B\,\omega_2, \quad h_3 = C\,\omega_3. \tag{24.14}$$

To secure this simplicity for all time, it is, in general, necessary to fix the triad in the body. But in a case of symmetry ($A = B$), a triad with one member along the axis of symmetry suffices. Then the triad is fixed neither in space nor in the body; note that in (24.14) the components of angular velocity are those of the body, not the triad.

25. Kinetic energy. The *kinetic energy* of a particle is $T = \frac{1}{2} m v^2$, where m is the mass and $\boldsymbol{v}$ the absolute velocity; for a system of particles the kinetic energy is

$$T = \sum_i \tfrac{1}{2} m_i v_i^2 = \tfrac{1}{2} \sum_i m_i \boldsymbol{v}_i \cdot \boldsymbol{v}_i. \tag{25.1}$$

Let $\boldsymbol{v}$ be the absolute velocity of the mass-centre of a system, $\boldsymbol{v}_i$ the absolute velocity of a particle of the system, and $\boldsymbol{v}'_i$ its velocity relative to the mass-centre. Then $\boldsymbol{v}_i = \boldsymbol{v} + \boldsymbol{v}'_i$, and (25.1) gives

$$T = \tfrac{1}{2} m v^2 + \sum_i \tfrac{1}{2} m_i v_i'^2, \tag{25.2}$$

since $\sum_i m_i \boldsymbol{v}'_i = 0$; here m is the total mass. This is the theorem of KÖNIG: *the kinetic energy of any system is the sum of two parts: (i) the absolute kinetic energy of a fictitious particle of mass m moving with the mass-centre, and (ii) the kinetic energy of the motion relative to the mass-centre.*

For a rigid body turning about a fixed point with angular velocity $\boldsymbol{\omega}$, the kinetic energy is

$$\left.\begin{aligned} T &= \tfrac{1}{2} \sum_i m_i (\boldsymbol{\omega} \times \boldsymbol{r}_i)^2 \\ &= \tfrac{1}{2} (A\,\omega_1^2 + B\,\omega_2^2 + C\,\omega_3^2 - 2F\,\omega_2\omega_3 - 2G\,\omega_3\omega_1 - 2H\,\omega_1\omega_2), \end{aligned}\right\} \tag{25.3}$$

where A, B, C, F, G, H are the moments and products of inertia. When principal axes are used, this reduces to

$$T = \tfrac{1}{2} (A\,\omega_1^2 + B\,\omega_2^2 + C\,\omega_3^2). \tag{25.4}$$

When the positions of the principal axes are described by the Eulerian angles (Sect. 11), by (19.4) the kinetic energy is

$$\left.\begin{aligned} T = \tfrac{1}{2} A (\dot\vartheta \sin\psi - \dot\varphi \sin\vartheta \cos\psi)^2 + \tfrac{1}{2} B (\dot\vartheta \cos\psi + \dot\varphi \sin\vartheta \sin\psi)^2 + \\ + \tfrac{1}{2} C (\dot\psi + \dot\varphi \cos\vartheta)^2. \end{aligned}\right\} \tag{25.5}$$

For rotation about a fixed axis, whether principal or not, we have

$$T = \tfrac{1}{2} I\,\omega^2 \tag{25.6}$$

where I is the moment of inertia about the axis.

In terms of the symmetric inertia matrix I_{rs} of (21.7), the kinetic energy is

$$T = T(\omega) = \tfrac{1}{2} I_{rs}\,\omega_r\,\omega_s, \tag{25.7}$$

with summation over 1, 2, 3 understood for repeated subscripts here and below, and by (24.13) we have

$$h_r = I_{rs}\,\omega_s, \qquad T = \tfrac{1}{2} h_r\,\omega_r. \tag{25.8}$$

In terms of the reciprocal matrix J_{rs}, satisfying $I_{rs} J_{rt} = \delta_{st}$, we have

$$\omega_r = J_{rs} h_s, \qquad T = T(h) = \tfrac{1}{2} J_{rs} h_r h_s. \tag{25.9}$$

Therefore

$$\frac{\partial T(\omega)}{\partial \omega_r} = h_r, \qquad \frac{\partial T(h)}{\partial h_r} = \omega_r. \tag{25.10}$$

For principal axes we have the formulae

$$\left.\begin{aligned} T &= \frac{1}{2} (A\,\omega_1^2 + B\,\omega_2^2 + C\,\omega_3^2) \\ &= \frac{1}{2} \left(\frac{h_1^2}{A} + \frac{h_2^2}{B} + \frac{h_3^2}{C} \right) \\ &= \frac{1}{2} (h_1\,\omega_1 + h_2\,\omega_2 + h_3\,\omega_3). \end{aligned}\right\} \tag{25.11}$$

26. Force systems. Consider forces F_i $(i=1, 2, \ldots, P)$ acting on P particles with position vectors r_i relative to a base-point O. The force F_i is regarded as made up of two parts: an *external* force F_i' and an *internal* force F_i'', the latter being the resultant of *reactions* exerted on the particle i by the other particles of the system.

We accept, as an hypothesis or axiom, NEWTON'S Third Law[1]: this states that the force exerted by particle i on particle j is equal and opposite to the force exerted by particle j on particle i, and that the two forces lie on the line joining the particles. This law, commonly called the *law of Action and Reaction*, is equivalently expressed by writing

$$F_i'' = \sum_{j=1}^{P} A_{ij}(r_j - r_i) \quad (i = 1, 2, \ldots, P), \tag{26.1}$$

where A_{ij} are scalar factors satisfying $A_{ij} = A_{ji}$.

Now, for any system of forces (F_i acting at r_i), the total force F and the total moment (or torque) G about the base-point O are respectively

$$F = \sum_{i=1}^{P} F_i, \qquad G = \sum_{i=1}^{P} r_i \times F_i. \tag{26.2}$$

The given system of forces is said to be *equipollent*[2] to a single force F acting at O and a couple G.

Note that F is a bound vector and G a free vector. If we change the base-point O, F remains unchanged but G changes; by suitable choice of O we can reduce the force system to a *wrench*, i.e. we can make $G = pF$, where p is a scalar factor, called the *pitch* of the wrench.

The vector pair (F, G) is called a *motor*[3] or a *torsor*[4]. It does not change if we slide the vectors F_i along their lines of action, and hence it is easily seen that the internal forces F_i'' contribute nothing to it. Therefore the total force and the total moment are given by

$$F = \sum_{i=1}^{P} F_i', \qquad G = \sum_{i=1}^{P} r_i \times F_i', \tag{26.3}$$

in which formulae only the external forces occur. This elimination of internal forces is of fundamental importance in Newtonian dynamics.

The work done by[5] a force F in a displacement δr of its point of application is $F \cdot \delta r$, and for a system of forces the work is

$$\delta W = \sum_i F_i \cdot \delta r_i. \tag{26.4}$$

If the forces act on a rigid body, and this body is given an infinitesimal displacement consisting of an infinitesimal translation δr_0 and an infinitesimal rotation $\delta\chi$ about a base-point, then

$$\delta r_i = \delta r_0 + \delta\chi \times r_i', \tag{26.5}$$

---

[1] "To every action there is always opposed an equal reaction: or, the mutual actions of two bodies upon each other are always equal, and directed to contrary parts". Sir ISAAC NEWTON'S Mathematical Principles of Natural Philisophy and his System of the World, MOTTE'S translation revised by F. CAJORI, p. 13. Berkeley: University of California Press 1946. Cf. Sect. 5 of the present article for a more general law, consistent with the homogeneity and isotropy of space.

[2] The word *equivalent* is often used, but it is misleading, since there is equivalence only if the particles on which the forces act form a rigid body.

[3] For references to motor symbolism, see Sect. 49.

[4] Cf. PÉRÈS [20], p. 9.

[5] The work done *against* the force is $-F \cdot \delta r$.

where $\boldsymbol{r}'_i$ is the position vector relative to the base-point. The work done in this displacement is

$$\left.\begin{aligned}\delta W &= \boldsymbol{F}\cdot\delta\boldsymbol{r}_0 + \sum_{i=1}^{P}\boldsymbol{F}_i\cdot(\delta\chi\times\boldsymbol{r}'_i)\\ &= \boldsymbol{F}\cdot\delta\boldsymbol{r}_0 + \boldsymbol{G}\cdot\delta\chi,\end{aligned}\right\}\tag{26.6}$$

in the notation of (26.2).

Given a plane Π passing through a point O, any infinitesimal rotation about O may be resolved into two infinitesimal rotations, one $(\delta_1\chi)$ about an axis perpendicular to Π and the other $(\delta_2\chi)$ about an axis in Π. If a couple is represented by a pair of equal opposite forces (magnitude P, distance apart p) in the plane Π, then the work done by it in an infinitesimal rotation is $\pm p\,P\,\delta_1\chi$, the sign depending on whether the senses of the couple and the rotation are the same or opposite.

If two equal and opposite forces $\boldsymbol{F}$, $-\boldsymbol{F}$, act at points $\boldsymbol{r}, \boldsymbol{r}'$, then the work done in arbitrary infinitesimal displacements is

$$\delta W = \boldsymbol{F}\cdot(\delta\boldsymbol{r} - \delta\boldsymbol{r}').\tag{26.7}$$

If, further, the forces act along the line joining the points, we may write $\boldsymbol{F} = \vartheta(\boldsymbol{r}-\boldsymbol{r}')$, where ϑ is some scalar factor, and (26.7) becomes

$$\delta W = \vartheta(\boldsymbol{r}-\boldsymbol{r}')\cdot\delta(\boldsymbol{r}-\boldsymbol{r}').\tag{26.8}$$

This vanishes if the displacements are such as to keep the distance $|\boldsymbol{r}-\boldsymbol{r}'|$ unchanged. Since by NEWTON's Third Law the reactions between any pair of particles of a system satisfy the above conditions on $\boldsymbol{F}$, it follows that no work is done by the reactions in a rigid body. Other cases of *workless reactions* are (a) reactions at smooth contacts, (b) reactions at rolling contacts (with no sliding). All such reactions disappear from those general equations of dynamics which are based on energy and work.

IV. Generalized coordinates.

27. Holonomic systems. Moving constraints. A configuration of a system composed of P particles may always be described by giving the $3P$ coordinates of the particles, but a smaller number suffices if these $3P$ coordinates are connected by equations of constraint. Thus, if the system is rigid and has a fixed point, three coordinates suffice (e.g. the Eulerian angles of Sect. 11). Any set of parameters which completely determine the configuration of a system are called *generalized coordinates*; their rates of change are *generalized velocities*.

Suppose that N generalized coordinates q_ϱ (and no smaller number) describe the configurations of system, and that it is possible to vary q_ϱ arbitrarily and independently without violating the constraints (such as rigidity); then we say that the system is *holonomic*, with N *degrees of freedom*.

The following are examples of holonomic systems, with the numbers of degrees of freedom indicated:

rigid body with fixed point (3),
free rigid body (6),
rigid body moving parallel to a plane (3),
rigid body in contact with a fixed plane (5).

A system may be subject to moving constraints (e.g. a particle may be constrained to move on a surface which itself moves in some prescribed manner). Then, to describe a configuration, we need the time t as well as the generalized coordinates q_ϱ, and the system is holonomic if arbitrary independent variations of q_ϱ and t do not violate the constraints. A system with fixed constraints is *scleronomic*; with moving constraints, *rheonomic*.

If (x_i, y_i, z_i) are the coordinates of the i-th particle of a system, relative to fixed axes $Oxyz$, the selection of generalized coordinates sets up equations of the form

$$x_i = x_i(q_1, q_2, \dots q_N, t), \quad y_i = y_i(q_1, q_2, \dots q_N, t), \quad z_i = z_i(q_1, q_2, \dots q_N, t), \tag{27.1}$$

t being absent for a scleronomic system[1].

The kinetic energy of a rheonomic system is

$$\left.\begin{aligned} T &= \tfrac{1}{2}\sum_{i=1}^{P} m_i(\dot{x}_i^2 + \dot{y}_i^2 + \dot{z}_i^2) \\ &= \tfrac{1}{2}\sum_{\varrho,\sigma=1}^{N} a_{\varrho\sigma}\dot{q}_\varrho\dot{q}_\sigma + \sum_{\varrho=1}^{N} a_\varrho\dot{q}_\varrho + a; \end{aligned}\right\} \tag{27.2}$$

here $a_{\varrho\sigma}$, a_ϱ and a are functions of $q_1, \dots q_N, t$. For a scleronomic system this reduces to

$$T = \tfrac{1}{2}\sum_{\varrho,\sigma=1}^{N} a_{\varrho\sigma}\dot{q}_\varrho\dot{q}_\sigma, \tag{27.3}$$

a positive-definite[2] quadratic form by virtue of the definition of T, the coefficients $a_{\varrho\sigma}$ being function of $q_1, \dots q_N$.

28. Non-holonomic systems. Consider a rigid lamina which can slide freely over a fixed plane; this is a scleronomic holonomic system with three degrees of freedom. But now suppose that a small sharp blade is fixed in the lamina, the blade being capable of motion only in the direction of its length. If (x, y) are the Cartesian coordinates of the blade and ϑ its inclination to the x-axis, then (x, y, ϑ) form a system of generalized coordinates for the lamina, but they are subject to the *non-integrable* relation

$$\frac{dy}{dx} = \tan\vartheta. \tag{28.1}$$

The number of generalized coordinates cannot be reduced below three, but these three coordinates are not freely variable.

We call a system *non-holonomic* when it is impossible to describe the configurations by generalized coordinates q_ϱ ($\varrho = 1, 2, \dots N$) and the time t, with q_ϱ and t freely and independently variable. In such cases there are present certain non-integrable[3] equations of constraint of the form

$$\sum_{\varrho=1}^{N} A_{c\varrho}\,dq_\varrho + A_c\,dt = 0, \quad (c = 1, 2, \dots M); \tag{28.2}$$

---

[1] But even for a system without moving constraints it may be convenient to use equations of the form (27.1). For example, to study the motion of a free rigid body (e.g. a rocket) relative to the earth (the motion of the latter being known), we might let $q_1, q_2, \dots q_6$ describe the configuration of the body relative to axes fixed in the earth; then the equations giving the coordinates of the particles of the body relative to fixed axes will be of the form (27.1), the time t entering through the earth's motion. From an analytic standpoint, it is sometimes convenient to use the word scleronomic when t is absent from (27.1) and rheonomic when it is present, without regard to the physical system under consideration.

[2] $T > 0$ unless $\dot{q}_1 = \cdots = \dot{q}_N = 0$.

[3] Pérès [*20*], p. 218, calls a system *semi-holonomic* if the equations of constraint are integrable, with constants of integration dependent on initial conditions. Mathematically, a semi-holonomic system is not very different from a holonomic system, since the integrated equations of constraint may be used to reduce the number of generalized coordinates.

here $A_{c\varrho}$ and A_c are functions of the variables (q, t). The system is said to have $N-M$ degrees of freedom.

The formulae (27.2) and (27.3) for kinetic energy are valid for non-holonomic systems.

Non-holonomicity usually occurs in systems with rolling contacts, the condition of rolling being the equality of the instantaneous velocities of the two particles at the point of contact, one particle belonging to each body[1]. The following examples illustrate non-holonomic constraints due to rolling.

α) Sphere rolling on a horizontal plane. Let $(\boldsymbol{I}, \boldsymbol{J}, \boldsymbol{K})$ be a fixed orthonormal triad with origin O in the given plane and $\boldsymbol{K}$ vertical (Fig. 9). As generalized coordinates we take $(x, y, \vartheta, \varphi, \psi)$ where (x, y) are the coordinates of the point of contact relative to axes Oxy drawn along $\boldsymbol{I}$ and $\boldsymbol{J}$, and $(\vartheta, \varphi, \psi)$ are the Eulerian angles (Sect. 11) which describe the position of an orthonormal triad $(\boldsymbol{i}, \boldsymbol{j}, \boldsymbol{k})$, fixed in the sphere, relative to $(\boldsymbol{I}, \boldsymbol{J}, \boldsymbol{K})$.

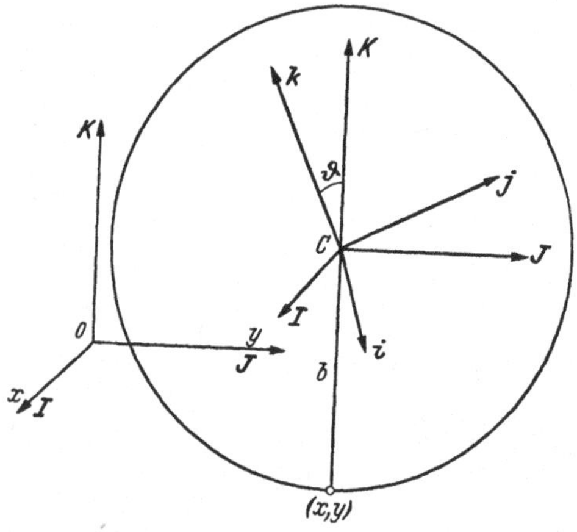

Fig. 9. Sphere rolling on plane.

Relative to the centre C of the sphere, the position vector of the point of contact is $-b\boldsymbol{K}$, where b is the radius of the sphere. Therefore, by (19.2), the velocity of that particle of the sphere which is instantaneously in contact with the plane is

$$\dot{x}\boldsymbol{I} + \dot{y}\boldsymbol{J} + \boldsymbol{\omega} \times (-b\boldsymbol{K}), \quad (28.3)$$

where $\boldsymbol{\omega}$ is the angular velocity of the sphere. Resolving the angular velocity along $(\boldsymbol{I}, \boldsymbol{J}, \boldsymbol{K})$, we have

$$\boldsymbol{\omega} = \Omega_1\boldsymbol{I} + \Omega_2\boldsymbol{J} + \Omega_3\boldsymbol{K}, \quad (28.4)$$

where the coefficients are expressed in terms of the Eulerian angles and their rates of change as in (19.5). Substituting (28.4) in (28.3), and equating the result to zero (condition of rolling), we get the following two non-integrable equations of constraint:

$$\left.\begin{aligned} \dot{x} - b(\dot{\vartheta}\cos\varphi + \dot{\psi}\sin\vartheta\sin\varphi) &= 0, \\ \dot{y} - b(\dot{\vartheta}\sin\varphi - \dot{\psi}\sin\vartheta\cos\varphi) &= 0. \end{aligned}\right\} \quad (28.5)$$

The sphere has $5-2=3$ degrees of freedom.

β) Circular disc rolling on a horizontal plane[2]. Let $(\boldsymbol{I}, \boldsymbol{J}, \boldsymbol{K})$ be a fixed orthonormal triad with origin O in the given plane and $\boldsymbol{K}$ vertical (Fig. 10). With these as axes, the position vector of the centre C of the disc is

$$\boldsymbol{r} = x\boldsymbol{I} + y\boldsymbol{J} + z\boldsymbol{K}. \quad (28.6)$$

Let $(\boldsymbol{i}, \boldsymbol{j}, \boldsymbol{k})$ by a second orthonormal triad centred at C, $\boldsymbol{i}$ being directed to the point of contact, $\boldsymbol{j}$ being horizontal, and $\boldsymbol{k}$ perpendicular to the plane of the disc.

[1] It is by no means always desirable to use generalized coordinates in the discussion of non-holonomic systems; Routh [22] II, pp. 165–205, applies direct methods with great elegance. For an extensive treatment of non-holonomic systems, with problems worked out in detail and consideration of non-linear constraints, see Hamel [11], pp. 464–507. See also Winkelmann and Grammel [29], pp. 434–440. For the dynamics of non-holonomic systems in the present article, see Sects. 46, 48, 85.

[2] For the dynamics of a rolling disc, see Appell [2] II, pp. 253–258.

Let ϑ be the inclination of $-\boldsymbol{i}$ to $\boldsymbol{K}$, φ the inclination to $\boldsymbol{I}$ of the tangent to the disc at the point of contact, and ψ the inclination to $\boldsymbol{i}$ of a radius fixed in the disc. Then (if b = radius of disc)

$$z = b\cos\vartheta, \qquad (28.7)$$

and $(x, y, \vartheta, \varphi, \psi)$ form a system of generalized coordinates.

The velocity of C is

$$\boldsymbol{v} = \dot{x}\boldsymbol{I} + \dot{y}\boldsymbol{J} - b\sin\vartheta\,\dot{\vartheta}\boldsymbol{K}, \qquad (28.8)$$

and the angular velocity of the disc is

$$\boldsymbol{\omega} = -\dot{\vartheta}\boldsymbol{j} + \dot{\varphi}\boldsymbol{K} + \dot{\psi}\boldsymbol{k}, \qquad (28.9)$$

as we see on increasing each of the angles in turn. Hence, by (19.2), the velocity of the particle of the disc instantaneously in contact with the plane is

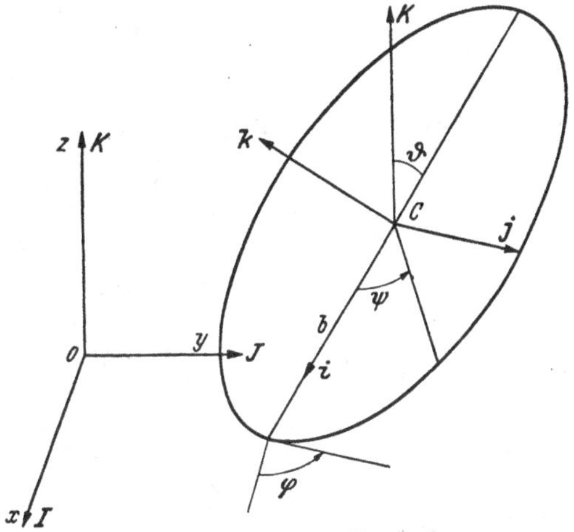

Fig. 10. Circular disc rolling on plane.

$$\boldsymbol{v} + \boldsymbol{\omega}\times b\,\boldsymbol{i} = \dot{x}\boldsymbol{I} + \dot{y}\boldsymbol{J} - b\sin\vartheta\,\dot{\vartheta}\boldsymbol{K} + b\dot{\vartheta}\boldsymbol{k} + b(\dot{\psi} + \dot{\varphi}\sin\vartheta)\boldsymbol{j}, \qquad (28.10)$$

since

$$\boldsymbol{K} = -\cos\vartheta\,\boldsymbol{i} + \sin\vartheta\,\boldsymbol{k}. \qquad (28.11)$$

We have

$$\left.\begin{aligned} \boldsymbol{j} &= \cos\varphi\,\boldsymbol{I} + \sin\varphi\,\boldsymbol{J},\\ \boldsymbol{k} &= \cos\vartheta\sin\varphi\,\boldsymbol{I} - \cos\vartheta\cos\varphi\,\boldsymbol{J} + \sin\vartheta\,\boldsymbol{K}, \end{aligned}\right\} \qquad (28.12)$$

and when these are substituted in (28.10), the condition of rolling gives the following two non-integrable equations of constraint:

$$\left.\begin{aligned} \dot{x} + b\cos\vartheta\sin\varphi\,\dot{\vartheta} + b\cos\varphi\,(\dot{\psi} + \sin\vartheta\,\dot{\varphi}) &= 0,\\ \dot{y} - b\cos\vartheta\cos\varphi\,\dot{\vartheta} + b\sin\varphi\,(\dot{\psi} + \sin\vartheta\,\dot{\varphi}) &= 0. \end{aligned}\right\} \qquad (28.13)$$

The disc has $5-2=3$ degrees of freedom.

γ) *Axle with pair of wheels which roll on a plane.* We suppose that each wheel is free to turn about the axle. (If they were both fixed to the axle we would have a holonomic system with one degree of freedom.) Let b be the radius of each wheel and $2c$ the length of the axle. We take generalized coordinates $(x, y, \varphi, \psi, \psi')$ as shown in Fig. 11. Then, computing the components of velocity, parallel and perpendicular to the axle, of the two particles instantaneously in contact with the plane, and then equating these components to zero to satisfy the condition of rolling, we get the following three

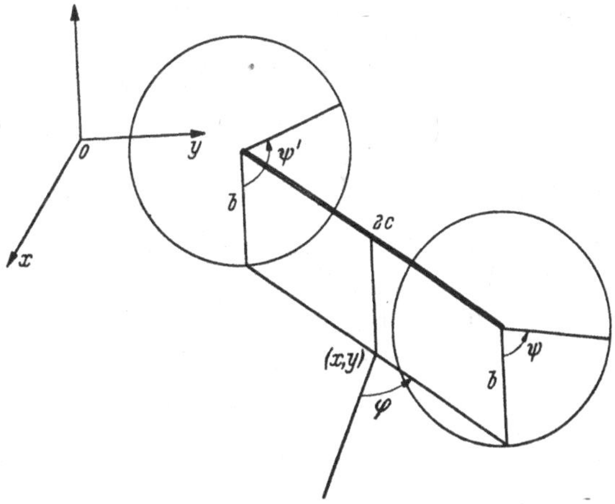

Fig. 11. Axle with pair of wheels which roll on a plane.

equations of constraint:

$$\left.\begin{aligned} \dot{x}\cos\varphi + \dot{y}\sin\varphi &= 0,\\ -\dot{x}\sin\varphi + \dot{y}\cos\varphi + c\dot{\varphi} + b\dot{\psi} &= 0,\\ -\dot{x}\sin\varphi + \dot{y}\cos\varphi - c\dot{\varphi} + b\dot{\psi}' &= 0. \end{aligned}\right\} \tag{28.14}$$

The last two yield the integrable combination

$$2c\dot{\varphi} + b(\dot{\psi} - \dot{\psi}') = 0. \tag{28.15}$$

If we are given initial values of the three angles, we have

$$2c\varphi = A - b(\psi - \psi'), \tag{28.16}$$

where A is a given constant; we may then take (x, y, ψ, ψ') as generalized coordinates, with the two non-integrable equations of constraint:

$$\left.\begin{aligned} \dot{x}\cos\varphi + \dot{y}\sin\varphi &= 0,\\ -\dot{x}\sin\varphi + \dot{y}\cos\varphi + \tfrac{1}{2}b(\dot{\psi} + \dot{\psi}') &= 0. \end{aligned}\right\} \tag{28.17}$$

The system has $4-2=2$ degrees of freedom.

29. Generalized forces. Work. Potential function. Consider a system of P particles, in general rheonomic and non-holonomic. Let forces with components (X_i^*, Y_i^*, Z_i^*) act on the particles with coordinates (x_i, y_i, z_i). Then, in any completely arbitrary set of infinitesimal displacements of the particles, all constraints being disregarded, the work done by these forces is

$$\delta W = \sum_{i=1}^{P} (X_i^*\,\delta x_i + Y_i^*\,\delta y_i + Z_i^*\,\delta z_i). \tag{29.1}$$

Now if (q_ϱ, t) describe the configurations of the system, we have equations of the form (27.1), and so the work done in the displacement corresponding to arbitrary variations $(\delta q_\varrho, \delta t)$ is

$$\delta W = \sum_{\varrho=1}^{N} Q_\varrho^*\,\delta q_\varrho + Q^*\,\delta t, \tag{29.2}$$

where

$$Q_\varrho^* = \sum_{i=1}^{P}\left(X_i^*\frac{\partial x_i}{\partial q_\varrho} + Y_i^*\frac{\partial y_i}{\partial q_\varrho} + Z_i^*\frac{\partial z_i}{\partial q_\varrho}\right), \tag{29.3}$$

$$Q^* = \sum_{i=1}^{P}\left(X_i^*\frac{\partial x_i}{\partial t} + Y_i^*\frac{\partial y_i}{\partial t} + Z_i^*\frac{\partial z_i}{\partial t}\right). \tag{29.4}$$

The quantities Q_ϱ^* are *generalized forces*; they are usually easier to calculate from (29.2) than from (29.3).

The total force (X_i^*, Y_i^*, Z_i^*) acting on a particle of the system can in general be split into two parts:

$$\left.\begin{aligned} (X_i, Y_i, Z_i) &= \text{given force}^1\text{, such as gravity,}\\ (X_i', Y_i', Z_i') &= \text{force of constraint.} \end{aligned}\right\} \tag{29.5}$$

Then the total generalized force Q_ϱ^* can be split into two parts:

$$Q_\varrho^* = Q_\varrho + Q_\varrho', \tag{29.6}$$

[1] Also called *applied* force.

so that Q_ϱ is the *given* generalized force and Q'_ϱ the generalized force of *constraint*; we suppose Q_ϱ to be known functions of $q_1, \ldots q_N, t, \dot{q}_1, \ldots \dot{q}_N$, and possibly of $\ddot{q}_1, \ldots \ddot{q}_N$ also.

The importance of this splitting is due to the fact that in many dynamical systems the forces of constraint are *workless*, by which we mean that these forces do no work in a displacement δq_ϱ which satisfies the instantaneous constraint (with $\delta t = 0$). This implies

$$\sum_{\varrho=1}^{N} Q'_\varrho \, \delta q_\varrho = 0, \tag{29.7}$$

so that the work done by all forces in this displacement is

$$\delta W = \sum_{\varrho=1}^{N} Q_\varrho \, \delta q_\varrho . \tag{29.8}$$

If there exists a function $V(q_1, \ldots q_N, t)$ such that

$$Q_\varrho = -\frac{\partial V}{\partial q_\varrho}, \tag{29.9}$$

so that by (29.8)

$$\delta W = -\delta V, \tag{29.10}$$

then V is the *potential function*[1], or *potential energy*, of the system. More generally, if there exists a function $V(q_1, \ldots q_N, t, \dot{q}_1, \ldots \dot{q}_N)$ such that

$$Q_\varrho = \frac{d}{dt}\frac{\partial V}{\partial \dot{q}_\varrho} - \frac{\partial V}{\partial q_\varrho}, \tag{29.11}$$

then V is an *extended potential function.*

C. Dynamics of a particle.

I. Equations of motion.

30. Basic equations. NEWTON's first two laws are embodied in the equation

$$\frac{d}{dt}(m\boldsymbol{v}) = \boldsymbol{F}, \tag{30.1}$$

where m is the mass of a particle, $\boldsymbol{v}$ its absolute velocity, and $\boldsymbol{F}$ the force acting on it. If m is constant (as we shall assume it to be), then (30.1) is equivalent to

$$m\boldsymbol{a} = \boldsymbol{F}, \tag{30.2}$$

where $\boldsymbol{a}$ is the absolute acceleration.

Let x^ϱ be curvilinear coordinates in absolute space (cf. Sects. 17, 18). Then, by (18.3), the equation of motion (30.2) may be written in the contravariant form

$$m a^\varrho \equiv m\left(\ddot{x}^\varrho + \left\{{\varrho \atop \mu\nu}\right\}\dot{x}^\mu \dot{x}^\nu\right) = F^\varrho. \tag{30.3}$$

Here F^ϱ is the contravariant force vector; the covariant force vector F_ϱ may be calculated from the invariant formula

$$\delta W = F_\varrho \, \delta x^\varrho, \tag{30.4}$$

[1] French writers prefer to speak of a *force function* U, where $U = -V$.

where δW is the work done by the force in an arbitrary displacement δx^ϱ, and F^ϱ is obtained from F_ϱ by the formula

$$F^\varrho = g^{\varrho\mu} F_\mu . \tag{30.5}$$

By (18.5) the covariant form of the equation of motion is[1]

$$m\, a_\varrho \equiv \frac{d}{dt}\frac{\partial T}{\partial \dot{x}^\varrho} - \frac{\partial T}{\partial x^\varrho} = F_\varrho , \tag{30.6}$$

where

$$T = \tfrac{1}{2} m\, g_{\varrho\sigma} \dot{x}^\varrho \dot{x}^\sigma , \tag{30.7}$$

the kinetic energy of the particle. If a potential function[2] $V(x, t)$ exists [cf. (29.9)], such that

$$F_\varrho = -\frac{\partial V}{\partial x^\varrho} , \tag{30.8}$$

or, more generally, an extended potential function[2] $V(x, \dot{x}, t)$ such that [cf. (29.11)]

$$F_\varrho = \frac{d}{dt}\frac{\partial V}{\partial \dot{x}^\varrho} - \frac{\partial V}{\partial x^\varrho} , \tag{30.9}$$

then (30.6) may be written

$$\frac{d}{dt}\frac{\partial L}{\partial \dot{x}^\varrho} - \frac{\partial L}{\partial x^\varrho} = 0 , \tag{30.10}$$

where L is the *Lagrangian function*

$$L = T - V . \tag{30.11}$$

For motion in a plane, the most convenient coordinates are usually either rectangular Cartesians (x, y), for which the equations of motion are

$$m\ddot{x} = X, \quad m\ddot{y} = Y , \tag{30.12}$$

or polar coordinates (r, ϑ), for which they read

$$m(\ddot{r} - r\dot{\vartheta}^2) = R, \quad m\frac{1}{r}\frac{d}{dt}(r^2\dot{\vartheta}) = \Theta , \tag{30.13}$$

where R is the radial component of force and Θ the transverse component.

In space, we may conveniently use cylindrical coordinates (18.8) or spherical polars (18.10).

31. Energy. Angular momentum. From (30.2) we deduce

$$\frac{dT}{dt} = \boldsymbol{F} \cdot \boldsymbol{v} , \tag{31.1}$$

so that the rate of increase of kinetic energy is equal to the rate of working (or *activity* or *power*) of the force $\boldsymbol{F}$. If there exists a potential function $V(x)$ (independent of t), then (31.1) gives the energy integral

$$T + V = E = \text{const} , \tag{31.2}$$

E being the total energy.

Also, by (30.2) we have

$$\boldsymbol{r} \times m\,\boldsymbol{a} = \boldsymbol{r} \times \boldsymbol{F} , \tag{31.3}$$

or

$$\frac{d\boldsymbol{h}}{dt} = \boldsymbol{G} , \tag{31.4}$$

[1] These are Lagrangian equations, as in (46.17).
[2] Here and later, x stands for the three quantities x^ϱ, and $\dot{x}$ for the three quantities $\dot{x}^\varrho$.

where $\boldsymbol{h}$ is the angular momentum about the origin O and $\boldsymbol{G}$ the moment of $\boldsymbol{F}$ about O. If the line of action of $\boldsymbol{F}$ passes through O, we have $\boldsymbol{G}=0$ and

$$\boldsymbol{h} = \text{const}. \tag{31.5}$$

This is the integral of angular momentum. In the case of a particle moving in a plane under the action of a force directed to, or from, the origin, it gives

$$r^2 \dot{\vartheta} = \text{const}, \tag{31.6}$$

from which it follows that the radius vector drawn from the origin to the particle traces out equal areas in equal times. It is useful to remember that

$$r^2 \dot{\vartheta} = p\, v, \tag{31.7}$$

where v is the speed and p the perpendicular dropped from the origin on the tangent to the orbit.

32. Moving frames of reference. Let S be a rigid body which has a prescribed motion relative to absolute space. This motion is described by selecting some base point O in S, giving the absolute velocity $\boldsymbol{v}_0(t)$ of O, and also giving the angular velocity $\boldsymbol{\omega}$ of S. We take S as a moving frame of reference. Then the absolute acceleration $\boldsymbol{a}$ of a particle, acted on by a force $\boldsymbol{F}$, may be broken down as in (20.10), and the equation of motion (30.2) may be written

$$m\,\boldsymbol{a}' = \boldsymbol{F} + \boldsymbol{F}_0 + \boldsymbol{F}_c + \boldsymbol{F}_t, \tag{32.1}$$

where m is the mass of the particle and

$$\left.\begin{aligned} \boldsymbol{F}_0 &= -m\,\boldsymbol{a}_0,\\ \boldsymbol{F}_c &= -m\,\boldsymbol{a}_c = -2m\,\boldsymbol{\omega}\times\boldsymbol{v}',\\ \boldsymbol{F}_t &= -m\,\boldsymbol{a}_t = -m\dot{\boldsymbol{\omega}}\times\boldsymbol{r}' - m\,\boldsymbol{\omega}\times(\boldsymbol{\omega}\times\boldsymbol{r}')\\ &= -m\dot{\boldsymbol{\omega}}\times\boldsymbol{r}' - m\,\boldsymbol{\omega}\,(\boldsymbol{\omega}\cdot\boldsymbol{r}') + m\,\omega^2\boldsymbol{r}'. \end{aligned}\right\} \tag{32.2}$$

In (32.1) $\boldsymbol{a}'$ is the relative acceleration, i.e. the acceleration as observed by an observer carried along with S, and we may say that, relative to a moving frame of reference, NEWTON's law of motion holds in the form

$$m\,\boldsymbol{a}' = \boldsymbol{F}', \tag{32.3}$$

where $\boldsymbol{F}'$ is made up of the real force $\boldsymbol{F}$ and the three *fictitious* forces $\boldsymbol{F}_0$, $\boldsymbol{F}_c$, $\boldsymbol{F}_t$; here $\boldsymbol{F}_0$ arises solely from the acceleration of the base point (and is present even if S is not rotating); $\boldsymbol{F}_c$ is the CORIOLIS *force*, a force of great importance in meteorology and indeed in all phenomena involving the rotation of the earth; $\boldsymbol{F}_t$ is of the nature of *centrifugal* force, although that term is properly applied only when $\boldsymbol{\omega}$ is constant, in which case we have [cf. (20.14)]

$$\boldsymbol{F}_t = m\,\boldsymbol{R}\,\omega^2. \tag{32.4}$$

If the motion of S is a uniform velocity, we have $\boldsymbol{a}_0 = \boldsymbol{\omega} = \boldsymbol{0}$, and (32.1) gives

$$m\,\boldsymbol{a}' = \boldsymbol{F}, \tag{32.5}$$

which means that such a frame is a Newtonian frame. This result is known as *Newtonian relativity*; starting with an absolute space S_0 and NEWTON's law of motion relative to S_0, we find that this law holds also for any frame S in uniform motion.

II. One-dimensional motions.

33. Simple harmonic oscillator. Damping. A simple harmonic oscillator is a particle which moves on a line under the influence of a restoring force directed to a point O on the line and proportional to the distance from O. If a frictional damping force, proportional to the velocity and opposing it, is added, and also a disturbing force, the equation of motion reads

$$\ddot{x} + 2\mu\dot{x} + p^2 x = X \tag{33.1}$$

where

$$\left.\begin{aligned} \text{restoring force} &= -m p^2 x, \\ \text{damping force} &= -2m\mu\dot{x}, \\ \text{disturbing force} &= m X, \end{aligned}\right\} \tag{33.2}$$

m being the mass of the particle. For the undamped undisturbed oscillator, the motion is given by

$$\left.\begin{aligned} \ddot{x} + p^2 x &= 0, \\ x &= a\cos(p t + \varepsilon), \end{aligned}\right\} \tag{33.3}$$

where

$$\begin{aligned} a &= \text{amplitude}^1, \\ \frac{2\pi}{p} &= \text{periodic time}, \\ \frac{p}{2\pi} &= \nu = \text{frequency}, \\ p &= 2\pi\nu = \omega = \text{circular frequency}, \\ p t + \varepsilon &= \varphi = \text{phase angle}, \\ \varepsilon &= \text{phase constant}. \end{aligned}$$

The oscillator is said to be "in the same phase" for two values of φ differing by a multiple of 2π. The solution (33.3) may also be written in complex form

$$x = A\,e^{ipt}, \tag{33.4}$$

where A is a complex amplitude, which includes the phase constant; the physical displacement x is the real part of (33.4).

If $X=0$, the general solution of the damped Eq. (33.1) is, in complex form,

$$x = A\,e^{n_1 t} + B\,e^{n_2 t}, \tag{33.5}$$

where n_1, n_2 are the roots of

$$n^2 + 2\mu n + p^2 = 0. \tag{33.6}$$

These roots may real or complex, giving non-oscillatory and oscillatory motions respectively[2].

When damping and disturbing forces are both present, the latter being a given function of t, the general solution of (33.1) is

$$x = A\,e^{n_1 t} + B\,e^{n_2 t} + \frac{1}{n_1 - n_2}\int\limits_0^t X(\tau)\left\{e^{n_1(t-\tau)} - e^{n_2(t-\tau)}\right\} d\tau. \tag{33.7}$$

---

[1] Sometimes $2a$ is called the amplitude, and a the semi-amplitude.

[2] The damping constant μ being positive, the real part of each root is necessarily negative; for details, see SYNGE and GRIFFITH [26], pp. 163–166.

For a sinusoidal disturbing force, we write

$$X(t) = X_0 e^{iqt}; \tag{33.8}$$

then (33.7) becomes

$$x = A' e^{n_1 t} + B' e^{n_2 t} + \frac{X_0 e^{iqt}}{n_1 - n_2}\left(\frac{1}{iq - n_1} - \frac{1}{iq - n_2}\right), \tag{33.9}$$

where A' and B' are new arbitrary constants. As $t \to \infty$, the first two terms die away, and we are left with the *forced oscillation*

$$x = \frac{X_0 e^{iqt}}{(iq - n_1)(iq - n_2)} = \frac{X_0 e^{iqt}}{p^2 - q^2 + 2i\mu q}. \tag{33.10}$$

The amplitude is large *(resonance)* if $(p - q)$ and μ are small, that is, when the frequency of the disturbing force is nearly equal to that of the undamped free oscillator and the damping is small[1].

34. Circular and cycloidal pendulums. Consider a particle of mass m moving under gravity on a smooth vertical circle of radius l (circular pendulum). If ϑ is the angular displacement from the downward vertical, the total energy is

$$\tfrac{1}{2} m l^2 \dot{\vartheta}^2 + m g l (1 - \cos\vartheta) = E = \text{const}, \tag{34.1}$$

and the equation of motion is

$$\ddot{\vartheta} + p^2 \sin\vartheta = 0, \quad p^2 = \frac{g}{l}. \tag{34.2}$$

For small amplitudes, this reduces to the Eq. (33.3) for the harmonic oscillator and the periodic time is

$$\tau = 2\pi \sqrt{\frac{l}{g}}. \tag{34.3}$$

In general, we get an oscillatory motion if

$$\omega_0 = |\dot{\vartheta}|_{\vartheta=0} < 2p, \tag{34.4}$$

and the motion is given by

$$\sin\tfrac{1}{2}\vartheta = \sin\tfrac{1}{2}\alpha \,\mathrm{sn}\, p(t - t_0), \tag{34.5}$$

where α is the maximum value of ϑ, so that

$$\sin\frac{1}{2}\alpha = \frac{1}{2}\frac{\omega_0}{p}, \tag{34.6}$$

t_0 is an arbitrary constant, and the Jacobian elliptic function sn has $\sin\frac{1}{2}\alpha$ for modulus[2].

The period is

$$\left.\begin{aligned} \tau &= \frac{4}{p}\int_0^{\frac{1}{2}\pi} \frac{d\varphi}{\sqrt{1 - k^2 \sin^2\varphi}} \\ &= \frac{2\pi}{p}\left[1 + \left(\frac{1}{2}\right)^2 k^2 + \left(\frac{1.3}{2.4}\right)^2 k^4 + \cdots\right] \end{aligned}\right\} \tag{34.7}$$

where $k = \sin\frac{1}{2}\alpha$; this gives, to order α^2,

$$\tau = 2\pi\sqrt{\frac{l}{g}}\left[1 + \frac{\alpha^2}{16}\right]. \tag{34.8}$$

---

[1] Cf. (104.19) for the generalization of (33.10) to an oscillating system with more degrees of freedom.

[2] Cf. WHITTAKER [28], p. 72; SYNGE and GRIFFITH [26], p. 371.

If $\omega_0 > 2p$, the motion is no longer oscillatory, the particle going round and round the circle. In this case the solution is[1]

$$\sin \tfrac{1}{2}\vartheta = \operatorname{sn} \frac{p}{k}(t - t_0), \tag{34.9}$$

where $k = 2p/\omega_0 < 1$, and the modulus of sn is k.

If $\omega_0 = 2p$, the particle reaches the highest point of the circle in an infinite time, the motion being given by[2]

$$\sin \tfrac{1}{2}\vartheta = \operatorname{Tan} p(t - t_0). \tag{34.10}$$

For motion under gravity on any smooth curve in a vertical plane, the equation of motion is

$$\ddot{s} + g\frac{dz}{ds} = 0, \tag{34.11}$$

where s is arc length and z height above some fixed level. If

$$z = k s^2, \tag{34.12}$$

we get

$$\ddot{s} + 2gks = 0, \tag{34.13}$$

the same form of equation as for a harmonic oscillator. The period is then independent of amplitude, and the curve (34.12), which is a cycloid, is called a *tautochrone* for the force of gravity[3].

III. Two-dimensional motions.

35. Projectiles. For a particle of mass m which moves in a uniform gravitational field, in a resisting medium with density depending only on height, the equations of motion are

$$\ddot{x} = -f(v, y)\frac{\dot{x}}{v}, \quad \ddot{y} = -g - f(v, y)\frac{\dot{y}}{v}. \tag{35.1}$$

Here the x-axis is horizontal, the y-axis is directed vertically upwards, and $mf(v, y)$ is the force of resistance, opposing the motion; v is the speed.

If $f = 0$, we get the elementary parabolic trajectory,

$$x = a + bt, \quad y = c + et - \tfrac{1}{2}gt^2, \tag{35.2}$$

where a, b, c, e are constants determined by the initial conditions[4]. If $f = kv$, the Eqs. (35.1) have simple exponential solutions[5]. Integration can also be carried out explicitly[6] for $f = k_0 + kv^n$, which contains, as a special case[7], resistance varying as the square of the speed ($f = kv^2$).

The integration of equations of the form (35.1) is a central problem in exterior ballistics, the function $f(v, y)$ being given numerically or by some empirical formula[8]. Numerical integration is used.

---

[1] Cf. WHITTAKER [*28*], p. 73.

[2] The hyperbolic functions sinh, cosh, tanh are printed Sin, Cos, Tan throughout this Encyclopedia.

[3] For tautochrones and brachistochrones (curves of quickest descent), see APPELL [*2*] I, pp. 478–489; MACMILLAN [*17*] I, pp. 322–329.

[4] For geometrical constructions connected with parabolic trajectories, see APPELL [*2*] I, p. 374; LAMB [*13*], p. 72; SYNGE and GRIFFITH [*26*], p. 151.

[5] Cf. SYNGE and GRIFFITH [*26*], p. 159.

[6] Cf. APPELL [*2*] I, p. 383; MACMILLAN [*17*] I, p. 256. This is the integrable case of LEGENDRE.

[7] Cf. SYNGE and GRIFFITH [*26*], p. 157.

[8] For the older approach to exterior ballistics, see P. CHARBONNIER: Traité de Balistique extérieure (2 vols.) (Paris: Doin & Gauthier-Villars 1921–1927); and for the modern approach, see E. J. MCSHANE, J. L. KELLEY and F. RENO: Exterior Ballistics (Denver: University Press 1953).

36. KEPLER problem[1]. In the KEPLER problem, a particle of mass m is attracted to (or repelled from) a fixed point O by a force which varies as the inverse square of its distance r from O. Let the inward component of this force be $m\mu/r^2$ (μ positive for attraction, negative for repulsion). By symmetry, the orbit is plane, and, in terms of polar coordinates (r, ϑ) in the plane of the orbit, the equations of motion are [cf. (30.13)]

$$\ddot{r} - r\dot{\vartheta}^2 = -\frac{\mu}{r^2}, \quad r^2\dot{\vartheta} = h, \tag{36.1}$$

where h is a constant, the angular momentum per unit mass [cf. (31.6)]. The kinetic and potential energies per unit mass are

$$T = \frac{1}{2}(\dot{r}^2 + r^2\dot{\vartheta}^2), \qquad V = -\frac{\mu}{r} \tag{36.2}$$

and we have the equation of energy

$$T + V = E, \tag{36.3}$$

where E is a constant, the total energy per unit mass.

Putting $u = 1/r$, and eliminating t from (36.1), we get

$$\frac{d^2u}{d\vartheta^2} + u = \frac{\mu}{h^2} \tag{36.4}$$

and hence

$$u = \frac{\mu}{h^2} + C\cos(\vartheta - \vartheta_0), \tag{36.5}$$

where C, ϑ_0 are constants of integration. From (36.3), C is determined in terms of E and h, and we can make $\vartheta_0 = 0$ by choice of the line $\vartheta = 0$; then the equation of the orbit may be written

$$u = \frac{1}{r} = \frac{1}{l}(1 + e\cos\vartheta), \tag{36.6}$$

where

$$l = \frac{h^2}{\mu}, \quad e = \sqrt{1 + \frac{2Eh^2}{\mu^2}}. \tag{36.7}$$

The orbit is a conic section of eccentricity e; it is an ellipse, parabola or hyperbola according as $E < 0$, $E = 0$, or $E > 0$, respectively, the centre of force being a focus of the conic.

If the force is repulsive, then $\mu < 0$, $E > 0$, and only the hyperbolic orbit occurs; it is that branch of the hyperbola which is convex towards O.

For an elliptical orbit, the semi-axis-major a and the eccentricity e determine the orbit; likewise E and h determine it. The relations between these constants are

$$\left.\begin{aligned} a &= -\frac{\mu}{2E}, & e &= \sqrt{1 + \frac{2Eh^2}{\mu^2}}, \\ E &= -\frac{\mu}{2a}, & h &= \sqrt{\mu a(1 - e^2)}. \end{aligned}\right\} \tag{36.8}$$

The speed v at distance r is given by

$$v^2 = \mu\left(\frac{2}{r} - \frac{1}{a}\right), \tag{36.9}$$

[1] For the solution of the KEPLER problem by means of the HAMILTON-JACOBI equation, see Sect. 78, or APPELL [2] I, 592–596. For the relativistic KEPLER problem, see Sect. 115. For a force of the form $r^{-2}\Phi(\vartheta)$, see APPELL [2] I, p. 408.

and the period is

$$\tau = \frac{2\pi a^2}{h}\sqrt{1-e^2} = 2\pi\sqrt{\frac{a^3}{\mu}} = \frac{\pi\mu}{\sqrt{-2E^3}}. \tag{36.10}$$

We cannot enter here into the details of elliptical (planetary) orbits, fundamental in celestial mechanics[1].

37. Central forces in general. For a particle of mass m repelled from a fixed point O by a force $mF(u)$, where $u=1/r$, the Eq. (30.13) give

$$\ddot{r} - r\dot{\vartheta}^2 = F, \quad r^2\dot{\vartheta} = h, \tag{37.1}$$

and hence

$$\frac{d^2u}{d\vartheta^2} + u = -\frac{F}{h^2u^2}. \tag{37.2}$$

The case of attraction is here included, F being then negative. A potential V exists, given by

$$V(u) = -\int_{r_0}^{r} F\,dr, \tag{37.3}$$

where r_0 is any constant, and the energy equation $T+V=E$ leads to

$$\left(\frac{du}{d\vartheta}\right)^2 + u^2 = \frac{2(E-V)}{h^2}, \tag{37.4}$$

which is, of course, a first integral of (37.2). This equation can be solved by a quadrature, being of the form[2]

$$\left(\frac{du}{d\vartheta}\right)^2 = f(u) = \frac{2(E-V)}{h^2} - u^2. \tag{37.5}$$

To relate the time t to the variables (u, ϑ), we have, by (37.1),

$$dt = \frac{r^2\,d\vartheta}{h} = \pm\frac{du}{h\,u^2\sqrt{f(u)}}, \tag{37.6}$$

the sign being chosen to make dt positive.

The *apsides* of an orbit are those points at which r is a maximum or a minimum; thus apsides occur for $u=u_1$, $u=u_2$, where

$$f(u_1) = 0, \quad f(u_2) = 0. \tag{37.7}$$

The apsidal angle is

$$\alpha = \int_{u=u_1}^{u=u_2} d\vartheta = \int_{u_1}^{u_2} \frac{du}{\sqrt{f(u)}}. \tag{37.8}$$

The whole orbit may be obtained from the part between two adjacent apsides, since the orbit is symmetric with respect to any apsidal radius. The whole orbit is contained between (and touches) two concentric circles, but in exceptional cases the radius of the inner circle may be zero and the radius of the outer circle infinite.

---

[1] For the anomalies, KEPLER's equation and LAMBERT's theorem, see APPELL [2] I, pp. 434–448; WHITTAKER [28], pp. 89–92; WINTNER [30], Chap. 4. Also MACMILLAN [17] I, pp. 278–292, where a repulsive force is included.

[2] An equation of this type is of frequent occurrence in dynamics, leading in general to periodic solutions. Elliptic functions may be discussed in terms of such an equation; cf. PÉRÈS [20], pp. 107–122, SYNGE and GRIFFITH [26], pp. 364–370.

Borrowing the words from astronomy, we may call any one of the apsides on the inner circle *perihelion*, and any one of the outer apsides *aphelion*.

When the force varies as r ($F = \varepsilon k^2 r$, $\varepsilon = \pm 1$), the motion is more simply discussed by means of rectangular Cartesians; we have

$$\ddot{x} = \varepsilon k^2 x, \quad \ddot{y} = \varepsilon k^2 y. \tag{37.9}$$

If $\varepsilon = -1$, the orbit is a central ellipse, with equations

$$\left.\begin{aligned} x &= A \cos kt + B \sin kt, \\ y &= C \cos kt + D \sin kt. \end{aligned}\right\} \tag{37.10}$$

If $\varepsilon = +1$, the solution is similarly expressed in hyperbolic functions; the orbit is a central hyperbola. In special cases (vanishing angular momentum) the orbit is a straight line through the origin; then we have a simple harmonic oscillator in the case $\varepsilon = -1$.

38. Stability of circular orbits. For a circular orbit of radius $r = 1/u$, we require

$$f(u) = 0, \quad f'(u) = 0. \tag{38.1}$$

The first condition is a consequence of (37.5), and the second follows from (37.2), which is equivalent to

$$\frac{d^2 u}{d\vartheta^2} = \frac{1}{2} f'(u). \tag{38.2}$$

Putting $u = u_0$ for the circular orbit, we write

$$u = u_0 + \xi \tag{38.3}$$

for a disturbed orbit (ξ is small), and obtain from (38.2) and (38.1), to the first order in ξ,

$$\frac{d^2 \xi}{d\vartheta^2} = \frac{1}{2} \xi f''(u_0). \tag{38.4}$$

This has sinusoidal solutions (giving stability) if and only if

$$f''(u_0) < 0. \tag{38.5}$$

Thus we have the following criterion for stability of a circular orbit of radius $r = 1/u$, described under a central attractive force:

$$\frac{1}{2} f''(u) = -1 - \frac{1}{h^2} \frac{d}{du}\left(\frac{F}{u^2}\right) < 0. \tag{38.6}$$

Here F is negative, the magnitude of the attractive force being $-mF$. But, by (37.2), applied to the circular orbit, we have

$$h^2 = -\frac{F}{u^3}, \tag{38.7}$$

and so the criterion for stability may be written

$$3F < u \frac{dF}{du} \tag{38.8}$$

or

$$3F + r \frac{dF}{dr} < 0. \tag{38.9}$$

If the attractive force varies as r^{-n}, there is stability if, and only if, $n < 3$.

39. Oscillations under gravity on a fixed surface[1]. Let S be a smooth fixed surface and O a point on it, the tangent plane at O being horizontal. Let $Oxyz$ be axes with Oz directed vertically upward, and Oxy in principal directions of curvature of S, so that its equation, near O, is

$$z = \frac{1}{2}\left(\frac{x^2}{R_1} + \frac{y^2}{R_2}\right), \tag{39.1}$$

approximately, R_1 and R_2 being the principal radii of curvature. If a particle of mass m moves on S, under the influence of gravity, its equations of motion are accurately

$$\ddot{x} = \lambda N, \quad \ddot{y} = \mu N, \quad \ddot{z} = \nu N - g, \tag{39.2}$$

where mN is the reaction of the surface and λ, μ, ν the direction cosines of the normal. To the first order in x, y, we have

$$\lambda = -\frac{x}{R_1}, \quad \mu = -\frac{y}{R_2}, \quad \nu = 1; \tag{39.3}$$

hence $N = g$ and

$$\ddot{x} + k_1^2 x = 0, \quad \ddot{y} + k_2^2 y = 0, \tag{39.4}$$

$$k_1^2 = \frac{g}{R_1}, \quad k_2^2 = \frac{g}{R_2}.$$

The solution is of the form

$$\left.\begin{aligned} x &= A\cos(k_1 t + B), \\ y &= C\cos(k_2 t + D). \end{aligned}\right\} \tag{39.5}$$

These curves in the xy-plane (called LISSAJOUS figures[2]) represent the composition of simple harmonic motions with different frequencies. They are closed curves if k_1/k_2 is rational; otherwise they fill[3] the whole rectangle $x^2 \leq A^2$, $y^2 \leq C^2$.

A *spherical pendulum* consists of a heavy particle attached to a fixed point by a light inextensible string. This is the same thing as a particle moving under gravity on a smooth fixed sphere. To investigate small oscillations, we put $R_1 = R_2$ in the above theory; the period is

$$\tau = 2\pi\sqrt{\frac{R_1}{g}}, \tag{39.6}$$

and the orbit (39.5) is a central ellipse.

Finite oscillations under gravity on a smooth sphere are much more complicated. Taking the origin at the centre of the sphere, with Oz directed vertically upwards, we have the four equations

$$\ddot{x} = -\frac{x}{a}N, \quad \ddot{y} = -\frac{y}{a}N, \quad \ddot{z} = -\frac{z}{a}N - g, \quad x^2 + y^2 + z^2 = a^2, \tag{39.7}$$

where a is the radius of the sphere. Alternatively, we may use conservation of energy (mE) and conservation of angular momentum (mh) about Oz. We obtain the equations[4]

$$\left.\begin{aligned} &(a^2 - z^2)\dot{\varphi} = h, \quad \dot{z}^2 = f(z), \\ &f(z) = \frac{2g}{a^2}\left[(z^2 - a^2)\left(z - \frac{E}{g}\right) - \frac{h^2}{2g}\right], \end{aligned}\right\} \tag{39.8}$$

---

[1] Cf. APPELL [2] I, pp. 510–541.

[2] For the generation of LISSAJOUS figures by means of BLACKBURN's pendulum, see LAMB [13], p. 80.

[3] Cf. APPELL [2] I, p. 522.

[4] For details, see APPELL [2] I, pp. 530–541; SYNGE and GRIFFITH [26], pp. 375–381.

(r, φ, z) being cylindrical coordinates. The three zeros z_1, z_2, z_3 of the cubic $f(z)$ are necessarily real, with $-a < z_1 < z_2 < a < z_3$, and the solution is expressible in terms of elliptic functions as follows:

$$\left.\begin{aligned} z - z_1 &= (z_2 - z_1)\,\mathrm{sn}^2 p\,(t - t_0)\,,\\ z_2 - z &= (z_2 - z_1)\,\mathrm{cn}^2 p\,(t - t_0)\,,\\ z_3 - z &= (z_3 - z_1)\,\mathrm{dn}^2 p\,(t - t_0)\,, \end{aligned}\right\} \tag{39.9}$$

where

$$p = \sqrt{\frac{g\,(z_3 - z_1)}{2a^2}}\,, \tag{39.10}$$

and the modulus of the elliptic functions is k, where

$$k^2 = \frac{z_2 - z_1}{z_3 - z_1}\,. \tag{39.11}$$

When the oscillations are small, the projection of the orbit on the xy-plane is a small ellipse which rotates through a small angle $3A/(4a^2)$ per orbital revolution, A being the area of the ellipse. This rotation, the sense of which is the same as the sense in which the orbit is described, is not to be confused with the FOUCAULT rotation (cf. Sect. 43).

IV. Three-dimensional motions.

40. Charged particle in electromagnetic field[1]. When a particle of mass m, carrying an electric charge e, moves in an electromagnetic field with electric vector $\boldsymbol{E}$ and magnetic vector $\boldsymbol{H}$, the equation of motion is[2]

$$m\boldsymbol{a} = e(\boldsymbol{E} + \boldsymbol{v}\times\boldsymbol{H})\,. \tag{40.1}$$

This gives

$$\frac{d}{dt}\left(\frac{1}{2}\,m v^2\right) = e\,\boldsymbol{v}\cdot\boldsymbol{E} \tag{40.2}$$

and hence, if $\boldsymbol{E}$ has a potential[3] ($\boldsymbol{E} = -\operatorname{grad} V$), the integral of energy

$$\tfrac{1}{2}\,m v^2 + e\,V = \text{const.} \tag{40.3}$$

If $\boldsymbol{E}$ and $\boldsymbol{H}$ are constant (uniform electromagnetic field), (40.1) is easily solved as follows. If $H = 0$, the trajectory is a parabola. If $H \neq 0$, we resolve $\boldsymbol{v}$ (obliquely) in the form

$$\boldsymbol{v} = v_1\boldsymbol{E} + v_2\boldsymbol{H} + v_3\boldsymbol{E}\times\boldsymbol{H}\,. \tag{40.4}$$

Substituting in (40.1) and separating the components, we obtain

$$\left.\begin{aligned} v_1 &= -A\,H\sin(kHt + B)\,,\\ v_2 &= C - \frac{\boldsymbol{E}\cdot\boldsymbol{H}}{H^2}\,\{kt - A\,H\sin(kHt + B)\}\,,\\ v_3 &= \frac{1}{H^2} - A\cos(kHt + B)\,, \end{aligned}\right\} \tag{40.5}$$

where $k = -e/m$, and A, B, C are constants of integration. The position vector of the particle, resolved as in (40.4), is then given by integration of (40.5).

---

[1] Cf. APPELL [2] I, p. 388.

[2] We take $\boldsymbol{E}$ measured in electrostatic units and $\boldsymbol{H}$ in electromagnetic units; if $\boldsymbol{H}$ is in electrostatic units, we must substitute $\boldsymbol{v}/c$ for $\boldsymbol{v}$, where c is the ratio of the units, i.e. the velocity of light. The relativistic equations have the same right-hand side as above; cf. (115.8).

[3] This is always the case for a field which is independent of the time.

If the electric and magnetic fields are both constant and orthogonal, we have $\boldsymbol{E}\cdot\boldsymbol{H}=0$, and hence $v_2=\text{const}$. Taking axes $Oxyz$ in the directions of $\boldsymbol{E}, \boldsymbol{H}, \boldsymbol{E}\times\boldsymbol{H}$, so that $\dot{x}=v_1 E$, $\dot{y}=v_2 H$, $\dot{z}=v_3 EH$, we have then

$$\left.\begin{aligned} x &= x_0 + A'\cos(kHt+B),\\ y &= y_0 + C't,\\ z &= z_0 + \frac{Et}{H} - A'\sin(kHt+B), \end{aligned}\right\} \tag{40.6}$$

where $A'=AE/k$, $C'=CH$. If $C=0$, this represents motion in a circle in a plane perpendicular to $\boldsymbol{H}$ with angular velocity kH, the centre of the circle moving in a straight line perpendicular to $\boldsymbol{E}$ and $\boldsymbol{H}$ with velocity E/H.

If $E=0$, the trajectory is a circular helix, with axis parallel to the magnetic field, described with azimuthal angular velocity He/m.

41. Axially symmetric electromagnetic fields. In a statical electromagnetic field, the electric potential V and the magnetic potential Ω are harmonic functions. If the field has the z-axis as axis of symmetry, these potentials can be expanded in power series:

$$\left.\begin{aligned} V &= V_0(z) + \tfrac{1}{2}R^2 V_1(z) + \tfrac{1}{24}R^4 V_2(z) + \cdots,\\ \Omega &= \Omega_0(z) + \tfrac{1}{2}R^2\Omega_1(z) + \tfrac{1}{24}R^4\Omega_2(z) + \cdots \end{aligned}\right\} \tag{41.1}$$

where $R^2=x^2+y^2$, and we find, from LAPLACE's equation,

$$V_0''(z) + 2V_1(z) = 0, \quad V_1''(z) + \tfrac{4}{3}V_2(z) = 0, \quad \ldots \tag{41.2}$$

with similar equations for Ω, so that the coefficients in (41.1) depend only on the axial potentials V_0, Ω_0. The equations of motion (40.1) of a charged particle, moving near the z-axis, can be treated approximately. By (40.3) we find $\dot{z}=w$, $w^2=2k(V_0-C)$, where $k=-e/m$ and C is a constant, and, on eliminating the time, we obtain as complex equation of the trajectory

$$\left.\begin{aligned} &\zeta'' + P\zeta' + Q\zeta = 0,\\ &P = \frac{w'}{w} + \frac{ik\Omega_0'}{w}, \qquad Q = \frac{1}{2}\left(\frac{w''}{w} + \frac{w'^2}{w^2}\right) + \frac{ik\Omega_0''}{2w},\\ &\zeta = x + iy. \end{aligned}\right\} \tag{41.3}$$

The prime indicates d/dz. Here P and Q are given functions of z. This second-order differential equation is basic in electron optics; the crude image-forming properties of an electron microscope depend on it[1]. To investigate aberrations, higher approximations must be used[2].

42. Motion relative to rotating earth[3]. Here orbital motion about the sun is neglected, and the earth is regarded as a rigid body rotating with angular velocity $\boldsymbol{\Omega}$. Let $Oxyz$ be rectangular axes (Fig. 12), with O a point on the earth's surface, Oz directed vertically upward (as determined by the plumb line), and Ox pointing south. Let $\boldsymbol{i}, \boldsymbol{j}, \boldsymbol{k}$ be an orthonormal triad in the directions of these axes. To discuss motion relative to the earth's surface, we use this frame of reference as if it were fixed, adding fictitious forces as in (32.1). Let λ be the latitude of O, and R the

[1] For some details, see SYNGE and GRIFFITH [22], pp. 387–403.

[2] For the principles of geometrical optics, see the article by A. MARÉCHAL in Vol. XXIV of this Encyclopedia, and for electron optics and electron microscopes see the articles by W. GLASER and S. LEISEGANG in Vol. XXXIII. See also W. GLASER: Grundlagen der Elektronenoptik. Vienna: Springer 1952.

[3] Cf. APPELL [2] II, pp. 289–292, or SYNGE and GRIFFITH [26], pp. 403–408 for details.

distance of O from the earth's axis. Then

$$\left.\begin{aligned}\boldsymbol{\Omega} &= -\Omega\cos\lambda\,\boldsymbol{i} + \Omega\sin\lambda\,\boldsymbol{k},\\ \boldsymbol{a}_0 &= -R\Omega^2\sin\lambda\,\boldsymbol{i} - R\Omega^2\cos\lambda\,\boldsymbol{k},\end{aligned}\right\}\tag{42.1}$$

and so, if $\boldsymbol{r}$, $\boldsymbol{v}$ now denote the position vector and velocity of a particle relative to $Oxyz$ and $\dot{\boldsymbol{v}}$ its relative acceleration, we have by (32.1)

$$\left.\begin{aligned}m\dot{\boldsymbol{v}} = \boldsymbol{F} + mR\Omega^2(\sin\lambda\,\boldsymbol{i} + \cos\lambda\,\boldsymbol{k}) - 2m\Omega(-\cos\lambda\,\boldsymbol{i} + \sin\lambda\,\boldsymbol{k})\times\boldsymbol{v}\\ - m\,\boldsymbol{\Omega}\times(\boldsymbol{\Omega}\times\boldsymbol{r});\end{aligned}\right\}\tag{42.2}$$

here $\boldsymbol{F}$ is the real force acting on the particle.

Let $\boldsymbol{F}_0$ be the force of gravity at O. Let g be defined by stating that mg is the tension in a plumb line suspending a particle of mass m at O. Then (42.2) are satisfied with

$$\boldsymbol{F} = \boldsymbol{F}_0 + mg\boldsymbol{k}, \quad \boldsymbol{v} = 0, \quad \boldsymbol{r} = 0, \tag{42.3}$$

and therefore we have

$$\boldsymbol{F}_0 + mg\boldsymbol{k} + mR\Omega^2(\sin\lambda\,\boldsymbol{i} + \cos\lambda\,\boldsymbol{k}) = 0. \tag{42.4}$$

We now subtract (42.4) from (42.2), omitting the last term in (42.2) on account of the smallness of Ω. This gives the equations of motion:

$$\left.\begin{aligned}m\ddot{x} &= X + 2m\Omega\sin\lambda\cdot\dot{y}\\ m\ddot{y} &= Y - 2m\Omega(\sin\lambda\cdot\dot{x} + \cos\lambda\cdot\dot{z})\\ m\ddot{z} &= Z - mg + 2m\Omega\cos\lambda\cdot\dot{y}\end{aligned}\right\}\tag{42.5}$$

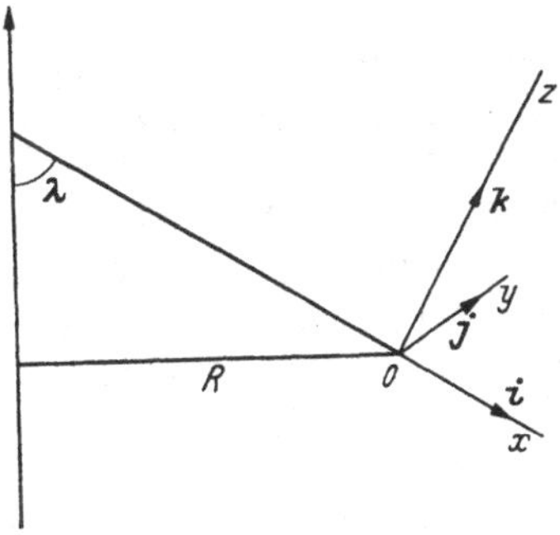

Fig. 12. Axes on rotating earth.

where (X, Y, Z) are the components of the difference between the total real force acting on the particle and the force of gravity on it in the position O.

In the case of a free particle (projectile in vacuo), we put $X = Y = Z = 0$, if we may neglect the variations of gravity with position. Neglecting Ω^2, we obtain, for a particle starting from the origin at $t = 0$ with velocity (u_0, v_0, w_0),

$$\left.\begin{aligned}x &= u_0 t + \Omega v_0 t^2\sin\lambda,\\ y &= v_0 t - \Omega t^2(u_0\sin\lambda + w_0\cos\lambda) + \tfrac{1}{3}\Omega g t^3\cos\lambda,\\ z &= w_0 t - \tfrac{1}{2}g t^2 + \Omega v_0 t^2\cos\lambda.\end{aligned}\right\}\tag{42.6}$$

For free fall from rest at height h, there is a deviation to the east of amount

$$\frac{1}{3}\Omega\cos\lambda\cdot 2h\sqrt{\frac{2h}{g}}. \tag{42.7}$$

For a projectile on a flat trajectory (w_0 small), the projection of the position vector on the horizontal plane Oxy grows at a constant rate and turns at a constant rate $-\Omega\sin\lambda$; this means a deviation to the right in the northern hemisphere, and to the left in the southern hemisphere (FEREL's law).

43. FOUCAULT's pendulum[1]. A particle of mass m is attached by a light string of length l to the point $(0, 0, l)$ (cf. Fig. 12). Then O is a position of equilibrium, and, when disturbed, the particle moves according to (42.5), in which

$$X = -\frac{x}{l}S, \quad Y = -\frac{y}{l}S, \quad Z = \frac{l-z}{l}S, \tag{43.1}$$

[1] Cf. APPELL [2] II, pp. 293—296; SYNGE and GRIFFITH [26], pp. 408—411.

where S is the tension in the string. For small disturbances, z is small of the second order, and we have

$$S = Z = mg - 2m\Omega\cos\lambda\cdot\dot{y} \tag{43.2}$$

by the last of (42.5); the other two equations give

$$\left.\begin{aligned}\ddot{x} - 2\Omega\sin\lambda\cdot\dot{y} + p^2 x = 0,\\ \ddot{y} + 2\Omega\sin\lambda\cdot\dot{x} + p^2 y = 0,\end{aligned}\right\} \tag{43.3}$$

where $p^2 = g/l$. Introducing $\zeta = x + iy$, we write the pair of equations as

$$\ddot{\zeta} + 2i\Omega\sin\lambda\,\dot{\zeta} + p^2\zeta = 0, \tag{43.4}$$

and the solution of this is (neglecting Ω^2)

$$\zeta = (A\,\mathrm{e}^{ipt} + B\,\mathrm{e}^{-ipt})\,\mathrm{e}^{-i\Omega t\sin\lambda}. \tag{43.5}$$

The first factor on the right represents motion in a fixed central ellipse; the effect of the second factor is to make this ellipse rotate with angular velocity $-\Omega\sin\lambda$, which is clockwise in the northern hemisphere and counterclockwise in the southern hemisphere.

This FOUCAULT rotation must not be confused with the areal effect for a spherical pendulum (Sect. 39). If the pendulum is started from a position of rest (by burning a supporting thread), the orbit would be a straight line if the earth were not rotating. Actually, it is elliptical, and the areal angular velocity is

$$\tfrac{3}{8}\alpha^2\Omega\sin\lambda,$$

where α is the initial angular amplitude. Since α is small, this is much less than the FOUCAULT rotation.

D. Dynamics of systems of particles and of rigid bodies.

I. Equations of motion.

44. Principles of linear and angular momentum. Consider a system of P particles with masses m_i $(i = 1, 2, \dots P)$ and position vectors $\boldsymbol{r}_i$ relative to an origin fixed in absolute space S_0. On these particles there act forces $\boldsymbol{F}_i$, which we split into external forces $\boldsymbol{F}_i'$ and internal forces $\boldsymbol{F}_i''$ as in Sect. 26. The equations of motion of the several particles are

$$m_i\ddot{\boldsymbol{r}}_i = \boldsymbol{F}_i = \boldsymbol{F}_i' + \boldsymbol{F}_i'' \quad (i = 1, \dots P). \tag{44.1}$$

Since $\sum_{i=1}^{P}\boldsymbol{F}_i'' = 0$, we deduce

$$\dot{\boldsymbol{M}} = \boldsymbol{F}, \tag{44.2}$$

where

$$\boldsymbol{M} = \sum_{i=1}^{P} m_i\dot{\boldsymbol{r}}_i, \qquad \boldsymbol{F} = \sum_{i=1}^{P}\boldsymbol{F}_i', \tag{44.3}$$

these being respectively the total linear momentum (Sect. 23) and the total external force (Sect. 26). Eq. (44.2) embodies the *principle of linear momentum: the rate of charge of linear momentum equals the total external force.*

We can also write (44.2) in the form

$$m\boldsymbol{a} = \boldsymbol{F}, \tag{44.4}$$

where m is the total mass of the system and $\boldsymbol{a}$ the acceleration of the mass-centre.

Since $\sum_{i=1}^{P} \boldsymbol{r}_i \times \boldsymbol{F}_i'' = 0$, (44.1) also gives

$$\dot{\boldsymbol{h}} = \boldsymbol{G}, \tag{44.5}$$

where

$$\boldsymbol{h} = \sum_{i=1}^{P} m_i \boldsymbol{r}_i \times \dot{\boldsymbol{r}}_i, \quad \boldsymbol{G} = \sum_{i=1}^{P} \boldsymbol{r}_i \times \boldsymbol{F}_i', \tag{44.6}$$

these being respectively the total angular momentum of the absolute motion about the origin, and the total moment of external force about the origin. Hence we have the *principle of angular momentum* in its first form: *the rate of change of angular momentum about a fixed point equals the total moment of external force about that point.*

Using (24.11), we easily deduce from (44.4) and (44.5) the equation

$$\dot{\boldsymbol{h}}^* = \boldsymbol{G}^*, \tag{44.7}$$

where $\boldsymbol{h}^*$ is the angular momentum about the mass-centre O^* of the motion relative to O^*, and $\boldsymbol{G}^*$ is the total moment of external force about O^*. This expresses the principle of angular momentum in its second form, the mass-centre taking the place of the fixed point.

The essence of the above work is the elimination of internal forces by means of NEWTON's Third Law[1]. The principles of linear and angular momentum hold for any Newtonian system. We can of course replace absolute space S_0 by any Newtonian frame S in uniform motion relative to S_0 (cf. Sect. 32).

45. D'ALEMBERT's principle. Energy. Let $\delta\boldsymbol{r}_i\ (i = 1, \ldots P)$ be any set of infinitesimal vectors. By (44.1) we have

$$\sum_{i=1}^{P} m_i \ddot{\boldsymbol{r}}_i \cdot \delta\boldsymbol{r}_i = \sum_{i=1}^{P} \boldsymbol{F}_i \cdot \delta\boldsymbol{r}_i = \delta W, \tag{45.1}$$

where δW is the work done by the forces $\boldsymbol{F}_i$ in the displacements $\delta\boldsymbol{r}_i$. These displacements are called *virtual* to distinguish them from the displacements actually occurring in the motion, viz. $d\boldsymbol{r}_i = \dot{\boldsymbol{r}}_i dt$.

To assimilate (45.1) to the principle of virtual work in statics, the vectors $m_i \ddot{\boldsymbol{r}}_i$ are called *effective forces* and these vectors reversed, viz. $-m_i \ddot{\boldsymbol{r}}_i$, are called *reversed effective forces* or *forces of inertia*. Then (45.1) may be stated in either of the two following forms (D'ALEMBERT's principle):

(i) *In any virtual displacement, the work done by the effective forces is equal to the work done by the actual forces.*

(ii) *In any virtual displacement, the total work done by the forces of inertia and the actual forces is zero.*

The importance of D'ALEMBERT's principle lies in two facts: (a) A set of *vector* equations (44.1) is replaced by a single *scalar* equation; (b) δW involves only those forces which do work in the displacements $\delta\boldsymbol{r}_i$. Hence, if the system has workless constraints (Sect. 26) and the displacements are consistent with them, the forces of constraint are not involved.

The principles of linear and angular momentum (Sect. 44) eliminate internal forces; D'ALEMBERT's principle eliminates workless reactions of constraint[2].

---

[1] Cf. Sect. 26. The axiom of homogeneity and isotropy of space (Sect. 5) would not suffice.

[2] For a discussion of D'ALEMBERT's principle, see APPELL [*2*] II, p. 303, and LANCZOS [*15*], p. 92. For applications to servo-mechanisms (asservissements), see APPELL [*2*] II, p. 403.

If we take the virtual displacements $\delta \boldsymbol{r}_i$ to be the actual displacements occurring in the motion ($\delta \boldsymbol{r}_i = \dot{\boldsymbol{r}}_i \, dt$), then (45.1) gives

$$\sum_{i=1}^{P} m_i \ddot{\boldsymbol{r}}_i \cdot \dot{\boldsymbol{r}}_i = \sum_{i=1}^{P} \boldsymbol{F}_i \cdot \dot{\boldsymbol{r}}_i = \dot{W}, \tag{45.2}$$

where $\dot{W}$ is the rate of working (or activity) of the forces; equivalently

$$\dot{T} = \dot{W}, \tag{45.3}$$

where T is the kinetic energy. *The rate of increase of kinetic energy is equal to the rate of working of the force acting on the system*; all working forces, external and internal, are to be included.

If the system is scleronomic, and a potential function V exists [cf. (29.9)], then (45.3) leads to the *equation of energy* or *integral of energy*

$$T + V = \text{const.} \tag{45.4}$$

This fundamental equation will be derived again in Sect. 46, with a more general form in (46.21).

46. LAGRANGE'S equations. Ignorable coordinates. *α) General theory.* Consider a system of P particles, as in Sects. 44 and 45, subject to constraints which, in general, we suppose to be rheonomic and non-holonomic. Let q_ϱ ($\varrho = 1, \ldots N$) be generalized coordinates, so that the position vectors of the particles may be written

$$\boldsymbol{r}_i = \boldsymbol{r}_i(q, t) \quad (i = 1, \ldots P); \tag{46.1}$$

let the equations of constraint be, as in (28.2),

$$\sum_{\varrho=1}^{N} A_{c\varrho} \, dq_\varrho + A_c \, dt = 0 \quad (c = 1, \ldots M), \tag{46.2}$$

the coefficients being given functions of the $N+1$ variables (q, t).

The velocities of the particles are

$$\dot{\boldsymbol{r}}_i = \sum_{\varrho=1}^{N} \frac{\partial \boldsymbol{r}_i}{\partial q_\varrho} \dot{q}_\varrho + \frac{\partial \boldsymbol{r}_i}{\partial t} \quad (i = 1, \ldots P), \tag{46.3}$$

so that the components of velocity are functions of the $2N+1$ variables $(q, \dot{q}, t)$. For the partial derivatives of these functions, we have

$$\frac{\partial \dot{\boldsymbol{r}}_i}{\partial \dot{q}_\varrho} = \frac{\partial \boldsymbol{r}_i}{\partial q_\varrho}, \qquad \frac{\partial \dot{\boldsymbol{r}}_i}{\partial q_\varrho} = \frac{d}{dt} \frac{\partial \boldsymbol{r}_i}{\partial q_\varrho} \quad (i = 1, \ldots P; \varrho = 1, \ldots N), \tag{46.4}$$

as is easily verified. The kinetic energy is

$$T = \tfrac{1}{2} \sum_{i=1}^{P} m_i \dot{\boldsymbol{r}}_i \cdot \dot{\boldsymbol{r}}_i, \tag{46.5}$$

a function of $(q, \dot{q}, t)$, and its partial derivatives are

$$\left.\begin{aligned} \frac{\partial T}{\partial q_\varrho} &= \sum_{i=1}^{P} m_i \dot{\boldsymbol{r}}_i \cdot \frac{\partial \dot{\boldsymbol{r}}_i}{\partial q_\varrho}, \\ \frac{\partial T}{\partial \dot{q}_\varrho} &= \sum_{i=1}^{P} m_i \dot{\boldsymbol{r}}_i \cdot \frac{\partial \dot{\boldsymbol{r}}_i}{\partial \dot{q}_\varrho} = \sum_{i=1}^{P} m_i \dot{\boldsymbol{r}}_i \cdot \frac{\partial \boldsymbol{r}_i}{\partial q_\varrho}. \end{aligned}\right\} \tag{46.6}$$

Hence, by (46.4),

$$\frac{d}{dt}\frac{\partial T}{\partial \dot{q}_\varrho} - \frac{\partial T}{\partial q_\varrho} = \sum_{i=1}^{P} m_i \ddot{\boldsymbol{r}}_i \cdot \frac{\partial \boldsymbol{r}_i}{\partial q_\varrho} \quad (\varrho = 1, \dots N). \tag{46.7}$$

Let δq_ϱ ($\varrho = 1, \dots, N$) be an arbitrary set of infinitesimals, and let $\delta \boldsymbol{r}_i$ be corresponding displacements of the particles of the system, obtained by differentiating (46.1) with t held fixed, so that

$$\delta \boldsymbol{r}_i = \sum_{\varrho=1}^{N} \frac{\partial \boldsymbol{r}_i}{\partial q_\varrho} \delta q_\varrho \quad (i = 1, \dots P). \tag{46.8}$$

Multiplying (46.7) by δq_ϱ and summing with respect to ϱ, we obtain

$$\sum_{\varrho=1}^{N} \left(\frac{d}{dt}\frac{\partial T}{\partial \dot{q}_\varrho} - \frac{\partial T}{\partial q_\varrho} \right) \delta q_\varrho = \sum_{i=1}^{P} m_i \ddot{\boldsymbol{r}}_i \cdot \delta \boldsymbol{r}_i. \tag{46.9}$$

Note that this is a purely kinematical result, no forces or equations of motion having been used; nor have the equations of constraint (46.2) been involved.

We now introduce the generalized force Q_ϱ^*, split into two parts as in (29.6):

$$Q_\varrho^* = Q_\varrho + Q_\varrho' \quad (\varrho = 1, \dots N), \tag{46.10}$$

where Q_ϱ is the given (or applied) force, and Q_ϱ' the force of constraint. We assume the constraint workless, in the sense that

$$\sum_{\varrho=1}^{N} Q_\varrho' \delta q_\varrho = 0 \tag{46.11}$$

for all values of δq_ϱ satisfying

$$\sum_{\varrho=1}^{N} A_{c\varrho} \delta q_\varrho = 0 \quad (c = 1, \dots M). \tag{46.12}$$

We turn now to D'ALEMBERT'S principle (45.1). Choose any δq_ϱ satisfying (46.12), and let $\delta \boldsymbol{r}_i$ be the corresponding displacements as in (46.8). We use (46.9) to change the first term in (45.1), and for the last term we have

$$\delta W = \sum_{\varrho=1}^{N} Q_\varrho^* \delta q_\varrho = \sum_{\varrho=1}^{N} Q_\varrho \delta q_\varrho, \tag{46.13}$$

in view of (46.11). Thus (45.1) gives

$$\sum_{\varrho=1}^{N} \left(\frac{d}{dt}\frac{\partial T}{\partial \dot{q}_\varrho} - \frac{\partial T}{\partial q_\varrho} - Q_\varrho \right) \delta q_\varrho = 0, \tag{46.14}$$

for all δq_ϱ satisfying (46.12).

From this last equation we obtain at once LAGRANGE'S *equations of motion for non-holonomic systems*:

$$\frac{d}{dt}\frac{\partial T}{\partial \dot{q}_\varrho} - \frac{\partial T}{\partial q_\varrho} = Q_\varrho + \sum_{c=1}^{M} \vartheta_c A_{c\varrho} \quad (\varrho = 1, \dots N) \tag{46.15}$$

where ϑ_c are undetermined multipliers. These equations are to be supplemented by the equations of constraint (46.2) in the form

$$\sum_{\varrho=1}^{N} A_{c\varrho} \dot{q}_\varrho + A_c = 0 \quad (c = 1, \dots M), \tag{46.16}$$

so that we have in all $N+M$ equations for the $N+M$ quantities (q_ϱ, ϑ_c).

We can write down the Eqs. (46.15) explicitly for any given system as soon as we given the form of the function T and the forms of the functions Q_ϱ; these last are most easily obtained in practice by calculating the work δW and using (46.13).

β) Holonomic systems. For a holonomic system we can take N to be the smallest possible number of generalized coordinates. The ϑ-terms disappear from (46.15), and we have LAGRANGE'S *equations of motion for a holonomic system*:

$$\frac{d}{dt}\frac{\partial T}{\partial \dot{q}_\varrho} - \frac{\partial T}{\partial q_\varrho} = Q_\varrho \quad (\varrho = 1, \ldots N). \tag{46.17}$$

But even though the system be holonomic, it is sometimes convenient to use more than the minimum number of coordinates; then the equations of motion are as in (46.15), supplemented with (integrable) equations of the form (46.16).

If a holonomic system possesses a potential function V as in (29.9) (in other words, a potential energy), or an extended potential function as in (29.11), the equations of motion (46.17) may be written

$$\frac{d}{dt}\frac{\partial L}{\partial \dot{q}_\varrho} - \frac{\partial L}{\partial q_\varrho} = 0 \quad (\varrho = 1, \ldots N), \tag{46.18}$$

where

$$L = T - V. \tag{46.19}$$

Here L is a function of the $2N+1$ variables $(q, \dot{q}, t)$; it is called the *Lagrangian function*, or the ***kinetic potential***. The peculiar virtue of (46.18) is that the equations of motion of a system can be written down once a single function is given. Also, if two different physical systems have Lagrangian functions of the same form, then they behave in the same way.

If we multiply (46.18) by $\dot{q}_\varrho$ and sum with respect to ϱ, the result may be rearranged to read

$$\frac{d}{dt}\left(\sum_{\varrho=1}^{N} \dot{q}_\varrho \frac{\partial L}{\partial \dot{q}_\varrho} - L\right) + \frac{\partial L}{\partial t} = 0. \tag{46.20}$$

If L does not depend explicitly on t (and this may happen even for a rheonomic system), we have $\partial L/\partial t = 0$; in that case (46.20) gives

$$\sum_{\varrho=1}^{N} \dot{q}_\varrho \frac{\partial L}{\partial \dot{q}_\varrho} - L = K, \tag{46.21}$$

a constant.

The kinetic energy is quadratic in the generalized velocities $(\dot{q}_\varrho)$, and we may write it

$$T = T_2 + T_1 + T_0, \tag{46.22}$$

where the subscripts indicate degrees of homogeneity in the generalized velocities. If $V = V(q)$, an ordinary potential function, application of EULER'S theorem for homogeneous functions to (46.21) gives

$$T_2 - T_0 + V = K. \tag{46.23}$$

If, further, the system is scleronomic, we have $T = T_2$, and (46.23) becomes

$$T + V = K, \tag{46.24}$$

the equation of energy or integral of energy, as in (45.4).

γ) Ignorable coordinates. Consider a holonomic system with Lagrangian L. If some one coordinate q_ϱ is absent from L, that coordinate is said to be *ignorable*[1], and the process described below is called the *ignoration* of coordinates.

If q_ϱ is ignorable, then the corresponding Lagrangian equation of motion (46.18) gives

$$\frac{\partial L}{\partial \dot{q}_\varrho} = c_\varrho, \tag{46.25}$$

a constant. This is a first integral of the equations of motion. If $q_1, \ldots, q_M$ are ignorable coordinates, there are M integrals like (46.25). Solving these equations, we obtain the velocities corresponding to the ignorable coordinates (that is, $\dot{q}_1, \ldots \dot{q}_M$) as functions of the other coordinates and velocities, of the time t, and of the constants $c_1, \ldots c_M$. The *Routhian function* R, defined as

$$R = L - \sum_{\varrho=1}^{M} \dot{q}_\varrho \frac{\partial L}{\partial \dot{q}_\varrho} = L - \sum_{\varrho=1}^{M} \dot{q}_\varrho c_\varrho, \tag{46.26}$$

can be expressed in the form

$$R = R(q_{M+1}, \ldots q_N, t, \dot{q}_{M+1}, \ldots \dot{q}_N, c_1, \ldots c_M). \tag{46.27}$$

This function, as we shall now show, may be used to replace L in the equations of motion.

Since the dynamical system may be thought of in any configuration at any time with any generalized velocities, the $2N - M + 1$ quantities

$$q_{M+1}, \ldots q_N, t, \dot{q}_1, \ldots \dot{q}_N \tag{46.28}$$

may be regarded as independently variable; equivalently, the $2N - M + 1$ quantities

$$q_{M+1}, \ldots q_N, t, c_1, \ldots c_M, \dot{q}_{M+1}, \ldots \dot{q}_N, \tag{46.29}$$

are independently variable. On applying such a variation, we get from (46.27) and (46.26), with (46.25),

$$\left.\begin{aligned} \delta R &= \sum_{\varrho=M+1}^{N} \frac{\partial R}{\partial q_\varrho} \delta q_\varrho + \frac{\partial R}{\partial t} \delta t + \sum_{\varrho=M+1}^{N} \frac{\partial R}{\partial \dot{q}_\varrho} \delta \dot{q}_\varrho + \sum_{\varrho=1}^{M} \frac{\partial R}{\partial c_\varrho} \delta c_\varrho \\ &= \sum_{\varrho=M+1}^{N} \frac{\partial L}{\partial q_\varrho} \delta q_\varrho + \frac{\partial L}{\partial t} \delta t + \sum_{\varrho=1}^{N} \frac{\partial L}{\partial \dot{q}_\varrho} \delta \dot{q}_\varrho - \sum_{\varrho=1}^{M} (\dot{q}_\varrho \delta c_\varrho + c_\varrho \delta \dot{q}_\varrho) \\ &= \sum_{\varrho=M+1}^{N} \frac{\partial L}{\partial q_\varrho} \delta q_\varrho + \frac{\partial L}{\partial t} \delta t + \sum_{\varrho=M+1}^{N} \frac{\partial L}{\partial \dot{q}_\varrho} \delta \dot{q}_\varrho - \sum_{\varrho=1}^{M} \dot{q}_\varrho \delta c_\varrho. \end{aligned}\right\} \tag{46.30}$$

Therefore, treating the variations of the quantities (46.29) as independent, we have

$$\frac{\partial R}{\partial q_\varrho} = \frac{\partial L}{\partial q_\varrho}, \quad \frac{\partial R}{\partial \dot{q}_\varrho} = \frac{\partial L}{\partial \dot{q}_\varrho} \quad (\varrho = M+1, \ldots N) \tag{46.31}$$

and

$$\frac{\partial R}{\partial t} = \frac{\partial L}{\partial t}, \quad \frac{\partial R}{\partial c_\varrho} = -\dot{q}_\varrho \quad (\varrho = 1, \ldots M). \tag{46.32}$$

---

[1] The words *kinosthenic* and *cyclic* are also used, particularly the latter, which is a pity, because *cyclic* may be needed in a topological sense; cf. Sect. 63. The word *ignorable* has been used in different senses: (i) absent from T, and (ii) absent from L; cf. GOLDSTEIN [7], p. 48; LANCZOS [*15*], p. 125.

Substituting from (46.31) in LAGRANGE's equations (46.18), we get equations of motion in the form

$$\frac{d}{dt}\frac{\partial R}{\partial \dot{q}_\varrho} - \frac{\partial R}{\partial q_\varrho} = 0 \quad (\varrho = M+1, \dots N). \tag{46.33}$$

The only unknowns in these equations are the $N-M$ non-ignorable coordinates q_ϱ. The equations contain the constants $c_1, \dots, c_M$. The original equations (46.18), being N equations of the second order, formed a system of differential equations of order $2N$; in (46.33) we have a system of order $2N-2M$, the Lagrangian form being preserved, with R in place of L. The passage from (46.18) to (46.33) is the process of ignoration of coordinates.

If (46.33) have been solved for the non-ignorable coordinates, the ignorable coordinates are given by

$$q_\varrho = -\int \frac{\partial R}{\partial c_\varrho}\, dt \quad (\varrho = 1, \dots M). \tag{46.34}$$

47. HAMILTON's equations. Consider any system with N degrees of freedom for which the motion is given by the Lagrangian equations (46.18), L being any function of the generalized coordinates q_ϱ, their derivatives $\dot{q}_\varrho$, and the time t. Define the *generalized momenta* p_ϱ by

$$p_\varrho = \frac{\partial L}{\partial \dot{q}_\varrho} \quad (\varrho = 1, \dots N). \tag{47.1}$$

If

$$\det \frac{\partial^2 L}{\partial \dot{q}_\varrho\, \partial \dot{q}_\sigma} \neq 0, \tag{47.2}$$

as is in general the case, (47.1) can be solved for the generalized velocities, so that we have

$$\dot{q}_\varrho = f_\varrho(q, t, p) \quad (\varrho = 1, \dots N). \tag{47.3}$$

Then L can be expressed as a function of the $2N+1$ variables (q, t, p), and so also can the *Hamiltonian function* H defined by

$$H(q, t, p) = \sum_{\varrho=1}^{N} p_\varrho \dot{q}_\varrho - L. \tag{47.4}$$

We now regard the $2N+1$ quantities (q, t, p) as independently variable, the generalized velocities being expressed in terms of them by (47.3). For an arbitrary variation we have

$$\left.\begin{aligned} \delta H &= \sum_\varrho \frac{\partial H}{\partial p_\varrho}\delta p_\varrho + \sum_\varrho \frac{\partial H}{\partial q_\varrho}\delta q_\varrho + \frac{\partial H}{\partial t}\delta t \\ &= \sum_\varrho \dot{q}_\varrho \delta p_\varrho + \sum_\varrho p_\varrho \delta \dot{q}_\varrho - \sum_\varrho \frac{\partial L}{\partial \dot{q}_\varrho}\delta \dot{q}_\varrho - \sum_\varrho \frac{\partial L}{\partial q_\varrho}\delta q_\varrho - \frac{\partial L}{\partial t}\delta t, \end{aligned}\right\} \tag{47.5}$$

all summations running $1, \dots, N$. By (47.1) the second and third summations on the right cancel, and we obtain the equations

$$\frac{\partial H}{\partial p_\varrho} = \dot{q}_\varrho, \qquad \frac{\partial H}{\partial q_\varrho} = -\frac{\partial L}{\partial q_\varrho}, \qquad \frac{\partial H}{\partial t} = -\frac{\partial L}{\partial t} \quad (\varrho = 1, \dots N). \tag{47.6}$$

On making use of (47.1), we can now write the Lagrangian equations (46.18) in the form

$$\dot{q}_\varrho = \frac{\partial H}{\partial p_\varrho}, \qquad \dot{p}_\varrho = -\frac{\partial H}{\partial q_\varrho} \quad (\varrho = 1, \dots N). \tag{47.7}$$

These are HAMILTON'S *equations of motion*; they are also called *canonical* equations. The passage from LAGRANGE'S equations to HAMILTON'S is a purely mathematical process, without reference to the original dynamical system; thus HAMILTON'S equations hold for any system provided LAGRANGE'S equations hold in the form (46.18); in particular, (47.7) are the equations of motion of any holonomic system (rheonomic or scleronomic), provided a potential function V, or an extended potential, exists.

For the rate of change of H we have, using (47.7),

$$\dot{H} = \sum_{\varrho=1}^{N}\left(\frac{\partial H}{\partial q_\varrho}\dot{q}_\varrho + \frac{\partial H}{\partial p_\varrho}\dot{p}_\varrho\right) + \frac{\partial H}{\partial t} = \frac{\partial H}{\partial t}. \tag{47.8}$$

Thus, if H does not involve t explicitly, so that $\partial H/\partial t = 0$, we have

$$H = \text{const}, \tag{47.9}$$

which may be called an integral of energy. If $T = T_2 + T_1 + T_0$ as in (46.22), and $V = V(q)$, then

$$H = \sum_{\varrho=1}^{N}\dot{q}_\varrho \frac{\partial L}{\partial \dot{q}_\varrho} - L = T_2 - T_0 + V, \tag{47.10}$$

and if $T = T_2$ (the case most commonly encountered in dynamics), then

$$H = T + V, \tag{47.11}$$

so that in this case *H is the total energy of the system, kinetic and potential.*

If, in addition to the forces taken care of by the potential function V (or the extended potential), forces Q_ϱ act on the system, the Lagrangian equations (46.18) are modified to read

$$\frac{d}{dt}\frac{\partial L}{\partial \dot{q}_\varrho} - \frac{\partial L}{\partial q_\varrho} = Q_\varrho \qquad (\varrho = 1, \dots N). \tag{47.12}$$

To convert these into Hamiltonian form, we note that (47.6) were obtained by purely mathematical manipulations, without reference to the equations of motion, and they are valid in the present case. It follows that the Hamiltonian equations (47.7) are modified to read

$$\dot{q}_\varrho = \frac{\partial H}{\partial p_\varrho}, \qquad \dot{p}_\varrho = -\frac{\partial H}{\partial q_\varrho} + Q_\varrho \qquad (\varrho = 1, \dots N). \tag{47.13}$$

48. APPELL'S equations[1]. For a system of P particles, the *energy of acceleration* is defined as

$$S = \tfrac{1}{2}\sum_{i=1}^{P} m\,\ddot{\boldsymbol{r}}_i\cdot\ddot{\boldsymbol{r}}_i, \tag{48.1}$$

accelerations replacing velocities in the definition of kinetic energy. If q_ϱ ($\varrho = 1, \dots N$) are generalized coordinates so that $\boldsymbol{r}_i = \boldsymbol{r}_i(q, t)$, then

$$\left.\begin{aligned} \dot{\boldsymbol{r}}_i &= \sum_{\varrho=1}^{N}\frac{\partial \boldsymbol{r}_i}{\partial q_\varrho}\dot{q}_\varrho + \frac{\partial \boldsymbol{r}_i}{\partial t},\\ \ddot{\boldsymbol{r}}_i &= \sum_{\varrho=1}^{N}\frac{\partial \boldsymbol{r}_i}{\partial q_\varrho}\ddot{q}_\varrho + \sum_{\varrho,\sigma=1}^{N}\frac{\partial^2 \boldsymbol{r}_i}{\partial q_\varrho\,\partial q_\sigma}\dot{q}_\varrho\dot{q}_\sigma + 2\sum_{\varrho=1}^{N}\frac{\partial^2 \boldsymbol{r}_i}{\partial q_\varrho\,\partial t}\dot{q}_\varrho + \frac{\partial^2 \boldsymbol{r}_i}{\partial t^2}\\ &\qquad (i = 1, \dots, P).\end{aligned}\right\} \tag{48.2}$$

[1] Cf. P. APPELL: Sur une forme générale des équations de la dynamique. Paris: Gauthier-Villars 1925; APPELL [2] II, pp. 388, 412, 498; NORDHEIM [18], p. 69; PÉRÈS [20], p. 219.

We can therefore write

$$S = S(q, \dot{q}, \ddot{q}, t). \tag{48.3}$$

This function of $3N+1$ quantities is APPELL's *function*; it is quadratic in the second derivatives $\ddot{q}_\varrho$, and its partial derivative with respect to one of them is [cf. (46.7)]

$$\left.\begin{aligned} \frac{\partial S}{\partial \ddot{q}_\varrho} = \sum_{i=1}^{P} m_i \ddot{\boldsymbol{r}}_i \cdot \frac{\partial \ddot{\boldsymbol{r}}_i}{\partial \ddot{q}_\varrho} = \sum_{i=1}^{P} m_i \ddot{\boldsymbol{r}}_i \cdot \frac{\partial \boldsymbol{r}_i}{\partial q_\varrho} = \frac{d}{dt}\frac{\partial T}{\partial \dot{q}_\varrho} - \frac{\partial T}{\partial q_\varrho} \\ (\varrho = 1, \dots N). \end{aligned}\right\} \tag{48.4}$$

Thus, if the system is holonomic and the q's form a system of coordinates of minimum number, LAGRANGE's equations (46.18) lead at once to APPELL's *equations of motion*

$$\frac{\partial S}{\partial \ddot{q}_\varrho} = Q_\varrho \quad (\varrho = 1, \dots N). \tag{48.5}$$

Suppose now that constraints, in general non-holonomic, are imposed by the equations

$$\sum_{\varrho=1}^{N} A_{c\varrho} \, dq_\varrho + A_c \, dt = 0 \quad (c = 1, \dots M) \tag{48.6}$$

as in (46.2), so that LAGRANGE's equations take the form (46.15). Using the purely kinematical result (48.4), we can express (46.14) as follows:

$$\sum_{\varrho=1}^{N} \frac{\partial S}{\partial \ddot{q}_\varrho} \delta q_\varrho = \sum_{\varrho=1}^{N} Q_\varrho \, \delta q_\varrho, \tag{48.7}$$

for all δq_ϱ satisfying

$$\sum_{\varrho=1}^{N} A_{c\varrho} \, \delta q_\varrho = 0 \quad (c = 1, \dots M). \tag{48.8}$$

By (48.6) we have

$$\sum_{\varrho=1}^{N} A_{c\varrho} \dot{q}_\varrho + A_c = 0 \quad (c = 1, \dots M), \tag{48.9}$$

and hence, differentiating with respect to t,

$$\sum_{\varrho=1}^{N} A_{c\varrho} \ddot{q}_\varrho + B_c(q, \dot{q}, t) = 0 \quad (c = 1, \dots M), \tag{48.10}$$

B_c being a function of the $2N+1$ variables indicated. These last equations may be used to express $\ddot{q}_1, \dots \ddot{q}_M$ in terms of the $3N-M+1$ quantities

$$q_1, \dots q_N, \dot{q}_1, \dots \dot{q}_N, \ddot{q}_{M+1} \dots, \ddot{q}_N, t,$$

and so we can write

$$\left.\begin{aligned} &S(q_1, \dots q_N, \dot{q}_1, \dots \dot{q}_N, \ddot{q}_1, \dots \ddot{q}_N, t) \\ &\quad = \overline{S}(q_1, \dots q_N, \dot{q}_1, \dots \dot{q}_N, \ddot{q}_{M+1}, \dots \ddot{q}_N, t). \end{aligned}\right\} \tag{48.11}$$

If we give variations $\delta\ddot{q}_1, \dots, \delta\ddot{q}_N$, arbitrary except for the conditions

$$\sum_{\varrho=1}^{N} A_{c\varrho} \, \delta\ddot{q}_\varrho = 0 \quad (c = 1, \dots M), \tag{48.12}$$

it follows from (48.10) and (48.11) that

$$\sum_{\varrho=1}^{N} \frac{\partial S}{\partial \ddot{q}_\varrho} \delta\ddot{q}_\varrho = \sum_{r=M+1}^{N} \frac{\partial \overline{S}}{\partial \ddot{q}_r} \delta\ddot{q}_r. \tag{48.13}$$

This is equivalent to saying that

$$\sum_{\varrho=1}^{N} \frac{\partial S}{\partial \ddot{q}_\varrho} \delta q_\varrho = \sum_{r=M+1}^{N} \frac{\partial \bar{S}}{\partial \ddot{q}_r} \delta q_r \tag{48.14}$$

for all variations $\delta q_1, \ldots, \delta q_N$ which satisfy

$$\sum_{\varrho=1}^{N} A_{c\varrho} \delta q_\varrho = 0 \quad (c = 1, \ldots M). \tag{48.15}$$

We now define $\bar{Q}_{M+1}, \ldots \bar{Q}_N$ by the condition that

$$\sum_{\varrho=1}^{N} Q_\varrho \delta q_\varrho = \sum_{r=M+1}^{N} \bar{Q}_r \delta q_r \tag{48.16}$$

for all variations satisfying (48.15). Then, by (48.14) and (48.16), we may write (48.7) in the form

$$\sum_{r=M+1}^{N} \frac{\partial \bar{S}}{\partial \ddot{q}_r} \delta q_r = \sum_{r=M+1}^{N} \bar{Q}_r \delta q_r, \tag{48.17}$$

and since these variations are arbitrary, we have

$$\frac{\partial \bar{S}}{\partial \ddot{q}_r} = \bar{Q}_r \quad (r = M+1, \ldots N), \tag{48.18}$$

which are APPELL's equations of motion in a form valid for non-holonomic systems[1].

49. Equations of motion of a rigid body. Consider a rigid body of mass m with mass-centre O and principal moments of inertia A, B, C at O. The four numbers m, A, B, C specify the body dynamically.

Let q_1, q_2, q_3 be generalized coordinates describing the position of O in absolute space S_0, and let q_1', q_2', q_3' be generalized coordinates describing the position of the body relative to O, i.e. specifying the directions of principal axes fixed in the body relative to axes fixed in space. By the theorem of KÖNIG (Sect. 25), the kinetic energy of the body may be written

$$T = T_0(q, \dot{q}) + T'(q', \dot{q}'), \tag{49.1}$$

where T_0 is the kinetic energy of a particle of mass m moving with O, and T' is the kinetic energy of the motion relative to O; these functions are of the forms

$$T_0 = \sum_{\varrho,\sigma=1}^{3} a_{\varrho\sigma}(q) \dot{q}_\varrho \dot{q}_\sigma, \qquad T' = \sum_{\varrho,\sigma=1}^{3} a'_{\varrho\sigma}(q') \dot{q}'_\varrho \dot{q}'_\sigma. \tag{49.2}$$

If the coordinates q are rectangular Cartesian coordinates (x, y, z), we have

$$T_0 = \tfrac{1}{2} m (\dot{x}^2 + \dot{y}^2 + \dot{z}^2), \tag{49.3}$$

and if the coordinates q' are the Eulerian angles $(\vartheta, \varphi, \psi)$, we have, as in (25.5),

$$\left.\begin{aligned} T' = \tfrac{1}{2} A (\dot{\vartheta} \sin\psi - \dot{\varphi} \sin\vartheta \cos\psi)^2 + \tfrac{1}{2} B (\dot{\vartheta} \cos\psi + \dot{\varphi} \sin\vartheta \sin\psi)^2 + \\ + \tfrac{1}{2} C (\dot{\psi} + \dot{\varphi} \cos\vartheta)^2. \end{aligned}\right\} \tag{49.4}$$

In the case of axial symmetry $(A = B)$, this simplifies to

$$T' = \tfrac{1}{2} A (\dot{\vartheta}^2 + \dot{\varphi}^2 \sin^2\vartheta) + \tfrac{1}{2} C (\dot{\psi} + \dot{\varphi} \cos\vartheta)^2. \tag{49.5}$$

[1] For the application of these equations to servo-mechanisms (asservissements), see APPELL [2] II, pp. 412–416.

Let Q_ϱ, Q'_ϱ $(\varrho = 1, 2, 3)$ be generalized forces such that the work done in an arbitrary displacement is

$$\delta W = \sum_{\varrho=1}^{3} Q_\varrho \, \delta q_\varrho + \sum_{\varrho=1}^{3} Q'_\varrho \, \delta q'_\varrho. \tag{49.6}$$

We have then the six Lagrangian equations of motion, as in (46.17),

$$\left.\begin{aligned} \frac{d}{dt}\frac{\partial T_0}{\partial \dot q_\varrho} - \frac{\partial T_0}{\partial q_\varrho} &= Q_\varrho, \\ \frac{d}{dt}\frac{\partial T'}{\partial \dot q'_\varrho} - \frac{\partial T'}{\partial q'_\varrho} &= Q'_\varrho \end{aligned}\right\} \quad (\varrho = 1, 2, 3). \tag{49.7}$$

The coordinates q are separated from the coordinates q' on the left hand sides, but this separation does not, in general, extend to the right hand sides; in other words, the problem of the motion of a rigid body does not, in general, split into two problems.

These Lagrangian equations hold for a rigid body without constraint. They hold also in the case of constraints, provided we include in Q_ϱ and Q'_ϱ contributions from the forces of constraint.

In discussing the motion of a rigid body, it is often more convenient to use the principles of linear and angular momentum instead of LAGRANGE'S equations. By (44.4) and (44.7) (dropping the asterisks), we have the two vector equations

$$m\boldsymbol{a} = \boldsymbol{F}, \tag{49.8}$$

$$\dot{\boldsymbol{h}} = \boldsymbol{G}, \tag{49.9}$$

where

$\boldsymbol{a}$ = acceleration of the mass-centre O,
$\boldsymbol{h}$ = angular momentum about O of the motion relative to O,
$\boldsymbol{F}$ = total external force,
$\boldsymbol{G}$ = total moment of external force about O.

In order to use these equations for the determination of the motion, we have to resolve them into components along some orthonormal triad; we can choose a triad fixed in absolute space, a triad fixed in the body, or neither of these. We shall later consider a triad fixed in absolute space; for practical purposes it is best to choose a moving triad which is a principal triad of inertia for the body.

But let us start by taking any arbitrary orthonormal triad $(\boldsymbol{i}, \boldsymbol{j}, \boldsymbol{k})$, rotating with angular velocity $\boldsymbol{\Omega}$. Resolving along this triad, we have

$$\left.\begin{aligned} \boldsymbol{h} &= h_1\boldsymbol{i} + h_2\boldsymbol{j} + h_3\boldsymbol{k}, \\ \boldsymbol{G} &= G_1\boldsymbol{i} + G_2\boldsymbol{j} + G_3\boldsymbol{k}, \\ \boldsymbol{\Omega} &= \Omega_1\boldsymbol{i} + \Omega_2\boldsymbol{j} + \Omega_3\boldsymbol{k}. \end{aligned}\right\} \tag{49.10}$$

Then, by (20.3), we can write (49.9) in the form

$$\frac{\delta \boldsymbol{h}}{\delta t} + \boldsymbol{\Omega} \times \boldsymbol{h} = \boldsymbol{G}, \tag{49.11}$$

or, explicitly,

$$\left.\begin{aligned} \dot h_1 - h_2\Omega_3 + h_3\Omega_2 &= G_1, \\ \dot h_2 - h_3\Omega_1 + h_1\Omega_3 &= G_2, \\ \dot h_3 - h_1\Omega_2 + h_2\Omega_1 &= G_3. \end{aligned}\right\} \tag{49.12}$$

Let us now choose $(\boldsymbol{i}, \boldsymbol{j}, \boldsymbol{k})$ be a principal triad of inertia at O, the corresponding moments of inertia being A, B, C. If $\boldsymbol{\omega}$ is the angular velocity of the body, we have then, by (24.14),

$$\left.\begin{aligned} \boldsymbol{\omega} &= \omega_1 \boldsymbol{i} + \omega_2 \boldsymbol{j} + \omega_3 \boldsymbol{k}, \\ \boldsymbol{h} &= A\,\omega_1 \boldsymbol{i} + B\,\omega_2 \boldsymbol{j} + C\,\omega_3 \boldsymbol{k}. \end{aligned}\right\} \tag{49.13}$$

There are three cases to consider.

α) Unsymmetrical body (A, B, and C all different). In this case the triad $(\boldsymbol{i}, \boldsymbol{j}, \boldsymbol{k})$, if it is to be principal, must be fixed in the body. Therefore $\boldsymbol{\Omega} = \boldsymbol{\omega}$, and (49.12) with (49.13) give us EULER'S *equations of motion for a rigid body*

$$\left.\begin{aligned} A\,\dot{\omega}_1 - (B - C)\,\omega_2\,\omega_3 &= G_1, \\ B\,\dot{\omega}_2 - (C - A)\,\omega_3\,\omega_1 &= G_2, \\ C\,\dot{\omega}_3 - (A - B)\,\omega_1\,\omega_2 &= G_3. \end{aligned}\right\} \tag{49.14}$$

β) Body with axis of symmetry $(A = B \neq C)$. Now (49.14) simplify to

$$\left.\begin{aligned} A\,\dot{\omega}_1 - (A - C)\,\omega_2\,\omega_3 &= G_1, \\ A\,\dot{\omega}_2 - (C - A)\,\omega_3\,\omega_1 &= G_2, \\ C\,\dot{\omega}_3 &= G_3. \end{aligned}\right\} \tag{49.15}$$

But we are no longer compelled to fix $(\boldsymbol{i}, \boldsymbol{j}, \boldsymbol{k})$ in the body in order to have a principal triad; it is sufficient to fix $\boldsymbol{k}$ in the body. Doing this, we have

$$\Omega_1 = \omega_1, \quad \Omega_2 = \omega_2, \tag{49.16}$$

Ω_3 remaining arbitrary, and (49.12) becomes

$$\left.\begin{aligned} A\,\dot{\omega}_1 - A\,\omega_2\,\Omega_3 + C\,\omega_3\,\omega_2 &= G_1, \\ A\,\dot{\omega}_2 + A\,\omega_1\,\Omega_3 - C\,\omega_3\,\omega_1 &= G_2, \\ C\,\dot{\omega}_3 &= G_3. \end{aligned}\right\} \tag{49.17}$$

The first two equations may be exhibited in complex form[1]:

$$\left.\begin{aligned} A\,\dot{\omega} + (A\,\Omega_3 - C\,\omega_3)\,i\,\omega &= \Gamma, \\ \omega = \omega_1 + i\,\omega_2, \quad \Gamma &= G_1 + i\,G_2. \end{aligned}\right\} \tag{49.18}$$

γ) Body with spherical symmetry $(A = B = C)$. Now (49.14) simplify to

$$A\,\dot{\omega}_1 = G_1, \quad A\,\dot{\omega}_2 = G_2, \quad A\,\dot{\omega}_3 = G_3, \tag{49.19}$$

for axes fixed in the body. In this case we can choose $\boldsymbol{\Omega}$ arbitrarily; if we choose $\boldsymbol{\Omega} = 0$, (49.12) gives (49.19), but with the components taken on axes with directions fixed in space.

Returning to the motion of the mass-centre as given by (49.8), we may use a moving triad $(\boldsymbol{i}, \boldsymbol{j}, \boldsymbol{k})$ here also. Let $\boldsymbol{v}$ be the absolute velocity of O, and let it and $\boldsymbol{F}$ be resolved along the triad, so that we have

$$\left.\begin{aligned} \boldsymbol{v} &= v_1 \boldsymbol{i} + v_2 \boldsymbol{j} + v_3 \boldsymbol{k}, \\ \boldsymbol{F} &= F_1 \boldsymbol{i} + F_2 \boldsymbol{j} + F_3 \boldsymbol{k}. \end{aligned}\right\} \tag{49.20}$$

[1] Useful in ballistics and other problems of stability; cf. K. L. NIELSEN and J. L. SYNGE: Quart. Appl. Math. **4**, 201 (1946); E. J. MCSHANE, J. L. KELLEY and F. V. RENO: Exterior Ballistics, p. 176 (Denver: University Press 1953); S. O'BRIEN and J. L. SYNGE: Proc. Roy. Irish Acad. A **56**, 23 (1954). The complex notation is also useful in the theory of KOWALEWSKI'S top (Sect. 56).

Then (49.8) may be written

$$m\left(\frac{\delta \boldsymbol{v}}{\delta t} + \boldsymbol{\Omega} \times \boldsymbol{v}\right) = \boldsymbol{F}, \tag{49.21}$$

or, explicitly,

$$\left.\begin{aligned} m(\dot{v}_1 - \Omega_3 v_2 + \Omega_2 v_3) &= F_1, \\ m(\dot{v}_2 - \Omega_1 v_3 + \Omega_3 v_1) &= F_2, \\ m(\dot{v}_3 - \Omega_2 v_1 + \Omega_1 v_2) &= F_3. \end{aligned}\right\} \tag{49.22}$$

In Case α, we put $\boldsymbol{\Omega} = \boldsymbol{\omega}$. Then we have in (49.14) and (49.22) six equations for the six components of $\boldsymbol{v}$ and $\boldsymbol{\omega}$. When these have been found as functions of t, the complete determination of the motion demands a further step. Assigning six generalized coordinates (q) to the body, we express the six components of $\boldsymbol{v}$ and $\boldsymbol{\omega}$ as functions of $(q, \dot{q})$; then, $\boldsymbol{v}$ and $\boldsymbol{\omega}$ being known, we have six differential equations of the first order to determine (q) as functions of t, and so complete the description of the motion.

Cases β and γ are treated similarly. In Case γ we may put $\boldsymbol{\Omega} = 0$, and then (49.22) read

$$m\dot{v}_1 = F_1, \quad m\dot{v}_2 = F_2, \quad m\dot{v}_3 = F_3. \tag{49.23}$$

The important case of *a rigid body turning about a fixed point* (see Sect. 55—57) is covered by the preceding theory. The body has three degrees of freedom, and its motion is determined by (44.5), which is formally identical with (49.9). But now $\boldsymbol{h}$ is the angular momentum about the fixed point, and $\boldsymbol{G}$ the total moment of forces about that point. With this change of interpretation, all the work from (49.10) to (49.19), inclusive, is applicable; but now A, B, C are principal moments of inertia at the fixed point, and not at the mass-centre.

In the case of a freely moving rigid body, explicit reference to the mass-centre is avoided in the motor symbolism (Motorrechnung) of STUDY and VON MISES[1].

50. Moving frames of reference[2]. Let S be a rigid body in some quite general prescribed motion. We take S as a frame of reference, and seek the equations of motion of a dynamical system relative to this frame. In Sect. 32 we treated this question for a single particle; now we take a dynamical system consisting of P particles, with scleronomic holonomic constraints, so that there exist generalized coordinates q_ϱ $(\varrho = 1, \dots N)$ determining the configuration of the system *relative to* S, these coordinates being freely variable without violation of the constraints.

For example, S might be the earth, having an orbital motion around the sun and also a rotation about its axis; the system might be a rigid body having one point attached to the earth; q_ϱ might be the three Eulerian angles relative to a triad fixed on the earth.

The plan is to reduce S to a Newtonian frame by the introduction of fictitious forces. As in (32.1), the equations of motion of the particles may be written

$$m_i \boldsymbol{a}'_i = \boldsymbol{F}_i + \boldsymbol{F}_{oi} + \boldsymbol{F}_{ci} + \boldsymbol{F}_{ti} \quad (i = 1, \dots P); \tag{50.1}$$

here $\boldsymbol{F}_i$ is the total real force on the particle (applied force + reaction of constraint), and the other three vectors on the right hand side are as in (32.2), but with a suffix i attached to $\boldsymbol{m}, \boldsymbol{r}', \boldsymbol{v}'$. Let δq_ϱ be an arbitrary variation of the generalized coordinates, and let $\delta \boldsymbol{r}'_i$ be the corresponding displacements of the particles

[1] R. VON MISES: Z. angew. Math. Mech. **4**, 155, 193 (1924). — FRANK [5] pp. 98—102. — C. B. BIEZENO: Handbuch der Physik, Vol. 5, pp. 247—250. Springer: Berlin 1927. — WINKELMANN and GRAMMEL [29] pp. 373—378. — L. BRAND, Chap. II of op. cit. in Sect. 12. — W. RAHER: Öst. Ing.-Arch. **9**, 55 (1954).

[2] For alternative methods of treating relative motion, see APPELL [2] II, pp. 360—376.

relative to S. Taking the scalar product of (50.1) by $\delta \boldsymbol{r}_i'$ and summing for i from 1 to P, we get

$$\sum_{i=1}^{P} m_i \boldsymbol{a}_i' \cdot \delta \boldsymbol{r}_i' = \sum_{i=1}^{P} \boldsymbol{F}_i \cdot \delta \boldsymbol{r}_i' + \sum_{i=1}^{P} \boldsymbol{F}_{oi} \cdot \delta \boldsymbol{r}_i' + \sum_{i=1}^{P} \boldsymbol{F}_{ci} \cdot \delta \boldsymbol{r}_i' + \sum_{i=1}^{P} \boldsymbol{F}_{ti} \cdot \delta \boldsymbol{r}_i'. \tag{50.1a}$$

Since $\boldsymbol{a}_i'$ is the acceleration of the particle relative to S, by the purely kinematical formula (46.9) we have

$$\sum_{i=1}^{P} m_i \boldsymbol{a}_i' \cdot \delta \boldsymbol{r}_i' = \sum_{\varrho=1}^{N} \left(\frac{d}{dt} \frac{\partial T'}{\partial \dot{q}_\varrho} - \frac{\partial T'}{\partial q_\varrho} \right) \delta q_\varrho, \tag{50.2}$$

where $T'(q, \dot{q})$ is the kinetic energy of the motion *relative to S*, i.e.

$$T'(q, \dot{q}) = \tfrac{1}{2} \sum_{i=1}^{P} m_i \dot{\boldsymbol{r}}_i' \cdot \dot{\boldsymbol{r}}_i' = \tfrac{1}{2} \sum_{i=1}^{P} m_i \boldsymbol{v}_i'^2. \tag{50.3}$$

To deal with the other summations in (50.1a), we first define generalized forces Q_ϱ by

$$\sum_{i=1}^{P} \boldsymbol{F}_i \cdot \delta \boldsymbol{r}_i' = \sum_{\varrho=1}^{N} Q_\varrho \, \delta q_\varrho, \tag{50.4}$$

this being the work done by the applied forces, the frame S being held fixed during the virtual displacement; the reactions of constraint do not appear in Q_ϱ. Next we define A_ϱ, B_ϱ, C_ϱ ($\varrho = 1, \ldots N$) by the equations. [cf. (32.2)]

$$\left. \begin{aligned} \sum_{i=1}^{P} \boldsymbol{F}_{oi} \cdot \delta \boldsymbol{r}_i' &= - \sum_{i=1}^{P} m_i \boldsymbol{a}_0 \cdot \delta \boldsymbol{r}_i' = \sum_{\varrho=1}^{N} A_\varrho \, \delta q_\varrho, \\ \sum_{i=1}^{P} \boldsymbol{F}_{ci} \cdot \delta \boldsymbol{r}_i' &= - 2 \sum_{i=1}^{P} m_i (\boldsymbol{\omega} \times \boldsymbol{v}_i') \cdot \delta \boldsymbol{r}_i' = \sum_{\varrho=1}^{N} B_\varrho \, \delta q_\varrho, \\ \sum_{i=1}^{P} \boldsymbol{F}_{ti} \cdot \delta \boldsymbol{r}_i' &= - \sum_{i=1}^{P} m_i (\dot{\boldsymbol{\omega}} \times \boldsymbol{r}_i') \cdot \delta \boldsymbol{r}_i' - \boldsymbol{\omega} \cdot \sum_{i=1}^{P} m_i (\boldsymbol{\omega} \cdot \boldsymbol{r}_i') \, \delta \boldsymbol{r}_i' + \\ &\quad + \omega^2 \sum_{i=1}^{P} m_i \boldsymbol{r}_i' \cdot \delta \boldsymbol{r}_i' = \sum_{\varrho=1}^{N} C_\varrho \, \delta q_\varrho, \end{aligned} \right\} \tag{50.5}$$

the variations δq_ϱ being arbitrary. In these expressions $\boldsymbol{a}_0$ and $\boldsymbol{\omega}$ are respectively the acceleration of the base-point O (fixed in S), and the angular velocity of S; they are given functions of t (since the motion of S is prescribed), and hence A_ϱ and C_ϱ are functions of (q, t), while B_ϱ is a function of $(q, \dot{q}, t)$.

After these preparations, we can use (50.1a) to obtain at once LAGRANGE'S equations for the motion relative to S in the form

$$\frac{d}{dt} \frac{\partial T'}{\partial \dot{q}_\varrho'} - \frac{\partial T'}{\partial q_\varrho'} = Q_\varrho + A_\varrho + B_\varrho + C_\varrho \quad (\varrho = 1, \ldots N), \tag{50.6}$$

the last three terms being fictitious generalized forces due to the motion of S.

II. Systems without constraints.

51. Two-body problem. Consider two particles with masses m_1, m_2, attracting or repelling one another with equal and opposite forces acting along the line joining the particles and depending only on the distance between them. Fig. 13 shows the case of repulsion.

Let $\boldsymbol{r}_1$, $\boldsymbol{r}_2$ be the position vectors of the particles relative to some fixed origin; let $\boldsymbol{P}$ be the force exerted by the first particle on the second. Then the equations of motion are

$$m_1 \ddot{\boldsymbol{r}}_1 = -\boldsymbol{P}, \quad m_2 \ddot{\boldsymbol{r}}_2 = \boldsymbol{P}. \tag{51.1}$$

The position vector of the second particle relative to the first is

$$\boldsymbol{r} = \boldsymbol{r}_2 - \boldsymbol{r}_1, \tag{51.2}$$

and from (51.1) we obtain

$$M \ddot{\boldsymbol{r}} = \boldsymbol{P}, \quad M = \frac{m_1 m_2}{m_1 + m_2}. \tag{51.3}$$

Thus, by abandoning the absolute frame of reference, and using an accelerated frame attached to the first particle, we reduce the two-body problem to the one-body problem, the mass of the second particle being fictitiously changed[1] in this process, but the force being unchanged. We can now apply to (51.3) the theory of central forces as developed in Sect. 37. But note that in (37.1) F is force per unit mass; in using (37.1) we are to put $F = P/M$, where $|P|$ is the magnitude of $\boldsymbol{P}$ and P is positive for repulsion and negative for attraction; for the potential V of (37.3) we have

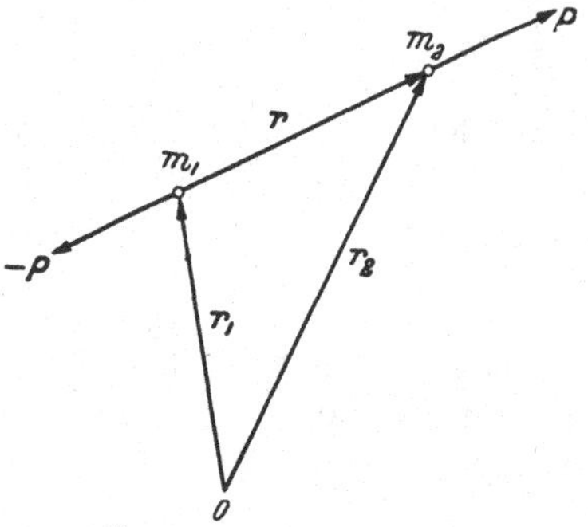

Fig. 13. Two-body problem.

$$V = -M^{-1} \int_{r_0}^{r} P\, dr. \tag{51.4}$$

We can also simplify the two-body problem by using a suitable Newtonian frame of reference. By (51.1) we have

$$m_1 \ddot{\boldsymbol{r}}_1 + m_2 \ddot{\boldsymbol{r}}_2 = 0, \tag{51.5}$$

and this tells us that the mass-centre is unaccelerated; a frame in which it is at rest is Newtonian. Using this frame, and taking the origin at the mass-centre, we have

$$m_1 \boldsymbol{r}_1 + m_2 \boldsymbol{r}_2 = 0. \tag{51.6}$$

It is then sufficient to work with only one of the Eq. (51.1). If the scalar law of repulsion or attraction is $P = P(r)$, the second of (51.1) may be written

$$m_2 \ddot{\boldsymbol{r}}_2 = \frac{\boldsymbol{r}_2}{r_2} P(r), \quad r = \frac{m_1 + m_2}{m_1} r_2. \tag{51.7}$$

Again we have a one-body problem. Now the mass is unchanged, but the law of force is altered. In the case of the inverse square law, we have $P = k/r^2$, and the motion relative to the mass-centre is given by

$$m_2 \ddot{\boldsymbol{r}}_2 = \left(\frac{m_1}{m_1 + m_2}\right)^2 \frac{k}{r_2^3} \boldsymbol{r}_2. \tag{51.8}$$

Viewed in a fixed frame of reference, the orbits of the two particles are twisted curves in space. It is much easier to study the motion in a moving frame, attached either to one particle as in (51.3), or to the mass-centre as in (51.7), for then the orbits are plane.

Note that the assumption regarding the interaction between the particles rules out magnetic interaction and electromagnetic retarded effects.

---

[1] The quantity M in (51.3) is called the *reduced mass*.

52. Capture and scattering[1]. Consider two particles which interact as in Sect. 51, the scalar force between them at distance r being $P(r)$, positive for repulsion and negative for attraction. We suppose that for large r this force goes to zero at least as fast as $r^{-1-\varepsilon}$ $(\varepsilon > 0)$, so that the potential V exists as in (51.4) with $r_0 = \infty$. For $t = -\infty$ the particles are infinitely distant from one another. They approach and interact. We are interested in the result of the encounter, i.e. the state of affairs at $t = +\infty$.

If $r \to 0$ as $t \to \infty$, or more generally if r is bounded as $t \to \infty$, we have *capture*. If $r \to \infty$ as $t \to \infty$, we have *scattering*. We wish to determine whether capture or scattering occurs in any encounter for which the initial conditions are prescribed, and, in the case of scattering, to find the directions in which the particles are scattered.

It is advisable to view any particular encounter in three frames of reference:

S_M, the *mass-centre frame*, in which the mass-centre is fixed.

S_R, the *relative frame*, in which the particle m_1 is fixed.

S_L, the *laboratory frame*, in which the particle m_1 is at rest for $t = -\infty$.

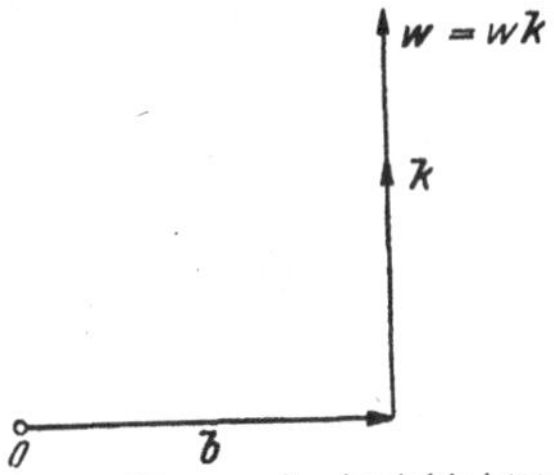

Fig. 14. Diagram showing initial state before an encounter.

We use parallel axes in all three frames. The frames S_M and S_L are Newtonian, but S_R is accelerated; however S_R is Newtonian for $t = -\infty$, and for $t = +\infty$ also if scattering takes place.

We shall denote initial velocities by $\boldsymbol{v}_1$, $\boldsymbol{v}_2$, indicating the frame of reference with M, R, or L, as in $(\boldsymbol{v}_1)_M$, $(\boldsymbol{v}_2)_R$. Then

$$\left.\begin{aligned} m_1(\boldsymbol{v}_1)_M + m_2(\boldsymbol{v}_2)_M = 0, \qquad (\boldsymbol{v}_1)_R = (\boldsymbol{v}_1)_L = 0, \\ (\boldsymbol{v}_2)_R = (\boldsymbol{v}_2)_L = (\boldsymbol{v}_2)_M - (\boldsymbol{v}_1)_M = \left(1 + \frac{m_2}{m_1}\right)(\boldsymbol{v}_2)_M, \end{aligned}\right\} \tag{52.1}$$

so that in all three frames $\boldsymbol{v}_2$ has the same direction.

In S_M the initial data provide a pair of infinite parallel lines (the asymptotes of the initial trajectories); in S_R and S_L we have a point (position of m_1) and an infinite line (the asymptote of the initial trajectory of m_2). A single diagram as in Fig. 14 serves to display the initial data of a given encounter, no matter which frame is used. Here $\boldsymbol{k}$ is a unit vector drawn on the line of $\boldsymbol{v}_2$. In S_R or S_L the point O is the initial position of m_1, and in S_M it is any point on the asymptote of the initial trajectory of m_1. The vector $\boldsymbol{b}$ is drawn from O to meet the line of $\boldsymbol{k}$ at right angles. It is the *impact vector* and its magnitude b is the *impact parameter* or *collision parameter*; it is in fact the shortest distance between the two asymptotes of the initial trajectories, viewed in any unaccelerated frame. The relative velocity appears in Fig. 14 as

$$\boldsymbol{w} = \boldsymbol{v}_2 - \boldsymbol{v}_1 = w\boldsymbol{k}; \tag{52.2}$$

it is of course the same for all the frames.

[1] Cf. CORBEN and STEHLE [3] pp. 86–90; GOLDSTEIN [7] pp. 81–89; D. BOHM: Quantum Theory, Chap. 21. New York: Prentice-Hall 1951. For details of encounters between particles obeying various particular laws of force, between smooth spheres and between rough spheres, see S. CHAPMAN and T. G. COWLING: The Mathematical Theory of Non-Uniform Gases, Chaps. 10, 11. Cambridge: University Press 1952. Cf. also H. GRAD: Comm. Pure Appl. Math. 2, 331 (1949) and his article on the kinetic theory of gases in Vol. XII of this Encyclopedia. For relativistic capture and scattering by a fixed centre, see J. L. SYNGE: Relativity: The Special Theory, p. 426. Amsterdam: North-Holland Publishing Co. 1956.

We deal with the encounter primarily in S_R, passing immediately to results in S_M, and with a little more complication to results in S_L.

In S_R the particle m_1 remains permanently at O in Fig. 14 and m_2 describes an orbit in the plane $(\boldsymbol{b}, \boldsymbol{k})$. By (37.5), (51.3) and (51.4), this orbit is determined by

$$\left(\frac{du}{d\vartheta}\right)^2 = f(u), \qquad f(u) = \frac{2(E-V)}{h^2} - u^2, \tag{52.3}$$

where

$$u = \frac{1}{r}, \qquad V(u) = \frac{m_1 + m_2}{m_1 m_2} \int\limits_r^\infty P(r)\, dr, \tag{52.4}$$

(r, ϑ) being polar coordinates in the plane of the motion. Here, in terms of the initial data,

$$E = \tfrac{1}{2} w^2, \qquad h = b w. \tag{52.5}$$

The time is given by [cf. (37.6)]

$$t = \frac{1}{h} \int^u \frac{du}{u^2 \sqrt{f(u)}}. \tag{52.6}$$

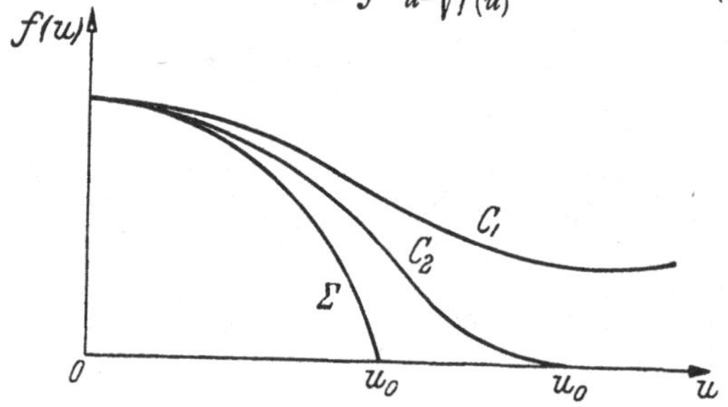

Fig. 15. Graphs of $f(u)$: capture for C_1 and C_2, scattering for Σ.

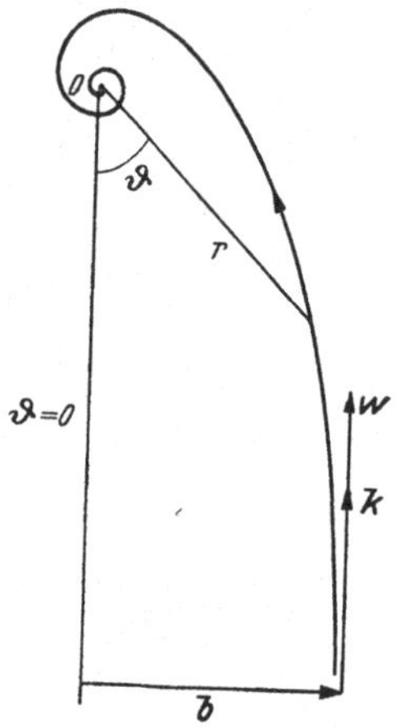

Fig. 16. Capture: $r \to 0$ as $t \to \infty$.

The outcome of the encounter depends entirely on the function $f(u)$, i.e. on the form of the function $V(u)$, on the masses of the particles, and on the two constants (b, w). We have

$$f(0) = \frac{2E}{h^2} = \frac{1}{b^2} > 0, \tag{52.7}$$

so that the graph of $f(u)$ starts above the u-axis (Fig. 15). If it does not meet that axis at all (curve C_1), then $f(u) > 0$ for all u, and the orbit spirals in to O, with capture resulting (Fig. 16). If the graph touches the u-axis at $u = u_0$ (C_2 in Fig. 15), there is an apse with apsidal distance $r = r_0 = 1/u_0$; but this apse is never attained, because $f(u)$ contains $(u - u_0)^2$ as a factor and the integral in (52.6) diverges; the result of the encounter is capture as shown in Fig. 17. If, finally, the graph cuts the u-axis at $u = u_0$ (curve Σ in Fig. 15), an apse occurs in finite time, and we have scattering as shown in Fig. 18.

Omitting head-on collisions, for which $b = 0$, it is clear that capture is impossible in the case of a repulsive force, since such a force bends the trajectory away from O.

Assuming now that scattering takes place (the force being either repulsive or attractive), we proceed to calculate the scattering angle χ_R as shown in Fig. 18, which shows also the base line $\vartheta = 0$, the two asymptotes of the orbit, and the apse A at which $du/d\vartheta = 0$. The apsidal distance is $OA = r_0 = 1/u_0$ and the apsidal angle, as shown, is

$$\chi = \int\limits_0^{u_0} \frac{du}{\sqrt{f(u)}}. \tag{52.8}$$

Since the orbit is symmetrical about the apsidal line OA, the scattering angle is

$$\chi_R = \pi - 2\alpha = \pi - 2\int_0^{u_0} \frac{du}{\sqrt{f(u)}}\,. \tag{52.9}$$

With the convention as to sign adopted here, we have $0 \leqq \chi_R \leqq \pi$ for repulsive scattering and $-\infty < \chi_R \leqq 0$ for attractive scattering. To calculate χ_R we have to evaluate an integral in which the upper limit is given by solving

$$f(u_0) = 0, \quad f(u) = \frac{1}{b^2} - \frac{2V(u)}{b^2 w^2} - u^2. \tag{52.10}$$

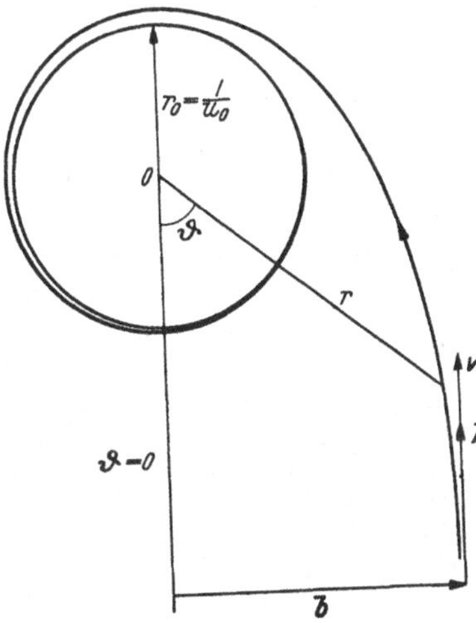

Fig. 17. Capture: $r \to r_0$ as $t \to \infty$.

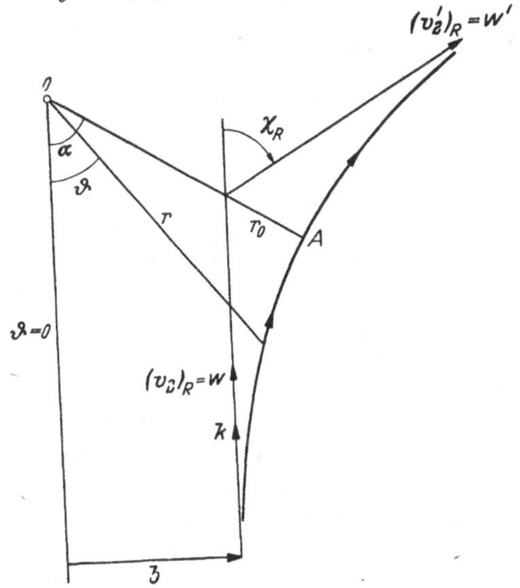

Fig. 18. Scattering in the relative frame S_R: the angle of scattering is χ_R.

The scattering angle χ_R is thus a function of the two basic parameters (b, w),

$$\chi_R = \chi_R(b, w), \tag{52.11}$$

the form of this function depending on the form of the function $V(u)$, obtained from the force function $P(r)$ by (52.4).

Let us denote final velocities with primes. By (51.3) we have in the frame S_R

$$\frac{d}{dt}\left(\frac{1}{2} M \dot{\boldsymbol{r}}^2\right) = \boldsymbol{P} \cdot \dot{\boldsymbol{r}} \tag{52.12}$$

and integration gives

$$(v_2'^2)_R = (v_2^2)_R. \tag{52.13}$$

Since

$$(v_1')_R = (v_1)_R = 0, \tag{52.14}$$

we have the invariant equation

$$w' = w, \tag{52.15}$$

true for all the frames; *the relative speed of the particles is unchanged by the encounter.* If we denote by $\boldsymbol{s}_R$ a unit vector in the direction of scattering, the final velocity may be written

$$(\boldsymbol{v}_2')_R = w\,\boldsymbol{s}_R. \tag{52.16}$$

We pass now to the frames S_M and S_L. They are Newtonian, and so momentum is conserved in them. Further, relative velocity is invariant under change of frame. Hence we have the equations

$$\left.\begin{aligned} m_2(\boldsymbol{v}_2')_M + m_1(\boldsymbol{v}_1')_M &= 0,\\ (\boldsymbol{v}_2')_M - (\boldsymbol{v}_1')_M &= \boldsymbol{w}' = w\,\boldsymbol{s}_R,\\ m_2(\boldsymbol{v}_2')_L + m_1(\boldsymbol{v}_1')_L &= m_2(\boldsymbol{v}_2)_L = m_2\boldsymbol{w},\\ (\boldsymbol{v}_2')_L - (\boldsymbol{v}_1')_L &= \boldsymbol{w}' = w\,\boldsymbol{s}_R, \end{aligned}\right\} \tag{52.17}$$

which give the final velocities in the form

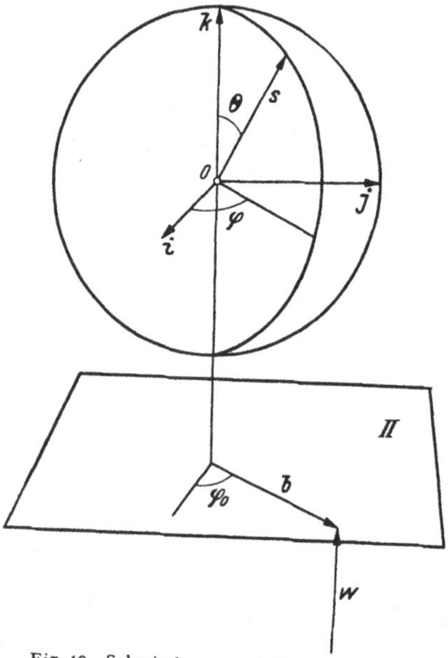

Fig. 19. Spherical representation of scattering.

$$\left.\begin{aligned} (m_1+m_2)(\boldsymbol{v}_1')_M &= -m_2 w\,\boldsymbol{s}_R,\\ (m_1+m_2)(\boldsymbol{v}_2')_M &= m_1 w\,\boldsymbol{s}_R,\\ (m_1+m_2)(\boldsymbol{v}_1')_L &= m_2\boldsymbol{w} - m_2 w\,\boldsymbol{s}_R,\\ (m_1+m_2)(\boldsymbol{v}_2')_L &= m_2\boldsymbol{w} + m_1 w\,\boldsymbol{s}_R. \end{aligned}\right\} \tag{52.18}$$

Therefore the unit vectors $\boldsymbol{s}_M$ and $\boldsymbol{s}_L$ in the direction of scattering [i.e. in the directions of $(\boldsymbol{v}_2')_M$ and $(\boldsymbol{v}_2')_L$ respectively] are

$$\left.\begin{aligned} \boldsymbol{s}_M &= \boldsymbol{s}_R,\\ \boldsymbol{s}_L &= \frac{m_2\boldsymbol{k} + m_1\boldsymbol{s}_R}{\sqrt{m_1^2 + m_2^2 + 2m_1 m_2\,\boldsymbol{k}\cdot\boldsymbol{s}_R}}, \end{aligned}\right\} \tag{52.19}$$

$\boldsymbol{k}$ being, as above, a unit vector in the direction of the initial relative velocity $\boldsymbol{w}$, which is the same direction as that of $(\boldsymbol{v}_2)_L$. We note that the directions of scattering are the same in S_R and S_M, and that the direction for S_L is given in terms of that for S_R by (52.19).

We now construct a spherical representation of scattering (Fig. 19), turning Fig. 14 into three-dimensional form. The vector $\boldsymbol{w}$ (initial relative velocity) is assigned, but we consider all impact vectors $\boldsymbol{b}$ perpendicular to it. For diagrammatic convenience we draw $\boldsymbol{b}$ in a displaced plane Π, and we construct, with centre O, a unit sphere with spherical polar coordinates (Θ, φ) on it. For given $\boldsymbol{b}$, the mechanism of scattering gives a unit scattering vector $\boldsymbol{s}$ and this appears as a point on the unit sphere. In fact, *the scattering maps the plane Π on the unit sphere*; the mapping is the same for S_R and S_M, but different for S_L.

In the frame S_R we have, in accordance with Fig. 18,

$$\left.\begin{aligned} \boldsymbol{s}_R &= \frac{\boldsymbol{b}}{b}\sin\chi_R + \boldsymbol{k}\cos\chi_R\\ &= \boldsymbol{i}\cos\varphi_0\sin\chi_R + \boldsymbol{j}\sin\varphi_0\sin\chi_R + \boldsymbol{k}\cos\chi_R, \end{aligned}\right\} \tag{52.20}$$

where φ_0 is the azimuth of $\boldsymbol{b}$ relative to unit vectors $\boldsymbol{i}, \boldsymbol{j}$, which, with $\boldsymbol{k}$, make up an orthogonal triad; thus the angles (Θ_R, φ_R) of $\boldsymbol{s}_R$ are given by

$$\left.\begin{aligned} \sin\Theta_R\cos\varphi_R &= \sin\chi_R\cos\varphi_0,\\ \sin\Theta_R\sin\varphi_R &= \sin\chi_R\sin\varphi_0,\\ \cos\Theta_R &= \cos\chi_R,\\ (0 \leq \Theta_R \leq \pi,\ &0 \leq \varphi_R < 2\pi). \end{aligned}\right\} \tag{52.21}$$

For repulsive scattering we have

$$\Theta_R = \chi_R, \qquad \varphi_R = \varphi_0, \tag{52.22}$$

and for attractive scattering (depending on the value of χ_R, which can in theory take any negative value) one of the following alternatives:

$$\left.\begin{aligned} \varphi_R &= \varphi_0, & \Theta_R &= \chi_R & &\text{modulo } 2\pi, \\ \varphi_R &= \pi + \varphi_0, & \Theta_R &= -\chi_R & &\text{modulo } 2\pi. \end{aligned}\right\} \tag{52.23}$$

Having thus obtained the angles (Θ_R, φ_R) for $\mathbf{s}_R$ [the angles (Θ_M, φ_M) are the same], we get the angles (Θ_L, φ_L) for $\mathbf{s}_L$ from (52.19). We have

$$\left.\begin{aligned} \sin\Theta_L \cos\varphi_L &= \beta \sin\Theta_R \cos\varphi_R, \\ \sin\Theta_L \sin\varphi_L &= \beta \sin\Theta_R \sin\varphi_R, \\ \cos\Theta_L &= \beta\left(\frac{m_2}{m_1} + \cos\Theta_R\right), \\ \beta^{-1} &= \sqrt{1 + \frac{m_2^2}{m_1^2} + 2\frac{m_2}{m_1}\cos\Theta_R}. \end{aligned}\right\} \tag{52.24}$$

Therefore

$$\left.\begin{aligned} \varphi_L &= \varphi_R, \\ \sin\Theta_L &= \beta\sin\Theta_R, \\ \tan\Theta_L &= \frac{\sin\Theta_R}{\frac{m_2}{m_1} + \cos\Theta_R}. \end{aligned}\right\} \tag{52.25}$$

The last of these gives Θ_L uniquely in the range $(0, \pi)$. We note the following formula:

$$\frac{\sin\Theta_R\, d\Theta_R}{\sin\Theta_L\, d\Theta_L} = \frac{\sin^3\Theta_R}{\sin^3\Theta_L}\,\frac{1}{1 + \frac{m_2}{m_1}\cos\Theta_R} = \frac{\left(1 + \frac{m_2^2}{m_1^2} + 2\frac{m_2}{m_1}\cos\Theta_R\right)^{\frac{3}{2}}}{1 + \frac{m_2}{m_1}\cos\Theta_R}. \tag{52.26}$$

Consider encounters with initial elements $(\boldsymbol{b}, \boldsymbol{w})$ as in Fig. 19, with the relative velocity $\boldsymbol{w}$ fixed and the impact vector $\boldsymbol{b}$ variable. We regard $\boldsymbol{b}$ as the position vector of a point in the plane Π; then to any point in Π there corresponds one of two results—capture or scattering. We define the *total cross section for capture* to be the area Π_c of Π corresponding to capture; Π_c may be zero, finite, or infinite.

The mechanism of scattering maps Π (except for the area Π_c) on the unit sphere by the scattering vectors $\mathbf{s}$, with one mapping for S_M, S_R and another for S_L. We define the *differential cross section for scattering into the solid angle* $d\Omega$ to be that area $d\Pi$ which is mapped on to the element $d\Omega$ of the unit sphere. If we define σ to be the mapping ratio

$$\sigma = \frac{d\Pi}{d\Omega}, \tag{52.27}$$

the differential cross section is

$$\sigma\, d\Omega = d\Pi. \tag{52.28}$$

We call σ the *density*; it is in fact a relative probability density over the unit sphere corresponding to uniform probability density over Π. There are of course two densities, $\sigma_M = \sigma_R$ and σ_L.

It is advisable to divide scattering into two categories: *zero-scattering* ($\Theta=0$) and *significant scattering* ($\Theta>0$). Zero-scattering can occur if, and only if, there is a cut-off of interaction, with $P(r)=0$ for $r>r_1$, say, so that if $b>r_1$ the two particles pass one another with straight trajectories.

We define the *total cross section for scattering* to be the area Π_s of Π corresponding to significant scattering; thus

$$\Pi_s = \int \sigma\, d\Omega, \tag{52.29}$$

an improper integral over the unit sphere with the point $\Theta=0$ excluded. If Π_0 is the area of Π corresponding to zero-scattering, then $\Pi_c+\Pi_s+\Pi_0$ is the total infinite area of Π; consequently at least one of Π_c, Π_s, Π_0 must be infinite.

The determination of the density σ is merely a question of finding the mapping ratio as in (52.27). The scattering vector is

$$\boldsymbol{s} = \boldsymbol{i}\sin\Theta\cos\varphi + \boldsymbol{j}\sin\Theta\sin\varphi + \boldsymbol{k}\cos\Theta, \tag{52.30}$$

and it traces out the solid angle

$$d\Omega = |\sin\Theta\, d\Theta\, d\varphi|. \tag{52.31}$$

Hence

$$\sigma = \frac{d\Pi}{d\Omega} = \frac{|b\, db\, d\varphi_0|}{|\sin\Theta\, d\Theta\, d\varphi|} = \frac{b}{\sin\Theta}\left|\frac{db}{d\Theta}\right|, \tag{52.32}$$

since $d\varphi_0 = d\varphi_R = d\varphi_L$.

In the frame S_R we know χ_R as a function of (b, w) [cf. (52.11)]. Let us solve for b as a function of (χ_R, w) and then substitute $\chi_R=\Theta_R$ for repulsive scattering, as in (52.22), and $\chi_R=\pm\Theta_R$ modulo 2π for attractive scattering, as in (52.23). Thus we get[1]

$$b = b(\Theta_R, w). \tag{52.33}$$

Then (52.32) gives for the density

$$\sigma_M = \sigma_R = \frac{b}{\sin\Theta_R}\left|\frac{db}{d\Theta_R}\right|, \tag{52.34}$$

the right hand side being a function of Θ_R and w; in this, as in the other differentiations, w is held fixed.

For the laboratory frame S_L, we get from (52.32)

$$\sigma_L = \frac{b}{\sin\Theta_L}\left|\frac{db}{d\Theta_L}\right| = \sigma_R\left|\frac{\sin\Theta_R\, d\Theta_R}{\sin\Theta_L\, d\Theta_L}\right|, \tag{52.35}$$

and by (52.26) this may be written

$$\left.\begin{aligned} \sigma_L &= \sigma_R \frac{\sin^3\Theta_R}{\sin^3\Theta_L}\frac{1}{\left|1+\frac{m_2}{m_1}\cos\Theta_R\right|} \\ &= \sigma_R \frac{\left(1+\frac{m_2^2}{m_1^2}+2\frac{m_2}{m_1}\cos\Theta_R\right)^{\frac{3}{2}}}{\left|1+\frac{m_2}{m_1}\cos\Theta_R\right|}. \end{aligned}\right\} \tag{52.36}$$

We would like to express σ_L explicitly as a function of Θ_L and w, but we cannot. The best plan is, for given w, to regard (52.25) and (52.36) as expressions for Θ_L and σ_L in terms of the parameter Θ_R. We can however use approximations when the mass-ratio m_2/m_1 is small, for then Θ_L and σ_L differ little from Θ_R and σ_R.

[1] This can be a multiple valued function in the case of attractive scattering.

In view of the symmetry of the mapping about the axis $\boldsymbol{k}$ in Fig. 19, it is often convenient to use the *differential cross section for scattering in the ring* $\Theta, \Theta+d\Theta$. This is the area of Π which is mapped on to the ring, and its value is

$$2\pi\sigma \sin\Theta\,|d\Theta| = 2\pi b\,|db|. \tag{52.37}$$

Here follow details for some special cases of scattering.

α) Smooth elastic spheres. A collision between two smooth elastic spheres may be regarded as an encounter between two particles with a cut-off at $r=D$, where D is the sum of radii of spheres; we take $V(u)=0$ for $u<1/D$ and $V(u)\to+\infty$ as $u\to 1/D$ from below. The solution of (52.10) is $u_0=1/D$, and by (52.9) the scattering angle is

$$\chi_R = \pi - 2\int\limits_0^{u_0} \frac{du}{\sqrt{b^{-2}-u^2}} = \pi - 2\arcsin\frac{b}{D}. \tag{52.38}$$

Thus $b=D\cos\frac{1}{2}\chi_R = D\cos\frac{1}{2}\Theta_R$, and by (52.34) the density is

$$\sigma_M = \sigma_R = \tfrac{1}{4}D^2, \tag{52.39}$$

which is independent of Θ_R and w. To get σ_L, we would use (52.25) and (52.36).

β) COULOMB *scattering.* Take

$$P(r) = \frac{\mu}{r^2}, \qquad V(u) = ku, \qquad k = \frac{m_1+m_2}{m_1 m_2}\,\mu \tag{52.40}$$

($k>0$ for repulsion, $k<0$ for attraction). Then by (52.10)

$$f(u) = \frac{1}{b^2} - \frac{2ku}{b^2w^2} - u^2 = (u_0-u)\,(u+u_1), \tag{52.41}$$

where

$$u_0 = \frac{1}{b^2w^2}\left(-k+\sqrt{k^2+b^2w^4}\right), \qquad u_1 = \frac{1}{b^2w^2}\left(k+\sqrt{k^2+b^2w^4}\right), \tag{52.42}$$

By (52.9) the scattering angle is

$$\chi_R = \pi - 2\int\limits_0^{u_0} \frac{du}{\sqrt{(u_0-u)\,(u+u_1)}} = \pi - 4\arctan\sqrt{\frac{u_0}{u_1}}, \tag{52.43}$$

so that for $k>0$ we have $0<\chi_R<\pi$, and for $k<0$ we have $-\pi<\chi_R<0$; thus $\Theta_R=|\chi|_R$ for both cases. By (52.42)

$$\sqrt{\frac{u_0}{u_1}} = \frac{\sqrt{k^2+b^2w^4}-k}{b\,w^2}, \tag{52.44}$$

and when we substitute this is (52.43) and solve, we get

$$bw^2 = k\cot\tfrac{1}{2}\chi_R = |k|\cot\tfrac{1}{2}\Theta_R. \tag{52.45}$$

By (52.34) the density is

$$\sigma_M = \sigma_R = \frac{b}{\sin\Theta_R}\left|\frac{\partial b}{\partial\Theta_R}\right| = \frac{k^2}{4w^4}\operatorname{cosec}^4\frac{1}{2}\Theta_R. \tag{52.46}$$

This is RUTHERFORD's scattering formula; it holds for both repulsive and attractive COULOMB fields. For the frame S_R the energy of the incident particle at infinity is $\frac{1}{2}m_2w^2$.

For the laboratory frame we have, by (52.25) and (52.36),

$$\left.\begin{aligned} \tan\Theta_L &= \frac{\sin\Theta_R}{\frac{m_2}{m_1}+\cos\Theta_R}, \\ \sigma_L &= \frac{k^2}{4w^4}\operatorname{cosec}^4\frac{1}{2}\Theta_R\,\frac{\left(1+\frac{m_2^2}{m_1^2}+2\frac{m_2}{m_1}\cos\Theta_R\right)^{\frac{3}{2}}}{\left|1+\frac{m_2}{m_1}\cos\Theta_R\right|}; \end{aligned}\right\} \tag{52.47}$$

thus Θ_L and σ_L are expressed in terms of the parameter Θ_R. For two particles of the same mass, we have $m_1=m_2$ and

$$\Theta_L=\frac{1}{2}\Theta_R, \qquad \sigma_L=\frac{k^2}{w^4}\cos\Theta_L\operatorname{cosec}^4\Theta_L. \tag{52.48}$$

γ) Inverse cube law. Take

$$P(r)=\frac{\mu}{r^3}, \qquad V(u)=\frac{1}{2}ku^2, \qquad k=\frac{m_1+m_2}{m_1 m_2}\mu \tag{52.49}$$

($k>0$ for repulsion, $k<0$ for attraction). By (52.10) we have

$$f(u)=\frac{1}{b^2}-u^2\left(\frac{k}{b^2w^2}+1\right). \tag{52.50}$$

Capture occurs if $k<-b^2w^2$. If k exceeds this value, there is scattering at the angle

$$\left.\begin{aligned} \chi_R &= \pi-2\int_0^{u_0}\frac{du}{\sqrt{f(u)}} \\ &= \pi\left(1-\frac{bw}{\sqrt{k+b^2w^2}}\right) \end{aligned}\right\} \quad \left(u_0=\frac{w}{\sqrt{k+b^2w^2}}\right). \tag{52.51}$$

In the case of repulsion we have $\Theta_R=\chi_R$ and

$$\frac{k}{b^2w^2}=\frac{\pi^2}{(\pi-\Theta_R)^2}-1; \tag{52.52}$$

hence

$$\sigma_M=\sigma_R=\frac{\pi^2 k}{w^2\sin\Theta_R}\cdot\frac{\pi-\Theta_R}{\Theta_R^2(2\pi-\Theta_R)^2}. \tag{52.53}$$

δ) Inverse fifth power[1]. Take

$$P(r)=\frac{\mu}{r^5}, \qquad V(u)=\frac{1}{4}ku^4, \qquad k=\frac{m_1+m_2}{m_1 m_2}\mu. \tag{52.54}$$

We have by (52.10)

$$f(u)=\frac{1}{b^2}-\frac{ku^4}{2b^2w^2}-u^2=\frac{k}{2b^2w^2}(u_0^2-u^2)(u^2+u_1^2) \tag{52.55}$$

where

$$\left.\begin{aligned} u_0^2 &= \frac{w}{k}\left(-b^2w+\sqrt{b^4w^2+2k}\right), \\ u_1^2 &= \frac{w}{k}\left(b^2w+\sqrt{b^4w^2+2k}\right). \end{aligned}\right\} \tag{52.56}$$

[1] Repulsion varying inversely as the fifth power of the distance is of importance in the kinetic theory of gases; cf. CHAPMAN and COWLING (op. cit. in preceding footnote) pp. 170 to 174, where a force varying as any power of r is discussed.

The scattering angle is given by the elliptic integral

$$\chi_R = \pi - \frac{4bw}{\sqrt{2k}} \int_0^{u_0} \frac{du}{\sqrt{(u_0^2 - u^2)(u^2 + u_1^2)}} . \tag{52.57}$$

53. *n*-body problem. The n-body problem is concerned with the motion of n particles which attract one another gravitationally according to the law of the inverse square. If $m_i (i=1, \dots n)$ are the masses of the particles, $\boldsymbol{r}_i$ their position vectors, and $\boldsymbol{r}_{ij} = -\boldsymbol{r}_{ji} = \boldsymbol{r}_i - \boldsymbol{r}_j$, then the equations of motion are

$$m_i \ddot{\boldsymbol{r}}_i = -G \sum_{j=1}^{n} \boldsymbol{r}_{ij} \frac{m_i m_j}{r_{ij}^3} \quad (i = 1, \dots n), \tag{53.1}$$

where G is the gravitational constant.

The system has three integrals of linear momentum and three integrals of angular momentum, contained in the vector formulae

$$\left.\begin{aligned} \sum_{i=1}^{n} m_i \dot{\boldsymbol{r}}_i &= \boldsymbol{M} = \text{const}, \\ \sum_{i=1}^{n} m_i \boldsymbol{r}_i \times \dot{\boldsymbol{r}}_i &= \boldsymbol{h} = \text{const}. \end{aligned}\right\} \tag{53.2}$$

There is also the integral of energy:

$$T + V = E = \text{const}, \tag{53.3}$$

where

$$T = \frac{1}{2} \sum_{i=1}^{n} m_i \dot{\boldsymbol{r}}_i \cdot \dot{\boldsymbol{r}}_i, \qquad V = -\sum_{\substack{i,j=1 \\ j>i}}^{n} \frac{G m_i m_j}{r_{ij}} . \tag{53.4}$$

The velocity of the mass-centre is constant, and we can, if we wish, use a frame of reference in which the mass-centre is permanently at rest.

Seven integrals of linear momentum, angular momentum, and energy exist also if the forces are of a more general type, provided they act in equal opposite pairs along the lines joining the particles and depend only on the mutual distances. The inverse square law is, however, definitely involved in JACOBI's equation, which reads

$$\frac{d^2 \Phi}{dt^2} = 2T + V, \tag{53.5}$$

where

$$\Phi = \tfrac{1}{2} \sum_{i=1}^{n} m_i r_i^2 . \tag{53.6}$$

This rather striking result is most easily proved[1] by applying HAMILTON's equations of motion (Sect. 47) to any system with N degrees of freedom having a Hamiltonian of the form

$$H(q, p) = T(p) + V(q), \tag{53.7}$$

where T is homogeneous of degree 2 in the generalized momenta (p) and V is homogeneous of degree -1 in the generalized coordinates (q); the Hamiltonian of the n-body problem is of this form. By virtue of the homogeneities, we have

$$\sum_{\varrho=1}^{N} \frac{\partial H}{\partial p_\varrho} p_\varrho = 2T, \qquad \sum_{\varrho=1}^{N} \frac{\partial H}{\partial q_\varrho} q_\varrho = -V. \tag{53.8}$$

[1] Cf. WHITTAKER [28], p. 342, for a different proof.

If we define Ψ by

$$\Psi = \sum_{\varrho=1}^{N} p_\varrho q_\varrho, \tag{53.9}$$

then, by HAMILTON's equations (47.7),

$$\frac{d\Psi}{dt} = \sum_{\varrho=1}^{N} \left(p_\varrho \frac{\partial H}{\partial p_\varrho} - \frac{\partial H}{\partial q_\varrho} q_\varrho \right) = 2T + V. \tag{53.10}$$

This is the general form of JACOBI's equation; in the n-body problem we have

$$\frac{d\Phi}{dt} = \sum_{i=1}^{n} m_i \dot{\boldsymbol{r}}_i \cdot \boldsymbol{r}_i = \sum_{\varrho=1}^{N} p_\varrho q_\varrho = \Psi. \tag{53.11}$$

and (53.10) gives (53.5).

The quantity

$$\Phi' = \frac{1}{2} \sum_{\substack{i,j=1 \\ j>i}}^{n} \frac{m_i m_j}{M} r_{ij}^2, \tag{53.12}$$

where M is the total mass of the system, is independent of the frame of reference, and it is easy to see that $\Phi' = \Phi$ when the origin is taken at the mass-centre. Hence JACOBI's equation may also be written in the form

$$\frac{d^2\Phi'}{dt^2} = 2T' + V, \tag{53.13}$$

where T is the kinetic energy relative to the mass-centre.

If $n=2$, we have the two-body problem (Sect. 51), which is easily solved. But for $n>2$ the problem is of great mathematical difficulty. The case $n=3$ (three-body problem) has been of particular interest to mathematicians and possesses a vast literature[1].

In the three-body problem there are 9 coordinates and 9 momenta, and the Hamiltonian equations of motion form a system of order 18. By means of the integrals (53.2) and (53.3) it is possible, by application of canonical transformations[2], to reduce the order from 18 to 6[3]; if the particles move in a plane, the reduction is from order 12 to order 4.

Although no general formal solution of the three-body problem is known, there exist special solutions known as LAGRANGE's particles[4], in which the configuration is a rigid line or triangle; these motions are as follows:

(a) The particles remain always on a straight line rotating with an arbitrary constant angular velocity, which determines the mutual distances of the particles.

(b) The triangle formed by the particles remains equilateral and of constant size, rotating in its plane with an arbitrary constant angular velocity, which determines the size of the triangle.

---

[1] For modern accounts, see WINTNER [*30*], Chap. 5, and C. L. SIEGEL: Vorlesungen über Himmelsmechanik. Berlin: Springer 1956. For current literature on the n-body problem, see the Subject Index of Mathematical Reviews under the heading "Astronomy: 3 and n-body problem"; about fourteen papers appear each year on the average.

[2] For canonical transformations, see Sects. 87, 91, 95 of the present Article.

[3] See WHITTAKER [*28*] Chap. 13; FRANK [*5*], p. 171; GRAMMEL [*8*], p. 346; G. D. BIRKHOFF: Dynamical Systems, Chap. 9. New York: American Mathematical Society 1927.

[4] Cf. WHITTAKER [*28*], p. 406; ROUTH [*22*] I, p. 232; C. CARATHÉODORY: Sitzgsber. Bayer. Akad. Wiss., Math.-nat. Abt. **1933**, 257 (Gesammelte mathematische Schriften, Bd. 2, p. 387. München: Beck 1955). For elementary solutions of the n-body problem, see HAMEL [*11*], pp. 449—464. For the stability of LAGRANGE's particles, see WHITTAKER [*28*], pp. 409—412, and GRAMMEL [*8*], pp. 370—372.

54. Periodic structures. Let particles, each of mass m, be attached at equal intervals along an infinite straight string, which is massless. If the particles execute small transverse oscillations, the displacements $y_p(t)$ satisfy the equations

$$\ddot{y}_p = a^2(y_{p+1} - 2y_p + y_{p-1}) \quad (p = 0, \pm 1, \pm 2, \ldots) \tag{54.1}$$

where $a^2 = S/(md)$, S = tension, d = separation of particles. Equations of the same form occur for longitudinal oscillations if there are elastic connections between the particles.

If the initial conditions are

$$y_p = \alpha_p, \quad \dot{y}_p = \beta_p \quad \text{for} \quad t = 0, \tag{54.2}$$

the solution of (54.1) is

$$y_p = \sum_{l=-\infty}^{\infty} \left[\alpha_{p+l} J_{2l}(2at) + \beta_{p+l} \int_0^t J_{2l}(2a\tau)\, d\tau\right], \tag{54.3}$$

where J_{2l} is the BESSEL function of order $2l$. Using the recurrence formulae for BESSEL functions[1], it is easy to verify this solution.

The above is the simplest example of a vibrating lattice, which may more generally consist of particles of several masses, and may be two-dimensional or three-dimensional, as in the crystal lattice of a solid body. The spatial periodicity of the system is an essential feature[2].

For a finite string with fixed ends, carrying n equal particles equally spaced, we have equations of motion as in (54.1), but now with end conditions:

$$\left.\begin{aligned} \ddot{y}_p &= a^2(y_{p+1} - 2y_p + y_{p-1}) \qquad (p = 1, \ldots n) \\ y_0 &= y_{n+1} = 0. \end{aligned}\right\} \tag{54.4}$$

To solve these equations, we substitute

$$y_p = \eta_p \cos(\omega t + \varepsilon) \quad (p = 0, 1, \ldots n+1), \tag{54.5}$$

where η_p, ω and ε are constants; then (54.4) become

$$\left.\begin{aligned} a^2\eta_{p+1} + (\omega^2 - 2a^2)\eta_p + a^2\eta_{p-1} &= 0 \quad (p = 1, \ldots n), \\ \eta_0 = \eta_{n+1} &= 0. \end{aligned}\right\} \tag{54.6}$$

This set of equations is satisfied by

$$\eta_p = \operatorname{Re} z_p \quad (p = 0, 1, \ldots n+1), \tag{54.7}$$

---

[1] A very convenient list of formulae for BESSEL functions is given in N. W. MCLACHLAN: BESSEL Functions for Engineers. Oxford: Clarendon Press 1934.

[2] For the vibrations of lattices, with an historical introduction and a discussion of electrical systems mathematically equivalent to the mechanical structures, see L. BRILLOUIN: Wave Propagation in Periodic Structures. New York and London: McGraw-Hill 1946. To supplement the history given by BRILLOUIN, it may be noted that HAMILTON worked intensively on this subject under the title "Dynamics of Light", but published only a brief account of his work; see W. R. HAMILTON: Mathematical Papers, Vol. 2, pp. 413–607. Cambridge: University Press 1940. HAMILTON obtained the formula (54.3) above by operational methods, the BESSEL functions appearing as integrals (op. cit., pp. 451, 576).

For loaded strings, chains of rods or gyrostats, and networks, see ROUTH [22] II, Chap. 9; for loaded strings and molecules, see CORBEN and STEHLE [3], Chap. 8. For the BORN-v. KARMAN theory of the specific heat of solids, see M. BLACKMAN, this Encyclopedia Vol. VII part 1, p. 330.

provided the complex z's satisfy

$$\left.\begin{aligned} a^2 z_{p+1} + (\omega^2 - 2a^2)\, z_p + a^2 z_{p-1} = 0 \quad (p = 1, \ldots n), \\ \operatorname{Re} z_0 = \operatorname{Re} z_{n+1} = 0. \end{aligned}\right\} \tag{54.8}$$

Choose $z_0 = -i\beta$, a pure imaginary, and write

$$z_p = -i\beta\, e^{ip\varphi} \quad (p = 0, 1, \ldots n+1). \tag{54.9}$$

Then all of (54.8) are satisfied provided only two equations are satisfied, viz.

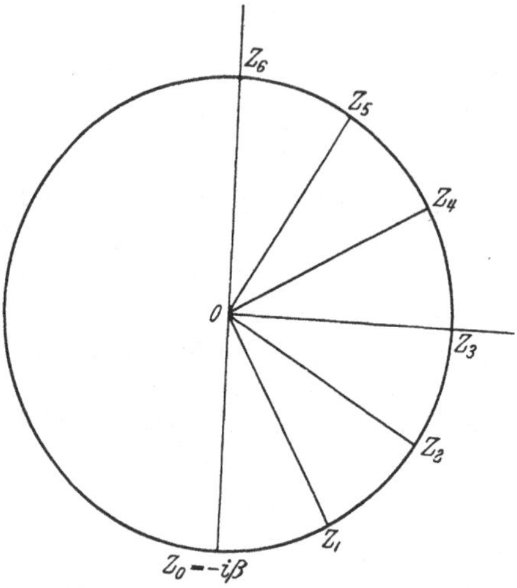

Fig. 20. Complex amplitudes for loaded string with fixed ends, drawn for the case of five particles ($n = 5$) vibrating in the fundamental mode ($r = 1$).

$$\left.\begin{aligned} a^2 e^{i\varphi} + (\omega^2 - 2a^2) + a^2 e^{-i\varphi} = 0, \\ \operatorname{Re} i\beta\, e^{i(n+1)\varphi} = 0. \end{aligned}\right\} \tag{54.10}$$

The second equation is satisfied if we give φ one of the values

$$\varphi_r = \frac{r\pi}{n+1} \quad (r = 1, \ldots n), \tag{54.11}$$

and the first of (54.10) is satisfied if, for $\varphi = \varphi_r$, ω has the value

$$\omega_r = 2a \sin \tfrac{1}{2}\varphi_r \quad (r = 1, \ldots n). \tag{54.12}$$

Thus the normal frequencies (or eigen frequencies) of the loaded string with fixed ends are

$$\left.\begin{aligned} \nu_r = \frac{\omega_r}{2\pi} = \frac{a}{\pi} \sin \frac{r\pi}{2(n+1)} \\ (r = 1, \ldots n), \end{aligned}\right\} \tag{54.13}$$

and the general vibration is given by the superposition of normal modes:

$$\left.\begin{aligned} y_p &= -\operatorname{Re} \sum_{r=1}^{n} i\beta_r \exp(ip\varphi_r) \cos(\omega_r t + \varepsilon_r) \\ &= \sum_{r=1}^{n} \beta_r \sin \frac{pr\pi}{n+1} \cos\left(2at \sin \frac{r\pi}{2(n+1)} + \varepsilon_r\right) \\ &\qquad (p = 1, \ldots n). \end{aligned}\right\} \tag{54.14}$$

The complex amplitudes z_p (54.9) may be displayed on a circle in the complex plane as in Fig. 20.

III. Rigid body with a fixed point.

55. Rigid body under no forces[1]. Consider a rigid body on which no external forces act. By (44.4) its mass-centre has a constant velocity, and by (44.7) the motion relative to the mass-centre satisfies

$$\dot{\boldsymbol{h}}^* = 0, \tag{55.1}$$

where $\boldsymbol{h}^*$ is the angular momentum about the mass-centre. Relative to the mass-centre, the body has three degrees of freedom, and the three scalar equations contained in (55.1) suffice to determine the motion.

[1] For analytical details and diagrams, see APPELL [2] II, pp. 164–195; MACMILLAN [17] II, pp. 192–216; ROUTH [22] II, Chap. 4; SYNGE and GRIFFITH [26], pp. 418–429; WHITTAKER [28], pp. 144–155; WINKELMANN and GRAMMEL [29], pp. 390–404; R. GRAMMEL: Der Kreisel, Bd. 1, pp. 121–164. 2nd. Edn.: Berlin: Springer 1950.

If external forces act but have no resultant moment about the mass-centre, the motion relative to the mass-centre is again given by (55.1). This situation arises when a rigid body moves in a uniform gravitational field; then the mass-centre moves in a parabola, but the motion relative to the mass-centre is uninfluenced by gravity.

If the rigid body is not free, but has a fixed point O about which it can turn freely, and if there act on the body no external forces except the reaction maintaining this constraint, then, as in (44.5), we have

$$\dot{\boldsymbol{h}} = 0, \tag{55.2}$$

$\boldsymbol{h}$ being the angular momentum about the fixed point.

The mathematical problems presented by (55.1) and (55.2) are identical, except for the fact that in (55.1) moments of inertia are to be taken relative to the mass-centre and in (55.2) they are to be taken relative to the fixed point. In the following discussion we shall deal with (55.2), with the body turning about a fixed point O; but the argument applies also to motion about the mass-centre in free motion.

Let $(\boldsymbol{i}, \boldsymbol{j}, \boldsymbol{k})$ be a principal orthonormal triad fixed in the body, and let $\boldsymbol{\omega}$ be the angular velocity of body and triad. Then

$$\boldsymbol{h} = A\,\omega_1 \boldsymbol{i} + B\,\omega_2 \boldsymbol{j} + C\,\omega_3 \boldsymbol{k}, \tag{55.3}$$

where A, B, C are the principal moments of inertia at the fixed point O. By (55.2) the vector $\boldsymbol{h}$ is fixed in space, and its magnitude h is a constant. We have then

$$A^2\omega_1^2 + B^2\omega_2^2 + C^2\omega_3^2 = h^2, \tag{55.4}$$

a constant. By (25.4) the kinetic energy T is given by

$$A\,\omega_1^2 + B\,\omega_2^2 + C\,\omega_3^2 = 2T, \tag{55.5}$$

and this is constant since the reaction of constraint does no work.

The motion may be given a vivid and simple description due to POINSOT[1]. The POINSOT ellipsoid, with equation

$$A x^2 + B y^2 + C z^2 = 2T, \tag{55.6}$$

is fixed in the body, and the motion is described by saying that this ellipsoid rolls on the invariable plane, which is the plane (fixed in space) drawn perpendicular to the fixed vector $\boldsymbol{h}$ at a distance $2T/h$ from O. The vector drawn from O to the point of contact is the angular velocity vector $\boldsymbol{\omega}$; the curves traced by this point of contact on the ellipsoid and the plane are called respectively the *polhode* and the *herpolhode*.

According to EULER's equations (49.14), the components of angular velocity satisfy

$$\left.\begin{aligned} A\,\dot{\omega}_1 - (B-C)\,\omega_2\omega_3 &= 0, \\ B\,\dot{\omega}_2 - (C-A)\,\omega_3\omega_1 &= 0, \\ C\,\dot{\omega}_3 - (A-B)\,\omega_1\omega_2 &= 0. \end{aligned}\right\} \tag{55.7}$$

The Eqs. (55.4) and (55.5) are integrals of these equations. Assuming the body unsymmetric, so that A, B and C are distinct, and choosing the triad $(\boldsymbol{i}, \boldsymbol{j}, \boldsymbol{k})$ so that $A > B > C$, we obtain an analytic solution of the problem as follows.

[1] L. POINSOT: Théorie nouvelle de la Rotation des corps. Paris: Bachelier 1851. This is interesting historically, because POINSOT revolted against the purely analytical approach to dynamics advocated by LAGRANGE.

The Eqs. (55.4) and (55.5) are solved for ω_1 and ω_3, and the solutions are substituted in the second of (55.7). This gives a differential equation for ω_2, of which the solution is an elliptic function. Two cases have to distinguished, according as h^2 is greater than or less than $2BT$. The solutions are as follows[1], expressed in terms of Jacobian elliptic functions of modulus k.

$h^2 > 2BT$:

$$\left.\begin{aligned} &\omega_1 = \alpha \operatorname{dn} p(t-t_0), \quad \omega_2 = \beta \operatorname{sn} p(t-t_0), \quad \omega_3 = \gamma \operatorname{cn} p(t-t_0), \\ &\beta = \sqrt{\frac{2AT-h^2}{B(A-B)}}, \quad p = \sqrt{\frac{(h^2-2CT)(A-B)}{ABC}}, \quad k = \sqrt{\frac{B-C}{A-B}\cdot\frac{2AT-h^2}{h^2-2CT}}. \end{aligned}\right\} \tag{55.8}$$

$h^2 < 2BT$:

$$\left.\begin{aligned} &\omega_1 = \alpha \operatorname{cn} p(t-t_0), \quad \omega_2 = \beta \operatorname{sn} p(t-t_0), \quad \omega_3 = \gamma \operatorname{dn} p(t-t_0), \\ &\beta = \sqrt{\frac{h^2-2CT}{B(B-C)}}, \quad p = \sqrt{\frac{(2AT-h^2)(B-C)}{ABC}}, \quad k = \sqrt{\frac{A-B}{B-C}\cdot\frac{h^2-2CT}{2AT-h^2}}. \end{aligned}\right\} \tag{55.9}$$

In both cases we have

$$\alpha = \sqrt{\frac{h^2-2CT}{A(A-C)}}, \quad \gamma = -\sqrt{\frac{2AT-h^2}{C(A-C)}}. \tag{55.10}$$

Once these components of angular velocity have been found, the description of the motion is completed by introducing the Eulerian angles ϑ, φ, ψ (Sect. 11) to describe the position of the triad $(\boldsymbol{i}, \boldsymbol{j}, \boldsymbol{k})$ relative to a fixed triad $(\boldsymbol{I}, \boldsymbol{J}, \boldsymbol{K})$. Choosing $\boldsymbol{K}$ in the direction of $\boldsymbol{h}$, one obtains ϑ and ψ from

$$\cos\vartheta = \frac{C\omega_3}{h}, \quad \tan\psi = -\frac{B\omega_2}{A\omega_1}, \tag{55.11}$$

and φ by a quadrature from

$$\sin\vartheta\,\dot\varphi = \omega_2\sin\psi - \omega_1\cos\psi. \tag{55.12}$$

In the above procedure we make use of the last row of (11.5) and (19.4).

The Eqs. (55.7) have special solutions in which any one of the three components of angular velocity is a constant and the other two components vanish. These correspond to steady rotations about the three principal axes.

To discuss the stability of these steady motions, we note that (55.4) and (55.5) may also be expressed as follows in terms of the components of $\boldsymbol{h}$ on $(\boldsymbol{i}, \boldsymbol{j}, \boldsymbol{k})$:

$$\left.\begin{aligned} &h_1^2 + h_2^2 + h_3^2 = h^2, \\ &\frac{h_1^2}{A} + \frac{h_2^2}{B} + \frac{h_3^2}{C} = 2T. \end{aligned}\right\} \tag{55.13}$$

Taking (h_1, h_2, h_3) as rectangular Cartesian coordinates in a representative space, we see that the steady rotations correspond to the points $(h, 0, 0)$, $(0, h, 0)$, $(0, 0, h)$. The Eqs. (55.13) restrict the representative point to a curve which is the intersection of a sphere and an ellipsoid, and by examining the forms of these curves it is easy to see that *steady rotations about the axes of greatest and least moments of inertia are stable, while a steady rotation about the axis of intermediate moment of inertia is unstable*[2].

---

[1] If $h^2 = 2BT$, the solution is exponential; cf. ROUTH [22] II, p. 120.

[2] The convenience of this representation is due to the fact that we have to deal with a sphere and an ellipsoid, whereas, if we stick to angular velocity, (55.4) and (55.5) provide two ellipsoids. For a discussion of the two approaches, with diagrams, see SCHAEFER [23], pp. 434–447.

If the body has an axis of inertial symmetry, so that $A = B \neq C$, the motion is greatly simplified. The POINSOT ellipsoid is now an ellipsoid of revolution, and the motion is described by saying that a right-circular polhode cone, fixed in the body, rolls on a right-circular herpolhode cone, fixed in space. The cases $A > C$ and $A < C$ have to be distinguished; in the former case the cones are outside one another, but in the latter case the polhode cone (or body cone) contains the herpolhode cone (or space cone)[1].

56. Spinning top. The toy spinning top is a solid of revolution which is set spinning about its axis of symmetry and placed in contact with a horizontal plane. The essential feature of this system is that we have a rigid body moving in contact with a fixed horizontal plane under the action of two forces, namely, the force of gravity acting at the mass-centre and the reaction at the point of contact. The contact may be regarded as smooth, in which case we have a holonomic system. Or it may be rough enough to prevent sliding; then the system is non-holonomic. Or it may be imperfectly rough, in which case the body slides or rolls according to circumstances[2].

α) Unsymmetrical top. In the usual mathematical idealization, the top is regarded as a rigid body with a fixed point O (the apex or vertex of the top); it moves under the influence of two forces, viz. the force of gravity acting at the mass-centre D and the reaction at O required to hold O fixed. The dynamical specification consists of seven numbers: the mass m, the principal moments of inertia A, B, C at O, and the coordinates ξ, η, ζ of D relative to the principal axes at O. The theory of such a top applies also to the motion about its mass-centre of a free rigid body acted on by forces equipollent to a single force with fixed magnitude and direction, acting at a point fixed in the body; with this interpretation, the theory has some significance in ballistics, the force being due to the resistance of the air.

A top is said to be *symmetrical* if

$$A = B, \quad \xi = \eta = 0, \tag{56.1}$$

so that OD is an axis of inertial symmetry. Otherwise the top is *unsymmetrical*.

To discuss the motion of an unsymmetrical top, we may use LAGRANGE'S equations with the kinetic energy expressed in terms of the Eulerian angles $(\vartheta, \varphi, \psi)$ as in (49.4). But we keep the physics of the problem better in mind by using EULER'S equations (49.14), which may be written

$$\left.\begin{aligned} A\,\dot{\omega}_1 - (B - C)\,\omega_2\,\omega_3 &= -\,m\,g\,(\eta\,K_3 - \zeta\,K_2),\\ B\,\dot{\omega}_2 - (C - A)\,\omega_3\,\omega_1 &= -\,m\,g\,(\zeta\,K_1 - \xi\,K_3),\\ C\,\dot{\omega}_3 - (A - B)\,\omega_1\,\omega_2 &= -\,m\,g\,(\xi\,K_2 - \eta\,K_1),\end{aligned}\right\} \tag{56.2}$$

where

$$\boldsymbol{K} = K_1\,\boldsymbol{i} + K_2\,\boldsymbol{j} + K_3\,\boldsymbol{k}, \tag{56.3}$$

---

[1] For description and diagrams, see SYNGE and GRIFFITH [26], p. 427.

[2] The best general reference for the treatment of these problems is ROUTH [22] II, Chap. 5. For the effects of friction, see also J. H. JELLETT: A Treatise on the Theory of Friction, Chaps. 5 and 8 (London: Macmillan 1872); and GRAMMEL, pp. 107–121 of op. cit. in Sect. 55. In a recent toy, the tippe-top, the body has a spherical base, and it turns over when set spinning with a sufficiently high angular velocity; for theory see C. M. BRAAMS: Physica, Haag 18, 497 (1952); A. D. FOKKER: Physica, Haag 18, 503 (1952); F. A. HARINGX: De Ingenieur 4, Technisch Wetenschappelijk Onderzoek 2 (1952); N. M. HUGENHOLTZ: Physica, Haag 18, 515 (1952); S. O'BRIEN and J. L. SYNGE: Proc. Roy. Irish Acad. A 56, 23 (1954); D. G. PARKYN: Math. Gaz. 40, 260 (1956).

a unit vector directed vertically upward (Fig. 21). Since $\boldsymbol{K}$ is fixed in direction, we have

$$0 = \dot{\boldsymbol{K}} = \frac{\delta \boldsymbol{K}}{\delta t} + \boldsymbol{\omega} \times \boldsymbol{K}, \tag{56.4}$$

or, explicitly,

$$\left.\begin{aligned} \dot{K}_1 + \omega_2 K_3 - \omega_3 K_2 &= 0, \\ \dot{K}_2 + \omega_3 K_1 - \omega_1 K_3 &= 0, \\ \dot{K}_3 + \omega_1 K_2 - \omega_2 K_1 &= 0. \end{aligned}\right\} \tag{56.5}$$

In (56.2) and (56.5) we have six differential equations of the first order for $\omega_1, \omega_2, \omega_3, K_1, K_2, K_3$, which quantities are expressible in terms of the three Eulerian angles and their first derivatives.

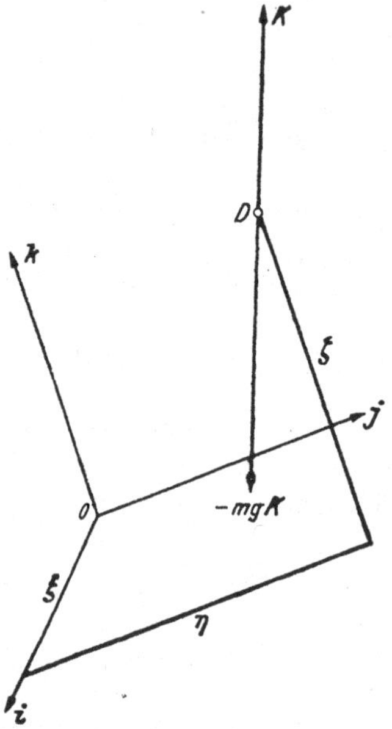

Fig. 21. The unsymmetrical top. The force of gravity is $-mg\boldsymbol{K}$.

These equations possess the following integrals, resulting from the constancy of total energy, the vanishing of the moment of gravity about the vertical through O, and the fact that $\boldsymbol{K}$ is a unit vector:

$$\left.\begin{aligned} &\tfrac{1}{2}(A\,\omega_1^2 + B\,\omega_2^2 + C\,\omega_3^2) + \\ &\quad + m g(\xi K_1 + \eta K_2 + \zeta K_3) = \text{const}, \\ &A\,\omega_1 K_1 + B\,\omega_2 K_2 + C\,\omega_3 K_3 = \text{const}, \\ &K_1^2 + K_2^2 + K_3^2 = 1. \end{aligned}\right\} \tag{56.6}$$

In certain special cases additional integrals exist. Of these, the most famous is KOWALEWSKI'S integral, which occurs in the case where

$$A = B = 2C, \quad \zeta = 0. \tag{56.7}$$

We make $\eta = 0$ by changing to new principal axes $(\boldsymbol{i}, \boldsymbol{j}, \boldsymbol{k})$. Then (56.2) and (56.5) give

$$\left.\begin{aligned} 2\dot{\omega} + i\,\omega\,\omega_3 &= i\,\beta K_3, \\ \dot{K} + i\,K\,\omega_3 &= i\,\omega K_3, \end{aligned}\right\} \tag{56.8}$$

where $\omega = \omega_1 + i\omega_2$, $K = K_1 + iK_2$, $\beta = mg\xi/C$. On eliminating K_3, we get

$$\frac{d}{dt}(\omega^2 - \beta K) + i\,\omega_3(\omega^2 - \beta K) = 0, \tag{56.9}$$

and hence

$$\frac{d}{dt}\log(\omega^2 - \beta K) = -i\,\omega_3. \tag{56.10}$$

On adding the complex conjugate we obtain the required integral

$$(\omega^2 - \beta K)(\overline{\omega}^2 - \beta \overline{K}) = \text{const}, \tag{56.11}$$

or

$$(\omega_1^2 - \omega_2^2 - \beta K_1)^2 + (2\omega_1\omega_2 - \beta K_2)^2 = \text{const}. \tag{56.12}$$

This integral, together with the first two of (56.6), gives us three equations in the Eulerian angles and their first derivatives; these equations can be solved in terms of hyperelliptic functions[1].

[1] The above argument follows ROUTH [*22*] II, pp. 159–161, or APPELL [*2*] II, pp. 209 to 211. For a treatment of KOWALEWSKI'S top by LAGRANGE'S equations, see WHITTAKER [*28*], pp. 164–167. These references may be consulted for other integrable cases of the unsymmetrical top. For much detailed work on the unsymmetrical top, including the use of HAMILTON'S equations, see HAMEL [*11*], pp. 407–449. See also GRAMMEL, pp. 164–214 of op. cit. in Sect. 55.

β) Symmetrical top: general motion. To deal with a symmetrical top, satisfying (56.1), it is convenient to use the two orthonormal triads $(\boldsymbol{i}, \boldsymbol{j}, \boldsymbol{k})$, $(\boldsymbol{I}, \boldsymbol{J}, \boldsymbol{K})$ shown in Fig. 22. $(\boldsymbol{I}, \boldsymbol{J}, \boldsymbol{K})$ is fixed in space, with $\boldsymbol{K}$ pointing vertically upward; $\boldsymbol{k}$ lies along the axis of symmetry OD and $\boldsymbol{j}$ is horizontal. The position of the triad $(\boldsymbol{i}, \boldsymbol{j}, \boldsymbol{k})$ is described by the co-latitude ϑ and the azimuthal angle φ shown in Fig. 22.

The angular velocities $\boldsymbol{\omega}$ and $\boldsymbol{\Omega}$ of the body and of the triad $(\boldsymbol{i}, \boldsymbol{j}, \boldsymbol{k})$ respectively are

$$\left.\begin{aligned} \boldsymbol{\omega} &= \omega_1 \boldsymbol{i} + \omega_2 \boldsymbol{j} + \omega_3 \boldsymbol{k}, \\ \boldsymbol{\Omega} &= \Omega_1 \boldsymbol{i} + \Omega_2 \boldsymbol{j} + \Omega_3 \boldsymbol{k}, \end{aligned}\right\} \tag{56.13}$$

where

$$\omega_1 = \Omega_1 = \sin\vartheta\,\dot{\varphi}, \qquad \omega_2 = \Omega_2 = -\dot{\vartheta}, \qquad \Omega_3 = \cos\vartheta\,\dot{\varphi}; \tag{56.14}$$

the angular momentum about O is

$$\boldsymbol{h} = A\,\omega_1 \boldsymbol{i} + A\,\omega_2 \boldsymbol{j} + C\,\omega_3 \boldsymbol{k}; \tag{56.15}$$

the moment of gravity about O is

$$\boldsymbol{G} = a\boldsymbol{k} \times (-m g \boldsymbol{K}) = -m g a \sin\vartheta\,\boldsymbol{j}, \tag{56.16}$$

where $OD = a$.

The equation of motion is

$$\dot{\boldsymbol{h}} = \boldsymbol{G}, \tag{56.17}$$

and this leads to three differential equations for ϑ, φ and ω_3. But it is convenient to proceed indirectly[1]. By (49.17) we have

$$\omega_3 = s, \tag{56.18}$$

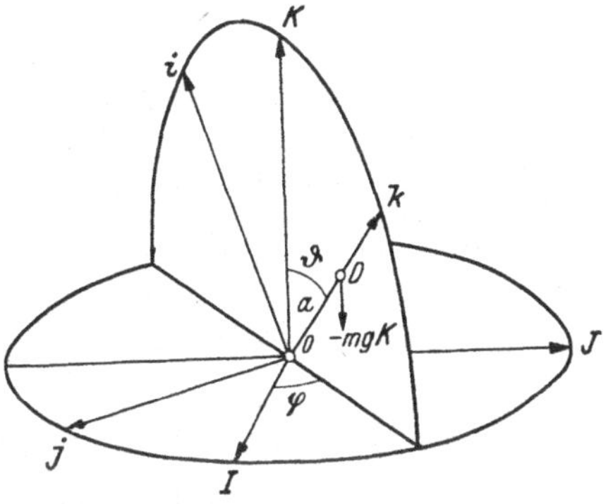

Fig. 22. Fixed triad $(\boldsymbol{I}, \boldsymbol{J}, \boldsymbol{K})$ and moving triad $(\boldsymbol{i}, \boldsymbol{j}, \boldsymbol{k})$ for symmetrical top.

a constant (the *spin* of the top). We have also

$$\left.\begin{aligned} \frac{d}{dt}(\boldsymbol{h}\cdot\boldsymbol{K}) &= \dot{\boldsymbol{h}}\cdot\boldsymbol{K} = \boldsymbol{G}\cdot\boldsymbol{K} = 0, \\ \boldsymbol{h}\cdot\boldsymbol{K} &= \alpha, \end{aligned}\right\} \tag{56.19}$$

a constant (the angular momentum about the axis $\boldsymbol{K}$), and we have the integral of energy

$$T + V = \tfrac{1}{2} A(\omega_1^2 + \omega_2^2) + \tfrac{1}{2} C\,\omega_3^2 + m g a \cos\vartheta = E, \tag{56.20}$$

a constant. Substituting from (56.14), (56.15), and (56.18) in (56.19) and (56.20), we get the following two equations for ϑ and φ:

$$\left.\begin{aligned} A \sin^2\vartheta\,\dot{\varphi} &= \alpha - \beta\cos\vartheta, \\ A(\dot{\vartheta}^2 + \sin^2\vartheta\,\dot{\varphi}^2) + \frac{\beta^2}{C} &= 2(E - m g a \cos\vartheta), \end{aligned}\right\} \tag{56.21}$$

where $\beta = Cs$. Eliminating $\dot{\varphi}$ and putting $\cos\vartheta = x$, we get the differential equation

$$\left.\begin{aligned} \dot{x}^2 &= f(x), \\ f(x) &= \frac{1}{A}\left(2E - \frac{\beta^2}{C} - 2m g a x\right)(1 - x^2) - \frac{(\alpha - \beta x)^2}{A^2}. \end{aligned}\right\} \tag{56.22}$$

---

[1] For a direct treatment of the symmetrical top by LAGRANGE's equations, see WHITTAKER [28], pp. 155–163, where both Eulerian angles and CAYLEY-KLEIN parameters are discussed. For a symmetrical top on a smooth plane, see WHITTAKER [28], pp. 163–164.

Since $f(x)$ is positive in the motion (except when $\dot{x}=0$), and $f(-1)<0$, $f(1)<0$, this cubic in x has three real zeros x_1, x_2, x_3, such that

$$-1<x_1<x_2<1<x_3, \tag{56.23}$$

special cases of equality being disregarded here. The variable x oscillates on the range $x_1 \leq x \leq x_2$, and the solution is

$$\cos\vartheta = x = x_1 + (x_2 - x_1)\,\mathrm{sn}^2\, n(t-t_0), \tag{56.24}$$

where

$$n = \sqrt{\frac{m g a (x_3 - x_1)}{2A}}, \quad k = \sqrt{\frac{x_2 - x_1}{x_3 - x_1}}, \tag{56.25}$$

k being the modulus of the Jacobian elliptic function sn. The azimuthal angle φ is given by

$$\dot{\varphi} = \frac{\alpha - \beta x}{A(1-x^2)}. \tag{56.26}$$

It is clear that $\dot{\varphi}$ has one sign throughout the motion if, and only if, α/β lies outside the range (x_1, x_2).

The motion is most clearly followed by tracing the path of the point $\boldsymbol{k}$ on the unit sphere; its polar coordinates are (ϑ, φ). This path is bounded by the two circles $x=x_1$ (above) and $x=x_2$ (below), and the path crosses itself if, and only if, $\dot{\varphi}$ changes sign during the motion[1].

γ) Symmetrical top: steady precession. Any motion of the top may be maintained by applying to it a torque $\boldsymbol{G}=\dot{\boldsymbol{h}}$, according to (56.17). Consider, with the notation of Fig. 22, the steady motion given by

$$\vartheta = \text{const}, \quad \varphi = pt, \quad \omega_3 = s, \tag{56.27}$$

where p and s are any constants. By (56.14) and (56.15) we have then

$$\left.\begin{aligned} \omega_1 = \Omega_1 = p\sin\vartheta, \quad \omega_2 = \Omega_2 = 0, \quad \Omega_3 = p\cos\vartheta, \\ \boldsymbol{h} = Ap\sin\vartheta\,\boldsymbol{i} + Cs\boldsymbol{k}, \\ \dot{\boldsymbol{h}} = \boldsymbol{\Omega}\times\boldsymbol{h} = p\sin\vartheta\,(Ap\cos\vartheta - Cs)\,\boldsymbol{j}. \end{aligned}\right\} \tag{56.28}$$

The torque required to maintain this motion is precisely the gravitational torque (56.16) provided that p and s satisfy the equation

$$Csp - Ap^2\cos\vartheta = mga. \tag{56.29}$$

This is the equation defining the steady motions of the symmetric top with its axis inclined at an angle ϑ to the vertical; s is the spin and p is the *precessional* angular velocity with which the axis of the top rotates about the vertical drawn through the vertex of the top.

Given any values of p and ϑ, a spin s can be found to satisfy (56.29). Conversely, given s and ϑ, (56.29) is satisfied by two real values of p, viz.

$$p = \frac{1}{2A\cos\vartheta}\left(Cs \pm \sqrt{C^2 s^2 - 4Amga\cos\vartheta}\right), \tag{56.30}$$

[1] For further details on the motion of a symmetrical top, see APPELL [2] II, pp. 197–209; MACMILLAN [17] II, pp. 216–249; ROUTH [22] II, Chap. 5; SYNGE and GRIFFITH [26], pp. 432–440; WINKELMANN and GRAMMEL [29], pp. 406–422. Reference may also be made to the classical treatise of F. KLEIN and A. SOMMERFELD: Über die Theorie des Kreisels. Leipzig: Teubner 1897–1910. For much detailed information about the theory of the top and gyroscopic applications, see R. GRAMMEL: Der Kreisel, 2 vols. 2nd. Edn.: Berlin: Springer 1950. See also A. GRAY: A Treatise on Gyrostatics and Rotational Motion. London: Macmillan 1918.

provided that

$$s^2 > \frac{4A\,m\,g\,a\cos\vartheta}{C^2}. \tag{56.31}$$

If s is large, one of these precessional angular velocities is small and the other is large; the small value is approximately

$$p = \frac{m\,g\,a}{C\,s}, \tag{56.32}$$

which is a very useful simple formula from which the spin can be computed from measurement of the slow precession.

δ) *Stability of a sleeping top.* A symmetrical top is said to be *sleeping* when it spins about its axis of symmetry with that axis vertical. In this motion, with

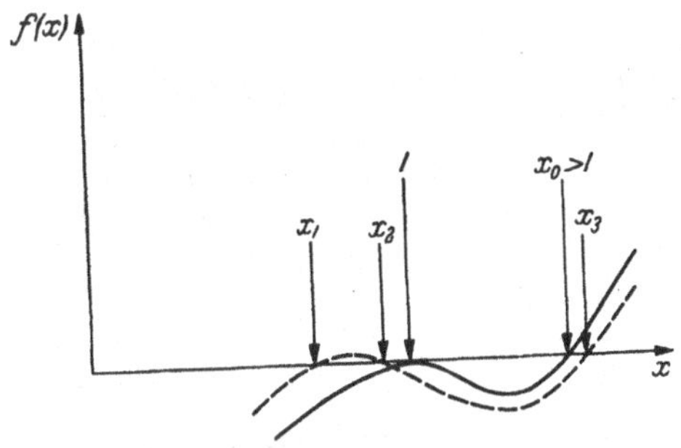

Fig. 23. Graph of the fundamental cubic for a stable sleeping top.

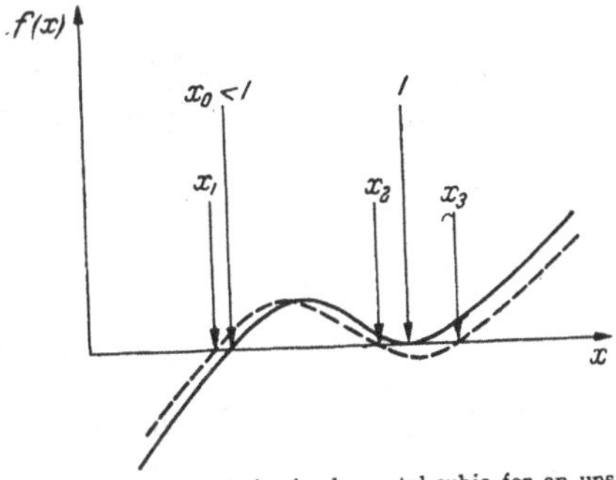

Fig. 24. Graph of the fundamental cubic for an unstable sleeping top.

spin s, the constants α, β, E which occur in the cubic $f(x)$ of (56.22) have the values

$$\alpha = \beta = C\,s, \qquad E = \tfrac{1}{2}C\,s^2 + m\,g\,a, \tag{56.33}$$

and the cubic is

$$f(x) = \frac{2m\,g\,a}{A}(1-x)^2\left(1 + x - \frac{C^2 s^2}{2A\,m\,g\,a}\right). \tag{56.34}$$

This has a double zero at $x = 1$ and a single zero at

$$x = x_0 = \frac{C^2 s^2}{2A\,m\,g\,a} - 1, \tag{56.35}$$

unless it happens that $x_0 = 1$, in which case there is a triple zero. Let us suppose that $x_0 \neq 1$. Then we have two cases: $x_0 > 1$ as shown in Fig. 23, and $x_0 < 1$ as shown in Fig. 24.

Any disturbed motion for which the constants α, β, E differ little from the values (56.33) will have its range of oscillation (x_1, x_2) controlled by a cubic function, as in (56.22), with a graph which differs little from the graph shown in Fig. 23 or 24, whichever applies to the undisturbed motion. The disturbed graph (indicated by the broken lines in the figures) will have zeros at (x_1, x_2), where $x_1 < x_2 < 1$ by (56.23). In the case of Fig. 23, these points are close to 1, and, in the case of Fig. 24, x_1 is close to x_0 and x_2 is close to 1. In the former case the range of oscillation in the disturbed motion is small, and in the latter case it is finite (from $x = x_0$ to $x = 1$ approximately). The former indicates stability, the latter instability.

By the same type of argument we see that $x_0 = 1$ gives stability, and so we have, as a necessary and sufficient condition for the stability of a sleeping top, $x_0 \geq 1$ or equivalently

$$s^2 \geq \frac{4Amga}{C^2}, \tag{56.36}$$

where s is the spin and (C, A) the axial and transverse moments of inertia at the vertex[1].

57. Gyroscopic stiffness. Gyrocompass. *α) Gyroscopic stiffness.* A *gyroscope* (or *gyrostat*) is a rigid body with an axis of symmetry, about which it is given a great angular velocity. Anyone who has handled a gyroscope knows that the spin imparts a sort of *stiffness* to it, so that it seems to resist efforts to change the direction of its axis. But this is only a crude muscular impression. A careful mathematical treatment is needed to elucidate the phenomenon, and three aspects of gyroscopic stiffness will now be discussed; in the first two, the symmetry of the body is not used.

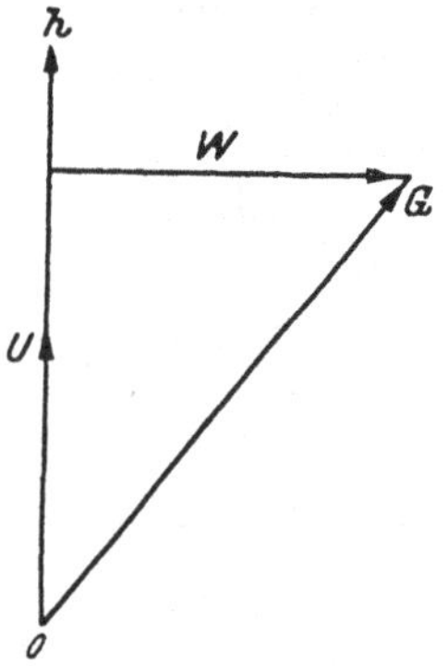

Fig. 25. Gyroscopic stiffness.

We consider a rigid body with a fixed point or a free rigid body; in the latter case we are concerned only with motion relative to the mass-centre. In either case the essential equation may be written [cf. (49.9)]

$$\dot{\boldsymbol{h}} = \boldsymbol{G}, \tag{57.1}$$

where $\boldsymbol{h}$ and $\boldsymbol{G}$ are the angular momentum and the applied torque about the fixed point or about the mass-centre, as the case may be.

As a first measure of gyroscopic stiffness, we consider the rate of change of the direction of the vector $\boldsymbol{h}$, which direction we describe by the unit vector $\boldsymbol{U} = \boldsymbol{h}/h$. Then

$$\dot{\boldsymbol{h}} = h\dot{\boldsymbol{U}} + \dot{h}\boldsymbol{U} = \boldsymbol{G}, \quad \boldsymbol{U}\cdot\dot{\boldsymbol{U}} = 0; \tag{57.2}$$

hence

$$\dot{h} = \boldsymbol{U}\cdot\boldsymbol{G}, \tag{57.3}$$

and (57.2) gives

$$\dot{\boldsymbol{U}} = \frac{\boldsymbol{W}}{h}, \tag{57.4}$$

where

$$\boldsymbol{W} = \boldsymbol{G} - \boldsymbol{U}(\boldsymbol{U}\cdot\boldsymbol{G}), \tag{57.5}$$

so that $-\boldsymbol{W}$ is the vector drawn from the extremity of $\boldsymbol{G}$ to meet $\boldsymbol{h}$ perpendicularly (Fig. 25). Now $\dot{\boldsymbol{U}}$ is the velocity of the point where the unit sphere is cut by the vector $\boldsymbol{h}$. From (57.4) and (57.5) we see that the speed of this point satisfies the inequality

$$|\dot{\boldsymbol{U}}| \leq \frac{|\boldsymbol{G}|}{h}, \tag{57.6}$$

so that the speed tends to zero as h tends to infinity. In this sense, *a large angular momentum imparts stiffness to the direction of the angular momentum vector.*

Secondly, let $(\boldsymbol{i}, \boldsymbol{j}, \boldsymbol{k})$ be a principal orthonormal triad fixed in the body. At $t = 0$, let the body be spinning about $\boldsymbol{k}$ with angular velocity s. A torque $\boldsymbol{G}$ is

---

[1] For a different treatment of the stability of the sleeping top, using LAGRANGE's equations, see WHITTAKER [*28*], p. 206.

applied. We consider the consequent motion of the representative point $\boldsymbol{k}$ on the unit sphere. For its velocity and acceleration at any instant we have

$$\dot{\boldsymbol{k}} = \boldsymbol{\omega} \times \boldsymbol{k}, \quad \ddot{\boldsymbol{k}} = \dot{\boldsymbol{\omega}} \times \boldsymbol{k} + \boldsymbol{\omega} \times (\boldsymbol{\omega} \times \boldsymbol{k}). \tag{57.7}$$

At $t=0$ we have

$$\omega_1 = \omega_2 = 0, \quad \omega_3 = s, \tag{57.8}$$

and so, by EULER's equations (49.14),

$$A\dot{\omega}_1 = G_1, \quad B\dot{\omega}_2 = G_2, \quad C\dot{\omega}_3 = G_3. \tag{57.9}$$

Thus at $t=0$ the velocity and acceleration are

$$\dot{\boldsymbol{k}} = 0, \quad \ddot{\boldsymbol{k}} = -\frac{G_1}{A}\boldsymbol{j} + \frac{G_2}{B}\boldsymbol{i}. \tag{57.10}$$

The spin s is not involved in this acceleration, which is in fact the same as when $s=0$. Thus *there is no gyroscopic stiffness as far as the acceleration of* $\boldsymbol{k}$ *is concerned.*

Thirdly, consider a body with axis of symmetry $\boldsymbol{k}$ and moments of inertia (A, A, C). Let $(\boldsymbol{I}, \boldsymbol{J}, \boldsymbol{K})$ be an orthonormal triad fixed in space, and let the body be made to move with spin $s\boldsymbol{k}$ and with $\boldsymbol{k}$ rotating in the plane $(\boldsymbol{K}, \boldsymbol{I})$ with (precessional) angular velocity $p\boldsymbol{J}$. Then its angular velocity and angular momentum are

$$\boldsymbol{\omega} = p\boldsymbol{J} + s\boldsymbol{k}, \quad \boldsymbol{h} = Ap\boldsymbol{J} + Cs\boldsymbol{k}, \tag{57.11}$$

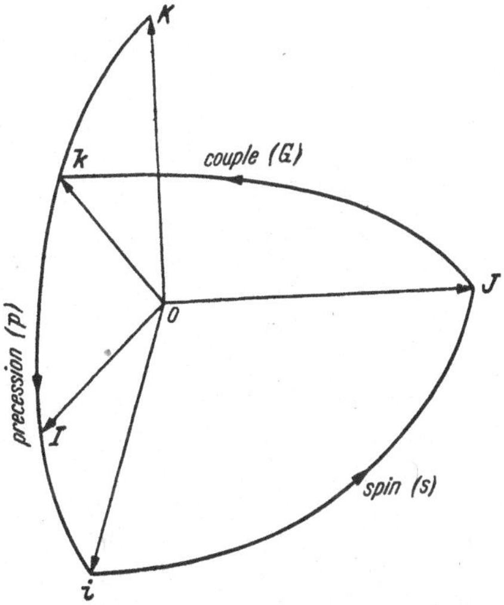

Fig. 26. Gyroscopic couple, precession, and spin.

and the torque required to maintain this motion is

$$\boldsymbol{G} = \dot{\boldsymbol{h}} = Cs\dot{\boldsymbol{k}} = Cs\boldsymbol{\omega} \times \boldsymbol{k} = Csp\boldsymbol{i}, \tag{57.12}$$

where $\boldsymbol{i}$ is such that $(\boldsymbol{i}, \boldsymbol{J}, \boldsymbol{k})$ form an orthonormal triad as in Fig. 26. This torque is known as the *gyroscopic couple*; since it is proportional to s, *the gyroscopic couple is a manifestation of gyroscopic stiffness.* Fig. 26 indicates, in the form of quadrants drawn on a unit sphere, the senses of the couple, the precession, and the spin.

β) Gyrocompass. Consider a rigid body with an axis of inertial symmetry at its mass-centre O, mounted so that it can turn freely about O, this point being attached to the surface of the rotating earth. This may be called a *free gyrocompass.* Its motion is determined by (57.1), $\boldsymbol{h}$ and $\boldsymbol{G}$ being measured relative to O. The torque $\boldsymbol{G}$ is due solely to the gravitational attraction of the earth. Supposing the earth uniform and spherical, the resultant of the gravitational forces on the particles of the body passes through the centre of the earth. But it does not in general pass through O, and so $\boldsymbol{G} \neq 0$. However, this torque is actually so small as to be negligible in practice. Putting $\boldsymbol{G}=0$, we have $\boldsymbol{h}=$ const, and the motion of the gyrocompass is the POINSOT motion of Sect. 55; the axis of symmetry of the gyrocompass rotates with constant precessional angular velocity about some line fixed in space, determined by the initial conditions. In fact, the free gyrocompass indicates to the terrestrial dweller a direction fixed in space.

Consider now the case where the axis of symmetry of the gyrocompass is compelled (by a workless constraint) to remain in a horizontal plane[1]. In Fig. 27, PQ is part of the axis of the earth (from south to north); $\boldsymbol{K}$ is a unit vector parallel to PQ; OQ is a horizontal line (south to north); λ is the latitude of O; $(\boldsymbol{i}, \boldsymbol{j}, \boldsymbol{k})$ is an orthonormal triad with $\boldsymbol{j}$ vertical and $\boldsymbol{k}$ along the axis of symmetry of the gyroscope; ϑ is the inclination of $\boldsymbol{k}$ to OQ (positive to the west); the circle in perspective represents a horizontal plane tangent to the earth's surface at O. The angular velocity of the gyroscope is compounded of (i) the angular velocity of the earth, $\Omega\boldsymbol{K}$, (ii) the angular velocity of the triad relative to the earth, $\dot{\vartheta}\boldsymbol{j}$, and (iii) the spin, $s_0\boldsymbol{k}$. Hence its angular momentum, relative to O, is

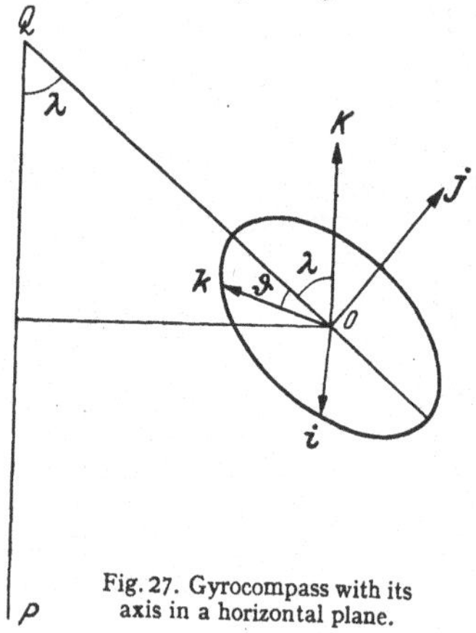

Fig. 27. Gyrocompass with its axis in a horizontal plane.

$$\left.\begin{aligned} \boldsymbol{h} = &- A\,\Omega \sin\vartheta \cos\lambda\,\boldsymbol{i} + \\ &+ A(\dot{\vartheta} + \Omega\sin\lambda)\,\boldsymbol{j} + Cs\boldsymbol{k}, \end{aligned}\right\} \tag{57.13}$$

where (A, A, C) are the moments of inertia at O and $s = s_0 + \Omega\cos\vartheta\cos\lambda$.

Neglecting, as earlier, any torque due to gravity, the torque $\boldsymbol{G}$ maintaining the axis in the horizontal plane has the direction of $\boldsymbol{i}$. The components of (57.1) in the direction of $\boldsymbol{k}$ and $\boldsymbol{j}$ give $s = \text{const}$ and

$$\left.\begin{aligned} &A\ddot{\vartheta} + Cs\,\Omega\cos\lambda\sin\vartheta - \\ &- A\,\Omega^2\sin\vartheta\cos\vartheta\cos^2\lambda = 0. \end{aligned}\right\} \tag{57.14}$$

This equation describes the oscillations of the gyrocompass about the south-north horizontal line OQ. If the term in Ω^2 is neglected, we get an equation of the form (34.2), which describes the finite oscillations of a circular pendulum. For small oscillations, the period is

$$\tau = 2\pi\sqrt{\frac{A}{Cs\,\Omega\cos\lambda}}\,. \tag{57.15}$$

Cases where the constraining plane is not horizontal, or where the point fixed to the earth's surface is not the mass-centre (the barygyroscope), are similarly treated[2].

IV. Impulsive motion.

58. Impulsive forces and moments. Impulsive work. LAGRANGE's equations. For a particle, the equation of motion (30.1) gives

$$m\boldsymbol{v}_2 - m\boldsymbol{v}_1 = \int_{t_1}^{t_2} \boldsymbol{F}\,dt, \tag{58.1}$$

the subscripts 1 and 2 referring to times t_1 and t_2 respectively. Proceeding to a limit in which the force $\boldsymbol{F}$ tends to infinity and the interval $t_2 - t_1$ tends to zero, we derive the concept of an *impulsive force* $\widehat{\boldsymbol{F}}$, such that its application to a particle produces a finite instantaneous change in velocity given by

$$m\,\Delta\boldsymbol{v} = \widehat{\boldsymbol{F}}. \tag{58.2}$$

---

[1] This is a mathematical simplification of the constraint used in practice; see LAMB [*14*], p. 144; TH. PÖSCHL: Handbuch der Physik, Vol. 5, pp. 543–552 (Berlin: Springer 1927); R. F. DEIMEL: Mechanics of the Gyroscope (New York: Macmillan 1929; reprinted by Dover Publications Inc.); A. L. RAWLINGS: The Theory of the Gyroscopic Compass (New York: Macmillan 1929); E. S. FERRY: Applied Gyrodynamics (New York: Wiley 1932); R. GRAMMEL: Der Kreisel, Bd. 2 (2nd. Edn.: Berlin: Springer 1950).

[2] Cf. APPELL [*2*] II, pp. 367–376.

For some mathematical purposes, impulsive forces may be treated in the same way as ordinary forces. We may speak of the *impulsive moment* $\boldsymbol{r}\times\hat{\boldsymbol{F}}$ of an impulsive force $\hat{\boldsymbol{F}}$ applied at the position $\boldsymbol{r}$; and we may speak of the *impulsive work* done by an impulsive force, defined by

$$\delta\hat{W}=\hat{\boldsymbol{F}}\cdot\delta\boldsymbol{r}. \tag{58.3}$$

We must of course remember that the addition of the word *impulsive* changes the nature of the entity involved; its physical dimensions are altered, being multiplied by time.

By integrating other equations of motion over a short range of time and proceeding to a limit as above, we derive impulsive principles from the principles of ordinary dynamics. Thus (44.2) leads to an impulsive law of linear momentum, expressed by

$$\Delta\boldsymbol{M}=\hat{\boldsymbol{F}}, \tag{58.4}$$

where $\hat{\boldsymbol{F}}$ is the sum of all external impulses; and (44.4) leads to

$$m\,\Delta\boldsymbol{v}=\hat{\boldsymbol{F}}, \tag{58.5}$$

where $\boldsymbol{v}$ is the velocity of the mass-centre of a system. Further, by (44.5) and (44.7), we have

$$\Delta\boldsymbol{h}=\hat{\boldsymbol{G}},\qquad \Delta\boldsymbol{h}^*=\hat{\boldsymbol{G}}^*, \tag{58.6}$$

the first applying to any fixed point and the second to the mass-centre; here $\hat{\boldsymbol{G}}$ and $\hat{\boldsymbol{G}}^*$ are impulsive moments.

In the case of energy, however, the transition from ordinary dynamics to impulsive dynamics cannot be effected in this simple way. By (45.3) the increase in the kinetic energy of a system is

$$\Delta T=\int_{t_1}^{t_2}\dot{W}\,dt=\int_{t_1}^{t_2}\sum_{i=1}^{P}\boldsymbol{F}_i\cdot\dot{\boldsymbol{r}}_i\,dt, \tag{58.7}$$

the summation being carried out over the P particles which form the system. But when we proceed to the limit, making the forces tend to infinity and the time-interval to zero, we get

$$\Delta T=\sum_{i=1}^{P}\hat{\boldsymbol{F}}_i\cdot\bar{\boldsymbol{v}}_i, \tag{58.8}$$

where $\bar{\boldsymbol{v}}_i$ are unknown mean values of the velocities. Although we may find it convenient to speak of impulsive work, we are not to connect it with increase in energy. The fact is that, in impulsive dynamics, mechanical energy may be converted into heat, and in that form has no place in Newtonian dynamics[1].

Consider now LAGRANGE's equations (46.17) for a holonomic system. Since the generalized velocities remain finite on proceeding to the limit, we get

$$\Delta\frac{\partial T}{\partial\dot{q}_\varrho}=\hat{Q}_\varrho\quad(\varrho=1,\dots N), \tag{58.9}$$

where $\hat{Q}_\varrho$ are generalized impulsive forces, which may be calculated in terms of impulsive work by the formula

$$\sum_{\varrho=1}^{N}\hat{Q}_\varrho\,\delta q_\varrho=\delta\hat{W}=\sum_{i=1}^{P}\hat{\boldsymbol{F}}_i\cdot\delta\boldsymbol{r}_i. \tag{58.10}$$

---

[1] In relativity, heat generated in a collision is taken care of within the framework of mechanics by an increase in proper mass; cf. Sect. 121 of this article and J. L. SYNGE: Relativity: the Special Theory, p. 184. Amsterdam: North-Holland Publishing Co. 1956.

Problems in impulsive dynamics are mathematically far simpler than problems of ordinary dynamics, because we have only algebraic equations instead of differential equations to solve.

59. Collisions. Coefficient of restitution. The theory of collisions has been largely inspired by ball-games, billiards in particular; at the same time this theory provides models for molecular encounters, with angular momentum taken into account[1].

Consider two rigid bodies, S_1 and S_2, moving in any manner. At a certain instant they find themselves in contact, and a continuation of their motions would cause their volumes to overlap. This overlapping is averted by the action of a pair of impulsive forces, equal and opposite, acting at the point of contact C (Fig. 28).

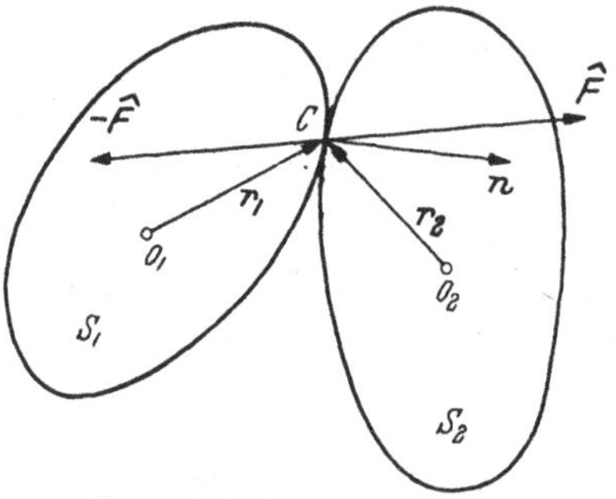

Fig. 28. Collision of rigid bodies.

Let O_1, O_2 be the mass-centres at the instant of collision and $\boldsymbol{r}_1$, $\boldsymbol{r}_2$ the position vectors of C relative to O_1, O_2 respectively. Let m_1, m_2 be the masses of the bodies. Let $\boldsymbol{v}_1$, $\boldsymbol{v}_2$ be the velocities of the mass-centres before collision and $\boldsymbol{h}_1$, $\boldsymbol{h}_2$ the angular momenta relative to the mass-centres, also before collision; let the same symbols marked with primes denote these quantities after collision. Let $-\hat{\boldsymbol{F}}$ and $\hat{\boldsymbol{F}}$ denote the impulsive forces acting on S_1 and S_2 respectively.

Then, as in (58.5) and (58.6), we have

$$\left.\begin{array}{ll} m_1(\boldsymbol{v}_1' - \boldsymbol{v}_1) = -\hat{\boldsymbol{F}}, & m_2(\boldsymbol{v}_2' - \boldsymbol{v}_2) = \hat{\boldsymbol{F}}, \\ \boldsymbol{h}_1' - \boldsymbol{h}_1 = -\boldsymbol{r}_1 \times \hat{\boldsymbol{F}}, & \boldsymbol{h}_2' - \boldsymbol{h}_2 = \boldsymbol{r}_2 \times \hat{\boldsymbol{F}}. \end{array}\right\} \tag{59.1}$$

Here $\boldsymbol{v}_1$, $\boldsymbol{v}_2$, $\boldsymbol{h}_1$, $\boldsymbol{h}_2$, $\boldsymbol{r}_1$, $\boldsymbol{r}_2$ are known, and we have 12 scalar equations for 15 unknowns, viz. the components of the 5 vectors

$$\boldsymbol{v}_1', \boldsymbol{v}_2', \boldsymbol{h}_1', \boldsymbol{h}_2', \hat{\boldsymbol{F}}. \tag{59.2}$$

Since the positions of the bodies are known, and also the moments of inertia, $\boldsymbol{h}_1'$ determines the angular velocity ω_1', and conversely; the same is true of $\boldsymbol{h}_2'$ and ω_2'. Thus, without further hypotheses, the problem of determining the motion resulting from collision has $15-12=3$ degrees of indeterminacy.

We assume now that the bodies are *smooth*[2], which means that the impulsive forces act at right angles to the common tangent plane, so that we have

$$\hat{\boldsymbol{F}} = \hat{F}\boldsymbol{n}, \quad \hat{F} > 0, \tag{59.3}$$

where $\boldsymbol{n}$ is the unit normal vector, drawn from S_1 into S_2. This assumption reduces the number of unknowns to 13, viz.

$$\boldsymbol{v}_1', \ \boldsymbol{v}_2', \ \boldsymbol{h}_1', \ \boldsymbol{h}_2', \ \hat{F}. \tag{59.4}$$

---

[1] In Sect. 52 the bodies in an encounter were particles without angular momentum.

[2] For a detailed treatment of collisions, rough or smooth, see HAMEL [*11*], pp. 395–402; ROUTH [*22*] I, pp. 257–266; G. BOULIGAND: Mécanique rationnelle, Chaps. 18, 19 (Paris: Vuibert 1954); TH. PÖSCHL: Handbuch der Physik, Vol. 6, pp. 503–525 (Berlin: Springer 1928). For a treatment in terms of Motorrechnung, see W. RAHER: Öst. Ing.-Arch. **9**, 55 (1954). The collision of two smooth elastic spheres was treated in a different way in Sect. 52 of the present article.

The remaining indeterminacy may be removed by assuming the conservation of energy,

$$T' = T. \tag{59.5}$$

Here T is known and T' can be expressed in terms of $\boldsymbol{v}_1', \boldsymbol{v}_2', \boldsymbol{h}_1', \boldsymbol{h}_2'$.

But we can follow a more general procedure which includes this as a particular case. Let $\boldsymbol{u}_1, \boldsymbol{u}_2$ be the velocities of the two particles in contact at C, so that

$$\left.\begin{aligned} \boldsymbol{u}_1 &= \boldsymbol{v}_1 + \boldsymbol{\omega}_1 \times \boldsymbol{r}_1, \\ \boldsymbol{u}_2 &= \boldsymbol{v}_2 + \boldsymbol{\omega}_2 \times \boldsymbol{r}_2, \end{aligned}\right\} \tag{59.6}$$

where $\boldsymbol{\omega}_1$ and $\boldsymbol{\omega}_2$ are the angular velocities of the bodies. Formulae of this form may be thought of as holding, not only at the beginning of the collision, but throughout the short period of its duration. We introduce the *speed of compression* c, defined by

$$c = \boldsymbol{n} \cdot (\boldsymbol{u}_1 - \boldsymbol{u}_2), \tag{59.7}$$

which is initially positive since the bodies tend to overlap. The whole collision is broken into (i) a *period of compression* with $c > 0$ and (ii) a *period of restitution* with $c < 0$. In the period of compression there act impulsive forces just sufficient to reduce c to zero. Denoting them by $(-I\boldsymbol{n}, I\boldsymbol{n})$, and writing $\boldsymbol{v}_1'', \boldsymbol{v}_2'', \ldots$ for quantities at the end of the period of compression, we have

$$\left.\begin{aligned} m_1(\boldsymbol{v}_1'' - \boldsymbol{v}_1) = -I\boldsymbol{n}, \quad & m_2(\boldsymbol{v}_2'' - \boldsymbol{v}_2) = I\boldsymbol{n}, \\ \boldsymbol{h}_1'' - \boldsymbol{h}_1 = -\boldsymbol{r}_1 \times I\boldsymbol{n}, \quad & \boldsymbol{h}_2'' - \boldsymbol{h}_2 = \boldsymbol{r}_2 \times I\boldsymbol{n}, \\ c'' = \boldsymbol{n} \cdot (\boldsymbol{u}_1'' & - \boldsymbol{u}_2'') = 0. \end{aligned}\right\} \tag{59.8}$$

The 13 scalar equations contained in (59.8) suffice to determine I and the state of motion at the end of the period of compression.

As for the period of restitution, it is assumed that the impulsive forces for this period are proportional to the impulsive forces during compression; the factor of proportionality, denoted by e, is called the *coefficient of restitution*. Its value ranges from $e = 0$ (*inelastic* collision) to $e = 1$ (*elastic* collision); collisions with intermediate values of e are called *semi-elastic*.

The final result of the collision is given by substituting

$$\hat{\boldsymbol{F}} = (1 + e)\, I\boldsymbol{n} \tag{59.9}$$

in (59.1), the value of I having been found from (59.8).

It can be shown that $e = 1$ implies $T' = T$ as in (59.5)

As an illustration of the coefficient of restitution, consider the collision of two smooth homogeneous spheres. In this case the impulsive forces have no moments about the mass-centres, and (59.8) reduce to

$$\left.\begin{aligned} m_1(\boldsymbol{v}_1'' - \boldsymbol{v}_1) = -I\boldsymbol{n}, \quad & m_2(\boldsymbol{v}_2'' - \boldsymbol{v}_2) = I\boldsymbol{n}, \\ \boldsymbol{n} \cdot (\boldsymbol{v}_1'' & - \boldsymbol{v}_2'') = 0. \end{aligned}\right\} \tag{59.10}$$

Hence we obtain

$$\left.\begin{aligned} I &= \frac{m_1 m_2}{m_1 + m_2}\, \boldsymbol{n} \cdot (\boldsymbol{v}_1 - \boldsymbol{v}_2) \\ &= \frac{m_1 m_2 c}{m_1 + m_2}, \end{aligned}\right\} \tag{59.11}$$

where c is the initial speed of compression. By (59.1) and (59.9) we have

$$m_1(\boldsymbol{v}_1' - \boldsymbol{v}_1) = -(1 + e)\, I\boldsymbol{n}, \quad m_2(\boldsymbol{v}_2' - \boldsymbol{v}_2) = (1 + e)\, I\boldsymbol{n}, \tag{59.12}$$

and therefore the velocities after collision are

$$\left.\begin{aligned} \boldsymbol{v}_1' &= \boldsymbol{v}_1 - \frac{m_2 c}{m_1 + m_2}(1+e)\,\boldsymbol{n}, \\ \boldsymbol{v}_2' &= \boldsymbol{v}_2 + \frac{m_1 c}{m_1 + m_2}(1+e)\,\boldsymbol{n}. \end{aligned}\right\} \tag{59.13}$$

The loss of kinetic energy is

$$T - T' = \frac{m_1 m_2 c^2}{2(m_1 + m_2)}(1 - e^2). \tag{59.14}$$

60. Minimal theorems in impulsive motion[1]. For a system of P particles, acted on by impulsive forces $\hat{\boldsymbol{F}}_i$, we have, by (58.2),

$$\sum m_i(\boldsymbol{v}_i' - \boldsymbol{v}_i)\cdot \boldsymbol{w}_i = \sum \hat{\boldsymbol{F}}_i \cdot \boldsymbol{w}_i, \tag{60.1}$$

where $\boldsymbol{w}_i$ are arbitrary vectors and $\boldsymbol{v}_i, \boldsymbol{v}_i'$ the velocities of the particles before and after the application of the impulsive forces. Here and below, summations are for $i = 1, \ldots P$.

The Eq. (60.1) may be regarded as a form of D'ALEMBERT's principle (Sect. 45), valid for impulsive motion.

The system may be subject to constraints, due to which certain particles are fixed or confined to smooth fixed curves or surfaces, or the distances between certain particles are kept constant (rigidity). Such constraints may persist through the application of the impulsive forces, or they may suddenly come into existence, or they may be suddenly abolished. In any case we can break down the impulsive forces into

$$\hat{\boldsymbol{F}}_i = \hat{\boldsymbol{P}}_i + \hat{\boldsymbol{R}}_i, \tag{60.2}$$

where $\hat{\boldsymbol{P}}_i$ are given or applied impulsive forces, and $\hat{\boldsymbol{R}}_i$ are impulsive forces of constraint. The latter satisfy

$$\sum \hat{\boldsymbol{R}}_i \cdot \boldsymbol{w}_i = 0, \tag{60.3}$$

if $\boldsymbol{w}_i$ are velocities satisfying the constraint.

α) CARNOT's theorem (first part). Theorem: If there are no applied impulsive forces, the sudden introduction of a constraint reduces the kinetic energy[2].

To prove this theorem, we choose $\boldsymbol{w}_i = \boldsymbol{v}_i'$ in (60.1); the right hand side vanishes, and we have

$$\sum m_i(\boldsymbol{v}_i' - \boldsymbol{v}_i)\cdot \boldsymbol{v}_i' = 0. \tag{60.4}$$

The loss of kinetic energy can therefore be expressed as a positive-definite expression:

$$\left.\begin{aligned} T - T' &= \tfrac{1}{2}\sum m_i \boldsymbol{v}_i \cdot \boldsymbol{v}_i - \tfrac{1}{2}\sum m_i \boldsymbol{v}_i' \cdot \boldsymbol{v}_i' \\ &= \tfrac{1}{2}\sum m_i(\boldsymbol{v}_i - \boldsymbol{v}_i')\cdot(\boldsymbol{v}_i - \boldsymbol{v}_i') > 0. \end{aligned}\right\} \tag{60.5}$$

β) CARNOT's theorem (second part). Theorem: Kinetic energy is increased when rigid bonds are broken by an explosion.

By an "explosion" we understand that impulsive forces operate between the particles of the system, in equal opposite pairs acting along the joining line, like action and reaction in NEWTON's Third Law. Although they occur in these balanced pairs, these are applied impulsive forces $\hat{\boldsymbol{P}}_i$, not impulsive forces of constraint $\hat{\boldsymbol{R}}_i$; the latter will be present in those bonds of rigidity which remain unbroken.

---

[1] Cf. ROUTH [22] I, pp. 298–304; APPELL [2] II, pp. 527–539.

[2] The kinetic energy is unchanged in the exceptional case where the introduction of the constraint changes no velocity. Exceptional cases like this are omitted in the enunciations and proofs.

For the same geometrical reason as that by which ordinary reactions in rigid bonds do no work, we have

$$\sum \hat{\boldsymbol{P}}_i \cdot \boldsymbol{v}_i = 0, \tag{60.6}$$

$\boldsymbol{v}_i$ being the velocities before the explosion; we have also

$$\sum \hat{\boldsymbol{R}}_i \cdot \boldsymbol{v}_i = 0, \tag{60.7}$$

and hence, on putting $\boldsymbol{w}_i = \boldsymbol{v}_i$ in (60.1),

$$\sum m_i (\boldsymbol{v}_i' - \boldsymbol{v}_i) \cdot \boldsymbol{v}_i = 0. \tag{60.8}$$

Consequently the gain in kinetic energy can be expressed as a positive-definite expression:

$$\left.\begin{aligned} T' - T &= \tfrac{1}{2} \sum m_i \boldsymbol{v}_i' \cdot \boldsymbol{v}_i' - \tfrac{1}{2} \sum m_i \boldsymbol{v}_i \cdot \boldsymbol{v}_i \\ &= \tfrac{1}{2} \sum m_i (\boldsymbol{v}_i' - \boldsymbol{v}_i) \cdot (\boldsymbol{v}_i' - \boldsymbol{v}_i) > 0. \end{aligned}\right\} \tag{60.9}$$

γ) KELVIN'S *theorem*. Theorem: If a system, initially at rest, is set in motion by applied impulsive forces acting on named particles of the system, these impulsive forces being such that the velocities of the named particles have prescribed values, then the kinetic energy is less than that of any hypothetical motion in which the constraints of the system are satisfied and the named particles have the prescribed velocities.

Let $\boldsymbol{v}_i'$ be the actual velocities and $\boldsymbol{v}_i''$ the velocities in the hypothetical motion, so that $\boldsymbol{v}_i' = \boldsymbol{v}_i''$ for the named particles. Then

$$\sum \hat{\boldsymbol{P}}_i \cdot \boldsymbol{w}_i = 0 \tag{60.10}$$

if $\boldsymbol{w}_i = \boldsymbol{v}_i' - \boldsymbol{v}_i''$, because $\hat{\boldsymbol{P}}_i = 0$ except for the named particles and $\boldsymbol{w}_i = 0$ for them. Further

$$\sum \hat{\boldsymbol{R}}_i \cdot \boldsymbol{w}_i = 0, \tag{60.11}$$

because both $\boldsymbol{v}_i'$ and $\boldsymbol{v}_i''$ satisfy the constraints. Putting $\boldsymbol{v}_i = 0$ in (60.1), since the system is initially at rest, we have then

$$\sum m_i \boldsymbol{v}_i' \cdot (\boldsymbol{v}_i' - \boldsymbol{v}_i'') = 0; \tag{60.12}$$

hence $T'' - T'$ can be expressed as a positive-definite expression:

$$\left.\begin{aligned} T'' - T' &= \tfrac{1}{2} \sum m_i \boldsymbol{v}_i'' \cdot \boldsymbol{v}_i'' - \tfrac{1}{2} \sum m_i \boldsymbol{v}_i' \cdot \boldsymbol{v}_i' \\ &= \tfrac{1}{2} \sum m_i (\boldsymbol{v}_i'' - \boldsymbol{v}_i') \cdot (\boldsymbol{v}_i'' - \boldsymbol{v}_i') > 0. \end{aligned}\right\} \tag{60.13}$$

δ) BERTRAND'S *theorem*. Theorem: A system in motion is acted on by given applied impulsive forces, and as a result its kinetic energy becomes T'. Then $T' > T''$, where T'' is the kinetic energy resulting from the application of the same impulsive forces to the same initial motion, but now subject to constraints consistent with that motion.

Let $\boldsymbol{v}_i$ be the initial velocities, $\boldsymbol{v}_i'$ the final velocities in the absence of the additional constraints, and $\boldsymbol{v}_i''$ the velocities in the presence of those constraints. If $\hat{\boldsymbol{R}}_i'$ and $\hat{\boldsymbol{R}}_i''$ are the impulsive forces of constraint in the two cases, we have

$$\sum \hat{\boldsymbol{R}}_i' \cdot \boldsymbol{v}_i'' = 0, \quad \sum \hat{\boldsymbol{R}}_i'' \cdot \boldsymbol{v}_i'' = 0, \tag{60.14}$$

since it is a question of *additional* constraints. Then, by (60.1), $\hat{\boldsymbol{P}}_i$ being the given applied impulsive forces, we have

$$\left.\begin{aligned} \sum m_i (\boldsymbol{v}_i' - \boldsymbol{v}_i) \cdot \boldsymbol{v}_i'' &= \hat{\boldsymbol{P}}_i \cdot \boldsymbol{v}_i'', \\ \sum m_i (\boldsymbol{v}_i'' - \boldsymbol{v}_i) \cdot \boldsymbol{v}_i'' &= \hat{\boldsymbol{P}}_i \cdot \boldsymbol{v}_i'', \end{aligned}\right\} \tag{60.15}$$

and so, by subtraction,

$$\sum m_i (\boldsymbol{v}_i' - \boldsymbol{v}_i'') \cdot \boldsymbol{v}_i'' = 0. \tag{60.16}$$

Therefore $T' - T''$ may be expressed as a positive-definite expression:

$$\left.\begin{aligned} T' - T'' &= \sum \tfrac{1}{2} m_i \boldsymbol{v}_i' \cdot \boldsymbol{v}_i' - \tfrac{1}{2} \sum m_i \boldsymbol{v}_i'' \cdot \boldsymbol{v}_i'' \\ &= \tfrac{1}{2} \sum m_i (\boldsymbol{v}_i' - \boldsymbol{v}_i'') \cdot (\boldsymbol{v}_i' - \boldsymbol{v}_i'') > 0. \end{aligned}\right\} \tag{60.17}$$

E. General dynamical theory.

I. Geometrical representations of dynamics.

61. The role of general dynamical theory. The most obvious goal of all dynamical theory is to solve dynamical problems which arise in physics or astronomy. Starting from a physical concept (Sect. 2), such as the solar system, we set up a corresponding mathematical concept, or mathematical model, and try to solve the differential equations belonging to that model.

But it is not altogether clear what we mean when we speak of *solving* a set of differential equations. True, a problem is regarded as solved when the coordinates of the particles of the model at time t have been expressed as simple functions of t and of those parameters which define the initial positions and velocities. But what are *simple functions*? We no longer regard a function $f(t)$ as a formal expression in t, but as a quantity determined by t, and it is impossible to draw a sharp line between functions which are simple and those which are not. If we drop the word *simple* and speak merely of *functions*, then every dynamical problem is solved as soon as it has been well stated, because the differential equations, with the initial conditions and the value of t, determine the coordinates at time t. This is not mathematical hair-splitting, but hard fact, for modern methods of electronic computation provide arithmetical solutions to dynamical problems to any desired degree of accuracy, the differential equations being replaced by difference equations. In ballistics, for example, this modern method has largely replaced the search for formulae representing the solution[1].

But, though precise definitions may elude us, there can be no doubt that the two-body problem has a simple solution, whereas the three-body problem has not. In the case of the two-body problem we have formulae involving parameters; we can change the values of those parameters, and so study what we may call the *mathematical structure* of the class of all solutions, with intellectual satisfaction and understanding. We can, moreover, form accurate vivid mental pictures of the behaviour of the two bodies, so that their motion becomes almost as real to us as the motion of a piece of machinery working before our eyes.

In the case of the three-body problem, a numerical solution, based on assigned values of the parameters involved, tells us how the bodies move under those given circumstances. But a single numerical solution does not reveal the mathematical structure of the problem, nor does a collection of such solutions. In this case, as in many others, we must seek an understanding of mathematical structure by examining the differential equations themselves.

But we can be more ambitious. We can aim, not at understanding the mathematical structure of some selected dynamical problem, but at understanding the mathematical structure of a class of problems so wide that we may regard the whole of dynamics as our objective. We shall concentrate our attention

[1] The historical development of the idea of an "unsolved" dynamical problem is discussed by WINTNER [*30*], p. 143.

on those systems to which LAGRANGE'S equations of motion, or HAMILTON'S, are applicable, but that includes a very wide range of problems indeed.

We recognize two purposes in the study of general methods in dynamics. First, the practical purpose, to increase our power in solving specific problems by developing standard techniques with a wide range of applicability. Secondly, the intellectual purpose, to understand the mathematical structure of dynamics. But it is not as simple as that. The development of quantum mechanics out of classical dynamics shows that, in the long run, an understanding of the mathematical structure of dynamics may have more practical results (i.e. results which increase our knowledge of the physical world) than a concentration on specific problems of the type for which dynamical methods were originally designed. With this in mind, the emphasis in the following account of general dynamical theory will be laid on mathematical structure, specific dynamical problems being regarded more in the nature of illustrations than as objectives in themselves.

We seek then to understand the mathematical structure of dynamics. But here we encounter a formidable difficulty, because *understanding* is a very personal matter. It is not a question of accepting this theorem or that, but of gaining an overall picture in which the details are subordinated to a central idea, and opinions are bound to differ widely as to the best central idea to choose. What is psychologically satisfying to one man may not suit another.

In elementary dynamics we all meet on common ground, for we share a vivid geometrical intuition of the motion of a particle, and to this intuition the formulae we use are subordinated. There exists, in fact, a triangle of mental links of this nature:

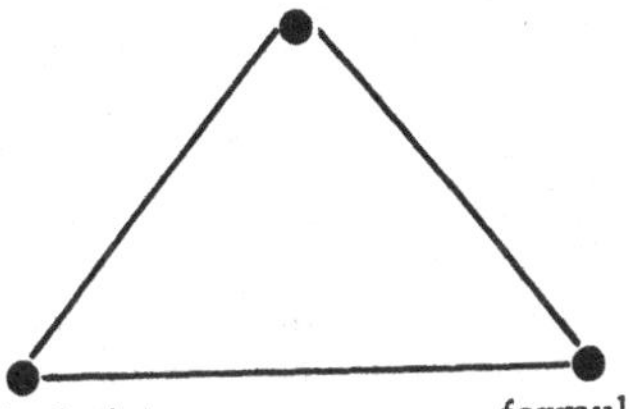

But when we pass on to more complicated dynamical systems, the geometrical picture becomes harder and harder to follow, and it is thrust into the background, so that dynamics becomes, to a great extent, a matter of formulae only. This is satisfactory for those who take pleasure in purely formal arguments, but for most of us the loss of geometrical intuition is a serious handicap. In the present article an attempt is made to give geometrical intuition its just place in general dynamical theory by the systematic use of representative spaces in which the motion of a representative point corresponds to the motion of a dynamical system[1].

[1] The geometrical approach to dynamics seems to have originated with H. HERTZ: Die Prinzipien der Mechanik in neuem Zusammenhang dargestellt (Leipzig 1894). An English translation by D. E. JONES and J. T. WALLEY (1899) has been republished recently: H. HERTZ: The Principles of Mechanics (New York: Dover Publications Inc. 1956). This contains an Introductory Essay by R. S. COHEN (with a bibliography) and a Preface by H. VON HELMHOLTZ. See also G. RICCI and T. LEVI-CIVITA: Méthodes de calcul absolu et leurs applications. Math. Ann. **54**, 125–201 (1901) and Paris: Blanchard 1923; J. W. GIBBS: Elementary Principles in Statistical Mechanics (Yale: Scribner 1902; see The Collected Works of J. W. GIBBS, Vol. 2. New York-London-Toronto: Longmans & Green 1928); J. L. SYNGE: Phil. Trans. Roy. Soc. Lond., Ser. A **226**, 31–106 (1926); L. BRILLOUIN: Les Tenseurs en mécanique et en élasticité (Paris: Masson 1938); LANCZOS [*15*]; PRANGE [*21*].

General dynamical theory occupies a curious position in physics. Historically, it has been suggested by, and developed in terms of, the Newtonian dynamics of particles and rigid bodies. But we feel an urgent need to give it a wider scope, presenting it as a consistent mathematical theory applicable to any physical system the behaviour of which can be expressed in Lagrangian or Hamiltonian form. There is a temptation to present it as pure mathematics, and the exposition which follows is a compromise. The argument is not of sufficient precision to satisfy a modern pure mathematician, but at the same time it does attempt to exhibit a mathematical structure independent (except for suggestion and explanation) of the preceding part of this article. Everything is based on a Lagrangian or a Hamiltonian, or an equivalent concept. Kinetic energy, so important in the direct physical applications of Newtonian dynamics, plays a minor part (in illustrative examples and in Sects. 84 and 85), and the Hamiltonian is not restricted to be a quadratic function of the generalized momenta, as it always is in Newtonian dynamics.

62. Representative spaces. The following representative spaces will be used to elucidate dynamical theory[1]:

Name	Symbol	Dimensionality	Coordinates
Configuration space	Q	N	q_ϱ
Space of events	QT	$N+1$	q_ϱ, t; or x_r
Momentum-energy space	PH	$N+1$	p_ϱ, H; or y_r
Phase space	QP	$2N$	q_ϱ, p_ϱ
Space of states	QTP	$2N+1$	q_ϱ, t, p_ϱ
Space of states and energy	$QTPH$	$2N+2$	$q_\varrho, t, p_\varrho, H$; or x_r, y_r

Here and throughout Part E, except where noted to the contrary, *small Greek suffixes* take the values $1, 2, \ldots N$, and *small Latin suffixes* the values $1, 2, \ldots N+1$, with the summation convention for a repeated suffix in each case.

The representative spaces are listed above in order of increasing dimensionality; they will be treated in a different, more convenient, order.

Our aim is to present dynamical theory in a fairly abstract way, so that the results may be applicable outside the range of traditional Newtonian dynamics. But, to keep our feet on the ground, let us briefly consider the above representative spaces relative to a Newtonian system (holonomic, and either scleronomic or rheonomic) with N degrees of freedom and generalized coordinates q_ϱ. The system possesses a Lagrangian $L(q, t, \dot{q})$ and moves in accordance with LAGRANGE'S equations of motion (Sect. 46); equivalently, it possesses a Hamiltonian $H(q, t, p)$ and moves in accordance with HAMILTON'S equations (Sect. 47). We shall call the system *conservative* if t is absent from H (or, equivalently, absent from L), so that we have

$$\frac{\partial H}{\partial t} = 0; \tag{62.1}$$

as in (47.9), this implies

$$H = E, \tag{62.2}$$

a constant of the motion. All motions for which E has a common value constitute an *isoenergetic* dynamics[2].

[1] There are other representative spaces which might be considered, such as the space of $3N$ dimensions, with coordinates $q_\varrho, \dot{q}_\varrho, p_\varrho$, used by P. A. M. DIRAC, Canad. J. Math. **2**, 129–148 (1950).

[2] A family of orbits which have the same constant energy is also called a *natural family*; cf. WHITTAKER [28], p. 387.

Of all the representative spaces, Q is the simplest. If the system consists of a single particle moving in ordinary space, then Q is ordinary space; and if the particle is constrained to move on a surface or curve, then Q is that surface or curve. However, the picture of the totality of trajectories is somewhat complicated, for a trajectory is not determined by a point in Q and a direction in Q (i.e. the ratios $dq_1 : dq_2 : \ldots dq_N$). For a conservative system, a single infinity of trajectories correspond to a given direction at a point (consider a particle in a gravitational field); for a non-conservative system, there is a double infinity of trajectories.

The totality of trajectories is easier to visualize in QT, in which a trajectory is determined by a point and a direction (i.e. the ratios $dq_1 : dq_2 : \ldots dq_N : dt$, which ratios are the generalized velocities). This applies whether the system is conservative or not. Moreover, the treatment of t on a parity with the coordinates q_ϱ makes QT a suitable background for relativistic dynamics.

The space PH is of rather secondary importance. It is useful when we deal with an encounter, in which a number of particles, initially in free independent motion, come under the influence of one another, and then separate with final motions free and independent. When the particles are moving freely and independently (i.e. before and after the encounter), the representative point maintains a fixed position in PH, and the effect of the encounter is to move this point from one such position to another.

The space QP is, thanks largely to the work of J. W. Gibbs on statistical mechanics, probably the best known of the spaces listed above. If the system is conservative, then the totality of trajectories appears as a congruence of curves in QP, one curve passing through each point. This is a satisfyingly simple picture, but it is complicated in the non-conservative case, for then there is a single infinity of trajectories through each point. Moreover, it is not well suited to relativity, for which t should be treated on a parity with q_ϱ.

In the space QTP the time t is treated on a parity with the coordinates q_ϱ and the momenta p_ϱ; the Hamiltonian $H(q, t, p)$ is a function of position in the space. The picture of the trajectories is simpler than in QP for a non-conservative system, for now we have a congruence of curves, one through each point. QTP differs from QP and $QTPH$ in having an odd dimensionality—an important difference from a mathematical standpoint.

The space $QTPH$ provides the most general approach to dynamics. In it t and H are treated on a parity with q_ϱ and p_ϱ, so that there is complete formal symmetry. The $2N+2$ coordinates fall into two groups, (q, t) and (p, H), the two groups being almost interchangeable in the dynamical theory. To preserve symmetry, a dynamics is best defined, not by giving H as a function of (q, t, p), but by writing down an energy equation involving, in general, all the $2N+2$ coordinates of $QTPH$. This equation defines a surface of $2N+1$ dimensions in $QTPH$, and the representative point is confined to this surface; but it is sometimes convenient to use an energy function instead of an energy equation, in order to deal with the whole space instead of merely with this surface.

The use of $QTPH$ is immediately suggested by Hamilton's optical method[1] in which all the coordinates in space are treated on a parity. The symmetrical approach to dynamics is indicated in Hamilton's calculus of principal

[1] W. R. Hamilton: Mathematical Papers I. Cambridge: University Press 1931. See footnote in Sect. 67.

relations[1], and in modern times it has been revived[2]. All theory developed in $QTPH$ can be immediately transferred to isoenergetic dynamics in QP by a mere reduction of dimensionality.

63. Topological remarks. As a simple illustration of a topological situation often encountered in a more complicated form, consider a particle moving on a circle with equation

$$\xi^2 + \eta^2 = a^2, \tag{63.1}$$

where (ξ, η) are rectangular Cartesian coordinates and a a constant. The configuration space Q is the circle itself, and we may assign a generalized coordinate q by writing

$$\xi = a\cos q, \quad \eta = a\sin q, \quad -\infty < q < \infty. \tag{63.2}$$

Then any value of q determines a configuration (a point of Q); but to any configuration (or point of Q) there corresponds an infinite set of values of q, separated by intervals of 2π. We may then speak of a representative space (say Q') which

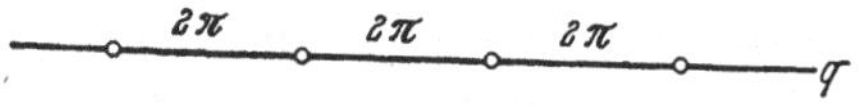

Fig. 29. The representative space Q' in which an infinite set of congruent points, with separations 2π in the cyclic coordinate q, correspond to a single configuration or point of the space Q.

is an infinite line with q measured off along it (Fig. 29); but we must always remember that the correspondence between configurations and the points of Q' is not 1:1. We may call q a *cyclic*[3] coordinate, and the increment in it which restores the configuration its *cyclic constant* (2π in the above case). The set of points in Q' obtained by increasing the cyclic coordinate by any multiple of the cyclic constant may be called a set of *congruent* points.

This many-valued representation is objectionable, but it is practical and commonly used. We can get a 1:1 correspondence between configurations and q-values by writing, instead of (63.2),

$$\xi = a\cos q, \quad \eta = a\sin q, \quad 0 \leq q < 2\pi. \tag{63.3}$$

But now the correspondence is discontinuous, for, by giving the particle a small displacement from the configuration $q=0$, in the sense of q decreasing, we find q suddenly increased by 2π.

Such discontinuous representations are even more objectionable than multiple valued representations, and we turn to the recognized topological plan. This involves the use of two coordinate systems, with an overlap. We assign coordinates q, q' as follows:

$$\left.\begin{aligned} \xi &= a\cos q, & \eta &= a\sin q, & -\alpha \leq q \leq \pi + \alpha,\\ \xi &= a\cos q', & \eta &= -a\sin q', & 0 \leq q' \leq \pi. \end{aligned}\right\} \tag{63.4}$$

In Fig. 30, the arc $EADBF$ is covered by q and $AECFB$ by q', with overlaps on AE and BF. In the overlaps we assign transformations

$$AE\colon\ q = -q', \quad BF\colon\ q - \pi = \pi - q'. \tag{63.5}$$

---

[1] W. R. HAMILTON: Mathematical Papers II, pp. 297–410. Cambridge: University Press 1940.

[2] P. A. M. DIRAC: Proc. Cambridge Phil. Soc. **29**, 389–400 (1933). — CORBEN and STEHLE [*3*], p. 298. — LANCZOS [*15*], p. 185.

[3] In many dynamical problems, cyclic coordinates are ignorable [cf. Sect. 46], and the word *cyclic* is often regarded as synonymous with *ignorable*; cf. GOLDSTEIN [*7*], p. 48. In this article we shall use the word *cyclic* only in the above topological sense.

If we take the particle round the circle in the counter-clockwise sense, starting from A, we use the coordinate q until we get into BF. Then we switch to q', and carry on with q' until we get into AE, when we switch back to q. In this way we have, outside the overlaps, a continuous 1:1 correspondence between the points of Q and the values of the appropriate coordinate (q or q'); in each overlap, we have a choice between two coordinates, and, when we leave an overlap, we leave it with that coordinate appropriate to the domain into which we are going. Although clumsier than the use of a cyclic coordinate, conceptually this is the most satisfactory method.

In ordinary dynamics we start with a physical system which we could, if need be, construct in the space of our experience, and topological questions concerning the configuration space Q may be answered by appeal to our intuition of ordinary space. No such intuition is available to us when we start to develop general dynamical theory; it must be built up on a mathematical basis, and, if our intuition is to be correct and useful, we should avoid purely formal arguments.

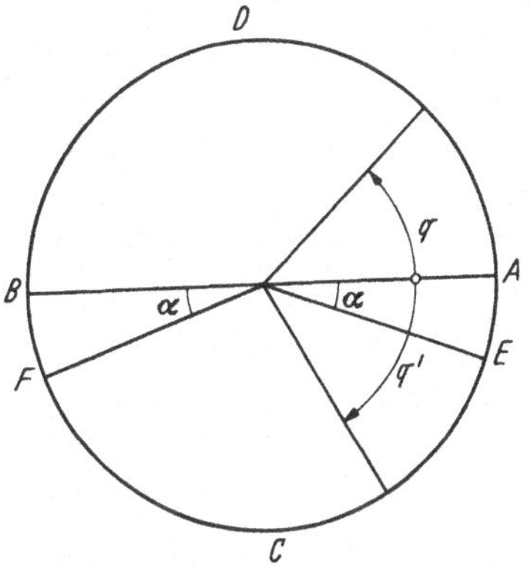

Fig. 30. Overlapping coordinate systems.

It is not enough to say that we shall discuss a dynamical system with generalized coordinates q_ϱ and a Lagrangian $L(q, t, \dot{q})$. The topology of the space Q should be prescribed. There are four ways of doing this:

(i) We may say that each of the coordinates q_ϱ ranges from $-\infty$ to $+\infty$, with a continuous 1:1 correspondence between configurations, or points of Q, and the sets of values $(q_1, q_2, \ldots)$. This is the simplest case (Euclidean topology). We need not stick to these coordinates; we can pass to others, provided that, in the domain of Q considered, the transformation is smooth and the Jacobian does not vanish.

(ii) We may regard Q as immersed in a Euclidean space of higher dimensionality. Then the topology of Q is intrinsic in the equations which define it as a subspace.

(iii) We may modify (i) by adding cyclic coordinates, thus replacing Q by a space Q', a point of Q corresponding to a set (in general multiply infinite) of congruent points in Q'.

(iv) We may introduce, not one coordinate system, but several, with overlapping domains and specified smooth[1] formulae of transformation in the overlaps.

The following topological terms are important.

A *circuit* is a closed curve in the representative space. It is *reducible* if it can be reduced to a single point by continuous transformation in the space; otherwise it is *irreducible*. Two circuits are *reconcilable* if they can be transformed into one another by continous transformation; if they cannot, they are *irreconcilable*.

All reducible circuits are reconcilable. Two irreducible circuits may be reconcilable or not; if irreconcilable, they are *independent*. A space possessing irreducible circuits is *multiply connected*; if it has none, it is *simply connected*. It is *doubly*[2] connected if there is one independent irreducible circuit; *triply* connected if there are two; and so on.

---

[1] In the space Q, we would demand smoothness of class C^2: $\partial^2 q'_\varrho / \partial q_\sigma \partial q_\tau$ continuous. In QP or $QTPH$, we would demand a canonical transformation (Sect. 87).

[2] Because there are then *two* essentially different ways of passing from one point to another.

The surface of a sphere is simply connected[1]; the circumference of a circle, or the surface of a cylinder, is doubly connected; the surface of a torus is triply connected.

If, for a cylinder, we treat the azimuthal angle φ as a cyclic coordinate, and measure z along the generators, a reducible circuit appears[2] as Γ in Fig. 31, and an irreducible circuit as Γ'.

Motion in a reducible circuit is called a *libration*; motion in an irreducible circuit is called a *rotation*.

For a rigid body moving in ordinary space, the configuration space is of six dimensions. Let (ξ, η, ζ) be the rectangular Cartesian coordinates of some base point selected in the body. The configuration space of the free body is the *product* of two 3-spaces; in the first of these the coordinates are (ξ, η, ζ), and it has Euclidean topology; a point in the second corresponds to a configuration of a rigid body with a fixed point. This second space is the *space of rotations*.

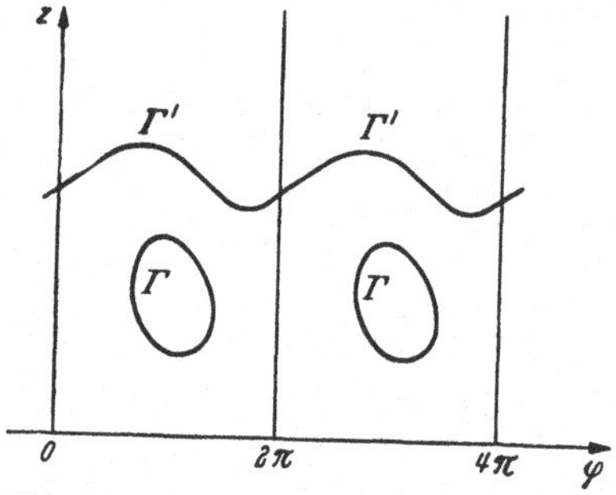

Fig. 31. The space Q' for a particle on a cylinder (φ cyclic); Γ is a reducible circuit, and Γ' irreducible.

The usual way of handling the space of rotations in dynamics is to use the Eulerian angles ϑ, φ, ψ of Sect. 11, treating φ and ψ as cyclic coordinates, so that we have a space Q' in which (if φ and ψ are regarded as rectangular Cartesians) the congruent points at the corners of an infinite array of cubes of edge 2π correspond to the same configuration. The configurations for which $\vartheta = 0$ and $\vartheta = \pi$ are omitted from this representation.

There is another way of dealing with the space of rotations. By Sect. 10, the points of this space are in continuous 1:1 correspondence with the *pairs* of diametrically opposed points on a hypersphere with equation

$$\lambda^2 + \mu^2 + \nu^2 + \varrho^2 = 1, \tag{63.6}$$

drawn in a Euclidean space of four dimensions. Thus the topology of the space Q for a rigid body with a fixed point is that of a hypersphere of elliptic type, in which opposite points are "identified" with one another (congruent points). This space is doubly connected, an irreducible circuit corresponding to a complete rotation of the body about any axis.

No treatment of dynamics, with due regard to topology, can be presented in small compass, and the above remarks are intended only to introduce the subject[3]. As long as we keep to a fairly small domain in the representative space ("dynamics in the small"), topological questions do not arise, for we may assume that small domain to have the simple topology of the interior of a Euclidean sphere with the appropriate number of dimensions. This article follows, in the main, the tradition of mathematical physics, in which topological questions are matters for ad hoc investigation in particular cases; they may be left in abeyance until we come to action-angle variables (Sects. 98 and 99).

[1] Nevertheless, for a particle moving on a sphere, there exists no coordinate system in continuous 1:1 correspondence with the points of the space Q. The common practice is to treat the azimuthal angle φ as a cyclic coordinate and exclude the poles, $\vartheta = 0$ and $\vartheta = \pi$.

[2] Only the simplest cases are illustrated; reducible and irreducible circuits may be much more complicated than those shown.

[3] In the extensive literature of modern topology, it is hard to find any treatment in which the physical interest is not heavily overshadowed by the rigour indispensable in a purely mathematical argument. But for a sprightly introduction to topology, see R. COURANT and H. ROBBINS: What is Mathematics? Chap. V. London-New York-Toronto: Oxford University Press 1948.

II. The space of events (QT).

64. The homogeneous Lagrangian $\Lambda(x, x')$ and the ordinary Lagrangian $L(q, t, \dot q)$. Consider the space of events QT, of $N+1$ dimensions and with coordinates $q_1, \dots q_N, t$. For notational reasons, and also with a view to relativistic applications, let us write

$$q_\varrho = x_\rho, \qquad t = x_{N+1}, \tag{64.1}$$

so that the coordinates in QT are x_r (see Sect. 62 for notation).

Let Γ be any curve in QT with equations

$$x_r = x_r(u). \tag{64.2}$$

We suppose these functions to be smooth (class C^2), and that all the derivatives x'_r $(=dx_r/du)$ do not vanish simultaneously for any value of u under consideration. The parameter u is of no particular importance; we keep in mind the whole class of parameters obtained from one of them by a C^2 transformation with non-vanishing positive first derivative, so that all the parameters increase together. It is the *curve* Γ in QT, with a certain sense on it, but not with any particular parametrization, that is the geometrical object now before us; it corresponds to a possible motion of the system.

We now introduce a *homogeneous Lagrangian*[1] *function*

$$\Lambda(x_1, \dots x_{N+1}, x'_1, \dots x'_{N+1}) \quad (x'_r = dx_r/du), \tag{64.3}$$

which we shall denote briefly by $\Lambda(x, x')$; we demand that this function shall be positive homogeneous of the first degree in the derivatives x'_r, so that we have

$$\Lambda(x, kx') = k\Lambda(x, x') \quad (k > 0), \tag{64.4}$$

and, by EULER's theorem for homogeneous functions,

$$x'_r \frac{\partial \Lambda}{\partial x'_r} = \Lambda. \tag{64.5}$$

As regards smoothness, we shall assume the existence and continuity of whatever derivatives we may require. Should piecewise discontinuities arise for consideration, they can be dealt with on the occasion.

In a region of overlapping coordinate systems, Λ transforms as an invariant in the sense of tensor calculus. If the two coordinate systems are x^*_r and x_r and the two Lagrangians are Λ^* and Λ, then

$$\Lambda^*(x^*, x^{*\prime}) = \Lambda(x, x'). \tag{64.6}$$

The *Lagrangian action*[2] along any directed curve Γ drawn from a point B^* (where $u = u_1$) to a point B (where $u = u_2 > u_1$) is defined to be

$$A_L(\Gamma) = \int_{u_1}^{u_2} \Lambda(x, x')\, du, \tag{64.7}$$

---

[1] The present discussion is intentionally kept on a somewhat abstract level, in order not to limit the scope of ultimate applications of the theory. The word "Lagrangian" forms a link with the more concrete theory of Sects. 46, and hence with physical concepts (Sect. 2).

[2] Unfortunately the simple word *action* is commonly used for $\int \frac{\partial L}{\partial \dot q_\varrho} dq_\varrho$ (cf. GOLDSTEIN [7], p. 228; WHITTAKER [28], p. 248), and there is no commonly accepted name for the more fundamental integral (64.7); it will be called *Lagrangian action* in this article.

so that $A_L(\Gamma)$ is a functional of the curve Γ. By reason of the homogeneity of Λ, $A_L(\Gamma)$ is independent of the parametrization. The *element of Lagrangian action* is

$$dA_L = \Lambda(x, x')\, du = \Lambda(x, dx). \tag{64.8}$$

There is, in general, no connection between the action from B^* to B and the action from B to B^*, even though the curve is the same in both cases.

By imposing an element of action $\Lambda(x, dx)$ on QT, we make it a FINSLER *space* in the language of geometry. If $\Lambda(x, dx)$ is the square root of a homogeneous quadratic form in the differentials, then QT is a *Riemannian space.*

The element of Lagrangian action (64.8) may be written more explicitly as

$$dA_L = \Lambda(x_1, \ldots x_N, x_{N+1}, x'_1, \ldots x'_N, x'_{N+1})\, du. \tag{64.9}$$

By reason of the homogeneity of Λ, dA_L is independent of the choice of the parameter u. Choose $u = t$; then, by (64.1),

$$dA_L = \Lambda(q_1, \ldots q_N, t, \dot{q}_1, \ldots \dot{q}_N, 1)\, dt. \tag{64.10}$$

Define the function $L(q, t, \dot{q})$ by

$$L(q_1, \ldots q_N, t, \dot{q}_1, \ldots \dot{q}_N) = \Lambda(q_1, \ldots q_N, t, \dot{q}_1, \ldots \dot{q}_N, 1). \tag{64.11}$$

Then the element of Lagrangian action is

$$dA_L = L(q, t, \dot{q})\, dt. \tag{64.12}$$

This function $L(q, t, \dot{q})$ we call the *ordinary Lagrangian function* (cf. Sect. 46).

The two functions, Λ (a function of $2N+2$ quantities, positive homogeneous of degree unity in the last $N+1$ of them) and L (a function of $2N+1$ quantities, with no such condition imposed), are equivalent to one another, in the sense that one determines the other. Given Λ, we get L by (64.11); given L, we get Λ by equating (64.9) and (64.12) and dividing by du. Thus

$$\left.\begin{aligned} &\Lambda(x_1, \ldots x_N, x_{N+1}, x'_1, \ldots x'_N, x'_{N+1}) \\ &\quad = L(q_1, \ldots q_N, t, \dot{q}_1, \ldots \dot{q}_N)\, t' \\ &\quad = L\left(x_1, \ldots x_N, x_{N+1}, \frac{x'_1}{x'_{N+1}}, \ldots \frac{x'_N}{x'_{N+1}}\right) x'_{N+1}, \end{aligned}\right\} \tag{64.13}$$

which has the required homogeneity.

To find the relationships between the partial derivatives of $\Lambda(x, x')$ and $L(q, t, \dot{q})$, we vary x_r and x'_r in (64.13). This gives

$$\frac{\partial \Lambda}{\partial x_r}\,\delta x_r + \frac{\partial \Lambda}{\partial x'_r}\,\delta x'_r = t'\left(\frac{\partial L}{\partial q_\varrho}\,\delta q_\varrho + \frac{\partial L}{\partial t}\,\delta t + \frac{\partial L}{\partial \dot{q}_\varrho}\,\delta \dot{q}_\varrho\right) + L\,\delta t'. \tag{64.14}$$

But

$$\dot{q}_\varrho = \frac{q'_\varrho}{t'}, \qquad \delta \dot{q}_\varrho = \frac{\delta q'_\varrho}{t'} - \frac{q'_\varrho\,\delta t'}{t'^2}. \tag{64.15}$$

Substituting this in (64.14), writing δx_ϱ for δq_ϱ, $\delta x'_\varrho$ for $\delta q'_\varrho$, $\delta x'_{N+1}$ for $\delta t'$, and equating the coefficients of the $2N+2$ independent differentials δx_r, $\delta x'_r$, we get

$$\left.\begin{aligned} &\frac{\partial \Lambda}{\partial x_\varrho} = t'\frac{\partial L}{\partial q_\varrho}, \qquad \frac{\partial \Lambda}{\partial x_{N+1}} = t'\frac{\partial L}{\partial t}, \\ &\frac{\partial \Lambda}{\partial x'_\varrho} = \frac{\partial L}{\partial \dot{q}_\varrho}, \qquad \frac{\partial \Lambda}{\partial x'_{N+1}} = L - \dot{q}_\varrho\frac{\partial L}{\partial \dot{q}_\varrho}. \end{aligned}\right\} \tag{64.16}$$

65. First form of HAMILTON's principle. LAGRANGE's equations of motion. Let Γ (Fig. 32) be any curve joining B^* to B. We vary it to a neighbouring curve Γ_1 with end points D^*, D. Whatever parameter u is used on Γ, we choose the parameter u on Γ_1 with the same end values, u_1, u_2. Then the variation of the Lagrangian action is

$$\left.\begin{aligned}\delta A_L &= A_L(\Gamma_1) - A_L(\Gamma)\\ &= \int_{u_1}^{u_2} \delta\Lambda\, du = \int_{u_1}^{u_2} \left(\frac{\partial\Lambda}{\partial x_r}\,\delta x_r + \frac{\partial\Lambda}{\partial x_r'}\,\delta x_r'\right) du.\end{aligned}\right\} \tag{65.1}$$

Integration by parts gives

$$\delta A_L = \left[\frac{\partial\Lambda}{\partial x_r'}\,\delta x_r\right]_{u=u_1}^{u=u_2} - \int_{u_1}^{u_2} \left(\frac{d}{du}\frac{\partial\Lambda}{\partial x_r'} - \frac{\partial\Lambda}{\partial x_r}\right)\delta x_r\, du. \tag{65.2}$$

Let us now restrict the varied curve Γ_1 by demanding that its end points coincide with B^* and B. The first term on the right hand side of (65.2) disappears. If $\delta A_L = 0$ for all variations of Γ to Γ_1, arbitrary except for the condition of fixed end points, then Γ must satisfy the EULER-LAGRANGE equations

$$\frac{d}{du}\frac{\partial\Lambda}{\partial x_r'} - \frac{\partial\Lambda}{\partial x_r} = 0. \tag{65.3}$$

We may refer to these equations as LAGRANGE's *equations of motion*, and to the curves satisfying them as *rays* or *trajectories*[1].

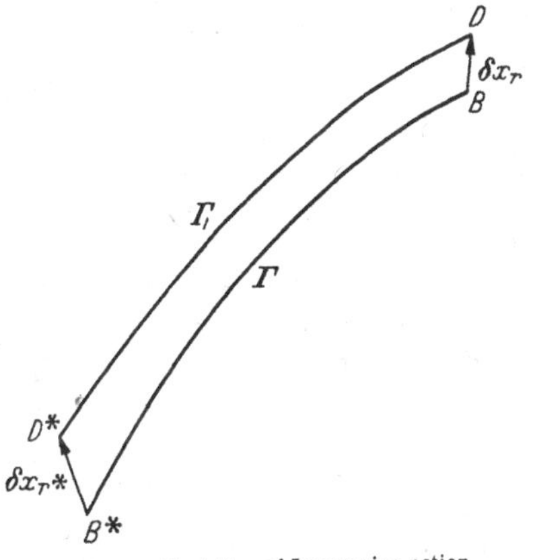

Fig. 32. Variation of Lagrangian action.

The variational equation

$$\delta \int \Lambda(x, x')\, du = 0, \tag{65.4}$$

for fixed end points, is equivalent to the set of differential equations (65.3). We call (65.4) the *first form of* HAMILTON's *principle*[2].

This principle can also be written

$$\delta \int L(q, t, \dot q)\, dt = 0, \tag{65.5}$$

for fixed end values of q_ϱ and t, and this leads at once to LAGRANGE's equations of motion in the form [cf. (46.18)]

$$\frac{d}{dt}\frac{\partial L}{\partial \dot q_\varrho} - \frac{\partial L}{\partial q_\varrho} = 0, \tag{65.6}$$

which are equivalent to (65.3). The expression occurring here is called the *Lagrangian derivative* of L.

[1] The word *trajectory* links the mathematical concept with the physical concepts of dynamics. The word *ray* links it with optics, and that may seem out of place in the present connection. But the wave mechanics of DE BROGLIE and SCHRÖDINGER has weakened the barrier between dynamics and optics. If we need the word *wave* in dynamics, the word *ray* enters naturally with it. Moreover, the present exposition of general dynamical theory owes as much to HAMILTON's method in optics as it owes to his method in dynamics, for in his optics all the coordinates were put on a parity, whereas in his dynamics the time was privileged.

[2] From a mathematical standpoint, (65.3) and (65.4) are different ways of saying the same thing. We seem to be indulging in a plethora of verbiage. But words are important, as indicated in Sect. 2. The differential equations (65.3) cannot be put into words which are more illuminating than the bare formulae. HAMILTON's principle can: it says that the Lagrangian action has a stationary value for a trajectory. These words are easier to link up with physical concepts than mathematical formulae are.

There are $N+1$ equations in (65.3) and only N in (65.6). However, (65.3) are identically related, for we have, remembering the homogeneity of Λ,

$$x'_r\left(\frac{d}{du}\frac{\partial\Lambda}{\partial x'_r}-\frac{\partial\Lambda}{\partial x_r}\right)=\frac{d}{du}\left(x'_r\frac{\partial\Lambda}{\partial x'_r}\right)-x''_r\frac{\partial\Lambda}{\partial x'_r}-x'_r\frac{\partial\Lambda}{\partial x_r}=\frac{d\Lambda}{du}-\frac{d\Lambda}{du}=0. \quad (65.7)$$

Provided the Eqs. (65.6) can be solved for $\ddot{q}_r$ (and we shall suppose that they can), these equations determine a ray in QT corresponding to assigned initial values of q_r, t and $\dot{q}_r$. Then, *through each point x_r of QT and in each direction (defined by the ratios $dx_1{:}dx_2{:}\ldots{:}dx_{N+1}$) there passes a unique ray or trajectory.*

We may use the term *Λ-dynamics* to refer to theory based on the variational Eq. (65.4) and its extremals (65.3); likewise the term *L-dynamics* for the variational Eq. (65.5) and the extremals (65.6). They are essentially equivalent to one another, L-dynamics being a form of Λ-dynamics in which t is treated in two capacities: a coordinate in QT and a parameter on a curve in QT. We include them both under the title *Lagrangian dynamics.*

66. Two illustrative examples. Lagrange's equations (65.6) form the link between the dynamical systems most commonly encountered and the present more abstract theory. This theory applies to all physical systems which behave in accordance with (65.6), whether those systems are truly dynamical or not. The system may consist of electrical circuits, with the generalized velocities corresponding to currents. In the truly dynamical domain, the present theory applies, by virtue of (46.18), to all holonomic systems for which the generalized forces are derivable from a potential function or an extended potential function. In such systems the kinetic energy is always quadratic in the generalized velocities, and so also is the Lagrangian $L\ (=T-V)$ when V is an ordinary potential. In the present general theory no such restriction is placed on the function $L(q, t, \dot{q})$, which is to be regarded as an arbitrary function of its $2N+1$ arguments.

To illustrate the general procedures, we shall keep before us the two following examples:

α) RS (relativistic system): We take for homogeneous Lagrangian

$$\Lambda(x, x')=\sqrt{b_{rs}\,x'_r x'_s}+A_r\,x'_r, \quad (66.1)$$

where $b_{rs}\,(=b_{sr})$ and A_r are functions of the x's. This is a generalization of the relativistic Lagrangian for a charged particle moving in a given electromagnetic field[1]. The homogeneous Lagrangian is simpler than the ordinary one, which, by (64.11), reads

$$L(q, t, \dot{q})=(b_{\varrho\sigma}\dot{q}_\varrho\dot{q}_\sigma+2b_{\varrho, N+1}\dot{q}_\varrho+b_{N+1, N+1})^{\frac{1}{2}}+A_\varrho\dot{q}_\varrho+A_{N+1}. \quad (66.2)$$

β) ODS (ordinary dynamical system): This is a holonomic scleronomic conservative system with an ordinary potential function, so that

$$L(q, \dot{q})=T(q, \dot{q})-V(q)=\tfrac{1}{2}\,a_{\varrho\sigma}\dot{q}_\varrho\dot{q}_\sigma-V, \quad (66.3)$$

the coefficients $a_{\varrho\sigma}\,(=a_{\sigma\varrho})$ and V being functions of the q's. Thus t is absent from L. In this case the homogeneous Lagrangian is, by (64.13),

$$\Lambda(x, x')=\tfrac{1}{2}\,a_{\varrho\sigma}\,x'_\varrho\,x'_\sigma/x'_{N+1}-V\,x'_{N+1}, \quad (66.4)$$

a little clumsier than $L(q, \dot{q})$.

[1] Cf. Sect. 114.

It is clear that the theory for RS will be simpler if we use $\Lambda(x, x')$, and the theory for ODS simpler if we use $L(q, \dot{q})$. HAMILTON's principle reads

$$\text{RS:} \quad \delta \int [(b_{rs}\, dx_r\, dx_s)^{\frac{1}{2}} + A_r\, dx_r] = 0, \tag{66.5}$$

$$\text{ODS:} \quad \delta \int (\tfrac{1}{2} a_{\varrho\sigma} \dot{q}_\varrho \dot{q}_\sigma - V)\, dt = 0. \tag{66.6}$$

For RS LAGRANGE's equations of motion read

$$\frac{d}{du}\left[\frac{b_{rs} x'_s}{\sqrt{b_{mn} x'_m x'_n}} + A_r\right] - \frac{\frac{\partial b_{st}}{\partial x_r} x'_s x'_t}{\sqrt{b_{mn} x'_m x'_n}} - \frac{\partial A_s}{\partial x_r} x'_s = 0, \tag{66.7}$$

and if we choose for u that parameter on the ray which makes

$$b_{mn} x'_m x'_n = 1, \tag{66.8}$$

they simplify to

$$\frac{d}{du}(b_{rs} x'_s) + \left(\frac{\partial A_r}{\partial x_s} - \frac{\partial A_s}{\partial x_r}\right) x'_s - \frac{\partial b_{st}}{\partial x_r} x'_s x'_t = 0, \tag{66.9}$$

or

$$b_{rs} x''_s + \left(\frac{\partial A_r}{\partial x_s} - \frac{\partial A_s}{\partial x_r}\right) x'_s + \left(\frac{\partial b_{rs}}{\partial x_t} - \frac{\partial b_{ts}}{\partial x_r}\right) x'_s x'_t = 0. \tag{66.10}$$

For ODS LAGRANGE's equations of motion read

$$\frac{d}{dt}\frac{\partial T}{\partial \dot{q}_\varrho} - \frac{\partial T}{\partial q_\varrho} = -\frac{\partial V}{\partial q_\varrho}, \tag{66.11}$$

or

$$a_{\varrho\sigma} \ddot{q}_\sigma + [\mu\nu, \varrho]\, \dot{q}_\mu \dot{q}_\nu = -\frac{\partial V}{\partial q_\varrho}, \tag{66.12}$$

or

$$\ddot{q}_\varrho + \left\{{\varrho \atop \mu\nu}\right\} \dot{q}_\mu \dot{q}_\nu = -a^{\varrho\sigma} \frac{\partial V}{\partial q_\sigma}, \tag{66.13}$$

where the CHRISTOFFEL symbols are

$$\left.\begin{aligned} [\mu\nu, \sigma] &= \frac{1}{2}\left(\frac{\partial a_{\mu\sigma}}{\partial q_\nu} + \frac{\partial a_{\nu\sigma}}{\partial q_\mu} - \frac{\partial a_{\mu\nu}}{\partial q_\sigma}\right), \\ \left\{{\varrho \atop \mu\nu}\right\} &= a^{\varrho\sigma}[\mu\nu, \sigma], \qquad a^{\varrho\sigma} a_{\mu\sigma} = \delta^\varrho_\mu. \end{aligned}\right\} \tag{66.14}$$

67. The energy equation $\Omega(x, y) = 0$ and the Hamiltonian $H(q, t, p)$. We now make a fresh start. We have before us a space QT of $N+1$ dimensions with coordinates x_r, expressed in alternative notation as

$$x_\varrho = q_\varrho, \qquad x_{N+1} = t. \tag{67.1}$$

Instead of imposing a Lagrangian $\Lambda(x, x')$ or $L(q, t, \dot{q})$ on QT, we impose an *energy equation*[1]

$$\Omega(x, y) = 0, \tag{67.2}$$

connecting the coordinates x_r of a point in QT with a vector y_r. This equation may be regarded geometrically as attaching to each point of QT a surface (of N dimensions) in an $(N+1)$-dimensional space tangent to QT, y_r being coordinates in that tangent space.

In a domain where two coordinate systems (x, x^*) overlap, y_r is to transform as a covariant vector:

$$y^*_r = y_s \frac{\partial x_s}{\partial x^*_r}. \tag{67.3}$$

[1] The reason for this name will appear at (67.8).

This makes

$$y_r^* dx_r^* = y_r dx_r, \tag{67.4}$$

so that this Pfaffian form is invariant.

For purposes of general theory it is often best to treat all the coordinates symmetrically, and then we leave (67.2) in this general implicit form. But we sometimes find it convenient to solve for y_{N+1}, so that the energy equation reads[1]

$$y_{N+1} + \omega(x_1, \dots x_N, x_{N+1}, y_1, \dots y_N) = 0. \tag{67.5}$$

We define p_ϱ and H by the equations

$$y_\varrho = p_\varrho, \quad y_{N+1} = -H \tag{67.6}$$

(note the minus sign). Then (67.2) expresses a relationship between the $2N+2$ quantities

$$q_\varrho, t, p_\varrho, H,$$

and (67.5) expresses H as a function of the others:

$$H = \omega(q_1, \dots q_N, t, p_1, \dots p_N), \tag{67.7}$$

or, if we like,

$$H = H(q, t, p). \tag{67.8}$$

We make a suggestive link with physical concepts (to be strengthened later) by calling H the *Hamiltonian* and y_r the *momentum-energy vector*; for brevity we may refer to y_r simply as *momenta*, if there is no danger of confusion. Since, for the commonest systems, the Hamiltonian is equal to the total energy, it seems appropriate to call (67.2) the energy equation, since it is equivalent to (67.8)[2].

68. Second form of HAMILTON'S principle. HAMILTON'S canonical equations of motion. Let Γ be any curve in QT joining a point B^* to a point B. We define the *Hamiltonian action* along Γ to be the integral

$$A_H(\Gamma) = \int y_r dx_r = \int (p_\varrho dq_\varrho - H\,dt), \tag{68.1}$$

the vector field y_r along Γ being assigned in any way consistent with the energy Eq. (67.2). Thus the Hamiltonian action is not a functional of Γ alone, but also of the assignment of y_r along Γ.

We now vary Γ as in Fig. 32 of Sect. 65, at the same time varying y_r consistently with (67.2), and obtain for the variation of Hamiltonian action

$$\left.\begin{aligned} \delta A_H &= \int (\delta y_r\, dx_r + y_r\, \delta\, dx_r) \\ &= [y_r\, \delta x_r] + \int (\delta y_r\, dx_r - \delta x_r\, dy_r). \end{aligned}\right\} \tag{68.2}$$

---

[1] The equation may have several roots, so that ω is multiple valued; in that case we concentrate on one of the values. We might of course solve for any of the y's, but for present purposes y_{N+1} is best.

[2] In this account of general dynamical theory, I follow the pattern of HAMILTON'S method in geometrical optics, which is essentially the same as his dynamical method, but more compact, because more symmetrical. To gain this compactness, one treats the coordinates q_ϱ on a parity with the time t, and the momenta p_ϱ on a parity with the negative of the Hamiltonian, $-H$. This symmetry is present, in thought if not in notation, in E. CARTAN, Leçons sur les invariants intégraux (Paris: Hermann 1922); one meets it also in LANCZOS [*15*], pp. 185—192, and in GOLDSTEIN [*7*], p. 243. LANCZOS uses the name *auxiliary condition* for (67.2) above. In HAMILTON'S optics, this equation is the *equation of the surface of components* and the quantities y_r are the *components of normal slowness*; cf. W. R. HAMILTON: Mathematical Papers, Vol. 1, pp. 291, 303 (Cambridge: University Press 1931). For modern general treatments of geometrical optics, see C. CARATHÉODORY: Geometrische Optik (Berlin: Springer 1937), or, following HAMILTON'S ideas more closely, J. L. SYNGE: J. Opt. Soc. Amer. **27**, 75—82 (1937).

Since $\Omega=0$ for Γ and for the varied curve, we have

$$\delta\Omega = \frac{\partial\Omega}{\partial x_r}\,\delta x_r + \frac{\partial\Omega}{\partial y_r}\,\delta y_r = 0. \tag{68.3}$$

If we hold the end points fixed, then (68.2) becomes

$$\delta A_H = \int (\delta y_r\, dx_r - \delta x_r\, dy_r). \tag{68.4}$$

In view of the side condition (68.3), a curve of stationary Hamiltonian action, i.e. a curve satisfying

$$\delta \int y_r\, dx_r = 0, \quad \Omega(x, y) = 0, \tag{68.5}$$

with fixed end points, satisfies the equations

$$dx_r = dw\,\frac{\partial\Omega}{\partial y_r}, \qquad dy_r = -\,dw\,\frac{\partial\Omega}{\partial x_r}, \tag{68.6}$$

where dw is an infinitesimal LAGRANGE multiplier. Hence the extremal satisfies the differential equations

$$\frac{dx_r}{dw} = \frac{\partial\Omega}{\partial y_r}, \qquad \frac{dy_r}{dw} = -\frac{\partial\Omega}{\partial x_r}. \tag{68.7}$$

We call (68.5) the *second form of* HAMILTON'S *principle* and (68.7) HAMILTON'S *equations of motion*. Equations of this form are called *canonical*.

A curve, with attached vectors y_r, satisfying this principle (or, equivalently, these differential equations), we call a *ray* or *trajectory*; the curve in QT and the associated vector field on it are described by equations of the form

$$x_r = x_r(w), \qquad y_r = y_r(w). \tag{68.8}$$

These functions are determined by (68.7) if initial values of the x's and the y's are assigned.

The parameter w in (68.7) is a special parameter in the sense that it cannot be changed once the function Ω has been assigned; for the element of Hamiltonian action is

$$dA_H = y_r\, dx_r = y_r\,\frac{dx_r}{dw}\,dw = y_r\,\frac{\partial\Omega}{\partial y_r}\,dw, \tag{68.9}$$

and this determines dw. But, of course, a *relationship* may be expressed by different *equations*, and if we change from an equation $\Omega=0$ to an equation $\Omega^*=0$, both expressing the same relationship, then the corresponding parameters satisfy

$$\frac{dw^*}{dw} = \frac{d\Omega}{d\Omega^*}. \tag{68.10}$$

We note that $\Omega=\text{const}$ is a direct formal consequence of the differential equations (68.7); for we have

$$\frac{d\Omega}{dw} = \frac{\partial\Omega}{\partial x_r}\frac{dx_r}{dw} + \frac{\partial\Omega}{\partial y_r}\frac{dy_r}{dw} = 0. \tag{68.11}$$

The theory presented here is fundamental and we shall express it in the unsymmetrical notation also. HAMILTON'S principle[1] (68.5) reads

$$\delta \int (p_\varrho\, dq_\varrho - H\, dt) = 0, \qquad H = H(q, t, p), \tag{68.12}$$

[1] This general form of HAMILTON'S principle is due to HELMHOLTZ; cf. LEVI-CIVITA and AMALDI [16] II$_2$, p. 559.

with fixed end points in QT. The general variation of this integral is

$$\left.\begin{aligned}\delta A_H &= \int (\delta p_\varrho\, dq + p_\varrho\, \delta\, dq_\varrho - \delta H\, dt - H\, \delta\, dt)\\ &= [p_\varrho\, \delta q_\varrho - H\, \delta t] + \int (\delta p_\varrho\, dq_\varrho - \delta q_\varrho\, dp_\varrho - \delta H\, dt + \delta t\, dH).\end{aligned}\right\} \quad (68.13)$$

But

$$\delta H = \frac{\partial H}{\partial q_\varrho}\,\delta q_\varrho + \frac{\partial H}{\partial t}\,\delta t + \frac{\partial H}{\partial p_\varrho}\,\delta p_\varrho, \quad (68.14)$$

and so

$$\left.\begin{aligned}\delta A_H = [p_\varrho\, \delta q_\varrho - H\, \delta t] + \int \Big\{\delta p_\varrho \Big(d q_\varrho - \frac{\partial H}{\partial p_\varrho}\, dt\Big) -\\ - \delta q_\varrho \Big(d p_\varrho + \frac{\partial H}{\partial q_\varrho}\, dt\Big) + \delta t \Big(dH - \frac{\partial H}{\partial t}\, dt\Big)\Big\}.\end{aligned}\right\} \quad (68.15)$$

The first term on the right hand side vanishes for fixed end points, the variations $\delta q_\varrho, \delta p_\varrho, \delta t$ remaining arbitrary on the rest of Γ. Hence the variational equation (68.12) leads to HAMILTON's equations of motion for the rays or trajectories in the form

$$\dot q_\varrho = \frac{\partial H}{\partial p_\varrho}, \qquad \dot p_\varrho = -\frac{\partial H}{\partial q_\varrho}, \quad (68.16)$$

agreeing with (47.7). Like (68.7), these equations are *canonical*. We get also

$$\frac{dH}{dt} = \frac{\partial H}{\partial t}. \quad (68.17)$$

This tells us that *H is constant along a ray or trajectory if t is absent from the function* $H(q, t, p)$ (conservative system, cf. Sect. 62).

We may use the term *Ω-dynamics* to indicate theory based on (68.5) and *H-dynamics* for theory based on (68.12). They are different ways of looking at the same thing, and we include them both under the title *Hamiltonian dynamics.*

The relative advantages and disadvantages of these two aspects of Hamiltonian dynamics are closely analogous to the relative advantages and disadvantages of expressing the equation of a surface in the two forms $f(x, y, z) = 0$ and $z = f(x, y)$; although, to improve the analogy, one should think of an even number of variables. Ω-dynamics seems preferable for general arguments in which it is desirable to put all the $2N + 2$ quantities on the same footing, whereas H-dynamics is in many ways preferable from an analytical standpoint Thus, the equations of motion (68.16) present themselves clearly as a system of order $2N$ ($2N$ equations, each of the first order), whereas in (68.7) we see a system apparently of order $2N + 2$. This latter order is reduced to $2N + 1$ by dividing all the equations by dx_{N+1}/dw, so that x_{N+1} (the time) becomes the independent variable, and the energy equation $\Omega(x, y) = 0$ implies the further reduction of order to $2N$. We shall return to the question of order in Sect. 91.

Comparing (68.7) with (68.16), we see that in H-dynamics the special parameter w is the time t. In Ω-dynamics, w has in general no simple physical meaning, but if we use the equation $\Omega(x, y) = 0$ to make $\Omega + 1$ homogeneous of degree unity in the y's, so that we have

$$y_r \frac{\partial \Omega}{\partial y_r} = y_r \frac{\partial}{\partial y_r}(\Omega + 1) = \Omega + 1 = 1, \quad (68.18)$$

then it follows from (68.9) that w is the Hamiltonian action A_H. In his optical work HAMILTON adopted this homogenization as standard procedure, but it will not be used in this article because it is more convenient to leave the form of the function $\Omega(x, y)$ unrestricted.

69. Equivalence of Lagrangian dynamics and Hamiltonian dynamics. We understand *Lagrangian dynamics* to be the theory of Sects. 64 and 65, based on a homogeneous Lagrangian $\Lambda(x, x')$ or an ordinary Lagrangian $L(q, t, \dot{q})$; and *Hamiltonian dynamics* to be that of Sects. 67 and 68, based on an energy equation $\Omega(x, y) = 0$ or a Hamiltonian $H(q, t, p)$. We shall show that these two dynamics are essentially equivalent, with some additional generality in Hamiltonian dynamics in the matter of the definition of the momentum-energy vector.

We shall establish the essential equivalence by setting up a correspondence

$$\Lambda(x, x') \leftrightarrow \Omega(x, y) = 0, \tag{69.1}$$

or equivalently

$$L(q, t, \dot{q}) \leftrightarrow H(q, t, p). \tag{69.2}$$

When this has been done, dynamics may be treated indifferently in terms of Λ or L or $\Omega = 0$ or H. The correspondence is established by demanding the equality of Lagrangian action and Hamiltonian action for an arbitrary curve in QT.

Let us start with an assigned homogeneous Lagrangian $\Lambda(x, x')$, and *define* y_r by

$$y_r = \frac{\partial \Lambda}{\partial x'_r}. \tag{69.3}$$

These partial derivatives are homogeneous of degree zero in the derivatives x'_r, and therefore involve only the N ratios $x'_1 : x'_2 : \ldots x'_{N+1}$, in addition of course to the coordinates x_r. Elimination of these ratios from the $N+1$ equations (69.3) gives an equation which we write[1]

$$\Omega(x, y) = 0. \tag{69.4}$$

Then along any curve with parameter u and with $x'_r = dx_r/du$, the element of Lagrangian action is, by (64.8),

$$d A_L = \Lambda(x, x')\, du = x'_r \frac{\partial \Lambda}{\partial x'_r} du = y_r\, dx_r. \tag{69.5}$$

By (68.1), this is the element of Hamiltonian action dA_H, corresponding to the energy equation (69.4). Actually, dA_H is more general than dA_L, because in it the momentum-energy vector y_r is restricted only by the energy equation (69.4), whereas in dA_L this momentum-energy vector is precisely specified for any curve by (69.3). However, if we vary a Hamiltonian ray or trajectory, keeping the ends fixed, then $\delta A_H = 0$ for all variations δy_r consistent with $\Omega = 0$, and therefore, in particular, for δy_r consistent with (69.3). Thus $\delta A_H = 0$ implies $\delta A_L = 0$, and this tells us that the Lagrangian rays coincide with the Hamiltonian rays.

The above procedure establishes the one-way correspondence

$$\Lambda(x, x') \to \Omega(x, y) = 0; \tag{69.6}$$

given a homogeneous Lagrangian, we obtain the energy equation by elimination from (69.3), as described. The equivalent one-way correspondence

$$L(q, t, \dot{q}) \to H(q, t, p) \tag{69.7}$$

[1] We assume that only one equation results from this elimination; there might be more Cf. P. A. M. Dirac: Canad. J. Math. **2**, 129–148 (1950).

may be obtained directly from the $N+1$ equations

$$p_\varrho = \frac{\partial L}{\partial \dot{q}_\varrho}, \qquad H = \dot{q}_\varrho \frac{\partial L}{\partial \dot{q}_\varrho} - L, \tag{69.8}$$

by eliminating the N quantities $\dot{q}_\varrho$ and expressing H as a function $H(q, t, p)$.

Let us now start with Hamiltonian dynamics, writing down an energy equation

$$\Omega(x, y) = 0. \tag{69.9}$$

On an arbitrary curve in QT, with equations $x_r = x_r(u)$, the momentum-energy vector y_r is to be regarded as arbitrary, except for this energy equation. But let us restrict the choice of y_r to what we shall call the *natural* momentum-energy by imposing the equations

$$\frac{dx_r}{du} = \vartheta \frac{\partial \Omega}{\partial y_r}, \tag{69.10}$$

where ϑ is an undetermined factor. We recognize here part of the equations of motion (68.7). After solving the $N+2$ equations in (69.9) and (69.10) for y_r and ϑ as functions of the x's and their derivatives $x'_r = dx_r/du$, we define the function Λ by

$$\Lambda(x, x') = y_r x'_r. \tag{69.11}$$

Then the element of Hamiltonian action may be written

$$dA_H = y_r dx_r = y_r x'_r du = \Lambda(x, x')\, du, \tag{69.12}$$

which is the element of Lagrangian action for the homogeneous Lagrangian $\Lambda(x, x')$, as in (64.8) (for the homogeneity of Λ see below).

Collecting the equations, we may say that the one-way correspondence

$$\Omega(x, y) = 0 \rightarrow \Lambda(x, x') \tag{69.13}$$

is obtained from the equations

$$x'_r = \vartheta \frac{\partial \Omega}{\partial y_r}, \qquad \Lambda = y_r x'_r, \qquad \Omega(x, y) = 0, \tag{69.14}$$

by eliminating ϑ and y_r and solving for Λ in terms of the remaining quantities. As for homogeneity, if these equations are satisfied by certain values $(\Lambda, x'_r, \vartheta)$, they are also satisfied by $(k\Lambda, k x'_r, k\vartheta)$ for any k, and this indicates that $\Lambda(x, x')$ is homogeneous of degree unity, and not merely positive homogeneous. However, the function Λ may have several branches, and multiplication by a negative k may involve a change of branch, as illustrated by an example in Sect. 70; only positive homogeneity is demanded in Sect. 64.

Similarly, the one-way correspondence

$$H(q, t, p) \rightarrow L(q, t, \dot{q}), \tag{69.15}$$

by which we pass from a given Hamiltonian to an equivalent Lagrangian, is obtained from the equations

$$\dot{q}_\varrho = \frac{\partial H}{\partial p_\varrho}, \qquad L = \dot{q}_\varrho p_\varrho - H; \tag{69.16}$$

we are to eliminate p_ϱ and express L in the form $L(q, t, \dot{q})$.

We have now established the essential equivalence of Lagrangian and Hamiltonian dynamics. The correspondences are illustrated in Sect. 70, and their geometrical significance is explained in Sect. 71.

70. Examples of LAGRANGE-HAMILTON correspondences. We consider the systems RS and ODS of Sect. 66.

For RS, we start with

$$\Lambda(x, x') = \sqrt{b_{rs} x'_r x'_s} + A_r x'_r. \tag{70.1}$$

Then (69.3) gives

$$y_r = \frac{\partial \Lambda}{\partial x'_\nu} = \frac{b_{rs} x'_s}{\sqrt{b_{mn} x'_m x'_n}} + A_r, \tag{70.2}$$

and elimination gives the energy equation

$$\Omega(x, y) \equiv \tfrac{1}{2}\left[b'^{rs}(y_r - A_r)(y_s - A_s) - 1\right] = 0, \tag{70.3}$$

where $b'^{rs} b_{rm} = \delta^s_m$; the factor $\tfrac{1}{2}$ is a mere notational convenience.

If we start from the energy equation (70.3), and seek to recover Λ, we write down (69.14), which read

$$x'_r = \vartheta\, b'^{rs}(y_s - A_s), \quad \Lambda = y_r x'_r, \quad \Omega(x, y) = 0. \tag{70.4}$$

Hence

$$\left.\begin{aligned} y_r - A_r &= \vartheta^{-1} b_{rs} x'_s, \\ 2\Omega &= \vartheta^{-2} b_{rs} x'_r x'_s - 1 = 0, \\ \vartheta &= \varepsilon \sqrt{b_{rs} x'_r x'_s}, \end{aligned}\right. \tag{70.5}$$

where $\varepsilon = \pm 1$ (all square roots are taken positive). Therefore

$$y_r = A_r + \frac{\varepsilon\, b_{rs} x'_s}{\sqrt{b_{mn} x'_m x'_n}}, \tag{70.6}$$

and we get the two Lagrangians

$$\left.\begin{aligned} \Lambda_+(x, x') &= A_r x'_r + \sqrt{b_{rs} x'_r x'_s}, \\ \Lambda_-(x, x') &= A_r x'_r - \sqrt{b_{rs} x'_r x'_s}. \end{aligned}\right\} \tag{70.7}$$

For $k > 0$, we have then

$$\Lambda_+(x, k x') = k \Lambda_+(x, x'), \quad \Lambda_-(x, k x') = k \Lambda_-(x, x'), \tag{70.8}$$

and for $k < 0$,

$$\Lambda_+(x, k x') = k \Lambda_-(x, x'), \quad \Lambda_-(x, k x') = k \Lambda_+(x, x'). \tag{70.9}$$

Thus the energy equation (70.3) yields two Lagrangians, both positive homogeneous of degree unity, but not homogeneous in the full sense, since the branches are interchanged in (70.9).

In ODS of Sect. 66, we start with the ordinary Lagrangian

$$L(q, \dot q) = \tfrac{1}{2} a_{\varrho\sigma} \dot q_\varrho \dot q_\sigma - V. \tag{70.10}$$

Then, as in (69.8),

$$p_\varrho = \frac{\partial L}{\partial \dot q_\varrho} = a_{\varrho\sigma} \dot q_\sigma, \quad \dot q_\varrho = a^{\varrho\sigma} p_\sigma, \tag{70.11}$$

and the Hamiltonian is

$$H(q, p) = \dot q_\varrho p_\varrho - L = \tfrac{1}{2} a^{\varrho\sigma} p_\varrho p_\sigma + V. \tag{70.12}$$

If we start from this Hamiltonian, then, as in (69.16),

$$\dot q_\varrho = \frac{\partial H}{\partial p_\varrho} = a^{\varrho\sigma} p_\sigma, \quad p_\varrho = a_{\varrho\sigma} \dot q_\sigma, \tag{70.13}$$

and we get

$$L(q, \dot q) = \dot q_\varrho p_\varrho - H = \tfrac{1}{2} a_{\varrho\sigma} \dot q_\varrho \dot q_\sigma - V, \tag{70.14}$$

thus recovering the Lagrangian (70.10).

71. Reciprocity theorem. The relationship between Lagrangian and Hamiltonian dynamics is simple when viewed geometrically[1].

Let $\Lambda(x, x')$ be a homogeneous Lagrangian and $\Omega(x, y)=0$ a corresponding energy equation, obtained by eliminating the ratios $x'_1 : x'_2 \ldots : x'_{N+1}$ from the equations

$$y_r = \frac{\partial \Lambda}{\partial x'_r}. \tag{71.1}$$

Consider a space Z_{N+1}, tangent to QT at x_r, with Euclidean metric and rectangular Cartesian coordinates z_r. In Z_{N+1} draw the unit sphere

$$S:\ z_r z_r = 1. \tag{71.2}$$

Consider the two surfaces (each of N dimensions) with the following equations:

$$\left.\begin{aligned} S_L&:\ \Lambda(x, z) - 1 = 0,\\ S_H&:\ \Omega(x, z) = 0.\end{aligned}\right\} \tag{71.3}$$

Here z_r are the current coordinates; the quantities x_r are mere constants, since our geometry is at present in the tangent space Z_{N+1}. We shall call S_L the *Lagrangian surface* and S_H the *Hamiltonian surface.*

The polar plane with respect to S of any point z^*_r has the equation

$$z_r z^*_r = 1, \tag{71.4}$$

and the polar reciprocal of S_L with respect to S is the envelope of these planes as z^*_r ranges over S_L. To find this envelope, we have the equations

$$\left.\begin{aligned} z_r\,\delta z^*_r = 0,&\qquad \frac{\partial \Lambda(x, z^*)}{\partial z^*_r}\,\delta z^*_r = 0,\\ z_r z^*_r = 1,&\qquad \Lambda(x, z^*) = 1.\end{aligned}\right\} \tag{71.5}$$

Hence

$$z_r = \varphi\,\frac{\partial \Lambda(x, z^*)}{\partial z^*_r}, \tag{71.6}$$

where φ is for the moment undetermined; but

$$\varphi = \varphi\,\Lambda(x, z^*) = \varphi\, z^*_r\,\frac{\partial \Lambda(x, z^*)}{\partial z^*_r} = z_r z^*_r = 1, \tag{71.7}$$

and so (71.6) reads

$$z_r = \frac{\partial \Lambda(x, z^*)}{\partial z^*_r}. \tag{71.8}$$

The polar reciprocal of S_L is now to be found by eliminating the ratios $z^*_1 : z^*_2 \ldots : z^*_{N+1}$ from these equations. But, on comparing with (71.1), we see that this is precisely the way in which the energy equation $\Omega=0$ is obtained. The polar reciprocal of S_L is therefore S_H, and since the operation of finding a polar reciprocal works the other way round, we have the *Reciprocity Theorem: The Hamiltonian surface is the polar reciprocal of the Lagrangian surface, and conversely.* This relationship is illustrated in Fig. 33.

[1] Cf. J. HADAMARD: Leçons sur le calcul des variations, Tome 1, pp. 75, 76, 96, 97 (Paris: Hermann 1910); C. CARATHÉODORY: Variationsrechnung und partielle Differentialgleichungen erster Ordnung pp. 243–248 (Leipzig u. Berlin: Teubner 1935). CARATHÉODORY calls S_L the *indicatrix* and S_H the *figuratrix*.

The equations
$$y_r = \frac{\partial \Lambda(x, x')}{\partial x'_r} \tag{71.9}$$
attach to any curve $x_r = x_r(u)$ a natural momentum-energy vector y_r; conversely, given a momentum-energy vector y_r, satisfying the energy equation, it determines the direction of a curve for which this momentum-energy vector is natural. In the process of finding the polar reciprocal of $\Omega(x, y) = 0$, with y_r as running coordinates, in the form $\Lambda(x, x') = 1$, with x'_r as running coordinates, we encounter the equations
$$x'_r = \vartheta \frac{\partial \Omega(x, y)}{\partial y_r}, \tag{71.10}$$
where ϑ is undetermined; these equations, like (71.9), establish a relationship between the vector y_r and the direction of a curve, i.e. the ratios of x'_r. From the reciprocity of the operation of taking the polar reciprocal, it follows that (71.9) and (7.10) are different ways of expressing a unique relationship between a direction in QT and the corresponding natural momentum-energy vector y_r. In (L, H)-notation, these equations give
$$p_\varrho = \frac{\partial L}{\partial \dot{q}_\varrho}, \qquad \dot{q}_\varrho = \frac{\partial H}{\partial p_\varrho}, \tag{71.11}$$
which are two equivalent ways of expressing a unique relationship between the velocity $\dot{q}_\varrho$ and the natural momentum p_ϱ.

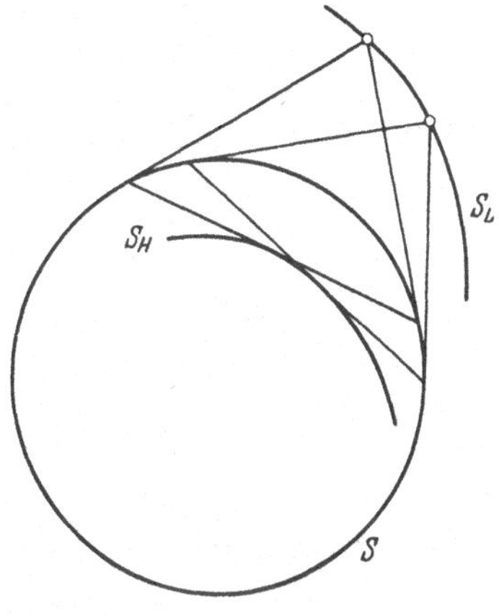

Fig. 33. Reciprocal relationship between the Lagrangian surface S_L and the Hamiltonian surface S_H.

To illustrate the above geometry, consider the systems RS and ODS of Sects. 66 and 70. In RS, take $b_{rs} = \delta_{rs}$ for simplicity. Then, by (70.1), S_L is a quadric and, by (70.3), S_H is a sphere with centre at A_r. As for ODS, by (66.4), S_L is a quadric through the origin with equation
$$\tfrac{1}{2} a_{\varrho\sigma} z_\varrho z_\sigma - V z_{N+1}^2 = z_{N+1}, \tag{71.12}$$
and, by (70.12), S_H is the quadric with equation
$$z_{N+1} + \tfrac{1}{2} a^{\varrho\sigma} z_\varrho z_\sigma + V = 0 \tag{71.13}$$
(actually an ellipsoid on account of the positive-definite character of kinetic energy). If ODS is a single particle moving in space, then S_L is a quadric of revolution and S_H a sphere.

72. HAMILTON's two-point characteristic or principal function[1]. The HAMILTON-JACOBI equation. Consider the space QT, of $N+1$ dimensions, and in it a Lagrangian or Hamiltonian dynamics; it does not matter which, on account of

[1] In HAMILTON's dynamics there was a distinction between the characteristic function and the principal function, a distinction still observed by many writers (cf. GOLDSTEIN [7], p. 283) who would regard the function used in the present exposition as the principal function, rather than the characteristic function, which is a somewhat less general concept. However, in this article we are following the optical method, in which the *characteristic* function has all the required generality, and any attempt to use the two words, characteristic and principal, in contrasted technical senses is likely to cause confusion. The word *characteristic* will be used here, the word *principal* being regarded as an equivalent alternative. It is rather unfortunate that there is still a third word, *eikonal*, invented by H. BRUNS in 1895, when he rediscovered HAMILTON's optical method in a somewhat specialized form; cf. W. R. HAMILTON: Mathematical Papers, Vol. 1, p. 488. Cambridge: University Press 1931. This word has found its way into dynamics; cf. NORDHEIM and FUES [*19*], p. 127, GOLDSTEIN [7], p. 311.

the correspondence set up in Sect. 69, and we may use indifferently a homogeneous Lagrangian $\Lambda(x, x')$ or an ordinary Lagrangian $L(q, t, \dot{q})$ as in Sects. 64 and 65, an energy equation $\Omega(x, y)=0$ or a Hamiltonian $H(q, t, p)$ as in Sects. 67 and 68.

Let Γ be a ray or trajectory joining a point B^* to a point B. We define the *two-point characteristic*[1] or *principal function* to be the Lagrangian or Hamiltonian action (they are equal) measured from B^* to B along this ray. We denote it by $S(B^*, B)$. It is a function of two points of QT. It may not exist for some choices of these points, there being no ray joining them. It may be single valued (one ray joining the points). Or it may be multiple valued (several rays joining the points). But these refinements will not concern us much. In the case of multiple values, we shall concentrate on one of the values.

The characteristic function is a function of $2N+2$ arguments, the coordinates of B^* and B, say, x_r^* and x_r. We write

$$\left.\begin{aligned} S(B^*, B) &= S(x^*, x) = S(x_1^*, \dots x_{N+1}^*, x_1, \dots x_{N+1}) \\ &= \int_{B^*}^{B} \Lambda(x, x')\, du = \int_{B^*}^{B} y_r\, dx_r, \end{aligned}\right\} \tag{72.1}$$

or

$$S(B^*, B) = \int_{B^*}^{B} L(q, t, \dot{q})\, dt = \int_{B^*}^{B} (p_\varrho\, dq_\varrho - H\, dt). \tag{72.2}$$

Here are some remarks about the characteristic function:

(i) The coordinate system for B^* need not be the same as the coordinate system for B. A domain of overlapping coordinates (cf. Sect. 63) may intervene. Or even if this should not occur, we might decide, for generality, to transform the coordinates for B^*, and perhaps those for B also, but independently. There is, then, a distinction between the notations $S(B^*, B)$ and $S(x^*, x)$, for the former indicates a function of two points (a number defined by those two points, irrespective of the coordinate systems employed), whereas the latter suggests a functional *form*, that form changing under transformation of coordinates. The quantity is an invariant (in the sense of tensor calculus) for independent transformations of the *two* systems of coordinates.

(ii) Strictly speaking, the transition from (72.1) and (72.2) requires that t shall increase monotonically from B^* to B. We may therefore prefer the more general form (72.1) in case we might wish to contemplate a system "moving backward in time", or in case, in some application of the general theory, the variable t did not correspond to the physical concept of time, but to something different.

(iii) In general, there is no connection between $S(B^*, B)$ and $S(B, B^*)$, since only the *positive* homogeneity of $\Lambda(x, x')$ is demanded. But if we have decided that B^* and B shall occur in one order only as far as the definition (72.1) is concerned (perhaps in the order of increasing t), then we are free to define $S(B, B^*)$ as we please, and it is usually convenient to define it as

$$S(B, B^*) = -S(B^*, B). \tag{72.3}$$

(iv) If B coincides with B^*, then at least one value of the function $S(B^*, B)$ vanishes. But this does not necessarily imply $S(x, x)=0$, because we may be using different coordinate systems for B^* and B.

[1] In Sect. 74 the *one-point* characteristic function will be introduced, and in Sect. 79 the *momentum-characteristic* and *mixed characteristics*.

In view of what has been said in (i) above, it is wise to use different functional symbols for the Lagrangians at B^* and B, and also for the energy equations:

$$\Lambda^*(x^*, x^{*\prime}), \quad \Lambda(x, x'), \quad \Omega^*(x^*, y^*) = 0, \quad \Omega(x, y) = 0. \tag{72.4}$$

We shall refer to B^* as the *initial* point and to B as the *final* point.

To find how $S(B^*, B)$ changes when the end points are changed, we turn to (68.2) and delete the integral, since we are dealing with a ray. We get

$$\delta S = y_r\,\delta x_r - y_r^*\,\delta x_r^*. \tag{72.5}$$

If the variations δx_r, δx_r^* are arbitrary and independent[1], this gives

$$\frac{\partial S}{\partial x_r} = y_r = \frac{\partial \Lambda}{\partial x_r'}, \qquad \frac{\partial S}{\partial x_r^*} = -y_r^* = -\frac{\partial \Lambda^*}{\partial x_r^{*\prime}}. \tag{72.6}$$

On account of the energy equations in (72.4), $S(x^*, x)$ satisfies the two partial differential equations

$$\Omega\left(x, \frac{\partial S}{\partial x}\right) = 0, \qquad \Omega^*\left(x^*, -\frac{\partial S}{\partial x^*}\right) = 0. \tag{72.7}$$

More explicitly, the first reads

$$\Omega\left(x_1, \ldots, x_{N+1}, \frac{\partial S}{\partial x_1}, \ldots \frac{\partial S}{\partial x_{N+1}}\right) = 0. \tag{72.8}$$

These are HAMILTON's *partial differential equations*; (72.8) is the HAMILTON-JACOBI equation[2].

If we hold the ray fixed and the point B on it, and slide B^* along the ray, the values of y_r (momentum-energy at B) are not changed. Hence, from the first of (72.6),

$$\frac{\partial^2 S}{\partial x_r\,\partial x_s^*}\,\delta x_s^* = 0, \tag{72.9}$$

and therefore the characteristic function satisfies the $(N+1)\times(N+1)$ determinantal equation

$$\det \frac{\partial^2 S}{\partial x_r\,\partial x_s^*} = 0. \tag{72.10}$$

Let us translate these results into the other notation, using

$$\left.\begin{aligned} x_\varrho &= q_\varrho, & x_{N+1} &= t, \\ y_\varrho &= p_\varrho, & y_{N+1} &= -H. \end{aligned}\right\} \tag{72.11}$$

By (72.5) we have

$$\delta S = p_\varrho\,\delta q_\varrho - H\,\delta t - p_\varrho^*\,\delta q_\varrho^* + H^*\,\delta t^*, \tag{72.12}$$

and by (72.6)

$$\left.\begin{aligned} \frac{\partial S}{\partial q_\varrho} &= p_\varrho, & \frac{\partial S}{\partial t} &= -H(q, t, p), \\ \frac{\partial S}{\partial q_\varrho^*} &= -p_\varrho^*, & \frac{\partial S}{\partial t^*} &= H^*(q^*, t^*, p^*); \end{aligned}\right\} \tag{72.13}$$

---

[1] This requires the existence of rays or trajectories joining all points in the neighbourhood of B^* to all points in the neighbourhood of B. In exceptional cases, when B^* and B are foci (or conjugate points), this requirement will not be met, and (72.6) will not hold.

[2] For JACOBI's adverse comments on HAMILTON's second equation, see C. G. J. JACOBI: Gesammelte Werke, Vol. IV, pp. 73, 74; and for some comments on this matter by A. W. CONWAY and A. J. MCCONNELL, see W. R. HAMILTON: Mathematical Papers, Vol. 2, pp. 613 to 621. Cambridge: University Press 1940.

the HAMILTON-JACOBI equation follows immediately in the form

$$\frac{\partial S}{\partial t} + H\left(q, t, \frac{\partial S}{\partial q}\right) = 0. \tag{72.14}$$

This is the unsymmetrical form (t privileged) of the general symmetric equation (72.8). It is in this form that it is chiefly used.

A certain confusion has existed with regard to the two-point characteristic function $S(x^*, x)$. It appeared to JACOBI that too much was demanded in (72.7), a *single* function being required to satisfy *two* partial differential equations. To clarify this point, let us set down the procedure by which the function $S(x^*, x)$ is determined. We are not at all concerned here with practical calculations leading to a *formula* for this function. Very few dynamical problems are "soluble" in that sense; as far as the mathematical structure of dynamics is concerned, it is *determinacy* of the operations that counts. It is in the sense of determinacy that we speak here of "solving" a set of ordinary differential equations, and of "finding" the characteristic function.

The function $S(x^*, x)$ is found by the following operations, starting from an energy equation $\Omega(x, y) = 0$. For simplicity, we shall use the same coordinate system for all the points involved.

(i) Choose a point x_r^* and a point x_r.

(ii) Choose y_r^* consistent with the equation

$$\Omega(x^*, y^*) = 0. \tag{72.15}$$

(iii) Solve the ordinary differential equations

$$\frac{dx_r}{dw} = \frac{\partial \Omega}{\partial y_r}, \qquad \frac{dy_r}{dw} = -\frac{\partial \Omega}{\partial x_r} \tag{72.16}$$

for the initial conditions $x_r = x_r^*$, $y_r = y_r^*$ for $w = 0$, so obtaining a ray Γ through x_r^*.

(iv) Giving y_r^* all values consistent with (72.15), obtain the set of all rays $\{\Gamma\}$ through x_r^*.

(v) Pick out that ray Γ of this set which passes through the point x_r chosen in (i). If no such ray exists, then $S(x^*, x)$ does not exist for the points chosen. If the ray does exist, then

$$S(x^*, x) = \int_{x^*}^{x} y_r \, dx_r, \tag{72.17}$$

the integral being taken along Γ.

The function $S(x^*, x)$ so obtained satisfies the two partial differential equations (72.7), with Ω^* replaced by Ω since we are using a single coordinate system.

73. Dynamics based on a chosen two-point characteristic function. In Sect. 72, the two-point characteristic function $S(x^*, x)$ was defined in terms of a Lagrangian $\Lambda(x, x')$ or an energy equation $\Omega(x, y) = 0$. We saw that it satisfies the determinantal equation

$$\det \frac{\partial^2 S}{\partial x_r \, \partial x_s^*} = 0. \tag{73.1}$$

In a sense, this is the most fundamental equation of Hamiltonian theory, since it holds in this one form for *all* dynamical systems, whereas the equation of energy changes its form when we change from system to system.

Note that the Eq. (73.1) is invariant under separate transformations of the variables x and the variables x^*. The quantities

$$\frac{\partial^2 S}{\partial x_r \partial x_s^*} \tag{73.2}$$

are, for fixed s, the components of a covariant vector with respect to transformations of the variables x; similarly, for fixed r, they are the components of a covariant vector with respect to transformations of the variables x^*. We shall assume the matrix (73.2) to be of rank N, this being the general case.

We shall now show that any chosen function $S(x^*, x)$, satisfying (73.1), may be used as a basis for dynamics[1]. Any such function being chosen, we write

$$y_r = \frac{\partial S(x^*, x)}{\partial x_r}, \qquad y_r^* = -\frac{\partial S(x^*, x)}{\partial x_r^*}. \tag{73.3}$$

Then, by virtue of (73.1), we can eliminate the variables x^* from the equations on the left and the variables x from the equations on the right, obtaining equations of the form

$$\Omega(x, y) = 0, \qquad \Omega^*(x^*, y^*) = 0. \tag{73.4}$$

These are the energy equations corresponding to the two-point characteristic function $S(x^*, x)$; they are different functions because we contemplate different coordinate systems for the initial and final points.

But we do not need an energy equation to obtain the rays or trajectories; we can get them directly from $S(x^*, x)$ as follows.

First, given an initial point x_r^* and an initial direction D^*, the ray or trajectory is given by

$$\frac{\partial^2 S(x^*, x)}{\partial x_r \partial x_s^*} d x_s^* = 0, \tag{73.5}$$

where dx_s^* correspond to the direction D^*. These equations enable us to express the variables x in terms of a single parameter.

Secondly, given two points, x_r^* and x_r, the ray or trajectory joining them is given by

$$\frac{\partial S(x^*, X)}{\partial X_r} + \frac{\partial S(X, x)}{\partial X_r} = 0, \tag{73.6}$$

where the variables X are current coordinates along the ray.

But it is perhaps most important to remark that, whether we start from an energy equation and derive the two-point characteristic function, or vice-versa, the equations

$$y_r^* = -\frac{\partial S(x^*, x)}{\partial x_r^*} \tag{73.7}$$

are the equations of a final ray or trajectory passing through the initial point x_r^* with initial momentum-energy y_r^*, consistent of course with the equation $\Omega^*(x^*, y^*) = 0$; in (73.7) the variables x are current coordinates.

74. Coherent systems of rays or trajectories. The one-point characteristic function. We assume an energy equation

$$\Omega(x, y) = 0, \tag{74.1}$$

and now use only one system of coordinates for initial and final points. The rays or trajectories satisfy

$$\frac{dx_r}{dw} = \frac{\partial \Omega}{\partial y_r}, \qquad \frac{dy_r}{dw} = -\frac{\partial \Omega}{\partial x_r}. \tag{74.2}$$

---

[1] HAMILTON gave this argument for optics; cf. Mathematical Papers, Vol. 1, p. 170. Cambridge: University Press 1931.

Consider a set of rays (Fig. 34) forming a subspace R of QT, or perhaps filling QT, in which case R is QT. Throughout R the first set of equations in (74.2) attach a vector field y_r to each point; equivalently and more explicitly in terms of the homogeneous Lagrangian,

$$y_r = \frac{\partial \Lambda(x, x')}{\partial x'_r}. \tag{74.3}$$

We say that the set of rays or trajectories forms a *coherent system*[1] if

$$\oint y_r \, dx_r = 0 \tag{74.4}$$

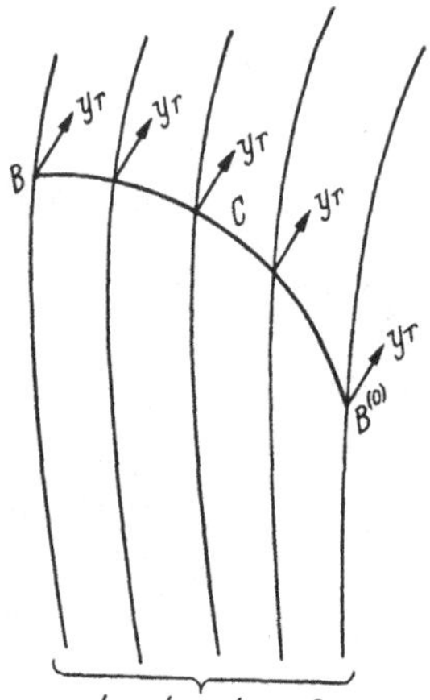

Fig. 34. The one-point characteristic function $U(B) = \int^B y_r dx_r$.

for every reducible circuit drawn in R. It is easy to see from (72.5) that, in particular, the totality of rays or trajectories drawn from a given point in QT form a coherent system.

For a coherent system, choose any point $B^{(0)}$ in R, and let B be any other point in R. Join $B^{(0)}$ to B by a curve C and write

$$U(B) = \int^B y_r \, dx_r, \tag{74.5}$$

the integral being taken along C. We suppress the dependence on $B^{(0)}$, because we keep this point fixed once for all. Note that, in (74.5), y_r is *not* the natural momentum-energy vector associated with C [cf. (69.10)], but the vector field given by (74.3) for the coherent system. By (74.4) the integral (74.5) has the same value for all reconcilable circuits in R. This means that, if R is simply connected, then U is a single-valued function of those coordinates which fix the position of B. If R is multiply connected, then U is a multiple-valued function. Let $C_1, C_2, \ldots C_m$ be a complete set of independent irreducible circuits. Then any two curves joining $B^{(0)}$ differ by a set of such circuits, and we have

$$U(B) = U_0(B) + n_1 J_1 + n_2 J_2 + \cdots + n_m J_m, \tag{74.6}$$

where $U_0(B)$ is one determination of U,

$$J_1 = \oint_{C_1} y_r \, dx_r, \ldots J_m = \oint_{C_m} y_r \, dx_r, \tag{74.7}$$

and $n_1, n_2, \ldots n_m$ are integers, positive, negative or zero[2].

If the rays or trajectories fill QT (forming a congruence of curves), then in (74.5) we can give arbitrary variations to the coordinates of B (say x_r). Then

$$y_r = \frac{\partial U(x)}{\partial x_r}, \tag{74.8}$$

or, equivalently,

$$p_\varrho = \frac{\partial U(q, t)}{\partial q_\varrho}, \qquad -H = \frac{\partial U(q, t)}{\partial t}, \tag{74.9}$$

and therefore, by (74.1), U satisfies the HAMILTON-JACOBI equation

$$\Omega\left(x, \frac{\partial U}{\partial x}\right) = 0, \tag{74.10}$$

---

[1] Or *family:* Cf. P. A. M. DIRAC: Canad. J. Math. **3**, 1–23 (1951), who remarks that "presumably the family has some deep significance in nature, not yet properly understood".

[2] The fact that $U(B)$ is multiple-valued when R is multiply connected is intimately connected with rules of quantization; cf. A. EINSTEIN: Verh. dtsch. phys. Ges. **19**, 82–92 (1917); J. L. SYNGE: Phys. Rev. **89**, 467–471 (1953). See also Sects. 98, 99; where the J's are action variables.

or, in the unsymmetrical notation,

$$\frac{\partial U}{\partial t} + H\left(q, t, \frac{\partial U}{\partial q}\right) = 0. \tag{74.11}$$

We call $U(B)$ the *one-point characteristic function* of the coherent system of rays or trajectories. True, it depends on the point $B^{(0)}$, but only trivially, for a change in $B^{(0)}$ merely adds a constant to $U(B)$.

75. Waves of constant action (Lagrangian or Hamiltonian)[1]. HUYGENS' construction. For a coherent system of rays or trajectories, as in Sect. 74, we define the *waves of constant action* (Lagrangian or Hamiltonian—they are the same) to be the loci[2]

$$U(B) = \text{const.} \tag{75.1}$$

The rays cut across the waves as in Fig. 35, which shows the waves W^*, W with points B^*, B on them, the curve joining these points being a ray Γ. The action along this ray is the same as the action along any other ray drawn from W^* to W, and is in fact equal to the integral $\int y_r\, dx_r$ taken along any curve in R drawn from W^* to W, y_r being the vector field defined by the coherent system as in (74.3).

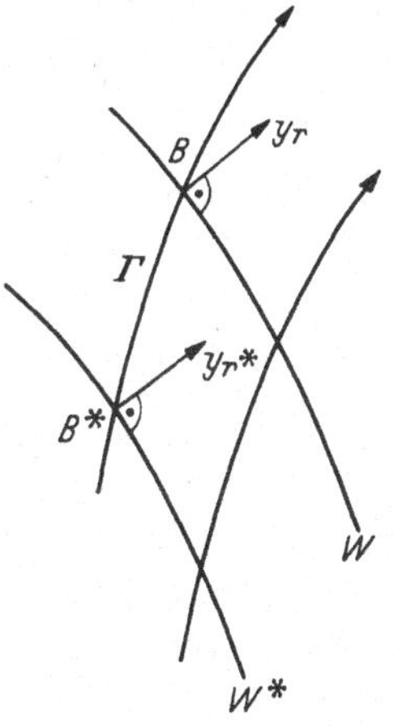

Fig. 35. Rays or trajectories in QT cutting across waves W^*, W.

In terms of the theory as here presented, it would be meaningless to ask whether the rays cut the waves orthogonally; we have no Riemannian metric in QT, and the concept of the orthogonality of a curve and a subspace is not invariant under coordinate transformations. This objection does not, however, apply to the momentum-energy vector y_r, since it is a covariant vector (to make $y_r\, dx_r$ invariant, dx_r being contravariant). This vector y_r is in fact orthogonal to the waves, in the sense that

$$y_r\, \delta x_r = 0 \tag{75.2}$$

for every infinitesimal displacement δx_r in a wave; this follows from (74.5), δU being zero for such a displacement.

The wave W may be generated from the wave W^* by HUYGENS' construction as follows (Fig. 36). Let B^* be any point on W^*. From B^* draw rays in all directions in QT and measure off on them an action

$$A = U(W) - U(W^*), \tag{75.3}$$

where these are the values of U on the two waves. This gives us an N-space, say V_N, with equation

$$S(x^*, x) = A, \tag{75.4}$$

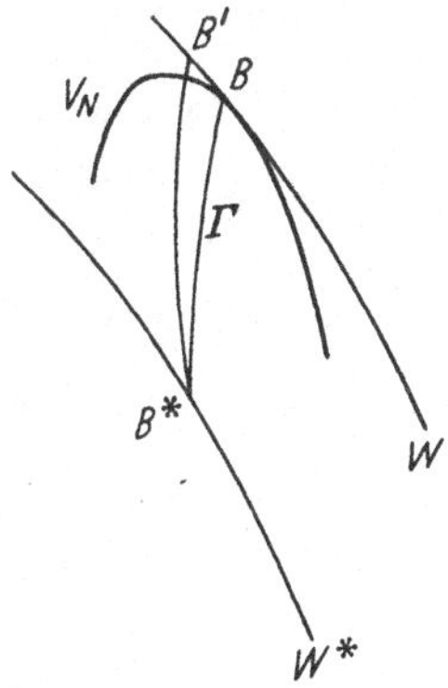

Fig. 36. HUYGENS' construction in QT.

where S is the two-point characteristic function. V_N is thus itself a wave, with B^* as source. In (75.4) the quantities x^* are fixed (coordinates of B^*), and the

---

[1] For waves in configuration space Q, see Sect. 81.

[2] In the language of the calculus of variations, they are *transversals*, the rays or trajectories being *extremals*. The condition of coherency (74.4) is called the condition of MAYER; it may be regarded in a sense as a condition of irrotationality as in hydrodynamics, cf. A. EINSTEIN: Sitzgsber. preuss. Akad. Wiss., phys.-math. Kl. **46**, 606—608 (1917). CARATHÉODORY, p. 249 of op. cit. in Sect. 71, calls the set of extremals and waves a *vollständige Figur*.

quantities x are current coordinates on V_N. We shall show that V_N touches W at that point B at which the ray Γ from B^* cuts W.

First, B must lie on V_N, because it is contained in the class of all points at action-distance A from B^*. Further, if we give an infinitesimal displacement δx_r to B, transferring it to a neighbouring position B' on W, the action for the ray joining B^* to B' exceeds A by $y_r \delta x_r$ [cf. (72.5)], and this vanishes by (75.2). Thus, to the first order, B' lies on V_N, and this establishes the tangency of V_N and W at the point B.

It is clear, then, that W is an envelope, in the space R of the coherent system, of the family of N-spaces (75.4), the value A being held fixed and B^* being allowed to range over the initial wave W^*. Since these N-spaces are themselves waves from sources on W^*, we have HUYGENS' construction.

We can, of course, regard the generation of one wave from another in this way as taking place in infinitesimal steps [make A infinitesimal in (75.4)]. Viewed in finite terms or infinitesimally, we have a *contact transformation* which establishes a correspondence between the points of the two waves, with tangent elements associated with the points; a tangent element here means the totality of infinitesimal vectors δx_r satisfying (75.2) for given y_r.

76. Determination of waves from initial data. Method of characteristic curves[1]. In the preceding work, the domain filled by the coherent system of rays was a subspace R of QT, or possibly QT itself. Let us now suppose that the rays fill QT, or a portion of it, so that they form a congruence. We seek to determine the waves from suitable initial data; we wish to solve for U the partial differential equation

$$\Omega(x, y) = 0, \quad y_r = \frac{\partial U}{\partial x_r}, \tag{76.1}$$

subject to initial conditions which assign to U the value U_0 (in general not constant) over a subspace Σ_M of QT, the dimensionality M being anything from zero (a point)[2] to N. The waves will then have the equations $U = \text{const}$, and U will be the one-point characteristic function of the coherent system rays or trajectories appropriate to the initial conditions.

The method is the method of characteristic curves, which curves in the present instance are the rays or trajectories. We write down the ordinary differential equations

$$\frac{dx_1}{\frac{\partial\Omega}{\partial y_1}} = \cdots = \frac{dx_{N+1}}{\frac{\partial\Omega}{\partial y_{N+1}}} = \frac{dy_1}{-\frac{\partial\Omega}{\partial x_1}} = \cdots = \frac{dy_{N+1}}{-\frac{\partial\Omega}{\partial x_{N+1}}}, \tag{76.2}$$

or equivalently, in terms of a parameter w,

$$\frac{dx_r}{dw} = \frac{\partial\Omega}{\partial y_r}, \quad \frac{dy_r}{dw} = -\frac{\partial\Omega}{\partial x_r}. \tag{76.3}$$

These equations determine a unique solution

$$x_r = x_r(w), \quad y_r = y_r(w) \tag{76.4}$$

if the values of x_r and y_r are assigned for $w = 0$.

---

[1] Cf. CARATHÉODORY, Chap. 3 of op. cit. in Sect. 71, also T. LEVI-CIVITA: Caratteristiche dei sistemi differenziali e propagazione ondosa (Bologna: Zanichelli 1931), or in French: Caractéristiques des systèmes différentiels et propagation des ondes (Paris: Alcan 1932).

[2] The case $M = 0$ is covered by Sect. 72, and may be disregarded in this work. The reader is reminded that QT is of $(N+1)$ dimensions and that Latin suffixes range from 1 to $(N+1)$; cf. Sect. 62.

Let B^*, with coordinates x_r^*, be any point on Σ_M (Fig. 37) and let y_r^* be chosen to satisfy

$$\Omega(x^*, y^*) = 0, \tag{76.5}$$

and also

$$y_r^* \, \delta x_r^* = \delta U_0 \tag{76.6}$$

for every infinitesimal displacement in Σ_M. It is impossible to enter here into all possibilities. The conditions (76.5) and (76.6) may be inconsistent, in which case no solution U of (76.1) and the initial conditions exists; and even if the conditions are consistent, certain degeneracies may occur. We shall merely follow a general argument, asserting that (76.5) and (76.6) contain $M+1$ conditions, and hence leave y_r^* with $N-M$ degrees of freedom. We use x_r^* and y_r^* as initial values for (76.3). From the first of (76.3), these values determine a direction in QT; there are ∞^{N-M} such directions at each point of Σ_M and there are ∞^M points on Σ_M, with the result that we get a congruence of curves (rays or trajectories) filling QT.

Let B, with coordinates x_r, be any point of QT. Through B draw the curve Γ belonging to the above congruence; let Γ cut Σ_M at B^*, with coordinates x_r^*. Then define $U(x)$ by

$$U(x) = U(x^*) + \int_{B^*}^{B} y_r \, dx_r, \tag{76.7}$$

the term $U(x^*)$ being the assigned value U_0 and the integration being along Γ. On varying B, and consequently B^*, we get

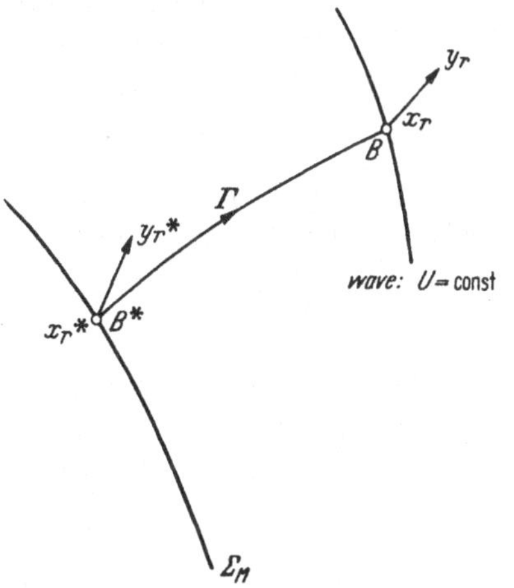

Fig. 37. Waves in QT obtained from initial data by the method of characteristic curves.

$$\delta U(x) = \delta U(x^*) + y_r \delta x_r - y_r^* \delta x_r^* + \int (\delta y_r \, dx_r - \delta x_r \, dy_r). \tag{76.8}$$

Now (76.3) imply $d\Omega/dw = 0$, and hence

$$\Omega(x, y) = 0, \tag{76.9}$$

on account of (76.5). Hence $\delta\Omega = 0$, and the integral in (76.8) vanishes; further, by (76.6), the equation reduces to

$$\delta U(x) = y_r \delta x_r. \tag{76.10}$$

Therefore

$$y_r = \frac{\partial U}{\partial x_r}, \tag{76.11}$$

and so, by (76.9), U satisfies the partial differential equation (76.1); by (76.7), it satisfies the initial condition. Thus the required solution has been found.

To revert to a point raised in Sect. 72, we are not so much interested here in finding a formula for $U(x)$ as in setting up a scheme which determines it, and indicates the conditions under which it may not exist.

77. JACOBI's complete integral of the HAMILTON-JACOBI equation. Suppose that we seek *all* the rays or trajectories, with their associated momentum-energy vectors, for a dynamical system with the energy equation

$$\Omega(x, y) = 0, \tag{77.1}$$

or equivalently

$$H = H(q, t, p). \tag{77.2}$$

The plan of JACOBI was not to attempt to integrate directly the ordinary differential equations of motion, but to work with the HAMILTON-JACOBI equation, which, corresponding to (77.2), reads

$$\frac{\partial U}{\partial t} + H\left(q, t, \frac{\partial U}{\partial q}\right) = 0. \tag{77.3}$$

He reduced the problem of motion to that of finding a *complete integral* of this equation of the form

$$U = J(q_1, \dots q_N, t, a_1, \dots a_N) + a_{N+1}. \tag{77.4}$$

Here the a's are arbitrary constants; a complete integral must contain $N+1$ of them, but one can be additive, since only derivatives of U occur in (77.3).

JACOBI's theorem asserts that, if b_ϱ are any constants, then the equations

$$b_\varrho = \frac{\partial J}{\partial a_\varrho} \tag{77.5}$$

determine the totality of all the rays or trajectories, and the equations

$$p_\varrho = \frac{\partial J}{\partial q_\varrho} \tag{77.6}$$

determine the associated momenta; or equivalently that the equations

$$\dot{q}_\varrho = \frac{\partial H}{\partial p_\varrho} \tag{77.7}$$

and

$$\dot{p}_\varrho = -\frac{\partial H}{\partial q_\varrho} \tag{77.8}$$

are satisfied by virtue of (77.5) and (77.6), and that (77.5), and (77.6) contain all the solutions of (77.7) and (77.8).

To prove this, we note first that (77.5) are N equations involving q_ϱ and t, and so determine functions $q_\varrho(t)$ for any values of the constants; further, that (77.6) then give associated p_ϱ. The derivatives of the q's and p's are obtained by differentiating (77.5) and (77.6) with respect to t; this gives

$$\frac{\partial^2 J}{\partial a_\varrho \partial q_\sigma} \dot{q}_\sigma + \frac{\partial^2 J}{\partial a_\varrho \partial t} = 0, \tag{77.9}$$

and

$$\dot{p}_\varrho = \frac{\partial^2 J}{\partial q_\varrho \partial q_\sigma} \dot{q}_\sigma + \frac{\partial^2 J}{\partial q_\varrho \partial t}. \tag{77.10}$$

Since (77.4) satisfies (77.3), we have

$$\frac{\partial J}{\partial t} + H\left(q, t, \frac{\partial J}{\partial q}\right) = 0 \tag{77.11}$$

for arbitrary values of the a's, and so

$$\frac{\partial^2 J}{\partial a_\varrho \partial t} + \frac{\partial H}{\partial p_\sigma} \frac{\partial^2 J}{\partial q_\sigma \partial a_\varrho} = 0. \tag{77.12}$$

Comparing this with (77.9), we see that (77.7) is satisfied. Now differentiate (77.11) with respect to q_ϱ, obtaining

$$\frac{\partial^2 J}{\partial q_\varrho \partial t} + \frac{\partial H}{\partial q_\varrho} + \frac{\partial H}{\partial p_\sigma} \frac{\partial^2 J}{\partial q_\sigma \partial q_\varrho} = 0. \tag{77.13}$$

Use (77.7) in this and compare with (77.10); we see that (77.8) is satisfied.

Since the number of constants (a_ϱ, b_ϱ) is $2N$, the same as the number of initial values (q_ϱ, p_ϱ) required to determine a solution of (77.7), (77.8), JACOBI's theorem is established. We may say that any dynamical problem is essentially solved if we can find a complete integral as in (77.4).

If a complete integral of the HAMILTON-JACOBI equation is given, as in (77.4), then the two-point characteristic function

$$S(q_1^*, \ldots q_N^*, t^*, q_1, \ldots q_N, t)$$

is obtained[1] by eliminating the N constants a_ϱ from the $N+1$ equations

$$\left.\begin{aligned} S &= J(q, t, a) - J(q^*, t^*, a), \\ \frac{\partial J(q, t, a)}{\partial a_\varrho} &= \frac{\partial J(q^*, t^*, a)}{\partial a_\varrho}. \end{aligned}\right\} \tag{77.14}$$

The connection between the complete integral and the two-point characteristic function is closely related to the fact that a complete integral of the partial differential equation

$$\Omega\left(x, \frac{\partial S}{\partial x}\right) = 0, \tag{77.15}$$

say $S(x^*, x)$, in which x_r^* are arbitrary constants, may also be regarded as *any* solution of the same equation if we take as independent variables the $2N+2$ quantities x_r and x_r^*, of which the second set do not appear explicitly in the equation. Whichever point of view we take, we are led to the fundamental determinantal equation

$$\det \frac{\partial^2 S}{\partial x_r \partial x_s^*} = 0, \tag{77.16}$$

as in (73.1).

JACOBI's theorem may be treated symmetrically as follows. Let $S(x^*, x)$ be a complete integral of

$$\Omega\left(x, \frac{\partial S}{\partial x}\right) = 0, \tag{77.17}$$

the quantities x_r^* being arbitrary constants. Then, defining y_r by

$$y_r = \frac{\partial S}{\partial x_r}, \tag{77.18}$$

we have

$$\frac{\partial \Omega}{\partial y_s} \frac{\partial^2 S}{\partial x_s \partial x_r^*} = 0, \tag{77.19}$$

and

$$\frac{\partial \Omega}{\partial x_r} + \frac{\partial \Omega}{\partial y_s} \frac{\partial^2 S}{\partial x_s \partial x_r} = 0. \tag{77.20}$$

Define y_r^* by

$$y_r^* = -\frac{\partial S}{\partial x_r^*}. \tag{77.21}$$

Now (77.19) implies the determinantal equation (77.16), and hence (77.21) implies a relationship between the quantities x^* and y^*, say

$$\Omega^*(x^*, y^*) = 0. \tag{77.22}$$

---

[1] Cf. A. W. CONWAY and A. J. MCCONNELL: Proc. Roy. Irish Acad. A **61**, 18–25 (1932); W. R. HAMILTON: Mathematical Papers, Vol. 2, pp. 613–621 (Cambridge: University Press 1940); also LANCZOS [*15*], p. 262.

If we give to these quantities any constant values consistent with this equation, then the equations

$$y_r = \frac{\partial S(x^*, x)}{\partial x_r}, \qquad y_r^* = -\frac{\partial S(x^*, x)}{\partial x_r^*} \tag{77.23}$$

define a curve $x_r(w)$, with associated $y_r(w)$, satisfying (for some parameter w)

$$\frac{dx_r}{dw} = \frac{\partial \Omega(x, y)}{\partial y_r}, \qquad \frac{dy_r}{dw} = -\frac{\partial \Omega(x, y)}{\partial x_r}. \tag{77.24}$$

To prove this, we differentiate (77.23) with respect to w, obtaining

$$\frac{dy_r}{dw} = \frac{\partial^2 S}{\partial x_r \partial x_s}\frac{dx_s}{dw}, \qquad 0 = \frac{\partial^2 S}{\partial x_r^* \partial x_s}\frac{dx_s}{dw}, \tag{77.25}$$

and the result follows on comparing these equations with (77.19) and (77.20). It is important to note that we do not actually need the Eq. (77.22); for we require only $2N+1$ equations to define a curve with associated vector y_r, and we can get these equations from (77.23) by omitting one of equations on the right, so that only N of the y^* are actually involved; they, and the constants x^*, may be given arbitrary values.

78. The practical use of JACOBI's theorem. Separation of variables. If, as often occurs in practice, the Hamiltonian does not contain the time explicitly (conservative system, cf. Sect. 62), we apply the procedure described in the preceding section by seeking a function J as in (77.4) of the form

$$J = -Et + K(q_1, \dots q_N, a_1, \dots a_{N-1}, E), \tag{78.1}$$

where $a_1, \dots a_{N-1}, E$ are arbitrary constants. Then, by (77.3), K is to satisfy

$$H\left(q_1, \dots q_N, \frac{\partial K}{\partial q_1}, \dots \frac{\partial K}{\partial q_N}\right) = E. \tag{78.2}$$

When such a complete integral has been found, the Eqs. (77.5), (77.6) give, for the trajectories, the equations

$$b_1 = \frac{\partial K}{\partial a_1}, \dots b_{N-1} = \frac{\partial K}{\partial a_{N-1}}, \quad b_N = -t + \frac{\partial K}{\partial E}, \tag{78.3}$$

and

$$p_1 = \frac{\partial K}{\partial q_1}, \dots p_{N-1} = \frac{\partial K}{\partial q_{N-1}}, \quad p_N = \frac{\partial K}{\partial q_N}. \tag{78.4}$$

The first $N-1$ equations in (78.3) determine a curve in the space Q, and the last gives the time t. H has a constant value in the motion (a consequence of $\partial H/\partial t = 0$), and this constant value is E.

The HAMILTON-JACOBI equation (78.2) may sometimes be solved by *separation of variables*. Let H be of the form

$$H(q, p) = \frac{H_1 + H_2 + \dots + H_N}{A_1 + A_2 + \dots + A_N}, \tag{78.5}$$

where H_1, A_1 are functions of q_1, p_1 only, H_2, A_2 functions of q_2, p_2 only, and so on. Then (78.2) may be written in the form

$$D_1 + D_2 + \dots + D_N = 0, \tag{78.6}$$

where

$$D_1 = H_1\left(q_1, \frac{\partial K}{\partial q_1}\right) - E\, A_1\left(q_1, \frac{\partial K}{\partial q_1}\right), \tag{78.7}$$

with similar expressions for $D_2, \dots D_N$. We satisfy (78.6) by taking

$$K = K_1(q_1, a_1, E) + K_2(q_2, a_2, E) + \cdots + K_N(q_N, a_N, E), \tag{78.8}$$

and making $K_1, K_2, \dots K_N$ satisfy

$$H_1\left(q_1, \frac{dK_1}{dq_1}\right) - E\,A_1\left(q_1, \frac{dK_1}{dq_1}\right) = a_1, \tag{78.9}$$

and similar equations, the quantities $a_1, a_2, \dots a_N$ being constants which are arbitrary except for the condition

$$a_1 + a_2 + \cdots + a_N = 0, \tag{78.10}$$

so that there are $N-1$ of them independent. Now (78.9) is an ordinary differential equation. When solved for dK_1/dq_1, it gives K_1 by a quadrature; similarly $K_2, \dots K_N$ are obtained by quadratures, and we have a complete integral of the HAMILTON-JACOBI equation (78.2) in the form (78.8), containing N arbitrary constants [we recall that an additive constant was thrown away in (77.4) in passing from U to J].

Systems of the LIOUVILLE type[1] are particular cases of (78.5); for them the kinetic and potential energies are of the forms

$$\left.\begin{aligned} T &= \tfrac{1}{2}(A_1 + \cdots + A_N)(B_1\dot{q}_1^2 + \cdots + B_N\dot{q}_N^2), \\ V &= \frac{V_1 + \cdots + V_N}{A_1 + \cdots + A_N}, \end{aligned}\right\} \tag{78.11}$$

where A_1, B_1, V_1 are functions of q_1 only, A_2, B_2, V_2 functions of q_2 only, and so on. The corresponding Hamiltonian is

$$H = \frac{\tfrac{1}{2}(p_1^2/B_1 + \cdots + p_N^2/B_N) + V_1 + \cdots + V_N}{A_1 + \cdots + A_N}, \tag{78.12}$$

and the ordinary differential equations as in (78.9), by which the problem is reduced to quadratures, now read

$$\left(\frac{dK_1}{dq_1}\right)^2 = 2B_1(E\,A_1 - V_1 + a_1), \dots \left(\frac{dK_N}{dq_N}\right)^2 = 2B_N(EA_N - V_N + a_N). \tag{78.13}$$

As a simple example, consider the KEPLER problem[2] (Sect. 36). For polar coordinates r, ϑ, we have the Lagrangian

$$L = T - V = \frac{1}{2}(\dot{r}^2 + r^2\dot{\vartheta}^2) + \frac{\mu}{r}, \tag{78.14}$$

and the momenta are

$$p_r = \frac{\partial L}{\partial \dot{r}} = \dot{r}, \qquad p_\vartheta = \frac{\partial L}{\partial \dot{\vartheta}} = r^2\dot{\vartheta}. \tag{78.15}$$

(We have taken the mass of the particle to be unity). The Hamiltonian is

$$\left.\begin{aligned} H &= \dot{r}\frac{\partial L}{\partial \dot{r}} + \dot{\vartheta}\frac{\partial L}{\partial \dot{\vartheta}} - L = T + V \\ &= \frac{1}{2}\left(p_r^2 + \frac{1}{r^2}p_\vartheta^2\right) - \frac{\mu}{r}, \end{aligned}\right\} \tag{78.16}$$

[1] For further details, and for the more general systems of STAECKEL, which also yield to the method of separation of variables, see APPELL [2] II, pp. 437–440; LEVI-CIVITA and AMALDI [16] II_2, pp. 415–424.

[2] Cf. CORBEN and STEHLE [3], pp. 251–257 for a detailed three-dimensional treatment, with consideration of the BOHR-SOMMERFELD quantum conditions. See also APPELL [2] I, pp. 592–596.

which is of the form (78.12). The HAMILTON-JACOBI equation (78.2) reads

$$\frac{1}{2}\left[\left(\frac{\partial K}{\partial r}\right)^2+\frac{1}{r^2}\left(\frac{\partial K}{\partial \vartheta}\right)^2\right]-\frac{\mu}{r}=E, \tag{78.17}$$

and we obtain a complete integral in the form

$$K=F(r,a,E)+a\vartheta, \tag{78.18}$$

where a and E are arbitrary constants, and F is given by a quadrature from

$$\left(\frac{dF}{dr}\right)^2=2E+\frac{2\mu}{r}-\frac{a^2}{r^2}. \tag{78.19}$$

By (78.3) the trajectories have the equations

$$b_1=\frac{dF}{\partial a}+\vartheta, \qquad b_2=-t+\frac{\partial F}{\partial E}. \tag{78.20}$$

It is obvious that the above method can be used when the potential is any function of r.

Separation of variables demands a special choice of coordinates. Thus, in the KEPLER problem, rectangular Cartesians would not do. In the problem of two centres of attraction, the variables may be separated (the system being reduced to LIOUVILLE type) by transforming from rectangular Cartesians (x, y) to elliptic coordinates (q_1, q_2) by the formulae

$$x=c\,\mathrm{Cos}\,q_1\cos q_2, \qquad y=c\,\mathrm{Sin}\,q_1\sin q_2, \tag{78.21}$$

the centres being at $x=\pm c$. If one attracting centre is removed to infinity, and its strength infinitely increased, we get in the limit the problem of a charged particle moving in a field which is the superposition of a uniform electric field on a COULOMB field (STARK effect)[1].

The relativistic KEPLER problem is treated in Sect. 116.

III. Momentum-energy space (*PH*).

79. The space *PH* and the momentum-energy characteristic function. In Part E II we took the $(N+1)$-dimensional space of events QT as the background for dynamical theory, using the coordinates x_r where[2]

$$x_\varrho=q_\varrho, \qquad x_{N+1}=t. \tag{79.1}$$

Let us now view dynamics in the $(N+1)$-dimensional *momentum-energy space PH* in which the coordinates are y_r where

$$y_\varrho=p_\varrho, \qquad y_{N+1}=-H. \tag{79.2}$$

We start from the beginning, assuming an energy equation

$$\Omega(x,y)=0, \tag{79.3}$$

or equivalently, on solving for y_{N+1},

$$H=H(q,t,p). \tag{79.4}$$

For any curve Γ in PH with equations $y_r=y_r(u)$, we define a new type of action by the integral

$$\tilde{A}=\int x_r\,dy_r, \tag{79.5}$$

[1] For a detailed treatment, see CORBEN and STEHLE [3], pp. 258–264; see also APPELL [2] I, pp. 602–607; GRAMMEL [8], p. 321; PÉRÈS [20], pp. 243, 244. For a Lagrangian treatment of the problem of two centres, see WHITTAKER [28], pp. 97–99.

[2] For notation, see Sect. 62.

where $x_r(u)$ are arbitrary except for (79.3); equivalently,

$$\tilde{A} = \int (q_\varrho \, dp_\varrho - t\, dH). \tag{79.6}$$

On varying Γ, we get

$$\delta \tilde{A} = [x_r \delta y_r] + \int (\delta x_r \, dy_r - \delta y_r \, dx_r). \tag{79.7}$$

If we hold the end points of Γ fixed (i.e. the end values of y_r), the variational equation

$$\delta \tilde{A} = \delta \int x_r \, dy_r = 0, \tag{79.8}$$

with the side condition (79.3), leads at once to the canonical equations

$$\frac{dx_r}{dw} = \frac{\partial \Omega}{\partial y_r}, \qquad \frac{dy_r}{dw} = -\frac{\partial \Omega}{\partial x_r}, \tag{79.9}$$

for some parameter w. These are the same equations as in (68.7), and we may call the curves satisfying them rays or trajectories as before.

It is clear then that dynamics in PH based on the energy equation (79.3) and the variational equation (79.8) is the same as the dynamics in QT based on the same energy equation and HAMILTON's principle (68.5), i.e.

$$\delta A = \delta \int y_r \, dx_r = 0, \tag{79.10}$$

for fixed end points in QT. The two actions are connected by

$$\left.\begin{aligned} A + \tilde{A} &= \int y_r \, dx_r + \int x_r \, dy_r \\ &= x_r y_r - x_r^* y_r^*, \end{aligned}\right\} \tag{79.11}$$

where (x^*, y^*) refer to the initial point and (x, y) to the final point. In QT we think of x_r as current coordinates on a curve and y_r as an associated vector field; in PH, we reverse the roles. There is a formal duality between the two representations; we could, for example, use the technique of Sect. 69 to define a "homogeneous Lagrangian" in PH as a function of the quantities y and y' (where $y'_r = dy_r/du$), and pass to an "ordinary Lagrangian", a function of p_ϱ, H and dp_ϱ/dH.

We now introduce in PH the *momentum-energy characteristic function*[1] $W(C^*, C)$ defined by

$$W(C^*, C) = \int x_r \, dy_r, \tag{79.12}$$

the integral being taken along the ray or trajectory joining the points C^* and C of PH. On varying these end points, we get from (79.7)

$$\delta W = x_r \delta y_r - x_r^* \delta y_r^*. \tag{79.13}$$

If the variations δy_r and δy_r^* can be taken arbitrarily (i.e. if there are rays joining arbitrarily varied positions of C^* and C), then

$$\frac{\partial W}{\partial y_r} = x_r, \qquad \frac{\partial W}{\partial y_r^*} = -x_r^*, \tag{79.14}$$

and hence, by (79.3), $W(y^*, y)$ satisfies the partial differential equations

$$\Omega\left(\frac{\partial W}{\partial y}, y\right) = 0, \qquad \Omega\left(-\frac{\partial W}{\partial y^*}, y^*\right) = 0. \tag{79.15}$$

The values of x_r^* and x_r define a ray or trajectory, and hence they determine y_r^* and y_r; conversely, the values of y_r^* and y_r in general define a ray or trajectory,

---

[1] In optics, this is HAMILTON's T-function, also known as angle-characteristic and angle-eikonal; it is the basis of the theory of the aberrations of optical instruments. It is denoted by W here to avoid confusion with kinetic energy.

and hence they determine x_r^* and x_r. The relationship between the two-point characteristic function $S(x^*, x)$ and the momentum-energy characteristic function $W(y^*, y)$ is

$$W(y^*, y) + S(x^*, x) = x_r y_r - x_r^* y_r^*. \tag{79.16}$$

In addition to these two characteristic functions, there are two other *mixed characteristic functions:*

$$F(x^*, y) = S(x^*, x) - x_r y_r, \tag{79.17}$$

and

$$G(x, y^*) = S(x^*, x) + x_r^* y_r^*. \tag{79.18}$$

If arbitrary variations are permissible, we have

$$\frac{\partial F}{\partial x_r^*} = -y_r^*, \qquad \frac{\partial F}{\partial y_r} = -x_r, \tag{79.19}$$

and

$$\frac{\partial G}{\partial x_r} = y_r, \qquad \frac{\partial G}{\partial y_r^*} = x_r^*. \tag{79.20}$$

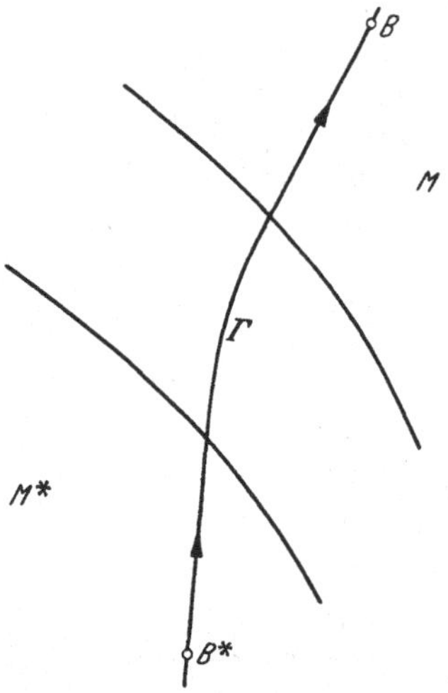

Fig. 38. A ray or trajectory in event-space QT with straight initial and final portions.

80. Encounters. We assumed, following (79.13), that C^* and C can be given arbitrary displacements δy_r^* and δy_r in PH, but there are important cases where this cannot be done. Let us consider a ray or trajectory Γ in QT, joining a point B^* to a point B (Fig. 38). Suppose that in a domain M^* of QT, containing B^*, the function $\Omega(x, y)$ is independent of its first set of arguments, the x's, and that the same is true in a domain M containing B. We then write the energy equation

$$\Omega^*(y^*) = 0 \quad \text{in} \quad M^*, \quad \Omega(y) = 0 \quad \text{in} \quad M. \tag{80.1}$$

By (79.9), the y's are constant along a ray in M^* or in M, and in fact the ray is a straight line in each of these domains of QT. It is clear from (80.1) that y_r^* and y_r cannot be given independent variations.

Solving (80.1), we get

$$y_{N+1}^* = -\omega^*(y_1^*, \ldots y_N^*), \qquad y_{N+1} = -\omega(y_1, \ldots y_N), \tag{80.2}$$

or equivalently

$$H^* = H^*(p^*), \qquad H = H(p). \tag{80.3}$$

We are in fact dealing with a system for which, initially and finally, the Hamiltonian depends only on the momenta, as is the case for a free particle or a set of free particles without interactions between them. Now (79.13) gives

$$\delta W = \left(x_\varrho - x_{N+1} \frac{\partial \omega}{\partial y_\varrho}\right) \delta y_\varrho - \left(x_\varrho^* - x_{N+1}^* \frac{\partial \omega^*}{\partial y_\varrho^*}\right) \delta v_\varrho^*, \tag{80.4}$$

which shows that W is a function only of the $2N$ quantities y_ϱ, y_ϱ^*. In the other notation, W is a function only of p_ϱ, p_ϱ^*, and (80.4) reads

$$\delta W = \left(q_\varrho - t \frac{\partial H}{\partial p_\varrho}\right) \delta p_\varrho - \left(q_\varrho^* - t^* \frac{\partial H^*}{\partial p_\varrho^*}\right) \delta p_\varrho^*. \tag{80.5}$$

Under these circumstances W is to be regarded as an arbitrary function of its $2N$ arguments, and is not required to satisfy any partial differential equation as in (79.15). Since p_ϱ, p_ϱ^* can be given arbitrary independent variations, (80.5) gives

$$\left.\begin{aligned} \frac{\partial W}{\partial p_\varrho^*} &= -q_\varrho^* + t^* \frac{\partial H^*}{\partial p_\varrho^*}, \\ \frac{\partial W}{\partial p_\varrho} &= q_\varrho - t\frac{\partial H}{\partial p_\varrho}. \end{aligned}\right\} \tag{80.6}$$

Regarding the function W as assigned, these are the equations of the initial and final rays or trajectories, the momenta p_ϱ, p_ϱ^* having constant values by virtue of the canonical equations (79.9). When viewed in QT, these initial and final rays are straight lines; when viewed in PH, they are mere *points*, each lying on a certain surface of N dimensions, as given in (80.1) or (80.3) (Fig. 39).

Consider now an encounter, as in the kinetic theory of gases, of a system of n particles which interact with one another, the generalized forces being derivable from a potential function, or an extended potential, so that the dynamics is Hamiltonian. Write $m_1 = m_2 = m_3$ for the mass of the first particle and q_1, q_2, q_3 for its rectangular Cartesian coordinates; write $m_4 = m_5 = m_6$ for the mass of the second particle and q_4, q_5, q_6 for its coordinates; and so on. Then the Lagrangian is

$$L = \tfrac{1}{2}\sum_{A=1}^{N} m_A \dot{q}_A^2 - V(q, \dot{q}), \tag{80.7}$$

where $N = 3n$. We suppose that V is a function of the positions and, perhaps, the velocities of the particles, vanishing when the distances between the particles tend to infinity.

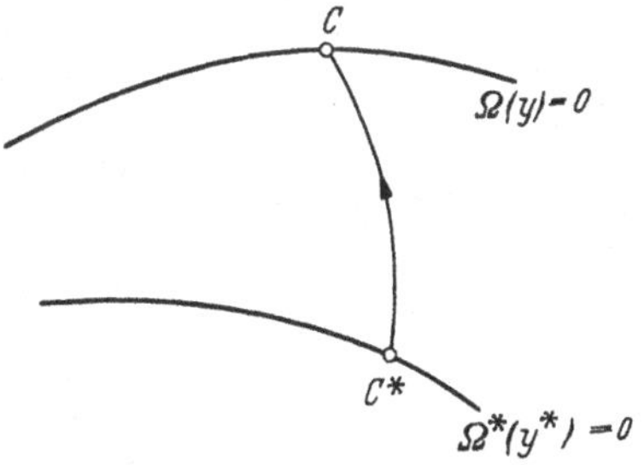

Fig. 39. Encounter viewed in momentum-energy space PH. Initially we have a fixed point C^* and finally a fixed point C.

Let the particles start at infinite distances from one another, approach, interact, and finally withdraw to infinite distances again. To avoid the awkward limits, $t \to \pm\infty$, let us suppose that the interactions are completely absent for $t < t_0^*$ and for $t > t_0$, being switched on and off as smoothly as we like. This means that we modify (80.7) by writing $V(q, t, \dot{q})$ for $V(q, \dot{q})$, with the understanding that $V = 0$ except for $t_0^* < t < t_0$. Then for the initial and final states we have the energy equations

$$\left.\begin{aligned} H^* &= \tfrac{1}{2}\sum_{A=1}^{N} m_A^{-1} p_A^{*2} \quad \text{for} \quad t^* < t_0^*, \\ H &= \tfrac{1}{2}\sum_{A=1}^{N} m_A^{-1} p_A^2 \quad \text{for} \quad t > t_0. \end{aligned}\right\} \tag{80.8}$$

We are now in a position to describe the effect of any possible Hamiltonian encounter in terms of a single function W of the arguments

$$p_1^*, \ldots p_N^*, \quad p_1, \ldots p_N, \tag{80.9}$$

in the sense that, if this characteristic function is assigned, then the initial and final trajectories are given by (80.6), the quantities (80.9) having arbitrary constant values, and the functions H^*, H being as in (80.8).

In view of the axiom of homogeneity and isotropy in Newtonian dynamics (Sect. 5), the function W is not completely arbitrary. If $\boldsymbol{p}^{*(1)}, \ldots \boldsymbol{p}^{*(n)}$ are the individual momentum vectors (in ordinary space) of the individual particles before the encounter, and $\boldsymbol{p}^{(1)}, \ldots \boldsymbol{p}^{(n)}$ the individual momentum vectors after the encounter, then W can involve the $6n$ components of these vectors only in the form of invariants under rigid body displacements. Thus, if there are only two particles, W must be a function of the nine scalar products (the last made a sum for symmetry)

$$\left.\begin{array}{c} \boldsymbol{p}^{(1)}\cdot\boldsymbol{p}^{(1)}, \quad \boldsymbol{p}^{(1)}\cdot\boldsymbol{p}^{(2)}, \quad \boldsymbol{p}^{(2)}\cdot\boldsymbol{p}^{(2)}, \\ \boldsymbol{p}^{*(1)}\cdot\boldsymbol{p}^{*(1)}, \quad \boldsymbol{p}^{*(1)}\cdot\boldsymbol{p}^{*(2)}, \quad \boldsymbol{p}^{*(2)}\cdot\boldsymbol{p}^{*(2)}, \\ \boldsymbol{p}^{(1)}\cdot\boldsymbol{p}^{*(1)}, \quad \boldsymbol{p}^{(2)}\cdot\boldsymbol{p}^{*(2)}, \\ \boldsymbol{p}^{(1)}\cdot\boldsymbol{p}^{*(2)} + \boldsymbol{p}^{(2)}\cdot\boldsymbol{p}^{*(1)}, \end{array}\right\} \tag{80.10}$$

or an equivalent set.

IV. Configuration space[1] (Q).

81. Reinterpretation of dynamics in the space Q. Rays and waves in a coherent system[2]. The space of configurations Q, in which the coordinates of a point are the N generalized coordinates q_ϱ of the dynamical system, gives the most natural geometrical representation; if the system consists of a single particle, the representative point in Q is identified with the position of the particle in ordinary space. All that has been said in Part E II about dynamics in QT can be reinterpreted in Q. A ray or trajectory, which was a curve in QT, now appears in Q as a moving point, the time t being definitely a parameter and not a coordinate; the coordinates q_ϱ and the associated momentum p_ϱ satisfy the canonical equations

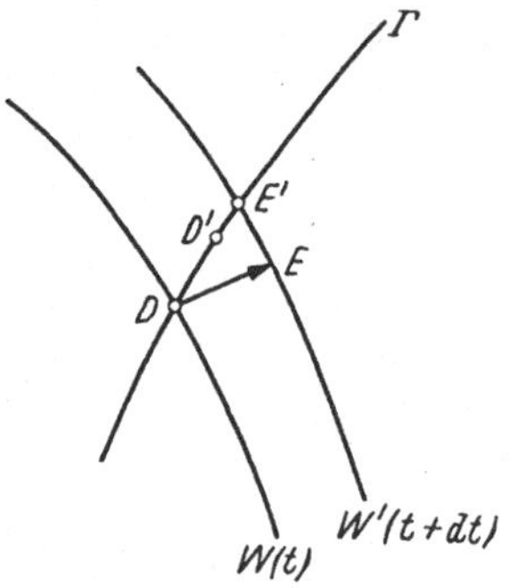

Fig. 40. A ray or trajectory Γ in Q and a moving wave of constant Lagrangian or Hamiltonian action. In time dt the ray displacement is DD' and the wave displacement DE'.

$$\dot{q}_\varrho = \frac{\partial H}{\partial p_\varrho}, \quad \dot{p}_\varrho = -\frac{\partial H}{\partial q_\varrho}. \tag{81.1}$$

But while Q may seem easier to think about than QT, the waves of constant action for a coherent system discussed in Sect. 75 present a rather complicated moving picture when viewed in Q. In QT the rays or trajectories are fixed curves and the waves fixed surfaces of N dimensions, as shown in Fig. 35 (p. 123); in Q the former are moving points and the latter moving surfaces of $N-1$ dimensions, with equations

$$U(q, t) = \text{const}, \tag{81.2}$$

where U is the one-point characteristic function. In Fig. 40, Γ is a ray or trajectory, and D, D' the positions on it of the representative point at times $t, t+dt$ respectively. W is the wave which passes through the point D at time t, and W' is the same wave at time $t+dt$. It is important to note that in general D' does not lie on W'; in other words, *the representative point does not ride on a wave.*

To explore the relation between the velocities of the representative point (the *ray velocity*) and the wave (the *wave velocity*), we note first that the ray velocity is $\dot{q}_\varrho$, but that, for a general Hamiltonian dynamical system, there is no way of converting this contravariant vector into an invariant speed. To investigate wave velocity, we take a point E on W', adjacent to D, and denote the displacement DE by δq_ϱ; then by (81.2), since it is a question of one moving wave, we have

$$\frac{\partial U}{\partial q_\varrho}\,\delta q_\varrho + \frac{\partial U}{\partial t}\,dt = 0, \tag{81.3}$$

---

[1] Often called q-space.

[2] Cf. LEVI-CIVITA and AMALDI [*16*] II_2, pp. 456–469. For the general theory of waves and characteristics, see T. LEVI-CIVITA, op. cit. in Sect. 76. For some interesting general remarks relative to wave mechanics, see T. LEVI-CIVITA: Bull. Amer. Math. Soc. **39**, 535–563 (1933).

or, by (74.9),

$$p_\varrho \delta q_\varrho - H\, dt = 0. \tag{81.4}$$

To follow the usual practice in finding wave velocity, we should take the displacement δq_ϱ along the normal to W at D. There is a normal, defined by $\partial U/\partial q_\varrho$ $(=p_\varrho)$, but this is a covariant vector, whereas δq_ϱ is contravariant, and we cannot (in any invariant sense) speak of them as having the same direction, without limiting the generality of the dynamical theory. The best we can do is to take δq_ϱ along the ray (taking E at E' in Fig. 40), so that

$$R\, \delta q_\varrho = d q_\varrho = \dot{q}_\varrho\, dt, \tag{81.5}$$

the proportionality factor R being describable as

$$R = \frac{\text{ray velocity}}{\text{wave velocity measured along ray}}. \tag{81.6}$$

Multiplying (81.5) by p_ϱ and using (81.4), we get

$$R = \frac{p_\varrho \dot{q}_\varrho}{H} = \frac{p_\varrho}{H} \frac{\partial H}{\partial p_\varrho}. \tag{81.7}$$

So far the argument has been of maximum generality. Consider now the ordinary dynamical system[1] (ODS) of Sects. 66 and 70; for it we have

$$\left.\begin{aligned} T = \tfrac{1}{2} a_{\varrho\sigma} \dot{q}_\varrho \dot{q}_\sigma, \quad L = T - V(q),\\ H = \tfrac{1}{2} a^{\varrho\sigma} p_\varrho p_\sigma + V. \end{aligned}\right\} \tag{81.8}$$

The tensor $a_{\varrho\sigma}$ provides us with an invariant kinematical line element ds in Q, defined by[2]

$$ds^2 = a_{\varrho\sigma}\, dq_\varrho\, dq_\sigma. \tag{81.9}$$

It is natural to define the ray speed v by

$$v^2 = \left(\frac{ds}{dt}\right)^2 = a_{\varrho\sigma} \dot{q}_\varrho \dot{q}_\sigma = 2T = 2(E - V), \tag{81.10}$$

where E is the constant[3] total energy $(E = H = T + V)$. In terms of the metric (81.9), the wave W has a contravariant unit normal n^ϱ given by

$$n^\varrho = \frac{a^{\varrho\sigma} \dfrac{\partial U}{\partial q_\sigma}}{\sqrt{a^{\mu\nu} \dfrac{\partial U}{\partial q_\mu} \dfrac{\partial U}{\partial q_\nu}}} = \frac{a^{\varrho\sigma} p_\sigma}{\sqrt{a^{\mu\nu} p_\mu p_\nu}} = \frac{a^{\varrho\sigma} p_\sigma}{\sqrt{2T}} = \frac{a^{\varrho\sigma} p_\sigma}{\sqrt{2(E - V)}}. \tag{81.11}$$

To find the normal velocity of wave propagation, we take E on the normal to W at D, so that

$$\delta q_\varrho = a^{\varrho\sigma} p_\sigma\, d\vartheta, \tag{81.12}$$

where $d\vartheta$ is an infinitesimal multiplier, and (81.4) gives

$$\frac{d\vartheta}{dt} = \frac{H}{a^{\varrho\sigma} p_\varrho p_\sigma}. \tag{81.13}$$

---

[1] The essential requirement here is the tensor $a_{\varrho\sigma}$, and this is available for rheonomic systems also [cf. (27.2)]; some slight modification of the statements is necessary.

[2] This kinematical line element is discussed more fully in Sect. 84.

[3] Constant, that is, for the ray or trajectory Γ; it is not necessary to take E constant throughout the coherent system. For the derivation of the formula (81.14) for the wave velocity from the HAMILTON-JACOBI equation, see E. SCHRÖDINGER: Ann. Physik (4) **79**, 489–527 (1926); Abhandlungen zur Wellenmechanik p. 494 (Leipzig: Johann Ambrosius Barth 1927). See also L. BRILLOUIN: Les Tenseurs en mécanique et en élasticité, Chap. VIII (Paris: Masson & Cie 1938), and GOLDSTEIN [7], p. 307.

Hence the normal velocity u of the wave is given by

$$\left.\begin{aligned} u^2 &= \frac{a_{\varrho\sigma}\,\delta q_\varrho\,\delta q_\sigma}{dt^2} = a^{\varrho\sigma}\,p_\varrho\,p_\sigma\left(\frac{d\vartheta}{dt}\right)^2 \\ &= \frac{H^2}{a^{\varrho\sigma}\,p_\varrho\,p_\sigma} = \frac{H^2}{2(H-V)} = \frac{E^2}{2(E-V)}\,. \end{aligned}\right\} \tag{81.14}$$

82. Isoenergetic dynamics in Q and its relation to general dynamics in QT. Consider a conservative dynamical system; the time t is absent from the Hamiltonian, so that we have

$$H = H(q, p)\,. \tag{82.1}$$

Then along any trajectory

$$H(q, p) = E\,. \tag{82.2}$$

We shall call E the *total energy*, since it is precisely that for ordinary dynamical systems.

We now study *isoenergetic dynamics* (cf. Sect. 62) in the space Q, by which we understand that E is a constant, not merely along each trajectory, but for the whole system of trajectories considered, and indeed for those varied curves which we have occasion to use. In fact, E is a constant built into the isoenergetic dynamics.

Let E be an assigned number, and let the system start from a point D^* of Q at time t^* with any initial momenta satisfying (82.2). Let Γ be the curve described in Q in accordance with the equations of motion. Then, if D is the position of the representative point on Γ at time t, the Hamiltonian action from D^* to D is, by (68.1),

$$A_H(\Gamma) = \int_\Gamma (p_\varrho\,dq_\varrho - H\,dt) = \int_\Gamma p_\varrho\,dq_\varrho - (t-t^*)\,E\,. \tag{82.3}$$

Consider any adjacent curve Γ_1 joining D^* and D, described in the same time interval (t^*, t) and having associated with it a vector field p_ϱ satisfying (82.2) and approximately the same as on Γ. Then

$$A_H(\Gamma_1) = \int_{\Gamma_1} p_\varrho\,dq_\varrho - (t-t^*)\,E\,,$$

and hence

$$\delta A_H = A_H(\Gamma_1) - A_H(\Gamma) = \delta\int p_\varrho\,dq_\varrho\,.$$

But by HAMILTON'S principle (68.12) we have $\delta A_H = 0$, and therefore the trajectory in Q satisfies (without reference to time) the variational equation and side condition

$$\delta\int p_\varrho\,dq_\varrho = 0, \qquad H(q, p) - E = 0\,, \tag{82.4}$$

the end points in Q being fixed.

If we now compare this isoenergetic dynamics in Q with the general dynamics in QT, based, as in (68.5), on the variational equation

$$\delta\int y_r\,dx_r = 0, \qquad \Omega(x, y) = 0\,, \tag{82.5}$$

we recognize at once the complete identity of the two dynamics, save for the trivial difference in dimensionality, N for (82.4) and $N+1$ for (82.5). The transition is effected as follows:

$$\left.\begin{aligned} x_r \to q_\varrho, \qquad y_r \to p_\varrho, \\ \Omega(x, y) = 0 \to H(q, p) - E = 0\,. \end{aligned}\right\} \tag{82.6}$$

All the theory developed in E II for general dynamics in QT is available for isoenergetic dynamics in Q, and we shall discuss some aspects of this in Sect. 83.

We note that the time t has disappeared from (82.4), so that, if we use nothing but (82.4), we can hardly expect the time to reappear. But it does. For if we apply to (82.4) the argument used at (68.5) to find the extremals, we get the equations

$$\frac{dq_\varrho}{dw} = \frac{\partial H}{\partial p_\varrho}, \qquad \frac{dp_\varrho}{dw} = -\frac{\partial H}{\partial q_\varrho}, \tag{82.7}$$

where w is some special parameter. But we know that the trajectories satisfy the canonical equations (68.16), that is

$$\frac{dq_\varrho}{dt} = \frac{\partial H}{\partial p_\varrho}, \qquad \frac{dp_\varrho}{dt} = -\frac{\partial H}{\partial q_\varrho}, \tag{82.8}$$

and, comparing these with (82.7), we see that, though we chased t out, it comes back as the special parameter in the canonical differential equations derived from (82.4).

If we have solved a dynamical problem in Q, obtaining a curve and a momentum field along it, we can use any one of (82.8) to find the time t; or we can use some derived equation, such as

$$dt = \frac{p_\varrho dq_\varrho}{p_\sigma \partial H/\partial p_\sigma}. \tag{82.9}$$

There is another way of keeping the time in isoenergetic dynamics which can be the cause of no little confusion. This method puts a parameter t on each varied curve according to the following plan. Given any arbitrary motion $q_\varrho = q_\varrho(t)$, not in general satisfying the canonical equations (82.8), there is a natural momentum [cf. (71.11)] $p_\varrho(t)$ given by

$$p_\varrho = \frac{\partial L}{\partial \dot q_\varrho}, \tag{82.10}$$

where L is the Lagrangian. By adjusting the parameter t we can satisfy $H(q, p) = E$; then we can state the variational equation (82.4) in a more restricted form as

$$\delta \int \frac{\partial L}{\partial \dot q_\varrho} \dot q_\varrho \, dt = 0, \qquad H\left(q, \frac{\partial L}{\partial \dot q}\right) - E = 0, \tag{82.11}$$

the variation being for fixed end points in Q, but *not* for fixed end times, the time t on each curve being obtained from the second equation in (82.11). This variational equation is more restricted than (82.4) because (82.10) is more restrictive than $H(q, p) = E$. It seems less confusing to work with the more general equation (82.4) and restore the time only on trajectories, as in (82.9) or in an equivalent way.

83. MAUPERTUIS action, the two-point characteristic function for isoenergetic systems, the homogeneous Lagrangian, and JACOBI's principle of least action. In isoenergetic dynamics in Q, we define the MAUPERTUIS[1] action as

$$\bar A = \int p_\varrho \, dq_\varrho, \tag{83.1}$$

where the p's and q's satisfy the energy equation

$$H(q, p) = E. \tag{83.2}$$

[1] Although it is the custom to use the single word *action* for $\bar A$ as in (83.1), (83.3) or (83.4) (cf. WHITTAKER [*28*], p. 248, GOLDSTEIN [*7*], p. 228), it seems desirable to have an adjective to distinguish this integral from the Lagrangian or Hamiltonian action of Sect. 64, 68. The name of MAUPERTUIS appears to be most commonly associated with the integral $\bar A$, particularly in the form (83.4) or $m\int v\, ds$ for a single particle, and will be used here as a distinguishing adjective, although historically it might be juster to call this the Eulerian action; cf. DUGAS, pp. 250—264, of op. cit. in Sect. 1.

If we further restrict p_ϱ to be natural momenta as in (82.10), we have

$$\bar{A} = \int \frac{\partial L}{\partial \dot{q}_\varrho} \dot{q}_\varrho \, dt. \tag{83.3}$$

For the system ODS of (81.8), we have

$$\bar{A} = 2 \int T \, dt. \tag{83.4}$$

Pursuing the analogy with general Hamiltonian dynamics in QT, we define, in isoenergetic dynamics in Q, a two-point characteristic function by

$$\bar{S}(D^*, D) = \int p_\varrho \, dq_\varrho, \tag{83.5}$$

the integral being taken along the ray or trajectory joining the point D^* to the point D. Variation of the end points gives

$$\delta \bar{S} = p_\varrho \, \delta q_\varrho - p_\varrho^* \, \delta q_\varrho^*, \tag{83.6}$$

and hence, if arbitrary variations are permissible, as is in general the case,

$$\frac{\partial \bar{S}}{\partial q_\varrho} = p_\varrho, \qquad \frac{\partial \bar{S}}{\partial q_\varrho^*} = -p_\varrho^*. \tag{83.7}$$

It follows from (83.2) that $\bar{S}$ satisfies the two partial differential equations

$$H\left(q, \frac{\partial \bar{S}}{\partial q}\right) = E, \qquad H\left(q^*, -\frac{\partial \bar{S}}{\partial q^*}\right) = E. \tag{83.8}$$

We recognize the HAMILTON-JACOBI equation in the form (78.2).

In the timeless theory of isoenergetic dynamics in Q, a coherent system of rays or trajectories appears as a set of fixed curves and the associated waves as a set of fixed surfaces, two waves being separated by a constant amount of MAUPERTUIS action.

Just as in QT we can pass from an energy equation $\Omega(x, y) = 0$ to a homogeneous Lagrangian $\Lambda(x, x')$ by the Eqs. (69.14), so in isoenergetic dynamics we can find a homogeneous Lagrangian $\bar{\Lambda}(q, q')$ (where $q'_\varrho = dq_\varrho/du$, u being any parameter) by eliminating ϑ and p_ϱ from the $N+2$ equations

$$q'_\varrho = \vartheta \frac{\partial H(q, p)}{\partial p_\varrho}, \qquad \bar{\Lambda} = p_\varrho q'_\varrho, \qquad H(q, p) - E = 0. \tag{83.9}$$

The rays or trajectories in Q then satisfy the variational equation

$$\delta \int \bar{\Lambda}(q, q') \, du = 0, \tag{83.10}$$

for fixed end points in Q.

Let us carry out this process for a system with the Lagrangian

$$L(q, \dot{q}) = \tfrac{1}{2} a_{\varrho\sigma} \dot{q}_\varrho \dot{q}_\sigma + a_\varrho \dot{q}_\varrho - V, \tag{83.11}$$

where $a_{\varrho\sigma} (= a_{\sigma\varrho})$, a_ϱ and V are functions of the q's only. (This is a little more general than the ODS of Sect. 66 and (81.8), for which $a_\varrho = 0$.) We first find the Hamiltonian as follows:

$$\left.\begin{aligned} p_\varrho &= \frac{\partial L}{\partial \dot{q}_\varrho} = a_{\varrho\sigma} \dot{q}_\sigma + a_\varrho, \qquad \dot{q}_\varrho = a^{\varrho\sigma} (p_\sigma - a_\sigma), \\ H(q, p) &= \dot{q}_\varrho \frac{\partial L}{\partial \dot{q}_\varrho} - L = \frac{1}{2} a_{\varrho\sigma} \dot{q}_\varrho \dot{q}_\sigma + V \\ &= \frac{1}{2} a^{\varrho\sigma} (p_\varrho - a_\varrho)(p_\sigma - a_\sigma) + V. \end{aligned}\right\} \tag{83.12}$$

Then we proceed by (83.9):

$$\left.\begin{aligned} q'_\varrho &= \vartheta\, a^{\varrho\sigma}(p_\sigma - a_\sigma), \quad p_\varrho - a_\varrho = \vartheta^{-1} a_{\varrho\sigma} q'_\sigma,\\ H(q,p) - E &= \tfrac{1}{2}\vartheta^{-2} a_{\varrho\sigma} q'_\varrho q'_\sigma + V - E = 0,\\ \vartheta^2 &= \frac{\frac{1}{2} a_{\varrho\sigma} q'_\varrho q'_\sigma}{E - V},\\ \bar{\Lambda} &= p_\varrho q'_\varrho = q'_\varrho(\vartheta^{-1} a_{\varrho\sigma} q'_\sigma + a_\varrho), \end{aligned}\right\} \tag{83.13}$$

and so we have the homogeneous Lagrangian for isoenergetic dynamics in Q:

$$\bar{\Lambda}(q, q') = \sqrt{2(E-V)}\sqrt{a_{\varrho\sigma} q'_\varrho q'_\sigma} + a_\varrho q'_\varrho. \tag{83.14}$$

The variational equation (83.10) may be written

$$\delta \int \left(\sqrt{2(E-V)}\sqrt{a_{\varrho\sigma}\, dq_\varrho\, dq_\sigma} + a_\varrho\, dq_\varrho\right) = 0. \tag{83.15}$$

If we put $a_\varrho = 0$, so that the system becomes ODS, and use the kinematical line element (81.9), this becomes

$$\delta \int \sqrt{(E-V)}\, ds = 0, \tag{83.16}$$

which is known as JACOBI's *principle of least action*[1].

This may be interpreted geometrically by saying that the trajectories are geodesics in the space Q if it is endowed with the Riemannian line element

$$ds_1^2 = (E-V)\, ds^2 = (E-V)\, a_{\varrho\sigma}\, dq_\varrho\, dq_\sigma, \tag{83.17}$$

which may be called the *action line element.* This metric is singular on the locus $V=E$, which corresponds to a state of instantaneous rest for the system, since $V=E$ implies $T=0$.

84. The kinematical line element. In this section and the next, we abandon Hamiltonian dynamics temporarily. Consider a dynamical system with kinetic energy

$$T = \tfrac{1}{2} g_{\varrho\sigma} \dot{q}^\varrho \dot{q}^\sigma. \tag{84.1}$$

Since this work is guided by the techniques of tensor calculus, we denote the coordinates by q^ϱ rather than q_ϱ, since dq^ϱ is a contravariant vector. We represent the system by a point in the space Q, which we endow with the kinematical line element[2]

$$ds^2 = 2T\, dt^2 = g_{\varrho\sigma}\, dq^\varrho\, dq^\sigma, \tag{84.2}$$

already used in (81.9) (we change a to g to avoid confusion with acceleration).

Before introducing forces, we consider kinematics. Any motion $q^\varrho = q^\varrho(t)$ defines a curve in Q, and at each point of the curve a contravariant *velocity vector*

$$v^\varrho = \dot{q}^\varrho, \tag{84.3}$$

which may also be written

$$v^\varrho = v\,\lambda^\varrho, \tag{84.4}$$

---

[1] For other derivations of this principle, see APPELL [2] II, p. 454; GOLDSTEIN [7], p. 232; PÉRÈS [20], pp. 229, 248. Written in the equivalent form $\delta\int T\,dt = 0$, with the side condition $\delta E = 0$, this is sometimes called HÖLDER's principle. This principle and HAMILTON's principle are both contained in the principle $\delta\int(2T - \lambda E)\,dt = 0$, with the side condition $\delta(E^{\lambda-1} dt^\lambda) = 0$, where λ is any constant; cf. E. STORCHI: Atti Accad. Naz. Lincei. Rend. Cl. Sci. Fis. Mat. Nat. (8) **14**, 771–778 (1953).

[2] This is almost the line element of HERTZ (pp. 55, 62 of op. cit. in Sect. 61); he divides by the total mass of the system, so as to give ds the dimensions of a length.

where

$$\lambda^\varrho = \frac{dq^\varrho}{ds}, \tag{84.5}$$

the unit tangent to the curve. There is also a contravariant *acceleration vector*

$$a^\varrho = \frac{\delta v^\varrho}{\delta t}, \tag{84.6}$$

the absolute derivative of v^ϱ, the absolute derivative of a vector field $V^\varrho(u)$ along a curve $q^\varrho(u)$ being[1]

$$\frac{\delta V^\varrho}{\delta u} = \frac{dV^\varrho}{du} + \left\{ {\varrho \atop \mu\nu} \right\} V^\mu \frac{dq^\nu}{du}. \tag{84.7}$$

Substitution from (84.4) in (84.6) gives

$$a^\varrho = \dot{v}\,\lambda^\varrho + \varkappa v^2 \nu^\varrho = v\frac{dv}{ds}\lambda^\varrho + \varkappa v^2 \nu^\varrho, \tag{84.8}$$

where ν^ϱ is the unit first normal to the curve of motion and $\varkappa$ the first curvature[2], defined by

$$\frac{\delta\lambda^\varrho}{\delta s} = \varkappa\nu^\varrho, \quad g_{\varrho\sigma}\nu^\varrho\nu^\sigma = 1, \quad \varkappa \geqq 0. \tag{84.9}$$

Thus the acceleration of the representative point can be resolved along the tangent and first normal in just the same way as in (18.2) for a moving particle.

Since we have a fundamental tensor $g_{\varrho\sigma}$ in Q, and its contravariant conjugate $g^{\varrho\sigma}$, we can pass from contravariant components to covariant, and vice versa. The covariant acceleration is

$$a_\varrho = g_{\varrho\sigma} a^\sigma = \frac{\delta v_\varrho}{\delta t} = \frac{d}{dt}\frac{\partial T}{\partial \dot{q}^\varrho} - \frac{\partial T}{\partial q^\varrho}. \tag{84.10}$$

Passing from kinematics to dynamics, we introduce the generalized force Q_ϱ, a covariant vector defined by

$$Q_\varrho\,\delta q^\varrho = \delta W, \tag{84.11}$$

δW being the work done in a displacement δq^ϱ (δW is an invariant). LAGRANGE'S equations of motion are

$$\frac{d}{dt}\frac{\partial T}{\partial \dot{q}^\varrho} - \frac{\partial T}{\partial q^\varrho} = Q_\varrho, \tag{84.12}$$

and these may be written

$$a_\varrho = Q_\varrho \quad \text{or} \quad a^\varrho = Q^\varrho, \tag{84.13}$$

where Q^ϱ is the contravariant force vector. In words: acceleration = force[3]. It should be noted that, while the physical dimensions of the several components of a vector depend on the choice of coordinates, the magnitude v of the velocity vector has the dimensions $[M^{\frac{1}{2}}LT^{-1}]$ and the magnitude a of the acceleration vector has the dimensions $[M^{\frac{1}{2}}LT^{-2}]$.

---

[1] For further details about absolute derivatives, see J. L. SYNGE and A. SCHILD: Tensor Calculus, pp. 47–51. Toronto: University Press 1952. For the CHRISTOFFEL symbols, see (18.4).

[2] We might call this *geometrical* curvature, to distinguish it from the *dynamical* curvature of Sect. 85. To within a constant factor, $\varkappa$ is the curvature of HERTZ (cf. pp. 74–77 of op. cit. in Sect. 61).

[3] If $Q_\varrho = 0$, then $a_\varrho = 0$, and hence, by (84.8), $\varkappa = 0$, so that the trajectory is a geodesic if no forces act.

We can form a geometrical picture of the problem of stability as in Fig. 41, Γ and Γ' being two adjacent trajectories. There are two simple ways of correlating points on them:

(i) isochronous correspondence, in which we correlate positions with the same value of t, the infinitesimal deviation vector being ξ^ϱ in Fig. 41;

(ii) normal correspondence, in which the infinitesimal deviation vector η^ϱ is orthogonal to Γ, so that

$$\eta^\varrho v_\varrho = 0. \tag{84.14}$$

Between the two types of deviation vector we have the relation

$$\xi^\varrho = \eta^\varrho + \vartheta v^\varrho, \tag{84.15}$$

in which, by virtue of (84.14),

$$\vartheta = \frac{\xi^\sigma v_\sigma}{v^2}. \tag{84.16}$$

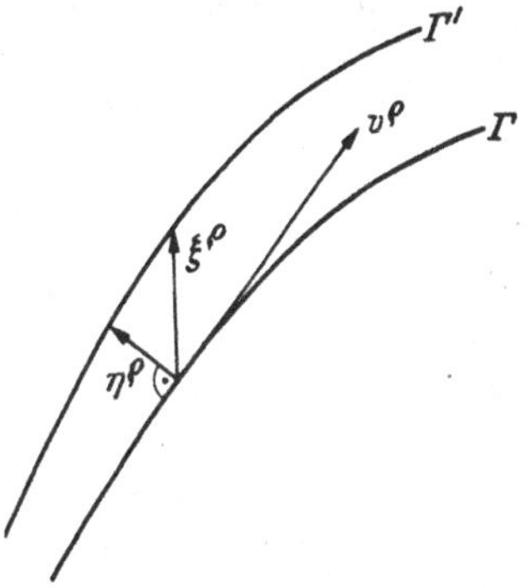

Fig. 41. Deviation of trajectories in the space Q, with the isochronous deviation vector ξ^ϱ and the normal deviation vector η^ϱ.

The isochronous vector ξ^ϱ satisfies the equation of deviation[1]

$$\frac{\delta^2 \xi^\varrho}{\delta t^2} + R^\varrho_{\cdot\sigma\mu\nu} v^\sigma \xi^\mu v^\nu = Q^\varrho_{|\sigma} \xi^\sigma \tag{84.17}$$

where $R^\varrho_{\cdot\sigma\mu\nu}$ is the curvature tensor of the space Q with metric (84.2),

$$R^\varrho_{\cdot\sigma\mu\nu} = \frac{\partial}{\partial q^\mu}\left\{{\varrho \atop \sigma\nu}\right\} - \frac{\partial}{\partial q^\nu}\left\{{\varrho \atop \sigma\mu}\right\} + \left\{{\tau \atop \sigma\nu}\right\}\left\{{\varrho \atop \tau\mu}\right\} - \left\{{\tau \atop \sigma\mu}\right\}\left\{{\varrho \atop \tau\nu}\right\}, \tag{84.18}$$

and $Q^\varrho_{|\sigma}$ is the covariant derivative of the contravariant force,

$$Q^\varrho_{|\sigma} = \frac{\partial Q^\varrho}{\partial q^\sigma} + \left\{{\varrho \atop \mu\sigma}\right\} Q^\mu. \tag{84.19}$$

The equation of deviation for normal correspondence is a little more complicated. Substitution from (84.15) in (84.17) gives, with the aid of (84.13),

$$\frac{\delta^2 \eta^\varrho}{\delta t^2} + \frac{d^2\vartheta}{dt^2} v^\varrho + 2\frac{d\vartheta}{dt} Q^\varrho + R^\varrho_{\cdot\sigma\mu\nu} v^\sigma \eta^\mu v^\nu = Q^\varrho_{|\sigma}\eta^\sigma; \tag{84.20}$$

with (84.14), we have then $N+1$ equations for ϑ and η^ϱ.

85. Least curvature. With the notation of Sect. 84, we define the *dynamical curvature* K of an arbitrary kinematical motion with acceleration a^ϱ, in the presence of given forces Q^ϱ, as the positive square root of

$$K^2 = g_{\varrho\sigma}(a^\varrho - Q^\varrho)(a^\sigma - Q^\sigma). \tag{85.1}$$

From the positive-definite character of kinetic energy, this expression is non-negative; $K=0$ if, and only if, the equations of motion (84.13) are satisfied.

We impose constraints, in general non-holonomic, given by [cf. (46.2)]

$$A_{c\varrho} v^\varrho + A_c = 0. \tag{85.2}$$

---

[1] Cf. J. L. Synge: Phil. Trans. Roy. Soc. Lond. A **226**, 31–106 (1926), and Tensorial Methods in Dynamics (University of Toronto Studies, Applied Mathematics Series No. 2; Toronto, University Press 1936). The latter contains a bibliography of work dealing with dynamics from the standpoint of Riemannian geometry.

The A's are functions of the coordinates and t, and the suffix c takes the values $1, 2, \dots M$, so that the constrained system has $N-M$ degrees of freedom. Differentiation gives

$$A_{c\varrho} a^\varrho + B_c = 0, \tag{85.3}$$

for any constrained motion, B_c being independent of the acceleration.

We now ask what acceleration a^ϱ minimizes K, a^ϱ being subject to (85.3). If we think of a^ϱ as rectangular Cartesian coordinates in a Euclidean space of N dimensions, this is equivalent to seeking the point of contact of an ellipsoid (85.1) (of given centre Q_ϱ and given shape) with a hyperplane represented by (85.3). Differentiating, we see that the minimizing a^ϱ satisfy

$$g_{\varrho\sigma}(a^\sigma - Q^\sigma) = \sum_{c=1}^{M} \vartheta_c A_{c\varrho}, \tag{85.4}$$

where ϑ_c are undetermined multipliers; but these are precisely the Lagrangian equations of motion (46.15), and we conclude that *the constrained trajectory (obeying the equations of motion) minimizes the dynamical curvature K, when it is compared with arbitrary constrained motions with the same position and velocity.*

For a system of particles, (85.1) reads, in ordinary notation,

$$K^2 = \sum m \left[\left(\ddot{x} - \frac{X}{m}\right)^2 + \left(\ddot{y} - \frac{Y}{m}\right)^2 + \left(\ddot{z} - \frac{Z}{m}\right)^2\right], \tag{85.5}$$

the summation being over the particles and (X, Y, Z) being the given force on a particle. With K in this form, the theorem $K =$ minimum is known as GAUSS's *theorem of least curvature or least constraint*[1].

For the more restricted case of scleronomic constraints,

$$A_{c\varrho} v^\varrho = 0, \tag{85.6}$$

with $A_{c\varrho}$ functions of the coordinates only, there is a theorem of minimum relative curvature, with *curvature* understood in the geometrical sense of Sect. 84. Consider

C, the unconstrained trajectory under given forces Q_ϱ,

C', an arbitrary constrained motion,

C'', the constrained trajectory under given forces Q_ϱ,

all three motions having a common configuration and velocity. As in (84.9) we define the first curvature vector of any curve by

$$\varkappa^\varrho = \frac{\delta \lambda^\varrho}{\delta s} = \varkappa \nu^\varrho, \tag{85.7}$$

and we define the curvatures of C' and C'' *relative* to C to be k' and k'', respectively, where

$$k'^2 = g_{\varrho\sigma}(\varkappa'^\varrho - \varkappa^\varrho)(\varkappa'^\sigma - \varkappa^\sigma), \qquad k''^2 = g_{\varrho\sigma}(\varkappa''^\varrho - \varkappa^\varrho)(\varkappa''^\sigma - \varkappa^\sigma). \tag{85.8}$$

Then

$$k'^2 - k''^2 = g_{\varrho\sigma}(\varkappa'^\varrho - \varkappa''^\varrho)(\varkappa'^\sigma - \varkappa''^\sigma) + 2 g_{\varrho\sigma}(\varkappa'^\varrho - \varkappa''^\varrho)(\varkappa''^\sigma - \varkappa^\sigma). \tag{85.9}$$

From the equations of motion of C and C'' we have, by (84.8) and (84.13),

$$\left.\begin{aligned} \varkappa^\varrho v^2 &= Q^\varrho - \dot{v}\lambda^\varrho = Q^\varrho - \lambda^\varrho Q^\sigma \lambda_\sigma, \\ \varkappa''^\varrho v^2 &= Q^\varrho - \lambda^\varrho Q^\sigma \lambda_\sigma + Q'^\varrho, \end{aligned}\right\} \tag{85.10}$$

where Q'^ϱ is the force of constraint, and hence

$$(\varkappa''^\varrho - \varkappa^\varrho)\, v^2 = Q'^\varrho; \tag{85.11}$$

---

[1] See APPELL [2] II, pp. 492–497; LANCZOS [15], pp. 106–110,; WHITTAKER [28], pp. 254–258. For a discussion of the principles of GAUSS, HERTZ and JOURDAIN, see NORDHEIM [18], pp. 62–69.

for the constrained motions C', C'', we have

$$A_{c\varrho}\lambda'^{\varrho} = 0, \qquad A_{c\varrho}\lambda''^{\varrho} = 0, \tag{85.12}$$

and differentiation gives, at the point of comparison,

$$A_{c\varrho}(\varkappa'^{\varrho} - \varkappa''^{\varrho}) = 0. \tag{85.13}$$

But by (46.15) or (85.4) the force of constraint is

$$Q'_{\varrho} = \sum_{c=1}^{M} \vartheta_c A_{c\varrho}, \tag{85.14}$$

where ϑ_c are undetermined multipliers, and it follows from (85.11) and (85.13) that the last expression in (85.9) vanishes. We have then

$$k'^2 - k''^2 = g_{\varrho\sigma}(\varkappa'^{\varrho} - \varkappa''^{\varrho})(\varkappa'^{\sigma} - \varkappa''^{\sigma}) \geq 0: \tag{85.15}$$

of all constrained motions with given position and velocity, the constrained trajectory has the smallest geometrical curvature relative to the unconstrained trajectory.

V. The space of states and energy (*QTPH*).

86. The energy surface und the energy function. Certain important aspects of dynamical theory are best illustrated by taking representative points in spaces of higher dimensionality than the $(N+1)$-dimensional space of events QT; these higher spaces are the $(2N+2)$-dimensional space of states and energy[1] $QTPH$, the $(2N+1)$-dimensional space of states[2] QTP, and the $2N$-dimensional phase space QP (as always, N denotes the number of degrees of freedom of the system). We shall now deal with $QTPH$, deferring QTP to Part E VI and QP to Part E VII. The theory developed for $QTPH$ can, as we shall see, be applied to QP by a simple change in notation, provided the system in QP is conservative ($\partial H/\partial t = 0$).

In $QTPH$ we take as coordinates[3] of a representative point the $(2N+2)$ quantities $q_{\varrho}, t, p_{\varrho}, H$, or, in the notation of (64.1) and (67.6)[4],

$$\left.\begin{aligned} x_{\varrho} &= q_{\varrho}, & x_{N+1} &= t, \\ y_{\varrho} &= p_{\varrho}, & y_{N+1} &= -H. \end{aligned}\right\} \tag{86.1}$$

The variables y_r are said to be *conjugate* to the variables x_r.

A dynamical system is defined by assigning a $(2N+1)$-dimensional *energy surface* in $QTPH$, and confining the representative point to this surface. In general we shall write the equation of the energy surface in the form

$$\Omega(x, y) = 0. \tag{86.2}$$

To a given surface there correspond an infinity of equations, each of which represents it. If we solve (86.2) for y_{N+1}, we may write the equation of the energy surface in the form

$$\Omega(x, y) = y_{N+1} + \omega(x_1, \ldots x_{N+1}, y_1, \ldots y_N) = 0, \tag{86.3}$$

---

[1] Called *extended phase space* by LANCZOS [*15*], p. 199.

[2] Called *l'espace des états* by E. CARTAN, Leçons sur les invariants intégraux, p. 4. Paris: Hermann 1922.

[3] We shall work in the small. In the large, overlapping coordinate systems may be required (cf. Sect. 63), with canonical transformations (cf. Sect. 87) in the overlaps.

[4] As in Sect. 62, Greek suffixes have the range $1, 2, \ldots N$ and small Latin suffixes the range $1, 2, \ldots N+1$, with the summation convention in both cases.

or, equivalently[1],

$$H = \omega(q, t, p). \tag{86.4}$$

But it is advisable to enlarge the scope of dynamics in $QTPH$ by using an *energy function* $\Omega(x, y)$ instead of merely an energy surface. To a given energy function there corresponds a unique energy surface with equation $\Omega(x, y) = 0$; to a given energy surface there correspond an infinity of energy functions.

We define *rays* or *trajectories* in $QTPH$ by the variational principle and side condition [cf. (68.5)]

$$\delta \int y_r \, dx_r = 0, \quad \Omega(x, y) = \text{const}, \tag{86.5}$$

the end values of x_r being fixed. Hence we obtain the canonical equations

$$\frac{dx_r}{dw} = \frac{\partial \Omega}{\partial y_r}, \quad \frac{dy_r}{dw} = -\frac{\partial \Omega}{\partial x_r}, \tag{86.6}$$

where w is a special parameter; these equations of course imply $\Omega = \text{const}$ along each ray or trajectory. The solutions of (86.6) fill $QTPH$ with a *natural congruence* of rays or trajectories, one through each point; some of these curves fill the energy surface $\Omega = 0$. Thus the totality of dynamical trajectories, including those off the energy surface, present a simpler geometrical picture than what we had in QT, where there was a trajectory passing through each point in each direction.

87. Canonical transformations. The bilinear invariant. If we apply an arbitrary transformation $(x, y) \to (x', y')$, the canonical equations (86.6) take the form

$$\frac{dx'_r}{dw} = X_r(x', y'), \quad \frac{dy'_r}{dw} = Y_r(x', y'). \tag{87.1}$$

These new equations will be in canonical form only if the right hand sides satisfy the conditions

$$\frac{\partial X_r}{\partial y'_s} = \frac{\partial X_s}{\partial y'_r}, \quad \frac{\partial Y_r}{\partial x'_s} = \frac{\partial Y_s}{\partial x'_r}, \quad \frac{\partial X_r}{\partial x'_r} + \frac{\partial Y_r}{\partial y'_r} = 0 \quad \text{(not summed)}. \tag{87.2}$$

For some purposes it is desirable to admit general transformations, but particular importance attaches to *canonical transformations*[2] $(x, y) \to (x', y')$ which preserve the canonical form of the equations of the rays or trajectories, i.e. those transformations under which the equations

$$\frac{dx_r}{dw} = \frac{\partial \Omega}{\partial y_r}, \quad \frac{dy_r}{dw} = -\frac{\partial \Omega}{\partial x_r} \tag{87.3}$$

transform into

$$\frac{dx'_r}{dw} = \frac{\partial \Omega}{\partial y'_r}, \quad \frac{dy'_r}{dw} = -\frac{\partial \Omega}{\partial x'_r}. \tag{87.4}$$

---

[1] In dealing with $QTPH$ it is confusing to write $H = H(q, t, p)$, because then H appears both as a coordinate and as a functional symbol. To illustrate this, in ordinary space a relation between x, y, z defines a surface. We can write the equation of this surface in the form $f(x, y, z) = 0$ [that corresponds to the general Eq. (86.2)] or in the form $z = F(x, y)$ [that corresponds to (86.4)]. It is confusing to write it in the form $z = z(x, y)$ if we have occasion to speak of points (x, y, z) which do not lie on the surface, because then we have $z \neq z(x, y)$.

[2] In the literature of dynamics, the terms *canonical transformations* and *contact transformations* are used almost synonymously. For remarks on usage, see GOLDSTEIN [*7*], p. 239; LANCZOS [*15*], uses the word *canonical* and WHITTAKER [*28*] uses *contact*. CORBEN and STEHLE [*3*], p. 302, use the expression *extended contact transformation* for a canonical transformation in $QTPH$. For the mathematical connection between canonical and contact transformations, see CARATHÉODORY, p. 107 of op. cit. in Sect. 71; H. TIETZ, this Encyclopedia, Vol. II, p. 193; WINTNER [*30*], p. 31. In the present article, only the expression *canonical transformation* will be used. For canonical transformations produced in the space QTP by the natural congruence, see Sect. 95; for canonical transformations in QP, see Sect. 96.

It is understood that under a canonical transformation (which we shall call CT for short) the special parameter w is to be unchanged and the energy function Ω is to be treated as an invariant in the sense of tensor calculus $[\Omega(x, y) = \Omega'(x', y')]$. We consider only non-singular (reversible) transformations.

Write

$$z = \begin{pmatrix} \boldsymbol{x} \\ \boldsymbol{y} \end{pmatrix}, \quad \delta z = \begin{pmatrix} \delta \boldsymbol{x} \\ \delta \boldsymbol{y} \end{pmatrix}, \quad z' = \begin{pmatrix} \boldsymbol{x}' \\ \boldsymbol{y}' \end{pmatrix}, \quad \delta z' = \begin{pmatrix} \delta \boldsymbol{x}' \\ \delta \boldsymbol{y}' \end{pmatrix}, \tag{87.5}$$

these being $(2N+2)\times 1$ matrices. Then any non-singular transformation gives, in differential form

$$\delta \boldsymbol{z} = \boldsymbol{J}\,\delta \boldsymbol{z}', \qquad \delta \boldsymbol{z}' = \boldsymbol{J}^{-1}\,\delta \boldsymbol{z}, \tag{87.6}$$

where $\boldsymbol{J}$ is the $(2N+2)\times(2N+2)$ Jacobian matrix

$$\boldsymbol{J} = \begin{pmatrix} \dfrac{\partial x_r}{\partial x'_m} & \dfrac{\partial x_r}{\partial y'_n} \\[2ex] \dfrac{\partial y_s}{\partial x'_m} & \dfrac{\partial y_s}{\partial y'_n} \end{pmatrix}, \qquad \det \boldsymbol{J} \neq 0. \tag{87.7}$$

Write

$$\boldsymbol{W} = \begin{pmatrix} \dfrac{\partial \Omega}{\partial x} \\[2ex] \dfrac{\partial \Omega}{\partial y} \end{pmatrix}, \qquad \boldsymbol{W}' = \begin{pmatrix} \dfrac{\partial \Omega}{\partial x'} \\[2ex] \dfrac{\partial \Omega}{\partial y'} \end{pmatrix}, \tag{87.8}$$

and use the tilde for transposition, turning a column matrix into a row matrix. From the existence of the invariant

$$\widetilde{\boldsymbol{W}}\,\delta \boldsymbol{z} = \delta\Omega = \widetilde{\boldsymbol{W}}'\,\delta \boldsymbol{z}' \tag{87.9}$$

or otherwise, we see that $\boldsymbol{W}$ transforms according to

$$\boldsymbol{W}' = \widetilde{\boldsymbol{J}}\boldsymbol{W}, \qquad \boldsymbol{W} = \widetilde{\boldsymbol{J}}^{-1}\boldsymbol{W}'. \tag{87.10}$$

We now introduce what is really the key to the algebra of CT, the skew-symmetric $(2N+2)\times(2N+2)$ numerical matrix

$$\boldsymbol{\Gamma} = \begin{pmatrix} \boldsymbol{0} & \boldsymbol{1} \\ -\boldsymbol{1} & \boldsymbol{0} \end{pmatrix}, \tag{87.11}$$

where $\boldsymbol{1}$ stands for the $(N+1)\times(N+1)$ unit matrix. We note that

$$\widetilde{\boldsymbol{\Gamma}} = -\boldsymbol{\Gamma}, \qquad \det \boldsymbol{\Gamma} = 1, \qquad \boldsymbol{\Gamma}^2 = -\boldsymbol{1}_{(2N+2)}, \qquad \boldsymbol{\Gamma}^{-1} = -\boldsymbol{\Gamma}. \tag{87.12}$$

In this notation the canonical equations (87.3) read

$$d\boldsymbol{z} = dw\cdot\boldsymbol{\Gamma}\boldsymbol{W}. \tag{87.13}$$

We have then, by (87.6) and (87.10),

$$d\boldsymbol{z} - dw\cdot\boldsymbol{\Gamma}\boldsymbol{W} = \boldsymbol{J}(d\boldsymbol{z}' - dw\cdot\boldsymbol{J}^{-1}\boldsymbol{\Gamma}\widetilde{\boldsymbol{J}}^{-1}\boldsymbol{W}'), \tag{87.14}$$

and it is clear that

$$\boldsymbol{J}^{-1}\boldsymbol{\Gamma}\widetilde{\boldsymbol{J}}^{-1} = \boldsymbol{\Gamma} \tag{87.15}$$

is a necessary and sufficient condition on $\boldsymbol{J}$ *in order that the transformation having* $\boldsymbol{J}$ *for Jacobian matrix should be a CT.* This condition may be written in the equivalent forms[1]

$$\boldsymbol{J}\boldsymbol{\Gamma}\tilde{\boldsymbol{J}}=\boldsymbol{\Gamma}, \quad \tilde{\boldsymbol{J}}\boldsymbol{\Gamma}\boldsymbol{J}=\boldsymbol{\Gamma}. \tag{87.16}$$

It follows from (87.16) that $\det \boldsymbol{J} = \pm 1$; we shall prove later in (88.29) that $\det \boldsymbol{J}=1$.

Let $\delta_1 \boldsymbol{z}$ and $\delta_2 \boldsymbol{z}$ be any independent variations; then

$$\delta_1 \tilde{\boldsymbol{z}} \cdot \boldsymbol{\Gamma} \cdot \delta_2 \boldsymbol{z} = (\delta_1 \boldsymbol{x}, \delta_1 \boldsymbol{y}) \begin{pmatrix} 0 & 1 \\ -1 & 0 \end{pmatrix} \begin{pmatrix} \delta_2 \boldsymbol{x} \\ \delta_2 \boldsymbol{y} \end{pmatrix} = \delta_1 x_r \, \delta_2 y_r - \delta_2 x_r \, \delta_1 y_r, \tag{87.17}$$

a bilinear form. Applying the CT (87.6), we obtain

$$\delta_1 \tilde{\boldsymbol{z}} \cdot \boldsymbol{\Gamma} \cdot \delta_2 \boldsymbol{z} = \delta_1 \boldsymbol{z}' \cdot \tilde{\boldsymbol{J}} \boldsymbol{\Gamma} \boldsymbol{J} \cdot \delta_2 \boldsymbol{z}' = \delta_1 \tilde{\boldsymbol{z}}' \cdot \boldsymbol{\Gamma} \cdot \delta_2 \boldsymbol{z}', \tag{87.18}$$

and so a CT has the *bilinear invariant*

$$\delta_1 x_r \, \delta_2 y_r - \delta_2 x_r \, \delta_1 y_r = \delta_1 x_r' \, \delta_2 y_r' - \delta_2 x_r' \, \delta_1 y_r'. \tag{87.19}$$

It is easy to see that (87.19) is a sufficient condition for CT. This invariant bilinear form may be regarded as the basis of CT in the same way as certain invariant quadratic forms ($dx^2+dy^2+dz^2$ and $dx^2+dy^2+dz^2-dt^2$) may be regarded as the bases of rigid body transformations in space and LORENTZ transformations in space-time, respectively. Just as these quadratic forms define the squares of invariant elements of length, so the bilinear form defines an invariant element of area

$$\{u, v\}\, du\, dv = \{u, v\}'\, du\, dv \tag{87.20}$$

on the 2-space with equations $x_r = x_r(u, v)$, $y_r = y_r(u, v)$ immersed in $QTPH$; here

$$\{u, v\} = \frac{\partial x_r}{\partial u}\frac{\partial y_r}{\partial v} - \frac{\partial x_r}{\partial v}\frac{\partial y_r}{\partial u}, \quad \{u, v\}' = \frac{\partial x_r'}{\partial u}\frac{\partial y_r'}{\partial v} - \frac{\partial x_r'}{\partial v}\frac{\partial y_r'}{\partial u}, \tag{87.21}$$

these being in fact LAGRANGE brackets (see Sect. 89).

Canonical transformations form a group. For they contain the identical transformation, to each CT $(x, y) \to (x', y')$ there corresponds the unique inverse CT $(x', y') \to (x, y)$, and from (87.16) or from the bilinear invariant it follows that a succession of two CT is itself a CT.

88. Generating functions. Let $(x, y) \to (x', y')$ be any non-singular transformation, not necessarily CT. Let A and B be any two points of $QTPH$ and C any curve joining them. Consider the integral

$$I(A, B; C) = \int (y_r\, dx_r - y_r'\, dx_r'), \tag{88.1}$$

taken along C from A to B. Here

$$dx_r' = \frac{\partial x_r'}{\partial x_s}\, dx_s + \frac{\partial x_r'}{\partial y_s}\, dy_s, \tag{88.2}$$

so that the integral is actually of the form

$$I(A, B; C) = \int (X_s\, dx_s + Y_s\, dy_s), \tag{88.3}$$

[1] Remember that we have imposed the condition $w' = w$; if we relax this, a necessary and sufficient condition for CT is $\boldsymbol{J}\boldsymbol{\Gamma}\tilde{\boldsymbol{J}} = \mu \boldsymbol{\Gamma}$, where μ is a scalar multiplier. A linear transformation $\boldsymbol{z} = \boldsymbol{J}\boldsymbol{z}'$ is called *symplectic* (or the matrix $\boldsymbol{J}$ is called *symplectic*) if $\boldsymbol{J}$ satisfies (87.16); cf. H. WEYL: The Classical Groups, Chap. 6 (2nd Edn. Princeton: University Press 1946). WINTNER [*30*], pp. 17, 29, 45, uses I to denote the matrix $\boldsymbol{\Gamma}$; SIEGEL (p. 9 of op. cit. in Sect. 53) uses $\mathfrak{J}$.

where

$$X_s = y_s - y'_r \frac{\partial x'_r}{\partial x_s}, \quad Y_s = - y'_r \frac{\partial x'_r}{\partial y_s}. \tag{88.4}$$

If we agree to keep A fixed once for all, we may denote the integral by $I(B;C)$. And if we make B coincide with A, so that C is a circuit, we may denote the integral by $I(C)$. Other similar obvious notations are used below.

Giving arbitrary variations to A, B and C, we get from (88.1), on integration by parts,

$$\left.\begin{aligned}\delta I(A, B; C) &= [y_r \delta x_r - y'_r \delta x'_r]_A^B \\ &+ \int [(dx_r \delta y_r - \delta x_r dy_r) - (dx'_r \delta y'_r - \delta x'_r dy'_r)].\end{aligned}\right\} \tag{88.5}$$

Suppose that $(x, y) \to (x', y')$ is CT. Then, on account of the bilinear invariant (87.19), the integral on the right hand side of (88.5) vanishes, and we have

$$\delta I(A, B; C) = [y_r \delta x_r - y'_r \delta x'_r]_A^B. \tag{88.6}$$

A number of consequences follow:

(i) Since the variation of I vanishes when A and B are held fixed, $I(A, B; C)$ has the same value for all reconcilable curves joining A and B; symbolically, $I(A, B; C) = I(A, B)$.

(ii) If A is held fixed, then $I(B; C) = I(B)$, a function of B only, multiple valued in the case of multiple connectivity; we have

$$y_r \delta x_r - y'_r \delta x'_r = \delta I(B) \tag{88.7}$$

for any arbitrary variation[1] of B.

(iii) If A and B coincide, so that C is a circuit and we can write $I(C)$ for the integral, then $\delta I(C) = 0$ for an arbitrary variation of the circuit. This implies that $I(C)$ has the same value for all reconcilable circuits, and $I(C) = 0$ for a reducible circuit. Equivalently,

$$\oint y_r dx_r = \oint y'_r dx'_r \tag{88.8}$$

for every reducible circuit; in words, *the circulation of action in a reducible circuit is invariant under CT*. In the case of an irreducible circuit, the effect of a CT is to increase or decrease the circulation by an amount which is the same for all reconcilable circuits.

By (88.7) we have this: if $(x, y) \to (x', y')$ is CT, then the Pfaffian

$$y_r dx_r - y'_r \delta x'_r = X_s \delta x_s + Y_s \delta y_s \tag{88.9}$$

is an exact differential. Let us prove the converse. Given a transformation $(x, y) \to (x', y')$ such that

$$y_r \delta x_r - y'_r \delta x'_r = \delta I(B), \tag{88.10}$$

where $I(B)$ is some function of (x, y), let us take a 2-space $x_r = x_r(u, v)$, $y_r = y_r(u, v)$, so that (x, y, x', y', I) are all functions of u and v. Then

$$y_r \frac{\partial x_r}{\partial v} - y'_r \frac{\partial x'_r}{\partial v} = \frac{\partial I}{\partial v}. \tag{88.11}$$

Differentiating with respect to u, interchanging u and v, and subtracting, we get

$$\{u, v\} = \{u, v\}' \tag{88.12}$$

[1] This means arbitrary variations of (x, y) or equivalently of (x', y'); the CT may be such that arbitrary independent variations cannot be given to (x, x'). This occurs in the case of MATHIEU transformations (see later).

in the notation of (87.21). This establishes the existence of the bilinear invariant, and hence the canonical character of $(x, y) \to (x', y')$.

We have now three tests for CT: (i) the symplectic test (87.16), based on the matrix $\boldsymbol{\Gamma}$, (ii) the bilinear test (87.19), and (iii) the exact differential test (88.10).

Canonical transformations may be generated as follows. Let $G_1(x, x')$ be an arbitrary function. If we define (y, y') by

$$y_r = \frac{\partial G_1(x, x')}{\partial x_r}, \qquad y'_r = -\frac{\partial G_1(x, x')}{\partial x'_r}, \tag{88.13}$$

then

$$y_r\,\delta x_r - y'_r\,\delta x'_r = \delta G_1(x, x'), \tag{88.14}$$

an exact differential in the space (x, x'). But this procedure does not necessarily give us a transformation $(x, y) \to (x', y')$, because we have no assurance that the Eq. (88.13) can be solved for (x, y) in terms of (x', y') or vice versa. To examine this, we differentiate (88.13), obtaining

$$\left.\begin{aligned} \delta y_r &= \frac{\partial^2 G_1}{\partial x_r\,\partial x_s}\,\delta x_s + \frac{\partial^2 G_1}{\partial x_r\,\partial x'_s}\,\delta x'_s, \\ \delta y'_r &= -\frac{\partial^2 G_1}{\partial x'_r\,\partial x_s}\,\delta x_s - \frac{\partial^2 G_1}{\partial x'_r\,\partial x'_s}\,\delta x'_s. \end{aligned}\right\} \tag{88.15}$$

Let us impose on $G_1(x, x')$ the condition

$$\det \frac{\partial^2 G_1}{\partial x_r\,\partial x'_s} \neq 0. \tag{88.16}$$

Then (88.15) can be solved for $(\delta x, \delta y)$ in terms of $(\delta x', \delta y')$ and vice versa. These solutions can be integrated, because, if we pass round a small circuit in the space of (x, x'), we are led round small circuits in the spaces of (x, y) and (x', y'). Hence we get a reversible transformation $(x, y) \to (x', y')$; *by virtue of (88.14), it is a CT.*

Thus, starting from a *generating function* $G_1(x, x')$, arbitrary except for the inequality (88.16), we obtain a CT from (88.13) or (88.14). This powerful method of generating CT does not, however, yield all CT. It does not yield those CT for which there exist one or more relationships[1] between the variables (x, x') or in particular the CT of MATHIEU[2] for which

$$y_r\,\delta x_r - y'_r\,\delta x'_r = 0. \tag{88.17}$$

Similarly, the alternative generating functions described below fail to yield certain special CT. But to understand the general theory of CT it seems advisable to neglect such special cases, and to suppose the CT under consideration such that any of the following sets of $(2N+2)$ variables forms a coordinate system in the space $QTPH$, in the sense that the variables in any set may be varied arbitrarily and independently:

$$(x, y), \quad (x', y'), \quad (x, x'), \quad (x, y'), \quad (y, x'), \quad (y, y').$$

The formula (88.14) may be written in the following equivalent forms:

$$y_r\,\delta x_r - y'_r\,\delta x'_r = \delta G_1, \tag{88.18a}$$
$$x_r\,\delta y_r - x'_r\,\delta y'_r = \delta G_2, \tag{88.18b}$$
$$y_r\,\delta x_r + x'_r\,\delta y'_r = \delta G_3, \tag{88.18c}$$
$$x_r\,\delta y_r + y'_r\,\delta x'_r = \delta G_4, \tag{88.18d}$$

[1] See WHITTAKER [*28*], p. 294.
[2] See WHITTAKER [*28*], p. 301.

where

$$\left.\begin{aligned} G_2 &= x_r y_r - x'_r y'_r - G_1, \\ G_3 &= x'_r y'_r + G_1, \\ G_4 &= x_r y_r - G_1. \end{aligned}\right\} \tag{88.19}$$

We have then four different ways of generating CT:

$$y_r = \frac{\partial G_1(x, x')}{\partial x_r}, \quad y'_r = -\frac{\partial G_1(x, x')}{\partial x'_r}, \tag{88.20a}$$

$$x_r = \frac{\partial G_2(y, y')}{\partial y_r}, \quad x'_r = -\frac{\partial G_2(y, y')}{\partial y'_r}, \tag{88.20b}$$

$$y_r = \frac{\partial G_3(x, y')}{\partial x_r}, \quad x'_r = \frac{\partial G_3(x, y')}{\partial y'_r}, \tag{88.20c}$$

$$x_r = \frac{\partial G_4(y, x')}{\partial y_r}, \quad y'_r = \frac{\partial G_4(y, x')}{\partial x'_r}. \tag{88.20d}$$

Any one of these formulae gives a CT, the generating function involved being arbitrary except for an inequality of the type (88.16).

Here are some particularly simple examples of CT. First, using (88.20a):

$$\left.\begin{aligned} G(x, x') &= x_r x'_r, & \det \frac{\partial^2 G}{\partial x_r \partial x'_s} &= 1, \\ y_r &= x'_r, & y'_r &= -x_r, \end{aligned}\right\} \tag{88.21}$$

so that the variables are interchanged, with a reversal of sign as indicated. Secondly, using (88.20c):

$$\left.\begin{aligned} G(x, y') &= x_r y'_r, & \det \frac{\partial^2 G}{\partial x_r \partial y'_s} &= 1, \\ y_r &= y'_r, & x'_r &= x_r, \end{aligned}\right\} \tag{88.22}$$

so that this is the identical transformation. Thirdly, using (88.20c) again, take

$$G(x, y') = f_r(x)\, y'_r, \quad \det \frac{\partial^2 G}{\partial x_r \partial y'_s} = \det \frac{\partial f_s}{\partial x_r} \neq 0, \tag{88.23}$$

the functions $f_s(x)$ being arbitrary except for this last condition. We get

$$y_r = \frac{\partial f_s}{\partial x_r} y'_s, \quad x'_r = f_r(x), \tag{88.24}$$

so that we have an arbitrary transformation $(x) \to (x')$ and y_r transforms like a covariant vector. This is an *extended point transformation*[1].

In (q, t, p, H) notation [cf. (86.1)], the CT (88.20) read as follows:

$$\left.\begin{aligned} &G_1 = G_1(q, t, q', t'): \\ &p_\varrho = \frac{\partial G_1}{\partial q_\varrho}, \quad -H = \frac{\partial G_1}{\partial t}, \quad p'_\varrho = -\frac{\partial G_1}{\partial q'_\varrho}, \quad -H' = -\frac{\partial G_1}{\partial t'}. \end{aligned}\right\} \tag{88.25a}$$

$$\left.\begin{aligned} &G_2 = G_2(p, H, p', H'): \\ &q_\varrho = \frac{\partial G_2}{\partial p_\varrho}, \quad t = -\frac{\partial G_2}{\partial H}, \quad q'_\varrho = -\frac{\partial G_2}{\partial p'_\varrho}, \quad t' = \frac{\partial G_2}{\partial H'}. \end{aligned}\right\} \tag{88.25b}$$

---

1 WHITTAKER [28], p. 293. GOLDSTEIN [7], p. 244, calls it simply a point transformation. These references may be consulted for further details about CT, with examples. See also WINTNER [30], p. 34 and CARATHÉODORY, pp. 78–102 of op. cit. in Sect. 71.

$$G_3 = G_3(q, t, p', H'):$$
$$\left.p_\varrho = \frac{\partial G_3}{\partial q_\varrho}, \quad -H = \frac{\partial G_3}{\partial t}, \quad q'_\varrho = \frac{\partial G_3}{\partial p'_\varrho}, \quad t' = -\frac{\partial G_3}{\partial H'}.\right\} \tag{88.25c}$$
$$G_4 = G_4(p, H, q', t'):$$
$$\left.q_\varrho = \frac{\partial G_4}{\partial p_\varrho}, \quad t = -\frac{\partial G_4}{\partial H}, \quad p'_\varrho = \frac{\partial G_4}{\partial q'_\varrho}, \quad -H' = \frac{\partial G_4}{\partial t'}.\right\} \tag{88.25d}$$

The following CT leave the time unchanged:

$$G_3(q, t, p', H') = -tH' + g(q, t, p'):$$
$$\left.p'_\varrho = \frac{\partial g}{\partial q_\varrho}, \quad H = H' - \frac{\partial g}{\partial t}, \quad q'_\varrho = \frac{\partial g}{\partial p'_\varrho}, \quad t' = t.\right\} \tag{88.26c}$$
$$G_4(p, H, q', t') = -Ht' + g(p, q', t'):$$
$$\left.q_\varrho = \frac{\partial g}{\partial p_\varrho}, \quad t = t', \quad p'_\varrho = \frac{\partial g}{\partial q'_\varrho}, \quad H' = H - \frac{\partial g}{\partial t'}.\right\} \tag{88.26d}$$

The following CT leave the Hamiltonian unchanged:

$$G_3(q, t, p', H') = -tH' + g(q, p', H'):$$
$$\left.p_\varrho = \frac{\partial g}{\partial q_\varrho}, \quad H = H', \quad q'_\varrho = \frac{\partial g}{\partial p'_\varrho}, \quad t' = t - \frac{\partial g}{\partial H'}.\right\} \tag{88.27c}$$
$$G_4(p, H, q', t') = -Ht' + g(p, H, q'):$$
$$\left.q_\varrho = \frac{\partial g}{\partial p_\varrho}, \quad t = t' - \frac{\partial g}{\partial H}, \quad p'_\varrho = \frac{\partial g}{\partial q'_\varrho}, \quad H' = H.\right\} \tag{88.27d}$$

The following CT leave both time and Hamiltonian unchanged:

$$G_3(q, t, p', H') = -tH' + g(q, p'):$$
$$\left.p_\varrho = \frac{\partial g}{\partial q_\varrho}, \quad H = H', \quad q'_\varrho = \frac{\partial g}{\partial p'_\varrho}, \quad t' = t.\right\} \tag{88.28c}$$
$$G_4(p, H, q', t') = -Ht' + g(p, q'):$$
$$\left.q_\varrho = \frac{\partial g}{\partial p_\varrho}, \quad t = t', \quad p'_\varrho = \frac{\partial g}{\partial q'_\varrho}, \quad H' = H.\right\} \tag{88.28d}$$

We shall now show[1] that a CT is unimodular in the sense that

$$\det \boldsymbol{J} = 1. \tag{88.29}$$

Take the CT to be generated as in (88.20c) and differentiate. In matrix notation we have

$$\delta \boldsymbol{y} = \boldsymbol{A}\,\delta\boldsymbol{x} + \boldsymbol{B}\,\delta\boldsymbol{y}', \quad \delta\boldsymbol{x}' = \tilde{\boldsymbol{B}}\,\delta\boldsymbol{x} + \boldsymbol{C}\,\delta\boldsymbol{y}', \tag{88.30}$$

where

$$\boldsymbol{A} = \left(\frac{\partial^2 G_3}{\partial x_r \partial x_s}\right), \quad \boldsymbol{B} = \left(\frac{\partial^2 G_3}{\partial x_r \partial y'_s}\right), \quad \boldsymbol{C} = \left(\frac{\partial^2 G_3}{\partial y'_r \partial y'_s}\right). \tag{88.31}$$

These equations may be written

$$\left.\begin{aligned} \delta\boldsymbol{y} - \boldsymbol{A}\,\delta\boldsymbol{x} &= \boldsymbol{B}\,\delta\boldsymbol{y}', \\ -\tilde{\boldsymbol{B}}\,\delta\boldsymbol{x} &= -\delta\boldsymbol{x}' + \boldsymbol{C}\,\delta\boldsymbol{y}', \end{aligned}\right\} \tag{88.32}$$

or

$$\begin{pmatrix} -\boldsymbol{A} & 1 \\ -\tilde{\boldsymbol{B}} & 0 \end{pmatrix} \begin{pmatrix} \delta\boldsymbol{x} \\ \delta\boldsymbol{y} \end{pmatrix} = \begin{pmatrix} 0 & \boldsymbol{B} \\ -1 & \boldsymbol{C} \end{pmatrix} \begin{pmatrix} \delta\boldsymbol{x}' \\ \delta\boldsymbol{y}' \end{pmatrix}. \tag{88.33}$$

---

[1] Cf. CARATHÉODORY, p. 92 of op. cit. in Sect. 71.

Comparing this with (87.6), we have

$$\begin{pmatrix} 0 & \boldsymbol{B} \\ -1 & \boldsymbol{C} \end{pmatrix} = \begin{pmatrix} -\boldsymbol{A} & 1 \\ -\boldsymbol{B} & 0 \end{pmatrix} \boldsymbol{J}, \tag{88.34}$$

and (88.29) follows on taking the determinants of the two sides.

89. Poisson brackets and Lagrange brackets in *QTPH*[1]. Let u, v be two functions of the $(2N+2)$ independent variables (x, y); the Poisson bracket $[u, v]$ is defined by

$$[u, v] = \frac{\partial u}{\partial x_r}\frac{\partial v}{\partial y_r} - \frac{\partial v}{\partial x_r}\frac{\partial u}{\partial y_r} = -[v, u]. \tag{89.1}$$

Let (x, y) be functions of the two independent variables u, v; the Lagrange bracket $\{u, v\}$ is defined as

$$\{u, v\} = \frac{\partial x_r}{\partial u}\frac{\partial y_r}{\partial v} - \frac{\partial x_r}{\partial v}\frac{\partial y_r}{\partial u} = -\{v, u\}. \tag{89.2}$$

In Sect. 97 we shall use the same notation with q_ϱ, p_ϱ substituted for x_r, y_r; the following results can be translated immediately from (x, y) to (q, p).

If u, v, w are any three functions of (x, y), then

$$[[u, v], w] + [[v, w], u] + [[w, u], v] = 0. \tag{89.3}$$

This Poisson-Jacobi identity is easy to prove by direct calculation[2].

In terms of the matrix $\boldsymbol{\Gamma}$ of (87.11), we have

$$[u, v] = \left(\frac{\partial u}{\partial x}, \frac{\partial u}{\partial y}\right) \boldsymbol{\Gamma} \begin{pmatrix} \dfrac{\partial v}{\partial x} \\ \dfrac{\partial v}{\partial y} \end{pmatrix}, \qquad \{u, v\} = \left(\frac{\partial x}{\partial u}, \frac{\partial y}{\partial u}\right) \boldsymbol{\Gamma} \begin{pmatrix} \dfrac{\partial x}{\partial v} \\ \dfrac{\partial y}{\partial v} \end{pmatrix}. \tag{89.4}$$

Under an arbitrary transformation $(x, y) \to (x', y')$, with u and v treated as invariants, we have the formulae of transformation [cf. (87.6) and (87.10)]

$$\begin{pmatrix} \dfrac{\partial x}{\partial u} \\ \dfrac{\partial y}{\partial u} \end{pmatrix} = \boldsymbol{J} \begin{pmatrix} \dfrac{\partial x'}{\partial u} \\ \dfrac{\partial y'}{\partial u} \end{pmatrix}, \qquad \begin{pmatrix} \dfrac{\partial u}{\partial x} \\ \dfrac{\partial u}{\partial y} \end{pmatrix} = \tilde{\boldsymbol{J}}^{-1} \begin{pmatrix} \dfrac{\partial u}{\partial x'} \\ \dfrac{\partial u}{\partial y'} \end{pmatrix}, \tag{89.5}$$

and the same equations with u replaced by v. Hence (89.4) gives

$$\left.\begin{aligned} [u, v] &= \left(\frac{\partial u}{\partial x'}, \frac{\partial u}{\partial y'}\right) \boldsymbol{J}^{-1} \boldsymbol{\Gamma} \tilde{\boldsymbol{J}}^{-1} \begin{pmatrix} \dfrac{\partial v}{\partial x'} \\ \dfrac{\partial v}{\partial y'} \end{pmatrix}, \\ \{u, v\} &= \left(\frac{\partial x'}{\partial u}, \frac{\partial y'}{\partial u}\right) \tilde{\boldsymbol{J}} \boldsymbol{\Gamma} \boldsymbol{J} \begin{pmatrix} \dfrac{\partial x'}{\partial v} \\ \dfrac{\partial y'}{\partial v} \end{pmatrix}. \end{aligned}\right\} \tag{89.6}$$

---

[1] It used to be the general custom to denote the Poisson bracket by (u, v) and the Lagrange bracket by $[u, v]$, cf. Whittaker [28], pp. 298, 299. That notation is used by H. Tietz, this Encyclopedia, Vol. II, pp. 194, 195. But in the application of classical dynamics to quantum theory, it has been found more convenient to denote the Poisson bracket by $[u, v]$; cf. P. A. M. Dirac: Quantum Mechanics, p. 94 (Oxford: Clarendon Press 1930). Following Goldstein [7], pp. 250, 252, we shall denote the Poisson bracket by $[u, v]$ and the Lagrange bracket by $\{u, v\}$.

[2] For an indirect proof, see Appell [2] II, p. 445; Nordheim and Fues [19], p. 107; Carathéodory, p. 55 op. cit. in Sect. 71.

If the transformation is CT, then, by (87.15) and (87.16), these equations become

$$[u, v] = [u, v]', \qquad \{u, v\} = \{u, v\}'; \tag{89.7}$$

the POISSON *and* LAGRANGE *brackets are invariant under CT.*

Returning to an arbitrary transformation $(x, y) \to (x', y')$, let u_A represent the variables $x'_1, \dots x'_{N+1}, y'_1, \dots y'_{N+1}$, capital Latin suffixes taking the values $1, 2, \dots 2N+2$, with the summation convention. We have then two skew-symmetric $(2N+2) \times (2N+2)$ matrices, a POISSON matrix $\boldsymbol{P}$ with elements $P_{AB} = [u_A, u_B]$ and a LAGRANGE matrix $\boldsymbol{L}$ with elements $L_{AB} = \{u_A, u_B\}$. The element AC of the product $\boldsymbol{LP}$ is

$$\left.\begin{aligned}(LP)_{AC} &= \{u_A, u_B\}[u_B, u_C] \\ &= \left(\frac{\partial x_r}{\partial u_A}\frac{\partial y_r}{\partial u_B} - \frac{\partial x_r}{\partial u_B}\frac{\partial y_r}{\partial u_A}\right)\left(\frac{\partial u_B}{\partial x_s}\frac{\partial u_C}{\partial y_s} - \frac{\partial u_C}{\partial x_s}\frac{\partial u_B}{\partial y_s}\right).\end{aligned}\right\} \tag{89.8}$$

Now

$$\frac{\partial x_r}{\partial u_B}\frac{\partial u_B}{\partial y_s} = \frac{\partial y_r}{\partial u_B}\frac{\partial u_B}{\partial x_s} = 0, \qquad \frac{\partial x_r}{\partial u_B}\frac{\partial u_B}{\partial x_s} = \frac{\partial y_r}{\partial u_B}\frac{\partial u_B}{\partial y_s} = \delta^r_s, \tag{89.9}$$

and hence

$$(LP)_{AC} = -\frac{\partial u_C}{\partial y_r}\frac{\partial y_r}{\partial u_A} - \frac{\partial u_C}{\partial x_r}\frac{\partial x_r}{\partial u_A} = -\delta^C_A. \tag{89.10}$$

In fact, we have

$$\boldsymbol{LP} = -1, \qquad \boldsymbol{L} = -\boldsymbol{P}^{-1}, \qquad \boldsymbol{P} = -\boldsymbol{L}^{-1}. \tag{89.11}$$

This relation between the POISSON and LAGRANGE matrices is true for an arbitrary transformation $(x, y) \to (x', y')$. Further we have, for arbitrary independent variations,

$$\left.\begin{aligned}(\delta_1\boldsymbol{x}', \delta_1\boldsymbol{y}')\,\boldsymbol{L}\begin{pmatrix}\delta_2\boldsymbol{x}'\\ \delta_2\boldsymbol{y}'\end{pmatrix} &= \delta_1 u_A\{u_A, u_B\}\,\delta_2 u_B \\ &= \left(\frac{\partial x_r}{\partial u_A}\frac{\partial y_r}{\partial u_B} - \frac{\partial x_r}{\partial u_B}\frac{\partial y_r}{\partial u_A}\right)\delta_1 u_A\,\delta_2 u_B \\ &= \delta_1 x_r\,\delta_2 y_r - \delta_2 x_r\,\delta_1 y_r \\ &= (\delta_1\boldsymbol{x}, \delta_1\boldsymbol{y})\,\boldsymbol{\Gamma}\begin{pmatrix}\delta_2\boldsymbol{x}\\ \delta_2\boldsymbol{y}\end{pmatrix} \\ &= (\delta_1\boldsymbol{x}', \delta_1\boldsymbol{y}')\,\tilde{\boldsymbol{J}}\boldsymbol{\Gamma}\boldsymbol{J}\begin{pmatrix}\delta_2\boldsymbol{x}'\\ \delta_2\boldsymbol{y}'\end{pmatrix}.\end{aligned}\right\} \tag{89.12}$$

Therefore, for an arbitrary transformation $(x, y) \to (x', y')$, the LAGRANGE matrix $\boldsymbol{L}$ is related to the Jacobian matrix $\boldsymbol{J}$ by

$$\boldsymbol{L} = \tilde{\boldsymbol{J}}\boldsymbol{\Gamma}\boldsymbol{J}. \tag{89.13}$$

For a CT, we have then by (87.16)

$$\boldsymbol{L} = \boldsymbol{\Gamma}, \qquad \boldsymbol{P} = -\boldsymbol{L}^{-1} = -\boldsymbol{\Gamma}^{-1} = \boldsymbol{\Gamma}. \tag{89.14}$$

In the above theory of POISSON and LAGRANGE brackets there has been no reference to an energy function Ω. We now introduce this, and consider a ray or trajectory satisfying the canonical equations (86.6). Let $F(x, y)$ be any function. Then as the representative point moves along the ray or trajectory, we have

$$\frac{dF}{dw} = \frac{\partial F}{\partial x_r}\frac{dx_r}{dw} + \frac{\partial F}{\partial y_r}\frac{dy_r}{dw} = \frac{\partial F}{\partial x_r}\frac{\partial \Omega}{\partial y_r} - \frac{\partial F}{\partial y_r}\frac{\partial \Omega}{\partial x_r} = [F, \Omega]. \tag{89.15}$$

In particular, the canonical equations themselves may be written

$$\frac{dx_r}{dw} = [x_r, \Omega], \quad \frac{dy_r}{dw} = [y_r, \Omega]. \tag{89.16}$$

Thus the POISSON brackets are intimately connected with the motion of a dynamical system.

It follows from (89.15) and the POISSON-JACOBI identity (89.3) that, for any two functions $f(x, y)$ and $F(x, y)$,

$$\frac{d}{dw}[f, F] = [[f, F], \Omega] = -[[F, \Omega], f] - [[\Omega, f], F]. \tag{89.17}$$

If f and F are constants of the motion, so that

$$\frac{df}{dw} = [f, \Omega] = 0, \quad \frac{dF}{dw} = [F, \Omega] = 0, \tag{89.18}$$

then it follows from (89.17) that the POISSON bracket $[f, F]$ is also a constant of the motion (POISSON's theorem[1]).

Let us now use the (q, t, p, H) notation, related to the (x, y) notation by (86.1), taking the energy function in the form

$$\Omega(x, y) = y_{N+1} + \omega(x_1, \ldots x_{N+1}, y_1, \ldots y_N) \tag{89.19}$$

as in (86.3), or, equivalently,

$$\Omega(x, y) = -H + \omega(q, t, p). \tag{89.20}$$

The canonical equations (86.6) read

$$\left.\begin{aligned} \frac{dq_\varrho}{dw} &= \frac{\partial\omega}{\partial p_\varrho}, & \frac{dp_\varrho}{dw} &= -\frac{\partial\omega}{\partial q_\varrho}, \\ \frac{dt}{dw} &= 1, & -\frac{dH}{dw} &= -\frac{\partial\omega}{\partial t}. \end{aligned}\right\} \tag{89.21}$$

Thus $t = w + \text{const}$, and we have

$$\frac{dq_\varrho}{dt} = \frac{\partial\omega}{\partial p_\varrho}, \quad \frac{dp_\varrho}{dt} = -\frac{\partial\omega}{\partial q_\varrho}, \quad \frac{dH}{dt} = \frac{\partial\omega}{\partial t}. \tag{89.22}$$

On the energy surface $\Omega = 0$, we may substitute H for ω, and the equations take the usual form

$$\frac{dq_\varrho}{dt} = \frac{\partial H}{\partial p_\varrho}, \quad \frac{dp_\varrho}{dt} = -\frac{\partial H}{\partial q_\varrho}, \quad \frac{dH}{dt} = \frac{\partial H}{\partial t}. \tag{89.23}$$

For any function $F(q, t, p, H)$ we have by (89.15)

$$\left.\begin{aligned} \frac{dF}{dt} &= \frac{\partial F}{\partial x_\varrho}\frac{\partial\Omega}{\partial y_\varrho} + \frac{\partial F}{\partial x_{N+1}}\frac{\partial\Omega}{\partial y_{N+1}} - \frac{\partial\Omega}{\partial x_\varrho}\frac{\partial F}{\partial y_\varrho} - \frac{\partial\Omega}{\partial x_{N+1}}\frac{\partial F}{\partial y_{N+1}} \\ &= \frac{\partial F}{\partial q_\varrho}\frac{\partial\omega}{\partial p_\varrho} + \frac{\partial F}{\partial t} - \frac{\partial\omega}{\partial q_\varrho}\frac{\partial F}{\partial p_\varrho} + \frac{\partial\omega}{\partial t}\frac{\partial F}{\partial H}. \end{aligned}\right\} \tag{89.24}$$

On the energy surface $\Omega = 0$ we may write H instead of ω, and obtain

$$\frac{dF}{dt} = \frac{\partial F}{\partial t} + \frac{\partial H}{\partial t}\frac{\partial F}{\partial H} + [F, H]_{qp}, \tag{89.25}$$

[1] Cf. APPELL [2] II, p. 447.

where

$$[F, H]_{qp} = \frac{\partial F}{\partial q_\varrho}\frac{\partial H}{\partial p_\varrho} - \frac{\partial H}{\partial q_\varrho}\frac{\partial F}{\partial p_\varrho}. \tag{89.26}$$

If $F = F(q, t, p)$, this becomes

$$\frac{dF}{dt} = \frac{\partial F}{\partial t} + [F, H]_{qp}, \tag{89.27}$$

and if $F = H$ it becomes simply

$$\frac{dH}{dt} = \frac{\partial H}{\partial t}. \tag{89.28}$$

90. Canonical transformations generated by the canonical equations. The basic relative integral invariant. The CT we have been considering are finite transformations. To get an infinitesimal CT, i.e. one near the identical transformation, we recall that the identical transformation was given in (88.22); accordingly, following the plan (88.20c), we introduce the generating function

$$G_3(x, y') = x_r y'_r + F(x, y')\, du, \tag{90.1}$$

the function F being arbitrary and du being an infinitesimal constant. This gives the CT

$$y_r = y'_r + du \cdot \frac{\partial F(x, y')}{\partial x_r}, \qquad x'_r = x_r + du \cdot \frac{\partial F(x, y')}{\partial y'_r}, \tag{90.2}$$

or, to the first order,

$$\left.\begin{aligned} dx_r &= x'_r - x_r = du \cdot \frac{\partial F(x, y)}{\partial y_r}, \\ dy_r &= y'_r - y_r = -du \cdot \frac{\partial F(x, y)}{\partial x_r}, \end{aligned}\right\} \tag{90.3}$$

in which we have replaced y' by y in the partial derivatives. If we write the canonical equations (86.6) in the form

$$dx_r = dw \cdot \frac{\partial \Omega(x, y)}{\partial y_r}, \qquad dy_r = -dw \cdot \frac{\partial \Omega(x, y)}{\partial x_r}, \tag{90.4}$$

and compare these with (90.3), we are led to say that *the canonical equations generate an infinitesimal CT*, the increment dw in the special parameter playing the part of the infinitesimal constant du, and the energy function Ω the part of the function F.

However, there is a change in point of view which may be a source of considerable confusion. Hitherto we have regarded a CT as a change of the labels attached to fixed points in the space $QTPH$, but we have regarded the canonical equations as the description of the motion of a representative point in $QTPH$ for some fixed coordinate system. This duality of interpretation is present in all transformation theory, and we face it by recognizing two alternative interpretations of a CT:

(i) We have an assembly of geometrical objects (points in $QTPH$) to which different sets of labels (x, y), (x', y') may be attached.

(ii) We have a Euclidean space E_{2N+2} with *one fixed set* of rectangular coordinate axes, and (x, y), (x', y') are the coordinates of two *different* points of E_{2N+2} relative to that set of coordinate axes.

According to the first point of view, a transformation $(x, y) \to (x', y')$ changes the labels on fixed points; according to the second, it moves the points, the space

E_{2N+2} being transformed into itself as a whole. If we put $F=\Omega$ and $du=dw$, there is complete formal agreement between (90.3) and (90.4); this common form can be interpreted geometrically in either of these two ways.

So far we have considered only infinitesimal CT generated by the canonical equations. In the interpretation (ii) above, we see all the points of E_{2N+2} given infinitesimal displacements, corresponding to some fixed infinitesimal value of dw. However, from the group property of CT it follows that a succession of infinitesimal CT is itself a CT, and we are led to the conclusion that, if we follow the points of E_{2N+2} along the rays or trajectories, with a common value of the finite increment Δw for them all, then the resulting transformation of E_{2N+2} into itself is a finite CT. We shall now show how a generating function for this finite CT may be constructed, the integration of the canonical equations of motion being assumed.

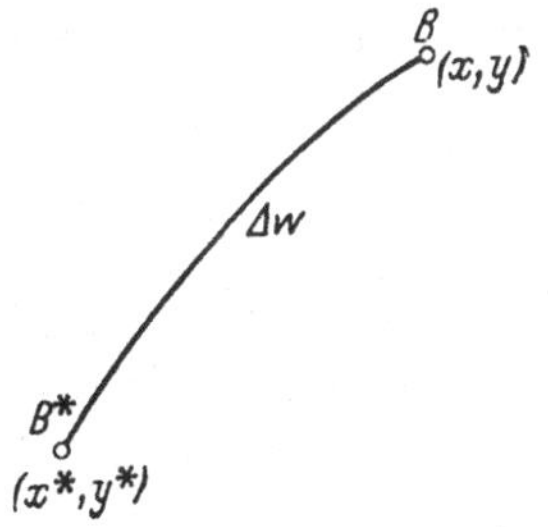

Fig. 42. Construction of a generating function $G(x^*, x, \Delta w)$ in $QTPH$.

For any curve C in $QTPH$ along which a monotone parameter u is assigned, the integral

$$G=\int\{y_r\,dx_r-\Omega(x,y)\,du\} \tag{90.5}$$

is meaningful. A general variation gives

$$\left.\begin{aligned}\delta G=&\,[y_r\,\delta x_r-\Omega\,\delta u]+\\&+\int(\delta y_r\,dx_r-\delta x_r\,dy_r-\delta\Omega\,du+d\Omega\,\delta u).\end{aligned}\right\} \tag{90.6}$$

Let us seek those curves C for which $\delta G=0$ when the variation is arbitrary except for the following conditions:

(i) The end values x_r^*, x_r are fixed.

(ii) The increment Δu in the parameter u along the curve is fixed.

We can put $\delta u=0$ in (90.6), and we get

$$\delta G=\int\left\{\delta x_r\left(-dy_r-\frac{\partial\Omega}{\partial x_r}du\right)+\delta y_r\left(dx_r-\frac{\partial\Omega}{\partial y_r}du\right)\right\}. \tag{90.7}$$

Therefore the required curves satisfy

$$\frac{dx_r}{du}=\frac{\partial\Omega}{\partial y_r},\qquad \frac{dy_r}{du}=-\frac{\partial\Omega}{\partial x_r}. \tag{90.8}$$

This tells us that these curves are rays or trajectories, and also that the parameter u on any one of them is the special parameter ($u=w$). Moreover, from the nature of the variational principle used here, it follows that the $2N+3$ quantities $(x^*, x, \Delta w)$, chosen arbitrarily, determine the value of an integral

$$G(x^*, x, \Delta w)=\int\{y_r\,dx_r-\Omega(x,y)\,dw\}, \tag{90.9}$$

calculated along a ray or trajectory (Fig. 42).

On giving arbitrary variations to the $2N+3$ quantities $(x^*, x, \Delta w)$, we get from (90.6)

$$\frac{\partial G}{\partial x_r}=y_r,\qquad \frac{\partial G}{\partial x_r^*}=-y_r^*, \tag{90.10}$$

and also

$$\frac{\partial G}{\partial\Delta w}=-\Omega(x,y)=-\Omega(x^*,y^*), \tag{90.11}$$

Ω being constant along the ray or trajectory by (90.8). We recognize in (90.10) the CT (88.20a), and so we conclude that *for any assigned value of Δw, the function $G(x^*, x, \Delta w)$, thus obtained by integrating along rays or trajectories, is the generating function of a finite CT which transforms the space $QTPH$ into itself.* We have, in fact, a one-parameter family of CT, with parameter Δw.

Assuming the integration of the canonical equations (90.8), the generating function is constructed in the following steps:

(i) Take any point B^* in $QTPH$ with coordinates (x^*, y^*).

(ii) Through B^* draw the ray or trajectory C satisfying (90.8) and proceed along it until the special parameter w is increased by an assigned amount Δw. Let $B(x, y)$ be the point so obtained. Then we have functional relationships

$$x_r = x_r(x^*, y^*, \Delta w), \qquad y_r = y_r(x^*, y^*, \Delta w). \tag{90.12}$$

(iii) Solve the first set of these equations, obtaining

$$y_r^* = y_r^*(x^*, x, \Delta w). \tag{90.13}$$

(iv) Calculate the integral (90.9) from B^* to B along C, (x, y) being functions of (x^*, y^*, w) of the form (90.12); thus G appears as a function of $(x^*, y^*, \Delta w)$.

(v) Substitute from (90.13) to obtain $G(x^*, x, \Delta w)$.

In (88.8) we established the invariance of the circulation of action under CT, a result capable of a dual interpretation according to the way in which we regard the CT. In (88.8) it was a question of changing the labels on fixed points. To get the other point of view, it is simplest to start all over again.

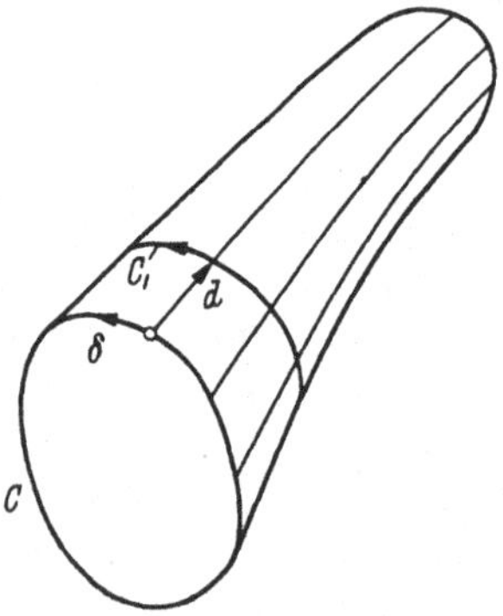

Fig. 43. A circuit C and a tube of the natural congruence in $QTPH$.

Fig. 43 shows a circuit C in $QTPH$ and a tube containing C, this tube consisting of rays or trajectories (part of the natural congruence). It is convenient to reserve d for a displacement along the natural congruence, so that

$$d x_r = d w \cdot \frac{\partial \Omega}{\partial y_r}, \qquad d y_r = - d w \cdot \frac{\partial \Omega}{\partial x_r}. \tag{90.14}$$

We shall use δ for a displacement along C, so that the circulation in C is

$$\varkappa(C) = \oint_C y_r \delta x_r. \tag{90.15}$$

If we displace C to C_1 along the natural congruence, we get

$$\varkappa(C_1) - \varkappa(C) = d\varkappa(C) = d \oint_C y_r \delta x_r = \oint_C (d y_r \delta x_r - d x_r \delta y_r), \tag{90.16}$$

or, by (90.14),

$$d\varkappa(C) = - \oint_C d w \, \delta\Omega, \tag{90.17}$$

in which the integration is with respect to $\delta\Omega$, dw being an infinitesimal scalar given along C.

If $dw = \text{const}$, we get

$$d\varkappa(C) = - d w \oint_C \delta\Omega = 0 \tag{90.18}$$

for a reducible circuit. Thus, *the circulation of action in a reducible circuit is unchanged by pushing that circuit along the natural congruence through a fixed infinitesimal increment in the special parameter, and, consequently, for a fixed finite increment also.*

As a second conclusion from (90.17), we note that if C is drawn on the energy surface $\Omega = 0$ (or, more generally, on $\Omega = \text{const}$), then $\delta\Omega = 0$ and hence $d\varkappa(C) = 0$. We are not compelled to make dw constant, and therefore *the circulation of action has a common value for all circuits (reducible or irreducible) drawn on the energy surface which can be deformed into one another by displacements along the natural congruence.*

In the (q, t, p, H) notation, the circulation in any circuit is

$$\varkappa(C) = \oint_C y_r\, \delta x_r = \oint_C (p_\varrho\, \delta q_\varrho - H\, \delta t). \tag{90.19}$$

For a circuit on the energy surface, H is assigned as a function of (q, t, p).

Since

$$\oint_C y_r\, \delta x_r = -\oint_C x_r\, \delta y_r, \tag{90.20}$$

the same results hold for a circulation defined as

$$\oint_C x_r\, \delta y_r = \oint_C (q_\varrho\, \delta p_\varrho - t\, \delta H). \tag{90.21}$$

The circulation of action is an example of a relative integral invariant. Integral invariants are defined as follows.

Let S_M be an M-dimensional subspace of $QTPH$, or possibly $QTPH$ itself, so that $1 \leq M \leq 2N+2$. Through S_M draw the natural congruence, and generate from S_M a set of subspaces $S_M(w)$ by measuring off along the rays or trajectories the same value of w from $w=0$ on S_M; thus $S_M = S_M(0)$. If an integral I taken over S_M is unchanged by this operation, i.e. if I for $S_M(w)$ is independent of w, then I is called an *integral invariant, absolute* if S_M is an open space and *relative* if it is closed (as for a circuit C). Integral invariants in the space QP are discussed in Sect. 98.

91. Transformation of the natural congruence to straight lines by solving the Hamilton-Jacobi equation. In Sect. 90 we recognized two different ways of looking at a transformation. Let us here take the view that we have a Euclidean space E_{2N+2} in which there are fixed axes, and a transformation $(x, y) \to (x', y')$ means the displacement of the points of this space to new positions. To avoid awkward questions about the topology of $QTPH$, we shall work in the small, confining our attention to a portion of $QTPH$ which has Euclidean topology.

From this point of view, the natural congruence of rays or trajectories, satisfying the canonical equations

$$\frac{dx_r}{dw} = \frac{\partial \Omega}{\partial y_r}, \qquad \frac{dy_r}{dw} = -\frac{\partial \Omega}{\partial x_r}, \tag{91.1}$$

appear as curves in E_{2N+2}, and the effect of a CT is to change those curves. We seek a CT $(x, y) \to (x', y')$ which transforms the natural congruence into a congruence of *parallel straight lines*.

Let $G(x, y')$ be any solution of the partial differential equation

$$\Omega\left(x, \frac{\partial G}{\partial x}\right) = y'_{N+1}, \tag{91.2}$$

considered in the domain of the $(2N+2)$ independent variables (x, y'), this solution being such that

$$\det \frac{\partial^2 G}{\partial x_r\, \partial y'_s} \neq 0. \tag{91.3}$$

Then, as in (88.20c), the equations

$$y_r = \frac{\partial G}{\partial x_r}, \qquad x'_r = \frac{\partial G}{\partial y'_r} \tag{91.4}$$

define a CT $(x, y) \to (x', y')$. For the energy function (always treated as an invariant in the sense of tensor calculus) we have

$$\Omega'(x', y') = \Omega(x, y) = \Omega\left(x, \frac{\partial G}{\partial x}\right) = y'_{N+1}. \tag{91.5}$$

Accordingly the new equations of the natural congruence read

$$\left.\begin{aligned} \frac{dx'_\varrho}{dw} &= \frac{\partial \Omega'}{\partial y'_\varrho} = 0, & \frac{dx'_{N+1}}{dw} &= \frac{\partial \Omega'}{\partial y'_{N+1}} = 1, \\ \frac{dy'_\varrho}{dw} &= -\frac{\partial \Omega'}{\partial x'_\varrho} = 0, & \frac{dy'_{N+1}}{dw} &= -\frac{\partial \Omega'}{\partial x'_{N+1}} = 0, \end{aligned}\right\} \tag{91.6}$$

and integration gives

$$\left.\begin{aligned} x'_\varrho &= a_\varrho, & x'_{N+1} &= w, \\ y'_\varrho &= b_\varrho, & y'_{N+1} &= k, \end{aligned}\right\} \tag{91.7}$$

where a_ϱ, b_ϱ, and k are $(2N+1)$ constants, the values of which depend on the particular ray or trajectory under consideration. We note that the special parameter w is equal to the coordinate x'_{N+1}, a trivial constant of integration having been dropped. Since, for arbitrary values of the constants, (91.7) represents a congruence of straight lines all parallel to the axis of x'_{N+1}, we have succeeded in transforming the natural congruence into a congruence of parallel straight lines. The energy surface $\Omega = 0$ has been transformed into the plane $y'_{N+1} = 0$.

By working in *QTPH* instead of in *QT*, we are able to present the theory of Sect. 77 in a more general way. The partial differential equation (91.2) is in fact the HAMILTON-JACOBI equation in a general form. To establish the connection, let us pass over into (q, t, p, H) notation, taking the energy function in the form

$$\Omega(x, y) = y_{N+1} + \omega(x_1, \ldots x_{N+1}, y_1, \ldots y_N). \tag{91.8}$$

Then by (86.1) the partial differential equation (91.2) becomes

$$\frac{\partial G}{\partial t} + H\left(q, t, \frac{\partial G}{\partial q}\right) = -H', \tag{91.9}$$

if we write H instead of ω as a functional symbol; we need a solution $G(q, t, p', H')$ such that

$$\begin{vmatrix} \dfrac{\partial^2 G}{\partial q_\varrho \partial p'_\sigma} & \dfrac{\partial^2 G}{\partial q_\varrho \partial H'} \\ \dfrac{\partial^2 G}{\partial t\, \partial p'_\sigma} & \dfrac{\partial^2 G}{\partial t\, \partial H'} \end{vmatrix} \neq 0. \tag{91.10}$$

We may regard (91.9) as a partial differential equation in the independent variables (q, t), the quantities (p', H') being constants. The first step towards integration is to put

$$G = -H't + U(q, t); \tag{91.11}$$

then U has to satisfy

$$\frac{\partial U}{\partial t} + H\left(q, t, \frac{\partial U}{\partial q}\right) = 0, \tag{91.12}$$

which is in fact the HAMILTON-JACOBI equation (77.3). We need a complete integral in order to carry out the transformation (91.4).

92. The order of the canonical equations, and its reduction by means of a first integral. We return to the question of the order of the equations of motion, raised near the end of Sect. 68.

Given an energy function $\Omega(x, y)$, the canonical equations

$$\frac{dx_r}{dw} = \frac{\partial \Omega}{\partial y_r}, \qquad \frac{dy_r}{dw} = -\frac{\partial \Omega}{\partial x_r} \tag{92.1}$$

form a system of order $2N+2$. If we multiply across by dw/dx_{N+1}, we get

$$\frac{dx_\varrho}{dx_{N+1}} = \frac{\partial\Omega/\partial y_\varrho}{\partial\Omega/\partial y_{N+1}}, \qquad \frac{dy_\varrho}{dx_{N+1}} = -\frac{\partial\Omega/\partial x_\varrho}{\partial\Omega/\partial y_{N+1}}, \tag{92.2}$$

and

$$\frac{dy_{N+1}}{dx_{N+1}} = -\frac{\partial\Omega/\partial x_{N+1}}{\partial\Omega/\partial y_{N+1}}. \tag{92.3}$$

This is a system of order $2N+1$; the independent variable x_{N+1} is contained in Ω.

We know that

$$\Omega(x, y) = c, \tag{92.4}$$

a constant along each trajectory. Let this equation be solved for y_{N+1}, so that we have

$$y_{N+1} = -\omega(x_1, \ldots x_{N+1}, y_1, \ldots y_N, c). \tag{92.5}$$

Substituting this in (92.2), we have a system of order $2N$, containing the constant c. If these equations are solved for $x_1, \ldots x_N, y_1, \ldots y_N$, then y_{N+1} is given by (92.5). Thus the canonical equations (92.1) are reducible to order $2N$; but the reduced Eqs. (92.2) are not in canonical form.

Suppose now that, instead of being given an energy function (which leads to a natural congruence filling $QTPH$), we are given an energy surface with equation

$$\Omega(x, y) = 0. \tag{92.6}$$

The trajectories are now confined to this surface. We have still the Eqs. (92.1) to (92.3) and we also have (92.4) and (92.5) with $c=0$. But now we are concerned with a *surface*, and the equation for that surface may be put into different forms. In fact, the functional form of Ω is not prescribed, and we are entitled to change it, so that the equation of the energy surface reads

$$\Omega(x, y) = y_{N+1} + \omega(x_1, \ldots x_{N+1}, y_1, \ldots y_N) = 0. \tag{92.7}$$

We have then

$$\frac{\partial\Omega}{\partial y_{N+1}} = 1, \tag{92.8}$$

and the Eqs. (92.2) become

$$\frac{dx_\varrho}{dx_{N+1}} = \frac{\partial\Omega}{\partial y_\varrho}, \qquad \frac{dy_\varrho}{dx_{N+1}} = -\frac{\partial\Omega}{\partial x_\varrho}. \tag{92.9}$$

This system is of order $2N$. *When we base dynamics on an energy surface in $QTPH$, the equations of motion can be reduced to order $2N$, with preservation of the canonical form, if we (i) write the equation of the energy surface in the form (92.7), and (ii) take x_{N+1} as parameter.* Note that the parameter x_{N+1} is now contained in Ω, whereas w was not contained in Ω in (92.1).

The standard translation from (x, y) to (q, t, p, H) is given as in (86.1) by

$$\left.\begin{aligned} x_\varrho &= q_\varrho, \quad & x_{N+1} &= t, \\ y_\varrho &= p_\varrho, \quad & y_{N+1} &= -H. \end{aligned}\right\} \tag{92.10}$$

But in view of the symmetry in the (x, y) notation, there is no necessity to stick to this translation; we are at liberty to permute the suffixes on x_r (making the same permutation for y_r). Thus, in (92.7), y_{N+1} need not necessarily stand for $-H$; it might stand for p_1, in which case the parameter in (92.9) would not be t, but q_1. This versatility of the (x, y) notation should never be forgotten.

In what follows we shall assume that an energy function is given. Consider a system which possesses a first integral $F(x, y)$, by which we mean that

$$\frac{dF}{dw} = [F, \Omega] = 0 \tag{92.11}$$

[cf. (89.15)], so that

$$F(x, y) = \text{const} \tag{92.12}$$

along each trajectory. We shall discuss the reduction of the order of the canonical equations (92.1) from $2N+2$ to $2N$ by means of this first integral[1], with preservation of the canonical form and of the special parameter w.

Let $G(x, y')$ be a solution of the partial differential equation

$$F\left(x, \frac{\partial G}{\partial x}\right) = y'_{N+1}, \tag{92.13}$$

satisfying

$$\det \frac{\partial^2 G}{\partial x_r \partial y'_s} \neq 0. \tag{92.14}$$

This procedure is, in fact similar to that of Sect. 91, the HAMILTON-JACOBI equation (91.2) being replaced by (92.13), and one might ask why we should spend our time on (92.13) when a solution of (91.2) solves the problem of motion. The answer is that the practical feasibility of getting an explicit solution depends very much on the complexity of the function involved (Ω and F respectively). It may well happen that F is much simpler than Ω.

Having solved (92.13), we apply the CT

$$y_r = \frac{\partial G}{\partial x_r}, \qquad x'_r = \frac{\partial G}{\partial y'_r}, \tag{92.15}$$

and the equations of motion transform to

$$\frac{dx'_r}{dw} = \frac{\partial \Omega'}{\partial y'_r}, \qquad \frac{dy'_r}{dw} = -\frac{\partial \Omega'}{\partial x'_r}, \tag{92.16}$$

the new energy function being

$$\Omega'(x', y') = \Omega(x, y). \tag{92.17}$$

We have then

$$\frac{\partial \Omega'}{\partial x'_{N+1}} = -\frac{dy'_{N+1}}{dw} = -\frac{d}{dw} F\left(x, \frac{\partial G}{\partial x}\right) = -\frac{d}{dw} F(x, y) = 0, \tag{92.18}$$

so that the variable x'_{N+1} is absent from Ω':

$$\Omega' = \Omega'(x'_1, \dots x'_N, y'_1, \dots y'_N, y'_{N+1}). \tag{92.19}$$

If, now, we select from (92.16) the $2N$ equations

$$\frac{dx'_\varrho}{dw} = \frac{\partial \Omega'}{\partial y'_\varrho}, \qquad \frac{dy'_\varrho}{dw} = -\frac{\partial \Omega'}{\partial x'_\varrho}, \tag{92.20}$$

we have a set of canonical equations of order $2N$, as required; the new energy function contains y'_{N+1} as a constant.

If we confine our attention to trajectories on the energy surface $\Omega = 0$, we can effect a further reduction of 2 in the order. In the new coordinates, the energy surface has the equation

$$\Omega'(x'_1, \dots x'_N, y'_1, \dots y'_{N+1}) = 0; \tag{92.21}$$

[1] The argument given here, being set in *QTPH*, has an appearance of greater generality than other treatments; cf. NORDHEIM and FUES [*19*], p. 115. See PRANGE [*21*], pp. 713–726, for a discussion, with considerable detail, of the simplification of a canonical system by knowledge of a first integral.

we solve this equation for one of the y's, say y'_N, and proceed as at (92.7), obtaining $2N-2$ canonical equations analogous to (92.9). Thus, using a first integral and an energy surface, the order is reducible to $2N-2$.

Some illustrative examples follow.

α) Reduction of order by ignoration of a coordinate or by the integral of energy in a conservative system. As will be seen, the work here is actually trivial, but it will serve to explain the method. Suppose that one of the coordinates, say x_{N+1}, is absent from $\Omega(x, y)$. In view of what has been said above about the symmetry of the notation, this may mean either that (i) the system has an ignorable coordinate (Sect. 46), or (ii) that the system is conservative [t is absent from $H(q, t, p)$]. Both cases are covered by the following argument.

Our assumption is

$$\frac{\partial \Omega}{\partial x_{N+1}} = 0, \tag{92.22}$$

and consequently we have the first integral

$$F(x, y) = y_{N+1} = \text{const}. \tag{92.23}$$

The partial differential equations (92.13) has the very simple form

$$\frac{\partial G}{\partial x_{N+1}} = y'_{N+1}, \tag{92.24}$$

and a suitable solution is

$$G(x, y') = x_r y'_r. \tag{92.25}$$

By (92.15) this gives the identical transformation

$$y_r = y'_r, \qquad x'_r = x_r, \tag{92.26}$$

and the problem of reduction in order to $2N$ is solved by merely picking out from the canonical equations the $2N$ equations

$$\frac{dx_\varrho}{dw} = \frac{\partial \Omega}{\partial y_\varrho}, \qquad \frac{dy_\varrho}{dw} = -\frac{\partial \Omega}{\partial x_\varrho}. \tag{92.27}$$

To complete the reduction to order $2N-2$ on the energy surface, we write the equation of that surface in the form

$$\Omega(x, y) = y_N + \omega(x_1, \ldots x_N, y_1, \ldots y_{N-1}, y_{N+1}) = 0. \tag{92.28}$$

Then (92.27) give, as in (92.9),

$$\left.\begin{aligned} \frac{dx_1}{dx_N} &= \frac{\partial \omega}{\partial y_1}, & \ldots \quad \frac{dx_{N-1}}{dx_N} &= \frac{\partial \omega}{\partial y_{N-1}}, \\ \frac{dy_1}{dx_N} &= -\frac{\partial \omega}{\partial x_1}, & \ldots \quad \frac{dy_{N-1}}{dx_N} &= -\frac{\partial \omega}{\partial x_{N-1}}. \end{aligned}\right\} \tag{92.29}$$

Let us translate this into (q, t, p, H) notation, using (92.10). We start with an energy function

$$\Omega(q, p, -H), \tag{92.30}$$

t being absent by hypothesis. We have the first integral

$$H = E, \tag{92.31}$$

and, as in (92.27), the equations of motion

$$\frac{dq_\varrho}{\partial w} = \frac{\partial \Omega}{\partial p_\varrho}, \qquad \frac{dp_\varrho}{dw} = -\frac{\partial \Omega}{\partial q_\varrho}. \tag{92.32}$$

We now take an energy surface $\Omega=0$, and write its equation in the new form

$$\Omega = p_N + \omega(q_1, \dots q_N, p_1, \dots p_{N-1}, -E) = 0, \tag{92.33}$$

the constant E having been substituted for H. Then, as in (92.29), we have the equations of motion

$$\left.\begin{aligned} \frac{dq_1}{dq_N} &= \frac{\partial\omega}{\partial p_1}, & \dots \quad \frac{dq_{N-1}}{dq_N} &= \frac{\partial\omega}{\partial p_{N-1}}, \\ \frac{dp_1}{dq_N} &= -\frac{\partial\omega}{\partial q_1}, & \dots \quad \frac{dp_{N-1}}{dq_N} &= -\frac{\partial\omega}{\partial q_{N-1}}. \end{aligned}\right\} \tag{92.34}$$

Here ω contains the independent variable q_N. *Thus, in a conservative system, we reduce*[1] *the order of the canonical equations to* $2N-2$ *by means of the integral of energy* $H=E$.

β) Reduction of order by means of an integral linear in the momenta. Suppose that a system has a first integral

$$y_1 + y_2 + y_3 = \text{const}. \tag{92.35}$$

It is convenient to modify the above plan, and use a generating function $G(x', y)$ and the CT

$$x_r = \frac{\partial G}{\partial y_r}, \quad y'_r = \frac{\partial G}{\partial x'_r}. \tag{92.36}$$

We seek G to satisfy

$$y_1 + y_2 + y_3 = y'_1 = \frac{\partial G}{\partial x'_1}. \tag{92.37}$$

A suitable solution is

$$G = x'_1(y_1 + y_2 + y_3) + x'_2 y_2 + \dots + x'_{N+1} y_{N+1}; \tag{92.38}$$

it satisfies

$$\det \frac{\partial^2 G}{\partial x'_r \partial y_s} \neq 0, \tag{92.39}$$

so that (92.36) gives a CT $(x, y) \to (x', y')$. Since y'_1 is a constant of the motion, $\Omega'(x', y')$ lacks x'_1, and the new equations of motion read

$$\left.\begin{aligned} \frac{dx'_2}{dw} &= \frac{\partial\Omega'}{\partial y'_2}, & \dots \quad \frac{dx'_{N+1}}{dw} &= \frac{\partial\Omega'}{\partial y'_{N+1}}, \\ \frac{dy'_2}{dw} &= -\frac{\partial\Omega'}{\partial x'_2}, & \dots \quad \frac{dy'_{N+1}}{dw} &= -\frac{\partial\Omega'}{\partial x'_{N+1}}, \end{aligned}\right\} \tag{92.40}$$

a system of order $2N$.

Integrals of the above type occur in the 3-body problem (Sect. 53); in that case we have three integrals of linear momentum,

$$\left.\begin{aligned} y_1 + y_4 + y_7 &= y'_1, \\ y_2 + y_5 + y_8 &= y'_2, \\ y_3 + y_6 + y_9 &= y'_3, \end{aligned}\right\} \tag{92.41}$$

the right hand sides being constants of the motion. The system has 9 degrees of freedom ($N=9$). By taking the generating function

$$\left.\begin{aligned} G(x', y) = {} & x'_1(y_1 + y_4 + y_7) + x'_2(y_2 + y_5 + y_8) + \\ & + x'_3(y_3 + y_6 + y_9) + x'_4 y_4 + \dots + x'_{10} y_{10}, \end{aligned}\right\} \tag{92.42}$$

[1] Cf. WHITTAKER [28], p. 313, for a different description of this reduction.

we eliminate x_1', x_2', x_3' from the transformed Ω, in which y_1', y_2', y_3' appear as constants. The canonical equations of the form (92.1) are thereby reduced from order 20 to $20-6=14$. But if we work in the space QP (Sect. 96), using an energy function $H(q, p)$ instead of $\Omega(x, y)$, the reduction in order is from 18 to 12[1].

γ) Reduction of order by means of an integral of angular momentum. Suppose that

$$F(x, y) = x_1 y_2 - x_2 y_1 = y_1', \tag{92.43}$$

a constant of the motion. According to (92.13) we are to solve the partial differential equation

$$x_1 \frac{\partial G}{\partial x_2} - x_2 \frac{\partial G}{\partial x_1} = y_1'. \tag{92.44}$$

A suitable solution is

$$G(x, y') = y_1'\left[y_2'(x_1^2 + x_2^2) + \arctan\frac{x_2}{x_1}\right] + x_3 y_3' + \cdots + x_{N+1} y_{N+1}'; \tag{92.45}$$

it satisfies the determinantal condition (92.14), and gives the CT

$$\left.\begin{aligned}
y_1 &= \frac{\partial G}{\partial x_1} = 2 y_1' y_2' x_1 - \frac{y_1' x_2}{x_1^2 + x_2^2}, \\
y_2 &= \frac{\partial G}{\partial x_2} = 2 y_1' y_2' x_2 + \frac{y_1' x_1}{x_1^2 + x_2^2}, \\
y_3 &= \frac{\partial G}{\partial x_3} = y_3', \\
&\cdots\cdots\cdots\cdots \\
x_1' &= \frac{\partial G}{\partial y_1'} = y_2'(x_1^2 + x_2^2) + \arctan\frac{x_2}{x_1}, \\
x_2' &= \frac{\partial G}{\partial y_2'} = y_1'(x_1^2 + x_2^2), \\
x_3' &= \frac{\partial G}{\partial y_3'} = x_3, \\
&\cdots\cdots\cdots\cdots
\end{aligned}\right\} \tag{92.46}$$

The variable x_1' is absent from the new energy function. The generating function (92.45) is reached by the use of polar coordinates, $x_1 = r\cos\vartheta$, $x_2 = r\sin\vartheta$.

VI. The space of states (QTP).

93. Circulation theorem. In the $(2N+1)$-dimensional space of states QTP the coordinates[2] of the representative point are q_ϱ, t, p_ϱ. The Hamiltonian H is not a coordinate, but a function of position in QTP:

$$H = H(q, t, p). \tag{93.1}$$

The canonical equations of motion are

$$\dot{q}_\varrho = \frac{\partial H}{\partial p_\varrho}, \quad \dot{p}_\varrho = -\frac{\partial H}{\partial q_\varrho}. \tag{93.2}$$

---

[1] For the reduction of the order of the equations of motion in the 3-body problem from 18 to 6, see WHITTAKER [28], pp. 340–351.

[2] We shall work in the small, and avoid reference to overlapping coordinate systems (cf. Sect. 63). For notation, see Sect. 62.

Written in the form

$$\frac{dq_1}{\frac{\partial H}{\partial p_1}} = \cdots = \frac{dq_N}{\frac{\partial H}{\partial p_N}} = \frac{dp_1}{-\frac{\partial H}{\partial q_1}} = \cdots = \frac{dp_N}{-\frac{\partial H}{\partial q_N}} = \frac{dt}{1}, \tag{93.3}$$

in order to put all the coordinates on a parity, these equations exhibit the natural congruence of trajectories in QTP, one curve passing through each point. We note that (93.2) imply

$$\frac{dH}{dt} = \frac{\partial H}{\partial t}. \tag{93.4}$$

To discuss circulation in QTP independently of the discussion in Sect. 90 for $QTPH$, we define the circulation in any circuit C to be

$$\varkappa(C) = \oint_C (p_\varrho \delta q_\varrho - H \delta t). \tag{93.5}$$

Giving to C any infinitesimal displacement d (not necessarily along the natural congruence), and integrating the varied expression by parts, we get

$$\left.\begin{aligned} d\varkappa(C) &= \oint_C (dp_\varrho \delta q_\varrho - dq_\varrho \delta p_\varrho - dH\delta t + dt\,\delta H) \\ &= \oint_C \Big[\delta q_\varrho\Big(dp_\varrho + \frac{\partial H}{\partial q_\varrho} dt\Big) + \delta p_\varrho\Big(-dq_\varrho + \frac{\partial H}{\partial p_\varrho} dt\Big) + \\ &\qquad + \delta t\Big(-dH + \frac{\partial H}{\partial t} dt\Big)\Big]. \end{aligned}\right\} \tag{93.6}$$

If, now, the displacement d is along the natural congruence, it follows from (93.3) and (93.4) that

$$d\varkappa(C) = 0. \tag{93.7}$$

On the other hand, if $d\varkappa(C) = 0$ for arbitrary C and for a displacement d taken along some congruence, then that congruence must satisfy (93.3) and (93.4), and is in consequence the natural congruence. In fact the circulation condition (93.7) is equivalent to the canonical equations (93.2).

94. Transformation of coordinates in QTP. The Pfaffian form. Let us make a general transformation

$$(q, t, p) \to (x), \tag{94.1}$$

where (x) stands for a set of $2N+1$ variables. We shall denote these variables by x_A, giving capital suffixes the range $1, 2, \ldots 2N+1$, with the summation convention in the case of a repeated suffix. We are in fact assigning completely arbitrary coordinates to the points of QTP.

Under the transformation (94.1) we get

$$p_\varrho \delta q_\varrho - H(q, t, p)\,\delta t = X_A \delta x_A, \tag{94.2}$$

where the X's are functions of the x's. The circulation in a circuit C is now

$$\varkappa(C) = \oint_C X_A \delta x_A, \tag{94.3}$$

and if we give C any displacement d, we get

$$d\varkappa(C) = \oint_C (dX_A \delta x_A - dx_A \delta X_A); \tag{94.4}$$

writing $\partial X_A/\partial x_B = X_{A,B}$, we change this to

$$\left.\begin{aligned} d\varkappa(C) &= \oint_C X_{A,B}(dx_B\,\delta x_A - dx_A\,\delta x_B) \\ &= \oint_C (X_{A,B} - X_{B,A})\,dx_B\,\delta x_A. \end{aligned}\right\} \tag{94.5}$$

If dx_A is along the natural congruence, then $d\varkappa(C) = 0$ for arbitrary C (cf. Sect. 93); therefore the trajectories must satisfy the equations

$$(X_{A,B} - X_{B,A})\,dx_B = 0. \tag{94.6}$$

This is the form taken by the canonical equations when we use a general coordinate system in QTP.

This important result can be seen otherwise; dx_A are the components of a contravariant vector with respect to arbitrary transformations (we might have written dx^A to emphasise that fact), and $X_{A,B} - X_{B,A}$ are the components of a skew-symmetric tensor; therefore (94.6) are vector equations, true for all choices of coordinates if true for one[1]. But if the coordinates x_A are chosen as follows:

$$\left.\begin{aligned} x_1 &= q_1, \dots\ x_N = q_N, \\ x_{N+1} &= p_1, \dots\ x_{2N} = p_N, \\ x_{2N+1} &= t, \end{aligned}\right\} \tag{94.7}$$

then, referring to the identity (94.2), we see that

$$\left.\begin{aligned} X_1 &= p_1, \dots\ X_N = p_N, \\ X_{N+1} &= 0, \dots\ X_{2N} = 0, \\ X_{2N+1} &= -H(q, t, p). \end{aligned}\right\} \tag{94.8}$$

On substituting these expressions in (94.6), we obtain the canonical equations (93.2), and therefore (94.6) represent the trajectories in all coordinate systems, since they represent them in the coordinate system (94.7).

In (94.6) we have $2N+1$ homogeneous equations in the $2N+1$ differentials; we know (quite apart from the preceding argument) that they are consistent because the matrix of the coefficients, being skew-symmetric and of odd order, is of rank $2N$ at most (in other words, its determinant vanishes). We have then a new approach to dynamics in QTP, based on a Pfaffian form[2]

$$X_A\,\delta x_A, \tag{94.9}$$

---

[1] It is interesting to compare (94.6) with (73.5), in which the total dimensionality is even, but the determinant of the coefficients vanishes by (73.1).

[2] For the theory of Pfaffian forms, see G. DARBOUX: Le Problème de PFAFF (Paris: Gauthier-Villars 1882); E. CARTAN: Leçons sur les invariants intégraux (Paris: Hermann 1922); E. GOURSAT: Leçons sur le problème de PFAFF (Paris: Hermann 1922); and for a very general treatment, J. A. SCHOUTEN and W. v. D. KULK: PFAFF's Problem and its Generalizations (Oxford: Clarendon Press 1949). See also WHITTAKER [28], p. 307 et seq. CARTAN's exterior multiplication and exterior differentiation (op. cit., pp. 52, 65) provide a highly condensed notation, but, as with all condensed notations, one needs much practice to use it with confidence; for the application of this notation to dynamics, see CARTAN (op. cit.); P. DEDECKER: Sur le théorème de la circulation de V. BJERKNES et la théorie des invariants intégraux, Inst. Roy. Météorol. Belg. Misc. no. 36, 1951; F. GALLISOT: Les formes extérieures en mécanique (Chartres: Durand 1954 — Thèse, Paris, published separately and in Tome 4, Ann. Inst. Fourier). See also E. CARTAN: Les systèmes différentiels extérieurs et leurs applications géométriques (Paris: Hermann 1945); TH. DE DONDER: Théorie des invariants intégraux (Paris: Gauthier-Villars 1927); W. ŚLEBODZIŃSKI: Formes extérieures et leurs applications (Warszawa: Państwowe Wydawnicto Naukowe 1954).

the X's being functions of the x's. Any such Pfaffian defines trajectories by the Eq. (94.6), and this is an invariant method, independent of any particular choice of the coordinates x_A.

If we have two Pfaffian forms,

$$X_A \delta x_A \text{ and } \quad X_A \delta x_A + \frac{\partial G(x)}{\partial x_A} \delta x_A, \tag{94.10}$$

where G is a function of position in QTP, then (94.6) gives the same trajectories for the two forms. In other words, *the trajectories are not changed if an exact differential is added to the Pfaffian form.*

Let (q, t, p) be canonical coordinates in QTP, so that the trajectories satisfy the equations

$$\frac{dq_\varrho}{dt} = \frac{\partial H}{\partial p_\varrho}, \quad \frac{dp_\varrho}{dt} = -\frac{\partial H}{\partial q_\varrho}. \tag{94.11}$$

The associated Pfaffian form is

$$p_\varrho \delta q_\varrho - H(q, t, p)\, \delta t. \tag{94.12}$$

Apply a transformation $(q, t, p) \to (q', t', p')$. In the transformed Pfaffian there will in general appear terms in $\delta q'_\varrho$, $\delta p'_\varrho$, $\delta t'$, and the coefficient of $\delta q'_\varrho$ will not in general be p'_ϱ. But suppose the transformation to be such that we can split off an exact differential from the transformed Pfaffian, leaving a Pfaffian in which there is no differential $\delta p'_\varrho$ and in which the coefficient of $\delta q'_\varrho$ is p'_ϱ. The coefficient of $\delta t'$ will be a function of (q', t', p'); write $-H'(q', t', p')$ for it. Then we have

$$p_\varrho \delta q_\varrho - H(q, t, p)\, \delta t = p'_\varrho \delta q'_\varrho - H'(q', t', p')\, \delta t' + \delta G, \tag{94.13}$$

δG being an exact differential, and the canonical Eqs. (94.11) transform into[1]

$$\frac{dq'_\varrho}{dt'} = \frac{\partial H'}{\partial p'_\varrho}, \quad \frac{dp'_\varrho}{dt'} = -\frac{\partial H'}{\partial q'_\varrho}. \tag{94.14}$$

95. Canonical transformations in *QTP*. If we want to make coordinate transformations in QTP which conserve the canonical form of the equations of motion, the best plan is to work with the Pfaffian form (94.12), canonical transformations as treated in Sect. 87 being suited to a space with an even number of dimensions. Phase space QP is of even dimensionality, and we shall return to canonical coordinate transformations in discussing QP in Part. E VII. However, before leaving QTP, we shall now see how the motion of the system provides us with a canonical transformation between two surfaces on each of which t is constant; these surfaces are suited to canonical transformations, being of even dimensionality.

Let us go back to the two-point characteristic function $S(x^*, x)$ of Sect. 72, which we now write

$$S = S(q^*, t^*, q, t). \tag{95.1}$$

Consider two surfaces $\boldsymbol{\Sigma}^*$, $\boldsymbol{\Sigma}$ in QTP, with $t = \text{const}$ on each of them. The natural congruence establishes a correspondence $E^* \leftrightarrow E$ between the points of these two surfaces, E^* and E being the points where the surfaces are cut by a single trajectory, as shown in Fig. 44. This must not be confused with Fig. 35 of Sect. 75, which shows, in the $(N+1)$-dimensional space QT, a set of trajectories forming a coherent system (an ∞^N set of curves); on the other hand, Fig. 44 shows representative curves taken from the congruence of *all* trajectories (an ∞^{2N} set of curves).

[1] For the connection of this with canonical transformations in the space QP, see Sect. 97.

By (72.12) we have

$$\delta S = p_\varrho \delta q_\varrho - H \delta t - p_\varrho^* \delta q_\varrho^* + H^* \delta t^*, \tag{95.2}$$

for arbitrary variations of E^* and E. The transformation $E^* \to E$ is therefore given by

$$p_\varrho = \frac{\partial S}{\partial q_\varrho}, \quad p_\varrho^* = -\frac{\partial S}{\partial q_\varrho^*}, \tag{95.3}$$

t^* and t being assigned the constant values which they have on Σ^* and Σ respectively. The transformation of H is given by

$$H = -\frac{\partial S}{\partial t}, \quad H^* = \frac{\partial S}{\partial t^*}. \tag{95.4}$$

We recognize (95.3) as a canonical transformation

$$(q^*, p^*) \to (q, p), \tag{95.5}$$

the quantities t^* and t entering as parameters in the generating function S. If we hold t^* fixed and let t vary, we have a continuous set of canonical transformations which transform the initial isochronous surface ($t = t^*$) into all subsequent isochronous surfaces.

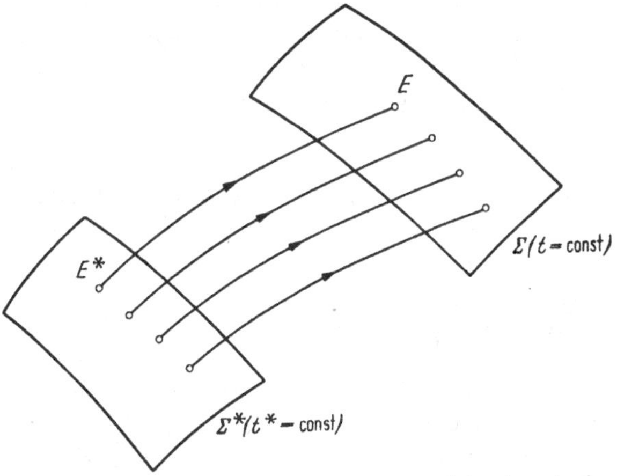

Fig. 44. Canonical transformation $E^* \to E$ produced by the natural congruence of trajectories in QTP.

We might also use the momentum-energy characteristic function $W(y^*, y)$ of Sect. 79 to generate a canonical transformation. Expressing it as

$$W = W(p^*, H^*, p, H), \tag{95.6}$$

the transformation is [cf. (79.14)]

$$q_\varrho = \frac{\partial W}{\partial p_\varrho}, \quad q_\varrho^* = -\frac{\partial W}{\partial p_\varrho^*}. \tag{95.7}$$

The surfaces which are thus canonically transformed into one another are surfaces on which H is constant. This transformation breaks down if $\partial H/\partial t = 0$, for then H is constant on each ray or trajectory, and so they cannot transform one surface $H = \text{const}$ into another.

VII. Phase space (QP).

96. Basic theory in QP for conservative systems. In Parts E II to VI a number of different representative spaces have been used in order to throw light on the mathematical structure of Hamiltonian dynamics; in spite of the variety of representations, they are all concerned with a single type of theory based on an assumed energy equation $\Omega(x, y) = 0$ or on a Hamiltonian $H(q, t, p)$. Of these spaces, QT, $QTPH$ and QTP are best suited for discussion of theory of the most general type; PH is of somewhat special interest in connection with encounters; Q is useful for conservative Hamiltonian systems [for which $H = H(q, p)$] and also for non-Hamiltonian dynamics.

We now come to the last of the representative spaces, the $2N$-dimensional phase space QP, in which the coordinates[1] of a point are q_ϱ, p_ϱ. Of all the representative spaces, QP is probably the most familiar to physicists, on account of its use in statistical mechanics[2].

In studying Hamiltonian dynamics in QP, the time t is to be regarded as a parameter and the Hamiltonian as a given function $H(q, t, p)$ of the variables (q, p) and of this parameter. The canonical equations

$$\dot{q}_\varrho = \frac{\partial H}{\partial p_\varrho}, \quad \dot{p}_\varrho = -\frac{\partial H}{\partial q_\varrho}, \tag{96.1}$$

describe the motion of the representative point in QP. The general case ($\partial H/\partial t \neq 0$) and the conservative case ($\partial H/\partial t = 0$) must be distinguished from one another. In the conservative case we have

$$H = H(q, p), \tag{96.2}$$

and we see QP filled with a natural congruence with equations

$$\frac{dq_1}{\partial H/\partial p_1} = \cdots = \frac{dq_N}{\partial H/\partial p_N} = \frac{dp_1}{-\partial H/\partial q_1} = \cdots = \frac{dp_N}{-\partial H/\partial q_N}, \tag{96.3}$$

each curve of the congruence (i.e. each trajectory) lying on a $(2N-1)$-dimensional surface with equation

$$H(q, p) = E, \tag{96.4}$$

where E is a constant. The corresponding picture for the case where $H = H(q, t, p)$ is much more complicated, because the direction of a trajectory at a point of QP depends on the value of t, so that there are ∞^1 possible directions at a point in QP, depending on the value of t.

For the rest of the present Section, only conservative systems are considered, with H as in (96.2).

Comparing (86.6) with (96.1), we see that *conservative dynamics in QP with Hamiltonian $H(q, p)$ is mathematically identical with dynamics in $QTPH$ with energy function $\Omega(x, y)$.* The only difference is one of notation.

This isomorphism is interesting because it links together two extremes of Hamiltonian dynamics. On the one hand, dynamics in $QTPH$ is as general as anything we could wish at present, both the time t and the Hamiltonian H being put on mathematical parity with (q, p), so that the theory is well suited for relativistic use. On the other hand, conservative dynamics in QP covers those problems with which we are most familiar in Newtonian dynamics, arising out of the motion of systems of particles and rigid bodies.

But although dynamics in $QTPH$ and conservative dynamics in QP are mathematically isomorphic, their physical interpretations are completely different, and it is advisable to translate briefly the theory developed for $QTPH$ into the form appropriate to conservative dynamics in QP.

---

[1] In the large, overlapping coordinate systems may be required (cf. Sect. 63), with canonical transformations in the overlaps.

[2] Cf. J. W. GIBBS: Elementary Principles in Statistical Mechanics (Yale: Scribner 1902); The Collected Works of J. WILLARD GIBBS, Vol. II (New York-London-Toronto: Longmans Green 1928). GIBBS (op. cit. pp. 6, 10) introduced the geometrical idea of phase space, possessing an invariant extension-in-phase given by $\int dp_1 \dots dp_N \, dq_1 \dots dq_N$. For a full and illuminating discussion of the work of GIBBS, see the articles by A. HAAS and P. S. EPSTEIN in A Commentary on the Scientific Writings of J. WILLARD GIBBS, Vol. II (New Haven: Yale University Press 1936), edited by A. HAAS. The space QP is also called μ-space (μ-Raum) or Γ-space (Γ-Raum) according to the way in which it is used in statistical theory; cf. A. MÜNSTER: Statistische Thermodynamik, pp. 27, 99 (Berlin-Göttingen-Heidelberg: Springer 1956).

The following table shows the correspondences[1]:

	Space of states and energy $QTPH$ of $2N+2$ dimensions.	Phase space QP of $2N$ dimensions.
Coordinates of point	x_r, y_r	q_ϱ, p_ϱ
Energy function	$\Omega(x, y)$	Hamiltonian $H(q, p)$
Canonical equations	$\frac{dx_r}{dw} = \frac{\partial \Omega}{\partial y_r}, \quad \frac{dy_r}{dw} = -\frac{\partial \Omega}{\partial x_r}$	$\frac{dq_\varrho}{dt} = \frac{\partial H}{\partial p_\varrho}, \quad \frac{dp_\varrho}{dt} = -\frac{\partial H}{\partial q_\varrho}$
Special parameter on trajectories	w	t
Circulation	$\oint_C y_r \, \delta x_r$	$\oint_C p_\varrho \, \delta q_\varrho$
Invariant bilinear form	$\delta_1 x_r \delta_2 y_r - \delta_2 x_r d_1 y_r$	$\delta_1 q_\varrho \delta_2 p_\varrho - \delta_2 q_\varrho \delta_1 p_\varrho$
POISSON brackets	$[u, v] = \frac{\partial u}{\partial x_r} \frac{\partial v}{\partial y_r} - \frac{\partial v}{\partial x_r} \frac{\partial u}{\partial y_r}$	$[u, v] = \frac{\partial u}{\partial q_\varrho} \frac{\partial v}{\partial p_\varrho} - \frac{\partial v}{\partial q_\varrho} \frac{\partial u}{\partial p_\varrho}$
LAGRANGE brackets	$\{u, v\} = \frac{\partial x_r}{\partial u} \frac{\partial y_r}{\partial v} - \frac{\partial x_r}{\partial v} \frac{\partial y_r}{\partial u}$	$\{u, v\} = \frac{\partial q_\varrho}{\partial u} \frac{\partial p_\varrho}{\partial v} - \frac{\partial q_\varrho}{\partial v} \frac{\partial p_\varrho}{\partial u}$

If we give a circuit C in QP a displacement (dq, dp) along the natural congruence, then, as in (90.17), the change in circulation is

$$d\varkappa(C) = d \oint_C p_\varrho \, \delta q_\varrho = -\oint_C dt \, \delta H. \tag{96.5}$$

Therefore the circulation is unchanged by the displacement if either

(i) the circuit C moves with the system, so that $dt = \text{const}$, or

(ii) the circuit C is drawn on an isoenergetic surface $H = \text{const}$.

The canonical transformations (88.20) are translated as follows:

$$p_\varrho = \frac{\partial G_1(q, q')}{\partial q_\varrho}, \quad p'_\varrho = -\frac{\partial G_1(q, q')}{\partial q'_\varrho}; \tag{96.6a}$$

$$q_\varrho = \frac{\partial G_2(p, p')}{\partial p_\varrho}, \quad q'_\varrho = -\frac{\partial G_2(p, p')}{\partial p'_\varrho}; \tag{96.6b}$$

$$p_\varrho = \frac{\partial G_3(q, p')}{\partial q_\varrho}, \quad q'_\varrho = \frac{\partial G_3(q, p')}{\partial p'_\varrho}; \tag{96.6c}$$

$$q_\varrho = \frac{\partial G_4(p, q')}{\partial p_\varrho}, \quad p'_\varrho = \frac{\partial G_4(p, q')}{\partial q'_\varrho}. \tag{96.6d}$$

Here the generating functions are arbitrary except for conditions of non-singularity such as

$$\det \frac{\partial^2 G_1(q, q')}{\partial q_\varrho \partial q'_\sigma} \neq 0. \tag{96.7}$$

The different generating functions yielding the same transformation are connected as in (88.19).

In $QTPH$, Ω transformed as an invariant and the special parameter w on the trajectories was unchanged under CT. Therefore, in QP, H transforms as an invariant and t is unchanged. The canonical equations (96.1) transform into

$$\dot{q}'_\varrho = \frac{\partial H'}{\partial p'_\varrho}, \quad \dot{p}'_\varrho = -\frac{\partial H'}{\partial q'_\varrho}, \quad H'(q', p') = H(q, p). \tag{96.8}$$

[1] For notation, see Sect. 62.

To transform the trajectories in QP into parallel straight lines we translate Sect. 91 as follows[1]. We find $G(q, p')$ to satisfy the HAMILTON-JACOBI equation

$$H\left(q, \frac{\partial G}{\partial q}\right) = p'_N, \qquad \det \frac{\partial^2 G}{\partial q_\varrho \partial p'_\sigma} \neq 0. \tag{96.9}$$

Then the CT

$$p_\varrho = \frac{\partial G(q, p')}{\partial q_\varrho}, \qquad q'_\varrho = \frac{\partial G(q, p')}{\partial p'_\varrho} \tag{96.10}$$

gives the Hamiltonian

$$H'(q', p') = H(q, p) = p'_N, \tag{96.11}$$

and the trajectories integrate into

$$\left.\begin{aligned} q'_1 = a_1, \ldots q'_{N-1} = a_{N-1}, \quad q'_N = t, \\ p'_1 = b_1, \ldots p'_{N-1} = b_{N-1}, \quad p'_N = E, \end{aligned}\right\} \tag{96.12}$$

where the a's, the b's and E are constants.

For a conservative system we can always depress the order of the canonical equations by 2 by means of the integral of energy

$$H(q, p) = E, \tag{96.13}$$

which is an immediate consequence of (96.1). This reduction was carried out in Sect. 92, and it suffices to note the technique here. We solve (96.13) for one of the momenta, p_N say, obtaining

$$p_N = -f(q_1, \ldots q_N, p_1, \ldots p_{N-1}, E); \tag{96.14}$$

then the reduced equations, of order $2N-2$, are as in (92.34), with ω replaced by f.

Given a first integral $F(q, p)$ (i.e. a constant of the motion), we reduce by 2 the order of the equations of motion as in Sect. 92, by solving

$$F\left(q, \frac{\partial G}{\partial q}\right) = p'_N, \qquad \det \frac{\partial^2 G}{\partial q_\varrho \partial p'_\sigma} \neq 0 \tag{96.15}$$

for $G(q, p')$ and using the CT (96.10). Since p'_N is constant in the motion, the new Hamiltonian $H'(q', p')$ lacks q'_N, and the new equations of motion read

$$\left.\begin{aligned} \dot{q}'_1 = \frac{\partial H'}{\partial p'_1} \quad \ldots \dot{q}'_{N-1} = \frac{\partial H'}{\partial p'_{N-1}}, \\ \dot{p}'_1 = -\frac{\partial H'}{\partial q'_1} \ldots \dot{p}'_{N-1} = -\frac{\partial H'}{\partial q'_{N-1}}. \end{aligned}\right\} \tag{96.16}$$

Having done this, we can effect a further reduction of 2 in the order by means of the integral of energy (96.13).

Note that solving the HAMILTON-JACOBI equation is equivalent to determining the motion. The solution of (96.15) may be a much simpler matter, for $F(q, p)$ may be a simple function like $p_1 + p_2 + p_3$ [cf. (92.35)].

97. Non-conservative systems. Canonical transformations in *QP*. POISSON brackets and LAGRANGE brackets[2]. We now turn to the general non-conservative system with Hamiltonian $H(q, t, p)$ involving the time, so that

$$\frac{\partial H}{\partial t} \neq 0. \tag{97.1}$$

[1] This formal argument is valid only in the small; cf. Sects. 63, 100.
[2] Cf. WHITTAKER [28], Chap. 11.

Since the energy function in $QTPH$ was $\Omega(x, y)$ and not $\Omega(x, y, w)$, we cannot apply the $QTPH$-theory by a simple reduction in dimensionality, as we did for a conservative system in Sect. 96. It is true that $QTPH$-theory is valid in all generality, but it is set in a space in which t is a coordinate, and we have now to demote t to the position of a mere parameter.

Consider a function $G(q, q', t)$ and the transformation $(q, p) \rightarrow (q', p')$ given by

$$p_\varrho = \frac{\partial G(q, q', t)}{\partial q_\varrho}, \quad p'_\varrho = -\frac{\partial G(q, q', t)}{\partial q'_\varrho}. \tag{97.2}$$

Then

$$p_\varrho \delta q_\varrho - p'_\varrho \delta q'_\varrho = \delta G - \frac{\partial G}{\partial t} \delta t, \tag{97.3}$$

or

$$p_\varrho \delta q_\varrho - H(q, t, p)\, \delta t = p'_\varrho \delta q'_\varrho - K(q', t, p')\, \delta t + \delta G, \tag{97.4}$$

where

$$K(q', t, p') = H(q, t, p) + \frac{\partial G(q, q', t)}{\partial t}. \tag{97.5}$$

Now G is a function of position in the space QTP [since we can solve (97.2) for q'_ϱ in terms of (q, t, p)], and we can apply the Pfaffian argument as in (94.13); this tells us that the canonical equations

$$\dot{q}_\varrho = \frac{\partial H}{\partial p_\varrho}, \quad \dot{p}_\varrho = -\frac{\partial H}{\partial q_\varrho}, \tag{97.6}$$

transform into

$$\dot{q}'_\varrho = \frac{\partial K}{\partial p'_\varrho}, \quad \dot{p}'_\varrho = -\frac{\partial K}{\partial q'_\varrho}, \tag{97.7}$$

the Hamiltonian being changed as in (97.5). Thus (97.2) is a canonical transformation (CT) in QP, the generating function G containing t as a parameter.

This treatment in QP is less general than the treatment given in Sect. 94 for QTP, because we have not in the present instance transformed the time. For time-preserving CT in $QTPH$, see (88.26).

Let u and v be any two functions of the $2N+1$ quantities (q, p, t), i.e. functions of position in QP and of the parameter t; their Poisson bracket is defined as

$$[u, v] = \frac{\partial u}{\partial q_\varrho}\frac{\partial v}{\partial p_\varrho} - \frac{\partial v}{\partial q_\varrho}\frac{\partial u}{\partial p_\varrho}. \tag{97.8}$$

As the representative point moves along a trajectory, the rate of change of any function $F(q, p, t)$ is

$$\left.\begin{aligned} \frac{dF}{dt} &= \frac{\partial F}{\partial t} + \frac{\partial F}{\partial q_\varrho}\dot{q}_\varrho + \frac{\partial F}{\partial p_\varrho}\dot{p}_\varrho \\ &= \frac{\partial F}{\partial t} + [F, H]. \end{aligned}\right\} \tag{97.9}$$

In particular we have, as an alternative expression of the canonical equations,

$$\dot{q}_\varrho = [q_\varrho, H], \quad \dot{p}_\varrho = [p_\varrho, H]. \tag{97.10}$$

We shall now show that *if $u(q, p, t)$ and $v(q, p, t)$ are two constants of the motion, then so also is their* Poisson *bracket* $[u, v]$.

We are given

$$\frac{du}{dt} = \frac{\partial u}{\partial t} + [u, H] = 0, \quad \frac{dv}{dt} = \frac{\partial v}{\partial t} + [v, H] = 0. \tag{97.11}$$

Now by (97.9) we have

$$\frac{d}{dt}[u,v] = \frac{\partial}{\partial t}[u,v] + [[u,v],H], \tag{97.12}$$

and in this equation we can change the last term by the POISSON-JACOBI identity [cf. (89.3)]

$$[[u,v],w] + [[v,w],u] + [[w,u],v] = 0,$$

so that, remembering the skew-symmetry of POISSON brackets,

$$\frac{d}{dt}[u,v] = \frac{\partial}{\partial t}[u,v] - [[v,H],u] + [[u,H],v]. \tag{97.13}$$

Applying (97.11), we get

$$\frac{d}{dt}[u,v] = \frac{\partial}{\partial t}[u,v] + \left[\frac{\partial v}{\partial t}, u\right] - \left[\frac{\partial u}{\partial t}, v\right] = 0, \tag{97.14}$$

which establishes the result.

Consider now an ∞^2 family of trajectories with equations

$$q_\varrho = q_\varrho(u,v,t), \qquad p_\varrho = p_\varrho(u,v,t), \tag{97.15}$$

where u and v are constant along each trajectory. Then the LAGRANGE bracket

$$\{u,v\} = \frac{\partial q_\varrho}{\partial u}\frac{\partial p_\varrho}{\partial v} - \frac{\partial q_\varrho}{\partial v}\frac{\partial p_\varrho}{\partial u} \tag{97.16}$$

is a function of u, v and t: we shall prove by direct calculation that *this* LAGRANGE *bracket is a constant of the motion*. We have

$$\frac{\partial q_\varrho}{\partial t} = \dot{q}_\varrho = \frac{\partial H}{\partial p_\varrho}, \qquad \frac{\partial p_\varrho}{\partial t} = \dot{p}_\varrho = -\frac{\partial H}{\partial q_\varrho}, \tag{97.17}$$

and these are functions of u, v and t. Then

$$\left.\begin{aligned}
\frac{d}{dt}\left(\frac{\partial q_\varrho}{\partial u}\frac{\partial p_\varrho}{\partial v}\right) &= \frac{\partial}{\partial t}\left(\frac{\partial q_\varrho}{\partial u}\frac{\partial p_\varrho}{\partial v}\right) = \frac{\partial}{\partial u}\left(\frac{\partial H}{\partial p_\varrho}\right)\frac{\partial p_\varrho}{\partial v} - \frac{\partial q_\varrho}{\partial u}\frac{\partial}{\partial v}\left(\frac{\partial H}{\partial q_\varrho}\right)\\
&= \frac{\partial^2 H}{\partial p_\varrho \partial q_\sigma}\frac{\partial q_\sigma}{\partial u}\frac{\partial p_\varrho}{\partial v} + \frac{\partial^2 H}{\partial p_\varrho \partial p_\sigma}\frac{\partial p_\sigma}{\partial u}\frac{\partial p_\varrho}{\partial v} -\\
&\quad - \frac{\partial q_\varrho}{\partial u}\frac{\partial^2 H}{\partial q_\varrho \partial q_\sigma}\frac{\partial q_\sigma}{\partial v} - \frac{\partial q_\varrho}{\partial u}\frac{\partial^2 H}{\partial q_\varrho \partial p_\sigma}\frac{\partial p_\sigma}{\partial v}\\
&= \frac{\partial^2 H}{\partial p_\varrho \partial p_\sigma}\frac{\partial p_\sigma}{\partial u}\frac{\partial p_\varrho}{\partial v} - \frac{\partial^2 H}{\partial q_\varrho \partial q_\sigma}\frac{\partial q_\varrho}{\partial u}\frac{\partial q_\sigma}{\partial v}.
\end{aligned}\right\} \tag{97.18}$$

Interchanging u and v and subtracting, we get

$$\frac{d}{dt}\{u,v\} = 0, \tag{97.19}$$

establishing the result.

98. Non-conservative systems. Absolute integral invariants in QP. LIOUVILLE'S theorem. We continue to consider a general system for which $H = H(q,t,p)$. Let capital suffixes $A, A_1, \ldots$ take the values $1, 2, \ldots 2M$ where $M \leq N$, N being, as always, the number of degrees of freedom of the system. Consider an ∞^{2M} family of trajectories with equations

$$q_\varrho = q_\varrho(u,t), \qquad p_\varrho = p_\varrho(u,t), \tag{98.1}$$

where u stands for a set of $2M$ quantities u_A which are constant along each trajectory.

For any fixed value of t, the Eqs. (98.1) define a surface of $2M$ dimensions immersed in QP. Let D be a domain in this surface, limited by bounds on the ranges of u_A. Introduce the following $2M\times 2M$ determinant, a function of the u's and of t:

$$\Delta_M(\varrho_1,\dots\varrho_M,\sigma_1,\dots\sigma_M)=\begin{vmatrix} \frac{\partial q_{\varrho_1}}{\partial u_1} & \frac{\partial q_{\varrho_1}}{\partial u_2} & \cdots & \frac{\partial q_{\varrho_1}}{\partial u_{2M}} \\ \frac{\partial q_{\varrho_2}}{\partial u_1} & \frac{\partial q_{\varrho_2}}{\partial u_2} & \cdots & \frac{\partial q_{\varrho_2}}{\partial u_{2M}} \\ \cdot & \cdot & \cdots & \cdot \\ \frac{\partial q_{\varrho_M}}{\partial u_1} & \frac{\partial q_{\varrho_M}}{\partial u_2} & \cdots & \frac{\partial q_{\varrho_M}}{\partial u_{2M}} \\ \frac{\partial p_{\sigma_1}}{\partial u_1} & \frac{\partial p_{\sigma_1}}{\partial u_2} & \cdots & \frac{\partial p_{\sigma_1}}{\partial u_{2M}} \\ \cdot & \cdot & \cdots & \cdot \\ \frac{\partial p_{\sigma_M}}{\partial u_1} & \frac{\partial p_{\sigma_M}}{\partial u_2} & \cdots & \frac{\partial p_{\sigma_M}}{\partial u_{2M}} \end{vmatrix}. \tag{98.2}$$

Here $\varrho_1,\dots\varrho_M,\sigma_1,\dots\sigma_M$ are any numbers in the range $1, 2, \dots N$. Then the integral

$$\int\limits_D \Delta_M(\varrho_1,\dots\varrho_M,\sigma_1,\dots\sigma_M)\,du_1\dots du_{2M} \tag{98.3}$$

is invariant in the sense that it has the same value no matter what parameters[1] u_A are used in D. Define Φ_M by

$$\Phi_M=\Delta_M(\varrho_1,\dots\varrho_M,\varrho_1,\dots\varrho_M), \tag{98.4}$$

with the summation convention operating; and define

$$I_M=\frac{1}{M!}\int\limits_D \Phi_M\,du_1\dots du_{2M}. \tag{98.5}$$

Its value is independent of the choice of parameters u_A in D.

Using the permutation symbol $\varepsilon_{A_1\dots A_{2M}}$, which is skew-symmetric in all its suffixes and equal to unity when they read $1, 2, \dots 2M$, we may write out the determinant (98.2) explicitly. But we need only Φ_M as in (98.4); it reads

$$\Phi_M=\varepsilon_{A_1\dots A_{2M}}\frac{\partial q_{\varrho_1}}{\partial u_{A_1}}\cdots\frac{\partial q_{\varrho_M}}{\partial u_{A_M}}\frac{\partial p_{\varrho_1}}{\partial u_{A_{M+1}}}\cdots\frac{\partial p_{\varrho_M}}{\partial u_{A_{2M}}}. \tag{98.6}$$

There is summation here for each ϱ on the range $1, \dots N$ and for each A on the range $1, \dots 2M$. In terms of LAGRANGE brackets, we have

$$\Phi_M=(\tfrac{1}{2})^M\,\varepsilon_{A_1\dots A_{2M}}\{u_{A_1},u_{A_{M+1}}\}\{u_{A_2},u_{A_{M+2}}\}\dots\{u_{A_M},u_{A_{2M}}\}. \tag{98.7}$$

Each LAGRANGE bracket is independent of t by (97.19). Therefore Φ_M is independent of t, and we conclude that *the integrals I_M, as given in (98.5) for $M=1, 2, \dots N$, are absolute*[2] *integral invariants.*

The case $M=N$ is of particular interest. The absolute integral invariant

$$I_N=\frac{1}{N!}\int\limits_D \Phi_N\,du_1\dots du_{2N} \tag{98.8}$$

is now an integral extended over a $2N$-dimensional portion D of the $2N$-dimensional phase space QP. This domain changes with t, the representative points

[1] Provided that their orientation is not changed; cf. H. D. BLOCK: Quart. Appl. Math. 12, 201–203 (1954).

[2] Called *absolute* (not *relative*) because D need not be a closed domain (e.g. a circuit).

being carried along according to the canonical equations (Fig. 45). For any chosen value of t, we may use (q, p) as coordinates in D, so that

$$u_1 = q_1, \dots u_N = q_N, \quad u_{N+1} = p_1, \dots u_{2N} = p_N. \tag{98.9}$$

Then (98.2) and (98.4) give

$$\left.\begin{aligned} \Delta_N(\varrho_1, \dots \varrho_N, \sigma_1, \dots \sigma_N) &= \varepsilon_{\varrho_1 \dots \varrho_N} \varepsilon_{\sigma_1 \dots \sigma_N}, \\ \Phi_N &= N!, \end{aligned}\right\} \tag{98.10}$$

and the integral invariant I_N becomes

$$I_N = \int_D dq_1 \dots dq_N \, dp_1 \dots dp_N = \int_D dq \, dp, \tag{98.11}$$

to use an abridged notation. Calling this integral the *volume*[1] of D, we have LIOUVILLE's theorem: *the volume of any portion of QP is conserved when the representative points which compose it move in accordance with the canonical equations.*

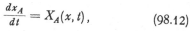

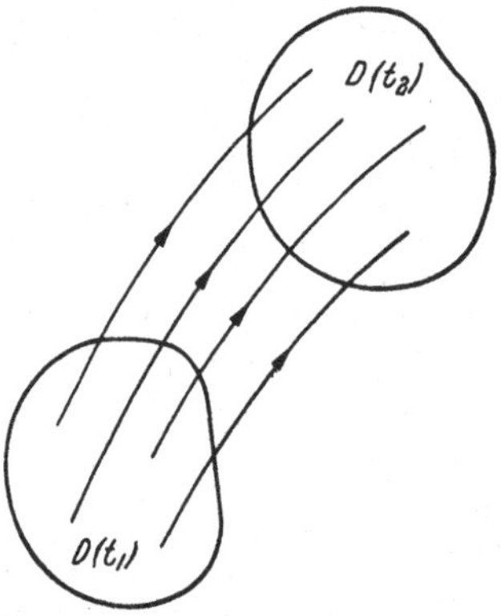

Fig. 45. Conservation of volume in QP (LIOUVILLE's theorem).

This result is so fundamental in statistical mechanics that we shall look at it in two other ways.

First, LIOUVILLE's theorem, as proved by him[2], is actually more general, neither the evenness of dimensionality of the space nor the canonical form of the equations of motion being required. Consider the equations

$$\frac{dx_A}{dt} = X_A(x, t), \tag{98.12}$$

where the right hand sides satisfy

$$\frac{\partial X_A}{\partial x_A} = 0, \tag{98.13}$$

the suffix A ranging $1, 2, \dots M$, with the summation convention. Instead of following LIOUVILLE, we may use a hydrodynamical argument. The Eqs. (98.12) define a velocity-field $v_A = X_A$ in an M-space in which x_A are taken as rectangular Cartesian coordinates, and, just as in ordinary hydrodynamics

$$\frac{\partial u}{\partial x} + \frac{\partial v}{\partial y} + \frac{\partial w}{\partial z}$$

is the expansion (rate of increase of volume per unit volume), so in this M-space $\partial v_A / \partial x_A$ is the expansion, volume being defined as $\int dx_1 \dots dx_M$. Then, by virtue of (98.13), volume is conserved. The details of proof can be filled in by expressing the rate of increase of a volume, moving according to (98.12), as an integral over the $(M-1)$-space bounding it, and applying GREEN's theorem. It is evident that (98.13) is true in particular if M is even and (98.12) are canonical.

The second alternative approach is through canonical transformations (CT). The essential point here is that the JACOBIAN of a CT is unity [cf. (88.29)], this being true even if a CT $(q, p) \to (q', p')$ contains t as a parameter. Two conclusions follow from this. First, without referring to motion at all, we recognize the integral I_N of (98.11) as a suitable definition of volume, since volume so defined has the same value for all coordinates (q, p) in QP obtained from one set

[1] GIBBS called it *extension-in-phase*; cf. footnote to Sect. 96.

[2] J. LIOUVILLE: J. de Math. **3**, 342 (1838); the famous theorem is a secondary result in his paper.

of coordinates by CT[1]. Secondly, volume is conserved in the motion because motion in accordance with the canonical equations consists of infinitesimal CT [cf. (90.4)].

In statistical mechanics[2] we consider a vast number n of identical Hamiltonian systems, differing only in their initial conditions. The superposition of these systems in the space QP gives an ensemble, a "fine dust" of representative points with a *probability density* $f(q, p, t)$ such that $n f\, dq\, dp$ is the number of representative points in the volume element $dq\, dp$ at time t. As the element $dq\, dp$ moves with the dust according to the canonical equations, its volume is conserved and also the number of representative points in it. Hence $df/dt = 0$, or, equivalently,

$$\frac{\partial f}{\partial t} + [f, H] = 0. \tag{98.14}$$

This is the fundamental partial differential equation to be satisfied by the density f, determining f for any t when f is given for $t = 0$.

99. Action-angle variables[3]. Action-angle variables were introduced by C. DELAUNAY for the discussion of astronomical perturbations[4]. Later, they were found to be admirably suited to the older form of quantum mechanics, for the BOHR-SOMMERFELD quantization consisted in making each action variable an integral multiple of PLANCK's constant h.

As treated below, the theory of action-angle variables depends on the separation of variables in the HAMILTON-JACOBI equation. Even though the space QP should have Euclidean topology, one or more of the separating variables may be cyclic (e.g. an azimuthal angle)[5]. However, the presence of cyclic coordinates is not an essential feature of the theory and their inclusion makes the discussion a little more complicated. Therefore it will be assumed that they are absent, essential modifications due to their presence being noted where necessary.

Let the Hamiltonian[6] $H(q, p)$ be such that the HAMILTON-JACOBI equation is of the separable type (Sect. 78). By this we mean that, in the $2N$-dimensional space of the variables (q, p'), the partial differential equation

$$H\left(q, \frac{\partial G}{\partial q}\right) = p'_N \tag{99.1}$$

has a solution of the form

$$G(q, p') = G_1(q_1, p') + G_2(q_2, p') + \cdots + G_N(q_N, p'), \tag{99.2}$$

p' standing for the N quantities p'_ϱ, and the determinantal condition

$$\det \frac{\partial^2 G}{\partial q_\varrho\, \partial p_\sigma} \neq 0, \tag{99.3}$$

being satisfied; in other words (99.2) is a complete integral.

---

[1] There is no definition of volume invariant under arbitrary transformations of (q, p).

[2] See R. H. FOWLER: Statistical Mechanics (Cambridge: University Press 1936); A. I. KHINCHIN: Mathematical Foundations of Statistical Mechanics (New York: Dover Publications, Inc. 1949); A. MÜNSTER: Statistische Thermodynamik (Berlin: Springer 1956); also the article by E. A. GUGGENHEIM in Vol. III, part 2 of this Encyclopedia.

[3] For various treatments of action-angle variables, with illustrative examples and reference to quantum conditions and adiabatic invariance, see M. BORN: The Mechanics of the Atom, Chap. 2 (London: Bell 1927); CORBEN and STEHLE [3], pp. 239–264; FUES [6]; GOLDSTEIN [7], pp. 288–307; LANCZOS [15], pp. 243–254; A. SOMMERFELD: Atomic Structure and Spectral Lines, Vol. 1, pp. 615–623 (translated from 5th. German Edn. by H. L. BROSE, London: Methuen 1934; 3rd Edn.).

[4] Cf. WHITTAKER [28], pp. 426, 431.

[5] Note that the word *cyclic* is used in a topological sense, and is not to be confused with *ignorable* (cf. Sect. 63).

[6] The system is assumed to be conservative, i.e. $\partial H/\partial t = 0$.

A canonical transformation (CT) is defined by

$$p_\varrho = \frac{\partial G(q, p')}{\partial q_\varrho}, \qquad q'_\varrho = \frac{\partial G(q, p')}{\partial p'_\varrho}. \tag{99.4}$$

We observe the effect of separation here; when written more explicitly, the first set of these equations read

$$p_1 = \frac{\partial G_1(q_1, p')}{\partial q_1}, \quad p_2 = \frac{\partial G_2(q_2, p')}{\partial q_2}, \quad \ldots \quad p_N = \frac{\partial G_N(q_N, p')}{\partial q_N}; \tag{99.5}$$

each equation contains only one p and the corresponding q, but of course all the quantities p'_ϱ are, in general, involved in each equation.

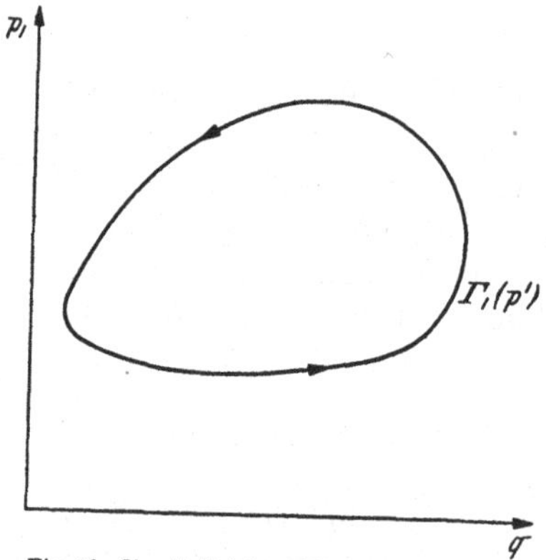

Fig. 46. Circuit $\Gamma_1(p')$ or $\Gamma_1(J)$ in the plane of the canonical variables q_1, p_1.

The new Hamiltonian is

$$H'(q', p') = H(q, p) = H\left(q, \frac{\partial G}{\partial q}\right) = p'_N, \tag{99.6}$$

and the new equations of motion are

$$\dot{q}'_\varrho = \frac{\partial H'}{\partial p'_\varrho}, \quad \dot{p}'_\varrho = -\frac{\partial H'}{\partial q'_\varrho}. \tag{99.7}$$

Therefore all the quantities (q', p') are constant along each trajectory, except q'_N; for it we have $\dot{q}'_N = 1$, so that $q'_N = t + \text{const.}$ We can write

$$p'_N = E, \tag{99.8}$$

E being the constant value of H on the trajectory. By the CT (99.4) we have transformed the trajectories into parallel straight lines, as in Sect. 96.

Consider a representative plane Π_1 in which q_1, p_1 are taken as rectangular Cartesian coordinates (Fig. 46). If the N quantities p'_ϱ are held fixed, the first equation in (99.5) defines a curve in Π_1; denote this curve by $\Gamma_1(p')$. Similarly the other equations in (99.5) define curves $\Gamma_2(p'), \ldots \Gamma_N(p')$ in representative planes $\Pi_2, \ldots \Pi_N$.

We now assume that these curves are all closed[1]; in each of the representative planes we have ∞^N circuits $\Gamma_\varrho(p')$, a set of circuits, one in each plane, being determined by the values of the N quantities p'_ϱ.

Define quantities J_ϱ by the formulae

$$J_1 = \oint_{\Gamma_1(p')} p_1\, dq_1, \quad \ldots \quad J_N = \oint_{\Gamma_N(p')} p_N\, dq_N, \tag{99.9}$$

these being in fact the "areas" contained within[2] the several circuits determined by the values p'_ϱ. Assuming that

$$\det \frac{\partial J_\varrho}{\partial p'_\sigma} \neq 0, \tag{99.10}$$

we have a two-way functional relationship $(J) \leftrightarrow (p')$, and we may express p'_ϱ as function of the J's:

$$p'_\varrho = p'_\varrho(J). \tag{99.11}$$

Substituting these functions in (99.1) and (99.2), we have a solution of

$$H\left(q, \frac{\partial G^*}{\partial q}\right) = p'_N(J) \tag{99.12}$$

[1] In the case of a cyclic coordinate, the circuit may appear like Γ' in Fig. 31, p. 104.
[2] Or *under*, in the case considered in preceding footnote.

in the $2N$-dimensional space of the variables (q, J), this solution being of the form (still separated)

$$G^*(q, J) = G_1^*(q_1, J) + G_2^*(q_2, J) + \cdots + G_N^*(q_N, J). \tag{99.13}$$

This function we now use as generating function for a CT $(q, p) \to (w, J)$, expressed by the formulae

$$p_\varrho = \frac{\partial G^*(q, J)}{\partial q_\varrho}, \qquad w_\varrho = \frac{\partial G^*(q, J)}{\partial J_\varrho}. \tag{99.14}$$

The quantities J_ϱ are called *action variables* and the quantities w_ϱ *angle variables.*

When action-angle variables are used, the Hamiltonian involves the action variables only; we write it $H^*(J)$. The canonical equations of motion read

$$\dot{J}_\varrho = -\frac{\partial H^*}{\partial w_\varrho} = 0, \qquad \dot{w}_\varrho = \frac{\partial H^*}{\partial J_\varrho}, \tag{99.15}$$

so that the J's are constant along each trajectory and the w's are given by

$$w_\varrho = \nu_\varrho t + \delta_\varrho, \tag{99.16}$$

where ν_ϱ and δ_ϱ are constants, the former being

$$\nu_\varrho = \frac{\partial H^*}{\partial J_\varrho}. \tag{99.17}$$

The constant value of H along the trajectory is

$$H = E = H^*(J), \tag{99.18}$$

the constant E being the total energy in ordinary dynamical systems.

100. The periodic property of angle variables. Let $\Gamma(J)$ be any circuit in the space QP such that all the action variables J_ϱ are constant along it. A position of the representative point B in QP determines positions of representative points B_ϱ in the planes Π_ϱ (Fig. 46), and when B is carried once round $\Gamma(J)$, these points B_ϱ are carried round the respective circuits $\Gamma_\varrho(J)$, perhaps several times over. Symbolically, we may write

$$\Gamma(J) = n_\varrho \Gamma_\varrho(J), \tag{100.1}$$

where the coefficients n_ϱ are integers (positive, negative or zero). In (100.1), and below, the summation convention operates as usual over the range $1, \ldots N$.

On carrying B round Γ, the generating function $G^*(q, J)$ is increased by

$$\Delta_\Gamma G^* = \oint\limits_{\Gamma(J)} \frac{\partial G^*}{\partial q_\varrho}\, dq_\varrho = \oint\limits_{\Gamma(J)} p_\varrho\, dq_\varrho. \tag{100.2}$$

To examine this final expression, we write out the first set of the transformation Eqs. (99.14):

$$p_1 = \frac{\partial G_1^*(q_1, J)}{\partial q_1}, \quad \ldots \quad p_N = \frac{\partial G_N^*(q_N, J)}{\partial q_N}. \tag{100.3}$$

The first of these equations sets up a connection between p_1 and q_1 which is the same for the point B on $\Gamma(J)$ and the corresponding point B_1 on $\Gamma_1(J)$; therefore, by the definition of J_1 in (99.9), we have

$$\oint\limits_{\Gamma(J)} p_1\, dq_1 = n_1 \oint\limits_{\Gamma_1(J)} p_1\, dq_1 = n_1 J_1. \tag{100.4}$$

Thus (100.2) gives

$$\Delta_\Gamma G^* = n_\varrho J_\varrho. \tag{100.5}$$

We see that, on the basis of assumptions explicit and implicit, the functions $G_1^*(q_1, J), \ldots G_N^*(q_N, J)$ are necessarily multiple valued functions of their q-arguments.

Now vary the action variables J_ϱ infinitesimally, the circuit $\Gamma(J)$ varying in consequence. The n's in (100.5), being integers, cannot change under this infinitesimal variation, and we get

$$\delta \Delta_\Gamma G^* = n_\varrho \delta J_\varrho. \tag{100.6}$$

On the other hand, from (100.2) we have

$$\delta \Delta_\Gamma G^* = \delta \oint_{\Gamma(J)} p_\varrho \, dq_\varrho = \oint_{\Gamma(J)} (\delta p_\varrho \, dq_\varrho - d\delta_\varrho \, dp_\varrho). \tag{100.7}$$

This last expression is the bilinear form invariant under CT (cf. Sect. 96), and therefore

$$\delta \Delta_\Gamma G^* = \oint_{\Gamma(J)} (\delta J_\varrho \, dw_\varrho - \delta w_\varrho \, dJ_\varrho). \tag{100.8}$$

But the J's are constants on both the varied and unvaried circuits Γ; therefore $dJ_\varrho = 0$ and $\delta J_\varrho = \text{const}$, and we have

$$\delta \Delta_\Gamma G^* = \delta J_\varrho \Delta_\Gamma w_\varrho, \tag{100.9}$$

where $\Delta_\Gamma w_\varrho$ is the increment in w_ϱ on passing once round $\Gamma(J)$. Comparing this last equation with (100.6), we may state the result: *On taking the representative point B once round any circuit $\Gamma(J)$ in QP for which the action variables J_ϱ all have fixed values, the increment in the angle variable w_ϱ is*

$$\Delta_\Gamma w_\varrho = n_\varrho, \tag{100.10}$$

where n_ϱ is the number of times the point B_ϱ goes round the circuit $\Gamma_\varrho(J)$ in the plane Π_ϱ.

In particular, if we hold all the quantities (q, J) fixed except q_1, then the point B_1 moves on the curve $\Gamma_1(J)$ in Π_1, and the points $B_2, \ldots B_N$ remain fixed. This causes the representative point B in QP to move on some curve, and when B_1 has completed a circuit of $\Gamma_1(J)$, B has completed some circuit $\Gamma_1^*(J)$ for which the numbers n_ϱ of (100.1) are

$$n_1 = 1, \quad n_2 = \cdots = n_N = 0. \tag{100.11}$$

Substituting in (100.10) and writing Γ_1^* for Γ, we have

$$\Delta_{\Gamma_1^*} w_1 = 1, \quad \Delta_{\Gamma_1^*} w_2 = 0, \quad \ldots \quad \Delta_{\Gamma_1^*} w_N = 0. \tag{100.12}$$

Hence more generally, with the notation interpreted as above,

$$\Delta_{\Gamma_\varrho^*} w_\sigma = \delta_{\varrho\sigma}. \tag{100.13}$$

Now let the second set of equations in (99.14) be solved for q_ϱ:

$$q_\varrho = q_\varrho(w, J). \tag{100.14}$$

If we fix all the quantities (w, J) except w_1, we leave one degree of freedom in the representative point B in QP, the p's being given to within that one degree of freedom by (100.3). Then B moves on some curve in QP, and the "projected" points B_ϱ move on the curves $\Gamma_\varrho(J)$. Let w_1 be increased continuously from 0 to 1, the other w's being held fixed as aforesaid. From inspection of (100.12) we conclude that, when this operation has been completed, the point B_1 will have gone once round $\Gamma_1(J)$ and the points $B_2, \dots B_N$ will have been restored to their original positions without having gone round their respective circuits. The same argument can be applied to increases of unity in each of the angle variables individually, and we conclude that *the functions (100.14) are periodic in each of the w's with period unity*[1].

We can therefore expand these functions in FOURIER series of the form

$$q_\varrho = \sum_{(n)} A_{\varrho;\, n_1, \dots n_N} e^{2\pi i (n_1 w_1 + \cdots + n_N w_N)}, \tag{100.15}$$

the summation running over all integer values of the n's (positive, negative and zero), and the A's being complex functions of the action variables J, such that a reversal in the signs of all the n's turns an A into its complex conjugate. Then, by (99.16), the motion of the system is given by

$$q_\varrho = \sum_{(n)} B_{\varrho;\, n_1, \dots n_N} e^{2\pi i (n_1 \nu_1 + \cdots + n_N \nu_N) t}, \tag{100.16}$$

the B's being functions of the J's. In this sense, the quantities ν_ϱ are "frequencies". If we know the function $H^*(J)$, they can be calculated at once from (99.17) by differentiating this function.

It has been assumed that the curves in the planes Π_ϱ defined by the Eqs. (100.3) (for fixed values of the J's) are closed, these closed curves being in fact the circuits $\Gamma_\varrho(J)$. This assumption does not at all imply that the motion of the system is periodic: we see from (100.16) that it is periodic if, and only if, the ratios of the frequencies ν_ϱ are rational numbers.

A system is called *degenerate* if the frequencies satisfy a relation of the form

$$s_1 \nu_1 + s_2 \nu_2 + \cdots + s_N \nu_N = 0, \tag{100.17}$$

where $s_1, \dots s_N$ are integers or zero, with at least two non-zero. Degeneracy occurs when the Hamiltonian $H^*(J)$ involves the action variables in certain ways which the following example illustrates. Suppose the Hamiltonian is of the form

$$\left.\begin{aligned} H^*(J) &= f(K, J_4, J_5, \dots J_N), \\ K &= m_1 J_1 + m_2 J_2 + m_3 J_3, \end{aligned}\right\} \tag{100.18}$$

where the m's are integers. Then

$$\nu_1 = \frac{\partial H^*}{\partial J_1} = m_1 \frac{\partial f}{\partial K}, \quad \nu_2 = m_2 \frac{\partial f}{\partial K}, \quad \nu_3 = m_3 \frac{\partial f}{\partial K}, \tag{100.19}$$

and we have a double degeneracy:

$$m_2 \nu_1 - m_1 \nu_2 = 0, \quad m_3 \nu_1 - m_1 \nu_3 = 0. \tag{100.20}$$

As indicated in Sect. 63, the exposition of general dynamical theory in this article is, from the standpoint of modern pure mathematics, on a rather low level of precision. As far as theory in the small is concerned, it would not be hard to make those additions which would make it precise, but action-angle variables take us out of the small into the large through the introduction of the circuits $\Gamma_\varrho(J)$. This raises topological questions of considerable complexity, not touched on here.

---

[1] A cyclic coordinate will not be periodic. It is increased by its cyclic constant, and (100.15) and (100.16) are modified by the addition of other terms; cf. FUES [6], p. 140, GOLDSTEIN [7], p. 295.

VIII. Small oscillations.

101. Reduction of energies to normal form. Normal modes and frequencies. Degeneracy. Consider a dynamical system with N generalized coordinates[1] q^ϱ and Lagrangian

$$L = T - V, \quad T = \tfrac{1}{2} a_{\varrho\sigma}(q)\,\dot{q}^\varrho \dot{q}^\sigma, \quad V = V(q), \tag{101.1}$$

the system moving in accordance with LAGRANGE's equations

$$\frac{d}{dt}\frac{\partial T}{\partial \dot{q}^\varrho} - \frac{\partial T}{\partial q^\varrho} = -\frac{\partial V}{\partial q^\varrho}. \tag{101.2}$$

This is the ordinary dynamical system (ODS) of Sect. 66.

We geometrize in configuration space Q. The representative point describes a trajectory in accordance with (101.2), the trajectory being determined by an initial point q^ϱ and an initial velocity $\dot{q}^\varrho$. If, at a certain point in Q we have

$$\frac{\partial V}{\partial q^\varrho} = 0, \tag{101.3}$$

then no motion results if the velocity vanishes; these N equations define *equilibrium configurations*, and we may expect in general to find a discrete set of such points in Q, since the number of equations is equal to the number of coordinates. We shall now discuss small oscillations about equilibrium.

By a change of coordinates, we can make the equilibrium point the origin O ($q^\varrho = 0$ there), and also make $V = 0$ at O, since potential energy is always undetermined to within an additive constant. Expanding $a_{\varrho\sigma}(q)$ and V in power series about O, the principal parts of T and V become

$$T = \tfrac{1}{2} a_{\varrho\sigma}\dot{q}^\varrho \dot{q}^\sigma, \quad V = \tfrac{1}{2} b_{\varrho\sigma} q^\varrho q^\sigma. \tag{101.4}$$

Here the coefficients are constants, and $a_{\varrho\sigma} = a_{\sigma\varrho}$, $b_{\varrho\sigma} = b_{\sigma\varrho}$. The equations of motion (101.2) now read

$$a_{\varrho\sigma}\ddot{q}^\sigma + b_{\varrho\sigma} q^\sigma = 0. \tag{101.5}$$

For the sake of mathematical clarity, it is wise to forget that we are dealing with an approximation, and regard the Eqs. (101.4) and (101.5) as defining our problem, with q_ϱ finite; the homogeneity of the system permits this.

The straightforward practical method of solving (101.5) is to substitute

$$q^\varrho = \alpha^\varrho e^{i\omega t}, \tag{101.6}$$

where α^ϱ are constant complex amplitude factors and ω a circular frequency. On eliminating the α's from (101.5), we get the *secular equation*

$$\det(a_{\varrho\sigma}\omega^2 - b_{\varrho\sigma}) = 0 \tag{101.7}$$

from which to determine the values of ω. Any positive root ω^2 gives a real ω; this is a *normal circular frequency*, and the corresponding *normal mode of oscillation* is given by the real part of (101.6), the amplitude factors being solutions of

$$(a_{\varrho\sigma}\omega^2 - b_{\varrho\sigma})\,\alpha^\sigma = 0. \tag{101.8}$$

The ratios of the α's are real, but they have an arbitrary common complex factor.

The above method is difficult to follow if the secular equation (101.7) has repeated roots, and a much deeper insight into the mathematical structure of

[1] To agree with tensor notation, we use superscripts here. See Sect. 62 for summation convention.

the problem presented by (101.4) and (101.5) is gained by starting afresh, using the geometry of the space Q. We assume (as is the case for all natural systems) that the kinetic energy is positive-definite. Then the quadratic form

$$A = a_{\varrho\sigma} q^\varrho q^\sigma \tag{101.9}$$

is also positive-definite, and there exists a linear homogeneous transformation $(q) \to (q')$ which makes[1]

$$A = q_1'^2 + q_2'^2 + \cdots + q_N'^2. \tag{101.10}$$

If we denote finite increments by Δ, the formula

$$D^2 = a_{\varrho\sigma} \Delta q^\varrho \Delta q^\sigma = \Delta q_1'^2 + \cdots + \Delta q_N'^2 \tag{101.11}$$

defines a finite Euclidean distance D between any two points of Q. (This is the integrated form of the kinematical line element of Sect. 84.) The kinetic energy is

$$T = \tfrac{1}{2} a_{\varrho\sigma} \dot{q}^\varrho \dot{q}^\sigma = \tfrac{1}{2} (\dot{q}_1'^2 + \dot{q}_2'^2 + \cdots + \dot{q}_N'^2). \tag{101.12}$$

We may now treat Q as a Euclidean N-space, q^ϱ being oblique Cartesian coordinates and q'_ϱ rectangular Cartesian coordinates. We are concerned with the geometrical form of the *equipotential surfaces*, which have the equations

$$B = b_{\varrho\sigma} q^\varrho q^\sigma = b'_{\varrho\sigma} q'_\varrho q'_\sigma = \text{const}, \tag{101.13}$$

$b'_{\varrho\sigma}$ being the new coefficients after application of the transformation.

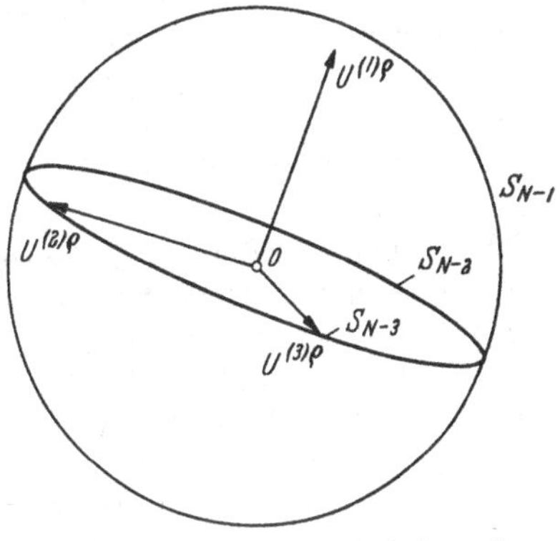

Fig. 47. Reduction to normal coordinates by maxima (the case where $N = 3$).

To investigate the principal axes of the equipotential surfaces, and to discover whether they are ellipsoidal or hyperboloidal in character, we proceed as follows. It is convenient to have before us at the same time expressions in both the coordinate systems (q) and (q'); the former will be written on the left, and the latter on the right. On the sphere S_{N-1} with equation

$$a_{\varrho\sigma} q^\varrho q^\sigma = 1 \quad \text{or} \quad q_1'^2 + q_2'^2 + \cdots + q_N'^2 = 1, \tag{101.14}$$

B is a function of position, and it attains a maximum value (say λ_1) at two or more points of S_{N-1}; let $U^{(1)\varrho}$ (or $U_\varrho'^{(1)}$) be the coordinates of such a point (Fig. 47).

Now intersect the sphere S_{N-1} by the plane orthogonal to this last vector, with equation

$$a_{\varrho\sigma} U^{(1)\varrho} q^\sigma = 0 \quad \text{or} \quad U_\varrho'^{(1)} q'_\varrho = 0, \tag{101.15}$$

thus obtaining a sphere of $N-2$ dimensions (say S_{N-2}). On S_{N-2} B attains a maximum (say λ_2) at two or more points; let $U^{(2)\varrho}$ (or $U_\varrho'^{(2)}$) be the coordinates of such a point. We have the orthogonality condition

$$a_{\varrho\sigma} U^{(1)\varrho} U^{(2)\sigma} = 0, \quad \text{or} \quad U_\varrho'^{(1)} U_\varrho'^{(2)} = 0. \tag{101.16}$$

Next we cut S_{N-2} by a plane orthogonal to $U^{(2)\varrho}$, obtaining a sphere S_{N-3}, and proceed as before. Carrying on this process, we arrive at a circle S_1 and finally at a pair of points S_0.

[1] It is convenient to denote the new coordinates by q'_ϱ (not q'^ϱ); for transformations conserving the form (101.10), there is no distinction between contravariant and covariant quantities.

In this way we get a set of N mutually orthogonal unit vectors, $U^{(\sigma)\varrho}$ or $U'^{(\sigma)}_{\varrho}$, with numbers λ_σ associated with them, these being the maxima of B under the conditions stated above. Then, by an orthogonal transformation,

$$q'_\varrho = A_{\varrho\sigma} q''_\sigma, \quad A_{\varrho\sigma} A_{\varrho\tau} = \delta_{\sigma\tau}, \tag{101.17}$$

we change to new rectangular Cartesian coordinates q''_ϱ with axes in the directions of the aforesaid orthogonal vectors, and it is easy to show, by reason of the maximal properties, that the form B lacks all product terms when expressed in the coordinates q''_ϱ.

Dropping the double primes on the final coordinates, we may state this result: *Given two quadratic forms, A and B, with A positive-definite, there exists a linear homogeneous transformation which turns A into a sum of squares and B into a form lacking product terms; equivalently, the kinetic and potential energies (101.4) can be transformed into*

$$\left.\begin{aligned} T &= \tfrac{1}{2}(\dot q_1^2 + \dot q_2^2 + \cdots + \dot q_N^2), \\ V &= \tfrac{1}{2}(\lambda_1 q_1^2 + \lambda_2 q_2^2 + \cdots + \lambda_N q_N^2). \end{aligned}\right\} \tag{101.18}$$

These final coordinates are *normal coordinates*. There is no implication in the above argument that the λ's are positive or that they are distinct; they are of course real, since the argument did not go outside the real domain[1].

Once the energies have been put into the normal form (101.18), the discussion of the motion is extremely simple, for the equations of motion (101.2) become

$$\ddot q_1 + \lambda_1 q_1 = 0, \quad \ldots \quad \ddot q_N + \lambda_N q_N = 0, \tag{101.19}$$

in which the variables are separated. Depending on the sign of the λ contained in it, the solution of any one of these equations is as follows:

$$\left.\begin{aligned} q &= a\cos\sqrt{\lambda}\,t + b\sin\sqrt{\lambda}\,t && \text{if } \lambda > 0, \\ q &= a t + b && \text{if } \lambda = 0, \\ q &= a\,\mathrm{Cos}\sqrt{-\lambda}\,t + b\,\mathrm{Sin}\sqrt{-\lambda}\,t && \text{if } \lambda < 0, \end{aligned}\right\} \tag{101.20}$$

where a and b are constants.

The equilibrium is *stable* under any one of the following equivalent conditions:

(i) All the λ's are positive.

(ii) $b_{\varrho\sigma} q^\varrho q^\sigma$ is a positive-definite form.

(iii) The potential energy V is a true minimum at the equilibrium configuration.

If any one of the λ's is zero or negative, the equilibrium is *unstable*. In the case of stability, the equipotential surfaces are ellipsoidal; in the case of instability, they are ellipsoidal (with V a maximum at the centre) or hyperboloidal or cylindrical.

Suppose the equilibrium stable, so that each normal coordinate varies sinusoidally as in the first of (101.20). Then a *normal mode* of oscillation is one in which only one normal coordinate oscillates, the others being zero, and the *normal frequencies* ν_ϱ and *normal circular frequencies* ω_ϱ are

$$\nu_\varrho = \frac{1}{2\pi}\sqrt{\lambda_\varrho}, \quad \omega_\varrho = \sqrt{\lambda_\varrho}. \tag{101.21}$$

---

[1] The transformation of the energies to normal form, as in (101.18), whether by means of maximal properties or otherwise, is basic in all thorough treatments of the theory of small oscillations. Cf. CORBEN and STEHLE [*3*], Chap. 8 (where there are a number of examples of systems with few and with many degrees of freedom); GOLDSTEIN [*7*], Chap. 10; WHITTAKER [*28*], Chap. 7.

If two or more of these frequencies coincide, the system is *degenerate*[1]. In a normal mode, the representative point in Q performs a harmonic oscillation on a straight line. These straight lines are the principal axes of the equipotential surfaces (101.13) when these surfaces are referred to the coordinate system (q'). In a non-degenerate system, these lines are fixed in direction; for a degenerate system, they are partially indeterminate, lying in a plane of two or more dimensions (according to the degree of degeneracy), and the normal mode can be performed on any one of these lines, the normal coordinates in that case being partially indeterminate. In a completely degenerate system, the direction of a normal mode of oscillation is completely arbitrary; in this case the equipotential surfaces are spheres in the coordinate system (q').

Under arbitrary initial conditions, the system performs a motion which is a superposition of all the normal modes; in general, the orbit in Q is a very complicated curve, and the motion is periodic only if the ratios of the normal frequencies are rational.

The roots of the equation

$$\det(a_{\varrho\sigma}\lambda - b_{\varrho\sigma}) = 0 \tag{101.22}$$

are invariant under linear transformations of the q's. For normal coordinates, as in (101.18), this equation becomes

$$(\lambda - \lambda_1)(\lambda - \lambda_2)\ldots(\lambda - \lambda_N) = 0. \tag{101.23}$$

Therefore $\lambda_1, \lambda_2, \ldots \lambda_N$ are the roots of (101.22); they are in fact the eigenvalues of the matrix $b_{\varrho\sigma}$ relative to the matrix $a_{\varrho\sigma}$. If λ is any one of these eigenvalues, the equations

$$\lambda\, a_{\varrho\sigma} U^\sigma = b_{\varrho\sigma} U^\sigma \tag{101.24}$$

define the corresponding eigenvectors for any system of coordinates. From the invariance of these equations it is easy to see, by using normal coordinates, that these eigenvectors point in the directions of the normal modes of oscillation. Since (101.22) is the same as (101.7), and (101.24) is the same as (101.8) (except for trivial change in notation), we recognize the mathematical significance of the simple substitution (101.6), which remains, when all is said, the most practical technique for dealing with vibration problems. A degeneracy is indicated by a multiple root of the secular equation (101.7).

102. The effect of constraints. To a system with kinetic and potential energies as in (101.4), let a constraint

$$A_\varrho q^\varrho = 0 \tag{102.1}$$

be applied, the A's being constants.

If we regard the energies in (101.4) as approximations, valid for small values of the velocities and coordinates, then (102.1) may be thought of as arising from any constraint which is independent of the time; it may even be non-holonomic, there being no distinction in a linear approximation between holonomic and non-holonomic.

As in (46.15), the equations of motion of the constrained system are

$$a_{\varrho\sigma}\ddot{q}^\sigma = -b_{\varrho\sigma}q^\sigma + \vartheta A_\varrho, \tag{102.2}$$

where ϑ is an undetermined multiplier. To investigate the motion, we substitute

$$q^\varrho = \alpha^\varrho e^{i\omega t} \tag{102.3}$$

[1] This word is overworked; cf. Sect. 100 for a different meaning.

in (102.1) and (102.2); on eliminating the α's and ϑ we get the following determinantal equation for the circular frequency ω:

$$\begin{vmatrix} a_{\varrho\sigma}\omega^2 - b_{\varrho\sigma} & A_\varrho \\ A_\sigma & 0 \end{vmatrix} = 0. \tag{102.4}$$

That is the practical plan. But to find the relationship between the frequencies of the unconstrained and the constrained systems, it is best to use normal coordinates of the unconstrained system, as in (101.18). Then, with λ written for ω^2, (102.4) becomes

$$\Delta(\lambda) = 0, \tag{102.5}$$

where

$$\Delta(\lambda) = -\begin{vmatrix} \lambda-\lambda_1 & 0 & \cdots & & A_1 \\ 0 & \lambda-\lambda_2 & \cdots & & A_2 \\ \cdot & \cdot & \cdot & \cdot & \cdot \\ 0 & 0 & \cdots & \lambda-\lambda_N & A_N \\ A_1 & A_2 & \cdots & A_N & 0 \end{vmatrix}, \tag{102.6}$$

the numbers A_ϱ being now the coefficients in the equation of constraint (102.1) when expressed in the normal coordinates. On expansion, we have

$$\left.\begin{aligned} \Delta(\lambda) = A_1^2(\lambda-\lambda_2)(\lambda-\lambda_3) & \ldots (\lambda-\lambda_N) + \\ + A_2^2(\lambda-\lambda_1)(\lambda-\lambda_3) & \ldots (\lambda-\lambda_N) + \\ + A_3^2(\lambda-\lambda_1)(\lambda-\lambda_2) & \ldots (\lambda-\lambda_N) + \\ \cdots\cdots\cdots & \cdots\cdots \\ + A_N^2(\lambda-\lambda_1)(\lambda-\lambda_2) & \ldots (\lambda-\lambda_{N-1}). \end{aligned}\right\} \tag{102.7}$$

Here $\lambda_1, \lambda_2, \ldots \lambda_N$ are the squares of the circular frequencies of the unconstrained system.

Suppose that the unconstrained system is non-degenerate; then, by a mere interchange of coordinates, we can arrange that

$$\lambda_1 < \lambda_2 < \cdots < \lambda_N. \tag{102.8}$$

Suppose, further, that none of the A's in (102.7) vanish. Then

$$\Delta(\lambda_N) > 0, \quad \Delta(\lambda_{N-1}) < 0, \quad \Delta(\lambda_{N-2}) > 0, \quad \ldots, \tag{102.9}$$

and therefore $\Delta(\lambda)$ has $N-1$ real zeros, separating the numbers $\lambda_1, \ldots \lambda_N$. Under these circumstances (i.e. in what we may call the general case) *the constrained frequencies separate the unconstrained frequencies.*

To allow for the possibility of one or more of the A's vanishing, this statement must be weakened to

$$\nu_1 \leqq \nu_1' \leqq \nu_2 \leqq \nu_2' \cdots \leqq \nu_{N-1} \leqq \nu_{N-1}' \leqq \nu_N, \tag{102.10}$$

where ν indicates an unconstrained frequency and ν' a constrained frequency. Degeneracy may be produced by constraint; in geometrical language, an ellipsoid possesses circular sections.

The effect of applying a constraint to a degenerate system is best illustrated by an example. Take $N=5$ and suppose

$$\lambda_1 < \lambda_2 = \lambda_3 = \lambda_4 < \lambda_5, \tag{102.11}$$

so that there is a triple degeneracy in the unconstrained system. Then (102.7) becomes

$$\left.\begin{aligned}\Delta(\lambda) = A_1^2(\lambda-\lambda_2)^3(\lambda-\lambda_5) + \\ +(A_2^2+A_3^2+A_4^2)(\lambda-\lambda_1)(\lambda-\lambda_2)^2(\lambda-\lambda_5)+\\ +A_5^2(\lambda-\lambda_1)(\lambda-\lambda_2)^3.\end{aligned}\right\} \tag{102.12}$$

Supposing that none of the A's vanish, the graph of $\Delta(\lambda)$ is now as in Fig. 48. We have

$$\Delta(\lambda_1)>0, \quad \Delta(\lambda_2)=0, \quad \Delta(\lambda_5)>0, \tag{102.13}$$

and, near $\lambda=\lambda_2$,

$$\Delta(\lambda)\sim(A_2^2+A_3^2+A_4^2)(\lambda_2-\lambda_1)(\lambda-\lambda_2)^2(\lambda_2-\lambda_5)<0. \tag{102.14}$$

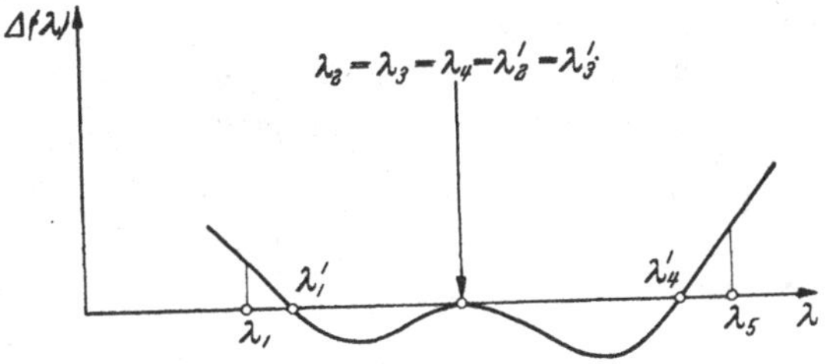

Fig. 48. Effect of a constraint on a degenerate system.

The constrained system has normal frequencies $\nu_1'<\nu_2'=\nu_3'<\nu_4'$, as indicated in Fig. 48 (pass from λ' to ν' by $4\pi^2\nu'^2=\lambda'$). The triple degeneracy has been reduced by the constraint to a double degeneracy.

A stable system remains stable under constraint, and an unstable system may be made stable by constraint.

For a system subject to one constraint, as discussed above, one of the coordinates may be eliminated, and $N-1$ new normal coordinates introduced. This procedure enables us to study the effect of additional constraints. In the general case where there is no degeneracy and the constraints are not specially chosen, we get successive separations as indicated below:

No constraint:	ν_1		ν_2		ν_3		...	ν_{N-1}		ν_N
One constraint:		ν_1'		ν_2'		ν_3'	...		ν_{N-1}'	
Two constraints:			ν_1''	ν_2''		ν_3''	...	ν_{N-1}''		
. . . .										
$N-2$ constraints:				$\nu_1^{(N-2)}$		$\nu_2^{(N-2)}$				
$N-1$ constraints:					$\nu_1^{(N-1)}$					

All these questions are, from a geometrical standpoint, questions about the lengths of the principal axes of plane sections of an ellipsoid in multidimensional Euclidean space; when so viewed, some of them can be answered very rapidly.

103. Dissipative systems. Gyroscopic stability. Consider the following set of N linear differential equations with real constant coefficients:

$$a_{\varrho\sigma}\ddot{q}^\sigma + c_{\varrho\sigma}\dot{q}^\sigma + b_{\varrho\sigma}q^\sigma = 0. \tag{103.1}$$

Such equations occur in the study of dissipative systems and of gyroscopic systems, and also in the theory of electrical circuits. We regard them here as exact equations, although in practice they may only be linear approximations to more complicated exact equations. We are interested in the stability of the solutions of (103.1), the case where $a_{\varrho\sigma}=a_{\sigma\varrho}$, $b_{\varrho\sigma}=b_{\sigma\varrho}$ and $c_{\varrho\sigma}=0$ having been already treated in Sect. 101.

Consider a system with kinetic and potential energies

$$T = \tfrac{1}{2} a_{\varrho\sigma} \dot{q}^\varrho \dot{q}^\sigma, \qquad V = \tfrac{1}{2} b_{\varrho\sigma} q^\varrho q^\sigma. \tag{103.2}$$

If, in addition to the generalized force $-\partial V/\partial q_\varrho$, a frictional (damping) force

$$Q_\varrho = -c_{\varrho\sigma} \dot{q}^\sigma \tag{103.3}$$

is applied, the equations of motion take the form (103.1). Hence we have

$$\frac{d}{dt}(T+V) = Q_\varrho \dot{q}^\varrho = -c_{\varrho\sigma} \dot{q}^\varrho \dot{q}^\sigma. \tag{103.4}$$

If we assume, as is natural, that the work done by the damping force is negative, then the quadratic form $c_{\varrho\sigma}\dot{q}^\varrho\dot{q}^\sigma$ is positive-definite, and the quantity $T+V$ decreases steadily. If the system is stable in the absence of damping, i.e. if V is positive-definite, then that stability is not destroyed by the damping. But if it is unstable without damping, we cannot tell whether the damping induces stability without recourse to a general argument as given below.

Equations of the form (103.1) also occur for a system with Lagrangian of the form

$$L = \tfrac{1}{2} a_{\varrho\sigma} \dot{q}^\varrho \dot{q}^\sigma + d_{\varrho\sigma} \dot{q}^\varrho q^\sigma - \tfrac{1}{2} b_{\varrho\sigma} q^\varrho q^\sigma. \tag{103.5}$$

Here we find

$$c_{\varrho\sigma} = d_{\varrho\sigma} - d_{\sigma\varrho}, \tag{103.6}$$

a skew-symmetric matrix. We meet Lagrangians of this form in rheonomic systems or as the result of the ignoration of coordinates, particularly in the case of gyroscopic systems; for that reason stability arising from the presence of the middle term in (103.1) is called *gyroscopic stability*[1].

We proceed to investigate the stability of the solutions of (103.1) by substituting

$$q^\varrho = \alpha^\varrho e^{st}, \tag{103.7}$$

where the α's and s are constants; on eliminating the former we obtain for s the determinantal equation

$$\Delta(s) = \det(a_{\varrho\sigma} s^2 + c_{\varrho\sigma} s + b_{\varrho\sigma}) = 0. \tag{103.8}$$

The criterion for stability is that every root of this equation should have a non-positive real part[2].

It is easy to establish the following result, already indicated in connection with (103.4): If $a_{\varrho\sigma} = a_{\sigma\varrho}$, $b_{\varrho\sigma} = b_{\sigma\varrho}$ (but $c_{\varrho\sigma} \neq c_{\sigma\varrho}$ in general), and if the three quadratic forms

$$A = a_{\varrho\sigma} q^\varrho q^\sigma, \qquad B = b_{\varrho\sigma} q^\varrho q^\sigma, \qquad C = c_{\varrho\sigma} q^\varrho q^\sigma \tag{103.9}$$

are all positive-definite, then the system is stable. To prove this, we note that the positive-definite character of B implies $\det b_{\varrho\sigma} \neq 0$, and therefore (103.8) has no zero root. Then, for any root s, we have

$$(s\, a_{\varrho\sigma} + c_{\varrho\sigma} + s^{-1} b_{\varrho\sigma}) \alpha^\sigma = 0 \tag{103.10}$$

for some non-vanishing (complex) vector α^σ; multiplying by the complex conjugate $\bar{\alpha}^\varrho$ and adding to the equation so obtained its complex conjugate, we get

$$(s + \bar{s}) A' + 2C' + (s^{-1} + \bar{s}^{-1}) B' = 0, \tag{103.11}$$

[1] See Sect. 105 for the Hamiltonian treatment of oscillations about steady motion.

[2] Overdamped systems, for which all the eigenvalues s are real and negative, have been discussed by R. J. Duffin, J. Rational Mech. Anal. 4, 221 (1955).

where $A' = a_{\varrho\sigma}\bar{\alpha}^\varrho\alpha^\sigma$, $B' = b_{\varrho\sigma}\bar{\alpha}^\varrho\alpha^\sigma$, $2C' = c_{\varrho\sigma}\bar{\alpha}^\varrho\alpha^\sigma + c_{\varrho\sigma}\alpha^\varrho\bar{\alpha}^\sigma$. (103.12)

This may be written

$$(s+\bar{s})\left(A' + \frac{B'}{s\bar{s}}\right) + 2C' = 0, \tag{103.13}$$

from which we obtain $s+\bar{s}<0$ (implying stability) since the positive-definite character of A, B, C implies that A', B', C' are all positive.

Let us return to the determinantal equation (103.8), placing no restriction on the matrices $a_{\varrho\sigma}$, $b_{\varrho\sigma}$, $c_{\varrho\sigma}$, except

$$\det a_{\varrho\sigma} > 0. \tag{103.14}$$

Then, on expanding the determinant, we have an algebraic equation of degree $2N$, the coefficient of s^{2N} being positive. We seek necessary and sufficient conditions that all its roots should have negative real parts (a slightly stronger condition than the requirement of stability, for which a zero real part would suffice).

In the argument which follows[1], the evenness of the degree of the equation plays no part, and it is convenient to write the equation

$$f(s) = a_0 s^n + a_1 s^{n-1} + \cdots = 0, \quad a_0 > 0. \tag{103.15}$$

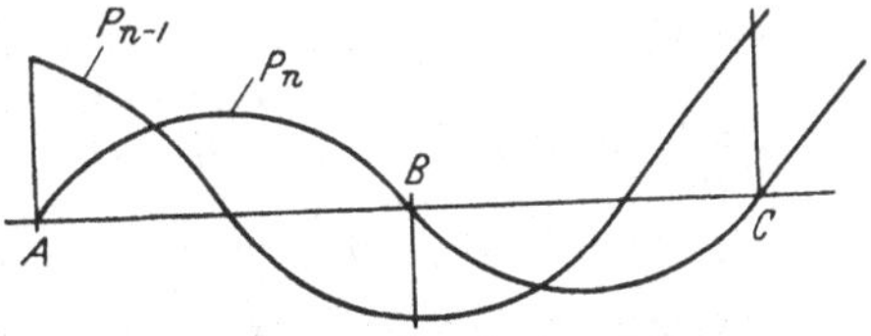

Fig. 49. Interlocked polynomials.

If, in the complex plane of s, we lead s along the imaginary axis from $-\infty$ to $+\infty$, then the increase in $\arg f(s)$ is precisely π times the number of roots of $f(s)=0$ with negative real parts. To use this fact, we write $s = iy$ and

$$f(s) = f(iy) = i^n (P_n - i P_{n-1}) \tag{103.16}$$

where

$$\left.\begin{aligned} P_n &= a_0 y^n - a_2 y^{n-2} + \cdots, \\ P_{n-1} &= a_1 y^{n-1} - a_3 y^{n-3} + \cdots. \end{aligned}\right\} \tag{103.17}$$

Thus, for s on the imaginary axis, we have

$$\arg f(s) = \frac{1}{2} n\pi - \arctan \frac{P_{n-1}}{P_n}, \tag{103.18}$$

and, if all the zeros of $f(s)$ have negative real parts, then $\arctan (P_{n-1}/P_n)$ decreases by $n\pi$ as y goes from $-\infty$ to $+\infty$. This occurs if, and only if, the polynomials (P_n, P_{n-1}) are *interlocked* in the following sense:

(i) All the zeros of $P_n(y)$ are real.

(ii) All the zeros of $P_{n-1}(y)$ are real and separate those of $P_n(y)$.

(iii) The relation between P_n and P_{n-1} is as shown in Fig. 49; P_{n-1} is positive at A and negative at B, these being successive zeros of P_n, with P_n positive between them. This relation is equivalent to the condition that P_n and P_{n-1} have the same sign in the limit $y \to +\infty$. (In Fig. 49 $\arctan P_{n-1}/P_n$ decreases by π as we go from A to B.)

Accordingly the question of the negative character of the real parts of the zeros of $f(s)$ is equivalent to the question of the interlocking of (P_n, P_{n-1}). To discuss this, we change the notation, writing $a_0 = A_n$, $a_1 = A_{n-1}$ and

$$\left.\begin{aligned} P_n &= A_n y^n + B_n y^{n-2} + \cdots, \\ P_{n-1} &= A_{n-1} y^{n-1} + B_{n-1} y^{n-3} + \cdots \end{aligned}\right\} \tag{103.19}$$

[1] E. J. ROUTH: Stability of a Given State of Motion, Chap. 3 (London: Macmillan 1877); PÉRÈS [20], p. 265; ROUTH [22] II, Chap. 6; WINKELMANN and GRAMMEL [29], p. 480.

By the following formulae we define a sequence of polynomials $P_{n-2}, P_{n-3}, \dots P_0$ and a sequence of numbers $A_{n-2}, A_{n-3}, \dots A_0$ (the coefficients of their highest powers):

$$\left.\begin{aligned} P_{n-2} &= A_{n-2} y^{n-2} + B_{n-2} y^{n-4} + \cdots = A_n y P_{n-1} - A_{n-1} P_n, \\ P_{n-3} &= A_{n-3} y^{n-3} + B_{n-3} y^{n-5} + \cdots = A_{n-1} y P_{n-2} - A_{n-2} P_{n-1}, \\ &\dots\dots\dots\dots\dots\dots \\ P_2 &= A_2 y^2 + B_2 = A_4 y P_3 - A_3 P_4, \\ P_1 &= A_1 y = A_3 y P_2 - A_2 P_3, \\ P_0 &= A_0 = A_2 y P_1 - A_1 P_2. \end{aligned}\right\} \tag{103.20}$$

As in (103.15), we take $A_n = a_0 > 0$. Suppose (P_n, P_{n-1}) interlocked, which implies $A_{n-1} > 0$. We shall prove that (P_{n-1}, P_{n-2}) are interlocked. To do this, we see from the first of (103.20) that, at any zero of P_{n-1}, P_{n-2} and P_n have opposite signs. Therefore P_{n-2} has all its zeros real, and they separate those of P_{n-1}. In the limit $y \to +\infty$, P_{n-2} has the same sign as it had at the greatest zero of P_{n-1}; that sign is the sign of $-P_n$ at that zero, and is positive (cf. Fig. 49). Therefore $A_{n-2} > 0$; (P_{n-1}, P_{n-2}) are interlocked.

Proceeding step by step in this way, we see that all the pairs (P_{n-2}, P_{n-3}), $(P_{n-3}, P_{n-4}), \dots (P_1, P_0)$ are interlocked[1], and (given $A_n > 0$) we conclude that the interlocking of (P_n, P_{n-1}) implies

$$A_{n-1} > 0, \quad A_{n-2} > 0, \quad \dots, \quad A_1 > 0, \quad A_0 > 0. \tag{103.21}$$

To prove the converse, we assume (103.21) with $A_n > 0$. For large y, the terms on the right in (103.20) are separately of higher order than the terms on the left; hence all the P's have the same sign in the limit $y \to +\infty$, and that sign is positive, the sign of A_0. Thus, by the last of (103.20), P_2 is positive at infinity and negative at $y = 0$, where $P_1 = 0$. Hence (P_2, P_1) are interlocked, and, working up the table step by step, we conclude that (P_n, P_{n-1}) are interlocked.

Thus (103.21) are necessary and sufficient conditions for the interlocking of (P_n, P_{n-1}), or, equivalently, for the negative character of the real parts of all the zeros of $f(s)$.

The quantities occurring in (103.21) may be expressed in determinantal form as follows, in terms of the coefficients in the formula (103.15) for $f(s)$[2]:

$$\left.\begin{aligned} &A_{n-1} = a_1, \quad A_{n-2} = \begin{vmatrix} a_1 & a_0 \\ a_3 & a_2 \end{vmatrix} \quad A_{n-3} = \begin{vmatrix} a_1 & a_0 & 0 \\ a_3 & a_2 & a_1 \\ a_5 & a_4 & a_3 \end{vmatrix}, \\ &\dots\dots\dots\dots\dots\dots \\ &A_0 = \begin{vmatrix} a_1 & a_0 & 0 & \dots & 0 \\ a_3 & a_2 & a_1 & \dots & 0 \\ \cdot & \cdot & \cdot & \cdot & \cdot \\ a_{2n-1} & a_{2n-2} & & \dots & a_n \end{vmatrix} \end{aligned}\right\} \tag{103.22}$$

with the understanding that $a_r = 0$ for $r > n$.

---

[1] In the case of (P_1, P_0), interlocking means merely that P_1 has the sign of P_0 as $y \to +\infty$.

[2] CH. HERMITE: Crelle's J. **52**, 39 (1850). — A. HURWITZ: Math. Ann. **46**, 273 (1895); cf. R. GRAMMEL: Der Kreisel, Bd. 1, p. 259. Berlin: Springer 1950.

104. Forced oscillations. Resonance. Operational methods. If, for a system moving in accordance with (103.1), all the roots of (103.8) have negative real parts, then, no matter what the initial conditions may be, the system ultimately tends to rest at the origin. In addition to the forces already represented in (103.1) we now supply a disturbing force $Q_\varrho(t)$, regarded as a given function of t, and so modify the equations of motion to read

$$a_{\varrho\sigma}\ddot{q}^\sigma + c_{\varrho\sigma}\dot{q}^\sigma + b_{\varrho\sigma}q^\sigma = Q_\varrho. \tag{104.1}$$

If the applied force is simple harmonic with circular frequency ω, we write

$$Q_\varrho = F_\varrho e^{i\omega t}. \tag{104.2}$$

After a long time, no matter what the initial conditions are, the system will tend to a *forced oscillation* given by

$$q^\varrho = \alpha^\varrho e^{i\omega t}, \tag{104.3}$$

the complex amplitude factors α^ϱ being found by substituting in (104.1); they must satisfy the equations

$$(-a_{\varrho\sigma}\omega^2 + i c_{\varrho\sigma}\omega + b_{\varrho\sigma})\alpha^\sigma = F_\varrho. \tag{104.4}$$

This is no eigenvalue problem; it is merely a question of solving a set of linear equations. But the eigenvalue problem represented by (103.8) is closely connected with the solution of (104.4), because the amplitude factors become large when the disturbing frequency is close to a natural frequency, or, more accurately, when $i\omega$ is close to one of the roots of (103.8). Then we have *resonance*.

Such problems are most compactly discussed by operational methods, and we shall show how to obtain the solution of (104.1) for a general disturbing force, not necessarily of the form (104.2). For initial conditions we take

$$q^\varrho = \pi^\varrho, \quad \dot{q}^\varrho = v^\varrho \quad \text{for} \quad t = 0. \tag{104.5}$$

Let I denote the operation[1] of integrating with respect to t:

$$I f(t) = \int_0^t f(\tau)\, d\tau. \tag{104.6}$$

Apply the operator I to (104.1) and use (104.5); this gives

$$a_{\varrho\sigma}(\dot{q}^\sigma - v^\sigma) + c_{\varrho\sigma}(q^\sigma - \pi^\sigma) + b_{\varrho\sigma} I q^\sigma = I Q_\varrho. \tag{104.7}$$

On repeating this operation, we get

$$A_{\varrho\sigma} q^\sigma = B_\varrho + I^2 Q_\varrho, \tag{104.8}$$

where

$$\left.\begin{aligned} A_{\varrho\sigma} &= a_{\varrho\sigma} + c_{\varrho\sigma} I + b_{\varrho\sigma} I^2, \\ B_\varrho &= a_{\varrho\sigma}\pi^\sigma + a_{\varrho\sigma} I v^\sigma + c_{\varrho\sigma} I \pi^\sigma. \end{aligned}\right\} \tag{104.9}$$

The Eq. (104.8) is equivalent to (104.1) and (104.5); if it is satisfied, they are, as we may see on differentiating it.

[1] For details of the operational method, see H. JEFFREYS and B. S. JEFFREYS: Mathematical Physics, Chap. 7 (Cambridge: University Press 1956; 3rd. Edn.). They use Q for the integration operator, here changed to I to avoid confusion with the generalized force. Much difficulty is caused in the operational method by the use of the HEAVISIDE operators p and p^{-1}, which are not commutative, and, although they lend some formal simplicity to the work, they will not be used here. For operational methods based on the LAPLACE transform, see R. V. CHURCHILL: Modern Operational Mathematics in Engineering (New York and London: McGraw-Hill 1944); N. W. MCLACHLAN: Modern Operational Calculus (London: Macmillan 1948); K. W. WAGNER: Operatorenrechnung (Leipzig: Johann Ambrosius Barth 1940) (lithoprinted Ann Arbor: Edwards 1944); I. N. SNEDDON: Fourier Transforms (New York: McGraw-Hill 1951), and this Encyclopedia, Vol. II, p. 251.

The essence of the operational method consists in treating the operator I as if it were a number, the validity of results obtained in this way being checked afterwards. We treat $A_{\varrho\sigma}$ as a matrix of numbers and define $D^{\varrho\sigma}$ as the cofactor of the element $A_{\varrho\sigma}$, so that

$$D^{\varrho\mu} A_{\varrho\sigma} = \delta^{\mu}_{\sigma} D, \tag{104.10}$$

where D is the determinant

$$D = \det (a_{\varrho\sigma} + c_{\varrho\sigma} I + b_{\varrho\sigma} I^2). \tag{104.11}$$

Multiplying (104.8) by $D^{\varrho\mu}$ and then dividing across by D, we get

$$q^{\mu} = \frac{1}{D} D^{\varrho\mu} (B_{\varrho} + I^2 Q_{\varrho}). \tag{104.12}$$

The general rule of operational calculus is to expand any rational fraction in I as a power series in I. This ensures the commutativity of two such fractional operators. Now D is a polynomial of degree $2N$ in I; it is related to the polynomial Δ of (103.8) by

$$D(I) = I^{2N} \Delta\left(\frac{1}{I}\right). \tag{104.13}$$

The fraction $1/D$ admits a unique power series expansion,

$$\frac{1}{D} = C_0 + C_1 I + C_2 I^2 + \cdots, \tag{104.14}$$

and if this is substituted in (104.12) we get on the right hand side a mathematically meaningful expression, provided that the infinite process converges. If it does, then (104.12) is the solution of the differential equations (104.1) with the initial conditions (104.5).

But although $1/D$ is basically interpreted as an infinite series of operations, it not actually necessary to use an infinite series in order to calculate the solution (104.12). For, by (104.13),

$$D = a_0 (1 - s_1 I)(1 - s_2 I) \ldots (1 - s_{2N} I), \tag{104.15}$$

where $a_0 = \det a_{\varrho\sigma}$ and $s_1, \ldots s_{2N}$ are the eigenvalues, satisfying (103.8). We can then make resolutions into partial fractions as follows:

$$\left.\begin{aligned} \frac{D^{\varrho\mu} B_{\varrho}}{D} &= \frac{K_1^{\mu}}{1 - s_1 I} + \frac{K_2^{\mu}}{1 - s_2 I} + \cdots + \frac{K_{2N}^{\mu}}{1 - s_{2N} I}, \\ \frac{D^{\varrho\mu} I}{D} &= \frac{L_1^{\varrho\mu}}{1 - s_1 I} + \frac{L_2^{\varrho\mu}}{1 - s_2 I} + \cdots + \frac{L_{2N}^{\varrho\mu}}{1 - s_{2N} I}, \end{aligned}\right\} \tag{104.16}$$

the numerators of the fractions on the left being polynomials of degree $2N-1$ in I; here the K's are constants depending on the initial data, and the L's are constants independent of the initial data, depending in fact only on the three matrices $a_{\varrho\sigma}$, $b_{\varrho\sigma}$, $c_{\varrho\sigma}$. The expressions (104.16) are now substituted in (104.12), and the following formulae[1] applied:

$$\left.\begin{aligned} \frac{1}{1 - \alpha I} 1 &= e^{\alpha t}, \\ \frac{I}{1 - \alpha I} f(t) &= \int_0^t f(\tau)\, e^{\alpha(t-\tau)}\, d\tau, \end{aligned}\right\} \tag{104.17}$$

[1] These formulae are easily established; cf. Jeffreys and Jeffreys, p. 233 of op. cit. in preceding footnote.

α being any constant. The solution (104.12) takes the form

$$\left.\begin{aligned} q^{\mu} &= K_1^{\mu} e^{s_1 t} + K_2^{\mu} e^{s_2 t} + \cdots + K_{2N}^{\mu} e^{s_{2N} t} + \\ &+ \int\limits_0^t [L_1^{\varrho\mu} e^{s_1(t-\tau)} + L_2^{\varrho\mu} e^{s_2(t-\tau)} + \cdots + L_{2N}^{\varrho\mu} e^{s_{2N}(t-\tau)}]\, Q_{\varrho}(\tau)\, d\tau . \end{aligned}\right\} \qquad (104.18)$$

The above result is very general, except for the assumption made implicitly at (104.16), that the eigenvalues are distinct; in cases of degeneracy the partial fraction expansions must be modified by the inclusion of fractions with higher powers in the denominators.

Let us now assume the disturbing force simple harmonic, as in (104.2). Further, let us assume that all the eigenvalues are distinct and have negative real parts. Then, omitting those terms which tend to zero as $t \to \infty$, we get from (104.18) the following expression for the forced oscillation due to the applied force (104.2):

$$q^{\mu} = \left[\frac{L_1^{\varrho\mu}}{i\omega - s_1} + \frac{L_2^{\varrho\mu}}{i\omega - s_2} + \cdots + \frac{L_{2N}^{\varrho\mu}}{i\omega - s_{2N}}\right] F_{\varrho}\, e^{i\omega t}. \qquad (104.19)$$

Since the frequency ω is here involved only where shown explicitly, the formula exhibits clearly the phenomenon of resonance[1].

105. Oscillations about steady motion or about a singular point in phase space (*QP*). Transformation of *H* to normal form. Consider a dynamical system with N degrees of freedom and Hamiltonian $H(q, p)$ from which some of the coordinates (q) are absent (ignorable coordinates). A *steady* motion is defined to be one in which the non-ignorable coordinates and the corresponding momenta are constant.

Let there be M ignorable coordinates q_A $(A = 1, 2, \ldots M)$, the non-ignorable coordinates being written $Q_\Gamma (\Gamma = 1, 2, \ldots N - M)$, with a similar notation for the momenta. Then the Hamiltonian may be written

$$H = H(Q, p, P), \qquad (105.1)$$

and the equations of motion are

$$\dot{q}_A = \frac{\partial H}{\partial p_A}, \qquad \dot{p}_A = -\frac{\partial H}{\partial q_A} = 0, \qquad (105.2)$$

$$\dot{Q}_\Gamma = \frac{\partial H}{\partial P_\Gamma}, \qquad \dot{P}_\Gamma = -\frac{\partial H}{\partial Q_\Gamma}. \qquad (105.3)$$

The conditions for steady motion are

$$\dot{Q}_\Gamma = 0, \qquad \dot{P}_\Gamma = 0. \qquad (105.4)$$

Combining these last conditions with (105.2), we get, for a steady motion,

$$q_A = \alpha_A t + \text{const}, \qquad p_A = \text{const}, \qquad (105.5)$$

where α_A are constants, so that the ignorable coordinates increase at constant rates and the corresponding momenta are constant. Combining (105.4) with (105.3), we get, for a steady motion,

$$\frac{\partial H}{\partial Q_\Gamma} = 0, \qquad \frac{\partial H}{\partial P_\Gamma} = 0. \qquad (105.6)$$

---

[1] Cf. (33.10) for the harmonic oscillator.

To obtain a steady motion, we have to satisfy these $2(N-M)$ equations by assigning appropriate values to the $2N-M$ constants (Q, p, P). Thus we may in general expect to find ∞^M steady motions, M being the number of ignorable coordinates.

If a system in steady motion is disturbed in such a way that the constants p_A are not changed, we may investigate the oscillations about steady motion by using the equations of motion (105.3), linearised by assuming (Q, P) to have values adjacent to the constant values which they have in the undisturbed steady motion.

Having thus formulated the problem of the oscillations about steady motion of a system possessing ignorable coordinates, we now present the same problem in a different way without reference to steady motions or ignorable coordinates.

Given a Hamiltonian $H(q, p)$, the canonical equations

$$\dot{q}_\varrho = \frac{\partial H}{\partial p_\varrho}, \quad \dot{p}_\varrho = -\frac{\partial H}{\partial q_\varrho} \tag{105.7}$$

define a direction of motion at every point in phase space (QP), except at those *singular* points where the $2N$ equations

$$\frac{\partial H}{\partial q_\varrho} = 0, \quad \frac{\partial H}{\partial p_\varrho} = 0 \tag{105.8}$$

are satisfied. Since there are $2N$ quantities (q, p), we may expect in general to find a finite number of singular points in QP. Each such point represents a complete history of the system, because the equations of motion (105.7) are satisfied by constant values of (q, p), provided these constant values satisfy (105.8).

We shall now investigate trajectories in QP adjacent to a singular point. On comparing (105.6) and (105.8), we recognize that such an investigation is at the same time an investigation of oscillations about a state of steady motion for a system with ignorable coordinates. The present approach has the advantage that we come at once to the heart of the matter.

The discussion which follows is closely connected with the matters treated in Sect. 103. But here we shall use Hamiltonian methods rather than Lagrangian. In the Lagrangian method we are restricted to transformations of the coordinates (q), the transformation of momenta (p) (if we bring in momenta at all) being a derived transformation; in the Hamiltonian method we can use canonical transformations (CT).

Singular points in QP are invariant under CT. For (105.8) are equivalent to $\delta H = 0$ for an arbitrary variation of position in QP, and this is invariant since H is invariant under CT.

Let $q_\varrho = a_\varrho$, $p_\varrho = b_\varrho$ be a singular point. The generating function

$$G(q, p') = (q_\varrho - a_\varrho)(p'_\varrho + b_\varrho) \tag{105.9}$$

gives the CT

$$p_\varrho = \frac{\partial G}{\partial q_\varrho} = p'_\varrho + b_\varrho, \quad q'_\varrho = \frac{\partial G}{\partial p'_\varrho} = q_\varrho - a_\varrho, \tag{105.10}$$

so that the singular point becomes $q'_\varrho = 0$, $p'_\varrho = 0$. We shall now use these new canonical variables, dropping the primes.

We assume $H(q, p)$ expansible in a power series about the singular point. The constant term in the expansion is ineffective in (105.7), and we drop it. Then, in view of (105.8), we have, to the second order inclusive,

$$H(q, p) = \tfrac{1}{2} A_{\varrho\sigma} q_\varrho q_\sigma + B_{\varrho\sigma} q_\varrho p_\sigma + \tfrac{1}{2} C_{\varrho\sigma} p_\varrho p_\sigma, \tag{105.11}$$

where the coefficients are constants.

For compactness, we introduce the notation of Sect. 87, with a slight modification of dimensions, since we are now in QP (of $2N$ dimensions) instead of in $QTPH$ (of $2N+2$ dimensions). We write

$$\boldsymbol{z} = \begin{pmatrix} \boldsymbol{q} \\ \boldsymbol{p} \end{pmatrix}; \tag{105.12}$$

then (105.11) may be written

$$2H = \tilde{\boldsymbol{z}} \boldsymbol{H} \boldsymbol{z}, \tag{105.13}$$

where $\boldsymbol{H}$ is a symmetric $2N \times 2N$ matrix ($\tilde{\boldsymbol{H}} = \boldsymbol{H}$). As in (87.11), we write

$$\boldsymbol{\Gamma} = \begin{pmatrix} \boldsymbol{0} & \boldsymbol{1} \\ -\boldsymbol{1} & \boldsymbol{0} \end{pmatrix}, \tag{105.14}$$

where $\boldsymbol{1}$ is now the unit $N \times N$ matrix, and note the properties

$$\tilde{\boldsymbol{\Gamma}} = -\boldsymbol{\Gamma}, \quad \boldsymbol{\Gamma}^{-1} = -\boldsymbol{\Gamma}, \quad \boldsymbol{\Gamma}^2 = -\boldsymbol{1}_{(2N)}, \quad \det \boldsymbol{\Gamma} = 1. \tag{105.15}$$

Then, as in (87.13), the equations of motion read

$$\dot{\boldsymbol{z}} = \boldsymbol{\Gamma} \boldsymbol{H} \boldsymbol{z}. \tag{105.16}$$

To find the nature of the motion, we seek to separate the variables in these equations, and this we do by applying a CT which reduces the matrix $\boldsymbol{H}$ to a simple (normal) form[1].

Consider the linear equations

$$\boldsymbol{\Gamma} \boldsymbol{H} \boldsymbol{z} = \lambda \boldsymbol{z}, \quad \text{or} \quad \boldsymbol{H} \boldsymbol{z} = -\lambda \boldsymbol{\Gamma} \boldsymbol{z}, \tag{105.17}$$

and the associated determinantal equation

$$\det(\boldsymbol{H} + \lambda \boldsymbol{\Gamma}) = 0, \tag{105.18}$$

which has $2N$ roots, real or complex, and not necessarily distinct. For any root λ we have (since $\det \boldsymbol{\Gamma} = 1$)

$$\det[\boldsymbol{\Gamma}(\boldsymbol{H} + \lambda \boldsymbol{\Gamma})\boldsymbol{\Gamma}] = 0. \tag{105.19}$$

Transposing this matrix and using (105.15), we obtain

$$\det(\boldsymbol{H} - \lambda \boldsymbol{\Gamma}) = 0, \tag{105.20}$$

and comparison with (105.18) shows that, if λ is a root, then so also is $-\lambda$. We can write the whole set of roots as $\pm\lambda_1, \ldots \pm\lambda_N$, and exhibit them in the form of a diagonal matrix

$$\boldsymbol{L} = \begin{pmatrix} \boldsymbol{L}_0 & \boldsymbol{0} \\ \boldsymbol{0} & -\boldsymbol{L}_0 \end{pmatrix}, \tag{105.21}$$

where $\boldsymbol{L}_0$ is the diagonal $N \times N$ matrix with elements $(\lambda_1, \ldots \lambda_N)$.

We shall assume that $\lambda_1, \ldots \lambda_N$ are distinct[2]. Then a set of solutions of (105.17) may be exhibited as a $2N \times 2N$ matrix $\boldsymbol{Z}$ such that

$$\boldsymbol{Z}^{-1} \boldsymbol{\Gamma} \boldsymbol{H} \boldsymbol{Z} = \boldsymbol{L}. \tag{105.22}$$

---

[1] Cf. WHITTAKER [28], pp. 427–429; C. LANCZOS, Ann. d. Phys. (5) **20**, 653–688; C. L. SIEGEL, pp. 76–80 of op. cit. in Sect. 53.

[2] The present argument applies only to this non-degenerate case. The case of repeated roots is covered by the WEIERSTRASS contour-integration treatment of stability of motion, as given by WHITTAKER [28], pp. 197–202.

However (and this is important) this equation does not determine $\boldsymbol{Z}$ uniquely. If $\boldsymbol{Z}$ is any solution-matrix, then so is $\boldsymbol{ZP}$, where $\boldsymbol{P}$ is any diagonal matrix; and if $\boldsymbol{Y}$ and $\boldsymbol{Z}$ are two solution-matrices, then

$$\boldsymbol{Z} = \boldsymbol{Y}\boldsymbol{P}, \tag{105.23}$$

where $\boldsymbol{P}$ is some diagonal matrix.

Let $\boldsymbol{Z}$ be any solution-matrix. Define $\boldsymbol{Y}$ by

$$\boldsymbol{Y} = -(\boldsymbol{\Gamma}\tilde{\boldsymbol{Z}}\boldsymbol{\Gamma})^{-1} = -\boldsymbol{\Gamma}\tilde{\boldsymbol{Z}}^{-1}\boldsymbol{\Gamma}. \tag{105.24}$$

Then

$$\boldsymbol{Y}^{-1}\boldsymbol{\Gamma}\boldsymbol{H}\boldsymbol{Y} = \boldsymbol{\Gamma}\tilde{\boldsymbol{Z}}\boldsymbol{\Gamma}\cdot\boldsymbol{\Gamma}\boldsymbol{H}\cdot\boldsymbol{\Gamma}\tilde{\boldsymbol{Z}}^{-1}\boldsymbol{\Gamma} = -\boldsymbol{\Gamma}\tilde{\boldsymbol{Z}}\boldsymbol{H}\boldsymbol{\Gamma}\tilde{\boldsymbol{Z}}^{-1}\boldsymbol{\Gamma}. \tag{105.25}$$

But, transposing (105.22), we have

$$\tilde{\boldsymbol{Z}}\boldsymbol{H}\boldsymbol{\Gamma}\tilde{\boldsymbol{Z}}^{-1} = -\boldsymbol{L}, \tag{105.26}$$

and so

$$\boldsymbol{Y}^{-1}\boldsymbol{\Gamma}\boldsymbol{H}\boldsymbol{Y} = \boldsymbol{\Gamma}\boldsymbol{L}\boldsymbol{\Gamma} = \boldsymbol{L}. \tag{105.27}$$

Thus $\boldsymbol{Y}$ is also a solution-matrix, and so is connected with $\boldsymbol{Z}$ by (105.23), where $\boldsymbol{P}$ is some diagonal matrix. Hence

$$\boldsymbol{P} = \boldsymbol{Y}^{-1}\boldsymbol{Z} = -\boldsymbol{\Gamma}\tilde{\boldsymbol{Z}}\boldsymbol{\Gamma}\boldsymbol{Z}, \quad \boldsymbol{\Gamma}\boldsymbol{P} = \tilde{\boldsymbol{Z}}\boldsymbol{\Gamma}\boldsymbol{Z}, \tag{105.28}$$

and, transposing, we get

$$\boldsymbol{P}\boldsymbol{\Gamma} = \boldsymbol{\Gamma}\boldsymbol{P}, \tag{105.29}$$

from which it follows that $\boldsymbol{P}$ is of the form

$$\boldsymbol{P} = \begin{pmatrix} \boldsymbol{P}_0 & 0 \\ 0 & \boldsymbol{P}_0 \end{pmatrix}, \tag{105.30}$$

where $\boldsymbol{P}_0$ is a diagonal $N \times N$ matrix.

In the class of solution-matrices satisfying (105.22) we now seek one (say $\boldsymbol{J}$) satisfying the symplectic condition [cf. (87.16)]

$$\tilde{\boldsymbol{J}}\boldsymbol{\Gamma}\boldsymbol{J} = \boldsymbol{\Gamma}, \tag{105.31}$$

and this we do by taking *any* solution matrix $\boldsymbol{Z}$, forming the corresponding diagonal $\boldsymbol{P}$ by (105.28), defining $\boldsymbol{D}$ by

$$\boldsymbol{D} = \begin{pmatrix} \boldsymbol{P}_0^{-1} & 0 \\ 0 & 1 \end{pmatrix}, \tag{105.32}$$

and writing

$$\boldsymbol{J} = \boldsymbol{Z}\boldsymbol{D}. \tag{105.33}$$

It is easy to verify that

$$\boldsymbol{D}\boldsymbol{\Gamma}\boldsymbol{D} = \boldsymbol{\Gamma}\boldsymbol{P}^{-1}, \tag{105.34}$$

and hence, using (105.28),

$$\tilde{\boldsymbol{J}}\boldsymbol{\Gamma}\boldsymbol{J} = \boldsymbol{D}\tilde{\boldsymbol{Z}}\boldsymbol{\Gamma}\boldsymbol{Z}\boldsymbol{D} = \boldsymbol{D}\boldsymbol{\Gamma}\boldsymbol{P}\boldsymbol{D} = \boldsymbol{D}\boldsymbol{\Gamma}\boldsymbol{D}\boldsymbol{P} = \boldsymbol{\Gamma}, \tag{105.35}$$

establishing (105.31).

With $\boldsymbol{J}$ defined by (105.33), we apply the CT

$$\boldsymbol{z} = \boldsymbol{J}\boldsymbol{z}', \tag{105.36}$$

and the Hamiltonian of (105.13) becomes

$$2H = \tilde{\boldsymbol{z}}'\boldsymbol{H}'\boldsymbol{z}', \quad \boldsymbol{H}' = \tilde{\boldsymbol{J}}\boldsymbol{H}\boldsymbol{J}. \tag{105.37}$$

Since $\boldsymbol{J}$ is a solution-matrix, (105.26) gives

$$\tilde{\boldsymbol{J}}\boldsymbol{H}\boldsymbol{\Gamma}\tilde{\boldsymbol{J}}^{-1} = -\boldsymbol{L}, \quad \tilde{\boldsymbol{J}}\boldsymbol{H} = \boldsymbol{L}\tilde{\boldsymbol{J}}\boldsymbol{\Gamma}, \tag{105.38}$$

and so

$$\boldsymbol{H}' = \boldsymbol{L}\tilde{\boldsymbol{J}}\boldsymbol{\Gamma}\boldsymbol{J} = \boldsymbol{L}\boldsymbol{\Gamma} = \begin{pmatrix} \boldsymbol{0} & \boldsymbol{L}_0 \\ \boldsymbol{L}_0 & \boldsymbol{0} \end{pmatrix}. \tag{105.39}$$

Thus by a CT *we are able to transform a quadratic Hamiltonian to the normal form*

$$H' = \lambda_1 q_1' p_1' + \lambda_2 q_2' p_2' + \cdots + \lambda_N q_N' p_N', \tag{105.40}$$

where the λ's are the roots of (105.18). The new coordinates (q', p') are in general complex.

If we apply the further CT with generating function

$$G(q', p'') = \sum_{\varrho=1}^{N} \left(q_\varrho' p_\varrho'' - \frac{1}{2} \frac{p_\varrho''^2}{\lambda_\varrho} - \frac{1}{4} \lambda_\varrho q_\varrho'^2 \right), \tag{105.41}$$

so that (without the summation convention)

$$p_\varrho' = \frac{\partial G}{\partial q_\varrho'} = p_\varrho'' - \frac{1}{2} \lambda_\varrho q_\varrho', \quad q_\varrho'' = \frac{\partial G}{\partial p_\varrho''} = q_\varrho' - \frac{p_\varrho''}{\lambda_\varrho}, \tag{105.42}$$

the Hamiltonian changes into an alternative normal form,

$$H'' = \tfrac{1}{2}(p_1''^2 + \cdots + p_N''^2 - \lambda_1^2 q_1''^2 - \cdots - \lambda_N^2 q_N''^2). \tag{105.43}$$

If we use the form (105.40), the equations of motion read

$$\dot{q}_1' = \frac{\partial H'}{\partial p_1'} = \lambda_1 q_1', \ldots, \quad \dot{p}_1' = -\frac{\partial H'}{\partial q_1'} = -\lambda_1 p_1', \ldots, \tag{105.44}$$

in which the variables are separated, and the motion is given by

$$q_1' = b_1 e^{\lambda_1 t}, \ldots, \quad p_1' = a_1 e^{-\lambda_1 t}, \ldots, \tag{105.45}$$

the constant coefficients depending on the initial conditions. It is clear that the motion given by these equations is stable if, and only if, all the λ's are pure imaginaries.

It is easy to show that if $H(q, p)$ is positive-definite, then all the roots of (105.18) are pure imaginaries, even in the degenerate case of repeated roots. Let λ, $\boldsymbol{z}$ (with $\boldsymbol{z} \neq \boldsymbol{0}$) be any solution of (105.17). Let $\bar{\boldsymbol{z}}$ be the complex conjugate of $\boldsymbol{z}$, and $\boldsymbol{z}^\dagger$ the transpose of $\bar{\boldsymbol{z}}$ (i.e. the row matrix). Then

$$\boldsymbol{z}^\dagger \boldsymbol{H} \boldsymbol{z} = -\lambda \boldsymbol{z}^\dagger \boldsymbol{\Gamma} \boldsymbol{z}. \tag{105.46}$$

From the positive-definite character of H, the left hand side is real and positive. Therefore $\lambda \neq 0$ (no zero roots) and

$$\boldsymbol{z}^\dagger \boldsymbol{\Gamma} \boldsymbol{z} \neq \boldsymbol{0}. \tag{105.47}$$

This quantity is easily seen to be a pure imaginary, and hence λ is a pure imaginary also.

We have the following result: *the motion of a system with homogeneous quadratic Hamiltonian H is stable if H is positive-definite*[1].

[1] We have proved it here only in the non-degenerate case of distinct eigenvalues; for the degenerate case, see preceding footnote

106. Perturbations. Consider two dynamical systems, S and S', with canonical variables (q, p), (q', p') and Hamiltonians $H(q, t, p)$, $H'(q', t, p')$ respectively. They are completely independent. Suppressing suffixes, we may write their equations of motion in the form

$$\left.\begin{aligned} S:\quad \dot{q} &= \frac{\partial H}{\partial p}, & \dot{p} &= -\frac{\partial H}{\partial q}; \\ S':\quad \dot{q}' &= \frac{\partial H'}{\partial p'}, & \dot{p}' &= -\frac{\partial H'}{\partial q'}. \end{aligned}\right\} \tag{106.1}$$

Now think of S and S' as forming a single system $S+S'$ with a Hamiltonian

$$H(q, t, p) + H'(q', t, p') + K(q, q', t, p, p'), \tag{106.2}$$

K being an *interaction* Hamiltonian. (As a very simple example, we might take S and S' to be two free particles and K to be the potential due to their mutual gravitational attraction.) The equations of motion of $S+S'$ are

$$S+S':\left\{\begin{aligned} \dot{q} &= \frac{\partial H}{\partial p} + \frac{\partial K}{\partial p}, & \dot{p} &= -\frac{\partial H}{\partial q} - \frac{\partial K}{\partial q}, \\ \dot{q}' &= \frac{\partial H'}{\partial p'} + \frac{\partial K}{\partial p'}, & \dot{p}' &= -\frac{\partial H'}{\partial q'} - \frac{\partial K}{\partial q'}. \end{aligned}\right\} \tag{106.3}$$

The effect of K is to *perturb* the original motions (106.1) of S and S'. If the derivatives of K are small, the values of $\dot{q}, \dot{p}, \dot{q}', \dot{p}'$ at a given point of the space QTP of $S+S'$ are nearly the same in the perturbed and unperturbed motions. But in a very long time significant changes in the motion may result; in that case we speak of *secular* changes.

For the system formed by the sun and the planets, the Hamiltonian may be written in the form

$$H = T(S) + \sum T(P) + \sum V(S\,P) + \sum V(P\,P'), \tag{106.4}$$

where $T(S)$ is the kinetic energy of the sun, $T(P)$ the kinetic energy of a planet, $V(S\,P)$ the mutual potential energy of the sun and a planet, and $V(P\,P')$ the mutual potential energy of two planets. To put this into perturbation form, we denote by S_0 a fictitious sun fixed at the origin and define $V'(S\,P)$ by

$$V'(S\,P) = V(S\,P) - V(S_0\,P). \tag{106.5}$$

Then the Hamiltonian (106.4) may be written

$$H = H(S) + \sum H(P) + K, \tag{106.6}$$

where

$$H(S) = T(S), \quad H(P) = T(P) + V(S_0\,P), \tag{106.7}$$

and K stands for the terms not otherwise accounted for. K is the perturbing Hamiltonian.

The unperturbed motion is known, for $H(S)$ corresponds to the free motion of a particle and $H(P)$ to the KEPLER problem. The practical significance of (106.6) lies in the fact that K is small; the actual motion of the solar system is a perturbation of a state of motion in which the sun is at rest, and the planets describe fixed elliptical orbits with the sun as focus, without mutual interaction.

The basic idea behind perturbation theory is this: starting at $t=t_1$, the motion up to $t=t_2$ of the complete system (including the perturbation) differs little from the unperturbed motion, provided we start the two motions from the same point in QTP-space and do not make the interval $t_2 - t_1$ long. Assuming the unperturbed motion known, the effect of the perturbation during such a finite interval can be found by approximate methods[1].

Without making any approximations based in smallness of the perturbing Hamiltonian, the general perturbation problem may be stated as follows: Given the general solution of the canonical equations

$$\dot{q}_\varrho = \frac{\partial H_0}{\partial p_\varrho}, \quad \dot{p}_\varrho = -\frac{\partial H_0}{\partial q_\varrho} \tag{106.8}$$

of unperturbed motion, it is required to set up a technique to find the motion for the Hamiltonian

$$H = H_0(q, t, p) + H_1(q, t, p), \tag{106.9}$$

H_1 being the perturbing Hamiltonian.

Let the solution of (106.8) be

$$q_\varrho = q_\varrho(c, t), \quad p_\varrho = p_\varrho(c, t), \tag{106.10}$$

where c stands for $2N$ arbitrary constants c_A, capital suffixes having the range $1, \ldots 2N$; these quantities are constant along each unperturbed trajectory. Solving (106.10), we have

$$c_A = c_A(q, t, p), \tag{106.11}$$

these $2N$ functions being determined by the form of the function $H_0(q, t, p)$.

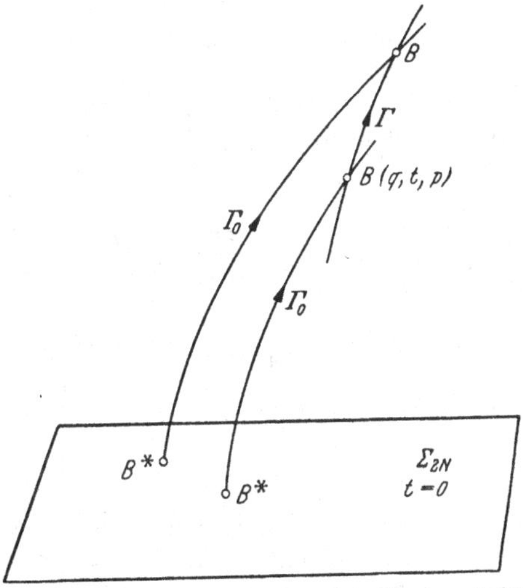

Fig. 50. Perturbation viewed in QTP. Γ_0 = unperturbed trajectories, Γ = perturbed trajectory.

Let us view the situation in QTP (Fig. 50)[2]. Σ_{2N} is the surface $t=0$, B is any point, and Γ_0 is the unperturbed trajectory through B, cutting Σ_{2N} at B^*, say. Since c_A are constant along Γ_0, we have at B^*

$$q_\varrho^* = q_\varrho(c, 0), \quad p_\varrho^* = p_\varrho(c, 0). \tag{106.12}$$

Thus c_A form a system of coordinates on Σ_{2N}; (c, t) form a system of coordinates in QTP, but not a canonical system in general. The unperturbed trajectories Γ_0 form a system of projection-lines by which a point B is projected into B^*, the corresponding values of c_A being given by (106.11).

Consider now a perturbed trajectory Γ. At B its direction differs from the direction of Γ_0, and as the representative point B traverses Γ, its projection B^* moves on Σ_{2N}. For that reason, the method given here is called the method of *variation of constants*, since c_A are constants for Γ_0 but not for Γ. The perturbation problem is reduced to the study of the way in which c_A vary with t as the representative point traverses Γ; if we knew this, then we would know Γ, its equations in the form

$$c_A = f_A(t) \tag{106.13}$$

determining a curve in QTP in the coordinate system (c, t).

---

[1] For a detailed treatment of perturbations, with use of action-angle variables, see Frank [5], pp. 127–156, and Fues [6].

[2] Remember that any set of canonical equations defines a congruence of curves in QTP, one curve through each point; cf. Sect. 93.

On Γ we have, by (97.9), c_A being the function (106.11),

$$\dot{c}_A = \frac{\partial c_A}{\partial t} + [c_A, H_0 + H_1], \tag{106.14}$$

and on Γ_0, since c_A is constant,

$$0 = \frac{\partial c_A}{\partial t} + [c_A, H_0]. \tag{106.15}$$

Hence, by subtraction, we have on Γ

$$\dot{c}_A = [c_A, H_1]. \tag{106.16}$$

By virtue of (106.10), or equivalently (106.11), the right hand side is a function of (c, t), and so we have here a set of $2N$ equations to determine the functions (106.13) and hence the perturbed motion.

The Eqs. (106.16) may be put in a different form. By (106.10) we can express H_1 as a function of (c, t):

$$H_1(q, t, p) = K(c, t). \tag{106.17}$$

Then

$$\frac{\partial H_1}{\partial q_\varrho} = \frac{\partial K}{\partial c_B}\frac{\partial c_B}{\partial q_\varrho}, \quad \frac{\partial H_1}{\partial p_\varrho} = \frac{\partial K}{\partial c_B}\frac{\partial c_B}{\partial p_\varrho}, \tag{106.18}$$

and

$$[c_A, H_1] = \frac{\partial c_A}{\partial q_\varrho}\frac{\partial H_1}{\partial p_\varrho} - \frac{\partial H_1}{\partial q_\varrho}\frac{\partial c_A}{\partial p_\varrho} = [c_A, c_B]\frac{\partial K}{\partial c_B}. \tag{106.19}$$

Thus (106.16) may be written

$$\dot{c}_A = [c_A, c_B]\frac{\partial K}{\partial c_B}. \tag{106.20}$$

So far everything is exact. But if the derivatives of H_1 are small, or equivalently the derivatives of K are small, then the right hand sides of (106.16) and (106.20) are small. The projected point B^* moves slowly over Σ_{2N}, and we may approximate to its motion in the finite interval (t_1, t_2) by substituting in these right hand sides the values of c_A for $t = t_1$ and integrating by quadratures[1].

F. Relativistic dynamics[2].

I. Minkowskian space-time and the laws of dynamics.

107. Lorentz transformations. *Small Latin suffixes* will now have the values 1, 2, 3, 4, and *small Greek suffixes* the values 1, 2, 3, with summation understood in either case for a repeated suffix.

Let x^r be real coordinates of an event in the 4-dimensional manifold of space-time; and let the separation ds between adjacent events be given by

$$ds^2 = \varepsilon\, g_{mn}\, dx^m dx^n, \tag{107.1}$$

[1] For perturbation theory from the standpoints of Lagrangian equations and contact transformations, see CORBEN and STEHLE [3], pp. 306–312.

[2] The basic laws of Newtonian and relativistic dynamics were contrasted in Sects. 4 and 5. Essential formulae are repeated here, but no more will be said about the peculiar difficulties surrounding relativistic *systems*. As general references for the special theory of relativity, the reader may consult O. COSTA DE BEAUREGARD: La Théorie de la Relativité restreinte (Paris: Masson & Cie. 1949); P. G. BERGMANN: Introduction to the Theory of Relativity (New York: Prentice-Hall 1942), and this Encyclopedia, Vol. IV; A. EINSTEIN: The Meaning of Relativity (5th Ed.; Princeton: University Press 1955); HALPERN [9]; M. VON LAUE: Die Relativitätstheorie, Bd. 1 (5th. Ed.; Braunschweig: Vieweg 1952); C. MØLLER: The Theory of Relativity (Oxford: Clarendon Press 1952); A. PAPAPETROU: Spezielle Relativitätstheorie (Berlin: VEB Deutscher Verlag der Wissenschaften 1955); J. L. SYNGE: Relativity: The Special Theory (Amsterdam: North-Holland Publishing Co. 1956).

where the coefficients are functions of the coordinates, and ε is an indicator chosen equal to $+1$ or -1 so as to make ds real.

In the special theory of relativity, with which alone we are concerned here, space-time is *flat*, which means that there exist real coordinates (x, y, z, t) such that

$$ds^2 = \varepsilon\,(dx^2 + dy^2 + dz^2 - c^2\,dt^2), \tag{107.2}$$

where c is a fundamental constant (the speed of light).

It is convenient to introduce Minkowskian coordinates, with "imaginary time", defined by

$$\left.\begin{aligned} x_1 &= x, & x_2 &= y, \\ x_3 &= z, & x_4 &= i\,c\,t; \end{aligned}\right\} \tag{107.3}$$

then (107.2) may be written compactly as

$$ds^2 = \varepsilon\,dx_r\,dx_r. \tag{107.4}$$

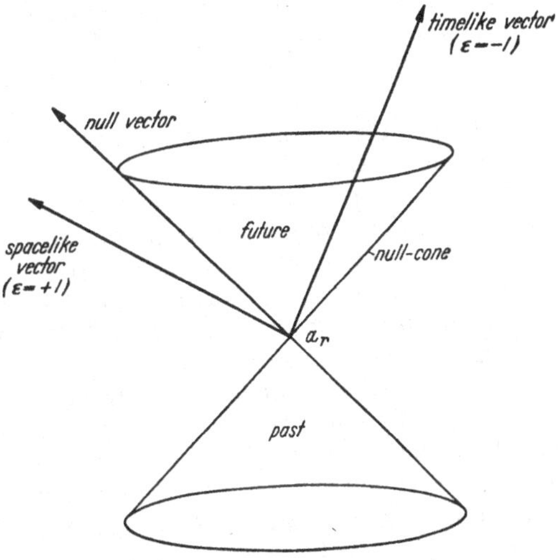

Fig. 51. Null-cone. Past and future. Timelike, spacelike and null vectors.

Minkowskian coordinates are used throughout this article; they have the great convenience that, for them, covariant components of vectors and tensors are the same as contravariant components, and we can write all vectors and tensors with subscripts, thus avoiding a notational complication. If the imaginary time x_4 should ever be a source of confusion, we can at once pass from Minkowskian coordinates x_r to real Cartesian coordinates x^r by writing $x_\varrho = x^\varrho$, $x_4 = i\,x^4$. We shall have occasion to pass to real coordinates in Sect. 111 in order to discuss a matter of sign.

The group of LORENTZ transformations consists of those transformations (necessarily linear) which conserve the quadratic form $dx_r\,dx_r$. Any such transformation is of the form

$$x'_r = A_{rs}\,x_s + B_r, \tag{107.5}$$

where the coefficients satisfy

$$A_{rs}\,A_{rt} = \delta_{st}. \tag{107.6}$$

Comparison with (9.6) shows that $\boldsymbol{A}$ is, formally, an orthogonal matrix, and this suggests that a LORENTZ transformation is a "rigid" displacement of space-time into itself. This is true, in a sense, and very important for the understanding of the LORENTZ transformation, but the presence of imaginary elements in $\boldsymbol{A}$ (due to the imaginary time) makes the geometry of the LORENTZ transformation essentially different from the geometry of orthogonal transformations of a 4-space with four real coordinates.

The *null-cone* drawn from any event a_r as vertex (Fig. 51) has the equation

$$(x_r - a_r)(x_r - a_r) = 0. \tag{107.7}$$

Events exterior to the null-cone satisfy

$$(x - a_r)(x_r - a_r) > 0, \tag{107.8}$$

and those interior to it satisfy

$$(x_r - a_r)(x_r - a_r) < 0. \tag{107.9}$$

The interior is divided into two parts, according as $(x_4 - a_4)/i$ is positive or negative; these two parts are respectively the *future* and the *past* with respect to the event a_r.

A displacement dx_r from a_r into the future or into the past is *timelike*, and for it $\varepsilon = -1$; a displacement on the null-cone is *null*, and one pointing outside the null-cone is *spacelike* ($\varepsilon = +1$).

The null-cone remains invariant under a LORENTZ transformation, and so do its interior and exterior regions. But future and past may be interchanged. However, for our purposes we reject LORENTZ transformations which cause this interchange.

By (107.6) the Jacobian J of a LORENTZ transformation is $J = \pm 1$; the transformation is *proper* if $J = +1$ and *improper* if $J = -1$.

It is a basic postulate of relativity that all the laws of dynamics should be invariant under proper future-preserving LORENTZ transformations. This is equivalent to saying that the laws are capable of geometrical construction in terms of the geometry of Minkowskian space-time (cf. Sect. 5).

It is further assumed that any displacement dx_r along the world line of a material particle is timelike. This is equivalent to saying that no particle can travel as fast as light (see Sect. 108 below).

The abstract concept of separation ds is given physical content by the assertion that, along the world line of a material particle, ds is a measure of *proper time*, i.e. the time recorded by a standard clock carried with the particle.

An observer is said to be *Galileian* (or to use a *Galileian frame of reference*) if the separation ds between any two events is expressible as in (107.2) or (107.4) in terms of his coordinates. When two Galileian observers, S and S', observe the same event, their observations are connected by a LORENTZ transformation. By cooperation between the two observers in the choice of space-axes, the LORENTZ transformation connecting the two observations may be put into the simple form

$$x' = \gamma(x - vt), \quad y' = y, \quad z' = z, \quad t' = \gamma\left(t - \frac{vx}{c^2}\right); \tag{107.10}$$

here

$$\gamma = \frac{1}{\sqrt{1 - \frac{v^2}{c^2}}}, \tag{107.11}$$

and v is the relative velocity of S and S'; more precisely, the velocity of S' relative to S is $(v, 0, 0)$, and the velocity of S relative to S' is $(-v, 0, 0)$.

108. Kinematics in space-time. 4-momentum. *α) Velocity of a point.* Consider a moving point (not necessarily a particle); the equations of its world line may be written

$$x_r = x_r(\chi), \tag{108.1}$$

where χ is a monotonic parameter. The 3-velocity of the point is

$$v_\varrho = \frac{dx_\varrho}{dt} = i c \frac{dx_\varrho}{dx_4}. \tag{108.2}$$

Hence the speed v, defined by

$$v = \sqrt{v_\varrho v_\varrho}, \tag{108.3}$$

is greater than, equal to, or less than c, according as the 4-vector $dx_r/d\chi$ is spacelike, null, or timelike.

For a material particle, this 4-vector is timelike, and we define the *4-velocity* to be the timelike unit vector

$$\lambda_r = \frac{dx_r}{ds}, \tag{108.4}$$

satisfying

$$\lambda_r \lambda_r = -1. \tag{108.5}$$

By (108.2) the relations between 3-velocity and 4-velocity are

$$\left.\begin{aligned} v_\varrho &= ic\frac{\lambda_\varrho}{\lambda_4},\\ \lambda_\varrho &= \frac{\gamma v_\varrho}{c}, \quad \lambda_4 = i\gamma,\\ \gamma &= \frac{1}{\sqrt{1-\frac{v^2}{c^2}}}. \end{aligned}\right\} \tag{108.6}$$

For a point moving faster than light, 4-velocity may be defined as in (108.4); we change -1 to $+1$ in (108.5), and the relations with 3-velocity are as in (108.6) except that γ is changed to γ^* where

$$\gamma^* = \frac{1}{\sqrt{\frac{v^2}{c^2}-1}}. \tag{108.7}$$

For a photon we assume that $dx_r/d\chi$ is a null vector. Thus the speed of a photon is c. It has no definable 4-velocity; we cannot use (108.4) because $ds=0$.

β) Velocity of a wave. An equation

$$F(x_1, x_2, x_3, x_4) = 0 \tag{108.8}$$

defines a 3-space in space-time. We may call it a *3-wave*. It is the history of a *2-wave*, the instantaneous 2-wave being given by putting $x_4 = \text{const}$ in (108.8).

The normal 3-velocity of the 2-wave is easily seen to be[1]

$$u_\varrho = -ic\frac{F_{,\varrho} F_{,4}}{F_{,\sigma} F_{,\sigma}}, \tag{108.9}$$

where $F_{,r} = \partial F/\partial x_r$. Hence the speed u of wave propagation satisfies

$$1 - \frac{c^2}{u^2} = \frac{F_{,r} F_{,r}}{(F_{,4})^2}. \tag{108.10}$$

Since $F_{,4}$ is pure imaginary, u is greater than, equal to, or less than c, according as the 4-vector $F_{,r}$ (which is the space-time normal to the 3-wave) is timelike, null, or spacelike.

γ) 4-acceleration. For a material particle the *4-acceleration* is the 4-vector

$$\frac{d\lambda_r}{ds} = \frac{d^2 x_r}{ds^2}. \tag{108.11}$$

It is orthogonal (in the space-time sense) to the 4-velocity, because (108.5) gives

$$\lambda_r \frac{d\lambda_r}{ds} = 0. \tag{108.12}$$

[1] Cf. SYNGE, p. 419 of op. cit. in Sect. 107 of this article.

The 3-acceleration, defined in the Newtonian way as

$$a_\varrho = \frac{d^2 x_\varrho}{dt^2}, \tag{108.13}$$

may be expressed as follows, since $\gamma\, ds = c\, dt$, $\lambda_4 = i\gamma$:

$$a_\varrho = \frac{c}{\gamma}\frac{d}{ds}\left(\frac{c}{\gamma}\lambda_\varrho\right) = -\frac{c^2}{\lambda_4}\frac{d}{ds}\left(\frac{\lambda_\varrho}{\lambda_4}\right). \tag{108.14}$$

δ) Composition of velocities. Let S and S' be two Galileian frames of reference with a LORENTZ transformation as in (107.10) connecting them. Let a particle move with velocity $(V, 0, 0)$ relative to S'. Then its velocity relative to S is $(u, 0, 0)$ where

$$u = \frac{v + V}{1 + \dfrac{vV}{c^2}}. \tag{108.15}$$

This law of composition of velocities may be expressed in terms of hyperbolic functions in the form

$$\psi = \varphi + \Phi \tag{108.16}$$

where

$$\operatorname{Tan}\psi = \frac{u}{c}, \qquad \operatorname{Tan}\varphi = \frac{v}{c}, \qquad \operatorname{Tan}\Phi = \frac{V}{c}. \tag{108.17}$$

ε) 4-momentum. With a particle we associate a number m, its *proper mass*, which is invariant under LORENTZ transformation. The 4-*momentum* of a particle is defined as

$$M_r = m\lambda_r, \tag{108.18}$$

where λ_r is the 4-velocity. All four components have the dimensions of a mass, but we can change to the dimensions of momentum or energy by inserting c or c^2 as a factor on the right hand side.

We define, with $\gamma = (1 - v^2/c^2)^{-\frac{1}{2}}$,

relative mass $= m\gamma$,
relative momentum (3-momentum) $= m\gamma v_\varrho$,
relative energy $= E = m\gamma c^2$,
proper energy $= E_0 = mc^2$,
relative kinetic energy $= T = mc^2(\gamma - 1)$.

On expansion in powers of v/c, we have

$$T = \frac{1}{2} m v^2 \left(1 + \frac{3}{4}\frac{v^2}{c^2} + \cdots\right), \tag{108.19}$$

the first term agreeing with the Newtonian value of kinetic energy.

The four components of M_r are expressible in terms of relative momentum and relative energy as follows:

$$\left.\begin{aligned} M_\varrho &= m\lambda_\varrho = \frac{m\gamma v_\varrho}{c}, \\ M_4 &= m\lambda_4 = im\gamma = \frac{iE}{c^2}. \end{aligned}\right\} \tag{108.20}$$

For a photon, the formula (108.18) fails, since $m = 0$ for a photon and the components of λ_r are infinite. We proceed as follows.

The equation
$$\varphi = \varphi_0 \cos \frac{2\pi\nu}{c}(n_\varrho x_\varrho - ct), \tag{108.21}$$
where $n_\varrho n_\varrho = 1$, represents plane waves of frequency ν advancing in the direction of the unit 3-vector n_ϱ with speed c. Putting
$$f_\varrho = \nu n_\varrho, \quad f_4 = i\nu, \tag{108.22}$$
we can write (108.21) in the form
$$\varphi = \varphi_0 \cos \frac{2\pi}{c} f_r x_r. \tag{108.23}$$

In order that the phase factor may be invariant under LORENTZ transformations, it is necessary and sufficient that f_r should transform as a 4-vector. Thus the four quantities in (108.22) are the components of a 4-vector, i.e. under the LORENTZ transformation (107.5) they transform according to
$$f'_r = A_{rs} f_s. \tag{108.24}$$

We use this *frequency 4-vector* f_r to define the 4-momentum M_r of a photon, writing
$$M_\varrho = \frac{h\nu}{c^2} n_\varrho, \quad M_4 = \frac{ih\nu}{c^2}, \tag{108.25}$$
where h is PLANCK's constant, ν is the frequency of the photon, and n_ϱ is a unit 3-vector in the direction of the photon's motion, so that the photon's 3-velocity is cn_ϱ.

The components of M_r have the dimensions of a mass. We define

relative momentum (3-momentum) of photon $= \dfrac{h\nu n_\varrho}{c}$,

relative energy of photon $= E = h\nu$.

Note that
$$\left.\begin{aligned} M_r M_r &= -m^2 \quad \text{for a material particle,} \\ M_r M_r &= 0 \quad \text{for a photon.} \end{aligned}\right\} \tag{108.26}$$

109. Equations of motion of a particle. If a 4-*force* X_r acts on a particle, we assume its equation of motion to be
$$\frac{d}{ds}(m\lambda_r) = X_r. \tag{109.1}$$
By (108.12) this leads to
$$\frac{dm}{ds} = -X_r \lambda_r, \tag{109.2}$$
and so the proper mass changes unless
$$X_r \lambda_r = 0, \tag{109.3}$$
which is the condition of orthogonality of the 4-force and the 4-velocity.

Using the relationship $\gamma\, ds = c\, dt$, we can write the first three of the Eq. (109.1) in the form
$$\frac{d}{dt}(m\gamma v_\varrho) = \frac{c^2 X_\varrho}{\gamma}, \tag{109.4}$$
a quasi-Newtonian formula giving the rate of change of relative momentum. The last of (109.1) gives
$$\frac{dE}{dt} = \frac{d}{dt}(m\gamma c^2) = -i\frac{c^3 X_4}{\gamma}, \tag{109.5}$$

a formula for the rate of change of relative energy. If (109.3) is satisfied, we have

$$-iX_4 = \frac{v_\varrho X_\varrho}{c}, \tag{109.6}$$

and hence

$$\frac{dE}{dt} = v_\varrho \frac{c^2 X_\varrho}{\gamma} = v_\varrho \frac{d}{dt}(m\gamma v_\varrho), \tag{109.7}$$

a formula connecting the rate of change of relative energy with the rate of change of relative momentum.

The orthogonality condition (109.3) is satisfied (and hence proper mass is conserved) if the 4-force depends on 4-velocity in the form

$$X_r = Y_{rs}\lambda_s, \tag{109.8}$$

where Y_{rs} is a skew-symmetric tensor, so that

$$Y_{rs} = -Y_{sr}. \tag{109.9}$$

This type of 4-force occurs in the case of a charged particle moving in an electromagnetic field (cf. Sects. 115, 116).

110. Lagrangian and Hamiltonian dynamics. Instead of laying down equations of motion of the form (109.1), we can base relativistic dynamics on a Lagrangian or Hamiltonian, using the methods of Part E, which are sufficiently general for relativistic application. Since we are dealing with a single particle, we put $N=3$; configuration space Q becomes the instantaneous space of a Galileian observer, and the space of events QT becomes space-time itself. The other representative spaces listed in Sect. 62 are also available for the description of relativistic dynamics. In particular, the 8-dimensional space $QTPH$ seems of interest, but we shall here use only QT and PH, with reference to Q for the physical interpretation of the formulae.

We shall use Minkowskian coordinates as in (107.3); we recall that small Latin suffixes range 1, 2, 3, 4 and small Greek suffixes range 1, 2, 3, with the summation convention. All general formulae can easily be translated into real curvilinear coordinates x^r as in (107.1).

Some essential formulae from Part E will now be restated, with a certain change in sign [cf. (110.4) and (110.8) below]; this change in sign is discussed in Sect. 111.

Consider any timelike curve in space-time with a parameter χ increasing from past to future. Let $\Lambda(x, x')$ be a given *invariant homogeneous*[1] *Lagrangian*, where $x'_r = dx_r/d\chi$. The Lagrangian action is

$$A_L = \int \Lambda(x, x')\, d\chi, \tag{110.1}$$

the value being independent of the choice of the parameter χ. The first form of HAMILTON's principle is [cf. (65.4)]

$$\delta \int \Lambda(x, x')\, d\chi = 0 \tag{110.2}$$

for fixed end events, and this gives the Lagrangian equations of motion

$$\frac{d}{d\chi}\frac{\partial \Lambda}{\partial x'_r} - \frac{\partial \Lambda}{\partial x_r} = 0, \tag{110.3}$$

identically related as in (65.7). We have then a Lagrangian relativistic dynamics based on a chosen Lagrangian; relativity appears only in the demand that the Lagrangian should be invariant under LORENTZ transformations.

[1] Positive homogeneous of degree 1 in x'_r.

Alternatively, we can put relativistic dynamics on a Hamiltonian basis. Let y_r be a *Hamiltonian 4-vector*[1] associated with an event x_r. As in (68.1), but with a change in sign[2], we define the Hamiltonian action along any curve in space-time to be

$$A_H = -\int y_r\, dx_r. \tag{110.4}$$

Let

$$\Omega(x, y) = 0 \tag{110.5}$$

be an invariant energy equation. The second form of HAMILTON's principle is [cf. (68.5)]

$$\delta \int y_r\, dx_r = 0, \quad \Omega(x, y) = 0. \tag{110.6}$$

This leads to canonical Hamiltonian equations of the form

$$\frac{dx_r}{dw} = \frac{\partial \Omega}{\partial y_r}, \quad \frac{dy_r}{dw} = -\frac{\partial \Omega}{\partial x_r}, \tag{110.7}$$

where w is a special parameter. Given an initial event x_r and an initial Hamiltonian 4-vector y_r satisfying (110.5), the Eqs. (110.7) determine a world line with a field of vectors y_r along it. This establishes a Hamiltonian relativistic dynamics, based on a chosen energy equation; relativity demands the invariance of that equation under LORENTZ transformations.

We have now three different ways of setting up relativistic dynamics. First, with a 4-force as in Sect. 109; secondly, by choosing a homogeneous Lagrangian $\Lambda(x, x')$; and thirdly, by choosing an energy equation $\Omega(x, y) = 0$.

As in Sect. 69, we reconcile these last two. Making a change in sign consequent on the change already made in (110.4), we pass from $\Lambda(x, x')$ to $\Omega(x, y) = 0$ by writing [cf. (69.3)]

$$y_r = -\frac{\partial \Lambda}{\partial x'_r}, \tag{110.8}$$

and eliminating the derivatives x'_r; and we pass from $\Omega(x, y) = 0$ to $\Lambda(x, x')$ by writing down [cf. (69.14)]

$$x'_r = \vartheta \frac{\partial \Omega}{\partial y_r}, \quad \Lambda = -y_r x'_r, \quad \Omega(x, y) = 0, \tag{110.9}$$

and eliminating y_r and ϑ. With this reconciliation, Lagrangian dynamics and Hamiltonian dynamics are one, and the common action is

$$A = \int \Lambda(x, x')\, d\chi = -\int y_r\, dx_r. \tag{110.10}$$

In Lagrangian dynamics there are two important special choices of the parameter χ, viz. $\chi = s$ and $\chi = t$. The former is good for general theory, being LORENTZ-invariant; the latter is good for making comparisons between relativistic dynamics and Newtonian dynamics.

If $\chi = s$, then

$$\Lambda(x, x')\, d\chi = \Lambda(x, dx) = \Lambda(x, \lambda)\, ds, \tag{110.11}$$

where $\lambda_r = dx_r/ds$, the 4-velocity. The action is

$$A = \int \Lambda(x, \lambda)\, ds, \tag{110.12}$$

---

[1] There is a risk of confusion of terminology in calling y_r the *momentum-energy 4-vector*, because it includes a contribution from the field acting on the particle [cf. (115.9)]. By analogy with HAMILTON's optics, y_r might be called the *slowness 4-vector*; cf. J. L. SYNGE: Geometrical Mechanics and DE BROGLIE Waves, p. 8. Cambridge: University Press 1954.

[2] It is desirable to arrange that action is positive when y_r and dx_r are both timelike vectors, pointing into the future; under those circumstances we have $y_r\, dx_r < 0$, and for that reason the minus sign is inserted in the definition (110.4). See also Sect. 111.

and LAGRANGE's equations read

$$\frac{d}{ds}\frac{\partial\Lambda}{\partial\lambda_r}-\frac{\partial\Lambda}{\partial x_r}=0, \tag{110.13}$$

with the special relation

$$\lambda_r\lambda_r=-1. \tag{110.14}$$

Some caution must be used. In carrying out the first partial differentiation in (110.13), we must employ a form of Λ homogeneous of the first degree in the 4-velocity. Should we at any time simplify Λ by means of (110.14), destroying the formal homogeneity, we must again use the same equation to restore that homogeneity before differentiating.

If $\chi=t$, we can write

$$\Lambda(x,x')\,d\chi=\Lambda(x,dx)=L\,dt, \tag{110.15}$$

the ordinary Lagrangian L, so defined, being a function of the seven quantities $x_\varrho, t, \dot{x}_\varrho$. The action is given by

$$A=\int L\,dt, \tag{110.16}$$

and LAGRANGE's equations have the familiar form

$$\frac{d}{dt}\frac{\partial L}{\partial\dot{x}_\varrho}-\frac{\partial L}{\partial x_\varrho}=0. \tag{110.17}$$

Note that the relativistic requirement is the LORENTZ-invariance of $L\,dt$, not of L itself; L is an invariant divided by the fourth component of a 4-vector, a curious requirement automatically satisfied if we generate L as in (110.15) from an invariant homogeneous Lagrangian.

In the Hamiltonian method, we may solve $\Omega(x,y)=0$ for y_4, obtaining

$$y_4-i\,\omega(x_1,x_2,x_3,t,y_1,y_2,y_3)=0 \tag{110.18}$$

(as the fourth component of a Minkowskian vector, y_4 is a pure imaginary). Define p_ϱ and H by

$$p_\varrho=y_\varrho,\qquad H=\frac{c\,y_4}{i}; \tag{110.19}$$

then (110.18) gives us the *Hamiltonian*

$$H=c\,\omega(x_1,x_2,x_3,t,p_1,p_2,p_3). \tag{110.20}$$

It is hard to find an unambiguous suggestive name for the 3-vector p_ϱ. In the case of a free particle (cf. Sect. 111), p_ϱ is the relative momentum or 3-momentum defined in Sect. 108. But when a charged particle moves in an electromagnetic field (cf. Sect. 115), the field contributes to p_ϱ. Probably *Hamiltonian 3-momentum* is the most satisfactory name for p_ϱ.

The action now reads

$$A=-\int y_r\,dx_r=\int(-p_\varrho\,dx_\varrho+H\,dt). \tag{110.21}$$

This agrees with the usual expression (68.1) for Hamiltonian action except for a reversal of sign; (110.21) gives a positive element of action for a particle instantaneous at rest ($dx_\varrho=0$), provided H is positive.

The canonical Hamiltonian equations have the familiar form

$$\dot{x}_\varrho=\frac{\partial H}{\partial p_\varrho},\qquad \dot{p}_\varrho=-\frac{\partial H}{\partial x_\varrho}. \tag{110.22}$$

Note that the relativistic requirement is that H should transform as the fourth component of a 4-vector, not that it should be a LORENTZ-invariant.

111. The free particle. In relativistic dynamics, the theory associated with a free particle is not trivial, for it serves as a link between physical concepts and the mathematical developments.

For a free particle with constant proper mass m, we choose the invariant homogeneous Lagrangian

$$\Lambda(x, x') = mc\sqrt{-x'_r x'_r}, \tag{111.1}$$

so that the element of action is

$$\Lambda(x, x')\, d\chi = mc\sqrt{-x'_r x'_r}\, d\chi = mc\sqrt{-dx_r dx_r} = mc\, ds = mc\sqrt{-\lambda_r \lambda_r}\, ds. \tag{111.2}$$

Note that this has the correct dimensions for action, $[ML^2T^{-1}]$.

LAGRANGE's equations (110.13) give

$$\frac{d\lambda_r}{ds} = 0, \tag{111.3}$$

so that the world line is straight.

For the Hamiltonian 4-vector we have, as in (110.8),

$$y_r = -\frac{\partial\Lambda}{\partial x'_r} = \frac{mc\, x'_r}{\sqrt{-x'_n x'_n}}, \tag{111.4}$$

and elimination gives the energy equation

$$2\Omega(x, y) = y_r y_r + m^2c^2 = 0, \tag{111.5}$$

from which the space-time coordinates are absent; the canonical equations (110.7) give

$$\frac{dx_r}{dw} = y_r, \quad \frac{dy_r}{dw} = 0. \tag{111.6}$$

We can write (111.4) in terms of 4-velocity:

$$y_r = mc\,\lambda_r. \tag{111.7}$$

Thus the Hamiltonian 4-vector is tangent to the world line in the case of a free particle; this is not generally true for a particle in a field. From (108.6), (110.19) and (111.7), we see that the Hamiltonian 3-momentum is

$$p_\varrho = m\gamma v_\varrho, \tag{111.8}$$

and we have also

$$y_4 = im\gamma c = \frac{iE}{c}, \tag{111.9}$$

where E is the relative energy, as in Sect. 108. By (110.19) we have then

$$H = E. \tag{111.10}$$

To find the form of the Hamiltonian function, we are to solve (111.5) for y_4, obtaining

$$y_4 = \pm i\sqrt{y_\varrho y_\varrho + m^2c^2}, \tag{111.11}$$

in which the $+$ sign is to be chosen on account of (111.9). Then, as in (110.20), the Hamiltonian is

$$H = c\sqrt{p_\varrho p_\varrho + m^2c^2}. \tag{111.12}$$

It remains to consider the ordinary Lagrangian for a free particle. Applying (110.15) to (111.2), we get

$$L\,dt = mc\,ds = \frac{mc^2}{\gamma}\,dt, \quad \gamma = \frac{1}{\sqrt{1-\frac{v^2}{c^2}}}, \tag{111.13}$$

so that

$$L = \frac{mc^2}{\gamma} = mc^2\sqrt{1-\frac{v^2}{c^2}} = mc^2 - \frac{1}{2}mv^2 + \cdots \tag{111.14}$$

the unwritten terms involving the square and higher powers of v/c.

This differs in three ways from the Lagrangian

$$L = T = \tfrac{1}{2}mv^2 \tag{111.15}$$

of a free particle in Newtonian dynamics. First, through the presence of the constant mc^2, representing proper energy; secondly, through the minus sign in $-\frac{1}{2}mv^2$; and thirdly, through the terms not written explicitly. If the Lagrangian is merely an integrand to be used in a variational equation for the determination of equations of motion, then the presence of mc^2 is trivial, since it gives the same contribution to $\int L\,dt$ for all the varied motions; nor is the sign in $-\frac{1}{2}mv^2$ significant, since $-L$ yields the same extremals as L. As for the unwritten terms, they represent the sort of relativistic correction one expects to find, tending to zero with v/c.

But if action itself is physically significant, then the difference between (111.14) and (111.15) is significant. For a particle at rest, (111.14) gives a positive action $mc^2\int dt$, whereas (111.15) gives zero. If we compare two particles, one at rest and the other moving, (111.14) ascribes the greater action (in a given time interval) to the particle at rest, whereas (111.15) ascribes the greater action to the moving particle.

Let us return to the change of sign introduced in (110.4) and consequentially in (110.8), and consider the definition of the Hamiltonian 4-vector when the coordinates are real.

If we use real coordinates x^r in space-time, we have to distinguish between covariant and contravariant vectors, the transition from the one to the other being effected by means of the fundamental tensor g_{mn} of (107.1). However, the geometry of space-time is not changed if we reverse the signs of all the quantities g_{mn}. Hence, when we diagonalize g_{mn} by using real Cartesian coordinates, there are two choices: we can take the diagonal form to be

$$(g_{mn}) = (+1, +1, +1, -1), \tag{111.16}$$

or we can take it to be

$$(\bar{g}_{mn}) = (-1, -1, -1, +1). \tag{111.17}$$

Some writers prefer the first, some the second. There is no physical difference between them, but (111.16) has the advantage that we can pass to the unit matrix (Minkowskian coordinates) by introducing *one* imaginary coordinate, whereas (111.17) requires *three* imaginary coordinates. Now, with real coordinates and g_{mn} and $\bar{g}_{mn}$ respectively as above, we would write (111.1) as

$$\Lambda(x, x') = mc\sqrt{-g_{rs}x^{r'}x^{s'}} = mc\sqrt{\bar{g}_{rs}x^{r'}x^{s'}}, \tag{111.18}$$

and partial differentiation gives in each case the same covariant vector:

$$\frac{\partial\Lambda}{\partial x^{r'}} = -mc\frac{g_{rs}x^{s'}}{\sqrt{-g_{mn}x^{m'}x^{n'}}} = mc\frac{\bar{g}_{rs}x^{s'}}{\sqrt{\bar{g}_{mn}x^{m'}x^{n'}}}. \tag{111.19}$$

This single covariant vector yields two (opposed) contravariant vectors, according as we use (111.16) or (111.17); they are, respectively,

$$\left.\begin{aligned} g^{rs}\frac{\partial \Lambda}{\partial x^{s'}} &= -\frac{m c x^{r'}}{\sqrt{-g_{mn} x^{m'} x^{n'}}}, \\ \bar{g}^{rs}\frac{\partial \Lambda}{\partial x^{s'}} &= \frac{m c x^{r'}}{\sqrt{\bar{g}_{mn} x^{m'} x^{n'}}}. \end{aligned}\right\} \tag{111.20}$$

The first of these points into the past, the second into the future. Now there are two possible definitions of the covariant Hamiltonian 4-vector, viz. $y_r = \pm\, \partial\Lambda/\partial x^{r'}$, and it is convenient to choose that sign which makes the corresponding contravariant 4-vector point into the future, at least in the case of a free particle. Thus we need two different definitions of y_r according as we use (111.16) or (111.17): they are

$$y_r = -\frac{\partial \Lambda}{\partial x^{r'}} \quad \text{for (111.16)}, \tag{111.21}$$

$$y_r = \frac{\partial \Lambda}{\partial x^{r'}} \quad \text{for (111.17)}. \tag{111.22}$$

In Minkowskian coordinates we do not have to distinguish between covariant and contravariant vectors; we have already seen that (111.21) is the appropriate definition, since (111.7) shows that y_r, so defined, points into the future.

112. The two-event characteristic function and the HAMILTON-JACOBI equation. The two-event characteristic function $S(x^*, x)$ is, as in (72.1) with a change of sign,

$$S(x^*, x) = -\int y_r\, dx_r, \tag{112.1}$$

integrated along the trajectory joining the two events. Hence

$$y_r = -\frac{\partial S}{\partial x_r}, \qquad y_r^* = \frac{\partial S}{\partial x_r^*}. \tag{112.2}$$

Substitution in the energy equation $\Omega(x, y) = 0$ gives the HAMILTON-JACOBI equation

$$\Omega\left(x, -\frac{\partial S}{\partial x}\right) = 0. \tag{112.3}$$

If, as in Sect. 77, we have a complete integral $S(x^*, x)$ of (112.3), then the Eqs. (112.2) determine the trajectories and associated Hamiltonian 4-vectors; in these equations we are to regard the quantities (x^*, y^*) as constants. Actually we need only six constants, since there are ∞^6 trajectories. If one of the (x^*) is taken additive to S, and the other three (x^*) and the first three of the (y^*) chosen arbitrarily, then the first three equations in the second set in (112.2) give a trajectory, and indeed all the trajectories, since six arbitrary constants are involved.

For a free particle as in Sect. 111, we have

$$S(x^*, x) = -y_r(x_r - x_r^*), \tag{112.4}$$

y_r being constant along the trajectory. Putting

$$s = \sqrt{-(x_r - x_r^*)(x_r - x_r^*)}, \tag{112.5}$$

we have, as in (111.7)

$$y_r = m c \lambda_r = \frac{m c (x_r - x_r^*)}{s}, \tag{112.6}$$

and hence

$$S(x^*, x) = m c \sqrt{-(x_r - x_r^*)(x_r - x_r^*)} = m c s. \tag{112.7}$$

II. Some special dynamical problems.

113. Hyperbolic motion. In Newtonian dynamics we can prescribe the motion of a particle, and calculate the force that gives this motion. Similarly, in relativity we can prescribe a world line and calculate the corresponding 4-force consistent with given constant proper mass of the particle, using the formula

$$X_r = m \frac{d\lambda_r}{ds}. \tag{113.1}$$

One of the simplest world lines is the pseudocircle with equations

$$x_1^2 + x_4^2 = a^2, \qquad x_2 = x_3 = 0, \tag{113.2}$$

where a is a constant. Since the first equation reads, in real coordinates,

$$x^2 - c^2 t^2 = a^2, \tag{113.3}$$

this is called *hyperbolic motion* (Fig. 52). The parametric equations of the world line are

$$x_1 = a \operatorname{Cos} \varphi, \qquad x_4 = i a \operatorname{Sin} \varphi, \tag{113.4}$$

and hence

$$ds^2 = -dx_1^2 - dx_4^2 = a^2 d\varphi^2, \tag{113.5}$$

so that the surviving components of the 4-velocity are

$$\lambda_1 = \frac{dx_1}{ds} = \operatorname{Sin}\varphi, \qquad \lambda_4 = \frac{dx_4}{ds} = i \operatorname{Cos}\varphi. \tag{113.6}$$

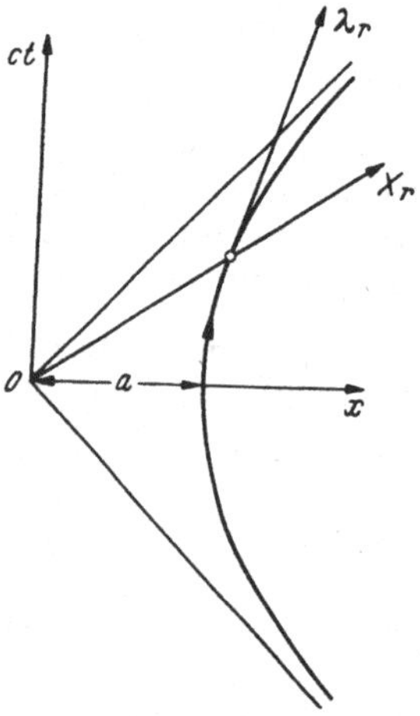

Fig. 52. Hyperbolic motion

By (113.1) the required 4-force is

$$\left.\begin{aligned} X_1 &= \frac{m}{a} \operatorname{Cos}\varphi = \frac{m x_1}{a^2}, \\ X_4 &= \frac{i m}{a} \operatorname{Sin}\varphi = \frac{m x_4}{a^2}, \\ X_2 &= X_3 = 0. \end{aligned}\right\} \tag{113.7}$$

Thus the 4-force is directed from the origin of space-time, with a magnitude proportional to the Minkowskian distance.

114. Particle in a potential field. Harmonic oscillator. Let $V(x_1, x_2, x_3)$ be a potential function, and let

$$\Lambda(x, x') = mc\sqrt{-x'_r x'_r} - \frac{iV x'_4}{c} \tag{114.1}$$

be the homogeneous Lagrangian for a particle of constant proper mass m moving in this field. This expression is not Lorentz-invariant; we are using a special frame of reference.

We can write

$$\Lambda(x, \lambda) = mc\sqrt{-\lambda_r \lambda_r} - \frac{iV\lambda_4}{c}, \tag{114.2}$$

and the corresponding ordinary Lagrangian L is given by

$$\left.\begin{aligned} L\,dt &= \Lambda(x, dx) = mc\,ds + V\,dt, \\ L &= \frac{mc^2}{\gamma} + V. \end{aligned}\right\} \tag{114.3}$$

By (110.13) the equations of motion are

$$\left.\begin{aligned}\frac{d}{ds}(-mc\lambda_\varrho)+\frac{i}{c}\frac{\partial V}{\partial x_\varrho}\lambda_4=0,\\ \frac{d}{ds}\left(-mc\lambda_4-\frac{iV}{c}\right)=0,\end{aligned}\right\}\tag{114.4}$$

so that

$$\left.\begin{aligned}\frac{d}{dt}(m\gamma v_\varrho)&=-\frac{\partial V}{\partial x_\varrho},\\ m\gamma c^2+V&=K,\end{aligned}\right\}\tag{114.5}$$

where K is a constant (constant of energy).

To discuss a harmonic oscillator, we consider motion along the x_1-axis with $V=\frac{1}{2}k^2x_1^2$. Let the particle start from the origin with $v=v_0$. Then, by the last of (114.5), we have

$$\frac{mc^2}{\sqrt{1-\frac{v^2}{c^2}}}=K-V,\qquad \frac{mc^2}{\sqrt{1-\frac{v_0^2}{c^2}}}=K.\tag{114.6}$$

The particle comes to rest at $x=a$ (we put $x_1=x$ for simplicity) where a is given by

$$mc^2=K-\tfrac{1}{2}k^2a^2,\tag{114.7}$$

and the constants v_0 and a are related by

$$\left.\begin{aligned}K=mc^2+\frac{1}{2}k^2a^2=\frac{mc^2}{\sqrt{1-\frac{v_0^2}{c^2}}},\\ \frac{1}{2}k^2a^2=mc^2(\gamma_0-1),\qquad \gamma_0=\frac{1}{\sqrt{1-\frac{v_0^2}{c^2}}}.\end{aligned}\right\}\tag{114.8}$$

By (114.6) we have

$$\frac{v}{c}=\sqrt{1-\frac{m^2c^4}{(K-V)^2}},\tag{114.9}$$

and so the period τ of the harmonic oscillator is given by

$$\left.\begin{aligned}\tau&=4\int_{x=0}^{x=a}dt=4\int_0^a\frac{dx}{v}\\ &=\frac{4}{c}\int_0^a\frac{(K-V)\,dx}{\sqrt{(K-V)^2-m^2c^4}}\\ &=\frac{2a}{c\varkappa}\int_0^{\frac{1}{2}\pi}\frac{(1+2\varkappa^2\cos^2\varphi)\,d\varphi}{\sqrt{1+\varkappa^2\cos^2\varphi}},\end{aligned}\right\}\tag{114.10}$$

where

$$\varkappa^2=\frac{k^2a^2}{4mc^2}.\tag{114.11}$$

The formula (114.10) is exact; expanding in powers of $\varkappa$, we get

$$\tau=2\pi\frac{\sqrt{m}}{k}\left(1+\frac{3}{16}\frac{k^2a^2}{mc^2}+\cdots\right),\tag{114.12}$$

in which the first term is the Newtonian period.

115. Charged particle in electromagnetic field. An electromagnetic field with electric vector E_ϱ and magnetic vector H_ϱ can be described by a skew-symmetric tensor F_{rs} where

$$\left.\begin{aligned} E_1 &= iF_{14}, & E_2 &= iF_{24}, & E_3 &= iF_{34}, \\ H_1 &= F_{23}, & H_2 &= F_{31}, & H_3 &= F_{12}; \end{aligned}\right\} \tag{115.1}$$

the corresponding 4-potential φ_r is such that

$$F_{rs} = \varphi_{s,r} - \varphi_{r,s}, \tag{115.2}$$

the commas denoting partial differentiation.

For a particle of constant proper mass m and charge e, moving in a given electromagnetic field[1], we take the homogeneous Lagrangian

$$\Lambda(x, x') = mc\sqrt{-x'_r x'_r} - \frac{e}{c}\varphi_r x'_r, \tag{115.3}$$

or, in terms of 4-velocity,

$$\Lambda(x, \lambda) = mc\sqrt{-\lambda_r \lambda_r} - \frac{e}{c}\varphi_r \lambda_r. \tag{115.4}$$

The corresponding ordinary Lagrangian is

$$L = \frac{mc^2}{\gamma} - \frac{e}{c}\varphi_\varrho \dot{x}_\varrho + V, \tag{115.5}$$

where

$$V = \frac{e\varphi_4}{i}, \tag{115.6}$$

the potential energy of the particle.

The equations of motion (110.13) give

$$m\frac{d\lambda_r}{ds} = \frac{e}{c^2}F_{rn}\lambda_n. \tag{115.7}$$

The term on the right is the LORENTZ ponderomotive force. This equation is of the form (109.1), and the condition (109.3) for constancy of proper mass is satisfied on account of the skew-symmetry of the electromagnetic tensor. These equations may also be expressed in the vector form [cf. (40.1), (40.2)]:

$$\left.\begin{aligned} &\frac{d}{dt}(m\gamma\boldsymbol{v}) = e\left(\boldsymbol{E} + \frac{\boldsymbol{v}}{c}\times\boldsymbol{H}\right), \\ &\frac{d}{dt}(m\gamma c^2) = e\,\boldsymbol{v}\cdot\boldsymbol{E}. \end{aligned}\right\} \tag{115.8}$$

To get the energy equation $\Omega(x, y) = 0$, we write the equation for the Hamiltonian 4-vector

$$y_r = -\frac{\partial\Lambda}{\partial\lambda_r} = mc\lambda_r + \frac{e}{c}\varphi_r, \tag{115.9}$$

and so obtain (since $\lambda_r\lambda_r = -1$)

$$2\Omega(x, y) = \left(y_r - \frac{e}{c}\varphi_r\right)\left(y_r - \frac{e}{c}\varphi_r\right) + m^2c^2 = 0. \tag{115.10}$$

The canonical equations are

$$\left.\begin{aligned} \frac{dx_r}{dw} &= \frac{\partial\Omega}{\partial y_r} = y_r - \frac{e}{c}\varphi_r, \\ \frac{dy_r}{dw} &= -\frac{\partial\Omega}{\partial x_r} = \frac{e}{c}\left(y_n - \frac{e}{c}\varphi_n\right)\varphi_{n,r}. \end{aligned}\right\} \tag{115.11}$$

[1] The units are electrostatic.

By (110.20) the Hamiltonian is [we solve (115.10) for y_4]

$$H = \frac{c y_4}{i} = V \pm c \sqrt{\left(p_\varrho - \frac{e}{c}\varphi_\varrho\right)\left(p_\varrho - \frac{e}{c}\varphi_\varrho\right) + m^2 c^2}. \tag{115.12}$$

The Hamiltonian 3-momentum is

$$p_\varrho = m\gamma v_\varrho + \frac{e}{c}\varphi_\varrho. \tag{115.13}$$

By (112.3) and (115.10), the HAMILTON-JACOBI equation is

$$\left(\frac{\partial S}{\partial x_r} + \frac{e}{c}\varphi_r\right)\left(\frac{\partial S}{\partial x_r} + \frac{e}{c}\varphi_r\right) + m^2 c^2 = 0. \tag{115.14}$$

In the case of an electrostatic field, we put $\varphi_\varrho = 0$, $e\varphi_4 = iV$; then we have

$$-\frac{1}{c^2}\left(\frac{\partial S}{\partial t} - V\right)^2 + \frac{\partial S}{\partial x_\varrho}\frac{\partial S}{\partial x_\varrho} + m^2 c^2 = 0. \tag{115.15}$$

In terms of a complete integral, the motion is given by (112.2).

116. Relativistic KEPLER problem. Consider a particle, of constant proper mass m and charge e, moving in the field of a charge e' of opposite sign, fixed at the origin. If e, e' are measured in Gaussian electrostatic units, the field and 4-potential are

$$\left.\begin{aligned} E_\varrho = i F_{\varrho 4} = \frac{e' x_\varrho}{r^3}, \quad & F_{\varrho\sigma} = 0, \\ \varphi_\varrho = 0, \quad & \varphi_4 = \frac{i e'}{r}. \end{aligned}\right\} \tag{116.1}$$

The equations of motion (115.8) give

$$\frac{d}{dt}(\gamma \boldsymbol{v}) = -\frac{k\boldsymbol{r}}{r^3}, \qquad \frac{d}{dt}(\gamma c^2) = -\frac{k}{r^3}\boldsymbol{v}\cdot\boldsymbol{r}, \tag{116.2}$$

where $\boldsymbol{r}$ is the position vector of the moving charge and

$$k = -\frac{e e'}{m} > 0. \tag{116.3}$$

It follows that the motion is plane, and if we use polar coordinates (r, ϑ) in the plane of the orbit, we get an integral of angular momentum

$$\gamma r^2 \dot{\vartheta} = A, \tag{116.4}$$

and an integral of energy

$$\gamma c^2 - \frac{k}{r} = W, \tag{116.5}$$

where A and W are constants. Putting $\varrho = 1/r$ and eliminating the time, the differential equation of the orbit comes out to be[1]

$$\frac{d^2\varrho}{d\vartheta^2} + \left(1 - \frac{k^2}{A^2 c^2}\right)\varrho = \frac{W k}{A^2 c^2}. \tag{116.6}$$

Assuming the coefficient of ϱ to be positive, and putting

$$p = \sqrt{1 - \frac{k^2}{A^2 c^2}}, \tag{116.7}$$

---

[1] Cf. BERGMANN, p. 218 of op. cit. in Sect. 107; SYNGE, p. 398 of op. cit. in Sect. 107.

we obtain the equation of the orbit in the form

$$\frac{1}{r} = \varrho = \frac{c}{A p^2}\left[\left(\frac{W^2}{c^4} - p^2\right)^{\frac{1}{2}} \cdot \cos(p\vartheta + C) + \frac{kW}{Ac^3}\right], \tag{116.8}$$

where C is a constant.

The condition for a finite orbit is

$$-c^2 < W < c^2. \tag{116.9}$$

The finite orbit may be regarded as a rotating ellipse with one focus at the origin, the advance of perihelion per revolution being

$$2\pi\left(\frac{1}{p} - 1\right). \tag{116.10}$$

If the velocity is small, this is approximately

$$\frac{\pi k^2}{A^2 c^2} = \frac{4\pi^3 a^2}{c^2 \tau^2 (1 - \varepsilon^2)}, \tag{116.11}$$

where a is the semi-axis major, τ the period, and ε the eccentricity; this is one sixth of the rotation given by the general theory of relativity for the motion of a planet in the gravitational field of the sun.

The solution of the relativistic KEPLER problem may also be obtained by separation of variables in the HAMILTON-JACOBI equation[1]. Transformed to polar coordinates (r, ϑ, φ), (115.15) reads

$$\left.\begin{aligned} &-\frac{1}{c^2}\left(\frac{\partial S}{\partial t} - V\right)^2 + \left(\frac{\partial S}{\partial r}\right)^2 + \frac{1}{r^2}\left(\frac{\partial S}{\partial \vartheta}\right)^2 + \\ &\qquad + \frac{1}{r^2 \sin^2\vartheta}\left(\frac{\partial S}{\partial \varphi}\right)^2 + m^2 c^2 = 0, \end{aligned}\right\} \tag{116.12}$$

where

$$V = \frac{ee'}{r}. \tag{116.13}$$

We get a complete integral by putting

$$S = a_1 + S_1(r, a_2, a_4) + S_2(\vartheta, a_2, a_3) + a_3\varphi + a_4 t, \tag{116.14}$$

the functions S_1 and S_2 being found by quadratures to satisfy the equations

$$\left.\begin{aligned} -\frac{1}{c^2}(a_4 - V)^2 + \left(\frac{\partial S_1}{\partial r}\right)^2 + m^2 c^2 &= \frac{a_2}{r^2}, \\ \left(\frac{\partial S_2}{\partial \vartheta}\right)^2 + \frac{a_3^2}{\sin^2\vartheta} &= -a_2. \end{aligned}\right\} \tag{116.15}$$

Here the a's are arbitrary constants. As in (112.2), the motion is given by

$$b_2 = \frac{\partial S_1}{\partial a_2} + \frac{\partial S_2}{\partial a_2}, \quad b_3 = \varphi + \frac{\partial S_2}{\partial a_3}, \quad b_4 = t + \frac{\partial S_1}{\partial a_4}, \tag{116.16}$$

where the b's are arbitrary constants. These three equations give r, ϑ, φ as functions of t and the six arbitrary constants $a_2, a_3, a_4, b_2, b_3, b_4$.

---

[1] Cf. A. SOMMERFELD, pp. 251–258 of op. cit. in Sect. 99, where the problem is approached in a slightly different way.

III. DE BROGLIE waves.

117. Coherent systems of trajectories in space-time and associated waves. The theory of Sects. 74 and 75 can be applied to space-time, with the alteration in sign indicated in Sect. 110 and discussed in Sect. 111. The essential steps are as follows.

We lay down an energy equation

$$\Omega(x, y) = 0. \tag{117.1}$$

For a free particle, this reads, as in (111.5),

$$2\Omega = y_r y_r + m^2 c^2 = 0; \tag{117.2}$$

for a charged particle in an electromagnetic field we have the more complicated Eq. (115.10); for a particle in a potential field, as in Sect. 114, we have

$$2\Omega = y_\varrho y_\varrho + \left(y_4 - \frac{iV}{c}\right)^2 + m^2 c^2 = 0. \tag{117.3}$$

The trajectories are given by the canonical equations

$$\frac{dx_r}{dw} = \frac{\partial\Omega}{\partial y_r}, \qquad \frac{dy_r}{dw} = -\frac{\partial\Omega}{dx_r}. \tag{117.4}$$

Picking out some set of trajectories, forming a subspace R of space-time, we associate with each event in R the Hamiltonian 4-vector y_r belonging to the trajectory passing through that event; this y_r may be found from the first set of equations in (117.4), or equivalently from the equations

$$y_r = -\frac{\partial\Lambda}{\partial x^{r\prime}}, \tag{117.5}$$

where $\Lambda(x, x')$ is the homogeneous Lagrangian corresponding to the energy equation (117.1). This set of trajectories forms a coherent system if

$$\oint y_r dx_r = 0 \tag{117.6}$$

for every reducible circuit in R. The one-event characteristic function for the coherent system is defined as

$$U(x) = -\int y_r dx_r, \tag{117.7}$$

the integral being taken along any curve in R, starting from some event which is fixed once for all and ending at the event x_r where U is evaluated.

Taking R to be a simply connected 4-dimensional region in space-time, and varying the event x_r, we have, as in (74.8) with a change of sign,

$$y_r = -\frac{\partial U}{\partial x_r}. \tag{117.8}$$

U satisfies the HAMILTON-JACOBI equation

$$\Omega\left(x, -\frac{\partial U}{\partial x}\right) = 0, \tag{117.9}$$

and the waves belonging to the coherent system have the equations

$$U(x) = \text{const}. \tag{117.10}$$

Note that the Hamiltonian 4-vector y_r is normal to the 3-wave in the space-time sense.

These waves are DE BROGLIE waves in the sense of geometrical mechanics.

118. Particle velocity and wave velocity. By (108.9) and (117.8), the normal velocity of propagation of the wave $U=\text{const}$ is

$$u_\varrho = -i\,c\,\frac{y_\varrho\, y_4}{y_\sigma\, y_\sigma}, \tag{118.1}$$

and, as in (108.10), the speed u satisfies

$$1-\frac{c^2}{u^2}=\frac{y_r y_r}{y_4^2}. \tag{118.2}$$

By (117.4) the particle velocity is

$$v_\varrho = i\,c\,\frac{d x_\varrho}{d x_4} = i\,c\,\frac{\partial\Omega/\partial y_\varrho}{\partial\Omega/\partial y_4}. \tag{118.3}$$

The connection between wave velocity u_ϱ and particle velocity v_ϱ is to be found (at any event x_r) by eliminating the four quantities y_r from the seven equations contained in (117.1), (118.1) and (118.3).

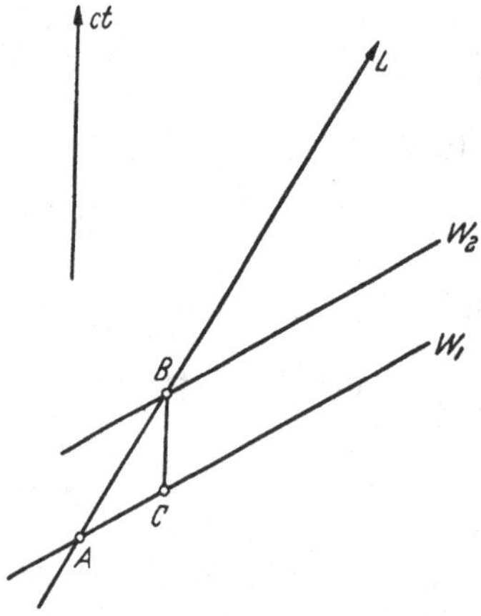

Fig. 53. DE BROGLIE waves with period BC/c.

For a free particle, with Ω as in (117.2), the particle velocity is, by (118.3),

$$v_\varrho = i\,c\,\frac{y_\varrho}{y_4}. \tag{118.4}$$

This has the same direction in space as the wave velocity u_ϱ given by (118.1); this direction is also that of the Hamiltonian 3-momentum p_ϱ. The relation between the two speeds (u for the wave and v for the particle) is

$$u\,v = u_\varrho\, v_\varrho = c^2, \tag{118.5}$$

the DE BROGLIE equation[1].

119. The DE BROGLIE wavelength and frequency. Let L be the world line of a particle in a coherent system (Fig. 53), and let W_1, W_2 be two DE BROGLIE 3-waves cutting L at the events A, B respectively. Let these waves be chosen so that the action along AB is h (PLANCK's constant).

Regarding h as infinitesimal, we have then

$$-\,y_r\, d x_r = h, \tag{119.1}$$

where y_r is the Hamiltonian 4-vector associated with L at A and dx_r is the displacement AB. If CB is drawn parallel to the time-axis, with C on W_1, then the period τ of the DE BROGLIE waves is

$$\tau = \frac{BC}{c}, \tag{119.2}$$

where BC is a Minkowskian separation.

By (117.8), y_r is orthogonal to the wave W_1. Therefore, since the displacement AB is the vector sum of the displacements AC and BC, we have from (119.1)

$$-\,y_r\, d\xi_r = h, \tag{119.3}$$

where $d\xi_r$ is the displacement CB. But this last displacement is parallel to the time-axis, and so

$$d\xi_\varrho = 0, \qquad d\xi_4 = i\,BC. \tag{119.4}$$

[1] For the more complicated relationship between the two velocities for a charged particle in an electromagnetic field, see SYNGE, p. 90 of op. cit. in Sect. 110.

Therefore (119.3) gives
$$BC = -\frac{h}{i y_4} \tag{119.5}$$
and
$$\tau = -\frac{h}{i c y_4}. \tag{119.6}$$

This expression for the period of the DE BROGLIE waves is approximate, since we took h to be infinitesimal.

In the case of a free particle, we have, as in (111.7),
$$y_4 = m c \lambda_4 = i m \gamma c, \tag{119.7}$$
and so the period τ and the frequency ν are such that
$$\tau = \frac{h}{m\gamma c^2}, \quad h\nu = m\gamma c^2. \tag{119.8}$$
Since the speed of the waves is $u = c^2/v$, the DE BROGLIE wavelength is
$$\lambda = u\tau = \frac{h}{m\gamma v}. \tag{119.9}$$
These expressions are accurate in the simplest case of all, viz. when the world lines of the coherent system are parallel and the waves are, in consequence, plane and parallel.

Passing back to the general case of a particle moving in accordance with any energy equation $\Omega(x, y) = 0$, or equivalently with a Hamiltonian H, we get from (110.19) and (119.6)
$$\tau = -\frac{h}{i c y_4} = \frac{h}{H}, \quad h\nu = H. \tag{119.10}$$

IV. Relativistic catastrophes.

120. Conservation of 4-momentum. In this discussion of catastrophes, the word *particle* includes both a material particle with 4-momentum
$$M_\varrho = \frac{m\gamma v_\varrho}{c}, \quad M_4 = i m \gamma = \frac{iE}{c^2} \tag{120.1}$$
and a photon with 4-momentum
$$M_\varrho = \frac{h\nu}{c^2} n_\varrho, \quad M_4 = \frac{ih\nu}{c^2}, \tag{120.2}$$
as in Sect. 108, all these components having the dimensions of a mass. By the word *catastrophe* we understand a collision or encounter of several particles, from which perhaps the number of particles emerging differs from the number coming in, or an explosion, in which one particle turns into several particles.

The basic assumption is the law of conservation of 4-momentum, which may be written
$$\sum M_r' = \sum M_r, \tag{120.3}$$
where the summation on the right extends over all the particles before the catastrophe, and the summation on the left over all the particles after the catastrophe; the two numbers need not be the same. This law may be written equivalently
$$\sum m'\gamma' v_\varrho' + \sum \frac{h\nu'}{c} n_\varrho' = \sum m\gamma v_\varrho + \sum \frac{h\nu}{c} n_\varrho, \tag{120.4}$$
$$\sum m'\gamma' c^2 + \sum h\nu' = \sum m\gamma c^2 + \sum h\nu, \tag{120.5}$$
where (120.4) expresses the conservation of relative momentum and (120.5) the conservation of relative energy. The laws are valid for all Galileian frames of reference.

Until we come to Sect. 123, there is no necessity to assume that the world lines of the participating particles intersect at a catastrophe; the conservation equation (120.3) makes no reference to the positions of the particles.

It is profitable to think of the 4-dimensional space PH in which the coordinates of a point are M_r and which has the same geometry as space-time, the only difference being that the origin of PH is a physically distinguishable point, whereas the origin of space-time is not. We have

$$M_r M_r = -m^2 \tag{120.6}$$

($m=0$ for a photon), and this provides the metric in PH. Viewed in this way, the Eq. (120.3) presents a very simple picture in PH, analogous to the polygon of forces in statics; a single resultant vector is broken down in two different ways, one corresponding to the initial state and the other to the final state. Fig. 54 shows a catastrophe with two initial particles and three final particles.

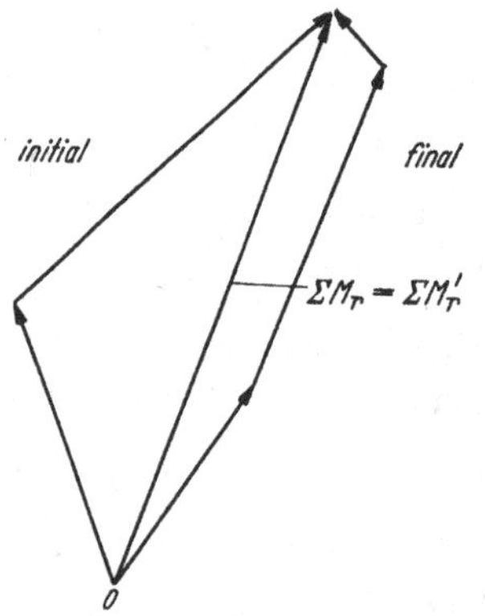

Fig. 54. Vector diagram of a catastrophe, drawn in the space PH, showing the conservation of 4-momentum.

Regarding the initial state as given, we have in (120.3) four equations to determine the final state. There may be supplementary conditions, the proper masses of the final particles being assigned, but, even with such conditions, there is only one case in which the final state is determinate, and that is when there is only *one* final particle. Its 4-momentum is given by

$$M'_r = \sum M_r, \tag{120.7}$$

and its proper mass by

$$m'^2 = -M'_r M'_r. \tag{120.8}$$

If there are *two* final particles, and we are told what their proper masses are, we have in all six equations, viz. four in (120.3) and two equations of the type (120.8); this leaves two degrees of freedom in the final state, the same as in the collision of two elastic spheres in Newtonian dynamics when the direction of the line of centres is unspecified.

If the material particles resulting from a catastrophe are brought to rest, without changing their proper masses, and any photons resulting from the catastrophe are allowed to escape, then there remains in hand a total relative energy $\sum m' c^2$. Thus the loss of relative energy is

$$Q = \sum h\nu + \sum m\gamma c^2 - \sum m' c^2. \tag{120.9}$$

If the catastrophe is the explosion of a single particle at rest, of proper mass m, then

$$Q = m c^2 - \sum m' c^2; \tag{120.10}$$

Q is the *binding energy* of the exploding particle (some writers prefer to use $-Q$).

The *mass-centre frame of reference*, or *reference system*, is that Galileian frame which has the total 4-momentum, $\sum M_r$, for time-axis. When this frame is used, the right hand side of (120.4) is zero. The use of this frame sometimes simplifies calculations, but of course it does not affect the question of determinacy in the result of the catastrophe.

121. Inelastic and elastic collisions. In a completely inelastic collision, all the incident particles coalesce into a single particle, with 4-momentum

$$M'_r = \sum M_r, \tag{121.1}$$

and proper mass m' given by

$$m'^2 = -M'_r M'_r. \tag{121.2}$$

If the number of incident particles is two, with proper masses $m, \tilde{m}$ and 4-momenta $M_r, \tilde{M}_r$, then the proper mass of the single final particle is given by

$$\left.\begin{aligned} m'^2 &= -(M_r + \tilde{M}_r)(M_r + \tilde{M}_r) \\ &= m^2 + \tilde{m}^2 - 2M_r\tilde{M}_r. \end{aligned}\right\} \tag{121.3}$$

If, in particular, the incident particles are photons with frequencies $\nu, \tilde{\nu}$ travelling in directions given by the unit vectors $n_\varrho, \tilde{n}_\varrho$, then we get

$$\left.\begin{aligned} m'^2 &= 2\frac{h^2\nu\tilde{\nu}}{c^4}(1 - n_\varrho\tilde{n}_\varrho) \\ &= 4\frac{h^2\nu\tilde{\nu}}{c^4}\sin^2\frac{1}{2}\vartheta, \end{aligned}\right\} \tag{121.4}$$

where ϑ is the angle between the lines of motion of the photons. This represents the creation of a single particle of matter from two photons. If we use the mass-centre reference system, we have $\tilde{\nu} = \nu$, $\vartheta = \pi$, and so

$$m'c^2 = 2h\nu, \tag{121.5}$$

as is indeed obvious, since the final particle must be at rest, with energy $m'c^2$, and the total energy of the photons is $2h\nu$.

In a completely inelastic collision of two material particles, moving with velocities $v_\varrho, \tilde{v}_\varrho$, (121.3) gives

$$m'^2 = m^2 + \tilde{m}^2 + 2m\tilde{m}\gamma\tilde{\gamma}\left(1 - \frac{v_\varrho\tilde{v}_\varrho}{c^2}\right), \tag{121.6}$$

where

$$\gamma = \frac{1}{\sqrt{1 - \frac{v^2}{c^2}}}, \quad \tilde{\gamma} = \frac{1}{\sqrt{1 - \frac{\tilde{v}^2}{c^2}}}. \tag{121.7}$$

It is easy to show that

$$\gamma\tilde{\gamma}\left(1 - \frac{v_\varrho\tilde{v}_\varrho}{c^2}\right) > 1 \tag{121.8}$$

(equality occurs only if $v_\varrho = \tilde{v}_\varrho$, in which case there is no collision), and hence

$$m' > m + \tilde{m}. \tag{121.9}$$

Thus total proper mass is always increased in a completely inelastic collision. The increase in total proper energy is

$$m'c^2 - mc^2 - \tilde{m}c^2; \tag{121.10}$$

if the speeds of the incident particles are small compared with c, it is easy to show that this increase is approximately equal to the heat generated in such a collision according to Newtonian dynamics (cf. Sects. 58, 59).

A completely inelastic collision of a material particle and a photon is similarly treated. This represents absorption of a photon by a material particle. Here also the total proper mass is increased; since the proper mass of the photon is zero, this means that we get a material particle with a greater proper mass than that of the incident material particle.

An elastic collision is characterized by two conditions:

(i) The number of particles is unchanged.

(ii) The proper mass of each particle is unchanged.

The result of an elastic collision between two material particles is very easy to describe in the mass-centre reference system: the speeds of the particles are unchanged, and they recede from the collision in opposite directions, the line of these directions being undetermined by the conservation law. The vector diagram in the space PH is as in Fig. 55. In the elastic collision of a material particle and a photon (the COMPTON effect, see Sect. 122), the frequency of the photon is unchanged when the mass-centre reference system is used.

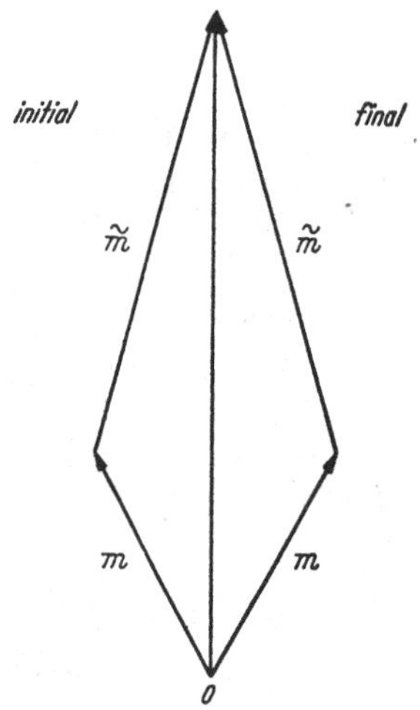

Fig. 55. Vector diagram of elastic collision.

122. COMPTON effect. An elastic collision between a material particle and a photon is called the COMPTON effect. As indicated in Sect. 120, there is a twofold indeterminary in the result of the collision.

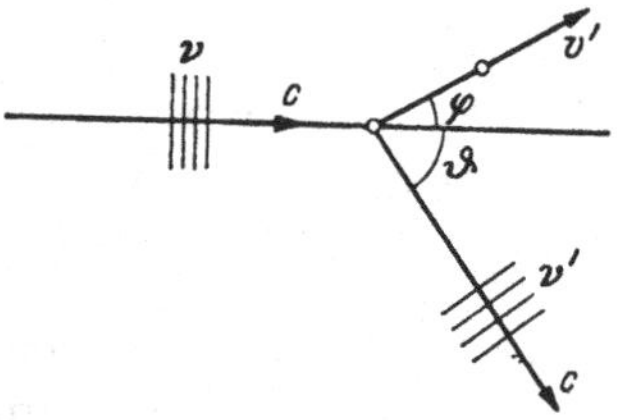

Fig. 56. COMPTON effect in the laboratory frame of reference.

To discuss the collision in a general Galileian frame of reference, we write M_r, M_r' for the 4-momenta of the material particle before and after the collision, and P_r, P_r' for the 4-momenta of the photon. We have the following equations:

$$\left.\begin{aligned} M_r' + P_r' &= M_r + P_r, \\ M_r' M_r' &= M_r M_r = -m^2, \\ P_r' P_r' &= P_r P_r = 0. \end{aligned}\right\} \tag{122.1}$$

These six equations contain all the information available for the determination of the eight quantities M_r', P_r'.

The description of the outcome of the collision depends on the frame of reference employed. The description is simplest for the mass-centre reference system (see Sect. 121), but in the usual description, as given below, one uses the laboratory frame, i.e. the frame in which the particle is initially at rest[1].

Fig. 56 shows the photon (with frequency ν) approaching from the left, and being scattered (with frequency ν') at an angle ϑ; the material particle recoils with speed v' at an angle φ as shown. From (120.4) and (120.5) it follows that the three lines of motion are coplanar and that

$$\left.\begin{aligned} m\gamma' v' \cos\varphi + \frac{h\nu'}{c}\cos\vartheta &= \frac{h\nu}{c}, \\ m\gamma' v' \sin\varphi - \frac{h\nu'}{c}\sin\vartheta &= 0, \\ m\gamma' c^2 + h\nu' &= m c^2 + h\nu. \end{aligned}\right\} \tag{122.2}$$

[1] See J. L. SYNGE, pp. 193–199 of op. cit. in Sect. 107 for further details and descriptions in other frames of reference.

Assuming ϑ to have any value (this is part of the indeterminacy of the problem), we find, after a little calculation,

$$\left.\begin{aligned} \tan\varphi &= \frac{\cot\frac{1}{2}\vartheta}{1+k}, \\ \frac{v'}{c} &= \frac{2k\sin\frac{1}{2}\vartheta\sqrt{1+k(2+k)\sin^2\frac{1}{2}\vartheta}}{1+2k(1+k)\sin^2\frac{1}{2}\vartheta}, \\ \gamma' &= \frac{1}{\sqrt{1-\frac{v'^2}{c^2}}} = \frac{1+2k(1+k)\sin^2\frac{1}{2}\vartheta}{1+2k\sin^2\frac{1}{2}\vartheta}, \\ \nu' &= \frac{\nu}{1+2k\sin^2\frac{1}{2}\vartheta}, \end{aligned}\right\} \quad (122.3)$$

where

$$k = \frac{h\nu}{mc^2}. \quad (122.4)$$

123. Angular momentum and mass-centre[1]. Let x_r be any event in the history of a particle, and let M_r be its 4-momentum at that event. Then the angular momentum of the particle at that event, relative to the origin of the space-time coordinates, is defined to be the skew-symmetric tensor

$$H_{rs} = x_r M_s - x_s M_r. \quad (123.1)$$

More generally, the angular momentum relative to an event a_r is defined to be

$$\left.\begin{aligned} H_{rs}(a) &= (x_r - a_r) M_s - (x_s - a_s) M_r \\ &= H_{rs} - a_r M_s + a_s M_r. \end{aligned}\right\} \quad (123.2)$$

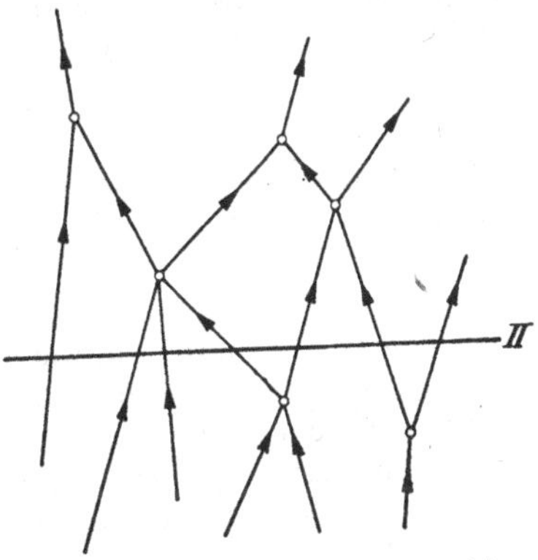

Fig. 57. World lines of a system of particles interacting at point catastrophes.

If the particle is free, then its world line is straight and M_r lies along it; in that case $H_{rs}(a)$ is independent of the particular event x_r chosen on the world line.

Consider now a system of particles, interacting with one another only by catastrophes in which the world lines intersect. Collisions may be elastic or inelastic; particles may combine and new particles may be formed; the particles may be material particles or photons. The essential condition is that 4-momentum should be conserved at each catastrophe, and that each catastrophe should take place at a single event. Fig. 57 illustrates the world lines of such a system.

Let Π be any spacelike 3-space. Let x_r be the event where a typical world line cuts Π, and let M_r be the corresponding 4-momentum. Then each world line cutting Π gives an angular momentum relative to the origin as in (123.1), and we get a total angular momentum for the system by adding the contributions of the several particles.

If we move Π in space-time, this total angular momentum certainly remains constant until Π crosses a catastrophe, because each particle is free between catastrophes. Further, there is no change in the total angular momentum when Π does cross a catastrophe, because only a single event is involved and the total 4-momentum of the particles involved is conserved. In fact, both the total 4-momentum and the total angular momentum of the world lines cutting Π are independent of the choice of Π; they are constants of the system.

[1] See the books by MøLLER and SYNGE cited in Sect. 107; also C. MøLLER: Ann. Inst. H. Poincaré **11**, 251–278 (1949) and Comm. Dublin Inst. Adv. Studies, Ser. A **1949**, No. 5.

Let us now use for the system the symbols previously employed for a single particle, writing

$$H_{rs} = \text{total angular momentum relative to the origin},$$
$$H_{rs}(a) = \text{total angular momentum relative to } a_r,$$
$$M_r = \text{total 4-momentum}.$$

We have as in (123.2), but now with an extended meaning,

$$H_{rs}(a) = H_{rs} - a_r M_s + a_s M_r. \tag{123.3}$$

Consider the four equations

$$H_{rs}(a)\, M_s = 0, \tag{123.4}$$

of which only three are independent on account of the skew-symmetry of $H_{rs}(a)$. Substitution from (123.3) gives

$$H_{rs} M_s - a_r M_s M_s + M_r a_s M_s = 0. \tag{123.5}$$

These equations locate a_r on a straight line in space-time, with equations

$$a_r = \frac{H_{rs} M_s}{M_n M_n} + \vartheta M_r, \tag{123.6}$$

where ϑ is a variable parameter. This is the history of the mass-centre of the system, *mass-centre* being so defined relativistically. This history is parallel to M_r.

If we use a frame of reference with time-axis parallel to M_r (this is the mass-centre reference system of Sect. 120), we have $M_\varrho = 0$; (123.6) gives a mass-centre fixed at the point with coordinates

$$a_\varrho = \frac{H_{\varrho 4}}{M_4}. \tag{123.7}$$

If, on the other hand, we leave the directions of the space-time axes arbitrary, but move the origin to a position on the history of the mass-centre, then (123.5) is satisfied by $a_r = 0$, and therefore

$$H_{rs} M_s = 0. \tag{123.8}$$

124. Particles with spin. The Eq. (123.8) suggests that the spin or intrinsic angular momentum of a particle with 4-momentum M_r should be represented by a skew-symmetric tensor H_{rs} satisfying the condition

$$H_{rs} M_s = 0. \tag{124.1}$$

Then the angular momentum of the particle about any event a_r will consist of two parts: an orbital angular momentum

$$(x_r - a_r) M_s - (x_s - a_s) M_r \tag{124.2}$$

and a spin angular momentum H_{rs}, satisfying (124.1) and independent of the choice of origin and of the event a_r.

Any skew-symmetric tensor can be represented by two 3-vectors, and we may describe the spin by the two 3-vectors, H_ϱ and H_ϱ^*, where

$$\left.\begin{aligned} H_1 &= i H_{14}, & H_2 &= i H_{24}, & H_3 &= i H_{34},\\ H_1^* &= H_{23}, & H_2^* &= H_{31}, & H_3^* &= H_{12}. \end{aligned}\right\} \tag{124.3}$$

The factor i is inserted here to yield a real 3-vector, $H_{\varrho 4}$ being a pure imaginary in Minkowskian coordinates.

The four equations contained in (124.1) give the vector equation

$$\boldsymbol{H} = \boldsymbol{H}^* \times \frac{\boldsymbol{v}}{c}, \tag{124.4}$$

where $\boldsymbol{v}$ is the velocity of the particle, and also the scalar equation

$$\boldsymbol{H} \cdot \boldsymbol{v} = 0, \tag{124.5}$$

which is of course a consequence of (124.4). By (124.4) the two spin vectors are perpendicular to one another.

If we use the rest frame of the particle, thus making $\boldsymbol{v}=0$, we have $\boldsymbol{H}=0$ by (124.4), and hence only one spin vector, $\boldsymbol{H}^*$.

The spin tensor yields a LORENTZ-invariant:

$$\Omega^2 = \tfrac{1}{2} H_{rs} H_{rs} = H^{*2} - H^2. \tag{124.6}$$

This expression is always positive, since we can make $H=0$ by choice of frame of reference, and hence Ω is real.

For a particle moving under the influence of a 4-force X_r and a torque Y_{rs} $(=-Y_{sr})$, equations of motion have been proposed[1], which read as follows in the notation of the present article:

$$\left.\begin{aligned} H_{rs}\lambda_s = 0, \quad & \lambda_r \lambda_r = -1, \\ M_r' = X_r, \quad & H_{rs}' = M_r \lambda_s - M_s \lambda_r + Y_{rs}; \end{aligned}\right\} \tag{124.7}$$

here λ_r is the 4-velocity and the prime means d/ds along the world line. The mass of the particle being defined by $m = -M_r \lambda_r$, the above equations imply

$$M_r = m\lambda_r + H_{rs}\lambda_s' + Y_{rs}\lambda_s. \tag{124.8}$$

If $X_r=0$ and $Y_{rs}=0$, then M_r is a constant 4-vector, and the orbit is a circle in that frame of reference for which $M_1 = M_2 = M_3 = 0$.

I wish to thank Professors C. LANCZOS and C. TRUESDELL for discussions and advice during the preparation of this Article, and Dr. L. BASS for assistance in proof-reading.

General references.

This is a selection of textbooks and articles on classical dynamics, with brief notes, each of which is intended to give some general indication of the nature of the contents.

[1] AMES, J. S., and F. D. MURNAGHAN: Theoretical Mechanics. Boston: Ginn 1929. — Vector analysis, including theory of screws. Kinematics. Dynamics of a particle and of a rigid body. Lagrangian and Hamiltonian equations. Variational principles. HAMILTON-JACOBI equation. POISSON brackets. Relativity.

[2] APPELL, P.: Traité de Mécanique rationnelle. Paris: Gauthier-Villars, Tome I, 1941 (6th. Edn.); Tome II, 1953 (6th Edn.). — A classical treatise, presenting the subject in detail and with great clarity. Sparing use of vector notation. Tome I deals with kinematics, statics and the dynamics of a particle. Tome II deals with systems, holonomic and non-holonomic, Lagrangian and Hamiltonian equations with associated general theory, shocks and percussions. Three further volumes deal with continuous media, rotating fluid masses and tensor calculus.

[3] CORBEN, H. C., and P. STEHLE: Classical Mechanics. New York: Wiley, and London: Chapman & Hall 1950. — A modern textbook, with emphasis placed on those parts of the subject most pertinent to quantum mechanics. Vector and matrix notation used. Hamiltonian theory, POISSON brackets, and contact transformations. Introduction to special theory of relativity.

---

[1] J. FRENKEL: Lehrbuch der Elektrodynamik, Bd. 1, p. 353. Berlin: Springer 1926. — M. MATHISSON: Acta Phys. Polon. **6**, 163, 218 (1937). — J. WEYSSENHOFF and A. RAABE: Acta Phys. Polon. **9**, 7, 19 (1947). — J. WEYSSENHOFF: Acta Phys. Polon. **9**, 26, 46 (1947) See also O. C. DE BEAUREGARD, p. 122 of op. cit. in Sect. 107.

[4] FINZI, B.: Meccanica Razionale. 2 volumes. Bologna: Zanichelli 1948. — A general textbook on mechanics, with some attention given to Lagrangian and Hamiltonian methods. Relativistic mechanics. Statistical mechanics.
[5] FRANK, PH.: Analytische Mechanik. Die Differential- und Integralgleichungen der Mechanik und Physik, Teil 2, pp. 1—176. Braunschweig: F. Vieweg & Sohn 1927. — Lagrangian and Hamiltonian equations, transformation theory, HAMILTON-JACOBI equation, action-angle variables, stability, rigid motions, perturbations.
[6] FUES, E.: Störungsrechnung. GEIGER-SCHEEL, Handbuch der Physik, Vol. V, pp. 131 to 177. Berlin: Springer 1927. — Multiply periodic motions, action-angle variables, degeneracy, adiabatic invariants, development in powers of parameter, secular disturbances, DELAUNAY's method, time-dependent disturbances.
[7] GOLDSTEIN, H.: Classical Mechanics. Cambridge, Mass.: Addison-Wesley 1950. — Emphasis on techniques required in quantum mechanics. Matrix and vector notations used. Special relativity. Hamiltonian equations, canonical transformations, small oscillations, introduction to Lagrangian and Hamiltonian formulations for continuous systems and fields.
[8] GRAMMEL, R.: Kinetik der Massenpunkte. GEIGER-SCHEEL, Handbuch der Physik, Vol. V, pp. 305—372. Berlin: Springer 1927. — Dynamics of a particle, free or constrained. Motion relative to rotating earth. Two-body problem. Three-body problem. Stability.
[9] HALPERN, O.: Relativitätsmechanik. GEIGER-SCHEEL, Handbuch der Physik, Vol. V, pp. 578—616. Berlin:Springer 1927. — Dynamics of a particle and of a continuum in special relativity. Light quantum. General relativity.
[10] HAMEL, G.: Die Axiome der Mechanik. GEIGER-SCHEEL, Handbuch der Physik, Vol. V, pp. 1—42. Berlin: Springer 1927. — Newtonian laws. Construction of mechanics from continuity hypothesis, from rigid body, from particles. Construction from Lagrangian and energetic principles. Non-classical forms of dynamics. Absence of contradictions.
[11] HAMEL, G.: Theoretische Mechanik. Berlin: Springer 1949. — A comprehensive textbook, with detailed treatments of rigid body, *n*-body problem and non-holonomic systems; 263 pages devoted to problems and their solutions.
[12] JUNG, G.: Geometrie der Massen. Encyklopädie der mathematischen Wissenschaften, Vol. IV.1, pp. 279—344. Leipzig: Teubner 1901—1908. — A specialized article on linear, quadratic and higher moments, with bibliography up to 1903.
[13] LAMB, H.: Dynamics. Cambridge: University Press 1929 (2nd Edn.). — Elementary textbook, without vector notation, noteworthy for direct simple treatment of plane problems.
[14] LAMB, H.: Higher Mechanics. Cambridge: University Press 1929 (2nd Edn.). — Sequel to [13]. Geometry of finite rotations, screws and wrenches. Motion of rigid body in space. Lagrangian and Hamiltonian equations. Vibrations.
[15] LANCZOS, C.: The Variational Principles of Mechanics. Toronto: University of Toronto Press 1949. — D'ALEMBERT's principle. Lagrangian and Hamiltonian equations. Canonical transformations. HAMILTON-JACOBI theory. Emphasis on the geometry of phase-space.
[16] LEVI-CIVITA, T., and U. AMALDI: Lezioni di Meccanica Razionale. Bologna: Zanichelli; Vol. I, 1923; Vol. II_1, 1926; Vol. II_2, 1927. — A comprehensive treatise, comparable to APPELL [2]. Vol. I deals with kinematics, geometry of masses, and statics. Vol. II_1, deals with dynamics of a particle, LAGRANGE's equations, stability of vibrations. Vol. II_2 deals with dynamics of a rigid body, Hamiltonian theory, variational principles, and impulsive motion.
[17] MACMILLAN, W. D.: Theoretical Mechanics. New York and London: McGraw-Hill; Vol. I, 1927; Vol. II, 1936. — Comprehensive textbook with much detail. Vol. I deals with orbits, ballistic trajectories, Lagrangian and Hamiltonian equations and variational principles for a particle. Vol. II deals with a rigid body, with fixed point or rolling, impulsive forces, Lagrangian and Hamiltonian methods in general, method of periodic solutions.
[18] NORDHEIM, L.: Die Prinzipe der Dynamik. GEIGER-SCHEEL, Handbuch der Physik, Vol. V, pp. 43—90. Springer: Berlin 1927. — Differential and integral principles, virtual work, D'ALEMBERT's principle, principles of GAUSS, HERTZ, HAMILTON, JACOBI.
[19] NORDHEIM, L., and E. FUES: Die HAMILTON-JACOBIsche Theorie der Dynamik. GEIGER-SCHEEL, Handbuch der Physik, Vol. V, pp. 91—130. Berlin: Springer 1927. — Canonical transformations, POISSON and LAGRANGE brackets, HAMILTON-JACOBI equation, eikonal.
[20] PÉRÈS, J.: Mécanique générale. Paris: Masson & Cie. 1953. — Compact textbook dealing with moments of inertia, non-holonomic constraints, virtual work, dynamics of a particle and of a rigid body, equations of LAGRANGE, APPELL and HAMILTON, HAMILTON-JACOBI equation, stability about equilibrium or steady motion. Shocks and percussions.

[21] PRANGE, G.: Die allgemeinen Integrationsmethoden der analytischen Mechanik. Encyklopädie der mathematischen Wissenschaften, Vol. IV.2, pp. 505–804. Leipzig: Teubner 1904–1935. — Detailed treatment of D'ALEMBERT's principle, LAGRANGE's equations, variational principles, variation of constants, HAMILTON's optics, characteristic function, HAMILTON-JACOBI equation, separation of variables, integral invariants, systematic integration of canonical systems, canonical transformations, substitution or generating functions, equivalent systems.

[22] ROUTH, E. J.: A Treatise on the Dynamics of a System of Rigid Bodies. London: Macmillan; Vol. I, 1897 (6th Edn.); Vol. II, 1905 (6th Edn.). German translation by A. SCHEPP, Leipzig 1898. Vol. II reprinted by Dover Publications, New York, 1955. — Although old-fashioned in notation, this book remains the most useful reference for miscellaneous special information on the dynamics of rigid bodies. Strings and membranes are also treated. The arrangement is unsystematic, but there is a good index.

[23] SCHAEFER, CL.: Einführung in die theoretische Physik, Bd. 1. Berlin: W. de Gruyter 1950 (5th Edn.). — The first 469 pp. of this book treat the dynamics of particles and rigid bodies in considerable detail, Lagrangian and Hamiltonian methods being included. The rest of the book (507 pp.) deals with the mechanics of continua.

[24] SCHOENFLIES, A., and M. GRÜBLER: Kinematik. Encyklopädie der mathematischen Wissenschaften, Vol. IV.1, pp. 190–278. Leipzig: Teubner 1901–1908. — Finite displacements, velocity, acceleration, linkages, mechanisms.

[25] STÄCKEL, P.: Elementare Dynamik der Punktsysteme und starren Körper. Encyklopädie der mathematischen Wissenschaften, Vol. IV.1, pp. 436–684. Leipzig: Teubner 1901–1908. — Direct treatment of the dynamics of particles and rigid bodies, with bibliography and detailed historical references. Gyroscopic motion handled rather fully, with diagrams.

[26] SYNGE, J. L., and B. A. GRIFFITH: Principles of Mechanics. New York-Toronto-London: McGraw-Hill 1959 (3rd Edn.). — Textbook of statics and dynamics up to gyroscopic theory. Motion of charged particles in axially symmetric electromagnetic field. Methods of LAGRANGE and HAMILTON. Vibrations. Introduction to relativity.

[27] VOSS, A.: Die Prinzipien der rationellen Mechanik. Encyklopädie der mathematischen Wissenschaften, Vol. IV.1, pp. 3–121. Leipzig: Teubner 1901–1908. — History and philosophy of mechanics, from GALILEO and NEWTON. Elimination of force by KELVIN and HERTZ. Principles of D'ALEMBERT, FOURIER, GAUSS, HAMILTON. Principle of energy.

[28] WHITTAKER, E. T.: A Treatise on the Analytical Dynamics of Particles and Rigid Bodies. Cambridge: University Press 1937 (4th Edn.). Reprinted by Dover Publications, New York 1944. The 2nd Edition (1917) translated into German by F. and K. MITTELSTEN-SCHEID (Berlin: Springer 1924). — Standard treatise, with systematic arrangement of material. More compact than APPELL [2] or LEVI-CIVITA and AMALDI [16], the usual problems of the dynamics of particles and rigid bodies being treated by Lagrangian methods, without vector notation or diagrams. The second half of the book deals with Hamiltonian systems, integral invariants, transformation theory, first integrals, three-body problem, theory of orbits.

[29] WINKELMANN, M., and R. GRAMMEL: Kinetik der starren Körper. GEIGER-SCHEEL, Handbuch der Physik, Vol. V, pp. 373–483. Berlin: Springer 1927. — The top, symmetric and unsymmetric, with many diagrams. Relative motion of a rigid body on the rotating earth. Systems of rigid bodies. Gyroscopic stability.

[30] WINTNER, A.: The Analytic Foundations of Celestial Mechanics. Princeton: University Press 1947. — The main theme is the n-body problem, but the book contains a compact critical treatment of Hamiltonian methods and canonical transformations, with interesting historical notes and references at the end.

The Classical Field Theories.

By

C. TRUESDELL and R. TOUPIN[1].

With 47 Figures.

With an Appendix on Invariants by

J. L. ERICKSEN.

A. The field viewpoint in classical physics.

1. Corpuscles and fields. Today matter is universally regarded as composed of molecules. Though molecules cannot be discerned by human senses, they may be defined precisely as the smallest portions of a material to exhibit certain of its distinguishing properties, and much of the behavior of individual molecules is predicted satisfactorily by known physical laws. Molecules in their turn are regarded as composed of atoms; these, of nuclei and electrons; and nuclei themselves as composed of certain elementary particles. The behavior of the elementary particles has been reduced, so far, but to a partial subservience to theory. Whether these elementary particles await analysis into still smaller corpuscles remains for the future.

Thus in the physics of today, corpuscles are supreme. It might seem mandatory, when we are to deal with extended matter and electricity, that we begin with the laws governing the elementary particles and derive from them, as mere corollaries, the laws governing apparently continuous bodies. Such a program is triply impractical:

A. The laws of the elementary particles are not yet fully established. Even such senior disciplines as quantum mechanics and general relativity remain open to possible basic revision and not yet satisfactorily interconnected.

B. The mathematical difficulties are at present insuperable. (Even on a lower level they remain: As is well known, the "proof" that a quantum-mechanical system may be replaced by a classical system in first approximation is defective.)

C. In such special cases as have actually been treated, the mathematical "approximations" committed in order to get to an answer are so drastic that the results obtained are not fair trials of what the basic laws may imply. When such a result appears in disaccord with experience, we are at a loss whether to assign the blame to the basic laws themselves or to the mathematical process used in the subsequent derivations.

[1] **Acknowledgment.** The authors are deeply indebted to Professor Dr. K. ZOLLER for thorough criticism of most of the manuscript and proofs. They are grateful also to Professors J. L. ERICKSEN and W. NOLL and to Dr. B. COLEMAN for help in certain passages.

During portions of the period of preparation of this treatise, TRUESDELL's work was supported by an ONR contract (1955 to 1956) the U. S. National Science Foundation (1956), the John Simon Guggenheim Memorial Foundation (1957), the Mathematics Research Center, U. S. Army, University of Wisconsin (1958), and the National Bureau of Standards (1959). During 1957 he was on sabbatical leave from Indiana University.

While all parts of this work have been discussed and revised jointly, Chaps. A to E and the first half of Chap. G were written by TRUESDELL; Chap. F and the second half of Chap. G, by TOUPIN.

But more than this, such a program even if successful would be illusory: (a) The future discovery of new entities within the present "elementary" particles would nullify any claim for such results as predictions from "basic" laws of physics. Indeed, *within any corpuscular view the possibility of an infinite regress is logically inevitable.*

(b) The details of the behavior of the corpuscles are extraneous to most mechanical and electromagnetic problems. Materials whose corpuscular structure is quite different may exhibit no perceptible difference of response to stress.

Avoiding illusory complications, it is possible to construct a direct theory of the *continuous field*, indefinitely divisible without losing any of its defining properties. The field may be the seat of motion, matter, force, energy, and electromagnetism. Theories expressed in terms of the field concept are called *phenomenological*, because they represent the immediate phenomena of experience, not attempting to explain them in terms of corpuscles or other inferred quantities.

The corpuscular theories and the field theories are mutually contradictory as direct models of nature[1]. The field is indefinitely divisible; the corpuscle is not. To mingle the terms and concepts appropriate to these two distinct representations of nature, while unfortunately a common practice, leads to confusion if not to error. For example, to speak of an element of volume in a gas as "a region large enough to contain many molecules but small enough to be used as an element of integration" is not only loose but also needless and bootless.

In a deeper sense, the continuous field and the assembly of corpuscles may be set into entire agreement. Adopting the viewpoint of statistical mechanics, consider a classical system of mass-points of any kind whatever, and assign a probability to its initial conditions. Extending a notable success by IRVING and KIRKWOOD[2], NOLL[3] has defined certain *phase averages* which he proved to satisfy exactly the laws of balance for a continuous field. This result, not a limit formula or approximation, is an *exact theorem* on distributions in phase space. Thus those who prefer to regard classical statistical mechanics as fundamental may nevertheless employ the field concept as exact in terms of *expected values*.

While sometimes the phenomenological approach is regarded as only approximate, the result just described shows that in representing matter as continuous rather than discrete we can in fact make no statement that is inconsistent with the statistical view of matter as composed of classical molecules, *so long as we confine attention to the exact and general theory of continuous media*[4].

This treatise presents the exact and general theory of the continuous field.

---

[1] The formal "derivations" of the field equations from the mass-point equations of mechanics given in many textbooks are illusory, such a derivation being impossible without added assumptions which are rendered superfluous by a direct approach to the continuum. The difficulty can be avoided by a formulation of the fundamental equations as Stieltjes integrals (cf. Sect. 201); in essence, this was done by EULER [1752, *2*, §§ 20–22].

[2] [1950, *12*]. Square brackets refer to the bibliography.

[3] [1955, *19*]. In NOLL's paper precise conditions of regularity for the density in phase are stated. The molecules, not restricted in variety, are supposed free of constraints but otherwise subject to arbitrary mutual and extrinsic forces. The expression for the resultant extrinsic force in general does not depend only on the extrinsic forces to which the molecules are subject; otherwise the agreement stated in the text above is unqualified. An extension to quantum-mechanical systems is given by IRVING and ZWANZIG [1951, *12*].

[4] That is, only the *general* equations expressing the balance of mass, momentum, and energy in the continuous field have been derived. There is no indication that any special theory of continuous bodies, such as the theory of perfect fluids, is consistent with statistical mechanics. In fact, a *simple* field theory seems to emerge only in approximation, and from a simple molecular picture an extremely complicated field theory results. Also, the exact agreement does not extend to thermodynamics, which from the statistical standpoint appears to be only an approximate theory.

2. Classical mass-points and classical fields. From the time of NEWTON until relatively recently, many natural scientists considered the mass-point the fundamental quantity of nature, or at least of mechanics. They believed that matter was composed of many very small particles obeying the laws of classical mechanics, and that, consequently, the behavior of gross matter could be predicted, in principle, to any desired accuracy, from a knowledge of the intermolecular forces. Thus continuum mechanics appears as an approximate or at best secondary theory within classical mechanics. While this tradition clings on in physics teaching today, it is not realistic. Aside from the as yet unconquered mathematical difficulties in putting this ideal program into practice, the program itself is out of keeping with modern views on matter. The smallest units of matter are no longer believed to obey the laws of Newtonian mechanics, except approximately and in circumstances rarely occurring in dense matter. Nevertheless, conditions in which the classical laws of momentum and energy fail perceptibly for *tangible portions of matter* are extremely rare if not altogether unknown.

To cite an example, no corpuscular theory based on Newtonian mechanics has produced formulae for the specific heats of solids which agree with experimental values. Nevertheless, there is not the slightest indication that a solid body when heated and set in motion fails, *as a body*, to obey the classical laws of balance of mass, momentum, and energy. In fact it is almost the rule that *Newtonian mechanics, while not appropriate to the corpuscles making up a body, agrees with experience when applied to the body as a whole*, except for certain phenomena of astronomical scale. Only paedagogical custom has hindered general realization that *as a physical theory, continuum mechanics is better than mass-point mechanics*[1].

Indeed, in physics it is inappropriate to lay down the laws of classical mechanics for small bodies, to which in general they do *not* apply, and thence to derive or state by analogy the corresponding laws for extended bodies, to which they *do* apply. Rather, the process should be reversed: *Classical mechanics is the mechanics of extended bodies*[2].

There remain certain special problems, particularly problems in celestial mechanics, ballistics, and mechanisms, where the mechanics of mass-points is accurate. As is shown in Sect. 167, problems of this kind are easily and reasonably regarded as special cases within continuum mechanics.

3. Experiments and axioms. It has become fashionable to present the foundations of theoretical physics in terms of experiments. Indeed, since physics is intended to predict numerous phenomena of nature from knowledge of a few, the preconception that a given physical discipline should be derivable from the results of certain basic experiments is most appealing. In fact, however, an experimental approach to mechanics and electromagnetism is not practical. The field, infinite in extent and indefinitely divisible, is by its very nature not measurable directly. The "experiments" sometimes used as the starting point for paedagogical treatments of field theories are *a posteriori* verifications at best; always unperformed and often unperformable, too often they are mere hoaxes. Moreover, they belie the true course by which the field theories have developed. *Experience* has been the guide, *thought* has been the creator[3]. Not only does any theory reduce and abstract experience, but also it overreaches it by extra

[1] Cf. TRUESDELL [1952, *22*, pp. 79–80].

[2] Cf. HAMEL [1908, *4*, p. 351]. Note also that from a theory of phase averages over systems governed by quantum mechanics IRVING and ZWANZIG [1951, *12*] infer the *classical* equations of balance of mass, momentum, and energy.

[3] Cf. e.g. DUGAS [1954, *5*], TRUESDELL [1956, *23*].

assumptions made for definiteness. Theory, in its turn, predicts the results of certain specific experiments. The body of theory furnishes the concepts and formulae by means of which experiment can be interpreted as in accord or disaccord with it. To *overturn* a theory by the results of experiment, we seek the aid of the theory itself; in terms of the theory, from experiment we may find agreement which develops confidence in the theory, but *establish* a theory by experiment we never can. Experiment, indeed, is a *necessary* adjunct to a physical theory; but it is an adjunct, not the master.

While most theoretical physicists seem to act in accord with the above views, they rarely admit to holding them. Therefore a fuller explanation, largely a paraphrase of a work by SOUTHWELL[1], is appended.

The "operational" system accepts as basic only quantities susceptible of direct measurement and, connecting them, laws which are to be tested by experiment. From these laws, logical inference is to derive a system shown by actual trial to keep contact with physical experience at every stage.

Apart from the practical limitations in checking any theoretical "law", there is a deeper objection against this view of physics, in that it rests on a circularity: No experiment can be interpreted without recourse to ideas in themselves part of the theory under examination. Similarly, no quantity can be measured in the absence of a theory explaining the experiment. Consider the measurement of "mass" by weighing or by impact; in the former case the law of falling bodies and in the latter case the law of conservation of momentum, both employing the concept of mass, are used to complete the measurement. If we seek to verify NEWTON'S "law" that "Every body continues in its state of rest, or of uniform motion in a right line, unless compelled to change that state by forces impressed upon it,"[2] we require a free body, unavailable because all bodies in the laboratory are subject to the earth's attraction. Indeed, we try to neutralize that attraction, as in "ATWOOD'S machine": The body is connected by a light string passing over a freely running pulley with a second body of equal weight, and it is found that, started with any initial velocity, the test body retains its velocity almost unchanged. Casting aside the small observed retardation, doubtless arising from friction, we still cannot accept this result as a proof of the "law" in question. The body found to move with substantially uniform speed and direction is *not* a free body, and without the principles of mechanics, themselves dependent upon the law we are supposedly establishing by experiment, we cannot justly assert that the forces present do in fact neutralize each other. More elaborate application of the principles of mechanics is required if we are to reason that the inertia of the pulley has no effect on the ideal experiment. Further, to estimate the "experimental error" in the real experiment, we require a hypothesis of friction and an application of the laws of mechanics both for the effect of this friction and for the partially counteracting effect of the inertia of the pulley.

Such difficulties are avoided by the postulational standpoint, according to which physics, as an abstract discipline, may employ any variables and any consistent initial assumptions or "laws" which are convenient. In construction of this mathematical system it is not necessary to maintain contact with experiment at every stage. The system is an *abstract model*, designed to represent some of the observed phenomena of the physical universe, but directly concerned only with ideal bodies. Some few of the properties of these ideal bodies are postulated; the numerous remainder is to be derived mathematically. Whether these derived properties correspond with physical observation is a separate question, to be decided by subsequent comparison with experiment. But the available tests apply only to the system as a whole: We cannot devise an experiment such as to verify any one of its assumptions apart from the rest.

Naturally it is possible to construct an ideal system without relevance to physics. However, since experience is the guide, entirely wrong physical theories have been rare. Rather, a well thought theory usually turns out to have relevance for certain physical phenomena but to be in error for others. Such is the case with the classical field theories. Their failures are well known and have provided the impetus for "modern" physics. Often their successes are forgotten. It is classical physics by which we grasp the world about us: the heavenly

---

[1] [1929, *9*].
[2] [1687, *1*, Lex 1].

motions, the winds and the tides, the terrestrial spin and the subterraneous tremors, prime movers and mechanisms, sound and flying, heat and light[1].

Thus the classical field theories have won indisputable *permanence* in the language by which we speak of nature. Whatever the future revisions of theories of the structure of matter, the place of the classical field theories will remain unchanged. This permanence, along with the difficulties mentioned in Sect. 1, makes necessary a *complete and independent presentation* of the foundations of the classical field theories. Being mathematical disciplines, they should be derived from *axioms*.

Indeed, as his sixth problem HILBERT[2] set the construction of a set of axioms, on the model of the axioms of geometry, for "those branches of physics where mathematics now plays a preponderant part; first among them are probability theory and mechanics." Like all of his problems concerning physical applications of mathematics, his proposal for mechanics has received little attention. The possibility that the future may revise the physics of small corpuscles does not reduce the need for axiomatic treatment of the field theories. Physics, like mathematics, may be constructed precisely at several different levels. The interconnection of the different levels, either exactly or by approximation or by addition of new axioms, then furnishes definite mathematical problems[3].

Having reached agreement that we should base the classical field theories on a set of axioms, we must now admit, ruefully, our inability to do so. In our opinion, none of the attempts to form such a system has been successful. Only in very recent years has an adequate set of axioms for pure mechanics, at last, been constructed by NOLL[4]. To present his development of the subject here would be premature, since a correspondingly clear and precise formulation of irreversible thermodynamics is not yet available. We regard the fully invariant formalism for electromagnetic theory given in our Chap. F as being essentially an axiomatization of the subject. NOLL and COLEMAN have disclosed to us the outline of what appears to be a satisfactory basis of general thermodynamics in deformable media. Thus there are grounds for expecting that HILBERT'S program will shortly be actualized.

Despite the lack of complete axiomatic formulation, the *general equations* governing the classical fields are known and universally accepted. The present article is devoted to a *formally precise* study of these general equations. Any future axiomatization, if successful, will necessarily lead to these same equations.

4. Mathematics and its physical interpretation. That a branch of theoretical physics is a mathematical science by no means implies its aim or interests to be those of pure mathematics. Rather, *the problems are set by the subject.* The developments must illumine the *physical aspects* of the theory, not necessarily in the narrower sense of prediction of numerical results for comparison with experimental measurement, but rather for the grasp and picture of the theory in relation to experience. In this spirit do we pursue our subject, *neither seeking nor avoiding* mathematical complexity[5].

---

[1] Cf. V. BJERKNES *et al.* [1933, *3*, Vorwort].

[2] [1900, *5*].

[3] In mathematics the economy of such independent constructions has long been realized. E.g., to construct the complex numbers we presume the properties of the real numbers given; for the real numbers, those of the integers; and for the integers, mathematical logic. To approach fluids in terms of nuclear physics is like treating functions of a complex variable with the apparatus of formal logic.

[4] [1957, *11*] [1958, *8*] [1959, *9*].

[5] The expression of such a program has been attributed to KELVIN and TAIT, but we are unable to trace the reference.

Some will reproach us with too much abstract and useless formalism. Not forgetting that such deprecation was bestowed upon WHITTAKER'S *Analytical Dynamics* half a century ago, we are confident that the reader of half a century hence will regard our compromise of the moment as erring rather toward insufficient use of the mathematical tools available.

Any mathematical theory of physics must idealize nature. That much of nature is left unrepresented in any one theory is obvious; less so, that theory may err in adding extra features not dictated by experience. For example, the infinity of space is itself a *purely mathematical concept*[1], and all theories erected within this space must share in the geometrical idealization already implied. Indeed, it is difficult to find any theory that does not contain infinities, and infinities, by definition, are immeasurable. While at one time certain theoretical statements were regarded as "laws" of physics, nowadays many theorists prefer to regard each theory as a *mathematical model*[2] of some aspect of nature.

In a sense, then, every theory is only "approximate" in respect to nature itself. This unavoidable defect in theory is often taken as a patent for "approximate" mathematics in the deductions from it. Indeed, while mathematics is generally understood to proceed by entirely logical processes, were the "derivations" in some of the accepted physical papers of today translated into common reasoning, they would fail to meet the logical standards of a competent historian or bibliographer. All too often is heard the alibi that since the theory itself is only approximate, the mathematics need be no better. In truth the opposite follows. Granted that the model represents but a part of nature, we are to find what such an ideal picture implies. A result strictly derived serves as a *test of the model*; a false result proves nothing but the failure of the theorist. To call an error by a sweeter name does not correct it. The oversimplification or extension afforded by the model is not error: The model, if well made, shows at least how the universe *might* behave, but logical errors bring us no closer to the reality of *any* universe. *In physical theory, mathematical rigor is of the essence.*

In this treatise we attempt to keep the argument rigorous. However, nothing is gained by laboring elementary details. We presume that the reader knows infinitesimal calculus, simple algebra, and tensor analysis; that he can supply for himself, without repetition on our part, conditions sufficient for interchanging differentiations, inversion of functions, expansions in power series, etc. Roughly speaking, our proportion of what is said to what is left unsaid is that which is customary in works on differential geometry.

5. Exact and approximate theories. While every theory is a model of nature, and thus not "exact" in relation to it, nevertheless there is a government among theories. A theory is tested by experiment, and a range of confidence in it is established. In this sense, a given theory is "good"; if the range of application is greater than another's, it is the "better" of the two. For example, the theory of the flow of viscous compressible fluids should suffice to predict definite results, fit for experimental test, concerning the propagation, absorption, and dispersion of sound in fluids. That such results have never been obtained is only from our lack of sufficient mathematics. Instead, a perturbation scheme has been used to infer equations governing "small" motions. The resulting acoustical theory is presumed to yield an "approximation" to the better but intractable theory of fluids.

Any given theory may be laid down as "exact". It is then a definite mathematical problem to discover the relation of its results to those derived from

[1] EULER [1736, *1*, § 8].
[2] HELMHOLTZ [1902, *4*, § 1].

other theories, considered as "approximate" in respect to it. Problems of this kind are important and difficult, indeed in most cases too difficult for the mathematics available today. We do not attempt to study them in this treatise. Neither do we present[1] the unjustified linearizations or formal schemes of perturbation which occupy much of the literature. Our scope is restricted to *exact treatment*.

6. Closed systems and armatures. In the nineteenth century there was a search for sets of physical laws which should include the maximum range of physical phenomena yet remain sufficiently specific to predict definite results in particular cases. The culmination of this trend came in the systems of JAUMANN and LOHR[2], in which an all-embracing set of equations governing mechanics, electrodynamics, chemical reactions, diffusion, heat transfer, electromechanical effects, etc. are postulated. Current knowledge of the structure of matter (cf. Sect. 1) has destroyed the *raison d'être* of such closed systems as well as rendering them impractical.

Rather, the classical field theories offer us *armatures* on which particular models of extended matter and electricity may be built. In this spirit, it is *inclusiveness* rather than particular problems that we seek here. For example, it is often claimed that in nature, if we look closely enough, only conservative forces occur; that such effects as friction are gross appearances resulting only from lack of knowledge of the underlying conservative process. But natural problems are not confined to those on the smallest or largest scale. The world about us, as we see it, must be mastered and controlled. Situations incompletely described are the rule, not the exception, and we must formulate good theories for these *limited aspects* of nature. Our object is a *general framework*[3] for such theories. The most general motions, the most general stresses, the most general flows of energy, and the most general electromagnetic fields, furnish the subject of this treatise.

7. Field equations and constitutive equations. *Motion, stress, energy, entropy,* and *electromagnetism* are the concepts upon which field theories are constructed. Certain laws of *conservation* or *balance* are laid down as relating these quantities in all cases. These basic principles, which are in integral form[4], in regions where

[1] An exception is Part e of Subchapter BI, which is included only so as to aid the reader in connecting the general theory with the results assumed in the ordinary theory of elasticity.

[2] [1911, *7*], [1918, *3*]; [1917, *5*].

[3] The field viewpoint is excellently apt to secure such generality, while to overcome the complications of corpuscular theories it is usual to make simplifying hypotheses which sharply lessen their scope. Cf. HELMHOLTZ [1902, *4*, § 2]. As was remarked by LAGRANGE [1788, *1*, Part 1, Sect. IV, § II, ¶ *9*] the field view has precisely the same mathematical advantage over the corpuscular view as the differential theory of curves over polygonal approximations.

[4] The view that all natural laws should be expressed by integrals is generally attributed to the Göttingen lectures of HILBERT. That jump conditions are not to be derived from smooth solutions was clearly understood by STOKES [1848, *4*, p. 353]: "... I wish the two subjects to be considered as quite distinct."

In recent years mathematicians have created various kinds of "generalized solutions", whereby, granted certain purely *analytic* presumptions as to the intended meaning of a problem formulated in terms of differential equations, discontinuous solutions may be inferred from continuous ones. We regard these approaches not only as demanding unnecessary mathematical apparatus but also as concealing the simple and immediate nature of physical laws. Neither in respect to rigor nor in any other regard do they offer advantages over HILBERT's program of stating physical laws in integral form. While the work of ZEMPLÉN [1905, *7*, p. 438] and HELLINGER [1914, *4*], which reflects HILBERT's influence, rests upon variational principles, the program of postulating *integral conservation laws*, which we follow here, seems first to have been laid down by KOTTLER [1922, *3* and *4*]. Cf. also CARTAN [1923, *1*, introd., §§ 8, 81], KOTCHINE [1926, *3*, § 3]. As remarked by VAN DANTZIG [1937, *11*], it is obvious that notions like differentiability can have no empirical basis at all. Since, on the contrary,

the variables change sufficiently smoothly are equivalent to differential *field equations*; at surfaces of discontinuity, to *jump conditions*.

The field equations and jump conditions form an underdetermined system, insufficient to yield specific answers unless further equations are supplied. Within the embracing concept of the balanced fields, it is possible to define *ideal materials*[1] by certain further conditions. These defining conditions are called *constitutive equations*. The most familiar constitutive equations, here expressed in words, are:

The distances between particles do not change (Rigid body).

The stress is hydrostatic (Perfect fluid).

The stress may be determined from the stretching alone (Viscous fluid, perfectly plastic body).

The stress may be determined from the strain alone (Perfectly elastic body).

The flux of energy is a linear function of the temperature gradient (Classical linear heat conduction).

The thermodynamic affinities are linear functions of the thermodynamic fluxes ("Irreversible thermodynamics").

The diffusion velocity of a constituent of a binary mixture is proportional to the gradient of its peculiar density (Classical linear mass diffusion).

The electric displacement is proportional to the electric field; the magnetic induction is proportional to the magnetic intensity (Classical linear electromagnetism).

The constitutive equations and field equations together, along with the jump conditions and boundary conditions, should lead to a definite theory, predicting specific answers to particular problems. For some of the special materials listed above, this definiteness has been proved through theorems of existence and uniqueness.

The present treatise is devoted to the *general principles of balance* alone. Thus we deal only with the *field equations and jump conditions*. Our last chapter mentions guiding principles by which rational constitutive equations may be formulated. The theories of certain particular ideal materials fill several later volumes of the Encyclopedia.

8. The nature and plan of this treatise. We present the *common foundation* of the field viewpoint[2]. We aim to provide the reader with a full panoply of *tools of research*, whereby he himself, put into possession not only of the latest discoveries but also of the profound but all too often forgotten achievements of previous generations, may set to work as a theorist.

This treatise is intended for the specialist, not the beginner. Necessarily it presents the foundations of the field theories, not as they appeared in the last century and linger on in the textbooks, nor as the experts in some other domains

---

such simple theories as the dynamics of perfect fluids and linear electromagnetism are known to furnish inadequate models of experience when the fields appearing are required to be everywhere continuous, no physical principle should be stated in differential form.

The older approach, still followed in many textbooks, either sets up more special postulates for discontinuities or employs an unrigorous limit process from the differentiable case.

[1] The program of mechanics was laid out by EULER [1752, *2*, § 19] (cf. also his remarks on rigid bodies and ideal fluids [1769, *1*, § 12]); of continuum mechanics in particular, by CAUCHY [1823, *1*], but the later emphasis on linear problems in very special theories caused it to be largely forgotten until it was stated anew by v. MISES [1930, *3*].

[2] The only single work attempting even a major part of our subject is the book of BRILL [1909, *2*].

may think they ought to be presented[1], but as they are cultivated by the specialists of today.

This treatise is organized as follows:

1. Kinematics, including conservation of mass, in Chaps. B and C.
2. Balance of momentum, Chap. D.
3. Balance of energy, including the thermodynamics of irreversible deformations, Chap. E.
4. Balance of electromagnetism, Chap. F.
5. Guiding principles for constitutive equations and examples of them, Chap. G.

For a more detailed plan, see the table of contents.

We interpret "classical" in the narrower sense, as confined to phenomena in *Euclidean three-dimensional space*[2] and governed by *Newtonian mechanical principles*. However, it would be crippling to maintain this restriction in dealing with electromagnetism. The four-dimensional viewpoint adopted in Chap. F necessitates further kinematical developments there and results in some duplication of subject. On the other hand, to have begun with a world-invariant formalism in the earlier chapters would have greatly lessened their direct usefulness to specialists in mechanics.

That over one half of the work is devoted to kinematics, the mathematical description of motion, is not malapropos. As the need for more and more general field theories has grown, the preliminary light which kinematics unencumbered by physical restrictions can provide, always appreciated by virtuosi of mechanics[3], has become a necessity. In presenting here as our Chaps. B and C the first general treatise on the kinematics of continua, we believe that we look toward the future course of the field theories.

9. Tradition. We have tried to supply full and correct attributions, not only for historical perspective but also in plain justice. If the name attached to many a proposition is but a small one, that is all the less reason that its owner should be pilled of what little he wrought by a no greater name of today, whose slight capacities are scarcely increased by wilful or heedless ignorance of what others have done. However, the multitude of detailed citations should not prevent the great names from emerging. Our subject is largely the creation of EULER and CAUCHY. If we present their results in forms often very different from the original, in return we have included many of their discoveries that have not previously found a place in expositions. Not only will these names be the most frequently encountered, but also their appearances are at the crucial theorems and definitions. Next come STOKES, HELMHOLTZ, KIRCHHOFF, KELVIN, MAXWELL and HUGONIOT. In the twentieth century, HADAMARD and HILBERT[4] continued and deepened the tradition. That no one later name is frequently cited does not indicate that the subject is dead. Rather, after a generation of

[1] Cf. KELVIN and TAIT [1867, *3*, Preface]: "... where we may appear to have rashly and needlessly interfered with methods and systems of proof in the present day generally accepted, we take the position of Restorers, not of Innovators."

[2] At the end of the treatise, references to relativistic and to non-Euclidean or higher-dimensional but non-relativistic theories are given in Special Bibliographies R and N, respectively.

[3] We follow the tradition of EULER, CAUCHY, and KELVIN. Cf. the remarks of HELMHOATZ [1858, *1*, Introd.], ZHUKOVSKI [1876, *7*], ST. VENANT [1880, *10*] and JAUMANN [1905, *2*, Introd.].

[4] While HILBERT published nothing relevant to our subject, his personal influence was widespread and continuing (cf. e.g. ZEMPLÉN [1905, *7*], HELLINGER [1914, *4*, footnote 6]), and the organization used in his Göttingen lectures [1907, *3* and *4*] has influenced ours. Cf. also footnote 4, p. 232.

quiescence, in very recent years it has experienced a revival in a form more compact and general, and, we believe, closer to nature.

10. General scheme of notation. We make frequent and essential use of notations and results given in the appendix, "Invariants", by J.L. ERICKSEN. Citations of that article are indicated by the prefix App. Thus "(App. 7.1)" refers to Eq. (7.1) of that appendix, and "Sect. App. 7" refers to its Sect. 7.

The following partial table explains the basis for selection of notations. It has not been possible to maintain this scheme without exception, nor to avoid use of the same symbols in different senses in widely separate passages.

Full-sized characters:

Italic letters A, a, ..., etc.: Scalars and the kernel indices of vectors and tensors. For distinctions, see Sects. App. 15, 14.

Bold-face letters $\boldsymbol{A}$, $\boldsymbol{a}$, $\boldsymbol{\mathfrak{A}}$ ...: Vectors, tensors, and matrices. For explanation, see Sect. App. 3.

Roman numerals I, II, III, $\overline{\text{II}}$, $\overline{\text{III}}$, etc.: Principal invariants and moments of second order tensors (Sect. App. 38).

Hebrew letters א, ב, ...: Scalar coefficients in expressions for isotropic functions (Sect. 299).

Russian capitals Д, Ц, ...: Other special scalar invariants.

Script letters $\mathscr{A}$, a, ...: Regions, surfaces and curves.

German letters $\mathfrak{A}$, $\mathfrak{a}$, ..., $\boldsymbol{\mathfrak{A}}$, $\boldsymbol{\mathfrak{a}}$, ...: Quantities defined by integrals.

Black-letter capitals $\mathfrak{R}$, $\mathfrak{W}$, ...: Similarity parameters.

Sans-serif capitals M, U, ..., M, U, ...: Dimensional units and dimensions (Sect. App. 7).

Greek letters Θ, ϑ, Φ, φ, ...: Angles, thermodynamic variables, world tensors.

Indices:

Italic and Greek indices K, k, Γ, γ, ...: Tensorial indices.

Roman indices s, t, ..., S, T, ...: Descriptive marks.

Bold-face indices: Tensors or matrices from which an invariant is constructed. For example, $\text{I}_{\boldsymbol{a}}$, $\text{II}_{\boldsymbol{a}}$, and $\text{III}_{\boldsymbol{a}}$ are the principal invariants of $\boldsymbol{a}$.

German indices $\mathfrak{a}$, $\mathfrak{A}$, ...: Enumerative indices, not indicating tensor character. The corresponding numbers are written in italics. Thus $b_{\mathfrak{a}}^{k}$ and $c_{m}^{\mathfrak{b}}$ stand for the sets of contravariant and covariant components of the vectors $\boldsymbol{b}_1, \boldsymbol{b}_2, \ldots, \boldsymbol{b}_t$ and $\boldsymbol{c}^1, \boldsymbol{c}^2, \ldots, \boldsymbol{c}^t$.

List of frequently used symbols

Symbol	Name	Place of definition or first occurrence
A	Abnormality of a vector field	(App. 30.1)
$\boldsymbol{A}$, A^k	Axis of finite rotation	(37.17)
$\boldsymbol{A}$, A_k	Magnetic potential	Sect. 276
$\boldsymbol{a}$, $a_{\gamma\delta}$, $\boldsymbol{A}$, $A_{\Gamma\Delta}$	First fundamental form of a surface	(App. 19.7)
$\boldsymbol{b}$, $b_{\gamma\delta}$, $\boldsymbol{B}$, $B_{\Gamma\Delta}$	Second fundamental form of a surface	(App. 21.5)
$\boldsymbol{B}$, B^k	Magnetic flux density	Sect. 274
$\boldsymbol{c}$	Position vector of the center of mass	(161.5)
$\boldsymbol{c}$, c_{km}	CAUCHY's deformation tensor	(26.1)
$\boldsymbol{C}$, C_{KM}	GREEN's deformation tensor	(26.2)
c	Velocity of light in the aether	Sect. 280
$c_{\mathfrak{A}}$	Absolute concentration of the constituent $\mathfrak{A}$	(158.4)
$\hat{c}_{\mathfrak{A}}$	Mass supply of the constituent $\mathfrak{A}$	(159.1)

Symbol	Name	Place of definition or first occurrence
$d_{(\boldsymbol n)}$	Stretching in the direction of $\boldsymbol n$	(82.5)
$d_{\mathfrak a}$	Principal stretching	Sect. 83
$\boldsymbol d$, d_{km}	Euler's stretching tensor	(82.2)
$\boldsymbol d_{\mathfrak a}$, $d^k_{\mathfrak a}$, $\boldsymbol d^{\mathfrak a}$, $\boldsymbol D_{\mathfrak a}$, $D^k_{\mathfrak a}$, $\boldsymbol D^{\mathfrak a}$, etc.	Directors and reciprocal directors of an oriented body	(61.1)
$\boldsymbol D$, D^k	Charge potential	Sect. 276
$e_{k_1 k_2 \dots k_n}$, $e^{k_1 k_2 \dots k_n}$	Absolute scalar permutation symbols	(App. 2.3)
$\boldsymbol e$, e_{km}, $\boldsymbol E$, E_{KM}	Strain tensors	(31.1)
$\tilde{\boldsymbol e}$, $\tilde e_{km}$, $\widetilde{\boldsymbol E}$, $\widetilde E_{KM}$	Elongation tensors	(31.6)
$\boldsymbol E$, E_k	Electric field	Sect. 274
$\boldsymbol f$, f^k	Assigned force field	(200.1)
$F^k{}_{mP}$, $F^{\mathfrak a}{}_{\mathfrak b P}$, etc.	Relative wryness of a strained oriented body	(61.14)
g	Determinant of metric tensor	Sect. App. 2
$\boldsymbol g$, g_{km}	Metric tensor	Sect. App. 2
g^k_K, g^K_k	Euclidean shifter	(App. 16.4)
$\boldsymbol g$, $g^{\Gamma\Delta}$	World space metric	Sect. 152
$\boldsymbol h$, h^k	Flux of energy	(241.1)
$\boldsymbol H$, H_k	Current potential	Sect. 276
$\boldsymbol i$, i^k, $\boldsymbol i[\Psi]$	Influx of ψ	(157.1)
j, J	Scalars defined by Jacobians	(16.5), (16.6)
$\boldsymbol J$, J^k	Current density	Sect. 274
$\boldsymbol l$, l_k, l_{km}	Assigned couple field	(200.3)
$\boldsymbol m_{(\boldsymbol n)}$	Couple-stress vector	(200.3)
$\boldsymbol m$, m^{kqp}	Couple-stress tensor	(203.5)
$\boldsymbol M$, M_k	Magnetization field	Sect. 283
$\boldsymbol n$	Unit normal to a surface, principal normal of a curve, etc.	
$\boldsymbol p$, p^k, $\boldsymbol P$, P^K	Position vector	(19.2)
$\dot{\boldsymbol p}\ (=\dot{\boldsymbol x})$	Velocity field	(67.2)
$\ddot{\boldsymbol p}\ (=\ddot{\boldsymbol x})$	Acceleration field	(98.1)
p	Total pressure in the hydrostatic special case	(204.2)
$\tilde p$	Mean pressure	(204.7)
$\hat{\boldsymbol p}_{\mathfrak A}$, $\hat p^k_{\mathfrak A}$	Supply of momentum for the constituent $\mathfrak A$	(215.2)
P	Stress power	(217.42)
$\boldsymbol P$, P^k	Polarization density	Sect. 283
Q	Charge density	Sect. 274
$\boldsymbol R$, $R^k{}_K$, $R^k{}_m$, etc.	Finite rotation tensor	(37.1)
$R^h{}_{kmp}$, $\boldsymbol R^{(\boldsymbol a)}$	Riemann tensor based on $\boldsymbol a$	(34.2)
$\widetilde{\boldsymbol R}$, $\widetilde R_{KM}$, $\tilde{\boldsymbol r}$, $\tilde r_{km}$, $\widetilde R_K$	Tensors and vectors of mean rotation	(31.6), (36.12)
s, $s[\Psi]$	Supply of ψ	(157.1)
t	Time	
$\boldsymbol t$	Unit tangent vector	(65.1)
$\boldsymbol t$, t_Γ	World covariant space normal	Sect. 152
$\boldsymbol t_{(\boldsymbol n)}$, $t^k_{(\boldsymbol n)}$	Stress vector	(200.1)
$\boldsymbol t$, t^{km}	Cauchy's stress tensor	(203.4)
$\boldsymbol T$, T^{kM}, T^{KM}, etc.	Piola-Kirchhoff stress tensors	(210.4), (210.9)
$T^\Gamma{}_\Delta$	World stress-momentum tensor	(211.3)
u_n	Speed of displacement of a surface	(74.4), (177.5)
$\boldsymbol u$, u_Γ	Tangential velocity of parametrization of a surface	(177.11)

Symbol	Name	Place of definition or first occurrence
U_{N}	Speed of propagation of a surface	(183.1)
U	Local speed of propagation of a surface	(183.4)
V	Velocity potential	(88.2)
V	Electric potential	Sect. 276
V^*	Acceleration potential	(109.1)
$\boldsymbol{w}$, w_{km}	CAUCHY's spin tensor	(86.1)
$\boldsymbol{w}$, w^k	Vorticity vector	(86.2)
$\boldsymbol{w}^*$, w^*_{km}	Spatial diffusion vector or tensor	(101.5)
$\boldsymbol{W}^*$, $W^*_{\alpha\beta}$	Material diffusion tensor	(101.12)
$\boldsymbol{W}$, $W^K{}_{MP}$, $W^{\mathfrak{a}}{}_{\mathfrak{b}P}$, etc.	Wryness of an oriented body	(61.7)
$\boldsymbol{x}$, x^k, $x^{I'}$	General co-ordinates of places or events	(14.1), (152.5)
$\boldsymbol{X}$, X^K	General symbols for particles	(14.1)
$x^k{}_{;K} = x^k{}_{,K} = \partial x^k / \partial X^K$, $X^K{}_{;k} = X^K{}_{,k} = \partial X^K / \partial x^k$	Deformation gradients	(17.1)
$\dot{\boldsymbol{x}}$, $\dot{x}^k$	Velocity field	(67.2)
$\ddot{\boldsymbol{x}}$, $\ddot{x}^k$	Acceleration field	(98.1)
$\boldsymbol{X}_{\mathfrak{A}}$	Particle of the constituent $\mathfrak{A}$	(158.1)
$\grave{\boldsymbol{x}}_{\mathfrak{A}}$, $\grave{x}^k_{\mathfrak{A}}$	Velocity of the constituent $\mathfrak{A}$	(158.2)
$\boldsymbol{z}$, z^k, z_k	Rectangular Cartesian co-ordinate of places	(13.1)
$\boldsymbol{Z}$, Z^K	Rectangular Cartesian initial co-ordinates of particles	(13.1)
$\widehat{km}$	Physical components of stress	(204.1)
$\left\{ {k \atop mp} \right\}$	Christoffel symbols	
$\lvert \boldsymbol{x}/\boldsymbol{X} \rvert$, $\lvert \boldsymbol{z}/\boldsymbol{Z} \rvert$	Jacobians	(16.1)
$d\boldsymbol{x}$, dx^k, $d\boldsymbol{X}$, dX^K	Elements of arc	(20.3)
$d\boldsymbol{a}$, da_k, da^{km}, $d\boldsymbol{A}$, etc.	Elements of area	(20.5), (20.8)
dv, dV	Elements of volume	(20.9)
A, [A]	Unit and dimension of action	Sect. 288
L, [L]	Unit and dimension of length	Sect. App. 8
M, [M]	Unit and dimension of mass	Sect. 155
Q, [Q]	Unit and dimension of charge	Sect. 270
T, [T]	Unit and dimension of time	Sect. 65
$\boldsymbol{\Theta}$, $[\Theta]$	Unit and dimension of temperature	(246.2)
Φ, $[\phi]$	Unit and dimension of magnetic flux	Sect. 270
$\mathfrak{A}$	Virtual work	(232.1)
$\boldsymbol{\mathfrak{D}}$, $\mathfrak{D}^k$	Potential of free charge	Sect. 283
$\mathfrak{E}$	Total internal energy	(240.1)
$\boldsymbol{\mathfrak{E}}$, $\boldsymbol{\mathfrak{E}}^{[\boldsymbol{a}]}$, $\mathfrak{E}_{km}$	EULER's tensor	(168.4)
$\boldsymbol{\mathfrak{E}}$, $\mathfrak{E}_k$	Electromotive intensity	Sect. 274
$\boldsymbol{\mathfrak{F}}$, $\mathfrak{F}_k$	Total force	(196.1)
$\boldsymbol{\mathfrak{H}}$, $\boldsymbol{\mathfrak{H}}^{[\boldsymbol{a}]}$, $\mathfrak{H}_k$, $\mathfrak{H}_{km}$	Total moment of momentum	(166.2)
$\boldsymbol{\mathfrak{H}}$, $\mathfrak{H}_k$	Potential of free current	Sect. 283
$\boldsymbol{\mathfrak{I}}$, $\boldsymbol{\mathfrak{I}}^{[\boldsymbol{a}]}$, $\mathfrak{I}_{km}$	Tensor of inertia	(168.4)
$\boldsymbol{\mathfrak{I}}$, $\mathfrak{I}^k$	Conduction current	Sect. 274
$\mathfrak{K}$	Kinetic energy	(94.1), (166.4)
$\boldsymbol{\mathfrak{L}}$, $\boldsymbol{\mathfrak{L}}^{[\boldsymbol{a}]}$, $\mathfrak{L}_k$, $\mathfrak{L}_{km}$	Total torque	(196.1)
$\mathfrak{M}$	Mass	(155.1)
$\boldsymbol{\mathfrak{P}}$, $\mathfrak{P}_k$	Total linear momentum	(166.1)

Symbol	Name	Place of definition or first occurrence
$\mathfrak{Q}$	Rate of non-mechanical working	(240.1)
$\mathfrak{W}$	Rate of mechanical working	(240.1)
$\mathfrak{W}_{(K)}$	K-th moment of vorticity	(118.1)
$\boldsymbol{\mathfrak{W}}_{\mathrm{K}}$	Kinematical vorticity number	(91.1)
$\mathcal{S}_k$	Oriented k-dimensional hypersurface	Sect. 268
$d\mathcal{S}_k$	Differential element of k-dimensional hypersurface	Sect. 268
$d\hat{\mathcal{S}}_k$	Dual of $d\mathcal{S}_k$	Sect. 268
α, α_Γ	Electromagnetic potential	Sect. 271
$\alpha_{\mathfrak{a}}$	Thermodynamic coefficient	(248.5)
$\beta_{\mathfrak{a}}$	Thermodynamic coefficient	(248.5)
γ	Ratio of specific heats	(249.7)
$\gamma_{(\boldsymbol{N},\boldsymbol{M})}$, $\gamma_{(\boldsymbol{n},\boldsymbol{m})}$	Shear	(25.11)
γ, $\gamma_{\Gamma\Omega}$	Lorentz-Minkowski metric	Sect. 280
$\boldsymbol{\Gamma}$, Γ^k_{mp}, Γ^K_{MP}, $\Gamma^\Delta_{\Phi\Xi}$	Affine connections	
$\delta_{(\boldsymbol{N})}$, $\delta_{(\boldsymbol{n})}$	Extension	(25.2)
$\delta_{\mathfrak{a}}$	Principal extension	Sect. 27
δ^k_m	Kronecker symbol	
Δ	Supply of entropy	(257.3)
$\boldsymbol{\Delta}$, $\Delta^{\Gamma\Phi}$	World tensor of stretching	(153.10)
ε	Specific internal energy	(241.2)
$\varepsilon_{\mathfrak{A}}$	Specific internal energy of the constituent $\mathfrak{A}$	Sect. 243
$\hat{\varepsilon}_{\mathfrak{A}}$	Supply of energy for the constituent $\mathfrak{A}$	(243.3)
$\varepsilon_{(\boldsymbol{N})}$, $\varepsilon_{(\boldsymbol{n})}$	Elongation	(25.9)
ε_0	Fundamental electromagnetic constant	Sect. 279
ζ	Specific free enthalpy (Gibbs function)	(251.1)
η	Specific entropy	(246.1)
$\eta_{\mathfrak{A}}$	Specific entropy of the constituent $\mathfrak{A}$	(254.1)
$\boldsymbol{\eta}$, $\eta^{\Gamma\Delta}$	Charge-current potential	Sect. 271
H	Total entropy	(257.1)
θ	Temperature	(247.1)
ϑ	Angle of finite rotation	(35.1)
$\varkappa_{\mathfrak{v}}$	Specific heat at constant substate	(249.4)
$\varkappa_{\boldsymbol{\tau}}$	Specific heat at constant tensions	(249.6)
$\lambda_{(\boldsymbol{N})}$, $\lambda_{(\boldsymbol{n})}$	Stretch	(25.1)
$\lambda_{\mathfrak{a}}$	Principal stretch	(27.3)
$\lambda_{\tau\mathfrak{a}}$, $\lambda_{\mathfrak{v}\mathfrak{a}}$	Latent heats	(250.2)
$\mu_{\mathfrak{A}}$	Chemical potential	(255.1)
μ_0	Fundamental electromagnetic constant	Sect. 279
$\nu_{\mathfrak{a}\mathfrak{b}}$	Thermodynamic coefficient	(248.5)
ν_Ω	4-vector normal to surface	Sect. 277
$\xi_{\mathfrak{a}\mathfrak{b}}$	Thermodynamic coefficient	(248.5)
π	Thermodynamic pressure	Sect. 247
$\boldsymbol{\pi}$, $\pi^{\Gamma\Delta}$	World density of polarization-magnetization	Sect. 283
$\varpi_{\mathfrak{a}}$	Thermodynamic coefficient	(247.9)
ϱ	Mass density	(155.4)
$\mathring{\varrho}$	Mass density in a reference configuration	Sect. 210
$\varrho_{\mathfrak{A}}$	Mass density of the constituent $\mathfrak{A}$	(158.3)
$\boldsymbol{\sigma}$, σ^Ω	Charge-current field	(270.1)
$\boldsymbol{\tau}$, $\tau_{\mathfrak{a}}$	Thermodynamic tension	(247 1)

Symbol	Name	Place of definition or first occurrence
v	Specific volume	(156.3)
$\boldsymbol{v}, v^{\Gamma}$	Absolute world velocity vector	(153.3)
$\boldsymbol{v}, v_{\alpha}$	Thermodynamic substate	(246.1)
$\boldsymbol{\Upsilon}, \Upsilon^{\Gamma\Delta}$	World potential of free charge-current	Sect. 283
$\phi_{\mathfrak{a}\mathfrak{b}}$	Thermodynamic coefficient	(247.9)
$\boldsymbol{\varphi}, \varphi_{\Gamma\Delta}$	Electromagnetic field	Sect. 270
ψ	Specific free energy	(251.1)
χ	Specific enthalpy	(251.1)
$\boldsymbol{\omega}, \omega_{km}$	Angular velocity of a frame or rigid motion	(143.2), (143.14)
$\boldsymbol{\Omega}, \Omega^{\Delta\Xi}$	World vorticity tensor	(154.5)
$\mathbf{1}$	Unit tensor or matrix	
Ψ	"Trident", i.e., arbitrary scalar, vector, or tensor	Sect. App. 3
$[\Psi]$	Jump of Ψ across a surface	(173.1)
-[Ψ]-	Impulse of Ψ at an instant	(194.13)
$\dot{\Psi}$	Material derivative of Ψ	(72.2)
$\frac{d_r}{dt}$	Co-rotational time flux	(148.7)
$\frac{d_c}{dt}$	Convected time flux	(150.5)
$\frac{\delta_d}{\delta t}$	Displacement derivative	(179.5)
curl	Curl of a k-vector	Sect. 268
div	Natural divergence (In earlier sections, "div" and "curl" are applied only to three-dimensional vectors)	Sect. 268
dual	Dual of a k-vector	Sect. 267
rot	Natural rotation of a k-vector	Sect. 268

11. Guide through this treatise. The chapters, and often even major subdivisions of the chapters, are largely independent of each other. We have tried to organize the ideas in such a way as to minimize cross-referencing. In most cases a reader with some experience in the subject will be able to start at any point he pleases.

Chap. B collects and organizes all the researches we have been able to find on the kinematics of continuous media. Because of its completeness, it contains much classic material found also in other works, but some parts of it deserve particular notice.

Subchapter I of Chap. B concerns the theory of a single deformation; that is, of finite strain and local rotation. Part f, concerning oriented bodies, presents apparatus for a type of physical theory as yet little studied but likely to be of future value. Subchapter II, which can be followed without reading its predecessor, is a treatise on the kinematics of continuous motions, i.e., deformations changing with time. Part e of this subchapter is directed especially toward interpretations in the flow of fluids and should be omitted by readers interested primarily in new theories; the essentials are given in the preceding parts. Part g, concerning relative motion, is conceptually the most important in the chapter and most apt to be enlightening in new studies of material behavior. Subchapter III concerns mass and momentum and includes general formulae of transformation under change of frame and also general solutions of the equation of continuity by means of stream functions of various kinds.

Chap. C contains the kinematical theory of slip surfaces, shock waves, and other kinds of surfaces of discontinuity. A general equation of balance, or "conservation law", was stated in Sect. 157 and was shown to be equivalent to a certain differential equation in regions of sufficient smoothness; at the end of Chap. C, this same equation of balance is applied to a surface of discontinuity and shown to be equivalent to a certain jump condition. Thus all the laws of classical physics may be derived by a uniform process from appropriate integral equations of balance.

Chap. D presents the laws of classical mechanics and the general theory of contact forces or stress, in terms of which mechanical theories of continuous media are formulated. Subchapter III gives many theorems on mean values of the stress as determined from boundary conditions when the stresses themselves are not uniquely determined. Subchapter IV is intended to be an exhaustive treatise on solutions of the equations of motion or equilibrium of a general continuum by means of stress functions. Subchapter V derives all the variational principles of mechanics that partake of any considerable generality.

The first part of Chap. E presents the general theory of energy in unexceptionable terms. The rest of the chapter concerns the more dubious subject of thermodynamics, set within field concepts. Entropy is taken as the primitive idea here, and emphasis is put upon exact formal properties of equations of state involving arbitrarily many variables. The rate of production of entropy is calculated.

A general theory of the motion of heterogeneous media may be collected from Sects. 158, 159, 215, 254, 255, 259, and 261 (cf. also Sect. 295).

Chap. F, on electromagnetism, begins by a study of a four-dimensional space-time in which no geometrical structure is presumed. In such a space-time the basic laws of conservation of charge and conservation of magnetic flux are then stated. These laws turn out to have an entirely general form, applicable alike in classical electromagnetism and in relativistic theories. A general law of conservation of energy-momentum, including as a special case the classical momentum principle developed in Chap. D, is formulated. Additional assumptions that characterize classical or relativistic electromagnetism are then analyzed.

Chap. G concerns constitutive equations defining particular materials. In contrast to the exhaustive earlier chapters, this one is selective. After a list of the principles used in forming constitutive equations, there follow brief sections on several classical or recent theories. These sections illustrate both the theoretical principles governing choiçe of constitutive equations and the immediate applicability of some of the general theory given in the preceding chapters.

B. Motion and mass.

12. Scope and plan of the chapter. This chapter presents the *kinematics of continuous media* in a Euclidean space of three dimensions. We treat those aspects which are fundamental, either for intuitive grasp or for solution of problems in the various field theories. We omit the older type of kinematical research, where motions preserving the similarity of certain classes of geometrical figures are analyzed; this work is the subject of special Bibliography K at the end of the treatise.

As to be expected of a general treatise on the kinematics of continua, this one is divided into three parts: Subchapter I presents the analysis of a single deformation, or *strain*; Subchapter II, of *motion*, which consists in a family of deformations continuously varying in time; Subchapter III, of *mass*.

The general theory is due primarily to EULER (1745—1766) and CAUCHY (1823—1841); important special concepts and results were added by D'ALEMBERT (1749), GREEN (1839), STOKES (1845), HELMHOLTZ (1858), KELVIN (1849—1869), E. and F. COSSERAT (1909), ZORAWSKI (1911), and many others.

While many expositions of the subject have been published, no other single work presents even a major part of the topics in this chapter[1].

I. Deformation.

a) Deformation gradients.

13. Geometrical axiom. The proper content of this subchapter begins at Sect. 15, the function of the first two sections being only to present some geometrical preliminaries.

Henceforth, except when the contrary is explicitly stated, we refer to Euclidean three-dimensional space, and we employ only real co-ordinates. Many of the results that follow hold in fact in Riemannian or even affine n-space and in complex co-ordinate systems; these generalizations will be obvious to those expert in geometry.

An exact and straightforward embodiment of the assertion that space is Euclidean is the existence of a *rectangular Cartesian co-ordinate system*[2], to which all points may be referred. One such system, called the *common frame*, is laid down at the beginning, other systems of reference being defined later in terms of it. When co-ordinates in the common frame are to be written out, instead of using Z_K and z_k we generally write

$$\boldsymbol{Z} = \boldsymbol{i}X + \boldsymbol{j}Y + \boldsymbol{k}Z, \qquad \boldsymbol{z} = \boldsymbol{i}x + \boldsymbol{j}y + \boldsymbol{k}z, \tag{13.1}$$

where $\boldsymbol{i}, \boldsymbol{j}, \boldsymbol{k}$ are unit co-ordinate vectors in the common frame.

While any one of the infinitely many possible rectangular Cartesian systems may be selected as "the" common frame, the choice is made once and for all. Thus, for example, we shall not use the notation (13.1) unless both points are referred to the same rectangular Cartesian system.

14. General co-ordinates. Invariant description. Duality. In Euclidean space it is permissible to restrict all considerations to rectangular Cartesian co-ordinates, to use absolute notations which eschew co-ordinate systems, or to employ preferred curvilinear nets. Any of these three styles suffices for proving general theorems; each shows peculiar advantages in certain special problems; and each has its passionate and exclusive devotees. We prefer to use two *independently selected general curvilinear co-ordinate systems*[3], one at $\boldsymbol{Z}$ and the other at $\boldsymbol{z}$. To help

---

[1] Surveys of certain parts are included in the following works, of which those distinguished by an asterisk possess the merit of originality, at least in part: KELVIN and TAIT* [1867, 3], ZHUKOVSKI* [1876, 7], E. and F. COSSERAT* [1896, 1] [1909, 5], JAUMANN [1905, 2], LOVE* [1906, 5, Appendix to Chap. I], CAFIERO [1906, 1], HEUN [1913, 4], ARIANO [1924, 1] [1925, 1] [1928, 1], L. BRILLOUIN [1925, 2] [1938, 2, Chap. X], PLATRIER [1936, 8], SIGNORINI [1943, 6], TONOLO [1943, 8], NOVOZHILOV* [1948, 18], GREEN and ZERNA [1950, 10] [1954, 7, Chap. II], MURNAGHAN [1951, 18], TRUESDELL* [1952, 21] [1953, 32] [1954, 24], MISICU [1953, 19], DEFRISE* [1953, 8], NOLL [1955, 18, Chap. I], DOYLE and ERICKSEN [1956, 5].

[2] We remind the reader of the notations explained in Sects. App. 2 and App. 3.

[3] This scheme was introduced by E. and F. COSSERAT [1896, 1, Chap. IV], put into tensorial form by MURNAGHAN [1937, 7, § 1], and developed by TRUESDELL [1952, 21, §§ 12—22] [1953, 32], DOYLE and ERICKSEN [1956, 5, § III], and TOUPIN [1956, 20, § 3]. GIBBS [1875, 1, p. 185] in using two rectangular Cartesian systems noted that "It is not necessary, nor always convenient, to regard these systems of axes as identical ...".

the reader find his way through the literature, we summarize some other schemes in Sect. 66B. Our formalism of course includes the special case when all points are referred to the common frame. Thus any reader unfamiliar with tensor analysis may follow most of the development by interpreting covariant derivatives simply as partial derivatives in a single rectangular system[1].

Let $\boldsymbol{Z}$ and $\boldsymbol{z}$ be points, and in the neighborhood of each let general curvilinear co-ordinates $\boldsymbol{X}$ and $\boldsymbol{x}$ be given by their equations of transformation from the common frame, as follows (Fig. 1):

$$\boldsymbol{X} = \boldsymbol{X}(\boldsymbol{Z}), \quad \boldsymbol{Z} = \boldsymbol{Z}(\boldsymbol{X}); \qquad \boldsymbol{x} = \boldsymbol{x}(\boldsymbol{z}), \quad \boldsymbol{z} = \boldsymbol{z}(\boldsymbol{x}). \tag{14.1}$$

Since the choices of co-ordinates employed at $\boldsymbol{X}$ and $\boldsymbol{x}$ should be entirely free and independent, the theory should be invariant under general changes of co-ordinates,

$$\boldsymbol{X}^* = \boldsymbol{X}^*(\boldsymbol{X}) \quad \text{and} \quad \boldsymbol{x}^* = \boldsymbol{x}^*(\boldsymbol{x}). \tag{14.2}$$

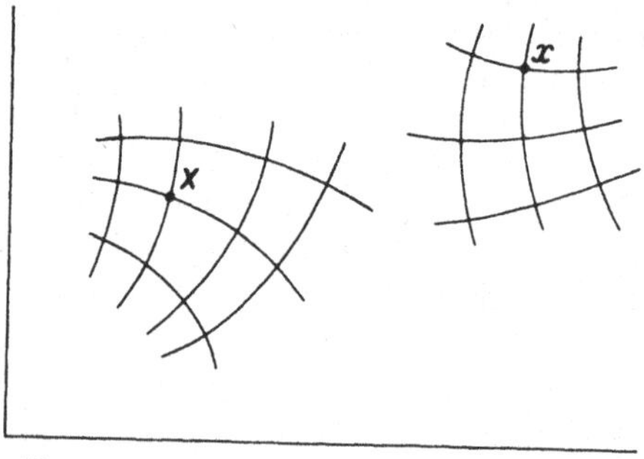

Fig. 1. Independent general curvilinear co-ordinates.

This suggests use of the *double tensors*[2] defined in Sect. App. 15, with the interpretation γ given there.

Throughout this subchapter, we shall describe conditions at $\boldsymbol{x}$ and $\boldsymbol{X}$ in symmetrical formalism, invariant with respect to choice of co-ordinates. Majuscule letters, in general, will refer to $\boldsymbol{X}$; minuscules, to $\boldsymbol{x}$. Often we shall give a *particular procedure* for constructing quantities $T^{K\ldots M k\ldots m}_{N\ldots P n\ldots p}$ in terms of information at two arbitrary points $\boldsymbol{X}$ and $\boldsymbol{x}$. Now suppose we interchange the roles of $\boldsymbol{X}$ and $\boldsymbol{x}$ systematically but otherwise follow the same procedure. We thus obtain dual quantities $t^{k\ldots m K\ldots M}_{n\ldots p N\ldots P}$. Our symmetrical notation achieves the economy expressed in the following, purely formal, ***principle of duality***[3]: *In any given equation, majuscules and minuscules may be interchanged.* In applying this rule, the interchange must be effected both in the kernel letters and in the indices. Henceforth, except in the most important cases, we shall not write down the duals. Also, for special reasons we shall later introduce special notations departing from the principle of duality, the most important of these being $\boldsymbol{u}$, $\boldsymbol{E}$, and $\boldsymbol{R}$.

For an easy example, consider the following prescription: Let $\boldsymbol{P}$ be the position vector from a fixed point to $\boldsymbol{X}$. The dual prescription is: Let $\boldsymbol{p}$ be the position vector from a fixed point to $\boldsymbol{x}$. The resulting components p^k obviously have no general relation to the components $P^k = g^k_K P^K$ of $\boldsymbol{P}$ shifted to $\boldsymbol{x}$. Similarly, if certain fields $\boldsymbol{T}$, $\boldsymbol{a}$, and $\boldsymbol{A}$ have been shown to satisfy the identity $T^k{}_K = a^k A_K$, then dual definitions will lead to the identity $t^K{}_k = A^K a_k$. The components $t^K{}_k$ of the double field $\boldsymbol{t}$, in general, will bear no relation to the corresponding converted components of $\boldsymbol{T}$; that is, $T^K{}_k = g^K_m g^M_k T^m{}_M \neq t^K{}_k$.

The line elements at $\boldsymbol{X}$ and $\boldsymbol{x}$ will be written

$$dS^2 = G_{KM}\, dX^K\, dX^M, \qquad ds^2 = g_{km}\, dx^k\, dx^m, \tag{14.3}$$

[1] Once and for all, however, we warn such a reader that *deformation is a general point transformation* (Sect. 15). Thus, willy nilly, he is employing the idea of general co-ordinates, for this idea is inherent in the theory of deformation.

[2] MICHAL [1947, *9*, Chap. XIV] was the first author to apply them to the theory of deformation. No earlier work was expressed in fully invariant form.

[3] In essence this principle was certainly known to CAUCHY and was used by FINGER [1892, *4*]. The nearest we have found to a formal statement is a remark of LE ROUX [1911, *9*].

where $\boldsymbol{G}$ and $\boldsymbol{g}$ are given in terms of the transformations (14.1) from the common frame to the arbitrarily selected co-ordinate systems by the usual formulae:

$$G_{KM} = g_{KM} = \delta_{PQ} \frac{\partial Z^P}{\partial X^K} \frac{\partial Z^Q}{\partial X^M}, \qquad g_{km} = \delta_{pq} \frac{\partial z^p}{\partial x^k} \frac{\partial z^q}{\partial x^m}. \tag{14.4}$$

The Riemann tensors R_{KMPQ} and r_{kmpq} based upon the components G_{KM} and g_{km} vanish identically. Since $\boldsymbol{X}$ and $\boldsymbol{x}$ are points in the same space, it is obvious that the components g_{km} and G_{KM} are components of the same metric tensor. That is, if $\boldsymbol{G}$ is the dual of $\boldsymbol{g}$, then $G_{KM} = g_{KM} = g^k_K\, g^m_M\, g_{km}$ and $G_{km} = g_{km}$, as is proved formally in Sect. App. 15. We are thus justified in avoiding the kernel index $\boldsymbol{G}$ entirely, as usually we will, but in some cases it helps to avoid ambiguity if we write G_{11}, G_{12}, etc. for the covariant components of the metric tensor in the co-ordinates at $\boldsymbol{X}$.

15. Deformation. This subchapter constructs the mathematical apparatus describing the deformation of a portion of matter from one configuration into another (Fig. 2). Let a typical point $\boldsymbol{Z}$ be carried into $\boldsymbol{z}$:

$$\boldsymbol{z} = \boldsymbol{z}(\boldsymbol{Z}), \qquad \boldsymbol{Z} = \boldsymbol{Z}(\boldsymbol{z}). \tag{15.1}$$

In explicit notation, (15.1) reads[1]

$$\left.\begin{aligned} x &= f(X, Y, Z), \text{ etc.,} \\ X &= F(x, y, z), \text{ etc.} \end{aligned}\right\} \tag{15.2}$$

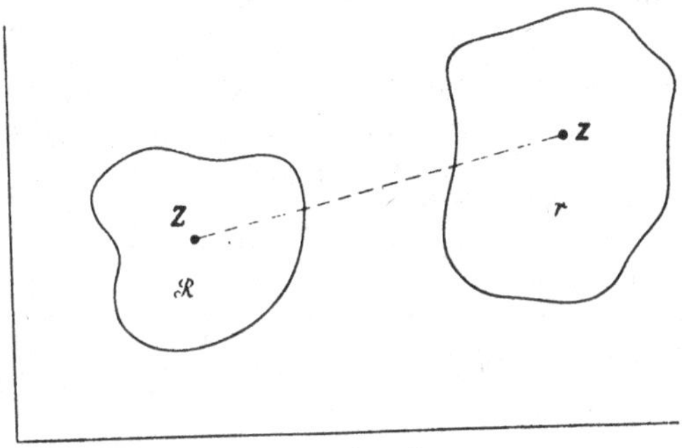

Fig. 2. Deformation.

This transformation will be called *the deformation*. Our object in this subchapter is to analyse the major properties of the deformation. With quantities associated with $\boldsymbol{Z}$ we shall by precise definitions set into correspondence certain quantities associated with $\boldsymbol{z}$, saying then that these quantities are *deformed* by (15.1).

As $\boldsymbol{Z}$ runs over a set of points $\mathscr{R}$, its image $\boldsymbol{z}$ under the mapping (15.1) runs over a set of points, say $\mathscr{r}$. In physical contexts the deformation of $\mathscr{R}$ into $\mathscr{r}$ will be produced by applying forces to the material. To promote physical interpretation we shall sometimes speak of $\mathscr{R}$ and $\mathscr{r}$ as the *undeformed* and the *deformed* material, respectively. More often, however, we shall prefer a symmetrical nomenclature for a mathematically symmetrical situation and shall speak accordingly of *the material about* $\boldsymbol{Z}$ and *the material about* $\boldsymbol{z}$.

16. Continuity. We now lay down the ***axiom of continuity:*** *Throughout* $\mathscr{R}$ *and* $\mathscr{r}$, *the deformation* $(15.1)_1$ *and its inverse* $(15.1)_2$ *are single-valued and as many times continuously differentiable as required.* Not only does this axiom cast aside deformations so irregular as to be useless in physics, but also it implies as a special case the ***permanence of matter:*** *No region of positive finite volume is deformed into one of zero or infinite volume.* For this permanence it is necessary that the Jacobians of (15.1) do not vanish; we shall not lose essential generality by assuming

$$\infty > |\boldsymbol{z}/\boldsymbol{Z}| > 0. \tag{16.1}$$

[1] This is the notation of EULER [1762, *1*] [1770, *1*, § 100]. The now more common notation a, b, c for X, Y, Z in this connection was introduced by LAGRANGE [1788, *1*, Part II, Sect. II, ¶ 4].

It follows also that (15.1) carries every region into a region, every surface into a surface, every curve into a curve. Another corollary of the axiom of continuity is the ***principle of impenetrability***[1]: *One portion of matter never penetrates within another.*

But in addition, the assumption of continuity excludes many singularities of physical interest. Such exclusion is legitimate because in the physical problems of the classical field theories singularities are confined to points, lines, or surfaces over an interval of time, or to instants: In a word, they are *isolated*. Thus they may be given special attention. Generally they fall under one of two types: boundary surfaces or wave surfaces whose specification is included at least in part in the definition of the problem, and singular points engendered by the differential equations of the particular material. The theory of the latter is outside the scope of the present treatise, which describes only general features common to all materials; the former are analysed in Chap. C. The assumption of continuity restricts the results of the present chapter to regions where phenomena are occurring smoothly. In any particular application, this assumption will generally be valid only in portions of the entire space under consideration.

For most of the theorems which follow, it is enough that the deformation possess two continuous derivatives; the reader whose taste favors weakening hypotheses of smoothness may easily satisfy it, though not sufficiently to include in the results of this chapter the essentially different discontinuous motions analysed in Chap. C.

We now generalize the terms and assumptions used in Sect. 15 and hitherto in this section so as to allow use of the general co-ordinate systems introduced in Sect. 14. We refer to $\boldsymbol{X}$ and $\boldsymbol{x}$ as co-ordinates in the *undeformed* and *deformed* material, respectively. From $(15.1)_1$ and $(14.1)_2$ we have $\boldsymbol{x}=\boldsymbol{x}(\boldsymbol{z}(\boldsymbol{Z}(\boldsymbol{X})))$, or, for short,

$$\boldsymbol{x}=\boldsymbol{x}(\boldsymbol{X}), \quad \boldsymbol{X}=\boldsymbol{X}(\boldsymbol{x}); \tag{16.2}$$

in co-ordinate form,

$$\left.\begin{aligned} x^k &= f^k(X^1, X^2, X^3), \quad k=1,2,3; \\ X^K &= F^K(X^1, X^2, X^3), \quad K=1,2,3. \end{aligned}\right\} \tag{16.3}$$

We employ only co-ordinate systems such that in the axiom of continuity we may replace $\boldsymbol{z}$ by $\boldsymbol{x}$ and $\boldsymbol{Z}$ by $\boldsymbol{X}$. In particular,

$$\infty > |\boldsymbol{x}/\boldsymbol{X}| > 0. \tag{16.4}$$

Of frequent use is the absolute scalar J whose numerical value is the Jacobian $|\boldsymbol{z}/\boldsymbol{Z}|$:

$$J \equiv \frac{\sqrt{\det g_{km}}}{\sqrt{\det g_{KM}}}\,|\boldsymbol{x}/\boldsymbol{X}| = |\boldsymbol{z}/\boldsymbol{Z}|. \tag{16.5}$$

By (16.4) follows

$$0 < J < \infty. \tag{16.6}$$

Writing j for the dual of J, we have $Jj=1$.

The bare Eqs. (16.2) and (14.2) are of the same form, but now we bring in the essential distinction wanted for the interpretation we have already given them: *The Euclidean metric is invariant under the co-ordinate transformations* (14.2) *but may be deformed arbitrarily by* (16.2). Thus the co-ordinates $\boldsymbol{X}$ and $\boldsymbol{X}^*$ are but different means of identifying the same point $\boldsymbol{Z}$ in the undeformed ma-

[1] This principle was emphasized in the earliest researches on deformable bodies, but the much stronger axiom of continuity is needed in order to derive most of the concrete results commonly used.

terial; the co-ordinates $\boldsymbol{x}$ and $\boldsymbol{x}^*$ but different names for the same point $\boldsymbol{z}$ in the deformed material. Two points $\boldsymbol{X}_1$ and $\boldsymbol{X}_2$ in the undeformed material are deformed by (16.2) into two points $\boldsymbol{x}_1 = \boldsymbol{x}(\boldsymbol{X}_1)$ and $\boldsymbol{x}_2 = \boldsymbol{x}(\boldsymbol{X}_2)$ whose distance from one another is in general different from the distance between $\boldsymbol{X}_1$ and $\boldsymbol{X}_2$. The co-ordinate invariance of (14.3), invariance with respect to the transformations (14.2), is a requirement imposed by the observer so as to assure himself that his own description of what happens is independent of chance circumstances[1]. Its consequences are rather trivial and may be read off from standard works on tensor algebra and tensor calculus. For example, the usual rules of tensor composition remain valid, a tensor whose components in one particular $\boldsymbol{x}$ system and one particular $\boldsymbol{X}$ system vanish is a zero tensor, etc.

17. Deformation gradients. The two sets of *deformation gradients* $x^k_{;K}$ and $X^K_{;k}$ are defined by

$$x^k_{;K} \equiv \frac{\partial x^k}{\partial X^K}, \quad X^K_{;k} \equiv \frac{\partial X^K}{\partial x^k}. \tag{17.1}$$

The deformation gradients are the fundamental quantities for the analysis of local properties of the deformation. Their physical components are *dimensionless*, as are all measures of relative configuration.

From (17.1) follows

$$x^k_{;K} X^K_{;m} = \delta^k_m, \quad X^K_{;k} x^k_{;M} = \delta^K_M. \tag{17.2}$$

Each of these formulae denotes a system of nine linear equations in nine unknowns, either the $x^k_{;K}$ or the $X^K_{;k}$. That a unique solution exists follows from (16.4). The solution by CRAMER's rule, expressing each $X^K_{;k}$ as a rational function of the nine $x^m_{;M}$, was first given by EULER[2]:

$$\left.\begin{aligned} X^K_{;k} &= \frac{\frac{1}{2}\varepsilon^{KMP}\varepsilon_{kmp}\, x^m_{;M}\, x^p_{;P}}{\frac{1}{6}\varepsilon^{QRS}\varepsilon_{qrs}\, x^q_{;Q}\, x^r_{;R}\, x^s_{;S}}, \\ &= \frac{\dfrac{\partial(x^m, x^p)}{\partial(X^M, X^P)}}{|\boldsymbol{x}/\boldsymbol{X}|}, \end{aligned}\right\} \tag{17.3}$$

where in the second form KMP and kmp are the same cyclic permutations of 1 2 3.

To calculate the derivatives of one set of deformation gradients with respect to the other, it is easiest to differentiate (17.2), thus obtaining

$$\frac{\partial x^k_{;M}}{\partial X^K_{;m}} X^M_{;p} = -x^k_{;M} \frac{\partial X^M_{;p}}{\partial X^K_{;m}} = -x^k_{;M}\, \delta^M_K\, \delta^m_p, \tag{17.4}$$

---

[1] Three classes of ordinary differential invariants present themselves: those with respect to the deformation (16.2), those with respect to the co-ordinate transformations $(14.2)_1$ at $\boldsymbol{X}$, and those with respect to the co-ordinate transformations $(14.2)_2$ at $\boldsymbol{x}$. A single quantity may enjoy a quite different invariant character in each of these three classes. For example, the element of volume dv at $\boldsymbol{x}$ is an absolute scalar with respect to changes of co-ordinates both at $\boldsymbol{x}$ and at $\boldsymbol{X}$, but by (20.9) below, dv and dV are the values at $\boldsymbol{x}$, and at $\boldsymbol{X}$, respectively, of a scalar capacity with respect to the deformation (16.2). Relations between the three classes are discussed by WUNDHEILER [1932, *14*] and HLAVATÝ [1933, *6*].

[2] [1762, *1*] [1766, *1*, § 36] [1770, *1*, §§ 105–111]. PIOLA [1836, *1*, § III, ¶ 28] noted the easy identity

$$|\boldsymbol{x}/\boldsymbol{X}|\, \delta^K_Q = \tfrac{1}{2}\varepsilon^{KMP}\varepsilon_{kmp}\, x^m_{;M}\, x^p_{;P}\, x^k_{;Q}.$$

whence by (17.2) follows [1]

$$\frac{\partial x^k_{;M}}{\partial X^K_{;m}} = \frac{\partial x^k_{;P}}{\partial X^M_{;m}} X^P_{;p}\, x^p_{;M} = -\, x^k_{;K}\, x^m_{;M}. \tag{17.5}$$

Hence

$$x^k_{;K}\,\delta^m_p = -\, X^M_{;p} \frac{\partial x^k_{;M}}{\partial X^K_{;m}}; \tag{17.6}$$

by contraction follows

$$x^k_{;K} = -\, X^M_{;m} \frac{\partial x^m_{;M}}{\partial X^K_{;k}}. \tag{17.7}$$

The usual formula for differentiating a determinant, combined with (17.3), yields JACOBI's identity [2]:

$$\frac{\partial |\boldsymbol{x}/\boldsymbol{X}|}{\partial x^k_{;K}} = X^K_{;k}\, |\boldsymbol{x}/\boldsymbol{X}|, \tag{17.8}$$

and the identity of EULER, PIOLA, and JACOBI [3]:

$$\frac{\partial (x^k_{;K}/|\boldsymbol{x}/\boldsymbol{X}|)}{\partial x^k} = 0. \tag{17.9}$$

The double vectors $x^k_{;K}$ and $X^K_{;k}$ are duals of one another and are *different* double fields. That is, $x^k_{;K} \neq g^k_M\, g^m_K\, X^M_{;m}$. The easiest way to see this is to refer both to the common frame, in which $g^k_M\, g^m_K\, X^M_{;m}$ has the value $\delta^k_M\, \delta^m_K\, \partial Z^M/\partial z^m$, while $x^k_{;K}$ has the value $\partial z^k/\partial Z^K$, the expression of which in terms of the gradients $\partial Z^M/\partial z^m$ is not linear but is of the form (17.3).

The most important property of the deformation gradients is their law of composition. Consider two successive deformations, the first from $\boldsymbol{X}$ to X, the second from X to $\boldsymbol{x}$. The chain rule of differential calculus asserts that

$$x^k_{;K} = x^k_{;\mathsf{K}}\, \mathsf{X}^{\mathsf{K}}_{;K}. \tag{17.10}$$

That is, *to obtain the matrix of deformation gradients $\boldsymbol{D}$ for the succession of two deformations whose matrices of gradients are $\boldsymbol{D}_1$ and $\boldsymbol{D}_2$, take the matrix product:*

$$\boldsymbol{D} = \boldsymbol{D}_2 \cdot \boldsymbol{D}_1. \tag{17.11}$$

This simple and universal formula is not restricted to any particular choice of co-ordinates. We observe that in general.

$$\boldsymbol{D}_2 \cdot \boldsymbol{D}_1 \neq \boldsymbol{D}_1 \cdot \boldsymbol{D}_2. \tag{17.12}$$

18. Double tensor equations in terms of the total covariant derivative. Differential identities concerning the deformation are easily put in terms of double tensors by using the total covariant derivative explained in Sect. App. 20. When a relation connecting double tensors has been derived in the common frame, replacing partial derivatives by corresponding total covariant derivatives suffices to infer an identity valid in all co-ordinate systems. Since the deformation $\boldsymbol{x} = \boldsymbol{x}(\boldsymbol{X})$ has a differentiable inverse $\boldsymbol{X} = \boldsymbol{X}(\boldsymbol{x})$, we may use the chain rule (App. 22.4). Thus the principle of duality stated in Sect. 14 applies to differential indices.

---

[1] TRUESDELL [1952, *21*, § 13].

[2] [1841, *3*, § 8].

[3] While (17.9) was first given by PIOLA [1825, *2*, § 253] [1848, *2*, Eq. (22)] and JACOBI [1844, *2*, § 2], to derive it we need only to combine EULER's Eq. (17.3) with his observation that $\partial(\varepsilon^{kmp} a_{,m} b_{,p})/\partial x^k = 0$ for any a and b [1770, *1*, §§ 26, 49]. PIOLA [1836, *1*, § III, ¶ 26] proved also that

$$\frac{\partial}{\partial x^k}(A^K\, x^k_{;K}/|\boldsymbol{x}/\boldsymbol{X}|) = \frac{\partial A^K}{\partial X^K}\Big/ |\boldsymbol{x}/\boldsymbol{X}|.$$

For example, consider the equation

$$(j\, x^k_{;K})_{;k} = 0. \tag{18.1}$$

where j is the dual of J as defined by (16.5). In the common frame, (18.1) assumes the same form as (17.9), which has already been derived. Therefore (18.1), being a double tensor equation, is valid in all co-ordinate systems[1]. By the principle of duality, we have also

$$(J\, X^K_{;k})_{;K} = 0. \tag{18.2}$$

19. Displacement vector. The *displacement vector* $\boldsymbol{u}$ has the components $\boldsymbol{z}-\boldsymbol{Z}$ in the common frame. These components are usually written

$$u \equiv x - X, \quad v \equiv y - Y, \quad w \equiv z - Z. \tag{19.1}$$

The dual of $\boldsymbol{u}$ is $-\boldsymbol{u}$. An invariant expression[2] for $\boldsymbol{u}$ is given by

$$u^K \equiv p^K - P^K, \quad u^k = p^k - P^k, \tag{19.2}$$

where $\boldsymbol{P}$ and $\boldsymbol{p}$ are the position vectors of $\boldsymbol{X}$ and $\boldsymbol{x}$, and $p^K \equiv g^K_k\, p^k$.

By differentiating $(19.2)_1$ and using (App. 20.4), (App. 22.4), and the fact that $P^K_{;M} = \delta^K_M$, $p^k_{;m} = \delta^k_{;m}$, we get[3]

$$\left.\begin{aligned} u^K_{;M} &= g^K_k\, p^k_{;M} - \delta^K_M = g^K_k\, p^k_{;m}\, x^m_{;M} - \delta^K_M, \\ &= g^K_m\, x^m_{;M} - \delta^K_M, \end{aligned}\right\} \tag{19.3}$$

so that

$$x^k_{;K} = g^k_M\,(\delta^M_K + u^M_{;K}). \tag{19.4}$$

These formulae express the deformation gradients $x^k_{;K}$ in terms of the *displacement gradients* $u^P_{;Q}$, and conversely. To relate the two sets of displacement gradients, $u^K_{;M}$ and $u^k_{;m}$, we have from (App. 20.4), (App. 22.6), and (19.4) the formula

$$\left.\begin{aligned} u^K_{;M} &= (g^K_k\, u^k)_{;M} = g^K_k\, u^k_{;m}\, x^m_{;M}, \\ &= g^K_k\, g^m_P\,(\delta^P_M + u^P_{;M})\, u^k_{;m}. \end{aligned}\right\} \tag{19.5}$$

More elaborate formulae giving any one component $u^K_{;M}$ explicitly in terms of the nine gradients $u^k_{;m}$ may be derived by putting the dual of (17.3) into (19.3) and then using the dual of (19.4). The formula (19.5) shows that while u^k and u^K are components of the same vector, $u^k_{;m}$ and $u^K_{;M}$ are not components of the same double tensor field; this illustrates the caution following (App. 18.6).

20. Elements of arc, surface, and volume. A curve $\mathscr{C}$ in $\mathscr{R}$ is deformed by (16.2) into a curve c in r. If $d\boldsymbol{X}$ is an *element of arc* along $\mathscr{C}$, from the rule for change of variable in integrals follows (Fig. 3)

$$\int_{\mathscr{C}} [\ldots]\, dX^K = \int_c [\ldots]\, X^K_{;k}\, dx^k, \tag{20.1}$$

---

[1] A formal derivation in general co-ordinates is given by DOYLE and ERICKSEN [1956, *5*, § III].

[2] We follow MICHAL [1947, *9*, pp. 72–73]. While the displacement vector figures largely in many of the older treatments and in approximate theories, KIRCHHOFF [1852, *1*, p. 762] remarked that in general its introduction serves only to lengthen and complicate the formulae in which it appears. His remark is borne out by the only partially successful attempt of DUPONT [1931, *5*, §§ 4–5, 8] to determine the invariant character of $\boldsymbol{u}$, but we believe that the present treatment achieves not only invariance but also simplicity.

[3] We follow TOUPIN [1956, *20*, § 4].

where the symbol $\int \ldots d \cdot$ is used in the usual sense of integral calculus. Thus to study the local properties of the deformation, it is appropriate to *define* the field $d\boldsymbol{x}$ in terms of the given field $d\boldsymbol{X}$ by the equation

$$dX^K = X^K_{;k}\, dx^k. \tag{20.2}$$

By (17.2), the unique solution $d\boldsymbol{x}$ is given by

$$dx^k = x^k_{;K}\, dX^K. \tag{20.3}$$

Upon these basic formulae, which are due to EULER[1], *rests all the remaining analysis of this subchapter.* It is essential that *$d\boldsymbol{x}$ and $d\boldsymbol{X}$ are not the same vector.* By the results of Sect. App. 16, the components of $d\boldsymbol{X}$ shifted to $\boldsymbol{x}$ are $g^k_K\, dX^K$; as is natural, these shifted components are determined from the displacement of the single point $\boldsymbol{X}$ into the single point $\boldsymbol{x}$ and are independent of the deformation of neighboring points. On the contrary, the field $d\boldsymbol{x}$ as given by (20.3) is a field *deformed* with the material, and to determine it we need to know the deformation in a neighborhood of $\boldsymbol{x}$. If $d\boldsymbol{X}$ is tangent to $\mathscr{C}$, then $d\boldsymbol{x}$ is tangent to c (Fig. 3), and the length of $d\boldsymbol{x}$ is adjusted in relation to the length of $d\boldsymbol{X}$ in such a way that the integral relation (20.1) holds. That is, the finite vector $d\boldsymbol{x}$ and the symbol $\int \ldots dx^k$ are related in just the same way as are the finite increment dx and the symbol $\int \ldots dx$ in integral calculus.

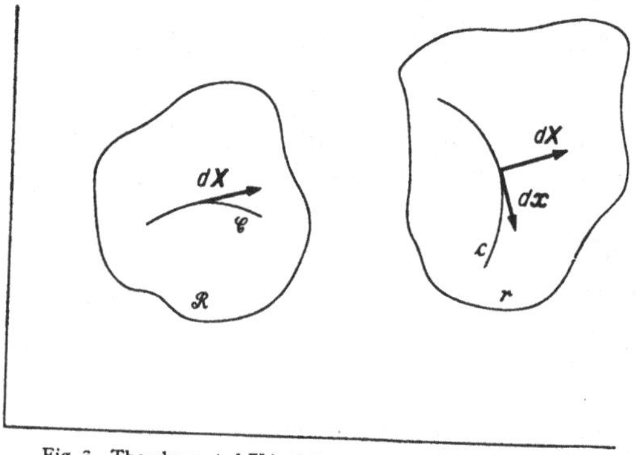

Fig. 3. The element $d\boldsymbol{X}$ is deformed into the element $d\boldsymbol{x}$.

The meaning of the displacement gradients $u^k_{;K}$ and $u^K_{;k}$ as measures of the local changes of length and angle is apparent from the relations

$$dx^k - g^k_K\, dX^K = u^k_{;K}\, dX^K, \quad g^K_k\, dx^k - dX^K = u^K_{;k}\, dx^k. \tag{20.4}$$

Simpler formal apparatus results from using the deformation gradients $x^k_{;K}$ and $X^K_{;k}$, as will be done in most of the rest of this work.

Let $\boldsymbol{X} = \boldsymbol{X}(L, M)$ represent a surface in the undeformed material. Then the element of area is the bivector[2] (cf. (App. 27.1))

$$dA^{KQ} \equiv 2\, \frac{\partial X^{[K}}{\partial L}\, \frac{\partial X^{Q]}}{\partial M}\, dL\, dM. \tag{20.5}$$

The deformed surface is $\boldsymbol{x} = \boldsymbol{x}(\boldsymbol{X}(L, M))$, with element of area

$$\left.\begin{aligned} da^{kq} &\equiv 2\, \frac{\partial x^{[k}}{\partial L}\, \frac{\partial x^{q]}}{\partial M}\, dL\, dM, \\ &= 2 x^{[k}_{;K} x^{q]}_{;Q}\, \frac{\partial X^K}{\partial L}\, \frac{\partial X^Q}{\partial M}\, dL\, dM = 2\, x^k_{;K}\, x^q_{;Q}\, \frac{\partial X^{[K}}{\partial L}\, \frac{\partial X^{Q]}}{\partial M}\, dL\, dM, \\ &= x^k_{;K}\, x^q_{;Q}\, dA^{KQ}. \end{aligned}\right\} \tag{20.6}$$

[1] [1762, *1*] [1766, *1*, § 36] [1770, *1*, §§ 105–111]. Cf. LAGRANGE [1762, *2*, § XLIV] [1788, *1*, Part II, Sect. 11, ¶ 5].

[2] As was noted by PIOLA [1848, *2*, ¶ 12], the squared magnitude of the element of area is a quadratic form in Jacobians:

$$(dA)^2 = \frac{1}{2}\, dA^{KP}\, dA_{KP} = e^K{}_{PQ}\, e_{KRS}\, \frac{\partial X^{[P}}{\partial L}\, \frac{\partial X^{Q]}}{\partial M}\, \frac{\partial X^{[R}}{\partial L}\, \frac{\partial X^{S]}}{\partial M}\, (dL\, dM)^2.$$

(Many works employ the equivalent covariant vector

$$d A_K = e_{KPQ} \frac{\partial X^P}{\partial L} \frac{\partial X^Q}{\partial M} dL\, dM = \frac{1}{2} e_{KPQ}\, dA^{PQ}; \tag{20.7}$$

in terms of it, (20.6) becomes the formula of NANSON[1]:

$$d a_k = J X^K_{;k} dA_K .) \tag{20.8}$$

The ratio of the magnitudes da/dA will be calculated in Sect. 29.

The similarity of (20.3) and (20.6) is immediate: They assert that the elements of arc and area are contravariant *with the deformation*[2]. The reasoning by which we derived the two formulae is intentionally somewhat different, for each proof illustrates a general argument which can be applied to either quantity, as well as to derive the formula of EULER[3] relating elements of volume in the deformed and undeformed material:

$$dv = J\, dV \tag{20.9}$$

where $dv \equiv \sqrt{\det g_{km}}\, dx^1\, dx^2\, dx^3$, $dV = \sqrt{\det g_{KM}}\, dX^1\, dX^2\, dX^3$. As CAUCHY[4] remarked, this relation shows that (16.6) is a mathematical expression of the principle of permanence of matter (Sect. 16).

At the beginning of this section we indicated two different ways of looking upon such differential transformation formulae as (20.2) and (20.6). A third way, which derives from EULER and still is commonly used in works by engineers and physicists, is to consider two "neighboring" points $\boldsymbol{X}_1$ and $\boldsymbol{X}_2$; setting $d\boldsymbol{X} = \boldsymbol{X}_2 - \boldsymbol{X}_1$, we say that $d\boldsymbol{X}$ is "small". $\boldsymbol{X}_1$ and $\boldsymbol{X}_2$ are displaced by (16.2) to $\boldsymbol{x}_1$ and $\boldsymbol{x}_2$, and by the theorem of mean value we get

$$x^k_2 - x^k_1 = x^k_{;K}\, dX^K + O(dX^2), \tag{20.10}$$

where the derivatives $x^k_{;K}$ are evaluated at $\boldsymbol{x}_1$. If we set $d\boldsymbol{x} = \boldsymbol{x}_2 - \boldsymbol{x}_1$, then (20.10) yields (20.3) in first approximation. While the language in the foregoing statements is loose, the argument is essentially rigorous. We prefer to avoid this approach not only because of the appearance of looseness but also because it is unnecessarily elaborate. The relation between $d\boldsymbol{X}$ and $d\boldsymbol{x}$ as so defined is not linear except in first approximation. For sufficiently smooth deformations, greater accuracy can be obtained if we replace (20.10) by

$$x^k_2 - x^k_1 = x^k_{;K}\, dX^K + \frac{1}{2} \frac{\partial^2 x^k}{\partial X^K \partial X^M} dX^K dX^M + O(dX^3), \tag{20.11}$$

---

[1] [1878, *8*].

[2] APPELL [1901, *2*] has remarked that if, corresponding to a field $\boldsymbol{A}$ in the material at $\boldsymbol{X}$, we assign a field $\boldsymbol{a}$ at $\boldsymbol{x}$ according to the *covariant* rule, *viz.*

$$a_k \equiv X^K_{;k} A_K,$$

then it follows that

$$a_{[k;m]} = X^K_{;k} X^M_{;m} A_{[K;M]}.$$

That is, the successive curls of a field transformed covariantly with the deformation also transform covariantly with the deformation.

[3] [1762, *1*] [1770, *1*, §§ 112–118].

[4] [1827, *5*, 1-st Part, Sect. 1, § 8]. Cf. HADAMARD [1903, *11*, ¶ 4]. From (20.9) we have for a sequence of deformations $dv = J_1 dV_1$, $dV_1 = J_2 dV$, and therefore $J = J_1 J_2$; this same result follows from (17.11) and the rule for the determinant of a product of matrices and may have motivated the formal proof given by PIOLA [1848, *2*, ¶ 45] in this connection, independently of the researches of JACOBI.

or by a full power series. Such more accurate expressions for the distance of neighboring points have rarely been used in physical problems[1]. Either of the two approaches used at the beginning of this section permits a rigorous treatment based on ideas just sufficient for the use to which the results are to be put. To avoid all possibility of confusion, we repeat that we choose to regard the differential element $d\boldsymbol{X}$ as an arbitrary (i.e. finite) contravariant vector at $\boldsymbol{X}$ and $d\boldsymbol{x}$ as the vector at $\boldsymbol{x}$ into which $d\boldsymbol{X}$ is deformed by the linear transformation (20.3).

Note added in proof. So as to construct a general theory of dislocations, KRÖNER[2] has replaced (20.3) by a linear relation which is not generally integrable:

$$d x^k = A^k_K d X^K \tag{20.12}$$

I.e., a deformation carrying $\boldsymbol{X}$ to $\boldsymbol{x}$ generally fails to exist. Metric concepts are not used; KRÖNER analyzes (20.12) in terms of an affine connection, generally with non-vanishing torsion, which is constructed from A^k_K, the essential tensor being defined in a fashion similar to (61.7) below.

Appendix to Part a). Geometrical representations of second-order tensors.

This appendix[3] concerns a subject that belongs rather to the theory of invariants than to the classical field theories but is traditionally included in expositions of elasticity and plasticity, in connection with which the results were first obtained. We feel compelled to include this material rather for completeness than for any real use we can see in it. For consecutive development of kinematical theory, the reader should turn at once to Sect. 25.

21. Quadrics of symmetric tensors. Corresponding to a symmetric second-order tensor $\boldsymbol{a}$ at a point $\boldsymbol{x}$, the central quadric

$$F(\boldsymbol{p}) = a_{km} p^k p^m = \pm K^2 = \text{const} \neq 0, \tag{21.1}$$

where $\boldsymbol{p}$ is the position vector issuing from $\boldsymbol{x}$, is called the *quadric of* $\boldsymbol{a}$. In (21.1), both signs are used when both yield real quadrics; otherwise, one chooses that sign which does. In the former case, the quadric is a pair of conjugate hyperboloids, their asymptotic cone[4] being the locus of points $\boldsymbol{p}$ such that $F(\boldsymbol{p}) = 0$; in the latter, it is an ellipsoid.

The tangent plane at $\boldsymbol{p}$ is the locus of points $\boldsymbol{p}^*$ such that

$$(\boldsymbol{p}^* - \boldsymbol{p}) \cdot \boldsymbol{n} = 0, \tag{21.2}$$

where $\boldsymbol{n}$ is any non-null vector proportional to the normal vector grad $F = 2\boldsymbol{a} \cdot \boldsymbol{p} = 2\boldsymbol{p} \cdot \boldsymbol{a}$. We set $\boldsymbol{n} = \boldsymbol{a} \cdot \boldsymbol{p}$ and use (21.1) to obtain from (21.2)

$$a_{km} p^{*k} p^m = a_{km} p^k p^m = \pm K^2. \tag{21.3}$$

In a rectangular Cartesian co-ordinate system z_k with origin at $\boldsymbol{x}$, the quadric intersects the r-th co-ordinate axis at the point $\boldsymbol{z}^0$ given by $p_m = \delta_{mr} z^0_r$ (r unsummed), and the tangent plane at this point intersects the s-th co-ordinate axis at the point $\boldsymbol{z}^{*0}$ given by $p^*_k = \delta_{ks} z^{*0}_s$

---

[1] PIOLA [1848, *2*, ¶¶ 73–74) wrote

$$(\Delta s)^2 \equiv g_{km}(u^k_2 - u^k_1)(u^m_2 - u^m_1),$$
$$= C_{KM}\, dX^K dX^M + C_{KMP}\, dX^K dX^M dX^P + C_{KMPQ}\, dX^K dX^M dX^P dX^Q + \cdots,$$

calculated C_{KM}, C_{KMP}, and C_{KMPQ} explicitly, and asserted that C_{KMP}, C_{KMPQ}, ... can be expressed in terms of the derivatives of C_{KM}, which is the same tensor as that defined by (26.2) below. Geometric interpretations of the second derivatives of the deformation have been found by LE ROUX [1911, *9*, *10*, *11*] [1913, *5*]. HADAMARD [1903, *11*, ¶¶ 57–60] considered deformations such that $x^k_{;K} = g^k_K$, $x^k_{;K_1 K_2 \dots K_{n-1}} = 0$, $x^k_{;K_1 K_2 \dots K_n} \neq 0$.

[2] [1960, *3*, §§ 4–6]. Cf. also KRÖNER and SEEGER [1959, *8*, §§ 1–2].

[3] We are indebted to J.L. ERICKSEN for assistance in preparing this appendix.

[4] FRITSCHE [1939, *8*] gives a simple geometric construction of the asymptotic directions of a plane tensor.

(s unsummed). Inserting these values of $\boldsymbol{p}^*$ and $\boldsymbol{p}$ in (21.3) yields

$$a_{rs} = \pm K^2/(z_r^{*0} z_s^0). \tag{21.4}$$

Therefore $a_{rs}=0$ if and only if one of these intercepts is infinite. When $r=s$, we have $z_r^{*0}=z_r^0=\pm p$, whence it follows that the *normal component of* $\boldsymbol{a}$ *for the direction of a co-ordinate axis is inversely proportional to the square of the magnitude of the parallel radius vector of the quadric*[1]. Of course, this direction may be assigned arbitrarily; another form of this invariant property follows directly from (21.1), since

$$a_{km} v^k v^m = p^{-2} a_{km} p^k p^m = \pm K^2/p^2, \tag{21.5}$$

where $\boldsymbol{v}$ is a unit vector parallel to $\boldsymbol{p}$. A corresponding result for any orthogonal shear component can be read off from (21.4).

Since $\boldsymbol{a}\cdot\boldsymbol{p}$ and $\boldsymbol{p}$ are parallel if and only if $\boldsymbol{p}$ is a proper vector of $\boldsymbol{a}$, the principal directions of $\boldsymbol{a}$ are the directions for which the position vector is normal to the quadric, which means they are *the directions of the principal axes of the quadric*. The squared magnitude of the position vector in such a direction is inversely proportional to the corresponding proper number a of $\boldsymbol{a}$, for if $a_{km} p^m = a\, p_k$, then

$$a_{km} p^k p^m = a\, p^2 = \pm K^2. \tag{21.6}$$

A necessary and sufficient condition that two proper numbers of $\boldsymbol{a}$ be equal is that the quadric $\boldsymbol{a}$ be a surface of revolution; that three be equal, a sphere. The latter property motivates the name *spherical* for symmetric tensors that are scalar multiples of $\mathbf{1}$.

The vector $\boldsymbol{n} = \pm\, \boldsymbol{a}\cdot\boldsymbol{p}/K^2$ is normal to the quadric at $\boldsymbol{p}$. For it, $\boldsymbol{p}\cdot\boldsymbol{n} = \pm\, a_{km} p^k p^m/K^2 = 1$, so its magnitude is the reciprocal of the distance from the center of the quadric to the tangent plane through $\boldsymbol{p}$. Let $\boldsymbol{p}^* \equiv \boldsymbol{n}$ be interpreted as a position vector issuing from the center of the quadric, and assume $\boldsymbol{a}^{-1}$ exists. As $\boldsymbol{p}$ sweeps out the quadric, $\boldsymbol{p}^*$ then sweeps out the *reciprocal* or *director quadric* of $\boldsymbol{a}$, coaxial with the former quadric[2] and satisfying the equation

$$\overset{-1}{a}{}^{km} p_k^* p_m^* = \overset{-1}{a}{}^{km} a_{kr} a_{ms} p^r p^s/K^4 = a_{rs} p^r p^s/K^4 = \pm 1/K^2. \tag{21.7}$$

The asymptotic cone of this quadric is the locus of normals to the asymptotic cone of the quadric of $\boldsymbol{a}$. It is called the *shear cone*[3] of $\boldsymbol{a}$. The normal component of $\boldsymbol{a}$ in any direction lying in this cone is zero.

$(21.7)_4$ may be given a different interpretation[4]: Interpret n_k as co-ordinates of $n-1$-dimensional planes in n dimensions[5], p^k as point coordinates, and take $n_k p^k - 1 = 0$ as the condition that a point $\boldsymbol{p}$ and plane $\boldsymbol{n}$ be incident. A tangent plane $\boldsymbol{n}$ and point of tangency $\boldsymbol{p}$ of the quadric of $\boldsymbol{a}$ are incident and related by $n_k = \pm\, a_{km} p^m/K^2$, whence it follows that $(21.7)_4$ with $\boldsymbol{p}^*$ replaced by $\boldsymbol{n}$ is the equation, in plane co-ordinates, of the quadric of $\boldsymbol{a}$, regarded as the envelope of its tangent planes. We always have in mind the former interpretation of (21.7) when we speak of the reciprocal or director quadric.

One can interpret $\boldsymbol{a}$ as a linear transformation, transforming any given vector $\boldsymbol{q}$ into the vector

$$\boldsymbol{p} = \boldsymbol{a}\cdot\boldsymbol{q}. \tag{21.8}$$

Geometrically, $\boldsymbol{q}$ and $\boldsymbol{p}$ are represented, respectively, by vectors parallel to the normal and radius vectors of the director quadric of $\boldsymbol{a}$. Equivalently, they are represented by vectors parallel, respectively, to the radius and normal vectors of the quadric of $\boldsymbol{a}$. From (21.8), the vectors $\boldsymbol{q}$ producing vectors $\boldsymbol{p}$ of given non-zero magnitude p satisfy

$$p^2 = \overset{2}{a}{}_{km} q^k q^m = \text{const}. \tag{21.9}$$

---

[1] CAUCHY [1827, *1*]. Much of the theory of quadric representations of symmetric tensors had been implicit though not distinctly stated in work of FRESNEL done in 1822 but not published until much later [1868, *6*, §§ 8–9] [1868, *7*, §§ 7–8].

[2] LAMÉ and CLAPEYRON [1833, *2*, pp. 486–508] introduced the reciprocal quadric of the stress tensor of a continuous medium, for which TODHUNTER and PEARSON [1886, *4*, p. 546] introduced the name "stress director". It was discussed further by WARREN [1862, *5*] and FINGER [1892, *3*, p. 1107].

[3] This cone was first discussed by LAMÉ and CLAPEYRON [1833, *2*, pp. 486–508]. Cf. LAMÉ [1852, *2*, § 23].

[4] FINGER [1892, *4*].

[5] Plane co-ordinates are discussed in various works on projective geometry, e.g., VEBLEN and YOUNG [1910, *9*, § 68].

Since the left member of (21.9) is non-negative, the quadric of $\boldsymbol{a}^2$ is always an ellipsoid. From (21.9) we have

$$\overset{2}{a}_{km} u^k u^m = p^2/q^2 = v^2, \tag{21.10}$$

where $\boldsymbol{u}$ is a unit vector parallel to $\boldsymbol{q}$. In other words, the length of the radius vector of the ellipsoid (21.9) is inversely proportional to the length of the image $\boldsymbol{v} = \boldsymbol{a} \cdot \boldsymbol{u}$ under the mapping (21.8) of a unit vector parallel to this radius vector. Since the proper vectors of $\boldsymbol{a}$ are proper vectors of $\boldsymbol{a}^2$, principal axes of the quadric of $\boldsymbol{a}$ are principal axes of the quadric[1] of $\boldsymbol{a}^2$.

Given the two quadrics

$$\overset{-1}{a}_{km} p^k p^m = \pm 1, \quad \overset{-2}{a}_{km} p^k p^m = 1, \tag{21.11}$$

one can easily construct the vector $\boldsymbol{p}_0 \equiv \boldsymbol{a} \cdot \boldsymbol{q}$, where $\boldsymbol{q}$ is any given unit vector[2]. As noted above, $\boldsymbol{q}$ must be normal to the quadric $(21.11)_1$ at the point whose radius vector has the direction of $\boldsymbol{p}_0$. Its magnitude is that of the parallel radius vector to the quadric $(21.11)_2$. To construct $\boldsymbol{p}_0$, one thus determines the point of tangency of a tangent plane to the quadric $(21.11)_1$ which has $\boldsymbol{q}$ as unit normal. To within sign, $\boldsymbol{p}_0$ is then determined as a radius vector of the quadric $(21.11)_2$ which is parallel to the line connecting this point of tangency to the common center of the quadrics. The ambiguity in sign is easily eliminated if one notes that, since $\boldsymbol{p}_0 \cdot \boldsymbol{q} = \boldsymbol{q} \cdot \boldsymbol{a} \cdot \boldsymbol{q} = \pm 1$, the angle determined by $\boldsymbol{p}_0$ and $\boldsymbol{q}$ is acute or obtuse according as the upper or lower sign holds in $(21.11)_1$. The transformation (21.8) maps orthogonal vectors $\boldsymbol{q}_1$ and $\boldsymbol{q}_2$ onto vectors $\boldsymbol{p}_1$ and $\boldsymbol{p}_2$ conjugate with respect to the quadric $(21.11)_2$, i.e.

$$\overset{-2}{a}_{km} p_1^k p_2^m = \overset{-1}{a}_{kr} p_1^k \overset{-1}{a}{}^r_m p_2^m = q_{1r} q_2^r = 0 \tag{21.12}$$

whenever $\boldsymbol{q}_1$ and $\boldsymbol{q}_2$ are orthogonal.

The quadric representation of a symmetric tensor $\boldsymbol{a}$, together with the extremal properties of the proper numbers obtained in Problem 1 of Sect. App. 46, enables us to describe the locus of directions for which the normal component has a given value K. First, it is necessary and sufficient that $a_1 \geqq K \geqq a_n$. Apart from the trivial case when $\boldsymbol{a}$ is spherical, this locus is a *cone*; while it is in general non-circular, it partakes of the symmetries of the quadric surfaces[3]. In particular, any possible value other than a_1 or a_n for the normal component is assumed for infinitely many directions. Any direction lying in the cone if projected through $\boldsymbol{x}$ or through any principal axis or principal plane yields another direction lying in the cone. In general, it seems easier to consider the trace of the cone upon the quadric of $\boldsymbol{a}$, normalized so that $a_{km} p^k p^m = \pm 1$. This trace is the curve of intersection of the quadric with a sphere of radius $|K|^{-\frac{1}{2}}$, provided that in the hyperbolic case only the part of the hyperbola corresponding to the normal components of the appropriate sign be considered. When $K = a_1$ or a_n, the cone degenerates to a single line. For simplicity, take $n = 3$; the curve then has eight congruent parts. When the quadric is a tri-axial, the curve is skew and non-degenerate; if $K \neq a_2$, it consists in two disjoint portions, while if $K = a_2$, it consists in two circles passing through the extremity of the intermediate principal axis. When the quadric is a quadric of revolution, the curve consists in a pair of circular normal sections.

22. Quadrics of asymmetric tensors[4]. Given any non-singular tensor $\boldsymbol{a}$, not necessarily symmetric, one can form the two symmetric tensors $\boldsymbol{a} \cdot \boldsymbol{a}'$ and $\boldsymbol{a}' \cdot \boldsymbol{a}$; since these tensors are positive definite (p. 841, footnote 5), their quadrics are ellipsoids. From the polar decomposition theorem (App. 43.4), together with $(\text{App. } 43.7)_1$, we see that these two ellipsoids congruent, or, equivalently, the proper numbers of $\boldsymbol{a}' \cdot \boldsymbol{a}$ and $\boldsymbol{a} \cdot \boldsymbol{a}'$ coincide, as follows are alternatively from SYLVESTER's theorem in Sect. App. 42.

The problem is now to determine $\boldsymbol{a}$ from the two ellipsoids, recognizing that such a determination cannot be unique. First, we select one of the rotations $\boldsymbol{o}$ that carry one congruent ellipsoid into the other, thus satisfying $(\text{App. } 43.7)_1$. Insertion of this $\boldsymbol{o}$ and the positive square root $\boldsymbol{s} \equiv (\boldsymbol{a} \cdot \boldsymbol{a}')^{\frac{1}{2}}$ into $(\text{App. } 43.4)_1$ then gives one determination of $\boldsymbol{a}$.

Let $\boldsymbol{a}'$ bé a second determination of $\boldsymbol{a}$; then by (App. 43.4) we have an orthogonal $\boldsymbol{o}'$ such that

$$\boldsymbol{a}' = \boldsymbol{o}' \cdot (\boldsymbol{a}' \cdot \boldsymbol{a})^{\frac{1}{2}} = (\boldsymbol{a} \cdot \boldsymbol{a}')^{\frac{1}{2}} \cdot \boldsymbol{o}'. \tag{22.1}$$

[1] The above facts were noted by CAUCHY [1827, *1*], who was first to discuss the quadric of $\boldsymbol{a}^2$.

[2] We follow LAMÉ and CLAPEYRON [1833, *2*], LAMÉ [1852, *2*, § 21].

[3] This was remarked by CAUCHY [1823, *1*] [1827, *2*]. The cone was described by KELVIN and TAIT [1867, *3*, §§ 166—168].

[4] That the quadrics of $\boldsymbol{a} \cdot \boldsymbol{a}'$ and $\boldsymbol{a}' \cdot \boldsymbol{a}$ are ellipsoids was noted by CAUCHY [1827, *2*]. The remaining analysis in this section is adapted from work of FINGER [1892, *4*].

If we set $\boldsymbol{o}_\backslash \equiv \boldsymbol{o}' \cdot \boldsymbol{o}^{-1}$, we verify that $\boldsymbol{o}_\backslash$ is orthogonal, and from (App. 43.4) and (22.1) we calculate that

$$\boldsymbol{a}' = \boldsymbol{o}_\backslash \cdot \boldsymbol{a}, \qquad \boldsymbol{o}_\backslash \cdot (\boldsymbol{a}\cdot\boldsymbol{a}')^{\frac{1}{2}} \cdot \boldsymbol{o}_\backslash^{-1} = (\boldsymbol{a}\cdot\boldsymbol{a}')^{\frac{1}{2}}. \tag{22.2}$$

Thus the second determination $\boldsymbol{a}'$ results from the first by a rotation under which the ellipsoid of $\boldsymbol{a}\cdot\boldsymbol{a}'$ is invariant. Conversely, if $\boldsymbol{a}$ is one determination and $\boldsymbol{o}_\backslash$ is any orthogonal matrix satisfying $(22.2)_2$, then $\boldsymbol{a}'$ as given by $(22.2)_1$ is another determination.

The foregoing results show that to within the ambiguity characterized by (22.2), any non-singular tensor $\boldsymbol{a}$ is characterized by two congruent ellipsoids, those of $\boldsymbol{a}\cdot\boldsymbol{a}'$ and $\boldsymbol{a}'\cdot\boldsymbol{a}$. It is equally well characterized by numerous other pairs of quadrics, e.g., those of $(\boldsymbol{a}\cdot\boldsymbol{a}')^{\frac{1}{2}}$ and $(\boldsymbol{a}'\cdot\boldsymbol{a})^{\frac{1}{2}}$. Cf. Sect. 32.

23. BOUSSINESQ'S construction. In the three-dimensional case, BOUSSINESQ[1] devised an elegant construction for the vector

$$\boldsymbol{u}_0 \equiv \boldsymbol{a}\cdot\boldsymbol{v}, \tag{23.1}$$

where $\boldsymbol{a}$ is a symmetric tensor with known proper numbers and vectors and $\boldsymbol{v}$ is any given unit vector. For convenience, we assume that the proper numbers of $\boldsymbol{a}$ are distinct and ordered as usual: $a_1 > a_2 > a_3$. Let

$$\boldsymbol{b} \equiv \boldsymbol{a} - \tfrac{1}{2}(a_1 + a_3)\,\boldsymbol{1}, \tag{23.2}$$

so that

$$\left.\begin{aligned} b_1 &= a_1 - \tfrac{1}{2}(a_1 + a_3) = \tfrac{1}{2}(a_1 - a_3),\\ b_2 &= a_2 - \tfrac{1}{2}(a_1 + a_3),\\ b_3 &= a_3 - \tfrac{1}{2}(a_1 + a_3) = \tfrac{1}{2}(a_3 - a_1) = -b_1. \end{aligned}\right\} \tag{23.3}$$

Clearly $b_1 > b_2 > b_3$, $b_1 > 0$. From (23.1) and (23.2) we have

$$\boldsymbol{u}_0 = \boldsymbol{u} + \tfrac{1}{2}(a_1 + a_3)\,\boldsymbol{v}, \qquad \boldsymbol{u} \equiv \boldsymbol{b}\cdot\boldsymbol{v}; \tag{23.4}$$

thus from given $\boldsymbol{u}$ and $\boldsymbol{v}$ one can construct $\boldsymbol{u}_0$. We thus concentrate on constructing $\boldsymbol{u}$ for given $\boldsymbol{v}$. Refer $\boldsymbol{b}$ to a rectangular Cartesian principal co-ordinate system z_k. In this system,

$$(v_1)^2 + (v_2)^2 + (v_3)^2 = 1$$

and

$$\left.\begin{aligned} u^2 \equiv \boldsymbol{u}\cdot\boldsymbol{u} = \boldsymbol{v}\cdot\boldsymbol{b}^2\cdot\boldsymbol{v} &= (b_1)^2\,[(v_1)^2 + (v_3)^2] + (b_2)^2\,(v_2)^2,\\ &= (b_1)^2 + [(b_2)^2 - (b_1)^2]\,(v_2)^2,\\ &= (b_1)^2 + [(b_2)^2 - (b_1)^2]\cos^2\phi, \end{aligned}\right\} \tag{23.5}$$

where ϕ is the angle determined by $\boldsymbol{v}$ and the proper vector of $\boldsymbol{b}$ corresponding to b_2, $\cos\phi \equiv v_2$. The angle ψ determined by $\boldsymbol{u}$ and this proper vector is given by

$$\left.\begin{aligned} \cos\psi = u_2/u &= b_2 v_2\,\{(b_1)^2 + [(b_2)^2 - (b_1)^2]\cos^2\phi\}^{-\frac{1}{2}},\\ &= b_2\cos\phi\,\{(b_1)^2 + [(b_2)^2 - (b_1)^2]\cos^2\phi\}^{-\frac{1}{2}}. \end{aligned}\right\} \tag{23.6}$$

Projecting $\boldsymbol{u}$ and $\boldsymbol{v}$ on a plane $z_2 = \text{const}$ yields the components $(b_1 v_1, 0, -b_1 v_3)$ and $(v_1, 0, v_3)$ respectively, so the projected vectors make equal angles with the z_1-axis, supplementary angles with the z_3-axis.

The above facts imply that $\boldsymbol{u}$ is given by the following construction: *In the plane of the proper vectors corresponding to the two larger proper numbers b_1 and b_2, draw a unit vector making the same angle ϕ with the proper vector corresponding to b_2 as does $\boldsymbol{v}$. In the same plane, draw a vector of magnitude $\{(b_1)^2 - [(b_2)^2 - (b_1)^2]\cos^2\phi\}^{\frac{1}{2}}$ which makes the angle ψ, given by* (23.6), *with the proper vector corresponding to b_2, and which lies on the same side of this proper vector as does the first vector. Rotate the two vectors thus constructed about the proper vector corresponding to b_2 through equal but opposite angles until the first vector coincides with $\boldsymbol{v}$. The second vector will then coincide with $\boldsymbol{u}$.*

24. MOHR'S mapping. For any symmetric tensor $\boldsymbol{a}$ in three dimensions, one has

$$X(\boldsymbol{v}) = a_{km} v^k v^m, \qquad Y^2(\boldsymbol{v}) = \overset{2}{a}_{km} v^k v^m - X^2, \qquad v^k v_k = 1, \tag{24.1}$$

where $X(\boldsymbol{v})$ and $Y(\boldsymbol{v})$ are, respectively, the normal component and the maximum shear component of $\boldsymbol{a}$ corresponding to the unit vector $\boldsymbol{v}$. Eqs. (24.1) define a mapping of the unit

---

[1] [1877, *1*]. An alternative procedure is given by GUEST [1939, *9*].

sphere $\boldsymbol{v}\cdot\boldsymbol{v}=1$ into the X-Y plane. Since $(24.1)_2$ is unaltered if Y is replaced by $-Y$, the image of the unit sphere is a region $\imath$ symmetric with respect to the X-axis. In rectangular Cartesian principal co-ordinates, (24.1) reads

$$\left.\begin{aligned} X &= a_1 (v_1)^2 + a_2 (v_2)^2 + a_3 (v_3)^2, \qquad Y^2 = (a_1)^2 (v_1)^2 + (a_2)^2 (v_2)^2 + (a_3)^2 (v_3)^2 - X^2, \\ &(v_1)^2 + (v_2)^2 + (v_3)^2 = 1, \end{aligned}\right\} \tag{24.2}$$

whence follows

$$\left.\begin{aligned} \left(X - \frac{a_2 + a_3}{2}\right)^2 + Y^2 &= (v_1)^2 (a_2 - a_1)(a_3 - a_1) + \left(\frac{a_2 - a_3}{2}\right)^2, \\ \left(X - \frac{a_3 + a_1}{2}\right)^2 + Y^2 &= (v_2)^2 (a_3 - a_2)(a_1 - a_2) + \left(\frac{a_3 - a_1}{2}\right)^2, \\ \left(X - \frac{a_1 + a_2}{2}\right)^2 + Y^2 &= (v_3)^2 (a_1 - a_3)(a_2 - a_3) + \left(\frac{a_1 - a_2}{2}\right)^2. \end{aligned}\right\} \tag{24.3}$$

The image of a curve on the sphere whose radius vector makes a constant angle with the proper vector of $\boldsymbol{a}$ corresponding to $a_{\mathfrak{b}}$, i.e., a curve $v_{\mathfrak{b}}=\text{const}$, is thus a circle with center on the X-axis at $X=\frac{1}{2}(a_{\mathfrak{c}}+a_{\mathfrak{d}})$, $(\mathfrak{b}, \mathfrak{c}, \mathfrak{d} \neq)$, whose squared radius is given by the right member of $(24.3)_{\mathfrak{b}}$. From $(24.2)_3$ and (24.3), the region $\imath$ consists of the points interior to the largest and exterior to the two smallest of the three circles given by

$$\left(X - \frac{a_{\mathfrak{b}} + a_{\mathfrak{c}}}{2}\right)^2 + Y^2 = \left(\frac{a_{\mathfrak{b}} - a_{\mathfrak{c}}}{2}\right)^2, \qquad \mathfrak{b} \neq \mathfrak{c} \tag{24.4}$$

points on the boundary being included in $\imath$. If $a_1=a_2=a_3$, $\imath$ reduces to the point $(a_1, 0)$. If two, but not three, proper values of $\boldsymbol{a}$ coincide, $\imath$ reduces to a circle. In all other cases, its area is non-zero. The circles (24.4)—more precisely, their centers—uniquely determine the proper numbers of $\boldsymbol{a}$. For two tensors $\boldsymbol{a}$ and $\boldsymbol{b}$ such that $\boldsymbol{b}=\boldsymbol{o}\cdot\boldsymbol{a}\cdot\boldsymbol{o}^{-1}$, where $\boldsymbol{o}$ is orthogonal, the corresponding regions $\imath$ coincide; conversely the boundary of $\imath$ determines $\boldsymbol{a}$ to within an orthogonal transformation.

Adding to $\boldsymbol{a}$ a tensor proportional to $\mathbf{1}$ does not alter Y but adds a constant to X. The effect is thus to translate the region $\imath$ parallel to the X-axis. Since X and Y are homogeneous of degree one in $\boldsymbol{a}$, multiplying $\boldsymbol{a}$ by any non-zero scalar factor effects a uniform expansion[1] of $\imath$. It is easy to show, conversely that, if the regions $\imath$ corresponding to two tensors $\boldsymbol{a}$ and $\boldsymbol{b}$ differ only by a translation parallel to the X-axis and a uniform expansion, then, assuming $\boldsymbol{a}\neq 0$, there exist scalars A and B and an orthogonal tensor $\boldsymbol{o}$ such that[2] $\boldsymbol{b}=A\,\boldsymbol{o}\cdot\boldsymbol{a}\cdot\boldsymbol{o}^{-1}+B\mathbf{1}$.

In any number of dimensions (24.1) defines a mapping of the unit sphere into the X-Y plane. In two dimensions, the unit circle $\boldsymbol{v}\cdot\boldsymbol{v}=1$ is mapped onto the circle

$$\left(X - \frac{a_1 + a_2}{2}\right)^2 + Y^2 = \left(\frac{a_1 - a_2}{2}\right)^2. \tag{24.5}$$

In this case

$$a_1 + a_2 = \mathrm{I}_{\boldsymbol{a}}, \qquad (a_1 - a_2)^2 = (a_1 + a_2)^2 - 4 a_1 a_2 = \mathrm{I}_{\boldsymbol{a}}^2 - 4\,\mathrm{II}_{\boldsymbol{a}}, \tag{24.6}$$

so the center and radius of the circle (24.5) are easily obtained from the components of $\boldsymbol{a}$ given in any co-ordinate system[3]. There seems to be no equally simple method of constructing the region $\imath$ in three dimensions from components given in an arbitrary co-ordinate system. There is an extensive literature concerned with applications of MOHR's mapping, much of it to theories of yield or rupture of solids[4].

By making transformations of the form $X^*=X^*(X, Y, a_1, a_2, a_3)$, $Y^*=Y^*(X, Y, a_1, a_2, a_3)$, one can obtain other plane representations of $\boldsymbol{a}$ which are essentially equivalent to MOHR's, but which may be more convenient for some purposes. CHARREAU[5] has considered the

---

[1] The facts thus far noted are contained in papers by MOHR [1882, *3*] [1900, *8*] [1914, *8*, pp. 192–235]. JUNG [1947, *7*] gives an earlier (1866) reference to CULMANN, which may contain some of the material given above. We have been unable to see CULMANN's work.

[2] KLOTTER [1933, *7*] discusses the relation between the maps of $\boldsymbol{a}$, its deviator ${}_0\boldsymbol{a}$, and the tensor $\boldsymbol{b}$ defined by $\boldsymbol{b}\equiv A\boldsymbol{a}+B\mathbf{1}$.

[3] Further discussion of the plane case is given, e.g., by NADAI [1950, *20*, Chap. 10], PRAGER and HODGE [1951, *20*, Chap. 5], DEWULF [1947, *4*], FADLE [1940, *10*], and WISE [1940, *18*].

[4] Cf. NADAI [1950, *20*, Chap. 15], TORRE [1946, *8*] [1951, *26*], BÖKER [1915, *1*], v. KÁRMÁN [1911, *8*] [1912, *5*] for discussion and further references.

[5] [1945, *1*].

case $X^* = X$, $Y^* = k(X^2 + Y^2)$, where k is a positive constant, Y^* then being proportional to the squared magnitude of the vector $\boldsymbol{a} \cdot \boldsymbol{v}$, CHARREAU and DUPONT[1] discuss the case

$$X^* = -Y^* + (X - a_3)(a_1 - a_2),$$

$$(a_2 - a_3) Y^* = (a_1 - a_2)\left\{\left(X - \frac{a_1 + a_2}{2}\right)^2 + Y^2 - \left(\frac{a_1 - a_3}{2}\right)^2\right\},$$

it being assumed that $a_1 > a_2 > a_3$. The latter transformation maps the unit sphere $\boldsymbol{v} \cdot \boldsymbol{v} = 1$ onto a triangular region in the X^*, Y^* plane.

b) Strain.

25. Stretch, extension, elongation, and shear. The change in length and relative direction occasioned by deformation is called, loosely, *strain*[2].

By (20.3), an element of arc $d\boldsymbol{X}$ at $\boldsymbol{X}$ is deformed into an element of arc $d\boldsymbol{x}$ at $\boldsymbol{x}$. Since (20.3) is linear and homogeneous, the ratio of lengths dx/dX is independent of the original length dX and hence for given displacement gradients is a function only of the direction of $d\boldsymbol{X}$. Let $\boldsymbol{N}$ be a unit vector along $d\boldsymbol{X}$ at $\boldsymbol{X}$. Then the *stretch* $\lambda_{(\boldsymbol{N})}$ *in the direction of* $\boldsymbol{N}$ is defined by

$$\lambda_{(\boldsymbol{N})} \equiv \frac{dx}{dX}, \tag{25.1}$$

while the *extension* $\delta_{(\boldsymbol{N})}$ is defined by

$$\delta_{(\boldsymbol{N})} \equiv \lambda_{(\boldsymbol{N})} - 1. \tag{25.2}$$

For the range of possible values of λ and δ, the axiom of continuity (Sect. 16) yields

$$0 < \lambda_{(\boldsymbol{N})} < \infty, \quad -1 < \delta_{(\boldsymbol{N})} < \infty. \tag{25.3}$$

In view of the remarks at the beginning of Sect. 20 we may interpret (25.1) by means of the length L of a finite arc $\mathscr{C}$ emanating from $\boldsymbol{X}$ and tangent there to $\boldsymbol{N}$, and the length $L + \Delta L$ of the arc c at $\boldsymbol{x}$ into which $\mathscr{C}$ is deformed. We have

$$\lambda_{(\boldsymbol{N})} = \lim_{L \to 0} \frac{L + \Delta L}{L}. \tag{25.4}$$

We agree *not* to dualize the concept of stretch. That is, letting $l + \Delta l$ be the length of an arc c' emanating from $\boldsymbol{x}$ and tangent there to $\boldsymbol{n}$, while l is the length of the arc $\mathscr{C}$ at $\boldsymbol{X}$ from which c' was deformed, we set

$$\lambda_{(\boldsymbol{n})} = \lim_{l \to 0} \frac{l + \Delta l}{l}. \tag{25.5}$$

If we choose c' as c, then $l = L$ and $\Delta L = \Delta l$, and we get

$$\lambda_{(\boldsymbol{N})} = \lambda_{(\boldsymbol{n})}. \tag{25.6}$$

This identification is not necessary, however, and it is convenient to specify stretch sometimes in terms of directions at $\boldsymbol{X}$, sometimes in terms of directions at $\boldsymbol{x}$. The important fact just established is that the *totality of stretches is the same*, whether the undeformed or the deformed material is used for reference.

Elements parallel to $\boldsymbol{N}$ are lengthened or shortened according as $\lambda_{(\boldsymbol{N})} > 1$ or $\lambda_{(\boldsymbol{N})} < 1$, or according as $\delta_{(\boldsymbol{N})}$ is positive or negative. Doubling the length corresponds to $\lambda = 2$, $\delta = 1$; halving the length, to $\lambda = \frac{1}{2}$, $\delta = -\frac{1}{2}$. Thus λ is

[1] [1945, *1*]; [1944, *5*].

[2] This term is due to RANKINE [1851, *1*, Sect. I, § 5], both in general and for the tensor $\boldsymbol{E}$ satisfying the relations (57.10).

multiplicatively symmetric with respect to lengthening and shortening but not additively so, while δ is not symmetric at all.

An additively symmetrical measure of strain $f(\lambda)$ must satisfy

$$f(1/\lambda) = -f(\lambda). \tag{25.7}$$

Among the infinitely many smooth functions that conform to this requirement are[1]

$$\left.\begin{aligned} f(\lambda) &= \log\lambda, & -\infty < f(\lambda) < +\infty, \\ f(\lambda) &= \frac{4}{\pi}\operatorname{arc\,tan}\lambda - 1, & -1 < f(\lambda) < +1. \end{aligned}\right\} \tag{25.8}$$

For doubling the length these measures have the respective values log 2 and $4/\pi$ arctan $2-1$. The difficulties which stand in the way of using such measures as these will be mentioned in Sect. 33. The stretch itself furnishes the most immediate measure of strain and is the basic concept in all serious studies of the subject.

For a given pair of points in a given deformation, there are in general infinitely many stretches, but they are not independent; as we shall see in Sect. 27, a properly selected set of three suffices to determine them all.

Returning to the definition of extension, if before calculating the ratio of lengths we project $d\boldsymbol{x}$ upon the direction of $d\boldsymbol{X}$, the result is called the *elongation* $\varepsilon_{(\boldsymbol{N})}$ in the direction of $d\boldsymbol{X}$:

$$\varepsilon_{(\boldsymbol{N})} \equiv \frac{d x_K\, dX^K}{(dX)^2} - 1. \tag{25.9}$$

Thus for elements which are turned through an obtuse angle the elongation is always negative. Elongation is not necessarily a measure of change of shape. For example, in a rigid rotation through a positive angle, every extension is zero, but every element not parallel to the axis suffers a negative elongation. More generally, from the definitions (25.1), (25.2), and (25.9) follows

$$-(2+\delta_{(\boldsymbol{N})}) \leqq \varepsilon_{(\boldsymbol{N})} \leqq \delta_{(\boldsymbol{N})}; \tag{25.10}$$

it is possible for $\varepsilon_{(\boldsymbol{N})}$ to assume any finite value. In the case of an element whose direction is unaltered by the deformation, we have $\varepsilon_{(\boldsymbol{N})} = \delta_{(\boldsymbol{N})}$. The condition $\varepsilon_{(\boldsymbol{N})} = -1$ is necessary and sufficient that the element be turned through a right angle, irrespective of its stretch. An identity connecting elongation and stretch is given as (35.3) below.

Let $d\boldsymbol{X}_1$ and $d\boldsymbol{X}_2$ be two elements at $\boldsymbol{X}$, and let them be deformed into the elements $d\boldsymbol{x}_1$ and $d\boldsymbol{x}_2$ at $\boldsymbol{x}$. Let unit tangents be $\boldsymbol{N}_1$, $\boldsymbol{N}_2$, $\boldsymbol{n}_1$, $\boldsymbol{n}_2$. If $\Theta_{(\boldsymbol{N}_1,\boldsymbol{N}_2)}$ is the angle between $d\boldsymbol{X}_1$ and $d\boldsymbol{X}_2$, while $\Theta_{(\boldsymbol{N}_1,\boldsymbol{N}_2)} - \gamma_{(\boldsymbol{N}_1,\boldsymbol{N}_2)}$ is the angle between $d\boldsymbol{x}_1$ and $d\boldsymbol{x}_2$, then $\gamma_{(\boldsymbol{N}_1,\boldsymbol{N}_2)}$, the *decrease* in angle, is the *shear* of the directions $\boldsymbol{N}_1$, $\boldsymbol{N}_2$. The concept of shear is not dualized. Thus with $\boldsymbol{n}_1$, $\boldsymbol{n}_2$ as defined we have

$$\gamma_{(\boldsymbol{N}_1,\boldsymbol{N}_2)} = \gamma_{(\boldsymbol{n}_1,\boldsymbol{n}_2)}. \tag{25.11}$$

The axiom of continuity forbids a shear equal to a right angle: $|\gamma| < \frac{1}{2}\pi$ for all pairs of directions.

From the linearity of (20.3) follow certain symmetries. First, the stretches of oppositely directed elements are equal:

$$\lambda_{(-\boldsymbol{N})} = \lambda_{(\boldsymbol{N})}. \tag{25.12}$$

---

[1] According to MEHMKE [1897, 5, § I], $(25.8)_1$ was used by IMBERT in 1880 to describe the extension of rubber. Cf. LUDWIK [1909, 7, Pt. 1, § I].

Second, the shears of oppositely directed pairs of elements are the same, and reversal of one of a pair of elements changes the sign of the shear:

$$\gamma_{(-\boldsymbol{N}_1,-\boldsymbol{N}_2)}=\gamma_{(\boldsymbol{N}_1,\boldsymbol{N}_2)},\qquad \gamma_{(-\boldsymbol{N}_1,\boldsymbol{N}_2)}=-\gamma_{(\boldsymbol{N}_1,\boldsymbol{N}_2)}. \tag{25.13}$$

In most cases there is thus no loss in generality if we take shear as an acute angle: $0\leqq\gamma<\frac{1}{2}\pi$.

26. The deformation tensors of CAUCHY and GREEN. The theory of finite strain is the creation of CAUCHY[1]. He observed that (20.2) put into the formula $dS^2=g_{KM}\,dX^K dX^M$ for the squared element of length at $\boldsymbol{X}$ gives

$$dS^2=c_{km}\,dx^k\,dx^m,\qquad\text{where}\qquad c_{km}\equiv g_{KM}\,X^K_{;k}\,X^M_{;m}. \tag{26.1}$$

$\boldsymbol{c}$ is *Cauchy's deformation tensor*[2]. As we shall see, all changes of length and angle are easily calculated from the values of the components c_{km}. The formulae dual to (26.1) are

$$ds^2=C_{KM}\,dX^K dX^M,\qquad C_{KM}\equiv g_{km}\,x^k_{;K}\,x^m_{;M}. \tag{26.2}$$

$\boldsymbol{C}$ is *Green's deformation tensor*[3]. Since $\boldsymbol{c}$ and $\boldsymbol{C}$ are metric tensors, their matrices are non-singular. In the common frame CAUCHY'S and GREEN'S tensors have the familiar forms[4]

$$\left.\begin{aligned}
C_{XX}&=\left(1+\frac{\partial u}{\partial X}\right)^2+\left(\frac{\partial v}{\partial X}\right)^2+\left(\frac{\partial w}{\partial X}\right)^2,\\
C_{XY}&=\left(1+\frac{\partial u}{\partial X}\right)\frac{\partial u}{\partial Y}+\frac{\partial v}{\partial X}\left(1+\frac{\partial v}{\partial Y}\right)+\frac{\partial w}{\partial X}\frac{\partial w}{\partial Y},\ \ldots,\\
c_{xx}&=\left(1-\frac{\partial u}{\partial x}\right)^2+\left(\frac{\partial v}{\partial x}\right)^2+\left(\frac{\partial w}{\partial x}\right)^2,\\
c_{xy}&=-\left(1-\frac{\partial u}{\partial x}\right)\frac{\partial u}{\partial y}-\frac{\partial v}{\partial x}\left(1-\frac{\partial v}{\partial y}\right)+\frac{\partial w}{\partial x}\frac{\partial w}{\partial y},\ \ldots,
\end{aligned}\right\} \tag{26.3}$$

where u, v, w are the components of the displacement vector (19.1). The tensors $\boldsymbol{c}$ and $\boldsymbol{C}$ are different tensors; i.e., $c_{km}\neq C_{km}$ in general. A formula relating $\boldsymbol{c}$ to $\boldsymbol{C}$ is given as (37.6) below.

Since $\boldsymbol{C}$ and $\boldsymbol{c}$ serve to measure all lengths, it is obvious that the stretches and shears can be expressed in terms of them. In fact, for the stretches in the directions of the unit vectors $\boldsymbol{N}$ and $\boldsymbol{n}$, by (25.4) and (25.5) we get

$$\lambda_{(\boldsymbol{N})}=\sqrt{C_{KM}\,N^K N^M},\qquad \lambda_{(\boldsymbol{n})}=\frac{1}{\sqrt{c_{km}\,n^k n^m}} \tag{26.4}$$

(the dissymetry in the definition of stretch accounts for the failure of duality here). Thus the *normal components* (Sect. App. 45) *of* $\boldsymbol{C}$ *and* $\boldsymbol{c}$ *in the directions* $\boldsymbol{N}$ *and* $\boldsymbol{n}$, *respectively, are the squares and reciprocal squares of the stretches in those directions.* If we let Λ_1 and λ_1 be the stretches and Δ_1 and δ_1 the extensions in

[1] [1823, *1*] [1827, *2*] [1841, *1*].

[2] [1827, *2*, Eqs. (10), (11)]. Cf. also LE ROUX [1911, *9*].

[3] While the tensor $\boldsymbol{C}$ appears formally in work by PIOLA [1836, *1*, Eq. (139)] [1848, *2'* ¶ 34], its components were first interpreted by GREEN [1841, *2*, pp. 295–296]. Cf. also CAUCHY [1841, *1*, § I, Eq. (15)].

[4] Corresponding explicit forms for cylindrical and revolution co-ordinates are written out by MAZZARELLA [1954, *15*].

the directions of the X^1 and x^1 co-ordinate curves, then[1]

$$\frac{C_{11}}{G_{11}} = (\Lambda_1)^2 = (1 + \Delta_1)^2, \quad \frac{c_{11}}{g_{11}} = \frac{1}{(\lambda_1)^2} = \frac{1}{(1+\delta_1)^2}. \tag{26.5}$$

Since there need be no connection between the choices of $\boldsymbol{x}$ co-ordinates and of $\boldsymbol{X}$ co-ordinates, there is no simple relation between Λ_1 and λ_1. When it is necessary to observe this distinction, we use the symbols Λ and Δ for stretch and extension of elements at $\boldsymbol{X}$.

The shear $\gamma_{(\boldsymbol{N}_1, \boldsymbol{N}_2)}$ of the directions $\boldsymbol{N}_1$, $\boldsymbol{N}_2$ may be calculated from

$$\left.\begin{aligned} \cos\Theta_{(\boldsymbol{N}_1,\boldsymbol{N}_2)} &= g_{KM} N_1^K N_2^M, \\ \cos(\Theta_{(\boldsymbol{N}_1,\boldsymbol{N}_2)} - \gamma_{(\boldsymbol{N}_1,\boldsymbol{N}_2)}) &= \frac{C_{KM} N_1^K N_2^M}{\sqrt{C_{PQ} N_1^P N_1^Q}\sqrt{C_{RS} N_2^R N_2^S}}, \\ &= \frac{1}{\lambda_{(\boldsymbol{N}_1)}\lambda_{(\boldsymbol{N}_2)}} C_{KM} N_1^K N_2^M. \end{aligned}\right\} \tag{26.6}$$

Thus the *shear component* (Sect. App. 45) of $\boldsymbol{C}$ for the directions $\boldsymbol{N}_1$ and $\boldsymbol{N}_2$ is insufficient to determine the shear of those directions. In fact, if we write S for the right-hand side of $(26.6)_3$, we obtain

$$\sin\gamma_{(\boldsymbol{N}_1,\boldsymbol{N}_2)} = S\sin\Theta_{(\boldsymbol{N}_1,\boldsymbol{N}_2)} - \sqrt{1-S^2}\cos\Theta_{(\boldsymbol{N}_1,\boldsymbol{N}_2)}. \tag{26.7}$$

From (26.6), *a necessary and sufficient condition that the shear of the directions $\boldsymbol{N}_1$, $\boldsymbol{N}_2$ be zero is that the ratio of the corresponding shear components of $\boldsymbol{C}$ and $\boldsymbol{g}$ equal the product of the stretches in the directions $\boldsymbol{N}_1$ and $\boldsymbol{N}_2$.* For orthogonal shears (26.7) reduces to

$$\sin\gamma_{(\boldsymbol{N}_1,\boldsymbol{N}_2)} = S \equiv \frac{C_{KM} N_1^K N_2^M}{\lambda_{(\boldsymbol{N}_1)}\lambda_{(\boldsymbol{N}_2)}}, \tag{26.8}$$

so that *the vanishing of an orthogonal shear component of $\boldsymbol{C}$ is necessary and sufficient for the vanishing of the corresponding orthogonal shear.* When applied to the directions of the X_1 and X_2 co-ordinate curves, (26.6) and its dual become

$$\left.\begin{aligned} \frac{C_{12}}{\sqrt{C_{11}}\sqrt{C_{22}}} &= \cos(\Theta_{12} - \Gamma_{12}), & \frac{G_{12}}{\sqrt{G_{11}}\sqrt{G_{22}}} &= \cos\Theta_{12}, \\ \frac{c_{12}}{\sqrt{c_{11}}\sqrt{c_{22}}} &= \cos\theta_{12}, & \frac{g_{12}}{\sqrt{g_{11}}\sqrt{g_{22}}} &= \cos(\theta_{12} - \gamma_{12}), \end{aligned}\right\} \tag{26.9}$$

where Γ_{12} is the shear of the directions tangent to the X^1 and X^2 co-ordinate curves at $\boldsymbol{X}$, while γ_{12} is the shear of the directions which are deformed into the tangents to the x^1 and x^2 curves at $\boldsymbol{x}$. Since the two systems of co-ordinates may be chosen independently at will, there is no simple relation between Γ_{12} and γ_{12}. In the orthogonal case, i.e., $\Theta_{12} = \frac{1}{2}\pi$ and $\theta_{12} - \gamma_{12} = \frac{1}{2}\pi$, (26.9) reduces to

$$\sin\Gamma_{12} = \frac{1}{\Lambda_1\Lambda_2}\cdot\frac{C_{12}}{\sqrt{G_{11}}\sqrt{G_{22}}}, \quad \sin\gamma_{12} = -\lambda_1\lambda_2\frac{c_{12}}{\sqrt{g_{11}}\sqrt{g_{22}}}. \tag{26.10}$$

From $(26.10)_1$ we see that *in the orthogonal case the shear component $|C_{12}|/\sqrt{G_{11}}\sqrt{G_{22}}$ constitutes an upper or a lower bound for $\sin\Gamma_{12}$ according as surface area in the X^3 co-ordinate surface be increased or decreased in deformation.*

[1] These interpretations and (26.10) are due to GREEN [1841, *2*, pp. 295–296]. Cf. E. and F. COSSERAT [1896, *1*, § 3]. While they may now be read off from familiar results in differential geometry, GREEN obtained them long before general co-ordinates were introduced in three dimensions, and in fact much of the theory of curvilinear co-ordinates grew out of continuum mechanics.

Shear is a property of a pair of directions. When these directions are selected as co-ordinate directions, it is obvious that the results obtained are not independent of the choice of co-ordinates. Shear is not an invariant concept unless the directions to which it refers are kept fixed. As we shall see in the next section, in any deformation at any points $\boldsymbol{X}$ and $\boldsymbol{x}$ it is always possible to find a system of orthogonal co-ordinates in which the co-ordinate shears all vanish. The maximum shears will be determined in Sect. 28.

A deformation is *rigid* if the distance between every pair of points is left unchanged. Necessary and sufficient that a deformation be rigid is that at each point

$$\boldsymbol{C} = \boldsymbol{c} = \boldsymbol{1}. \tag{26.11}$$

At a single point where (26.11) is satisfied we shall say the deformation is *locally rigid*; when no confusion is likely, the qualification "locally" may be omitted.

In order for every portion of a curve $\mathscr{C}$ to be carried into a corresponding portion of a curve c of the same length, it is necessary and sufficient that the stretch in the direction of the tangent to $\mathscr{C}$ be 1 at each point. A differential equation for inextended curves has been derived and discussed by CASTOLDI[1].

27. The strain ellipsoids and the principal stretches. I: Geometrical treatment. CAUCHY created his representation of symmetric tensors by quadric surfaces, which has been explained in Sect. 21, as a means of visualizing strain, inertia, and stress. The quadrics of $\boldsymbol{C}$ and $\boldsymbol{c}$ we shall call the *strain ellipsoids* at $\boldsymbol{X}$ and $\boldsymbol{x}$, respectively[2]. That these quadrics are indeed ellipsoids may be proved in many ways. The simplest proof consists in noting that by the axiom of continuity, there exists a sphere about $\boldsymbol{x}$ which contains the image of every sufficiently small sphere about $\boldsymbol{X}$, and the only quadric contained in a sphere is an ellipsoid. This proof[3] involves the undesirable complications mentioned at the end of Sect. 20.

An algebraic proof was given in Sect. App. 43. For a formal yet geometric approach, consider the differential elements $d\boldsymbol{X}$ which sweep out a sphere of radius K at $\boldsymbol{X}$, and hold K fixed. By (26.1) these $d\boldsymbol{X}$ are deformed into differential elements $d\boldsymbol{x}$ which sweep out at $\boldsymbol{x}$ the quadric surface

$$K^2 = g_{KM}\, dX^K\, dX^M = c_{km}\, dx^k\, dx^m. \tag{27.1}$$

Since $\boldsymbol{g}$ is a Euclidean metric tensor, the first form is positive definite, and hence the quadric of the second form is an ellipsoid.

Select a vector $d\boldsymbol{X}_1$ at $\boldsymbol{X}$, and let $d\boldsymbol{X}_2$ be any vector in the plane perpendicular to $d\boldsymbol{X}_1$, so that $g_{KM}\, dX_1^K\, dX_2^M = 0$. By (20.2) and $(26.1)_2$ follows

$$0 = g_{KM}\, X^K_{;k}\, dx_1^k\, X^M_{;m}\, dx_2^m = c_{km}\, dx_1^k\, dx_2^m. \tag{27.2}$$

Now the gradient vector of the strain ellipsoid (27.1) at the terminus of $d\boldsymbol{x}_1$ is $2c_{km}\, dx_1^m$. Since (27.2) asserts that $d\boldsymbol{x}_2$ is perpendicular to this gradient, we have derived ***Cauchy's first fundamental theorem***[4]: *An element of arc and its normal plane in an infinitesimal sphere at* $\boldsymbol{X}$ *are deformed into an element of arc and its conjugate plane in the strain ellipsoid at* $\boldsymbol{x}$. This theorem is often phrased

---

[1] [1950, *3*].

[2] The quadric of $\boldsymbol{C}$ was called the *strain ellipsoid* by KELVIN and TAIT [1867, *3*, § 160]; the quadric of $\boldsymbol{c}$, the *reciprocal strain ellipsoid* by LOVE [1906, *5*, § 6].

[3] This and other arguments based on geometrical considerations regarding infinitesimals were given by ST. VENANT [1864, *4*] [1880, *10*].

[4] [1828, *3*, Ths. I and II]. Let $\mathscr{P}$ be the plane element normal to $d\boldsymbol{X}$, and let it be deformed into p; according to GALLI [1933, *5*], for small deformation the normal to p subtends equal angles with $d\boldsymbol{x}$ and $d\boldsymbol{X}$, but this relation does not hold for arbitrary strains.

as follows: *Perpendicular diameters of an infinitesimal sphere at* $\boldsymbol{X}$ *are deformed into conjugate diameters of the strain ellipsoid at* $\boldsymbol{x}$, *and conversely* (Fig. 4). Cf. also the presentation in Sect. 21.

Now a quadric has three diameters that are perpendicular to their conjugate planes, these diameters being its axes. The axes of the strain ellipsoid at $\boldsymbol{x}$ and $\boldsymbol{X}$ are called the *principal axes of strain* at $\boldsymbol{x}$ and $\boldsymbol{X}$, respectively. Thus it follows from CAUCHY's theorem *that there exists an orthogonal triad at* $\boldsymbol{X}$ *which is deformed into an orthogonal triad*[1] *at* $\boldsymbol{x}$. It is but a change of words to say that there exists at $\boldsymbol{x}$ an orthogonal triad which results from the deformation of a certain orthogonal triad at $\boldsymbol{X}$. Therefore, by the property dual to the foregoing, the original orthogonal triad at $\boldsymbol{X}$ is a set of principal axes of strain at $\boldsymbol{X}$. Thus follows another corollary of CAUCHY's theorem: *The deformation rotates the principal axes of strain at* $\boldsymbol{X}$ *into the principal axes of strain at* $\boldsymbol{x}$ (Fig. 5).

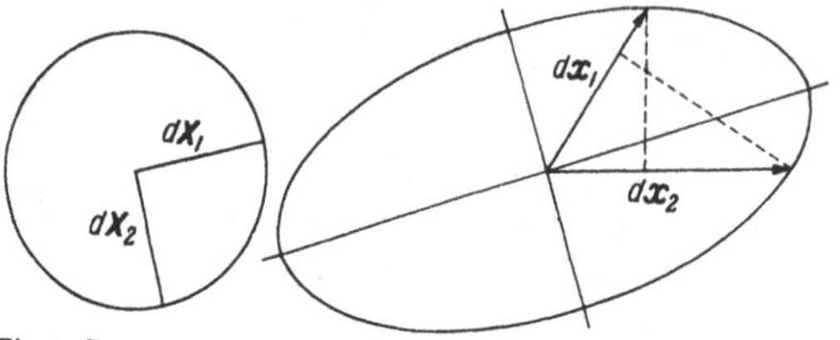

Fig. 4. Perpendicular diameters of a sphere at $\boldsymbol{X}$ are deformed into conjugate diameters of an ellipsoid at $\boldsymbol{x}$.

For derivation of the last corollary, we have tacitly assumed that the lengths of the axes of the strain ellipsoid at $\boldsymbol{x}$ are all unequal, so that the principal axes of strain are uniquely determined. If the ellipsoid degenerates into a spheroid or a sphere, an infinite number of orthogonal triads are deformed into orthogonal triads. Any one of these may be selected and called "the" principal axes of strain. In this case, so that the last corollary will remain correct as stated we take for "the" axes at $\boldsymbol{x}$ those directions into which "the" axes at $\boldsymbol{X}$ are deformed.

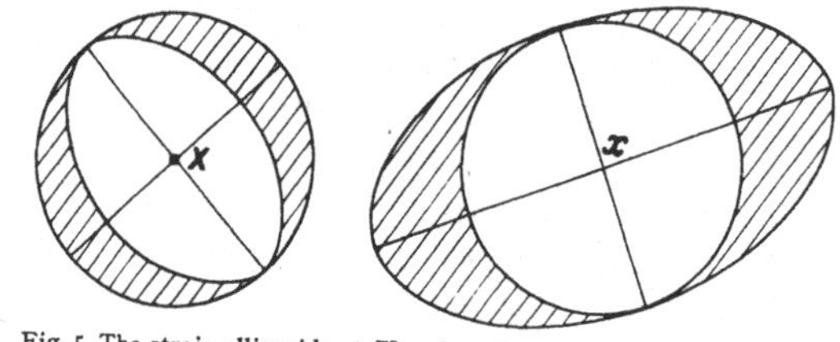

Fig. 5. The strain ellipsoids at $\boldsymbol{X}$ and $\boldsymbol{x}$, showing the rotation of the principal axes of strain.

The stretch in a given direction was defined in Sect. 25. Since the diameters of an infinitesimal sphere at $\boldsymbol{X}$ are deformed into diameters of the strain ellipsoid at $\boldsymbol{x}$, *the ratios of corresponding diameters are exactly the stretches*. That is, the stretches in the several directions at a point vary as the distance from the point to the surface of the ellipsoid. Thus, as CAUCHY[2] remarked, *the stretches are distributed symmetrically about the principal axes*: in particular, any stretch that is not a principal stretch is experienced either in no direction or in infinitely many directions. Cf. the remarks at the end of Sect. 21.

Similarly, the ratios of diameters of a sphere at $\boldsymbol{x}$ to corresponding diameters of the strain ellipsoid at $\boldsymbol{X}$ are the reciprocals of the stretches. Thus the longest axis of the ellipsoid at $\boldsymbol{X}$ is rotated into the shortest axis of the ellipsoid at $\boldsymbol{x}$, and if for one of the ellipsoids there is an axis of intermediate length, it is deformed into one of intermediate length for the other. Hence the strain ellipsoid at $\boldsymbol{X}$ degenerates to a spheroid or a sphere if and only if the strain ellipsoid at $\boldsymbol{x}$ does so. The interpretation of stretches as proportional to diameters of an ellip-

[1] By applying WEINGARTEN's conditions (App. 48.3), TONOLO [1949, *32*] has obtained necessary and sufficient conditions that the orthogonal triple of plane elements so determined envelop a triply orthogonal system of surfaces.

[2] [1823, *1*] [1827, *2*].

soid yields at once ***Cauchy's second fundamental theorem***[1]: *At any point X there exists a direction in which the stretch is not less than in any other direction; a second, perpendicular to it, in which the stretch is not greater than in any other direction. The stretches in these two directions and in a mutually perpendicular third direction are uniquely defined pure numbers, the* principal stretches $\lambda_1, \lambda_2, \lambda_3$, *satisfying*

$$\lambda_1 \geqq \lambda_2 \geqq \lambda_3. \tag{27.3}$$

If $\lambda_1 > \lambda_2 > \lambda_3$, the stretch λ_2 is a minimax. The lengths of the axes of the strain ellipsoid at x stand in the ratio $\lambda_1 : \lambda_2 : \lambda_3$ to one another; the lengths of the corresponding axes of the strain ellipsoid at X, in the reciprocal ratios. The directions in which $\lambda_1, \lambda_2, \lambda_3$ are the stretches are the principal directions of strain at X; alternatively, they are the principal directions of strain at x.

The principal stretches are the most important quantities connected with the strain. Their magnitudes determine the range of stretches, since every stretch λ must satisfy $\lambda_1 \geqq \lambda \geqq \lambda_3$. If we know the attitudes of the principal axes of strain, either at x or at X, the values of the three principal stretches enable us to construct the strain ellipsoids and hence to determine the stretch in general as a function of direction.

The extensions $\delta_\mathfrak{a}$ corresponding to the principal stretches $\lambda_\mathfrak{a}$ are called the *principal extensions*.

For a deformation to be locally rigid it is necessary and sufficient that

$$\lambda_\mathfrak{a} = 1, \quad \mathfrak{a} = 1, 2, 3. \tag{27.4}$$

In a rigid displacement, the strain quadrics are spheres, but the converse is not true. Indeed, suppose the quadric at X be a sphere. Then all stretches are equal, and $C_{KM} = \lambda^2 g_{KM}$, so that from (26.6) follows[2] $\gamma_{(N_1, N_2)} = 0$. Thus the deformation is conformal. Again the converse is not true, since an inversion is conformal but need not produce equal stretches in all elements[3].

28. The strain ellipsoids and the principal stretches, II: Algebraic treatment. While the geometric proofs given in the foregoing section are rigorous, some readers will prefer algebraic demonstrations, which in any case bring with them formal apparatus useful later as well as some additional results. We preserve unity of treatment by making no reference to any of the theorems already derived by geometric means. However, we leave to the reader the geometrical interpretation of the formulae we now derive.

Since by its definition $(26.1)_2$ the tensor $\boldsymbol{c}$ is real and symmetric, we may apply to it a now celebrated theorem of CAUCHY, first proved in this very connection[4] (cf. Sect. App. 37): *The proper numbers $c_\mathfrak{a}$ are real; proper vectors corresponding to distinct proper numbers are orthogonal; and there exists a rectangular Cartesian frame such that at x*

$$\|c^k_m\| = \left\| \begin{matrix} c_1 & 0 & 0 \\ \cdot & c_2 & 0 \\ \cdot & \cdot & c_3 \end{matrix} \right\|. \tag{28.1}$$

Since $\boldsymbol{c}$ is a Euclidean metric tensor, we have $c_{kk} > 0$ in all co-ordinate systems, and hence by the theorem just stated it follows in particular that $c_\mathfrak{a} > 0$. The

[1] [1841, *1*, § I] (in part, [1823, *1*] [1827, *2*]).

[2] CISOTTI [1944, *3*, § 3].

[3] On the basis of theorems of CAUCHY and LIOUVILLE, SIGNORINI [1943, *6*, ¶¶ 15—16] develops the properties of conformal strains.

[4] And in connection with the tensor of inertia, see Sect. 168 [1828, *1*, pp. 15—17] [1841, *1*, Th. VIII].

directions of the co-ordinate axes in a system for which (28.1) holds are principal directions of strain at $\boldsymbol{x}$. Dual statements hold for $\boldsymbol{C}$, whose proper numbers are $C_{\mathfrak{a}}$.

Let[1] the vectors $\boldsymbol{n}_{\mathfrak{a}}$ be an orthogonal unit triad in the directions of the principal axes of strain at $\boldsymbol{x}$, so that

$$c^k_m n^m_{\mathfrak{a}} = c_{\mathfrak{a}} n^k_{\mathfrak{a}}, \quad \text{or} \quad c_{km} n^m_{\mathfrak{a}} = c_{\mathfrak{a}} g_{km} n^m_{\mathfrak{a}}, \tag{28.2}$$

while $g_{km} n^k_{\mathfrak{a}} n^m_{\mathfrak{b}} = \delta_{\mathfrak{a}\mathfrak{b}}$. Set

$$N^K_{\mathfrak{a}} \equiv X^K_{;k} n^k_{\mathfrak{a}} / \sqrt{c_{\mathfrak{a}}}. \tag{28.3}$$

Then by $(26.2)_2$, (17.2), and (28.2) follows

$$\left.\begin{aligned} C_{KM} N^M_{\mathfrak{a}} &= g_{km} x^k_{;K} x^m_{;M} X^M_{;p} n^p_{\mathfrak{a}} / \sqrt{c_{\mathfrak{a}}}, \\ &= g_{km} x^k_{;K} n^m_{\mathfrak{a}} / \sqrt{c_{\mathfrak{a}}} = (c_{\mathfrak{a}})^{-1} c_{km} n^m_{\mathfrak{a}} x^k_{;K} / \sqrt{c_{\mathfrak{a}}}, \\ &= (c_{\mathfrak{a}})^{-1} g_{MP} X^M_{;k} X^P_{;m} x^k_{;K} n^m_{\mathfrak{a}} / \sqrt{c_{\mathfrak{a}}} = (c_{\mathfrak{a}})^{-1} g_{KM} X^M_{;k} n^k_{\mathfrak{a}} / \sqrt{c_{\mathfrak{a}}}, \\ &= (c_{\mathfrak{a}})^{-1} g_{KM} N^M_{\mathfrak{a}} \end{aligned}\right\} \tag{28.4}$$

and

$$\left.\begin{aligned} g_{KM} N^K_{\mathfrak{a}} N^M_{\mathfrak{b}} &= g_{KM} X^K_{;k} X^M_{;m} n^k_{\mathfrak{a}} n^m_{\mathfrak{b}} / \sqrt{c_{\mathfrak{a}} c_{\mathfrak{b}}}, \\ &= c_{km} n^k_{\mathfrak{a}} n^m_{\mathfrak{b}} / \sqrt{c_{\mathfrak{a}} c_{\mathfrak{b}}} = c_{\mathfrak{a}} g_{km} n^k_{\mathfrak{a}} n^k_{\mathfrak{b}} / \sqrt{c_{\mathfrak{a}} c_{\mathfrak{b}}}, \\ &= \delta_{\mathfrak{a}\mathfrak{b}}. \end{aligned}\right\} \tag{28.5}$$

The results (28.4) and (28.5) assert that the transformation (28.3) carries the orthogonal triad of unit proper vectors $\boldsymbol{n}_{\mathfrak{a}}$ of $\boldsymbol{c}$ with proper numbers $c_{\mathfrak{a}}$ into an orthogonal triad of unit proper vectors $\boldsymbol{N}_{\mathfrak{a}}$ of $\boldsymbol{C}$ with proper numbers $C_{\mathfrak{a}}$ given by

$$C_{\mathfrak{a}} = \frac{1}{c_{\mathfrak{a}}}. \tag{28.6}$$

The formula (28.3) sets the principal directions of strain at $\boldsymbol{x}$ into one-to-one correspondence with the principal directions of strain at $\boldsymbol{X}$, the dual formula being

$$n^k_{\mathfrak{a}} = x^k_{;K} N^K_{\mathfrak{a}} / \sqrt{C_{\mathfrak{a}}} = x^k_{;K} N^k_{\mathfrak{a}} \sqrt{c_{\mathfrak{a}}}. \tag{28.7}$$

Since the principal stretches $\lambda_{\mathfrak{a}}$ are defined as the stretches in the principal directions of strain, by (28.1) and (26.4) we get

$$C_{\mathfrak{a}} = \lambda^2_{\mathfrak{a}}, \quad c_{\mathfrak{a}} = \lambda^{-2}_{\mathfrak{a}}, \tag{28.8}$$

incidentally verifying (28.6). Hence by (25.3) follows $0 < C_{\mathfrak{a}} < \infty$, $0 < c_{\mathfrak{a}} < \infty$.

The tensors $\boldsymbol{C}$ and $\boldsymbol{c}$ may be characterized as the *unique symmetric tensors whose principal axes are the principal axes of strain at $\boldsymbol{X}$ and $\boldsymbol{x}$ and whose proper numbers are the squares and the reciprocal squares of the principal stretches, respectively.*

Let $\boldsymbol{N}$ be a unit vector at $\boldsymbol{X}$; then its components in a rectangular Cartesian system whose axes are the principal axes of strain at $\boldsymbol{X}$ are $\cos(\boldsymbol{N}\mathfrak{a})$, where $(\boldsymbol{N}\mathfrak{a})$ is the angle between $\boldsymbol{N}$ and the $\mathfrak{a}$-th principal axis. For the stretch in the direction of $\boldsymbol{N}$, $(28.8)_1$ and the dual of (28.1) put into $(26.4)_1$ yield

$$\lambda_{(\boldsymbol{N})} = \sqrt{\sum_{\mathfrak{a}=1}^{3} \lambda^2_{\mathfrak{a}} \cos^2(\boldsymbol{N}\mathfrak{a})}. \tag{28.9}$$

Similarly

$$\lambda_{(\boldsymbol{n})} = \frac{1}{\sqrt{\sum_{\mathfrak{a}=1}^{3} \frac{\cos^2(\boldsymbol{n}\mathfrak{a})}{\lambda^2_{\mathfrak{a}}}}}, \tag{28.10}$$

[1] We follow TOUPIN's arrangement of the argument [1953, *32*, § 14].

but it is essential to notice that whether or not $\boldsymbol{n}$ is taken in the direction into which the direction of $\boldsymbol{N}$ is deformed, the angles $(\boldsymbol{N}\mathfrak{a})$ and $(\boldsymbol{n}\mathfrak{a})$ are in general different.

By $(26.6)_2$, the orthogonal shear for the directions of the unit vectors $\boldsymbol{N}_1$ and $\boldsymbol{N}_2$ is given by

$$\sin\gamma_{(\boldsymbol{N}_1,\boldsymbol{N}_2)} = C_{KM}\, U^K V^M, \tag{28.11}$$

where $\boldsymbol{U}$ and $\boldsymbol{V}$ are parallel to $\boldsymbol{N}_1$ and $\boldsymbol{N}_2$ and are subject to the restrictions $C_{KM} U^K U^M = 1$, $C_{KM} V^K V^M = 1$. Therefore the extreme values of the shears, which may be called the *principal shears*, are given at once by the solution of Problem 5 in Sect. App. 46: *The pairs of directions in which the orthogonal shears are extreme are the same as those in which the shear components of* $\boldsymbol{C}$ *are extreme, namely, the pairs of directions at* $\boldsymbol{X}$ *that are normal to one principal axis of* $\boldsymbol{C}$ *and bisect the angles between the other two.* The extreme values of the shear components of $\boldsymbol{C}$ are $\pm\frac{1}{2}(C_1 - C_3)$, $\pm\frac{1}{2}(C_2 - C_3)$, $\pm\frac{1}{2}(C_1 - C_2)$. The extreme values of $\sin\Gamma_{(\boldsymbol{N}_1,\boldsymbol{N}_2)}$ are $\pm(C_1 - C_3)/(C_1 + C_3)$, $\pm(C_2 - C_3)/(C_2 + C_3)$, $\pm(C_1 - C_2)/(C_1 + C_2)$. As it should be, this last statement is self-dual. Recalling the convention of Sect. 25 that shears are always acute angles, for the *principal shears* $\gamma_{\mathfrak{a}}$ we have

$$\gamma_2 = \operatorname{Arc\,sin}\frac{\lambda_1^2 - \lambda_3^2}{\lambda_1^2 + \lambda_3^2}, \quad \gamma_3 = \operatorname{Arc\,sin}\frac{\lambda_1^2 - \lambda_2^2}{\lambda_1^2 + \lambda_2^2}, \quad \gamma_1 = \operatorname{Arc\,sin}\frac{\lambda_2^2 - \lambda_3^2}{\lambda_2^2 + \lambda_3^2}, \tag{28.12}$$

where γ_2 is the greatest.

29. Further formulae relating to the deformation tensors. Since $\boldsymbol{c}$ and $\boldsymbol{C}$ are positive definite, they have unique inverse tensors $\boldsymbol{c}^{-1}$ and $\boldsymbol{C}^{-1}$. From $(26.1)_2$ and $(26.2)_2$ it is easy to verify that[1]

$$\overset{-1}{c}{}^k{}_m = x^{k;K} x_{m;K}, \qquad \overset{-1}{C}{}^K{}_M = X^{K;k} X_{M;k}, \tag{29.1}$$

where we are using the notation of associated components. To interpret $\boldsymbol{C}^{-1}$, we calculate the magnitude of the element of area $d\boldsymbol{a}$ at $\boldsymbol{x}$ in terms of the components dA_K at $\boldsymbol{X}$. By using (20.8), (30.5), and $(29.1)_2$ we obtain

$$\left.\begin{aligned}(da)^2 &= J^2 g^{km} X^K_{;m} X^M_{;m}\, dA_K\, dA_M,\\ &= \mathrm{III}_{\boldsymbol{C}}\, \overset{-1}{C}{}^{KM}\, dA_K\, dA_M.\end{aligned}\right\} \tag{29.2}$$

Comparing this result with $(26.2)_1$ shows that *the tensor* $\mathrm{III}_{\boldsymbol{C}}\,\boldsymbol{C}^{-1}$ *measures the ratios of areas in the same sense that the tensor* $\boldsymbol{C}$ *measures the ratios of lengths*[2].

The principle of duality enables us to assert an analogous interpretation for $\mathrm{III}_{\boldsymbol{c}}\,\boldsymbol{c}^{-1}$.

If, in analogy to the theory of deformation of lengths, we were to construct a theory of deformation of areas, the formula (29.2) and its dual would enable us to proceed in steps parallel to those we have based upon (26.1) and (26.2). This fact enables us to assert a **second principle of duality:** *In any proposition concerning change of lengths and of angles between elements of length expressed in terms of the tensors* $\boldsymbol{C}$ *and* $\boldsymbol{c}$, *a valid proposition results if we replace* "length", $\boldsymbol{C}$, *and* $\boldsymbol{c}$ *by* "area", $\mathrm{III}_{\boldsymbol{C}}\,\boldsymbol{C}^{-1}$, *and* $\mathrm{III}_{\boldsymbol{c}}\,\boldsymbol{c}^{-1}$, *and conversely.* In particular, since $\boldsymbol{C}^{-1}$ and $\boldsymbol{C}$ are co-axial, the extremal changes of area occur in elements normal

---

[1] The tensor $\boldsymbol{C}^{-1}$ was introduced by PIOLA [1833, *3*, § 5] [1836, *1*, Eqs. (142), (143), (152)]. FINGER [1894, *4*, Eqs. (12), (31)] introduced $\mathrm{III}_{\boldsymbol{C}}\,\boldsymbol{C}^{-1}$ and $\boldsymbol{c}^{-1}$. Cf. also [1892, *4*].

[2] TONOLO [1943, *8*, § V.4] derived a formula which is essentially $(29.2)_1$; by substituting in it the dual of (37.8), he observed that the rotation tensor cancels out, and hence the ratio da/dA can be calculated from the components C_{KM}, as is immediate from our result $(29.2)_2$. In this paragraph and the next two we follow TRUESDELL [1958, *10*].

to the principal axes of strain, and the greatest (least) change of area occurs in the element normal to the axis of least (greatest) stretch. In fact, the extremal changes of area are $\lambda_{\mathfrak{a}}\lambda_{\mathfrak{b}}$, $\mathfrak{a}\neq\mathfrak{b}$, satisfying $\lambda_2\lambda_3\leqq\lambda_3\lambda_1\leqq\lambda_1\lambda_2$.

In the notation of associated components, $(26.1)_2$ and $(26.2)_2$ read

$$c^k_m = X^{K;k} X_{K;m}, \qquad C^K_M = x^{k;K} x_{k;M}. \tag{29.3}$$

From this result and from (29.1) we have[1]

$$\left.\begin{aligned} &\frac{\partial c_{km}}{\partial X^K_{;p}} = \delta^p_k X_{K;m} + \delta^p_m X_{K;k}, \qquad \frac{\partial \overset{-1}{c}{}^{km}}{\partial x^p_{;K}} = \delta^m_p x^{k;K} + \delta^k_p x^{m;K},\\ &c_{km;p} = X_{K;k} X^K_{;mp} + X_{K;m} X^K_{;kp},\\ &\overset{-1}{c}{}^{km}_{;K} = x^{k;M} x^m_{;MK} + x^{m;M} x^k_{;MK}. \end{aligned}\right\} \tag{29.4}$$

From $(29.1)_1$ and $(29.3)_2$ we get

$$\overset{-2}{c}{}^k_m = \overset{-1}{c}{}^k_p \overset{-1}{c}{}^p_m = C^{KM} x^k_{;K} x_{m;M}. \tag{29.5}$$

But by (App. 38.10) we have

$$\mathrm{III}_{\boldsymbol{c}}\boldsymbol{c}^{-2} = \boldsymbol{c}^{-2}\cdot \mathrm{III}_{\boldsymbol{c}}\,\mathbf{1} = \boldsymbol{c}^{-2}\cdot[\boldsymbol{c}^3 - \mathrm{I}_{\boldsymbol{c}}\boldsymbol{c}^2 + \mathrm{II}_{\boldsymbol{c}}\boldsymbol{c}]. \tag{29.6}$$

Combining (29.5) and (29.6) yields an identity due to FINGER[2]:

$$\mathrm{III}_{\boldsymbol{c}} C^{KM} x^k_{;K} x_{m;M} = c^k_m - \mathrm{I}_{\boldsymbol{c}}\delta^k_m + \mathrm{II}_{\boldsymbol{c}} \overset{-1}{c}{}^k_m. \tag{29.7}$$

TRUESDELL[3] noted that each of the deformation tensors may be expressed as the trace of a matrix of derivatives of one set of deformation gradients with respect to one another. The formulae are immediate consequences of (17.5):

$$c^k_m = -\frac{\partial X^{K;k}}{\partial x^{m;K}}, \qquad \overset{-1}{c}{}^k_m = -\frac{\partial x^{k;K}}{\partial X^{K;m}}, \tag{29.8}$$

and their duals.

By the theorem of SPOTTISWOODE given in Sect. App. 40, we may order the proper numbers $\overset{-1}{c}_{\mathfrak{a}}$ and $\overset{-1}{C}_{\mathfrak{a}}$ of $\boldsymbol{c}^{-1}$ and $\boldsymbol{C}^{-1}$ in such a way that

$$\overset{-1}{c}_{\mathfrak{a}} = (c_{\mathfrak{a}})^{-1} = \lambda^2_{\mathfrak{a}}, \qquad \overset{-1}{C}_{\mathfrak{a}} = (C_{\mathfrak{a}})^{-1} = \lambda^{-2}_{\mathfrak{a}}. \tag{29.9}$$

The tensor $\boldsymbol{c}^{-1}$ is the deformation measure most commonly used in modern work on finite deformation, since (1) it is an ordinary symmetric tensor whose principal axes are the principal axes of strain in the deformed material (this distinguishes it from $x^k_{;K}$, which is a non-trivially double field with 9 independent components rather than 6, and from $\boldsymbol{C}$, whose principal axes are the principal axes of strain in the undeformed material), (2) it is a quadratic polynomial in derivatives with respect to points in the undeformed material (this distinguishes it from $\boldsymbol{c}$), and (3) its proper numbers are the squares of principal stretches (this also distinguishes it from $\boldsymbol{c}$).

30. The strain invariants. The principal invariants $\mathrm{I}_{\boldsymbol{a}}$, $\mathrm{II}_{\boldsymbol{a}}$, $\mathrm{III}_{\boldsymbol{a}}$ of a tensor $\boldsymbol{a}$ have been discussed in Sect. App. 38. By (28.8) and (29.9) we have the following

---

[1] Cf. BONVICINI [1932, *3*] [1935, *2*].

[2] In the form given by FINGER [1894, *4*, Eq. (34)], the invariants of $\boldsymbol{c}$ are replaced by those of $\boldsymbol{C}$ through (30.2).

[3] [1952, *20*, § 14].

expressions for the principal invariants of $\boldsymbol{C}$, $\boldsymbol{c}^{-1}$, $\boldsymbol{c}$, and $\boldsymbol{C}^{-1}$ in terms of the principal stretches λ_a and principal extensions δ_a:

$$\left.\begin{aligned}
\mathrm{I}_{\boldsymbol{C}} &= \mathrm{I}_{\boldsymbol{c}^{-1}} = \lambda_1^2 + \lambda_2^2 + \lambda_3^2 = (1+\delta_1)^2 + (1+\delta_2)^2 + (1+\delta_3)^2,\\
\mathrm{II}_{\boldsymbol{C}} &= \mathrm{II}_{\boldsymbol{c}^{-1}} = \lambda_1^2\lambda_2^2 + \lambda_2^2\lambda_3^2 + \lambda_3^2\lambda_1^2,\\
&= (1+\delta_1)^2(1+\delta_2)^2 + (1+\delta_2)^2(1+\delta_3)^2 + (1+\delta_3)^2(1+\delta_1)^2,\\
\mathrm{III}_{\boldsymbol{C}} &= \mathrm{III}_{\boldsymbol{c}^{-1}} = \lambda_1^2\lambda_2^2\lambda_3^2 = (1+\delta_1)^2(1+\delta_2)^2(1+\delta_3)^2,\\
\mathrm{I}_{\boldsymbol{c}} &= \mathrm{I}_{\boldsymbol{C}^{-1}} = \frac{1}{\lambda_1^2} + \frac{1}{\lambda_2^2} + \frac{1}{\lambda_3^2},\\
\mathrm{II}_{\boldsymbol{c}} &= \mathrm{II}_{\boldsymbol{C}^{-1}} = \frac{1}{\lambda_1^2\lambda_2^2} + \frac{1}{\lambda_2^2\lambda_3^2} + \frac{1}{\lambda_3^2\lambda_1^2},\\
\mathrm{III}_{\boldsymbol{c}} &= \mathrm{III}_{\boldsymbol{C}^{-1}} = \frac{1}{\lambda_1^2\lambda_2^2\lambda_3^2}.
\end{aligned}\right\} \tag{30.1}$$

Hence follow the ***fundamental identities***[1]:

$$\mathrm{I}_{\boldsymbol{c}} = \frac{\mathrm{II}_{\boldsymbol{C}}}{\mathrm{III}_{\boldsymbol{C}}}, \qquad \mathrm{II}_{\boldsymbol{c}} = \frac{\mathrm{I}_{\boldsymbol{C}}}{\mathrm{III}_{\boldsymbol{C}}}, \qquad \mathrm{III}_{\boldsymbol{c}} = \frac{1}{\mathrm{III}_{\boldsymbol{C}}}. \tag{30.2}$$

As a corollary of these formulae and of the representation theorem for isotropic scalar functions[2] we have the ***fundamental lemma***: *An absolute scalar function of any one of the tensors* $\boldsymbol{c}$, $\boldsymbol{c}^{-1}$, $\boldsymbol{C}$, $\boldsymbol{C}^{-1}$ *equals an absolute scalar function of any other.*

From (30.1) follows

$$\infty > \mathrm{I} > 0, \qquad \infty > \mathrm{II} > 0, \qquad \infty > \mathrm{III} > 0, \tag{30.3}$$

where the invariants are calculated from any of the measures $\boldsymbol{C}$, $\boldsymbol{c}$, $\boldsymbol{C}^{-1}$, $\boldsymbol{c}^{-1}$, or from $\boldsymbol{C}^n$ or $\boldsymbol{c}^n$, for any n. Moreover, since the principal invariants determine unique values of the proper numbers (Sect. App. 38), *corresponding to any assigned values of* $\mathrm{I}_{\boldsymbol{C}}$, $\mathrm{II}_{\boldsymbol{C}}$, *and* $\mathrm{III}_{\boldsymbol{C}}$ *satisfying* (30.3) *and to any assigned principal directions at* $\boldsymbol{X}$, *there exists a unique deformation tensor* $\boldsymbol{C}$. In particular the conditions

$$\mathrm{I} = \mathrm{II} = 3, \qquad \mathrm{III} = 1, \tag{30.4}$$

for any of the tensors $\boldsymbol{C}^n$ or $\boldsymbol{c}^n$, are *necessary and sufficient for a rigid deformation.*

When the stretches are great, $\mathrm{I}_{\boldsymbol{C}}$, $\mathrm{II}_{\boldsymbol{C}}$, and $\mathrm{III}_{\boldsymbol{C}}$ are also very large; when the stretches are small, $\mathrm{I}_{\boldsymbol{C}}$, $\mathrm{II}_{\boldsymbol{C}}$, and $\mathrm{III}_{\boldsymbol{C}}$ are much smaller. The invariants of $\boldsymbol{c}$ show a reciprocal behavior. E.g., if the deformation doubles all lengths, then $\mathrm{I}_{\boldsymbol{C}}$, $\mathrm{II}_{\boldsymbol{C}}$, and $\mathrm{III}_{\boldsymbol{C}}$ assume the values 12, 48, 64; $\mathrm{I}_{\boldsymbol{c}}$, $\mathrm{II}_{\boldsymbol{c}}$, and $\mathrm{III}_{\boldsymbol{c}}$, $\frac{3}{4}$, $\frac{3}{16}$, and $\frac{1}{64}$. The axiom of continuity is violated at a point where any of the principal invariants equals 0 or ∞. If $\mathrm{I}_{\boldsymbol{C}} = 0$, all lengths are annulled; if $\mathrm{I}_{\boldsymbol{C}} = \infty$, at least one stretch is infinite.

---

1 Doubtless these identities were known to CAUCHY. FINGER [1894, *4*, Eq. $(28)_5$] noted $(30.2)_1$ but apparently did not realize its importance. $(30.2)_2$ follows from $(30.2)_1$ by the principle of duality. Cf. also [1892, *4*]. $(30.2)_3$ is really obvious, but apparently was first given, in the more elaborate form $(31.5)_{6,7}$, by ALMANSI [1911, *2*, § 2] and HAMEL [1912, *4*, §§ 364 to 365]. In the form (31.5), all three identies are given an elaborate derivation by MURNAGHAN [1937, *7*, Appendix]. Apparently RIVLIN [1948, *27*, § 3] was the first to state them in the simple and immediate form (30.2). As has been observed by SIGNORINI [1943, *6*, ¶ 13], since $\boldsymbol{C}$ and $\boldsymbol{c}^{-1}$ have the same proper numbers, (30.2) is no more than a rewriting of the well known relation between the invariants of a matrix and of its inverse, immediate from SPOTTISWOODE's theorem.

2 Since $f(\boldsymbol{O}\,\boldsymbol{c}\,\boldsymbol{O}^{-1}) = f(\boldsymbol{c})$ for any orthogonal transformation $\boldsymbol{O}$, $f(\boldsymbol{c}) = g(c_1, c_2, c_3)$, where the function g is a symmetric function of its three arguments. Hence $g(c_1, c_2, c_3) = h(\mathrm{I}_{\boldsymbol{c}}, \mathrm{II}_{\boldsymbol{c}}, \mathrm{III}_{\boldsymbol{c}})$.

For any deformation, we derive by (26.2), the rule for the determinant of a product of matrices, and (16.5) the identity

$$\mathrm{III}_{\boldsymbol{C}} = \det C^K_M = \det g_{pq} \det g^{PQ} (\det x^k_{;K})^2 = J^2. \tag{30.5}$$

Hence by (20.9) follow CAUCHY's formulae[1]

$$dv = \sqrt{\mathrm{III}_{\boldsymbol{C}}}\, dV, \quad dV = \sqrt{\mathrm{III}_{\boldsymbol{c}}}\, dv. \tag{30.6}$$

From these results and the inequalities (App. 39.10)$_{3,4}$ we derive bounds for the change of volume in terms of the first invariants of $\boldsymbol{c}$ and $\boldsymbol{C}$:

$$\left(\frac{1}{3}\,\mathrm{I}_{\boldsymbol{c}}\right)^{-\frac{3}{2}} \leqq \frac{dv}{dV} \leqq \left(\frac{1}{3}\,\mathrm{I}_{\boldsymbol{C}}\right)^{\frac{3}{2}}, \tag{30.7}$$

with equality holding if and only if all three principal stretches are equal.

It is of some use to express the principal strain invariants in terms of the elementary symmetric function of the principal extensions, which we shall denote by I_δ, II_δ, and III_δ. Then[2]

$$\left.\begin{aligned} \mathrm{I}_{\boldsymbol{C}} - 3 &= 2\mathrm{I}_\delta + \mathrm{I}_\delta^2 - 2\mathrm{II}_\delta, \\ \mathrm{II}_{\boldsymbol{C}} - 3 &= 4\mathrm{I}_\delta + 2\mathrm{I}_\delta^2 + 2\mathrm{I}_\delta \mathrm{II}_\delta - 6\mathrm{III}_\delta + \mathrm{II}_\delta^2 - 2\mathrm{I}_\delta \mathrm{III}_\delta, \\ \sqrt{\mathrm{III}_{\boldsymbol{C}}} - 1 &= \frac{dv - dV}{dV} = \mathrm{I}_\delta + \mathrm{II}_\delta + \mathrm{III}_\delta. \end{aligned}\right\} \tag{30.8}$$

As a measure of *strain magnitude* we suggest

$$\mathcal{K} \equiv \sqrt{\frac{1}{2}\,(\mathrm{I}_{\boldsymbol{C}} + \mathrm{I}_{\boldsymbol{c}}) - 3} = \sqrt{\frac{1}{2}\sum_{\mathfrak{a}=1}^{3}\left(\lambda_{\mathfrak{a}} - \frac{1}{\lambda_{\mathfrak{a}}}\right)^2}. \tag{30.9}$$

This measure vanishes if and only if the deformation is rigid, and it increases if the absolute value of any one extension increases while the others are held constant; moreover, lengthening and shortening are measured symmetrically. A simple measure of *shear intensity* is given by

$$\mathfrak{Z} \equiv \sqrt{\sum_{\mathfrak{a}=1}^{3} \sin^2 \gamma_{\mathfrak{a}}}. \tag{30.10}$$

By (28.12), it is easy to express this measure in terms of the principal invariants of $\boldsymbol{C}$, but the result is complicated. The octahedral invariant $\mathrm{II}_{\boldsymbol{C}}$, defined by (App. 38.11), is a measure of the intensity of the shear components of $\boldsymbol{C}$, but for large strain this bears no simple relation to $\mathfrak{Z}$.

31. The classical strain tensors and elongation tensors. Much of the older work on finite strain employs one or other of the *strain tensors*[3] $\boldsymbol{E}$ and $\boldsymbol{e}$, defined by

$$2\boldsymbol{E} \equiv \boldsymbol{C} - \boldsymbol{1}, \quad 2\boldsymbol{e} \equiv \boldsymbol{1} - \boldsymbol{c}. \tag{31.1}$$

So as to reflect the dissymmetry in the definition of stretch (Sect. 25), the definition (31.1)$_2$ takes $-\boldsymbol{e}$ as the dual of $\boldsymbol{E}$.

---

[1] [1827, *2*, Eq. (28)].

[2] The formula for the ratio of volumes was noted by BOUSSINESQ [1912, *1*, § 15].

[3] $\boldsymbol{E}$ was introduced by GREEN [1841, *2*, p. 29] and ST. VENANT [1844, *3*] [1847, *3*, § 2] and is probably the commonest strain measure even today. ALMANSI [1911, *1*, § 2] and HAMEL [1912, *4*, § 363] introduced $\boldsymbol{e}$, which has figured largely in the Italian literature and more recently has become popular with British authors, who often attribute it to COKER and FILON [1931, *3*, § 3.06]. Cf. also L. BRILLOUIN [1925, *2*, § 4] [1938, *2*, Chap. X, § VII] and MURNAGHAN [1937, *7*, § 1].

To interpret $\boldsymbol{E}$, note that we may put (26.6) into the form[1]

$$\left.\begin{aligned}\lambda_{(\boldsymbol{N}_1)}\lambda_{(\boldsymbol{N}_2)}\cos(\Theta_{(\boldsymbol{N}_1,\boldsymbol{N}_2)}-\gamma_{(\boldsymbol{N}_1,\boldsymbol{N}_2)})-\cos\Theta_{(\boldsymbol{N}_1,\boldsymbol{N}_2)}\\ =C_{KM}N_1^K N_2^M-\cos\Theta_{(\boldsymbol{N}_1,\boldsymbol{N}_2)},\\ =2E_{KM}N_1^K N_2^M.\end{aligned}\right\}\tag{31.2}$$

In terms of the gradients of the displacement vector, by (19.4) we have[2]

$$E_M^K=u^{(K}{}_{;M)}+\tfrac{1}{2}u^{P;K}u_{P;M},\qquad e_m^k=u^{(k}{}_{;m)}-\tfrac{1}{2}u^{p;k}u_{p;m},\tag{31.3}$$

the latter being the dual of the former since $-\boldsymbol{u}$ is the dual of $\boldsymbol{u}$. The vanishing of either $\boldsymbol{E}$ or $\boldsymbol{e}$ is necessary and sufficient for a rigid displacement. The normal components of $\boldsymbol{E}$ and $\boldsymbol{e}$ are called *normal strains*; the shear components, *shear strains*.

The principal strains E_a and e_a are related to the principal stretches λ_a and principal extensions δ_a by

$$\left.\begin{aligned}2E_a&=C_a-1=\lambda_a^2-1=(1+\delta_a)^2-1=2\delta_a+\delta_a^2,\\ 2e_a&=1-c_a=1-\lambda_a^{-2}=1-(1+\delta_a)^{-2}=2E_a(1+\delta_a)^2;\end{aligned}\right\}\tag{31.4}$$

hence by (25.3) follows $-\frac{1}{2}<E_a<\infty$, $-\infty<e_a<\frac{1}{2}$.

Formulae expressing the principal invariants of $\boldsymbol{E}$ and $\boldsymbol{e}$ in terms of those of $\boldsymbol{c}$, $\boldsymbol{C}$, etc., follow by (30.1). For example[3],

$$\left.\begin{aligned}\mathrm{I}_{\boldsymbol{c}}&=\mathrm{III}_{\boldsymbol{c}}\mathrm{II}_{\boldsymbol{c}^{-1}}=3-2\mathrm{I}_{\boldsymbol{e}},\\ \mathrm{II}_{\boldsymbol{c}}&=\mathrm{III}_{\boldsymbol{c}}\mathrm{I}_{\boldsymbol{c}^{-1}}=3-4\mathrm{I}_{\boldsymbol{e}}+4\mathrm{II}_{\boldsymbol{e}},\\ (dv/dV)^2&=1+2\mathrm{I}_{\boldsymbol{E}}+4\mathrm{II}_{\boldsymbol{E}}+8\mathrm{III}_{\boldsymbol{E}}=\mathrm{III}_{\boldsymbol{C}},\\ &=(1-2\mathrm{I}_{\boldsymbol{e}}+4\mathrm{II}_{\boldsymbol{e}}-8\mathrm{III}_{\boldsymbol{e}})^{-1}=\mathrm{III}^{-1},\end{aligned}\right\}\tag{31.5}$$

where in the last sequence (30.6) has been used. From these formulae and (30.3) may be read off[4] inequalities satisfied by invariants of $\boldsymbol{E}$ and $\boldsymbol{e}$.

We may split the displacement gradients $u_{K;M}$ and $u_{k;m}$ into symmetric and skew-symmetric parts:

$$\left.\begin{aligned}u_{K;M}&=\widetilde{E}_{KM}+\widetilde{R}_{KM},\quad &\widetilde{E}_{KM}&\equiv u_{(K;M)},\quad &\widetilde{R}_{KM}&\equiv u_{[K;M]},\\ u_{k;m}&=\tilde{e}_{km}+\tilde{r}_{km},\quad &\tilde{e}_{km}&\equiv u_{(k;m)},\quad &\tilde{r}_{km}&\equiv u_{[k;m]}.\end{aligned}\right\}\tag{31.6}$$

Since $-\boldsymbol{u}$ is the dual of $\boldsymbol{u}$, it follows that $-\tilde{\boldsymbol{e}}$ and $-\tilde{\boldsymbol{r}}$ are the duals of $\widetilde{\boldsymbol{E}}$ and $\widetilde{\boldsymbol{R}}$.

To interpret $\widetilde{\boldsymbol{E}}$, begin by noting that from (19.4) and (20.3) we have

$$dx^K=g_k^K\,dx^k=(\delta_M^K+u^K_{;M})\,dX^M,\tag{31.7}$$

so that by $(31.6)_1$ follows

$$dx_K=g_{KM}\,g_k^M\,dx^k=(g_{KM}+\widetilde{E}_{KM}+\widetilde{R}_{KM})\,dX^M.\tag{31.8}$$

---

[1] Cf. CISOTTI [1944, *3*, § 1].

[2] $u^{(K}{}_{;M)}$ stands for $g^{KP}u_{(P;M)}$, etc. In the formulation given here, both the derivation of (31.2) and the proof that $\boldsymbol{E}$ is a tensor with respect to general transformations of co-ordinates are trivial. The result seems to be equivalent to that given in very elaborate form by E. and F. COSSERAT [1896, *1*, § 41], but perhaps it is toward this same end that L. BRILLOUIN [1928, *2*, § 4] [1925, *2*, § 2] [1938, *2*, Chap. X, § 6] and DUPONT [1931, *6*] [1931, *5*, §§ 4–5] struggle with power series expansions (cf. also TONOLO [1943, *8*, § I.2]).

[3] For references see footnote 1, p. 265.

[4] SIGNORINI [1943, *6*, ¶ 9].

Hence by (25.9) we find for the elongation in the direction of $d\boldsymbol{X}$

$$\varepsilon_{(\boldsymbol{N})} = \widetilde{E}_{KM} N^K N^M: \tag{31.9}$$

The normal component of $\widetilde{\boldsymbol{E}}$ in the direction of $\boldsymbol{N}$ is the elongation in that direction. The tensor $\widetilde{\boldsymbol{E}}$, introduced by LOVE[1], is called the *elongation tensor.* Its quadric (Sect. 21) is the *elongation quadric.* Dual results hold for the dual tensor, $-\tilde{\boldsymbol{e}}$.

The two elongation tensors $\widetilde{\boldsymbol{E}}$ and $\tilde{\boldsymbol{e}}$ are not strain measures (Sect. 32). That this is so is obvious from the remarks in Sect. 25: In general, the proper numbers of $\widetilde{\boldsymbol{E}}$ and $\tilde{\boldsymbol{e}}$ are not functions of the principal stretches only.

A geometrical interpretation for $\widetilde{\boldsymbol{R}}$ will be given in Sect. 36, and in Sect. 38 it will be proved that $\widetilde{\boldsymbol{R}} = 0$ is a necessary but not sufficient condition for a pure strain, defined in Sect. 35.

By (31.3) we may express $\boldsymbol{E}$ as a function of $\widetilde{\boldsymbol{E}}$ and $\widetilde{\boldsymbol{R}}$:

$$E^K_M = \widetilde{E}^K_M + \tfrac{1}{2}(\widetilde{E}^{KP}\widetilde{E}_{MP} - \widetilde{E}^{KP}\widetilde{R}_{MP} - \widetilde{R}^{KP}\widetilde{E}_{MP} + \widetilde{R}^{KP}\widetilde{R}_{MP}). \tag{31.10}$$

From this formula we see that $\widetilde{\boldsymbol{E}} = 0$ does not generally imply $\boldsymbol{E} = 0$; in other words, if the elongations in three independent directions, even in three mutually perpendicular directions, are zero, it does not follow that the deformation is rigid. A non-linear condition for rigid deformation is obtained by equating to zero the right-hand side of (31.10).

The elongation tensor $\widetilde{\boldsymbol{E}}$ is a simple linear combination of displacement gradients, but it is not a strain measure. The strain tensor $\boldsymbol{E}$ is a strain measure, but it is a complicated quadratic function of the displacement gradients. By using $\boldsymbol{E}$ or $\boldsymbol{e}$ rather than $\boldsymbol{C}$ or $\boldsymbol{c}$, much of the literature on finite strain succeeds in making simple problems appear complicated.

32. The equivalence of strain measures. Many measures of strain have been proposed, and there is a large literature (cf. the appendix to the next section) claiming to show that some one is better than all the others. While for a particular problem a particular choice of strain measure may be helpful, in general such contentions are futile. As WEISSENBERG (1946), REINER, and RICHTER[2] have had the merit to observe, since the stretch of an arbitrary element can be calculated from the orientation of the element with respect to either set of axes of strain and from the values of the principal stretches, *any* measure sufficient to determine the directions of the principal axes of strain and the magnitudes of the principal stretches may be employed and is fully general.

Now there are two sets of principal axes, those at $\boldsymbol{X}$ and those at $\boldsymbol{x}$. From some quantities, such as the nine deformation gradients $x^k_{;K}$, both these sets of axes may be determined. From others, such as the six C_{KM}, the axes at $\boldsymbol{X}$ may be determined but not those at $\boldsymbol{x}$. From yet others, such as the six c_{km}, we may find the axes at $\boldsymbol{x}$ but not those at $\boldsymbol{X}$. Quantities of the last two types are called *strain measures* at $\boldsymbol{X}$ and at $\boldsymbol{x}$, respectively, while quantities of the first type are measures of both strain and rotation. From the foregoing remarks and the representation theorem for isotropic tensor functions[3] follows the ***theorem of equivalence***: *Any uniquely invertible isotropic second order tensor function of* $\boldsymbol{c}$ *is a strain measure at* $\boldsymbol{x}$; *of* $\boldsymbol{C}$, *at* $\boldsymbol{X}$.

Also from the general theory of isotropic functions it follows that the two sets of strain measures obtained by the foregoing theorem consist in symmetric tensor fields, those of the former set having as their principal directions the principal axes of strain in the deformed material, while those of the latter set have as their

[1] [1892, *8*, §§ 8–9].

[2] [1949, *36*]; [1948, *23*, § 3]; [1949, *26*, § 3].

[3] While the assertion has a long history, the first general proof is due to RIVLIN and ERICKSEN [1955, *21*, § 29]. For a simpler proof, see § 59 of *Mathematical principles of classical fluid mechanics* by J. SERRIN, this Encyclopedia, Vol. VIII/1.

principal directions the principal axes of strain in the undeformed material. Examples for the deformed material are $\boldsymbol{c}$, $\boldsymbol{c}^{-1}$, and $\boldsymbol{e}$; for the undeformed material, $\boldsymbol{C}$, $\boldsymbol{C}^{-1}$, and $\boldsymbol{E}$.

It is possible to describe strain correctly by a measure which is not a tensor, but there can hardly be any advantage, and attempts of this kind have usually led to confusion if not disaster.

It is obvious that *a description of strain in terms of a strain measure at $\boldsymbol{x}$ is not in general equivalent to one in terms of a strain measure at $\boldsymbol{X}$.* In certain special situations the two descriptions may become equivalent, e.g. for phenomena in certain isotropic materials (the much abused term *isotropic* will be explained in Sect. 293 γ), or when the two sets of principal directions happen to coincide.

33. Certain particular strain measures. Many students prefer to use a tensor whose proper numbers reduce for small extensions to the principal extensions δ_a themselves. By (31.4), $\boldsymbol{E}$ and $\boldsymbol{e}$ are such tensors, but so also is

$$\frac{1}{2K}\left(1-\boldsymbol{c}^K\right), \qquad K \neq 0, \tag{33.1}$$

where we employ the K-th power of a tensor as defined by (App. 41.2). Thus this requirement does not define a unique measure. The measures $\boldsymbol{C}^{\frac{1}{2}}$ and $\boldsymbol{c}^{-\frac{1}{2}}$ are attractive in that their proper numbers are *exactly* the principal stretches, but fractional powers are difficult to use in practice, since the components of such a tensor referred to co-ordinates other than principal are in general complicated infinite series in the displacement gradients. For example

$$\left.\begin{aligned} \boldsymbol{c}^{-\frac{1}{2}} &= (1-2\boldsymbol{e})^{-\frac{1}{2}} = \sum_{n=0}^{\infty} \frac{(2n-1)!!}{n!}\, \boldsymbol{e}^n, \\ &= \sum_{n=0}^{\infty} \frac{(2n-1)!!}{2^n n!}\, (1-\boldsymbol{c})^n. \end{aligned}\right\} \tag{33.2}$$

By the theorem at the beginning of Sect. App. 42, the series converges if $\lambda_3 > 2^{-\frac{1}{2}}$, diverges if $\lambda_1 < 2^{-\frac{1}{2}}$.

Similar objections apply to transcendental measures such as

$$\left.\begin{aligned} \boldsymbol{H} &\equiv \tfrac{1}{2}\log \boldsymbol{C} = \tfrac{1}{2}\log(1+2\boldsymbol{E}), \\ \boldsymbol{h} &\equiv -\tfrac{1}{2}\log \boldsymbol{c} = -\tfrac{1}{2}\log(1-2\boldsymbol{e}), \end{aligned}\right\} \tag{33.3}$$

where the definitions, which may be achieved by analytic continuation of the series for the logarithms, have a sense[1] for all deformations since $\boldsymbol{c}$ and $\boldsymbol{C}$ are positive definite (Sect. 28). Since $h_a = H_a = \log \lambda_a$, we have

$$\mathrm{I}_{\boldsymbol{h}} = \mathrm{I}_{\boldsymbol{H}} = \log \sqrt{\mathrm{III}_{\boldsymbol{C}}} = \log \frac{dv}{dV}, \tag{33.4}$$

and accordingly the deviators of $\boldsymbol{H}$ and $\boldsymbol{h}$ are zero for a uniform dilation (Sect. 43) and thus may serve as distorsion tensors measuring change of shape apart from change of volume[2]. While logarithmic measures of strain are a favorite in

---

[1] Rather than using the concepts of Sect. App. 42, it is preferable, in analogy to (App. 41.2), to define $\boldsymbol{H}$ as the unique symmetric tensor whose principal directions are the principal axes of strain at $\boldsymbol{X}$ and whose proper numbers are $\log \lambda_a$.

[2] REINER [1948, *23*, § 7]. Cf. also the next footnote.

one-dimensional or semi-qualitative treatments [1], they have never been successfully applied in general.

Such simplicity for certain problems as may result from a particular strain measure is bought at the cost of complexity for other problems. In a Euclidean space, distances are measured by a quadratic form, and attempt to elude this fact is unlikely to succeed. Most modern work uses the metric tensors $\boldsymbol{c}$ and $\boldsymbol{C}$ or their reciprocals $\boldsymbol{c}^{-1}$ and $\boldsymbol{C}^{-1}$.

33A. Appendix. History of the theory of strain. *1. Early work.* The idea of strain, that is, of *relative* rather than absolute change of configuration, seems to begin with the definition of what we here call extension, viz

$$\text{extension} \equiv \frac{\text{change in length}}{\text{original length}},$$

which was introduced by Beeckman (1630) and James Bernoulli (1705); the latter understood that a law relating stress to extension characterizes a material, while formulae relating force to change in length, favored by empiricists such as Hooke (1675), can refer only to a particular specimen. A detailed history has been written by Truesdell [1960, *4*, §§ 3, 8, 13, 20]. The history of one-dimensional strain measures given by Mehmke [1897, *5*] is misleadingly incomplete; the account of the early work by Todhunter and Pearson [1886, *4*, Chap. I] is inaccurate as well.

2. Infinitesimal strain. The theory of small strain in a three-dimensional medium was created by Euler (1750–1770), who dealt with time rates. For displacement, the theory of small strain was fully elaborated by Cauchy (1822–1840), who obtained it by specialization from his general theory of finite strain. References to this work will be given in the course of our exposition of it in Part e of this subchapter. While Cauchy was certainly influenced by the researches of Fresnel, it is Cauchy's special achievement to have disengaged and developed individually the concepts of strain, stress, and elasticity, which in Fresnel's writing appear to be for the most part taken for granted. The work on special problems concerning deformable solids from the time of Galileo (1638) through the researches of Hooke (1675), Leibniz (1685), James Bernoulli (1691–1705), Parent (1713), Euler (1720–1776), Daniel Bernoulli (1734–1766), and Coulomb (1773–1784), even including Navier's derivation of the general equations of linear elasticity (1821), suffers from lack of a definite and explicit concept of strain as distinct from displacement but comprising extension, dilation, bending, shearing, and all other special deformations. A history of this work to 1788 is given by Truesdell [1960, *4*].

3. Finite strain. References to all important work on three-dimensional finite strain have been given already in connection with the theory itself. Nothing was done prior to Cauchy's time, and little has been added by the extensive subsequent literature. Cauchy himself acknowledged, if somewhat vaguely [1841, *1*, second sentence], a debt to earlier work in differential geometry. The most relevant part is provided by Euler's and Gauss's work on the line element on a surface and in particular the theory of applicable surfaces; however, these developments, the history of which has been written by Speiser [1955, *23*] [1956, *19*], by no means sufficed to make the theory of strain evident, especially since the now familiar formulae analogous to (26.3) for a curved surface embedded in Euclidean space, while discovered by Euler, were not published until much later [1862, *3*].

3a. Non-tensorial measures. Green in his first formulation [1839, *1*, p. 249] laid down a rectangular system at $\boldsymbol{X}$ and took the extensions and changes of mutual angle suffered by the co-ordinate lines as measures of strain. Kelvin remarked that the usual strain measures are unsymmetrical in this sense, that each normal component is a function of the extension in but a single direction, while each shear component, being essentially a change of angle, depends upon the extensions in a pair of directions; he constructed a symmetrical specifica-

[1] Cf. Ludwik [1909, *7*, Pt. 1, § 1], Hencky [1928, *4*, § 1] [1929, *2*] [1929, *3*, § 2], Weissenberg [1935, *10*, pp. 59–60], Biezeno and Grammel [1939, *2*, Chap. 1, ¶ 15], Reiner [1948, *23*]. None of this work is unequivocal as regards general finite strain, for which the definitions (33.3) were given by Murnaghan [1941, *3*, p. 127] and Richter [1948, *24*, § 2]. Later Richter [1949, *26*, § 3] worked out various special properties of $\boldsymbol{h}$ and $\boldsymbol{H}$. Noticing that the condition of vanishing in uniform dilation does not determine a unique strain measure, Richter proposed a set of axioms, including a superposition principle for coaxial stretches, and showed that there are at $\boldsymbol{x}$ and $\boldsymbol{X}$ unique distorsion tensors [1949, *26*, § 4] which satisfy them. This corrects an earlier attempt by Moufang [1947, *10*]; cf. [1948, *25*]. Richter's distorsion tensors are complicated algebraic functions of $\boldsymbol{e}$ and $\boldsymbol{E}$, respectively.

tion of strain in terms of the extensions of the edges of a tetrahedron [1877, *5*, Chap. VIII, Ex. 1] [1902, *5*]. DORN and LATTER [1948, *8*] used the logarithms of the principal stretches and the cosines of the angles. SWAINGER [1947, *15* and *16*] [1948, *29* and *30*] [1949, *30*] [1950, *29* and *30*] [1954, *23*] claims to define a new linear strain measure, embellished with polemics against the classical strain measures and the authors who use them; his views have been criticized by GORDON [1950, *8*], TRUESDELL [1952, *21*, § 15[1] and § 17[1]] [1955, *29*], and RICHTER [1955, *20*], among others, and in his numerous publications we have been unable to find anywhere a prescription for calculating his strain measure from a given displacement (but see the table below). The authors favoring logarithmic measures such as $(25.8)_1$ and (33.3) usually call them "natural"; this term is applied by YOSHIMURA [1953, *36*] to $-\log\tan\frac{1}{2}(\frac{1}{2}\pi-\gamma)$, where γ is the shear.

3b. Table of tensor measures.

Author	Definition	Special Property
SIGNORINI [1930, *6*, § 9]. (Most of SIGNORINI's work employs $\boldsymbol{E}$ or $\boldsymbol{e}$)	$\frac{1}{2}(\boldsymbol{c}^{-1}-\boldsymbol{1})$	The case $K=-1$ in (33.1)
BIOT [1939, *3*, § 1] [1939, *4*, p. 118] [1939, *5*, p. 108] [1940, *3*, § 1]	$\boldsymbol{C}^{\frac{1}{2}}-\boldsymbol{1}$	The unique strain measure at $\boldsymbol{X}$ whose proper numbers are δ_a
MURNAGHAN [1941, *3*, pp. 127—128] (in reference to BIOT), RICHTER [1948, *24*, § 2]	$\boldsymbol{c}^{-\frac{1}{2}}$	The unique strain measure at $\boldsymbol{x}$ whose proper numbers are λ_a
SWAINGER according to HERSHEY [1952, *11*] and REINER [1954, *20*, § 4]	$\boldsymbol{1}-\boldsymbol{c}^{-\frac{1}{2}}$	The unique strain measure at $\boldsymbol{x}$ whose proper numbers are $-\delta_a$
MOONEY [1948, *15*, pp. 435—436]	$(\mathrm{III}_{\boldsymbol{c}})^{\frac{1}{6}}\boldsymbol{c}^{-\frac{1}{2}}$	$\mathrm{III}=1$
OLDROYD [1950, *23*, § 6]	$(\mathrm{III}_{\boldsymbol{c}})^{-\frac{1}{2}}\boldsymbol{c}$	$dV=\mathrm{III}\,dv$

See also footnote 1, p. 256 and the formulae for logarithmic measures in Sect. 33.

REINER [1948, *23*, §§ 3, 5, 7] [1954, *20*] and HERSHEY [1952, *11*] present reviews of various definitions of strain and give comparative numerical values in simple cases. HANIN and REINER [1956, *11*] work out the forms of the coefficients in formulae expressing one of the above measures as an isotropic function of another (cf. Sect. 33).

3c. Other discussions of strain. KILCHEVSKI [1938, *5*, §§ 3—7] regards the components of the metric tensors at $\boldsymbol{x}$ and at $\boldsymbol{X}$ as generalized anholonomic components of the same tensor. Setting $B^{KM}_{km}\equiv\frac{1}{3}g^{KM}g_{km}$, we have $g_{km}=B^{KM}_{km}g_{KM}$, etc. He introduces a very general measure of deformation which includes as special cases both the deformation tensor $\boldsymbol{c}$ and the stretching tensor $\boldsymbol{d}$ (Sect. 82). HENCKY [1949, *13*] on the basis of a discussion of a special kind of finite deformation of a finite element criticizes all strain measures which depend only on the deformation gradients and obtains formulae of the "projective" type $\Delta x^k=A^k_K\Delta X^K$, in which A^k_K depends on the shape of the original element and on a certain arbitrary vector; in passing to the limit of infinitesimal volume he gets a formula different from (20.3), but LODE [1954, *13*, § 3.3] points out an error in his work. LODE himself prefers to use $x^k_{;K}$ and rediscovers such formulae as (20.8).

A correct theory of affine type is mentioned at the end of Sect. 20; cf. also Sect. 61.

34. Conditions of compatibility. Given the deformation (16.2), it is a straightforward matter to calculate the six components C_{KM} by $(26.2)_2$ or to calculate any other measure of strain. Conversely, we may ask how to calculate the deformation when the components C_{KM} are given as functions of $\boldsymbol{X}$. To solve this problem we have to integrate a system of six partial differential equations in three unknowns:

$$g_{km}\,x^k_{;K}\,x^m_{;M}=f_{KM}(X^1,X^2,X^3)\,. \tag{34.1}$$

Such a system, being overdetermined, in general admits no solution unless the assigned functions f_{KM} satisfy a condition of integrability.

In the present case, by $(26.2)_1$ C_{KM} is a metric tensor of Euclidean three-dimensional space. According to a well known theorem asserted by RIEMANN, a symmetric tensor a_{km} is a metric tensor for Euclidean space if and only if it is a non-singular positive definite tensor such that the Riemann-Christoffel tensor $R^{(a)}_{kmps}$ formed from it vanish identically[1]. Various forms of this tensor in the co-ordinates y^k are

$$\left.\begin{aligned} R^{(a)}_{kmpq} &= a_{kr}\left[\frac{\partial}{\partial y^p}\left\{{r \atop m\,q}\right\} - \frac{\partial}{\partial y^q}\left\{{r \atop m\,p}\right\} + \left\{{s \atop m\,q}\right\}\left\{{r \atop s\,p}\right\} - \left\{{s \atop m\,p}\right\}\left\{{r \atop s\,q}\right\}\right],\\ &= \frac{1}{2}\left(\frac{\partial^2 a_{kq}}{\partial y^m \partial y^p} - \frac{\partial^2 a_{kp}}{\partial y^m \partial y^q} + \frac{\partial^2 a_{mp}}{\partial y^k \partial y^q} - \frac{\partial^2 a_{mq}}{\partial y^k \partial y^p}\right) + \\ &\quad + a_{rs}\left(\left\{{r \atop k\,q}\right\}\left\{{s \atop m\,p}\right\} - \left\{{r \atop k\,p}\right\}\left\{{s \atop m\,q}\right\}\right),\\ &= -\frac{1}{2}\frac{\partial^2 a_{rs}}{\partial y^t \partial y^u}\delta^{rt}_{km}\delta^{su}_{pq} + a_{rs}\left(\left\{{r \atop k\,q}\right\}\left\{{s \atop m\,p}\right\} - \left\{{r \atop k\,p}\right\}\left\{{s \atop m\,q}\right\}\right),\\ &= \frac{\partial[k, m q]}{\partial y^p} - \frac{\partial[k m, p]}{\partial y^q} + \left\{{r \atop k\,q}\right\}[r, m p] - \left\{{r \atop k\,p}\right\}[r, m q], \end{aligned}\right\} \tag{34.2}$$

where

$$[p, k m] \equiv \frac{1}{2}\left(\frac{\partial a_{kp}}{\partial y^m} + \frac{\partial a_{pm}}{\partial y^k} - \frac{\partial a_{km}}{\partial y^p}\right), \qquad \left\{{p \atop k\,m}\right\} \equiv \overset{-1}{a}{}^{pq}[q, k m]. \tag{34.3}$$

Hence

$$R^{(c)}_{kmpq} = 0, \qquad R^{(C)}_{KMPQ} = 0, \tag{34.4}$$

where in the former equations $\boldsymbol{y} = \boldsymbol{x}$, in the latter, $\boldsymbol{y} = \boldsymbol{X}$. There are six algebraically independent non-identically vanishing components of R_{kmpq} in a space of three dimensions. Either of the two sets of Eqs. (34.4) is referred to as the *conditions of compatibility*.

Many equivalent forms have been found, of which we record only that given by GRAIFF[2]:

$$\left.\begin{aligned} &E_{KM,PQ} + E_{PQ,KM} - E_{KP,MQ} - E_{MQ,KP} + \\ &\quad + \overset{-1}{C}{}^{RS}[(E_{MR,K} + E_{KR,M} - E_{KM,R})(E_{PS,Q} + E_{QS,P} - E_{PQ,S}) - \\ &\quad - (E_{MR,Q} + E_{QR,M} - E_{MQ,R})(E_{PS,K} + E_{KS,P} - E_{PK,S})] = 0, \end{aligned}\right\} \tag{34.4a}$$

where the covariant differentiation is based upon g_{JL}.

---

[1] This theorem was asserted rather vaguely in RIEMANN's second Habilitationsschrift (1854) [1868, *13*, §§ II 2, II 4, III 1]; in his Paris prize essay (1861) [1876, *5*, Pars secunda], RIEMANN proved necessity and asserted that sufficiency is not difficult to prove. Priority in publication belongs to CHRISTOFFEL [1869, *1*, §§ 1–6] [1869, *2*] and LIPSCHITZ [1869, *3*, § 7] [1870, *2*, Part I, § 8]. For a modern proof, cf. e.g. VEBLEN [1927, *8*, § V 4]. Although geometry and mechanics were closer together in the last century than today, it was long before students of elasticity theory (even including some great geometers) recognized the connection between the conditions of compatibility and the equivalence problem of Riemannian geometry.

[2] [1958, *3*, Eq. (10)]. References for the linearized case will be given in Sect. 57. The general case has been treated by MANVILLE [1904, *5*], MARCOLONGO [1905, *4*], RIQUIER [1905, *5*], CAFIERO [1906, *1*, Chap. I, § 3], CRUDELI [1911, *3*], BURGATTI [1914, *1*, § 3], SIGNORINI [1930, *4*, § 2] [1942, *11*, p. 60] [1943, *6*, Chap. 1, ¶ 20], KILCHEVSKI [1938, *5*, § 18] (including a four-dimensional generalization), TONOLO [1943, *8*, § IV], SETH [1944, *11*], ZELMANOV [1948, *39*, Eqs. (6), (7)], NOVOZHILOV [1948, *18*, § 39], OLDROYD [1950, *23*, § 5], GREEN and ZERNA [1950, *10*, § 4], ZERNA [1950, *36*, § 4], SEUGLING [1950, *26*], LODGE [1951, *15*], GALIMOV [1951, *9*], PLATRIER [1953, *21*, *22*], KOPPE [1956, *14*]. PADOVA [1889, *7*] was the first to recognize the geometrical nature of the problem, for the linearized case (cf. also the works of CAFIERO and CRUDELI just cited).

In some problems, it is useful to measure the strain with regard to different configurations for different particles[1]. The duals of Eqs. (34.4a) are satisfied, but (34.4a) themselves need not be, and the tensor $R^{(C)}_{KMPQ}$ may be taken as a measure of the *incompatibility* of the reference configurations for different particles.

The conditions (34.4) are not independent since $\boldsymbol{R}^{(a)}$, in virtue of its definition (34.2), satisfies the identities of BIANCHI[2],

$$R^{(a)h}{}_{mpk,q} + R^{(a)h}{}_{mkq,p} + R^{(a)h}{}_{mqp,k} = 0. \tag{34.5}$$

The implications of these identities in respect to solution of (34.4) are not known; a simpler analogous problem which has been solved will be mentioned later on in this section.

If $\boldsymbol{x}_1(\boldsymbol{X})$ and $\boldsymbol{x}_2(\boldsymbol{X})$ are any pair of solutions of (34.4), the difference $\boldsymbol{p}_1 - \boldsymbol{p}_2$ of the corresponding position vectors is a rigid displacement.

In a problem formulated entirely in terms of the deformation, the conditions of compatibility need not be regarded, since they are satisfied automatically in virtue of the definitions of $\boldsymbol{c}$ and $\boldsymbol{C}$. It is only in a problem where the components of $\boldsymbol{c}$ or of $\boldsymbol{C}$ are themselves taken as basic unknowns that the conditions of compatibility become additional equations to be solved. For such a problem any additional conditions imposed on $\boldsymbol{c}$ or on $\boldsymbol{C}$ must be compatible with (34.4).

Consider the quantities R^*_{kpmn} defined by

$$\left.\begin{aligned} R^*_{kpmn} &\equiv \tfrac{1}{2}\left(a_{kn,mp} + a_{pm,kn} - a_{km,np} - a_{pn,km}\right) + \\ &\quad + \overset{-1}{a}{}^{rs}\left(A_{pmr}A_{kns} - A_{pnr}A_{kms}\right), \\ A_{kpm} &\equiv \tfrac{1}{2}\left(a_{km,p} + a_{pm,k} - a_{kp,m}\right), \end{aligned}\right\} \tag{34.6}$$

where the comma denotes the covariant derivative based upon the metric tensor g_{km}. From this definition, the quantities R^*_{kpmn} form components of a tensor field; in a Euclidean space, we may choose a rectangular Cartesian co-ordinate system and by inspection of $(34.2)_2$ conclude that $\boldsymbol{R}^* = \boldsymbol{R}$ in this co-ordinate system and hence in all co-ordinate systems. Thus in writing the conditions of compatibility $\boldsymbol{R}^{(c)} = 0$, we may if we please replace partial derivatives by covariant derivatives[3].

We may pose an analogous question in regard to the elongation tensor $\tilde{\boldsymbol{E}}$. In this case, the problem is a linear one: To find necessary and sufficient conditions that given a symmetric tensor $\tilde{\boldsymbol{E}}$ there exist a vector $\boldsymbol{u}$ such that

$$\tilde{E}_{KM} = u_{(K;M)}. \tag{34.7}$$

---

[1] The intended application is to bodies initially stressed though free from applied loads. The stress may then be thought of as arising in response to deformation from natural states which are different for different particles. There is no natural state for the body as a whole. This idea was first put forward, it seems, by ECKART [1948, *10*, §§ 1–2], though his discussion is obscured by remarks about cutting a body apart and by failure to take account of the fact that in a deformation in Euclidean space the conditions of compatibility need not be satisfied at points where the axiom of continuity (Sect. 16) is not satisfied. Cf. the criticism by TRUESDELL [1952, *21*, § 18[4]]. While the idea that yield may be represented by a deformation from a hypothetical non-Euclidean space was expressed by KONDO [1949, *15*] [1950, *14–16*] [1954, *10*] and by BILBY, BULLOUGH and SMITH [1955, *2*, §§ 2–3], the first clear statement and formalism is that of KRÖNER and SEEGER [1959, *8*, §§ 1–2].

[2] E.g., EISENHART [1926, *1*, § 26]. For generalizations, cf. SCHOUTEN [1954, *21*, Chap. III, § 5].

[3] ERICKSEN [1955, *6*].

The desired condition is easily shown to have the following equivalent forms[1]:

$$\left.\begin{aligned} \delta^{RS}_{KM}\,\delta^{TU}_{PQ}\,\widetilde{E}_{RT;\,SU}=0, \qquad \varepsilon^{KMP}\,\varepsilon^{RST}\,\widetilde{E}_{MS;\,PT}=0, \\ \widetilde{E}_{KN;\,MP}+\widetilde{E}_{PM;\,KN}-\widetilde{E}_{KM;\,NP}-\widetilde{E}_{PN;\,KM}=0, \end{aligned}\right\} \tag{34.8}$$

the second of which is valid only in three-dimensional space, while the others remain true in n-space.

There are six linearly independent, not identically vanishing conditions (34.8). While they resemble in some ways the conditions (34.4), in principle they are of a different kind, for they are conditions of integrability for the differential system (34.7) when (34.4) holds, i.e., *when the space is assumed to be Euclidean.* Conditions of this kind are discussed from a more general standpoint in Sect. 84.

If we set

$$J^{(\widetilde{E})}_{KPMN}\equiv\widetilde{E}_{KN;\,MP}+\widetilde{E}_{PM;\,KN}-\widetilde{E}_{KM;\,NP}-\widetilde{E}_{PN;\,KM}\,, \tag{34.9}$$

then we have identically

$$J^{(\widetilde{E})}_{KPMN;\,R}+J^{(\widetilde{E})}_{KPNR;\,M}+J^{(\widetilde{E})}_{KPRM;\,N}=0, \tag{34.10}$$

a relation formally analogous to the Bianchi identities (34.5) but presupposing that the space is flat. The conditions of compatibility (34.8) may be written in the form $J^{(\widetilde{E})}_{KPMN}=0$. WASHIZU[2] has shown that in virtue of the identities (34.10) and GREEN's transformation, the conditions (34.8) may be divided into two sets of three, viz., $J^{(\widetilde{E})}_{1213}=J^{(\widetilde{E})}_{2321}=J^{(\widetilde{E})}_{3132}=0$ and $J^{(\widetilde{E})}_{1212}=J^{(\widetilde{E})}_{2323}=J^{(\widetilde{E})}_{3131}=0$, such that if both sets are satisfied upon the boundary of a region, then the vanishing of either set in the interior implies the vanishing of both sets. Thus the degree of redundancy that the identities (34.10) imply in the conditions of compatibility (34.8) is rendered definite.

If $\widetilde{\boldsymbol{E}}$ satisfies (34.8), and if ${}^0u_K$ is any solution of $\widetilde{E}_{KM}=u_{(K;\,M)}$, then the most general solution is

$$u_K={}^0u_K+P^M\widetilde{R}_{MK}+B_K\,, \tag{34.11}$$

where $\boldsymbol{P}$ is any vector satisfying $P^M_{;Q}=\delta^M_Q$, $\widetilde{\boldsymbol{R}}$ is any skew-symmetric tensor satisfying $\widetilde{R}_{KM;\,Q}=0$, and $\boldsymbol{B}$ is any vector satisfying $B_{K;\,M}=0$. That is, the displacement corresponding to a given elongation tensor is indeterminate to within a vector having the same form as an infinitesimal rigid displacement.

c) Rotation.

35. Fundamental theorem. The results of CAUCHY presented in Sects. 25 to 27 show that the stretch of every element is determined by either of the strain ellipsoids, providing the orientations of its axes in the common frame be specified. Both the stretch and the change in direction of every element are known if either ellipsoid and both sets of principal axes of strain are specified. Hence follows the ***fundamental theorem***: *The deformation at any point may be regarded as resulting from a translation, a rigid rotation of the principal axes of strain, and stretches along these axes.* The translation, rotation, and stretch may be applied in any order, but their tensorial measures are not independent of this order.

This theorem, while obvious from CAUCHY's work, was not stated by him[3]; not only did he wait fourteen years after completing the theory of strain in all

[1] References to proofs are given in Sect. 57.

[2] [1958, *12*].

[3] It is given rather vaguely by KELVIN and TAIT [1867, *3*, § 182], and apparently LOVE [1892, *8*, § 10] was the first to assert it explicitly.

detail before giving a theory of rotation (Sect. 36 below), but also his measure of rotation is not one of those that come at once to mind on stating the fundamental theorem. The theory of finite rotation has always presented singular difficulty, although the essential idea is simple. Perhaps the trouble lies in its being a truly Euclidean concept: Often those least desirous of generality are also least successful in grasping the special peculiarities afforded by a special case.

The angle ϑ through which the element $d\boldsymbol{X}$ is rotated by the deformation follows at once from (20.3):

$$\left.\begin{aligned} \cos\vartheta &= \frac{g_{KM}\, dX^K g^M_k\, dx^k}{dX\, dx}, \\ &= \frac{g_{KM} g^M_k\, x^k_{;P}\, dX^K dX^P}{\sqrt{g_{RS}\, dX^R dX^S}\sqrt{C_{TU}\, dX^T dX^U}}, \end{aligned}\right\} \tag{35.1}$$

with the convention $0 \leqq \vartheta \leqq \pi$. In particular, we may choose $d\boldsymbol{X}$ as a unit vector $\boldsymbol{N}$, obtaining

$$\left.\begin{aligned} \cos\vartheta &= \frac{1}{\lambda_{(\boldsymbol{N})}}\, g_{KM} g^M_k\, x^k_{;P} N^K N^P, \\ &= \frac{1}{\lambda_{(\boldsymbol{n})}}\, g_{km} g^m_K X^K_{;p} n^k n^p. \end{aligned}\right\} \tag{35.2}$$

Stretch, extension, rotation, and elongation (Sect. 25) are related thus[1]:

$$\varepsilon = \lambda \cos\vartheta - 1 = \delta \cos\vartheta - 2\sin^2 \tfrac{1}{2}\vartheta, \tag{35.3}$$

this being but another form of (35.2).

If we let $\boldsymbol{n}_\mathfrak{a}$ be a proper vector of $\boldsymbol{c}$, we get for the angle $\vartheta_\mathfrak{a}$ through which the deformation has turned it

$$\cos\vartheta_\mathfrak{a} = c_{km} g^m_K X^K_{;p} n^k_\mathfrak{a} n^p_\mathfrak{a}. \tag{35.4}$$

These three angles specify the rotation, since the $\mathfrak{a}$-th axis of strain at $\boldsymbol{x}$ lies on a cone of vertex angle $\vartheta_\mathfrak{a}$ about the $\mathfrak{a}$-th axis of strain at $\boldsymbol{X}$ shifted to $\boldsymbol{x}$.

If the three angles $\vartheta_\mathfrak{a}$ vanish, or, what is the same thing, if any two of them vanish, the rotation itself is said to *vanish*, and the deformation is called a *pure strain*.

It is important to realize that while by definition the principal axes of strain are not rotated in a pure strain, it by no means follows that no linear elements suffer rotation. This is grasped most easily through an example. Suppose two tri-axial ellipsoids have axes mutually parallel but of lengths in pairwise different ratios. Then a linear transformation of one into the other changes the orientation of every radius vector not lying along one of the axes.

An algebraic criterion for invariant directions will be given in Sect. 38.

36. The mean rotation tensor of CAUCHY and NOVOZHILOV. CAUCHY[2] took as a measure of rotation the mean values of the angles through which all elements in each of three perpendicular planes are turned. Refer all quantities to the common frame. Let $\boldsymbol{N}_X$ be a unit vector perpendicular to the X-axis at $\boldsymbol{Z}$, so that in terms of a real angle Φ we may write

$$\boldsymbol{N}_X = \boldsymbol{j}\cos\Phi + \boldsymbol{k}\sin\Phi. \tag{36.1}$$

[1] BOUSSINESQ [1877, *1*, § 3] described this formula and gave further discussion of the rotation by means of the apparatus of Sect. 23.

[2] [1841, *1*, Th. IV].

CAUCHY's departure from the ideas of Sect. 35 was to consider not ϑ as given by (35.2) but rather the angle ϑ_X between $\boldsymbol{N}_X$ and the projection of $\boldsymbol{n}_X$ upon the Y-Z plane. The slope of this projection is given by

$$\tan\varphi = \frac{z,_Y\cos\Phi + z,_Z\sin\Phi}{y,_Y\cos\Phi + y,_Z\sin\Phi}, \tag{36.2}$$

and

$$\vartheta_X \equiv \varphi - \Phi. \tag{36.3}$$

Therefore (36.2) asserts that

$$\frac{\tan\vartheta_X + \tan\Phi}{1-\tan\vartheta_X\tan\Phi} = \frac{z,_Y\cos\Phi + z,_Z\sin\Phi}{y,_Y\cos\Phi + y,_Z\sin\Phi}, \tag{36.4}$$

whence follows

$$\left.\begin{aligned}\tan\vartheta_X &= \frac{z,_Y\cos^2\Phi + (z,_Z - y,_Y)\sin\Phi\cos\Phi - y,_Z\sin^2\Phi}{y,_Y\cos^2\Phi + (y,_Z + z,_Y)\sin\Phi\cos\Phi - z,_Z\sin^2\Phi},\\ &= \frac{-\tilde{R}_{YZ} + \tilde{E}_{YZ}\cos 2\Phi + \frac{1}{2}(\tilde{E}_{ZZ} - \tilde{E}_{YY})\sin 2\Phi}{1 + \tilde{E}_{YY}\cos^2\Phi + \tilde{E}_{ZZ}\sin^2\Phi + \tilde{E}_{ZY}\sin 2\Phi},\end{aligned}\right\} \tag{36.5}$$

where $\tilde{\boldsymbol{R}}$ and $\tilde{\boldsymbol{E}}$ are defined by (31.6). For the determination of ϑ_X, the x-component of the deformation is not used.

The expression (36.5) is periodic of period π in Φ, reflecting the fact that oppositely directed elements suffer equal rotations. From (36.5), ϑ_X as an angle satisfying $0 \leqq \vartheta_X \leqq \pi$ is uniquely determined *with an important exception*: $\vartheta_X = 0$ and $\vartheta_X = \pi$ are indistinguishable.

CAUCHY's measure of rotation about the X-axis is

$$\chi_X \equiv \frac{1}{2\pi}\int_0^{2\pi}\vartheta_X(\Phi)\,d\Phi = \frac{1}{\pi}\int_0^{\pi}\vartheta_X(\Phi)\,d\Phi. \tag{36.6}$$

However, CAUCHY failed to notice that because of its equivocal treatment of $\vartheta_X = 0$ and $\vartheta_X = \pi$ his formula (36.5) does not always determine ϑ_X uniquely and hence is not sufficient to calculate χ_X. To remedy this difficulty, we may easily replace (36.5) by a formula for $\cos\vartheta_X$. However, despite the elegance of CAUCHY's concept, his measure of rotation has never been used. First, the angles χ_X, χ_Y, χ_Z do not form a vector field[1]. Second, the values of the mean rotations about three perpendicular axes do not suffice to determine the rotations of the individual elements at $\boldsymbol{Z}$.

NOVOZHILOV[2] has most happily modified CAUCHY's definition by putting

$$\tan\tau_X \equiv \frac{1}{2\pi}\int_0^{2\pi}\tan\vartheta_X(\Phi)\,d\Phi \tag{36.7}$$

in place of (36.6). The integration can now be performed explicitly; from $(36.5)_2$ it is easy to see that in fact

$$\left.\begin{aligned}\tan\tau_X &= -\frac{\tilde{R}_{YZ}}{2\pi}\int_0^{2\pi}\frac{d\Phi}{1+\tilde{E}_{YY}\cos^2\Phi + \tilde{E}_{ZZ}\sin^2\Phi + \tilde{E}_{ZY}\sin 2\Phi},\\ &= \frac{-\tilde{R}_{YZ}}{\sqrt{(1+\tilde{E}_{YY})(1+\tilde{E}_{ZZ}) - \tilde{E}_{ZY}^2}}.\end{aligned}\right\} \tag{36.8}$$

[1] Their law of transformation was calculated by CAUCHY [1841, *1*, § I, Eq. (37)].
[2] [1948, *18*, § 7].

This elegant formula of NOVOZHILOV shows that $\tilde{R}_{KM}$ is a measure of the *mean rotation* suffered by elements in the K-M plane[1].

While we have used a special co-ordinate system, our results are easily put in invariant form. Let $X^1 = \text{const}$ be a surface at $\boldsymbol{X}$, let X^2 and X^3 be co-ordinates on that surface, and consider only transformations of these surface co-ordinates. Then, with dummy indices restricted to the range 2, 3, (36.5) reads

$$\tan \vartheta_{X^1} = \frac{-\tilde{R}_{23} + e_{KM} N^K \tilde{E}^M_P N^P}{1 + \tilde{E}_{QS} N^Q N^S}, \tag{36.9}$$

while (36.8) reads

$$\tan \tau_{X^1} = -\frac{\tilde{R}_{23}}{\sqrt{1 + \mathrm{I}_{{}_1\tilde{\boldsymbol{E}}} + \mathrm{II}_{{}_1\tilde{\boldsymbol{E}}}}}, \tag{36.10}$$

where ${}_1\tilde{\boldsymbol{E}}$ is the two-dimensional tensor obtained from $\tilde{\boldsymbol{E}}$ by suppressing all components having the index 1. The formula (36.10) justifies our calling $\tilde{\boldsymbol{R}}$ the *mean rotation tensor*.

Since $\tilde{\boldsymbol{R}}$ is a tensor, if it vanishes in some one co-ordinate system, it vanishes in all. In view of (36.10), therefore, if the mean rotations of the elements in three perpendicular planes at a point be 0 or π radians, the mean rotation of the elements in every plane at that point is 0 or π radians.

As is indicated by the two possibilities in the foregoing statement, $\tilde{\boldsymbol{R}}$, *while indeed a measure of mean rotation, is not a measure of rotation alone.* The easiest way to see this is to consider a rigid rotation through π radians: $x = -X$, $y = -Y$, $z = Z$. In such a rotation, $\tilde{\boldsymbol{R}} = 0$. Since $\tilde{\boldsymbol{R}} = 0$ if and only if there exists a *displacement potential*[2] U:

$$u_K = U_{;K}, \tag{36.11}$$

we may call deformations for which $\tilde{\boldsymbol{R}} = 0$ *potential deformations.* These will be characterized in Sect. 38.

If we replace NOVOZHILOV's measure of rotation by CAUCHY's or by the mean of $\cos \vartheta_X$, we get a measure of rotation only, but there is no simple expression for these measures in terms of $\tilde{\boldsymbol{R}}$ and $\tilde{\boldsymbol{E}}$.

Since $\tilde{\boldsymbol{R}}$ is a skew-symmetric tensor of second order, we may always replace it by an axial vector, which we choose to define as follows:

$$\tilde{\boldsymbol{R}} \equiv \tfrac{1}{2} \operatorname{curl} \boldsymbol{u}. \tag{36.12}$$

The context will make clear whether the vector or the tensor is intended by the symbol $\tilde{\boldsymbol{R}}$.

37. Algebraic proof of the fundamental theorem. The fundamental theorem of Sect. 35 is but an interpretation of the polar decomposition theorem in Sect. App. 43. Historically, however, the former was much the earlier and in fact gave rise to the latter. It is illuminating to give a different proof of the fundamental theorem,

---

[1] As was observed by KELVIN and TAIT [1867, *3*, § 190], KELVIN's transformation (Sect. App. 28) yields an obvious but not very illuminating connection between $\tilde{\boldsymbol{R}}$ and the mean rotation around a closed circuit:

$$\oint_{\mathscr{C}} u_K \, dX^K = \int_{\mathscr{S}} \tilde{R}_{KM} \, dA^{KM}.$$

The interpretation of $\tilde{\boldsymbol{R}}$ given by M. BRILLOUIN [1891, *1*, § 1] is faulty, being based on an error in calculation.

[2] LOVE [1892, *8*, § 12].

based indeed on the same ideas as the proof of FINGER, the COSSERATS, and AUTONNE given in Sect. App. 43, but using the explicit formulae[1] of Sect. 28.

Let $\boldsymbol{N}_\mathfrak{a}$ be an orthogonal triad at $\boldsymbol{X}$, $\boldsymbol{n}_\mathfrak{a}$ an orthogonal triad at $\boldsymbol{x}$. If we shift the $\boldsymbol{N}_\mathfrak{a}$ to $\boldsymbol{x}$, we can then exhibit a unique orthogonal tensor (cf. Sect. App. 43) $\boldsymbol{R}$ which rotates the shifted triad $\boldsymbol{N}_\mathfrak{a}$ into the triad $\boldsymbol{n}_\mathfrak{a}$:

$$n^k_\mathfrak{a} = R^k{}_m g^m_K N^K_\mathfrak{a} = R^k{}_K N^K_\mathfrak{a}, \quad N^K_\mathfrak{a} = g^K_m \overset{-1}{R}{}^m{}_k n^k_\mathfrak{a} = \overset{-1}{R}{}^K{}_k n^k_\mathfrak{a}, \tag{37.1}$$

where the intermediate components $R^k{}_K = R^k{}_m g^m_K$ represent the translation from $\boldsymbol{X}$ to $\boldsymbol{x}$ followed by the rotation $\boldsymbol{R}$. In terms of the reciprocal triads $\boldsymbol{N}^\mathfrak{a}$ and $\boldsymbol{n}^\mathfrak{a}$, we easily see that $\boldsymbol{R}$ and $\boldsymbol{R}^{-1}$ assume the forms

$$R^k{}_K = n^k_\mathfrak{a} N^\mathfrak{a}_K, \quad \overset{-1}{R}{}^K{}_k = N^K_\mathfrak{a} n^\mathfrak{a}_k, \tag{37.2}$$

where we employ the summation convention for diagonally repeated German indices. The duals of $(37.1)_1$ are also formulae for the $\boldsymbol{N}_\mathfrak{a}$ in terms of the $\boldsymbol{n}_\mathfrak{a}$. Comparison of them with $(37.1)_2$ shows that $\boldsymbol{R}^{-1}$ is the dual of $\boldsymbol{R}$, as is geometrically obvious, and yields the formulae

$$N^K_\mathfrak{a} = \overset{-1}{R}{}^K{}_M g^M_k n^k_\mathfrak{a}, \quad n^k_\mathfrak{a} = g^k_M R^M{}_K N^K_\mathfrak{a}. \tag{37.3}$$

For the rotation of the reciprocal triads we have

$$n^\mathfrak{a}_k = R_k{}^K N^\mathfrak{a}_K, \quad R_k{}^K \equiv g^{KM} g_{km} R^m{}_M = n^\mathfrak{a}_k N^K_\mathfrak{a}. \tag{37.4}$$

Thus far the triads $\boldsymbol{n}_\mathfrak{a}$ and $\boldsymbol{N}_\mathfrak{a}$ are arbitrary. Henceforth we restrict them to be directed along the principal axes of strain at $\boldsymbol{x}$ and at $\boldsymbol{X}$. There follows the *necessary and sufficient condition for a pure strain*:

$$\boldsymbol{R} = \boldsymbol{1}. \tag{37.5}$$

Co-ordinate forms of this condition are $R^k{}_m = \delta^k_m$, $R_{km} = g_{km}$, $R^k{}_K = g^k_K$, etc.

There is an important formula which with some justice may be called **FINGER's theorem**[2]:

$$\overset{-n}{c}{}_m{}^k = R^k{}_K \overset{n}{C}{}^K_M \overset{-1}{R}{}^M{}_m, \tag{37.6}$$

whereby the n-th powers of $\boldsymbol{c}^{-1}$ and $\boldsymbol{C}$ are related. To prove it, we need only observe that by (28.6) and the theorems on matrices given in Sect. App. 37, the tensors on the two sides of the equation are symmetric tensors having the same proper numbers, while by the definition of $\boldsymbol{R}$ the principal axes of the two tensors coincide. A more formal proof can be constructed from $(\text{App. } 37.12)_3$, (37.1), and (37.4); also, the result may be read off from (App. 41.2).

We turn now to a proof of the fundamental theorem. By the dual of (28.3) and (37.1) follows

$$\left.\begin{aligned} x^k_{;K} &= x^k_{;M}\,\delta^M_K = x^k_{;M} N^M_\mathfrak{a} N^\mathfrak{a}_K = \sqrt{C_\mathfrak{a}}\, n^k_\mathfrak{a} N^\mathfrak{a}_K, \\ &= R^k{}_M \sqrt{C_\mathfrak{a}}\, N^M_\mathfrak{a} N^\mathfrak{a}_K. \end{aligned}\right\} \tag{37.7}$$

---

[1] We follow TOUPIN [1953, *32*, pp. 595–597] [1956, *20*, § 4].

[2] FINGER [1892, *4*] studied the properties of $\boldsymbol{a}\cdot\boldsymbol{a}'$ and $\boldsymbol{a}'\cdot\boldsymbol{a}$, where $\boldsymbol{a}$ is any non-singular matrix. Most of his results have been stated in Sect. 22. If we identify $\boldsymbol{a}$ with $\|x^k_{;K}\|$, then one of his formulae is (37.6) with $n=1$. The theorem was given by RICHTER [1952, *17*, § 2] for $n=\frac{1}{2}$; by NOLL [1955, *18*, § 2b] for $n=1$; and by TOUPIN [1956, *20*, § 4] for the general case.

By (App. 37.12)$_3$, this is equivalent to

$$x^k_{;K} = R^k{}_M \overset{\frac{1}{2}}{C}{}^M_K = g^k_P R^P{}_M \overset{\frac{1}{2}}{C}{}^M_K = R^k{}_m g^m_M \overset{\frac{1}{2}}{C}{}^M_K . \tag{37.8}$$

If in this we substitute the dual of (37.6) with $n = -\frac{1}{2}$, we get

$$x^k_{;K} = \overset{-\frac{1}{2}}{c}{}^k_m R^m{}_K = \overset{-\frac{1}{2}}{c}{}^k_m R^m{}_k g^k_K = \overset{-\frac{1}{2}}{c}{}^k_m g^m_M R^M{}_K . \tag{37.9}$$

Now in a *pure strain*, by (37.5) we get from (37.8) and (37.9) the formulae

$$x^k_{;K} = g^k{}_M \overset{\frac{1}{2}}{C}{}^M_K = \overset{-\frac{1}{2}}{c}{}^k_m g^m_K ; \tag{37.10}$$

in a *translation*, which is a rigid pure strain, by (26.11) follows

$$x^k_{;K} = g^k_K ; \tag{37.11}$$

while in a more general *rigid displacement*, by (26.11) we get from (37.8) and (37.9) the formulae

$$x^k_{;K} = g^k_M R^M{}_K = R^k{}_m g^m_K . \tag{37.12}$$

In view of these special cases and (17.11), we see that the four formulae (37.8)$_2$, (37.8)$_3$, (37.9)$_2$, and (37.9)$_3$ constitute statements of the ***fundamental theorem***. For example, (37.8)$_2$ asserts that to obtain the set of nine $x^k_{;K}$ we may begin with stretches λ_a along the principal axes of strain at $\boldsymbol{X}$, then rotate those axes into the principal directions of strain at $\boldsymbol{x}$, then translate from $\boldsymbol{X}$ to $\boldsymbol{x}$. The remaining three formulae express similar decompositions in different orders. The dual formulae give corresponding expressions for $X^K_{;k}$.

From (37.8)$_1$ follows

$$R^k{}_K = x^k_{;M} \overset{-\frac{1}{2}}{C}{}^M_K , \tag{37.13}$$

which may be regarded as an expression for the rotation tensor as an infinite series in the displacement gradients. The formulae (37.2) are much simpler, but they cannot be used until both sets of principal axes of strain have been calculated.

By (19.3), equivalent to (37.8) is

$$u_{K;M} = R_K{}^Q \overset{\frac{1}{2}}{C}_{QM} - g_{KM} . \tag{37.14}$$

Hence we may connect the elongation tensor $\widetilde{\boldsymbol{E}}$ and the mean rotation tensor $\widetilde{\boldsymbol{R}}$ with the rotation tensor $\boldsymbol{R}$ and the deformation tensor $\boldsymbol{C}$:

$$\left.\begin{aligned} \widetilde{R}_{KM} = R_{[K}{}^P \overset{\frac{1}{2}}{C}_{M]P} , \qquad \widetilde{E}_{KM} = R_{(K}{}^P \overset{\frac{1}{2}}{C}_{M)P} - g_{KM} , \\ R_{KM} = (g_{KP} + \widetilde{E}_{KP} + \widetilde{R}_{KP}) \overset{-\frac{1}{2}}{C}{}^P_M . \end{aligned}\right\} \tag{37.15}$$

In a potential deformation (Sect. 36), by combining (37.15)$_1$ with (37.15)$_2$ we get

$$\widetilde{E}^K_M = R^K{}_P \overset{\frac{1}{2}}{C}{}^P_M - \delta^K_M . \tag{37.16}$$

No such simple formula of composition holds when $\widetilde{\boldsymbol{R}} \neq 0$.

For the general expression of a spatial rotation in terms of angles or other parameters, the reader is referred to standard treatises on kinematics. Here we mention only EULER's theorem that every rotation about a point may be regarded as a rotation about a line through the point and hence may be characterized by a unit vector and an angle. In the language of matrices, this theorem asserts that $\|R^K{}_M\|$ if not the identity matrix possesses a real unit vector invariant, its *axis* $\boldsymbol{A}$, uniquely determined up to sign, and a real scalar invariant, its *angle* ϑ, uniquely determined up to a convention of sign and quadrant. An expression for $\boldsymbol{R}$ in terms of these two invariants is[1]

$$R^K{}_M = \cos\vartheta\, \delta^K_M + (1-\cos\vartheta)\, A^K A_M + \sin\vartheta\, e^K{}_{MP} A^P. \tag{37.17}$$

If we write $\boldsymbol{R} = \boldsymbol{R}(\boldsymbol{A}, \vartheta)$, then for the inverse rotation we have

$$\boldsymbol{R}^{-1} = \boldsymbol{R}(\boldsymbol{A}, -\vartheta): \tag{37.18}$$

In other words, the duals of $\boldsymbol{A}$ and ϑ are $\boldsymbol{A}$ and $-\vartheta$. Also

$$R^K{}_M A^M = A^K, \quad R^K{}_K = R^k{}_k = 1 + 2\cos\vartheta, \quad R_{[KM]} = \sin\vartheta\, e_{KMP} A^P. \tag{37.19}$$

This last formula asserts that the axial vector $\frac{1}{2}\boldsymbol{R}_\times$ points along the axis of rotation and has magnitude $\sin\vartheta$. Alternative necessary and sufficient conditions for pure strain are

$$R^K{}_K = 3, \quad R_{[KM]} = 0, \quad \vartheta = 0. \tag{37.20}$$

Perhaps preferable to ϑ or $R^K{}_K$ as a scalar measure of rotation is

$$\mathfrak{R} \equiv \sin^2 \tfrac{1}{2}\vartheta = \tfrac{1}{4}(3 - R^K{}_K), \tag{37.21}$$

since $\mathfrak{R}$ vanishes if and only if the deformation is a pure strain, and for deformations which are not pure strains it satisfies $0 < \mathfrak{R} \leqq 1$, with $\mathfrak{R} = 1$ corresponding to a rotation through angle π.

In summary, *every deformation carrying a neighborhood of* $\boldsymbol{X}$ *into a neighborhood of* $\boldsymbol{x}$ *is specified locally by the translation from* $\boldsymbol{X}$ *to* $\boldsymbol{x}$, *the axis* $\boldsymbol{A}$, *the angle of rotation* ϑ, *either of the two sets of principal axes of strain, and the three principal stretches* λ_a.

A study of further matters concerned with the separation of strain from rotation has been made by SIGNORINI[2].

38. Invariant directions. A different question, first considered by KELVIN and TAIT[3], is to find the particular elements $d\boldsymbol{X}$ which suffer no rotation. It is easiest to start with (20.3), which yields as the condition for such an invariant element

$$g^K_k\, dx^k = g^K_k\, x^k{}_{;M}\, dX^M = b\, dX^K, \tag{38.1}$$

where b is a real factor of proportionality. Thus it is necessary and sufficient that b satisfy

$$\det\left[(b-1)\,\delta^K_M - u^K_{;M}\right] = 0, \tag{38.2}$$

where we have used (19.4). This is a real cubic. Since it has at least one real root, we obtain the ***theorem of Kelvin and Tait***: *In any deformation, at least one direction is left unaltered.* By (16.4), $b \neq 0$. Let D be the discriminant of (38.2). Then if $D > 0$, there are three distinct invariant directions[4]; if $D < 0$, there is one and only one; if $D = 0$, there may be one, two, three, or an infinite number. In Sect. 78 we shall solve a formally analogous question where the

[1] FINGER [1892, *3*, p. 1121] gave formulae for $\boldsymbol{A}$ and ϑ in terms of the $\boldsymbol{N}_a$ and $\boldsymbol{n}_a$. Cf. also SIGNORINI [1943, *6*, ¶ 11].

[2] [1943, *6*, ¶¶ 20–28]. Cf. also the theorem of GRIOLI in Sect. 42.

[3] [1867, *3*, § 181]. WARREN [1861, *2*, § II] had asserted that there are three invariant directions in all cases, but he did not take up the question of their reality.

[4] WARREN [1868, *16*] asserted a geometrical relation between these three directions and the principal directions of strain, but the result appears to be valid only in small deformations.

results are easier to interpret. Here we are content to remark that in the case $D>0$ the three invariant directions need not be mutually orthogonal, but if they are, then *a fortiori* they are principal axes of strain at $\boldsymbol{X}$ and at $\boldsymbol{x}$. Cf. the theorem of KELVIN and TAIT presented in Sect. App. 37.

By $(31.6)_1$, the tensor whose proper numbers satisfy (38.2) is symmetric if and only if $\tilde{\boldsymbol{R}}=0$. If $\tilde{\boldsymbol{R}}=0$, (38.1) becomes identical with the equation for the principal elongations. Therefore $\tilde{\boldsymbol{R}}=0$ is a necessary and sufficient condition that the principal axes of strain be invariant *as lines*. Alternatively, *a necessary and sufficient condition that the principal axes of strain be deformed into themselves is that they coincide with the principal axes of elongation*; equivalently, *that there exist an orthogonal triple of invariant directions*[1]. Thus, in particular, *the condition*

$$\tilde{\boldsymbol{R}}=0 \tag{38.3}$$

is necessary for pure strain. But it is not sufficient[2]. For in (38.1) it is possible that $b<0$; i.e., what we have called an invariant direction includes the possibility of a reversal of sense. Suppose now (38.3) holds. Then, as already established, the principal axes of strain are invariant and coincide with the principal axes of elongation. By (16.4), it is impossible that just one or all three principal axes may be reversed in sense, but the case of two reversals is possible and corresponds to a rotation through angle π about one principal axis[3]. Since when $\tilde{\boldsymbol{R}}=0$ the roots $b_\mathfrak{a}$ of (38.2) are related to the principal elongations through $b_\mathfrak{a}-1=\varepsilon_\mathfrak{a}$, we see that if $\tilde{\boldsymbol{R}}=0$ then either $\varepsilon_\mathfrak{a}>-1$ for $\mathfrak{a}=1, 2, 3$ or else $\varepsilon_1>-1$, $\varepsilon_2<-1$, $\varepsilon_3<-1$, where $\varepsilon_1\geqq\varepsilon_2\geqq\varepsilon_3$. Hence we have the following **theorem**: *In a potential deformation which is not a pure strain, the axis of rotation is the axis of greatest elongation and the angle of rotation is a straight angle.* The axis of greatest elongation may be either the axis of least stretch or the axis of greatest stretch. In terms of the measure $\mathfrak{R}$ defined by (37.21), we have the following **criterion**: *if $\tilde{\boldsymbol{R}}=0$, then either $\mathfrak{R}=0$ or $\mathfrak{R}=1$. The former case is a pure strain; the latter, a non-pure potential deformation in which the axis $\boldsymbol{A}$ points in the direction of greatest elongation. Conversely, $\mathfrak{R}=0$ implies $\tilde{\boldsymbol{R}}=0$, but in general $\mathfrak{R}=1$ does not imply $\tilde{\boldsymbol{R}}=0$.* If we refer $R^k{}_m$ and $R^K{}_M$ to the principal axes of strain, for a pure strain they both reduce to the unit matrix, while for a non-pure potential deformation they both reduce to one of the forms diag $(1, -1, -1)$, diag $(-1, 1, -1)$, diag $(-1, -1, 1)$.

Our considerations here are all local. By the axiom of continuity Sect. 16 it follows that if $\tilde{\boldsymbol{R}}=0$ along a curve, upon a surface, or throughout a region, then in the *entirety* of the said manifold the deformation is a pure strain or a non-pure potential deformation according as it is one or the other at some one point of the manifold.

---

[1] DARBOUX [1901, *3*] [1910, *5*, §§ 319–330] found necessary and sufficient conditions that $\tilde{\boldsymbol{E}}$ have a triply orthogonal set of isostatic surfaces when $\tilde{\boldsymbol{R}}=0$. Cf. Sect. App. 48. He proved also that corresponding to any triply orthogonal family of surfaces there exist infinitely many potential deformations. TONOLO [1952, *20*] has replaced DARBOUX's complicated analysis by a short and elegant demonstration.

[2] Almost the entire literature on finite strain (e.g. LOVE [1927, *6*, § 33]) follows KELVIN and TAIT [1867, *3*, § 183] in asserting erroneously that (38.3) is a sufficient condition for pure strain. Apparently the error and its correction were first noted by SIGNORINI [1943, *6*, ¶ 14].

[3] Our convention regarding the definition of principal axes when there are two or more equal principal stretches (Sect. 27) makes this case no exception to our statement here.

In a potential deformation, (31.10) reduces to

$$E^K_M = \tilde{E}^K_M + \tfrac{1}{2}\tilde{E}^{KP}\tilde{E}_{PM}, \quad \text{or} \quad \boldsymbol{E} = \tilde{\boldsymbol{E}} + \tfrac{1}{2}\tilde{\boldsymbol{E}}^2. \tag{38.4}$$

Thus in a potential deformation $\boldsymbol{E}$ is an isotropic function[1] of $\tilde{\boldsymbol{E}}$. Moreover, from (38.4) we get

$$\varepsilon_a = \delta_a \quad \text{or} \quad -(2+\delta_a), \tag{38.5}$$

where for a pure strain the former alternative holds in all three cases, while for a non-pure potential deformation the former alternative holds only for the elongation along the axis.

For a potential deformation, from $(37.15)_1$ we get

$$\overset{\frac{1}{2}}{C}_{KM} = \overset{-1}{R}_K{}^P \overset{\frac{1}{2}}{C}_{PQ} R_M{}^Q. \tag{38.6}$$

The rotation $\boldsymbol{R}$ therefore leaves the quadric of $\boldsymbol{C}^{\frac{1}{2}}$, and hence also the strain ellipsoid, invariant; this gives a second proof[2] of the theorem above.

A different condition for invariant directions follows from the theorem at the beginning of Sect. App. 37: *If a deformation with distinct principal stretches is the succession of two pure strains, then there are three invariant directions.*

39. Composition of strains. For two successive deformations, the first from $\boldsymbol{X}$ to X, the second from X to $\boldsymbol{x}$, we have the simple formula (17.10) for composition of the deformation gradients. No such simple result is valid for the corresponding strains. We now seek to relate the Cauchy tensors for these two deformations to the Cauchy tensor for the deformation from $\boldsymbol{X}$ to $\boldsymbol{x}$. Equivalently, we compare the two measures of strain at $\boldsymbol{x}$ from the two different unstrained states X and $\boldsymbol{X}$. By $(26.1)_2$ we get[3]

$$\left.\begin{aligned} c_{km} &= g_{KM} X^K_{;k} X^M_{;m} = g_{KM} X^K_{;\mathsf{K}} X^M_{;\mathsf{M}} \mathsf{X}^{\mathsf{K}}_{;k} \mathsf{X}^{\mathsf{M}}_{;m}, \\ &= {}_1c_{\mathsf{KM}} \mathsf{X}^{\mathsf{K}}_{;k} \mathsf{X}^{\mathsf{M}}_{;m}, \end{aligned}\right\} \tag{39.1}$$

where $X^K_{;\mathsf{K}} \equiv \partial X^K/\partial \mathsf{X}^{\mathsf{K}}$, $\mathsf{X}^{\mathsf{K}}_{;k} \equiv \partial \mathsf{X}^{\mathsf{K}}/\partial x^k$, and ${}_1\boldsymbol{c}$ is the Cauchy tensor measuring the strain from $\boldsymbol{X}$ to X. By the dual of (37.8) follows

$$c_{km} = {}_1c_{\mathsf{KM}} \overset{-1}{R}{}^{\mathsf{K}}{}_p \overset{-1}{R}{}^{\mathsf{M}}{}_q \, {}_2\overset{\frac{1}{2}}{c}{}^p_k \, {}_2\overset{\frac{1}{2}}{c}{}^q_m, \tag{39.2}$$

where ${}_2\boldsymbol{c}$ is the Cauchy tensor measuring the strain from X to $\boldsymbol{x}$. This is the general law of composition of strains, or of change of strain reference. As would be expected, to calculate the strain from $\boldsymbol{X}$ to $\boldsymbol{x}$ a knowledge of the strains from $\boldsymbol{X}$ to X and from X to $\boldsymbol{x}$ is insufficient. In addition, knowledge of the rotation from X to $\boldsymbol{x}$ is necessary.

In case the displacement from X to $\boldsymbol{x}$ is a pure strain, (39.2) becomes

$$c_{km} = {}_1c_{\mathsf{KM}} \, g^{\mathsf{K}}_p \, g^{\mathsf{M}}_q \, {}_2\overset{\frac{1}{2}}{c}{}^p_k \, {}_2\overset{\frac{1}{2}}{c}{}^q_m = {}_1c_{pq} \, {}_2\overset{\frac{1}{2}}{c}{}^p_k \, \overset{\frac{1}{2}}{c}{}^q_m, \tag{39.3}$$

which is not the ordinary matrix law of composition, being of the form $\boldsymbol{c} = {}_2\boldsymbol{c}^{\frac{1}{2}} \cdot {}_1\boldsymbol{c} \cdot {}_2\boldsymbol{c}^{\frac{1}{2}}$. In the special case when $\mathsf{Z}^{\mathsf{K}} = \lambda \delta^{\mathsf{K}}_K Z^K$, so that the strain from $\boldsymbol{Z}$ to

[1] In a work written in 1956 but not published, TOUPIN has shown that in the case when the principal stretches are all distinct or all equal, $\tilde{\boldsymbol{R}} = 0$ is necessary and sufficient that every principal direction of $\tilde{\boldsymbol{E}}$ be a principal direction of $\boldsymbol{E}$. In the case when two principal stretches are equal, the strain ellipsoid is a spheroid, and rotating it about its axis of symmetry produces an equal elongation in all elements in its equatorial plane. Therefore the elongation quadric is also a quadric of revolution with the same axis. If the angle of rotation is not 0 or π, we have $\tilde{\boldsymbol{R}} \neq 0$, yet every principal axis of $\boldsymbol{E}$ is also a principal axis of $\tilde{\boldsymbol{E}}$. This theorem was stated by TOUPIN without proof.

[2] Given in the work of TOUPIN mentioned in the preceding footnote.

[3] The dual is given by LOVE [1892, *8*, § 10].

Z is a uniform dilation (cf. Sect. 43), we get the obvious relation

$$_2\mathbf{c} = \lambda^2 \mathbf{c}; \tag{39.4}$$

hence

$$\mathrm{I}_{_2\mathbf{c}} = \lambda^2 I_{\mathbf{c}}, \quad \mathrm{II}_{_2\mathbf{c}} = \lambda^4 \mathrm{II}_{\mathbf{c}}, \quad \mathrm{III}_{_2\mathbf{c}} = \lambda^6 \mathrm{III}_{\mathbf{c}}. \tag{39.5}$$

The multiplicative resolution (39.2) gives in general the simplest expression for a given strain in terms of composite strains and rotations. For some problems, however, a more complicated additive resolution is preferable. To obtain it, we begin with the dual of $(39.1)_3$:

$$\left.\begin{aligned} C_{KM} &= {}_2C_{\mathsf{KM}} X^{\mathsf{K}}_{;K} X^{\mathsf{M}}_{;M} = (g_{\mathsf{KM}} + 2\,{}_2E_{\mathsf{KM}}) X^{\mathsf{K}}_{;K} X^{\mathsf{M}}_{;M}, \\ &= {}_1C_{KM} + 2\,{}_2E_{\mathsf{KM}} X^{\mathsf{K}}_{;K} X^{\mathsf{M}}_{;M}. \end{aligned}\right\} \tag{39.6}$$

In this we substitute the formula that results from replacing $\boldsymbol{x}$ by X in (19.4), so obtaining[1]

$$\left.\begin{aligned} E_{KM} = {}_1E_{KM} + {}_2E_{KM} + 2\,{}_1u_{P;(K}\,{}_2E^P_{M)} \\ + {}_2E^{PQ}\,{}_1u_{P;K}\,{}_1u_{Q;M}. \end{aligned}\right\} \tag{39.7}$$

The apparent simplicity of this formula is somewhat deceiving, since all tensors relating to either of the deformations are shifted to the point $\boldsymbol{X}$. As would be expected of a result dual to (39.2), this one calculates the total strain in terms of the second strain, the first strain, and the first rotation. A more complicated result follows from (39.7) if we replace ${}_1E_{KM}$ and ${}_1u_{K;M}$ by their expressions in terms of ${}_1\tilde{E}_{KM}$ and ${}_1\tilde{R}_{KM}$.

By setting ${}_1\boldsymbol{E}=0$ in (39.7) we get an expression for the change in the components of strain induced by a superposed rigid deformation. From (39.6) it follows that $\mathbf{c} = {}_1\mathbf{c}$ if and only if ${}_2\boldsymbol{E}=0$; this is another way of saying that *the components of* $\mathbf{c}$ *characterize the deformation to within a rigid deformation.*

d) Special deformations.

40. Isochoric deformations. There are four important special types of deformation defined by conditions referring neither to a particular direction nor to a special geometrical configuration. The first three, *rigid* deformation, *pure* strain, and *potential* deformation, were defined in Sects. 26, 35, and 36. The fourth, *isochoric* deformation, is defined by the condition that volumes be unaltered. By (20.9), (16.5), (30.6), and $(30.8)_4$, we get as alternative necessary and sufficient local conditions for an isochoric deformation

$$|z/\boldsymbol{Z}| = 1, \quad J = 1, \quad \mathrm{III}_{\mathbf{c}} = 1, \quad \mathrm{III}_{\mathbf{c}^{-1}} = 1, \quad \mathrm{I}_{\delta} + \mathrm{II}_{\delta} + \mathrm{III}_{\delta} = 0, \tag{40.1}$$

and many other forms are easily found.

In view of $(40.1)_3$, by specialization from the fundamental theorem on isotropic scalar functions we conclude that *in an isochoric deformation, an isotropic scalar function of* $\mathbf{c}$ *is equal to a function of* $\mathrm{I}_{\mathbf{c}}$ *and* $\mathrm{II}_{\mathbf{c}}$ *only*. These two invariants satisfy a number of simple relations. First, (30.2) becomes

$$\mathrm{I}_{\mathbf{c}} = \mathrm{II}_{\mathbf{C}}, \quad \mathrm{II}_{\mathbf{c}} = \mathrm{I}_{\mathbf{C}}. \tag{40.2}$$

---

[1] Cf. SIGNORINI [1943, *6*, ¶ 8].

Second, since $\mathrm{III}_C = \lambda_1^2\lambda_2^2\lambda_3^2 = 1$, one of the principal stretches, say λ_3, can always be eliminated in terms of the others. For example, from (30.1) follows

$$\mathrm{I}_C = \mathrm{I}_{c^{-1}} = \lambda_1^2 + \lambda_2^2 + \frac{1}{\lambda_1^2\lambda_2^2}. \tag{40.3}$$

From this formula and (40.2) we conclude

$$\mathrm{I}_c,\ \mathrm{II}_c,\ \mathrm{I}_{c^{-1}},\ \mathrm{II}_{c^{-1}} \geqq 3, \tag{40.4}$$

where the lower bound 3 is assumed in a rigid deformation. This is to be contrasted with the general case, where all that can be inferred is that the principal invariants are positive.

In an isochoric deformation, from putting $(40.1)_5$ into $(30.8)_1$ and $(30.8)_2$ we derive

$$\left.\begin{aligned}\mathrm{II}_c - 3 = \mathrm{I}_{c^{-1}} - 3 &= -4\,\mathrm{II}_\delta - 2\,\mathrm{III}_\delta + (\mathrm{II}_\delta + \mathrm{III}_\delta)^2,\\ \mathrm{I}_c - 3 = \mathrm{II}_{c^{-1}} - 3 &= -4\,\mathrm{II}_\delta - 10\,\mathrm{III}_\delta + \mathrm{II}_\delta^2 + 4\,\mathrm{II}_\delta\,\mathrm{III}_\delta + 2\,\mathrm{III}_\delta^2.\end{aligned}\right\} \tag{40.5}$$

On the one hand, these identities put into (40.4) yield inequalities connecting the principal extensions. On the other hand, if we set $\delta^2 \equiv \sum_{a=1}^{3} \delta_a^2$, then as $\delta \to 0$ we have from (40.5)

$$\mathrm{I}_c - 3 = O(\delta^2), \qquad \mathrm{II}_c - 3 = O(\delta^2). \tag{40.6}$$

This, as RIVLIN[1] has remarked, is to be contrasted with the general case, where all that can be concluded is that $\mathrm{I}_c - 3$, $\mathrm{II}_c - 3$, and $\mathrm{III}_c - 1$ are $O(\delta)$.

41. Plane strain[2]. A deformation is said to be a *plane strain* if there exists a family of parallel planes which are individually preserved, while the family of lines normal to these planes is preserved as a family. If we choose the preserved planes as rectangular co-ordinate planes and the remaining co-ordinate surfaces as cylinders normal to these planes, the deformation assumes the form

$$x^k = x^k(X^1, X^2), \qquad k = 1, 2, \qquad z = Z = x^3 = X^3. \tag{41.1}$$

To visualize the strain and discuss it, it suffices to restrict attention to points in the $z = 0$ plane, to replace the strain ellipsoids there by their elliptical sections by that plane, etc.

The Cauchy and Green tensors and the rotation tensor of a plane strain have matrices of the type

$$\left\|\begin{matrix}\cdot & \cdot & 0\\ \cdot & \cdot & 0\\ 0 & 0 & 1\end{matrix}\right\|. \tag{41.2}$$

Hence the z-direction at a point is both a principal axis of strain and the axis of rotation, so that the angle of rotation is the common angle through which the two principal axes lying in the $z = 0$ plane are turned.

A necessary and sufficient condition that one principal stretch be 1, and hence a necessary but not sufficient condition for a plane strain, is that the principal invariants of any one of the tensors $\boldsymbol{c}$, $\boldsymbol{C}$, $\boldsymbol{c}^{-1}$ and $\boldsymbol{C}^{-1}$ satisfy (App. 38.8). Since $\mathrm{III}_C = \lambda_1^2\lambda_2^2$, by $(40.1)_3$ it follows that *a plane strain is isochoric if and only if the principal stretches are reciprocal to one another*. By (App. 38.8), a necessary but not sufficient condition for plane isochoric strain is $\mathrm{I}_C = \mathrm{II}_C$. By $(40.2)_1$ follows then $\mathrm{I}_c = \mathrm{I}_C$: *In a plane isochoric strain, each principal invariant of any one of*

[1] [1951, *22*, § 21].

[2] More detailed analysis is given by SIGNORINI [1943, *6*, ¶ 31].

the tensors $\mathbf{c}$, $\boldsymbol{C}$, $\mathbf{c}^{-1}$, *and* $\boldsymbol{C}^{-1}$ *equals the corresponding principal invariant of any other.*

A deformation is said to be a *generalized plane strain*[1] when (41.1) is replaced by

$$x^k = x^k(X^1, X^2), \qquad k = 1, 2, \qquad z = f(Z). \tag{41.3}$$

The planes $Z = \text{const}$ are preserved as a family, but not necessarily individually. For example, there may be a uniform stretch λ normal to these planes, so that $z = \lambda Z$. A number of properties of plane strain carry over to the generalized case. For example, the z-direction is again both a principal axis of strain and the axis of rotation. The principal stretch in this direction is $\lambda = f'(Z)$. Since λ_1 and λ_2 are independent of Z, and since $\mathrm{III}_{\boldsymbol{C}} = \lambda_1^2 \lambda_2^2 \lambda^2$, it follows that *a generalized plane strain is isochoric if and only if* $\lambda = \text{const}$ *and* $\lambda_1 \lambda_2 = \lambda^{-1}$.

42. Homogeneous strain. A deformation is called a *homogeneous strain* when every straight line is deformed into a straight line. Equivalently, every ellipse (including the circular special case) is deformed into an ellipse, or every ellipsoid is deformed into an ellipsoid. A formal and invariant local condition for homogeneous strain is

$$x^k_{;KM} = 0; \tag{42.1}$$

of the many equivalent forms of this condition we note only

$$\mathfrak{R}_{;K} = 0, \qquad A^K_{;M} = 0, \qquad C_{KM;P} = 0. \tag{42.2}$$

The most convenient condition is that in the common frame the deformation (15.1) has the form

$$\boldsymbol{z} = \boldsymbol{D} \cdot \boldsymbol{Z} + \boldsymbol{B}, \qquad \boldsymbol{Z} = \boldsymbol{D}^{-1} \cdot (\boldsymbol{z} - \boldsymbol{B}), \tag{42.3}$$

where matrix notation is used and where $\boldsymbol{D}$ is a real matrix with positive determinant and $\boldsymbol{B}$ is a real constant vector. When referred to the common frame, all the measures of deformation and rotation (such as $\mathbf{c}$, $\boldsymbol{C}$, $\boldsymbol{R}$, $\mathfrak{R}$, $\boldsymbol{A}$, $\tilde{\boldsymbol{R}}$, $\tilde{\boldsymbol{E}}$, etc.) are constants. There is no loss in generality in supposing the origin of the common frame so chosen that $\boldsymbol{B} = 0$.

Homogeneous strain was introduced and studied exhaustively by KELVIN and TAIT[2]. For homogeneous strain, the reservations expressed in Sect. 20 become unnecessary, for finite line segments are deformed according to the same law as are differential elements in a general deformation.

By uniformly extending initially circular or rectangular frames across which marked cloth or rubber sheets had previously been stretched, WEISSENBERG[3] has obtained the beautiful illustrations of homogeneous strains which are shown here as Fig. 6.

By (20.10), we may choose to regard any deformation satisfying the axiom of continuity as a homogeneous strain, with error $O(dX^2)$. Thus for general deformations all properties which depend only on $x^k_{;K}$ may be developed by application of appropriate theorems on homogeneous strain, and many authors attack the subject in this way. We follow the reverse procedure to this extent, that we leave to the reader the specialization of general results already obtained to the case of homogeneous strain.

The following sections develop the properties of certain particularly important types of homogeneous strain and certain of their generalizations. The co-ordinate system used is always rectangular Cartesian.

---

[1] The case $z = \lambda Z$ was introduced by LOVE [1906, *5*, § 94].

[2] [1867, *3*, §§ 155–189]. Further geometrical properties were discussed by ISÉ [1890, *5*, §§ II–III]; algebraic treatments were given by METZLER [1894, *6*] and CAFIERO [1906, *1*, Chap. II].

[3] [1935, *10*, pp. 85–87] [1949, *37*].

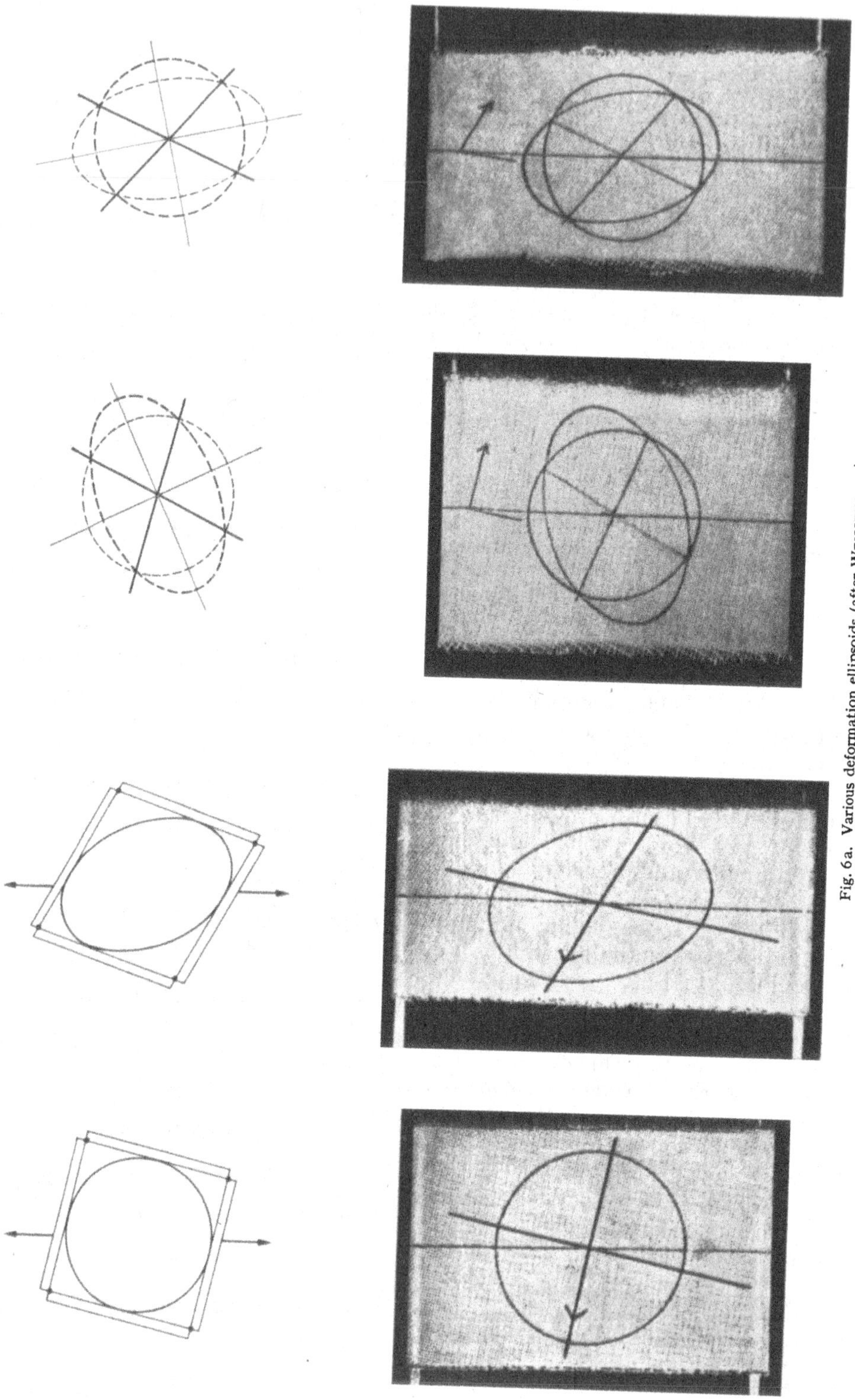

Fig. 6a. Various deformation ellipsoids (after Weissenberg).

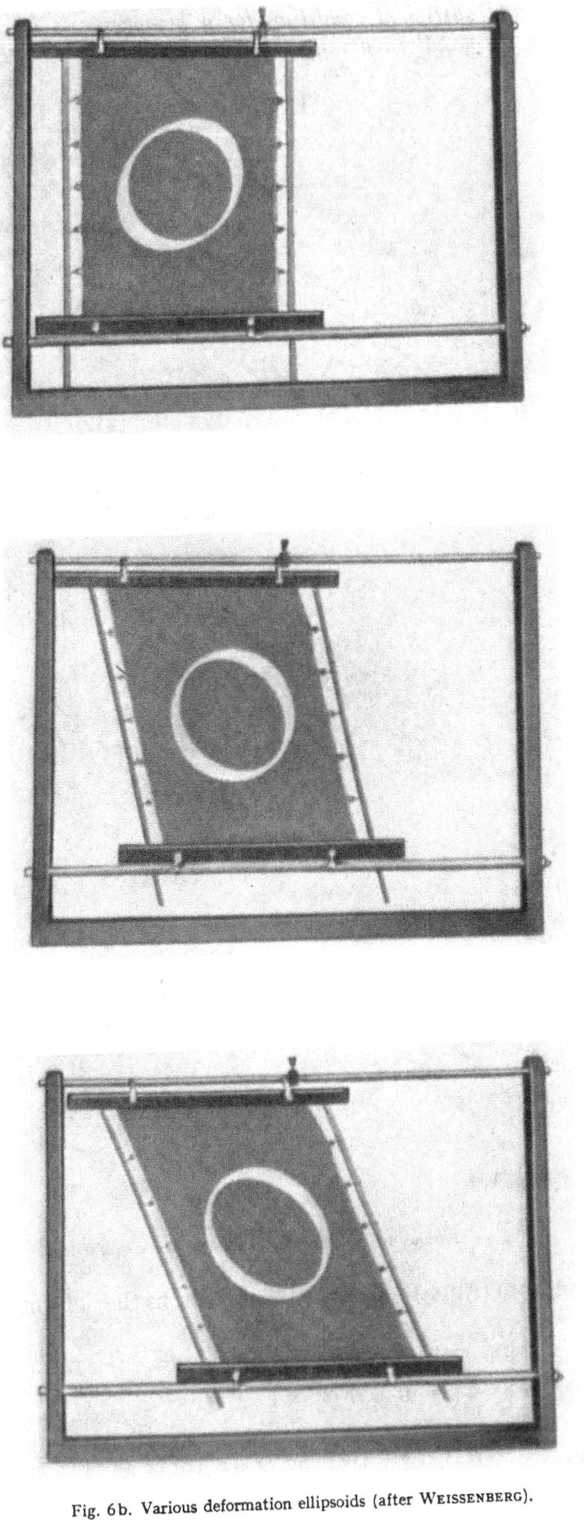

Fig. 6b. Various deformation ellipsoids (after WEISSENBERG).

A necessary and sufficient condition for a homogeneous isochoric strain to be plane is that the principal invariants of any one of the tensors $\mathbf{c}$, $\boldsymbol{C}$, $\mathbf{c}^{-1}$, *and* $\boldsymbol{C}^{-1}$ *satisfy*

$$\mathrm{I} = \mathrm{II}. \tag{42.4}$$

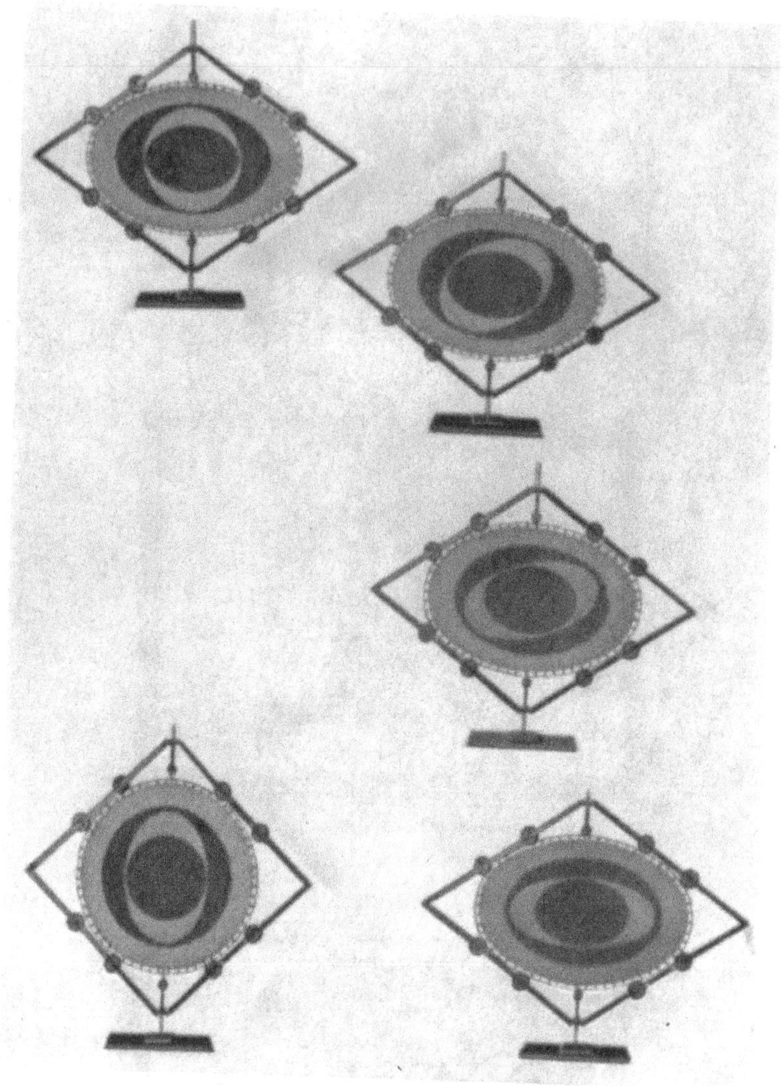

Fig. 6c. Various deformation ellipsoids (after WEISSENBERG).

This is immediate from (App. 38.8), since by hypothesis the deformation is a homogeneous strain.

Since $\boldsymbol{D}$ in (42.3) is the matrix of deformation gradients, by (37.8) and (37.9) we have

$$\boldsymbol{D} = \boldsymbol{R} \cdot \boldsymbol{C}^{\frac{1}{2}} = \mathbf{c}^{-\frac{1}{2}} \cdot \boldsymbol{R}, \tag{42.5}$$

where $\boldsymbol{R}$ is the rotation tensor defined by (37.1). GRIOLI[1] has obtained a theorem of approximation which characterizes $\boldsymbol{R}$. Let $\boldsymbol{z} = \boldsymbol{z}(\boldsymbol{Z})$ and $\boldsymbol{z}^* = \boldsymbol{z}^*(\boldsymbol{Z})$ be any two homogeneous defor-

[1] [1940, *12*].

mations, and define their *deviation* $d(\boldsymbol{z}^*, \boldsymbol{z})$ by

$$d(\boldsymbol{z}^*, \boldsymbol{z}) \equiv \int_{\mathscr{V}_a} |\boldsymbol{z}^* - \boldsymbol{z}|^2 \, dV, \tag{42.6}$$

where $\mathscr{V}_a$ is a sphere of fixed radius a about $\boldsymbol{Z}$. Given $\boldsymbol{z}$, we now seek to determine $\boldsymbol{z}^*$ as a *rigid* deformation minimizing $d(\boldsymbol{z}^*, \boldsymbol{z})$. Since $\boldsymbol{z}^* = \boldsymbol{R}^* \cdot \boldsymbol{Z} + \boldsymbol{B}^*$, where $\boldsymbol{R}^*$ is a rotation

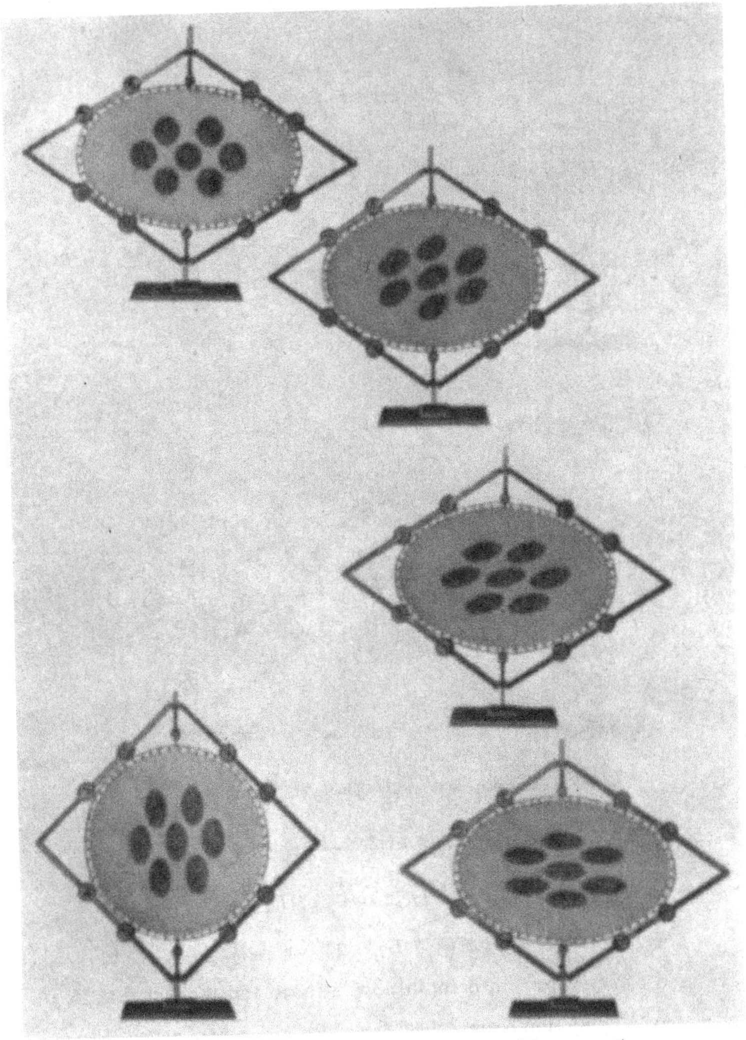

Fig. 6d. Various deformation ellipsoids (after WEISSENBERG).

matrix, we have

$$\left.\begin{aligned} d(\boldsymbol{z}^*, \boldsymbol{z}) &= \int_{\mathscr{V}_a} |(\boldsymbol{B}^* - \boldsymbol{B}) + (\boldsymbol{R}^* - \boldsymbol{R} \cdot \boldsymbol{C}^{\frac{1}{2}}) \cdot \boldsymbol{Z}|^2 \, dV, \\ &= \int_{\mathscr{V}_a} |\boldsymbol{R} \cdot \{\boldsymbol{R}^{-1} \cdot (\boldsymbol{B}^* - \boldsymbol{B}) + (\boldsymbol{R}^{-1} \cdot \boldsymbol{R}^* - \boldsymbol{C}^{\frac{1}{2}}) \cdot \boldsymbol{Z}\}|^2 \, dV, \\ &= \int_{\mathscr{V}_a} |\boldsymbol{B}' + (\boldsymbol{R}' - \boldsymbol{C}^{\frac{1}{2}}) \cdot \boldsymbol{Z}|^2 \, dV, \end{aligned}\right\} \tag{42.7}$$

where

$$\boldsymbol{B}' \equiv \boldsymbol{R}^{-1} \cdot (\boldsymbol{B}^* - \boldsymbol{B}), \qquad \boldsymbol{R}' = \boldsymbol{R}^{-1} \cdot \boldsymbol{R}^*, \tag{42.8}$$

and where the factor $\boldsymbol{R}$, since it does not affect the length being integrated, was dropped in passing from $(42.7)_2$ to $(42.7)_3$. Hence

$$\left.\begin{aligned} d(\boldsymbol{z}^*, \boldsymbol{z}) &= \tfrac{4}{3}\pi a^3 \boldsymbol{B}'^2 + 2\boldsymbol{B}' \cdot (\boldsymbol{R}' - \boldsymbol{C}^{\frac{1}{2}}) \cdot \int_{\mathscr{V}_a} \boldsymbol{Z}\, dV + \\ &+ (\boldsymbol{R}' - \boldsymbol{C}^{\frac{1}{2}}) \cdot \Big(\int_{\mathscr{V}_a} \boldsymbol{Z}\boldsymbol{Z}\, dV\Big) \cdot (\boldsymbol{R}' - \boldsymbol{C}^{\frac{1}{2}})'. \end{aligned}\right\} \tag{42.9}$$

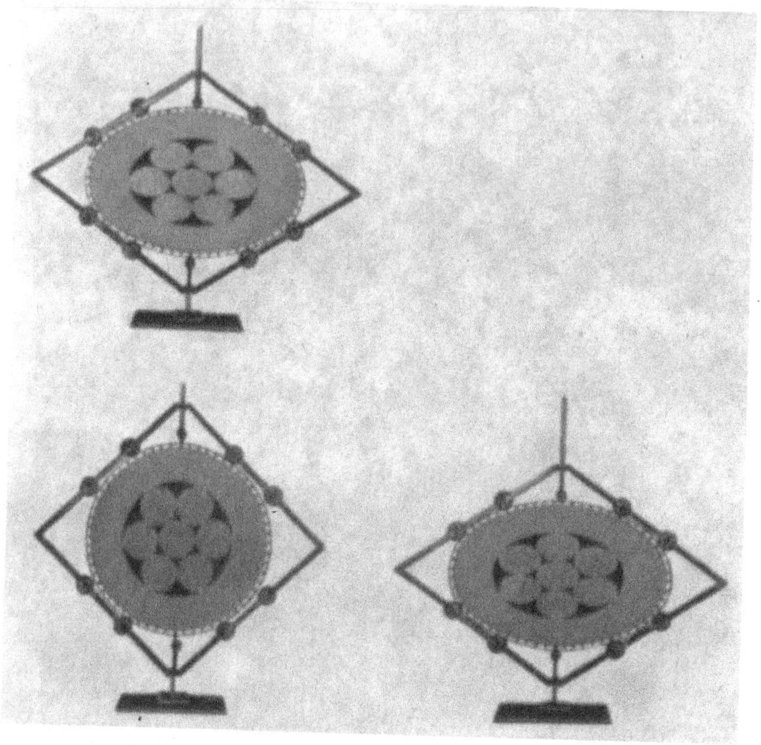

Fig. 6e. Various deformation ellipsoids (after WEISSENBERG).

Since $\mathscr{V}_a$ is a sphere, by choosing the origin of $\boldsymbol{Z}$ at its center we get

$$\left.\begin{aligned} d(\boldsymbol{z}^*, \boldsymbol{z}) &= \tfrac{4}{3}\pi a^3 \boldsymbol{B}'^2 + \tfrac{4}{15}\pi a^5 (R'_{km} - \overset{\frac{1}{2}}{C}_{km})(R'_{km} - \overset{\frac{1}{2}}{C}_{km}), \\ &= \tfrac{4}{3}\pi a^3 [\boldsymbol{B}'^2 + \tfrac{1}{5} a^2 (3 + \mathrm{I}_{\boldsymbol{C}} - 2 R'_{km} \overset{\frac{1}{2}}{C}_{km})], \end{aligned}\right\} \tag{42.10}$$

where we have used BORCHARDT's theorem in Sect. App. 40 and the fact that $(\boldsymbol{R}')^{-1} = (\boldsymbol{R}')'$. Since $\boldsymbol{R}'$ is a matrix of direction cosines, we have $|R'_{km}| \leqq 1$, and hence $|R'_{km} \overset{\frac{1}{2}}{C}_{km}| \leqq \mathrm{I}_{\boldsymbol{C}^{\frac{1}{2}}}$, with equality holding if and only if $\boldsymbol{R}' = \boldsymbol{1}$. Since $3 + \mathrm{I}_{\boldsymbol{C}} - 2\mathrm{I}_{\boldsymbol{C}^{\frac{1}{2}}} = \overline{\mathrm{II}}_\delta \equiv \sum_{\mathfrak{a}=1}^{3} \delta_{\mathfrak{a}}^2$, from (42.10) we conclude that

$$d(\boldsymbol{z}^*, \boldsymbol{z}) \geqq \tfrac{4}{15}\pi a^5 \overline{\mathrm{II}}_\delta, \tag{42.11}$$

where equality holds if and only if $\boldsymbol{B}' = 0$, $\boldsymbol{R}' = \boldsymbol{1}$. These conditions, by (42.8), are equivalent to $\boldsymbol{B}^* = \boldsymbol{B}$, $\boldsymbol{R}^* = \boldsymbol{R}$. We have proved **Grioli's theorem**: *Let a given homogeneous strain be decomposed into a translation, rotation, and a pure strain; then the translation and the rotation are precisely those defining the rigid deformation whose deviation from the given strain is the least possible.*

43. Uniform dilation. A pure homogeneous deformation in which the principal stretches have a common value, λ, is called a *uniform dilation*. For such a deformation we have

$$\left.\begin{aligned} &\boldsymbol{D}=\lambda\mathbf{1}, \quad \boldsymbol{c}=\boldsymbol{C}^{-1}=\frac{1}{\lambda^2}\mathbf{1}, \quad \boldsymbol{C}=\boldsymbol{c}^{-1}=\lambda^2\mathbf{1},\\ &\mathrm{I}_{\boldsymbol{C}}=3\lambda^2,\ \mathrm{II}_{\boldsymbol{C}}=3\lambda^4, \qquad \mathrm{III}_{\boldsymbol{C}}=\lambda^6, \end{aligned}\right\} \tag{43.1}$$

where $0<\lambda<\infty$. An invariant necessary and sufficient condition that a pure homogeneous deformation be a uniform dilation is[1]

$$\left(\frac{1}{3}\,\mathrm{I}\right)^3=\left(\frac{1}{3}\,\mathrm{II}\right)^{\frac{3}{2}}=\mathrm{III}, \tag{43.2}$$

where the invariants are calculated from any one strain measure. This is so because (43.2) is necessary and sufficient that the strain measure have equal proper numbers.

44. Simple extension. A pure homogeneous deformation in which two but not three principal stretches are equal is called a *simple extension*. In a rectangular co-ordinate system whose axes are the principal axes of strain, we have

$$\left.\begin{aligned} &\boldsymbol{D}=\begin{Vmatrix} B\lambda & 0 & 0\\ 0 & B\lambda & 0\\ 0 & 0 & \lambda\end{Vmatrix}, \qquad \begin{aligned}&0<B<\infty,\\ &0<\lambda<\infty,\\ &B\neq 1,\end{aligned}\\ &\boldsymbol{c}=\boldsymbol{C}^{-1}=\begin{Vmatrix} \frac{1}{B^2\lambda^2} & 0 & 0\\ 0 & \frac{1}{B^2\lambda^2} & 0\\ 0 & 0 & \frac{1}{\lambda^2}\end{Vmatrix}, \quad \boldsymbol{c}^{-1}=\boldsymbol{C}=\begin{Vmatrix} B^2\lambda^2 & 0 & 0\\ 0 & B^2\lambda^2 & 0\\ 0 & 0 & \lambda^2\end{Vmatrix},\\ &\mathrm{I}_{\boldsymbol{C}}=\lambda^2(1+2B^2), \quad \mathrm{II}_{\boldsymbol{C}}=B^2\lambda^4(2+B^2), \quad \mathrm{III}_{\boldsymbol{C}}=B^4\lambda^6. \end{aligned}\right\} \tag{44.1}$$

The direction in which the stretch is λ is the *axis* of the extension, here taken as the axis of z and Z. The factor B is related to the *transverse contraction ratio* ν through the definition

$$\nu\equiv-\frac{\delta_x}{\delta_z}=\frac{1-B\lambda}{\lambda-1}, \qquad B=\frac{1}{\lambda}(1+\nu)-\nu. \tag{44.2}$$

The case $B=1$, or $\nu=-1$, which would lead to the results of Sect. 43, is excluded by definition. According as $\nu>0$ or $\nu<0$, the cross sections normal to the axis wane or wax as the length along the axis increases, while if $\nu=0$, lengths normal to the axis are unchanged. The extension is isochoric if and only if $B=\lambda^{-\frac{3}{2}}$, in which case $(44.2)_2$ becomes

$$\nu=\frac{1-\lambda^{-\frac{1}{2}}}{\lambda-1}=\frac{1}{2}+O(\delta_z) \tag{44.3}$$

as $\delta_z\to 0$.

Some authors[2] restrict the term *simple extension* to the case when $\nu=0$. Such a pure homogeneous strain may be characterized by the necessary and sufficient conditions[3] $\mathrm{II}_{\boldsymbol{E}}=\mathrm{III}_{\boldsymbol{E}}=0$, the stretch being determined by $\mathrm{I}_{\boldsymbol{E}}=\lambda^2-1$. A corresponding invariant condition for the case when ν does not necessarily vanish is more elaborate[4]:

$$18\,\mathrm{I}\,\mathrm{II}\,\mathrm{III}-4\,\mathrm{I}^3\,\mathrm{III}+\mathrm{I}^2\,\mathrm{II}^2-4\,\mathrm{II}^3-27\,\mathrm{III}^2=0, \qquad \mathrm{I}^2\neq 3\,\mathrm{II}, \tag{44.4}$$

[1] A more elaborate but equivalent condition was given by SIGNORINI [1930, *6*, § 10].
[2] E.g., LOVE [1892, *8*, § 3].
[3] CAFIERO [1906, *1*, Chap. II, § 16].
[4] A more complicated but equivalent condition is given by SIGNORINI [1930, *6*, § 10].

where the invariants may be calculated from any strain measure. To see that this condition is necessary and sufficient, recall that the matrix of any strain measure may be reduced to diagonal form by a suitable choice of co-ordinates (Sect. 28). From the definition of a strain measure (Sect. 32), for a pure homogeneous deformation to be a simple extension it is necessary and sufficient that exactly two proper numbers of the strain measure be equal. But $(44.4)_1$ asserts that the discriminant of the cubic equation for the proper numbers vanishes, while $(44.4)_2$ asserts that the case of Sect. 43 is excluded. Q.E.D.

It is easy to consider a more general homogeneous extension in which all three stretches are unequal. By the fundamental theorem of Sect. 27, however, any pure homogeneous strain if referred to principal axes is of this form, and hence this case, being special only in choice of co-ordinate system but not intrinsically, is of no interest.

A deformation, not necessarily homogeneous, is a *pure extension*[1] if a certain triply orthogonal system of planes is deformed into itself. Referred to principal axes, such a deformation assumes the form $x=f(X)$, $y=g(Y)$, $z=h(Z)$, and hence $\lambda_x=f'$, $\lambda_y=g'$, $\lambda_z=h'$, whence follow formulae similar to (44.1). In particular, $\mathrm{III}_{\boldsymbol{C}}=(f'g'h')^2$, and hence by obtaining the general solution of $f'g'h'=1$ we conclude that *an isochoric pure extension is necessarily homogeneous.*

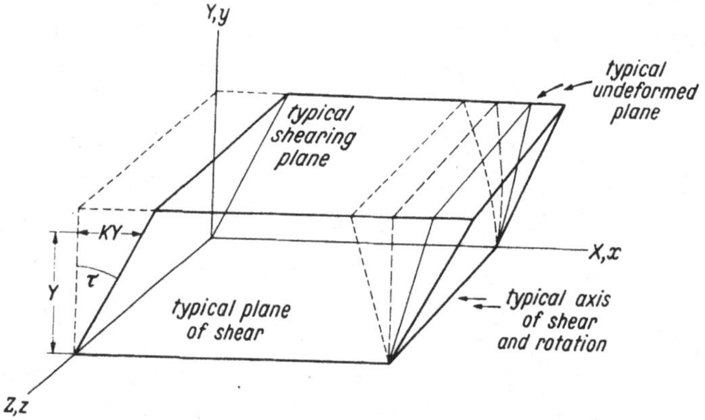

Fig. 7. Simple shear.

45. Simple shear. Consider a homogeneous strain such that

1. Two orthogonal families of parallel planes are individually preserved.
2. The lines normal to one family remain normal to that family.
3. The lines common to both families are not stretched.

Such a homogeneous strain is called a *simple shear*; the planes whose normals remain normals are called the *planes of shear*, while those of the second family are called the *shearing planes* (Fig. 7). It is obvious that the lines common to both families are individually preserved. The direction normal to a plane of shear is the *axis* of the shear. Since simple shear is the most important and instructive of all special deformations[2], we shall analyse it in detail.

[1] TRUESDELL [1952, *21*, § 42B].

[2] The concept of shear received currency relatively late in the history of continuum mechanics. While in 1770 EULER had briefly and clearly explained and calculated not only the time rate of extension but also the time rate of shearing (for the results and reference, see Sects. 82–82A below), his work on this subject escaped notice until 1955. According to TODHUNTER and PEARSON [1886, *4*, § 726 and Appendix, Note A (6)], the first clear discussion of shear in reference to solids was given by VICAT (1831), earlier remarks being due to YOUNG (1807). The first analysis of finite shear was given by KELVIN and TAIT [1867, *3*, §§ 169 to 176]. SIGNORINI [1943, *6*, ¶ 12] approaches simple shear through a preliminary characterization of all deformations that leave a single plane point-by-point invariant.

Ignoring, as usual, a possible translation, we let the planes of shear be the planes $Z=\text{const}$; the shearing planes, $Y=\text{const}$. Hence $z=Z$, $y=Y$, and $x=MX+KY+LZ$. The condition that the Z-direction be preserved yields $L=0$, while the condition that the stretch in the x-direction be 1 yields $M=1$. Thus

$$\left.\begin{array}{c} x=X+KY, \qquad y=Y, \qquad z=Z, \qquad -\infty<K<\infty, \\ \boldsymbol{D}=\begin{Vmatrix} 1 & K & 0 \\ 0 & 1 & 0 \\ 0 & 0 & 1 \end{Vmatrix}. \end{array}\right\} \tag{45.1}$$

The $X=\text{const}$ planes are rotated rigidly about their lines of intersection with the $Y=0$ plane through the *angle of shear*[1], τ, which is related to the *amount*, K, of the shear by

$$\tau \equiv \operatorname{arc\,tan} K. \tag{45.2}$$

For definiteness, we suppose $K>0$ henceforth. To visualize the shear, imagine that each of the planes $Y=\text{const}$ be slid in the X-direction by an amount proportional to its distance from the $Y=0$ plane. Such a displacement is easily approximated with a pack of cards. The stretch suffered by lines in the Y-direction is $\sqrt{1+K^2}$. Either from the definition of simple shear or from the fact that $\det \boldsymbol{D}=1$ it is obvious that simple shear is an isochoric deformation.

The point $(-\frac{1}{2}K, 1, 0)$ is carried into the point $(+\frac{1}{2}K, 1, 0)$. Hence the length of the line whose slope angle is $\frac{1}{2}\pi+\operatorname{arc\,tan}\frac{1}{2}K$ in the $Z=0$ plane in the undeformed material is unaltered. It follows that in any given simple shear there are two families of parallel planes that are transported rigidly, i.e., *undeformed*; both these are normal to the planes of shear, one set being the shearing planes themselves and the other the planes subtending the angle $\frac{1}{2}\pi+\operatorname{arc\,tan}\frac{1}{2}K$ from these planes. The Z-direction is of course a principal direction of strain. The other principal directions of strain are easily found, since by the symmetrical distribution of stretches and the extremal property of the principal directions (Sect. 27) it follows that the principal directions bisect the angles between the undeformed planes. In the undeformed body, one of these families of planes is inclined at the angle $\frac{1}{4}\pi+\frac{1}{2}\operatorname{Arc\,tan}\frac{1}{2}K$ to the shearing planes, and in these planes lie the elements suffering greatest stretch. Hence and from (45.1) the magnitudes of the two principal stretches in the plane of shearing are easily calculated:

$$\left.\begin{aligned} \lambda_1^2 &= 1+\tfrac{1}{2}K^2+K\sqrt{1+\tfrac{1}{4}K^2}, \\ \lambda_2^2 &= 1/\lambda_1^2 = 1+\tfrac{1}{2}K^2-K\sqrt{1+\tfrac{1}{4}K^2}. \end{aligned}\right\} \tag{45.3}$$

For small shears we have $\lambda_1=1+\frac{1}{2}K+O(K^2)$, $\lambda_2=1-\frac{1}{2}K+O(K^2)$, while the principal axes approach the bisectors of the co-ordinate axes.

The shear experienced by the pair of elements $(1, 0, 0)$ and $(0, 1, 0)$ is τ, the angle of shear, but this is not the greatest shear. As follows from the general theory in Sect. 28, the greatest orthogonal shear is experienced by the elements bisecting the principal axes of strain in the plane of shear. One of these elements is inclined at angle $\frac{1}{2}\operatorname{arc\,tan}\frac{1}{2}K$ to the shearing planes. By applying (45.3) to (28.12) we conclude that the magnitude of the greatest shear, γ, is given by

$$\tan\gamma = K\sqrt{1+\tfrac{1}{4}K^2}. \tag{45.4}$$

Hence for small shears $\gamma=K+O(K^3)=\tau+O(\tau^3)$.

[1] Some authors (e.g. LOVE [1892, *8*, § 3]) use the term *angle of shear* for the angle through which an unstretched fibre in a plane of shear but not in a shearing plane is turned. As shown below, this angle is $2\operatorname{Arc\,tan}\frac{1}{2}K=2\vartheta$.

The principal axes of strain in the planes of shear in the deformed material are the lines into which the two previously determined axes are deformed. They may be calculated by the same construction, starting from the deformed rather than the undeformed positions of the lines whose length is unaltered. Hence the principal axis of greatest stretch in the deformed material subtends the angle $\frac{1}{4}\pi - \frac{1}{2}\arctan\frac{1}{2}K$ from the shearing planes. The axis of rotation is parallel to the Z axis, and the angle of rotation is given by

$$\vartheta = \arctan\tfrac{1}{2}K = \arctan\left(\tfrac{1}{2}\tan\tau\right). \tag{45.5}$$

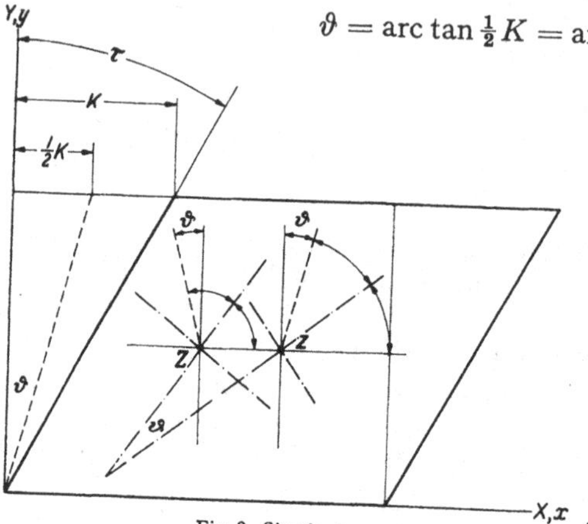

Fig. 8. Simple shear.

For small shear, $\vartheta = \frac{1}{2}\tau + O(\tau^3) = \frac{1}{2}K + O(K^3)$. In terms of the rotation, we may write (45.3) in the form

$$\left.\begin{aligned} \lambda_1 &= \sec\vartheta + \tan\vartheta, \\ \lambda_2 &= \sec\vartheta - \tan\vartheta. \end{aligned}\right\} \tag{45.6}$$

Hence[1]

$$\delta_1 - \delta_2 = \lambda_1 - \lambda_2 = 2\tan\vartheta = K. \tag{45.7}$$

The amount of shear is the difference of the principal extensions.

A homogeneous isochoric pure strain is sometimes called a *pure shear*[2]. The results just established are equivalent to resolution of a given simple shear into a pure shear followed by or preceded by a rotation about an axis normal to its plane.

The features of shear just discussed are illustrated in Fig. 8, which is accurately drawn for a shear angle $\frac{1}{6}\pi$. The portion of material shown is, before deformation, the intersection of a unit cube with a typical plane of shear; the particle $\boldsymbol{Z}$ selected for illustration is at the center of this area.

The foregoing geometric treatment, which follows KELVIN and TAIT, is elegantly simple. For solution of mechanical problems, however, it is more convenient to have at hand the explicit formulae for various tensors introduced earlier in the chapter. The Cauchy and Green tensors and their reciprocals have the values

$$\left.\begin{aligned} &\boldsymbol{c} = \begin{Vmatrix} 1 & -K & 0 \\ \cdot & 1+K^2 & 0 \\ \cdot & \cdot & 1 \end{Vmatrix}, \quad \boldsymbol{c}^{-1} = \begin{Vmatrix} 1+K^2 & K & 0 \\ \cdot & 1 & 0 \\ \cdot & \cdot & 1 \end{Vmatrix}, \\ &\boldsymbol{C} = \begin{Vmatrix} 1 & K & 0 \\ \cdot & 1+K^2 & 0 \\ \cdot & \cdot & 1 \end{Vmatrix}, \quad \boldsymbol{C}^{-1} = \begin{Vmatrix} 1+K^2 & -K & 0 \\ \cdot & 1 & 0 \\ \cdot & \cdot & 1 \end{Vmatrix}, \\ &\mathrm{I}_{\boldsymbol{C}} = \mathrm{II}_{\boldsymbol{C}} = 3 + K^2, \quad \mathrm{III}_{\boldsymbol{C}} = 1. \end{aligned}\right\} \tag{45.8}$$

The proper numbers of all four deformation tensors are 1 and λ_1^2 and λ_2^2 as given by (45.3). The unit proper vectors of $\boldsymbol{C}$ corresponding to these proper numbers, in the reverse order, are

$$\boldsymbol{N}_1, \boldsymbol{N}_2 = \frac{\boldsymbol{i} + \boldsymbol{j}\left(\frac{1}{2}K \pm \sqrt{1+\frac{1}{4}K^2}\right)}{\sqrt{2 + \frac{1}{2}K^2 \pm K\sqrt{1+\frac{1}{4}K^2}}}, \quad \boldsymbol{N}_3 = \boldsymbol{k}, \tag{45.9}$$

[1] CAFIERO [1906, *1*, Chap. II, § 16].
[2] LOVE [1906, *5*, § 2].

while the corresponding unit proper vectors of $\boldsymbol{c}$ are

$$\boldsymbol{n}_1, \boldsymbol{n}_2 = \frac{\boldsymbol{i} + \boldsymbol{j}\left(-\frac{1}{2}K \pm \sqrt{1 + \frac{1}{4}K^2}\right)}{\sqrt{2 + \frac{1}{2}K^2 \mp K\sqrt{1 + \frac{1}{4}K^2}}}, \qquad \boldsymbol{n}_3 = \boldsymbol{k}. \tag{45.10}$$

Since we are employing the common frame, there is no need to distinguish the different types of components of $\boldsymbol{R}$, and by substituting (45.9) and (45.10) into (37.2) we get

$$\boldsymbol{R} = \|R_M{}^m\| = \left\| \begin{matrix} \frac{1}{\sqrt{1+\frac{1}{4}K^2}} & \frac{\frac{1}{2}K}{\sqrt{1+\frac{1}{4}K^2}} & 0 \\ \frac{-\frac{1}{2}K}{\sqrt{1+\frac{1}{4}K^2}} & \frac{1}{\sqrt{1+\frac{1}{4}K^2}} & 0 \\ 0 & 0 & 1 \end{matrix} \right\|. \tag{45.11}$$

where the row index is M and the column index is m. Hence the axis of rotation is given by $\boldsymbol{A} = \pm \boldsymbol{k}$, while by (37.21) the angle and measure of rotation are given by

$$\cos\vartheta = \frac{1}{\sqrt{1+\frac{1}{4}K^2}}, \qquad \mathfrak{R} = \frac{1}{2}\left(1 - \frac{1}{\sqrt{1+\frac{1}{4}K^2}}\right). \tag{45.12}$$

This result for ϑ agrees with (45.5) and also follows directly from (45.9) and (45.10) and the fact that $\cos\vartheta = n_1^M N_{1M} = n_2^M N_{2M}$.

By (45.8), for the strain tensors $\boldsymbol{E}$ and $\boldsymbol{e}$ we have

$$\left.\begin{aligned} \boldsymbol{e} = \left\| \begin{matrix} 0 & \frac{1}{2}K & 0 \\ \cdot & -\frac{1}{2}K^2 & 0 \\ \cdot & \cdot & 0 \end{matrix} \right\|, \quad \boldsymbol{E} = \left\| \begin{matrix} 0 & \frac{1}{2}K & 0 \\ \cdot & \frac{1}{2}K^2 & 0 \\ \cdot & \cdot & 0 \end{matrix} \right\|, \\ \mathrm{I}_{\boldsymbol{e}} = -\mathrm{I}_{\boldsymbol{E}} = 2\,\mathrm{II}_{\boldsymbol{e}} = 2\,\mathrm{II}_{\boldsymbol{E}} = -\tfrac{1}{2}K^2, \end{aligned}\right\} \tag{45.13}$$

while for the tensors of elongation we have

$$\tilde{\boldsymbol{e}} = \tilde{\boldsymbol{E}} = \left\| \begin{matrix} 0 & \frac{1}{2}K & 0 \\ \cdot & 0 & 0 \\ \cdot & \cdot & 0 \end{matrix} \right\|. \tag{45.14}$$

Hence, as is obvious, the elongations of elements in the co-ordinate directions are all zero. The directions of elements suffering greatest and least elongation lie in the plane of shear and are inclined at angle $\frac{1}{4}\pi$ to the shearing planes, the extremal elongations being $\pm\frac{1}{2}K$. The axial vectors of mean rotation are given by

$$\tilde{\boldsymbol{r}} = \tilde{\boldsymbol{R}} = \tfrac{1}{2}\operatorname{curl}\boldsymbol{u} = -\tfrac{1}{2}K\boldsymbol{k} = -\tan\vartheta\,\boldsymbol{k}. \tag{45.15}$$

Of course this vector and the axis $\boldsymbol{A}$ are parallel, but only for small rotation does the magnitude of this vector equal the angle of rotation (45.5). This result is to be contrasted with the formula for the cross of $\boldsymbol{R}$:

$$\tfrac{1}{2}\boldsymbol{R}_\times = -\sin\vartheta\,\boldsymbol{k}, \tag{45.16}$$

which follows from the general formula $(37.19)_4$ and the fact that we may choose $-\boldsymbol{k}$ as the axis.

The concept of simple shear is not invariant, since it singles out a particular rotation. We may approach an invariant characterization as follows: *A homogeneous isochoric strain may be regarded as simple shear followed by or preceded by a rotation about the axis of shear if and only if*[1]

$$\mathrm{I}_C = \mathrm{II}_C; \tag{45.17}$$

equivalently, if and only if it is plane. The amount of shear is $\sqrt{\mathrm{I}_C - 3}$. The equivalence of the two conditions follows from (42.4). The theorem itself is geometrically obvious. For an algebraic proof, we need only notice that in any plane isochoric strain, as was remarked in Sect. 41, we have $\lambda_1 = 1/\lambda_2$, and hence by (45.6) and (45.5) follow uniquely determined values of ϑ and K.

The foregoing theorem is particularly useful when applied to deformations which are not homogeneous, as follows: *At a point where* $\mathrm{III}_C = 1$ *and* $\mathrm{I}_C = \mathrm{II}_C$, *the measures of strain and rotation have the same values as for a simple shear of amount* $K = \sqrt{\mathrm{I}_C - 3}$, *in the plane normal to the principal direction whose stretch is* 1.

46. Composition of shears and other special homogeneous strains. By (17.11), a succession of homogeneous strains may be regarded as a single homogeneous strain, but the result depends on the order in which the strains are taken, as follows from (17.12). However, no such rule as (17.11) holds when the matrices $\boldsymbol{D}$ are restricted to special subclasses such as pure strains, simple extensions, or simple shears. That is, while homogeneous strains constitute a group, none of the previously defined special types, apart from uniform dilation, form subgroups. This fact suggests that these special deformations may be used to build up a general homogeneous deformation. Two resolutions of this kind are stated now without proof[2]:

1. Any homogeneous strain may be produced by the succession of a simple shear, a simple extension (without transverse contraction) normal to the plane of shear, a uniform dilation, and a rotation.

2. Any homogeneous strain may be produced by the succession of three simple shears in mutually perpendicular planes, a uniform dilation, and a rotation.

As is indicated by No. 2, the result of two successive shears will in general be complicated[3]. However, in the special cases when the planes of the two shears are perpendicular and one or two families of planes are individually preserved in each deformation, the formulae are simple enough to be worth noting. First[4], if both the $Y = \mathrm{const}$ and the $Z = \mathrm{const}$ planes are preserved, we get

$$\left.\begin{aligned} \boldsymbol{D} &= \begin{Vmatrix} 1 & K_1 & K_2 \\ 0 & 1 & 0 \\ 0 & 0 & 1 \end{Vmatrix} = \begin{Vmatrix} 1 & K_1 & 0 \\ 0 & 1 & 0 \\ 0 & 0 & 1 \end{Vmatrix} \cdot \begin{Vmatrix} 1 & 0 & K_2 \\ 0 & 1 & 0 \\ 0 & 0 & 1 \end{Vmatrix}, \\ &= \boldsymbol{D}_1 \cdot \boldsymbol{D}_2 = \boldsymbol{D}_2 \cdot \boldsymbol{D}_1. \end{aligned}\right\} \tag{46.1}$$

Second[5], if only the $Y = \mathrm{const}$ planes are individually preserved, it follows that

$$\left.\begin{aligned} \boldsymbol{D} &= \begin{Vmatrix} 1 & K_1 & 0 \\ 0 & 1 & 0 \\ 0 & K_2 & 1 \end{Vmatrix} = \begin{Vmatrix} 1 & K_1 & 0 \\ 0 & 1 & 0 \\ 0 & 0 & 1 \end{Vmatrix} \cdot \begin{Vmatrix} 1 & 0 & 0 \\ 0 & 1 & 0 \\ 0 & K_2 & 1 \end{Vmatrix}, \\ &= \boldsymbol{D}_1 \cdot \boldsymbol{D}_2 = \boldsymbol{D}_2 \cdot \boldsymbol{D}_1. \end{aligned}\right\} \tag{46.2}$$

---

[1] LOVE [1892, *8*, § 7].

[2] The proofs are easy from geometrical considerations (cf. KELVIN and TAIT [1867, *3*, §§ 178–185]) or by theorems on the factorization of matrices. An analysis of shears into products of reflections has been given by WILSON [1907, *8*].

[3] Cf. LOVE [1892, *8*, Note A].

[4] E. and F. COSSERAT [1896, *1*, § 7].

[5] TRUESDELL [1952, *21*, § 42G].

For these two cases we get, respectively,

$$C = \begin{Vmatrix} 1 & K_1 & K_2 \\ \cdot & 1+K_1^2 & K_1 K_2 \\ \cdot & \cdot & 1+K_2^2 \end{Vmatrix}, \quad \begin{Vmatrix} 1 & K_1 & 0 \\ \cdot & 1+K_1^2+K_2^2 & K_2 \\ \cdot & \cdot & 1 \end{Vmatrix}. \tag{46.3}$$

Hence in both cases

$$I_C = II_C = 3 + K_1^2 + K_2^2, \quad III_C = 1. \tag{46.4}$$

By the criterion given at the end of Sect. 45, each of these composite shears may be regarded as a simple shear of amount[1] $K = \sqrt{K_1^2 + K_2^2}$. The principal axes of the resultant shear are of course different in the two cases. The planes of shearing are the planes normal to the proper vectors corresponding to the proper number 1. For the former, this gives $K_2 Y = K_1 Z$; for the latter, $K_2 X = K_1 Z$.

47. Generalized shear. In the deformations[2]

$$x = X f(Y) + g(Y), \qquad y = h(Y), \qquad z = k(Z) + l(Y), \tag{47.1}$$

the planes $Y =$ const are slid parallel to each other, but the direction and amount of the slide varies arbitrarily from plane to plane. At the same time, there are stretches in the X, Y, and Z directions; in particular, $\lambda_X = f(Y)$ and $\lambda_Z = k'(Z) = k'(k^{-1}(z - l(Y)))$, so that in a given plane $Y =$ const the stretch λ_Z depends only on z, and the stretch λ_X is constant. This class of deformations reduces to a special kind of generalized plane strain if $l(Y) = 0$; it includes simple extension, simple shear, and pure shear as special cases. We readily find that

$$\left.\begin{aligned} C &= \begin{Vmatrix} f^2 & (Xf' + g')f & 0 \\ \cdot & (Xf'+g')^2 + h'^2 + l'^2 & k'l' \\ \cdot & \cdot & k'^2 \end{Vmatrix}, \\ C^{-1} &= \begin{Vmatrix} f^2 + \left(\frac{x-g}{f} f' + g'\right)^2 & \left(\frac{x-g}{f} f' + g'\right) h' & \left(\frac{x-g}{f} f' + g'\right) l' \\ \cdot & h'^2 & h' l' \\ \cdot & \cdot & l'^2 + k'^2 \end{Vmatrix}, \\ I_C &= f^2 + h'^2 + k'^2 + l'^2 + (Xf' + g')^2, \\ II_C &= f^2(h'^2 + k'^2 + l'^2) + k'^2[(Xf'+g')^2 + h'^2], \\ III_C &= f^2 h'^2 k'^2. \end{aligned}\right\} \tag{47.2}$$

Hence such a deformation is isochoric if and only if $k' = \text{const} = \lambda$, say, while $f(Y)h'(Y) = 1/\lambda$.

Another generalization of simple shear, given by

$$x = X + f(Y), \quad y = Y + g(Z), \quad z = Z + h(X), \tag{47.3}$$

---

[1] For the former case, this was observed by LOVE [1892, *8*, § 10(iv)].

[2] This class of deformations combines the features of (46.2) and of a class considered by ADKINS [1955, *1*, § 3].

may be regarded as the succession of three variable shears in mutually perpendicular planes[1]. For this case we get

$$\left.\begin{aligned} &\boldsymbol{C}=\left\|\begin{matrix}1+h'^2 & f' & h'\\ \cdot & 1+f'^2 & g'\\ \cdot & \cdot & 1+g'^2,\end{matrix}\right\|, \quad \boldsymbol{c}^{-1}=\left\|\begin{matrix}1+f'^2 & f' & h'\\ \cdot & 1+g'^2 & g'\\ \cdot & \cdot & 1+h'^2\end{matrix}\right\|,\\ &\mathrm{I}_{\boldsymbol{C}}=3+f'^2+g'^2+h'^2,\\ &\mathrm{II}_{\boldsymbol{C}}=3+f'^2+g'^2+h'^2+f'^2g'^2+g'^2h'^2+h'^2f'^2,\\ &\mathrm{III}_{\boldsymbol{C}}=(1+f'g'h')^2.\end{aligned}\right\} \tag{47.4}$$

Thus the deformation (47.3) is isochoric if and only if at least one of the functions f, g, h reduces to a constant; that is, if and only if at least one of the three constituent shears is a simple shear. By (45.17), it is locally a simple shear if and only if at least two of the three constituent shears are simple shears.

SIGNORINI[2] has determined the most general deformation in which all components of $\boldsymbol{E}$ except a single shear component vanish.

48. Simple torsion. When a cylinder is twisted uniformly along its length, we get the isochoric deformation

$$r=R,\quad \theta=\Theta+KZ,\quad z=Z, \tag{48.1}$$

where R, Θ, Z and r, θ, z are cylindrical polar co-ordinates referred to the same system. Torsion deserves detailed analysis because it is the most important simple deformation which is neither a homogeneous strain nor a plane strain. The constant K is the *twist per unit length.* Since $G_{11}=1$, $G_{22}=R^2$; $G_{33}=1$, $g_{11}=1$, $g_{22}=r^2$, $g_{33}=1$, while all other components G_{MP} and g_{mp} vanish, from $(26.2)_2$ and $(29.1)_2$ we get

$$\left.\begin{aligned} &\|C^M_P\|=\left\|\begin{matrix}1 & 0 & 0\\ 0 & 1 & K\\ 0 & KR^2 & 1+K^2R^2\end{matrix}\right\|, \quad \left\|\overset{-1}{c}{}^k_p\right\|=\left\|\begin{matrix}1 & 0 & 0\\ 0 & 1+K^2r^2 & K\\ 0 & Kr^2 & 1\end{matrix}\right\|,\\ &\mathrm{I}_{\boldsymbol{C}}=\mathrm{II}_{\boldsymbol{C}}=3+K^2R^2,\quad \mathrm{III}_{\boldsymbol{C}}=1.\end{aligned}\right\} \tag{48.2}$$

The physical components of $\boldsymbol{C}$ and $\boldsymbol{c}$ are precisely equal to their counterparts for simple shear as given by (45.8), provided we take the plane of shear as the tangent plane to the cylinder $R=$const and the shearing planes as the planes $Z=$const, at the same time replacing K by KR. In other words, simple torsion may be regarded as effecting on each cylinder $R=$const, when cut along a generator and developed upon a plane, a simple shear of magnitude KR. Thus a complete analysis of torsion may be read off from the results obtained in Sect. 45. The formulae so gotten for $\boldsymbol{R}$ and other tensors are expressed in terms of physical components rather than tensor components. The angle of rotation is given by

$$\vartheta=\arctan\tfrac{1}{2}KR, \tag{48.3}$$

the principal stretches are 1, $\sec\vartheta+\tan\vartheta$, $\sec\vartheta-\tan\vartheta$, etc.

---

[1] This is not the succession of the three shears

$$\begin{aligned} X_1&=X+f(Y) & X_2&=X_1 & x&=X_2\\ Y_1&=Y & Y_2&=Y_1+g(Z_1) & y&=Y_2\\ Z_1&=Z & Z_2&=Z_1 & z&=Z_2+h(X_2),\end{aligned}$$

their composition being the isochoric deformation $x=X+f(Y)$, $y=Y+g(Z)$, $z=Z+h(X+f(Y))$.

[2] [1943, *6*, ¶ 30].

49. Generalized torsion, shear, and inflation of a circular cylinder. A family of deformations of torsional type is given by[1]

$$r = f(R), \qquad \theta = g(\Theta) + h(R)\,k(Z), \qquad z = l(Z) + m(\Theta) + n(R), \tag{49.1}$$

where the co-ordinates are chosen as in Sect. 48. If this deformation is considered throughout an entire cylinder, rather than merely a sector, g and m must be periodic of period 2π. The function f represents a radial expansion of the cylinders $R=$ const; g, an annular expansion; k, a torsion of magnitude varying from cylinder to cylinder, as moderated by h; l, an extension along the axis; m, a shearing of the generators of the cylinders; and n, a shearing of the cylinders along their generators. We readily calculate

$$\left.\begin{aligned}
\|C^T_U\| &= \left\|\begin{matrix} f'^2 + f^2 h'^2 k^2 + n'^2 & f^2 g' h' k + m' n' & f^2 h' k' h k + l' n' \\ (f^2 g' h' k + m' n')/R^2 & (f^2 g'^2 + m'^2)/R^2 & (f^2 g' h k' + l' m')/R^2 \\ f^2 h' k' h k + l' n' & f^2 g' h k' + l' m' & f^2 h^2 k'^2 + l'^2 \end{matrix}\right\|, \\
\|\overset{-1}{c}{}^{t}_{u}\| &= \left\|\begin{matrix} f'^2 & r^2 f' h' k & f' n' \\ f' h' k & r^2 (h'^2 k^2 + g'^2/R^2 + h^2 k'^2) & g' m'/R^2 + h k' l' + h' k n' \\ f' n' & r^2 (g' m'/R^2 + h k' l' + h' k n') & l'^2 + m'^2/R^2 + n'^2 \end{matrix}\right\|, \\
\mathrm{I}_C &= f^2 (h'^2 k^2 + g'^2/R^2 + h^2 k'^2) + f'^2 + l'^2 + m'^2/R^2 + n'^2, \\
\mathrm{II}_C &= f^2 [(h k' n' - h' k l')^2 + \{(h k' m' - g' l')^2 + (g' n' - m' k h')^2\}/R^2] + \\
&\quad + f'^2 [f^2 h^2 k'^2 + l'^2 + (f^2 g'^2 + m'^2)/R^2, \\
\mathrm{III}_C &= f^2 f'^2 (g' l' - m' h k')^2/R^2.
\end{aligned}\right\} \tag{49.2}$$

To find the isochoric subclass of these deformations, we set $\mathrm{III}_C = 1$ and solve the functional differential equation so obtained. If $h=$ const, the result is

$$r = \sqrt{A R^2 + B}, \quad \theta = C\Theta + DZ, \quad z = E\Theta + FZ + n(R), \quad A(CF - DE) = \pm 1; \tag{49.3}$$

if $h' g' k' l' m' \neq 0$, the result is

$$\left.\begin{aligned}
r &= \sqrt{\int_0^R \frac{2 A v\, dv}{h(v) + G} + B}, \\
\theta &= C\Theta + K + h(R)(DZ + L), \\
z &= E\Theta + FZ + n(R), \\
A C F &= \pm G, \qquad A D E = \mp 1;
\end{aligned}\right\} \tag{49.4}$$

if $h' \neq 0$, $g' = 0$, the result is

$$\left.\begin{aligned}
r &= \sqrt{\int_0^R \frac{2 A v\, dv}{h(v)} + B}, \\
\theta &= K + h(R)(DZ + L), \qquad z = l(Z) + E\Theta + n(R);
\end{aligned}\right\} \tag{49.5}$$

if $h' \neq 0$, $l' \neq 0$, the result is

$$\left.\begin{aligned}
r &= \sqrt{\int_0^R \frac{2 A v\, dv}{h(v)} + B}, \\
\theta &= g(\Theta) + h(R)(DZ + L), \qquad z = E\Theta + n(R);
\end{aligned}\right\} \tag{49.6}$$

[1] This family includes one introduced by ERICKSEN and RIVLIN [1954, *6*, §§ 8–10].

if $h' \neq 0$, $m' = 0$, the result is

$$r = \sqrt{A R^2 + B}, \qquad \theta = C\Theta + h(R)\,k(Z), \qquad z = FZ + n(R); \tag{49.7}$$

if $h' \neq 0$, $k' = 0$, the result is

$$r = \sqrt{A R^2 + B}, \qquad \theta = C\Theta + h(R), \qquad z = FZ + m(\Theta) + n(R), \tag{49.8}$$

where in all cases insignificant additive constants have been dropped.

Numerous interesting cases are included, of which we mention but a few. If $A < 0$, r is a decreasing function of R in $(49.3)_1$; such deformations represent the eversion of a hollow cylinder. For a cylinder which is subjected first to longitudinal stretch λ and then to twist τ, put $A = \lambda^{-1}$, $B = 0$, $C = 1$, $D = \tau/\lambda$, $E = 0$, $F = \lambda$, $n = 0$. If $A = \Theta_0/(2\pi K^2)$, $C = 2\pi/\Theta_0$, $F = K^2$, $B = D = E = n = 0$, the deformation represents the closing of a circular cylindrical shell of wedge angle Θ_0 so as to form a complete cylinder, furnishing an example of the *distorsioni* introduced by WEINGARTEN and VOLTERRA[1]. If $A = C = F = 1$, $B = D = E = 0$, then (49.2) reduces to the forms included in (45.8), up to a permutation of co-ordinates, if K is replaced by n'. Thus the result of a shearing along the generators is locally the same as a simple shear of amount n' in the planes $\Theta = \text{const}$. A similar remark holds for the case $f' = g' = l' = 1$, $k = 1$, $m = n = 0$, except that physical components must be used, the amount of shear is $r h'$, and the planes of shearing are the $Z = \text{const}$ planes.

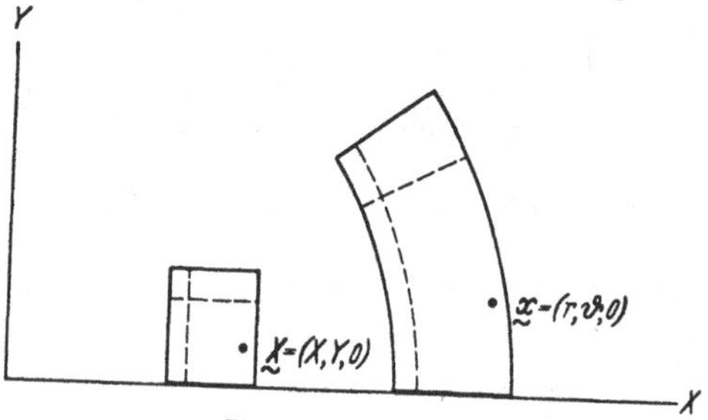

Fig. 9. Bending of a block.

50. Bending of a block into a cylindrical wedge.

As our first example of the advantage of the use of different co-ordinate systems at $\boldsymbol{X}$ and $\boldsymbol{x}$, we consider the deformation[2]

$$\left.\begin{aligned} r &= f(X), \quad \theta = g(Y), \\ z &= h(Z), \end{aligned}\right\} \tag{50.1}$$

where for the most applications it is desirable, but not necessary, to regard the rectangular Cartesian X, Y, Z system and the cylindrical polar r, θ, z system as having a common origin and a common z-Z axis. The family of planes $Z = \text{const}$ is preserved; the $Y = \text{const}$ planes are deformed into planes intersecting along the z axis; the planes $X = \text{const}$ are deformed into concentric circular cylinders whose axis is the Z-z axis. The deformation is most easily visualized by restricting attention to a block whose faces are X, Y, Z planes and which does not include the origin (Fig. 9).

Since $G_{KM} = \delta_{KM}$, while $g_{11} = g_{33} = 1$, $g_{22} = r^2$, $g_{km} = 0$ if $k \neq m$, by straightforward calculation from $(26.2)_2$ and $(29.1)_1$ we get

$$\left.\begin{aligned} &\|\overset{-1}{c}{}^k_m\| = \|C^K_M\| = \left\|\begin{matrix} f'^2 & 0 & 0 \\ \cdot & r^2 g'^2 & 0 \\ \cdot & \cdot & h'^2 \end{matrix}\right\|, \\ &\mathrm{I}_C = f'^2 + r^2 g'^2 + h'^2, \qquad \mathrm{II}_C = f'^2 h'^2 + r^2 g'^2 (f'^2 + h'^2), \\ &\mathrm{III}_C = r^2 f'^2 g'^2 h'^2. \end{aligned}\right\} \tag{50.2}$$

To interpret the formulae just obtained, note that corresponding components of $\boldsymbol{c}^{-1}$ and $\boldsymbol{C}$ have the same numerical values for a given particle; of course, if we regard $\boldsymbol{c}^{-1}$ as a function of $\boldsymbol{x}$ and $\boldsymbol{C}$ as a function of $\boldsymbol{X}$, the functions are different. For example, $\overset{-1}{c}{}^\theta_\theta = C^Y_Y$, but if we take $\boldsymbol{c}^{-1}$ as a function of $\boldsymbol{x}$ and $\boldsymbol{C}$ as a function of $\boldsymbol{X}$, we have $\overset{-1}{c}{}^\theta_\theta = [r\,g'(g^{-1}(\theta))]^2 = C^Y_Y = [f(X)\,g'(Y)]^2$.

[1] [1901, *15*]; [1905, *6*].

[2] We follow TRUESDELL [1952, *21*, § 42I]. SIGNORINI [1943, *6*, § 29] approaches the case $z = Z$ through conditions of integrability for the pure strain.

Since (50.2) is in diagonal form, the directions of the co-ordinate curves both at $\boldsymbol{X}$ and $\boldsymbol{x}$ are principal directions of strain, and the principal stretches are f', rg', h'. The contravariant components of the unit proper vectors, expressed in the appropriate co-ordinates, are given by

$$\left.\begin{aligned} &\boldsymbol{N}_1 = (1, 0, 0), \quad &&\boldsymbol{N}_2 = (0, 1, 0), \quad &&\boldsymbol{N}_3 = (0, 0, 1), \\ &\boldsymbol{n}_1 = (1, 0, 0), \quad &&\boldsymbol{n}_2 = \left(0, \tfrac{1}{r}, 0\right), \quad &&\boldsymbol{n}_3 = (0, 0, 1), \end{aligned}\right\} \tag{50.3}$$

whence follows by $(37.2)_1$

$$\|R^k{}_K\| = \begin{Vmatrix} 1 & 0 & 0 \\ \cdot & r^{-1} & 0 \\ \cdot & \cdot & 1 \end{Vmatrix}. \tag{50.4}$$

To calculate $R^k{}_m$ from this result by the formula $R^k{}_m = g^K_m R^k{}_K$, we need the values of the shifter g^K_k, given by (App. 17.2) for the case when the co-ordinate systems at $\boldsymbol{X}$ and at $\boldsymbol{x}$ have a common origin and a common Z-z axis. Hence follows

$$\|R^k{}_m\| = \begin{Vmatrix} \cos\theta & -r\sin\theta & 0 \\ \frac{1}{r}\sin\theta & \cos\theta & 0 \\ 0 & 0 & 1 \end{Vmatrix}, \tag{50.5}$$

so that the axis of rotation is the z axis, and by (37.21) we get

$$\cos\vartheta = \cos\theta, \quad \text{or} \quad \vartheta = \pm\theta: \tag{50.6}$$

Apart from a possible reflection, the angle of rotation is the azimuth angle. The result is obvious, but we have given the foregoing formulae in illustration of the general apparatus of Sect. 37.

The isochoric subclass of (50.1) is

$$r = \sqrt{2AX + B}, \qquad \theta = CY, \qquad z = DZ, \qquad ACD = \pm 1, \tag{50.7}$$

where two constants of integration have been adjusted so as to center the deformed block with respect to the co-ordinates at $\boldsymbol{X}$.

If we wish to allow shearing of the cylindrical surfaces, we may generalize (50.1) by

$$r = f(X), \qquad \theta = g(Y) + k(X), \qquad z = h(Z) + l(X), \tag{50.8}$$

resulting in

$$\left.\begin{aligned} &\|\overset{-1}{c}{}^k_m\| = \begin{Vmatrix} f'^2 & r^2 f' k' & f' l' \\ f' k' & r^2(g'^2 + k'^2) & k' l' \\ f' l' & r^2 k' l' & h'^2 + l'^2 \end{Vmatrix}, \quad \|C^K_M\| = \begin{Vmatrix} f'^2 + f^2 k'^2 + l'^2 & f^2 g' k' & h' l' \\ \cdot & f^2 g'^2 & 0 \\ \cdot & \cdot & h'^2 \end{Vmatrix}, \\ &\mathrm{I}_C = f'^2 + r^2(g'^2 + k'^2) + h'^2 + l'^2, \\ &\mathrm{II}_C = f'^2 h'^2 + r^2[g'^2(f'^2 + h'^2 + l'^2) + h'^2 k'^2], \\ &\mathrm{III}_C = r^2 f'^2 g'^2 h'^2. \end{aligned}\right\} \tag{50.9}$$

The isochoric subclass of these deformations is obtained by adding to the second members of (50.7) the functions 0, $k(X)$, $l(X)$.

51. Bending of a block into a spherical wedge. If we let the co-ordinates at $\boldsymbol{x}$ be spherical polar, φ being the azimuth angle, we get in analogy to (50.8)

$$r = f(X), \qquad \theta = g(Y) + k(X), \qquad \varphi = h(Z) + l(X). \tag{51.1}$$

Analogous calculations yield

$$\left.\begin{aligned}
\|\overset{-1}{c}{}^k_m\| &= \begin{Vmatrix} f'^2 & r^2 f' k' & r^2\sin^2\theta\, f' l' \\ f' k' & r^2(g'^2+k'^2) & r^2\sin^2\theta\, k' l' \\ f' l' & r^2 k' l' & r^2\sin^2\theta\,(h'^2+l'^2) \end{Vmatrix},\\
\|C^K_M\| &= \begin{Vmatrix} f'^2+r^2k'^2+r^2\sin^2\theta\, l'^2 & r^2 g' k' & r^2\sin^2\theta\, h' l' \\ \cdot & r^2 g'^2 & 0 \\ \cdot & \cdot & r^2\sin^2\theta\, h'^2 \end{Vmatrix},\\
\mathrm{I}_C &= f'^2 + r^2(g'^2+k'^2) + r^2\sin^2\theta\,(h'^2+l'^2),\\
\mathrm{II}_C &= r^4\sin^2\theta\,(g'^2h'^2+h'^2k'^2+g'^2l'^2) + r^2\sin^2\theta\, h'^2 f'^2 + r^2 f'^2 g'^2,\\
\mathrm{III}_C &= r^4\sin^2\theta\, f'^2 g'^2 h'^2.
\end{aligned}\right\} \qquad (51.2)$$

When the shearing functions k and l vanish, the co-ordinate axes are principal axes of strain, and counterparts of the results at the beginning of Sect. 50 are easily obtained.

For the isochoric case, it is easy to show that h' and k must be constants. Without loss of generality we can then set $k=0$, $h'=H$, and obtain

$$r=\sqrt[3]{A\,X+B}, \qquad \cos\theta = \pm\frac{3}{A\,H}\,Y+C, \qquad \varphi = H\,Z + l(X). \qquad (51.3)$$

52. Inflation, torsion, and shear of a sphere. Let the co-ordinates at $\boldsymbol{x}$ and at $\boldsymbol{X}$ be spherical polar, and suppose

$$r=f(R), \qquad \theta = g(\Theta)+h(R)\,k(\Phi), \qquad \varphi = l(\Phi). \qquad (52.1)$$

Then

$$\left.\begin{aligned}
\|\overset{-1}{c}{}^k_m\| &= \begin{Vmatrix} f'^2 & r^2 f' h' k & 0 \\ f' h' k & r^2\left(h'^2k^2+g'^2/R^2+\dfrac{h^2k'^2}{R^2\sin^2\Theta}\right) & \dfrac{r^2\sin^2\theta}{R^2\sin^2\Theta}\,h\,k'\,l' \\ 0 & \dfrac{r^2}{R^2\sin^2\Theta}\,h\,k'\,l' & \dfrac{r^2\sin^2\theta}{R^2\sin^2\Theta}\,l'^2 \end{Vmatrix},\\
\|C^K_M\| &= \begin{Vmatrix} f'^2+r^2h'^2k^2 & r^2 g' h' k & r^2 h\,h'\,k\,k' \\ \dfrac{r^2 g' h' k}{R^2} & \dfrac{r^2 g'^2}{R^2} & \dfrac{r^2 g' h k'}{R^2} \\ \dfrac{r^2 h\,h'\,k\,k'}{R^2\sin^2\Theta} & \dfrac{r^2 g' h k'}{R^2\sin^2\Theta} & \dfrac{r^2h^2k'^2+r^2\sin^2\theta\, l'^2}{R^2\sin^2\Theta} \end{Vmatrix},\\
\mathrm{I}_C &= f'^2 + r^2h'^2k^2 + \frac{r^2 g'^2}{R^2} + \frac{r^2(h^2k'^2+\sin^2\theta\, l'^2)}{R^2\sin^2\Theta},\\
\mathrm{II}_C &= \frac{r^2 f'^2 g'^2}{R^2} + \frac{r^2\left[l'^2\sin^2\theta\,(f'^2+r^2\{h'^2k^2+g'^2/R^2\}) + f'^2h^2k'^2\right]}{R^2\sin^2\Theta},\\
\mathrm{III}_C &= \frac{r^4\sin^2\theta}{R^4\sin^2\Theta}\, f'^2 g'^2 l'^2.
\end{aligned}\right\} \qquad (52.2)$$

The isochoric subclass in general does not appear to be easy to determine unless h and k are constants. Taking them as zero yields for the isochoric subclass

$$r=\sqrt[3]{A\,R^3+B}, \qquad \theta = \pm\,\mathrm{arc}\cos\left(\frac{\cos\Theta}{A\,L}\right), \qquad \varphi = L\,\Phi, \qquad (52.3)$$

where insignificant constants have been dropped. For this subclass, as also more generally whenever h and k are constants, the co-ordinate directions are principal axes of strain. Even though the two sets of co-ordinate surfaces coincide, it does not then follow that the deformation is a pure strain. In fact, it is evident that the principal axes of strain experience non-zero rotation unless

$$g(\Theta) = \Theta, \quad l(\Phi) = \Phi.$$

Since f may be a decreasing function of R, eversion of a spherical shell is included as a special case in (52.1) and (52.3).

e) Small deformation[1].

53. The meaning of "small". The looseness with which the term "small" is often used has led to much confusion and some error. In this work, unless the contrary is stated explicitly, by "A is small when B is small" we mean "$B \to 0$ implies $A \to 0$". By

$$\text{"}A(y_1, y_2, \ldots, y_n) \approx B(y_1, y_2, \ldots, y_n) \quad \text{when} \quad y_1, \ldots, y_n \quad \text{are small"} \tag{53.1}$$

we mean "A and B are analytic functions of $y_1, \ldots, y_n$ at $y_1 = y_2 = \cdots = y_n = 0$, and the expansions for A and B in powers of the y_k have the same non-identically zero leading terms". (As is well known, successive applications of this definition may lead to seemingly inconsistent results[2].)

While a more general definition could easily be formulated, this one suffices for unequivocal and correct statements of results of the type used by authors who neglect terms because they are "small".

We must warn the reader however, that it is unjustified to assume that the sequences of deformations standing behind the limit processes used in the following purely kinematical statements are compatible with any particular theory of continuous bodies. For example, we sometimes consider extension and rotation as *independently* variable, but in any particular theory of bodies a sequence of deformations is produced by a sequence of loads, and whether it is possible to choose the latter in such a way as to leave the local rotation fixed but vary the extensions is dubious.

It does not seem possible to give a precise meaning to the concept of "numerically small" often encountered in physical works. For example, if by "$x \ll 1$" we agree to mean "$x \leqq 10^{-6}$", then certainly $10^6 \cdot 10^{-6}$ is not small. However, when this concept of smallness is used, the users invariably manipulate it in a fashion justified only if the sum of *any* finite number of small quantities is small, and a constant multiple of any small quantity is small.

54. Small principal extensions[3]. When the three principal extensions δ_a are small, by (31.4) we have

$$E_a \approx e_a \approx \delta_a. \tag{54.1}$$

From (30.8) and (App. 39.11) it follows that every one of the components E_{KM} and e_{km} is small. By (26.4) and (31.1), the extension in every direction is small,

---

[1] We do not have space to include the large body of results on infinitesimal deformation of surfaces.

[2] For example, if $f = y^3$ and $g = y$, f and g are small when y is small, and also y is small if either f or g is small. However, if we set

$$v = f^3 + f g^2 + f g^3 = y^5 + y^6 + y^9,$$

we obtain

$$\begin{aligned} v &\approx f^3 = y^9 \text{ when } g \text{ is small} \\ &\approx f g^2 + f g^3 = y^5 + y^6 \text{ when } f \text{ is small} \\ &\approx f^3 + f g^2 = y^5 + y^9 \text{ when } \textit{both } f \text{ and } g \text{ are small} \\ &\approx y^5 \text{ when } y \text{ is small.} \end{aligned}$$

Thus to assume both f and g small is not the same thing as to assume y small.

[3] While the results of this section are due to CAUCHY [1827, 2], he did not distinguish the case considered here from the more restricitve one considered in Sect. 56.

and
$$\frac{E_{KK}}{g_{KK}} \approx \Delta_{KK}, \quad \frac{e_{kk}}{g_{kk}} \approx \delta_{kk}. \tag{54.2}$$

From (26.9) and (31.1) we see that the shears are small, and after some manipulation we derive the formula

$$\Gamma_{KM} \approx \frac{2E_{KM}}{\sqrt{g_{KK}}\sqrt{g_{MM}}} \operatorname{cosec} \Theta_{KM} - \left(\frac{E_{KK}}{g_{KK}} + \frac{E_{MM}}{g_{MM}}\right) \cot \Theta_{KM}, \quad K \neq M. \tag{54.3}$$

As we have said, from (54.1) it follows that the shears are small. It would not be sufficient to replace (54.1) by the assumption that the extensions in some non-principal orthogonal triad of directions are small. To see this, note that by (26.5) and (31.1) the vanishing of the extensions in three orthogonal directions yields $\mathrm{I}_{\boldsymbol{E}}=0$ but imposes no bounds on $\mathrm{II}_{\boldsymbol{E}}$ and $\mathrm{III}_{\boldsymbol{E}}$. Hence the δ_a may be arbitrarily large. To describe the case considered in this section without reference to the principal directions we should have to speak of *small extensions and shears.*

For an orthogonal system of co-ordinates at $\boldsymbol{X}$ it follows that $E_1^1 \approx$ the extension in the X^1 direction, $2(\sqrt{g_{11}}/\sqrt{g_{22}})E_2^1 \approx$ the shear of the X^1 and X^2 directions. In other words, for the physical components $E\langle KM\rangle$ we have

$$\|E\langle KM\rangle\| \approx \left\| \begin{matrix} \Delta_{11} & \frac{1}{2}\Gamma_{12} & \frac{1}{2}\Gamma_{13} \\ \cdot & \Delta_{22} & \frac{1}{2}\Gamma_{23} \\ \cdot & \cdot & \Delta_{33} \end{matrix} \right\| : \tag{54.4}$$

In an orthogonal co-ordinate system, the physical components of $\boldsymbol{E}$ *for small extensions equal the corresponding extensions and halves of the shears.* The dual formula holds for $\boldsymbol{e}$, but there is in general no simple relation between the two sets of extensions and shears.

From (54.1), $(31.5)_5$ and $(31.5)_7$ we get[1]

$$\frac{dv - dV}{dV} \approx \mathrm{I}_{\boldsymbol{E}} \approx \mathrm{I}_{\boldsymbol{e}}: \tag{54.5}$$

For small extensions, the first invariants of $\boldsymbol{E}$ *and* $\boldsymbol{e}$ *equal the relative increment of volume.* The fundamental identities (30.2) now assume the forms

$$\mathrm{I}_{\boldsymbol{E}} \approx \mathrm{I}_{\boldsymbol{e}} \approx \mathrm{I}_{\delta}, \quad \mathrm{II}_{\boldsymbol{E}} \approx \mathrm{II}_{\boldsymbol{e}} \approx \mathrm{II}_{\delta}, \quad \mathrm{III}_{\boldsymbol{E}} \approx \mathrm{III}_{\boldsymbol{e}} \approx \mathrm{III}_{\delta}, \tag{54.6}$$

while the fundamental lemma of Sect. 30 may be expressed thus: Any scalar function of $\boldsymbol{E}$ equals *the same* scalar function of $\boldsymbol{e}$. Also from (30.8) follows[2]

$$\mathrm{I}_{\boldsymbol{C}} \approx 3 + 2\mathrm{I}_{\delta}, \quad \mathrm{II}_{\boldsymbol{C}} \approx 3 + 4\mathrm{I}_{\delta}, \quad \mathrm{III}_{\boldsymbol{C}} \approx 1 + 2\mathrm{I}_{\delta}. \tag{54.7}$$

It is only in the case of small extensions that the tensors $\boldsymbol{E}$ and $\boldsymbol{e}$ are convenient measures of strain.

From (31.1) follows a formula which in itself includes much of what we have just derived:

$$\boldsymbol{C}^{\frac{1}{2}} \approx \boldsymbol{1} + \boldsymbol{E}. \tag{54.8}$$

By (37.15) we now get simple expressions for the measures of rotation and elongation:

$$\left.\begin{aligned} \widetilde{R}_{KM} &\approx R_{[KM]} + R_{[K}{}^{P} E_{M]P}, \\ \widetilde{E}_{KM} &\approx R_{(KM)} + R_{(K}{}^{P} E_{M)P} - g_{KM}, \\ R^{K}{}_{M} &\approx \delta^K_M + \widetilde{R}^{K}{}_{M} + \widetilde{E}^K_M - E^K_M - E^P_M u^K_{;P}, \end{aligned}\right\} \tag{54.9}$$

[1] CAUCHY [1827, 2, Eq. (33)].

[2] Conditions of compatibility (Sect. 34) appropriate to the case of small extension have been discussed by CASTOLDI [1954, 2].

where we have retained terms of order 1 in the extensions. If we revert to our usual usage of "small", we get simply

$$\left.\begin{aligned} \tilde{R}_{KM} &\approx R_{[KM]}, & \tilde{E}_{KM} &\approx R_{(KM)} - g_{KM}, \\ R^K{}_M &\approx \delta^K_M + u^K_{;M}, & \overset{-1}{R}{}^K{}_M &\approx \delta^K_M - u^K_{;M}. \end{aligned}\right\} \tag{54.10}$$

Note that for these simple formulae to be true, the rotations need not be small. E.g., in any rigid rotation (54.10) holds, with $\approx$ replaced by $=$.

55. Small rotation[1]. The term "small rotation" shall refer to the case when the rotation ϑ of *every* element is small. By (35.3) follows[2]

$$\varepsilon = \delta - \tfrac{1}{2}(1+\delta)\,\vartheta^2 + (1+\delta)\,O(\vartheta^4). \tag{55.1}$$

Also, from (36.7) and (36.8) it follows that τ_X is small and is given by

$$\tau_X \approx \frac{-\tilde{R}_{YZ}}{\sqrt{(1+\tilde{E}_{YY})(1+\tilde{E}_{ZZ}) - \tilde{E}^2_{YZ}}}. \tag{55.2}$$

Consequently $\tilde{\boldsymbol{R}}$ is small when the rotation is small. By the results of Sect. 38, then, a small potential deformation is a pure strain; alternatively: *For the case of small rotation, a necessary and sufficient condition for pure strain is* $\tilde{\boldsymbol{R}} = 0$.

From (37.17) follows

$$R^K{}_M \approx \delta^K_M + \vartheta\, e^K{}_{MP} A^P, \qquad R_{[KM]} \approx \vartheta\, e_{KMP} A^P, \tag{55.3}$$

from which is plain the vectorial character of small rotations:

$${}_1R^K{}_P\, {}_2R^P{}_M \approx \delta^K_M + e^K{}_{MP}(\vartheta_1 A^P_1 + \vartheta_2 A^P_2). \tag{55.4}$$

Also, comparison of $(55.3)_1$ with $(55.3)_2$ shows that the rotation tensor $\boldsymbol{R}$ is completely characterized by its cross; writing $\boldsymbol{R} \equiv \tfrac{1}{2}\boldsymbol{R}_\times$, we have

$$R^K \approx \vartheta\, A^K. \tag{55.5}$$

Extending the convention of Sect. 36, we now use the same symbol $\boldsymbol{R}$ to denote at will either the rotation tensor or its cross; the context will show which one is intended.

56. Small extensions and small rotation. When both the rotations and the extensions are small, we may combine the results of the last two sections. From (55.1) it follows that the elongations are small, and hence the components $\tilde{E}_{KM}$ are small. By (55.2) and (36.12) we get

$$\tau_X \approx -\tilde{R}_{YZ} = \tilde{R}_X: \tag{56.1}$$

For small extension and rotation, the length of the projection of the vector $\tilde{\boldsymbol{R}}$ *upon a given direction is the mean rotation about that direction*[3]. While $\tilde{\boldsymbol{R}}$ is a small

---

[1] That it is possible in interesting cases of deformation for the rotation to be numerically large when the extension is numerically small was noticed by St. Venant [1844, *3*] [1847, *3*], who mentioned the following examples: a thin sheet may be bent back so that its two ends touch, a long slender shaft may be twisted through several diameters. Cf. Poincaré [1892, *11*, § 2], Almansi [1917, *1*, § 3], Kappus [1939, *11*, § 4, footnote]. It is possible to exhibit families of such deformations in which $\boldsymbol{E} \to 0$ while $\tilde{\boldsymbol{E}}$ remains non-zero. See [1944, *4*], where Cisotti, using a previously derived result [1932, *2*], has calculated $\tilde{\boldsymbol{E}}$ for the case of a rigid displacement, when $\boldsymbol{E} = 0$. Cf. also Castoldi [1954, *3*].

[2] Novozhilov [1948, *18*, Eq. (1.110)].

[3] Cauchy [1841, *1*, Ths. 5, 6, 7].

vector representing mean rotations and $\boldsymbol{R}$ as given by (55.5) is a small vector representing rotation of the axes of strain, the two vectors are distinct unless further assumptions are made.

Recalling our definition of "small" in Sect. 53, we see that in the present case (31.10) reduces to[1]

$$E^K_M \approx \tilde{E}^K_M + \tfrac{1}{2}(-\tilde{E}^{KP}\tilde{R}_{MP} - \tilde{R}^{KP}\tilde{E}_{MP} + \tilde{R}^{KP}\tilde{R}_{MP}), \tag{56.2}$$

while (55.1) becomes

$$\varepsilon \approx \delta - \tfrac{1}{2}\vartheta^2. \tag{56.3}$$

57. Small displacement gradients. If the extensions and rotations are small, it follows that the gradients $u_{K;M}$ are small. However, according to the definition in Sect. 53, to suppose conversely that the $u_{K;M}$ are small is not the same thing. In fact, if we retain only terms of lowest order in expansions in powers of the $u_{K;M}$ we get not only all the results of Sects. 54 to 56 but also in place of (56.2) simply[2]

$$\boldsymbol{E} \approx \tilde{\boldsymbol{E}}, \quad \boldsymbol{e} \approx \tilde{\boldsymbol{e}}: \tag{57.1}$$

When the displacement gradients are small, the strain and elongation tensors are equal, their physical components being related to the extensions and shears by (54.4).

By observing (57.1) and (56.1), we see that the invariant identity

$$u_{K;M} = \tilde{E}_{KM} + \tilde{R}_{KM} \tag{57.2}$$

now expresses an *additive decomposition of any displacement into a pure strain and a mean rotation.* This is ***Cauchy's resolution of a displacement with small gradients***[3]. For an alternative interpretation based on the rotation of the principal axes, we may calculate the rotation $\boldsymbol{R}$ from $(54.9)_3$, by (57.1) now obtaining[4]

$$R^K{}_M \approx \delta^K_M + \tilde{R}^K{}_M, \quad \overset{-1}{R}{}^k{}_m \approx \delta^k_m - \tilde{r}^k{}_m. \tag{57.3}$$

Hence

$$R^K{}_M \tilde{R}^M \approx \tilde{R}^K + \tilde{R}^K{}_M \tilde{R}^M \approx \tilde{R}^K. \tag{57.4}$$

This last equation states that the axis of rotation is a unit vector in the direction of $\tilde{\boldsymbol{R}}$, i.e., $A^K = \pm \tilde{R}^K/\tilde{R}$. If we attempt to calculate the angle of rotation from (57.3), we get only 0. Instead, we go back to $(54.9)_3$, obtaining

$$R^K{}_K \approx 3 + \tilde{E}^K_K - E^K_K \approx 3 - \tfrac{1}{2}\tilde{R}^{KM}\tilde{R}_{KM}, \tag{57.5}$$

where the last step follows by (31.10). By $(37.21)_2$ and (36.12) we get

$$\vartheta = \pm \tilde{R}: \tag{57.6}$$

When the displacement gradients are small, the vector $\tilde{\boldsymbol{R}}$ points along the axis of rotation and is of magnitude equal to the angle of rotation, in radians[5].

---

[1] We do not follow the argument of NOVOZHILOV [1948, *18*, § 14], according to which the two terms with minus signs are dropped. We note also that NOVOZHILOV's equation (1.105) is false.

[2] CAUCHY [1827, *2*, Eq. (41)].

[3] Implied by [1841, *1*, Part II], but not explicitly stated.

[4] To obtain $(54.10)_{3,4}$, we held the rotation fixed and let the extensions approach zero. The apparently contradictory formula (57.3) results from the limit $u_{K;M} \to 0$, implying that both the extensions and the rotations approach zero together and at certain relative rates implied by their relation to the quantities $u_{K;M}$. This illustrates the remarks in Sect. 53[1].

[5] E. and F. COSSERAT [1896, *1*, § 11]. Other proofs are given by BOUSSINESQ [1912, *1*, §§ 9–10] and ODONE [1933, *9*].

A different interpretation of $\tilde{\boldsymbol{R}}$, due to CAUCHY[1], is more conveniently presented in Sect. 86.

Also by (57.1) and (34.8) it follows at once[2] that *when the displacement gradients are small, the conditions of compatibility assume* ST. VENANT'S *form:*

$$\delta^{RS}_{KM}\,\delta^{TU}_{PQ}\,E_{RT;SU} = 0. \tag{57.7}$$

These equations possess an extensive literature[3].

Since the co-ordinates at $\boldsymbol{X}$ and at $\boldsymbol{x}$ are independently selected, no simple relation between $\boldsymbol{E}$ and $\boldsymbol{e}$ can be expected. From (19.5), however, we conclude that if the $u^k{}_{;m}$ are small, so are the $u^K{}_{;M}$, and conversely, and moreover

$$u^K_{;M} \approx g^K_k\, g^m_M\, u^k_{;m}. \tag{57.8}$$

That is, *when the displacement gradients are small, $u_{k;m}$ and $u_{K;M}$ are components of the same tensor at $\boldsymbol{x}$ and at $\boldsymbol{X}$, respectively*, and consequently the same holds of $\tilde{\boldsymbol{R}}$ and $\tilde{\boldsymbol{r}}$, $\tilde{\boldsymbol{E}}$ and $\tilde{\boldsymbol{e}}$:

$$\tilde{R}_{KM} \approx g^k_K\, g^m_M\, \tilde{r}_{km}, \quad \tilde{E}_{KM} \approx g^k_K\, g^m_M\, \tilde{e}_{km}. \tag{57.9}$$

By (57.1), analogous relations hold between the R_{KM} and the R_{km}, the E_{KM} and the e_{km}. As a corollary, it follows that *for small displacement gradients, corresponding components of $\boldsymbol{R}+1$, $\tilde{\boldsymbol{R}}$, and $\tilde{\boldsymbol{r}}$, and of $\boldsymbol{E}$, $\boldsymbol{e}$, $\tilde{\boldsymbol{E}}$, and $\tilde{\boldsymbol{e}}$ in the common frame are numerically equal:*

$$\left.\begin{aligned} 1+R_{XX} &\approx 1+R_{xx} \approx \tilde{R}_{XX} \approx \tilde{r}_{xx},\\ R_{XY} &\approx R_{xy} \approx \tilde{R}_{XY} \approx \tilde{r}_{xy}, \ldots,\\ E_{XX} &\approx e_{xx} \approx \tilde{E}_{XX} \approx \tilde{e}_{xx},\\ E_{XY} &\approx e_{xy} \approx \tilde{E}_{XY} \approx \tilde{e}_{xy}, \ldots. \end{aligned}\right\} \tag{57.10}$$

For a curvilinear co-ordinate system, no such result as (57.10) holds without further assumptions.

---

[1] [1841, *1*, Ths. 5—9].

[2] It follows, that is, if we add to our definition of "small" in Sect. 53 the convention that the derivative of a small quantity is small. Such a convention may be achieved by restricting all comparisons to specially selected sequences of functions, e.g. by a perturbation series such as that of Sect. 59. While we follow the custom of writers on elasticity in presenting the conditions of compatibility in this way, we feel compelled to point out that we do not share the confidence many authors put in these formal procedures. The mathematical questions at issue are these. Given a family of displacement vectors ${}^B\boldsymbol{u}$ depending upon a parameter B in such a way that ${}^Bu_{K;M}\to 0$ as $B\to 0$, is it true that $R^{(C)}_{KMPQ} \sim \delta^{RS}_{KM}\,\delta^{TU}_{PQ}\,\tilde{E}_{RT;SU}$ as $B\to 0$? Conversely, if $\tilde{E}_{KM}$ satisfies (34.8), let $\boldsymbol{u}$ be the corresponding displacement; then does there exist a family ${}^B\boldsymbol{u}$ which for each B satisfies (34.4) and also satisfies ${}^B\boldsymbol{u}\sim\boldsymbol{u}$ as $B\to 0$?

CASTOLDI [1954, *2*] derives forms of the conditions of compatibility for the case of small extensions and for the case of small extensions with small gradients of the extensions. For the latter case he obtains precisely (57.7); hence there is a vector $\boldsymbol{v}$ such that $\tilde{E}_{KM} = v_{(K;M)}$, and he finds that this vector $\boldsymbol{v}$ is either the displacement vector $\boldsymbol{u}$ or else $v_K = u_K + \frac{1}{4}(u^M u_M)_{;K}$.

[3] They were given by KIRCHHOFF [1859, *2*, §§ 1—2], but without a clear statement of their meaning, which was first explained by ST. VENANT [1864, *3*, § 32]. The classical arguments do not deal with the question raised in the preceding footnote but in fact concern the necessity and sufficiency of (34.8) as conditions on the elongation tensor $\tilde{\boldsymbol{E}}$; cf. BOUSSINESQ [1871, *2*, § I], KIRCHHOFF [1876, *2*, Vorl. 27, § 4], BELTRAMI [1886, *1*, Note at end] [1889, *2*], PADOVA [1889, *7*], E. and F. COSSERAT [1896, *1*, § 13], CESARO [1906, *2*], VOLTERRA [1907, *7*, Chap. I]. Explicit forms of (57.7) in curvilinear co-ordinates are given by ODQVIST [1937, *8*] and VLASOV [1944, *12*].

The foregoing theorem is a special case of a general principle for replacement of the derivatives $\partial/\partial X^K$ by the derivatives $\partial/\partial x^k$. By (19.4) we have in general

$$\frac{\partial f}{\partial X^K} = \frac{\partial f}{\partial x^k} x^k_{;K} = \frac{\partial f}{\partial x^k} g^k_M (\delta^M_K + u^M_{;K}). \tag{57.11}$$

Hence if the displacement gradients are small, $\partial f/\partial X^K$ and $\partial f/\partial x^k$ are components of the same quantity at $\boldsymbol{X}$ and at $\boldsymbol{x}$:

$$\frac{\partial f}{\partial X^K} \approx g^k_K \frac{\partial f}{\partial x^k}. \tag{57.12}$$

The most important property of the case of small displacement gradients is the superposability of a succession of strains and rotations. Beginning with strain, we consider first the more general case analysed in Sect. 39, in which an arbitrary deformation from $\boldsymbol{X}$ to $\overline{\boldsymbol{X}}$ is followed by a second from $\overline{\boldsymbol{X}}$ to $\boldsymbol{x}$, yielding the exact formula (39.6). When the second displacement has small gradients, it is immediate from $(39.6)_3$ and (57.1) that[1]

$$E_{KM} \approx {}_1E_{KM} + {}_2\widetilde{E}_{\overline{K}\,\overline{M}} \overline{X}^{\overline{K}}_{;K} \overline{X}^{\overline{M}}_{;M}. \tag{57.13}$$

If the gradients of the first displacement are small, independently of those of the second, by (39.7) follows

$$E_{KM} \approx \widetilde{E}_{KM} \approx {}_1\widetilde{E}_{KM} + {}_2\widetilde{E}_{KM} + 2\,{}_1u_{P;(K}\,{}_2\widetilde{E}^P_{M)}, \tag{57.14}$$

while if the gradients of both displacements, taken together, are small, we get

$$\boldsymbol{E} \approx \widetilde{\boldsymbol{E}} \approx {}_1\boldsymbol{E} + {}_2\boldsymbol{E} \approx {}_1\widetilde{\boldsymbol{E}} + {}_2\widetilde{\boldsymbol{E}}. \tag{57.15}$$

Analogous formulae for the mean rotation $\widetilde{\boldsymbol{R}}$ follow easily, and we may assert ***Cauchy's superposition theorem***[2]: *In the common frame, the extensions, shears, and mean rotations corresponding to the succession of two displacements whose gradients are small may be obtained by adding together the extensions, shears, and mean rotations corresponding to each individual displacement.*

58. Small displacement. If we refer the co-ordinates at $\boldsymbol{x}$ and $\boldsymbol{X}$ to the same system (e.g., the common frame), for continuous f we have

$$f(\boldsymbol{x}) \approx f(\boldsymbol{X}) \tag{58.1}$$

for small displacements, the dimensionless error being of the order $u^k(\log f)_{,k}$. That is, *for small displacement, any continuous function of $\boldsymbol{x}$ may be taken as the same function of $\boldsymbol{X}$, and conversely*, provided the co-ordinate systems at $\boldsymbol{x}$ and $\boldsymbol{X}$ be the same[3].

For small displacement we have

$$g^k_K \approx \delta^k_K, \tag{58.2}$$

again in a single co-ordinate system. Hence when both the displacement and the displacement gradients are small, (57.11) yields $\partial f/\partial X^1 \approx \partial f/\partial x^1$, and (57.9) shows that the tensors measuring strain, rotation, and elongation at $\boldsymbol{X}$ are numerically equal, component by component, to their counterparts at $\boldsymbol{x}$.

[1] TRUESDELL [1952, *21*, § *19*]. An apparently more complicated result is obtained by CISOTTI [1944, *4*] for the case when the first deformation is rigid.

[2] While not stated by CAUCHY, this theorem is obvious from his work [1841, *1*, Part II].

[3] Otherwise, we shall not have $x^1 \to X^1$ as $\boldsymbol{u} \to 0$. Note also that the above statement does not apply to the displacement vector when considered as a function of *both* $\boldsymbol{x}$ and $\boldsymbol{X}$, but does apply to it when considered as a function of $\boldsymbol{x}$ only or of $\boldsymbol{X}$ only.

59. Perturbation series. A concept of "small" more restricted than that used in the previous sections was formalized by E. and F. COSSERAT[1] and has been used by many authors subsequently. We consider a family of displacement vectors ${}^B\boldsymbol{u}$ given by the formal power series

$$ {}^B\boldsymbol{u} = \sum_{n=0}^{\infty} B^n \boldsymbol{u}_n, \tag{59.1}$$

where the functions $\boldsymbol{u}_n$ are given and kept fixed and B is a parameter to which no meaning need be attached. Various powers of B occur in the various measures of deformation and rotation calculated from (59.1). In each of these, the terms multiplied by B^n are said to be "of order n". In such a scheme it is impossible, for example, to have mean rotation $\tilde{\boldsymbol{R}}$ and elongation $\tilde{\boldsymbol{E}}$ which are not of the same order. Moreover, the first, second, third, ..., derivatives of the n-th order approximation to ${}^B\boldsymbol{u}$ cannot decrease in order of smallness. Such a scheme, therefore, cannot include a displacement small in the sense used in earlier sections yet yielding strains which are not small, though it is easy to construct examples of families of displacements depending on a parameter in such a way that as the displacement vanishes the accompanying strain does not. The use of perturbation series restricts the class of functions which are considered in the limit process which always stands behind the concept of "small". By so doing, it facilitates formal calculation and perhaps also the proving of approximation theorems, but at the same time it limits the breadth and usefulness of results dependent upon it.

f) Oriented bodies.

60. Physical motivation for the theory of oriented bodies. All the foregoing analysis in this subchapter has concerned the changes of length and direction undergone by an arbitrary differential element $d\boldsymbol{X}$ in a deformation from $\boldsymbol{X}$ to $\boldsymbol{x}$. In some physical problems more than this much information is needed. The additional data most commonly relevant are the fates of certain *preferred directions*.

First, in some physical materials as they come to hand physical anisotropy is observable. For example, the stress required to effect a given extension in an elastic crystal differs with the direction in which the extension is to occur. For the description of the deformation of crystals, then, a mathematical model should carry not only the particle $\boldsymbol{X}$ to the place $\boldsymbol{x}$ but also directions at $\boldsymbol{X}$ into directions at $\boldsymbol{x}$. In fact, all relevant analysis for an arbitrary element at $\boldsymbol{X}$ has already been given. For the intended interpretation, however, the directions selected at $\boldsymbol{X}$ will not be arbitrary, and a dual formalism is not appropriate. All that is needed is a notation suited for distinguishing directions given at $\boldsymbol{X}$ by some rule known *a priori*.

It is a different matter, however, with *thin bodies*. Thin bodies are mathematical models for physical objects having one or two dimensions much smaller than the third. For definiteness, consider a circular cylinder. The object of a theory of thin cylinders is to produce directly the same predictions as would result from a corresponding three-dimensional theory in the limit as the diameter of the cylinder approaches zero while the resultant loads and the resistance of the material to deformation approach finite and non-vanishing limits. For results of the kind wished, obviously it is not enough to represent the rod only

[1] [1896, *1*, § 8]. Equivalently, SIGNORINI [1943, *6*, § 18] systematically calculated the n-th derivatives at $B=0$ of quantities associated with a family of displacements ${}^B\boldsymbol{u}$.

as a straight line which may be lengthened and bent, since such a theory does not describe the *twisting* of the cylinder about its axis. A proper one-dimensional theory, then, cannot be restricted to point deformations, but must also include directions which can suffer rotations independent of the deformations of the points with which they are associated[1]. An adequate theory of thin bodies cannot be merely the intrinsic deformation theory for a Riemannian space of one or two dimensions, but rather must consider such spaces imbedded in a Euclidean space of three dimensions.

In the eighteenth century special theories of rods and shells were proposed with little or no reference to phenomena in three dimensions. At the beginning of the nineteenth century, after general three-dimensional theories were accepted in certain domains of mechanics, CAUCHY and POISSON sought to obtain theories for thin bodies by averaging over a cross-section the results from a three-dimensional theory and then letting the cross-sectional area approach zero. From this point of view, no special geometrical apparatus is necessary for the description of thin bodies. The twisting of a circular cylinder, for example, may be measured in a three-dimensional description by the parameter K introduced in Sect. 48; if we let the diameter of the cylinder approach zero, the value of K remains unchanged and emerges as the measure of twist in the limiting formulae appropriate to the cylinder represented as but a line. It has turned out, however, that in the case of most if not all of the specific theories of materials in common use, the behavior of solutions for a body of thickness a in the neighborhood of $a=0+$ is highly singular[2]. While the three-dimensional viewpoint is certainly the more fundamental, the mathematical problems to which it has given rise are among the most difficult in analysis. The majority of modern studies of thin bodies pretending to adopt the three-dimensional approach in fact fail to establish any rigorous limit formulae but instead rest on concealed assumptions equivalent to those for an independent theory. In any case, independent theories of thin bodies are of interest and currency sufficient to justify discussion of their foundations.

Returning once more to three-dimensional bodies, even for them we may discern possible interpretations for an arbitrarily oriented medium. Suppose, for example, that the dielectric polarization vector is to be taken into account in the deformation of a crystal. This vector may be thought of as attached to a particle and suffering deformation, a deformation, however, which is not in general the same as that of a material element such as an axis of aeolotropy. In a similar way, a continuum model in which directions attached to a particle suffer deformation differing from that of material elements may be equivalent, as far as gross behavior is concerned, to a molecular model whose molecules have internal structure.

That physical bodies should be presented as assemblies not only of points but also of *directions associated with the points*, in brief, as *oriented bodies*, was suggested by DUHEM[3]. Theories based on this idea were constructed by E. and F. COSSERAT[4], but in the half century since their profound work was published,

[1] This was first remarked explicitly, though not very clearly, by ST. VENANT [1843, *3*, § 2].

[2] Cf. E. and F. COSSERAT [1907, *1*] [1908, *2*] [1909, *5*, § 4]. The typical phenomenon is evanescence or confluence of boundary conditions in the limit. Of course the remark and those immediately following do not apply to *exact* averagings such as that presented in Sect. 213.

[3] [1893, *1*, Chap. II].

[4] [1907, *1*] [1909, *5*].

scant attention has been given to it[1]. Sects. 61 to 64 present the theory of oriented bodies in the general and invariant form given to it by ERICKSEN and TRUESDELL[2].

61. The oriented bodies of E. and F. COSSERAT[3]. To the point $\boldsymbol{X}$ assign a set of $\mathfrak{p}$ vectors $\boldsymbol{D}_{\mathfrak{a}}(\boldsymbol{X})$, $\mathfrak{a}=1, 2, \ldots, \mathfrak{p}$, the *directors* of the body at $\boldsymbol{X}$. According as $\boldsymbol{X}$ ranges over a region of positive volume, a surface, or a curve, we may speak of a *solid*, a *shell*, or a *rod*. By a *deformation* of the body we shall mean a transformation carrying $\boldsymbol{X}$ into $\boldsymbol{x}$ and the directors $\boldsymbol{D}_{\mathfrak{a}}$ at $\boldsymbol{X}$ into directors $\boldsymbol{d}_{\mathfrak{a}}$ at $\boldsymbol{x}$. In equations,

$$\boldsymbol{x}=\boldsymbol{x}(\boldsymbol{X}), \quad \boldsymbol{d}_{\mathfrak{a}}=\boldsymbol{d}_{\mathfrak{a}}(\boldsymbol{D}_{\mathfrak{b}})=\boldsymbol{d}_{\mathfrak{a}}(\boldsymbol{X}); \tag{61.1}$$

in geometrical terms, a deformation consists in a displacement of the points and *independent* rotations and stretches of the directors. In the special case when $d_{\mathfrak{a}}^{k}=x^{k}_{;K}D^{K}_{\mathfrak{a}}$, the directors are material elements, and their presence adds nothing to the description already given in this chapter. In the special case when $d^{k}_{\mathfrak{a}}=g^{k}_{K}D^{K}_{\mathfrak{a}}$, the directors are invariable elements and add nothing to what has been explained above[4]. In general, the directors are *neither material nor invariable*; i.e., $d^{k}_{\mathfrak{a}}\neq x^{k}_{;K}D^{K}_{\mathfrak{a}}$ and $d^{k}_{\mathfrak{a}}\neq g^{k}_{K}D^{K}_{\mathfrak{a}}$.

In an oriented body, strain and rotation are defined from $(61.1)_1$ as for the ordinary bodies considered earlier in this subchapter. What is new is the relation between the directors. For the full possibility of interpretation mentioned at the end of Sect. 60, it should be possible to use an arbitrarily large number of directors. However, an appropriate description of strain has never been constructed in this degree of generality. Therefore we add the assumption that there are but *three* directors $\boldsymbol{D}_{\mathfrak{a}}$, and that these are linearly independent. This number suffices for theories of rods, shells, and anisotropic solids.

Let the *reciprocal directors* $\boldsymbol{D}^{\mathfrak{a}}$ be defined as the triad reciprocal to $\boldsymbol{D}_{\mathfrak{a}}$, so that

$$D^{K}_{\mathfrak{a}}D^{\mathfrak{a}}_{M}=\delta^{K}_{M}, \quad D^{K}_{\mathfrak{a}}D^{\mathfrak{b}}_{K}=\delta^{\mathfrak{a}}_{\mathfrak{b}}, \quad D^{K}_{\mathfrak{a}}=g^{KM}G_{\mathfrak{a}\mathfrak{b}}D^{\mathfrak{b}}_{M}, \tag{61.2}$$

where

$$G_{\mathfrak{a}\mathfrak{b}}\equiv D^{K}_{\mathfrak{a}}D^{M}_{\mathfrak{b}}g_{KM}. \tag{61.3}$$

Note that

$$\det G_{\mathfrak{a}\mathfrak{b}}=\det(D_{\mathfrak{a}}D_{\mathfrak{b}}\cos\vartheta_{\mathfrak{a}\mathfrak{b}})=D^{2}_{1}D^{2}_{2}D^{2}_{3}\det\cos\vartheta_{\mathfrak{a}\mathfrak{b}} \tag{61.4}$$

where $\vartheta_{\mathfrak{a}\mathfrak{b}}$ is the angle between $D_{\mathfrak{a}}$ and $D_{\mathfrak{b}}$. When the directors form an orthogonal unit triad, we thus obtain $\det G_{\mathfrak{a}\mathfrak{b}}=1$.

The director triads may be used to define anholonomic components. For example, if we set

$$X^{\mathfrak{a}}_{k}\equiv D^{\mathfrak{a}}_{K}X^{K}_{;k}, \tag{61.5}$$

then $X^{K}_{;k}=D^{K}_{\mathfrak{a}}X^{\mathfrak{a}}_{k}$, and from $(26.1)_2$ follows

$$c_{km}=G_{\mathfrak{a}\mathfrak{b}}X^{\mathfrak{a}}_{k}X^{\mathfrak{b}}_{m}, \quad etc. \tag{61.6}$$

Now set[5]

$$W^{K}{}_{MP}\equiv D^{K}_{\mathfrak{a};P}D^{\mathfrak{a}}_{M}=-D^{K}_{\mathfrak{a}}D^{\mathfrak{a}}_{M;P}, \tag{61.7}$$

---

[1] There is an exposition by SUDRIA [1935, *8*]. In his § 9, SUDRIA notes an error in the COSSERATS argument and gives a different proof of invariance. We do not follow either the COSSERATS or SUDRIA in detail, and we do not present the material on time rates given in SUDRIA's Chap. II.

[2] [1958, *1*, Part II].

[3] E. and F. COSSERAT [1909, *5*, §§ 48–50], ERICKSEN and TRUESDELL [1958, *1*, §§ 5–6]. Cf. also the discussion of infinitesimal strain by GÜNTHER [1958, *4*, § 2].

[4] Selecting three orthogonal and invariable directors, we may use them as a fixed anholonomic frame and so obtain invariant forms for the results given in many of the older treatments of finite strain, where all quantities are referred to the common frame.

[5] A definition of this kind is the starting point of the theory of non-integrable strain constructed by KRÖNER [1960, *3*, § 4, Eq. (13)].

where we use the definition (App. 20.2). Hence

$$D^K_{\mathfrak{a};M} = W^K{}_{PM}\, D^P_{\mathfrak{a}}. \tag{61.8}$$

From (61.7) and (61.2) we find that

$$2\, W_{(KM)P} = G_{\mathfrak{a}\mathfrak{b};P}\, D^{\mathfrak{a}}_K\, D^{\mathfrak{b}}_M. \tag{61.9}$$

Therefore $G_{\mathfrak{a}\mathfrak{b}}=\text{const}$ is equivalent to

$$W_{KMP} = -\, W_{MKP}. \tag{61.10}$$

In this case, also, if we transform the $\boldsymbol{D}_{\mathfrak{a}}$ at all points by the same orthogonal transformation, the components $W^K{}_{MP}$ are invariant. From these results it follows that if the lengths of the directors and the angles between them are fixed, as is the case, for example, if the directors are chosen as an orthogonal unit triad, and if the point $\boldsymbol{X}$ is made to traverse the X^P co-ordinate curve at unit speed, the quantities W_{KMP} are the components of angular velocity of the director frame $\boldsymbol{D}_{\mathfrak{a}}$ carried by $\boldsymbol{X}$. We do not need to use (61.10), and therefore we do not impose the restriction $G_{\mathfrak{a}\mathfrak{b}}=\text{const}$, but this special case serves to motivate our calling $W^K{}_{MP}$ the *wryness* of the director frame in the undeformed material.

Dual results hold for the deformed material, but the dual wryness tensor, since it refers only to the relative configurations of the director frames at different points in the deformed material, does not afford a comparison between the deformed and undeformed conditions. What we wish, in the kinematic terms used above, are generalizations of the angular velocities of the director frame at $\boldsymbol{x}$ relative to those of the director frame at $\boldsymbol{X}$ when $\boldsymbol{x}$ traverses the curve into which the path of $\boldsymbol{X}$ is deformed. To this end, introduce the *relative wryness* at $\boldsymbol{x}$:

$$F^k{}_{mP} \equiv d^k_{\mathfrak{a};P}\, d^{\mathfrak{a}}_m = -\, d^k_{\mathfrak{a}}\, d^{\mathfrak{a}}_{m;P}, \quad d^k_{\mathfrak{a};K} = F^k{}_{mK}\, d^m_{\mathfrak{a}}. \tag{61.11}$$

Here, however, we encounter the quantities $d^k_{\mathfrak{a};K}$, the *director gradients*, which appear in the theory of deformation of oriented bodies along with the deformation gradients $x^k_{;K}$ as primary local variables. In general there are 27 director gradients; when the directors form an orthogonal unit triad, only 9 director gradients are independent.

Set

$$A^k{}_K \equiv d^k_{\mathfrak{a}}\, D^{\mathfrak{a}}_K, \qquad a^K{}_k \equiv D^K_{\mathfrak{a}}\, d^{\mathfrak{a}}_k; \tag{61.12}$$

then

$$d^k_{\mathfrak{a}} = A^k{}_K\, D^K_{\mathfrak{a}}, \qquad D^K_{\mathfrak{a}} = a^K{}_k\, d^k_{\mathfrak{a}}, \qquad \boldsymbol{a} = \boldsymbol{A}^{-1}. \tag{61.13}$$

From (61.11) follows

$$F^k{}_{mP} = A^k{}_{K;P}\, a^K{}_m + A^k{}_K\, W^K{}_{MP}\, a^M{}_m. \tag{61.14}$$

A deformation of an oriented body is *rigid* if not only $\boldsymbol{C}=\boldsymbol{1}$ but also the directors at $\boldsymbol{x}$ may be obtained from those at $\boldsymbol{X}$ by a uniform orthogonal transformation. In this case the tensor $\boldsymbol{A}$ defined by (61.12) is a covariantly constant orthogonal tensor, the first term on the right-hand side of (61.14) vanishes, and we see that $\boldsymbol{F}$ is *orthogonally equivalent to* $\boldsymbol{W}$.

To make use of these results, we consider the anholonomic components of $\boldsymbol{W}$ with respect to the directors at $\boldsymbol{X}$, the anholonomic components of $\boldsymbol{F}$ with respect to the directors at $\boldsymbol{x}$:

$$W_{\mathfrak{a}\mathfrak{b}P} \equiv W_{KMP}\, D^K_{\mathfrak{a}}\, D^M_{\mathfrak{b}} = D^K_{\mathfrak{a}}\, D_{\mathfrak{b}K;P}, \quad F_{\mathfrak{a}\mathfrak{b}P} \equiv F_{kmP}\, d^k_{\mathfrak{a}}\, d^m_{\mathfrak{b}} = d^k_{\mathfrak{a}}\, d_{\mathfrak{b}k;P}, \tag{61.15}$$

so that by (61.8) and $(61.11)_3$ we have

$$D_{\mathfrak{b}K;M} = W_{\mathfrak{a}\mathfrak{b}M}\, D^{\mathfrak{a}}_K, \qquad d_{\mathfrak{b}k;M} = F_{\mathfrak{a}\mathfrak{b}M}\, d^{\mathfrak{a}}_k. \tag{61.16}$$

If we put $g_{\mathfrak{a}\mathfrak{b}} \equiv g_{km}\, d^k_{\mathfrak{a}}\, d^m_{\mathfrak{b}}$, then (61.14) assumes the form[1]

$$F_{\mathfrak{a}\mathfrak{b}P} = A^k{}_{K;P}\, D^K_{\mathfrak{b}}\, d_{\mathfrak{a}k} + g_{\mathfrak{a}\mathfrak{c}}\, G^{\mathfrak{e}\mathfrak{c}} W_{\mathfrak{e}\mathfrak{b}P}. \tag{61.17}$$

Also $2F_{(\mathfrak{a}\mathfrak{b})P} = g_{\mathfrak{a}\mathfrak{b};P}$.

Now $\boldsymbol{A}$ is an orthogonal tensor if and only if $g_{\mathfrak{a}\mathfrak{b}} = G_{\mathfrak{a}\mathfrak{b}}$. When $\boldsymbol{A}$ is a uniform orthogonal tensor, (61.17) reduces to

$$F_{\mathfrak{a}\mathfrak{b}P} = W_{\mathfrak{a}\mathfrak{b}P}. \tag{61.18}$$

Conversely, suppose (61.18) holds. From (61.15) follows

$$g_{km}\, d^k_{\mathfrak{a}}\, d^m_{\mathfrak{b};P} = g_{KM}\, D^K_{\mathfrak{a}}\, D^M_{\mathfrak{b};P}; \tag{61.19}$$

Since this holds for all choices of $\mathfrak{a}$ and $\mathfrak{b}$, we deduce

$$g_{\mathfrak{a}\mathfrak{b};P} = G_{\mathfrak{a}\mathfrak{b};P}. \tag{61.20}$$

Hence

$$g_{\mathfrak{a}\mathfrak{b}} = G_{\mathfrak{a}\mathfrak{b}} + K_{\mathfrak{a}\mathfrak{b}}, \tag{61.21}$$

where the $K_{\mathfrak{a}\mathfrak{b}}$ are constants of integration representing constant differences of length and angle. Thus (61.18) asserts that the lengths and angles of the two sets of directors differ by constants, so that the tensor $\boldsymbol{A}$ is orthogonal everywhere if it is orthogonal at one point. From the foregoing analysis we conclude that *necessary and sufficient conditions for rigid deformation of an oriented body are*

$$\left.\begin{array}{l} 1.\ C_{KM} = g_{KM}, \\ 2.\ F_{\mathfrak{a}\mathfrak{b}K} = W_{\mathfrak{a}\mathfrak{b}K}, \\ 3.\ \textit{At some one point, } \boldsymbol{A} \textit{ is orthogonal.} \end{array}\right\} \tag{61.22}$$

The tensor $\boldsymbol{W}$ thus appears as analogous to the metric tensor $\boldsymbol{g}$, while $\boldsymbol{F}$ is analogous to GREEN's deformation tensor, $\boldsymbol{C}$. However, we must bear in mind that while $(61.22)_1$ is a general tensorial condition, $(61.22)_2$ is not, since the anholonomic components $F_{\mathfrak{a}\mathfrak{b}M}$ and $W_{\mathfrak{a}\mathfrak{b}M}$ are calculated with respect to different frames.

The relative wryness $\boldsymbol{F}$ is thus a measure of *strain of orientation*, as contrasted with the *strain of position* analysed earlier in this chapter. Just as the C_{KM} are certain quadratic combinations of the deformation gradients $x^k_{;K}$, the $F_{\mathfrak{a}\mathfrak{b}M}$ are certain linear combinations of the director gradients $d^{\mathfrak{c}}_{k;K}$. In general, the numbers of independent quantities $x^k_{;K}$, C_{KM}, $d^{\mathfrak{a}}_{k;K}$, and $F_{\mathfrak{a}\mathfrak{b}M}$ are, respectively, 9, 6, 27, 27.

It would be possible, in analogy to Sect. 32, to define and characterize a general measure of strain of orientation. One such measure is given by

$$J_{\mathfrak{a}\mathfrak{b}M} \equiv F_{\mathfrak{a}\mathfrak{b}M} - W_{\mathfrak{a}\mathfrak{b}M}, \tag{61.23}$$

---

[1] Since the anholonomic components of $\boldsymbol{W}$ and $\boldsymbol{F}$ are defined with respect to *different* anholonomic frames, the usual rules for manipulating anholonomic components do not always apply. For our purposes the particular choices (61.15) for the anholonomic components are essential. For example, if instead we set

$$W^{\mathfrak{a}}{}_{\mathfrak{b}P} \equiv W^K{}_{MP}\, D^{\mathfrak{a}}_K\, D^M_{\mathfrak{b}}, \quad F^{\mathfrak{a}}{}_{\mathfrak{b}P} \equiv F^k{}_{MP}\, d^{\mathfrak{a}}_k\, d^m_{\mathfrak{b}},$$

we obtain from (61.14)

$$F^{\mathfrak{a}}{}_{\mathfrak{b}P} = A^k{}_{K;P}\, d^{\mathfrak{a}}_k\, D^K_{\mathfrak{b}} + W^{\mathfrak{a}}{}_{\mathfrak{b}P}.$$

Therefore a necessary and sufficient condition for $\boldsymbol{A}$ to be covariantly constant is

$$F^{\mathfrak{a}}{}_{\mathfrak{b}P} = W^{\mathfrak{a}}{}_{\mathfrak{b}P},$$

but this is not at all the same as condition (61.18).

corresponding to GREEN's tensor $\boldsymbol{E}$ (Sect. 31). Necessary and sufficient for a rigid deformation are the conditions

$$\boldsymbol{E}=0, \quad \boldsymbol{J}=0, \tag{61.24}$$

along with the orthogonality of $\boldsymbol{A}$ at one point. We do not take up the question of characterizing all possible measures of strain of orientation.

A principle of duality analogous to that of Sect. 14 may be formulated. The reader will easily construct analogues of $\boldsymbol{c}$, $\boldsymbol{e}$, etc.

Not pausing to explore the structure whose traits have just been presented, we note only a formula for the gradient of relative wryness. By differentiating (61.11) we get

$$\left.\begin{aligned} F^k{}_{mM;K} &= d^k_{\mathfrak{a};MK}\, d^{\mathfrak{a}}_m + d^k_{\mathfrak{a};M}\, d^{\mathfrak{a}}_{m;K}, \\ &= d^k_{\mathfrak{a};MK}\, d^{\mathfrak{a}}_m - F^k{}_{pM} F^p{}_{mK}. \end{aligned}\right\} \tag{61.25}$$

Hence

$$\left.\begin{aligned} F_{\mathfrak{a}\mathfrak{b}M;K} &= F_{kmM;K}\, d^k_{\mathfrak{a}} d^m_{\mathfrak{b}} + F_{kmM}\,(d^k_{\mathfrak{a};K}\, d^m_{\mathfrak{b}} + d^k_{\mathfrak{a}}\, d^m_{\mathfrak{b};K}), \\ &= d_{\mathfrak{b}k;MK}\, d^k_{\mathfrak{a}} + g^{\mathfrak{c}\mathfrak{e}}(F_{\mathfrak{a}\mathfrak{e}M} F_{\mathfrak{c}\mathfrak{b}K} + F_{\mathfrak{e}\mathfrak{b}M} F_{\mathfrak{c}\mathfrak{a}K} - F_{\mathfrak{a}\mathfrak{c}M} F_{\mathfrak{e}\mathfrak{b}K}). \end{aligned}\right\} \tag{61.26}$$

There is no doubt that with the apparatus constructed here much of the work on special theories of oriented media could be unified and correlated. In the following sections we present only the simplest and most immediate applications.

In view of what we have explained already, the theory of deformation of an oriented body can be organized as follows:

I. Strain of position
  A. Intrinsic theory
  B. Imbedding theory
II. Strain of orientation.

The intrinsic theory, based on relations between the fundamental tensors $\boldsymbol{g}$ and $\boldsymbol{C}$, for three-dimensional bodies has been studied in detail earlier in the chapter. For bodies of one and two dimensions it is somewhat simpler but not essentially different. While curved rods and shells are not Euclidean spaces, the analysis based on the mapping (16.2) and the formulae (26.1) is valid in a Riemannian space of any number of dimensions[1]. However, the conditions of compatibility given in Sect. 34 are no longer appropriate, since the curvature of a rod or shell may be changed arbitrarily by deformation. The theory of finite rotation given in Part c continues to be meaningful, but only when the rod or shell is regarded from the Euclidean three-dimensional space in which it is imbedded. In treating the imbedding theory, however, it is fitter to use the concepts of differential geometry: The curvature and torsion of a curve, the second differential form of a surface. In the above program, then, Part I A has been given in principle already, while Part I B can be gotten with little labor from geometry, so in what follows the major attention will be given to Part II.

62. Anisotropic solids. To represent anisotropic solids, the directors are chosen as material elements[2] and hence are related by (20.3):

$$d^k_{\mathfrak{a}} \equiv x^k_{;K}\, D^K_{\mathfrak{a}}, \qquad x^k_{;K} = d^k_{\mathfrak{a}}\, D^{\mathfrak{a}}_K, \qquad X^K_{;k} = D^K_{\mathfrak{a}}\, d^{\mathfrak{a}}_k. \tag{62.1}$$

Hence by (26.2) and (29.1) follows

$$g_{\mathfrak{a}\mathfrak{b}} = C_{KM}\, D^K_{\mathfrak{a}} D^M_{\mathfrak{b}}, \qquad C_{KM} = g_{\mathfrak{a}\mathfrak{b}}\, D^{\mathfrak{a}}_K\, D^{\mathfrak{b}}_M, \qquad \overset{-1}{c}{}^{km} = G^{\mathfrak{a}\mathfrak{b}}\, d^k_{\mathfrak{a}}\, d^m_{\mathfrak{b}}, \tag{62.2}$$

[1] PADOVA [1890, 8].
[2] E.g. ERICKSEN and RIVLIN [1954, 6, § 2].

so that the anholonomic components of $\boldsymbol{E}$ are given by

$$E_{\mathfrak{a}\mathfrak{b}} \equiv \tfrac{1}{2}(g_{\mathfrak{a}\mathfrak{b}} - G_{\mathfrak{a}\mathfrak{b}}). \tag{62.3}$$

Also

$$\mathrm{I}_{\boldsymbol{C}} = g_{\mathfrak{a}\mathfrak{b}} G^{\mathfrak{a}\mathfrak{b}}, \qquad \mathrm{I}_{\boldsymbol{c}} = g^{\mathfrak{a}\mathfrak{b}} G_{\mathfrak{a}\mathfrak{b}}, \qquad \mathrm{III}_{\boldsymbol{C}} = (\det g_{\mathfrak{a}\mathfrak{b}})(\det G^{\mathfrak{a}\mathfrak{b}}). \tag{62.4}$$

When the $\boldsymbol{D}_{\mathfrak{a}}$ form an orthogonal unit triad, we have $G_{\mathfrak{a}\mathfrak{b}} = G^{\mathfrak{a}\mathfrak{b}} = \delta_{\mathfrak{a}\mathfrak{b}}$, and the foregoing formulae simplify. E.g., from $(62.2)_3$ and (62.4) we get

$$\overset{-1}{c}{}^{km} = \sum_{\mathfrak{a}=1}^{3} d_{\mathfrak{a}}^{k} d_{\mathfrak{a}}^{m}, \qquad \mathrm{I}_{\boldsymbol{C}} = \operatorname{trace} g_{\mathfrak{a}\mathfrak{b}}, \qquad \mathrm{I}_{\boldsymbol{c}} = \operatorname{trace} g^{\mathfrak{a}\mathfrak{b}}, \qquad \mathrm{III}_{\boldsymbol{C}} = \det g_{\mathfrak{a}\mathfrak{b}}. \tag{62.5}$$

In the application to anisotropic solids, assignment of the directors $\boldsymbol{D}_{\mathfrak{a}}$ results from *a priori* knowledge concerning the nature of the undeformed body. For a different interpretation of this same formalism, we may select the directors as tangent to the co-ordinate curves. The resulting formulae in terms of anholonomic components are then identifiable with the results usually derived by use of convected co-ordinate systems (Sects. 66B, 150).

The case when the directors are material is a trivial one from the viewpoint of the general theory in Sect. 61, since the strain of orientation is then not independent of the strain of position. Although the wryness is not needed for applications to anisotropic solids, it is interesting to specialize its theory to the present case. By comparing (62.1) and (61.12) we obtain $A^k{}_K = x^k_{;K}$, $a^K{}_k = X^K_{;k}$, and (61.14) becomes

$$F^k{}_{mP} = x^k_{;KP} X^K_{;m} + x^k_{;K} X^M_{;m} W^K{}_{MP}. \tag{62.5}$$

In homogeneous strain, the first term on the right-hand side is zero, and we see that $\boldsymbol{F}$ and $\boldsymbol{W}$ are the components of the same tensor with respect to the deformation. More generally, Eq. (62.5) shows that when the present formalism is applied to material directors, we may construct in terms of it an interpretation of the second derivatives of the deformation.

63. Rods. Let a curve $\mathscr{C}$ in the undeformed material be given by the parametric equations

$$X^K = X^K(S), \qquad K = 1, 2, 3, \tag{63.1}$$

where it is convenient to regard S as arc length. In a *deformation* $\mathscr{C}$ is mapped onto a curve c given by

$$x^k = x^k(s), \qquad k = 1, 2, 3, \tag{63.2}$$

where

$$s = s(S). \tag{63.3}$$

The *stretch* λ is given by

$$\lambda^2 \equiv \frac{ds^2}{dS^2} = s'^2 = C_{KM} T^K T^M, \tag{63.4}$$

where $\boldsymbol{T}$ is the unit tangent to $\mathscr{C}$. This completes the intrinsic theory of strain of position.

For the imbedding theory of strain of position, we remind the reader that if $\mathscr{C}$ and c have the same curvature and torsion as functions of S, then one may be brought into point-by-point coincidence with the other by means of suitable rigid motion. Thus a complete description of the strain of position, as far as the imbedding theory is concerned, is given by the curvatures and torsions of c and $\mathscr{C}$ as functions of S.

We come now to the strain of orientation. As a preliminary, we note that (63.4) illustrates two alternatives that we shall always have before us. If the deformation is specified only by the relations (63.1), (63.2), and (63.3), we have

(63.4)$_1$ at once, but the tensor $\boldsymbol{C}$ occuring in (63.4)$_3$ has no meaning. If, on the other hand, we regard $\mathscr{C}$ as a single material line in a three-dimensional body subject to the deformation (16.2), then $\mathscr{C}$ may be calculated from (26.2)$_2$, and (63.4)$_3$ emerges as a special case from (26.4).

For example, we can set

$$\hat{A}^{\cdots}_{\cdots} \equiv A^{\cdots}_{\cdots;K} \frac{dX^K}{dS}. \tag{63.5}$$

For a double field $A^{\cdots}_{\cdots}(S, s)$, by putting the definition (App. 20.2) into (63.5) we get

$$\hat{A}^{\cdots}_{\cdots} = \frac{\partial A^{\cdots}_{\cdots}}{\partial S} + \frac{\partial A^{\cdots}_{\cdots}}{\partial s}\frac{ds}{dS} + \left\{\begin{matrix} \cdot \\ \cdot\; K\end{matrix}\right\} A^{\cdots}_{\cdots}\frac{dX^K}{dS} + \cdots + \left\{\begin{matrix} \cdot \\ \cdot\; k\end{matrix}\right\} A^{\cdots}_{\cdots}\frac{dx^k}{ds}\frac{ds}{dS} + \cdots, \tag{63.6}$$

and in this formula appear only quantities which can be calculated from the equations of the curves $\mathscr{C}$ and c and the deformation $s = s(S)$. For a strictly one-dimensional approach, Eq. (63.6) rather than (63.5) should be taken as the *definition* of $\hat{\boldsymbol{A}}$; as far as results are concerned, it makes no difference.

In conformity with this notation, set

$$\hat{X}^K \equiv \frac{dX^K}{dS}, \qquad \hat{x}^k \equiv \frac{dx^k}{ds}\frac{ds}{dS}, \tag{63.7}$$

where X^K and x^k are given parametrically by (63.1) and (63.2). Note also that for a scalar $F(S)$ we have simply $\hat{F} = dF/dS$.

We begin by constructing a differential description of the undeformed rod $\mathscr{C}$. We assign directors $\boldsymbol{D}_{\mathfrak{a}}$ to the points of $\mathscr{C}$ and set

$$T^{\mathfrak{a}} \equiv D^{\mathfrak{a}}_K \hat{X}^K, \quad G_{\mathfrak{a}\mathfrak{b}} \equiv g_{KM} D^K_{\mathfrak{a}} D^M_{\mathfrak{b}}, \quad W^{\mathfrak{a}}{}_{\mathfrak{b}} \equiv D^{\mathfrak{a}}_K \hat{D}^K_{\mathfrak{b}}, \tag{63.8}$$

so that

$$\hat{X}^K = D^K_{\mathfrak{a}} T^{\mathfrak{a}}, \quad \hat{D}^K_{\mathfrak{a}} = D^K_{\mathfrak{b}} W^{\mathfrak{b}}{}_{\mathfrak{a}}, \quad 2W_{(\mathfrak{a}\mathfrak{b})} = \hat{G}_{\mathfrak{a}\mathfrak{b}}, \quad G_{\mathfrak{a}\mathfrak{b}} T^{\mathfrak{a}} T^{\mathfrak{b}} = 1. \tag{63.9}$$

The scalars $T^{\mathfrak{a}}$, anholonomic components of the unit tangent, are the lengths of the projections of the unit tangent upon the reciprocal directors; the $G_{\mathfrak{a}\mathfrak{b}}$ prescribe the lengths and mutual angles of the directors; the tensor $\boldsymbol{W}$, whose anholonomic components are the rates of change of the directors resolved along the directions of the reciprocal directors, is the *wryness* of the directors along $\mathscr{C}$. By (63.9)$_3$ we see that if the $G_{\mathfrak{a}\mathfrak{b}}$ are constant along $\mathscr{C}$, and in that case only, $\boldsymbol{W}$ may be regarded as an angular velocity as explained following (61.10). Sometimes it is preferable to use tensor components rather than anholonomic components:

$$W^K{}_M = \hat{D}^K_{\mathfrak{a}} D^{\mathfrak{a}}_M = D^K_{\mathfrak{b}} W^{\mathfrak{b}}{}_{\mathfrak{a}} D^{\mathfrak{a}}_M, \quad \hat{D}^K_{\mathfrak{a}} = W^K{}_M D^M_{\mathfrak{a}}, \quad W^{\mathfrak{a}}{}_{\mathfrak{b}} = D^{\mathfrak{a}}_K W^K{}_M D^M_{\mathfrak{b}}. \tag{63.10}$$

(If $\mathscr{C}$ is thought of as imbedded in a three-dimensional oriented body, the wryness $W^K{}_M$ along $\mathscr{C}$ is related to the wryness of the body, viz. $W^K{}_{MP}$, as follows:

$$W^K{}_M = W^K{}_{MP} \hat{X}^P.) \tag{63.11}$$

Suppose the quantities $T^{\mathfrak{a}}$, $W^{\mathfrak{a}}{}_{\mathfrak{b}}$, and $G_{\mathfrak{a}\mathfrak{b}} = G_{\mathfrak{b}\mathfrak{a}}$ be given as any functions of S such that (63.9)$_3$ and (63.9)$_4$ are satisfied and such that $\boldsymbol{G}$ is positive definite. Eqs. (63.9)$_{1,2}$ may then be regarded as a differential system for determining $X^K(S)$ and $D^K_{\mathfrak{a}}(S)$. From (63.9)$_2$, any solution satisfies

$$\widehat{G_{KM} D^K_{\mathfrak{a}} D^M_{\mathfrak{b}}} = 2 G_{KM} D^K_{\mathfrak{c}} D^M_{(\mathfrak{a}} W^{\mathfrak{c}}{}_{\mathfrak{b})}. \tag{63.12}$$

We may write (63.9)$_3$ in the form $\hat{G}_{\mathfrak{a}\mathfrak{b}} = 2G_{\mathfrak{c}(\mathfrak{a}} W^{\mathfrak{c}}{}_{\mathfrak{b})}$; comparing this result with (63.12) shows that $G_{KM} D^K_{\mathfrak{a}} D^M_{\mathfrak{b}}$ and $G_{\mathfrak{a}\mathfrak{b}}$ satisfy the same first order differential system. By uniqueness of solutions of such systems, $(63.8)_2$ will hold for all S if it holds for one value of S. Given a solution $X^K(S)$, $D^K_{\mathfrak{a}}(S)$ of the system $(63.9)_{1\,2}$, suppose $(63.8)_2$ is satisfied; by $(63.9)_4$ follows

$$\hat{X}^K \hat{X}_K = G_{\mathfrak{a}\mathfrak{b}}\, T^{\mathfrak{a}}\, T^{\mathfrak{b}} = 1, \tag{63.13}$$

so the solution represents an oriented rod with S as arc length. It is easily shown that two sets of initial data consistent with $(63.8)_2$ are related by a rigid motion, perhaps combined with a reflection, and that any rigid motion combined with a reflection carries any solution of the system $(63.9)_{1\,2}$ into another. Uniqueness of solutions of this system thus implies that *assignment of $T^{\mathfrak{a}}$, $W^{\mathfrak{a}}{}_{\mathfrak{b}}$, and $G_{\mathfrak{a}\mathfrak{b}}$ determines $\mathscr{C}$ and its directors to within a rigid motion combined with a reflection.*

In the special case when the directors are chosen as the unit tangent, principal normal, and binormal to $\mathscr{C}$, oriented so as to form a right-handed system, the Serret-Frenet formulae assert that

$$\| W^{\mathfrak{a}}{}_{\mathfrak{b}} \| = \begin{Vmatrix} 0 & \varkappa & 0 \\ -\varkappa & 0 & \tau \\ 0 & -\tau & 0 \end{Vmatrix}, \tag{63.14}$$

where $\varkappa$ and τ are the curvature and the torsion of $\mathscr{C}$. No such result holds for a general triad of directors or would be of interest here, since we seek not the properties of the curve itself but rather of a set of axes attached to the curve but turning independently of it.

For another special case, consider a single unit director $\boldsymbol{D}$ which is normal to the curve $\mathscr{C}$, and let it subtend an angle $\varphi(S)$ with the principal normal $\boldsymbol{N}$, measured positively away from the binormal, so that

$$D^K = N^K \cos\varphi + B^K \sin\varphi, \tag{63.15}$$

where $\boldsymbol{B}$ is the binormal. The quantity $\hat{\varphi}$ is the *twist* of $\boldsymbol{D}$ with respect to $\mathscr{C}$. If both the curve $\mathscr{C}$ and the director $\boldsymbol{D}$ are subjected to the same orthogonal transformation, $\hat{\varphi}$ is unchanged. Since $D_K N^K = \cos\varphi$, we have

$$\hat{D}^K N_K + D_K \hat{N}^K = -\sin\varphi\,\hat{\varphi}; \tag{63.16}$$

by $(63.10)_3$ and the Serret-Frenet formula for $\hat{\boldsymbol{N}}$ follows

$$W^K{}_M D^M N_K + D_K(-\varkappa T^K + \tau B^K) = -\sin\varphi\,\hat{\varphi}. \tag{63.17}$$

Substitution of (63.15) in this result yields

$$W_{(K\,M)} N^K N^M \cos\varphi + (W_{K\,M} N^K B^M + \tau + \hat{\varphi}) \sin\varphi = 0. \tag{63.18}$$

Thus far in this paragraph we have spoken of but a single director $\boldsymbol{D}$, leaving arbitrary the choice of the remaining two directors required for definition of the wryness $\boldsymbol{W}$ by $(63.8)_3$. Since we are here interested only in the single director $\boldsymbol{D}$, and since this is a unit vector, there is no loss of generality in selecting the other two directors as of fixed length and subtending fixed angles with $\boldsymbol{D}$ and with each other. By $(63.9)_3$ follows $W_{(KM)} = 0$, and thus, provided $\varphi \neq 0$, from (63.18) we obtain

$$\hat{\varphi} = W_{MK} N^K B^M - \tau. \tag{63.19}$$

That is, *the twist is the excess of $W_{MK} N^K B^M$ over the torsion.* The scalar $W_{MK} N^K B^M$ is itself an anholonomic component of $\boldsymbol{W}$ with respect to the principal frame of $\mathscr{C}$. If we let $\boldsymbol{D}^*$ be a unit vector such that $\boldsymbol{D}^*$, $\boldsymbol{D}$, and $\boldsymbol{T}$ form a right-handed unit triad, since $\boldsymbol{D}^{*K} = -N^K \sin\varphi + B^K \cos\varphi$, then $W_{KM} B^K N^M = D^{*K} W_{KM} D^M$, so that by (63.10)₃ and (63.19) follows

$$\hat{\varphi} = D_K^* \hat{D}^K - \tau. \qquad (63.20)$$

This formula is the basis of the classical attempts to construct a theory of strain for bent and twisted rods (for references, see the appendix at the end of this section).

Since the rotation of $\boldsymbol{D}$ may be prescribed in any smooth way as we traverse $\mathscr{C}$, there can be no general connection between the twists of two different directors. In the applications of the theory to elasticity, it is customary to think of a rod as a line furnished with normal *cross-sections*; these are represented, for the purposes of the theory, only by their principal axes of geometrical inertia and perhaps one or two constants such as their geometrical moments of inertia about these axes. The *twist of the unstrained* rod is then defined as the twist of either of these axes relative to $\mathscr{C}$. Since these axes are normal to one another, a unique twist is obtained.

Now consider the deformation of $\mathscr{C}$ into c, defined not only by (63.3) but also by a relation setting directors $\boldsymbol{d}_\mathfrak{a}$ along c into correspondence with the directors $\boldsymbol{D}_\mathfrak{a}$ along $\mathscr{C}$:

$$\boldsymbol{d}_\mathfrak{a} = \boldsymbol{d}_\mathfrak{a}(s, \boldsymbol{D}_1, \boldsymbol{D}_2, \boldsymbol{D}_3). \qquad (63.21)$$

Pursuing the interpretation mentioned above, we might choose for one director $\boldsymbol{d}$ a unit vector along the direction into which one of the directors $\boldsymbol{D}$, considered as material, is deformed in the three-dimensional deformation experienced by the finite rod whose strain our one-dimensional theory is intended to represent. Such a director, $\boldsymbol{d}_0$, would not in general be orthogonal to c. In practice, $\boldsymbol{d}$ is selected as the normal to the unit tangent $\boldsymbol{t}$ which lies in the plane of $\boldsymbol{d}_0$ and $\boldsymbol{t}$. The triple orthogonal directions so determined are called the *principal torsion-flexure axes* in the deformed rod. The twist in the deformed rod is then defined as the twist of $\boldsymbol{d}$ relative to c.

Obviously this classical procedure is motivated only by formal simplicity and furnishes an inadequate description of the strain of a rod. In the first place, twist is defined unsymmetrically with respect to the strained and unstrained rods. While the twist of the unstrained rod is uniquely determined, two different twists can be obtained for the strained rod, depending on which of the principal axes of inertia of the cross-section is selected for $\boldsymbol{D}$ in the unstrained rod, and by the above remarks, these two twists are in general entirely unrelated to one another. Moreover, the insistence that $\boldsymbol{d}$ be normal to c is merely artificial and does not represent any geometrical requirement.

For a more precise description, we need only adapt to c what has already been done for $\mathscr{C}$. Since the operation "^" defined by (63.6) is a derivative with respect to S, not s, the analysis, while strictly parallel, is not merely dual to that given above. We set

$$t^\mathfrak{a} \equiv d_k^\mathfrak{a} \hat{x}^k, \quad C_{\mathfrak{a}\mathfrak{b}} \equiv g_{km} d_\mathfrak{a}^k d_\mathfrak{b}^m, \quad F^\mathfrak{a}{}_\mathfrak{b} \equiv d_k^\mathfrak{a} \hat{d}_\mathfrak{b}^k. \qquad (63.22)$$

The $C_{\mathfrak{a}\mathfrak{b}}$ are the *components of deformation*; $F^\mathfrak{a}{}_\mathfrak{b}$ is the *relative wryness* of c.

By analogy with what was said earlier in connection with $\mathscr{C}$ it follows that if $t^\mathfrak{a}$, $C_{\mathfrak{a}\mathfrak{b}}$, and $F^\mathfrak{a}{}_\mathfrak{b}$ are given functions of S subject to the conditions that $C_{\mathfrak{a}\mathfrak{b}}$ be symmetric and positive definite and that

$$\hat{C}_{\mathfrak{a}\mathfrak{b}} = 2 C_{\mathfrak{c}(\mathfrak{a}} F^\mathfrak{c}{}_{\mathfrak{b})}, \qquad (63.23)$$

the rod c is determined to within a rigid motion combined with a reflection, its arc length s being obtained by integrating the equation

$$\lambda^2 = \frac{d s^2}{d S^2} = \hat{x}_k \hat{x}^k = C_{\mathfrak{a}\mathfrak{b}} t^{\mathfrak{a}} t^{\mathfrak{b}}, \tag{63.24}$$

λ being the stretch. What has been shown is summarized in the following ***fundamental theorem on the strain of a rod***: *Given a rod $\mathscr{C}$ with arc length S and directors $\boldsymbol{D}_{\mathfrak{a}}$, prescription of the eighteen scalars $t^{\mathfrak{a}}$, $C_{\mathfrak{a}\mathfrak{b}}$, and $F^{\mathfrak{a}}{}_{\mathfrak{b}}$ as functions of S subject to the aforementioned conditions determines a second rod c with arc length s and directors $d_{\mathfrak{a}}(s)$, uniquely to within a rigid motion combined with a reflection.* In other words, the quantities $t^{\mathfrak{a}}$, $C_{\mathfrak{a}\mathfrak{b}}$, and $F^{\mathfrak{a}}{}_{\mathfrak{b}}$ furnish *a complete differential description of the strain of a rod.* It suffices to specify only the twelve scalars $t^{\mathfrak{a}}$, $C_{\mathfrak{a}\mathfrak{b}}$, and $C_{\mathfrak{a}[\mathfrak{b}} F^{\mathfrak{a}}{}_{\mathfrak{c}]}$ since one can then calculate $F^{\mathfrak{a}}{}_{\mathfrak{b}}$ using (63.23).

The apparatus constructed is very general, enough so to include what would be regarded physically as an anisotropic rod. To represent a physically isotropic rod, let $\boldsymbol{D}_1$ be the unit tangent to $\mathscr{C}$ and $\boldsymbol{d}_1$ the unit tangent to c. In this case, $T^1 = t^1 = 1$, $T^2 = T^3 = t^2 = t^3 = 0$. The remaining two directors, both along $\mathscr{C}$ and along c, may be assigned arbitrarily.

Again we consider a general director frame. To determine the *finite rotation* of a rod, it is necessary to regard the rod from the Euclidean space in which it is imbedded. So as to motivate the definition shortly to be given, we first consider the case when $\mathscr{C}$ is a material curve in a three-dimensional body and the directors $\boldsymbol{D}_{\mathfrak{a}}$ are material elements, this being the case that one-dimensional theories of rods are intended to idealize. Then, as remarked in Sect. 62, in (61.13) we have $A^k{}_K = x^k_{;K}$, $a^K{}_k = X^K_{;k}$. By the theory given in Sect. 37, we can decompose $x^k{}_{;K}$ uniquely into a stretch and a rotation, and the rotation so determined is the rotation of the principal axes of strain at $\boldsymbol{X}$ into the principal axes of strain at $\boldsymbol{x}$. In a purely one-dimensional theory, we do not have available the $x^k_{;K}$, but from the two sets of directors we can define $A^k{}_K$ by (61.12), and, since the linear independence of the directors assures us that $A^k{}_K$ is non-singular, from the polar decomposition theorem (Sect. App. 43) we can write

$$A^k{}_K = O^k{}_M P^M_K = p^k_m O^m{}_K, \tag{63.25}$$

where $\boldsymbol{O}$ is an orthogonal double tensor and $\boldsymbol{P}$ and $\boldsymbol{p}$ are positive definite symmetric tensors. The tensor $\boldsymbol{O}$ then represents the *local rotation* of the rod. What we have proved regarding it is summarized as follows: *Given two rods with directors $\boldsymbol{D}_{\mathfrak{a}}$ and $\boldsymbol{d}_{\mathfrak{a}}$, a unique local rotation is defined by* (63.25); *the rotation is independent of the choice of the directors to this extent, that if we imagine and then leave fixed a deformation of a three-dimensional body in which $\mathscr{C}$ is a material line and the given directors $\boldsymbol{D}_{\mathfrak{a}}$ are carried materially into the given directors $\boldsymbol{d}_{\mathfrak{a}}$, then we may choose as directors any other sets of linearly independent material vectors and obtain the same rotation.* When the directors are chosen as above, $(63.22)_2$ becomes

$$C_{\mathfrak{a}\mathfrak{b}} = C_{KM} D^K_{\mathfrak{a}} D^M_{\mathfrak{b}}: \tag{63.26}$$

the quantities $C_{\mathfrak{a}\mathfrak{b}}$ are anholonomic components of the three-dimensional deformation tensor $\boldsymbol{C}$ with respect to the directors of $\mathscr{C}$.

63 A. Appendix. History of the description of strain in a rod. The concept of twist may be traced back to St. Venant [1843, *3*, ¶ 2], who called it "gauchissement". He obtained (63.20) [1845, *3*], having given an approximation to it earlier [1843, *3*, ¶ 15]. Binet [1844, *1*] claimed that he had introduced the concept of twist shortly after 1815. The introduction of the principal torsion-flexure axes is due to St. Venant [1843, *3*, ¶ 2]. Other attempts to formulate a general theory of strain in a rod were made by Kirchhoff [1859, *2*, § 2]

[1876, *2*, Vorl. 28, § 2] and CLEBSCH [1862, *2*, §§ 48–49, 55]. All this early work employs more or less hidden approximations and is difficult to follow with confidence. The first straightforward analysis is that of LOVE [1893, *5*, § 233].

Our Sect. 63 follows ERICKSEN and TRUESDELL [1958, *1*, §§ 7–14]; their treatment is freely adapted from that of E. and F. COSSERAT [1909, *5*, § 21] (cf. also [1908, *2*]), whose considerations, in our opinion rather fragmentary, are restricted to the common frame.

The only previous analysis approaching the generality of Sect. 63 is that of HAY [1942, *8*, §§ 2–3]. In effect, he takes the $\boldsymbol{D}_\mathfrak{a}$ as any unit triad such that $\boldsymbol{D}_1$ is the unit tangent to $\mathscr{C}$, and he chooses co-ordinate systems such that not only $D^K_\mathfrak{a} = \delta^K_\mathfrak{a}$ but also $d^k_\mathfrak{a} = \delta^k_\mathfrak{a}$. Hence (61.12) yields $A^k{}_K = \delta^k_K$, $a^K{}_k = \delta^K_k$, but it must be remembered that the co-ordinates at $\boldsymbol{x}$ are not generally orthogonal. In these co-ordinates we have from $(63.22)_2$

$$C_{\mathfrak{a}\mathfrak{b}} = g_{km}\, d^k_\mathfrak{a}\, d^m_\mathfrak{b} = g_{km}\, \delta^k_\mathfrak{a}\, \delta^m_\mathfrak{b}: \tag{63 A.1}$$

thus the components of the metric tensor at $\boldsymbol{x}$ are numerically equal to the components of stretch, this being a generalization of HAY's Eq. (3.9). From $(63.22)_3$ we get in these co-ordinates

$$\left.\begin{aligned} F^\mathfrak{a}{}_\mathfrak{b} &= d^\mathfrak{a}_k \left[\frac{d d^k_\mathfrak{b}}{ds} + \left\{ {k \atop m\,p} \right\} d^m_\mathfrak{b} \frac{dx^p}{ds}\right] \frac{ds}{dS}, \\ &= \delta^\mathfrak{a}_k \left\{ {k \atop m\,1} \right\} \delta^m_\mathfrak{b}\, \lambda, \\ &= \delta^\mathfrak{a}_k\, \delta^m_\mathfrak{b}\, g^{kp} \left(\frac{\partial g_{1[p}}{\partial x^{m]}} + \frac{1}{2} \frac{\partial g_{pm}}{\partial x^1} \right) \lambda, \end{aligned}\right\} \tag{63 A.2}$$

where the rod c is taken as the x^1-curve with $x^1 = s$. The skew-symmetric part of this equation is essentially HAY's Eq. (3.5); hence our relative wryness $\boldsymbol{F}$ includes and generalizes HAY's rotation vector $\boldsymbol{\omega}$.

64. Shells[1].

Let the undeformed surface $\mathscr{S}$ be given by

$$X^K = X^K(V^1, V^2) = X^K(V^\Xi), \tag{64.1}$$

the deformed surface s by

$$x^k = x^k(v^1, v^2) = x^k(v^\xi), \tag{64.2}$$

where, as in the rest of this section, Latin indices run from 1 to 3, Greek indices from 1 to 2. The deformation is specified by the functional forms of the right-hand sides of (64.1) and (64.2), augmented by

$$\boldsymbol{v} = \boldsymbol{v}(\boldsymbol{V}). \tag{64.3}$$

For full generality, it is sufficient but not necessary to take

$$v^\xi = \delta^\xi_\Xi\, V^\Xi. \tag{64.4}$$

The surface metric tensors are given by

$$\left.\begin{aligned} dS^2 &= A_{\Delta\Xi}\, dV^\Delta dV^\Xi = A_{\delta\xi}\, dv^\delta dv^\xi, \\ ds^2 &= a_{\delta\xi}\, dv^\delta dv^\xi = a_{\Delta\Xi}\, dV^\Delta dV^\Xi, \end{aligned}\right\} \tag{64.5}$$

where

$$A_{\delta\xi} = A_{\Delta\Xi}\, V^\Delta_{;\delta}\, V^\Xi_{;\xi}, \qquad a_{\Delta\Xi} = a_{\delta\xi}\, v^\delta_{;\Delta}\, v^\xi_{;\Xi}, \tag{64.6}$$

and where ";" denotes a partial derivative. If $\boldsymbol{A}(\boldsymbol{V})$ and $\boldsymbol{a}(\boldsymbol{v})$ are known, and if the parameter transformation (64.3) is given, we may calculate for any material element the changes of length and angle occasioned by the deformation. Also, we may calculate the total curvatures R_{1212} of $\mathscr{S}$ and s. Corresponding to any assigned real symmetric tensor $\boldsymbol{a}(\boldsymbol{v})$, with $\boldsymbol{A}(\boldsymbol{V})$ and the relation (64.3) regarded

[1] The general theory of strain of position for a surface was discussed by USAI [1909, *9*]. Strain of orientation was considered but scarcely analysed by E. and F. COSSERAT [1908, *1*] [1909, *5*, §§ 30–32]. Our treatment follows ERICKSEN and TRUESDELL [1958, *1*, §§ 15–20].

as given, there exists a surface $\mathfrak{s}$ into which $\mathscr{S}$ is deformed, and any two such surfaces $\mathfrak{s}_1$ and $\mathfrak{s}_2$ are applicable. In principle, this completes the intrinsic theory of strain of position of a shell.

In terms of the three-dimensional equations (64.1) and (64.2), by (App. 19.7) and (App. 21.5) we may calculate the fundamental forms $\boldsymbol{A}$, $\boldsymbol{B}$, and $\boldsymbol{a}$, $\boldsymbol{b}$:

$$A_{\Delta\Xi} = g_{KM} X^K_{;\Delta} X^M_{;\Xi}, \qquad a_{\delta\xi} = g_{km} x^k_{;\delta} x^m_{;\xi}. \tag{64.7}$$

$$B_{\Delta\Xi} = N_K X^K_{;\Delta\Xi}, \qquad b_{\delta\xi} = n_k x^k_{;\delta\xi},$$

where $\boldsymbol{N}$ and $\boldsymbol{n}$ are the unit normals to $\mathscr{S}$ and $\mathfrak{s}$, and where the total covariant derivatives are defined as in Sect. App. 20. We have

$$B_{\Delta\Xi} dV^\Delta dV^\Xi = B_{\delta\xi} dv^\delta dv^\xi, \qquad b_{\delta\xi} dv^\delta dv^\xi = b_{\Delta\Xi} dV^\Delta dV^\Xi, \tag{64.8}$$

where

$$B_{\delta\xi} = B_{\Delta\Xi} V^\Delta_{;\delta} V^\Xi_{;\xi}, \qquad b_{\Delta\Xi} = b_{\delta\xi} v^\delta_{;\Delta} v^\xi_{;\Xi}. \tag{64.9}$$

The tensors $\boldsymbol{a}$ and $\boldsymbol{b}$ are related by (App. 21.8), the Gauss and Mainardi-Codazzi equations; $\boldsymbol{A}$ and $\boldsymbol{B}$, by the duals of those equations. (In applying here the results of Sect. App. 21, we are to replace Greek majuscules by Greek minuscules throughout.)

When $\mathscr{S}$ and a parameter mapping (64.3) are given, assignment of arbitrary symmetric tensors $\boldsymbol{a}$ and $\boldsymbol{b}$ satisfying (App. 21.8) determines a surface $\mathfrak{s}$ into which $\mathscr{S}$ is deformed, uniquely to within a rigid motion[1]. The surface $\mathfrak{s}$ is obtained by solving (App. 21.7) with x^k and n^k assigned at one point, and $\mathscr{S}$ is obtained by solving the dual equations. In principle, this completes the imbedding theory of strain of position.

In approximate theories of elastic shells it is customary to take the changes of principal curvature as measures of strain. To do this in the exact theory would introduce complications analogous to those resulting from using the displacement vector and the strain tensors in finite strain of three-dimensional bodies.

The theory of strain of orientation for a shell may be constructed in analogy to what was done for three-dimensional bodies in Sect. 61. At the points $\boldsymbol{X}$ of the undeformed shell $\mathscr{S}$, assign a set of three directors $\boldsymbol{D}_\mathfrak{a}$, and put

$$X^\mathfrak{a}_\Delta \equiv D^\mathfrak{a}_K X^K_{;\Delta}, \qquad G_{\mathfrak{a}\mathfrak{b}} \equiv G_{KM} D^K_\mathfrak{a} D^M_\mathfrak{b}, \qquad W^\mathfrak{a}{}_{\mathfrak{b}\Delta} \equiv D^\mathfrak{a}_K D^K_{\mathfrak{b};\Delta}, \tag{64.10}$$

so that

$$X^K_{;\Delta} = D^K_\mathfrak{a} X^\mathfrak{a}_\Delta, \qquad D^K_{\mathfrak{a};\Delta} = D^K_\mathfrak{b} W^\mathfrak{b}{}_{\mathfrak{a}\Delta}. \tag{64.11}$$

Then by $(64.7)_1$ follows

$$A_{\Delta\Xi} = G_{\mathfrak{a}\mathfrak{b}} X^\mathfrak{a}_\Delta X^\mathfrak{b}_\Xi, \tag{64.12}$$

while by $(64.7)_3$ follows

$$B_{\Delta\Xi} = N_K D^K_\mathfrak{a} (X^\mathfrak{a}_{\Delta;\Xi} + W^\mathfrak{a}{}_{\mathfrak{b}\Xi} X^\mathfrak{b}_\Delta). \tag{64.13}$$

From the relation (64.11),

$$X^K_{;\Xi\Delta} = D^K_\mathfrak{b} (X^\mathfrak{b}_{\Xi;\Delta} + W^\mathfrak{b}{}_{\mathfrak{a}\Delta} X^\mathfrak{a}_\Xi), \tag{64.14}$$

$$D^K_{\mathfrak{a};\Xi\Delta} = D^K_\mathfrak{c} (W^\mathfrak{c}{}_{\mathfrak{b}\Delta} W^\mathfrak{b}{}_{\mathfrak{a}\Xi} + W^\mathfrak{c}{}_{\mathfrak{a}\Xi;\Delta}). \tag{64.15}$$

We also have the integrabi ity conditions

$$\frac{\partial^2 X^K}{\partial V^{[\Delta} \partial V^{\Xi]}} = X^K_{;[\Delta\Xi]} = 0, \qquad \frac{\partial^2 D^K_\mathfrak{a}}{\partial V^{[\Delta} \partial V^{\Xi]}} = D^K_{\mathfrak{a};[\Delta\Xi]} = 0, \tag{64.16}$$

[1] E.g. EISENHART [1940, 9, § 39].

whence follows

$$\frac{\partial X^{\mathfrak{b}}_{[\Delta}}{\partial V^{\Xi]}} + W^{\mathfrak{b}}{}_{\mathfrak{a}[\Xi} X^{\mathfrak{a}}{}_{\Delta]} = 0, \tag{64.17}$$

$$W^{\mathfrak{c}}{}_{\mathfrak{b}\Xi} W^{\mathfrak{b}}{}_{\mathfrak{a}\Delta} - W^{\mathfrak{c}}{}_{\mathfrak{b}\Delta} W^{\mathfrak{b}}{}_{\mathfrak{a}\Xi} + 2\frac{\partial W^{\mathfrak{c}}{}_{\mathfrak{a}[\Delta}}{\partial V^{\Xi]}} = 0. \tag{64.18}$$

From $(64.10)_2$ and $(64.11)_2$

$$G_{\mathfrak{a}\mathfrak{b};\Delta} = 2W_{(\mathfrak{a}\mathfrak{b})\Delta}. \tag{64.19}$$

Eqs. (64.11) may be regarded as a differential system for the equation $\boldsymbol{X}(\boldsymbol{V})$ of the shell and for the assignment of directors upon it. When $X^{\mathfrak{a}}_{\Delta}$, $G_{\mathfrak{a}\mathfrak{b}} = G_{\mathfrak{b}\mathfrak{a}}$, and $W^{\mathfrak{b}}{}_{\mathfrak{a}\Delta}$ are prescribed as functions of V^{Δ} subject to the conditions (64.17) to (64.19), the system (64.11) is completely integrable. Locally there will then exist a unique solution $\boldsymbol{X}(\boldsymbol{V})$ and $\boldsymbol{D}_{\mathfrak{a}}(\boldsymbol{V})$ taking on prescribed values at one value $\boldsymbol{V}_0$ of $\boldsymbol{V}$. Provided that not all the quantities $X^{\mathfrak{a}}_{[\Delta} X^{\mathfrak{b}}_{\Xi]}$ vanish, that $G_{\mathfrak{a}\mathfrak{b}}$ be positive definite and that the prescription of $\boldsymbol{D}_{\mathfrak{a}}(\boldsymbol{V}^0)$ be consistent with $(64.10)_2$, the solution represents an oriented shell and the relation $(64.10)_1$ holds. Using the uniqueness of solutions, it is a simple matter to show that *$X^{\mathfrak{b}}_{\Delta}$, $G_{\mathfrak{a}\mathfrak{b}}$, and $W^{\mathfrak{b}}{}_{\mathfrak{a}\Delta}$ determine $\mathscr{S}$ and its directors to within a rigid displacement combined with a reflection.*

At the points $\boldsymbol{x}$ of the deformed shell s, assign the directors $\boldsymbol{d}_{\mathfrak{a}}$, and, as suggested by (63.22), put

$$\left.\begin{aligned} x^{\mathfrak{a}}_{\Delta} &\equiv d^{\mathfrak{a}}_{k}\, x^{k}_{;\Delta} = d^{\mathfrak{a}}_{k}\, x^{k}_{;\delta}\, v^{\delta}_{;\Delta},\\ C_{\mathfrak{a}\mathfrak{b}} &\equiv g_{km}\, d^{k}_{\mathfrak{a}}\, d^{m}_{\mathfrak{b}},\\ F^{\mathfrak{a}}{}_{\mathfrak{b}\Delta} &\equiv d^{\mathfrak{a}}_{k}\, d^{k}_{\mathfrak{b};\Delta} = d^{\mathfrak{a}}_{k}\, d^{k}_{\mathfrak{b};\delta}\, v^{\delta}_{;\Delta}, \end{aligned}\right\} \tag{64.20}$$

so that

$$x^{k}_{;\delta} = d^{k}_{\mathfrak{a}}\, x^{\mathfrak{a}}_{\Delta}\, V^{\Delta}_{;\delta}, \qquad d^{k}_{\mathfrak{b};\delta} = d^{k}_{\mathfrak{a}}\, F^{\mathfrak{a}}{}_{\mathfrak{b}\Delta}\, V^{\Delta}_{;\delta}. \tag{64.21}$$

The $C_{\mathfrak{a}\mathfrak{b}}$ are the *components of deformation*; $F^{\mathfrak{a}}{}_{\mathfrak{b}\Delta}$ is the *relative wryness* of s; the $x^{\mathfrak{a}}_{\Delta}$ are certain tangent vectors. Then

$$\left.\begin{aligned} a_{\delta\xi} &= C_{\mathfrak{a}\mathfrak{b}}\, x^{\mathfrak{a}}_{\Delta}\, x^{\mathfrak{b}}_{\Xi}\, V^{\Delta}_{;\delta}\, V^{\Xi}_{;\xi},\\ b_{\delta\xi} &= n_k\, d^{k}_{\mathfrak{a}} \left[F^{\mathfrak{a}}{}_{\mathfrak{b}\Xi}\, V^{\Xi}_{;\xi}\, x^{\mathfrak{b}}_{\Delta}\, V^{\Delta}_{;\delta} + x^{\mathfrak{a}}_{\Delta;\Xi}\, V^{\Delta}_{;\delta}\, V^{\Xi}_{;\xi} + x^{\mathfrak{a}}_{\Delta}\, V^{\Delta}_{;\delta\xi}\right]. \end{aligned}\right\} \tag{64.22}$$

Thus we see that knowledge of $\boldsymbol{v}$, $x^{\mathfrak{a}}_{\Delta}$, $C_{\mathfrak{a}\mathfrak{b}}$, and $F^{\mathfrak{a}}{}_{\mathfrak{b}\Delta}$ and $d^{k}_{\mathfrak{a}}$ as functions of $\boldsymbol{V}$ suffices to determine the first and second forms of the deformed shell s.

By analogy with (64.17) to (64.19), we have

$$\frac{\partial x^{\mathfrak{b}}_{[\Delta}}{\partial V^{\Xi]}} + F^{\mathfrak{b}}{}_{\mathfrak{a}[\Xi} X^{\mathfrak{a}}_{\Delta]} = 0, \tag{64.23}$$

$$F^{\mathfrak{c}}{}_{\mathfrak{b}\Xi} F^{\mathfrak{b}}{}_{\mathfrak{a}\Delta} - F^{\mathfrak{c}}{}_{\mathfrak{b}\Delta} F^{\mathfrak{b}}{}_{\mathfrak{a}\Xi} + 2\frac{\partial F^{\mathfrak{c}}{}_{\mathfrak{a}[\Delta}}{\partial V^{\Xi]}} = 0, \tag{64.24}$$

and

$$C_{\mathfrak{a}\mathfrak{b};\Delta} = 2C_{\mathfrak{c}(\mathfrak{a}} F^{\mathfrak{c}}{}_{\mathfrak{b})\Delta}. \tag{64.25}$$

Further, when $\boldsymbol{v}$, $F^{\mathfrak{a}}{}_{\mathfrak{b}\Xi}$, $x^{\mathfrak{a}}_{\Delta}$, and $C_{\mathfrak{a}\mathfrak{b}}$ are given, subject to the requirements (64.23) to (64.25), $|v^{\xi}{}_{;\Xi}| \neq 0$, and the conditions that $C_{\mathfrak{a}\mathfrak{b}}$ be symmetric and positive definite and that not all the quantities $x^{\mathfrak{a}}_{[\Delta} x^{\mathfrak{b}}_{\Xi]}$ vanish, the differential system (64.21) determines an oriented shell s to within a rigid motion combined with a reflection, and (64.20) is satisfied.

What has been shown, then, is summarized in the following ***fundamental theorem on the strain of a shell***: *Given a shell $\mathscr{S}$ with fundamental forms $\boldsymbol{A}(\boldsymbol{V})$ and $\boldsymbol{B}(\boldsymbol{V})$ and with directors $\boldsymbol{D}_{\mathfrak{a}}$, prescription of the 32 quantities v^{ξ}, $F^{\mathfrak{a}}{}_{\mathfrak{b}\Xi}$, $x^{\mathfrak{a}}_{\Delta}$, and $C_{\mathfrak{a}\mathfrak{b}}$*

as functions of $\boldsymbol{V}$ *subject to the aforementioned conditions determines a second shell* $\mathfrak{s}$ *with equation* $\boldsymbol{x}=\boldsymbol{x}(\boldsymbol{v})$ *and directors* $\boldsymbol{d}_{\mathfrak{a}}(\boldsymbol{v})$, *uniquely to within a rigid displacement combined with a reflection.* In other words, the quantities v^{ξ}, $F^{\mathfrak{a}}{}_{\mathfrak{b}\Xi}$, $x^{\mathfrak{a}}_{\Delta}$, and $C_{\mathfrak{a}\mathfrak{b}}$ furnish a *complete description of the strain of a shell.*

A theory of rotation is easily constructed by analogy to what was done for rods in Sect. 63.

We now consider resolution of these results in terms of normal and tangential components. Since $X^K_{;1}$, $X^K_{;2}$, and N^K are three linearly independent vectors, we may write

$$D^K_{\mathfrak{a}} = D^{\Delta}_{\mathfrak{a}} X^K_{;\Delta} + D_{\mathfrak{a}} N^K, \tag{64.26}$$

where

$$D^{\Delta}_{\mathfrak{a}} \equiv A^{\Delta\Xi} D^K_{\mathfrak{a}} X_{K;\Xi}, \qquad D_{\mathfrak{a}} = D^K_{\mathfrak{a}} N_K. \tag{64.27}$$

Then by (App. 21.6) follows

$$D^K_{\mathfrak{a};\Delta} = (D^{\Xi}_{\mathfrak{a};\Delta} - D_{\mathfrak{a}} B^{\Xi}_{\Delta}) X^K_{;\Xi} + (D^{\Xi}_{\mathfrak{a}} B_{\Xi\Delta} + D_{\mathfrak{a};\Delta}) N^K, \tag{64.28}$$

whence

$$N_K D^K_{\mathfrak{a};\Delta} = D^{\Xi}_{\mathfrak{a}} B_{\Xi\Delta} + D_{\mathfrak{a};\Delta}, \tag{64.29}$$

$$X_{K;\Xi} D^K_{\mathfrak{a};\Delta} = D_{\mathfrak{a}\Xi;\Delta} - D_{\mathfrak{a}} B_{\Xi\Delta}. \tag{64.30}$$

From the fact that $B_{\Delta\Xi}$ and $A_{\Delta\Xi}$ determine $\boldsymbol{X}(\boldsymbol{V})$ to within a rigid motion and that $D^k_{\mathfrak{a}}(\boldsymbol{V})$ is uniquely determined by $D^{\Delta}_{\mathfrak{a}}$ and $D_{\mathfrak{a}}$ when $\boldsymbol{X}(\boldsymbol{V})$ is known, it follows that $B_{\Delta\Xi}$, $A_{\Delta\Xi}$, $D^{\Delta}_{\mathfrak{a}}$, and $D_{\mathfrak{a}}$ determine $\mathscr{S}$ and its directors to within a rigid motion. Other than the duals of (App. 21.8), there are no compatibility conditions to be satisfied by these quantities.

It involves no restriction to require that $\mathfrak{s}$ and one director not tangent to $\mathfrak{s}$, say $\boldsymbol{d}_1$, be material with respect to an unspecified three-dimensional deformation. That is, if $\boldsymbol{x}(\boldsymbol{v})$, $\boldsymbol{d}_1(\boldsymbol{v})$, $\boldsymbol{v}(\boldsymbol{V})$, $\boldsymbol{X}(\boldsymbol{V})$, and $\boldsymbol{D}_1(\boldsymbol{V})$ are given subject to the condition that $\boldsymbol{d}_1(\boldsymbol{D}_1)$ be not tangent to $\mathfrak{s}(\mathscr{S})$, there will exist infinitely many mappings $\boldsymbol{x}(\boldsymbol{X})$ such that

$$\boldsymbol{x}(\boldsymbol{v}) = \boldsymbol{x}(\boldsymbol{X}(\boldsymbol{V}(\boldsymbol{v}))), \qquad \boldsymbol{X}(\boldsymbol{V}) = \boldsymbol{X}(\boldsymbol{x}(\boldsymbol{v}(\boldsymbol{V}))) \tag{64.31}$$

and

$$d^k_1 = x^k_{;K} D^K_1, \tag{64.32}$$

the quantities $x^k_{;K}$ being uniquely determined as functions of $\boldsymbol{V}$ by (64.32) and

$$x^k_{;\delta} = x^k_{;K} X^K_{;\Delta} V^{\Delta}_{;\delta}. \tag{64.33}$$

Since the remaining directors can be assigned arbitrarily, they will not necessarily be material with respect to this deformation.

The formalism just given is easily specialized so as to give results depending on particular choices of co-ordinates for the strain of position. We consider here only the most general form of the usual approach, that given by SYNGE and CHIEN[1]. The intended interpretation is that $\mathfrak{s}$ and $\boldsymbol{d}_1$ are material, so that (64.31) to (64.33) apply, with $\boldsymbol{D}_1$ varying with the deformation $\boldsymbol{x}(\boldsymbol{X})$ in such a way that $\boldsymbol{d}_1=\boldsymbol{n}$. Choose co-ordinates and parameters such that

$$\boldsymbol{x}(\boldsymbol{X}) = \boldsymbol{X}, \qquad \boldsymbol{X}(\boldsymbol{V}) = (V^1, V^2, 0), \qquad \boldsymbol{x}(\boldsymbol{v}) = (v^1, v^2, 0), \qquad g_{k3} = \delta_{k3}. \tag{64.34}$$

Then

$$d^k_1 = n^k = n_k = \delta^k_K D^K_1 = \delta^k_3, \qquad N_M = \delta_{M3}(G^{33})^{-\frac{1}{2}}, \tag{64.35}$$

and

$$A_{\Xi\Delta} = G_{KM} \delta^K_{\Xi} \delta^M_{\Delta}, \qquad a_{\xi\delta} = g_{km} \delta^k_{\xi} \delta^m_{\delta}. \tag{64.36}$$

---

[1] [1941, 9]. Our formula (64.38) is essentially their formula (69). Cf. also CHIEN [1944, 2, Eq. (6.13)].

From (64.7)$_3$, (64.34)$_2$, (64.35)$_5$, and (64.36)$_1$ we have[1]

$$\left.\begin{aligned} 2B_{\Xi\Delta} &= 2N_K\left(\frac{\partial^2 X^K}{\partial V^\Xi \partial V^\Delta} + \left\{ {K \atop M\,P} \right\} X^M_{;\Xi} X^P_{;\Delta} - \left\{ {\Lambda \atop \Xi\Delta} \right\} X^K_{;\Lambda}\right), \\ &= 2(G^{33})^{-\frac{1}{2}} \left\{ {3 \atop J\,K} \right\} \delta^J_\Xi \delta^K_\Delta, \\ &= (G^{33})^{\frac{1}{2}} \left(\frac{\partial G_{3J}}{\partial V^\Delta}\delta^J_\Xi + \frac{\partial G_{K3}}{\partial V^\Xi}\delta^K_\Delta - \frac{\partial G_{JK}}{\partial X^3}\delta^J_\Xi\delta^K_\Delta\right) + \\ &\quad + (G^{33})^{-\frac{1}{2}} G^{3K}\delta^\Lambda_K \left(\frac{\partial A_{\Xi\Lambda}}{\partial V^\Delta} + \frac{\partial A_{\Delta\Lambda}}{\partial V^\Xi} - \frac{\partial A_{\Xi\Delta}}{\partial V^\Lambda}\right), \end{aligned}\right\} \tag{64.37}$$

which we may write in the form

$$\left.\begin{aligned} \frac{\partial G_{JK}}{\partial X^3}\delta^J_\Xi\delta^K_\Delta &= \frac{\partial G_{J3}}{\partial V^\Delta}\delta^J_\Xi + \frac{\partial G_{J3}}{\partial V^\Xi}\delta^J_\Delta + (G^{33})^{-1} G^{3K}\delta^\Lambda_K \times \\ &\quad \times \left(\frac{\partial A_{\Xi\Lambda}}{\partial V^\Delta} + \frac{\partial A_{\Delta\Lambda}}{\partial V^\Xi} - \frac{\partial A_{\Xi\Delta}}{\partial V^\Lambda}\right) - 2(G^{33})^{-\frac{1}{2}} B_{\Xi\Delta}. \end{aligned}\right\} \tag{64.38}$$

From (64.27) and (64.35),

$$D_{1\Delta} = G_{K3}\delta^K_\Delta, \qquad D_1 = (G^{33})^{-\frac{1}{2}}. \tag{64.39}$$

Now, using (64.34)$_1$, (64.35)$_1$, and (64.39), we obtain

$$\left.\begin{aligned} 2X_{K;\Xi} D^K_{1;\Delta} &= 2G_{JK} X^J_{;\Xi}\left(\frac{\partial D^K_1}{\partial V^\Delta} + \left\{ {K \atop M\,P} \right\} X^M_{;\Delta} D^P_1\right), \\ &= 2G_{JK}\delta^J_\Xi \left\{ {K \atop M\,P} \right\} \delta^M_\Delta \delta^P_3, \\ &= \frac{\partial G_{JK}}{\partial X^3}\delta^J_\Xi\delta^K_\Delta + \frac{\partial G_{J3}}{\partial V^\Delta}\delta^J_\Xi - \frac{\partial G_{K3}}{\partial V^\Xi}\delta^K_\Delta. \end{aligned}\right\} \tag{64.40}$$

From (64.30), (64.39), and (64.40), we have

$$\left.\begin{aligned} 2X_{K;(\Xi} D^K_{1;\Delta)} &= \frac{\partial G_{JK}}{\partial X^3}\delta^J_\Xi\delta^K_\Delta = D_{1\Xi;\Delta} + D_{1\Delta;\Xi} - 2D_1 B_{\Xi\Delta}, \\ &= \frac{\partial D_{1\Xi}}{\partial V^\Delta} + \frac{\partial D_{1\Delta}}{\partial V^\Xi} - 2D_{1\Lambda}\left\{ {\Lambda \atop \Xi\Delta} \right\} - 2D_1 B_{\Xi\Delta}, \\ &= \frac{\partial G_{J3}}{\partial V^\Delta}\delta^J_\Xi + \frac{\partial G_{J3}}{\partial V^\Xi}\delta^J_\Delta - G_{J3}\delta^J_\Lambda A^{\Lambda\Sigma}\left(\frac{\partial A_{\Xi\Sigma}}{\partial V^\Delta} + \frac{\partial A_{\Delta\Sigma}}{\partial V^\Xi} - \frac{\partial A_{\Xi\Delta}}{\partial V^\Sigma}\right) \\ &\quad - 2(G^{33})^{-\frac{1}{2}} B_{\Xi\Delta}, \end{aligned}\right\} \tag{64.41}$$

which is similar to, but not identical with (64.38). The apparent discrepancy is resolved by noting that the equations

$$A^{\Xi\Delta} A_{\Delta\Lambda} = \delta^\Xi_\Lambda, \qquad 0 = \delta^3_\Delta = G^{3K} G_{KJ}\delta^J_\Delta = G^{33} G_{3J}\delta^J_\Delta + G^{3K}\delta^\Xi_K A_{\Xi\Delta} \tag{64.42}$$

imply that

$$(G^{33})^{-1} G^{3K}\delta^\Xi_K = -G_{3J}\delta^J_\Delta A^{\Delta\Xi}. \tag{64.43}$$

For s, we have by analogy with (64.38) together with (64.34)$_{1,4}$

$$2b_{\alpha\beta} = -\frac{\partial g_{km}}{\partial x^3}\delta^k_\alpha\delta^m_\beta = -\frac{\partial g_{km}}{\partial X^3}\delta^k_\alpha\delta^j_\beta. \tag{64.44}$$

SYNGE and CHIEN find it sufficient to introduce nine measures of strain. The set $a_{\alpha\beta}$, $b_{\alpha\beta}$, $D_{1\Delta}$, and D_1 is equivalent to that which they use. These quantities

[1] Note that $G^{13} \neq 0$, $G^{23} \neq 0$ in general.

satisfy three compatibility conditions, which may be taken as the duals[1] of (App. 21.8), and when $\mathcal{S}$ is given, they determine $\boldsymbol{D}_l$ uniquely and $\boldsymbol{s}$ to within a rigid motion. With the conventions adopted above, part of the problem involves determining a tensor g_{km} consistent with the conditions $(64.36)_2$, (64.44), and the fact that the Riemann tensor based on g_{km} must vanish; such complexity is the price one pays for eliminating the functions $\boldsymbol{x}(\boldsymbol{v})$ as unknowns.

II. Motion.

a) Velocity.

65. Continuous motion. In the common frame (Sect. 13), consider a transformation

$$\boldsymbol{z} = \boldsymbol{z}(\boldsymbol{Z}, t). \tag{65.1}$$

The real parameter t is to be identified with the *time*; it is assigned a physical unit T distinct from that of length, L. For each fixed time t, (65.1) is a deformation of the type considered in Sect. 15. Such a one-parameter family of deformations is called a *motion*.

The co-ordinates $\boldsymbol{Z}$ we think of as assigned once and for all to a given point in the material. Since they are the co-ordinates of the points at an arbitrary initial time, $t=t_0$, they serve for all time as names for the *particles* of the material[2]. The co-ordinates $\boldsymbol{z}$, on the other hand, we think of as assigned once and for all to a point in the Euclidean space where material bodies reside. They are the names of *places*. The motion (65.1) chronicles the places $\boldsymbol{z}$ occupied by the particle $\boldsymbol{Z}$ in the course of time[3].

To the assumptions already made in Sect. 16, now appropriate for each fixed t, we add counterparts for differentiation of (65.1) with respect to time[3]: ***Axiom of continuity.*** *The motion* (65.1) *and its inverse,*

$$\boldsymbol{Z} = \boldsymbol{Z}(\boldsymbol{z}, t), \tag{65.2}$$

possess continuous partial derivatives of as many orders as needed.

For most developments, partial derivatives of the first two or three orders suffice[4].

As a result of the axiom of continuity, during the motion a region remains always a region, a surface remains a surface, a line remains a line. Two particles once distinct remain ever so, and a body never splits asunder. In consequence of this severe restriction, many motions of physical interest are excluded from the formalism presented in this chapter. Among these are impacts, sliding, rolling, shocks, and the opening and sealing of cavities.

What is really essential is an axiom expressing the ***impenetrability of matter:*** While a body may split in two, or two disjoint bodies may coalesce,

---

[1] These equations are equivalent to the three given by SYNGE and CHIEN [1941, *9*, Eqs. (38), (39)].

[2] For physical interpretation, it is not only useless but in fact misleading and generally incorrect to attempt to identify what throughout this treatise we call *particles* with anything of a similar name in the molecular or atomic theories of physical matter. Cf. Sect. 2.

[3] A principle of duality extending that of Sect. 14 is easy to work out; cf. DEUKER [1941, *1*, § IV]. In such a scheme, e.g., one can introduce as dual to the ordinary velocity $\partial \boldsymbol{z}/\partial t$ the quantity $\partial \boldsymbol{Z}/\partial t$, which is the rate at which particles move by a given place. We avoid such dual quantities as they do not promote the intended application, for which space and the matter occupying it are conceptually different.

[4] In the case when the second spatial derivatives of Eqs. (65.1) and (65.2) do not necessarily exist, LICHTENSTEIN [1929, *5*] has formulated a condition sufficient that the ratio of the distances between two distinct particles at different times remain bounded above and below, uniformly over a finite region and in a finite interval of time.

it never happens that the particles of two initially disjoint bodies intermingle. Mathematically expressed, this axiom requires that at any fixed time the dimension of the set of points upon which (65.1) and (65.2) fail to be single-valued and continuous is always less than three[1]: in a word, that all singularities are *isolated*.

This restriction allows us conveniently to divide kinematics into two parts, one of which, given in the present chapter, concerns continuous motions, while the theory of singular surfaces is given in the next.

A different formalism is required to describe the diffusion of the several constituents of a mixture; this is presented in Sect. 159.

66. Material and spatial co-ordinates. As in Sect. 14, we may introduce general co-ordinates. It is possible to select a different co-ordinate system at each time t, or, equivalently, to view the motion in terms of a co-ordinate system in motion with respect to the common frame. To do so is in some problems very convenient, but it introduces formal complications which we wish to avoid now, deferring them until Part g of this subchapter. Here we restrict attention to a *single, fixed co-ordinate system* $\boldsymbol{x}$. That is, the motion assumes the form

$$\boldsymbol{x} = \boldsymbol{x}(\boldsymbol{X}, t), \qquad \boldsymbol{X} = \boldsymbol{X}(\boldsymbol{x}, t), \tag{66.1}$$

where

$$\boldsymbol{x} = \boldsymbol{f}(\boldsymbol{z}), \qquad \boldsymbol{X} = \boldsymbol{f}(\boldsymbol{Z}), \tag{66.2}$$

the function $\boldsymbol{f}$ being the same function in both $(66.2)_1$ and $(66.2)_2$.

Insofar as we have occasion to apply them, the results of Subchapter I are available at each fixed instant t. To indicate the restricted class of co-ordinates, we replace X^K by X^α. The variables X^α and t are the *material variables*, the X^α being the *material co-ordinates*. The x^k and t are the *spatial variables*, the x^k being the *spatial co-ordinates*. Problems for which the X^α and t are taken as independent variables are said to be set in the *material description*; the x^k and t, the *spatial description* (cf. the first appendix to this section).

The remarks following (16.6) should be reread in the present connection. In particular, for purposes of interpretation we shall never describe properties of (66.1) as if it were a co-ordinate transformation. (For some other viewpoints, see the second appendix to this section.) Most of the results concerning motion are statements of *instantaneous conditions* at $\boldsymbol{x}$. Such statements will be tensorial equations of the form $t^{k\ldots m}_{p\ldots q} = 0$, invariant with respect to changes of stationary spatial co-ordinates and thus valid independently of our assumptions (66.2) regarding the co-ordinates at $\boldsymbol{X}$.

The material description is an immediate extension of the scheme used in the mechanics of mass-points, where the paths of the several distinct masses are traced, while the spatial description has no counterpart in elementary mechanics. As DIRICHLET[2] remarked, the spatial description would seem inappropriate to problems concerning a finite free portion of matter, since the region of space to be occupied, and hence the range of the independent variables x^k, is not known in advance. Nevertheless, the mathematical difficulties which accompany the material description have limited its use to two special ends: problems in one dimension, and the proof of general theorems. For the latter it is particularly convenient, besides being the more fundamental, and we shall use it much in this work. In any problem of the mechanics of continua it is always possible, though not always convenient, to use the material description exclusively[3].

---

[1] The formulation given by HADAMARD [1903, *11*, ¶ 46] is that (65.2) shall be single-valued: Two distinct particles never occupy the same place. Since it excludes the junction or welding of two bodies, this formulation is too restrictive.

[2] [1860, *1*, Introd.].

[3] Cf. LODGE [1951, *16*].

Since $\boldsymbol{X}$ is the initial co-ordinate of the particle now at the place $\boldsymbol{x}$, a dependence upon the choice of the initial instant is implied, and $(66.1)_1$ should properly be written

$$\boldsymbol{x} = \boldsymbol{f}(\boldsymbol{X}, t_0, t). \tag{66.3}$$

Let $\boldsymbol{x}_1$ and $\boldsymbol{x}_2$ be the places occupied by $\boldsymbol{X}$ at the times t_1 and t_2:

$$\boldsymbol{x}_1 \equiv \boldsymbol{f}(\boldsymbol{X}, t_0, t_1), \quad \boldsymbol{x}_2 \equiv \boldsymbol{f}(\boldsymbol{X}, t_0, t_2). \tag{66.4}$$

It is equally possible to select $\boldsymbol{x}_1$ as an identifying label for the particle $\boldsymbol{X}$, and since (66.3) is to hold for *all* particles and *all* times, we must have then

$$\boldsymbol{x}_2 = \boldsymbol{f}(\boldsymbol{x}_1, t_1, t_2). \tag{66.5}$$

Comparison of (66.4) and (66.5) yields the functional equation

$$\boldsymbol{f}(\boldsymbol{f}(\boldsymbol{X}, t_0, t_1), t_1, t) = \boldsymbol{f}(\boldsymbol{X}, t_0, t), \tag{66.6}$$

which must be satisfied by the relation $\boldsymbol{f}$ defining a motion[1]. By means of (66.6) it is possible to prove the invariance of kinematical quantities with respect to choice of the initial time.

Our special choice of co-ordinates, according to which the X^α are the values of the x^k when $t = t_0$, implies

$$\left.\begin{aligned} &X^\alpha = \delta^\alpha_k \, x^k|_{t=t_0}, \\ &x^k(X^1, X^2, X^3, t_0) = \delta^k_\alpha X^\alpha, \\ &\left.\frac{\partial x^k}{\partial X^\alpha}\right|_{t=t_0} = x^k_{;\alpha}|_{t=t_0} = \delta^k_\alpha, \qquad \left.\frac{\partial X^\alpha}{\partial x^k}\right|_{t=t_0} = X^\alpha_{;k}|_{t=t_0} = \delta^\alpha_k. \end{aligned}\right\} \tag{66.7}$$

More generally, if $t^{k\ldots m}_{p\ldots q}$ is a tensor defined at $\boldsymbol{x}$ for all t, we put

$$t^{\alpha\ldots\beta}_{\gamma\ldots\delta} \equiv \delta^\alpha_k \ldots \delta^\beta_m \, \delta^p_\gamma \ldots \delta^q_\delta \, t^{k\ldots m}_{p\ldots q} \tag{66.8}$$

for its values at $\boldsymbol{X}$ when $t = t_0$. In many cases our results will hold independently of these formulae and hence when the material co-ordinates are any independent smooth designations for the particles, not necessarily their initial positions[2].

66 A. Appendix. History of the field description of motion. The first attempt to discuss any local features of the motion of a continuous medium in more than one dimension occurs in an isolated passage by D. BERNOULLI [1738, *1*, § 11, ¶ 4], but he failed to introduce the velocity field. The first author to attempt to do so was EULER [1745, *2*, Chap. II, Satz 1, Anm. 3], who used intrinsic co-ordinates based on the stream fillets; the apparatus is far from complete. The velocity field for plane and for rotationally symmetric motions, in a fixed co-ordinate system, was employed and developed by D'ALEMBERT (1749) [1752, *1*, § 43]. The fully general spatial description is due to EULER (1752) [1757, *2*, §§ 1–21] [1761, *2*, §§ 1–41].

After a partial attempt in 1751 [1767, *1*, § 15], EULER formulated the material description in 1760 [1762, *1*] [1766, *1*, § 36] [1770, *1*, §§ 100–118].

The origin of DIRICHLET's erroneous and commonly followed terminology, "Eulerian" for *spatial* and "Lagrangian" for *material*, was explained by RIEMANN, published by HANKEL [1861, *1*, § 1], and corrected in detail by TRUESDELL [1954, *24*, § 14[2]].

The detailed history of field theories to 1788 has been written by TRUESDELL [1954, *25*] [1955, *26*] [1960, *4*].

66 B. Appendix. Other schemes of co-ordinates and notation. Many textbooks present elementary topics in kinematics in one of the systems of vectorial and polyadic notation; the best is that of GIBBS and WILSON [1909, *6*], but it is impractical for the more elaborate parts of the subject. Only the books of JAUMANN [1905, *2*] and SPIELREIN [1916, *5*] and the article of HEUN [1913, *4*] begin to embrace the generality required today. The direct concepts and notations of the BOURBAKI school, at bottom little different from those of GIBBS, are explained and employed successfully by NOLL [1955, *18*, Chap. 1].

Some of the older writers on finite deformation restricted the co-ordinate surfaces at $\boldsymbol{x}$ to be the material surfaces which at $t = t_0$ were co-ordinate surfaces at $\boldsymbol{X}$. Various properties

1 MOISIL [1942, *10*].
2 Cf. HILL [1881, *3*, § IV], DEUKER [1941, *1*, §§ II–IV], OLDROYD [1950, *22*, §§ 2–3].

of such *convected co-ordinates* have been formalized by Hencky [1925, *7*], Zelmanov [1948, *39*], Gleyzal [1949, *12*], Oldroyd [1950, *22*, §§ 2–3], Green and Zerna [1950, *10*, §§ 2–3] [1954, *7*, §§ 2.1–2.2], and Lodge [1951, *16*], and these co-ordinates are now the favorite of British authors. Some studies employing them regard (66.1) as a transformation of co-ordinates in a space without an invariant metric; others introduce as fixed spatial co-ordinate surfaces the loci of the convected co-ordinate surfaces at time t. Insofar as time rates other than the velocity are not involved, convected co-ordinates are included as a special case of the scheme used in the present work, and the results of using them may be obtained by setting $t=0$ in our formulae. For time rates in a convected system, see Sect. 150.

67. Velocity[1]. The *velocity* $\dot{\boldsymbol{x}}$ is the rate of change of position for a given particle:

$$\dot{x}^k \equiv \frac{\partial x^k}{\partial t}, \tag{67.1}$$

where $\boldsymbol{X}$ is held constant in the motion $(66.1)_1$. Hence the $\dot{x}^k$ are the contravariant components of a vector, whose magnitude $\dot{x}$ is the *speed:* $\dot{x} = \sqrt{\dot{x}^k \dot{x}_k}$. (In the common frame, it is customary to set $u \equiv \dot{x}$, $v \equiv \dot{y}$, $w \equiv \dot{z}$, but we shall not use this notation.) Since the origin of the position vector $\boldsymbol{p}$ is fixed, we have

$$\dot{\boldsymbol{p}} = \dot{\boldsymbol{x}}, \tag{67.2}$$

and in passages where the direct vectorial significance rather than the tensorial components of the formula is uppermost, we shall prefer to write $\dot{\boldsymbol{p}}$ rather than $\dot{\boldsymbol{x}}$.

From (67.1) and $(66.1)_1$ follows

$$\dot{\boldsymbol{x}} = \dot{\boldsymbol{x}}(\boldsymbol{X}, t): \tag{67.3}$$

The velocity emerges as a function of time for a given particle. However, we may eliminate $\boldsymbol{X}$ by $(66.1)_2$, so obtaining the velocity as a function of time for a given place:

$$\dot{\boldsymbol{x}} = \dot{\boldsymbol{x}}(\boldsymbol{X}(\boldsymbol{x}, t), t) = \dot{\boldsymbol{x}}(\boldsymbol{x}, t). \tag{67.4}$$

The functional dependence (67.3) is appropriate to the material description; (67.4), to the spatial.

A point where $\dot{\boldsymbol{x}} = 0$ is called a *stagnation point.*

A motion such that the velocity field does not change in time at any given place is *steady.* For such a motion, (67.4) reduces to the form

$$\dot{\boldsymbol{x}} = \dot{\boldsymbol{x}}(\boldsymbol{x}). \tag{67.5}$$

More generally, any quantity which is a function of place only is said to be *steady.*

Sometimes it is convenient to define steadiness not in the x, y, z system but in another appropriate to an observer moving at speed V in the x direction. Then a steady quantity f must satisfy

$$f(x, y, z, t) = g(x - Vt, y, z), \tag{67.6}$$

and hence

$$\frac{\partial f}{\partial t} = -V \frac{\partial f}{\partial x}. \tag{67.7}$$

The general problem of the invariance of steady motion with respect to change of observer is discussed in Sect. 146.

68. Geometrically restricted motions. Various special kinds of motion are defined by special properties of the velocity field.

---

[1] In the early hydrodynamics of Newton, the Bernoullis, D'Alembert, etc., and in Euler's treatments published prior to 1760, the velocity was taken as a primitive concept rather than being defined in terms of particle motion.

If the velocity is constant over each member of a family of parallel planes, and normal to those planes, the motion is *lineal*[1]

$$\dot{x}^1 = \dot{x}^1(x^1, t), \quad \dot{x}^2 = \dot{x}^3 = 0, \tag{68.1}$$

where the above-mentioned planes are the surfaces $x^1 = \text{const}$. In a *pseudo-lineal motion of the first kind*, (68.1) is generalized by

$$\dot{x}^1 = \dot{x}^1(\boldsymbol{x}, t), \quad \dot{x}^2 = \dot{x}^3 = 0; \tag{68.2}$$

in a *pseudo-lineal motion of the second kind*, by

$$\dot{\boldsymbol{x}} = \dot{\boldsymbol{x}}(x^1, t). \tag{68.3}$$

A motion is said to be *plane*[2] if its velocity field is a plane field and hence representable in the form (App. 33.6) with the surfaces $x^3 = \text{const}$ being a family of parallel planes. In kinematical terms, along every line in a certain direction the velocity is constant and normal to that direction. If either of the two requirements (App. 33.6)$_{1,2}$ or (App. 33.6)$_3$ is removed, but the other retained, the motion is called *pseudo-plane*. For a *pseudo-plane motion of the first kind*[3],

$$\dot{x}^k = \dot{x}^k(x^1, x^2, x^3, t), \quad k = 1, 2, \quad \dot{x}^3 = 0; \tag{68.4}$$

for a *pseudo-plane motion of the second kind*,

$$\dot{x}^k = \dot{x}^k(x^1, x^2, t), \quad k = 1, 2, 3. \tag{68.5}$$

A motion is said to be *rotationally symmetric* if its velocity field is a rotationally symmetric field and hence representable in the form (App. 33.6) with the surfaces $x^3 = \text{const}$ being a family of co-axial planes.

In plane and rotationally symmetric motion, we shall follow the custom, obviously general enough, of restricting attention to a single plane, $x^3 = 0$, say, which may be called the *plane of motion*. For interpretation, however, it is necessary to add a third dimension, this being achieved for a plane motion by translating the plane of motion normal to itself; for a rotationally symmetric motion, by rotating the plane of motion about the axis.

The idea of plane motion is easily generalized. The plane of motion may be replaced by any curved surface, $x^3 = 0$, upon which x^1 and x^2 may be any curvilinear co-ordinates. The equations $\dot{x}^k = \dot{x}^k(x^1, x^2)$, $k = 1, 2$, now describe a strictly two-dimensional motion. To generate a three-dimensional motion in a region of space near the surface $x^3 = 0$, choose a co-ordinate system in which the surface $x^3 = \text{const}$ are surfaces parallel to $x^3 = 0$, while the surfaces $x^1 = \text{const}$ and $x^2 = \text{const}$ are those swept out by the normals to $x^3 = 0$ along the x^1 and x^2 co-ordinate curves upon it.

Properties of some of these motions will be developed in Sect. 160.

69. Velocity at a boundary. A *boundary* is a surface which the material does not cross. A necessary but not sufficient condition that $\mathscr{s}$ be a boundary is that

---

[1] In the period 1650 to 1750 most studies of fluid motion rested upon the "hypothesis of parallel sections", which supposes that (68.1) holds approximately even in tubes of varying cross-section.

[2] Both plane and rotationally symmetric motions first appear fairly explicitly in the work of D'ALEMBERT [1752, *1*, § 73; §§ 43–48]. Plane motion is more or less implied in some parts of earlier hydrodynamic studies by NEWTON, D. BERNOULLI, EULER, and other writers.

[3] The terminology follows BERKER [1936, *2*, §§ 14–17], who defines also similar generalizations of rotationally symmetric motion.

upon it the normal component of velocity be equal to the normal velocity v_n of $\mathfrak{s}$:

$$\dot{x}_n \equiv \dot{x}^k n_k = \dot{p}_n = v_n. \tag{69.1}$$

For the theory, v_n is regarded as assigned. If

$$v_n = 0 \quad \text{on} \quad \mathfrak{s}, \tag{69.2}$$

the boundary $\mathfrak{s}$ is *stationary* in the frame considered; otherwise, it is *moving*. A boundary surface may be a rigid surface or a deforming surface.

If not merely the normal component but rather the velocity vector itself is prescribed on a surface $\mathfrak{s}$, the material is said to *adhere* to $\mathfrak{s}$:

$$\dot{\boldsymbol{x}} = \dot{\boldsymbol{p}} = \boldsymbol{v} \quad \text{on} \quad \mathfrak{s}. \tag{69.3}$$

If

$$\boldsymbol{v} = 0 \quad \text{on} \quad \mathfrak{s}, \tag{69.4}$$

then $\mathfrak{s}$ is a stationary boundary to which the material adheres. When the boundary $\mathfrak{s}$ is in rigid rotation at angular velocity $\boldsymbol{\omega}$, we have $\boldsymbol{v} = \boldsymbol{b} + \boldsymbol{\omega} \times \boldsymbol{p}$ and hence (69.1) and (69.3) become, respectively,

$$\dot{x}_n = \dot{p}_n = b_n + \boldsymbol{n} \cdot \boldsymbol{\omega} \times \boldsymbol{p}, \tag{69.5}$$

$$\dot{\boldsymbol{x}} = \dot{\boldsymbol{p}} = \boldsymbol{b} + \boldsymbol{\omega} \times \boldsymbol{p}. \tag{69.6}$$

70. Path lines, stream lines, and streak lines. The curve in space traversed by $\boldsymbol{X}$ as t varies is the *path line*[1] of $\boldsymbol{X}$. The vector lines (Sect. App. 29) of the field $\dot{\boldsymbol{x}}$ at time t are the *stream lines*. The *streak line* through $\boldsymbol{x}$ at time t is the locus at time t of all particles which at any time, past or future, will occupy or have occupied the place $\boldsymbol{x}$. In equations, we have

Path line of the particle $\boldsymbol{X}$:

$$\boldsymbol{x} = \boldsymbol{x}(\boldsymbol{X}, t), \quad \boldsymbol{X} \text{ fixed}, \quad -\infty < t < \infty; \tag{70.1}$$

also, the integral curve of the system

$$d x^k = \dot{x}^k d t \tag{70.2}$$

which passes through $\boldsymbol{X}$ at $t = t_0$.

Stream lines at time t:

Integral curves

$$f_1(\boldsymbol{x}, t) = 0, \quad f_2(\boldsymbol{x}, t) = 0 \tag{70.3}$$

of the system

$$d x^1 : d x^2 : d x^3 = \dot{x}^1 : \dot{x}^2 : \dot{x}^3, \quad t = \text{const}. \tag{70.4}$$

Streak lines:

Write the motion (66.1) in the form $\boldsymbol{x} = \boldsymbol{f}(\boldsymbol{X}, t)$, $\boldsymbol{X} = \boldsymbol{F}(\boldsymbol{x}, t)$. Then the streak line through $\boldsymbol{x}$ at time t is given parametrically by the locus of x, where

$$\mathsf{x} = \boldsymbol{f}(\boldsymbol{F}(\boldsymbol{x}, t'), t), \quad -\infty < t' < \infty. \tag{70.5}$$

At a given place $\boldsymbol{x}$ and time t, the stream line through $\boldsymbol{x}$, the path line of the particle occupying $\boldsymbol{x}$, and the streak line through $\boldsymbol{x}$ all have a common tangent. When the motion is steady, all three curves coincide, but in general for unsteady motions they are distinct.

From the assumption of Sect. 65, a stream line never crosses itself nor ends, except possibly at a stagnation point. The possible singular behavior of stream lines may be read

[1] D'ALEMBERT [1752, *1*, §§ 36–39], EULER [1757, *2*, § 67]. Cf. also EULER [1745, *2*, Chap. II, Satz 1, Anm. 3].

off from well known theorems on differential systems of the type (70.4). By relaxing the assumption of Sect. 65 so as to permit more liberal behavior of $\dot{\boldsymbol{x}}$ at isolated points, more elaborate singularities for the stream lines may be obtained[1]. Since the stream lines are determined at a fixed instant, such singularities may be generated or destroyed in the course of time. The path lines and streak lines of an unsteady motion may cross themselves or double back upon themselves.

To render trajectories in the motion of a fluid visible, the commonest technique, first mentioned by LEONARDO DA VINCI (1452–1519)[2], is to cast small discernible foreign objects upon or within it, the result being kept most easily by a photograph. A long time exposure of a fluid in which a single such object has been injected records the path line of a particle. An instantaneous exposure of a fluid into which the objects are being injected continuously at one place shows a portion of the streak line through that place at that time. A short time exposure of a fluid onto whose surface many objects have just previously been dropped shows the direction field of $\dot{\boldsymbol{x}}$, whence the stream lines follow by a graphical integration[3].

A vector sheet (Sect. App. 29) of the velocity field is a *stream sheet*. Any congruence of stream lines sweeps out a stream sheet[4].

From (70.2) and the general theory of ordinary differential equations, we see that any non-vanishing field $\boldsymbol{v}$ which satisfies a Lipschitz condition may be regarded as the velocity field of a motion, at least in a sufficiently small region. In particular, any differentiable congruence of curves may serve as stream lines for infinitely many different motions of a continuous medium. Thus the vector lines of the electric field $\boldsymbol{E}$ or the magnetic intensity $\boldsymbol{B}$ may be regarded as stream lines; indeed, the whole of the theory of vector fields may be visualized in terms of motions of a continuous medium. It has seemed to us more economical to follow the opposite course, referring to the Appendix, "Invariants", for most properties of motions of continuous media that are obvious interpretations of general theorems on vector fields.

71. Examples of line systems. For a mathematical example, consider the plane motion whose spatial description is

$$\dot{x} = \frac{x}{1+t}, \quad \dot{y} = 1. \tag{71.1}$$

By integrating (70.4) for this case we get

$$y + C_2 = (1+t)\log\left|\frac{x}{C_1}\right| \tag{71.2}$$

as an equation for the stream lines; equivalently, a parametric equation for the stream line through $\boldsymbol{x}_0$ at time t is

$$x = x_0\, e^{\frac{\tau}{1+t}}, \quad y = y_0 + \tau. \tag{71.3}$$

---

[1] ZHUKOVSKI [1876, 7, Gl. II]. For the centrally important case of homogeneous motion (Sect. 142), these singularities were classified into 19 types and illustrated by POLUBARINOVA [1929, *6*]. Cf. also STEPHANSEN [1903, *16*], HESSELBERG and SVERDRUP [1914, *5*], DIETZIUS [1918, *2*].

[2] [1889, *4*, MS F, ff. 34v, 44r].

[3] Following a vague hint of FARADAY [1831, *1*], ROEVER [1914, *10*] constructed a machine for visualizing stream lines. It consists in two slotted wheels rotating with different angular speeds about different axes. ROEVER analysed only special cases and did not give the design of the slots to represent a given stream field. According to LAMPE [1922, *5*], a similar idea was employed by DOVE in 1859 to 1860. We make no attempt to discuss the numerous physical analogies for stream lines, since they presuppose constitutive equations.

[4] RIABOUCHINSKY [1938, *11*] has suggested the use of two independent families of stream sheets, $F_1 = \text{const}$ and $F_2 = \text{const}$, along with a third family of independent functions G so as to obtain a representation $\dot{\boldsymbol{x}} = \dot{\boldsymbol{x}}(G, F_1, F_2, t)$. Cf. Sect. 136.

By integrating (70.2), we get the material description:

$$x = X(1+t), \quad y = Y + t \quad (t_0 = 0). \tag{71.4}$$

By eliminating t from (71.4) we get the path line of the particle $\boldsymbol{X}$:

$$x - X\,y = X(1 - Y). \tag{71.5}$$

Each particle moves in a straight line at constant speed, but the stream lines change in time according to (71.2). To get the streak line through x, y when $t=0$ we need only hold x, y fixed in (71.5):

$$X\,Y - X(y+1) + x = 0. \tag{71.6}$$

This is a hyperbola. To get the streak line through x, y at time t, we first invert (71.4) at time t':

$$X = \frac{x}{1+t'}, \quad Y = y - t'. \tag{71.7}$$

This gives the particle which occupies the place x, y at time t'. The place x, y occupied by this particle at time t follows from (71.4):

$$\mathsf{x} = x\,\frac{1+t}{1+t'}, \quad \mathsf{y} = y + t - t', \tag{71.8}$$

or the curve

$$\mathsf{x}\mathsf{y} - \mathsf{x}\,(1 + y + t) + x(1+t) = 0, \tag{71.9}$$

again a hyperbola, including (71.6) as the special case $t=0$. These results are illustrated in Fig. 10.

A physically more typical and mathematically more difficult example of plane motion was worked out by MAXWELL[1]. In the plane of motion, let r, θ be polar co-ordinates and let the contravariant velocity components $\dot{r}$, $\dot{\theta}$ be given by

$$\dot{r} = -V\left(1 - \frac{a^2}{r^2}\right)\cos\theta, \quad r\dot{\theta} = V\left(1 + \frac{a^2}{r^2}\right)\sin\theta, \tag{71.10}$$

where V and a are constants. For the stream lines, integration of (70.4) yields

$$\left(1 - \frac{a^2}{r^2}\right) r\sin\theta = \text{const}. \tag{71.11}$$

Hence $r=a$ is a stream line, and as $r\to\infty$ we have $\dot{x}\to -V$, $\dot{y}\to 0$, so that (71.10) represents a steady motion at speed V past a circular cylinder. The stream lines (71.11) are drawn in Fig. 11.

Now consider this same motion as apparent to an observer for whom the material is at rest at ∞, the cylinder in motion with the velocity V along the x axis from $-\infty$ to $+\infty$. The appropriate velocity field is obtained by adding

[1] [1870, 5]. The analysis had been initiated by RANKINE [1864, 2, § 18], who determined the radius of curvature of the path lines corresponding to certain fields of stream lines which he called "oögenous neoïds". In obtaining the special case cited in the next footnote, RANKINE observed that the corresponding path line is a looped elastica and gave a sketch of its form.

The path lines for flow within an ellipsoid were obtained by KELVIN [1885, 8] (according to LARMOR [note, p. 197 of the 1910 reprint of [1885, 8]], the analysis had been given in substance earlier, probably by FERRERS, in the Cambridge examinations); for flow past a sphere and past two parallel rectilinear vortices, by RIECKE [1888, 8]; for flow around or within certain rotating prisms, by MORTON [1913, 7]. Cf. also MORTON and VINT [1915, 4]. These problems have been considered afresh by DARWIN [1953, 7]. All these flows are examples of isochoric irrotational motion, a general class which will be defined in Sect. 88.

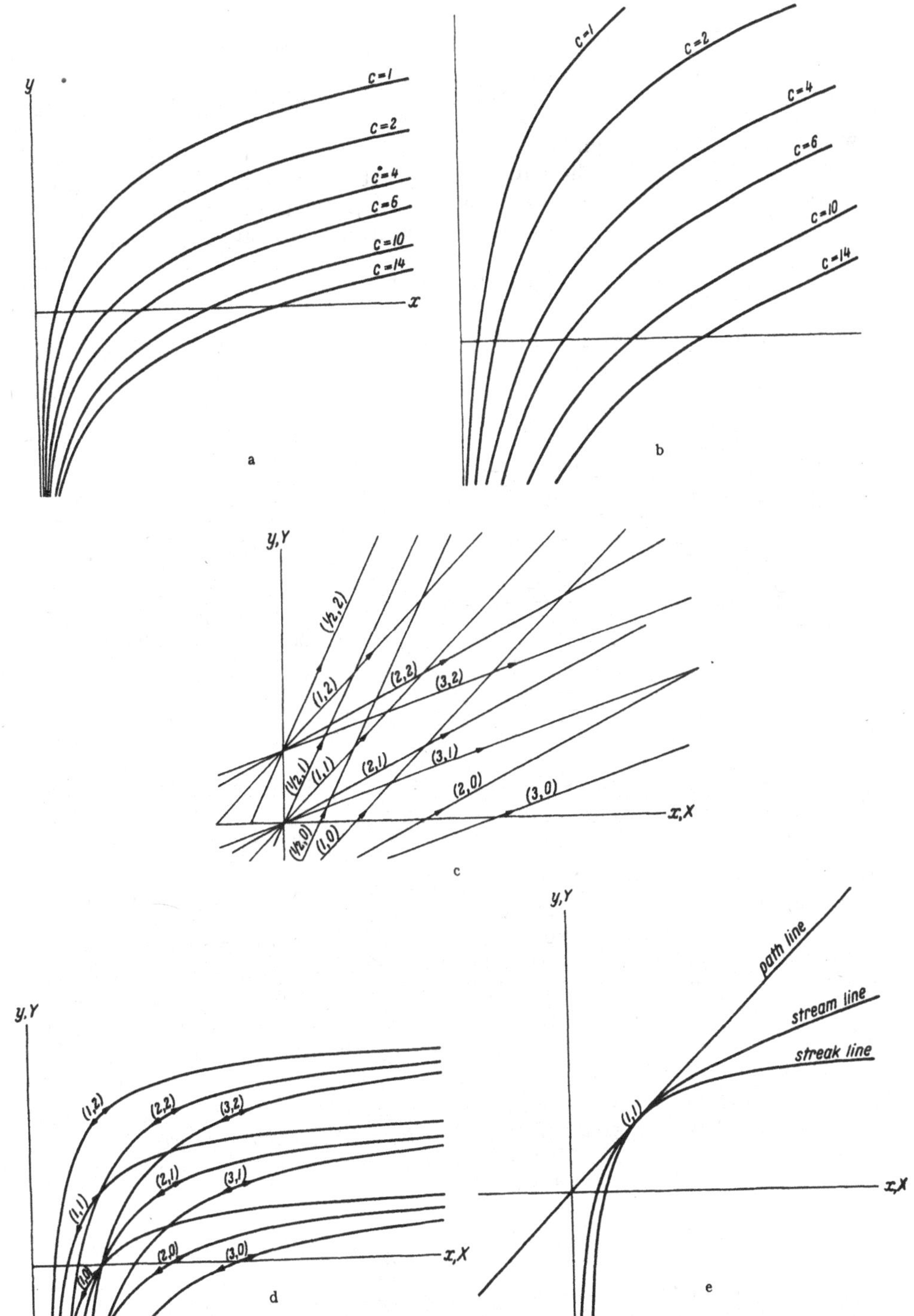

Fig. 10a—e. (a) Stream lines of the velocity field (71.1) at $t=0$, with $C_2=0$ and C_1 as indicated. (b) Stream lines of the velocity field (71.1) at $t=1$, with $C_2=0$ and C_1 as indicated. (c) Path lines of the particles (X, Y) for the velocity field (71.1). (d) Initial streak lines, i.e., loci at $t=0$ of all particles ever occupying the place (x, y), for the velocity field (71.1). (e) Path line of the particle (1, 1); stream line at $t=0$ and initial streak line through the place (1, 1), for the velocity field (71.1).

the components $\dot{r} = V\cos\theta$, $r\dot{\theta} = -V\sin\theta$, to (71.10), so that[1]

$$\dot{r} = V\frac{a^2}{r^2}\cos\theta, \quad r\dot{\theta} = V\frac{a^2}{r^2}\sin\theta, \tag{71.12}$$

where, as before, the origin of r and θ is the center of the circular cross-section, now moving. To sketch the motion, note that the concentric circles $r = \text{const}$ are curves of equal speed, while, since $\dot{y}/\dot{x} \doteq \tan 2\theta$, the angle of the velocity vector at $\boldsymbol{x}$ is twice the azimuth angle. The stream lines are the circles

$$\frac{\sin\theta}{r} = \text{const}, \tag{71.13}$$

shown in Fig. 12. This stream field is said to be that induced by a *doublet*. The circle $r = a$ is no longer a stream line. The motion is not steady in the frame of

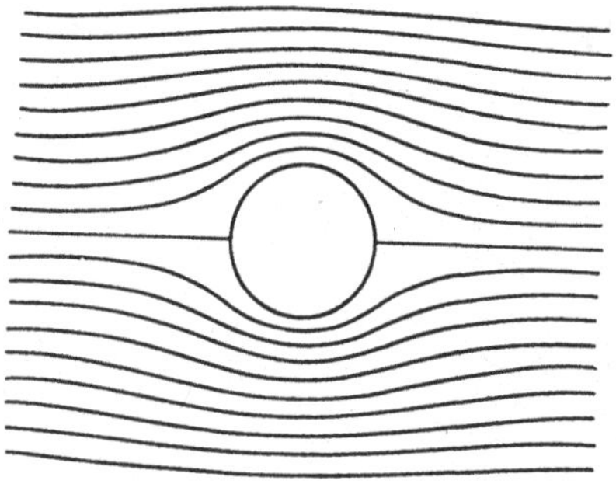

Fig. 11. Stream lines of steady isochoric potential flow past a circular cylinder.

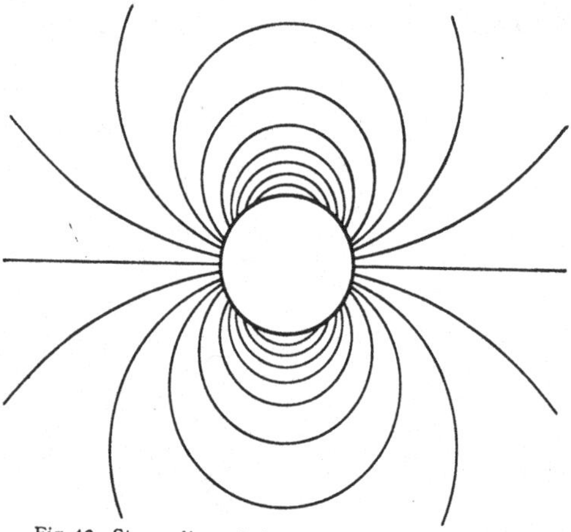

Fig. 12. Stream lines of the same flow as apparent to an observer for whom the material is at rest at ∞.

reference presently employed. At a later instant, the cylinder will have moved, and the stream field will be similar to the present one but centered upon the new position of the cylinder. Note the entire difference between Figs. 11 and 12, both representing the same motion, but the former as apparent to an observer for whom the cylinder is stationary, the latter to an observer for whom the material is stationary at ∞. In Part g of this subchapter we shall determine in full generality the apparent differences induced by the motion of an observer.

Our problem is to obtain the path lines corresponding to (71.12). This was done somewhat indirectly by MAXWELL; we shall follow a direct attack by DARWIN[2], which has in common with MAXWELL's analysis the use of r, θ in their previous meanings as radial distance and azimuth from the center of the moving cylinder. Thus by (71.11) we get

$$\left(1 - \frac{a^2}{r^2}\right) y = Y, \quad \left(1 - \frac{a^2}{y_{\mathrm{m}}^2}\right) y_{\mathrm{m}} = Y, \tag{71.14}$$

where Y is the material co-ordinate and where for given Y the quantity y_{m} is the maximum value of y, namely, the value of y when the cylinder is abreast of the particle. From $(71.14)_2$ we have $2(y_{\mathrm{m}} - Y) = \sqrt{Y^2 + 4a^2}$ if $Y \geq 0$, whence

[1] RANKINE [1864, *2*, Eq. (17)].
[2] [1953, *7*, § 2].

$y_m > a$ if $Y > 0$, expressing the fact that each particle moves aside sufficiently to let the cylinder pass by. Equivalent to (71.12) is the system

$$\dot{x} = V\frac{a^2}{r^2}\cos 2\theta, \quad \dot{y} = V\frac{a^2}{r^2}\sin 2\theta. \tag{71.15}$$

From $(71.10)_2$ and $(71.14)_1$ follows

$$\dot{\theta} = \frac{V_y}{r^2}\left(1+\frac{a^2}{r^2}\right) = \frac{V}{r^2}\sqrt{Y^2+4a^2\sin^2\theta}, \tag{71.16}$$

and hence from (71.15)

$$\frac{dx}{d\theta} = \frac{a^2\cos 2\theta}{\sqrt{Y^2+4a^2\sin^2\theta}}, \quad \frac{dy}{d\theta} = \frac{a^2\sin 2\theta}{\sqrt{Y^2+4a^2\sin^2\theta}}, \tag{71.17}$$

where Y is kept constant. As t varies from $-\infty$ to $+\infty$, the parameter θ varies from 0 to π. In the notation of Jacobian elliptic functions, put

$$k = \frac{2a}{\sqrt{Y^2+4a^2}}, \quad \cos\theta = -\operatorname{sn} v, \tag{71.18}$$

so that the range of the parameter v is $-K(k)$ to $+K(k)$. By integrating (71.17) we get the equations of the paths[1] in terms of the parameter v:

$$\left.\begin{aligned} y(v) &= Y + \frac{a}{k}(\operatorname{dn} v - k'), \\ x(v) &= X + \frac{a}{k}\left[\left(1-\frac{1}{2}k^2\right)(v+K) - E(\operatorname{am} v) - E\right]. \end{aligned}\right\} \tag{71.19}$$

To get the corresponding times, we observe directly from (71.15) and (71.16) that

$$\left.\begin{aligned} \frac{d}{dt}(x - y\cot\theta) &= -\dot{x}\sec 2\theta + y\csc^2\theta\,\dot{\theta}, \\ &= -V\frac{a^2}{r^2} + V\left(1+\frac{a^2}{r^2}\right) = V; \end{aligned}\right\} \tag{71.20}$$

hence

$$Vt = x - y\cot\theta, \tag{71.21}$$

where $t=0$ corresponds to $x=0$, $\theta=\frac{1}{2}\pi$, $y=y_m$.

Thus the cylinder drives the particles ahead and, except for those situate upon its course, aside. In so doing it causes each particle not directly before or behind it to traverse a loop which is symmetrical about a line normal to the path of the cylinder. The maximum normal displacement d_m, which occurs when the cylinder is abreast of the particle, follows from $(71.14)_2$:

$$d_m \equiv |y_m - Y| = \tfrac{1}{2}\left(\sqrt{4a^2+Y^2} - |Y|\right). \tag{71.22}$$

The greatest such displacement, $d_m = a$, results from the limit $Y\to 0$, showing that the particles nearest the cylinder's path are those most pushed aside, yet in amount barely sufficient to let the cylinder by, while if $Y=0$ the particle is carried straight along ahead or behind the cylinder. Each path is a loop. For very distant particles, it is approximately a small circle of radius $a^2/2Y$. When the cylinder is gone past, each particle has suffered a permanent forward displacement d_f, the amount of which may be read off from $(71.19)_2$:

$$d_f \equiv x(K) - X = \frac{2a}{k}\left[\left(1-\frac{1}{2}k^2\right)K - E\right], \tag{71.23}$$

[1] As was shown by Rankine [1864, *2*, § 18], the curvature of the path line is $4(y-\frac{1}{2}Y)/a^2$. A very simple expression for the path in terms of arc length was obtained by Havelock [1913, *3*]. Cf. also Milne-Thomson [1938, *9*, § 9.21].

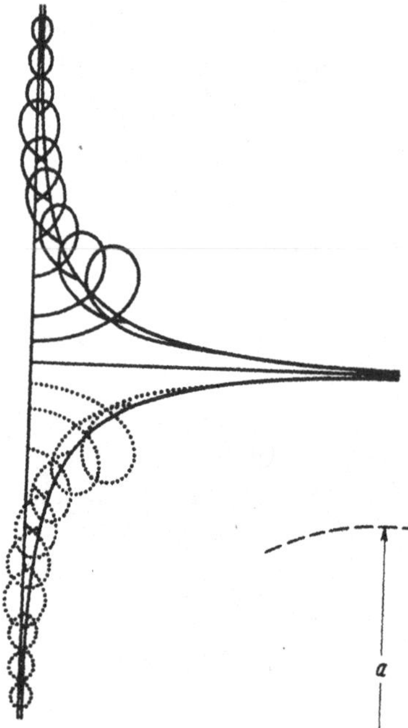

Fig. 13a.

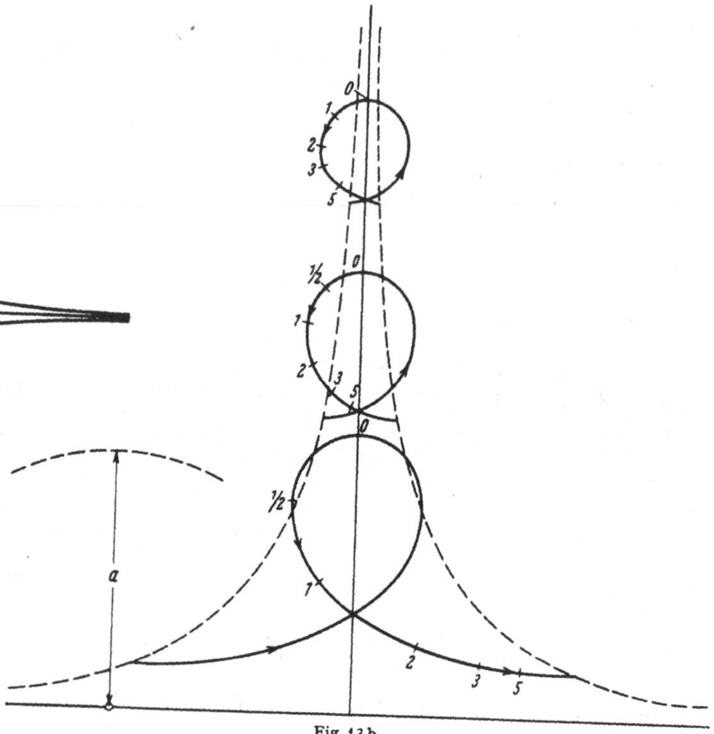

Fig. 13b.

Fig. 13a. MAXWELL's sketch of the paths of the particles which at $t = -\infty$ were situated upon a line perpendicular to the direction of motion.

Fig. 13b. DARWIN's accurate drawing of particle paths. The broken lines at left and right are the loci at $t = \mp\infty$ of the particles which are abreast of the cylinder as its center crosses the central line of the drawing. The numbers indicate the times of passage to the points marked, in units in which the cylinder moves a distance a.

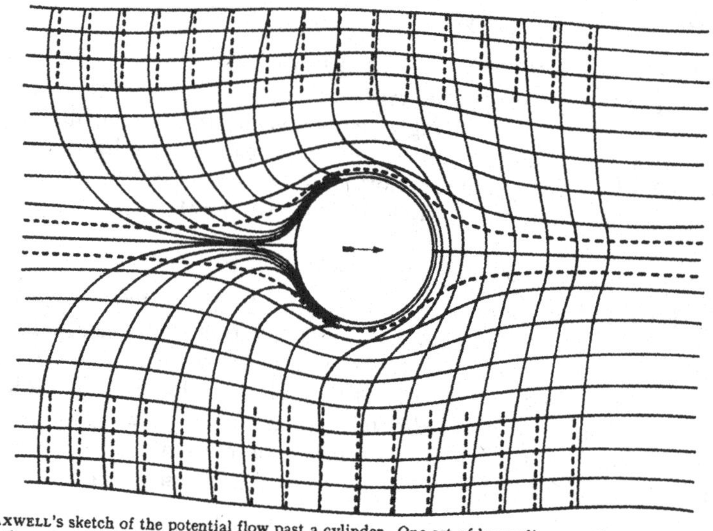

Fig. 13c. MAXWELL's sketch of the potential flow past a cylinder. One set of heavy lines are the stream lines of Fig. 11. The other set are the present loci of the particles which at time $t = -\infty$ were situated upon perpendiculars.

and for particles near the path of the cylinder this is approximately $a\,[\log(8a/Y) - 2]$.

MAXWELL's and DARWIN's drawings of this motion are reproduced as Fig. 13.

b) Material systems.

72. Material derivative. In the common frame, let $A^{\cdots}_{\cdots}$ be a tensor expressed as a function of $\boldsymbol{Z}$ and t only. Then the partial derivative $\partial A^{\cdots}_{\cdots}/\partial t$ is the rate of change of $A^{\cdots}_{\cdots}$ apparent to an observer situate upon the moving particle $\boldsymbol{Z}$. For general co-ordinates $\boldsymbol{X}$, we write

$$\frac{DA^{\cdots}_{\cdots}}{Dt} \equiv \frac{\partial A^{\cdots}_{\cdots}(\boldsymbol{X},t)}{\partial t}\bigg|_{\boldsymbol{X}=\text{const}}, \tag{72.1}$$

but this formula does not in general define a tensor. Now consider the double tensor $\boldsymbol{A}=\boldsymbol{A}(\boldsymbol{x},\boldsymbol{X},t)$, $\boldsymbol{x}$ and $\boldsymbol{X}$ being general co-ordinates. Then the tensor $dA^{\cdots}_{\cdots}/dt$, or $\dot{A}^{\cdots}_{\cdots}$, defined by

$$\left.\begin{aligned}\frac{dA^{\alpha\ldots\beta k\ldots m}_{\gamma\ldots\delta n\ldots p}}{dt} \equiv \dot{A}^{\cdots}_{\cdots} &\equiv \frac{\partial A^{\cdots}_{\cdots}}{\partial t} + \left[\frac{\partial A^{\cdots}_{\cdots}}{\partial x^q} + \begin{Bmatrix} k \\ q\,r\end{Bmatrix} A^{\alpha\ldots\beta r\ldots m}_{\gamma\ldots\delta n\ldots p} + \cdots - \right.\\ &\qquad \left. - \begin{Bmatrix} r \\ p\,q\end{Bmatrix} A^{\alpha\ldots\beta k\ldots m}_{\gamma\ldots\delta r\ldots p} - \cdots\right]\dot{x}^q,\\ &= \frac{\partial A^{\cdots}_{\cdots}}{\partial t} + A^{\cdots}_{\cdots,q}\dot{x}^q,\end{aligned}\right\} \tag{72.2}$$

is the *material derivative*[1] of $A^{\cdots}_{\cdots}$. In (72.2), $A^{\cdots}_{\cdots}$ is a function of $\boldsymbol{x}$, $\boldsymbol{X}$, and t, and $\partial/\partial t$ is executed with both $\boldsymbol{x}$ and $\boldsymbol{X}$ held constant, while $\partial/\partial x^q$ is executed with t, all x^k except x^q, and $\boldsymbol{X}$ all held constant. The field $\dot{\boldsymbol{x}}$ is regarded as given. If we suppose $\boldsymbol{x}$ and $\boldsymbol{X}$ related through a motion $\boldsymbol{x}=\boldsymbol{x}(\boldsymbol{X},t)$, $\boldsymbol{X}=\boldsymbol{X}(\boldsymbol{x},t)$, with $\dot{\boldsymbol{x}}$ obtained from (67.1), then (72.2) is an invariant time derivative whose value is independent of whether $\boldsymbol{x}$ be replaced by $\boldsymbol{x}(\boldsymbol{X},t)$ in some or all of its occurrences in $A^{\cdots}_{\cdots}$. There are two major special cases. First, when the *material description* is employed, we have $\boldsymbol{A}=\boldsymbol{A}(\boldsymbol{X},t)$, and (72.2) becomes

$$\left.\begin{aligned}\frac{dA^{\alpha\ldots\beta k\ldots m}_{\gamma\ldots\delta n\ldots p}}{dt} = \dot{A}^{\alpha\ldots\beta k\ldots m}_{\gamma\ldots\delta n\ldots p} &= \frac{DA^{\alpha\ldots\beta k\ldots m}_{\gamma\ldots\delta n\ldots p}}{Dt} + \left[\begin{Bmatrix} k \\ q\,r\end{Bmatrix} A^{\alpha\ldots\beta r\ldots m}_{\gamma\ldots\delta n\ldots p} + \cdots - \right.\\ &\qquad\left. - \begin{Bmatrix} r \\ n\,q\end{Bmatrix} A^{\alpha\ldots\beta k\ldots m}_{\gamma\ldots\delta r\ldots p} - \cdots\right]\dot{x}^q;\end{aligned}\right\} \tag{72.3}$$

in this case, $\dot{\boldsymbol{A}}$ is the intrinsic derivative of $\boldsymbol{A}$ with respect to the field $\dot{\boldsymbol{x}}$. Second, when the *spatial description* is employed, we have $\boldsymbol{A}=\boldsymbol{A}(\boldsymbol{x},t)$, and (72.2) becomes

$$\frac{dA^{\cdots}_{\cdots}}{dt} = \dot{A}^{\cdots}_{\cdots} = \frac{\partial A^{\cdots}_{\cdots}}{\partial t} + A^{\cdots}_{\cdots,q}\dot{x}^q, \quad \text{or} \quad \dot{\boldsymbol{A}} = \frac{\partial \boldsymbol{A}}{\partial t} + \dot{\boldsymbol{x}}\cdot\operatorname{grad}\boldsymbol{A}, \tag{72.4}$$

where

$$\frac{\partial A^{\cdots}_{\cdots}}{\partial t} \equiv \frac{\partial A^{\cdots}_{\cdots}(\boldsymbol{x},t)}{\partial t}\bigg|_{\boldsymbol{x}=\text{const}}. \tag{72.5}$$

The essential term $A^{\cdots}_{\cdots,q}\dot{x}^q$ in (72.4) is called the *convection* of $\boldsymbol{A}$. Its non-linearity gives rise to many of the celebrated difficulties of hydrodynamics.

We have introduced two notations, d/dt and a superposed dot, to stand for the material derivative. Except for rather complicated expressions, we shall

---

[1] The concept of material derivative and the formula $(72.2)_3$ in the common frame were used implicitly by Euler [1770, *1*, § 6] and Lagrange [1783, *1*, §§ 10–11] [1788, *1*, Part II, § 11, ¶¶ 11–12] and were formalized by Cousin [1786, *1*, § 1], Piola [1836, *1*, § VII, ¶ 77] [1848, *2*, ¶ 14], and Stokes [1845, *4*, § 5] [1851, *2*, § 49], with the notations $\partial/\partial t$, $'$, and D/Dt, respectively. In the classical literature general results are usually derived in the common frame, where there is no distinction between the operators denoted in the present work by D/Dt and d/dt, but in the cases concerning rates of change along a curved path the classical notation D/Dt often really stands for what we here call d/dt.

prefer to use the dot, whose extent of application will be indicated by an overbar rather than parentheses, thus:

$$\left.\begin{aligned}\overline{\dot{A^\alpha_{,k}}} &\equiv \frac{d}{dt}\left(A^\alpha_{,k}\right), \quad \dot{A}^\alpha_{,k} \equiv \left(\frac{dA^\alpha}{dt}\right)_{,k},\\ \overline{\dot{A^\alpha_k b^k}}\, c_m &\equiv \frac{d}{dt}\left(A^\alpha_k b^k\right) c_m, \quad \text{etc.}\end{aligned}\right\} \tag{72.6}$$

Since

$$\frac{\partial}{\partial t} A^{\cdots}_{\cdots,k} = \left(\frac{\partial A^{\cdots}_{\cdots}}{\partial t}\right)_{,k}, \tag{72.7}$$

from (72.2) we get the commutation rule

$$\frac{d}{dt} A^{\cdots}_{\cdots,k} - \left(\frac{dA^{\cdots}_{\cdots}}{dt}\right)_{,k} = -A^{\cdots}_{\cdots,m}\,\dot{x}^m_{,k}. \tag{72.8}$$

This refers to the partial covariant derivative with respect to the spatial co-ordinate x^k.

No simple rule holds, in general, for the partial covariant derivative with respect to the material co-ordinate X^α. Recalling that in (72.2) $\partial/\partial t$ is executed with both $\boldsymbol{x}$ and $\boldsymbol{X}$ held constant, we have for the total covariant derivative (App. 20.2) the following commutation rule:

$$\dot{A}^{\cdots}_{\cdots;\alpha} = \overline{\dot{A^{\cdots}_{\cdots;\alpha}}}\,. \tag{72.9}$$

This holds for double fields $\boldsymbol{A}(\boldsymbol{x}, \boldsymbol{X}, t)$, but in carrying out the operation on the left we must regard $\boldsymbol{A}$ as a function of $\boldsymbol{X}$ and t only, and in carrying out the operation on the right we must regard $\dot{\boldsymbol{x}}$ as a function of $\boldsymbol{X}$ and t only. To prove this rule, it suffices to establish it in the common frame. Verification in general co-ordinates is possible but more elaborate, requiring use of a commutation rule for the derivatives of the Christoffel symbols which follows from the vanishing of the Riemann tensor.

Also, from (App. 20.3), (App. 20.4), and the fact that the co-ordinate system is stationary, we get

$$\dot{g}_{km} = 0, \quad \dot{g}_{\alpha\beta} = 0, \quad \dot{g}^k_\alpha = 0, \quad \text{etc.} \tag{72.10}$$

In problems of motion, it is unnecessary and often inconvenient to consider double fields $A^{\cdots}_{\cdots}(\boldsymbol{x}, \boldsymbol{X}, t)$ as such. Henceforth we adopt the **convention** that *either the material or the spatial description will be used, but not a mixture of both.* That is, the tensor $A^{\cdots}_{\cdots}$ will always be assumed a function of $\boldsymbol{X}$ and t, or a function of $\boldsymbol{x}$ and t, but not a function of $\boldsymbol{x}$, $\boldsymbol{X}$, and t. This occasions no loss in generality and results in economy of description. E.g. when we write $A^m_{\alpha,\beta k}$ we think of A^m_α as a function of $\boldsymbol{X}$ and t to get $A^m_{\alpha,\beta}$, then of $A^m_{\alpha,\beta}$ as a function of $\boldsymbol{x}$ and t to get $(A^m_{\alpha,\beta})_{,k}$. The two different partial time derivatives are then distinguished by the notations (72.1) and (72.5). Also we shall employ only the partial covariant derivatives "$,k$" and "$,\alpha$" rather than the total covariant derivative, and for uniformity of notation we set

$$x^k_{,\alpha} \equiv x^k_{;\alpha} \equiv \frac{\partial x^k}{\partial X^\alpha}, \qquad X^\alpha_{,k} \equiv X^\alpha_{;k} \equiv \frac{\partial X^\alpha}{\partial x^k}, \tag{72.11}$$

where t is held constant.

Under the foregoing conventions, the commutation rules (72.7) and (72.8) hold *a fortiori*, but (72.9) does not generally hold if the total covariant derivative is replaced by the ordinary covariant derivative[1]. However, for a purely material field viewed in the material description we have

$$\dot{A}^{\alpha\ldots\beta}_{\gamma\ldots\delta,\varepsilon} = \overline{\dot{A^{\alpha\ldots\beta}_{\gamma\ldots\delta,\varepsilon}}}\,, \tag{72.12}$$

[1] TRUESDELL [1954, *24*, § 18].

as follows from (72.3), whence it is equally plain that this rule cannot be extended to non-trivially double fields: $\dot{A}^{\alpha\ldots\beta\, k\ldots m}_{\gamma\ldots\delta\, p\ldots q,\varepsilon} \neq \overline{A^{\alpha\ldots\beta\, k\ldots m}_{\gamma\ldots\delta\, p\ldots q,\varepsilon}}^{\boldsymbol{\cdot}}$, in contradistinction to (72.9).

If $\dot{\boldsymbol{A}} = 0$, the quantity $\boldsymbol{A}$ is said to be *substantially constant*. I.e., a substantially constant quantity is a function of $\boldsymbol{X}$ only.

The material derivative is sometimes called the "substantial derivative".

In a *steady* motion, if $\dot{\boldsymbol{A}} = 0$ we obtain from (72.2), (67.5) and (72.7)

$$\left.\begin{aligned} 0 &= \frac{\partial \dot{A}^{\cdots}_{\cdots}}{\partial t} = \frac{\partial}{\partial t}\left[\frac{\partial A^{\cdots}_{\cdots}}{\partial t} + A^{\cdots}_{\cdots,k}\dot{x}^k\right] = \frac{\partial}{\partial t}\left(\frac{\partial A^{\cdots}_{\cdots}}{\partial t}\right) + \left(\frac{\partial A^{\cdots}_{\cdots}}{\partial t}\right)_{,k}\dot{x}^k, \\ &= \overline{\frac{\partial A^{\cdots}_{\cdots}}{\partial t}}^{\boldsymbol{\cdot}}. \end{aligned}\right\} \tag{72.13}$$

That is, *in a steady motion the local rate of change of a substantially constant function is also substantially constant.*

73. Material systems. A manifold consisting of a set of particles is said to be *material*. The set of places occupied by a material manifold at time t is its *configuration* at time t.

A curve

$$X^\alpha = X^\alpha(S) \tag{73.1}$$

is a *material line*; its configuration at time t is given by

$$x^k(S, t) = x^k(X^1(S),\ X^2(S),\ X^3(S), t), \tag{73.2}$$

where $\boldsymbol{x} = \boldsymbol{x}(\boldsymbol{X}, t)$ is the motion $(66.1)_1$.

Similarly, a *material surface* is defined by

$$X^\alpha = X^\alpha(L, M), \tag{73.3}$$

or, alternatively,

$$F(\boldsymbol{X}) = 0; \tag{73.4}$$

the corresponding configurations are given by

$$\boldsymbol{x} = \boldsymbol{x}(\boldsymbol{X}(L, M), t), \quad F(\boldsymbol{X}(\boldsymbol{x}, t)) = 0. \tag{73.5}$$

A *material volume* is a region of particles.

A single particle is sometimes called a *material point*.

74. Material surfaces. From (73.4), (72.3), and (72.4) follows ***Lagrange's criterion***[1]: *A necessary and sufficient condition for the surface $f(\boldsymbol{x}, t) = 0$ to be material is*

$$\dot{f} = \frac{\partial f}{\partial t} + \dot{\boldsymbol{x}} \cdot \operatorname{grad} f = \frac{\partial f}{\partial t} + f_{,k}\dot{x}^k = 0. \tag{74.1}$$

By constructing a geometrical interpretation[2] for $\dot{f}$, we can make (74.1) entirely plausible from the spatial standpoint. Let a typical point upon a moving surface $f(\boldsymbol{x}, t) = 0$ be endowed with a certain velocity $\boldsymbol{u}$, not necessarily the velocity $\dot{\boldsymbol{x}}$ of the particle instantaneously occupying that point. If, continually observing this point on the surface as it moves, we differentiate the equation

[1] [1783, *1*, §§ 10–11] [1788, *1*, Part II, § 11, ¶ 12]. LAGRANGE's proof of sufficiency rests on the method of characteristics.

[2] KELVIN [1848, *5*].

$f=0$ with respect to time, we obtain

$$\frac{\partial f}{\partial t}+f_{,k}u^k=0. \tag{74.2}$$

Now the length of the projection of $\boldsymbol{u}$ onto the normal to $f=0$ is given by

$$u_n=\frac{u^k f_{,k}}{\sqrt{f_{,m}f^{,m}}}. \tag{74.3}$$

Hence from (74.2) follows

$$u_n=-\frac{\frac{\partial f}{\partial t}}{\sqrt{f_{,m}f^{,m}}}. \tag{74.4}$$

But the length of the projection of the particle velocity $\dot{\boldsymbol{x}}$ onto the normal to $f=0$ is

$$\dot{x}_n=\dot{x}^k\frac{f_{,k}}{\sqrt{f_{,m}f^{,m}}}. \tag{74.5}$$

Comparing (74.4), (74.5), and (72.4) yields

$$\dot{f}=\sqrt{f_{,m}f^{,m}}\,(\dot{x}_n-u_n), \tag{74.6}$$

an assertion that *at a point upon the surface $f=0$, $\dot{f}$ is proportional to the speed of propagation, relative to the surface, of the particle instantaneously situate thereon.* Consequently LAGRANGE's criterion (74.1) is a statement that at a given instant the normal component of the particle motion relative to the surface $f=0$ is zero.

The ideas suggested here will be generalized in Sects. 177 and 182 to 183.

In the spatial description, when the velocity field $\dot{\boldsymbol{x}}$ is a given smooth function of place and time, it does not always yield a continuous motion in the sense of Sect. 65. The motion (66.1) is to be found by an integration (cf. Sect. 70) and may possess singularities. For example, it may not always be possible, throughout all parts of space where $\dot{\boldsymbol{x}}$ is defined, to determine corresponding particles $\boldsymbol{X}$. Little is known about such singularities. An isolated result was asserted by SOMOFF[1]: *A surface which envelopes path-lines is also an envelope of successive spatial configurations of a single material surface.* The proof of SOMOFF is purely formal. First, a necessary condition that in the motion $\boldsymbol{x}=\boldsymbol{x}(\boldsymbol{X},t)$ the surface $F(\boldsymbol{X})=0$ be an envelope of paths is that there exist non-zero solutions dX^α of the system

$$x^k_{,\alpha}\,dX^\alpha+\dot{x}^k\,dt=0,\qquad F_{,\alpha}\,dX^\alpha=0. \tag{74.7}$$

Hence

$$\dot{x}^k\,\varepsilon_{kmp}\,\varepsilon^{\alpha\beta\gamma}\,x^m_{,\beta}\,x^p_{,\gamma}\,F_{,\alpha}=0. \tag{74.8}$$

By (17.3) follows

$$\dot{x}^k\,X^\alpha_{,k}\,F_{,\alpha}\,|\boldsymbol{x}/\boldsymbol{X}|=0. \tag{74.9}$$

On the other hand, let $F(\boldsymbol{X})=0$ be a material surface. Its configurations in space are then $F(\boldsymbol{X}(\boldsymbol{x},t))=0$, so that a condition for these to possess an evelope is

$$F_{,\alpha}X^\alpha_{,k}\dot{x}^k=0. \tag{74.10}$$

If $|\boldsymbol{x}/\boldsymbol{X}|\neq 0$, (74.9) and (74.10) are identical. This completes the argument of SOMOFF. It is inadequate in that it consists merely in showing that a necessary condition for one type of singular surface is necessary also for another. Sufficient conditions do not appear easy to establish.

75. Material vector lines. If a curve is given by the equations $F(\boldsymbol{x},t)=0$, $G(\boldsymbol{x},t)=0$, for that curve to be a material line it is obviously sufficient, but not necessary, that $\dot{F}=0$ and $\dot{G}=0$. A more convenient criterion can be expressed in terms of the representation (73.1), or of any field of tangents to the

[1] [1885, 7].

curve. Equivalently we may consider a field $\mathbf{c}(\boldsymbol{x}, t)$ and find *a necessary and sufficient condition that the vector lines of* $\mathbf{c}$ *be material lines.*

At the instant t, let c be a vector line of $\mathbf{c}$, and for an interval about t let $\boldsymbol{x}=\boldsymbol{x}(L, t')$ be the material line $\mathscr{C}$ which coincides with c at the time t. Since $\mathscr{C}$ is a material line, we have by (72.9)

$$\overline{c^{[k}\frac{\partial x^{m]}}{\partial L}}^{\bullet} = \dot{c}^{[k}\frac{\partial x^{m]}}{\partial L} + c^{[k}\dot{x}^{m]}_{,p}\frac{\partial x^p}{\partial L}. \tag{75.1}$$

By (App. 29.1)$_2$, if $\mathscr{C}$ is to remain a vector line of $\mathbf{c}$ we must have

$$\overline{c^{[k}\frac{\partial x^{m]}}{\partial L}}^{\bullet} = 0. \tag{75.2}$$

Since $\mathscr{C}$ is a vector line of $\mathbf{c}$ at the instant t, we have at that instant

$$\frac{\partial x^k}{\partial L} = a\, c^k, \tag{75.3}$$

where a is a scalar. Hence, combining (75.1), (75.2), and (75.3), we conclude that a necessary condition for $\mathscr{C}$, once being a vector line of $\mathbf{c}$, to remain so, is

$$c^{[k}(\dot{c}^{m]} - \dot{x}^{m]}_{,p}\, c^p) = 0. \tag{75.4}$$

This is the ***Helmholtz-Zorawski criterion***[1]. To see that it is sufficient, note that if it is satisfied then $c^{[k}\partial x^{m]}/\partial L$ is a quantity initially zero whose time derivative, when $\boldsymbol{X}$ is kept constant, is also zero.

Equivalent to (75.4) is

$$c^{[k}\left(\overline{a\, c^{m]}}^{\bullet} - a\,\dot{x}^{m]}_{,p}\, c^p\right) = 0, \quad \text{or} \quad \mathbf{c}\times\left(\overline{a\mathbf{c}}^{\bullet} - a\,\mathbf{c}\cdot\operatorname{grad}\dot{\boldsymbol{p}}\right) = 0, \tag{75.5}$$

where a is any non-vanishing quantity, not necessarily constant. This form indicates that the result is invariant when $\mathbf{c}$ is replaced by $a\mathbf{c}$, as is geometrically obvious. Also equivalent to (75.4) is

$$\mathbf{c}\times\left[\frac{\partial\mathbf{c}}{\partial t} + \operatorname{curl}(\mathbf{c}\times\dot{\boldsymbol{p}}) + \dot{\boldsymbol{p}}\operatorname{div}\mathbf{c}\right] = 0. \tag{75.6}$$

By putting $\mathbf{c}=\dot{\boldsymbol{p}}$ in (75.6) we get

$$\dot{x}^{[k}\frac{\partial\dot{x}^{m]}}{\partial t} = 0, \quad \dot{\boldsymbol{p}}\times\frac{\partial\dot{\boldsymbol{p}}}{\partial t} = 0, \quad \text{or, equivalently,} \quad \frac{\partial\dot{\boldsymbol{p}}}{\partial t} = C(\boldsymbol{x}, t)\,\dot{\boldsymbol{p}}, \tag{75.7}$$

as necessary and sufficient that the stream lines be material, whence follows the theorem that *the stream lines and the path lines coincide if and only if they are steady* (cf. Sect. 70). Motions in which (75.7) is satisfied are called motions with *steady stream lines.*

76. Material derivative of the elements of arc, surface, and volume. The fundamental lemma for this section is

$$\overline{x^k_{,\alpha}}^{\bullet} = \dot{x}^k_{,\alpha} = \dot{x}^k_{,m}\, x^m_{,\alpha}. \tag{76.1}$$

[1] It was implicit in the analysis of HELMHOLTZ [1858, *1*, § 2] and NANSON [1874, *5*] but was first explicitly stated by ZORAWSKI [1900, *11*]. The present proof is that of TRUESDELL and PRIM [1947, *18*] [1950, *24*]. Others were given by LAMPARIELLO [1938, *6*] and CASTOLDI [1950, *4*].

To prove it, we write the sequence of identities

$$\left.\begin{aligned} \overline{x^k_{,\alpha}}^{\boldsymbol{\cdot}} &= \frac{d}{dt}\left(\frac{\partial x^k}{\partial X^\alpha}\right) = \frac{D}{Dt}\left(\frac{\partial x^k}{\partial X^\alpha}\right) + \left\{ {k \atop p\,m} \right\} \frac{\partial x^p}{\partial X^\alpha} \dot{x}^m, \\ &= \frac{\partial \dot{x}^k}{\partial X^\alpha} + \left\{ {k \atop p\,m} \right\} \frac{\partial x^p}{\partial X^\alpha} \dot{x}^m = \dot{x}^k_{;\alpha} = \dot{x}^k_{,\alpha}, \\ &= \left[\frac{\partial \dot{x}^k}{\partial x^p} + \left\{ {k \atop p\,m} \right\} \dot{x}^m\right] \frac{\partial x^p}{\partial X^\alpha} = \dot{x}^k_{,m}\, x^m_{,\alpha}, \end{aligned}\right\} \tag{76.2}$$

which follow from (72.3), (72.1), (App. 20.2), our convention in Sect. 72, and (App. 22.4).

By differentiating the identity $x^k_{,\alpha} X^\alpha_{,m} = \delta^k_m$ and using (76.2) we get also

$$\overline{X^\alpha_{,k}}^{\boldsymbol{\cdot}} = -X^\alpha_{,m} X^\beta_{,k} \dot{x}^m_{;\beta} = -X^\alpha_{,m} \dot{x}^m_{,k}. \tag{76.3}$$

From (20.3) and (76.1) we have[1]

$$\overline{dx^k}^{\boldsymbol{\cdot}} = \dot{x}^k_{,m} x^m_{,\alpha} dX^\alpha = \dot{x}^k_{,m} dx^m = d\dot{x}^k. \tag{76.4}$$

From $(20.6)_4$ follows similarly

$$\overline{da^{km}}^{\boldsymbol{\cdot}} = -2\dot{x}^{[k}_{,p}\, da^{m]p}; \tag{76.5}$$

equivalently[2],

$$\overline{da_k}^{\boldsymbol{\cdot}} = -\dot{x}^m_{,k}\, da_m + \dot{x}^m_{,m}\, da_k. \tag{76.6}$$

We set

$$\mathrm{I}_{\boldsymbol{d}} \equiv \dot{x}^k_{,k} = \operatorname{div} \dot{\boldsymbol{p}} \tag{76.7}$$

(the notation $\mathrm{I}_{\boldsymbol{d}}$ will be motivated in Sect. 83). The rule for differentiating a determinant gives

$$\left.\begin{aligned} \overline{|\boldsymbol{x}/\boldsymbol{X}|}^{\boldsymbol{\cdot}} &= \overline{x^k_{,\alpha}}^{\boldsymbol{\cdot}} X^\alpha_{,k} |\boldsymbol{x}/\boldsymbol{X}| = \dot{x}^k_{,m} x^m_{,\alpha} X^\alpha_{,k} |\boldsymbol{x}/\boldsymbol{X}|, \\ &= \mathrm{I}_{\boldsymbol{d}} |\boldsymbol{x}/\boldsymbol{X}|, \end{aligned}\right\} \tag{76.8}$$

where we have used (76.1). Equivalent formulae are

$$\dot{J} = \mathrm{I}_{\boldsymbol{d}} J, \quad \overline{dv}^{\boldsymbol{\cdot}} = \mathrm{I}_{\boldsymbol{d}}\, dv = \dot{x}^k_{,k}\, dv; \tag{76.9}$$

the former follows from (16.5) and $(72.10)_1$, the latter, from (20.9). The quantity $\mathrm{I}_{\boldsymbol{d}}$ is called the *expansion* and is of fundamental importance. Each of the results (76.8) and (76.9) as well as the equivalent forms

$$\mathrm{I}_{\boldsymbol{d}} = \dot{x}^k_{,k} = \operatorname{div} \dot{\boldsymbol{p}} = \overline{\log J}^{\boldsymbol{\cdot}} = \overline{\log dv}^{\boldsymbol{\cdot}} \tag{76.10}$$

is called ***Euler's expansion formula***[3]. An equivalent form involving an arbitrary function B such that $\dot{B} = 0$ is the ***spatial equation of continuity***:

$$\overline{\log B j}^{\boldsymbol{\cdot}} + \mathrm{I}_{\boldsymbol{d}} = 0, \quad \text{or} \quad \frac{\partial (B j)}{\partial t} + \operatorname{div}(B j \dot{\boldsymbol{p}}) = 0. \tag{76.11}$$

Any quantity of the form Bj, where $\dot{B} = 0$, is called a *density* for the motion. The properties of such functions are developed in Subchapter III.

[1] Euler [1761, 2, §§ 13, 55] [1757, 2, § 11].
[2] Lamb [1877, 4].
[3] [1757, 2, §§ 10—15] [1770, 1, § 14] [1862, 4, § 156].

Since $\operatorname{div}(\dot{\boldsymbol{p}}\boldsymbol{p}) = \dot{\boldsymbol{p}} + \boldsymbol{p}\,\mathrm{I}_{\boldsymbol{d}}$, by (App. 26.1)$_2$ we have[1]

$$\int_v \dot{\boldsymbol{p}}\,dv = \oint_s d\boldsymbol{a}\cdot\dot{\boldsymbol{p}}\boldsymbol{p} - \int_v \boldsymbol{p}\,\mathrm{I}_{\boldsymbol{d}}\,dv. \tag{76.12}$$

This identity, an alternative form of (76.10), expresses the total velocity in a region by means of the normal velocity $\dot{p}_n$ on the boundary and the expansion $\mathrm{I}_{\boldsymbol{d}}$ at interior points.

The formulae of this section make it plain that from the velocity gradient tensor $\dot{x}^k_{,m}$ the material rates of change of all elements of arc, surface, and volume are determined. Pursuit of this idea furnishes the subject of Part c of this subchapter.

77. Isochoric motion; steady motion with steady density. A motion such that the volume occupied by any material region is unaltered, however that region may change its shape in the course of time, is *isochoric* (cf. Sect. 40). By (76.9) follows as a local and instantaneous condition for isochoric motion ***Euler's criterion***[2]:

$$\mathrm{I}_{\boldsymbol{d}} \equiv \dot{x}^k_{,k} = \operatorname{div}\dot{\boldsymbol{p}} = 0. \tag{77.1}$$

That is, *a motion is isochoric if and only if its velocity field is solenoidal.* From the theorem of HELMHOLTZ in Sect. App. 31 we conclude that *a motion is isochoric if and only if, at any given instant, the strength of each stream tube is the same at all of its cross sections.* This assertion was taken, in one form or another, as the defining expression of the principle of continuity in hydrodynamic researches before 1752[3] and is still in frequent use in engineering treatments. Of course, it is but a special case of the defining property of a solenoidal field, which in the present connection assumes the following form: *A motion is isochoric if and only if the flux of velocity out of every reducible closed surface is zero*:

$$\oint_s d\boldsymbol{a}\cdot\dot{\boldsymbol{p}} = 0. \tag{77.2}$$

Since steady motion is defined by (67.5) alone, in a steady motion the quantity j need not be steady. However, by (76.10) the quantity $\dot{\overline{\log j}}$ is steady.

Moreover, if B is any steady quantity which is constant on each stream line, we have $\dot{B} = 0$, and also

$$\operatorname{div}(j B\dot{\boldsymbol{p}}) = j\dot{\boldsymbol{p}}\cdot\operatorname{grad} B + B\operatorname{div}(j\dot{\boldsymbol{p}}) = 0. \tag{77.3}$$

From this formula and (76.11) we see that *in a steady motion, by assigning the quantity B a constant value for each stream line we obtain a steady density jB.* Since such a choice of B is not necessary, we distinguish the case when it is made by speaking of a *steady motion with steady density.*

In two important cases, then, the velocity is proportional to a solenoidal field:

1. *Isochoric* motion, in which $\dot{\boldsymbol{p}}$ is solenoidal,
2. *Steady motion with steady density*, in which $Bj\dot{\boldsymbol{p}}$ is solenoidal.

For these two classes of motion, we may read off from the results in Sects. App. 31 and App. 32 a sequence of special properties. These properties are extremely important, and it is only for economy that we refrain from repeating them here.

---

[1] MUNK [1936, *7*]; more obscurely in [1922, *6*].

[2] [1761, *2*, § 36]. Special cases are due to D'ALEMBERT [1752, *1*, §§ 45, 73].

[3] E.g. NEWTON [1687, *1*, Lib. II, Prop. XXXVI], D. BERNOULLI [1738, *1*, Chap. III, § 2]. Perhaps the earliest recorded statement is that of LEONARDO DA VINCI [1923, *5*, Book VIII, §§ 37–43].

In particular, we emphasize (App. 31.12), which gives the total squared speed in the isochoric case, the total of $j^2 B^2 \dot{p}^2$ in the steady case.

In an isochoric motion with steady stream lines, we may take the divergence of $(75.7)_3$, by (77.1) thus obtaining

$$0 = C_{,k} \dot{x}^k. \tag{77.4}$$

This is a statement that $C = \text{const}$ on each stream line at each fixed time. But if we write $\dot{x}^k = \dot{x}\, d^k$, where $\boldsymbol{d}$ is the field of unit tangent to the stream lines, we get

$$\frac{\partial \dot{x}^k}{\partial t} = \frac{\partial \dot{x}}{\partial t} d^k, \tag{77.5}$$

since $\boldsymbol{d}$ is steady. By $(75.7)_2$, therefore,

$$C \dot{x}\, d^k = \frac{\partial \dot{x}}{\partial t} d^k, \tag{77.6}$$

whence

$$\frac{\partial \log \dot{x}}{\partial t} = C. \tag{77.7}$$

By (77.4) follows a theorem of CAUCHY[1]: *In an isochoric motion with steady stream lines, $\partial \log \dot{x} / \partial t$ is constant on each stream line at each instant.* The converse is not generally true. In fact, it is easy to show that in a motion with steady stream lines a necessary and sufficient condition that C be constant along each stream line at each fixed time is that $I_{\boldsymbol{d}}/\dot{x}$ be steady.

78. Invariant directions. If the element $d\boldsymbol{x}$ is to be instantaneously constant in direction, we must have

$$\overline{dx^m}^{\boldsymbol{\cdot}} = k\, dx^m, \tag{78.1}$$

where k is a real factor of proportionality, which may be zero. By $(76.4)_2$ follows

$$(\dot{x}^m_{,p} - k\, \delta^m_p)\, dx^p = 0, \tag{78.2}$$

and hence

$$\det (k\, \delta^m_p - \dot{x}^m_{,p}) = 0. \tag{78.3}$$

The analogy to (38.2) is immediate, and the analysis proceeds in the same way as in Sect. 38.

First, since (78.3) is a real cubic, we have at once the ***theorem of Bertrand***[2]: *In any motion, at each point there is at least one real direction suffering no instantaneous rotation.* Second, let D be the discriminant of (78.3). Recalling from Sect. App. 37 that a complex proper number k cannot yield a real solution $d\boldsymbol{x}$ of (78.2), and that to distinct real proper numbers there correspond linearly independent real solutions $d\boldsymbol{x}$, from the theory of cubic equations we obtain Cases 1 and 2 of the following:

1. *If $D > 0$, there are three and only three invariant directions.*
2. *If $D < 0$, there is one and only one invariant direction.*
3. *If $D = 0$, there may be one, two, three, or an infinite number of invariant directions.*

The additional possibilities in Case 3 are most easily seen by example. The condition $D = 0$ is requisite and sufficient for multiple proper numbers. In the rectilinear shearing motion $\dot{x} = 0$, $\dot{y} = x$, $\dot{z} = 0$, there is but the single proper number $k = 0$, thrice repeated, and any material straight line initially parallel to the y-z plane is carried ever parallel to itself. In the motion $\dot{x} = 0$, $\dot{y} = x$, $\dot{z} = z$

[1] [1823, *2*].
[2] [1867, *1*].

the proper numbers are $k=0, 1$, the former being double, and the only lines not being rotated are those parallel to the y axis or to the z axis.

For D we have the following expression[1]:

$$D = -18 I_{\boldsymbol{d}} \text{Ю}\, \text{Б} - 4 I_{\boldsymbol{d}}^2 \text{Б} + I_{\boldsymbol{d}}^2 \text{Ю}^2 - 4\text{Ю}^3 - 27\text{Б}^2, \tag{78.4}$$

where

$$\text{Ю} \equiv \tfrac{1}{2} \delta^{k m}_{n p} \dot{x}^n_{,k} \dot{x}^p_{,m}, \qquad \text{Б} \equiv \det \dot{x}^k_{,m}. \tag{78.5}$$

While interpretation of the sign of D is difficult, Ericksen[2] has observed that the problem is simplified by introducing the deviator, $v^k_{\ m} \equiv \dot{x}^k_{\ ,m} - \frac{1}{3} I_{\boldsymbol{d}} \delta^k_m$. From (78.2) it is plain that $d\boldsymbol{x}$ is invariant with respect to $\dot{x}^k_{\ ,m}$ if and only if it is invariant with respect to $v^k_{\ m}$, the proper numbers for the latter being $k - \frac{1}{3} I_{\boldsymbol{d}}$, where k is a root of (78.3). If we let D_0 be the discriminant for $v^k_{\ m}$, then by (78.4) follows

$$D_0 = -4\text{Ю}_0^3 - 27\text{Б}_0^2, \tag{78.6}$$

where Ю_0 and Б_0 are formed from $v^k_{\ m}$ as are Ю and Б from $\dot{x}^k_{,m}$. Hence there are three distinct proper numbers, at least one multiple proper number, or one and only one real proper number according as

$$-\tfrac{1}{3} \text{Ю}_0 \gtreqless (\tfrac{1}{2} \text{Б}_0)^{\frac{2}{3}}. \tag{78.7}$$

In particular, $\text{Ю}_0 > 0$ is sufficient that $D_0 < 0$ and hence that there be one and only one invariant direction, while $\text{Ю}_0 < 0$ is necessary in order that $D_0 > 0$ and hence that there be three distinct invariant directions.

A kinematical interpretation for the sign of Ю will be given in Sect. 91.

Since $\dot{x}_{k,m}$ is generally not symmetric, the algebraic theorem of Kelvin and Tait (Sect. App. 37) implies that in the case when three distinct invariant directions exist, they are not usually orthogonal.

A condition that an instantaneously invariant direction shall remain invariant has been found by Zorawski[3].

In order that there exist an element of area which is instantaneously constant[4], we must have $\dot{\overline{da^k}} = 0$; by (76.6) follows

$$\dot{x}^m_{,k} \, d a_m = \mathrm{I}_{\boldsymbol{d}} \, d a_k, \tag{78.8}$$

a statement that $\mathrm{I}_{\boldsymbol{d}}$ is a proper number of $\dot{x}^m_{,k}$. From the characteristic equation of $\dot{x}^m_{,k}$ follows the necessary and sufficient condition

$$\mathrm{I}_{\boldsymbol{d}} \text{Ю} = \text{Б}. \tag{78.9}$$

79. Kinematics of line integrals. Definitions and conventions regarding line, surface, and volume integrals are given in Sects. App. 23 to App. 25 and Sect. 73. Let $\mathscr{C}$ be a material line. Then for the rate of change of the flow of Ψ along $\mathscr{C}$ we have

$$\frac{d}{dt} \int_{\mathscr{C}} \Psi \, dx^k = \int_{\mathscr{C}} \dot{\overline{\Psi \, dx^k}}, \tag{79.1}$$

---

[1] Truesdell [1954, *24*, § 22].

[2] [1955, *5*].

[3] [1900, *12*].

[4] Beltrami [1871, *1*, § 9] found also conditions for (1) lineal elements suffering pure rotation, (2) plane elements all directions in which are experiencing rates of change normal to one direction, (3) plane elements suffering no instantaneous rotation. These last form a tetrahedron whose edges, possibly imaginary, are the directions determined by (78.2).

since an equation for $\mathscr{C}$ is $\boldsymbol{X}=\boldsymbol{X}(L)$ and thus the integral has fixed limits in the material description. Hence by (76.4) follows

$$\left.\begin{aligned}\frac{d}{dt}\int_{\mathscr{C}}\Psi\,dx^k &= \int_{\mathscr{C}}(\dot{\Psi}\,dx^k+\Psi\,\overline{dx^k}^{\boldsymbol{\cdot}}),\\ &= \int_{\mathscr{C}}(\dot{\Psi}\,dx^k+\Psi\,\dot{x}^k_{,m}\,dx^m),\\ &= \int_{\mathscr{C}}(\dot{\Psi}\,dx^k+\Psi\,d\dot{x}^k),\end{aligned}\right\}\tag{79.2}$$

results implicit in the analysis of KELVIN[1].

On the other hand, for a spatial line c we have

$$\frac{\partial}{\partial t}\int_c \Psi\,dx^k=\int_c\frac{\partial\Psi}{\partial t}\,dx^k.\tag{79.3}$$

Choosing the spatial line c which is the configuration at time t of the material line $\mathscr{C}$, from (79.2), (79.3), and (72.4) we obtain

$$\frac{d}{dt}\int_{\mathscr{C}}\Psi\,dx^k-\frac{\partial}{\partial t}\int_c\Psi\,dx^k=\int_c(\Psi_{,m}\dot{x}^m\,dx^k+\Psi\,\dot{x}^k_{,m}\,dx^m).\tag{79.4}$$

80. Kinematics of surface integrals. For a spatial surface s we have

$$\frac{\partial}{\partial t}\int_s\Psi\,da^{km}=\int_s\frac{\partial\Psi}{\partial t}\,da^{km},\tag{80.1}$$

but for a material surface $\mathscr{S}$ a more elaborate formula is necessary. By (76.5) we get

$$\frac{d}{dt}\int_{\mathscr{S}}\Psi\,da^{km}=\int_{\mathscr{S}}(\dot{\Psi}\,da^{km}-2\Psi\,\dot{x}^{[k}_{,p}\,da^{m]p}),\tag{80.2}$$

or, equivalently,

$$\frac{d}{dt}\int_{\mathscr{S}}\Psi\,da_k=\int_{\mathscr{S}}[\dot{\Psi}\,da_k+\Psi(-\dot{x}^m_{,k}\,da_m+\dot{x}^m_{,m}\,da_k)].\tag{80.3}$$

When Ψ is a vector field $\boldsymbol{c}$, this becomes a formula[2] for the rate of change of the flux of $\boldsymbol{c}$ through $\mathscr{S}$:

$$\frac{d}{dt}\int_{\mathscr{S}}d\boldsymbol{a}\cdot\boldsymbol{c}=\int_{\mathscr{S}}d\boldsymbol{a}\cdot(\dot{\boldsymbol{c}}-\boldsymbol{c}\cdot\operatorname{grad}\dot{\boldsymbol{p}}+\boldsymbol{c}\operatorname{div}\dot{\boldsymbol{p}}).\tag{80.4}$$

A generalization is given in Sect. 277.

From (80.4) follows ***Zorawski's criterion***: *In order that the flux of* $\boldsymbol{c}$ *across every material surface remain constant in time, it is necessary and sufficient that*

$$\dot{c}^k-\dot{x}^k_{,m}c^m+\dot{x}^m_{,m}c^k=0.\tag{80.5}$$

Equivalent forms are

$$\overline{AJc^k}^{\boldsymbol{\cdot}}-AJ\dot{x}^k_{,m}c^m=0,\qquad \frac{\partial\boldsymbol{c}}{\partial t}+\operatorname{curl}(\boldsymbol{c}\times\dot{\boldsymbol{p}})+\dot{\boldsymbol{p}}\operatorname{div}\boldsymbol{c}=0;\tag{80.6}$$

---

[1] [1869, *7*, § 59(a)]. The general formula is given by JAUMANN [1905, *2*, § 383] and SPIELREIN [1916, *5*, § 29].

[2] Cf. ZORAWSKI [1900, *11*], JAUMANN [1905, *2*, § 383], ABRAHAM [1909, *1*, § 2], SPIELREIN [1916, *5*, § 29].

in the former, which is a consequence of $(76.9)_1$, A is any substantially constant quantity.

By comparing (80.5) with (75.4), we conclude that *in order for the flux of* $\mathbf{c}$ *across every material surface to be constant in time, it is necessary but not sufficient that the vector lines of* $\mathbf{c}$ *be material.* When (80.5) holds, the vector tubes of $\mathbf{c}$ are material tubes whose strength (Sect. App. 29) at any given cross-section remains constant as the motion proceeds.

81. Kinematics of volume integrals. The transport theorem. For a stationary volume v we have

$$\frac{\partial}{\partial t}\int_v \Psi\, dv = \int_v \frac{\partial \Psi}{\partial t}\, dv. \tag{81.1}$$

For a material volume $\mathscr{V}$, by $(76.9)_3$, (72.4), and $(\text{App. } 26.1)_2$ we get[1]

$$\left.\begin{aligned}\frac{d}{dt}\int_{\mathscr{V}} \Psi\, dv &= \int_{\mathscr{V}} (\dot{\Psi} + \Psi\, \dot{x}^k_{,k})\, dv,\\ &= \int_{\mathscr{V}} \left[\frac{\partial \Psi}{\partial t} + (\Psi\, \dot{x}^k)_{,k}\right] dv,\\ &= \int_{\mathscr{V}} \frac{\partial \Psi}{\partial t}\, dv + \oint_{\mathscr{S}} \Psi\, \dot{x}^k\, da_k.\end{aligned}\right\} \tag{81.2}$$

Choosing the spatial volume v which at time t is the configuration of the material volume $\mathscr{V}$, we derive the ***transport theorem***[2]:

$$\frac{d}{dt}\int_{\mathscr{V}} \Psi\, dv = \frac{\partial}{\partial t}\int_v \Psi\, dv + \oint_s \Psi\, \dot{x}^k\, da_k = \int_v \frac{\partial \Psi}{\partial t}\, dv + \oint_s \Psi\, \dot{x}_n\, da. \tag{81.3}$$

Thus *the rate of change of the total* Ψ *over a material volume* $\mathscr{V}$ *equals the rate of change of the total* Ψ *over the fixed volume* v *which is the instantaneous configuration of* $\mathscr{V}$, *plus the flux of* $\dot{\boldsymbol{x}}\,\Psi$ *out of the bounding surface.* The transport theorem is really an alternative statement of EULER's expansion formula (76.10).

The formula (81.3) expresses, for a given Ψ, a time derivative of an integral taken over a volume whose bounding surface is in motion at an arbitrary velocity $\dot{\boldsymbol{x}}$. To consider a volume $v(t)$ bounded by a surface $s(t)$ moving at a different velocity $\boldsymbol{u}$, we need only imagine fictitious particles whose velocity is $\boldsymbol{u}$. The result, then, is

$$\frac{d_{\boldsymbol{u}}}{dt}\int_{v(t)} \Psi\, dv = \int_v \frac{\partial \Psi}{\partial t}\, dv + \oint_s \Psi\, u^k\, da_k, \tag{81.4}$$

where $d_{\boldsymbol{u}}/dt$ indicates that the volume of integration is material with respect to the velocity $\boldsymbol{u}$. By taking $\boldsymbol{u}=0$ or $\boldsymbol{u}=\dot{\boldsymbol{x}}$, we recover (81.1) or (81.3).

c) Stretching and spin.

82. The stretching tensor of EULER. Let $d\boldsymbol{x}_1$ and $d\boldsymbol{x}_2$ be material elements. Then by (72.10) and (76.4) we have

$$\left.\begin{aligned}\overline{g_{km}\, dx^k_1\, dx^m_2}^{\,\cdot} &= g_{km}(d\dot{x}^k_1\, dx^m_2 + dx^k_1\, d\dot{x}^m_2),\\ &= g_{km}(\dot{x}^k_{,p}\, dx^p_1\, dx^m_2 + \dot{x}^m_{,p}\, dx^k_1\, dx^p_2),\\ &= 2 d_{km}\, dx^k_1\, dx^m_2,\end{aligned}\right\} \tag{82.1}$$

---

[1] REYNOLDS [1903, *15*, § 14], JAUMANN [1905, *2*, § 383], SPIELREIN [1916, *5*, § 29].

[2] Asserted by REYNOLDS [1903, *15*, § 14], proved by SPIELREIN [1916, *5*, § 29], who gave numerous alternative forms and corollaries.

where $\boldsymbol{d}$ is the symmetric *stretching tensor* or *rate of deformation tensor* of EULER:

$$d_{km} \equiv \dot{x}_{(k,m)}. \tag{82.2}$$

By (82.1) it is plain that the tensor $\boldsymbol{d}$ serves to measure the instantaneous rates of change of length and angle of material elements in the moving material.

First, if $d\boldsymbol{x}_1 = d\boldsymbol{x}_2$, (82.1) reduces to

$$\dot{\overline{ds^2}} = 2\, d_{km}\, dx^k\, dx^m, \tag{82.3}$$

or

$$\dot{\overline{\log ds}} = \frac{d_{km}\, dx^k\, dx^m}{g_{pq}\, dx^p\, dx^q}. \tag{82.4}$$

If we write

$$d_{(\boldsymbol{n})} \equiv \lim_{l\to 0} \frac{\dot{l}}{l} = \lim_{l\to 0} \dot{\overline{\log l}} \tag{82.5}$$

for the *stretching* [cf. Sect. 25, especially (25.5)] of a material line of length l emanating from $\boldsymbol{x}$ with unit tangent $\boldsymbol{n}$, then we have from (82.4)

$$d_{(\boldsymbol{n})} = d_{km}\, n^k\, n^m. \tag{82.6}$$

In particular, if $\boldsymbol{n}$ points along the x^1 co-ordinate curve, we have

$$d_{(\boldsymbol{n})} = \frac{d_{11}}{g_{11}}. \tag{82.7}$$

Second, if $\mathfrak{G}$ is the angle between $d\boldsymbol{x}_1$ and $d\boldsymbol{x}_2$, from (82.1) we get

$$-\sin\mathfrak{G}\,\dot{\mathfrak{G}} = 2\, d_{km}\, n_1^k\, n_2^m - \cos\mathfrak{G}\,(d_{(\boldsymbol{n}_1)} + d_{(\boldsymbol{n}_2)}). \tag{82.8}$$

For $\mathfrak{G}=0$ this reduces to (82.4). If $\mathfrak{G} \neq 0$, (82.8) gives us the *shearing*, $-\dot{\mathfrak{G}}$, of the directions of $d\boldsymbol{x}_1$ and $d\boldsymbol{x}_2$. For the case of orthogonal directions we get

$$-\tfrac{1}{2}\dot{\mathfrak{G}}_{12} = d_{km}\, n_1^k\, n_2^m. \tag{82.9}$$

We have shown, then, that *in an orthogonal co-ordinate system the physical components of $\boldsymbol{d}$ equal the stretchings and the halves of the shearings in the co-ordinate directions*:

$$\|d\langle km\rangle\| = \left\| \begin{matrix} d_{(1)} & -\tfrac{1}{2}\dot{\mathfrak{G}}_{12} & -\tfrac{1}{2}\dot{\mathfrak{G}}_{13} \\ \cdot & d_{(2)} & -\tfrac{1}{2}\dot{\mathfrak{G}}_{23} \\ \cdot & \cdot & d_{(3)} \end{matrix} \right\|. \tag{82.10}$$

The physical dimension of $\boldsymbol{d}$ is $[T^{-1}]$; thus $\boldsymbol{d}$ is a *pure rate*.

The linearity of $\boldsymbol{d}$ in $\dot{\boldsymbol{x}}$ is important: *The stretchings and orthogonal shearings corresponding to the vector sum of two velocity fields are the sums of the stretchings and shearings of the two constituent fields.*

The results just given, as well as those to follow in this part of the subchapter, could be derived by reinterpretation of the formulae concerning small deformation given in Part e of Subchapter I. Because of the definition of "small" there employed, there would be no loss of rigor in such a treatment. It seems to us clearer, however, to make the theory of rates of change entirely independent of the theory of finite strain. In so doing, in this section we have given a compact, modern version of the original argument of EULER. The rather elaborate exact connection between stretching and rate of finite strain will be presented in Sect. 95.

The rate of change of a material element of area is given by (76.6) and hence is not determined by the stretching alone. For the rate of change of the magnitude da, however, we readily calculate[1] from (76.6), (82.2), and (82.6)

$$\dot{\overline{\log da}} = \mathrm{I}_{\boldsymbol{d}} - d_{(\boldsymbol{da})}. \tag{82.11}$$

82A. Appendix. History of the theory of stretching and shearing. All the analysis just given, except for the last equation, is due to EULER [1770, *1*, §§ 9–12]. His form of the argument, based on consideration of the infinitesimal displacement $\dot{\boldsymbol{x}}\,dt$, is often given in engineering texts today. In reference to infinitesimal strain (cf. Sect. 33A and Sect. 57), the material was re-created by CAUCHY [1823, *1*] [1827, *2*] [1828, *3*], who added the contents of Sect. 83. Cf. also PIOLA [1836, *1*, § VII, ¶¶ 70–71]. All this was discussed afresh in terms of rates by STOKES [1845, *4*, § 2], whose kinematical description is more immediate. The formula (82.4) in a less symmetrical form was derived by PIOLA [1848, *2*, ¶ 15] in connection with rods. From this, as well as some later work, we must conclude that the equivalence of the theory of *small deformation* to that of *rates of change* was not as plain as now it seems; the difficulty, perhaps, lay in a certain vagueness in the concept of "small" (cf. Sect. 53).

The explicit equation (82.3), in terms of rates, is due to BELTRAMI [1871, *1*, § 4] and DURRANDE [1871, *4*].

Further properties of $\boldsymbol{d}$ were derived by KLEITZ [1872, *2*, § 8] [1873, *4*, §§ 30–32, 38–40]. He analysed the ellipsoid of $\boldsymbol{d}^{-2}$ and the properties of the vector $d_{km}n^m$, where $\boldsymbol{n}$ is a unit normal. His results have been presented in more general form in Sect. App. 46. He constructed also a geometrical interpretation for the shear components of $\boldsymbol{d}$ in terms of certain curves associated with the stream lines.

A formula for $\boldsymbol{d}$ according to the material description, which follows at once by substituting $\dot{x}_{k,m} = \dot{x}_{k,\alpha} X^{\alpha}_{,m}$ and (17.3) into (82.2), was written out by DUHEM [1903, *3*].

83. The principal stretchings, the invariants of stretching, and the quadric of stretching. Since d_{km} is a real symmetric tensor, it has many properties which may be read off from results in Sects. App. 37 to App. 39, Sect. App. 46 and Sect. 21. Its real and orthogonal principal axes are the *axes of stretching*; its real proper numbers $d_\mathfrak{a}$ are the *principal stretchings*, which may be distinct or multiple, positive, negative, or zero. Therefore the quadric of $\boldsymbol{d}$, which is called the *quadric of stretching*[2], may be an ellipsoid or a hyperboloid or any of their degenerate cases. If it is a hyperboloid, elements along its asymptotic cone are suffering no extension, and the cone divides elements which are being shortened from elements which are being lengthened. If $d_\mathfrak{a} > 0$ for $\mathfrak{a} = 1, 2, 3$, the quadric is an ellipsoid and all elements are waxing; if $d_\mathfrak{a} < 0$, the quadric is again an ellipsoid, but all elements are shrinking.

The stretchings along the principal axes are the extremal stretchings. For two orthogonal directions suffering extremal stretching, the shearing is zero. The orthogonal directions suffering extremal shearings are those bisecting the angles between two of the axes of stretching and normal to a third, while the amounts of these extremal shearings are $\pm(d_\mathfrak{a} - d_\mathfrak{b})$.

The principal invariants of $\boldsymbol{d}$ are called the *invariants of stretching* and are written as $\mathrm{I}_{\boldsymbol{d}}$, $\mathrm{II}_{\boldsymbol{d}}$, $\mathrm{III}_{\boldsymbol{d}}$. Since $\mathrm{I}_{\boldsymbol{d}} = 0$ in an isochoric motion, from (App. 39.7) we obtain $\mathrm{II}_{\boldsymbol{d}} \leqq 0$. In any motion, the invariants $\sqrt{\overline{\mathrm{II}}_{\boldsymbol{d}}}$ and $\sqrt{\mathrm{II}_{\boldsymbol{d}}}$ are measures of the *intensity of stretching* and of the *intensity of shearing*, respectively.

${}_0\boldsymbol{d}$, the *deviator* of $\boldsymbol{d}$ as defined by (App. 38.12) with $\boldsymbol{d} = \boldsymbol{a}$, is often used in works on plasticity. In the sense of the maximal decomposition (App. 43.1), ${}_0\boldsymbol{d}$ may be regarded as a measure of instantaneous change of shape without change of volume. By (82.9), a necessary condition for every orthogonal shearing to be zero is that $d_{km}n_1^k n_2^m = 0$ whenever $g_{km}n_1^k n_2^m = 0$. Hence $\boldsymbol{d} = \alpha\,\mathbf{1}$, where α is

[1] This result is due in principle to PIOLA [1825, *2*, § 236] [1848, *2*, ¶ 15].

[2] A very detailed study of the quadric of stretching was made by ZHUKOVSKI [1876, *5*, Gl. I].

a scalar; hence $\boldsymbol{d} = \frac{1}{3} \mathrm{I}_{\boldsymbol{d}} \mathbf{1}$; equivalently, ${}_0\boldsymbol{d} = 0$. If ${}_0\boldsymbol{d} = 0$, it follows by (82.8) that $\dot{\mathfrak{S}} = 0$ if $\mathfrak{S} \neq 0$; that is, every shearing is zero. We have shown, then, that *a necessary and sufficient condition for all shearings to vanish is that* ${}_0\boldsymbol{d} = 0$. Accordingly ${}_0\boldsymbol{d}$ is often called the *distortion tensor*. The principal axes and the principal shear components of ${}_0\boldsymbol{d}$ coincide with their counterparts for $\boldsymbol{d}$; however, in general ${}_0\boldsymbol{d}$ is not the stretching tensor of any motion[1]. Since the proper numbers of a deviator cannot be of one sign, the quadric of ${}_0\boldsymbol{d}$ is always a hyperboloid or one of its degenerate cases[2]. Further properties of ${}_0\boldsymbol{d}$ may be read off from results in Sect. App. 38.

When $\boldsymbol{d}$ is spherical, or, equivalently, when the three principal stretchings are equal, the motion is a *dilatation*. Equivalent invariant conditions are

$$(\tfrac{1}{3} \mathrm{I}_{\boldsymbol{d}})^3 = (\tfrac{1}{3} \mathrm{II}_{\boldsymbol{d}})^{\frac{3}{2}} = \mathrm{III}_{\boldsymbol{d}} \quad \text{and} \quad {}_0\boldsymbol{d} = 0. \tag{83.1}$$

The most general dilatation will be determined in Sect. 142.

PAILLOUX[3] has derived the condition

$$\mathrm{III}_{\boldsymbol{d}} = 0 \tag{83.2}$$

as necessary and sufficient that there exist a material surface whose spatial configuration is instantaneously invariant. The direction in which the stretching is zero is the normal to the surface.

84. Rigid motion. A motion is *rigid* if all material lengths remain unchanged by it. From (82.3) follows as a local and instantaneous necessary and sufficient condition for rigid motion ***Euler's criterion***[4]:

$$d_{km} = 0. \tag{84.1}$$

If (84.1) holds throughout a region, its general solution is[5]

$$\dot{x}_k = \omega_{km} p^m + \dot{c}_k, \tag{84.2}$$

where $\dot{\boldsymbol{c}}$ and $\boldsymbol{\omega}$ depend only upon t, where $\boldsymbol{\omega}$ is skew symmetric, and where $p^k_{,m} = \delta^k_m$. Therefore $\boldsymbol{p}$ is the position vector of $\boldsymbol{x}$ with respect to a certain origin and $\boldsymbol{\omega}$ is the angular velocity tensor of a certain frame. Since $\dot{\boldsymbol{x}} - \dot{\boldsymbol{c}}$ is perpendicular both to $\boldsymbol{\omega}$ and to $\boldsymbol{p}$, $\dot{\boldsymbol{c}}$ is the linear velocity of the origin from which $\boldsymbol{p}$ is directed, and with respect to the frame whose angular velocity is $\boldsymbol{\omega}$ the velocity of the material is zero. The origin is arbitrary, but $\boldsymbol{\omega}$ is uniquely determined by (84.2). That is, all the possible frames determined by (84.2) have the same angular velocity. We may think of these frames as rigidly attached to the material, and we call $\boldsymbol{\omega}$ the *angular velocity of the motion*. (Cf. the general treatment in Sect. 143.). If $\boldsymbol{\omega} = 0$, the rigid motion is a *translation*.

A rigid motion is isochoric, since $\dot{x}^k_{,k} = 0$. In a rigid motion, the invariants Ю and Б defined by (78.5) have the values Ю $= \omega^2$, Б $= 0$, ω being the angular speed, and hence by the theorem of Sect. 78 it follows that in a rigid motion which is not a translation there is one and only one invariant direction. The foregoing statements are geometrically trivial and are made here only so as to illustrate the general criteria established earlier.

For a general motion, six functions d_{km} assigned arbitrarily cannot in general serve as the components of stretching. Rather, in order that there exist a field

[1] In order that ${}_0\boldsymbol{d}$ be the stretching tensor of a motion, by (84.3) it is necessary and sufficient that $\mathrm{I}_{\boldsymbol{d}} g_{km}$ be the stretching tensor of a motion. At the end of Sect. 142 it is shown that only special forms for $\mathrm{I}_{\boldsymbol{d}}$ are possible.

[2] BELTRAMI [1871, *1*, § 3].

[3] [1938, *10*].

[4] [1761, *2*, §§ 75–77] [1770, *1*, § 13]. A generalization of this result in differential geometry was obtained by KILLING [1892, *6*, p. 167].

[5] EULER [1770, *1*, § 13].

$\dot{\boldsymbol{x}}$ such that (82.2) holds, we have the condition of integrability[1],

$$\delta^{km}_{pq}\,\delta^{tu}_{rs}\,d_{kt,mu}=0. \tag{84.3}$$

If (84.3) is satisfied, let two velocity fields corresponding to $\boldsymbol{d}$ be $\dot{\boldsymbol{x}}_1$ and $\dot{\boldsymbol{x}}_2$; for their vector difference $\dot{\boldsymbol{x}}^*\equiv\dot{\boldsymbol{x}}_1-\dot{\boldsymbol{x}}_2$, we calculate the stretching tensor $\boldsymbol{d}^*$ and by the linearity of (84.2) conclude that $\boldsymbol{d}^*=0$; therefore, *the vector difference of two velocities having the same stretching tensor is a rigid motion.* In view of (84.2), this furnishes proof for the assertion (34.11).

The problem of determining corresponding conditions in more general spaces of n dimensions is a difficult one, not yet fully solved. Given a tensor d_{km}, we are to find necessary and sufficient conditions that there exist a vector c_k such that

$$d_{km}=c_{(k,m)} \tag{84.4}$$

where, as usual,

$$c_{k,m}\equiv\frac{\partial c_k}{\partial x^m}-\Gamma^p_{km}\,c_p. \tag{84.5}$$

For a given field c_k, let d_{km} be defined by (84.4), and put $w_{km}\equiv c_{[k,m]}$. For a symmetric connection Γ we have the identities

$$\left.\begin{aligned} w_{km,p}+w_{mp,k}+w_{pk,m}&=0,\\ c_{k,mp}-c_{k,pm}&=c_q R^q{}_{kmp},\\ h_{km,ps}-h_{km,sp}&=h_{kq}R^q{}_{mps}+h_{qm}R^q{}_{kps}.\end{aligned}\right\} \tag{84.6}$$

Since $c_{k,mp}=d_{km,p}+w_{km,p}$, by forming $c_{k,mp}-c_{k,pm}$ and using $(84.6)_{1,2}$ we find that

$$w_{mp,k}=d_{km,p}-d_{kp,m}-R^q{}_{kmp}\,c_q. \tag{84.7}$$

Now forming $w_{mp,ks}-w_{mp,sk}$ and using $(84.6)_3$, we obtain the identity[2]

$$\left.\begin{aligned} &d_{km,ps}-d_{kp,ms}-d_{sm,pk}+d_{sp,mk}+\\ &\quad+(R^q{}_{smp,k}-R^q{}_{kmp,s})\,c_q+R^q{}_{smp}\,d_{qk}-R^q{}_{kmp}\,d_{qs}-\\ &\quad-R^q{}_{kmp}\,w_{qs}+R^q{}_{smp}\,w_{qk}+R^q{}_{pks}\,w_{qm}-R^q{}_{mks}\,w_{qp}=0.\end{aligned}\right\} \tag{84.8}$$

For a flat space, (84.8) reduces to (84.3). In more general spaces, we are to derive from (84.8) and from the geometry of the space further identities, by means of which c_q and w_{km} are to be eliminated.

PALATINI[3] has indicated how the calculation can be initiated in the case of a metric space. First, we replace sums of the type $R^q\ldots a_q$ by corresponding sums $R_q\cdots a^q$, then use the relation $R_{kmps}=R_{pskm}$ and the Bianchi identities (34.5) to obtain $(R^q{}_{smp,k}-R^q{}_{kmp,s})\,c_q=-R_{mpsk,q}\,c^q$. Thus (84.8) becomes

$$\left.\begin{aligned} &\delta^{rq}_{kp}\,\delta^{tu}_{ms}\,d_{rt,qu}-R_{mpsk,q}\,c^q-R_{mpsq}\,d^q_k+R_{mpkq}\,d^q_s+\\ &\quad+R_{mpkq}\,w^q{}_s-R_{mpsq}\,w^q{}_k-R_{kspq}\,w^q{}_m+R_{ksmq}\,w^q{}_p=0.\end{aligned}\right\} \tag{84.9}$$

Raising the index k and then contracting upon k and m yields[4]

$$\mathrm{I}_{,ps}+d_{ps}{}^{,k}{}_{,k}-2d^q_{(p,s)q}-2d^q_{(p}R_{s)q}-2w^q_{(p}R_{s)q}-R_{ps,q}\,c^q=0, \tag{84.10}$$

---

[1] ZORAWSKI [1911, *14*, § 4]. Cf. (34.8). A geometrical interpretation for these conditions in terms of the shearing required to render compatible a given tensor $\boldsymbol{d}$ has been constructed by L. FINZI [1956, *7*, § 7].

[2] RICCI [1888, *7*, Eq. (20)].

[3] [1943, *3*]. The method was sufficiently indicated by his earlier analysis of a space of constant curvature [1916, *4*]. Cf. also ANDRUETTO [1932, *1* and *2*], AGOSTINELLI [1933, *1*], GRAIFF [1958, *3*, §§ 6–9].

[4] This identity was given in a special co-ordinate system by PALATINI [1934, *3*]; an equivalent general form is due to GRAIFF [1958, *3*, Eq. (22)].

where $(84.6)_3$ has been used, where $R_{pq} \equiv R^h{}_{pqh}$, and where we have written I for I_d. Raising the index p in (84.10) and the contracting upon p and s yields[1]

$$I^{,p}{}_{,p} - d^{kp}{}_{,kp} - R_{pq}\, d^{pq} = R_{,q}\, c^q, \tag{84.11}$$

where $R \equiv R^k_k$.

Eqs. (84.10) are to be regarded as a system of $\frac{1}{2}n(n+1)$ non-homogeneous linear equations for the $\frac{1}{2}n(n+1)$ unknowns c^q and $w^k{}_s$ in terms of the tensor $\boldsymbol{d}$ and its second derivatives. Substituting the result into (84.9) yields, in principle, a set of differential equations for $\boldsymbol{d}$ alone[2]. The geometrical meaning of the condition of solubility expressed in terms of a determinant Formel from the components R_{pq} and $R_{pq,s}$ is not known; of course, it is not satisfied in a flat space.

More generally, directly from (84.4) we see that the nature of the solution depends very strongly upon the geometry of the space considered. In spaces admitting a group of "motions", i.e., instantaneously rigid velocity fields, the solution of (84.4), if compatible, is determinate only to within such a motion. In spaces where instantaneously rigid motions are not possible, the system (84.4) will determine $\boldsymbol{c}$ uniquely from a given $\boldsymbol{d}$, if compatible. GRAIFF[3] has initiated a discussion of the problem in terms of the principal invariants of the tensor R_{pq}, according as these are linearly independent and non-constant or not. The problem must still be regarded as open.

In a space of constant curvature, we have $n(n-1)\, R^q{}_{kmp} = R(g_{km}\,\delta^q_p - g_{kp}\,\delta^q_m)$, where $n R_{km} = R g_{km}$, $R = \text{const} = R^k_k$. The terms involving $\boldsymbol{c}$ and $\boldsymbol{w}$ in (84.9) vanish. The resulting conditions, easily seen to be sufficient as well as necessary, are[4]

$$\delta^{rq}_{kp}\,\delta^{tu}_{ms}\left(d_{rt,qu} - \frac{R}{n(n-1)}\, d_{rt}\, g_{qu}\right) = 0. \tag{84.12}$$

When $n = 2$, there is but one linearly independent non-vanishing component of (84.12), which may be written in the equivalent forms

$$\left.\begin{aligned} e^{\alpha\beta} e^{\gamma\delta}\, d_{\alpha\gamma,\beta\delta} + K\,\mathrm{I} &= 0,\\ \mathrm{I}^{,\alpha}{}_{,\alpha} - d^{\alpha\beta}{}_{,\alpha\beta} + K\,\mathrm{I} &= 0, \end{aligned}\right\} \tag{84.13}$$

where K is the total curvature, $K = -\frac{1}{2}R$.

More generally, when $n = 2$ and K is arbitrary, we have $R_{\alpha\beta\gamma\delta} = K\, e_{\alpha\beta}\, e_{\gamma\delta}$ and $R_{\alpha\beta} = -K a_{\alpha\beta}$, where $\boldsymbol{a}$ is the metric tensor. Again the terms containing $\boldsymbol{w}$ in (84.9) are annulled, and we have

$$e^{\alpha\beta} e^{\gamma\delta}\, d_{\alpha\gamma,\beta\delta} + K\,\mathrm{I} = -2K_{,\alpha}\, c^\alpha. \tag{84.14}$$

Further differentiations are needed to eliminate $\boldsymbol{c}$. B. FINZI[5] has shown that for a surface applicable upon a surface of revolution the final condition assumes the form

$$(K_{,\gamma}\, K^{,\gamma}\, a^{\alpha\beta} + K^{,\alpha} K^{,\beta})\, d_{\alpha\beta} + K\, K^{,\gamma}\, a^{\alpha\beta}\, d_{\alpha\beta,\gamma} + e^{\alpha\beta} e^{\gamma\delta}\, K^{,\varepsilon}\, d_{\beta\delta,\alpha\gamma\varepsilon} = 0. \tag{84.15}$$

For a general surface, he obtained a system of three conditions of compatibility, each being a differential equation of fourth order. TRUESDELL[6] showed indirectly that they must be equivalent to a single condition of fifth order. By a method of infinitesimal variation applied to a complete set of differential invariants of the surface in question, GRAIFF[7] has given a definitive treatment of the problem. She has found two identities relating the quantities occurring in B. FINZI's conditions, but she has not effected the elimination explicitly. She has characterized the case when the least possible order of the single scalar condition is 4; in this case, as for a surface of revolution, the lines $K = \text{const}$ are geodesic parallels, but $K^{,\alpha}{}_{,\alpha}$ is not a function of K only.

GRAIFF[8] has discussed also the various possibilities when $n = 3$ and $n = 4$.

---

[1] GRAIFF [1958, *3*, Eq. (23)].

[2] This seems to be a correct substitute for the method of PALATINI [1934, *3*], who notes that it is possible to choose co-ordinates so that at a fixed point (84.10) with $p = s$ becomes a system of n linear equations for the n components c^q but fails to note that the solution of this system is not sufficient, in general, to obtain the derivatives of c^q and so to determine w^{km}.

[3] [1958, *3*, § 10].

[4] PALATINI [1916, *4*] [1934, *3*], ANDRUETTO [1932, *2*], FINZI [1934, *2*] including a simplification when $n = 3$.

[5] [1930, *1*].

[6] [1957, *17*]; the reasoning is presented in Sect. 229.

[7] [1957, *5*].

[8] [1958, *3*, §§ 13–14].

Enlightenment is cast upon these results by the analysis in Sect. 229. For another approach to the conditions of compatibility, see Sect. 234.

85. Relative spin. Let $d\boldsymbol{x}$ be a material element, and let $\boldsymbol{n}$ be a unit vector fixed in space. Then the angle $\varphi_{(d\boldsymbol{x},\boldsymbol{n})}$ between these two vectors is given to within a multiple of 2π by

$$\cos\varphi_{(d\boldsymbol{x},\boldsymbol{n})} = \frac{g_{km}\, d x^k n^m}{\sqrt{g_{np}\, d x^n d x^p}} \tag{85.1}$$

and the right-hand rule. By (72.10) and (76.4) follows

$$\left.\begin{aligned} -\sin\varphi\,\dot{\varphi} &= \frac{g_{km}\, d\dot{x}^k n^m}{\sqrt{g_{np}\, d x^n d x^p}} - \cos\varphi\, d_{(d\boldsymbol{x})}, \\ &= \dot{x}_{k,m} n^k N^m - \cos\varphi\, d_{(\boldsymbol{N})}, \end{aligned}\right\} \tag{85.2}$$

where $\boldsymbol{N}$ is the unit vector $d\boldsymbol{x}/dx$. The angular rate $\dot{\varphi}$ is the *spin* of $d\boldsymbol{x}$ relative to the fixed direction $\boldsymbol{n}$. In particular, for the spin relative to an orthogonal direction we have

$$-\dot{\varphi}_{(\boldsymbol{N},\boldsymbol{n})} = \dot{x}_{k,m} n^k N^m. \tag{85.3}$$

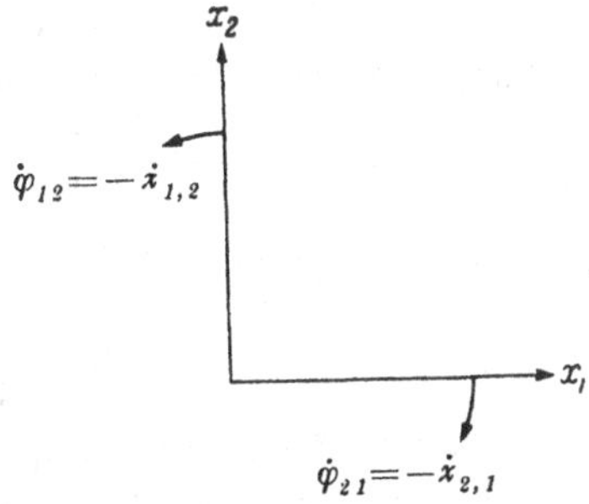

Fig. 14. Relative spin.

The elegant formula (85.3) enables us to give a geometric interpretation for the shear components $\dot{x}_{m,k}$, $k \neq m$, of the velocity gradient in an orthogonal co-ordinate system. For let $\boldsymbol{N}$ and $\boldsymbol{n}$ point along the m and k co-ordinate curves respectively, and for the corresponding $\dot{\varphi}$ write $\dot{\varphi}_{\mathrm{km}}$; then

$$-\dot{\varphi}_{\mathrm{km}} = \frac{\dot{x}_{k,m}}{\sqrt{g_{mm}}\sqrt{g_{kk}}}: \tag{85.4}$$

for orthogonal co-ordinates, $\dot{x}_{1,2}/\sqrt{g_{11}}\sqrt{g_{22}}$ *is the rate at which a material element instantaneously pointing along the* x^2 *axis is turning toward the* x^1 *axis*[1].

From this result and (82.9) we se that $-\dot{\mathfrak{G}}_{12} = -\dot{\varphi}_{12} - \dot{\varphi}_{21}$, as is obvious: the shearing of two orthogonal directions equals the negative sum of their spins relative to fixed orthogonal directions (Fig. 14). In interpreting this result, due attention must be paid to the convention of sign. In a right-handed orthogonal system, then, $-\dot{\mathfrak{G}}_{12}$ is the *excess* of the rate of *right-handed* rotation of an element instantaneously pointing along the x^2 axis over that of one pointing along the x^1 axis.

If we put $\varphi = 0$ in (85.2), we recover (82.6).

86. The spin tensor of CAUCHY. Set

$$w_{km} \equiv \dot{x}_{[k,m]} = \partial\dot{x}_{[k}/\partial x^{m]}. \tag{86.1}$$

With the interpretation of $\dot{x}_{k,m}$ as proportional to the relative spin, constructed in the previous section, we are able to identify the components w_{km} in the orthogonal case as *the halves of the differences of the relative spins of the elements in the co-ordinate directions.* Here again due caution regarding the convention of sign is necessary. For a right-handed system, the quantity $-w_{12}/\sqrt{g_{11}}\sqrt{g_{22}}$ is one half the *sum* of the rates of *right-handed* rotation of elements in the x^1 direction and the x^2 direction.

$\boldsymbol{w}$, like $\boldsymbol{d}$, is a pure rate, having the physical dimension $[T^{-1}]$.

[1] This result is implied though not actually stated by CAUCHY [1841, *1*, Th. IX].

The axial vector

$$w^k \equiv - e^{kmq} w_{mq} = e^{kmq} \dot{x}_{q,m}, \quad \text{or} \quad \boldsymbol{w} = \operatorname{curl} \dot{\boldsymbol{p}}, \tag{86.2}$$

is the *vorticity vector;* its direction is the *axis* of spin, and its magnitude w is the *vorticity magnitude.* We have

$$w = \sqrt{w_k w^k} = \sqrt{2 w_{km} w^{km}}. \tag{86.3}$$

From the vectorial character of $\boldsymbol{w}$ follows, as another wording for the proposition demonstrated above, the ***first theorem of Cauchy***[1]: *The length of the projection of $\boldsymbol{w}$ upon any given direction is the sum of the rates of right-handed rotation about that direction of elements in any two directions perpendicular to it and to each other.*

The theorem just proved contains a statement of invariance: The component of vorticity is the sum of rates associated with *any* two perpendicular directions in a plane normal to the component. This invariance may be investigated by means of the *skew projections* introduced in Sect. App. 4. Let Greek majuscule indices run from 1 to 2, and write ${}_3w_{\Delta\Xi}$ for the skew projection of $\dot{x}_{k,m}$ onto the x^3 direction. Then (App. 4.2) yields

$$\| {}_3w^{\Xi}_{\Delta} \| = \frac{1}{\sqrt{\det g_{\Phi\Psi}}} \begin{Vmatrix} \dot{x}_{1,2} & -\dot{x}_{1,1} \\ \dot{x}_{2,2} & -\dot{x}_{2,1} \end{Vmatrix}, \tag{86.4}$$

while from (App. 4.3) follows

$${}_3w^{\Delta}_{\Delta} = \frac{2 w_{12}}{\sqrt{\det g_{\Phi\Psi}}} = \frac{\sqrt{\det g_{km}}}{\sqrt{\det g_{\Phi\Psi}}} w^3. \tag{86.5}$$

The result is of interest mainly for the case when the x^3 direction is normal to the other two co-ordinate directions. Then

$${}_3w^{\Delta}_{\Delta} = w\langle 3 \rangle. \tag{86.6}$$

That is, *the physical component of the vorticity vector in a given direction is the trace of the skew projection of the velocity gradient onto the plane normal to that direction.* Moreover, since the skew projection ${}_3w^{\Xi}_{\Delta}$ is a plane tensor, its symmetric part may be represented by a quadric, as usual, and this quadric is a conic section. By (86.4) and (85.4), the normal components of ${}_3w^{\Xi}_{\Delta}$ are proportional to the relative spins of elements normal to the x^3 direction. Hence we obtain the ***second theorem of Cauchy***[2]: *The rates of right-handed rotation of elements normal to the x^3 direction are inversely proportional to the square roots of the lengths of the corresponding radius vectors to the conic section*

$${}_3w^{\Xi}_{\Delta} n^{\Delta} n_{\Xi} = \text{const}. \tag{86.7}$$

Let the constant be given any non-zero value which renders the locus real. There are then three possibilities. (1), the conic is an ellipse; the rates of rotation of the elements in the directions of its axes are minimal and maximal rates, and all elements normal to the x^3 direction are rotating in the same sense. (2), the conic is a pair of parallel straight lines; again all elements are rotating in the same sense, except that the elements in the direction of the lines are not rotating at all. (3), the conic is a pair of conjugate hyperbolae; its asymptotes divide the plane normal to the x^3 direction into two portions, elements in one of which are rotating in one sense, while elements in the other are rotating in the opposite sense;

[1] [1841, *1*, Th. IX].

[2] [1841, *1*, Th. VII]. Cf. also BERTRAND [1867, *1*].

elements in the directions of the asymptotes are suffering no rotation whatever, while the rates of rotation of elements parallel to the axes are maximal and minimal rates (in absolute value, both are maximal). Reinterpretation of (86.6) now yields the ***third theorem of Cauchy***[1]: *The length of the projection of the vorticity vector upon a given direction equals the sum of the greatest and least rates of right-handed rotation of normal elements.*

Returning to (86.4), we use rectangular Cartesian co-ordinates and observe that by the tensor law of transformation we have

$$\dot{x}^*_{2,1} = \dot{x}_{2,1}\cos^2\psi - \dot{x}_{1,1}\cos\psi\sin\psi + \dot{x}_{2,2}\sin\psi\cos\psi - \dot{x}_{1,2}\sin^2\psi, \tag{86.8}$$

where ψ is the angle between the x^1 and x^{1*} axes. This formula expresses the relative orthogonal spin of an element subtending an angle ψ with the x^1 axis. The mean of all these spins is thus

$$\frac{1}{2\pi}\int_0^{2\pi} \dot{x}^*_{2,1}\, d\psi = \frac{1}{2}(\dot{x}_{2,1} - \dot{x}_{1,2}) = -w_{12} = \frac{1}{2} w_3. \tag{86.9}$$

This is the ***fourth theorem of Cauchy***[2]: *The length of the projection of the vorticity vector upon a direction is twice the mean of the rates of right-handed rotation of all elements perpendicular to that direction.* Cf. Sect. 36 and Sect. 56.

From (84.2) it is immediate that *the spin tensor of a rigid motion is its angular velocity.* In making this observation, STOKES[3] suggested that in general the spin tensor of a deforming medium could be regarded, at each place and time, as a *local angular velocity.* STOKES's interpretation was extended by BELTRAMI[4]. Imagine a vanishingly small massy element whose center of mass is at $\boldsymbol{x}$ and whose principal axes of inertia are the principal axes of stretching at $\boldsymbol{x}$ at time t; let this element suddenly be solidified into a rigid body, and at the same time all the surrounding material shorn away; then this rigid body will continue to rotate indefinitely with angular velocity $\boldsymbol{w}$.

An elegant reformulation of STOKES's result, free of dynamical concepts, is due to GOSIEWSKI[5]. Since material elements along the principal axes of extension are instantaneously suffering no rotation with respect to one another, their motion as lines, no account being taken of the motion of particles along them, is instantaneously rigid; hence *the spin is the angular velocity of the principal axes of extension,* in the ordinary sense of rigid motions. A formal analysis equivalent to the above argument was also presented by GOSIEWSKI, who noted that by (76.4) the rate of change of a unit vector $\boldsymbol{n}$ parallel to $d\boldsymbol{x}$ is given by[6]

$$\left.\begin{aligned} \dot{n}^k &= \frac{\overline{d x^k}^{\,\cdot}}{d x} - \frac{\overline{d x}^{\,\cdot}}{(d x)^2}\, d x^k = \frac{1}{d x}\left[d\,\dot{x}^k - d_{(\boldsymbol{n})}\, d x^k\right],\\ &= (w^k{}_m + d^k_m - d_{(\boldsymbol{n})}\,\delta^k_m)\, n^m. \end{aligned}\right\} \tag{86.10}$$

---

[1] [1841, *1*, Th. IX].

[2] [1841, *1*, Ths. V, VI, VII].

[3] [1845, *4*, § 2]. A similar idea had been put forward by CORIOLIS [1835, *1*, pp. 100–101] in connection with systems of mass-points.

[4] [1871, *1*, § 10]. Cf. HADAMARD [1903, *11*, ¶ 63]. STOKES had considered only a spherical element.

[5] [1890, *4*, §§ 2–3]. APPELL [1921, *1*, § 706] attributed the result to BOUSSINESQ. It was asserted also by LEVY [1890, *7*, § 5]. A somewhat obscure treatment based on the equations of relative motion (Sect. 143) was given by INGRAM [1924, *5*]. We do not follow the analysis of VALCOVICI [1946, *9*], who, on the ground that the material lines which at a given instant coincide with the principal axes of extension no longer do so after an infinitesimal time, objects to $\boldsymbol{w}$ as a measure of rotation and proposes a substitute.

[6] GOSIEWSKI [1890, *4*, § 8], ZORAWSKI [1901, *17*, § 2].

If $\boldsymbol{n}$ is a proper vector of $\boldsymbol{d}$, then $(d^k_m - d_{(\boldsymbol{n})}\delta^k_m)\, n^m = 0$, and hence (86.10) reduces to

$$\dot{n}^k = w^k{}_m n^m, \tag{86.11}$$

which is an analytical statement of the result to be proved.

To get a formula for $\boldsymbol{w}$ in the material description, we begin with the evident

$$w_{km} = -X^{\alpha}_{,[m}\dot{x}_{k],\alpha}. \tag{86.12}$$

From (17.3) it then follows that[1]

$$J w^k = e^{\alpha\beta\gamma} x^m_{,\beta} x^k_{,\gamma} \dot{x}_{m,\alpha}. \tag{86.13}$$

The vorticity magnitude is related to other scalar invariants of the velocity gradient by identities which follow at once from (App. 38.15):

$$\left.\begin{aligned} \mathrm{IO} &= \mathrm{II}_{\boldsymbol{d}} + \tfrac{1}{4} w^2 = -\tfrac{1}{2}\overline{\mathrm{II}}_{\boldsymbol{d}} + \tfrac{1}{2}\mathrm{I}^2_{\boldsymbol{d}} + \tfrac{1}{4} w^2, \\ \dot{x}^k_{,m}\dot{x}^m_{,k} &= \overline{\mathrm{II}}_{\boldsymbol{d}} - \tfrac{1}{2} w^2, \end{aligned}\right\} \tag{86.14}$$

where IO is defined by $(78.5)_1$. Also

$$4 w_k{}^m w_m{}^p d^k{}_p = -\mathrm{I}_{\boldsymbol{d}} w^2 + d_{km} w^k w^m. \tag{86.15}$$

86 A. Appendix. History of the theory of spin. The components w_{km} occur in the earliest analyses of the velocity field, having been introduced in special cases by D'ALEMBERT (1749) [1752, *1*, §§ 45–49] and generally by EULER (1752–1755) [1761, *2*, §§ 46–47], whose theorems concerning spin will be presented below. LAGRANGE (1760) [1762, *3*, Chap. XLII] and CAUCHY [1827, *5*, 2nd Part, § 1, Subsect. 4] were the first to introduce single letters to stand for the components.

All this early work is purely formal and somewhat mystifying. While in 1770 EULER presented all the apparatus necessary to obtain what we have called CAUCHY's first theorem in Sect. 86, he did not take the final step, and the components w_{km}, which appear fluently in eighteenth-century works on hydrodynamics, remained uninterpreted symbols.

The beginning of the theory is the proof by MACCULLAGH in 1839 that the components of the curl satisfy the vectorial law of transformation. Before this work was published [1848, *1*, § II, Lemma 2], however, CAUCHY, more than a decade after he had completed in all detail his theory of strain, constructed the theory of rotation. MACCULLAGH's result is included in CAUCHY's great paper [1841, *1*, § II, Eq. 12] from which we have reproduced the major theorems in the foregoing section.

The innovations of STOKES, HELMHOLTZ, and KELVIN will appear in the following sections.

The terminology of the subject, always a stumbling block, gave rise to a controversy between BERTRAND and HELMHOLTZ [1867, *1*] [1868, *2–4*, *8–10*]. The term *spin* is due to CLIFFORD [1878, *2*, Bk. II, Chap. II, pp. 122–123, Bk. III, Chap. II, pp. 193–194]; *vorticity*, to LAMB [1916, *3*, Preface and § 30].

87. Circulation. KELVIN's transformation (App. 28.1) when applied to the velocity field reads

$$\oint_c \dot{x}_k\, dx^k = \int_s w_k\, da^k, \quad \text{or} \quad \oint_c d\boldsymbol{x}\cdot\dot{\boldsymbol{p}} = \int_s d\boldsymbol{a}\cdot\boldsymbol{w}: \tag{87.1}$$

The mean value of the tangential component of velocity around a closed circuit, multiplied by the length of the circuit, equals the flux of vorticity through any surface entirely bounded by the circuit. It was in this connection that KELVIN's transformation was discovered (Sect. App. 28). The line integral $\oint d\boldsymbol{x}\cdot\dot{\boldsymbol{p}}$ is the *circulation* of c.

Let w_n be the length of the projection of $\boldsymbol{w}$ onto the normal $\boldsymbol{n}$ to a surface s at the point $\boldsymbol{x}$. Letting $\mathfrak{s}$ be the area of s, we have by (87.1)

$$\oint_c \dot{x}_k\, dx^k = \mathfrak{s}\, w_n + o(\mathfrak{s}) \tag{87.2}$$

[1] BELTRAMI [1871, *1*, § 6]. BONVICINI [1932, *3*] resolved $\boldsymbol{w}$ with respect to the principal axes of finite strain.

as c is shrunk down to the point $\boldsymbol{x}$. Hence

$$w_n = \lim_{\mathfrak{s}\to 0} \frac{\oint_c \dot{x}_k\, dx^k}{\mathfrak{s}}. \tag{87.3}$$

That is, *if $\mathfrak{s}$ is the area inclosed by a circuit c upon a surface s, then the ultimate ratio of the circulation of c to the area $\mathfrak{s}$ as c is shrunk down to a point P equals the component of vorticity in the direction normal to s at P.* Hence the direction of the vorticity vector at P is perpendicular to the plane in which the circulation per unit area is greatest, and the vorticity magnitude is the circulation per unit area in that plane[1].

The striking interpretation just presented for the vorticity vector is of a quality essentially different from those of the previous section, which involved the rotations of material lines and hence the fact that $\dot{\boldsymbol{x}}$ is the velocity of a motion. Interpretations in terms of the circulation, on the other hand, are but especially vivid statements of the properties of the curl of any vector field (cf. footnote 1, p. 277).

88. Irrotational motions. A motion for which the spin vanishes is *irrotational*[2]:

$$w_{km} = 0, \quad \text{or} \quad \boldsymbol{w} = 0. \tag{88.1}$$

Motions of this kind form the subject of most of classical hydrodynamics. Motions not satisfying (88.1) are called *rotational.*

In the terminology of Sect. App. 33, *a motion is irrotational if and only if its velocity field is lamellar,* and (88.1) is necessary and sufficient that $\dot{x}_k\, dx^k$ be the exact differential of a *velocity potential* V:

$$\dot{x}_k = -V_{,k} \quad \text{or} \quad \dot{\boldsymbol{p}} = -\operatorname{grad} V. \tag{88.2}$$

The velocity potential is a discovery of EULER (1752)[3].

In simply connected regions of irrotational motion the velocity potential is single-valued; its application in multiply connected regions, where it is generally a cyclic function, was initiated by HELMHOLTZ and KELVIN[4]. In the former case, we have

$$-\int_{\boldsymbol{x}_1}^{\boldsymbol{x}_2} \dot{x}_k\, dx^k = \int_{\boldsymbol{x}_1}^{\boldsymbol{x}_2} dV = V(\boldsymbol{x}_2) - V(\boldsymbol{x}_1). \tag{88.3}$$

If we call $\int_c \dot{x}_k dx^k$ the *flow* along c, then (88.3) asserts that *in a simply connected region of irrotational motion, the value of the velocity potential V at a point P is the negative of the flow along any curve connecting P with some arbitrary fixed point where V is assigned the value zero.* Taking $\boldsymbol{x}_1$ and $\boldsymbol{x}_2$ as the same point and the closed circuit of integration as lying within a simply connected region, we find the right-hand side of (88.3) to be zero and hence derive ***Kelvin's kinematical***

---

[1] TRICOMI [1934, *9*] has constructed an interpretation of w_n in terms of the angular velocity of a ring in the plane normal to $\boldsymbol{n}$.

[2] The term is due to KELVIN [1869, *7*, §§ 59–60]; earlier authors, from 1757 onward, had spoken of these motions as "those in which $\dot{x}\, dx + \dot{y} dy + \dot{z}\, dz$ is an exact differential". Nowadays irrotational motions are often called *potential flows.*

[3] [1761, *2*, §§ 46–48, 50, 55–56] [1757, *2*, §§ 26–27]. While D'ALEMBERT [1752, *1*, §§ 59, 86] based his theory of plane and rotationally symmetric fluid motions on the contention that $\dot{x}_k dx^k$ is exact, he did not make direct use of the velocity potential, whose existence follows from this assumption. The term *Geschwindigkeitspotential* is due to HELMHOLTZ [1858, *1*, Introd.].

[4] [1858, *1*, §§ 5–6]; [1869, *7*, §§ 60(s)–64]. An elaborate analytical treatment is given by LICHTENSTEIN [1929, *4*, Chap. 3, § 4].

theorem[1]: *A motion is irrotational if and only if the circulation about every reducible circuit is zero.* In case the region in question is multiply connected, let it be rendered simply connected by the imposition of suitable barriers. Then from two reconcileable but not necessarily reducible circuits may be formed a single reducible circuit (Fig. 15) by adding one connecting path, to be traversed in opposite senses, upon each barrier crossed by the circuits. By applying KELVIN's kinematical theorem to the resulting single reducible circuit and noting that the net flow along the twice traversed paths on the barriers is zero, we conclude that *a motion is irrotational if and only if the circulations about any two reconcileable circuits are equal.* More generally, if we let $C_{\mathfrak{t}}$, $\mathfrak{t}=1, 2, \ldots, \mathfrak{p}$, be the circulations about any $\mathfrak{p}$ circuits such that none is reconcileable with any circuit formed by the connection of any number of the others by barriers, then (88.3) must be replaced by

$$-\int_{\boldsymbol{x}_1}^{\boldsymbol{x}_2} \dot{x}_k\,dx^k = V(\boldsymbol{x}_2) - V(\boldsymbol{x}_1) + \sum_{\mathfrak{t}=1}^{\mathfrak{p}} n_{\mathfrak{t}} C_{\mathfrak{t}}, \qquad (88.4)$$

where the $n_{\mathfrak{t}}$ are integers. The $C_{\mathfrak{t}}$ are a set of *cyclic constants* of the motion. The cyclic constants are not uniquely defined, but any member of one possible set may be expressed as a linear combination, with integral coefficients, of the members of any other possible set. The coefficients $n_{\mathfrak{t}}$ in the sum on the right-hand side of (88.4) depend on the path of integration in the integral on the left-hand side. By choosing $\boldsymbol{x}_1 = \boldsymbol{x}_2$, we conclude that *any two determinations of the velocity potential at a point differ by a linear integral combination of the cyclic constants of the motion.*

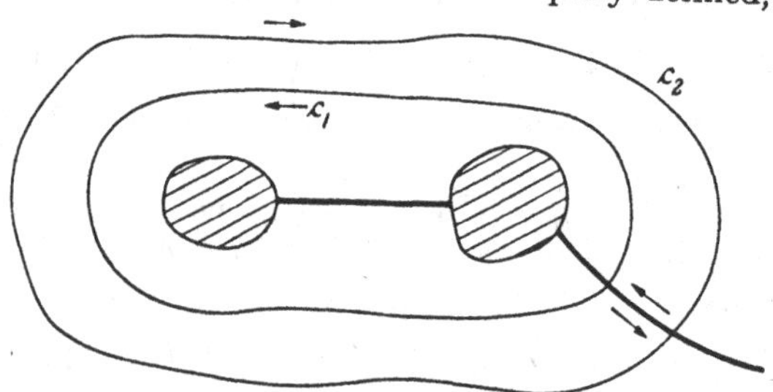

Fig. 15. Circulation in a multiply connected region.

Since $\dot{x}_{k,m}$ is symmetric if and only if $w_{km}=0$, by applying the algebraic theorems of CAUCHY and KELVIN and TAIT in Sect. App. 37 we conclude that *a necessary and sufficient condition for a motion to be instantaneously irrotational at a point is that at that point there exist three mutually orthogonal directions suffering no instantaneous rotation* (cf. Sect. 78). This result follows also from CAUCHY's conics of rotations (Sect. 86): *A necessary and sufficient condition that a motion in which $\dot{x}_{k,m} \neq 0$ be instantaneously irrotational at a point is that the conics of rotations in three different planes through the point be equilateral hyperbolae.* The proof lies in the fact that the asymptotes of a hyperbolic conic of rotations are suffering no rotation, so that if these are orthogonal, the sum of the rates of rotation of two orthogonal directions in the plane of the conic is zero, and hence by the first theorem of CAUCHY (Sect. 86) the component of vorticity normal to that plane is zero.

That in an irrotational motion there are at least three mutually orthogonal invariant directions was established above, but a more specific statement holds. Since $\dot{x}_{k,m}$ is symmetric, its proper vectors are mutually orthogonal if its proper numbers are distinct. By the analysis in Sect. 78 it follows that *in an irrotational motion there are at each point either three or an infinite number of instantaneously invariant directions, according as the three principal stretchings are distinct or not.*

[1] [1869, *7*, §§ 59(e), 59(f)]. The result follows immediately from analysis given by HANKEL [1861, *1*, § 8] but was not stated by him.

Comparison of the above theorem with the fact that in a rigid motion which is not a translation there is but one invariant direction (Sect. 84) yields the following characterization: *A motion is locally a translation if and only if it is irrotational and rigid*[1].

By (88.2) we have for the derivative of V in the direction of the velocity vector

$$\frac{dV}{ds} = \frac{\dot{x}^k}{\dot{x}} V_{,k} = -\dot{x} \leqq 0. \tag{88.5}$$

Hence *the velocity potential can never increase in the direction of motion along a stream line, and in steady irrotational motion a particle always moves toward a region of lower velocity potential. A stagnation point is a stationary point for the velocity potential along the stream line on which it occurs.* Thus a closed stream line lying wholly in a simply connected region of irrotational motion can never exist. In these circumstances it follows also from the continuity of the velocity field that the stream lines in a simply connected region cannot return arbitrarily near to themselves, as in general they may in a rotational motion[2].

By (88.5) and (72.4) we have

$$\frac{\partial V}{\partial t} - \dot{V} = \dot{x}^2 \geqq 0. \tag{88.6}$$

Hence the squared speed is the excess of the local over the material time derivative of the velocity potential, and in particular the rate of increase of the velocity potential as apparent to an observer moving with a particle can never exceed the rate apparent to a fixed observer.

In an irrotational motion, from (88.2) and (82.2) we have[3]

$$d_{km} = -V_{,km}. \tag{88.7}$$

Hence in particular

$$\mathrm{I}_{\boldsymbol{d}} = -V^{,k}{}_{,k} = -\nabla^2 V. \tag{88.8}$$

By (77.1) we derive **Euler's theorem on irrotational motions**[4]: *An irrotational motion is isochoric if and only if*

$$\nabla^2 V = 0. \tag{88.9}$$

Thus all properties of isochoric irrotational motions follow from potential theory, and, conversely, every result of potential theory has a kinematical interpretation in terms of isochoric irrotational motions.

From (88.8) we see more generally that V is subharmonic, harmonic, or superharmonic according as $\mathrm{I}_{\boldsymbol{d}} \leqq 0$, $\mathrm{I}_{\boldsymbol{d}} = 0$, or $\mathrm{I}_{\boldsymbol{d}} \geqq 0$. Therefore *in the interior of a region where material volumes are not decreasing, the velocity potential cannot suffer a minimum; not increasing, a maximum.*

Gosiewski[5] has noted a connection between irrotational motions and a special property of spin. If a principal axis of $\boldsymbol{d}$ is *invariant*, then a unit vector $\boldsymbol{n}$ along it must satisfy both

$$(d_{km} + w_{km})\, n^m = C\, n_k \quad \text{and} \quad d_{km}\, n^m = d_{(\boldsymbol{n})}\, n_k. \tag{88.10}$$

Hence

$$w_{km}\, n^m = (C - d_{(\boldsymbol{n})})\, n_k; \tag{88.11}$$

[1] Euler [1761, *2*, § 77].

[2] Cf. Hadamard [1903, *11*, ¶ 67].

[3] Further properties of the stretching tensor of an irrotational motion have been discussed by Reiff [1887, *4*].

[4] [1761, *2*, § 67] [1770, *1*, § 93]. The Eq. (88.9) is often erroneously associated with the name of Laplace, who was three years old when Euler read the paper deriving it in this connection and obtaining all its polynomial solutions.

[5] [1890, *4*, §§ 4—5].

since $w_{km} = -w_{mk}$ implies $w_{km} n^k n^m = 0$, it follows from (88.11) that $C - d_{(\boldsymbol{n})} = 0$, and putting this result back into (88.11) yields

$$w_{km} n^m = 0. \tag{88.12}$$

From this formula we may read off the **theorem of Gosiewski**: *If a principal axis of stretching is invariant, either it must coincide with the direction of the vorticity vector or the motion is irrotational; if two principal axes of stretching are invariant, the motion is irrotational; in an irrotational motion, all three axes of stretching are invariant.* The last part of the theorem is obvious from the last interpretation of the spin in Sect. 86.

89. Two central examples: rectilinear shearing and rectilinear vortex. The general nature of stretching, shearing, and spin is greatly illuminated by two special classes of motion.

A *rectilinear shearing*[1] is a steady motion in which the paths of the particles are parallel straight lines and the speed of each particle is constant in time. Thus in a suitable rectangular Cartesian system we have

$$\dot{x} = f(y, z), \quad \dot{y} = 0, \quad \dot{z} = 0. \tag{89.1}$$

In fact, by (68.2) and (77.1) we see that a pseudo-lineal motion of the first kind is isochoric if and only if it is a rectilinear shearing. The stretching and spin are given by

$$\left.\begin{aligned} \boldsymbol{d} &= \begin{Vmatrix} 0 & \tfrac{1}{2} f_{,y} & \tfrac{1}{2} f_{,z} \\ \cdot & 0 & 0 \\ \cdot & \cdot & 0 \end{Vmatrix}, \\ \boldsymbol{w} &= \begin{Vmatrix} 0 & \tfrac{1}{2} f_{,y} & \tfrac{1}{2} f_{,z} \\ \cdot & 0 & 0 \\ \cdot & \cdot & \cdot \end{Vmatrix}. \end{aligned}\right\} \tag{89.2}$$

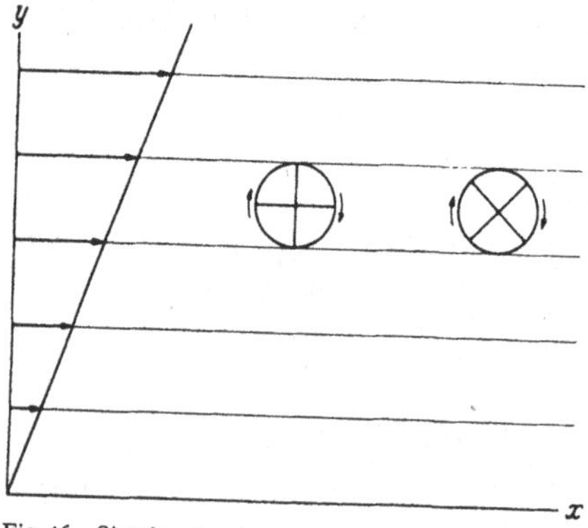

Fig. 16. Simple shearing, a rotational motion with straight stream lines.

Elements parallel or normal to the direction of motion are not being stretched. Elements parallel to the direction of motion remain so and thus never change in length. Elements normal to the direction of motion, however, in general are being rotated[2], and thus at a later instant are no longer normal; hence their exemption from stretching is only temporary.

Entirely typical is the plane homogeneous case, called *simple shearing* (Fig. 16):

$$\dot{x} = K y, \quad \dot{y} = 0, \quad \dot{z} = 0. \tag{89.3}$$

For this special case the quadric of stretching is the hyperbolic cylinder

$$K x y = \text{const}, \tag{89.4}$$

the principal axes of stretching are the z axis and the bisectors of the x and y axes, the principal stretchings are $\pm\frac{1}{2}K$, the maximum orthogonal shears are experienced by elements in the x and y co-ordinate directions, the axis of spin is the z axis, and both the vorticity and the maximum shearing have the value K.

This example is instructive[3] in that although every particle travels in a straight line at uniform speed, in general the spin is not zero. Rotational motion

---

[1] While this class of motions was introduced and analysed by EULER [1757, *2*, §§ 48—49], it is usually named after COUETTE or POISEUILLE.

[2] ST. VENANT [1869, *6*, § 6, footnote].

[3] It was used by BERTRAND [1867, *1*] as a basis for objection to HELMHOLTZ's theory of vorticity. Cf. [1868, *1*].

of a continuum is thus different in nature from rotation of a rigid body. The particles need not circle about in orbits like planets. Rather, if we allow a small object marked with a cross to be carried along in a motion of a continuum, and if we suppose the object partakes of the *tangential* motion, so that its rate of turning is a measure of the *circulation*[1], then in the rotational case the arms of the cross will turn, while in the irrotational case they will not. Spin is a *local* property of a motion.

Shearing, like shear (Sect. 45) has been defined by a particular set of geometrical objects. To replace it by an invariant concept, we observe that at any given point in order for there to exist a co-ordinate system in which $\boldsymbol{d}$ assumes the form

$$\boldsymbol{d} = \left\| \begin{matrix} 0 & \tfrac{1}{2}K & 0 \\ \cdot & 0 & 0 \\ \cdot & \cdot & 0 \end{matrix} \right\|, \tag{89.5}$$

it is necessary and sufficient that $d_1 = -d_3$, $d_2 = 0$. But this is not enough, since in (89.3) the spin is parallel to the second principal axis of stretching, and its magnitude satisfies $w = 2d_1$. Therefore, *in order that a motion may be regarded locally as a simple shearing, it is necessary and sufficient that:*

$$\left.\begin{array}{l} 1.\ \mathrm{I}_{\boldsymbol{d}} = \mathrm{III}_{\boldsymbol{d}} = 0; \\ 2.\ \boldsymbol{w} \textit{ is parallel to the principal axis of stretch along which the stretching is zero;} \\ 3.\ w = \sqrt{-4\,\mathrm{II}_{\boldsymbol{d}}} = \textit{the amount of shearing.} \end{array}\right\} \tag{89.6}$$

For some purposes it is preferable to have a definition independent of the spin. We shall say that a motion is a *shearing* if $(89.6)_1$ holds. An equivalent condition is

$$d_1 = -d_3, \quad d_2 = 0. \tag{89.7}$$

From (89.2) it is plain that the rectilinear shearings (89.1) are included as a special case. Also, any shearing may be regarded locally as a simple shearing superposed upon a rigid rotation.

A *rectilinear vortex*[2] is a motion in which the paths of the particles are circles, whose planes are parallel and whose centers lie upon a straight line, while the speeds of the particles upon a given circle are the same at any one time. Thus in a suitable cylindrical polar system we have

$$\dot{r} = 0, \quad \dot{\theta} = f(r, z, t), \quad \dot{z} = 0. \tag{89.8}$$

The circulation of the circle $r = \text{const}$, $z = \text{const}$ is given by

$$\int_0^{2\pi} r\,\dot{\theta} \cdot r\,d\theta = 2\pi r^2 \dot{\theta} = 2\pi r^2 \omega, \tag{89.9}$$

the physical components of the vorticity vector by

$$w\langle r\rangle = -r\,\partial\dot{\theta}/\partial z, \quad w\langle\theta\rangle = 0, \quad w\langle z\rangle = r^{-1}\,\partial(r^2\dot{\theta})/\partial r. \tag{89.10}$$

---

[1] The arms of the cross must *not* be confused with material elements, since even in an irrotational motion material elements are generally rotating (cf. Sects. 78, 86, 88).

[2] The idea of a vortex goes back to DESCARTES or earlier. The mathematical theory was initiated, rather incompletely if not inaccurately, by NEWTON [1687, *1*, Bk. II, § IX], D. BERNOULLI [1738, *1*, § 11] and D'ALEMBERT [1744, *1*, Bk. III, Chap. IV]. The theory as given here is due to EULER [1757, *2*, §§ 30–33] [1757, *3*, §§ 57–61] [1770, *1*, §§ 75–87]. See also Sect. 98.

If $\dot{\theta}=0$ at $r=0$, (89.9) and (89.10) are related through (87.1). A rectilinear vortex is a pseudo-plane motion of the first kind; typical is the plane special case when $\dot{\theta}=\dot{\theta}(r,t)$.

Whatever be the function $\dot{\theta}$, so long as it is not zero when $r>0$, the particles circle about in orbits. Nevertheless, the motion need not be rotational. As EULER observed, the spin vanishes if and only if

$$r^2\dot{\theta}=f(t). \tag{89.11}$$

In this case, the circulations (89.9) all have the same value, $2\pi f(t)$, called the *strength* of the *irrotational vortex*. Whether or not the motion be irrotational, or indeed the value of the vorticity altogether, in this example has nothing to do with the paths followed by the particles. Again, it is a local matter. Fig. 17 represents the irrotational and rotational cases schematically by means of an object marked with a cross[1].

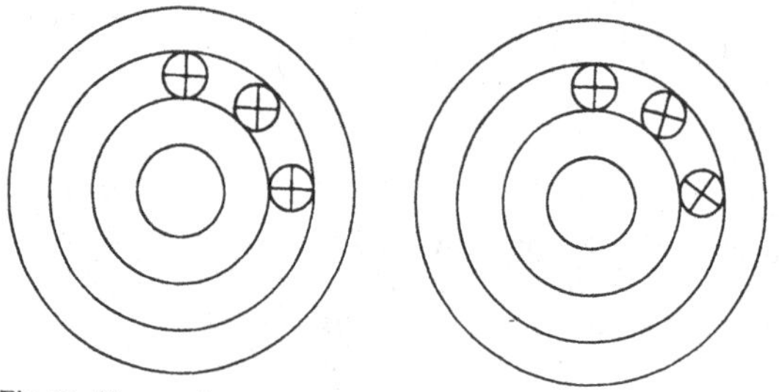

Fig. 17. Two motions with the same stream lines. At the left, an irrotational vortex; at the right, a rotational one.

To describe these two special motions, we have used the spatial description. Corresponding material descriptions are, respectively,

$$x=X+F(Y,Z)\,t,\qquad y=Y,\qquad z=Z, \tag{89.12}$$

$$r=R,\quad \theta=\Theta+\int F(R,Z,t)\,dt,\quad z=Z. \tag{89.13}$$

90. The fundamental decomposition of CAUCHY and STOKES. Since

$$\dot{x}_{k,m}=d_{km}+w_{km}, \tag{90.1}$$

and since $\dot{x}_{k,m}=0$ defines a motion of uniform translation, from the results of EULER and CAUCHY presented in Sect. 82 and Sect. 86 we may read off the following ***fundamental theorem***, first explicitly stated by STOKES[2]: *An arbitrary instantaneous state of motion may be resolved at each point into a uniform translation, three in general unequal stretchings along mutually perpendicular axes suffering no shearing, and a rigid rotation of these axes.*

The most important implication of this theorem is the simplest: *At any point, a given motion may be decomposed uniquely into a translation, an irrotational motion, and a rigid rotation.*

Other decompositions will be mentioned in Sect. 92 and Sect. 142.

As noted by ZORAWSKI[3], the several identities connecting the gradients of $\boldsymbol{d}$ and $\boldsymbol{w}$ follow from the fact that $\dot{x}_{k,mp}=\dot{x}_{k,pm}$. The simplest of these is

$$w_{kp,m}=d_{km,p}-d_{pm,k}. \tag{90.2}$$

Various other relations follow by simple combinations of this identity and the symmetry conditions $d_{km}=d_{mk}$, $w_{km}=-w_{mk}$; in particular, these in turn imply $\dot{x}_{k,pm}=\dot{x}_{k,mp}$.

---

[1] Recall the caution expressed in footnote 1, p. 361.

[2] [1845, *4*, § 2]. The result is equivalent to the theorem of CAUCHY presented in Sect. 57. APPELL [1903, *2*, §§ 1–5] studied further properties of $\dot{x}_{k,m}$. On the basis of arguments we do not follow, HENCKY [1949, *13*] replaced $\dot{x}_{k,m}$ by $\dot{x}_{k,m}+c_k\dot{x}_m$, where $\boldsymbol{c}$ is an arbitrary vector, and discussed a resolution analogous to (90.1).

[3] [1911, *14*, § 4]. Cf. the more general result (84.7) of RICCI.

91. The kinematical vorticity number. Since there is a unique local decomposition of any motion into stretching and spin, it is natural to seek a measure of the *rotational quality* or *rotationality* of a given motion. This question, the answer to which has important applications in hydrodynamics, was first attacked by LEVI-CIVITA[1], but the measure he suggested, $\int \boldsymbol{w}\, dt$, is obviously unsatisfactory. TRUESDELL[2] has introduced the *kinematical vorticity number*:

$$\mathfrak{W}_{\mathrm{K}} \equiv \sqrt{\frac{w_{km} w^{km}}{d_{np} d^{np}}} = \sqrt{\frac{-\mathrm{II}_{\boldsymbol{w}}}{\mathrm{II}_{\boldsymbol{d}}}} = \frac{w}{\sqrt{2\,\overline{\mathrm{II}}_{\boldsymbol{d}}}} = \frac{w}{\sqrt{2(d_1^2+d_2^2+d_3^2)}}, \tag{91.1}$$

the ratio of the magnitude of $\boldsymbol{w}$ to the intensity of $\boldsymbol{d}$. $\mathfrak{W}_{\mathrm{K}}$ is dimensionless; when $\mathfrak{W}_{\mathrm{K}}$ is large, vorticity predominates over deformation, and when $\mathfrak{W}_{\mathrm{K}}$ is small, the motion is nearly irrotational; $\mathfrak{W}_{\mathrm{K}}=0$ is a necessary and sufficient condition that a motion which is not a translation be irrotational, while $\mathfrak{W}_{\mathrm{K}}=\infty$ is a necessary and sufficient condition that a motion which is not a translation be rigid. Thus all motions which are not translations are assigned a numerical degree of rotation between 0 and ∞, with the rigid rotations appearing as the most rotational of all.

All that we have just said would apply equally well had we taken $k\,\mathfrak{W}_{\mathrm{K}}$, where k is any dimensionless constant, as a measure. To normalize such a measure, we might select a particular rotational non-rigid motion and assign it a measure 1. We prefer, however, to approach the normalization geometrically. A motion in which all the components d_{km} reduce to zero except for one shearing component, say d_{xy}, may be described entirely in terms of two angular rates. One of these, as usual, is the local angular velocity, $\omega=\frac{1}{2}w$; for the other, χ, select the rate at which elements in the x and y co-ordinate directions are turning away from each other, i.e., $\chi=d_{xy}$. Then

$$k\,\mathfrak{W}_{\mathrm{K}} = k\,\frac{2\omega}{\sqrt{4\chi^2}} = k\,\frac{\omega}{|\chi|}. \tag{91.2}$$

Thus if we take $k=1$, the measure $k\,\mathfrak{W}_{\mathrm{K}}$ becomes, for this class of motions, just the ratio of the two characterizing angular rates. Some arbitrariness remains in the choice of this particular pair of angular rates; perhaps the final justification for taking $k=1$, as we shall do, is the symmetry and simplicity of the resulting formulae.

It is easy to verify that

$$\left.\begin{aligned} \mathfrak{W}_{\mathrm{K}} &= \left(1+\frac{2\dot{x}^m_{,n}\dot{x}^n_{,m}}{w^2}\right)^{-\frac{1}{2}},\\ &= \frac{w}{\sqrt{2\mathrm{I}_{\boldsymbol{d}}^2-4\,\mathrm{II}_{\boldsymbol{d}}}} = \frac{1}{\sqrt{1+\dfrac{2\mathrm{I}_{\boldsymbol{d}}^2-4\mathrm{IO}}{w^2}}}, \end{aligned}\right\} \tag{91.3}$$

where (86.14) has been used. Thus we get the following alternative *necessary and sufficient conditions that* $\mathfrak{W}_{\mathrm{K}}=1$:

$$w^2 = 2\,\overline{\mathrm{II}}_{\boldsymbol{d}}, \qquad \dot{x}^m_{,n}\dot{x}^n_{,m}=0, \qquad 4\,\mathrm{II}_{\boldsymbol{d}}+w^2=2\,\mathrm{I}_{\boldsymbol{d}}^2, \qquad 2\mathrm{IO}=\mathrm{I}_{\boldsymbol{d}}^2. \tag{91.4}$$

It is obvious that the second of these is satisfied by (89.1); hence *in a rectilinear shearing*, $\mathfrak{W}_{\mathrm{K}}=1$.

[1] [1940, *13*].
[2] [1953, *31*].

More generally, from (91.3) we may assert that $\mathfrak{W}_K \lesseqgtr 1$ *according as*

$$\dot{x}^m_{,n}\dot{x}^n_{,m} \gtreqless 0; \quad \textit{equivalently,} \quad \mathrm{I}^2_{\boldsymbol{d}} \gtreqless 2\mathrm{II}. \tag{91.5}$$

These are conditions on the proper numbers of the velocity gradient. In fact, we may express $(91.5)_1$ as follows: *If the proper numbers of $\dot{x}^m_{,n}$ are real, then $\mathfrak{W}_K < 1$; if the proper numbers are a, $b \pm ic$, then $\mathfrak{W}_K \lesseqgtr 1$ according as $c^2 \lesseqgtr a^2 + b^2$.* This result is not very illuminating, since the proper numbers of $\dot{x}^m_{,n}$ do not have an immediate kinematical significance.

Further properties of $\mathfrak{W}_K$, including other forms of the condition for the sign of $\mathfrak{W}_K - 1$, will be obtained in Sect. 102.

Now consider the deviator v_{km} of $\dot{x}_{k,m}$, already mentioned in Sect. 78, and let a subscript 0 distinguish quantities calculated from it. Since $\boldsymbol{w}_0 = \boldsymbol{w}$, by (App. 39.6) and $(91.1)_3$ follows[1]

$$\mathfrak{W}_{K0} \geqq \mathfrak{W}_K, \tag{91.6}$$

where equality holds if and only if the motion is isochoric or irrotational or both.

We can now put some of the conditions in Sect. 78 regarding invariant directions into a clearer form. We have

$$\left.\begin{aligned} \mathrm{II}_0 &\equiv -\frac{1}{2} v_{km} v^{mk} = \frac{1}{2}\left(w_{km} w^{km} - d_{0km} d_0^{km}\right), \\ &= \frac{1}{4} w^2 \left(1 - \frac{1}{\mathfrak{W}^2_{K0}}\right). \end{aligned}\right\} \tag{91.7}$$

Therefore $\mathrm{II}_0 < 0$ if and only if $\mathfrak{W}_{K0} < 1$; also, by (91.6), $\mathfrak{W}_K > 1$ is sufficient that $\mathrm{II}_0 > 0$. From the remarks following (78.7) we derive a result of ERICKSEN[2]: *If $\mathfrak{W}_K > 1$, or if $\mathfrak{W}_{K0} = 1$ and $\mathrm{III}_0 \neq 0$, there is one and only one invariant direction; for there to be three invariant directions, it is necessary that $\mathfrak{W}_{K0} < 1$.* Rigid motions are included in the first alternative, irrotational motions in the second. For a simple shearing (Sect. 89) we have $\mathfrak{W}_{K0} = \mathfrak{W}_K = 1$, $\mathrm{III}_0 = 0$, and as noted in Sect. 78 there are infinitely many invariant directions.

92. Sources and vortices. A decomposition which differs from that of Sect. 90 in that it is cumulative rather than local is expressed by the existence of STOKES's potentials (App. 36.1) for the velocity:

$$\dot{x}_k = -S_{,k} + e_{kmq} v^{q,m}, \quad \text{or} \quad \dot{\boldsymbol{p}} = -\operatorname{grad} S + \operatorname{curl} \boldsymbol{v}. \tag{92.1}$$

The theory of this representation, like that presented in Sect. 87, is purely geometrical. We have $\nabla^2 S = -\mathrm{I}_{\boldsymbol{d}}$, $\nabla^2 \boldsymbol{v} = -\boldsymbol{w}$, and (App. 36.2) asserts that *in a finite region, a velocity field is uniquely determined by its values on the boundary and on any surfaces of discontinuity, by its cyclic constants, and by the values of its expansion and vorticity at interior points.* In particular, with the special choice of potentials (App. 36.2), the normal velocity $\dot{p}_n$ on $\mathscr{S}$ contributes to the scalar potential S, while the tangential velocity $\dot{\boldsymbol{p}}_t$ contributes only to the vector potential $\boldsymbol{v}$. It is not possible to assign $\mathrm{I}_{\boldsymbol{d}}$, $\boldsymbol{w}$, $\dot{p}_n$, and $\dot{\boldsymbol{p}}_t$ arbitrarily, however. For example, GREEN's transformation $(\text{App. } 26.1)_{2,3}$ yields

$$\int_v \mathrm{I}_{\boldsymbol{d}}\, dv = \oint_s d\boldsymbol{a} \cdot \dot{\boldsymbol{p}}, \qquad \int_v \boldsymbol{w}\, dv = \oint_s d\boldsymbol{a} \times \dot{\boldsymbol{p}}, \tag{92.2}$$

as necessary conditions to be satisfied by the four quantities occurring in (App. 36.2). In Sect. 116 we shall see that (92.2) is not always sufficient.

[1] ERICKSEN [1955, *5*].
[2] [1955, *5*].

The general theory of *sources* and *vortices* is given in works on potential theory. We may interpret (92.1) and (App. 36.2) with $\boldsymbol{c} = \dot{\boldsymbol{p}}$ as asserting that *any motion may be expressed as the sum of an isochoric irrotational motion and a motion induced by continuously distributed sources and vortices.*

93. Total squared speed. I. Estimates. The results of the foregoing section imply that any quantity associated with a motion in a region may be calculated from the values of the vorticity and expansion in the interior and of $\dot{\boldsymbol{p}}$ upon the boundary. As a specimen, we now reckon upper bounds for the total squared speed.

First, a formula generalizing (App. 31.12) is

$$\mathfrak{J} \equiv \int_v \dot{p}^2 dv = \oint_s d\boldsymbol{a} \cdot (\dot{p}^2 \boldsymbol{p} - 2\dot{\boldsymbol{p}}\,\dot{\boldsymbol{p}} \cdot \boldsymbol{p}) + 2\int_v \boldsymbol{p} \cdot (\boldsymbol{w} \times \dot{\boldsymbol{p}} + \dot{\boldsymbol{p}}\,\mathrm{I}_{\boldsymbol{d}})\,dv, \qquad (93.1)$$

as follows by setting $\boldsymbol{b} = \boldsymbol{c} = \dot{\boldsymbol{p}}$ and $K = 1$ in (App. 26.2), then contracting the resulting identity. Since the integrand of the volume integral contains $\dot{\boldsymbol{p}}$ as well as $\boldsymbol{w}$ and $\mathrm{I}_{\boldsymbol{d}}$, we do not yet have a result of the type desired. However, from (93.1) we can derive an inequality free of interior values of $\dot{\boldsymbol{p}}$, as follows. First,

$$\left|\oint_s d\boldsymbol{a} \cdot (\dot{p}^2 \boldsymbol{p} - 2\dot{\boldsymbol{p}}\,\dot{\boldsymbol{p}} \cdot \boldsymbol{p})\right| \leq K_1 \dot{p}_{\mathrm{m}}^2 + 2K_2 \dot{p}_{\mathrm{m}} \dot{p}_{n\mathrm{m}} \qquad (93.2)$$

where

$$\left.\begin{aligned} \dot{p}_{\mathrm{m}} &\equiv \operatorname{Max} \dot{p} \text{ on } s, & \dot{p}_{n\mathrm{m}} &\equiv \operatorname{Max} \dot{p}_n \text{ on } s,\\ K_1 &\equiv \oint_s |d\boldsymbol{a} \cdot \boldsymbol{p}|, & K_2 &\equiv \oint_s da\, p. \end{aligned}\right\} \qquad (93.3)$$

We note that

$$K_1 \leq K_2 \leq \tfrac{1}{2} D\,\mathfrak{S}, \qquad (93.4)$$

where $\mathfrak{S}$ is the area of s and where to derive the last inequality we have chosen the origin of $\boldsymbol{p}$ at the center of the smallest sphere whose interior contains s, D being the diameter of that sphere. Also

$$|\boldsymbol{p} \cdot (\boldsymbol{w} \times \dot{\boldsymbol{p}}) + \dot{\boldsymbol{p}}\,\mathrm{I}_{\boldsymbol{d}}| \leq p\,\dot{p}\,(w + |\mathrm{I}_{\boldsymbol{d}}|); \qquad (93.5)$$

therefore, the Cauchy-Schwarz inequality implies that[1]

$$\left.\begin{aligned} \left|\int_v \boldsymbol{p} \cdot (\boldsymbol{w} \times \dot{\boldsymbol{p}} + \dot{\boldsymbol{p}}\,\mathrm{I}_{\boldsymbol{d}})\,dv\right|^2 &\leq \mathfrak{J} \int_v p^2 (w + |\mathrm{I}_{\boldsymbol{d}}|)^2\,dv,\\ &\leq (\tfrac{1}{2} D)^2\,\mathfrak{J} \int_v (w + |\mathrm{I}_{\boldsymbol{d}}|)^2\,dv. \end{aligned}\right\} \qquad (93.6)$$

From (93.1), (93.2), (93.4), and (93.6) we get

$$\mathfrak{J} \leq \tfrac{1}{2} D\,\mathfrak{S}\,(\dot{p}_{\mathrm{m}}^2 + 2\dot{p}_{\mathrm{m}} \dot{p}_{n\mathrm{m}}) + D\sqrt{\mathfrak{J}}\,\sqrt{\int_v (w + |\mathrm{I}_{\boldsymbol{d}}|)^2\,dv}. \qquad (93.7)$$

Set

$$A \equiv \tfrac{1}{2} D\,\mathfrak{S}\,(\dot{p}_{\mathrm{m}}^2 + 2\dot{p}_{\mathrm{m}} \dot{p}_{n\mathrm{m}}), \qquad B \equiv D\sqrt{\int_v (w + |\mathrm{I}_{\boldsymbol{d}}|)^2\,dv}. \qquad (93.8)$$

Then (93.7) yields

$$\mathfrak{J} = \int_v \dot{p}^2\,dv \leq \tfrac{1}{2}\left(B^2 + 2A + B\sqrt{B^2 + 4A}\right). \qquad (93.9)$$

Thus the total squared speed in a finite region may be estimated from

1. The diameter of the region and the area of its boundary;
2. The maximum speed on the boundary;
3. The total $(w + |\mathrm{I}_{\boldsymbol{d}}|)^2$.

[1] Inequalities of this type were first obtained by Kampé de Feriet [1946, *6*]; we follow and generalize the work of Berker [1949, *2*, § 3].

Several special cases are interesting. First, if $\boldsymbol{w} = \mathrm{I}_{\boldsymbol{d}} = 0$, we have an isochoric irrotational motion, and (93.9) reduces to $\int_{\mathscr{v}} \dot{p}^2 dv \leqq A$, a result which is also an immediate consequence of the fact that the maximum speed and the maximum of each velocity component of such a motion always occur on the boundary (cf. Sect. 121). Second, if $\mathscr{s}$ is a stationary boundary of the motion, A reduces to the value $\frac{1}{2} D \mathfrak{S} \dot{p}_{\mathrm{m}}^2$. Third, if $\mathscr{s}$ is a stationary boundary to which the material adheres, (93.9) reduces to $\int_{\mathscr{v}} \dot{p}^2 dv \leqq B^2$, or

$$\int_{\mathscr{v}} \dot{p}^2 dv \leqq D^2 \int_{\mathscr{v}} (w + |\mathrm{I}_{\boldsymbol{d}}|)^2 dv. \tag{93.10}$$

If $w = \mathrm{I}_{\boldsymbol{d}} = 0$, we conclude that $\dot{p} = 0$: *In a bounded stationary domain to which the material adheres, there exists no isochoric irrotational motion other than a state of rest.* Cf. the generalizations and alternative developments of this result in Sects. 111 and 113.

We now deduce some stronger results for the case of a simply connected domain, where the explicit formulae (App. 36.2) are available. First we note the identity

$$\left.\begin{aligned} \dot{p}^2 &= \dot{\boldsymbol{p}} \cdot [-\operatorname{grad} S + \operatorname{curl} \boldsymbol{v}], \\ &= -\operatorname{div}(\dot{\boldsymbol{p}}\, S) + S\, \mathrm{I}_{\boldsymbol{d}} - \operatorname{div}(\dot{\boldsymbol{p}} \times \boldsymbol{v}) + \boldsymbol{v} \cdot \boldsymbol{w}, \end{aligned}\right\} \tag{93.11}$$

where the second step is possible because S is single-valued. By (App. 36.2) follows

$$\left.\begin{aligned} \dot{p}^2 = \operatorname{div}(\boldsymbol{v} \times \dot{\boldsymbol{p}} - \dot{\boldsymbol{p}}\, S) + \frac{\mathrm{I}_{\boldsymbol{d}}}{4\pi}\left[\int_{\mathscr{v}} \frac{\mathrm{I}_{\boldsymbol{d}}\, dv}{d} - \oint_{\mathscr{s}} \frac{d\boldsymbol{a} \cdot \dot{\boldsymbol{p}}}{d}\right] + \\ + \frac{\boldsymbol{w}}{4\pi} \cdot \left[\int_{\mathscr{v}} \frac{\boldsymbol{w}\, dv}{d} - \oint_{\mathscr{s}} \frac{d\boldsymbol{a} \times \dot{\boldsymbol{p}}}{d}\right]. \end{aligned}\right\} \tag{93.12}$$

We now integrate this identity over the volume $\mathscr{v}$. If we use primes to distinguish quantities evaluated at a second running point while unprimed quantities are understood to be evaluated at the running point occurring in the integrations already indicated in (93.12), we obtain

$$\left.\begin{aligned} \int_{\mathscr{v}} \dot{p}^2 dv = \frac{1}{4\pi} \int_{\mathscr{v}} \int_{\mathscr{v}} \frac{\mathrm{I}_{\boldsymbol{d}}\, \mathrm{I}'_{\boldsymbol{d}} + \boldsymbol{w} \cdot \boldsymbol{w}'}{d}\, dv\, dv' + \oint_{\mathscr{s}} d\boldsymbol{a} \cdot (\boldsymbol{v} \times \dot{\boldsymbol{p}} - \dot{\boldsymbol{p}}\, S) - \\ - \frac{1}{4\pi} \int_{\mathscr{v}} \oint_{\mathscr{s}} \frac{d\boldsymbol{a} \cdot \dot{\boldsymbol{p}}\, \mathrm{I}'_{\boldsymbol{d}} + (d\boldsymbol{a} \times \dot{\boldsymbol{p}}) \cdot \boldsymbol{w}'}{d}\, dv'. \end{aligned}\right\} \tag{93.13}$$

Under various circumstances the second and third terms on the right-hand side are zero. Such is the case, for example, if $\mathscr{s}$ is a fixed boundary to which the material adheres, or if $\mathscr{v}$ is an infinite region and $\dot{p} = o(r^{-2})$ at ∞. We then obtain the elegant formula of HELMHOLTZ and A. FÖPPL[1]:

$$\int_{\mathscr{v}} \dot{p}^2 dv = \frac{1}{4\pi} \int_{\mathscr{v}} \int_{\mathscr{v}} \frac{\mathrm{I}_{\boldsymbol{d}}\, \mathrm{I}'_{\boldsymbol{d}} + \boldsymbol{w} \cdot \boldsymbol{w}'}{d}\, dv\, dv'. \tag{93.14}$$

[1] [1858, *1*, § 4] (in the case $\mathrm{I}_{\boldsymbol{d}} = 0$); [1897, *4*, §§ 32–33].

We now determine a bound for the right-hand side of (93.14) in the case of a finite region[1]. From the Cauchy-Schwarz inequality we get

$$\left(\int_v\int_v \frac{\mathrm{I}_{\boldsymbol{d}}\mathrm{I}'_{\boldsymbol{d}}}{d}\,dv\,dv'\right)^2 \leqq (4\pi C)^2 \int_v\int_v (\mathrm{I}_{\boldsymbol{d}}\mathrm{I}'_{\boldsymbol{d}})^2\,dv\,dv' = (4\pi C)^2\left(\int_v \mathrm{I}^2_{\boldsymbol{d}}\,dv\right)^2, \tag{93.15}$$

where

$$(4\pi C)^2 \equiv \int_v\int_v \frac{dv\,dv'}{d^2}. \tag{93.16}$$

Since $(\boldsymbol{w}\cdot\boldsymbol{w}')^2 \leqq w^2 w'^2$, we similarly obtain

$$\left(\int_v\int_v \frac{\boldsymbol{w}\cdot\boldsymbol{w}'}{d}\,dv\,dv'\right)^2 \leqq (4\pi C)^2\left(\int_v w^2\,dv\right)^2. \tag{93.17}$$

Thus for a finite simply connected region to whose boundary the material adheres we get the elegant bound

$$\int_v \dot{p}^2\,dv \leqq C\int_v (\mathrm{I}^2_{\boldsymbol{d}} + w^2)\,dv, \tag{93.18}$$

to be compared with (93.10).

The constant C depends only upon the domain v. We now obtain an estimate[2] for C in terms of the diameter D that was used above. Clearly

$$(4\pi C)^2 \leqq \int_{v_0}\int_{v_0} \frac{dv\,dv'}{d^2} = \int_{v_0} G\,dv, \tag{93.19}$$

where v_0 is the smallest sphere that contains v, and

$$G \equiv \int_{v_0} \frac{dv}{d^2}. \tag{93.20}$$

Let R_1 be the distance from the argument point of G to the center of the sphere v_0, and introduce polar co-ordinates with axis along the line connecting these two points. Then

$$\left.\begin{aligned} G &= \int_{v_0} \frac{r^2\sin\theta\,d\theta\,d\varphi\,dr}{r^2 + R_1^2 - 2rR_1\cos\theta},\\ &= \pi\frac{R^2 - R_1^2}{R_1}\log\frac{R+R_1}{R-R_1} + 2\pi R, \end{aligned}\right\} \tag{93.21}$$

where $R \equiv \frac{1}{2}D$. A further quadrature yields $\int_{v_0} G\,dv = \frac{1}{4}\pi^2 D^4$, and hence[3]

$$C \leqq \tfrac{1}{8}D^2. \tag{93.22}$$

Comparison with (93.10) shows that the second estimate is much sharper than the first. However, the first estimate holds without restriction on the connectivity of the domain.

---

[1] Berker [1949, *2*, §§ 4–5].

[2] Berker [1949, *2*, §§ 4–5].

[3] For the plane case, Kampé de Feriet [1946, *6*] obtained a plane analogue of (93.18) with

$$(2\pi C)^2 = \int_a\int_a \left(\log\frac{1}{d}\right)^2 da\,da'.$$

94. Total squared speed. II. Extremal theorems. A simple and profound theorem of KELVIN[1] characterizes irrotational motions as minimizing the total squared speed. Indeed, let us define the *kinetic energy*[2] $\mathfrak{K}$ of a motion as

$$2\mathfrak{K} \equiv \int_{\mathscr{v}} A j \dot{x}^2 \, dv, \tag{94.1}$$

where A is any substantially constant function not identically zero. I.e., Aj is a density for the motion (Sect. 76). We now compare the kinetic energy $\mathfrak{K}'$ of the motion $\dot{\boldsymbol{x}}'$ with the kinetic energy $\mathfrak{K}$ of an irrotational motion, $\dot{x}_k = -V_{,k}$. If V is single-valued, we have

$$\left.\begin{aligned} A j \dot{x}'^2 &= A j [\dot{x}^2 - 2(\dot{x}'^k - \dot{x}^k) V_{,k} + (\dot{\boldsymbol{x}}' - \dot{\boldsymbol{x}})^2], \\ &= A j \dot{x}^2 - 2[(\dot{x}'^k - \dot{x}^k) A j V]_{,k} + \\ &\quad + 2V[A j (\dot{x}'^k - \dot{x}^k)]_{,k} + A j (\dot{\boldsymbol{x}}' - \dot{\boldsymbol{x}})^2. \end{aligned}\right\} \tag{94.2}$$

By (76.11), the third term vanishes in two cases: when both $\dot{\boldsymbol{x}}$ and $\dot{\boldsymbol{x}}'$ are isochoric, or when both motions are steady and have the same steady density Aj. In these cases, integration over a volume $\mathscr{v}$ and use of (App. 26.1)$_2$ yields

$$\mathfrak{K}' = \mathfrak{K} + \tfrac{1}{2} \int_{\mathscr{v}} A j (\dot{\boldsymbol{x}}' - \dot{\boldsymbol{x}})^2 \, dv - \oint_{\mathscr{s}} d\boldsymbol{a} \cdot (\dot{\boldsymbol{x}}' - \dot{\boldsymbol{x}}) A j V. \tag{94.3}$$

The surface integral vanishes if both motions are assumed to have the same normal flow $Aj(\dot{x}' - \dot{x})_n$ through the bounding surface $\mathscr{s}$. If we assume $A > 0$, the second term in (94.3) is positive unless $\dot{\boldsymbol{x}}' = \dot{\boldsymbol{x}}$. Rewording this result yields ***Kelvin's theorem of minimum energy:*** *Let the normal flow at each point of the bounding surface of a simply connected region be assigned; then among all isochoric motions, or among all steady motions with an assigned positive steady density, the least possible kinetic energy is attained if and only if the motion is irrotational.* For multiply connected regions the theorem still holds, provided fixed values be assigned to the fluxes through a set of barriers rendering the region simply connected[3].

A variational theorem which is a partial converse of KELVIN's theorem has been obtained by PRATELLI[4]. Consider the velocity field as given in terms of its Stokes potentials (92.1). We vary the potentials S and $\boldsymbol{v}$, obtaining $\delta \dot{\boldsymbol{p}} = -\operatorname{grad} \delta S + \operatorname{curl} \delta \boldsymbol{v}$. We assume the region simply connected, so that S is single-valued, and we set

$$\mathfrak{X} \equiv \int_{\mathscr{v}} [\tfrac{1}{2} \dot{p}^2 - S B] \, dv \tag{94.4}$$

and restrict the variations to those leaving the quantity B unaltered. Then by (App. 26.1)$_2$ and easy vectorial identities follows

$$\left.\begin{aligned} \delta \mathfrak{X} &= \int_{\mathscr{v}} [\dot{\boldsymbol{p}} \cdot \delta \dot{\boldsymbol{p}} - B \delta S] \, dv, \\ &= \int_{\mathscr{v}} [\dot{\boldsymbol{p}} \cdot (-\operatorname{grad} \delta S + \operatorname{curl} \delta \boldsymbol{v}) - B \delta S] \, dv, \\ &= \int_{\mathscr{v}} [-\operatorname{div}(\dot{\boldsymbol{p}} \delta S + \dot{\boldsymbol{p}} \times \delta \boldsymbol{v}) + (\mathrm{I}_{\boldsymbol{d}} - B) \delta S + \boldsymbol{w} \cdot \delta \boldsymbol{v}] \, dv, \\ &= -\oint_{\mathscr{s}} d\boldsymbol{a} \cdot (\dot{\boldsymbol{p}} \delta S + \dot{\boldsymbol{p}} \times \delta \boldsymbol{v}) + \int_{\mathscr{v}} [(\mathrm{I}_{\boldsymbol{d}} - B) \delta S + \boldsymbol{w} \cdot \delta \boldsymbol{v}] \, dv. \end{aligned}\right\} \tag{94.5}$$

---

[1] [1849, *3*]. A variational formulation is given by PRATELLI [1953, *24*, §§ 1—4], who notes also that of all isochoric motions in a tube with assigned discharge, the irrotational motion has the least total squared speed.

[2] The general theory of the kinetic energy is given in Sects. 166ff.

[3] LAMB [1895, *2*, § 55], but the result is really immediate from KELVIN's theory of cyclic potentials. A rigorous proof, based on topological theorems, of a result slightly stronger in that the motions considered for comparison may have any density not greater than that of the irrotational motion, is given by HAYES [1960, *2*].

[4] [1953, *24*, §§ 6—7].

If we restrict the variations so that $\delta S = 0$ and $\delta \boldsymbol{v} = 0$ on $\mathfrak{s}$, we get as equivalent to $\delta \mathfrak{X} = 0$ the conditions

$$\boldsymbol{w} = 0, \quad \mathrm{I}_{\boldsymbol{d}} = B. \tag{94.6}$$

That is, *in order to render $\mathfrak{X}$ an extreme when the scalar and vector potentials of the velocity are subjected to variations vanishing on the boundary but otherwise arbitrary, it is necessary and sufficient that the motion be irrotational and that its expansion be the quantity B.* In the case when only S is varied, from (94.5) follows only the latter, not the former of (94.6). That is, *of all motions having the same vorticity, those such that $\mathrm{I}_{\boldsymbol{d}} = B$ give an extreme value to $\mathfrak{X}$ when the scalar potential is subjected to variations which vanish on the boundary but are otherwise arbitrary.* These results may be regarded as variational theorems for the equation of continuity (76.11). In the isochoric case, we may set $B = 0$ and reduce $2\mathfrak{X}$ to the total squared speed; the condition $(94.6)_2$ then becomes $\mathrm{I}_{\boldsymbol{d}} = 0$.

As has been remarked by SERRIN[1], DIRICHLET's principle may also be regarded as a converse of KELVIN's theorem, as follows. Let h be given; corresponding to any single-valued function V', set

$$\mathfrak{J}' \equiv -\tfrac{1}{2}\int_{\mathfrak{v}} (\operatorname{grad} V')^2\, dv + \oint_{\mathfrak{s}} V' h\, da. \tag{94.7}$$

Let $\mathfrak{J}$ be the value of $\mathfrak{J}'$ when $V' = V$, the potential of the unique isochoric irrotational motion in $\mathfrak{v}$ such that $\dot{p}_n = -\partial V/\partial n = h$ on $\mathfrak{s}$. Since $\nabla^2 V = 0$, from (94.7) it follows by use of GREEN's transformation that

$$\mathfrak{J} = \mathfrak{J}' + \tfrac{1}{2}\int_{\mathfrak{v}} [\operatorname{grad}(V - V')]^2\, dv. \tag{94.7}$$

Hence $\mathfrak{J} > \mathfrak{J}'$ unless $V = V' + \text{const}$. That is, *the potential of the isochoric irrotational motion in $\mathfrak{v}$ such that $\dot{p}_n = h$ on $\mathfrak{s}$ renders $\mathfrak{J}$ a maximum.*

For an isochoric motion, we have $j = 1$, and we may take $A = 1$. In the notation used from Sect. 155 onward, this is equivalent to taking $\varrho = Aj = \text{const.}$, so that comparison of (169.7) with (94.7) yields

$$\varrho\,\mathfrak{J} = -\mathfrak{K} + 2\mathfrak{K} = \mathfrak{K}; \tag{94.8}$$

i.e., *the maximum value of $\varrho\,\mathfrak{J}$ equals the minimum value of $\mathfrak{K}$, namely, the kinetic energy of the isochoric irrotational motion which is characterized as the unique solution of each variational problem.*

95. Rate of strain from an initial configuration. Our purpose now is to connect the concept of finite *stretch*, formulated in Sect. 25, with the *stretching* introduced in Sect. 82. In terms of motion, if the material element given by $d\boldsymbol{X}$ at time $t = 0$ is carried into $d\boldsymbol{x}$ at time t, then its stretch at time t is the ratio of its present length to its original length:

$$\lambda \equiv \frac{dx}{dX}. \tag{95.1}$$

As repeatedly emphasized in Subchapter I, stretch is a function of *two different configurations* of the material. Not so with stretching, which refers only to the present configuration:

$$d \equiv \frac{d}{dt} \log dx \bigg|_{d\boldsymbol{x} = d\boldsymbol{X}}. \tag{95.2}$$

But, since for a material element we have $\dot{\overline{dX}} = 0$, from (95.1) follows

$$\dot{\lambda} = \frac{\dot{\overline{dx}}}{dX} = \lambda d, \quad \text{or} \quad d = \dot{\overline{\log \lambda}}. \tag{95.3}$$

Thus, in general, *the stretching of a material element is its rate of stretch per unit stretch.* If we put $t = 0$ in (95.3), we get $\lambda = 1$, and hence

$$\dot{\lambda}|_{t=0} = d: \tag{95.4}$$

[1] Sect. 24 of *Mathematical Principles of Fluid Mechanics*, This Encyclopedia, Vol. VIII/1.

The stretching of an element is its rate of stretch with respect to its present configuration.

Despite their truth, the foregoing simple remarks are of little use[1]. The stretch of an element is calculated from its components by means of the components of the tensors $\boldsymbol{C}$ and $\boldsymbol{c}$, and the manner in which these components change with time is not obvious.

For the tensor $\boldsymbol{C}$, the calculation is easy, since by (26.2)$_1$ and the fact that $\dot{\overline{dX^\alpha}} = 0$ we get

$$\dot{\overline{ds^2}} = \dot{C}_{\alpha\beta}\, dX^\alpha dX^\beta. \tag{95.5}$$

From (82.3) and (20.3), on the other hand, we get

$$\left.\begin{aligned}\dot{\overline{ds^2}} &= 2 d_{km}\, dx^k\, dx^m\\ &= 2 d_{km}\, x^k_{,\alpha}\, x^m_{,\beta}\, dX^\alpha\, dX^\beta.\end{aligned}\right\} \tag{95.6}$$

Since $d\boldsymbol{X}$ is arbitrary, comparison of (95.5) and (95.6) yields[2]

$$\dot{C}_{\alpha\beta} = 2 d_{km}\, x^k_{,\alpha}\, x^m_{,\beta} = 2\dot{E}_{\alpha\beta}. \tag{95.7}$$

Similarly, differentiating (26.1)$_1$ yields

$$0 = \dot{c}_{km}\, dx^k dx^m + c_{km}(\dot{x}^k_{,n}\, dx^n\, dx^m + dx^k\, \dot{x}^m_{,n}\, dx^n), \tag{95.8}$$

where we have used (76.4)$_2$. Since dx^k is arbitrary, we get

$$\dot{c}_{km} = -2 c_{n(k}\, \dot{x}^n_{,m)}, \tag{95.9}$$

or, equivalently[3],

$$\dot{e}_{km} = d_{km} - 2 e_{n(k}\, \dot{x}^n_{,m)}. \tag{95.10}$$

From (95.10) and (95.7)$_2$ we get

$$\dot{e}_{km}|_{t=0} = d_{km}, \qquad \dot{E}_{\alpha\beta}|_{t=0} = \delta^k_\alpha\, \delta^m_\beta\, d_{km} = d_{\alpha\beta}, \tag{95.11}$$

as expected from (95.4). However, the rates $\dot{\boldsymbol{e}}$ and $\dot{\boldsymbol{E}}$ in a strained configuration are quite different from $\boldsymbol{d}$ and from one another. This is most easily seen by setting $\boldsymbol{d} = 0$ in (95.10) and (95.7), yielding

$$\dot{E}_{\alpha\beta} = 0, \qquad \dot{e}_{km} = -2 e_{n(k}\, \dot{x}^n_{,m)}. \tag{95.12}$$

Thus in an instantaneously rigid motion the components of $\boldsymbol{E}$ remain constant, but those of $\boldsymbol{e}$ generally change. This difference reflects an inconvenience of the

---

[1] Further results of this kind are noted by REINER [1948, *23*, § 7] and TRUESDELL [1952, *21*, § 22].

[2] E. and F. COSSERAT [1896, *1*, § 15, Eq. (2)]. Cf. also DUHEM [1904, *1*, Part I, Chap. I, § 1, Eq. (30)]. An equivalent but more complicated variational formula had been given by PIOLA [1833, *3*, § 6] [1848, *2*, ¶ 36]; in effect, he also calculated $\dot{\overline{C^{-1}}}$. BONVICINI [1932, *3*] calculated $\dot{\overline{C^{\frac{1}{2}}}}$. A particularly interesting equivalent form was obtained by DEUKER [1941, *1*, § VI] in terms of the velocity $\underset{\sim}{X}_\alpha$ of space relative to the material (cf. Sect. 65): $\dot{E}_{\alpha\beta} = -\underset{\sim}{X}_{(\alpha,\beta)}$. Comparison with (82.3) shows that the rate of strain tensor $\dot{\boldsymbol{E}}$ is the negative of the stretching tensor for the motion of space relative to the body. This is an example of the principle of duality mentioned in Sect. 65. As a corollary, $\underset{\sim}{X}_{(\alpha,\beta)} = 0$ is necessary and sufficient for rigid motion.

[3] HENCKY [1929, *3*, § 2], MURNAGHAN [1937, *7*, p. 243]. BONVICINI [1935, *2*] calculated $\dot{\overline{c^{-1}}}$ and resolved it with respect to the principal axes of $\boldsymbol{c}$.

spatial description: In a rigid motion of a strained material, the quantity $ds^2 - dS^2$, and hence also the scalar $e_{km} dx^k dx^m$, remains constant for each particle, but the individual components e_{km} generally change because to an observer moving with a particle the spatial co-ordinate frame appears to rotate as places move by.

These same results, along with their counterparts for the rotation[1], may be obtained by differentiating materially the fundamental decomposition $(37.9)_2$ By (72.10) and (76.1) we thus obtain

$$\left.\begin{aligned} \dot{x}_{k,m} &= g_{kr} g^p_\alpha X^\alpha_{,m} \overline{\overset{-\frac{1}{2}}{c}{}^r_q R^q{}_p}^{\,\boldsymbol{\cdot}}, \\ &= g^p_\alpha X^\alpha_{,m} \left(\dot{\overline{\overset{-\frac{1}{2}}{c}}}_{kq} R^q{}_p + \overset{-\frac{1}{2}}{c}_{kq} \dot{R}^q{}_p \right). \end{aligned}\right\} \qquad (95.13)$$

If we evaluate this formula at $t = 0$, we get

$$\dot{x}_{k,m} = \dot{\overline{\overset{-\frac{1}{2}}{c}}}_{km}\Big|_{t=0} + \dot{R}_{km}\Big|_{t=0}. \qquad (95.14)$$

Now $\boldsymbol{c}^{-\frac{1}{2}}$ is symmetric and $\boldsymbol{R}$ is orthogonal; by differentiating (B) of footnote 3, p. 841, we see that $\dot{R}_{km} = -\dot{R}_{mk}$ when $t = 0$. Since the decomposition of a matrix into symmetric and skew-symmetric parts is unique, by comparison with (90.1) we conclude that

$$d_{km} = \dot{\overline{\overset{-\frac{1}{2}}{c}}}_{km}\Big|_{t=0}, \qquad w_{km} = \dot{R}_{km}\Big|_{t=0}. \qquad (95.15)$$

Since $\dot{\overset{n}{\boldsymbol{c}}} = n \dot{\boldsymbol{c}}$ at $t = 0$, for any power n, we see that $(95.15)_1$ is but another form of $(95.11)_1$. The content of $(95.15)_2$ is essentially that of $(57.3)_1$.

From these results it is plain that there is in general no simple connection between vorticity and rate of change of finite rotation. For example, if we set

$$Q_{\alpha\beta} \equiv w_{km} x^k_{,\alpha} x^m_{,\beta}, \qquad (95.16)$$

there is no analogue of (95.7) connecting $\dot{\boldsymbol{R}}$ with $\boldsymbol{Q}$, except at $t = 0$. From (37.13), it is easy to calculate a formula for $\dot{R}_{\alpha\beta}$, but the result is a complicated expression involving the components of $\boldsymbol{w}$, $\boldsymbol{d}$, and $\boldsymbol{R}$ as well as the deformation gradients.

By (95.16) and $(95.7)_1$ we have

$$\dot{E}_{\alpha\beta} + Q_{\alpha\beta} = \dot{x}_{k,m} x^k_{,\alpha} x^m_{,\beta} = \dot{x}_{k,\beta} \dot{x}^k_{,\alpha}. \qquad (95.17)$$

Differentiating this equation materially, we obtain[2]

$$\left.\begin{aligned} \ddot{E}_{\alpha\beta} &= -\dot{Q}_{\alpha\beta} + \ddot{x}_{k,\beta} x^k_{,\alpha} + \dot{x}_{k,\alpha} \dot{x}^k_{,\beta}, \\ &= -\dot{Q}_{\alpha\beta} + \ddot{x}_{k,m} x^k_{,\alpha} x^m_{,\beta} + (\dot{E}_{\alpha\gamma} + Q_{\alpha\gamma})(\dot{E}_{\beta\delta} + Q_{\beta\delta}) X^\gamma_{,k} X^{\delta,k}, \\ &= -\dot{Q}_{\alpha\beta} + \ddot{x}_{k,m} x^k_{,\alpha} x^m_{,\beta} + (\dot{E}_{\alpha\gamma} + Q_{\alpha\gamma})(\dot{E}_{\beta\delta} + Q_{\beta\delta}) \overset{-1}{C}{}^{\gamma\delta}, \end{aligned}\right\} \qquad (95.18)$$

where we have used (95.17) and $(29.1)_2$.

Since the conditions of compatibility (Sect. 34) may be expressed in the form $R^{(\mathbf{c})}_{kmnp} = 0$, it would be valuable to calculate the explicit form of $\dot{R}^{(\mathbf{c})}_{kmnp}$. This seems to be an elaborate matter. However, the work is simplified when $R^{(\mathbf{c})}_{kmnp} = 0$; the result gives a condition that an instantaneously possible strain shall remain compatible. To calculate it, we choose co-ordinates such that at the instant $t = 0$ we have $c_{km} = \delta_{km}$. Then we shall have $c_{km} = \delta_{km} +$

[1] NOLL [1955, *18*, § 2d]. $(95.15)_2$ is to some extent foreshadowed by PIOLA [1836, *1*, Eq. (188)], but his results are not correct in general.

[2] This result, in a convected co-ordinate system, is obtained by ZELMANOV [1948, *39*, Eq. (5)].

$t\dot{c}_{km}(0)+o(t)$. For these co-ordinates, by the same argument as given in Sect. 57 we get

$$\dot{R}^{(\mathbf{c})}_{kmpq}=\delta^{rs}_{km}\,\delta^{uv}_{pq}\,\frac{\partial^2\dot{c}_{ru}}{\partial x^s\,\partial x^v}. \tag{95.19}$$

To put this result in invariant form, we replace partial derivatives by covariant derivatives based on $\mathbf{c}$. Denoting such derivatives by a double bar, we thus obtain[1]

$$\dot{R}^{(\mathbf{c})}_{kmpq}=\delta^{rs}_{km}\,\delta^{uv}_{pq}\,\dot{c}_{ru\|sv}\quad\text{when}\quad R^{(\mathbf{c})}_{kmpq}=0. \tag{95.20}$$

96. Rates with respect to a varying configuration of reference. Setting aside the restrictions laid down in Sect. 66, as in Sect. 14 we now consider two arbitrary configurations X^K and x^k. To describe the motion of the medium, we retain the equations

$$x^k=x^k(X^\alpha,t), \tag{96.1}$$

where X^α are material co-ordinates, but to this we adjoin a second motion

$$X^K=X^K(X^\alpha,t). \tag{96.2}$$

That is, to set X^K and x^k into correspondence at time t, we employ the particles X^α as parameters. The second mapping (96.2) gives a *reference configuration*[2] for the particles at time t. In the special case when (96.2) reduces to identity, we regain the description used hitherto in this subchapter. In the general case, the relation obtained by eliminating the X^α between (96.1) and (96.2) gives the motion of the material with respect to a *varying reference configuration*. The particles need never acturally occupy the reference configuration.

The field $\dot{\boldsymbol{X}}$ defined by

$$\dot{X}^K\equiv\left.\frac{\partial X^K}{\partial t}\right|_{X^\alpha=\text{const}}, \tag{96.3}$$

the velocity of the reference configuration with respect to the particles, may be called the *infixity* of the reference configuration. In the earlier parts of this subchapter, the infixity was zero. For the material derivative of a double field $\boldsymbol{A}(\boldsymbol{x},\boldsymbol{X},t)$, we set

$$\dot{A}^{\cdots}_{\cdots}\equiv\frac{\partial A^{\cdots}_{\cdots}}{\partial t}+A^{\cdots}_{\cdots,k}\,\dot{x}^k+A^{\cdots}_{\cdots,K}\,\dot{X}^K, \tag{96.4}$$

where $\partial/\partial t$ is executed with both the x^k and the X^K constant, and where the other derivatives indicated are covariant partial derivatives. Comparison with $(72.2)_3$ shows that when the infixity vanishes, (96.4) reduces to the material derivative used previously in this subchapter.

We easily see from (96.3) and (96.4) that

$$\overline{dX^K}^{\,\boldsymbol{\cdot}}=\dot{X}^K_{,\alpha}\,dX^\alpha=\dot{X}^K_{,M}\,dX^M. \tag{96.5}$$

[cf. (76.4)]. Hence if we assign to the reference configuration a metric tensor G_{KM}, which need not be Euclidean, by analysis very like that leading to (82.3) we obtain

$$\overline{dS^2}^{\,\boldsymbol{\cdot}}=2\,S^*_{KM}\,dX^K\,dX^M=2\,s_{km}\,dx^k\,dx^m, \tag{96.6}$$

where the *slippage tensors*[3] $\boldsymbol{S}^*$ and $\boldsymbol{s}$ are given by

$$S^*_{KM}=\frac{1}{2}\,\frac{\partial G_{KM}}{\partial t}+\dot{X}_{(K,M)},\qquad s_{km}=S^*_{KM}\,X^K_{,k}\,X^M_{,m}. \tag{96.7}$$

[1] This result seems to be included in one of Zelmanov [1948, *39*, Eq. (10)]. Cf. Sect. 150.

[2] The first to use such a reference configuration was Piola [1836, *1*, § III, ¶ 32] [1848, *2*, ¶ 42], but in his work it seems to provide only a detour.

[3] The analysis here generalizes that of Truesdell [1952, *21*, § 22], which was based on work of Eckart [1948, *10*, § 3].

The slippage tensors are thus measures of the stretching of the reference configuration relative to the material. From (96.7) it is plain that the components of the slippage tensors may not be assigned arbitrarily, being subject to a condition of integrability dependent upon the curvature tensor of G_{KM}.

By differentiating $(26.1)_2$ we obtain from $(96.6)_2$

$$2 s_{km} = \dot{c}_{km} + 2 c_{n(k} \dot{x}^n_{,m)}; \tag{96.8}$$

this is ECKART's[1] generalization of (95.9). If we multiply (96.8) by $\overset{-1}{c}{}^{km}$, we obtain

$$2 \overset{-1}{c}{}^{km} s_{km} = \overset{-1}{c}{}^{km} \dot{c}_{km} + 2 \dot{x}^k_{,k}. \tag{96.9}$$

By (76.7) and $(30.6)_2$ follows

$$\mathrm{I}_{\boldsymbol{d}} = \overset{-1}{c}{}^{km} s_{km} - \overline{\log \sqrt{\mathrm{III}_{\boldsymbol{c}}}}^{\,\bullet} = \overset{-1}{c}{}^{km} s_{km} + \overline{\log \frac{dv}{dV}}^{\,\bullet}. \tag{96.10}$$

The special case when $\boldsymbol{s} = 0$ yields a new proof of EULER's expansion formula $(76.9)_2$. Using this special case to eliminate dv, we finally obtain

$$\overset{-1}{c}{}^{km} s_{km} = \overline{\log dV}^{\,\bullet}. \tag{96.11}$$

When the slippage vanishes, this result reduces to $0 = 0$.

A material counterpart of (96.8) may be obtained by the principle of duality (Sect. 14). If we compare (96.6) with (82.3), we see that the dual of $\boldsymbol{s}$ is not $\boldsymbol{S}^*$ but rather the tensor $\boldsymbol{S}$ given by

$$S_{KM} = d_{km} x^k_{,K} x^m_{,M}. \tag{96.12}$$

Hence the dual of (96.8) is[2]

$$2 d_{km} x^k_{,K} x^m_{,M} = \dot{C}_{KM} + 2 C_{N(K} \dot{X}^N_{,M)}. \tag{96.13}$$

This generalizes $(95.7)_1$.

Comparison of (96.8) with (96.13) shows that $\dot{\boldsymbol{c}}$ depends on the slippage, the actual strain, and the stretching and spin of the medium, while $\dot{\boldsymbol{C}}$ depends on the slippage, the actual strain and rotation, and the rotation of the reference configuration relative to the material.

For a simple example, consider the motion

$$x = X e^{-t}, \quad y = Y e^{\frac{1}{2}t}, \quad z = Z e^{\frac{1}{2}t}, \tag{96.14}$$

representing a slab of material pressed isochorically and homogeneously between parallel plates $X = \text{const}$. Agree to measure strain not with respect to the initial configuration X, Y, Z but rather with respect to an intermediate state, say, half way between the initial and actual states. Writing A, B, C for the co-ordinates X^K, set

$$\left.\begin{aligned} A &\equiv X - \tfrac{1}{2}(X - x) = \tfrac{1}{2}(X + x) = \tfrac{1}{2} X (e^{-t} + 1) = \tfrac{1}{2} x (1 + e^{t}), \\ B &\equiv Y + \tfrac{1}{2}(y - Y) = \tfrac{1}{2}(Y + y) = \tfrac{1}{2} Y (e^{\frac{1}{2}t} + 1) = \tfrac{1}{2} y (1 + e^{-\frac{1}{2}t}), \\ C &\equiv Z + \tfrac{1}{2}(z - Z) = \tfrac{1}{2}(Z + z) = \tfrac{1}{2} Z (e^{\frac{1}{2}t} + 1) = \tfrac{1}{2} z (1 + e^{-\frac{1}{2}t}). \end{aligned}\right\} \tag{96.15}$$

For the element of squared arc dS^2 at A, B, C we have

$$\left.\begin{aligned} 4\, dS^2 &= 4\,(dA^2 + dB^2 + dC^2), \\ &= (e^{-t} + 1)^2 dX^2 + (e^{\frac{1}{2}t} + 1)^2 (dY^2 + dZ^2). \end{aligned}\right\} \tag{96.16}$$

[1] [1948, *10*, § 3].
[2] The result was derived in a more lengthy way by TRUESDELL [1952, *21*, § 22].

Hence

$$\left.\begin{aligned}\overline{4\,dS^2}^{\boldsymbol{\cdot}} &= -2e^{-t}(e^{-t}+1)\,dX^2 + e^{\frac{1}{2}t}(e^{\frac{1}{2}t}+1)\,(dY^2+dZ^2),\\ &= -2(1+e^t)\,dx^2 + (1+e^{-\frac{1}{2}t})\,(dy^2+dz^2).\end{aligned}\right\} \tag{96.17}$$

Therefore the slippage tensor $\boldsymbol{s}$ has the components

$$8\,\boldsymbol{s} = \left\| \begin{matrix} -2\,(1+e^t) & 0 & 0\\ \cdot & 1+e^{-\frac{1}{2}t} & 0\\ \cdot & \cdot & 1+e^{-\frac{1}{2}t}\end{matrix} \right\| = \left\| \begin{matrix} -2\left(1+\frac{X}{x}\right) & 0 & 0\\ \cdot & 1+\frac{Y}{y} & 0\\ \cdot & \cdot & 1+\frac{Z}{z}\end{matrix} \right\| \tag{96.18}$$

A different choice of state with respect to which strain is to be measured would of course have led to a different slippage tensor.

97. Variational formulae. We consider a family of deformations $\boldsymbol{x}(\boldsymbol{X}, t, Q)$, where Q is a parameter, and we suppose this family so adjusted that $\boldsymbol{x}(\boldsymbol{X}, t, 0)$ is a given motion. The *first variation* of the motion is defined by

$$\delta \boldsymbol{x} \equiv \left.\frac{\partial \boldsymbol{x}}{\partial Q}\right|_{Q=0,\ \boldsymbol{X}=\text{const},\ t=\text{const}}. \tag{97.1}$$

If $f = f(x^k, x^k_{;\alpha}, x^k_{;\alpha\beta}, \ldots, t)$, we define the first variation of f by

$$\delta f \equiv \left.\frac{\partial f}{\partial Q}\right|_{Q=0,\ \boldsymbol{X}=\text{const},\ t=\text{const}}. \tag{97.2}$$

Thus the symbol of variation, δ, commutes with differentiations[1] with respect to the material variables X^α and t:

$$\delta \dot{x}^k = \dot{\overline{\delta x^k}}, \qquad \delta \frac{\partial f}{\partial X^\alpha} = \frac{\partial\, \delta f}{\partial X^\alpha}, \qquad \text{etc.} \tag{97.3}$$

Also, the metric tensor is not varied. Thus, for example,

$$\ddot{x}_k\, \delta x^k = \dot{\overline{\dot{x}_k\, \delta x^k}} - \delta\left(\tfrac{1}{2}\dot{x}^2\right). \tag{97.4}$$

Variations, as is clear from their definition, are no more than derivatives with respect to a parameter. Many of the formulae already derived can be put into variational form by inspection. The only point of difference is that variations are always executed at a fixed time.

d) Acceleration.

98. The acceleration field. The *acceleration* $\ddot{\boldsymbol{x}}$ is the rate of change of velocity as apparent to an observer situate upon a moving particle of the medium:

$$\left.\begin{aligned}\ddot{x}^k &\equiv \frac{D\dot{x}^k}{Dt} + \left\{ {k \atop pm} \right\} \dot{x}^p \dot{x}^m,\\ &= \frac{\partial \dot{x}^k}{\partial t} + \dot{x}^k_{\ ,m}\, \dot{x}^m,\end{aligned}\right\} \tag{98.1}$$

where the former expression, which follows from (72.3), is appropriate to the material description, the latter, which follows from (72.4), to the spatial[2]. In the spatial form, the first term is the *local* acceleration, since it represents the

[1] When the co-ordinate system is varied along with the displacement, these results no longer hold. Cf. footnote 1, p. 604.

[2] Special cases of $(98.1)_2$ are due to D'ALEMBERT [1752, *1*, §§ 43–44, 73, 86], the general expression to EULER [1761, *2*, §§ 40–41, 56] [1757, *2*, § 19].

change in the velocity field apparent to an observer fixed in the spatial co-ordinate system; in a steady motion, it vanishes. The second term is the *convective acceleration*, for it represents the acceleration which in a steady motion coinciding with $\dot{\boldsymbol{x}}$ at time t is necessary so as to direct the particles along their appointed paths at the required speed.

Since the basic idea of classical mechanics is that *a priori* knowledge regarding the acceleration in certain co-ordinate systems is available, detailed study of the acceleration field is useful. The acceleration, like all other concepts in this chapter, is purely kinematical, and the analysis we give is not restricted to any particular type of material or co-ordinate frame.

99. Various formulae for the acceleration. The form $(98.1)_1$ is not different from that appropriate to the kinematics of a single moving point. Therefore all formulae for the acceleration derived in punctual kinematics are valid as well in the theory of the acceleration field of a medium. We recapitulate some of these.

First, let $\boldsymbol{t}$ be the unit tangent to the path line, and let $\boldsymbol{n}_1$ and $\boldsymbol{n}_2$ be any pair of unit vectors normal to $\boldsymbol{t}$ and to each other, ordered so that $\boldsymbol{t}, \boldsymbol{n}_1, \boldsymbol{n}_2$ form a right-handed triad. Then if d/ds denotes the directional derivative along the path of a particle, we have

$$\left.\begin{aligned} \frac{d\boldsymbol{t}}{ds} &= \varkappa_1 \boldsymbol{n}_1 + \varkappa_2 \boldsymbol{n}_2, \\ \frac{d\boldsymbol{n}_1}{ds} &= -\varkappa_1 \boldsymbol{t} + \tau \boldsymbol{n}_2, \qquad \frac{d\boldsymbol{n}_2}{ds} = -\varkappa_2 \boldsymbol{t} - \tau \boldsymbol{n}_1, \end{aligned}\right\} \tag{99.1}$$

where the coefficients, which are defined by these resolutions, are the curvature-torsion components, and the curvature $\varkappa$ satisfies $\varkappa^2 = \varkappa_1^2 + \varkappa_2^2$. (Cf. the more general theory of wryness presented in Sect. 61.) Hence

$$\left.\begin{aligned} \frac{d\dot{\boldsymbol{x}}}{ds} &= \frac{d}{ds}(\dot{x}\,\boldsymbol{t}) = \frac{d\dot{x}}{ds}\boldsymbol{t} + \dot{x}(\varkappa_1 \boldsymbol{n}_1 + \varkappa_2 \boldsymbol{n}_2), \\ \frac{d^2\dot{\boldsymbol{x}}}{ds^2} &= \left(\frac{d^2\dot{x}}{ds^2} - \dot{x}\varkappa^2\right)\boldsymbol{t} + \left(2\frac{d\dot{x}}{ds}\varkappa_1 + \dot{x}\frac{d\varkappa_1}{ds} - \dot{x}\varkappa_2\tau\right)\boldsymbol{n}_1 + \\ &\quad + \left(2\frac{d\dot{x}}{ds}\varkappa_2 + \dot{x}\frac{d\varkappa_2}{ds} + \dot{x}\varkappa_1\tau\right)\boldsymbol{n}_2, \end{aligned}\right\} \tag{99.2}$$

and[1]

$$\ddot{\boldsymbol{x}} = \dot{x}\frac{d\dot{\boldsymbol{x}}}{ds} = \frac{d}{ds}\left(\frac{1}{2}\dot{x}^2\right)\boldsymbol{t} + \dot{x}^2(\varkappa_1 \boldsymbol{n}_1 + \varkappa_2 \boldsymbol{n}_2). \tag{99.3}$$

Also[2]

$$\left.\begin{aligned} \varkappa_1 \dot{x} &= \boldsymbol{n}_1 \cdot \frac{d\dot{\boldsymbol{x}}}{ds} = \boldsymbol{n}_2 \cdot \boldsymbol{t} \times \frac{d\dot{\boldsymbol{x}}}{ds}, \\ \varkappa_2 \dot{x} &= \boldsymbol{n}_2 \cdot \frac{d\dot{\boldsymbol{x}}}{ds} = -\boldsymbol{n}_1 \cdot \boldsymbol{t} \times \frac{d\dot{\boldsymbol{x}}}{ds}, \\ \dot{x}^2\left(\varkappa^2\tau + \varkappa_1\frac{d\varkappa_2}{ds} - \varkappa_2\frac{d\varkappa_1}{ds}\right) &= \boldsymbol{t} \cdot \frac{d\dot{\boldsymbol{x}}}{ds} \times \frac{d^2\dot{\boldsymbol{x}}}{ds^2} = \frac{1}{\dot{x}}\boldsymbol{t} \cdot \frac{d\dot{\boldsymbol{x}}}{ds} \times \frac{d\ddot{\boldsymbol{x}}}{ds}. \end{aligned}\right\} \tag{99.4}$$

These formulae show how to determine $\varkappa_1$, $\varkappa_2$, and τ from $\dot{\boldsymbol{x}}$, $d\dot{\boldsymbol{x}}/ds$, and $d\ddot{\boldsymbol{x}}/ds$, once the vectors $\boldsymbol{n}_1$ and $\boldsymbol{n}_2$ are selected. If we choose $\boldsymbol{n}_1$ and $\boldsymbol{n}_2$ as the principal normal $\boldsymbol{n}$ and the binormal $\boldsymbol{b}$, then $\varkappa_2 = 0$, $\varkappa_1 = \varkappa$, and $\tau =$ the torsion. Among

[1] EULER [1736, *1*, §§ 802–809]. The organization of this material in the text follows TRUESDELL [1960, *5*, §§ 2–3].

[2] These formulae seem to be more or less well known. $(99.4)_5$ generalizes a result of DUNCAN [1957, *3*, Eq. (17)].

the simplifications that accrue in the above results are[1]

$$\left.\begin{aligned} \dot{x}^2\varkappa\boldsymbol{b} &= \dot{\boldsymbol{x}}\times\frac{d\dot{\boldsymbol{x}}}{ds} = \boldsymbol{t}\times\ddot{\boldsymbol{x}}, \qquad \varkappa\dot{x}^3 = \boldsymbol{b}\cdot\dot{\boldsymbol{x}}\times\ddot{\boldsymbol{x}},\\ \ddot{\boldsymbol{x}} &= \frac{d}{ds}\left(\frac{1}{2}\dot{x}^2\right)\boldsymbol{t} + \dot{x}^2\varkappa\boldsymbol{n},\\ \dot{x}^2\varkappa\tau &= -\boldsymbol{n}\cdot\dot{\boldsymbol{x}}\times\frac{d^2\dot{\boldsymbol{x}}}{ds^2} = -\boldsymbol{n}\cdot\boldsymbol{t}\times\frac{d\ddot{\boldsymbol{x}}}{ds}. \end{aligned}\right\} \tag{99.5}$$

These formulae show that the directions of the velocity and acceleration determine that of the binormal, and that the normal component of the acceleration is along the principal normal.

The foregoing results are immediately applicable in the kinematics of a continuum, but care must be taken in identifying the derivatives along the path with partial derivatives of the kinematic fields. We have

$$\frac{d}{dt} = \dot{x}\frac{d}{ds} = \frac{\partial}{\partial t} + \dot{x}\frac{\partial}{\partial s}, \tag{99.6}$$

where $\partial/\partial t$ and $\partial/\partial s$ are spatial derivatives of the usual kind. Hence $(99.5)_4$ becomes[2]

$$\ddot{\boldsymbol{x}} = \left[\frac{\partial\dot{x}}{\partial t} + \frac{\partial}{\partial s}\left(\frac{1}{2}\dot{x}^2\right)\right]\boldsymbol{t} + \dot{x}^2\varkappa\boldsymbol{b}. \tag{99.7}$$

If we wish a resolution along the principal triad $\boldsymbol{t}$, $\boldsymbol{n}_s$, $\boldsymbol{b}_s$ of the stream lines rather than of the path lines, we have in place of $(99.1)_1$

$$\frac{\partial\boldsymbol{t}}{\partial s} = \varkappa_s\boldsymbol{n}_s, \tag{99.8}$$

and in place of (99.7) we have[3]

$$\left.\begin{aligned} \ddot{\boldsymbol{x}} &= \frac{d\dot{\boldsymbol{x}}}{dt} = \frac{\partial\dot{\boldsymbol{x}}}{\partial t} + \dot{x}\frac{\partial}{\partial s}(\dot{x}\boldsymbol{t}),\\ &= \left[\frac{\partial\dot{x}}{\partial t} + \frac{\partial}{\partial s}\left(\frac{1}{2}\dot{x}^2\right)\right]\boldsymbol{t} + \left[\frac{\partial\dot{x}_{n_s}}{\partial t} + \dot{x}^2\varkappa_s\right]\boldsymbol{n}_s + \frac{\partial\dot{x}_{b_s}}{\partial t}\boldsymbol{b}_s, \end{aligned}\right\} \tag{99.9}$$

where $\partial\dot{x}_{n_s}/\partial t$ is the time derivative of the length of the projection of $\dot{\boldsymbol{x}}(\boldsymbol{x}, t)$ onto the fixed direction instantaneously coinciding with $\boldsymbol{n}_s$, etc. It is plain that (99.9) is too complicated to be useful except in the case of a motion with steady stream-lines, in which case $\boldsymbol{n}_s = \boldsymbol{n}$, $\boldsymbol{b}_s = \boldsymbol{b}$, $\varkappa_s = \varkappa$, and (99.9) reduces to the form (99.7)

Explicit formulae for $\ddot{\boldsymbol{x}}$ in curvilinear co-ordinates are immediate from (98.1) and the explicit form of the Christoffel symbols[4]. For example, in a cylindrical system r, θ, z we have the contravariant components

$$\ddot{r} = \frac{D\dot{r}}{Dt} - r\dot{\theta}^2, \qquad \ddot{\theta} = \frac{D\dot{\theta}}{Dt} + \frac{2\dot{r}\dot{\theta}}{r}, \qquad \ddot{z} = \frac{D\dot{z}}{Dt}. \tag{99.10}$$

Since $(99.10)_2$ is equivalent to $r^2\ddot{\theta} = D(r^2\dot{\theta})/Dt$, or, in physical components, $r\ddot{x}\langle\theta\rangle = D(r\dot{x}\langle\theta\rangle)/Dt$, we conclude that *in order for the azimuthal acceleration to vanish, it is necessary*

[1] $(99.5)_4$ is classical. The remaining Eqs. (99.5) were given by TRUESDELL [1958, *11*, § 2]; of course all of them are immediate consequences of the Serret-Frenet formulae.

[2] Derivations restricted to the case of steady motion and based upon $(98.1)_2$ were constructed by TOUCHE [1893, *7*] and COBURN [1952, *2*].

[3] v. MISES [1909, *8*, § 1.4]. Cf. also CAUCHY [1823, *2*].

[4] GILBERT [1890, *2*] expressed the components of $\ddot{\boldsymbol{x}}$ in terms of the radii of curvature of triply orthogonal surfaces.

and sufficient that the projection of the radius vector upon a plane normal to the axis sweep out equal areas in equal times[1].

Set

$$2k \equiv \dot{x}^2 = g_{km} \dot{x}^k \dot{x}^m. \tag{99.11}$$

The importance of the scalar k, the *specific kinetic energy*, will appear later. From a well known formula[2] for the acceleration of a single moving point we obtain

$$\ddot{x}^m = g^{mn} \left[\frac{D}{Dt} \frac{\partial k}{\partial \dot{x}^n} - \frac{\partial k}{\partial x^n} \right], \tag{99.12}$$

where D/Dt is the partial derivative with $\boldsymbol{X}$ and $\boldsymbol{g}$ held constant. Moreover, if we introduce moving co-ordinates $\boldsymbol{x}^*$ by the transformation $\boldsymbol{x} = \boldsymbol{x}(\boldsymbol{x}^*, t)$, so that in the $\boldsymbol{x}^*$ system $2k$ is a quadratic but not necessarily homogeneous form, then (99.12) with $\boldsymbol{x}$ replaced by $\boldsymbol{x}^*$ gives the components of acceleration in a *stationary* system whose co-ordinate curves coincide at time t with those of the moving $\boldsymbol{x}^*$ system.

Now we turn to forms of the acceleration which depend essentially on there being a field of motion, not just a single particle in motion.

The most important is especially simple. From the D'Alembert-Euler formula $(98.1)_2$ we get

$$\left.\begin{aligned} \ddot{x}_k &= \frac{\partial \dot{x}_k}{\partial t} + (\dot{x}_{k,m} - \dot{x}_{m,k}) \dot{x}^m + \dot{x}_{m,k} \dot{x}^m, \\ &= \frac{\partial \dot{x}_k}{\partial t} + 2 w_{km} \dot{x}^m + \left(\frac{1}{2} \dot{x}^2\right)_{,k}, \end{aligned}\right\} \tag{99.13}$$

a result due to LAGRANGE[3]. The modification of $(98.1)_2$ may seem trivial, but it is not. The D'Alembert-Euler formula distinguishes the local acceleration from the convective, but it does not distinguish the roles of stretching and spin. The Lagrange formula singles out a part of the acceleration, namely $2 w_{km} \dot{x}^m$, associated with the spin alone. Of course both the formulae under discussion are valid in spaces of any number of dimensions. In the case of a three-dimensional space, however, LAGRANGE's formula has a particular advantage resting on the fact that the spin may be represented by the vorticity *vector*, and its effect on the acceleration appears through the cross product of vorticity and velocity, so that $(99.13)_2$ becomes

$$\ddot{\boldsymbol{p}} = \frac{\partial \dot{\boldsymbol{p}}}{\partial t} + \boldsymbol{w} \times \dot{\boldsymbol{p}} + \operatorname{grad} \frac{1}{2} \dot{p}^2. \tag{99.14}$$

Thus while the D'Alembert-Euler formula $(98.1)_2$ expresses the acceleration in terms of two vectors and one tensor of the second order, LAGRANGE's formula (99.14) shows that *in a space of three dimensions* the acceleration is expressed in terms of *four vectors*: the local acceleration $\partial \dot{\boldsymbol{p}}/\partial t$, the vorticity vector $\boldsymbol{w}$, the velocity vector $\dot{\boldsymbol{p}}$, and the gradient of the specific kinetic energy. Thus every property of the acceleration in a space of three dimensions may be stated and visualized in terms of lines, tubes, and the other apparatus associated with vector fields[4]. Most of the special properties of the vorticity of fluid motions

---

[1] Cf. SVANBERG [1841, *5*, § 3].

[2] E.g. WHITTAKER [1904, *8*, § 28]. The essential idea involved may be traced back to LAGRANGE.

[3] [1783, *1*, § 14] [1788, *1*, Part II, ¶ 12]. A special case had been implied by D'ALEMBERT [1761, *1*, § XI], the general case by EULER [1757, *3*, § 14].

[4] It appears possible to fail to appreciate this fact. Cf. McVITTIE [1955, *17*].

are closely connected with this *characterizing property of three-dimensional space.* Some of these properties will be presented in Part e of this Subchapter. For reasons to appear there, the vector $\boldsymbol{w}\times\dot{\boldsymbol{p}}$ will be called the *Lamb vector.*

Since

$$(\dot{x}^k \dot{x}^m)_{,m} = \dot{x}^k_{,m} \dot{x}^m + \dot{x}^k \mathrm{I}_{\boldsymbol{d}}, \tag{99.15}$$

we have from $(98.1)_2$ a formula of BELTRAMI[1]:

$$\ddot{x}^k = \frac{\partial \dot{x}^k}{\partial t} - \mathrm{I}_{\boldsymbol{d}} \dot{x}^k + (\dot{x}^k \dot{x}^m)_{,m}. \tag{99.16}$$

This formula has the advantage of expressing all space derivatives of velocity only through divergences (cf. Sect. 157), and of bringing out the effect of the expansion. Pursuing the latter of these ideas, we notice that for any scalar a we have from (99.16)

$$a\ddot{x}^k = \frac{\partial}{\partial t}(a\dot{x}^k) + (a\dot{x}^k\dot{x}^m)_{,m} - (\dot{a} + a\,\mathrm{I}_{\boldsymbol{d}})\,\dot{x}^k. \tag{99.17}$$

In the case when $a = jB$, where $\dot{B}=0$, by the equation of continuity (76.11) we obtain an important result of GREENHILL[2]:

$$jB\ddot{x}^k = \frac{\partial}{\partial t}(jB\dot{x}^k) + (jB\dot{x}^k\dot{x}^m)_{,m}. \tag{99.18}$$

Finally, let $\dot{\boldsymbol{x}}$ be expressed in terms of its Monge potentials (App. 35.2):

$$\dot{x}_k = H_{,k} + F G_{,k}, \quad \text{or} \quad \dot{\boldsymbol{p}} = \operatorname{grad} H + F \operatorname{grad} G. \tag{99.19}$$

We note first that[3]

$$w_{km} = F_{,[m} G_{,k]}, \quad \boldsymbol{w} = \operatorname{grad} F \times \operatorname{grad} G \tag{99.20}$$

Hence

$$\dot{\boldsymbol{p}} \cdot \boldsymbol{w} = \dot{x}_k w^k = \frac{1}{\sqrt{\det g_{km}}} \cdot \frac{\partial(F, G, H)}{\partial(x^1, x^2, x^3)}. \tag{99.21}$$

Also

$$\left.\begin{aligned} \frac{\partial \dot{x}_k}{\partial t} &= \left(\frac{\partial H}{\partial t}\right)_{,k} + \frac{\partial F}{\partial t} G_{,k} + F\left(\frac{\partial G}{\partial t}\right)_{,k}, \\ &= \left(\frac{\partial H}{\partial t} + F\frac{\partial G}{\partial t}\right)_{,k} + \frac{\partial F}{\partial t} G_{,k} - \frac{\partial G}{\partial t} F_{,k}. \end{aligned}\right\} \tag{99.22}$$

From (99.20) and (99.22) follows

$$\frac{\partial \dot{x}_k}{\partial t} + 2 w_{km}\dot{x}^m = \left(\frac{\partial H}{\partial t} + F\frac{\partial G}{\partial t}\right)_{,k} + \dot{F} G_{,k} - \dot{G} F_{,k}. \tag{99.23}$$

From this result and $(99.13)_2$ we derive the formula of DUHEM[4]:

$$\ddot{x}_k = \left(\frac{1}{2}\dot{x}^2 + \frac{\partial H}{\partial t} + F\frac{\partial G}{\partial t}\right)_{,k} + \dot{F} G_{,k} - \dot{G} F_{,k}. \tag{99.24}$$

100. Measures of acceleration. From $(98.1)_2$ and (99.14) we may form dimensionless measures of the relative magnitudes of different portions of the acceleration.

---

[1] [1889, *1*, Eq. (5a)].

[2] While not actually stated, this formula follows at once from equations given in [1875, *2*, § 22]; the analysis there was reproduced by BASSET [1888, *1*, § 21].

[3] MORERA [1889, *6*].

[4] [1901, *4*, § 2]. Cf. also LAMB [1906, *4*, § 16].

First, the *dynamical vorticity number*[1] is given by

$$\mathfrak{W}_D \equiv \frac{|\boldsymbol{w} \times \dot{\boldsymbol{p}}|}{\left|\frac{\partial \dot{\boldsymbol{p}}}{\partial t} + \operatorname{grad} \frac{1}{2} \dot{p}^2\right|}; \tag{100.1}$$

it is measure of the relative importance of the Lamb vector with respect to the remainder of the acceleration. Except for motions such that $\partial \dot{\boldsymbol{p}}/\partial t + \operatorname{grad} \frac{1}{2} \dot{p}^2 = 0$, the condition $\mathfrak{W}_D = 0$ is *necessary and sufficient for irrotational motion*. In a steady rigid motion which is not a translation, $\mathfrak{W}_D = 2$. In a motion with zero acceleration, $\mathfrak{W}_D = 1$. In steady motions with uniform speed, we have $\mathfrak{W}_D = \infty$, provided the Lamb vector be not zero. For small values of $\mathfrak{W}_D$, we have

$$\ddot{\boldsymbol{p}} \approx \frac{\partial \dot{\boldsymbol{p}}}{\partial t} + \operatorname{grad} \frac{1}{2} \dot{p}^2 \tag{100.2}$$

with an error of magnitude $\mathfrak{W}_D |\partial \dot{\boldsymbol{p}}/\partial t + \operatorname{grad} \frac{1}{2} \dot{p}^2|$. Other properties of $\mathfrak{W}_D$ have been developed by TRUESDELL.

SZEBEHELY'S *number*[2], a measure of unsteadiness, is defined by

$$\mathfrak{S} \equiv \frac{\left|\frac{\partial \dot{\boldsymbol{p}}}{\partial t}\right|}{|\dot{\boldsymbol{p}} \cdot \operatorname{grad} \dot{\boldsymbol{p}}|}. \tag{100.3}$$

It measures the relative importance of the convective and local accelerations. In a steady accelerationless motion, it is indeterminate. We have $\mathfrak{S} = \infty$ if and only if the motion is unsteady and the convective acceleration is zero; $\mathfrak{S} = 0$ if and only if the motion is steady and the convective acceleration not zero. For an unsteady accelerationless motion, $\mathfrak{S} = 1$. For small values of $\mathfrak{S}$, we have

$$\ddot{x}^k \approx \dot{x}^k{}_{,m} \dot{x}^m, \quad \text{or } \ddot{\boldsymbol{p}} \approx \dot{\boldsymbol{p}} \cdot \operatorname{grad} \boldsymbol{p}, \tag{100.4}$$

with an error of magnitude $\mathfrak{S} |\dot{\boldsymbol{p}} \cdot \operatorname{grad} \dot{\boldsymbol{p}}|$, while for large values of $\mathfrak{S}$ we have

$$\ddot{x}_k \approx \frac{\partial \dot{x}^k}{\partial t}, \quad \text{or } \ddot{\boldsymbol{p}} \approx \frac{\partial \dot{\boldsymbol{p}}}{\partial t}, \tag{100.5}$$

with an error of magnitude $|\partial \dot{\boldsymbol{p}}/\partial t|/\mathfrak{S}$. Motions in which (100.5) is regarded as correct are often said to be *slow*. We note that in an unsteady slow motion we have $\mathfrak{W}_D \approx 0$; in an isochoric slow motion, $\mathfrak{W}_K \approx 1$ (cf. Sect. 91). Further properties of $\mathfrak{S}$ have been developed by SZEBEHELY.

To illustrate the use of the measures $\mathfrak{W}_K, \mathfrak{W}_D$, and $\mathfrak{S}$, consider the following special case of the rectilinear vortex (89.8):

$$\dot{r} = 0, \quad \dot{\theta} = \frac{K}{2\pi r^2}\left(1 - e^{-\frac{r^2}{4\nu t}}\right), \quad \dot{z} = 0, \tag{100.6}$$

where ν is a positive constant. The interest in this particular case lies in its being a dynamically possible motion for a viscous incompressible fluid of kinematic viscosity ν. As $t \to 0^+$, the velocity field (100.6) approaches that of an irrotational vortex of strength K, so that (100.6) represents the decay of such a vortex due to the action of viscosity. Put $\lambda \equiv r^2/(4\nu t)$.

---

[1] TRUESDELL [1953, *31*, § 13].

[2] SZEBEHELY [1952, *19*] [1953, *30*] mentions that (100.3) was suggested by analogy to (100.1), at that time unpublished.

Then from (91.1), (100.1), and (100.3) we calculate

$$\mathfrak{W}_K = \frac{1}{\frac{1}{\lambda}(e^\lambda - 1) - 1}, \tag{100.7}$$

$$\mathfrak{W}_D = \left[\frac{16\pi^2 \nu^2}{K^2} \cdot \frac{\lambda^2 e^{2\lambda}}{(e^\lambda - 1)^2} + \left(1 - \frac{e^\lambda - 1}{2\lambda}\right)^2\right]^{-\frac{1}{2}}, \tag{100.8}$$

$$\mathfrak{S} = \frac{8\pi\nu}{K} \frac{\lambda^2 e^{-\lambda}}{(1 - e^{-\lambda})^2}. \tag{100.9}$$

To interpret these results we let λ decrease from ∞ to 0, representing the effects either of the increase of time at a fixed place or the decrease of r at a fixed time. As would be expected, at $\lambda = \infty$ all three measures vanish. As λ decreases, both $\mathfrak{W}_K$ and $\mathfrak{S}$ increase steadily, until at $\lambda = 0$ we have $\mathfrak{W}_K = \infty$ and $\mathfrak{S} = 8\pi\nu/K$. *At a fixed place, then, the motion becomes more and more rotational and more and more unsteady as time goes on.* The measure $\mathfrak{W}_D$ vanishes at $\lambda = \infty$; at $\lambda = 0$ it approaches the value $\left(\frac{16\pi^2\nu^2}{K^2} + \frac{9}{4}\right)^{-\frac{1}{2}}$.

101. Convection and diffusion of stretching and spin. We show now that the material rates of change of stretching and spin are expressible in terms of the gradient of the acceleration. While it is easy to derive the formulae by calculating the material derivatives of $\boldsymbol{d}$ and $\boldsymbol{w}$, we prefer to follow CARSTOIU[1] in beginning with expressions for finite increments along the path of a particle.

Since

$$\dot{\overline{\dot{x}_{k,\beta}\, x^k_{,\alpha}}} = \ddot{x}_{k,\beta}\, x^k_{,\alpha} + \dot{x}_{k,\beta}\, \dot{x}^k_{,\alpha}, \tag{101.1}$$

we have

$$\dot{\overline{\dot{x}_{k,m}\, x^k_{,\alpha}\, x^m_{,\beta}}} = (\ddot{x}_{k,m} + \dot{x}_{p,m}\, \dot{x}^p_{,k})\, x^k_{,\alpha}\, x^m_{,\beta}. \tag{101.2}$$

Integrating along the path of a particle, we obtain the ***basic identity***:

$$\dot{x}_{k,m} = \left[\dot{x}_{\alpha,\beta} + \int_0^t (\ddot{x}_{n,p} + \dot{x}_{q,p}\, \dot{x}^q_{,n})\, x^n_{,\alpha}\, x^p_{,\beta}\, dt\right] X^\alpha_{,k}\, X^\beta_{,m}, \tag{101.3}$$

where $\dot{x}_{\alpha,\beta}$ stands for the initial value of $\dot{x}_{k,m}$ in accord with the convention of Sect. 66.

The first term on the right, $\dot{x}_{\alpha,\beta}\, X^\alpha_{,k}\, X^\beta_{,m}$, expresses the mere transport of the initial velocity gradient to the present location of the particle $\boldsymbol{X}$, while the remainder calculates the cumulative contributions made by the values of the acceleration, its gradients, the gradients of the speed, and the deformation gradients at each point on the path of the particle $\boldsymbol{X}$. We say that the mechanism of change associated with the former is *convection*; with the latter, *diffusion*[2].

The equation that follows by taking the material derivative of (101.3) is

$$\dot{\overline{\dot{x}_{k,m}}} = \ddot{x}_{k,m} - \dot{x}_{k,p}\, \dot{x}^p_{,m}. \tag{101.4}$$

It is easier, perhaps, to establish this result directly by differentiating $(98.1)_2$. In the terminology of the preceding paragraph, the first and second terms on the right of (101.4) are the *diffusive* and *convective* rates of change, respectively. Thus *the gradient of the acceleration field is the diffusive rate of change of the velocity gradient.*

By a somewhat elaborate calculation it is possible also to derive (101.3) as the integral of (101.4) along the path of a particle.

---

[1] [1954, *1*, § 1].

[2] The terms were defined in this manner by TRUESDELL [1948, *36*], who derived them from JAFFÉ [1921, *3*].

As suggested by (90.1), we now split $\ddot{x}_{k,m}$ into its symmetric and skew symmetric parts:

$$d^*_{km} \equiv \ddot{x}_{(k,m)}, \quad w^*_{km} \equiv \ddot{x}_{[k,m]}, \quad \ddot{x}_{k,m} = d^*_{km} + w^*_{km}. \tag{101.5}$$

The skew symmetric part of (101.4) is BELTRAMI's equation for the rate of change of spin[1]:

$$\dot{w}_{km} = w^*_{km} + 2\, d^p_{[k} w_{m]p}, \tag{101.6}$$

or, equivalently,

$$\left.\begin{aligned} \dot{w}^k &= w^{*k} + w^m \dot{x}^k_{,m} - w^k \mathrm{I}_{\boldsymbol{d}}, \\ \overline{J w^k}^{\bullet}/J &= w^{*k} + w^m \dot{x}^k_{,m}, \quad \overline{\boldsymbol{w}\, d\boldsymbol{v}}^{\bullet} = \boldsymbol{w}^* d\boldsymbol{v} + \boldsymbol{w}\, d\boldsymbol{v} \cdot \operatorname{grad} \dot{\boldsymbol{p}}. \end{aligned}\right\} \tag{101.7}$$

Hence the rate of change of vorticity magnitude is given by[2]

$$\left.\begin{aligned} \tfrac{1}{2}\, \overline{J^2 w^2}^{\bullet}/J^2 &= w^*_k w^k + d_{km} w^k w^m, \quad \text{or} \\ \tfrac{1}{2}\, \overline{w^2 (dv)^2}^{\bullet} &= (\boldsymbol{w}^* \cdot \boldsymbol{w} + \boldsymbol{w} \cdot \boldsymbol{d} \cdot \boldsymbol{w})\, (dv)^2. \end{aligned}\right\} \tag{101.8}$$

The general solution[3] of BELTRAMI's equation (101.6) is the skew symmetric part of (101.3):

$$w_{km} = \left[W_{\alpha\beta} + \int_0^t w^*_{np}\, x^n_{,\alpha}\, x^p_{,\beta}\, dt\right] X^\alpha_{,k} X^\beta_{,m}; \tag{101.9}$$

equivalently,

$$Q_{\alpha\beta} = W_{\alpha\beta} + \int_0^t w^*_{np}\, x^n_{,\alpha}\, x^p_{,\beta}\, dt, \tag{101.10}$$

where $\boldsymbol{Q}$ is defined by Eq. (95.16), and where we write $W_{\alpha\beta}$ rather than $w_{\alpha\beta}$ for the initial value of w_{km} [cf. the convention (66.8)]; another equivalent form is[4]

$$w^k = \left[W^\alpha + \int_0^t W^\alpha_* \, dt\right] x^k_{,\alpha}/J, \tag{101.11}$$

where

$$W^\alpha_* \equiv e^{\alpha\beta\gamma} (\ddot{x}_k\, x^k_{,\gamma})_{,\beta} = J\, X^\alpha_{,k}\, e^{kmp}\, \ddot{x}_{p,m} = J\, X^\alpha_{,k}\, w^{*k}. \tag{101.12}$$

The vectors $\boldsymbol{w}^*$ and $\boldsymbol{W}_*$ are called the *spatial diffusion vector* and the *material diffusion vector*, respectively.

Turning now to the symmetric part of (101.3), we obtain[5]

$$d_{km} = \left[D_{\alpha\beta} + \int_0^t (d^*_{np} + \dot{x}_{q,p}\, \dot{x}^q_{,n})\, x^n_{,\alpha}\, x^p_{,\beta}\, dt\right] X^\alpha_{,k} X^\beta_{,m}. \tag{101.13}$$

Comparison of (101.13) with (101.9) shows that the diffusion of stretching is a more complicated process than the diffusion of spin. The diffusion of stretching results not only from gradients of acceleration but also from the existing spin and stretching at each point on the path of the particle.

---

[1] [1871, *1*, § 6]. For earlier special cases, see Sect. 130.

[2] APPELL [1903, *1*]. Special cases were obtained by WARREN [1870, *8*] and v. KÁRMÁN [1937, *3*].

[3] TRUESDELL [1948, *36*]. Special cases are due to APPELL [1917, *3*, § 1] [1921, *1*, § 814], VILLAT [1930, *8*, pp. 10–12], BOGGIO [1935, *1*], CARSTOIU [1946, *2*], and TRUESDELL [1948, *37*]. For further discussion, cf. LICHTENSTEIN [1925, *10*, Chap. I, § 5] [1927, *5*, Chap. 1, § 2] [1929, *4*, Chap. 10, §§ 1–2]. For earlier work along this line, see Sect. 134. A derivation in material co-ordinates is given by TRUESDELL [1954, *24*, §§ 84–85].

[4] TRUESDELL [1954, *24*, § 84].

[5] CARSTOIU [1954, *1*, § 6]; the main step was given in [1953, *3*].

The differential equation whose general solution is (101.13) was given by APPELL[1]:

$$\dot{d}_{km} = d^*_{km} - \dot{x}^p_{,(k}\dot{x}_{m),p}, \tag{101.14}$$

as follows at once by taking the symmetric part of (101.4).

Various other properties of the acceleration gradient and its invariants have been worked out by APPELL and CARSTOIU[2]. The divergence of the acceleration, $\ddot{x}^k_{,k}$, offers particular interest and will be studied in the next section.

102. The divergence of the acceleration[3]. From (101.14), (App. 38.15), and $(86.14)_1$ follows[4]

$$\left.\begin{aligned}\operatorname{div}\ddot{\boldsymbol{p}} = \ddot{x}^k_{,k} = d^{*k}_{\ k} = \dot{\mathrm{I}}_{\boldsymbol{d}} + \dot{x}^k_{,m}\dot{x}^m_{,k} &= \dot{\mathrm{I}}_{\boldsymbol{d}} + \overline{\mathrm{II}}_{\boldsymbol{d}} - \tfrac{1}{2}w^2,\\ &= \dot{\mathrm{I}}_{\boldsymbol{d}} + \mathrm{I}^2_{\boldsymbol{d}} - 2\mathrm{II}_{\boldsymbol{d}} - \tfrac{1}{2}w^2 = \dot{\mathrm{I}}_{\boldsymbol{d}} + \mathrm{I}^2_{\boldsymbol{d}} - 2K.\end{aligned}\right\} \tag{102.1}$$

Since $\overline{\mathrm{II}}_{\boldsymbol{d}} \geqq 0$, as corollaries follow $\ddot{x}^k_{,k} < 0$ in a rigid motion which is not a translation; in an isochoric motion,

$$\ddot{x}^k_{,k} \geqq -\tfrac{1}{2}w^2, \tag{102.2}$$

where equality holds if and only if the motion is rigid; and in an isochoric irrotational motion

$$\ddot{x}^k_{,k} \geqq 0, \tag{102.3}$$

where again equality holds if and only if the motion is rigid.

These results may be put into a more general and illuminating form by noting that from $(102.1)_2$ and (91.1) we have

$$\mathfrak{W}_{\mathrm{K}} = \left[1 + 2\,\frac{\ddot{x}^k_{,k} - \dot{\mathrm{I}}_{\boldsymbol{d}}}{w^2}\right]^{-\frac{1}{2}}. \tag{102.4}$$

Hence $\mathfrak{W}_{\mathrm{K}} \lesseqgtr 1$ according as $\ddot{x}^k_{,k} \gtreqless \dot{\mathrm{I}}_{\boldsymbol{d}}$. In particular, $\mathfrak{W}_{\mathrm{K}} = 1$ in an isochoric rotational accelerationless motion.

Let us now write S^* and $\boldsymbol{v}^*$ for the scalar and vector potentials (App. 36.1) of the acceleration:

$$\ddot{x}_k = -S^*_{,k} + e_{kmq}v^{*q,m}, \quad \ddot{\boldsymbol{p}} = -\operatorname{grad} S^* + \operatorname{curl}\boldsymbol{v}^*. \tag{102.5}$$

Then

$$\operatorname{div}\ddot{\boldsymbol{p}} = \ddot{x}^k_{,k} = -\nabla^2 S^*, \tag{102.6}$$

and from (102.4) follows

$$\mathfrak{W}_{\mathrm{K}} = \left[1 - 2\,\frac{\nabla^2 S^* + \dot{\mathrm{I}}_{\boldsymbol{d}}}{w^2}\right]^{-\frac{1}{2}}, \tag{102.7}$$

a result which is remarkable in that it expresses $\mathfrak{W}_{\mathrm{K}}$ in terms of three very simple scalars. From it we see that $\mathfrak{W}_{\mathrm{K}} \lesseqgtr 1$ according as $\dot{\mathrm{I}}_{\boldsymbol{d}} \lesseqgtr -\nabla^2 S^*$.

Finally, from (99.13) we calculate

$$\left.\begin{aligned}\ddot{x}^k_{,k} &= \frac{\partial \mathrm{I}_{\boldsymbol{d}}}{\partial t} + \nabla^2\left(\frac{1}{2}\dot{x}^2\right) + 2\left(w^k_{\ m}\dot{x}^m\right)_{,k},\\ &= \frac{\partial \mathrm{I}_{\boldsymbol{d}}}{\partial t} + \nabla^2\left(\frac{1}{2}\dot{x}^2\right) + 2w^k_{\ m,k}\dot{x}^m - w^2,\\ &= \frac{\partial \mathrm{I}_{\boldsymbol{d}}}{\partial t} + \nabla^2\left(\frac{1}{2}\dot{x}^2\right) + e_{kpm}w^{m,p}\dot{x}^k - w^2,\\ &= \dot{\mathrm{I}}_{\boldsymbol{d}} + \nabla^2\left(\frac{1}{2}\dot{x}^2\right) - \dot{x}^k_{,km}\dot{x}^m - w^2.\end{aligned}\right\} \tag{102.8}$$

---

[1] [1903, *1*]. An equivalent result of more elaborate form was given earlier by PIOLA [1836, *1*, Eq. (214)].

[2] [1903, *1*] [1903, *2*, § 6] [1917, *2*]; [1946, *3*] [1954, *1*, Chap. II].

[3] This organization of more or less well known results was given by TRUESDELL [1954, *24*, § 41].

[4] LIPSCHITZ [1875, *3*], ZHUKOVSKI [1876, *7*, § 28], APPELL [1903, *1*].

If we put $I_d = 0$ and $\boldsymbol{w} = 0$ in $(102.8)_1$, by (102.3) we conclude that $\nabla^2 \dot{x}^2 \geqq 0$; by the maximum principle for subharmonic functions follows ***Kelvin's theorem***[1]: *In a region of isochoric irrotational motion the greatest speed cannot occur at an interior point.* A broad generalization of this result will be given in Sect. 121.

Motions such that $\ddot{x}^k_{,k} = 0$ have been studied by Zhukovski[2].

103. The higher accelerations. The N-th kinestate. The $N-1^{\text{st}}$ *acceleration* $\overset{(N)}{\boldsymbol{x}}$ is defined by

$$\overset{(N)}{x}{}^k = \frac{d^N x^k}{d t^N}. \tag{103.1}$$

Hence $\overset{(0)}{\boldsymbol{x}} = \boldsymbol{x}$, $\overset{(1)}{\boldsymbol{x}} = \dot{\boldsymbol{x}}$, $\overset{(2)}{\boldsymbol{x}} = \ddot{\boldsymbol{x}}$, and also

$$\overset{(N+1)}{x}{}^k = \frac{\partial \overset{(N)}{x}{}^k}{\partial t} + \overset{(N)}{x}{}^k_{,m}\, \dot{x}^m = \left[\frac{\partial}{\partial t} + \dot{x}^m (\;)_{,m}\right]^N \dot{x}^k. \tag{103.2}$$

Thus the N-th acceleration is a combination of derivatives of the velocity field up to the N-th order.

The set of spatial fields

$$\dot{x}^k_{,m},\; \ddot{x}^k_{,m},\; \ldots,\; \overset{(N)}{x}{}^k_{,m} \tag{103.3}$$

may be called the N-th *kinestate*. If we adjoin the double field $x^k_{;\alpha}$, we may call the resulting set the N-th *referential kinestate*.

These kinestates, sometimes augmented by higher derivatives up to a specified order, are the stuff of which the most recent models in continuum mechanics are wrought. As yet, little is known about their kinematical significance and properties. The next section is devoted to the relation between the N-th derivatives of material elements. As will appear, these are not obtained by simple analogy to the stretching and spin tensors, which suggests the notations

$$d^{(N)}_{km} \equiv \overset{(N)}{x}{}_{(k,m)}, \qquad w^{(N)}_{km} \equiv \overset{(N)}{x}{}_{[k,m]}, \quad \text{so that} \quad \overset{(N)}{x}{}_{k,m} = d^{(N)}_{km} + w^{(N)}_{km} \tag{103.4}$$

and $\boldsymbol{d}^{(1)} = \boldsymbol{d}$, $\boldsymbol{d}^{(2)} = \boldsymbol{d}^*$, $\boldsymbol{w}^{(1)} = \boldsymbol{w}$, $\boldsymbol{w}^{(2)} = \boldsymbol{w}^*$.

104. Higher rates of change of arc length and angle[3]. Using a superposed (N) to stand for d^N/dt^N, we see as in Sect. 76 that

$$\overset{(N)}{\overline{dx^k}} = \overset{(N)}{x}{}^k_{,m}\, dx^m. \tag{104.1}$$

Hence there is a tensor $\boldsymbol{A}^{(N)}$ such that

$$\overset{(N)}{\overline{d s^2}} = A^{(N)}_{pq}\, dx^p\, dx^q. \tag{104.2}$$

$\boldsymbol{A}^{(N)}$ may be called the N-th *Rivlin-Ericksen tensor*, since Rivlin and Ericksen showed the importance of these tensors in continuum mechanics.

First we obtain a recurrence relation[4] for $\boldsymbol{A}^{(N)}$. Since

$$\left.\begin{aligned} \overset{(N+1)}{\overline{d s^2}} &= \dot{\overline{A^{(N)}_{pq}\, dx^p\, dx^q}}, \\ &= \dot{A}^{(N)}_{pq}\, dx^p\, dx^q + A^{(N)}_{pq}\, \dot{\overline{dx^p}}\, dx^q + A^{(N)}_{pq}\, dx^p\, \dot{\overline{dx^q}}, \end{aligned}\right\} \tag{104.3}$$

---

[1] An equivalent statement, in another connection, was made without proof by Kelvin [1850, *3*, p. 509] and is now a well known theorem of potential theory.

[2] [1876, *6*, § 33].

[3] We do not attempt to summarize the intensive study of the differential invariants of motion given by Zorawski [1911, *14*] [1912, *8*].

[4] Dupont [1931, *5*, §§ 2—3].

by (76.4) follows

$$A^{(N+1)}_{pq} = \dot{A}^{(N)}_{pq} + 2\,A^{(N)}_{m(p}\,\dot{x}^m_{,q)}. \tag{104.4}$$

It is easily possible, however, to get an explicit formula[1] for $\boldsymbol{A}^{(N)}$. We have

$$\left.\begin{aligned} \overset{(N)}{\overline{ds^2}} &= \overset{(N)}{\overline{g_{km}\,dx^k\,dx^m}} = g_{km}\,\overset{(N)}{\overline{dx^k\,dx^m}},\\ &= g_{km}\sum_{K=0}^{N}\binom{N}{K}\overset{(N-K)}{\overline{dx^k}}\;\overset{(K)}{\overline{dx^m}},\\ &= g_{km}\sum_{K=0}^{N}\binom{N}{K}\overset{(N-K)}{x^k_{,p}}\;\overset{(K)}{x^m_{,q}}\,dx^p\,dx^q, \end{aligned}\right\} \tag{104.5}$$

and hence

$$A^{(N)}_{pq} = 2\,d^{(N)}_{pq} + \sum_{K=1}^{N-1}\binom{N}{K}\overset{(N-K)}{x_{m,p}}\;\overset{(K)}{x^m_{,q}}. \tag{104.6}$$

But also, directly from $(26.2)_1$ we have

$$\overset{(N)}{\overline{ds^2}} = \overset{(N)}{C}_{\alpha\beta}\,dX^\alpha\,dX^\beta = \overset{(N)}{C}_{\alpha\beta}\,X^\alpha_{;k}\,X^\beta_{;m}\,dx^k\,dx^m. \tag{104.7}$$

Comparison with (104.2) shows that

$$A^{(N)}_{pq} = \overset{(N)}{C}_{\alpha\beta}\,X^\alpha_{;p}\,X^\beta_{;q}; \tag{104.8}$$

hence[2]

$$\boldsymbol{A}^{(N)} = \overset{(N)}{\boldsymbol{C}}\Big|_{t=0}, \tag{104.9}$$

although the definition (104.2) makes it clear that $\boldsymbol{A}^{(N)}$ is independent of the choice of the initial state, so that (104.9) furnishes only one possible interpretation.

NOLL[3] approaches the whole problem of higher rates of change by repeated differentiations of the polar decomposition theorem. By $(37.8)_3$ and (72.9) we have

$$\left.\begin{aligned} \overset{(N)}{x}_{k,m} &= \overset{(N)}{x}_{k;\alpha}\,X^\alpha_{;m} = X^\alpha_{;m}\,\overset{(N)}{\overline{x_{k;\alpha}}} = X^\alpha_{;m}\,g^q_\beta\,\overset{(N)}{\overline{R_{kq}\,C^{\frac12\,\beta}_{\ \ \alpha}}},\\ &= X^\alpha_{;m}\,g^q_\beta\sum_{K=0}^{N}\binom{N}{K}\overset{(N-K)}{R}_{kq}\,\overset{(K)}{C}{}^{\frac12\,\beta}_{\ \ \alpha},\\ &= \sum_{K=0}^{N}\binom{N}{K}W^{(N-K)}_{kq}\,D^{(K)\,q}_{m}, \end{aligned}\right\} \tag{104.10}$$

where the N-th *stretching* and the N-th *spin*, $\boldsymbol{D}^{(N)}$ and $\boldsymbol{W}^{(N)}$, are given by the definitions

$$D^{(N)\,q}_{m} \equiv X^\alpha_{;m}\,g^q_\beta\,\overset{(N)}{C}{}^{\frac12\,\beta}_{\ \ \alpha}, \qquad W^{(N)}_{kq} \equiv \overset{(N)}{R}_{kq}. \tag{104.11}$$

While the foregoing formulae are valid for any choice of the initial configuration, different tensors $\boldsymbol{D}^{(N)}$ and $\boldsymbol{W}^{(N)}$ may result from different choices. The simplest

[1] RIVLIN and ERICKSEN [1955, *21*, § 10]. A somewhat different method was used by OLDROYD [1950, *22*, § 3]. A formula for $\ddot{\lambda}_a$ had been obtained by BONVICINI [1932, *3*].

[2] NOLL [1958, *8*, Eq. (8.9)] uses this property as the definition of $\boldsymbol{A}^{(N)}$. What is essentially the same result had been obtained by DUPONT [1931, *5*, §§ 2–3], using power series expansions.

[3] [1958, *8*, §§ 8–9].

choice is furnished by taking $t=0$, so that[1]

$$\boldsymbol{D}^{(N)} = \overset{(N)}{\boldsymbol{C}^{\frac{1}{2}}}\Big|_{t=0}, \qquad \boldsymbol{W}^{(N)} = \overset{(N)}{\boldsymbol{R}}\Big|_{t=0}. \tag{104.12}$$

In particular, this yields

$$\boldsymbol{D}^{(0)} = \boldsymbol{W}^{(0)} = 1, \qquad \boldsymbol{D}^{(1)} = \boldsymbol{d} = \boldsymbol{d}^{(1)} = \tfrac{1}{2}\boldsymbol{A}^{(1)}, \qquad \boldsymbol{W}^{(1)} = \boldsymbol{w} = \boldsymbol{w}^{(1)} \tag{104.13}$$

[cf. (95.15)], and from (144.10) we now get the simple identity

$$W^{(N)}_{km} = \overset{(N)}{x}_{k,m} - D^{(N)}_{km} - \sum_{K=1}^{N-1}\binom{N}{K} W^{(N-K)}_{kq}\, D^{(K)q}_{m}. \tag{104.14}$$

Moreover, differentiating the identity $\boldsymbol{C} = (\boldsymbol{C}^{\frac{1}{2}})^2$ and then using $(104.12)_2$ and (104.9) yields

$$A^{(N)}_{pq} = 2D^{(N)}_{pq} + \sum_{K=1}^{N-1}\binom{N}{K} D^{(N-K)}_{pm}\, D^{(K)m}_{q}. \tag{104.15}$$

In view of (104.6), it follows that $\boldsymbol{D}^{(N)}$ may be calculated recursively as a polynomial in the N-th kinestate. By (104.13), the same may be said of $\boldsymbol{W}^{(N)}$. While $\boldsymbol{D}^{(N)}$ and $\boldsymbol{A}^{(N)}$ are symmetric tensors, $\boldsymbol{W}^{(K)}$ is not skew symmetric in general; by (103.4), $\boldsymbol{D}^{(N)} \neq \boldsymbol{d}^{(N)}$, $\boldsymbol{W}^{(N)} \neq \boldsymbol{w}^{(N)}$ in general if $N>1$. In particular,

$$\left.\begin{aligned} \tfrac{1}{2}A^{(2)}_{pq} &= d^{(2)}_{pq} + \dot{x}_{m,p}\,\dot{x}^{m}_{,q}, \\ D^{(2)}_{pq} &= \tfrac{1}{2}A^{(2)}_{pq} - d_{pm}\,d^{m}_{q} = d^{(2)}_{pq} + \dot{x}_{m,p}\,\dot{x}^{m}_{,q} - d_{pm}\,d^{m}_{q} \\ W^{(2)}_{pq} &= w^{(2)}_{pq} - d_{pm}\,w^{m}{}_{q} - w_{pm}\,d^{m}_{q} + w_{pm}\,w^{m}{}_{q}. \end{aligned}\right\} \tag{104.16}$$

The tensors introduced here are only a few of the many in terms of which a correct and complete description of the higher rates of change of length and angle may be expressed. The possible variety corresponds to that for strain mentioned in Sect. 32.

Further properties of the tensors $\boldsymbol{A}^{(N)}$ are developed in Sects. 144 and 150.

Since the displacement vector u^{α} is given by the formal power series

$$u^{\alpha} = \sum_{N=1}^{\infty} u^{\alpha}_{(N)}, \qquad u^{\alpha}_{(N)} = \frac{t^N}{N!}\overset{(N)}{p}{}^{\alpha}\Big|_{t=0}, \tag{104.17}$$

one may calculate a series for $\overset{(N)}{x}{}^{m}_{,q}$; substituting this series into (104.6) and then putting the result into the formal series equivalent to (104.9) yields a series[2] for $C_{\alpha\beta}$ in terms of the derivatives of the $u^{\alpha}_{(N)}$.

e) Special developments concerning vorticity.

e I) The vorticity field.

105. Vorticity and circulation. Throughout this part of the subchapter, in the main, we employ direct vectorial notations. Our subject is the *vorticity vector* $\boldsymbol{w}$, defined by $(86.2)_3$.

---

[1] Had we started from (37.9) instead of (37.8), we should have obtained a similar result except that the order of the factors $\boldsymbol{W}^{(N-K)}$ and $\boldsymbol{D}^{(K)}$ in $(104.10)_6$ would be interchanged, and we should have to set

$$\boldsymbol{D}^{(N)} \equiv \overset{(N)}{\boldsymbol{C}^{-\frac{1}{2}}}\Big|_{t=0}$$

instead of $(104.12)_1$. The tensors so obtained are the same as those given in the text above for $N=0$ or 1 but not for larger values of N.

[2] The beginning of this elaborate calculation is given by DUPONT [1931, *5*, §§ 4–5].

KELVIN's transformation $(87.1)_2$ may be read as follows[1]: *Let $\mathfrak{s}$ be any surface lying upon the boundary of a closed region in which the velocity $\dot{\boldsymbol{p}}$ is continuous and the vorticity $\boldsymbol{w}$ is piecewise continuous; then the flux of vorticity through $\mathfrak{s}$ equals the circulation around its bounding circuit $\mathfrak{c}$.* Thus the net rotation of the finite circuit $\mathfrak{c}$ equals the sum of the strengths of all the vortices it embraces[2]. It is important that the vorticity need not be defined on both sides of the surface $\mathfrak{s}$: (87.1) can be applied not only to interior surfaces, but also to bounding surfaces.

As a first application of KELVIN's transformation we shall now determine the nature of the vorticity on a stationary surface to which the material adheres, so that the boundary condition for the velocity field is (69.4). By applying $(87.1)_2$ to any closed circuit lying wholly upon the boundary we obtain

$$\int_{\mathfrak{s}} d\boldsymbol{a} \cdot \boldsymbol{w} = 0 \tag{105.1}$$

for the included area $\mathfrak{s}$; hence, since $\mathfrak{s}$ may be shrunk down to a point, it follows that

$$d\boldsymbol{a} \cdot \boldsymbol{w} = 0: \tag{105.2}$$

Upon a stationary surface to which the material adheres, the vorticity is zero or tangent to the surface[3]. Thus all particles rotate, if at all, about axes tangential to the wall. Conversely, *if at each point of a surface the vorticity vanishes or is tangential to the surface, then the material adheres to the surface.*

We may express the abnormality of the velocity field in terms of the vorticity, since by $(\text{App. }30.1)_1$ we have

$$A(\dot{\boldsymbol{p}}) = \frac{\dot{\boldsymbol{p}}}{\dot{p}} \cdot \operatorname{curl} \frac{\dot{\boldsymbol{p}}}{\dot{p}} = \frac{\dot{\boldsymbol{p}} \cdot \boldsymbol{w}}{\dot{p}^2} = \frac{w_{\dot{\boldsymbol{p}}}}{\dot{p}}, \tag{105.3}$$

where $w_{\dot{\boldsymbol{p}}}$ is the length of the projection of $\boldsymbol{w}$ upon the direction of $\dot{\boldsymbol{p}}$. That is, *the component of vorticity in the direction of the velocity equals the product of the speed by the abnormality of the velocity field.*

106. Vortex lines and vortex tubes. The vector lines of the vorticity field are called *vortex lines*, its vector surfaces are called *vortex surfaces*, and its vector tubes are called *vortex tubes*. These concepts, which were introduced by HELMHOLTZ[4], have proved to be of central importance.

Since $\operatorname{div} \boldsymbol{w} = 0$, *the vorticity field is solenoidal.* Herein lies one reason that the vorticity field is often simpler than the velocity field from which it is derived. From the Helmholtz characterization of solenoidal fields (Sect. App. 31) we derive ***Helmholtz's first vorticity theorem**: The strength of a vortex tube is the*

[1] KELVIN [1869, *7*, §§ 59–60], BELTRAMI [1871, *1*, § 12].

[2] PASCAL [1920, *2*,] [1921, *5*] proposed to take $\oint_{\mathfrak{s}} \dot{\boldsymbol{p}} \times d\boldsymbol{a}$, which he called the *superficial circulation* of $\mathfrak{s}$, as a measure of rotation. He showed that a motion is irrotational if and only if the superficial circulations of every two reconcileable surfaces are equal; equivalently, if and only if the superficial circulation of every reducible surface is zero. For the case when $\mathfrak{s}$ is the complete boundary of a region $\mathfrak{v}$ of continuous motion, by $(\text{App. }26.1)_3$ we see that $-\oint_{\mathfrak{s}} \dot{\boldsymbol{p}} \times d\boldsymbol{a} = \int_{\mathfrak{v}} \boldsymbol{w}\, dv =$ the total vorticity in $\mathfrak{v}$. Cf. Sect. 118. The discussion of superficial circulation by various authors in C. R. Acad. Sci., Paris **176** and **177** (1923) seems to consist mainly in mutual misunderstandings. As remarked by NOAILLON [1924, *11*], the presence of singularities in potential flows will generally prevent the superficial circulation about a submerged body from vanishing.

[3] This result is well known, but we have not been able to trace its origin. It was proved by LICHTENSTEIN [1929, *4*, Chap. 5, § 16]. Other constraints imposed on the values of $\dot{x}_{k,m}$ by various types of boundary conditions were derived by BJØRGUM [1955, *3*].

[4] [1858, *1*, § 2]. KELVIN [1869, *7*, §§ 60(i), 60(m)] used the terms "axial lines" and "axial sheets" for vortex lines and vortex surfaces, respectively. For vortex surfaces in general, see POINCARÉ [1893, *6*, § 12].

same at all cross-sections[1]. By applying $(87.1)_2$ we may at once put HELMHOLTZ'S theorem into a form derived by KELVIN: *The circulations taken in the same sense about any two reconcileable circuits lying upon a given vortex tube are equal.* In the foregoing proof we have assumed, as mentioned at the outset of this work, that the velocity field is twice continuously differentiable, and in particular that the vorticity field is once continuously differentiable. Stated thus, however, the result holds subject to rather weaker hypotheses[2], as we shall now discover by following the proof given by KELVIN[3].

First, *the circulation around any circuit lying wholly upon a vortex surface and reducible upon it is zero*[4], for we may choose for the surface in KELVIN'S transformation $(87.1)_2$ the inclosed portion of the vortex surface, upon which $\boldsymbol{w}$ is normal to $d\boldsymbol{a}$, so that

$$0 = \oint_c d\boldsymbol{x} \cdot \dot{\boldsymbol{p}}. \tag{106.1}$$

Alternatively, *let two points lying upon the same vortex surface be connected by two curves c_1 and c_2 reconcileable upon it; then the flow along c_1 equals the flow along c_2.* Now for the validity of KELVIN'S transformation, and hence also for the validity of this result, the vorticity $\boldsymbol{w}$ need not even be continuous: It suffices that the velocity field $\dot{\boldsymbol{p}}$ be continuous and that the vortex surface be part of the boundary of a closed region in which $\boldsymbol{w}$ is piecewise continuous (Sect. App. 28).

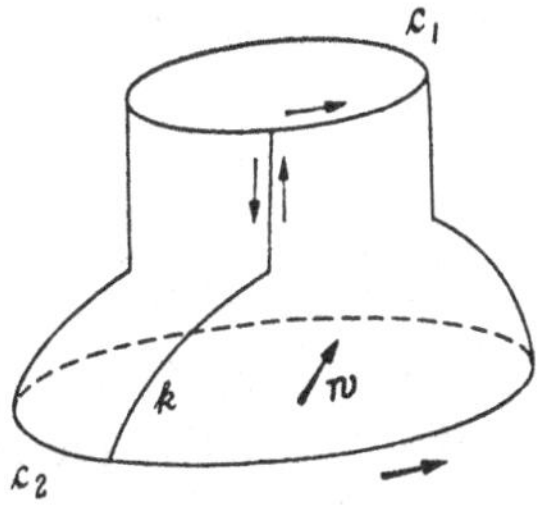

Fig. 18. Diagram for proof of KELVIN'S extension of HELMHOLTZ'S first theorem.

Thus (106.1) holds for circuits upon vortex surfaces which suffer sharp but isolated bends. Consider now two simple circuits c_1 and c_2 lying upon a vortex tube, reconcileable upon it but not reducible upon it (Fig. 18). The vortex tube itself need not be reducible. Let the two circuits be connected by a barrier k traversed in opposite senses, so that $c \equiv c_1 + k - c_2 - k$ is a single closed circuit lying upon the surface of the vortex tube and inclosing a simply connected portion of it. To this circuit our previous result (106.1) applies:

$$0 = \oint_c d\boldsymbol{x} \cdot \dot{\boldsymbol{p}} = \oint_{c_1} + \int_k - \oint_{c_2} - \int_k, \tag{106.2}$$

whence follows

$$\oint_{c_1} d\boldsymbol{x} \cdot \dot{\boldsymbol{p}} = \oint_{c_2} d\boldsymbol{x} \cdot \dot{\boldsymbol{p}}. \tag{106.3}$$

The result is easily extended to reconcileable circuits which are not simple. We have thus established ***Kelvin's extension*** of HELMHOLTZ'S first vorticity theorem: *Let there be given a vortex tube in a motion where the velocity field is continuous and the vorticity field is piecewise continuous in a closed region, of whose boundary the vortex tube forms a part; then the circulations about any two circuits lying wholly upon this vortex tube and reconcileable upon it are equal.*

107. Criteria for permanence and constant strength of the vortex tubes. In order to investigate the possibility that the vortex lines be material lines, we

---

[1] [1858, *1*, § 2]. HELMHOLTZ erroneously concluded that vortex lines either form closed curves or else end upon a boundary; see footnote 2, p. 820. Closed vortex lines, while uncommon, are of great interest in the theory of vortices.

[2] LAMB [1895, *2*, § 145].

[3] [1869, *7*, § 60 (l)]. An alternative proof, based on direct analysis of the discontinuity of $\boldsymbol{w}$ across a surface, was given by WEINGARTEN [1901, *16*].

[4] POINCARÉ [1893, *6*, § 12].

put $\boldsymbol{c}=\boldsymbol{w}$ in the HELMHOLTZ-ZORAWSKI criterion (75.6), so obtaining

$$\boldsymbol{w}\times\left[\frac{\partial\boldsymbol{w}}{\partial t}+\operatorname{curl}(\boldsymbol{w}\times\dot{\boldsymbol{p}})\right]=0. \tag{107.1}$$

But taking the curl of LAGRANGE's acceleration formula (99.14) yields

$$\frac{\partial\boldsymbol{w}}{\partial t}+\operatorname{curl}(\boldsymbol{w}\times\dot{\boldsymbol{p}})=\operatorname{curl}\ddot{\boldsymbol{p}}\equiv\boldsymbol{w}^*, \tag{107.2}$$

for any motion. Hence (107.1) becomes

$$\boldsymbol{w}\times\boldsymbol{w}^*=0: \tag{107.3}$$

A necessary and sufficient condition that the vortex lines be material lines is that the acceleration be lamellar or that its curl be parallel to the vorticity.

By putting $\boldsymbol{c}=\boldsymbol{w}$ in ZORAWSKI's criterion $(80.6)_2$ we similarly obtain

$$\boldsymbol{w}^*=0: \tag{107.4}$$

A necessary and sufficient condition that the strengths of all vortex tubes remain constant in time is that the acceleration be lamellar; this condition is also sufficient that the vortex tubes be material tubes[1].

108. Circulation preserving motions. The Helmholtz theorems. The condition just derived, *viz.*

$$\boldsymbol{w}^*=0, \tag{108.1}$$

is of great importance. Both D'ALEMBERT (1749)[2] and EULER (1752—1755)[3] derived it as a consequence of the dynamical equations for barotropic motions of perfect[4] fluids subject to conservative extrinsic force, and we shall call it the ***D'Alembert-Euler condition***. It is a statement that *the acceleration field is lamellar*. Its meaning emerges when we put $\psi=\dot{x}_k$ in $(79.2)_3$, obtaining

$$\frac{d}{dt}\oint_{\mathscr{C}}d\boldsymbol{x}\cdot\dot{\boldsymbol{p}}=\oint_{\mathscr{C}}d\boldsymbol{x}\cdot\ddot{\boldsymbol{p}}; \tag{108.2}$$

alternatively, we may interpret the second theorem of Sect. 107 in the light of KELVIN's transformation: *The D'Alembert-Euler condition is necessary and sufficient that the circulation of every material circuit remain constant in time*[5]. Classical hydrodynamics is characterized by this one basic statement, and all the main theorems of that subject are consequences of it alone[6]. Motions in which the circulation of material circuits does not change in time we shall call *circulation preserving motions*. They will be frequent subject of remark and illustration in the rest of this part of the subchapter.

By (101.12), equivalent to (108.1) is the ***Hankel-Appell condition***[7]

$$\boldsymbol{W}_*=0. \tag{108.3}$$

---

[1] POINCARÉ [1893, *6*, §§ 5—6, 150—151], ZORAWSKI [1900, *11*], JAUMANN [1905, *2*, § 386], VESSIOT [1911, *12*, § 4]. Earlier but partially incorrect statements were given by MÜLLER [1878, *7*] and LEVY [1890, *7*, § 10].

[2] [1752, *1*, § 86] [1761, *1*, §§ 10, 15].

[3] [1761, *2*, § 58] [1757, *2*, § 35]. EULER noted the connection of this condition with BERNOULLI's equation (see Sect. 120).

[4] It was remarked by BRANDES [1806, *1*, § 150, footnote] that in a theory where friction is taken into account it will generally follow that $\boldsymbol{w}^*\neq 0$; this is borne out by the various theories of fluid friction proposed later.

[5] The sufficiency was proved by HANKEL [1861, *1*, § 8] and KELVIN [1869, *7*, § 59(e)].

[6] LEVY [1890, *7*, § 7].

[7] In connection with perfect fluids, the condition $\dot{\boldsymbol{W}}_*=0$ was derived by HANKEL [1861, *1*, § 6]; the condition (108.3), by APPELL [1897, *1*, § 2]. An equivalent but more complicated formula in material co-ordinates had been obtained by LAGRANGE [1762, *2*, § XLIV].

This form of the condition for circulation preserving motion is appropriate to the material description.

We may now reformulate the results of Sect. 107 in a fashion so closely connected with the hydrodynamical theorems of HELMHOLTZ (1858)[1] that, although they are purely kinematical, we shall call them the ***second and third vorticity theorems of Helmholtz***: (2) *In a circulation preserving motion the vortex lines are material lines*, and (3) *in a motion such that the vortex lines are material, in order that the strengths of all vortex tubes remain constant in time it is necessary and sufficient that the motion be circulation preserving*. The Helmholtz theorems furnish a vivid and full picture of the general character of circulation preserving motions, whose theory may thus be said to be, in a certain sense, closed. Most motions that take place in mechanical theories fail to be circulation preserving, however, and a major objective is to clarify the manner in which a general motion departs from circulation preserving character, or equivalently, to describe the mechanism by which the vortex tubes turn away from coincident material tubes and change their strength—that is to say, to generalize the Helmholtz theorems. This research is presented in Part e IV.

The stream-line patterns possible in circulation preserving motions are of a restricted type; a geometrical characterization of those possible in the steady case was given by ZHUKOVSKI[2].

109. The acceleration potential. A function V^* such that

$$\ddot{x}_k = -V^*_{,k}, \quad \text{or} \quad \ddot{\boldsymbol{p}} = -\operatorname{grad} V^* \tag{109.1}$$

is called an *acceleration potential*[3]. The function V^* may be single-valued or many-valued. By a theorem in Sect. App. 33 we may reformulate the D'Alembert-Euler condition (108.1) in the following way: *A motion is circulation preserving if and only if it possesses an acceleration potential.*

An evident example of a circulation preserving motion is an irrotational motion. By putting (88.2) into (99.14) we obtain

$$\ddot{\boldsymbol{p}} = -\operatorname{grad}\left[\frac{\partial V}{\partial t} - \frac{1}{2}(\operatorname{grad} V)^2\right]; \tag{109.2}$$

comparison with (109.1) shows that if a meaningless function of time only is absorbed into V, we shall have[4]

$$V^* = \frac{\partial V}{\partial t} - \frac{1}{2}(\operatorname{grad} V)^2. \tag{109.3}$$

Another example is furnished by the rectilinear vortices (89.8), which are circulation preserving if and only if they are steady and strictly plane; i.e., $\dot{\theta} = \dot{\theta}(r)$. In this case[5]

$$V^* = \int r\,\dot{\theta}^2\,dr. \tag{109.4}$$

---

[1] [1858, *1*, § 2]. The validity of the Helmholtz theorems was extended by KELVIN [1869, 7, §§ 59(d), 60(f)–60(i)]. Cf. NANSON [1874, *5*]. That these theorems are essentially kinematical was remarked by SCHÜTZ [1895, *5*]. According to LAMB [1895, *2*, § 143], LARMOR observed that HELMHOLTZ's own proofs of the second and third theorems are open to the same objection as that raised by STOKES against certain faulty proofs of the velocity potential theorem; for discussion of the question at issue, see TRUESDELL [1954, *24*, §§ 104–107].

[2] [1876, *7*, §§ 30–32]. Cf. the unsuccessful attempts of TOUCHE [1895, *6*] [1897, *8*]. For still more special cases in which formal solutions are available, see Sects. 110, 111, 161. Proof of local existence and uniqueness of steady isochoric circulation preserving motion is approached through consideration of Eq. (108.1) by GODAL [1958, *2*, § 2].

[3] The acceleration potential is a discovery of EULER [1757, *2*, § 35] [1770, *1*, § 42]. Cf. also PIOLA [1836, *1*, Eq. (219)]. Its importance was stressed by LEVY [1890, *7*, § 7]. Cf. POINCARÉ [1893, *6*, § 6], APPELL [1921, *1*, §§ 729, 753, *et passim*].

[4] EULER [1761, *2*, §§ 79–80]. Cf. VESSIOT [1911, *12*, § 4], CARSTOIU [1946, *3*, § 3].

[5] EULER [1757, *2*, §§ 30–33] [1757, *3*, §§ 57–61].

110. Complex-lamellar motions. A motion is said to be *complex-lamellar*[1] if the velocity field $\dot{\boldsymbol{p}}$ is complex-lamellar (Sect. App. 33). In this case (99.19) reduces to

$$\dot{\boldsymbol{p}} = F \operatorname{grad} G; \tag{110.1}$$

equivalently

$$\boldsymbol{w} \cdot \dot{\boldsymbol{p}} = 0. \tag{110.2}$$

Hence *a rotational motion is complex-lamellar if and only if its vortex lines are orthogonal to its stream lines.* Equivalently, *a motion is complex-lamellar if and only if its stream lines possess normal surfaces.* Both plane motions and rotationally symmetric motions (Sect. 68) are complex-lamellar.

Complex-lamellar motions possess some of the distinguishing properties of irrotational motions[2]. The function $-G$ is somewhat analogous to the potential function V: The formulae

$$-F\frac{\partial G}{\partial s} = -\dot{x} \leqq 0, \quad -F\left(\frac{\partial G}{\partial t} - \dot{G}\right) = \dot{x}^2 \geqq 0, \tag{110.3}$$

generalize (88.5) and (88.6). Thus the conclusions regarding the stream lines of an irrotational motion derived from these two formulae in Sect. 88 carry over to a region of complex-lamellar motion where F is of one sign. Now by the continuity of the velocity field, F can be of opposite sign at two points upon a stream line only if there is an intermediate point where $F=0$, and by (110.1) this point is a stagnation point. Consequently *in a region of complex-lamellar motion without stagnation points, the conclusions of Sect. 88 regarding the speed and the stream line pattern of an irrotational motion continue to hold, provided $-G$ be substituted for V if $F>0$, G for V if $F<0$.*

The condition of complex-lamellar motion is purely geometric, being no more than a statement that the stream lines possess orthogonal surfaces. Important properties of complex-lamellar motion may be read off from results in Sect. App. 33. For example, in an isochoric complex-lamellar motion, the speed is constant on each stream line if and only if the normal surfaces are minimal.

There have been many studies of the geometry of stream fields. In most cases the work is unnecessarily complicated by superfluous physical considerations. Indeed, most such results are purely kinematical in nature and may be derived by kinematical reasoning from kinematical assumptions alone. The prime example is the entire classical theory of vorticity for circulation preserving motions, which is included in Part IV of this subchapter. This theory follows in its entirety from the assumption that the acceleration field is lamellar[3] (cf. Sects. 109–110). Similarly, the classical theory of inviscid fluids in the more general case when there is no acceleration potential may be characterized by the statement that *the acceleration field is the sum of a given field and a complex-lamellar field*[4].

To illustrate the above we shall give a single example: *In a motion with steady straight stream lines, the acceleration is complex-lamellar if and only if the motion is complex-lamellar.* For proof, note only that by (99.9) the acceleration and velocity fields are necessarily parallel;

---

[1] These motions were first discussed explicitly by EARNSHAW [1837, *1*, §§ 2–5].

[2] TRUESDELL [1954, *24*, § 50]. Certain geometric conditions on the stream lines according as they do or do not possess normal surfaces were obtained by KLEITZ [1873, *4*, Chap. V].

[3] LEVY [1890, *7*, § 7], CARSTOIU [1942, *2*].

[4] EULER [1757, *3*, § 17]. This is the content of EULER's dynamical equation $\ddot{x}_k = -p_{,k}/\varrho + f_k$. Conditions of integrability for the velocity corresponding to a complex-lamellar acceleration and subject to additional conditions relevant for gas dynamics have been obtained by FRIEDMANN [1916, *2*] [1925, *6*] and BERKER [1956, *1*].

therefore, the one can have normal surfaces if and only if the other does. The result itself has been stated and derived in a more elaborate way in the hydrodynamical literature[1].

111. Steady vortex lines. Steady vorticity. *The vortex lines are steady* if and only if

$$\frac{\partial \boldsymbol{w}}{\partial t} \times \boldsymbol{w} = 0. \tag{111.1}$$

For the vortex lines to be steady it is of course sufficient, but not necessary, that the *vorticity itself be steady:*

$$\frac{\partial \boldsymbol{w}}{\partial t} = 0. \tag{111.2}$$

From (111.1) follows

$$\frac{\partial \boldsymbol{w}}{\partial t} = C\,\boldsymbol{w}, \tag{111.3}$$

where C is a scalar quantity. Thus we have

$$\operatorname{curl}\frac{\partial \dot{\boldsymbol{p}}}{\partial t} = \operatorname{curl}(C\,\dot{\boldsymbol{p}}) - \operatorname{grad} C \times \dot{\boldsymbol{p}}. \tag{111.4}$$

Taking the divergence of this equation yields

$$0 = \operatorname{div}(\operatorname{grad} C \times \dot{\boldsymbol{p}}) = -\operatorname{grad} C \cdot \boldsymbol{w}. \tag{111.5}$$

Hence it follows that *in a motion where the vortex lines are steady but the vorticity is not steady, the surfaces upon which $\partial \boldsymbol{w}/\partial t$ bears a constant ratio to $\boldsymbol{w}$ are vortex surfaces.*

Following the analysis of MASOTTI[2], we may characterize flows with steady vorticity by noting that (111.2) is equivalent to

$$\operatorname{curl}\frac{\partial \dot{\boldsymbol{p}}}{\partial t} = 0. \tag{111.6}$$

Hence (111.2) is satisfied if and only if there exists a scalar $U(\boldsymbol{x}, t)$ such that[3]

$$\frac{\partial \dot{\boldsymbol{p}}}{\partial t} = \operatorname{grad} U; \tag{111.7}$$

that is, the local acceleration is lamellar, and thus by integration we obtain

$$\dot{\boldsymbol{p}}(\boldsymbol{x}, t) = \operatorname{grad}\left[\int U(\boldsymbol{x}, t)\,dt\right] + \boldsymbol{u}(\boldsymbol{x}): \tag{111.8}$$

A necessary and sufficient condition that a motion have steady vorticity is that the velocity be the sum of a lamellar field and a steady field. An irrotational motion furnishes a special case, with $U = -\partial V/\partial t$, $\boldsymbol{u} = 0$. Henceforward in this part of the subchapter the symbol U will be employed only in the sense indicated by (111.7).

For any motion with steady vorticity we may put (111.7) into LAGRANGE'S acceleration formula (99.14), obtaining

$$\ddot{\boldsymbol{p}} = \boldsymbol{w} \times \dot{\boldsymbol{p}} + \operatorname{grad}\left(\tfrac{1}{2}\dot{p}^2 + U\right), \tag{111.9}$$

a result whose significance will appear in Part e IV.

---

[1] BYUSHGENS [1948, *3*, § 4.3], COBURN [1952, *2*].

[2] [1927, *7*, § 2].

[3] By (79.3), the condition (111.7) with single-valued U is necessary and sufficient that the circulation about every *spatial* circuit remain constant in time.

More generally, from (107.2), which holds in any motion, and from the condition (107.4) we conclude that *a motion with steady vorticity is circulation preserving if and only if the Lamb vector is lamellar:*

$$\operatorname{curl}(\boldsymbol{w}\times\dot{\boldsymbol{p}})=0. \tag{111.10}$$

If we substitute (110.1) into (111.10), a rather complicated condition results. For the case when the normal surfaces are minimal (cf. Sect. App. 33), PRIM[1] has used this condition to derive the following theorem: *In a steady isochoric circulation preserving complex-lamellar motion such that the speed is constant along each stream line, it is possible to parametrize the normal surfaces* $G=\text{const}$ *in such a way that* $\nabla^2 G=0$. By (88.9) it follows that the totality of stream-line patterns for this class of motions is exhausted by the special case when "circulation preserving" is replaced by "irrotational". The irrotational case has been characterized in a celebrated and difficult analysis by HAMEL[2].

112. Screw motions. In a rotational motion with vanishing Lamb vector

$$\boldsymbol{w}\times\dot{\boldsymbol{p}}=0,\qquad \boldsymbol{w}\neq 0, \tag{112.1}$$

the velocity field is a screw field (Sect. App. 34); such a motion will be called a *screw motion.*

Screw motions and complex-lamellar motions are mutually exclusive types, and each shares some of the properties of irrotational motions, which, by the definitions employed in the present work, are included as special cases of the latter but not of the former. In an irrotational motion there are no vortex lines, the stream lines are normal to the equipotential surfaces, and the convective acceleration may be determined from the speed alone. In a complex-lamellar motion the vortex lines are normal to the stream lines, the stream lines are endowed with normal surfaces, but the acceleration possesses no particularly simple quality. On the other hand, in a screw motion, just as in an irrotational motion, (99.14) reduces to

$$\ddot{\boldsymbol{p}}=\frac{\partial\dot{\boldsymbol{p}}}{\partial t}+\operatorname{grad}\frac{1}{2}\dot{p}^2; \tag{112.2}$$

the vortex lines coincide with the stream lines, and a congruence of surfaces normal to the stream lines does not exist.

From (112.2) in the steady case follows the ***first Gromeka-Beltrami theorem***[3]: *Any steady screw motion is circulation preserving, its acceleration potential being*

$$V^*=-\tfrac{1}{2}\dot{p}^2+\text{const}. \tag{112.3}$$

Fro a steady screw motion with steady density (Sect. 77), we have $\operatorname{div}(\jmath B\dot{\boldsymbol{p}})=0$, where $B=\text{const}$ along each stream line, so we may put $m=\jmath B$ in (App. 34.5); the first corollary following (App. 34.6) then yields the ***second Gromeka-Beltrami theorem***: *In a steady screw motion*

$$\frac{w}{\jmath\dot{p}}=\text{const} \tag{112.4}$$

on each stream line.

Finally, BELTRAMI showed in effect that a *circulation preserving screw motion is necessarily steady*. We give the elegant proof of BJØRGUM[4]. First, it follows immediately from (107.2) and (108.1) that *in a circulation preserving screw motion, the vorticity is steady*. Since by $(\text{App. }34.4)_2$ the abnormality A of a screw field is

[1] [1948, *22*, §§ 4–5] [1952, *16*, Chap. V, § B].
[2] [1937, *2*]. Cf. also HOWARD [1953, *13*].
[3] References for this sequence of theorems are given in Sect. App. 34.
[4] [1951, *2*, Sect. 2.5]. Cf. also BALLABH [1940, *1*, § 6] [1948, *1*, § 4].

determined by the vorticity field alone, A is steady if $\boldsymbol{w}$ is steady. But $\dot{\boldsymbol{p}} = \boldsymbol{w}/A$, and hence *a screw motion is steady if and only if its vorticity is steady.* Several of the foregoing statements may be combined in a single ***theorem of Beltrami***: *A screw motion is circulation preserving if and only if it satisfies the following two equivalent conditions: It is steady, or its vorticity is steady.*

With $\boldsymbol{c} = \dot{\boldsymbol{p}}$, the condition (App. 31.14) is satisfied by a screw motion. The theorem following (App. 31.12) thus implies that *an isochoric screw motion adhering to fixed boundaries and at rest at* ∞, *in the sense* (App. 31.13), *cannot exist*[1]; *the only irrotational motion possible under these conditions is a state of rest.*

Certain restrictions on the stream-line pattern of a screw motion have been noticed by MORERA and TRUESDELL[2].

113. Circumstances when an irrotational motion is impossible. Following the line of thought suggested by the last theorem, we can determine various cases when an irrotational motion is incompatible with the boundary conditions of Sect. 69.

We begin with the ***theorem of Kelvin and Helmholtz***[3]: *In a simply connected region, an irrotational motion such that* $j V \frac{dV}{dn} = \bar{o}(p^{-2})$ *in any infinite portion, while all finite boundaries are stationary, if either steady or isochoric is a state of rest.*

Proof. Since the domain is simply connected, any irrotational motion within it must be possessed of a single-valued velocity potential V. For an isochoric motion we may put $\Psi = V$, $\boldsymbol{c} = \dot{\boldsymbol{p}}$ in (App. 31.5), thus obtaining

$$\int_v \dot{p}^2 \, dv = -\oint_s d\boldsymbol{a} \cdot \dot{\boldsymbol{p}} \, V = 0, \tag{113.1}$$

where the vanishing of the surface integral follows from (69.2) and the assumed order condition at ∞. From (113.1) we conclude that $\dot{\boldsymbol{p}} = 0$. For a *steady* motion we may put $\Psi = V$, $\boldsymbol{c} = j\dot{\boldsymbol{p}}$ in (App. 31.5), thus obtaining

$$\int_v j \, \dot{\boldsymbol{p}}^2 \, dv = -\oint_s d\boldsymbol{a} \cdot \dot{\boldsymbol{p}} \, j \, V = 0. \tag{113.2}$$

Since $j > 0$, the theorem again follows, Q.E.D.

We note the following corollaries. First, for a motion in the interior of a finite region the condition at ∞ becomes superfluous, and the theorem holds unconditionally: *There is no isochoric or steady irrotational motion, other than a state of rest, within a finite simply connected region with stationary boundary.* Second, if the region of motion is such that $\dot{\boldsymbol{p}} \to 0$ uniformly at ∞, then for the isochoric case a theorem on isolated singularities of harmonic functions implies that the three components $\dot{p}^k$ in the common frame are single-valued harmonic functions analytic at ∞. This fact together with the condition that $\oint_s d\boldsymbol{a} \cdot \dot{\boldsymbol{p}} = 0$ for a sufficiently large sphere yields $\dot{p}^k = O(p^{-2})$, whence follows $V = O(p^{-1})$. Thus $V \frac{dV}{dn} = O(p^{-3}) = o(p^{-2})$, so at ∞ the order condition is satisfied. That is: *In an isochoric irrotational motion in an infinite region such that all bounding surfaces are stationary, if the velocity field vanishes uniformly at infinity and if no material is supplied at infinity, then the motion is a state of rest.*

---

[1] TRUESDELL [1951, *29*].
[2] [1889, *6*]; [1954, *24*, § 52].
[3] [1849, *3*]; [1858, *1*, § 1].

That the foregoing result cannot be extended to multiply connected regions may be seen from the counter example of the irrotational vortex (89.11), which is irrotational in the doubly connected domain $0<r_1\leqq r\leqq r_2$, $|z|\leqq z_0$. The reason the above proof fails to hold for multiply connected regions is that then V is many-valued, so that the substitution $\psi=V$ in (App. 31.5) is no longer permissible. That when the motion is not isochoric the restriction to steady motion is essential may be seen from the counter example of the oscillating motion

$$\left.\begin{aligned} \dot{j} &= l[1+q\sin kt\sin x], \quad 0<q<1,\\ \dot{x} &= J\,l\,q\,k\cos kt\cos x, \quad \dot{y}=\dot{z}=0, \end{aligned}\right\} \tag{113.3}$$

which is irrotational in the finite simply connected domain $|x|\leqq\frac{1}{2}\pi$, $|y|\leqq a$, $|z|\leqq b$, upon whose boundaries it satisfies (69.2)

In the case of boundaries where the condition of *adherence* (69.4) is to hold, irrotational motion is often impossible. A theorem to this effect, for the isochoric case, is included in that proved at the end of Sect. 112. Indeed, it was asserted by DUHEM[1] that the adherence condition and irrotational motion are usually incompatible, but this proof is not convincing, and his statement cannot be true in general, since the oscillating motion given in spherical co-ordinates r, θ, ϕ by

$$\left.\begin{aligned} \dot{j} &= \frac{1}{r^2}[1+q\sin kt\sin r], \quad 0<q<1,\\ \dot{r} &= \frac{Jlqk}{r^2}\cos kt\cos r, \quad \dot{\theta}=0, \quad \dot{\phi}=0, \end{aligned}\right\} \tag{113.4}$$

is a motion which is continuous and irrotational within the spherical shell $\frac{1}{2}\pi\leqq r\leqq\frac{2}{3}\pi$, to whose stationary boundaries the material adheres. However, for the *isochoric* case there are theorems asserting the incompatibility of irrotational motion with adherence to *any finite surface*, however small. We begin with the ***theorem of Kirchhoff***[2]: *The only isochoric irrotational motion in which the material adheres to a finite stationary surface, however small, is a state of rest.*

Proof: By (69.4), the velocity potential V is a harmonic function all of whose first derivatives with respect to the space variables vanish upon the finite surface. Upon that finite surface, consequently, the function itself is constant and its normal derivative vanishes. The only such harmonic function is a constant. Q.E.D.

As an immediate corollary it follows that the only isochoric irrotational motion in which the material adheres to a finite surface in rigid translation is itself a rigid translation. The results just stated and proved refer to the state of motion at any one instant.

Following the analysis of SUPINO[3], let us now consider the case when the material adheres to surfaces suffering rigid rotation at angular velocity $\boldsymbol{\omega}(t)$. If we refer an irrotational motion whose velocity potential is V to a co-ordinate frame which is at rest with respect to these surfaces, and whose origin coincides with the origin of a system with respect to which the motion is irrotational, for the velocity with respect to the moving frame we obtain

$$\dot{\boldsymbol{p}}' = -\boldsymbol{\omega}\times\boldsymbol{p} - \operatorname{grad} V, \tag{113.5}$$

[1] [1901, *7*, 5e partie, Chap. II, § 1].

[2] While the proof of KIRCHHOFF [1876, *2*, Vorl. 16, § 6] is not rigorous, the result is true, being in fact one way of phrasing a now well known theorem of potential theory.

[3] [1949, *29*].

or equivalently

$$\dot{x}' = 2\omega_z y - P_{,x}, \quad \dot{y}' = 2\omega_x z - P_{,y}, \quad \dot{z}' = 2\omega_y x - P_{,z}, \tag{113.6}$$

where P, given by

$$P \equiv V + x\omega_y z + y\omega_z x + z\omega_x y, \tag{113.7}$$

is also harmonic. At any given instant we may orient the axes in such a way that $\omega_x = \omega_y = 0$, $\omega_z = \omega$, so that (113.6) reduces to

$$\dot{x}' = 2\omega y - P_{,x}, \quad \dot{y}' = -P_{,y}, \quad \dot{z}' = -P_{,z}. \tag{113.8}$$

Since the adherence condition (69.4) now assumes the form $\dot{\boldsymbol{p}}' = 0$, the harmonic function P must satisfy the boundary conditions

$$P_{,x} = 2\omega y, \quad P_{,y} = 0, \quad P_{,z} = 0. \tag{113.9}$$

We suppose $\omega \neq 0$ so as to exclude the case treated in the previous paragraph.

Consider first a motion within finite rigid boundary surfaces, on each point of which (113.9) applies. Then $\dot{y}'$ and $\dot{z}'$ are harmonic functions vanishing upon a closed bounding surface, and hence vanishing identically. Thus $P_{,y} = 0$, $P_{,z} = 0$; since $\nabla^2 P = 0$ it follows that $P = ax + b$. Comparing this result with $(113.9)_1$ yields $2\omega y = a$, but since $\omega \neq 0$ this condition can be satisfied only upon a single plane $y = \text{const}$, a type of boundary not included in the hypothesis. Hence no such motion exists. Second, consider a motion exterior to finite rigid surfaces, on each point of which (113.9) applies. Let us call *regular* a velocity field which, relative to the surfaces in question, approaches a definite limit at ∞ and possesses a gradient which is $O(p^{-2})$. Then by the uniqueness of solution to the exterior Dirichlet problem the functions $\dot{y}'$ and $\dot{z}'$ must vanish identically, and again there is no such regular motion. These results constitute the following ***first theorem of Supino***: *No continuous isochoric irrotational motion of a material adhering to boundary surfaces and completely filling a finite domain whose boundaries form a rigid system exists. There exists no such motion with a regular velocity field filling an infinite domain exterior to a rigid system of surfaces to which the material adheres.*

In the statement of the above theorem the term "rigid system" denotes a set of rigid surfaces rigidly attached to one another, so that all are endowed with the same angular velocity. Various irrotational motions of materials adhering to rigid bounding surfaces can indeed exist if these are in relative motion, or if the motion relative to them is not regular at ∞. A simple example is the plane irrotational vortex already cited, for the stream lines are the rigid concentric circles $r = r_0$, $z = z_0$, rotating at angular speeds $\omega = K r^{-2}$. Another example is furnished by the rectilinear motion $\dot{x}' = 2\omega y$, $\dot{y}' = 0$, $\dot{z}' = 0$, or, equivalently, $\dot{x} = \omega y$, $\dot{y} = \omega x$, where the material adheres to the plane $y = 0$, which is rotating at angular speed ω about the z-axis.

In general, however, as SUPINO points out, the conditions $(113.9)_2$ and $(113.9)_3$ are themselves sufficient to determine P, and hence are not likely to be compatible with $(113.9)_1$ except in degenerate cases. Another way of putting this same thing is to say that if there exists a harmonic function P satisfying (113.9) on any finite surface, however small, then by the theorem of KIRCHHOFF mentioned above it is unique in all space to within an additive constant. Expressed in kinematical terms, this result becomes the following ***second theorem of Supino***: *Suppose there exist an isochoric irrotational motion of a material adhering to a certain rigid boundary surface rotating at a certain angular velocity Then there is no other such motion adhering to any finite portion of that surface, however* *small, rotating at the same angular velocity.* Applying this theorem to the second example noted in the preceding paragraph shows that the only such motion of a material adhering to a finite plane area rotating about an axis in its plane is the motion induced by the rigid rotation of the entire plane. Similarly, from the first example we conclude that the only such motion of a material adhering to a finite portion of a circular cylinder rotating about its axis is the irrotational vortex motion induced by the rotation of the entire cylinder[1].

114. Superposable motions[2]. Two velocity fields $\dot{\boldsymbol{p}}_1$ and $\dot{\boldsymbol{p}}_2$ belonging to a certain class are said to be *superposable* if their vector sum, $\dot{\boldsymbol{p}}_1+\dot{\boldsymbol{p}}_2$, belongs to the same class. If $2\dot{\boldsymbol{p}}$ belongs to the same class as $\dot{\boldsymbol{p}}$, then $\dot{\boldsymbol{p}}$ is said to be *self-superposable.*

Consider first the class of circulation preserving motions. Since from (107.2) we have for all motion[3]

$$\operatorname{curl}\dot{\overline{(\dot{\boldsymbol{p}}_1+\dot{\boldsymbol{p}}_2)}} = \operatorname{curl}\ddot{\boldsymbol{p}}_1 + \operatorname{curl}\ddot{\boldsymbol{p}}_2 + \operatorname{curl}(\boldsymbol{w}_1\times\dot{\boldsymbol{p}}_2+\boldsymbol{w}_2\times\dot{\boldsymbol{p}}_1) \tag{114.1}$$

by (108.1) follows the condition for superposability:

$$\operatorname{curl}(\boldsymbol{w}_1\times\dot{\boldsymbol{p}}_2+\boldsymbol{w}_2\times\dot{\boldsymbol{p}}_1)=0. \tag{114.2}$$

It is trivial to remark that any two irrotational motions or any two screw motions having the same abnormality are superposable, and hence that any irrotational or steady screw motion is self-superposable. In the irrotational case, from (88.2) it is immediate that *the velocity potential for the combined motion is the sum of the velocity potentials of the two original motions*[4]. Also from (111.10) it follows that *any circulation preserving motion with steady vorticity is self-superposable.* It is not true in general however that an unsteady circulation preserving motion is self-superposable, that any two screw motions are superposable, or that an irrotational motion is superposable upon a circulation preserving motion.

Second, consider the class of motions whose acceleration is complex-lamellar. From (114.1) and (App. 33.5)$_1$ follows the necessary and sufficient condition

$$\ddot{\boldsymbol{p}}_1\cdot\operatorname{curl}\ddot{\boldsymbol{p}}_2+\ddot{\boldsymbol{p}}_2\cdot\operatorname{curl}\ddot{\boldsymbol{p}}_1+(\ddot{\boldsymbol{p}}_1+\ddot{\boldsymbol{p}}_2)\cdot\operatorname{curl}(\boldsymbol{w}_1\times\dot{\boldsymbol{p}}_2+\boldsymbol{w}_2\times\dot{\boldsymbol{p}}_1)=0. \tag{114.3}$$

For the self-superposable case this becomes

$$\ddot{\boldsymbol{p}}\cdot\operatorname{curl}(\boldsymbol{w}\times\dot{\boldsymbol{p}})=0, \tag{114.4}$$

since $\ddot{\boldsymbol{p}}\cdot\operatorname{curl}\ddot{\boldsymbol{p}}=0$; equivalently,

$$\ddot{\boldsymbol{p}}\cdot\frac{\partial\boldsymbol{w}}{\partial t}=0. \tag{114.5}$$

Thus *a motion with complex-lamellar acceleration is self-superposable if and only of the local rate of change of vorticity is normal to the acceleration.*

e II) Vorticity averages.

115. Intensity balance. The decomposition theorem of CAUCHY and STOKES (Sect. 90) resolves the local and instantaneous motion into stretching and spin. We now consider spatial averages of scalar measures of these portions. With

[1] SUPINO shows further that these two motions are the only possible plane isochoric irrotational motions in which the material adheres to any finite rigid surface.

[2] The following analysis, suggested by earlier work of BALLABH [1940, *1* and *2*], STRANG [1948, *28*], and ERGUN [1949, *6*], was given by TRUESDELL [1954, *24*, § 95].

[3] The dot over the bar indicates, of course, the material derivative based upon the total velocity, $\dot{\boldsymbol{p}}_1+\dot{\boldsymbol{p}}_2$.

[4] STOKES [1844, *4*, § 5].

IO defined by $(78.5)_1$, it is easy to show that

$$\operatorname{div}(\dot{\boldsymbol{p}}\cdot\operatorname{grad}\dot{\boldsymbol{p}} - \mathrm{I}_{\boldsymbol{d}}\dot{\boldsymbol{p}}) = -2\mathrm{IO} = -2\mathrm{II}_{\boldsymbol{d}} - \tfrac{1}{2}w^2, \tag{115.1}$$

where the second step follows by $(86.14)_1$. Hence

$$\left.\begin{aligned}\int_v \mathrm{IO}\,dv &= -\frac{1}{2}\oint_s d\boldsymbol{a}\cdot(\dot{\boldsymbol{p}}\cdot\operatorname{grad}\dot{\boldsymbol{p}} - \mathrm{I}_{\boldsymbol{d}}\dot{\boldsymbol{p}}),\\ &= -\frac{1}{2}\oint_s d\boldsymbol{a}\cdot\left(\ddot{\boldsymbol{p}} - \frac{\partial\dot{\boldsymbol{p}}}{\partial t} - \mathrm{I}_{\boldsymbol{d}}\dot{\boldsymbol{p}}\right).\end{aligned}\right\} \tag{115.2}$$

Hence follows the ***theorem of intensity balance***[1]: *If all finite boundaries are stationary, and if upon them the material adheres without slipping, while in any portion of the material extending to* ∞ *the condition*

$$(\dot{\boldsymbol{p}}\cdot\operatorname{grad}\dot{\boldsymbol{p}} - \mathrm{I}_{\boldsymbol{d}}\dot{\boldsymbol{p}})_n = \bar{o}(p^{-2}) \tag{115.3}$$

is satisfied, then the average value of IO *over the entire motion is zero; equivalently,*

$$\int_v (4\mathrm{II}_{\boldsymbol{d}} + w^2)\,dv = 0. \tag{115.4}$$

We note several corollaries. (1) *For a motion of the type described in the theorem to be rotational it is necessary that there exist within it a region where* $\mathrm{II}_{\boldsymbol{d}}<0$. (2) *In an irrotational motion satisfying the hypotheses of the theorem, the average value of* $\mathrm{II}_{\boldsymbol{d}}$ *is zero.* (3) *In an isochoric motion of the type described in the theorem the average value of the squared vorticity must equal twice the average value of the squared intensity of deformation:*

$$\int_v w^2\,dv = 2\int_v \overline{\mathrm{II}}_{\boldsymbol{d}}\,dv. \tag{115.5}$$

From the third corollary, which is a consequence of $(86.14)_2$ and (77.1), follows another proof of the theorem on the impossibility of irrotational adhering motions in a finite domain (Sect. 113).

116. Linear balance. The previous section demonstrated that in a broad class of motions an average balance between the second deformation invariant and the squared magnitude of the vorticity is maintained. We shall now establish two simpler but kinematically less informative relations of balance connecting the vorticity vector $\boldsymbol{w}$ and the expansion $\mathrm{I}_{\boldsymbol{d}}$.

Let $\boldsymbol{h}$ be any single-valued harmonic gradient: $\boldsymbol{h}=\operatorname{grad}Q$, $\nabla^2 Q=0$. In (App. 26.2), put $K=0$, $\boldsymbol{b}=\dot{\boldsymbol{p}}$, $\boldsymbol{c}=\boldsymbol{h}$. Then there results

$$\oint_s [d\boldsymbol{a}\cdot(\dot{\boldsymbol{p}}\,\boldsymbol{h} + \boldsymbol{h}\,\dot{\boldsymbol{p}}) - d\boldsymbol{a}\,\dot{\boldsymbol{p}}\cdot\boldsymbol{h}] = \int_v [\boldsymbol{h}\,\mathrm{I}_{\boldsymbol{d}} - \boldsymbol{h}\times\boldsymbol{w}]\,dv. \tag{116.1}$$

By formulating conditions sufficient for the vanishing of the surface integral we obtain the ***first linear balance theorem***[2]: *Let* $\boldsymbol{h}$ *be a single-valued harmonic gradient, and let* $\mathrm{I}_{\boldsymbol{d}}$ *be the expansion and* $\boldsymbol{w}$ *the vorticity of a motion in a region* v *such that*

1. *Each finite boundary is stationary, and upon it the material adheres without slipping;*

---

[1] TRUESDELL [1950, *34*]. A geometrical construction for the space average of IO had previously been invented by BILIMOVITCH [1948, *2*] and has been discussed further [1950, *34*] [1953, *1*]. From these results follows a characterization of those motions in which $\mathrm{IO}=0$; this class had been studied previously by HAMEL [1936, *4*, §§ 2–3].

[2] TRUESDELL [1951, *30*]. The special case $I_{\boldsymbol{d}}=0$ for a finite domain was given by BERKER [1949, *1*, Th. IV].

2. *In any portion of $\mathfrak{v}$ which extends to ∞,*

$$(\dot{\boldsymbol{p}}\,\boldsymbol{h} + \boldsymbol{h}\,\dot{\boldsymbol{p}})_n = \bar{o}\,(p^{-2}), \quad \dot{\boldsymbol{p}}\cdot\boldsymbol{h} = \bar{o}\,(p^{-2});$$

then

$$\int_{\mathfrak{v}} [\boldsymbol{h}\,\mathrm{I}_{\boldsymbol{d}} - \boldsymbol{h}\times\boldsymbol{w}]\,dv = 0. \tag{116.3}$$

A simple condition sufficient for (116.2) is $h\dot{p} = o\,(p^{-2})$.

Let $\boldsymbol{f}$ be a field whose curl is a harmonic gradient: $\mathrm{curl}\,\boldsymbol{f} = \mathrm{grad}\,F$, $\nabla^2 F = 0$, the harmonic function F being single-valued. Then we have

$$\left.\begin{aligned} \mathrm{div}\,(F\,\dot{\boldsymbol{p}}) &= F\,\mathrm{I}_{\boldsymbol{d}} + \dot{\boldsymbol{p}}\cdot\mathrm{curl}\,\boldsymbol{f},\\ &= F\,\mathrm{I}_{\boldsymbol{d}} + \mathrm{div}\,(\boldsymbol{f}\times\dot{\boldsymbol{p}}) + \boldsymbol{f}\cdot\boldsymbol{w}. \end{aligned}\right\} \tag{116.4}$$

Hence by GREEN'S transformation follows

$$\oint_{\mathfrak{s}} d\boldsymbol{a}\cdot[F\,\dot{\boldsymbol{p}} + \dot{\boldsymbol{p}}\times\boldsymbol{f}] = \int_{\mathfrak{v}} [F\,\mathrm{I}_{\boldsymbol{d}} + \boldsymbol{f}\cdot\boldsymbol{w}]\,dv. \tag{116.5}$$

By formulating conditions sufficient for the vanishing of the surface integral we obtain the ***second linear balance theorem***[1]: *Let F be a single-valued harmonic gradient, and let $\mathrm{curl}\,\boldsymbol{f} = \mathrm{grad}\,F$; let $\mathrm{I}_{\boldsymbol{d}}$ be the expansion and $\boldsymbol{w}$ the vorticity of a motion in a region $\mathfrak{v}$ such that*

1. *Each finite boundary is stationary, and upon it the material adheres without slipping;*

2. *In any portion of $\mathfrak{v}$ which extends to ∞*

$$(F\,\dot{\boldsymbol{p}} + \dot{\boldsymbol{p}}\times\boldsymbol{f})_n = \bar{o}\,(p^{-2}); \tag{116.6}$$

then

$$\int_{\mathfrak{v}} [F\,\mathrm{I}_{\boldsymbol{d}} + \boldsymbol{f}\cdot\boldsymbol{w}]\,dv = 0. \tag{116.7}$$

A simple condition sufficient for (116.6) is $F\,\dot{p}_n = o\,(r^{-2})$, $(\dot{\boldsymbol{p}}\times\boldsymbol{f})_n = o\,(r^{-2})$.

An immediate corollary of (116.7), following from the choice $\boldsymbol{f} = 0$, $F = 1$, is the vanishing of the total expansion:

$$\int_{\mathfrak{v}} \mathrm{I}_{\boldsymbol{d}}\,dv = 0. \tag{116.8}$$

Putting $\boldsymbol{h} = \mathrm{const}$ in (116.3) and employing (116.8) yields

$$\boldsymbol{h}\times\int_{\mathfrak{v}} \boldsymbol{w}\,dv = 0, \tag{116.9}$$

whence, since $\boldsymbol{h}$ is arbitrary, follows

$$\int_{\mathfrak{v}} \boldsymbol{w}\,dv = 0: \tag{116.10}$$

In a motion such that both the linear balance theorems hold, the total vorticity vanishes. We shall discover a broad generalization of this result in Sect. 118.

For the case of a motion in a finite simply connected domain, it is possible to show[2] that either (116.7) by itself or (116.3) combined with (105.2) is sufficient that assigned functions $\mathrm{I}_{\boldsymbol{d}}$ and $\boldsymbol{w}$ be the expansion and the vorticity of the motion of a material adhering to the boundary of $\mathfrak{v}$. That (116.3) is not by itself sufficient is plain from the example $\mathrm{I}_{\boldsymbol{d}} = 0$, $\boldsymbol{w} = \mathrm{grad}\,p^{-1}$ when the bounding surface $\mathfrak{s}$

---

[1] TRUESDELL [1951, *31*]. The special case $I_{\boldsymbol{d}} = 0$ for a finite domain is given by BERKER [1949, *1*, Th. I].

[2] TRUESDELL [1951, *30* and *31*]. The results were asserted also by VAN DEN DUNGEN [1951, *37*] and SYNGE [1951, *25*], but their proofs are incomplete.

is a pair of concentric spheres, since then we have

$$\left.\begin{aligned}\int_v \boldsymbol{h}\times\boldsymbol{w}\,dv &= \int_v \boldsymbol{h}\times\operatorname{grad} p^{-1}\,dv = \int_v \operatorname{curl}(Q\operatorname{grad} p^{-1})\,dv,\\ &= \oint_s d\boldsymbol{a}\times Q\operatorname{grad} p^{-1} = 0.\end{aligned}\right\} \tag{116.11}$$

For this special case, then, the condition (116.3) is satisfied, but upon the spherical boundaries we have $w_n = \frac{\partial p^{-1}}{\partial p} = -p^{-2} \neq 0$, whence by (105.2) it follows that the material cannot adhere to the boundary.

When applied to the special case of a plane motion, both (116.3) and (116.7) yield[1]

$$\int_s (G\,\mathrm{I}_{\boldsymbol{d}} - H\,w)\,da = 0, \tag{116.12}$$

where H and G are any pair of conjugate harmonic functions. When s is finite and simply connected, (116.12) is sufficient that $\mathrm{I}_{\boldsymbol{d}}$ and w be the expansion and the vorticity of a material adhering to the boundary.

117. The theorems of LAMB, POINCARÉ, J. J. THOMSON, and BJØRGUM. Leaving behind the connections between vorticity and expansion, we shall now learn the regularities inherent in the distribution of rotation alone by establishing theorems concerning the average value of the vorticity itself. The first such results discovered concern only isochoric motions, upon which we now fix our attention. Since in an isochoric motion the velocity field is solenoidal, we may put $\boldsymbol{c} = \dot{\boldsymbol{p}}$ in (App. 31.7), so obtaining a simple formula for the total Lamb vector. This formula shows that for *an isochoric motion in a finite stationary domain, the average value of* $\boldsymbol{w}\times\dot{\boldsymbol{p}}$ *is determined by the speed on the bounding surfaces; in the case when the speed is constant on each closed boundary surface, the average value of* $\boldsymbol{w}\times\dot{\boldsymbol{p}}$ *is zero.* More generally, by replacing $\boldsymbol{c}$ by $\dot{\boldsymbol{p}}$ in the italicized statement following (App. 31.7), we obtain ***Lamb's vorticity average theorem***[2]: *In an isochoric motion, if all finite boundaries are stationary and the material adheres to them, while in any portion of the material extending to infinity* $\dot{p}\,\dot{p}_n = \bar{o}(p^{-2})$, $\dot{p}^2 = \bar{o}(p^{-2})$, *then*

$$\int_v \boldsymbol{w}\times\dot{\boldsymbol{p}}\,dv = 0; \tag{117.1}$$

that is, the average value of the Lamb vector is zero. The property which defines the class of irrotational and screw motions, i.e., $\boldsymbol{w}\times\dot{\boldsymbol{p}} = 0$, thus has been shown to hold *on the average* for a much greater class of motions. A simple sufficient condition at ∞ is $\dot{p} = o(p^{-1})$.

By putting $\boldsymbol{c} = \dot{\boldsymbol{p}}$ in (App. 31.16) we similarly obtain a formula for the total moment of the Lamb vector; hence follows ***Poincaré's vorticity average theorem***[3]: *In an isochoric motion, if all finite boundaries are stationary and the material adheres to them, while in any portion of the material extending to infinity* $\boldsymbol{p}\times\dot{\boldsymbol{p}}\,\dot{p}_n = \bar{o}(p^{-2})$, $\dot{p}^2\boldsymbol{p}_t = \bar{o}(p^{-2})$, *then*

$$\int_v \boldsymbol{p}\times(\boldsymbol{w}\times\dot{\boldsymbol{p}})\,dv = 0. \tag{117.2}$$

A simple sufficient condition at ∞ is $\dot{p} = o(p^{-\frac{3}{2}})$.

Three other vorticity theorems may be obtained by putting $\boldsymbol{c} = \dot{\boldsymbol{p}}$ in (App. 31.23), (App. 31.25), and (App. 31.21). The surface integrals in these formulae vanish when

---

[1] SYNGE [1950, *31*]. Special cases had been obtained earlier by HAMEL [1911, *6*, p. 266] SYNGE [1936, *9*, § 2], and KAMPÉ DE FERIET [1946, *6*] [1947, *8*].

[2] [1879, *2*, § 136].

[3] [1893, *6*, § 115].

the velocity is tangent to $\mathfrak{s}$, and in the case of an infinite region, when $\dot{p}_n = o(p^{-3})$, $o(p^{-4})$, $o(p^{-4})$, respectively. The interest in these formulae[1] lies in the fact that when the motion is isochoric they express the momentum and the moment of momentum of the material in $\mathfrak{v}$ in terms of the vorticity and the normal component of velocity upon the boundary (cf. Sect. 166).

Finally, consider the case when the Monge representation (99.19) is valid over an entire motion. For the abnormality (App. 30.1)$_1$ we obtain

$$\left.\begin{aligned} \dot{p}^2 A = \dot{\boldsymbol{p}} \cdot \boldsymbol{w} &= \operatorname{grad} H \cdot \operatorname{grad} F \times \operatorname{grad} G, \\ &= \operatorname{div}(H \operatorname{grad} F \times \operatorname{grad} G), \\ &= \operatorname{div}(H \boldsymbol{w}), \end{aligned}\right\} \tag{117.3}$$

where the last step follows by (99.20). Integration over the volume yields the ***vorticity average theorem of Bjørgum***[2]: *When H is single-valued,*

$$\int_{\mathfrak{v}} \dot{p}^2 A\, dv = \oint_{\mathfrak{s}} d\boldsymbol{a} \cdot \boldsymbol{w} H. \tag{117.4}$$

Hence in a region on all whose finite boundaries $\boldsymbol{w}$ is tangential, while in any portion extending to ∞ $H w_n = \bar{o}(p^{-2})$, we have

$$\int_{\mathfrak{v}} \dot{p}^2 A\, dv = 0; \tag{117.5}$$

in such a region, then, if the motion is not complex-lamellar, the abnormality A must assume both positive and negative values. In particular, the theorem holds for a motion in a finite domain to whose walls the material adheres[3].

An interesting formula for the total squared vorticity follows from the substitution $\boldsymbol{c} = \boldsymbol{w}$ in (App. 31.12), permissible because $\boldsymbol{w}$ is solenoidal[4]. This formula expresses the total squared vorticity in terms of the boundary values of $\boldsymbol{w}$ and interior values of $\operatorname{curl} \boldsymbol{w} \times \boldsymbol{w}$. In particular, if $\boldsymbol{w}$ vanishes on the boundary and $\operatorname{curl} \boldsymbol{w} \times \boldsymbol{w}$ vanishes in the interior, it follows that $\boldsymbol{w}$ vanishes everywhere.

118. The vorticity moment theorem. Since the vorticity field is solenoidal, we may put $\boldsymbol{c} = \boldsymbol{w}$ in (App. 31.3), thus obtaining[5]

$$\mathfrak{W}_{(K)} \equiv \int_{\mathfrak{v}} \{\boldsymbol{p}^{(K)} \boldsymbol{w}\}\, dv = \oint_{\mathfrak{s}} d\boldsymbol{a} \cdot \boldsymbol{w} \boldsymbol{p}^{(K+1)}; \tag{118.1}$$

---

[1] MOREAU [1948, *16*] [1949, *20*] [1950, *18*] [1952, *13*, §§ 14—22] has studied the rates of change of (App. 31.23) and (App. 31.25) in an isochoric motion, expressing them in terms of the resultant force and moment of force of any non-conservative extrinsic forces. Proofs of essentially equivalent but purely kinematical formulae based on the decomposition (124.1) were given by TRUESDELL [1951, *28*, § 8]. MOREAU noted also some alternative forms of (App. 31.25) which are easily obtained by contracting (App. 31.3) in the case $K = 2$. He emphasized the application of his results to a limitless fluid, all but a finite interior part of which is in irrotational or circulation preserving motion. In this connection we should beware of the extremely strong order conditions at ∞ required in order to get simple results, order conditions, indeed, which possibly may never be satisfied except in trivial cases.

[2] [1951, *2*, § 6.6].

[3] It is worth noting that on the right-hand side in (117.4) H may be replaced by F, which is necessarily single-valued, or by G, assumed single-valued, provided $\boldsymbol{w}$ be replaced by $\operatorname{grad} G \times \operatorname{grad} H$ or $\operatorname{grad} H \times \operatorname{grad} F$, respectively. Thus in case any of one of the functions F, G, H vanishes on a closed surface, the result (117.5) follows independently of any further hypothesis regarding $\boldsymbol{w}$.

[4] TRUESDELL [1951, *29*].

[5] The case $K = 0$ was apparently known to A. FÖPPL [1897, *4*, §§ 4, 32] and was stated by MUNK [1941, *4*, Eq. (7)]. The general formula and its consequences were derived by TRUESDELL [1949, *35*] [1951, *28*, § 11]; the invariance of the result with respect to change of the origin of the position vector was checked in [1954, *24*, § 65].

All the moments $\mathfrak{W}_{(K)}$ *of the vorticity field over any region* v *are independent of conditions at interior points, being completely determined by the normal component of vorticity upon the boundary* s. This purely kinematical statement indicates the predominant effect of boundaries upon vorticity. It may be regarded as asserting that the familiar hydrodynamical theorem that vorticity cannot be generated in the interior of a homogeneous viscous liquid subject to conservative extrinsic force, but must be diffused inward from the boundaries, continues to hold for arbitrary continuous media, provided it be expressed in terms of the *average* rather than the local vorticity.

Now upon a stationary boundary to which the material adheres, by (105.2) the normal component of vorticity vanishes, and thus *in any motion bounded by finite stationary walls to which the material adheres, all moments* $\mathfrak{W}_{(K)}$ *vanish.* The case $K=0$ has already been derived as (116.10). By formulating more general conditions sufficient for the vanishing of the surface integral on the right-hand side of (118.1), we obtain the following ***vorticity moment theorem***: *Subject to the boundary conditions*

$$w_n = 0 \quad \text{or} \quad w_n = \bar{o}\,(p^{-K-3}), \tag{118.2}$$

the first $K+1$ *moments of vorticity vanish*[1].

For the special case of *a motion enclosed by finite stationary boundaries to which the material adheres*, our several investigations in Sects. 112 to 113 and Sects. 115 to 118 have revealed a high degree of regularity: *The motion is almost certainly rotational (if isochoric, certainly rotational), the intensity balance theorem holds, the linear balance theorems hold, and all the moments* $\mathfrak{W}_{(K)}$ *are zero.* It is to be borne in mind that no restriction regarding the connectivity has been presupposed, and indeed the main interest here is in the case of multiply connected regions.

119. A general vorticity average. Since $\boldsymbol{w}$ is solenoidal, we may put $\boldsymbol{c}=\boldsymbol{w}$ in (App. 31.5) and obtain

$$\int_v \boldsymbol{w}\cdot \operatorname{grad}\Psi\, dv = \oint_s d\boldsymbol{a}\cdot \boldsymbol{w}\Psi\,. \tag{119.1}$$

Thus[2] *for any twice continuously differentiable quantity* Ψ, *the total* $\boldsymbol{w}\cdot\operatorname{grad}\Psi$ *in a region is completely determined by the values of* Ψ *and of the normal component of vorticity* w_n *upon the boundary.* The vorticity moment theorem (118.1) is the special $\Psi = \boldsymbol{p}^{(K+1)}$ in (119.1).

Another special case has been obtained by Berker[3]. First, put $\Psi = B$ and write $\boldsymbol{f} \equiv \operatorname{grad} B$ in (119.1):

$$\int_v \boldsymbol{w}\cdot\boldsymbol{f}\, dv = \oint_s d\boldsymbol{a}\cdot\boldsymbol{w}B\,. \tag{119.2}$$

By formulating conditions sufficient for the vanishing of the surface integral we obtain the following vorticity average theorem: *Given a motion in a region* v *such that upon any finite boundaries the normal component of vorticity vanishes, let B be a twice continuously differentiable scalar and put* $\boldsymbol{f}\equiv\operatorname{grad} B$; *if in any portion of* v *extending to* ∞ *we have* $Bw_n = \bar{o}\,(p^{-2})$, *then*

$$\int_v \boldsymbol{w}\cdot\boldsymbol{f}\, dv = 0\,. \tag{119.3}$$

The conditions of this theorem are satisfied by the material within a closed vortex tube of any continuous motion; by the entire material of any continuous motion in a bounded domain upon whose boundaries $\boldsymbol{w}=0$; and for any such motion within finite stationary boundaries

[1] A special case of the case $K=0$ was given by A. Föppl [1897, *4*, §§ 4, 32].

[2] Truesdell [1954, *24*, § 66].

[3] [1949, *1*, Th. II].

to which the material adheres without slipping. Putting $\boldsymbol{f} = \boldsymbol{p} = \operatorname{grad} \frac{1}{2} \dot{p}^2$, we obtain

$$\int_v \boldsymbol{p} \cdot \boldsymbol{w} \, dv = 0, \tag{119.4}$$

a result which follows equally by taking the scalar of the equation $\mathfrak{W}_{(1)} = 0$.

The form of the result (119.3) coincides with that of the special case $\mathrm{I}_{\boldsymbol{d}} = 0$ of (116.7) but the conditions under which the two statements hold are somewhat different. For the validity of (119.3) it is required that $\operatorname{curl} \boldsymbol{f} = 0$, but the motion need not be isochoric, while for the special case of (116.7) valid in an isochoric motion it is sufficient that $\operatorname{curl} \operatorname{curl} \boldsymbol{f} = 0$.

The completeness of the set of theorems of vorticity average has been investigated by HOWARD[1]. He has shown that, within a certain algebraic class, for general motions there are no additional theorems beyond those obtained by linear combination of the results in Sect. 118; for isochoric motions, there are in addition not only those obtained in Sect. 117 but also

$$\int_v \boldsymbol{\beta} \cdot \boldsymbol{w} \, dv = \oint_s d\boldsymbol{a} \cdot (\alpha \dot{\boldsymbol{x}} + \dot{\boldsymbol{x}} \times \boldsymbol{\beta}), \tag{119.5}$$

where $\operatorname{curl} \boldsymbol{\beta} = \operatorname{grad} \alpha$.

It is possible to obtain further vorticity average theorems only by extending the algebraic classes considered by HOWARD (e.g., the results of Sect. 116 and some of those of Sect. 119 are not included), or by narrowing the class of motions.

e III) Bernoullian theorems.

120. The nature of Bernoullian theorems. For an irrotational motion, the speed is related to the acceleration potential by EULER's formula (109.3), which we now write in the form

$$V^* + \frac{1}{2} \dot{p}^2 - \frac{\partial V}{\partial t} = 0. \tag{120.1}$$

This formula gives the essential content of what in classical hydrodynamics is called ***Bernoulli's theorem***. Since it would be inappropriate here to tarry over hydrodynamical details[2], we rest content with the remark that (120.1) serves to suggest for rotational motions a search for properties of the *specific kinetic energy* $\frac{1}{2}\dot{p}^2$ or the *specific motive energy* $S^* + \frac{1}{2}\dot{p}^2$, where S^* is defined by (102.5). In such relations, naturally, $\frac{1}{2}\dot{p}^2$ or $S^* + \frac{1}{2}\dot{p}^2$ should be determined by something less than a knowledge of the velocity field as a function of place and time.

The basic formulae on which the analysis rests are, first, the Stokes resolution (102.5) of the acceleration field in terms of its scalar potential S^* and vector potential $\boldsymbol{v}^*$, and second, LAGRANGE's formula (99.14) for the acceleration, in which $\boldsymbol{w} \times \dot{\boldsymbol{p}}$, the *Lamb vector*, is of central interest.

Using these two formulae, we shall find classes of motions for which $\frac{1}{2}\dot{p}^2$ or $S^* + \frac{1}{2}\dot{p}^2$ has particularly simple properties.

121. Poisson equations for the specific energy. By (102.6) any of the expressions for $\operatorname{div} \ddot{\boldsymbol{p}}$ given in Sect. 102 may be written in the form $\nabla^2 S^* = \cdots$. All results of this type express *properties common to all motions whose accelerations have the same scalar potential*, whatever the vector potential may be.

First, from $(102.8)_1$ we have[3]

$$\left.\begin{aligned} \nabla^2\left(S^* + \frac{1}{2}\dot{p}^2\right) &= -\frac{\partial \mathrm{I}_{\boldsymbol{d}}}{\partial t} + \operatorname{div}(\dot{\boldsymbol{p}} \times \boldsymbol{w}), \\ &= -\frac{\partial \mathrm{I}_{\boldsymbol{d}}}{\partial t} + w^2 - \dot{\boldsymbol{p}} \cdot \operatorname{curl} \boldsymbol{w}. \end{aligned}\right\} \tag{121.1}$$

---

[1] [1957, *8*].

[2] The connection of this formulation with that usual in hydrodynamics is explained by TRUESDELL [1954, *24*, § 68]. For a history of the researches of DANIEL BERNOULLI, JOHN I BERNOULLI, and EULER in this connection, see TRUESDELL [1954, *25*, Parts IV and VI].

[3] Special cases are due to BOBYLEW [1873, *1*, § 4], FORSYTH [1879, *1*, p. 139], CRAIG [1880, *5*, pp. 223–225] [1880, *7*, p. 274], and ROWLAND [1880, *9*, p. 268].

As a corollary it follows that *in any motion with steady expansion, in order that the specific motive energy be harmonic:*

$$\nabla^2 (S^* + \tfrac{1}{2}\dot{p}^2) = 0, \tag{121.2}$$

it is necessary and sufficient that the Lamb vector be solenoidal. In particular, (121.2) *holds in any irrotational or screw motion in which the expansion is steady.* As a further corollary it follows that *in an irrotational or screw motion filling all space, if the expansion is steady and if the specific motive energy $S^* + \frac{1}{2}\dot{p}^2$ is uniformly bounded, then the classical Bernoulli theorem holds*[1]:

$$S^* + \tfrac{1}{2}\dot{p}^2 = f(t). \tag{121.3}$$

This result extends (120.1) besides viewing it from a different aspect.

From (121.1) may be read off conditions that $S^* + \frac{1}{2}\dot{p}^2$ be subharmonic or superharmonic, and hence follow extremal theorems for $S^* + \frac{1}{2}\dot{p}^2$. Unfortunately it does not seem easy to put in kinematically informative terms the conditions so obtained. We note[2] that *if the expansion nowhere increases and if the vectors $\dot{\boldsymbol{p}}$ and* curl $\boldsymbol{w}$ *nowhere subtend an acute angle then $\nabla^2 S^* \geqq 0$, and hence $S^* + \frac{1}{2}\dot{p}^2$ cannot experience a maximum at an interior point.*

For S^* itself, however, elegantly simple results may be obtained. We begin with the identity[2]

$$-\nabla^2 S^* = \dot{\mathrm{I}}_{\boldsymbol{d}} + \overline{\mathrm{II}}_{\boldsymbol{d}} - \tfrac{1}{2} w^2, \tag{121.4}$$

which is immediate from $(102.1)_4$ and (102.6). For the special case $\dot{\mathrm{I}}_{\boldsymbol{d}} = 0$, Hamel[3] has remarked that since $w^2 \geqq 0$, $\overline{\mathrm{II}}_{\boldsymbol{d}} \geqq 0$, we may obtain bounds for $\nabla^2 S^*$, viz., in a motion in which *the expansion remains constant for each particle we have*

$$-\overline{\mathrm{II}}_{\boldsymbol{d}} \leqq \nabla^2 S^* \leqq \tfrac{1}{2} w^2. \tag{121.5}$$

This result, which holds *a fortiori* for isochoric motions, may be expressed in terms of the vorticity number $\mathfrak{W}_{\mathrm{K}}$ of Sect. 91:

$$\frac{\nabla^2 S^*}{\overline{\mathrm{II}}_{\boldsymbol{d}}} \leqq \mathfrak{W}_{\mathrm{K}}^2, \qquad \frac{\nabla^2 S^*}{\frac{1}{2} w^2} \geqq -\frac{1}{\mathfrak{W}_{\mathrm{K}}^2}. \tag{121.6}$$

More generally, either from (121.4) or from (102.7) we have

$$-\nabla^2 S^* = \dot{\mathrm{I}}_{\boldsymbol{d}} + \overline{\mathrm{II}}_{\boldsymbol{d}} (1 - \mathfrak{W}_{\mathrm{K}}^2), \tag{121.7}$$

inspection of which yields the following results: *In a non-rigid motion, sufficient conditions for S^* to be superharmonic, harmonic, or subharmonic, respectively, are $\dot{\mathrm{I}}_{\boldsymbol{d}} \geqq 0$ and $\mathfrak{W}_{\mathrm{K}} \leqq 1$; $\dot{\mathrm{I}}_{\boldsymbol{d}} = 0$ and $\mathfrak{W}_{\mathrm{K}} = 1$; $\dot{\mathrm{I}}_{\boldsymbol{d}} \leqq 0$ and $\mathfrak{W}_{\mathrm{K}} \geqq 1$.* In the special case when $\dot{\mathrm{I}}_{\boldsymbol{d}} = 0$ the value of $\mathfrak{W}_{\mathrm{K}}$ becomes the sole criterion of the character of S^*. Hence follow conclusions regarding various types of motion: first, the ***maximum theorem for motions in which $\mathfrak{W}_{\mathrm{K}} \geqq 1$***: *Given a motion in which the expansion experienced by a particle does not increase ($\dot{\mathrm{I}}_{\boldsymbol{d}} \leqq 0$), the greatest value of the scalar potential S^* in a region where $\mathfrak{W}_{\mathrm{K}} \geqq 1$ cannot be attained in the interior but must be attained on the boundary*; and, second, the ***minimum theorem for motions in***

[1] Truesdell [1954, *24*, § 70].

[2] Special cases were given by Rowland [1880, *9*, p. 267], Gosiewski [1890, *3*, §§ 3–4], Lichtenstein [1929, *4*, Chap. 10, § 6], and Lagally [1937, *4*].

[3] [1936, *4*, § 1]. As had been observed by Gosiewski [1890, *3*, § 8], from (121.4) it follows that the condition

$$-\nabla^2 S^* = \overline{\mathrm{II}}_{\boldsymbol{d}} - \tfrac{1}{2} w^2$$

is necessary and sufficient that a material volume once in isochoric motion remain ever in isochoric motion.

which $\mathfrak{W}_K \leqq 1$: *Given a motion in which the expansion experienced by a particle does not decrease* ($\dot{\mathrm{I}}_{\boldsymbol{d}} \geqq 0$), *the least value of* S^* *in a region where* $\mathfrak{W}_K \leqq 1$ *cannot be attained in the interior, but must be attained on the boundary*[1]. The former theorem includes the case of rigid motion ($\mathfrak{W}_K = \infty$, $\mathrm{I}_{\boldsymbol{d}} = 0$); the latter, the case of irrotational motion ($\mathfrak{W}_K = 0$); both include the case $\mathrm{I}_{\boldsymbol{d}} = 0$, $\mathfrak{W}_K = 1$, which was mentioned from another point of view in Sect. 91.

For irrotational or screw motions, an analogous result can be obtained for the speed $\dot{p}$, generalizing KELVIN's theorem in Sect. 102. By (121.7) and (121.1) when $\boldsymbol{w} \times \dot{\boldsymbol{p}} = 0$ we get

$$\left.\begin{aligned} \nabla^2 \frac{1}{2} \dot{p}^2 &= \dot{\mathrm{I}}_{\boldsymbol{d}} - \frac{\partial \mathrm{I}_{\boldsymbol{d}}}{\partial t} + \overline{\mathrm{II}}_{\boldsymbol{d}} (1 - \mathfrak{W}_{K}^2), \\ &= \dot{\boldsymbol{p}} \cdot \operatorname{grad} \mathrm{I}_{\boldsymbol{d}} + \overline{\mathrm{II}}_{\boldsymbol{d}} (1 - \mathfrak{W}_{K}^2). \end{aligned}\right\} \tag{121.8}$$

By formulating conditions sufficient that the right-hand side have a given sign, we derive the ***theorem of maximum speed for irrotational or screw motions***[2]: *In a region of irrotational or screw motion such that* $\mathfrak{W}_K \leqq 1$ ($\mathfrak{W}_K = 1$) ($\mathfrak{W}_K \geqq 1$), *while the expansion* $\mathrm{I}_{\boldsymbol{d}}$ *does not decrease (change) (increase) in the direction of motion along a stream line, then at an interior point the speed cannot experience a maximum (maximum or minimum) (minimum).* In the irrotational case, of course, $\mathfrak{W}_K = 0$, so the results concerning minimum speed do not apply. A corollary is that in a region of isochoric screw motion where $\mathfrak{W}_K \geqq 1$ there can be no stagnation point.

By applying POISSON's integral from the theory of the potential, it is possible from the Poisson equations given above to write down various expressions for S^* or for $S^* + \frac{1}{2}\dot{p}^2$ as a volume integral[3].

122. Lamb planes and Lamb surfaces. In a motion which is neither an irrotational nor a screw motion the vorticity $\boldsymbol{w}$ and velocity $\dot{\boldsymbol{p}}$ differ in direction, except possibly at certain singular points, lines, or surfaces, and hence at each regular point determine the *Lamb plane*, whose normal is parallel to the Lamb vector $\boldsymbol{w} \times \dot{\boldsymbol{p}}$. A necessary and sufficient condition for the existence of *Lamb surfaces*[4], which are simultaneously vortex surfaces and stream surfaces, is that the Lamb vector $\boldsymbol{w} \times \dot{\boldsymbol{p}}$ be complex-lamellar and non-vanishing. By the Euler-Kelvin criterion (App. 33.5) it is then necessary and sufficient that

$$\boldsymbol{w} \times \dot{\boldsymbol{p}} \cdot \operatorname{curl} (\boldsymbol{w} \times \dot{\boldsymbol{p}}) = 0, \qquad \boldsymbol{w} \times \dot{\boldsymbol{p}} \neq 0, \tag{122.1}$$

a condition which may be put into the form

$$\boldsymbol{w} \times \dot{\boldsymbol{p}} \cdot (\dot{\boldsymbol{p}} \cdot \operatorname{grad} \boldsymbol{w} - \boldsymbol{w} \cdot \operatorname{grad} \dot{\boldsymbol{p}}) = 0. \tag{122.2}$$

By eliminating $\boldsymbol{w} \times \dot{\boldsymbol{p}}$ between (122.1) and LAGRANGE's acceleration formula (99.14), we may obtain a form of the condition for the existence for Lamb surfaces which though lacking the symmetry of (122.2) is nevertheless easier to apply, viz.

$$\boldsymbol{w} \times \dot{\boldsymbol{p}} \cdot \left(\boldsymbol{w}^* - \frac{\partial \boldsymbol{w}}{\partial t} \right) = 0. \tag{122.3}$$

---

[1] These broad generalizations of results of BOULIGAND [1927, *1* and *2*] and of HAMEL [1936, *4*, § 1] were given by TRUESDELL [1953, *31*, § 10].

[2] TRUESDELL [1953, *34*].

[3] BOBYLEW [1873, *1*, § 4], FORSYTH [1879, *1*, p. 139], CRAIG [1880, *5*, pp. 223–225] [1880, *7*, p. 276], TRUESDELL [1954, *24*, § 71].

[4] These surfaces were introduced by LAMB [1878, *5*] [1879, *2*, § 145]. Cf. POINCARÉ [1893, *6*, §§ 22–24], APPELL [1921, *1*, § 762]. The name "Bernoulli surfaces" was proposed by CALDONAZZO [1924, *2*] [1925, *3*, § 2] in a somewhat different sense.

By the d'Alembert-Euler condition (108.1) it follows then that *Lamb surfaces exist in any circulation preserving motion with steady vorticity.*

Equivalently, for the existence of Lamb surfaces it is necessary and sufficient that there exist a non-constant scalar B and a non-vanishing scalar C such that

$$\boldsymbol{w} \times \dot{\boldsymbol{p}} = C \operatorname{grad} B. \tag{122.4}$$

The surfaces $B = \text{const}$ are the Lamb surfaces, and B must satisfy the differential system

$$(\boldsymbol{w} \times \dot{\boldsymbol{p}}) \times \operatorname{grad} B = 0. \tag{122.5}$$

For a very simple example of Lamb surfaces in motions which need not be circulation preserving, consider the case when the convective acceleration vanishes:

$$\ddot{\boldsymbol{p}} = \frac{\partial \dot{\boldsymbol{p}}}{\partial t}. \tag{122.6}$$

Then by (99.14) follows

$$\boldsymbol{w} \times \dot{\boldsymbol{p}} = -\operatorname{grad} \tfrac{1}{2} \dot{p}^2. \tag{122.7}$$

Thus the surfaces of constant speed are Lamb surfaces. Moreover, by (99.7) it follows that if the stream lines are steady, they are straight, and the speed is constant along them at each instant[1]. Thus the Lamb surfaces are ruled by the stream lines.

Since both stream and vortex lines lie upon the Lamb surfaces (if these exist), these surfaces together with the stream surfaces and vortex surfaces normal to them form three one-parameter families of surfaces, and hence serve to define a natural curvilinear co-ordinate system[2].

The following statements are but immediate applications of a classical theorem of DARBOUX[3].

(a) *In a complex-lamellar motion with Lamb surfaces, the vorticity is complex-lamellar if and only if the vortex lines are lines of curvature both on the Lamb surfaces and on the surfaces normal to the velocity.*

(b) *In a motion in which both the velocity and the vorticity are complex-lamellar, Lamb surfaces exist if and only if the stream lines are lines of curvature on the surfaces normal to the vorticity, while the vortex lines are lines of curvature on the surfaces normal to the velocity.*

(c) *In a motion with Lamb surfaces and with complex-lamellar vorticity, the motion itself is complex-lamellar if and only if the stream lines are lines of curvature both on the Lamb surfaces and on the surfaces normal to the vorticity.*

In a steady motion, the Lamb surfaces, since they are stream surfaces, are material surfaces (Sect. 74), and since they are also vortex surfaces, it follows that *in any steady motion such that* curl $\ddot{\boldsymbol{p}}$ *is zero or normal to the Lamb vector there exist stationary surfaces which are both stream surfaces and vortex surfaces.*

123. The line integral Bernoulli theorem. If $\ddot{p}_t$ denotes the component of acceleration along any direction $\boldsymbol{t}$ which lies in the Lamb plane, and d/ds_t denotes the directional derivative in that direction, then from (111.9), which is valid only in motions with steady vorticity, it follows that[4]

$$\ddot{p}_t = \frac{d}{d s_t}\left(\frac{1}{2}\dot{p}^2 + U\right). \tag{123.1}$$

A curve which is everywhere tangent to the Lamb planes is a *Lamb curve* c_L. Both stream lines and vortex lines are Lamb curves, and in a motion where Lamb surfaces exist any curve lying wholly upon some one of them is a Lamb curve. If we integrate (123.1) along a Lamb curve we obtain the ***line integral***

[1] CASTOLDI [1953, *5*].

[2] CRAIG [1881, *2*, pp. 5–6] used as co-ordinate surfaces the Lamb surfaces (which he incorrectly assumed always to exist), any independent family of stream surfaces, and any independent family of vortex surfaces.

[3] [1866, *1*, ¶ 15].

[4] FABRI [1894, *3*].

Bernoulli theorem:

$$(\tfrac{1}{2}\dot{p}^2 + U)|_{c_L} = \int_{c_L} d\boldsymbol{x} \cdot \ddot{\boldsymbol{p}}: \tag{123.2}$$

In a motion with steady vorticity, the flow of acceleration along a Lamb curve at any instant equals the difference of the values of $\frac{1}{2}\dot{p}^2 + U$ *at the two ends of the curve.* In particular, *in a motion with steady vorticity the circulation of the acceleration around a closed Lamb curve is zero.* In a steady motion, since $U = \text{const}$ the line integral Bernoulli theorem gives a direct connection between speed and acceleration. For any particular dynamical model the acceleration is expressed in terms of other quantities (cf. Chap. G), and (123.2) then shows directly the effect of these quantities upon the speed of flow[1].

124. The curvilinear Bernoulli theorem. Now in general it is possible to express any vector, and in particular the acceleration $\ddot{\boldsymbol{p}}$, as the sum of a gradient plus a second field in an infinite number of ways:

$$\ddot{\boldsymbol{p}} = -\operatorname{grad} Q + \ddot{\boldsymbol{p}}^*. \tag{124.1}$$

The function Q may be the Stokes scalar potential S^*, the negative of the Monge potential H^*, or some other function. We shall assume that $Q \neq \text{const}$, so that (124.1) really expresses a decomposition of the acceleration field, and we shall assume further that at least one possible choice of Q be such that[2] $\ddot{\boldsymbol{p}}^* \neq \boldsymbol{w} \times \dot{\boldsymbol{p}}$. We have

$$\boldsymbol{w}^* = \operatorname{curl} \ddot{\boldsymbol{p}} = \operatorname{curl} \ddot{\boldsymbol{p}}^*. \tag{124.2}$$

As suggested by the results in Sect. 101, we shall call $\ddot{\boldsymbol{p}}^*$ the *diffusive acceleration*, taking care to recall ever that this field is determined only to within an arbitrary gradient. The numerous theorems to follow in whose statements reference to the diffusive acceleration occurs may be divided into two classes. Those which essentially employ only curl $\ddot{\boldsymbol{p}}^*$ are single statements, but those which employ $\ddot{\boldsymbol{p}}^*$ itself are really an infinity of statements, one for each admissible choice of $\ddot{\boldsymbol{p}}^*$.

In a motion with steady vorticity, comparison of (124.1) with (111.9) yields

$$\operatorname{grad}(Q + U + \tfrac{1}{2}\dot{p}^2) = \dot{\boldsymbol{p}} \times \boldsymbol{w} + \ddot{\boldsymbol{p}}^* \neq 0. \tag{124.3}$$

Suppose $\dot{\boldsymbol{p}} \times \boldsymbol{w} \neq 0$, and at each point let $\boldsymbol{t}$ be a vector determined by the intersection of the Lamb plane with the plane normal to $\ddot{\boldsymbol{p}}^*$. Then by taking the dot product of (124.3) with $\boldsymbol{t}$ we obtain

$$\boldsymbol{t} \cdot \operatorname{grad}(Q + U + \tfrac{1}{2}\dot{p}^2) = 0. \tag{124.4}$$

Now the field $\boldsymbol{t}$ is a tangent field for a certain congruence of Lamb curves, determined by the condition that they be normal to the field $\ddot{\boldsymbol{p}}^*$. From (124.4) follows then the ***curvilinear Bernoulli theorem***[3]: *In a motion with steady vorticity, let the curves* c_L *be the Lamb curves normal to the diffusive acceleration field. Then*

$$Q + U + \tfrac{1}{2}\dot{p}^2 = f(c_L, t); \tag{124.5}$$

that is, along any one of these curves at any one instant the expression on the left has a constant value. It is possible that some admissible diffusive acceleration

---

[1] An example was given by CARSTOIU [1947, *3*, Chap. VI, § 3].

[2] From a kinematical point of view the foregoing statements are trivial. Dynamically, however, a medium is defined by specifying the acceleration, and thus some one decomposition may have particular physical significance.

[3] TRUESDELL [1954, *24*, § 74]. Special cases involving the determination of particular Lamb curves in the motion of viscous fluids were given earlier by SBRANA [1931, *9*], CASTOLDI [1948, *6*], and TRUESDELL [1950, *33*].

field $\ddot{\boldsymbol{p}}^*$ be parallel to the Lamb vector $\boldsymbol{w}\times\dot{\boldsymbol{p}}$ but unequal to it. In this case the result (124.5) holds for any Lamb curve.

In the special case of steady motion, the curvilinear Bernoulli theorem (124.5) assumes a simpler form:

$$Q+\tfrac{1}{2}\dot{p}^2=f(c_{\mathrm{L}}). \tag{124.6}$$

Thus upon each of the curves c_{L} there is a finite least upper bound $\overline{Q}(c_{\mathrm{L}})$ for Q, attained (if at all) at and only at a stagnation point:

$$\overline{Q}(c_{\mathrm{L}})=f(c_{\mathrm{L}}), \tag{124.7}$$

so that (124.6) becomes

$$Q+\tfrac{1}{2}\dot{p}^2=\overline{Q}. \tag{124.8}$$

If further there is a finite greatest lower bound $\underline{Q}(c_{\mathrm{L}})$ for Q on some particular Lamb curve c_{L}, then on that same curve *there must be a finite upper bound $\bar{\dot{p}}$ for the speed:*

$$\tfrac{1}{2}\bar{\dot{p}}^2=\overline{Q}-\underline{Q}. \tag{124.9}$$

An equivalent form for (124.6) then is

$$\tfrac{1}{2}(\bar{\dot{p}}^2-\dot{p}^2)=Q-\underline{Q}. \tag{124.10}$$

From these last results follow the principal applications of BERNOULLI'S theorem in hydrodynamics. One of these, for example, consists in the observation that if $Q=\text{const}$ upon one of the curves, then the speed also must be constant upon that curve.

125. The superficial Bernoulli theorem. We consider now a motion in which Lamb surfaces exist and in which also the vorticity is steady. By inserting (122.4) and (124.1) into (111.9) we then obtain

$$\operatorname{grad}(Q+U+\tfrac{1}{2}\dot{p}^2)=\ddot{\boldsymbol{p}}^*-C\operatorname{grad}B. \tag{125.1}$$

If further

$$\ddot{\boldsymbol{p}}^*=E\operatorname{grad}B, \tag{125.2}$$

then

$$\operatorname{grad}(Q+U+\tfrac{1}{2}\dot{p}^2)=(E-C)\operatorname{grad}B, \tag{125.3}$$

whence it follows that $Q+U+\frac{1}{2}\dot{p}^2$ is constant upon each of the Lamb surfaces $B=\text{const}$. We may state this result as the ***superficial Bernoulli theorem***[1]: *In a motion where Lamb surfaces s_{L} exist and where the vorticity is steady, if it is possible to find a diffusive acceleration field $\ddot{\boldsymbol{p}}^*$ which is zero or normal to the Lamb surfaces, then*

$$Q+U+\tfrac{1}{2}\dot{p}^2=f(s_{\mathrm{L}},t): \tag{125.4}$$

That is, the expression on the left is constant upon each of the Lamb surfaces at each instant.

Consider now the case of a circulation preserving motion with steady vorticity. As was shown in Sect. 122, Lamb surfaces do indeed exist. By (109.1) we may put $Q=V^*$, $\ddot{\boldsymbol{p}}^*=0$ in (124.1), and hence (124.3) becomes

$$\dot{\boldsymbol{p}}\times\boldsymbol{w}=\operatorname{grad}(U+V^*+\tfrac{1}{2}\dot{p}^2). \tag{125.5}$$

[1] TRUESDELL [1954, *24*, § 75]. In any motion, it is possible to *define* certain surfaces by the condition $Q+U+\frac{1}{2}\dot{p}^2=\text{const}$, but these do not generally enjoy particularly simple kinematical properties; cf. CASTOLDI [1953, *4*].

The foregoing theorem applies, and (125.4) follows. Conversely, suppose that (125.5) holds; from LAGRANGE'S acceleration formula (99.14) we then obtain

$$\ddot{\boldsymbol{p}} = \frac{\partial \dot{\boldsymbol{p}}}{\partial t} - \operatorname{grad}(U + V^*), \tag{125.6}$$

and hence $\boldsymbol{w}^* = \partial \boldsymbol{w}/\partial t$. Thus $\boldsymbol{w}^* = 0$ if and only if $\partial \boldsymbol{w}/\partial t = 0$. In summary of the foregoing analysis we may state then that *in a circulation preserving motion with steady vorticity, Lamb surfaces* $\mathfrak{s}_L$ *exist, and*

$$U + V^* + \tfrac{1}{2}\dot{p}^2 = f(\mathfrak{s}_L, t). \tag{125.7}$$

Conversely, in a motion such that Lamb surfaces exist and (125.5) *holds, the motion is circulation preserving if and only if the vorticity is steady.* As a corollary follows the ***Lamb characterization of steady circulation preserving motion***[1]: *For a circulation preserving motion to be steady it is both necessary and sufficient that Lamb surfaces exist and that*

$$\left|\frac{d}{dn}\left(V^* + \frac{1}{2}\dot{p}^2\right)\right| = |\boldsymbol{w} \times \dot{\boldsymbol{p}}|, \tag{125.8}$$

where d/dn denotes differentiation in a direction normal to the Lamb surface. The formula (125.8) is but an alternative expression for (125.5) in the special case of steady motion. The statement that (125.7) actually holds in a circulation preserving motion with steady vorticity we may call the ***Lamb-Masotti***[2] ***form of Bernoulli's theorem.*** In particular, $U + V^* + \frac{1}{2}\dot{p}^2$ is a function of time only along each stream line[3], along each vortex line[4], and along any Lamb curve[5].

Under conditions sufficient for the validity of (125.7), the finite upper and lower bounds discussed in Sect. 124 exist more generally and refer to a whole Lamb surface, rather than to a single Lamb curve.

126. The spatial Bernoulli theorem. In any irrotational motion, steady or not, we have the formula of EULER (120.1), and in any steady screw motion we have (112.3). Recalling that in an unsteady irrotational motion $U = -\partial V/\partial t$, we may write both these results in a single formula, obtaining the ***spatial Bernoulli theorem***: *In any irrotational motion and in any steady screw motion*

[1] [1878, *5*] [1879, *2*, § 145]. A special case had been discovered previously by COTTERILL [1876, *2*]. The derivation above is based upon that of BASSET [1888, *1*, § 39]. Cf. also CLEBSCH [1857, *1*, § 5], CRAIG [1880, *5*, p. 220] [1880, *6*, pp. 344–347]. POINCARÉ [1893, *6*, §§ 23–24].

[2] [1927, *7*, § 3]. A Bernoulli theorem for a special class of steady circulation preserving complex-lamellar motions is obtained by OSWATITSCH [1956, *17*] [his Eq. (4) is only a special solution of his Eq. (3)].

[3] In a sense, the hydraulic statements of D. and J. BERNOULLI may be regarded as assertions pertaining to a single stream line. As a hydrodynamical theorem for plane steady flow, the "Bernoulli equation for the stream lines" was first derived by EULER in 1751 [1767, *1*, §§ 18–20] by a remarkable analysis in a partially material description. The result in three-dimensional steady flow is also EULER'S [1757, *3*, §§ 50–52], and in [1757, *3*, §§ 58 to 59] he showed that specialization of his general analytical expressions yields

$$V^* + \tfrac{1}{2}\dot{p}^2 = \int r\,\dot{\theta}^2\,dr + \tfrac{1}{2}r^2\dot{\theta}^2 = \int r\,\omega^2\,dr + \tfrac{1}{2}r^2\omega^2$$

for the specific energy of motion of a stream line of a plane steady vortex (Sect. 89), as is evident from first principles. STOKES regarded the general theorem as well known [1842, *4*, p. 1]. Cf. also CLEBSCH [1857, *1*, § 5], BASSET [1888, *1*, § 39].

[4] It was stated incorrectly by CISOTTI [1923, *3*, § 2] that these are the only curves upon which $V^* + \frac{1}{2}\dot{p}^2$ is constant in a steady motion of this type; the error was corrected by SEGRE [1923, *4*, § 4, footnote].

[5] Certain Lamb curves occurring in certain special motions are remarked by POPOV [1951, *19*].

the acceleration potential S^* satisfies

$$U + S^* + \tfrac{1}{2}\dot{p}^2 = 0, \tag{126.1}$$

where in the case of steady motion U reduces to a constant. This result is a slight generalization of the classical Bernoulli-Euler theorem discussed in Sect. 120.

Under conditions sufficient for the validity of (126.1), the finite upper and lower bounds discussed in Sect. 124 exist more generally and refer to the whole motion rather than to a particular Lamb curve.

Conversely, suppose that in a circulation preserving motion with steady vorticity the spatial theorem (126.1) holds. Since by (111.9) and (109.1) we have

$$\operatorname{grad}(\tfrac{1}{2}\dot{p}^2 + U + S^*) + \boldsymbol{w}\times\dot{\boldsymbol{p}} = 0, \tag{126.2}$$

from (126.1) it follows that

$$\boldsymbol{w}\times\dot{\boldsymbol{p}} = 0. \tag{126.3}$$

We have thus proved the ***characterization of Stokes***[1]; *In a circulation preserving motion with steady vorticity, in order that the spatial Bernoulli theorem hold it is both necessary and sufficient that the motion be an irrotational motion or a steady screw motion.*

e IV) Convection and diffusion of vorticity.

127. Aim and plan of the analysis of convection and diffusion of vorticity. In Sect. 101 we have given the essential formulae governing the convection and diffusion of stretching and spin. In the case of the spin tensor, or, equivalently, the vorticity vector, the distinction represented by the two corresponding portions of (101.9) and (101.11) is striking. Motions in which there is no diffusion of vorticity are characterized as *circulation preserving*; alternatively, the D'Alembert-Euler condition (108.1) or the equivalent Hankel-Appell condition (108.3) is necessary and sufficient for circulation preserving motion. The Helmholtz theorems (Sect. 108) furnish yet another defining condition. Circulation preserving motions occur in classical hydrodynamics and in a sense may be said to characterize it. They have been studied with intensity and perseverance for two centuries, and though a large mathematical structure has grown up about them, their secrets are far from exhausted, and interesting new researches concerning them continue to appear.

The viewpoint we adopt is the most general. Insofar as possible, in each case we shall outline analysis applicable to *all continuous motions*, thereafter particularizing the result to the circulation preserving case. However, so as to point up the enlightening simplicity of circulation preserving motions, we begin by reversing this order and following a line of argument appropriate only when the circulation of each material circuit is constant in time.

128. KELVIN's proofs of the Helmholtz theorems[2]. To prove the second Helmholtz theorem, we compute the circulation around a material circuit $\mathscr{C}$ which at time t_1 lies entirely upon a given vortex surface $\mathscr{W}$ and is reducible upon it. By (106.1), the circulation about $\mathscr{C}$ is zero at time t_1. At time t_2 the particles initially comprising the vortex surface $\mathscr{W}$ constitute a new surface $\boldsymbol{w}$, which we do not know to be a vortex surface. Upon it lies c, the present locus of the

[1] STOKES [1842, *4*, pp. 2—3] thus demonstrated the formula (126.3) for steady motion but erroneously concluded therefrom that $\boldsymbol{w} = 0$. See Sect. App. 34.

[2] KELVIN [1869, *7*, §§ 60(f)—60(i); §§ 59(d), 60(q)]. The third theorem had been proved previously in essentially the same way by HANKEL [1861, *1*, § 9], who had used a different method for the more difficult second theorem [ibid., § 11].

particles comprising $\mathscr{C}$. Since the motion is circulation preserving, the circulation about c is zero, and hence by KELVIN's transformation (87.1) we have

$$0 = \oint_c d\boldsymbol{x} \cdot \dot{\boldsymbol{p}} = \int_s d\boldsymbol{a} \cdot \boldsymbol{w}, \tag{128.1}$$

where s is the portion of w inclosed by c. But $\mathscr{C}$ is an arbitrary circuit upon $\mathscr{W}$; by the continuity of motion, it follows that c is an arbitrary circuit upon w; therefore s is an arbitrary portion of w, and hence, finally, (128.1) implies that $d\boldsymbol{a} \cdot \boldsymbol{w} = 0$ upon w. Consequently w is a vortex surface. Since vortex lines are the curves of intersection of vortex surfaces, and since all vortex surfaces are material surfaces, the vortex lines are material lines. Q.E.D.

To prove the third Helmholtz theorem, we recall that by KELVIN's transformation the strength of a vortex tube equals the circulation about any curve once embracing it, and from HELMHOLTZ's second theorem, in a circulation preserving motion a material curve once lying upon a vortex tube always lies upon the same vortex tube. Since the circulation about this curve is constant during the motion, the strength of the vortex tube is constant during the motion. Conversely, if the vortex lines are material and the strengths of the vortex tubes constant in time, it follows by KELVIN's transformation that the circulation about any circuit once embracing a vortex tube is constant in time, but since any circuit defines a vortex tube, the motion is circulation preserving. Q.E.D.

129. Some aspects of the diffusion of vorticity. The vivid geometrical picture at the basis of the arguments of the last section is available only in the circulation preserving case. To find the effect of diffusion upon the circulation and the vortex lines, we must revert to formulae.

First, let $\mathscr{C}$ be a material line, not necessarily a closed circuit, and from $(79.2)_2$ calculate the rate of change of $\oint_{\mathscr{C}} d\boldsymbol{x} \cdot \dot{\boldsymbol{p}}$, the *flow* along it, obtaining[1]

$$\left.\begin{aligned} \frac{d}{dt}\int_{\mathscr{C}} d\boldsymbol{x} \cdot \dot{\boldsymbol{p}} &= \int_{\mathscr{C}} d\boldsymbol{x} \cdot \left[\ddot{\boldsymbol{p}} + \operatorname{grad} \frac{1}{2} \dot{p}^2\right], \\ &= \int_{\mathscr{C}} d\boldsymbol{x} \cdot \ddot{\boldsymbol{p}} + \frac{1}{2} \dot{p}^2 \bigg|_{\mathscr{C}}. \end{aligned}\right\} \tag{129.1}$$

When $\mathscr{C}$ is a closed circuit, we have simply

$$\frac{d}{dt}\oint_{\mathscr{C}} d\boldsymbol{x} \cdot \dot{\boldsymbol{p}} = \int_{\mathscr{C}} d\boldsymbol{x} \cdot \ddot{\boldsymbol{p}}: \tag{129.2}$$

The material rate of change of circulation equals the circulation of the acceleration. By KELVIN's transformation, an equivalent formula is[2]

$$\frac{d}{dt}\oint_{\mathscr{C}} d\boldsymbol{x} \cdot \dot{\boldsymbol{p}} = \frac{d}{dt}\int_{\mathscr{S}} d\boldsymbol{a} \cdot \boldsymbol{w} = \int_{\mathscr{S}} d\boldsymbol{a} \cdot \operatorname{curl} \ddot{\boldsymbol{p}}, \tag{129.3}$$

where $\mathscr{S}$ is a surface bounded by $\mathscr{C}$.

When the curve $\mathscr{C}$ is instantaneously a Lamb curve in a motion with steady vorticity, we may eliminate either the line integral or the speed from (129.1) and (123.2), thus obtaining

$$\left.\begin{aligned} \frac{d}{dt}\int_{\mathscr{C}_L} d\boldsymbol{x} \cdot \dot{\boldsymbol{p}} &= 2\int_{\mathscr{C}_L} d\boldsymbol{x} \cdot \ddot{\boldsymbol{p}} - U\Big|_{\mathscr{C}_L}, \\ &= (\dot{p}^2 + U)\Big|_{\mathscr{C}_L}. \end{aligned}\right\} \tag{129.4}$$

[1] KELVIN [1869, 7, § 59(c)], BELTRAMI [1871, 1, § 12].
[2] FÖPPL [1897, 4, § 31].

Two major special cases follow: The circulation about a *closed material Lamb curve* is constant, and in a *steady* motion the rate of change of the flow along a material Lamb curve is the difference between the squared speed at the two ends. These last results become self-evident when the Lamb curve is a stream line. Another interesting special case may be obtained by choosing as the Lamb curve a vortex line: *The circulation about a closed material vortex line in a steady motion is constant in time.* In interpreting this result one must take care to recollect first that closed material vortex lines generally do not exist, and second that in any case they are not generally steady.

We turn now to the change in flux of vorticity occasioned by diffusion. From (101.11) and (20.8) follows

$$\boldsymbol{w}\cdot d\boldsymbol{a} - \boldsymbol{W}\cdot d\boldsymbol{A} = \int_0^t \boldsymbol{W}_* \, dt \cdot d\boldsymbol{A}. \tag{129.5}$$

Thus the vector $\boldsymbol{W}_*$ is the time rate of change of flux of vorticity per unit initial area. It is this result, perhaps, which places the diffusion of vorticity in its clearest setting. KIRCHHOFF's proof[1] of the third Helmholtz theorem (Sects. 108, 128) is equivalent to annulling the material diffusion vector $\boldsymbol{W}_*$ in (129.5), whence follows

$$\boldsymbol{w}\cdot d\boldsymbol{a} = \boldsymbol{W}\cdot d\boldsymbol{A}, \tag{129.6}$$

an elegant formula which might be put into words as follows: *The specific flux of vorticity for a particle does not change.*

To study the effect of diffusion on the vortex lines, from (101.9) we calculate

$$\left.\begin{aligned} w_{km}\, dx^m &= \left[W_{\alpha\beta} + \int_0^t w^*_{pq}\, x^p_{,\alpha}\, x^q_{,\beta}\, dt\right] X^\alpha_{,k}\, X^\beta_{,m}\, dx^m, \\ &= X^\alpha_{,k}\left[W_{\alpha\beta} + \int_0^t w^*_{pq}\, x^p_{,\alpha}\, x^q_{,\beta}\, dt\right] dX^\beta; \end{aligned}\right\} \tag{129.7}$$

equivalently,

$$d\boldsymbol{x}\times\boldsymbol{w} = \operatorname{grad}\boldsymbol{X}\cdot\left[d\boldsymbol{X}\times\left(\boldsymbol{W} + \int_0^t \boldsymbol{W}_*\, dt\right)\right]. \tag{129.8}$$

KIRCHHOFF's proof of the second Helmholtz theorem is equivalent to the observation that in the circulation preserving case, when (108.3) holds, (129.8) yields

$$d\boldsymbol{x}\times\boldsymbol{w} = 0 \quad \text{if and only if} \quad d\boldsymbol{X}\times\boldsymbol{W} = 0: \tag{129.9}$$

A material line initially a vortex line remains ever a vortex line. For general motions, (129.8) gives a quantitative measure of how for a material line which at time 0 was a vortex line has been turned away from the vortex line at time t by diffusion.

The formulae (129.5) and (129.8) were obtained by TRUESDELL[2], who used them to formulate conditions that a material line be a vortex line at two different instants, or that a particular material surface element carry the same flux of vorticity at two different instants.

For steady circulation preserving motion, a measure of convection in terms of stream tubes, vortex tubes, and the cells into which they decussate space has been constructed by ERTEL and KÖHLER[3].

As is well known from the researches of POINCARÉ and CARTAN[4], virtually all the most important properties of the circulation preserving case may be generalized to motions in

---

[1] More precisely, both KIRCHHOFF [1876, *2*, Vorlesung 15, § 3] and STOKES (note added in 1883 reprint of [1848, *3*]) gave indications of proofs of the second and third Helmholtz theorems following from CAUCHY's formula (134.1). The details were worked out by APPELL [1897, *1*, § 8]. The history of the question has been written by TRUESDELL [1954, *24*, § 86[1]].

[2] [1948, *36*]. Related additional results were given by CARSTOIU [1947, *3*, Chap. IV] and TRUESDELL [1954, *24*, §§ 86—88].

[3] [1949, *9*]. A different proof of a more precise statement is given by TRUESDELL [1954, *24*, § 99]. Cf. also Sect. 163.

[4] [1922, *2*].

a space of any number of dimensions, provided those motions satisfy a variational principle of the form

$$\delta \int_{t_1}^{t_2} B\, dt = 0, \tag{129.10}$$

where B is a *linear* differential form[1], the variations being zero at $t=t_1$ and $t=t_2$. Eq. (129.10) would seem to be a severe restriction, but in fact a method of BATEMAN[2] enables us to cast any continuous motion into a form derivable from such a principle. Indeed, *Bateman's function* is given by

$$B \equiv g_{km} \dot{x}^k \dot{y}^m - a_k y^k, \tag{129.11}$$

where $\boldsymbol{y}$ is an auxiliary vector field to be varied along with $\boldsymbol{x}$ and where $\boldsymbol{a}$ is a vector field to be taken as given, not varied. Regard $(\boldsymbol{x}, \boldsymbol{y})$ as a 6-vector. Then the Euler equations of (129.10) with B given by (129.11) are

$$\frac{d}{dt}\frac{\partial B}{\partial \dot{x}^k} - \frac{\partial B}{\partial x^k} = 0, \qquad \frac{d}{dt}\frac{\partial B}{\partial \dot{y}^k} - \frac{\partial B}{\partial y^k} = 0, \tag{129.12}$$

where g is held constant; equivalently,

$$\ddot{y}_k - a_{k,m} y^m = 0, \qquad \ddot{x}_k = a_k. \tag{129.13}$$

From $(129.13)_2$ it follows that if we take $\boldsymbol{a}$ as the acceleration of a given motion, then the variational principle (129.10) yields in the 6-dimensional $(\boldsymbol{x}, \boldsymbol{y})$ space trajectories whose projections onto the $\boldsymbol{x}$ space are the paths of the particles in the given motion. The vector $\boldsymbol{y}$ has no obvious physical significance; its presence makes for some awkwardness, as will appear.

DEDECKER[3] has used BATEMAN's principle to study the diffusion of vorticity. If we write $(\boldsymbol{u}, \boldsymbol{v})$ for the 6-dimensional velocity and $\mathscr{X}$ for a curve which is material under the 6-dimensional motion, POINCARÉ's theorem asserts that

$$\frac{d}{dt}\oint_{\mathscr{X}} u_k\, dx^k + \frac{d}{dt}\int_{\mathscr{X}} v_k\, dy^k = 0. \tag{129.14}$$

Since both the six-dimensional and the three-dimensional motions are uniquely determined by the initial values of the corresponding velocities at a given point, by choosing $\boldsymbol{u}(\boldsymbol{x}, \boldsymbol{y}, t_1) = \dot{\boldsymbol{x}}(\boldsymbol{x}, t_1)$ we obtain $\boldsymbol{u}(\boldsymbol{x}, \boldsymbol{y}, t) = \dot{\boldsymbol{x}}(\boldsymbol{x}, t)$. We choose also $\boldsymbol{v}(\boldsymbol{x}, \boldsymbol{y}, t_1) = \dot{\boldsymbol{x}}(\boldsymbol{y}, t_1)$, but this does not have any obvious consequence when $t \neq t_1$. At $t=t_1$, however, if $\mathscr{X}$ lies in the hyperplane $\boldsymbol{x}=\boldsymbol{y}$, it follows that $\overline{dx^k}|_{\mathscr{X}} = \overline{dy^k}|_{\mathscr{X}} = \overline{dx^k}|_{\mathscr{C}}$, where $\mathscr{C}$ is the projection of $\mathscr{X}$ onto $\boldsymbol{x}$-space. Hence, at $t=t_1$,

$$\frac{d}{dt}\oint_{\mathscr{X}} u_k\, dx^k = \frac{d}{dt}\int_{\mathscr{C}} \dot{x}_k\, dx^k = -\frac{d}{dt}\int_{\mathscr{X}} v_k\, dy^k, \tag{129.15}$$

where the last equality follows from (129.14). Both the right-hand and the left-hand member of (129.15) can be calculated from (129.13); by each calculation, as would be expected, the result is (129.2). That no *new formulae* emerge from use of this deeper approach is not surprising. Its interest lies in showing that *diffusion of vorticity may be regarded as resulting from projection onto a three-dimensional space of a six-dimensional circulation preserving motion*. Indeed, it is always possible to render a dissipative process conservative by adjoining extra quantities—be they extra dimensions or only simple absorbers—which drain off the excesses and supply the defects of the original process. Little illumination results from such extensions, since it is the dissipative process itself that commands our interest (cf. Sect. 6).

---

[1] Cf. the expositions of CARTAN [1922, *2*, §§ 20—25] and DEDECKER [1951, *3*, §§ 9—11], who derive the theorems of HELMHOLTZ and KELVIN on this basis.

[2] [1931, *2*]. We modify slightly the application given by DEDECKER [1951, *3*, § 21].

[3] [1951, *3*, §§ 20—22]. In [1951, *4*] DEDECKER considers circulations about arbitrary circuits in space-time.

130. The Euler-Ertel theorem. In the circulation preserving case, BELTRAMI'S equation (101.7) reduces to the ***D'Alembert-Euler vorticity equation*** [1], three equivalent forms of which are

$$\overline{J\boldsymbol{w}}^{\cdot} = J\boldsymbol{w}\cdot\operatorname{grad}\dot{\boldsymbol{p}}, \quad \overline{\boldsymbol{w}\,dv}^{\cdot} = \boldsymbol{w}\,dv\cdot\operatorname{grad}\dot{\boldsymbol{p}}, \quad \dot{\boldsymbol{w}} = \boldsymbol{w}\cdot\operatorname{grad}\dot{\boldsymbol{p}} - \boldsymbol{w}\operatorname{div}\dot{\boldsymbol{p}}. \tag{130.1}$$

The second of these shows that $\boldsymbol{w}\,dv$, the vorticity carried by an element of volume, is transported as an absolute vector [cf. Eq. $(134.1)_3$].

ERTEL[2] has introduced an arbitrary quantity into (101.7):

$$\overline{J\boldsymbol{w}\cdot\operatorname{grad}\Psi}^{\cdot} - J\boldsymbol{w}\cdot\operatorname{grad}\dot{\Psi} = J\boldsymbol{w}^*\cdot\operatorname{grad}\Psi, \tag{130.2}$$

a result which follows easily from (101.7) and (72.8). Substitution of various quantities for Ψ yields various different vorticity theorems, (101.7) itself being the case $\Psi = \boldsymbol{p}$. If Ψ is a scalar such that $\dot{\Psi} = 0$, we obtain the ***generalized Euler-Ertel conservation theorem*** [3]: *If it is possible to find a substantially constant function Ψ such that*

$$\boldsymbol{w}^*\cdot\operatorname{grad}\Psi = 0, \tag{130.3}$$

then

$$J\boldsymbol{w}\cdot\operatorname{grad}\Psi = \text{const} \quad \text{for each particle}; \tag{130.4}$$

equivalently,

$$J w\frac{d\Psi}{dw} = \text{const} \quad \text{for each particle}, \tag{130.5}$$

where d/dw denotes the directional derivative along the vortex line.

By the d'Alembert-Euler condition (108.1), in a circulation preserving motion the requirement (130.3) is satisfied by all functions Ψ, while in a motion which is not circulation preserving, for a scalar function Ψ it states that $\boldsymbol{w}^*$ shall be tangent to the surfaces $\Psi = \text{const}$.

For any Ψ, in a circulation preserving motion we have from (130.2) simply

$$\overline{J\boldsymbol{w}\cdot\operatorname{grad}\Psi}^{\cdot} = J\boldsymbol{w}\cdot\operatorname{grad}\dot{\Psi}. \tag{130.6}$$

We may call this result the ***Ertel commutation formula***, since it shows that d/dt commutes with $J\boldsymbol{w}\cdot\operatorname{grad}$ for circulation preserving motions.

Conversely, the Ertel formula (130.6) leads to various characterizations of convection. For (130.6) to hold for general Ψ is obviously sufficient to ensure $\boldsymbol{w}^* = 0$, since (130.1) is then included as a special case. But we may restrict Ψ considerably, since for any given Ψ (130.6) is equivalent to

$$\boldsymbol{w}^*\cdot\operatorname{grad}\Psi = 0. \tag{130.7}$$

Thence to conclude that $\boldsymbol{w}^* = 0$, all we need is to be able to select at each point scalars Ψ having gradients parallel to three linearly independent directions. Thus

---

[1] The analysis of D'ALEMBERT [1752, *1*, § 48] is restricted to special classes of motions. EULER derived (130.1) in [1761, *2*, § 59]. Though well known in the eighteenth century, it was forgotten and was rediscovered in the next by PIOLA, STOKES, and HELMHOLTZ, usually being named after the last.

DELVAL [1950, *6*, § 5] has attempted to formulate (130.1) as a principle of least compulsion (cf. Sect. 237); his analysis was modified by TRUESDELL [1954, *24*, § 94[1]]. Another variational principle was proposed by MOREAU [1952, *13*, § 7c].

[2] The original work of ERTEL [1942, *4*–*7*] was restricted to special cases. (130.2) was derived by TRUESDELL [1951, *33*]; a slightly different form, by ERTEL [1955, *7*].

[3] The analysis of EULER [1757, *3*, § 55] was put into modern notation by TRUESDELL [1954, *25*, Part XIII]. Both EULER and ERTEL considered only the circulation preserving case, in which (130.3) is satisfied for all Ψ.

it follows[1] that *if there exists a class C of scalars B such that for each of three linearly independent directions $\mathbf{e}_1, \mathbf{e}_2, \mathbf{e}_3$ at each point the gradient of some one B is parallel to $\mathbf{e}_1$, another to $\mathbf{e}_2$, and a third to $\mathbf{e}_3$, such that for all $B \in C$ ERTEL's eqution holds:*

$$\overline{J\boldsymbol{w}\cdot\operatorname{grad} B}^{\bullet} = J\boldsymbol{w}\cdot\operatorname{grad}\dot{B}, \tag{130.8}$$

then the motion is circulation preserving, and (130.6) *holds for all* Ψ. If $\dot{B} = \text{const}$ is a vortex surface, (130.8) implies that $B = \text{const}$ is also a vortex surface. If $B = \text{const}$ is a material surface, (130.8) asserts that $\overline{J\boldsymbol{w}\cdot\operatorname{grad} B}^{\bullet} = 0$ for all t; thus if $J\boldsymbol{w}\cdot\operatorname{grad} B$ vanishes at $t = 0$, it vanishes for all t, so we have a new proof of the second theorem of HELMHOLTZ (Sects. 108, 128).

An interesting special case of the Ertel theorem (130.6) is obtained by choosing $\Psi \equiv E$, where

$$E \equiv \int_0^t \left(\tfrac{1}{2}\dot{p}^2 - V^*\right) dt, \tag{130.9}$$

where the integration is to be carried out for a fixed particle $\boldsymbol{X}$. Then the Ertel theorem yields

$$\overline{J\boldsymbol{w}\cdot\operatorname{grad} E}^{\bullet} = J\boldsymbol{w}\cdot\operatorname{grad}\left(\tfrac{1}{2}\dot{p}^2 - V^*\right). \tag{130.10}$$

By this result and (130.1) we have

$$\overline{J\boldsymbol{w}\cdot(\dot{\boldsymbol{p}} - \operatorname{grad} E)}^{\bullet} = J\boldsymbol{w}\cdot\left[\operatorname{grad}\dot{\boldsymbol{p}}\cdot\dot{\boldsymbol{p}} + \ddot{\boldsymbol{p}} - \operatorname{grad}\left(\tfrac{1}{2}\dot{p}^2 - V^*\right)\right] = 0. \tag{130.11}$$

Integrating this equation from 0 to t yields the ***convection theorem of Ertel and Rossby***[2]:

$$J\boldsymbol{w}\cdot(\dot{\boldsymbol{p}} - \operatorname{grad} E) = \boldsymbol{W}\cdot\dot{\boldsymbol{P}}, \tag{130.12}$$

where $\dot{\boldsymbol{P}}$ stands for the initial value of $\dot{\boldsymbol{p}}$, in accord with the conventions of Sect. 66.

An elegant vorticity theorem of the same type as (130.2) but expressed in terms of surface integrals was noted more recently by ERTEL[3]:

$$\frac{d}{dt}\int_{\mathscr{S}} d\boldsymbol{a}\cdot\boldsymbol{w}\Psi - \int_{\mathscr{S}} d\boldsymbol{a}\cdot\boldsymbol{w}\dot{\Psi} = \int_{\mathscr{S}} d\boldsymbol{a}\cdot\boldsymbol{w}^*\Psi. \tag{130.13}$$

In the circulation preserving case this becomes a commutation formula for flux integrals; the case when $\dot{\Psi} = 0$ shows that to multiply $\boldsymbol{w}$ by a substantially constant function Ψ yields a quantity whose rate of diffusion is $\boldsymbol{w}^*\Psi$. Derivation of (130.13) is easy, either from (80.3) or by integrating (130.2).

131. Diffusion in mean. There are very broad circumstances in which there is *no average diffusion of vorticity*. By this we mean that the average values of certain rates of change in regions of certain types can be calculated from the velocity field alone, *independently of the diffusion vector* $\boldsymbol{w}^*$. These averages, then, have *the same value in any motion of the type considered, whether circulation preserving or not.*

First, noting that

$$\boldsymbol{w}^*\cdot\operatorname{grad}\Psi = \operatorname{div}(\ddot{\boldsymbol{p}}\times\operatorname{grad}\dot{\Psi}), \tag{131.1}$$

---

[1] TRUESDELL [1954, *24*, § 98]. However, some conclusions drawn from this result by TRUESDELL are not justified.

[2] [1949, *7* and *8*].

[3] [1955, *8*].

we may integrate (130.2) over a volume and obtain[1]

$$\left.\begin{aligned}\int_v [\overline{J\,\boldsymbol{w}\cdot\operatorname{grad}\Psi}^{\bullet}/J - \boldsymbol{w}\cdot\operatorname{grad}\dot{\Psi}]\,dv &= \oint_s d\boldsymbol{a}\cdot(\ddot{\boldsymbol{p}}\times\operatorname{grad}\Psi),\\ &= \oint_s (d\boldsymbol{a}\times\ddot{\boldsymbol{p}})\cdot\operatorname{grad}\Psi.\end{aligned}\right\}\tag{131.2}$$

By formulating conditions sufficient for the vanishing of the surface integral we conclude that *in a region such that on all finite boundaries* $\ddot{\boldsymbol{p}}_t = 0$, *while in any part extending to infinity* $(\ddot{\boldsymbol{p}}\times\operatorname{grad}\Psi)_n = \bar{o}(p^{-2})$, *then the Ertel commutation formula* (130.6) *holds in mean over the whole motion*:

$$\int_v [\overline{J\,\boldsymbol{w}\cdot\operatorname{grad}\Psi}^{\bullet}/J - \boldsymbol{w}\cdot\operatorname{grad}\dot{\Psi}]\,dv = 0,\tag{131.3}$$

or

$$\frac{d}{dt}\int_{\mathscr{V}} \boldsymbol{w}\cdot\operatorname{grad}\Psi\,dv = \int_v \boldsymbol{w}\cdot\operatorname{grad}\dot{\Psi}\,dv.$$

In particular, the result holds for any motion of a material confined within finite walls to which it adheres. In this sense, then, we may say that in the type of volume specified by the conditions, there is no mean diffusion of vorticity[2]. In the conditions stated, it is clear from (131.1) that in the conditions sufficient for the validity of (131.3) the true acceleration $\ddot{\boldsymbol{p}}$ may be replaced by the diffusive acceleration $\ddot{\boldsymbol{p}}^*$ in accord with (124.2)

We now calculate the rate of change of the moments of vorticity $\mathfrak{W}_{(K)}$, defined by $(118.1)_1$. For a material volume $\mathscr{V}$ we have from (80.3)

$$\left.\begin{aligned}\dot{\mathfrak{W}}_{(K)} = \oint_{\mathscr{S}} \big(d\boldsymbol{a}\cdot[\boldsymbol{w}\{\boldsymbol{p}^{(K)}\dot{\boldsymbol{p}}\} + \dot{\boldsymbol{w}}\,\boldsymbol{p}^{(K+1)} + \boldsymbol{w}\,\boldsymbol{p}^{(K+1)}\operatorname{div}\dot{\boldsymbol{p}}]\\ - [\operatorname{grad}\dot{\boldsymbol{p}}\cdot d\boldsymbol{a}]\cdot\boldsymbol{w}\,\boldsymbol{p}^{(K+1)}\big).\end{aligned}\right\}\tag{131.4}$$

Upon substituting for $\dot{\boldsymbol{w}}$ from Beltrami's formula $(101.7)_3$, we find that by a happy circumstance two of the terms so introduced cancel the last two terms in (131.4), which becomes simply

$$\dot{\mathfrak{W}}_{(K)} = \oint_{\mathscr{S}} d\boldsymbol{a}\cdot\boldsymbol{w}\{\boldsymbol{p}^{(K)}\dot{\boldsymbol{p}}\} + \oint_{\mathscr{S}} d\boldsymbol{a}\cdot\boldsymbol{w}^*\,\boldsymbol{p}^{(K+1)}.\tag{131.5}$$

By $(\text{App. } 26.1)_1$ and the fact that $\operatorname{div}\boldsymbol{w}^* = 0$ follows

$$\dot{\mathfrak{W}}_{(K)} = \oint_{\mathscr{S}} d\boldsymbol{a}\cdot\boldsymbol{w}\{\boldsymbol{p}^{(K)}\dot{\boldsymbol{p}}\} + \int_{\mathscr{V}} \{\boldsymbol{p}^{(K)}\boldsymbol{w}^*\}\,dv.\tag{131.6}$$

Now by the transport theorem (81.3) we have

$$\dot{\mathfrak{W}}_{(K)} = \frac{\partial\mathfrak{W}_{(K)}}{\partial t} + \oint_{\mathscr{S}} d\boldsymbol{a}\cdot\dot{\boldsymbol{p}}\{\boldsymbol{p}^{(K)}\boldsymbol{w}\}.\tag{131.7}$$

From $(\text{App. } 26.1)_1$ we get the identities

$$\left.\begin{aligned}0 &= \int_{\mathscr{V}} \operatorname{div}\operatorname{curl}(\boldsymbol{p}^{(K)}\ddot{\boldsymbol{p}})\,dv,\\ &= \oint_{\mathscr{S}} d\boldsymbol{a}\cdot\operatorname{curl}(\boldsymbol{p}^{(K)}\ddot{\boldsymbol{p}}),\\ &= -\oint_{\mathscr{S}} \{\boldsymbol{p}^{(K)}(d\boldsymbol{a}\times\ddot{\boldsymbol{p}})\} + \oint_{\mathscr{S}} d\boldsymbol{a}\cdot\boldsymbol{w}^*\,\boldsymbol{p}^{(K)}.\end{aligned}\right\}\tag{131.8}$$

---

[1] Truesdell [1951, *34*].

[2] The special case $\Psi = \boldsymbol{p}$ was observed by Truesdell [1948, *34*].

By combining this result and (131.5) with (131.7) we finally obtain[1]

$$\frac{\partial \mathfrak{W}_{(K)}}{\partial t} = \oint_{\mathscr{S}} d\boldsymbol{a}\cdot[\boldsymbol{w}\{\boldsymbol{p}^{(K)}\dot{\boldsymbol{p}}\} - \dot{\boldsymbol{p}}\{\boldsymbol{p}^{(K)}\boldsymbol{w}\}] + \oint_{\mathscr{S}} \{\boldsymbol{p}^{(K+1)}(d\boldsymbol{a}\times\ddot{\boldsymbol{p}})\}. \quad (131.9)$$

The vorticity moment theorem of Sect. 118 states conditions sufficient that $\mathfrak{W}_{(0)}, \mathfrak{W}_{(1)}, \ldots, \mathfrak{W}_{(K)}$ shall vanish. From (131.9) we may infer weaker conditions sufficient that these moments remain constant in time: *Subject to the boundary condition*

$$\left.\begin{array}{ccc} w_n = 0, & \ddot{p}_t = 0 & \text{or} \\ \dot{\boldsymbol{p}}\dot{w}_n = \bar{o}(p^{-K-2}), & \boldsymbol{w}\dot{p}_n = \bar{o}(p^{-K-2}), & \ddot{\boldsymbol{p}}_t = \bar{o}(p^{-K-3}), \end{array}\right\} \quad (131.10)$$

the first $K+1$ moments of vorticity are constant in time. In the case when all finite boundaries are stationary and the material adheres without slipping, we have $\ddot{\boldsymbol{p}}_t = 0$ and $w_n = 0$, so that the only conditions of the theorem which remain to be considered are the order conditions at infinity. In the conditions as stated, it is clear from (131.6) and (131.8) that the true acceleration $\ddot{\boldsymbol{p}}$ may be replaced by the diffusive acceleration $\ddot{\boldsymbol{p}}^*$, in accord with (124.2).

132. The generalized convection vector. Complex-screw motions. The properties of motions in which vorticity is transported by convection only are relatively simple and easy to picture. Kinematical analysis of more general motions usually seeks conditions such that some of the properties of circulation preserving motions can be carried over or adjusted to more complicated circumstances. The theorems of mean value in the previous section are examples. Another example is the following theorem of APPELL[2], which we state without proof: *Given any family of lines, furnished with continuously turning tangents, which in a given motion $\dot{\boldsymbol{p}}$ are material lines, there exists a continuously differentiable field $\boldsymbol{v}$ whose circulation about any material circuit is constant and whose vortex lines are the given material lines.*

We now construct apparatus for a more fruitful generalization. The underlying idea[3] consists in introducing a class of vector fields proportional to the velocity:

$$\boldsymbol{v}_C \equiv \frac{\dot{\boldsymbol{p}}}{v_0}, \quad (132.1)$$

where v_0 is any non-vanishing substantially constant scalar:

$$\dot{v}_0 = 0, \quad v_0 \neq 0. \quad (132.2)$$

Any such field $\boldsymbol{v}_C$ we shall call a *generalized convection vector*, and v_0 we shall call the *defining parameter*. We introduce also the curl of the convection vector:

$$\boldsymbol{w}_C \equiv \operatorname{curl} \boldsymbol{v}_C. \quad (132.3)$$

Then we have identically

$$\boldsymbol{w} = \operatorname{curl} v_0 \boldsymbol{v}_C = v_0 \boldsymbol{w}_C + \operatorname{grad} v_0 \times \boldsymbol{v}_C, \quad (132.4)$$

[1] The special case $K=0$ was derived by TRUESDELL [1948, *32*], generalizing an earlier analysis of JAFFÉ [1921, *3*]. The general formula was given by TRUESDELL [1951, *28*, § 12]. Cf. also HOWARD [1957, *8*, § V].

[2] The analysis of APPELL is in the material description [1899, *1*, §§ 1–10]; a shorter spatial proof was given by TRUESDELL [1954, *24*, § 89]. Another proof is given by DROBOT and RYBARSKI [1959, *4*, § III.4].

[3] Due to HICKS, GUENTHER and WASSERMAN [1947, *6*, Introd.] and extended by TRUESDELL [1951, *32*] [1952, *23*, § 8]. A different class of convection vectors was considered by HICKS [1949, *14*]. The definition (132.1) is motivated by work of CROCCO [1936, *3*], although he considered only the trivial case when $v_0 = \text{const}$; also the introduction of m in (133.1) is suggested by certain special results of CROCCO.

and hence quite independently of the condition (132.2) we obtain

$$v_0^2 \boldsymbol{v}_C \cdot \boldsymbol{w}_C = \dot{\boldsymbol{p}} \cdot \boldsymbol{w}. \tag{132.5}$$

By (App. 33.5) it follows that *the generalized convection vector is complex-lamellar if and only if the motion is complex-lamellar.*

The researches of NEMÉNYI and PRIM[1] have drawn attention to motions in which

$$\boldsymbol{v}_C \times \boldsymbol{w}_C = 0, \tag{132.6}$$

the possibility $\boldsymbol{w}_C = 0$ not being excluded. The special case $v_0 = 1$ is an irrotational or screw motion, and the class of motions satisfying the generalization (132.6) will be called *complex-screw motions.* The remainder of this section presents results equivalent to those of NEMÉNYI and PRIM concerning this interesting type of motion.

We first establish the connection between complex-screw motions and irrotational or ordinary screw motions. By (132.4) and $(132.2)_1$ we obtain

$$\dot{\boldsymbol{p}} \times \boldsymbol{w} = v_0^2 \boldsymbol{v}_C \times \boldsymbol{w}_C + \dot{p}^2 \operatorname{grad} \log v_0 + \dot{\boldsymbol{p}} \frac{\partial \log v_0}{\partial t}. \tag{132.7}$$

From this identity it is obvious that a *complex-screw motion in which the defining parameter is either uniform or steady is an irrotational or screw motion if and only if the defining parameter is both uniform and steady.* This result implies broadly that the class of complex-screw motions is more extensive than that of irrotational and screw motions. We may calculate the angle ψ between the vortex line and the stream line in the following way[2]. If $\boldsymbol{w}_C \times \boldsymbol{v}_C = 0$, the two summands on the right-hand side of (132.4) are perpendicular, so that

$$\left.\begin{aligned} w^2 &= v_0^2 w_C^2 + (\operatorname{grad} \log v_0 \times \dot{\boldsymbol{p}})^2, \\ &= v_0^2 w_C^2 + (\operatorname{grad} \log v_0)^2 \dot{p}^2 - (\dot{\boldsymbol{p}} \cdot \operatorname{grad} \log v_0)^2, \\ &= v_0^2 w_C^2 + (\operatorname{grad} \log v_0)^2 \dot{p}^2 - \left(\frac{\partial \log v_0}{\partial t}\right)^2. \end{aligned}\right\} \tag{132.8}$$

Simultaneously (132.7) becomes

$$\dot{\boldsymbol{p}} \times \boldsymbol{w} = \dot{p}^2 \operatorname{grad} \log v_0 + \dot{\boldsymbol{p}} \frac{\partial \log v_0}{\partial t}; \tag{132.9}$$

in view of $(132.2)_1$, the square of this equation is

$$(\dot{\boldsymbol{p}} \times \boldsymbol{w})^2 = \dot{p}^2 \left[\dot{p}^2 (\operatorname{grad} \log v_0)^2 - \left(\frac{\partial \log v_0}{\partial t}\right)^2\right]. \tag{132.10}$$

The angle ψ is then obtained by combining (132.8) and (132.10):

$$\csc \psi = \frac{\dot{p}\, w}{|\dot{\boldsymbol{p}} \times \boldsymbol{w}|} = \left(1 + \frac{v_0^2 w_C^2}{\dot{p}^2 (\operatorname{grad} \log v_0)^2 - \left(\frac{\partial \log v_0}{\partial t}\right)^2}\right), \tag{132.11}$$

whence follows

$$\tan^2 \psi = \frac{v_C^2 (\operatorname{grad} \log v_0)^2 - \left[\frac{\partial}{\partial t}\left(\frac{1}{v_0}\right)\right]^2}{w_C^2}. \tag{132.12}$$

By comparing (132.9) with (122.4) we conclude that *in a complex-screw motion whose defining parameter is steady but not uniform, the surfaces upon which that parameter is constant are Lamb surfaces.*

[1] [1948, *17*, § 5] [1949, *21* and *25*] [1952, *16*, Chap. V, Sect. C].
[2] We follow and correct the analysis of TRUESDELL [1954, *24*, § 90].

By putting (132.9) into LAGRANGE'S acceleration formula (99.14) we obtain

$$\left.\begin{aligned}\ddot{\boldsymbol p} &= \frac{\partial \dot{\boldsymbol p}}{\partial t} - \dot p^2 \operatorname{grad}\log v_0 - \dot{\boldsymbol p}\,\frac{\partial \log v_0}{\partial t} + \operatorname{grad}\tfrac{1}{2}\dot p^2,\\ &= v_0\frac{\partial \boldsymbol v_{\mathrm C}}{\partial t} + v_0^2 \operatorname{grad}\tfrac{1}{2} v_{\mathrm C}^2 .\end{aligned}\right\} \qquad (132.13)$$

Hence *in a complex-screw motion whose convection vector is steady the acceleration is complex-lamellar, its normal surfaces being the surfaces of constant magnitude of the convection vector.*

Taking the curl of (132.13) yields

$$\operatorname{curl}\ddot{\boldsymbol p} = \operatorname{grad} v_0 \times \frac{\partial \boldsymbol v_{\mathrm C}}{\partial t} + v_0\frac{\partial \boldsymbol w_{\mathrm C}}{\partial t} + \frac{1}{2}\operatorname{grad} v_0^2 \times \operatorname{grad} v_{\mathrm C}^2 . \qquad (132.14)$$

On the assumption that the local time derivatives are zero, we now consider in turn the situations that annul the remaining term on the right. First, if v_0 is uniform, then from $\boldsymbol w_{\mathrm C}\times\boldsymbol v_{\mathrm C}=0$ it follows that $\boldsymbol w\times\dot{\boldsymbol p}=0$. Second, we may have $v_{\mathrm C}^2=\text{const}$. Third, if the surfaces $v_0=\text{const}$ coincide with the surfaces $v_{\mathrm C}=\text{const}$, these in turn are surfaces of constant speed. In summary of these results we state that *a complex-screw motion whose convection vector is steady is a circulation preserving motion if and only if*

a) *it is an irrotational or screw motion, or*

b) *its convection vector is of uniform magnitude, or*

c) *at each fixed time, the defining parameter is a function of the speed alone.*

When the motion itself is steady, we may apply the theorems stated just before and just after (132.13) to replace c) by

c′) *the surfaces of constant speed are Lamb surfaces, and the acceleration is normal to them.*

Although, as indicated by these results, complex-screw motions usually fail to be circulation preserving, yet in some types of such motions the mechanism of diffusion operates in a fashion closely analogous to convection, as will be shown now.

133. Generalized convection theorems. Consider first a steady complex-screw motion, and let m be any solution of

$$\operatorname{div}(m\boldsymbol v_{\mathrm C}) = 0 . \qquad (133.1)$$

Noting that vector sheets of $\boldsymbol v_{\mathrm C}$ are stream surfaces, we apply the theorem of GROMEKA and BELTRAMI derived in Sect. App. 34 and obtain the ***convection theorem for steady complex-screw motions***: *The surfaces*

$$\frac{w_{\mathrm C}}{m v_{\mathrm C}} = \text{const} \qquad (133.2)$$

are stream surfaces; in particular, (133.2) *holds on each stream line.* The analogy to convection is immediate, and the result generalizes the second Gromeka-Beltrami theorem (112.4).

The corresponding generalized convection theorem for complex-lamellar motions lies deeper[1]. For any function F we readily obtain the identity

$$\left.\begin{aligned}\operatorname{curl}\left[F\left(\frac{1}{v_0}\frac{\partial \boldsymbol v_{\mathrm C}}{\partial t} + \boldsymbol w_{\mathrm C}\times\boldsymbol v_{\mathrm C}\right)\right] &= \operatorname{grad}\frac{F}{v_0}\times\frac{\partial \boldsymbol v_{\mathrm C}}{\partial t} - \frac{\boldsymbol w_{\mathrm C}}{v_0}\frac{\partial F}{\partial t} + \frac{1}{v_0}\frac{\partial F\boldsymbol w_{\mathrm C}}{\partial t} + {}\\ &+ \boldsymbol v_{\mathrm C}\cdot\operatorname{grad}F\boldsymbol w_{\mathrm C} - F\boldsymbol w_{\mathrm C}\cdot\operatorname{grad}\boldsymbol v_{\mathrm C} + F\boldsymbol w_{\mathrm C}\operatorname{div}\boldsymbol v_{\mathrm C} - \boldsymbol v_{\mathrm C}\operatorname{div}(F\boldsymbol w_{\mathrm C}) .\end{aligned}\right\} \qquad (133.3)$$

[1] TRUESDELL [1951, *32*].

Now let F be chosen as a solution of

$$\operatorname{grad}\frac{F}{v_0}\times\frac{\partial \boldsymbol{v}_C}{\partial t}=\frac{\boldsymbol{w}_C}{v_0}\frac{\partial F}{\partial t}, \tag{133.4}$$

and let M^{1/v_0} be any permissible density for $\boldsymbol{v}_C$, i.e., let M be any solution of

$$\frac{1}{v_0}\,\dot{\overline{\log M}}=-\operatorname{div}\boldsymbol{v}_C, \tag{133.5}$$

or equivalently

$$\frac{1}{v_0}\frac{\partial M}{\partial t}+\operatorname{div}(M\boldsymbol{v}_C)=0. \tag{133.6}$$

Our identity (133.3) then reduces to

$$\left.\begin{aligned}&\frac{1}{M}\operatorname{curl}\left[F\left(\frac{1}{v_0}\frac{\partial \boldsymbol{v}_C}{\partial t}+\boldsymbol{w}_C\times\boldsymbol{v}_C\right)\right]\\&\qquad=\frac{1}{v_0}\frac{d}{dt}\left(\frac{F\boldsymbol{w}_C}{M}\right)-\frac{F\boldsymbol{w}_C}{M}\cdot\operatorname{grad}\boldsymbol{v}_C-\frac{\boldsymbol{v}_C}{M}\operatorname{div}(F\boldsymbol{w}_C),\end{aligned}\right\} \tag{133.7}$$

a generalization of BELTRAMI's diffusion equation (101.7), to which it reduces when $v_0=1$, $F=1$, $M=J^{-1}$.

We now form the scalar product of (133.7) with $F\boldsymbol{w}_C/M$, thus obtaining a generalization of (101.8):

$$\left.\begin{aligned}\frac{1}{2v_0}\frac{d}{dt}\left(\frac{F\boldsymbol{w}_C}{M}\right)^2&=\frac{F\boldsymbol{w}_C}{M}\cdot\operatorname{grad}\boldsymbol{v}_C\cdot\frac{F\boldsymbol{w}_C}{M}+\frac{F}{M^2}\boldsymbol{v}_C\cdot\boldsymbol{w}_C\,\boldsymbol{w}_C\cdot\operatorname{grad}F+\\&\quad+\frac{F\boldsymbol{w}_C}{M^2}\cdot\operatorname{curl}\left[F\left(\frac{1}{v_0}\frac{\partial \boldsymbol{v}_C}{\partial t}+\boldsymbol{w}_C\times\boldsymbol{v}_C\right)\right].\end{aligned}\right\} \tag{133.8}$$

By (132.5), the assumption that the motion is complex-lamellar, now employed for the first time, annuls the second term on the right in (133.8). Let us assume further that the third term is zero:

$$\boldsymbol{w}_C\cdot\operatorname{curl}\left[F\left(\frac{1}{v_0}\frac{\partial \boldsymbol{v}_C}{\partial t}+\boldsymbol{w}_C\times\boldsymbol{v}_C\right)\right]=0. \tag{133.9}$$

We then obtain the equation

$$\frac{1}{2v_0}\frac{d}{dt}\left(\frac{F\boldsymbol{w}_C}{M}\right)^2=\frac{F\boldsymbol{w}_C}{M}\cdot\operatorname{grad}\boldsymbol{v}_C\cdot\frac{F\boldsymbol{w}_C}{M}. \tag{133.10}$$

We now impose the further requirement that the vortex lines be steady. Then it is possible to choose stationary co-ordinates at a single point[1] in such a way that the x^1 co-ordinate curve is tangent to the vortex line at the point in question:

$$\left.\begin{aligned}&ds^2=h^2(dx^1)^2+g_{22}(dx^2)^2+g_{33}(dx^3)^2,\\&(w)^2=w^1w_1,\qquad \dot{x}^1=0,\qquad \frac{\partial h}{\partial t}=0.\end{aligned}\right\} \tag{133.11}$$

From (133.10) we have then

$$\frac{1}{v_0}\frac{d}{dt}\log\frac{F w_C}{M}=v^1_{C,1}=\frac{\partial v^1_C}{\partial x^1}+\left\{{1\atop 1\,k}\right\}v^k_C, \tag{133.12}$$

[1] The neighboring x^1 co-ordinate curves need not be vortex lines. Thus (133.11) does not impose any restriction on the class of motions considered.

whence follows, since $v_C^1 = 0$,

$$\left.\begin{aligned}\frac{d}{dt}\log\frac{F w_C}{M} &= \begin{Bmatrix}1\\1\,2\end{Bmatrix}\dot{x}^2 + \begin{Bmatrix}1\\1\,3\end{Bmatrix}\dot{x}^3,\\ &= \frac{\partial\log h}{\partial x^2}\dot{x}^2 + \frac{\partial\log h}{\partial x^3}\dot{x}^3,\\ &= \frac{\partial\log h}{\partial x^1}\dot{x}^1 + \frac{\partial\log h}{\partial x^2}\dot{x}^2 + \frac{\partial\log h}{\partial x^3}\dot{x}^3 + \frac{\partial\log h}{\partial t},\\ &= \dot{\overline{\log h}}.\end{aligned}\right\} \tag{133.13}$$

Hence

$$\frac{d}{dt}\left(\frac{F w_C}{h M}\right) = 0. \tag{133.14}$$

The equation just derived expresses a simple conservation law. Going back and collecting the assumptions we have made to derive it, we obtain the ***generalized vorticity convection theorem for complex-lamellar motions***: *Given a complex-lamellar motion with steady vortex lines, let the x^1 co-ordinate curve at the point in question be tangent to the vortex line and let $dx^1/ds = h^{-1}$; let v_0 be any substantially constant function, let $\boldsymbol{v}_C$ and $\boldsymbol{w}_C$ be defined by*

$$\boldsymbol{v}_C \equiv \frac{\dot{\boldsymbol{p}}}{v_0}, \qquad \boldsymbol{w}_C \equiv \operatorname{curl}\boldsymbol{v}_C; \tag{133.1), (133.3}$$

let F be any solution of

$$\operatorname{grad}\frac{F}{v_0}\times\frac{\partial\boldsymbol{v}_C}{\partial t} = \frac{\boldsymbol{w}_C}{v_0}\,\frac{\partial F}{\partial t}; \tag{133.4}$$

and let M be any solution of

$$\frac{1}{v_0}\,\frac{\partial M}{\partial t} + \operatorname{div}(M\boldsymbol{v}_C) = 0; \tag{133.6}$$

if further, it is possible to find among the functions v_0 and F satisfying these conditions a pair such that also

$$\boldsymbol{w}_C\cdot\operatorname{curl}\left[F\left(\frac{1}{v_0}\,\frac{\partial\boldsymbol{v}_C}{\partial t} + \boldsymbol{w}_C\times\boldsymbol{v}_C\right)\right] = 0, \tag{133.9}$$

then

$$\frac{F w_C}{h M} = \text{const} \tag{133.15}$$

for each particle. Of the conditions of this theorem only the requirement of complex-lamellar motion with steady vortex lines and the one Eq. (133.9) are truly restrictive, the others being rather in the nature of definitions of the class of admissible functions v_0, F, and M.

For the simplest special case, put $v_0 = 1$, $F = 1$. Then $\boldsymbol{v}_C = \dot{\boldsymbol{p}}$, and we may satisfy the condition (133.6) by the choice $M = J^{-1}$. By (107.2) the condition (133.9) reduces to

$$\boldsymbol{w}\cdot\boldsymbol{w}^* = 0. \tag{133.16}$$

Hence it follows from the theorem above that *in a complex-lamellar motion with steady vortex lines, in order that*

$$\frac{J w}{h} = \text{const} \tag{133.17}$$

for each particle it is necessary and sufficient that the motion be circulation preserving or that the diffusion vector be normal to the vorticity. In particular the theorem applies both to plane motions and to rotationally symmetric motions. For the former, we take $h=1$ and obtain ***D'Alembert's vorticity theorem***[1]: *In order that a plane motion be circulation preserving it is necessary and sufficient that*

$$J\,w = \text{const} \tag{133.18}$$

for each particle. For rotationally symmetric motion, we have $h=r$, where r is the distance from the axis of symmetry. Hence follows ***Svanberg's vorticity theorem***[2]: *In order that a rotationally symmetric motion be circulation preserving, it is necessary and sufficient that*

$$\frac{J w}{r} = \text{const} \tag{133.19}$$

for each particle.

It was these theorems, stating that $J w$ or $J w/r$ is carried by the motion as if it were some native property of the particles, that originally motivated the name "convection" for the process of transfer of vorticity in a circulation preserving motion.

Special cases of (133.15) appropriate to certain gas motions which are not circulation preserving have been developed by TRUESDELL[3].

134. Characterizations of convection, I. CAUCHY's formula.

The most useful characterization of convection, the celebrated ***Cauchy vorticity formula***[4], follows at once from (101.9) or (101.11):

$$w_{km} = X^{\alpha}_{,k} X^{\beta}_{,m} W_{\alpha\beta}, \quad \text{or} \quad J\,w^k = W^{\alpha} x^k_{,\alpha}, \quad \text{or} \quad w^k\,dv = W^{\alpha}\,dV\,x^k_{,\alpha}. \tag{134.1}$$

$(134.1)_1$ is the tensor law of transformation for the covariant components of spin, when the motion is regarded as change of co-ordinates. That is, *in order for the motion to be circulation preserving it is necessary and sufficient that vorticity be convected.* The main consequences of (134.1) have already been given in generalized form in Sect. 129. Here we add only the corollary that $\boldsymbol{w}=0$ if and only if $\boldsymbol{W}=0$. This is the famous ***velocity potential theorem*** of LAGRANGE and CAUCHY[5]: *In a circulation preserving motion, a particle once in irrotational motion is always in irrotational motion.* Thus a portion of material for which a velocity potential exists moves about and carries this property with it, but the part of space which it originally occupied may in the course of time come to be occupied by material which did not originally possess the property, and which therefore cannot have acquired it. This theorem may be proved in several other ways. We leave it to the reader to see how proofs may be constructed from the results given in Sect. 108, Sect. 130, and the two sections to follow.

---

[1] D'ALEMBERT [1761, *1*, § XIII] treated only steady isochoric motion. Cf. also STOKES [1842, *4*, Eq. (10)], HELMHOLTZ [1858, *1*, § 5], LAMB [1878, *5*]. Sufficiency is not proved by these authors.

[2] [1841, *5*, § 4]. Cf. STOKES [1842, *4*, Eq. (21)], HELMHOLTZ [1858, *1*, § 6], LAMB [1878, *5*]. Sufficiency is not proved by these authors.

[3] [1952, *23*, §§ 9–10 [1954, *24*, § 92]. This work was motivated by earlier results of CROCCO [1936, *3*] and PRIM [1948, *21*] [1952, *16*, Eqs. (65), (194)].

[4] [1827, *5*, 1e partie, Sect. 1, ¶ 4]. CAUCHY's analysis is elaborate and does not make evident that $(134.1)_2$ is sufficient as well as necessary. Cf. CRUDELI [1918, *1*, § I]. That the form $(134.1)_1$ is a necessary and sufficient condition for circulation preserving motion in n-dimensional spaces was observed by DE DONDER [1912, *2*, § 5].

[5] The statement and proof of LAGRANGE [1783, *1*, §§ 17–19] [1788, *1*, Part II, § 11, ¶ 16 – 17] are faulty. They were corrected by CAUCHY, loc. cit. An extensive literature presents the proofs, misunderstandings, and controversies that have arisen in this connection: POISSON [1831, *2*, ¶ 73] [1833, *4*, § 654], POWER [1842, *3*], STOKES [1845, *4*, §§ 10–13] [1846, *1*, § I] [1848, *3*], ST. VENANT [1869, *6*], BRESSE [1880, *3* and *4*], BOUSSINESQ [1880, *1* and *2*], POINCARÉ [1893, *6*, § 152], HADAMARD [1901, *8*, ¶ 4, footnote]. Critical reviews are given by DUHEM [1901, *7*, Part 5] and TRUESDELL [1954, *24*, §§ 104–107].

135. Characterizations of convection, II. The transformation of WEBER. CAUCHY's formula (134.1) is a first integral of the kinematical equations. We now seek a corresponding statement in terms of the velocity field itself. For any motion we have

$$\left.\begin{aligned} x^k_{,\alpha}\ddot{x}_k &= \overline{x^k_{,\alpha}\dot{x}_k}^{\,\bullet} - \dot{x}^k_{,\alpha}\dot{x}_k, \\ &= \overline{x^k_{,\alpha}\dot{x}_k}^{\,\bullet} - (\tfrac{1}{2}\dot{x}^2)_{,\alpha}. \end{aligned}\right\} \qquad (135.1)$$

From (109.1) follows

$$x^k_{,\alpha}\ddot{x}_k = -V^*_{,\alpha}. \qquad (135.2)$$

By combining (135.1) and (135.2) we get

$$\overline{x^k_{,\alpha}\dot{x}_k}^{\,\bullet} = (\tfrac{1}{2}\dot{x}^2 - V^*)_{,\alpha}. \qquad (135.3)$$

Intetration along the path of a particle then yields

$$\int_0^t (\tfrac{1}{2}\dot{x}^2 - V^*)_{,\alpha}\,dt = x^k_{,\alpha}\dot{x}_k - \dot{x}_\alpha. \qquad (135.4)$$

where $\dot{x}_\alpha$ stands for the initial value of $\dot{x}_k$ in accord with the convention (66.8). If we introduce the function E defined by (130.9), this last result may be written

$$E_{,\alpha} = x^k_{,\alpha}\dot{x}_k - \dot{x}_\alpha. \qquad (135.5)$$

which is ***Weber's transformation***[1]. It expresses the condition for circulation preserving motion in an integrated material form. It asserts that the actual velocity $\dot{\boldsymbol{x}}$ exceeds the mere transport of the initial velocity $\dot{\boldsymbol{X}}$ by the gradient of E, where E is the accumulated difference of the specific kinetic energy and the acceleration potential along the path traversed.

For the case of steady motion, an ingenious modification of WEBER's transformation has been constructed by ERTEL[2]. Instead of (130.9) write

$$E(t,t') = \int_{t'}^t (\tfrac{1}{2}\dot{x}^2 - V^*)\,dt, \qquad (135.6)$$

where t' may vary arbitrarily from one particle to another: $t' = t'(\boldsymbol{X})$. Denote by primes functions evaluated at time t'. In particular, write

$$\boldsymbol{x}' \equiv \boldsymbol{x}(\boldsymbol{X}, t'(\boldsymbol{X})) \qquad (135.7)$$

for the place occupied by the particle $\boldsymbol{X}$ at time t', and write $\dot{\boldsymbol{x}}'$ for its velocity. Differentiating (135.7) yields

$$x'^k_{,\alpha} = (x^k_{,\alpha})' + \dot{x}'^k t'_{,\alpha}, \qquad (135.8)$$

whence

$$x'^k_{,\alpha}\dot{x}'_k = (x^k_{,\alpha}\dot{x}_k)' + \dot{x}'^2 t'_{,\alpha}. \qquad (135.9)$$

But

$$\left.\begin{aligned} E_{,\alpha} &= \int_{t'}^t (\tfrac{1}{2}\dot{x}^2 - V^*)_{,\alpha}\,dt - (\tfrac{1}{2}\dot{x}^2 - V^*)' t'_{,\alpha}, \\ &= \int_{t'}^t (\tfrac{1}{2}\dot{x}^2 - V^*)_{,\alpha}\,dt - (\tfrac{1}{2}\dot{x}^2 + V^*)\,(t-t')_{,\alpha} - \dot{x}'^2 t'_{,\alpha}, \end{aligned}\right\} \qquad (135.10)$$

where we have used the fact that in steady motion $\frac{1}{2}\dot{x}^2 + V^*$ is constant on each stream line (Sect. 125). Hence by (135.9) follows

$$E_{,\alpha} = \int_{t'}^t (\tfrac{1}{2}\dot{x}^2 - V^*)_{,\alpha}\,dt - (\tfrac{1}{2}\dot{x}^2 + V^*)\,(t-t')_{,\alpha} - x'^k_{,\alpha}\dot{x}'_k + (x^k_{,\alpha}\dot{x}_k)'. \qquad (135.11)$$

---

[1] [1868, *17*, § 2]. The proof given here is that of APPELL [1897, *1*, § 7].
[2] [1952, *4*].

Integrating (135.3) from t' to t yields

$$\int_{t'}^{t} (\tfrac{1}{2}\dot{x}^2 - V^*)_{,\alpha}\, dt = x^k_{,\alpha}\dot{x}_k - (x^k_{,\alpha}\dot{x}_k)', \tag{135.12}$$

whence by (135.11) follows

$$E_{,\alpha} = x^k_{,\alpha}\dot{x}_k - (\tfrac{1}{2}\dot{x}^2 + V^*)(t - t')_{,\alpha} - (x'^k_{,\alpha}\dot{x}'_k), \tag{135.13}$$

generalizing (135.5). Multiplying this result by $X^\alpha_{,k}$ yields finally

$$\dot{x}_k = E_{,k} + (\tfrac{1}{2}\dot{x}^2 + V^*)(t - t')_{,k} + x'_{k,m}\dot{x}'^m. \tag{135.14}$$

The next step is to formulate conditions sufficient that the last term shall vanish. First, $a^m x'_{k,m} = 0$ for some non-vanishing $\boldsymbol{a}$ if and only if $x'_{k,m}$ is singular. This is the case for all $\boldsymbol{X}$ if and only if all the places $\boldsymbol{x}'(\boldsymbol{X})$ lie upon a certain surface; that is, if and only if $t'(\boldsymbol{X})$ is such that the particle $\boldsymbol{X}$ crosses that surface at the time $t'(\boldsymbol{X})$. If it is possible to choose $t'(\boldsymbol{X})$ in this way, then we shall have $\dot{x}'^m \dot{x}_{k,m} = 0$ if and only if $\dot{\boldsymbol{x}}'$ is normal to the surface. Summarizing these results, we obtain ***Ertel's potential theorem***: *In a steady circulation preserving motion such that every stream line is normal to a certain surface, let $t - t'$ be the time which the particle now at $\boldsymbol{x}$ has travelled since crossing that surface. Then a set of Monge potentials for the velocity is*[1]

$$H = E \equiv \int_{t'}^{t} (\tfrac{1}{2}\dot{x}^2 - V^*)\, dt, \quad F = \tfrac{1}{2}\dot{x}^2 + V^*, \quad G = t - t'. \tag{135.15}$$

Put

$$S \equiv E + (\tfrac{1}{2}\dot{x}^2 + V^*)(t - t'). \tag{135.16}$$

Then evidently when (135.15) is valid we have the alternative set

$$H = S, \quad F = t' - t, \quad G = \tfrac{1}{2}\dot{x}^2 + V^*. \tag{135.17}$$

Now by (135.6) and the fact that $\frac{1}{2}\dot{x}^2 + V^*$ is constant in time for each particle we see that

$$S = \int_{t'}^{t} \dot{x}^2\, dt = \int_0^x dx^m \dot{x}_m = \int_0^x dx\, p_{\dot{\boldsymbol{x}}}, \tag{135.18}$$

where x is the arc length along the stream line, measured from its point of intersection with the normal surface. The right-hand side of (135.18) is simply the flow along the stream line (Sect. 129). Then the Monge resolution asserted by (135.17) is equivalent to

$$\dot{x}_k = \Big(\int_0^x dx^m \dot{x}_m\Big)_{,k} - (t - t')(\tfrac{1}{2}\dot{x}^2 + V^*)_{,k}. \tag{135.19}$$

These Monge potentials H, F, G have simple kinematical interpretations: flow along the stream line, time for traversing the stream line, and specific motive energy of the stream line.

Now consider the special case when the stream lines in a certain region are all closed curves, of length, say, l, where l varies in general from one to another. Since the motion is steady, the time required for a particle to traverse each stream line once is the same for all particles upon the stream line, and may be called its *period*, T. If we replace x by $x + l$ and t by $t + T$, $\dot{\boldsymbol{x}}$ must remain unchanged, so that by (138.19)

$$\dot{x}_k = \Big(\int_0^{x+l} dx^m \dot{x}_m\Big)_{,k} - (t + T - t')(\tfrac{1}{2}\dot{x}^2 + V^*)_{,k}. \tag{135.20}$$

[1] When the stream lines possess more than one orthogonal surface, FORTAK [1953, *10*] has shown that the representation (135.15) is independent of the choice of surface.

Subtracting (135.19) yields

$$0 = \left(\int_x^{x+l} dx^m \dot{x}_m\right)_{,k} - T\left(\tfrac{1}{2}\dot{x}^2 + V^*\right)_{,k}. \tag{135.21}$$

But $\int_x^{x+l} dx^m \dot{x}_m = \oint dx^m \dot{x}_m$, the circulation around the stream line. By (135.21), this quantity if not constant is a function of $\frac{1}{2}\dot{x}^2 + V^*$, and hence is constant on each Lamb surface (Sect. 125). Moreover, from (135.21) follows

$$T = \frac{d(\oint dx^m \dot{x}_m)}{d(\frac{1}{2}\dot{x}^2 + V^*)}, \tag{135.22}$$

provided that $\oint dx^m \dot{x}_m \neq \text{const}$. We have thus obtained the elegant ***Ertel theorem on circulating motions***[1]: *In a steady rotational circulation preserving motion whose stream lines are closed curves intersecting a certain surface orthogonally, the circulation of the stream lines is a function of their energy $\frac{1}{2}\dot{x}^2 + V^*$ only. Hence it is the same for all stream lines lying upon the same Lamb surface. Moreover, the periods of the stream lines are given by* (135.22).

To verify ERTEL's theorem in the case of a steady plane rotational simple vortex (Sect. 89), for which any azimuthal plane may be taken as the normal surface, by (89.8) and the formula in footnote 3, p. 408, we have

$$T = \frac{d(\oint dx^m \dot{x}_m)}{d(\frac{1}{2}\dot{x}^2 + V^*)} = \frac{\frac{d(\oint dx^m \dot{x}_m)}{dr}}{\frac{d(\frac{1}{2}\dot{x}^2 + V^*)}{dr}} = \frac{2\pi(2r\omega + r^2\omega')}{2r\omega^2 + r^2\omega\omega'} = \frac{2\pi}{\omega}, \tag{135.23}$$

as is evident.

136. Characterizations of convection, III. The transformation of CLEBSCH. We now seek a spatial counterpart of the analysis of the last section. We consider the Monge potentials of the velocity, (99.19). In any motion we have (99.20), and the surfaces $F = \text{const}$, $G = \text{const}$ are vortex surfaces. Since in a circulation preserving motion the second Helmholtz theorem asserts that the vortex lines are material, it is natural to conjecture, and we shall now accordingly prove, that it is possible to choose the functions F and G in such a way that the surfaces $F = \text{const}$, $G = \text{const}$ are also material surfaces, that is, in such a way that

$$\dot{F} = 0, \quad \dot{G} = 0. \tag{136.1}$$

Let $F^0(\boldsymbol{X})$ and $G^0(\boldsymbol{X})$ be any pair of continuously differentiable functions satisfying (99.20) at $t = 0$. First, as was stated in Sect. App. 32, for one of the functions, say G, in (99.20) at time t we may select any function such that the surfaces $G = \text{const}$ are vortex surfaces, and thus a permissible choice is given by

$$G(\boldsymbol{x}, t) = G^0(\boldsymbol{X}(\boldsymbol{x}, t)), \tag{136.2}$$

whereby $(136.1)_2$ is satisfied. By (App. 32.4), the two functions F^0 and F satisfy the relations

$$F^0 = \int \frac{W_{\alpha\beta} E^\alpha dX^\beta}{E^\gamma G^0_{,\gamma}}, \quad F = \int \frac{w_{km} e^k dx^m}{e^p G_{,p}}, \tag{136.3}$$

where $\boldsymbol{E}$ and $\boldsymbol{e}$ are arbitrary continuous fields, which we may select independently and at pleasure, subject only to the restriction that the denominators in (136.3) be non-vanishing in the regions considered, and in each integral the path of integration is the same, viz., any curve lying wholly upon one of the surfaces $G = G^0 = \text{const}$. By employing CAUCHY's vorticity formula (134.1) and (20.2)

[1] [1950, 7]. Our proof is taken from [1952, 4, § V].

we obtain

$$\left.\begin{aligned}\frac{w_{km}\, e^k\, d x^m}{e^p\, G_{,p}} &= \frac{W_{\alpha\beta}\, X^\alpha_{,k}\, X^\beta_{,m}\, e^k\, d x^m}{e^p\, G_{,\gamma}\, X^\gamma_{,p}}\,,\\ &= \frac{(X^\alpha_{,k}\, e^k)\, W_{\alpha\beta}\, dX^\beta}{(X^\gamma_{,p}\, e^p)\, G_{,\gamma}}\,.\end{aligned}\right\} \tag{136.4}$$

Hence if we make the choice $E^\alpha \equiv X^\alpha_{,k}\, e^k$, then comparison of $(136.3)_2$ with $(136.3)_1$ yields

$$F(\boldsymbol{x}, t) = F^0(\boldsymbol{X})\,, \tag{136.5}$$

so that $(136.1)_1$ also is satisfied. The possibility of such a choice of the functions F and G was first proved by CLEBSCH[1] in a different way.

With the aid of (136.1) we are able to reduce DUHEM's acceleration formula (99.24):

$$\ddot{x}_k = \left(\frac{1}{2}\,\dot{x}^2 + \frac{\partial H}{\partial t} + F\,\frac{\partial G}{\partial t}\right)_{,k}\,. \tag{136.6}$$

If we absorb into H a function of time only, whose value, by (99.19), cannot affect the velocity, comparison of (136.6) with (109.1) yields

$$V^* + \frac{1}{2}\,\dot{x}^2 + \frac{\partial H}{\partial t} + F\,\frac{\partial G}{\partial t} = 0\,. \tag{136.7}$$

This result may be called ***Clebsch's transformation***[2]. It expresses the condition for circulation preserving motion in integrated form and wholly in terms of the spatial variables. According to the definition of Sect. 120, CLEBSCH's transformation is a Bernoullian theorem, but unlike other results of this type it is not limited in its validity to motions with steady vorticity.

Another form of CLEBSCH's transformation may be obtained by noticing that from $(136.1)_2$ and (99.19) we have

$$\frac{\partial G}{\partial t} = -\,\dot{\boldsymbol{p}} \cdot \operatorname{grad} G = -\,[\operatorname{grad} H + F \operatorname{grad} G] \cdot \operatorname{grad} G\,, \tag{136.8}$$

while by squaring (99.19) follows

$$\tfrac{1}{2}\,\dot{p}^2 = \tfrac{1}{2}\,(\operatorname{grad} H)^2 + F \operatorname{grad} G \cdot \operatorname{grad} H + \tfrac{1}{2}\,F^2\,(\operatorname{grad} G)^2\,. \tag{136.9}$$

Putting these two results into (136.7) yields[3]

$$V^* + \frac{\partial H}{\partial t} + \frac{1}{2}\,(\operatorname{grad} H)^2 - \frac{1}{2}\,F^2\,(\operatorname{grad} G)^2 = 0\,. \tag{136.10}$$

A part of the apparent simplicity of CLEBSCH's transformation is illusory, since in steady motion the functions H and G generally cannot be steady. In fact, comparison of (136.6) with LAGRANGE's acceleration formula (99.14) in the case of steady motion yields

$$\boldsymbol{w} \times \dot{\boldsymbol{p}} = \operatorname{grad}\left[\frac{\partial H}{\partial t} + F\,\frac{\partial G}{\partial t}\right]. \tag{136.11}$$

---

[1] [1859, *1*, § 3]. Other proofs were given by HANKEL [1861, *1*, § VI], HILL [1881, *3*, § III and pp. 168–210], BASSET [1888, *1*, §§ 33–35], LAMB [1906, *4*, § 166], APPELL [1921, *1*, § 799], and TRUESDELL [1954, *24*, § 101]. The last of these is the one reproduced here.

[2] CLEBSCH [1857, *1*, § 5] derived an equivalent formula from the equations for perfect fluids. See also BELTRAMI [1871, *1*, § 13], HILL [1881, *3*, § III].

[3] This formula, suggested by similar but generally false equations of CHALLIS [1842, *1*] and CRAIG [1881, *2*, p. 12], was derived by TRUESDELL [1954, *24*, § 101].

Hence H and G can be steady if and only if the motion is an irrotational or screw motion, in which case CLEBSCH's transformation (136.7) reduces to the same form[1] as does the spatial Bernoulli theorem (120.1).

Finally we consider the more general case in which the functions F and G are not subjected to (136.1). By taking the curl of DUHEM's acceleration formula (99.24) and comparing the result with the d'Alembert-Euler condition (108.1), we then obtain

$$\operatorname{grad}\dot{F}\times\operatorname{grad}G=\operatorname{grad}\dot{G}\times\operatorname{grad}F \tag{136.12}$$

as a necessary and sufficient condition to be satisfied by the Monge potentials of the velocity in order that the motion be circulation preserving. That is, the vectors $\operatorname{grad}\dot{G}$, $\operatorname{grad}\dot{F}$, $\operatorname{grad}G$, and $\operatorname{grad}F$ must all lie in the plane perpendicular to the vorticity $\boldsymbol{w}$, and the area inclosed by the parallelogram whose edges are $\operatorname{grad}\dot{G}$ and $\operatorname{grad}F$ must equal the area enclosed by that whose edges are $\operatorname{grad}\dot{F}$ and $\operatorname{grad}G$.

Without direct use of the foregoing, we introduce the abbreviation

$$W\equiv V^*+\frac{1}{2}\dot{p}^2+\frac{\partial H}{\partial t}+F\frac{\partial G}{\partial t}, \tag{136.13}$$

form $\operatorname{grad}W$, and simplify the result with the aid of (99.24) and (109.1) so as to obtain[2]

$$\operatorname{grad}W=\dot{G}\operatorname{grad}F-\dot{F}\operatorname{grad}G. \tag{136.14}$$

Hence

$$\frac{\partial(W,F,G)}{\partial(x,y,z)}=\operatorname{grad}W\cdot\operatorname{grad}F\times\operatorname{grad}G=0. \tag{136.15}$$

Since F and G are necessarily independent, from this result follows the equation of DUHEM[3]:

$$W=W(F,G,t). \tag{136.16}$$

Hence

$$\operatorname{grad}W=\frac{\partial W}{\partial F}\operatorname{grad}F+\frac{\partial W}{\partial G}\operatorname{grad}G. \tag{136.17}$$

Comparison with (136.14) now yields the Hamiltonian equations of STUART and LAMB[4]:

$$\frac{\partial W}{\partial F}=\dot{G},\qquad \frac{\partial W}{\partial G}=-\dot{F}. \tag{136.18}$$

Given the acceleration potential V^* of a circulation preserving motion, that F, G, W, H satisfy (136.12), (136.13), (136.16), and (136.18) is plainly not sufficient for F, G, H to be Monge potentials of any given velocity field, since H is indeterminate up to an arbitrary steady function. The contribution of any H in (99.19), however, is only an irrotational motion. Thus a solution F, G, W, H always yields a circulation preserving motion. CLEBSCH's transformation (136.7) results from use of the particular solution $W=0$.

---

[1] This same result follows also from the fact that the surfaces $G=$ const are material vortex surfaces, and hence in a steady motion in order to be themselves steady they must be Lamb surfaces.

[2] DUHEM [1901, *4*, § 2].

[3] [1901, *4*, § 3].

[4] The result was first published by LAMB [1906, *4*, § 166]; later [1924, *9*, § 167] he attributed it to STUART. DUHEM [1901, *4*, §§ 4–7] discussed the general theory of the integration of the system (136.1), (136.7) by analogy to JACOBI's method in analytical dynamics. MASUDA [1953, *17*] uses the Hamilton-Jacobi theory to obtain some of the theorems given above. Such general methods, failing to take account of the first integral (134.1) or its various equivalents, lead to an elaborate and awkward treatment.

The reader may verify that the special potentials (135.15) and (135.17) satisfy (136.12) and (136.18).

Generalizations of the transformations of CLEBSCH and WEBER to arbitrary motions are easy to work out but not illuminating[1].

136 A. Appendix. Variational form of CLEBSCH'S transformation. CLEBSCH[2] observed that for the isochoric case, use of the apparatus of Sect. 136 leads to a simple variational principle. The three functions H, F, and G are to be varied so as to yield, through (99.19), a velocity field $\dot{x}$ which is *isochoric and circulation preserving*. The formula (136.7) is taken as a *definition* of a function V^*, not yet known to be an acceleration potential, in terms of F, G, H. Then

$$\left.\begin{aligned} -\delta V^* &= \dot{x}^k\,\delta\dot{x}_k + \frac{\partial\,\delta H}{\partial t} + F\frac{\partial\,\delta G}{\partial t} + \frac{\partial G}{\partial t}\,\delta F,\\ &= [\delta H_{,k} + G_{,k}\,\delta F + F\,\delta G_{,k}]\,\dot{x}^k + \frac{\partial\,\delta H}{\partial t} + F\frac{\partial\,\delta G}{\partial t} + \frac{\partial G}{\partial t}\,\delta F,\\ &= [(\delta H + F\,\delta G)\,\dot{x}^k]_{,k} + \frac{\partial}{\partial t}(\delta H + F\,\delta G) - (\delta H + F\,\delta G)\,\dot{x}^k_{,k} - \dot{F}\,\delta G + \dot{G}\,\delta F.\end{aligned}\right\}\quad(136\text{A}.1)$$

Integrating over a fixed volume $\boldsymbol{v}$ yields

$$\left.\begin{aligned} -\delta\int_{t_1}^{t_2} dt\int_{\boldsymbol{v}} V^*\,dv &= \int_{t_1}^{t_2}\left[\oint_{\boldsymbol{s}}(\delta H + F\,\delta G)\,\dot{x}^k\,da_k\right]dt + \int_{\boldsymbol{v}}(\delta H + F\,\delta G)\Big|_{t_1}^{t_2} dv\,+\\ &\quad + \int_{t_1}^{t_2}\left[\int_{\boldsymbol{v}}\{\dot{G}\,\delta F - \dot{F}\,\delta G - (\delta H + F\,\delta G)\,\dot{x}^k_{,k}\}\,dv\right]dt.\end{aligned}\right\}\quad(136\text{A}.2)$$

We now impose the condition that δG and δH vanish throughout $\boldsymbol{v}$ at $t = t_1$ and at $t = t_2$ as well as vanishing on $\boldsymbol{s}$ at all times. Then from (136A.2) it follows that

$$\delta\int_{t_1}^{t_2} dt\int_{\boldsymbol{v}} V^*\,dv = 0 \qquad (136\text{A}.3)$$

is equivalent to the conditions $\dot{x}^k_{,k} = 0$, $\dot{F} = 0$, $\dot{G} = 0$ in $\boldsymbol{v}$. The first, by (77.1), is the condition of isochoric motion; the second and third show that (136.12) is satisfied and hence that the motion is circulation preserving. By CLEBSCH's transformation (136.7), the function V^* satisfying (136.7) is in fact an acceleration potential.

137. The Euler-Clebsch reduction of steady circulation preserving motion. After establishing a representation of the form (App. 32.5) for a solenoidal field, EULER[3] applied it to the velocity field of a steady motion. Writing

$$\dot{\boldsymbol{p}} = J\,C(A, B)\,\text{grad}\,A \times \text{grad}\,B, \qquad (137.1)$$

we present EULER's determination of conditions on A, B, C in order that the motion be circulation preserving. First we have the identity

$$\left.\begin{aligned} \boldsymbol{w}\times\dot{\boldsymbol{p}}\cdot d\boldsymbol{x} &= \boldsymbol{w}\cdot(J\,C\,\text{grad}\,A\times\text{grad}\,B)\times d\boldsymbol{x},\\ &= J\,C\,\boldsymbol{w}\cdot[(\text{grad}\,A\cdot d\boldsymbol{x})\,\text{grad}\,B - (\text{grad}\,B\cdot d\boldsymbol{x})\,\text{grad}\,A],\\ &= J\,C\,\boldsymbol{w}\cdot[dA\,\text{grad}\,B - dB\,\text{grad}\,A).\end{aligned}\right\}\quad(137.2)$$

Using (125.5) as the condition for circulation preserving motion, in the steady case we write it as

$$-d\,(V^* + \tfrac{1}{2}\dot{p}^2) = \boldsymbol{w}\times\dot{\boldsymbol{p}}\cdot d\boldsymbol{x}. \qquad (137.3)$$

---

[1] APPELL [1917, *3*, § 2] [1921, *1*, § 815], CARSTOIU [1947, *2*, § 2], TRUESDELL [1954, *24*, § 103].

[2] [1859, *1*, § 3]. Cf. also BASSET [1888, *1*, § 34].

[3] [1757, *3*, §§ 54–56]. His analysis, which was rediscovered by BASSET [1888, *1*, §§ 39 to 40] and v. MISES [1909, *8*, § 4.3], was put into modern terms by TRUESDELL [1954, *24*, § 98].

Comparison of (137.2) and (137.3) yields EULER's formulae

$$\left.\begin{aligned}\frac{\partial(V^*+\frac{1}{2}\dot{p}^2)}{\partial A} &= -J\,C\,\boldsymbol{w}\cdot\operatorname{grad} B = -J\,C\,w\,\frac{dB}{dw},\\ \frac{\partial(V^*+\frac{1}{2}\dot{p}^2)}{\partial B} &= J\,C\,\boldsymbol{w}\cdot\operatorname{grad} A = J\,C\,w\,\frac{dA}{dw},\end{aligned}\right\} \tag{137.4}$$

whence follows the condition of integrability

$$\frac{\partial}{\partial A}(J\,C\,\boldsymbol{w}\cdot\operatorname{grad} A) + \frac{\partial}{\partial B}(J\,C\,\boldsymbol{w}\cdot\operatorname{grad} B) = 0, \tag{137.5}$$

where $\boldsymbol{w}$ is to be thought of as expressed by $\boldsymbol{w} = \operatorname{curl}(J\,C\operatorname{grad} A \times \operatorname{grad} B)$.

We may use this apparatus to determine conditions that the validity of (130.5) for two families of surfaces, $\Psi = A = \text{const}$ and $B = \text{const}$, be sufficient to ensure that a given velocity field $\dot{\boldsymbol{p}}$ be circulation preserving. We assume C determined by (137.1), and we assume that A, B, and C satisfy (137.5). Then there exists a function $F(A, B)$ such that (137.4) is satisfied with F replacing $V^* + \frac{1}{2}\dot{p}^2$, and hence by the identity (137.3) follows

$$-\operatorname{grad} F = \boldsymbol{w} \times \dot{\boldsymbol{p}}. \tag{137.6}$$

Hence by (99.14)

$$-\operatorname{grad}(F - \tfrac{1}{2}\dot{p}^2) = \boldsymbol{w} \times \dot{\boldsymbol{p}} + \operatorname{grad}\tfrac{1}{2}\dot{p}^2 = \ddot{\boldsymbol{p}}. \tag{137.7}$$

Therefore the motion is circulation preserving with $F - \frac{1}{2}\dot{p}^2$ as an acceleration potential.

In summary of these results, we have EULER's characterization of steady convection: *In a steady motion, let it be possible to find two independent families of stream sheets* $A = \text{const}$, $B = \text{const}$, *such that* (130.5) *hold for each; with* C *determined by* (137.1), *if* (137.5) *holds also then it follows that the motion is circulation preserving. Conversely, in a steady circulation preserving motion* (130.5) *and* (137.5) *hold for any two substantially constant functions* A, B.

Taking $C = 1$, as is always possible, CLEBSCH found an alternative form for (137.5) in the case when the co-ordinates are rectangular Cartesian. Since in any steady motion $\dot{z}/J$ is a function of the six components $A_{,k}$ and $B_{,k}$, we have

$$\left.\begin{aligned}\frac{\partial}{\partial A_{,k}}\left(\frac{1}{2}\,\dot{z}^2/J^2\right) &= \frac{1}{2}\,\frac{\partial}{\partial A_{,k}}\,[\varepsilon_{mnp}\,A_{,n}\,B_{,p}\,\varepsilon_{mqr}\,A_{,q}\,B_{,r}],\\ &= \varepsilon_{mkn}\,B_{,n}\,\dot{z}_m,\end{aligned}\right\} \tag{137.8}$$

so that

$$\left.\begin{aligned}\left[\frac{\partial}{\partial A_{,k}}\left(\frac{1}{2}\,\dot{z}^2/J^2\right)\right]_{,k} &= \varepsilon_{kmn}\,B_{,n}\,\dot{z}_{m,k} = -\boldsymbol{w}\cdot\operatorname{grad} B,\\ \left[\frac{\partial}{\partial B_{,k}}\left(\frac{1}{2}\,\dot{z}^2/J^2\right)\right]_{,k} &= \boldsymbol{w}\cdot\operatorname{grad} A.\end{aligned}\right\} \tag{137.9}$$

Hence (137.5) may be written

$$\frac{\partial}{\partial B}\left[J\,\frac{\partial}{\partial A_{,k}}\left(\frac{1}{2}\,\dot{z}^2/J^2\right)\right]_{,k} = \frac{\partial}{\partial A}\left[J\,\frac{\partial}{\partial B_{,k}}\left(\frac{1}{2}\,\dot{z}^2/J^2\right)\right]_{,k}. \tag{137.10}$$

In the isochoric case, the problem of determining all steady circulation preserving motions is thus reduced in principle to the solution of a single differential equation for the speed.

These results may be used to derive a variational principle of CLEBSCH[1] for the isochoric case. Again taking $C = 1$, we vary A and B. Since, as established above, $V^* + \frac{1}{2}\dot{z}^2$ is a function of A and B only, we have

$$\delta\int_v\left(V^* + \frac{1}{2}\,\dot{z}^2\right)dv = \int_v\left[\frac{\partial}{\partial A}\left(V^* + \frac{1}{2}\,\dot{z}^2\right)\delta A + \frac{\partial}{\partial B}\left(V^* + \frac{1}{2}\,\dot{z}^2\right)\delta B\right]dv; \tag{137.11}$$

[1] [1857, *1*, §§ 3–4]. Our derivation follows BASSET [1888, *1*, § 41].

since $\dot{z}$ is a function of $A_{,k}$ and $B_{,k}$ only, we have

$$\left.\begin{aligned}\delta\int_v \frac{1}{2}\dot{z}^2\,dv &= \int_v\left[\frac{\partial\frac{1}{2}\dot{z}^2}{\partial A_{,k}}\,\delta A_{,k} + \frac{\partial\frac{1}{2}\dot{z}^2}{\partial B_{,k}}\,\delta B_{,k}\right]dv,\\ &= \int_v\left[\left(\frac{\partial\frac{1}{2}\dot{z}^2}{\partial A_{,k}}\,\delta A + \frac{\partial\frac{1}{2}\dot{z}^2}{\partial B_{,k}}\,\delta B\right)_{,k} - \delta A\left(\frac{\partial\frac{1}{2}\dot{z}^2}{\partial A_{,k}}\right)_{,k} - \delta B\left(\frac{\partial\frac{1}{2}\dot{z}^2}{\partial B_{,k}}\right)_{,k}\right]dv,\\ &= \oint_s\left(\frac{\partial\frac{1}{2}\dot{z}^2}{\partial A_{,k}}\,\delta A + \frac{\partial\frac{1}{2}\dot{z}^2}{\partial B_{,k}}\,\delta B\right)da_k -\\ &\quad - \int_v\left[\left(\frac{\partial\frac{1}{2}\dot{z}^2}{\partial A_{,k}}\right)_{,k}\delta A + \left(\frac{\partial\frac{1}{2}\dot{z}^2}{\partial B_{,k}}\right)_{,k}\delta B\right]dv.\end{aligned}\right\}\tag{137.12}$$

We restrict the variations so that $\delta A = 0$ and $\delta B = 0$ on s. Adding (137.11) to (137.12) then yields

$$\left.\begin{aligned}\delta\int_v (V^* + \dot{z}^2)\,dv = \int_v\Big\{&\left[\frac{\partial}{\partial A}\left(V^* + \frac{1}{2}\dot{z}^2\right) - \left(\frac{\partial\frac{1}{2}\dot{z}^2}{\partial A_{,k}}\right)_{,k}\right]\delta A +\\ &+\left[\frac{\partial}{\partial B}\left(V^* + \frac{1}{2}\dot{z}^2\right) - \left(\frac{\partial\frac{1}{2}\dot{z}^2}{\partial B_{,k}}\right)_{,k}\right]\delta B\Big\}\,dv.\end{aligned}\right\}\tag{137.13}$$

By the identity (137.9), it follows that if there exists a function V^* such that

$$\delta\int_v (V^* + \dot{z}^2)\,dv = 0 \tag{137.14}$$

when A and B are varied as stated, $\dot{z}$ being given by (137.1) with $JC = 1$, then the differential system (137.4) is satisfied. Hence (137.14) is a necessary and sufficient condition that a given solenoidal field be the velocity field of a steady isochoric circulation preserving motion.

138. Appell's theorem on the stretching of vortex lines.

138. Appell's theorem on the stretching of vortex lines. In any rotational motion, for any material element of arc $d\boldsymbol{x}$ we have

$$\left.\begin{aligned}\frac{d}{dt}\left(\frac{dx}{Jw}\right) &= \frac{1}{2}\left[\frac{1}{Jw\,dx}\,\dot{\overline{dx^2}} - \frac{dx}{(Jw)^3}\,\dot{\overline{(Jw)^2}}\right],\\ &= \frac{1}{Jw}\left[\frac{d_{km}\,dx^k\,dx^m}{dx} - \frac{dx\,(w_k^*\,w^k + d_{km}\,w^k\,w^m)}{w^2}\right],\end{aligned}\right\}\tag{138.1}$$

where we have used (82.3) and (101.8). In any motion such that the vortex lines are material, we may choose $d\boldsymbol{x}$ as tangent to the vorticity, so that

$$dx\,w^k = dx^k\,w. \tag{138.2}$$

In this case, (138.1) becomes

$$\frac{d}{dt}\left(\frac{dx}{Jw}\right) = -\frac{w_k^*\,w^k}{Jw^3}\,dx. \tag{138.3}$$

The vortex lines have been assumed material; by (107.3) it follows that $w_k^* w^k = 0$ if and only if $\boldsymbol{w}^* = 0$. This yields ***Appell's theorem***[1]: *In a motion such that the vortex lines are material, in order that*

$$\frac{dx}{Jw} = \text{const} \tag{138.4}$$

[1] Following indications by APPELL [1921, *1*, § 760], this theorem was stated and proved by TRUESDELL [1954, *24*, § 97], who showed also [ibid., § 96] that it is equivalent to a theorem of LAMB [1885, *5*].

for each material element $d\boldsymbol{x}$ in the direction of the vorticity, it is necessary and sufficient that the motion be circulation preserving. If $\mathscr{C}$ is a finite portion of a vortex line, then (138.4) is equivalent to

$$\int_{\mathscr{C}} \frac{dx}{Jw} = \text{const.} \tag{138.5}$$

Like the circulation itself, the integral (138.5) is an integral invariant. While circulation is calculated for any closed material circuit, the invariant (138.5) is calculated for any part of a vortex line. The theorem implies that *if Jw increases during the course of the motion, the vortex lines are stretched.* As a corollary, *in order that $Jw = \text{const}$ for each particle in a circulation preserving motion, it is necessary and sufficient that the length of each vortex line remain constant*[1]. This result is an interesting supplement to D'ALEMBERT'S vorticity theorem (133.18).

Let $\boldsymbol{n}$ be a unit vector. Then

$$dx_{\boldsymbol{n}} \equiv d\boldsymbol{x}\cdot\boldsymbol{n} = J\boldsymbol{w}\cdot\boldsymbol{n}\frac{dx}{Jw} = Jw_{\boldsymbol{n}}\frac{dx}{Jw}, \tag{138.6}$$

where again $d\boldsymbol{x}$ is a material element parallel to the vorticity. Then (138.4) is equivalent to the statement that for a given particle $dx_{\boldsymbol{n}} \propto Jw_{\boldsymbol{n}}$. In particular, we may choose $\boldsymbol{n}$ as the unit normal to a material surface and obtain ***v. Mises' theorem***[2]: *In a circulation preserving motion, let a particle $\boldsymbol{X}$ be infinitely near to a material surface $\mathscr{S}$; then the distance of $\boldsymbol{X}$ from $\mathscr{S}$ is proportional to the component of $J\boldsymbol{w}$ normal to $\mathscr{S}$.* Since, as will be shown in Sect. 184, a bounding surface is always material, v. MISES was able to use this theorem to discuss the behavior of the vorticity near a wall.

f) Further special motions.

139. Summary of special motions previously defined. In the foregoing sections of this subchapter numerous special types of motion have been defined. These will be listed, along with their formally simplest defining equations and references to our earlier discussion of them.

First, there are classes defined by a *geometrical* condition:

1. *Lineal* motions (Sect. 68):

$$\dot{x} = \dot{x}(x,t),\quad \dot{y} = 0,\quad \dot{z} = 0,\quad j = j(x,t). \tag{139.1}$$

2. *Pseudo-lineal* motions of the *first* kind (Sect. 68):

$$\dot{x} = \dot{x}(x,y,z,t),\quad \dot{y} = 0,\quad \dot{z} = 0,\quad j = j(x,y,z,t). \tag{139.2}$$

3. *Pseudo-lineal* motions of the *second* kind (Sect. 68):

$$\dot{x} = \dot{x}(x,t),\quad \dot{y} = \dot{y}(x,t),\quad \dot{z} = \dot{z}(x,t),\quad j = j(x,t). \tag{139.3}$$

4. *Plane* motions (Sect. 68):

$$\dot{x} = \dot{x}(x,y,t),\quad \dot{y} = \dot{y}(x,y,t),\quad \dot{z} = 0,\quad j = j(x,y,t). \tag{139.4}$$

5. *Pseudo-plane* motions of the *first* kind (Sect. 68):

$$\dot{x} = \dot{x}(x,y,z,t),\quad \dot{y} = \dot{y}(x,y,z,t),\quad \dot{z} = 0,\quad = j(x,y,z,t). \tag{139.5}$$

---

[1] Cf. the more special result of CARSTOIU [1946, *4*].

[2] [1909, *8*, § 2.5].

6. *Pseudo-plane* motions of the *second* kind (Sect. 68):

$$\dot{x} = \dot{x}(x, y, t), \quad \dot{y} = \dot{y}(x, y, t), \quad \dot{z} = \dot{z}(x, y, t), \quad j = j(x, y, t). \tag{139.6}$$

7. *Rotationally symmetric* motions (Sect. 68):

$$\dot{r} = \dot{r}(r, z, t), \quad \dot{z} = \dot{z}(r, z, t), \quad \dot{\theta} = 0, \quad j = j(r, z, t). \tag{139.7}$$

Second, there *kinematical* classes. First among these are those defined by a condition of steadiness:

8. *Steady* motion (Sect. 67):

$$\dot{\boldsymbol{x}} = \dot{\boldsymbol{x}}(\boldsymbol{x}). \tag{139.8}$$

9. *Steady* motion with *steady density* (Sect. 77):

$$\dot{\boldsymbol{x}} = \dot{\boldsymbol{x}}(\boldsymbol{x}), \quad B j = f(\boldsymbol{x}), \quad \dot{B} = 0. \tag{139.9}$$

10. Motion with *steady stream lines* (Sect. 75):

$$\dot{\boldsymbol{p}} \times \frac{\partial \dot{\boldsymbol{p}}}{\partial t} = 0. \tag{139.10}$$

Then there are motions defined by conditions on the *stretching*:

11. *Rigid* motion (Sect. 84):

$$\boldsymbol{d} = 0. \tag{139.11}$$

12. *Isochoric* motion (Sect. 77):

$$\mathrm{I}_{\boldsymbol{d}} = 0. \tag{139.12}$$

13. *Dilatation* (Sect. 83):

$$d_1 = d_2 = d_3 = d. \tag{139.13}$$

14. *Shearing* (Sect. 89):

$$d_1 = -d_3, \quad d_2 = 0. \tag{139.14}$$

Finally, there are motions defined by conditions on the *spin*:

15. *Irrotational* motion (Sect. 88):

$$\boldsymbol{w} = 0. \tag{139.15}$$

16. *Complex-lamellar* motion (Sect. 110):

$$\dot{\boldsymbol{p}} \cdot \boldsymbol{w} = 0. \tag{139.16}$$

17. *Screw* motion (Sect. 112):

$$\boldsymbol{w} \times \dot{\boldsymbol{p}} = 0, \quad \boldsymbol{w} \neq 0. \tag{139.17}$$

18. *Complex-screw* motion (Sect. 132):

$$\left.\begin{aligned} \boldsymbol{v}_{\mathrm{C}} &\equiv \dot{\boldsymbol{p}}/v_0, \quad \dot{v}_0 = 0, \\ \boldsymbol{w}_{\mathrm{C}} &\equiv \operatorname{curl} \boldsymbol{v}_{\mathrm{C}}, \quad \boldsymbol{w}_{\mathrm{C}} \times \boldsymbol{v}_{\mathrm{C}} = 0. \end{aligned}\right\} \tag{139.18}$$

19. Motion with *steady vorticity* (Sect. 111):

$$\frac{\partial \boldsymbol{w}}{\partial t} = 0. \tag{139.19}$$

20. *Circulation preserving* motion (Sect. 108):

$$\boldsymbol{w}^* = 0. \tag{139.20}$$

Also, Sect. 89 concerns two very particular motions: *rectilinear shearing*, which includes *simple shearing* as a special case and itself a special pseudo-lineal motion of the first kind, and the *rectilinear vortex*, which includes the *irrotational vortex* as a special case.

The next three sections describe additional special classes of motions.

140. D'Alembert motions. Motions such that

$$\dot{\boldsymbol{p}}(\boldsymbol{x}, t) = T(t)\, \boldsymbol{v}(\boldsymbol{x}) \tag{140.1}$$

were introduced by D'ALEMBERT[1]. These motions have steady stream lines, they are isochoric if and only if $\boldsymbol{v}$ is solenoidal, and their kinematical vorticity numbers $\mathfrak{W}_K$ are steady.

Since

$$\ddot{\boldsymbol{p}} = T' \boldsymbol{v} + T^2 \boldsymbol{v} \cdot \operatorname{grad} \boldsymbol{v}, \tag{140.2}$$

we have

$$\boldsymbol{w}^* \equiv \operatorname{curl} \ddot{\boldsymbol{p}} = T' \operatorname{curl} \boldsymbol{v} + T^2 \operatorname{curl}(\operatorname{curl} \boldsymbol{v} \times \boldsymbol{v}). \tag{140.3}$$

If $T'/T^2 \neq$ const, then (139.20) cannot be satisfied unless $\boldsymbol{w} = 0$. Thus follows a theorem of D'ALEMBERT and LAGRANGE[2]: *Apart from the case when*

$$T = \frac{1}{At + B}, \qquad A, B = \text{const}, \qquad AB \neq 0, \tag{140.4}$$

a D'Alembert motion is circulation preserving if and only if it is irrotational. In the case when (140.4) holds, the condition for circulation preserving motion is that $A \operatorname{curl} \boldsymbol{v} = \operatorname{curl}(\operatorname{curl} \boldsymbol{v} \times \boldsymbol{v})$; if $A = 0$, this is merely the condition that $\boldsymbol{v}$ itself be the velocity field of a circulation preserving motion. Again, if $\boldsymbol{v}$ is the velocity field of a rotational circulation preserving motion, by (140.3) the D'Alembert motion (140.1) is circulation preserving if and only if $T =$ const. From (140.1) and (100.3), SZEBEHELY's measure of unsteadiness assumes the form

$$\mathfrak{S} = \frac{|T'|}{T^2} \cdot \frac{v}{|\boldsymbol{v} \cdot \operatorname{grad} \boldsymbol{v}|}. \tag{140.5}$$

If $v \neq 0$ and $|\boldsymbol{v} \cdot \operatorname{grad} \boldsymbol{v}| \neq 0$, the condition (140.4) emerges as necessary and sufficient that $\mathfrak{S}$ be steady. The theorem of D'ALEMBERT and LAGRANGE may thus be expressed in terms of the steadiness of $\mathfrak{S}$.

141. Accelerationless motions. If

$$\ddot{\boldsymbol{x}} = 0, \tag{141.1}$$

then every particle $\boldsymbol{X}$ travels in a straight line at a uniform velocity $\dot{\boldsymbol{x}}(\boldsymbol{X})$. Despite the apparent simplicity of this fact, accelerationless motions have never been characterized in a form really manageable in the spatial description. Indeed, a *steady* accelerationless motion, typified by the rectilinear shearing motions (Sect. 89), is rather trivial, but when the stream lines are not steady, they are not straight, and the straight path lines may cross each other in a bewildering variety of ways. A relatively simple example was given in Sect. 71.

[1] [1752, *1*, § 148].

[2] The faulty statement and proof by D'ALEMBERT [1761, *1*, § 10] were corrected by LAGRANGE [1762, *3*, § 52].

Following CALDONAZZO[1], we derive a spatial functional equation for these motions. In the common frame the material description of an accelerationless motion is

$$z = \dot{z}(Z)\,t + Z. \tag{141.2}$$

Since $\dot{z} = f(Z)$, we may put (141.2) into the form

$$\dot{z} = f(z - \dot{z}\,t). \tag{141.3}$$

Conversely, if (141.3) holds, put $y \equiv z - \dot{z}t$; then

$$\ddot{z}^k = f^k_{,m}\dot{y}^m = f^k_{,m}(\dot{z}^m - \dot{z}^m - \ddot{z}^m t), \tag{141.4}$$

or $(\delta^k_m + t f^k_{,m})\,\ddot{z}^m = 0$. Since $\delta^k_m + t f^k_{,m}$ is singular for at most three values of t, it follows that $\ddot{z} = 0$. Therefore, *the functional equation* (141.3) *is necessary and sufficient for accelerationless motion.*

Returning to the material description (141.2), we now calculate the ratio of the elements of volume at time t and 0, by (20.9) obtaining[2]

$$\begin{aligned} dv/dV = J &= \det z_{k,\alpha} \\ &= \det(\delta_{k\alpha} + t\,\dot{z}_{k,\alpha}), \\ &= 1 + \mathrm{I}_A t + \mathrm{II}_A t^2 + \mathrm{III}_A t^3, \end{aligned} \tag{141.5}$$

where we have used the expansion of the secular determinant, and where $A \equiv \|\dot{z}_{k,\alpha}\|$. If we choose the present instant as the initial instant, $\dot{z}_{k,\alpha}$ becomes the spatial velocity gradient matrix, and for the invariants defined by (78.5) we get the interpretations

$$\text{Ю} = \frac{1}{2}\cdot\frac{1}{dV}\left.\frac{d^2(dv)}{dt^2}\right|_{t=0}, \qquad \text{Б} = \frac{1}{6}\cdot\frac{1}{dV}\left.\frac{d^3(dv)}{dt^3}\right|_{t=0}. \tag{141.6}$$

For accelerationless motions, these results extend $(76.9)_2$.

Given an accelerationless motion with velocity field $\dot{z}(Z)$, in order that it be *isochoric* we derive from $(141.5)_3$ the necessary and sufficient conditions

$$\mathrm{I}_A = 0, \qquad \mathrm{II}_A = 0, \qquad \mathrm{III}_A = 0, \tag{141.7}$$

which were obtained by MERLIN[3], who noted that (141.7) is an assertion that the hodograph is degenerate. If the hodograph is a single point, the motion is a uniform translation. MERLIN has characterized the curves and surfaces which may serve as hodographs and in terms of them has shown how to construct all isochoric accelerationless motions.

Other properties of accelerationless motions have been noted in Sect. 100 and Sect. 102.

---

[1] [1947, *1*, § 3].

[2] The argument is due in principle to EULER [1761, *2*, §§ 27–35]; cf. LIPSCHITZ [1875, *3*], MERLIN [1938, *8*, § 2].

Generalizing this idea, KOSTIUK [1936, *6*] has considered the analytic motion

$$z = Z + \sum_{k=0}^{\infty} \frac{\overset{(k)}{z}(Z, t)}{(k+1)!}\, T^{k+1}$$

and has calculated explicitly the coefficients A_k in the series

$$|z/Z| = \sum_{k=0}^{\infty} A_k T^k,$$

the first ones being $A_0 = 1$, $A_1 = \mathrm{I}_d$, and has shown by calculation at length that

$$(k+1)\,A_{k+1} = \dot{A}_k + A_1 A_k.$$

[3] [1938, *8*, § 2]. Cf. also [1937, *6*].

142. Homogeneous motions. A motion is *homogeneous* if in the common frame it assumes the form

$$\dot{\boldsymbol{z}} = \boldsymbol{a}(t) \cdot \boldsymbol{z} + \boldsymbol{b}(t). \tag{142.1}$$

The analogy to homogeneous strain (Sect. 42) is immediate, but most of the results which can be read off from this analogy are of little interest. Much of the work on geometrical kinematics listed in Special Bibliography K concerns the homogeneous case; cf. also the papers cited in footnote 1, p. 331. Since any continuous motion may be approximated in the neighborhood of any one point by an appropriate homogeneous motion, this special class is of central importance. However, we have preferred to develop general properties of general motions, and we leave to the reader their applications to this simplest of cases.

In a homogeneous motion, from (142.1) we get

$$d_{km} = a_{(km)}, \qquad w_{km} = a_{[km]} \tag{142.2}$$

for the tensors of stretching and spin. It is only an application of the fundamental theorem of Sect. 90 to assert that any homogeneous motion may be regarded, at any given instant, as a rigid rotation superposed upon an irrotational homogeneous motion of stretching along three mutually perpendicular axes. In a suitable co-ordinate system, the motion of stretching assumes the form

$$\left.\begin{aligned}
\|\dot{\boldsymbol{z}}\| &= \begin{Vmatrix} d_1 & 0 & 0 \\ \cdot & d_2 & 0 \\ \cdot & \cdot & d_3 \end{Vmatrix} \cdot \|\boldsymbol{z}\|, \\
&= \tfrac{1}{3}(d_1 + d_2 + d_3) \begin{Vmatrix} 1 & 0 & 0 \\ \cdot & 1 & 0 \\ \cdot & \cdot & 1 \end{Vmatrix} \cdot \|\boldsymbol{z}\| + \\
&\quad + \tfrac{1}{3}(d_1 + d_2 - 2d_3) \begin{Vmatrix} 1 & 0 & 0 \\ \cdot & 0 & 0 \\ \cdot & \cdot & -1 \end{Vmatrix} \cdot \|\boldsymbol{z}\| + \\
&\quad + \tfrac{1}{3}(d_1 + d_3 - 2d_2) \begin{Vmatrix} 1 & 0 & 0 \\ \cdot & -1 & 0 \\ \cdot & \cdot & 0 \end{Vmatrix} \cdot \|\boldsymbol{z}\|.
\end{aligned}\right\} \tag{142.3}$$

Now simple shearing is a special case of homogeneous motion. From (89.6) we easily conclude that *a plane homogeneous motion is isochoric if and only if it is a simple shearing superposed upon a rigid rotation about an axis normal to the plane of shearing*. Therefore, apart from a rigid rotation, the motions represented by the second and third matrices in (142.3) are simple shearings in the planes normal to the second and third principal axes of stretching, respectively. The first matrix represents a dilatation (139.13). One of several ways of expressing the foregoing decomposition, due to STOKES[1], is: *Any homogeneous motion may be regarded as composed of a rigid rotation, a uniform dilatation, and simple shearings in two of the planes normal to the principal axes of stretching.* Cf. also Sect. 46, No. 2.

Accelerationless homogeneous motions have been characterized by TRUESDELL[2]. By using $(98.1)_2$ to calculate $\ddot{\boldsymbol{z}}$ from (142.1), we get the conditions

$$\dot{\boldsymbol{a}} + \boldsymbol{a}^2 = 0, \qquad \dot{\boldsymbol{b}} + \boldsymbol{a} \cdot \boldsymbol{b} = 0. \tag{142.4}$$

[1] [1845, *4*, § 2].
[2] [1955, *27*, § 2].

In the steady case these conditions reduce to

$$\boldsymbol{a}^2 = 0, \quad \boldsymbol{a}\cdot\boldsymbol{b} = 0. \tag{142.5}$$

In the unsteady case, the general solutions $\boldsymbol{a}$ and $\boldsymbol{b}$ may be obtained from the algebraic system

$$(1+\boldsymbol{A}t)\cdot\boldsymbol{a} = \boldsymbol{A}, \quad (1+\boldsymbol{A}t)\cdot\boldsymbol{b} = \boldsymbol{B}, \tag{142.6}$$

where $\boldsymbol{A}$ and $\boldsymbol{B}$ are the initial values of $\boldsymbol{a}$ and $\boldsymbol{b}$. If $\boldsymbol{A}$ has positive (negative) proper numbers, write the largest (smallest) as $-1/t_-(-1/t_+)$; otherwise, write $t_- = -\infty$ $(t_+ = +\infty)$. Then a unique solution of (142.6) exists and is a differentiable function of t in the interval $t_- < t < t_+$. Moreover, if m is the largest among the absolute values of all the proper numbers of $\boldsymbol{A}$, then $\boldsymbol{a}$ and $\boldsymbol{b}$ are analytic functions of t when $|t| < 1/m$. Thus in general these motions do not remain continuous indefinitely but develop singularities after a finite time, determined by the initial velocities and velocity gradients.

A simple example is given by

$$\boldsymbol{a} = k \begin{Vmatrix} -\sigma & 0 & 0 \\ \cdot & -\sigma & 0 \\ \cdot & \cdot & -1 \end{Vmatrix}, \quad \boldsymbol{b} = 0, \tag{142.7}$$

where in order to satisfy (142.6) we must have

$$k(t) = \frac{k_0}{1+k_0 t}, \quad \sigma(t) = \sigma_0 \frac{1+k_0 t}{1-k_0\sigma_0 t}. \tag{142.8}$$

In this motion, a rectangular block with edges parallel to the co-ordinate planes is extended along the z-direction at the rate $k(t)$, contracted transversely in the ratio $\sigma(t)$. The stream lines in the $y=0$ plane are the curves

$$z\, x^{1/\sigma} = \text{const}. \tag{142.9}$$

The ratio of the volume v at time t to the initial volume V is

$$v/V = (1+k_0 t)(1-k_0\sigma_0 t)^2, \tag{142.10}$$

whence appears that in the cases when the motion develops a singularity, the volume is reduced to zero. This motion is illustrated in Fig. 19.

For an accelerationless homogeneous motion both (141.2) and (142.1) hold. Hence

$$\boldsymbol{z} = (\boldsymbol{a}\cdot\boldsymbol{z} + \boldsymbol{b})\,t + \boldsymbol{Z}. \tag{142.11}$$

We multiply each side by $1+\boldsymbol{A}t$ and use (142.6) to simplify the result, so obtaining

$$(1+\boldsymbol{A}t)\cdot\boldsymbol{z} = \boldsymbol{A}\cdot\boldsymbol{z}t + \boldsymbol{B}t + (1+\boldsymbol{A}t)\cdot\boldsymbol{Z}. \tag{142.12}$$

Hence

$$\boldsymbol{z} = (\boldsymbol{A}t+1)\cdot\boldsymbol{Z} + \boldsymbol{B}t. \tag{142.13}$$

This gives $\dot{\boldsymbol{z}} = \boldsymbol{A}\cdot\boldsymbol{Z} + \boldsymbol{B}$, so that $\dot{\boldsymbol{z}}$ is linear in $\boldsymbol{Z}$ as well as in $\boldsymbol{z}$. There is no essential loss in generality if we set $\boldsymbol{B}=0$, since its presence indicates only a superposed uniform translation. If $\boldsymbol{B}=0$, the point $\boldsymbol{Z}=0$ remains fixed, and if we consider the parallelepiped whose vertices are the points $0, \boldsymbol{Z}_1, \boldsymbol{Z}_2, \boldsymbol{Z}_3$ at time $t=0$, the material volume so defined is the parallelepiped whose vertices at time t are $0, \boldsymbol{z}_1, \boldsymbol{z}_2, \boldsymbol{z}_3$. We may now interpret (141.5) as giving the ratio of the volumes of these finite parallelepipeds[1].

[1] TRUESDELL [1953, 33].

In a homogeneous motion, the principal axes of stretching and the principal stretchings themselves are uniform in space at each time. If we relinquish the latter of these requirements, we seek motions such that in a certain rectangular Cartesian frame we have

$$d_{km} = 0 \quad \text{if} \quad k \neq m. \tag{142.14}$$

These three conditions are necessary and sufficient for the existence of functions $F_1(x, y, z, t)$, $F_2(x, y, z, t)$, $F_3(x, y, z, t)$ such that

$$\left.\begin{aligned} &\dot{x} = \frac{\partial F_1}{\partial x}, \quad \dot{y} = \frac{\partial F_2}{\partial y}, \quad \dot{z} = \frac{\partial F_3}{\partial z}, \\ &\frac{\partial^2}{\partial x\,\partial z}(F_1 + F_3) = 0, \quad \frac{\partial^2}{\partial y\,\partial x}(F_2 + F_1) = 0, \quad \frac{\partial^2}{\partial z\,\partial y}(F_3 + F_2) = 0. \end{aligned}\right\} \tag{142.15}$$

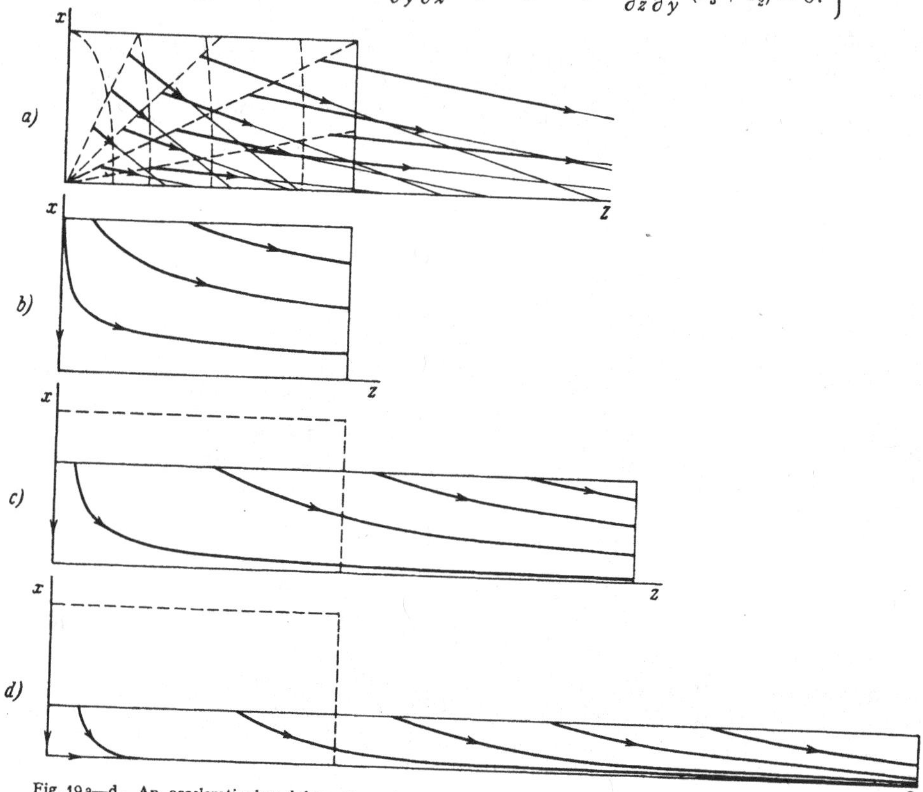

Fig. 19a—d.. An accelerationless deformation of a rectangular block. a) The undeformed block. The arrows are the velocity vectors of the particles; the light lines are their paths; the dashed straight lines are loci of particles whose velocities are parallel; the dashed ellipses are loci of particles having the same speed. b) Stream lines when $k_0 t = 0$. c) Stream lines when $k_0 t = 1$. d) Stream lines when $k_0 t = 2$.

The general solution of the latter three equations when put into the former three yields

$$\left.\begin{aligned} \dot{x} &= \frac{\partial}{\partial x}[G_3(x, y) - G_2(x, z)], \\ \dot{y} &= \frac{\partial}{\partial y}[G_1(y, z) - G_3(x, y)], \\ \dot{z} &= \frac{\partial}{\partial z}[G_2(x, z) - G_1(y, z)], \end{aligned}\right\} \tag{142.16}$$

where, as henceforth, the functions occurring may depend also on t. The principal stretchings are

$$\frac{\partial^2}{\partial x^2}(G_3 - G_2), \quad \frac{\partial^2}{\partial y^2}(G_1 - G_3), \quad \frac{\partial^2}{\partial z^2}(G_3 - G_1). \tag{142.17}$$

For the isochoric subclass, the additional condition imposed is

$$\frac{\partial^2 F_1}{\partial x^2} + \frac{\partial^2 F_2}{\partial y^2} + \frac{\partial^2 F_3}{\partial z^2} = 0. \tag{142.18}$$

Consider first the possibility $\mathrm{III}_{\boldsymbol{d}} = 0$, equivalent to, say, $\partial^2 F_3/\partial z^2 = 0$. From (142.15) and (142.18) it follows quickly that

$$\frac{\partial^2 F_1}{\partial x^2} = \frac{\partial^2 F_1}{\partial y^2}, \qquad \frac{\partial^3 F_1}{\partial x\,\partial z^2} = 0. \tag{142.19}$$

Hence $(142.15)_{1,2,3}$ must reduce to

$$\left.\begin{aligned} \dot{x} &= f(x+y) + g(x-y) + C x z, \\ \dot{y} &= -f(x+y) + g(x-y) - C y z, \\ \dot{z} &= -\tfrac{1}{2} C (x^2 - y^2), \end{aligned}\right\} \tag{142.20}$$

where the functions f and g and the constant C are arbitrary. This result was obtained by PRAGER[1], who determined also the more restricted motions in which $\mathrm{III}_{\boldsymbol{d}} \neq 0$.

While a dilatation may be characterized by specializing the preceding results, it is easier[2] to start by substituting (139.13) into (84.3), whence it is immediate that $\boldsymbol{d}$ is a linear function, and therefore the velocity is given to within a rigid motion by

$$\dot{\boldsymbol{z}} = (\boldsymbol{a}\cdot\boldsymbol{z} + k)\,\boldsymbol{z} - \tfrac{1}{2} z^2 \boldsymbol{a}, \tag{142.21}$$

where the vector $\boldsymbol{a}$ and the scalar k are functions of time only.

g) Relative motion.

143. Motion relative to a rigid rotating frame. Let $\boldsymbol{i}, \boldsymbol{j}, \boldsymbol{k}$ and $\boldsymbol{i}', \boldsymbol{j}', \boldsymbol{k}'$ be unit vectors along the co-ordinate directions of two rectangular Cartesian frames, and let $\boldsymbol{b}$ be the position vector of the origin of the primed frame relative to the unprimed frame. Then the position vectors $\boldsymbol{p}$ and $\boldsymbol{p}'$ of a given point relative to the two origins are related by

$$\boldsymbol{p} = \boldsymbol{b} + \boldsymbol{p}' \tag{143.1}$$

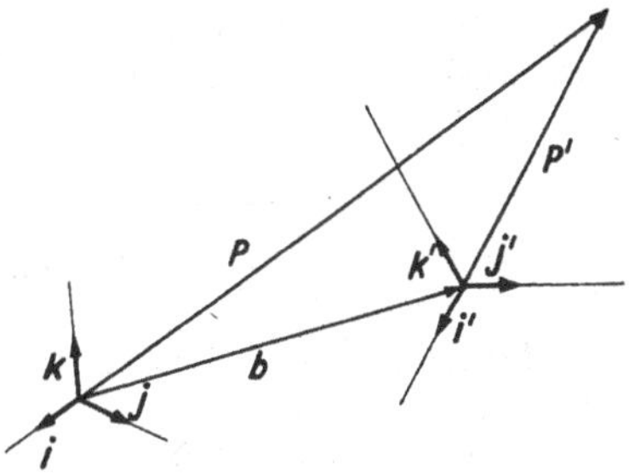

Fig. 20. Position vectors $\boldsymbol{p}$ and $\boldsymbol{p}'$ of the same point with respect to two different frames.

(Fig. 20). If the primed frame is in motion relative to the unprimed, the vectors $\boldsymbol{i}', \boldsymbol{j}', \boldsymbol{k}'$ are assigned functions of t, and $d\boldsymbol{i}'/dt$, $d\boldsymbol{j}'/dt$, $d\boldsymbol{k}'/dt$ serve to specify the rate of rotation. The axial vector $\boldsymbol{\omega}$ defined by

$$\boldsymbol{\omega} \equiv \boldsymbol{i}'\,\frac{d\boldsymbol{j}'}{dt}\cdot\boldsymbol{k}' + \boldsymbol{j}'\,\frac{d\boldsymbol{k}'}{dt}\cdot\boldsymbol{i}' + \boldsymbol{k}'\,\frac{d\boldsymbol{i}'}{dt}\cdot\boldsymbol{j}' \tag{143.2}$$

is the *angular velocity*[3] of the primed frame with respect to the unprimed. If $\dot{\boldsymbol{p}}$ and $\dot{\boldsymbol{p}}'$ are the velocities of a given particle relative to these two frames, it is easy to calculate directly that

$$\dot{\boldsymbol{p}} = \dot{\boldsymbol{b}} + \boldsymbol{\omega}\times\boldsymbol{p}' + \dot{\boldsymbol{p}}'. \tag{143.3}$$

---

[1] [1953, *23*].

[2] CAUCHY [1829, *3*].

[3] Recognition of the vectorial character of the angular velocity is due to EULER [1765, *3*, § 5]. It is sometimes more convenient to replace $\boldsymbol{i}', \boldsymbol{j}', \boldsymbol{k}'$ by any rigid triad of independent vectors $\boldsymbol{a}, \boldsymbol{b}, \boldsymbol{c}$. Then

$$\boldsymbol{\omega} = \frac{\boldsymbol{a}\left(\frac{d\boldsymbol{b}}{dt}\cdot\boldsymbol{c}\right) + \boldsymbol{b}\left(\frac{d\boldsymbol{c}}{dt}\cdot\boldsymbol{a}\right) + \boldsymbol{c}\left(\frac{d\boldsymbol{a}}{dt}\cdot\boldsymbol{b}\right)}{\boldsymbol{a}\cdot\boldsymbol{b}\times\boldsymbol{c}}.$$

In fact, if $\boldsymbol{b}=0$ we have the operator identity

$$\frac{d}{dt}=\boldsymbol{\omega}\times+\frac{d'}{dt}. \tag{143.4}$$

In particular

$$\begin{aligned}\ddot{\boldsymbol{p}}=\frac{d\dot{\boldsymbol{p}}}{dt}&=\boldsymbol{\omega}\times\dot{\boldsymbol{p}}+\frac{d'\dot{\boldsymbol{p}}}{dt},\\&=\boldsymbol{\omega}\times[\boldsymbol{\omega}\times\boldsymbol{p}'+\dot{\boldsymbol{p}}']+\frac{d'}{dt}[\boldsymbol{\omega}\times\boldsymbol{p}'+\dot{\boldsymbol{p}}'],\\&=\boldsymbol{\omega}\times(\boldsymbol{\omega}\times\boldsymbol{p}')+2\boldsymbol{\omega}\times\dot{\boldsymbol{p}}'+\dot{\boldsymbol{\omega}}\times\boldsymbol{p}'+\ddot{\boldsymbol{p}}'.\end{aligned} \tag{143.5}$$

For unrestricted $\boldsymbol{b}$, this is to be replaced by[1]

$$\ddot{\boldsymbol{p}}=\ddot{\boldsymbol{b}}+\dot{\boldsymbol{\omega}}\times\boldsymbol{p}'+\boldsymbol{\omega}\times(\boldsymbol{\omega}\times\boldsymbol{p}')+2\boldsymbol{\omega}\times\dot{\boldsymbol{p}}'+\ddot{\boldsymbol{p}}', \tag{143.6}$$

relating the accelerations $\ddot{\boldsymbol{p}}$ and $\ddot{\boldsymbol{p}}'$. For the higher accelerations, similarly[2]

$$\overset{(N)}{\boldsymbol{p}}=\overset{(N)}{\boldsymbol{b}}+\left(\boldsymbol{\omega}\times+\frac{d'}{dt}\right)^{N}\boldsymbol{p}. \tag{143.7}$$

The formula (143.6) separates into easily identifiable parts the apparent acceleration in the unprimed frame arising from the rotation of the primed frame. The terms $\dot{\boldsymbol{\omega}}\times\boldsymbol{p}'$ and $2\boldsymbol{\omega}\times\dot{\boldsymbol{p}}'$ are traditionally named after EULER and CORIOLIS, respectively, while $\boldsymbol{\omega}\times(\boldsymbol{\omega}\times\boldsymbol{p}')$ is the *centripetal acceleration*. This last is always a lamellar field, since

$$\boldsymbol{\omega}\times(\boldsymbol{\omega}\times\boldsymbol{p})=\operatorname{grad}\left[\tfrac{1}{2}(\boldsymbol{\omega}\cdot\boldsymbol{p})^2-\tfrac{1}{2}\omega^2p^2\right]. \tag{143.8}$$

To see if the Coriolis acceleration may be lamellar, we observe that

$$\operatorname{curl}'(2\boldsymbol{\omega}\times\dot{\boldsymbol{p}}')=-2\boldsymbol{\omega}\cdot\operatorname{grad}\dot{\boldsymbol{p}}'+2\boldsymbol{\omega}\operatorname{div}\dot{\boldsymbol{p}}'. \tag{143.9}$$

For the right-hand side to vanish it is sufficient, though not necessary, that the velocity $\dot{\boldsymbol{p}}'$ be that of a plane isochoric motion in a plane normal to $\boldsymbol{\omega}$. In this case we have[3]

$$2\boldsymbol{\omega}\times\dot{\boldsymbol{p}}'=-\operatorname{grad}'2\omega Q', \tag{143.10}$$

where Q' is a stream function satisfying an equation of the type (161.2). From $(130.1)_3$ we see that a more general sufficient condition is that the vorticity of the primed motion be parallel to $\boldsymbol{\omega}$ and be substantially constant.

Since $\operatorname{curl}'(\dot{\boldsymbol{\omega}}\times\boldsymbol{p}')=2\dot{\boldsymbol{\omega}}$, the Euler acceleration is lamellar if and only if it vanishes.

We now give a formal derivation of (143.3), (143.6), and certain further results, using rectangular Cartesian co-ordinates and a formalism valid in a Euclidean space of any dimension[4]. The equations corresponding to (143.1) are

$$z_k=A_{kk'}z'_{k'}+b_k,\qquad z'_{k'}=A_{kk'}(z_k-b_k), \tag{143.11}$$

[1] An attempt of CLAIRAUT [1745, *1*, Art. I, § IV] to calculate relative acceleration in a special case is faulty from lack of the term corresponding to $2\boldsymbol{\omega}\times\dot{\boldsymbol{p}}'$. Correct results involving special cases in which $2\boldsymbol{\omega}\times\dot{\boldsymbol{p}}'\neq 0$, employing special units and angular variables appropriate to hydraulic machines, were first obtained by EULER [1752, *3*, Probl. 1] [1753, *1*, §§ 9–10] [1756, *1*, § 35], but when he came to use rectangular Cartesian co-ordinates, he obtained a result [1757, *3*, § 37] which is erroneous in having $\boldsymbol{\omega}\times\dot{\boldsymbol{p}}'$ rather than $2\boldsymbol{\omega}\times\dot{\boldsymbol{p}}'$. The complete and correct result (143.6) was obtained, if rather obscurely, by CORIOLIS [1835, *2*].

[2] There is a literature on this subject: LEVY [1878, *6*], GILBERT [1878, *3*] [1884, *1*] [1888, *5*] [1889, *3*], LAISANT [1878, *4*], RIVLIN and ERICKSEN [1955, *21*, § 3]. Some of these authors obtain resolutions of $\overset{(N)}{\boldsymbol{p}}$ along the principal directions of $\dot{\boldsymbol{p}}$ or in orthogonal curvilinear co-ordinates.

[3] TAYLOR [1916, *6*, pp. 100–102].

[4] ZORAWSKI [1911, *13*, § 1] [1911, *14*, § 1].

where we employ the summation convention for Cartesian tensors, where $\boldsymbol{A} = \boldsymbol{A}(t)$, $\boldsymbol{b} = \boldsymbol{b}(t)$, and

$$A_{kk'} A_{km'} = \delta_{k'm'}, \quad A_{kk'} A_{mk'} = \delta_{km}. \tag{143.12}$$

Differentiating (143.12) yields

$$A_{kk'} \dot{A}_{km'} + \dot{A}_{kk'} A_{km'} = 0, \quad A_{kk'} \dot{A}_{mk'} + \dot{A}_{kk'} A_{mk'} = 0. \tag{143.13}$$

The definition corresponding to (143.2) is

$$\omega_{km} \equiv \dot{A}_{kk'} A_{mk'}, \quad \omega'_{k'm'} = \dot{A}_{kk'} A_{km'}, \tag{143.14}$$

whence by (143.12) and (143.13) follows

$$\omega_{km} = -\omega_{mk}, \quad \omega'_{k'm'} = -\omega'_{m'k'}, \quad \omega'_{k'm'} = -A_{kk'} A_{mm'} \omega_{km}. \tag{143.15}$$

In dyadic form, the third of these relations reads $\boldsymbol{\omega}' = -\boldsymbol{\omega}$: The angular velocity of the unprimed frame relative to the primed is equal in magnitude but opposite in sign to that of the primed frame relative to the unprimed.

We note also that from (143.14)

$$\dot{A}_{kk'} \dot{A}_{mk'} = \omega_{kq} \omega_{mq}, \quad \dot{\omega}_{km} = \ddot{A}_{kk'} A_{mk'} + \omega_{kq} \omega_{mq}. \tag{143.16}$$

In fact

$$\left.\begin{aligned} 2\overset{(N)}{\omega}_{km} &= 2\,\overset{(N)}{\overline{\dot{A}_{kk'} A_{mk'}}} = \overset{(N)}{\overline{\dot{A}_{kk'} A_{mk'} - A_{kk'} \dot{A}_{mk'}}}, \\ &= \sum_{p=0}^{N} \binom{N}{p} \Big[\overset{(N+1-p)}{A_{kk'}}\, \overset{(p)}{A_{mk'}} - \overset{(p)}{A_{kk'}}\, \overset{(N+1-p)}{A_{mk'}}\Big], \end{aligned}\right\} \tag{143.17}$$

where each summand in brackets is skew symmetric in k and m. Similarly,

$$\left.\begin{aligned} \overset{(N+1)}{A_{kk'}} &= \overset{(N)}{\overline{\dot{A}_{kk'}}} = \overset{(N)}{\overline{\dot{A}_{km'} A_{mk'} A_{mm'}}}, \\ &= \overset{(N)}{\overline{\omega_{km} A_{mk'}}}, \\ &= \sum_{p=0}^{N} \binom{N}{p} \overset{(p)}{\omega}_{km}\, \overset{(N-p)}{A_{mk'}}. \end{aligned}\right\} \tag{143.18}$$

Thus $\overset{(N)}{\boldsymbol{\omega}}$ is determined by $\boldsymbol{A}, \dot{\boldsymbol{A}}, \ldots, \overset{(N+1)}{\boldsymbol{A}}$, while $\overset{(N)}{\boldsymbol{A}}$ is determined by $\boldsymbol{A}, \boldsymbol{\omega}, \dot{\boldsymbol{\omega}}, \ldots, \overset{(N)}{\boldsymbol{\omega}}$. Indeed, from (143.18) it follows that

$$\overset{(N+1)}{A_{kk'}} = \psi^{(N)}_{km} A_{mk'}, \tag{143.19}$$

where $\psi^{(N)}_{km}$ satisfies the recurrence relation

$$\psi^{(N+1)}_{km} = \overset{(N+1)}{\omega}_{km} + \sum_{p=0}^{N} \binom{N+1}{p} \overset{(p)}{\omega}_{kq} \psi^{(N-p)}_{qm}. \tag{143.20}$$

Thus

$$\left.\begin{aligned} \psi^{(0)}_{km} &= \omega_{km}, \\ \psi^{(1)}_{km} &= \dot{\omega}_{km} + \omega_{kq} \omega_{qm}, \\ \psi^{(2)}_{km} &= \ddot{\omega}_{km} + 2\dot{\omega}_{kq} \omega_{qm} + \omega_{kq} \dot{\omega}_{qm} + \omega_{kq} \omega_{qr} \omega_{rm}, \quad \text{etc.} \end{aligned}\right\} \tag{143.21}$$

From (143.19) follows

$$\overset{(p+1)}{A_{kk'}}\, \overset{(q+1)}{A_{mk'}} = \psi^{(p)}_{kr} \psi^{(q)}_{mr}, \tag{143.22}$$

generalizing $(143.16)_1$. Putting this result into (143.17) yields an explicit formula for $\overset{(N)}{\omega}$ in terms of the ψ's:

$$2\overset{(N)^{\cdot}}{\omega}_{km} = \sum_{p=0}^{N} \binom{N}{p} \left[\psi_{kr}^{(N-p)} \psi_{mr}^{(p-1)} - \psi_{kr}^{(p-1)} \psi_{mr}^{(N-p)}\right]. \tag{143.23}$$

Thus the quantites $\psi^{(0)}, \psi^{(1)}, \ldots, \psi^{(N)}$ are determined by the quantities $\omega, \dot{\omega}, \ldots, \overset{(N)}{\omega}$, and conversely.

Differentiation of (143.11) with respect to time, followed by use of $(143.14)_1$, yields

$$\left.\begin{aligned} \dot{z}_k &= \dot{b}_k + \dot{A}_{kk'} z'_{k'} + A_{kk'} \dot{z}'_{k'}, \\ &= \dot{b}_k + \omega_{km} A_{mk'} z'_{k'} + A_{kk'} \dot{z}'_{k'}, \end{aligned}\right\} \tag{143.24}$$

a co-ordinate expression for (143.3). Differentiation of $(143.24)_1$ with respect to time, followed by use of $(143.16)_2$ and $(143.16)_1$, yields

$$\left.\begin{aligned} \ddot{z}_k &= \ddot{b}_k + \ddot{A}_{kk'} z'_{k'} + 2\dot{A}_{kk'} \dot{z}'_{k'} + A_{kk'} \ddot{z}'_{k'}, \\ &= \ddot{b}_k + (\dot{\omega}_{km} - \omega_{kq}\omega_{mq}) A_{mk'} z'_{k'} + 2\omega_{km} A_{mk'} \dot{z}'_{k'} + A_{kk'} \ddot{z}'_{k'}, \end{aligned}\right\} \tag{143.25}$$

a co-ordinate expression for (143.6). Similarly,

$$\left.\begin{aligned} \overset{(N)}{z}_k &= \overset{(N)}{b}_k + \sum_{p=0}^{N} \binom{N}{p} \overset{(N-p)}{A}_{kk'} \overset{(p)}{z}{}'_{k'}, \\ &= \overset{(N)}{b}_k + A_{kk'} \overset{(N)}{z}{}'_{k'} + \sum_{p=0}^{N-1} \binom{N}{p} \psi_{km}^{(N-p-1)} A_{mk'} \overset{(p)}{z}{}'_{k'}, \end{aligned}\right\} \tag{143.26}$$

where ψ is defined by the recurrence relation (143.20). This result is an explicit form for (143.7).

144. Invariance of stretching and spin. The analysis thus far in Part g refers to a single particle. Whether it be isolated or a participant in a continuous medium makes no difference. We now obtain results peculiar to the kinematics of a continuous body. If we differentiate $(143.24)_2$ with respect to z_m, we obtain

$$\left.\begin{aligned} \dot{z}_{k,m} &= A_{kk'} \frac{\partial \dot{z}'_{k'}}{\partial z_m} + \omega_{kq} A_{qk'} \frac{\partial z'_{k'}}{\partial z^m}, \\ &= A_{kk'} A_{mm'} \dot{z}'_{k',m'} + \omega_{km}, \end{aligned}\right\} \tag{144.1}$$

where we have used $(143.11)_2$ and $(143.12)_2$. Taking symmetric and skew symmetric parts yields

$$\left.\begin{aligned} d_{km} &= A_{kk'} A_{mm'} d'_{k'm'}, \\ w_{km} &= A_{kk'} A_{mm'} w'_{k'm'} + \omega_{km}. \end{aligned}\right\} \tag{144.2}$$

Thus the stretching tensors $\boldsymbol{d}$ and $\boldsymbol{d}'$ are related by the usual tensor law of transformation, independent of the relative motion of the two co-ordinate frames, while the spin tensors $\boldsymbol{w}$ and $\boldsymbol{w}'$ differ by the angular velocity $\boldsymbol{\omega}$. In dyadic form,

$$\boldsymbol{d} = \boldsymbol{d}', \qquad \boldsymbol{w} = \boldsymbol{w}' + \boldsymbol{\omega}. \tag{144.3}$$

These formulae embody the ***theorem of Zorawski***[1]: *For a given motion, observers in two rigid frames moving arbitrarily with respect to one another perceive the same stretchings, but spins which differ by the relative angular velocity of the frames.*

[1] [1911, *14*, § 2]. Cf. also McVittie [1949, *18*, § 3].

Stated in geometric terms, the result is obvious. However, the phrase "the same" must be interpreted in the tensor sense, given explicitly by (144.2). In the case of scalar invariants, however, "the same" means "numerically equal": *The principal stretchings and all scalar invariants of stretching have the same value for all observers.*

The formula $(144.3)_2$ casts additional light upon STOKES' interpretation of the spin as a local angular velocity (Sect. 86).

In view of the definition (143.14), equivalent to (144.2) is the relation

$$\dot{A}_{kk'} = w_{km} A_{mk'} - A_{km'} w'_{m'k'}. \tag{144.4}$$

The extension of (144.2) to the case of independently selected curvilinear co-ordinate systems is immediate and easy[1], so long as these systems be obtainable by a *time-independent* transformation from the z_k and $z_{k'}$ systems, respectively. The phrase "all observers" above refers to observers employing such co-ordinate systems. If, however, the co-ordinate surfaces themselves are in non-rigid motion with respect to the primed and unprimed frames, a more elaborate treatment is required and will be given below in Sects. 152 to 154.

An important generalization of the theorem of ZORAWSKI asserts that the Rivlin-Ericksen tensors $\boldsymbol{A}^{(N)}$ and the higher stretchings $\boldsymbol{D}^{(N)}$ are the same for all observers[2]:

$$\boldsymbol{A}^{(N)} = \boldsymbol{A}^{(N)'}, \quad \boldsymbol{D}^{(N)} = \boldsymbol{D}^{(N)'}. \tag{144.5}$$

This result is really clear from (104.9) and (104.12) but may also be established directly.

145. Invariance of diffusion of vorticity and circulation. A relation between the acceleration gradients in the two frames may be obtained by differentiating $(143.25)_2$:

$$\ddot{z}_{k,m} = \dot{\omega}_{km} - \omega_{kq}\omega_{mq} + 2\omega_{kq} A_{qk'} A_{mm'} \dot{z}'_{k',m'} + A_{kk'} A_{mm'} \ddot{z}'_{k',m'}. \tag{145.1}$$

The skew symmetric part of this equation may be written symbolically in the form[3]

$$\boldsymbol{w}^* - \boldsymbol{w}^{*\prime} = 2[\dot{\boldsymbol{\omega}} - \boldsymbol{\omega}\cdot \operatorname{grad}' \dot{\boldsymbol{p}}' + \boldsymbol{\omega} \operatorname{div}' \dot{\boldsymbol{p}}']. \tag{145.2}$$

Comparison with $(130.1)_3$ yields the following ***theorem***: *The difference between the rates of diffusion of vorticity as apparent to two different observers equals twice the excess of the rate of change of angular velocity over the rate of convection of angular velocity,* where the terms "convection" and "diffusion" are used as in Sect. 101 and Sect. 129.

From (145.2) it is apparent that *a motion which is circulation preserving with respect to one observer generally fails to be so with respect to another*[4].

An analysis equivalent to the foregoing but formally simpler may be constructed in terms of the circulation. First, from $(144.3)_2$ we have[5]

$$\int_{\mathfrak{s}} d\boldsymbol{a}\cdot\boldsymbol{w} = \int_{\mathfrak{s}} d\boldsymbol{a}'\cdot\boldsymbol{w}' + 2\boldsymbol{\omega}\cdot\int_{\mathfrak{s}} d\boldsymbol{a}', \tag{145.3}$$

since at any given instant $d\boldsymbol{a} = d\boldsymbol{a}'$. By KELVIN's transformation $(87.1)_2$ follows

$$\oint_{\mathfrak{c}} d\boldsymbol{x}\cdot\dot{\boldsymbol{p}} = \oint_{\mathfrak{c}} d\boldsymbol{x}'\cdot\dot{\boldsymbol{p}}' + 2\omega\,\mathfrak{S}_{\perp} \tag{145.4}$$

---

[1] Cf. BOBYLEW [1885, *1*].

[2] For $\boldsymbol{A}^{(N)}$, this was shown by RIVLIN and ERICKSEN [1955, *21*, Eq. (10.15)]; for $\boldsymbol{D}^{(N)}$, it is implied though not explicitly stated by NOLL [1958, *8*, § 9].

[3] An equivalent result was obtained by v. MISES [1909, *8*, § 2.6].

[4] Some special circumstances in which the circulation preserving property is invariant have been worked out by TRUESDELL [1954, *24*, § 47].

[5] The approach was suggested by ZORAWSKI [1911, *14*, § 2], but he did not obtain the geometric interpretation (145.4).

where $\mathfrak{S}_\perp \equiv \int_{\mathfrak{s}} d\boldsymbol{a} \cdot \boldsymbol{n}$, $\boldsymbol{n}$ being the unit vector in the direction of the angular velocity. Thus *the apparent circulation induced by a rotating frame equals twice the angular speed times the area inclosed by the projection of the circuit upon a plane normal to the axis of rotation.*

Now let $\mathscr{C}$ be a material circuit, and differentiate (145.4) materially, obtaining

$$\frac{d}{dt}\oint_{\mathscr{C}} d\boldsymbol{x}\cdot\dot{\boldsymbol{p}} = \frac{d'}{dt}\oint_{\mathscr{C}} d\boldsymbol{x}'\cdot\dot{\boldsymbol{p}}' + 2\frac{d'}{dt}(\omega\mathfrak{S}_\perp). \tag{145.5}$$

This elegant result[1] is often applied to the study of atmospheric circulations.

Formulae similar to (145.1) are easily obtained for the higher accelerations. Differentiating (143.26) with respect to z_m yields

$$\left.\begin{aligned}\overset{(N)}{z}_{k,m} &= \sum_{p=0}^{N}\binom{N}{p}\overset{(N-p)}{A}_{kk'} A_{mm'}\overset{(p)}{z'}_{k',m'},\\ &= \psi_{km}^{(N-1)} + A_{kk'} A_{mm'}\overset{(N)}{z'}_{k',m'} + \sum_{p=1}^{N-1}\binom{N}{p}\psi_{kq}^{(N-p-1)} A_{qk'} A_{mm'}\overset{(p)}{z'}_{k',m'},\end{aligned}\right\} \tag{145.6}$$

where we have used (143.19). By (143.20) we get

$$\psi_{(km)}^{(N)} = f\big(\omega, \dot{\omega}, \ldots, \overset{(N-1)}{\omega}\big), \quad \psi_{[km]}^{(N)} = \overset{(N)}{\omega}_{km} + g\big(\omega, \dot{\omega}, \ldots, \overset{(N-1)}{\omega}\big). \tag{145.7}$$

Therefore the symmetric and skew symmetric parts of (145.6) may be written in the symbolic forms

$$\left.\begin{aligned}\boldsymbol{d}^{(N)} &= \boldsymbol{d}^{(N)\prime} + F\big(\omega, \ldots, \overset{(N-2)}{\omega}, \boldsymbol{d}', \ldots, \boldsymbol{d}^{(N-1)\prime}, \boldsymbol{w}', \ldots, \boldsymbol{w}^{(N-1)\prime}\big),\\ \boldsymbol{w}^{(N)} &= \boldsymbol{w}^{(N)\prime} + \overset{(N-1)}{\omega} + G\big(\omega, \ldots, \overset{(N-2)}{\omega}, \boldsymbol{d}', \ldots, \boldsymbol{d}^{(N-1)\prime}, \boldsymbol{w}', \ldots, \boldsymbol{w}^{(N-1)\prime}\big).\end{aligned}\right\} \tag{145.8}$$

146. Condition for steady motion. The definition (67.5) of steady motion refers to a particular frame and plainly is not invariant with respect to rotating co-ordinates. Following analysis given by ZORAWSKI[2], we now seek to determine whether or not there exists a rotating frame in which a particular given motion is steady. While a solution to this problem is not known, it is worthwhile to carry the analysis as far as possible.

First, if $f = f(\boldsymbol{z}, t)$, set $f'(\boldsymbol{z}', t) \equiv f(\boldsymbol{z}(\boldsymbol{z}', t), t)$, where $\boldsymbol{z}(\boldsymbol{z}', t)$ is the function defined by the right-hand side of $(143.11)_1$; then for the partial time derivative of f' with $\boldsymbol{z}'$ held constant we have[3]

$$\left.\begin{aligned}\frac{\partial' f'}{\partial t} &= \frac{\partial f}{\partial t} + \frac{\partial z_k}{\partial t} f_{,k},\\ &= \frac{\partial f}{\partial t} + (\dot{A}_{kk'} z'_{k'} + \dot{b}_k) f_{,k},\\ &= \frac{\partial f}{\partial t} + (\omega_{km} z_m + a_k) f_{,k},\end{aligned}\right\} \tag{146.1}$$

where we have used both $(143.11)_1$ and $(143.11)_2$ and have put

$$a_k \equiv \dot{b}_k - \omega_{km} b_m. \tag{146.2}$$

---

[1] In the case $\boldsymbol{\omega} =$ const, it was regarded as well known by POINCARÉ [1893, *6*, § 158] and was rediscovered by V. BJERKNES [1902, *1*]; a formal derivation using (145.2) was constructed by TRUESDELL [1954, *24*, § 81].

[2] [1911, *13*]. Cf. also DE DONDER [1912, *2*].

[3] A vectorial form of (146.1) was derived by JARRE [1951, *13*, § 3].

We now differentiate $(143.24)_1$ partially with respect to t, holding z' constant. By (146.1) follows

$$\frac{\partial \dot{z}_k}{\partial t} + (\omega_{mq} z_q + a_m)\, \dot{z}_{k,m} = \ddot{b}_k + \ddot{A}_{kk'} z'_{k'} + \dot{A}_{kk'} \dot{z}'_{k'} + A_{kk'} \frac{\partial' \dot{z}_{k'}}{\partial t}. \tag{146.3}$$

Use of several of the formulae of Sect. 143, including the inverse of (143.24), allows us to put this result into the form

$$\frac{\partial \dot{z}_k}{\partial t} + (\omega_{mq} z_q + a_m)\, \dot{z}_{k,m} - \dot{a}_k - \dot{\omega}_{km} z_m - \omega_{km} \dot{z}_m = A_{kk'} \frac{\partial' \dot{z}'_{k'}}{\partial t}. \tag{146.4}$$

In order, then, that there exist a frame in which the motion is steady, it is necessary and sufficient that

$$\frac{\partial \dot{z}_k}{\partial t} + (\omega_{mq} z_q + a_m)\, \dot{z}_{k,m} - \dot{a}_k - \dot{\omega}_{km} z_m - \omega_{km} \dot{z}_m = 0. \tag{146.5}$$

This condition is expressed entirely in terms of the given velocity field $\dot{\boldsymbol{z}}(\boldsymbol{z}, t)$.

It is evident that only in most exceptional cases will the differential equations (146.5) admit a solution $\boldsymbol{\omega}(t)$, $\boldsymbol{a}(t)$. Conditions of integrability for (146.5) are not known[1].

147. The general problem of invariance with respect to moving observers. From (143.24), (143.25), and (145.1) we see that some quantities of central interest in kinematics do not transform as tensors with respect to change of frame. In the cases considered, a quantity was defined by a fixed rule in each of two arbitrarily selected frames, after which the connection between the two quantities was established.

A different problem must be considered now. Euclidean space demands the study of rigid motion, and under such a motion some important kinds of statement fail of invariance; unavoidably, we encounter the notion of a *preferred frame*. In Sect. 196 we shall give a dynamical specification of certain frames in reference to the laws of classical mechanics; in the sections following now we construct certain kinematical classes.

Given a class F_0 of preferred frames such that a certain group $\mathfrak{G}_0$ of linear time-dependent transformations (143.11) carries this class into itself, consider a quantity Ψ_0 defined for each member of F_0 and transforming as a tensor field under the group $\mathfrak{G}_0$. Let F be the class of all rigid frames, and let $\mathfrak{G}$ be the group of linear time-dependent transformations among them[2]. Then $\mathfrak{G}_0$ is a subgroup of $\mathfrak{G}$. Thus it is in order to define a quantity Ψ such that

1. In F_0, $\Psi = \Psi_0$,
2. Ψ is a tensor under $\mathfrak{G}$.

---

[1] Much of ZORAWSKI's paper consists in an attempt to express such conditions in kinematical form, but the attempt is a failure. Though more elaborately stated, his result is equivalent to the obvious fact that in order for the motion to be steady, it is necessary that there exist a frame in which the quadric of stretching and the vorticity vector are steady.

[2] The group property of these transformations may be established as follows. Writing $(143.11)_1$ and a second transformation of the same kind in matrix form, viz.

$$\boldsymbol{z}' = \boldsymbol{A} \cdot (\boldsymbol{z} - \boldsymbol{b}), \quad \boldsymbol{z}'' = \boldsymbol{A}^* \cdot (\boldsymbol{z}' - \boldsymbol{b}'),$$

where $\boldsymbol{A}$ and $\boldsymbol{A}^*$ are orthogonal, we obtain

$$\boldsymbol{z}'' = \boldsymbol{A}_* \cdot (\boldsymbol{z} - \boldsymbol{b}_*), \quad \text{where} \quad \boldsymbol{A}_* = \boldsymbol{A}^* \cdot \boldsymbol{A}, \quad \boldsymbol{b}_* = \boldsymbol{b} + \boldsymbol{A}^{-1} \cdot \boldsymbol{b}'.$$

Therefore the succession of two such transformation is of the same kind. That the inverse is of the same kind is immediate from $(143.11)_2$. Any transformation of $\mathfrak{G}$ is determined by specification of the image of some one frame; conversely, any such specification determines an element of the group.

In fact, ψ_0 need not be a tensor, but may satisfy any definite law of transformations under $\mathfrak{G}_0$; we then demand that it satisfy a corresponding law under $\mathfrak{G}$.

Our problem, then, is one of *extension* of a field defined in special circumstances so as to get a field defined more generally. We remark first that the definition of ψ in a general frame will generally involve the form of the transformation to that frame from a special one[1]. Second, when ψ_0 is a tensor the extension is always possible and the result is unique; however, it may disagree with the result of using other and sometimes desirable rules for defining ψ. Third, if ψ_0 is not a tensor, there is no assurance that a solution to the problem exists or is unique, since there is ambiguity in specifying a corresponding law under $\mathfrak{G}$.

An important class F_0 is the class of frames at rest with respect to a given frame. The corresponding group $\mathfrak{G}_0$ consists in those transformations (143.11) for which $\boldsymbol{A}=\text{const}$, $\boldsymbol{b}=\text{const}$. Other classes F_0 of preferred frames which are of kinematical interest are defined in terms of a particular motion: by the principal directions of its stream lines, by its principal axes of stretching, etc. The motion itself may then be referred to a frame based upon these directions. Preferred frames of kinematical interest may also be suggested by the geometry of the underlying space. Of course, use of a given rule in different preferred frames will generally lead to different invariants.

The problem we have set is purely kinematical. Its usefulness will appear in Sect. 293, where we discuss constitutive equations for materials. These equations should be invariant with respect to motion of the observer. By what we have said above, such invariance will imply the use of a preferred frame, to be chosen on the basis of some intrinsic property of the motion or of the material.

148. Co-rotational frames. Following the motion of a single particle $\boldsymbol{Z}$, we may choose a moving frame with origin $\boldsymbol{z}'=0$ at the place $\boldsymbol{z}$ occupied by $\boldsymbol{Z}$ at time t, and such that both $\dot{\boldsymbol{z}}'=0$ and $\boldsymbol{w}'=0$ at $\boldsymbol{z}'=0$. Such a frame, whose origin is always at $\boldsymbol{Z}$ and whose angular velocity tensor is exactly the spin tensor $\boldsymbol{w}$ at $\boldsymbol{Z}$, may be called *co-rotational*[2]. It is only a rewording of a familiar proposition in rigid kinematics to assert that *any two co-rotational frames for the same particle are related by an orthogonal transformation which is constant in time*[3].

The foregoing statement empowers us to refer to a quantity as a "tensor in a co-rotational frame", without specifying which co-rotational frame we select, so long as we continue to fix attention on a single particle. Taking up the problem of extension set in Sect. 147, we let F_0 be the class of co-rotational frames, and we first consider quantities which transform as tensors under $\mathfrak{G}_0$.

Still more specifically, let ψ be a tensor under the entire group $\mathfrak{G}$:

$$\psi'_{k'\ldots m'} = A_{kk'}\ldots A_{mm'}\,\psi_{k\ldots m}. \tag{148.1}$$

Differentiating with respect to t with $\boldsymbol{Z}$ held constant yields

$$\dot{\psi}'_{k'\ldots m'} = A_{kk'}\ldots A_{mm'}\,\dot{\psi}_{k\ldots m} + \dot{A}_{kk'}\ldots A_{mm'}\,\psi_{k\ldots m} + \cdots + A_{kk'}\ldots \dot{A}_{mm'}\,\psi_{k\ldots m}. \tag{148.2}$$

[1] Thus our problem differs somewhat from the problem of extension discussed e.g. by VEBLEN [1927, *8*, Chap. VI, §§ 5, 13–17].

[2] These frames were introduced by ZAREMBA [1903, *20* and *21*, §§ 2–4] [1937, *12*, § 2] and were employed by HENCKY [1929, *3*, § 2] and FROMM [1933, *3*, §§ I, VAb]; in elaborating ZAREMBA's ideas, THOMAS [1955, *24*, § 2] [1955, *25*, § 3] used the term "kinematically preferred co-ordinate systems".

[3] Treatments in the present notation have been given by ZORAWSKI [1911, *13*, § 3], THOMAS [1955, *25*, § 3], MARGULIES [1956, *15*].

Thus Ψ is *not* a tensor under $\mathfrak{G}$. However, if we restrict attention to any subgroup whose members are suffering relative translation only, by (143.14) we have $\dot{A}=0$, and hence $\dot{\Psi}$ is a tensor under this subgroup. In particular, from the italicized statement above it follows that this holds for the subgroup $\mathfrak{G}_0$ of transformations among co-rotational frames. We now define the *co-rotational time flux* of Ψ, $d_r\Psi/dt$, as that tensor under $\mathfrak{G}$ which reduces to $\dot{\Psi}$ at the origin of a co-rotational frame. Since at the origin of a co-rotational frame we have $\partial\Psi/\partial t=\dot{\Psi}$ we might equally well have used $\partial\Psi/\partial t$ in setting up the definition.

To calculate the form of the operator d_r/dt in any frame, let primes refer to a co-rotational frame. Then (144.4) yields

$$\dot{A}_{kk'}=w_{km}A_{mk'}. \tag{148.3}$$

By substitution in (148.2) follows

$$\left.\begin{aligned}\dot{\Psi}'_{k'\ldots m'}=&A_{kk'}\ldots A_{mm'}\dot{\Psi}_{k\ldots m}+\\&+w_{kq}A_{qk'}\ldots A_{mm'}\Psi_{k\ldots m}+\cdots+A_{kk'}\ldots w_{mq}A_{qm'}\Psi_{k\ldots m}.\end{aligned}\right\} \tag{148.4}$$

Hence

$$\left.\begin{aligned}\frac{d_r\Psi_{k\ldots m}}{dt}&\equiv A_{kk'}\ldots A_{mm'}\dot{\Psi}'_{k'\ldots m'},\\&=\dot{\Psi}_{k\ldots m}-w_{kq}\Psi_{q\ldots m}-\cdots-w_{mq}\Psi_{k\ldots q}.\end{aligned}\right\} \tag{148.5}$$

This is the result proved by Zaremba for tensors of the second order. To verify that $d_r\Psi/dt$ is indeed a tensor, we need only observe that since $\boldsymbol{w}'=0$, on the left-hand side of (148.4) we may replace $\dot{\Psi}'_{k'\ldots m'}$ by $\dot{\Psi}'_{k'\ldots m'}-w'_{k'q'}\Psi'_{q'\ldots m'}-\cdots-w'_{m'q'}\Psi'_{k'\ldots q'}$.

It is easy to show that in a co-rotational frame we have

$$\frac{d_r^N\Psi'_{k'\ldots m'}}{dt^N}=\overset{(N)}{\Psi}{}'_{k'\ldots m'}. \tag{148.6}$$

That is, N successive applications of d_r/dt yields the same result as does extension of d^N/dt^N, or of $\partial^N/\partial t^N$.

If we introduce general curvilinear co-ordinates, so long as these be *stationary*, there is only one tensor which reduces to (148.5) in a rectangular Cartesian system. This tensor is given by

$$\frac{d_r\Psi^{k\ldots m}_{p\ldots q}}{dt}=\dot{\Psi}^{k\ldots m}_{p\ldots q}-w^k{}_r\Psi^{r\ldots m}_{p\ldots q}-\cdots-w^m{}_r\Psi^{k\ldots r}_{p\ldots q}-w_p{}^r\Psi^{k\ldots m}_{r\ldots q}-\cdots-w_q{}^r\Psi^{k\ldots m}_{p\ldots r}. \tag{148.7}$$

For the metric tensor g_{km}, by (72.10) this gives

$$\frac{d_r g_{km}}{dt}=-w_k{}^q g_{qm}-w_m{}^q g_{kq}=0. \tag{148.8}$$

Hence *raising and lowering of indices commutes with d_r/dt.*

Since $\dot{\boldsymbol{x}}'=0$ and $\boldsymbol{w}'=0$ at the origin of the co-rotational system, it is an immediate consequence of the definition of d_r/dt that

$$\frac{d_r^N\dot{x}^k}{dt^N}=0,\quad \frac{d_r^N w_{km}}{dt^N}=0. \tag{148.9}$$

This does not follow, however, from (148.7). In fact (148.7) was derived on *the assumption that Ψ itself is a tensor under $\mathfrak{G}$*, while neither $\dot{\boldsymbol{x}}$ nor $\boldsymbol{w}$ enjoys this property, being subject rather to the transformation laws (143.24) and $(144.2)_2$. However, as we have said, the process of extension may be applied to quantities which transform as tensors only under the subgroup $\mathfrak{G}_0$, and this is the case for $\dot{\boldsymbol{x}}$ and $\boldsymbol{w}$. If we carry out the process of extension for $\dot{\boldsymbol{x}}$, we find that we are led through just the same steps as were used to derive (143.24), except that $\dot{\boldsymbol{b}}=0$, $\boldsymbol{z}'=0$, and $\dot{\boldsymbol{z}}'=0$, and this yields $(148.9)_1$ for the case $N=1$. Similarly, it is possible to derive $(148.9)_2$ by differentiating $(144.2)_2$ and evaluating the result at the origin of a co-rotational system. From $(148.9)_2$ we get[1]

$$\left.\begin{aligned}\frac{d_r}{dt}\dot{x}_{k,m} &= \frac{d_r}{dt}(d_{km}+w_{km}) = \frac{d_r d_{km}}{dt},\\ &= \dot{d}_{km} + w_k{}^q d_{qm} + w_m{}^q d_{qk},\end{aligned}\right\} \tag{148.10}$$

where, since $(144.2)_1$ asserts that $\boldsymbol{d}$ is a tensor under $\mathfrak{G}$, the last step follows from (148.7). This formula illustrates the fact that for tensors under $\mathfrak{G}_0$ the operation d_r/dt need not commute with the covariant derivative. By $(148.9)_1$, $(d_r\dot{x}_k/dt)_{,m}=0$.

Further properties of the extensions of a tensor, defined as those tensors under $\mathfrak{G}$ which reduce to $\partial^N\psi/\partial z'_{k'}\,\partial z'_{m'}\dots\partial t\dots\partial t$ at the origin of a co-rotational system, have been noticed by THOMAS[2].

149. Irrotational frames.

149. Irrotational frames. We shall call a frame *irrotational of order* N_1+N_2 if at *two* times $t_\mathfrak{a}$, $\mathfrak{a}=1, 2$, its orientations with respect to a common frame are assigned and also for a particle at the origin we have

$$\boldsymbol{w}^{(r)\prime}=0, \quad r=1,2,\dots,N_\mathfrak{a}, \tag{149.1}$$

where $\boldsymbol{w}^{(r)\prime}$ is defined by $(103.4)_2$. For the case when $N_2=0$, these frames were introduced by RIVLIN and ERICKSEN[3], who selected the orientation at time t_1 as that of the fixed spatial frame, the orientation at time t_2 as that of the spatial axes of finite strain for the deformation from $\boldsymbol{x}(\boldsymbol{X},t_1)$ to $\boldsymbol{x}(\boldsymbol{X},t_2)$.

Using (145.8), we may prove by induction that to satisfy the requirement (149.1) we need only assign certain values to $\omega, \dot{\omega}, \dots, \overset{(N_\mathfrak{a}-1)}{\omega}$ at the times $t_\mathfrak{a}$. The existence of irrotational frames is then a consequence of the following theorem of NOLL[4]:

There are infinitely many transformations (143.11) *such that*

1. $\boldsymbol{A}(t_\mathfrak{a})=\boldsymbol{A}_\mathfrak{a}$, *where* $\boldsymbol{A}_1$ *and* $\boldsymbol{A}_2$ *are arbitrarily assigned rotation matrices, both proper or both improper;*

2. $\boldsymbol{b}(t_\mathfrak{a})=\boldsymbol{b}_\mathfrak{a}$, *where* $\boldsymbol{b}_1$ *and* $\boldsymbol{b}_2$ *are arbitrary; and*

3. $\omega(t_\mathfrak{a}), \dot{\omega}(t_\mathfrak{a}), \dots, \overset{(N_\mathfrak{a})}{\omega}(t_\mathfrak{a})$ *are assigned arbitrarily.*

Passing to the proof, we note that to satisfy Condition 2 is trivial, and that without loss of generality we may assume the $\boldsymbol{A}_\mathfrak{a}$ both proper. Further, it is possible to express the general proper rotation $\boldsymbol{A}$ as an analytic function $\boldsymbol{A}(\lambda,\mu,\nu)$ of three parameters λ,μ,ν in such a way that $\boldsymbol{A}(0,0,0)=\boldsymbol{1}$; e.g., we may employ EULER's angles. Specifying $\boldsymbol{A}$ is thus equivalent to specifying three functions $\lambda(t)$, $\mu(t)$, $\nu(t)$. From $(143.14)_1$ we have, writing $\boldsymbol{A}'$ for the transpose of $\boldsymbol{A}$,

$$\omega = \dot{\boldsymbol{A}}\cdot\boldsymbol{A}' = \dot{\lambda}\boldsymbol{A}_{,\lambda}\cdot\boldsymbol{A}' + \dot{\mu}\boldsymbol{A}_{,\mu}\cdot\boldsymbol{A}' + \dot{\nu}\boldsymbol{A}_{,\nu}\cdot\boldsymbol{A}', \tag{149.2}$$

so that ω and $\dot{\lambda},\dot{\mu},\dot{\nu}$ determine each other uniquely, as is geometrically evident. Moreover, ω vanishes if and only if $\dot{\lambda},\dot{\mu}$, and $\dot{\nu}$ vanish. By induction it may be

[1] THOMAS [1955, *25*, § 6].
[2] [1955, *25*, § 7].
[3] [1955, *21*, §§ 5, 15].
[4] Not previously published.

shown that the set of quantities $\boldsymbol{A}, \boldsymbol{\omega}, \dot{\boldsymbol{\omega}}, \ldots, \overset{(N)}{\boldsymbol{\omega}}$ is uniquely determined by the set of quantities $\lambda, \mu, \nu, \dot{\lambda}, \dot{\mu}, \dot{\nu}, \ldots, \overset{(N+1)}{\lambda}, \overset{(N+1)}{\mu}, \overset{(N+1)}{\nu}$, and conversely; also $\overset{(k)}{\boldsymbol{\omega}} = 0$ for $k=0, 1, \ldots, N$ if and only if $\overset{(k+1)}{\lambda} = \overset{(k+1)}{\mu} = \overset{(k+1)}{\nu} = 0$ for $k=0, 1, \ldots, N$. The problem is thus reduced to finding three functions $\lambda(t), \mu(t), \nu(t)$ such that $\overset{(k)}{\lambda}(t_\mathfrak{a}), \overset{(k)}{\mu}(t_\mathfrak{a}), \overset{(k)}{\nu}(t_\mathfrak{a})$, $k=0, 1, \ldots, N_\mathfrak{a}+1$, are prescribed. There are infinitely many such functions; e.g., it is possible to determine them as polynomials of degree N_1+N_2+1. Q.E.D.

We may determine the class of *all* frames satisfying Conditions 1, 2, 3 as follows. Let $\boldsymbol{z}' = \boldsymbol{A}(\lambda, \mu, \nu) \cdot (\boldsymbol{z} - \boldsymbol{b})$ be a transformation to any one frame, determined as above. Let $\boldsymbol{b}^*(t)$ be any vector such that $\boldsymbol{b}^*(t_\mathfrak{a}) = 0$, and let λ^*, μ^*, ν^* be any three functions such that $\overset{(k)}{\lambda}(t_\mathfrak{a}) = \overset{(k)}{\mu}(t_\mathfrak{a}) = \overset{(k)}{\nu}(t_\mathfrak{a}) = 0$ for $k=0, 1, \ldots, N_\mathfrak{a}+1$. Then any frame of the type sought may be obtained by a transformation of the form

$$\boldsymbol{z}' = \boldsymbol{A}(\lambda+\lambda^*, \mu+\mu^*, \nu+\nu^*) \cdot [\boldsymbol{z} - (\boldsymbol{b}+\boldsymbol{b}^*)]. \tag{149.3}$$

Since $\overset{(k)}{\boldsymbol{A}} = \overset{(k)}{\boldsymbol{1}}$ for $k=0, 1, \ldots, N_\mathfrak{a}+1$, if and only if $\overset{(k)}{\lambda} = \overset{(k)}{\mu} = \overset{(k)}{\nu} = 0$ for $k=0, 1, \ldots, N_\mathfrak{a}+1$, transformations carrying one of the frames sought into another are of the form

$$\boldsymbol{z}' = \boldsymbol{A}(\lambda^*, \mu^*, \nu^*) \cdot (\boldsymbol{z} - \boldsymbol{b}^*), \tag{149.4}$$

and these form a group.

When $N_2 = 0$, it is possible to indicate a simple solution. Writing $N \equiv N_1$, from (143.14) we see that we are to find a solution of

$$\overline{\overset{(r)}{\dot{A}_{kk'} A_{mk'}}}\Big|_{t=t_1} = \Omega^{(r)}_{km}, \qquad r = 0, 1, \ldots, N, \tag{149.5}$$

where $\boldsymbol{\Omega}^{(r)} \equiv \overset{(r)}{\boldsymbol{\omega}}(t_1)$; moreover, the solution is to satisfy (143.12). There is a polynomial of degree N satisfying (149.5), namely

$$\dot{A}_{kk'} A_{mk'} = \sum_{r=0}^{N} \Omega^{(r)}_{km} \frac{(t-t_1)^r}{r!}. \tag{149.6}$$

Since $\Omega^{(r)}_{km} = -\Omega^{(r)}_{mk}$, any solution $\boldsymbol{A}$ of (149.6) satisfies $\dot{\overline{A_{kk'} A_{mk'}}} = 0$; hence a solution satisfying (143.12) at one time satisfies it at all times. For such a solution, (149.6) is equivalent to

$$\dot{A}_{kk'} = \sum_{r=0}^{N} \Omega^{(r)}_{km} A_{mk'} \frac{(t-t_1)^r}{r!}. \tag{149.7}$$

This is a linear differential system for $\boldsymbol{A}$; it has polynomial coefficients and therefore a unique solution $\boldsymbol{A}^*(t)$ such that $\boldsymbol{A}^*(t_1) = \boldsymbol{A}_1$. The matrix $\boldsymbol{A}(t)$ given by

$$\boldsymbol{A}(t) = \boldsymbol{A}^* \cdot \exp\left\{\left(\frac{t-t_1}{t_2-t_1}\right)^{N+1} \log[(\boldsymbol{A}^*)^{-1} \cdot \boldsymbol{A}_2]\right\} \tag{149.8}$$

then satisfies Condition 1 as well as Condition 3 with $N_2 = 0$.

An invariant time flux based upon time differentiation in an irrotational frame has been constructed by COTTER and RIVLIN[1]. As might be expected from the difficulties mentioned above, they do not attempt the extension process directly. Rather, they appear to lay down the following requirements for the M-th time flux $d_\mathrm{i}^M \Psi / dt^M$ of an absolute tensor Ψ:

1. $d_\mathrm{i}^M \Psi / dt^M$ shall be a tensor under $\mathfrak{G}$.

2. In any irrotational frame with origin at the point in question and with $N \geqq M$, $d_\mathrm{i}^M \Psi / dt^M$ shall reduce to a linear combination of the material derivatives $\overset{(p)}{\Psi}$ with $p = 1, 2, \ldots, M$.

---

[1] [1955, 4].

These requirements do not determine $d_i^M \Psi/dt^M$ uniquely. The particular choice made by COTTER and RIVLIN, in the special case they consider, coincides with the result to be obtained by a different method in the section following.

150. Convected time flux. If we select three independent families of material surfaces, whose equations may be taken as $X^\alpha = \text{const}$, their configurations at time t define a *convected co-ordinate system* for the given motion[1]. Setting $x^k = \delta^k_\alpha X^\alpha$, we thus describe the motion in terms of a specially selected moving co-ordinate system. It is usual to define *body tensor fields* as tensors under transformations of the X^α and *spatial tensor fields* as tensors under transformations of the x^k. It is possible, in any given case, to associate a pair of such fields in such a way that their components are respectively equal. In such a case, the motion of the co-ordinate system x^k will generally prevent the time derivatives of the two fields from being equal.

In the convected co-ordinate system, lengths are determined from a varying squared element of arc:

$$d s^2 = G_{\alpha\beta}(\boldsymbol{X}, t)\, dX^\alpha dX^\beta. \tag{150.1}$$

$\boldsymbol{G}$ is related to the tensors denoted by $\boldsymbol{g}$ and $\boldsymbol{C}$ in Sects. 14 and 26: At time t_0 the components of $\boldsymbol{G}$ coincide with the components of $\boldsymbol{g}$, while at time t the components of $\boldsymbol{G}$ coincide with the components of $\boldsymbol{C}$. These remarks suggest how the formalism of convected co-ordinates may be used to treat the subjects developed in the foregoing sections by other and in our opinion preferable methods. Our purpose in introducing convected co-ordinates here is for the following special application.

Let Ψ be a tensor field of weight W under the group $\mathfrak{G}$ of *all* co-ordinate transformations, including transformations to moving co-ordinates. Then the material derivative, $\dot{\Psi}$, is not generally a tensor under $\mathfrak{G}$. However, since the convected co-ordinates X^α for a particle $\boldsymbol{X}$ do not change in time, $\dot{\Psi}$ is a tensor under the subgroup $\mathfrak{G}_0$ of transformations of convected co-ordinates. (The formal proofs of these statements are analogous to those given at the beginning of Sect. 148.) The *convected time flux* $d_c\Psi/dt$ of Ψ is defined as that tensor under $\mathfrak{G}$ which reduces to $\dot{\Psi}$ in a convected co-ordinate system. To calculate $d_c\Psi/dt$, we follow the method of OLDROYD[2].

---

[1] General references for such co-ordinates were given in Sect. 66B. Cf. also the remark of ZAREMBA [1903, *21*, pp. 619—620].

[2] [1950, *22*, § 3a]. COTTER and RIVLIN [1955, *4*] approach the convected time flux in a special case by a method analogous to that we have used to calculate the tensors $\boldsymbol{k}^{(N)}$ in Sect. 104. It is easy to generalize their method so as to yield (150.6).

The idea of convected time flux may be traced back to a very special case given by CAUCHY [1829, *4*]; in effect, he defined $\tilde{t}_{km}$ by the requirement that $\overline{t_{km}\, da^k dx^m}^{\,\cdot} = \tilde{t}_{km}\, da^k dx^m$, where $\boldsymbol{t}$ is an absolute tensor and $d\boldsymbol{a}$ is an element of area; the result is the same as that of taking $\boldsymbol{t}$ to be a tensor of weight 1 in the first place. Cf. TRUESDELL [1953, *32*, § 55 bis].

HELMHOLTZ [1892, *5*, § 5] in formulating a type of variation suitable for Hamiltonian principles in spatial co-ordinates was led by some rather mysterious steps to put

$$\delta\dot{\boldsymbol{p}} = \frac{\partial\,\delta\boldsymbol{x}}{\partial t} + \operatorname{curl}(\dot{\boldsymbol{p}}\times\delta\boldsymbol{x}) + \dot{\boldsymbol{p}}\operatorname{div}\delta\boldsymbol{x} - \delta\boldsymbol{x}\operatorname{div}\dot{\boldsymbol{p}}.$$

By the same vectorial identities that lead from $(98.1)_2$ to (99.14), this formula is equivalent to

$$\begin{aligned}\delta\dot{\boldsymbol{p}} &= \frac{\partial\,\delta\boldsymbol{x}}{\partial t} + \dot{\boldsymbol{p}}\cdot\operatorname{grad}\delta\boldsymbol{x} - \delta\boldsymbol{x}\cdot\operatorname{grad}\dot{\boldsymbol{p}},\\ &= \dot{\overline{\delta\boldsymbol{x}}} - \delta\boldsymbol{x}\cdot\operatorname{grad}\dot{\boldsymbol{p}},\end{aligned}$$

as was remarked by C. NEUMANN [1897, *7*], whose derivation is equally mysterious. By (150.6), however, this last result is a statement that $\delta\dot{\boldsymbol{p}}$ is to be calculated in a convected co-ordinate system. Cf. the next footnote.

Since ψ is a tensor under $\mathfrak{G}$, we may calculate its components $\psi^{k\ldots m}_{p\ldots q}$ in any fixed system x^k in terms of its components $\psi^{\alpha\ldots\beta}_{\gamma\ldots\delta}$ in any convected system by the usual law of transformation:

$$\psi^{\alpha\ldots\beta}_{\gamma\ldots\delta}\frac{\partial x^k}{\partial X^\alpha}\cdots\frac{\partial x^m}{\partial X^\beta}=|\boldsymbol{x}/\boldsymbol{X}|^W\psi^{k\ldots m}_{p\ldots q}\frac{\partial x^p}{\partial X^\gamma}\cdots\frac{\partial x^q}{\partial X^\delta}. \tag{150.2}$$

For a given particle $\boldsymbol{X}$, the co-ordinates X^α in any one convected co-ordinate system do not change. Letting D/Dt denote the partial derivative with $\boldsymbol{X}$ held constant, as in Sect. 72, by operating with D/Dt on (150.2) we get

$$\left.\begin{aligned}\frac{D\psi^{\alpha\ldots\beta}_{\gamma\ldots\delta}}{Dt}\frac{\partial x^k}{\partial X^\alpha}\cdots\frac{\partial x^m}{\partial X^\beta}+\psi^{\alpha\ldots\beta}_{\gamma\ldots\delta}\left[\frac{\partial\dot{x}^k}{\partial X^\alpha}\cdots\frac{\partial x^m}{\partial X^\beta}+\cdots+\frac{\partial x^k}{\partial X^\alpha}\cdots\frac{\partial\dot{x}^m}{\partial X^\beta}\right]\\ =|\boldsymbol{x}/\boldsymbol{X}|^W\left\{\left[\frac{\partial\psi^{k\ldots m}_{p\ldots q}}{\partial t}+\frac{\partial\psi^{k\ldots m}_{p\ldots q}}{\partial x^s}\dot{x}^s\right]\frac{\partial x^p}{\partial X^\gamma}\cdots\frac{\partial x^q}{\partial X^\delta}+\right.\\ +\psi^{k\ldots m}_{p\ldots q}\left[\frac{\partial\dot{x}^p}{\partial X^\gamma}\cdots\frac{\partial x^q}{\partial X^\delta}+\cdots+\frac{\partial x^p}{\partial X^\gamma}\cdots\frac{\partial\dot{x}^q}{\partial X^\delta}\right]+\\ \left.+W\frac{\partial\dot{x}^s}{\partial x^s}\psi^{k\ldots m}_{p\ldots q}\frac{\partial x^p}{\partial X^\gamma}\cdots\frac{\partial x^q}{\partial X^\delta}\right\},\end{aligned}\right\} \tag{150.3}$$

where we have used $\dot{x}^k=Dx^k/Dt$ [this being (67.1) expressed in the present notation] and

$$\frac{D}{Dt}\frac{\partial x^k}{\partial X^\alpha}=\frac{\partial}{\partial X^\alpha}\frac{Dx^k}{Dt}=\frac{\partial\dot{x}^k}{\partial X^\alpha}=\frac{\partial\dot{x}^k}{\partial x^m}\frac{\partial x^m}{\partial X^\alpha}. \tag{150.4}$$

Putting this result into (150.3) shows that if we set

$$\left.\begin{aligned}\frac{d_c\psi^{k\ldots m}_{p\ldots q}}{dt}\equiv\frac{\partial\psi^{k\ldots m}_{p\ldots q}}{\partial t}+\frac{\partial\psi^{k\ldots m}_{p\ldots q}}{\partial x^s}\dot{x}^s+\psi^{k\ldots m}_{s\ldots q}\frac{\partial\dot{x}^s}{\partial x^p}+\cdots+\psi^{k\ldots m}_{p\ldots s}\frac{\partial\dot{x}^s}{\partial x^q}-\\ -\psi^{s\ldots m}_{p\ldots q}\frac{\partial\dot{x}^k}{\partial x^s}-\cdots-\psi^{k\ldots s}_{p\ldots q}\frac{\partial\dot{x}^m}{\partial x^s}+W\psi^{k\ldots m}_{p\ldots q}\frac{\partial\dot{x}^s}{\partial x^s},\end{aligned}\right\} \tag{150.5}$$

then these quantities furnish the components in the x^k system of the tensor whose components in the X^α system are $D\psi^{\alpha\ldots\beta}_{\gamma\ldots\delta}/Dt$. By (72.3), in the previous sentence we may replace $D\psi/Dt$ by $\dot{\psi}$, since $\psi^{\alpha\ldots\beta}_{\gamma\ldots\delta}$ is a body tensor field. Thus the right-hand side of (150.5) gives the desired convected time flux.

The so-called *Lie derivative*[1] of ψ with respect to $\boldsymbol{v}$ is given by

$$\left.\begin{aligned}\underset{\boldsymbol{v}}{\pounds}\psi^{k\ldots m}_{p\ldots q}\equiv\partial_s\psi^{k\ldots m}_{p\ldots q}v^s+\psi^{k\ldots m}_{s\ldots q}\partial_p v^s+\cdots+\psi^{k\ldots m}_{p\ldots s}\partial_q v^s-\\ -\psi^{s\ldots m}_{p\ldots q}\partial_s v^k-\cdots-\psi^{k\ldots s}_{p\ldots q}\partial_s v^m+W\psi^{k\ldots m}_{p\ldots q}\partial_s v^s.\end{aligned}\right\} \tag{150.6}$$

Thus (150.5) may be written in the form

$$\frac{d_c\psi}{dt}=\frac{\partial\psi}{\partial t}+\underset{\dot{\boldsymbol{x}}}{\pounds}\psi, \tag{150.7}$$

showing that the Lie derivative is the special case of the convected time flux that results when ψ is steady. In Sect. 153 we shall see that a four-dimensional

[1] SLEBODZINSKI [1931, *10*], VAN DANTZIG [1932, *13*, § 3], SCHOUTEN and VAN KAMPEN [1934, *7*, § 1], SCHOUTEN and STRUIK [1935, *4*, § 12]. A very general exposition of the nature and properties of this derivative is given by SCHOUTEN [1954, *21*, Chap. III, § 10]. The German verb associated with a convected co-ordinate system is "mitschleppen"; a virtual transliteration occurs in some works ostensibly written in English. WUNDHEILER [1932, *14*, § 2] introduced a differential of the type

$$\delta v^k=dv^k+\Gamma^k_{mp}v^m dx^p+\Lambda^k_m v^m dt,$$

which transforms as a contravariant vector under time-dependent transformations of co-ordinates.

treatment enabling identification of convected time flux with a more general Lie derivative is possible also for unsteady motions.

In works on geometry the invariant character of $\underset{v}{\pounds}\Psi$ is established in considerable generality. Our treatment above shows that the right-hand side of (150.5) is a tensor under transformations from a convected to a fixed co-ordinate system, but it is not obviously a tensor under transformations of fixed systems. However, the fixed system has not been restricted in any way, and we may choose it as rectangular Cartesian. In this case (150.5) may be written

$$\left.\begin{aligned}\frac{\partial_c \Psi^{k\ldots m}_{p\ldots q}}{\partial t} = \dot{\Psi}^{k\ldots m}_{p\ldots q} + \Psi^{k\ldots m}_{s\ldots q}\,\dot{x}^s_{,p} + \cdots + \Psi^{k\ldots m}_{p\ldots s}\,\dot{x}^s_{,q} - \Psi^{s\ldots m}_{p\ldots q}\,\dot{x}^k_{,s} - \cdots \\ - \Psi^{k\ldots s}_{p\ldots q}\,\dot{x}^m_{,s} + W\,\Psi^{k\ldots m}_{p\ldots q}\,\dot{x}^s_{,s}.\end{aligned}\right\} \tag{150.8}$$

The right-hand side of (150.8) is plainly a tensor under all co-ordinate transformations. Being established for a single choice of a fixed co-ordinate system, (150.8) is thus valid in all fixed systems. Alternatively, it is possible to transform (150.8) into (150.5) by writing out the explicit forms of the covariant derivatives. For actual calculation, (150.5) is preferable.

If we apply (150.8) to the metric tensor g_{km} itself, we get[1]

$$\frac{1}{2}\,\frac{d_c g_{km}}{dt} = \frac{1}{2}\,(g_{pm}\,\dot{x}^p_{,k} + g_{kp}\,\dot{x}^p_{,m}) = \dot{x}_{(k,m)} = d_{km}: \tag{150.9}$$

The convected time flux of the metric tensor is twice the stretching tensor, and only in a rigid motion do we have $d_c\boldsymbol{g}/dt = 0$. This is really obvious from the definition of the convected time flux. However, it has the important consequence that *at a point where Ψ does not vanish, d_c/dt commutes with raising and lowering of indices of Ψ if and only if the motion is rigid.* In fact, since $a_k = g_{km}a^m$, we have identically $d_c(a_k - g_{km}a^m)/dt = 0$, and hence

$$\left.\begin{aligned}\frac{d_c a_k}{dt} - g_{km}\frac{d_c a^m}{dt} &= \frac{d_c g_{km}}{dt}\,a^m = 2d_{km}\,a^m,\\ \frac{d_c a^k}{dt} - g^{km}\frac{d_c a_m}{dt} &= -\,2d^{km}\,a_m,\end{aligned}\right\} \tag{150.10}$$

specimens of the rules which hold for general Ψ.

In the derivation of (150.5) and (150.8) we have used the tensor law of transformation (150.2) from a convected to a fixed co-ordinate system. This might seem to imply that our results are applicable only to quantities Ψ which transform as tensors under *the motion itself*. This is not so. We have employed (150.2) only at a single instant, and the formulae used in passing from (150.3) to (150.5) need hold only at that instant. At this one instant, we may choose any convected and fixed co-ordinates we please; for example, we may choose the two sets of co-ordinate surfaces as instantaneously identical, as is done by many authors. We thus obtain (150.5) without the necessity of assuming that Ψ enjoys any particular transformation property under the motion itself.

It is interesting, however, to investigate properties under the entire motion. If $d_c a^k/dt = 0$ and $d_c b_k/dt = 0$ over an interval of time, we have

$$\dot{a}^k - a^m\,\dot{x}^k_{,m} = 0, \qquad \dot{b}_k + b_m\,\dot{x}^m_{,k} = 0. \tag{150.11}$$

[1] HENCKY [1925, 7]. ZELMANOV [1948, *39*, Eq. (10)] has calculated the flux of the Riemann tensor; the terms involving the rotation tensor in his result seem to us to cancel one another.

The general solutions of these differential equations are

$$a^k = x^k_{,\alpha} A^\alpha, \qquad b_k = X^\alpha_{,k} B_\alpha, \tag{150.12}$$

where $\boldsymbol{A}$ and $\boldsymbol{B}$ are arbitrary. Thus $d_c a^k/dt = 0$ is equivalent to the assertion that a^k is carried by the motion as a contravariant vector, or material line segment; $d_c b_k/dt = 0$, that b_k is carried as a covariant vector, or plane area.

Since, as just established, we have $d_c(dx^k)/dt = 0$, it follows that

$$\overset{(N)}{ds^2} = \frac{d_c^N (ds^2)}{dt^N} = \frac{d_c^N g_{km}}{dt^N} dx^k dx^m, \tag{150.13}$$

where the first step is a consequence of the fact that d_c/dt reduces to the material derivative when applied to a scalar. Comparison with (104.2) yields[1]

$$A^{(N)}_{pq} = \frac{d_c^N g_{pq}}{dt^N}. \tag{150.14}$$

Generalizing (150.9), this formula offers a new interpretation for the Rivlin-Ericksen tensors $\boldsymbol{A}^{(N)}$.

151. Contrast of the various time fluxes. The three time fluxes, d_r/dt, d_i/dt, and d_c/dt, are in general different from one another. Since an explicit form for d_i/dt has not been calculated, we omit it from the discussion. Comparison of (148.7) and (150.8) for an absolute tensor $\Psi^{\cdots}_{\cdots}$ yields

$$\left(\frac{d_c}{dt} - \frac{d_r}{dt}\right) \Psi^{k\ldots m}_{p\ldots q} = d^s_p \Psi^{k\ldots m}_{s\ldots q} + \cdots + d^s_q \Psi^{k\ldots m}_{p\ldots s} - d^k_s \Psi^{s\ldots m}_{p\ldots q} - \cdots - d^m_s \Psi^{k\ldots s}_{p\ldots q}. \tag{151.1}$$

Consequently, *the co-rotational and convected time fluxes equal one another if:*

1. *Ψ is a scalar; or*
2. *$\Psi = 0$ at the place and time in question; or*
3. *The motion is locally and instantaneously rigid.*

In cases 1 and 2 we have

$$\dot{\Psi} = \frac{d_c \Psi}{dt} = \frac{d_r \Psi}{dt}. \tag{151.2}$$

Returning to Sect. 147, we see that a flux satisfying Requirement 2 is indeterminate to within an arbitrary tensor which vanishes when $\boldsymbol{d} = 0$. The problem of constructing the most general invariant time flux is then equivalent to finding the most general tensor under $\mathfrak{G}$ which vanishes in a rigid motion. We do not know of any solution to this problem.

Invariant time fluxes are required for the formation of correct constitutive equations for materials (Chap. G). Special circumstances may make one or another flux appear preferable. From a purely kinematical standpoint, however, the most that can be done is to make the meaning of each flux clear. $d_c\Psi/dt$ is a measure of the instantaneous departure of the rate of change of Ψ from the value it would assume if Ψ were carried as a tensor under the motion itself. $d_r\Psi/dt$ and $d_i\Psi/dt$ are, in different senses, the rates of change of Ψ as apparent to an observer whose frame of reference is carried by the medium and turns with it.

It is clear that any fluxes defined in terms of the motion of the material will depend upon $\boldsymbol{w}$ when referred to a common frame. The co-rotational flux $d_r\Psi/dt$ is thus the simplest possible, in that it depends on $\boldsymbol{w}$ only, not on $\boldsymbol{d}$ or on higher

[1] Oldroyd [1950, *22*, § 3a].

derivatives of $\dot{\boldsymbol{x}}$. The most striking advantage of $d_r\Psi/dt$ is expressed by (148.8) and the statement following it. Since in particular problems in Euclidean space there is usually no circumstance indicating contravariant or covariant character, rather than merely tensorial character, for the quantities occurring, a flux operator that commutes with the raising and lowering of indices seems most natural. Finally, the simple way in which extension of non-tensorial quantities from their values in a co-rotational frame may be achieved is a further recommendation for $d_r\Psi/dt$.

152. World invariant kinematics. In the preceding sections of this subchapter we have studied various properties of a motion within the mathematical framework of 3-dimensional tensor analysis. We now present essentially the same considerations within a unified formalism of 4-dimensional tensors[1]. As preparations we introduce the following notions from geometry.

α) Some geometrical preliminaries. An n-dimensional space S is a set of points $\boldsymbol{p}$ such that to every point there corresponds a subset $\eta(\boldsymbol{p})$ containing $\boldsymbol{p}$ which can be placed in one-to-one correspondence with the ordered sets of n real numbers $x^\Gamma=(x^1, x^2, \ldots, x^n)$ lying in some interval $x_0^\Gamma - h^\Gamma < x^\Gamma < x_0^\Gamma + h^\Gamma$, $h^\Gamma > 0$ and such that $\boldsymbol{p}$ corresponds to x_0^Γ. A co-ordinate transformation $x^{\Gamma^*} = x^{\Gamma^*}(\boldsymbol{x})$, $x^\Gamma = x^\Gamma(\boldsymbol{x}^*)$ is regarded as replacing the one-to-one correspondence $\boldsymbol{p} \leftrightarrow (x^\Gamma)$ by another one, $\boldsymbol{p} \leftrightarrow (x^{\Gamma^*})$. The *geometry of the space* S may sometimes be determined by a group $\mathfrak{G}$ of *allowable co-ordinate transformations*. Elements of the group $\mathfrak{G}$ relate the *preferred co-ordinate systems* of the space S. A *field (figure)* in the space S is represented by an ordered set of functions $\boldsymbol{\Phi} = (\Phi_1(\boldsymbol{x}), \Phi_2(\boldsymbol{x}), \ldots, \Phi_{\mathfrak{N}}(\boldsymbol{x}))$ having an assigned law of transformation under $\mathfrak{G}$. The representation $\Phi_m(\boldsymbol{x})$ in any one preferred co-ordinate system x^Γ and the co-ordinate transformation $x^{\Gamma^*} = x^{\Gamma^*}(\boldsymbol{x})$ uniquely determine the representation $\Phi_{m^*}(\boldsymbol{x}^*)$ of the field $\boldsymbol{\Phi}$ in every other preferred co-ordinate system x^{Γ^*}. One also considers spaces S characterized by a group of *allowable* co-ordinate transformations $\mathfrak{G}$ *and* one or more fields $\boldsymbol{\Phi}$. By choosing different groups $\mathfrak{G}$ and different assigned fields $\boldsymbol{\Phi}$, we obtain various familiar examples of geometric spaces. Thus ordinary Euclidean metric space corresponds to letting $\mathfrak{G}$ be the group of 3-dimensional orthogonal transformations $\mathfrak{G}_{\mathfrak{O}}$, where with this choice for $\mathfrak{G}$ the set of fields $\boldsymbol{\Phi}$ is empty. However, we can also characterize ordinary 3-dimensional Euclidean metric space by the group $\mathfrak{G}_{\mathfrak{A}}$ of unrestricted analytic transformations provided we let $\boldsymbol{\Phi}$ be an analytic Euclidean metric tensor field $g_{km}(\boldsymbol{x})$. The two spaces so defined are regarded as equivalent.

Curved Riemannian spaces, affinely connected spaces, conformal spaces, etc., correspond to various other choices of the pair of objects $(\mathfrak{G}, \boldsymbol{\Phi})$. The foregoing example of Euclidean metric space shows that different choices of $(\mathfrak{G}, \boldsymbol{\Phi})$ may define spaces regarded as equivalent.

An *invariant* of a field $\boldsymbol{\Psi}$ in a space S with group $\mathfrak{G}$ is a property possessed in common by each of its representations $\Psi_m(\boldsymbol{x})$. Joint invariants of a set of fields $\boldsymbol{\Psi}_1, \boldsymbol{\Psi}_2, \ldots, \boldsymbol{\Psi}_{\mathfrak{N}}$ are invariant properties of the fields held singly and jointly. The *geometry of a field* $\boldsymbol{\Psi}$ in a space S with assigned fields $\boldsymbol{\Phi}$ is the theory of the joint invariants of $\boldsymbol{\Psi}$ and $\boldsymbol{\Phi}$.

---

[1] CARTAN [1923, *1*, §§ 7, 15—17] initiated a 4-dimensional treatment of classical kinematics which has been extended and developed by DEFRISE [1953, *8*]. The same class of problems has been approached by McVITTIE [1949, *18*], via a detour through the kinematics of special relativity theory. DE DONDER [1931, *4*, § 3] and KILCHEVSKI [1938, *5*, § 7] have constructed a 4-dimensionally invariant formalism for finite strain. Here we follow TOUPIN [1958, *9*].

β) Euclidean and Galilean space-time. Consider now the 4-dimensional spaces $S_{\mathfrak{E}}$ and $S_{\mathfrak{G}}$ defined by the following groups of preferred co-ordinate transformations[1]:

I. Euclidean space-time $S_{\mathfrak{E}}$ and the group $\mathfrak{G}_{\mathfrak{E}}$ of Euclidean transformations:

$$\left.\begin{aligned} z^{k'} &= A^{k'}_{m}(z^4)\, z^m + b^{k'}(z^4)\,, \\ z^{4'} &= z^4 + \text{const} \end{aligned}\right\} \tag{152.1}$$

where $A^{k'}_{m}$ and $d^{k'}$ are analytic functions of z^4 and $A^{k'}_{m}$ is an orthogonal matrix.

II. Galilean space-time $S_{\mathfrak{E}}$ and the group $\mathfrak{G}_{\mathfrak{G}}$ of Galilean transformations:

$$\left.\begin{aligned} z^{k'} &= A^{k'}_{m}\, z^m + u^{k'} z^4 + \text{const}\,, \\ z^{4'} &= z^4 + \text{const}\,, \end{aligned}\right\} \tag{152.2}$$

where $A^{k'}_{m}$ is a constant orthogonal matrix and the $u^{k'}$ are constants.

A point $\boldsymbol{z}$ in either of these 4-dimensional spaces is called an *event.*

Let a *motion of a material medium in Euclidean or Galilean space-time* be defined by a set of three scalar functions $Z^K(\boldsymbol{z})$ such that the matrix $\overset{-1}{C}{}^{KL}(\boldsymbol{z})$, where

$$\overset{-1}{C}{}^{KL} \equiv \delta^{ij}\, \partial_i Z^K\, \partial_j Z^L\,, \quad \partial_i Z^K \equiv \frac{\partial Z^K}{\partial z^i}\,, \tag{152.3}$$

is positive definite, this definition being consistent with $(29.1)_2$.

The geometry of a motion in $S_{\mathfrak{E}}$ or $S_{\mathfrak{G}}$ will be called *Euclidean kinematics* and *Galilean kinematics*, respectively. Since $\mathfrak{G}_{\mathfrak{G}}$ is a subgroup of $\mathfrak{G}_{\mathfrak{E}}$ and a motion in either space-time is defined in the same way, it is clear that any Euclidean invariant of a motion is also a Galilean invariant of a motion but that the converse is not necessarily true.

γ) General co-ordinates in Euclidean and Galilean space-time. A primary objective in this 4-dimensional treatment of classical kinematics is the systematic study of the invariants of a motion under the Euclidean and Galilean groups expressed in terms of curvilinear and possibly deforming co-ordinates. Such co-ordinate systems are related to the z^Γ in (152.1) and (152.2) by transformations of the general form

$$\left.\begin{aligned} x^i &= x^i(z^i, z^4)\,, \\ x^4 &= z^4 + \text{const}\,, \end{aligned}\right\} \tag{152.4}$$

where the $x^i(\boldsymbol{z})$ are unrestricted analytic functions of the four co-ordinates $\boldsymbol{z}$. The set of all transformations having the form (152.4) constitutes a group which is a subgroup of the group $\mathfrak{G}_{\mathfrak{A}}$ of *unrestricted analytic transformations on all four co-ordinates of events.* General co-ordinates in $S_{\mathfrak{E}}$ or $S_{\mathfrak{G}}$ will be denoted by x^Γ, and a typical element of $\mathfrak{G}_{\mathfrak{A}}$ is written in the form

$$x^{\Gamma'} = x^{\Gamma'}(\boldsymbol{x})\,, \qquad x^\Gamma = x^\Gamma(\boldsymbol{x}')\,. \tag{152.5}$$

We shall develop here a formalism for Euclidean and Galilean kinematics which is invariant under $\mathfrak{G}_{\mathfrak{A}}$. Since the transformations (152.4) relating general curvilinear and deforming co-ordinates form a subgroup of the more general transformations $\mathfrak{G}_{\mathfrak{A}}$, the classical problem of introducing curvilinear and deforming

[1] For the remainder of this subchapter, Latin indices will range over the three values 1, 2, and 3. Greek indices will range over the four values 1, 2, 3, and 4.

co-ordinates will then be solved. The concepts needed to construct such a formalism for kinematics are embodied in ***Klein's principle***[1]:

If in any space with group $\mathfrak{G}_1$ the subgroup $\mathfrak{G}_2$ is introduced, consisting in all transformations which leave a figure (field) Φ_1 invariant, then the geometry of a figure Φ_2 with respect to $\mathfrak{G}_2$ is identical with the geometry of the set of figures (Φ_1, Φ_2) with respect to $\mathfrak{G}_1$.

Let us illustrate the application of KLEIN's principle we intend to make by the following familiar example. Suppose we have given a tensor field $f^{k\cdots}_{m\cdots}$ in ordinary 3-dimensional Euclidean metric space where $\mathfrak{G}_2$ is the orthogonal group, i.e., $f^{k\cdots}_{m\cdots}$ is a Cartesian tensor. Let $\mathfrak{G}_1$ be the group of general analytic co-ordinate transformations in 3 dimensions. $\mathfrak{G}_2$ is a subgroup of $\mathfrak{G}_1$. Let $g_{km}(\boldsymbol{x})$ be an absolute symmetric positive definite tensor field under $\mathfrak{G}_1$ such that its Riemann curvature tensor vanishes. Then in the space with group $\mathfrak{G}_1$ we know that there exist preferred co-ordinate systems z^k such that $g_{km}(\boldsymbol{z}) = \delta_{km}$. Furthermore, any such pair of preferred co-ordinate systems are related to each other by an orthogonal transformation. Thus the group $\mathfrak{G}_2$ can be defined as the subgroup of $\mathfrak{G}_1$ which leaves the *canonical form* δ_{ij} of the Euclidean metric field g_{ij} invariant. Let $\varrho^{k\cdots}_{m\cdots}(\boldsymbol{x})$ be any field in the space with group $\mathfrak{G}_1$ having any law of transformation under $\mathfrak{G}_1$ such that

$$\varrho^{k\cdots}_{m\cdots}(\boldsymbol{z}) = f^{k\cdots}_{m\cdots}(\boldsymbol{z}) \tag{152.6}$$

in every preferred co-ordinate system in which $g_{ij} = \delta_{ij}$. According to KLEIN's principle, the theory of the invariants of the field $f^{k\cdots}_{m\cdots}$ under the group $\mathfrak{G}_2$ is identical with the theory of the joint invariants of the fields $(\varrho^{k\cdots}_{m\cdots}, g_{pq})$ under the group $\mathfrak{G}_1$ of general analytic transformations or under any group containing $\mathfrak{G}_2$ as a subgroup.

With these ideas in mind, consider the group $\mathfrak{G}_{\mathfrak{A}}$ of general analytic transformations of the four co-ordinates x^Γ of events. Our objective is to define two sets of fields $\{\Phi\}_{\mathfrak{E}}$ and $\{\Phi\}_{\mathfrak{G}}$ having assigned transformation laws under $\mathfrak{G}_{\mathfrak{A}}$ such that (1) there exists a subclass of preferred co-ordinates z^Γ in which the fields $\{\Phi\}_{\mathfrak{E}}$ and $\{\Phi\}_{\mathfrak{G}}$ assume certain canonical forms, and (2) the subgroups $\mathfrak{G}_{\mathfrak{E}}$ and $\mathfrak{G}_{\mathfrak{G}}$ consist in all the transformations of $\mathfrak{G}_{\mathfrak{A}}$ which leave invariant the canonical forms of the sets of fields $\{\Phi\}_{\mathfrak{E}}$ and $\{\Phi\}_{\mathfrak{G}}$, respectively. Once we determine such a set of fields, we invoke KLEIN's principle and give new but equivalent definitions of Euclidean and Galilean kinematics. That is, the study of the invariants of a motion under the restricted groups $\mathfrak{G}_{\mathfrak{E}}$ and $\mathfrak{G}_{\mathfrak{G}}$ can then be replaced by the equivalent theory of the joint invariants of the combined sets of fields $(Z^K(x^\Gamma), \{\Phi\}_{\mathfrak{E}})$ and $(Z^K(x^\Gamma), \{\Phi\}_{\mathfrak{G}})$ under the group $\mathfrak{G}_{\mathfrak{A}}$.

Case I. Euclidean space-time. Let $t_\Gamma(\boldsymbol{x}) \not\equiv 0$ be an absolute covariant vector field under $\mathfrak{G}_{\mathfrak{A}}$ such that

$$\partial_{[\Gamma} t_{\Delta]} = 0. \tag{152.7}$$

Let $g^{\Gamma\Delta}(\boldsymbol{x})$ be a symmetric contravariant singular absolute tensor field under $\mathfrak{G}_{\mathfrak{A}}$ such that

$$g^{\Gamma\Delta} t_\Delta = 0, \quad g^{\Gamma\Delta} v_\Gamma v_\Delta > 0, \tag{152.8}$$

for all $v_\Gamma \not\equiv 0$ and not parallel to t_Δ. The condition (152.7) is necessary and sufficient for the existence of an absolute scalar field $t(\boldsymbol{x})$ such that

$$t_\Gamma = \partial_\Gamma t. \tag{152.9}$$

The field $t(\boldsymbol{x})$ is uniquely determined to within an additive constant. Let us introduce $t(\boldsymbol{x})$ as a co-ordinate surface by setting $z^4 = t$ and impose on the non-

[1] SCHOUTEN [1954, *21*, p. 65].

vanishing components $g^{km}(x^p, z^4)$ in such a system of co-ordinates the conditions

$$R^k{}_{mpq}(\boldsymbol{g}) = 0. \tag{152.10}$$

These conditions are necessary and sufficient that we be able to make a further transformation

$$z^k = z^k(x^p, z^4)$$

of the first three co-ordinates such that in the co-ordinate system z^Γ the fields $g^{\Gamma\Delta}$ and t_Γ have the *canonical form*

$$\boldsymbol{g} = \begin{bmatrix} \delta^{rs} & 0 \\ 0 & 0 \end{bmatrix}, \quad \boldsymbol{t} = (0, 0, 0, 1). \tag{152.11}$$

Applying the assumed tensor law of transformation to these canonical forms, we then see that *the Euclidean group of transformations* (152.1) *is the subgroup of* $\mathfrak{G}_{\mathfrak{A}}$ *which leaves these canonical forms invariant.*

Therefore, Euclidean kinematics is the theory of the joint invariants under the group $\mathfrak{G}_{\mathfrak{A}}$ *of the set of fields*

$$Z^K(\boldsymbol{x}), \quad g^{\Gamma\Delta}(\boldsymbol{x}), \quad t_\Gamma(\boldsymbol{x}), \tag{152.12}$$

where the definition (152.3) of $\overset{-1}{\boldsymbol{C}}$, invariant under $\mathfrak{G}_{\mathfrak{E}}$, is replaced by the definition

$$\overset{-1}{C}{}^{KL} \equiv g^{\Gamma\Delta}\, \partial_\Gamma Z^K\, \partial_\Delta Z^L, \tag{152.13}$$

invariant under $\mathfrak{G}_{\mathfrak{A}}$.

A co-ordinate system in Euclidean space-time such that (152.11) holds will be called a *Euclidean frame*. We shall call $g^{\Gamma\Delta}$ the *space metric*; we shall call t_Γ the *covariant space normal*; and we identify t with the *time*.

Case II. Galilean space-time. Let $\Gamma^\Theta_{\Gamma\Delta}(\boldsymbol{x})$ be a symmetric affine connection under $\mathfrak{G}_{\mathfrak{A}}$. We assume that $\boldsymbol{\Gamma}$ is a flat or integrable connection so that

$$R^\Gamma{}_{\Theta\Lambda\Delta}(\boldsymbol{\Gamma}) = 0. \tag{152.14}$$

Let $g^{\Gamma\Delta}$ and t_Γ be tensors under $\mathfrak{G}_{\mathfrak{A}}$ having the same properties assigned to these fields as in Case I above, but which in addition satisfy the conditions

$$\left.\begin{aligned} \underset{\Gamma}{\nabla}_\Theta g^{\Phi\Delta} &= \partial_\Theta g^{\Phi\Delta} + \Gamma^\Phi_{\Theta\Lambda} g^{\Lambda\Delta} + \Gamma^\Delta_{\Theta\Lambda} g^{\Phi\Lambda} = 0, \\ \underset{\Gamma}{\nabla}_\Delta t_\Phi &= \partial_\Delta t_\Phi - \Gamma^\Theta_{\Phi\Delta} t_\Theta = 0, \end{aligned}\right\} \tag{152.15}$$

jointly with the connection $\Gamma^\Theta_{\Phi\Delta}$. That is, the covariant derivatives of $g^{\Phi\Delta}$ and t_Γ based on the connection $\Gamma^\Phi_{\Theta\Lambda}$ vanish identically.

From (152.14) follows the existence of preferred co-ordinate systems in which all of the components of the connection vanish[1]. Any two such systems are related by a *linear* transformation. It follows from (152.15) that, in any of the co-ordinate systems in which the connection vanishes, the components of $g^{\Gamma\Delta}$ and t_Π are constants. Set $z^4 = t(\boldsymbol{x})$. This will be a linear transformation leaving the connection zero, and $\boldsymbol{t}$ will assume its canonical form

$$\boldsymbol{t} = (0, 0, 0, 1). \tag{152.16}$$

From (152.8) it follows that $g^{\Gamma\Delta}$ is reduced to the form

$$\boldsymbol{g} = \begin{bmatrix} g^{km} & 0 \\ 0 & 0 \end{bmatrix}, \tag{152.17}$$

[1] Eisenhart [1927, 4, § 29].

where g^{km} is a constant symmetric positive definite matrix. Thus by a further linear transformation of the first three co-ordinates not involving z^4, preserving the condition (152.16) and the vanishing of the connection, we can reduce $\boldsymbol{g}$ to its canonical form $(152.11)_1$. It is then an easy matter to verify that the Galilean group (152.2) is the subgroup of $\mathfrak{G}_{\mathfrak{E}}$ which leaves invariant the canonical forms (152.11) and $\Gamma^{\Phi}_{\Delta\Lambda} = 0$.

Therefore, Galilean kinematics is the theory of the joint invariants under $\mathfrak{G}_{\mathfrak{A}}$ of the set of fields

$$Z^K(\boldsymbol{x}), \quad g^{\Gamma\Delta}(\boldsymbol{x}), \quad t_\Theta(\boldsymbol{x}), \quad \Gamma^{\Theta}_{\Phi\Delta}(\boldsymbol{x}), \tag{152.18}$$

where the Z^K again satisfy the invariant condition (152.15). The preferred co-ordinate systems in Galilean space-time in which we have (152.11) and $\Gamma^{\Delta}_{\Phi\Delta} = 0$ will be called *Galilean frames*, and $\Gamma^{\Delta}_{\Phi\Delta}$ will be called the *Galilean connection*. Galilean frames may be identified with the inertial frames of classical mechanics (cf. Sect. 196).

Thus we have succeeded in formulating Euclidean and Galilean kinematics as theories of the joint invariants of a motion and a suitable set of fields under the group $\mathfrak{G}_{\mathfrak{A}}$ of unrestricted general transformations of the co-ordinates in space-time. Quantities transforming as a tensor under $\mathfrak{G}_{\mathfrak{A}}$ will be called *world tensors*. An affine connection under $\mathfrak{G}_{\mathfrak{A}}$ will be called a *world affine connection*. The Galilean connection is a world affine connection, $g^{\Gamma\Delta}$ and t_Γ are world tensors, and the $Z^K(\boldsymbol{x})$ are world scalars.

We shall also consider the theory of the joint invariants of an arbitrary world tensor field $\Psi^{\Gamma\dots}_{\Delta\dots}$ and a motion as part of the subject matter of kinematics. In regard to this aspect of the theory, we shall be especially interested in quantities like the stress tensor t^{km} of mechanics (cf. Sect. 203) which are required to transform as the components of a 3-dimensional Cartesian tensor under the restricted group $\mathfrak{G}_{\mathfrak{E}}$. We shall represent quantities of this type in the 4-dimensional formalism by *contravariant* world tensors $\Psi^{\Gamma\Delta\dots}$ which satisfy the invariant conditions

$$\Psi^{\Gamma\Delta\dots} t_\Gamma = \Psi^{\Gamma\Delta\dots} t_\Delta = \dots = 0. \tag{152.19}$$

Contravariant world tensors satisfying (152.19) will be called *space tensors*. We easily verify that, in every Euclidean or Galilean frame, each component of a space tensor with any index equal to 4 vanishes, and the non-vanishing components have the transformation law

$$\Psi^{k'l'} = A^{k'}{}_k A^{l'}{}_l \dots \Psi^{kl}, \tag{152.20}$$

under the restricted groups $\mathfrak{G}_{\mathfrak{E}}$ and $\mathfrak{G}_{\mathfrak{G}}$. Thus, according to KLEIN's principle, *the theory of the joint invariants of Ψ^{ij} and a motion under the restricted groups $\mathfrak{G}_{\mathfrak{E}}$ and $\mathfrak{G}_{\mathfrak{G}}$ is equivalent to the theory of the joint invariants of $\Psi^{\Gamma\Delta}$ and the lists of fields* (152.14) *and* (152.18), *respectively.*

153. World invariant Euclidean kinematics. In this section we consider certain of the invariants of a motion in Euclidean space-time. As previously mentioned, since $\mathfrak{G}_{\mathfrak{G}}$ is a subgroup of $\mathfrak{G}_{\mathfrak{E}}$, any invariant of a motion in Euclidean space-time is also an invariant of a motion in Galilean space-time, so that all of the results and considerations of this section apply equally well to motions in Galilean space-time.

α) World velocity field of a motion. Consider the world scalar of weight 1 defined by

$$\mathfrak{D} \equiv \frac{1}{3!} \varepsilon^{\Gamma\Delta\Lambda\Sigma} \partial_\Gamma Z^K \partial_\Delta Z^L \partial_\Lambda Z^M \partial_\Sigma t \, \varepsilon_{KLM}, \tag{153.1}$$

and the world vector of weight 1 defined by

$$\mathfrak{v}^\Gamma \equiv \frac{1}{3!}\, \varepsilon^{\Delta\Lambda\Sigma\Gamma}\, \partial_\Delta Z^K\, \partial_\Lambda Z^L\, \partial_\Sigma Z^M\, \varepsilon_{KLM}. \tag{153.2}$$

In a Euclidean frame, $\mathfrak{D} = \det \partial_k Z^K = \pm\sqrt{\det \overset{-1}{C}} \neq 0$. Since the law of transformation for $\mathfrak{D}$ is $\mathfrak{D}^* = |\boldsymbol{x}^*/\boldsymbol{x}|^{-1}\mathfrak{D}$ and $|\boldsymbol{x}^*/\boldsymbol{x}|$ is never zero, $\mathfrak{D} \neq 0$ in any co-ordinate system. Thus we can define the absolute world contravariant vector field

$$v^\Gamma = \frac{\mathfrak{v}^\Gamma}{\mathfrak{D}} \tag{153.3}$$

called the *world velocity vector of the motion*. The form which any world tensor or other type of world invariant takes in every Euclidean or Galilean frame will be called its *canonical form*. The canonical form of the world velocity vector v^Γ is

$$v^\Gamma = (v^k, 1). \tag{153.4}$$

Since $\mathfrak{D}$ is never zero, we can always solve for any system of general co-ordinates x^Γ in terms of the material co-ordinates Z^K and the time $T = t(\boldsymbol{x})$. Thus we always have relations of the form

$$x^\Gamma = x^\Gamma(Z^K, T). \tag{153.5}$$

In terms of these relations, we have

$$v^\Gamma = \frac{\partial x^\Gamma}{\partial T}. \tag{153.6}$$

The result (153.5) serves to promote the geometric interpretation of a motion as a congruence of lines in space-time which are nowhere tangent to the surfaces $t(\boldsymbol{x}) = T = \text{const}$. Such a surface is called an *instantaneous space*. The material co-ordinates Z^K serve as names for the lines of the congruence, and T is an admissible parameter whose value is never stationary as one moves along a line of the congruence. It is clear that the first three components of the world velocity in a Euclidean frame coincide with the usual definition (67.1) of the velocity vector of a continuous medium.

β) Invariant definition of a rigid motion. A motion in Euclidean or Galilean space-time is called *rigid* if and only if

$$\dot{\overset{-1}{C}}{}^{KL} \equiv v^\Gamma \partial_\Gamma \overset{-1}{C}{}^{KL} = 0. \tag{153.7}$$

To see that this definition corresponds to the customary one, note that from (153.1) and the fact that $\mathfrak{D} \neq 0$, we can always introduce the Z^K and $T = t(\boldsymbol{x})$ as a system of co-ordinates in space-time. The components of the space metric $\boldsymbol{g}$ in such a system of *convected co-ordinates* will have the values $g^{K4} = g^{44} = 0$ and g^{KL} where

$$g^{KL} = \overset{-1}{C}{}^{KL} \quad \text{(convected co-ordinates)}. \tag{153.8}$$

The result (153.8) follows immediately from the definition of the $\overset{-1}{C}{}^{KL}$ in a general system of co-ordinates. Let $g_{KL}(Z^K, T)$ denote the inverse of g^{KL}. The distance between two neighboring material points Z^K and $Z^K + dZ^K$ at time T is given by

$$dS^2 = g_{KL}\, dZ^K\, dZ^L = C_{KL}\, dZ^K\, dZ^L. \tag{153.9}$$

Thus a motion is rigid if and only if $\dot{\overline{dS}} = 0$ for every pair of neighboring material points.

γ) World tensor of stretching. Consider the contravariant absolute world tensor $\boldsymbol{\Delta}$ defined by[1]

$$\left.\begin{aligned}\Delta^{\Gamma\Sigma} &\equiv -\tfrac{1}{2}\underset{v}{\pounds}\, g^{\Gamma\Sigma},\\ &= -\tfrac{1}{2}\left(v^{\Lambda}\partial_{\Lambda} g^{\Gamma\Sigma} - g^{\Lambda\Sigma}\partial_{\Lambda} v^{\Gamma} - g^{\Lambda\Gamma}\partial_{\Lambda} v^{\Sigma}\right),\end{aligned}\right\} \tag{153.10}$$

where $\underset{v}{\pounds}$ denotes the Lie derivative (150.6) with respect to the world velocity vector v^{Γ}. One sees by inspection that the Lie derivative of any space tensor is a space tensor. The canonical form of the space tensor $\boldsymbol{\Delta}$ is

$$\boldsymbol{\Delta} = \begin{bmatrix} d^{km} & 0 \\ 0 & 0 \end{bmatrix}, \qquad d^{km} \equiv \tfrac{1}{2}\left(\partial_k v^m + \partial_m v^k\right). \tag{153.11}$$

The quantities d^{km} are the familiar Cartesian components of the stretching tensor (82.2). We call $\boldsymbol{\Delta}$ the *world tensor of stretching.*

The field g^{-1} defined by

$$g^{-1} \equiv \frac{1}{3!}\,\varepsilon_{\Gamma\Psi\Lambda\Xi}\,\varepsilon_{\Sigma\Omega\Pi\Phi}\, g^{\Psi\Omega} g^{\Lambda\Pi} g^{\Xi\Phi} v^{\Gamma} v^{\Sigma} \tag{153.12}$$

is a world scalar of weight -2 having the constant value $g^{-1}=1$ in every Euclidean frame. The familiar absolute scalar invariants of the stretching tensor d^{km} are given in world invariant form by the formulae

$$\left.\begin{aligned}\mathrm{I}_{\boldsymbol{d}} &= \frac{g}{2}\,\varepsilon_{\Gamma\Psi\Lambda\Xi}\,\varepsilon_{\Sigma\Omega\Pi\Phi}\, g^{\Psi\Omega} g^{\Lambda\Pi} \Delta^{\Gamma\Sigma} v^{\Xi} v^{\Phi},\\ \mathrm{II}_{\boldsymbol{d}} &= \frac{g}{2!}\,\varepsilon_{\Gamma\Psi\Lambda\Xi}\,\varepsilon_{\Sigma\Omega\Pi\Phi}\, g^{\Psi\Omega} \Delta^{\Lambda\Pi} \Delta^{\Gamma\Sigma} v^{\Xi} v^{\Phi},\\ \mathrm{III}_{\boldsymbol{d}} &= \frac{g}{3!}\,\varepsilon_{\Gamma\Psi\Lambda\Xi}\,\varepsilon_{\Sigma\Omega\Pi\Phi}\, \Delta^{\Psi\Omega} \Delta^{\Lambda\Pi} \Delta^{\Gamma\Sigma} v^{\Xi} v^{\Phi}.\end{aligned}\right\} \tag{153.13}$$

The canonical form of these scalars (cf. Sect. 83) is

$$\mathrm{I}_{\boldsymbol{d}} = \delta_{km} d^{km}, \qquad \mathrm{II}_{\boldsymbol{d}} = \frac{1}{2!}\,\varepsilon_{klm}\,\varepsilon_{kpq}\, d^{lp} d^{mq}, \qquad \mathrm{III}_{\boldsymbol{d}} = \det d^{km}. \tag{153.14}$$

Comparing the definitions of the Lie derivative with respect to v^{Γ} of a world space tensor and the convected time flux (150.5) of a Cartesian tensor under $\mathfrak{G}_{\mathfrak{E}}$, we see that

$$\underset{v}{\pounds}\, \psi^{ij\ldots} \doteq \frac{d_c \psi^{ij\ldots}}{dt}, \tag{153.15}$$

where, as henceforth, the symbol "$\doteq$" indicates that equality holds in a Euclidean frame and, in general, only in such a frame.

From this result and from (150.6) we see that in the spatial description, where the velocity field is basic and is regarded as given, the convected time flux $d_c\psi/dt$ defines an *infinitesimal transformation* of ψ. As was remarked by ZORAWSKI[2], the central problem of the kinematics of the spatial description is to characterize all invariants of this infinitesimal transformation.

δ) World affine connections defined by a motion. The structure of Euclidean space-time implied by the underlying group $\mathfrak{G}_{\mathfrak{E}}$ of Euclidean transformations does not of itself lead to an affinely connected space in the four-dimensional sense as in the case of the linear Galilean group. However, as was shown by DEFRISE[3], given the basic tensors $g^{\Xi\Sigma}$ and t_{Ξ} of Euclidean space-time *and* a

[1] WUNDHEILER [1932, *15*, § 8].

[2] The analysis of ZORAWSKI [1900, *11* and *12*] [1911, *13* and *14*] [1912, *8*] refers only to steady motion described in the common frame.

[3] [1953, *8*].

world velocity vector v^{Ξ}, we can construct from these quantities and their partial derivatives a set of quantities $\Omega^{\Xi}_{\Sigma\Phi}$ having the law of transformation of a world affine connection under $\mathfrak{G}_{\mathfrak{A}}$. The method of constructing the components of $\Omega^{\Xi}_{\Sigma\Phi}$ from the $g^{\Xi\Sigma}$, t_{Ξ}, and v^{Ξ} is analogous to the method of constructing the Christoffel symbols based on a non-singular symmetric absolute tensor of rank two, familiar in Riemannian geometry. Let $\underset{\Omega}{\nabla}_{\Phi}$ denote covariant differentiation based on the world affine connection $\Omega^{\Pi}_{\Xi\Sigma}$. The equations used by DEFRISE to determine the components $\Omega^{\Pi}_{\Xi\Sigma}$ are equivalent to the following:

$$\underset{\Omega}{\nabla}_{\Xi} v^{\Lambda} = 0, \quad \underset{\Omega}{\nabla}_{\Phi} g^{\Xi\Sigma} = -2\Delta^{\Xi\Sigma} t_{\Phi}, \quad \underset{\Omega}{\nabla}_{\Sigma} t_{\Xi} = 0. \tag{153.16}$$

These equations have a unique solution for all 64 components of the world connection $\Omega^{\Pi}_{\Xi\Sigma}$. The canonical form of these quantities is

$$\Omega^{i}_{jk} = 0, \quad \Omega^{4}_{\Xi\Sigma} = 0, \quad \Omega^{i}_{4j} = -\partial_j v^i, \quad \Omega^{i}_{44} = -\frac{\partial v^i}{\partial z^4} + v^j \partial_j v^i. \tag{153.17}$$

As a variant of DEFRISE's procedure for determining a world affine connection in terms of a motion in Euclidean space-time, we can solve the equations

$$\underset{\Lambda}{\nabla}_{\Phi} g^{\Xi\Sigma} = 0, \quad v^{\Sigma} \underset{\Lambda}{\nabla}_{\Sigma} v^{\Xi} = 0, \quad \underset{\Lambda}{\nabla}_{\Sigma} t_{\Xi} = 0, \quad \underset{\Lambda}{\nabla}_{\Phi} v^{[\Xi} g^{\Sigma]\Phi} = 0, \tag{153.18}$$

for the components $\Lambda^{\Pi}_{\Xi\Sigma}$ of a different symmetric world affine connection. The canonical forms of the components of the connection $\Lambda^{\Pi}_{\Xi\Sigma}$ are

$$\Lambda^{p}_{jk} = 0, \quad \Lambda^{4}_{\Xi\Sigma} = 0, \quad \Lambda^{p}_{4s} = (w^{ps} - d^{ps})\, v^s - \frac{\partial v^p}{\partial z^4}, \tag{153.19}$$

where $w^{ps} \equiv \frac{1}{2}(\partial_s v^p - \partial_p v^s)$ are the components of vorticity.

It is an immediate consequence of the law of transformation for affine connections that the difference between two affine connections is an absolute tensor. In the case of the two world affine connections $\Omega^{\Pi}_{\Xi\Sigma}$ and $\Lambda^{\Pi}_{\Xi\Sigma}$, this difference turns out to be

$$U^{\Pi}_{\Xi\Sigma} \equiv \Lambda^{\Pi}_{\Xi\Sigma} - \Omega^{\Pi}_{\Xi\Sigma} = p_{\Xi\Phi} \Delta^{\Phi\Pi} t_{\Sigma} + p_{\Sigma\Phi} \Delta^{\Phi\Pi} t_{\Xi}, \tag{153.20}$$

where $p_{\Xi\Phi}$ is the absolute tensor which has the canonical form

$$\boldsymbol{p} = \begin{bmatrix} \delta_{km} & -v^k \\ -v_m & v^p v^p \end{bmatrix}. \tag{153.21}$$

ε) *Invariant time fluxes in the 4-dimensional formalism.* We now note an interesting relation between the covariant derivatives of world space tensors based on the two connections $\Omega^{\Pi}_{\Xi\Sigma}$ and $\Lambda^{\Pi}_{\Xi\Sigma}$ and the two invariant time fluxes d_c/dt and d_r/dt defined in Sects. 150 and 148. Let $\Psi^{\Xi\Sigma\ldots}$ be a space tensor, and consider the space tensors given by

$$v^{\Phi} \underset{\Omega}{\nabla}_{\Phi} \Psi^{\Xi\Sigma\ldots} \quad \text{and} \quad v^{\Phi} \underset{\Lambda}{\nabla}_{\Phi} \Psi^{\Xi\Sigma\ldots}. \tag{153.22}$$

As one easily verifies from the canonical forms (153.17) and (153.19), in a Euclidean frame we have

$$v^{\Phi} \underset{\Omega}{\nabla}_{\Phi} \Psi^{ij\ldots} \doteq \frac{d_c}{dt} \Psi^{ij\ldots}, \qquad v^{\Phi} \underset{\Lambda}{\nabla}_{\Phi} \Psi^{ij\ldots} \doteq \frac{d_r}{dt} \Psi^{ij\ldots}. \tag{153.23}$$

In an arbitrary system of space-time co-ordinates we have

$$v^{\Phi} \underset{\Lambda}{\nabla}_{\Phi} \Psi^{\Xi\Sigma\ldots} - v^{\Phi} \underset{\Omega}{\nabla}_{\Phi} \Psi^{\Xi\Sigma\ldots} = v^{\Phi} U^{\Xi}_{\Phi\Pi} \Psi^{\Pi\Sigma\ldots} + v^{\Phi} U^{\Sigma}_{\Phi\Pi} \Psi^{\Xi\Pi\ldots} + \cdots \tag{153.24}$$

which generalizes the relation (151.1) contrasting the invariant time fluxes d_c/dt and d_r/dt.

If one adds any symmetric absolute tensor $S^{\Pi}_{\Xi\Sigma}=S^{\Pi}_{\Sigma\Xi}$ to the world affine connection determined by either of the sets of Eqs. (153.16) or (153.18), one obtains yet another symmetric world affine connection. Many tensors $S^{\Pi}_{\Xi\Sigma}$ can be defined in terms of the basic set $g^{\Xi\Sigma}$, t_Ξ, and v^Ξ. Thus, many world affine connections can be constructed from these quantities and their partial derivatives, and the absolute derivatives of world tensors can be defined in terms of them. The question as to which of these methods of absolute differentiation in Euclidean space-time defined in terms of a motion in that space is the most natural or useful is analogous to that raised in Sect. 151 concerning invariant time fluxes.

ζ) *Convected co-ordinates.* In a convected system of co-ordinates $x^\Xi=(Z^K, T)$, the world velocity vector and the covariant space normal have the *convected form*

$$\boldsymbol{v}=(0,0,0,1), \qquad \boldsymbol{t}=(0,0,0,1), \tag{153.25}$$

and, as noted previously, the space metric has the form

$$\boldsymbol{g}=\begin{bmatrix} \overset{-1}{C}{}^{KL} & 0 \\ 0 & 0 \end{bmatrix}. \tag{153.26}$$

Thus it is a simple matter to determine the convected form of the various world tensors $\Delta^{\Xi\Sigma}$, v^Φ, $\Psi^{\Xi\Sigma}$, etc., which have been introduced. From (153.10) it follows immediately that, in a convected system of co-ordinates,

$$\Delta^{\Xi\Sigma}=\begin{bmatrix} -\dfrac{1}{2}\dfrac{\partial g^{KL}}{\partial T} & 0 \\ 0 & 0 \end{bmatrix}. \tag{153.27}$$

The convected form of the Lie derivative of any space tensor is given by

$$\underset{v}{£}\,\Psi^{4\Sigma\ldots}=\underset{v}{£}\,\Psi^{\Xi 4\ldots}=0, \qquad \underset{v}{£}\,\Psi^{KL\ldots}=\frac{\partial\Psi^{KL\ldots}}{\partial T}. \tag{153.28}$$

The convected forms of the world affine connections $\Omega^{\Pi}_{\Xi\Sigma}$ and $\Lambda^{\Pi}_{\Xi\Sigma}$ are

$$\left.\begin{aligned} &\Omega^K_{44}=0, \quad \Omega^K_{LM}=\left\{\begin{matrix}K\\LM\end{matrix}\right\}_C, \quad \Omega^4_{\Xi\Sigma}=0, \quad \Omega^K_{4L}=0,\\ &\Lambda^K_{44}=0, \quad \Lambda^K_{LM}=\left\{\begin{matrix}K\\LM\end{matrix}\right\}_C, \quad \Lambda^4_{\Xi\Sigma}=0, \quad \Lambda^K_{4L}=-\frac{1}{2}C_{LM}\frac{\partial \overset{-1}{C}{}^{MK}}{\partial T}, \end{aligned}\right\} \tag{153.29}$$

where $\left\{\begin{matrix}K\\LM\end{matrix}\right\}_C$ denotes the Christoffel symbols based on the C_{KL}. In Sect. 150 it was shown that the convected time flux of a 3-dimensional tensor under $\mathfrak{G}_{\mathfrak{C}}$ reduces to $\partial\psi^{KL\ldots}/\partial T$ in convected co-ordinates. In fact, this was made the basis of its definition. It follows from (153.28) and (153.29) that the conditional equalities (153.15) and (153.23) holding in Euclidean frames may be extended to the case of convected co-ordinates. However, for the absolute derivative $(153.22)_2$ of a space tensor in convected co-ordinates we have

$$\left.\begin{aligned} &v^\Phi \underset{\Lambda}{\nabla}_\Phi\, \psi^{4\Sigma\ldots}=v^\Phi \underset{\Lambda}{\nabla}_\Phi\, \psi^{\Xi 4\ldots}=0,\\ &v^\Phi \underset{\Lambda}{\nabla}_\Phi\, \psi^{KL\ldots}=\frac{\partial\psi^{KL\ldots}}{\partial T}-\frac{1}{2}C_{MN}\frac{\partial g^{MK}}{\partial T}\psi^{NL}-\frac{1}{2}C_{MN}\frac{\partial g^{ML}}{\partial T}\psi^{KN}-\cdots. \end{aligned}\right\} \tag{153.30}$$

If the 3-dimensional tensor components $\psi^{ij\ldots}$ occurring in the definition (148.5) of the co-rotational time flux are transformed as the components of a *contra-*

variant tensor under the group of transformations (152.4) relating general curvilinear and deforming co-ordinates with the time held fixed, we can also extend the conditional equality $(153.23)_2$ to these more general types of co-ordinate systems. In this case, $(153.30)_3$ is a formula for the components of the co-rotational time flux referred to a convected system of co-ordinates.

154. World invariant Galilean kinematics. In the case of Galilean kinematics, we study the invariants of the fields (152.18) and the joint invariants of a given world tensor $\Psi^{\Xi\Sigma\ldots}$ and these fields. In Galilean space-time, as opposed to Euclidean space-time, we have given *a priori* an integrable or flat world affine connection $\Gamma^{\Phi}_{\Xi\Sigma}$ with which to construct additional invariants of a motion and other fields defined in the space.

α) *The world acceleration vector and the vorticity tensor.* Let $\underset{\Gamma}{\nabla}_{\Xi}$ denote covariant differentiation based on the Galilean connection $\Gamma^{\Phi}_{\Xi\Sigma}$. Then in terms of the world velocity field of a motion in Galilean space-time, we define the *absolute world acceleration vector*

$$a^{\Xi} \equiv v^{\Phi} \underset{\Gamma}{\nabla}_{\Phi} v^{\Xi}. \tag{154.1}$$

In a Galilean frame the acceleration vector has the canonical form

$$a^{\Gamma} = \left(\frac{\partial v^k}{\partial z^4} + v^p \,\partial_p v^k, 0\right). \tag{154.2}$$

Thus the first three components of the acceleration vector in a Galilean frame are equal to the usual expressions (98.1) for the acceleration. Moreover, since a^{Ξ} is a space tensor, they have the customary 3-dimensional vector law of transformation under $\mathfrak{G}_{\mathfrak{G}}$. The *world velocity gradients* are defined by $\underset{\Gamma}{\nabla}_{\Xi} v^{\Sigma}$. Consider the symmetric and skew-symmetric parts of the world tensor $W^{\Xi\Sigma}$ defined by

$$W^{\Xi\Sigma} \equiv g^{\Xi\Phi} \underset{\Gamma}{\nabla}_{\Phi} v^{\Sigma}. \tag{154.3}$$

These are given by

$$W^{(\Xi\Sigma)} = \Delta^{\Xi\Sigma}, \qquad W^{[\Xi\Sigma]} = \Omega^{\Xi\Sigma}, \qquad W^{\Xi\Sigma} = \Delta^{\Xi\Sigma} + \Omega^{\Xi\Sigma}. \tag{154.4}$$

We have met with several invariant definitions of $\Delta^{\Xi\Sigma}$ already, but $\Omega^{\Xi\Sigma}$ is a new tensor quantity having the canonical form

$$\boldsymbol{\Omega} = \begin{bmatrix} -w^{ij} & 0 \\ 0 & 0 \end{bmatrix}, \qquad w^{ij} = \tfrac{1}{2}(\partial_j v^i - \partial_i v^j). \tag{154.5}$$

Since w^{ij} is the classical vorticity tensor (86.1), we call $\boldsymbol{\Omega}$ the *absolute world vorticity tensor.* Now the world tensor of stretching Δ was defined independently of the Galilean connection, but this independence is not immediately apparent from (154.3) and (154.4). It may be seen from

$$W^{(\Xi\Sigma)} = g^{\Phi(\Xi}\,\partial_{\Phi} v^{\Sigma)} + g^{\Phi(\Xi}\,\Gamma^{\Sigma)}_{\Phi\Pi} v^{\Pi}, \tag{154.6}$$

since by $(152.15)_1$ we have

$$2 g^{\Phi(\Xi}\,\Gamma^{\Sigma)}_{\Phi\Pi} = -\,\partial_{\Pi} g^{\Xi\Sigma}. \tag{154.7}$$

The components of the world vorticity tensor $\boldsymbol{\Omega}$ cannot in this same way be expressed solely in terms of the fields $\boldsymbol{v}, \boldsymbol{g}, \boldsymbol{t}$ and their derivatives. That is, vorticity is not a Euclidean invariant of a motion. This is clear from an intuitive point of view since vorticity measures a rate of rotation, which does not have an absolute significance under the Euclidean group but does have an absolute significance under the smaller Galilean group.

β) Curvilinear, rotating, and deforming co-ordinate systems. Consider the class of co-ordinate systems related to a Galilean frame by a co-ordinate transformation of the general form,

$$x^{k'} = x^{k'}(z^i, z^4), \qquad x^{4'} = z^4 + \text{const}. \tag{154.8}$$

In general, the co-ordinate system $x^{k'}$ will be curvilinear, rotating, accelerated and deforming. The tensors $\boldsymbol{g}$ and $\boldsymbol{t}$ in such a system of co-ordinates will have the general form

$$\boldsymbol{g} = \begin{bmatrix} g^{k'm'}(x^{n'}, x^{4'}) & 0 \\ 0 & 0 \end{bmatrix}, \qquad \boldsymbol{t} = (0, 0, 0, 1). \tag{154.9}$$

The components of the Galilean connection $\boldsymbol{\Gamma}$ can be written in the form

$$\left.\begin{aligned} \Gamma^{k'}_{m'n'} &= \begin{Bmatrix} k' \\ m'\,n' \end{Bmatrix}_{\boldsymbol{g}}, \qquad \Gamma^{4}_{\Xi'\Sigma'} = 0, \\ \Gamma^{k'}_{m'4'} &= -\frac{\partial u^{k'}}{\partial x^{m'}} - \begin{Bmatrix} k' \\ m'\,n' \end{Bmatrix}_{\boldsymbol{g}} u^{n'} \equiv -u^{k'}{}_{,m'}, \\ \Gamma^{k'}_{4'4'} &= -\frac{\partial u^{k'}}{\partial x^{4'}} + u^{k'}{}_{,m'} u^{m'}, \end{aligned}\right\} \tag{154.10}$$

where $\begin{Bmatrix} k' \\ l'\,m' \end{Bmatrix}_{\boldsymbol{g}}$ are the Christoffel symbols based on the $g_{k'm'}$ and $u^{k'}$ is the velocity of a fixed point in the Galilean frame $\boldsymbol{z}$ relative to the deforming co-ordinate system $\boldsymbol{x}'$. The $u^{k'}$ are defined in terms of the co-ordinate transformation (154.8) by

$$u^{k'} \equiv \frac{\partial x^{k'}}{\partial z^4}. \tag{154.11}$$

If the transformation (154.8) has the Euclidean form (143.11), then

$$\left.\begin{aligned} \boldsymbol{g} &= \begin{bmatrix} \delta^{kn} & 0 \\ 0 & 0 \end{bmatrix}, \qquad \boldsymbol{t} = (0, 0, 0, 1) \\ u^{k'} &= \dot{A}_{kk'} A_{km'} x^{m'} - A_{kk'} \dot{b}_k, \\ &= \omega'_{k'm'} x^{m'} - A_{kk'} \dot{b}_k. \end{aligned}\right\} \tag{154.12}$$

The components of $\boldsymbol{\Gamma}$ in this rotating and accelerated Cartesian frame have the values

$$\left.\begin{aligned} &\Gamma^{k'}_{m'n'} = 0, \qquad \Gamma^{4}_{\Xi'\Sigma'} = 0, \qquad \Gamma^{k'}_{m'4'} = -\omega'_{k'm'}, \\ &\Gamma^{k'}_{4'4'} = (-\dot{\omega}'_{k'n'} + \omega'_{k'm'}\omega'_{m'n'}) x^{n'} + A_{kk'} \ddot{b}_k. \end{aligned}\right\} \tag{154.13}$$

The components of the acceleration vector in a general rotating, accelerating and deforming co-ordinate system are obtained simply by substituting the components of the Galilean connection (154.10) into the invariant definition (154.1). In this manner, we get

$$\left.\begin{aligned} a^4 &= 0, \\ a^{k'} &= \frac{\partial v^{k'}}{\partial x^{4'}} + v^{k'}{}_{,j'} v^{j'} - \left(\frac{\partial u^{k'}}{\partial z^{4'}} + u^{k'}{}_{,j} u^{j'}\right) - 2u^{k'}{}_{,j}(v^{j'} - u^{j'}). \end{aligned}\right\} \tag{154.14}$$

If the non-Galilean frame is Cartesian, so that (154.13) applies, the expression for $a^{k'}$ reduces to

$$a^{k'} = \frac{\partial v^{k'}}{\partial x^4} + \frac{\partial v^{k'}}{\partial x^{m'}} v^{m'} - 2\omega'_{k'm'} v^{m'} - (\dot{\omega}'_{k'm'} - \omega'_{k'n'}\omega_{n'm'}) x^{m'} + A_{kk'} \ddot{b}_k. \tag{154.15}$$

Since a^{Ξ} is a space vector, we also have $a^{k'} = \frac{\partial x^{k'}}{\partial z^m} a^m$. Therefore, (154.15) is the formula of CORIOLIS (143.6) expressed in component form, and (154.14) is a generalization of the Coriolis formula to the case of curvilinear and deforming co-ordinate systems.

We pause to remark that the formulae (154.9) and (154.10) achieve in a fully general yet explicit form all that is attempted by older treatments of the subject, since in all those treatments[1] the equations of transformation to the general, deforming co-ordinate system from a "fixed" rectangular Cartesian system are considered as given.

In an arbitrary curvilinear and deforming co-ordinate system, we have (154.9) and (154.10), and the world tensor of stretching and world vorticity tensor assume the forms

$$\left.\begin{aligned} &\Delta^{\Xi' 4'} = \Omega^{\Xi' 4'} = 0, \\ &\Delta^{k' m'} = d^{k' m'} + \Gamma^{(k'}_{4 n'} g^{m') n'} = d^{k' m'} - \frac{1}{2}\frac{\partial g^{k' m'}}{\partial z^4}, \\ &\Omega^{k' m'} = \omega^{k' m'} - u^{[k'}{}_{,n'} g^{m'] n'}. \end{aligned}\right\} \tag{154.16}$$

If the $\boldsymbol{x}'$ frame is Cartesian, these formulae reduce to

$$\Delta^{k' m'} = d^{k' m'}, \qquad \Omega^{k' m'} = \omega^{k' m'} - \omega'^{k' m'},$$

affording a new proof of ZORAWSKI's theorem (144.3). Eq. (154.16) is the extension of ZORAWSKI's formulae to the case of curvilinear and deforming co-ordinate systems.

γ) Invariant equations for the rate of change of stretching and spin. Let a superposed dot denote the *world material derivative* defined by

$$\dot{\Psi}^{\Xi \Sigma \ldots} \equiv v^{\Phi} \underset{\Gamma}{\nabla_{\Phi}} \Psi^{\Xi \Sigma \ldots}. \tag{154.18}$$

If Ψ is a space tensor, so also is its material derivative, and in a Galilean frame we have

$$\dot{\Psi}^{4 \Xi \ldots} = \dot{\Psi}^{\Sigma 4 \ldots} = 0, \qquad \dot{\Psi}^{k m \ldots} = \frac{\partial \Psi^{k m \ldots}}{\partial z^4} + v^n \partial_n \Psi^{k m \ldots}; \tag{154.19}$$

thus the definition (154.18) affords an invariant generalization of (72.4). With this notation, we have $a^{\Xi} = \dot{v}^{\Xi}$. Consider the world material derivative of the tensor $\boldsymbol{W}$ defined in (154.3). We have

$$\left.\begin{aligned} \dot{W}^{\Sigma \Xi} &= \overline{\underset{\Gamma}{\nabla_{\Phi}} v^{\Xi} g^{\Sigma \Phi}}^{\boldsymbol{\cdot}} = \underset{\Gamma}{\nabla_{\Phi}} \dot{v}^{\Xi} g^{\Sigma \Phi} - \underset{\Gamma}{\nabla_{\Pi}} v^{\Xi} \underset{\Gamma}{\nabla_{\Phi}} v^{\Pi} g^{\Sigma \Phi} \\ &= \underset{\Gamma}{\nabla_{\Phi}} a^{\Xi} g^{\Sigma \Phi} - \underset{\Gamma}{\nabla_{\Pi}} v^{\Xi} \underset{\Gamma}{\nabla_{\Phi}} v^{\Pi} g^{\Sigma \Phi}. \end{aligned}\right\} \tag{154.20}$$

Let $D^{\Xi \Sigma}$ and $Q^{\Xi \Sigma}$ denote the symmetric and antisymmetric parts of the *acceleration gradients* $\underset{\Gamma}{\nabla_{\Phi}} a^{\Xi} g^{\Sigma \Phi}$. Taking the symmetric and antisymmetric parts of (154.20) then yields

$$\left.\begin{aligned} \dot{\Delta}^{\Xi \Sigma} &= \dot{W}^{(\Xi \Sigma)} = D^{\Xi \Sigma} - \underset{\Gamma}{\nabla_{\Pi}} v^{(\Xi} g^{\Sigma) \Phi} \underset{\Gamma}{\nabla_{\Phi}} v^{\Pi}, \\ \dot{\Omega}^{\Xi \Sigma} &= \dot{W}^{[\Xi \Sigma]} = Q^{\Xi \Sigma} - \underset{\Gamma}{\nabla_{\Pi}} v^{[\Xi} g^{\Sigma] \Phi} \underset{\Gamma}{\nabla_{\Phi}} v^{\Pi}. \end{aligned}\right\} \tag{154.21}$$

The classical problem of finding the equation for the rate of change of vorticity in a rotating, deforming co-ordinate system is thus solved by introducing the expressions (154.9) and (154.10) into $(154.21)_2$[2].

---

[1] E.g., that of McVITTIE [1949, *18*].

[2] This equation generalizes one derived by McVITTIE [1949, *18*, Eq. (6.8)] by different means. There is an extensive earlier literature devoted to special cases, mostly in meteorological contexts.

III. Mass.

a) Definition of mass.

155. The meaning of mass. In classical mechanics, each body is assigned a *mass*, a positive real number expressing the *quantity of matter* in the body, according to the requirements:

1. *The mass of a whole body is the sum of the masses of its parts;*

2. *The mass of a body never changes, no matter how that body is moved, accelerated, or deformed;*

3. *The mass of a body is not in general determined by its size.*

These requirements are translated into mathematical form as follows:

1. *Mass is a measure.*

2. *Mass is invariant under motion.*

3. *Mass bears a physical dimension* [M], *independent of* [L] *and* [T].

We discuss these in reverse order.

No. 3. The dimension of mass. This is the only property of mass that distinguishes it from other measures, such as the probability in phase of statistical mechanics or any other kind of probability or measure. While the dimensional independence of mass has definite and essential consequences in certain parts of mechanics, notably in the theory of modelling, for the developments presented in this treatise no use is made of it.

No. 2. The conservation of mass. In classical mechanics, the quantity of matter in a body generally does not change, while in more recent physical theories there are laws governing the change of mass. These physical principles must not close our eyes to the simple fact that it is *always* possible to define an invariant measure. Indeed, let the motion in the interval of time from t_0 to t be regarded as a transformation $T_{t_0}^t$ which carries the particle $\boldsymbol{X}$ to the place $\boldsymbol{x} = T_{t_0}^t \boldsymbol{X}$, and the set of particles $\mathcal{S}$ into the set of places $\mathfrak{s} = T_{t_0}^t \mathcal{S}$. From the axiom of continuity of motion, if $\mathcal{S}$ is measurable then $\mathfrak{s}$ is measurable. Assign any measure $\mathfrak{M}(\mathcal{S})$ to the measurable sets of particles; then the definition

$$\mathfrak{M}(\mathfrak{s}, t) \equiv \mathfrak{M}(T_{t_0}^t \mathcal{S}, t) \equiv \mathfrak{M}(\mathcal{S}) \tag{155.1}$$

induces a measure of sets of places, and this measure is invariant in time for a given set of particles. That is, *the measure of the set of places occupied by the set of particles $\mathcal{S}$ is the same at each time.* This trivial construction is valid in the greatest generality. For any sort of motion, whether or not a possible one for a body obeying the laws of classical mechanics, we may have conservation of mass *by definition.*

When we assert that in classical mechanics mass is conserved but in relativistic mechanics it is not conserved, we mean that the foregoing construction is useful in classical mechanics, not useful or at least not appropriate in relativistic mechanics. The distinction is a physical one. In relativistic mechanics, it is equally possible to define an invariant mass, but such a mass does not enter simply into the dynamical equations of the theory and does not correspond to the physical idea of quantity of matter; it is not fit to be compared with the result of an experiment designed to measure quantity of matter.

Even in classical mechanics there are cases when conservation of mass is not appropriate. The constituents of a mixture undergoing chemical reactions may lose or gain mass, though the mass of the mixture is constant. A formalism for such exchanges of mass will be presented in Sect. 158. Another example is furnished by the motion of a burning body in a theory which does not take account of the motion of the combustion products. That in both these cases the apparent changes of mass result from confining attention to only a part of all the matter "really" present does not reduce their cogency as examples: Mechanics must be general enough to admit models for limited situations (cf. Sect. 6).

An invariant measure is at our disposal: Conservation of mass results not from an axiom but from a *definition*, a definition we may wish to use, or may not. Kinematics is neither more nor less general after the introduction of mass. This subchapter presents topics in kinematics whose usefulness is connected with mass. It would be possible, though less interesting and less fruitful, to develop all this material without mentioning mass.

No. 1. Mass as a measure. A positive, finite mass $\mathfrak{M}(\boldsymbol{X})$ may be assigned to the particle $\boldsymbol{X}$. In this case, the mathematical statement of Property 1 becomes

1_{disc}. *The measure $\mathfrak{M}(\boldsymbol{X})$ is discrete,*

and the particle $\boldsymbol{X}$ is called a *mass-point*[1]. The mass of any finite number of mass points is finite, but the mass of an infinite number of like mass-points is always infinite. Therefore, if we are to represent finite physical bodies by a model consisting in an aggregate of mass-points having only finitely many different masses, only a finite number of mass-points may be used, and thus necessarily *almost all of space is void of matter.*

In the model of matter as continuous, to which this treatise is devoted, finite mass is assigned not to individual particles but only to sets of particles having positive volume. Moreover, if we have an infinite decreasing sequence of bodies whose common part, if any, has no volume, the masses of this sequence of bodies must approach zero. In mathematical terms,

1_{cont}. *The measure $\mathfrak{M}(\mathscr{S})$ is an absolutely continuous function of volume.*

This requirement is more stringent than the physical motivation No. 1 suggests. The difficulty is typical of infinite models (cf. Sect. 4); it is just the same as a classical difficulty in the experiential foundation of probability theory; thus we pass over it here. The requirement of absolute continuity modifies No. 3 to this extent, that *a body of zero volume has zero mass.*

Requirements 1_{disc} and 1_{cont} may be combined so as to permit mass-points and continuous masses simultaneously:

1_{mixed}. *Let $\mathfrak{M}(\mathscr{S})$ be the sum of an absolutely continuous measure and a discrete measure defined over a finite number of points.*

For application to particular cases, 1_{mixed} may be the most convenient. More general definitions of mass are suggested by measure theory, but these do not seem useful in mechanics. According to the field viewpoint, indeed, 1_{mixed} is unnecessarily general, since, as we shall see in Sect. 167, the concept of mass-point emerges derivatively in the theory of continuous bodies.

Integration over mass. Since mass is a measure, a Lebesgue-Stieltjes integration may be defined over it. By the fundamental theorem on absolutely continuous measures, from 1_{mixed} follows

$$\int_{\mathscr{S}} f\, d\mathfrak{M} = \int_{\mathscr{V}} \varrho f\, dV + \sum_{\mathfrak{a}=1}^{n} \mathfrak{M}_{\mathfrak{a}} f_{\mathfrak{a}}, \tag{155.2}$$

where $f_{\mathfrak{a}} \equiv f(\boldsymbol{X}_{\mathfrak{a}})$, where $\mathfrak{M}_{\mathfrak{a}}$ is the mass assigned to $\boldsymbol{X}_{\mathfrak{a}}$, and where ϱ, the *mass density*, satisfies

$$\varrho = \operatorname*{Lim}_{\mathfrak{V}\to 0} \frac{\mathfrak{M}(\mathscr{S})}{\mathfrak{V}(\mathscr{S})}, \qquad \dim \varrho = [M L^{-3}], \tag{155.3}$$

[1] EULER [1736, *1*, §§ 98, 117, 134]. The concept of "body" used by NEWTON and other earlier authors was vague.

$\mathfrak{M}(\mathscr{S})$ being the mass of the set $\mathscr{S}$ whose volume is $\mathfrak{V}(\mathscr{S})$. One way of asserting (155.2) is to write

$$d\mathfrak{M} = \varrho\, dV, \tag{155.4}$$

except at the points bearing discrete masses.

By definition, ϱ is an *absolute scalar* under co-ordinate transformations, since dV, as defined in the line following (20.9), is the absolute scalar element of volume rather than the scalar capacity $dX^1\, dX^2\, dX^3$.

While the theory of measure is needed to justify rigorous deductions from (155.2), an equivalent idea was used regularly by EULER, LAGRANGE, and other savants of their time. In results based on (155.2), theorems for systems of mass points emerge by taking $\varrho = 0$; for continuous media, by taking $\mathfrak{M}_a = 0$. We prefer to follow the latter course, adopting 1_{cont} rather than 1_{mixed}, and confining our attention to continuous bodies alone. Especially in view of the results to be presented in Sect. 167, this seems conceptually the cleanest course. However, the reader who desires to employ the mixed model may do so by trivial modification of the analysis in the following sections.

The concept of body. Thus far we have used the term "material" to refer to any set of particles. We now introduce the term *body* to describe a set $\mathscr{S}$ of particles such that:

1. *$\mathscr{S}$ has positive mass.*
2. *No subset of $\mathscr{S}$ which has positive volume has zero mass.*

This makes *body* synonymous with *portion of matter occupying a finite non-zero volume* and suggests a model of the universe consisting in certain bodies moving through a massless aether.

A finite portion of a body is itself a body, if its volume is positive. The requirement of absolute continuity asserts that a set of volume zero cannot be a body, but leaves it possible for an infinite volume to have finite mass, or for a finite volume to have infinite mass. In the former case, it is necessary that $\varrho \to 0$ at ∞ except possibly on a set of points of volume 0; in the latter, that $\varrho \to \infty$ at one or more points. It is also possible that $\varrho = 0$ over a set of volume zero within a body. Points where $\varrho \to \infty$ usually require special attention for other reasons, so we classify them as *singularities* and strengthen the axiom of continuity (Sect. 65) so as to exclude them:

$$0 \leqq \varrho < \infty. \tag{155.5}$$

Points where $\varrho = 0$ occasion no difficulty.

From this additional requirement it results that *every finite body has finite mass.*

For one-dimensional and two-dimensional systems, the *arc density* α and the *surface density* σ of invariant measures have the dimensions $[ML^{-1}]$ and $[ML^{-2}]$, respectively, and (155.4) is to be replaced by

$$d\mathfrak{M} = \alpha\, dX \quad \text{or} \quad d\mathfrak{M} = \sigma\, dA. \tag{155.6}$$

156. Equations expressing conservation of mass. The foregoing section has made it clear that conservation of mass is no more than a definition. Adopting the field viewpoint as expressed in the requirement 1_{cont}, we may say that an *initial density* $\varrho_0(\boldsymbol{X})$ is assigned as an arbitrary integrable function of $\boldsymbol{X}$ at time t_0. The requirement of conservation of mass then determines the density $\varrho(\boldsymbol{X}, t)$ associated with the particle $\boldsymbol{X}$ at time t. By (155.4), therefore,

$$\varrho\, dv = \varrho_0\, dV. \tag{156.1}$$

By (20.9) follow forms of the ***material equation of continuity***[1]:

$$\varrho J = \varrho_0, \quad \varrho = \varrho_0 j. \tag{156.2}$$

All four symbols occuring in these formulae are absolute scalars under transformations of spatial or material co-ordinates. Other material forms of the equation of continuity may be obtained by eliminating J or j between (156.2) and various other relations derived in Subchapter I.

From (156.2) it follows that in a continuous motion the density $\varrho(\boldsymbol{X}, t)$ of the particle $\boldsymbol{X}$ is either always zero, always finite and positive, or always infinite.

The *specific volume* v is defined by

$$v \equiv \frac{1}{\varrho}. \tag{156.3}$$

Equivalent to (156.1) and (156.2) are

$$dv/dV = v/v_0, \quad v j = v_0, \quad v = v_0 J. \tag{156.4}$$

In an isochoric motion, $\varrho = \varrho_0$: The density of each particle remains constant throughout the motion. A *homochoric* motion, defined by $\varrho_0 = \text{const}$, is an isochoric motion in which the density remains uniform in space and time.

Since $\varrho_0 = \varrho_0(\boldsymbol{X})$, we may set $B = \varrho_0$ in (76.11) and by $(156.2)_2$ obtain the ***spatial continuity equation*** of D'ALEMBERT and EULER[2] in the forms

$$\frac{\partial \varrho}{\partial t} + (\varrho \dot{x}^k)_{,k} = 0, \quad \overline{\log \varrho}^{\,\cdot} + \mathrm{I}_{\boldsymbol{d}} = 0, \quad \mathrm{I}_{\boldsymbol{d}} = \overline{\log v}^{\,\cdot}, \quad \text{etc.} \tag{156.5}$$

From analysis given in Sect. 77 it follows that *in any steady motion, it is possible to assign a density ϱ such that $\partial \varrho / \partial t = 0$.* This motivates the term *steady motion with steady density*, already introduced. For such a density, $(156.5)_1$ becomes

$$(\varrho \dot{x}^k)_{,k} = 0, \quad \text{or} \quad \operatorname{div}(\varrho \dot{\boldsymbol{p}}) = 0. \tag{156.6}$$

In general, the vector $\varrho \dot{\boldsymbol{x}}$ is called the *mass velocity* or *mass flow*. According to the result established, *in a steady motion with steady density the mass flow is solenoidal.* For such motions a sequence of theorems may be read off from the results of Sects. App. 31 to App. 32.

Two important identities are immediate consequences of the definition of ϱ. First, by setting $B = \varrho_0$ in (99.18) we obtain

$$\varrho \ddot{x}^k = \frac{d}{dt}(\varrho \dot{x}^k) + (\varrho \dot{x}^k \dot{x}^m)_{,m}, \tag{156.7}$$

a result which will be interpreted in Sect. 170. Second, since (156.1) implies $\overline{\varrho\, dv}^{\,\cdot} = 0$, for a material volume $\mathscr{V}$ we have

$$\frac{d}{dt}\int_{\mathscr{V}} \varrho F\, dv = \int_{\mathscr{V}} \varrho \dot{F}\, dv, \quad \frac{d}{dt}\int_{\mathscr{V}} F\, d\mathfrak{M} = \int_{\mathscr{V}} \dot{F}\, d\mathfrak{M}. \tag{156.8}$$

These simple formulae will be used frequently.

---

[1] EULER [1762, *1*] [1766, *1*, §§ 6, 35] [1770, *1*, §§ 123–129]. PIOLA [1836, *1*, § V, Eq. 158)] obtained the form

$$\left(\frac{\varrho}{\varrho_0} x^k_{;K}\right)_{;k} = 0,$$

which follows from (156.2) and (18.1).

[2] D'ALEMBERT [1752, *1*, § 116] obtained the special case for steady rotationally symmetric motion; the general equation is due to EULER [1757, *2*, §§ 16–17]. References for the isochoric special case were cited in Sect. 77.

For one-dimensional and two-dimensional systems, the counterparts of (156.1) are[1]

$$\alpha\, dx = \alpha_0\, dX, \quad \sigma\, da = \sigma_0\, dA; \tag{156.9}$$

by (82.5) and (82.11), the counterparts of (156.5) are[2]

$$\overline{\log \alpha}^{\bullet} + d_{(\boldsymbol{n})} = 0, \quad \overline{\log \sigma}^{\bullet} + \mathrm{I}_{\boldsymbol{d}} - d_{(\boldsymbol{n})} = 0, \tag{156.10}$$

where for the former $\boldsymbol{n}$ is the unit tangent, for the latter, the unit normal.

157. The general balance[3]. Let ϱ be the density of a not necessarily invariant mass, and let Ψ be any quantity. For a material volume $\mathscr{V}$, set

$$\frac{d}{dt}\int_{\mathscr{V}} \varrho \Psi\, dv = -\oint_{\mathscr{S}} d\boldsymbol{a} \cdot \boldsymbol{i}[\Psi] + \int_{\mathscr{V}} \varrho\, s[\Psi]\, dv. \tag{157.1}$$

This relation we call the ***general balance*** or ***general conservation law***; $\boldsymbol{i}[\Psi]$ is the *influx* of Ψ through $\mathscr{S}$, while $s[\Psi]$ is the *supply* of Ψ within $\mathscr{V}$. The general balance may serve as a *definition* of any one of the three quantities $\dot{\Psi}$, $\boldsymbol{i}[\Psi]$, or $s[\Psi]$ in terms of the other two.

Despite the tautologism of the general conservation law, it is useful because in many cases we have *a priori* knowledge of $\boldsymbol{i}$ or s, or we can derive special forms for these quantities from information about Ψ. Since we may always define appropriate $\boldsymbol{i}$ and s for any given Ψ, we may say that *all quantities* Ψ *may be balanced*, by definition. This usage is not conventional, but it should help to avoid current ambiguity of the term "conservation", which sometimes refers to equations of the type (157.1), sometimes to the stricter cases defined below.

From (157.1), $(81.2)_1$, and $(\text{App. } 26.1)_2$ follows

$$\int_{\mathscr{V}} \{\varrho \dot{\Psi} + \Psi (\dot{\varrho} + \varrho \mathrm{I}_{\boldsymbol{d}})\}\, dv = \int_{\mathscr{V}} (-\operatorname{div} \boldsymbol{i} + \varrho s)\, dv. \tag{157.2}$$

Since this holds for all sufficiently regular volumes $\mathscr{V}$, however small, a classical argument yields the ***differential form of the general balance***: *In a region where* $\dot{\Psi}$, Ψ, ϱ, $\dot{\varrho}$, $\mathrm{I}_{\boldsymbol{d}}$, $\operatorname{div} \boldsymbol{i}$, *and* s *are continuous,*

$$\varrho \dot{\Psi} + \Psi (\dot{\varrho} + \varrho \mathrm{I}_{\boldsymbol{d}}) = -\operatorname{div} \boldsymbol{i} + \varrho s. \tag{157.3}$$

A corresponding formulae at a surface of discontinuity will be given in Sect. 193.

Take $\Psi = 1$ and $\boldsymbol{i} = 0$; then (157.3) becomes

$$\dot{\varrho} + \varrho \mathrm{I}_{\boldsymbol{d}} = \varrho s. \tag{157.4}$$

This is the form of the equation of continuity appropriate when mass is being supplied at a rate s per unit mass in the interior[4].

Henceforth we restrict attention to the case when mass is conserved. By (156.5), (157.3) reduces to

$$\varrho \dot{\Psi} = -\operatorname{div} \boldsymbol{i} + \varrho s, \tag{157.5}$$

as follows also by applying (156.8) to (157.1). If Ψ is a tensor of any order, by the convention of Sect. App. 23 we obtain the tensor equation

$$\varrho \dot{\Psi}^{\cdots}_{\cdots} = -i^{\cdots k}_{\cdots, k} + \varrho s^{\cdots}_{\cdots}, \tag{157.6}$$

[1] PIOLA [1825, *2*, § 178] [1848, *2*, ¶¶ 11–12].
[2] PIOLA [1825, *2*, §§ 202, 237] [1848, *2*, ¶ 15].
[3] Cf. e.g. PRIGOGINE [1947, *12*, Chap. VII, § 2], GRAD [1952, *7*, § 4].
[4] ARRIGHI [1933, *2*].

valid in all co-ordinate systems. Another form follows by use of $(156.5)_2$:

$$\varrho s^{\dots}_{\dots} = \frac{\partial}{\partial t}(\varrho \Psi^{\dots}_{\dots}) + (\varrho \Psi^{\dots}_{\dots} \dot{x}^k + i^{\dots k}_{\dots})_{,k}. \tag{157.7}$$

Both from (157.1) and from (157.5) it is plain that for any given Ψ we may take $\boldsymbol{i}=0$ or $s=0$, as we please. This result may be called ***the equivalence of surface and volume sources***: *To secure balance of* Ψ, *it is sufficient to replace*

1. *Any influx* $\boldsymbol{i}$ *by a supply* $-\varrho^{-1}\,\mathrm{div}\,\boldsymbol{i}$, *or*
2. *Any supply* s *by an influx* $\boldsymbol{i}$ *such that* $-\mathrm{div}\,\boldsymbol{i}=\varrho s$.

Given any s satisfying a Hölder condition, there are infinitely many $\boldsymbol{i}$ satisfying Requirement 2, such an $\boldsymbol{i}$ being indeterminate to within an arbitrary solenoidal field. (In cases when discontinuities of $\boldsymbol{i}$ or surface sources are present, this principle may fail. Cf. Sect. 193.)

By the principle just established, it is always possible to define infinitely many fields $\boldsymbol{i}[\Psi]$ in such a way that[1]

$$\varrho \dot{\Psi} = -\,\mathrm{div}\,\boldsymbol{i}, \quad \text{or, if we prefer,} \quad \frac{\partial}{\partial t}(\varrho\Psi) = -\,\mathrm{div}\,\boldsymbol{i}. \tag{157.8}$$

With any such a choice of $\boldsymbol{i}$, we shall say that Ψ is *locally* conserved[2].

Finally, when the total $\varrho\Psi$ is invariant for a given material volume $\mathscr{V}$,

$$\frac{d}{dt}\int\limits_{\mathscr{V}} \varrho\Psi\, dv = 0, \tag{157.9}$$

we shall say that *the total* $\varrho\Psi$ *in* $\mathscr{V}$ *is conserved*[3]. This is consistent with our earlier statement that mass is conserved. In fact, in order that the total $\varrho\Psi$ be conserved for *every* $\mathscr{V}$, by (156.8) it is necessary and sufficient that $\dot{\Psi}=0$. The typical cases of conservation are not so universal, referring rather to bodies within boundaries of certain type or otherwise in special circumstances.

In Chap. F we shall use a somewhat different definition of "conservation law", given in Sects. 269 to 270.

158. Kinematics of diffusion in a heterogeneous medium. In order to treat motion of physical mixtures possibly undergoing chemical changes, Fick and Stefan[4] suggested that each place $\boldsymbol{x}$ may be regarded as occupied simultaneously by several different particles $\boldsymbol{X}_{\mathfrak{A}}$, $\mathfrak{A}=1, 2, \dots, \mathfrak{K}$, one for each constituent $\mathfrak{A}$. The mixture is thus represented as a *superposition* of $\mathfrak{K}$ continuous media, each of which follows its own *individual motion*

$$x^k = f^k_{\mathfrak{A}}(\boldsymbol{X}_{\mathfrak{A}}, t). \tag{158.1}$$

Henceforth media whose motion is described by (158.1) will be called *heterogeneous* if $\mathfrak{K}>1$; if $\mathfrak{K}=1$, they will be called *simple*.

For heterogeneous media, German capital subscripts are used to distinguish quantities associated with the individual motions. The *individual velocity* $\dot{\boldsymbol{x}}_{\mathfrak{A}}$ is

---

[1] Ferrari [1913, *1*, § 2].

[2] For the case $\Psi=\frac{1}{2}\dot{x}^2$, Wien [1892, *14*] discussed the form taken by certain choices of $\boldsymbol{i}$ in certain particular field theories. Cf. also Mattioli [1914, *7*].

[3] The term "conservation law" is used in a different sense by Osborn [1954, *17*], who restricts it to a formula of the type $\sum\limits_{k=1}^{n} \partial a^k/\partial x^k = 0$; he attacks the problem of determining all such laws which follow from a given set of partial differential equations.

[4] [1855, *1*]; [1871, *6*]. Cf. Hilbert [1907, *3*, pp. 43–47] [1907, *4*, pp. 42–45]. Reynolds [1903, *15*, § 35] employed a similar mathematical resolution of turbulent flow.

defined by

$$\dot{x}^k_{\mathfrak{A}} \equiv \left.\frac{\partial x^k}{\partial t}\right|_{X_{\mathfrak{A}}=\text{const}} . \tag{158.2}$$

The spatial description of the motion consists in the $\mathfrak{K}$ functional relations $\dot{\boldsymbol{x}}_{\mathfrak{A}} = \dot{\boldsymbol{x}}_{\mathfrak{A}}(\boldsymbol{x}, t)$. Each constitutent has its *individual density* $\varrho_{\mathfrak{A}}$, the total density ϱ being given by

$$\varrho \equiv \sum_{\mathfrak{A}=1}^{\mathfrak{K}} \varrho_{\mathfrak{A}} . \tag{158.3}$$

The absolute *concentration*[1] $c_{\mathfrak{A}}$ of the constituent $\mathfrak{A}$ is defined by

$$c_{\mathfrak{A}} \equiv \varrho_{\mathfrak{A}}/\varrho , \tag{158.4}$$

so that (158.3) is equivalent to

$$\sum_{\mathfrak{A}=1}^{\mathfrak{K}} c_{\mathfrak{A}} = 1 . \tag{158.5}$$

The *mean velocity* $\dot{\boldsymbol{x}}$ of the mixture is defined by the requirement that the total mass flow be the sum of the individual mass flows:

$$\varrho \dot{\boldsymbol{x}} \equiv \sum_{\mathfrak{A}=1}^{\mathfrak{K}} \varrho_{\mathfrak{A}} \dot{\boldsymbol{x}}_{\mathfrak{A}}, \quad \text{or} \quad \dot{\boldsymbol{x}} = \sum_{\mathfrak{A}=1}^{\mathfrak{K}} c_{\mathfrak{A}} \dot{\boldsymbol{x}}_{\mathfrak{A}} . \tag{158.6}$$

The *particles* $\boldsymbol{X}$ of the mixture may be defined by integrating (70.2); such particles, in general, bear no simple relation to the particles $\boldsymbol{X}_{\mathfrak{A}}$ of the constitutents.

The *diffusion velocity* or *peculiar velocity* of the constituent $\mathfrak{A}$ is its velocity relative to the mean:

$$\boldsymbol{u}_{\mathfrak{A}} \equiv \dot{\boldsymbol{x}}_{\mathfrak{A}} - \dot{\boldsymbol{x}} . \tag{158.7}$$

By (158.6) follows

$$\sum_{\mathfrak{A}=1}^{\mathfrak{K}} \varrho_{\mathfrak{A}} \boldsymbol{u}_{\mathfrak{A}} = 0, \quad \sum_{\mathfrak{A}=1}^{\mathfrak{K}} c_{\mathfrak{A}} \boldsymbol{u}_{\mathfrak{A}} = 0 . \tag{158.8}$$

That is to say, the mean velocity has been defined in such a way that the total mass flow of the diffusive motions is zero.

We now introduce two different material derivatives $\dot{\Psi}$ and $\overset{\mathfrak{A}}{\dot{\Psi}}$; the former, which coincides with that used for simple media, follows the mean motion, while the latter follows the individual motion of the constituent $\mathfrak{A}$:

$$\dot{\Psi} \equiv \frac{\partial \Psi}{\partial t} + \Psi_{,k} \dot{x}^k, \quad \overset{\mathfrak{A}}{\dot{\Psi}} = \frac{\partial \Psi}{\partial t} + \Psi_{,k} \dot{x}^k_{\mathfrak{A}} . \tag{158.9}$$

Hence

$$\overset{\mathfrak{A}}{\dot{\Psi}} - \dot{\Psi} = \Psi_{,k} u^k_{\mathfrak{A}} , \tag{158.10}$$

so that the two derivatives coincide, in the case when Ψ is a non-constant scalar, if and only if the diffusion velocity of the constituent $\mathfrak{A}$ is tangent to the surface $\Psi = \text{const}$.

From (158.7) to (158.9) we derive the following identity:

$$\left.\begin{aligned} \sum_{\mathfrak{A}=1}^{\mathfrak{K}} c_{\mathfrak{A}} \overset{\mathfrak{A}}{\overline{\tfrac{1}{2} u^2_{\mathfrak{A}}}} &= \sum_{\mathfrak{A}=1}^{\mathfrak{K}} c_{\mathfrak{A}} u_{\mathfrak{A}k} \left[\ddot{x}^k_{\mathfrak{A}} - \frac{\partial \dot{x}^k}{\partial t} - \dot{x}^k_{,m} \dot{x}^m_{\mathfrak{A}} \right], \\ &= \sum_{\mathfrak{A}=1}^{\mathfrak{K}} c_{\mathfrak{A}} u_{\mathfrak{A}k} \left[\ddot{x}^k_{\mathfrak{A}} - \dot{x}^k_{,m} (\dot{x}^m + u^m_{\mathfrak{A}}) \right], \\ &= \sum_{\mathfrak{A}=1}^{\mathfrak{K}} c_{\mathfrak{A}} u_{\mathfrak{A}k} \ddot{x}^k_{\mathfrak{A}} - \sum_{\mathfrak{A}=1}^{\mathfrak{K}} c_{\mathfrak{A}} u^k_{\mathfrak{A}} u^m_{\mathfrak{A}} d_{km} . \end{aligned}\right\} \tag{158.11}$$

[1] This usage is customary in works by physicists; chemists often call $\varrho_{\mathfrak{A}}$ the "concentration" and $c_{\mathfrak{A}}$ the "mass fraction".

For any function $\Psi_{\mathfrak{A}}$, from (158.9) it follows that

$$\left.\begin{aligned}\varrho_{\mathfrak{A}}\dot{\Psi}_{\mathfrak{A}} &= \frac{\partial}{\partial t}(\varrho_{\mathfrak{A}}\Psi_{\mathfrak{A}}) + (\varrho_{\mathfrak{A}}\Psi_{\mathfrak{A}}\dot{x}^k_{\mathfrak{A}})_{,k} - \Psi_{\mathfrak{A}}\left[\frac{\partial\varrho_{\mathfrak{A}}}{\partial t} + (\varrho_{\mathfrak{A}}\dot{x}^k_{\mathfrak{A}})_{,k}\right],\\ &= \frac{\partial}{\partial t}(\varrho_{\mathfrak{A}}\Psi_{\mathfrak{A}}) + (\varrho_{\mathfrak{A}}\Psi_{\mathfrak{A}}\dot{x}^k)_{,k} + (\varrho_{\mathfrak{A}}\Psi_{\mathfrak{A}}u^k_{\mathfrak{A}})_{,k} - \Psi_{\mathfrak{A}}\left[\frac{\partial\varrho_{\mathfrak{A}}}{\partial t} + (\varrho_{\mathfrak{A}}\dot{x}^k_{\mathfrak{A}})_{,k}\right].\end{aligned}\right\} \quad (158.12)$$

We set

$$\Psi \equiv \sum_{\mathfrak{A}=1}^{\mathfrak{K}} c_{\mathfrak{A}}\Psi_{\mathfrak{A}} \qquad (158.13)$$

and by summing (158.12) from $\mathfrak{A}=1$ to $\mathfrak{A}=\mathfrak{K}$ obtain the following ***fundamental identity***[1]:

$$\left.\begin{aligned}\sum_{\mathfrak{A}=1}^{\mathfrak{K}} c_{\mathfrak{A}}\dot{\Psi}_{\mathfrak{A}} &= \dot{\Psi} + \frac{\Psi}{\varrho}\left[\frac{\partial\varrho}{\partial t} + (\varrho\dot{x}^k)_{,k}\right] + \frac{1}{\varrho}\sum_{\mathfrak{A}=1}^{\mathfrak{K}}(\varrho_{\mathfrak{A}}\Psi_{\mathfrak{A}}u^k_{\mathfrak{A}})_{,k} - \\ &\quad - \sum_{\mathfrak{A}=1}^{\mathfrak{K}}\frac{\Psi_{\mathfrak{A}}}{\varrho}\left[\frac{\partial\varrho_{\mathfrak{A}}}{\partial t} + (\varrho_{\mathfrak{A}}\dot{x}^k_{\mathfrak{A}})_{,k}\right].\end{aligned}\right\} \quad (158.14)$$

Upon this identity, which relates the material derivative of the mean value (158.13) to the mean value of the material derivatives, are founded all our proofs of equations of balance in a heterogeneous medium.

In the foregoing analysis and in our later developments concerning heterogeneous media we employ the definition (158.6), according to which the mean motion of the mixture is regarded as the velocity of the center of mass of the several constituents (cf. Sect. 165 below). This mean velocity $\dot{\boldsymbol{x}}$ may be called the *barycentric velocity*. A more general mean velocity $\tilde{\boldsymbol{x}}$ is defined by[2]

$$\tilde{\boldsymbol{x}} \equiv \frac{\sum_{\mathfrak{A}=1}^{\mathfrak{K}} w_{\mathfrak{A}}\dot{\boldsymbol{x}}_{\mathfrak{A}}}{\sum_{\mathfrak{A}=1}^{\mathfrak{K}} w_{\mathfrak{A}}}, \qquad (158.15)$$

where the weight factors $w_{\mathfrak{A}}$ are any real numbers such that the denominator is not zero. The corresponding diffusion velocity is defined by

$$\tilde{\boldsymbol{u}}_{\mathfrak{A}} \equiv \dot{\boldsymbol{x}}_{\mathfrak{A}} - \tilde{\boldsymbol{x}} = \boldsymbol{u}_{\mathfrak{A}} + \dot{\boldsymbol{x}} - \tilde{\boldsymbol{x}}. \qquad (158.16)$$

Formulae related to the motion of the constituents are generally more complicated when expressed in terms of (158.15) and (158.16) rather than $\dot{\boldsymbol{x}}$ and $\boldsymbol{u}_{\mathfrak{A}}$. For example, we have

$$\sum_{\mathfrak{A}=1}^{\mathfrak{K}} \tilde{u}^k_{\mathfrak{A}} g_{\mathfrak{A}k} = \sum_{\mathfrak{A}=1}^{\mathfrak{K}} u^k_{\mathfrak{A}} g_{\mathfrak{A}k} + (\dot{x}^k - \tilde{x}^k)\sum_{\mathfrak{A}=1}^{\mathfrak{K}} g_{\mathfrak{A}k}. \qquad (158.17)$$

Hence in order that

$$\sum_{\mathfrak{A}=1}^{\mathfrak{K}} \tilde{u}^k_{\mathfrak{A}} g_{\mathfrak{A}k} = \sum_{\mathfrak{A}=1}^{\mathfrak{K}} u^k_{\mathfrak{A}} g_{\mathfrak{A}k}, \qquad (158.18)$$

it is necessary and sufficient that $\sum_{\mathfrak{A}=1}^{\mathfrak{K}} \boldsymbol{g}_{\mathfrak{A}}$ be orthogonal to $\dot{\boldsymbol{x}} - \tilde{\boldsymbol{x}}$; in particular, it is sufficient that

$$\sum_{\mathfrak{A}=1}^{\mathfrak{K}} \boldsymbol{g}_{\mathfrak{A}} = 0. \qquad (158.19)$$

The condition (158.19) is of interest in that it restricts the quantities $\boldsymbol{g}_{\mathfrak{A}}$, not the definition of $\tilde{\boldsymbol{x}}$, and in consequence holds for any $\tilde{\boldsymbol{x}}$. Moreover, if the $\boldsymbol{g}_{\mathfrak{A}}$ are given, (158.17) shows that

---

[1] TRUESDELL [1957, *16*, § 3]. The reader will recognize a certain analogy to MAXWELL'S equation of transfer in the kinetic theory of gases. The present identity, however, is purely kinematical; not only does it embody no special hypotheses regarding the material, but also it is independent even of the general conservation principles of mechanics.

[2] DE GROOT [1952, *3*, §§ 47–48]. Cf. also PRIGOGINE [1947, *12*, Chap. VII, § 3].

there are infinitely many definitions of mean velocity such that (158.18) holds, since all that is necessary is to set $\tilde{\boldsymbol{x}}$ equal to the sum of $\dot{\boldsymbol{x}}$ and any vector orthogonal to $\sum_{\mathfrak{A}=1}^{\mathfrak{K}} \boldsymbol{g}_{\mathfrak{A}}$.

159. Conservation of mass in a heterogeneous medium. In the course of time, mass may be exchanged among the constituents, but the total mass is conserved. The formal structure expressing the balance of the several masses has been worked out by many authors[1]. The *mass supply* for the constituent $\mathfrak{A}$ is the pure rate $\hat{c}_{\mathfrak{A}}$ defined by[2]

$$\varrho\,\hat{c}_{\mathfrak{A}} \equiv \frac{\partial \varrho_{\mathfrak{A}}}{\partial t} + (\varrho_{\mathfrak{A}}\,\dot{x}^k_{\mathfrak{A}})_{,k}. \tag{159.1}$$

From (158.9) follows

$$\varrho\,\dot{c}_{\mathfrak{A}} = \varrho\,\hat{c}_{\mathfrak{A}} - c_{\mathfrak{A}}(\varrho\,u^k_{\mathfrak{A}})_{,k}, \qquad \varrho\,\dot{c}_{\mathfrak{A}} = \varrho\,\hat{c}_{\mathfrak{A}} - (\varrho_{\mathfrak{A}}\,u^k_{\mathfrak{A}})_{,k}. \tag{159.2}$$

These relations show that only diffusion or the creation of mass, not convection, can alter the concentrations at $\boldsymbol{X}$ or at $\boldsymbol{X}_{\mathfrak{A}}$. In an *inert mixture*, Eqs. (159.2) are satisfied by the vanishing of each term: $\hat{c}_{\mathfrak{A}}=0$, $\boldsymbol{u}_{\mathfrak{A}}=0$, and $c_{\mathfrak{A}}=c_{\mathfrak{A}}(\boldsymbol{X})$. That is, each constituent flows with the same velocity, and the concentration of each constituent remains constant for each particle of the mean motion, and thus such a particle may be regarded as a perpetual union of fixed particles from each constituent.

Returning to general motions, in (158.13) and (158.14) put $\Psi_{\mathfrak{A}}=1$. By (158.5) we thus obtain the identity

$$\frac{\partial \varrho}{\partial t} + (\varrho\,\dot{x}^k)_{,k} = \varrho \sum_{\mathfrak{A}=1}^{\mathfrak{K}} \hat{c}_{\mathfrak{A}}, \tag{159.3}$$

whence we read off the following **theorem**: *In order that the mean motion of a heterogeneous medium obey the ordinary equation of continuity* (156.5)$_1$, *it is necessary and sufficient that*

$$\sum_{\mathfrak{A}=1}^{\mathfrak{K}} \hat{c}_{\mathfrak{A}} = 0. \tag{159.4}$$

The condition (159.4) asserts that the total mass supply is zero. Thus the ordinary equation of continuity and (159.4) are equivalent statements of the *conservation of total mass in the mixture.*

When we adopt (159.4), as we do henceforth, we may use it and (159.1) to reduce the fundamental identity (158.14) to the following simpler form:

$$\sum_{\mathfrak{A}=1}^{\mathfrak{K}} c_{\mathfrak{A}}\,\dot{\Psi}_{\mathfrak{A}} = \dot{\Psi} + \frac{1}{\varrho}\sum_{\mathfrak{A}=1}^{\mathfrak{K}} (\varrho_{\mathfrak{A}}\,\Psi_{\mathfrak{A}}\,u^k_{\mathfrak{A}})_{,k} - \sum_{\mathfrak{A}=1}^{\mathfrak{K}} \hat{c}_{\mathfrak{A}}\,\Psi_{\mathfrak{A}}. \tag{159.5}$$

Putting $\Psi_{\mathfrak{A}}=\dot{x}^k_{\mathfrak{A}}$ in (159.5), from (158.6) we derive the relation

$$\sum_{\mathfrak{A}=1}^{\mathfrak{K}} c_{\mathfrak{A}}\,\ddot{x}^k_{\mathfrak{A}} = \ddot{x}^k + \frac{1}{\varrho}\sum_{\mathfrak{A}=1}^{\mathfrak{K}} (\varrho_{\mathfrak{A}}\,u^k_{\mathfrak{A}}\,u^m_{\mathfrak{A}})_{,m} - \sum_{\mathfrak{A}=1}^{\mathfrak{K}} \hat{c}_{\mathfrak{A}}\,u^k_{\mathfrak{A}}, \tag{159.6}$$

connecting the acceleration of the mean motion with the mean of the individual accelerations. That nothing simpler than (159.6) is to be expected follows from the remark put after (158.6).

---

[1] JAUMANN [1911, *7*, §§ I, VIII], LOHR [1917, *5*, §§ 5, 7–8, 12, 15] MEISSNER [1938, *7*, § III], BATEMAN [1939, *1*], ECKART [1940, *8*, p. 271] MEIXNER [1941, *2*, § 3], VERSCHAFFELT [1942, *13*, §§ 12–14] [1942, *14*, §§ 4–5] [1951, *38*, §§ 7–13]. We follow the argument as arranged by TRUESDELL [1957, *16*, § 4].

[2] REYNOLDS [1903, *15*, §§ 13, 21, 36], JAUMANN [1911, *7*, §§ IV, VIII], HEUN [1913, *4*, § 24c].

159A. Appendix. Chemical interpretation for the balance of mass in a heterogeneous medium[1]. For definiteness, we shall employ freely the terms *atomic* and *compound*. These names, however, serve only as a conceptual guide, and are not to be confused with the units occurring in discrete models of matter. For us, *atomic* will denote a constituent whose mass is indestructible; *compound*, a constituent whose mass may vary with circumstances and is the sum of masses of constituents whose mass is permanent. The formal structure we present is easily adapted to a great variety of different physical situations: to chemical reactions in which atoms unite to form molecules, to the dissociation of a pure substance into ions variously distinguished—in short, to any phenomenon governed by simple laws of combination and separation.

Of the $\mathfrak{K}$ constituents, let the first $\mathfrak{P}$ be atomic; the remainder, compound. Then the *atomic weights* and *molecular weights* $M_{\mathfrak{A}}$ satisfy

$$M_{\mathfrak{A}} = \sum_{\mathfrak{B}=1}^{\mathfrak{P}} n_{\mathfrak{A}\mathfrak{B}} M_{\mathfrak{B}}, \quad \mathfrak{A} = \mathfrak{P}+1, \mathfrak{P}+2, \ldots, \mathfrak{K}, \tag{159A.1}$$

where, in the strictly chemical interpretation, $n_{\mathfrak{A}\mathfrak{B}}$ is a non-negative integer expressing the number of atoms of the atomic substance $\mathfrak{B}$ in one molecule of the compound $\mathfrak{A}$. In the formal analysis, no use is made of this restriction, which may be dropped for more general interpretations. If we set $n_{\mathfrak{A}\mathfrak{B}} \equiv \delta_{\mathfrak{A}\mathfrak{B}}$ when $\mathfrak{A} \leq \mathfrak{P}$, while $n_{\mathfrak{A}\mathfrak{B}} \equiv 0$ when $\mathfrak{B} > \mathfrak{P}$, we may replace (159A.1) by

$$M_{\mathfrak{A}} = \sum_{\mathfrak{B}=1}^{\mathfrak{K}} n_{\mathfrak{A}\mathfrak{B}} M_{\mathfrak{B}}, \quad \mathfrak{A} = 1, 2, \ldots, \mathfrak{K}. \tag{159A.2}$$

The indestructibility of the atomic substances is expressed by the postulate

$$\sum_{\mathfrak{A}=1}^{\mathfrak{K}} n_{\mathfrak{A}\mathfrak{B}} \hat{c}_{\mathfrak{A}} / M_{\mathfrak{A}} = 0, \tag{159A.3}$$

$\hat{c}_{\mathfrak{A}}$ being defined by (159.1). If we multiply this relation by $M_{\mathfrak{B}}$ and sum on $\mathfrak{B}$, by (159A.2) we derive (159.4), which we have already seen to be equivalent to the conservation of total mass.

The postulate (159A.3) implies a pregnant identity connecting an arbitrary function $g_{\mathfrak{A}}$ to $g_{\mathfrak{A}}^*$, the difference between $g_{\mathfrak{A}}$ and its equivalent in atomic substances:

$$g_{\mathfrak{A}}^* \equiv g_{\mathfrak{A}} - \sum_{\mathfrak{B}=1}^{\mathfrak{K}} n_{\mathfrak{A}\mathfrak{B}} \frac{M_{\mathfrak{B}}}{M_{\mathfrak{A}}} g_{\mathfrak{B}}, \tag{159A.4}$$

implying that $g_{\mathfrak{A}}^* = 0$ for $\mathfrak{A} \leq \mathfrak{P}$. Then from (159A.3) follows

$$\sum_{\mathfrak{A}=1}^{\mathfrak{K}} \hat{c}_{\mathfrak{A}} g_{\mathfrak{A}} = \sum_{\mathfrak{A}=\mathfrak{P}+1}^{\mathfrak{K}} \hat{c}_{\mathfrak{A}} g_{\mathfrak{A}}^*. \tag{159A.5}$$

This identity asserts that the total rate of production of the quantities $g_{\mathfrak{A}}$ through creation of mass equals the total rate of production of the quantities $g_{\mathfrak{A}}^*$ for the compound substances only. The case $g_{\mathfrak{A}} = 1$ again yields (159.4).

In the study of chemical reactions, it is customary to divide the mass supply $\hat{c}_{\mathfrak{A}}$ of the constituent $\mathfrak{A}$ into portions contributed by the $\mathfrak{k}$ reactions taking place[2]:

$$\varrho \hat{c}_{\mathfrak{A}} = \sum_{\mathfrak{a}=1}^{\mathfrak{k}} N_{\mathfrak{A}\mathfrak{a}} J_{\mathfrak{a}}, \tag{159A.6}$$

where $J_{\mathfrak{a}}$ is the *reaction rate* of the $\mathfrak{a}$-th reaction, and where the pure numbers $N_{\mathfrak{A}\mathfrak{a}}$ may be called *stoechiometric coefficients*. The reaction rates are assumed to be variable independ-

---

[1] Not attempting to trace the origin of the ideas used here in the early chemical literature, we follow ECKART [1940, *8*]. In a celebrated work which has given rise to the enormous literature on the phenomenological theory of molecular relaxation, EINSTEIN [1920, *1*] applied this structure to study reaction rates in a chemically pure gas, some of whose molecules are dissociated. Unfortunately the loose language in many of the more recent studies of relaxation phenomena might mislead the unwary reader into regarding the analysis as belonging to the kinetic theory, while in fact almost all of it employs field concepts, even if not very accurately.

[2] DE DONDER [1927, *3*, pp. 10–11] [1938, *3*]. The reaction rates are often written in the form $J_{\mathfrak{a}} = d\xi_{\mathfrak{a}}/dt$, where $\xi_{\mathfrak{a}}$ is called "the degree of advancement" of the $\mathfrak{a}$-th reaction. RAW and YOURGRAU [1956, *18*] propose to call $dJ_{\mathfrak{a}}/dt$ the "acceleration" of the $\mathfrak{a}$-th reaction.

ently. Hence the condition (159.4) for conservation of total mass in the mixture becomes

$$\sum_{\mathfrak{A}=1}^{\mathfrak{K}} N_{\mathfrak{A}\mathfrak{a}} = 0, \qquad \mathfrak{a} = 1, 2, \ldots, \mathfrak{k}. \tag{159A.7}$$

If we replace (159.4) by the stronger condition (159A.3), then (159A.7) in turn is to be replaced by

$$\sum_{\mathfrak{A}=1}^{\mathfrak{K}} n_{\mathfrak{A}\mathfrak{B}} N_{\mathfrak{A}\mathfrak{a}}/M_{\mathfrak{a}} = 0, \qquad \mathfrak{a} = 1, 2, \ldots, \mathfrak{k}. \tag{159A.8}$$

b) Solution of the equation of continuity.

160. The need for a solution of the equation of continuity. The spatial equation $(156.5)_1$ is in effect the result of differentiating the material equation (156.1) with respect to time[1]. Thus (156.1) is the general solution of the spatial equation of continuity[2]. However, (156.1) is expressed within the material description, and in order to render it explicit we must know the motion. A spatial form, for application of which the motion itself need not be given, is preferable[3].

161. Plane motion and related cases. We begin with the special case of an *isochoric plane or pseudo-plane motion*, defined by (139.5) or (139.6). Then $(156.5)_2$ becomes

$$\frac{\partial \dot{x}}{\partial x} + \frac{\partial \dot{y}}{\partial y} = 0. \tag{161.1}$$

This equation is a necessary and sufficient condition that there exist a function $Q(x, y, z, t)$ such that

$$\dot{x} = -\frac{\partial Q}{\partial y}, \qquad \dot{y} = \frac{\partial Q}{\partial x}. \tag{161.2}$$

Since $dQ = \dot{y}\, dx - \dot{x}\, dy$, the curves $Q = \text{const}$ satisfy (70.4) and hence are stream lines. The function Q is ***D'Alembert's stream function***[4]. To interpret it, let $\boldsymbol{x}$ vary along a curve c in the x-y plane. The vector $d\boldsymbol{x}_n = -\boldsymbol{i}\, dy + \boldsymbol{j}\, dx$ is normal to c and oriented by the right-hand rule. Since $dQ = \dot{y}\, dx - \dot{x}\, dy = \dot{\boldsymbol{x}} \cdot d\boldsymbol{x}_n$, it follows that for two points $\boldsymbol{x}_1$ and $\boldsymbol{x}_2$ lying in a simply connected region in the same plane $z = \text{const}$, the difference $Q(\boldsymbol{x}_1, t) - Q(\boldsymbol{x}_2, t)$ is the flow per unit height across any curve c connecting them[5]. The independence of the flow from the choice of the curve c is a statement that the motion is isochoric.

The classical argument just given establishes the single-valuedness of the stream function Q in a simply connected region, but in fact the result holds in much greater generality[6]. All that is required is to show that the line integral $\int_{\boldsymbol{x}_1}^{\boldsymbol{x}_2} \dot{\boldsymbol{x}} \cdot d\boldsymbol{x}_n$, or $\int_{\boldsymbol{x}_1}^{\boldsymbol{x}_2} \dot{x}_n\, d x_n$, is independent of path for any pair of points $\boldsymbol{x}_1, \boldsymbol{x}_2$

[1] A formal proof was given by LAGRANGE [1762, *3*, § 51].

[2] A formal proof was given by EULER [1770, *1*, § 112].

[3] Most of the stream functions whose properties are presented in the following sections are included in the listing of KRZYWOBLOCKI [1958, *6*]. Interpretations of the equation of continuity or of similar relations satisfied by the acceleration as equalities among certain areas or volumes have been constructed by POMPEIU [1929, *7*], VALCOVICI [1933, *12*], JACOB [1944, *6*, § 3], CARSTOIU [1944, *1*] [1948, *4*] [1954, *1*, §§ 9–12], BILIMOVITCH [1948, *2*] [1953, *1*], and TRUESDELL [1950, *34*].

[4] [1761, *1*, § 12]. A more general type of stream function for pseudo-plane motions had been introduced by EULER [1757, *3*, § 62], but he did not note the specially simple properties of Q.

[5] RANKINE [1864, *2*, § 2].

[6] NOLL [1957, *12*].

interior to the motion. For this, it is sufficient that *the boundary consist in a finite number of closed curves moving rigidly*, and, in the case of an infinite region, that there be *no flux out of the portion of the region of flow that lies within a sufficiently large circle*. To prove this, we begin noting that the condition satisfied upon the finite boundaries is (69.1); since v_n is the normal component of the velocity of a rigid motion, we have $\oint_{\ell} v_n \, d x_{\boldsymbol{n}} = 0$ for each closed rigid bounding curve ℓ. Let c be any simple circuit, not necessarily reducible, in the interior. Then the regions of flow interior and exterior to c are bounded by c and by a curve b, finite or infinite, such that $\oint_{b} \dot{x}_n \, d x_{\boldsymbol{n}} = 0$. Since the condition of isochoric motion is

$$0 = \oint_{c} \dot{x}_n \, d x_{\boldsymbol{n}} - \oint_{b} \dot{x}_n \, d x_{\boldsymbol{n}}, \tag{161.3}$$

it follows that

$$\oint_{c} \dot{x}_n \, d x_{\boldsymbol{n}} = 0 \tag{161.4}$$

for all circuits c, whether reducible or not. Q.E.D.

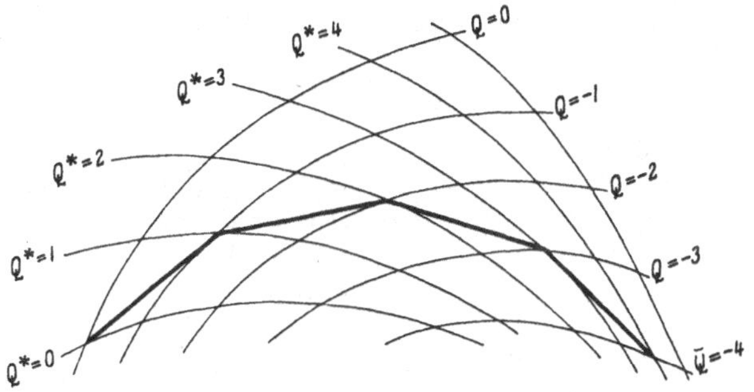

Fig. 21. MAXWELL's construction for stream lines.

Any family of plane curves $Q(x, y, t) = \text{const}$ may be the stream lines of a plane isochoric motion. Given the stream function Q and Q^* of two such motions, $Q + Q^*$ is thus also a stream function; if we sketch the stream lines $Q = 0, \pm 1, \pm 2, \ldots$ and $Q^* = 0, \pm 1, \pm 2, \ldots$, in any system of units, stream lines corresponding to $Q + Q^*$ may be sketched approximately by connecting appropriate points of intersection of the two sets $Q = \text{const}$ and $Q^* = \text{const}$ (Fig. 21)[1].

From (161.2) we have at once[2]

$$\nabla^2 Q = w_z, \tag{161.5}$$

where $\nabla^2 = \partial^2/\partial x^2 + \partial^2/\partial y^2$. Hence in an irrotational motion Q is a plane harmonic function[3]:

$$\nabla^2 Q = 0. \tag{161.6}$$

More generally, D'ALEMBERT's vorticity theorem (Sect. 133) asserts that a necessary and sufficient condition for a plane isochoric motion to be circulation preserving is that w be constant for each particle. From (161.5) we see that an equivalent statement is[4]

$$\dot{\overline{\nabla^2 Q}} = 0; \tag{161.7}$$

[1] According to RANKINE [1864, *2*, Appendix], this observation is due to MAXWELL.
[2] This was implied by D'ALEMBERT [1761, *1*, § 12].
[3] D'ALEMBERT [1768, *1*, § II, ¶ 7].
[4] This result has been misinterpreted by MEISSEL [1855, *3*].

in the steady case, w is constant on each stream line, and equivalent forms of (161.7) are[1]

$$\nabla^2 Q = f(Q), \qquad \frac{\partial(\nabla^2 Q, Q)}{\partial(x, y)} = 0. \tag{161.8}$$

Now we seek *a necessary and sufficient condition that a given family of curves* $F(x, y) = \text{const}$ *be possible stream lines of a steady plane isochoric circulation preserving motion.* First, this is equivalent to the existence of a function Q that satisfies $(161.8)_2$ and also has the property $Q = Q(F)$. That is

$$Q''(F)\,\frac{\partial(F, |\operatorname{grad} F|^2)}{\partial(x, y)} + Q'(F)\,\frac{\partial(F, \nabla^2 F)}{\partial(x, y)} = 0. \tag{161.9}$$

Now Q''/Q' is a function of F; hence it is constant on each stream line. For such a Q to exist, then, it is necessary and sufficient that

$$\frac{\partial(F, \nabla^2 F)}{\partial(x, y)} \Big/ \frac{\partial(F, |\operatorname{grad} F|^2)}{\partial(x, y)} = G(F). \tag{161.10}$$

This criterion is due to STOKES[2]. For the irrotational special case, (161.10) is to be replaced by

$$\nabla^2 F / |\operatorname{grad} F|^2 = G(F). \tag{161.11}$$

The condition (161.8) results from elimination of w from (161.5) by means of the condition of steady plane circulation preserving motion. As was noticed by HILL[3] we may instead eliminate Q, even in the unsteady case. The condition $\dot{w} = 0$ may be written

$$\frac{\partial Q}{\partial x}\frac{\partial w}{\partial y} - \frac{\partial Q}{\partial y}\frac{\partial w}{\partial x} + \frac{\partial w}{\partial t} = 0. \tag{161.12}$$

If $w = w(x, y, t)$ is known, this is a linear partial differential equation for Q; its characteristics satisfy

$$\frac{dx}{\frac{\partial w}{\partial y}} = \frac{dy}{-\frac{\partial w}{\partial x}} = \frac{dQ}{\frac{\partial w}{\partial t}}. \tag{161.13}$$

Therefore the characteristic curves are given by

$$w(x, y, t) = \text{const}, \qquad Q - \int \frac{\frac{\partial w}{\partial t}}{\frac{\partial w}{\partial y}}\,dx = \text{const}, \tag{161.14}$$

where in the second equation y is supposed eliminated by aid of the first. The general solution of (161.12) is

$$F\left(w,\ Q - \int \frac{\frac{\partial w}{\partial t}}{\frac{\partial w}{\partial y}}\,dx,\ t\right) = 0, \tag{161.15}$$

or

$$Q = \int \frac{\frac{\partial w}{\partial t}}{\frac{\partial w}{\partial y}}\,dx + G(w, t). \tag{161.16}$$

---

[1] LAGRANGE [1783, *1*, § 21], STOKES [1842, *4*].

[2] [1842, *4*].

[3] [1885, *2*, § I].

Substitution in (161.5) yields

$$\Delta^2\left[\int \frac{\frac{\partial w}{\partial t}}{\frac{\partial w}{\partial y}}\, d x + G(w, t)\right] = w, \tag{161.17}$$

a differential functional equation for w alone.

For a *steady plane motion with steady density*, $(156.5)_1$ becomes

$$\frac{\partial}{\partial x}(\varrho \dot{x}) + \frac{\partial}{\partial y}(\varrho \dot{y}) = 0. \tag{161.18}$$

Hence there exists a function Q' such that[1]

$$\varrho \dot{x} = -\frac{\partial Q'}{\partial y}, \qquad \varrho \dot{y} = \frac{\partial Q'}{\partial x}. \tag{161.19}$$

The interpretation of Q' as a measure of mass flow is immediate from the interpretation of Q as a measure of flow. Q' is related to the vorticity through the equation

$$\Delta^2 Q' = \varrho w + \frac{\partial Q'}{\partial x}\frac{\partial \log \varrho}{\partial x} + \frac{\partial Q'}{\partial y}\frac{\partial \log \varrho}{\partial y}. \tag{161.20}$$

Neither Q nor Q' is a generalization of the other. In a steady plane homochoric motion, both Q and Q' exist, and $Q' = \varrho Q$. However, in a steady plane motion in which the density is not uniform, while both Q' and Q exist, they may be related by an arbitrary functional relation $F(Q, Q') = 0$. For example, in a motion such that $\dot{x} = V = \text{const}$, $\dot{y} = 0$, $\varrho = \varrho(y)$, we have $Q = -Vy$, $Q' = -V\int_0^y \varrho(u)\, du$.

For *pseudo-lineal motions*, defined by (139.2) and (139.3), the continuity equation $(156.5)_1$ becomes

$$\frac{\partial \varrho}{\partial t} + \frac{\partial}{\partial x}(\varrho \dot{x}) = 0. \tag{161.21}$$

Hence there exists a function Q such that[2]

$$\varrho = -\frac{\partial Q}{\partial x}, \qquad \varrho \dot{x} = \frac{\partial Q}{\partial t}. \tag{161.22}$$

162. Stream functions for two-dimensional motions. The equation of continuity (156.5) may be written in the explicit forms

$$\sqrt{g}\,\dot{\varrho} + \varrho \frac{\partial}{\partial x^k}(\sqrt{g}\,\dot{x}^k) = 0, \qquad \sqrt{g}\,\frac{\partial \varrho}{\partial t} + \frac{\partial}{\partial x^k}(\sqrt{g}\,\varrho \dot{x}^k) = 0. \tag{162.1}$$

These suggest various extensions of the results given in the foregoing section.

In the isochoric case and in the case of steady motion with steady density, respectively, suppose

$$\frac{\partial}{\partial x^3}(\sqrt{g}\,\dot{x}^3) = 0, \qquad \frac{\partial}{\partial x^3}(\sqrt{g}\,\varrho \dot{x}^3) = 0. \tag{162.2}$$

---

[1] Crocco [1936, *3*]. We do not follow the analysis of Birkeland [1916, *1*], who claims to derive a result of the form (161.2) even when the expansion is not zero.

[2] W. Kirchhoff [1930, *2*]. An equivalent result had been given by Euler [1757, *3*, §§ 48–49].

These conditions are necessary and sufficient for the existence of *stream functions* $Q(x^1, x^2, x^3, t)$ and $Q'(x^1, x^2, x^3, t)$ such that

$$\sqrt{g}\,\dot{x}^1 = -\frac{\partial Q}{\partial x^2}, \quad \sqrt{g}\,\dot{x}^2 = \frac{\partial Q}{\partial x^1}; \quad \sqrt{g}\,\varrho\,\dot{x}^1 = -\frac{\partial Q'}{\partial x^2}, \quad \sqrt{g}\,\varrho\,\dot{x}^2 = \frac{\partial Q'}{\partial x^1}, \tag{162.3}$$

respectively[1]. The factor $\sqrt{g}$ in general depends upon x^3 as well well as on x^1 and x^2. In the special case when Q or Q' does not depend upon x^3, knowledge of the motion in any one surface $x^3=$const suffices for construction of the motion in any other surface $x^3=$const, since the factor $\sqrt{g}$ is specified by the choice of co-ordinates. In this case, let $F=F(x^1, x^2, t)$, and we have by (162.3)

$$-\dot{x}^k F_{,k} = \frac{1}{\sqrt{g}} \frac{\partial(F, Q)}{\partial(x^1, x^2)} \quad \text{or} \quad \frac{1}{\varrho\sqrt{g}} \frac{\partial(F, Q')}{\partial(x^1, x^2)}. \tag{162.4}$$

Hence the curves $F=$const, $x^3=$const are stream lines if and only if $F=F(Q)$ or $F=F(Q')$, respectively. In particular, the curves $x^3=$const, $Q=$const or $Q'=$const are stream lines.

The simplest assumption leading to (162.2) is $\dot{x}^3=0$. In this case, a particle once upon a given surface $x^3=$const remains ever upon it. Such motions are strictly two-dimensional and may be approached alternatively as motions in a curved space of two dimensions. If the co-ordinates are chosen so that $g_{3k}=\delta_{3k}$, then, since $\sqrt{g}\,w^3 = \dot{x}_{2,1} - \dot{x}_{1,2}$, we have, respectively,

$$w^3 = g^{km} Q_{,km} \quad \text{or} \quad g^{km}\left(\frac{1}{\varrho} Q'_{,k}\right)_{,m}. \tag{162.5}$$

When the surfaces $x^3=$const are *parallel surfaces*, the interpretation is particularly simple. For example, if they are concentric spheres of radius r, we may take x^1, x^2, x^3 as the polar co-ordinates θ, ϕ, r; then $(162.3)_1$ and $(162.3)_2$ become

$$r \sin\theta\, \dot{x}\langle\theta\rangle = -\frac{\partial Q}{\partial \phi}, \quad r\,\dot{x}\langle\phi\rangle = \frac{\partial Q}{\partial \theta}, \tag{162.6}$$

where $\dot{x}\langle\theta\rangle$ and $\dot{x}\langle\phi\rangle$ are the physical components of velocity. This case was considered in detail by ZERMELO.

In a *rotationally symmetric motion* (Sect. 68) the surfaces $x^3=$const are planes, and we may take x^1, x^2, x^3 as the cylindrical co-ordinates r, z, θ. Then $(162.3)_1$ and $(162.3)_2$ become[2]

$$r\dot{r} = -\frac{\partial Q}{\partial z}, \quad r\dot{z} = \frac{\partial Q}{\partial r}; \tag{162.7}$$

an explicit form for $(162.5)_1$ in the rotationally symmetric case is

$$\frac{\partial^2 Q}{\partial r^2} - \frac{1}{r}\frac{\partial Q}{\partial r} + \frac{\partial^2 Q}{\partial z^2} = r^2 w^3 = r\,w\langle 3\rangle. \tag{162.8}$$

By SVANBERG's vorticity theorem (Sect. 133), it follows from (162.8) that a necessary and sufficient condition for a steady isochoric rotationally symmetric motion to be circulation preserving is that Q satisfy

$$\Sigma[Q] \equiv \frac{1}{r^2}\left[\frac{\partial^2 Q}{\partial r^2} - \frac{1}{r}\frac{\partial Q}{\partial r} + \frac{\partial^2 Q}{\partial z^2}\right] = f(Q); \tag{162.9}$$

---

[1] In essence, we follow ZERMELO [1902, *10*, Chap. I, §§ 2–4], who used the intrinsic two-dimensional approach mentioned below. In this spirit ZERMELO obtained two-dimensional counterparts of some of the classical theorems on potential motion and circulation preserving motion. Cf. SBRANA [1934, *6*].

[2] STOKES [1842, *4*]. Cf. BASSET [1888, *1*, § 306], SAMPSON [1891, *5*]. The results that follow by specialization of $(162.3)_3$ and $(162.3)_4$ in this case were noted by CROCCO [1936, *3*]. Formulae appropriate to a rotating cylindrical polar system were obtained by v. MISES [1909, *8*, § 4.4].

that is,

$$\frac{\partial(\Sigma[Q], Q)}{\partial(r, z)} = 0. \tag{162.10}$$

If $Q = Q(F)$, substitution into (162.10) yields

$$Q' \frac{\partial(\Sigma[F], F)}{\partial(r, z)} + Q'' \frac{\partial(|\operatorname{grad} F|^2 r^{-2}, F)}{\partial(r, z)} = 0. \tag{162.11}$$

Hence the curves $F = \text{const}$ may be stream lines of a steady isochoric rotationally symmetric circulation preserving motion if and only if

$$\frac{\partial(F, \Sigma[F])}{\partial(r, z)} \Big/ \frac{\partial(F, |\operatorname{grad} F|^2 r^{-2})}{\partial(r, z)} = G(F) \tag{162.12}$$

along them[1].

163. Stream functions for certain three-dimensional motions. If the motion is isochoric or steady, the field $\dot{\boldsymbol{p}}$ or the field $\varrho\dot{\boldsymbol{p}}$ is solenoidal; hence by (App. 32.7) and (App. 32.9) we have in the former case

$$\dot{\boldsymbol{p}} = C(A, B) \operatorname{grad} A \times \operatorname{grad} B = \operatorname{curl} \boldsymbol{v}; \tag{163.1}$$

in the latter

$$\varrho\dot{\boldsymbol{p}} = C'(A', B') \operatorname{grad} A' \times \operatorname{grad} B' = \operatorname{curl} \boldsymbol{v}'. \tag{163.2}$$

The reader may verify that all the foregoing results in this part of the subchapter except (161.22) are included as special cases of these four formulae[2].

In EULER's representation $(163.1)_1$ the potentials A and B may be selected as any independent functions such that the surfaces $A = \text{const}$ and $B = \text{const}$ are stream surfaces[3]; a similar statement may be made about $(163.2)_1$. By suitable parametrization of these surfaces, we may take $C = 1$.

A kinematical interpretation analogous to that in Sect. 161 is then easily obtained[4], for if $\boldsymbol{x} = \boldsymbol{x}(\boldsymbol{v})$ is a parametric representation of a closed surface $\mathfrak{s}$, for the mass flux out of $\mathfrak{s}$ we find from $(163.2)_1$

$$\left.\begin{aligned} \oint_{\mathfrak{s}} \varrho\dot{\boldsymbol{p}} \cdot d\boldsymbol{a} &= \oint_{\mathfrak{s}} (\operatorname{grad} A' \times \operatorname{grad} B') \cdot \left(\frac{\partial \boldsymbol{x}}{\partial v^1} \times \frac{\partial \boldsymbol{x}}{\partial v^2}\right) dv^1\, dv^2, \\ &= \oint_{\mathfrak{s}} \frac{\partial(A', B')}{\partial(v^1, v^2)}\, dv^1\, dv^2, \\ &= \int_{\mathfrak{s}'} dA'\, dB', \end{aligned}\right\} \tag{163.3}$$

where it is assumed that it is possible to select the area $\mathfrak{s}'$ in a plane with co-ordinates A', B' such that $\mathfrak{s}'$ is mapped onto $\mathfrak{s}$ by the transformation $A' = A'(\boldsymbol{x}(\boldsymbol{v}))$, $B' = B'(\boldsymbol{x}(\boldsymbol{v}))$. Under these assumptions, then, we may decussate space into tubes bounded by the surfaces $A' = 0, \pm 1, \pm 2, \ldots$, $B' = 0, \pm 1, \pm 2, \ldots$, and the number of these unit tubes crossing $\mathfrak{s}$ equals the mass flux out of $\mathfrak{s}$.

The representations $(163.1)_1$ and $(163.2)_1$ are awkward to use because they are non-linear. While STOKES's representations $(163.1)_2$ and $(163.2)_2$ are linear, the components of $\boldsymbol{v}$ have no simple kinematical interpretation. Thus each of the alternatives in (163.1) and (163.2) possesses one, but not both, of the properties which make the stream function of a plane motion convenient.

---

[1] STOKES [1842, *4*]. Cf. also CLEBSCH [1857, *1*, § 7]. HILL [1885, *2*, § II] gave analysis parallel to (161.12) to (161.17) for the rotationally symmetric case.

[2] Cf. ABRAHAM [1901, *1*, § 7].

[3] A kinematical interpretation of A and B is given by PRATELLI [1953, *24*, § 5]. EULER's use of $(163.2)_1$ has been discussed in Sect. 137. Cf. also the sources cited in footnote 2, p. 823.

[4] ERTEL and KÖHLER [1949, *9*], GIESE [1951, *10*].

In the representations $(163.1)_2$ and $(163.2)_2$, one of the three components of the vector potentials $\boldsymbol{v}$ and $\boldsymbol{v}'$ may be eliminated. For example, if $\boldsymbol{v}$ is any vector potential of $\dot{\boldsymbol{p}}$, so that

$$\sqrt{g}\,\dot{x}^k = \varepsilon^{kmp}\,v_{p,m} = \varepsilon^{kmp}\,(\partial v_p/\partial x^m), \quad \text{set}$$

$$\left.\begin{aligned} L &\equiv -v_2 + \int \frac{\partial v_3}{\partial x^2}\,dx^3 + F(x^1, x^2),\\ M &\equiv v_1 - \int \frac{\partial v_3}{\partial x^1}\,dx^3 + G(x^1, x^2). \end{aligned}\right\} \tag{163.4}$$

Then by differentiation follows[1]

$$\frac{\partial L}{\partial x^3} = \sqrt{g}\,\dot{x}^1, \quad \frac{\partial M}{\partial x^3} = \sqrt{g}\,\dot{x}^2, \quad \frac{\partial L}{\partial x^1} + \frac{\partial M}{\partial x^2} = -\sqrt{g}\,\dot{x}^3, \tag{163.5}$$

provided $\partial F/\partial x^1 + \partial G/\partial x^2 = 0$. Thus the velocity $\dot{\boldsymbol{x}}$ may always be expressed in terms of two potentials.

164. The equation of continuity in space-time; its general solution. Let $\boldsymbol{v}$ be the world velocity vector (153.3) of a motion, and consider the contravariant world vector density ϱ^Ω defined by

$$\varrho^\Omega \equiv \sqrt{g}\,\varrho\,v^\Omega, \tag{164.1}$$

where ϱ is the mass density and g is the world scalar density defined by (153.12). The equation

$$\frac{\partial \varrho^\Omega}{\partial x^\Omega} = \frac{\partial}{\partial x^\Omega}\,(\sqrt{g}\,\varrho\,v^\Omega) = 0 \tag{164.2}$$

is invariant under general transformations of the four space-time coordinates. In every Euclidean frame (cf. Sect. 152), we have

$$\varrho = (\varrho\,\dot{z}^k, \varrho), \tag{164.3}$$

so that (164.2) reduces to

$$\frac{\partial \varrho}{\partial t} + \frac{\partial}{\partial z^k}\,(\varrho\,\dot{z}^k) = 0, \tag{164.4}$$

which is the Cartesian co-ordinate form of $(156.5)_1$. Thus (164.2) is a world tensor form of the continuity equation. It can also be written in the alternative world tensor form

$$\dot{\varrho} + \varrho\,\mathrm{I}_d = 0, \tag{164.5}$$

where the absolute world scalar invariant I_d is given by $(153.13)_1$, and $\dot{\varrho} = v^\Omega \dfrac{\partial \varrho}{\partial x^\Omega}$ in accord with (154.18).

Equations of the type (164.2) were solved by EULER[2] for spaces of any number of dimensions. For the four-dimensional case, his solution is

$$\sqrt{g}\,\varrho\,v^\Gamma = \varepsilon^{\Gamma\Lambda\Delta\Sigma}\,F_{,\Lambda}\,G_{,\Delta}\,H_{,\Sigma}; \tag{164.6}$$

since, trivially, $\dot{F} = F_{,\Gamma}\,v^\Gamma = 0$, the function $F(\boldsymbol{x}, t)$ is substantially constant, as are G and H. Conversely, (164.6) yields a solution of (164.2) when F, G, and H

[1] That these formulae furnish a solution was noted by LAGRANGE [1783, *1*, § 13] for the case of rectangular Cartesian co-ordinates; our proof of sufficiency is adapted from one given by GEIS [1956, *8*, § 3] under an unnecessary restriction. Note that (163.5) yields the contravariant components, not the physical components of velocity, in arbitrary curvilinear co-ordinates.

[2] [1770, *1*, §§ 44–49]. Cf. the treatment of this and related problems by FINZI and PASTORI [1949, *11*, Chap. IV, § 7]; cf. also our Subchapter D IV.

are any substantially constant functions. This result is fairly obvious, since it amounts to a statement that the most general velocity field may be obtained by mapping vectorially onto $\boldsymbol{x}$-t space a field of vectors in $\boldsymbol{X}$ space tangent to the curves of intersection of three appropriate families in surfaces which are stationary and hence consist always in the same particles. The mass in $\mathscr{V}$ is given by[1]

$$\left.\begin{aligned}\mathfrak{M} &= \int_{\mathscr{V}} \varrho\, dv = \int_{\mathscr{V}} \frac{\partial(F, G, H)}{\partial(x, y, z)}\, dx\, dy\, dz,\\ &= \int dF\, dG\, dH\\ &= (F_2 - F_1)(G_2 - G_1)(H_2 - H_1),\end{aligned}\right\} \tag{164.7}$$

provided $\mathscr{V}$ is small enough that the mapping $x, y, z \leftrightarrow F, G, H$ be one-to-one. This generalizes (163.3).

Euler's solution (164.6) includes as special cases all those given above in Sects. 161 to 163, and the interpretation just obtained likewise includes the foregoing.

It is to be noted that the derivatives occurring in (164.6) are ordinary partial derivatives, even though the result is valid for arbitrary co-ordinates in space-time. However, it is valid only locally, and its non-linearity makes it difficult to use with profit.

c) Momentum.

Throughout this part of the subchapter, the co-ordinate system is assumed to be rectangular Cartesian. Generalization to curvilinear co-ordinates may be achieved by the method of Sect. App. 17.

165. Center of mass. The *center of mass* of body $\mathscr{V}$ is the point whose position vector $\boldsymbol{c}$ is defined by[2]

$$\mathfrak{M}\boldsymbol{c} \equiv \int_{\mathscr{V}} \boldsymbol{p}\, d\mathfrak{M} = \int_{\mathscr{V}} \varrho\, \boldsymbol{p}\, dv. \tag{165.1}$$

In a frame such that the origin of the position vector $\boldsymbol{p}'$ is at the center of mass, we have

$$\int_{\mathscr{V}} \boldsymbol{p}'\, d\mathfrak{M} = 0. \tag{165.2}$$

Hence a body may not lie entirely on one side of any plane through its center of mass[3].

The center of mass of a deformable body generally moves about within the body in the course of time. In order for the center of mass to reside in a particle and move with it, it is necessary and sufficient that the velocity of the motion, $\dot{\boldsymbol{p}}$, at the center of mass shall equal $\dot{\boldsymbol{c}}$ as calculated from (165.1). Any homogeneous motion (142.1) has this property[4].

166. Momentum, moment of momentum, and kinetic energy. The *momentum* $\mathfrak{P}$, sometimes called the *linear momentum* or *impulse* or *inertia*, of the body $\mathscr{V}$ is defined by

$$\mathfrak{P}_k \equiv \int_{\mathscr{V}} \dot{z}_k\, d\mathfrak{M}, \quad \mathfrak{P} \equiv \int_{\mathscr{V}} \dot{\boldsymbol{p}}\, d\mathfrak{M}. \tag{166.1}$$

---

[1] Yih [1957, *19*, § VII].

[2] In researches done in 1758—1759, Euler [1765, *2*, §§ 7—10] [1765, *1*, §§ 285—287], after criticizing loose ideas on the "center of gravity" of earlier writers, introduced (165.1) and the name *center of mass* as well as an alternative, "center of inertia", and established the vectorial invariance of the definition [1765, *2*, §§ 11—16].

[3] (165.2) and the above inference from it were given by Euler [1765, *1*, § 286] 1765, *2*, § 18], the latter somewhat obscurely.

[4] Levi-Civita [1935, *3*]. We do not follow the analysis of Schwerdtfeger [1935, *5*], who claims a broad extension of this result.

The *moment of momentum* $\mathfrak{H}^{[*p]}$, sometimes called the *angular momentum*, of the body $\mathscr{V}$ with respect to the point $*z$ is defined by

$$\mathfrak{H}_{km}^{[*p]} \equiv \int_{\mathscr{V}} (z_{[k} - *z_{[k}) (\dot{z}_{m]} - *\dot{z}_{m]}) \, d\mathfrak{M}. \tag{166.2}$$

The axial vector $\mathfrak{H}^{[*p]}$ corresponding to the skew-symmetric tensor $\mathfrak{H}_{km}^{[*p]}$ is called by the same names and is given by

$$\mathfrak{H}^{[*p]} \equiv \int_{\mathscr{V}} (\boldsymbol{p} - *\boldsymbol{p}) \times (\dot{\boldsymbol{p}} - *\dot{\boldsymbol{p}}) \, d\mathfrak{M}. \tag{166.3}$$

Sometimes, but not always, there is no loss in generality in taking $*\boldsymbol{p} = 0$. The superscript $*\boldsymbol{p}$ is omitted in cases where no confusion should result.

The *kinetic energy* $\mathfrak{K}$ of the body $\mathscr{V}$ is defined by

$$\mathfrak{K} \equiv \tfrac{1}{2} \int \dot{p}^2 \, d\mathfrak{M}. \tag{166.4}$$

[Cf. (94.1).] $2\mathfrak{K}$, or sometimes also $\mathfrak{K}$, is called the *vis viva* or *live force* of the body.

The concepts of momentum, moment of momentum, and kinetic energy of a finite body are peculiar to Euclidean space and are the stuff of which classical mechanics is made. They deserve the most minute analysis[1].

In non-Euclidean spaces, the lack of an invariant finite parallel transport makes it impossible to define such volume integrals meaningfully[2] —at least, impossible without introducing new concepts not present in Euclidean mechanics. Since all physical experience necessarily concerns *finite bodies*, the impossibility of forming geometrically invariant measures of the inertia and energy of such bodies reduces mechanics in general spaces to an uneasy formalism. Two approaches are used: (1) Euclidean arguments are applied to infinitely small volumes, leading to formal analogues of all local equations in classical mechanics, or (2) the non-Euclidean space is supposed imbedded in a Euclidean space of higher dimension, where the usual measures and laws of mechanics are valid, thus inducing certain laws in the non-Euclidean subspace. The results of these two approaches are not the same. For the former, see Sect. 238; examples of the latter are given in Sects. 212 to 214. From the standpoint of mere postulation, there is no error in either of these approaches. For the conceptual foundations, they are both inadequate in that they use Euclidean concepts as a crutch. (Of course, the objection to the latter method is not relevant to the theories of rods and shells, where the Euclidean imbedding space is the usual three-dimensional one.)

In the more general case when discrete as well as distributed masses are present, by (155.2) we get

$$\left.\begin{aligned}
\mathfrak{P} &= \int_{\mathscr{V}} \varrho \dot{\boldsymbol{p}} \, dv + \sum_{\mathfrak{a}=1}^{n} \mathfrak{M}_{\mathfrak{a}} \dot{\boldsymbol{p}}_{\mathfrak{a}}, \\
\mathfrak{H} &= \int_{\mathscr{V}} (\boldsymbol{p} - *\boldsymbol{p}) \times (\dot{\boldsymbol{p}} - *\dot{\boldsymbol{p}}) \varrho \, dv + \sum_{\mathfrak{a}=1}^{n} (\boldsymbol{p}_{\mathfrak{a}} - *\boldsymbol{p}) \times (\dot{\boldsymbol{p}}_{\mathfrak{a}} - *\dot{\boldsymbol{p}}) \mathfrak{M}_{\mathfrak{a}}, \\
2\mathfrak{K} &= \int_{\mathscr{V}} \varrho \dot{p}^2 \, dv + \sum_{\mathfrak{a}=1}^{n} \mathfrak{M}_{\mathfrak{a}} \dot{p}_{\mathfrak{a}}^2.
\end{aligned}\right\} \tag{166.5}$$

In particular, for a single mass-point we have

$$\mathfrak{P} = \mathfrak{M} \dot{\boldsymbol{p}}, \quad \mathfrak{H} = \mathfrak{M} (\boldsymbol{p} - *\boldsymbol{p}) \times (\dot{\boldsymbol{p}} - *\dot{\boldsymbol{p}}), \quad 2\mathfrak{K} = \mathfrak{M} \dot{p}^2. \tag{166.6}$$

The quantities $\mathfrak{P}$, $\mathfrak{H}$, and $\mathfrak{K}$ are special cases of general moments of the type

$$\left.\begin{aligned}
\mathfrak{M}_{k_1 \ldots k_n m} &\equiv \int_{\mathscr{V}} z_{k_1} \cdots z_{k_n} \dot{z}_m \, d\mathfrak{M}, \\
\mathfrak{K}_{k_1 \ldots k_n m q} &\equiv \int_{\mathscr{V}} z_{k_1} \cdots z_{k_n} \dot{z}_m \dot{z}_q \, d\mathfrak{M},
\end{aligned}\right\} \tag{166.7}$$

---

[1] The history of these quantities has never been written; on the basis of the information known to us, we rest content with stating that the concept of linear momentum seems to originate in the middle ages, while kinetic energy and moment of momentum were introduced by LEIBNIZ and EULER, respectively. It is certain that the *general* definitions of all three, and some of their major properties, first appear in the later writings of EULER.

[2] Cf. VAN DANTZIG [1934, *10*, Part IV, § 1].

etc., where we have set $^*z=0$. In this notation

$$\mathfrak{H}_{km}=\mathfrak{M}_{[km]}, \quad 2\mathfrak{K}=\mathfrak{K}_{kk}. \tag{166.8}$$

While the introduction of the higher moments is suggested by the occurrence of analogous quantities in statistical mechanics and in the kinetic theory of gases, their dynamical significance is as yet unknown, and their properties remain to be learned. Dynamical laws concerning them will be proposed in Sect. 205 and Sect. 232.

The quantities $\mathfrak{P}$, $\mathfrak{H}$, and $\mathfrak{K}$ are defined in a *particular frame*. Their invariance under change of frame will be investigated in the two following sections.

167. Theorems on the center of mass. If we take the volume $\mathscr{V}$ as material and differentiate (165.1) materially, from (166.1), $(81.2)_1$, and (76.9), whether or not mass is conserved, we get

$$\left.\begin{aligned}\mathfrak{P}&=\dot{\overline{\mathfrak{M}\boldsymbol{c}}}-\int_{\mathscr{V}}\boldsymbol{p}\,(\dot{\overline{\log\varrho}}+\mathrm{I}_{\boldsymbol{d}})\,d\mathfrak{M},\\ &=\mathfrak{M}\,\dot{\boldsymbol{c}}+\int_{\mathscr{V}}(\boldsymbol{c}-\boldsymbol{p})\,(\dot{\overline{\log\varrho}}+\mathrm{I}_{\boldsymbol{d}})\,d\mathfrak{M}.\end{aligned}\right\} \tag{167.1}$$

When mass is conserved, by $(156.5)_2$ the volume integral vanishes, and we get the ***fundamental theorem on the center of mass***[1]: *The momentum of any body, no matter what deformation it be undergoing, equals the momentum of a mass-point located at the center of mass and having the same mass as the body.* In particular, in a frame such that the origin is constantly at the center of mass, the momentum vanishes.

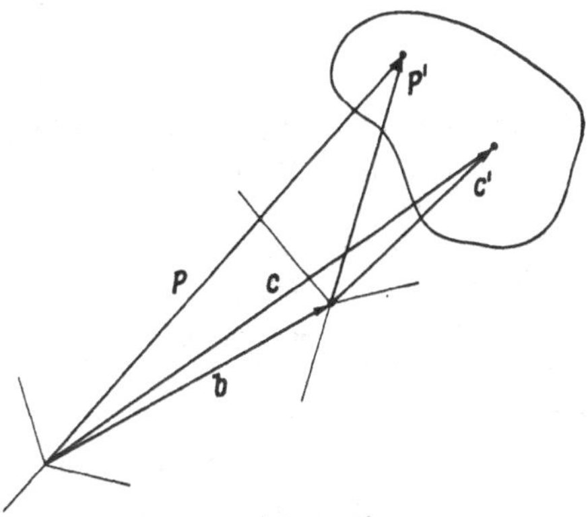

Fig. 22. Center of mass.

The moment of momentum (166.3), being a function only of differences of position, is invariant with respect to change of frame, so long as the primed frame be not rotating:

$$\left.\begin{aligned}\mathfrak{H}=\mathfrak{H}'\equiv\int_{\mathscr{V}}(\boldsymbol{p}'-{}^*\boldsymbol{p}')\times\\ \times(\dot{\boldsymbol{p}}'-{}^*\dot{\boldsymbol{p}}')\,d\mathfrak{M},\end{aligned}\right\} \tag{167.2}$$

it being understood that $^*\boldsymbol{p}$ and $^*\boldsymbol{p}'$ are the position vectors of the *same point* in the two frames. Sometimes, however[2], it is preferable to consider the relation between the moments of momentum with respect to the origins of the two frames:

$$\mathfrak{H}^{[O]}\equiv\int_{\mathscr{V}}\boldsymbol{p}\times\dot{\boldsymbol{p}}\,d\mathfrak{M}, \quad \mathfrak{H}'^{[O']}\equiv\int_{\mathscr{V}}\boldsymbol{p}'\times\dot{\boldsymbol{p}}'\,d\mathfrak{M}. \tag{167.3}$$

Putting $\boldsymbol{p}=\boldsymbol{b}+\boldsymbol{p}'$, we obtain

$$\mathfrak{H}^{[O]}=\mathfrak{M}\,[\boldsymbol{b}\times\dot{\boldsymbol{b}}+\boldsymbol{b}\times\dot{\boldsymbol{c}}'-\dot{\boldsymbol{b}}\times\boldsymbol{c}']+\mathfrak{H}'^{[O']}, \tag{167.4}$$

where $\boldsymbol{c}'$ is the position vector of the center of mass in the primed frame (Fig. 22). When there is a mere static shift of origin, $\dot{\boldsymbol{b}}=0$, and we get

$$\mathfrak{H}^{[O]}=\boldsymbol{b}\times\mathfrak{P}'+\mathfrak{H}'^{[O']}=\boldsymbol{b}\times\mathfrak{P}+\mathfrak{H}^{[O']}, \tag{167.5}$$

[1] Kelvin and Tait [1867, *3*, § 230]. A closely related, much older statement is given in § 196 below.

[2] Painlevé [1895, *4*, 2e Leçon].

showing that the difference of the moments of momentum with respect to two different points is simply $\boldsymbol{b}\times\mathfrak{P}$ (or $\boldsymbol{b}\times\mathfrak{P}'$), where $\boldsymbol{b}$ is the vector joining the two[1]. When the origin of the primed frame is taken as the center of mass, (167.4) reduces to

$$\mathfrak{H}^{[O]} = \boldsymbol{c}\times\mathfrak{M}\,\dot{\boldsymbol{c}} + \mathfrak{H}'^{[O']}: \tag{167.6}$$

The moment of momentum of any body equals the sum of the moment of momentum of a mass-point having the same mass as the body and located at the center of mass, plus the moment of momentum with respect to the center of mass. From (167.4) it is plain that no such simple decomposition is valid if the relative moment of momentum is taken with respect to a point other than the center of mass. In order for (167.6) to hold, it is necessary and sufficient that

$$\boldsymbol{b}\times\dot{\boldsymbol{c}}' = \dot{\boldsymbol{b}}\times\boldsymbol{c}'. \tag{167.7}$$

Similarly, when

$$2\,\mathfrak{K}' \equiv \int_{\mathscr{V}} \dot{p}'^2\, d\mathfrak{M}, \tag{167.8}$$

we have

$$\mathfrak{K} = \tfrac{1}{2}\,\mathfrak{M}\,\dot{b}^2 + \mathfrak{M}\,\dot{\boldsymbol{b}}\cdot\boldsymbol{c}' + \mathfrak{K}'. \tag{167.9}$$

If the origin of the primed system is taken at the center of mass, this becomes

$$\mathfrak{K} = \tfrac{1}{2}\,\mathfrak{M}\,\dot{c}^2 + \mathfrak{K}', \tag{167.10}$$

whence follows $\mathfrak{K} \geqq \mathfrak{K}'$, where $\mathfrak{K} = \mathfrak{K}'$ if and only if the center of mass is stationary. This same formula may be used to compare the relative kinetic energies with respect to the center of mass and any other point. Hence follow the ***theorems of König***[2]: *If we compare the kinetic energies calculated in a class of frames not in relative rotation, then*

1. *The relative kinetic energy with respect to the center of mass is the least possible;*
2. *The kinetic energy of any body is the sum of the kinetic energy of a mass-point having the same mass as the body and located at the center of mass, plus the kinetic energy relative to the center of mass.*

The foregoing theorems enable us to give a rational position to the classical mechanics of point masses without recourse to the strict concept of mass-point (Sect. 155) or any other mention of the infinitely small: *We may consider a body of mass $\mathfrak{M}$, whatever its size, to be in motion as a mass-point of mass $\mathfrak{M}$, located at its center of mass, if*

(1) *Our knowledge is sufficient to determine the motion of its center of mass, and*
(2) *We do not require knowledge of its moment of momentum $\mathfrak{H}'$ and kinetic energy $\mathfrak{K}'$ relative to the center of mass.*

According to this view, the mechanics of mass-points appears not as the fundamental descipline of mechanics but rather as a degenerate special case when we have but an incomplete description of the actual motion of an extended body. Since we have not said that it is an infinitely small body, or a geometrical

---

[1] In (167.5) the symbol $\mathfrak{H}^{[O']}$ stands for the moment of momentum taken with respect to the origin of the primed frame but calculated in the unprimed frame. Since the frames are not in relative rotation, we have $\dot{\boldsymbol{p}} = \dot{\boldsymbol{b}} + \dot{\boldsymbol{p}}'$, and hence

$$\mathfrak{H}^{[O']} \equiv \int_{\mathscr{V}} (\boldsymbol{p}-\boldsymbol{b})\times(\dot{\boldsymbol{p}}-\dot{\boldsymbol{b}})\, d\mathfrak{M} = \mathfrak{H}'^{[O']}.$$

[2] [1751, *1*, pp. 172—173]. Cf. Masotti [1932, *9*].

point, we suffer no conceptual revulsion in treating the earth as a mass-point on occasion; we are aware of its extension, its spin, its geometrical irregularity, and the motions of its winds and tides, which in some problems may be the focus of our interest, but for its motion with respect to the sun they are of little or no importance.

168. Momentum, moment of momentum, and energy in a rotating frame. We generalize the formulae of the last section by allowing the primed frame to be in rotation with respect to the unprimed frame.

By substituting (143.3) into (166.1) we get

$$\mathfrak{P} = \mathfrak{M}(\dot{\boldsymbol{b}} + \dot{\boldsymbol{c}}' + \boldsymbol{\omega}\times\boldsymbol{c}') = \mathfrak{M}\,\dot{\boldsymbol{c}}, \tag{168.1}$$

in conformity with (167.1).

By substituting (143.3) into (166.3) we get

$$\mathfrak{H} = \mathfrak{H}' + \mathfrak{H}_{\text{rel}}^{[*p]} \tag{168.2}$$

where

$$\mathfrak{H}_{\text{rel}}^{[a]} \equiv \int_{\mathscr{V}} (\boldsymbol{p}-\boldsymbol{a})\times[\boldsymbol{\omega}\times(\boldsymbol{p}-\boldsymbol{a})]\,d\mathfrak{M}. \tag{168.3}$$

The relation (168.2) generalizes $(167.2)_1$; the corresponding generalization of (167.4) is elaborate but easy to work out. Since $\boldsymbol{p}-\boldsymbol{a}=\boldsymbol{p}'-\boldsymbol{a}'$, primed quantities may be used to calculate $\mathfrak{H}_{\text{rel}}^{[*p]}$, but the angular velocity $\boldsymbol{\omega}$ is that of the primed frame with respect to the unprimed. To discuss the form of $\mathfrak{H}_{\text{rel}}^{[a]}$, we introduce EULER'S *tensor* $\mathfrak{E}^{[a]}$ and the *tensor of inertia* $\mathfrak{J}^{[a]}$ by the definitions[1]

$$\left.\begin{aligned} \mathfrak{E}^{[a]} &\equiv \int_{\mathscr{V}} (\boldsymbol{p}-\boldsymbol{a})(\boldsymbol{p}-\boldsymbol{a})\,d\mathfrak{M}, \\ \mathfrak{J}^{[a]} &\equiv 1\int_{\mathscr{V}} (\boldsymbol{p}-\boldsymbol{a})^2\,d\mathfrak{M} - \mathfrak{E}^{[a]}; \end{aligned}\right\} \tag{168.4}$$

---

[1] The great researches of EULER on these tensors deserve special notice because they are never cited in reference to the much later work on second-order tensors and matrices they in part anticipated. In 1741 or earlier, EULER introduced the name *moments of inertia* for the diagonal components of $\mathfrak{J}$; these quantities were already familiar. In [1765, *2*, §§ 26—29] he derived the tensor law of transformation for them. In [ibid., § 34] he defined the *principal axes* by the extremal properties of the moments of inertia, and he gave these axes the name they have retained ever after; in [ibid., § 37] he derived the characteristic equation of $\mathfrak{J}$, which he proved to have three real roots, to which correspond at least three mutually perpendicular principal axes [ibid., §§ 38—42]. For these axes, the tensor $\mathfrak{J}$ assumes diagonal form, and the corresponding components are named *principal moments of inertia;* any moment of inertia is expressed as a linear combination of these [ibid., §§ 43—45]. If the principal moments are distinct, there are only three principal axes; if two principal moments coincide, any axis normal to the one corresponding to the third moment is a principal axis; if all three principal moments are equal, every axis is a principal axis [ibid., §§ 46—47]. These results, obtained in 1758—1759, are presented even more clearly and fully in [1765, *1*, Chap. 5]. Cf. also [1776, *2*, §§ 33—34].

The principal axes themselves have a short prior history. In [1746, *1*, § 77] EULER had found that in order for a lamina to rotate freely about an axis, the two products of inertia with respect to that axis must vanish. In [1752, *2*, § 48], EULER had introduced the full tensor of inertia and had shown that permanent rotation of a rigid body is possible only about an axis such that the appropriate products of inertia vanish [ibid., §§ 14—16, 53], but he had not determined the existence of such axes. SEGNER [1755, *1*, pp. 16—30], who acknowledged having seen EULER'S paper [ibid., p. 15], then proved at length that at least three such axes exist and are mutually orthogonal, etc.; in particular, he derived the characteristic cubic satisfied by the principal moments of inertia.

these symmetric tensors are associated with the body $\mathscr{V}$ and with a particular point $\boldsymbol{a}$. We find that

$$\left.\begin{aligned}\mathfrak{H}^{[\boldsymbol{a}]}_{\text{rel}} &= \int_{\mathscr{V}} \{(\boldsymbol{p}-\boldsymbol{a})^2\,\boldsymbol{\omega} - [\boldsymbol{\omega}\cdot(\boldsymbol{p}-\boldsymbol{a})]\,(\boldsymbol{p}-\boldsymbol{a})\}\,d\mathfrak{M},\\ &= \boldsymbol{\omega}\cdot\mathfrak{J}^{[\boldsymbol{a}]} = \mathfrak{J}^{[\boldsymbol{a}]}\cdot\boldsymbol{\omega}.\end{aligned}\right\} \tag{168.5}$$

In component form, (168.5) may be written

$$\mathfrak{H}^{[\boldsymbol{a}]}_{k m} = \varepsilon_{k m p}\,\varepsilon_{q r s}\,\omega_{r s}\,\mathfrak{J}^{[\boldsymbol{a}]}_{q p}. \tag{168.6}$$

Since $\mathfrak{J}^{[\boldsymbol{a}]}_{m k} = \mathfrak{J}^{[\boldsymbol{a}]}_{k m}$, there exist three real mutually orthogonal principal directions for $\mathfrak{J}^{[\boldsymbol{a}]}$; these are the *principal axes of inertia*. From (168.4) the normal components of $\mathfrak{J}^{[\boldsymbol{a}]}$ are $\int_{\mathscr{V}} [(y-a_y)^2 + (z-a_z)^2]\,d\mathfrak{M}$, etc., and hence the proper numbers of $\mathfrak{J}^{[\boldsymbol{a}]}$ are positive if the body has positive volume. Thus the quadric of $\mathfrak{J}^{[\boldsymbol{a}]}$ is an ellipsoid, the *ellipsoid of inertia*[1]. The normal components of $\mathfrak{J}^{[\boldsymbol{a}]}$ are called *moments of inertia*; the shear components, *products of inertia*.

The tensor of inertia, its principal axes, and its proper numbers are all defined in terms of a particular point. The importance of the formula $\mathfrak{H}^{[\boldsymbol{a}]}_{\text{rel}} = \mathfrak{J}^{[\boldsymbol{a}]}\cdot\boldsymbol{\omega}$ lies in its linearity in $\boldsymbol{\omega}$: Once a point $\boldsymbol{a}$ is selected, the tensor $\mathfrak{J}^{[\boldsymbol{a}]}$ can be evaluated from a knowledge only of the shape of the body $\mathscr{V}$ and of the distribution of density within it; $\mathfrak{J}^{[\boldsymbol{a}]}$ depends on the orientation of the primed frame only through the tensor law, and it is independent of the angular velocity of the primed frame. In the special case when the body is *at rest with respect to the primed frame*, we have $\mathfrak{H}'=0$, and (168.2) shows that the entire moment of momentum of the body is $\mathfrak{H}^{[{}^*\boldsymbol{p}]}_{\text{rel}}$. In the case of *rigid motion*, it is possible to choose a primed frame of this kind, and in this frame $\mathfrak{J}^{[{}^*\boldsymbol{p}]}$ is constant in time; $\boldsymbol{\omega}$ is the angular velocity of the body (Sects. 84, 86, 143), and the geometrical meaning of (168.3) is simple.

To compare the moments of momentum $\mathfrak{H}^{[\boldsymbol{O}]}$ and $\mathfrak{H}'^{[\boldsymbol{O}']}$ with respect to the two origins, by substituting (143.3) into $(167.3)_1$ we easily calculate

$$\mathfrak{H}^{[\boldsymbol{O}]} = \mathfrak{M}\,[\boldsymbol{b}\times\dot{\boldsymbol{b}} + \boldsymbol{b}\times(\boldsymbol{\omega}\times\boldsymbol{c}') + \boldsymbol{b}\times\dot{\boldsymbol{c}}' + \boldsymbol{c}'\times\dot{\boldsymbol{b}}] + \mathfrak{H}^{[\boldsymbol{O}']}_{\text{rel}} + \mathfrak{H}'^{[\boldsymbol{O}']}. \tag{168.7}$$

If the origin of the primed frame is the center of mass, three of the four punctual terms vanish; also in the case when the origin is stationary there is a corresponding simplification. In a fully general motion, (168.5) remains valid, but since $\mathfrak{H}'^{[\boldsymbol{O}']}$ generally does not vanish and the components of $\mathfrak{J}^{[\boldsymbol{O}']}$ are continually changing, the result is seldom useful.

By substituting (143.3) into (166.4) we get

$$\mathfrak{K} = \mathfrak{M}\,[\tfrac{1}{2}\dot{b}^2 + \dot{\boldsymbol{b}}\cdot\dot{\boldsymbol{c}}' + \boldsymbol{\omega}\cdot\boldsymbol{c}'\times\dot{\boldsymbol{b}}] + \boldsymbol{\omega}\cdot\mathfrak{H}'^{[\boldsymbol{O}']} + \mathfrak{K}^{[\boldsymbol{O}']}_{\text{rel}} + \mathfrak{K}' \tag{168.8}$$

where

$$2\,\mathfrak{K}^{[\boldsymbol{O}']}_{\text{rel}} \equiv \boldsymbol{\omega}\cdot\mathfrak{J}^{[\boldsymbol{O}']}\cdot\boldsymbol{\omega} = \mathfrak{H}^{[\boldsymbol{O}']}_{\text{rel}}\cdot\boldsymbol{\omega} = \boldsymbol{\omega}\cdot\mathfrak{H}^{[\boldsymbol{O}']}_{\text{rel}}. \tag{168.9}$$

169. Interpretation of certain previous results as theorems on momentum and energy. Various formulae derived earlier may be regarded as expressions for momentum and energy. First, in a *homochoric* motion we have

$$\mathfrak{P}/\varrho = \int_{\mathscr{V}} \dot{\boldsymbol{p}}\,dv, \qquad \mathfrak{H}^{[\boldsymbol{O}]}/\varrho = \int_{\mathscr{V}} \boldsymbol{p}\times\dot{\boldsymbol{p}}\,dv, \qquad 2\,\mathfrak{K}/\varrho = \int_{\mathscr{V}} \dot{p}^2\,dv, \tag{169.1}$$

and since $\dot{\boldsymbol{p}}$ is solenoidal, we may apply the results of Sect. App. 31. By (App. 31.23), (App. 31.25), (App. 31.21), and (App. 31.12) we thus get formulae

[1] CAUCHY [1827, *3*].

relating $\mathfrak{P}$, $\mathfrak{H}$, and $\mathfrak{K}$ to certain vorticity averages:

$$\left.\begin{aligned} 2\mathfrak{P}/\varrho &= \int_{\mathscr{V}} \boldsymbol{p}\times\boldsymbol{w}\,dv + \oint_{\mathscr{s}} (d\boldsymbol{a}\times\dot{\boldsymbol{p}})\times\boldsymbol{p},\\ 2\mathfrak{H}^{[O]}/\varrho &= -\int_{\mathscr{V}} p^2\boldsymbol{w}\,dv + \oint_{\mathscr{s}} d\boldsymbol{a}\times p^2\dot{\boldsymbol{p}},\\ 3\mathfrak{H}^{[O]}/\varrho &= \int_{\mathscr{V}} \boldsymbol{p}\times(\boldsymbol{p}\times\boldsymbol{w})\,dv - \oint_{\mathscr{s}} [(d\boldsymbol{a}\times\dot{\boldsymbol{p}})\times\boldsymbol{p}]\times\boldsymbol{p},\\ \mathfrak{K}/\varrho &= \int_{\mathscr{V}} \boldsymbol{p}\cdot\boldsymbol{w}\times\dot{\boldsymbol{p}}\,dv + \oint_{\mathscr{s}} d\boldsymbol{a}\cdot[\tfrac{1}{2}\dot{p}^2\boldsymbol{p} - \dot{\boldsymbol{p}}\,\dot{\boldsymbol{p}}\cdot\boldsymbol{p}]. \end{aligned}\right\} \tag{169.2}$$

In the formulae for $\mathfrak{P}$ and $\mathfrak{H}^{[O]}$, the surface integrals vanish when the velocity is normal to $\mathscr{s}$. In an isochoric irrotational motion, the volume integrals all vanish, and thus $\mathfrak{P}$, $\mathfrak{H}^{[O]}$, and $\mathfrak{K}$ are determinate from the boundary velocities alone.

In a *steady motion with steady density*, $\varrho\dot{\boldsymbol{p}}$ is solenoidal, and in just the same way we derive

$$\left.\begin{aligned} 2\mathfrak{P} &= \int_{\mathscr{V}} \boldsymbol{p}\times\operatorname{curl}(\varrho\dot{\boldsymbol{p}})\,dv + \oint_{\mathscr{s}} [(d\boldsymbol{a}\times\varrho\dot{\boldsymbol{p}})\times\boldsymbol{p}],\\ 2\mathfrak{H}^{[O]} &= -\int_{\mathscr{V}} p^2\operatorname{curl}(\varrho\dot{\boldsymbol{p}})\,dv + \oint_{\mathscr{s}} d\boldsymbol{a}\times\varrho p^2\dot{\boldsymbol{p}},\\ 3\mathfrak{H}^{[O]} &= \int_{\mathscr{V}} \boldsymbol{p}\times[\boldsymbol{p}\times\operatorname{curl}(\varrho\dot{\boldsymbol{p}})]\,dv - \oint_{\mathscr{s}} \{[(d\boldsymbol{a}\times\varrho\dot{\boldsymbol{p}})\times\boldsymbol{p}]\times\boldsymbol{p}\}. \end{aligned}\right\} \tag{169.3}$$

However, if we apply (App. 31.12) to $\varrho\dot{\boldsymbol{p}}$ we get an expression for $\int_{\mathscr{V}} \varrho^2\dot{p}^2\,dv$ rather than for $2\mathfrak{K}$ itself. The formula

$$2\mathfrak{K} = \oint_{\mathscr{s}} d\boldsymbol{a}\cdot\dot{\boldsymbol{p}}(\varrho\dot{\boldsymbol{p}}\cdot\boldsymbol{p}) - \int_{\mathscr{V}} \boldsymbol{p}\cdot\operatorname{div}(\varrho\dot{\boldsymbol{p}}\dot{\boldsymbol{p}})\,dv, \tag{169.4}$$

valid for all types of motion, is easy to verify directly. In this formula the surface integral vanishes when $\mathscr{s}$ is a fixed boundary.

It is easy to generalize (169.2) and (169.3) by expressions valid for any motion, but these involve various combinations of the derivatives of ϱ which render them difficult to apply. For an exception, consider the simpler formula (76.12), which in a homochoric motion yields

$$\mathfrak{P}/\varrho = \oint_{\mathscr{s}} d\boldsymbol{a}\cdot\dot{\boldsymbol{p}}\boldsymbol{p}; \tag{169.5}$$

for a general motion, a corresponding formula is

$$\left.\begin{aligned} \mathfrak{P} &= \oint_{\mathscr{s}} d\boldsymbol{a}\cdot\varrho\dot{\boldsymbol{p}}\boldsymbol{p} - \int_{\mathscr{V}} \boldsymbol{p}\operatorname{div}(\varrho\dot{\boldsymbol{p}})\,dv,\\ &= \oint_{\mathscr{s}} d\boldsymbol{a}\cdot\varrho\dot{\boldsymbol{p}}\boldsymbol{p} + \int_{\mathscr{V}} \boldsymbol{p}\frac{\partial\varrho}{\partial t}\,dv. \end{aligned}\right\} \tag{169.6}$$

Hence we see that *in a motion with steady density, if all finite boundaries are stationary and if in any infinite regions* $\varrho\dot{\boldsymbol{p}}\boldsymbol{p} = \bar{o}(p^{-2})$, *then the momentum is zero*. In particular, it is sufficient that $\varrho\dot{p} = o(p^{-3})$.

For a steady irrotational motion, whether or not it is isochoric, in (App. 31.5) we may put $\boldsymbol{c} = \varrho\dot{\boldsymbol{p}}$, $\Psi = V$, where V is the velocity potential (Sect. 88), and obtain[1]

$$2\mathfrak{K} = \oint_{\mathscr{s}} \varrho V\frac{dV}{dn}\,da \tag{169.7}$$

[1] Kelvin [1849, *3*, § 7].

for the case when the region bounded by $\mathfrak{s}$ is simply connected. In the multiply connected case, appropriate multiples of the mass fluxes through the barriers sufficient to render the region simply connected must be added to the result of using (169.7) with any particular determination of the cyclic function V.

For plane homochoric motion, by applying AMPÈRE's transformation to the pair of functions $Q\,\partial Q/\partial x$, $Q\,\partial Q/\partial y$ and then using (161.2) and (161.5), we obtain

$$2\,\mathfrak{K}^0/\varrho = \oint_c Q\,\frac{dQ}{dn}\,ds - \int_{\mathfrak{s}} Q\,w\,da, \tag{169.8}$$

where $\mathfrak{K}^0$ is the kinetic energy per unit height normal to the plane of motion. Various other formulae for $\mathfrak{K}$ or $\mathfrak{K}/\varrho$ in various circumstances may be derived from the representation given in Sects. 161 to 164.

The kinetic energy $\mathfrak{K}$ was defined by (94.1), which is formally equivalent to (166.4), and KELVIN's fundamental theorem of minimum energy was given in Sect. 94. In some cases the total squared speed, to which $\mathfrak{K}/\varrho$ reduces for homochoric motion, has properties more illuminating than those of $\mathfrak{K}$ itself; the general formula (93.1) as well as estimates in terms of vorticity and expansion were given in Sect. 93.

Since the density of $\mathfrak{K}$, per unit mass, is $\frac{1}{2}\dot{p}^2$, or $\frac{1}{2}\dot{x}^2$, every theorem concerning the speed is an energy theorem. Indeed, we have called $\frac{1}{2}\dot{x}^2$ the *specific kinetic energy* and have obtained many of its properties in Part e III of Subchapter II. Note also that the results cited in footnote 2, p. 822, may be regarded as energy theorems.

170. Rate of change of momentum, moment of momentum, and kinetic energy. By applying (156.8) to the definitions (166.1), (166.3), and (166.4) we obtain[1], for any body $\mathscr{V}$,

$$\dot{\boldsymbol{p}} = \int_{\mathscr{V}} \ddot{\boldsymbol{p}}\,d\mathfrak{M}, \quad \dot{\mathfrak{H}} = \int_{\mathscr{V}} (\boldsymbol{p} - {}^*\boldsymbol{p})\times(\ddot{\boldsymbol{p}} - {}^*\ddot{\boldsymbol{p}})\,d\mathfrak{M}, \quad \dot{\mathfrak{K}} = \int_{\mathscr{V}} \dot{\boldsymbol{p}}\cdot\ddot{\boldsymbol{p}}\,d\mathfrak{M}. \tag{170.1}$$

For the higher moments (166.7), formulae of such simplicity do not generally hold. For example, with ${}^*\boldsymbol{z} = 0$ we get

$$\dot{\mathfrak{M}}_{km} = \int_{\mathscr{V}} (\dot{z}_k\,\dot{z}_m + z_k\,\ddot{z}_m)\,d\mathfrak{M}, \tag{170.2}$$

whence $(170.1)_2$ follows by taking the skew-symmetric part.

If in the general transport theorem (81.3) we substitute successively $\Psi = \varrho\dot{\boldsymbol{p}}$, $\Psi = \boldsymbol{p}\times\varrho\dot{\boldsymbol{p}}$, $\Psi = \frac{1}{2}\varrho\dot{p}^2$, we obtain

$$\left.\begin{aligned}
\dot{\mathfrak{P}} &= \frac{\partial\mathfrak{P}}{\partial t} + \oint_{\mathfrak{s}} d\boldsymbol{a}\cdot\varrho\,\dot{\boldsymbol{p}}\,\dot{\boldsymbol{p}},\\
\dot{\mathfrak{H}}^{[O]} &= \frac{\partial\mathfrak{H}^{[O]}}{\partial t} + \oint_{\mathfrak{s}} d\boldsymbol{a}\cdot\dot{\boldsymbol{p}}\,\boldsymbol{p}\times\varrho\,\dot{\boldsymbol{p}} = \frac{\partial\mathfrak{H}^{[O]}}{\partial t} + \oint_{\mathfrak{s}} \boldsymbol{p}\times(\varrho\,\dot{\boldsymbol{p}}\,\dot{\boldsymbol{p}}\cdot d\boldsymbol{a}),\\
\dot{\mathfrak{K}} &= \frac{\partial\mathfrak{K}}{\partial t} + \oint_{\mathfrak{s}} d\boldsymbol{a}\cdot\dot{\boldsymbol{p}}\cdot\frac{1}{2}\varrho\,\dot{p}^2.
\end{aligned}\right\} \tag{170.3}$$

These formulae express the rate of change of $\mathfrak{P}$, $\mathfrak{H}^{[O]}$, and $\mathfrak{K}$ for a given body as the sum of a local or apparent rate and an appropriate flux through the bound-

[1] $(170.1)_3$ was noticed by STOKES [1851, *2*, § 49]. $(170.1)_1$ and $(170.1)_2$, in principle, are still older, but the classical writers, beginning with EULER (1776), were content to regard the right-hand sides *a priori* as measures of force and moment. As purely kinematical formulae, apparently $(170.1)_{1,2}$ were first derived by v. MISES [1909, *8*, § 9], who obtained also $(170.3)_1$. The straightforward treatment given here derives from CISOTTI [1917, *4*, §§ 3–4]. Cf. also SERINI [1941, *6*].

ing surface[1]. From them we conclude that *in a steady motion with steady density, if no material enters or leaves v, then the momentum, the moment of momentum, and the kinetic energy of the material occupying v are conserved.* The truth of this statement is obvious, but special cases of it are sometimes proved at length.

In the formulae for $\dot{\mathfrak{P}}$ and $\dot{\mathfrak{H}}^{[O]}$, the flux is proportional to $\varrho\dot{\boldsymbol{p}}\dot{\boldsymbol{p}}$. This important tensor, whose contravariant components are $\varrho\dot{x}^k\dot{x}^m$, is called the *momentum transfer*. It has appeared already in the identities (99.18) and (156.7), which may be used for an alternative derivation of $(170.3)_1$ and $(170.3)_2$; those identities assert that the divergence of the momentum transfer may be regarded as the supply or source strength for creation of momentum per unit mass, to be added to the apparent rate of change of momentum in order to yield the total rate experienced by a moving particle. Cf. also Sect. 207.

We record a variational formula which follows at once from (97.4):

$$\int_{\mathscr{V}} \ddot{x}_k\,\delta x^k\,d\mathfrak{M} = \frac{d}{dt}\int_{\mathscr{V}} \dot{x}_k\,\delta x^k\,d\mathfrak{M} - \delta\mathfrak{K}, \tag{170.4}$$

where the variation of density is so adjusted that $\delta\,d\mathfrak{M} = \delta(\varrho\,dv) = 0$.

We now calculate the apparent rate of change of momentum[2] and energy due to a difference of observers. Differentiating (168.1), (168.2), and (168.8) with respect to time yields

$$\left.\begin{aligned}
\dot{\mathfrak{P}} &= \mathfrak{M}\,[\ddot{\boldsymbol{b}} + \ddot{\boldsymbol{c}}' + \dot{\boldsymbol{\omega}}\times\boldsymbol{c}' + 2\boldsymbol{\omega}\times\dot{\boldsymbol{c}}' + \boldsymbol{\omega}\times(\boldsymbol{\omega}\times\boldsymbol{c}')] = \mathfrak{M}\,\ddot{\boldsymbol{c}},\\
\dot{\mathfrak{H}} &= \dot{\mathfrak{H}}' + \dot{\mathfrak{H}}^{[*p]}_{\mathrm{rel}} + \boldsymbol{\omega}\times\mathfrak{H}',\\
\dot{\mathfrak{K}} &= \mathfrak{M}\,[\dot{\boldsymbol{b}}\cdot\ddot{\boldsymbol{b}} + \ddot{\boldsymbol{b}}\cdot\dot{\boldsymbol{c}}' + \dot{\boldsymbol{b}}\cdot(\ddot{\boldsymbol{c}}' + \boldsymbol{\omega}\times\dot{\boldsymbol{c}}') +\\
&\quad + \dot{\boldsymbol{\omega}}\cdot\boldsymbol{c}'\times\dot{\boldsymbol{b}} + \boldsymbol{\omega}\cdot(\dot{\boldsymbol{c}}' + \boldsymbol{\omega}\times\boldsymbol{c}')\times\dot{\boldsymbol{b}} + \boldsymbol{\omega}\cdot\boldsymbol{c}'\times\ddot{\boldsymbol{b}}] +\\
&\quad + \dot{\boldsymbol{\omega}}\cdot\mathfrak{H}'^{[O']} + \boldsymbol{\omega}\cdot(\dot{\mathfrak{H}}'^{[O']} + \boldsymbol{\omega}\times\mathfrak{H}'^{[O']}) + \dot{\mathfrak{K}}^{[O']}_{\mathrm{rel}} + \dot{\mathfrak{K}}',
\end{aligned}\right\} \tag{170.5}$$

where a dot superposed upon a primed quantity stands for d'/dt. The simplicity of $(170.5)_3$ is somewhat deceiving, since $\dot{\mathfrak{H}}^{[*p]}$ stands for $d\,\mathfrak{H}^{[*p]}/dt$, not $d'\,\mathfrak{H}^{[*p]}/dt$. The relation between $\dot{\mathfrak{H}}^{[O]}$ and $\dot{\mathfrak{H}}'^{[O']}$ is still more elaborate since in general the two origins are in motion with respect to one another[3]. By $(170.1)_2$, the two quantities to be compared are given by

$$\dot{\mathfrak{H}}^{[O]} = \int_{\mathscr{V}} \boldsymbol{p}\times\ddot{\boldsymbol{p}}\,d\mathfrak{M}, \qquad \dot{\mathfrak{H}}'^{[O']} = \int_{\mathscr{V}} \boldsymbol{p}'\times\ddot{\boldsymbol{p}}'\,d\mathfrak{M}. \tag{170.6}$$

Using (143.6) in $(170.6)_1$, we get

$$\dot{\mathfrak{H}}^{[O]} = \int_{\mathscr{V}} (\boldsymbol{b} + \boldsymbol{p}')\times(\ddot{\boldsymbol{b}} + \dot{\boldsymbol{\omega}}\times\boldsymbol{p}' + 2\boldsymbol{\omega}\times\dot{\boldsymbol{p}}' + \boldsymbol{\omega}\times(\boldsymbol{\omega}\times\boldsymbol{p}') + \ddot{\boldsymbol{p}}')\,d\mathfrak{M}. \tag{170.7}$$

Before evaluating all the terms, we note first that if $\mathfrak{J}'^{[O']}$ is the tensor of inertia with respect to the origin of the primed frame, for its material rate of change as apparent to an observer in the primed frame we have by (168.4)

$$\dot{\mathfrak{J}}'^{[O']} = 2\,\mathbf{1}\int_{\mathscr{V}} \boldsymbol{p}'\cdot\dot{\boldsymbol{p}}'\,d\mathfrak{M} - \int_{\mathscr{V}} (\boldsymbol{p}'\dot{\boldsymbol{p}}' + \dot{\boldsymbol{p}}'\boldsymbol{p}')\,d\mathfrak{M}. \tag{170.8}$$

---

[1] It is easily possible to study the material rate of change of other quantities associated with the mass flow $\varrho\dot{\boldsymbol{p}}$. For example, V. BJERKNES [1898, *1*, § 18] calculated the rate of change of the circulation of $\varrho\dot{\boldsymbol{p}}$ about a material circuit, but the result is not illuminating.

[2] v. MISES [1909, *8*, § 9.2] constructed an interpretation for the relative rate of change of momentum in a stream tube.

[3] Cf. PAINLEVÉ [1895, *4*, 2e Leçon], HUNTINGTON [1914, *6*], KELLOGG [1924, *8*].

Hence

$$\left.\begin{aligned}\int_{\mathscr{V}} \boldsymbol{p}'\times(2\boldsymbol{\omega}\times\dot{\boldsymbol{p}}')\,d\mathfrak{M} &= 2\boldsymbol{\omega}\cdot\int_{\mathscr{V}}[\mathbf{1}(\boldsymbol{p}'\cdot\dot{\boldsymbol{p}}')-\boldsymbol{p}'\dot{\boldsymbol{p}}']\,d\mathfrak{M},\\ &= \boldsymbol{\omega}\cdot\int_{\mathscr{V}}[2\,\mathbf{1}(\boldsymbol{p}'\cdot\dot{\boldsymbol{p}}')-(\boldsymbol{p}'\dot{\boldsymbol{p}}'+\dot{\boldsymbol{p}}'\boldsymbol{p}')+\\ &\qquad+(\dot{\boldsymbol{p}}'\boldsymbol{p}'-\boldsymbol{p}'\dot{\boldsymbol{p}}')]\,d\mathfrak{M},\\ &= \boldsymbol{\omega}\cdot\dot{\mathfrak{J}}'^{[O']}+\boldsymbol{\omega}\times\mathfrak{H}'^{[O']}.\end{aligned}\right\}\tag{170.9}$$

Also

$$\left.\begin{aligned}\int_{\mathscr{V}} \boldsymbol{p}'\times[\boldsymbol{\omega}\times(\boldsymbol{\omega}\times\boldsymbol{p}')]\,d\mathfrak{M} &= -\int_{\mathscr{V}}(\boldsymbol{p}'\cdot\boldsymbol{\omega})(\boldsymbol{\omega}\times\boldsymbol{p}')\,d\mathfrak{M},\\ &= -\boldsymbol{\omega}\times\int_{\mathscr{V}}\boldsymbol{p}'\boldsymbol{p}'\,d\mathfrak{M}\cdot\boldsymbol{\omega},\\ &= \boldsymbol{\omega}\times\mathfrak{J}'^{[O']}\cdot\boldsymbol{\omega}.\end{aligned}\right\}\tag{170.10}$$

Using (170.9) and (170.10) in (170.7) yields[1]

$$\dot{\mathfrak{H}}^{[O]} = \dot{\mathfrak{H}}'^{[O']} - \mathfrak{L}_{\mathrm{c}} - \mathfrak{L}_{\mathrm{r}},\tag{170.11}$$

where

$$\left.\begin{aligned}-\mathfrak{L}_{\mathrm{c}} &\equiv \boldsymbol{b}\times\mathfrak{M}[\ddot{\boldsymbol{b}}+\dot{\boldsymbol{\omega}}\times\boldsymbol{c}'+2\boldsymbol{\omega}\times\dot{\boldsymbol{c}}'+\boldsymbol{\omega}\times(\boldsymbol{\omega}\times\boldsymbol{c}')+\ddot{\boldsymbol{c}}']+\boldsymbol{c}'\times\mathfrak{M}\ddot{\boldsymbol{b}},\\ &= \boldsymbol{b}\times\mathfrak{M}\ddot{\boldsymbol{c}}+\boldsymbol{c}'\times\mathfrak{M}\ddot{\boldsymbol{b}},\\ -\mathfrak{L}_{\mathrm{r}} &\equiv \dot{\boldsymbol{\omega}}\cdot\mathfrak{J}'^{[O']}+\boldsymbol{\omega}\cdot\dot{\mathfrak{J}}'^{[O']}+\boldsymbol{\omega}\times\mathfrak{H}'^{[O']}+\boldsymbol{\omega}\times\mathfrak{J}'^{[O']}\cdot\boldsymbol{\omega}.\end{aligned}\right\}\tag{170.12}$$

The simple formula $(170.5)_2$ compares for two different observers the rates of change of moment of momentum calculated with respect to the *same point*, which may be moving in any way. (170.11), on the other hand, compares the rate of change of moment of momentum when each observer calculates it with respect to his *own frame and origin*. In (170.12), all dots on primed quantities indicate time rates as apparent in the primed frame. Thus an observer in the primed frame by combining appropriate quantities he himself observes may calculate the rate $\dot{\mathfrak{H}}^{[O]}$ apparent to an observer in the unprimed frame. In rigid motion, it is possible to choose the primed frame rigidly attached to the body, and then we have both $\dot{\mathfrak{J}}'^{[O']}=0$ and $\mathfrak{H}'^{[O']}=0$, so that $\mathfrak{L}_{\mathrm{r}}$ assumes the classical simple form used in rigid dynamics (cf. Sect. 294). Further simple cases will be mentioned in Sect. 197.

The formula $(170.1)_2$ has been used by Nyborg[2] to obtain an interpretation for the spatial diffusion tensor $\boldsymbol{w}^*$ defined by $(101.5)_2$. Writing quantities evaluated at the center of mass with a superscript $\boldsymbol{c}$, we have by (165.1)

$$\left.\begin{aligned}\int_{\mathscr{V}}(z_k-c_k)\,\ddot{z}_m\,d\mathfrak{M} &= \int_{\mathscr{V}}(z_k-c_k)\,[\ddot{z}^{\boldsymbol{c}}_m+(z_q-c_q)\,\ddot{z}^{\boldsymbol{c}}_{m,q}+O(r^2)]\,d\mathfrak{M},\\ &= \mathfrak{E}_{kq}\,\ddot{z}^{\boldsymbol{c}}_{m,q}+O(r^6),\end{aligned}\right\}\tag{170.13}$$

where r is the diameter of $\mathscr{V}$ and where $\mathfrak{E}$ is the Euler tensor defined by $(168.4)_1$, with $\boldsymbol{a}=\boldsymbol{c}$. Hence

$$\ddot{z}^{\boldsymbol{c}}_{m,q}\lim_{r\to0}\frac{\mathfrak{E}_{qk}}{r^5} = \lim_{r\to0}\frac{\int_{\mathscr{V}}(z_k-c_k)\,\ddot{z}_m\,d\mathfrak{M}}{r^5}.\tag{170.14}$$

If the ellipsoid of inertia is a sphere, we have $\mathfrak{E}_{mk}=\frac{1}{3}\mathfrak{J}\,\delta_{mk}$, where $\mathfrak{J}$ is the polar moment of inertia about $\boldsymbol{c}$. In this case (170.14) reduces to

$$\ddot{z}^{\boldsymbol{c}}_{m,k} = \lim_{r\to0}\frac{\int_{\mathscr{V}}(z_k-c_k)\,\ddot{z}_m\,d\mathfrak{M}}{\frac{1}{3}\mathfrak{J}};\tag{170.15}$$

---

[1] In principle, this calculation is due to Euler [1765, *3*, §§ 16—24] [1776. *3*, §§ 31—34].

[2] [1953, *20*]. Cf. Truesdell [1956, *22*]. The analysis presented above is more general than that in the two sources cited.

hence, by $(170.1)_2$,

$$w^*_{km} = \operatorname{Lim}_{r\to 0} \frac{\dot{\mathfrak{H}}_{km}}{\frac{1}{3}\mathfrak{J}}, \tag{170.16}$$

where we have omitted the superscript $\mathbf{c}$. This result asserts that *the curl of the acceleration at $\boldsymbol{x}$ is proportional to the rate of change of moment of momentum experienced by a vanishingly small sphere centered*[1] *at $\boldsymbol{x}$.*

171. Galilean invariance. We now determine the class of frames in which the rate of change of momentum of every body is the same as its rate of change of momentum in one given frame. Since $(170.5)_1$ may be written

$$\dot{\mathfrak{P}} = \mathfrak{M}[\ddot{\boldsymbol{b}} + \dot{\boldsymbol{\omega}}\times\boldsymbol{c}' + 2\boldsymbol{\omega}\times\dot{\boldsymbol{c}}' + \boldsymbol{\omega}\times(\boldsymbol{\omega}\times\boldsymbol{c}')] + \dot{\mathfrak{P}}', \tag{171.1}$$

in order that $\dot{\mathfrak{P}} = \dot{\mathfrak{P}}'$ for arbitrary $\mathfrak{M}$ and arbitrary $\boldsymbol{c}'$, i.e., for all bodies, whatever their masses and locations, we must have $\ddot{\boldsymbol{b}} = 0$ and $\boldsymbol{\omega} = 0$. This result is sometimes called the ***Galilean principle of relativity***: *In order that every body shall appear to have the same rate of change of momentum to two observers, it is necessary and sufficient that their frames be in relative translation at constant velocity.* That rate of change of momentum fails to be invariant under rotation or acceleration of the frame of the observer is a very old remark, whose history we do not attempt to relate here.

Given any frame, the totality of frames in uniform translation with respect to it constitutes its *Galilean class*; quantities invariant under change of frame within this subgroup of the group of rigid motions are called *Galilean invariants*. By (167.2), $\mathfrak{H}$ is a Galilean invariant, and so is $\dot{\mathfrak{H}}$. Of course $\mathfrak{H}^{[\boldsymbol{O}]}$, $\dot{\mathfrak{H}}^{[\boldsymbol{O}]}$, $\mathfrak{K}$, and $\dot{\mathfrak{K}}$ are not Galilean invariants, and also steadiness of motion (Sects. 67, 146) fails of Galilean invariance. Cf. the general treatment of the kinematical basis of Galilean invariance in Sect. 154.

C. Singular surfaces and Waves.

172. Scope and plan of the chapter. We organize and describe those properties of surfaces of discontinuity, such as vortex sheets, shock waves, and acceleration waves, as are common to all media. First we derive conditions that hold at any given instant and express the fact that the discontinuity is spread out smoothly over a surface, not isolated at a point or a line. The second subchapter presents a general differential description of moving surfaces. Then we prove kinematical conditions expressing the persistence of a surface of discontinuity. The fourth subchapter classifies the various kinds of discontinuities associated with the motion of a material and proves simple theorems characterizing them. Finally, we obtain the general form taken on by a conservation law when applied at a surface of discontinuity.

Most of the major ideas of the subject derive from the work of CHRISTOFFEL (1877) and HUGONIOT (1885), extended in the classical treatise of HADAMARD (1899—1900)[2].

---

[1] The variation of density within the sphere need not be considered, since the distance of the center of the sphere from its center of mass is $O(r^2)$ and hence may be neglected in the interpretation of (170.16).

[2] [1903, *11*, Chap. II]. There are also the expositions of ZEMPLÉN [1905, *9*] and LICHTENSTEIN [1929, *4*, Chap. 6]; the latter labors the analytical side but is confined to a narrower range than the older works. Some major additions are contained in a paper by KOTCHINE [1926, *3*]. Recent studies of discontinuities have emphasized calculation of solutions in particular theories of materials and have not extended the general theory. There is no modern treatise on the subject.

Surfaces at which the derivatives of the velocity are discontinuous first appear in the acoustical researches of EULER (1764—1765); the possibility that the velocity itself may be discontinuous was first remarked by STOKES (1848).

I. Geometry of singular surfaces.

173. Definition of a singular surface. Intentionally we begin our analysis on a very general plane; the application to the three-dimensional Euclidean space, which furnishes our main interest in this work, will follow shortly.

Consider a regular surface $\mathscr{s}$ which is the common boundary of two regions $\mathscr{R}^+$ and $\mathscr{R}^-$ in any real space (Fig. 23). Let $\psi(\boldsymbol{x})$ be a function which is continuous in the interiors of $\mathscr{R}^+$ and $\mathscr{R}^-$ and which approaches definite limit values ψ^+ and ψ^- as $\boldsymbol{x}$ approaches a point $\boldsymbol{x}_0$ on $\mathscr{s}$ while remaining within $\mathscr{R}^+$ and $\mathscr{R}^-$, respectively. At $\boldsymbol{x}_0$, ψ need not be defined. The *jump* of ψ across $\mathscr{s}$ at $\boldsymbol{x}_0$ is denoted by[1]

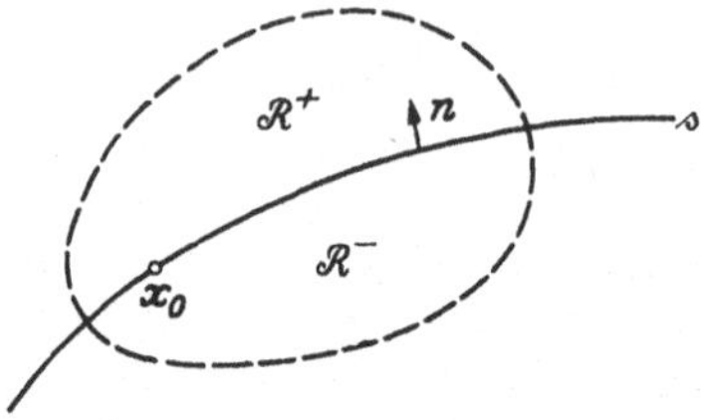

Fig. 23. Singular surface.

$$[\psi] \equiv \psi^+ - \psi^- \qquad (173.1)$$

[cf. (App. 36.3)]. The sign of the jump is a matter of convention, but, since all our considerations are local, it will occasion no trouble. The quantity $[\psi]$ is a function of position upon $\mathscr{s}$.

When (173.1) is applied to a tensor $\boldsymbol{T}$, the jump $[\boldsymbol{T}]$ is a tensor defined as a function of position upon $\mathscr{s}$. If $[\boldsymbol{T}]$ is normal to $\mathscr{s}$, the discontinuity of $\boldsymbol{T}$ is said to be *longitudinal*; if tangent to $\mathscr{s}$, *transversal*[2]. In a metric space, the jump of any tensor may be resolved uniquely into longitudinal and transversal components, such a resolution being given for three-dimensional vectors[3] by an identity such as (App. 30.6).

If $[\psi] \neq 0$, the surface $\mathscr{s}$ is said to be *singular* with respect ψ.

174. HADAMARD's lemma. The entire theory of singular surfaces rests upon ***Hadamard's lemma***[4]: *Let ψ be defined and continuously differentiable in the interior of a region $\mathscr{R}^+$ with smooth boundary $\mathscr{s}$, and let ψ and $\partial_k\psi$ approach finite limits ψ^+ and $\partial_k\psi^+$ as $\mathscr{s}$ is approached upon paths interior to $\mathscr{R}^+$. Let $\boldsymbol{x}=\boldsymbol{x}(l)$ be a smooth curve upon $\mathscr{s}$, and assume that ψ^+ is differentiable on this path. Then*

$$\frac{d\psi^+}{dl} = \partial_k \psi^+ \frac{dx^k}{dl}. \qquad (174.1)$$

In other words, the theorem of the total differential holds for the limiting values as $\mathscr{s}$ is approached *from one side only*. The function ψ need not be defined upon

[1] The notation was introduced by CHRISTOFFEL [1877, *2*, § 6].

[2] HADAMARD [1901, *8*, § 3], [1903, *11*, ¶ 115].

[3] In some three-dimensional applications it is convenient to use the notations of EMDE [1915, *2*, § 1] (cf. SPIELREIN [1916, *5*, §§ 10, 26]):

$$\operatorname{Grad} \boldsymbol{u} \equiv \boldsymbol{n}[\boldsymbol{u}],$$
$$\operatorname{Div} \boldsymbol{u} \equiv \boldsymbol{n}\cdot[\boldsymbol{u}],$$
$$\operatorname{Curl} \boldsymbol{u} \equiv \boldsymbol{n}\times[\boldsymbol{u}],$$

where $\boldsymbol{n}$ is the unit normal pointing into $\mathscr{R}^+$. Thus $\operatorname{Div}\boldsymbol{u}$ and $\operatorname{Curl}\boldsymbol{u}$ are the longitudinal and transversal jumps of $\boldsymbol{u}$, and the resolution of $[\boldsymbol{u}]$ is

$$[\boldsymbol{u}] = \boldsymbol{n}\operatorname{Div}\boldsymbol{u} - \boldsymbol{n}\times\operatorname{Curl}\boldsymbol{u}.$$

While $\operatorname{Grad}\boldsymbol{u}$, $\operatorname{Div}\boldsymbol{u}$, and $\operatorname{Curl}\boldsymbol{u}$ are sometimes called the "surface gradient, divergence, and curl" of the discontinuous field $\boldsymbol{u}$, they must not be confused with the differential invariants of a field defined intrinsically on the surface. EMDE's notations are motivated by (App. 36.2).

[4] HADAMARD [1903, *11*, ¶ 72] gives two proofs, the first of which is that reproduced above. The lemma was used tacitly by earlier authors. A more elaborate proof is given by LICHTENSTEIN [1929, *4*, Chap. 1, § 9].

the other side of $\mathscr{s}$. If it is, and if the corresponding limiting values Ψ^- and $\partial_k \Psi^-$ exist and have the required smoothness, a similar result holds for them, but $d\Psi^-/dl$ is in general unrelated to $d\Psi^+/dl$.

To prove (174.1), we need only remark that for two sufficiently near points on $\mathscr{s}$, of the polygonal paths employed in the classical proof applicable to interior points at least one is interior to $\mathscr{R}^+$, so the classical argument may be applied to it. For the case of two dimensions, the two paths are indicated in Fig. 24. The diagram is merely heuristic: HADAMARD'S lemma is a proposition in differential calculus, independent of any geometry that may hold in the spaces where we choose to apply it.

In particular, in an affine space it may be applied to the individual components $T^{\cdots}_{\cdots}$ of a tensor field. By adding to each side of (174.1) in this case suitable expressions made up from connection symbols and from $T^{\cdots}_{\cdots}$, we thus obtain

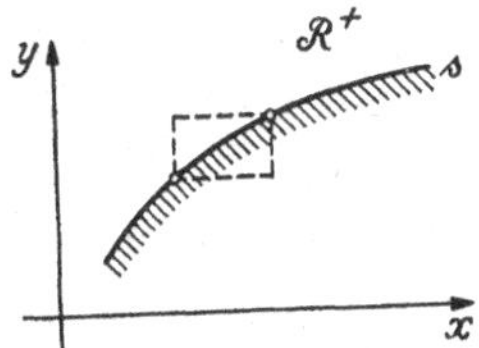

Fig. 24. Diagram for proof of HADAMARD'S lemma.

$$\tilde{T}^{\cdots +}_{\cdots} \equiv T^{\cdots +}_{\cdots ; \Gamma} \frac{dv^\Gamma}{dl} = T^{\cdots +}_{\cdots , k} \frac{dx^k}{dl} \tag{174.2}$$

where $\tilde{T}^{\cdots +}_{\cdots}$, the intrinsic derivative of $T^{\cdots +}_{\cdots}$ along the path $\boldsymbol{x} = \boldsymbol{x}(\boldsymbol{v}(l))$ on $\mathscr{s}$, is defined as the quantity to the right of the identity symbol, and where $T^{\cdots +}_{\cdots , k}$ stands for the limiting value of the covariant derivative $T^{\cdots}_{\cdots , k}$ as the point in question on $\mathscr{s}$ is approached from within $\mathscr{R}^+$.

175. Superficial and geometrical conditions of compatibility. We now apply HADAMARD'S lemma to singular surfaces as defined in Sect. 173. When, hereinafter, we derive a condition concerning the derivatives of a quantity Ψ, we presume without further explicit statement that all derivatives of Ψ of orders up to and including the one considered exist and are continuous in $\mathscr{R}^+$ and $\mathscr{R}^-$ and approach definite limits at $\boldsymbol{x}_0$ along paths lying wholly in $\mathscr{R}^+$ and wholly in $\mathscr{R}^-$. These limit values are assumed continuously differentiable functions of position on $\mathscr{s}$. Thus HADAMARD'S lemma (174.1) may be applied to the limiting values on *each side* of the singular surface $\mathscr{s}$:

$$\frac{d\Psi^+}{dl} = \partial_k \Psi^+ \frac{dx^k}{dl}, \quad \frac{d\Psi^-}{dl} = \partial_k \Psi^- \frac{dx^k}{dl}. \tag{175.1}$$

Subtracting the second of these equations from the first yields

$$\frac{d}{dl}[\Psi] = [\partial_k \Psi] \frac{dx^k}{dl} = \left[\partial_k \Psi \frac{dx^k}{dl}\right]. \tag{175.2}$$

The entire differential theory of singular surfaces grows from applications of this formula, which asserts that *the jump of a tangential derivative is the tangential derivative of the jump*. Since the values of Ψ in $\mathscr{R}^+$ and $\mathscr{R}^-$ are in general entirely unrelated to one another, the limiting values of the normal derivatives of Ψ on the two sides of the singular surface need be connected in no way:

$$\left[\frac{d\Psi}{dn}\right] \text{ is unrestricted.} \tag{175.3}$$

To express the fact that the discontinuity is *spread out smoothly over a surface* of exactly $n-m$ dimensions, not isolated upon a surface of lower dimension nor varying abruptly from one part of $\mathscr{s}$ to another, we need only apply (175.2) to $n-m$ *independent families of curves* upon $\mathscr{s}$. If we choose these as co-ordinate curves $v^\Gamma = \text{const}$, we have

$$\partial_\Gamma [\Psi] = [\partial_k \Psi]\, x^k_{;\Gamma}. \tag{175.4}$$

Equations (175.3) and (175.4) are the *superficial conditions of compatibility*. Like HADAMARD's lemma, from which they follow at once, they presuppose no geometry in the space.

In the important special case when Ψ is continuous, or, more generally, when $[\Psi] = \text{const}$, from (175.4) we obtain

$$[\partial_k \Psi]\, x^k_{;\Gamma} = 0. \tag{175.5}$$

Thus $[\partial_k \Psi]$ lies in the manifold normal to $\mathscr{s}$. If the dimension of $\mathscr{s}$ is but one less than that of the space, there is but one linearly independent normal vector, which may be selected as that given by taking $n - N = 1$ in (App. 19.4). Therefore

$$[\partial_k \Psi] = B\nu_k, \tag{175.6}$$

where B is a factor of proportionality.

We have intentionally kept the argument on the highest level of generality. As soon as geometrical structure is introduced, more explicit formulae become possible. In an affine space, for the jump of a tensor Ψ we have the following important identity valid when Ψ is continuous or when Ψ is a scalar and $[\Psi] = \text{const}$:

$$[\Psi_{,k}] = [\partial_k \Psi], \tag{175.7}$$

since $\Psi_{,k} - \partial_k \Psi$ is a continuous quantity. In a metric space, we may introduce the unit normal $\boldsymbol{n}$ and evaluate explicitly the factor B in (175.6), so obtaining ***Maxwell's theorem***[1]: *When* $[\Psi] = 0$, *or when* Ψ *is a scalar and* $[\Psi] = \text{const}$, *then*

$$[\Psi_{,k}] = B n_k, \qquad B = [n^k \partial_k \Psi] = [n^k \Psi_{,k}] = \left[\frac{d\Psi}{dn}\right]. \tag{175.8}$$

We now consider applications to vector fields in the space of three dimensions. Generalization to tensor fields in n-dimensional spaces is immediate and is left to the reader.

If we take $\Psi = \boldsymbol{c}$, a vector, (175.8) becomes

$$[c_{k,m}] = B_k n_m, \qquad B_k = [n^m c_{k,m}] = \left[\frac{d\boldsymbol{c}}{dn}\right]. \tag{175.9}$$

Hence

$$[\operatorname{curl} \boldsymbol{c}] = \boldsymbol{n} \times \boldsymbol{B}, \quad [\operatorname{div} \boldsymbol{c}] = \boldsymbol{n} \cdot \boldsymbol{B}, \quad \boldsymbol{B} = \boldsymbol{n}[\operatorname{div} \boldsymbol{c}] - \boldsymbol{n} \times [\operatorname{curl} \boldsymbol{c}]. \tag{175.10}$$

These formulae embody ***Weingarten's first theorem***[2]: *The longitudinal and transversal jumps of the gradient of a continuous vector are the jumps of its divergence and curl, respectively.* It follows as a corollary that the only possible jump in the gradient of a continuous lamellar field is normal; of a continuous solenoidal field, transversal.

Results concerning the jump of the field itself can be inferred by assuming a representation in terms of a continuous potential. First, suppose we have a *lamellar* field $\boldsymbol{c} = -\operatorname{grad} P$ with a potential P [cf. (App. 33.2)] which is continuous or suffers a constant jump on the singular surface. It is but reinterpretation of MAXWELL's theorem (175.8) to say that *only the normal component* c_n can suffer a jump; the tangential component must be continuous[3]. Second, suppose we

[1] [1873, *5*, § 78] [1881, *4*, § 78a]. Cf. also WEINGARTEN [1901, *15*]. HADAMARD [1903, *11*, ¶ 73] called the result "the identical conditions". MAXWELL seems to have been the first to derive compatibility conditions by differentiating a jump relation along a path lying on a surface.

[2] [1901, *15*].

[3] MAXWELL [1873, *5*, § 78] [1881, *4*, § 78a], FERRARIS [1897, *2*, ¶ 42]. This seems to be the content of the obscure statements and proofs of BROCA [1899, *2*], [1900, *2*].

have a *solenoidal* field $\boldsymbol{c} = \operatorname{curl} \boldsymbol{v}$ with a solenoidal vector potential $\boldsymbol{v}$ which is continuous or has a constant jump on the surface. From (175.10) we have at once $[\boldsymbol{c}] = [\operatorname{curl} \boldsymbol{v}] = \boldsymbol{n} \times \boldsymbol{B}$, and therefore $[c_n] = 0$. That is, *only the tangential component* $\boldsymbol{n} \times \boldsymbol{c}$ can suffer a jump; the normal component must be continuous[1]. These results are used frequently in electromagnetism.

In a metric space, the more general conditions (175.4) and (175.3), which do not presuppose $[\Psi] = \text{const}$, can be included in a single formula, which we now derive for the case of a surface in three-dimensional space. We simply multiply both sides of (175.4) by $a^{\Delta\Gamma} x^m_{;\Delta}$, use the identity (App. 21.4)$_1$, and obtain

$$\left.\begin{aligned}[\partial_k \Psi] &= [n^m \partial_m \Psi]\, n_k + g_{km} a^{\Delta\Gamma} x^m_{;\Delta} \partial_\Gamma [\Psi], \\ [\Psi_{,k}] &= [n^m \Psi_{,m}]\, n_k + g_{km} a^{\Delta\Gamma} x^m_{;\Delta} [\Psi]_{;\Gamma},\end{aligned}\right\} \tag{175.11}$$

where for the second form Ψ is assumed to be a tensor, and the semicolon denotes the total covariant derivative defined in Sect. App. 20. These *geometrical conditions of compatibility*[2] are no more than formal alternative expressions of the validity of (175.4) for two independent families of curves upon $\mathscr{s}$. Indeed, taking the scalar product of (175.11)$_1$ first by n^k and then by $x^k_{;\Gamma}$ yields (175.3) and (175.4).

An easy calculation using (175.11) and (App. 21.3)$_2$ yields the following identity for the magnitudes of the jumps:

$$[\Psi_{,k}]\,[\Psi^{,k}] = [n^k \Psi_{,k}]^2 + [\Psi]_{;\Gamma} [\Psi]^{;\Gamma}, \tag{175.12}$$

which might easily have been predicted from our remarks above concerning the nature of the geometrical conditions. In the case when Ψ is continuous, by comparing (175.12) with (175.8) we obtain

$$[\Psi_{,k}]\,[\Psi^{,k}] = B^2, \quad \text{or} \quad |B| = |[\Psi_{,k}]|. \tag{175.13}$$

While we have written these last two results in notations appropriate only when Ψ is a scalar, results of the same kind hold when Ψ is a tensor of any order.

There is a curious special condition of compatibility which is more easily derived directly than by application of the above results and those in the next section. For any vector field $\boldsymbol{c}$ in $\mathscr{R}^+$, we begin with the identity

$$\left.\begin{aligned} d c_k = \{c_{(k,r)} - (p^q - p^q_{(1)})\,(c_{(k,r),q} - c_{(q,r),k})\}\, d x^k + \\ + d\,\{c_{[k,q]}\,(p^q - p^q_{(1)})\},\end{aligned}\right\} \tag{175.14}$$

where $\boldsymbol{p}$ and $\boldsymbol{p}_{(1)}$ are the position vectors of a variable and a fixed point on the curve along which $d\boldsymbol{c}$ is calculated, so that $d\boldsymbol{p} = d\boldsymbol{x}$, $d\boldsymbol{p}_{(1)} = 0$. By HADAMARD's lemma, this identity still holds for a curve on the + side of $\mathscr{s}$, provided all the covariant derivatives be interpreted as limit values from the + side. We now consider the special case when $c_{(k,m)}$ and $c_{(k,m),p}$ are continuous across $\mathscr{s}$, while c_k and $c_{[k,m]}$ may suffer jump discontinuities. By writing an identity of the type (175.14) for each side of the surface and subtracting the results, we obtain

$$d[c_k] = d\,\{[c_{[k,q]}]\,(p^q - p^q_{(1)})\}. \tag{175.15}$$

Integrating from some point $\boldsymbol{p}_{(0)}$ to $\boldsymbol{p}_{(1)}$ on $\mathscr{s}$, and then dropping the subscript (1), we obtain

$$\left.\begin{aligned}[c_k] &= [c_k]_0 + [c_{[k,q]}]_0\,(p^q - p^q_{(0)}), \\ [\boldsymbol{c}] &= [\boldsymbol{c}]_0 + \tfrac{1}{2}[\operatorname{curl} \boldsymbol{c}]_0 \times (\boldsymbol{p} - \boldsymbol{p}_{(0)}).\end{aligned}\right\} \tag{175.16}$$

---

[1] FERRARIS [1897, *2*, ¶ 55], with an inadequate proof. A result of this kind had been given by HELMHOLTZ [1858, *1*, § 4].

[2] As has been remarked by KOTCHINE [1926, *3*, § 1] these conditions are included, if not very obviously, in much more general ones given by COULON [1902, *3*, §§ 3, 8, 46]. They were rediscovered by THOMAS [1957, *15*, § 3]. The elegant formal proof in the text is due to KANWAL [1958, *5*, § 5].

These formulae express **Weingarten's second theorem**[1]: *The discontinuity of a vector field* **c** *across a surface where* $c_{(k,m)}$ *and* $c_{(k,m),p}$ *are continuous is determined, as a function of position upon the surface, by its value and the value of* [curl **c**] *at any one point.* The theorem arose in the context of small displacement and small strain (Sect. 58). So interpreted, it asserts that a displacement corresponding to continuous strain and continuous strain gradient can suffer at most a discontinuity corresponding to a rigid motion of the material on one side of σ with respect to that on the other.

The more general problem of characterizing those discontinuities of a vector **c** across which $c_{(k,m)}$ is continuous is easily settled[2]. By $(175.11)_2$ we have

$$[c_{(k,m)}] = [n^p n_{(k} c_{m),p}] + x^p_{;\Delta} a^{\Delta\Gamma} g_{p(k}[c_{m)}]_{;\Gamma}. \tag{175.17}$$

By $(\text{App. }21.3)_2$ it follows that

$$\left.\begin{aligned} &[n^k n^m c_{(k,m)}] \quad \text{is unrestricted},\\ &2[c_{(k,m)}]\, n^k x^m_{;\Delta} = [n^k c_{m,k}]\, x^m_{;\Delta} + n^k [c_k]_{;\Delta},\\ &[c_{(k,m)}]\, x^k_{;\Delta} x^m_{;\Lambda} = [c_k]_{;(\Delta}\, x^k_{;\Lambda)}. \end{aligned}\right\} \tag{175.18}$$

These formulae express the resolution of (175.17) into normal and tangential components. Therefore, necessary and sufficient conditions that $[c_{(k,m)}] = 0$ are

$$\left.\begin{aligned} [n^k n^m c_{(k,m)}] &= 0,\\ [n^k c_{m,k}]\, x^m_{;\Delta} + n^k [c_k]_{;\Delta} &= 0,\\ [c_k]_{;(\Delta}\, x^k_{;\Lambda)} &= 0. \end{aligned}\right\} \tag{175.19}$$

These conditions do not suffice for the truth of WEINGARTEN's result (175.16), which presumes that $[c_{(k,m),q}] = 0$ as well.

176. Iterated geometrical conditions of compatibility. Since the geometrical conditions of compatibility (175.11) are mere identities resolving the jump of an arbitrary derivative into the jump of the normal derivative and the tangential derivatives of the jump of the function, it is clear that iteration will yield an expression for the jump of the second derivatives, $[\partial_k \partial_m \Psi]$, in terms of the values of

$$[n^p n^q \partial_p \partial_q \Psi], \quad [n^p \partial_p \Psi], \quad \partial_\Gamma [n^p \partial_p \Psi], \quad [\Psi], \quad \partial_\Gamma [\Psi], \quad \partial_\Gamma \partial_\Delta [\Psi],$$

and of the geometry of the singular surface. An analogous result holds for the derivatives of any order. This was perceived by HADAMARD[3] and applied in special cases; the general reduction for the second derivatives in a Euclidean space of three dimensions was worked out by THOMAS[4], whose analysis we present now. It is possible, but cumbrous, to carry through the work using partial derivatives, but an elegant form results if we suppose Ψ to be a tensor and employ covariant derivatives throughout.

Writing

$$\left.\begin{aligned} &A \equiv [\Psi], \quad A_k \equiv [\Psi_{,k}],\\ &B \equiv A_k n^k = [n^k \Psi_{,k}], \quad B_k \equiv [\Psi_{,km}]\, n^m,\\ &C \equiv B_k n^k = [n^k n^m \Psi_{,km}], \end{aligned}\right\} \tag{176.1}$$

---

[1] [1901, *15*]. The proof is a simplified version of that given by CESARO [1906, *2*]. The original treatment and those in textbooks do not make it sufficiently clear that $c_{(k,m),q}$ is assumed to be continuous. As may be seen from the formulae of the next section, it is only this assumption that makes so definite a conclusion possible.

[2] The problem is related to one solved by SOMIGLIANA [1914, *11*, § II], who considered the case when $c_{(k,m)}$ is the strain tensor of a linearly elastic equilibrated body. The continuity of the stress vector, as required by (205.5), implies that $(175.19)_{1,2}$ are satisfied, so that only $(175.19)_3$ remains, and this is SOMIGLIANA's result.

[3] [1903, *11*, ¶¶ 119—120].

[4] [1957, *15*, § 5]. Much more general results are given in schematic form by COULON [1902, *3*, §§ 4, 46].

from $(175.11)_2$ we have

$$\left.\begin{aligned}[\Psi_{,k}] &= B n_k + g_{km} a^{\Delta\Gamma} x^m_{;\Delta} A_{;\Gamma} = A_k,\\ [\Psi_{,km}] &= B_m n_k + g_{kp} a^{\Delta\Gamma} x^p_{;\Delta} A_{m;\Gamma}.\end{aligned}\right\} \tag{176.2}$$

Since the left-hand side of $(176.2)_2$ is symmetric in the indices k and m, it follows that

$$B_m n_k + g_{kp} a^{\Delta\Gamma} x^p_{;\Delta} A_{m;\Gamma} = B_k n_m + g_{mp} a^{\Delta\Gamma} x^p_{;\Delta} A_{k;\Gamma}. \tag{176.3}$$

Taking the scalar product of this equation by $\boldsymbol{n}$ yields

$$B_m = C n_m + g_{mp} a^{\Delta\Gamma} x^p_{;\Delta} n^k A_{k;\Gamma}, \tag{176.4}$$

where we have used $(\text{App.}\,21.3)_2$ and $(176.1)_6$.

From $(176.1)_2$ and $(176.2)_1$ we have at once

$$A_{k;\Gamma} = (B n_k + g_{km} a^{\Delta\Lambda} x^m_{;\Delta} A_{;\Lambda})_{;\Gamma} = (B n_k)_{;\Gamma} + g_{km} a^{\Delta\Lambda} (x^m_{;\Delta} A_{;\Lambda})_{;\Gamma}. \tag{176.5}$$

By use of $(\text{App.}\,21.6)_{1,2}$ follows

$$A_{k;\Gamma} = n_k B_{;\Gamma} - g_{kp} b^\Delta_\Gamma x^p_{;\Delta} B + g_{km} a^{\Delta\Lambda} x^m_{;\Delta} A_{;\Lambda\Gamma} + n_k b^\Delta_\Gamma A_{;\Delta}. \tag{176.6}$$

Hence by use of $(\text{App.}\,21.3)_2$ we have

$$n^k A_{k;\Gamma} = B_{;\Gamma} + b^\Delta_\Gamma A_{;\Delta}. \tag{176.7}$$

Substituting from (176.7) into (176.4) and then putting the result and (176.6) into $(176.2)_2$, we obtain THOMAS's *iterated geometrical condition of compatibility:*

$$\left.\begin{aligned}[\Psi_{,km}] &= C n_k n_m + 2 n_{(k} g_{m)p} a^{\Delta\Gamma} x^p_{;\Delta} (B_{;\Gamma} + b_{\Gamma\Sigma} a^{\Sigma\Lambda} A_{;\Lambda}) +\\ &\quad + g_{kp} g_{mq} x^p_{;\Delta} x^q_{;\Gamma} a^{\Delta\Lambda} a^{\Gamma\Sigma} (A_{;(\Sigma\Lambda)} - b_{\Sigma\Lambda} B),\\ &= C n_k n_m + 2 n_{(k} x_{m);}{}^{\Gamma} (B_{;\Gamma} + b^\Delta_\Gamma A_{;\Delta}) + x_{(k;}{}^{\Gamma} x_{m);}{}^{\Lambda} (A_{;(\Gamma\Lambda)} - b_{\Gamma\Lambda} B).\end{aligned}\right\} \tag{176.8}$$

The scalar product of $(176.8)_2$ by g^{km}, simplified by use of $(\text{App.}\,21.33)_2$, (App. 19.7), and $(\text{App.}\,21.10)_6$, becomes

$$[\Psi^{,k}{}_{,k}] = C + A^{;\Gamma}{}_{;\Gamma} - \overline{K} B = \left[\frac{d^2\Psi}{dn^2}\right] + [\Psi]^{;\Gamma}{}_{;\Gamma} - \overline{K}\left[\frac{d\Psi}{dn}\right], \tag{176.9}$$

which may be set side by side with (175.12).

While the simple condition of compatibility (175.11), or $(176.2)_1$, expresses the 3 jumps $[\Psi_{,k}]$ in terms of the 3 quantities B and $A_{;\Gamma}$ defined on the surface, the iterated condition (176.8) is in a measure redundant, since it expresses the 6 jumps $[\Psi_{,km}]$ in terms of the 9 surface quantities C, B, $B_{;\Gamma}$, $A_{;\Lambda}$, and $A_{;(\Sigma\Phi)}$. Since, however, the 3 quantities $A_{;\Lambda}$ and B are determined by $A_{;(\Sigma\Phi)}$ and $B_{;\Gamma}$, there are but 6 independent jumps, as expected. Conditions of this kind, being pure identities, merely rearrange the variables. Only when they are used in connection with some further hypothesis, such that some particular quantity is continuous, may fruit be gotten from them.

In the important special case when both Ψ and $\Psi_{,k}$ are continuous[1], so that $A = B = 0$, (176.8) reduces to the form[2]

$$\left.\begin{aligned}[\Psi_{,km}] &= [\partial_k \partial_m \Psi] = C n_k n_m,\\ C &= [n^p n^q \Psi_{,pq}] = [n^p n^q \partial_p \partial_q \Psi] = [\Psi_{,p}{}^{p}].\end{aligned}\right\} \tag{176.10}$$

[1] If we suppose merely that $\Psi_{,k}$ is continuous and $[\Psi]_{;\Gamma} = 0$, we obtain

$$[\Psi_{,km}] = C n_k n_m, \quad C = [n^k n^m \Psi_{,km}],$$

but not all the other forms included in (176.10) remain necessarily valid.

[2] HADAMARD [1901, *8*, § 1], [1903, *11*, ¶ 74].

More generally[1], if Ψ and its derivatives of orders $1, 2, \ldots, p-1$ are continuous, we have

$$\left.\begin{aligned}[\Psi_{,k_1 k_2 \ldots k_p}] &= [\partial_{k_1} \partial_{k_2} \ldots \partial_{k_p} \Psi], \\ &= [n^{m_1} n^{m_2} \ldots n^{m_p} \partial_{m_1} \partial_{m_2} \ldots \partial_{m_p} \Psi]\, n_{k_1} n_{k_2} \ldots n_{k_p}, \\ &= [n^{m_1} n^{m_2} \ldots n^{m_p} \Psi_{,m_1 m_2 \ldots m_p}]\, n_{k_1} n_{k_2} \ldots n_{k_p}, \\ &= \begin{cases} [\Psi^{,m_1}{}_{,m_1}{}^{,m_2}{}_{,m_2} \ldots {}^{,m_{p/2}}{}_{,m_{p/2}}]\, n_{k_1} n_{k_2} \ldots n_{k_p} & \text{if } p \text{ is even} \\ [n^q \Psi^{,m_1}{}_{,m_1}{}^{,m_2}{}_{,m_2} \ldots {}^{,m_{(p-1)/2}}{}_{q}]\, n_{k_1} n_{k_2} \ldots n_{k_p} & \text{if } p \text{ is odd.} \end{cases}\end{aligned}\right\} \quad (176.11)$$

That is, the jumps in all the $p!$ derivatives of p-th order are determined uniquely by the jump in the completely normal p-th derivative and by the unit normal $\boldsymbol{n}$. By using an argument such as that leading to (175.8), we see that (176.11) holds in a metric space of any dimension; a corresponding generalization of (175.6) holds in any kind of space, irrespective of what geometrical structure it may have.

A second iteration would enable us to derive a general resolution of $[\Psi_{,kmp}]$ in the Euclidean three-dimensional case, but the formal complexities encountered in deriving (176.8) render the details of such an analysis forbidding and the result too complicated to be useful.

II. The motion of surfaces.

177. The speed of displacement and the normal velocity of a moving surface. Consider a family of surfaces given by

$$\boldsymbol{x} = \boldsymbol{x}(\boldsymbol{V}, t), \qquad (177.1)$$

where $\boldsymbol{V}$ stands for a pair of surface parameters V^Δ identifying what we shall call a *surface point*. $\boldsymbol{V}$, in general, is not to be confused with a material particle of any motion that may be occurring; indeed, the considerations in this section should be regarded as independent of the motion of substances, although as an aid to visualization it is often convenient to picture the moving surface as consisting of identifiable particles. The representation (177.1) gives the places $\boldsymbol{x}$ occupied by the surface point $\boldsymbol{V}$ as the time t progresses; thus it describes the motion of a surface. The *velocity* of the surface point $\boldsymbol{V}$ is defined by

$$\boldsymbol{u} = \left.\frac{\partial \boldsymbol{x}}{\partial t}\right|_{\boldsymbol{V}=\text{const}}. \qquad (177.2)$$

If we eliminate the parameters $\boldsymbol{V}$, we may write (177.1) in the form

$$f(\boldsymbol{x}, t) = 0. \qquad (177.3)$$

Conversely, however, from a spatial representation (177.3) it is not possible to calculate a unique form (177.1). This is easy to understand: Given a moving surface, there are infinitely many ways of identifying the points on its successive configurations in such a way that all those configurations are swept out smoothly by the surface points constituting any one of them.

Supposing, now, that we have any one parametrization (177.1), by differentiating (177.3) with respect to t we get

$$\frac{\partial f}{\partial t} + \boldsymbol{u} \cdot \operatorname{grad} f = 0. \qquad (177.4)$$

[1] COULON [1902, *3*, § 46], HADAMARD [1903, *11*, ¶ 74]. Contrary to the implication of THOMAS [1957, *15*, § 1], the analysis of HADAMARD is not restricted to any special choice of co-ordinates.

Writing $\boldsymbol{n}$ for the unit normal to the surface, by (177.4) we have

$$u_n = \boldsymbol{u} \cdot \boldsymbol{n} = \boldsymbol{u} \cdot \frac{\operatorname{grad} f}{|\operatorname{grad} f|} = - \frac{\frac{\partial f}{\partial t}}{\sqrt{f_{,k} f^{,k}}}; \tag{177.5}$$

since the right member is determined by the spatial equation (177.3) alone, it is independent of our choice of the parametrization (177.1). That is, *all possible velocities* $\boldsymbol{u}$ *of the moving surface have the same normal component* u_n, which is called the *speed of displacement* of the surface[1]. Cf. Sect. 74.

For some purposes it is convenient to make the particular choice of surface points implied by requiring $\boldsymbol{u}$ to be normal to the surface;

$$\boldsymbol{u} = u_n \boldsymbol{n}. \tag{177.6}$$

This velocity will be called the *normal velocity* of the surface. The identification of surface points may be visualized by erecting normal vectors of magnitude $u_n\, dt$ from each point on the configuration of the surface at some one time t; the termini of these vectors then sweep out the configuration at time $t + dt$. When $u_n = f(t)$, the surfaces so generated are parallel surfaces.

Suppose now that we have any parametrization

$$\boldsymbol{x} = \boldsymbol{x}(\boldsymbol{v}, t) \tag{177.7}$$

which is consistent with (177.3). Then both of these equations may be used simultaneously, so that (177.3) becomes $f(\boldsymbol{x}(\boldsymbol{v}, t), t) = 0$. Differentiation with respect to t yields

$$\frac{\partial f}{\partial t} + f_{,k} \frac{\partial x^k}{\partial t} = 0. \tag{177.8}$$

From (177.5) it follows that

$$u_n = n_k \frac{\partial x^k}{\partial t}. \tag{177.9}$$

Conversely, (177.5) follows from (177.9), so that these two equations furnish equivalent definitions of the speed of displacement according as the representation (177.3) or (177.7) for the surface is preferred.

The parameter $\boldsymbol{v}$ in (177.7) may, but need not, be identified with what was called a surface point $\boldsymbol{V}$ above. In any case, given a particular parametrization (177.7), an observer moving with the velocity (177.6), which we have called the normal velocity of the surface, will encounter points on the surface (177.7) having surface co-ordinates $\boldsymbol{v}$ which vary in time. Their rates of change u^Γ, which we shall call the *tangential velocity* of the parametrization, may be calculated[2]. Such a velocity must satisfy

$$\frac{\partial x^k}{\partial t} + u^\Gamma x^k_{;\Gamma} = u_n n^k. \tag{177.10}$$

Taking the scalar product first by n_k and then by $g_{km} x^m_{;\Delta}$ yields (177.9) and[3]

$$u_\Gamma = - g_{km} \frac{\partial x^k}{\partial t} x^m_{;\Gamma}. \tag{177.11}$$

[1] This quantity was introduced by STOKES [1848, *4*, p. 353], who called it "the speed of propagation"; cf. also KELVIN [1848, *5*]; for general surfaces it first appears, unnamed, in the work of CHRISTOFFEL [1877, *2*, § 1]; also HUGONIOT [1885, *3*, p. 1120] first called it "vitesse de propagation" but immediately thereafter [1885, *4*, p. 1231] distinguished "deux *vitesses de propagation*", the other being that we consider in Sect. 183. The term "vitesse de déplacement de l'onde" was introduced by HADAMARD [1901, *8*, § 1], [1903, *11*, ¶ 100].

[2] THOMAS [1957, *15*, Eqs. (52) to (54)].

[3] Cf. (177.2). Eq. (177.10) thus furnishes a resolution of the particular choice of surface velocity specified by (177.6).

Conversely, if we choose to define u_n and u_Γ by (177.9) and (177.11), we have by (App. 21.4)$_1$

$$\left.\begin{aligned}\frac{\partial x^k}{\partial t}+u^\Gamma x^k_{;\Gamma}&=\frac{\partial x^k}{\partial t}-a^{\Gamma\Delta}g_{pm}\frac{\partial x^p}{\partial t}x^m_{;\Delta}x^k_{;\Gamma}\\&=\frac{\partial x^k}{\partial t}-g_{pm}\frac{\partial x^p}{\partial t}(g^{mk}-n^m n^k),\end{aligned}\right\}\tag{177.12}$$

whence (177.10) follows. A necessary and sufficient condition that the parameter $\boldsymbol{v}$ correspond to a surface point $\boldsymbol{V}$ for the normal motion (177.6) is $u_\Gamma=0$.

178. Differential description of a moving surface[1]. Supposing that $\boldsymbol{a}$, $\boldsymbol{b}$, u_n, and u^Γ are given as functions of $\boldsymbol{v}$ and t, we ask if there exists a representation (177.7) such that these quantities belong to it. In other words, in order to define a moving surface does it suffice to assign as functions of t and the parameters $\boldsymbol{v}$ its first and second forms, its speed of displacement, and the tangential velocity of the parametrization? The answer, in general, is negative. In addition to the Eqs. (App. 21.8) of MAINARDI-CODAZZI and GAUSS, it is necessary that other conditions of compatibility be satisfied.

First we differentiate (177.11), obtaining

$$u_{\Gamma;\Delta}=-g_{km}\left(\frac{\partial x^k_{;\Delta}}{\partial t}x^m_{;\Gamma}+\frac{\partial x^k}{\partial t}x^m_{;\Gamma\Delta}\right).\tag{178.1}$$

By (App. 19.7) and (App. 21.6)$_1$ it follows that

$$\frac{\partial a_{\Gamma\Delta}}{\partial t}+2u_{(\Gamma;\Delta)}=-2u_n b_{\Gamma\Delta}.\tag{178.2}$$

From the definition of the total covariant derivative it is easy to verify that

$$\left(\frac{\partial x^k_{;\Gamma}}{\partial t}\right)_{;\Delta}=\frac{\partial x^k_{;\Gamma\Delta}}{\partial t}+x^k_{;\Lambda}\frac{\partial}{\partial t}\left\{\begin{matrix}\Lambda\\\Delta\Gamma\end{matrix}\right\}.\tag{178.3}$$

By (App. 19.6)$_1$ we then have

$$n_k\left(\frac{\partial x^k_{;\Gamma}}{\partial t}\right)_{;\Delta}=n_k\frac{\partial x^k_{;\Gamma\Delta}}{\partial t}.\tag{178.4}$$

We now differentiate (177.9) and by use of (178.4), (App. 21.6)$_2$, (App. 21.7)$_1$, (App. 21.6)$_1$, (177.11), (177.9), and (App. 19.7) obtain

$$\left.\begin{aligned}u_{n;\Gamma\Delta}&=n_{k;\Gamma\Delta}\frac{\partial x^k}{\partial t}+2n_{k;(\Gamma}\frac{\partial x^k_{;\Delta)}}{\partial t}+n_k\frac{\partial x^k_{;\Gamma\Delta}}{\partial t},\\&=-g_{km}\frac{\partial x^k}{\partial t}(b^\Lambda_{\Gamma;\Delta}x^m_{;\Lambda}+b^\Lambda_\Gamma b_{\Lambda\Delta}n^m)-\\&\quad-2g_{km}x^m_{;\Lambda}b^\Lambda_{(\Gamma}\frac{\partial x^k_{;\Delta)}}{\partial t}+n_k\frac{\partial}{\partial t}(n^k b_{\Gamma\Delta}),\\&=u_\Lambda b^\Lambda_{\Gamma;\Delta}-u_n b^\Lambda_\Gamma b_{\Lambda\Delta}-b^\Lambda_{(\Gamma}\frac{\partial a_{\Delta)\Lambda}}{\partial t}+\frac{\partial b_{\Gamma\Delta}}{\partial t}.\end{aligned}\right\}\tag{178.5}$$

From (App. 21.8)$_1$ and (178.2) it follows that

$$\frac{\partial b_{\Gamma\Delta}}{\partial t}+u^\Lambda b_{\Gamma\Delta;\Lambda}+b^\Lambda{}_{(\Gamma}u_{\Delta);\Lambda}=u_{n;\Gamma\Delta}-u_n b^\Lambda_\Gamma b_{\Lambda\Delta}.\tag{178.6}$$

[1] The results (178.2) and (178.6), though not the proofs given here, were disclosed to us by J.L. ERICKSEN.

Equations (178.2) and (178.6) are *conditions of compatibility* to be satisfied by $\boldsymbol{a}$, $\boldsymbol{b}$, u_n, and u^Γ. ERICKSEN has shown that conversely, if the quantities $\boldsymbol{a}, \boldsymbol{b}, u_n$, and u^Γ satisfy (App. 21.8), (178.2), and (178.6), then they are derivable from a relation of the form (177.7) with an assigned spatial metric $\boldsymbol{g}$; that is, the conditions of compatibility here derived are also *sufficient* for the existence of a moving surface. Therefore any other condition satisfied by $\boldsymbol{a}$, $\boldsymbol{b}$, u_n, and u^Γ will be a consequence of the relations already derived.

179. The displacement derivative. Given a function $F(\boldsymbol{v}, t)$ defined upon the moving surface, its rate of change $\delta F/\delta t$ as apparent to an observer moving with the normal velocity (177.6) of the surface is[1]

$$\frac{\delta F}{\delta t} = \frac{\partial F}{\partial t} + u^\Gamma \partial_\Gamma F, \tag{179.1}$$

where u^Γ is the tangential velocity of the parametrization, given by (177.11). Suppose that $G(\boldsymbol{x}, t)$ be a function such that on the surface $\mathscr{s}$ we have $G(\boldsymbol{x}(\boldsymbol{v},t)t) = F(\boldsymbol{v}, t)$. Then

$$\frac{\partial F}{\partial t} = \frac{\partial G}{\partial t} + \frac{\partial x^k}{\partial t} \partial_k G, \qquad \partial_\Gamma F = x^k_{;\Gamma} \partial_k G. \tag{179.2}$$

Hence by (177.10) we have

$$\left.\begin{aligned} \frac{\delta G}{\delta t} \equiv \frac{\delta F}{\delta t} &= \frac{\partial G}{\partial t} + \left(\frac{\partial x^k}{\partial t} + u^\Gamma x^k_{;\Gamma}\right) \partial_k G, \\ &= \frac{\partial G}{\partial t} + u_n n^k \partial_k G. \end{aligned}\right\} \tag{179.3}$$

If F and G are tensors, $\delta F/\delta t$ and $\delta G/\delta t$ as defined by (179.1) and $(179.3)_3$ generally fail to be tensors. For a double tensor $\Psi^{k\ldots m\,\Gamma\ldots\Delta}_{p\ldots q\,\Lambda\ldots\Sigma}(\boldsymbol{x}, \boldsymbol{v}, t)$, we define the *displacement derivative* $\delta_d \Psi/\delta t$ as that double tensor under the group of transformations $\boldsymbol{x}^* = \boldsymbol{x}^*(\boldsymbol{x})$, $\boldsymbol{v}^* = \boldsymbol{v}^*(\boldsymbol{v}, t)$ which reduces to $\delta\Psi/\delta t$ when the spatial co-ordinates are rectangular Cartesian and the tangential velocity u^Γ vanishes. To calculate $\delta_d \Psi/\delta t$, we first introduce the "Lie derivative" $\underset{\boldsymbol{u}}{\pounds}\Psi$ when any spatial indices of Ψ or dependence of Ψ upon $\boldsymbol{x}$ is ignored, viz.

$$\left.\begin{aligned} \underset{\boldsymbol{u}}{\pounds}\Psi^{k\ldots m\,\Gamma\ldots\Delta}_{p\ldots q\,\Lambda\ldots\Sigma} &= u^\Phi \partial_\Phi \Psi^{k\ldots m\,\Gamma\ldots\Delta}_{p\ldots q\,\Lambda\ldots\Sigma} - \Psi^{k\ldots m\,\Phi\ldots\Delta}_{p\ldots q\,\Lambda\ldots\Sigma} \partial_\Phi u^\Gamma - \cdots + \\ &\quad + \Psi^{k\ldots m\,\Gamma\ldots\Delta}_{p\ldots q\,\Phi\ldots\Sigma} \partial_\Lambda u^\Phi + \cdots, \\ &= u^\Phi \Psi^{k\ldots m\,\Gamma\ldots\Delta}_{p\ldots q\,\Lambda\ldots\Sigma,\Phi} - \Psi^{k\ldots m\,\Phi\ldots\Delta}_{p\ldots q\,\Lambda\ldots\Sigma} u^\Gamma{}_{,\Phi} - \cdots + \\ &\quad + \Psi^{k\ldots m\,\Gamma\ldots\Delta}_{p\ldots q\,\Phi\ldots\Gamma} u^\Phi{}_{,\Lambda} + \cdots. \end{aligned}\right\} \tag{179.4}$$

Then we have

$$\frac{\delta_d \Psi}{\delta t} = \frac{\partial \Psi}{\partial t} + \Psi_{,k} u_n n^k + \underset{\boldsymbol{u}}{\pounds}\Psi, \tag{179.5}$$

where both $\boldsymbol{x}$ and $\boldsymbol{v}$ are held constant when $\partial\Psi/\partial t$ is calculated. For this formula to be meaningful, it is not necessary to use any equation $\boldsymbol{x} = \boldsymbol{x}(\boldsymbol{v}, t)$ for the surface whose normal and tangential velocities are $u_n \boldsymbol{n}$ and u^Γ. To verify its correctness, however, let us eliminate $\boldsymbol{x}$; by (177.10) we have

$$\left.\begin{aligned} \frac{\delta_d \Psi}{\delta t} &= \frac{\partial \Psi}{\partial t} + \Psi_{,k}\left(\frac{\partial x^k}{\partial t} + u^\Gamma x^k_{;\Gamma}\right) + \underset{\boldsymbol{u}}{\pounds}\Psi, \\ &= \frac{\partial \Psi}{\partial t} + \frac{\partial x^k}{\partial t}\partial_k \Psi + \frac{\partial x^k}{\partial t}\left[\Psi^{q\ldots}_{\ldots}\left\{{p \atop k\,q}\right\} + \cdots\right] + \underset{\boldsymbol{u}}{\pounds}\Psi + u^\Gamma x^k_{;\Gamma}\Psi_{,k}, \\ &= \left.\frac{\partial \Psi}{\partial t}\right|_{\boldsymbol{v}=\text{const}} + \underset{\boldsymbol{u}}{\tilde{\pounds}}\Psi + \frac{\partial x^k}{\partial t}\left[\Psi^{q\ldots}_{\ldots}\left\{{p \atop k\,q}\right\} + \cdots\right], \end{aligned}\right\} \tag{179.6}$$

[1] This definition, stated in words by HAYES [1957, 7, p. 595], seems to give the sense intended also by THOMAS [1957, *15*, § 4].

where $\tilde{\pounds}_{\boldsymbol{u}} \Psi$ is obtained by replacing "$,_\Phi$" by "$;_\Phi$" in (179.4)₂. When $u^\Gamma = 0$ and the spatial co-ordinates are rectangular Cartesian, by (179.6)₃ and (179.1) we have

$$\frac{\delta_d \Psi}{\delta t} = \left.\frac{\partial \Psi}{\partial t}\right|_{\boldsymbol{v}=\text{const}} = \frac{\delta \Psi}{\delta t}. \tag{179.7}$$

Since the right-hand side of (179.5) is a double tensor under transformations $\boldsymbol{x}^* = \boldsymbol{x}^*(\boldsymbol{x})$, $\boldsymbol{v}^* = \boldsymbol{v}^*(\boldsymbol{v}, t)$, it furnishes the required expression for the displacement derivative in all co-ordinates systems.

For a spatial tensor $\Psi^{k\ldots m}_{p\ldots q}(\boldsymbol{x}, t)$, (179.5) reduces to

$$\frac{\delta_d \Psi}{\delta t} = \frac{\partial \Psi}{\partial t} + \Psi_{,k}\, u_n\, n^k, \tag{179.8}$$

and it is this form that is most useful.

From (179.8) we have at once

$$\frac{\delta_d g_{km}}{\delta t} = 0; \tag{179.9}$$

hence raising and lowering of spatial indices commutes with $\delta_d/\delta t$. Not so, however with surface indices, for the conditions of compatibility (178.2) and (178.6) assume the forms

$$\left.\begin{aligned} \frac{\delta_d a_{\Gamma\Delta}}{\delta t} &= -2u_n\, b_{\Gamma\Delta}, \\ \frac{\delta_d b_{\Gamma\Delta}}{\delta t} &= u_{n;\Gamma\Delta} - u_n\, b_\Gamma^\Lambda\, b_{\Lambda\Delta}. \end{aligned}\right\} \tag{179.10}$$

By differentiating the relation $a_{\Gamma\Delta}\, a^{\Delta\Lambda} = \delta_\Gamma^\Lambda$ we see that (179.10)₁ is equivalent to

$$\frac{\delta_d a^{\Gamma\Delta}}{\delta t} = +2u_n\, b^{\Gamma\Delta}. \tag{179.11}$$

Also

$$\frac{\delta_d a}{\delta t} = -2u_n\, a\, \overline{K}, \tag{179.12}$$

whence it follows that $\delta_d a/\delta t = 0$ is a necessary and sufficient condition that a moving surface be and remain a minimal surface.

From (179.11) and (179.10)₂ we have

$$\left.\begin{aligned} \frac{\delta_d b_\Delta^\Gamma}{\delta t} &= u_n{}^{;\Gamma}{}_{;\Delta} + u_n\, b^{\Gamma\Lambda}\, b_{\Lambda\Delta}, \\ \frac{\delta_d b^{\Gamma\Delta}}{\delta t} &= u_n{}^{;\Gamma\Delta} + 3u_n\, b^{\Gamma\Lambda}\, b_\Lambda^\Delta. \end{aligned}\right\} \tag{179.13}$$

From these results and (App. 21.10) it is easy to show that

$$\left.\begin{aligned} \frac{\delta_d \overline{K}}{\delta t} &= u_n(\overline{K}^2 - 2K) + u_{n;\Gamma}{}^{;\Gamma}, \\ \frac{\delta_d K}{\delta t} &= u_n\, K\, \overline{K} + (\overline{K}\, a^{\Gamma\Delta} - b^{\Gamma\Delta})\, u_{n;\Gamma\Delta}, \\ \frac{\delta_d b}{\delta t} &= -u_n\, a\, K\, \overline{K} + a(\overline{K}\, a^{\Gamma\Delta} - b^{\Gamma\Delta})\, u_{n;\Gamma\Delta}. \end{aligned}\right\} \tag{179.14}$$

We now calculate[1] $\dfrac{\delta_d n^k}{\delta t}$. From the relations (App. 19.6)₁ and (App. 21.3)₁ it follows that

$$n^k \frac{\partial n_k}{\partial t} = 0, \qquad n_{k;\Gamma}\, n^k = 0, \qquad \frac{\partial n_k}{\partial t}\, x^k_{;\Gamma} = -n_k \frac{\partial x^k_{;\Gamma}}{\partial t}. \tag{179.15}$$

[1] The result is stated by HAYES [1957, 7, Eq. (22)] without proof; a proof different from ours is given by THOMAS [1957, 15, Eq. (60)].

By differentiating (177.9) we obtain

$$\left.\begin{aligned} u_{n;\Gamma} &= n_{k;\Gamma}\frac{\partial x^k}{\partial t} + n_k\frac{\partial x^k_{;\Gamma}}{\partial t}, \\ &= -u^\Delta x_{k;\Delta} n^k_{;\Gamma} - x^k_{;\Gamma}\frac{\partial n_k}{\partial t}, \end{aligned}\right\} \tag{179.16}$$

where we have used (177.10), (App. 19.6)$_1$, and (179.15)$_3$. Since (177.9) presumes that $\boldsymbol{x}=\boldsymbol{x}(\boldsymbol{v},t)$, the time derivative $\partial/\partial t$ is taken on the understanding that $\boldsymbol{x}$ is eliminated. Hence

$$\left.\begin{aligned} a^{\Delta\Gamma} x^k_{;\Delta} u_{n;\Gamma} &= u^\Delta g_{qp} x^q_{;\Delta} x^p_{;\Sigma} b^\Sigma_\Gamma a^{\Gamma\Lambda} x^k_{;\Lambda} - (g^{kq} - n^k n^q)\frac{\partial n_q}{\partial t}, \\ &= u^\Delta b^\Lambda_\Delta x^k_{;\Lambda} - \frac{\partial n^k}{\partial t}, \end{aligned}\right\} \tag{179.17}$$

where we have used (App. 21.6)$_2$, (App. 21.4)$_1$, (App. 19.7), and (179.15)$_1$. By (App. 21.6)$_2$ it follows that

$$\frac{\partial n^k}{\partial t} + u^\Delta n^k_{;\Delta} = -a^{\Gamma\Delta} x^k_{;\Gamma} u_{n;\Delta}, \tag{179.18}$$

where, as stated above, $\boldsymbol{x}$ has been eliminated before $\partial n^k/\partial t$ is calculated. By use of (179.6)$_2$ we thus obtain the desired formula:

$$\frac{\delta_d n^k}{\delta t} = -a^{\Gamma\Delta} x^k_{;\Gamma} u_{n;\Delta}. \tag{179.19}$$

For an alternative derivation, we may suppose that $u^\Gamma=0$ and the spatial co-ordinates are rectangular Cartesian; most of the terms in the above calculations are then absent, and we quickly obtain a formula recognizable as the appropriate special case of (179.19), which is a tensorial equation.

From (179.19) and (App. 19.7) we have

$$x^k_{;\Gamma}\frac{\delta_d n_k}{\delta t} = -u_{n;\Gamma}, \tag{179.20}$$

a result that might have been expected directly from the definitions, since it asserts that the length of the projection of the displacement derivative of the normal onto a given direction on the surface is just the negative of the gradient of the speed of displacement in that direction. In particular, a necessary and sufficient condition for parallel propagation is $u_n = f(t)$, as was already remarked in Sect. 177.

III. Kinematics of singular surfaces.

180. Kinematical condition of compatibility. We now consider a moving surface $\mathscr{s}(t)$ which divides a varying region $\mathscr{R}^+(t)$ from another, $\mathscr{R}^-(t)$. The moving surface is assumed to satisfy the conditions stated in Sect. 177 and at each instant t to be a singular surface with respect to a quantity Ψ, as defined in Sect. 173; the conditions laid down for Ψ are now supposed to hold for each t. Assuming also that the limiting values Ψ^+ and Ψ^- are continuously differentiable functions of t in $\mathscr{R}^+$ and $\mathscr{R}^-$, respectively, we derive a condition that the discontinuity in Ψ *persists in time* rather than appearing and disappearing at some particular instant.

In a general space this temporal persistency is expressed by superficial conditions analogous to those discussed in Sect. 175 but applied when one more dimension, that of t, is added both to the surface and to the space in which it lies, i.e., we need only regard $f(\boldsymbol{x}, t) = 0$ as a single surface in the $\boldsymbol{x}$-t-space. In this degree of generality, nothing new results.

In a metric space, however, the existence of a definite speed of displacement u_n for the moving surface makes possible results of a more concrete kind. The essential step, again, is furnished by HADAMARD's lemma (174.1), but this time we apply it to a particular tangential path on the n-dimensional surface $f(\boldsymbol{x}, t) = 0$ in the $n+1$-dimensional $\boldsymbol{x}$-t-space, namely, the path tangent to the vector $u_n \boldsymbol{n}, 1$. The derivative[1] on the $+$ side of the surface is the quantity $\delta\Psi^+/\delta t$ as defined by (179.1). Thus

$$\frac{\delta\Psi^+}{\delta t} = \left(\frac{\partial\Psi}{\partial t}\right)^+ + u_n n^k \partial_k \Psi^+, \tag{180.1}$$

where, as in previous formulae, $\partial_k \Psi^+ \equiv (\partial_k \Psi)^+$. Writing a similar equation for the other side of the surface and subtracting the result from (180.1), we obtain the ***kinematical condition of compatibility***[2]:

$$\left[\frac{\partial\Psi}{\partial t}\right] = -u_n [n^k \partial_k \Psi] + \frac{\delta}{\delta t}[\Psi]. \tag{180.2}$$

The jumps occurring on the right-hand side are those occurring also in the geometrical conditions (175.11). Thus *the jumps of the derivatives $\partial_k \Psi$ and $\partial\Psi/\partial t$ across a persistent singular surface are determined by the quantities u_n, $[n^k \partial_k \Psi]$, and $[\Psi]$.*

A condition equivalent to (180.2) but expressed in terms of tensors is easy to obtain by using the displacement derivative (179.5). Considering a spatial tensor field $\Psi^{k\ldots m}_{p\ldots q}$, we have

$$\left[\frac{\partial\Psi}{\partial t}\right] = -u_n [n^k \Psi_{,k}] + \frac{\delta_d}{\delta t}[\Psi]. \tag{180.3}$$

This is so because (1) it is a tensorial equation and (2) when the spatial co-ordinates are rectangular Cartesian, it reduces to (180.2).

In the important special case when Ψ is continuous, (180.2) reduces to the form

$$\left[\frac{\partial\Psi}{\partial t}\right] = -u_n [n^k \partial_k \Psi] = -u_n [n^k \Psi_{,k}]. \tag{180.4}$$

Thus, in particular, across a stationary surface that is singular with respect to $\partial_k \Psi$ but not with respect to Ψ, the time derivative $\partial\Psi/\partial t$ is continuous, as is evident also directly from the definitions.

We may write (175.8) and (180.4) as the system

$$[\Psi_{,k}] = B n_k, \quad \left[\frac{\partial\Psi}{\partial t}\right] = -u_n B. \tag{180.5}$$

---

[1] There are n independent paths at any one point of $\sigma(t)$, but we use only a particular one. If all n are employed, we may derive the geometrical and kinematical conditions of compatibility simultaneously, as is done in a special case by HADAMARD [1903, *11*, ¶ 97] and more generally by COULON [1902, *3*, § 46]. We prefer separate treatment of the two sets of conditions so as to separate the underlying ideas. In many cases useful in continuum mechanics, the geometrical conditions are satisfied when the kinematical is not; e.g., at the instant a portion of material splits in two, or two parts are joined together.

[2] The essential content of this condition seems to be contained in the "phoronomic conditions" of CHRISTOFFEL [1877, *2*, § 7], but these are not easy to use, and certainly the general concept of *compatibility* is due to HUGONIOT. First [1885, *3*, p. 1119] he used "propagation" to mean compatibility, but soon thereafter [1887, *1*, §§ 3, 5] he introduced the terms "compatibilité" and "conditions de compatibilité". HADAMARD [1903, *11*, ¶ 97] and LOVE [1904, *4*, § 7] obtained the system (180.5); HADAMARD's term is "conditions de compatibilité cinématique". The full condition (180.2) is implied by the results of COULON [1902, *3*, § 46]; cf. also KOTCHINE [1926, *3*, § 1]. Our presentation follows THOMAS [1957, *15*, § 4].

Of all the forms of the conditions of compatibility, it is these, which presume Ψ itself to be continuous, that are most often used. If we square $(180.5)_2$ and use $(175.13)_1$, we obtain[1]

$$\left[\frac{\partial \Psi}{\partial t}\right]^2 = u_n^2 B^2 = u_n^2 [\Psi_{,k}][\Psi^{,k}], \tag{180.6}$$

whereby the magnitude of the speed of displacement is shown to be the quotient of the magnitude of the jump in $\partial\Psi/\partial t$ by the magnitude of the jump in $\Psi_{,k}$. This last result is expressed in terms and notations appropriate to the case when Ψ is a scalar, but generalization is easy.

Since HUGONIOT'S time[2] it has been stated that a singular surface upon which the kinematical condition of compatibility does not hold will instantly split into two or more singular surfaces or will become singular with respect to a different quantity, such as a derivative of Ψ. To substantiate a statement of this kind, the theory of some particular material is needed. In the generality maintained here, all that can be said is that the singular surface will not persist.

181. Iterated kinematical conditions of compatibility. Amplifying the notations (176.1), set

$$A' \equiv \left[\frac{\partial \Psi}{\partial t}\right], \qquad B' \equiv \left[n^k \frac{\partial \Psi_{,k}}{\partial t}\right]. \tag{181.1}$$

We may then write (180.3) in the form

$$A' = -u_n B + \frac{\delta_d A}{\delta t}. \tag{181.2}$$

By replacing Ψ by $\partial\Psi/\partial t$ in $(176.2)_1$ and (181.2) we obtain

$$\left.\begin{aligned}\left[\frac{\partial \Psi_{,k}}{\partial t}\right] &= B' n_k + g_{km} a^{\Delta\Gamma} x^m_{;\Delta} A'_{;\Gamma},\\ \left[\frac{\partial^2 \Psi}{\partial t^2}\right] &= -u_n B' + \frac{\delta_d A}{\delta t}.\end{aligned}\right\} \tag{181.3}$$

By (181.2), the quantity A' is already expressed in terms of A and B. We now obtain a like expression for B'. To this end, we differentiate $(176.1)_3$ and use $(176.2)_1$, obtaining

$$\frac{\delta_d B}{\delta t} = g_{km} a^{\Delta\Gamma} x^m_{;\Delta} A_{;\Gamma} \frac{\delta_d n^k}{\delta t} + n^k \frac{\delta_d [\Psi_{,k}]}{\delta t}. \tag{181.4}$$

Now by another application of the argument leading to (180.1) we can show that

$$\frac{\delta}{\delta t}(\partial_k \Psi^+) = \left(\frac{\partial}{\partial t}\partial_k \Psi\right)^+ + u_n n^m \partial_m \partial_k \Psi^+, \tag{181.5}$$

since the various derivatives are assumed continuous in $\mathscr{R}^+$; consistently with our practice, we have put $\partial_m \partial_k \Psi^+ \equiv (\partial_m \partial_k \Psi)^+$. Hence

$$\frac{\delta_d}{\delta t}[\Psi_{,k}] = \left[\frac{\partial \Psi_{,k}}{\partial t}\right] + u_n n^m [\Psi_{,mk}]. \tag{181.6}$$

Substitution of this result into (181.4) yields a result which when simplified by use of $(176.1)_6$, $(181.3)_1$ and $(\text{App. }21.3)_2$ becomes

$$\left.\begin{aligned}B' &= -u_n C + \frac{\delta_d B}{\delta t} - g_{km} \frac{\delta_d n^k}{\delta t} a^{\Delta\Gamma} x^m_{;\Delta} A_{;\Gamma},\\ &= -u_n C + \frac{\delta_d B}{\delta t} + u_n{}^{;\Gamma} A_{;\Gamma},\end{aligned}\right\} \tag{181.7}$$

where the latter form follows by use of (179.19).

---

[1] DUHEM [1900, *3*].
[2] [1887, *1*, § 14]. Cf. HADAMARD [1903, *11*, ¶ 108].

The identities (181.3), with A' and B' replaced by right-hand sides of (181.2) and (181.7), are THOMAS's *iterated kinematical conditions of compatibility*[1].

In the case when Ψ is continuous, we have $A=0$, and the conditions (181.3) reduce to the simpler forms

$$\left.\begin{aligned} \left[\frac{\partial \Psi_{,k}}{\partial t}\right] &= \left(-u_n C + \frac{\delta_{\mathrm{d}} B}{\delta t}\right) n_k - g_{km} a^{\Delta\Gamma} x^m_{;\Delta} (u_n B)_{;\Gamma}, \\ \left[\frac{\partial^2 \Psi}{\partial t^2}\right] &= u_n^2 C - 2u_n \frac{\delta_{\mathrm{d}} B}{\delta t} - B \frac{\delta_{\mathrm{d}} u_n}{\delta t}. \end{aligned}\right\} \quad (181.8)$$

When not only Ψ but also $\partial\Psi/\partial t$ and $\partial_k\Psi$ are continuous, the results are still simpler, for (181.8) and (176.10) yield[2]

$$\left.\begin{aligned} [\partial_k \partial_m \Psi] &= [\Psi_{,km}] = C n_k n_m, \\ \left[\frac{\partial \partial_k \Psi}{\partial t}\right] &= \left[\frac{\partial \Psi_{,k}}{\partial t}\right] = -u_n C n_k, \\ \left[\frac{\partial^2 \Psi}{\partial t^2}\right] &= u_n^2 C, \end{aligned}\right\} \quad (181.9)$$

where

$$C = [n^k n^m \partial_k \partial_m \Psi] = [n^k n^m \Psi_{,km}] = [\Psi_{,k}{}^{,k}]. \quad (181.10)$$

More generally[3], if Ψ and all its derivatives of orders $1, 2, \ldots, p-1$ are continuous, we have

$$\left.\begin{aligned} \left[\partial_{k_1} \partial_{k_2} \ldots \partial_{k_s} \frac{\partial^{p-s}}{\partial t^{p-s}} \Psi\right] &= \left[\frac{\partial^{p-s}}{\partial t^{p-s}} \Psi_{,k_1 k_2 \ldots k_s}\right] \\ &= (-u_n)^{p-s} [n^{m_1} n^{m_2} \ldots n^{m_p} \partial_{m_1} \partial_{m_2} \ldots \partial_{m_p} \Psi] n_{k_1} n_{k_2} \ldots n_{k_s}, \\ &= (-u_n)^{p-s} [n^{m_1} n^{m_2} \ldots n^{m_p} \Psi_{,m_1 m_2 \ldots m_p}] n_{k_1} n_{k_2} \ldots n_{k_s}, \end{aligned}\right\} \quad (181.11)$$

the result being valid in any metric space. In particular, choosing $s=0$ we have

$$(-u_n)^p = \frac{\left[\dfrac{\partial^p \Psi}{\partial t^p}\right]}{[n^{m_1} n^{m_2} \ldots n^{m_p} \partial_{m_1} \partial_{m_2} \ldots \partial_{m_p} \Psi]}, \quad (181.12)$$

interpretation of which yields the **Hugoniot-Duhem theorem**[4]: *The speed of displacement of a singular surface across which Ψ and its derivatives of orders $1, 2, \ldots, p-1$ are continuous but at least one p-th derivative of Ψ is discontinuous is determined up to sign by the ratio of the jump of $\partial^p\Psi/\partial t^p$ to that of the fully normal p-th derivative, $d^p\Psi/dn^p$.*

IV. Singular surfaces associated with a motion.

182. Material and spatial representations of a surface. So far in this chapter our considerations have been independent of the motion of any material medium. We now suppose that a medium consisting of particles $\boldsymbol{X}$ is in motion through the space of places $\boldsymbol{x}$ according to (66.1). For the time being, we shall assume

[1] [1957, *15*, § 6]. THOMAS obtains also an alternative form for $\partial_\Gamma A'$; see his Eq. (51) as corrected. As regards the history of these conditions, remarks similar to those at the beginning of Sect. 176 may be made.

[2] HUGONIOT [1885, *4*, p. 1231], HADAMARD [1901, *8*, § 2].

[3] DUHEM [1901, *6*], COULON [1902, *3*, § 46], HADAMARD [1903, *11*, ¶ 97].

[4] Given in the special cases $p=1$ and $p=2$ by HUGONIOT [1885, *3*, p. 1120] [1885, *4*, p. 1231], in general by DUHEM [1900, *3*] [1901, *7*, Part II, Chap. II, § 2], but expressed by means of Laplacians (cf. (176.11) and (181.10)) rather than normal derivatives.

that the functions occurring in (66.1) are single-valued and continuous; modifications appropriate to motions suffering discontinuities will be given in Sect. 185. We consider a surface $\mathfrak{s}(t)$, given by a representation of the form (177.3), and we set

$$F(\boldsymbol{X}, t) \equiv f(\boldsymbol{x}(\boldsymbol{X}, t), t), \quad \text{so that} \quad f(\boldsymbol{x}, t) \equiv F(\boldsymbol{X}(\boldsymbol{x}, t), t), \tag{182.1}$$

identically in $\boldsymbol{x}$, $\boldsymbol{X}$, and t. Alternative representations of the moving surface are thus[1]

$$f(\boldsymbol{x}, t) = 0, \quad F(\boldsymbol{X}, t) = 0. \tag{182.2}$$

In the latter representation, which we denote by $\mathcal{S}(t)$, we may conceive the particles as stationary and the surface $\mathcal{S}(t)$ moving amongst them, being occupied by a different set of particles at each time t. The two representatives (182.2) are the duals of one another in the sense of Sect. 14. The analysis given earlier in this chapter did not presuppose any particular choice of co-ordinates, so long as they be independent of time, and is equally applicable to both representations. It is easier to visualize in terms of the spatial variables $\boldsymbol{x}$, t, and from now on we agree to regard all the foregoing equations as so expressed; where we wish to employ a material counterpart, we shall invoke the principle of duality.

Thus in the special case when $(182.2)_1$ reduces to the form

$$f(\boldsymbol{x}) = 0, \tag{182.3}$$

we shall say that the surface $\mathfrak{s}$ is *stationary*; when $(182.2)_2$ reduces to the form

$$F(\boldsymbol{X}) = 0, \tag{182.4}$$

that $\mathcal{S}$ is *material*[2] [cf. (73.4)]. In the former case, the surface consists always of the same places; in the latter, of the same particles.

Although $(182.2)_1$ and $(182.2)_2$ are but different means of representing the same phenomenon, the two surfaces so defined are, in general, entirely different from one another geometrically. The surface $f(\boldsymbol{x}, t) = 0$ is a surface in the space of places, while the surface $F(\boldsymbol{X}, t) = 0$ is the locus, in the space of particles, of the *initial* positions of the particles $\boldsymbol{X}$ that are situate upon the surface $f(\boldsymbol{x}, t) = 0$ at time t. Such connections as there are must be established by use of the transformations (182.1). For example, from the assumption that $f(\boldsymbol{x}, t) = 0$ has a continuous normal it follows that $F(\boldsymbol{X}, t) = 0$ also has a continuous normal[3]. We assume, in fact, that (182.2) are sufficiently smooth as to permit any number of differentiations and functional inversions. The theory we construct is local.

Some aspects of the foregoing theory, while not losing their validity, lose their intuitive appeal when applied to the material variables. The shape of $\mathcal{S}(t)$, including its first and second differential forms, and its unit normal, have no immediate interpretation, for they do not correspond to any geometrical properties that an observer of a singular surface in space would perceive.

The material representation, rather, is of the nature of a *diagram* for the moving surface. It is only one of many such diagrams, for by choice of the initial instant, or of the co-ordinates or parameters $\boldsymbol{X}$ corresponding to given initial positions, the particular functional form that results from $(182.1)_1$ will differ[4].

As an example of the above remarks we consider the unit normals $\boldsymbol{n}$ and $\boldsymbol{N}$ to $f = 0$ and $F = 0$, given by

$$n_k = \frac{f_{,k}}{\sqrt{f_{,m} f^{,m}}}, \quad N_\alpha = \frac{F_{,\alpha}}{\sqrt{F_{,\beta} F^{,\beta}}}, \tag{182.5}$$

[1] HUGONIOT [1885, *4*, p. 1231].
[2] French: *stationnaire*.
[3] This is proved under weaker assumptions by LICHTENSTEIN [1929, *4*, Chap. 6, § 1].
[4] The effect of such changes is discussed somewhat by HADAMARD [1903, *11*, ¶¶ 79 to 84].

where, as usual, $f^{,m} = g^{mq} f_{,q}$ and $F^{,\beta} = g^{\beta\gamma} F_{,\gamma}$. The former has an immediate geometric significance as a unit vector normal to the surface $\mathfrak{s}(t)$ in the space of places $\boldsymbol{x}$. The latter, dual to the former, applies only to one of many possible diagrams for the moving surface in various possible spaces of particles $\boldsymbol{X}$ and has no immediate interpretation.

Since $F_{,\alpha} = f_{,k}\, x^k_{;\alpha}$, by the dual of (17.3) we have[1]

$$F_{,\alpha} F^{,\alpha} = |\boldsymbol{x}/\boldsymbol{X}|^2 g^{\beta\gamma} \cdot \tfrac{1}{2} \varepsilon^{krp} \varepsilon_{\beta\delta\varepsilon} f_{,k} X^\delta_{;r} X^\varepsilon_{;p} \cdot \tfrac{1}{2} \varepsilon^{msq} \varepsilon_{\gamma\zeta\eta} f_{,m} X^\zeta_{;s} X^\eta_{;q}, \tag{182.6}$$

a result which in a common frame assumes the form

$$\sqrt{F_{,\alpha} F^{,\alpha}} = |\boldsymbol{z}/\boldsymbol{Z}| \sqrt{\left(\frac{\partial(f, Y, Z)}{\partial(x, y, z)}\right)^2 + \left(\frac{\partial(X, f, Z)}{\partial(x, y, z)}\right)^2 + \left(\frac{\partial(X, Y, f)}{\partial(x, y, z)}\right)^2}. \tag{182.7}$$

For the unit normals themselves we have the relations

$$n_k = \frac{F_{,\alpha} X^\alpha_{;k}}{\sqrt{f_{,m} f^{,m}}} = N_\alpha X^\alpha_{;k} \cdot \frac{\sqrt{F_{,\beta} F^{,\beta}}}{\sqrt{f_{,m} f^{,m}}}, \tag{182.8}$$

where $\sqrt{F_{,\beta} F^{,\beta}}$ and $X^\alpha_{;k}$ are thought of as expressed in terms of spatial gradients by means of (182.6) and (17.3), respectively.

It is a natural requirement that the moving surface $\mathfrak{s}(t)$ shall have a continuous and non-vanishing gradient vector $f_{,k}$. Such a requirement if put upon the gradient vector $F_{,\alpha}$ of $\mathscr{S}$ has no immediate appeal. From (182.7), we see that $F_{,\alpha}$ if continuous can vanish at a point if and only if either the radical or the Jacobian on the right-hand side vanishes. For the radical to vanish in a neighborhood, it is necessary and sufficient that f be functionally independent of X, Y, Z: that is, by (182.1), that $F = F(t)$, and such an equation cannot represent a surface. Thus $F_{,\alpha}$ vanishes only with the Jacobian $|\boldsymbol{z}/\boldsymbol{Z}|$. By the Axiom of Continuity in Sect. 65, it follows that for a material diagram $F(\boldsymbol{X}, t) = 0$ obtained from a surface $f(\boldsymbol{x}, t) = 0$ in a continuous motion, $F_{,\alpha}$ does not vanish except possibly at isolated points or lines.

183. Speeds of propagation. Waves.

183. Speeds of propagation. Waves. The dual of the speed of displacement, defined by (177.5), is the *speed of propagation*[2] U_N:

$$U_N \equiv -\frac{\frac{\partial F}{\partial t}}{\sqrt{F_{,\alpha} F^{,\alpha}}}. \tag{183.1}$$

This speed is a measure of the rate at which the moving surface $\mathscr{S}(t)$ traverses the material. In particular, a necessary and sufficient condition that $F = 0$ be a material surface in an interval of time is that $U_N = 0$ throughout the interval.

A surface that is singular with respect to some quantity and that has a non-zero speed of propagation is said to be a *propagating* singular surface or *wave*[3].

Now it is evident that the value of U_N for a given surface $f(\boldsymbol{x}, t) = 0$ in a given motion, unless $U_N = 0$, depends in general upon the choice of the instant regarded as the time $t = 0$ for the motion of each particle[4]. Thus there are infinitely many different speeds of propagation. Often it is most convenient to take the instant $t = 0$ as the present instant. Then $X^\alpha = \delta^\alpha_k x^k$, $\partial/\partial X^\alpha = \delta^\alpha_k \partial/\partial x^k$, and the functional forms of F and f, at this one instant and *qua* functions of $\boldsymbol{x}$ or $\boldsymbol{X}$, are the same, but of course the time rates calculated with $\boldsymbol{x}$ held constant do not generally coincide with those calculated when $\boldsymbol{X}$ is held constant. In particular, with this choice of $\boldsymbol{X}$ the speed of propagation U_N will be written as U and called the *local speed of propagation*[5] of the surface. This speed, which

---

[1] HADAMARD [1903, *11*, ¶ 82, footnote], TRUESDELL [1951, *35*].

[2] This quantity, for a general surface, first appears in the work of CHRISTOFFEL [1877, *2*, § 1]; it is the second of the "deux *vitesses de propagation*" introduced by HUGONIOT [1885, *4*, p. 1231]. Cf. also HADAMARD [1901, *8*, § 1] [1903, *11*, ¶ 98].

[3] HADAMARD [1901, *8*, § 1] [1903, *11*, ¶ 91].

[4] HUGONIOT [1887, *2*, Part 2, § 7], HADAMARD [1903, *11*, ¶ 99].

[5] HUGONIOT [1885, *3*, p. 1120], "vitesse de propagation rapportée au fluide lui-même". RANKINE [1870, *6*, § 2] in dealing with a one-dimensional case called U "the linear velocity of advance of the wave".

is *the normal speed of the surface with respect to the particles instantaneously situate upon it*, is related to the speed of displacement as follows:

$$U \frac{\partial f}{\partial t} = u_n \frac{\partial F}{\partial t} \,. \tag{183.2}$$

More generally, we have from (182.5), (183.1), (177.5), and (74.1) the alternative forms

$$\left.\begin{aligned} U_N \sqrt{F_{,\alpha} F^{,\alpha}} &= u_n \sqrt{f_{,k} f^{,k}} - \dot{x}^k f_{,k} \\ &= (u_n - \dot{x}_n) \sqrt{f_{,k} f^{,k}} \\ &= -\dot{f} \end{aligned}\right\} \tag{183.3}$$

(cf. (74.6)). If we choose the present configuration as the initial one, $(183.3)_3$ becomes

$$U = -\frac{\dot{f}}{\sqrt{f_{,k} f^{,k}}} \,, \tag{183.4}$$

while $(183.3)_2$ reduces to the more elegant form[1]

$$U = u_n - \dot{x}_n \,, \tag{183.5}$$

expressing the evident fact that the normal speed at which the particles now comprising $\mathscr{s}(t)$ are leaving it is the excess of their normal speed over the normal speed of the surface.

184. Boundaries. Recalling that a body $\mathscr{B}$ is a set of particles $\boldsymbol{X}$ having positive mass, we define its boundary[2] at time t as the set of places $\boldsymbol{x}$ whose every neighborhood contains two places distinct from $\boldsymbol{x}$, one of which is occupied by a particle of $\mathscr{B}$ and one is not. In kinematical terms, the bounding surface is adjacent to $\mathscr{B}$ but not crossed by any particle of $\mathscr{B}$. In general, it is a moving and deforming surface.

While an axiom of continuity was laid down in Sect. 65, we now replace it by the weaker requirement that the motion (66.1) be a topological transformation, i.e., a transformation that puts open sets in a region of $\boldsymbol{X}$-t-space into one-to-one correspondence with open sets in $\boldsymbol{x}$-t-space. In particular, for each fixed t the transformation of $\boldsymbol{X}$ into $\boldsymbol{x}$ will then be topological. Hence the boundary of a set in $\boldsymbol{X}$-space is mapped, at each t, into the boundary of the corresponding set in $\boldsymbol{x}$-space. Therefore, *the boundary surface of every body in a topological motion is a material surface*[3]. Conversely, any material surface permanently divides the material (if there is any) on one side from that on the other, and thus consists in boundary points of the bodies (if any) upon each side.

From these results and LAGRANGE's criterion in Sect. 74, it follows that *in a continuous motion, a necessary and sufficient condition that a surface $f=0$ be a portion of the boundary of the material (if any) instantaneously lying upon either side of it is*

$$\dot{f} = 0 \,. \tag{184.1}$$

When the motion fails to be topological, these results hold no longer, and boundaries may be instantly created or destroyed. General transformations

---

[1] HADAMARD [1903, *11*, ¶ 100] but given in effect by HUGONIOT [1885, *3*, p. 1120] in a sentence which is confusingly mispunctuated.

[2] Some properties of boundaries have been given in Sect. 69.

[3] HADAMARD [1903, *11*, ¶ 48] uses the differentiability of the motion to prove this result. The result was asserted by LAGRANGE [1783, *1*, §§ 10–11] (cf. Sect. 74), but his discussion of a bounding surface is insufficient.

are not of interest in field theories. There are important cases, however, when motions fail to be topological at one certain instant only, or upon certain isolated surfaces, curves or points.

An example of the former is furnished by the *fracture* or *welding* of a solid (Fig. 25) or by the *formation or coalescence of drops* in a fluid. In these cases,

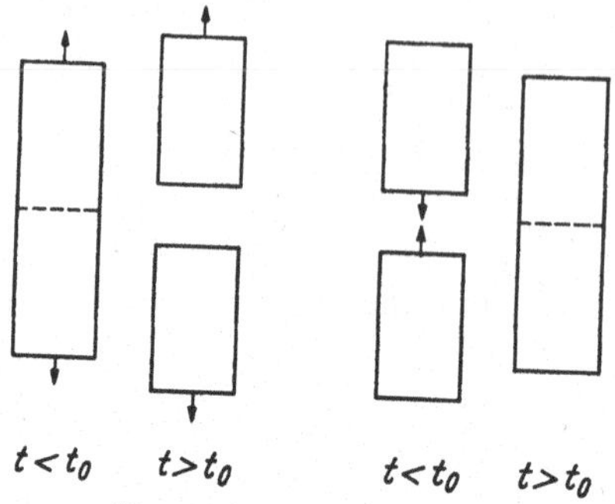

Fig. 25. Fracture and welding.

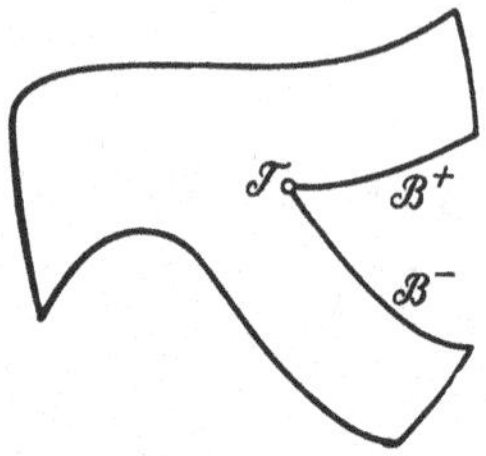

Fig. 26. Tear.

interior particles suddenly find themselves on the boundary of two disjoint motions, or boundary particles suddenly become interior ones. A still more complicated singularity is a *tear* such as that shown in Fig. 26, where the singular line $\mathscr{T}$ is propagating into the material, splitting each particle $\boldsymbol{X}$ in its progress into two particles $\boldsymbol{X}^+$ and $\boldsymbol{X}^-$, one of which stays upon the boundary $\mathscr{B}^+$ and the other upon the boundary $\mathscr{B}^-$. Singularities located upon lines seem not to have been studied from a general viewpoint.

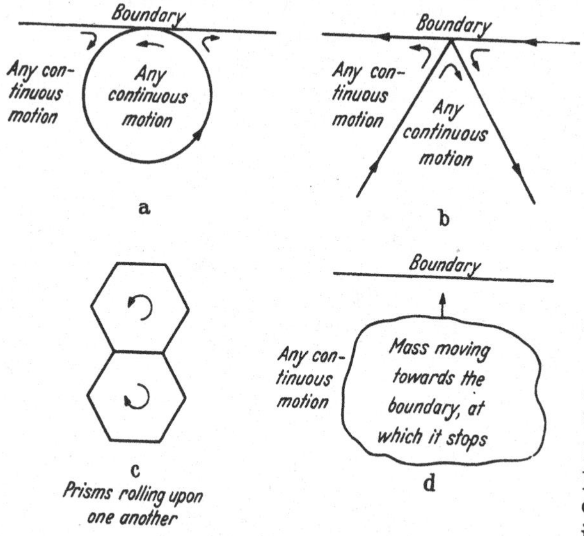

Fig. 27 a—d. Examples of non-topological motions.

The latter type of non-topological motion allows us to represent motions in which particles enter and depart from a boundary surface, possibly quite smoothly along tangential paths[1]. Let Fig. 27(a) represent, say, a rigid cylinder or a fluid vortex spinning just below the plane free surface of a fluid and tangent at the top, and let the region outside the cylinder be endowed with any topological motion such that the cylinder and the plane constitute its boundary[2]. The combined motion fails to be topological at the line of tangency, and we may say that particles on the cylindrical stream surface continually rise into the plane boundary and fall away from it again. In this example, however, as in all others formed by piecing together topological motions upon portions of their boundaries, the

[1] POISSON [1831, *2*, § 12] [1833, *4*, § 652].
[2] This example and that in Fig. 27(d) were given by KELVIN [1848, *5*].

complete boundary, consisting of the union of the constituent boundaries, is again a material surface[1], though a material surface on which the motion fails to be topological. Other examples are shown in Fig. 27.

In the case of motions that fail to be continuous in the sense defined in Sect. 65, the condition (184.1) is *neither necessary nor sufficient* that $f=0$ be a material surface or consist in boundary points of the material, if any, that it instantaneously separates.

From (183.3) it is plain that $U_N=0$, for all choices of the initial configuration, is equivalent to $\dot{f}=0$ provided $F_{,\alpha}F^{,\alpha}\neq 0$. Now $F_{,\alpha}F^{,\alpha}$ is not determined by the instantaneous shape of the moving surface alone, but is influenced also by the motion. The two effects are in some measure separated in the identity (182.7). The quantity under the root sign on the right assumes the values 0 or ∞ if and only if the equation $F(\boldsymbol{X}, t)=0$ reduces to the form $F(t)=0$, which does not represent a surface. Thus $F_{,\alpha}F^{,\alpha}$, for a surface $F=0$, is singular only with $|\boldsymbol{z}/\boldsymbol{Z}|$; if the Axiom of Continuity in Sect. 65 holds, from (183.3) we thus read off a formal proof of LAGRANGE' criterion. More generally, by (156.2) we have

$$\sqrt{F_{,\alpha}F^{,\alpha}} \propto \frac{\varrho_0}{\varrho}, \quad \text{and hence} \quad U_N\,\varrho_0 \propto \varrho\,\dot{f}. \tag{184.2}$$

We consider particles such that $\varrho_0\neq 0, \infty$. Then from $(184.2)_2$ we conclude that[1]:

1. *If $\varrho=\infty$ upon the surface $f=0$ and $U_N\neq\infty$, then* (184.1) *is satisfied, but the surface may or may not be material.*

2. *If $\varrho=0$ upon the surface $f=0$ over an interval of time, then the surface is material, whether or not* (184.1) *is satisfied. The condition* (184.1) *remains sufficient, but not necessary, that $f=0$ be a material surface.*

3. *Both in Case 1 and in Case 2, the condition* (184.1) *is necessary, but not sufficient, that the surface $f=0$ be an admissible boundary.*

The case when $\varrho=\infty$ upon $f=0$ is illustrated by the following example[1]:

$$x = X - c\,t, \qquad y = (Y^{\frac{1}{3}} - k\,t)^3, \qquad z = Z. \tag{184.3}$$

Since

$$\dot{x} = -c, \qquad \dot{y} = -3k(Y^{\frac{1}{3}} - k\,t)^2 = -3k\,y^{\frac{2}{3}}, \qquad \dot{z} = 0, \tag{184.4}$$

the velocity field is plane, single-valued, steady, and irrotational[2], and the stream lines are the similar cubical parabolas

$$y = \left[\frac{k}{c}(x - X) + Y^{\frac{1}{3}}\right]^3, \qquad z = Z, \tag{184.5}$$

which cross the $y=0$ plane tangentially (Fig. 28). The density is given by

$$\frac{\varrho_0}{\varrho} = \frac{\partial(x, y, z)}{\partial(X, Y, Z)} = \left(\frac{y}{Y}\right)^{\frac{2}{3}}, \tag{184.6}$$

so that $\varrho=\infty$ upon the plane $y=0$. For this plane we have the equations

$$f = y = F = (Y^{\frac{1}{3}} - k\,t)^3 = 0, \tag{184.7}$$

and hence

$$\dot{f} = \dot{y} = -3k y^{\frac{2}{3}} = 0. \tag{184.8}$$

Although (184.1) is satisfied, the plane $y=0$ is neither a material surface nor a boundary, since the particles are continuously crossing it. Neither is $y=0$, while indeed a persistent surface of discontinuity, a singular surface in the sense defined in Sect. 180, since no jump discontinuity occurs across it.

---

[1] TRUESDELL [1951, *35*].

[2] Hence this motion, despite its artificial appearance, is dynamically possible (in theory) for an ideal gas subject to no extrinsic force; the motion is to be conceived as occurring in a channel bounded by cylinders erected upon two of the curves shown in Fig. 28.

The speed of displacement, u_n, is zero, since the plane $y=0$ is stationary. The speed of propagation U_N is given by

$$U_N = 3k\,Y^{\frac{2}{3}} = 3k^3 t^2. \tag{184.9}$$

Thus the surface $y=0$ is suffering propagation with respect to all the particles except those instantaneously situate upon it. The local speed of propagation is zero, $U=0$, but for no other choice of the initial instant or of material variables can U_N vanish. These observations are also evident from Fig. 28.

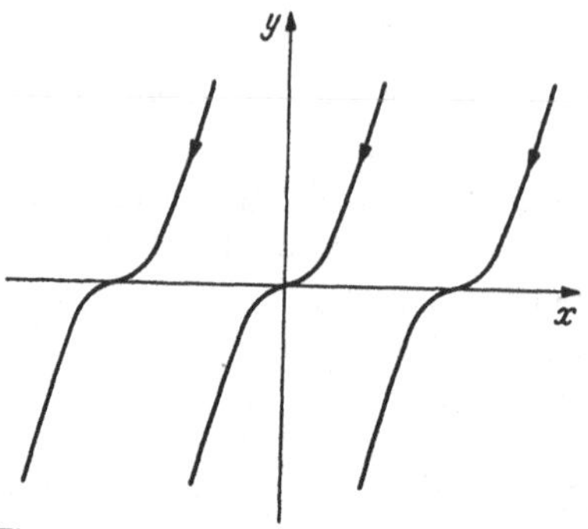

Fig. 28. The plane $y=0$ is not a material surface even though $\dot{y}=0$ upon it.

The nature of the discontinuity of the motion is made clearer if we calculate the distance d at time t between the two particles X, Y, Z and $X, (Y^{\frac{1}{3}}+D^{\frac{1}{3}})^3, Z$, where $D>0$. From (184.3) we have

$$d = D + 3(Y^{\frac{1}{3}} - kt)\,D^{\frac{2}{3}} + 3(Y^{\frac{1}{3}} - kt)^2\,D^{\frac{1}{3}}. \tag{184.10}$$

Thus D is the distance between the two particles at the instant they cross the plane $y=0$. From (184.10) it follows that no matter how small is D, there exists a time t_0 such that both after $t=t_0$ and before $t=-t_0$ the distance d is arbitrarily large. Thus no material volume remains bounded.

185. Slip surfaces, dislocations, vortex sheets, shock waves. The first kind of singular surface to be used in continuum mechanics was the *slip surface* of HELMHOLTZ[1]. On such a surface, the inverse motion $\boldsymbol{X}=\boldsymbol{X}(\boldsymbol{x},t)$ is discontinuous. Each place $\boldsymbol{x}$ is simultaneously occupied by two particles, $\boldsymbol{X}^+$ and $\boldsymbol{X}^-$. Two different masses thus slip past one another without penetration. The surface $f(\boldsymbol{x},t)=0$ is a material boundary of each motion.

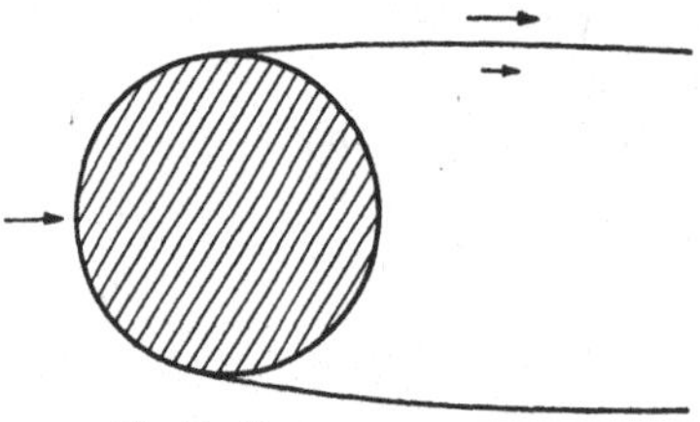

Fig. 29. Vortex sheet of order 0.

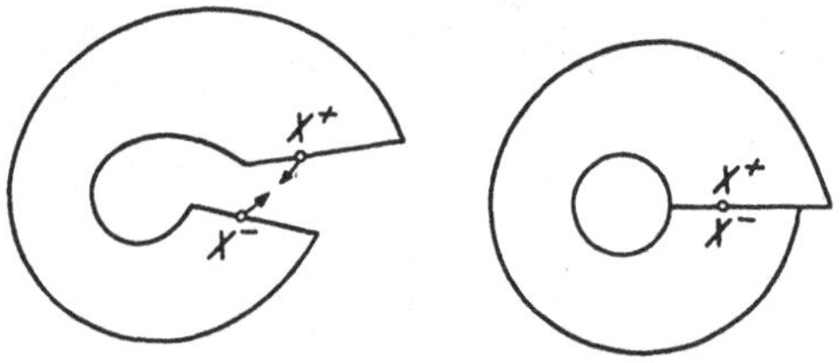

Fig. 30. Dislocation.

A surface across which the velocity suffers a transversal discontinuity,

$$[\dot{\boldsymbol{x}}] \neq 0, \quad [\dot{x}_n] = 0, \tag{185.1}$$

is called a *vortex sheet.*

Slip surfaces are vortex sheets; sometimes they are called vortex sheets of order 0. Such vortex sheets are easily constructed by placing adjacent to one another two motions having a common boundary. Unless it happens that $[\dot{\boldsymbol{x}}]=0$, that boundary will be a vortex sheet of the composite motion. Thus, alternatively, a slip surface may be regarded as a material surface on which the functions occurring in $(66.1)_1$ are double-valued (Fig. 29).

The *dislocations* of VOLTERRA[2] are singular surfaces intended to represent the deformation corresponding to removal or insertion of one mass within another, or the welding of boundaries (Fig. 30). There results a surface upon which $\boldsymbol{X}(\boldsymbol{x},t)$ is double-valued. Such surfaces need not be material; they may propagate, and they may bear any kind of discontinuity in the velocity.

[1] [1858, *1*, § 4], „diskontinuierliche Flüssigkeitsbewegung".
[2] [1905, *6*], "distorsioni". Cf. the example given in Sect. 49.

A *shock surface*[1] is one across which the normal velocity is discontinuous:

$$[\dot{x}_n] \neq 0. \tag{185.2}$$

Dislocations may be shock surfaces.

Little of a general nature may be said regarding such discontinuities, especially since they may represent removal or insertion of material.

The considerations of Sect. 182 regarding the material diagram must now be modified, since to a single spatial surface $f(\boldsymbol{x}, t) = 0$ there correspond *two distinct diagrams* $F^+ = 0$, $F^- = 0$, viz.,

$$0 = f(\boldsymbol{x}, t) \equiv F^+(\boldsymbol{X}^+(\boldsymbol{x}, t), t) = F^-(\boldsymbol{X}^-(\boldsymbol{x}, t), t), \tag{185.3}$$

the functions $\boldsymbol{X}^+(\boldsymbol{x}, t)$ and $\boldsymbol{X}^-(\boldsymbol{x}, t)$ being the two inverse functions to the single-valued equation $\boldsymbol{x} = \boldsymbol{x}(\boldsymbol{X}, t)$ defining the motion of the medium. *Thus the principle of duality fails to hold for these singularities.*

Moreover, we cannot apply the conditions of compatibility given in Sects. 175 and 180, since the function $\boldsymbol{x}(\boldsymbol{X}, t)$ is not defined, in general, upon both sides of either one of the diagrams $F^+ = 0$, $F^- = 0$ in the space of particles. HADAMARD'S lemma still holds, however, and may be used to derive some meager information[2].

For example, we can calculate the normal $\boldsymbol{n}$ in terms of the normals $\boldsymbol{N}^+$ and $\boldsymbol{N}^-$, using (17.3) and (182.8) with appropriate limiting values from one side only. Thus we may put the condition $(185.1)_2$ for a vortex sheet into the following form[3]:

$$[\dot{x}^k]\, \varepsilon_{kmp}\, \varepsilon^{\alpha\beta\gamma} F^{\pm}_{,\alpha}\, x^{m\pm}_{,\beta}\, x^{p\pm}_{,\gamma} = 0. \tag{185.4}$$

Each of the diagrams $F^+ = 0$ and $F^- = 0$ has its own speed of propagation, U_N^+ and U_N^-, given by (183.1) applied to F^+ and to F^-. Each of these speeds of propagation has the indeterminacy described in Sect. 183. By using (185.3) we may repeat the analysis leading to (183.3) and so obtain

$$U_N^+ \frac{\sqrt{F^+_{,\alpha} F^{+,\alpha}}}{\sqrt{f_{,k} f^{,k}}} + \dot{x}_n^+ = u_n = U_N^- \frac{\sqrt{F^-_{,\alpha} F^{-,\alpha}}}{\sqrt{f_{,k} f^{,k}}} + \dot{x}_n^-. \tag{185.5}$$

For this identity to hold, the derivatives $x^k_{,\alpha}$ and $\dot{x}^k$ need not exist upon the singular surface; by HADAMARD'S lemma, limit values from the $+$ side of $F^+ = 0$ and the $-$ side of $F^- = 0$, respectively, are employed throughout, it being assumed, as usual, that these limit values are continuously differentiable functions of position upon the surfaces.

This relation connects the two different speeds of propagation, whatever be the choices of the initial configurations corresponding to the two diagrams $F^+ = 0$ and $F^- = 0$. We are at liberty to choose the present configuration as the initial one, both for the particles on the $+$ side of $F^+ = 0$ and for those on the $-$ side of $F^- = 0$. By (185.5), the corresponding local speeds of propagation, U^+ and U^-, satisfy the relations[4]

$$U^+ = u_n - \dot{x}_n^+, \quad U^- = u_n - \dot{x}_n^-, \quad [U] = -[\dot{x}_n]: \tag{185.6}$$

---

[1] RIEMANN [1860, *4*, §§ 4–5], „Verdichtungsstoß"; cf. STOKES [1848, *4*], "a surface of discontinuity".

[2] E.g., the counterpart of (175.5) is

$$(x^k_{,\alpha} X^{\alpha}_{,\Delta})^+ = (x^k_{,\alpha} X^{\alpha}_{,\Delta})^-,$$

where the curves $v^\Gamma = \text{const}$ on $F^+ = 0$ and $F^- = 0$ are selected by identification of the points $\boldsymbol{X}^+$ and $\boldsymbol{X}^-$ that occupy the same place $\boldsymbol{x}$. It does not seem possible to make any use of this relation.

[3] BJERKNES *et al.* [1933, *3*, § 18].

[4] CHRISTOFFEL [1877, *2*, § 1], HADAMARD [1903, *11*, ¶ 102].

The jump in the speed of propagation of a singular surface is the negative of the jump in the normal velocity of the material, and a necessary and sufficient condition for the speed of propagation to be continuous is that the normal speed of the material be continuous, i.e., that the singular surface not be a shock surface.

From (185.6) we read off the following theorems concerning shock surfaces and vortex sheets, subject to the assumption that $\boldsymbol{x}(\boldsymbol{X}, t)$ be single-valued and continuous, though $\boldsymbol{X}(\boldsymbol{x}, t)$, in general, is double-valued:

1. *The local speed of propagation cannot be continuous across a shock surface.*

2. *There are no material shock surfaces,* i.e., *shock surfaces are always waves*[1].

3. *The local speed of propagation of a vortex sheet is always continuous. In particular, a vortex sheet which is material with respect to the particles on one side is also material with respect to those on the other.*

In the case of singular surfaces where the motion itself, or the velocity, is singular, the dual of the kinematical condition of compatibility (180.3) must be modified. Retracing the steps leading to that condition, we first write the dual of (180.1) for paths on each of the two material diagrams that may represent the singular surface $f(\boldsymbol{x}, t) = 0$, thus obtaining

$$\frac{\delta_+ \Psi^+}{\delta t} = \dot{\Psi}^+ + U_N^+ N^\alpha \partial_\alpha \Psi^+, \qquad \frac{\delta_- \Psi^-}{\delta t} = \dot{\Psi}^- + U_N^- N^\beta \partial_\beta \Psi^-, \tag{185.7}$$

where $\delta_+/\delta t$ and $\delta_-/\delta t$ as defined with respect to the normal velocities of the two diagrams in the space of particles $\boldsymbol{X}$. Even when the two diagrams coincide, the speeds of propagation appropriate to the two sides are in general different, as already noted. We now choose the present configuration as the initial configuration, both for the + and for the − sides, and subtract the second of (185.7) from the first. Thus follows[2]

$$\left.\begin{aligned} [\dot{\Psi}] &= \left[\frac{\delta \Psi}{\delta t}\right] - [U n^k \partial_k \Psi], \\ &= \left[\frac{\delta \Psi}{\delta t}\right] - U^+[n^k \partial_k \Psi] - [U]\, n^k \partial_k \Psi^-, \\ &= \left[\frac{\delta \Psi}{\delta t}\right] - U^+[n^k \partial_k \Psi] + [\dot{x}_n]\, n^k \partial_k \Psi^-, \end{aligned}\right\} \tag{185.8}$$

where we have used (185.6). In these formulae it must be remembered that on the two sides of the surface $\dot{\Psi}$ is calculated on the basis of $\dot{\boldsymbol{x}}^+$ and $\dot{\boldsymbol{x}}^-$, while the normal velocities used in calculating $\delta/\delta t$ are $U^+ \boldsymbol{n}$ and $U^- \boldsymbol{n}$, respectively.

---

[1] A shock surface may be material with respect to the particles on one side but not with respect to those on both sides.

[2] A result of this kind is asserted by KOTCHINE [1926, *3*, § 1], but his statement has $\delta[\Psi]/\delta t$ rather than $[\delta\Psi/\delta t]$ and thus is incorrect unless the operator $\delta/\delta t$, referred to the material diagram, is continuous, i.e., unless the singular surface is a vortex sheet.

For a check on (185.8), put $\Psi = \dot{x}^k$. The dual of (179.3) yields

$$\frac{\delta_+ x^k}{\delta t} = \dot{x}^{k+} + U_N^+ N^\alpha x^{k+}_{,\alpha}.$$

When the present configuration on the + side is taken as the initial state, this formula reduces to

$$\frac{\delta_+ x^k}{\delta t} = \dot{x}^{k+} + U^+ n^k;$$

hence

$$\left[\frac{\delta x^k}{\delta t}\right] = [\dot{x}^k] + [U]\, n^k.$$

Therefore (185.8) is satisfied.

When $\dot{\boldsymbol{x}}$ is continuous, (185.8) reduces to the dual of (180.2). Another special case will be discussed in the next section.

186. Material vortex sheets. In the case of a material singularity, irrespective of whether or not $\boldsymbol{x}(\boldsymbol{X}, t)$ is continuous, we have $U^+ = U^- = 0$, and in virtue of the duals of (179.1) we may reduce (185.8) to the form

$$[\dot{\Psi}] = \frac{d_+\Psi^+}{dt} - \frac{d_-\Psi^-}{dt}, \tag{186.1}$$

where d_+/dt and d_-/dt are the material derivatives calculated with $\boldsymbol{X}^+$ and $\boldsymbol{X}^-$ held constant, respectively. Hence[1]

$$\left.\begin{aligned}
[\dot{\Psi}] &= \frac{d_+[\Psi]}{dt} + [\dot{x}^k]\,\Psi^-_{,k}, \\
&= \frac{d_-[\Psi]}{dt} + [\dot{x}^k]\,\Psi^+_{,k}, \\
&= \frac{1}{2}\left(\frac{d_+}{dt} + \frac{d_-}{dt}\right)[\Psi] + \frac{1}{2}[\dot{x}^k]\,(\Psi^+_{,k} + \Psi^-_{,k}).
\end{aligned}\right\} \tag{186.2}$$

It was suggested by HELMHOLTZ[2] that the *velocity of the vortex sheet* be defined as the mean of the velocities on each side:

$$\boldsymbol{u} \equiv \tfrac{1}{2}(\dot{\boldsymbol{x}}^+ + \dot{\boldsymbol{x}}^-). \tag{186.3}$$

Only when the sheet is steady is this velocity tangent to it. Writing $\bar{d}/dt$ for the material derivative following this velocity $\boldsymbol{u}$, we have

$$\frac{\bar{d}}{dt} = \frac{1}{2}\left(\frac{d_+}{dt} + \frac{d_-}{dt}\right), \tag{186.4}$$

and (186.2) becomes

$$\left.\begin{aligned}
[\dot{\Psi}] &= \frac{\bar{d}[\Psi]}{dt} + \frac{1}{2}[\dot{x}^k]\,(\Psi^+_{,k} + \Psi^-_{,k}), \\
&= \frac{\bar{d}[\Psi]}{dt} + [\dot{\boldsymbol{x}}]\cdot\overline{\operatorname{grad}\Psi},
\end{aligned}\right\} \tag{186.5}$$

where we write

$$\overline{\operatorname{grad}\Psi} \equiv \tfrac{1}{2}(\operatorname{grad}\Psi)^+ + \tfrac{1}{2}(\operatorname{grad}\Psi)^-, \tag{186.6}$$

etc. Eqs. (186.1) and (186.5) are alternative forms of the *kinematical condition of compatibility for material singular surfaces*.

In particular, if Ψ and $\operatorname{grad}\Psi$ are continuous, from (186.5) we have

$$[\dot{\Psi}] = [\dot{\boldsymbol{x}}]\cdot\operatorname{grad}\Psi. \tag{186.7}$$

Since the singular surface has been assumed material, we have $d_+f/dt = d_-f/dt = 0$, and hence $[\dot{f}] = 0$. Putting $\Psi = f$ in (186.7) yields

$$[\dot{\boldsymbol{x}}]\cdot\operatorname{grad} f = 0. \tag{186.8}$$

Hence it follows that *if the velocity is not continuous across a singular surface that is persistently material with respect to the motion on each side, that surface is necessarily a vortex sheet*[3]. Thus we have another proof that *shock surfaces are always waves*.

[1] The first of these forms is given by KOTCHINE [1926, *3*, § 4].
[2] [1858, *1*, § 4].
[3] HADAMARD [1903, *11*, ¶ 94].

For any material derivative, we have identically $\dot{\overline{f_{,k}}} = \dot{f}_{,k} - \dot{x}^m{}_{,k} f_{,m}$. Since $d_+ f/dt = 0$, this equation yields $d_+ f_{,k}/dt = -\dot{x}^{m+}{}_{,k} f_{,m}$, and hence

$$\frac{\bar{d} f_{,k}}{dt} = -\frac{1}{2}(\dot{x}^{m+}{}_{,k} + \dot{x}^{m-}{}_{,k}) f_{,m}, \tag{186.9}$$

or

$$\frac{\bar{d}\,\mathrm{grad}\, f}{dt} = -\overline{\mathrm{grad}\,\dot{\boldsymbol{x}}} \cdot \mathrm{grad}\, f. \tag{186.10}$$

Therefore,

$$\frac{\bar{d}\boldsymbol{n}}{dt} = -\overline{\mathrm{grad}}\,\dot{\boldsymbol{x}} \cdot \boldsymbol{n} + \boldsymbol{n}(\boldsymbol{n} \cdot \overline{\mathrm{grad}\,\dot{\boldsymbol{x}}} \cdot \boldsymbol{n}). \tag{186.11}$$

This result shows that the manner in which the unit normal to the material singular surface changes as viewed by an observer moving with the mean velocity $\boldsymbol{u}$ is definitely determined.

The same is true of the jump in the acceleration, for if we put $\Psi = \dot{\boldsymbol{x}}$ in (186.5), we obtain

$$\left.\begin{aligned} [\ddot{x}_k] &= \frac{\bar{d}[\dot{x}_k]}{dt} + \frac{1}{2}[\dot{x}^m](x^+_{k,m} + x^-_{k,m}), \\ [\ddot{\boldsymbol{x}}] &= \frac{\bar{d}[\dot{\boldsymbol{x}}]}{dt} + [\dot{\boldsymbol{x}}] \cdot \overline{\mathrm{grad}\,\dot{\boldsymbol{x}}}, \end{aligned}\right\} \tag{186.12}$$

a result which bears a formal similarity to $(98.1)_2$ for the acceleration of a continuous motion. (186.12) is an iterated kinematical condition of compatibility for the special case of a material singular surface. It is a simple corollary that $[\dot{\boldsymbol{x}}] = 0$ implies $[\ddot{\boldsymbol{x}}] = 0$; that is, *if the velocity is continuous across a material singular surface, so is the acceleration*[1].

We now interpret (186.12) more closely by resolving it in directions normal and tangential to the surface. From $(186.12)_1$ and (186.9) we have

$$\left.\begin{aligned} f_{,k}[\ddot{x}^k] &= f_{,k}\frac{\bar{d}[\dot{x}^k]}{dt} + \frac{1}{2}[\dot{x}^m](\dot{x}^{k+}_{,m} + \dot{x}^{k-}_{,m}) f_{,k}, \\ &= f_{,k}\frac{\bar{d}[\dot{x}^k]}{dt} - [\dot{x}^k]\frac{\bar{d} f_{,k}}{dt}. \end{aligned}\right\} \tag{186.13}$$

By (186.8), we may write this result in the alternative forms[2]

$$\left.\begin{aligned} f_{,k}[\ddot{x}^k] &= -2[\dot{x}^k]\frac{\bar{d} f_{,k}}{dt} = +2 f_{,k}\frac{\bar{d}[\dot{x}^k]}{dt}, \\ [\ddot{x}_n] &= 2\boldsymbol{n} \cdot \frac{\bar{d}[\dot{\boldsymbol{x}}]}{dt}, \end{aligned}\right\} \tag{186.14}$$

whence it follows that the normal acceleration is continuous if and only if the time rate of change of the velocity as apparent to an observer moving with the mean velocity is tangent to the surface.

The more difficult reduction of the tangential component of (186.12) has been achieved by MOREAU[3]. By $(186.12)_2$ and (186.11) we have

[1] HADAMARD [1903, *11*, ¶ 94] has noted an interesting variant: By differentiating twice the equation $\dot{f} = 0$, we conclude that $[\ddot{x}_n] = 0$ if $[\dot{\boldsymbol{x}}] = 0$. That is, if *at some particular instant the velocity is continuous across a material singular surface, then at that instant the acceleration may suffer a transversal jump but not a longitudinal one.* The italicized result in the text above (due also to HADAMARD) refers to *persistent* continuity of $\dot{\boldsymbol{x}}$.

[2] KOTCHINE [1926, *3*, § 4].

[3] [1949, *19*] [1952, *13*, § 8c].

$$\left.\begin{aligned}\frac{\bar d}{dt}(\boldsymbol n\times[\dot{\boldsymbol x}]) &= [\dot{\boldsymbol x}]\times(\overline{\operatorname{grad}\dot{\boldsymbol x}}\cdot\boldsymbol n)-\boldsymbol n\times([\dot{\boldsymbol x}]\cdot\overline{\operatorname{grad}\dot{\boldsymbol x}})+\\ &\quad+(\boldsymbol n\times[\dot{\boldsymbol x}])(\boldsymbol n\cdot\overline{\operatorname{grad}\dot{\boldsymbol x}}\cdot\boldsymbol n)+\boldsymbol n\times[\ddot{\boldsymbol x}],\\ &=(\boldsymbol n\times[\dot{\boldsymbol x}])\cdot\overline{\operatorname{grad}\dot{\boldsymbol x}}+[\dot{\boldsymbol x}](\boldsymbol n\cdot\overline{\boldsymbol w})-\\ &\quad-(\boldsymbol n\times[\dot{\boldsymbol x}])(\overline{\operatorname{div}\dot{\boldsymbol x}}-\boldsymbol n\cdot\overline{\operatorname{grad}\dot{\boldsymbol x}}\cdot\boldsymbol n)+\boldsymbol n\times[\ddot{\boldsymbol x}],\end{aligned}\right\}\quad(186.15)$$

where we have used vector identities and (186.8), and where $\overline{\boldsymbol w}\equiv\overline{\operatorname{curl}\dot{\boldsymbol x}}=\frac12(\boldsymbol w^++\boldsymbol w^-)$, the mean of the vorticities on the two sides. Now by (App. 21.4)$_1$ we have

$$\left.\begin{aligned}\overline{\operatorname{div}\dot{\boldsymbol x}}-\boldsymbol n\cdot\overline{\operatorname{grad}\dot{\boldsymbol x}}\cdot\boldsymbol n &= \tfrac12(g^{km}-n^kn^m)(\dot x^+_{k,m}+\dot x^-_{k,m})\\ &=\tfrac12 a^{\Gamma\Delta}x^k_{,\Gamma}x^m_{,\Delta}(\dot x^+_{k,m}+\dot x^-_{k,m}).\end{aligned}\right\}\quad(186.16)$$

Thus the quantity on the left-hand side is the divergence, calculated intrinsically upon the surface, of the projection of the mean velocity onto the surface. If, therefore, we imagine on the surface itself a fictitious motion with the velocity field $\boldsymbol n\times\boldsymbol u$, an element of area da that is carried by this motion will change according to the formula

$$\frac{\bar d\,da}{dt}=(\overline{\operatorname{div}\dot{\boldsymbol x}}-\boldsymbol n\cdot\overline{\operatorname{grad}\dot{\boldsymbol x}}\cdot\boldsymbol n)\,da\quad(186.17)$$

[cf. (76.6), which holds in any metric space]. Putting (186.17) into (186.15) yields

$$\left.\begin{aligned}\frac{\bar d}{dt}(\boldsymbol n\times[\dot{\boldsymbol x}]\,da) &= \boldsymbol n\times[\ddot{\boldsymbol x}]\,da+(\boldsymbol n\times[\dot{\boldsymbol x}]\,da)\cdot\overline{\operatorname{grad}\dot{\boldsymbol x}}+[\dot{\boldsymbol x}](\boldsymbol n\cdot\overline{\boldsymbol w})\,da\\ &=\boldsymbol n[\ddot{\boldsymbol x}]\,da+\boldsymbol n\times[\dot{\boldsymbol x}]\,da\cdot\overline{\operatorname{grad}\dot{\boldsymbol x}}-(\boldsymbol n\cdot\overline{\boldsymbol w})\,\boldsymbol n\times(\boldsymbol n\times[\dot{\boldsymbol x}]\,da).\end{aligned}\right\}\quad(186.18)$$

This is Moreau's result. Its significance is easier to assess when we use Emde's notation (footnote 3, on p. 492): $\boldsymbol W\equiv\operatorname{Curl}\dot{\boldsymbol x}$, $\boldsymbol W^*\equiv\operatorname{Curl}\ddot{\boldsymbol x}$, for then we have

$$\frac{\bar d(\boldsymbol W da)}{dt}=\boldsymbol W^*da+\boldsymbol W da\cdot\overline{\operatorname{grad}\dot{\boldsymbol x}}-(\boldsymbol n\cdot\overline{\boldsymbol w})\,\boldsymbol n\times\boldsymbol W da.\quad(186.19)$$

Except for the last term, this equation has the same form as Beltrami's vorticity Eq. (101.7)$_3$, establishing an analogy with between the convection and diffusion of the vorticity $\boldsymbol w\,dv$ of a material element of volume in a continuous motion and the transport of surface vorticity $\boldsymbol W\,da$ in an element of area which is material with respect to the mean motion on the vortex sheet. In many cases of vortex sheets in fluid mechanics, the spatial vorticity $\boldsymbol w$ vanishes or is tangent to the sheet on each side; in such cases, $\boldsymbol n\cdot\overline{\boldsymbol w}=0$, and the analogy becomes precise. The analogue of the circulation preserving case is $\boldsymbol W^*=0$. For this case, as Moreau remarks, superficial analogues of all the classical theorems such as the Helmholtz theorems on spatially circulation preserving motions exist.

187. General classification of singular surfaces. In Sect. 173 a singular surface was defined with respect to an arbitrary quantity Ψ. Duhem[1] proposed to regard all quantities associated with a motion as functions $\Psi(\boldsymbol X,t)$ of the *material variables* $\boldsymbol X, t$ and to define the *order* of a singular surface with respect to Ψ as the order of the derivative $\left(\frac{D}{Dt}\right)^q\partial_{\alpha_1}\partial_{\alpha_2}\dots\partial_{\alpha_p}\Psi$ of lowest order $p+q$ suffering a non-zero jump upon the surface. Here, as in all that follows, we assume that in regions $\mathscr R^+$ and $\mathscr R^-$ on each side of the singular surface $\mathscr S(t)$ in

[1] [1900, 3].

the space of material variables $\boldsymbol{X}$, the function $\Psi(\boldsymbol{X}, t)$ and all its derivatives up to the highest order considered exist and are continuously differentiable functions of $\boldsymbol{X}$ and t, while on $\mathscr{S}(t)$ they approach definite limits which are continuously differentiable functions of position. Thus we may apply the geometrical and kinematical conditions of compatibility (175.11) and (180.2) and use differential manipulations freely.

There is no compelling reason to allow only discontinuities of this special type. Jump discontinuities upon surfaces are not the only ones that occur in physical problems; e.g. in Sect. 184 we have examined a simple and otherwise smooth motion in which the Jacobian $|\boldsymbol{x}/\boldsymbol{X}|$ increases steadily to ∞ as a certain surface is crossed and decreases steadily thereafter. Boundaries, studied in Sect. 184, and slip surfaces, dislocations, and tears, studied or mentioned in Sects. 185 to 186, are excluded as not being defined by sufficiently smooth jump discontinuities in functions of the material variables. Singularities at isolated lines or points are common; some of these are described in works on potential theory. In the case of jump discontinuities on surfaces, there is no *a priori* ground to expect that the limit values on each side of the surface be continuously differentiable on the surface, as we have assumed. The reasons for considering here only singularities of this kind are, first, that for more general singularities other than those analyzed above, scarcely any definite results are known except in very particular cases, and, second, that singular surfaces of the above types are frequently found useful in special theories of materials.

Clearly the definition of the order of a singular surface may be expressed alternatively in terms of the covariant derivatives $\overset{(q)}{\Psi}_{;\alpha_1\alpha_2\ldots\alpha_p}$. No modification in the results of Sects. 175 to 176 and 180 to 181 is needed to allow us to substitute double tensors of the type $T^{k\ldots m\,\alpha\ldots\beta}_{p\ldots q\,\gamma\ldots\delta}$ in the various jump conditions.

Many of the singularities of greatest interest are included in the case when

$$\Psi = \boldsymbol{x}(\boldsymbol{X}, t), \tag{187.1}$$

i.e., are surfaces across which the motion itself, or one of its derivatives, is discontinuous. By the *order*[1] of a singular surface henceforth we shall mean, unless some other quantity is mentioned explicitly, that we are taking $\Psi = \boldsymbol{x}$. Thus surfaces across which at least one of the functional relations (66.1) defining the motion itself is discontinuous are singularities of zero order; those across which some of the derivatives $\dot{x}^k$ and $x^k_{;\alpha}$ are discontinuous are of first order, etc.

In the classification of LICHTENSTEIN[2], the definition of the order is based upon the derivatives of the velocity field, $\dot{\boldsymbol{x}}$. Since $\dot{x}^k_{,m} = \dot{x}^k_{;\alpha} X^\alpha_{;m}$, a singular surface of order p in LICHTENSTEIN's scheme is also one of order p in that of DUHEM and HADAMARD, but the converse does not hold, for it is possible that a gradient such as $x^k_{;\alpha}$ may be discontinuous without there being any discontinuity in $\dot{x}^k$, etc. Thus the DUHEM-HADAMARD scheme includes a greater variety of singularities. Most researches on singular surfaces in fluids follow LICHTENSTEIN's classification, since in hydrodynamics it is possible largely to avoid consideration of the material variables. In retaining the DUHEM-HADAMARD classification we recognize its more fundamental scope[3] and its necessity in contexts such as the theory of

---

[1] HADAMARD [1901, *8*, § 1] [1903, *11*, ¶ 75].

[2] LICHTENSTEIN [1929, *4*, Chap. 6, ¶ 2].

[3] E.g., LICHTENSTEIN [1929, *4*, Chap. 6, ¶ 2] concludes that there are no material singularities of first or second order. While this is true according to his definitions, it is true only because those definitions offer no possibility of considering discontinuities in $x^k_{,\alpha}$ and $x^k_{,\alpha\beta}$, unaccompanied by discontinuities in time derivatives of $\boldsymbol{x}$. This example illustrates the insufficiency of LICHTENSTEIN's scheme.

elasticity, but we take care to derive from it, among other consequences, the spatial formulae that have found use in hydrodynamics.

In the following sections we find the kinematical properties of singular surfaces of finite order[1].

At a singular surface of order 0, the motion $\boldsymbol{x} = \boldsymbol{x}(\boldsymbol{X}, t)$ suffers a jump discontinuity. This must be interpreted as stating that the particles $\boldsymbol{X}$ upon the singular surface at time t are simultaneously occupying two places $\boldsymbol{x}^+$ and $\boldsymbol{x}^-$ or jump instantaneously from $\boldsymbol{x}^-$ to $\boldsymbol{x}^+$. Such discontinuities have not been found useful in field theories up to the present time. Therefore, in what follows, we study singular surfaces of orders 1 and greater.

188. Material singular surfaces. Material vortex sheets have been studied in Sect. 186. The results derived there remain valid for material singular surfaces of all orders. For a singularity of order 1 or greater, $\boldsymbol{X}(\boldsymbol{x}, t)$ is continuous, $d_+/dt = d_-/dt$, and (186.1) reduces to

$$[\dot{\Psi}] = \dot{\overline{[\Psi]}}. \tag{188.1}$$

This is *the general kinematical condition of compatibility for material singularities of order greater than* 0.

Its major use is to show that $[\Psi] = 0$ implies $[\dot{\Psi}] = 0$: Continuity of Ψ implies continuity of $\dot{\Psi}$. In other words, *the derivative of lowest order that is discontinuous across a material singularity is always a purely spatial derivative,* never a time derivative[2]. This is the dual of the theorem stated just after (180.4).

In particular, across a material singularity of first order, since $\boldsymbol{x}(\boldsymbol{X}, t)$ is continuous, so is $\dot{\boldsymbol{x}}$. That is, not only shock surfaces but also *vortex sheets of first order are waves,* while for the material vortex sheets described in Sect. 186 it is impossible that the motion itself be continuous across them. Vortex sheets are thus divided into two distinct categories: those of order 0, which are material, and those of order 1, which propagate. Across a material singularity of first order, $\dot{\boldsymbol{x}}$ is continuous, but at least one of the deformation gradients $x^k_{,\alpha}$ suffers a discontinuity.

189. Singular surfaces of order 1: Shock waves and propagating vortex sheets. For a singular surface of order 1, we put $\Psi = x^k$ in the duals of (180.5) and obtain[3]

$$[x^k_{,\alpha}] = s^k N_\alpha, \quad s^k = [N^\beta x^k_{,\beta}], \quad [\dot{x}^k] = -U_N s^k. \tag{189.1}$$

---

[1] A singular surface of infinite order is defined by HADAMARD [1903, *11*, ¶ 76] as one such that on each side, the function occurring in (66.1) are different analytic functions, yet all their derivatives are continuous across the surface. Such singularities seem not to have been studied. They offer interesting possibilities. For example, a one-dimensional motion starting very smoothly from rest at $t = 0$ is furnished by $x = X$ for $t \leqq 0$, $x = X + f(X)\, e^{-c/t^2}$ for $t > 0$, where $c > 0$.

In works on physics we often encounter discontinuous solutions regarded as limits of continuous ones; cf., e.g., the definition of a discontinuity given by MAXWELL [1873, *5*, §§ 7–8], [1881, *4*, § 8]. Perhaps seeking to justify such a treatment, LICHTENSTEIN [1929, *4*, Chap. 6, §§ 6–7] proves that any motion with a singular surface of order 1 or 2 may be obtained as a limit of analytic motions. Such theorems, however, do not reflect the physical situation. In electromagnetic theory, for example, physicists are wont to regard the solution of a problem for a material whose dielectric constant changes abruptly from K_1 to K_2 upon a certain surface as the limit of the solutions of "the same" problem for a dielectric such that K varies smoothly from K_1 to K_2 in a thin layer containing the surface. In gas dynamics, a flow of an ideal gas with a shock wave is regarded as the limit of solutions of a corresponding boundary value problem for viscous, thermally conducting fluids as the viscosity and thermal conductivity tend to zero. Problems of this kind presuppose some definite theory of materials and have no meaning in the generality of the present treatise. Even in relatively simple definite cases, they offer the highest mathematical difficulty. Cf. Sect. 5.

[2] HADAMARD [1903, *11*, ¶¶ 93, 104].

[3] HADAMARD [1903, *11*, ¶ 101].

The vector $\boldsymbol{s}$ is the *singularity vector*; while $(189.1)_2$ shows it to be parallel to the jump of velocity, its magnitude varies with the choice of the initial state and thus does not furnish a measure of the strength of the singularity. Rather, guided by the result given in Sect. 188, it is convenient to divide singular surfaces of order 1 into two classes:

1. Material singularities, which affect only the deformation gradients $x^k_{,\alpha}$.
2. Waves, including both shock waves and propagating vortex sheets.

For the former, the choice of the initial state is of prime importance. For the latter, it is not, and the nature of the waves is best specified in terms of the jump of velocity itself, $[\dot{\boldsymbol{x}}]$, which may be arbitrary both in direction and in magnitude. Indeed, if we adopt a strictly spatial standpoint, we may say the *only geometrical and kinematical requirement is that discontinuities in velocity be propagated*, both the amount of the discontinuity and the speed of propagation being arbitrary. Even here the adherents of a strictly spatial standpoint are closing their eyes to one of the phenomena occurring, since from (189.1) it follows that *a jump in velocity is impossible unless it is accompanied by jumps in the deformation gradients* $x^k_{,\alpha}$.

The results derived in Sect. 185, since they presume less regularity than is here assumed, remain valid for singularities of order 1. In particular, the speed of propagation satisfies (185.6), and the first and second theorems derived from it remain relevant. The third theorem is irrelevant, since, as just shown, *vortex sheets of first order are always waves.*

Since the functions occurring in (66.1) are continuous and single-valued by hypothesis, there is only one material diagram corresponding to a given initial state, and the principle of duality now holds. However, it is necessary to proceed with caution, since the fact that x^k_{α} experiences a jump on σ implies that if we regard a succession of initial states $\boldsymbol{X}$ corresponding to different initial times t_0, these states jump discontinuously at $t_0 = t \pm 0$. In particular, choosing the states on the $+$ and $-$ sides of the singular surfaces leads to different local speeds of propagation, U^+ and U^-, satisfying (185.6), as already stated. Writing the corresponding vectors $\boldsymbol{s}$ as $\boldsymbol{s}^+$ and $\boldsymbol{s}^-$, we have

$$[\dot{\boldsymbol{x}}] = -U^+ \boldsymbol{s}^+ = -U^- \boldsymbol{s}^-; \tag{189.2}$$

hence

$$[U \boldsymbol{s}] = 0. \tag{189.3}$$

By (185.6) follows

$$u_n[\boldsymbol{s}] = [\dot{x}_n \boldsymbol{s}]. \tag{189.4}$$

In the case of a vortex sheet, this relation reduces to

$$u_n = \dot{x}_n; \tag{189.5}$$

in the case of a stationary shock wave, to

$$[\dot{x}_n \boldsymbol{s}] = 0. \tag{189.6}$$

The nature of first order shock waves is illustrated by the very simple example[1] furnished by the one-dimensional motion defined by the equations $y = Y$, $z = Z$, and

$$x = \begin{cases} X + 2vt & \text{when } X \leqq 0, \quad -\infty < vt \leqq -\tfrac{1}{2}X, \\ \tfrac{1}{2}X + vt & \text{when } X \leqq 0, \quad -\tfrac{1}{2}X \leqq vt < \infty, \\ X + vt & \text{when } X \geqq 0, \quad -X \leqq vt < \infty, \\ 2X + 2vt & \text{when } X \geqq 0, \quad -\infty < vt \leqq -X, \end{cases} \tag{189.7}$$

[1] LICHTENSTEIN [1929, *4*, Chap. 6, § 8].

$v \neq 0$. This motion is defined over all space, and the equation $x = x(X, t)$ is continuous for every value of X and t. The stationary surface $x = 0$ is a shock wave. The particles constituting the material plane $X = \text{const}$ occupy a spatial plane at all times. They travel with the uniform speed $\dot{x} = 2v$ until they encounter the shock wave; their speed then drops to $\dot{x} = v$. With respect to the particles X it has not yet reached, a material diagram of the shock wave is given by

$$F \equiv X + 2vt = 0; \tag{189.8}$$

with respect to those it has passed, by

$$F \equiv X + vt = 0. \tag{189.9}$$

The speed of propagation is $U^- = -2v$ in the former case, $U^+ = -v$ in the latter. In both cases, $\boldsymbol{N} = (1, 0, 0)$; for the former case, $\boldsymbol{s} = \boldsymbol{s}^- = (\frac{1}{2}, 0, 0)$, while for the latter, $\boldsymbol{s} = \boldsymbol{s}^+ = (1, 0, 0)$.

This example is specially simple in that conditions are homogeneous on each side of the shock wave, so that only two different possibilities for $\boldsymbol{s}$ arise naturally; as always, there are in fact infinitely many[1].

The two planes $X = X_0$ and $X = X_0 + D$, where $D > 0$, remain a distance D apart until the latter encounters the shock. After the shock has passed, their distance apart is $\frac{1}{2}D$. The shock thus effects not only a sudden drop in velocity but also a sudden condensation. We now prove that this is representative of the general case.

To determine the jump experienced by a volume element through which a wave of first order passes, we calculate the jump in the Jacobian $\sqrt{g}\,J/\sqrt{G} = \boldsymbol{x}/\boldsymbol{X}$. By $(189.1)_1$ we have

$$\left.\begin{aligned} 6\,|\boldsymbol{x}/\boldsymbol{X}|^+ &= \varepsilon_{kqp}\,\varepsilon^{\alpha\beta\gamma}\,x^{k+}_{,\alpha}\,x^{q+}_{,\beta}\,x^{p+}_{,\gamma}, \\ &= \varepsilon_{kmp}\,\varepsilon^{\alpha\beta\gamma}\,(x^{k-}_{,\alpha} + s^k N_\alpha)\,(x^{m-}_{,\beta} + s^m N_\beta)\,(x^{p-}_{,\gamma} + s^p N_\gamma), \\ &= 6\,|\boldsymbol{x}/\boldsymbol{X}|^- + 3\,\varepsilon_{kmp}\,\varepsilon^{\alpha\beta\gamma}\,x^{k-}_{,\alpha}\,x^{m-}_{,\beta}\,s^p N_\gamma. \end{aligned}\right\} \tag{189.10}$$

By the dual of (17.3) follows

$$\frac{J^+}{J^-} = 1 + s^p N_\gamma X^{\gamma-}_{,p}. \tag{189.11}$$

Now we may apply (182.8) and $(189.1)_3$, writing U_N^- to recall that the derivatives $X^{\alpha-}_{,k}$ are used in the calculation[2]; using also $(185.5)_2$ we thus obtain

$$\left.\begin{aligned} \frac{J^+}{J^-} &= 1 + s^p n_p \frac{\sqrt{f_{,k} f^{,k}}}{\sqrt{F_{,\alpha} F^{,\alpha}}} = 1 - [\dot{x}_n]\,\frac{\sqrt{f_{,k} f^{,k}}}{U_N^- \sqrt{F_{,\alpha} F^{,\alpha}}}, \\ &= 1 - \frac{\dot{x}_n^+ - \dot{x}_n^-}{u_n - \dot{x}_n^-} \end{aligned}\right\} \tag{189.12}$$

or, by (185.6),

$$\frac{dv^+}{dv^-} = \frac{J^+}{J^-} = \frac{\dot{x}_n^+ - u_n}{\dot{x}_n^- - u_n} = \frac{U^+}{U^-}. \tag{189.13}$$

Thus the passage of a shock wave of first order[3] causes an abrupt change of volume, the ratio being that of the local speeds of propagation.

---

[1] We are not obliged to use initial positions or rectangular co-ordinates as material co-ordinates.

[2] The calculation may be shortened by taking the state on the − side of the surface as the initial state from the beginning.

[3] The qualification, "of first order", refers to the possibility of shocks in which $\boldsymbol{x}(\boldsymbol{X}, t)$ is discontinuous. About such shocks, mentioned in Sect. 185, little definite can be said.

This result is usually presented in terms of the density ϱ; from (156.2) then follows the ***Stokes-Christoffel condition:***

$$[\varrho\, U] = 0 \quad \text{or} \quad [\varrho(\dot{x}_n - u_n)] = 0. \tag{189.14}$$

These forms are of frequent use[1].

From (189.14) we see that *shock waves of first order are impossible in an isochoric motion, and that the passage of a vortex sheet of first order leaves the volume unchanged.*

By combining $(185.6)_3$ and $(189.14)_1$ we obtain the identity

$$\varrho^{\pm}\, U^{\pm}[\dot{x}_n] = -[\varrho\, U^2] = (U^-)^2 \frac{\varrho^-}{\varrho^+}[\varrho]. \tag{189.15}$$

Since, by hypothesis, $J > 0$, from (189.13) it follows that $\dot{x}_n - u_n$ is of the same sign on each side of the singularity. That is, if to an observer situate upon the surface the material seems to approach on one side, it departs upon the opposite side. The side on which the material approaches is called the *front* side, and the sign — will be assigned to it; the other side is the *rear.* The *strength* δ of a shock wave of first order is defined as

$$\delta \equiv \frac{[\varrho]}{\varrho^-} = \frac{\varrho^+}{\varrho^-} - 1 = \frac{[\dot{x}_n]}{U^+}. \tag{189.16}$$

If $\delta > 0$, the shock effects *condensation*; in the contrary case, *rarefaction*[2].

By putting $\Psi = x^k$ in the duals of (176.8) and (181.8) we may calculate the forms of the jumps of $x^k_{,\alpha\beta}$, $\dot{x}^k_{,\alpha}$, and $\ddot{x}^k$ across a shock wave. The results are complicated. We shall give them only for the quantity of greatest interest, the acceleration $\ddot{\boldsymbol{x}}$. Putting

$$c^k \equiv [N^\alpha N^\beta x^k_{,\alpha\beta}], \tag{189.17}$$

from (181.8) and $(189.1)_{2,3}$ we obtain

$$\left.\begin{aligned} [\ddot{x}^k] &= U_N^2\, c^k - 2\, U_N \frac{\delta_{\mathrm{d}}\, s^k}{\delta t} - s^k \frac{\delta_{\mathrm{d}}\, U_N}{\delta t}, \\ &= U_N^2\, c^k + 2 \frac{\delta_{\mathrm{d}}[\dot{x}^k]}{\delta t} - [\dot{x}^k] \frac{\delta_{\mathrm{d}} \log U_N}{\delta t}, \end{aligned}\right\} \tag{189.18}$$

where the displacement derivative $\delta_{\mathrm{d}}/\delta t$ is defined in terms of the motion of the same material diagram $F(\boldsymbol{X}, t) = 0$ as is used to calculate the speed of propagation, U_N. Since the quantity c^k, the jump of the fully normal second spatial derivatives, is essentially arbitrary, there is no immediate interpretation for the result (189.18).

Indeed, since the jump in velocity is arbitrary, so are the jumps in its derivatives, and there is no restriction in general upon $[\dot{x}_{k,m}]$ or $[\ddot{x}_k]$. It may be convenient, nevertheless, to resolve these jumps by means of the identities (175.11) and (180.3). The results are[3]

$$\left.\begin{aligned} [\dot{x}_{k,m}] &= [n^p \dot{x}_{k,p}]\, n_m + g_{mp}\, a^{\Delta\Gamma} x^p_{;\Delta}\, [\dot{x}_k]_{;\Gamma}, \\ \left[\frac{\partial \dot{x}_k}{\partial t}\right] &= -\, u_n [n^p \dot{x}_{k,p}] + \frac{\delta_{\mathrm{d}}[\dot{x}_k]}{\delta t}, \end{aligned}\right\} \tag{189.19}$$

[1] The direct proof given in the text simplifies and generalizes that of KOTCHINE [1926, *3*, § 3, ¶ 1]; the customary proof, resting upon an integral form of the principle of conservation of mass, will be given in Sect. 193. For one-dimensional motion, the result is due to STOKES [1848, *4*, Eq. (2)], RIEMANN [1860, *4*, § 5], and RANKINE [1870, *6*, §§ 2—4], the general case, to CHRISTOFFEL [1877, *2*, § 1] and JOUGUET [1901, *9*]. The classical treatment is that of HADAMARD [1903, *11*, ¶¶ 109—110).

[2] There is no kinematical or dynamical reason why rarefaction shocks cannot occur, although special thermodynamic conditions, including those usually assumed in gas dynamics, may forbid them.

[3] The somewhat different results obtained by HADAMARD [1903, *11*, ¶¶ 112, 113 bis, 119—120] are equally inconclusive.

where the displacement derivative $\delta_d/\delta t$ now refers to the spatial equation of the singularity, $f(\boldsymbol{x}, t) = 0$.

190. Singular surfaces of order 2: Acceleration waves. For singular surfaces of order 2, the analysis is simpler, since the speed of propagation U_N is continuous.

Substituting $\Psi = x^k$ into the dual of (176.10) yields

$$\left.\begin{aligned} [x^k_{,\alpha\beta}] &= [\partial_\beta x^k_{,\alpha}] = [\partial_\alpha x^k_{,\beta}] = s^k N_\alpha N_\beta, \\ s^k &= [N^\gamma N^\delta \partial_\delta x^k_{,\gamma}] = [N^\gamma N^\delta x^k_{,\gamma\delta}]. \end{aligned}\right\} \tag{190.1}$$

From the dual of (181.9) we have similarly[1]

$$[\dot{x}^k_{,\alpha}] = -U_N s^k N_\alpha, \quad [\ddot{x}^k] = U_N^2 s^k. \tag{190.2}$$

These formulae show that a singular surface of order 2 is completely determined by a vector $\boldsymbol{s}$ and the speed of propagation, U_N. In particular, material discontinuities of second order affect only the derivatives $x^k_{,\alpha\beta}$, while *discontinuities in the acceleration and in the velocity gradient are necessarily propagated*, and conversely, every wave of second order carries jumps in the velocity gradient and the acceleration. Waves of second order are therefore called *acceleration waves*.

Since $X^\alpha_{,m}$ is continuous, by $(190.2)_1$ and (182.8) we have

$$\left.\begin{aligned} [\dot{x}^k_{,m}] &= -U_N s^k N_\alpha X^\alpha{}_{,m}, \\ &= -U_N s^k \frac{\sqrt{f_{,p} f_{,p}}}{\sqrt{F_{,\beta} F_{,\beta}}} n_m. \end{aligned}\right\} \tag{190.3}$$

The left-hand side is independent of the choice of the initial state; therefore so also is the coefficient of $\boldsymbol{n}$ upon the right-hand side. Thus if we put

$$U s_0{}^k \equiv U_N s^k \frac{\sqrt{f_{,p} f_{,p}}}{\sqrt{F_{,\beta} F_{,\beta}}}, \tag{190.4}$$

the vector $U\boldsymbol{s}_0$ is independent of the choice of the initial state. We may choose to regard U as the local speed of propagation and $\boldsymbol{s}_0$ as the value of $\boldsymbol{s}$ corresponding to a choice of the material co-ordinates as being equal to the spatial co-ordinates at the instant the wave passes. In this notation, (190.3) and $(190.2)_2$ become[2]

$$[\dot{x}^k_{,m}] = -U s_0^k n_m, \quad [\ddot{x}^k] = U^2 s_0^k. \tag{190.5}$$

From (190.5) we have a number of corollaries. First,

$$\left.\begin{aligned} -U[n^m \dot{x}^k_{,m}] &= [\ddot{x}^k], \\ \left[\frac{\partial \dot{x}^k}{\partial t}\right] &= [\ddot{x}^k] - [\dot{x}^k{}_{,m}]\,\dot{x}^m, \\ &= U^2 s_0^k + U s_0^k \dot{x}_n, \\ &= U u_n s_0^k, \end{aligned}\right\} \tag{190.6}$$

where we have used (183.5). Thus the local acceleration is continuous across an acceleration wave if and only if the wave is stationary. Second,

$$\left.\begin{aligned} [\mathrm{I}_d] &= [\operatorname{div} \dot{\boldsymbol{x}}] = -U \boldsymbol{s}_0 \cdot \boldsymbol{n}, \\ [\boldsymbol{w}] &= [\operatorname{curl} \dot{\boldsymbol{x}}] = -U \boldsymbol{s}_0 \times \boldsymbol{n}; \end{aligned}\right\} \tag{190.7}$$

---

[1] Eqs. (190.1) to (190.2) are due essentially to HUGONIOT [1885, *4*, p. 1231] [1887, *2*, Part II, § 9)]. Cf. HADAMARD [1901, *8*, § 2] [1903, *11*, ¶ 102].

[2] These results and (190.6) were given by HUGONIOT [1885, *3*, p. 1120] [1887, *2*, Part I, § 9] in forms with $U s_0^k$ eliminated.

interpretation of these identities yields **Hadamard's theorem**[1]: *A longitudinal acceleration wave carries a jump in the expansion but leaves the vorticity unchanged, while a transverse acceleration wave carries a jump in the vorticity but does not affect the expansion.*

By $(156.5)_2$, we may put $(190.7)_2$ into the alternative form

$$[\overline{\log \varrho}^{\,\cdot}] = U \boldsymbol{s}_0 \cdot \boldsymbol{n} = \frac{1}{U}[\ddot{x}_n]. \tag{190.8}$$

Since $\log \varrho$ is continuous across an acceleration wave, by putting $\Psi = \log \varrho$ in the dual of $(180.5)_2$ we have

$$[\overline{\log \varrho}^{\,\cdot}] = -U\left[\frac{d \log \varrho}{dn}\right]. \tag{190.9}$$

By (190.8) follows the important relation

$$[\ddot{x}_n] = -U^2\left[\frac{d \log \varrho}{dn}\right]. \tag{190.10}$$

191. Singular surfaces of higher order[2]. For a singular surface of order p, the results are easy generalizations of those for the case when $p=2$.

Putting $\Psi = x^k$ into the dual of (176.11) yields

$$\left.\begin{aligned} [x^k{}_{,\alpha_1\alpha_2\ldots\alpha_p}] &= [\partial_{\alpha_1}\partial_{\alpha_2}\ldots\partial_{\alpha_p}x^k] = s^k N_{\alpha_1}N_{\alpha_2}\ldots N_{\alpha_p},\\ s^k &= [N^{\beta_1}N^{\beta_2}\ldots N^{\beta_p}\partial_{\beta_1}\partial_{\beta_2}\ldots\partial_{\beta_p}x^k], \end{aligned}\right\} \tag{191.1}$$

while the dual of (181.11) yields

$$\left.\begin{aligned} [\dot{x}^k{}_{,\alpha_1\alpha_2\ldots\alpha_{p-1}}] &= -U_N s^k N_{\alpha_1}N_{\alpha_2}\ldots N_{\alpha_{p-1}},\\ [\ddot{x}^k{}_{,\alpha_1\alpha_2\ldots\alpha_{p-2}}] &= U_N^2 s^k N_{\alpha_1}N_{\alpha_2}\ldots N_{\alpha_{p-2}},\\ &\vdots\\ [\overset{(p-1)}{x}{}^k{}_{,\alpha_1}] &= (-U_N)^{p-1}s^k N_{\alpha_1},\\ [\overset{(p)}{x}{}^k] &= (-U_N)^p s^k = (-U)^p s_0^k. \end{aligned}\right\} \tag{191.2}$$

Thus a singularity of order p carries a jump in the p-th acceleration if and only if it is a wave.

Now by (156.5) we have

$$-\overline{\log \varrho}^{\,\cdot} = \dot{x}^k{}_{,\alpha}X^\alpha{}_{,k} = \dot{x}^k{}_{,k},$$

so that

$$\tag{191.3}$$

$$-\overline{(\log \varrho)}^{(h)}{}_{,\alpha_1\alpha_2\ldots\alpha_{p-1-h}} = \overset{(h)}{x}{}^k{}_{,\alpha\alpha_1\alpha_2\ldots\alpha_{p-1-h}}X^\alpha{}_{,k} + \cdots, \tag{191.4}$$

where the dots stand for a polynomial in derivatives of orders less than p. By (191.2) we thus obtain

$$\left.\begin{aligned} -[\overline{(\log \varrho)}^{(h)}{}_{,\alpha_1\alpha_2\ldots\alpha_{p-1-h}}] &= (-U_N)^h N_{\alpha_1}N_{\alpha_2}\ldots N_{\alpha_{p-1-h}}N_\alpha X^\alpha{}_{,k}s^k,\\ h &= 0, 1, \ldots, p-1. \end{aligned}\right\} \tag{191.5}$$

Choosing the initial state as that on the singular surface yields

$$-[\overline{(\log \varrho)}^{(h)}{}_{,k_1k_2\ldots k_{p-1-h}}] = (-U)^h s_{0n} n_{k_1}n_{k_2}\ldots n_{k_{p-1-h}}. \tag{191.6}$$

---

[1] HADAMARD [1903, *11*, ¶¶ 111–115]; announced in part in [1901, *8*, § 3]. In part, these results are equivalent to WEINGARTEN's formulae (175.10) and are foreshadowed by a theorem of HUGONIOT [1887, *2*, Part I, § 12].

[2] HADAMARD [1903, *11*, ¶¶ 88, 103, 111–111 bis].

In particular, from this result and $(191.2)_4$ we have

$$[\overset{(p-1)}{(\log \varrho)}] = \frac{1}{U} [\overset{(p)}{x}_n]. \tag{191.7}$$

This relation enables us to attach a meaning to the sign of a discontinuity. [Cf. (189.16) and (190.8).] If the propagating singular surface of p-th order carries with it an increase in $\overset{(p-1)}{\log \varrho}$, it may be said to be a *compressive* wave; in the contrary case *expansive*. From (191.7) we see that a wave is compressive or expansive according as it brings an increase or decrease in the normal component $\overset{(p)}{x}_n$ of the p-th acceleration[1]. In particular, *in an isochoric motion, acceleration waves of all orders are necessarily transversal*, and, conversely, *material singularities and transverse waves of all orders leave the density and all its derivatives continuous.*

In the case of a surface which is singular with respect to Ψ and also a singular surface of order 2 or greater with respect to the motion itself, the principle of duality when applied to (180.3) yields

$$[\dot{\Psi}] = -U_N[N^\alpha \Psi_{,\alpha}] + \frac{\delta_d}{\delta t}[\Psi], \tag{191.8}$$

where the displacement derivative $\delta_d/\delta t$ is defined in terms of the motion of the material diagram $F(\boldsymbol{X}, t) = 0$. This result follows at once because, corresponding to any selected initial state, there is a unique speed of propagation U_N. [The modification necessary for a singularity of first order has been given as (185.8).] We now give an alternative derivation[2] of (191.8), based upon the spatial equations. From $(176.2)_1$ we have

$$\left.\begin{aligned} \dot{x}^k[\Psi_{,k}] &= B\dot{x}_n + g_{km} a^{\Delta\Gamma} \dot{x}^k x^m_{;\Delta} A_{;\Gamma}, \\ &= B\dot{x}_n - U^\Gamma A_{;\Gamma}, \end{aligned}\right\} \tag{191.9}$$

where U^Γ is the dual of u_Γ, given by (177.11). From (181.2) we have

$$\left[\frac{\partial \Psi}{\partial t}\right] = -u_n B + \frac{\delta_d A}{\delta t}, \tag{191.10}$$

where $\delta_d/\delta t$ is the displacement derivative defined with respect to the spatial surface $f(\boldsymbol{x}, t) = 0$. By adding together (191.9) and (191.10) and then using (183.5) and the dual of (179.5), we obtain the result (191.8) in the case when the present configuration is chosen as the initial one.

V. Discontinuous equations of balance.

192. The transport theorem for a region containing a singular surface. Consider a material volume $\mathscr{V}$ within which there occurs a surface $\mathscr{s}(t)$ that is a persistent singular surface with respect to a quantity Ψ and possibly also with respect to $\dot{\boldsymbol{x}}$. The singular surface, assumed smooth, may be in motion with any speed of displacement, u_n. Our purpose is to generalize the transport theorem (81.3) so as to be applicable to such a volume.

---

[1] This analysis is due to HADAMARD [1903, *11*, ¶ 117], who gives also [ibid., ¶ 116] a more general definition, applicable whether or not the kinematical conditions of compatibility hold; this definition leads to the sign of $[\overset{(p)}{x}_n]$ as the criterion. HADAMARD's use of "comprimant" and "dilatant" is the inverse of ours.

[2] Suggested by KOTCHINE [1926, *3*, § 1].

While we consider a finite volume $\mathscr{V}$, we are interested only in the case when it is arbitrarily small and arbitrarily smooth. Thus it is sufficiently general to assume that $\mathscr{s}$ divides $\mathscr{V}$ into two regions $\mathscr{R}^+$ and $\mathscr{R}^-$ in which Ψ and $\dot{\boldsymbol{x}}$ are continuously differentiable (Fig. 31) and divides the material boundary $\mathscr{S}$ into two parts $\mathscr{S}^+$ and $\mathscr{S}^-$. In general, the regions and surfaces $\mathscr{R}^+$, $\mathscr{R}^-$, $\mathscr{S}^+$, $\mathscr{S}^-$ fail to be material. We now define fields $\boldsymbol{u}^+$ and $\boldsymbol{u}^-$ as follows

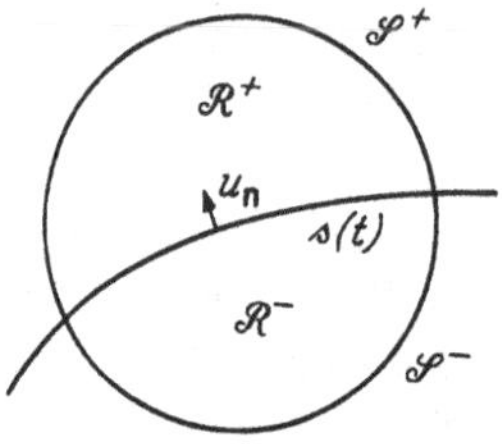

Fig. 31. Diagram for proof of the transport theorem for a region divided by a singular surface.

$$\left.\begin{aligned}\boldsymbol{u}^+ &\equiv \begin{cases}\dot{\boldsymbol{x}} \text{ on } \mathscr{S}^+,\\ u_n \boldsymbol{n} \text{ on } \mathscr{s},\end{cases}\\ \boldsymbol{u}^- &\equiv \begin{cases}\dot{\boldsymbol{x}} \text{ on } \mathscr{S}^-,\\ u_n \boldsymbol{n} \text{ on } \mathscr{s}.\end{cases}\end{aligned}\right\} \tag{192.1}$$

Since $\mathscr{s}$ is a persistent common boundary of $\mathscr{R}^+$ and $\mathscr{R}^-$, we have

$$\frac{d}{dt}\int_{\mathscr{V}} \Psi\, dv = \frac{d_{\boldsymbol{u}^+}}{dt}\int_{\mathscr{R}^+} \Psi\, dv + \frac{d_{\boldsymbol{u}^-}}{dt}\int_{\mathscr{R}^-} \Psi\, dv, \tag{192.2}$$

where $d_{\boldsymbol{u}^+}/dt$ indicates that we are to take the time derivative of the integral over a region that instantaneously coincides with $\mathscr{R}^+$ and is material with respect to the field $\boldsymbol{u}^+$, while $d_{\boldsymbol{u}^-}/dt$ is analogously defined. To each of the integrals on the right, since Ψ and $\dot{\boldsymbol{x}}$ are continuously differentiable in $\mathscr{R}^+$ and $\mathscr{R}^-$ and approach continuous limits on the entire boundaries $\mathscr{S}^+ + \mathscr{s}$ and $\mathscr{S}^- + \mathscr{s}$, we may apply the usual transport theorem (81.4), obtaining

$$\left.\begin{aligned}\frac{d_{\boldsymbol{u}^+}}{dt}\int_{\mathscr{R}^+} \Psi\, dv &= \int_{\mathscr{R}^+} \frac{\partial \Psi}{\partial t}\, dv + \int_{\mathscr{S}^+} \Psi\, \dot{x}_n\, da - \int_{\mathscr{s}} \Psi^+ u_n\, da,\\ \frac{d_{\boldsymbol{u}^-}}{dt}\int_{\mathscr{R}^-} \Psi\, dv &= \int_{\mathscr{R}^-} \frac{\partial \Psi}{\partial t}\, dv + \int_{\mathscr{S}^-} \Psi\, \dot{x}_n\, da + \int_{\mathscr{s}} \Psi^- u_n\, da.\end{aligned}\right\} \tag{192.3}$$

By substituting these results into (192.2) we obtain[1]

$$\frac{d}{dt}\int_{\mathscr{V}} \Psi\, dv = \int_{\mathscr{V}} \frac{\partial \Psi}{\partial t}\, dv + \oint_{\mathscr{S}} \Psi\, \dot{x}_n\, da - \int_{\mathscr{s}} [\Psi]\, u_n\, da. \tag{192.4}$$

This is the desired generalization of (81.3).

193. The general balance at a surface of discontinuity. Let it be supposed that a general equation of balance of the form (157.1) holds for an arbitrary material volume, whether or not there is a singular surface within it. Then by applying (192.4) to a sufficiently small material volume $\mathscr{V}$ containing a singular surface $\mathscr{s}$ we obtain

$$\int_{\mathscr{V}} \frac{\partial(\varrho\Psi)}{\partial t}\, dv + \oint_{\mathscr{S}} \varrho\Psi\, \dot{x}_n\, da - \int_{\mathscr{s}} [\varrho\Psi]\, u_n\, da = -\oint_{\mathscr{S}} i_n\, da + \int_{\mathscr{V}} \varrho\, s\, dv. \tag{193.1}$$

We now assume that in the neighborhood of $\mathscr{s}$, the quantities $\frac{\partial(\varrho\Psi)}{\partial t}$ and ϱs are bounded, while on each side of $\mathscr{s}$ the quantities $\varrho\Psi$, $\dot{x}_n$, and i_n approach limits that are continuous functions of position. Under these conditions, we let

[1] THOMAS [1949, *31*, § 18].

$\mathscr{S}^+$ and $\mathscr{S}^-$ shrink down to $\mathfrak{s}$ (Fig. 32), so that the volume of $\mathscr{R}^+ + \mathscr{R}^-$ vanishes while the area of $\mathfrak{s}$ remains finite in the limit. The volume integrals then vanish in the limit, and (193.1) becomes

$$\int_{\mathfrak{s}} \left([\varrho\,\Psi\,\dot{x}_n] - [\varrho\,\Psi]\,u_n + [i_n]\right) da = 0. \tag{193.2}$$

Since this equation holds for any area on $\mathfrak{s}$, and since the integrand, by hypothesis, is continuous, it must vanish. By using (185.6) we thus obtain ***Kotchine's theorem***[1]: *At a singular surface, the general balance* (157.1) *is equivalent to the condition*

$$[\varrho\,\Psi\,U] - [i_n] = 0. \tag{193.3}$$

In this result, the source ϱs of Ψ plays no part. This phenomenon does not contradict the principle of the equivalence of surface and volume sources, stated in Sect. 157, since a condition for replacing the influx $\boldsymbol{i}$ by an equivalent supply $\varrho s = -i^k{}_{,k}$ is that $\boldsymbol{i}$ be continuously differentiable throughout the interior of $\mathscr{V}$, and in this case it follows that $\boldsymbol{i}$ is continuous on $\mathfrak{s}$, so that $[i_n] = 0$.

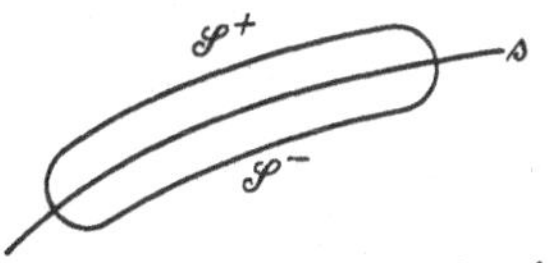

Fig. 32. Diagram for derivation of the general balance at a singular surface.

Here we give only one example of the use of Kotchine's theorem. We assume that the principle of conservation of mass, in the form stated in Sect. 155, holds also for material volumes containing a singular surface. Then we may put $\Psi = 1$ and $\boldsymbol{i}[\Psi] = 0$. Substituting these assumptions into (193.3) yields the same result (189.14) that was derived from different concepts in Sect. 189.

It follows that we may always express (193.3) in the form

$$\varrho^{\pm}\,U^{\pm}[\Psi] - [i_n] = 0. \tag{193.4}$$

The two derivations we have given for (189.14) must not be regarded as equivalent, for they rest upon different definitions of what is meant by conservation of mass at a shock wave. The derivation given in Sect. 189 employs the identity (189.12), which is a corollary of the geometrical and kinematical conditions of compatibility; conservation of mass is then taken to mean that the density ϱ is so adjusted that $[\varrho\,dv] = 0$, this being a natural extension of the postulate (156.1). In the present section we have postulated instead that for any material volume $\mathscr{V}$, irrespective of whether or not it contains a shock wave, $\int_{\mathscr{V}} \varrho\,dv$ remains constant in time. That the two postulates, while consistent, are not equivalent, is made yet more clear by observing that the proof in this section goes through in just the same way if there is a bounded mass supply ϱs and a continuous mass influx $\boldsymbol{i}$, in the sense used in Sect. 157. In these circumstances the usual continuity equation is replaced by

$$\dot{\varrho} + \varrho\,\dot{x}^k{}_{,k} = -i^k{}_{,k} + \varrho s \tag{193.5}$$

at interior points, but on the shock wave the jump condition (189.14) still holds. Thus (189.14), while a necessary and sufficient condition that $\varrho\,dv$ remain constant in crossing the shock, is only a necessary but not sufficient condition that mass be conserved in every material volume, however small, through which the shock is propagated.

194. Impulsive balance. The general balance (157.1) may be integrated with respect to t to yield

$$\int_{\mathfrak{v}_{t_2}} \varrho\,\Psi\,dv - \int_{\mathfrak{v}_{t_1}} \varrho\,\Psi\,dv = \int_{t_1}^{t_2} \left\{-\oint_{\mathfrak{s}_t} d\boldsymbol{a}\cdot\boldsymbol{i} + \int_{\mathfrak{v}_t} \varrho\,s\,dv\right\} dt, \tag{194.1}$$

where $\mathfrak{v}_t$ is the spatial volume occupied at time t by the material volume $\mathscr{V}$ and where $\mathfrak{s}_t$ is the bounding surface of $\mathfrak{v}_t$. When Ψ is discontinuous at a certain

[1] Kotchine [1926, *3*, § 3]. Special cases due to earlier authors will be given in Sects. 205 and 241. Kotchine called (193.3) "les conditions dynamiques de compatibilité", following the terminology used earlier in a special case by Zemplén [1905, *7*, p. 438].

instant t, where $t_1 \leqq t \leqq t_2$, the form (157.1) loses meaning, but (194.1), with the time integral regarded as a Stieltjes integral, may then be taken as a postulate replacing it. That is,

$$\int_{v_{t_2}} \varrho \Psi \, dv - \int_{v_{t_1}} \varrho \Psi \, dv = \int_{t_1}^{t_2} \Big\{ - \oint_{s_t} d\boldsymbol{a} \cdot d\boldsymbol{i} + \int_{v_t} \varrho \, ds \, dv \Big\} . \tag{194.2}$$

This form may be regarded as a *fully general statement of any law of balance.*

If Ψ suffers a jump discontinuity as a function of t, so that $\Psi(\boldsymbol{x}, t+0)$ and $\Psi(\boldsymbol{x}, t-0)$ are continuously differentiable functions of $\boldsymbol{x}$, the quantity Ψ is said to experience an *impulse.* We write

$$[\Psi] \equiv \Psi(\boldsymbol{x}, t+0) - \Psi(\boldsymbol{x}, t-0) . \tag{194.3}$$

From (194.2), by the rule for differentiating a Stieltjes integral, it follows that

$$\Big[\int_{v_t} \varrho \Psi \, dv \Big] = - \oint_{s_t} d\boldsymbol{a} \cdot \boldsymbol{k} + \int_{v_t} \varrho \, h \, dv , \tag{194.4}$$

where $\boldsymbol{k}[\Psi]$ is an integrable field defined upon s_t and $h[\Psi]$ is an integrable quantity defined over v_t. We may call $\boldsymbol{k}$ and h the *impulsive influx* and the *impulsive supply* of Ψ. The remarks made in Sect. 157 are applicable here, *mutatis mutandis.*

Now how the particles $\boldsymbol{X}$ are regarded as behaving in an impulse is only a matter of convention. It is all one whether we choose to regard all the particles as jumping instantly to different places, where the velocity is different, or as staying fast but instantly gaining a new velocity. Henceforth we lay down the convention that *in an impulse, the particles stand still.* Therefore

$$[\boldsymbol{x}] = 0 , \quad [dv] = 0 ; \tag{194.5}$$

hence

$$\Big[\int_{v_t} \varrho \Psi \, dv \Big] = \int_{v_t} [\varrho \Psi] \, dv . \tag{194.6}$$

If we assume further that mass is not impulsively created, we have

$$[\varrho \, dv] = 0 . \tag{194.7}$$

From $(194.5)_2$ it follows that

$$[\varrho] = 0 , \tag{194.8}$$

and hence

$$\Big[\int_{v_t} \varrho \Psi \, dv \Big] = \int_{v_t} \varrho [\Psi] \, dv \tag{194.9}$$

[cf. (156.8)].

Combining (194.9) and (194.4) yields

$$\int_{v_t} \varrho [\Psi] \, dv = - \oint_{s_t} d\boldsymbol{a} \cdot \boldsymbol{k} + \int_{v_t} \varrho \, h \, dv , \tag{194.10}$$

this being the impulsive analogue of the general balance (157.1). By applying GREEN's transformation to (194.10), when $\varrho [\Psi]$, k^p, $k^q{}_{,q}$, and ϱh are assumed continuous, we obtain the *general impulsive balance*[1]:

$$\varrho \{\!\!\{\psi\}\!\!\} = -\operatorname{div} \boldsymbol{k} + \varrho \boldsymbol{h}, \quad \text{or} \quad \varrho \{\!\!\{\psi^{\cdots}_{\cdots}\}\!\!\} = -k^{\cdots q}_{\cdots, q} + \varrho h^{\cdots}_{\cdots}. \tag{194.11}$$

The analogy to (157.5) and (157.6) is immediate.

194A. Appendix to Chapter C: Remarks on other definitions of waves and discontinuities. In particular theories of materials two other definitions of waves and/or discontinuities frequently appear: *characteristic manifolds* and *harmonic oscillations*.

Given a particular and determinate system of partial differential equations which is linear in the highest derivatives occurring, prescription of the values of the derivatives of lower order upon a surface element of appropriate dimension generally determines unique values for all the highest derivatives. Particular surface elements on which this is not so are called *characteristic* elements. If the characteristic elements envelope a surface, this surface may serve as the common boundary of two solutions that are otherwise independent of one another. Thus characteristic surfaces resemble singular surfaces.

Now the general principles of physics are always expressed by underdetermined systems. A determinate system results only by the adoption of constitutive equations defining a particular material (cf. Sect. 7). Thus it is clear that the theory of characteristics cannot be a consequence of the general principles of physics. It is possible to develop the theory of characteristics in very general terms, but this theory, independent of *all* physical principles, becomes a part of pure mathematics. On both counts, it would be inappropriate to present this theory here.

It should be remarked, however, that in the various common theories of fluids and elastic or plastic bodies, definite relations between characteristic surfaces and singular surfaces exist. In some theories, singular surfaces are always characteristic surfaces; in others, never; in still others, sometimes. Some theories in which singular surfaces are possible have no real characteristic surfaces. Characteristic surfaces need not be, and in general are not, singular surfaces. This is further evidence that the theory of characteristics is special in respect to the principles of physics, since singular surfaces, as we have seen, may be envisaged in the greatest generality and are governed by purely kinematical restrictions in addition to dynamical and energetic ones to be derived in Sects. 205 and 241, besides such others as may result from particular constitutive equations.

The situation is somewhat similar with respect to *infinitesimal oscillations*. These are obtained by assuming, in the differential equations of a particular theory, that the acceleration $\ddot{\boldsymbol{x}}$ may justifiably be replaced by $\partial^2 \boldsymbol{u}/\partial t^2$, where $\boldsymbol{u}$ is the displacement at $\boldsymbol{x}, t$; such motions are *slow* in the sense defined in Sect. 100. The equations, by one means or another, are then replaced "approximately" by linear ones, so that

$$\frac{\partial^2 \boldsymbol{u}}{\partial t^2} = L(\boldsymbol{u}), \tag{194A.1}$$

where $L(\boldsymbol{u})$ is a linear differential expression in space derivatives. It may then be expected that there are solutions of the form $\boldsymbol{u} = \boldsymbol{s} \cos \omega t$ satisfying

$$L(\boldsymbol{s}) + \omega^2 \boldsymbol{s} = 0. \tag{194A.2}$$

From boundary conditions result limitations on the possible values of the circular frequency ω, implying that these infinitesimal motions propagate at definite speeds. In many theories, these speeds turn out to be exactly the same as those possible for certain kinds of propagating singular surfaces and/or characteristic surfaces. This is the case in regard to large classes of motions of elastic solids and perfect fluids, the common speeds being called *speeds of sound*; also in respect to electromagnetic theories, where they are called *speeds of light*. No explanation of this remarkable agreement is known[2].

---

[1] In principle, this result is due to KOTCHINE [1926, *3*, § 3], who considered a more general singular surface in space-time, allowing for the possibility of an impulse combined with a singular surface in space, an impulse confined to a line, etc. Since such complexities have not yet appeared in applications, and since KOTCHINE's greater generality is purchased with the expense of an intricate proof, we have thought it sufficient to consider spatial singular surfaces and volume impulses separately in Sect. 193 and the present section, taking advantage of the almost trivial proofs which thus become possible.

[2] Of course, in the *particular* classical theories to which the text refers, the explanation is almost trivial, since the operator $L(\boldsymbol{u})$ is totally hyperbolic and hence can be put into the normal form

$$L(\boldsymbol{u}) = c^2 (\nabla^2 \boldsymbol{u} + \cdots),$$

D. Stress.

195. Scope and plan of the chapter. This chapter presents the general theory of momentum and moment of momentum, or, equivalently, of *force* and *torque*, in classical mechanics.

Subchapter I introduces *Euler's laws* of mechanics; these integral relations express for an entire body, whether solid, superficial, lineal, or punctual in form, the balance of momentum and of moment of momentum. This formulation emphasizes the properties of classical mechanics distinguishing it from other physical theories: Space is presumed to be Euclidean, allowing finite parallel transport of vectors, and the statements of the mechanical laws distinguish the particular class of Euclidean frames called inertial frames.

The rest of the chapter is an exhaustive exposition of the properties of the *stress tensor*, the characteristic dynamical quantity of the mechanics of continua. In Subchapter II the stress principle is stated, whence follow the basic differential equations, called *Cauchy's laws of motion*, as well as jump conditions at singular surfaces. The next subchapter concerns applications of these laws to determine consumption of power, mean values of the stress, etc. Subchapter IV is a treatise on general solutions of Cauchy's laws by means of *stress functions*. The last subchapter explains alternative *variational principles*.

After major but special preliminary work by James Bernoulli (1691—1704), John Bernoulli (1742), and Euler (1749—1771), the general theory was created by Cauchy (1823—1829); important special results were obtained by Piola (1835—1845), Signorini (1930—1933), and B. Finzi (1934).

While every textbook on elasticity or plasticity makes some show of presenting the general theory of stress, no reasonably comprehensive exposition has ever been published before[1].

---

where $c = c(\boldsymbol{x})$ and where the dots stand for an expression involving derivatives of at most the first order. For a surface across which $\boldsymbol{u}$ and its first derivatives are continuous, from (194A.1) we then have

$$\left[\frac{\partial^2 \boldsymbol{u}}{\partial t^2}\right] = c^2[\nabla^2 \boldsymbol{u}].$$

As was remarked by Hugoniot [1887, *2*, Prem. Part., § 7] and, Duhem [1891, *3*, Ch. IX § 11], comparison with $(181.9)_5$ and $(181.10)_3$ shows at once that

$$c^2 = u_n^2:$$

The speed of displacement is $\pm c$. The same result follows also by the method of characteristics. In these theories, the speeds of displacement and propagation are exactly or approximately equal.

In the method of infinitesimal oscillations, a solution of the spatial equation

$$L(\boldsymbol{s}) = c^2(\nabla^2 \boldsymbol{s} + \cdots) = -c^2 \boldsymbol{s}/(4\pi^2 l^2)$$

is said to be a *wave-form* of *wave-length* l. While the terms are motivated by the one-dimensional case when $L(s) = c^2 s''$, the nature and existence of wave forms is no trivial matter. If such solutions exist, from (194A.2) it follows that $c^2 = \omega^2 l^2/(4\pi^2)$; hence $u_n = \pm\omega l/(2\pi) = \pm\nu l$. Q.E.D. While for the simpler theories, which rest upon the wave equation itself, the method of infinitesimal osillations yields many easily soluble cases, for more complicated theories it does not seem to be practicable, but the method of characteristics and the method of singular surfaces lead easily to precise results.

The theory of singular surfaces for (194A.1) was initiated by Hugoniot [1886, *3*].

[1] Even the treatises of the last century are astonishingly superficial on this subject. A reliable brief account of some of the essentials in our Subchapter II is given by Love [1906, *5*, Chap. II]. Much of the material in our Subchapter V may be found in the article of Hellinger [1914, *4*]. No more than isolated details from the contents of our Subchapters III and IV are given in any other expository work.

I. The balance of momentum.

196. The laws of classical mechanics. Without attempting an axiomatic presentation[1] (cf. Sect. 3) of the subject, we lay down statements summarizing the *practice* of classical mechanics. To the undefined objects that occur in kinematics, namely, particle, position, time, and mass, are added two more: the *force* $\mathfrak{F}$ and the *torque* $\mathfrak{L}$ acting upon bodies. These quantities are vector measures, defined over bodies and their measurable parts:

$$\mathfrak{F} = \int_{\mathscr{V}} d\mathfrak{F}, \quad \mathfrak{L} = \int_{\mathscr{V}} (\boldsymbol{p} \times d\mathfrak{F} + d\mathfrak{L}). \tag{196.1}$$

$\mathfrak{F}$ is an absolute vector, $\mathfrak{L}$ an axial vector; these vectors are known *a priori* and are *the same for all observers*. In their prescription consists the specification of a particular dynamical problem; this prescription is equally available and invariant to all observers, whatever their relative motion. When the measures occurring in (196.1) are discrete, they are appropriate to mass points; when continuous, to continuous media; the fully general case represents a partly continuous, partly discrete body subject to loads concentrated at points, lines, and surfaces as well as distributed continuously throughout its interior.

The torque $\mathfrak{L}$ consists in two parts: the *moment* $\int \boldsymbol{p} \times d\mathfrak{F}$ of the force distribution taken with respect to the origin of the position vector $\boldsymbol{p}$, and the *couple* $\int d\mathfrak{L}$. Thus, properly, the torque should be called the *torque with respect to the origin*, and when the distinction of origins is important, we shall write $\mathfrak{L}^{[O]}$, $\mathfrak{L}^{[O']}$, etc. To consider the effect of change of origin, we set $\boldsymbol{p} = \boldsymbol{b} + \boldsymbol{p}'$ and conclude from $(196.1)_{1,2}$ that

$$\mathfrak{L}^{[O]} = \boldsymbol{b} \times \mathfrak{F} + \mathfrak{L}^{[O']}, \tag{196.2}$$

so that the difference of the torques with respect to O and O' is the moment of the total force, thought of as concentrated at O', with respect to O.

Let $\mathfrak{P}$ be the momentum and $\mathfrak{H}$ the moment of momentum, with respect to the origin of the position vector $\boldsymbol{p}$, of the body $\mathscr{V}$. Then the postulated ***laws of classical mechanics***, or ***Euler's laws***[2], assert that *there exists a frame in which*

$$\mathfrak{F} = \dot{\mathfrak{P}}, \quad \mathfrak{L} = \dot{\mathfrak{H}}, \tag{196.3}$$

for every body $\mathscr{V}$. These two equations express the *balance of momentum* and the *balance of moment of momentum*, respectively. It is important to recall that since each measurable part of a body is also a body, these equations apply not only to the body as a whole but also to every part, no matter how the body be subdivided.

The pair of quantities $\dot{\mathfrak{P}}$, $\dot{\mathfrak{H}}$ in any frame, constitute the *reaction* of the body, while $\mathfrak{F}$, $\mathfrak{L}$ constitute the *load*. EULER's laws may be read as a statement that *in an inertial frame, the reaction of any body is equal to the load acting upon it.* In Sect. 197 we shall see that (196.3) does not hold in frames which are not inertial.

---

[1] However, our treatment has been influenced by the system of axioms given by NOLL [1957, *11*] [1958, *8*] [1959, *9*]. For older work, see Special Bibliography P at the end of the treatise.

[2] EULER [1776, *2*, §§ 26–28] (while the memoir concerns rigid bodies, Eqs. (196.3) are expressly stated to hold for arbitrary bodies). Earlier [1752, *2*, §§ 20–22], EULER had given a local formula equivalent to $(196.3)_1$. Cf. also KIRCHHOFF [1876, *2*, Eilfte Vorl., § 1]. It should be, but unfortunately it is not, unnecessary to comment that the laws of NEWTON [1687, *1*] are neither unequivocally stated nor sufficiently general to serve as a foundation for continuum mechanics. See the scholion at the end of this section.

From the postulate (196.3) it follows that force and torque have the physical dimensions

$$\dim \mathfrak{F} = [M L T^{-2}], \quad \dim \mathfrak{L} = [M L^2 T^{-2}]. \tag{196.4}$$

The balance of moment of momentum should properly be written $\mathfrak{L}^{[O]} = \dot{\mathfrak{H}}^{[O]}$ to indicate the role of the origin $\boldsymbol{O}$. To establish its invariance with respect to change of origin, we note from (196.2) and (167.5) that

$$\mathfrak{L}^{[O]} - \dot{\mathfrak{H}}^{[O]} = \boldsymbol{b} \times \mathfrak{F} + \mathfrak{L}^{[O']} - \boldsymbol{b} \times \dot{\mathfrak{P}} - \dot{\mathfrak{H}}^{[O']} = \mathfrak{L}^{[O']} - \dot{\mathfrak{H}}^{[O']}, \tag{196.5}$$

by $(196.3)_1$. It follows then that *when linear momentum is balanced, the balance of moment of momentum with respect to one origin implies the balance of moment of momentum with respect to all origins.* In this statement it is understood that all "origins" are *stationary* in the frame in which the laws (196.3) hold. Thus the choice of origin is immaterial, and we shall generally omit the superscript $\boldsymbol{O}$.

The static case of (196.3) may be read as follows: In order for a body to remain motionless in an inertial frame, it is necessary and sufficient that the total force and torque acting upon every part of it vanish. To express the general case in statical terms, we may say that the rates $-\dot{\mathfrak{P}}$ and $-\dot{\mathfrak{H}}$ are the *effective force* and *effective torque* arising from motion. This statement may be called the *Euler-D'Alembert principle*[1].

From Eqs. (196.3) and (167.1), mass being assumed invariant, we have

$$\mathfrak{F} = \mathfrak{M} \ddot{\boldsymbol{c}}: \tag{196.6}$$

The center of mass of any body moves as a mass-point, of mass equal to the mass of the body and subject to the total force acting on the body[2]. No correspondingly simple statement holds exactly for the effect of the torque, but from NYBORG's theorem (170.16) we may assert that in first approximation the torque is proportional to the curl of the acceleration.

The particular frame in which the laws (196.3) hold is called an *inertial frame.* The class of inertial frames will be delimited in Sect. 197. Thus the physical assumption basic to classical dynamics is that there exists a special frame in which observers may *prescribe* the rates of change of momentum and moment of momentum. The nature of this prescription is difficult to render definite in a fashion broad enough to cover all cases occurring in practice yet narrow enough to delimit the subject. Usually it consists in a relation, possibly elaborate, between the forces and torques acting on a body and the motion it experiences. Further progress is made by distinguishing various types of forces and the prescriptions available for them. The development of continuum mechanics within this framework is taken up in Sect. 199. Here we append some explanations regarding other views of mechanics. We remark also that much of what

[1] This principle, only a rewording of EULER's laws (196.3), while sometimes called "D'ALEMBERT's principle" is different in spirit and in detail from the principle asserted by D'ALEMBERT [1743, *2*, § 50]: Force is only a name for the product of acceleration by mass, and the system of forces corresponding to all the constraints is a system in static equilibrium. The idea that motion gives rise to an effective force equal to the negative of the product of mass by acceleration and that both the resultant and the static moment of the augmented forces must vanish may be traced back through special cases occurring in the early work of EULER and DANIEL BERNOULLI to a fundamental paper of JAMES BERNOULLI (1703). For still a different "D'ALEMBERT's principle", see Sect. 232.

[2] Special cases were proved on the basis of various specializing hypotheses by NEWTON [1687, *1*, Coroll. IIII to The Laws of Motion], D'ALEMBERT [1743, *2*, § 66], and EULER [1765, *1*, §§ 280, 290]. According to EULER [1765, *3*, § II], the general law (196.6) "is proved in mechanics", but the earliest general proof we have found is that of LAGRANGE [1788, *1*, Seconde Partie, 3me Sect., ¶ 3].

would seem the insuperable difficulty of identifying an inertial frame is removed in the practice of continuum mechanics by requiring the form of *constitutive equations* (Sect. 7) to be *the same in all frames*, whether or not inertial (cf. Sect. 293ζ).

196A. Scholion. *α) On inertial frames.* The difficulty of identifying an inertial frame by experiment is well known. This difficulty, however, is not reflected in the formalism of the subject. Rather, it arises from the fact that in practice force is generally measured by acceleration; that is, by the balance of linear momentum in the form (196.6). If this is the case, in order to determine the force one must first, and independently, verify that his frame is inertial, which results in a circularity. This is a problem of interpretation, not of formalism, and need not concern us here.

Were mechanics restricted to dynamical problems, the circularity could be avoided by regarding (196.3) as the *definitions* of force and torque, as was advocated by D'ALEMBERT and KIRCHHOFF. An inertial frame is then one in which, for example, a small massy sphere fixed to the end of a light horizontal spring suffers no acceleration beyond that proportional to the extension of the spring. This view is supported by the recognition that much of what is called dynamics in paedagogical treatments consists in memorized vocabulary. For example, when the student is told to find the motion of three pendulums connected by springs, he is expected to remember that "pendulum" means a point constrained to move on a fixed sphere with an additional constant three-dimensional acceleration; that "spring" means an acceleration proportional to the distance between two points. Unless the contrary is stated explicitly, the observer is assumed situate in an inertial frame. In other words, the student acquires a set of terms which, while based on abstraction from physical experience, are in fact *prescriptions of accelerations* in a preferred frame. In continuum mechanics, the *a priori* knowledge concerning the forces may be more elaborate. "Perfect fluid" means "the stress tensor is spherical"; ideal materials may be defined by integro-differential equations connecting the acceleration with other quantities, etc. But in all cases, the *practice* of dynamics begins with an *a priori* statement regarding the force and torque, or, equivalently, the mass density and the acceleration, in an inertial frame.

Against this view stands the whole science of statics, wherein the discourse is always of forces that produce no acceleration. In a mechanics general enough to include statics must occur not only the resultant force and torque that enter the Eqs. (196.3) but also systems of forces exerted by the parts of a body upon each other.

β) On mutual forces and Newton's third law. In the more extensive set of assumptions required for statics[1], to each *part* $\mathscr{B}$ of a body is associated a vector measure $\int d\,\mathfrak{F}_{\mathscr{B}}$, and for each measurable set $\mathscr{C}$ within the part $\mathscr{B}$ we call

$$\mathfrak{F}_{\mathscr{B}}(\mathscr{C}) \equiv \int_{\mathscr{C}} d\,\mathfrak{F}_{\mathscr{B}} \tag{196A.1}$$

the *portion* of the force over $\mathscr{B}$ that is exerted on $\mathscr{C}$. What we have above called force, the quantity that enters the fundamental law $(196.3)_1$, is then $\mathfrak{F}_{\mathscr{B}}(\mathscr{B})$, the *resultant force* on $\mathscr{B}$. The broader concept (196A.1) enables us to speak of cases when the resultant force on $\mathscr{B}$ vanishes, or is balanced by rate of change of momentum, though the portion of force over $\mathscr{B}$ exerted upon a part $\mathscr{C}$ does not vanish or is not balanced. A similar extension is made for torque.

If $\mathscr{B}_1$ and $\mathscr{B}_2$ are two disjoint parts of a body, then the *resultant mutual force* $\mathfrak{F}_{\mathscr{B}_1\mathscr{B}_2}$ exerted on $\mathscr{B}_1$ by $\mathscr{B}_2$ is given by the definition

$$\mathfrak{F}_{\mathscr{B}_1\mathscr{B}_2} = \mathfrak{F}_{\mathscr{B}_1}(\mathscr{B}_1) - \mathfrak{F}_{\mathscr{B}_1\cup\mathscr{B}_2}(\mathscr{B}_1), \tag{196A.2}$$

that is, the resultant force on $\mathscr{B}_1$ less the portion of the force over $\mathscr{B}_1$ and $\mathscr{B}_2$ together that is exerted on $\mathscr{B}_1$.

On the basis of this definition, we may show that when there are no surface, line, or point forces acting within either of the disjoint parts $\mathscr{B}_1$ and $\mathscr{B}_2$, then it is a consequence of the balance of linear momentum that

$$\mathfrak{F}_{\mathscr{B}_1\mathscr{B}_2} = -\mathfrak{F}_{\mathscr{B}_2\mathscr{B}_1}: \tag{196A.3}$$

The sum of the mutual forces exerted by any two parts of a body on each other is zero. This property of mutual forces is sometimes called the *principle of action and reaction*, or *Newton's third law*[2].

---

[1] NOLL [1957, *11*, § 2].

[2] The name is merely traditional, as the statement of NEWTON [1687, *1*, Lex 3] is so vague as to be only suggestive, while subsequent writers apply this name sometimes to (196A.3), sometimes to (196A.8), and sometimes to broader ideas but tenuously related to the formal science of mechanics. The precise statement and proof given above are due to NOLL [1957, *11*, § 6].

Indeed, from the definition (196 A.2) we have

$$\mathfrak{F}_{\mathscr{B}_1 \mathscr{B}_2} + \mathfrak{F}_{\mathscr{B}_2 \mathscr{B}_1} = \mathfrak{F}_{\mathscr{B}_1}(\mathscr{B}_1) + \mathfrak{F}_{\mathscr{B}_2}(\mathscr{B}_2) - \mathfrak{F}_{\mathscr{B}_1 \cup \mathscr{B}_2}(\mathscr{B}_1) - \mathfrak{F}_{\mathscr{B}_1 \cup \mathscr{B}_2}(\mathscr{B}_2). \quad (196\text{A}.4)$$

The above-specified absence of singular loads enables us to replace the last two terms by $\mathfrak{F}_{\mathscr{B}_1 \cup \mathscr{B}_2}(\mathscr{B}_1 \cup \mathscr{B}_2)$. Each term on the right-hand side is now a resultant force, so that the balance of linear momentum $(196.3)_1$ may be applied to each. Therefore

$$\mathfrak{F}_{\mathscr{B}_1 \mathscr{B}_2} + \mathfrak{F}_{\mathscr{B}_2 \mathscr{B}_1} = \dot{\mathfrak{P}}_1 + \dot{\mathfrak{P}}_2 - \dot{\overline{\mathfrak{P}_1 + \mathfrak{P}_2}} = 0. \quad (196\text{A}.5)$$

Q.E.D.

For the mutual force between mass-points not subject to a couple, the balance of moment of momentum $(196.3)_2$ implies that *the mutual force is directed along the line joining the two points*[1]. To see this, let the external and mutual forces, respectively, be $\mathfrak{F}_a$ and $\mathfrak{F}_{ab}$, and write the balance of linear momentum for each mass-point separately:

$$\mathfrak{F}_1 + \mathfrak{F}_{12} = \mathfrak{M}_1 \ddot{\boldsymbol{p}}_1, \quad \mathfrak{F}_2 + \mathfrak{F}_{21} = \mathfrak{M}_2 \ddot{\boldsymbol{p}}_2. \quad (196\text{A}.6)$$

Since only the resultant force on each part enters the balance of moment of momentum, for the body consisting of the two mass-points we have

$$\left.\begin{aligned} \boldsymbol{p}_1 \times \mathfrak{F}_1 + \boldsymbol{p}_2 \times \mathfrak{F}_2 &= \boldsymbol{p}_1 \times \mathfrak{M}_1 \ddot{\boldsymbol{p}}_1 + \boldsymbol{p}_2 \times \mathfrak{M}_2 \ddot{\boldsymbol{p}}_2, \\ &= \boldsymbol{p}_1 \times (\mathfrak{F}_1 + \mathfrak{F}_{12}) + \boldsymbol{p}_2 \times (\mathfrak{F}_2 + \mathfrak{F}_{21}), \\ &= \boldsymbol{p}_1 \times \mathfrak{F}_1 + \boldsymbol{p}_2 + \mathfrak{F}_2 + (\boldsymbol{p}_1 - \boldsymbol{p}_2) \times \mathfrak{F}_{12}, \end{aligned}\right\} \quad (196\text{A}.7)$$

by Eqs. (196 A.6) and (196 A.3). Hence follows

$$(\boldsymbol{p}_1 - \boldsymbol{p}_2) \times \mathfrak{F}_{12} = 0. \quad (196\text{A}.8)$$

Q.E.D.

It is thus shown that "NEWTON's laws" for mass-points, with the third law being taken as an assertion that the mutual forces between pairs of mass-points are pairwise equilibrated and central, follow from EULER's fundamental equations (196.3). It is obvious that the converse problem of derivation of EULER's laws (196.3) from NEWTON's laws cannot even be stated meaningfully without severe restrictions. Nevertheless, the mass-point is so ingrained in paedagogical repetition that even today many textbooks strive to foster an illusion that NEWTON's laws suffice as a basis for mechanics.

The special nature of the usual argument is best seen by presenting it. The forces are all taken as *absolutely continuous functions of mass* and are restricted to be external forces $\boldsymbol{f}_P$ and mutual forces $\boldsymbol{f}_{PQ}$ only, where P and Q refer to the mass-bearing points. That is, the resultant force and torque are given by

$$\left.\begin{aligned} \mathfrak{F} &= \int_{\mathscr{V}} \boldsymbol{f}_P \, d\mathfrak{M}_P + \int_{\mathscr{V}} \int_{\mathscr{V}} \boldsymbol{f}_{PQ} \, d\mathfrak{M}_Q \, d\mathfrak{M}_P, \\ \mathfrak{L} &= \int_{\mathscr{V}} \boldsymbol{p}_P \times \boldsymbol{f}_P \, d\mathfrak{M}_P + \int_{\mathscr{V}} \left[\boldsymbol{p}_P \times \int_{\mathscr{V}} \boldsymbol{f}_{PQ} \, d\mathfrak{M}_Q\right] d\mathfrak{M}_P. \end{aligned}\right\} \quad (196\text{A}.9)$$

If (196 A.3) holds, the first double integral vanishes; if in addition (196 A.8) holds, the second double integral vanishes, and $(196.3)_2$ follows from $(196.3)_1$.

In this argument, the masses may be discrete or continuous. However, the forms (196 A.9) are not sufficiently general for continuum mechanics, where, while the consideration of mutual forces is unusual, the occurrence of *surface forces*, not depending continuously upon the distribution of mass, is the rule. The basic law $(196.3)_2$ of balance of moment of momentum assumes for continuum mechanics a form far different from that of NEWTON's third law for mass-points, as will be seen in Sects. 203 and 205.

γ) Classification of forces and torques. The simpler problems of mechanics concern prescriptions of $\mathfrak{F}$ and $\mathfrak{L}$ which do not depend explicitly upon the velocity $\dot{\boldsymbol{p}}$, the acceleration $\ddot{\boldsymbol{p}}$, or the higher accelerations $\overset{(N)}{\boldsymbol{p}}$. Forces and torques of this kind are called *frictionless*, while reactions depending upon $\dot{\boldsymbol{p}}$ are called *frictional*. There have been many attempts to relegate frictional reactions to mere appearances or approximations; particularly in statistical mechanics, it is often sought to show that they arise from an incomplete or approximate description of problems governed in fact by frictionless reactions. These efforts have found only limited success. In any case, in the practice of continuum mechanics many of the most interesting theories concern frictional forces. In some of the more recent theories, reactions depending upon the accelerations $\ddot{\boldsymbol{p}}, \ldots, \overset{(N)}{\boldsymbol{p}}$ occur; such reactions may be called *hyperfrictional*.

[1] NOLL [1957, *11*, § 10].

The foregoing classification of reactions follows conventional notions but does not deserve to be perpetuated in continuum mechanics, where $\mathfrak{F}$ and $\mathfrak{L}$ are specified in terms of the stress tensor $\boldsymbol{t}$ (cf. Sect. 200), for which in turn a functional dependence is prescribed. If, for example, that functional dependence is of the form $\dot{\boldsymbol{t}}=\boldsymbol{f}(\dot{\boldsymbol{p}})$, it is not legitimate to infer that $\mathfrak{F}$ is prescribed in terms of $\boldsymbol{p}$, or of $\dot{\boldsymbol{p}}$, and the conventional classification cannot be applied.

197. Apparent forces and torques. In an inertial frame, EULER's laws (196.3) assert that the load $\mathfrak{F}$, $\mathfrak{L}$ on a body equals its reaction $\dot{\mathfrak{P}}$, $\dot{\mathfrak{H}}$. Since, as already asserted, $\mathfrak{F}$ and $\mathfrak{L}$ are the same for all observers, while $\dot{\mathfrak{P}}$ and $\dot{\mathfrak{H}}$ do not enjoy this invariance, it follows that EULER's laws are not an invariant statement of the principles of mechanics. Indeed, by the results of Sect. 171 we may express the ***Galilean principle of relativity*** as follows: *The reaction of each body is the same for two observers if and only if their frames belong to the same Galilean class; in particular, the class of inertial frames is a Galilean class.*

Letting $\dot{\mathfrak{P}}'$, $\dot{\mathfrak{H}}'^{[O']}$ be the reaction in an arbitrary frame, by $(196.3)_1$ and (171.1) we have

$$\dot{\mathfrak{P}}'=\mathfrak{F}-\mathfrak{M}(\ddot{\mathbf{c}}-\ddot{\mathbf{c}}')=\mathfrak{F}+\mathfrak{F}_{\mathrm{t}}+\mathfrak{F}_{\mathrm{r}} \tag{197.1}$$

where

$$-\mathfrak{F}_{\mathrm{t}}\equiv\mathfrak{M}\,\ddot{\boldsymbol{b}},\qquad -\mathfrak{F}_{\mathrm{r}}\equiv\mathfrak{M}\,[\dot{\boldsymbol{\omega}}\times\mathbf{c}'+2\boldsymbol{\omega}\times\dot{\mathbf{c}}'+\boldsymbol{\omega}\times(\boldsymbol{\omega}\times\mathbf{c}')]. \tag{197.2}$$

By $(196.3)_2$, (196.2), (170.11), and (170.12) follows

$$\left.\begin{aligned}\dot{\mathfrak{H}}'^{[O']}&=\mathfrak{L}^{[O]}+\mathfrak{L}_{\mathrm{c}}+\mathfrak{L}_{\mathrm{r}},\\&=\mathfrak{L}^{[O']}-(\mathfrak{L}^{[O]}-\mathfrak{L}^{[O']})+\mathfrak{L}_{\mathrm{c}}+\mathfrak{L}_{\mathrm{r}},\\&=\mathfrak{L}^{[O']}+\boldsymbol{b}\times\mathfrak{F}+\mathfrak{L}_{\mathrm{c}}+\mathfrak{L}_{\mathrm{r}},\\&=\mathfrak{L}^{[O']}+\mathfrak{L}_{\mathrm{t}}+\mathfrak{L}_{\mathrm{r}},\end{aligned}\right\} \tag{197.3}$$

where

$$\left.\begin{aligned}-\mathfrak{L}_{\mathrm{t}}&\equiv-\boldsymbol{b}\times\mathfrak{F}-\mathfrak{L}_{\mathrm{c}}=\mathbf{c}'\times\mathfrak{M}\,\ddot{\boldsymbol{b}}=-\mathbf{c}'\times\mathfrak{F}_{\mathrm{t}},\\-\mathfrak{L}_{\mathrm{r}}&\equiv\dot{\boldsymbol{\omega}}\cdot\mathfrak{J}'^{[O']}+\boldsymbol{\omega}\cdot\dot{\mathfrak{J}}'^{[O']}+\boldsymbol{\omega}\times\mathfrak{H}'^{[O']}+\boldsymbol{\omega}\times\mathfrak{J}'^{[O']}\cdot\boldsymbol{\omega}.\end{aligned}\right\} \tag{197.4}$$

Thus if we were to regard the primed frame as one in which the laws of mechanics (196.3) hold, we should have to say that the body is subject not only to the real load $\mathfrak{F}$, $\mathfrak{L}$, but also to the *apparent force*[1] $\mathfrak{F}_{\mathrm{t}}+\mathfrak{F}_{\mathrm{r}}$ and the *apparent torque* $\mathfrak{L}_{\mathrm{t}}+\mathfrak{L}_{\mathrm{r}}$. In all these formulae, dots on primed quantities indicate time rates as apparent to an observer in the primed frame.

The minus signs occur in (197.2) and (197.4) because we follow the tradition of using the position vector $\boldsymbol{b}$ of O' with respect to the unprimed frame, and the angular velocity of the primed frame with respect to the unprimed frame. Introduction of the corresponding quantities with respect to the primed frame, viz $\boldsymbol{b}'\equiv-\boldsymbol{b}$ and $\boldsymbol{\omega}'\equiv-\boldsymbol{\omega}$, so that $\ddot{\boldsymbol{b}}'=-\ddot{\boldsymbol{b}}$, $\dot{\boldsymbol{\omega}}'=-\dot{\boldsymbol{\omega}}$, removes the minus signs except from the centripetal terms $\boldsymbol{\omega}'\times(\boldsymbol{\omega}'\times\mathbf{c}')$ and $\boldsymbol{\omega}'\times\mathfrak{J}'^{[O']}\cdot\boldsymbol{\omega}'$.

In the apparent reaction the subscripts t and r distinguish the portions arising from translation and rotation, respectively. It is only a rephrasing of the Galilean principle of relativity to remark that when $\boldsymbol{\omega}=0$, a necessary and sufficient condition that the apparent reaction be zero is $\ddot{\boldsymbol{b}}=0$. More generally, to an observer in the primed frame $\mathfrak{F}_{\mathrm{t}}$ and $\mathfrak{L}_{\mathrm{t}}$ equal, respectively, the momentum and the moment of momentum about the origin of a mass-point of mass $\mathfrak{M}$, located

[1] A verbal *statement* of the principle of apparent forces was given by CLAIRAUT [1745, *1*, Art. I, § I]; clearer statements are due to EULER [1752, *3*, Probl. 1] [1756, *1*, § 21] [1757, *3*, §§ 33–34] (and many other papers). While EULER used the principle correctly to solve special problems, notably in his analysis of fluid motion in a turbine and in his theory of rigid bodies, the correct general *equations* rest upon the full theorem of CORIOLIS. Cf. Sect. 143.

at the center of mass, and enjoying the same instantaneous acceleration as does the origin of the unprimed frame. In order that $\mathfrak{L}_t = 0$, by $(197.4)_1$ it is necessary and sufficient that to an observer at the origin of the primed frame the accelerations of the origin of the unprimed frame appear to be in the direction of the radius vector to the center of mass[1]. It is sufficient, but not necessary that the origin of one frame be unaccelerated with respect to the other, or that the origin of the primed frame be at the center of mass. When the motion is rigid and the primed frame is rigidly attached to the body, the second and third summands in $(197.4)_2$ vanish.

198. Impulse. The laws (196.3) are suitable only to motions in which the velocity exists and is integrable. There are cases when we wish to regard a motion as started or altered instantaneously. There are then, generally, impulsive discontinuities $-[\mathfrak{P}]-$ and $-[\mathfrak{H}]-$ in $\mathfrak{P}$ and $\mathfrak{H}$, where the symbol $-[\;]-$ is defined by (194.3). In analogy to (196.3), for an impulse suffered at time t we write

$$\mathfrak{R} \equiv -[\mathfrak{P}]- = \int_{\mathscr{V}} -[\dot{\boldsymbol{p}}]- \, d\mathfrak{M}, \qquad \mathfrak{S} \equiv -[\mathfrak{H}]- \equiv \int_{\mathscr{V}} \boldsymbol{p} \times -[\dot{\boldsymbol{p}}]- \, d\mathfrak{M}. \tag{198.1}$$

Regarding the impulses $\mathfrak{R}$ and $\mathfrak{S}$, we make an assumption analogous to the balance of momentum, namely, that we have *a priori* knowledge of them in an inertial frame. In particular, we assume that $\mathfrak{R} = 0$ and $\mathfrak{S} = 0$ except at a finite number of instants; at instants when $\mathfrak{R}$ and $\mathfrak{S}$ do not both vanish, an impulse is said to *occur*. We have

$$\dim \mathfrak{R} = \dim \mathfrak{P} = [M L T^{-1}], \qquad \dim \mathfrak{S} = \dim \mathfrak{H} = [M L^2 T^{-1}]. \tag{198.2}$$

Moreover, since by (143.3) follows $-[\dot{\boldsymbol{p}}]- = -[\dot{\boldsymbol{p}}']-$, *the laws of impulse are valid for all observers.* In all the foregoing statements it is assumed that the material itself does not jump about: $-[\boldsymbol{p}]- = 0$.

We may include both (198.1) and (196.3) in the same equations by writing

$$\mathfrak{P}(t) - \mathfrak{P}(t') = \int_{t'}^{t} \mathfrak{F}\, dt, \qquad \mathfrak{H}(t) - \mathfrak{H}(t') = \int_{t'}^{t} \mathfrak{L}\, dt \tag{198.3}$$

and interpreting the integration in the Stieltjes sense.

Since impulse consists in outright prescription of momentum and moment of momentum, there is little to be said regarding it.

II. The stress principle.

199. Extrinsic, mutual, and contact loads. There are three types of forces and torques:

1. *Extrinsic loads.* These arise at least in part from outside the body and are regarded as acting upon the particles comprising the body.

2. *Mutual loads.* These arise within the body and are regarded as acting upon pairs of particles. They were defined and discussed in Sect. 196 Aβ.

3. *Contact loads.* These loads are not assignable as functions of position but are to be imagined as acting upon bounding surfaces, lines, or points in such a way as to be equipollent to the loading exerted by one portion of the material upon another.

Examples of extrinsic loads are the fields of uniform gravity, electrostatics, or polarisation; of mutual loads, Newtonian gravitation or the forces between

[1] Cf. the references cited in Sect. 170.

charges; of contact loads, the stress field, to which the remainder of this chapter is devoted.

In their effect upon motion the three types of loads are indiscernible, as follows from (196.3). For example, if two mass-points are subject only to their mutual gravitation, it is equally permissible to regard them as moving subject to specially adjusted time-dependent extrinsic forces. The distinction of forces, as may be expected from the *a priori* nature of force itself, classifies our *knowledge* of the circumstances of a body. Generally, a problem which is simple and hence natural when put in terms of contact loads becomes intricate and artificial if put in terms of mutual or extrinsic loads.

200. The stress principle of EULER and CAUCHY[1]. The *stress principle* specifies the nature of the contact load[2]. It asserts that upon any imagined closed diaphragm $\mathscr{S}$ within a body there exists a field of *stress vectors* $\boldsymbol{t}_{(\boldsymbol{n})}$, where $\boldsymbol{n}$ is the unit normal; this field is equipollent to the loads exerted by the material outside upon the material inside, or, in case $\mathscr{S}$ is a boundary, to the applied loads. If we let $\boldsymbol{f}$ be the field of extrinsic and mutual forces per unit mass, or, as we shall say henceforth, the *assigned force*, the stress principle assumes the form[3]

$$\mathfrak{F} = \oint_{\mathscr{S}} \boldsymbol{t}_{(\boldsymbol{n})}\, da + \int_{\mathscr{V}} \boldsymbol{f}\, d\mathfrak{M} = \frac{d}{dt}\int_{\mathscr{V}} \dot{\boldsymbol{p}}\, d\mathfrak{M} = \dot{\mathfrak{P}}, \tag{200.1}$$

$$\mathfrak{L} = \oint_{\mathscr{S}} \boldsymbol{p}\times\boldsymbol{t}_{(\boldsymbol{n})}\, da + \int_{\mathscr{V}} \boldsymbol{p}\times\boldsymbol{f}\, d\mathfrak{M} = \frac{d}{dt}\int_{\mathscr{V}} \boldsymbol{p}\times\dot{\boldsymbol{p}}\, d\mathfrak{M} = \dot{\mathfrak{H}}, \tag{200.2}$$

where for convenience we have repeated the content of (196.3). Since $\boldsymbol{t}_{(\boldsymbol{n})}$ represents the effect of the material outside on the material inside, it is a *tension* if it subtends an acute angle with the unit outward normal $\boldsymbol{n}$; if an obtuse angle, a *pressure*.

The projection of $\boldsymbol{t}_{(\boldsymbol{n})}$ upon $\boldsymbol{n}$ is called the *normal stress* or *normal tension* on the surface element normal to $\boldsymbol{n}$; the projection of $\boldsymbol{t}_{(\boldsymbol{n})}$ upon the plane normal to $\boldsymbol{n}$, the *shearing stress* on the element. If the normal stress is zero, the element is said to be subject to *pure shearing stress*; if the shearing stress is zero, to *pure*

---

[1] The notion of stress arose in special cases in theories of flexible, elastic, and fluid bodies. The general concept and mathematical theory are due to CAUCHY [1823, *1*] [1827, *1*, pp. 60—61].

The following principal steps may be extracted from the detailed historical accounts of TRUESDELL [1954, *25*, Parts II and IV] [1960, *4*, §§ 7, 11, 14, 19, 58, 59, 64]; cf. also [1956, *21*]. That the action of one part of a flexible line upon its neighbor is equipollent to a tangent vector at the junction, after being tacitly assumed in special cases by GALILEIO (1638) and PARDIES (1673), was given general mathematical form by JAMES BERNOULLI (1691—1704) and HERMANN (1716). The need for interior shear stress (Sect. 204) in a terminally loaded beam, while seen by PARENT (1713), was first analysed by COULOMB [1776, *1*, § V]. That the action of a portion of fluid in a tube upon a neighboring portion is equipollent to a vector normal to the plane face separating them is the basic assumption of JOHN BERNOULLI'S hydraulics [1743, *1*, Part I, § 1] and is fully elaborated in EULER'S hydraulic researches (1749—1752). These led to EULER'S three-dimensional hydrodynamics, where the diaphragm is of arbitrary form but the stress vector is assumed to be normal to it [1757, *1*, §§ 6—9]. The mathematical theory of resultant shear on a plane deformable line was given by EULER [1771, *2*, §§ 1—11, 35—40].

CAUCHY'S general theory is obtained by adopting the common features and discarding the special aspects of the foregoing theories.

The history of the term "stress", which was introduced by RANKINE [1856, *1*, § 2], has been written by MUIRHEAD [1901, *10*].

[2] A general theory of contact forces is constructed by NOLL [1958, *8*, § 6].

[3] KIRCHHOFF [1876, *2*, Vorl. 11, §§ 1—2]. CAUCHY had used equivalent local postulates.

normal stress. If $\boldsymbol{t}_{(\boldsymbol{n})}$ is always parallel to $\boldsymbol{n}$ and of magnitude independent of $\boldsymbol{n}$, so that $\boldsymbol{t}_{(\boldsymbol{n})} = -P\,\boldsymbol{n}$, where P is a scalar independent of $\boldsymbol{n}$, the stress is *hydrostatic*.

In (200.2), all torques arise from forces. To remove this restriction of generality we allow in addition a field of assigned *couples*[1] $\boldsymbol{l}$ and a *couple stress*[2] $\boldsymbol{m}_{(\boldsymbol{n})}$, so that (200.2) is replaced by

$$\mathfrak{L} = \oint_{\mathscr{S}} (\boldsymbol{m}_{(\boldsymbol{n})} + \boldsymbol{p}\times\boldsymbol{t}_{(\boldsymbol{n})})\,da + \int_{\mathscr{V}} (\boldsymbol{l} + \boldsymbol{p}\times\boldsymbol{f})\,d\mathfrak{M}. \tag{200.3}$$

Most work in continuum mechanics concerns the *non-polar case*, defined as that in which

$$\boldsymbol{l} = 0, \quad \boldsymbol{m} = 0. \tag{200.4}$$

201. Concentrated loads. Eqs. (200.1) to (200.3) presume the entire mechanical action to arise from fields which are continuous, or at least integrable, over the interior or the surface of the body. Sometimes in addition we wish to take account of *concentrated loads* $\boldsymbol{f}_{\mathfrak{a}}$ and $\boldsymbol{l}_{\mathfrak{a}}$, acting at $\mathfrak{q}$ individual points $\boldsymbol{p}_{\mathfrak{a}}$. The corresponding extensions of (200.1) and (200.3), included within the scheme of Sect. 196, are

$$\left.\begin{aligned}
\mathfrak{F} &= \oint_{\mathscr{S}} \boldsymbol{t}_{(\boldsymbol{n})}\,da + \int_{\mathscr{V}} \boldsymbol{f}\,d\mathfrak{M} + \sum_{\mathfrak{a}=1}^{\mathfrak{q}} \boldsymbol{f}_{\mathfrak{a}},\\
\mathfrak{L} &= \oint_{\mathscr{S}} (\boldsymbol{m}_{(\boldsymbol{n})} + \boldsymbol{p}\times\boldsymbol{t}_{(\boldsymbol{n})})\,da + \int_{\mathscr{V}} (\boldsymbol{l} + \boldsymbol{p}\times\boldsymbol{f})\,d\mathfrak{M} + \sum_{\mathfrak{a}=1}^{\mathfrak{q}} (\boldsymbol{l}_{\mathfrak{a}} + \boldsymbol{p}_{\mathfrak{a}}\times\boldsymbol{f}_{\mathfrak{a}}).
\end{aligned}\right\} \tag{201.1}$$

Usually the points $\boldsymbol{p}_{\mathfrak{a}}$ are either within $\mathscr{V}$ or on its surface $\mathscr{S}$. It is left to the reader to generalize (201.1) to so as to allow distributions of assigned loads upon particular curves or surfaces.

While concentrated loads are hardly in the spirit of continuum mechanics, they provide a convenient idealization and are frequently used. Often they are regarded as limiting cases when the area or volume on which a distributed load acts is shrunk down to a point. In the theory of any particular ideal material, it is an analytical question, and often a difficult one, to determine whether solutions possessing singularities of this type exist and are the limits, unique or not, of certain families of regular solutions. Such problems are outside our scope. Let it suffice that any such singular solutions in order to be acceptable must be consistent with the general laws of mechanics (196.3), where $\mathfrak{F}$ and $\mathfrak{L}$ are given by (201.1). To this limited extent, there is no difficulty in including concentrated loads without special distinction, simply by writing (200.1) and (200.3) in terms of Stieltjes integrals. General theorems following from these laws are of two kinds. For results concerning *finite portions of matter*, the device mentioned allows us to include concentrated loads with no extra effort. However, difficulties arise in the proofs of *local* theorems, to which much of this chapter is devoted. When a concentrated load occurs, the point at which it acts is a singularity, and the behavior of the stress vector nearby cannot be rendered definite without additional assumptions, such as those appropriate to special theories of materials.

---

[1] MAXWELL [1873, *5*, § 641]; cf. LARMOR [1892, *7*], COMBEBIAC [1902, *2*]. DUHEM [1904, *1*, Part I, Chap. II, §§ I–V] [1911, *4*, Chap. XIV] obtained couples by supposing the mutual forces not Newtonian.

[2] VOIGT [1887, *5*, Chap. I, § 2], [1895, *7*]. Cf. E. and F. COSSERAT [1909, *5*, § 53], HEUN [1913, *4*, § 21a], GÜNTHER [1958, *4*, § 3]. The necessity of couple stresses in certain physical circumstances is emphasized by KRÖNER [1960, *3*, § 2].

We therefore lay down the following *convention regarding concentrated loads*:

1. For local theorems, we assume there are *no concentrated loads* in a neighborhood of the point in question.

2. For integral theorems, concentrated loads may be present, but we refrain from writing down the obvious additional terms that arise when (200.1) and (200.3) are replaced by (201.1). In particular examples, the reader will supply these terms as needed.

202. The reaction on bounding surfaces. If we combine (200.1) and (200.3) with $(170.3)_{1,2}$, we obtain[1]

$$\left.\begin{aligned}\mathfrak{F}_{\mathscr{S}} &\equiv \oint_{\mathscr{S}} \boldsymbol{t}_{(\boldsymbol{n})}\, da = \int_{\mathscr{V}} \frac{\partial}{\partial t}(\varrho\,\dot{\boldsymbol{p}})\, dv + \oint_{\mathscr{S}} \varrho\,\dot{\boldsymbol{p}}\,\dot{\boldsymbol{p}}\cdot d\boldsymbol{a} - \int_{\mathscr{V}} \boldsymbol{f}\, d\mathfrak{M},\\ \mathfrak{L}_{\mathscr{S}} &\equiv \oint_{\mathscr{S}} (\boldsymbol{m}_{(\boldsymbol{n})} + \boldsymbol{p}\times\boldsymbol{t}_{(\boldsymbol{n})})\, da,\\ &= \int_{\mathscr{V}} \frac{\partial}{\partial t}(\boldsymbol{p}\times\varrho\,\dot{\boldsymbol{p}})\, dv + \oint_{\mathscr{S}} \boldsymbol{p}\times\varrho\,\dot{\boldsymbol{p}}\,\dot{\boldsymbol{p}}\cdot d\boldsymbol{a} - \int_{\mathscr{V}} (\boldsymbol{l} + \boldsymbol{p}\times\boldsymbol{f})\, d\mathfrak{M}.\end{aligned}\right\} \tag{202.1}$$

Thus $\mathfrak{F}_{\mathscr{S}}$ and $\mathfrak{L}_{\mathscr{S}}$, the force and torque exerted by the material outside $\mathscr{S}$ upon that inside, may be calculated from the fields $\frac{\partial(\varrho\dot{\boldsymbol{p}})}{\partial t}$, $\frac{\partial \boldsymbol{p}}{\partial t}\times\varrho\,\dot{\boldsymbol{p}}$, $\boldsymbol{f}$ and $\boldsymbol{l}$ within $\mathscr{V}$ and the field $\varrho\,\dot{\boldsymbol{p}}\,\dot{\boldsymbol{p}}$ on $\mathscr{S}$.

First, in steady motion within a stationary vessel of any form, by (69.1) we get from (202.1)

$$\begin{aligned}\mathfrak{F}_{\mathscr{S}} &= -\int_{\mathscr{V}} \boldsymbol{f}\, d\mathfrak{M},\\ \mathfrak{L}_{\mathscr{S}} &= -\int_{\mathscr{V}} (\boldsymbol{l} + \boldsymbol{p}\times\boldsymbol{f})\, d\mathfrak{M}.\end{aligned} \tag{202.2}$$

Thus *the steady motion of a material filling a closed stationary vessel has no reaction upon the vessel*[2].

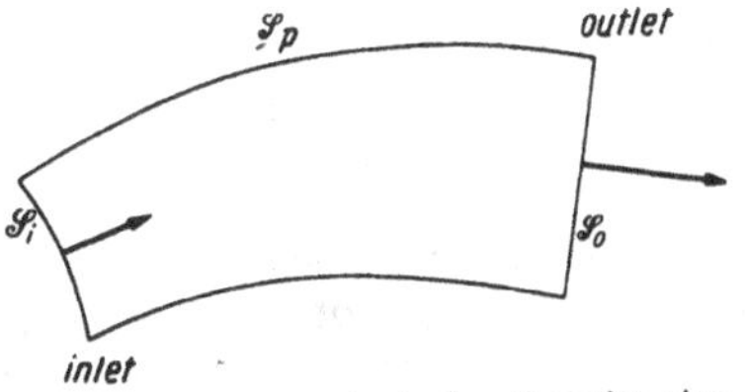

Fig. 33. Material in steady flow through a pipe.

For a less trivial example[3], consider a material which is being forced in steady flow through a stationary pipe (Fig. 33). Suppose that $\boldsymbol{l}=0$ and that $\boldsymbol{f}$ is the field of uniform gravity. If we take the normal to the inlet cross section $\mathscr{S}_{\text{i}}$ inward, retaining that on the outlet cross section $\mathscr{S}_{\text{o}}$ as outward, then (202.1) reduces to

$$\left.\begin{aligned}\mathfrak{F}_{\mathscr{S}} &= \int_{\mathscr{S}_{\text{o}}} \varrho\,\dot{\boldsymbol{p}}\,\dot{\boldsymbol{p}}\cdot d\boldsymbol{a} - \int_{\mathscr{S}_{\text{i}}} \varrho\,\dot{\boldsymbol{p}}\,\dot{\boldsymbol{p}}\cdot d\boldsymbol{a} - \mathfrak{W},\\ \mathfrak{L}_{\mathscr{S}} &= \int_{\mathscr{S}_{\text{o}}} \boldsymbol{p}\times\varrho\,\dot{\boldsymbol{p}}\,\dot{\boldsymbol{p}}\cdot d\boldsymbol{a} - \int_{\mathscr{S}_{\text{i}}} \boldsymbol{p}\times\varrho\,\dot{\boldsymbol{p}}\,\dot{\boldsymbol{p}}\cdot d\boldsymbol{a} - \mathfrak{M}\,\boldsymbol{c}\times\mathfrak{W},\end{aligned}\right\} \tag{202.3}$$

where $\mathfrak{W}$ is the weight of the material in the pipe. Thus, no matter what the shape of the pipe or the nature of the material in it, from a knowledge of $\mathfrak{W}$ and of the density and velocity of the material *at the inlet and outlet only*, we can determine the total reaction on the material instantaneously occupying the pipe. Writing $\mathscr{S}_{\text{p}}$ for the surface constituting the pipe itself, by (202.3) and (202.1)

---

[1] In principle, these formulae are old. Apparently the first general discussion is that of v. MISES [1909, *8*, § 9]. A clear treatment is given by CISOTTI [1917, *4*, §§ 1–4]. Cf. LELLI [1925, *8* and *9*], MÜLLER [1933, *8*].

[2] CISOTTI [1917, *4*, § 7].

[3] CISOTTI [1917, *4*, § 8].

we obtain also

$$\left.\begin{aligned}\int_{\mathscr{S}_p} \boldsymbol{t}_{(\boldsymbol{n})}\, d a &= \int_{\mathscr{S}_1} (\boldsymbol{t}_{(\boldsymbol{n})}\, d a - \varrho\, \dot{\boldsymbol{p}}\, \dot{\boldsymbol{p}} \cdot d\boldsymbol{a}) - \int_{\mathscr{S}_0} (\boldsymbol{t}_{(\boldsymbol{n})}\, d a - \varrho\, \dot{\boldsymbol{p}}\, \dot{\boldsymbol{p}} \cdot d\boldsymbol{a}) - \mathfrak{W},\\ \int_{\mathscr{S}_p} (\boldsymbol{m}_{(\boldsymbol{n})} + \boldsymbol{p}\times\boldsymbol{t}_{(\boldsymbol{n})})\, d a &= \int_{\mathscr{S}_1} [\boldsymbol{m}_{(\boldsymbol{n})}\, d a + \boldsymbol{p}\times(\boldsymbol{t}_{(\boldsymbol{n})}\, d a - \varrho\, \dot{\boldsymbol{p}}\, \dot{\boldsymbol{p}} \cdot d\boldsymbol{a}) - {}\\ &\quad - \int_{\mathscr{S}_0} [\boldsymbol{m}_{(\boldsymbol{n})}\, d a + \boldsymbol{p}\times(\boldsymbol{t}_{(\boldsymbol{n})}\, d a - \varrho\, \dot{\boldsymbol{p}}\, \dot{\boldsymbol{p}} \cdot d\boldsymbol{a}) - \mathfrak{M}\, \boldsymbol{c}\times\mathfrak{W}.\end{aligned}\right\}\quad (202.4)$$

Thus a knowledge of $\mathfrak{W}$ and ϱ, $\dot{\boldsymbol{p}}$, $\boldsymbol{t}_{(\boldsymbol{n})}$, and $\boldsymbol{m}_{(\boldsymbol{n})}$ at the inlet and the outlet alone suffices to determine the reaction the pipe exerts upon the material.

Consider next a single finite closed rigid boundary submerged and fixed in a continuous medium, and let $\mathscr{S}_c$ be an imagined *control surface* sufficiently large as to include all of $\mathscr{S}$ (Fig. 34). For simplicity, take $\boldsymbol{f}=0$, $\boldsymbol{l}=0$. So as to calculate the reaction exerted by the rigid obstacle on the material, we apply (202.1) to the material between $\mathscr{S}$ and $\mathscr{S}_c$, thus obtaining for the force and torque exerted *by the obstacle* the expressions[1]

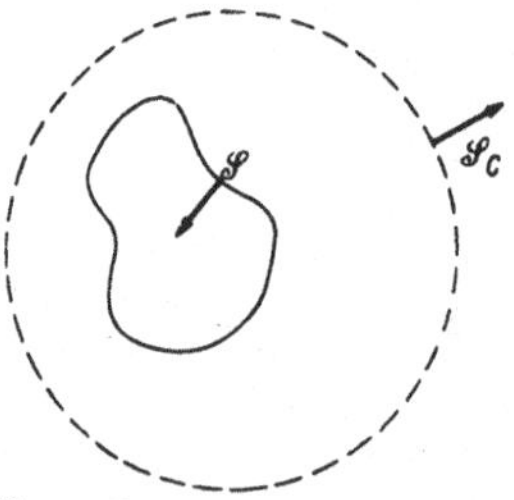

Fig. 34. Reaction on a submerged rigid object.

$$\left.\begin{aligned}\mathfrak{F} &= \frac{\partial}{\partial t}\int_{\mathscr{v}} \dot{\boldsymbol{p}}\, d\mathfrak{M} + \oint_{\mathscr{S}_c} \varrho\, \dot{\boldsymbol{p}}\, \dot{\boldsymbol{p}} \cdot d\boldsymbol{a} - \oint_{\mathscr{S}_c} \boldsymbol{t}_{(\boldsymbol{n})}\, d a,\\ \mathfrak{L} &= \frac{\partial}{\partial t}\int_{\mathscr{v}} \boldsymbol{p}\times\dot{\boldsymbol{p}}\, d\mathfrak{M} + \oint_{\mathscr{S}_c} \boldsymbol{p}\times\varrho\, \dot{\boldsymbol{p}}\, \dot{\boldsymbol{p}} \cdot d\boldsymbol{a} - {}\\ &\quad - \oint_{\mathscr{S}_c} \boldsymbol{p}\times\boldsymbol{t}_{(\boldsymbol{n})}\, d a.\end{aligned}\right\}\quad (202.5)$$

The first terms represent the force and torque arising from local changes in velocity. In steady motion, they vanish, and (202.5) shows that then *the reaction of a submerged body may be determined from the values of the stress vector, the velocity, and the density on a control surface far from the body.*

To calculate this reaction in the steady case, let $\boldsymbol{V}$ be any constant vector; then by mere algebraic identity we have

$$\left.\begin{aligned}\oint_{\mathscr{S}_c} \varrho\, \dot{\boldsymbol{p}}\, \dot{\boldsymbol{p}} \cdot d\boldsymbol{a} &= \oint_{\mathscr{S}_c} \varrho\, (\dot{\boldsymbol{p}} - \boldsymbol{V})(\dot{\boldsymbol{p}} - \boldsymbol{V}) \cdot d\boldsymbol{a} + {}\\ &\quad + \oint_{\mathscr{S}_c} \varrho\, (\dot{\boldsymbol{p}} - \boldsymbol{V})\, d\boldsymbol{a} \cdot \boldsymbol{V} + \boldsymbol{V} \oint_{\mathscr{S}_c} \varrho\, \dot{\boldsymbol{p}} \cdot d\boldsymbol{a}.\end{aligned}\right\}\quad (202.6)$$

Since $\varrho\,\dot{\boldsymbol{p}}$ is solenoidal in a steady motion, and since $\dot{\boldsymbol{p}}$ is normal to the fixed boundary $\mathscr{S}$, application of GREEN's transformation to the integral of $\operatorname{div}(\varrho\,\dot{\boldsymbol{p}})$ over the region between $\mathscr{S}$ and $\mathscr{S}_c$ shows that the last summand vanishes. Thus $(202.5)_1$ may be written

$$\mathfrak{F} = \oint_{\mathscr{S}_c} \varrho\, (\dot{\boldsymbol{p}} - \boldsymbol{V})(\dot{\boldsymbol{p}} - \boldsymbol{V}) \cdot d\boldsymbol{a} + \oint_{\mathscr{S}_c} \varrho\, (\dot{\boldsymbol{p}} - \boldsymbol{V})\, d\boldsymbol{a} \cdot \boldsymbol{V} - \oint_{\mathscr{S}_c} (\boldsymbol{t}_{(\boldsymbol{n})} + P\,\boldsymbol{n})\, d a, \quad (202.7)$$

where P is an arbitrary constant. This formula may be applied to the case when at great distances from the obstacle the material is moving at a uniform velocity $\boldsymbol{V}$ and the stress is hydrostatic. Indeed, from (202.7) we read off the

[1] For the classical special case of isochoric irrotational motion subject to hydrostatic pressure, see e.g. MILNE-THOMSON [1938, *9*, §§ 17.10—17.51] and the more general result of HASKIND [1956, *12*].

general form of the **Euler-D'Alembert paradox**[1]: *Let a stationary rigid body be immersed in an infinite material in steady motion past it; if there exist constants* $\boldsymbol{V}$ *and* P *such that*

$$\varrho(\dot{\boldsymbol{p}}-\boldsymbol{V})(\dot{\boldsymbol{p}}-\boldsymbol{V})=\bar{o}(p^{-2}),\quad \varrho(\dot{\boldsymbol{p}}-\boldsymbol{V})=\bar{o}(p^{-2}),\quad \boldsymbol{t}_{(\boldsymbol{n})}+P\boldsymbol{n}=\bar{o}(p^{-2}),\qquad(202.8)$$

then the obstacle exerts no force on the material. Simple sufficient conditions are

$$\dot{\boldsymbol{p}}-\boldsymbol{V}=o(p^{-2}),\quad \varrho=O(1),\quad \boldsymbol{t}_{(\boldsymbol{n})}+P\boldsymbol{n}=o(p^{-2}).\qquad(202.9)$$

The result is extremely general: If the velocity approaches its constant limit at ∞ sufficiently quickly, and if the stress vector approaches a uniform pressure at ∞ sufficiently quickly, the submerged body exerts no force. However, the "paradox" has limited application, as the order conditions (202.8), which are *the very essence*[2] of the assumptions underlying it, are rarely fulfilled in particular theories of materials. The classical example is homochoric irrotational flow of a perfect fluid, where (202.9) follows easily from theorems of potential theory. It is often asserted that in fact it is the adherence of materials to solid boundaries that accounts for the resistance to steady motion in real fluids. This may be so, but it does not enter the present argument directly. Indeed, if the stronger condition (69.4) were to replace (69.2) here, the most we could expect would be the annulling of further surface integrals, not the addition of extra terms. Rather, so far as is now known, taking account of adherence in conjunction with any dissipative mechanism has the effect of transmitting stronger disturbances to ∞, sufficient to violate the order condition (202.8).

There are various generalizations. If the material is confined by a stationary canal, from (202.6) we see that instead of integrating over all of $\mathscr{S}_c$, for the first two integrals in (202.7) we need consider only the cross sections of the canal at great distances, and again $(202.8)_{1,2}$ are sufficient to make these integrals vanish in the limit. To make the third integral vanish also, in addition to $(202.8)_3$ we supply an appropriate assumption regarding the value of the stress vector on the canal walls[3]. If there are surfaces on which $\dot{\boldsymbol{p}}$ is discontinuous, provided they do not extend to infinity, the result still holds[4]; if, however, there are infinite shocks or slip surfaces, the resultant force in general is not zero[5].

To calculate the torque, we note first that

$$\left.\begin{aligned}d\boldsymbol{a}\cdot\dot{\boldsymbol{p}}(\boldsymbol{p}\times\varrho\,\dot{\boldsymbol{p}})&=d\boldsymbol{a}\cdot(\dot{\boldsymbol{p}}-\boldsymbol{V})[\boldsymbol{p}\times\varrho(\dot{\boldsymbol{p}}-\boldsymbol{V})]+\\&\quad+(d\boldsymbol{a}\cdot\boldsymbol{V})[\boldsymbol{p}\times\varrho(\dot{\boldsymbol{p}}-\boldsymbol{V})]+d\boldsymbol{a}\cdot\varrho\,\dot{\boldsymbol{p}}\,\boldsymbol{p}\times\boldsymbol{V}.\end{aligned}\right\}\qquad(202.10)$$

Now by GREEN's transformation and (69.2) and (156.6) we have

$$\int_{\mathscr{S}_c}d\boldsymbol{a}\cdot\varrho\,\dot{\boldsymbol{p}}\,\boldsymbol{p}=\int_{\mathscr{V}}\operatorname{div}(\varrho\,\dot{\boldsymbol{p}}\,\boldsymbol{p})\,dv=\int_{\mathscr{V}}\dot{\boldsymbol{p}}\,d\mathfrak{M}=\mathfrak{P}\qquad(202.11)$$

---

[1] The result was asserted and proved correctly by EULER [1745, *2*, Satz I, Anm. 3] for steady plane flow of a perfect fluid, on the assumption that the stream lines straighten out at ∞. D'ALEMBERT [1752, *1*, §§ 66–69] rediscovered or appropriated it; later [1768, *2*, § I] he reasserted it in sensational terms and proved it by an argument assuming the body to consist in eight congruent parts. It has given rise to misunderstandings lasting over centuries. For the hydrodynamical case, a correct proof based on integral transformation was given by CISOTTI [1906, *3*]; we present, essentially, the reformulation by BOGGIO [1910, *2*]. A variant argument based partly on the energy balance was suggested by DUHEM [1914, *2*] and worked out by B. FINZI [1926, *2*].

[2] In the older treatments this is glossed over, but it is made clear by CISOTTI [1917, *4*, § 9].

[3] CISOTTI [1909, *4*].

[4] DUHEM [1914, *2* and *3*], PICARD [1914, *9*], MANARINI [1948, *14*] [1949, *17*].

[5] Examples were given by HELMHOLTZ and others; a general discussion is presented by JOUGUET and ROY [1924, *7*].

where $\mathfrak{P}$ is the momentum of the material between $\mathscr{S}$ and $\mathscr{S}_c$. Substituting (202.10) and (202.11) into $(202.5)_2$ yields

$$\left.\begin{aligned}\mathfrak{L} = \oint_{\mathscr{S}_c} \boldsymbol{p}\times\varrho(\dot{\boldsymbol{p}}-\boldsymbol{V})(\dot{\boldsymbol{p}}-\boldsymbol{V})\cdot d\boldsymbol{a} + \boldsymbol{V}\cdot\int_{\mathscr{S}_c} d\boldsymbol{a}\,\boldsymbol{p}\times\varrho(\dot{\boldsymbol{p}}-\boldsymbol{V}) + \\ + \mathfrak{P}\times\boldsymbol{V} - \oint_{\mathscr{S}_c} d\,a\,\boldsymbol{p}\times(\boldsymbol{t}_{(\boldsymbol{n})} + P\,\boldsymbol{n}).\end{aligned}\right\} \quad (202.12)$$

Therefore if we strengthen (202.9) to read

$$\dot{\boldsymbol{p}} - \boldsymbol{V} = o(p^{-3}), \qquad \varrho = O(1), \qquad \boldsymbol{t}_{(\boldsymbol{n})} + P\boldsymbol{n} = o(p^{-3}), \qquad (202.13)$$

we conclude not only the Euler-D'Alembert paradox but also[1]

$$\mathfrak{L} = \mathfrak{P}_\infty \times \boldsymbol{V}, \qquad (202.14)$$

where $\mathfrak{P}_\infty$ is the relative momentum[2] of the material exterior to the obstacle. Under the severe conditions (202.13), then, a stationary rigid body will generally exert a torque perpendicular to the direction of motion. In certain cases of symmetry we shall have $\mathfrak{P}_\infty = 0$; the body then exerts no torque.

CISOTTI[3] extended these results to the case of an obstacle spinning at angular velocity $\boldsymbol{\omega}$, provided the motion *relative to the obstacle* be steady. He found that then

$$\mathfrak{F} = \boldsymbol{\omega}\times\mathfrak{P}_\infty; \qquad (202.15)$$

the expression for $\mathfrak{L}$ is more complicated.

All our results are expressed in terms of the reaction of the obstacle *on the material*. That this is equal in magnitude and opposite in direction to the reaction of the material on the obstacle is to be expected and follows from CAUCHY's lemma, to be proved in the next section.

203. The stress tensor[4]. The field of stress vectors $\boldsymbol{t}_{(\boldsymbol{n})}$ is not an ordinary vector field. Rather, since the stress vectors across two different surfaces through the same point are generally different, at any given time $\boldsymbol{t}_{(\boldsymbol{n})}$ is a function both of the position vector $\boldsymbol{p}$ and of the direction $\boldsymbol{n}$. The first problem is to delimit the class of $\boldsymbol{t}_{(\boldsymbol{n})}$ for fixed $\boldsymbol{p}$ as $\boldsymbol{n}$ varies.

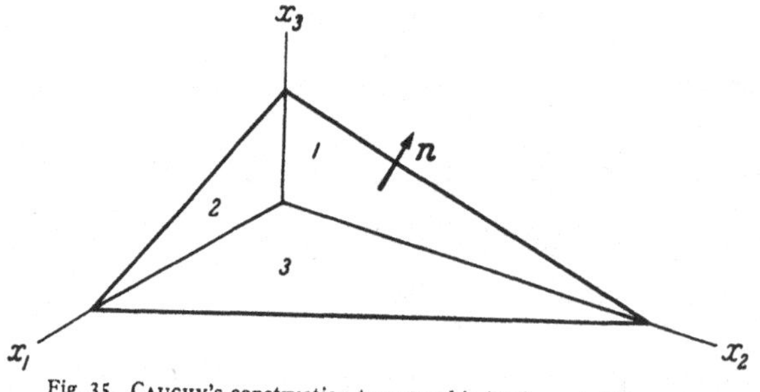

Fig. 35. CAUCHY's construction to prove his fundamental theorem.

We apply (200.1) to a tetrahedron (Fig. 35), three sides of which are mutually orthogonal, the fourth having outward unit normal $\boldsymbol{n}$. Let the altitude of the tetrahedron be h; the area of the inclined face, $\mathfrak{s}$; the projections of $\boldsymbol{n}$ onto the orthogonal faces, n_1, n_2, and n_3, so that the areas of the inclined faces are $\mathfrak{s}n_1$, $\mathfrak{s}n_2$, and $\mathfrak{s}n_3$. Assume that the fields $\varrho\ddot{\boldsymbol{p}}$ and $\varrho\boldsymbol{f}$ are

[1] CISOTTI [1910, *4*], BOGGIO [1910, *2*].

[2] I.e., $\mathfrak{P}_\infty \equiv \int(\dot{\boldsymbol{p}}-\boldsymbol{V})\,d\mathfrak{M}$, where by $(202.13)_1$ the integral, taken over the infinite space exterior to the obstacle, is convergent.

[3] [1910, *4*]; a vectorial derivation was given by BOGGIO [1910, *2*].

[4] The idea and the results here are due to CAUCHY [1823, *1*] [1827, *1*]. In a sense, the fundamental theorem (203.4) is contained in a memoir written by FRESNEL in 1822 but not published until much later [1868, *7*, §§ 3—4]; however, FRESNEL did not disentangle stress in general from purely elastic stress. Cf. also HOPKINS [1847, *1*, § 2].

bounded, that $\boldsymbol{t}_{(\boldsymbol{n})}$ is a continuous function both of $\boldsymbol{p}$ and of $\boldsymbol{n}$. We may then estimate the volume integrals in (200.1) and apply the theorem of mean value to the surface integral:

$$\mathfrak{s}[n_1 \boldsymbol{t}_1^* + n_2 \boldsymbol{t}_2^* + n_3 \boldsymbol{t}_3^* + \boldsymbol{t}_{(\boldsymbol{n})}^*] + h\mathfrak{s}K = 0, \tag{203.1}$$

where K is a bound and where $\boldsymbol{t}_{(\boldsymbol{n})}^*$ and $\boldsymbol{t}_a^*$ are the stress vectors at certain points upon the outsides of the respective faces. We cancel $\mathfrak{s}$ and let h tend to zero, so obtaining

$$\boldsymbol{t}_{(\boldsymbol{n})} = -(\boldsymbol{t}_1 n_1 + \boldsymbol{t}_2 n_2 + \boldsymbol{t}_3 n_3), \tag{203.2}$$

where all stress vectors are evaluated at the vertex of the tetrahedron. From their definitions, the quantities $\boldsymbol{t}_1, \boldsymbol{t}_2, \boldsymbol{t}_3$ do not depend upon $\boldsymbol{n}$.

In the case when $n_1 = 1$, $n_2 = n_3 = 0$, we have $\boldsymbol{t}_1 = \boldsymbol{t}_{(-\boldsymbol{n})}$, since the outward normal to the face 1 is $-\boldsymbol{n}$. Although the construction of the tetrahedron fails for this case, we may nevertheless suppose $n_1 \to 1$, $n_2 \to 0$, $n_3 \to 0$, and by the assumed continuity of $\boldsymbol{t}_{(\boldsymbol{n})}$ as a function of $\boldsymbol{n}$ infer from (203.2) that

$$\boldsymbol{t}_{(\boldsymbol{n})} = -\boldsymbol{t}_{(-\boldsymbol{n})}. \tag{203.3}$$

This is ***Cauchy's lemma***: *The stress vectors acting upon opposite sides of the same surface at a given point are equal in magnitude and opposite in direction.*

Now select a rectangular Cartesian co-ordinate system whose planes are the three orthogonal faces. With the convention that t_{km} is the k-th component of the stress vector acting upon the *positive* side of the plane $z_m =$ const, by (203.3) we infer that $-t_{1x} = t_{xx}$, $-t_{1y} = t_{yx}$, $-t_{1z} = t_{zx}$, etc., and hence (203.2) becomes $t_{(\boldsymbol{n})k} = t_{km} n_m$, where the quantities t_{km} are independent of $\boldsymbol{n}$. By the quotient law of tensors[1], the quantities t_{km} form a Cartesian tensor of second order. The tensor equation

$$t_{(\boldsymbol{n})}^k = t^{km} n_m, \quad \text{or} \quad t_{(\boldsymbol{n})}^k\, da = t^{km}\, da_m, \tag{203.4}$$

having been established in a special co-ordinate system, is valid in all co-ordinate systems. This proves ***Cauchy's fundamental theorem***: *From the stress vectors acting across three mutually perpendicular planes at a point, all stress vectors at that point are determinate; they are given by* (203.4) *as linear functions of the stress tensor* t^{km}.

Application of a parallel argument to (200.3) yields $\boldsymbol{m}_{(\boldsymbol{n})} = -\boldsymbol{m}_{(-\boldsymbol{n})}$, $m_{(\boldsymbol{n})}^k = m^{kq} n_q$; in terms of the equivalent skew-symmetric tensor $m_{(\boldsymbol{n})}^{kq}$, we have

$$m_{(\boldsymbol{n})}^{kq} = m^{kqp} n_p, \quad m^{kqp} = -m^{qkp}. \tag{203.5}$$

In the foregoing analysis, the point under consideration was an interior point. However, the argument may easily be adjusted to the case of a boundary point where the bounding surface has a continuous tangent plane. Thus if a body is subject to loading upon its surface $\mathfrak{s}$, the corresponding *boundary condition* is

$$t_{(\boldsymbol{n})}^k = t^{km} n_m = s^k, \quad m_{(\boldsymbol{n})}^{kr} = m^{krp} n_p = q^{kr}, \tag{203.6}$$

where $\boldsymbol{s}$ and $\boldsymbol{q}$ are prescribed functions of position upon $\mathfrak{s}$. Moreover, if $\mathfrak{s}$ is a surface of discontinuity separating two regions in which $\boldsymbol{t}_{(\boldsymbol{n})}$ and $\boldsymbol{m}_{(\boldsymbol{n})}$ are continuous, the argument may be applied separately on each side, again yielding (203.4) and (203.5), though the components t^{km} and m^{kpq} will fail to be continuous across $\mathfrak{s}$. Thus CAUCHY's fundamental theorem holds in all cases of interest.

[1] At this point CAUCHY [1829, *1*] made a fresh appeal to (200.1).

204. Further properties of the stress tensor. The *physical components of stress*, $t\langle km\rangle$ are regularly written[1] as $\widehat{km}$. We have

$$\dim \widehat{km} = [M L^{-1} T^{-2}]. \tag{204.1}$$

The quantity $\widehat{km}$ is the k component of tension per unit area acting upon the side of the surface $x^m = \text{const}$ which faces in the direction of increasing x^m. The normal components $\widehat{kk}$ are called *normal stresses*; the shear components $\widehat{km}$, $k \neq m$, are called *shearing stresses*[2]. These terms are consistent with those introduced in Sect. 200, but not identical with them.

By (203.4) and the definition given in Sect. 200, the stress is *hydrostatic* if and only if for all $\boldsymbol{n}$ we have $t^k{}_m n^m = -p n^k$, where p is a scalar independent of $\boldsymbol{n}$. First we relinquish this last condition and allow p possibly to depend upon $\boldsymbol{n}$. Then we must have $(t^k{}_m + p\,\delta^k_m) n^m = 0$ for all unit vectors $\boldsymbol{n}$. Choosing $\boldsymbol{n} = (1, 0, 0)$, we conclude that $t^k{}_1 + p(1, 0, 0)\,\delta^k_1 = 0$, etc., so that in a certain rectangular Cartesian co-ordinate system, $\|t^k{}_m\|$ is diagonal. Therefore $\|t^k{}_m\|$ is diagonal in all co-ordinate systems; i.e., $\boldsymbol{t}$ is spherical. Rephrasing these remarks, we obtain a theorem of CAUCHY[3]: *The stress is hydrostatic if and only if every element is free of shearing stress; in this case*

$$t^k{}_m = -p\,\delta^k_m, \tag{204.2}$$

where p is a scalar independent of $\boldsymbol{n}$. For any stress tensor which is symmetric, it is of course possible to find at each point three particular surface elements free of shearing stress, this being but a restatement of the diagonalization theorem in Sect. App. 37.

In hydrostatic stress, all stress vectors at a point have the same magnitude, but the converse does not hold. For if all stress vectors have equal magnitude, then $g_{kq}\, t^{km}\, t^{qp}\, n_m\, n_p$ is independent of $\boldsymbol{n}$, subject to the condition $g^{kq} n_k n_q = 1$. Introducing a multiplier M and setting equal to zero the coefficients of the resulting quadratic form which must vanish identically, we get

$$t^{qm}\, t_{qp} = M\,\delta^m_p. \tag{204.3}$$

Eliminating M by taking the trace of this equation, we obtain

$$t^{qk}\, t_{qm} = \tfrac{1}{3}\, t^{sp}\, t_{sp}\, \delta^k_m, \tag{204.4}$$

a homogeneous quadratic condition to be satisfied by $\boldsymbol{t}$ as necessary and sufficient that all stress vectors have the same magnitude. It asserts that the product of $\boldsymbol{t}$ by its transpose shall be a spherical tensor.

However, for the normal stresses on all elements to be equal, $t^{km} n_m n_k$ must be independent of $\boldsymbol{n}$ for all units vectors $\boldsymbol{n}$, and this leads at once to (204.2). That is, *the stress is hydrostatic if and only if all elements suffer equal normal stress.* Also, *the stress is hydrostatic if and only if every stress component $t^k{}_m$ transforms as a scalar*[4].

Instead of the stress tensor $\boldsymbol{t}$, sometimes the *pressure tensor*, defined by

$$\boldsymbol{p} \equiv -\boldsymbol{t}, \tag{204.5}$$

is employed. Also, if p is any scalar whatever, a corresponding *extra stress* may defined by

$$\boldsymbol{v} \equiv p\,\mathbf{1} + \boldsymbol{t}. \tag{204.6}$$

[1] PEARSON [1886, *4*, § 610].

[2] The first special studies of shearing stresses were made by COULOMB [1776, *1*, § V] and by HOPKINS [1847, *1*, §§ 4–5].

[3] [1827, *1*].

[4] PIOLA [1848, *2*, ¶ 62].

The *mean pressure* is the scalar defined by

$$\tilde{p} \equiv -\tfrac{1}{3} t^k{}_k = -\tfrac{1}{3} \mathrm{I}_{\boldsymbol{t}}. \tag{204.7}$$

In the case when $p = \tilde{p}$, the tensor $\boldsymbol{v}$ is the *stress deviator*[1], $\boldsymbol{v} = {}_0\boldsymbol{t}$. The decomposition (204.6) is then maximal in the sense of Sect. App. 43.

When there exist three perpendicular elements subject to pure shearing stress (Sect. 200), the stress is described as a *state of pure shearing stress*. By the results at the beginning of this section, for a state of pure shearing stress it is necessary and sufficient that in some rectangular Cartesian co-ordinate system $\boldsymbol{t}$ assume the form (App. 44.3) with $a = 0$. Thus a *necessary and sufficient condition for a state of pure shear is that the mean pressure vanish*: $\tilde{p} = 0$. Therefore the maximal decomposition $\boldsymbol{t} = -\tilde{p}\,\mathbf{1} + {}_0\boldsymbol{t}$ splits an arbitrary state of stress uniquely into a hydrostatic pressure and a state of pure shearing stress. Further results concerning pure shearing stress will be given in Sect. 209.

When $\boldsymbol{t}$ is a plane tensor, the state of stress is said to be *plane*. In plane stress, a family of parallel planes is free of traction.

205. Cauchy's laws of motion. If in (200.1) and (200.3) we replace $\boldsymbol{t}_{(\boldsymbol{n})}$ and $\boldsymbol{m}_{(\boldsymbol{n})}$ by their expressions (203.4) and (203.5)$_1$, we obtain

$$\left.\begin{aligned} \frac{d}{dt}\int_{\mathscr{V}} \dot{z}_k \, d\mathfrak{M} &= \dot{\mathfrak{P}}_k = \mathfrak{F}_k = \oint_{\mathscr{S}} t_{km}\, da_m + \int_{\mathscr{V}} f_k \, d\mathfrak{M}, \\ \frac{d}{dt}\int_{\mathscr{V}} z_{[k}\dot{z}_{p]}\, d\mathfrak{M} &= \dot{\mathfrak{H}}_{kp} = \mathfrak{L}_{kp} = \oint_{\mathscr{S}} (m_{kpq} + z_{[k} t_{p]q})\, da_q + \int_{\mathscr{V}} (l_{kp} + z_{[k} f_{p]})\, d\mathfrak{M}. \end{aligned}\right\} \tag{205.1}$$

These equations are of the form (157.1) and hence are equations of balance. The tensor $-\boldsymbol{t}$ is the influx of momentum per unit mass (i.e., velocity), the vector $\boldsymbol{f}$ is the supply of momentum, etc.

By applying (157.6) to (205.1)$_1$ we conclude that in regions where $\boldsymbol{t}$ is continuously differentiable and $\varrho\ddot{\boldsymbol{x}}$ and $\varrho\boldsymbol{f}$ are continuous, *a necessary and sufficient condition for the balance of linear momentum*, expressed in general co-ordinates, is

$$\varrho \ddot{x}^k = t^{km}{}_{,m} + \varrho f^k. \tag{205.2}$$

This is ***Cauchy's first law of motion***[2].

Similarly, by application of (193.3) to (205.1)$_1$ we conclude that at a surface across which $\boldsymbol{t}$, ϱ, and $\dot{\boldsymbol{x}}$ may be discontinuous, but in whose vicinity these quantities and $\partial\dot{\boldsymbol{x}}/\partial t$ and $\varrho\boldsymbol{f}$ remain bounded, we have[3]

$$n_m[t^{km}] + [\varrho U \dot{x}^k] = 0, \quad \text{or} \quad [\boldsymbol{t}_{(\boldsymbol{n})} + \varrho U \dot{\boldsymbol{x}}] = 0, \tag{205.3}$$

[1] KLEITZ [1873, *4*, § 23].

[2] [1827, *4*] [1828, *2*, Eq. (25)], the special case when $\boldsymbol{t}$ is spherical being due to EULER [1757, *2*, § 21]. That (205.2) is sufficient as well as necessary for the balance of momentum was proved by POISSON [1829, *5*, § I, ¶ 11] and KELVIN and TAIT [1867, *3*, § 698]. The Cartesian tensorial invariance of (205.2) was verified by PIOLA [1848, *2*, ¶¶ 55, 58]. The derivation given above (i.e., in Sect. 157) is due to KIRCHHOFF [1876, *2*, Vorl. 11, § 2]; cf. also DONATI [1888, *4*, pp. 347–349], LOVE [1892, *8*, § 14].

[3] For the case when $\boldsymbol{t}$ is spherical, hydrodynamical special cases were obtained by STOKES [1848, *4*, Eq. (3)], RIEMANN [1860, *4*, § 5], RANKINE [1870, *6*, §§ 3–5], CHRISTOFFEL [1877, *2*, § 1], HUGONIOT [1887, *1*, § 140] and JOUGUET [1901, *9*]; for general $\boldsymbol{t}$ but linearized acceleration, by CHRISTOFFEL [1877, *3*, § 4, Eq. (δ)]; the special case for finite elastic deformation, by ZEMPLÉN [1905, *7*, p. 448] [1905, *9*, § 3], who approached it via HAMILTON's principle in the form given in Sect. 236A; the general result is due to KOTCHINE [1926, *3*, Eq. (18)].

where U is the local speed of propagation of the singular surface. By (189.14) and (183.4) we have the equivalent forms

$$[\boldsymbol{t}_{(\boldsymbol{n})}] + \varrho^{\pm} U^{\pm}[\dot{\boldsymbol{x}}] = 0, \qquad f_{,m}[t^{km}] - \varrho^{\pm} \dot{f}^{\pm}[\dot{x}^k] = 0, \tag{205.4}$$

where $f=0$ is an equation of the surface of discontinuity, and where either the $+$ mark or the $-$ mark is to be used throughout. When $U=0$ or $[\dot{\boldsymbol{x}}]=0$, these formulae reduce to

$$[\boldsymbol{t}_{(\boldsymbol{n})}] = 0: \tag{205.5}$$

Across a surface of discontinuity which is not a shock, the stress vector is continuous[1].

Resolving (205.3) into components normal and tangential to the singular surface yields

$$[t_{nn} - \varrho U^2] = 0, \qquad [\boldsymbol{t}_{(\boldsymbol{n})t}] + \varrho^{\pm} U^{\pm}[\dot{\boldsymbol{x}}_t] = 0, \tag{205.6}$$

where (185.6) has been used. At a point where $\boldsymbol{t}$ is hydrostatic, from $(205.6)_2$ we have $\varrho^{\pm} U^{\pm}[\dot{\boldsymbol{x}}_t] = 0$, showing that either the tangential velocity is continuous, or the singular surface is material, or both. From (189.15) and $(205.6)_1$ we obtain[2]

$$(U^-)^2 = -\frac{\varrho^+}{\varrho^-}\cdot\frac{[t_{nn}]}{[\varrho]}; \qquad \text{or} \qquad U^+ U^- = -\frac{[t_{nn}]}{[\varrho]}. \tag{205.7}$$

Thus *the speeds of propagation of a shock wave are determined by the jump of the normal tension and by the values of the density on each side of the shock*, independently of all thermodynamic principles or constitutive equations. This result shows also that at a condensation shock, the normal pressure increases.

Second, we apply (157.6) to $(205.1)_2$, obtaining

$$\varrho z_{[k}\ddot{z}_{p]} = (m_{kpq} + z_{[k} t_{p]q})_{,q} + \varrho (l_{kp} + z_{[k} f_{p]}). \tag{205.8}$$

Hence

$$m_{kpq,q} - t_{[kp]} + \varrho l_{kp} + z_{[k}(t_{p]q,q} + \varrho f_{p]} - \varrho \ddot{z}_{p]}) = 0. \tag{205.9}$$

By CAUCHY's first law (205.2), the term in parentheses vanishes. The resulting formula, put into general co-ordinates, is[3]

$$m^{kpq}{}_{,q} + \varrho l^{kp} = t^{[kp]}. \tag{205.10}$$

In the non-polar case, this becomes $t^{[kp]}=0$, expressing ***Cauchy's second law of motion***[4]: *When there are no assigned couples and no couple-stresses, a necessary and sufficient condition for the balance of moment of momentum in a body where linear momentum is balanced is*[5]

$$t^{km} = t^{mk}; \tag{205.11}$$

i.e., *the stress tensor is symmetric.*

---

[1] Announced by POISSON [1829, *5*, § I, ¶ 10] [1831, *2*, § 25], proved by CAUCHY [1843, *1*] and HAUGHTON [1849, *1*, p. 176] [1855, *2*, § I]. Note that (205.5) holds at all vortex sheets (Sects. 185, 186).

[2] KOTCHINE [1926, *3*, Eq. (20)], the one-dimensional hydrostatic case being a celebrated result of RANKINE [1870, *6*, Eq. (6)] and HUGONIOT [1887, *2*, §§ 144, 149].

[3] For references, see footnotes 1 and 2, p. 538. GRAD [1952, *7*, § 4], motivated by a model in which the molecules have internal degrees of freedom, adds to the ordinary moment of momentum $z_{[k}\dot{z}_{m]}$ an internal moment of momentum μ_{km}, resulting in the additional term $\varrho\dot{\mu}^{kp}$ on the left-hand side of (205.10). Cf. also ERICKSEN [1960, *1*, § 2].

[4] [1827, *1*, Th. II]. This law had been discovered but not published by FRESNEL in 1822 [1868, *7*, § 5]. That (205.11) and (205.2) imply (200.2) was proved by POISSON [1829, *5*, § I, ¶ 12]. The observation that the classical form (205.11) involves simplifying hypotheses and is not a *general* statement of balance of moment of momentum was made by HAUGHTON [1855, *2*, p. 107]. The German literature persists in attributing credit to BOLTZMANN here.

[5] Inspection of (205.10) shows clearly conditions under which the stress tensor is symmetric and conditions under which it is not. We do not cite the recent literature propagating misunderstandings on this classical subject.

Let $\boldsymbol{n}$ and $\boldsymbol{n}'$ be two unit vectors at a given point. By (203.4) and (205.11) we have

$$\boldsymbol{t}_{(\boldsymbol{n})}\cdot\boldsymbol{n}' = t^{km} n_m n'_k = t^{mk} n_m n'_k = \boldsymbol{t}_{(\boldsymbol{n}')}\cdot\boldsymbol{n}, \tag{205.12}$$

and conversely, if we have (203.4), then (205.11) is the special case of (205.12) when $\boldsymbol{n}$ and $\boldsymbol{n}'$ are perpendicular. This is ***Cauchy's reciprocal theorem***[1]: *In the second law, the symmetry of the stress tensor is equivalent to the condition that each of two stress vectors at a point has an equal projection upon the normal to the surface on which the other acts* (Fig. 36). Extension of this result to the case when (205.10) rather than (205.11) holds is easy and uninteresting.

By application of (193.3) to $(205.1)_2$ we get the following condition at a surface across which $\boldsymbol{m}$, $\boldsymbol{t}$, ϱ, and $\dot{\boldsymbol{x}}$ may be discontinuous, but $\partial\dot{\boldsymbol{z}}/\partial t$, $\boldsymbol{f}$, and $\boldsymbol{l}$ remain bounded:

$$\left.\begin{aligned} n_q[m_{kpq}] + z_{[k}([t_{p]q} n_q + \\ + \varrho U \dot{z}_{p]}]) = 0. \end{aligned}\right\} \tag{205.13}$$

By (205.3), the term in parentheses vanishes, and (205.13), put into general coordinates, becomes

$$n_q[m^{kpq}] = 0, \quad \text{or} \quad [\boldsymbol{m}_{(\boldsymbol{n})}] = 0: \tag{205.14}$$

the couple-stress vector is continuous.

If $\boldsymbol{m} = 0$, this condition is satisfied automatically: *When there is no couple-stress, balance of linear momentum at a surface of discontinuity implies balance of moment of momentum as well.* In particular, this holds for the non-polar case.

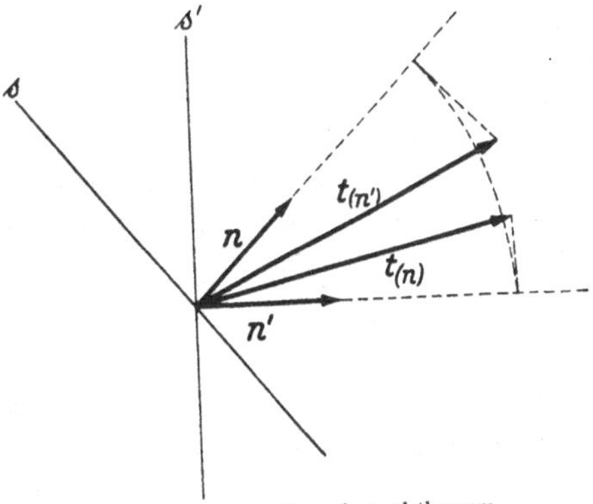

Fig. 36. CAUCHY's reciprocal theorem.

The stress vector $\boldsymbol{t}_{(\boldsymbol{n})}$ is not uniquely defined[2] by the requirement given here, being indeterminate to within an arbitrary vector function of place and direction such that its vector sum and vector moment over any closed surface vanishes, By (205.2), (205.5), and (205.11), the corresponding stress tensor ${}^0\boldsymbol{t}$ must satisfy

$${}^0t^{km}{}_{,m} = 0, \qquad {}^0t^{km} = {}^0t^{mk}, \qquad [{}^0t^{km}]\, n_m = 0. \tag{205.15}$$

Such a stress tensor we shall call a *null stress*. Any dynamically correct statement regarding the stress of an arbitrary continuum must remain invariant when an arbitrary null stress is added to the stress tensor. All results derived in the foregoing paragraphs satisfy this principle of invariance: *Any motion dynamically possible subject to a stress tensor* $\boldsymbol{t}$ *is also possible subject to* $\boldsymbol{t} + {}^0\boldsymbol{t}$*, where* ${}^0\boldsymbol{t}$ *is an arbitrary null stress*[3].

Eqs. (205.2) and (205.11) [or (205.10)] in regions of sufficient smoothness and Eq. (205.3) [supplemented, if necessary, by (205.14)] at a surface of discontinuity are the ***fundamental equations of continuum mechanics.***

In the general case, by (205.2) and (205.10) we have

$$t^{(kp)}{}_{,p} = t^{kp}{}_{,p} - t^{[kp]}{}_{,p} = \varrho \ddot{x}^k - \varrho f^k - m^{kpq}{}_{,qp} - (\varrho l^{kp})_{,p}. \tag{205.16}$$

---

[1] [1829, *2*].

[2] POINCARÉ [1892, *11*, § 38]. POINCARÉ objected also that the existence of a field of stress vectors is not self-evident, but such an objection could be raised almost anywhere in theoretical physics.

[3] This statement refers only to general continua; if $\boldsymbol{t}$ satisfies the equations of some particular theory of materials, then $\boldsymbol{t} + {}^0\boldsymbol{t}$ generally fails to be a solution.

or

$$\varrho \ddot{x}^k = t^{(k p)}{}_{,p} + \varrho f^k + m^{k p q}{}_{,q p} + (\varrho l^{k p})_{,p}. \tag{205.17}$$

This equation, which shows that the symmetric part of the stress tensor obeys an equation of the same form as CAUCHY's first law but with suitably supplemented forces, may be taken along with (205.10) as one of the two fundamental equations of motion.

A more general scheme of mechanics might be set up in terms of the moments $(166.7)_1$, with basic equations as follows:

$$\left.\begin{aligned}\mathfrak{F}_{k_1\ldots k_n q} &= \dot{\mathfrak{M}}_{k_1\ldots k_n q},\\ &= \oint_{\mathscr{S}} t_{k_1\ldots k_n q}\, da + \int_{\mathscr{V}} f_{k_1\ldots f_n q}\, d\mathfrak{M}.\end{aligned}\right\} \tag{205.18}$$

The corresponding boundary conditions and differential equations are, respectively,

$$\left.\begin{aligned} t_{k_1\ldots k_n q} &= t_{k_1\ldots k_n q p}\, n_p,\\ \varrho\, \overline{z_{k_1}\ldots z_{k_n} \dot{z}_q} &= t_{k_1\ldots k_n q p,\,p} + \varrho f_{k_1\ldots k_n q}.\end{aligned}\right\} \tag{205.19}$$

They may be recast by writing $t_{k_1\ldots k_n q}$ as the sum of the n-th moments of all stresses of lower order plus a tensor of excess; e.g., in the notation introduced above,

$$t_{[k q] p} = z_{[k}\, t_{q] p} + m_{k q p}, \qquad f_{[k p]} = z_{[k} f_{p]} + l_{k p}. \tag{205.20}$$

The assigned moments $f_{k_1\ldots k_n q}$ are subject to a law of transformation such as to render (205.20) invariant with respect to change of the origin of co-ordinates.

205 A. Appendix. Notations for stress. The following table, shortened from that of PEARSON [1886, *4*, § 610], presents the notations for stress found in the classical works and in some cases still in use today. The blocks stand for the matrix of t_{km} in rectangular Cartesian co-ordinates, or possibly for physical components in curvilinear co-ordinates.

CAUCHY's earlier work	POISSON	CORIOLIS, CAUCHY's later work, ST.-VENANT, MAXWELL	LAMÉ
$\begin{matrix} A & F & E\\ F & B & D\\ E & D & C\end{matrix}$	$\begin{matrix} P_3 & Q_3 & R_3\\ P_2 & Q_2 & R_2\\ P_1 & Q_1 & R_1\end{matrix}$	$\begin{matrix} p_{xx} & p_{xy} & p_{xz}\\ p_{yx} & p_{yy} & p_{yz}\\ p_{zx} & p_{zy} & p_{zz}\end{matrix}$	$\begin{matrix} N_1 & T_3 & T_2\\ T_3 & N_2 & T_1\\ T_2 & T_1 & N_3\end{matrix}$
F. NEUMANN, KIRCHHOFF, RIEMANN	KELVIN	CLEBSCH	PEARSON
$\begin{matrix} X_x & X_y & X_z\\ Y_x & Y_y & Y_z\\ Z_x & Z_y & Z_z\end{matrix}$	$\begin{matrix} P & V & T\\ V & Q & S\\ T & S & R\end{matrix}$	$\begin{matrix} t_{11} & t_{12} & t_{13}\\ t_{21} & t_{22} & t_{23}\\ t_{31} & t_{32} & t_{33}\end{matrix}$	$\begin{matrix} \widehat{xx} & \widehat{xy} & \widehat{xz}\\ \widehat{yx} & \widehat{yy} & \widehat{yz}\\ \widehat{zx} & \widehat{zy} & \widehat{zz}\end{matrix}$

German engineering works usually employ N_1, N_x, or σ_x for t_{xx} and T_{xy}, T_z, τ_{xy}, or τ_z for t_{xy}, etc. French authors usually follow LAMÉ, though some adopt the more luminous notation of CORIOLIS. The definitions sometimes differ in sign. The notation t_{km} may be remembered by the word *tension*; p_{km}, defined by (204.5), by the word *pressure*.

In this work we always use PEARSON's notation $\widehat{km}$, already introduced in Sect. 204, when employing physical components in orthogonal curvilinear co-ordinates. By (App. 14.6) and (App. 14.7), CAUCHY's first law (205.2) then assumes the explicit form

$$\left.\begin{aligned}\varrho \ddot{x}\langle k\rangle = \varrho f\langle k\rangle + \sum_{m=1}^{3}\Bigg\{&\frac{1}{\sqrt{g}}\frac{\partial}{\partial x^m}\left(\sqrt{\frac{g}{g_{mm}}}\,\widehat{km}\right) +\\ &+ \frac{\widehat{mk}}{\sqrt{g_{kk}}}\cdot\frac{1}{\sqrt{g_{mm}}}\frac{\partial\sqrt{g_{kk}}}{\partial x^m} - \frac{\widehat{mm}}{\sqrt{g_{mm}}}\cdot\frac{1}{\sqrt{g_{kk}}}\frac{\partial\sqrt{g_{mm}}}{\partial x^k}\Bigg\}\end{aligned}\right\} \tag{205 A.1}$$

(LAMÉ [1841, *4*, §§ VII—VIII] [1859, *3*, § 14, Eqs. (14) (15)]; cf. MORERA [1885, *6*], ANDRUETTO [1931, *1*]; for general curvilinear co-ordinates, cf. RICCI and LEVI-CIVITA [1901, *14*, Chap. 6, § 3], TONOLO [1930, *7*]).

206. Stress impulse. For the laws of impulse in continuum mechanics, we apply to (198.1) the general results given in Sect. 194. The influx of impulse is the negative of the *stress-impulse* tensor $\boldsymbol{i}$; the supply of impulse is a vector $\boldsymbol{s}$; and as the counterpart of CAUCHY's first law (205.2) we get from (194.11)

$$\{\varrho \dot{x}^k\} = i^{km}{}_{,m} + \varrho s^k, \tag{206.1}$$

while the balance of moment of momentum assumes exactly the form (205.10) with $\boldsymbol{i}$ replacing $\boldsymbol{t}$ and with appropriate new symbols replacing $\boldsymbol{m}$ and $\boldsymbol{l}$.

In the special case when $\boldsymbol{i}$ is spherical and $\boldsymbol{s}=0$, the covariant form of (206.1) is

$$\{\varrho \dot{x}_k\} = i_{,k}. \tag{206.2}$$

Hence follows trivially a celebrated theorem of LAGRANGE and CAUCHY[1]: *The mass flow produced by impulsive hydrostatic pressure is lamellar; in particular, a homochoric motion initiated by impulsive hydrostatic pressure is irrotational in the first instant.*

More generally, by comparing (205.2) and (206.1) we derive the following *analogy: The correspondences*

$$\{\varrho \dot{\boldsymbol{x}}\} \leftrightarrow \varrho \ddot{\boldsymbol{x}}, \quad \boldsymbol{i} \leftrightarrow \boldsymbol{t}, \quad \boldsymbol{s} \leftrightarrow \boldsymbol{f} \tag{206.3}$$

at a given instant enable us to construct a dynamically possible impulse, stress-impulse, and supply of impulse from a dynamically possible acceleration, stress, and force, and conversely.

207. The equivalence of stress, extrinsic loads, and transfer of momentum. As follows from the remarks in Sects. 157 and 199, in any given problem the stress may be replaced by equipollent extrinsic forces, or the extrinsic and mutual forces by a stress. Indeed, CAUCHY's first law (205.2) asserts that *the resultant force exerted by the stress is $t^{km}{}_{,m}$ per unit volume*, while (205.10) asserts that *the stress and the couple-stress exert a resultant couple of magnitude $m^{kpq}{}_{,q} - t^{[kp]}$ per unit volume.*

Conversely, to replace mutual and extrinsic loads by stresses we must integrate $t^{km}{}_{,m} = \varrho f^k$ with f^k given. Any solution of this equation may be added to the stress $\boldsymbol{t}$ in (205.2), resulting in a combined stress under which the material moves as if subject to no extrinsic or mutual force. In general, to calculate an equivalent stress is elaborate, but it turns out to be simple in the special case when $\varrho \boldsymbol{f}$ is lamellar:

$$\varrho f_k = - V_{,k}. \tag{207.1}$$

For then

$$t_k{}^m{}_{,m} + \varrho f_k = t_k{}^m{}_{,m} - V_{,k} = (t_k{}^m - \delta_k^m V)_{,m}, \tag{207.2}$$

so that *when the assigned force per unit volume is lamellar, it is equipollent to an appropriate hydrostatic pressure*[2].

This equivalence of various loads is generally artificial and useless, however. The course of the discipline of mechanics is to *prescribe* functional dependences for $\boldsymbol{f}$ and $\boldsymbol{t}$ (cf. Sect. 7). With a single prescription for $\boldsymbol{t}$, a host of different motions

---

[1] The statement and proof of LAGRANGE [1783, *1*, § 20], motivated perhaps by earlier remarks of D'ALEMBERT [1752, *1*, §§ 50—51], were corrected by CAUCHY [1827, *5*, Part 2, Sect. 1, §§ 4—5].

[2] In effect, this observation is due to EULER [1757, *1*, §§ 25—31].

of the body, corresponding to different initial and boundary conditions, are possible; for each of these different motions, a *different* equivalent $\boldsymbol{f}$ will result.

There is a simple and natural equivalence of stress to transfer of momentum, obtained[1] by substituting (156.7) into CAUCHY's first law (205.2):

$$\frac{\partial}{\partial t}(\varrho \dot{x}^k) = (t^{km} - \varrho \dot{x}^k \dot{x}^m)_{,m} + \varrho f^k. \qquad (207.3)$$

This equation asserts that if the tensor $-\varrho \dot{x}^k \dot{x}^m$ is added to t^{km}, the true rate of change of momentum may be replaced by the local apparent rate $\partial(\varrho \dot{\boldsymbol{x}})/\partial t$; therefore $-\varrho \dot{\boldsymbol{x}} \dot{\boldsymbol{x}}$ is called the tensor of ***apparent stress due to transfer of momentum***[2]. It is the negative of the momentum transfer, which we have discussed in Sect. 170.

Very little is known regarding the transformation of systems of stresses appropriate to one problem into those appropriate to another. An elegant result of this kind is due to OLDROYD and THOMAS[3] and NOLL[4]. We present NOLL's form of the argument. For an incompressible medium, we consider two velocity fields $\dot{\boldsymbol{x}}_1$ and $\dot{\boldsymbol{x}}_2$ which differ only by a rigid rotation at constant angular velocity $\boldsymbol{\omega}$ (cf. Sect. 143). We assume further that, apart from arbitrary hydrostatic pressures $p_1 \mathbf{1}$ and $p_2 \mathbf{1}$, the stress tensors in the two motions are the same functions of place and time[5]:

$$p_1 \mathbf{1} + \boldsymbol{t}_1 = p_2 \mathbf{1} + \boldsymbol{t}_2. \qquad (207.4)$$

Let the frame in which the velocity is $\dot{\boldsymbol{x}}_1$ be inertial, so that CAUCHY's first law (205.2) is satisfied. By the results in Sect. 197, the second motion, if it is to be dynamically possible, must then satisfy an equation of the same form but including the apparent forces arising from the rotation with respect to an inertial frame. Thus we have

$$\left.\begin{aligned} \varrho \ddot{\boldsymbol{x}}_1 &= \operatorname{div} \boldsymbol{t}_1 + \varrho \boldsymbol{f}, \\ \varrho \ddot{\boldsymbol{x}}_1 &= \operatorname{div} \boldsymbol{t}_2 + \varrho (\boldsymbol{f} + \boldsymbol{g}), \end{aligned}\right\} \qquad (207.5)$$

where $-\boldsymbol{g}$ is the sum of the CORIOLIS and centrifugal accelerations. Subtracting $(207.5)_1$ from $(207.5)_2$ and using (207.4), we see that

$$\varrho \boldsymbol{g} = \operatorname{grad}(p_2 - p_1). \qquad (207.6)$$

In order that the motion $\dot{\boldsymbol{x}}_2$ be dynamically possible, it is thus necessary and sufficient that $\boldsymbol{g}$ possess a single-valued potential. From the results in Sect. 143, a sufficient condition for $\boldsymbol{g}$ to be lamellar is that the motion be plane and that $\boldsymbol{\omega}$ be normal to the plane of motion[6]. Conditions sufficient to ensure single-valuedness of the function Q' in (143.10) have been stated in Sect. 161. Directly from (207.6) we see that the reaction per unit height exerted upon any cylinder perpendicular to the plane of motion and of cross-sectional area $\mathfrak{A}$ differs, in the two cases, only by the reaction of a hydrostatic pressure equalling the potential of the centrifugal and CORIOLIS accelerations. The moment is thus zero, and the resultant force is easily shown to be that requisite to impel a mass-point of mass $\varrho \mathfrak{A}$ located at the centroid to move with the velocity of the centroid. Alternatively, the difference of the forces is that which would be exerted upon the region occupied by the cylinder if it were filled with the surrounding material. Hence *if a rigid homogeneous cylinder immersed in an incompressible substance of the same density as the cylinder is made to undertake any motion in a plane normal to its generators, then a uniform rotation of the whole system about any axis parallel to the generators will not alter the motion of the cylinder relative to the surrounding substance.* This interesting discovery of G. I. TAYLOR, in the special case of a fluid, has been verified by experiments.

---

[1] Given for the hydrostatic case by GREENHILL [1875, *2*, § 22], for the general case by MATTIOLI [1914, *7*, § 2].

[2] While equations of the form (207.3) in the case when $\boldsymbol{t}$ is hydrostatic were given by earlier writers, the tensor $\varrho \dot{\boldsymbol{x}} \dot{\boldsymbol{x}}$ is often named after REYNOLDS; it was discussed by LORENTZ [1907, *5*, § 11].

[3] [1956, *16*].

[4] [1957, *12*].

[5] There is reason to believe that all material constitutive equations for incompressible media must satisfy this requirement; cf. NOLL [1955, *18*, §§ 4, 10].

[6] This result, for a perfect fluid, and also the result below for the force exerted upon an immersed cylinder, are due to TAYLOR [1916, *6*, pp. 100–104]. Cases of intermediate generality are treated by DEAN [1954, *4*] and JAIN [1955, *15*].

208. Principal stresses and stress invariants. In the non-polar case, CAUCHY'S second law (205.11) asserts that $\boldsymbol{t}$ is symmetric. By the first theorem in Sect. App. 37, it has real proper numbers, called the *principal stresses*[1], and real orthogonal principal directions, which define the *principal axes of stress*. Elements normal to the principal axes of stress are free of shearing stress, being subject to normal tension or pressure according to the sign of the corresponding principal stress; these are extremal values of the normal stresses. The scalar *invariants of stress*, $\mathrm{I}_{\boldsymbol{t}}$, $\mathrm{II}_{\boldsymbol{t}}$, $\mathrm{III}_{\boldsymbol{t}}$, are symmetric functions of the principal stresses.

When $t_2 = t_3 = 0$, $t_1 \neq 0$, the state of stress is called *simple tension*, and the principal direction corresponding to t_1 is called the *axis* of the tension. In the engineering literature, when one and only one principal stress vanishes, the state of stress is often called *bi-axial*; similarly, if no principal stress vanishes, the state of stress is *tri-axial*.

A state of non-vanishing plane symmetric pure shearing stress (Sect. 200) is called *simple shearing stress*; in a suitable rectangular Cartesian system

$$\|t_{km}\| = \left\| \begin{matrix} 0 & t_{xy} & 0 \\ \cdot & 0 & 0 \\ \cdot & \cdot & 0 \end{matrix} \right\|, \qquad t_{xy} \neq 0; \tag{208.1}$$

an invariant condition is

$$\mathrm{I}_{\boldsymbol{t}} = 0, \qquad \mathrm{III}_{\boldsymbol{t}} = 0, \tag{208.2}$$

so that $t_{xy} = t_1 = -t_2 = \sqrt{-\mathrm{II}_{\boldsymbol{t}}}$, and the principal axes of stress lying in the plane of stress bisect the angle between the elements suffering pure shearing stress.

In the non-polar case, the condition (204.4) that all stress vectors have equal magnitude reduces to $(t_1)^2 = (t_2)^2 = (t_3)^2$, as is obvious.

An exhaustive study of the general state of stress at a point was made by KLEITZ[2]. Some of his results, as well as the fundamental theorem of COULOMB and HOPKINS on the maximum shearing stress, have been given in more general form in Sect. App. 46. In particular, (App. 46.14) shows that the magnitude of the maximum shearing stress for a pair of directions, one of which is kept fixed as the other swings perpendicularly about it, is a function only of the differences of the principal stresses, and hence is independent of the mean pressure. For the overall maximum and minimum shears, this independence holds *a fortiori*.

From results of BOUSSINESQ[3] given in more general form in Sect. 23 it follows that among the planes through the principal axis corresponding to the principal stress t_2 when $t_1 \geqq t_2 \geqq t_3$ occur

1. The planes across which the magnitude of the stress vector is greatest and least;
2. The planes across which the magnitude of the normal stress is greatest and least;
3. The planes on which the magnitude of the shearing stress is greatest;
4. The planes across which the angle subtended by the stress vector is greatest and least.

Since theorems of this kind are no more than verbal adjustments of theorems on an arbitrary symmetric matrix, for economy we refer the reader to Sects. App. 37 — App. 50. Perhaps the most important function of the present section is to give warning that these results follow only from the assumption that there are no extrinsic or mutual couples or couple stresses.

---

[1] The theory is due to FRESNEL (1821—1822) [1868, *5*, § 28] [1868, *6*, §§ 1, 8—9] [1868, *7*, § 1].

[2] [1872, *2*, § 6] [1873, *4*, Chap. II].

[3] [1877, *1*, § 2].

209. Geometrical representations of stress. Geometrical representations for the stress tensor and stress vector may be read off from the results of Sects. 21 to 24. For the non-polar case, the following quadrics were introduced by CAUCHY and by LAMÉ and CLAPEYRON[1]:

1. The quadric of $\boldsymbol{t}$, called *Cauchy's stress quadric*. The normal stress across any plane through its center is inversely proportional to the squared length of that radius vector of the quadric which is normal to the plane.
2. The quadric of $\boldsymbol{t}^{-1}$, called *Lamé's stress director quadric*. The radius vector from the center to any point of the surface is in the direction of the stress vector across a plane parallel to the tangent plane at the point.
3. The quadric of $\boldsymbol{t}^2$, called *Cauchy's stress ellipsoid*. The central radius vector in any direction is inversely proportional to the magnitude of the stress vector across the plane at right angles to that direction.
4. The quadric of $\boldsymbol{t}^{-2}$, called *Lamé's stress ellipsoid*. The magnitude of the stress vector across any plane is proportional to the central perpendicular on the parallel tangent plane of the ellipsoid.

The asymptotic cone of LAMÉ's stress director quadric, called *Lamé's cone of shearing stress*, is the locus of elements subject to pure shearing stress. When it is real, it separates the planes across which the normal traction is a tension from those across which it is a pressure. When the cone is imaginary, the normal traction across all planes is a tension or a pressure according as $\mathrm{I}_{\boldsymbol{t}}>0$ or $\mathrm{I}_{\boldsymbol{t}}<0$. For there to be a state of pure shearing stress (Sect. 204), it is necessary and sufficient that this cone be rectangular. There are then infinitely many orthogonal triples of elements subject to pure shearing stress.

Further analysis of the stress quadrics and of curves upon them has been given by PACELLA[2].

The circle diagrams of MOHR (Sect. 24) originated in connection with stress and are frequently used by engineers.

In the non-polar case, $\boldsymbol{t}$ has three mutually orthogonal families of real principal trajectories (Sect. App. 47), called the *stress trajectories*[3]. The problem of isostatic surfaces (Sect. App. 48) arose in connection with stress. When an isostatic surface exists for the stress tensor, it is a surface free of shearing stress. If there are three families of isostatic surfaces, they may be used as an orthogonal curvilinear co-ordinate system; when referred to these co-ordinates, by (App. 47.4) CAUCHY's first law (205 A.1) reduces to the form[4]

$$\varrho\, \ddot{x}\langle k\rangle = \varrho\, f\langle k\rangle + \frac{1}{\sqrt{g_{kk}}}\frac{\partial t_k}{\partial x^k} + \frac{t_k - t_m}{\varrho_{kq}} - \frac{t_k - t_q}{\varrho_{km}}, \tag{209.1}$$

where the three indices k, m, q are unequal, and where the ϱ_{km} are the principal radii of curvature of the surface $x^k = \text{const}$. While these equations have limited correctness in three dimensions, their specialization to plane stress is always valid and often useful. In view of the results in Sect. 208, in the case when $\ddot{\boldsymbol{x}}=0$ and $\boldsymbol{f}=0$ these formulae express the rate of variation of a given principal stress along its trajectory in terms of two principal shearing stresses. Further results concerning the lines and sheets associated with the stress field may be read off from the analysis in Sects. App. 47 to App. 50.

---

[1] See Sect. 21 for references. In 1821–1822 FRESNEL [1868, *5*, § 28] [1868, *6*, §§ 1–7] [1868, *7*, §§ 1–4] constructed but did not publish a theory of elasticity based on postulating the ellipsoidal distribution of stress and the coincidence of the principal axes of stress and of infinitesimal strain.

For plane stress, a method of representing both the magnitude and the direction of the principal stress fields on a single diagram was constructed by TESAR [1933, *11*].

[2] [1948, *19*].

[3] While the existence of these trajectories is well known, we have been unable to trace their history or to find literature concerning them except in very special cases.

[4] LAMÉ [1841, *4*, § XI] [1859, *3*, § 149]. Cf. also WARREN [1864, *5*], MORERA [1885, *6*, § 3].

Other trajectories can be associated with a state of stress. Rather arbitrarily, VOLTERRA[1] chose to consider the trajectories of the field of stress vectors across the elements normal to $\dot{\boldsymbol{x}}$. These trajectories, which he called *lines of tension*, are indeterminate for static problems.

210. The equations of motion expressed in terms of a reference state. We now give forms of CAUCHY'S laws in which the independent positional variable is not the place $\boldsymbol{x}$ where the stress is experienced but rather is a reference point $\boldsymbol{X}$ functionally related to $\boldsymbol{x}$ through (66.1) [or (16.2)]. The apparatus of Subchapter BI is at our disposal. Thus far we have considered t^{km} as a function of $\boldsymbol{x}$ and t only; for such a tensor field, the definition (App. 20.2) yields $t^{km}{}_{,m}=t^{km}{}_{;m}$. In this section we prefer to consider $\boldsymbol{t}$ as a double tensor field. CAUCHY'S laws (205.2) and (205.11) then assume the forms

$$\varrho\ddot{x}^k = t^{km}{}_{;m}+\varrho f^k, \qquad t^{km}=t^{mk}. \tag{210.1}$$

Since $t^{km}{}_{;m}=t^{km}{}_{;M}X^M_{;m}$, by substituting (17.8) into $(210.1)_1$ we obtain[2]

$$\tilde{\varrho}\ddot{x}^k = t^{km}{}_{;M}\,\partial J/\partial x^m{}_{;M}+\tilde{\varrho}f^k, \tag{210.2}$$

where $\tilde{\varrho}\equiv\varrho J$. When the differentiations are carried out, the first term on the right becomes a sum of three Jacobians[3]:

$$\tilde{\varrho}\ddot{x}^k = \tfrac{1}{2}\,e^{QMN}\,e_{qmn}\,t^{kq}{}_{;Q}\,x^m_{;M}\,x^n_{;N}+\tilde{\varrho}f^k. \tag{210.3}$$

PIOLA[4] introduced the double vector T^{kK}:

$$T^{kK}\equiv J\,t^{km}X^K_{;m}, \qquad t^{km}=j\,T^{kM}x^m{}_{;M}. \tag{210.4}$$

From $(203.4)_2$ and (20.8) we have

$$t^k_{(\boldsymbol{n})}\,da = T^{kK}\,dA_K. \tag{210.5}$$

Hence $\|T^{kK}\|$ gives the stress at $\boldsymbol{x}$ measured per unit area at $\boldsymbol{X}$; the quantity T^{kK} is the component along the X^K co-ordinate of the component of the stress vector along the x^k co-ordinate, multiplied by the ratio of the area at $\boldsymbol{x}$ to the area at $\boldsymbol{X}$. Thus the quantities T^{kK}, sometimes called *pseudo-stresses*, are awkward to interpret. Moreover, in terms of T^{kK} CAUCHY'S second law $(210.1)_2$ assumes the elaborate form[5]

$$T^{kK}x^m{}_{;K}=T^{mK}x^k_{;K}. \tag{210.6}$$

By $(210.4)_1$ and (18.2) we have

$$\left.\begin{aligned} T^{kK}{}_{;K} &= (J X^K_{;m})_{;K}\,t^{km}+J X^K_{;m}\,t^{km}{}_{;K}, \\ &= J\,t^{km}{}_{;m}. \end{aligned}\right\} \tag{210.7}$$

---

[1] [1899, *4*]. VOLTERRA calculated the flux of mechanical energy across vector tubes of this field.

[2] BOUSSINESQ [1872, *1*, § I, Eq. (3)]. The result may be read off from a transformation given by CLEBSCH [1857, *1*, § 2]. Cf. E. and F. COSSERAT [1896, *1*, § 17].

[3] Due essentially to E. and F. COSSERAT [1896, *1*, § 43, Eq. (117)]; cf. TRUESDELL [1952, *21*, § 26]. The result in the hydrostatic case was given by EULER [1770, *1*, § 119].

[4] [1833, *3*, Eq. (42)] [1836, *1*, Eq. (73)] [1848, *2*, ¶¶ 34–38]. Cf. also KIRCHHOFF [1852, *1*, pp. 763–764], C. NEUMANN [1860, *3*, §§ 2, 4–5], E. and F. COSSERAT [1896, *1*, § 15, Eq. (33)]. KIRCHHOFF remarked that the matrix $\|T^{kK}\|$ is not generally symmetric, as indeed is manifest from the present notation. Although POINCARÉ [1892, *11*, § 40] gave a clear explanation of this fact, based upon (210.5) and the observation that three orthogonal elements $d\boldsymbol{a}_\alpha$ at $\boldsymbol{x}$ do not in general arise from three orthogonal elements $d\boldsymbol{A}_\alpha$ at $\boldsymbol{X}$ by (20.8), nevertheless his elaborate, confusing, and unnecessarily restricted manner of stating [1892, *10*, § 35] that for infinitesimal displacement gradients the matrix $\|T^{kK}\|$ is approximately symmetric gave rise to an unnecessary discussion of this "paradox" and an incorrect notion that it is connected with the presence of initial stress (E. and F. COSSERAT [1896, *1*, § 26], COLLINET [1924, *3*], JOUGUET [1924, *6*]).

[5] The corresponding form of the more general law (205.10) is obtained by E. and F. COSSERAT [1909, *5*, § 53].

Substituting into $(210.1)_1$ yields an expression[1] for CAUCHY's first law in terms of T^{kK}:

$$\tilde{\varrho}\,\ddot{x}^k = T^{kK}{}_{;K} + \tilde{\varrho}\,f^k. \tag{210.8}$$

PIOLA[2] introduced also the tensor T^{KM} defined by

$$T^{KM} \equiv x^K{}_{;k}\,T^{kM} = J\,X^K_{;k}\,X^M_{;m}\,t^{km}. \tag{210.9}$$

In terms of it, CAUCHY's laws $(210.1)_{1,2}$ assume the forms[3]

$$\tilde{\varrho}\,\ddot{x}^k = (x^k_{;K}\,T^{KM})_{;M} + \tilde{\varrho}\,f^k, \qquad T^{KM} = T^{MK}. \tag{210.10}$$

While the expression of the first law is simple in terms of T^{kK}, the second law is complicated; for T^{KM}, the reverse holds. There is no simple and exact form[4] with $\boldsymbol{X}$ as independent variable.

211. CAUCHY's laws in space-time; the stress-momentum tensor. The results of Sect. 205 express the balance of momentum and moment of momentum in an inertial frame. Apparent forces and torques on a body were calculated in Sect. 197, but for the local equations in a frame which is not inertial it is easier to use the transformations (143.3) and (143.6). Since neither $\ddot{\boldsymbol{x}}$ nor $\dot{\boldsymbol{x}}$ occurs in (205.10) or (205.14), we see that *the equations for local balance of moment of momentum are valid for all observers.* This follows only because, as emphasized in Sect. 205, linear momentum is assumed already in balance and the quantities $\boldsymbol{m}$ and $\boldsymbol{l}$ represent couples only, not torques combined with forces. In the condition (205.3) we find not $\ddot{\boldsymbol{x}}$ but $\dot{\boldsymbol{x}}$; however, from (143.3) follows $[\dot{\boldsymbol{x}}] = [\dot{\boldsymbol{x}}']$, and hence *the condition for conservation of linear momentum at a surface of discontinuity, when written in terms of the velocity of propagation of the surface relative to the material, is valid for all observers.* Thus among all the dynamical laws, only the differential equation (205.2) for balance of linear momentum is affected by the apparent forces.

To obtain a form of CAUCHY's first law valid in a rotating frame, we have only to replace $\ddot{\boldsymbol{x}}$ by the expression on the right-hand side of (143.6). The result shows that in order to calculate the balance of linear momentum in an arbitrary frame, in addition to the metric tensor we must know the linear acceleration $\ddot{\boldsymbol{b}}$ and the angular velocity $\boldsymbol{\omega}$ of that frame with respect to an inertial frame. From the discussion of Sect. 197 it is clear that such dependence on the observer's motion relative to an inertial frame is unavoidable.

---

[1] PIOLA [1833, *3*, Eqs. (22), (29)] [1836, *1*, Eq. (56)] [1848, *2*, ¶ 36, Eq. (26)].

[2] [1833, *3*, Eq. (45)] [1836, *1*, Eq. (132)]. Cf. also KIRCHHOFF [1852, *1*, p. 767], E. and F. COSSERAT [1896, *1*, § 15, Eq. (31)]. L. BRILLOUIN [1928, *2*, § 11] [1925, *2*, § 7] [1938, *2*, Chap. X, § X] and RIVAUD [1944, *10*] have remarked that (210.9) is a statement that the quantities T^{KM} and t^{km} are the components at $\boldsymbol{X}$ and at $\boldsymbol{x}$ of a tensor density under the deformation (16.2). Since (210.9) is a mere definition, motivated only by the relative simplicity of some of the resulting forms of CAUCHY's laws, this observation does not seem to have mechanical significance.

[3] PIOLA [1833, *3*, Eqs. (33), (35)]. Cf. also SIGNORINI [1930, *5*, § 4] [1943, *6*, Chap. II, § 4], TOLOTTI [1943, *7*].

[4] Other forms are given by SIGNORINI [1930, *5*, § 5] [1930, *6*, § 2], ZELMANOV [1948, *39*, Eq. (4)], and CASTOLDI [1948, *5*, § 7]. The form given by DEUKER [1941, *1*, Eq. (8.7)] was shown to be false by TRUESDELL [1952, *21*, § 26[12]]. The forms given by PLATRIER [1948, *20*, Eq. (13)] and GLEYZAL [1949, *12*, Eq. (5)] also seem dubious. Various approximate forms of (210.10) or (210.8) occur in the literature; e.g., NOVOZHILOV [1948, *18*, §§ 21–22]. Explicit forms for (210.8) in orthogonal curvilinear co-ordinates are worked out by YOSHIMURA [1953, *37*].

However, it is possible to obtain a more elegant formalism[1] for the balance of momentum by using the ideas and methods of Sect. 152. First we recall that the stress tensor $\boldsymbol{t}$ was introduced through use of the balance of linear momentum in an inertial frame; the contravariant components t^{km} occurring in (203.4) therefore are defined in *all inertial co-ordinate systems*. We now introduce the *agreement* that what we shall mean by the stress components t^{km} in *any* frame obtained from an inertial frame by a time-preserving transformation (154.8) is

$$t^{k'm'} \equiv \frac{\partial x^{k'}}{\partial x^p} \frac{\partial x^{m'}}{\partial x^q} t^{pq}. \tag{211.1}$$

That is, in the terms used in Sect. 152, the components t^{km} may be regarded as the non-vanishing components of a space tensor $t^{\Omega\Delta}$ having the canonical form

$$\| t^{\Omega\Delta} \| = \begin{Vmatrix} t^{km} & 0 \\ 0 & 0 \end{Vmatrix} \tag{211.2}$$

in any frame related to an inertial frame by a time-preserving transformation. The agreement (211.1) is consistent with the intuitive notion that assignable forces (including the stress vector[2]) are invariant under the Euclidean group of transformations (152.1). The action of one part of the material upon another is thus assumed to be the same to all observers.

From (211.2) it follows that the quantities $T^{\Omega\Delta}$ defined by

$$T^{\Omega\Delta} \equiv t^{\Omega\Delta} - \varrho\, v^\Omega v^\Delta, \tag{211.3}$$

where ϱ is the mass density, taken as a world scalar, and where $\boldsymbol{v}$ is the world velocity (153.6), also constitute an absolute contravariant world tensor, which we shall call the *stress-momentum tensor*. The name is suggested by the result $(207)_3$, since in a Euclidean frame

$$\left.\begin{aligned} T^{km} &= t^{km} - \varrho\, \dot z^k \dot z^m, \\ T^{4k} &= T^{k4} = -\varrho\, \dot z^k, \\ T^{44} &= -\varrho. \end{aligned}\right\} \tag{211.4}$$

Similarly, we define a space vector $\boldsymbol{F}$ such that in every Euclidean frame $F^\Omega = (f^k, 0)$.

Now consider the world tensor equation

$$\underset{\Gamma}{\nabla_\Phi} T^{\Omega\Phi} + \varrho F^\Omega = 0, \tag{211.5}$$

where $\underset{\Gamma}{\nabla_\Phi}$ denotes the covariant derivative based on the Galilean connection $\boldsymbol{\Gamma}$. (Cf. Sect. 152.) In every Galilean (inertial) frame, (211.5) with $\Omega = k$ is equivalent to CAUCHY's first law; the equation that results from putting $\Omega = 4$ is the equation of continuity $(156.5)_1$. Thus *Eq.* (211.5) *expresses the balance of mass and of linear momentum in world-invariant form.*

From the world-invariant form (164.2) of the continuity equation and from the definition (154.1) of the world acceleration vector, it follows that an alternative form of (211.5) is

$$\underset{\Gamma}{\nabla_\Phi} t^{\Omega\Phi} + \varrho F^\Omega = \varrho\, a^\Omega. \tag{211.6}$$

---

[1] Special cases have often been noted; e.g. LEVI-CIVITA [1928, *5*, pp. 67–81], FINZI [1934, *2*], KILCHEVSKI [1936, *5*, §§ 1–2] [1938, *5*, Part I, § 10], PAILLOUX [1947, *11*], MANARINI [1948, *13*], ARZHANIKH [1952, *1*]. Our approach differs basically from that of CARTAN [1923, *1*, §§ 15–17], who makes the connection Γ^k_{mp} depend upon the dynamical law.

[2] This concept is developed and emphasized as a postulate by NOLL [1957, *11*, § 9].

By substituting the components of the Galilean connection (154.10) into (211.6), we easily obtain an *explicit form* for CAUCHY's first law in an arbitrary rotating and deforming co-ordinate system[1]:

$$t^{km}{}_{,m} + \varrho f^k = \varrho \left\{ \ddot{x}^k - \left(\frac{\partial u^k}{\partial t} + u^k_{,m} u^m \right) - 2 u^k_{,m} (\dot{x}^m - u^m) \right\}, \qquad (211.7)$$

where $u^k \equiv \dfrac{\partial x^k(z^m, t)}{\partial t}$, $\boldsymbol{x} = \boldsymbol{x}(\boldsymbol{z}, t)$ being the transformation giving the general co-ordinates x^k in terms of the co-ordinates z^k in a Galilean reference frame (cf. Sect. 154), where $\ddot{x}^k$ is the acceleration as apparent to an observer in the $\boldsymbol{x}$-system, and where the comma denotes covariant differentiation based upon the time-dependent metric $\boldsymbol{g}(\boldsymbol{x}, t)$ in the rotating, deforming $\boldsymbol{x}$-system.

212. Stress and couple resultants for shells. I. Direct theory. From the dynamical standpoint[2], a shell may be regarded in one of two ways: as a surface $\mathfrak{s}$, or as a region between two surfaces $\mathfrak{s}_1$ and $\mathfrak{s}_2$. In both cases, the shell is subject to normal forces as well as to tangential forces, and therefore its theory is that of a surface or body in three-dimensional space, not a merely intrinsic theory. To follow consistently the second view, in which the shell is regarded as a region in space, we must *derive* from the general theory of three-dimensional stress the nature of the forces and couples acting upon the shell. To follow consistently the first view, we cannot use the momentum principle and the stress principle as stated in Sect. 200, since the forms given there are appropriate only to bodies of non-zero volume. Rather, for the first approach we must *postulate* new forms of the stress principle and the momentum principle. In the older work[3] on shells as models for solid bodies the two approaches are often confused, while theories of shells as two-dimensional models for soap films, water bells, etc., are so specialized as hardly to be typical. We present the two alternatives independently, beginning with the first.

Consider a portion of a surface $\mathfrak{s}$ bounded by a circuit c, and let this surface be in *equilibrium* subject to forces $\boldsymbol{F}$ and couples $\boldsymbol{L}$ per unit area. $\boldsymbol{F}$ and $\boldsymbol{L}$ are three-dimensional vectors which may point in any direction in space. We *postulate*[4] a ***stress principle for shells***: The action of the part of the shell outside c

---

[1] Inability to read Ukrainian prevents us from following in detail the related work of KILCHEVSKI [1936, *5*, §§ 1–4] [1938, *5*, Part I, § 10]; like the corresponding result of McVITTIE [1949, *18*, § 4], it seems to rest upon the unnecessary and in general unjustified assumption that there is a four-dimensional metric. The meteorological literature abounds in special cases obtained by laborious transformation.

[2] Just as the theory of stress in three-dimensional bodies is independent of kinematics, so also for the dynamics of shells we do not need to mention the theory of deformation given in Sect. 64.

[3] While the dynamical equations are implicit in the pioneer work of LOVE [1888, *6*, ¶ 8], he did not disengage them from special elastic hypotheses and approximations, and they were first given in the forms (212.13), (212.14), (212.20), in special co-ordinates, by LAMB [1890, *6*, § 4]. Later, LOVE [1893, *5*, § 339] remarked that LAMB's dynamical equations are valid for finite deformations if referred to the strained shell. Among the numerous repetitions of essentially the same argument as LAMB's we cite only the efficient vectorial derivation of E. REISSNER [1941, *5*, §§ 3–4].

Fundamentally sharper reasoning has been applied in the two special cases when 1. there are no cross forces or moments, the shell being then called a *membrane*, or 2. the surface is plane, the shell being then called a *plate*. For the former, a geometrically intrinsic theory is easy; for the latter, the trivial geometry makes a rigorous treatment easy. The history of membranes and plates, which goes back to the eighteenth century, we make no attempt to trace; cf. TRUESDELL [1960, *4*, §§ 48–49].

[4] Our treatment follows ERICKSEN and TRUESDELL [1958, *1*, §§ 24–26], who shortened the argument of SYNGE and CHIEN [1941, *9*, pp. 104–111]. SYNGE and CHIEN were the first to obtain the equations of equilibrium in the fully general form (212.6).

on the part inside is equipollent to a field of *stress resultant vectors* $\mathbf{S}_{(\mathbf{n})}$ and *couple resultant tensors* $\mathbf{M}_{(\mathbf{n})}$ located on c. The subscript $\mathbf{n}$ refers to the unit outward normal to c; of course, $\mathbf{n}$ is a vector defined intrinsically in s, but $\mathbf{S}_{(\mathbf{n})}$ and $\mathbf{M}_{(\mathbf{n})}$ are three-dimensional fields. In analogy to (200.1) and (200.3), a mathematical expression of this postulate is

$$\left.\begin{aligned} \oint_c \mathbf{S}_{(\mathbf{n})}\, ds + \int_s \mathbf{F}\, da &= 0, \\ \oint_c (\mathbf{M}_{(\mathbf{n})} + \mathbf{p}\times\mathbf{S}_{(\mathbf{n})})\, ds + \int_s (\mathbf{L} + \mathbf{p}\times\mathbf{F})\, da &= 0, \end{aligned}\right\} \tag{212.1}$$

where ds is arc length along c and $\mathbf{p}$ is the three-dimensional position vector. According to the convention of Sect. App. 13 these vector integrals are understood to be written in *rectangular Cartesian co-ordinates*.

The argument in Sect. 203 can be adapted to two dimensions if we replace the tetrahedron by a curvilinear triangle on s. The results, analogous to (203.3), (203.4), and (203.5) are

$$\mathbf{S}_{(\mathbf{n})} = -\mathbf{S}_{(-\mathbf{n})}, \quad \mathbf{M}_{(\mathbf{n})} = -\mathbf{M}_{(-\mathbf{n})}, \tag{212.2}$$

$$S^k_{(\mathbf{n})} = S^{k\delta} n_\delta, \qquad M^k_{(\mathbf{n})} = M^{k\delta} n_\delta. \tag{212.3}$$

In (212.3) the quantities n_δ are covariant components of the unit normal to c in *any curvilinear co-ordinate system* v^1, v^2 on s. Greek minuscule indices run from 1 to 2 in this section and the next one. By hypothesis, $\mathbf{S}_{(\mathbf{n})}$ is an absolute vector and $\mathbf{M}_{(\mathbf{n})}$ is an axial vector, for given $\mathbf{n}$; while to derive (212.3) we employed rectangular Cartesian co-ordinates, the results are tensorial equations within the scheme of double tensors of Sect. App. 15, and hence are valid in all co-ordinate systems. The double tensors $S^{k\delta}$ and $M^{k\delta}$ are the fields of *stress resultants* and *stress couples*. We have

$$\left.\begin{aligned} \dim \mathbf{F} &= [M L^{-1} T^{-2}], & \dim \mathbf{L} &= [M T^{-2}], \\ \dim \mathbf{S} &= [M T^{-2}], & \dim \mathbf{M} &= [M L T^{-2}]. \end{aligned}\right\} \tag{212.4}$$

There are six components $S^{k\delta}$ and six components $M^{k\delta}$; the latter we may sometimes wish to replace by the components of the equivalent skew-symmetric tensor $M^{kp\delta}$. In the classical treatments of the theory of shells the vectors $\mathbf{L}$ and $\mathbf{M}_{(\mathbf{n})}$ are assumed tangent to s; this assumption, which is analogous to that defining the non-polar case in three dimensions, reduces the number of non-vanishing components $M^{k\delta}$ from six to four.

Again we suppose the space co-ordinates rectangular Cartesian, we consider $\mathbf{S}$ and $\mathbf{M}$ as functions of $\mathbf{v}$ only, and we substitute (212.3) into (212.1). By reasoning strictly parallel to that in Sect. 205 we obtain as analogues of (205.2) and (205.10) the differential equations

$$\left.\begin{aligned} S^{k\delta}{}_{,\delta} + F^k &= 0, \\ \overline{M}^{kp\delta}{}_{,\delta} + z^{[k}{}_{,\delta}\, S^{p]\delta} + \overline{L}^{kp} &= 0, \end{aligned}\right\} \tag{212.5}$$

where $\overline{M}^{kp\delta}$ and $\overline{L}^{kp}$ are absolute alternating tensors equivalent to the axial vectors $M^{k\delta}$ and L^k. In these equations the subscript comma indicates covariant differentiation with respect to the surface metric $\mathbf{a}$, except that $z^k_{,\delta} \equiv \partial z^k/\partial v^\delta$, where $\mathbf{z} = \mathbf{z}(\mathbf{v})$ is a rectangular Cartesian equation of the surface s. Now consider the equations

$$\left.\begin{aligned} S^{k\delta}{}_{;\delta} + F^k &= 0, \\ \overline{M}^{kp\delta}{}_{;\delta} + x^{[k}{}_{;\delta}\, S^{p]\delta} + \overline{L}^{kp} &= 0, \end{aligned}\right\} \tag{212.6}$$

where the space co-ordinates and the surface co-ordinates are *arbitrary independently selected general curvilinear systems*, where $\boldsymbol{x}=\boldsymbol{x}(\boldsymbol{v})$ is an equation of $\mathfrak{s}$ referred to these two systems, where $x^k_{;\delta}\equiv \partial x^k/\partial v^\delta$, and where other occurrences of the semi-colon indicate the total covariant derivative (App. 20.2). These equations are in double tensor form; when the space co-ordinates are rectangular Cartesian, they reduce to (212.5), which have been proved valid for such co-ordinate systems; therefore (212.6) are the *general differential equations of equilibrium for shells*. We may continue to regard $\boldsymbol{S}$ and $\boldsymbol{M}$ as functions of $\boldsymbol{v}$ only, or we may consider them to be functions of $\boldsymbol{x}$ also, as we please.

The elegant simplicity of this derivation should not conceal the complexity of the result. When $(212.6)_1$ is written out, it assumes the form

$$\frac{\partial S^{k\delta}}{\partial v^\delta}+\left\{ {k \atop m\,p} \right\} x^p_{;\delta}\, S^{m\delta}+\left\{ {\xi \atop \delta\,\xi} \right\} S^{k\delta}+F^k=0, \tag{212.7}$$

where $\left\{ {k \atop m\,p} \right\}$ and $\left\{ {\xi \atop \delta\,\sigma} \right\}$ are Christoffel symbols based upon the space metric $\boldsymbol{g}$ and the surface metric $\boldsymbol{a}$, respectively, the two metrics being related as usual: $a_{\delta\xi}=g_{km}\, x^k_{;\delta}\, x^m_{;\xi}$. When $(212.6)_2$ is written explicitly in terms of the axial vector $M^{k\delta}$, it assumes the form

$$\frac{\partial M^{k\delta}}{\partial v^\delta}+\left\{ {k \atop m\,p} \right\} x^p_{;\delta}\, M^{m\delta}+\left\{ {\xi \atop \delta\,\xi} \right\} M^{k\delta}+e^k_{mp}\, x^m_{;\delta}\, S^{p\delta}+L^k=0. \tag{212.8}$$

These formulae involve doubly contravariant tensors considered as functions of $\boldsymbol{v}$ only. In terms of physical components, or of components allowed to depend on $\boldsymbol{x}$ as well, they would take on still more complicated forms.

To assume $\boldsymbol{L}$ and $\boldsymbol{M}$ tangent to $\mathfrak{s}$ is equivalent to assuming the existence of surface fields L^δ and $M^{\xi\delta}$ such that

$$L^k=x^k_{;\delta}\, L^\delta,\quad M^{k\delta}=x^k_{;\xi}\, M^{\xi\delta}. \tag{212.9}$$

To obtain equations of motion instead of equations of equilibrium, we may assign momentum to $\mathfrak{s}$ and express its rate of change in terms of a surface density A^k, then replace F^k by F^k-A^k in $(212.6)_1$. However, the vector $\boldsymbol{A}$ does not bear any simple relation to the accelerations of points on $\mathfrak{s}$, and we prefer to postpone determining the effect of inertial force until the general discussion of relations between three-dimensional and surface variables in Sect. 213.

We now resolve all quantities into components normal and tangent to $\mathfrak{s}$:

$$\left.\begin{aligned} F^k&=F^\delta\, x^k_{;\delta}+F N^k, & L^k&=L^\delta\, x^k_{;\delta}+L N^k,\\ S^{k\delta}&=S^{\gamma\delta}\, x^k_{;\gamma}+S^\delta N^k, & M^{k\delta}&=M^{\gamma\delta}\, x^k_{;\gamma}+M^\delta N^k, \end{aligned}\right\} \tag{212.10}$$

where $\boldsymbol{N}$ is the unit normal to $\mathfrak{s}$. The following table connects the components occurring in (212.10) with the terms usually employed in shell theory:

F, L = *normal components of specific applied force and couple.*
F^δ, L^δ = *specific applied force and couple tangent to the shell.*
S^δ = *cross force resultant.*
$S^{\gamma\delta}$ = *membrane stress resultant.*
M^δ = *cross moment resultant.*
$M^{\gamma\delta}$ = *couple resultant.*

According to the usual assumptions (212.9), $L=0$ and $M^\delta=0$. The normal and shear components of $S^{\gamma\delta}$ in an orthogonal co-ordinate system are called *normal* and *shear* membrane stress resultants; the normal and shear components of $M^{\gamma\delta}$ in such a system are called *bending* and *twisting* couple resultants, respectively.

To express the equations of equilibrium in terms of tangential and normal components[1], we use the identities (App. 21.6), (App. 21.2), and (App. 21.4)$_2$ to obtain from (212.10)$_3$ the following resolution:

$$S^{k\delta}{}_{;\delta} = (S^{\gamma\delta}{}_{;\delta} - a^{\sigma\gamma} b_{\sigma\delta} S^{\delta})\, x^k_{;\gamma} + (S^{\delta}_{;\delta} + b_{\gamma\delta} S^{\gamma\delta})\, N^k. \tag{212.11}$$

Substituting this result and (212.10)$_1$ into (212.6)$_1$ yields

$$(S^{\gamma\delta}{}_{;\delta} - a^{\sigma\gamma} b_{\sigma\delta} S^{\delta} + F^{\delta})\, x^k_{;\gamma} + (S^{\delta}{}_{;\delta} + b_{\gamma\delta} S^{\gamma\delta} + F)\, N^k = 0. \tag{212.12}$$

Taking the scalar product of this equation by $\boldsymbol{N}$ yields as the condition for equilibrium of normal forces

$$S^{\delta}{}_{;\delta} + b_{\gamma\delta} S^{\gamma\delta} + F = 0; \tag{212.13}$$

taking the vector product by $\boldsymbol{N}$, the condition for equilibrium of tangential forces:

$$S^{\gamma\delta}{}_{;\delta} - a^{\gamma\sigma} b_{\sigma\delta} S^{\delta} + F^{\gamma} = 0. \tag{212.14}$$

Similar resolution of (212.6)$_2$ yields as the condition for equilibrium of bending moments[2]

$$M^{\delta}_{;\delta} + b_{\gamma\delta} M^{\gamma\delta} + e_{\gamma\delta} S^{\gamma\delta} + L = 0; \tag{212.15}$$

for twisting moments,

$$M^{\gamma\delta}{}_{;\delta} - a^{\gamma\sigma} b_{\sigma\delta} M^{\delta} + a^{\gamma\delta} e_{\delta\sigma} S^{\sigma} + L^{\gamma} = 0. \tag{212.16}$$

In these formulae, it is legitimate and natural to regard all fields as functions of $\boldsymbol{v}$ only and to interpret "$;\delta$" as covariant differentiation based on $\boldsymbol{a}$. Under the usual assumption $L=0$, $M^{\delta}=0$, the total number of independent components of $\boldsymbol{S}$ and $\boldsymbol{M}$ is reduced from 12 to 10, the second term in (212.16) vanishes, while (212.15) reduces to an algebraic equation expressing the difference of shear resultants $S^{[12]}$ as a linear combination of the four couple resultants $M^{\gamma\delta}$.

In works on shell theory it is customary to use in place of $\boldsymbol{M}$ the dual tensor $\boldsymbol{B}$, defined as follows:

$$B^{\alpha\beta} = a^{\alpha\gamma} e_{\gamma\delta} M^{\delta\beta}, \qquad M^{\alpha\beta} = -a^{\alpha\gamma} e_{\gamma\delta} B^{\delta\beta}, \tag{212.17}$$

so that the physical components of these tensors in an orthogonal co-ordinate system satisfy the relations

$$M\langle 11\rangle = -B\langle 21\rangle, \quad M\langle 12\rangle = -B\langle 22\rangle, \quad M\langle 21\rangle = B\langle 11\rangle, \quad M\langle 22\rangle = B\langle 12\rangle. \tag{212.18}$$

Eqs. (212.15) and (212.16) representing the balance of moments may be expressed in terms of $\boldsymbol{B}$:

$$\left.\begin{aligned} M^{\delta}_{;\delta} - e_{\gamma\delta} b^{\gamma}_{\sigma} B^{\delta\sigma} + e_{\gamma\delta} S^{\gamma\delta} + L &= 0,\\ B^{\nu\delta}{}_{;\delta} - a^{\nu\alpha} e_{\alpha\gamma} b^{\gamma}_{\delta} M^{\delta} - S^{\nu} + C^{\nu} &= 0, \end{aligned}\right\} \tag{212.19}$$

where $C^{\nu} = a^{\nu\alpha} e_{\alpha\gamma} L^{\gamma}$; when the classical assumption (212.9) is adopted, these equations reduce to

$$S^{[\gamma\delta]} = -B^{\sigma[\gamma} b^{\delta]}_{\sigma}, \qquad B^{\nu\delta}{}_{;\delta} - S^{\nu} + C^{\nu} = 0, \tag{212.20}$$

the former of which is especially interesting because it shows the tensors $\boldsymbol{S}$ and $\boldsymbol{B}$ to be non-symmetric except in special circumstances.

---

[1] This resolution was effected by SYNGE and CHIEN [1941, *9*, p. 109] by use of the special co-ordinate system we explain in Sect. 213.

[2] The fully general equations (212.15) and (212.16) were given by E. and F. COSSERAT [1909, *5*, §§ 35–37], who derived also forms in material co-ordinates. Cf. also HEUN [1913, *4*, § 20].

The dynamics of shells is not a mere two-dimensional analogue of the dynamics of three-dimensional bodies (cf. Sect. 238). As is evident from the ideas used to formulate it and from the occurrence of the second fundamental form $\boldsymbol{b}$ in every one of its equations, shell theory concerns properties of two-dimensional idealizations of three-dimensional bodies. For an intrinsic analogue to the three-dimensional case, we should have to have a tensor $F^{\xi\delta}$ satisfying

$$F^{\xi\delta}{}_{,\delta} + G^{\xi} = 0, \tag{212.21}$$

where $\boldsymbol{G}$ is assigned. From (212.14), we see that an equation of this form emerges from shell theory if the cross force S^{γ} is known. In particular, when $S^{\gamma} = 0$ we obtain from (212.14) and (212.13) the equations

$$S^{\xi\delta}{}_{,\delta} + F^{\xi} = 0, \quad b_{\xi\delta} S^{\delta\xi} + F = 0. \tag{212.22}$$

When, in addition, the couple resultants and assigned couples are zero, (212.15) reduces to

$$S^{\xi\delta} = S^{\delta\xi}. \tag{212.23}$$

Eqs. (212.22) and (212.23) are said to describe a *state of membrane stress*. The four membrane stress resultants satisfy a system of two linear partial differential equations and two linear algebraic equations with coefficients which are determined by the surface s. There exists an extensive theory of integration of this determined system. See also Sect. 229.

213. Stress and couple resultants for shells. II. Derivation from three-dimensional theory. If we choose to regard a shell as a portion of material between two surfaces s_1 and s_2, the theory of equilibrium and motion of a shell is derivable as a consequence of the three-dimensional theory. According to this view, the stress and couple resultants, instead of being introduced through a postulated two-dimensional stress principle as in Sect. 212, should be defined in terms of the three-dimensional stress tensor $\boldsymbol{t}$. Moreover, a shell need not now be a body by itself: All results we are now going to derive hold equally for a shell which is but a part of a three-dimensional body, though this interpretation is unlikely to be useful.

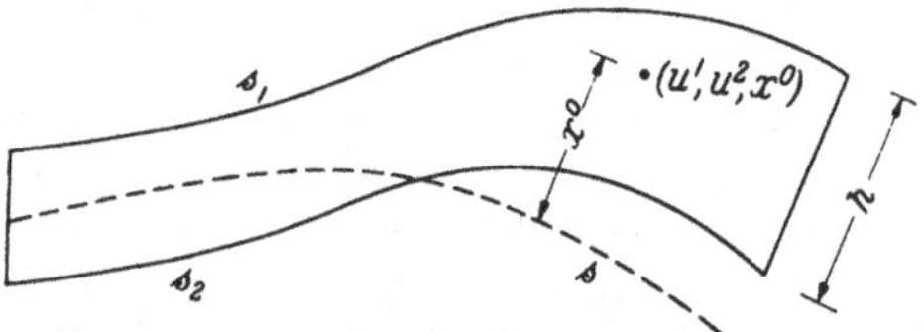

Fig. 37. Shell regarded as a three-dimensional body.

The resultants $S^{k\delta}$ and $M^{k\delta}$ are defined by the condition that their action upon a curve c lying on a reference surface s shall be equipollent to the action of the three-dimensional stress tensor t^{km} upon a finite surface $h(c)$ intersecting s along c. E.g.,

$$\int_c S^{k\delta} n_\delta \, ds = \int_{h(c)} t^{km} n_m^* \, da, \tag{213.1}$$

where $\boldsymbol{n}^*$ is the unit outward normal to $h(c)$, and where our usual convention regarding integrals of vectors in curvilinear co-ordinates is understood (Sect. App. 17). For each curve c, the surface $h(c)$ is to be fixed once and for all, subject to the understanding that to a curve which is a part of c there corresponds a surface which is a part of h. The requirement (213.1) then defines $S^{k\delta}$ uniquely.

In the practice of shell theory, h is always taken as a surface swept out by the normals to s along c, and it is always assumed that the surfaces s_1 and s_2 are given by equations $x^0 = h_1(\boldsymbol{v})$ and $x^0 = h_2(\boldsymbol{v})$, where x^0 is the normal distance from s and where v^1, v^2 are curvilinear co-ordinates upon s (Fig. 37). The quantity $|h_2 - h_1|$ is then the *thickness* of the shell at the point $\boldsymbol{v}$ on s. In many applications the two surfaces are supposed given by the equations $x^0 = \pm h(\boldsymbol{v})$; in this case s is called the *middle surface* of the shell. For the general theory, however, no such restriction is necessary, and the reference surface s need not even lie within the shell.

It is natural to use a co-ordinate system[1] in which one family of co-ordinate surfaces consists in the surfaces $x^0 = \text{const}$, which are parallel to $\mathscr{s}$. If $\boldsymbol{a}(\boldsymbol{v})$ and $\boldsymbol{b}(\boldsymbol{v})$ are the fundamental tensors of $\mathscr{s}$, then the spatial metric $\boldsymbol{g}(\boldsymbol{v}, x^0)$ assumes the form

$$\left.\begin{aligned} g_{00} &= g^{00} = 1, \quad g_{0\alpha} = g^{0\alpha} = 0, \\ g_{\alpha\beta} &= a_{\alpha\beta} + 2x^0 b_{\alpha\beta} + (x^0)^2 b_{\alpha\gamma} b^\gamma_\beta; \end{aligned}\right\} \tag{213.2}$$

i.e., the superficial components of $\boldsymbol{g}$ are of the form $\boldsymbol{g} = \boldsymbol{a}\cdot(\boldsymbol{1} + x^0\boldsymbol{b})^2$. In most of the older researches, the lines of curvature on $\mathscr{s}$ are chosen as co-ordinate curves, so that

$$g_{11} = a_{11}(1 + x^0 K_1)^2 = \frac{1}{g^{11}}, \quad g_{12} = 0, \quad g_{22} \equiv a_{22}(1 + x^0 K_2)^2 = \frac{1}{g^{22}}, \tag{213.3}$$

where K_1 and K_2 are the principal curvatures of $\mathscr{s}$. From (213.2) we have

$$\left.\begin{aligned} \sqrt{g} &= \sqrt{a}\, K^*, \\ \text{where}^2 \quad K^* &\equiv 1 + x^0 \overline{K} + (x^0)^2 K = (1 + x^0 K_1)(1 + x^0 K_2), \\ &= 1 + x^0 b^\alpha_\alpha + (x^0)^2 \frac{b}{a}, \end{aligned}\right\} \tag{213.4}$$

and also

$$\left.\begin{aligned} &\begin{Bmatrix} k \\ 0\,0 \end{Bmatrix} = \begin{Bmatrix} 0 \\ k\,0 \end{Bmatrix} = 0, \\ &\begin{Bmatrix} 0 \\ \alpha\beta \end{Bmatrix} = [\alpha\beta, 0] = -b_{\alpha\gamma}(\delta^\gamma_\beta + x^0 b^\gamma_\beta), \\ &\begin{Bmatrix} \alpha \\ 0\,\beta \end{Bmatrix} = g^{\alpha\lambda} b_{\beta\gamma}(\delta^\gamma_\lambda + x^0 b^\gamma_\lambda), \\ K^* &\begin{Bmatrix} k \\ k\,\alpha \end{Bmatrix} = K^* \begin{Bmatrix} \beta \\ \beta\,\alpha \end{Bmatrix}_{\boldsymbol{a}} + \frac{\partial K^*}{\partial x^\alpha}, \quad K^* \frac{\partial \log\sqrt{g}}{\partial x^0} = \frac{\partial K^*}{\partial x^0}. \end{aligned}\right\} \tag{213.5}$$

With the above choice of surfaces and co-ordinates, the unit normal $\boldsymbol{n}^*$ to $\mathscr{b}$ at $(\boldsymbol{v}, x^0)$ is related to the unit normal $\boldsymbol{n}$ to the generating curve $\mathscr{c}$ of $\mathscr{b}$ at $(\boldsymbol{v}, 0)$ as follows:

$$n^*_0 = 0, \quad n^*_\alpha = n_\gamma(\delta^\gamma_\alpha + x^0 b^\gamma_\alpha), \tag{213.6}$$

while $da = K^*\, ds\, dx^0$. From these facts and (213.1) we obtain the relations

$$\left.\begin{aligned} S^{\alpha\beta} &= \int_{h_1}^{h_2} t^{\beta\gamma}(\delta^\alpha_\gamma + x^0 b^\alpha_\gamma) K^*\, dx^0, \\ S^\alpha &= -\int_{h_1}^{h_2} t^{\alpha 0} K^*\, dx^0, \\ B^{\alpha\beta} &= -\int_{h_1}^{h_2} x^0 t^{\beta\gamma}(\delta^\alpha_\gamma + x^0 b^\alpha_\gamma) K^*\, dx^0. \end{aligned}\right\} \tag{213.7}$$

These definitions make the stress principle for shells, as stated in Sect. 212, a corollary of the three-dimensional stress principle, stated in Sect. 200. Thus it is legitimate to use the same notations as in Sect. 212.

From (213.7) it is plain that the quantities $S^{\alpha\beta}$, S^α, and $B^{\alpha\beta}$ transform as surface tensors of the indicated variance. The physical components of these

---

[1] Special co-ordinate systems of this kind were used in many of the older researches on the theory of shells. General formulations were given by SYNGE and CHIEN [1941, *9*, p. 109], ZERNA [1949, *39*, § 2], GREEN and ZERNA [1950, *9*, § 2], and others.

[2] $K^* > 0$, since the parameters $\boldsymbol{v}$, x^0 are not admissible as co-ordinates if $(1 + x^0 K_1) \leqq 0$ or $(1 + x^0 K_2) \leqq 0$.

tensors are the quantities usually employed in shell theory; e.g., if we make the special choice of co-ordinates leading to (213.3), we have[1]

$$\left.\begin{aligned}
S\langle 11\rangle &= a_{11}\,S^{11} = a_{11}\int\limits_{h_1}^{h_2} t^{11}\,(1+x^0 K_1)^2\,(1+x^0 K_2)\,dx^0\,,\\
&= \int\limits_{h_1}^{h_2} \widehat{11}\,(1+x^0 K_2)\,dx^0\,,\\
S\langle 12\rangle &= \int\limits_{h_1}^{h_2} \widehat{21}\,(1+x^0 K_1)\,dx^0\,,\\
S\langle 21\rangle &= \int\limits_{h_1}^{h_2} \widehat{12}\,(1+x^0 K_2)\,dx^0\,,\\
S\langle 1\rangle &= -\int\limits_{h_1}^{h_2} \widehat{01}\,(1+x^0 K_2)\,dx^0\,,\\
B\langle 11\rangle &= -\int\limits_{h_1}^{h_2} x^0\,\widehat{11}\,(1+x^0 K_2)\,dx^0\,,\\
B\langle 12\rangle &= -\int\limits_{h_1}^{h_2} x^0\,\widehat{21}\,(1+x^0 K_1)\,dx^0\,,
\end{aligned}\right\}\qquad (213.8)$$

etc., where, as usual, $\widehat{km}$ is the matrix of physical components of $\boldsymbol{t}$.

We now find the reflection of CAUCHY's second law, in its narrower form (205.11), upon the stress resultants and couples. Integrating the algebraic identity

$$\left.\begin{aligned}
e_{\alpha\beta}\,t^{\beta\gamma}\,(\delta^\alpha_\gamma + x^0\,b^\alpha_\gamma) &= x^0\,e_{\alpha\beta}\,b^\alpha_\gamma\,t^{\beta\gamma}\,,\\
&= -\,x^0\,e_{\alpha\beta}\,b^\beta_\delta\,t^{\delta\gamma}\,(\delta^\alpha_\gamma + x^0\,b^\alpha_\gamma)
\end{aligned}\right\}\qquad (213.9)$$

across the shell and expressing the result in terms of the definitions (213.7), we obtain a formula identical to $(212.20)_1$. This should not be surprising, since CAUCHY's second law is itself a condition for the balance of moments. The interesting thing about this result is its indication that *the classical assumptions* (212.9) *of shell theory are consistent with the three-dimensional non-polar case*, for it is the absence of three-dimensional applied couples and couple-stress that CAUCHY's second law in the narrower form (205.11) asserts.

We now derive Eqs. (212.13) and (212.14) for a shell by integrating the corresponding components of CAUCHY's first law (205.2), thus showing that the balance of momentum for a shell as a whole is a consequence of the balance of momentum of the three-dimensional body with which we identify it, as indeed is physically plain[2]. Since there are some formal difficulties in using general co-ordinates on the surface $\mathfrak{s}$, we shall use the special metric components (213.3), one advantage of which is the form assumed by the Mainardi-Codazzi identities, viz.,

$$\frac{\partial\sqrt{g_{11}}}{\partial v^2} = \frac{\partial}{\partial v^2}\left[\sqrt{a_{11}}\,(1+x^0 K_1)\right] = (1+x^0 K_2)\,\frac{\partial\sqrt{a_{11}}}{\partial v^2}\,, \qquad (213.10)$$

and a similar formula obtained by interchanging 1 and 2, *1* and *2*.

[1] The curvature factors were mentioned by LAMB [1890, *6*, § 2] and were used in special cases by BASSET [1890, *1*, §§ 5, 18]; the full set (213.8) was given by LOVE [1893, *5*, § 399], and the general equations (213.7) were formulated by GREEN and ZERNA [1950, *9*, § 3]. For discussion of the mechanical significance of the resultants, cf. ZERNA [1949, *39*, §§ 3—4]. Definitions not obviously equivalent to these were given by KILCHEVSKI [1938, *5*, Part II, § 3].

[2] Our derivation follows that of NOVOZHILOV [1943, *4*] and of NOVOZHILOV and FINKELSTEIN [1943, *6*, §§ 1, 4], which is more general than that given independently by TRUESDELL [1945, *6*, § 8]. Derivations in general co-ordinates have been sketched by CHIEN [1948, *7*] and GREEN and ZERNA [1950, *9*, § 3].

As the normal component of CAUCHY's first law (205.2), from (205 A.1) we have

$$\left.\begin{aligned}&\frac{1}{\sqrt{a_{11}}\sqrt{a_{22}}}\left\{\frac{\partial}{\partial v^1}\left[\sqrt{a_{22}}\,(1+x^0K_2)\,\widehat{01}\right]+\frac{\partial}{\partial v^2}\left[\sqrt{a_{11}}\,(1+x^0K_1)\,\widehat{02}\right]\right\}-\\&\qquad-K_1(1+x^0K_2)\,\widehat{11}-K_2(1+x^0K_1)\,\widehat{22}+\\&\qquad+\frac{\partial}{\partial x^0}(K^*\,\widehat{00})+\varrho K^*(f\langle 0\rangle-\ddot{x}\langle 0\rangle)=0.\end{aligned}\right\}\quad(213.11)$$

Integrating this equation from $x^0=h_1$ to $x^0=h_2$ and taking account of (213.8) yields

$$\left.\begin{aligned}&\frac{1}{\sqrt{a_{11}}\sqrt{a_{22}}}\left\{\frac{\partial}{\partial v^1}\left(\sqrt{a_{22}}\,S\langle 1\rangle\right)+\frac{\partial}{\partial v^2}\left(\sqrt{a_{11}}\,S\langle 2\rangle\right)\right\}+\\&\qquad+K_1\,S\langle 11\rangle+K_2\,S\langle 22\rangle+F=0,\end{aligned}\right\}\quad(213.12)$$

providing we put

$$F=-\int_{h_1}^{h_2}\varrho K^*(f\langle 0\rangle-\ddot{x}\langle 0\rangle)\,\partial x^0-K^*\,\widehat{00}\Big|_{h_1}^{h_2}.\quad(213.13)$$

The result (213.12) is identical in form with (212.13), in the co-ordinates employed and in terms of physical components.

As the tangential component of CAUCHY's first law (205.2) corresponding to v^1, from (205 A.1) we have

$$\left.\begin{aligned}&\frac{1}{\sqrt{a_{11}}\sqrt{a_{22}}}\left\{\frac{\partial}{\partial v^1}\left[\sqrt{a_{22}}\,(1+x^0K_2)\,\widehat{11}\right]+\frac{\partial}{\partial v^2}\left[\sqrt{a_{11}}\,(1+x^0K_1)\,\widehat{12}\right]\right\}+\\&+\frac{1}{\sqrt{a_{22}}}\,\frac{\partial\log\sqrt{a_{11}}}{\partial v^2}\,(1+x^0K_2)\,\widehat{21}-\frac{1}{\sqrt{a_{11}}}\,\frac{\partial\log\sqrt{a_{22}}}{\partial v^1}\,(1+x^0K_1)\,\widehat{22}+\\&+K_1(1+x^0K_2)\,\widehat{01}+\frac{\partial}{\partial x^0}(K^*\,\widehat{01})+\varrho K^*(f\langle 1\rangle-\ddot{x}\langle 1\rangle)=0,\end{aligned}\right\}\quad(213.14)$$

where (213.10) has been used. Integrating this equation from $x^0=h_1$ to $x^0=h_2$ and taking account of (213.8) yields

$$\left.\begin{aligned}&\frac{1}{\sqrt{a_{11}}\sqrt{a_{22}}}\left\{\frac{\partial}{\partial v^1}\left(\sqrt{a_{22}}\,S\langle 11\rangle\right)+\frac{\partial}{\partial v^2}\left(\sqrt{a_{11}}\,S\langle 12\rangle\right)\right\}+\\&\quad+\frac{1}{\sqrt{a_{22}}}\,\frac{\partial\log\sqrt{a_{11}}}{\partial v^2}\,S\langle 21\rangle-\frac{1}{\sqrt{a_{11}}}\,\frac{\partial\log\sqrt{a_{22}}}{\partial v^1}\,S\langle 22\rangle-\\&\quad-K_1\,S\langle 1\rangle+F\langle 1\rangle=0,\end{aligned}\right\}\quad(213.15)$$

provided we put

$$F\langle 1\rangle=\int_{h_1}^{h_2}\varrho K^*(f\langle 1\rangle-\ddot{x}\langle 1\rangle)\,dx^0+K^*\,\widehat{01}\Big|_{h_1}^{h_2}.\quad(213.16)$$

The result (213.15) is identical in form with (212.14), in the co-ordinates employed and in terms of physical components, when $\gamma=1$.

To complete the identification of the results obtained by integration of the three-dimensional equations with those of the direct shell theory of Sect. 212, we need to replace (213.13) and (213.16) by invariant formulae valid in general co-ordinates on the surface $\mathfrak{s}$, as follows:

$$\left.\begin{aligned}F&=-\int_{h_1}^{h_2}\varrho K^*(f^0-\ddot{x}^0)\,dx^0-K^*\,t^{00}\Big|_{h_1}^{h_2},\\F^\alpha&=\int_{h_1}^{h_2}\varrho K^*(\delta^\alpha_\gamma+x^0\,b^\alpha_\gamma)\,(f^\gamma-\ddot{x}^\gamma)\,dx^0+K^*(\delta^\alpha_\gamma+x^0\,b^\alpha_\gamma)\,t^{0\gamma}\Big|_{h_1}^{h_2}.\end{aligned}\right\}\quad(213.17)$$

Thus far, except in deriving (213.9) and its direct consequence $(212.20)_1$, we have not assumed the three-dimensional stress tensor to be symmetric. To obtain the equation of moments, however, we rest content with the non-polar case and simply multiply the three-dimensional equation (213.14) by x^0, then integrate across the thickness of the shell[1]. Thus follows

$$\left.\begin{aligned}&\frac{1}{\sqrt{a_{11}}\sqrt{a_{22}}}\left\{\frac{\partial}{\partial v^1}\left(\sqrt{a_{22}}\,B\langle 11\rangle\right)+\frac{\partial}{\partial v^2}\left(\sqrt{a_{11}}\,B\langle 12\rangle\right)\right\}+\\&+\frac{1}{\sqrt{a_{22}}}\frac{\partial\log\sqrt{a_{11}}}{\partial v^2}B\langle 21\rangle-\frac{1}{\sqrt{a_{11}}}\frac{\partial\log\sqrt{a_{22}}}{\partial v^1}B\langle 22\rangle-\\&-S\langle 1\rangle+C\langle 1\rangle=0,\end{aligned}\right\}\qquad(213.18)$$

where, in general co-ordinates upon s, we have put

$$C^\alpha=-\int_{h_1}^{h_2}\varrho K^*x^0(\delta^\alpha_\gamma+x^0 b^\alpha_\gamma)(f^\gamma-\ddot{x}^\gamma)\,dx^0-K^*x^0(\delta^\alpha_\gamma+x^0 b^\alpha_\gamma)\,t^{0\gamma}\Big|_{h_1}^{h_2}.\qquad(213.19)$$

The result (213.18) is identical in form with $(212.20)_2$, in the co-ordinates employed and in terms of physical components, when $\alpha=1$.

In comparing the results of this section with those in the preceding one, we must understand that it is impossible to prove that the quantities entering the two systems are identical, since the difference of basic assumptions and definitions in the two cases makes a statement of isomorphism the best that can be hoped for. What we have shown is that no error can result if we choose to regard the surface fields $S^{\alpha\beta}$, S^γ, and $B^{\delta\varepsilon}$, defined in terms of the three-dimensional stress tensor $\boldsymbol{t}$ by (213.7), as equivalent to the fields denoted by the same symbols in Sect. 212, where they were defined in terms of the double tensors $\boldsymbol{S}$ and $\boldsymbol{M}$, introduced *a priori*.

If we accept this identification, then the results of this section show how the equilibrium theory of the previous section can be generalized to the case of motion. In (213.17) and (213.19) appears not only the assigned forces $\boldsymbol{f}$ but also the acceleration $\ddot{\boldsymbol{x}}$. From (213.17) we see that the effective force of inertia, per unit area and surface mass on s, is not necessarily the acceleration of any particle on s; rather, at a given point P on s it is a certain weighted mean of the accelerations at all points on the normal to s through P. Moreover, inertial forces occur also in (213.19) and hence affect the balance of moment of momentum—an unusual phenomenon in mechanics. Finally, even in the static case the effective surface loads $\boldsymbol{F}$ and $\boldsymbol{C}$ which enter the equations of equilibrium for shells are not merely the vector differences of the loads in the interior, but rather are weighted averages and differences, influenced by the thickness and the curvature of the shell as well as by the forces and couples applied.

214. Stress and couple resultants for rods[2]. If we consider a rod simply as a curve c which may be the seat of dynamical actions, by considerations anal-

[1] In the general case, the definitions $(213.7)_3$ are no longer adequate, since additional resultant couples are produced by the couple stress $\boldsymbol{m}$. Also, instead of simply multiplying the equations of linear momentum by x^0 and then integrating, we should integrate (205.10).

[2] For the plane case, the stress principle for rods and the appropriate special cases of (214.1), (214.2) and (214.7), independent of any hypothesis regarding the constitution of the material, were first given by EULER [1771, *2*, §§ 1–11, 35–40] [1776, *4*, § 17]. For the history of the earlier special theories of rods and flexible lines by PARDIES, JAMES BERNOULLI, and others cf. TRUESDELL [1959, *8*, §§ 2–3, 7–14, 20–21, 25].

ST. VENANT [1843, *3*, ¶ 3] was the first to remark that six equations are needed to express the equilibrium of rods which are twisted as well as bent, but he did not succeed in obtaining them without special simplifying hypotheses. The general equations were given in principle, but very obscurely, by KIRCHHOFF [1859, *2*, § 3], explicitly by CLEBSCH [1862, *2*, § 50]. These

ogous to those at the beginning of Sect. 212 we are led to postulate a ***stress principle for rods***: At each point on a rod, the action of the material to one side upon the material to the other is equipollent to that of a *stress resultant* vector $\boldsymbol{S}$ and a *couple resultant* $\boldsymbol{M}$. These quantities have the physical dimensions $[M\,L\,T^{-2}]$ and $[M\,L^2\,T^{-2}]$, respectively. Properly, we should define them as acting on the opposite sides, $+$ and $-$, of a cut through the rod; then analogously to (212.2) and (212.3), (203.3) we have

$$\boldsymbol{S}_{(+)} = -\boldsymbol{S}_{(-)}, \quad \boldsymbol{M}_{(+)} = -\boldsymbol{M}_{(-)} \tag{214.1}$$

as the first consequences of the principle of equilibrium. Dropping the subscripts $+$ and $-$ but adopting an appropriate convention of sign, as the definitive condition of equilibrium we obtain

$$\left.\begin{aligned} \check{\boldsymbol{S}} + \boldsymbol{F} &= 0, \\ \check{\boldsymbol{M}} + \overline{\boldsymbol{p}\times\boldsymbol{S}}^{\vee} + \boldsymbol{L} + \boldsymbol{p}\times\boldsymbol{F} &= 0, \end{aligned}\right\} \tag{214.2}$$

where "$\vee$" is the dual of the intrinsic derivative defined by (63.6), where $\boldsymbol{p}$ is the position vector with respect to a fixed origin, and where $\boldsymbol{F}$ and $\boldsymbol{L}$ are the applied force and couple, per unit length. By substituting $(214.2)_1$ into $(214.2)_2$ we obtain

$$\check{\boldsymbol{M}} + \boldsymbol{t}\times\boldsymbol{S} + \boldsymbol{L} = 0, \tag{214.3}$$

where $\boldsymbol{t}$ is the unit tangent to the rod $\mathscr{c}$.

A rod such that $\boldsymbol{M}=0$ if $\boldsymbol{L}=0$ is said to be *perfectly flexible*; such rods are often called *strings*. By (214.3), a necessary and sufficient condition for perfect flexibility is that the stress resultant $\boldsymbol{S}$ always be tangent to the rod.

Since the two Eqs. $(214.2)_1$ and (214.3) are in vectorial form, they are valid in an arbitrary curvilinear co-ordinate system. It is customary, however, to refer them to a particular frame defined with respect to the rod $\mathscr{c}$. Retaining full generality at the start, in the scheme of Sect. 61 let us assign any three linearly independent directors and reciprocal directors $\boldsymbol{d}_\mathfrak{a}$ and $\boldsymbol{d}^\mathfrak{a}$ to $\mathscr{c}$. With S^k and F^k as the contravariant components of $\boldsymbol{S}$ and $\boldsymbol{F}$, M_k and L_k as the covariant components of $\boldsymbol{M}$ and $\boldsymbol{L}$, in general curvilinear co-ordinates, we define corresponding anholonomic components:

$$\left.\begin{aligned} S^\mathfrak{a} &\equiv d^\mathfrak{a}_k S^k, & M_\mathfrak{a} &\equiv d^k_\mathfrak{a} M_k, \\ F^\mathfrak{a} &\equiv d^\mathfrak{a}_k F^k, & L_\mathfrak{a} &\equiv d^k_\mathfrak{a} L_k. \end{aligned}\right\} \tag{214.4}$$

By $(214.2)_1$ and the result *dual*[1] to the reciprocal of $(63.10)_3$, we have

$$\left.\begin{aligned} \frac{d S^\mathfrak{a}}{d s} = \check{S}^\mathfrak{a} &= d^\mathfrak{a}_k \check{S}^k + \check{d}^\mathfrak{a}_k S^k, \\ &= -d^\mathfrak{a}_k F^k - d^\mathfrak{a}_m w^m{}_k S^k, \end{aligned}\right\} \tag{214.5}$$

---

and other early treatments are difficult to follow, sometimes imparting the impression that some approximation is made. E.g. LOVE [1906, *5*, § 254] says "the extension of the central line may be disregarded". In fact, as was noted by BASSET [1895, *1*, § 2] (cf. also [1892, *1*, § 4]), Eqs. (214.7), analogous to CAUCHY's laws, are exact when referred to the actual position of the rod; just as in the three-dimensional theory, no question of approximation appears unless we attempt to refer the equations to a configuration assumed by the rod prior to its being loaded by the forces under which it is in equilibrium. The derivation given in the text is that of ERICKSEN and TRUESDELL [1958, *1*, §§ 21–23], patterned on earlier work of E. and F. COSSERAT [1909, *5*, § 10] and HEUN [1913, *4*, § 19].

[1] Note that $\boldsymbol{w}$ is *not* the $\boldsymbol{F}$ of Sect. 63 but rather the dual of $\boldsymbol{W}$; neither is it to be confused with the vorticity vector, which is denoted by the same kernel index.

where $\boldsymbol{w}$ is the wryness of the directors along $\boldsymbol{c}$. Similarly

$$\frac{dM_{\mathfrak{a}}}{ds} = \breve{M}_{\mathfrak{a}} = - d^k_{\mathfrak{a}} (e_{kpq} t^p S^q + L_k) + d^m_{\mathfrak{a}} w^k{}_m M_k. \tag{214.6}$$

Hence the statical equations in anholonomic components are[1]

$$\left.\begin{aligned} &\frac{dS^{\mathfrak{a}}}{ds} + w^{\mathfrak{a}}{}_{\mathfrak{b}} S^{\mathfrak{b}} + F^{\mathfrak{a}} = 0, \\ &\frac{dM_{\mathfrak{a}}}{ds} - w^{\mathfrak{b}}{}_{\mathfrak{a}} M_{\mathfrak{b}} + e_{\mathfrak{abc}} t^{\mathfrak{b}} S^{\mathfrak{c}} + L_{\mathfrak{a}} = 0. \end{aligned}\right\} \tag{214.7}$$

Thus far the director frame has been arbitrary. We now require it to be a unit orthogonal triad such that $\boldsymbol{d}_1 = \boldsymbol{t}$, the unit tangent. By the dual of $(63.9)_3$, the wryness $\boldsymbol{w}$ then satisfies $w_{\mathfrak{ab}} = - w_{\mathfrak{ba}}$ and may be interpreted as an angular velocity[2]. The component S^1, which is the projection of $\boldsymbol{S}$ onto the tangent to $\boldsymbol{c}$, is called the specific *tension* in the rod; the components S^2 and S^3, the specific *shearing forces*; M^1, the specific *twisting couple*; M^2 and M^3, the specific *bending couples*. This special choice of directors, while not simplifying $(214.7)_1$, implies that $t^1 = 1$, $t^2 = t^3 = 0$ and hence reduces the three components of $(214.7)_2$ to the following explicit forms:

$$\left.\begin{aligned} &\frac{dM_1}{ds} - w^{\mathfrak{b}}{}_1 M_{\mathfrak{b}} + L_1 = 0, \\ &\frac{dM_2}{ds} - w^{\mathfrak{b}}{}_2 M_{\mathfrak{b}} - S^3 + L_2 = 0, \\ &\frac{dM_3}{ds} - w^{\mathfrak{b}}{}_3 M_{\mathfrak{b}} + S^2 + L_3 = 0. \end{aligned}\right\} \tag{214.8}$$

In analogy to the reasoning in Sect. 213, it should be possible to derive the equations for rods by integrating the three-dimensional equations or the equations for shells; using power series expansions, GREEN[3] has given a derivation of the former type.

If $\boldsymbol{F} = 0$, from $(214.2)_1$ it follows that $\boldsymbol{S} = \text{const}$; (214.3) then becomes a statement that $\boldsymbol{M}$ is a prescribed function of s. Since $\boldsymbol{M}$ is an axial vector, such a statement is a formal analogue of (196.3), the general condition for balance of moment of momentum, provided s and t are made to correspond. Such a correspondence may be carried further by taking the director frame of the rod as an orthogonal unit triad, so that the wryness $\boldsymbol{w}$ becomes the analogue of the angular velocity $\boldsymbol{\omega}$. This observation forms the basis of KIRCHHOFF's

---

[1] Since by definition

$$e_{\mathfrak{abc}} = + \sqrt{g}\, \varepsilon_{kpq} d^k_{\mathfrak{a}} d^p_{\mathfrak{b}} d^q_{\mathfrak{c}},$$

we have

$$e_{\mathfrak{abc}} = + \sqrt{g}\, \varepsilon_{\mathfrak{abc}} \det d^k_{\mathfrak{e}}.$$

Now

$$g (\det d^k_{\mathfrak{e}})^2 = \det g_{km} d^k_{\mathfrak{e}} d^m_{\mathfrak{f}} = \det g_{\mathfrak{ef}};$$

hence

$$e_{\mathfrak{abc}} = \pm \sqrt{\det g_{\mathfrak{ef}}}\, \varepsilon_{\mathfrak{abc}},$$

where the sign is to be selected so as to agree with that of $\det d^k_{\mathfrak{a}}$. The quantity $\det g_{\mathfrak{ef}}$ is evaluated by the dual of (61.4). In particular, for a right-handed unit orthogonal triad we have

$$e_{\mathfrak{abc}} = \varepsilon_{\mathfrak{abc}}.$$

[2] The classical notation for the component w_{23} is τ or $-\tau$; the other two independent components are written as $\pm\varkappa$ and $\pm\varkappa'$. It is important to remember that in the exact theory all these quantities refer to the *loaded* rod. Cf. the footnotes on p. 565.

[3] [1959, *7*]. Various earlier authors, e.g. KIRCHHOFF [1876, *2*, Vorl. 28, § 5] and LOVE [1906, *5*, § 254], had given definitions of the stress resultants and couple resultants in terms of three-dimensional stresses, but their subsequent arguments rest on unnecessary and unrigorous limit processes rather than exact integration such as that given for shells in Sect. 213. The method of power series expansion was initiated by HAY [1942, *8*, § 6].

celebrated analogy between the motion of a rigid body and the deflection of an elastic rod[1]. The result as usually presented takes on an appearance of greater complexity because (214.8) rather than (214.3) is used as the starting point.

215. Partial stresses in a heterogeneous medium[2]. To discuss the transfer of momentum in a mixture, we employ the formalism of Sects. 158 to 159. Each constituent $\mathfrak{A}$ is regarded as being subject to *partial stress* $\boldsymbol{t}_{\mathfrak{A}}$ whose action upon any imagined closed diaphragm is equipollent to the action of all constituents exterior to the diaphragm upon the material of the constituent $\mathfrak{A}$ within the diaphragm. The total stress $\boldsymbol{t}$ is the sum of these partial stresses plus the apparent stresses arising from diffusion[3]:

$$\boldsymbol{t} \equiv \sum_{\mathfrak{A}=1}^{\mathfrak{K}} (\boldsymbol{t}_{\mathfrak{A}} - \varrho_{\mathfrak{A}} \boldsymbol{u}_{\mathfrak{A}} \boldsymbol{u}_{\mathfrak{A}}). \tag{215.1}$$

The momentum of the constituent $\mathfrak{A}$ need not be balanced by itself, as momentum may be transferred from one constituent to another. We define the *supply of momentum* $\hat{\boldsymbol{p}}_{\mathfrak{A}}$ of the constituent $\mathfrak{A}$ by

$$\varrho\, \hat{p}^k_{\mathfrak{A}} \equiv \varrho_{\mathfrak{A}} (\ddot{x}^k_{\mathfrak{A}} - f^k_{\mathfrak{A}}) - t^{km}_{\mathfrak{A},\,m}, \tag{215.2}$$

where $\boldsymbol{f}_{\mathfrak{A}}$ is the applied force per unit mass acting upon the constituent $\mathfrak{A}$. Thus $\hat{\boldsymbol{p}}_{\mathfrak{A}} = 0$ is a necessary and sufficient condition that the linear momentum of the constituent $\mathfrak{A}$ be in balance by itself. Summing (215.2) over all constituents, by (159.6) and (215.1) we obtain the identity

$$\varrho \sum_{\mathfrak{A}=1}^{\mathfrak{K}} (\hat{p}^k_{\mathfrak{A}} + \hat{c}_{\mathfrak{A}} u^k_{\mathfrak{A}}) = \varrho (\ddot{x}^k - f^k) - t^{km}{}_{,m}, \tag{215.3}$$

---

[1] [1859, *2*, § 3]. In the classical theory of elastic rods, $\boldsymbol{M}$ is a linear function of $\boldsymbol{w}$, and in the theory of rigid motions, $\mathfrak{H}$ is a linear function of $\boldsymbol{\omega}$; thus the analogy can be extended by setting the tensor of elastic moduli into correspondence with the tensor of inertia. A generalization is given by E. and F. COSSERAT [1907, *2*].

[2] Equations of the type (215.2) with special forms for $\hat{\boldsymbol{p}}_{\mathfrak{A}}$ arise in MAXWELL's kinetic theory of gas mixtures; for detailed references, see the next two footnotes. The first attempts at a continuum theory were given by DUHEM [1893, *1*, Chap. II], REYNOLDS [1903, *15*, § 38], and JAUMANN [1911, *7*, § V]. In another work [1893, *2*, Part I, Chap. VI], DUHEM derived from a variational principle the special case of (215.2) appropriate to mixtures of perfect fluids when $\hat{c}_{\mathfrak{A}} = 0$; the quantities $\hat{\boldsymbol{p}}_{\mathfrak{A}}$, which appear as multipliers, satisfy the appropriate special case of (215.5); hence DUHEM derived (205.2), again in the special case. Eq. (215.2) with $\hat{\boldsymbol{p}}_{\mathfrak{A}} = 0$ was given by LEAF [1946, *7*, Eq. (11)], but he did not derive from it any conclusion regarding the mean motion. We do not follow the argument by which PRIGOGINE [1947, *12*, Chap. VIII, § 2] claims to infer CAUCHY's first law for a mixture. The definition (215.1) and the consequent (215.7) are in agreement with classical results from the kinetic theory of mixtures of monatomic gases but appeared only much later in the phenomenological theory; while NACHBAR, WILLIAMS and PENNER [1957, *10*, § IV] attribute them to unpublished lectures of v. KÁRMÁN (1950–1951), they were published by PRIGOGINE and MAZUR [1951, *21*, § 3] [1951, *17*, § 2], who introduced the definition (215.2), assumed CAUCHY's first law (205.2), and derived the condition (215.5). Our treatment follows the rediscovery and completion of their results by TRUESDELL [1957, *16*, § 6]. Cf. also the special theories of diffusion cited in Sect. 295.

[3] The diffusion velocities $\boldsymbol{u}_{\mathfrak{A}}$ are defined by (158.7). Motivation for regarding $-\varrho_{\mathfrak{A}} \boldsymbol{u}_{\mathfrak{A}} \boldsymbol{u}_{\mathfrak{A}}$ as a stress was given in Sect. 207. Cf. ECKART [1940, *8*, p. 271]: "There is now the possibility that $\boldsymbol{t}$ may also depend on the $\boldsymbol{u}_{\mathfrak{A}}$, but this complication may be ignored." In fact, (215.1) is fully consistent with the kinetic theory of monatomic gas mixtures as created by MAXWELL [1867, *2*]. There, however, it is customary to define all quantities in terms of molecular motion relative to $\dot{\boldsymbol{x}}$, not $\dot{\boldsymbol{x}}_{\mathfrak{A}}$; thus the partial pressure tensor $p_{\mathfrak{A}}$ of CHAPMAN and COWLING [1939, *6*, Eq. (2.5.11)] and of HIRSCHFELDER, CURTISS and BIRD [1954, *9*, Eq. (7.2.22)] is to be identified as a special case of our $-\boldsymbol{t}_{\mathfrak{A}} + \varrho_{\mathfrak{A}} u_{\mathfrak{A}} u_{\mathfrak{A}}$, not of our $-\boldsymbol{t}_{\mathfrak{A}}$.

where

$$\boldsymbol{f} = \sum_{\mathfrak{A}=1}^{\mathfrak{K}} c_{\mathfrak{A}} \boldsymbol{f}_{\mathfrak{A}}. \tag{215.4}$$

Therefore *a necessary and sufficient condition that Cauchy's first law* (205.2) *shall hold for the mixture is*[1]

$$\sum_{\mathfrak{A}=1}^{\mathfrak{K}} (\hat{p}^k_{\mathfrak{A}} + \hat{c}_{\mathfrak{A}} u^k_{\mathfrak{A}}) = 0. \tag{215.5}$$

This condition asserts that momentum supplied by unbalanced inertial forces of the several constituents plus momentum supplied through the creation of constituent diffusing masses shall add up to zero; in other words, the total momentum of the mixture is conserved.

If we wish to separate the effect of diffusion, we define the *interior part* $\boldsymbol{t}_{\mathrm{I}}$ of the stress by

$$\boldsymbol{t}_{\mathrm{I}} \equiv \sum_{\mathfrak{A}=1}^{\mathfrak{K}} \boldsymbol{t}_{\mathfrak{A}}. \tag{215.6}$$

By (215.1), CAUCHY's first law (205.2) then assumes the form

$$\varrho \ddot{x}^k = t^{km}_{\mathrm{I},m} - \sum_{\mathfrak{A}=1}^{\mathfrak{K}} (\varrho_{\mathfrak{A}} u^k_{\mathfrak{A}} u^m_{\mathfrak{A}})_{,m} + \varrho f^k. \tag{215.7}$$

In the case when the several constituents are not subject to applied couples, $\boldsymbol{t}_{\mathfrak{A}}$ is symmetric; by (215.1) it follows that CAUCHY's second law in its usual form (205.11) is necessary and sufficient for balance of moment of momentum when linear momentum is already in balance. Alternatively, $t^{km}_{\mathrm{I}} = t^{mk}_{\mathrm{I}}$. For non-symmetric partial stresses, it would be possible to construct a theory of balance of moment of momentum in steps parallel to those above for linear momentum, but we refrain from doing so. An interesting problem would be the characterization of cases when $\boldsymbol{t}$ is symmetric even though the $\boldsymbol{t}_{\mathfrak{A}}$ are not.

III. Applications of Cauchy's laws.

216. A general theorem of mean value. If Ψ is any continuously differentiable function, from CAUCHY's first law (205.2) alone we have

$$\left.\begin{aligned}(\Psi t_{km})_{,m} &= t_{km} \Psi_{,m} + \Psi t_{km,m}, \\ &= t_{km} \Psi_{,m} + \varrho \Psi (\ddot{z}_k - f_k),\end{aligned}\right\} \tag{216.1}$$

where we are using rectangular Cartesian co-ordinates. Integrating this identity over a volume $\mathfrak{v}$, by GREEN's transformation we obtain

$$\oint_{\mathfrak{s}} \Psi t_{km} da_m = \oint_{\mathfrak{s}} \Psi t_{(\boldsymbol{n})} da = \int_{\mathfrak{v}} [t_{km} \Psi_{,m} + \varrho \Psi (\ddot{z}_k - f_k)] dv. \tag{216.2}$$

This simple ***theorem of stress means***[2], in principle the same as KELVIN's identity (119.1), has many important consequences. To derive it we have assumed the

[1] Under the special assumptions used in the kinetic theory of monatomic gas mixtures, including $\hat{c}_{\mathfrak{A}} = 0$, this equation is implied by the work of MAXWELL [1867, *2*] but does not appear explicitly; in the notation of CHAPMAN and COWLING [1939, *6*, § 8.1], it assumes the form $\sum_{\mathfrak{A}=1}^{\mathfrak{K}} n_{\mathfrak{A}} \Delta m_{\mathfrak{A}} \bar{C}_{\mathfrak{A}} = 0$.

[2] SIGNORINI [1933, *10*, § 1]. MAXWELL [1870, *4*, pp. 194–195] had obtained formulae for the mean values of $\mathrm{I}_{\boldsymbol{t}}$ and $\mathrm{II}_{\boldsymbol{t}}$ in plane stress.

stress itself to be continuously differentiable, but, as follows from (205.5), the presence of surfaces of discontinuity other than shock fronts does not invalidate it. Concentrated loads are included, according to the convention in Sect. 201.

The use of (216.2) is two-fold: (1) to derive general theorems; (2) in the static case, when it becomes

$$\int_v t_{km} \Psi_{,m} dv = \oint_s \Psi t_{(n)k} da + \int_v \varrho \Psi f_k dv, \tag{216.3}$$

it serves to give the mean value of $t_{km} \Psi_{,m}$ over a body in terms of assigned functions, namely, the mean of $\Psi \boldsymbol{f}$ over the body and the mean of $\Psi \boldsymbol{t}_{(n)}$ over the boundary.

Putting $\Psi = 1$ in (216.2) yields (200.1); since at bottom only (200.1) has been used to derive it, (216.2) is thus *equivalent* to the balance of linear momentum for a continuous medium.

Put $\Psi = z_m$. Then (216.2) yields a formula of FINGER that will be discussed in Sect. 219, viz.:

$$\oint_s z_m t_{kq} da_q = \int_v [t_{km} + \varrho z_m (\ddot{z}_k - f_k)] dv. \tag{216.4}$$

Taking the skew-symmetric part yields[1]

$$\oint_s z_{[m} t_{k]q} da_q = \int_v [t_{[km]} + \varrho z_{[m} (\ddot{z}_{k]} - f_{k]})] dv, \tag{216.5}$$

an expression for the torque exerted by the stress vectors acting upon s. This furnishes a new proof that when linear momentum is balanced, $\boldsymbol{t}$ is symmetric if and only if all torques are the moments of forces [cf. (205.10)].

The next sections present some applications of the theorem of stress means[2].

217. Theorem of power expended[3]. In (216.2), put $\Psi = \dot{z}_k$ and write the result in general co-ordinates. By $(170.1)_3$, there emerges an expression[4] for the rate of change of kinetic energy $\mathfrak{K}$ in the material volume $\mathscr{V}$:

$$\dot{\mathfrak{K}} = \int_{\mathscr{V}} f_k \dot{x}^k d\mathfrak{M} + \oint_{\mathscr{S}} t_{(n)k} \dot{x}^k da - \int_{\mathscr{V}} t^{km} \dot{x}_{k,m} dv. \tag{217.1}$$

Thus the change of kinetic energy may be thought of as threefold:

1. Increase at the rate work is done by the assigned force $\boldsymbol{f}$.
2. Increase at the rate work is done by the stress vector on the bounding surface.
3. Decrease at the rate $t^{km} \dot{x}_{k,m}$ per unit volume in the interior.

By the transport theorem (81.3), an alternative form[5] is

$$\frac{\partial \mathfrak{K}}{\partial t} = \oint_s \left(t_k{}^m - \frac{1}{2} \varrho \dot{x}^2 \delta_k^m \right) \dot{x}^k da_m + \int_v (\varrho f_k \dot{x}^k - t^{km} \dot{x}_{k,m}) dv. \tag{217.2}$$

Both of these results also follow easily from the identities

$$\left.\begin{aligned} \varrho \overline{\tfrac{1}{2} \dot{x}^2}^{\,\cdot} &= \varrho \ddot{x}^k \dot{x}_k = (t^{km} \dot{x}_k)_{,m} - t^{km} \dot{x}_{k,m} + \varrho f^k \dot{x}_k, \\ \frac{\partial}{\partial t}\left(\frac{1}{2} \varrho \dot{x}^2\right) &= \left(t^{km} \dot{x}_k - \frac{1}{2} \varrho \dot{x}^2 \dot{x}^m\right)_{,m} - t^{km} \dot{x}_{k,m} + \varrho f_k \dot{x}^k. \end{aligned}\right\} \tag{217.3}$$

[1] RAYLEIGH [1900, *9*].
[2] The approach is that of CISOTTI [1940, *6*] [1942, *3*].
[3] A general discussion of the definitions of work and power is given by BEGHIN [1951, *1*].
[4] STOKES [1851, *2*, § 49], UMOV [1874, *6*, § 3]. Cf. also v. MISES [1909, *8*, § 10].
[5] UMOV [1874, *6*, § 5].

There have been many attempts[1] to identify one or another of the flux terms as "the" flux of mechanical energy, but such attempts are futile for the general reason given in Sect. 157. Our interpretation above does not share this defect because it refers simultaneously to all the terms in the equation.

The important scalar[2]

$$P \equiv t^{km}\dot{x}_{k,m}, \tag{217.4}$$

which figures in all these expressions, is called the *stress power*; it is a measure of the dissipation of kinetic energy incident upon stretching and spin of the medium. By (90.1) we have

$$P = t^{(km)} d_{km} + t^{[km]} w_{km}: \tag{217.5}$$

The symmetric part of the stress does work in stretching the body, while the skew-symmetric part does work in spinning the elements of the body[3]. In view of the generalized form (205.10) of CAUCHY's second law, we have

$$P = t^{(kp)} d_{kp} + (m^{kpq}{}_{,q} + \varrho\, l^{kp})\, w_{kp}. \tag{217.6}$$

In particular, when $\boldsymbol{m}=0$ the work of spinning depends solely on the assigned couples $\boldsymbol{l}$.

If we decompose the stress according to (204.6), where p is any scalar, we get

$$P = -p\,\mathrm{I}_{\boldsymbol{d}} + v^{(km)} d_{km} + t^{[km]} w_{km}. \tag{217.7}$$

In terms of the deviators ${}_0\boldsymbol{d}$ and ${}_0\boldsymbol{t}$, we have

$$\left.\begin{aligned} P &= -\tilde{p}\,\mathrm{I}_{\boldsymbol{d}} + t^{(km)}\,{}_0d_{km} + t^{[km]} w_{km},\\ &= -\tilde{p}\,\mathrm{I}_{\boldsymbol{d}} + {}_0t^{(km)} d_{km} + {}_0t^{[km]} w_{km},\\ &= -\tilde{p}\,\mathrm{I}_{\boldsymbol{d}} + {}_0t^{(km)}\,{}_0d_{km} + {}_0t^{[km]} w_{km}, \end{aligned}\right\} \tag{217.8}$$

where $\tilde{p}$ is the mean pressure (204.7).

Since the three terms in this expression vanish, respectively, when there is no expansion, no isochoric stretching, and no spin, it is often stated that this decomposition separates the work done into a part arising from change of volume, a part from change of shape, and a part from rotation. While this statement seems to do no harm, it is difficult to attach meaning to it, since such a decomposition is not unique. Indeed, let $F(\boldsymbol{t}, \boldsymbol{d})$ be any scalar function of $\boldsymbol{t}$ and $\boldsymbol{d}$ which vanishes *both* when $I_{\boldsymbol{d}}=0$ and when ${}_0\boldsymbol{d}=0$; e.g., $\mathrm{I}_{\boldsymbol{d}}\,\mathrm{II}_{{}_0\boldsymbol{d}}$. From (217.8) we have

$$P = [-\tilde{p}\,\mathrm{I}_{\boldsymbol{d}} + F(\boldsymbol{t},\boldsymbol{d})] + [t^{(km)}\,{}_0d_{km} - F(\boldsymbol{t},\boldsymbol{d})] + t^{[km]} w_{km}, \tag{217.9}$$

and the sentence following (217.8) applies equally to this decomposition. Neither can we impose the stronger requirement that the respective rates vanish *only* in the special classes of motion mentioned, since this is not so even for (217.8): e.g. the second term vanishes when $\boldsymbol{t}$ is spherical, no matter what the motion. In order to secure uniqueness for the decomposition (217.8) we may add the requirement of linearity in the velocity gradient, but there appears to be no physical motive for such a requirement. Alternatively, we may find significance for (217.8) indirectly in the maximal decomposition (App. 43.1).

The stress power P is easily expressed in terms of a reference state. From (217.4), $(210.4)_2$, and (20.9) follows[4]

$$\left.\begin{aligned} P\,dv &= t^{km}\dot{x}_{k,m}\,dv = j\,T^{kM} x^m_{;M}\dot{x}_{k,m}\,dv,\\ &= T^{kM}\dot{x}_{k;M}\,dV = T^{KM} x^k_{;K} x^m_{;M}\dot{x}_{k,m}\,dV. \end{aligned}\right\} \tag{217.10}$$

---

[1] E.g. WIEN [1892, *14*], VOLTERRA [1899, *3*], MATTIOLI [1914, *7*, § 5], HEINRICH [1952, *10*] [1955, *12*].

[2] STOKES [1851, *2*, § 49].

[3] VOIGT [1887, *5*, Chap. I, § 3], COMBEBIAC [1902, *2*].

[4] KIRCHHOFF [1852, *1*, p. 771].

When the variables $\boldsymbol{X}$ are material, by $(95.7)_2$ and (95.16) we may put (217.10) into the form[1]

$$P\,dv = (T^{(\alpha\beta)}\,\dot{E}_{\alpha\beta} + T^{[\alpha\beta]}\,Q_{\alpha\beta})\,dV_0, \tag{217.11}$$

reducing in the non-polar case to[2]

$$P\,dv = t^{km}\,d_{km}\,dv = T^{\alpha\beta}\,\dot{E}_{\alpha\beta}\,dV_0. \tag{217.12}$$

218. Potential energy. The theorem of power expended takes on a simpler form in the special case when the field of extrinsic and mutual forces is steady and lamellar[3]:

$$f_k = -U_{,k}, \qquad U = U(\boldsymbol{x}). \tag{218.1}$$

If we set

$$\mathfrak{U} \equiv \int_{\mathscr{V}} U\,d\mathfrak{M}, \tag{218.2}$$

then for a material volume $\mathscr{V}$ we have by $(156.8)_2$ and (72.4)

$$\dot{\mathfrak{U}} = \int_{\mathscr{V}} \dot{U}\,d\mathfrak{M} = \int_{\mathscr{V}} U_{,k}\,\dot{x}^k\,d\mathfrak{M} = -\int_{\mathscr{V}} f_k\,\dot{x}^k\,d\mathfrak{M}. \tag{218.3}$$

Thus the rate at which the forces $\boldsymbol{f}$ do work is the time derivative of a function $-\mathfrak{U}$ of the set of particles comprising the body. In this case $\mathfrak{U}$ is called the *potential energy*; force fields satisfying (218.1) are sometimes called *conservative*[4]. Substituting (218.3) into (217.1) yields

$$\dot{\mathfrak{K}} + \dot{\mathfrak{U}} = \oint_{\mathscr{S}} t_{(\boldsymbol{n})k}\,\dot{x}^k\,da - \int_{\mathscr{V}} t^{km}\,\dot{x}_{k,m}\,dv. \tag{218.4}$$

The rate of change of the sum of the kinetic and potential energies is thus expressible entirely in terms of the work done by the stresses.

The assumption (218.1) has produced a result having the form of an equation of balance (157.1). Further terms may be expressed as time derivatives of set functions by adding further assumptions. We divide the total stress $\boldsymbol{t}$ into two parts,

$$\boldsymbol{t} = {}_{\mathrm{E}}\boldsymbol{t} + {}_{\mathrm{D}}\boldsymbol{t}, \tag{218.5}$$

such that the former, which we call the *elastic* stress, is derivable from a *strain energy* $\sigma(x^k_{,\alpha})$ according to the formula

$${}_{\mathrm{E}}T_k^{\ \alpha} = \varrho_0\,\frac{\partial\sigma}{\partial x^k_{,\alpha}}, \tag{218.6}$$

where the X^α are material co-ordinates and where ${}_{\mathrm{E}}T_k^{\ \alpha}$ is PIOLA's double vector $(210.4)_1$. The remaining stress, ${}_{\mathrm{D}}\boldsymbol{t}$, will be called the *dissipative* stress. For the stress power (217.4), by $(217.10)_3$, $(76.2)_7$, and $(156.2)_2$ we obtain

$$\left.\begin{aligned} P = ({}_{\mathrm{E}}t^{km} + {}_{\mathrm{D}}t^{km})\,\dot{x}_{k,m} &= \frac{\varrho}{\varrho_0}\,{}_{\mathrm{E}}T_k^{\ \alpha}\,\dot{\overline{x^k_{,\alpha}}} + {}_{\mathrm{D}}t^{km}\,\dot{x}_{k,m},\\ &= \varrho\,\dot{\sigma} + {}_{\mathrm{D}}t^{km}\,\dot{x}_{k,m}. \end{aligned}\right\} \tag{218.7}$$

The *total strain energy* $\mathfrak{S}$ is defined by

$$\mathfrak{S} \equiv \int_{\mathscr{V}} \varrho\,\sigma\,dv. \tag{218.8}$$

---

[1] E. and F. COSSERAT [1909, *5*, §§ 51–52, 54–55].
[2] E. and F. COSSERAT [1896, *1*, § 15, Eq. (30)].
[3] In the homochoric case, (218.1) is equivalent to (207.1), with $V = \varrho U$.
[4] The concepts used here arose in the early development of mass-point mechanics. In continuum mechanics they are rather obvious and unimportant.

From (218.7) and (218.4) follows

$$\dot{\mathfrak{K}} + \dot{\mathfrak{U}} + \dot{\mathfrak{S}} = \oint_{\mathscr{S}} t_{(n)k}\, \dot{x}^k \, da - \int_{\mathscr{V}} {}_{\mathrm{D}}t^{km} \dot{x}_{k,m} \, dv. \tag{218.9}$$

While the decomposition (218.5) is always trivially possible in infinitely many ways, the dissipative stress ${}_{\mathrm{D}}\boldsymbol{t}$ will generally possess no specially simple property. To obtain a conservation law as at the end of Sect. 157, we formulate assumptions such that the right-hand side of (218.9) vanishes, so deriving the ***first theorem of conservation of mechanical energy***[1]: *Assume that*

1. *The total stress is the sum of an elastic stress given by* (218.6) *and a dissipative stress;*

2. *The dissipative stress does no total work; and*

3. *The total stress vector on the boundary is normal to the velocity there; then*

$$\mathfrak{K} + \mathfrak{U} + \mathfrak{S} = \text{const.} \tag{218.10}$$

In the still more special case when

$$\sigma(x^k{}_{,\alpha}) = \sigma(J) = \sigma(v/v_0), \tag{218.11}$$

from (218.6) and (17.8) we have[2]

$$\left.\begin{aligned} {}_{\mathrm{E}}T_m{}^{\alpha} &= \varrho_0 \frac{\partial\sigma}{\partial(v/v_0)} \frac{\partial J}{\partial x^m{}_{,\alpha}} = \frac{\partial\sigma}{\partial v} \frac{\partial J}{\partial x^m{}_{,\alpha}}, \\ &= \frac{\partial\sigma}{\partial v} J X^{\alpha}{}_{,m}; \end{aligned}\right\} \tag{218.12}$$

hence

$$ {}_{\mathrm{E}}t^k{}_m = -\pi\,\delta^k_m, \quad \text{where} \quad \pi = -\frac{\partial\sigma}{\partial v} = \pi(v): \tag{218.13}$$

The elastic stress is hydrostatic. Conversely, if $\boldsymbol{t} = -\pi\mathbf{1}$ and $\pi = \pi(v)$, we may infer (218.11). In this case, since $\sigma = -\int \pi\, dv$, for the total strain energy (218.8) we may write

$$\mathfrak{S} = -\int_{\mathscr{V}} \left(\int \pi\, dv\right) dv. \tag{218.14}$$

As we remarked in Sect. 217, in an isochoric motion a hydrostatic pressure does no work. Combining this observation with the result above, we get a ***second theorem of conservation of mechanical energy***[3]: *In the first theorem, replace 1 by*

$1'_{\mathrm{a}}$. *The total stress is the sum of a hydrostatic pressure and a dissipative stress; and*

$1'_{\mathrm{b}}$. *Either the motion is isochoric or the pressure is a function of the density only; retain assumptions 2 and 3; then* (218.10) *follows, with the total strain energy being* 0 *in the isochoric alternative while given by* (218.14) *otherwise.*

For a stationary region on whose boundary $\boldsymbol{t}_{(n)}$ is normal, the third condition for both theorems is satisfied in virtue of (69.2). Also, it is easy to formulate assumptions under which the surface integral in (218.4) or (218.9) may be expressed as the time derivative of another surface integral. For example, if we apply the stress boundary condition (203.6), assume that the surface load $\boldsymbol{s}$

---

[1] Suggested by the somewhat vague analysis of GREEN [1839, *1*, pp. 248–250] and KELVIN [1855, *4*, § 187], given in essentially the above form by LOVE [1906, *5*, § 125].

[2] HADAMARD [1903, *11*, ¶ 265].

[3] Due in principle to HELMHOLTZ [1858, *1*, § 4], though it may be traced back to D'ALEMBERT in restricted cases.

satisfies $s_k = -B_{,k}$ where $B = B(\boldsymbol{x})$; then when $\mathscr{S}$ is stationary we have

$$\oint_{\mathscr{S}} t_{(\boldsymbol{n})k}\, \dot{x}^k\, da = \oint_{\mathscr{S}} s_k\, \dot{x}^k\, da = -\dot{\mathfrak{B}}, \quad \text{where } \mathfrak{B} \equiv \oint_{\mathscr{S}} B\, da. \tag{218.15}$$

In this case, from (218.4) follows

$$\dot{\mathfrak{K}} + \dot{\mathfrak{U}} + \dot{\mathfrak{B}} = -\int_{\mathscr{V}} P\, dv. \tag{218.16}$$

We leave it to the reader to formulate conservation theorems appropriate to this case.

While the requirements 1 and $1'_a$ may always be satisfied trivially, and while there are many cases where requirement 3 is relevant, the other requirements are satisfied only in restricted elastic and hydrodynamic situations. Our purpose in giving these theorems here is to make it clear that such a conservation law as (218.10) is *not* to be expected in any typical situation in continuum mechanics, where dissipation of energy is the rule, not the exception.

219. The virial theorem[1]. Put

$$\mathfrak{Y}_{mk} \equiv \int_{\mathscr{V}} z_m\, \dot{z}_k\, d\mathfrak{M}, \quad 2\,\mathfrak{K}_{mk} \equiv \int_{\mathscr{V}} \dot{z}_m\, \dot{z}_k\, d\mathfrak{M}, \tag{219.1}$$

the quantity $-2\mathfrak{K}_{mk}$ being the total apparent stress due to transfer of momentum (Sect. 207). Then (216.4) may be written

$$\dot{\mathfrak{Y}}_{mk} = 2\,\mathfrak{K}_{mk} + \oint_{\mathscr{S}} z_m\, t_{kq}\, da_q + \int_{\mathscr{V}} (\varrho\, z_m\, f_k - t_{km})\, dv. \tag{219.2}$$

The skew-symmetric part of this equation was considered in Sect. 216. To interpret the symmetric part, notice that

$$\mathfrak{Y}_{(mk)} = \tfrac{1}{2}\, \dot{\mathfrak{E}}_{mk}, \tag{219.3}$$

where $\mathfrak{E}$ is EULER's tensor $(168.4)_1$ with $\boldsymbol{a} = 0$. From (219.2) follows

$$\tfrac{1}{2}\, \ddot{\mathfrak{E}}_{mk} = 2\,\mathfrak{K}_{mk} + \oint_{\mathscr{S}} z_{(m}\, t_{k)q}\, da_q + \int_{\mathscr{V}} (\varrho\, z_{(m}\, f_{k)} - t_{(km)})\, dv, \tag{219.4}$$

the trace of this equation being[2]

$$\tfrac{1}{2}\, \ddot{\mathfrak{E}} = 2\,\mathfrak{K} + \oint_{\mathscr{S}} z_m\, t_{mq}\, da_q + \int_{\mathscr{V}} (\varrho\, z_m\, f_m + 3\,\tilde{p})\, dv, \tag{219.5}$$

where we write $\mathfrak{E}$ for $\mathfrak{E}_{kk}$, the polar moment of inertia of the body about the origin, where $\mathfrak{K}$ is the kinetic energy and where $\tilde{p}$ is the mean pressure (204.7).

The typical application of these results is to obtain time means by integration with respect to t, often under the added assumption that certain terms are periodic.

---

[1] The quantity $\sum_{\mathfrak{a}} \boldsymbol{F}_{\mathfrak{a}} \cdot \boldsymbol{p}_{\mathfrak{a}}$ was introduced into statics by MÖBIUS [1837, *3*, § 123] and studied by SCHWEINS [1849, *2*] [1854, *2*], who called it the „Fliehmoment" of the forces. Its introduction in the dynamics of mass-points is due to JACOBI [1837, *2*, § 6] [1866, *2*, Vierte Vorl.]. Cf. also LIPSCHITZ [1866, *3*] [1872, *3*], CLAUSIUS [1870, *1*], VILLARCEAU [1872, *4*]. Here we follow a generalization due to FINGER [1897, *3*, § I] and elaborated by PARKER [1954, *18*, §§ 1, 3]. Cf. also the hint of MAXWELL [1874, *2*, p. 410].

[2] CISOTTI [1923, *2*, § 3] [1940, *6*, § 6] [1942, *3*].

220. SIGNORINI's theory of stress means. A more useful application of FINGER's virial formula (216.4) has been found by SIGNORINI[1]. Set

$$\mathfrak{v}\, a_{km} \equiv \oint_{s} z_m t_{kq}\, d a_q - \int_{v} z_m (\ddot{z}_k - f_k)\, d\mathfrak{M}, \tag{220.1}$$

where $\mathfrak{v}$ is the volume of v. If we use a superposed bar to denote a mean value over the body, (216.4) is equivalent to

$$\overline{t_{km}} = a_{km}. \tag{220.2}$$

In the static case, the quantities a_{km} may be calculated from the *applied loads* only, so that (220.2) furnishes the *mean stresses directly in terms of known quantities.* Simple as is the reasoning used to derive (220.2), the result is important: While CAUCHY's first law (205.2) in itself constitutes an underdetermined system and is thus insufficient to yield unique values for the stresses, its corollary (220.2)

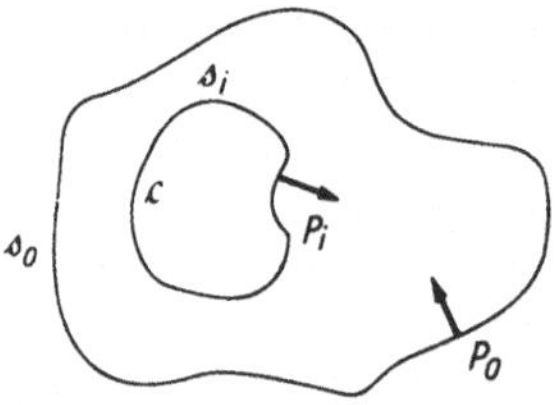

Fig. 38. Body subject to internal and external normal pressure.

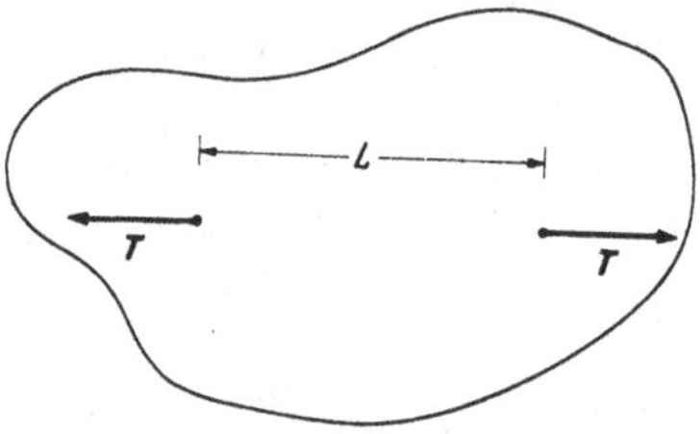

Fig. 39. Body subject to tensile load.

enables us to calculate the mean stress uniquely and *independently of the physical constitution of the material.*

We now give SIGNORINI's examples, all for the static case.

1. *Torque.* If $\boldsymbol{m} = 0$ on s and $\boldsymbol{l} = 0$ in v, $a_{[km]}$ is the torque acting on v. If $a_{[km]} = 0$, (220.2) yields $\overline{t_{km}} = \overline{t_{mk}}$. This is consistent with CAUCHY's second law (205.11), which holds under the stronger assumption that $\boldsymbol{m} = 0$ throughout v. If $a_{km} = 0$, the load on v is said to be *astatic*. From (220.2), a necessary and sufficient condition for astatic load on v is

$$\overline{t_{km}} = 0. \tag{220.3}$$

2. *Hydrostatic pressure.* Suppose v is the region between a surface s_o subject to hydrostatic pressure p_o and a surface s_i subject to hydrostatic pressure p_i (Fig. 38), and suppose $\boldsymbol{f} = 0$. Then if we write $\mathfrak{c}$ for the volume of the cavity, from (220.2) follows

$$-\overline{t_{km}} = \left[p_o + \frac{\mathfrak{c}}{\mathfrak{v}}(p_o - p_i)\right] \delta_{km}. \tag{220.4}$$

This shows that *hydrostatic loading always gives rise to a stress system which is hydrostatic in mean*[2]. Moreover, if $p_o \geqq p_i$ and $p_o > 0$, the mean normal stress is a pressure.

3. *Tensile loading.* Let a body be subject to two equilibrated concentrated forces $\boldsymbol{T}$ and $-\boldsymbol{T}$, acting at points a distance L apart (Fig. 39). Choosing the

[1] [1932, *13*, §§ 1–2]. In [1939, *11*], SIGNORINI applied these results to the motion of rigid bodies.

[2] NARDINI [1952, *14*] has shown that when a body is subject to equal loads applied at the vertices of a regular polyhedron and directed toward its center, the mean stress is hydrostatic. He has obtained a similar result for loads applied at the vertices of a regular polygon.

axis of z_1 parallel to $\boldsymbol{T}$, from (220.2) we get

$$\overline{t_{11}} = \frac{L\,T}{\mathfrak{v}}, \tag{220.5}$$

while all other $\overline{t_{km}}$ vanish. Thus *tensile loading gives rise to simple tension in mean.*

4. *Body rotating steadily about a principal axis of inertia.* Consider a body in steady rotation at angular speed ω about the z_1-axis, so that $\ddot{z}_1 = 0$, $\ddot{z}_2 = -\omega^2 z_2$, $\ddot{z}_3 = -\omega^2 z_3$. In order that the force and torque acting on the body be zero, we must have

$$\left.\begin{aligned} &\int_{\mathfrak{v}} z_k\, d\,\mathfrak{M} = 0 \\ \text{and}\quad &\int_{\mathfrak{v}} z_1 z_k\, d\,\mathfrak{M} = 0, \\ &k = 2, 3; \end{aligned}\right\} \tag{220.6}$$

that is, the axis must pass through the center of mass and coincide with a principal axis of inertia. From (220.2) we get

$$\left.\begin{aligned} &\overline{t_{kk}} = \frac{\omega^2\,\mathfrak{E}_{kk}}{\mathfrak{v}}, \\ &k = 2, 3 \text{ (unsummed)}, \end{aligned}\right\} \tag{220.7}$$

where $\mathfrak{E}_{km}$ is EULER's tensor $(168.4)_1$ with $\boldsymbol{a} = 0$, and all other $\overline{t_{km}}$ vanish.

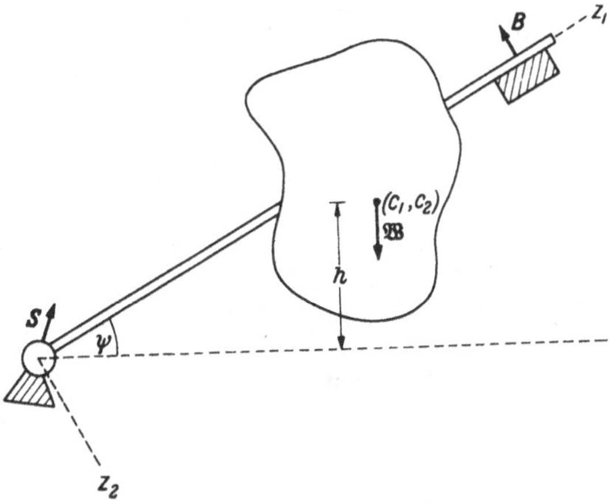

Fig. 40. Heavy solid on an axle.

5. *Heavy solid on an axle*[1]. Consider a heavy solid on an axle which makes an angle ψ with the horizontal, being supported by a hinge at one end, a bearing at the other (Fig. 40). Choose the z_1-axis along the axle, the z_2-axis normal to it, the origin at the hinge, and the plane of z_1-z_2 vertical. The loads are equilibrated; the reaction $\boldsymbol{S}$ of the hinge, being located at the origin, makes no contribution to (220.1); the reaction $\boldsymbol{B}$ of the bearing, being directed along z_2 at a point where $z_2 = z_3 = 0$, contributes only to a_{21}, but need not be mentioned in the non-polar case, since then $a_{21} = a_{12}$, and a_{12} can be calculated without knowledge of $\boldsymbol{B}$. By (165.1), from (220.2) we calculate

$$\mathfrak{v}\,\overline{t_{11}} = -\mathfrak{W}\,c_1 \sin\psi, \quad \mathfrak{v}\,\overline{t_{12}} = -\mathfrak{W}\,c_2 \sin\psi = \mathfrak{v}\,\overline{t_{21}}, \quad \mathfrak{v}\,\overline{t_{22}} = \mathfrak{W}\,c_2 \cos\psi, \tag{220.8}$$

where $\boldsymbol{c}$ is the vector from the hinge to the center of mass and $\mathfrak{W}$ is the weight of the body and axle. All other mean stresses vanish. For the mean value of the mean pressure $\bar{p}$, we have

$$3\bar{\bar{p}} = -\overline{t_{kk}} = \mathfrak{W}\,(c_1 \sin\psi - c_2 \cos\psi)/\mathfrak{v} = \mathfrak{W}\,h/\mathfrak{v}, \tag{220.9}$$

where h is the height of $\boldsymbol{c}$ above the hinge. Thus $\bar{\bar{p}}$ is a pressure or a tension according as $\boldsymbol{c}$ is above or below the hinge. When the axle is horizontal and $c_2 \neq 0$, t_{22} is the only stress which does not vanish in mean; when the axle is vertical and the center of mass lies upon it, we have the case of a body balanced upon a single point, and the resulting expression for $\overline{t_{11}}$, the only non-vanishing mean, agrees with (220.5).

[1] This example and the next are due to TEDONE [1942, *13*, §§ 2—3].

6. *Pendulous body.* Consider a body swinging rigidly about a fixed point (Fig. 41) located at distance L from its center of mass. So that the motion be possible, we assume the support pin is parallel to a principal axis of inertia, while the motion takes place in a plane containing the other two; we take these axes as co-ordinate axes. Since the co-ordinates of the center of mass are $c_1 = L$, $c_2 = 0$, the only contribution made by the weight $\mathfrak{W}$ to the $\mathfrak{v} a_{km}$ is made to $\mathfrak{v} a_{11}$ and is of amount $\mathfrak{W} L \cos\phi$, where ϕ is the angle of inclination to the vertical. From (143.6), for the components of acceleration in an inertial frame instantaneously coinciding with the co-ordinate frame we get $-\dot{\phi}^2 z_1 - \ddot{\phi} z_2$, $-\dot{\phi}^2 z_2 + \ddot{\phi} z_1$, 0. From (220.2) then follows

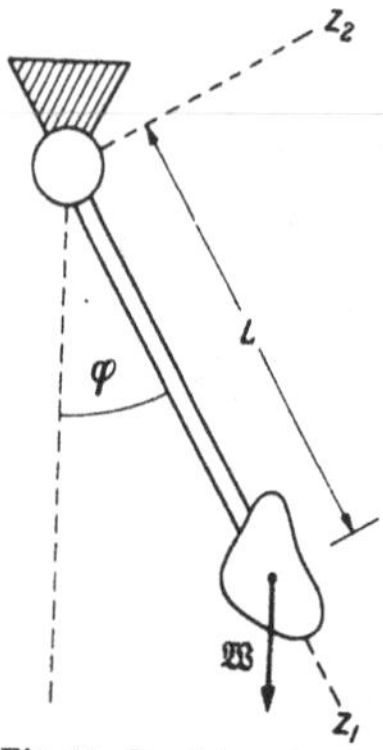

Fig. 41. Pendulous body.

$$\left.\begin{array}{lll} \mathfrak{v}\overline{t_{11}} = \mathfrak{W} L \cos\phi + \dot{\phi}^2 \mathfrak{E}_{11}, & \mathfrak{v}\overline{t_{22}} = \dot{\phi}^2 \mathfrak{E}_{22}, & \\ \mathfrak{v}\overline{t_{12}} = \ddot{\phi}\, \mathfrak{E}_{22}, & \mathfrak{v}\overline{t_{21}} = -\ddot{\phi}\, \mathfrak{E}_{11}, & \overline{t_{3k}} = 0. \end{array}\right\} \tag{220.10}$$

That $\overline{t_{12}} \neq \overline{t_{21}}$ except when $\mathfrak{E}_{11} = \mathfrak{E}_{22} = 0$ or when the angular acceleration vanishes is because a pendulous body not devoid of rotary inertia must be provided with a restraining torque from the support if it is to execute an accelerated rotation.

221. Moments of stress. Setting

$$\mathfrak{v} b_{pqr} \equiv \oint_s z_p z_q t_{rm}\, d a_m - \int_v z_p z_q (\ddot{z}_r - f_r)\, d\mathfrak{M} = \mathfrak{v} b_{qpr}, \tag{221.1}$$

in (216.2) we put $\Psi = z_p z_q$ and obtain[1]

$$\overline{t_{rp} z_q} + \overline{t_{rq} z_p} = b_{pqr}. \tag{221.2}$$

Therefore

$$\overline{t_{(pq)} z_r} + \overline{t_{[pr]} z_q} + \overline{t_{[qr]} z_p} = \tfrac{1}{2}\left(b_{qrp} + b_{prq} - b_{pqr}\right) \equiv c_{pqr}. \tag{221.3}$$

Henceforth we consider only the non-polar case. From (221.3) follows then

$$\overline{t_{pq} z_r} = c_{pqr}, \tag{221.4}$$

so that the second moments of the load determine the mean values of all the first moments of the stresses.

Since $b_{pqr} = c_{rpq} + c_{rqp}$ in the non-polar case, the first moments of the stress determine the second moments of the load. It is impossible that this one-to-one correspondence between moments of the load and moments of the stress can continue indefinitely, for if it did, the stress would be determinate from the load, while in fact Cauchy's laws form an underdetermined system. Indeed, we have $\overline{t_{pq}} = a_{pq}$ from Cauchy's first law alone; to obtain (221.4), we use the second law as well, reducing the number of independent stress moments $\overline{t_{pq} z_r}$ from 27 to 18, the number of independent load moments. For higher moments, the number of independent load moments of a given order is much less than the number of independent stress moments of the next order, and thus the full set of load moments is insufficient to determine all the stress moments.

To calculate the higher moments[2], set

$$\mathfrak{v}\, b^{(r)}_{\mathfrak{a}\mathfrak{b}\mathfrak{c}} \equiv \oint_s z_1^{\mathfrak{a}} z_2^{\mathfrak{b}} z_3^{\mathfrak{c}}\, t_{rm}\, d a_m - \int_v z_1^{\mathfrak{a}} z_2^{\mathfrak{b}} z_3^{\mathfrak{c}} (\ddot{z}_r - f_r)\, d\mathfrak{M}. \tag{221.5}$$

[1] Signorini [1933, *10*, § 2]. A special study of the b_{pq} is made in [1932, *12*].
[2] Grioli [1953, *12*, § 1] [1952, *8*, § 1].

The $b^{(r)}_{\mathfrak{abc}}$ for which $\mathfrak{a}+\mathfrak{b}+\mathfrak{c}=\mathfrak{n}$ are the load moments of order $\mathfrak{n}$, being $\frac{3}{2}(\mathfrak{n}+1)\times(\mathfrak{n}+2)$ in number; for $\mathfrak{n}=1$ they reduce to the a_{rp}; for $\mathfrak{n}=2$, to the b_{pqr}. By putting $\Psi=z_1^{\mathfrak{a}}z_2^{\mathfrak{b}}z_3^{\mathfrak{c}}$ in (216.2), we obtain an equation satisfied by the $3\mathfrak{n}(\mathfrak{n}+1)$ stress moments of order $\mathfrak{n}$:

$$\mathfrak{a}\,\overline{t_{r1}z_1^{\mathfrak{a}-1}z_2^{\mathfrak{b}}z_3^{\mathfrak{c}}}+\mathfrak{b}\,\overline{t_{r2}z_1^{\mathfrak{a}}z_2^{\mathfrak{b}-1}z_3^{\mathfrak{c}}}+\mathfrak{c}\,\overline{t_{r3}z_1^{\mathfrak{a}}z_2^{\mathfrak{b}}z_3^{\mathfrak{c}-1}}=\mathfrak{b}^{(r)}_{\mathfrak{abc}},\tag{221.6}$$

on the understanding that any term in which the exponent "-1" appears is to be annulled. From the choice $\mathfrak{a}=\mathfrak{n}$, $\mathfrak{b}=\mathfrak{c}=0$, we get

$$\overline{t_{rs}z_s^{\mathfrak{n}-1}}=\frac{1}{\mathfrak{n}}\,p^{(r)}_{s\mathfrak{n}}\qquad (s \text{ unsummed, } \mathfrak{n}\geqq 1),\tag{221.7}$$

where $p^{(r)}_{s\mathfrak{n}}$ denotes $b^{(r)}_{\mathfrak{abc}}$ with exponent $\mathfrak{n}$ for z_s and with the other two exponents taken as 0. From the choice $\mathfrak{a}=\mathfrak{n}-1$, $\mathfrak{b}=1$, $\mathfrak{c}=0$, we get

$$(\mathfrak{n}-1)\,\overline{t_{rk}z_k^{\mathfrak{n}-2}z_s}+\overline{t_{rs}z_k^{\mathfrak{n}-1}}=q^{(r)}_{ks\mathfrak{n}}\qquad (k \text{ unsummed}),\tag{221.8}$$

where $q^{(r)}_{ks\mathfrak{n}}$ denotes $b^{(r)}_{\mathfrak{abc}}$ with $\mathfrak{n}-1$ for the exponent of z_k, 1 for the exponent of z_s, and 0 for the third exponent. When $k=r$, (221.8) and (221.7) yield in the non-polar case

$$\overline{t_{rr}z_r^{\mathfrak{n}-2}z_s}=\frac{1}{\mathfrak{n}-1}\left[q^{(r)}_{rs\mathfrak{n}}-\frac{1}{\mathfrak{n}}\,p^{(s)}_{r\mathfrak{n}}\right]\qquad (r \text{ unsummed}).\tag{221.9}$$

From (221.7) and (221.9) we see that when $\mathfrak{n}\geqq 3$, the values of the load moments of order $\mathfrak{n}$ determine unique values for at least 15 of the stress moments of order $\mathfrak{n}$ in the non-polar case.

222. Estimates for the maximum stress. SIGNORINI was the first to observe that a lower bound for the maximum stress is determined by the loading. Here we present generalizations and extensions of his work by GRIOLI[1].

Considering only the non-polar case, write

$$\left.\begin{aligned} &t_1\equiv t_{11},\quad t_2\equiv t_{22},\quad t_3\equiv t_{33},\\ &t_4\equiv t_{23}=t_{32},\quad t_5\equiv t_{31}=t_{13},\quad t_6\equiv t_{12}=t_{21};\end{aligned}\right\}\tag{222.1}$$

let $Q_{\mathfrak{A}}$, $\mathfrak{A}=0, 1, \ldots, \mathfrak{m}$ be a set of functions orthogonal over $\mathscr{v}$, with mean norms $\mathfrak{M}^2_{\mathfrak{A}}$ given by $\mathfrak{v}\,\mathfrak{M}^2_{\mathfrak{A}}\equiv\int_{\mathscr{v}} Q^2_{\mathfrak{A}}\,dv$; let $k_{\mathfrak{bc}}$, $\mathfrak{b},\mathfrak{c}=1, 2, \ldots, 6$ be the symmetric coefficients of any constant positive semi-definite form; and let $C_{\mathfrak{bA}}$, $\mathfrak{b}=1, 2, \ldots, 6$, $\mathfrak{A}=0, 1, \ldots, \mathfrak{m}$, be any constants. Then

$$\left.\begin{aligned} 0&\leqq\frac{1}{\mathfrak{v}}\int_{\mathscr{v}}\sum_{\mathfrak{b},\mathfrak{c},\mathfrak{A},\mathfrak{B}}k_{\mathfrak{bc}}(t_{\mathfrak{b}}-C_{\mathfrak{bA}}Q_{\mathfrak{A}})\,(t_{\mathfrak{c}}-C_{\mathfrak{cB}}Q_{\mathfrak{B}})\,dv,\\ &=\sum_{\mathfrak{b},\mathfrak{c}}k_{\mathfrak{bc}}\,\overline{t_{\mathfrak{b}}t_{\mathfrak{c}}}+\sum_{\mathfrak{b},\mathfrak{c},\mathfrak{A}}k_{\mathfrak{bc}}\,C_{\mathfrak{bA}}(\mathfrak{M}^2_{\mathfrak{A}}C_{\mathfrak{cA}}-2\,\overline{Q_{\mathfrak{A}}t_{\mathfrak{c}}})\,.\end{aligned}\right\}\tag{222.2}$$

Choosing the constants $C_{\mathfrak{cA}}$ so that

$$C_{\mathfrak{cA}}=\overline{Q_{\mathfrak{A}}t_{\mathfrak{c}}}/\mathfrak{M}^2_{\mathfrak{A}},\tag{222.3}$$

from (222.2) we infer

$$\sum_{\mathfrak{b},\mathfrak{c}}k_{\mathfrak{bc}}\,\overline{t_{\mathfrak{b}}t_{\mathfrak{c}}}\geqq\sum_{\mathfrak{b},\mathfrak{c},\mathfrak{A}}k_{\mathfrak{bc}}\,\overline{Q_{\mathfrak{A}}t_{\mathfrak{b}}}\;\overline{Q_{\mathfrak{A}}t_{\mathfrak{c}}}/\mathfrak{M}^2_{\mathfrak{A}}.\tag{222.4}$$

If the form whose coefficients are $k_{\mathfrak{bc}}$ is positive definite, equality holds in (222.4) if and only if

$$t_{\mathfrak{b}}=\sum_{\mathfrak{A}}\overline{Q_{\mathfrak{A}}t_{\mathfrak{b}}}\,Q_{\mathfrak{A}}/\mathfrak{M}^2_{\mathfrak{A}}.\tag{222.5}$$

[1] [1953, *12*, § 2] [1952, *8*, §§ 3—5]. In [1952, *9*, §§ 1—3] GRIOLI applies these results to plates of arbitrary thickness.

This is in itself an interesting observation: For all states of stress yielding given means $\overline{Q_\mathfrak{A} t_\mathfrak{b}}$, that given by (222.5) minimizes $\sum_{\mathfrak{b},\mathfrak{c}} k_{\mathfrak{b}\mathfrak{c}} \overline{t_\mathfrak{b} t_\mathfrak{c}}$ for any positive definite form.

The inequality (222.4) may be used to estimate the values of the stresses if the means $\overline{Q_\mathfrak{A} t_\mathfrak{b}}$ are known. To this end, in particular cases, we employ the results of Sects. 220 and 221. For example, if we choose the origin at the centroid and the co-ordinate axes along the principal axes of geometrical inertia of v, so that $\int_v z_k \, dv = 0$, $\int_v z_k z_m \, dv = 0$ if $k \neq m$, then the functions $1, z_1, z_2, z_3$ form an orthogonal set, with $\mathfrak{M}_0^2 = 1$, $\mathfrak{v}\, \mathfrak{M}_k^2 = \int_v z_k^2 \, dv$. The values of $\overline{Q_\mathfrak{A} t_\mathfrak{b}}$ are given by (220.2) and (221.4). If we introduce for the quantities a_{pq} and c_{pqr} a notation parallel to (222.1), then (222.4) yields ***Signorini's inequality***[1]:

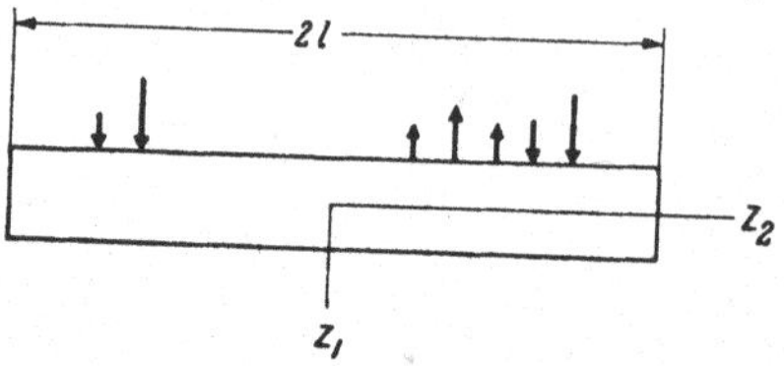

Fig. 42. Normally loaded beam.

$$\left.\begin{aligned} &\sum_{\mathfrak{b},\mathfrak{c}} k_{\mathfrak{b}\mathfrak{c}} \overline{t_\mathfrak{b} t_\mathfrak{c}} \\ &\geq \sum_{\mathfrak{b},\mathfrak{c}} k_{\mathfrak{b}\mathfrak{c}} \Big(a_\mathfrak{b} a_\mathfrak{c} + \sum_{e=1}^{3} c_{\mathfrak{b}e} c_{\mathfrak{c}e} / \mathfrak{M}_e^2\Big). \end{aligned}\right\} \tag{222.6}$$

Examples. 1. Taking one of the $k_{\mathfrak{b}\mathfrak{c}}$ as 1 and all other $k_{\mathfrak{b}\mathfrak{c}}$ as 0 yields

$$t_\mathfrak{b}^2|_{\max} \geq \overline{t_\mathfrak{b}^2} \geq a_\mathfrak{b}^2 + \sum_{e=1}^{3} \frac{c_{\mathfrak{b}e}^2}{\mathfrak{M}_e^2}. \tag{222.7}$$

2. In order that the stress satisfy the condition

$$\sum_{\mathfrak{b},\mathfrak{c}} k_{\mathfrak{b}\mathfrak{c}} t_\mathfrak{b} t_\mathfrak{c} \leq K, \qquad K = \text{const} \tag{222.8}$$

the loads must satisfy

$$\sum_{\mathfrak{b},\mathfrak{c}} k_{\mathfrak{b}\mathfrak{c}} \Big(a_\mathfrak{b} a_\mathfrak{c} + \sum_{e=1}^{3} c_{\mathfrak{b}e} c_{\mathfrak{c}e} / \mathfrak{M}_e^2\Big) \leq K^2. \tag{222.9}$$

SIGNORINI[2] has applied (222.7) to the equilibrium of a beam which is a horizontal rectangular block in its deformed state when supported at its two ends, loaded by its own weight and by arbitrary forces acting upon its upper and lower faces. Choosing co-ordinates so that the axis of z_1 points vertically downward through the center of mass, while the axis of z_2 is horizontal (Fig. 42), we have $f_2 = f_3 = 0$, $f_1 = g$; on the surface, the z_2 and z_3 components of load vanish, while the z_1 component may be arbitrary. From (220.1) and (221.1) we get at once $a_{22} = 0$, $c_{222} = 0$, $c_{223} = 0$, $c_{221} = -\frac{1}{2} b_{221}$, and hence by (222.7) follows

$$|t_2|_{\max} \geq \frac{|c_{221}|}{\mathfrak{M}_1} = \frac{|b_{221}|}{2c}, \tag{222.10}$$

where c is the radius of gyration of the cross-section. We now evaluate b_{221}. If we write $p(z_2)$ for the density of total load per unit length along the beam, taking account of the contribution of the two supports, which equilibrate the total load on the beam, from (221.1) we get

$$-\mathfrak{v}\, b_{221} = l^2 \int_{-l}^{l} p(s)\, ds - \int_{-l}^{l} s^2 p(s)\, ds. \tag{222.11}$$

The total bending moment acting at z_2 in virtue of the reaction at the end $z_2 = -l$ and of the load on the part of the beam to the left-hand side of z_2 is given by

$$m(z_2) = -(l + z_2) \cdot \frac{1}{2l} \int_{-l}^{l} (l - s)\, p(s)\, ds + \int_{-l}^{z_2} (z_2 - s)\, p(s)\, ds. \tag{222.12}$$

[1] [1933, *10*, § 4].
[2] [1941, *7*]. The result is generalized somewhat in [1954, *22*].

If we write m for the mean value of $m(z_2)$, i.e. $2lm \equiv \int_{-l}^{l} m(s)\,ds$, it is easy to show from (222.12) and (222.11) that

$$\mathfrak{A}\, l\, b_{221} = 4m, \tag{222.13}$$

where $\mathfrak{A}$ is the cross-sectional area. Thus (222.10) may be written in the form

$$|t_2|_{\max} \geqq \frac{|m|}{\mathfrak{A}\, c}. \tag{222.14}$$

Various special theories of the strength of beams assume formulae similar to (222.14); it furnishes a general bound with which such theories must agree if they are to be statically possible.

GRIOLI has discussed the use of (222.4) for choices of the $Q_{\mathfrak{A}}$ suggested by the results in Sect. 221.

The inequality (222.7) can be generalized easily, since the same choice of the $k_{\mathfrak{b}\mathfrak{c}}$ with general $Q_{\mathfrak{A}}$ in (222.4) yields

$$t_{\mathfrak{b}}^2|_{\max} \geqq \overline{t_{\mathfrak{b}}^2} \geqq \sum_{\mathfrak{A}} \frac{(\overline{Q_{\mathfrak{A}}\, t_{\mathfrak{b}}})^2}{\mathfrak{M}_{\mathfrak{A}}^2}. \tag{222.15}$$

Other estimates for the maximum stress[1] follow immediately from (221.7) and (221.9):

$$\left.\begin{aligned} |t_{rs}|_{\max} &\geqq \frac{\mathfrak{V}\,|p_{s\mathfrak{n}}^{(r)}|}{\mathfrak{n}\int_v |z_s^{\mathfrak{n}-1}|\,dv}, \\ |t_{rr}|_{\max} &\geqq \frac{\mathfrak{V}\,|q_{s\mathfrak{n}}^{(r)} - p_{s\mathfrak{n}}^{(r)}/\mathfrak{n}|}{(\mathfrak{n}-1)\int_v |z_r^{\mathfrak{n}-2}\, z_s|\,dv}. \end{aligned}\right\} \tag{222.16}$$

From the identity

$$\sum_{\mathfrak{A}} k_{\mathfrak{A}} \int_v Q_{\mathfrak{A}}\, t_{rs}\, dv = \mathfrak{V} \sum_{\mathfrak{A}} k_{\mathfrak{A}}\, \overline{Q_{\mathfrak{A}}\, t_{rs}}, \tag{222.17}$$

where the $k_{\mathfrak{A}}$ are arbitrary constants, follows

$$|t_{rs}|_{\max} \geqq \mathfrak{V}\, \frac{\sum_{\mathfrak{A}} |k_{\mathfrak{A}}\, \overline{Q_{\mathfrak{A}}\, t_{rs}}|}{\sum_{\mathfrak{B}} \int_v |k_{\mathfrak{B}}\, Q_{\mathfrak{B}}|\,dv}. \tag{222.18}$$

GRIOLI[2] has shown that in each case there exists a particular choice of the constants $k_{\mathfrak{A}}$ rendering the bound (222.18) sharper than (222.15). In fact, with r and s held fixed, put

$$k_{\mathfrak{A}} = \overline{Q_{\mathfrak{A}}\, t_{rs}}/\mathfrak{M}_{\mathfrak{A}}^2. \tag{222.19}$$

Then by the Schwarz inequality and the orthogonality of the $Q_{\mathfrak{A}}$ follows

$$\left.\begin{aligned} \mathfrak{V}^2 \left(\frac{\sum_{\mathfrak{A}} |k_{\mathfrak{A}}\, \overline{Q_{\mathfrak{A}}\, t_{rs}}|}{\int_v \sum_{\mathfrak{B}} |k_{\mathfrak{B}}\, Q_{\mathfrak{B}}|\,dv}\right)^2 &= \mathfrak{V}^2 \left[\frac{\sum_{\mathfrak{A}} (\overline{Q_{\mathfrak{A}}\, t_{rs}})^2/\mathfrak{M}_{\mathfrak{A}}^2}{\int_v \sum_{\mathfrak{B}} |\overline{Q_{\mathfrak{B}}\, t_{rs}} Q_{\mathfrak{B}}|\,dv/\mathfrak{M}_{\mathfrak{B}}^2}\right]^2, \\ &\geqq \mathfrak{V}\, \frac{\left[\sum_{\mathfrak{A}} (\overline{Q_{\mathfrak{A}}\, t_{rs}})^2/M_{\mathfrak{A}}^2\right]^2}{\int_v \left(\sum_{\mathfrak{B}} \frac{\overline{Q_{\mathfrak{B}}\, t_{rs}}\, Q_{\mathfrak{B}}}{\mathfrak{M}_{\mathfrak{B}}^2}\right)^2 dv}, \\ &= \mathfrak{V}\, \frac{\left[\sum_{\mathfrak{A}} (\overline{Q_{\mathfrak{A}}\, t_{rs}})^2/\mathfrak{M}_{\mathfrak{A}}^2\right]^2}{\int_v \sum_{\mathfrak{B}} \frac{\overline{Q_{\mathfrak{B}}\, t_{rs}}\, Q_{\mathfrak{B}}}{\mathfrak{M}_{\mathfrak{B}}^2} \sum_{\mathfrak{C}} \frac{\overline{Q_{\mathfrak{C}}\, t_{rs}}\, Q_{\mathfrak{C}}}{\mathfrak{M}_{\mathfrak{C}}^2}\, dv}, \\ &= \mathfrak{V}\, \frac{\left[\sum_{\mathfrak{A}} (\overline{Q_{\mathfrak{A}}\, t_{rs}})^2/\mathfrak{M}_{\mathfrak{A}}^2\right]^2}{\sum_{\mathfrak{B}} \frac{(\overline{Q_{\mathfrak{B}}\, t_{rs}})^2}{\mathfrak{M}_{\mathfrak{B}}^4} \int_v Q_{\mathfrak{B}}^2\, dv}, \\ &= \sum_{\mathfrak{A}} (\overline{Q_{\mathfrak{A}}\, t_{rs}})^2/\mathfrak{M}_{\mathfrak{A}}^2. \end{aligned}\right\} \tag{222.20}$$

[1] GRIOLI [1953, *12*, § 2] [1952, *8*, § 2].
[2] [1955, *10*, § 1].

Since the left-hand side is the square of the bound given by (222.18), while the last right-hand side is the bound given by (222.15), it is established that with the choice of constants given by (222.19), the bound (222.18) is the best so far obtained. In order to obtain a best possible bound of the type (222.18), we should have to choose the $k_{\mathfrak{A}}$ so as to maximize the right-hand side.

GRIOLI[1] has applied (222.18) to the estimation of the maximum longitudinal stress in a cylinder of arbitrary cross-section.

BRESSAN[2] has observed that if we take Ψ as a vector $\boldsymbol{b}$ in (216.2), in the static and non-polar case follows

$$\overline{t^{km} b_{k,m}} = \overline{\sum_{\mathfrak{a}} t_{\mathfrak{a}} b_{\mathfrak{a}}} = N \equiv \frac{1}{\mathfrak{V}} \left[\oint_{\mathfrak{s}} b_k t^{km} \, d a_m + \int_{\mathfrak{v}} \varrho f_k b^k \, dv \right], \qquad (222.21)$$

where a notation analogous to (222.1) is used. Let $b^{(\mathfrak{q})}$ be the value of $b_{\mathfrak{q}}$ when $\boldsymbol{b}$ is a solution of the compatible system of five linear partial differential equations

$$b_{\mathfrak{a}} = 0 \quad \text{for} \quad \mathfrak{a} \neq \mathfrak{q}, \qquad (222.22)$$

and $N^{(\mathfrak{q})}$ the corresponding value of N. Then (222.21) yields $\overline{t_{\mathfrak{q}} b^{(\mathfrak{q})}} = N^{(\mathfrak{q})}$, and hence

$$|t_{\mathfrak{q}}|_{\max} \geqq \frac{N^{(\mathfrak{q})}}{|\overline{b^{(\mathfrak{q})}}|}. \qquad (222.23)$$

For a particular solution in cylindrical polar co-ordinates, BRESSAN has worked out a full set of such bounds explicitly.

223. Statically determinate problems. In the practice of mechanics a problem is said to be *statically determinate* if the state of stress may be determined without knowledge of the displacement or motion. Since, as will appear in Subchapter IV, it is always possible to solve the equations of motion in terms of arbitrary functions, and since arbitrary functions often occur in what are called statically determinate solutions, it does not seem that the term has any precise meaning[3]. We shall use it as a title for the following collection of problems arising from *direct hypotheses regarding the stress*, independently of any possible constitutive equations.

First, a system of stresses whose rectangular Cartesian components are constant is said to be *uniform*. Obviously *a uniform stress is a null stress* (Sect. 205). Also, from (205.2) we see that *a body subject to uniform stress, or to any null stress, moves as if its particles were free mass-points subject to force* $\boldsymbol{f}$ *per unit mass.*

The next simplest case is that when the stress depends linearly on the rectangular Cartesian co-ordinates[4]: From (205.2) it follows at once that $\varrho(\ddot{\boldsymbol{x}} - \boldsymbol{f})$ must be a constant vector field. Assuming this condition to be satisfied, we use (207.2) to reduce the general problem to one in which

$$t_{km,m} = 0 \quad \text{in} \quad \mathfrak{v}, \qquad t_{km} n_m = s_k \quad \text{on} \quad \mathfrak{s}. \qquad (223.1)$$

That is, we are to find the most general linear null stress such that the stress vector assumes prescribed values on $\mathfrak{s}$. Now if we write

$$t_{km} = A_{km} + C_{kmq} z_q, \qquad (223.2)$$

and if we take the origin at the centroid and direct the co-ordinate axes along the principal axes of geometrical inertia, we have

$$a_{km} = \overline{t_{km}} = A_{km}, \qquad c_{pqr} = \overline{t_{pq} z_r} = C_{pqr} \mathfrak{J}_r \quad \text{(unsummed)} \qquad (223.3)$$

[1] [1955, *10*, § 2]. We do not reproduce his interesting result, because we cannot verify his initial formulae $(11)_{2,3}$.

[2] [1956, *2*].

[3] The same may be said of the convenient German expression „in geschlossener Form".

[4] SIGNORINI [1933, *10*, §§ 6—7]. Examples are given in §§ 8—9.

where $\mathfrak{I}_r \equiv \int\limits_v z_r^2\, dv$. Hence in the linear case, the coefficients are determined by the means and first moments of the stress, and vice-versa. By (223.2), the conditions (223.1) become

$$c_{k\,m\,m}/\mathfrak{I}_m = 0 \quad \text{in} \quad v, \qquad (a_{k\,m} + c_{k\,m\,q}\, z_q/\mathfrak{I}_q)\, n_m = s_k \quad \text{on} \quad s. \tag{223.4}$$

The former of these is equivalent to

$$b_{m\,m\,k}/\mathfrak{I}_m = 0. \tag{223.5}$$

Conversely, if (223.5) and $(223.4)_2$ are satisfied, the linear stress defined by (223.3) and (223.2) will be a null stress such that the stress vector on s is $\mathbf{s}$. Thus the conditions $(223.4)_2$ and (223.5), along with the condition $\varrho(\ddot{\mathbf{x}} - \mathbf{f}) = \text{const}$, constitute *necessary and sufficient conditions to be satisfied by the loads in order that a linear solution to Cauchy's laws, with a prescribed load on the boundary, may exist.*

In Sect. 208 we have derived a condition that all stress vectors at a point have the same magnitude. The hydrostatic special case, $\mathbf{t} = -p\,\mathbf{1}$ and $\ddot{\mathbf{x}} = 0$, is statically determinate (cf. Sect. 208). The corresponding problem for plane stress yields $t_3 = 0$ and $(t_1)^2 = (t_2)^2$ and is statically determinate even in the non-hydrostatic alternative[1], since $t_1 = -t_2$ implies $t^1_1 = -t^2_2$ in all co-ordinate systems in the plane of stress, so that the equations of equilibrium become

$$t^1_{1,1} + t^2_{1,2} = \varrho f_1, \qquad t^1_{2,1} - t^1_{1,2} = \varrho f_2, \tag{223.6}$$

with $(1 - g^{12} g_{12})\, t^1_2 = (g^{11} g_{21} - g^{12} g_{22})\, t^1_1 + g^{11} g_{22}\, t^2_1$.

Statically determinate problems also follow from assumptions regarding the stress trajectories. For example, from (209.1) we see that in plane stress with $\mathbf{f} - \ddot{\mathbf{x}} = 0$, if one family of stress trajectories consists in straight lines, the other principal stress does not vary along its trajectories[2]. In this same case, the angle ϕ in (App. 6.3) is of course constant along the straight trajectories.

TRUESDELL[3] has found the most general stresses compatible with the contravariant velocity components given in cylindrical co-ordinates r, θ, z by

$$\dot{r} = r\,R(t), \qquad \dot{\theta} = z\,A(t), \qquad \dot{z} = z\,Z(t); \tag{223.7}$$

this motion is a simple type of torsion, expansion, and extension of a circular cylinder. For the physical components of $\boldsymbol{d}$ we have

$$\boldsymbol{d} = \begin{Vmatrix} R & 0 & 0 \\ \cdot & R & \tfrac{1}{2} r A \\ \cdot & \cdot & Z \end{Vmatrix}. \tag{223.8}$$

By solving $(156.5)_2$ we obtain

$$\varrho = \varrho_0\, S(t), \qquad S \equiv \exp\left[-\int (2R + Z)\, dt\right]. \tag{223.9}$$

The contravariant components of acceleration are

$$\ddot{r} = r[\dot{R} + R^2 - z^2 A^2], \qquad \ddot{\theta} = z[\dot{A} + A(2R + Z)], \qquad \ddot{z} = z[\dot{Z} + Z^2]. \tag{223.10}$$

We restrict attention to stresses such that

$$\frac{\partial\,\widehat{km}}{\partial \theta} = 0, \qquad \widehat{rz} = 0, \tag{223.11}$$

[1] THEODORESCO [1937, *9*].
[2] HEYMANS [1924, *4*].
[3] [1955, *28*, § 11]

and we put $f=0$; the dynamical equations (205 A.1), in cylindrical co-ordinates, become

$$\left.\begin{aligned} \frac{\partial \widehat{r r}}{\partial r}+\frac{\widehat{r r}-\widehat{\theta \theta}}{r} &= \varrho_0\, S\, r\,[\dot{R}+R^2-z^2 A^2],\\ \frac{\partial \widehat{r \theta}}{\partial r}+\frac{\partial \widehat{\theta z}}{\partial z}+\frac{2 \widehat{r \theta}}{r} &= \varrho_0\, S\, r\, z\,[\dot{A}+A(2R+Z)],\\ \frac{\partial \widehat{z z}}{\partial z} &= \varrho_0\, S\, z\,[\dot{Z}+Z^2]. \end{aligned}\right\} \tag{223.12}$$

The general solution is

$$\left.\begin{aligned} \widehat{\theta z} &= -\frac{1}{r^2}\frac{\partial}{\partial r}\left(r^2 \int \widehat{r \theta}\, d z\right)+\frac{1}{2}\varrho_0\, r\, z^2\, S\,[\dot{A}+A(2R+Z)]+Q,\\ \widehat{\theta \theta} &= \frac{\partial}{\partial r}(r\, \widehat{r r})+\varrho_0\, r^2\, S\,[z^2 A^2-\dot{R}-R^2],\\ \widehat{z z} &= \frac{1}{2}\varrho_0\, z^2\, S\,[\dot{Z}+Z^2]+P, \end{aligned}\right\} \tag{223.13}$$

where P and Q are arbitrary functions of r, t. The terms proportional to ϱ_0 express the effect of the inertia of the material.

The most interesting statically determinate problems arise in the theory of shells, where a state of membrane stress leads to the determined system (212.22), (212.23). The general solution of these equations will be developed in Sect. 229.

IV. General solutions of the equations of motion.

224. Steady plane problems. So as to make clear the approach to stress functions in general, we begin with the simplest case, when the motion is steady and plane, the stress is plane[1], and the assigned forces satisfy (207.1). Then in rectangular Cartesian co-ordinates CAUCHY'S first law (205.2) assumes the form

$$\left.\begin{aligned} (t_{xx}-\varrho\dot{x}^2-V)_{,x}+(t_{xy}-\varrho\dot{x}\dot{y})_{,y}&=0,\\ (t_{yx}-\varrho\dot{y}\dot{x})_{,x}+(t_{yy}-\varrho\dot{y}^2-V)_{,y}&=0. \end{aligned}\right\} \tag{224.1}$$

Since each of these formulae is a condition of integrability for a differential form, the pair is equivalent to the existence of functions F and G such that

$$\left.\begin{aligned} t_{xx}-\varrho\dot{x}^2-V&=F_{,y}, & t_{xy}-\varrho\dot{x}\dot{y}&=-F_{,x},\\ t_{yx}-\varrho\dot{y}\dot{x}&=G_{,y}, & t_{yy}-\varrho\dot{y}^2-V&=-G_{,x}. \end{aligned}\right\} \tag{224.2}$$

When the stress is not symmetric, no simplification is possible, but in the non-polar case, to which this subchapter is restricted, the left-hand sides of $(224.2)_2$ and $(224.2)_3$ are equal, so that

$$F_{,x}+G_{,y}=0. \tag{224.3}$$

This, in turn, is a condition of integrability for the existence of a function A such that

$$F=-A_{,y}, \quad G=A_{,x}. \tag{224.4}$$

Substitution in (224.2) yields

$$t_{xx}-\varrho\dot{x}^2-V=-A_{,yy}, \quad t_{yy}-\varrho\dot{y}^2-V=-A_{,xx}, \quad t_{xy}-\varrho\dot{x}\dot{y}=A_{,xy}. \tag{224.5}$$

When $\boldsymbol{t}-\varrho\dot{\boldsymbol{x}}\dot{\boldsymbol{x}}-V\boldsymbol{1}$ is twice continuously differentiable, the theorem on exact differentials implies that the existence of a function A satisfying (224.5) is

[1] These results hold also when $t_{zz}\neq 0$ but $t_{zz,z}=0$, as for example when the stress is a hydrostatic pressure independent of z.

necessary and sufficient that (224.1) hold. Accordingly, (224.5) gives the ***general solution*** of the equations of motion for the case considered. The function A is the celebrated *stress function* of AIRY[1]. Our argument, since it rests upon the theorem of the exact differential, implies that A is single-valued in simply connected regions, generally multivalued in multiply connected regions[2].

Since a unit normal to the element of arc $\boldsymbol{i}\,dx+\boldsymbol{j}\,dy$ is $-\boldsymbol{i}\,dy/ds+\boldsymbol{j}\,dx/ds$, from (203.4) and (224.5) we obtain the components of the stress vector across the arc in the forms

$$\left.\begin{aligned} t_{(\boldsymbol{n})x}-\varrho\dot{x}\dot{p}_n+V\frac{dy}{ds}&=A_{,yy}\frac{dy}{ds}+A_{,xy}\frac{dx}{ds}=\frac{dA_{,y}}{ds},\\ t_{(\boldsymbol{n})y}-\varrho\dot{y}\dot{p}_n-V\frac{dx}{ds}&=-A_{,yx}\frac{dy}{dx}-A_{,xx}\frac{dx}{ds}=-\frac{dA_{,x}}{ds}.\end{aligned}\right\}\quad(224.6)$$

Therefore the normal stress and shear stress on the element of arc are given by

$$\left.\begin{aligned} t_n-\varrho\dot{p}_n^2-V&=-\frac{dA_{,y}}{ds}\frac{dy}{ds}-\frac{dA_{,x}}{ds}\frac{dx}{ds},\\ &=-\frac{d}{ds}\left(A_{,x}\frac{dx}{ds}+A_{,y}\frac{dy}{ds}\right)+A_{,x}\frac{d}{ds}\left(\frac{dx}{ds}\right)+A_{,y}\frac{d}{ds}\left(\frac{dy}{ds}\right),\\ &=-\frac{d^2A}{ds^2}+\varkappa\left(-A_{,x}\frac{dy}{ds}+A_{,y}\frac{dx}{ds}\right),\\ &=-\frac{d^2A}{ds^2}+\varkappa\frac{dA}{dn},\\ t_t-\varrho\dot{p}_n\dot{p}_t&=\frac{d^2A}{ds\,dn}+\varkappa\frac{dA}{ds},\end{aligned}\right\}\quad(224.7)$$

where $\dot{p}_n$ and $\dot{p}_t$ are the normal and tangential components of velocity. These results are due to MICHELL[3], for the case of equilibrium. If the element is a stationary boundary, the momentum transfer vanishes upon it, and (224.7) yields the stress vector directly.

To find the dynamical significance of the intermediate function F, we integrate along a curve c from $\boldsymbol{x}_1$ to $\boldsymbol{x}_2$, obtaining

$$\left.\begin{aligned} F|_{\boldsymbol{x}_2}-F|_{\boldsymbol{x}_1}&=\int_c dF=\int_c[(-t_{xy}+\varrho\dot{x}\dot{y})\,dx+(t_{xx}-\varrho\dot{x}^2-V)\,dy],\\ &=\int_c[\boldsymbol{i}\,(t_{xx}-\varrho\dot{x}^2-V)+\boldsymbol{j}\,(t_{xy}-\varrho\dot{x}\dot{y})]\,[\boldsymbol{i}\,dy-\boldsymbol{j}\,dx].\end{aligned}\right\}\quad(224.8)$$

If we include the apparent stress due to assigned force and to transfer of momentum, by (203.4) the integrand is the x-component of the stress vector acting upon a cylinder of unit height based upon the curve c. Thus the difference of

---

[1] AIRY [1863, *1*] considered only the case when $\varrho\dot{\boldsymbol{x}}\dot{\boldsymbol{x}}-V\boldsymbol{1}=0$, and he did not prove necessity; the generality of the solution was asserted by MAXWELL [1870, *4*, pp. 192–193]. The first fully satisfactory treatment for the case of equilibrium was given by MICHELL [1900, *6*]; among other things, he included V [ibid., p. 100] and gave an explicit form of (224.5) in orthogonal curvilinear co-ordinates [ibid., p. 111]. E.R. NEUMANN [1907, *6*, § 1] obtained (224.5) for motion in which $\boldsymbol{t}$ is hydrostatic and $V=0$. Cf. also BRAHTZ [1934, *1*], BATEMAN [1936, *1*, § 2], CROCCO [1950, *5*]. VOIGT [1882, *4*, pp. 297–298] remarked that (224.1) continues to hold when all components of stress are constant in the z-direction; in this case, which is often appropriate to torsion of a cylinder, the z-component of CAUCHY's first law yields a function W such that $t_{yz}-\varrho\dot{y}\dot{z}=W_{,x}$, $t_{xz}-\varrho\dot{x}\dot{z}=-W_{,y}$, when $(V+\varrho\dot{z}^2)_{,z}=0$.

[2] After making this observation, MICHELL [1900, *6*, p. 103ff.] determined the nature of this multi-valuedness for the case of linear elasticity theory.

[3] [1900, *6*, p. 110]. An incorrect formula of this kind was given by NEUMANN [1907, *6*, § 3].

the values F at $\boldsymbol{x}_1$ and $\boldsymbol{x}_2$ equals the x-component of the force[1], including inertial force, acting upon c. When c is a stationary boundary, the momentum transfer terms contribute nothing to the integral, which then yields the force alone. A similar interpretation holds for G.

To interpret[2] the function A, we see that for a curve c connecting $\boldsymbol{x}_2$ to $\boldsymbol{x}_1$ we have

$$\left.\begin{aligned} A|_{\boldsymbol{x}_2} - A|_{\boldsymbol{x}_1} &= \int_c dA = \int_c (A_{,x}\,dx + A_{,y}\,dy) = \int_c (G\,dx - F\,dy),\\ &= (x_2 - x_1)\, G|_{\boldsymbol{x}_2} - (y_2 - y_1)\, F|_{\boldsymbol{x}_2} + \\ &\quad + \int_c \{(x - x_1)\left[(t_{yy} - \varrho \dot{y}^2 - V)\,dx - (t_{xy} - \varrho \dot{x}\dot{y})\,dy\right] + \\ &\quad + (y - y_1)\left[(t_{xx} - \varrho \dot{x}^2 - V)\,dy - (t_{xy} - \varrho \dot{x}\dot{y})\,dx\right]\},\\ &= (x_2 - x_1)\, G|_{\boldsymbol{x}_2} - (y_2 - y_1)\, F|_{\boldsymbol{x}_2} + \mathfrak{L} \end{aligned}\right\} \tag{224.9}$$

where $\mathfrak{L}$ is the torque about $\boldsymbol{x}_1$ exerted by the stress, including the apparent stress due to rate of change of momentum and to applied force, acting upon the right-hand side of c, thought of as directed from $\boldsymbol{x}_1$ toward $\boldsymbol{x}_2$.

While we have used rectangular Cartesian co-ordinates, the extension of (224.5) to arbitrary curvilinear co-ordinates is immediate[3]:

$$t^{km} - \varrho \dot{x}^k \dot{x}^m - V g^{km} = - e^{kp} e^{mq} A_{,pq}, \tag{224.10}$$

where g^{km} is the contravariant metric tensor in the plane, e^{kp} is the absolute alternating tensor, and "," denotes covariant differentiation based on $\boldsymbol{g}$. Since (224.10) is a tensor equation, and since it reduces to (224.5) when the co-ordinates are rectangular Cartesian, it is valid in all co-ordinate systems.

225. Generalized lineal motion. Consider a motion in which $\boldsymbol{t}, \dot{\boldsymbol{x}}, \varrho$, and V all depend only upon x and t. Using (161.22) and the x-component of (205.2), we readily infer the existence of a function U such that[4]

$$\varrho = U_{,xx}, \quad \dot{x} = -\frac{U_{,xt}}{U_{,xx}}, \quad t_{xx} = V + \frac{\dfrac{\partial(U_{,t}, U_{,x})}{\partial(x, t)}}{U_{,xx}}. \tag{225.1}$$

The second of these equations may be written in the form

$$\dot{x} = -\frac{(U_{,x})_{,t}}{(U_{,x})_{,x}}; \tag{225.2}$$

thus the velocity equals the slope of the curve $U_{,x} = \text{const}$ in the t-x plane. The remaining components of (205.2) yield the existence of functions V and W such that

$$\left.\begin{aligned} \dot{y} &= V_{,x}/U_{,xx}, \quad & t_{yx} &= \frac{\partial(V, U_{,x})}{\partial(t, x)}\Big/U_{,xx},\\ \dot{z} &= W_{,x}/U_{,xx}, \quad & t_{zx} &= \frac{\partial(W, U_{,x})}{\partial(t, x)}\Big/U_{,xx}. \end{aligned}\right\} \tag{225.3}$$

In the lineal case, $V = W = 0$.

---

[1] This result is due to MAXWELL [1870, *4*, pp. 192–193], who considered only the static case; he called the vector F, G "the diagram of stress". His derivation was criticized and corrected by MICHELL [1900, *6*, p. 107]. In rediscovering this result for the case of hydrostatic stress, BATEMAN [1938, *1*, § 1] called F and G the "drag and lift functions".

[2] PHILLIPS [1934, *4*] based his proof of the existence of AIRY's function on this interpretation. Cf. also SOBRERO [1935, *7*]. An energetic interpretation for A and for stress functions of all types considered below has been given by L. FINZI [1956, *7*, § 6].

[3] B. FINZI [1934, *2*, § 1].

[4] The result was worked out by W. KIRCHHOFF [1930, *2*, § 1], following a suggestion of E. R. NEUMANN [1907, *6*, § 6]. It is rediscovered by MCVITTIE [1953, *18*, § 3].

226. Conventions for the remaining general solutions. All general solutions are obtained by essentially the same reasoning as in Sect. 224, although the details may be more elaborate. At bottom, we are to integrate[1]

$$s^{km}{}_{,m} = 0, \qquad s^{km} = s^{mk}; \tag{226.1}$$

that is, to find the most general symmetric null stress. It is the additional condition of symmetry that makes the problem interesting and different from that solved in Sects. 161 to 164. The variations in the answers from case to case arise because of different numbers of dimensions and different metrics on which the covariant differentiation is based.

For the Euclidean case, whatever the number of dimensions, the entire analysis consists in repeated use of the fact that a continuously differentiable tensor whose divergence vanishes may be expressed as the curl of another tensor. The *tensor potentials* so determined exist subject to various conditions that may be found in the literature on the theory of the potential.

We do not remind the reader again that when all co-ordinates are space co-ordinates we may take s^{km} as $t^{km} - \varrho \dot{x}^k \dot{x}^m$ with an additional term $-V g^{km}$ when $\boldsymbol{f}$ satisfies (207.1). For general $\boldsymbol{f}$, the linearity of CAUCHY's laws enables the general solution to be gotten by adding to any particular solution the general null stress.

The problem of finding a particular integral is relatively trivial in a flat space. In a convex region, quadratures suffice. More generally, for any Riemannian space, let $\boldsymbol{u}$ be any solution of the linear differential equation

$$u^k{}_{,m}{}^{,m} - R^k_m u^m + \varrho f^k = 0, \tag{226.2}$$

where R^k_m is the Ricci tensor. Then it is easy to verify that a particular solution ${}_{\mathrm{P}}s^{km}$ of the system $s^{km}{}_{,m} + \varrho f^k = 0$, $s^{km} = s^{mk}$ is given by[2]

$${}_{\mathrm{P}}s^{km} = 2u^{(k,m)} - u^q{}_{,q}\, g^{km}. \tag{226.3}$$

We now turn our attention to determining the most general null stress[3].

227. General solution for a flat space. We suppose $\boldsymbol{s}$ to be twice continuously differentiable. In order that $s^{km}{}_{,m} = 0$ in a flat space of any number of dimensions, it is necessary and sufficient that there exist a tensor $\boldsymbol{b}$ such that[4]

$$s^{km} = b^{kmp}{}_{,p}, \quad \text{where} \quad b^{kmp} = -b^{kpm}. \tag{227.1}$$

The condition $s^{km} = s^{mk}$ may now be written in the form

$$(b^{kmp} - b^{mkp})_{,p} = 0. \tag{227.2}$$

---

[1] Cf. the treatment of this class of problems by FINZI and PASTORI [1949, *11*, Chap. IV, § 7].

[2] This result is suggested by the special case for flat spaces given by SCHAEFER [1953, *28*, § 5].

[3] Most of the material in the rest of this subchapter is taken from a work of TRUESDELL [1959, *11*].

[4] As observed by GWYTHER [1913, *2*], for a stress tensor which is not symmetric the analysis breaks off at this point. In the more general viewpoint allowing couple stress as well as ordinary stress, all differential conditions of equilibrium are prescriptions of divergences, as shown at the end of Sect. 205. Stress functions for the system (205.2), (205.10) are given by GÜNTHER [1958, *4*, § 3].

This condition, in turn, is equivalent to the existence of a tensor $\boldsymbol{c}$ such that

$$b^{kmp} - b^{mkp} = c^{kmpq}{}_{,q}, \tag{227.3}$$

where

$$c^{kmpq} = -c^{kmqp} = -c^{mkpq}. \tag{227.4}$$

Therefore

$$2b^{kmp} = (c^{pkmq} + c^{mpqk} + c^{kmpq})_{,q}, \tag{227.5}$$

so that (227.1) becomes

$$s^{km} = h^{pkmq}{}_{,pq}, \tag{227.6}$$

where

$$\left.\begin{aligned} h^{pkmq} &\equiv \tfrac{1}{2}(c^{pkmq} + c^{mqpk}), \\ h^{pkmq} &= -h^{kpmq} = -h^{pkqm} = h^{mqpk}. \end{aligned}\right\} \tag{227.7}$$

The foregoing derivation, given by DORN and SCHILD[1], shows that (227.6) furnishes the ***general solution of Cauchy's laws*** in a flat space of any dimension.

In the three-dimensional case, set

$$a_{rs} \equiv \tfrac{1}{4}\, e_{rpk}\, e_{sqm}\, h^{pkmq}, \tag{227.8}$$

so that

$$h^{pkmq} = e^{krp}\, e^{msq}\, a_{rs}, \qquad a_{rs} = a_{sr}. \tag{227.9}$$

Then (227.6) becomes the ***Gwyther-Finzi general solution***[2]:

$$s^{km} = e^{krp}\, e^{msq}\, a_{rs,pq}. \tag{227.10}$$

B. FINZI noticed that the tensor $\boldsymbol{a}$ is indeterminate[3] to within an arbitrary symmetric tensor $\boldsymbol{a}^0$ satisfying $e^{krp}\, e^{msq}\, a^0_{rs,pq} = 0$. As we have mentioned in

[1] [1956, *4*]. The basic idea was suggested by BELTRAMI [1892, *2*]. Cf. also MORERA [1892, *10*].

A variant of this derivation had been given earlier by GÜNTHER [1954, *8*, § 1]. He begins by introducing the skew-symmetric dual tensor of fourth order,

$$T_{kmpq} = e_{kmn}\, e_{pqs}\, T^{ns}, \tag{*}$$

which he interprets as a "transversal stress tensor". An explicit solution for T_{kmpq} in terms of stress functions is simply obtained. In DORN and SCHILD's proof, the duals appear in (227.8). While GÜNTHER's solution for the *dual* tensor is indeed valid, as he says, in a flat space of any dimension, to derive from it the ordinary stress tensor we must use the inverse of (*) and hence presume the number of dimensions to be three. Of course GÜNTHER's proof can be adjusted to the n-dimensional case also, but DORN and SCHILD's proof is equally simple and natural in all cases.

[2] While the Cartesian tensor form of (227.10) is almost obvious from an equation of KLEIN and WIEGHARDT [1905, *3*, Eq. (33)], they did not infer it. GWYTHER [1912, *3*] obtained (227.10) in orthogonal curvilinear co-ordinates, writing out the special cases appropriate to rectangular Cartesian, cylindrical polar, and spherical polar co-ordinates (cf. also [1911, *5*]). His steps are such as to imply the necessity of the result; the sufficiency is immediate. B. FINZI [1934, *2*, § 3] observed that (227.10) yields a symmetric tensor satisfying $s^{km}{}_{,m} = 0$; for a proof of completeness he was content to refer to the previously known fact that the special cases (227.12) and (227.13) are complete. While KRÖNER [1954, *11*, § 1] does not prove completeness, he begins from an invariant decomposition of symmetric tensors that might be made the basis of a rigorous proof.

[3] This fact is used by PERETTI [1949, *23*] to show that it is possible to choose $\boldsymbol{a}$ in such a way that from its components may be obtained simple expressions for the resultant force and moment of the stresses on a surface. Cf. also BLOKH [1950, *2*]. GÜNTHER [1954, *8*, § 3] gives simple expressions for the resultant force and torque on a body in terms of integrals of stress functions around particular curves. SCHAEFER [1955, *22*] discusses the nature of null stress on this basis. In a later work, SCHAEFER [1959, *10*] interprets the components of $\boldsymbol{a}$ as dynamical actions upon the bounding surface, conceived as being that of a plate and of a slab simultaneously.

Sects. 34 and 84, such a tensor is of the form $a^0_{mr} = b_{(m,r)}$, where $\boldsymbol{b}$ is an arbitrary vector. For a given $\boldsymbol{a}$, in rectangular Cartesian co-ordinates we may choose $\boldsymbol{b}$ so as to satisfy one or the other of the conditions

$$\left.\begin{aligned} b_{(m,r)} &= -a_{mr}, \quad r \neq m \\ b_{m,m} &= -a_{mm} \quad \text{(unsummed)}. \end{aligned}\right\} \tag{227.11}$$

These two choices of $\boldsymbol{b}$ show that for rectangular Cartesian co-ordinates there is no loss in generality in assuming in the first place that $\boldsymbol{a}$ is diagonal, or that the diagonal components of $\boldsymbol{a}$ are zero. The former alternative yields

$$s_{xx} = a^2_{,zz} + a^3_{,yy}, \qquad s_{xy} = -a^3_{,yx}, \qquad \text{etc.} \tag{227.12}$$

with $a^1 \equiv a_{xx}$, $a^2 \equiv a_{yy}$, $a^3 \equiv a_{zz}$; the latter alternative yields

$$s_{xx} = -2a^4_{,yz}, \qquad s_{xy} = (a^4_{,x} + a^5_{,y} - a^6_{,z})_{,z}, \qquad \text{etc.} \tag{227.13}$$

with $a^4 \equiv a_{23}$, $a^5 \equiv a_{31}$, $a^6 \equiv a_{12}$. These two forms of the general solution were obtained by MAXWELL and MORERA[1], respectively. As follows from the special choices of $\boldsymbol{a}$ made to derive them, these special forms are not invariant under transformations even of rectangular Cartesian co-ordinates. The explicit form for (227.10) in rectangular Cartesian co-ordinates, with no restrictions on the six potentials, may be obtained by adding together the right-hand sides of (227.12) and (227.13). Other special choices of the potentials are possible[2], but it by no means follows that a solution obtained by imposing three *arbitrary* conditions on the six potentials a_{km} remains complete[3].

An attempt to adjust the stress functions so as to satisfy the stress boundary condition $(203.6)_1$ on a rectangular parallelepiped has been made by FILONENKO-BORODICH[4].

228. Two applications: the rotationally symmetric case, and plane unsteady motion. If we write (227.10) explicitly in cylindrical polar co-ordinates, at the same time supposing that all derivatives with respect to the azimuth angle are zero, we

---

[1] [1868, *12*], [1870, *4*]; [1892, *9*]. Using results given by BELTRAMI [1892, *2*], MORERA [1892, *10*] modified his derivation so as to yield (227.12) alternatively to (227.13). Cf. also GWYTHER [1913, *2*]. A literature devoted mainly to rediscovery of known results regarding this subject has arisen recently; cf. KUZMIN [1945, *3*], WEBER [1948, *38*], MORINAGA and NÔNO [1950, *19*], SCHAEFER [1953, *28*], LANGHAAR and STIPPES [1954, *12*], ORNSTEIN [1954, *16*].

[2] Cf. MORINAGA and NÔNO [1950, *19*, § 3]. BLOKH [1950, *2*] lists the essentially different forms which result from such special choices: 5 in rectangular Cartesian co-ordinates, 20 in general co-ordinates, 18 in cylindrical co-ordinates, 10 in cylindrical co-ordinates for rotationally symmetric problems, 19 in spherical co-ordinates.

[3] What reductions are possible is not obvious. In writings on stress functions there is a deplorable custom of inferring completeness by merely counting the number of arbitrary functions. Apart from the logical gap in such inference, its danger is illustrated by the solution written down without proof of completeness by PRATELLI [1953, *25*, § 1]:

$$s^{km} = e^{kpq} e^{mrs} (F_{,pr} g_{qs} + H_{,ps} g_{qr} + K_{,qs} g_{pr}). \tag{A}$$

While indeed a solution for any choice of the three arbitrary potentials F, H, K, it is not general. In fact, by using the expression for the product $\varepsilon^{kpq}\varepsilon^{mrs}$ as a determinant of δ^r_n's, we may put (A) into the form

$$s^{km} = P^{,q}{}_{,q}\, g^{km} - P^{,km}, \tag{B}$$

where $P \equiv F - H + K$, and it is easy to exhibit solutions of $s^{km}{}_{,m} = 0$ which cannot be expressed in the form (B).

[4] [1951, *7* and *8*] [1957, *4*]. The work rests on trigonometric series or special functional forms.

obtain[1]

$$\left.\begin{aligned}
\widehat{rr} &= a^2_{,zz} + \frac{1}{r} a^3_{,r} - \frac{2}{r} a^5_{,z},\\
\widehat{\theta\theta} &= a^3_{,rr} + a^2_{,zz} - 2a^5_{,rz},\\
\widehat{zz} &= a^2_{,rr} + \frac{2}{r} a^2_{,r} - \frac{1}{r} a^1_{,r},\\
\widehat{rz} &= -\left[a^2_{,r} + \frac{1}{r} a^2 - \frac{1}{r} a^1\right]_{,z},\\
\widehat{r\theta} &= \left[a^4_{,r} - \frac{1}{r} a^4 - a^6_{,z}\right]_{,z},\\
\widehat{\theta z} &= -a^4_{,rr} - \frac{1}{r} a^4_{,r} + \frac{1}{r^2} a^4 + a^6_{,zr} + \frac{2}{r} a^6_{,z},
\end{aligned}\right\} \tag{228.1}$$

where commas denote partial derivatives, and where we have set $a^1 \equiv a_{rr}$, $a^2 \equiv a_{\theta\theta}/r^2$, $a^3 \equiv a_{zz}$, $a^4 \equiv a_{\theta z}/r$, $a^5 \equiv a_{zr}$, $a^6 \equiv a_{r\theta}/r$. The potentials a^1, a^2, a^3, and a^5 occur only in the first four members of Eq. (228.1); the potentials a^4 and a^6, only in the last two. Rotationally symmetric stress distributions in which $\widehat{r\theta}=0$, $\widehat{\theta z}=0$ are often called *torsionless*; the most general stress of this kind is obtained by setting $a^4=0$, $a^6=0$ in (228.1). In any case, the six potentials may be reduced to three in a variety of ways[2]. For example, if we set

$$\left.\begin{aligned}
L_{,r} &\equiv a^2_{,r} + \frac{1}{r} a^2 - \frac{1}{r} a^1,\\
M &\equiv a^2_{,zz} + \frac{1}{r} a^3_{,r} - \frac{2}{r} a^5_{,z} - L_{,zz},
\end{aligned}\right\} \tag{228.2}$$

then the first four members of (228.1) become[3]

$$\left.\begin{aligned}
\widehat{rr} &= L_{,zz} + M, & \widehat{\theta\theta} &= (rM)_{,r} + L_{,zz},\\
\widehat{zz} &= L_{,rr} + \frac{1}{r} L_{,r}, & \widehat{rz} &= -L_{,zr},
\end{aligned}\right\} \tag{228.3}$$

furnishing the *general solution for torsionless rotationally symmetric stress*. Similarly, if we put

$$W \equiv -r^2\left[a^4_{,r} - \frac{1}{r} a^4 - a^6_{,z}\right], \tag{228.4}$$

the last two members of (228.1) become[4]

$$\widehat{r\theta} = -\frac{1}{r^2} W_{,z}, \qquad \widehat{\theta z} = \frac{1}{r^2} W_{,r}, \tag{228.5}$$

furnishing the *general solution for purely torsional stress*.

A more interesting application, resting upon a device to be used more strikingly in Sect. 229, begins with the observation that in the case of plane motion,

[1] BRDIČKA [1957, *2*, § 6]. A more symmetrical expression is given by MARGUERRE [1955, *16*], but his potentials are connected by a condition of compatibility, and there is no proof of completeness.

[2] Cf. the work of BLOKH [1950, *2*].

[3] This solution, whose completeness is easy to prove also directly from the equations of equilibrium, was obtained by LOVE [1906, *5*, § 188]. Variants are given by BRDIČKA [1957, *2*, § 4].

[4] This is another form of the solution of VOIGT mentioned in footnote 1, p. 583. Cf. also MICHELL [1900, *7*, p. 133], PHILLIPS [1934, *4*].

Eqs. (211.5) with $\boldsymbol{F}=0$ are of the form $s^{km}{}_{,m}=0$ in a flat space of three dimensions with rectangular Cartesian co-ordinates x, y, t. Restricting attention to these special co-ordinates, we apply the solution (227.10). After time differentiations and time components are written explicitly, the result turns out to be in tensorial form under transformation of general co-ordinates in the plane:

$$\left.\begin{aligned} -\varrho &= s^{33} = e^{\delta\sigma} e^{\varphi\psi} a_{\delta\varphi,\sigma\psi}, \\ -\varrho \dot{x}^{\gamma} &= s^{\gamma 3} = e^{\gamma\delta} e^{\varphi\psi} (a'_{\delta\varphi,\psi} - a_{3\varphi,\delta\psi}), \\ t^{\gamma\lambda} - \varrho \dot{x}^{\gamma} \dot{x}^{\lambda} &= s^{\gamma\lambda} = e^{\gamma\varphi} e^{\lambda\psi} (a_{33,\varphi\psi} + a''_{\varphi\psi}) - (e^{\gamma\varphi} e^{\lambda\psi} + e^{\gamma\psi} e^{\lambda\varphi})\, a'_{3\psi,\varphi}, \end{aligned}\right\} \tag{228.6}$$

where a prime denotes $\partial/\partial t$ and where the range of indices is 1, 2. In the steady case, this reduces to AIRY's solution (224.10) with $A=-a_{33}$. The six potentials may be reduced to three in various ways. For example, if we choose $\boldsymbol{a}$ to be diagonal in a particular rectangular Cartesian co-ordinate system, we obtain[1]

$$\left.\begin{aligned} &-\varrho = a^{1}_{,yy} + a^{2}_{,xx}, \quad \varrho\dot{x} = a^{2}_{,xt}, \quad \varrho\dot{y} = a^{1}_{,yt}, \\ &t_{xx} - \varrho\dot{x}^2 = -A_{,yy} + a^{2}_{,tt}, \quad t_{yy} - \varrho\dot{y}^2 = -A_{,xx} + a^{1}_{,tt}, \\ &t_{xy} - \varrho\dot{x}\dot{y} = A_{,xy}, \end{aligned}\right\} \tag{228.7}$$

generalizing AIRY's solution to the case of arbitrary plane motion.

229. Stress functions for membranes. From the result at the end of Sect. 212, the problem of equilibrium of a membrane leads to a system of the form (226.1); explicitly, we are to find the most general symmetric tensor $s^{\delta\xi}$ satisfying

$$\frac{\partial s^{\delta\xi}}{\partial u^{\xi}} + \left\{ {\delta \atop \lambda\,\xi} \right\} s^{\lambda\xi} + \left\{ {\xi \atop \lambda\,\xi} \right\} s^{\delta\lambda} = 0, \tag{229.1}$$

where the $\left\{ {\delta \atop \lambda\,\xi} \right\}$ are Christoffel symbols based on the positive definite surface metric tensor $a_{\delta\xi}$, and the range of indices is 1, 2.

There have been several attacks upon this intrinsic problem, which turns out to be more difficult than those considered in the preceding sections because for curved spaces an analogue of STOKES's representation (App. 32.9) of a solenoidal field, which yields (227.1) in a Euclidean space of arbitrary dimension, is not known. A different approach is needed. STORCHI[2] has obtained a solution in geodesic co-ordinates by a direct and laborious reduction. At the present writing, a general invariant solution is not yet known, but we present what results are available.

The similarity in form between the conditions of compatibility (84.3) and the general solution (227.10) has been remarked for half a century. This simi-

---

[1] A similar but not obviously identical solution is obtained by KILCHEVSKI [1953, *14*].

[2] [1950, *28*].

STORCHI [1950, *27*] observed that if for an arbitrary surface we choose co-ordinates x, y so that $ds^2 = \lambda(dx^2+dy^2)$, then a solution is furnished by the formulae $\lambda^2 s^{xx} = \partial^2 A/\partial y^2$, $\lambda^2 s^{xy} = \lambda^2 s^{yx} = -\partial^2 A/\partial x\,\partial y$, $\lambda^2 s^{yy} = \partial^2 A/\partial x^2$, provided $\partial^2 A/\partial x^2 + \partial^2 A/\partial y^2 = 0$. For such solutions the mean pressure vanishes: $\lambda(s^{xx}+s^{yy}) = s^{\delta}_{\delta} = 0$, but it is not shown that all solutions such that $s^{\delta}_{\delta}=0$ are included.

STORCHI treated the case of a surface of revolution in [1949, *28*]; in [1952, *18*] [1953, *29*], the case when a general solution involving derivatives of orders no higher than the fourth is possible. Cf. the counterpart expressed in terms of the conditions of compatibility which we have mentioned in Sect. 84. A solution for minimal surfaces is initiated by COLONNETTI [1956, *3*]. We do not discuss the older solutions for special cases defined by conditions of inextensibility.

For Riemannian spaces of 2, 3, and 4 dimensions, a special solution containing an arbitrary function of the total curvature is obtained by STORCHI [1957, *13*].

larity may be used to yield a direct solution[1] of (226.1) based upon the classical principle of virtual work. While that principle will be developed in Sect. 232, here we need only a special case which may be verified at once, namely, that $\boldsymbol{s}$ satisfies (226.1) if and only if

$$\int_{v} s^{km} d_{km} \, dv = 0 \tag{229.2}$$

for all fields $\boldsymbol{d}$ such that $d_{km} = c_{(k,m)}$ for some vector $\boldsymbol{c}$ vanishing upon the boundary. Here, as henceforth in this section, we systematically neglect surface integrals as having no effect on the differential equations we seek; then the condition (229.2) follows at once from GREEN's transformation and hence is valid in any space where covariant differentiation is defined. Now suppose the conditions of compatibility (Sect. 84) for the system $d_{km} = c_{(k,m)}$ have the explicit form

$$0 = \alpha^{km} d_{km} + \beta^{kmr} d_{km,r} + \gamma^{kmrs} d_{km,rs} + \delta^{kmrst} d_{km,rst} + \cdots, \tag{229.3}$$

where the tensors $\alpha, \beta, \ldots$, are symmetric in the first two indices. Introducing a multiplier A, we replace (229.2) by

$$\int_{v} [s^{km} d_{km} - A(\alpha^{km} d_{km} + \beta^{kmr} d_{km,r} + \cdots)] \, dv = 0, \tag{229.4}$$

where the variation of $\boldsymbol{d}$ is now unrestricted. Equivalently,

$$\int [s^{km} - \alpha^{km} A + (\beta^{kmr} A)_{,r} - \cdots] \, d_{km} \, dv = 0. \tag{229.5}$$

Hence for any $\boldsymbol{s}$ satisfying (226.1) there exists a function A such that

$$s^{km} = \alpha^{km} A - (\beta^{kmr} A)_{,r} + (\gamma^{kmrs} A)_{,rs} - \cdots. \tag{229.6}$$

In this general solution, the stress function A appears as a multiplier for the constant expressing the compatibility of the field $\boldsymbol{d}$ with virtual displacements in the space considered.

To apply to the two-dimensional case the local result obtained, consider first a surface of constant Gaussian curvature K. By comparing (229.3) with (84.13) we have

$$\alpha^{\delta\xi} = K a^{\delta\xi}, \quad \beta^{\delta\xi\sigma} = 0, \quad \gamma^{\delta\xi\sigma\varphi} = e^{\delta\sigma} e^{\xi\varphi}. \tag{229.7}$$

Substitution into (229.6) yields the *general solution for surfaces of constant curvature*[2]:

$$s^{\delta\xi} = e^{\delta\sigma} e^{\xi\varphi} A_{,\sigma\varphi} + K A a^{\delta\xi}. \tag{229.8}$$

Second, consider a surface applicable upon a surface of revolution. By comparing (229.3) with (84.15) we have

$$\left.\begin{aligned} \alpha^{\delta\xi} &= K_{,\lambda} K^{,\lambda} a^{\delta\xi} + K^{,\delta} K^{,\xi}, & \beta^{\delta\xi\sigma} &= K K^{,\sigma} a^{\delta\xi}, \\ \gamma^{\delta\xi\sigma\varphi} &= 0, & \delta^{\delta\xi\sigma\varphi\psi} &= -e^{\sigma(\delta} e^{\xi)\varphi} K^{,\psi}. \end{aligned}\right\} \tag{229.9}$$

[1] Given by TRUESDELL [1957, *17*], revising work of L. FINZI to be described in Sect. 224. Earlier SCHAEFER [1953, *28*, § 4] had introduced multipliers in just the same way, concluding that "Jeder Verträglichkeitsbedingung ist eine Spannungsfunktion zugeordnet", but his presentation employs results of a kind valid only in flat spaces, and he did not mention any further possibilities. GÜNTHER [1954, *8*, § 2] had in effect noted the method and had remarked that the tensor of stress functions in a three-dimensional flat space may be interpreted as Lagrangean multipliers expressing the reactions against the geometrical constraints but had concluded that "no new point of view results".

[2] B. FINZI [1934, *2*, § 2] verified that (229.8) satisfies (226.1) when K is constant but did not prove the completeness of this solution; his result is rediscovered by LANGHAAR [1953, *15*]. B. FINZI [1934, *2*, §§ 4, 7] conjectured also corresponding general solutions for spaces of three and four dimensions with constant curvature; TRUESDELL [1959, *11*, § 12] establishes their completeness by this method.

Substitution into (229.6) yields the *general solution for a surface applicable upon a surface of revolution of non-constant curvature*[1]:

$$\left.\begin{aligned} s^{\delta\xi} = {} & [-K K_{,\lambda}^{;\lambda} a^{\delta\xi} + K^{,\delta} K^{,\xi} + e^{\sigma(\delta} e^{\xi)\lambda} K^{\varphi}_{,\lambda\sigma\varphi}] A + \\ & + [-K K^{,\lambda} a^{\delta\xi} + e^{\sigma(\delta} e^{\xi)\psi} K^{,\lambda}{}_{,\psi\sigma} + 2 e^{\lambda(\delta} e^{\xi)\varphi} K^{,\psi}{}_{,\varphi\psi}] A_{,\lambda} + \\ & + [2 e^{\lambda(\delta} e^{\xi)\psi} K^{,\sigma}{}_{,\psi} + e^{\sigma(\delta} e^{\xi)\lambda} K_{,\gamma}^{;\gamma}] A_{,\lambda\sigma} + e^{\sigma(\delta} e^{\xi)\lambda} K^{,\varphi} A_{,\lambda\sigma\varphi}. \end{aligned}\right\} \quad (229.10)$$

If a condition of compatibility of the form (229.3) were available for a general surface, we should be able to write down the general solution of (229.1) in terms of a single stress function. Since STORCHI obtained such a solution, in geodesic co-ordinates, we know that a condition of compatibility of the form (229.3) exists and involves derivatives of $\boldsymbol{d}$ up to the fifth order; however, the result cited at the end of Sect. 84 is a system of three equations of fourth order, connected by two identities. By ingenious formal devices, GRAIFF[2] has exhibited a general tensorial solution in terms of derivatives of order up to 4 of two arbitrary scalar functions. She has inferred also that one of the two may always be set equal to zero, but in order to justify this inference, it would be necessary to investigate the conditions of solubility of certain differential equations occurring in her work.

In spaces of dimension greater than two, the conditions of compatibility will be tensorial rather than scalar equations. For a space of three dimensions, tensors α^{uvkm}, β^{uvkmr}, ..., replace the tensors α^{km}, β^{kmr}, ..., in (229.3). In place of a scalar multiplier A we then have a tensor A_{uv} in (229.4), and hence also in (229.6). The reader may easily verify[3] that use of (84.3) thus yields the solution (227.10). In the general case, if the conditions of compatibility have the form

$$L(\boldsymbol{d}) = 0 \quad (229.10\text{a})$$

where L is a symmetric tensorial linear differential operator, then the general solution of (226.1) is

$$\boldsymbol{s} = \bar{L}(\boldsymbol{A}), \quad (229.10\text{b})$$

where $\bar{L}$ is the operator adjoint to L.

When the solution of (229.1) is known, we may solve $(212.22)_1$ by quadratures; substitution of the result into $(212.22)_2$ then yields a single linear differential equation, of fifth order in the general case, to be satisfied by the single stress function.

A different approach was presented by PUCHER[4]; we now follow a formal derivation of his result by L. FINZI[5]. The essential differences from the foregoing are two-fold. (1) both of the Eqs. (212.22) are used from the start, so that the purely intrinsic problem is never solved, and (2) special co-ordinates are employed, yielding an extraordinarily simple result of the same form as AIRY's solution (224.5), except that in general the stress function must satisfy a linear differential equation of second order. The first step is to replace $\boldsymbol{s}$ by a corresponding tensor density,

$$w^{\delta\xi} \equiv \sqrt{a}\, s^{\delta\xi}, \quad (229.11)$$

so that $(212.22)_1$ assumes the form

$$\frac{\partial w^{\delta\xi}}{\partial u^{\xi}} + \left\{\begin{matrix}\delta \\ \lambda\,\xi\end{matrix}\right\} w^{\lambda\xi} + \sqrt{a}\, F^{\delta} = 0. \quad (229.12)$$

---

[1] Given in geodesic co-ordinates, with a proof of completeness, by STORCHI; in the above form, with an imperfect proof of completeness, by L. FINZI.

[2] [1959, *5*, §§ 4 to 5].

[3] Also, using the tensorial conditions of compatibility for a general two-dimensional surface, by this method we may obtain a general solution of (229.1) expressed in terms of fourth derivatives of a tensor A_{uv}, but, as mentioned above, the ultimate solution should rather be expressed in terms of fifth derivatives of a single stress function.

[4] [1934, *5*] [1939, *12*].

[5] [1955, *9*].

Let a rectangular Cartesian equation of $\mathfrak{s}$ be $z=z(x, y)$, and on $\mathfrak{s}$ itself choose x and y as curvilinear co-ordinates; this amounts to referring $\mathfrak{s}$ to special co-ordinate curves so chosen that their projection upon some plane is a rectangular Cartesian net. In these co-ordinates, which we write as x^δ, we have

$$\left\{ {\delta \atop \lambda\,\xi} \right\} = z_{,\varphi}\, z_{,\delta\xi}\, a^{\varphi\delta}, \qquad \sqrt{a}\, b_{\delta\xi} = z_{,\delta\xi}, \tag{229.13}$$

where the comma now indicates a partial derivative. Eq. $(212.22)_2$ thus assumes the form

$$z_{,\delta\xi}\, w^{\delta\xi} = \sqrt{a}\,(F_x z_{,x} + F_y z_{,y} - F_z), \tag{229.14}$$

where F_x, F_y, F_z are the rectangular Cartesian components of the surface load $\boldsymbol{F}$. Putting (229.13) and (229.14) into (229.12) yields after some manipulation[1]

$$w^{11}{}_{,x} + w^{12}{}_{,y} + \sqrt{a}\, F_x = 0, \qquad w^{21}{}_{,x} + w^{22}{}_{,y} + \sqrt{a}\, F_y = 0. \tag{229.15}$$

Since this result is of the same form as (224.1), with more general $\boldsymbol{F}$ but with $\varrho\ddot{\boldsymbol{x}} = 0$, by (224.5) we read off the general solution

$$w^{11} = -P_{,yy} - \int \sqrt{a}\, F_x\, dx, \quad w^{22} = -P_{,xx} - \int \sqrt{a}\, F_y\, dy, \quad w^{12} = w^{21} = P_{,xy}. \tag{229.16}$$

Substituting these results into (229.14) yields

$$\left.\begin{aligned} &z_{,xx} P_{,yy} + z_{,yy} P_{,xx} - 2 z_{,xy} P_{,xy} \\ &\quad = - z_{,xx} \int \sqrt{a}\, F_x\, dx - z_{,yy} \int \sqrt{a}\, F_y\, dy - \sqrt{a}\,(F_x z_{,x} + F_y z_{,y} - F_z). \end{aligned}\right\} \tag{229.17}$$

This is *Pucher's general solution*. When the membrane is a plane, Eq. (229.17) for PUCHER's function is satisfied automatically, and comparison with (224.5) shows that PUCHER's solution reduces to AIRY's.

In the special case when $\mathfrak{s}$ is a surface of revolution with meridian $z = z(r)$ in an azimuthal plane $\theta = \text{const}$, PUCHER obtained the specialization[2]

$$\left.\begin{aligned} w\langle r r\rangle &= -\frac{1}{\cos\phi}\left[\frac{1}{r^2} P_{,\theta\theta} + \frac{1}{r} P_{,r} + A\right], \\ w\langle\theta\,\theta\rangle &= -\cos\phi\,[P_{,rr} + B], \\ w\langle r\,\theta\rangle &= \left[\frac{1}{r} P_{,\theta}\right]_{,r}, \\ \frac{z''}{r}\left(\frac{1}{r} P_{,\theta\theta} + P_{,r}\right) &+ \frac{z'}{r} P_{,rr} = Z - (A z')_{,r} - \frac{z'}{r} A, \end{aligned}\right\} \tag{229.18}$$

where

$$\tan\phi = z', \qquad r A = \int \left[R + \frac{1}{r} \int \Theta\, r\, d\theta\right] r\, dr, \qquad B = \int \Theta\, r\, d\theta, \tag{229.19}$$

R, Θ and Z being the physical components of $\boldsymbol{F}$.

230. General solution for unsteady motion in three dimensions. For a flat space of n dimensions, we have the solution (227.6), whose completeness was established in Sect. 227. When $n = 4$, letting Greek capital indices have the range 1, 2, 3, 4, we may put

$$A_{\Sigma\Lambda\Phi\Psi} \equiv \tfrac{1}{16}\, e_{\Omega\Gamma\Sigma\Lambda}\, e_{\Delta\Xi\Phi\Psi}\, h^{\Omega\Gamma\Delta\Xi}, \tag{230.1}$$

so that

$$h^{\Gamma\Lambda\Delta\Xi} = e^{\Lambda\Gamma\Sigma\Phi}\, e^{\Delta\Xi\Psi\Omega}\, A_{\Sigma\Phi\Psi\Omega}. \tag{230.2}$$

---

[1] Note that $F^\delta = a^{1\delta} F_x + a^{2\delta} F_y + (a^{1\delta} z_{,x} + a^{2\delta} z_{,y})\, F_z$.

[2] Further properties of the function P and of Eq. (229.17) have been worked out by PUCHER [1939, *12*] and L. FINZI [1955, *9*].

From (227.6) it follows that[1]

$$s^{\Gamma\Delta} = e^{\Gamma\Lambda\Sigma\Phi} e^{\Delta\Xi\Psi\Omega} A_{\Sigma\Phi\Psi\Omega,\Lambda\Xi}, \tag{230.3}$$

where (227.7) imply the following conditions of symmetry for $\boldsymbol{A}$:

$$A_{\Sigma\Phi\Psi\Omega} = -A_{\Phi\Sigma\Psi\Omega} = -A_{\Sigma\Phi\Omega\Psi}, \quad A_{\Sigma\Phi\Psi\Omega} = A_{\Psi\Omega\Sigma\Phi}. \tag{230.4}$$

Inspection of (230.3) shows that only the second of these sets of symmetry conditions is essential; the first set may be abandoned without impairing the solution. The tensor $\boldsymbol{A}$ is indeterminate to within a tensor $\boldsymbol{A}^0$ such that

$$e^{\Gamma\Lambda\Sigma\Phi} e^{\Delta\Xi\Psi\Omega} A^0_{\Sigma\Phi\Psi\Omega,\Lambda\Sigma} = 0.$$

The most general such tensor is a linear combination of tensors of the type $B_{\Sigma\Phi\Psi,\Omega}$; to satisfy the essential symmetry condition $(230.4)_3$, we may choose $B_{\Sigma\Phi\Psi,\Omega} + B_{\Psi\Omega\Sigma,\Phi}$; if, finally, we wish to satisfy also the condition $(230.4)_{1,2}$, we have

$$\left.\begin{aligned} A^0_{\Sigma\Phi\Psi\Omega} &= 4B_{[\Sigma\Phi][\Psi,\Omega]} + 4B_{[\Psi\Omega][\Sigma,\Phi]}, \\ &= B_{\Sigma\Phi\Psi,\Omega} - B_{\Sigma\Phi\Omega,\Psi} - B_{\Phi\Sigma\Psi,\Omega} + B_{\Phi\Sigma\Omega,\Psi} + \\ &\quad + B_{\Psi\Omega\Sigma,\Phi} - B_{\Psi\Omega\Phi,\Sigma} - B_{\Omega\Psi\Sigma,\Phi} + B_{\Omega\Psi\Phi,\Sigma}. \end{aligned}\right\} \tag{230.5}$$

Since classical space-time need not be regarded as a flat four-dimensional space, these formulae do not necessarily have invariant significance for it (but cf. Sect. 153). However, *in rectangular Cartesian co-ordinates in an inertial frame* (211.5) with $\boldsymbol{F}=0$ do reduce to the form

$$T^{\Gamma\Delta}{}_{,\Delta} = 0, \qquad T^{\Gamma\Delta} = T^{\Delta\Gamma}. \tag{230.6}$$

For these special co-ordinates, then, the solution (230.3) is general. In this solution, we may write time differentiations and time components explicitly. The resulting formulae, derived in rectangular co-ordinates, turn out to be of tensorial form under transformations of the space co-ordinates alone. These formulae, valid for all curvilinear co-ordinate systems in an inertial frame, are[2]:

$$\left.\begin{aligned} -\varrho &= s^{44} = e^{smn} e^{pqr} A_{mnqr,sp}, \\ -\varrho \dot{x}^k &= s^{k4} = -e^{ksm} e^{pqr} (A'_{smqr,p} + 2A_{4mqr,sp}) \\ t^{km} - \varrho \dot{x}^k \dot{x}^m &= s^{km} = 4e^{kpq} e^{msn} A_{4p4s,qn} + \\ &\quad + 2(e^{kpq} e^{msn} + e^{ksn} e^{mpq}) A'_{4npq,s} + e^{kpq} e^{msn} A''_{pqsn}, \end{aligned}\right\} \tag{230.7}$$

---

[1] B. FINZI [1934, *2*, § 5] wrote down (230.3) and symmetry conditions consisting in (230.4) and a further requirement which we do not verify; he was content to infer completeness by counting the number of assignable arbitrary functions. A somewhat involved proof was given by MORINAGA and NÔNO [1950, *19*, § 4]; we do not follow the argument whereby they claim [ibid., § 5] to establish the alternative form

$$s^{\Gamma\Delta} = e^{\Gamma\Lambda\Phi\Psi} e^{\Delta\Xi\Sigma}{}_{\Psi} A_{\Phi\Xi,\Lambda\Sigma};$$

they give corresponding results in n dimensions.

[2] B. FINZI [1934, *2*, § 6]. In rectangular Cartesian co-ordinates, this result is rediscovered by ARZHANIKH [1952, *1*] (while he uses 21 potentials, obviously one may be eliminated). KILCHEVSKI [1953, *14*] observes that if $R^{\Gamma}{}_{\Delta}$ is the contracted Riemann tensor based on the Riemannian metric tensor $G_{\Gamma\Delta}$, then in any Riemannian 4-space the quantities $T^{\Gamma\Delta} \equiv R^{\Gamma\Delta} - \frac{1}{2} G^{\Gamma\Delta} R^{\Phi}{}_{\Phi}$ satisfy $T^{\Gamma\Delta}{}_{,\Delta} = 0$. Putting $G_{\Gamma\Delta} = \delta_{\Gamma\Delta} + \varepsilon H_{\Gamma\Delta}$, he calculates $T^{\Gamma\Delta} = \varepsilon Q^{\Gamma\Delta} + O(\varepsilon^2)$; hence follows $\partial Q^{\Gamma\Delta}/\partial x^{\Delta} = 0$, so that $Q^{\Gamma\Delta}$, in rectangular Cartesian co-ordinates, gives a solution of the type presented in the text above. A similar approach, involving a detour through relativity theory, is presented by MCVITTIE [1953, *18*, § 2]; that his solution is not general is remarked by WHITHAM [1954, *26*], who obtains what appears to be a special case of FINZI's solution in a special co-ordinate system. Rediscoveries of other special cases are made by MILNE-THOMSON [1957, *9*] and by BLANKFIELD and MCVITTIE [1959, *1* and *2*].

where primes denote $\partial/\partial t$, and where the symmetries of the tensors A_{pqmn}, A_{4npq}, and A_{4p4q} may be read off from (230.4). This is ***Finzi's general solution*** of the equations of balance of mass and momentum in an inertial frame. In the case of equilibrium, this solution reduces to (227.10) with $4A_{4p4m}=a_{pm}$.

In a particular co-ordinate system, by means of (230.5) we may impose 14 conditions upon the 20 independent potentials occurring in the solution (230.7). For example, we are tempted to take A_{4p4m} as diagonal, A_{4mqr} as zero, and A_{mnqr} as zero when $m \neq q$ and $n \neq r$. We have been unable to prove that it is possible to choose $\boldsymbol{B}$ in (230.5) in such a way as to justify this special choice of $\boldsymbol{A}$; if it is legitimate, then, writing $A^1 \equiv 4A_{4141}, \ldots, A^4 \equiv 2A_{2323}, \ldots$, in this case we reduce (230.3) to the form

$$\left.\begin{aligned} -\varrho = A^4_{,xx} + A^5_{,yy} + A^6_{,zz}, \quad \varrho\dot{x} = A^4_{,xt}, \ldots, \\ t_{xx} - \varrho\dot{x}^2 = A^2_{,zz} + A^3_{,yy} + A^4_{,tt}, \ldots, \quad t_{xy} - \varrho\dot{x}\dot{y} = -A^3_{,xy}, \ldots, \end{aligned}\right\} \tag{230.8}$$

extending MAXWELL's formulae (227.12) so as to yield a simple yet general solution in terms of six potentials.

When the field theories were discovered in the eighteenth century, solutions in arbitrary functions such as those presented in this subchapter were sought earnestly, but, for the most part, sought in vain. In the nineteenth century, researches on partial differential equations turned away from such general solutions so as to concentrate upon boundary-value problems. When, in the twentieth century, the general solutions were at last obtained, scarce attention was paid to them, and to this day they remain virtually unknown. Though so far they have been used but rarely, they might turn out to be illuminating in studies of underdetermined systems, where the conventional viewpoint of partial differential equations has gained little.

V. Variational principles.

231. On the qualities of variational principles. In favor of variational principles as expressions of physical laws it is commonly alleged[1]:

1. They are statements about a system as a whole, rather than the parts that it comprises.

2. Since they refer to the extremum of a scalar, they are invariant, and may be used to derive the special forms appropriate to any particular description.

3. They imply boundary conditions and jump conditions as well as differential equations.

4. They automatically include the effects of constraints, without requiring that the corresponding reactions be known.

5. They have heuristic value for suggesting generalizations[2].

No. 2 is outmoded, now that the principles of tensor analysis offer us a simpler and more direct method, used throughout this treatise, for obtaining invariant statements. No. 3 is shared by the direct statement of physical laws in *integral form* as *equations of balance*[3], as shown by the development of continuum mechanics given earlier in this chapter (cf. especially Sects. 203 and 205). Moreover, the boundary conditions emerging from a variational principle depend upon what boundary integrals, if any, are included in the statement of the principle, and the selection of these boundary integrals is not always dictated by the physical idea which the variational principle is assumed to express. No. 4 is a somewhat dubious blessing, since only a special kind of constraints is included

[1] Cf. HELLINGER [1914, *4*, § 1], who speaks of "the pregnant brevity".

[2] A sixth, the use of direct variational methods to calculate or prove existence of solutions, may be added in cases where a determinate system is considered but in the present treatise, devoted to underdetermined systems, is not relevant.

[3] Cf. footnote 4, p. 232.

in each case, namely, those constraints having no effect on the quantity being varied. In mechanics, these are typically constraints which do no work. Not all constraints are of this kind, and for those which are not, the variational approach requires as direct a statement as does any other. No. 5, while having a basis in the physics of this century, is largely an expression of taste.

There remains only No. 1, along with the elegance that variational principles sometimes exhibit. Both these are reduced if not annulled when the variational principle itself is awkward of unnatural. This is usually the case in continuum mechanics. The lines of thought which have led to beautiful variational statements for systems of mass-points have been applied in continuum mechanics also, but only rarely are the results beautiful or useful.

For completeness, we now present variational principles and related topics, but we regard them as derivative and subservient to the principles of mechanics already developed. In particular, *no variational principle has ever been shown to yield Cauchy's fundamental theorem* (203.4) in its basic sense as asserting that existence of the stress vector implies the existence of the stress tensor[1]. For the statements henceforth we take the stress tensor rather than the stress vector as given. Our purpose, in each case, is to learn the role of the effective force of the stress, $t^{km}{}_{,m}$, in modifying the classical theorems concerning mass-points[2].

Our presentation concerns solely the formal problem[3] of setting up expressions such that the vanishing of their first variation is equivalent to CAUCHY's laws. Analytical questions and, except in Sect. 235, the existence of minima are not discussed.

232. Virtual work and the Lagrange-D'Alembert principle. The principle of *virtual work*, which when the reaction of inertia is included may be called *the Lagrange-D'Alembert principle*, is the oldest general variational form of the equations of mechanics. We begin by following but extending the traditional development[4] of the principle. Some variants are presented afterward.

Corresponding to a sequence of independent variations $\delta x_k, \delta x_{km}, \delta x_{kmp}, \ldots$, that is, a set of arbitrary covariant fields, we define the *virtual work* done on a

[1] The derivation given by HELLINGER [1914, *4*, § 3a] fails through *petitio principi*, since the stress components appear in the original variational principle. We do not understand the remark attributed to CARATHÉODORY by MÜLLER and TIMPE [1906, *6*, footnote 32]. Existence of the stress tensor can be proved from variational principles which assume the existence of an internal energy having a special functional form. Such results are presented in Sects. 232A, 236, and 262.

[2] It is possible simply to transcribe the theorems for systems of mass-points, written as Stieltjes integrals over material volumes, and then add $t^{km}{}_{,m}$ to the contravariant force vector. Cf. EICHENWALD [1939, *7*].

[3] We do not attempt to discuss variational principles from the axiomatic or conceptual standpoint. For the difficult question of the contrast between the momentum principle and the Lagrange-D'Alembert principle, cf. HAMEL [1908, *4*, Kap. 2, §§ 1, 3].

[4] HAUGHTON [1855, *2*, pp. 99–100], KIRCHHOFF [1876, *2*, Eilfte Vorlesung, § 5], BELTRAMI [1881, *1*, pp. 385–388], HELLINGER [1914, *4*, §§ 3a–b]. A variant, bringing in explicitly the dependence of general curvilinear co-ordinates on rectangular Cartesian co-ordinates, was given by MORERA [1885, *6*, § 1]. We do not list the many sources that formulate a principle of virtual work for special systems such as rods or membranes, nor do we trace the origin of the principle for continuous media in general through the studies of LAGRANGE on perfect fluids, of GREEN and KELVIN on elastic bodies. The first treatment valid for general continua is that of PIOLA (1833), to be given below. The "general formula" of LAGRANGE [1788, *1*, Seconde Partie, Seconde Sect., ¶ 7] is valid only for systems of mass-points and certain other special systems. For D'ALEMBERT's principle, see above, p. 532, footnote 1.

All the foregoing references concern only the classical case, where the terms in δx_{km}, $\delta x_{kmp}, \ldots$, are absent from (232.1). The general theory given here is suggested by an intermediate case due to HELLINGER [1914, *4*, § 4b].

body $\mathscr{V}$ as the linear form[1]

$$\left.\begin{aligned}\mathfrak{A} \equiv \oint_{\mathscr{S}} [s^k \delta x_k + s^{km} \delta x_{km} + s^{kmp} \delta x_{kmp} + \cdots] \, da + \\ + \int_{\mathscr{V}} \{\varrho [f^k \delta x_k + f^{km} \delta x_{km} + f^{kmp} \delta x_{kmp} + \cdots] - \\ - [t^{km} \delta x_{k,m} + t^{kmp} \delta x_{km,p} + \cdots]\} \, dv.\end{aligned}\right\} \tag{232.1}$$

The quantities $s^k, s^{km}, \ldots, \varrho f^k, \varrho f^{km}, \ldots, t^{km}, t^{kmp}, \ldots$ are defined simply as the *coefficients* of the form. By GREEN's transformation follows

$$\left.\begin{aligned}\mathfrak{A} = \oint_{\mathscr{S}} [(s^k - t^{km} n_n) \delta x_k + (s^{km} - t^{kmp} n_p) \delta x_{km} + \cdots] \, da + \\ + \int_{\mathscr{V}} [(\varrho f^k + t^{km}{}_{,m}) \delta x_k + (\varrho f^{km} + t^{kmp}{}_{,p}) \delta x_{km} + \cdots] \, dv.\end{aligned}\right\} \tag{232.2}$$

The ***principle of virtual work*** is the assertion

$$\int_{\mathscr{V}} \varrho [\ddot{x}^k \delta x_k + \overline{p^k \dot{x}^m}^{\cdot} \delta x_{km} + \overline{p^k p^m \dot{x}^p}^{\cdot} \delta x_{kmp} + \cdots] \, dv = \mathfrak{A}, \tag{232.3}$$

$\boldsymbol{p}$ being, as usual, the position vector from an arbitrary origin. Eq. (232.3) is to hold for all variations consistent with the constraints. For the time being, we leave aside the effect of constraints and assume that the virtual fields may be varied arbitrarily, Then, by (232.2), the principle of virtual work is equivalent[2] to the system (205.19), with boundary conditions $s^k = t^{km} n_m$, etc.

The classical special case of (232.3) is

$$\int_{\mathscr{V}} \varrho \ddot{x}^k \delta x_k = \mathfrak{A} \equiv \oint_{\mathscr{S}} s^k \delta x_k \, da + \int_{\mathscr{V}} [\varrho f^k \delta x_k - t^{km} \delta x_{k,m}] \, dv. \tag{232.4}$$

This special case, for unconstrained variations, is equivalent to CAUCHY's first law (205.2).

We now consider some alternative variational formulations for one or both of CAUCHY's laws.

If we restrict the variations considered, we may avoid using the stress components directly in the principle of virtual work. Set

$$\mathfrak{W} \equiv \oint_{\mathscr{S}} s^k \delta x_k \, da + \int_{\mathscr{V}} \varrho f^k \delta x_k \, dv. \tag{232.5}$$

The ***theorem of Piola***[3] asserts that the condition

$$\int_{\mathscr{V}} \varrho \ddot{x}^k \delta x_k \, dv = \mathfrak{W} \tag{232.6}$$

---

[1] Here and the sequel we set $\delta x_{k,m} \equiv (\delta x_k)_{,m}$, etc.

[2] For the non-polar case, PIOLA [1848, *2*, ¶ 48] and HELLINGER [1914, *4*, § 3d] proved the equivalence by adjusting the variations so that (232.2) reduces to (200.1). We prefer the simpler argument (232.3) ↔ (205.19) ↔ (205.20). The result we have established shows also that the common claim that symmetry of the stress tensor does not follow from the principle of virtual work is misleading. Indeed, it does not follow *naturally*. The principle must be adjusted so as to imply that the virtual work of the torques is exactly the virtual work of the moments of corresponding forces. As follows from (205.21), this may be done by adding *a priori* the assumption that the coefficients in (232.1) satisfy $t_{[km]p} = p_{[k} t_{m]p}$, $f_{[kp]} = p_{[k} f_{p]}$, and only skew-symmetric δx_{km} need be considered.

[3] The pioneer work of PIOLA [1833, *3*] [1848, *2*, ¶¶ 34–38, 46–50] is somewhat involved. First, PIOLA used the material variables, and his condition of rigidity is $\delta C_{KM} = 0$ or $\delta \overset{-1}{C}_{KM} = 0$, so that the outcome is (210.8) or (210.10) rather than (205.2); (205.11) and (205.2) are then proved by transformation. Second, he seemed loth to confess that his principle employed rigid *virtual* displacements; instead, he claimed to establish it first for rigid bodies only. In the former work, he promised to remove the restriction in a later memoir; in the latter, he claimed to do so by use of an intermediate reference state. He was also the first to derive the stress boundary conditions from a variational principle [1848, *2*, ¶ 52], and he formulated an analogous variational principle for one-dimensional and two-dimensional systems [1848, *2*, Chap. VII].

for virtual translations is equivalent to Cauchy's first law (205.2); *for rigid virtual displacements, to Cauchy's second law* (205.11) *as well.* In these statements, a *virtual translation* is a field $\delta \boldsymbol{x}$ such that $\delta x_{k,m}=0$, while a *rigid virtual displacement* is a field $\delta \boldsymbol{x}$ such that $\delta x_{(k,m)}=0$. To prove PIOLA's theorem we first set up the nine side conditions $\delta x_{k,m}=0$, and we write $-t^{km}$ for the corresponding *multipliers*[1]. Then (232.6) is equivalent to

$$\int_{\mathscr{V}} \varrho \ddot{x}^k \delta x_k \, dv = \oint_{\mathscr{S}} s^k \delta x_k \, da + \int_{\mathscr{V}} (\varrho f^k \delta x_k - t^{km} \delta x_{k,m}) \, dv \tag{232.7}$$

for arbitrary variations $\delta \boldsymbol{x}$. By applying GREEN's transformation, we derive both (203.6) and (205.2); conversely, these latter imply (232.6). To derive the second statement in PIOLA's theorem, we set up the six side conditions $\delta x_{(k,m)}=0$, corresponding to which there is a symmetric tensor of multipliers, $-t^{km}$, and the proof proceeds as before.

An elegant modification of PIOLA's theorem, not using multipliers but again introducing the stresses as coefficients, is due to MURNAGHAN[2]. In (232.5) we add terms representing the virtual work of the assigned couples, and then we add and subtract terms containing $\boldsymbol{t}$ and $\boldsymbol{m}$:

$$\left.\begin{aligned}
\mathfrak{W} &\equiv \oint_{\mathscr{S}} [s^k \delta x_k - s^{kp} \delta x_{k,p}] \, da + \int_{\mathscr{V}} \varrho [f^k \delta x_k - l^{kp} \delta x_{k,p}] \, dv, \\
&= \oint_{\mathscr{S}} [(s^k da - t^{kq} da_q) \delta x_k - (s^{kp} da - m^{kpq} da_q) \delta x_{k,p}] + \\
&\quad + \oint_{\mathscr{S}} [t^{kq} \delta x_k - m^{kpq} \delta x_{k,p}] \, da_q + \int_{\mathscr{V}} \varrho [f^k \delta x_k - l^{kp} \delta x_{k,p}] \, dv, \\
&= \oint_{\mathscr{S}} [(s^k da - t^{kq} da_q) \delta x_k - (s^{kp} da - m^{kpq} da_q) \delta x_{k,p}] + \\
&\quad + \int_{\mathscr{V}} [(\varrho f^k + t^{kq}{}_{,q}) \delta x_k - (\varrho l^{kp} + m^{kpq}{}_{,q}) \delta x_{k,p}] \, dv + \\
&\quad + \int_{\mathscr{V}} [t^{kq} \delta x_{k,q} - m^{kpq} \delta x_{k,pq}] \, dv,
\end{aligned}\right\} \tag{232.8}$$

where $s^{kp}=-s^{pk}$, $l^{kp}=-l^{pk}$, $m^{kpq}=-m^{pkq}$. We now assert the principle of virtual work in the form (232.6), except that $\mathfrak{W}$ is given by (232.8) rather than by (232.5), and again *only rigid virtual displacements* $\delta \boldsymbol{x}$ *are allowed.*

First we consider only the subclass of virtual translations. For these, we must have

$$\int_{\mathscr{V}} \varrho \ddot{x}^k \delta x_k \, dv = \oint_{\mathscr{S}} (s^k da - t^{kq} da_q) \delta x_k + \int_{\mathscr{V}} (\varrho f^k + t^{kq}{}_{,q}) \delta x_k \, dv, \tag{232.9}$$

a result equivalent to both (203.6) and (205.2). In other words, *for Cauchy's first law and the associated boundary condition to hold it is necessary and sufficient that the virtual work done in any virtual translation equal the virtual work of the inertial force.* We now assume that this condition is satisfied[3]. Then from (232.8)

[1] Following the views of LAGRANGE for incompressible perfect fluids, PIOLA regarded it as a great merit of the method that nothing need be said about the stresses in advance; rather, their appearance is compelled by the variational process alone, once one accepts the principle (232.6) with appropriately restricted variations. Whether or not this is a natural approach to mechanics is a matter of taste. We share the opinion of the Lagrange-Piola method expressed by TODHUNTER [1886, *4*, § 764]; cf. also Sect. 231.

[2] MURNAGHAN [1937, *7*, § 2] considers only the non-polar case. The additional terms used above were suggested by HELLINGER [1914, *4*, § 4a].

[3] In the treatment based on (232.1), all variations δx_k, δx_{km}, ..., are *independent*, and the members of the system (205.19) follow independently of one another in any order we please. In the present treatment, the variations $\delta x_{k,m}$ being defined as $(\delta x_k)_{,m}$, are determined from the variations δx_k, and in order to derive the balance of moment of momentum we must first assume (in effect) that linear momentum is already in balance.

PIOLA [1848, *2*, ¶¶ 71—76] in effect considered a variational principle similar to (232.3) with $\delta x_{km\ldots p}$ replaced by $\delta x_{k,m\ldots p}$, and for rigid virtual displacements he showed that such a principle implies CAUCHY's first law, with the stresses being combinations of the multipliers of all orders.

follows

$$\left.\begin{aligned}\mathfrak{W} &= \int_{\mathscr{V}} \varrho\, \ddot{x}^k\, \delta x_k\, dv - \oint_{\mathscr{S}} (s^{kp}\, da - m^{kpq}\, da_q)\, \delta x_{[k,p]} - \\ &- \int_{\mathscr{V}} (\varrho\, l^{kp} + m^{kpq}{}_{,q})\, \delta x_{[k,p]}\, dv + \\ &+ \int_{\mathscr{V}} (t^{(kq)}\, \delta x_{(k,q)} + t^{[kq]}\, \delta x_{[k,q]} - m^{kpq}\, \delta x_{[k,p]q})\, dv,\end{aligned}\right\} \tag{232.10}$$

and our principle asserts that $\mathfrak{W} = \int_{\mathscr{V}} \varrho\, \ddot{x}^k\, \delta x_k\, dv$ for all variations $\delta \boldsymbol{x}$ such that $\delta x_{(k,m)} = 0$. With the further condition $\delta x_{[k,p]m} = 0$, we are still able to give arbitrary constant values to the three independent components of the field $\delta x_{[k,m]}$, and this suffices to derive (203.5) and (205.10). In other words, *for the general form of Cauchy's second law and the associated boundary condition to hold, supposing the first law holds, it is necessary and sufficient that the virtual work done in any virtual rotation be zero.* When both of CAUCHY's laws and the associated boundary conditions hold, (232.10) reduces to

$$\mathfrak{W} - \int_{\mathscr{V}} \varrho\, \ddot{x}^k\, \delta x_k\, dv = \int_{\mathscr{V}} (t^{(kq)}\, \delta x_{(k,q)} - m^{kpq}\, \delta x_{[k,p]q})\, dv. \tag{232.11}$$

In the non-polar case, $\mathfrak{W} - \int_{\mathscr{V}} \varrho\, \ddot{x}^k\, \delta x_k\, dv$ is analogous to the total stress power, but in the general case this analogy does not hold, as is plain from (217.6). An advantage of the formulation we have just given is that the symmetry of the stress tensor in the non-polar case follows easily and naturally if we simply omit in (232.8) the terms involving $\boldsymbol{l}$ and $\boldsymbol{m}$.

The variational forms of the principle of virtual work are now traditional, but it is equally possible to put it directly in terms of the velocity field, replacing δx_k by *virtual velocities* v_k throughout. A virtual velocity, like a virtual displacement, is at bottom no more than an arbitrary vector field. For example, a rewording of some of the above results in terms of virtual velocities is: *The actual motion of a continuous medium is such as to make the total rate of working of the stress on the boundary and the effective force* $\boldsymbol{f} - \ddot{\boldsymbol{x}}$ *in the interior zero in an arbitrary virtual translation; in the actual motion, for the non-polar case, this rate of working is the total stress power.*

An interesting application of the principle of virtual work was given in Sect. 229.

232A. Appendix. Virtual work in the case when there is a strain energy. We consider the special case when there exists a function $\sigma(x^k_{;K}, X^M)$ such that[1]

$$\mathfrak{A} = \oint_{\mathscr{S}} s_k\, \delta x^k\, da + \int_{\mathscr{V}} \varrho\, f_k\, \delta x^k\, dv - \delta \int_{\mathscr{V}} \varrho\, \sigma\, dv, \tag{232A.1}$$

where ϱ is varied in such a way that $\delta(\varrho\, dv) = 0$. Since by (156.1) and GREEN's transformation follows[2]

$$\left.\begin{aligned}\delta \int_{\mathscr{V}} \varrho\, \sigma\, dv &= \int_{\mathscr{V}} \varrho\, \delta\sigma\, dv = \int_{\tilde{\mathscr{V}}} \tilde{\varrho}\, \delta\sigma\, dV, \\ &= \int_{\tilde{\mathscr{V}}} \tilde{\varrho}\, \frac{\partial \sigma}{\partial x^k_{;K}}\, \delta x^k_{;K}\, dV, \\ &= \oint_{\tilde{\mathscr{S}}} \tilde{\varrho}\, \frac{\partial \sigma}{\partial x^k_{;K}}\, \delta x^k\, dA_K - \int_{\tilde{\mathscr{V}}} \left(\tilde{\varrho}\, \frac{\partial \sigma}{\partial x^k_{;K}}\right)_{;K} \delta x^k\, dV,\end{aligned}\right\} \tag{232A.2}$$

where tildes indicate quantities appropriate to the reference configuration, in the case when there are no constraints we derive from $(232.4)_1$

$$s_k\, da = T_k^K\, dA_K \;\text{ on } \mathscr{S}, \quad \tilde{\varrho}\, \ddot{x}_k = T^K_{k;K} + \tilde{\varrho}\, f_k \;\text{ in } \mathscr{V}, \tag{232A.3}$$

[1] In principle, the analysis is due to GREEN [1839, *1*, pp. 253—256]; his work was corrected and extended by HAUGHTON [1849, *1*, p. 152] and KIRCHHOFF [1850, *2*, § 1].

[2] Since $x^k_{;K} \equiv \partial x^k / \partial x^K$, by the rule given in Sect. 97 we have $\delta x^k{}_{;K} = (\delta x^k)_{;K}$. In the variations, $\boldsymbol{X}$ is kept constant.

where

$$T_k^{\ K} \equiv \tilde{\varrho}\, \frac{\partial \sigma}{\partial x^k{}_{;K}}\,. \tag{232A.4}$$

Equations (232 A.3) have the same form as (210.5) and (210.8). The foregoing analysis signifies that *when the inner part of the virtual work is the variation of a volume integral whose density is a function of the deformation gradients only, the existence of the stress tensor follows directly from the principle of virtual work; moreover, the stresses are then purely elastic*, in the sense defined in Sect. 218.

If we assume that the assigned force $\boldsymbol{f}$ is conservative in the sense of (218.1), and if we adopt the convention that in the varied motion the assigned surface force $\boldsymbol{s}\, da$ is the same as at corresponding points on the actual motion, (232 A.1) may be written as a simple variation:

$$\mathfrak{A} = \delta \mathfrak{T}, \quad \text{where} \quad -\mathfrak{T} \equiv -\oint_{\mathscr{S}} s_k u^k\, da + \int_{\mathscr{V}} \varrho\, \tau\, dv, \tag{232A.5}$$

$\boldsymbol{u}$ is the displacement vector, and τ is the total potential, $U + \sigma$. There is an evident analogy to the work theorem (218.9), for which, however, it was assumed that the reference configuration $\boldsymbol{X}$ was material, while for the present development no such restriction is necessary.

If we add the assumption that (232 A.4) may be inverted[1] to yield $x^k_{;K}$ as a function of $\boldsymbol{X}$ or $\boldsymbol{x}$, t, and $T_m^{\ M}$, by the usual device of a contact transformation we may exhibit this function. Indeed, put

$$\varkappa \equiv \sigma - \frac{1}{\tilde{\varrho}}\, T_k^{\ K} x^k_{;K}\,. \tag{232A.6}$$

Then for $t = \text{const}$ we have by (232 A.4)

$$\left.\begin{aligned} d\varkappa &= d\sigma - x^k_{;K}\, d(T_k^{\ K}/\tilde{\varrho}) - (T_k^{\ K}/\tilde{\varrho})\, d x^k_{;K}, \\ &= \frac{\partial \sigma}{\partial X^K}\, dX^K - x^k_{;K}\, d(T_k^{\ K}/\tilde{\varrho}), \end{aligned}\right\} \tag{232A.7}$$

so that with σ considered as a function of the variables X^M, $x^k_{;K}$ and with $\varkappa$ considered as a function of the variables X^M, $T_k^{\ K}/\tilde{\varrho}$, we have

$$\frac{\partial x^k}{\partial X^K} = x^k_{;K} = -\frac{\partial \varkappa}{\partial (T_k^{\ K}/\tilde{\varrho})}\,, \qquad \frac{\partial \varkappa}{\partial X^K} = \frac{\partial \sigma}{\partial X^K}\,. \tag{232A.8}$$

In what follows now, we consider the special case of (232 A.8) when

$$x^k_{;K} = \frac{\partial \mu}{\partial T_k^{\ K}}\,, \qquad \mu = \mu(T_k^{\ K}, t)\,. \tag{232A.9}$$

Then the *theorem of Hellinger and E. Reissner*[2] asserts that the *variational principle*

$$\int_{\tilde{\mathscr{V}}} \tilde{\varrho}\,(\ddot{x}_k - f_k)\, \delta x^k\, dV = \delta \Big[\int_{\tilde{\mathscr{V}}} (\mu - T_k^{\ K} x^k_{;K})\, dV + \int_{\tilde{\mathscr{S}}_1} s_k u^k\, dA + \int_{\tilde{\mathscr{S}}_2} T_k^{\ K} (u^k - v^k)\, dA_K \Big], \tag{232A.10}$$

where both *the deformation* $\boldsymbol{x}(\boldsymbol{X}, t)$ *and the stress components* $T_k^{\ K}$ *are varied independently, is equivalent to Cauchy's first law and* (232 A.9) *in* $\tilde{\mathscr{V}}$, *to the stress boundary condition* (203.6)$_1$ *on the part* $\tilde{\mathscr{S}}_1$ *of the boundary, and to the displacement boundary condition* $\boldsymbol{u} = \boldsymbol{v}$ *on the remaining part* $\tilde{\mathscr{S}}_2$. To establish this result, we execute the variations on the right-hand side of (232 A.10)

---

[1] For the variational theorem which follows now, it is essential that this inversion be possible not merely locally but in the large. Most writers on the subject apparently fail to appreciate that for typical non-quadratic functional forms for σ, this is most unlikely.

[2] HELLINGER [1914, *4*, § 7e] considered μ as depending on $\boldsymbol{x}$ as well as $T_k^{\ K}$, thus including the potential energy as in (232 A.5). REISSNER [1953, *27*] adjusted the boundary integrals, which were omitted by HELLINGER, so as to be appropriate to mixed boundary conditions. He considered only the special case when (232A.9) is replaced by $E_{KM} = \partial\mu/\partial T^{KM}$; effectively, CAUCHY's second law is assumed. Cf. also MANACORDA [1954, *14*].

obtaining

$$\left.\begin{aligned}&\int_{\tilde{\mathscr{V}}}\left[\left(\frac{\partial\mu}{\partial T_k{}^K}-x^k_{;K}\right)\delta T_k{}^K-T_k{}^K\,\delta x^k_{;K}\right]dV+\\&\quad+\int_{\tilde{\mathscr{S}}_1}s_k\,\delta x^k\,dA+\int_{\tilde{\mathscr{S}}_2}[\delta T_k{}^K(u^k-v^k)+T_k{}^K\,\delta x^k]\,dA_K\\&=\int_{\tilde{\mathscr{V}}}\left[\left(\frac{\partial\mu}{\partial T_k{}^K}-x^k_{;K}\right)\delta T_k{}^K+T^K_{k;K}\,\delta x^k\right]dV-\oint_{\tilde{\mathscr{S}}_1+\tilde{\mathscr{S}}_2}T_k{}^K\,\delta x^k\,dA_K+\\&\quad+\int_{\tilde{\mathscr{S}}_1}s_k\,\delta x^k\,dA+\int_{\tilde{\mathscr{S}}_2}[\delta T_k{}^K(u^k-v^k)+T_k{}^K\,\delta x^k]\,dA_K,\\&=\int_{\tilde{\mathscr{V}}}\left[\left(\frac{\partial\mu}{\partial T_k{}^K}-x^k_{;K}\right)\delta T_k{}^K+T^K_{k;K}\,\delta x^k\right]dV+\\&\quad+\int_{\tilde{\mathscr{S}}_1}(s_k\,dA-T_k{}^K\,dA_K)\,\delta x^k+\int_{\tilde{\mathscr{S}}_2}(u^k-v^k)\,\delta T_k{}^K\,dA_K.\end{aligned}\right\}\quad(232\text{A}.11)$$

Substituting this last form for the right-hand side of (232A.10) gives a formula from which the desired result follows at once.

If in the HELLINGER-REISSNER theorem we suppose that only the stress, not the displacement, is varied, we obtain a variational principle yielding only (232A.9), not CAUCHY'S first law. This principle may be used to select from the class of equilibrated stress systems one in which there is a strain energy[1]. If, on the other hand, we vary only the displacement, then $\delta\mu=0$, and by omitting μ from (232A.10) we get the principle of virtual work in its classical form, to be studied further in Sect. 236A.

233. Virtual work in a constrained body. When a body is subject to internal constraints, *providing these do no work*, the virtual work is still given by (232.1), but the variations $\delta\boldsymbol{x}$ at neighboring points are no longer fully arbitrary, and from (232.3) we cannot conclude CAUCHY'S first law without delay.

Consider first the special case when the constraint is that of incompressibility. We now restrict the variations $\delta\boldsymbol{x}$ so as to comply with this constraint: $(\delta x^k)_{,k}=0$. Therefore

$$-\int_{\mathscr{V}}p\,(\delta x^k)_{,k}\,dv=0, \tag{233.1}$$

where we have introduced a multiplier $-p$ corresponding to the constraint[2], and we may replace (232.4) by

$$\left.\begin{aligned}0&=\oint_{\mathscr{S}}(s^k-t^{km}n_m)\,\delta x_k\,da+\int_{\mathscr{V}}[t^{km}{}_{,m}+\varrho(f^k-\ddot{x}^k)]\,\delta x_k\,dv-\int_{\mathscr{V}}p\,(\delta x^k)_{,k}\,dv,\\&=\oint_{\mathscr{S}}(s^k-t^{km}n_m+p\,g^{km}n_m)\,\delta x_k\,da+\int_{\mathscr{V}}[t^{km}{}_{,m}+\varrho(f^k-\ddot{x}^k)-g^{km}p_{,m}]\,\delta x_k\,dv,\end{aligned}\right\}\quad(233.2)$$

with unrestricted $\delta\boldsymbol{x}$. We infer (203.6) and (205.2) as before, *except that the stress tensor* $\boldsymbol{t}$ *is now replaced by* $\boldsymbol{t}-p\boldsymbol{1}$, where p is arbitrary.

The difference between CAUCHY'S first law and the result just obtained is easy to understand. CAUCHY'S law refers to the *entire stress*, whatever its origin. The stress entering the principle of virtual work *does not include any stress required to maintain the constraints, so long as such stresses do no work.* Naturally, then, results obtained from the principle of virtual work are *indeterminate* to within

[1] This principle is usually named after MENABREA and CASTIGLIANO, who formulated it in terms of frameworks. For the case of finite strain, cf. B. FINZI [1940, *11*] and LANGHAAR [1953, *16*].

[2] This procedure, in essence, derives from LAGRANGE [1762, *3*, § 40]. Surface constraints are consdered by FERRARESE [1958, *1A*].

whatever stress may exist without doing work in constrained motions of the type considered. Indeed, from (217.7) we see that the stress power of a non-zero hydrostatic pressure is zero if and only if the motion is isochoric.

General constraints dependent on the velocity gradient may be taken into account in just the same way. Let the tensorial equations of constraint be

$$ {}_{\mathfrak{a}}c^{k\ldots n}_{p\ldots q}(\dot{x}^r_{,s}, t) = 0, \qquad \mathfrak{a} = 1, 2, \ldots, \mathfrak{m}, \tag{233.3}$$

where the orders of the tensors ${}_{\mathfrak{a}}\boldsymbol{c}$ need not be the same. Consider only those variations $\delta\boldsymbol{x}$ satisfying

$$\sum_{\mathfrak{a}=1}^{\mathfrak{m}} \int_{\mathscr{V}} {}_{\mathfrak{a}}p^{n\ldots q}_{k\ldots m} \frac{\partial\, {}_{\mathfrak{a}}c^{k\ldots m}_{n\ldots q}}{\partial \dot{x}^r_{,s}} (\delta x^r)_{,s}\, dv = 0, \tag{233.4}$$

where ${}_{\mathfrak{a}}p^{n\ldots q}_{k\ldots m}$ is a tensor of multipliers. The same procedure leads to boundary conditions and equations of motion in which t^{km} is replaced by

$$t^{km} + g^{kr} \sum_{\mathfrak{a}=1}^{\mathfrak{m}} {}_{\mathfrak{a}}p^{n\ldots q}_{u\ldots s} \frac{\partial\, {}_{\mathfrak{a}}c^{u\ldots s}_{n\ldots q}}{\partial \dot{x}^r_{,m}}. \tag{233.5}$$

The fact that in general there exist no motions satisfying the fully general constraints (233.3) does not invalidate the procedure.

Sometimes there are constraints depending on the deformation gradients $x^k_{;K}$ from a reference state[1]. For simplicity, consider a single equation of the form

$$c(x^k_{;K}, t) = 0. \tag{233.6}$$

The variations are now subject to the constraint

$$\frac{\partial c}{\partial x^k_{;K}} (\delta x^k)_{;K} = 0. \tag{233.7}$$

To apply this side condition, it is convenient to transform (232.2) by introducing the variables appropriate to the description in terms of a reference state. By (210.4), (210.6), and (20.9), in the classical non-polar case we get

$$\mathfrak{A} = \oint_{\mathscr{S}} (s^k da - T^{kK} dA_K)\, \delta x_k + \int_{\mathscr{V}} (T^{kK}{}_{;K} + \tilde{\varrho} f^k)\, \delta x_k\, dV. \tag{233.8}$$

Now introducing a multiplier p corresponding to the constraint (233.7) and proceeding as before, we obtain the boundary condition corresponding to (210.5) and the equations of motion in PIOLA's form (210.8), except that in both the double vector T^{kK} is replaced by

$$T^{kK} - p\, g^{km} \frac{\partial c}{\partial x^m{}_{;K}}. \tag{233.9}$$

Equivalently, the stress tensor $\boldsymbol{t}$ is replaced by

$$t^{km} - j\, p\, x^m_{;K}\, g^{kq} \frac{\partial c}{\partial x^q{}_{;K}}. \tag{233.10}$$

From the results (233.5) and (233.10) it is evident that general constraints such as (233.3) or (233.6) yield a non-symmetric contribution to the stress[2]. As is shown by the example of isochoric motion, certain special constraints may be maintained by symmetric stresses.

---

[1] POINCARÉ [1889, *8*, § 152] [1892, *11*, § 33]. Cf. HELLINGER [1914, *4*, § 4c], who considers constraints involving higher derivatives. ERICKSEN and RIVLIN [1954, *6*, § 4] work out the explicit form of the result which is implied because c is a scalar.

[2] This was remarked by ERICKSEN and RIVLIN [1954, *6*, § 3].

All the constraints considered in this section are of the type called *holonomic*[1]. Non-holonomic constraints will be mentioned in Sect. 237.

234. Converse of the principle of virtual work. DORN and SCHILD[2] have formulated an elegant ***converse to the principle of virtual work*** in the static and non-polar case: *Suppose that, corresponding to any vector* $\boldsymbol{u}$ *given on* $\mathscr{S}$, *there exists a symmetric tensor* $\mathbf{c}$ *such that*

$$\int_{\mathscr{V}} {}^0t^{km}\, c_{km}\, dv = \oint_{\mathscr{S}} {}^0t^{km}\, u_k\, da_m \tag{234.1}$$

for all null stresses ${}^0\boldsymbol{t}$; *then the vector field* $\boldsymbol{u}$ *can be extended throughout* $\mathscr{V}$ *in such a way that*

$$c_{km} = u_{(k,m)}. \tag{234.2}$$

In other words, if $\boldsymbol{u} = \dot{\boldsymbol{x}}$ it follows that in fact $\mathbf{c} = \boldsymbol{d}$. The statement of the theorem is limited to null stresses only for simplicity, the extension to general equilibrated stresses being easy, but for motion rather than equilibrium the formulation is awkward. The proof of the theorem is divided into two stages: (a) we show that $\boldsymbol{c}$ satisfies the conditions of compatibility for $\boldsymbol{d}$, and thereafter (b) we prove (234.2).

Proof of (a). By (227.10), we may write the hypothesis (234.1) in the form

$$\int_{\mathscr{V}} e^{krs}\, e^{qmn}\, a_{rm,sn}\, c_{kq}\, dv = \oint_{\mathscr{S}} e^{krs}\, e^{qmn}\, a_{rm,sn}\, u_k\, da_q, \tag{234.3}$$

where $\boldsymbol{a}$ is an arbitrary symmetric tensor. Two applications of GREEN's transformation put (234.3) into the form

$$\left.\begin{aligned} &\int_{\mathscr{V}} e^{krs}\, e^{qmn}\, c_{kq,sn}\, a_{rm}\, dv \\ &\quad = \oint_{\mathscr{S}} e^{krs}\, e^{qmn}\, \{c_{kq,n}\, a_{rm}\, da_s - c_{kq}\, a_{rm,s}\, da_n + a_{rm,sn}\, u_k\, da_q\}. \end{aligned}\right\} \tag{234.4}$$

We choose $\boldsymbol{a}$ as zero outside a small region surrounding a given point, thus annulling the surface integral on the right; since $\boldsymbol{a}$ may be arbitrary, to within requirements of smoothness, we conclude that $e^{krs}\, e^{qmn}\, c_{kq,sn}$, since it is symmetric, must vanish. By the remarks following (84.3), or by those at the end of Sect. 34, there exists a vector $\boldsymbol{v}$ such that

$$c_{km} = v_{(k,m)} \quad \text{in} \quad \mathscr{V}. \tag{234.5}$$

Proof of (b). From (234.5), GREEN's transformation, and the fact that ${}^0\boldsymbol{t}$ is a null stress, we have

$$\int_{\mathscr{V}} {}^0t^{km}\, c_{km} = \int_{\mathscr{V}} {}^0t^{km}\, v_{(k,m)}\, dv = \oint_{\mathscr{S}} {}^0t^{km}\, v_k\, da_m. \tag{234.6}$$

Comparison with (234.1) yields

$$\oint_{\mathscr{S}} {}^0t^{km}\, (u_k - v_k)\, da_m = 0 \tag{234.7}$$

for arbitrary null stresses ${}^0\boldsymbol{t}$. This condition asserts that the virtual work of an arbitrary equilibrated stress in the virtual displacement $u_k - v_k$ is zero. Hence the motion $\boldsymbol{u} - \boldsymbol{v}$ is rigid. That is, there exist constants ω_{km} and b_k such that

[1] HELLINGER [1914, *4*, § 4c] discusses also holonomic constraints applied only on surfaces or lines, as well as constraints which are inequalities.

[2] [1956, *4*]. The statement (*a*) and its proof are due to LOCATELLI [1940, *15* and *16*], who considered also the case when the assigned force does not vanish.

$\omega_{km} = -\omega_{mk}$ and

$$u_k = v_k + \omega_{km} z_m + b_k \tag{234.8}$$

on $\mathscr{S}$. Now $\boldsymbol{v}$ is defined through $\mathscr{V}$. Therefore we may take (234.8) as extending the definition of $\boldsymbol{u}$ throughout $\mathscr{V}$. Since $u_{(k,m)} = v_{(k,m)}$, (234.2) follows from (234.5). Q.E.D.

L. FINZI[1] has remarked that the process leading from (234.1) to the conditions of compatibility for (234.2) is entirely general and may be applied in any space where covariant differentiation is defined. For example, let the general null stress have the form

$${}_0t^{km} = a^{km} A + b^{kmr} A_{,r} + c^{kmrs} A_{,rs} + \cdots. \tag{234.9}$$

Substitution in (234.3) and proceeding as in the lines following yields

$$0 = a^{km} c_{km} - (b^{kmr} c_{km})_{,r} + (c^{kmrs} c_{km})_{,rs} - \cdots \tag{234.10}$$

as the conditions of compatibility. The reader may use this result to derive (84.5) from (229.8), and to derive (84.4) from (229.10). The process is easily modified so as to apply to the case when the general null stress is given in terms of a tensor of stress functions A_{km} or A_{kmpq}, *etc.*

235. Principle of minimum stress intensity. Given a symmetric tensor field s_{km} in $\mathscr{V}$, we shall say it is given a *potential increment* when it is replaced by $s_{km} + b_{(k,m)}$, where $\boldsymbol{b}$ is a vector field which vanishes upon the boundary $\mathscr{S}$ of $\mathscr{V}$. If $s^{km}{}_{,m} = 0$, we have

$$[s_{km} + b_{(k,m)}]\,[s^{km} + b^{(k,m)}] = s_{km} s^{km} + (2 b_k s^{km})_{,m} + b_{(k,m)} b^{(k,m)}. \tag{235.1}$$

By GREEN's transformation follows

$$\left.\begin{aligned} \int_{\mathscr{V}} [s_{km} + b_{(k,m)}]\,[s^{km} + b^{(k,m)}]\,dv &= \int_{\mathscr{V}} [s_{km} s^{km} + b_{(k,m)} b^{(k,m)}]\,dv, \\ &\geqq \int_{\mathscr{V}} s_{km} s^{km}\,dv, \end{aligned}\right\} \tag{235.2}$$

where equality holds if and only if $b_{(k,m)} = 0$. The analysis is valid in a Riemannian space of any number of dimensions, so long as the metric be positive definite.

We apply the foregoing result to a body subject to no assigned force, both in the three-dimensional and the four-dimensional cases, so obtaining PRATELLI's ***theorems of minimum stress intensity***[2]:

1. *In steady motion, the total intensity of a stress tensor* $\mathbf{t} - \varrho \dot{\boldsymbol{x}} \dot{\boldsymbol{x}}$ *satisfying Cauchy's laws is a minimum with respect to potential increments.*

2. *In a rectangular Cartesian co-ordinate system in an inertial frame, the total intensity of a world stress-momentum tensor (211.2) satisfying the equations of continuity and momentum balance is a minimum with respect to potential increments.*

From the analysis, it is evident that no result of this kind can be expected to hold for general variations of stress.

236. Lagrangian and Hamiltonian principles. In this section and the next we derive principles related to (232.4). Thus we are restricting attention to principles equivalent to CAUCHY's first law.

---

[1] [1956, 7, § 10]. Indeed, the statement above may easily be inferred from the work of LOCATELLI [1940, *15* and *16*].

[2] [1953, *25*]. PRATELLI, who uses variational calculus, does not obtain an absolute minimum in space-time because he uses the energy-momentum tensor of special relativity rather than the classical tensor (211.2).

By (170.4), we may write (232.4) in the form

$$\frac{d}{dt}\int_{\mathscr{V}} \varrho \dot{x}_k \delta x^k \, dv = \delta\mathfrak{K} + \mathfrak{A}, \tag{236.1}$$

where $\mathfrak{K}$ is the kinetic energy and where the variation of density satisfies $\delta(\varrho\, dv) = 0$. Eq. (236.1), which is merely a rewriting of the principle of virtual work, is called the *Lagrangian central equation*[1].

Integrating (236.1) from $t = t_1$ to $t = t_2$, we impose the condition that $\delta \boldsymbol{x} = 0$ at $t = t_1$ and at $t = t_2$ so as to obtain ***Hamilton's principle***[2]:

$$\int_{t_1}^{t_2} (\delta\mathfrak{K} + \mathfrak{A})\, dt = 0, \tag{236.2}$$

und conversely, by the identity (236.1), if (236.2) holds for every pair of times t_1 and t_2, (232.4) for all admissible $\delta\boldsymbol{x}$ follows.

By further restriction of the variations, it is easy to derive the *principle of least action* from (236.2).

The elegant and useful forms that these principles assume for systems of mass-points do not carry over to general continuum mechanics.

236A. Appendix. HAMILTON'S principle in the case when there is a strain energy. In the special case when the formula (232A.5) holds, HAMILTON'S principle (236.2) may be written in the more familiar form

$$\delta \int_{t_1}^{t_2} \mathfrak{L}\, dt = 0, \tag{236A.1}$$

where

$$\mathfrak{L} \equiv \int_{\mathscr{V}} \lambda\, dV + \oint_{\mathscr{S}} s_k u^k\, da, \qquad \lambda \equiv \varrho_0 [\tfrac{1}{2}\dot{x}^2 - \tau]. \tag{236A.2}$$

Here the tildes have been dropped, for the reference configuration must be a fixed one. While the equivalence of this variational principle to CAUCHY'S first law, the associated stress boundary condition, and the stress-strain relations (232A.4) follows from results already given in Sect. 232A, (236A.1) is our only example of a variational principle of the conventional simple type to which the Euler formulae may be applied, so we give an independent verification[3]. Recalling that in the material description $\dot{x}^k = \partial x^k/\partial t$, we think of the time integral of the volume integral in (236A.1) as an integral over a four-dimensional manifold, and to this integral we apply the usual rules of the calculus of variations. Hence, when there are no constraints, the spatial differential equations following from (236A.1) are

$$\left.\begin{aligned} 0 &= \frac{d}{dt}\left(\frac{\partial\lambda}{\partial \dot{x}^k}\right) + \frac{\partial}{\partial X^\alpha}\left(\frac{\partial\lambda}{\partial x^k_{;\alpha}}\right) - \frac{\partial\lambda}{\partial x^k}, \\ &= \varrho_0 \ddot{x}_k - \left(\varrho_0 \frac{\partial\tau}{\partial x^k_{;\alpha}}\right)_{;\alpha} - (\varrho_0 \tau)_{,k}, \end{aligned}\right\} \tag{236A.3}$$

equivalent to CAUCHY'S first law when there is a strain energy and a potential energy.

[1] For continuum mechanics, this equation was introduced in a somewhat more general form by HEUN [1913, *4*, § 21] and HELLINGER [1914, *4*, § 5b]. These authors consider also an alternative form in which the time as well as the path is varied.

[2] According to HELLINGER [1914, *4*, § 5b], the extension of HAMILTON'S principle to continuous media was first obtained by WALTER [1868, *15*]; it was given also by KIRCHHOFF [1876, *2*, Vorl. 11, § 5].

[3] KIRCHHOFF [1859, *2*, § 1]. While in [1852, *1*] it was assumed that the strain energy is a quadratic function, as far as concerns the variational formulation this restriction was not used.

The second variational principle of CLEBSCH, which we have presented in a purely kinematical form in Sect. 137, may be interpreted as a special case of HAMILTON'S principle. ZEMPLÉN [1905, *7*] [1905, *9*, § 3] used (236.1) to obtain (236A.3), (210.5), and (205.3). The general formula has been rediscovered by DE DONDER and VAN DEN DUNGEN [1949, *5*] and by E. HÖLDER [1950, *11*, §§ 1—3], who discuss alternative forms.

In the above formulation of HAMILTON's principle, ϱ_0 is taken as an assigned function of $\boldsymbol{X}$, and any additional parameters upon which the function τ may depend are kept fixed in the variation. We may set $\varrho_0 = \varrho J$ by (156.2), and then, as has been remarked by HERIVEL[1], we may vary ϱ and the additional parameters, at the same time setting up as side conditions the continuity equation and the constancy of the additional parameters for each particle. The result is unchanged. It is also possible, though more elaborate, to formulate HAMILTON's principle in terms of the spatial variables and to vary the velocity field rather than the displacement[2].

237. The principle of extreme compulsion[3]. In the principle of virtual work (232.4), the variation $\delta\boldsymbol{x}$ may be selected as any field satisfying the constraints, presumed holonomic. Precisely, as we have seen in Sect. 233, this means that if there are constraints[4]

$$_{\mathfrak{a}}C(\boldsymbol{x}, \boldsymbol{X}, x^k_{;K}, x^k_{;KM}, \ldots, t) = 0, \tag{237.1}$$

then $\delta\boldsymbol{x}$ ranges over the class of vectors $\boldsymbol{v}$ satisfying

$$\frac{\partial_{\mathfrak{a}}C}{\partial x^k} v^k + \frac{\partial_{\mathfrak{a}}C}{\partial x^k_{;K}} v^k_{;K} + \frac{\partial_{\mathfrak{a}}C}{\partial x^k_{;KM}} v^k_{;KM} + \cdots = 0. \tag{237.2}$$

For the actual motion, differentiating (237.1) materially yields

$$\frac{\partial_{\mathfrak{a}}C}{\partial x^k} \dot{x}^k + \frac{\partial_{\mathfrak{a}}C}{\partial x^k_{;K}} \dot{x}^k_{;K} + \frac{\partial_{\mathfrak{a}}C}{\partial x^k_{;KM}} \dot{x}^k_{;KM} + \cdots + \frac{\partial_{\mathfrak{a}}C}{\partial t} = 0, \tag{237.3}$$

$$\frac{\partial_{\mathfrak{a}}C}{\partial x^k} \ddot{x}^k + \frac{\partial_{\mathfrak{a}}C}{\partial x^k_{;K}} \ddot{x}^k_{;K} + \frac{\partial_{\mathfrak{a}}C}{\partial x^k_{;KM}} \ddot{x}^k_{;KM} + \cdots + F = 0, \tag{237.4}$$

where F is a function of $\boldsymbol{x}, \boldsymbol{X}, x^k_{;K}, x^k_{;KM}, \ldots, \dot{x}^k, \dot{x}^k_{;K}, \ldots$. This identity suggests a means of constructing variational fields which conform to the constraints. Consider the class of variations such that, at each fixed instant, the varied motion has the same displacement and the same velocity as the original motion, but not necesarily the same acceleration:

$$\delta\boldsymbol{x} = 0, \quad \delta\dot{\boldsymbol{x}} = 0. \tag{237.5}$$

For such variations, by (237.1) we have $\delta(\partial_{\mathfrak{a}}C/\partial x^k) = 0, \ldots$, and $\delta F = 0$. From (237.4) follows

$$\frac{\partial_{\mathfrak{a}}C}{\partial x^k} \delta\ddot{x}^k + \frac{\partial_{\mathfrak{a}}C}{\partial x^k_{;K}} \delta\ddot{x}^k_{;K} + \cdots = 0. \tag{237.6}$$

This is an equation of the form (237.2). What we have proved is that if $\delta\boldsymbol{x}$ is any variation satisfying (237.5), then $\delta\ddot{\boldsymbol{x}}$ is an admissible virtual displacement field. We may therefore replace $\delta\boldsymbol{x}$ by $\delta\ddot{\boldsymbol{x}}$ in (232.4), obtaining

$$\int_{\mathscr{V}} \varrho \ddot{x}_k \delta\ddot{x}^k dv = \oint_{\mathscr{S}} s_k \delta\ddot{x}^k da + \int_{\mathscr{V}} [\varrho f_k \delta\ddot{x}^k - t^{km} \delta\ddot{x}_{k,m}] dv. \tag{237.7}$$

Hence, supposing $\delta(\varrho\, dv) = 0$, we get

$$\delta \int_{\mathscr{V}} \tfrac{1}{2} \varrho \ddot{x}^2 dv = \oint_{\mathscr{S}} s_k \delta\ddot{x}^k da + \int_{\mathscr{V}} [\varrho f_k \delta\ddot{x}^k - t^{km} \delta\ddot{x}_{k,m}] dv. \tag{237.8}$$

---

[1] [1955, *13*, § 1].

[2] Unsatisfactory special cases are given by ECKART [1938, *4*, § 3] and HERIVEL [1955, *13*, § 2]; their result is corrected by LIN in a work not yet published. We make no attempt to cite the large hydrodynamical literature on special variational principles, some of which is discussed by SERRIN, Sect. 15 of The Mathematical Principles of Fluid Mechanics, this Encyclopedia, Vol. VIII/1.

[3] For systems of mass-points, this principle is associated with the names of GAUSS, LIPSCHITZ, GIBBS, and APPELL. For continuum mechanics, we follow the development of BRILL [1909, *2*, § 18] and HELLINGER [1914, *4*, § 5c].

[4] The tensorial character of the equations of constraint is irrelevant here.

If we add the convention that at points and times in the varied motions the same force $\boldsymbol{f}$ is assigned as at the corresponding points in the actual motion, so that $\delta\boldsymbol{f}=0$, we may replace (237.8) by

$$\delta\int_{\mathscr{V}}\tfrac{1}{2}\varrho(\ddot{\boldsymbol{x}}-\boldsymbol{f})^2\,dv=\oint_{\mathscr{S}}s_k\,\delta\ddot{x}^k\,da-\int_{\mathscr{V}}t^{km}\,\delta\ddot{x}_{k,m}\,dv. \tag{237.9}$$

The integral whose variation stands on the left is the total squared effective force, or the *compulsion*, of the motion, and the variational equation (237.9) is the *principle of extreme compulsion*[1].

The principle may be extended so as to hold for *non-holonomic* constraints of the more general form

$${}_{\mathfrak{a}}C(\boldsymbol{x},\boldsymbol{X},x^k_{;K},x^k_{;KM},\ldots,\dot{x}^k,\dot{x}^k_{;K},\ldots,t)=0. \tag{237.10}$$

For the actual motion under such a constraint, we have

$$\frac{\partial\,{}_{\mathfrak{a}}C}{\partial\dot{x}^k}\ddot{x}^k+\frac{\partial\,{}_{\mathfrak{a}}C}{\partial\dot{x}^k_{;K}}\ddot{x}^k_{;K}+\cdots+\frac{\partial\,{}_{\mathfrak{a}}C}{\partial x^k}\dot{x}^k+\cdots+\frac{\partial\,{}_{\mathfrak{a}}C}{\partial t}=0. \tag{237.11}$$

In order that varied motions satisfying (237.5) be compatible with (237.11), it is necessary and sufficient that

$$\frac{\partial\,{}_{\mathfrak{a}}C}{\partial\dot{x}^k}\delta\ddot{x}^k+\frac{\partial\,{}_{\mathfrak{a}}C}{\partial\dot{x}^k_{;K}}\delta\ddot{x}^k_{;K}+\cdots=0. \tag{237.12}$$

Thus for constraints of the type (237.10) we may still infer the principle of extreme compulsion, but the variations besides satisfying (237.5) are subjected to (237.12) as an additional condition.

238. Remarks on mechanics in generalized spaces. There have been many discussions of mechanical principles appropriate to non-Euclidean spaces[2]. Except for relativistic mechanics, which is outside the scope of this treatise, these developments seem to consist mainly in observations that certain parts of the theory do not require Euclidean three-dimensional space, but may be carried over bodily to more general ambients.

For example[3], while the momentum principle in the form (196.3), since it requires that the integrals of vector fields over bodies enjoy invariance, is restricted to Euclidean spaces, or at least to spaces with distant parallelism, CAUCHY'S laws (205.2) and (205.11) are meaningful in any space where covariant differentiation may be defined[4]. Sometimes they are derived for Riemannian spaces by assuming that the momentum principle applies to infinitely small volumes, but in essence such a derivation is no more than a direct postulation of the desired result. There are infinitely many possible "laws" of mechanics in generalized spaces if we demand no more than that they be invariant and intrinsic equations which in the Euclidean case reduce to CAUCHY'S laws. For example, if $R^k{}_m$ is the contracted curvature tensor of an affine space, we may replace $\ddot{x}^k$ by $\ddot{x}^k+KR^k{}_m\ddot{x}^m$ in (205.2), and for all values of K the resulting equation reduces in Euclidean spaces to the usual form of CAUCHY'S first law.

Another formal generalization, the analogue of common practice in mass-point mechanics, replaces $\mathfrak{K}$ in (236.1) or in (236.2) by

$$\mathfrak{K}\equiv\tfrac{1}{2}\int_{\mathscr{V}}\varrho\,a_{km}\dot{x}^k\dot{x}^m\,dv, \tag{238.1}$$

[1] As if English vocabulary were insufficient to supply two different words to translate „Nebenbedingung" and „Zwang", (237.9) is often called "the principle of least constraint".

[2] See Special Bibliography M at the end of this treatise.

[3] VAN DANTZIG [1934, *10*, Part IV, § 1].

[4] At bottom, this is the content of BELTRAMI'S observation [1881, *1*, p. 389] that Eq. (232.4) is meaningful and may be applied in curved Riemannian spaces.

where $a_{km}\, d\dot{x}^k\, d\dot{x}^m$ is an arbitrary quadratic form, not necessarily reducible to a sum of squares, and not necessarily the metric tensor of space.

A third generalization[1], more interesting from the mechanical point of view, replaces the principle of virtual work (232.4), after converting it as in (232A.2) to an expression in terms of a material reference state X^α, by a four-dimensional equation

$$\int_{t_1}^{t_2} dt \left\{ \oint_{\mathscr{S}} s_k\, \delta x^k\, da + \int_{\mathscr{V}_0} \left[\varrho_0 \left(f_k\, \delta x^k + b_k\, \dot{\overline{\delta x^k}} \right) - T_k^{\alpha}\, \delta x^k_{;\alpha} \right] dv \right\} = 0, \qquad (238.2)$$

putting the *impulsive coefficients* b_k on a par with force and stress as coefficients in a linear variational form. As in HAMILTON'S principle, it is supposed that $\delta \boldsymbol{x} = 0$ at $t = t_1$ and at $t = t_2$. Equivalent to (238.2), in the case when there are no constraints, are the equations

$$\left. \begin{aligned} s_k\, da &= T_k^{\alpha}\, dA_\alpha && \text{on } \mathscr{S}, \\ \varrho_0\, \dot{b}_k &= \varrho_0 f_k + T_k^{\alpha}{}_{;\alpha} && \text{in } \mathscr{V}. \end{aligned} \right\} \qquad (238.3)$$

The case when $\boldsymbol{b} = \dot{\boldsymbol{x}}$ gives the classical equations in the form (210.5), (210.8); the case when $b_k = a_{km}\dot{x}^m$ gives the generalization described in the paragraph preceding. It is easy to extend (238.2) to a mechanics of moments which generalizes that following from (232.3).

E. Energy and entropy.

239. Scope and plan of the chapter. We attempt to collect here everything of a *general* nature concerning the balance of energy in continuous media and the mathematical properties of entropy. There can be no doubt of the relevance of the first subchapter, which defines the specific internal energy and derives differential equations and jump conditions expressing the balance of total energy in such a way as to reflect the interconvertibility of heat and work.

The second subchapter concerns the more special and more dubious subject of thermodynamics. The specific entropy is regarded as a static defining parameter entering a caloric equation of state which is supposed to regulate the specific internal energy, regardless of deformation and motion. The formal content of this subchapter coincides with that customarily said to describe "reversible" processes in a substance obeying an equation of state depending upon any finite number of parameters, except that our considerations are phrased in terms of a particle in a general continuum and that we do not take up the special thermodynamic properties distinguishing fluids from solids. The subchapter ends by deriving a differential equation for the production of entropy and by discussing relations between the rate of change of total entropy and inequalities governing local changes of entropy.

The last subchapter considers several definitions of equilibrium and mentions connections between stability of equilibrium and inequalities restricting thermodynamic equations and quantities.

In historical origin the balance of energy and the theory of the equation of state, logically independent and in fact relevant to different levels of physical

---

[1] Due to HELLINGER [1914, *4*, § 5d], generalizing a form proposed by E. and F. COSSERAT [1909, *5*, §§ 61–67, 76–80].

generality, are intertwined not only with each other but also with a third concept, that heat and temperature are mean manifestations of molecular motion. To CARNOT (1824–1832) is due not only a general understanding and statement of the equivalence of heat and work, formulated independently by JOULE (1843) and WATERSTON (1843), but also the concept of entropy. While owing much to numerous earlier researches, the first thermodynamic writing to achieve a modicum of clarity is that of GIBBS (1873–1875), upon which our treatment is based, though we do not enter many of the details and applications he developed, nor do we fail to elaborate the mathematical structure rendered possible and natural by his approach.

There exists no other comparably general and complete exposition of the material given in this chapter[1].

I. The balance of energy.

240. The law of energy balance. In Sect. 217 we have seen that kinetic energy $\mathfrak{K}$ generally fails to be conserved. Using the apparatus of Sect. 157, we could set up an influx and a supply of kinetic energy and so obtain a general equation of balance for it[2]. Such a balance, while not incorrect, does not lead to a fruitful theory, because it fails to reflect the physical principle of *interconvertibility of heat and mechanical work*. This principle, which is too broad and too vague for us to attempt a general formulation, as well as too old for us to include its history[3] in this treatise, suggests that any equation of energy balance should contain terms which can be identified with *non-mechanical* transfer of energy. These terms may, but need not be dependent upon changes of *temperature*. To keep full generality, we refrain from defining temperature specifically until the next subchapter, but its effects remain in our minds to motivate the introduction of the *non-mechanical power*, $\mathfrak{Q}$.

At the same time, the special theorems on conservation of energy in Sect. 218, in the circumstances when they hold, should not remain outside the general for-

---

[1] We have been unable to derive much help from any of the numerous treatises except that of PARTINGTON [1949, *22*, Sect. II], distinguished for its concise and clear statements of practice and for its critical references.

[2] Such is the approach of MATTIOLI [1914, *7*].

[3] That heat is a mode of motion was widely believed in the eighteenth century, and both EULER (1729, 1782) and DANIEL BERNOULLI (1738) constructed kinetic molecular models in which temperature may be identified with the kinetic energy of the molecules. The *general* and phenomenological principle, independent of a molecular interpretation, is more recent. That it was known to CARNOT by 1832 is proved by his notes [1878, *1*], which calculate the mechanical equivalent of heat and project the porous plug experiment. The contention of CLAUSIUS, still reproduced in textbooks, that CARNOT's celebrated treatise [1824, *1*] obtained correct results from an incorrect axiom, is shown to be false, as PARTINGTON [1949, *22*, Part II, § 33, last footnote] has remarked, by the presentation of LIPPMAN [1889, *5*, pp. 76–78]; to render CARNOT's work in accord with later views, translate "calorique" as "entropy" (cf. CALLENDER [1910, *3*, §§ 16, 20, 22], LAMER [1949, *16*]), but is unlikely that anyone would grasp the principle of conservation of energy from reading CARNOT's treatise.

In our opinion, the first clear statements to be published are those of JOULE [1843, *2*] [1845, *1* and *2*] [1847, *2*] and WATERSTON [1843, *5*], the latter's being more restricted in scope because derived exclusively from a kinetic molecular model. Forerunners are MOHR [1837, *4*], SEGUIN [1839, *2*, Chap. VII, § 1], and J.R. MAYER [1842, *2*]. The history of the "first law of thermodynamics" is discussed by JOULE [1864, *1*], TAIT [1868, *14*, Introd. and Chap. I] [1876, *6*, §§ II, III, and Introd. to 2nd ed.], CHERBULIEZ [1871, *3*], HELMHOLTZ [1882, *1*], MACH [1896, *3*, pp. 238–268], SARTON [1929, *8*], EPSTEIN [1937, *1*, § 11], BOYER [1943, *1*]; the most nearly complete account is given by PARTINGTON [1949, *22*, Part II, §§ 10–12].

malism. The existence of a strain energy in some cases suggests that in the most general motions we introduce an *internal energy* $\mathfrak{E}$, an additive set function such that the *total energy* $\mathfrak{K}+\mathfrak{E}$ is balanced.

The fundamental ***energy balance*** is then[1]

$$\dot{\overline{\mathfrak{K}+\mathfrak{E}}}=\mathfrak{W}+\mathfrak{Q}, \tag{240.1}$$

where $\mathfrak{W}$ is the *mechanical power* or total rate of working of the mechanical actions upon the body. This basic law, sometimes called the "first law of thermodynamics"[2], is to be set alongside the laws of momentum (196.3).

241. The equation of energy balance for continuous media. For a continuum, the mechanical power $\mathfrak{W}$ is the rate of working of the stress vector and the couple stress on the boundary, plus the rate of working of the assigned forces and couples in the interior (cf. Sects. 217, 232); the non-mechanical power $\mathfrak{Q}$, as to be expected from Sect. 157, is expressed in terms of an *efflux of energy*[3] $\boldsymbol{h}$ and a *supply of energy* q. Thus (240.1) becomes[4]

$$\left.\begin{aligned}\dot{\mathfrak{K}}+\dot{\mathfrak{E}}=\oint_{\mathscr{S}}(t^{pr}\dot{x}_p-m^{pqr}w_{pq})\,da_r+\int_{\mathscr{V}}(f^p\dot{x}_q-l^{pq}w_{pq})\,d\mathfrak{M}+\\+\oint_{\mathscr{S}}h^p\,da_p+\int_{\mathscr{V}}q\,d\mathfrak{M}.\end{aligned}\right\} \tag{241.1}$$

The internal energy $\mathfrak{E}$, being an additive set function[5], may be expressed in terms of a *specific internal energy* ε:

$$\mathfrak{E}=\int_{\mathscr{V}}\varepsilon\,d\mathfrak{M}. \tag{241.2}$$

Thus (241.1) is an equation of balance (157.1). In regions where $\boldsymbol{t}, \dot{\boldsymbol{x}}, \boldsymbol{m}, \boldsymbol{w}$, and $\boldsymbol{h}$ are continuously differentiable, while $\ddot{\boldsymbol{x}}, \dot{\varepsilon}, \boldsymbol{f}, \boldsymbol{l}$, and q are continuous, we may apply (157.6) and obtain

$$\left.\begin{aligned}\varrho\ddot{x}^p\dot{x}_p+\varrho\dot{\varepsilon}=t^{pr}{}_{,r}\dot{x}_p+t^{(pq)}d_{pq}+t^{[pq]}w_{pq}-\\-m^{pqr}{}_{,r}w_{pq}-m^{pqr}w_{pq,r}+\\+\varrho f^p\dot{x}_p-\varrho l^{pq}w_{pq}+h^p{}_{,p}+\varrho q,\end{aligned}\right\} \tag{241.3}$$

where we have used (217.6). By Cauchy's laws (205.2) and (205.10), this reduces to

$$\varrho\dot{\varepsilon}=t^{(pq)}d_{pq}-m^{pqr}w_{pq,r}+h^p{}_{,p}+\varrho q, \tag{241.4}$$

---

[1] The first statement approaching this degree of generality was given by Duhem [1892, *3*, Chap. III, § 3].

[2] Since the same title is often conferred upon far more restricted assertions, we prefer to eschew it, especially since (240.1) is a *fundamental balance*, on a par with the balances of momentum and electromagnetism, while the other "laws" of thermodynamics are more special in nature. See Sect. 244.

[3] The introduction of this vector, not necessarily restricted by any constitutive equation relating it to the temperature, is due to Stokes [1851, *3*].

[4] The minus signs put before the rates of working of the couples are required by the conventions expressed by (App. 3.3) and the remarks following it. For example, $\boldsymbol{l}\cdot\frac{1}{2}\boldsymbol{w}=-l^{pq}w_{pq}$. Most treatments take $\boldsymbol{h}$ with the opposite sign so as to represent an influx rather than an efflux.

[5] It is possible, of course, to consider a more general case when there are surface distributions and isolated sources of energy.

the differential equation of ***energy balance***[1]. It is equivalent to the general principle of energy balance (241.1) when the above-stated requirements of smoothness are satisfied and when in addition *the balance of momentum and of moment of momentum is assumed*. It is important to notice that the stress power $t^{km}\dot{x}_{k,m}$, which figures in the work theorem (217.1), does not give all of the mechanical contribution to the increase of internal energy except in the non-polar case.

At a surface of discontinuity, by (193.4) we obtain from (241.1) the jump condition[2]

$$\varrho^{\pm} U^{\pm}[\varepsilon + \tfrac{1}{2}\dot{x}^2] + n_r[t^{pr}\dot{x}_p - m^{pqr} w_{pq} + h^r] = 0. \tag{241.5}$$

It is natural to try to simplify the result by using also the conditions (205.3), (205.4) expressing balance of momentum, but no very neat result emerges. In the non-polar case, the term involving the vorticity drops out, and the term preceding it becomes $[\boldsymbol{t}_{(n)} \cdot \dot{\boldsymbol{x}}]$. Resolving $\dot{\boldsymbol{x}}$ and $\boldsymbol{t}_{(n)}$ into components normal and tangential to the singular surface, by use of (185.6) and (189.14) we have

$$\left.\begin{aligned} [\boldsymbol{t}_{(n)} \cdot \dot{\boldsymbol{x}}] &= [t_{nn}\dot{x}_n] + [\boldsymbol{t}_{(n)t} \cdot \dot{\boldsymbol{x}}_t], \\ [t_{nn}\dot{x}_n] &= u_n[t_{nn}] - \varrho^{\pm} U^{\pm}[t_{nn} v], \\ [\tfrac{1}{2}\dot{x}^2] &= [\tfrac{1}{2}U^2] + u_n[\dot{x}_n] + [\tfrac{1}{2}\dot{x}_t^2]. \end{aligned}\right\} \tag{241.6}$$

Substituting these results into (241.5) and use of $(205.6)_1$ yields

$$\varrho^{\pm} U^{\pm}[\varepsilon - t_{nn} v + \tfrac{1}{2}U^2 + \tfrac{1}{2}\dot{x}_t^2] + [\boldsymbol{t}_{(n)t} \cdot \dot{\boldsymbol{x}}_t] + [h_n] = 0. \tag{241.7}$$

In fact, $(205.6)_2$ has not been used, but it does not seem to make possible any real simplification of (241.7) without further assumptions.

Two special cases are commonly met. First, if the surface of discontinuity is neither a slip surface nor a shock front, and if the vorticity is continuous or there is no couple stress, (241.5) reduces to FOURIER's condition[3]

$$[h_n] = 0: \tag{241.8}$$

The normal flux of energy is continuous. Second, if $\boldsymbol{t} = -p\,\boldsymbol{1}$ and there is neither couple stress nor flux of energy, (241.7) reduces to the form[4]

$$[\varepsilon + p v + \tfrac{1}{2}U^2] = 0 \tag{241.9}$$

---

[1] That the material derivative should replace the partial time derivative in the equation of heat conduction for a moving medium was observed by FOURIER [1833, *1*, Eq. (3)]; cf. OBERBECK [1879, *3*, § 2]. The principle and scheme of the derivation of (241.4) are due to KIRCHHOFF [1868, *11*, § 1], but his result is limited to infinitesimal motion of a perfect gas; later [1894, *5*, Elfte Vorl., § 3] he removed the former restriction. That use of a differential equation expressing balance of energy is necessary except in specially simple circumstance was first emphasized by DUHEM [1891, *3*, Vol. I, Livre II, Chap. III, § 5], [1901, *5*], [1901, *7*, Part I, Chap. 1, § 7], [1903, *5*], who called it "la relation supplémentaire"; fairly general cases of (241.4) were derived also by C. NEUMANN [1894, *7*, § 4], VAN LERBERGHE [1933, *13*], MEIXNER [1941, *2*, § 3] [1942, *9*, § 2], STEWART [1942, *12*], TRUESDELL [1948, *35*], and others. A theory of the energy and entropy of a continuum based on the existence of extrinsic and mutual potential energies was constructed by DUHEM [1893, *1*, Chap. II] [1911, *4*, Chap. XIV]. WEISSENBERG [1931, *11*, § A c II] [1935, *10*, pp. 52, 138–141] divided the internal energy into "free" and "bound" portions.

[2] A fairly general case is given by KOTCHINE [1926, *3*, Eq. (21)]. A special result of this kind is obtained by MARSHAK [1958, *7*, § 1].

[3] [1822, *1*, § 146].

[4] That the balance of energy is in general a condition independent of those of mass and momentum has long been evident in purely thermal problems, but the first mechanical context in which this fact was noted is the shock wave in a gas. The analysis of STOKES [1848, *4*] presumed tacitly that the theorem of conservation of mechanical energy holding on each

when the discontinuity is a shock[1], while when the discontinuity is not a shock, no restriction upon ε results. In all these jump conditions it is assumed that no surface sources of energy lie upon the surface of discontinuity.

To write the equation of energy balance in four-dimensionally invariant form is easy. Considering only the non-polar case, we make the following agreements: ε and q shall be absolute world scalars, $\boldsymbol{h}$ shall be represented by an absolute space tensor (cf. Sect. 153) h^Ω such that $h^\Omega = (h^k, 0)$ in every Euclidean frame. Let $t^{\Omega\Delta}$ be the space tensor (211.2), let g be the world scalar (153.12), and let $\Delta^{\Phi\Psi}$ be the world tensor of stretching (153.10). Put

$$D_{\Phi\Psi} = p_{\Phi\Theta} p_{\Psi\Omega} \Delta^{\Theta\Omega}, \tag{241.10}$$

where $p_{\Omega\Delta}$ is the world tensor (153.21). The world tensor equation

$$\varrho v^\Omega \frac{\partial \varepsilon}{\partial x^\Omega} = \varrho \dot{\varepsilon} = t^{\Omega\Delta} D_{\Omega\Delta} + \frac{1}{\sqrt{g}} \frac{\partial}{\partial x^\Omega} (\sqrt{g}\, h^\Omega) + \varrho q, \tag{241.11}$$

where $\mathfrak{v}$ is the world velocity vector, reduces to the non-polar case of (241.4) in every Euclidean frame. Therefore, (241.11) is a *world tensor form of the equation of energy balance*[2]. Note that (241.11) does not presume the existence of a world connection. It is a Euclidean invariant equation.

241 A. Appendix: Relation between strain energy and internal energy. The cases we interpolate here are far too special to deserve a place now, but we pause to note them so as to give the reader who is accustomed to a more conventional approach to energetic problems a hold on its place in our development.

We restrict attention to the non-polar case. Considering the general energy equation (241.4), we search for assumptions such that terms on the right-hand side may be written in the form $\varrho \dot{\Psi}$, where Ψ is some function of the variables encountered in earlier developments. Alternatively, we seek to express the stress-power P in terms of other scalars. Since the field $\boldsymbol{f}$ does not enter (241.4), to assume the existence of a potential energy, which simplified the equation of mechanical energy in Sect. 218, will have no effect here. However, the special assumptions (218.5) and (218.6) yield

$$\varrho(\dot{\varepsilon} - \dot{\sigma}) = {}_D t^{pq} d_{pq} + h^p_{,p} + \varrho q, \tag{241 A.1}$$

as follows at once from (218.7). In particular, if the right-hand side is zero, we get $\varepsilon = \sigma + f(\boldsymbol{X})$, showing that when there is no dissipative stress, no flux of energy, and no supply of energy, the internal energy and the strain energy differ by a constant for each particle.

Further formal simplification results when we assume[3]

$${}_D t^{pq} = \frac{\partial D}{\partial d_{pq}}, \tag{241 A.2}$$

---

side of the shock ensures conservation of energy also at the shock. This is not so. That a distinct additional condition is needed was first seen by RANKINE [1870, *6*, §§ 7—9], who obtained a special case of (241.9) for lineal motion, as did HUGONIOT [1887, *1*, § 149]. According to a note added in the 1883 reprint of [1848, *4*], KELVIN and RAYLEIGH were aware of the matter. Cf. the discussion by HADAMARD [1903, *11*, ¶ 209]. A special case for general motion was derived by JOUGUET [1901, *9*] and discussed by DUHEM [1901, *7*, Part 2, Chap. I, §§ 7—8] and ZEMPLÉN [1905, *9*, § 8]. A fairly general case was obtained by COURANT and FRIEDRICHS [1948, *8*, §§ 54, 118].

We are unable to follow the physical arguments sometimes adduced to infer (241.9) directly from remarks concerning the enthalpy. Indeed, if $p = \pi$, the thermodynamic pressure (Sect. 247) in a fluid obeying a caloric equation of state of the form $\varepsilon = \varepsilon(\eta, v)$, then (241.9) assumes the form $[\chi + \frac{1}{2}U^2] = 0$, but in more general circumstances (241.7) bears no apparent relation to the enthalpy defined by $(251.1)_3$. Indeed, to derive (241.7) no thermodynamic formalism is used, and in the degree of generality maintained here, the thermodynamic tensions need not be defined, and hence an enthalpy need not exist.

[1] Note that the condition $[\dot{\boldsymbol{x}}_t] = 0$, which follows from $(205.6)_2$ in the present case, has been used here; cf. the remark after (241.7).

[2] Contrary to the implication of MCVITTIE [1949, *18*, § 7], solution of this problem requires no commitment as to the form of the energy equation in relativistic theories. MCVITTIE obtains a special case of (241.11), in which ε has a special functional form.

[3] RAYLEIGH [1873, *6*, § II].

where D is a homogeneous function of degree r in the components d_{pq}, since (241 A.1) then becomes

$$\varrho(\dot{\varepsilon} - \dot{\sigma}) = h^p_{,p} + \varrho q + r D. \tag{241 A.3}$$

This result is an equation of balance (Sect. 157): If we set $\lambda \equiv \varepsilon - \sigma$, then $\boldsymbol{h}$ is the efflux of λ and $\varrho q + rD$ is the supply of λ. The function D is called a *dissipation function*. This same result, with $D \equiv 0$, holds *a fortiori* when the dissipative stresses vanish.

Thus it is clear that energy equations of the type encountered in classical elasticity, hydrodynamics, and thermal conduction, while consistent with the general scheme of energy balance, result only from the special assumptions peculiar to those disciplines and would be false guides in any general approach to energetic theory.

242. Energy impulse. Equations for energy impulse may be gained by considerations parallel to those of Sect. 198 and Sect. 206, using the results of Sect. 194. Apparently, however, this has never been done, and when we set about it, we find an unexpected complication. Suppose a stress impulse given by the tensor i^{pr} on a boundary $\mathscr{S}$ gives rise to a velocity impulse $\{\dot{x}_p\}$. Then there certainly results an energy impulse on $\mathscr{S}$, but it is not $\oint_{\mathscr{S}} i^{pr} \{\dot{x}_p\} \, d a_r$. Indeed, since $i^{pr} d a_r$ is a *difference* of momenta, while $\{\dot{x}_p\}$ is a difference of velocities, the difference of energies resulting will not in general bear any simple relation either to $\boldsymbol{i}$ or to $\{\dot{\boldsymbol{x}}\}$.

Thus the causes of energy impulse in general we cannot easily separate into a purely energetic portion plus a portion arising from the impulses of velocity, momentum, and stress. In full generality we have

$$\{\varrho(\varepsilon + \tfrac{1}{2}\dot{x}^2)\} = k^p_{,p} + \varrho n, \tag{242.1}$$

where $\boldsymbol{k}$ and n are the *total* influx and the *total* supply of energy impulse, but for the reasons given above we cannot generally write $\boldsymbol{k}$ and n partially in terms of $\boldsymbol{i}$, $\boldsymbol{s}$, and $\{\dot{\boldsymbol{x}}\}$, and thus analysis parallel to that leading from (241.3) to (241.4) is not possible.

An exception is the case when the motion is generated from *rest*, for then we have $\{\dot{\boldsymbol{x}}\} = \dot{\boldsymbol{x}}$, $\{\varrho \dot{\boldsymbol{x}}\} = \varrho \dot{\boldsymbol{x}}$, and in the non-polar case

$$k^p = \tfrac{1}{2} i^{qp} \dot{x}_q + {}^0k^p, \qquad n = \tfrac{1}{2} s^q \dot{x}_q + {}^0n, \tag{242.2}$$

where ${}^0\boldsymbol{k}$ and ${}^0n$ are the non-mechanical influx and supply of energy impulse. In this case (242.1) may be reduced by use of (206.1), so that

$$\{\varrho \varepsilon\} = \tfrac{1}{2} i^{qp} \dot{x}_{q,p} + {}^0k^p_{,p} + \varrho\, {}^0n. \tag{242.3}$$

243. Balance of energy in a heterogeneous medium[1]. To discuss the transfer of energy in a mixture, we employ the formalism of Sect. 158, and we proceed

[1] Equations of the type (243.1), (243.2), and (243.3), with special forms for $\varepsilon_{\mathfrak{A}} \hat{\varepsilon}$ and $_{\mathfrak{A}}$ arise in MAXWELL's kinetic theory of mixtures of monatomic gases [1867, *2*, pp. 47–49]. There, however, it is customary to define all quantities in terms of molecular motion relative to $\dot{\boldsymbol{x}}$, not to $\dot{\boldsymbol{x}}_{\mathfrak{A}}$; thus the partial heat flux vector $\boldsymbol{q}_{\mathfrak{A}}$ of HIRSCHFELDER, CURTISS and BIRD [1954, *9*, Eq. (7.2–25)] is not to be identified with our $-\boldsymbol{h}_{\mathfrak{A}}$ but rather, subject to the special assumptions of the kinetic theory, with the negative of the whole expression in brackets on the right-hand side of (243.2). Our $\varrho \hat{\varepsilon}_{\mathfrak{A}}$ includes as a special case what CHAPMAN and COWLING [1939, *6*, § 8.1] denote by $n_{\mathfrak{A}} \Delta \overline{E}_{\mathfrak{A}} - \overline{\boldsymbol{C}}_{\mathfrak{A}} \cdot n_{\mathfrak{A}} \Delta m_{\mathfrak{A}} \overline{\boldsymbol{C}}_{\mathfrak{A}}$; our $\varrho \varepsilon$, as defined by (243.1), includes their $n \overline{E}$, our Eq. (243.6) corresponds to the sum of their equations $\sum_{\mathfrak{A}=1}^{\mathfrak{K}} n_{\mathfrak{A}} \Delta m_{\mathfrak{A}} \overline{\boldsymbol{C}}_{\mathfrak{A}} = 0$ and $\sum_{\mathfrak{A}=1}^{\mathfrak{K}} n_{\mathfrak{A}} \Delta \overline{E}_{\mathfrak{A}} = 0$, etc.

The continuum theory, while more general and simpler in concept, developed only later, imperfectly, and apparently in oblivion of what had been done long before in the kinetic theory. Differential equations for the internal energy of a mixture of continua have been given

as in Sects. 159 and 215. Each constituent $\mathfrak{A}$ has its own *partial internal energy* $\varepsilon_{\mathfrak{A}}$ and is subject to flux of energy $\boldsymbol{h}_{\mathfrak{A}}$ and supply of energy $q_{\mathfrak{A}}$. The total internal energy ε is the sum of the partial internal energies plus the kinetic energies of diffusion:

$$\varepsilon \equiv \sum_{\mathfrak{A}=1}^{\mathfrak{K}} c_{\mathfrak{A}} \left(\varepsilon_{\mathfrak{A}} + \tfrac{1}{2} u_{\mathfrak{A}}^2\right). \tag{243.1}$$

The total flux of energy $\boldsymbol{h}$ arises from three sources: The constituent non-mechanical fluxes of energy $\boldsymbol{h}_{\mathfrak{A}}$, the rates of working of the partial stresses against diffusion, and the fluxes of total constituent energies by diffusion:

$$h^k \equiv \sum_{\mathfrak{A}=1}^{\mathfrak{K}} \left[h_{\mathfrak{A}}^k + t_{\mathfrak{A}}^{km} u_{\mathfrak{A}\,m} - \varrho_{\mathfrak{A}} \left(\varepsilon_{\mathfrak{A}} + \tfrac{1}{2} u_{\mathfrak{A}}^2\right) u_{\mathfrak{A}}^k\right]. \tag{243.2}$$

The total energy of the constituent $\mathfrak{A}$ need not be balanced in itself, as energy may be transferred from one constituent to another. Thus, restricting attention to the non-polar case, so that $t_{\mathfrak{A}}^{km} = t_{\mathfrak{A}}^{mk}$, we define the *supply of energy* $\hat{\varepsilon}_{\mathfrak{A}}$ by

$$\varrho\, \hat{\varepsilon}_{\mathfrak{A}} \equiv \varrho_{\mathfrak{A}} \left(\dot{\varepsilon}_{\mathfrak{A}} - q_{\mathfrak{A}}\right) - t_{\mathfrak{A}}^{km} \dot{x}_{\mathfrak{A}\,k,m} - h_{\mathfrak{A},k}^k; \tag{243.3}$$

the condition $\hat{\varepsilon}_{\mathfrak{A}} = 0$ is then necessary and sufficient that the energy of the constituent $\mathfrak{A}$ be in balance by itself.

---

for various special cases and under various special hypotheses by REYNOLDS [1903, *15*, § 39], JAUMANN [1911, *7*, § IV], HEUN [1913, *4*, § 24c], LOHR [1917, *5*, Eqs. (108), (109)] [1924, *10*], MEIXNER VAN MIEGHEM [1935, *9*, § 2], MEISSNER [1938, *7*, § 3], ECKART [1940, *8*, p. 272], MEIXNER [1941, *2*, Eq. (12)] [1943, *2*, Eq. (2,8)], VERSCHAFFELT [1942, *14*, §§ 15—16] [1942, *15*, §§ 7—8], PRIGOGINE [1947, *12*, Chap. VIII, § 3], KIRKWOOD and CRAWFORD [1952, *12*, Eqs. (12) and (18)], and later writers. These authors write down their differential equations essentially by inspection, without deriving them from the equations governing the constituents, and without unequivocal specification of the total internal energy in terms of constituent energies. ECKART [1940, *8*, p. 271] hinted at the definition (243.1) but in the end neglected the kinetic energies of diffusion; they are included by KIRKWOOD and CRAWFORD [1952, *12*]. An equation of the form (243.3) with $\hat{\varepsilon}_{\mathfrak{A}} = 0$ was proposed by LEAF [1946, *7*, Eq. (14)]; from the kinetic theory it is clear that such an assumption is usually false. Thermodynamic writers often pass over mechanical aspects rather cavalierly; typically (e.g. DE GROOT [1952, *3*, § 44]) they replace or supplement the mechanical power term $t_{\mathrm{I}}^{km} d_{km}$ by expressions containing some or all of the thermodynamic power terms in (255.15). Thus it is not surprizing that the results do not always agree with one another and do not contain all the terms in (243.9) or any counterpart of (243.6). A discussion similar to ours is given by PRIGOGINE and MAZUR [1951, *21*, § 3] [1951, *17*, § 2], but their definitions differ from (243.1) and (243.2) in including terms depending on potential energy and in employing the velocities $\dot{\boldsymbol{x}}_{\mathfrak{A}}$ rather than the diffusion velocities $\boldsymbol{u}_{\mathfrak{A}}$; thus their definitions of ε and $\boldsymbol{h}$ do not reduce to ε_I and $\boldsymbol{h}_I$ when $\mathfrak{K} = 1$. Also, they do not derive any counterpart of the condition (243.6), but it may be implied in the equation they write down for the balance of potential energy. Cf. also the special case considered by NACHBAR, WILLIAMS and PENNER [1957, *10*, § V].

Our treatment follows TRUESDELL [1957, *16*, § 7]. In view of the divergences among thermodynamic writers, it appears necessary to emphasize that all results in the text stand in detailed consistency with their counterparts in the kinetic theory. In particular, that the correct energy equation for the mixture should be of just the same form as that for a simple medium, viz., (243.7), has long been known; cf., e.g., the careful derivation of HIRSCHFELDER, CURTISS and BIRD [1954, *9*, Eqs. (7.2—49) and (7.6—7)]. The difference in results obtained by thermodynamic writers arises only partly from their apparent use of ε_{I} rather than ε (cf. our analysis in Sect. 259) but seems to be inherently unresolvable because of their failure to define $\boldsymbol{t}$ and $\boldsymbol{h}$ in terms of quantities associated with the constituents.

Summing (243.3) on $\mathfrak{A}$, we obtain

$$\left.\begin{aligned}
\varrho\sum_{\mathfrak{A}=1}^{\mathfrak{K}}\hat{\varepsilon}_{\mathfrak{A}} &= \sum_{\mathfrak{A}=1}^{\mathfrak{K}}\{\varrho_{\mathfrak{A}}(\dot{\varepsilon}_{\mathfrak{A}}+\overline{\tfrac{1}{2}u_{\mathfrak{A}}^2}^{\,\prime}) - \varrho_{\mathfrak{A}}u_{\mathfrak{A}k}\ddot{x}_{\mathfrak{A}}^k - \varrho_{\mathfrak{A}}q_{\mathfrak{A}} - \\
&\quad - (t_{\mathfrak{A}}^{km} - \varrho_{\mathfrak{A}}u_{\mathfrak{A}}^k u_{\mathfrak{A}}^m)\dot{x}_{k,m} - \\
&\quad - [h_{\mathfrak{A}}^k + t_{\mathfrak{A}}^{km}u_{\mathfrak{A}m} - \varrho_{\mathfrak{A}}(\varepsilon_{\mathfrak{A}}+\tfrac{1}{2}u_{\mathfrak{A}}^2)u_{\mathfrak{A}}^k]_{,k} + \\
&\quad + t_{\mathfrak{A},k}^{km}u_{\mathfrak{A}m} - [\varrho_{\mathfrak{A}}(\varepsilon_{\mathfrak{A}}+\tfrac{1}{2}u_{\mathfrak{A}}^2)u_{\mathfrak{A}}^k]_{,k}\},\\
&= \varrho\dot{\varepsilon} - t^{km}\dot{x}_{k,m} - h^k{}_{,k} - \\
&\quad - \sum_{\mathfrak{A}=1}^{\mathfrak{K}}\{\varrho\hat{c}_{\mathfrak{A}}(\varepsilon_{\mathfrak{A}}+\tfrac{1}{2}u_{\mathfrak{A}}^2) + u_{\mathfrak{A}k}(\varrho_{\mathfrak{A}}\ddot{x}_{\mathfrak{A}}^k - t_{\mathfrak{A},m}^{km}) + \varrho_{\mathfrak{A}}q_{\mathfrak{A}}\},\\
&= \varrho\varepsilon - t^{km}d_{km} - h^k{}_{,k} - \varrho q - \\
&\quad - \varrho\sum_{\mathfrak{A}=1}^{\mathfrak{K}}[\hat{c}_{\mathfrak{A}}(\varepsilon_{\mathfrak{A}}+\tfrac{1}{2}u_{\mathfrak{A}}^2) + \hat{p}_{\mathfrak{A}}^k u_{\mathfrak{A}k}],
\end{aligned}\right\}\tag{243.4}$$

where

$$q \equiv \sum_{\mathfrak{A}=1}^{\mathfrak{K}} c_{\mathfrak{A}}(q_{\mathfrak{A}} + f_{\mathfrak{A}}^k u_{\mathfrak{A}k}).\tag{243.5}$$

To derive (243.4), at the first equality we have used only algebraic rearrangement, (158.7), and (158.11); at the second equality we have used (243.1), the fundamental identity (159.5), and (215.1) and (243.2); the last equality follows by (215.2). From (243.4) we see that a necessary and sufficient condition for the non-polar case of (241.4) to hold for the mixture is

$$\sum_{\mathfrak{A}=1}^{\mathfrak{K}}[\hat{\varepsilon}_{\mathfrak{A}} + \hat{p}_{\mathfrak{A}}^k u_{\mathfrak{A}k} + \hat{c}_{\mathfrak{A}}(\varepsilon_{\mathfrak{A}}+\tfrac{1}{2}u_{\mathfrak{A}}^2)] = 0.\tag{243.6}$$

This result, analogous to (159.4) and (215.5), asserts that the energy supplied by an excess internal energy rate, plus the energy supplied by the work of the excess inertial forces against diffusion, plus the energy supplied by the creation of mass, must add up to zero for the mixture.

We have shown that for the non-polar case the condition (243.6), along with the definitions (243.1), (243.2) and (243.5), leads to an energy equation of the form

$$\varrho\dot{\varepsilon} = t^{km}d_{km} + h^k{}_{,k} + \varrho q\tag{243.7}$$

for the mixture. For later use we require this same equation put in terms of the *inner parts* ε_{I}, h_{I}, and q_{I} of ε, $\boldsymbol{h}$, and q:

$$\varepsilon_{\mathrm{I}} \equiv \sum_{\mathfrak{A}=1}^{\mathfrak{K}} c_{\mathfrak{A}}\varepsilon_{\mathfrak{A}},\qquad \boldsymbol{h}_{\mathrm{I}} \equiv \sum_{\mathfrak{A}=1}^{\mathfrak{K}}\boldsymbol{h}_{\mathfrak{A}},\qquad q_{\mathrm{I}} \equiv \sum_{\mathfrak{A}=1}^{\mathfrak{K}} c_{\mathfrak{A}}q_{\mathfrak{A}}.\tag{243.8}$$

Either by transforming (243.7) or by summing (243.3) on $\mathfrak{A}$ and then using (159.5), (215.7), and (243.6), we obtain

$$\left.\begin{aligned}
\varrho\dot{\varepsilon}_{\mathrm{I}} &= t_{\mathrm{I}}^{km}d_{km} + \sum_{\mathfrak{A}=1}^{\mathfrak{K}} t_{\mathfrak{A}}^{km}u_{\mathfrak{A}k,m} + \Big(h_{\mathrm{I}}^k - \sum_{\mathfrak{A}=1}^{\mathfrak{K}}\varrho_{\mathfrak{A}}\varepsilon_{\mathfrak{A}}u_{\mathfrak{A}}^k\Big)_{,k} + \\
&\quad + \varrho q_{\mathrm{I}} - \varrho\sum_{\mathfrak{A}=1}^{\mathfrak{k}}[\hat{p}_{\mathfrak{A}}^k u_{\mathfrak{A}k} + \hat{c}_{\mathfrak{A}}\cdot\tfrac{1}{2}u_{\mathfrak{A}}^2],
\end{aligned}\right\}\tag{243.9}$$

where $\boldsymbol{t}_{\mathrm{I}}$ is defined by (215.6).

244. Remarks on the general energy balance. The interconvertibility of heat and mechanical work is expressed only indirectly in (241.4) and (241.6), or, for that matter, in the basic equation (240.1). In the interpretation, the variables $\mathfrak{Q}$, $\boldsymbol{h}$, q, and ε are to be associated at least in part with *thermal phenomena*, yet they are measured in *mechanical units*. In fact,

$$\left.\begin{aligned} &\dim \mathfrak{Q} = \dim \mathfrak{W} = [M L^2 T^{-3}], \\ &\dim \varepsilon = \dim \dot{x}^2 = [L^2 T^{-2}], \\ &\dim \boldsymbol{h} = [M T^{-3}], \quad \dim q = [L^2 T^{-3}]. \end{aligned}\right\} \tag{244.1}$$

In order that $\boldsymbol{h}$ may be connected with a temperature gradient, for example, the mechanical equivalent of heat must be used. Only in its tacit assumption, necessary for its intended applications, that *all thermo-energetic phenomena may be measured in mechanical units*, does the energy balance bring in any new physical idea beyond those used in pure mechanics.

Indeed, with an equation such as (241.4) we appear to be further from solving any problems concerning energy than we were in Sect. 217. To secure balance of energy, we have introduced three new quantities ε, $\boldsymbol{h}$, and q, and only one condition connecting them. In any given motion, (241.4) may be used for each particle $\boldsymbol{X}$ as a definition of ε to within an additive constant, as a definition of $\boldsymbol{h}$ to within an arbitrary solenoidal field, or as a definition of q, when two of these quantities are assigned arbitrarily. But this is not the way in which (241.4) is used in practice. Rather, its new variables correspond to physical ideas, and the simple structure set up in this subchapter serves as a framework (cf. Sect. 6) within which more special considerations concerning changes of energy may conveniently be expressed. Some of these special assumptions will be discussed in the remainder of the chapter.

II. Entropy.

a) The caloric equation of state.

245. Thermostatics and thermodynamics. The reader who has no preconception of thermodynamics may pass over this section, entering at once into the theory in Sect. 246.

1. *The classical difficulties. Thermostatics*, which even now is usually called *thermodynamics*, has an unfortunate history and an unfortunate tradition. As compared with the older science of mechanics and the younger science of electromagnetism, its mathematical structure is meager. Though claims for its breadth of application are often extravagant, the examples from which its principles usually are inferred are most special, and extensive mathematical developments based on fundamental equations, such as typify mechanics and electromagnetism, are wanting. The logical standards acceptable in thermostatics fail to meet the criteria of most other branches of physics; books and papers concerning it contain a high proportion of descriptive matter to equations and results. The obscurity of its concepts is witnessed by the many attempts, made alike by engineers, physicists, and mathematicians and continuing today in greater number, to reformulate them and to set the house of thermostatics in order.

The difficulty of the subject lies partly in its task of comparing different equilibria without describing the intermediate states whereby bodies may reach equilibrium. At the outset, the reader is told to imagine a system changing so slowly as to be in equilibrium at all times, for such paradoxical "quasistatic processes" are to furnish the main subject of the theory. The critical student must long have realized that some kind of linearization is involved; in the shadows behind classical thermostatics must stand a better theory including motion and change of motion, and from this theory the classical quasistatic process should result when kinetic energy, diffusion, and inhomogeneity are neglected. In addition to quasistatic or "reversible" processes, the classical theory attempts to deal with certain "irreversible" processes, but rather than describing their course in its formal structure, it merely lays down prohibitions regarding their outcomes[1]. Here too the traditional obscurities arise from an *incomplete description*[2] of the phenomena the theory is envisioned as representing.

2. *"Irreversible" thermodynamics.* Both these difficulties are cleared by a simple expedient: The basic equations of classical thermostatics are applied to elements of volume in a moving material, or in a mixture of materials. This device of a *local* thermostatic state[3], when employed in conjunction with the general principle of energy given in Subchapter I, leads to a theory in which equilibrium is but a special case of motion and change, and through which many processes considered "irreversible" in thermostatics are described easily and naturally, at least in principle. This theory, usually called "irreversible thermodynamics", we prefer to call simply *thermodynamics*[4].

That so great a difference in scope can result from transferring long familiar ideas to a local scale should not be surprizing. It is typical of the success of the *field viewpoint* (cf. Sect. 2). Similarly, problems of fluid motion which could be treated only grossly by the mechanics of "bodies" (i.e., mass-points) were solved successfully 200 years ago by hydrodynamics, which applied the principles of mechanics to the continuous field. What is surprizing is that, granted occasional exceptions, the dynamical theory of heat should have remained for nearly a century in a stage of development analogous to the theory of mass-points in mechanics.

Even though both theories employ parallel concepts on different scales, beyond the first few steps we cannot expect simply to read off results for the *thermodynamic field* from counterparts in thermostatics, any more than hydrodynamics can be read off from mass-point mechanics.

3. *The concept of entropy.* Much of the wordiness of the traditional presentation grows from its insistence on justifying the basic assumptions by experience, and in particular on developing the concept of entropy in terms of heat and

---

[1] Cf. the remark of PARTINGTON [1949, *22*, Sect. II, § 51]: "The thermodynamics of irreversible processes is entirely qualitative and of little interest in physical chemistry." This remark applies to the traditional view, to which the text alludes, but not to the more recent studies mentioned in footnote 1, p. 618, and in Sect. 306.

[2] Cf. the remarks of DUHEM [1904, *2*]: "La Thermodynamique ne possède pas de moyens qui suffisent à mettre complètement en équations le mouvement des systèmes qu'elle étudie...", etc. Hence arise the peculiar difficulties of the theory of thermodynamic stability (Sects. 264 and 265).

[3] The earliest examples are cited in Sect. 248 in connection with the thermal equation of state for gas flows. The first systematic treatments of the energy and entropy fields in a deformable medium were given by JAUMANN [1911, *7*] [1918, *3*] and LOHR [1917, *5*] [1924, *10*]; their work is difficult to study because its main object, the explanation of a set of linear constitutive equations intended to describe all physical phenomena known to the authors, has lost what interest it may have had (cf. Sect. 6), and from the maze of calculation, which is highly condensed despite its length, the reader can scarcely disengage the physical principles. It is clear, however, that JAUMANN and LOHR deserve great credit for realizing the nature and importance of the production of entropy and for being the first to derive differential equations for it. Cf. also the early exposition of DE DONDER [1931, *4*, § 6].

[4] A description of "irreversible thermodynamics" in classical terms would be: Volume elements are assumed to suffer only reversible changes, possibly resulting in irreversible changes for the body as a whole. Such terms can be rather confusing, as when PRIGOGINE [1947, *12*, Chap. IX, § 1] interprets (255.1) as an assertion that the mean motion of a mixture does not produce entropy and is thus a "reversible phenomenon", even if accompanied by viscosity, diffusion, and the conduction of heat. We prefer to cleave to equations and eschew verbalisms.

temperature[1]. In the more highly developed parts of theoretical physics, such discussions do not ordinarily form a part of a treatise on the theory itself, but belong rather to works on the physical foundations and on the connection between theory and experiment. While it is true that the physics laboratory does not contain an entropy meter, the concept of entropy is not more difficult than some others, such as electric displacement[2]; even temperature and mass prove elusive to critical inquiry.

A glance at the *equations* of the theory, once the preliminary words are past, shows that *thermodynamics is the science of entropy*. This is true even more of recent works on irreversible processes than of the classics on thermostatics.

4. *The nature and scope of our presentation*. An axiomatic development, deriving entropy from heat and temperature, would be desirable, but in our opinion there exists no acceptable treatment of this kind[3]. As in the other domains presented in this treatise, we are content to explain the *formal structure* of the theory *as it is practised* (cf. Sects. 3, 196). Thus we take *entropy* as the primitive concept in terms of which thermodynamics is constructed[4]. Surely, any future axiomatic treatment if successful will lead to the same *equations* as those from which our presentation begins. For readers who prefer arguments concerning steam engines, we cite the original memoirs on thermostatics[5]. Neither do we

---

[1] Hence results an apologetic tone in many recent works on "irreversible thermodynamics", which often find it necessary to discuss whether or not it is "meaningful" to speak of temperature and entropy for systems not in equilibrium. Often included are arguments from statistical mechanics. While a *rigorous* development of equations governing entropy and temperature from *general* statistical mechanics would be most illuminating (cf. Sect. 1), all attempts thus far rest on formal approximation procedures in the kinetic theory of monatomic gases or on the theory of small perturbations from statistical equilibrium, so that their validity is confined *a fortiori* to physical situations far less general than those which the results they claim to derive are intended to represent. In any case, it is not right to single out thermodynamics as the only branch of physics where such arguments are in order. Rather, the development of field theories from statistical theories constitutes a general program of inquiry (cf. Sect. 5). Such a program is outside the scope of the present treatise, the purpose of which is to explain *the field theories as such*.

[2] The parallel is good. Entropy cannot be measured except in terms of other quantities, such as energy and temperature; the same is true of dielectric displacement, but the constitutive assumption $\mathfrak{D} = \varepsilon \boldsymbol{E}$ (cf. Sect. 308) is so common that we are often led to regard $\mathfrak{D}$ as closer to experience than in reality it is.

[3] Special mention must be made of the celebrated work of CARATHÉODORY [1909, *3*], [1925, *4*]; cf. BORN [1921, *2*], EHRENFEST-AFANASSJEWA [1925, *5*], LANDÉ [1926, *4*], MIMURA [1931, *7* and *8*], IWATSUKI and MIMURA [1932, *7*], JARDETSKY [1939, *9*], WHAPLES [1952, *24*], FÉNYES [1952, *6*]. CARATHÉODORY succeeded in deriving the concepts of absolute temperature and of entropy from a suitable formalization of the idea of *equilibrium* and the assumption that for any state, there is an arbitrarily near state that the system cannot reach without work's being expended. Despite the mathematical elegance and success of this approach, we cannot regard it as fundamental for a theory intended to describe arbitrary changes of energy, where thermal equilibrium is as little to be expected as is mechanical equilibrium in dynamics. For thermodynamics, it is not equilibrium that is basic, but *entropy production*. Cf. also the criticisms expressed by LEAF [1944, *8*, p. 94].

[4] This method is due to GIBBS [1873, *2*, p. 2, footnote] [1873, *3*, p. 31] [1875, *1*, pp. 56, 63] who did not attempt to justify it; it was recognized at once by MAXWELL [1875, *4*, p. 195] and was adopted by HILBERT [1907, *4*, pp. 435—438]. Among modern authors who follow it, we cite ECKART [1940, *8*] and MEIXNER [1941, *2*, § 3] [1943, *2*, § 2].

[5] The traditional theory was developed by CLAPEYRON [1834, *1*], KELVIN [1849, *4*] [1853, *3*], CLAUSIUS [1850, *1*] [1854, *1*] [1862, *1*] [1865, *1*], REECH [1853, *2*], RANKINE [1853, *1*], and F. NEUMANN [1950, *21*] (deriving from 1854 to 1855 or earlier). A particularly careful discussion of the dozens of assumptions, mostly tacit, on which the traditional development rests was given by DUHEM [1893, *3*]. A variant system, largely unpublished until 1928, was devised by WATERSTON from 1843 onward; it is recommended and developed by HALDANE [1928, *3*, Chap. II]. Other variants or extensions are proposed by BRØNSTED [1940, *4* and *5*] and LEAF [1944, *7*]. Brief histories of thermostatics are given by MACH

consider it the province of any treatise on mathematical theory to explain how measurements testing the theory are to be made. Finally, and also in defiance of the tradition of the subject, we do not consider it any more appropriate here than in the theory of stress to work out elementary examples based on special assumptions. This treatise presents the *general theory of the entropy field*[1].

The theory of entropy, like any other theory, has limitations. That there are many physical occurrences to which it does not apply may be taken for granted. To justify its inclusion in this treatise, it is enough that there are many physical situations to which it *does* apply, and even in the rather limited special cases now being explored its relevance seems to be waxing.

246. Entropy[2]. At a given place and time in a body, let there be given $\mathfrak{k}$ parameters v_α which are regarded as influencing the internal energy ε. The assignment of these parameters[3] is made *a priori*; their totality is the *thermodynamic substate*. The physical dimensions of the v_α are made up of mechanical and electromagnetic units but are otherwise arbitrary. The most familiar case is when the thermodynamic substate consists in a single scalar parameter, the specific volume v. In another common example, the v_α are the nine deformation gradients $x^k_{;\alpha}$ from a material reference state. In still another, they are the densities or the concentrations of the constituents of a mixture. For all general

[1896, *3*, pp. 269—301], Duhem [1903, *10*], Callender [1910, *3*, §§ 20—23], and Partington [1949, *25*, Part II, §§ 23—37] [1952, *15*].

In the system of Duhem [1893, *1*, Eqs. (43 bis) and (56)], [1911, *4*, Chap. XIV, §§ 1—2], the caloric equation of state (246.1) is regarded as only approximate; Duhem adds a mutual internal energy resulting from the interaction of all pairs of elements of mass.

The system of Fink will be described in Sect. 253.

Here we mention the system of Reik [1953, *26*] [1954, *19*], which replaces the traditional "second law" by an axiom governing the time change of entropy. This axiom seems to us to be a constitutive relation, generalizing the conventional linear ones (cf. the next footnote), and thus we do not attempt to present Reik's theory here. We recognize the difficulty of drawing a firm distinction: Eq. (246.1), on which the rest of the chapter is founded, is itself, in a strict view, a constitutive equation, and the chapter should stop after Sect. 244.

[1] The majority of recent studies on "irreversible thermodynamics" rest on the special constitutive assumption (cf. Sect. 7) that the affinities are linear functions of the fluxes. In the present treatise, these special developments would be as inappropriate as classical linear elasticity. The literature is too extensive to cite, but we mention the expositions of Prigogine [1947, *12*], Haase [1951, *11*], de Groot [1952, *3*] and Meixner and Reik, this Encyclopedia, Vol. III/2. Although special cases of such linear relations are old, apparently the first proposal of a general theory was made by de Donder [1938, *3*]. While great emphasis is currently laid upon the so-called "Onsager relations", we are unable to see in them, at least for the present, anything more than an indication of a special choice of variables; cf. Coleman and Truesdell [1960, *1A*]. Future analysis may show that they result by linearization from some as yet undiscovered general principle of invariance.

For the now generally accepted approach to the theory of the entropy field, see the summary by de Groot [1953, *9*].

[2] Since the developments of Sects. 246—249 are parallel to those of classical thermostatics, we do not cite references beyond those for Sect. 245, except to remark that the early papers concern only the case $\mathfrak{k} = 1$, corresponding developments for equations of state with arbitrarily many variables having been given by Gibbs [1875, *1*], Schiller [1879, *4*] [1894, *8*], Helmholtz [1882, *2*, § 1], Duhem [1886, *2*, Part II, Chap. II] [1894, *2*] [1891, *2*], and Oumoff [1895, *3*]. Some of the results of Gibbs are special in that they rest upon a condition of homogeneity (Sect. 260). The general view of the subject is due to Helmholtz: „Der Zustand des Systems sei durch θ und eine Anzahl von passend gewählten Parametern p_a vollständig bestimmt." Some of the identities are derived anew and interpreted in linear thermo-elastic contexts by Ting and Li [1957, *14*].

That $\boldsymbol{X}$ may enter the equations of state, so that the thermodynamic behavior of one particle may differ from that of another, was noted by Hugoniot [1885, *4*] and emphasized by Duhem [1901, *7*, Part II, Chap. IV, § 1].

[3] In classical thermostatics they are often divided into two classes, called "extensive" and "intensive", but this distinction is not necessary here. See, however, Sect. 260.

developments, the parameters $v_\mathfrak{a}$ are left unspecified. They are tensor fields of arbitrary order, functions of place $\boldsymbol{x}$ and time t, or, if we prefer, functions of time for each particle $\boldsymbol{X}$.

In following the more recent custom of the subject, where, except in the case of fluids, the variables $v_\mathfrak{a}$ are arbitrary, we feel compelled to caution the reader that the physical meaning of the results is likewise left uncertain. Results depending only on the possibility of differentiation are in the main rather insensitive to the choice of variables, but results following from inversion of functional relations are applicable, in most cases, only to *particular choices* of the variables $v_\mathfrak{a}$ in any given physical system. Our aim here is to present certain mathematical features common to all the simpler thermodynamic theories. Compared to the contents of the other chapters of this treatise, the matter here is not concrete; in a satisfactory treatment, the variables $v_\mathfrak{a}$ should be identified with *definite physical quantities*, as they always were in the studies of GIBBS.

We have said that the thermodynamic substate is regarded as influencing ε. The basic assumption of thermodynamics is: *The substate plus a single further dimensionally independent scalar parameter suffices to determine ε, independently of time, place, motion, and stress.* That is, we assume that it is possible to assign *a priori* a function f such that

$$\varepsilon = f(\eta, v_1, v_2, \ldots, v_\mathfrak{t}, \boldsymbol{X}) = \varepsilon(\eta, \boldsymbol{v}, \boldsymbol{X}). \tag{246.1}$$

The parameter η is called the *specific entropy*[1]. Its physical dimension, postulated to be independent of $[M]$, $[L]$, $[T]$ and the electromagnetic units, is traditionally left unnamed, but the dimension given by

$$[\Theta] \equiv \frac{\dim \varepsilon}{\dim \eta} = \frac{[L^2 T^{-2}]}{\dim \eta} \tag{246.2}$$

is called the dimension of *temperature*.

In any given motion, of course we have $\varepsilon = g(\boldsymbol{X}, t)$. In a different motion, a like functional relation of different form, $\varepsilon = h(\boldsymbol{X}, t)$, will hold. The first implication of the postulate (246.1) is that we can determine ε *without knowing the particular motion occurring, and without regard to the time.* In other words, the value of the internal energy can be ascertained from information which is *static* and *universal*. This information consists in

1. *The value of the substate, $\boldsymbol{v}$.*
2. *The value of the entropy, η.*
3. *The functional form of the relation* (246.1).

Thus the role of entropy is that of a *specifying parameter*. The mechanical and electromagnetic information expressed by the substate is in itself not enough, but, for any given substance, the value of the one additional quantity η suffices to yield the internal energy of each particle, whatever the motion it is undergoing or has undergone[2]. Adjoining the entropy η to the substate $\boldsymbol{v}$, we obtain the

---

[1] Entropy is to be identified with the "calorique" of CARNOT (1824). While used by others, notably by RANKINE (1854), its distinction from the caloric of CLAPEYRON and earlier writers was emphasized by CLAUSIUS, who invented the name [1865, *1*, § 14]. That thermodynamics is, at bottom, the science of entropy was first made clear by the researches of GIBBS (1875).

[2] This striking property of entropy results only from the field viewpoint. Typically of field theories, the greater generality obtained by introducing a field which may vary from point to point is gained at the cost of closeness to experiment. HADAMARD [1903, *11*, ¶ 107] remarked that experimental verification of equations of state for large masses in equilibrium gives no indication whatever that local equations of state hold in deforming media. As in any other field theory, the experimental justification must result indirectly by comparing the solution of specific problems with measurements.

thermodynamic state: η, $\boldsymbol{v}$. The relation (246.1) is the *caloric equation of state*. Choice of the form of f in (246.1) defines different *thermodynamic substances*. When $\boldsymbol{X}$ does not appear in (246.1), the substance is *thermodynamically homogeneous*. In the present treatise, the form of the caloric equation of state is left arbitrary, except for some inequalities to be laid down in Sects. 263 and 265.

The meaning of (246.1) is most easily seen from *Gibbs' diagram*[1], where ε is represented by a height over the space of thermodynamic states. For the case when $\mathfrak{k}=1$, such a diagram is shown in Fig. 43. The relation (246.1) is then represented by an *energy surface*, whose form is fixed once and for all for each substance considered. Suppose we have a motion in which a particle $\boldsymbol{X}$ is carried from a state v, η at time t to a second state v^*, η^* at time t^*. During the motion, ε, v, and η will vary: $\varepsilon=\varepsilon(t)$, $v=v(t)$, $\eta=\eta(t)$. By (246.1), the motion in space is mapped onto *a curve on the energy surface*, and the changes in energy and thermodynamic state during the motion must be consistent with this fact. Conversely, however, the energy surface does not determine the motion, and many different motions may be mapped onto the same curve on the energy surface. Moreover, if two motions carry the particle from the state v, η to the state v^*, η^* by different paths on the energy surface and over different intervals of time, the *energy increments are the same*, being determined by the form of the energy surface.

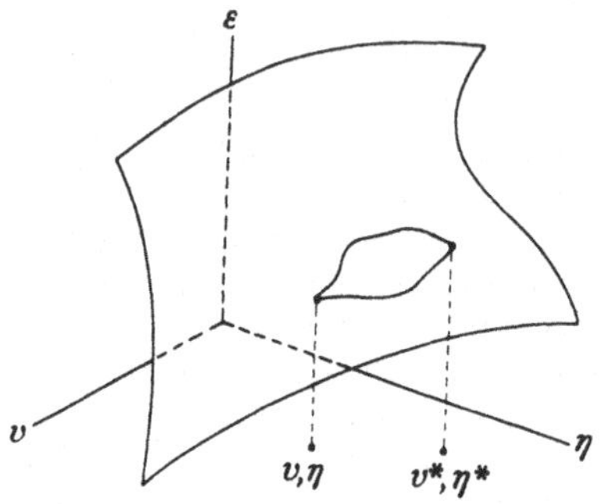

Fig. 43. GIBBS' diagram.

The *dimensional independence* of η furnishes the avenue by which thermal phenomena are to be connected with mechanical and electromagnetic actions. Were it not for this dimensional independence, η would be but another parameter in the set of $v_\mathfrak{a}$. An essential part of the physical content of the caloric equation of state is that *thermal information must be added to mechanical and electromagnetic information if we are to determine the internal energy without knowing the motion.*

Since ε and η are dimensionally independent, any relation between them must involve constants ε_0 and η_0 having the same dimensions as ε and η, respectively. Thus the caloric equation of state (246.1) must really be of the form

$$\frac{\varepsilon}{\varepsilon_0}=f\left(\frac{\eta}{\eta_0}, \boldsymbol{v}, \boldsymbol{X}\right). \tag{246.3}$$

Dimensional invariance imposes some restriction upon the manner in which $\boldsymbol{v}$ enters (246.3) also, but until the physical dimensions of the $v_\mathfrak{a}$ have been rendered explicit, nothing definite can be concluded. Even after the introduction of reduced variables, it is not clear what geometric structure, if any, is to be adduced for the thermodynamic space of points ε-η-$\boldsymbol{v}$. In particular, there is no reason to think of it as metric. But until the space itself is fully defined, we are at a loss to know what properties of the energy surface can have physical significance for a given substance, apart from the accidents of its representation[2]. Invariance requirements for thermodynamics have never been established.

---

[1] The method of GIBBS [1873, *3*, pp. 33–34] was adopted and made known by MAXWELL [1875, *4*, pp. 195–208].

[2] Cf. the remarks of GIBBS [1873, *3*, pp. 34–35] and L. BRILLOUIN [1938, *2*, Chap. I, §§ VIII–IX].

247. Temperature and thermodynamic tensions. The *temperature* θ and the *thermodynamic tensions* $\tau_{\mathfrak{a}}$ are defined from (246.1) by

$$\theta \equiv \frac{\partial \varepsilon}{\partial \eta}, \qquad \tau_{\mathfrak{a}} \equiv \frac{\partial \varepsilon}{\partial v_{\mathfrak{a}}}. \tag{247.1}$$

Hence for any change whatever in the thermodynamic state of a given particle $\boldsymbol{X}$ we have[1]

$$d\varepsilon = \theta\, d\eta + \sum_{\mathfrak{a}=1}^{\mathfrak{k}} \tau_{\mathfrak{a}}\, d v_{\mathfrak{a}}. \tag{247.2}$$

The temperature and the tensions are the slopes of the curves of intersection of the energy surface with planes parallel to the co-ordinate planes in the diagram of Sect. 246. The temperature measures the sensitivity of energy to changes in entropy; the tensions, to changes in the corresponding parameters. When v_1 is the specific volume, $-\tau_1$ is called the *thermodynamic pressure*, π. In the case of a homogeneous mixture, when the substate includes both the total volume and the masses of the constituents, and when the caloric equation of state is taken as referring to the whole mass, then the tension $\tau_{\mathfrak{a}}$ corresponding to the mass of the constituent $\mathfrak{B}$ is called its *chemical potential*[2], $\mu^{\star}_{\mathfrak{B}}$ (cf. Sect. 260). When $v_{\mathfrak{a}}$ is the deformation gradient $x^k_{;\alpha}$, we may set ${}_{\varepsilon}T_k^{\alpha} \equiv \tau_{\mathfrak{a}}/\varrho_0$ and obtain a double vector of elastic stress analogous to (218.6). See Sects. 255 to 256A for development of the properties of these coefficients.

In all cases, from (247.1) and (246.2) follow

$$\dim \theta = [\Theta], \qquad \dim \tau_{\mathfrak{a}} = \frac{[L^2 T^{-2}]}{\dim v_{\mathfrak{a}}}. \tag{247.3}$$

Thus temperature is dimensionally independent from the thermodynamic tensions.

Also, from (247.1) and (246.1) it is immediate that

$$\theta = \theta(\eta, \boldsymbol{v}), \qquad \tau_{\mathfrak{a}} = \tau_{\mathfrak{a}}(\eta, \boldsymbol{v}): \tag{247.4}$$

The temperature and the tensions are functions of the thermodynamic state[3].

Since (247.2), being the result of differentiating a scalar relation with fixed $\boldsymbol{X}$, is valid for all paths on the energy surface for a given particle, as a special case it is valid along the curve onto which is mapped the actual motion of the particle $\boldsymbol{X}$. Hence

$$\dot{\varepsilon} = \theta \dot{\eta} + \sum_{\mathfrak{a}=1}^{\mathfrak{k}} \tau_{\mathfrak{a}} \dot{v}_{\mathfrak{a}}. \tag{247.5}$$

---

[1] A special equation of this kind was derived by GIBBS in his first work [1873, *2*, Eq. (4)] but is taken as the starting point of his later work [1873, *3*, Eq. (1)] [1875, *1*, Eq. (12)] and hence is often called "the GIBBS equation". For references to related earlier work, see Sect. 248.

[2] GIBBS [1875, *1*, p. 63].

[3] The temperature is easier to interpret physically than is the entropy, and for this reason most treatments of thermodynamics prefer to take the temperature as a primitive concept and then introduce the entropy as a defined concept. We have remarked upon this in Sect. 245, No. 4, and further remarks are given in Sect. 250. Our reasons for preferring the present course are two: (1) it is formally much simpler, and (2) in our opinion no existing treatment along classical lines is logically clear. A possible formal alternative would be to take free energy ψ rather than internal energy ε as primitive, define η through $(251.4)_3$, then define ε through $(251.1)_2$, but to us this seems indirect and more difficult to motivate. This same criticism applies to the work of DUHEM [1893, *1*] [1911, *4*] who always started from the free enthalpy ζ, taken as a function of θ and the substate.

For arbitrary changes, however, we have

$$\left.\begin{aligned}\frac{\partial\varepsilon}{\partial t} &= \theta\frac{\partial\eta}{\partial t} + \sum_{\mathfrak{a}=1}^{\mathfrak{k}} \tau_{\mathfrak{a}}\frac{\partial v_{\mathfrak{a}}}{\partial t} + \frac{\partial\varepsilon}{\partial X^\alpha}\frac{\partial X^\alpha}{\partial t},\\ \varepsilon_{,k} &= \theta\eta_{,k} + \sum_{\mathfrak{a}=1}^{\mathfrak{k}} \tau_{\mathfrak{a}} v_{\mathfrak{a},k} + \frac{\partial\varepsilon}{\partial X^\alpha} X^\alpha_{,k}.\end{aligned}\right\} \tag{247.6}$$

In a homogeneous material, the last terms in these expressions vanish.

We lay down an *assumption of regularity*: All thermodynamic functional relations are differentiable as many times as needed and are invertible to yield any one variable as a function of the others. The caution stated when the substate $\boldsymbol{v}$ was introduced in Sect. 246 must be borne constantly in mind: a strong restriction not only on the functional forms admissible for the caloric equation of state (246.1) but also on the choice of the variables $v_{\mathfrak{a}}$ is implied; in particular, it is thus assumed that various partial derivatives occurring are of one sign. While some related inequalities will be derived in Sect. 265, an adequate analysis of the nature of equations of state and of the singularities they may possess is not available.

We lay down also a *notation* for partial derivatives: A subscript denotes the variables held constant, and

$$\left(\frac{\partial f}{\partial v_{\mathfrak{a}}}\right)_{\mathfrak{v}\mathfrak{a}} \equiv \left(\frac{\partial f}{\partial v_{\mathfrak{a}}}\right)_{v_1, v_2, \ldots, v_{\mathfrak{a}-1}, v_{\mathfrak{a}+1}, \ldots, v_{\mathfrak{k}}}. \tag{247.7}$$

Moreover, we agree to hold $\boldsymbol{X}$ constant in all differentiations unless the contrary is noted explicitly. In this notation, (247.1) reads

$$\theta \equiv \left(\frac{\partial\varepsilon}{\partial\eta}\right)_{\mathfrak{v}}, \qquad \tau_{\mathfrak{a}} \equiv \left(\frac{\partial\varepsilon}{\partial v_{\mathfrak{a}}}\right)_{\eta,\mathfrak{v}\mathfrak{a}}. \tag{247.8}$$

In all such expressions it is understood that the quantity being differentiated is regarded as a function *only* of $\boldsymbol{X}$ and of the variables actually written in the denominator and the subscripts.

Rates of change subject to the condition $\eta = \text{const}$ are called *isentropic*. Thus the thermodynamic tensions are the rates of change of the internal energy in isentropic changes of the substate. Other isentropic rates are defined by[1]

$$\phi_{\mathfrak{a}\mathfrak{b}} \equiv \left(\frac{\partial\tau_{\mathfrak{a}}}{\partial v_{\mathfrak{b}}}\right)_{\eta,\mathfrak{v}\mathfrak{b}}, \qquad \varpi_{\mathfrak{a}} \equiv \left(\frac{\partial\tau_{\mathfrak{a}}}{\partial\eta}\right)_{\mathfrak{v}}. \tag{247.9}$$

As a consequence of the assumption of smoothness, we may invert (246.1) and obtain $\eta = \eta(\varepsilon, \boldsymbol{v})$. Hence

$$\left.\begin{aligned} d\eta &= \left(\frac{\partial\eta}{\partial\varepsilon}\right)_{\mathfrak{v}} d\varepsilon + \sum_{\mathfrak{a}=1}^{\mathfrak{k}} \left(\frac{\partial\eta}{\partial v_{\mathfrak{a}}}\right)_{\varepsilon} dv_{\mathfrak{a}},\\ &= \left(\frac{\partial\eta}{\partial\varepsilon}\right)_{\mathfrak{v}} \left[\theta\, d\eta + \sum_{\mathfrak{a}=1}^{\mathfrak{k}} \tau_{\mathfrak{a}}\, dv_{\mathfrak{a}}\right] + \sum_{\mathfrak{a}=1}^{\mathfrak{k}} \left(\frac{\partial\eta}{\partial v_{\mathfrak{a}}}\right)_{\varepsilon,\mathfrak{v}\mathfrak{a}} dv_{\mathfrak{a}}.\end{aligned}\right\} \tag{247.10}$$

Equating coefficients of differentials yields

$$\frac{1}{\theta} = \left(\frac{\partial\eta}{\partial\varepsilon}\right)_{\mathfrak{v}}, \qquad \tau_{\mathfrak{a}} = -\left(\frac{\partial\eta}{\partial v_{\mathfrak{a}}}\right)_{\varepsilon,\mathfrak{v}\mathfrak{a}}. \tag{247.11}$$

---

[1] When $v_1 = v$, we have $v^2\phi_{11} = (\partial\pi/\partial\varrho)_\eta = V_0^2$, where V_0 is the Laplacean speed of sound. Cf. (297.13).

248. Thermal equations of state. From the assumption of smoothness it follows also that we may invert $(247.1)_1$ and obtain

$$\eta = \eta(\theta, \boldsymbol{v}). \tag{248.1}$$

Substituting this into (246.1) yields a functional relation of the form

$$\varepsilon = \varepsilon(\theta, \boldsymbol{v}). \tag{248.2}$$

But also we may substitute (248.1) into $(247.4)_2$ and obtain

$$\tau_{\mathfrak{a}} = \tau_{\mathfrak{a}}(\theta, \boldsymbol{v}). \tag{248.3}$$

Similarly,

$$v_{\mathfrak{a}} = v_{\mathfrak{a}}(\theta, \boldsymbol{\tau}). \tag{248.4}$$

The relations (248.3) and (248.4) are the ***thermal equations of state***[1]. They are conveniently represented by surfaces over the subspace θ–$\boldsymbol{v}$; a set of such surfaces may be called *Euler's diagrams* for the particle or the subspace[2].

The following coefficients may be calculated from the thermal equations of state:

$$\alpha_{\mathfrak{a}} \equiv \left(\frac{\partial v_{\mathfrak{a}}}{\partial \theta}\right)_{\boldsymbol{\tau}}, \quad \beta_{\mathfrak{a}} \equiv \left(\frac{\partial \tau_{\mathfrak{a}}}{\partial \theta}\right)_{\boldsymbol{v}}, \quad \xi_{\mathfrak{a}\mathfrak{b}} \equiv \left(\frac{\partial \tau_{\mathfrak{a}}}{\partial v_{\mathfrak{b}}}\right)_{\theta, v^{\mathfrak{b}}}, \quad \nu_{\mathfrak{a}\mathfrak{b}} \equiv \left(\frac{\partial v_{\mathfrak{a}}}{\partial \tau_{\mathfrak{b}}}\right)_{\theta, \tau^{\mathfrak{b}}}, \tag{248.5}$$

so that

$$d v_{\mathfrak{a}} = \sum_{\mathfrak{b}=1}^{\mathfrak{k}} \nu_{\mathfrak{a}\mathfrak{b}}\, d\tau_{\mathfrak{b}} + \alpha_{\mathfrak{a}}\, d\theta, \qquad d\tau_{\mathfrak{a}} = \sum_{\mathfrak{b}=1}^{\mathfrak{k}} \xi_{\mathfrak{a}\mathfrak{b}}\, d v_{\mathfrak{b}} + \beta_{\mathfrak{a}}\, d\theta. \tag{248.6}$$

Since the coefficients (248.5) are rates of change of measurable quantities, they are useful for inferring the forms of the thermal equations of state by experiment. When v_1 is the specific volume, α_1/v_1 is called the *coefficient of thermal expansion* or *isobaric compressibility*, $-\nu_{11}/v_1$ is called the *isothermal compressibility*, and $-\beta_1$ is called the *pressure coefficient*. The coefficients (248.5) are related by the identities

$$\left.\begin{aligned} &\beta_{\mathfrak{a}} + \sum_{\mathfrak{b}=1}^{\mathfrak{k}} \xi_{\mathfrak{a}\mathfrak{b}}\alpha_{\mathfrak{b}} = 0, \qquad && \alpha_{\mathfrak{a}} + \sum_{\mathfrak{b}=1}^{\mathfrak{k}} \nu_{\mathfrak{a}\mathfrak{b}}\beta_{\mathfrak{b}} = 0, \\ &\sum_{\mathfrak{c}=1}^{\mathfrak{k}} \nu_{\mathfrak{a}\mathfrak{c}}\xi_{\mathfrak{c}\mathfrak{b}} = \delta_{\mathfrak{a}\mathfrak{b}}, && \sum_{\mathfrak{c}=1}^{\mathfrak{k}} \xi_{\mathfrak{a}\mathfrak{c}}\nu_{\mathfrak{c}\mathfrak{b}} = \delta_{\mathfrak{a}\mathfrak{b}}. \end{aligned}\right\} \tag{248.7}$$

As compared with the caloric equation of state, the thermal equations of state offer the advantage of connecting easily measurable quantities, the disadvantage of being insufficient, despite their number, to determine all the thermodynamic properties of the material. The latter statement will be proved in Sect. 251.

In a theory in which entropy is not regarded as a primitive, the thermal equations of state are often taken as the first postulate, and from them an argument leading to the caloric equation of state is constructed. The *heat increment* $\langle Q \rangle$, the excess of the increment of internal energy over thermostatic work is defined by[3]

$$\langle Q \rangle \equiv d\varepsilon - \sum_{\mathfrak{a}=1}^{\mathfrak{k}} \tau_{\mathfrak{a}}\, d v_{\mathfrak{a}}. \tag{248.8}$$

---

[1] For a perfect gas in equilibrium, the thermal equation of state $\pi v = A(\theta - \theta_0)$ was established by combining the experimental results of BOYLE (1662) and of AMONTONS (1699), rediscovered many times subsequently. This equation, along with the more general equation $\pi = f(\theta, v)$ and other of its special cases, was used regularly by EULER [1745, *2*, Chap. I, Laws 3, 4, 5] [1757, *1*, §§ 17–18] [1757, *2*, § 21] [1757, *3*, §§ 29–31] [1764, *1*] [1769, *1*, §§ 24–30, 90–108] [1771, *1*, Chap. V], but did not reappear in field mechanics until the work of KIRCHHOFF [1868, *11*], after which it quickly became universal for studies of gas flows.

[2] EULER [1769, *1*, § 28]. Cf. also GIBBS [1873, *2*].

[3] Special equations of this form appear as derived results in the work of CLAUSIUS [1850, *1*, Eq. IIa] [1854, *1*, Eq. (2)].

$\langle Q\rangle$ is a differential form in the variables $\boldsymbol{v}$, θ, since both ε and the $\tau_{\mathfrak{a}}$, according to this approach, are assumed to be known functions of these variables. In general, this differential form is not integrable. Special arguments and added assumptions are required in order to conclude that $\langle Q\rangle$ is integrable and that an integrating factor is $1/\theta$. Under these assumptions, entropy is defined by integrating

$$d\eta = \frac{\langle Q\rangle}{\theta} = \frac{1}{\theta}\left[d\varepsilon - \sum_{\mathfrak{a}=1}^{\mathfrak{k}} \tau_{\mathfrak{a}}\, d v_{\mathfrak{a}}\right], \tag{248.9}$$

identical with (247.2), and the consequent theory is the same as that we have discussed in Sects. 247 to 248 and will develop further in the rest of this chapter.

It is also possible to consider a theory in which the thermal equations of state hold but there is no entropy or at least no caloric equation of state.

249. Thermodynamic paths. Specific heats, I. Treatment based on entropy[1]. For a given particle $\boldsymbol{X}$ a *thermodynamic path* is a path $\eta=\eta(\lambda)$, $\boldsymbol{v}=\boldsymbol{v}(\lambda)$ in the space of thermodynamic states η, $\boldsymbol{v}$. On any path P, at a given point, any differentiable function g of the thermodynamic state has a differential, which we may write as dg_{P}. *Isentropic* paths are defined by $\eta=\text{const}$ (Sect. 247); *isothermal* paths, by $\theta=\text{const}$.

The *specific heat* $\varkappa_{\mathrm{P}}$ on the path P is defined by

$$\varkappa_{\mathrm{P}} \equiv \theta\,\frac{d\eta_{\mathrm{P}}}{d\theta_{\mathrm{P}}}. \tag{249.1}$$

For an assigned path, the ratios of differentials $d v_{\mathfrak{a}\mathrm{P}}/d\eta_{\mathrm{P}}$ are assigned, and from (249.1), by use of $(247.4)_1$, results a definite value of the specific heat. We have $\dim \varkappa_{\mathrm{P}} = \dim \eta = [L^2\, T^{-2}\, \Theta^{-1}]$.

By (247.2) we may write

$$\varkappa_{\mathrm{P}} = \frac{d\varepsilon_{\mathrm{P}} - \sum_{\mathfrak{a}=1}^{\mathfrak{k}} \tau_{\mathfrak{a}}\, d v_{\mathfrak{a}\mathrm{P}}}{d\theta_{\mathrm{P}}}. \tag{249.2}$$

From the assumption of regularity in Sect. 247, we may regard ε as a function of θ and $\boldsymbol{\upsilon}$, and thus (249.2) becomes

$$\begin{aligned}\varkappa_{\mathrm{P}} &= \frac{\left(\frac{\partial\varepsilon}{\partial\theta}\right)_{\boldsymbol{\upsilon}} d\theta_{\mathrm{P}} + \sum_{\mathfrak{a}=1}^{\mathfrak{k}}\left[\left(\frac{\partial\varepsilon}{\partial v_{\mathfrak{a}}}\right)_{\theta,\,\boldsymbol{\upsilon}^{\mathfrak{a}}} d v_{\mathfrak{a}\mathrm{P}} - \tau_{\mathfrak{a}}\, d v_{\mathfrak{a}\mathrm{P}}\right]}{d\theta_{\mathrm{P}}},\\ &= \left(\frac{\partial\varepsilon}{\partial\theta}\right)_{\boldsymbol{\upsilon}} + \sum_{\mathfrak{a}=1}^{\mathfrak{k}}\left[\left(\frac{\partial\varepsilon}{\partial v_{\mathfrak{a}}}\right)_{\theta,\,\boldsymbol{\upsilon}^{\mathfrak{a}}} - \tau_{\mathfrak{a}}\right]\frac{d v_{\mathfrak{a}\mathrm{P}}}{d\theta_{\mathrm{P}}}.\end{aligned} \tag{249.3}$$

In particular, if the path is one on which $\boldsymbol{\upsilon}=\text{const}$, $\varkappa_{\mathrm{P}}$ is called the *specific heat at constant substate* and is written $\varkappa_{\boldsymbol{\upsilon}}$. For this quantity, (249.3) yields[2]

$$\varkappa_{\boldsymbol{\upsilon}} \equiv \theta\left(\frac{\partial\eta}{\partial\theta}\right)_{\boldsymbol{\upsilon}} = \left(\frac{\partial\varepsilon}{\partial\theta}\right)_{\boldsymbol{\upsilon}}. \tag{249.4}$$

Hence (249.3) becomes

$$\varkappa_{\mathrm{P}} = \varkappa_{\boldsymbol{\upsilon}} + \sum_{\mathfrak{a}=1}^{\mathfrak{k}}\left[\left(\frac{\partial\varepsilon}{\partial v_{\mathfrak{a}}}\right)_{\theta,\,\boldsymbol{\upsilon}^{\mathfrak{a}}} - \tau_{\mathfrak{a}}\right]\frac{d v_{\mathfrak{a}\mathrm{P}}}{d\theta_{\mathrm{P}}}. \tag{249.5}$$

From the assumption of regularity, we may regard $v_{\mathfrak{a}}$ as expressed in the form (248.4). Writing $\varkappa_{\boldsymbol{\tau}}$ for the *specific heat at constant tensions*[3], by holding $\boldsymbol{\tau}$

---

[1] REECH [1853, *2*, Chap. IV], for the case of a fluid.

[2] CLAUSIUS [1854, *1*, p. 486] for the case $\mathfrak{k}=1$, HELMHOLTZ [1882, *2*, § 1] for the general case.

[3] DUHEM [1894, *2*, Chap. IV, § 7].

constant in (249.5) we obtain

$$\varkappa_\tau - \varkappa_\upsilon = \sum_{\mathfrak{a}=1}^{\mathfrak{k}} \left[\left(\frac{\partial \varepsilon}{\partial v_\mathfrak{a}} \right)_{\theta, \upsilon^\mathfrak{a}} - \tau_\mathfrak{a} \right] \alpha_\mathfrak{a}, \tag{249.6}$$

where $\alpha_\mathfrak{a}$ is defined by $(248.5)_1$.

The ratio of specific heats,

$$\gamma \equiv \frac{\varkappa_\tau}{\varkappa_\upsilon}, \tag{249.7}$$

is an important dimensionless scalar. It is connected with the coefficients $\phi_{\mathfrak{a}\mathfrak{b}}$ and $\xi_{\mathfrak{a}\mathfrak{b}}$ by interesting identities which will be derived in Sect. 252.

250. Specific heats and latent heats, II. M. BRILLOUIN's general theory. The theory of specific heats is much older than thermostatics, since it can be founded directly upon (248.8) and does not require the concept of entropy, provided not only the substate $\boldsymbol{\upsilon}$ but also the tensions $\boldsymbol{\tau}$ be taken as quantities known *a priori*. Indeed, with no assumed functional relations connecting the variables, we may set[1]

$$\varkappa \equiv \frac{\dot{\varepsilon} - \sum_{\mathfrak{a}=1}^{\mathfrak{k}} \tau_\mathfrak{a} \dot{v}_\mathfrak{a}}{\dot{\theta}} = \frac{\langle Q \rangle}{d\theta}, \tag{250.1}$$

where $\langle Q \rangle$ is the differential form (248.8). Similarly, the *latent heats* $\lambda_{\tau_\mathfrak{a}}$ and $\lambda_{v_\mathfrak{a}}$ are defined by

$$\lambda_{\tau_\mathfrak{a}} \equiv \frac{\dot{\varepsilon} - \sum_{\mathfrak{b}=1}^{\mathfrak{k}} \tau_\mathfrak{b} \dot{v}_\mathfrak{b}}{\dot{\tau}_\mathfrak{a}} = \frac{\langle Q \rangle}{d\tau_\mathfrak{a}}, \quad \lambda_{v_\mathfrak{a}} \equiv \frac{\dot{\varepsilon} - \sum_{\mathfrak{b}=1}^{\mathfrak{k}} \tau_\mathfrak{b} \dot{v}_\mathfrak{b}}{\dot{v}_\mathfrak{a}} = \frac{\langle Q \rangle}{d v_\mathfrak{a}}. \tag{250.2}$$

In general, such specific heats and latent heats will be functions of time even for a given path. Definitions equivalent to these were used long prior to the concept of entropy and the theory of thermostatics. While they serve to record the results of measurements, they are too general to be the basis of a mathematical treatment.

M. BRILLOUIN[2] proposed assumptions which, while far more general than those used in the preceding sections, are yet definite enough that the major classical formal properties of specific and latent heats remain valid. In order to represent the possibility of permanent deformation, he suggested discarding even the thermal equations of state, replacing them by the differential forms (248.6), where the coefficients $\nu_{\mathfrak{a}\mathfrak{b}}, \alpha_\mathfrak{a}, \xi_{\mathfrak{a}\mathfrak{b}}, \beta_\mathfrak{a}$ are given functions of $\boldsymbol{v}$, $\boldsymbol{\tau}$, and θ. The resulting differential forms in all $2\mathfrak{k}+1$ variables $\boldsymbol{v}, \boldsymbol{\tau}, \theta$ are not necessarily integrable. The identities (248.7) remain valid, but the interpretations (248.5) are correct only for the integrable case.

The idea is easiest to picture in the case of three variables, $dv = \nu(v, \tau, \theta)\, d\tau + \alpha(v, \tau, \theta)\, d\theta$. Consider a closed loop in the τ–θ plane, described by $\tau = \tau(t)$, $\theta = \theta(t)$. As the system is carried through this closed cycle, the point τ, θ, v traverses a curve on the right cylinder whose base is the loop. This curve is obtained by integrating

$$\dot{v} = \nu(v, \tau(t), \theta(t))\, \dot{\tau}(t) + \alpha(v, \tau(t), \theta(t))\, \dot{\theta}(t). \tag{250.3}$$

---

[1] An attempt to study specific heats in something approaching this degree of generality was made by M. BRILLOUIN [1888, 2], who discussed possible restrictions on the dependence of $\varkappa$ on the substate.

[2] [1888, 3, §§ 3, 8–9]. Our treatment is somewhat more compact and general than BRILLOUIN's. BRILLOUIN also studied a generalized entropy $\eta(\boldsymbol{v}, \boldsymbol{\tau}, \theta)$ [1888, 3, §§ 12–25]. The criticism of BRILLOUIN's theory expressed by DUHEM [1896, 2, Introd. to Part I] refers to this generalized entropy, not to the developments presented above. DUHEM [ibid., Chap. I, § 1] proposes a generalized free enthalpy.

If the argument v does not actually appear in ν and α, as is the case when there is a thermal equation of state, this curve is closed, and v returns to its initial value when the loop is traversed; indeed, the only possible curve is given by the intersection of the cylinder with EULER's diagram. If the form $(248.6)_1$ is not integrable, however, v will not generally return to its initial value, and after the system is carried through the closed cycle of values of θ and τ, a change in v will result. But all these changes are reversible to the extent that if a path may be traversed in one sense, it may also be traversed in the opposite sense, since the form $(248.6)_1$ is linear and homogeneous.

We now add the assumption that an internal energy exists, but we replace (248.8) by allowing the more general assumed functional relation

$$\varepsilon = \varepsilon(\boldsymbol{v}, \boldsymbol{\tau}, \theta). \tag{250.4}$$

Then from $(248.6)_1$ follows

$$\left.\begin{aligned}
\langle Q\rangle &= \sum_{\mathfrak{a}=1}^{\mathfrak{f}}\left(\frac{\partial\varepsilon}{\partial v_{\mathfrak{a}}}\,d v_{\mathfrak{a}} + \frac{\partial\varepsilon}{\partial\tau_{\mathfrak{a}}}\,d\tau_{\mathfrak{a}}\right) + \frac{\partial\varepsilon}{\partial\theta}\,d\theta - \sum_{\mathfrak{a}=1}^{\mathfrak{f}}\tau_{\mathfrak{a}}\,d v_{\mathfrak{a}},\\
&= \sum_{\mathfrak{a}=1}^{\mathfrak{f}}\left[\left(\frac{\partial\varepsilon}{\partial v_{\mathfrak{a}}} - \tau_{\mathfrak{a}}\right) d v_{\mathfrak{a}} + \frac{\partial\varepsilon}{\partial\tau_{\mathfrak{a}}}\,d\tau_{\mathfrak{a}}\right] + \frac{\partial\varepsilon}{\partial\theta}\,d\theta,\\
&= \sum_{\mathfrak{a}=1}^{\mathfrak{f}}\left[\left(\frac{\partial\varepsilon}{\partial v_{\mathfrak{a}}} - \tau_{\mathfrak{a}} + B_{\mathfrak{a}}\right) d v_{\mathfrak{a}} + \left(\frac{\partial\varepsilon}{\partial\tau_{\mathfrak{a}}} - \sum_{\mathfrak{b}=1}^{\mathfrak{f}} B_{\mathfrak{b}}\,\nu_{\mathfrak{b}\mathfrak{a}}\right) d\tau_{\mathfrak{a}}\right] + \\
&\qquad + \left(\frac{\partial\varepsilon}{\partial\theta} - \sum_{\mathfrak{a}=1}^{\mathfrak{f}} B_{\mathfrak{a}}\,\alpha_{\mathfrak{a}}\right) d\theta,\\
&= \sum_{\mathfrak{a}=1}^{\mathfrak{f}}\left[\left(\frac{\partial\varepsilon}{\partial v_{\mathfrak{a}}} - \tau_{\mathfrak{a}} - \sum_{\mathfrak{b}=1}^{\mathfrak{f}} C_{\mathfrak{b}}\,\xi_{\mathfrak{b}\mathfrak{a}}\right) d v_{\mathfrak{a}} + \left(\frac{\partial\varepsilon}{\partial\tau_{\mathfrak{a}}} + C_{\mathfrak{a}}\right) d\tau_{\mathfrak{a}}\right] + \\
&\qquad + \left(\frac{\partial\varepsilon}{\partial\theta} - \sum_{\mathfrak{a}=1}^{\mathfrak{f}} C_{\mathfrak{a}}\,\beta_{\mathfrak{a}}\right) d\theta,
\end{aligned}\right\} \tag{250.5}$$

where the quantities $B_{\mathfrak{a}}$ and $C_{\mathfrak{a}}$ are arbitrary. With the choices[1] $B_{\mathfrak{a}} = \tau_{\mathfrak{a}} - \frac{\partial\varepsilon}{\partial v_{\mathfrak{a}}}$, $C_{\mathfrak{a}} = -\frac{\partial\varepsilon}{\partial\tau_{\mathfrak{a}}}$, we get

$$\left.\begin{aligned}
\langle Q\rangle &= \sum_{\mathfrak{a}=1}^{\mathfrak{f}}\lambda_{\tau_{\mathfrak{a}}}\,d\tau_{\mathfrak{a}} + \varkappa_{\tau}\,d\theta,\\
&= \sum_{\mathfrak{a}=1}^{\mathfrak{f}}\lambda_{v_{\mathfrak{a}}}\,d v_{\mathfrak{a}} + \varkappa_{v}\,d\theta,
\end{aligned}\right\} \tag{250.6}$$

where the latent heats and specific heats are given by

$$\left.\begin{aligned}
\lambda_{\tau_{\mathfrak{a}}} &= \frac{\partial\varepsilon}{\partial\tau_{\mathfrak{a}}} - \sum_{\mathfrak{b}=1}^{\mathfrak{f}}\left(\tau_{\mathfrak{b}} - \frac{\partial\varepsilon}{\partial v_{\mathfrak{b}}}\right)\nu_{\mathfrak{b}\mathfrak{a}}, &\quad \varkappa_{\tau} &= \frac{\partial\varepsilon}{\partial\theta} - \sum_{\mathfrak{a}=1}^{\mathfrak{f}}\alpha_{\mathfrak{a}}\left(\tau_{\mathfrak{a}} - \frac{\partial\varepsilon}{\partial v_{\mathfrak{a}}}\right),\\
\lambda_{v_{\mathfrak{a}}} &= \frac{\partial\varepsilon}{\partial v_{\mathfrak{a}}} - \tau_{\mathfrak{a}} + \sum_{\mathfrak{b}=1}^{\mathfrak{f}}\frac{\partial\varepsilon}{\partial\tau_{\mathfrak{b}}}\,\xi_{\mathfrak{b}\mathfrak{a}}, &\quad \varkappa_{v} &= \frac{\partial\varepsilon}{\partial\theta} + \sum_{\mathfrak{a}=1}^{\mathfrak{f}}\beta_{\mathfrak{a}}\,\frac{\partial\varepsilon}{\partial\tau_{\mathfrak{a}}}.
\end{aligned}\right\} \tag{250.7}$$

Hence

$$\varkappa_{\tau} - \varkappa_{v} = \sum_{\mathfrak{a}=1}^{\mathfrak{f}}\left[\alpha_{\mathfrak{a}}\left(\frac{\partial\varepsilon}{\partial v_{\mathfrak{a}}} - \tau_{\mathfrak{a}}\right) - \beta_{\mathfrak{a}}\,\frac{\partial\varepsilon}{\partial\tau_{\mathfrak{a}}}\right], \tag{250.8}$$

---

[1] By other choices of the $B_{\mathfrak{a}}$ or $C_{\mathfrak{a}}$ it is also possible to eliminate $d\theta$ and any $\mathfrak{f}-1$ of the $d v_{\mathfrak{a}}$ and $d\tau_{\mathfrak{a}}$, but the result is not illuminating except in the special case $\mathfrak{f}=1$, for which it is given at the end of this section.

a result which reduces to (249.6) when (250.5) is replaced by the more special assumption (248.8). From (250.7), (250.8), and (248.7) we obtain

$$\varkappa_\tau - \varkappa_v = \sum_{\mathfrak{a}=1}^{\mathfrak{k}} \alpha_{\mathfrak{a}} \lambda_{v_{\mathfrak{a}}} = - \sum_{\mathfrak{a}=1}^{\mathfrak{k}} \beta_{\mathfrak{a}} \lambda_{\tau_{\mathfrak{a}}}. \tag{250.9}$$

These identities express the difference of specific heats as bilinear forms in the thermal coefficients $\alpha_{\mathfrak{a}}, \beta_{\mathfrak{a}}$ and the latent heats $\lambda_{\tau_{\mathfrak{a}}}, \lambda_{v_{\mathfrak{a}}}$. *Their form is precisely the same for Brillouin's theory as for the classical special case when there are thermal equations of state.* Thus if one specific heat is known, the other may be calculated directly from the measurable quantities defined as coefficients in the forms (248.6) and (250.6). Moreover, the relation (250.9) does not serve as a test for the existence of a thermal equation of state.

When $\mathfrak{k}=1$, we write β for β_1, α for α_1, etc., and obtain $\varkappa_\tau - \varkappa_v = -\beta\lambda_\tau = \alpha\lambda_v$, so that *the latent heats are proportional to the differences of specific heats*, and (250.6) become

$$\langle Q\rangle = -\frac{1}{\beta}(\varkappa_\tau - \varkappa_v)\,d\tau + \varkappa_\tau\,d\theta = \frac{1}{\alpha}(\varkappa_\tau - \varkappa_v)\,dv + \varkappa_v\,d\theta, \tag{250.10}$$

with $\beta + \alpha\xi = 0$, $\alpha + \beta\nu = 0$, $\xi\nu = 1$. Also we have the alternative form

$$\left.\begin{aligned}\langle Q\rangle &= \left(\frac{\partial\varepsilon}{\partial v} - \tau + \frac{1}{\alpha}\frac{\partial\varepsilon}{\partial\theta}\right)dv + \left(\frac{\partial\varepsilon}{\partial\tau} + \frac{1}{\beta}\frac{\partial\varepsilon}{\partial\theta}\right)d\tau,\\ &= \frac{\varkappa_\tau}{\alpha}\,dv + \frac{\varkappa_v}{\beta}\,d\tau.\end{aligned}\right\} \tag{250.11}$$

Therefore

$$\gamma = \frac{\varkappa_\tau}{\varkappa_v} = -\frac{\alpha}{\beta}\left.\frac{d\tau}{dv}\right|_{\langle Q\rangle = 0}. \tag{250.12}$$

Thus when $\mathfrak{k}=1$ the ratio of specific heats may be calculated from α, β, and the ratio $d\tau/dv$ in a process where $\langle Q\rangle = 0$, or, equivalently, all energy changes are balanced by thermostatic work. Such a process generalizes the notion of "isentropic path" introduced in Sect. 249.

251. Thermodynamic potentials and transformations. We return to the classical theory based upon the caloric equation of state (246.1). The *thermodynamic potentials*[1] are named and defined as follows:

$$\left.\begin{aligned}&\textit{internal energy:} && \varepsilon,\\ &\textit{free energy:} && \psi \equiv \varepsilon - \eta\theta,\\ &\textit{enthalpy:} && \chi \equiv \varepsilon - \sum_{\mathfrak{a}=1}^{\mathfrak{k}} \tau_{\mathfrak{a}} v_{\mathfrak{a}},\\ &\textit{free enthalpy:} && \zeta \equiv \chi - \eta\theta = \varepsilon - \eta\theta - \sum_{\mathfrak{a}=1}^{\mathfrak{k}} \tau_{\mathfrak{a}} v_{\mathfrak{a}}.\end{aligned}\right\} \tag{251.1}$$

They are related through the identity

$$\varepsilon - \psi + \zeta - \chi = 0. \tag{251.2}$$

Each of the potentials has certain specially simple properties. From (247.2) and (251.1) it follows that

$$\left.\begin{aligned}d\varepsilon &= \theta\,d\eta + \sum_{\mathfrak{a}=1}^{\mathfrak{k}} \tau_{\mathfrak{a}}\,dv_{\mathfrak{a}}, & d\psi &= -\eta\,d\theta + \sum_{\mathfrak{a}=1}^{\mathfrak{k}} \tau_{\mathfrak{a}}\,dv_{\mathfrak{a}},\\ d\chi &= \theta\,d\eta - \sum_{\mathfrak{a}=1}^{\mathfrak{k}} v_{\mathfrak{a}}\,d\tau_{\mathfrak{a}}, & d\zeta &= -\eta\,d\theta - \sum_{\mathfrak{a}=1}^{\mathfrak{k}} v_{\mathfrak{a}}\,d\tau_{\mathfrak{a}}.\end{aligned}\right\} \tag{251.3}$$

[1] Proportional functions were introduced by MASSIEU [1869, *4* and *5*] [1876, *3*, §§ I and IV], who was motivated by the desire to express all thermodynamic properties in terms of functions of θ, $\mathfrak{v}$ and of θ, $\boldsymbol{\tau}$. Cf. Eqs. (251.7). The theory was elaborated by GIBBS [1875, *1*, p. 87]. Cf. also HELMHOLTZ [1882, *2*, § 1]. Further comments on the interpretations of the potentials are made by NATANSON [1891, *4*] and TREVOR [1897, *9*].

Hence

$$\left.\begin{aligned}
\theta &= \left(\frac{\partial\varepsilon}{\partial\eta}\right)_{\mathfrak{v}}, & \tau_{\mathfrak{a}} &= \left(\frac{\partial\varepsilon}{\partial v_{\mathfrak{a}}}\right)_{\eta,\,\mathfrak{v}^{\mathfrak{a}}},\\
\eta &= -\left(\frac{\partial\psi}{\partial\theta}\right)_{\mathfrak{v}}, & \tau_{\mathfrak{a}} &= \left(\frac{\partial\psi}{\partial v_{\mathfrak{a}}}\right)_{\theta,\,\mathfrak{v}^{\mathfrak{a}}},\\
\theta &= \left(\frac{\partial\chi}{\partial\eta}\right)_{\boldsymbol{\tau}}, & v_{\mathfrak{a}} &= -\left(\frac{\partial\chi}{\partial\tau_{\mathfrak{a}}}\right)_{\eta,\,\boldsymbol{\tau}^{\mathfrak{a}}},\\
\eta &= -\left(\frac{\partial\zeta}{\partial\theta}\right)_{\boldsymbol{\tau}}, & v_{\mathfrak{a}} &= -\left(\frac{\partial\zeta}{\partial\tau_{\mathfrak{a}}}\right)_{\theta,\,\boldsymbol{\tau}^{\mathfrak{a}}},
\end{aligned}\right\} \tag{251.4}$$

the first two of these having already appeared as (247.1). From (251.1) and (251.4) follow the relations

$$\left.\begin{aligned}
\psi-\varepsilon &= \theta\left(\frac{\partial\psi}{\partial\theta}\right)_{\mathfrak{v}}, \quad \chi-\varepsilon = \sum_{\mathfrak{a}=1}^{\mathfrak{k}} \tau_{\mathfrak{a}}\left(\frac{\partial\chi}{\partial\tau_{\mathfrak{a}}}\right)_{\eta,\,\boldsymbol{\tau}^{\mathfrak{a}}},\\
\zeta-\varepsilon &= \theta\left(\frac{\partial\zeta}{\partial\theta}\right)_{\boldsymbol{\tau}} + \sum_{\mathfrak{a}=1}^{\mathfrak{k}} \tau_{\mathfrak{a}}\left(\frac{\partial\zeta}{\partial\tau_{\mathfrak{a}}}\right)_{\theta,\,\boldsymbol{\tau}^{\mathfrak{a}}}.
\end{aligned}\right\} \tag{251.5}$$

From (251.3) it appears that we may use any one of the alternative sets of independent variables

$$\eta,\boldsymbol{v};\quad \theta,\boldsymbol{v};\quad \eta,\boldsymbol{\tau};\quad \theta,\boldsymbol{\tau} \tag{251.6}$$

and yet be able to calculate all energy changes. The functions appropriate to the four cases are the four potentials $\varepsilon, \psi, \chi, \zeta$. In particular, $(251.4)_3$ shows that from the equations $\psi=\psi(\theta,\boldsymbol{v})$ and $\zeta=\zeta(\theta,\boldsymbol{\tau})$, the entropy may be calculated as a change of free energy per unit temperature. The name *free energy* is appropriate for ψ because, as follows from $(251.3)_2$, it is the portion of the energy available for doing work at constant temperature. Similarly, by $(251.3)_3$ it follows that the *enthalpy* χ is the portion of the energy which can be released as heat when the thermodynamic tensions are kept constant.

When $\mathfrak{k}=1$, the free enthalpy is called the "Gibbs function" or "thermodynamic potential". When $\mathfrak{k}>1$, the usage of Gibbs, followed in most texts today, was somewhat different from ours. In the case when v_1 is the specific volume v, Gibbs set

$$\chi=\varepsilon+\pi v,\quad \zeta=\chi-\eta\vartheta=\varepsilon+\pi v-\eta\vartheta, \tag{251.6A}$$

even when $\mathfrak{k}>1$; i.e., in the enthalpy he left out of account the energy corresponding to any parameter other than v_1. This results in a different physical interpretation for χ and ζ if $\mathfrak{k}>1$; in particular, Gibbs' function ζ is useful in situations where the temperature is controlled and a uniform hydrostatic pressure is maintained while other thermostatic parameters vary. In adopting here the generalized enthalpy given by $(251.1)_3$, not only do we seek to exploit the neater mathematical development which will be seen below, but also we recognize that in the general case in continuum mechanics the thermostatic pressure does not exist or does not enjoy much importance.

An equation giving one thermodynamic variable as a function of one of the four sets (251.6) is said to be a ***fundamental equation***[1] if from it all thermodynamic variables that are not its arguments may be calculated by partial differentiation, functional inversion, and algebraic operations. In this definition, the set of "thermodynamic variables" is $v_{\mathfrak{a}}, \tau_{\mathfrak{a}}, \eta, \theta, \varepsilon$. Fundamental equations are

$$\varepsilon=\varepsilon(\eta,\boldsymbol{v}),\quad \psi=\psi(\theta,\boldsymbol{v}),\quad \chi=\chi(\eta,\boldsymbol{\tau}),\quad \zeta=\zeta(\theta,\boldsymbol{\tau}). \tag{251.7}$$

[1] Gibbs [1875, *1*, pp. 85–92].

As will be shown in Sect. 253, an example of a thermodynamic relation which is not a fundamental equation is a thermal equation of state such as $\tau_{\mathfrak{a}}=\tau_{\mathfrak{a}}(\theta, \boldsymbol{v})$.

There is a formal analogy between the equations appropriate to one potential and those appropriate to another. Examples will be given in Sect. 252. For the case when $\mathfrak{k}=1$, HAYES[1] has constructed a systematic and exhaustive method of permutation to obtain them all.

252. Thermodynamic identities. From the assumption of smoothness in Sect. 247, we may write down the following conditions of integrability[2] for the four differential forms (251.3):

$$\left.\begin{aligned}
&\left(\frac{\partial \theta}{\partial v_{\mathfrak{a}}}\right)_{\eta, \boldsymbol{v}^{\mathfrak{a}}}=\left(\frac{\partial \tau_{\mathfrak{a}}}{\partial \eta}\right)_{\boldsymbol{v}}=\varpi_{\mathfrak{a}}, &&-\left(\frac{\partial \eta}{\partial v_{\mathfrak{a}}}\right)_{\theta, \boldsymbol{v}^{\mathfrak{a}}}=\left(\frac{\partial \tau_{\mathfrak{a}}}{\partial \theta}\right)_{\boldsymbol{v}}=\beta_{\mathfrak{a}},\\
&\left(\frac{\partial \theta}{\partial \tau_{\mathfrak{a}}}\right)_{\eta, \boldsymbol{\tau}^{\mathfrak{a}}}=-\left(\frac{\partial v_{\mathfrak{a}}}{\partial \eta}\right)_{\boldsymbol{\tau}}, &&\left(\frac{\partial \eta}{\partial \tau_{\mathfrak{a}}}\right)_{\theta, \boldsymbol{\tau}^{\mathfrak{a}}}=\left(\frac{\partial v_{\mathfrak{a}}}{\partial \theta}\right)_{\boldsymbol{\tau}}=\alpha_{\mathfrak{a}},\\
&\phi_{\mathfrak{a}\mathfrak{b}}=\phi_{\mathfrak{b}\mathfrak{a}}, &&\xi_{\mathfrak{a}\mathfrak{b}}=\xi_{\mathfrak{b}\mathfrak{a}},\\
&\left(\frac{\partial v_{\mathfrak{a}}}{\partial \tau_{\mathfrak{b}}}\right)_{\eta, \boldsymbol{\tau}^{\mathfrak{a}\mathfrak{b}}}=\left(\frac{\partial v_{\mathfrak{b}}}{\partial \tau_{\mathfrak{a}}}\right)_{\eta, \boldsymbol{\tau}^{\mathfrak{a}\mathfrak{b}}}, &&\nu_{\mathfrak{a}\mathfrak{b}}=\nu_{\mathfrak{b}\mathfrak{a}},
\end{aligned}\right\} \tag{252.1}$$

where $\boldsymbol{\tau}^{\mathfrak{a}\mathfrak{b}}$ stands for the set of all the τ's except $\tau_{\mathfrak{a}}$ and $\tau_{\mathfrak{b}}$. These identities are called *Maxwell's relations* or *reciprocal relations*. The expressions for $\alpha_{\mathfrak{a}}$ and $\beta_{\mathfrak{a}}$ are of particular interest in showing that certain rates of change of entropy can be inferred from the thermal equations of state.

When $\mathfrak{k}=1$, the above identities are easily expressed in terms of Jacobians. For example, $(252.1)_1$ implies that

$$\frac{\partial(\theta, \eta)}{\partial(v, \eta)}=\frac{\partial(\tau, v)}{\partial(\eta, v)}; \tag{252.2}$$

hence[3]

$$\frac{\partial(\tau, v)}{\partial(\eta, \theta)}=1. \tag{252.3}$$

This may be regarded as a summary of all the Maxwell relations when $\mathfrak{k}=1$, since the same procedure applied to any one of the identities (252.1) yields this same end product, which asserts that a mapping from the plane of η, θ to the plane of τ, v is area-preserving.

Again by the assumption of smoothness, we consider both η and ε as functions of θ and $\boldsymbol{v}$, by (247.2) obtaining

$$\left.\begin{aligned}
\theta\, d\eta &= d\varepsilon-\sum_{\mathfrak{a}=1}^{\mathfrak{k}} \tau_{\mathfrak{a}}\, d v_{\mathfrak{a}},\\
&=\left(\frac{\partial \varepsilon}{\partial \theta}\right)_{\boldsymbol{v}} d\theta+\sum_{\mathfrak{a}=1}^{\mathfrak{k}}\left[\left(\frac{\partial \varepsilon}{\partial v_{\mathfrak{a}}}\right)_{\theta, \boldsymbol{v}^{\mathfrak{a}}}-\tau_{\mathfrak{a}}\right] d v_{\mathfrak{a}}.
\end{aligned}\right\} \tag{252.4}$$

---

[1] [1946, *5*]. Earlier SHAW [1935, *6*] had given tables based on rather laborious calculation. It has long been noted that the relations (251.1) are contact transformations. CORBEN [1949, *3*] and FÉNYES [1952, *5*] have constructed an analogy to the dynamics of mass-points, whereby each dynamical formula enables us to write down a thermodynamic identity, certain new thermodynamical quantities being thus suggested.

[2] The results, when $\mathfrak{k}=1$, are due to MAXWELL [1871, *5*, Chap. IX] and, for the general case, to GIBBS [1875, *1*, Eq. (272)] and DUHEM [1886, *2*, Part II, Chap. II, Eq. (82)]. Cf. also MILLER [1897, *6*]. It was NATANSON [1891, *4*, § I] who first remarked that they are conditions of integrability.

[3] Allowing for the fact that CLAPEYRON did not distinguish properly between entropy and caloric, we may see nevertheless that he gave the essential content of this relation and of $(252.1)_6$ [1834, *1*, § V].

Hence we again obtain $(249.4)_2$, but also[1]

$$\theta\left(\frac{\partial\eta}{\partial v_{\mathfrak{a}}}\right)_{\theta,\,v^{\mathfrak{a}}}=\left(\frac{\partial\varepsilon}{\partial v_{\mathfrak{a}}}\right)_{\theta,\,v^{\mathfrak{a}}}-\tau_{\mathfrak{a}}. \tag{252.5}$$

This last result enables us to express the latent heats in terms of entropy changes. Indeed, specializing the formulae $(250.7)_{3,1}$ to the case when $\varepsilon=\varepsilon(\boldsymbol{v},\theta)$, we obtain

$$\lambda_{v_{\mathfrak{a}}}=\theta\left(\frac{\partial\eta}{\partial v_{\mathfrak{a}}}\right)_{\theta,\,v^{\mathfrak{a}}},\qquad \lambda_{\tau_{\mathfrak{a}}}=\theta\sum_{\mathfrak{b}=1}^{\mathfrak{k}}\left(\frac{\partial\eta}{\partial v_{\mathfrak{b}}}\right)_{\theta,\,v^{\mathfrak{b}}}\nu_{\mathfrak{b}\mathfrak{a}}=\sum_{\mathfrak{b}=1}^{\mathfrak{k}}\lambda_{v_{\mathfrak{b}}}\nu_{\mathfrak{b}\mathfrak{a}}. \tag{252.6}$$

By the Maxwell relation $(252.1)_4$, these results may be put into the forms[2]

$$\lambda_{v_{\mathfrak{a}}}=-\theta\beta_{\mathfrak{a}},\qquad \lambda_{\tau_{\mathfrak{a}}}=-\theta\sum_{\mathfrak{b}=1}^{\mathfrak{k}}\beta_{\mathfrak{b}}\nu_{\mathfrak{b}\mathfrak{a}}. \tag{252.7}$$

While the last formulae express the latent heats in terms only of the quantities occuring in the differential forms (248.6), the presence of the factor θ indicates that the differential properties of the entropy have been used, and in fact test of the relations (252.7) serves as an experimental check on the integrability of the form $\langle Q\rangle/\theta$.

By differentiating $(252.7)_4$ and using (252.5), (249.4), and $(252.1)_1$ we obtain the following identity of CLAUSIUS[3]:

$$\left(\frac{\partial\lambda_{v_{\mathfrak{a}}}}{\partial\theta}\right)_{\boldsymbol{v}}-\left(\frac{\partial\varkappa_{\boldsymbol{v}}}{\partial v_{\mathfrak{a}}}\right)_{\theta,\,v^{\mathfrak{a}}}=-\beta_{\mathfrak{a}}. \tag{252.8}$$

From $(252.7)_1$ and (250.9) it follows that

$$\varkappa_{\boldsymbol{\tau}}-\varkappa_{\boldsymbol{v}}=-\theta\sum_{\mathfrak{a}=1}^{\mathfrak{k}}\alpha_{\mathfrak{a}}\beta_{\mathfrak{a}}, \tag{252.9}$$

whereby the difference $\varkappa_{\boldsymbol{\tau}}-\varkappa_{\boldsymbol{v}}$ is given in terms of quantities obtainable from the thermal equations of state. For alternative forms, we use (249.7), $(248.7)_{1,2}$, and (252.9) to obtain

$$\gamma-1=\frac{\theta}{\varkappa_{\boldsymbol{v}}}\sum_{\mathfrak{a},\mathfrak{b}=1}\xi_{\mathfrak{a}\mathfrak{b}}\alpha_{\mathfrak{a}}\alpha_{\mathfrak{b}}=\frac{\theta}{\varkappa_{\boldsymbol{v}}}\sum_{\mathfrak{a},\mathfrak{b}=1}^{\mathfrak{k}}\nu_{\mathfrak{a}\mathfrak{b}}\beta_{\mathfrak{a}}\beta_{\mathfrak{b}}. \tag{252.10}$$

We now address ourselves to the task of finding an explicit relation between the coefficients $\phi_{\mathfrak{a}\mathfrak{b}}$ and $\varpi_{\mathfrak{c}}$, defined from the caloric equation of state by (247.9), and the coefficients $\xi_{\mathfrak{a}\mathfrak{b}}$ and $\beta_{\mathfrak{c}}$, defined from the thermal equations of state by $(248.5)_3$. First we write out the two Hessian matrices whose individual components we seek to relate:

$$\left.\begin{aligned}\left\|\frac{\partial(\boldsymbol{\tau},\theta)}{\partial(\boldsymbol{v},\eta)}\right\|&=\left\|\begin{matrix}\phi_{\mathfrak{a}\mathfrak{b}} & \varpi_{\mathfrak{c}}\\ \varpi_{\mathfrak{c}} & \theta/\varkappa_{\boldsymbol{v}}\end{matrix}\right\|,\\ \left\|\frac{\partial(\boldsymbol{v},\eta)}{\partial(\boldsymbol{\tau},\theta)}\right\|&=\left\|\begin{matrix}\nu_{\mathfrak{a}\mathfrak{b}} & \alpha_{\mathfrak{c}}\\ \alpha_{\mathfrak{c}} & \varkappa_{\boldsymbol{\tau}}/\theta\end{matrix}\right\|,\end{aligned}\right\} \tag{252.11}$$

where we have employed only the definitions (247.9), $(248.5)_{1,4}$, and (249.1). The product of the two matrices on the left-hand side is the unit matrix; there-

[1] CLAUSIUS [1854, *1*, p. 486], when $\mathfrak{k}=1$.

[2] The first of these relations, for the case when $\mathfrak{k}=1$, is due to CLAPEYRON [1834, *1*, § IV], with the proviso noted in footnote 3, p. 629. It was obtained from the present theory by CLAUSIUS, first for the case of a perfect gas [1850, *1*, Eq. IV] and then more generally [1854, *1*, Eq. (13a)].

[3] [1854, *1*, Eq. (3)], given earlier for the case of a perfect gas [1850, *1*, Eq. II].

fore, so also is the product of those on the right-hand side. Thus

$$\left.\begin{aligned} \sum_{\mathfrak{c}=1}^{\mathfrak{k}} \phi_{\mathfrak{a}\mathfrak{c}} \nu_{\mathfrak{c}\mathfrak{b}} + \varpi_{\mathfrak{a}} \alpha_{\mathfrak{b}} &= \delta_{\mathfrak{a}\mathfrak{b}}, \\ \sum_{\mathfrak{b}=1}^{\mathfrak{k}} \varpi_{\mathfrak{b}} \nu_{\mathfrak{b}\mathfrak{a}} + \theta \alpha_{\mathfrak{a}}/\varkappa_{v} &= 0, \\ \sum_{\mathfrak{b}=1}^{\mathfrak{k}} \phi_{\mathfrak{a}\mathfrak{b}} \alpha_{\mathfrak{b}} + \varpi_{\mathfrak{a}} \varkappa_{\tau}/\theta &= 0, \\ \sum_{\mathfrak{a}=1}^{\mathfrak{k}} \varpi_{\mathfrak{a}} \alpha_{\mathfrak{a}} + \gamma &= 1, \end{aligned}\right\} \tag{252.12}$$

where for the last identity we have used the definition (249.7).

From $(252.12)_4$ we have

$$\gamma - 1 = -\sum_{\mathfrak{a}=1}^{\mathfrak{k}} \varpi_{\mathfrak{a}} \alpha_{\mathfrak{a}}. \tag{252.13}$$

Multiplying $(252.12)_3$ by $\alpha_{\mathfrak{a}}$ and summing yields

$$\varkappa_{\tau}(\gamma - 1) = \theta \sum_{\mathfrak{a},\mathfrak{b}=1}^{\mathfrak{k}} \phi_{\mathfrak{a}\mathfrak{b}} \alpha_{\mathfrak{a}} \alpha_{\mathfrak{b}}, \tag{252.14}$$

where (252.13) has been used, while a similar process applied to $(252.12)_2$ yields

$$\gamma - 1 = \frac{\varkappa_{v}}{\theta} \sum_{\mathfrak{a},\mathfrak{b}=1}^{\mathfrak{k}} \nu_{\mathfrak{a}\mathfrak{b}} \varpi_{\mathfrak{a}} \varpi_{\mathfrak{b}}. \tag{252.15}$$

Similarly, from $(252.12)_3$ and $(252.12)_4$ we have

$$\gamma - 1 = \frac{\theta}{\varkappa_{\tau}} \sum_{\mathfrak{a},\mathfrak{b}=1}^{\mathfrak{k}} \overset{-1}{\phi}_{\mathfrak{a}\mathfrak{b}} \varpi_{\mathfrak{a}} \varpi_{\mathfrak{b}}. \tag{252.16}$$

These formulae for $\gamma - 1$ are to be set alongside (252.9) and (252.10).

Coming now to $(252.12)_1$, we multiply by $\xi_{\mathfrak{b}\mathfrak{e}}$ and sum on $\mathfrak{b}$; simplifying the result by use of $(248.7)_{1,3}$ and $(252.1)_9$ yields

$$\phi_{\mathfrak{a}\mathfrak{b}} = \xi_{\mathfrak{a}\mathfrak{b}} + \varpi_{\mathfrak{a}} \beta_{\mathfrak{b}}. \tag{252.17}$$

From this identity and $(252.1)_{8,9}$ we obtain a reciprocity theorem:

$$\varpi_{\mathfrak{a}} \beta_{\mathfrak{b}} = \varpi_{\mathfrak{b}} \beta_{\mathfrak{a}}. \tag{252.18}$$

If we write

$$\phi \equiv \det \phi_{\mathfrak{a}\mathfrak{b}}, \qquad \xi \equiv \det \xi_{\mathfrak{a}\mathfrak{b}}, \qquad \nu \equiv \det \nu_{\mathfrak{a}\mathfrak{b}}, \tag{252.19}$$

then taking the determinant of $(252.12)_1$ yields

$$\left.\begin{aligned} \phi \nu = \frac{\phi}{\xi} &= \det(\delta_{\mathfrak{a}\mathfrak{b}} - \varpi_{\mathfrak{a}} \alpha_{\mathfrak{b}}), \\ &= 1 - \sum_{\mathfrak{a}=1}^{\mathfrak{k}} \varpi_{\mathfrak{a}} \alpha_{\mathfrak{a}} \end{aligned}\right\} \tag{252.20}$$

[cf. the steps used in deriving (189.11)]. By (252.13) it follows that[1]

$$\phi = \gamma \xi. \tag{252.21}$$

---

[1] For the case when $\mathfrak{k} = 1$, the identity (252.21) becomes $\gamma = \left(\frac{\partial \pi}{\partial v}\right)_{\eta} \Big/ \left(\frac{\partial \pi}{\partial v}\right)_{\theta}$, which may be interpreted as a statement that in a perfect fluid, the ratio of the isentropic and isothermal speeds of sound is always γ, whatever be the form of the equation of state $\varepsilon = \varepsilon(\eta, v)$ (cf. Sect. 297). While this well known result is often attributed to REECH [1853, *2*, p. 414], he did not derive or state it, although he gave related expressions for $\varkappa_{\pi}$ and $\varkappa_{v}$. A different generalization is given by DUHEM [1903, *10*].

Looking back at $(252.11)_1$, we take the determinant of each side and expand the right-hand member according to minors of elements in the bottom row:

$$\left.\begin{aligned}\frac{\partial(\mathfrak{r},\theta)}{\partial(\mathfrak{v},\eta)} &= \frac{\theta\phi}{\varkappa_{\mathfrak{v}}} - \phi\sum_{\mathfrak{a},\mathfrak{b}=1}^{\mathfrak{k}} \phi_{\mathfrak{a}\mathfrak{b}}^{-1}\,\varpi_{\mathfrak{a}}\varpi_{\mathfrak{b}},\\ &= \frac{\theta\phi}{\varkappa_{\mathfrak{r}}}\left[\gamma-(\gamma-1)\right],\\ &= \frac{\theta\phi}{\varkappa_{\mathfrak{r}}} = \frac{\theta\phi\varkappa_{\mathfrak{v}}}{\gamma} = \theta\,\xi\,\varkappa_{\mathfrak{v}},\end{aligned}\right\}\tag{252.22}$$

where we have used (252.16) and (252.21).

The identities just derived enable information about the caloric equation of state to be inferred from properties of the thermal equations of state, which are more easily accessible to measurement. Essential use will be made of them in Sect. 265 in connection with the theory of stability.

Returning to the system composed of (252.5) and $(249.4)_2$, as its condition of compatibility we derive the following *necessary and sufficient condition for the existence of entropy* as an integral of the form $\langle Q\rangle/\theta = d\eta$:

$$\left(\frac{\partial\varepsilon}{\partial v_{\mathfrak{a}}}\right)_{\theta,v^{\mathfrak{a}}} = \tau_{\mathfrak{a}} - \theta\left(\frac{\partial\tau_{\mathfrak{a}}}{\partial\theta}\right)_{\mathfrak{v}} = -\,\theta^2\left(\frac{\partial(\tau_{\mathfrak{a}}/\theta)}{\partial\theta}\right)_{\mathfrak{v}}.\tag{252.23}$$

Many treatments of thermostatics contain arguments rendering plausible the validity of (252.23) as a postulate, thus enabling the concept of entropy to be derived from those of temperature and an internal energy assumed given by an equation of the form (248.2), supplemented by thermal equations of state (248.3).

While we have developed only the identities following from use of ε as a potential, the same processes lead to corresponding identities if any one of the functions ψ,χ,ζ is used as the starting point. For physical applications, the function ζ is often the best suited.

For a concrete example, consider the van der Waals gas, defined by a caloric equation of state (246.1) given parametrically as follows:

$$\varepsilon = \int c(u)\,du - \frac{a}{v},\qquad \eta = \int\frac{c(u)}{u}\,du + R\log(v - {}_0v),\qquad v > {}_0v\tag{252.24}$$

where a, R, and ${}_0v$ are positive constants, and where $c(u)$ is an arbitrary function. From $(247.7)_1$ and (249.4) we obtain at once

$$\theta = u,\qquad \varkappa_v = c(u) = c(\theta),\tag{252.25}$$

whereby physical interpretation is attached to the parameter u and the function $c(u)$. From $(247.7)_2$ and (252.25) follows the thermal equation of state:

$$\pi = -\tau = \frac{R\theta}{v - {}_0v} - \frac{a}{v^2}.\tag{252.26}$$

By (248.5), (249.11), (252.10), and (252.13) we have

$$\left.\begin{aligned}&\alpha = \frac{R}{\pi - \dfrac{a}{v^2} + \dfrac{2a\,{}_0v}{v^3}}, && \beta = -\frac{R}{v - {}_0v},\\ &\xi = \frac{\pi + \dfrac{a}{v^2}}{v - {}_0v} - \frac{2a}{v^3}, && \nu = \frac{1}{\xi},\\ &\phi = \gamma\,\xi, && \varpi = -\frac{\gamma-1}{\alpha},\\ &(\gamma-1)\,c(\theta) = \varkappa_\pi - \varkappa_v = \frac{R}{1 - \dfrac{2a(v-{}_0v)^2}{R\theta v^3}}. &&\end{aligned}\right\}\tag{252.27}$$

The conditions of ultrastability, as shown in Sect. 265, are $\gamma > 1$ and $\varkappa_v > 0$, where the former is equivalent, alternatively, to $\nu > 0$ and $\xi > 0$. Hence with $c(\theta)$ assumed positive, a state is ultrastable for this substance if and only if $R\theta > 2a(v - {}_0v)^2 v^{-3}$; neutral stability occurs when "$>$" is replaced by "$=$".

When $a = 0$ and ${}_0v = 0$, the van der Waals gas reduces to the ideal gas, which is a special case of the ideal materials to be defined in the next section. For the ideal gas, if $c(\theta) = \text{const}$, u is easily eliminated between $(252.24)_1$ and $(252.24)_2$, so that

$$\varepsilon = \varkappa_v \theta = \varkappa_v \, e^{\frac{\eta - \eta_0}{\varkappa_v}} \left(\frac{v_0}{v}\right)^{\gamma - 1}, \tag{252.28}$$

where v_0 and η_0 are arbitrary constants connected with the zero of ε and the unit of θ. All states are ultrastable for the ideal gas if $\theta > 0$, provided, of course, that $\varkappa_v > 0$.

253. Thermodynamic degeneracy. When $\mathfrak{k}+1$ is the greatest number of independent variables occurring in any equation of state, either caloric or thermal, but in one of them less than $\mathfrak{k}+1$ are present, the material is said to be *degenerate*. The most familiar example of a degenerate substance is one for which the equation of state $\varepsilon = \varepsilon(\theta, \boldsymbol{v})$ reduces to $\varepsilon = \varepsilon(\theta)$. Such a substance is called *ideal*. From (252.23) we read off as a necessary and sufficient condition for an ideal material

$$\tau_{\mathfrak{a}} = \theta f_{\mathfrak{a}}(\boldsymbol{v}) = \theta \beta_{\mathfrak{a}} = -\lambda_{v_{\mathfrak{a}}}, \tag{253.1}$$

where the steps follow by $(248.5)_2$ and $(252.7)_1$. This example establishes the contention just following (251.7), since thermal equations of state of the form (253.1) do not yield the functional form of the caloric equation of state (246.1), but only the restriction

$$f\left(\varepsilon, \left(\frac{\partial \varepsilon}{\partial \eta}\right)_{\boldsymbol{v}}\right) = 0. \tag{253.2}$$

For an ideal material, inversion of (253.1) yields $v_{\mathfrak{a}} = g_{\mathfrak{a}}(\boldsymbol{\tau}/\theta)$ and hence by (248.5)

$$-\theta \alpha_{\mathfrak{a}} = \sum_{\mathfrak{b}=1}^{\mathfrak{k}} \frac{\tau_{\mathfrak{b}}}{\theta} \frac{\partial g_{\mathfrak{a}}}{\partial(\tau_{\mathfrak{b}}/\theta)} = h_{\mathfrak{a}}(\boldsymbol{v}). \tag{253.3}$$

Substitution of these results into (252.9) yields

$$\varkappa_\tau - \varkappa_v = -\sum_{\mathfrak{a}=1}^{\mathfrak{k}} \alpha_{\mathfrak{a}} \tau_{\mathfrak{a}} = \sum_{\mathfrak{a}=1}^{\mathfrak{k}} f_{\mathfrak{a}}(\boldsymbol{v}) \, h_{\mathfrak{a}}(\boldsymbol{v}) = F(\boldsymbol{v}). \tag{253.4}$$

Thus for an ideal material the difference of specific heats, besides enjoying the special expression $(253.4)_1$, is a function of the substate only[1]. For an ideal material the identity $(248.7)_1$ becomes a differential equation for determining $\beta_{\mathfrak{a}}$ when all the functions $-\theta\alpha_{\mathfrak{b}}$ or $h_{\mathfrak{b}}(\boldsymbol{v})$ are known:

$$\theta \sum_{\mathfrak{b}=1}^{\mathfrak{k}} \frac{\partial \beta_{\mathfrak{a}}}{\partial v_{\mathfrak{b}}} \alpha_{\mathfrak{b}} + \beta_{\mathfrak{a}} = 0. \tag{253.5}$$

A second familiar degenerate material is one in which the thermodynamic tensions are determinate from the thermodynamic substate alone, and conversely.

[1] For an ideal gas, $\tau v = -R\theta$, so that $\alpha = -R/\tau$ and $\beta = -R/v$, and (253.4) reduces to $\varkappa_\tau - \varkappa_v = R$ [cf. $(252.27)_8$]. While this celebrated relation is often named after MAYER, the only possibly related specific statement we have been able to find in his paper [1842, *2*, p. 240] is the unproved assertion, "... findet man die Senkung einer ein Gas comprimirenden Quecksilbersäule gleich der durch die Compression entbundenen Wärmemenge ...". Even the generous interpretation of MACH [1896, *3*, pp. 247–250], who finds MAYER „so unzweifelhaft klar ..., daß ein Mißverständnis *nicht möglich* ist," does not impute to him the relation in question, which seems to be due in fact to CLAUSIUS [1850, *1*, Eq. (10a)].

Such a material, characterized by the relations

$$\tau_{\mathfrak{a}} = \tau_{\mathfrak{a}}(\boldsymbol{v}), \quad v_{\mathfrak{a}} = v_{\mathfrak{a}}(\boldsymbol{\tau}), \tag{253.6}$$

is called *piezotropic*. From $(252.1)_1$ follows the necessary and sufficient condition $\theta = \theta(\eta)$. In other words, the caloric equation of state (246.1) in the piezotropic case reduces to[1]

$$\varepsilon = \varepsilon_\theta(\eta) + \varepsilon_v(\boldsymbol{v}), \tag{253.7}$$

so that the internal energy may be split into a thermal portion ε_θ and a substantial portion ε_v. For the former, by (249.4) we obtain

$$\varepsilon_\theta = \int \varkappa_v \, d\theta, \tag{253.8}$$

where $\varkappa_v = \varkappa_v(\theta)$. By (248.5) and (247.9) we obtain as alternative necessary and sufficient conditions for a piezotropic material

$$\alpha_{\mathfrak{a}} = 0, \quad \beta_{\mathfrak{a}} = 0, \quad \varpi_{\mathfrak{a}} = 0 \quad \xi_{\mathfrak{a}\mathfrak{b}} = \phi_{\mathfrak{a}\mathfrak{b}}. \tag{253.9}$$

From (249.10), or alternatively from (252.10) or any of Eqs. (252.13) to (252.16), follows

$$\gamma = 1, \tag{253.10}$$

this condition being necessary but not sufficient unless $\mathfrak{k} = 1$.

An interesting question regarding apparent degeneracy has been raised by FINK[2]. In all the formal developments of thermodynamics it is assumed that the $\mathfrak{k}$ parameters $v_{\mathfrak{a}}$ are sufficient to describe the physical body under consideration. If this is not so, the theory may yet be sufficient to describe some, though not all, of the physical behavior of the body. To see the effect of neglecting the additional parameters necessary for a complete description, we consider the theory based on

$$\varepsilon = f(\eta, v_1, v_2, \ldots, v_{\mathfrak{k}}, v_{\mathfrak{k}+1}, \ldots, v_{\mathfrak{k}+\mathfrak{m}}) \tag{253.11}$$

as exact, and we compare the exact results with those which result from treating the material as if it were degenerate:

$$\varepsilon = f(\eta, v_1, v_2, \ldots, v_{\mathfrak{k}}, A_1, \ldots, A_{\mathfrak{m}}), \tag{253.12}$$

where $A_1, \ldots, A_{\mathfrak{m}}$ are constants. If f did not in fact depend on its last $\mathfrak{m}$ arguments, the material would be really degenerate and the results obtained from (253.12) exact. Since, by assumption, ε does depend on these arguments, quantities such as θ and $\tau_1, \ldots, \tau_{\mathfrak{k}}$ calculated from (253.12) will depend on the constants $A_1, \ldots, A_{\mathfrak{m}}$. That is, different behavior as an apparently degenerate material may correspond to different values of the unknown constants $A_1, \ldots, A_{\mathfrak{m}}$.

254. Heterogeneous media I. Compatibility of an equation of state for the mixture with equations of state for the constituents. We now divide the parameters $\boldsymbol{v}$ into two groups: the set of individual densities $\varrho_{\mathfrak{A}}$ and the set of additional parameters $\omega_1, \ldots, \omega_{\mathfrak{k}}$, independent of the densities. We assume that each constituent has its own *partial entropy* $\eta_{\mathfrak{A}}$ and its own *caloric equation of state*

$$\varepsilon_{\mathfrak{A}} = \varepsilon_{\mathfrak{A}}(\eta_{\mathfrak{A}}, \varrho_1, \varrho_2, \ldots, \varrho_{\mathfrak{K}}, \omega), \quad \mathfrak{A} = 1, 2, \ldots, \mathfrak{K}. \tag{254.1}$$

We now define the *total specific inner energy* ε_I and the *total specific entropy* η of the mixture:

$$\varepsilon_I \equiv \sum_{\mathfrak{A}=1}^{\mathfrak{K}} c_{\mathfrak{A}} \varepsilon_{\mathfrak{A}}, \quad \eta \equiv \sum_{\mathfrak{A}=1}^{\mathfrak{K}} c_{\mathfrak{A}} \eta_{\mathfrak{A}}. \tag{254.2}$$

---

[1] COURANT and FRIEDRICHS [1948, *8*, § 3].

[2] [1947, *5*] [1948, *11*] [1949, *10*] [1951, *6*]. Most of FINK's work considers the effect of adding one additional parameter in the usual thermostatic theory for fluids so as to describe metastable states, etc.

By (254.1) and the definition $(254.2)_1$, which was given earlier as $(243.8)_1$, ε_{I} is defined as a function of all the $\eta_{\mathfrak{A}}$, all the $\varrho_{\mathfrak{A}}$, and the set $\boldsymbol{\omega}$. It is then a definite mathematical problem to find conditions under which ε_{I} depends on the $\eta_{\mathfrak{A}}$ only through the linear combination η defined by $(254.2)_2$. Thus no thermodynamic arguments, but only straightforward use of the theory of functional dependence, is required to determine whether or not the caloric equations of state (254.1) for the constituents imply a *caloric equation for the mixture*:

$$\varepsilon_{\mathrm{I}} = \varepsilon_{\mathrm{I}}(\eta, \varrho_1, \ldots, \varrho_{\mathfrak{K}}, \omega). \tag{254.3}$$

Indeed, a necessary and sufficient condition for this functional dependence is the vanishing of the following Jacobians:

$$\left.\begin{array}{l} \dfrac{\partial(\varepsilon_{\mathrm{I}}, \eta, \varrho_1, \ldots, \varrho_{\mathfrak{K}}, \omega_1, \ldots, \omega_{\mathfrak{k}})}{\partial(\eta_1, \eta_2, \varrho_1, \ldots, \varrho_{\mathfrak{K}}, \omega_1, \ldots, \omega_{\mathfrak{k}})}, \ldots, \\[2ex] \dfrac{\partial(\varepsilon_{\mathrm{I}}, \eta, \varrho_1, \ldots, \varrho_{\mathfrak{K}}, \omega_1, \ldots, \omega_{\mathfrak{k}})}{\partial(\eta_1, \eta_2, \eta_3, \varrho_2, \ldots, \varrho_{\mathfrak{K}}, \omega_1, \ldots, \omega_{\mathfrak{k}})}, \ldots, \\ \quad\vdots \\ \dfrac{\partial(\varepsilon_{\mathrm{I}}, \eta, \varrho_1, \ldots, \varrho_{\mathfrak{K}}, \omega_1, \ldots, \omega_{\mathfrak{k}})}{\partial(\eta_1, \eta_2, \varrho_1, \ldots, \varrho_{\mathfrak{K}}, \eta_3, \ldots, \omega_{\mathfrak{k}})}, \ldots. \end{array}\right\} \tag{254.4}$$

Since the variables $\eta_1, \ldots, \eta_{\mathfrak{K}}, \varrho_1, \ldots, \varrho_{\mathfrak{K}}, \omega_1, \ldots, \omega_{\mathfrak{k}}$ are independent, the only members of the above set of determinants that are not zero by definition are those of the first line, which reduce to

$$\frac{\partial(\varepsilon_{\mathrm{I}}, \eta)}{\partial(\eta_1, \eta_2)}, \ldots, \tag{254.5}$$

where $\varrho_1, \ldots, \varrho_{\mathfrak{K}}, \omega_1, \ldots, \omega_{\mathfrak{k}}$ are held constant. If we set

$$\theta_{\mathfrak{A}} \equiv \frac{\partial \varepsilon_{\mathfrak{A}}}{\partial \eta_{\mathfrak{A}}} \tag{254.6}$$

for the *temperature of the constituent* $\mathfrak{A}$, the determinants (254.5) may be expressed in the forms

$$\left.\begin{array}{ll} c_1 c_2 (\theta_1 - \theta_2), & c_1 c_3 (\theta_1 - \theta_3), \ldots, \\ & c_2 c_3 (\theta_2 - \theta_3), \ldots. \end{array}\right\} \tag{254.7}$$

Therefore we have the following local **theorem**[1]: *If each constituent of a mixture has its own caloric equation of state* (254.1), *and if the total inner energy* ε_{I} *and total entropy* η *of the mixture are defined by* (254.2), *then in order that there be a caloric equation of state* (254.3), *it is necessary and sufficient that at each place and time the constituents fall into two categories*:

1. $\mathfrak{L}$ *constituents have the same temperature*:

$$\theta_{\mathfrak{A}_1} = \theta_{\mathfrak{A}_2} = \cdots \theta_{\mathfrak{A}_{\mathfrak{L}}} = \theta. \tag{254.8}$$

---

[1] TRUESDELL [1957, *16*, § 10]. The *necessity* of (254.8) and (254.9) is easily established alternatively as follows: from $(254.2)_2$ and (254.3) we have

$$\left(\frac{\partial \varepsilon_{\mathrm{I}}}{\partial \eta_{\mathfrak{A}}}\right)_{\boldsymbol{\rho}, \boldsymbol{\omega}} = c_{\mathfrak{A}} \left(\frac{\partial \varepsilon_{\mathrm{I}}}{\partial \eta}\right)_{\boldsymbol{\rho}, \boldsymbol{\omega}},$$

but by $(254.2)_1$

$$\left(\frac{\partial \varepsilon_{\mathrm{I}}}{\partial \eta_{\mathfrak{A}}}\right)_{\boldsymbol{\rho}, \boldsymbol{\omega}} = c_{\mathfrak{A}} \left(\frac{\partial \varepsilon_{\mathfrak{A}}}{\partial \eta_{\mathfrak{A}}}\right)_{\boldsymbol{\rho}, \boldsymbol{\omega}}.$$

Equating these two results shows that $\theta_{\mathfrak{A}} = \theta$ if $c_{\mathfrak{A}} \neq 0$.

2. *The remaining* $\mathfrak{R}-\mathfrak{L}$ *constituents are not present:*

$$c_{\mathfrak{A}_{\mathfrak{L}+1}} = c_{\mathfrak{A}_{\mathfrak{L}+2}} = \cdots c_{\mathfrak{A}_{\mathfrak{R}}} = 0. \tag{254.9}$$

The form of the caloric equation (254.3) is thus *derived*; it depends not only on the form of the caloric equations (254.1) for the individual constituents but also on the particular variety defined by the conditions (254.8) and (254.9).

255. Heterogeneous media II. Explicit form of the Gibbs equation. Although it is only a matter of functional elimination to determine the caloric equation of state according to the results just preceding, we cannot expect to find its explicit form in general. However, in the practice of thermodynamics for continuous media Eq. (254.3) is never used other than through its material derivative following the mean motion:

$$\dot{\varepsilon}_{\mathrm{I}} = \theta\dot{\eta} - \pi\dot{v} + \sum_{\mathfrak{A}=1}^{\mathfrak{R}} \mu_{\mathfrak{A}}\dot{c}_{\mathfrak{A}} + \sum_{\mathfrak{a}=1}^{\mathfrak{t}} \sigma_{\mathfrak{a}}\dot{\omega}_{\mathfrak{a}}. \tag{255.1}$$

Of course, this equation, often called the *Gibbs equation*, is a special case of the differential relation (247.2); the differentials are taken along the mean motion of the mixture, and the parameters $v_{\mathfrak{a}}$ are rendered partly explicit as $v, c_1, \ldots, c_{\mathfrak{R}}$, where the $c_{\mathfrak{A}}$ satisfy (158.5). The *thermodynamic pressure* π, the *chemical potentials*[1] $\mu_{\mathfrak{A}}$, and the remaining tensions $\sigma_{\mathfrak{a}}$ are defined by

$$\pi \equiv -\left(\frac{\partial\varepsilon_{\mathrm{I}}}{\partial v}\right)_{\eta,\,\boldsymbol{c},\,\boldsymbol{\omega}}, \quad \mu_{\mathfrak{A}} \equiv \left(\frac{\partial\varepsilon_{\mathrm{I}}}{\partial c_{\mathfrak{A}}}\right)_{\eta,\,v,\,\boldsymbol{c}^{\mathfrak{A}},\,\boldsymbol{\omega}} + f(\eta, v, \boldsymbol{c}, \boldsymbol{\omega}), \quad \sigma_{\mathfrak{a}} \equiv \left(\frac{\partial\varepsilon_{\mathrm{I}}}{\partial\omega_{\mathfrak{a}}}\right)_{\eta,\,v,\,\boldsymbol{c},\,\boldsymbol{\omega}^{\mathfrak{a}}}. \tag{255.2}$$

The arbitrary function f in the definition of $\mu_{\mathfrak{A}}$ reflects the indeterminacy corresponding to the infinitely many possible functional forms for dependence of ε_{I} upon the variables $c_{\mathfrak{A}}$, which are related by the condition (158.5). A possible method of resolving this indeterminacy is to regard the $\mathfrak{R}$-th constituent as the "solvent", so that ε_{I} depends only upon $c_1, c_2, \ldots, c_{\mathfrak{R}-1}$, not upon $c_{\mathfrak{R}}$; the function f in $(255.2)_2$ must then be taken as 0, yielding unique values for $\mu_1, \mu_2, \ldots, \mu_{\mathfrak{R}-1}$, but $\mu_{\mathfrak{R}}$ is not defined.

Alternatively, (255.1) may be written in the form

$$\left.\begin{aligned}\dot{\varepsilon}_{\mathrm{I}} &= \theta\dot{\eta} - \Big(\pi - \sum_{\mathfrak{A}=1}^{\mathfrak{R}} \varrho_{\mathfrak{A}}\mu_{\mathfrak{A}}\Big)\dot{v} + \sum_{\mathfrak{A}=1}^{\mathfrak{R}} v\mu_{\mathfrak{A}}\dot{\varrho}_{\mathfrak{A}} + \sum_{\mathfrak{a}=1}^{\mathfrak{t}} \sigma_{\mathfrak{a}}\dot{\omega}_{\mathfrak{a}},\\ &= \theta\dot{\eta} + \sum_{\mathfrak{A}=1}^{\mathfrak{R}} v\Big[\mu_{\mathfrak{A}} + v\Big(\pi - \sum_{\mathfrak{B}=1}^{\mathfrak{R}} \varrho_{\mathfrak{B}}\mu_{\mathfrak{B}}\Big)\Big]\dot{\varrho}_{\mathfrak{A}} + \sum_{\mathfrak{a}=1}^{\mathfrak{t}} \sigma_{\mathfrak{a}}\dot{\omega}_{\mathfrak{a}};\end{aligned}\right\} \tag{255.3}$$

that is, the chemical potentials are related to the tensions corresponding to the individual densities by the following equation:

$$\left(\frac{\partial\varepsilon_{\mathrm{I}}}{\partial\varrho_{\mathfrak{A}}}\right)_{\eta,\,\boldsymbol{\varrho}^{\mathfrak{A}},\,\boldsymbol{\omega}} = v\Big[\mu_{\mathfrak{A}} + v\Big(\pi - \sum_{\mathfrak{B}=1}^{\mathfrak{R}} \varrho_{\mathfrak{B}}\mu_{\mathfrak{B}}\Big)\Big]. \tag{255.4}$$

The quantity on the left-hand side is a uniquely defined thermodynamic coefficient; the identity is valid for all possible choices of $\mu_{\mathfrak{A}}$. Properties of these equations may be derived from the developments given in Sects. 246 to 252. However, there is no need to follow the general custom[2] of assuming the validity of

[1] This kind of chemical potential, $\mu_{\mathfrak{A}}$, is used in works on irreversible thermodynamics; it differs from the chemical potential $\mu_{\mathfrak{A}}^{\star}$ of GIBBS, defined in Sects. 247 and 260, which is used by chemists in studies of homogeneous mixtures in equilibrium; cf. (260.12).

[2] E.g. ECKART [1940, *8*], MEIXNER [1943, *2*, § 2], PRIGOGINE [1947, *12*, Chap. IX, § 1], DE GROOT [1952, *3*, §§ 3, 43] and virtually all other works on "irreversible thermodynamics".

the basic relation (255.1); rather, as suggested by the results in the last section, that equation should be *derived*, under the assumption that all constituents have the same temperature.

Indeed, the explicit form of the coefficients occurring in the Gibbs equation (255.1) may be calculated[1] in terms of the corresponding coefficients for the Eqs. (254.1). We set

$$v_{\mathfrak{A}} \equiv \frac{1}{\varrho_{\mathfrak{A}}}, \qquad \tau_{\mathfrak{A}\mathfrak{B}} \equiv \frac{\partial \varepsilon_{\mathfrak{A}}}{\partial v_{\mathfrak{B}}}, \qquad \sigma_{\mathfrak{A}\mathfrak{a}} \equiv \frac{\partial \varepsilon_{\mathfrak{A}}}{\partial \omega_{\mathfrak{a}}}, \tag{255.5}$$

and from (254.1) obtain

$$d\varepsilon_{\mathfrak{A}} = \theta_{\mathfrak{A}}\, d\eta_{\mathfrak{A}} + \sum_{\mathfrak{B}=1}^{\mathfrak{K}} \tau_{\mathfrak{A}\mathfrak{B}}\, dv_{\mathfrak{B}} + \sum_{\mathfrak{a}=1}^{\mathfrak{k}} \sigma_{\mathfrak{A}\mathfrak{a}}\, d\omega_{\mathfrak{a}}, \tag{255.6}$$

for arbitrary values of the differentials. Hence in particular

$$\left.\begin{aligned}
\varepsilon_{\mathfrak{A},k} &= \theta_{\mathfrak{A}}\, \eta_{\mathfrak{A},k} + \sum_{\mathfrak{B}=1}^{\mathfrak{K}} \tau_{\mathfrak{A}\mathfrak{B}}\, v_{\mathfrak{B},k} + \sum_{\mathfrak{a}=1}^{\mathfrak{k}} \sigma_{\mathfrak{A}\mathfrak{a}}\, \omega_{\mathfrak{a},k},\\
\acute{\varepsilon}_{\mathfrak{A}} &= \theta_{\mathfrak{A}}\, \acute{\eta}_{\mathfrak{A}} + \sum_{\mathfrak{B}=1}^{\mathfrak{K}} \tau_{\mathfrak{A}\mathfrak{B}}\, \acute{v}_{\mathfrak{B}} + \sum_{\mathfrak{a}=1}^{\mathfrak{k}} \sigma_{\mathfrak{A}\mathfrak{a}}\, \acute{\omega}_{\mathfrak{a}},\\
&= \theta_{\mathfrak{A}}\, \acute{\eta}_{\mathfrak{A}} + \sum_{\mathfrak{B}=1}^{\mathfrak{K}} \tau_{\mathfrak{A}\mathfrak{B}} (\dot{v}_{\mathfrak{B}} + v_{\mathfrak{B},k} u^k_{\mathfrak{A}}) + \sum_{\mathfrak{a}=1}^{\mathfrak{k}} \sigma_{\mathfrak{A}\mathfrak{a}} (\dot{\omega}_{\mathfrak{a}} + \omega_{\mathfrak{a},k} u^k_{\mathfrak{A}}),
\end{aligned}\right\} \tag{255.7}$$

where we have used (158.10). We assume

$$\theta_{\mathfrak{A}} = \theta_{\mathfrak{B}} = \theta \quad \text{for} \quad \mathfrak{A} = 1, 2, \ldots, \mathfrak{K}. \tag{255.8}$$

Putting

$$\tau_{\mathfrak{B}} \equiv \sum_{\mathfrak{A}=1}^{\mathfrak{K}} c_{\mathfrak{A}}\, \tau_{\mathfrak{A}\mathfrak{B}}, \qquad \sigma_{\mathfrak{a}} \equiv \sum_{\mathfrak{A}=1}^{\mathfrak{K}} c_{\mathfrak{A}}\, \sigma_{\mathfrak{A}\mathfrak{a}}, \tag{255.9}$$

we sum $(255.7)_2$ with respect to $\mathfrak{A}$; from the fundamental identity (159.5) we obtain

$$\left.\begin{aligned}
\dot{\varepsilon}_{\mathrm{I}} = \theta\dot{\eta} + \sum_{\mathfrak{B}=1}^{\mathfrak{K}} \tau_{\mathfrak{B}}\, \dot{v}_{\mathfrak{B}} + \sum_{\mathfrak{a}=1}^{\mathfrak{k}} \sigma_{\mathfrak{a}}\, \dot{\omega}_{\mathfrak{a}} + \sum_{\mathfrak{A}=1}^{\mathfrak{K}} \psi_{\mathfrak{A}} \left[\hat{c}_{\mathfrak{A}} - \frac{1}{\varrho} (\varrho_{\mathfrak{A}} u^k_{\mathfrak{A}})_{,k}\right] -\\
- \sum_{\mathfrak{A}=1}^{\mathfrak{K}} c_{\mathfrak{A}} u^k_{\mathfrak{A}} \left[\varepsilon_{\mathfrak{A},k} - \theta\, \eta_{\mathfrak{A},k} - \sum_{\mathfrak{B}=1}^{\mathfrak{K}} \tau_{\mathfrak{A}\mathfrak{B}}\, v_{\mathfrak{B},k} - \sum_{\mathfrak{a}=1}^{\mathfrak{k}} \sigma_{\mathfrak{A}\mathfrak{a}}\, \omega_{\mathfrak{a},k}\right],
\end{aligned}\right\} \tag{255.10}$$

where $\psi_{\mathfrak{A}} \equiv \varepsilon_{\mathfrak{A}} - \theta\eta_{\mathfrak{A}}$, the free energy of the constituent $\mathfrak{A}$. From $(255.7)_1$ we see that the last term in brackets in (255.10) is zero. From $(159.2)_2$ and (255.10) we thus obtain

$$\dot{\varepsilon}_{\mathrm{I}} = \theta\dot{\eta} + \sum_{\mathfrak{B}=1}^{\mathfrak{K}} \tau_{\mathfrak{B}}\, \dot{v}_{\mathfrak{B}} + \sum_{\mathfrak{a}=1}^{\mathfrak{k}} \sigma_{\mathfrak{a}}\, \dot{\omega}_{\mathfrak{a}} + \sum_{\mathfrak{A}=1}^{\mathfrak{K}} \psi_{\mathfrak{A}}\, \dot{c}_{\mathfrak{A}}. \tag{255.11}$$

Since $\dot{v}_{\mathfrak{B}} = \overline{v/c_{\mathfrak{B}}}^{\,\boldsymbol{\cdot}} = \dot{v}/c_{\mathfrak{B}} - \dot{c}_{\mathfrak{B}}/(c^2_{\mathfrak{B}}\varrho)$, from (255.11) we conclude a result having precisely the form of the Gibbs equation (255.1), with $\sigma_{\mathfrak{a}}$ given by (255.9), pro-

[1] The theorem given at the end of this section extends to general mixtures a result long used, without derivation, for mixtures of perfect gases; cf., e.g., DUHEM [1893, *2*, Part I, Chap. I, Eq. (15)], ECKART [1940, *8*, Eq. (39)]. For a special case, the derivation was given by LEAF [1946, *7*, pp. 750–751], but his argument seems obscured by other equations and unnecessary side issues. We follow TRUESDELL [1957, *16*, § 11].

vided we set

$$\left.\begin{aligned}\pi &= -\sum_{\mathfrak{A}=1}^{\mathfrak{K}} \frac{\tau_{\mathfrak{A}}}{c_{\mathfrak{A}}} = -\sum_{\mathfrak{A},\mathfrak{B}=1}^{\mathfrak{K}} \frac{c_{\mathfrak{B}}}{c_{\mathfrak{A}}} \left(\frac{\partial \varepsilon_{\mathfrak{B}}}{\partial v_{\mathfrak{A}}}\right)_{\eta_{\mathfrak{B}},\, v^{\mathfrak{a}},\, \omega}, \\ \mu_{\mathfrak{A}} &= \psi_{\mathfrak{A}} - \frac{\tau_{\mathfrak{A}} v_{\mathfrak{A}}}{c_{\mathfrak{A}}}, \\ &= \psi_{\mathfrak{A}} - \frac{v_{\mathfrak{A}}}{c_{\mathfrak{A}}} \sum_{\mathfrak{B}=1}^{\mathfrak{K}} c_{\mathfrak{B}} \left(\frac{\partial \varepsilon_{\mathfrak{B}}}{\partial v_{\mathfrak{A}}}\right)_{\eta_{\mathfrak{B}},\, v^{\mathfrak{a}},\, \omega}.\end{aligned}\right\} \tag{255.12}$$

Thus the thermodynamic pressure π and the chemical potentials $\mu_{\mathfrak{A}}$ are evaluated in terms of the caloric equations of state of the constituents. The generality of (254.1), however, may be excessive; if we are content with the special case when

$$\varepsilon_{\mathfrak{A}} = \varepsilon_{\mathfrak{A}}(\eta_{\mathfrak{A}}, \varrho_{\mathfrak{A}}, \omega), \tag{255.13}$$

then the evaluations (255.12) simplify:

$$\left.\begin{aligned}\pi &= \sum_{\mathfrak{A}=1}^{\mathfrak{K}} \pi_{\mathfrak{A}} \quad \text{where} \quad \pi_{\mathfrak{A}} = -\frac{\partial \varepsilon_{\mathfrak{A}}}{\partial v_{\mathfrak{A}}}, \\ \mu_{\mathfrak{A}} &= \psi_{\mathfrak{A}} + v_{\mathfrak{A}} \pi_{\mathfrak{A}} = \zeta^0_{\mathfrak{A}},\end{aligned}\right\} \tag{255.14}$$

where we may call $\zeta^0_{\mathfrak{A}}$ the *reduced free enthalpy*[1] of the constituent $\mathfrak{A}$. We have thus established the following **theorem**: *In a mixture whose several constituents have the same temperature at each place and time and obey caloric equations of state of the form* (255.13), *for the mean motion there is a Gibbs equation in which*

1. *The total pressure is the sum of the partial pressures, and*

2. *The chemical potential of the constituent $\mathfrak{A}$ in the mixture equals the individual reduced free enthalpy of the constituent $\mathfrak{A}$.*

In $(255.12)_3$ and $(255.14)_3$, the potentials $\mu_{\mathfrak{A}}$ must be understood as sharing the indeterminacy already expressed in $(255.2)_3$.

In the later work we prefer to replace the Gibbs equation (255.1) by the form

$$\varrho \dot{\varepsilon}_{\mathrm{I}} = \varrho \theta \dot{\eta} - \varrho \pi \dot{v} + \sum_{\mathfrak{A}=1}^{\mathfrak{K}} [\varrho_{\mathfrak{A}} \mu_{\mathfrak{A},k} u^k_{\mathfrak{A}} - (\varrho_{\mathfrak{A}} \mu_{\mathfrak{A}} u^k_{\mathfrak{A}})_{,k} + \varrho \mu_{\mathfrak{A}} \hat{c}_{\mathfrak{A}}] + \varrho \sum_{\mathfrak{a}=1}^{\mathfrak{k}} \sigma_{\mathfrak{a}} \dot{\omega}_{\mathfrak{a}}, \tag{255.15}$$

which follows by $(159.2)_2$.

b) The production of entropy

256. Differential equation for the specific entropy in a simple medium. Thus far, the considerations of this subchapter have been independent from those of the preceding. We now combine the two. First confining attention to simple media, we write the equation of energy balance (241.4) in the form

$$\varrho \dot{\varepsilon} = P_{\mathrm{E}} + Q_{\mathrm{E}}, \tag{256.1}$$

where P_{E} and Q_{E} are the *external power* and *external non-mechanical supply of energy*:

$$P_{\mathrm{E}} \equiv t^{(pq)} d_{pq} - m^{pqr} w_{pq,r}, \qquad Q_{\mathrm{E}} \equiv h^p_{\ ,p} + \varrho q. \tag{256.2}$$

In the non-polar case, P_{E} reduces to the stress power, P. Analogously, we write (247.5) in the form

$$\varrho \dot{\varepsilon} = P_{\mathrm{I}} + Q_{\mathrm{I}}, \tag{256.3}$$

[1] The free enthalpy $\zeta_{\mathfrak{A}}$ is obtained by subtracting from $\zeta^0_{\mathfrak{A}}$ the quantity $\sum_{\mathfrak{a}=1}^{\mathfrak{k}} \omega_{\mathfrak{a}} \sigma_{\mathfrak{A}\mathfrak{a}}$, in accord with $(251.1)_4$.

where
$$P_{\mathrm{I}} \equiv \varrho \sum_{\mathfrak{a}=1}^{\mathfrak{k}} \tau_{\mathfrak{a}} \dot{v}_{\mathfrak{a}}, \quad Q_{\mathrm{I}} \equiv \varrho \theta \dot{\eta}, \tag{256.4}$$

the subscript I standing for *inner*. The work done by the thermodynamic tensions is said to be *recoverable*; thus P_{I} is the rate of recoverable working, per unit volume. Elimination of $\dot{\varepsilon}$ between (256.1) and (256.3) yields
$$P_{\mathrm{E}} - P_{\mathrm{I}} + Q_{\mathrm{E}} - Q_{\mathrm{I}} = 0, \tag{256.5}$$

an equation stating that the difference of external and inner working cancels the difference of external and inner supplies of non-mechanical energy.

By $(256.4)_2$, (256.5) may be regarded as an equation for *production of specific entropy*:
$$\varrho \theta \dot{\eta} = P_{\mathrm{E}} - P_{\mathrm{I}} + Q_{\mathrm{E}}. \tag{256.6}$$

Thus only the *surplus* of external work over inner work affects the entropy, while the same non-mechanical factors contribute to the entropy as to the energy. In all cases, entropy rate = (energy rate)/(temperature).

By (256.3) and (251.3) we may introduce P_{E} or Q_{E} or both into the differential equations governing the thermodynamic potentials:
$$\left.\begin{aligned} \varrho \dot{\psi} &= P_{\mathrm{I}} - \varrho \eta \dot{\theta} = P_{\mathrm{E}} + Q_{\mathrm{E}} - Q_{\mathrm{I}} - \varrho \eta \dot{\theta}, \\ \varrho \dot{\chi} &= P_{\mathrm{E}} - P_{\mathrm{I}} - \varrho \sum_{\mathfrak{a}=1}^{\mathfrak{k}} v_{\mathfrak{a}} \dot{\tau}_{\mathfrak{a}} + Q_{\mathrm{E}}, \\ \varrho \dot{\zeta} &= P_{\mathrm{E}} - P_{\mathrm{I}} - \varrho \sum_{\mathfrak{a}=1}^{\mathfrak{k}} v_{\mathfrak{a}} \dot{\tau}_{\mathfrak{a}} + Q_{\mathrm{E}} - Q_{\mathrm{I}} - \varrho \eta \dot{\theta}. \end{aligned}\right\} \tag{256.7}$$

These equations add further motivation for the names "free energy" and "enthalpy". If the non-mechanical terms in $(256.7)_1$ balance one another, $Q_{\mathrm{E}} = Q_{\mathrm{I}} + \varrho \eta \dot{\theta}$, then all external work is converted to free energy. If the mechanical terms in $(256.7)_2$ balance one another, $P_{\mathrm{E}} = P_{\mathrm{I}} + \varrho \sum_{\mathfrak{a}=1}^{\mathfrak{k}} v_{\mathfrak{a}} \dot{\tau}_{\mathfrak{a}}$, then all external non-mechanical supply of energy is converted into enthalpy.

Returning to (256.6), we proceed to express the surplus of work in the form
$$P_{\mathrm{E}} - P_{\mathrm{I}} = \varrho \sum_{\mathfrak{a}=1}^{\mathfrak{k}} \gamma_{\mathfrak{a}} \dot{v}_{\mathfrak{a}}. \tag{256.8}$$

To do so, we note that by (90.1) and (72.9) we have
$$\left.\begin{aligned} P_{\mathrm{E}} &= g_{pr} t^{(rq)} \dot{x}^p_{,q} - m_p{}^{qr} \dot{x}^p{}_{,qr}, \\ &= g_{pr} t^{(rq)} \dot{x}^p_{;\alpha} X^\alpha{}_{;q} - m_p{}^{qr} (\dot{x}^p_{;\alpha} X^\alpha_{;q})_{;r}, \\ &= (g_{pr} t^{(rq)} X^\alpha_{;q} - m_p{}^{qr} X^\alpha_{;qr}) \dot{\overline{x^p_{;\alpha}}} - m_p{}^{qr} X^\alpha_{;q} X^\beta_{;r} \dot{\overline{x^p_{;\alpha\beta}}}. \end{aligned}\right\} \tag{256.9}$$

We now agree to choose the first 9 parameters $v_{\mathfrak{a}}$ as the $x^p_{;\alpha}$, the next 18 as the 18 independent $x^p_{;\alpha\beta}$, in some fixed order. This is not a loss in generality, since in materials such that ε does not depend on some or all of these variables, we may take the corresponding $\tau_{\mathfrak{a}}$ as zero. With this choice of parameters $v_{\mathfrak{a}}$, we obtain (256.8) by putting
$$\gamma_{\mathfrak{a}} = \left\{\begin{array}{ll} \dfrac{1}{\varrho} (g_{pr} t^{(rq)} X^\alpha_{;q} - m_p{}^{qr} X^\alpha_{;qr}) - \left(\dfrac{\partial \varepsilon}{\partial x^p_{;\alpha}}\right)_\eta, & \mathfrak{a} = 1, 2, \ldots, 9, \\ -\dfrac{1}{\varrho} m_p{}^{qr} X^\alpha_{;q} X^\beta_{;r} - \left(\dfrac{\partial \varepsilon}{\partial x^p_{;\alpha\beta}}\right)_\eta, & \mathfrak{a} = 10, 11, \ldots, 27, \\ -\left(\dfrac{\partial \varepsilon}{\partial v_{\mathfrak{a}}}\right)_\eta & \mathfrak{a} = 28, 29, \ldots, \mathfrak{k}. \end{array}\right\} \tag{256.10}$$

The first 27 γ_a represent the excesses of the total symmetric stress and total couple stress over the parts of the stress and couple stress which do recoverable work. Equivalently, we may write[1]

$$\left.\begin{aligned} t^{(pq)} &= \varrho\, g^{pr}\left[\left(\frac{\partial\varepsilon}{\partial x^r_{;\alpha}}\right)_\eta x^q_{;\alpha} + \left(\frac{\partial\varepsilon}{\partial x^r_{;\alpha\beta}}\right)_\eta x^q_{;\alpha\beta}\right] + {}_{\mathrm{D}}t^{(pq)}, \\ m_p{}^{qr} &= -\varrho\left(\frac{\partial\varepsilon}{\partial x^p_{;\alpha\beta}}\right)_\eta x^q_{;\alpha}\, x^r_{;\beta} + {}_{\mathrm{D}}m_p{}^{qr}, \end{aligned}\right\} \tag{256.11}$$

where the second result has been used to simplify the first, and conclude that

$$P_{\mathrm{E}} - P_{\mathrm{I}} = {}_{\mathrm{D}}t^{(pq)}\, d_{pq} - {}_{\mathrm{D}}m^{pqr}\, w_{pq,r} - \varrho \sum_{a=28}^{t} \tau_a \dot{v}_a, \tag{256.12}$$

so that *only the dissipative parts* ${}_{\mathrm{D}}\boldsymbol{t}$ *and* ${}_{\mathrm{D}}\boldsymbol{m}$ *of* $\boldsymbol{t}$ *and* $\boldsymbol{m}$ *contribute to the entropy*[2].

The definition of ${}_{\mathrm{D}}\boldsymbol{t}$ implied by (256.11) coincides with that given in Sect. 218 only in certain special cases, neither being strictly more general than the other. To render explicit the definitions used here, we rewrite (256.10) by setting

$$\gamma_a \equiv \left\{\begin{matrix} \gamma_p^\alpha & \text{for} & a = 1, 2, \ldots, 9 \\ \gamma_p^{\alpha\beta} & \text{for} & a = 10, 11, \ldots, 27 \end{matrix}\right\} \tag{256.13}$$

so as to indicate the tensorial character of the parameters γ_a. Then from $(256.10)_2$ and $(256.11)_2$ we have

$${}_{\mathrm{D}}m_p{}^{qr} = -\varrho\, \gamma_p^{\alpha\beta}\, x^q_{;\alpha}\, x^n_{;\beta}; \tag{256.14}$$

by $(256.10)_1$ and $(256.11)_1$ follows

$${}_{\mathrm{D}}t^{(pq)} = \varrho\left(\gamma^{\alpha p}\, x^q_{;\alpha} + \gamma^{\alpha\beta p}\, x^q_{;\alpha\beta}\right). \tag{256.15}$$

In particular, these results hold when the stresses doing recoverable work are hydrostatic, since this is the special case when $(256.11)_1$ reduces to[3]

$$t^p_q = -\pi\,\delta^p_q + {}_{\mathrm{D}}t^p_q, \qquad \pi = -(\partial\varepsilon/\partial v)_\eta \tag{256.16}$$

[cf. (218.11) and the developments following from it].

---

[1] In principle, this result is due to HELLINGER [1914, *4*, § 7b], generalizing an analysis of E. and F. COSSERAT [1909, *5*, §§ 53–55].

[2] Since it is possible that the entire stress be dissipative, a better statement is: *Any portion of the stress that does recoverable work makes no contribution to the entropy.* Also, as shown in Sect. 256A, there may be constraints such as to prevent certain dissipative stresses from contributing to the entropy.

[3] Here we take occasion to warn the reader against a celebrated pitfall that continues to trip writers on hydrodynamics: In (256.16), π is not just any hydrostatic pressure but exactly the *thermodynamic pressure*, related to the caloric equation of state by $\pi \equiv -(\partial\varepsilon/\partial v)_\eta$. If $\boldsymbol{t} = -p\,\mathbf{1}$, then $\boldsymbol{t} = -\pi\,\mathbf{1} - (p-\pi)\,\mathbf{1}$, so that in (256.5)

$${}_{\mathrm{D}}\boldsymbol{t} = -(p-\pi)\,\mathbf{1}. \tag{A}$$

In words: *When the stress is hydrostatic, the excess of the total pressure over the thermodynamic pressure is dissipative.* More generally, for a substance with caloric equation of state $\varepsilon = \varepsilon(\eta, v, \boldsymbol{X})$, we have

$$\boldsymbol{t} = -\pi\,\mathbf{1} + {}_{\mathrm{D}}\boldsymbol{t}. \tag{B}$$

If we choose to write the total stress in the form (204.6), then *the extra stress* $\boldsymbol{v}$ *is not necessarily the dissipative stress.* In fact

$${}_{\mathrm{D}}\boldsymbol{t} = -(p-\pi)\,\mathbf{1} + \boldsymbol{v}. \tag{C}$$

Choosing p as the mean pressure $\bar{p}$ and $\boldsymbol{v}$ as the stress deviator ${}_0\boldsymbol{t}$ does not alter or simplify this result. In an isochoric motion, an arbitrary hydrostatic pressure does no work (cf.

From $(256.11)_1$, only the symmetric part of the stress doing recoverable work is determined. This is consistent with (205.10), which yields the skew symmetric part of $\boldsymbol{t}$ when $\boldsymbol{m}$ is known and $\boldsymbol{l}$ is assigned[1].

256A. Appendix. The fully recoverable case. So as to point up the special and restricted character of many discussions of the subject, we insert here a characterization of the case when all work is recoverable. By definition,

$$P_{\mathrm{E}} = P_{\mathrm{I}}. \tag{256A.1}$$

From (256.5) follows the necessary and sufficient condition[2]

$$Q_{\mathrm{E}} = Q_{\mathrm{I}} = \varrho\,\theta\,\dot{\eta}. \tag{256A.2}$$

For (256A.2) to hold it is sufficient, but not necessary, that the process be isentropic and $Q_{\mathrm{E}}=0$. From (256.7) we have the alternative necessary and sufficient conditions

$$\varrho\dot{\psi} = P_{\mathrm{E}} - \varrho\eta\dot{\theta}, \quad \varrho\dot{\chi} = -\varrho\sum_{\mathfrak{a}=1}^{\mathfrak{t}} v_{\mathfrak{a}}\dot{\tau}_{\mathfrak{a}} + Q_{\mathrm{E}}. \tag{256A.3}$$

The fully recoverable case is characterized by the splitting[3] of the equation of energy into the two Eqs. (256A.1) and (256A.2). Eq. (256A.1), or an equivalent form, may be used to *relate the stress to the thermodynamic tensions*; Eq. (256A.2) remains for the control of such thermal phenomena as may be occurring.

If there are *no constraints*, either kinematical or thermodynamic, then from (256.8) we conclude that $\gamma_{\mathfrak{a}}=0$; this implies that we may put $\mathrm{D}\boldsymbol{t}=0$ and $\mathrm{D}\boldsymbol{m}=0$ in (256.11) and obtain explicit formulae for $t^{(pq)}$ and $m_p{}^{qr}$ in terms of the derivatives of ε at constant entropy. Such formulae are called *stress-strain relations*. When ε does not depend upon the $x^p_{;\alpha\beta}$, we may write the result in the more familiar form[4]

$$T_k^{\alpha} = \varrho_0\left(\frac{\partial\varepsilon}{\partial x^k_{;\alpha}}\right)_{\eta}, \tag{256A.4}$$

where T_k^{α} is Piola's double vector defined by $(210.4)_1$.

If there are constraints, (256A.1) no longer necessarily implies $\mathrm{D}\boldsymbol{t}=0$ and $\mathrm{D}\boldsymbol{m}=0$. When the constraints are kinematical, the appropriate modifications are easily read off from results in Sect. 233. For general thermodynamic constraints, the proper stress-strain relations are not known. However, if the constraint is $\theta=\text{const}$, from $(256\text{A}.3)_1$ we obtain

$$\varrho\dot{\psi} = P_{\mathrm{E}}. \tag{256A.5}$$

Sect. 217); hence in an *incompressible* fluid, for which $\pi\equiv 0$, the hydrostatic pressure may be taken as any convenient scalar without affecting the balance of energy.

Among the few hydrodynamic writers who treat the distinction between π and $\tilde{p}$ correctly may be cited Duhem [1901, *7*, Part I, Chap. I, § 8], Zaremba [1903, *18*, pp. 385, 390, 402], Hilbert [1907, *3*, pp. 220–222], and Lamb [1932, *8*, § 325, footnote, § 358].

The idea illustrated here is: *The concept of recoverable work is subservient to the existence of a caloric equation of state*, and only the tensions derived from such an equation do work which makes no contribution to the entropy.

[1] This explains why thermodynamic treatments of elasticity never prove the symmetry of the stress tensor, since even when $\boldsymbol{m}=0$ the condition $t^{[kq]}=0$ follows only if $\boldsymbol{l}=0$, and the assigned couple field $\boldsymbol{l}$ does not enter the formalism of thermodynamics. This illustrates the remark in Sect. 262 that a satisfactory energetics fully independent of mechanical considerations has never been attained.

[2] The form we give to the argument extends a suggestion by Toupin and Ericksen [1952, *21*, § 33]. The argument itself, usually couched in verbal evasions designed to lend it a specious generality, is old and common. Cf. Voigt [1889, *9*, pp. 943–949] [1895, *8*, Pt. III, Chap. I, §§ 6–9] [1910, *10*, §§ 277, 381–382, 389, 392, 394], Müller and Timpe [1906, *6*, § 5c], Love [1927, *6*, § 62], L. Brillouin [1938, *2*, Chap. X, § VIII], Signorini [1949, *27*, Chap. I, ¶¶ 1–2]. Cf. the discussion by Zoller [1959, *12*].

[3] For the hydrodynamical special case, the distinction was emphasized by V. Bjerknes et al. [1933, *3*, Vorwort and § 24].

[4] Kirchhoff [1852, *1*, p. 772]. A list of equivalent forms is given by Truesdell [1952, *21*, §§ 39–40].

From (256.9), provided there are no kinematical constraints, we thus obtain for the *isothermal case* stress-strain relations of the same form as those described in the preceding paragraph for the isentropic case, except that ψ replaces ε, and the derivatives are taken with θ held constant. In this case there are in general non-zero dissipative stresses ${}_D\boldsymbol{t}$ and ${}_D\boldsymbol{m}$, but they do no work in motions compatible with the constraint $\theta = \text{const}$.

The foregoing results characterize the classical theory of finite elastic strain, as well as the theory of perfect compressible fluids, included as the special case when ε depends on the nine $x^k_{;\alpha}$ only through J [cf. (218.11) to (218.13)]. In the non-polar case, when $\boldsymbol{t}$ is symmetric and $\boldsymbol{m} = 0$, comparison with (218.6) enables us to assert that *if all work is recoverable, there exists a strain energy σ if either of the following additional conditions hold:*

1. *There are no constraints, either kinematical or thermodynamic; or*
2. *There are no kinematical constraints, and the motion is isothermal.*

The strain energies for these two cases are:

1. $\sigma = \varrho_0 \varepsilon$,
2. $\sigma = \varrho_0 \psi$.

In both these cases, the work theorem (218.10) holds when the stress vector on the boundary is normal to the velocity there.

It is noteworthy that the basic assumption for *both cases* is that in any possible deformation of the material, *all work is recoverable*[1]. Also noteworthy is the characterizing quality of the results, which show, for example, that if ε does not depend on the $x^k_{;\alpha}$ or the $x^k_{;\alpha\beta}$, and if there are no constraints, then only in the trivial case when $\boldsymbol{t} = 0$ and $\boldsymbol{m} = 0$ can the material experience a deformation in which all the work done is recoverable.

257. Production of total entropy[2].

The *total entropy* H of the body $\mathscr{V}$ is given by the definition[3]

$$\mathsf{H} \equiv \int_{\mathscr{V}} \eta \, d\mathfrak{M}. \tag{257.1}$$

The occurrence of θ on the left-hand side of (256.6) makes the equation for production of total entropy assume a form strikingly different from that for the production of specific entropy. Since (256.6) may be written in the form

$$\varrho \dot{\eta} = \frac{P_E - P_I}{\theta} + \left(\frac{h^p}{\theta}\right)_{,p} + \frac{h^p \theta_{,p}}{\theta^2} + \frac{\varrho q}{\theta}, \tag{257.2}$$

by (257.1) and $(156.8)_1$ we have

$$\dot{\mathsf{H}} - \oint_{\mathscr{S}} \frac{h^p \, da_p}{\theta} = \int_{\mathscr{V}} \left(\Delta + \frac{\varrho q}{\theta}\right) dv, \tag{257.3}$$

---

[1] The criticism of TRUESDELL [1952, *21*, § 33] is directed toward this assumption, which is usually not stated in discussions of Case 2.

In evaluating the theory, it is essential to recall that its *object is to relate the total stress $\boldsymbol{t}$ to the thermodynamic tensions τ_a*. Were we content, as are most writers on thermodynamics, to accept the τ_a as the actual tensions in the material, there would be no problem, since the τ_a may always be calculated from $(247.8)_2$ or from $(251.4)_4$.

[2] Our presentation in this section and the next follows ECKART [1940, *7*]. Cf. also MEIXNER [1941, *2*, §§ 2–3] [1943, *3*, §§ 3–4]. Earlier authors were inclined to select one or another quantity bearing the dimension of [entropy]/[time] and on the basis of some physical argument call it the "irreversible" production of entropy; this arbitrary procedure has been criticized by DE GROOT [1952, *3*, §§ 1, 82]. However, the criticism is not applicable to the work of TOLMAN and FINE [1948, *31*], who in essence follow ECKART's procedure and on the basis of (257.3) call Δ the "irreversible" rate of production of entropy, thereafter deriving other equations in which Δ occurs, e.g.

$$\varrho \theta \dot{\eta} = h^p_{,p} - h^p (\log \theta)_{,p} + \theta \Delta + \varrho q.$$

Some alternative forms involving the coefficients (248.5) are discussed by HUNT [1955, *14*, § 1c, d].

[3] There is always a question as to how the total entropy of a system should be defined. The definition (257.1), asserting that what we call the total entropy of a body is the sum of the entropies of the several elements of mass, is unequivocal and is that customarily introduced in continuum mechanics. We do not attempt to decide whether this definition is consistent with others, such as those used in statistical mechanics.

where

$$\theta\Delta \equiv P_{\rm E} - P_{\rm I} + \frac{h^p \theta_{,p}}{\theta} = \varrho \sum_{a=1}^{t} \gamma_a \dot{v}_a + h^p (\log\theta)_{,p}, \tag{257.4}$$

γ_a being given by (256.10). Eq. (257.3) for ***production of total entropy*** is an equation of balance of the type (157.1).

Unlike our earlier examples of equations of balance, (257.3) is not set down by definition[1]. Rather, it is *derived* from the assumptions expressed by previous equations, and the quantities occurring in it have meanings in earlier associations. From (257.3) we see that total entropy may be regarded as flowing into a body through its boundary at the rate $-\boldsymbol{h}/\theta$, where $-\boldsymbol{h}$ is the influx of non-mechanical energy. The supply of entropy is $\Delta/\varrho + q/\theta$; in particular, by (257.4), a non-zero flow of non-mechanical energy creates entropy only when it is in the presence of a temperature gradient. Both the supply of non-mechanical energy and the excess of total working over recoverable working contribute to the total entropy in just the same way as to the specific entropy. In summary, *total entropy is changed by the same factors that change energy, with two modifications*:

1. *Entropy rate = (energy rate)/(temperature), and*

2. *In addition to the sources just stated, there is also an effective supply of energy of amount* $h^p(\log\theta)_{,p}$.

The supply of entropy Δ is given by the bilinear form (257.4). The terms occurring in this form are the building blocks of recent theories of particular thermodynamic phenomena[2]. It has grown customary to name the two terms in each summand a *thermodynamic force* or *affinity* and a corresponding *thermodynamic flux*. For example, we may set up the table:

Affinity	Corresponding flux
$(1/\theta)_{,k}$	$-h^k$
$\dot{v}_a$	$\varrho\gamma_a/\theta$

There seems to be no compelling reason for assigning any one term to one category or to the other, and usage varies[3]. In any case, the terms entering Δ are not uniquely determined, since, as was remarked in Sect. 157, in an equation of balance it is trivially possible to shift any term from the surface integral into the volume integral at will, and conversely, though not always explicitly, to shift any term from volume integral into the surface integral. Thus we are unable to see any physical significance in the interpretation of any one term in (257.3) unless accompanied by interpretation of all the other terms.

258. The entropy inequality[4]. It is a matter of experience that a substance at uniform temperature and free from sources of heat may consume mechanical work but cannot give it out. That is, whatever work is not recoverable is lost,

[1] A formally analogous treatment can be applied to the calculation of the rate of change of $\int_{\mathscr{V}} v_a\, d\mathfrak{M}$ but has no interest. Indeed, as is plain from the manner in which entropy was introduced in Sect. 246, its properties differ from those of the parameters v_a only in physical dimension.

[2] Cf. footnote 1, p. 618.

[3] Cf. the references cited in Sect. 259.

[4] The postulate (258.3), which sometimes shares with (247.2) the name *second law of thermodynamics*, for the case when $\boldsymbol{h} = 0$ and $q = 0$ is due to CLAUSIUS [1854, *1*, p. 152] [1862, *1*, § 1] [1865, *1*, §§ 1, 14–17]; the surface integral was added by DUHEM [1901, *7*, Part. I, Chap. 1, § 6].

not created. Similarly, in a body at rest and subject to no sources of heat, the flow of heat is from the hotter to the colder parts, not *vice versa*. The two observations, abstracted and generalized, may be put as follows:

$$\left.\begin{array}{ll} 1. & P_{\mathrm{E}} - P_{\mathrm{I}} \geqq 0 \text{ when } q = 0 \text{ and } \theta = \text{const}, \\ 2. & h^p \theta_{,p} \geqq 0 \text{ when } q = 0 \text{ and } P_{\mathrm{E}} - P_{\mathrm{I}} = 0. \end{array}\right\} \tag{258.1}$$

In both these cases, by (257.4) we conclude that

$$\theta \Delta \geqq 0. \tag{258.2}$$

When $\theta > 0$, by (257.3) we see that (258.2) is equivalent to

$$\dot{\mathsf{H}} - \oint_{\mathscr{S}} \frac{h^p \, da_p}{\theta} \geqq \int_{\mathscr{V}} \frac{\varrho q}{\theta} \, dv. \tag{258.3}$$

Guided by these special cases, we might set up (258.3), or the equivalent condition (258.2), as a general ***postulate of irreversibility***[1]. Unlike the previous assumptions of the field theories, it is an inequality rather than an equation. It asserts a *trend in time* for various processes.

The most familiar of these is an *adiabatic* process, defined by the condition that non-mechanical energy flow neither in nor out through the boundary, nor be created or destroyed within the body: $\boldsymbol{h} = 0$ on $\mathscr{S}$, $q = 0$ in $\mathscr{V}$. Then (258.3) yields $\dot{\mathsf{H}} \geqq 0$: *In an adiabatic process, the total entropy cannot decrease*[2].

If (258.3) is to hold for all bodies, it is equivalent to the local condition (258.2) restricting the sum of products of affinities by fluxes (cf. Sect. 257). Whether or not this condition may be broken down into a statement that separate parts, such as $h^p \theta_{,p}$, are to be *severally* non-negative depends on whether or not the body is susceptible of independent variation of the parameters v_a and θ or of sets of these parameters, and whether or not the partial sums selected are scalars under appropriate transformations[3].

---

[1] We are aware of the unsatisfactory nature of (258.3) in that $\boldsymbol{h}$ and q are not uniquely defined. However, since q is to include the possibility of arbitrarily assignable sources and sinks of heat, we cannot restrict its sign or its value.

[2] Notice that the specific entropy η is not necessarily non-decreasing at all places and times in an adiabatic process. Cf. MEISSNER [1938, *7*, § 11].

[3] Here we touch on a great mystery of the subject. Writers on irreversible thermodynamics appear to select partial sums at will and then demand that each one be non-negative. Certainly, however, an arbitrary choice is not justified: E.g., it is not necessary that $h^1 \theta_{,1} \geqq 0$, and it would make no sense to require it, since $h^1 \theta_{,1}$ is not scalar under co-ordinate transformations. In dealing with the more complicated situation to be considered in the next section, PRIGOGINE and MAZUR [1951, *21*, § 3d] demand that certain terms in Δ be separately non-negative, "tout couplage entre quantités de caractère tensoriel différent étant interdit ..." but the meaning of this assertion is not clear to us. Cf. also KIRKWOOD and CRAWFORD [1952, *12*, p. 1050]: "We must treat scalars, vectors and tensors separately, for entities of different tensorial character cannot interact (CURIE's theorem)." In the publication of CURIE [1894, *1*] sometimes cited in this connection we are unable to find anything relevant. Interactions between quantities of different tensorial orders are well known in the kinetic theory of gases and are illustrated in Sect. 307. Possibly what the writers on "irreversible thermodynamics" mean to describe is the separation of effects that follows by *linearization* of isotropic functions (cf. Sects. 293η and 307). Also we stumble again over the most serious gap in the fundamentals of thermodynamics, that the appropriate group of transformations of the thermodynamic variables and the invariance to be required are not known. Cf. footnote 2, p. 620.

If we raise (258.3) to the level of a general postulate of mechanics, then it is to be applied also to regions containing surfaces of discontinuity. At such a surface, provided $\partial(\varrho\eta)/\partial t$ and $\varrho q/\theta$ be bounded, it is equivalent to[1]

$$\varrho^{\pm} U^{\pm}[\eta] + \left[\frac{h_n}{\theta}\right] \geqq 0. \tag{258.4}$$

259. Production of entropy in a heterogeneous medium[2]. Confining attention to the non-polar case, we eliminate $\dot{\varepsilon}_{\mathrm{I}}$ between (243.9) and (255.15), obtaining

$$\varrho\theta\dot{\eta} = h^k_{*,k} + \sum_{\mathfrak{A}=1}^{\mathfrak{K}} (\varrho_{\mathfrak{A}}\mu_{\mathfrak{A}} u^k_{\mathfrak{A}})_{,k} - \sum_{\mathfrak{A}=1}^{\mathfrak{K}} \varrho_{\mathfrak{A}}\mu_{\mathfrak{A},k} u^k_{\mathfrak{A}} + D, \tag{259.1}$$

where

$$\left.\begin{aligned} h^k_* &= h^k_{\mathrm{I}} - \sum_{\mathfrak{A}=1}^{\mathfrak{K}} \varrho_{\mathfrak{A}}\varepsilon_{\mathfrak{A}} u^k_{\mathfrak{A}},\\ D &\equiv t^{km}_{\mathrm{I}} d_{km} + \pi d^k_k - \varrho\sum_{\mathfrak{b}=1}^{\mathfrak{t}} \sigma_{\mathfrak{b}}\dot{\omega}_{\mathfrak{b}} + \sum_{\mathfrak{A}=1}^{\mathfrak{K}} t^{km}_{\mathfrak{A}} u_{\mathfrak{A}k,m} - \\ &\quad - \varrho\sum_{\mathfrak{A}=1}^{\mathfrak{K}} \left[\hat{p}^k_{\mathfrak{A}} u_{\mathfrak{A}k} + \hat{c}_{\mathfrak{A}}(\mu_{\mathfrak{A}} + \tfrac{1}{2}u^2_{\mathfrak{A}})\right] + \varrho q_{\mathrm{I}}. \end{aligned}\right\} \tag{259.2}$$

Therefore

$$\varrho\dot{\eta} - \left(\frac{s^k}{\theta}\right)_{,k} = \Delta + \frac{\varrho q_{\mathrm{I}}}{\theta}, \tag{259.3}$$

---

[1] In the case when $\boldsymbol{h}=0$, this famous condition was introduced into gas dynamics by JOUGUET [1901, *9*] [1904, *3*, § 2] and ZEMPLÉN [1905, *8*] (cf. also HADAMARD [1905, *1*], RAYLEIGH [1910, *8*, pp. 590—591]), where it is used to prove that shocks of rarefaction are impossible, since it may be shown to follow from the conditions of stability (Sect. 265) and from 205.7) that sgn $[\eta]$ = sgn $[\varrho]$. For proof, see Sects. 55 to 56 of the article by SERRIN, Mathematical Principles of Classical Fluid Mechanics, this Encyclopedia, Vol. VIII/1.

[2] This subject has been discussed by numerous authors, e.g. ECKART [1940, *8*], MEIXNER [1943, *2*, § 2] [1943, *3*, § 4], PRIGOGINE and MAZUR [1951, *21*, § 3] [1951, *17*, § 2]. Our treatment follows TRUESDELL [1957, *16*, § 12].

As mentioned in Sect. 243, detailed comparison of our results with those of thermodynamic writers is not possible, since they employ an equation for balance of energy obtained by intuition rather than derivation. An exception is furnished by the treatment of HIRSCHFELDER, CURTISS and BIRD [1954, *9*, §§ 7.6a, 7.6b, and 11.1]; if we presume that their phenomenological variables are to be understood as including as special cases the corresponding quantities they define explicitly for the kinetic theory of gas mixtures, then our treatment and theirs are in entire agreement up to the point where they introduce a caloric equation of state for the mixture. Instead of our (255.1) they write an equation of similar form containing $\dot{\varepsilon}$ rather than $\dot{\varepsilon}_{\mathrm{I}}$ on the left-hand side. It seems to us that such an equation cannot be exact: the total kinetic energy of diffusion cannot be a function of static parameters alone, since any diffusion velocities, at a given place and instant, are compatible with any values of the local thermodynamic state. From our point of view, HIRSCHFELDER, CURTISS and BIRD neglect the quadratic term in (243.1) when they calculate (255.1), although they do not neglect it when deriving (243.7). This observation accounts entirely for the difference between their equation for production of total entropy and our Eq. (259.4).

Perhaps motivated by results in the approximate kinetic theory of diffusion, numerous writers on irreversible thermodynamics (e.g. PRIGOGINE [1949, *24*, § 1]) recommend use of a "reduced" heat flux which in our notation is given by

$$-\left[\boldsymbol{h} + \sum_{\mathfrak{A}=1}^{\mathfrak{K}} \chi_{\mathfrak{A}}\varrho_{\mathfrak{A}}\boldsymbol{u}_{\mathfrak{A}}\right].$$

From (243.2) and $(243.8)_2$ we see that this reduced heat flux is precisely our $-\boldsymbol{h}_{\mathrm{I}}$ if we assume (as in theories of diffusion) that $\boldsymbol{t}_{\mathfrak{A}} = -\pi_{\mathfrak{A}}\mathbf{1}$ and if we neglect the kinetic energy of diffusion.

where

$$\left.\begin{aligned}
s^k &\equiv h_*^k + \sum_{\mathfrak{A}=1}^{\mathfrak{K}} \varrho_{\mathfrak{A}} \mu_{\mathfrak{A}} u_{\mathfrak{A}}^k, \\
&= h_{\mathrm{I}}^k + \sum_{\mathfrak{A}=1}^{\mathfrak{K}} \varrho_{\mathfrak{A}} (\mu_{\mathfrak{A}} - \varepsilon_{\mathfrak{A}}) u_{\mathfrak{A}}^k, \\
\theta \Delta &\equiv h_*^k (\log\theta)_{,k} - \theta \sum_{\mathfrak{A}=1}^{\mathfrak{K}} \varrho_{\mathfrak{A}} (\mu_{\mathfrak{A}}/\theta)_{,k} u_{\mathfrak{A}}^k + D - \varrho q_{\mathrm{I}}, \\
&= \Big(h_{\mathrm{I}}^k - \sum_{\mathfrak{A}=1}^{\mathfrak{K}} \varrho_{\mathfrak{A}} \varepsilon_{\mathfrak{A}} u_{\mathfrak{A}}^k\Big) (\log\theta)_{,k} + t_{\mathrm{I}}^{km} d_{km} + \pi d_k^k - \varrho \sum_{\mathfrak{b}=1}^{\mathfrak{k}} \sigma_{\mathfrak{b}} \dot{\omega}_{\mathfrak{b}} + \\
&\quad + \sum_{\mathfrak{A}=1}^{\mathfrak{K}} \big[t_{\mathfrak{A}}^{km} u_{\mathfrak{A}k,m} - \varrho_{\mathfrak{A}} \theta (\mu_{\mathfrak{A}}/\theta)_{,k} u_{\mathfrak{A}}^k - \varrho \hat{p}_{\mathfrak{A}}^k u_{\mathfrak{A}k} - \varrho \hat{c}_{\mathfrak{A}} (\mu_{\mathfrak{A}} + \tfrac{1}{2} u_{\mathfrak{A}}^2)\big].
\end{aligned}\right\} \tag{259.4}$$

The form of Δ suggests the following possible division of the variables into affinities and fluxes:

Affinity	Corresponding flux
$(1/\theta)_{,k}$	$-\Big[h_{\mathrm{I}}^k - \sum_{\mathfrak{A}=1}^{\mathfrak{K}} \varrho_{\mathfrak{A}} \varepsilon_{\mathfrak{A}} u_{\mathfrak{A}}^k\Big]$
d_{km}	$\frac{1}{\theta}\,[t_{\mathrm{I}}^{km} + \pi g^{km}]$
$\dot{\omega}_{\mathfrak{b}}$	$-\varrho \sigma_{\mathfrak{b}}/\theta$
$u_{\mathfrak{A}k,m}$	$t_{\mathfrak{A}}^{km}/\theta$
$u_{\mathfrak{A}}^k$	$-[\varrho_{\mathfrak{A}} (\mu_{\mathfrak{A}}/\theta)_{,k} + \varrho \hat{p}_{\mathfrak{A}k}/\theta]$
$\varrho \hat{c}_{\mathfrak{A}}$	$-\frac{1}{\theta}\Big[\mu_{\mathfrak{A}} + \frac{1}{2} u_{\mathfrak{A}}^2\Big]$

To the doubts mentioned at the end of Sect. 257 must be added the remark that various alternative forms and regroupings of terms in (259.4) itself are obviously possible and lead to different selection of affinities and fluxes[1]. To mention the simplest of examples, if as in Sect. 257 we choose some of the parameters ω_a as the $x_{;\alpha}^k$, the second and third lines in the above table coalesce into the single line

$$\dot{\omega}_a \qquad \gamma_a,$$

as has already appeared in our treatment of a simple medium. Also, part of the flux on the first line, since it is proportional to $\boldsymbol{u}_{\mathfrak{A}}$, could equally well be combined with that on the next-to-last line.

For heterogeneous media the entropy inequality is assumed to hold in the form (258.3), or, equivalently, (258.2). It is customary, though there appears to be no solid reason for it, to infer that various sums occurring in Δ are separately

[1] We share the view of ECKART [1940, *8*] that the classification is arbitrary. However, most thermodynamic writers disagree with this view and with each other's choices. The point has been discussed from a physical standpoint by MEIXNER [1942, *9*, Zusatz bei der Korrektur] and by DE GROOT [1952, *3*, §§ 2, 9–13, 18]. MEIXNER [1943, *2*, § 3] has also investigated the invariance of the classification under linear transformation, but only in the case when the affinities and fluxes are assumed linearly related. Cf. also PRIGOGINE [1947, *12*, Chap. IX, § 2], DE GROOT [1952, *3*, §§ 29, 44, 52, 78]. MEIXNER and REIK in § 5 of their article, "Thermodynamik der irreversiblen Prozesse", this Encyclopedia, Vol. III/2, notice the infinitely many possible ways of rearranging and regrouping terms in (259.3); they suggest that it should be done in such a way as to render the source of entropy non-negative in all circumstances, and they assert that the only possibility of satisfying this requirement is given by the particular form they adopt.

non-negative. For example, ECKART[1] concluded that $\sum_{\mathfrak{A}=1}^{\mathfrak{K}} \hat{c}_{\mathfrak{A}} \mu_{\mathfrak{A}} \leqq 0$; by (159A.5), we have equivalently

$$\sum_{\mathfrak{A}=\mathfrak{P}+1}^{\mathfrak{K}} \hat{c}_{\mathfrak{A}} \mu_{\mathfrak{A}}^{*} \leqq 0. \tag{259.5}$$

In particular, if only one compound $\mathfrak{A}$ is being created, it is forming or dissociating according as its potential difference $\mu_{\mathfrak{A}}^{*}$ is negative or positive. Thus when but a single compound substance can result, the reaction proceeds so as to reduce its potential difference to zero. ECKART concluded also that from his special cases of (259.3) and (258.3) that

$$c_{\mathfrak{A}} \mu_{\mathfrak{A},k} u_{\mathfrak{A}}^{k} \leqq 0; \tag{259.6}$$

that is, the diffusion current for the substance $\mathfrak{A}$ carries it toward a region of lower chemical potential. Interesting as are such inequalities, it is not justified at present to regard them as derived from any general principle.

III. Equilibrium.

There are several different ideas of the meaning of "equilibrium"; when put into mathematical form, they lead to conditions that are not generally equivalent to one another. These conditions occur frequently in the literature of irreversible thermodynamics, but their interrelations have been studied only subject to the assumption of linear constitutive equations. We confine the following sections to the general definitions.

260. Differential condition for homogeneity. Thus far we have not needed to restrict the variables $v_{\mathfrak{a}}$ specifying the sub-state. Now, however, we presume that they are the densities of additive set functions. The set functions themselves are called *extensive variables*[2]. When the thermodynamic state is *homogeneous*, which here is taken to mean *uniform*, throughout a body, we have then

$$\eta = \mathsf{H}/\mathfrak{M}, \quad \varepsilon = \mathfrak{E}/\mathfrak{M}, \quad v_{\mathfrak{a}} = \Upsilon_{\mathfrak{a}}/\mathfrak{M}, \quad \text{where} \quad \Upsilon_{\mathfrak{a}} \equiv \int_{\mathscr{V}} v_{\mathfrak{a}}\, d\mathfrak{M},$$

and (246.1) may be written in the equivalent form

$$\mathfrak{E} = \mathfrak{M}\,\varepsilon\left(\frac{\mathsf{H}}{\mathfrak{M}}, \frac{\Upsilon}{\mathfrak{M}}\right) = \mathfrak{E}(\mathsf{H}, \Upsilon, \mathfrak{M}), \tag{260.1}$$

---

[1] [1940, *8*, esp. p. 924]. The quantity $A = \varrho \sum_{\mathfrak{A}=1}^{\mathfrak{K}} \hat{c}_{\mathfrak{A}} \mu_{\mathfrak{A}}$ was called the "affinity" or "chemical affinity" by DE DONDER and has been studied intensively in connection with chemical reactions; cf. DE DONDER [1927, *3*] [1929, *1*] [1932, *5*], DUPONT [1932, *6*]. If we use (159A.6), we may write the chemical affinity A in the form

$$A = \sum_{\mathfrak{a}=1}^{\mathfrak{l}} J_{\mathfrak{a}} A_{\mathfrak{a}}, \quad A_{\mathfrak{a}} = \sum_{\mathfrak{A}=1}^{\mathfrak{K}} N_{\mathfrak{A}\mathfrak{a}} \mu_{\mathfrak{A}},$$

where $\mathfrak{l}$ is the total number of chemical reactions that may occur. The quantity $A_{\mathfrak{a}}$ is the chemical affinity of the reaction $\mathfrak{a}$. Further developments follow by assuming that $A_{\mathfrak{a}}$ is an assignable function of the thermodynamic state, etc. The literature of this subject is obfuscated by the habit of writing all rates as time derivatives of otherwise undefined quantities.

[2] The distinction between extensive and intensive variables is due to MAXWELL [1876, *4*]; the former kind, which he called *magnitudes*, "represents a physical quantity, the value of which, for a material system, is the sum of its values for the parts of the system", while the latter "denote the intensity of certain physical properties of the substance". The defining property of intensive variables is expressed by (260.2) and hence is an alternative statement of the homogeneity of (260.1) when the densities of the extensive variables are uniform. In general, then, "intensive variable" is no more than another name for "field".

where the function on the right-hand side is homogeneous of degree 1 in all its arguments. Since by (247.1) we have

$$\frac{\partial \mathfrak{E}}{\partial \mathsf{H}} = \frac{\partial (\mathfrak{M}\,\varepsilon)}{\partial (\mathfrak{M}\,\eta)} = \frac{\partial \varepsilon}{\partial \eta} = \theta, \qquad \frac{\partial \mathfrak{E}}{\partial \varUpsilon_{\mathfrak{a}}} = \frac{\partial \varepsilon}{\partial v_{\mathfrak{a}}} = \tau_{\mathfrak{a}}, \tag{260.2}$$

the temperature and thermodynamic tensions are independent of the amount of material present. Such variables are called *intensive*. The Euler differential equation expressing the homogeneity of (260.1) is[1]

$$\left.\begin{aligned} \mathfrak{E} &= \theta\,\mathsf{H} + \sum_{1=\mathfrak{a}}^{\mathfrak{k}} \tau_{\mathfrak{a}}\,\varUpsilon_{\mathfrak{a}} + \mu\,\mathfrak{M}, \\ \varepsilon &= \theta\,\eta + \sum_{\mathfrak{a}=1}^{\mathfrak{k}} \tau_{\mathfrak{a}}\,v_{\mathfrak{a}} + \mu, \end{aligned}\right\} \tag{260.3}$$

where

$$\mu \equiv \frac{\partial \mathfrak{E}}{\partial \mathfrak{M}}. \tag{260.4}$$

By $(251.1)_4$ we may write (260.3) in the form

$$\zeta = \mu, \tag{260.5}$$

expressing succinctly the relation between the caloric equation of state for the density, ε, and that for the total energy, $\mathfrak{E}$ [cf. the analogous result $(255.14)_3$].

While (260.3) is the basic expression of homogeneity, other forms are more commonly encountered. First, from (260.2) and (260.3) we have

$$d\mathfrak{E} = \theta\,d\mathsf{H} + \sum_{\mathfrak{a}=1}^{\mathfrak{k}} \tau_{\mathfrak{a}}\,d\,\varUpsilon_{\mathfrak{a}} + \mu\,d\,\mathfrak{M} \tag{260.6}$$

as the counterpart of (247.2). If we subtract this from the differential of $(260.3)_1$, we obtain the *Gibbs-Duhem equation*[2]:

$$0 = \mathsf{H}\,d\theta + \sum_{\mathfrak{a}=1}^{\mathfrak{k}} \varUpsilon_{\mathfrak{a}}\,d\tau_{\mathfrak{a}} + \mathfrak{M}\,d\mu. \tag{260.7}$$

For an alternative derivation, we need only multiply $(251.3)_4$ by $\mathfrak{M}$ and then take note of (260.5).

In the usual applications the $v_{\mathfrak{a}}$ are supposed to consist in the $\mathfrak{K}$ masses $\mathfrak{M}_{\mathfrak{A}}$ of the constituents of a mixture and in a further set $\boldsymbol{\Omega}$ of extensive variables $\Omega_{\mathfrak{a}}$ independent of the partial masses. These quantities, in the case of a homogeneous system, are related to the densities used in Sect. 254 and 255 as follows:

$$\left.\begin{aligned} \mathfrak{M}_{\mathfrak{A}} &\equiv \int_{\mathscr{v}} c_{\mathfrak{A}}\,d\,\mathfrak{M} = c_{\mathfrak{A}}\,\mathfrak{M}, \quad \sum_{\mathfrak{A}=1}^{\mathfrak{K}} \mathfrak{M}_{\mathfrak{A}} = \mathfrak{M}, \\ \Omega_{\mathfrak{a}} &\equiv \int_{\mathscr{v}} \omega_{\mathfrak{a}}\,d\,\mathfrak{M} = \omega_{\mathfrak{a}}\,\mathfrak{M}. \end{aligned}\right\} \tag{260.8}$$

From (260.3) we have

$$\mathfrak{E}_{\mathrm{I}} = \theta\,\mathsf{H} + \sum_{\mathfrak{a}=1}^{\mathfrak{k}} \sigma_{\mathfrak{a}}\,\Omega_{\mathfrak{a}} + \sum_{\mathfrak{A}=1}^{\mathfrak{K}} \mu_{\mathfrak{A}}\,\mathfrak{M}_{\mathfrak{A}} + \mu\,\mathfrak{M}, \tag{260.9}$$

where

$$\left.\begin{aligned} \mu_{\mathfrak{A}} &= f + \left(\frac{\partial \mathfrak{E}_{\mathrm{I}}}{\partial \mathfrak{M}_{\mathfrak{A}}}\right)_{\mathsf{H},\,\mathfrak{M}^{\mathfrak{A}},\,\boldsymbol{\Omega},\,\mathfrak{M}} = f + \left(\frac{\partial \varepsilon_{\mathrm{I}}}{\partial c_{\mathfrak{A}}}\right)_{\eta,\,\mathbf{c}^{\mathfrak{A}},\,\boldsymbol{\omega},\,v} \\ \mu &= -f + \left(\frac{\partial \mathfrak{E}_{\mathrm{I}}}{\partial \mathfrak{M}}\right)_{\mathsf{H},\,\mathfrak{M},\,\boldsymbol{\Omega}}, \end{aligned}\right\} \tag{260.10}$$

---

[1] GIBBS [1875, *1*, Eq. (54)].

[2] GIBBS [1875, *1*, Eq. (97)], DUHEM [1886, *2*, Part II, Chap. II, Eq. (81)].

and where we have written $\mathfrak{E}_{\mathrm{I}}$ instead of $\mathfrak{E}$ as a reminder that a mixture is being considered (cf. Sects. 254 and 255). The indeterminacy represented by the occurrence of $f(\mathsf{H}, \mathfrak{M}, \boldsymbol{\Omega})$ is the same as that already encountered in (255.2). So as to eliminate this indeterminacy, we may agree to regard $\mathfrak{E}_{\mathrm{I}}$ as a function of $\mathfrak{M}_1, \mathfrak{M}_2, \ldots, \mathfrak{M}_{\mathfrak{K}}$ but not also of $\mathfrak{M}$; then we may set

$$\mu_{\mathfrak{A}}^{\star} \equiv \left(\frac{\partial \mathfrak{E}_{\mathrm{I}}}{\partial \mathfrak{M}_{\mathfrak{A}}}\right)_{\mathsf{H}, \mathfrak{M}^{\mathfrak{A}}, \boldsymbol{\Omega}}. \tag{260.11}$$

It is these chemical potentials $\mu_{\mathfrak{A}}^{\star}$ which generally are used in works on chemical equilibrium. They have the disadvantage of not being easily interpretable in terms of densities. However, for any f in (260.10), by (260.4), $(260.6)_3$, and (260.8) we obtain

$$\mu_{\mathfrak{A}}^{\star} = \mu_{\mathfrak{A}} + \mu, \tag{260.12}$$

while (260.6) and (260.7) become, respectively,

$$\left.\begin{aligned} d\,\mathfrak{E}_{\mathrm{I}} &= \theta\, d\,\mathsf{H} + \sum_{\mathfrak{a}=1}^{\mathfrak{k}} \sigma_{\mathfrak{a}}\, d\Omega_{\mathfrak{a}} + \sum_{\mathfrak{A}=1}^{\mathfrak{K}} \mu_{\mathfrak{A}}^{\star}\, d\,\mathfrak{M}_{\mathfrak{A}}, \\ 0 &= \mathsf{H}\, d\,\theta + \sum_{\mathfrak{a}=1}^{\mathfrak{k}} \Omega_{\mathfrak{a}}\, d\sigma_{\mathfrak{a}} + \sum_{\mathfrak{A}=1}^{\mathfrak{K}} \mathfrak{M}_{\mathfrak{A}}\, d\mu_{\mathfrak{A}}^{\star}. \end{aligned}\right\} \tag{260.13}$$

The latter relation may be written in terms of densities:

$$0 = \eta\, d\theta + \sum_{\mathfrak{a}=1}^{\mathfrak{k}} \omega_{\mathfrak{a}}\, d\sigma_{\mathfrak{a}} + \sum_{\mathfrak{A}=1}^{\mathfrak{K}} c_{\mathfrak{A}}\, d\mu_{\mathfrak{A}}^{\star}, \tag{260.14}$$

but it is important to note that the differentials of the $\mu_{\mathfrak{A}}^{\star}$, not generally the $\mu_{\mathfrak{A}}$, occur here. However, the custom among thermodynamical writers of expressing all relations in differential form can lead to confusion, since the differentials $d c_{\mathfrak{A}}$ are not independent. From (158.5) it follows that

$$\sum_{\mathfrak{A}=1}^{\mathfrak{K}} d c_{\mathfrak{A}} = 0. \tag{260.15}$$

By (260.12), then,

$$\sum_{\mathfrak{A}=1}^{\mathfrak{K}} \mu_{\mathfrak{A}}\, d c_{\mathfrak{A}} = \sum_{\mathfrak{A}=1}^{\mathfrak{K}} \mu_{\mathfrak{A}}^{\star}\, d c_{\mathfrak{A}}. \tag{260.16}$$

Thus the Gibbs equation $(260.13)_1$, for homogeneous conditions, may be written in *either* of the forms

$$\left.\begin{aligned} d\varepsilon_{\mathrm{I}} &= \theta\, d\eta + \sum_{\mathfrak{a}=1}^{\mathfrak{k}} \sigma_{\mathfrak{a}}\, d\omega_{\mathfrak{a}} + \sum_{\mathfrak{A}=1}^{\mathfrak{K}} \mu_{\mathfrak{A}}^{\star}\, d c_{\mathfrak{A}}, \\ &= \theta\, d\eta + \sum_{\mathfrak{a}=1}^{\mathfrak{k}} \sigma_{\mathfrak{a}}\, d\omega_{\mathfrak{a}} + \sum_{\mathfrak{A}=1}^{\mathfrak{K}} \mu_{\mathfrak{A}}\, d c_{\mathfrak{A}}, \end{aligned}\right\} \tag{260.17}$$

as well as in other forms, for variations in which the total mass is kept constant.

It is customary to take one of the parameters $\Omega_{\mathfrak{a}}$ as the volume, $\mathfrak{V}$. The Gibbs-Duhem relation (260.14) then assumes the form

$$0 = \eta\, d\theta - v\, d\pi + \sum_{\mathfrak{A}=1}^{\mathfrak{K}} c_{\mathfrak{A}}\, d\mu_{\mathfrak{A}}^{\star} + \sum_{\mathfrak{a}=1}^{\mathfrak{k}} \omega_{\mathfrak{a}}\, d\sigma_{\mathfrak{a}}. \tag{260.18}$$

While writers on "irreversible thermodynamics" sometimes use the relations of this section in problems concerning deformation, we are unable to find any solid ground for ascribing any relevance to them except in equilibrium. Consequences of (260.13) are

$$\sum_{\mathfrak{A}=1}^{\mathfrak{K}} c_{\mathfrak{A}} \left(\frac{\partial \mu_{\mathfrak{A}}^{\star}}{\partial c_{\mathfrak{B}}}\right)_{\theta,\pi,c^{\mathfrak{B}},\sigma} = 0, \qquad v_{\mathfrak{A}} = \left(\frac{\partial \mu_{\mathfrak{A}}^{\star}}{\partial \pi}\right)_{\theta,c,\sigma}. \tag{260.19}$$

Another condition of equilibrium, apparently independent of the foregoing, is the *principle of detailed balancing*, asserting that in thermostatic equilibrium each chemical reaction is individually in equilibrium. In terms of the reaction rates $J_{\mathfrak{a}}$ occurring in (159A.6), this postulate asserts that $\eta = \text{const}$ and $\boldsymbol{v} = \text{const}$ implies $J_{\mathfrak{a}} = 0$ for all $\mathfrak{a}$.

261. Mechanical equilibrium. By definition, a system is in mechanical equilibrium if $\ddot{\boldsymbol{x}} = 0$ in an inertial system. For simple media, no special thermodynamic consequences result. For heterogeneous media, however, by (215.4) and (205.2) we have

$$t^{km}{}_{,m} + \sum_{\mathfrak{A}=1}^{\mathfrak{K}} \varrho_{\mathfrak{A}} f_{\mathfrak{A}}^{k} = 0. \tag{261.1}$$

If now we add assumptions[1] sufficient to validate the Gibbs-Duhem equation (260.18), from its two consequences (260.19) we obtain

$$\left.\begin{aligned}
\sum_{\mathfrak{A}=1}^{\mathfrak{K}} \varrho_{\mathfrak{A}} (f_{\mathfrak{A}k} - \mu_{\mathfrak{A},k}^{\star}) &= - t_{k,m}^{m} - \varrho \sum_{\mathfrak{A}=1}^{\mathfrak{K}} c_{\mathfrak{A}} \Bigg[\sum_{\mathfrak{B}=1}^{\mathfrak{K}} \left(\frac{\partial \mu_{\mathfrak{A}}^{\star}}{\partial c_{\mathfrak{B}}}\right)_{\theta,\pi,c^{\mathfrak{B}},\sigma} c_{\mathfrak{B},k} + \\
&\quad + \left(\frac{\partial \mu_{\mathfrak{A}}^{\star}}{\partial \theta}\right)_{\pi,c,\sigma} \theta_{,k} + \left(\frac{\partial \mu_{\mathfrak{A}}^{\star}}{\partial \pi}\right)_{c,\theta,\sigma} \pi_{,k} + \sum_{\mathfrak{b}=1}^{\mathfrak{k}} \left(\frac{\partial \mu_{\mathfrak{A}}^{\star}}{\partial \sigma_{\mathfrak{b}}}\right)_{\theta,\pi,c,\sigma^{\mathfrak{b}}} \sigma_{\mathfrak{b},k} \Bigg], \\
&= - (t_{k}^{m} + \pi \delta_{k}^{m})_{,m} - \varrho \sum_{\mathfrak{A}=1}^{\mathfrak{K}} c_{\mathfrak{A}} \Bigg[\left(\frac{\partial \mu_{\mathfrak{A}}^{\star}}{\partial \theta}\right)_{\pi,c,\sigma} \theta_{,k} + \sum_{\mathfrak{b}=1}^{\mathfrak{k}} \left(\frac{\partial \mu_{\mathfrak{A}}^{\star}}{\partial \sigma_{\mathfrak{b}}}\right)_{\theta,\pi,c,\sigma^{\mathfrak{b}}} \sigma_{\mathfrak{b},k} \Bigg].
\end{aligned}\right\} \tag{261.2}$$

Further drastic assumptions are required in order to get a simple result. If $\theta = \text{const}$ and if the parameters $\boldsymbol{\omega}$ do not appear or if $\sigma_{\mathfrak{b}} = \text{const}$ for all $\mathfrak{b}$, and if further $t_{k}^{m} = -\pi \delta_{k}^{m}$, then (261.2) yields

$$\sum_{\mathfrak{A}=1}^{\mathfrak{K}} \varrho_{\mathfrak{A}} (f_{\mathfrak{A}k} - \mu_{\mathfrak{A},k}^{\star}) = 0. \tag{261.3}$$

In this case, by (158.19), we may apply (158.18) to show that

$$\sum_{\mathfrak{A}=1}^{\mathfrak{K}} \varrho_{\mathfrak{A}} (f_{\mathfrak{A}k} - \mu_{\mathfrak{A},k}^{\star}) \tilde{u}_{\mathfrak{A}}^{k} = \sum_{\mathfrak{A}=1}^{\mathfrak{K}} \varrho_{\mathfrak{A}} (f_{\mathfrak{A}k} - \mu_{\mathfrak{A},k}^{\star}) u_{\mathfrak{A}}^{k} \tag{261.4}$$

for all definitions of the diffusion velocity $\tilde{\boldsymbol{u}}_{\mathfrak{A}}$.

It is also possible to reorganize the terms in (261.2) in a fashion analogous to that used in Sect. 256.

According to ECKART[2], it should be possible to arrange the affinities and fluxes (Sect. 259) in such a way that the vanishing of the ones is a definition of equilibrium; the vanishing of the others should then be a provable criterion of equilibrium. This interesting idea does not serve to classify the quantities uniquely.

262. Variational condition of equilibrium. Toward the end of the nineteenth century arose a tendency to regard all gross phenomena as essentially thermodynamical, whence followed attempts to construct a system of energetics including mechanics as a subsidiary part[3]. The strength and the weakness of this

[1] The idea was suggested by PRIGOGINE [1947, *12*, Chap. IX, § 3]; we generalize the treatment of DE GROOT [1952, *3*, § 47].

[2] [1940, *8*, p. 270], "*D*-factors" and "*C*-factors".

[3] Cf. DUHEM [1911, *4*], JAUMANN [1911, *7*] [1918, *3*], LOHR [1917, *5*] [1924, *10*] [1940, *17*] [1948, *12*].

approach are illustrated in the special case when the substate $\boldsymbol{v}$ in (246.1) consists in the 9 deformation gradients $x^k_{;K}$ from a reference state $\boldsymbol{X}$. The objective is to give a *purely thermodynamic definition of stress*, without reference to mechanical considerations, and to derive CAUCHY'S laws (Sect. 205) as special cases of a general criterion of equilibrium[1].

This condition is bipartite. First, the principle of virtual work in the form (232.6) when $\ddot{\boldsymbol{x}}=0$—*i.e.*, $\mathfrak{W}=0$—along with the postulate

$$\mathfrak{W} = \delta \int_{\mathscr{V}} \varepsilon \, d\mathfrak{M}, \tag{262.1}$$

is assumed. Second, thermal equilibrium is defined by

$$\delta \mathsf{H} = 0. \tag{262.2}$$

All variations are assumed to respect the conservation of mass. From (262.1), (232.5), (247.2), and (156.1) follows

$$\left.\begin{aligned} \int_{\mathscr{V}} \tilde{\varrho}\,[\theta\,\delta\eta + \sum_{\mathfrak{a}=1}^{\mathfrak{k}} \tau_{\mathfrak{a}}\,\delta v_{\mathfrak{a}}]\,dV &= \oint_{\mathscr{S}} s^k\,\delta x_k\,da + \int_{\mathscr{V}} \tilde{\varrho}\,f^k\,\delta x_k\,dV, \\ \int_{\mathscr{V}} \tilde{\varrho}\,\delta\eta\,dV &= 0. \end{aligned}\right\} \tag{262.3}$$

The second condition is equivalent to the requirement that $\eta=\text{const}$ in the variations; in this case, or in the alternative case that (262.3) is used as a side condition and the variations are executed with $\theta=\text{const}$, the former condition becomes

$$\int_{\mathscr{V}} \tilde{\varrho} \sum_{\mathfrak{a}=1}^{\mathfrak{k}} \tau_{\mathfrak{a}}\,\delta v_{\mathfrak{a}}\,dV = \oint_{\mathscr{S}} s^k\,\delta x_k\,da + \int_{\mathscr{V}} \tilde{\varrho}\,f^k\,\delta x_k\,dV. \tag{262.4}$$

Further progress cannot be made unless we specify the relation between the $\delta v_{\mathfrak{a}}$ and the δx_k. If all variations are independent, we get

$$\tau_{\mathfrak{a}} = 0, \qquad s^k = 0, \qquad f^k = 0 \tag{262.5}$$

as the conditions of equilibrium for this trivial case. However, if the $v_{\mathfrak{a}}$ are in fact the $x^k_{;K}$, as we agreed to assume, then we write (262.3) in the form

$$\int_{\mathscr{V}} \tilde{\varrho}\left(\frac{\partial\varepsilon}{\partial x^k_{;K}}\right)\delta x^k_{;K}\,dV = \oint_{\mathscr{S}} s^k\,\delta x_k\,da + \int_{\mathscr{V}} \tilde{\varrho}\,f^k\,\delta x_k\,dV. \tag{262.6}$$

Since $\delta x^k_{;K} = (\delta x^k)_{;K}$, by GREEN'S transformation we thus obtain

$$0 = \oint_{\mathscr{S}} \left(s_k\,da - \tilde{\varrho}\,\frac{\partial\varepsilon}{\partial x^k_{;K}}\,dA_K\right)\delta x^k + \int_{\mathscr{V}} \left[\tilde{\varrho}\,f^k + \left(\tilde{\varrho}\,\frac{\partial\varepsilon}{\partial x^k_{;K}}\right)_{;K}\right]\delta x_k\,dV. \tag{262.7}$$

If there are no kinematical constraints, it follows that

$$\left.\begin{aligned} s_k\,da &= \tilde{\varrho}\,\frac{\partial\varepsilon}{\partial x^k_{;K}}\,dA_k \qquad \text{on } \mathscr{S} \\ 0 &= \left(\tilde{\varrho}\,\frac{\partial\varepsilon}{\partial x^k_{;K}}\right)_{;K} + \tilde{\varrho}\,f^k \quad \text{in } \mathscr{V}. \end{aligned}\right\} \tag{262.8}$$

---

[1] The analysis we present is due to GIBBS [1875, *1*, pp. 184–190]. The theory based on the more special assumptions $\varepsilon = \varepsilon(\boldsymbol{E}, \boldsymbol{X})$ was initiated by GREEN [1839, *1*, pp. 248–255] [1841, *2*, pp. 298–300] and developed by KIRCHHOFF [1850, *2*, § 1] [1852, *1*, pp. 770–772] and KELVIN [1855, *4*] [1856, *2*, Chaps. XIII, XIV] [1863, *2*, §§ 61–67] [1867, *3*, § 673] and App. C. §§ (c)–(d)]. Cf. Sect. 232. Further attempts along this line have never been successful. The usual approach is to assume ε depends on various variables at hand. For example, VAN MIEGHEM [1935, *9*, § 3] sets up the equation $\varepsilon = \varepsilon(\pi, \theta, A, \tilde{\boldsymbol{E}})$, where A is the affinity and $\tilde{\boldsymbol{E}}$ is defined by $(31.6)_2$, as supposedly appropriate for a viscous fluid.

By (210.5) and (210.8), the quantities defined by

$$F_k^{\ K} = \tilde{\varrho} \frac{\partial \varepsilon}{\partial x^k_{;K}} \tag{262.9}$$

in virtue of (262.1) and (262.2) satisfy the same equations as those imposed on PIOLA's double vector $T_k^{\ K}$ in virtue of the conditions of equilibrium in continuum mechanics. *Thus it is not inconsistent with mechanical principles if we set*

$$T_k^{\ K} = F_k^{\ K} \tag{262.10}$$

and replace the laws of statics by the conditions (262.1) *and* (262.2) *of thermodynamic equilibrium.* At bottom, this analysis rests on the same idea as that in Sect. 256A.

If there are additional parameters v_a beyond the $x^k_{;K}$, the corresponding τ_a must vanish in equilibrium, provided there are no constraints. The details here are akin to those in Sect. 256.

Despite the elegance of the foregoing analysis, it cannot be accepted. Indeed, from a *special* thermodynamic assumption, CAUCHY's first law has been derived. That CAUCHY's *second* law cannot be derived by such methods is plain from its general form (205.10) and from footnote 2, p. 596. In fact, in general the stress $\boldsymbol{t}$ derived from (262.10) is not symmetric unless some additional assumption, such as that ε depends on the $x^k_{;K}$ only through $\boldsymbol{E}$, is added.

But there is a deeper objection. The equations of mechanics describe a wider range of phenomena than do the principles of thermodynamics. No one will contest the principles of balance of mass, momentum, and energy, but the existence of a caloric equation of state is an assumption of a more special kind[1]. CAUCHY's laws are valid for all sorts of continuous media, but in essence the thermodynamic method just explained is restricted to perfectly elastic bodies[2]. If there are viscous or plastic stresses, they must be dragged in by extra assumptions having nothing to do with thermodynamic principles[3]. Finally, to obtain equations of motion it is customary to apply the Euler-D'Alembert principle to the equations of equilibrium, and this is at bottom no different from assuming the balance of momentum to start with.

263. Inequalities restricting the equations of state, I. "Absolute" temperature. Thus far in our formal structure the temperature θ has been taken as defined by $(247.1)_1$ with no other restriction than that the equations of state be such as to permit any number of differentiations and functional inversions. For the validity of one or two of the formulae it has been tacitly supposed that $\theta \neq 0$, and at one point in connection with the entropy inequality (258.3) we have assumed that $\theta > 0$. This last requirement is customarily imposed throughout the subject. When the caloric equation of state (246.1) is so restricted that

$$\theta > 0, \quad \inf \theta = 0 \tag{263.1}$$

---

[1] This is borne out also by general statistical mechanics, whence, as shown by NOLL [1955, *19*], the field Eqs. (156.5), (205.2), (205.11), and (241.4) follow in the greatest generality, but the existence of thermodynamical equations for cases other than equilibrium remains in doubt. In the kinetic theory of monatomic gases, there is a thermal equation of state in all circumstances, but a caloric equation of state is valid only in conditions sufficiently near to equilibrium.

[2] Within this limitation, COLEMAN and NOLL [1959, *3*] have replaced the analysis in this section by a rigorous development based upon a genuine minimum principle; they obtain not only the classical stress-strain relations but also full conditions of stability.

[3] Cf. the method of DUHEM [1901, *7*, Chap. I, §§ 1–5] [1903, *4* and *5*] [1904, *1*, Chap II. § V] [1911, *4*, Chap. XIV, § 3], who was a devotee of the thermodynamic approach. Only the details, not the principles, of more recent allegedly thermodynamic treatments are different.

for all allowable values of the thermodynamic state η, $\mathfrak{v}$, then θ is said to be an *absolute temperature*.

It does not seem to be possible within the concepts of thermodynamics to give any clear idea of what absolute temperature is[1]. We may, however, rephrase the assumption as follows: the caloric equation of state (246.1) is such that θ has *a finite lower bound*. By $(247.1)_1$ this is equivalent to

$$\theta_0 \equiv \inf \left(\frac{\partial \varepsilon}{\partial \eta}\right)_{\mathfrak{v}} > -\infty. \tag{263.2}$$

For a given function $\varepsilon(\eta, \mathfrak{v})$, put

$$\varepsilon' \equiv \varepsilon - \theta_0 \eta. \tag{263.3}$$

Then

$$\inf \left(\frac{\partial \varepsilon'}{\partial \eta}\right)_{\mathfrak{v}} = 0. \tag{263.4}$$

While, as stated in Sect. 245, the requirements of invariance to which the caloric equation of state is subject are not clear, it seems that ε' as an internal energy function is physically equivalent to ε; by (263.4), if θ is defined from ε' rather than from ε, we obtain (263.1). Thus for an internal energy function satisfying (263.2) it is possible to find a physically equivalent energy function such that the temperature is absolute.

The requirement (263.1) has a simple physical interpretation: Addition of heat to a body increases its internal energy. This is so because η, while itself not a measure of heat, is to be interpreted as a quantity which increases or decreases according as heat is being supplied or drawn off. Pursuing this same interpretation suggests also the additional restriction

$$\lim_{\eta \to \infty} \varepsilon = \infty: \tag{263.5}$$

If, at a fixed substate $\mathfrak{v}$, heat is added indefinitely, the internal energy also increases indefinitely.

264. Stability of equilibrium. At the conclusion of a paper on various forms of the "second law of thermodynamics", CLAUSIUS[2] asserted, „Die Energie der Welt ist konstant. Die Entropie der Welt strebt einem Maximum zu." GIBBS[3] replaced this claim of a trend in time by a *definition of thermostatic stability* for an isolated system:

$$(\Delta \mathsf{H})_{\mathfrak{E}} \leqq 0, \tag{264.1}$$

where inequality refers to *stable*, equality to *neutral* equilibrium. The symbol Δ indicates a finite increment, not necessarily a differential, consistent with whatever thermostatic constraints there may be beyond the assignment of a fixed energy $\mathfrak{E}$. GIBBS inferred that (264.1) is equivalent to

$$(\Delta \mathfrak{E})_{\mathsf{H}} \geqq 0: \tag{264.2}$$

The energy of an isolated system in stable equilibrium is the maximum compatible with its entropy. The argument of GIBBS rests upon a third but unstated idea,

[1] The traditional developments rest upon concepts taken from outside thermodynamics: either that temperature is to be identified, or at least connected, with the energy or the mean energy of a mechanical model, or that temperature obeys FOURIER's law of heat conduction (Sect. 296 below). That (263.1) amounts to a postulated restriction upon the caloric equation of state seems first to have been made clear by HELMHOLTZ [1882, *2*, § 1].

The criticism in the text, however, does not apply to the method of CARATHÉODORY; cf. the references cited in footnote 3, p. 617.

[2] [1865, *1*, § 17].

[3] [1875, *1*, pp. 56–57]. Earlier ideas of GIBBS on stability will be mentioned in the next section.

that it is always possible to produce changes such that $\Delta \mathsf{H}$ and $\Delta \mathfrak{E}$ are of one sign. The numerous applications of these criteria by GIBBS reflect but do not specify a broad concept of admissible thermostatic variation.

GIBBS drew many interesting conclusions from (264.1) and (264.2). E.g., from $(251.1)_2$ we have, when $\theta = \text{const}$,

$$(\Delta \Psi)_\theta = \Delta \mathfrak{E} - \theta \Delta \mathsf{H}, \tag{264.3}$$

where Ψ is the total free energy. From (264.1) and (264.2) we have then

$$(\Delta \Psi)_{\theta, \mathfrak{E}} \geqq 0, \quad (\Delta \Psi)_{\theta, \mathsf{H}} \geqq 0. \tag{264.4}$$

GIBBS[1] inferrred, more generally, that

$$(\Delta \Psi)_\theta \geqq 0: \tag{264.5}$$

In stable equilibrium, the free energy of an isolated system is the least possible among all states at the same temperature. GIBBS inferred also that

$$(\Delta \mathsf{Z})_{\theta, \boldsymbol{\tau}} \geqq 0, \tag{264.6}$$

where Z is the total thermodynamic potential analogous to $(251.6\text{A})_2$.

The arguments adduced by GIBBS are non-mathematical; in particular, he does not define "isolated" and does not specify precisely what variations are allowed[2].

There is a conceptually different approach, due, apparently, to PLANCK[3]. The differential equation (247.2), equivalent to the existence of a caloric equation of state (246.1), is replaced by a postulated differential inequality:

$$\theta\, d\eta \geqq d\varepsilon - \sum_{a=1}^{\mathfrak{k}} \tau_a\, d v_a, \tag{264.7}$$

where, of course, the differentials are no longer restricted to a path on the energy surface; in fact, they may be replaced by finite increments, it being understood that the values of θ and of τ_a are those for the state from which the increments are taken. As in Sect. 250, the quantities τ_a and θ are supposed given *a priori*, since (247.1) can no longer be used to define them. When equality holds in (264.7), we have the results deduced earlier in this chapter. But, more generally, (264.7) is to be regarded, not as a criterion for equilibrium but as a statement forbidding certain non-equilibrium states; sometimes taken as an expression of the entropy inequality, it bears no obvious relation to the form (258.3) adopted in this work. Neither is its relation to the Gibbsian variations clear, but it enables us to derive easily results of the type GIBBS obtained.

Specifically, we apply (264.7) to three kinds of differential changes:

1. $\theta = \text{const}$, $\boldsymbol{v} = \text{const}$; then

$$(d\psi)_{\theta, \boldsymbol{v}} = d\varepsilon - \theta\, d\eta \leqq \sum_{a=1}^{\mathfrak{k}} \tau_a\, d v_a = 0. \tag{264.8}$$

2. $\eta = \text{const}$, $\boldsymbol{\tau} = \text{const}$; then

$$(d\chi)_{\eta, \boldsymbol{\tau}} = d\varepsilon - \sum_{a=1}^{\mathfrak{k}} \tau_a\, d v_a \leqq \theta\, d\eta = 0. \tag{264.9}$$

3. $\theta = \text{const}$, $\boldsymbol{\tau} = \text{const}$; then

$$(d\zeta)_{\theta, \boldsymbol{\tau}} = d\varepsilon - \theta\, d\eta - \sum_{a=1}^{\mathfrak{k}} \tau_a\, d v_a \leqq 0. \tag{264.10}$$

---

[1] [1875, *1*, pp. 89–91].

[2] Cf. the criticism of DUHEM [1904, *2*].

[3] [1887, *3*, § I].

That is, in the three conditions stated, the only possible changes are such as to decrease ψ, χ, ζ, respectively. These results are suggestive of GIBBS' criteria (264.5) and (264.6). In addition to the difference of concepts used, we note especially that in (264.5) no assumption is made regarding the substate, while in (264.8) it is held constant.

In a work in progress, which they have kindly disclosed to us, COLEMAN and NOLL have taken up the ideas of GIBBS and replaced them by more concrete mathematical definitions. They restrict attention to the case of a fluid mixture as discussed in Sect. 255, the caloric equation of state being

$$\varepsilon = \varepsilon(\eta, v, c_1, \ldots, c_{\mathfrak{R}-1}). \tag{264.11}$$

This was the case to which GIBBS generally referred. In effect, COLEMAN and NOLL define the class of all possible variations of an isolated system as the set of all configurations leaving fixed the total mass and the positions of the particles on the boundary, while conserving mass in such chemical reactions as may take place. This definition they render mathematically explicit. They assume that (263.1) and (263.5) hold, these being restrictions upon the form of the function on the right-hand side of (246.11). They then replace GIBBS' criteria (264.1) and (264.2) by more general ones referring to stable equilibrium of an isolated system as a whole:

$$\mathsf{H}|_{\mathfrak{E}=\text{const}} = \max, \tag{264.12}$$

$$\mathfrak{E}|_{\mathsf{H}=\text{const}} = \min, \tag{264.13}$$

where H and $\mathfrak{E}$ are the total entropy and total energy of the system, and where the variations considered are those defined above, subject to the additional restriction indicated in each case. COLEMAN and NOLL then give a strict proof that (264.12) and (264.13) are equivalent to each other.

265. Inequalities restricting the equations of state, II. Consequences of the stability of equilibrium. GIBBS saw that if the equations of state are such as to ensure that every equilibrium state is stable, they must satisfy certain inequalities, and these he obtained. Because of the vagueness of his definition of stability, however, his arguments are not convincing.

In the work mentioned in the foregoing section, COLEMAN and NOLL establish the validity of GIBBS' inequalities by use of the precisely formulated criterion (264.13). Let us write ω for the local thermostatic state $\eta, v, c_1, c_2, \ldots, c_{\mathfrak{R}-1}$, so that the specific energy is given by $\varepsilon(\omega)$. Restricting attention to global states that are homogeneous (Sect. 260), which were those GIBBS customarily considered, COLEMAN and NOLL prove the following elegant theorem: In order for the homogeneous state characterized by ω to be one of stable equilibrium, it is necessary and sufficient that

$$\varepsilon(\omega) < a'\,\varepsilon(\omega') + a''\,\varepsilon(\omega'') \tag{265.1}$$

for all pairs (a', a'') and (ω', ω'') such that

$$\left.\begin{aligned} a' + a'' = 1, \quad 0 < a' < 1, \quad 0 < a'' < 1, \\ \omega = a'\,\omega' + a''\,\omega'', \end{aligned}\right\} \tag{265.2}$$

ω' and ω'' being states. The condition (265.1) asserts a kind of convexity of the energy function. While ω is a homogeneous state, ω' and ω'' are not; the energy at the homogeneous state ω is thus compared with those of all non-homogeneous states that may arise from internal fluctuations.

From this theorem, the inequalities of GIBBS follow as special cases. To see this, consider a function $f(\omega)$ of a single variable which is subject to a condition of the type (265.1), and take $a' = a'' = \frac{1}{2}$, $\omega' = \omega - h$, $\omega'' = \omega + h$. Then it follows from (265.1) that

$$f(\omega + h) - f(\omega) - [f(\omega) - f(\omega - h)] > 0. \tag{265.3}$$

When f is twice differentiable, (265.3) implies that $f''(\omega) \geqq 0$. Conversely, the condition $f''(x) > 0$ is sufficient, but not necessary, for convexity. Likewise the more general condition (265.1) implies that

$$\left\| \frac{\partial \sigma_{\mathfrak{a}}}{\partial \omega_{\mathfrak{b}}} \right\| \quad \text{is positive semi-definite,} \tag{265.4}$$

where we have put

$$\sigma_{\mathfrak{a}} \equiv \frac{\partial \varepsilon}{\partial \omega_{\mathfrak{a}}}. \tag{265.5}$$

This is the result inferred by GIBBS[1].

Suppose that for a given mixture there exists a convex set of local states $\boldsymbol{\omega}$ such that the homogeneous state corresponding to each is a state of stable equilibrium. It follows then from (265.1) and (265.2) that for all states $\boldsymbol{\omega}'$, $\boldsymbol{\omega}''$ in the set we have

$$\left.\begin{aligned} &\varepsilon(\boldsymbol{\omega}'') - \varepsilon(\boldsymbol{\omega}') - (\eta'' - \eta')\,\varepsilon_\eta(\boldsymbol{\omega}') - \\ &\quad - (v'' - v')\,\varepsilon_v(\boldsymbol{\omega}') - \sum_{\mathfrak{A}=1}^{\mathfrak{K}-1} (c''_{\mathfrak{A}} - c'_{\mathfrak{A}})\,\varepsilon_{c_{\mathfrak{A}}}(\boldsymbol{\omega}') > 0, \end{aligned}\right\} \tag{265.5 A}$$

[1] [1875, 1, pp. 111–112]. In his earlier study of simple fluids [1873, 2, p. 29] GIBBS had written, "The condition of stability requires that, when the pressure is constant, the temperature shall increase with the heat received,—therefore, with the entropy. ...It also requires that, when there is no transmission of heat, the pressure should increase as the volume diminishes ..." I.e., the "condition of stability" is

$$\left(\frac{\partial \theta}{\partial \eta}\right)_\pi \geqq 0, \qquad \left(\frac{\partial \pi}{\partial v}\right)_\eta \leqq 0$$

(equivalently,

$$\varkappa_\pi \geqq 0, \qquad \phi \geqq 0),$$

neutral stability being defined as the case when equality holds instead of inequality. GIBBS then discussed the implications of these inequalities upon the topological properties of the Euler diagram. He remarked that $(\partial \pi / \partial \eta)_v$ can have any sign, and discussed "the effect of heat as increasing or diminishing the pressure when the volume is maintained constant".

The only significant subsequent work we have found in the general theory of thermodynamic stability is a long study by DUHEM [1893, 2, Part I, Chaps. IV and V]. DUHEM considered a system held at fixed temperature subject to constant hydrostatic pressure on its boundaries. His definition of stability, accordingly, differs from that of GIBBS, as does his choice of variables. COLEMAN and NOLL show, however, that the conditions on the free energy function which DUHEM derived as *necessary* for the stability of a homogeneous global state, including $(\partial \pi / \partial \varrho)_{\theta, \boldsymbol{c}} > 0$, are consequences of (265.1). DUHEM's methods did not enable him to derive (265.8); nevertheless, he knew that this equation holds in various applications, so he adopted it, calling it "the postulate of HELMHOLTZ".

DUHEM criticized certain inequalities of GIBBS similar to $(265.7)_3$ [1875, 1, Ineqs. (167), (168), (169)]: "une des rares inexactitudes"; also [1893, 2, Part I, Chap. V, § V] [1894, 2, Chap. IV, especially §§ 5 and 7] [1898, 2] [1911, 8, Chap. XVI, § 9]. Such results are very sensitive to the choice of variables, and GIBBS is obscure in the detail of argument as well as reluctant to state just what is being varied. COLEMAN and NOLL have straightened the whole matter out; they find that DUHEM's criticism is well taken if his interpretation of what GIBBS meant to say is correct, but they find another interpretation in which GIBBS' inequalities are correct.

In summary, there is no conflict between the results of GIBBS and those of DUHEM, which complement one another; all their results and many more of like nature are stated unequivocally and derived precisely in the forthcoming work of COLEMAN and NOLL.

where the subscripts to ε denote partial derivatives. It is clear that (265.5 A), for the special variables considered here, is a mathematical statement of the physical notions behind (264.7). Thus the approach based upon (264.7), when made clear, amounts to an asserting as a postulate the main result GIBBS sought to derive.

For the case of a fluid obeying a caloric equation of state of the form (264.11) —the case to which COLEMAN and NOLL's analysis applies—we have[1] $\sigma = (\theta, -\pi, \mu_1^{\backprime}, \mu_2^{\backprime}, \ldots, \mu_{\mathfrak{R}-1}^{\backprime})$, and the matrix occurring in (265.4) assumes the form

$$\left\| \begin{array}{ccccc} \dfrac{\partial\theta}{\partial\eta} & \dfrac{\partial\theta}{\partial v} & \dfrac{\partial\theta}{\partial c_1} & \cdots & \dfrac{\partial\theta}{\partial c_{\mathfrak{R}-1}} \\ -\dfrac{\partial\pi}{\partial\eta} & -\dfrac{\partial\pi}{\partial v} & -\dfrac{\partial\pi}{\partial c_1} & \cdots & -\dfrac{\partial\pi}{\partial c_{\mathfrak{R}-1}} \\ \dfrac{\partial\mu_1^{\backprime}}{\partial\eta} & \cdots & \cdots & \cdots & \cdots \\ \vdots & \cdots & \cdots & \cdots & \dfrac{\partial\mu_{\mathfrak{R}-1}^{\backprime}}{\partial c_{\mathfrak{R}-1}} \end{array} \right\|, \tag{265.6}$$

where all quantities are taken as functions of $\eta, v, c_1, \ldots, c_{\mathfrak{R}-1}$. In order that this matrix be positive semi-definite, it is necessary that each principal minor be non-negative. In particular,

$$\left(\frac{\partial\theta}{\partial\eta}\right)_{v,\mathbf{c}} \geqq 0, \quad \left(\frac{\partial\pi}{\partial\varrho}\right)_{\eta,\mathbf{c}} \geqq 0, \quad \left(\frac{\partial\mu_{\mathfrak{a}}^{\backprime}}{\partial c_{\mathfrak{a}}}\right)_{\eta,v,\mathbf{c}^{\mathfrak{a}}} \geqq 0. \tag{265.7}$$

The first of these, by (249.1), may be written in the form[2]

$$\varkappa_v \geqq 0; \tag{265.8}$$

that is, in order to increase the temperature of a body held at constant substate, it is necessary to supply heat. The second of the inequalities (265.7) may be interpreted as a statement that the Laplacian speed of sound is real [cf. footnote 1, p. 631 and (297.13)]. But also all principal minors of (265.6) must be non-negative; in particular,

$$\left.\frac{\partial(\pi,\theta)}{\partial(\eta,v)}\right|_{\mathbf{c}} \geqq 0. \tag{265.9}$$

Up to now we have identified the variables constituting the substate, conformably to the caution stated in Sect. 246. For more general thermostatic systems, it is clear that (265.4) cannot generally hold. For if $\omega_{\mathfrak{b}}$ is a suitable parameter for describing the state of a system, so also are $-\omega_{\mathfrak{b}}$ and $1/\omega_{\mathfrak{b}}$; use of $-\omega_{\mathfrak{b}}$, leaving all other $\omega_{\mathfrak{a}}$ unchanged, would change the sign of $\partial\sigma_{\mathfrak{a}}/\partial\omega_{\mathfrak{b}}$, and use of $1/\omega_{\mathfrak{b}}$ would change the sign of $\partial\sigma_{\mathfrak{b}}/\partial\omega_{\mathfrak{b}}$. However, it seems likely that for *some* admissible choice of the $\omega_{\mathfrak{b}}$ in any system, (265.4) will result, and its simplest consequence,

$$\frac{\partial\sigma_{\mathfrak{a}}}{\partial\omega_{\mathfrak{a}}} \geqq 0, \tag{265.10}$$

has often been taken as a condition for stability.

---

[1] Note that $\mu_{\mathfrak{A}}^{\backprime} = \mu_{\mathfrak{A}} - \mu_{\mathfrak{R}}$, $\mathfrak{A} = 1, 2, \ldots, \mathfrak{R} - 1$, where $\mu_{\mathfrak{A}}$ is the chemical potential defined by $(255.2)_2$. In virtue of (158.5) we have

$$\sum_{\mathfrak{A}=1}^{\mathfrak{R}} \mu_{\mathfrak{A}} \dot{c}_{\mathfrak{A}} = \sum_{\mathfrak{A}=1}^{\mathfrak{R}-1} \mu_{\mathfrak{A}}^{\backprime} \dot{c}_{\mathfrak{A}}.$$

[2] GIBBS [1875, *1*, Eq. (166)], HELMHOLTZ [1882, *2*, § 1].

We now translate the condition of stability (265.1) for homogeneous states of isolated systems into forms more accessible to experimental test. We have seen that (265.4) is a necessary but not sufficient condition for stability; if we replace "positive semi-definite" in (265.4) by "positive definite", we obtain a condition which is sufficient but not necessary. We agree to consider this stronger condition only[1]. Also we shall rule out the possibility of thermodynamic degeneracy (Sect. 253), and we exclude from consideration states such that $\varpi_{\mathfrak{a}}=0$ or $\alpha_{\mathfrak{a}}=0$, for any $\mathfrak{a}$. Also we adopt (263.1). A system satisfying these conditions will be called *ultrastable.*

By a well known theorem on matrices, from (252.11) we see that a necessary and sufficient condition for ultrastability is that $\|\phi_{\mathfrak{a}\mathfrak{b}}\|$ be positive definite and $\partial(\tau, \theta)/\partial(v, \eta) > 0$. From the former statement it follows that $\phi > 0$. By $(252.22)_3$ we conclude that the latter statement, when the former is satisfied, is equivalent to $\varkappa_\tau > 0$. Hence we have the

First necessary and sufficient condition for ultrastability:

$$\|\phi_{\mathfrak{a}\mathfrak{b}}\| \text{ is positive definite, } \varkappa_\tau > 0. \tag{265.11}$$

Proceeding in just the same way from $(252.11)_2$ and using $(242.22)_5$, yields the

Second necessary and sufficient condition for ultrastability[2]:

$$\|\xi_{\mathfrak{a}\mathfrak{b}}\| \text{ is positive definite, } \varkappa_v > 0. \tag{265.12}$$

In this condition, $\xi_{\mathfrak{a}\mathfrak{b}}$ may be replaced by $\nu_{\mathfrak{a}\mathfrak{b}}$.

From (265.11) and (252.14) or (252.16), or, alternatively, from (265.12) and (252.15), it follows that $\gamma > 1$. Conversely, if $\gamma > 1$ and $\|\phi_{\mathfrak{a}\mathfrak{b}}\|$ is positive definite, from (252.14) or (252.16) it follows that $\varkappa_\tau > 0$, whence by (265.11) the equilibrium is ultrastable. Thus we have proved the

Third necessary and sufficient condition for ultrastability:

$$\|\phi_{\mathfrak{a}\mathfrak{b}}\| \text{ is positive definite, } \gamma > 1. \tag{265.13}$$

In just the same way, from (265.12) and (252.15) we prove the

Fourth necessary and sufficient condition for ultrastability:

$$\|\xi_{\mathfrak{a}\mathfrak{b}}\| \text{ is positive definite, } \gamma > 1. \tag{265.14}$$

The stability of thermodynamically degenerate substances requires a separate analysis. The particular kind of degeneracy which defines an ideal material (Sect. 253) does not affect the arguments given above, so the conditions of stability (265.11) to (265.14) remain applicable. For piezotropic substances, by (253.9) and (253.10) it is easy to see directly from (252.11) that a necessary and sufficient condition for ultrastability is

$$\|\nu_{\mathfrak{a}\mathfrak{b}}\| \text{ is positive definite, } \varkappa_v > 0 \text{ or } \varkappa_\tau > 0. \tag{265.15}$$

We do not attempt to find conditions appropriate to other degenerate materials or to investigate the conditions of neutral stability.

---

[1] This was done by SAUREL [1904, *6* and *7*], who phrased GIBBS' considerations more formally in terms of differentials. Our text, in effect, replaces the Gibbs-Saurel arguments by more efficient and precise mathematical analysis. It appears that the subtle logical difference between stability and ultrastability was recognized by GIBBS, though he did not emphasize it; in the work of DUHEM, it is completely obscured, and a reader with modern standards of rigor will need to apply some labor if he is to disentangle DUHEM's arguments.

[2] The condition $(265.12)_1$, when $\mathfrak{f} = 1$, is traditionally used in analysis of the stability of a VAN DER WAALS gas, as was mentioned in Sect. 252.

An important consequence of (265.12) has been derived by EPSTEIN[1]. Consider two tensions, $\sigma_\mathfrak{a}$ and $\sigma_\mathfrak{b}$; then $\sigma_\mathfrak{a} = \sigma_\mathfrak{a}(\omega_\mathfrak{a}, \omega_\mathfrak{b}, \omega^{\mathfrak{a}\mathfrak{b}})$, $\sigma_\mathfrak{b} = \sigma_\mathfrak{b}(\omega_\mathfrak{a}, \omega_\mathfrak{b}, \omega^{\mathfrak{a}\mathfrak{b}})$, where $\omega^{\mathfrak{a}\mathfrak{b}}$ stands for all $\omega_\mathfrak{a}$'s except $\omega_\mathfrak{a}$ and $\omega_\mathfrak{b}$. The Maxwell relations (252.1) may all be expressed in the form

$$\left(\frac{\partial \sigma_\mathfrak{a}}{\partial \omega_\mathfrak{b}}\right)_{\omega^\mathfrak{b}} = \left(\frac{\partial \sigma_\mathfrak{b}}{\partial \omega_\mathfrak{a}}\right)_{\omega^\mathfrak{a}}. \tag{265.16}$$

Inverting the relation giving $\sigma_\mathfrak{b}$ yields $\omega_\mathfrak{b} = \omega_\mathfrak{b}(\sigma_\mathfrak{b}, \omega_\mathfrak{a}, \omega^{\mathfrak{a}\mathfrak{b}})$ so that $\sigma_\mathfrak{a} = \sigma_\mathfrak{a}(\omega_\mathfrak{a}, \omega_\mathfrak{b}(\sigma_\mathfrak{b}, \omega_\mathfrak{a}, \omega^{\mathfrak{a}\mathfrak{b}}), \omega^{\mathfrak{a}\mathfrak{b}})$. Differentiating this relation yields

$$\left(\frac{\partial \sigma_\mathfrak{a}}{\partial \omega_\mathfrak{a}}\right)_{\sigma_\mathfrak{b},\, \omega^{\mathfrak{a}\mathfrak{b}}} = \left(\frac{\partial \sigma_\mathfrak{a}}{\partial \omega_\mathfrak{a}}\right)_{\omega^\mathfrak{a}} + \left(\frac{\partial \sigma_\mathfrak{a}}{\partial \omega_\mathfrak{b}}\right)_{\omega^\mathfrak{b}} \left(\frac{\partial \omega_\mathfrak{b}}{\partial \omega_\mathfrak{a}}\right)_{\sigma_\mathfrak{b},\, \omega^{\mathfrak{a}\mathfrak{b}}}. \tag{265.17}$$

But

$$\left.\begin{aligned} 0 &= \left(\frac{\partial \omega_\mathfrak{b}}{\partial \omega_\mathfrak{a}}\right)_{\omega^{\mathfrak{a}\mathfrak{b}}} = \left(\frac{\partial \omega_\mathfrak{b}}{\partial \omega_\mathfrak{a}}\right)_{\sigma_\mathfrak{b},\, \omega^{\mathfrak{a}\mathfrak{b}}} + \left(\frac{\partial \omega_\mathfrak{b}}{\partial \sigma_\mathfrak{b}}\right)_{\omega^\mathfrak{b}} \left(\frac{\partial \sigma_\mathfrak{b}}{\partial \omega_\mathfrak{a}}\right)_{\omega^\mathfrak{a}}, \\ &= \left(\frac{\partial \omega_\mathfrak{b}}{\partial \omega_\mathfrak{a}}\right)_{\sigma_\mathfrak{b},\, \omega^{\mathfrak{a}\mathfrak{b}}} + \left(\frac{\partial \omega_\mathfrak{b}}{\partial \sigma_\mathfrak{b}}\right)_{\omega^\mathfrak{b}} \left(\frac{\partial \sigma_\mathfrak{a}}{\partial \omega_\mathfrak{b}}\right)_{\omega^\mathfrak{b}}, \end{aligned}\right\} \tag{265.18}$$

where the second step follows by (265.16). Substituting this result into (265.17) yields the identity

$$\left(\frac{\partial \sigma_\mathfrak{a}}{\partial \omega_\mathfrak{a}}\right)_{\sigma_\mathfrak{b},\, \omega^{\mathfrak{a}\mathfrak{b}}} = \left(\frac{\partial \sigma_\mathfrak{a}}{\partial \omega_\mathfrak{a}}\right)_{\omega^\mathfrak{a}} - \left[\left(\frac{\partial \sigma_\mathfrak{a}}{\partial \omega_\mathfrak{b}}\right)_{\omega^\mathfrak{b}}\right]^2 \left(\frac{\partial \omega_\mathfrak{b}}{\partial \sigma_\mathfrak{b}}\right)_{\omega^\mathfrak{b}}. \tag{265.19}$$

By (265.12) we see that the term subtracted is non-negative; hence

$$\left(\frac{\partial \sigma_\mathfrak{a}}{\partial \omega_\mathfrak{a}}\right)_{\sigma_\mathfrak{b},\, \omega^{\mathfrak{a}\mathfrak{b}}} \leqq \left(\frac{\partial \sigma_\mathfrak{a}}{\partial \omega_\mathfrak{a}}\right)_{\omega^\mathfrak{a}}; \tag{265.20}$$

equivalently,

$$\left(\frac{\partial \omega_\mathfrak{a}}{\partial \sigma_\mathfrak{a}}\right)_{\sigma_\mathfrak{b},\, \omega^{\mathfrak{a}\mathfrak{b}}} \geqq \left(\frac{\partial \omega_\mathfrak{a}}{\partial \sigma_\mathfrak{a}}\right)_{\omega^\mathfrak{a}}. \tag{265.21}$$

The foregoing analysis is due to EPSTEIN, who calls (265.20) and (265.21) the *restricted Le Chatelier-Braun principle*. For the case of a simple fluid, the relation just derived is equivalent to $\gamma \geqq 1$.

Whether all inequalities called "conditions of stability" in the literature are consequences of GIBBS' condition (265.4) is not certain.

Further inequalities to be satisfied by the equations of state have been discussed[2], but not definitively.

It is to be noted that equations of state need not satisfy the conditions of stability for all values of the thermostatic state. Rather, the conditions serve to distinguish stable states from unstable ones. In a theory where mechanical phenomena are of primary interest, it may be natural to seek and impose a require-

---

[1] [1937, *1*, § 143].

[2] Cf. WEYL [1949, *38*, § 3], who treats only the case when $\mathfrak{f} = 1$, and who proposes also the postulate $\left(\frac{\partial^2 v}{\partial \tau^2}\right)_\eta > 0$, i.e., $\left(\frac{\partial \phi}{\partial \tau}\right)_\eta < 0$, and derives from it some further inequalities. WEYL notes also that from (252.3) it follows that $\left(\frac{\partial \eta}{\partial \tau}\right)_\eta$ and $\left(\frac{\partial \eta}{\partial v}\right)_\tau$ cannot vanish simultaneously; hence (265.19) implies that $\left(\frac{\partial \eta}{\partial \tau}\right)_v$ and $\left(\frac{\partial \eta}{\partial v}\right)_\tau$ are of one sign. COURANT and FRIEDRICHS [1948, *8*, § 2] propose the stronger inequality $\left(\frac{\partial^2 \tau}{\partial \varrho^2}\right)_\eta \leqq 0$. For the effect of these inequalities in gas dynamics, see § 37 and § 56 of SERRIN's article, Mathematical Principles of Classical Fluid Mechanics, This Encyclopedia, Vol. VIII/1.

ment of universal stability, but in a theory aiming to determine criteria of stability of equilibrium it is more natural to include the theoretical possibility of unstable states[1].

F. Charge and Magnetic Flux.

I. Introduction.

266. The scope of this chapter. Classical mechanics is founded on the principles of conservation or balance of mass, momentum, moment of momentum, and energy. Alongside these mechanical principles we now set up two further principles of conservation as a basis for the theory of electromagnetism. These are, the *conservation of charge* and the *conservation of magnetic flux*. Subchapter II formulates these principles and deduces and interprets some of their consequences. In Subchapter III we introduce an additional postulate, the *aether relations*, and in Subchapter IV we consider the mechanics of the electromagnetic and charge-current fields. The sequence of hypotheses and the order of logical development that we adopt here depart from the traditional treatments. A principal objective is to isolate those aspects of the theory which are independent of the assigned geometry of space-time from those whose formulation and interpretation depend on or imply a particular space-time geometry[2]. For example, we regard the conservation of charge as a physical or intuitive concept logically independent of the concepts of rigid rods, uniform clocks, and inertial frames, and we have chosen to express this law in a mathematical form likewise independent of the representation of these extraneous entities.

The development of electricity and magnetism is too closely connected with special cases and special phenomena for detailed historical references as in the previous chapters to be practicable here. The central importance of MAXWELL'S work is well known; in FARADAY and KELVIN he had major predecessors, and the classical theory is in part the creation of his successors, HERTZ and LORENTZ. An excellent history has been written by WHITTAKER[3].

267. Antisymmetric tensors. Since the only quantities occurring in the conservation laws of electromagnetic theory are antisymmetric tensors, we introduce some relevant specific terminology and notation.

Let

$$x^{r'} = x^{r'}(\boldsymbol{x}) \tag{267.1}$$

denote an element of the group of analytic transformations of the co-ordinates of an n-dimensional space. Let

$$U' = U^{-1}\, U \tag{267.2}$$

denote a transformation of the *unit*[4] U. As U runs over the real numbers, (267.2) generates the *group of unit transformations* on the unit U.

[1] E.g., in gas dynamics the van der Waals equation may be used for the entire density range only so long as the temperature exceeds the critical temperature, while in the theory of liquefaction it is useful to consider it also below this temperature.

[2] Our development and ordering is similar to that of KOTTLER [1922, *4*]. For a history of the mathematics and physics leading to KOTTLER'S formulation of the basic equations of electromagnetic theory, cf. WHITTAKER [1953, *35*, p. 192]. This earlier work includes that of HARGREAVES [1908, *5*], BATEMAN [1910, *1*], and MURNAGHAN [1921, *4*]. Cf. also the remarks of WEYL [1921, *6*, § 17], [1950, *35*, § 17]. KOTTLER'S ideas were taken up by the Dutch school of geometers and mathematical physicists and culminated in the series of papers by VAN DANTZIG [1934, *11*, ¶ 12] [1937, *10* and *11*]. Cf. also SCHOUTEN and HAANTJES [1934, *8*]. Shortcomings of the metric viewpoint even in strictly mechanical situations are emphasized by KRÖNER [1960, *3*, § 18].

[3] [1951, *39*] [1953, *35*].

[4] See Sect. App. 9 for a discussion of units and unit transformations.

A tensor field under the group of analytic co-ordinate transformations having absolute dimension $[U]$ is a set of functions of the co-ordinates $\boldsymbol{x}$ whose law of transformation has the general form

$$f^{m'n'\ldots}_{r'\ldots}(\boldsymbol{x}', U') = |(\boldsymbol{x}'/\boldsymbol{x})|^{-w} \operatorname{sgn}(\boldsymbol{x}'/\boldsymbol{x})^{p}\, U \frac{\partial x^{m'}}{\partial x^{m}} \frac{\partial x^{n'}}{\partial x^{n}} \cdots \frac{\partial x^{r}}{\partial x^{r'}}, \cdots f^{mn\ldots}_{r\ldots}(\boldsymbol{x}, U), \quad (267.3)$$

where $(\boldsymbol{x}'/\boldsymbol{x})$ denotes the Jacobian of the co-ordinate transformation. If $p=w=0$, $\boldsymbol{f}$ is called an *absolute tensor*. If $p=0$, $w \neq 0$, $\boldsymbol{f}$ is called a *relative tensor of weight* w. If $p=1$, $w=0$, $\boldsymbol{f}$ is called an *axial tensor*, and if $p=1$, $w \neq 0$, $\boldsymbol{f}$ is called an *axial relative tensor of weight* w.

A covariant or contravariant completely antisymmetric tensor of rank k will be called a *k-vector*. We can restrict the value of k to be less than or equal to n since k-vectors such that $k>n$ vanish identically. 0-vectors are called *scalars* and 1-vectors are called *vectors*.

We shall be concerned primarily with the following special types of k-vectors:

1. Absolute covariant or contravariant k-vectors.
2. Contravariant relative k-vectors of weight $+1$.
3. Covariant relative k-vectors of weight -1.
4. Covariant and contravariant axial k-vectors.
5. Contravariant axial relative k-vectors of weight $+1$.
6. Covariant axial relative k-vectors of weight -1.

Tensors of types 2 and 3 will be called *k-vector densities*. Tensors of types 5 and 6 will be called *axial k-vector densities*.

If the group of co-ordinate transformations consists solely of unimodular transformations, $(\boldsymbol{x}'/\boldsymbol{x})=+1$, then the transformation laws of absolute k-vectors, axial k-vectors, k-vector densities and axial k-vector densities coalesce. If the group consists solely of transformations with positive Jacobian, then the transformation laws of absolute k-vectors and axial k-vectors coalesce, as do the transformation laws of k-vector densities and axial k-vector densities.

In the following definitions, $\boldsymbol{F}$ and $\boldsymbol{Y}$ stand for k-vectors; they may be absolute, axial, densities, or axial densities, but their variance is specified by their indices. The *dot product* of a contravariant k-vector $\boldsymbol{F}$ and a covariant m-vector $\boldsymbol{Y}$, $k \leq m$, is defined by (cf. Sect. App. 3)

$$(\boldsymbol{F}\cdot\boldsymbol{Y})_{r_1 r_2 \ldots r_{m-k}} \equiv \frac{1}{k!} F^{s_1 s_2 \ldots s_k}\, Y_{s_1 s_2 \ldots s_k r_1 r_2 \ldots r_{m-k}} \quad (267.4)$$

$$(\boldsymbol{Y}\cdot\boldsymbol{F})_{r_1 r_2 \ldots r_{m-k}} \equiv \frac{1}{k!} Y_{r_1 r_2 \ldots r_{m-k} s_1 s_2 \ldots s_k}\, F^{s_1 s_2 \ldots s_k}. \quad (267.5)$$

Let $\varepsilon^{r_1 r_2 \ldots r_n}$ and $\varepsilon_{r_1 r_2 \ldots r_n}$ denote the permutation symbols such that $\varepsilon^{12\ldots n} = \varepsilon_{12\ldots n} = +1$; these symbols define axial n-vector densities. The *duals*, dual $\boldsymbol{F}$ and dual $\boldsymbol{Y}$, of covariant and contravariant k-vectors $\boldsymbol{F}$ and $\boldsymbol{Y}$ are defined by

$$(\text{dual}\,\boldsymbol{F})^{r_1 r_2 \ldots r_{n-k}} \equiv \frac{1}{k!} \varepsilon^{r_1 r_2 \ldots r_{n-k} s_1 \ldots s_k} F_{s_1 s_2 \ldots s_k}, \quad \text{or} \quad \text{dual}\,\boldsymbol{F} = \boldsymbol{\varepsilon}\cdot\boldsymbol{F}, \quad (267.6)$$

$$(\text{dual}\,\boldsymbol{Y})_{r_1 r_2 \ldots r_{n-k}} \equiv \frac{1}{k!} Y^{s_1 s_2 \ldots s_k}\, \varepsilon_{s_1 s_2 \ldots s_k r_1 \ldots r_{n-k}}, \quad \text{or} \quad \text{dual}\,\boldsymbol{Y} = \boldsymbol{Y}\cdot\boldsymbol{\varepsilon}. \quad (267.7)$$

These definitions imply that for any type of k-vector we have

$$\text{dual dual}\,\boldsymbol{F} = \boldsymbol{F}. \quad (267.8)$$

The cross product of a covariant (contravariant) k-vector $\boldsymbol{F}$ and a covariant (contravariant) m-vector $\boldsymbol{Y}$, $k+m\leqq n$, is defined by

$$(\boldsymbol{F}\times\boldsymbol{Y})^{r_1 r_2\ldots r_{n-k-m}} \equiv \frac{1}{k!\,m!}\,\varepsilon^{r_1 r_2\ldots r_{n-k-m} s_1\ldots s_k t_1\ldots t_m} F_{s_1\ldots s_k} Y_{t_1\ldots t_m} \qquad (267.9)$$

$$(\boldsymbol{F}\times\boldsymbol{Y})_{r_1 r_2\ldots r_{n-k-m}} \equiv \frac{1}{k!\,m!}\,F^{s_1\ldots s_k} Y^{t_1\ldots t_m}\,\varepsilon_{s_1\ldots s_k t_1\ldots t_m r_1 r_2\ldots r_{n-k-m}}\,. \qquad (267.10)$$

These definitions generalize that for the cross product of a vector and a tensor given in Sect. App. 3. We have the useful identities relating the cross product, dot product, and dual:

$$\boldsymbol{F}\times\boldsymbol{Y} = (\operatorname{dual}\boldsymbol{Y})\cdot\boldsymbol{F}, \qquad (\boldsymbol{F} \text{ and } \boldsymbol{Y} \text{ covariant}), \qquad (267.11)$$

$$\boldsymbol{F}\times\boldsymbol{Y} = \boldsymbol{Y}\cdot\operatorname{dual}\boldsymbol{F}, \qquad (\boldsymbol{F} \text{ and } \boldsymbol{Y} \text{ contravariant}). \qquad (267.12)$$

A constant tensor field whose components have the same values in all co-ordinate systems related by a group of transformations $\mathcal{G}$ is called a *constant invariant tensor of the group* $\mathcal{G}$. The axial n-vector densities $\varepsilon^{r_1 r_2\ldots r_n}$ and $\varepsilon_{r_1 r_2\ldots r_n}$ are constant invariant tensors of the full group of general analytic transformations. The mixed Kronecker deltas $\delta^{r_1 r_2\ldots r_k}_{s_1 s_2\ldots s_k}$ are examples of absolute constant invariant tensors of the group of analytic transformations. The covariant Kronecker delta δ_{rs} is a constant invariant tensor of the orthogonal group. Since the fundamental axial n-vector densities are constant invariant tensors of the group of general analytic transformations, the identity (267.8) establishes a one-to-one correspondence between k-vectors and their dual $(n-k)$-vectors which is *invariant* under the group of general analytic co-ordinate transformations. This implies that the transformation laws of a k-vector and its $(n-k)$-vector dual are indistinguishable. For example, suppose that $f_1, f_2, \ldots, f_6$ is a set of six functions of four co-ordinates whose transformation law is suitably defined by setting

$$f_1 = a_{12},\; f_2 = a_{13},\; f_3 = a_{14},\; f_4 = a_{23},\; f_5 = a_{24},\; f_6 = a_{34},$$

where $\boldsymbol{a}$ is an absolute covariant 2-vector. The same transformation law for the f's is obtained by setting

$$f_1 = (\operatorname{dual}\boldsymbol{a})^{34},\quad f_2 = (\operatorname{dual}\boldsymbol{a})^{42},\quad f_3 = (\operatorname{dual}\boldsymbol{a})^{23},$$
$$f_4 = (\operatorname{dual}\boldsymbol{a})^{14},\quad f_5 = (\operatorname{dual}\boldsymbol{a})^{31},\quad f_6 = (\operatorname{dual}\boldsymbol{a})^{12}.$$

The point we make is analogous to the better known fact that, under the orthogonal group, the transformation laws of covariant and contravariant tensors are indistinguishable. The similarity of the two situations can be made even more apparent by noting that a contravariant index may be raised or lowered by the Kronecker deltas δ_{rs} or δ^{rs} and that this process of association is invariant under the orthogonal group because the covariant and contravariant Kronecker deltas are constant invariant tensors of that group.

268. Invariant integral and differential equations independent of a metric or connection. Let ∂_r stand for $\partial/\partial x^r$. If $\boldsymbol{F}$ is an absolute or axial k-vector field, the quantities defined by

$$(\operatorname{rot}\boldsymbol{F})_{r s_1 s_2\ldots s_k} = (k+1)\,\partial_{[r} F_{s_1 s_2\ldots s_k]} \qquad (268.1)$$

transform as the components of an absolute or axial $(k+1)$-vector field under general transformations of the co-ordinates. Similarly, if $\boldsymbol{Y}$ is a contravariant k-vector density or axial k-vector density, the quantities defined by

$$(\operatorname{div}\boldsymbol{Y})^{r_1 r_2\ldots r_{k-1}} = \partial_s Y^{r_1 r_2\ldots r_{k-1} s} \qquad (268.2)$$

transform as the components of a $(k-1)$-vector density or an axial $(k-1)$-vector density. The $(k+1)$-vector field defined in (268.1) is called the *natural rotation* of the field $\boldsymbol{F}$, and the $(k-1)$-vector field defined in (268.2) is called the *natural divergence* of the field $\boldsymbol{Y}$. Note that the natural rotation is defined only for covariant absolute or axial k-vectors and not for covariant k-vector densities or contravariant k-vectors of any type. Similarly, the natural divergence is defined only for contravariant densities or axial densities. The dual of the natural rotation of a covariant k-vector is a contravariant $(n-k-1)$-vector called the *curl* of $\boldsymbol{F}$.

$$\operatorname{curl} \boldsymbol{F} = \operatorname{dual} \operatorname{rot} \boldsymbol{F}. \tag{268.3}$$

We have the following identity relating the divergence, curl, and dual of a k-vector:

$$\operatorname{curl} \boldsymbol{F} = \operatorname{div} \operatorname{dual} \boldsymbol{F}. \tag{268.4}$$

Consider the oriented k-dimensional surfaces in an n-dimensional space admitting a parametric representation

$$x^r = x^r(u^1, u^2, \ldots, u^k) \tag{268.5}$$

by piecewise continuously differentiable functions of the parameters u^a, $a=1, 2, \ldots, k$. We denote such a surface (hypersurface) by $\mathscr{S}_k$. If the parameters $\boldsymbol{u}$ are transformed by continuously differentiable one-to-one parameter transformations with positive Jacobian

$$u^{a'} = u^{a'}(\boldsymbol{u}), \qquad (\boldsymbol{u}'/\boldsymbol{u}) > 0, \tag{268.6}$$

we obtain another admissible parametrization of the surface $\mathscr{S}_t$ given by

$$x^r = x^r(\boldsymbol{u}') = x^r(\boldsymbol{u}(\boldsymbol{u}')). \tag{268.7}$$

A *k-dimensional circuit* is an $\mathscr{S}_k$ topologically equivalent to the complete boundary of a $(k+1)$-dimensional interval.

Let $\mathfrak{M}[\mathscr{S}_k, U]$ denote a quantity defined for every $\mathscr{S}_k$. We may think of $\mathfrak{M}[\mathscr{S}_k, U]$ as a physical quantity having the dimension $[U]$ such that for every k-dimensional set $\mathscr{S}_k$ of events or points in space a value is assigned to it in principle. Thus, for example, we may think of $\mathfrak{M}$ as the total charge in a given spatial region or as the total charge which has passed through a 2-dimensional surface in space in a given interval of time. In the general case, we assume that $\mathfrak{M}$ is an *additive set function*. More specifically, we shall assume that $\mathfrak{M}$ is expressible in the form

$$\mathfrak{M}[\mathscr{S}_k, U] = \int_{\mathscr{S}_k} m\left(\boldsymbol{x}(\boldsymbol{u}), \frac{\partial x^r}{\partial u^a}\right) du^1\, du^2 \ldots du^k, \tag{268.8}$$

where the integrand is a polynomial in the vectors $\dfrac{\partial x^r}{\partial u^a}$.

Theorem: *If $\mathfrak{M}[\mathscr{S}_k, U]$ has the transformation law*

$$\mathfrak{M}'[\mathscr{S}_k, U'] = U\,\mathfrak{M}[\mathscr{S}_k, U] \tag{268.9}$$

under independent transformations of the unit U, general analytic transformations of the co-ordinates $\boldsymbol{x}$, and transformations of the parameters $\boldsymbol{u}$, the set function $\mathfrak{M}[\mathscr{S}_k, U]$ is expressible in the form

$$\mathfrak{M} = \frac{1}{k!}\int m_{r_1 r_2 \ldots r_k}\, d\mathscr{S}_k^{r_1 r_2 \ldots r_k} = \int \boldsymbol{m} \cdot d\mathscr{S}_k, \tag{268.10}$$

where $\boldsymbol{m}$ *is an absolute covariant k-vector field, of absolute dimension* $[U]$, *independent of the vectors* $\frac{\partial x^r}{\partial u^a}$, *and where* $d\mathscr{S}_k$ *is the absolute contravariant k-vector defined by*

$$d\mathscr{S}_k^{r_1 r_2 \ldots r_k} \equiv k! \frac{\partial x^{[r_1}}{\partial u^1} \frac{\partial x^{r_2}}{\partial u^2} \cdots \frac{\partial x^{r_k]}}{\partial u^k} du^1 du^2 \ldots du^k. \tag{268.11}$$

Proof: First consider the invariance of $\mathfrak{M}$ under the subgroup of unimodular parameter transformations: $(\boldsymbol{u}'/\boldsymbol{u}) = +1$. Under parameter transformations with the co-ordinates held fixed, the quantities $\partial x^r/\partial u^a$ transform as a set of n absolute covariant vectors in a k-dimensional space. A known theorem of classical invariant theory[1] states that every invariant polynomial function of a set of n covariant vectors in k dimensions: $(n \geqq k)$ under the group of unimodular transformations is reducible to a polynomial in the $\binom{n}{k}$ determinants of the vectors taken k at a time. In the present application of this theorem we conclude that the integrand of (268.8) must reduce to a polynomial in the $\binom{n}{k}$ variables

$$D^{r_1 r_2 \ldots r_k} = k! \frac{\partial x^{[r_1}}{\partial u^1} \frac{\partial x^{r_2}}{\partial u^2} \cdots \frac{\partial x^{r_k]}}{\partial u^k}. \tag{268.12}$$

The coefficients in the polynomial are at most functions of the co-ordinates $x^r(\boldsymbol{u})$. Under parameter transformations with Jacobian not equal to 1, the variables $\boldsymbol{D}$ of (268.12) transform as relative scalars of weight 1, while the coefficients of these quantities transform as absolute scalars. The invariance of $\mathfrak{M}$ under this larger group allows us to conclude that the integrand must reduce to a linear homogeneous function of the variables $\boldsymbol{D}$. Hence we can write $\mathfrak{M}$ in the form (268.10). The invariance of $\mathfrak{M}$ under general analytic transformations of the co-ordinates and the quotient rule of tensor algebra allow us finally to conclude that the coefficients $m_{r_1 r_2 \ldots r_k}$ must transform as a covariant tensor. Only the antisymmetric part of the tensor of coefficients contributes to the transvected sum in (268.10); hence, there is no loss in generality by assuming that $\boldsymbol{m}$ is a k-vector. On transforming the unit U, we see that the absolute dimension of $\boldsymbol{m}$ must be $[U]$ provided we assume that the absolute dimension of $d\mathscr{S}_k$ is $[1]$. As explained in Sect. App. 9, co-ordinates and parameters will always be assigned the absolute dimension $[1]$. The physical dimension of $d\mathscr{S}_k$ may differ from $[1]$, and the physical dimension of $\boldsymbol{m}$ may differ from the absolute dimension of $\mathfrak{M}$; but this problem will be treated in some detail later in Sect. 277.

Corollary: *If* $\mathfrak{M}[\mathscr{S}_k, U]$ *is an axial scalar with the transformation law*

$$\mathfrak{M}[\mathscr{S}_k, U'] = \operatorname{sgn}(\boldsymbol{x}'/\boldsymbol{x})\, U\, \mathfrak{M}[\mathscr{S}_k, U], \tag{268.13}$$

then it is expressible in the form

$$\mathfrak{M}[\mathscr{S}_k, U] = \int \boldsymbol{m} \cdot d\mathscr{S}_k, \tag{268.14}$$

where $\boldsymbol{m}(\boldsymbol{x}(\boldsymbol{u}))$ *is an axial k-vector.*

The set function $\mathfrak{M}[\mathscr{S}_k, U]$ can be written in the dual form

$$\mathfrak{M}[\mathscr{S}_k, U] = (-1)^{k(n-k)} \int (\operatorname{dual} \boldsymbol{m}) \cdot d\hat{\mathscr{S}}_k, \tag{268.15}$$

where $d\hat{\mathscr{S}}_k \equiv \operatorname{dual} d\mathscr{S}_k$. In an odd-dimensional space, the factor $(-1)^{k(n-k)} = +1$ for every value of k; however, in even-dimensional spaces, this factor alternates in sign as k runs over the integers.

[1] WEYL [1946, *10*, p. 45].

Let $\mathcal{S}_k$ be a circuit given by $x^r = x^r(u^1, u^2, \ldots, u^k)$, and let it form the complete boundary of $\mathcal{S}_{k+1}$, given by $x^r = x^r(v^1, v^2, \ldots, v^{k+1})$; if the vectors $w^r, \frac{\partial x^r}{\partial u^1} \frac{\partial x^r}{\partial u^2} \cdots \frac{\partial x^r}{\partial u^k}$ and the vectors $\frac{\partial x^r}{\partial v^1}, \frac{\partial x^r}{\partial v^2} \cdots \frac{\partial x^r}{\partial v^{k+1}}$ have the same orientation, where $\boldsymbol{w}$ is a vector directed out of $\mathcal{S}_{k+1}$, we have (Sect. App. 21)

$$\oint \boldsymbol{m} \cdot d\mathcal{S}_k = \int \operatorname{rot} \boldsymbol{m} \cdot d\mathcal{S}_{k+1}, \tag{268.16}$$

where $\boldsymbol{m}$ is a continuously differentiable absolute or axial covariant k-vector field in $\mathcal{S}_{k+1}$. The dual form of this fundamental integral identity is[1]

$$(-)^{n+1} \oint (\operatorname{dual} \boldsymbol{m}) \cdot d\hat{\mathcal{S}}_k = \int \operatorname{div} \operatorname{dual} \boldsymbol{m} \cdot d\hat{\mathcal{S}}_{k+1} \tag{268.17}$$

Substituting from the identity (268.4) we can write (268.17) in the alternative form

$$(-)^{n+1} \oint (\operatorname{dual} \boldsymbol{m}) \cdot d\hat{\mathcal{S}}_k = \int \operatorname{curl} \boldsymbol{m} \cdot d\hat{\mathcal{S}}_{k+1}. \tag{268.18}$$

269. Conservative k-vector fields. In mechanics, if we are given a force field $\boldsymbol{f}$ such that

$$\oint f_r \, dx^r = 0 \tag{269.1}$$

for every circuit, the field $\boldsymbol{f}$ is said to be *conservative*. Vector fields satisfying (269.1) play an important role in many physical theories. In other contexts, vector fields satisfying (269.1) are called *lamellar* (see Sect. App. 33). Here in this chapter we prefer to use the name "conservative" owing to the connection we shall establish between the conservation laws of charge and magnetic flux and k-vector fields satisfying equations bearing a formal resemblance to (269.1). If a vector field satisfies (269.1), there exists an infinity of scalar fields h such that

$$h(\boldsymbol{x}_2) - h(\boldsymbol{x}_1) = \int_{\boldsymbol{x}_1}^{\boldsymbol{x}_2} \boldsymbol{f} \cdot d\boldsymbol{x}, \tag{269.2}$$

where $\boldsymbol{x}_1$ and $\boldsymbol{x}_2$ are the end points of the curve $\boldsymbol{x}(u)$. At a point where $\boldsymbol{f}$ is continuous, we have

$$f_r = \partial_r h. \tag{269.3}$$

A scalar field h with these properties is called a *potential* of the conservative field $\boldsymbol{f}$. The potential h is not uniquely determined by the field $\boldsymbol{f}$ and these requirements. If h is any one field satisfying (269.2), then so also is the field h' given by

$$h' = h + \text{const}. \tag{269.4}$$

We shall now extend the above definitions to k-vector fields in n dimensions.

Let $\boldsymbol{F}$ denote an absolute or axial covariant k-vector field and $\boldsymbol{Y}$ an absolute or axial covariant $(k+1)$-vector field such that

$$\oint \boldsymbol{F} \cdot d\mathcal{S}_k = \int \boldsymbol{Y} \cdot d\mathcal{S}_{k+1} \tag{269.5}$$

for every k-dimensional circuit. We then call $\boldsymbol{Y}$ the *source* of $\boldsymbol{F}$. A source-free k-vector field $\boldsymbol{F}$, i.e., one for which $\boldsymbol{Y}=0$, is called a *conservative k-vector field.* The integral theorem (268.16) states that if $\boldsymbol{F}$ is continuously differentiable, its

[1] The factor $(-)^{n+1}$ appears here and not in the corresponding formula of SCHOUTEN [1954, *21*, Chap. II, Eq. (8.14)] because our definitions (267.6) (267.7) of the dual tensors differ slightly from his [ibid., Chap. I, Eq. (7.15)]. If we had followed SCHOUTEN's definitions, we should have had to insert a factor $(-)^{n+1}$ before the right-hand side of the identity (268.4).

source $\boldsymbol{Y}$ is given by

$$\boldsymbol{Y} = \operatorname{rot} \boldsymbol{F}. \tag{269.6}$$

Now let $\boldsymbol{F}$ denote a conservative k-vector field so that

$$\oint \boldsymbol{F} \cdot d\mathscr{S}_k = 0. \tag{269.7}$$

If $\boldsymbol{F}$ has continuous derivatives of all orders up to $p > 0$, there exists at least one $(k-1)$-vector field $\boldsymbol{K}$ having continuous derivatives of all orders up to $p+1$ such that

$$\int \boldsymbol{F} \cdot d\mathscr{S}_k = \oint \boldsymbol{K} \cdot d\mathscr{S}_{k-1} \tag{269.8}$$

for every $\mathscr{S}_k$, where $\mathscr{S}_{k-1}$ is the complete boundary of $\mathscr{S}_k$.[1] A field $\boldsymbol{K}$ satisfying (269.8) will be called a *potential* of the conservative field $\boldsymbol{F}$. The field $\boldsymbol{K}$ is not uniquely determined by the field $\boldsymbol{F}$ and (269.8). Given any potential field $\boldsymbol{K}$ of $\boldsymbol{F}$, we obtain any other possible potential by adding to $\boldsymbol{K}$ an appropriate conservative $(k-1)$-vector field. For example, if $\boldsymbol{G}$ is any twice differentiable $(k-2)$-vector field, then $\boldsymbol{K}'$ given by

$$\boldsymbol{K}' = \boldsymbol{K} + \operatorname{rot} \boldsymbol{G} \tag{269.9}$$

is an alternative potential of the field $\boldsymbol{F}$. The group of transformations relating the potentials of a given conservative field will be called the *group of potential transformations*.

II. The conservation of charge and magnetic flux.

270. The electromagnetic and charge-current fields and the Maxwell-Bateman laws. *Space-time* is a 4-dimensional space. A point in this space will be called an *event*, whose four real co-ordinates we denote by x^Ω, $\Omega = 1, 2, 3, 4$[2]. Quantities transforming as a tensor under the group of general analytic transformations of the $\boldsymbol{x}$ will be called *world tensors* (cf. Sect. 152).

We assign an additive set function $\mathfrak{C}[\mathscr{S}_3, \mathsf{Q}]$ called the *charge* to every $\mathscr{S}_3$ in space-time. We assume that the charge is expressible in the form

$$\mathfrak{C}[\mathscr{S}_3, \mathsf{Q}] = \int \boldsymbol{\sigma} \cdot d\hat{\mathscr{S}}_3 = \int \sigma^\Omega \, d\hat{\mathscr{S}}_{3\Omega}, \tag{270.1}$$

where $\boldsymbol{\sigma}$ is the *charge-current field* and where Q is the *unit of charge*. It follows from (268.15) that we can also write $\mathfrak{C}[\mathscr{S}_3, \mathsf{Q}]$ in the dual form

$$\mathfrak{C}[\mathscr{S}_3, \mathsf{Q}] = -\int (\operatorname{dual} \boldsymbol{\sigma}) \cdot d\mathscr{S}_3. \tag{270.2}$$

The charge $\mathfrak{C}[\mathscr{S}_3, \mathsf{Q}]$ is assumed to transform as an axial scalar having absolute dimension $[Q]$. Therefore, it follows by the corollary (268.14) that dual $\boldsymbol{\sigma}$ is an axial 3-vector having absolute dimension $[Q]$. Thus $\boldsymbol{\sigma}$ is a vector density having absolute dimension $[Q]$.[3] Under the group of analytic transformations of the co-ordinates $\boldsymbol{x}$ and the group of transformations

$$\mathsf{Q}' = Q^{-1} \mathsf{Q} \tag{270.3}$$

[1] Throughout this chapter we assume that the underlying space is topologically equivalent to the Euclidean space of the same dimension. For such spaces, the theorem embodied in (269.8) follows from WHITNEY's lemma [1957, *18*, § 25].

[2] Henceforth in this chapter Greek indices will range over the four values 1, 2, 3, and 4.

[3] Our reasons for assuming that the charge transforms as an axial scalar instead of an absolute scalar will be explained in Sect. 283. In Sect. 285 we introduce a more general expression for the charge.

of the unit of charge Q, σ has the transformation law

$$\sigma^{\Omega'} = Q\,|(\boldsymbol{x}'/\boldsymbol{x})|^{-1} \frac{\partial x^{\Omega}}{\partial x^{\Omega}} \sigma^{\Omega}. \tag{270.4}$$

We now postulate the ***law of conservation of charge***: *the world scalar invariant integral* $\mathfrak{C}[\mathscr{S}_3, Q]$ *vanishes for every 3-dimensional circuit,*

$$\oint \sigma \cdot d\hat{\mathscr{S}}_3 = 0. \tag{270.5}$$

In a similar fashion we assign to every $\mathscr{S}_2$ in space-time a world invariant additive set function $\mathfrak{F}[\mathscr{S}_2, \Phi]$ called the *magnetic flux* and assume that magnetic flux is expressible in the form

$$\mathfrak{F}[\mathscr{S}_2, \Phi] = \int \varphi \cdot d\mathscr{S}_2 = \tfrac{1}{2} \int \varphi_{\Omega\Delta}\, d\mathscr{S}_2^{\Omega\Delta}, \tag{270.6}$$

where φ is the *electromagnetic field* and where Φ is the *unit of magnetic flux*. The magnetic flux and the electromagnetic field have the absolute dimension $[\phi]$. Under the group of analytic coordinate transformations of the $\boldsymbol{x}$ and the transformations

$$\Phi' = \phi^{-1}\,\Phi \tag{270.7}$$

of the unit of magnetic flux, the electromagnetic field has the transformation law

$$\varphi_{\Omega'\Delta'} = \phi\, \frac{\partial x^{\Omega}}{\partial x^{\Omega'}} \frac{\partial x^{\Delta}}{\partial x^{\Delta'}} \varphi_{\Omega\Delta}. \tag{270.8}$$

We now postulate the ***law of conservation of magnetic flux***: *the world scalar invariant* $\mathfrak{F}[\mathscr{S}_2, \Phi]$ *vanishes for every 2-dimensional circuit:*

$$\oint \varphi \cdot d\mathscr{S}_2 = 0. \tag{270.9}$$

Eqs. (270.5) and (270.9) are the *cornerstones of electromagnetic theory*. The laws of nature embodied in these postulates are perhaps the most lasting achievements of the classical theory of electromagnetism. For the moment we regard the charge-current and electromagnetic fields as independent. In Subchapter III we shall see the manner in which they are related to one another by an additional hypothesis. Also, we shall subsequently give a more general mathematical expression for the law of conservation of charge. The intended generalization, however, involves no new physical idea, and we prefer now to consider the simpler Eqs. (270.5) and (270.9), independently of any further physical assumptions or increased mathematical generality. When we are on more familiar ground, these generalizations may be added with less difficulty and abstractness.

The world invariant integral equations (270.5) and (270.9) were deduced by BATEMAN[1], who took as a starting point the differential equations commonly referred to as MAXWELL'S equations. Consistently with the program of Sect. 7, we prefer, following KOTTLER[2], to announce our basic premises in the stronger form of the integral equations (270.5) and (270.9). The acceptance of these ***Maxwell-Bateman laws*** may be motivated on the grounds of simplicity and the general intuitive notion of conservation. As we shall see, the customary *field equations* and *boundary conditions* of electromagnetic theory follow as consequences of these integral equations if we supply appropriate assumptions regarding the continuity of the fields and the nature of the co-ordinates $\boldsymbol{x}$.

[1] [1910, *1*]. BATEMAN cites the earlier work of HARGREAVES [1908, *5*] on invariant integral forms.

[2] [1922, *4*].

We trust that our rather abstract postulation of the laws of conservation of charge and magnetic flux will not discourage the reader seeking more concrete and familiar results. Our main reason for this approach is to emphasize the independence of the conservation laws from any geometry of space-time. The ideas of conservation as formulated here have, in a certain sense, topological significance, transcending both the intuition and the mathematics of length, time, and angles.

This "metrical independence" of the conservation laws has been noted by VAN DANTZIG[1]. As he remarks, the concept of metric and the measurement of lengths, angles, and time intervals is perhaps one of the most sophisticated and complex aspects of any physical theory. Furthermore, is not the intuitive notion of conservation of charge, for example, quite independent of measurements of length and time? We view the conservation of charge and magnetic flux as independent of ideas like inertial frames, rigid rods, absolute or uniform time, Lorentz transformations, Galilean transformations, etc., and hence as deserving an independent mathematical expression.

Since there is a one-to-one correspondence between k-vectors and their duals, the same physical quantity or physical law involving only k-vectors may be expressed in equivalent dual forms. It is a matter of taste and convenience which representation is used. Usually the representation involving the fewer number of indices is to be preferred. Thus the charge-current field is usually represented by a contravariant vector density rather than by the dual covariant axial 3-vector. As far as the electromagnetic field is concerned, both $\boldsymbol{\varphi}$ and dual $\boldsymbol{\varphi}$ are of rank 2, and it is a matter of convention that we tend to favor the covariant representation. However, for some purposes, a degree of formal simplification is obtained by using one or the other representation, and in what follows we do not restrict ourselves to any one particular choice.

271. Electromagnetic and charge-current potentials. If the electromagnetic and charge-current fields are continuous, the Maxwell-Bateman laws (270.5) and (270.9) are sufficient conditions for the existence of continuously differentiable fields $\boldsymbol{\alpha}$ and $\boldsymbol{\eta}$ such that

$$\int \boldsymbol{\sigma}\cdot d\hat{\mathscr{S}}_3 = -\oint \boldsymbol{\eta}\cdot d\hat{\mathscr{S}}_2, \tag{271.1}$$

$$\int \boldsymbol{\varphi}\cdot d\mathscr{S}_2 = \oint \boldsymbol{\alpha}\cdot d\mathscr{S}_1. \tag{271.2}$$

The integral theorems (268.16) and (268.17) lead to the local relations

$$\boldsymbol{\sigma} = \operatorname{div}\boldsymbol{\eta}, \qquad \boldsymbol{\varphi} = \operatorname{rot}\boldsymbol{\alpha}. \tag{271.3}$$

In terms of components, we have

$$\sigma^{\Omega} = \partial_{\Delta}\,\eta^{\Omega\Delta}, \qquad \varphi_{\Omega\Delta} = 2\,\partial_{[\Omega}\,\alpha_{\Delta]}. \tag{271.4}$$

A contravariant 2-vector density $\boldsymbol{\eta}$ satisfying (271.1) will be called a *charge-current potential*, and a covariant absolute vector $\boldsymbol{\alpha}$ satisfying (271.2) will be called an *electromagnetic potential*.

The existence of charge-current potentials $\boldsymbol{\eta}$ is a consequence of the law of conservation of charge. In Subchapter III we shall subsequently introduce the aether relations fixing a particular charge-current potential in terms of the electromagnetic field. Our procedure here is not unlike a procedure sometimes adopted in mechanics, where the stress $t^r{}_s$ may be introduced as a solution of the equations

$$\int \varrho\, f_r\, dv = \oint t^s{}_r\, da_s,$$

$$\int \varrho\, z_{[r}\, f_{s]}\, dv = \oint z_{[r}\, t^p{}_{s]}\, da_p,$$

where the force field $\varrho\boldsymbol{f}$ is regarded as prescribed (cf. Sect. 203). These equations do not uniquely determine the stress field $\boldsymbol{t}$ since addition of a null stress, i.e., any symmetric tensor

[1] [1934, *11*].

satisfying $\partial_r t'_s = 0$, to a given solution of these equations yields another solution (cf. the remarks at the end of Sect. 205). Nevertheless, theories of continuum mechanics generally employ constitutive equations for the stress which are not invariant under the process of adding a divergence-free symmetric tensor. For the moment we emphasize that the existence of an infinity of distinct charge-current potentials follows from the law of conservation of charge. Similarly, conservation of magnetic flux is a sufficient condition for the existence of an infinity of distinct electromagnetic potentials provided we assume rather weak continuity properties for the electromagnetic field. The group of transformations of the electromagnetic potentials has been called the *group of gauge transformations*[1]. It is customary to render the electromagnetic potential unique by imposing upon it additional restrictions in the form of boundary conditions, continuity requirements, and algebraic or differential relations amongst its components. Some remarks on these conditions are given in Sect. 276.

272. Field equations and boundary conditions. If the electromagnetic and charge-current fields are continuously differentiable in a region, by applying the classical argument based on (268.16) to (268.17) (cf. Sect. 157), from the Maxwell-Bateman laws (270.5) and (270.9) we derive the *field equations*

$$\operatorname{div} \boldsymbol{\sigma} = 0, \tag{272.1}$$

$$\operatorname{curl} \boldsymbol{\varphi} = 0 = \operatorname{dual} \operatorname{rot} \boldsymbol{\varphi}. \tag{272.2}$$

In terms of components, we have

$$\partial_\Omega \sigma^\Omega = 0, \tag{272.3}$$

$$\varepsilon^{\Omega\Delta\Psi\Theta} \partial_\Delta \varphi_{\Psi\Theta} = 0. \tag{272.4}$$

Let

$$\Sigma(\boldsymbol{x}) = 0 \tag{272.5}$$

be the equation of a 3-dimensional surface in space-time dividing a region $\mathscr{R}$ into two regions $\mathscr{R}^+$ and $\mathscr{R}^-$. We suppose the electromagnetic and charge-current fields continuous in the closure of $\mathscr{R}^+$ and $\mathscr{R}^-$ but possibly discontinuous at $\Sigma = 0$. We may think of the surface $\Sigma = 0$ as the set of events representing the history of a 2-dimensional surface in space across which the electromagnetic and charge-current fields have finite jumps. Applying the basic integral laws (270.5) and (270.9) to circuits divided by the surface $\Sigma = 0$, we conclude that the discontinuities $[\operatorname{dual} \boldsymbol{\varphi}]$ and $[\boldsymbol{\sigma}]$ are such as to satisfy the restrictions

$$\left.\begin{aligned} [\operatorname{dual} \boldsymbol{\varphi}] \cdot \operatorname{grad} \Sigma &= 0, \\ [\boldsymbol{\sigma}] \cdot \operatorname{grad} \Sigma &= 0. \end{aligned}\right\} \tag{272.6}$$

In terms of components, these equations read

$$\left.\begin{aligned} (\operatorname{dual} \boldsymbol{\varphi})^{\Omega\Delta} \partial_\Delta \Sigma &= 0, \\ \sigma^\Omega \partial_\Omega \Sigma &= 0. \end{aligned}\right\} \tag{272.7}$$

These are the *electromagnetic* and *charge-current boundary conditions*. They imply that the most general discontinuities in the electromagnetic and charge-current fields allowed by the conservation laws are expressible in the form

$$[\operatorname{dual} \boldsymbol{\varphi}] = \boldsymbol{\beta} \times \operatorname{grad} \Sigma, \tag{272.8}$$

$$[\boldsymbol{\sigma}] = \boldsymbol{\omega} \times \operatorname{grad} \Sigma, \tag{272.9}$$

or, in terms of components,

$$\left.\begin{aligned} [(\operatorname{dual} \boldsymbol{\varphi})^{\Omega\Delta}] &= \varepsilon^{\Omega\Delta\Psi\Theta} \beta_\Psi \partial_\Theta \Sigma, \\ [\sigma^\Omega] &= \tfrac{1}{2} \varepsilon^{\Omega\Delta\Psi\Theta} \omega_{\Delta\Psi} \partial_\Theta \Sigma, \end{aligned}\right\} \tag{272.10}$$

where $\boldsymbol{\beta}$ and $\boldsymbol{\omega}$ are arbitrary fields defined on $\Sigma = 0$.

---

[1] Bergmann [1942, *1*, p. 115].

273. The geometry of space-time, reduction of k-vectors, and the units of length and time. So as to obtain equations and results familiar from conventional treatments of electromagnetic theory, we now introduce a space-time geometry. In Sect. 152 we studied two such geometries based on the idea of the underlying Euclidean and Galilean groups of transformations. We showed that the class of Euclidean frames could be defined as the class of co-ordinate systems in space-time for which the space metric $\boldsymbol{g}(\boldsymbol{x})$ and the covariant space normal $\boldsymbol{t}(\boldsymbol{x})$ assume the canonical forms

$$\boldsymbol{g} = \begin{bmatrix} \delta^{rs} & 0 \\ 0 & 0 \end{bmatrix}, \quad \boldsymbol{t} = (0, 0, 0, 1). \tag{273.1}$$

The subclass of these frames for which the Galilean connection $\boldsymbol{\Gamma}(\boldsymbol{x}) = 0$ were identified as the inertial frames of classical mechanics. The question of inertial frames in electromagnetic theory will be taken up in Subchapter III, where we shall discuss the Lorentz invariance of the aether relations, Lorentz frames, and the relation between these entities and their Galilean counterparts. For the remainder of this Subchapter it suffices that we consider the entire class of rectangular Cartesian co-ordinate systems in space-time characterized by the canonical forms (273.1). In addition to the co-ordinate transformations considered in Sect. 152, we shall here consider transformations of the *units of length and time*

$$\mathsf{L}' = \mathsf{L}^{-1}\,\mathsf{L}, \qquad \mathsf{T}' = \mathsf{T}^{-1}\,\mathsf{T}. \tag{273.2}$$

The space metric $\boldsymbol{g}$ will be assigned the absolute dimension $[\mathsf{L}^{-2}]$, and the co-variant space normal $\boldsymbol{t}$ will be assigned the absolute dimension $[\mathsf{T}]$. Thus, under transformations of the co-ordinates $\boldsymbol{x}$ and the units of length and time, the world tensors $\boldsymbol{g}$ and $\boldsymbol{t}$ have the transformation laws

$$\left.\begin{aligned} g^{\Omega'\Delta'}(\boldsymbol{x}', \mathsf{L}') &= \mathsf{L}^{-2}\,\frac{\partial x^{\Omega'}}{\partial x^{\Omega}}\,\frac{\partial x^{\Delta'}}{\partial x^{\Delta}}\,g^{\Omega\Delta}(\boldsymbol{x},\mathsf{L}), \\ t_{\Omega'}(\boldsymbol{x}', \mathsf{T}) &= \mathsf{T}\,\frac{\partial x^{\Omega}}{\partial x^{\Omega'}}\,t_{\Omega}(\boldsymbol{x}, \mathsf{T}). \end{aligned}\right\} \tag{273.3}$$

Recall that $t_{\Omega} = \partial_{\Omega} t$, where $t(\boldsymbol{x})$ is the time. The absolute dimension of t is $[\mathsf{T}]$.

If $\boldsymbol{g}$ and $\boldsymbol{t}$ are to retain their canonical form (273.1) when the units of length and time are changed, the co-ordinates belonging to the two systems of units must be related by a transformation of the form

$$\left.\begin{aligned} z^{r'}(\mathsf{L}') &= \mathsf{L}\,[A^{r'}{}_{r}(t)\, z^{r}(\mathsf{L}) + d^{r'}(t)], \\ z^{4'}(\mathsf{T}') &= t' = \mathsf{T}\, z^{4}(\mathsf{T}) + \text{const} = \mathsf{T}\, t + \text{const}, \end{aligned}\right\} \tag{273.4}$$

where $\boldsymbol{A}$ is an orthogonal matrix. We shall denote the co-ordinates $\boldsymbol{x}$ by $\boldsymbol{z}$ and t when (273.1) holds. That is, $\boldsymbol{z}$ denotes rectangular Cartesian spatial co-ordinates and $t = x^4$ denotes the time. Recall that in Sect. 152 we called a co-ordinate system for which we have the canonical forms (273.1), a *Euclidean frame*.

Let $\boldsymbol{F}$ denote a covariant world k-vector. Let us denote the components of $\boldsymbol{F}$ referred to a Euclidean frame by

$$F_{r_1 r_2 \ldots r_k} = a_{r_1 r_2 \ldots r_k}, \qquad F_{r_1 r_2 \ldots r_{k-1} 4} = b_{r_1 r_2 \ldots r_{k-1}}. \tag{273.5}$$

All the components of $\boldsymbol{F}$ are determined from these particular components. With reference to the decomposition (273.5) we shall use the notation

$$\boldsymbol{F} = (\boldsymbol{a}, \boldsymbol{b}). \tag{273.6}$$

For a contravariant world k-vector $\boldsymbol{Y}$ we write

$$\boldsymbol{Y} = (\boldsymbol{c}, \boldsymbol{d}), \tag{273.7}$$

where

$$c^{r_1 r_2 \ldots r_k} = Y^{r_1 r_2 \ldots r_k}, \quad d^{r_1 r_2 \ldots r_{k-1}} = Y^{4 r_1 r_2 \ldots r_{k-1}}. \tag{273.8}$$

Note the different position of the index 4 in (273.5) and (273.8).

If $\boldsymbol{F}$ is an absolute covariant k-vector of absolute dimension $[U]$, the transformation laws of its components $\boldsymbol{a}$ and $\boldsymbol{b}$ can be put in the form

$$\left.\begin{aligned} a_{r'_1 r'_2 \ldots r'_k} &= U L^{-k} A^{s_1}_{r'_1} A^{s_2}_{r'_2} \ldots A^{s_k}_{r'_k} a_{s_1 s_2 \ldots s_k} \\ b_{r'_1 r'_2 \ldots r'_{k-1}} &= U L^{-k+1} T^{-1} A^{s_1}_{r'_1} A^{s_2}_{r'_2} \ldots A^{s_{k-1}}_{r'_{k-1}} (b_{s_1 s_2 \ldots s_{k-1}} + a_{s_1 \ldots s_{k-1} t} u^t), \end{aligned}\right\} \tag{273.9}$$

where $u^t \equiv \partial z^t / \partial z^{4\prime}$. Thus the quantities $\boldsymbol{a}$ transform as a 3-dimensional tensor; but unless $\boldsymbol{a} = 0$, the quantities $\boldsymbol{b}$ do not.

For contravariant absolute k-vectors, the components $\boldsymbol{c}$ and $\boldsymbol{d}$ in (273.8) have the transformation laws

$$\left.\begin{aligned} c^{r'_1 r'_2 \ldots r'_k} &= U L^k A^{r'_1}_{s_1} A^{r'_2}_{s_2} \ldots A^{r'_k}_{s_k} (c^{s_1 \ldots s_k} - k u^{[s_1} d^{s_2 s_3 \ldots s_k]}), \\ d^{r'_1 r'_2 \ldots r'_{k-1}} &= U L^{k-1} T A^{r'_1}_{s_1} \ldots A^{r'_2}_{s_2} A^{r'_{k-1}}_{s_{k-1}} d^{s_1 s_2 \ldots s_{k-1}}. \end{aligned}\right\} \tag{273.10}$$

Similar considerations apply in the case of covariant and contravariant axial k-vectors, k-vector densities, and axial k-vector densities. For example, if $\boldsymbol{Y}$ is a contravariant k-vector density we have

$$\left.\begin{aligned} c^{r'_1 r'_2 \ldots r'_k} &= U L^{k-3} T^{-1} A^{r'_1}_{s_1} A^{r'_2}_{s_2} \ldots A^{r'_k}_{s_k} (c^{s_1 \ldots s_k} - k u^{[s_1} d^{s_2 \ldots s_k]}), \\ d^{r'_1 r'_2 \ldots r'_{k-1}} &= U L^{k-4} A^{r'_1}_{s_1} A^{r'_2}_{s_2} \ldots A^{r'_{k-1}}_{s_{k-1}} d^{s_1 s_2 \ldots s_{k-1}}. \end{aligned}\right\} \tag{273.11}$$

For axial k-vectors and axial k-vector densities a factor, $\det \boldsymbol{A} = \pm 1$, will occur in the transformation laws of the Euclidean components.

The rectangular Cartesian co-ordinates $\boldsymbol{z}(L)$ of a Euclidean frame based on the unit of length L are assigned the physical dimension $[L]$ (cf. Sect. App. 9) and, consistent with (273.4), $z^4(T)$ is assigned the physical dimension $[T]$. The physical dimensions of the components $\boldsymbol{a}$, $\boldsymbol{b}$, $\boldsymbol{c}$, etc., of world tensors referred to a Euclidean frame are determined by the absolute dimension of the corresponding world tensor, its variance, and its weight by writing the transformation laws for its components in the forms (273.9), (273.10), (273.11), etc. Thus if $\boldsymbol{F}$ is an axial or absolute covariant k-vector of absolute dimension $[U]$, then

$$\left.\begin{aligned} \text{phys. dim.}\, \boldsymbol{a} &= [U L^{-k}], \\ \text{phys. dim.}\, \boldsymbol{b} &= [U L^{-k+1} T^{-1}]. \end{aligned}\right\} \tag{273.12}$$

If $\boldsymbol{Y}$ is an absolute or axial contravariant k-vector, then

$$\left.\begin{aligned} \text{phys. dim.}\, \boldsymbol{c} &= [U L^k], \\ \text{phys. dim.}\, \boldsymbol{d} &= [U L^{k-1} T]. \end{aligned}\right\} \tag{273.13}$$

If $\boldsymbol{F}$ is a covariant density or axial density, then

$$\left.\begin{aligned} \text{phys. dim.}\, \boldsymbol{a} &= [U L^{-k+3} T], \\ \text{phys. dim.}\, \boldsymbol{b} &= [U L^{-k+4}]. \end{aligned}\right\} \tag{273.14}$$

Finally, if $\boldsymbol{Y}$ is a contravariant density or axial density, then

$$\left.\begin{aligned} \text{phys. dim.}\, \boldsymbol{c} &= [U L^{k-3} T^{-1}], \\ \text{phys. dim.}\, \boldsymbol{d} &= [U L^{k-4}]. \end{aligned}\right\} \tag{273.15}$$

Consider a world covariant k-vector or axial k-vector $\boldsymbol{F}$ and a world covariant $(k-1)$-vector or axial $(k-1)$-vector $\boldsymbol{G}$ such that

$$\boldsymbol{F} = \operatorname{rot} \boldsymbol{G}. \tag{273.16}$$

Let

$$\boldsymbol{F} = (\boldsymbol{a}, \boldsymbol{b}), \quad \boldsymbol{G} = (\boldsymbol{g}, \boldsymbol{h}). \tag{273.17}$$

We shall then have

$$\boldsymbol{a} = \operatorname{rot} \boldsymbol{g}, \qquad \boldsymbol{b} = \operatorname{rot} \boldsymbol{h} + (-1)^{k-1} \frac{\partial \boldsymbol{g}}{\partial t}, \tag{273.18}$$

or, equivalently,

$$\operatorname{dual} \boldsymbol{a} = \operatorname{curl} \boldsymbol{g}, \quad \operatorname{dual} \boldsymbol{b} = \operatorname{curl} \boldsymbol{h} + (-1)^{k-1} \frac{\partial (\operatorname{dual} \boldsymbol{g})}{\partial t}. \tag{273.19}$$

274. The charge density, current density, electric field, and magnetic flux density. If we assume the geometry of space-time to be Euclidean in the sense of Sect. 152, then we can apply the considerations of Sect. 273 to a reduction of the charge-current and electromagnetic fields into four distinct fields. Refer the electromagnetic field $\boldsymbol{\varphi}$ and the charge-current field to a Euclidean frame, and set

$$\boldsymbol{\varphi} = (\operatorname{dual} \boldsymbol{B}, \boldsymbol{E}), \quad \sigma = (\boldsymbol{J}, Q), \tag{274.1}$$

where $\boldsymbol{B}$ is called the *density of magnetic flux*, $\boldsymbol{E}$ the *electric field*, $\boldsymbol{J}$ the *current density*, and Q the *charge density*. From the rules for determining the physical dimensions of these quantities given in Sect. 273 we get immediately,

$$\left.\begin{aligned} \text{phys. dim. } \boldsymbol{B} &= [\phi\, L^{-2}], \\ \text{phys. dim. } \boldsymbol{E} &= [\phi\, L^{-1}\, T^{-1}], \\ \text{phys. dim. } \boldsymbol{J} &= [Q\, L^{-2}\, T^{-1}], \\ \text{phys. dim. } Q &= [Q\, L^{-3}]. \end{aligned}\right\} \tag{274.2}$$

Let v^Ω be the world velocity field of a motion as defined in Sect. 152. In terms of $\boldsymbol{v}$, the electromagnetic field, and the charge-current field, we can define the space tensors

$$\mathfrak{E}^\Omega \equiv g^{\Omega\Delta} \varphi_{\Delta\Theta} v^\Theta, \quad \mathfrak{J}^\Omega \equiv \sigma^\Omega - v^\Omega \sigma^\Delta t_\Delta. \tag{274.3}$$

Recall that a space tensor was defined in Sect. 152 as a contravariant world tensor for which $F^{\Omega\Delta\cdots} t_\Omega = F^{\Delta\Omega\cdots} t_\Omega = \cdots = 0$. A k-vector which is a space tensor has the representation

$$\boldsymbol{F} = (\boldsymbol{c}, 0), \tag{274.4}$$

so that, according to (273.10) and (273.11), $\boldsymbol{c}$ transforms as a 3-dimensional Cartesian tensor. If we refer the components of $\mathfrak{E}$ and $\mathfrak{J}$ to a Euclidean frame, we get

$$\left.\begin{aligned} \mathfrak{E} &= (\boldsymbol{E} + \boldsymbol{v}\times\boldsymbol{B}, 0) = (\mathfrak{E}, 0), \\ \mathfrak{J} &= (\boldsymbol{J} - Q\,\boldsymbol{v}, 0) = (\mathfrak{J}, 0), \end{aligned}\right\} \tag{274.5}$$

where $\mathfrak{E} \equiv \boldsymbol{E} + \boldsymbol{v}\times\boldsymbol{B}$ is called the *electromotive intensity* at a point moving with the particles of the motion, and $\mathfrak{J} \equiv \boldsymbol{J} - Q\boldsymbol{v}$ is called the *conduction current* relative to the particles of the motion. The electromotive intensity and the conduction current of a motion transform as vectors under time-dependent transformations of the spatial coordinates. However, one should note that the electric field and current density do not transform as vectors under transformations between co-ordinate frames in relative motion. Thus, if the electric field and current density vanish in one Euclidean frame, they need not necessarily vanish in every Euclidean frame. A distribution of charge in one frame will constitute a current density in a frame in relative motion. Similarly, a density of magnetic flux in one frame will be interpreted as an electric field in a frame in relative motion. However, if the density of magnetic flux and the density of charge vanish in one Euclidean frame, they vanish in every Euclidean frame.

275. The 3-dimensional integral form of the laws of conservation of charge and magnetic flux. Consider first the 3-dimensional circuits in space-time formed by 3-dimensional tubes (Sect. App. 29) of the world velocity field $\boldsymbol{v}$ defined by a given motion closed on either end by surfaces lying in the instantaneous spaces $t(\boldsymbol{x}) = t_1$ and $t(\boldsymbol{x}) = t_2$. (See Fig. 44.) If we apply the law of conservation of charge (270.5) to such a circuit and refer all quantities to a Euclidean frame, we get

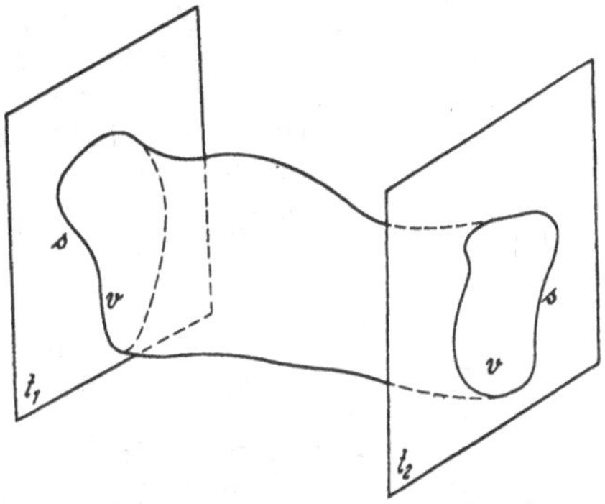

Fig. 44. Tube of integration in Eq. (275.1).

$$\int_v Q\,dv\Big]_{t_1}^{t_2} + \int_{t_1}^{t_2} dt \oint_{\mathscr{S}} \mathfrak{J}\cdot d\boldsymbol{a} = 0, \qquad (275.1)$$

where v is a spatial region moving with the particles of the motion and $\mathscr{S}$ is its complete boundary. Set $\Delta t = t_2 - t_1$ and consider the limit

$$\lim_{\Delta t\to 0} \frac{1}{\Delta t}\left\{\int Q\,dv\Big]_{t_1}^{t_2} + \int_{t_1}^{t_2} dt \oint \mathfrak{J}\cdot d\boldsymbol{a}\right\} = 0. \qquad (275.2)$$

If the limits of the two terms exist separately, we get

$$\frac{d}{dt}\int Q\,dv + \oint \mathfrak{J}\cdot d\boldsymbol{a} = 0, \qquad (275.3)$$

which has the traditional form of an equation of balance when sources are excluded (cf. Sect. 157) and puts the law of conservation of charge in terms easy to understand. The conduction current relative to the particles[1] of the motion is the efflux of charge out of the moving region through its boundary.

Consider next the 2-dimensional circuits in space-time formed by segments of 2-dimensional tubes of the world velocity field closed on either end by surfaces lying in the instantaneous spaces $t(\boldsymbol{x}) = t_1$ and $t(\boldsymbol{x}) = t_2$. If we apply the law of conservation of magnetic flux (270.9) to such a circuit and refer all quantities to a Euclidean frame, we get

$$\int_{\mathscr{S}} \boldsymbol{B}\cdot d\boldsymbol{a}\Big]_{t_1}^{t_2} + \int_{t_1}^{t_2} dt \oint_{\mathscr{C}} \mathfrak{E}\cdot d\boldsymbol{x} = 0, \qquad (275.4)$$

where $\mathscr{S}$ is a 2-dimensional surface in space moving with the particles of the motion and $\mathscr{C}$ is its complete boundary. Applying a limit argument as in (275.2), we then get

$$\frac{d}{dt}\int \boldsymbol{B}\cdot d\boldsymbol{a} + \oint \mathfrak{E}\cdot d\boldsymbol{x} = 0, \qquad (275.5)$$

---

[1] These particles are defined mathematically by integration of the given velocity field and are a convenient device for visualizing it; they need not be mass-bearing.

which is the traditional form of *Faraday's law of induction for moving circuits.* If we apply (270.9) to a 2-dimensional circuit which lies in the surface $t(\boldsymbol{x}) = \text{const}$ (Fig. 45) we get the condition

$$\oint \boldsymbol{B} \cdot d\boldsymbol{a} = 0. \tag{275.6}$$

A visualization of FARADAY'S law of induction is obtained by introducing the notion of *lines of magnetic flux* or *lines of induction.* The number of lines of magnetic flux threading a closed curve in space is measured by the integral $\int \boldsymbol{B} \cdot d\boldsymbol{a}$, where $\mathscr{S}$ is any surface having the curve for its boundary. Eq. (275.6) states that this measure is independent of the choice of $\mathscr{S}$ and depends only on the curve. FARADAY'S law of induction states that the time rate of change of the total number of lines of magnetic flux threading a moving circuit is measurde by the negative line integral of the electromotive intensity around the moving circuit.

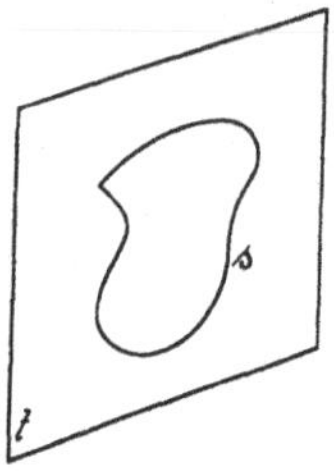

Fig. 45. Surface of integration in Eq. (275.6).

Since we have (275.3), (275.5), and (275.6) for any motion, the corresponding laws for circuits which are at rest in some Euclidean frame follow as a special case by setting $v^r = 0$.

276. The 3-dimensional integral form of the potential equations. Let the components of the electromagnetic potential and the charge-current potential in a Euclidean frame be denoted by

$$\boldsymbol{\alpha} = (\boldsymbol{A}, -V), \qquad \eta = (\text{dual}\, \boldsymbol{H}, \boldsymbol{D}), \tag{276.1}$$

where $\boldsymbol{A}$ is called the *magnetic potential,* V the *electric potential,* $\boldsymbol{H}$ the *current potential,* and $\boldsymbol{D}$ the *charge potential*[1].

If we apply the world invariant equations (271.1) and (271.2) to circuits in space-time constructed in a manner similar to those used to obtain (275.3), (275.5), and (275.6), we obtain the equations

$$\int \mathfrak{J} \cdot d\boldsymbol{a} = \oint (\boldsymbol{H} + \boldsymbol{D} \times \boldsymbol{v}) \cdot d\boldsymbol{a} - \frac{d}{dt} \int \boldsymbol{D} \cdot d\boldsymbol{a}, \tag{276.2}$$

$$\int Q\, dv = \oint \boldsymbol{D} \cdot d\boldsymbol{a}, \tag{276.3}$$

$$\int \boldsymbol{B} \cdot d\boldsymbol{a} = \oint \boldsymbol{A} \cdot d\boldsymbol{a}, \tag{276.4}$$

$$\int_{\boldsymbol{z}_1}^{\boldsymbol{z}_2} \mathfrak{E} \cdot d\boldsymbol{a} = -\frac{d}{dt} \int \boldsymbol{A} \cdot d\boldsymbol{a} + (\boldsymbol{A} \cdot \boldsymbol{v} - V)\Big]_{\boldsymbol{z}_1}^{\boldsymbol{z}_2}, \tag{276.5}$$

where $\boldsymbol{z}_1$ and $\boldsymbol{z}_2$ denote the end points of the moving curve $\boldsymbol{c}$.

Eq. (276.2) will be called the *current equation,* and (276.3) will be called the *charge equation*[2].

The potential equations for stationary circuits follow simply by setting $v^r = 0$. For the classical theory, (276.4) and (276.5) are less used than are the current and charge equations, although the introduction of magnetic and electric potentials and their relative importance

---

[1] The axial vector $\boldsymbol{H}$ is generally called the "magnetic field intensity" and the vector density $\boldsymbol{D}$ the "electric displacement". Since the charge-current field and the electromagnetic field have not been related to one another in any way, this customary terminology would be inappropriate for our purposes here. The conventional names for the magnetic potential $\boldsymbol{A}$ and the electric potential V of the electromagnetic field are appropriate and suggestive, and we have assigned names to $\boldsymbol{H}$ and $\boldsymbol{D}$ on the basis of their similar relation to the charge-current field. The terms "magnetic field intensity" and "electric displacement" will take on their proper connotation after we introduce the aether relations in Subchapter III.

[2] Eq. (276.2) is HERTZ's form of the current equation for moving circuits. HERTZ [1900, *4*, Chap. XIV], WHITTAKER [1951, *39*, pp. 329—331].

is a matter of debate. The charge and current potentials $\boldsymbol{H}$ and $\boldsymbol{D}$ play a more fundamental role in the classical theory than do the magnetic and electric potentials $\boldsymbol{A}$ and V. MIE and DIRAC[1] have proposed theories of electromagnetism in which the four potentials $\boldsymbol{H}, \boldsymbol{D}, \boldsymbol{A}$, and V enter the theoretical structure on a more equal footing. However, in most treatments of the classical theory, the magnetic and electric potentials are introduced as auxiliary fields, determined only to within a gauge transformation. Additional restrictions are often imposed in the form of boundary conditions, continuity requirements or algebraic and differential relations between the four fields A_1, A_2, A_3, and V. For example, many authors introduce the "gauge condition" $V=0$, while others impose the "Lorentz gauge condition" $\operatorname{div}\boldsymbol{A}+c^{-2}\frac{\partial V}{\partial t}=0$. MAXWELL[2] imposed the condition $\operatorname{div}\boldsymbol{A}=0$ and called $\boldsymbol{A}$ the *electromagnetic momentum*. Perhaps a suitably "gauged" electromagnetic potential and the ideas of MIE and DIRAC will serve as the basis of a better theory of electromagnetism. We do not take up these questions, resting content with our choice of charge and magnetic flux as the fundamental quantities entering the equations of conservation.

277. Time derivatives of scalar integrals over moving curves, surfaces, and regions. The electromagnetic equations for moving circuits involve the time derivatives of scalar integrals having the form

$$\mathfrak{I}=\frac{1}{k!}\int F_{r_1 r_2\ldots r_k}\,d\mathscr{S}_k^{r_1 r_2\ldots r_k}=\int \boldsymbol{F}\cdot d\mathscr{S}_k=\int(\operatorname{dual}\boldsymbol{F})\cdot d\hat{\mathscr{S}}_k, \tag{277.1}$$

where $\boldsymbol{F}$ is a 3-dimensional k-vector ($k=0, 1, 2$, or 3) and $\mathscr{S}_k$ is a spatial region, surface, or curve moving with the particles of a motion with velocity field v^r in a Euclidean frame. Now the motion can be presented in the form $z^r=z^r(Z^K,t)$ where the Z^K are material co-ordinates. In general, the k-vector field $\boldsymbol{F}$ depends explicitly on the time t as well as the spatial co-ordinates $\boldsymbol{z}$. The surface $\mathscr{S}_k$ is given by parametric equations

$$z^r=z^r(u^1,u^2,\ldots,u^k,t)=z^r(Z^K(\boldsymbol{u}),t). \tag{277.2}$$

The limits of integration (277.1) correspond to fixed values of the $\boldsymbol{Z}$ and $\boldsymbol{u}$ independent of the time t. Thus we can write

$$\frac{d\mathfrak{I}}{dt}=\frac{1}{k!}\int\frac{\partial F_{K_1K_2\ldots K_k}}{\partial t}\,d\mathscr{S}_k^{K_1K_2\ldots K_k} \tag{277.3}$$

provided that the field $\boldsymbol{F}$ and the motion $\boldsymbol{z}(\boldsymbol{Z},t)$ are continuously differentiable. In (277.3), the $F_{K_1K_2\ldots K_k}$ are defined by

$$F_{K_1K_2\ldots K_k}\equiv\frac{\partial z^{r_1}}{\partial Z^{K_1}}\frac{\partial z^{r_2}}{\partial Z^{K_2}}\cdots\frac{\partial z^{r_k}}{\partial Z^{K_k}}F_{r_1r_2\ldots r_k}, \tag{277.4}$$

and $d\mathscr{S}_k^{K_1K_2\ldots K_k}$ is the corresponding transform of $dS_k^{r_1r_2\ldots r_k}$. If follows from (277.3) and the results in Sect. 150 that $d\mathfrak{I}/dt$ can be written in the form

$$\frac{d\mathfrak{I}}{dt}=\frac{1}{k!}\int\frac{d_c}{dt}F_{r_1r_2\ldots r_k}\,d\mathscr{S}_k^{r_1r_2\ldots r_k}, \tag{277.5}$$

where $d_c\boldsymbol{F}/dt$ denotes the convected time-flux of the k-vector field $\boldsymbol{F}$.

We can also write $d\mathfrak{I}/dt$ in the dual form

$$\frac{d\mathfrak{I}}{dt}=\int\frac{d_c}{dt}(\operatorname{dual}\boldsymbol{F})\cdot d\hat{\mathscr{S}}_k. \tag{277.6}$$

Now the convected time-flux of an absolute or axial k-vector field has the particularly simple form

$$\frac{d_c}{dt}\boldsymbol{F}=\frac{\partial\boldsymbol{F}}{\partial t}+\boldsymbol{v}\cdot\operatorname{rot}\boldsymbol{F}+\operatorname{rot}(\boldsymbol{v}\cdot\boldsymbol{F})=\frac{\partial\boldsymbol{F}}{\partial t}+\operatorname{curl}\boldsymbol{F}\times\boldsymbol{v}+\operatorname{rot}(\boldsymbol{v}\cdot\boldsymbol{F}). \tag{277.7}$$

[1] [1912, 6]; [1951, 5].
[2] [1881, 4, § 618].

Similarly, the convected time-flux of a contravariant k-vector density or axial density has the simple form

$$\frac{d_c}{dt}(\operatorname{dual}\boldsymbol{F}) = \operatorname{dual}(\operatorname{div}\operatorname{dual}\boldsymbol{F}\times\boldsymbol{v}) + \operatorname{curl}(\operatorname{dual}\boldsymbol{F}\times\boldsymbol{v}). \tag{277.8}$$

One can easily verify that

$$\operatorname{dual}\frac{d_c\boldsymbol{F}}{dt} = \frac{d_c}{dt}\operatorname{dual}\boldsymbol{F}. \tag{277.9}$$

Let us set $\hat{\boldsymbol{F}} \equiv \operatorname{dual}\boldsymbol{F}$. Eqs. (277.7) and (277.8) written in terms of components are

$$\frac{d_c}{dt}F_{r_1 r_2\ldots r_k} = \frac{\partial F_{r_1 r_2\ldots r_k}}{\partial t} + (k+1)\,\partial_{[s}F_{r_1 r_2\ldots r_k]}v^s + k\,\partial_{[r_1}(F_{|s|r_2 r_3\ldots r_k]}v^s), \tag{277.10}$$

$$\left.\begin{aligned}\frac{d_c}{dt}\hat{F}^{r_1 r_2\ldots r_p} = \frac{\partial\hat{F}^{r_1 r_2\ldots r_p}}{\partial t} + p\,\partial_s\hat{F}^{[r_1 r_2\ldots r_{p-1}|s|}v^{r_p]} + (p+1)\,\partial_s(\hat{F}^{[r_1 r_2\ldots r_p}v^{s]}),\\ p+k=3.\end{aligned}\right\} \tag{277.11}$$

The foregoing integral formulae and vector indentities are useful for the study and interpretation of the electromagnetic equations for moving circuits. In what follows we shall also need to consider a moving surface given in the form of an equation

$$\Sigma(\boldsymbol{z}, t) = 0, \tag{277.12}$$

where $\boldsymbol{z}$ denotes the rectangular Cartesian spatial co-ordinates of a Euclidean frame, and t denotes the time. The vector $\boldsymbol{n}$ defined by

$$n_r \equiv \frac{\partial_r\Sigma}{\sqrt{\partial_s\Sigma\,\partial_s\Sigma}} \tag{277.13}$$

is called the *instantaneous unit normal* to the moving surface $\Sigma=0$, and the quantity s defined by

$$u_n \equiv -\frac{\dfrac{\partial\Sigma}{\partial t}}{\sqrt{\partial_s\Sigma\,\partial_s\Sigma}}, \tag{277.14}$$

is called the *speed* (cf. (177.5)) of the surface $\Sigma=0$ relative to the reference frame $(\boldsymbol{z}, t)$. The quantities $\boldsymbol{n}$ and u_n may be regarded as the Euclidean components of the world vector

$$\nu_\Omega \equiv \frac{\partial_\Omega\Sigma}{\sqrt{g^{\Delta\Theta}\partial_\Delta\Sigma\,\partial_\Theta\Sigma}}. \tag{277.15}$$

That is, in the notation of Sect. 273

$$\boldsymbol{\nu} = (\boldsymbol{n}, -u_n). \tag{277.16}$$

278. Field equations and boundary conditions in 3-dimensional form. The 3-dimensional form of the field equations and boundary conditions may be obtained in two ways. One method is to work with the 3-dimensional integral equations (275.3), (275.5), (275.6), (276.2) to (276.5). By assuming the fields in these equations continuously differentiable, one can use the results of Sect. 277 to express the time derivatives of integrals over moving surfaces, curves and regions in terms of the convected time-flux. Then, by appropriate application of the fundamental integral theorem, line integrals can be transformed into surface integrals, surface integrals into volume integrals, etc. Various terms cancel and, by the usual type of limit arguments, certain local conditions in the form of partial differential equations follow from the integral equations. Boundary conditions in 3-dimensional form follow by applying these same integral equations to appropriate circuits divided by a surface of possible discontinuity. A simpler and more direct method applies the reduction formulae derived in Sect. 273 to the world tensor field equations and boundary conditions already obtained. By this latter method we obtain as an immediate consequence of the field equations (272.1) and (272.2):

$$\operatorname{div}\boldsymbol{J} + \frac{\partial Q}{\partial t} = 0, \tag{278.1}$$

$$\operatorname{curl}\boldsymbol{E} + \frac{\partial \boldsymbol{B}}{\partial t} = 0, \tag{278.2}$$

$$\operatorname{div}\boldsymbol{B} = 0. \tag{278.3}$$

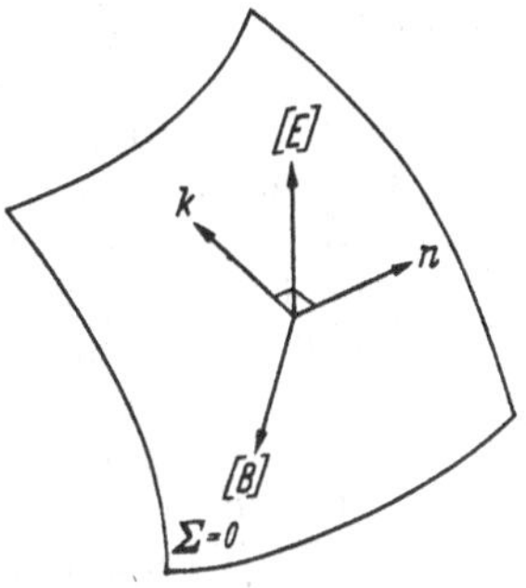

Fig. 46. Geometry of an electromagnetic discontinuity.

The field equations (271.3) yield the relations

$$Q = \operatorname{div}\boldsymbol{D}, \tag{278.4}$$

$$\boldsymbol{J} = \operatorname{curl}\boldsymbol{H} - \frac{\partial \boldsymbol{D}}{\partial t}, \tag{278.5}$$

$$\boldsymbol{B} = \operatorname{curl}\boldsymbol{A}, \tag{278.6}$$

$$\boldsymbol{E} = -\frac{\partial \boldsymbol{A}}{\partial t} - \operatorname{grad} V. \tag{278.7}$$

Eq. (278.1) is the differential or local form of the law of conservation of charge, and (278.2) is the differential form of FARADAY'S law of magnetic induction.

The 3-dimensional forms of the boundary conditions follow easily from the world tensor equations (272.6) and the definitions (277.13) and (277.14) of the unit normal $\boldsymbol{n}$ and the speed u_n of a moving surface of discontinuity[1].

$$\boldsymbol{n}\times[\boldsymbol{E}] - u_n[\boldsymbol{B}] = 0, \tag{278.8}$$

$$[\boldsymbol{B}]\cdot\boldsymbol{n} = 0, \tag{278.9}$$

$$[\boldsymbol{J}]\cdot\boldsymbol{n} - u_n[Q] = 0. \tag{278.10}$$

Eq. (278.8) implies that the most general discontinuities in the magnetic flux density and the electric field allowed by the law of conservation of magnetic flux are expressible in the form

$$[\boldsymbol{E}] = f\boldsymbol{n} - u_n\boldsymbol{k}, \tag{278.11}$$

$$[\boldsymbol{B}] = \boldsymbol{k}\times\boldsymbol{n}, \tag{278.12}$$

where the fields f and $\boldsymbol{k}$ are arbitrary fields defined on the surface of discontinuity. We see that conservation of magnetic flux restricts considerably the geometry of any electromagnetic discontinuity. Eq. (278.9) demands that any discontinuity in the magnetic flux density be transversal. For a stationary surface of discontinuity, we have $u_n = 0$, and from Eq. (278.8) it follows that the discontinuity in the electric field is longitudinal. If the surface of discontinuity is moving, the discontinuity in the magnetic flux density must be normal to the plane determined by the discontinuity in the electric field and the normal $\boldsymbol{n}$ (see Fig. 46).

III. The Maxwell-Lorentz aether relations.

279. The 3-dimensional form of the aether relations. In all that precedes, we have treated the electromagnetic and charge-current fields as independent. We now introduce a fundamental relation between these fields by postulating the ***Maxwell-Lorentz aether relations.***

[1] LUNEBERG [1944, *9*, p. 21] has obtained these boundary conditions holding at a moving surface of discontinuity in the electromagnetic field by other means.

So as to introduce the aether relations in their simplest and most familiar form, we shall begin with a statement of them in terms of the vector fields $\boldsymbol{E}$, $\boldsymbol{B}$, $\boldsymbol{D}$, and $\boldsymbol{H}$. First of all, we assume the existence of at least one Euclidean frame in which the Maxwell-Lorentz aether relations

$$\left.\begin{aligned} \boldsymbol{D} &= \varepsilon_0 \boldsymbol{E}, \\ \boldsymbol{H} &= \frac{1}{\mu_0} \boldsymbol{B}, \end{aligned}\right\} \tag{279.1}$$

are valid. The quantities ε_0 and μ_0 are fundamental constants depending only on the units of length, time, charge, and magnetic flux. That is, we assume the existence of at least one Euclidean frame in which a charge potential $\boldsymbol{D}$ is proportional to the electric field $\boldsymbol{E}$ and in which a current potential $\boldsymbol{H}$ is proportional to the magnetic flux density $\boldsymbol{B}$. Now it is easy to show that if the Maxwell-Lorentz aether relations (279.1) hold in one Euclidean frame, they cannot hold generally in every Euclidean frame. To see this, simply consider the transformation laws of the four fields $\boldsymbol{E}$, $\boldsymbol{B}$, $\boldsymbol{H}$, and $\boldsymbol{D}$ as have been determined in Sect. 273. These can be written in the form

$$\left.\begin{aligned} \boldsymbol{D}' &= \boldsymbol{D}, \\ \boldsymbol{E}' &= \boldsymbol{E} + \boldsymbol{u} \times \boldsymbol{B}, \\ \boldsymbol{B}' &= \boldsymbol{B}, \\ \boldsymbol{H}' &= \boldsymbol{H} - \boldsymbol{u} \times \boldsymbol{D}, \end{aligned}\right\} \tag{279.2}$$

from which it is immediately plain that the aether relations cannot hold generally and simultaneously in Euclidean frames in relative motion. Thus *the Maxwell-Lorentz aether relations are not Euclidean nor even Galilean invariants.* If we adopt the aether relations we see that the equations of electromagnetic theory will not have the same form in every Euclidean frame, or more importantly, they will not have the same form in every Galilean frame, as do the equations of classical mechanics. Thus it appeared, from the classical point of view, that if the aether relations could be verified in some selected and preferred inertial frame, then by a simultaneous application of the laws of mechanics and electromagnetic theory one should be able to detect the relative motion of inertial frames. To put it otherwise, the aether relations determine a preferred class of Euclidean frames all at rest relative to one another. It may also be assumed for our purposes here that this preferred class of frames are inertial or Galilean. In Galilean frames which are in motion relative to this preferred class, the aether relations do not generally hold. This gives rise in a natural way to the notion of an "aether". That is, we may think of the preferred inertial frames for which the aether relations are valid as the class of inertial frames in which the "aether" is at rest. In all other inertial frames, there will be an "aether wind", controverting the validity of the simple relations (279.1).

Matter resides in an aether characterized by the relations (279.1). These relations are not to be regarded as constitutive equations for matter. We may think of them as constitutive relations for the aether. Allow us to remind the reader already familiar with the classical theory of dielectric and magnetic materials that the aether relations apply only to the charge and current potentials of the resultant charge and current distribution of *all kinds* of charge and current. Charge and current distributions arising from the polarization and magnetization of a material medium are to be included in the charge-current field σ, and $(\boldsymbol{H}, \boldsymbol{D})$ is a resultant potential.

We shall enlarge on this point in Sect. 283. Here it suffices that our intuitive motivation for adopting the aether relations (279.1) as valid both inside and outside of "matter" is based on the guiding principle used by LORENTZ [1915, 3]. In LORENTZ's theory, matter is regarded as a collection of charged point particles which exist and move through an aether or vacuum whose properties are unaffected by the presence or motion of the particles. Here we treat matter as a continuous medium but carry over the Lorentz hypothesis that the aether relations are unaffected by the presence of matter. The effects of polarization and magnetization in the Lorentz electron theory were determined by arguments based on a particle model of a dielectric and magnetic material medium. A treatment of these effects regarding matter as a continuous medium is given in Sect. 283.

At the same time, we give warning that the aether relations represent an assumption not adopted in every existing theory of electromagnetism[1], whereas the conservation laws of charge and magnetic flux are, to our knowledge, common to all. Thus the considerations of this subchapter are of a rather special nature, and we have attempted to present but one point of view. The principal objective of the chapter as a whole is to formulate and develop the conservation laws of electromagnetic theory. These laws hold independently of the aether relations[2], and the contents of this subchapter do not in any way restrict the considerations in Subchapter II.

280. The world tensor form of the Maxwell-Lorentz aether relations. Let the constant c^2 be defined by

$$c^2 \equiv \frac{1}{\varepsilon_0 \mu_0}. \tag{280.1}$$

The fundamental nature of this constant will become apparent as we proceed; it is called the square of the *speed of light in vacuum*. Consider a space-time co-ordinate system (z, t) for which the aether relations are valid. As we have seen, from the classical view of space-time geometry the frame (z, t) will be one of a restricted subclass of all Galilean and Euclidean space-time frames. Consider a world contravariant, absolute, symmetric tensor of rank two whose components in the frame (z, t) have the particular values

$$\|\overset{-1}{\gamma}{}^{\Omega\Delta}\| \doteq \begin{bmatrix} \delta^{rs} & 0 \\ 0 & -\frac{1}{c^2} \end{bmatrix}. \tag{280.2}$$

Now the components of a tensor in a general system of co-ordinates are determined uniquely by its transformation law and its components in any one co-ordinate system, so that (280.2) determines the components of $\overset{-1}{\gamma}$ in every space-time co-ordinate system. The determinant of the inverse γ of $\overset{-1}{\gamma}$ is given by

$$\det \gamma = -c^2 \tag{280.3}$$

in the frame (z, t) and transforms as a scalar density of weight 2 under general transformations of the co-ordinates. Now consider the world tensor equation

$$\eta^{\Omega\Delta} = \sqrt{\frac{\varepsilon_0}{\mu_0}}\,(-\det\gamma)^{\frac{1}{2}}\,\overset{-1}{\gamma}{}^{\Omega\Psi}\,\overset{-1}{\gamma}{}^{\Delta\Theta}\,\varphi_{\Psi\Theta}. \tag{280.4}$$

By the quotient rule of tensor algebra, we easily verify that (280.4) is indeed a tensor equation; i.e., the rank and weight of both sides of the equation agree.

---

[1] ABRAHAM [1909, 1] reviews several of the points of view regarding the aether relations in material media.

[2] A similar distinction was made in Chap. E, where developments resting on an equation of state were separated from the general theory of energy. Cf. the remarks at the end of footnote 5, pp. 617—618.

Since it is a tensor equation, it will be satisfied in every co-ordinate system if it is satisfied in one co-ordinate system. In the frame $(\boldsymbol{z}, t)$ where the $\overset{-1}{\gamma}$ have the values (280.2), we can verify that the tensor equation (280.4) is satisfied if the Maxwell-Lorentz aether relations (279.1) are satisfied. Thus we call (280.4) the *world tensor form of the Maxwell-Lorentz aether relations.*

For some purposes it proves convenient to write (280.4) in a somewhat different form. Consider the world contravariant tensor of weight $\frac{1}{2}$ defined by

$$\overset{-1}{\mathfrak{G}}{}^{\Omega\Delta} \equiv \frac{\overset{-1}{\gamma}{}^{\Omega\Delta}}{\sqrt{-\det\overset{-1}{\gamma}}}. \tag{280.5}$$

The determinant of $\overset{-1}{\mathfrak{G}}$ is an absolute world scalar having the value -1 in every co-ordinate system. We find that the aether relations can be put in the alternative invariant form

$$\eta^{\Omega\Delta} = \sqrt{\frac{\varepsilon_0}{\mu_0}}\, \overset{-1}{\mathfrak{G}}{}^{\Omega\Psi}\, \overset{-1}{\mathfrak{G}}{}^{\Delta\Theta}\, \varphi_{\Psi\Theta}. \tag{280.6}$$

In the frames where $\overset{-1}{\gamma}$ has the special form (280.2), $\overset{-1}{\mathfrak{G}}$ will have the form

$$\overset{-1}{\mathfrak{G}} = \sqrt{c}\begin{bmatrix} \delta^{rs} & 0 \\ 0 & -\frac{1}{c^2} \end{bmatrix}. \tag{280.7}$$

281. Dimensional transformations and the aether relations. The absolute dimension of the world tensor fields σ, φ, η, and $\boldsymbol{\alpha}$ are given by

$$\left.\begin{aligned} &\text{abs. dim. } \sigma = [Q], \quad &&\text{abs. dim. } \varphi = [\phi], \\ &\text{abs. dim. } \eta = [Q], \quad &&\text{abs. dim. } \boldsymbol{\alpha} = [\phi]. \end{aligned}\right\} \tag{281.1}$$

From these dimensions and the rules of Sect. 273 follow the physical dimensions of the fields $\boldsymbol{B}$, $\boldsymbol{E}$, $\boldsymbol{J}$, Q, $\boldsymbol{D}$, $\boldsymbol{H}$, $\boldsymbol{A}$, and V. Some of these have been stated elsewhere already, but we shall list all of them now for easy reference:

$$\left.\begin{aligned} &\text{phys. dim. } \boldsymbol{B} = [\phi\, L^{-2}], \quad &&\text{phys. dim. } \boldsymbol{E} = [\phi\, L^{-1}\, T^{-1}], \\ &\text{phys. dim. } \boldsymbol{J} = [Q\, L^{-2}\, T^{-1}], \quad &&\text{phys. dim. } Q = [Q\, L^{-3}], \\ &\text{phys. dim. } \boldsymbol{D} = [Q\, L^{-2}], \quad &&\text{phys. dim. } \boldsymbol{H} = [Q\, L^{-1}\, T^{-1}], \\ &\text{phys. dim. } \boldsymbol{A} = [\phi\, L^{-1}], \quad &&\text{phys. dim. } V = [\phi\, T^{-1}]. \end{aligned}\right\} \tag{281.2}$$

Thus, in order that the aether relations be invariant under independent transformations of the units of length, time, charge, and magnetic flux, we must have

$$\text{phys. dim. } \varepsilon_0 = [\phi^{-1}\, Q\, L^{-1}\, T], \quad \text{phys. dim. } \mu_0 = [\phi\, Q^{-1}\, L^{-1}\, T]. \tag{281.3}$$

The constant c defined in (280.1) has, therefore, the dimension

$$\text{phys. dim. } c = [L\, T^{-1}]. \tag{281.4}$$

The dimension of the constant $\sqrt{\frac{\varepsilon_0}{\mu_0}}$ occurring in the world invariant forms of the aether relations (280.4) and (280.6) is given by

$$\text{phys. dim. } \sqrt{\frac{\varepsilon_0}{\mu_0}} = [Q\, \phi^{-1}]. \tag{281.5}$$

The absolute dimension and physical dimension of absolute world scalars and constants are equal. From (281.5) and (280.6) we see that dimensional invariance of (280.6) requires that

$$\text{abs. dim. } \mathfrak{G} = 1, \tag{281.6}$$

which is consistent with the fact that the determinant of this tensor has the value -1 in all co-ordinate systems and unit systems. Dimensional invariance of the canonical form (280.2) requires that

$$\text{abs. dim. } \overset{-1}{\gamma} = [L^{-2}]. \tag{281.7}$$

Of course, this requires that the absolute dimension of the inverse γ be $[L^2]$.

When magnetic flux, charge, length, and time are measured in units of

$$Q = 1 \text{ Coulomb}, \quad \Phi = 1 \text{ Weber}, \quad L = 1 \text{ Meter}, \quad T = 1 \text{ Second}, \tag{281.8}$$

the fundamental constants in the aether relations have the values[1]:

$$\left.\begin{aligned}
\varepsilon_0 &= 8.854 \times 10^{-12}\, \frac{\text{Coulomb-Second}}{\text{Weber-Meter}},\\
\mu_0 &= 1.257 \times 10^{-6}\, \frac{\text{Weber-Second}}{\text{Coulomb-Meter}},\\
c &= 2.998 \times 10^{8}\, \frac{\text{Meter}}{\text{Second}},\\
\sqrt{\frac{\varepsilon_0}{\mu_0}} &= 2.654 \times 10^{-3}\, \frac{\text{Coulomb}}{\text{Weber}}.
\end{aligned}\right\} \tag{281.9}$$

282. The Lorentz invariance of the Maxwell-Faraday aether relations, Lorentz transformations, and Lorentz frames. In Sect. 152 we showed that the Galilean (inertial) frames of classical mechanics could be characterized as the preferred co-ordinates in space-time for which the world tensors $g^{\Omega\Delta}$, t_Δ, and the Galilean connection $\boldsymbol{\Gamma}$ assumed the canonical forms

$$\boldsymbol{g} = \begin{bmatrix} \delta^{rs} & 0 \\ 0 & 0 \end{bmatrix}, \quad \boldsymbol{t} = (0, 0, 0, 1), \quad \boldsymbol{\Gamma} = 0. \tag{282.1}$$

The co-ordinates of any two such frames must be related by a Galilean transformation having the form

$$\left.\begin{aligned}
z^{r'} &= A^{r'}{}_{r}\, z^r + u^{r'} z^4 + \text{const},\\
z^{4'} &= z^4 + \text{const},
\end{aligned}\right\} \tag{282.2}$$

where $\boldsymbol{A}$ is an orthogonal matrix and the $u^{r'}$ are constants representing the relative velocity of the two Galilean frames. If we also consider transformations of the units of length and time and assign $\boldsymbol{g}$ the absolute dimension $[L^{-2}]$ and $\boldsymbol{t}$ the absolute dimension $[T]$, the co-ordinates of a Galilean frame based on the units of length L and T are related to the co-ordinates of a Galilean frame based on the units L' and T' by a transformation of the form

$$\left.\begin{aligned}
z^{r'}(L') &= L\left(A^{r'}{}_{r}\, z^r(L) + u^{r'} z^4(T) + \text{const}\right),\\
z^{4'}(T') &= T\left(z^{4'}(T) + \text{const}\right).
\end{aligned}\right\} \tag{282.3}$$

So as to distinguish these more general transformations from those having the special form (282.2), we shall call (282.3) a *generalized Galilean transformation.*

Following the same general procedure outlined above for the case of Galilean space-time, let us consider the co-ordinate transformations and space-time frames

[1] These values are quoted from STRATTON [1941, *8*, p. 601].

defined by the canonical form (280.2) of the *Lorentz tensor* $\overset{-1}{\gamma}$. A co-ordinate transformation which leaves the Lorentz tensor invariant must satisfy the equations

$$\overset{-1}{\gamma}{}^{\Omega'\Delta'}(L', T') = L^{-2}\frac{\partial x^{\Omega'}}{\partial x^{\Omega}}\frac{\partial x^{\Delta'}}{\partial x^{\Delta}}\overset{-1}{\gamma}{}^{\Omega\Delta}(L, T), \tag{282.4}$$

where

$$\overset{-1}{\gamma}(L, T) = \begin{bmatrix}\delta^{rs} & 0\\ 0 & -\frac{1}{c^2}\end{bmatrix}, \quad \overset{-1}{\gamma}{}'(L', T') = \begin{bmatrix}\delta^{rs} & 0\\ 0 & -\frac{T^2}{c^2 L^2}\end{bmatrix}. \tag{282.5}$$

Since the $\overset{-1}{\gamma}$ and $\overset{-1}{\gamma}{}'$ are constants, it follows that the co-ordinate transformation must be linear. A coordinate transformation satisfying the (282.4) will be called a *generalized Lorentz transformation*. One can deduce the general form of these transformations by multiplying the well known special Lorentz transformation[1],

$$z_1' = \frac{z_1 - u z_4}{\sqrt{1-\frac{u^2}{c^2}}}, \quad z_2' = z_2, \quad z_3' = z_3, \quad z_4' = \frac{z_4 - \frac{u}{c^2} z_1}{\sqrt{1-\frac{u^2}{c^2}}}, \tag{282.6}$$

by an orthogonal transformation of the spatial co-ordinates, simple extensions of the co-ordinates corresponding to transformations of L and T, and by time inversions. The generalized Lorentz transformations have the form

$$\left.\begin{aligned} z^{r'}(L') &= L A^{r'}{}_r\{[\delta^r{}_s + (\zeta - 1)\,u^{-2} u^r u_s]\, z^s(L) - \zeta\, u^r z^4(T)\},\\ z^{4'}(T') &= \pm T\zeta\left\{z^4(T) - \frac{1}{c^2} u_r z^r(L)\right\}, \end{aligned}\right\} \tag{282.7}$$

where $\boldsymbol{A}$ is an orthogonal matrix, $\zeta \equiv \left(1-\frac{u^2}{c^2}\right)^{-\frac{1}{2}}$, and $u^2 \equiv u_r u^r < c^2$ is the *squared relative speed of the two Lorentz frames.*

Various subgroups of the group of generalized Lorentz transformations have received special attention and special names. If we keep the units of length and time fixed, so that $L = T = 1$ in (282.7), we get what has been called the *extended Lorentz group*[2]. If we transform the units of length and time by the same factor so that $L = T$ and c is invariant, we get what BATEMAN[3] has called the *group of spherical wave transformations*. This subgroup is also called the *conformal Lorentz group*. BATEMAN noted that the aether relations are invariant under the group of spherical wave transformations. If we require that $\det \boldsymbol{A} = +1$ and disallow time inversions corresponding to the minus sign in $(282.7)_2$, we get the *proper* Lorentz transformations. On writing the aether relations in the form (280.6), we see that they are invariant under the conformal Lorentz group.

A Lorentz transformation with $u/c \ll 1$ approximates a Galilean transformation. In special relativity theory, the notion of a stationary aether is abandoned. The inertial frames of relativistic mechanics are identified with the Lorentz frames. The equations and definitions of mechanics are revised so as to be Lorentz invariant rather than Galilean invariant as in classical mechanics.

The world invariant form of the aether relations (280.6) provides us with our principal motivation for assuming that the charge $\mathfrak{C}[(\mathscr{S}_3, Q)]$ transforms as an axial scalar rather than as an absolute scalar under general transformations of the space-time coordinates. With magnetic flux an absolute scalar and charge an axial scalar as we have assumed, the transformation law of the tensor density $\overset{-1}{\mathfrak{G}}$ is

$$\overset{-1}{\mathfrak{G}}{}^{\Omega\Delta} = |(\boldsymbol{x}'/\boldsymbol{x})|^{-\frac{1}{2}}\frac{\partial x^{\Omega'}}{\partial x^{\Omega}}\frac{\partial x^{\Delta'}}{\partial x^{\Delta}}\overset{-1}{\mathfrak{G}}{}^{\Omega\Delta}. \tag{282.8}$$

Had we assumed that both charge and magnetic flux were absolute scalars, the charge-current potential η would have been an axial density. The transformation law (280.8) for the $\overset{-1}{\mathfrak{G}}$

[1] See, for example, BERGMANN [1942, *1*].
[2] CORSON [1953, *6*, Chap. I].
[3] BATEMAN [1910, *1*].

would then have involved the square root of the Jacobian rather than the square root of the absolute value of the Jacobian. Thus, in some coordinate systems, the coefficients $\overset{-1}{\mathfrak{G}}$ would have been imaginary or complex-valued. We have considered this undesirable. Another motivation for assuming charge to be an axial scalar is that the current density $\boldsymbol{J}$ and the charge density Q, being tensor densities under time-independent transformations of the spatial co-ordinates when this assumption is adopted, transform as absolute tensors under the group of time independent orthogonal transformations of the spatial co-ordinates. This is the transformation usually assumed for these quantities in traditional treatments of electromagnetic theory. The distinction between axial tensors and absolute tensors and the distinction between axial densities and densities is made necessary only because we find the consideration of improper co-ordinate transformations to be of some major concern in many physical theories. It is for this reason that we did not restrict ourselves at the outset to the group of proper co-ordinate transformations.

283. Polarization and magnetization. α) *The principle of Ampère and Lorentz.* *Polarization* and *magnetization* are auxiliary fields introduced into the general theory so as to serve in formulating constitutive relations for special types of materials called *dielectrics* and *magnets* or *magnetic materials.* In the classical theory of dielectrics and magnets, surfaces across which the electromagnetic properties of the medium change abruptly are important for the description of many electromagnetic phenomena. These surfaces are most often associated with surfaces of discontinuity in the polarization and magnetization. Surfaces of discontinuity in the polarization and magnetization are in turn associated with surface distributions of charge and current. The point of view we adopt here is that *charge and current are the fundamental entities* while polarization and magnetization are simply auxiliary fields introduced as mathematical devices providing a convenient description of special distributions of charge and current in special types of materials. This may be called the *principle of Ampère and Lorentz*[1]. Thus, if we anticipate a treatment of surfaces of discontinuity in the polarization and magnetization, we must first introduce surface distributions of charge and current into the conservation law of charge. For mathematical simplicity, at the outset we did not burden the reader with this additional complication. The fundamental physical idea remains the same, conservation of charge, but now we give a slightly more general mathematical expression for the measure of charge. The mathematician will readily infer a general mathematical statement for the physical idea of conservation in terms of additive set functions.

β) *Surface distributions of charge and current.* Let $\Sigma(\boldsymbol{x})=0$ be the equation of a 3-dimensional surface in space-time representing the history of a 2-dimensional surface in space bearing a surface distribution of charge and current. We now replace (270.1) by the more general assumption that the charge is expressible in the form

$$\mathfrak{C}[\mathscr{S}_3, Q] = \int \sigma \cdot d\mathscr{S}_3 + \int_{\mathscr{S}_3\cap\Sigma} \omega \cdot d\mathscr{S}_2^*, \tag{283.1}$$

where σ is the *volume density of charge-current* and the world contravariant 2-vector density ω is the *surface density of charge-current.* The field ω is defined only at events corresponding to the surface $\Sigma(\boldsymbol{x})=0$. In (283.1), $\mathscr{S}_3\cap\Sigma$ denotes the intersection of the arbitrary 3-dimensional surface $\mathscr{S}_3$ and the special 3-dimensional surface $\Sigma(\boldsymbol{x})=0$. Here we have assumed that, if the intersection is not empty, $\mathscr{S}_3\cap\Sigma$ is a 2-dimensional surface and $d\mathscr{S}_2^*$ is its differential element.

Generalizing (270.5), we now write the law of conservation of charge in the form

$$\oint \sigma \cdot d\mathscr{S}_3 + \int \omega \cdot d\mathscr{S}_2^* = 0. \tag{283.2}$$

[1] Whittaker [1951, *39*, p. 88, pp. 393–400].

We know that, for all surfaces $\mathcal{S}_3$ not intersecting $\Sigma=0$, as a consequence of (283.2) there exist potential fields $\eta_{\mathscr{V}}$ such that

$$\int \sigma \cdot d\mathcal{S}_3 = -\oint \eta_{\mathscr{V}} \cdot d\mathcal{S}_2. \tag{283.3}$$

We can also construct fields $\eta_{\mathcal{S}}$ such that

$$\oint \eta_{\mathcal{S}} \cdot d\mathcal{S}_2 = 0 \tag{283.4}$$

for every circuit not intersecting $\Sigma(\boldsymbol{x})=0$, and such that

$$\oint \eta_{\mathcal{S}} \cdot d\mathcal{S}_2 = -\int \omega \cdot d\mathcal{S}_2^* \tag{283.5}$$

for every circuit intersecting $\Sigma=0$. Therefore, we can write the charge-current potential equation in the form

$$\int \sigma \cdot d\mathcal{S}_3 + \int_{\mathcal{S}_3 \cap \Sigma} \omega \cdot d\mathcal{S}_2^* = -\oint \eta \cdot d\mathcal{S}_2, \tag{283.6}$$

where $\eta = \eta_{\mathscr{V}} + \eta_{\mathcal{S}}$. Eq. (283.6) generalizes (271.1). If the fields are continuously differentiable except perhaps at $\Sigma(\boldsymbol{x})=0$, at every event not on $\Sigma(\boldsymbol{x})=0$ we have

$$\sigma = \operatorname{div} \eta. \tag{283.7}$$

However, at $\Sigma(\boldsymbol{x})=0$, the potential η must have a discontinuity consistent with the condition

$$[\eta] \cdot \operatorname{grad} \Sigma = -\omega \cdot \operatorname{grad} \Sigma, \tag{283.8}$$

or, in component form

$$[\eta^{\Omega\Delta}]\, \partial_\Delta \Sigma = -\omega^{\Omega\Delta}\, \partial_\Delta \Sigma. \tag{283.9}$$

The aether relations (280.4) or (280.6) are postulated to hold between the electromagnetic field $\boldsymbol{\varphi}$ and the generalized potential η of (283.6). Thus, the electromagnetic field is discontinuous at a surface bearing a distribution of charge and current.

The 3-dimensional form of the law of conservation of charge and the charge and current equations taking into account a surface distribution of charge and current can be obtained from the world invariant equations (283.2) and (283.6) by introducing a Euclidean frame. The resulting equations have an interesting but complicated structure. Since we do not make use of these equations here, we leave to the reader the task of deriving them.

Having established these preliminary results on surface distributions of charge and current, we are in a position to give a fairly general treatment of polarization and magnetization in moving and deforming material media, allowing that these fields may be discontinuous at special surfaces.

γ) Polarization charge and current and magnetization current. As an introduction to the more difficult dynamical case, we note first some simpler and better known results of the static theory of dielectrics and magnets. These latter theories are concerned with the determination of the electric and magnetic fields in regions of space containing dielectric and magnetic materials at rest.

It is customary in the electrostatic theory of dielectrics to divide the charge distribution into two types: (1) the *bound charge* or *polarization charge*; (2) the *free charge*[1]. The polarization charge contained in a 3-dimensional spatial region $\mathscr{v}$ is given by

$$\mathfrak{C}[\mathscr{v}, Q] = -\oint \boldsymbol{P} \cdot d\boldsymbol{s} \tag{283.10}$$

[1] The polarization charge is also called the "induced" charge, and the free charge is also called the "real" or "true" charge. See, for example, STRATTON [1941, *8*, Sect. 3.13, p. 183].

where $\mathscr{s}$ is the complete boundary of $\mathscr{v}$ and $\boldsymbol{P}$ is the *density of polarization.* One generally considers problems where $\boldsymbol{P}$ is a continuously differentiable field except at special surfaces where it suffers a jump or discontinuity. Excepting points on these surfaces, we can use the integral theorem to transform the surface integral (283.10) to a volume integral. Thus, if $\mathscr{v}$ is a region in which the polarization field is everywhere continuously differentiable, we have

$$\mathfrak{C}[\mathscr{v}, Q] = -\int \operatorname{div} \boldsymbol{P}\, dv. \tag{283.11}$$

The scalar field $-\operatorname{div} \boldsymbol{P}$ may be called the *volume density of polarization charge.* If the region $\mathscr{v}$ is divided into two regions $\mathscr{v}^+$ and $\mathscr{v}^-$ by a surface $\mathscr{s}^*$ across which $\boldsymbol{P}$ suffers a jump $[\boldsymbol{P}]$, and if $\boldsymbol{P}$ is continuously differentiable in the closure of the regions $\mathscr{v}^+$ and $\mathscr{v}^-$, we can write (283.10) in the form

$$\mathfrak{C}[\mathscr{v}, Q] = -\int \operatorname{div} \boldsymbol{P}\, dv - \int_{\mathscr{v}\cap\mathscr{s}^*} [\boldsymbol{P}] \cdot d\boldsymbol{a}^*. \tag{283.12}$$

The differential element $d\boldsymbol{a}^*$ can be written in the form $\boldsymbol{n}\, da$, where da is the element of area, and $\boldsymbol{n}$ is the unit normal to $\mathscr{s}^*$. The scalar $[\boldsymbol{P}] \cdot \boldsymbol{n}$ thus plays the role of a surface density of charge. The fields $-\operatorname{div} \boldsymbol{P}$ and $[\boldsymbol{P}] \cdot \boldsymbol{n}$ are called the *Poisson-Kelvin equivalent charge distributions of a polarized medium*[1].

Consider next the case when the polarization $\boldsymbol{P}$ depends upon the time. The time rate of change of the polarization charge contained in a stationary region $\mathscr{v}$ is given by

$$\frac{d}{dt}\mathfrak{C}[\mathscr{v}, Q] = -\oint \frac{\partial \boldsymbol{P}}{\partial t} \cdot d\boldsymbol{a} \tag{283.13}$$

provided the polarization field be sufficiently smooth as to allow interchanging the operations of integration and differentiation. Thus, in this simple case, it is reasonable to call $\partial \boldsymbol{P}/\partial t$ the *current of polarization.* In the more general case when the region $\mathscr{v}$ is moving with the particles of a motion, we have

$$\frac{d}{dt}\mathfrak{C}[\mathscr{v}, Q] = -\oint \frac{d_c \boldsymbol{P}}{dt} \cdot d\boldsymbol{a}. \tag{283.14}$$

Thus, by analogy with (275.3), the convected time flux $d_c\boldsymbol{P}/dt$ will be called the *polarization current relative to the particles of the motion.*

In the case of magnetic materials, it is known[2] that the magnetic field of a stationary magnet characterized by a magnetization field $\boldsymbol{M}$ is equivalent to the magnetic field of a volume distribution of magnetization current $\boldsymbol{J}_{\mathfrak{M}}$ and a surface distribution of magnetization current $\boldsymbol{K}_{\mathfrak{M}}$ on surfaces of discontinuity of the field $\boldsymbol{M}$ such that

$$\int_{\mathscr{s}} \boldsymbol{J}_{\mathfrak{M}} \cdot d\boldsymbol{a} + \int_{\mathscr{s}\cap\mathscr{s}^*} \boldsymbol{K}_{\mathfrak{M}} \cdot d\boldsymbol{x}^* = \oint_{c} \boldsymbol{M} \cdot d\boldsymbol{x} \tag{283.15}$$

where c is the complete boundary of $\mathscr{s}$ and c^* is the curve of intersection of the surface of discontinuity $\mathscr{s}^*$ and $\mathscr{s}$. Thus we have

$$\boldsymbol{J}_{\mathfrak{M}} = \operatorname{curl} \boldsymbol{M}, \qquad \boldsymbol{K}_{\mathfrak{M}} = [\boldsymbol{M}]. \tag{283.16}$$

The differential element $d\boldsymbol{x}^*$ can be written in the form $d\boldsymbol{x}^* = \boldsymbol{\tau}\, ds$, where ds is the differential element of length along the curve $\mathscr{s} \cap \mathscr{s}^*$ and $\boldsymbol{\tau}$ is the unit tangent vector to this curve. Let $\boldsymbol{n}$ denote the unit normal to the surface of discontinuity

---

[1] SMITH-WHITE [1951, *24*].
[2] STRATTON [1941, *8*, Sect. 4.10, p. 242].

s^* (see Fig. 47). We can then introduce the unit vector $\boldsymbol{m}$ through the relation $\boldsymbol{\tau}=\boldsymbol{n}\times\boldsymbol{m}$ and write the line integral over c^* in (283.15) in the alternative form

$$\int_{s\cap s^*} \boldsymbol{K}_{\mathfrak{M}}\cdot d\boldsymbol{x}^* = \int_{s\cap s^*} \overline{\boldsymbol{K}}_{\mathfrak{M}}\cdot \boldsymbol{m}\, d s, \tag{283.17}$$

where

$$\overline{\boldsymbol{K}}_{\mathfrak{M}} = \boldsymbol{K}_{\mathfrak{M}}\times\boldsymbol{n} = [\boldsymbol{M}]\times\boldsymbol{n}. \tag{283.18}$$

The surface density of magnetization current is usually defined as the vector field $\overline{\boldsymbol{K}}_{\mathfrak{M}}$ rather than the field $\boldsymbol{K}_{\mathfrak{M}}$. The fields curl $\boldsymbol{M}$ and $[\boldsymbol{M}]$ or $[\boldsymbol{M}]\times\boldsymbol{n}$ may be called the *equivalent Amperian currents of magnetized matter.*

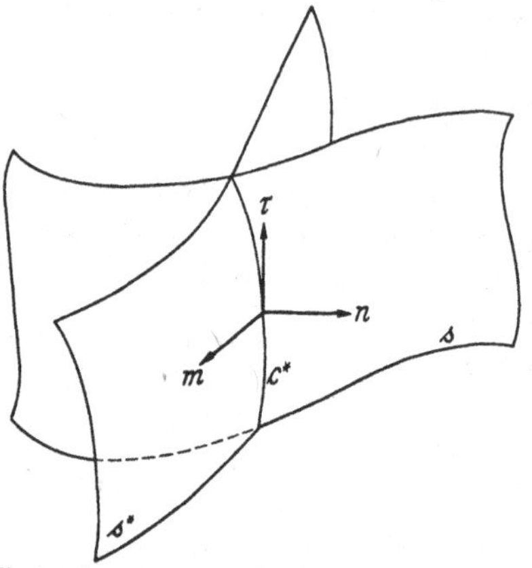

Fig. 47. Geometry of a surface of discontinuity in the magnetization.

Eq. (274.5) states that the current $\boldsymbol{J}$ is, in the general case, the sum of the conduction current $\mathfrak{J}$ relative to the particles of a motion and the "convection current" $Q\boldsymbol{v}$.

$$\boldsymbol{J} = \mathfrak{J} + Q\boldsymbol{v}. \tag{283.19}$$

Now we have seen that a polarized material medium has an equivalent volume distribution of charge $-\operatorname{div}\boldsymbol{P}$, and we have seen that the convected time flux of the polarization field $d_c\boldsymbol{P}/dt$ is analogous to the conduction current. We are thus led to define the *current of polarization* in a moving medium by

$$\boldsymbol{J}_{\mathfrak{P}} = \frac{d_c\boldsymbol{P}}{dt} - \operatorname{div}\boldsymbol{P}\,\boldsymbol{v}. \tag{283.20}$$

We shall call the term $-\operatorname{div}\boldsymbol{P}\,\boldsymbol{v}$, the *current of dielectric convection*[1]. If we expand the convected time-flux using the definitions (277.10) to (277.11) we get

$$\boldsymbol{J}_{\mathfrak{P}} = \frac{d\boldsymbol{P}}{dt} + \operatorname{curl}(\boldsymbol{P}\times\boldsymbol{v}). \tag{283.21}$$

In the general case of a polarized and magnetized moving material medium in which there is a distribution of free charge $Q_{\mathfrak{F}}$ and free current, $\boldsymbol{J}_{\mathfrak{F}}$, the resultant or total volume density of charge and current $\boldsymbol{J}$ and Q is given by

$$\left.\begin{aligned} Q &= Q_{\mathfrak{F}} - \operatorname{div}\boldsymbol{P},\\ \boldsymbol{J} &= \boldsymbol{J}_{\mathfrak{F}} + \frac{d_c\boldsymbol{P}}{dt} - \operatorname{div}\boldsymbol{P}\boldsymbol{v} + \operatorname{curl}\boldsymbol{M}.\end{aligned}\right\} \tag{283.22}$$

If we substitute these expressions for Q and $\boldsymbol{J}$ into the charge and current equations (278.4) and (278.5), we obtain LORENTZ's form of these equations for moving polarized and magnetized material media[2].

δ) World tensor representation of polarization and magnetization charge-current. Consider the world contravariant 2-vector density $\boldsymbol{\pi}$ whose components in a Euclidean frame are given by

$$\boldsymbol{\pi} = (\operatorname{dual}\boldsymbol{M} + \operatorname{dual}(\boldsymbol{P}\times\boldsymbol{v}), -\boldsymbol{P}). \tag{283.23}$$

[1] According to WHITTAKER [1951, *39*, p. 398], the term curl $\boldsymbol{P}\times\boldsymbol{v}$ is called the current of dielectric convection. We prefer the arrangement of terms and definitions used here because the analogous terms in (283.19) and (283.20) have similar transformation properties, while the analogous terms in (283.19) and (283.21) do not.

[2] WHITTAKER [1951, *39*, p. 398].

The components of the natural divergence of π in a Euclidean frame are given by

$$\sigma_{\mathfrak{B}} = \operatorname{div} \pi = (J_{\mathfrak{P}} + J_{\mathfrak{M}}, -\operatorname{div} P). \tag{283.24}$$

The world vector density $\sigma_{\mathfrak{B}}$ may be called the *volume density of bound charge-current.* The distribution of charge and current represented by $\sigma_{\mathfrak{B}}$ describes the effects of polarization and magnetization in the moving medium. An invariant decomposition of the magnetization-polarization world tensor π is obtained by introducing the *world polarization and magnetization densities* defined in terms of π by

$$\left.\begin{aligned} \mathfrak{P}^{\Omega} &\equiv \pi^{\Omega \Delta} t_{\Delta}, \\ \mathfrak{M}^{\Omega \Delta} &\equiv \pi^{\Delta \Omega} - 2\mathfrak{P}^{[\Omega} v^{\Delta]}. \end{aligned}\right\} \tag{283.25}$$

One sees that $\mathfrak{P}$ and $\mathfrak{M}$ are space tensors, so that their vanishing in one Euclidean frame implies their vanishing in every Euclidean frame. From the definitions (283.23) to (283.25), we have

$$\pi = \mathfrak{M} + \operatorname{dual} \mathfrak{P} \times v \tag{283.26}$$

$$\mathfrak{M} = (\operatorname{dual} M, 0), \quad \mathfrak{P} = (P, 0). \tag{283.27}$$

If π is discontinuous across a surface $\Sigma(x) = 0$ in space-time, we introduce a surface distribution of bound charge-current represented by a world contravariant 2-vector density $\omega_{\mathfrak{B}}$ such that

$$\omega_{\mathfrak{B}} = -[\pi]. \tag{283.28}$$

From (283.24) and (283.28) we get

$$\mathfrak{C}_{\mathfrak{B}}[\mathscr{S}_3, Q] = \int \sigma_{\mathfrak{B}}\, d\mathscr{S}_3 + \int_{\mathscr{S}_3 \cap \Sigma} \omega_{\mathfrak{B}} \cdot d\mathscr{S}_2^* = \oint \pi \cdot d\mathscr{S}_2. \tag{283.29}$$

Comparing this equation with (283.6), we see that the polarization-magnetization field π is a potential of a conservative distribution of bound charge. If we assume that the free charge is representable in the form

$$\mathfrak{C}_{\mathfrak{F}}[\mathscr{S}_3, Q] = \int \sigma_{\mathfrak{F}} \cdot d\mathscr{S}_3, \tag{283.30}$$

we get for the resultant charge

$$\left.\begin{aligned} \mathfrak{C}[\mathscr{S}_3, Q] &= \mathfrak{C}_{\mathfrak{B}}[\mathscr{S}_3, Q] + \mathfrak{C}_{\mathfrak{F}}[\mathscr{S}_3, Q] \\ &= \int (\sigma_{\mathfrak{F}} + \sigma_{\mathfrak{B}}) \cdot d\mathscr{S}_3 + \int \omega_{\mathfrak{B}} \cdot d\mathscr{S}_2^* = -\oint \eta \cdot d\mathscr{S}_2, \end{aligned}\right\} \tag{283.31}$$

where the *resultant potential* η is assumed to be related to the electromagnetic field by the fundamental aether relations (280.6). Substituting (283.29) in (283.31) we get the result

$$\mathfrak{C}_F[\mathscr{S}_3, Q] = \int \sigma_F \cdot d\mathscr{S}_3 = -\oint \Upsilon \cdot d\mathscr{S}_2, \tag{283.32}$$

where

$$\Upsilon \equiv \eta - \pi = (\operatorname{dual}(H - M - P \times v), D + P). \tag{283.33}$$

To state this result in words, we can say that, in polarizable and magnetizable media, there exists a world potential Υ for that portion of the charge and current distribution which has been characterized as free. Let us introduce the special notation

$$\Upsilon = (\operatorname{dual} \mathfrak{H}, \mathfrak{D}) \tag{283.34}$$

for the components of Υ in a Euclidean frame. According to (283.33) $\mathfrak{H}$ and $\mathfrak{D}$ are given by

$$\left.\begin{aligned} \mathfrak{H} &= H - M - P \times v, \\ \mathfrak{D} &= D + P. \end{aligned}\right\} \tag{283.35}$$

It is important that we distinguish between the fields $(\boldsymbol{D}, \boldsymbol{H})$ and $(\mathfrak{D}, \mathfrak{H})$ in polarizable and magnetizable materials. The characteristic feature of the resultant potentials $\boldsymbol{D}$ and $\boldsymbol{H}$ not shared by the partial potentials $\mathfrak{D}$ and $\mathfrak{H}$ is that they are related to the electric field $\boldsymbol{E}$ and the magnetic flux density $\boldsymbol{B}$ by the simple aether relations (279.1) *even in the interior of dielectrics and magnetic materials in motion.* In theories of dielectric and magnetic materials, it is customary to prescribe a constitutive relation between the polarization and magnetization fields and the electromagnetic field. This constitutive relation may be as general as a functional relationship where the polarization and magnetization at a given instant depend on the history of the electromagnetic field experienced by the polarized and magnetized particle plus other variables such as strain and temperature which describe the state of the material medium. Thus we see that the relationship between the partial potentials $\mathfrak{D}$ and $\mathfrak{H}$ and the electromagnetic field may be quite complicated and is not the same in all materials and for all motions. If we regard the charge-current distribution as prescribed, the aether relations have the effect of imposing certain additional conditions on the electromagnetic field other than the law of conservation of magnetic flux. In a polarized and magnetized material medium, these additional conditions are to be obtained by substituting $\varepsilon_0 \boldsymbol{E}$ for $\boldsymbol{D}$ and $\frac{1}{\mu_0}\boldsymbol{B}$ for $\boldsymbol{H}$ in (283.32). We are not to substitute $\varepsilon_0 \boldsymbol{E}$ and $\frac{1}{\mu_0}\boldsymbol{B}$ for the partial potentials $\mathfrak{D}$ and $\mathfrak{H}$. If the material medium is at rest in a Euclidean frame which is simultaneously a Lorentz frame, and if the constitutive relation for the polarization and magnetization is of the extremely simple type

$$\boldsymbol{P} = \varepsilon_0 \chi \boldsymbol{E}, \qquad \boldsymbol{M} = 0, \tag{283.36}$$

where χ is a constant, we have the relations

$$\mathfrak{D} = \varepsilon_0 K \boldsymbol{E}, \qquad \mathfrak{H} = \frac{1}{\mu_0}\boldsymbol{B}, \tag{283.37}$$

where $K = 1 + \chi$. A material characterized by these simple constitutive relations may be called a non-magnetic, linear, isotropic, rigid dielectric. If this same material is set in motion with respect to the reference frame for which the aether relations are valid, the constitutive relations (283.36) are replaced by

$$\boldsymbol{P} = \varepsilon_0 \chi (\boldsymbol{E} + \boldsymbol{v} \times \boldsymbol{B}) = \varepsilon_0 \mathfrak{E}, \qquad \boldsymbol{M} = 0. \tag{283.38}$$

Note that for a moving medium of this simple type, the relations between the partial potentials $\mathfrak{D}$ and $\mathfrak{H}$ are rather complicated. We have

$$\left.\begin{aligned} \mathfrak{D} &= \varepsilon_0 (K\boldsymbol{E} + \chi \boldsymbol{v} \times \boldsymbol{B}), \\ \mathfrak{H} &= \frac{1}{\mu_0}\left(\boldsymbol{B} + \frac{\chi}{c^2}\boldsymbol{v} \times (\boldsymbol{E} + \boldsymbol{v} \times \boldsymbol{B})\right). \end{aligned}\right\} \tag{283.39}$$

The point emphasized by LORENTZ in his electron theory is that the aether relations between the resultant potentials $\boldsymbol{D}$ and $\boldsymbol{H}$ and the electric field $\boldsymbol{E}$ and magnetic flux density $\boldsymbol{B}$ are unaffected by the presence of electrons or their motion, but it is clear that the simple linear relations (283.37) between $\mathfrak{D}$, $\mathfrak{H}$, $\boldsymbol{E}$ and $\boldsymbol{B}$ fail to hold in a moving medium of this type[1]. Finally, we caution the reader regarding notation. In most elementary and even advanced texts on

[1] The constitutive relations (283.38) for the moving medium of this simple type will be discussed further in Sect. 308, where we shall also consider their relativistic or Lorentz invariant generalization.

electromagnetic theory, a clear distinction is not made between the partial potentials $\mathfrak{D}$ and $\mathfrak{H}$ and the resultant potentials $\boldsymbol{D}$ and $\boldsymbol{H}$ in discussion of polarizable and magnetizable media. However, it is only when one attempts to treat the problem of dielectrics and magnets set in motion that serious difficulties arise from confusing these fields.

The world tensor form of the field equations and boundary conditions following from (283.32) is

$$\operatorname{div} \boldsymbol{\Upsilon} = \sigma_{\mathfrak{F}}, \tag{283.40}$$

$$[\boldsymbol{\Upsilon}] \cdot \operatorname{grad} \Sigma = 0, \tag{283.41}$$

where (283.41) must hold at an arbitrary surface $\Sigma(\boldsymbol{x}) = 0$ provided we exclude surface distributions of free charge and current.

If we refer all quantities to a Euclidean frame which is simultaneously a Lorentz frame, these equations take the 3-dimensional tensor form

$$\varrho_{\mathfrak{F}} - \operatorname{div} \boldsymbol{P} = \varepsilon_0 \operatorname{div} \boldsymbol{E}, \tag{283.42}$$

$$\boldsymbol{J}_{\mathfrak{F}} + \operatorname{curl} \boldsymbol{M} - \operatorname{div} \boldsymbol{P}\boldsymbol{v} + \frac{d_c \boldsymbol{P}}{dt} = \frac{1}{\mu_0} \operatorname{curl} \boldsymbol{B} - \varepsilon_0 \frac{\partial \boldsymbol{E}}{\partial t}, \tag{283.43}$$

$$[\varepsilon_0 \boldsymbol{E} + \boldsymbol{P}] \cdot \boldsymbol{n} = 0, \tag{283.44}$$

$$\boldsymbol{n} \times \left[\frac{1}{\mu_0} \boldsymbol{B} - \boldsymbol{M} - \boldsymbol{P} \times \boldsymbol{v}\right] + u_n [\varepsilon_0 \boldsymbol{E} + \boldsymbol{P}] = 0, \tag{283.45}$$

where $\boldsymbol{n}$ is the unit normal and u_n is the speed of the moving surface of discontinuity.

IV. Conservation of energy and momentum.

284. Electromagnetic energy, momentum, stress, and energy flux. If the polarization and magnetization vanish, there is general agreement that the *density of electromagnetic energy* U_e and the *density of electromagnetic momentum* $\mathfrak{P}$ are to be defined as follows:

$$U_e \equiv \boldsymbol{D} \cdot \boldsymbol{E} + \boldsymbol{B} \cdot \boldsymbol{H}, \tag{284.1}$$

$$\mathfrak{P} = \boldsymbol{D} \times \boldsymbol{B}. \tag{284.2}$$

We assume that the reference frame is one for which we have the aether relations (279.1) so that,

$$U_e = \varepsilon_0 E^2 + \frac{1}{\mu_0} B^2, \tag{284.3}$$

$$\mathfrak{P} = \varepsilon_0 \boldsymbol{E} \times \boldsymbol{B}, \tag{284.4}$$

expressing the energy and momentum in terms of the electromagnetic field. In any case, independently of any hypothesis regarding polarization and magnetization, from these definitions and from the field equations (278.1) through (278.7) we derive the following *identities:*

$$\frac{d}{dt} \int U_e \, dv + \oint (\boldsymbol{E} \times \boldsymbol{H}) \cdot d\boldsymbol{a} = - \int \boldsymbol{J} \cdot \boldsymbol{E} \, dv, \tag{284.5}$$

$$\frac{d}{dt} \int \mathfrak{P}_r \, dv - \oint m^k{}_r \, da_k = - \int [\varrho E_r + (\boldsymbol{J} \times \boldsymbol{B})_r] \, dv, \tag{284.6}$$

where $m^k{}_r$ is the *electromagnetic stress tensor,*

$$m^k{}_r \equiv D^k E_r + B^k H_r - \tfrac{1}{2} \delta^k_r (\boldsymbol{D} \cdot \boldsymbol{E} + \boldsymbol{B} \cdot \boldsymbol{H}). \tag{284.7}$$

We record the caution that proof of the identities (284.5) and (284.6) depends in an essential way on the aether relations $\boldsymbol{D}=\varepsilon_0\boldsymbol{E}$ and $\boldsymbol{H}=\frac{1}{\mu_0}\boldsymbol{B}$.

These identities have the form of *equations of balance* (Sect. 157). The scalar field $-\boldsymbol{J}\cdot\boldsymbol{E}$ is the supply of electromagnetic energy. Alternatively, $+\boldsymbol{J}\cdot\boldsymbol{E}$ may be regarded as the rate at which electromagnetic energy is converted into other forms of energy; while our development in this chapter has so far been purely electromagnetic, we expect such other forms of energy to include kinetic energy and internal energy. Similarly, the vector field $Q\boldsymbol{E}+\boldsymbol{J}\times\boldsymbol{B}$, called the *Lorentz force*, represents the rate at which electromagnetic momentum is transformed into mechanical momentum. The vector $\boldsymbol{E}\times\boldsymbol{H}$, called the *Poynting vector*, is the influx of electromagnetic energy. The electromagnetic stress tensor is the efflux of electromagnetic momentum; moreover, it is symmetric[1]:

$$m^{[r}{}_{s]}=0. \tag{284.8}$$

The formal and dimensional analogy between the identities (284.5) and (284.6) and the principles of balance of energy and linear momentum in continuum mechanics is immediate. However, the results here are identities, formal consequences of the conservation laws and the aether relations of electromagnetism, while the corresponding principles in mechanics are postulates. This difference and this similarity provide the motive for the connection between mechanics and electromagnetism that is the subject of the rest of this chapter.

Other definitions of electromagnetic momentum, energy, stress, and energy flux are sometimes made in the case of polarized and magnetized material media. For example, if the electromagnetic energy density $\overline{U}_e$ is defined as a solution of the equation

$$\frac{\partial \overline{U}_e}{\partial t}=\frac{\partial \mathfrak{D}}{\partial t}\cdot\boldsymbol{E}+\frac{\partial \boldsymbol{B}}{\partial t}\cdot\mathfrak{H}, \tag{284.9}$$

where the fields $\mathfrak{D}$ and $\mathfrak{H}$ are the partial potentials of the free charge and current as defined in Sect. 283, we obtain an identity of the form

$$\frac{d}{dt}\int \overline{U}_e\,dv+\oint(\boldsymbol{E}\times\boldsymbol{H})\cdot d\boldsymbol{a}=-\int \boldsymbol{J}_{\mathfrak{F}}\cdot\boldsymbol{E}\,dv, \tag{284.10}$$

where $\boldsymbol{J}_{\mathfrak{F}}$ is the density of free current. Similarly, suppose one retains the definition of electromagnetic momentum (284.2), but defines electromagnetic stress to be

$$\overline{m}^r{}_s\equiv\mathfrak{D}^r E_s+B^r H_s-\tfrac{1}{2}\delta^r_s(\mathfrak{D}\cdot\boldsymbol{E}+\boldsymbol{B}\cdot\boldsymbol{H}); \tag{284.11}$$

we shall then have an identity of the form

$$\frac{d}{dt}\int \mathfrak{P}_k\,dv-\oint\overline{m}^r{}_k\,da_r=-\int[Q_{\mathfrak{F}}E_k+P^r\partial_r E_k+(\boldsymbol{J}\times\boldsymbol{B})_k]\,dv. \tag{284.12}$$

The antisymmetric part of the stress tensor $\overline{m}^r{}_k$, given by

$$\overline{m}^{[r}{}_{k]}=P^{[r}E_{k]}, \tag{284.13}$$

represents a torque exerted by the electric field on the polarized matter.

Still other definitions have been proposed. The problem of what electromagnetic quantities are to be identified or related to the essentially mechanical concepts of stress, energy, momentum, and energy flux in polarized and magne-

[1] Recall that our reference frame is rectangular Cartesian and that Cartesian covariant and contravariant indices are equivalent, so that (284.8) is an invariant condition under the group of orthogonal transformations. The indices of Cartesian tensors are usually written all as subscripts or all as superscripts; however, the position of the indices we use here follows quite naturally from the way these Cartesian tensors are related to world tensors and is a more suggestive scheme of notation.

tized material media, the subject of extensive debate for more than half a century, is reviewed by PAULI[1]. As we shall attempt to show here, this problem is not one to be solved independently of consideration of the properties of the material medium. Neither are we to regard identities such as (284.5), (284.6), (284.10), and (284.12) as conservation laws. These equations place no new restrictions on the electromagnetic and charge-current fields but are satisfied identically as consequences of the aether relations and the laws of conservation of charge and magnetic flux. Conservation of energy and momentum are conditions placed on the total energy and momentum of all types: mechanical, caloric, electromagnetic and any other forms of energy and momentum one may choose to define. G. GYORGYI[2] has recently demonstrated the equivalence of various definitions of the electromagnetic stress, energy, momentum, and energy flux provided the mechanical counterparts of these quantities be suitably adjusted. Additional and special assumptions classifying or dissecting momentum, energy, stress, etc., into various types are necessary in the formulation of constitutive relations but in no way affect what we shall call the ***conservation laws of energy and momentum***. The following treatment aims primarily to *formulate the conservation laws of energy and momentum independently of any resolution of stress, energy, momentum, and energy flux*. Continuing the program initiated in Sect. 270, we give to the laws of conservation of energy and momentum a statement also independent of any geometry of space-time. Such a general statement once obtained, the resulting equations expressing conservation of energy and momentum are susceptible to the introduction of Galilean, Lorentzian, or other space-time geometry and to such resolution of the energy, momentum, stress, and energy flux as may be appropriate for the formulation of special constitutive relations appropriate to various materials. Rather than beginning with an abstract postulational approach as we did in treating conservation of charge and magnetic flux and then proceeding to obtain more familiar results by adding restrictive assumptions and a space-time geometry, here we shall begin with equations and concepts familiar from a traditional point of view and proceed thence to the more abstract and general conclusions.

285. Mechanical energy, momentum, stress, and energy flux. Let b_r denote the *density of mechanical momentum*, $t^r{}_s$ the *mechanical stress*, U_m the *density of mechanical energy* and h^r the *heat flux* or *extra flux of energy*[3].

The equations of *balance of mechanical momentum and energy* may be written in the form

$$\frac{d}{dt}\int_{\mathscr{V}} b_r\,dv = \oint_{\mathscr{S}} t^s{}_r\,da_s + \int_{\mathscr{V}} \varrho f_r\,dv, \tag{285.1}$$

$$\frac{d}{dt}\int_{\mathscr{V}} U_\mathrm{m}\,dv - \oint_{\mathscr{S}} (t^r{}_s\dot{z}^s + h^r)\,da_r = \int_{\mathscr{V}} (q + \varrho f_r \dot{z}^r)\,dv, \tag{285.2}$$

where $\mathscr{V}$ is a spatial region moving with the particles of the motion and $\mathscr{S}$ is its boundary. Cf. (200.1) and (241.1). The extrinsic force $\boldsymbol{f}$ is the supply of mechani-

---

[1] [1920, *3*].

[2] [1955, *11*].

[3] The density of mechanical momentum $\boldsymbol{b}$ in classical mechanics is assumed proportional to the velocity field $\dot{\boldsymbol{z}}$, $\boldsymbol{b} = \varrho\dot{\boldsymbol{z}}$, where ϱ is the *mass density*. Also, in classical mechanics, the density of mechanical energy U_m is given by $U_\mathrm{m} = \varrho(\varepsilon + \frac{1}{2}\dot{z}^2)$, where ε is the *internal energy per unit of mass* and $\frac{1}{2}\dot{z}^2$ is the *kinetic energy per unit of mass*. Cf., however, Sect. 238.

We assume that the 3-dimensional tensors occurring in these definitions are Cartesian tensors under time-independent transformations of the spatial co-ordinates. Thus the vertical placement of the indices does not represent any difference of transformation law. Our choices for these positions anticipate a subsequent world tensor representation.

cal momentum. The quantity $q+\boldsymbol{f}\cdot\dot{\boldsymbol{z}}$ is the supply of mechanical energy; it consists of two parts, the latter, $\boldsymbol{f}\cdot\dot{\boldsymbol{z}}$, being the rate of working of the extrinsic force and the former, q, the supply of internal energy. If we write (285.1) and (285.2) in terms of regions and surfaces stationary with respect to the reference frame we get

$$\frac{\partial}{\partial t}\int b_r\,dv - \oint (t^s{}_r - b_r\dot{z}^s)\,da_s = \int \varrho f_r\,dv, \tag{285.3}$$

$$\frac{\partial}{\partial t}\int U_{\mathrm{m}}\,dv - \oint (t^s{}_r\dot{z}^r + h^s - U_{\mathrm{m}}\dot{z}^s)\,da_s = \int (q+\varrho f_r\dot{z}^r)\,dv. \tag{285.4}$$

The terms $b_r\dot{z}^s$ and $U_{\mathrm{m}}\dot{z}^s$ appearing in the surface integrals of this latter form of the equations of balance are called the *convective flux of momentum* and *energy*, respectively.

In classical mechanics, we also have the equation of *balance of the moment of momentum* or angular momentum. For regions of integration moving with the particles of a motion, this equation takes the form

$$\frac{d}{dt}\int z^{[r}b_{s]}\,dv - \oint (z^{[r}t^k{}_{s]} + m^{kr}{}_s)\,da_k = \int (z^{[r}f_{s]} + l^r{}_s)\,dv, \tag{285.5}$$

where the assigned couples $l^r{}_s$ may be interpreted as the rate of supply of mechanical angular momentum in excess of the torque exerted by the assigned force. The couple stress $m^{kr}{}_s$ may be interpreted as a flux of mechanical angular momentum in excess of the torque exerted by the mechanical stress. Cf. $(200.1)_2$.

286. The interaction of matter and the electromagnetic field. Here we consider only materials for which the couple stress $m^{kr}{}_s$ vanishes, and we neglect all supplies of mechanical momentum and internal energy other than those arising from the interaction of the material with an electromagnetic field. In particular, we neglect the interaction of the material with a gravitational field.

A *law of interaction* of matter with the electromagnetic field is prescribed by assuming particular forms for the fields $\boldsymbol{f}$, q, and $\boldsymbol{l}$ as functions or functionals of the electromagnetic variables. As preliminaries to the general treatment in the next section, we consider two special laws. First, suppose

$$\varrho\boldsymbol{f} = Q\boldsymbol{E} + \boldsymbol{J}\times\boldsymbol{B}, \tag{286.1}$$

$$q = \mathfrak{J}\cdot\mathfrak{E}, \tag{286.2}$$

$$\boldsymbol{l} = 0. \tag{286.3}$$

These definitions for $\boldsymbol{f}$ and q imply the identity

$$q + \varrho\boldsymbol{f}\cdot\boldsymbol{v} = \boldsymbol{J}\cdot\boldsymbol{E}. \tag{286.4}$$

Substituting the expressions (286.1) through (286.4) into the mechanical equations of balance (285.1), (285.2), and (285.4) and eliminating the charge and current fields by means of the identities (284.5) and (284.6), we get the equations

$$\frac{d}{dt}\int [b_r + (\boldsymbol{D}\times\boldsymbol{B})_r]\,dv - \oint [t^k{}_r + m^k{}_r + (\boldsymbol{D}\times\boldsymbol{B})_r\dot{z}^k]\,da_k = 0, \tag{286.5}$$

$$\frac{d}{dt}\int (U_{\mathrm{m}} + U_{\mathrm{e}})\,dv - \oint [t^k{}_r v^r + h^k - (\boldsymbol{E}\times\boldsymbol{H})^k + U_{\mathrm{m}}\dot{z}^k]\,da_k = 0, \tag{286.6}$$

$$\frac{d}{dt}\int [z^{[k}b_{r]} + z^{[k}(\boldsymbol{D}\times\boldsymbol{B}_{r]})]\,dv - \oint [z^{[k}t^n{}_{r]} + z^{[k}m^n{}_{r]} + z^{[k}(\boldsymbol{D}\times\boldsymbol{B})_{r]}\dot{z}^n]\,da_n = 0. \tag{286.7}$$

As a second example, set

$$\varrho f_r = P^k \partial_k E_r + (\boldsymbol{J}\times\boldsymbol{B})_r + Q_{\mathfrak{F}} E_r, \tag{286.8}$$

$$q = \mathfrak{J}\cdot\mathfrak{E}, \tag{286.9}$$

$$l^k{}_r = P^{[k} E_{r]}. \tag{286.10}$$

Substituting these expressions into the mechanical equations of balance (285.1), (285.2), and (285.4) and eliminating the charge and current by means of the identities (284.8) and (284.10), we obtain

$$\frac{d}{dt}\int [b_r + (\boldsymbol{D}\times\boldsymbol{B})_r]\, dv - \oint [t^k{}_r + \overline{m}^k{}_r + (\boldsymbol{D}\times\boldsymbol{B})_r \dot{z}^k]\, d a_k = 0, \tag{286.11}$$

$$\frac{d}{dt}\int (U_{\mathrm{m}} + U_{\mathrm{e}})\, dv - \oint [t^k{}_r \dot{z}^r + h^k - (\boldsymbol{E}\times\boldsymbol{H})^k + U_{\mathrm{m}} \dot{z}^k]\, d a_k = 0, \tag{286.12}$$

$$\frac{d}{dt}\int [z^{[k} b_{r]} + z_{[k} (\boldsymbol{D}\times\boldsymbol{B})_{r]}]\, dv - \oint [z^{[k} t^n{}_{r]} + z^{[k} m^n{}_{r]} + z^{[k} (\boldsymbol{D}\times\boldsymbol{B})_{r]} \dot{z}^n]\, d a_n = 0, \tag{286.13}$$

where $\overline{m}^k{}_r$ is given by (284.11).

The assumptions (286.1) to (286.3) and (286.8) to (286.10) embody different laws of interaction, examples illustrative of special theories of matter interacting with an electromagnetic field. The system of equations consisting in the conservation laws of charge and magnetic flux, the aether relations, and the equations of balance of energy and momentum is an underdetermined system. In any particular theory of materials, there are relations connecting the mechanical stress $\boldsymbol{t}$ and the energy flux $\boldsymbol{h}$ to the motion and to other variables characterizing the state of the medium. Now it is clear that for a given material, these relations cannot all be the same if the systems (286.5) to (286.7) and (286.11) to (286.13) are to be equivalent. For example, if momentum is balanced, (286.7) implies that

$$t^{[r}{}_{s]} + m^{[r}{}_{s]} = 0; \tag{286.14}$$

whereas, (286.13) implies that

$$t^{[r}{}_{s]} + \overline{m}^{[r}{}_{s]} = 0. \tag{286.15}$$

Thus, in the first case, the mechanical stress is symmetric since $\boldsymbol{m}$ is symmetric. In the second case, the mechanical stress is asymmetric, in general, since the electromagnetic stress $\overline{\boldsymbol{m}}$ has an antisymmetric part given by (284.3). If it is assumed that the mechanical stress is a function of the local state of the material, as for example in elasticity theory, the functional dependence must be consistent with (286.14) or (286.15) depending on our choice for the law of interaction. However, both of these special models lead to results having a property in common. Indeed, since (285.1) and (285.2) are equations of balance, substitution of special forms for the supplies $\boldsymbol{f}$, q, and $\boldsymbol{l}$ will of course yield results again having the form of equations of balance. What is important here is that in both cases use of the conservation laws of electromagnetism, through the agency of their corollaries (284.5) and (284.6), has led to equations of balance in which the volume integral is missing, i.e., *the supplies are zero.* That is, both cases lead to *local* conservation laws (Sect. 157). Conversely, but always subject to the assumed electromagnetic equations, these local conservation laws are equivalent to the balance of momentum, moment of momentum, and energy. These results suggest then that there is a *general* formulation of the principles of Sect. 285 in the form of local conservation laws, and this we seek.

The complexity of the Eqs. (286.5) to (286.7) and (286.11) to (286.13) is due in part to special assumptions regarding the resolution of energy, momentum, stress, and energy flux into various types. *The mechanical momentum and internal energy and the electromagnetic momentum and energy are not separately conserved* but are transformable into one another. The rate at which this transformation takes place for a given material is specified by the law of interaction of the matter with the electromagnetic field, characteristic of that material. Conservation of energy and momentum surely are laws of nature holding independently of the rates at which energy and momentum are converted from one form to another. Our principal objective in Subchapter II was to put the conservation laws of electromagnetic theory into a form as pure and devoid of extraneous and special assumptions as possible without losing contact with familiar concepts and mathematics. The review of familiar ideas and equations given in this section and the two foregoing should provide sufficient background and motivation for the following treatment of the laws of conservation of energy and momentum in a correspondingly pure form.

287. Conservation of energy and momentum in a preferred inertial frame. The systems (286.5) to (286.7) and (286.11) to (286.13) each constitute a set of four equations having the general form

$$\oint T_\mu^\Omega \, d\mathscr{S}_{3\Omega} = 0, \qquad \Omega, \mu = 1, 2, 3, 4. \tag{287.1}$$

That is, each equation in each of these systems of four equations may be regarded as having been obtained from an equation of the form (287.1) with a fixed value of the index μ if the quantities $T_\mu^\Omega(\boldsymbol{z}, t)$ are appropriately defined, subject to sufficient continuity of all the fields. For example, the energy equation (286.6) can be put in the form (287.1) with $\mu = 4$ by integrating (286.6) with respect to the time between the limits t_1 and t_2, and making the following definition of the quantities T_4^Ω:

$$T_4^\Omega = (t^r{}_s \dot{z}^s + h^r - (\boldsymbol{E}\times\boldsymbol{H})^r - U_{\mathrm{m}} \dot{z}^r, \quad - U_{\mathrm{m}} - U_{\mathrm{e}}). \tag{287.2}$$

Similarly, the three momentum equations (286.5) can be written in the form (287.1) by setting

$$T_k^\Omega = (- t^r{}_k - m^r{}_k + b_k \dot{z}^r, \quad b_k + (\boldsymbol{D}\times\boldsymbol{B})_k). \tag{287.3}$$

In the case of (286.11) we must set

$$T_k^\Omega = (- t^r{}_k - \overline{m}^r{}_k + b_k \dot{z}^r, \quad b_k + (\boldsymbol{D}\times\boldsymbol{B})_k). \tag{287.4}$$

In either case, assuming that the momentum equations are satisfied, the angular momentum equations are equivalent to the conditions

$$T^{[r}{}_{k]} = 0, \tag{287.5}$$

provided the mechanical momentum $\boldsymbol{b}$ is parallel to $\dot{\boldsymbol{z}}$ as is assumed in classical mechanics.

That is, we have observed that in all theories of electromagnetodynamics of continuous media equations of the form (287.1) and (287.5), with T_μ^Ω suitably identified, are valid and are equivalent to the principles of linear momentum and energy and to the principle of moment of momentum, respectively. In these theories, then, we may *replace* the more familiar mechanical principles of Sect. 285 by the conservation law (287.1) and the symmetry condition (287.5). Thus we are led to assume, independently of any resolution of momentum, energy, stress, and energy flux into various types and independently of any law of interaction between matter and the electromagnetic field, the existence of a stress-energy-momentum tensor T_μ^Ω satisfying (287.1) and (287.5). To these equations we have been led through an electromagnetic context. They are not to be regarded, however, as being restricted to electrodynamical phenomena. Rather, they encompass and extend the principles of continuum mechanics as explained in Subchapters DII and EI. We shall call (287.1), in the special coordinates there employed, the *law of conservation of energy and momentum*; the symmetry condition (287.5), the *law of conservation of angular momentum.*

The extent to which these principles may replace the usual principles of mechanics is not determined. In view of the principle of equivalence of surface and volume sources (Sect. 157), it would seem that a formal reconciliation is always possible when the couple stress is zero, but such a reconciliation would sometimes involve redefinition of the mechanical stress tensor and in many cases would scarcely be natural or useful. In a fully general scheme, extending that embodied in (205.18), a non-zero right-hand side would have to be supplied for (287.1), and (287.5) would be replaced by a system of tensor equations of the form (287.1) but involving tensors of increasingly high order.

288. World invariant form of the law of conservation of energy and momentum. We have thus far simply introduced an array or matrix of sixteen quantities T_μ^Ω satisfying (287.1) and (287.5). The coordinates z and t in these equations have a definite and special interpretation. They are the coordinates of a space-time frame for which the aether relations hold in the canonical form (279.1) and which is simultaneously an inertial frame from the point of view of classical mechanics. As was discussed in Sect. 279, any two frames so restricted must be related by a time-independent orthogonal transformation of the spatial co-ordinates. Under this group of transformations, the quantities T_j^i transform as a tensor of rank 2, the quantities T_4^i and T_i^4 transform as vectors, and the quantity T_4^4 is a scalar. We wish to extend this transformation law to the full group of unrestricted analytic transformations of the four space-time co-ordinates $\boldsymbol{x}$. We can retain a tensor law of transformation for the quantities T_μ^Ω under this larger group of transformations without introducing auxiliary fields or special assumptions regarding the geometry of space-time provided we assume that T_μ^Ω is a 2-event world tensor field with the transformation law[1]

$$T_{\mu'}^{\Omega'}(\boldsymbol{x}_1', \boldsymbol{x}_2') = \mathsf{A}\,|(\boldsymbol{x}_1'/\boldsymbol{x}_1)|^{-1} \left.\frac{\partial x^{\Omega'}}{\partial x^\Omega}\right|_{\boldsymbol{x}_1} \left.\frac{\partial x^\mu}{\partial x^{\mu'}}\right|_{\boldsymbol{x}_2} T_\mu^\Omega(\boldsymbol{x}_1, \boldsymbol{x}_2), \tag{288.1}$$

under general analytic transformations of the co-ordinates and transformations $\mathsf{A}' = \mathsf{A}^{-1}\mathsf{A}$ of the *unit of action*. A unit of action is related to the units of energy and time by $\mathsf{A} = \mathsf{E}\,\mathsf{T}$.

If the two events $\boldsymbol{x}_1$ and $\boldsymbol{x}_2$ are always referred to the same co-ordinate system, the transformation law of the 2-event tensor field $\boldsymbol{T}$ under any subgroup of *linear* transformations (e.g. Galilean or Lorentzian) of the co-ordinates is indistinguishable from the transformation law of an ordinary single-event mixed tensor field of rank 2 and weight 1. Hence, the extended law of transformation (288.1) for the quantities T_μ^Ω under the group of general transformations of the co-ordinates is in agreement with the previously assigned transformation law of the quantities under the subgroup of time independent orthogonal transformations of the spatial co-ordinates. Moreover, our principal reason for assuming the transformation law (288.1) for the T_μ^Ω is that the integral equation

$$\oint T_\mu^\Omega \, d\mathscr{S}_{3\Omega} = 0 \tag{288.2}$$

is a world tensor equation holding not only in the preferred inertial frames, but in *an arbitrary space-time co-ordinate system.* Before proceeding, we should

[1] See Sect. App. 15 for a general discussion of double tensor fields.

point out that the symmetry condition $T^{[i}{}_{j]}=0$ holds only in the preferred inertial frames and is obviously is not a world tensor invariant condition.

We have been led to assume a world tensor equation (288.2) and the transformation law (288.1) for the quantities $\boldsymbol{T}$ by a process of *abstraction* and *extension* of the classical equations and transformation law. Clearly, such a process is not unique if all that we demand is that Eq. (288.2) and the transformation law (288.1) reduce to the classical equation and transformation law when the co-ordinates are specialized. We have chosen this extension of the classical transformation law largely because of its simplicity and the fact that (288.2) is a world tensor equation not involving special assumptions regarding the geometry of space-time. A more abstract formulation of the law of conservation of energy and momentum can be obtained by first postulating equation (288.2) and the transformation law (288.1). Then, by introducing a number of special assumptions regarding the 2-event field $\boldsymbol{T}$, the geometry of space-time, and the resolution of stress, energy, momentum, and energy flux into mechanical and electromagnetic components, we can derive the classical energy and momentum equations by a process of *specialization* and *reduction*. We shall now outline briefly such a treatment of energy and momentum paralleling our treatment of charge and magnetic flux in Subchapter II.

We assign a world vector additive set function $\mathfrak{W}[\mathscr{S}_3, \boldsymbol{x}, A]$ of energy and momentum to every $\mathscr{S}_3$ in space-time. The set function $\mathfrak{W}$ has the absolute dimension $[A]$ where A is the unit of action. The value of $\mathfrak{W}$ in a general system of co-ordinates may depend on the co-ordinates of a *reference event* $\boldsymbol{x}$. We shall call $\mathfrak{W}$ the *energy-momentum vector* of the set $\mathscr{S}_3$ referred to the event $\boldsymbol{x}$. The components of $\mathfrak{W}$ in a preferred inertial frame are denoted by

$$\mathfrak{W} = (\mathfrak{P}, \mathfrak{U}), \tag{288.3}$$

where $\mathfrak{P}$ is the *resultant momentum of the set* $\mathscr{S}_3$ and $\mathfrak{U}$ is the *total energy of the set* $\mathscr{S}_3$. It follows from the rules given in Sect. 273 that

$$\text{phys. dim. } \mathfrak{P} = [A\, L^{-1}], \tag{288.4}$$

$$\text{phys. dim. } \mathfrak{U} = [A\, T^{-1}], \tag{288.5}$$

in agreement with the customary relations between the units of action, length, time, momentum, and energy.

Next we assume that $\mathfrak{W}$ is expressible in the form

$$\mathfrak{W}_\mu[\mathscr{S}_3, \boldsymbol{x}, A] = \int T_\mu^\Omega(\boldsymbol{x}_1, \boldsymbol{x}_2)\, d\mathscr{S}_{3\Omega}, \tag{288.6}$$

where $\boldsymbol{T}$ is a 2-event world tensor field having the transformation law (288.1).

We postulate the ***law of conservation of energy and momentum:*** the world vector $\mathfrak{W}$ vanishes for every circuit,

$$\oint \boldsymbol{T}\cdot d\mathscr{S}_3 = 0. \tag{288.7}$$

We call $\boldsymbol{T}$ the *stress-energy-momentum tensor*.

From the point of view of world geometry, a preferred inertial frame can be defined invariantly as a space-time co-ordinate system for which the Lorentz tensor $\overset{-1}{\gamma}$ has the canonical form (280.2) *and* for which the classical space normal $\boldsymbol{t}$ has its canonical form $\boldsymbol{t}=(0, 0, 0, 1)$. We have interpreted such a reference frame as an inertial frame in which the "aether" is at rest; these frames have intrinsic *mechanical* and *electromagnetic* significance. As a characteristic feature of the *classical* theory of energy and momentum we postulate that if the co-ordinates of $\boldsymbol{x}_1$ and $\boldsymbol{x}_2$ are referred to a common preferred inertial frame, the stress-energy-momentum-tensor $T_\mu^\Omega(\boldsymbol{x}_1, \boldsymbol{x}_2)$ is independent of the co-ordinates $\boldsymbol{x}_2$. Thus, in such a frame, the components of the energy-momentum vector $\mathfrak{W}[\mathscr{S}_3, \boldsymbol{x}_2, \mathsf{A}]$ are independent of $\boldsymbol{x}_2$ and depend only on $\mathscr{S}_3$ and the unit of action. Since

$$\oint \mathfrak{W}[\mathscr{S}_3, \boldsymbol{x}, \mathsf{A}] \cdot d\mathscr{S}_1 = 0 \tag{288.8}$$

is a world tensor equation holding trivially in every preferred inertial frame, it is satisfied by the components of $\mathfrak{W}$ in a general system of co-ordinates.

The field equations and boundary conditions following from the law of conservation of energy and momentum are

$$\partial_\Omega T_\mu^\Omega = 0, \tag{288.9}$$

$$[T_\mu^\Omega]\, \partial_\Omega \Sigma = 0. \tag{288.10}$$

Substituting the special forms (287.2) and (287.3) for the $\boldsymbol{T}$ into (288.9), we find that these four equations are equivalent to the system

$$\varrho \ddot{z}_k = \partial_r t^r{}_k + Q E_k + (\boldsymbol{J} \times \boldsymbol{B})_k, \tag{288.11}$$

$$\varrho \dot{\varepsilon} = t^r{}_k \partial_r \dot{z}^k + \mathfrak{J} \cdot \mathfrak{E} + \operatorname{div} \boldsymbol{h}, \tag{288.12}$$

where we have used the full complement of assumptions embodied in the conservation laws of charge and magnetic flux, the aether relations, and the continuity equation [cf. (156.5)]:

$$\frac{\partial \varrho}{\partial t} + \operatorname{div}(\varrho \dot{\boldsymbol{z}}) = 0. \tag{288.13}$$

The jump conditions (288.10) holding at a moving surface of discontinuity $\Sigma(\boldsymbol{x}, t) = 0$ encompass as a special case and extend the jump conditions (205.3) and (241.5) to the case of electromagnetodynamical problems where discontinuities in the electromagnetic stress, energy, and momentum must also be considered.

The ease with which the field equations and boundary conditions (288.9) and (288.10) follow from the world invariant form of the law of conservation of energy and momentum is perhaps the best recommendation for a 4-dimensional formalism as applied to classical mechanics. The world invariant formulation of the energy and momentum equations given in this section may be regarded simply as a mathematical device designed to illuminate the intrinsic *structure* of the classical equations and to calculate explicitly their form in an arbitrary coordinate system. Such a procedure in no way alters or impairs their essential physical and mathematical content but suggests generalizations of classical procedure and allows an easier comparison of the classical and relativistic theories.

289. Conservation of angular momentum, special relativity, and the principle of equivalence of momentum and energy flux. In Sect. 287 we have seen that the classical law of conservation of angular momentum is embodied in the equations

$$T^{[k}{}_{r]} = 0, \tag{289.1}$$

where, from the point of view of world tensors, the $T^k{}_r$ are nine of the components of the stress-energy-momentum tensor $\boldsymbol{T}$ referred to an inertial frame in which the aether is at rest. Now consider the world tensor

$$T^{\Omega\mu} \equiv \overset{-1}{\gamma}{}^{\mu\nu}\, T_\nu^\Omega. \tag{289.2}$$

In a preferred inertial frame we have

$$T^{[kr]} = 0. \tag{289.3}$$

However, in classical mechanics, the momentum density and the energy flux are so constituted that we do not generally have

$$T^{4r} = T^{r4}. \tag{289.4}$$

This equation is equivalent to the statement that

$$\text{momentum density} = \frac{1}{c^2}\,\text{energy flux}. \tag{289.5}$$

If, on the other hand, in the classical theory based on the resolutions (287.2) and (287.3), we assume that the mechanical momentum, energy, heat flux, and stress vanish, (289.4) reduces to the equation

$$\boldsymbol{D}\times\boldsymbol{B} = \frac{1}{c^2}\,\boldsymbol{E}\times\boldsymbol{H}, \tag{289.6}$$

which is satisfied by virtue of the aether relations. That is, electromagnetic momentum is *defined* as electromagnetic energy flux divided by the fundamental constant c^2. In special relativity theory, there is an equivalence between momentum and energy flux and between mass and energy. The classical relation (289.6) between electromagnetic momentum and electromagnetic energy flux is extended to *all* forms of momentum and energy flux. Momentum in special relativity theory is *defined*[1] as energy flux divided by c^2. The constitutive equations and definitions of mechanics are revised so as to be Lorentz invariant. Thus, in special relativity theory, we have the symmetry conditions

$$T^{[\mu\nu]} = 0, \tag{289.7}$$

holding in every Lorentz frame. The symmetry of a single-point tensor field is invariant under transformations of the co-ordinates. The Lorentz transformations are a linear subgroup of the group of general transformations of the space-time co-ordinates. Under this group of linear transformations, our 2-event stress-energy-momentum tensor has a transformation law indistinguishable from the single-event tensor law of transformation assumed in relativity theory. Hence, the symmetry condition (289.7) is a Lorentz invariant condition. Interpreted from the point of view of classical mechanics, the symmetry condition (289.7) of relativity mechanics embodies the notion of conservation of angular momentum *and* the equivalence of momentum and energy flux.

290. The dimensions of energy and momentum. The absolute dimension of the field $\boldsymbol{T}$ has been called the dimension of action.

$$\text{abs. dim. } \boldsymbol{T} = [A]. \tag{290.1}$$

The physical dimensions of stress, energy, momentum, and energy flux are determined by the following rule. Refer the co-ordinates of both events on which $\boldsymbol{T}$ depends to a common preferred inertial frame based on a unit of length L and a unit of time T. The components of $\boldsymbol{T}$ reduce to functions of a single event in such a frame and transform as Cartesian tensors under the group of transformations relating any pair of such frames. If we transform the co-ordinates

[1] WHITTAKER [1951, *39*, Vol. II, p. 54], PLANCK [1908, *6*]. This unified definition of momentum was PLANCK's method of expressing the relativistic equivalence of mass and energy.

and the units of length, time, and action simultaneously, we obtain from (288.1)

$$T^{r'}_{k'} = \mathsf{L}^{-3}\,\mathsf{A}\,\mathsf{T}^{-1} A^{r'}{}_{r} A^{k}{}_{k'} T^{r}_{k}, \tag{290.2}$$

$$T^{r'}_{4'} = \mathsf{L}^{-2}\,\mathsf{A}\, A^{r'}{}_{r} T^{r}_{4}, \tag{290.3}$$

$$T^{4'}_{r'} = \mathsf{L}^{-4}\,\mathsf{A}\, A^{r}{}_{r'} T^{4}_{r}, \tag{290.4}$$

$$T^{4'}_{4'} = \mathsf{L}^{-3}\,\mathsf{A}\,\mathsf{T}^{-1} T^{4}_{4}. \tag{290.5}$$

It is customary to introduce the unit of mass as an independent unit instead of the unit of action. In terms of the unit of mass, the physical dimensions of stress, momentum, energy flux, and energy density as determined by the foregoing transformation formulae can be obtain by substituting

$$[\mathsf{A}] = [\mathsf{M}\,\mathsf{L}^2\,\mathsf{T}^{-1}]. \tag{290.6}$$

Also, it is not customary to regard the units of action, charge, and magnetic flux as independent but to relate them according to

$$\mathsf{A} = \mathsf{Q}\,\Phi. \tag{290.7}$$

Thus, for example, transformations of the unit of magnetic flux are generally regarded as fixed in terms of the transformations of M, L, T, and Q by the relation

$$\Phi = \mathsf{Q}^{-1}\,\mathsf{M}\,\mathsf{L}^2\,\mathsf{T}^{-1}. \tag{290.8}$$

This is consistent with our having called $\boldsymbol{D}\times\boldsymbol{B}$ a momentum density and $\boldsymbol{E}\times\boldsymbol{H}$ a flux of energy without introducing any additional dimensional constants of proportionality. Thus we have

$$\text{phys. dim. } \boldsymbol{D}\times\boldsymbol{B} = [\mathsf{M}\,\mathsf{L}^{-2}\,\mathsf{T}^{-1}], \tag{290.9}$$

$$\text{phys. dim. } \boldsymbol{E}\times\boldsymbol{H} = [\mathsf{M}\,\mathsf{T}^{-3}], \tag{290.10}$$

in agreement with the customary dimensions of momentum per unit volume, and energy per unit area per unit time.

291. Conclusions. The principal result contained in this chapter is a world tensor invariant formulation of the laws of conservation of charge, magnetic flux, energy, and momentum, independent of special assumptions as to the geometry of space-time.

In Subchapter IV, we motivated the law of conservation of energy and momentum by considering the classical energy and momentum equations of a continuous medium interacting with an electromagnetic field. We assumed that the stress-energy-momentum tensor T^{Ω}_{μ} was a 2-event world tensor field. While the charge-current potential and the electromagnetic field are connected through the aether relations, no general relation between the stress-energy-momentum tensor and the other fields has been proposed here. The basic content of Subchapter IV is thus independent of the two preceding. Rather, the conservation law of energy and momentum as formulated here is to be regarded as encompassing and generalizing the purely mechanical considerations of Subchapters D II and E I; it also includes and extends the various purely electromagnetic proposals for conservation of energy and momentum.

The conservation laws of charge, magnetic flux, momentum, energy and angular momentum constitute an underdetermined system of equations. They must be supplemented by relations between the conservative fields. The aether relations are an example of such relations. The many determinate special theories encompassed by the conservation laws is evidence of the different special ideas of material behavior consistent with them. The conservation laws provide a general framework common to all electrodynamic theories, classical and relativistic, with which we are familiar. Special and general relativity theory entail modification of our concepts of time, mass, energy, momentum, the aether relations, and inertial frames, but have yet to alter in principle either the formal or the intuitive aspects of the laws of conservation of charge, magnetic flux, energy, and momentum.

G. Constitutive Equations.

I. Generalities.

292. The nature of constitutive equations. In Sect. 7 we explained that the field equations and jump conditions express the general principles of mechanics, thermodynamics, and electromagnetism, while constitutive equations define *ideal materials*, which are mathematical models of particular classes of materials encountered in nature. The preceding treatise has developed in detail the properties of motions and of the fundamental physical principles of balance, which imply both field equations and jump conditions.

In this final chapter we illuminate the concept of constitutive equation by stating general principles and adducing common examples.

293. Principles to be used in forming constitutive equations. It should be needless to remark that while from the mathematical standpoint a constitutive equation is a postulate or a definition, the first guide is physical experience, perhaps fortified by experimental data. However, it is rarely if ever possible to determine all the basic equations of a theory by physical experience alone. Every theory abstracts and simplifies the natural phenomena it is intended to describe (cf. Sect. 4). Supposing that the theorist has assembled the facts of experience he wishes to use in defining an ideal material, we now list the mathematical principles he may call to his aid when he attempts to formulate definite constitutive equations.

We know of no ideal material for which all these principles have been demonstrated to hold, although for the simpler classical theories it is generally believed that they do.

α) Consistency. Any constitutive equation must be consistent with the general principles of balance of mass, momentum, energy, charge, and magnetic flux. This is obvious and easy to say, but to test it in a special case may be difficult.

β) Co-ordinate invariance. Constitutive equations must be stated by a rule which holds equally in all inertial co-ordinate systems, at any fixed time. Otherwise, a mere change of description would imply a different response in the material. Such a rule may be achieved, usually trivially, by stating the equations either in tensorial form or by the aid of direct notations not employing co-ordinates at all[1].

γ) Isotropy or aeolotropy. Materials exhibiting no preferred directions of response are said to be *isotropic*. There are differences of opinion as to how this somewhat vague concept should be rendered mathematically precise. In the common phrase "isotropic material" we are unable to discern any meaning. Only *after* a kind of material has been defined by a particular constitutive equa-

[1] However, the literature of hydraulics abounds in "power laws" which are not invariant, as was pointed out by KLEITZ [1873, *4*, § 22] in a work dating from 1856; the same may be said of rheology.

tion does it become possible to state an unequivocal concept of isotropy. Pursuant to this idea, NOLL[1] defines the *isotropy group* of a material as the group of transformations of the material coordinates which leave the constitutive equations invariant. The nature of this group specifies the symmetries of the material. A material is then isotropic if its isotropy group is the full orthogonal group. Various kinds of *aeolotropy*, which is a symmetry with respect to certain preferred directions, are shown by materials whose isotropy groups are proper subgroups of the orthogonal group, or are other groups.

While these definitions reflect broader ideas, up to this time they have been rendered concrete only in pure mechanics, since the general constitutive equations expressing energetic and electromagnetic response are not yet known. In theories not limited to purely mechanical phenomena it is usual to define isotropy in respect to the variables $A, B, C, \ldots$ entering a constitutive equation by requiring the functional dependence of $A, B, C, \ldots$ to be isotropic in the mathematical sense. A material which is isotropic in respect to one property, e.g. stress and strain, need not be so in respect to another, e.g., electric displacement and electric field. There is a large current literature[2] concerning means of expressing isotropy and aeolotropy, but there is no adequate survey of the field as yet.

δ) Just setting. Constitutive equations connecting a given set of variables should be such that, when combined with all the principles of balance affecting these same variables, there should result a unique solution corresponding to appropriate initial and boundary data, and a solution that depends continuously on that data. This principle can rarely be used. With great mathematical labor, it has been proved to hold only in the simplest classes of boundary-value problems in the simplest classical theories[3]. At best, it shifts the problem to another domain, that of finding adequate initial and boundary conditions[4].

ε) Dimensional invariance. It is essential that included in each constitutive equation should be a full list of all the dimensionally independent moduli or *material constants* upon which the response of the material may depend. This requirement has often been neglected in recent work, although it is a commonly accepted principle of physics, known at least since GALILEO's day, that any fully stated physical result is dimensionally invariant. (It is tacitly understood that dimensionless moduli need not be listed, since they cannot be specified until a particular functional form is stated.)

The use of dimensional invariance to classify constitutive equations and in some cases to exclude inappropriate terms was initiated by TRUESDELL[5]. The tool used is the classical π-theorem, a precise statement and simple rigorous proof of which was recently given by BRAND[6].

---

[1] [1948, *8*, §§ 19–20].

[2] The basic principles derive from work of CAUCHY. Numerous articles on the subject have appeared in the Journal and Archive for Rational Mechanics and Analysis, 1952 to date.

[3] The honored custom of verifying that the number of equations equals the number of unknown functions seems to bring comfort.

[4] In many cases the *traditional* setting is known to fail to lead to a unique solution. E.g., in buckling problems the specification of the loads on the boundary is not sufficient to insure a unique solution. Such problems may nevertheless be regarded as well set if we amplify either the boundary conditions or the constitutive relations. E.g., in the case of buckling we may include as a part of the constitutive equations the requirement that the elastic energy shall be a minimum with respect to compatible deformations; a unique solution results. Still better, the solution for static buckling may be regarded as the limit of a uniquely soluble dynamic problem.

[5] [1947, *17*] [1948, *32* and *33*] [1949, *33* and *34*] [1950, *32*, §§ 7–8] [1951, *27*, §§ 22, 25] [1952, *22*] [1952, *21*, 1§ 47, 62–65, 67–69, 74–75] [1955, *28*, § 1].

[6] [1957, *1*].

ζ) Material indifference. The most important and the most frequently used correct idea for formulating constitutive equations is that *the response of a material is independent of the observer.* In proposing the first of all constitutive equations for deformable materials, namely, the law of linear elasticity, HOOKE[1] gave the first dim hint toward this principle in the suggestion that by carrying a spring scale to the bottom of a deep mine or to the top of a mountain, the change of gravity could be measured. A more striking example is furnished by use of a spring, attached to the center of a horizontal, uniformly rotating table, to measure the centrifugal force acting upon a terminal mass. The assumption implicit in these measurements is that the force exerted by the spring in response to a given elongation is *independent* of the observer, being the same to an observer moving with the table as to one standing upon the floor.

Another example is furnished by FOURIER's law of heat conduction, to be discussed in Sect. 296 below. According to this law, the flow of heat is proportional to the temperature gradient. Since both of these quantities are independent of the motion of the observer, in order that FOURIER's theory satisfy the principle of material indifference the constant of proportionality, or heat conductivity, must also be so invariant and hence is an absolute scalar under general transformations of space-time.

This invariance or ***material indifference*** has nothing to do with the co-ordinate invariance described in Subsection β, above. Indeed, if we conceive forces and motions directly, without the intermediary of co-ordinates and components, the requirement of material indifference is not thereby satisfied automatically[2]. Neither can it be achieved by a thoughtless four-dimensional formulation, for the precise meaning of "observer" in classical mechanics, where the time is a preferred co-ordinate, must be stated in terms of time-dependent orthogonal transformations, not of more general ones.

That this requirement has a fundamental physical meaning is shown by the fact that *the laws of motion themselves do not enjoy invariance with respect to the observer.* "Apparent" forces and torques, in general, are needed to reconcile the descriptions of mechanical phenomena given by two observers in relative motion (Sect. 197). The principle of material indifference states that these are the *only* mechanical effects of the motion. For example, the deflection of the spring on the rotating table is supposed to be proportional to the *entire* force acting parallel to the spring, and thus, since the end is free, to measure precisely the force measurable by an observer at rest upon the table, this force being the "apparent" force that accompanies the rotation of the observer's frame with respect to one in which the spring would suffer no extension[3].

All the classical linear theories, and some of the non-linear ones, satisfy the principle of material indifference trivially and automatically. For example, since mutual distances are invariant under change of observer, any theory in which the stress is determined only by mutual distances and differences or derivatives of mutual distances or relative velocities of particles exhibits material indifference. This is the case with most theories of elasticity and fluid motion[4].

[1] [1678, *1*].

[2] Cf. THIRRING [1929, *10*, § 6]: „Es sei ausdrücklich darauf hingewiesen, daß Unabhängigkeit von der *Koordinatenwahl* durchaus nicht mit Unabhängigkeit vom *Bezugssystem* zu verwechseln ist."

[3] If the spring is not idealized as massless, a simple correction is made for the centrifugal force acting upon the mass of the spring itself, but otherwise the response of the spring is unchanged.

[4] It is important not to confuse material indifference with any kind of tensorial invariance. Some constitutive equations relate quantities which transform as tensors under change of

But as soon as time rates of stressing come into consideration, this invariance holds no longer. The problem was first faced and solved correctly, in a special case, by CAUCHY[1]; the idea was expressed more clearly and given a somewhat different form by ZAREMBA[2]. While several recent studies have attacked the problem in one way or another, a general mathematical statement, for purely mechanical theories, was first achieved by NOLL[3]. Since this principle has not yet been formulated in a scope broad enough to cover all situations envisaged in this treatise, and since in the purely mechanical case it will be presented in the article, "The Non-linear Field Theories of Mechanics" (Vol. VIII, Part 2), we do not discuss it further here, though we give an example of its use in Sect. 298, below.

η) Equipresence. In the most general physical situations, mass, motion, energy, and electromagnetism are simultaneously present. In the classical theories, the variables describing these phenomena are divided more or less arbitrarily into classes, the members of each of which are supposed to influence only each other, not the members of the other classes. Stress is coupled with strain or stretching, flux of energy with temperature gradient, electric displacement with electric field strength, magnetic intensity with magnetic induction.

Such simple divisions of phenomena, perhaps lingering remnants of old views on "causes" and "effects", suffice to describe a great range of physical experience, but, as is well known, the members of the different classes often react upon each other, and there are also various special theories of energetico-mechanical, electro-energetic, and electro-mechanical interactions. Here the process of the theorists has been conservative. They have maintained as much of the old separation of variables as possible, relinquishing just enough of it to include the bare existence of a particular phenomenon of interaction.

In truth, the separation in itself is unnatural and unjustified by physical principle. Resulting only from the gradual discovery of individual phenomena, it reflects the old opinions that break physics up into compartments. The present treatise is conceived in the belief that the classical field theories, if rightly understood, describe most of the gross physical phenomena of nature. Constitutive equations, then, should not artificially divert these theories into disjoint channels.

This same view, for energetico-mechanical effects alone, results also from statistical models of continuous matter. There, stress and flux of energy appear as gross or mean expressions of a purely mechanical process. The separation

---

frame; e.g., the theories of fluids discussed in Sects. 298–299 relate $\boldsymbol{t}$ to $\boldsymbol{d}$, both of which, by the results in Sects. 144 and 211, are indeed tensors under change of frame. The principle of material indifference requires that any relation between such tensors be a tensor relation under change of frame. But some variables occurring in constitutive relations are not tensors under change of frame, the vorticity $\boldsymbol{w}$ and the deformation tensor $\boldsymbol{C}$ being examples. The classical theory of finite elastic strain (Sect. 302) relates $\boldsymbol{t}$ to $\boldsymbol{C}$; its constitutive equations satisfy the principle of material indifference, but the elastic coefficients do not generally transform as tensors under change of frame.

[1] [1829, *4*].

[2] [1903, *20* and *21*] [1937, *12*, Chap. I, § 2].

[3] [1955, *18*, § 4], "isotropy of space"; [1957, *11*, § 9] [1959, *9*, § 7] [1958, *8*, § 11], "principle of objectivity". A noteworthy earlier attempt is given by OLDROYD [1950, *22*]: "The form of the completely general equations must be restricted by the requirement that the equations describe properties independent of the frame of reference ... *Moreover, only those tensor quantities need be considered which have a significance for the material element independent of its motion as a whole in space.*" OLDROYD's process is to write all constitutive equations in convected co-ordinates, extension to spatial systems then being achieved by imposing tensorial invariance. Cf. LODGE [1951, *16*]. While this procedure leads to correct results, it is not the only one possible, nor is OLDROYD's formulation of the problem unequivocal. The principles of RIVLIN and ERICKSEN [1955, *21*, § 11 ff], of THOMAS [1955, *24* and *25*], and of COTTER and RIVLIN [1955, *4*], which likewise rest upon use of special co-ordinate systems, are special cases of NOLL's.

of stress as arising from deformation and flux of energy from changes of temperature emerges, not without reason, as a *first approximation*; but on a finer scale, it is illusory. Any attempt to draw precise conclusions from statistical theories show that the classical separation of phenomena is artificial and unphysical[1].

For energetico-mechanical phenomena, a principle denying any fundamental separation was formulated by TRUESDELL[2]: *The stress and the flux of energy depend upon the same variables.* Here we propose the following more general ***principle of equipresence****: A variable present as an independent variable in one constitutive equation should be so present in all.*

Let it not be thought that this principle would invalidate the classical separate theories in the cases for which they are intended, or that no separation of effects remains possible. Quite the reverse: The various principles of invariance, stated above, when brought to bear upon a general constitutive equation have the effect of *restricting* the manner in which a particular variable, such as the spin tensor or the temperature gradient, may occur. The classical separations may always be expected, in one form or another, for small changes—not as assumptions, but as *proven consequences* of invariance requirements. The principle of equipresence states, in effect, that no restrictions *beyond* those of invariance are to be imposed in constitutive equations. It may be regarded as a natural extension of OCKHAM'S razor as restated by NEWTON[3]: "We are to admit no more causes of natural things than such as are both true and sufficient to explain their appearances, for nature is simple and affects not the pomp of superfluous causes." This more general approach has the added value of showing in what way the classical separations fail to hold when interactions actually occur. We illustrate these remarks in Sect. 307, below.

II. Examples of kinematical constitutive equations.

294. Rigid bodies. A body is *rigid* if it is susceptible of rigid motions only: The distance between any two particles of the body is constant in time. The motion of the body is then completely determined when the position of one particle and the orientation of an appropriate rigid frame with origin at that particle are given as functions of time. For any body, rigid or not, the motion of the center of mass $\boldsymbol{c}$ is determined from the total assigned force by integration of (196.6). This fact is of little or no use when deformation occurs, since the center of mass generally moves about and fails to reside in any one particle, but the center of mass of a rigid body is stationary within it; more precisely, in any frame with respect to which the body is at rest, the center of mass $\boldsymbol{c}$ is also at rest. Thus we may say that the motion of the center of mass determines, or, more properly, constitutes the translatory motion of the rigid body.

The kinematical statement that there exists a frame with respect to which the body is permanently at rest may be expressed in terms of the tensor of inertia and the moment of momentum as follows: *In a suitably selected frame, denoted by a prime, we have*

$$\dot{\mathfrak{J}}'^{[\boldsymbol{O}']} = 0, \quad \mathfrak{H}'^{[\boldsymbol{O}']} = 0 \tag{294.1}$$

for all t. These are the *constitutive equations of a rigid body*. Substituting them into the general equations of balance of moment of momentum $(197.3)_4$ and

[1] M. BRILLOUIN [1900, *1*, § 37], supported by many later results.
[2] [1949, *34*, § 19] [1951, *27*, § 19], "BRILLOUIN'S Principle".
[3] [1687, *1*, Lib. III, Hypoth. I].

(197.4) yields ***Euler's equations for a rigid body***[1]:

$$\dot{\omega}\cdot\mathfrak{J}'^{[O']}+\omega\times\mathfrak{J}'^{[O']}\cdot\omega=\mathfrak{L}^{[O']}-c'\times\mathfrak{M}\ddot{b}. \tag{294.2}$$

Here $\mathfrak{L}^{[O']}$ is the applied torque with respect to O'. The last term on the right-hand side vanishes if O' is the center of mass: $c'=0$; it vanishes also if there is some one particle of the body which moves at uniform velocity in an inertial frame and if O' is chosen at that particle: $\ddot{b}=0$. In these cases and in any others when $c'\times\ddot{b}$ is known, *the angular velocity ω of any frame rigidly attached to the body is determined to within an arbitrary initial value by the applied torque, the mass and the translatory motion of the body, and the tensor of inertia $\mathfrak{J}'^{[O']}$ with respect to the same frame.* This conclusion follows from the existence and uniqueness of solution to (294.2), which is a non-linear differential equation of first order for ω.

Thus the purely *kinematical* assumption that the motion is persistently rigid renders it simply determinable. This is no wonder, for the infinite number of degrees of freedom generally present in a body of finite mass and volume have been reduced to six. While the laws of motion (196.3) are linear in an inertial frame, EULER's equations (294.2) are non-linear, since they state the balance of moment of momentum in a non-inertial frame, the coupling between these frames being that expressed by the principles of apparent torques in Sect. 197.

In a rigid body, we are at liberty to imagine internal stress and flux of energy and their respective balances, but since the motion is determinate without these considerations, we invoke OCKHAM's razor[2] to excise them.

A theorem of balance of mechanical energy follows from (294.2). Choosing the origin of the primed frame at c, by (168.8) and (168.9) we obtain

$$2\mathfrak{K}=\mathfrak{M}\dot{c}^2+\omega\cdot\mathfrak{J}^{[O']}\cdot\omega. \tag{294.3}$$

By differentiating this result and using (196.6) and (294.2) we obtain

$$\dot{\mathfrak{K}}=\mathfrak{F}\cdot\dot{c}+\mathfrak{L}'^{[O']}\cdot\omega. \tag{294.4}$$

This equation asserts that the entire rate of working of the applied force $\mathfrak{F}$ in producing the translatory motion and of the applied torque $\mathfrak{L}'^{[O']}$ in producing the rotation is converted into kinetic energy. Assumptions analogous to those discussed in Sect. 218 suffice to derive from (294.4) a law of conservation of total energy.

The theory of rigid bodies is presented in detail in the article by SYNGE in this volume.

295. Diffusion of mass in a mixture. A simple theory of diffusion of mass in a non-uniform binary mixture was proposed by FICK[3]: The rate of diffusion of

---

[1] The theory is usually and justly attributed to EULER [1765, *1*]; it has a long prior and a short but important subsequent history, which has never been adequately traced.

[2] Adopted by NEWTON [1687, *1*] as the first of the "Hypotheses", called in later editions Rules of Philosophizing", set at the head of Book III. On p. 704, we have quoted in full the statement in the edition of 1687.

Here we mention that the usual derivation of EULER's equations given in textbooks must also be excised by OCKHAM's razor. That derivation rests on an argument, summarized in Sect. 196A, showing that if all the mutual forces between the particles of a body are central and pairwise cancelling, balance of momentum implies balance of moment of momentum. In a rigid body, by definition, the mutual forces never manifest themselves in any way except to maintain rigidity. Rigidity has already been assumed, and it suffices; to hypothecate mutual forces is to multiply causes. In any case, the derivation is absurd from the standpoint of modern physics, which does not represent the smallest portions of any kind of body as stationary centers of force.

[3] FICK [1855, *1*, pp. 65–66] gave as his only physical basis an assertion that diffusion of matter in a binary mixture at uniform total density and pressure seems analogous to flow of heat according to FOURIER's theory (Sect. 296).

mass of the constituent $\mathfrak{A}$ is proportional to the gradient of the density of that constituent and is such as to tend to restore uniformity. In equations,

$$\varrho_{\mathfrak{A}} u_{\mathfrak{A}k} = -D \varrho_{\mathfrak{A},k}, \quad D > 0, \quad \mathfrak{A} = 1, 2, \tag{295.1}$$

where the coefficient D, the *diffusivity*, has the physical dimensions $[L^2 T^{-1}]$. FICK assumed that $\varrho_{,k} = 0$, as indeed is necessary in order that (295.1) be compatible with $(158.8)_1$. Substituting (295.1) into $(159.2)_2$, an expression of the conservation of mass, we have the ***diffusion equation:***

$$\dot{c}_{\mathfrak{A}} = (D c_{\mathfrak{A},k})^{,k} + \hat{c}_{\mathfrak{A}}, \tag{295.2}$$

where (159.4) is assumed to hold. This equation is of the same form as the isotropic case of FOURIER's equation (296.10).

FICK's law was proposed and has been applied under various specializing hypotheses, typically that $\hat{c}_{\mathfrak{A}} = 0$, that there is no mean motion, and that the pressure is constant. Also, only a binary mixture was envisioned, and while formally (295.1) is meaningful for an arbitrary number of constituents, it does not allow for their possible mutual actions. While there have been numerous studies[1]

---

[1] The main features of a general theory of diffusion might have been found in the kinetic theory of gases; indeed, MAXWELL [1860, *2*, Prop. XVIII] [1867, *2*, Eq. (76)] emphasized the fact that diffusion arises from ***equal and opposite forces*** proportional to the differences of velocities, so that (295.5) is virtually suggested by him. However the later development of the formal theory (e.g. [1939, *6*, § 8.4]), emphasizing special features of the binary case, tended to obscure the simple mechanical idea. The effort of HELLUND [1940, *13*] to calculate the basic equations for multi-constituent mixtures according to the kinetic theory did not lead to clear results.

The general continuum theory based upon (295.5) and (295.8) may fairly be attributed to STEFAN [1871, *6*, Eqs. (3)], who gave the special case of (295.6) appropriate to a ternary mixture of perfect fluids, but his work seems to have attracted no attention, and many inferior attempts were published later. DUHEM [1893, *2*, Chap. VI, § I] in effect noted that a constitutive equation for a fluid mixture should specify the functional form of $\hat{\boldsymbol{p}}_{\mathfrak{A}}$, but instead of perceiving a connection with FICK's law, he proposed to determine the $\hat{\boldsymbol{p}}_{\mathfrak{A}}$ by assuming the mixture to be such that all constituents may participate in a common infinitely small isentropic motion. We do not understand the meanings of all the terms in the theories of diffusion proposed by JAUMANN [1911, *7*, § XXXVI] [1918, *3*, §§ 136–141, 150–154] and LOHR [1917, *5*, §§ 5, 7, 12, 15].

A thermodynamic theory of diffusion, including thermal diffusion and the diffusion-thermal effect, seems first to have been proposed by ECKART [1940, *8*, Eq. (49)]. Cf. the earlier and more special theory of ONSAGER and FUOSS [1932, *10*, § 4.12]. A more fully elaborated theory of the same kind, adopting the "Onsager relations", was proposed independently by MEIXNER [1941, *2*, § 5] [1943, *2*, § 2] [1943, *3*, § 4]. The later literature in this field does not always take pains to recognize all the conditions laid down by MEIXNER but otherwise seems to diverge from his work only in minor points; cf. LAMM [1944, *7*], LEAF [1946, *7*, Eq. (48)], PRIGOGINE [1947, *12*, Chap. 10, §§ 1 and 5, Chap. 11, § 3], DE GROOT [1952, *3*, §§ 45–46], KIRKWOOD and CRAWFORD [1952, *12*], HIRSCHFELDER, CURTISS and BIRD [1954, *9*, § 11.2d]. There is also a more primitive theory of ONSAGER [1945, *4*, pp. 242–247], resting upon an inverse of (295.4); apparently only special circumstances were envisioned, but it is not clear what they are. In the thermodynamic theories, little if any use is made of mechanical concepts and principles.

Meanwhile, clear results from the kinetic theory approximations for a multi-constituent mixture had been obtained by COWLING [1945, *2*, Eq. (34)]; since his $\boldsymbol{d}_{\mathfrak{A}}$ is approximately (though not exactly) proportional to our $\hat{\boldsymbol{p}}_{\mathfrak{A}}$, his result is equivalent to the inverse of a special case of (295.5). His analysis was generalized by CURTISS and HIRSCHFELDER [1949, *4*, Eqs. (19), (20), (21)] so as to include thermal diffusion; after allowance is made for the approximations of the kinetic theory, their result is seen to be a special case of (295.5). In the kinetic theory the relation (295.8) is valid in first approximation even if $\mathfrak{K} > 2$ but has not been shown to hold in a more accurate treatment.

The theory of STEFAN was proposed anew by SCHLÜTER [1950, *25*, Eq. (3)], [1951, *23*, Eqs. (1) to (3)] and by JOHNSON [1951, *14*]; the latter gave an argument indicating that $F_{\mathfrak{B}\mathfrak{A}}/\varrho_{\mathfrak{B}}\varrho_{\mathfrak{A}}$ is roughly independent of the densities.

of more general cases, we prefer to follow an independent argument[1] toward a linear theory of diffusion which reduces to FICK's law under the conditions in which it is usually applied.

First, we note that by means of (158.8), we may express (295.1) alternatively in the form

$$\varrho_{\mathfrak{A},k} = \frac{1}{2D}(\varrho_{\mathfrak{B}}\, u_{\mathfrak{B}k} - \varrho_{\mathfrak{A}}\, u_{\mathfrak{A}k}) = \frac{\varrho_{\mathfrak{A}}\,\varrho_{\mathfrak{B}}}{\varrho D}(u_{\mathfrak{B}k} - u_{\mathfrak{A}k}) = \frac{\varrho_{\mathfrak{A}}\,\varrho_{\mathfrak{B}}}{\varrho D}(\dot{x}_{\mathfrak{B}k} - \dot{x}_{\mathfrak{A}k}), \qquad (295.3)$$

still for a binary mixture. This symmetrical expression indicates that it is an excess of the mass flow of one constituent over another's that increases the density gradient of the latter. Thus for a general mixture we might have

$$\varrho_{\mathfrak{A},k} = \sum_{\mathfrak{B}=1}^{\mathfrak{K}} \gamma_{\mathfrak{B}\mathfrak{A}}(\varrho_{\mathfrak{A}}\, u_{\mathfrak{B}k} - \varrho_{\mathfrak{A}}\, u_{\mathfrak{B}k}) = \sum_{\mathfrak{B}=1}^{\mathfrak{K}} F_{\mathfrak{B}\mathfrak{A}}(\dot{x}_{\mathfrak{B}k} - \dot{x}_{\mathfrak{A}k}). \qquad (295.4)$$

But since the right-hand side of this relation is equal to a linear combination of the excess of the momenta of all constituents above the momentum of the constituent $\mathfrak{A}$, a still more natural idea of diffusion is expressed by the more general constitutive equation

$$\varrho\,\hat{\boldsymbol{p}}_{\mathfrak{A}} = \sum_{\mathfrak{B}=1}^{\mathfrak{K}} F_{\mathfrak{B}\mathfrak{A}}(\dot{\boldsymbol{x}}_{\mathfrak{B}} - \dot{\boldsymbol{x}}_{\mathfrak{A}}), \qquad (295.5)$$

where $\hat{\boldsymbol{p}}_{\mathfrak{A}}$ is the supply of momentum, defined by (215.2) and restricted by (215.5). In this relation the purely kinematical view expressed in FICK's law (295.1) is replaced by a dynamical one: Diffusion gives rise to a *force* tending to restore uniformity. The dynamical equations corresponding to (295.5) follow at once from (215.2):

$$\varrho_{\mathfrak{A}}(\ddot{x}^k_{\mathfrak{A}} - f^k_{\mathfrak{A}}) - t^{km}_{\mathfrak{A},m} = \sum_{\mathfrak{B}=1}^{\mathfrak{K}} F_{\mathfrak{B}\mathfrak{A}}(\dot{x}^k_{\mathfrak{B}} - \dot{x}^k_{\mathfrak{A}}). \qquad (295.6)$$

Since $\hat{c}_{\mathfrak{A}} = 0$, in order that (295.5) be consistent with the balance of total momentum in the mixture, expressed by (215.5), it is necessary and sufficient that

$$\sum_{\mathfrak{A}=1}^{\mathfrak{K}} (F_{\mathfrak{B}\mathfrak{A}} - F_{\mathfrak{A}\mathfrak{B}}) = 0. \qquad (295.7)$$

When $\mathfrak{K} = 2$, this result is equivalent to[2]

$$F_{\mathfrak{A}\mathfrak{B}} = F_{\mathfrak{B}\mathfrak{A}}, \qquad (295.8)$$

---

[1] The ideas given here are presented more fully by TRUESDELL [1960, *6*].

[2] This condition, for the binary case, originated in the work of MAXWELL [1860, *2*, Prop. XVIII], to whom is due the appeal to "NEWTON's third law" which JOHNSON [1951, *14*] phrased more generally: "The force exerted by species $\mathfrak{A}$ on species $\mathfrak{B}$ is equal in magnitude and opposite in direction to that exerted by $\mathfrak{B}$ on $\mathfrak{A}$ ...." The argument in the text above, basically different in that it rests upon the principle of linear momentum, was first suggested by STEFAN [1871, *6*, p. 74]: „Da die ... Kräfte aus Wechselwirkungen zwischen den im Element $dx\,dy\,dz$ zur selben Zeit befindlichen Teilchen des ersten und zweiten Gases entspringen, so ändern sie die Bewegung des Schwerpunktes dieses Elementes nicht." Indeed, this establishes (295.8) in the binary case, but if $\mathfrak{K} > 2$, only (295.7) follows. For the ternary case, STEFAN assumed (295.8) without analysis or comment. On the other hand, there is twofold insufficiency in the argument of MAXWELL and JOHNSON, for in order to invoke "NEWTON's third law" we have to know that the diffusive term represents the entire force arising from the mutual forces of the two species, which it generally does not, and, secondly, the influence of species $\mathfrak{C}$ upon the mutual action of species $\mathfrak{A}$ and $\mathfrak{B}$ is left out of account. We feel the true content of the Maxwell-Johnson argument does not yield a proof but merely suggests the supplementary assumption (1) in the text above, stating that species $\mathfrak{C}$ has no effect upon the relative diffusion of species $\mathfrak{A}$ and $\mathfrak{B}$.

The relations (295.8) for arbitrary $\mathfrak{K}$ may well be named "STEFAN's relations". The "Onsager reciprocity relations" for pure diffusion refer to a different set of coefficients. LAMM [1954, *11A*], who considered STEFAN's relations to be "self-evident for physical reasons", showed that they are equivalent to ONSAGER's relations when $\mathfrak{K} = 3$; for general $\mathfrak{K}$, this is established by TRUESDELL [1960, *6*].

but if $\mathfrak{K} > 2$, such a relation of symmetry does not hold without further assumptions. It may be shown to be sufficient that (1) $F_{\mathfrak{A}\mathfrak{B}}$ shall not depend upon $\varrho_{\mathfrak{C}}$ if $\mathfrak{C} \neq \mathfrak{A}$ or $\mathfrak{B}$ and (2) $F_{\mathfrak{A}\mathfrak{B}} = 0$ if $\varrho_{\mathfrak{A}} = 0$.

To satisfy the additional requirement (243.6), expressing balance of total energy in the mixture, it suffices to assume the additional constitutive relation

$$\varrho\,\hat{\varepsilon}_{\mathfrak{A}} = \tfrac{1}{2} \sum_{\mathfrak{B}=1}^{\mathfrak{K}} F_{\mathfrak{B}\mathfrak{A}} (\dot{\boldsymbol{x}}_{\mathfrak{B}} - \dot{\boldsymbol{x}}_{\mathfrak{A}})^2, \tag{295.9}$$

as is easily verified. Thus a definite supply of energy is associated with the diffusive supply of momentum. Such supplies of energy, though dynamically required, have never before appeared in a theory of diffusion, perhaps because the diffusive motions are considered small enough that quadratic terms are negligible.

To see the relation between (295.5) and FICK's law[1], we consider the assumptions usual in connection with this latter. The applied forces are supposed to balance the accelerations, $\ddot{x}^k_{\mathfrak{A}} = f^k_{\mathfrak{A}}$, and $t^{km}_{\mathfrak{A}}$ reduces to a hydrostatic pressure: $t^m_{\mathfrak{A}k} = -p_{\mathfrak{A}}\,\delta^m_k$. Then (295.6) becomes

$$p_{\mathfrak{A},k} = \sum_{\mathfrak{B}=1}^{\mathfrak{K}} F_{\mathfrak{B}\mathfrak{A}} (u_{\mathfrak{B}k} - u_{\mathfrak{A}k}). \tag{295.10}$$

If, further, $p_{\mathfrak{A}} \propto \varrho_{\mathfrak{A}}$, as is the case in a gas mixture at uniform temperature, then (295.10) reduces to (295.3), which, for a binary mixture, has already been shown to be equivalent to FICK's law.

A precise mathematical study of the motion of mixtures and of constitutive equations appropriate to them has never been attempted. We venture the following proposal of constitutive equations for a general class of diffusive motions:

$$\left.\begin{aligned} \hat{c}_{\mathfrak{A}} &= 0, \\ \hat{\boldsymbol{p}}_{\mathfrak{A}} &= \boldsymbol{F}_{\mathfrak{A}} (\dot{\boldsymbol{x}}_{\mathfrak{B}} - \dot{\boldsymbol{x}}_{\mathfrak{A}}), \\ \hat{\varepsilon}_{\mathfrak{A}} &= G_{\mathfrak{A}} (\dot{\boldsymbol{x}}_{\mathfrak{B}} - \dot{\boldsymbol{x}}_{\mathfrak{A}}), \end{aligned}\right\} \tag{295.11}$$

where $\boldsymbol{F}_{\mathfrak{A}}$ and $G_{\mathfrak{A}}$ are functions of the indicated arguments, $\mathfrak{B} = 1, 2, \ldots, \mathfrak{K}$. These relations assert that each constituent mass is separately conserved but that the momentum and internal energy of the constituent $\mathfrak{A}$ changes in response to the diffusive motion of each other constituent in respect to it. The functions entering (295.11) are subject to the requirements (215.5) and (243.6), which here take the forms

$$\sum_{\mathfrak{A}=1}^{\mathfrak{K}} \boldsymbol{F}_{\mathfrak{A}} = 0, \quad \sum_{\mathfrak{A}=1}^{\mathfrak{K}} (G_{\mathfrak{A}} + \dot{\boldsymbol{x}}_{\mathfrak{A}} \cdot \boldsymbol{F}_{\mathfrak{A}}) = 0, \tag{295.12}$$

as well as to restrictions following from the fact that $\hat{\boldsymbol{p}}_{\mathfrak{A}}$ is a vector and $\hat{\varepsilon}_{\mathfrak{A}}$ is a scalar. Nothing is presumed, however, regarding the material constitutive equations for the constituents, which may be perfect fluids, viscous fluids, etc.

If chemical reactions are occurring, we must generally assume that $\hat{c}_{\mathfrak{A}} \neq 0$. The usual constitutive equations for chemical reactions have already been given as Eqs. (159A.6). The reaction rates J_a occurring there are to be restricted by use of such information as is available concerning the substances participating in the reactions[2].

---

[1] Our steps reverse those given for a special case by JOHNSON and HULBERT [1950, *13*, Eqs. (1) to (3)].

[2] Cf. ECKART [1940, *8*, p. 274], MEIXNER [1943, *2*, § 2] [1943, *3*, § 4].

III. Example of an energetic constitutive equation.

296. FOURIER's law of heat conduction. Let it be supposed that the flux of energy h^k is a linear function of the temperature gradient $\theta_{,m}$. Then[1]

$$h^k = \varkappa^{km}\theta_{,m}. \tag{296.1}$$

While the quantities $\varkappa^{km}$ are introduced simply as coefficients in a linear relation in some particular co-ordinate system, their law of transformation is easily inferred. Since h^k, by hypothesis, is a Euclidean contravariant tensor, while θ is a scalar, from the quotient law of tensors it follows that $\varkappa^{km}$ is a contravariant tensor of second order. Moreover, if as in Sect. 241 we put $h^\Gamma = (h^k, 0)$ in Euclidean frames and take θ as a world scalar, by defining $\varkappa^{\Gamma\Delta}$ as the world tensor such that

$$\varkappa^{\Gamma\Delta} = \begin{Vmatrix} \varkappa^{km} & 0 \\ 0 & 0 \end{Vmatrix} \tag{296.2}$$

in every Euclidean frame, we may put (296.1) into world-invariant form:

$$h^\Gamma = \varkappa^{\Gamma\Delta}\theta_{,\Delta}. \tag{296.3}$$

It is unusual to be able to ascertain so easily a world-invariant constitutive relation. The principle of material indifference (Sect. 293) is satisfied if $\boldsymbol{\varkappa}$ is assumed independent of all kinematic variables[2], although it may depend upon the temperature and other scalars.

The tensor $\boldsymbol{\varkappa}$ is the *thermal conductivity*.

A body is *thermally isotropic* if (296.1) is an isotropic relation; in this case, the tensor $\boldsymbol{\varkappa}$ is an isotropic tensor; hence $\varkappa^{km} = \varkappa g^{km}$, where $\varkappa$ is a scalar, so that (296.1) reduces to the form[3]

$$h_k = \varkappa\,\theta_{,k}. \tag{296.4}$$

Since the sign of $h^k\theta_{,k}$ depends upon $\boldsymbol{\varkappa}$ and $\theta_{,p}$ alone, application of the consequence $(258.1)_2$ of the entropy inequality to (296.1) and (296.2) shows that

$$\varkappa^{km} \text{ is positive semi-definite or } \varkappa \geqq 0, \tag{296.5}$$

respectively, unless there are thermal constraints limiting the range of $\theta_{,k}$.

Substitution of this result into (241.7) does not yield much enlightenment in general. When, however, assumptions sufficient to derive (241.8) are added, we obtain

$$[n_k \varkappa^{km}\theta_{,m}] = 0, \tag{296.6}$$

where $\boldsymbol{n}$ is the unit normal to the discontinuity. This is the condition appropriate to the interface of two materials having different thermal conductivities, provided, of course, that the surface not be the seat of sources of energy. In the isotropic case, we thus obtain[4]

$$\left[\varkappa \frac{d\theta}{dn}\right] = 0. \tag{296.7}$$

---

[1] DUHAMEL [1832, *1*, pp. 361–366].

[2] However, it is not impossible for $\boldsymbol{\varkappa}$ to depend upon the velocity gradient. E.g., if $\varkappa^{\Gamma\Lambda} = C\,\Delta^{\Gamma\Lambda}$, where $\Delta^{\Gamma\Lambda}$ is the world tensor of stretching, the principle of material indifference is satisfied. Cf. Sect. 307 below.

[3] FOURIER [1822, *1*, § 98]. There is a formal similarity to NEWTON's law for the radiative cooling through a boundary.

[4] FOURIER [1822, *1*, § 146].

When the temperature itself is continuous, we may put $\psi = \theta$ in (175.8); substituting the result into (296.6) and rearranging yields

$$n_m n_k \varkappa_+^{km} \left[\frac{d\theta}{dn}\right] + [n_k \varkappa^{km}]\, \theta_{,m}^- = 0. \tag{296.8}$$

By (296.5), the coefficient of the first jump is positive unless $\boldsymbol{\varkappa}_+ = 0$. Thus we obtain the following result: Apart from the case when the conductivity vanishes on one side of a surface on which θ is continuous, the thermal flux (296.1) cannot suffer a discontinuity unless also the normal conductivity $n_k \varkappa^{km}$, or, in the isotropic case, the conductivity $\varkappa$, is discontinuous.

Substitution of (296.1) into the general equation (241.4) of energy balance makes it possible to determine ε when the stress, the motion, the temperature, and the supply of energy are known. This is not, however, the classical approach to problems of heat conduction, which rests, if often rather obscurely, upon additional assumptions leading to the completely recoverable case described in Sect. 256A. By (256A.2), $(256.2)_2$, and (296.1) we have

$$\varrho\, \theta\, \dot{\eta} = (\varkappa^{km}\, \theta_{,m})_{,k} + \varrho\, q. \tag{296.9}$$

If it is further supposed that the material is *ideal*[1] in the sense defined in Sect. 253, so that $\eta = \eta(\theta, \boldsymbol{X})$, then (296.9) reduces to ***Fourier's equation of thermal conduction***[2]:

$$c\, \dot{\theta} = \frac{1}{\varrho} (\varkappa^{km}\, \theta_{,m})_{,k} + q, \tag{296.10}$$

c being a known function of θ and possibly also of $\boldsymbol{X}$. When the velocity field $\dot{\boldsymbol{x}}$, the density ϱ, and the supply of energy q are known, (296.10) is a partial differential equation for the temperature. In particular, in a material at rest we have[3]

$$c\, \frac{\partial \theta}{\partial t} = \frac{1}{\varrho} (\varkappa^{km}\, \theta_{,m})_{,k} + q. \tag{296.11}$$

The foregoing derivation may seem more laborious than that to which the reader is accustomed. This is so because the classical argument, deriving in part from researches prior to the establishment of the general principles of thermomechanics, condenses all the assumptions into a single one (essentially, that all thermal flux arises solely by transference of a caloric energy which is proportional to the temperature), with no connection to other domains of physics.

IV. Examples of mechanical constitutive equations.

297. Perfect fluids. A continuous medium is a *perfect fluid* if it can support no shearing stress and no couple stress. By the theorem of CAUCHY given in Sect. 204, the stress tensor $\boldsymbol{t}$ is then in all circumstances hydrostatic, $\boldsymbol{t} = -p\mathbf{1}$, and from CAUCHY's first law of motion (205.2) we have ***Euler's dynamical equation***[4]:

$$\varrho\, \ddot{\boldsymbol{x}} = -\operatorname{grad} p + \varrho \boldsymbol{f}. \tag{297.1}$$

CAUCHY's second law (205.11) is satisfied automatically; in other words, balance of linear momentum in a perfect fluid implies balance of moment of momentum,

---

[1] This assumption is made in a disguised form by VOIGT [1889, *9*, pp. 743–759] [1895, *8*, Part III, Chap. I, § 6–9] [1910, *10*, §§ 277, 381–382, 389, 392].

[2] FOURIER [1833, *1*, Eq. (3)], for the isotropic case.

[3] FOURIER [1822, *1*, §§ 127–128], for the isotropic case.

[4] [1757, *2*, §§ 20–21]. The theory had a long earlier history in special cases.

so long as there be no extrinsic couples, while if extrinsic couples are present, the perfect fluid is incompatible with the principles of mechanics (cf. Sect. 293α).

From (205.4) we have at once ***Christoffel's condition***[1]:

$$[p]\,\boldsymbol{n} + \varrho^{\pm} U^{\pm} [\dot{\boldsymbol{x}}] = 0; \tag{297.2}$$

hence[2]

$$[p] + \varrho^{\pm} U^{\pm} [\dot{x}_n] = 0, \qquad \varrho^{\pm} U^{\pm} [\dot{\boldsymbol{x}}_t] = 0. \tag{297.3}$$

We distinguish two cases. First, assume that $U^+ = U^- = 0$, or that $[\dot{x}_n] = 0$, or that both these conditions hold; then $(297.3)_1$ reduces to

$$[p] = 0: \tag{297.4}$$

Across a singular surface which is material and across all waves other than shock waves, the pressure is continuous. Second, assume $\varrho^+ U^+ \neq 0$. Then $(297.3)_2$ yields

$$[\dot{\boldsymbol{x}}_t] = 0: \tag{297.5}$$

In a perfect fluid, shock waves are always longitudinal. This last result can be extended to waves of higher order, as follows. By (297.4), we may put $\Psi = p$ in (175.8) and conclude that [grad p] is parallel to $\boldsymbol{n}$, and hence, by (297.1), on the assumption that $\boldsymbol{f}$ is continuous, it follows that $[\ddot{\boldsymbol{x}}]$ is parallel to $\boldsymbol{n}$. Therefore, *the tangential acceleration is continuous across all material singularities and all waves other than shock waves*[3], and, in particular, acceleration waves are always longitudinal. For waves of third order, a parallel argument applied to the gradient of (297.1) shows that $\overset{(3)}{\boldsymbol{x}}$ is parallel to $\boldsymbol{n}$. By induction, then, *a perfect fluid admits only longitudinal waves*[4]. By (191.2) and (191.7), a wave of k-th order carries jumps of $\overset{(k-1)}{x_n}$, div $\overset{(k-1)}{\boldsymbol{x}}$, $\overset{(k-1)}{\overline{\log \varrho}}$, and $d^{k-1} p/d n^{k-1}$, but leaves unaffected $\overset{(k-1)}{\boldsymbol{x}_t}$, curl $\overset{(k-1)}{\boldsymbol{x}}$, and all tangential derivatives of p up to the $k-1$-st order. Finally, in *an isochoric motion of a perfect fluid, wave propagation of any kind is impossible*[5], the only compatible discontinuities being material.

Note that all the foregoing results follow at once from the definition of a perfect fluid and from the principles of kinematics and mechanics; none of the usual specializing assumptions have been invoked.

Returning to consideration of (297.1), assume that field $\boldsymbol{f}$ of assigned force be lamellar; the mechanical significance of this assumption has been presented in Sect. 218. From (297.1) we have then

$$\boldsymbol{w}^* \equiv \operatorname{curl} \ddot{\boldsymbol{x}} = \operatorname{grad} p \times \operatorname{grad} \frac{1}{\varrho}. \tag{297.6}$$

By interpreting this formula in the light of results given in Sect. 108, we derive ***Kelvin's theorem***[6]: *A flow of a perfect fluid subject to lamellar assigned force is circulation preserving if and only if there exists a functional relation*

$$f(p, \varrho, t) = 0; \tag{297.7}$$

---

[1] [1877, *2*, § 1].

[2] Eqs. (297.3) form a part of the set called "the Rankine-Hugoniot equations". Further references are given in Sect. 205 and in footnote 2, p. 713.

[3] HADAMARD [1903, *11*, ¶ 244]. MOREAU [1952, *13*, § 8c] notes that the condition $\boldsymbol{n} \times [\ddot{\boldsymbol{x}}] = 0$, or Curl $\ddot{\boldsymbol{x}} = 0$ on the surface, is analogous to the D'ALEMBERT-EULER condition curl $\ddot{\boldsymbol{x}} = 0$ in space, and thus he derives a superficial analogue of KELVIN's theorem; cf. his results presented in Sect. 186, but note that, in contrast to KELVIN's theorem, barotropic flow is not required.

[4] Proved for acceleration waves by HUGONIOT [1887, *2*, Part I, § 12], for shock waves and waves of higher order by HADAMARD [1903, *11*, ¶ 242] and DUHEM [1901, *7*, Part II, Chap. IV, § 2].

[5] HADAMARD [1903, *11*, ¶¶ 245–246], DUHEM [1901, *7*, Part II, Chap. I, § 10].

[6] [1869, *7*, § 59(d)].

alternatively, if and only if, for each fixed time, the pressure is constant, or the density is constant, or the surfaces $p=\text{const}$ coincide with the surfaces $\varrho=\text{const}$. Isochoric flows, in which each particle has the same density, are included as a special case.

KELVIN's theorem is regarded as the fundamental theorem of classical hydrodynamics. Flows satisfying (297.7) are called *barotropic*[1].

A perfect fluid may be such that all its flows are barotropic; this is the case for homogeneous incompressible fluids, for which $\varrho=\varrho_0=\text{const}$ in space and time, and for piezotropic fluids, for which there is an equation of state of the form $p=f(\varrho)$ (discussed more generally in Sect. 253). But these conditions are merely sufficient, not necessary, for barotropic flow. For example, in a fluid having equations of state $p=F(\varrho,\theta)=G(\varrho,\eta)$, special conditions may lead to an isothermal flow, in which $\theta=\text{const}$, or to an isentropic flow, in which $\eta=\text{const}$; any such flow, obviously, is barotropic, but the functional form of f in (297.7) depends upon the particular conditions giving rise to the flow.

While various thermodynamic circumstances may determine whether or not a particular flow be barotropic, KELVIN's theorem shows that these circumstances, once it has been ascertained that an equation of the form (297.7) does in fact hold, may be largely disregarded. For when (297.7) holds, all the numerous theorems appropriate to circulation preserving motion may be applied: the Helmholtz vorticity theorems (Sect. 108), Bernoullian theorems of various kinds (Sects. 125 to 126), with the acceleration-potential V^* being given by[2]

$$V^*=\int\frac{dp}{\varrho}+v+g(t),\qquad \boldsymbol{f}=-\operatorname{grad}v, \tag{297.8}$$

HELMHOLTZ's theorem of conservation of energy (Sect. 218)—indeed, all the *general* theorems of classical hydrodynamics follow at once from the circulation preserving property, as we have taken pains to show in Sects. 105 to 138. The theory of irrotational motion is included as an important special case. When we combine HADAMARD's theorem of Sect. 190 with the result proved above that acceleration waves are necessarily longitudinal, we see that *the passage of waves other than shock waves leaves the circulation unchanged*, and, in particular, the Lagrange-Cauchy theorem on the permanence of irrotational flow (Sect. 134) is unimpaired by the passage of waves other than shock waves[3].

In the case of a barotropic flow that is neither isochoric nor isobaric, we may write (297.1) in the form

$$\ddot{\boldsymbol{x}}=-c^2\operatorname{grad}(\log\varrho)+\boldsymbol{f},\qquad c^2\equiv\frac{\partial p}{\partial\varrho}. \tag{297.9}$$

Hence at an acceleration wave we have

$$[\ddot{x}_n]=-c^2\left[\frac{d\log\varrho}{dn}\right]. \tag{297.10}$$

Comparison of this formula with the kinematical identity (190.10) yields $U^2=c^2$. This is ***Hugoniot's theorem***[4]: *The speed of propagation of acceleration waves is c.*

[1] BJERKNES *et al.* [1933, *3*, §§ 1, 24].

[2] EULER [1757, *2*, § 35] [1770, *1*, § 42].

[3] Hinted by HUGONIOT [1887, *2*, Part I, § 12], proved explicitly by HADAMARD [1903, *10*, ¶¶ 254–255].

[4] [1885, *3*] [1887, *2*, Part I, § 10]. As remarked in [1885, *4*] [1887, *2*, Part II, §§ 11–12], the result still holds if (297.7) is replaced by $p=f(\varrho,t,\boldsymbol{X})$; while HUGONIOT used a formulation in material variables to establish this more general result, it also follows easily by noting that $(297.9)_1$ is modified only by the addition of $-\frac{1}{\varrho}\frac{\partial p}{\partial X^\alpha}X^\alpha_{,k}$, which is continuous across an acceleration wave.

Similar analysis applied to the results in Sect. 191 shows that $U^2=c^2$ for waves of higher order. In view of $(297.9)_2$, these results may be interpreted as follows[1]: *Given a barotropic flow in which neither the pressure nor the density is uniform, a necessary and sufficient condition for wave propagation to be possible is that the pressure be an increasing function of the density; this being so, waves of all orders greater than* 1 *propagate with the unique speed c.* Since c is the common speed of propagation of so many kinds of waves, it is called the *speed of sound.*

The speed of propagation of a shock wave satisfies (205.7), which reduces when $\boldsymbol{t}=-p\mathbf{1}$ to the celebrated ***Rankine-Hugoniot equation***[2]

$$(U^-)^2=\frac{\varrho^+}{\varrho^-}\frac{[p]}{[\varrho]} \quad \text{or} \quad U^+U^-=\frac{[p]}{[\varrho]}. \tag{297.11}$$

The formal similarity of these results to $(297.9)_2$ is deceiving, for these are purely dynamical and do not presume the existence of a barotropic relation (297.7) on either side of the shock. If there is such a relation, it need not have the same form on each side of the shock; i.e., $[p]=f^+(\varrho^+)-f^-(\varrho^-)$, where the functions f^+ and f^- need not be the same. In order to conclude that a shock of given strength propagates at a definite speed, it is necessary to supply further assumptions.

Flows in which no relation of the form (297.7) holds are called *baroclinic*[3]. The theory of these flows is more complicated and cannot be approached in such generality. We consider only the commonest special case. It is assumed that the fluid obeys a caloric equation of state of the form $\varepsilon=f(\eta, v)$ relating specific internal energy ε to specific entropy η and specific volume $v=1/\varrho$, and that the pressure p in EULER's equation (297.1) is the thermodynamic pressure: $p=\pi=-\tau$, where τ is given by $(247.1)_2$. Then ${}_{\mathrm{D}}\boldsymbol{t}=0$ and $\boldsymbol{m}=0$ in (256.11), and (256.12) yields $P_{\mathrm{E}}=P_{\mathrm{I}}$. It is customary to put $\boldsymbol{h}=0$ and $q=0$; by $(256.2)_2$, $Q_{\mathrm{E}}=0$. In physical terms, it has been assumed that all work done is fully recoverable and that there is neither heat flux nor any source of heat. From the balance of energy as expressed by (256.6) it then follows that

$$\dot{\eta}=0: \tag{297.12}$$

The entropy of each particle remains constant, where we have assumed that $\theta\neq 0$.

If there exists a surface $\eta=$const which is traversed by every particle, as may be presumed to be the case in a gas flow originating from a quiet reservoir, it follows that η is uniform, and we fall back upon a barotropic case, but one of a particular kind: Since the barotropic relation (297.7) which holds in these conditions is obtained by putting $\eta=$const in the equation of state $p=p(\varrho,\eta)$, the speed of sound as given by (297.9) now has the value

$$U^2=c^2=\left(\frac{\partial p}{\partial \varrho}\right)_\eta, \tag{297.13}$$

commonly associated with the name of LAPLACE.

If the fluid is assumed to conduct heat, however, a very different result follows. By putting FOURIER's law (296.4) into $(256.2)_2$, we obtain $Q_{\mathrm{E}}\neq 0$ in general, and (297.12) does not hold. We consider an acceleration wave across which p, θ, and the thermal conductivity $\varkappa$ are continuous. By the result derived from (296.6), $[\operatorname{grad}\theta]=0$. By using the thermal equation of state $p=p(\varrho,\theta)$ in

[1] HADAMARD [1903, *10*, ¶ 242], DUHEM [1901, *7*, Part II, Chap. IV, § 2].

[2] [1870, *6*, Eq. (6)] (lineal motion); [1887, *2*, §§ 144, 149].

[3] BJERKNES *et al.* [1933, *3*, §§ 1, 24].

(297.1) we have

$$\ddot{\boldsymbol{x}} = -\left(\frac{\partial p}{\partial \varrho}\right)_\theta \operatorname{grad}\log\varrho - \left(\frac{\partial p}{\partial \theta}\right)_\varrho \operatorname{grad}\theta + \boldsymbol{f}. \tag{297.14}$$

Hence, in the present circumstances,

$$[\ddot{\boldsymbol{x}}] = -\left(\frac{\partial p}{\partial \varrho}\right)_\theta [\operatorname{grad}\log\varrho], \tag{297.15}$$

although the flow, which is *not* assumed isothermal, is generally baroclinic. Comparison with the kinematical identity (190.10) yields[1]

$$U^2 = \left(\frac{\partial p}{\partial \varrho}\right)_\theta. \tag{297.16}$$

This is the value of the speed of sound first obtained in a very special case by NEWTON; it is thus seen to be the speed of propagation of acceleration waves in a perfect fluid having a thermal equation of state $p = p(\varrho, \theta)$ and a non-vanishing thermal conductivity $\varkappa$, no matter how small[2].

Notice that in order to derive (297.13) and (297.16) we have used appropriate forms of the balance of energy and appropriate equations of state. The theory is thus a thermo-mechanical one.

While (297.12) holds in every region free of shocks, the constant value of η it implies is different upon each side of the shock. That this is so follows from the balance of energy in the form (241.9), or, equivalently,

$$[\chi + \tfrac{1}{2}U^2] = 0. \tag{297.17}$$

Since the functional forms of the equations of state $\chi = \chi(\eta, v)$ and $\pi = p = \pi(\eta, v)$ are assumed unchanged by any circumstance, (297.17) may be used along with (297.11) to calculate the relative jumps of η, U and p as functions of the strength δ, where $\delta \equiv [\varrho]/\varrho^- = -[v]/v^+$. When the equation of state is restricted by suitable inequalities (cf. Sect. 265, esp. footnote 1, p. 656), it can be shown that the function $U^- = f(\delta)$ is monotone increasing, that $f(0) = c$, where c is the speed of sound given by (297.13), that $\delta \leqq \delta_{\max}$, where $\delta_{\max}$ is a number determined by the equation of state, and that $f(\delta_{\max}) = \infty$. That is, all shocks are propagated at supersonic speed relative to the fluid ahead and subsonic speed relative to the fluid behind; infinitely weak shocks travel at nearly sonic speed, and the stronger the shock, the faster it travels; but only a certain ultimate strength is possible in any given fluid, no matter how fast the shock[3].

When the flow ahead of the shock satisfies not only (297.12) but also the stronger condition $\eta = \text{const}$, the flow is barotropic and hence circulation preserving. The passage of a shock wave which is curved or oblique to the flow causes

[1] DUHEM [1901, *7*, Part II, Chap. IV, § 2], KOTCHINE [1926, *3*, § 3].

[2] As $\varkappa \to 0$, the speed given by (297.16) does not approach that given by (297.13), and when $\varkappa = 0$, no wave can travel at the speed given by (297.16) except in the very special circumstances when $\theta = \text{const}$. This difference ceases to seem paradoxical when we recall that the conduction of heat is a dissipative mechanism, and that when the second main dissipative mechanism, that of viscosity, is included in the mathematical model, both the kinds of waves considered here become impossible, as is shown in Sect. 298. The limit behavior is discussed by DUHEM [1901, *7*, Part III, §§ 2–3] and by SERRIN, § 57 of Mathematical Principles of Classical Fluid Mechanics, this Encyclopedia, Vol. VIII. Part 1.

[3] The first analysis of this kind is that of STOKES [1848, *4*, p. 355], faulty from failure to take account of the balance of energy. For a perfect gas, the full results are due to HUGONIOT [1887, *2*, §§ 154, 161–163]; cf. the summary of VIEILLE [1900, *10*, p. 185]. The general theory of weak shocks had been given earlier by CHRISTOFFEL [1877, *2*, §§ 2–5]. An exposition for general tri-variate fluids is given by SERRIN, Sect. 56 of the article just cited.

a jump of η which is not uniform, thus rendering the flow baroclinic and destroying the circulation preserving property, as well as inducing a jump of vorticity[1].

Further general theorems of gas dynamics follow from further assumptions; typically, that the flow is steady, that $\boldsymbol{f}=0$, etc.[2].

298. Linearly viscous fluids. To consider a substance which in equilibrium has the same behavior as that predicted by EULER's equation (297.1), viz.

$$\operatorname{grad} p = \varrho \boldsymbol{f}, \tag{298.1}$$

yet when in motion can support appropriate shearing stresses, assume that the stress tensor $\boldsymbol{t}$ be a linear function of the velocity $\dot{\boldsymbol{x}}$ and the velocity gradient. By (90.1) we may write

$$\boldsymbol{t} = \boldsymbol{g}(\dot{\boldsymbol{x}}, \boldsymbol{w}, \boldsymbol{d}), \tag{298.2}$$

where $\boldsymbol{g}$ is a linear function. The constitutive equations (298.2) define a *linearly viscous fluid*, it being supposed also that there is no couple stress: $\boldsymbol{m}=0$.

We now apply the principle of material indifference (Sect. 293 θ). The constitutive equation (298.2) is to have the same form for all observers. To an observer in a co-rotational frame, $\dot{\boldsymbol{x}}=0$ and $\boldsymbol{w}=0$; for him, (298.2) reduces to a relation giving $\boldsymbol{t}$ as a function of $\boldsymbol{d}$ alone, and therefore, since both $\boldsymbol{t}$ and $\boldsymbol{d}$ transform independently of $\dot{\boldsymbol{x}}$ and $\boldsymbol{w}$ [cf. (144.3) and (211.1)], it must reduce to such a relation for all observers. I.e.[3],

$$\boldsymbol{t} = \boldsymbol{f}(\boldsymbol{d}). \tag{298.3}$$

Now consider frames whose axes coincide with the principal directions of $\boldsymbol{d}$ (Sects. 82 and 83), so that (298.3) becomes

$$t_{km} = \mathfrak{f}_{km}(d_1, d_2, d_3). \tag{298.4}$$

For a definite assignment of the undirected axes different assignments of the positive senses yield four different positively oriented frames, any one of which may be obtained from any other by a rotation through a straight angle about one axis. Under such rotations, it follows from (82.6) that $d_\mathfrak{a}$ is invariant. By the principle of material indifference, then, $\mathfrak{f}_{km}$ in (298.4) is invariant under these rotations: $\mathfrak{f}^*_{km}=\mathfrak{f}_{km}$, say. But $\boldsymbol{t}$ transforms according to the tensor law; in particular, under rotation through a straight angle about the k-axis we have $t^*_{km}=-t_{km}$ for $k\neq m$. Comparing these two results yields $t_{km}=0$ if $k\neq m$. Thus *the principal axes of stretching are also principal axes of stress.* Since further orthogonal transformations may permute the $d_\mathfrak{a}$ in any way, the principal stress $t_\mathfrak{b}$ is a symmetric function of them. Hence the relation (298.3) reduces to one giving

[1] Indicated by HADAMARD [1903, *10*, Note III]. A modern exposition is given by SERRIN, § 54 of the article, just cited.

Despite the above-stated facts, the speed of sound behind the shock is still given by (297.13), since, as remarked by TRUESDELL [1951, *36*], this follows from (297.12) and the more general theorem proved in footnote 4 on p. 712.

[2] A development of these principles is given by TRUESDELL [1952, *23*].

[3] This equation, in an inertial frame, along with $\boldsymbol{f}(0) = -p\mathbf{1}$, was taken as the definition of a fluid by STOKES [1845, *4*, § 1].

$\boldsymbol{t}$ as an isotropic function of $\boldsymbol{d}$. In other words, *all fluids included in the definition* (298.2) *are necessarily isotropic*[1].

The most general linear isotropic function $\boldsymbol{t}$ of a symmetric second-order tensor $\boldsymbol{d}$ may be written in the form[2] of the ***Navier-Poisson law***:

$$\left.\begin{aligned} \boldsymbol{t} &= -p\boldsymbol{1} + \lambda \mathrm{I}_{\boldsymbol{d}} \boldsymbol{1} + 2\mu \boldsymbol{d}, \\ t^k_m &= -p\delta^k_m + \lambda d^q_q \delta^k_m + 2\mu d^k_m, \end{aligned}\right\} \tag{298.5}$$

where we have used the requirement that $\boldsymbol{t} = -p\boldsymbol{1}$ when $\boldsymbol{d} = 0$. From (298.5) it follows that $\boldsymbol{t}$ is symmetric. Therefore CAUCHY'S second law in the form (205.11) is satisfied automatically. In other words, it is a consequence of the principle of material indifference that *a fluid cannot support extrinsic couples; when there are no extrinsic couples, balance of moment of momentum follows from balance of momentum*[3]. Substitution of (298.5) into CAUCHY'S first law (205.2) yields a system of three differential equations known, at least when subjected to further simplifying assumptions, as "the Navier-Stokes equations". If the pressure p and the three components $\dot{x}^k$ of the velocity are regarded as unknowns and the coefficients λ, μ are regarded as given, the system is still underdetermined, but since the means of rendering it determinate are essentially the same as those used in the theory of perfect fluids (Sect. 297), they will not be discussed here.

Since the physical dimensions of the coefficients λ and μ in (298.5) are $[M L^{-1} T^{-1}]$, it is plain that our definition (298.2) violates the principle stated in Sect. 293 ε. In rectification, we replace (298.2) by the complete ***definition of a linearly viscous fluid***:

$$\boldsymbol{t} = \boldsymbol{f}(\dot{\boldsymbol{x}}, \boldsymbol{w}, \boldsymbol{d}, p, \theta, \theta_0, \mu_0), \tag{298.6}$$

where, as before, $\boldsymbol{f}$ is a linear function of $\dot{\boldsymbol{x}}$, $\boldsymbol{w}$, and $\boldsymbol{d}$ and a continuous function of its remaining arguments, and where θ_0 and μ_0 are material constants such that

$$\text{phys. dim}\,\theta_0 = [\Theta], \qquad \text{phys. dim.}\,\mu_0 = [M L^{-1} T^{-1}]. \tag{298.7}$$

The presence of the first of these constants makes it possible for the fluid's response to deformation to vary with the temperature. Motivation for introducing the second constant may be found in simple experiments on the resistance of fluids in viscometers of various kinds. Since it would be possible, still within the framework of pure mechanics, to lay down a relation like (298.6) but involving four

---

[1] Sometimes encountered are "anisotropic fluids" defined by constitutive relations of the type $t^k_m = C^{kp}_{mq} d^q_p + D^k_m$, where $\boldsymbol{C}$ and $\boldsymbol{D}$ are general tensors of the orders indicated. Such an equation does not generally satisfy the principle of material indifference. Given such a relation in an inertial frame, for example, transformation of $\boldsymbol{t}$ and $\boldsymbol{d}$ by the appropriate laws to a non-inertial frame yields a relation of the same form except that the components of $\boldsymbol{C}$ and $\boldsymbol{D}$ in the non-inertial frame depend upon the time t, violating the original postulate (298.2).

In order to obtain a properly invariant theory of anisotropic fluids, it is necessary to modify (298.2) by introduction of some vector or vectors specifying preferred directions. Cf. the oriented bodies studied in Sects. 60–64 and the theory recently proposed by ERICKSEN [1960, *1*].

[2] The simplest case of this law was proposed by NEWTON [1687, *1*, Lib. II, Chap. IX]. For incompressible fluids, equivalent dynamical equations were obtained from a molecular model by NAVIER [1821, *1*] [1822, *2*] [1825, *1*] [1827, *6*]. The continuum theory of CAUCHY [1823, *1*] [1828, *2*, § III, Eqs. (95), (96)] lacks the term $-p\delta^k_m$. The general formula was obtained by POISSON [1831, *2*, ¶¶ 60–63] from a molecular model. The continuum theory, subject to an unjustified but easily removed specialization, is due to ST. VENANT [1843, *4*] and STOKES [1845, *4*, §§ 3–5].

[3] The reader is to recall that in Sect. 205 symmetry of $\boldsymbol{t}$ was proved equivalent to balance of moment of momentum only under the assumptions that $\boldsymbol{l} = 0$, that $\boldsymbol{m} = 0$, and that linear momentum was balanced.

rather than two dimensionally independent material constants, our definition (298.6) implies not only the kinematical restrictions we have already demonstrated but also dimensional ones, which we now proceed to determine.

First, the same reasoning as given above suffices to reduce (298.6) to an equation of the form[1]

$$\boldsymbol{t} = \boldsymbol{f}(\boldsymbol{d}, p, \theta, \theta_0, \mu_0). \tag{298.8}$$

The dimensional matrix of the 11 quantities appearing in any one component of (298.8) is

	L	T	M	Θ	
t_{km}	−1	−2	1	0	(k and m fixed)
p	−1	−2	1	0	
$\boldsymbol{d}$	0	−1	0	0	(6 rows)
μ_0	−1	−1	1	0	
θ	0	0	0	1	
θ_0	0	0	0	1	

The first 2 rows are alike; so are the next 6 and the last 2; the ninth may be obtained by subtracting the third from the first; thus the rank is at most 3; in fact, it is exactly 3. By the π-theorem, (298.8) is equivalent to a relation among $11-3=8$ dimensionless ratios formed from the quantities entering it. A possible set of such ratios is t_{km}/p, $\mu_0 \boldsymbol{d}/p$, θ/θ_0. Hence (298.8) is equivalent to a relation of the form[2]

$$\boldsymbol{t}/p = \boldsymbol{g}(\mu_0 \boldsymbol{d}/p, \theta/\theta_0), \tag{298.9}$$

where the function $\boldsymbol{g}$ is a dimensionless function depending linearly upon $\mu_0 \boldsymbol{d}/p$ and continuously upon θ/θ_0. Thus the requirement of dimensional invariance implies that p may enter the dynamical equations only in a strikingly restricted way. Re-examining the argument used to reduce (298.3), we see that the presence of additional scalar arguments does not affect it. Hence we again obtain (298.5), except that now the coefficients λ and μ are shown to have restricted functional forms:

$$\lambda = \mu_0 f_1(\theta/\theta_0), \quad \mu = \mu_0 f_2(\theta/\theta_0), \tag{298.10}$$

the functions f_1 and f_2 being dimensionless. While, as shown by (298.9), it has been tacitly assumed that $p \neq 0$ in the region considered, the assumed continuity of $\boldsymbol{f}$ as a function of p in (298.6) allows this restriction to be removed by inspection of the final result.

The material coefficients λ and μ are the *viscosities* of the fluid. The dimensional constant μ_0 is a parameter which, in any system of units selected, may be assigned a numerical value representing the amount of shearing stress or other resistance to a given stretching that a particular physical fluid may offer. The dimensional constant θ_0 and the dimensionles functions f_1 and f_2 are parameters which enable representation of a viscous response varying with temperature. From their definition via (298.6), the viscosities are *independent*[3] *of all kinematical*

---

[1] In the following argument, we tacitly employ rectangular Cartesian co-ordinates throughout, so that all tensor components bear the physical dimensions shown in the table. The final results are in tensorial form and hence valid in all co-ordinates.

[2] TRUESDELL [1949, *33*, §§ 3, 6] [1950, *32*, §§ 4, 7] [1952, *21*, §§ 63–65] [1952, *22*, p. 90].

[3] To speak of a "frequency-dependent viscosity" or a "non-linear viscosity", as is not uncommon in the literature of ultrasonics and rheology, is a misleading way of saying that the linear law (298.5) does *not* hold but some particular (and usually imperfectly specified) non-linear law does..

variables; according to the results (298.10), the viscosities are also *independent of the pressure*[1], being functions of the temperature only.

When $\lambda = \mu = 0$, the dynamical equations for viscous fluids reduce to EULER's equation (297.1) for perfect fluids. Thus perfect fluids are often called *inviscid*.

That, irrespective of the values of λ and μ, the equations of the theory of viscous fluids reduce to the *same* statical equation (298.1) when $\boldsymbol{d}=0$, is a standard example to show that the commonest statement of "D'ALEMBERT's principle" is false[2]; for this reason, the portion $\lambda \mathrm{I}_{\boldsymbol{d}} \mathbf{1} + 2\mu \boldsymbol{d}$ of the stress is regarded as arising from *internal friction*.

Since

$$t^k_m = 2\mu\, d^k_m \quad \text{when} \quad k \neq m, \tag{298.11}$$

the quantity μ is the ratio of shear stress to the corresponding shearing (Sect. 82) of any two orthogonal elements; hence it is called the *shear viscosity*.

The mean pressure (204.7) is given by

$$\tilde{p} - p = -\left(\lambda + \tfrac{2}{3}\mu\right) d^k_k = \left(\lambda + \tfrac{2}{3}\mu\right) \overline{\log \varrho}^{\,\bullet}. \tag{298.12}$$

Thus $\lambda + \frac{2}{3}\mu$, the *bulk viscosity*, is the ratio of the excess of the mean of the three normal pressures over the static pressure to the rate of condensation. To render this last interpretation definite, it is best to distinguish two cases. First, for an *incompressible* fluid we have $\dot{\varrho} = 0$, and from (298.12) follows $p = \tilde{p}$ in all circumstances. Thus, for an incompressible fluid, the pressure p occurring in the constitutive relation (298.5) is always the mean pressure, and λ drops out of all equations. Second, for a *compressible* fluid, in accord with the agreement that (298.1) holds in equilibrium, we take p in (298.5) to be the same pressure as would hold in equilibrium under the same thermodynamic conditions, i.e., $p = \pi =$ $\pi(\varrho, \theta)$ *as given by the thermal equation of state*. This pressure is then *determined*, independently of the motion, as soon as ϱ and θ are known. (298.12) relates this pressure to the mean of the pressures actually exerted upon three perpendicular planes at the point in question. (A pressure-measuring device, in general, measures some component of the stress tensor; in a viscous fluid, it is not justifiable to identify either p or $\tilde{p}$ with the results of measurement, a theory of the flow near the instrument being required in order to interpret the experimental values in terms of the variables occurring in the theory.)

[1] I.e., to obtain viscosities which depend also upon the pressure, it is necessary to start with a definition more general than (298.6): A constant or scalar bearing the dimension of time or stress must be included, as is done in the following section. Cf. the discussion of this point by TRUESDELL [1950, *30*, §§ 4, 7, 11] [1952, *21*, §§ 62–63].

[2] I.e., to obtain "the" equations of motion from the statical equations, for a given system, add the "inertia force" $-\ddot{\boldsymbol{x}}$ to the assigned force $\boldsymbol{f}$ per unit mass. If (298.1) are the statical equations, then this form of "D'ALEMBERT's principle" is to be supplemented by adding to the "inertial force" any "frictional forces" that may be present. "Frictional forces" are then defined as any forces arising from motion other than inertial force. Putting all these definitions together leads to the conclusion that D'ALEMBERT's principle asserts that equations of motion follow from statical equations by supplying inertial force plus such other forces as may arise in conjunction with motion. Thus D'ALEMBERT's principle appears to have no content at all.

To the reader who finds this confusing we remark that it was not by oversight that from the list of guiding principles in Sect. 293 we omitted "D'ALEMBERT's principle", for we consider it to be either trivial or false in the usual statements in this context. (This does not affect the validity of the different "D'ALEMBERT-LAGRANGE principle" given in Sect. 232.)

A correct statement, revealing the limited but useful validity of this form of the principle, is as follows: *Given the statical equations for a material, constitutive equations for a dynamically possible material result if* $\boldsymbol{f}$ *is replaced by* $\boldsymbol{f} - \ddot{\boldsymbol{x}}$. *This special kind of material, being only one of the infinitely many that share the same statical properties, is called "perfect"*. This definition applies to several classical theories.

The stress power (217.4) assumes the form

$$P = P_{\mathrm{E}} = t^{km} d_{km} = -p\, d^k_k + \lambda (d^k_k)^2 + 2\mu\, d^k_m\, d^m_k\,, \tag{298.13}$$

while by the above agreement regarding p we have from $(256.4)_1$

$$P_{\mathrm{I}} = -p\, d^k_k. \tag{298.14}$$

Hence Eq. (256.6) for production of total entropy becomes

$$\varrho\,\theta\,\dot{\eta} = \Phi + h^p_{,p} + \varrho\, q, \tag{298.15}$$

where

$$\Phi = \lambda (d^k_k)^2 + 2\mu\, d^k_m\, d^m_k\,. \tag{298.16}$$

[It is plain that $\frac{1}{2}\Phi$ is a dissipation function in the sense of (241 A.2).] If we adopt the corollary $(258.1)_1$ of the entropy inequality, we conclude that Φ as given by (298.16) must be a positive semi-definite quadratic form. An easy analysis of (298.16) shows that this is the case if and only if[1]

$$\mu \geqq 0, \quad 3\lambda + 2\mu \geqq 0. \tag{298.17}$$

These results have immediate mechanical interpretations. By (298.11), $(298.17)_1$ asserts that *the shear stress always opposes the shearing.* By (298.12), $(298.17)_2$ asserts that *in order to produce condensation (expansion), a mean pressure not less (not greater) than that required to maintain equilibrium at the same density and temperature must be applied.* These interpretations, showing that the effect of the viscous stress $\boldsymbol{t} + p\,\mathbf{1}$ as given by (298.5) is always to *resist* change of shape or bulk, reinforce the view that the viscous stress is of the nature of frictional resistance.

The quantity Φ in (298.15) is the viscous dissipation of energy per unit volume. As (298.15) shows, this energy goes into the increase of entropy, or is carried off by the flux $-\boldsymbol{h}$, or is drawn away by sinks $-q$. It is customary to render the theory more definite by setting $q=0$ and adopting Fourier's law (296.4). The resulting equations furnish an example, somewhat more typical than the perfect non-conducting gas (Sect. 297), of a fully *thermomechanical theory*, in which the basic principles of mechanics, energetics, and thermodynamics are employed.

The presence of viscosity has the effect of rendering propagation of most kinds of waves impossible. We consider here only the simplest case, that of an acceleration wave, across which p, $\dot{\boldsymbol{x}}$, and ϱ are continuous. By substituting (298.5) into the dynamical conditions (205.5) we obtain

$$\lambda\, n_k\, [\dot{x}^q_{,q}] + \mu\, n^m\, [\dot{x}_{m,k}] + \mu\, n^m\, [\dot{x}_{k,m}] = 0, \tag{298.18}$$

where it is assumed that λ, μ, and $\varrho\boldsymbol{f}$ are continuous. Writing $s^k \equiv -U s^k_0$ in $(190.5)_1$ and substituting the result into (298.18) yields

$$(\lambda + \mu)\, n^m s_m n_k + \mu\, s_k = 0. \tag{298.19}$$

Taking the scalar product of this equation by $\boldsymbol{n}$, we have

$$(\lambda + 2\mu)\, n^m s_m = 0. \tag{298.20}$$

It is a consequence of (298.17) that in order for $\lambda + 2\mu = 0$ to hold, it is necessary and sufficient that $\lambda = 0$ and $\mu = 0$. Hence by (298.20) we conclude that in a viscous fluid $n^m s_m = 0$. Putting this result back into (298.19) yields $\mu s_k = 0$, and hence $s_k = 0$. What has been proved[2] is that *the instantaneous existence of*

[1] Duhem [1901, *7*, Part I, Chap. 1, § 3], Stokes [Note, pp. 136–137 of the 1901 reprint of [1851, *2*]].

[2] Kotchine [1926, *3*, § 3], but the result is really included in an earlier one of Duhem [1901, *7*, Part II, Chap. III], who uses a different terminology.

a surface upon which $\dot{x}$ *and* p *are continuous but* $\dot{x}_{k,m}$ *suffers a jump discontinuity is incompatible with the law of linear viscosity* (298.5).

We recall that kinematical analysis alone (Sect. 190) shows that a discontinuity in $\dot{x}_{k,m}$ in order to persist must be propagated as an acceleration wave. In Sect. 297 we showed that acceleration waves are propagated in a barotropic flow of a perfect fluid at the speed c, determined by the barotropic relation $p = f(\varrho)$. Any thermal conductivity, however small, generally renders the flow baroclinic and results in the Newtonian speed of propagation (297.16) instead of the Laplacian speed (297.13) that holds when there is a non-degenerate equation of state $p = p(\varrho, \eta)$ and when there is neither flux nor supply of energy. Now we have seen that in fluid with non-vanishing viscosity, however small, such a surface cannot exist at even at a single instant. This last result is included in a general theorem of DUHEM[1]: *In a linearly viscous fluid, no waves of order greater than 1 are possible.*

Since viscous and thermally conducting fluids are regarded as a refined model, superior to the perfect fluid, for the same physical materials, a device must be found whereby wave propagation in some sort may occur. Moreover, perfect fluids are formally a special case of viscous fluids; thus, properties of perfect fluids must be reflected in corresponding properties of viscous fluids. The appropriate device is the *quasi-wave* of DUHEM, a thin layer in which some of the variables suffer rapid but nevertheless continuous changes. The solutions containing waves that occur in perfect fluids are to be regarded as limit cases of an appropriate solution of "the same" problem for viscous fluids, in the limit as λ, μ, and $\varkappa$ approach 0, and hence the layer becomes arbitrarily thin. The rate at which $\varkappa \to 0$, relative to those at which $\lambda \to 0$ and $\mu \to 0$, influences the results. A full analysis of the plane quasi-wave has been given by GILBARG (1951)[2].

A similar difficulty arises in connection with the boundary conditions. For viscous fluids it is customary to impose the condition (69.3) representing adherence of the fluid to the boundary. For perfect fluids, solutions satisfying this condition generally fail to exist, and only the weaker condition (69.1) is employed. In many cases the solutions afforded by the two theories are sensibly the same throughout most of the region occupied by the fluid but differ only in a thin *boundary layer* near solid objects. The existence of this small region of difference, while perhaps effecting little alteration of the gross appearance of the flow, can yield results which are dynamically of a different kind; e.g., the force exerted by the fluid on an obstacle will generally be very different, as is plain from the results given in Sect. 202.

The purpose of the foregoing remarks is to point up a major instance of the ideas sketched in the second paragraph of Sect. 5.

299. Non-linearly viscous fluids. A simple and now familiar theory of non-linear viscosity is obtained by taking (298.2), but without the restriction that $\boldsymbol{f}$ be linear, as the definition of a fluid. The analysis at the beginning of Sect. 298 made no use of the linearity of $\boldsymbol{f}$; thus it follows, in full generality, that (298.2) must reduce to a form giving $\boldsymbol{t}$ as an isotropic function of $\boldsymbol{d}$. By the representation theorem for such functions we thus have

$$\boldsymbol{t} = -p\,\mathbf{1} + \aleph_0\,\mathbf{1} + \aleph_1\,\boldsymbol{d} + \aleph_2\,\boldsymbol{d}^2, \tag{299.1}$$

[1] [1901, 6] [1901, 7, Part II, Chap. III]. DUHEM asserted also that shock waves are not possible in a viscous fluid, but this result holds only subject to some qualification. Cf. Sect. 54 of the article by SERRIN, Mathematical Principles of Classical Fluid Mechanics, this Encyclopedia, Vol. VIII, Part 1.

[2] For an exposition, see § 57 of SERRIN's article, just cited.

where

$$\aleph_0, \aleph_1, \aleph_2 = f(\mathrm{I}_{\boldsymbol{d}}, \mathrm{II}_{\boldsymbol{d}}, \mathrm{III}_{\boldsymbol{d}}), \quad \aleph_0(0, 0, 0) = 0. \tag{299.2}$$

This reduction was first given by REINER[1]. In an incompressible fluid, there is no loss in generality in taking $\aleph_0 = 0$. The scalar coefficient functions generalize the classical viscosities occuring in the linear law (298.5). If we set $2\mu \equiv \aleph_1(0, 0, 0)$ and $\lambda \equiv \partial \aleph_0 / \partial \mathrm{I}_{\boldsymbol{d}}$ when $\boldsymbol{d} = 0$, then (298.5) results from (299.1) by linearization. The coefficient $\aleph_2$, the *cross-viscosity*, gives rise to effects of a type not present in the linear theory.

For the dimensional analysis, we may proceed just as in Sect. 298, but instead we prefer to replace (298.6) by the more general defining equation

$$\boldsymbol{t} = \boldsymbol{f}(\dot{\boldsymbol{x}}, \boldsymbol{w}, \boldsymbol{d}, p, \theta, \theta_0, \mu_0, t_0) \tag{299.3}$$

where t_0, the *natural time*, is a material constant such that

$$\text{phys. dim. } t_0 = [T]. \tag{299.4}$$

Use of the principle of material indifference and the π-theorem reduces (299.3) to the form[2]

$$t_0 \boldsymbol{t}/\mu_0 = \boldsymbol{g}(t_0 \boldsymbol{d}, t_0 p/\mu, \theta/\theta_0), \tag{299.5}$$

where, as shown already, the dimensionless function $\boldsymbol{g}$ is an isotropic function of its first argument. Thus in (299.1) the coefficients have the more explicit forms

$$\left.\begin{aligned} \aleph_\Gamma &= \frac{\mu_0}{t_0} \cdot t_0^{\Gamma-1} \beth_\Gamma, \quad (\Gamma \text{ unsummed}) \\ \beth_\Gamma &= f_\Gamma(t_0 \mathrm{I}_{\boldsymbol{d}}, t_0^2 \mathrm{II}_{\boldsymbol{d}}, t_0^3 \mathrm{III}_{\boldsymbol{d}}, t_0 p/\mu_0, \theta/\theta_0), \end{aligned}\right\} \tag{299.6}$$

where the functions f_Γ are dimensionless functions of their five dimensionless scalar arguments.

When the definition of a fluid is narrowed, as it was in Sect. 298, so as to exclude the time constant t_0, the forms (299.6) must be replaced by others, as follows[3]:

$$\left.\begin{aligned} \aleph_\Gamma &= p \left(\frac{\mu_0}{p}\right)^{\Gamma-1} \daleth_\Gamma, \\ \daleth_\Gamma &= f_\Gamma\left(\frac{\mu_0}{p} \mathrm{I}_{\boldsymbol{d}}, \frac{\mu_0^2}{p^2} \mathrm{II}_{\boldsymbol{d}}, \frac{\mu_0^3}{p^3} \mathrm{III}_{\boldsymbol{d}}, \theta/\theta_0\right). \end{aligned}\right\} \tag{299.7}$$

The theory based upon (299.7) is easily seen to be a special case of that based upon (299.6). The specialization, however, is one of consequence. The theory devoid of a time constant is nearer to the classical linear theory in that it possesses no more dimensionally independent material constants. A possible parameter governing dynamical similarity is the *truncation number* $\mathfrak{I}$, given by[4]

$$\mathfrak{I} \equiv \frac{\sqrt{\mathrm{II}_{\boldsymbol{d}}}}{p/\mu}. \tag{299.8}$$

---

[1] [1945, *5*, § 4]; cf. the treatment of the incompressible case by RIVLIN [1947, *13*] [1948, *24*]. A simple rigorous proof is given in Sect. 59 of the article by SERRIN, just cited.

[2] TRUESDELL [1950, *32*, § 11] [1952, *21*, §§ 68–69] [1952, *22*, pp. 89–90].

[3] TRUESDELL [1949, *33*, § 4] [1950, *32*, § 5], "Stokesian fluid".

[4] TRUESDELL [1949, *33*, § 7] [1950, *32*, § 8]. It was stated erroneously by TRUESDELL that $\mathfrak{I}$ must be controlled independently of the classical scaling parameters. In fact, in two geometrically similar flows having the same Euler and Reynolds numbers the two truncation numbers (299.8) are necessarily also the same. The only further scaling parameters required for Stokesian non-linear viscosity are such dimensionless material constants as are used to specify the functions f_Γ in (299.7). No such remark applies to (299.9).

A condition for approximating the constitutive relation (299.1), supposed differentiable with respect to $\boldsymbol{d}$, by the linear relation (298.5), is $\mathfrak{J} \ll 1$; that is, the intensity of stretching (Sect. 83) shall be small in respect to the ratio of pressure to linear viscosity. While the parameter (299.8) may be introduced in the more general theory defined by (299.6), it does not suffice; for dynamical similarity it is necessary to control also a parameter such as

$$\mathfrak{J} \equiv t_0 \sqrt{\mathrm{II}_{\boldsymbol{d}}}\,. \tag{299.9}$$

The criterion for truncation is now that the product of the intensity of stretching by the natural time of the fluid shall be small. The difference between the two theories is reflected even in the linear case, for in the theory based upon (299.7) the linear viscosities λ and μ, as we have seen in Sect. 298, are independent of the pressure, while in the theory based upon (299.6) this need not be so, for they may depend upon to p/μ_0. Further, it can be shown that the stress of the "Stokesian" theory is of a type that occurs also in results following from the kinetic theory of monatomic gases, while the more general theory based upon (299.6) is generally in contradiction with the kinetic theory, in which there is no time constant[1].

A summary of existing knowledge of the theories of non-linear viscosity is given in "The Non-linear Field Theories of Mechanics", Vol. VIII, Part 2 of this Encyclopedia.

300. Perfectly plastic bodies. The theory of perfectly plastic bodies is intended to describe an *elastic* rather than viscous flow in response to stretching and hence, while adopting the constitutive relation (298.3), relinquishes the requirement of consistency with (298.1). Thus we have at once in the linear case[2]

$$t^k_m = \lambda\, d^q_q\, \delta^k_m + 2\mu\, d^k_m, \tag{300.1}$$

but λ and μ are *not* material constants, being rather functions of $\boldsymbol{d}$ to be determined by additional conditions.

Writing (300.1) in terms of the deviators of $\boldsymbol{t}$ and $\boldsymbol{d}$ (cf. App. 38.12), we have

$$\left.\begin{aligned} -3\tilde{p} &= \alpha\, \mathrm{I}_{\boldsymbol{d}}, \qquad \alpha \equiv 3\lambda + 2\mu, \\ {}_0\boldsymbol{t} &= 2\mu\, {}_0\boldsymbol{d}. \end{aligned}\right\} \tag{300.2}$$

The additional conditions imposed as an essential part of the theory are (1) the mean pressure is a function of the dilatation only, and (2) some scalar invariant of ${}_0\boldsymbol{t}$ vanishes. These conditions are said to represent *plastic flow* or *yield*. The first amounts to replacing $(300.2)_1$ by a general relation $f(\tilde{p}, \mathrm{I}_{\boldsymbol{d}}) = 0$ and is a law of compressibility. Often it is replaced by the condition of incompressibility, $\mathrm{I}_{\boldsymbol{d}} = 0$; in this case $\tilde{p}$ becomes an additional unknown function. The second, characteristic of the theory, may be put in the form

$$Y\left(\frac{\overline{\mathrm{II}}_{{}_0\boldsymbol{t}}}{\sigma^2}, \frac{\mathrm{III}_{{}_0\boldsymbol{t}}}{\sigma^3}\right) = 0, \tag{300.3}$$

where σ is an elastic modulus called the *yield stress* of the material and where the *yield function* Y is dimensionless. This condition represents a material which responds, or at least responds in the manner indicated by (300.2), only when stresses of a certain kind reach an appropriately large value, while the state of plastic flow is assumed such as to maintain this value unaltered.

---

[1] The "relaxation time" in a Maxwellian gas is not a material constant such as t_0, being in fact μ/p, a function of temperature and pressure.

[2] CAUCHY [1823, *1*] [1828, *2*, § III, Eqs. (95), (96)] gave these formulae as appropriate for a "soft" body but stated that λ and μ are constants.

A more general theory for incompressible substances[1] is obtained by replacing (300.2) by a relation of the form

$$2\mu\, d^k_m = \sigma^2 \frac{\partial P}{\partial t^m_k}, \tag{300.4}$$

where the dimensionless *plastic potential* $P(\boldsymbol{t}/\sigma)$ is subject to the condition

$$\delta^k_m \frac{\partial P}{\partial t^m_k} = 0, \tag{300.5}$$

so that (300.4) is consistent with the condition of incompressibility, $\mathrm{I}_{\boldsymbol{d}} = 0$. For isotropic materials, P is taken as a function of $\overline{\mathrm{II}}_{\mathring{\boldsymbol{t}}}/\sigma^2$ and $\mathrm{III}_{\mathring{\boldsymbol{t}}}/\sigma^3$ only.

While (300.4) might seem to determine $\boldsymbol{d}$ uniquely when $\boldsymbol{t}$ is known, this is not so. Indeed, $\mu\boldsymbol{d}/\sigma$ is so determined; conversely, if P is a sufficiently smooth function, $\boldsymbol{t}/\sigma$ is determined as a function of $\mu\boldsymbol{d}/\sigma$. Substituting this function into the yield condition (300.3), in the isotropic case we obtain a functional relation of the form

$$f\left(\frac{\mu^2}{\sigma^2}\overline{\mathrm{II}}_{\boldsymbol{d}},\ \frac{\mu^3}{\sigma^3}\mathrm{III}_{\boldsymbol{d}}\right) = 0. \tag{300.6}$$

Assuming this can be solved for μ/σ, we see that

$$\mu = \frac{\sigma}{\sqrt{\overline{\mathrm{II}}_{\boldsymbol{d}}}}\, f\left(\frac{\sqrt[3]{\mathrm{III}_{\boldsymbol{d}}}}{\sqrt{\overline{\mathrm{II}}_{\boldsymbol{d}}}}\right). \tag{300.7}$$

Thus the factor μ occuring in (300.4) is determined, not assignable.

Comparing a formula of the type (300.4), μ being eliminated by means of (300.7), with the relation (299.1) for a non-linearly viscous fluid, we see that a still more general theory of *perfectly soft bodies*, including both viscous fluids and perfectly plastic bodies, results by allowing the coefficients $\aleph_0$, $\aleph_1$, and $\aleph_2$ to be discontinuous at $\boldsymbol{d} = 0$ and by leaving the physical dimensions of the moduli unrestricted. The principal difference between the theories of viscosity and plasticity arises from the different physical dimensions of the material constants. While λ and μ are viscosities, having the dimension $[M\, L^{-1}\, T^{-1}]$, σ is a stress or elasticity, so that phys. dim $\sigma = [M\, L^{-1}\, T^{-2}]$.

The theory of perfectly plastic bodies, as defined by $(300.2)_2$ and (300.3), is due to St. Venant, Levy, and v. Mises[2]. The most commonly employed yield condition is that of Maxwell and v. Mises[3], viz., $\mathrm{II}_{\mathring{\boldsymbol{t}}} = K\sigma^2$, where K is a constant (cf. the alternative forms of $\mathrm{II}_{\mathring{\boldsymbol{t}}}$ given in Sect. App. 38); this amounts to taking $f = \text{const}$ in (300.7). In this theory $P = Y$.

As appears at once by confronting (300.1) with Cauchy's first law (205.2), the theory of perfectly plastic bodies is a dynamical theory. Nevertheless, almost all of its large literature treats it as if it were statical. Thus, as far as the exact theory is concerned, very little is known regarding it. A survey of the mathematical developments as practised by current specialists in the field is given in the article by Freudenthal and Geiringer in Vol. VI of this Encyclopedia.

301. Linearly elastic bodies. The simplest kind of elastic or springy body is one such that *the stress arises solely in response to such change of shape as the body has undergone from its "natural" or unstressed state.* Considering the strain to be very small, by the results in Sect. 57 we may take the tensor $\tilde{\boldsymbol{e}}$ as a measure of it, and we assume the stress depends linearly upon it. Moreover, $\boldsymbol{t} = 0$ if $\tilde{\boldsymbol{e}} = 0$.

[1] Geiringer [1953, *11*, § 3].
[2] [1870, *7*]; [1870, *3*]; [1913, *6*].
[3] [1937, *5*, pp. 32—33] (written in 1856), [1913, *6*, §§ 2—3].

There results the constitutive equation

$$t^{mk} = C^{kmpq}\, e_{pq}, \qquad C^{kmpq} = C^{mkpq} = C^{kmqp}, \tag{301.1}$$

where for simpler writing we have dropped the tilde from $\tilde{e}$. A linearly elastic body is *isotropic* if the relation (301.1) is an isotropic one. For an isotropic body, (301.1) reduces to ***Cauchy's law***[1]:

$$t^k_m = \lambda\, e^q_q\, \delta^k_m + 2\mu\, e^k_m. \tag{301.2}$$

To the extent that $\boldsymbol{e}$ is an approximate measure of the mutual distances of particles, the principle of material indifference is satisfied by (301.1) and (301.2). If (57.1) is interpreted strictly, however, the principle of material indifference is violated, although (301.1) and (301.2) are invariant under infinitesimal rigid time-independent displacements. Linear elasticity theory does not represent exactly the kind of behavior possible in any real material. Rather, it is to be regarded as a mathematical approximation to the properly invariant theories described in the two following sections.

The *elasticities* λ, μ, and C^{kmpq} in (301.1) and (301.2) are material constants or functions of the temperature or entropy; their physical dimensions are those of stress, $[M\, L^{-1}\, T^{-2}]$, and they bear no physical connection with the mathematically analogous viscosities appearing in (298.5). The static theory is a linear one: Uniformly doubled displacements always result from uniformly doubled loads, and, more generally, from displacements $\boldsymbol{u}'$, $\boldsymbol{u}''$ corresponding to stresses $\boldsymbol{t}'$, $\boldsymbol{t}''$, assigned forces $\boldsymbol{f}'$, $\boldsymbol{f}''$, and assigned surface loads $\boldsymbol{t}'_{(\boldsymbol{n})}$, $\boldsymbol{t}''_{(\boldsymbol{n})}$ we may construct a displacement $\boldsymbol{u} \equiv \boldsymbol{u}' - \boldsymbol{u}''$ answering to the stress $\boldsymbol{t} = \boldsymbol{t}' - \boldsymbol{t}''$, force $\boldsymbol{f} = \boldsymbol{f}' - \boldsymbol{f}''$, and surface load $\boldsymbol{t}_{(\boldsymbol{n})} = \boldsymbol{t}'_{(\boldsymbol{n})} - \boldsymbol{t}''_{(\boldsymbol{n})}$.

A more restricted concept of elasticity was proposed by GREEN[2]: *The work done by the stress in a deformation depends only upon the strain and is recoverable work.* The former part of the assertion may be expressed thus:

$$\tfrac{1}{2}\, t^k_m\, e^k_m = \Sigma(\boldsymbol{e}), \tag{301.3}$$

the function Σ being the *stored energy*. The latter part may be used to derive a special case of (301.1), but we defer the argument until the next section, where a more general definition is considered. Here we simply combine (301.1) and (301.3), obtaining

$$\Sigma = \tfrac{1}{2}\, C^{k\ \,p}_{\ m\ q}\, e^q_p\, e^m_k = \tfrac{1}{2}\, G^{k\ \,p}_{\ m\ q}\, e^q_p\, e^m_k, \tag{301.4}$$

where

$$G^{k\ \,p}_{\ m\ q} \equiv \tfrac{1}{2}\left(C^{k\ \,p}_{\ m\ q} + C^{p\ \,k}_{\ q\ m}\right), \qquad G^{k\ \,p}_{\ m\ q} = G_m^{\ kp}{}_q = G^k{}_{mq}{}^p = G^{p\ \,k}_{\ q\ m}. \tag{301.5}$$

Hence

$$t^k_m = G^{k\ \,p}_{\ m\ q}\, e^q_p = \frac{\partial \Sigma}{\partial e^m_k}. \tag{301.6}$$

Thus GREEN's theory allows but the 21 independent elasticities $G^{k\ \,p}_{\ m\ q}$ in place of the 36 independent elasticities $C^{k\ \,p}_{\ m\ q}$ of CAUCHY's theory. In GREEN's theory we have two alternative ways of defining isotropy: either in the same way as in CAUCHY's theory, or by requiring that Σ be a scalar invariant of $\boldsymbol{e}$. By the

[1] Eqs. (301.1) and (301.2) are commonly referred to as "HOOKE's law". After a long prior history in very special cases, (301.2) with $\lambda = \mu$ was derived from a molecular model by NAVIER [1823, *3*] [1827, *7*]; more generally, by POISSON [1829, *5*, § 7]. The basic ideas of the continuum theory were constructed in 1821, the year of NAVIER's earliest work, by FRESNEL [1866, *4*, pp. LXXVIII–LXXXI]. The general continuum theory is due to CAUCHY [1823, *1*] [1828, *2*, Eqs. (67), (70)] [1829, *4*, Eqs. (7), (8)] [1830, *1*]. Cf. also POISSON [1831, *2*, ¶ 23, Eq. (10)].

[2] [1839, *1*, p. 249] [1841, *2*, pp. 295–296]. Some special cases were known long before.

representation theorem for isotropic scalar functions, we see that both definitions yield the same result, viz.,

$$\Sigma = \tfrac{1}{2}(\lambda + 2\mu)\, \mathrm{I}_{\boldsymbol{e}}^2 - 2\mu\, \mathrm{II}_{\boldsymbol{e}}, \tag{301.7}$$

and hence *for isotropic bodies Green's and Cauchy's definitions of elasticity lead to the same theory.*

We shall call *hyperelastic* a body answering to GREEN's theory, based upon the use of Σ as a stress potential according to $(301.6)_2$; this theory allows some remarkable deductions.

First we establish the uniqueness of solution[1] to boundary-value problems of equilibrium where the stress vector and the displacement are prescribed upon disjoint surfaces $\mathscr{s}_1$ and $\mathscr{s}_2$, respectively, such that the closure of $\mathscr{s}_1 + \mathscr{s}_2$ is the complete boundary of a finite body $\mathscr{v}$. Consider two solutions $\boldsymbol{u}'$ and $\boldsymbol{u}''$ corresponding to the same assigned loads and boundary values. Form a solution $\boldsymbol{u} \equiv \boldsymbol{u}' - \boldsymbol{u}''$ as indicated above; for this solution we have $\boldsymbol{f} = 0$ in $\mathscr{v}$, $\boldsymbol{t}_{(\boldsymbol{n})} = 0$ on $\mathscr{s}_1$, and $\boldsymbol{u} = 0$ on $\mathscr{s}_2$. Substitution of $(301.6)_2$ into CAUCHY's first law (205.2) yields

$$\left(\frac{\partial \Sigma}{\partial e^m_k}\right)_{,k} = 0. \tag{301.8}$$

Hence

$$\left.\begin{aligned} 0 &= \int_{\mathscr{v}} u^m \left(\frac{\partial \Sigma}{\partial e^m_k}\right)_{,k} dv = \int_{\mathscr{v}} \left[\left(u^m \frac{\partial \Sigma}{\partial e^m_k}\right)_{,k} - u^m{}_{,k} \frac{\partial \Sigma}{\partial e^m_k}\right] dv, \\ &= \oint_{\mathscr{s}} u^m\, t_{(\boldsymbol{n})m}\, da - \int_{\mathscr{v}} e^m_k \frac{\partial \Sigma}{\partial e^m_k}\, dv \\ &= -2 \int_{\mathscr{v}} \Sigma\, dv, \end{aligned}\right\} \tag{301.9}$$

where we have used EULER's theorem on homogeneous functions as well as the fact that on $\mathscr{s}$ either $\boldsymbol{u}$ or $\boldsymbol{t}_{(\boldsymbol{n})}$ vanishes. Now if $\Sigma(\boldsymbol{e})$ is of one sign for all values of $\boldsymbol{e}$, it follows from (301.9) that $\Sigma = 0$ in $\mathscr{v}$. Looking back at (301.4), we see that *if Σ is a definite quadratic form (whether positive or negative), the general boundary-value problem of static linear hyperelasticity cannot have two distinct solutions.* What has been proved is that $\boldsymbol{e} = 0$; the strains is thus unique, and from the results in Sect. 57 it follows that the displacement $\boldsymbol{u}$ is determined uniquely to within an infinitesimal rigid displacement. This degree of indeterminacy is inherent in the linear theory of elasticity and is to be understood in all statements concerning it, except in cases where this indeterminacy is removed by specification of the displacement on the boundary.

There is physical reason to require that Σ be a positive definite form, for then in any given small strain from an unstressed state, the stress must do positive work. This idea seems to be related to, but is not identical with, the requirement $(258.1)_1$ following from the entropy inequality, which is expressed rigorously in terms of time rates rather than displacements. In the isotropic case, Σ is positive definite if and only if

$$\mu > 0, \quad 3\lambda + 2\mu > 0. \tag{301.10}$$

There is a remarkable principle enabling us, in the case of equilibrium subject to given surface displacements and vanishing assigned force in the interior, to select among all kinematically possible deformations that one which is consistent with the theory of hyperelasticity, a positive definite stored energy function

[1] KIRCHHOFF [1859, 2, § 1] [1876, 2, Vorl. 27, § 2].

being assigned[1]: *The displacement that satisfies the equations of equilibrium as well as the conditions at the bounding surface yields a smaller value for the total stored energy than does any other displacement satisfying the same conditions at the bounding surface.*

To prove this theorem, we write $\boldsymbol{e}$ for the strain leading to a stress that satisfies the conditions of equilibrium, and $\boldsymbol{e}+\boldsymbol{e}'$ for some other strain, where it is assumed that $\boldsymbol{u}'=0$ on the boundary $\mathfrak{s}$. Then

$$\int_{\mathfrak{v}} \Sigma(\boldsymbol{e}+\boldsymbol{e}')\,dv = \int_{\mathfrak{v}} \Sigma(\boldsymbol{e})\,dv + \int_{\mathfrak{v}} \left(e'^{\,m}_{k}\frac{\partial\Sigma(\boldsymbol{e})}{\partial e^{m}_{k}} + \Sigma(\boldsymbol{e}')\right)dv, \tag{301.11}$$

since $\Sigma(\boldsymbol{e})$ is a homogeneous quadratic form in the components e^{k}_{m}. Now

$$\left.\begin{aligned}\int_{\mathfrak{v}} e'^{\,m}_{k}\frac{\partial\Sigma(\boldsymbol{e})}{\partial e^{m}_{k}}\,dv &= \int_{\mathfrak{v}} u'_{(k,m)}\,t^{km}\,dv,\\ &= \int_{\mathfrak{v}} [(u'_k\,t^{km})_{,m} - u'_k\,t^{km}{}_{,m}]\,dv,\\ &= \oint_{\mathfrak{s}} u'_k\,t^{k}_{(\boldsymbol{n})}\,da = 0,\end{aligned}\right\} \tag{301.12}$$

where we have used (301.6) and (205.2) and the fact that $\boldsymbol{u}'=0$ on $\mathfrak{s}$. Substitution of (301.12) into (301.11) yields

$$\int_{\mathfrak{v}} \Sigma(\boldsymbol{e}+\boldsymbol{e}')\,dv = \int_{\mathfrak{v}} \Sigma(\boldsymbol{e})\,dv + \int_{\mathfrak{v}} \Sigma(\boldsymbol{e}')\,dv. \tag{301.13}$$

This identity shows that the energy stored by a deformation corresponding to the vector sum of two displacements, one of which leads to an equilibrated elastic stress and both of which have the same values on the boundary, is the sum of the energies of the constituent displacements. Since the stored energy is assumed to be a positive definite form, the above-stated theorem of minimum energy follows immediately.

The two theorems just proved are representative of the many that are known[2] in this classical subject, the theory of which has been brought to a state of analytical completeness second only to that of the theory of the potential[3].

Having given some consideration to static theorems, we turn now to the propagation of waves.

For a body of continuous constant elasticity $\boldsymbol{C}$, putting $(301.1)_1$ into (205.2) yields

$$[\varrho\,\ddot{x}^{k}] = C^{kmpq}[u_{p,qm}], \tag{301.14}$$

where we have supposed $\varrho\boldsymbol{f}$ to be continuous. In linear elasticity theory we have $[u_{p,qm}] = g^{\alpha}_{q}\,g^{\beta}_{m}\,[x_{p,\alpha\beta}]$. By applying the general identities (190.1) and (190.2) for an acceleration wave, when the present configuration is taken as the initial one, from (301.14) we thus obtain

$$\varrho\,U^2 s^k = C^{kmpq}\,n_q\,n_m\,s_p, \tag{301.15}$$

or

$$(C_k{}^{mpq}\,n_q\,n_m - \varrho\,U^2\,\delta^p_k)\,s_p = 0. \tag{301.16}$$

---

1 KELVIN [1863, *2*, § 62] took the assertion as "the elementary condition of stable equilibrium"; in this sense, that of a postulated variational principle, its history may be traced back to an idea of DANIEL BERNOULLI in respect to elastic bands (1738). As a proved theorem of linear three-dimensional elasticity, it seems first to have been given by LOVE [1906, *5*, § 119].

2 A masterly exposition of some of them is given by LOVE [1927, *6*, Chap. VII].

3 Despite this fact, there exists no general exposition of the theory from a rigorous mathematical standpoint.

Thus in order for an acceleration wave with normal $\boldsymbol{n}$ to exist and propagate, the jump $\mathbf{s}$ which it carries must be a proper vector of $C_k{}^{mpq} n_q n_m$ corresponding to the proper number ϱU^2. For a body such that the work of the stress in any deformation is positive, as is the case for a hyperelastic body with positive definite stored energy, the tensor $C_k{}^{mpq} n_q n_m$ is positive definite, its quadric being called *Fresnel's ellipsoid* for the direction $\boldsymbol{n}$; therefore all proper numbers ϱU^2 are positive, and therefore all possible speeds U are real. In the general case, then, *in any linearly elastic body such that the work of the stress is positive for arbitrary deformations, a wave with given normal* $\mathbf{n}$ *may carry a discontinuity of the acceleration parallel to any one of three uniquely determined, mutually orthogonal directions, and corresponding to each of these directions there is a speed of propagation determined uniquely by the elasticities of the material and by* $\boldsymbol{n}$.

When the proper numbers ϱU^2 are not distinct, the above conclusion must be modified, as is seen most easily by considering the isotropic case, for then (301.14) assumes the more special form

$$[\varrho \ddot{x}_k] = (\lambda + \mu)\,[u^p{}_{,pk}] + \mu\,[u_{k,p}^{,p}], \tag{301.17}$$

so that for an acceleration wave we have

$$\varrho U^2 s_k = (\lambda + \mu)\, s^p n_p n_k + \mu s_k, \tag{301.18}$$

specializing (301.15). Taking the scalar and vector products of this equation by $\boldsymbol{n}$ yields

$$\left.\begin{aligned} \{\varrho U^2 - (\lambda + 2\mu)\}\, \mathbf{s}\cdot\boldsymbol{n} &= 0,\\ \{\varrho U^2 - \mu\}\, \mathbf{s}\times\boldsymbol{n} &= 0. \end{aligned}\right\} \tag{301.19}$$

If $\mathbf{s}\cdot\boldsymbol{n} \neq 0$, the first equation yields $\varrho U^2 = \lambda + 2\mu$, and the second, if we exclude the case when $\lambda + \mu = 0$, yields $\mathbf{s}\times\boldsymbol{n} = 0$. If $\mathbf{s}\cdot\boldsymbol{n} = 0$ but $\mathbf{s}\times\boldsymbol{n} \neq 0$, the second equation yields $\varrho U^2 = \mu$. Summarizing these results, we see that *in an isotropic linearly elastic body for which* $\lambda + \mu \neq 0$, *a necessary and sufficient condition that acceleration waves be propagated at positive speeds is* $\lambda + 2\mu > 0$, $\mu > 0$. *This condition is satisfied when the stored energy is positive definite. Two kinds of acceleration waves are possible: longitudinal waves, whose speed of propagation is given by*

$$U^2 = \frac{\lambda + 2\mu}{\varrho}, \tag{301.20}$$

and transverse waves, for which

$$U^2 = \frac{\mu}{\varrho}. \tag{301.21}$$

In view of the kinematical interpretation furnished by HADAMARD'S theorem in Sect. 190, the longitudinal waves are called *expansion* waves or *irrotational* waves, while the transverse waves are called *equivoluminal* waves or *shear* waves. The foregoing results, which are due to CHRISTOFFEL and HUGONIOT[1], illustrate the far-reaching effect of isotropy: instead of three speeds of propagation, for an isotropic body there are only two, but instead of there being only three possible directions for the discontinuity, there are infinitely many, though the possible directions are still far from arbitrary.

302. The rotationally elastic aether. The quest for a mechanical theory of light as a vibration of an elastic medium attracted the attention of many of the

[1] CHRISTOFFEL [1877, *3*] obtained really all of the above results and more, but he did not present them very clearly, nor did he recognize as such the isotropic case, for which HUGONIOT [1886, *3*] gave a very simple treatment. Our proof is essentially that of HADAMARD [1903, *11*, ¶¶ 260, 267—268] and DUHEM [1904, *1*, Part IV, Chap. I, § V].

illustrious physicists of the nineteenth century[1]. The most successful of these mechanical theories as applied to the deduction of the laws of transmission, reflection, and refraction of light in transparent media was devised by MACCULLAGH in 1839[2]. Some thirty years after the publication of MACCULLAGH's theory, FITZGERALD[3] applied the new electromagnetic theory of MAXWELL to derive the laws of reflection and refraction of light at the interface between two dissimilar dielectric media. In this memoir, FITZGERALD called attention to the formal analogy between the equations of the new theory and the old equations of MACCULLAGH's theory. This seems remarkable, for the physical ideas underlying the two theories are totally different. The constitutive equations of MACCULLAGH's "rotationally elastic aether" do not conform to the principle of material indifference known to be satisfied by ordinary elastic materials. In all other regards, however, the theory is based on the usual equations of mechanics, specialized somewhat by linearization.

α) *MacCullagh's equations and boundary conditions.* As a starting point we consider the equations of motion (205.2) and boundary conditions (205.5) at a stationary surface of discontinuity which is not a shock:

$$t^{mk}{}_{,k} = \varrho \ddot{x}^m, \qquad [t^{mk}]\, n_k = 0. \tag{302.1}$$

Now contrary to the theory of ordinary elastic materials, where the stress arises solely in response to changes in shape, MACCULLAGH supposed the aether to be a medium in which the stress, while completely insensitive to changes in shape, arises only in response to rotations about its relaxed state. This implies the existence of a preferred class of reference frames. We may regard these preferred frames as the inertial frames of classical mechanics. The aether in its relaxed state will be at rest or in uniform translatory motion with respect to one of these frames. Let $u^k(\boldsymbol{x}, t)$ denote the displacement vector of the medium taken in this sense. Then the quantities

$$w_{rs} \equiv u_{[r,s]} \equiv \tfrac{1}{2}\left(u_{r,s} - u_{s,r}\right) \tag{302.2}$$

measure an infinitesimal rotation of the medium (cf. Sect. 57). Thus MACCULLAGH assumed that for small rotations of the aether medium the stress is given by

$$t^{rs} = A^{rspq}\, w_{pq}. \tag{302.3}$$

Although MACCULLAGH considered the more difficult case of crystalline media, we shall here treat only the isotropic case, where A^{rspq} is an isotropic tensor. It follows that, in this case, the relations (302.3) reduce to

$$t_{rs} = K\, w_{rs}. \tag{302.4}$$

The constant K is the *gyrostatic rigidity*. The aether is assumed to pervade all ordinary material media and to have the same density ϱ in all materials. However, the gyrostatic rigidity of the aether is assumed to have a different value in materials with different indices of refraction. Thus at the interface between two dissimilar isotropic media of this simple type we have $[\varrho] = 0$, $[K] \neq 0$. Moreover, at such an interface, the displacement $\boldsymbol{u}$ is assumed continuous, $[\boldsymbol{u}] = 0$. Thus it follows from the results of Sect. 175 that if $\boldsymbol{u}$ is differentiable in each of the adjoining media with differentiable limit values for $\partial \boldsymbol{u}/\partial t$ and grad $\boldsymbol{u}$ on each side of the

---

[1] WHITTAKER [1951, *39*] has given us a detailed and fascinating account of the evolution of these theories and their interrelations.

[2] [1848, *1*].

[3] [1880, *8*].

interface, then

$$\left[\frac{\partial \boldsymbol{u}}{\partial t}\right] = 0, \quad [\operatorname{curl} \boldsymbol{u}] \cdot \boldsymbol{n} = 0. \tag{302.5}$$

Substituting the constitutive relation (302.4) into (302.1), linearizing the acceleration with respect to $\boldsymbol{u}$ and its derivatives, and collecting our assumptions thus far, we have

$$\left.\begin{aligned} K \operatorname{curl}\operatorname{curl} \boldsymbol{u} + \varrho \frac{\partial^2 \boldsymbol{u}}{\partial t^2} &= 0, \\ [K \operatorname{curl} \boldsymbol{u}] \times \boldsymbol{n} &= 0, \\ \left[\varrho \frac{\partial \boldsymbol{u}}{\partial t}\right] = 0, \quad [\operatorname{curl} \boldsymbol{u}] \cdot \boldsymbol{n} &= 0. \end{aligned}\right\} \tag{302.6}$$

These are the basic equations of MACCULLAGH's theory of light.

β) Fitzgerald's analogy. MAXWELL's constitutive equations for a non-magnetic, linear, rigid, homogeneous, isotropic stationary dielectric were discussed briefly in Sect. 283 and are treated in detail in Sect. 308. For such media we have

$$\left.\begin{aligned} \mathfrak{D} &= \varepsilon \boldsymbol{E}, \\ \mathfrak{H} &= \frac{1}{\mu_0} \boldsymbol{B}, \end{aligned}\right\} \tag{302.7}$$

where the dielectric constant ε is a measure of the ease of electric polarization of the medium. If we substitute these relations into the general electromagnetic equations (278.2), (278.3), (278.8), (278.9), and (283.42) to (283.45), we obtain the system

$$\left.\begin{aligned} \operatorname{curl} \boldsymbol{E} + \frac{\partial \boldsymbol{B}}{\partial t} = 0, \quad \operatorname{div} \boldsymbol{B} = 0, \\ [\boldsymbol{E}] \times \boldsymbol{n} = 0, \quad [\boldsymbol{B}] = 0, \quad [\varepsilon \boldsymbol{E}] \cdot \boldsymbol{n} = 0, \end{aligned}\right\} \tag{302.8}$$

$$\frac{1}{\mu_0} \operatorname{curl} \boldsymbol{B} - \varepsilon \frac{\partial \boldsymbol{E}}{\partial t} = 0, \quad \operatorname{div} \boldsymbol{E} = 0. \tag{302.9}$$

These equations, generalized to the case of crystalline media, formed the basis of FITZGERALD's derivation of the laws of reflection and refraction of light from the Maxwell theory. We see that, leaving aside the question of physical interpretation of the symbols, if we make the substitutions

$$\left.\begin{aligned} K \operatorname{curl} \boldsymbol{u} \Rightarrow \alpha \boldsymbol{E}, \quad \varrho \frac{\partial \boldsymbol{u}}{\partial t} \Rightarrow \alpha \boldsymbol{B}, \\ K \Rightarrow \beta/\varepsilon, \quad \varrho \Rightarrow \beta \mu_0, \end{aligned}\right\} \tag{302.10}$$

in (302.6), we obtain (302.8) and (302.9). The quantities α and β in (302.10) are arbitrary non-vanishing constants. The last two Maxwell equations (302.9) are a consequence of the identities $\operatorname{div}\operatorname{curl} \boldsymbol{u} = 0$, $\operatorname{curl} \frac{\partial \boldsymbol{u}}{\partial t} - \frac{\partial}{\partial t} \operatorname{curl} \boldsymbol{u} = 0$. This mathematical equivalence between the electromagnetic equations in ideal media characterized by the constitutive equations (302.7) and MACCULLAGH's equations was first perceived by FITZGERALD[1].

As remarked by WHITTAKER[2], "...there can be no doubt that MACCULLAGH really solved the problem of devising a medium whose vibrations, calculated

---

[1] [1880, *8*]. The analogy between MACCULLAGH's equations and MAXWELL's equations based on the replacements (302.10) is discussed briefly by SOMMERFELD in his lectures [1947, *14*]. He also considers a second set of replacements which renders the equations equivalent. HEAVISIDE [1893, *4*, Chap. III] gives an elaborate account of electro-mechanical analogies based on the Maxwell equations of dielectrics, conductors, etc., and the mechanical equations of elastic solids, viscous fluids, etc.

[2] [1951, *39*, p. 144].

in accordance with the correct laws of dynamics, should have the same properties as the vibrations of light".

303. Finitely elastic bodies. The classical theory of finite elastic strain generalizes the concepts of linear elasticity by considering a body in which the stress arises solely in response to the difference of the present shape from that in an unstressed "natural state". An apparently still more general notion is included in the constitutive equation

$$t^{km} = f^{km}(x^q{}_{,\alpha}), \tag{303.1}$$

where $x^q{}_{,\alpha} \equiv \partial x^q / \partial X^\alpha$ and where $\boldsymbol{x} = \boldsymbol{x}(\boldsymbol{X})$ is the deformation from the natural state to the present configuration. The prominence of the natural state is extreme: neither any intermediate state nor any event in the stress and strain history need be considered in ascertaining the stress. We may say that the body exhibits a perfect memory of its natural state and is entirely oblivious to every other except the present one.

The principle of material indifference (Sect. 293 θ) makes it possible to reduce (303.1) to the form[1]

$$\boldsymbol{T} = \boldsymbol{f}(\boldsymbol{C}), \tag{303.2}$$

where $\boldsymbol{C}$ is the deformation tensor defined by (26.2) and where $\boldsymbol{T}$ is PIOLA's stress tensor, connected with $\boldsymbol{t}$ through (210.9).

Although both $\boldsymbol{C}$ and $\boldsymbol{T}$ transform as double tensors under time-independent changes of material and spatial co-ordinates, they do not transform as tensors under change of frame. Thus the principle of material indifference does *not* force the relation (303.2) to be an isotropic one; indeed, any relation of the form (303.2) satisfies that principle. Herein lies the explanation of why the classical theory of elasticity includes anisotropic as well as isotropic behavior, in contrast to the theories of fluids described in Sects. 298 and 299.

An elastic body is *isotropic* if (303.2) reduces to an isotropic relation. The principles given in Sect. 32 suffice to show that in this case, alternatively, $\boldsymbol{t}$ may be regarded as an isotropic function of $\boldsymbol{c}$, allowing a statement of the law of elasticity in terms of spatial tensors only[2]:

$$\begin{aligned} \boldsymbol{t} &= \aleph_0 \boldsymbol{1} + \aleph_1 \boldsymbol{c} + \aleph_2 \boldsymbol{c}^2 \\ &= \beth_0 \boldsymbol{1} + \beth_1 \boldsymbol{c} + \beth_2 \boldsymbol{c}^{-1}, \end{aligned} \tag{303.3}$$

where the coefficients $\aleph_\Gamma$ and $\beth_\Gamma$ are scalar functions of $\boldsymbol{c}$, and hence may be taken as functions of $\mathrm{I}_{\boldsymbol{c}}$, $\mathrm{II}_{\boldsymbol{c}}$, $\mathrm{III}_{\boldsymbol{c}}$ or as functions of $\mathrm{I}_{\boldsymbol{c}}$, $\mathrm{I}_{\boldsymbol{c}^{-1}}$ and $\mathrm{III}_{\boldsymbol{c}}$ or $\mathrm{III}_{\boldsymbol{c}^{-1}}$, etc.

Thus far we have followed CAUCHY's concept of elasticity. GREEN's concept, which may be stated exactly as in Sect. 301, leads to a more restricted theory, as we shall see now. The basic assumption of *finite hyperelasticity* is that there exists a stored energy Σ, a scalar function of the material co-ordinates $\boldsymbol{X}$ and of the deformation gradients $x^k{}_{,\alpha}$, such that all the work of the stress is recoverable. We have then, by hypothesis,

$$\frac{\varrho}{\varrho_0} \dot{\Sigma} = t^{km} d_{km}, \tag{303.4}$$

where the inessential factor ϱ/ϱ_0 is added for conformity with standard usage. The principle of material indifference requires that in fact $\Sigma = \Sigma(\boldsymbol{X}, \boldsymbol{C})$. A classical argument, which has been given in three different forms in Sects. 218, 232A,

[1] NOLL [1955, *18*, § 15a]. A weaker theorem had been proved under stronger hypotheses by earlier writers.

[2] REINER [1948, *23*, §§ 1–2].

and 256A, enables us to conclude that[1]

$$T_k^\alpha = \frac{\partial \Sigma}{\partial x^k_{,\alpha}}, \qquad t^{km} = 2\,\frac{\varrho}{\varrho_0}\,\frac{\partial \Sigma}{\partial C_{\alpha\beta}}\, x^k_{,\alpha}\, x^m_{,\beta}, \qquad T^\alpha_\beta = \frac{\partial \Sigma}{\partial E^\beta_\alpha}, \tag{303.5}$$

etc. These stress-strain relations furnish a special case of (303.2).

Within hyperelasticity, a body is *isotropic* if its stored energy is an isotropic function of $\boldsymbol{C}$. From the fundamental lemma in Sect. 30, it follows that, alternatively, $\Sigma = \Sigma(\boldsymbol{c}) = \Sigma(\mathrm{I}_{\boldsymbol{c}^{-1}}, \mathrm{II}_{\boldsymbol{c}^{-1}}, \mathrm{III}_{\boldsymbol{c}^{-1}})$. It is easy to show by using (App. 38.16) that for isotropic bodies any one of Eqs. (303.5) reduces to the form given by FINGER[2]:

$$\left.\begin{aligned} t^k_m = \frac{2\varrho}{\varrho_0}\Big[\Big(&\mathrm{II}_{\boldsymbol{c}^{-1}}\frac{\partial \Sigma}{\partial \mathrm{II}_{\boldsymbol{c}^{-1}}} + \mathrm{III}_{\boldsymbol{c}^{-1}}\frac{\partial \Sigma}{\partial \mathrm{III}_{\boldsymbol{c}^{-1}}}\Big)\delta^k_m + \\ &+ \frac{\partial \Sigma}{\partial \mathrm{I}_{\boldsymbol{c}^{-1}}}\, \overset{-1}{c}{}^k_m - \mathrm{III}_{\boldsymbol{c}^{-1}}\frac{\partial \Sigma}{\partial \mathrm{II}_{\boldsymbol{c}^{-1}}}\, c^k_m\Big]. \end{aligned}\right\} \tag{303.6}$$

Equivalently, the principal axes of stress coincide with the principal axes of strain at $\boldsymbol{x}$, and the principal stresses t_a are related to the principal stretches λ_a as follows[3]:

$$\lambda\frac{t_a}{\lambda_a} = \frac{\partial \Sigma}{\partial \lambda_a}, \qquad \lambda \equiv \lambda_1\lambda_2\lambda_3, \qquad a = 1, 2, 3. \tag{303.7}$$

The definition (303.1) is incomplete in that the dimensional moduli are not specified; it should be amplified to read

$$t^{km} = f^{km}(x^q_{,\alpha}, \mu), \tag{303.8}$$

where

$$\text{phys. dim. } \mu = [M\,L^{-1}\,T^{-2}]. \tag{303.9}$$

Thus, as in the theory of perfectly plastic solids, there is but a single independent dimensional material constant, and it has the dimensions of stress. By the π-theorem we see at once that (303.8) must reduce to $t^{km}/\mu = h^{km}(x^q_{,\alpha})$, and hence (303.2) may be replaced by

$$\frac{\boldsymbol{T}}{\mu} = \boldsymbol{g}(\boldsymbol{C}), \tag{303.10}$$

where $\boldsymbol{g}$ is dimensionless. Corresponding modifications are easily made in all equations of the theory.

After a quiescence of half a century, the theory of finite elastic strain has undergone remarkable development in the past decade. See "The Non-linear Field Theories of Mechanics" in Vol. VIII, Part 2 of this Encyclopedia.

304. Hypo-elasticity. A different generalization of the classical linear theory of elasticity is obtained by regarding it as describing approximately a material not necessarily having any natural state but rather experiencing a *stress increment arising in response to the rate of strain from the immediately preceding state.* Thus no finite memory is ascribed to the material. From the results at the beginning of Sect. 95 it is plain that the appropriate tensor measuring the rate of strain is the stretching, $\boldsymbol{d}$. From the principle of material indifference in Sect. 293 θ we see that one of the various time fluxes introduced in Sects. 149 to 151 may be used

---

[1] The argument is due in principle to GREEN [1839, *1*, p. 249] [1841, *2*, pp. 295–296] whose analysis was corrected by KELVIN [1863, *2*, §§ 51–57]. All essential arguments occur in the treatment of a special case by KIRCHHOFF [1852, *1*, pp. 770–772]. The many known equivalent forms are summarized by TRUESDELL [1952, *21*, §§ 39–40].

[2] [1894, *4*, Eq. (35)].

[3] KÖTTER [1910, *6*, § 1)], ALMANSI [1911, *1*, § 7].

to measure the rate of stress in a properly invariant way, and that the constitutive equations must be sufficiently general as to render immaterial the choice of a particular flux. We select the co-rotational flux $d_r\boldsymbol{t}/dt$ as defined by (148.7); allowing the magnitude of the present stress $\boldsymbol{t}$ to moderate the response of the material to stretching, we write the constitutive equation in the form

$$\frac{d_r \boldsymbol{t}}{dt} = \boldsymbol{f}(\boldsymbol{t}, \boldsymbol{d}, \mu), \tag{304.1}$$

where $\mu = \text{const}$ and

$$\text{phys. dim. } \mu = [M\,L^{-1}\,T^{-2}]. \tag{304.2}$$

This last requirement asserts that all material moduli shall be *elasticities*, just as in the theories of linear or finite elasticity or of perfectly plastic materials. Fluid behavior, to the extent that it is accompanied by effects of viscosity or relaxation, is thus explicitly and intentionally excluded. (Effects of temperature differences are easily taken into account if desired but are here omitted for simplicity.)

Application of dimensional analysis shows that $\boldsymbol{t}$ may enter (304.1) only in the ratio $\boldsymbol{t}/\mu$ and that $\boldsymbol{f}$, assumed continuous at $\boldsymbol{d} = 0$, must be linear in $\boldsymbol{d}$:

$$\frac{d_r s^{km}}{dt} = K^{kmpq}\, d_{pq}, \qquad s^{km} \equiv t^{km}/2\mu, \tag{304.3}$$

where $\boldsymbol{K}$ is a function of $\mathbf{s}$. Since $d_r\mathbf{s}/dt$ and $\boldsymbol{d}$ are both tensors under change of frame (Sects. 144 and 148), $\boldsymbol{K}$ must also be such a tensor; consequently $\boldsymbol{K}$ is an isotropic tensor function of $\mathbf{s}$. By a representation theorem due to RIVLIN and ERICKSEN[1] it follows that the most general constitutive equation satisfying the hypotheses (304.1) and (304.2) is

$$\left.\begin{aligned} \frac{d_r \mathbf{s}}{dt} = {} & \aleph_0\, \mathrm{I}_{\boldsymbol{d}}\, \mathbf{1} + \aleph_1\, \boldsymbol{d} + \aleph_2\, \mathrm{I}_{\boldsymbol{d}}\, \mathbf{s} + \aleph_3\, M\, \mathbf{1} + \\ & + \tfrac{1}{2}\aleph_4\, (\boldsymbol{d}\,\mathbf{s} + \mathbf{s}\,\boldsymbol{d}) + \aleph_5\, \mathrm{I}_{\boldsymbol{d}}\, \mathbf{s}^2 + \aleph_6\, M\mathbf{s} + \\ & + \aleph_7\, N\,\mathbf{1} + \tfrac{1}{2}\aleph_8\, (\boldsymbol{d}\,\mathbf{s}^2 + \mathbf{s}^2\boldsymbol{d}) + \aleph_9\, M\,\mathbf{s}^2 + \\ & + \aleph_{10}\, N\mathbf{s} + \aleph_{11}\, N\mathbf{s}^2, \end{aligned}\right\} \tag{304.4}$$

where

$$M \equiv s^k_m\, d^m_k, \qquad N \equiv s^k_m\, s^m_p\, d^p_k, \tag{304.5}$$

$$\aleph_\Gamma = \aleph_\Gamma(\mathrm{I}_{\mathbf{s}}, \mathrm{II}_{\mathbf{s}}, \mathrm{III}_{\mathbf{s}}) \qquad \Gamma = 1, 2, \ldots 11. \tag{304.6}$$

The theory based upon these constitutive equations is called *hypo-elasticity*[2].

In keeping with the expressed aim of elastic rather than viscous response, Eqs. (304.4) are invariant under a change of the time scale. While the stress at a given time generally depends upon the manner in which the load has been applied during previous instants, it is thus independent of the actual speed of deformation.

The exact theory is a fully dynamical one. However, Eqs. (304.4) reduce to those of the linear theory of elasticity (Sect. 301) under the assumptions usually made in formulating that theory. Moreover, every isotropic elastic body is also hypo-elastic[3]. This result, however, must not be regarded as reducing hypo-elasticity to elasticity, for the converse is generally false, and in particular the simpler special cases of (304.4) are not elastic cases. Finally, it has been shown that most of the common "incremental" theories of plasticity other than that of

[1] [1955, *21*, §§ 40 and 33]. A result of the same form had been deduced from a too restrictive definition by TRUESDELL [1951, *27*, § 26].

[2] TRUESDELL [1953, *32*, § 56 (revised)] [1955, *27* and *28*].

[3] NOLL [1955, *18*, § 15b].

perfectly plastic materials are included as special cases of hypo-elasticity, providing they be first corrected so as to satisfy the principle of material indifference[1].

In much of the foregoing discussion, it has been assumed that the body is initially unstressed, but this assumption is not at all necessary in hypo-elasticity. Suppose that in an interval dt a body subject to initial stress $\boldsymbol{s}_0$ undergoes a displacement having small gradients in the sense defined in Sect. 57. Then we have $w^k{}_r\, dt \approx \tilde{R}^k{}_r$ and $d_{pq}\, dt \approx \tilde{e}_{pq}$, where $\tilde{\boldsymbol{R}}$ and $\tilde{\boldsymbol{e}}$ are the tensors of infinitesimal rotation and strain for the displacement considered. Also $\dot{s}^{km} dt \approx s^{km} - s_0^{km}$. From (304.3) and (148.7) we have then

$$s^{km} \approx s_0^{km} + \tilde{R}^k{}_r\, s_0^{rm} + \tilde{R}^m{}_r\, s_0^{kr} + C^{kmpq}\, \tilde{e}_{pq}. \tag{304.7}$$

The essential content of these equations, which define a theory of small deformation of an initially stressed body, was given by CAUCHY[2]. It reflects the fundamentally greater generality of hypo-elasticity in comparison to elasticity. While, as is plain from (303.6), the most general theory of finite elasticity is not capable of representing an elastically isotropic body which in its natural state suffers any but hydrostatic stress, in hypo-elasticity there is, in general, no natural state, and the stress at any given instant may be arbitrary. It was CAUCHY'S theory of initially stressed bodies, a special case of that defined by (304.7), that suggested the theory of hypo-elasticity. Hypo-elasticity may be regarded as a theory in which relations of CAUCHY'S type are applied in each time interval dt.

305. Visco-elastic and accumulative theories. Studies of elasticity and viscosity, according to the classical theories presented in Sects. 298 and 301, made it natural to attempt to combine the two kinds of phenomena within a single theory. Two ideas immediately present themselves:

1. *The total stress is the sum of an elastic stress arising from the strain and a viscous stress arising from the stretching,* and

2. *The total rate of strain is the sum of an elastic rate arising from the rate of stressing and a viscous rate arising from the stress.*

Schematically, the two alternatives may be written

$$t = f(e) + g(\dot{e}) \quad \text{and} \quad \dot{e} = h(\dot{t}) + k(t). \tag{305.1}$$

Both imply necessarily the existence of material constants having the physical dimensions of elasticity and of viscosity; hence there is a modulus having the physical dimensions of time, and relaxation effects may be expected. The former alternative was proposed by O.-E. MEYER, VOIGT, and DUHEM[3]; the latter, by MAXWELL, NATANSON, and ZAREMBA[4]. Theories of this type, including generalizations obtained by allowing time derivatives up to arbitrarily high order to occur on the each side of $(305.1)_1$ or $(305.1)_2$, are called *visco-elastic*. If $\boldsymbol{e}$ is the infinitesimal strain tensor, the constitutive relations are

$$\sum_{k=0}^{n} A_k \left(\frac{\partial}{\partial t}\right)^k \boldsymbol{t} = \sum_{k=0}^{m} B_k \left(\frac{\partial}{\partial t}\right)^k \boldsymbol{e}, \tag{305.2}$$

[1] GREEN [1956, *9* and *10*]. Perfectly plastic materials are included formally by a limit process.

[2] [1829, *4*, Eqs. (36), (37)].

[3] [1874, *3* and *5*] [1875, *4*]; [1889, *10*] [1892, *12* and *13*] [1910, *10*, §§ 395–396]; [1903, *6–9*] [1904, *1*, Part I, Chap. II, and Part IV, Chaps. II–III].

[4] [1867, *2*, pp. 30–31]; [1901, *11–13*] [1902, *7–9*] [1903, *12–14*]; [1903, *17–20*] [1937, *12*].

where the coefficients A_k and B_k are usually assumed constant. These theories possess a large recent literature, mostly in one-dimensional and/or linearized contexts.

Such linear theories generally violate the principle of material indifference, as was first noticed by ZAREMBA, who was led to his form of the principle in this context. His is the only early attempt in viscoelasticity to yield a theory that is a mechanically possible one for unrestricted motions.

A general and properly invariant theory of the MEYER-VOIGT type, allowing the stress to depend upon the first spatial derivatives of the displacement and of the accelerations of all orders, has been achieved by RIVLIN and ERICKSEN[1]. The Maxwell-Zaremba theory is generalized by NOLL's theory of *hygrosteric materials*[2], which are defined by (304.1) without requiring $\boldsymbol{f}$ to be linear in $\boldsymbol{d}$, and by a more general theory sketched by COTTER and RIVLIN[3].

BOLTZMANN and VOLTERRA[4] proposed a theory in which the stress is determined by the entire sequence of strains undergone by the body in the past. The material is represented as having a weaker memory for older experiences; the stress is obtained by integrating the strains from $-\infty$ to t, with a suitable damping or "memory" function. The linear theory, sometimes called "hereditary" but more fitly named *accumulative*, has a considerable literature.

It is still more general to allow the stress to be determined in any way by the strain history, i.e.,

$$\boldsymbol{t} = \overset{t}{\underset{-\infty}{F}} [x^k_{,\alpha}], \tag{305.3}$$

where F denotes a functional. This extremely general and natural concept of material behavior has been put into properly invariant from by NOLL[5] and by GREEN, RIVLIN, and SPENCER[6] (1956). The latter authors determined conditions under which the functional in (305.3) may be replaced by a sum of integrals of VOLTERRA's type or by a finite combination of time derivatives as in the theory of RIVLIN and ERICKSEN. NOLL's method, which rests directly upon properties of invariance as the definitions of particular materials, is presented in "The Non-linear Field Theories of Mechanics", This Encyclopedia, Vol. VIII, Part 2.

V. Examples of thermo-mechanical constitutive equations.

306. Irreversible thermodynamics. While even the classical theory of isotropic viscous and thermally conducting fluids furnishes an example where both mechanical and energetic principles must be used in order to solve any definite problem, the constitutive equations themselves, namely, (296.1) and (298.5), embody separate principles, one being purely mechanical and the other, purely energetic.

The various theories of "irreversible thermodynamics" attempt to describe the numerous physical phenomena which cannot be separated as belonging to one or the other category to the exclusion of the other. Such phenomena occur especially in heterogeneous substances. Current practice concentrates upon the production of entropy owing to such interactions and supposes that the "affinities" and "fluxes" which enter into the production of entropy depend functionally, and usually linearly, upon one another.

---

[1] [1955, *21*].
[2] [1955, *18*, §§ 6–9].
[3] [1955, *4*].
[4] [1874, *1*]; [1909, *10* and *11*] [1930, *9*].
[5] [1957, *11*] [1958, *8*].
[6] [1957, *6*] [1959, *6*].

In Sects. 257 and 259 we have explained the concepts in terms of which such theories are constructed. Since we do not consider that the problem of dynamical and energetic invariance is yet sufficiently understood, we rest content with citing expositions of the subject as it is currently received[1].

307. The "Maxwellian" fluid. As our sole example of the use of the principle of equipresence (Sect. 293 η), we mention a fully thermomechanical theory of fluids based on two constitutive assumptions:

1. Both the stress and the flux of energy depend upon the spatial and temporal derivatives of the thermodynamic state and of the velocity, of all orders.

2. The constitutive relations involve material coefficients having the physical dimensions of viscosity, thermal conductivity, and temperature, but no others.

These constitutive relations define the *Maxwellian fluid* of TRUESDELL[2].

While this theory is known[3] to stand in need of revision so as to be rendered properly invariant, the general forms of the leading terms in the expansions of the stress and flux of energy in powers of the viscosity give an idea of what is to be expected. From the first of the defining conditions, expressing an application of the principle of equipresence, it might be thought that the expansions for the stress $\boldsymbol{t}$ and flux of energy $\boldsymbol{h}$ would be very similar. However, the different tensorial characters of $\boldsymbol{t}$ and $\boldsymbol{h}$, combined with requirements of dimensional invariance, force the the counterparts of terms present in the expansion of $\boldsymbol{t}$ to be absent from the expansion of $\boldsymbol{h}$, and conversely.

To avoid long formulae we summarize the forms of the terms of orders 0, 1, 2 in words. The stress $\boldsymbol{t}$ is the sum of

T1. The terms in the linear law of fluid viscosity (298.5).

T2. The quadratic viscous terms according to the theory of RIVLIN and ERICKSEN mentioned in Sect. 305.

T3. The most general linear isotropic function of $p_{,k}\,p_{,m}$.

T4. The most general linear isotropic tensor function of $p_{,(k}\theta_{,m)}$.

T5. The most general linear isotropic tensor function of $\theta_{,k}\,\theta_{,m}$.

T6. The most general linear isotropic tensor function of $p_{,km}$.

T7. The most general linear isotropic tensor function of $\theta_{,km}$.

Similarly, the flux of energy $\boldsymbol{h}$ is the sum of

H1. A linear term,

$$h_k = \varkappa\,\theta_{,k} + \alpha\,\frac{\varkappa\theta}{p}\,p_{,k}, \tag{307.1}$$

where α is dimensionless, and

H2. The most general linear isotropic vector function of

$$p_{,k}, \quad \theta_{,m}, \quad \text{and} \quad \dot{x}_{p,q}.$$

H3. The most general linear isotropic vector function of $\dot{x}_{p,qr}$.

It is most striking that in these results the separation of effects follows from considerations of invariance alone. For example, despite the much more general definition of a fluid used here, the linear terms T1 in the stress are exactly the same as occur both in the classical linear theory and in the simpler non-linear

---

1 PRIGOGINE [1947, *12*], DE GROOT [1952, *3*], J. MEIXNER and H. REIK in Vol. III, Part 2 of this Encyclopedia.

2 [1949, *34*, §§ 2–4] [1951, *27*, §§ 19–21]. We have altered slightly TRUESDELL'S definition and results.

3 Cf. the remarks of TRUESDELL [1956, *13*, § 16].

"Stokesian" theory (Sect. 299). While FOURIER's law (296.4) is slightly generalized by (307.1), the added term is of a thermodynamic type, parallel to the classical thermal one, and in the linear approximation the effects of deformation *cannot* enter the constitutive equation for $\boldsymbol{h}$. Similar, though more elaborate, separations of effects are seen in the quadratic terms T2 to T7 and H2 to H3. These quadratic terms allow for interactions of a *definite*, not arbitrary kind.

Restrictions such as these, which follow from principles of invariance alone, are only now coming to be studied.

VI. Electromagnetic constitutive equations.

308. The Maxwellian dielectric. The problem of formulating constitutive equations for moving and deforming material media is one of the most difficult and controversial in electromagnetic theory. We shall illustrate some of the relevant ideas by treating a simple example.

α) *The Euclidean invariant constitutive equations of a Maxwellian dielectric.* In classical electromagnetic theory as refined by LORENTZ we have the aether relations (cf. Sect. 279), which can be written in the 4-dimensional tensor form

$$\eta^{\Omega\Delta} = \sqrt{\frac{\varepsilon_0}{\mu_0}}\,(-\det\gamma)^{\frac{1}{2}}\,\overset{-1}{\gamma}{}^{\Omega\Psi}\,\overset{-1}{\gamma}{}^{\Delta\Theta}\,\varphi_{\Psi\Theta} \tag{308.1}$$

or the 3-dimensional vector form

$$\boldsymbol{D} = \varepsilon_0\boldsymbol{E}, \qquad \boldsymbol{H} = \frac{1}{\mu_0}\boldsymbol{B}. \tag{308.2}$$

According to the Lorentz point of view, these relations hold in all material media, moving or stationary, but it must be recalled that $\boldsymbol{D}$ and $\boldsymbol{H}$ are the charge and current potentials for the *total* charge including that due to polarization and magnetization of the material medium.

We define the ideal material called a *Maxwellian dielectric* by the relations

$$\left.\begin{aligned} \boldsymbol{P} &= \varepsilon_0\chi\boldsymbol{E}, \qquad \chi = \text{const}\\ \boldsymbol{M} &= 0 \end{aligned}\right\} \tag{308.3}$$

for the polarization $\boldsymbol{P}$ and the magnetization $\boldsymbol{M}$, provided the medium is at rest in a Euclidean frame which is simultaneously a Lorentz frame. Recall that such a frame is one for which we have the canonical forms (cf. Sects. 280 and 152)

$$\overset{-1}{\gamma} = \begin{bmatrix} \delta_{rs} & 0 \\ 0 & -\frac{1}{c^2} \end{bmatrix}, \quad \boldsymbol{g} = \begin{bmatrix} \delta_{rs} & 0 \\ 0 & 0 \end{bmatrix}, \quad \boldsymbol{t} = (0,0,0,1), \quad c^2 = \frac{1}{\varepsilon_0\mu_0}. \tag{308.4}$$

If we introduce the potential $\boldsymbol{\Upsilon} = (\mathfrak{H}, \mathfrak{D})$ of free charge and current (cf. Sect. 283) in a polarizable and magnetizable medium, the constitutive relations (308.3) can be put in the 4-dimensional form

$$\Upsilon^{\Omega\Delta} = \sqrt{\frac{\varepsilon}{\mu_0}}\,(-\det\boldsymbol{\varkappa})^{\frac{1}{2}}\,\overset{-1}{\varkappa}{}^{\Omega\Psi}\,\overset{-1}{\varkappa}{}^{\Delta\Theta}\,\varphi_{\Psi\Theta} \tag{308.5}$$

with

$$\overset{-1}{\boldsymbol{\varkappa}} = \begin{bmatrix} \delta_{rs} & 0 \\ 0 & -\varepsilon\mu_0 \end{bmatrix}, \qquad \varepsilon \equiv \varepsilon_0 K, \qquad K \equiv 1+\chi \tag{308.6}$$

or in the 3-dimensional vector form

$$\mathfrak{D} = \varepsilon\boldsymbol{E}, \qquad \mathfrak{H} = \frac{1}{\mu_0}\boldsymbol{B}. \tag{308.7}$$

To see how these relations are generalized to the case of moving media, it is easiest to employ the world tensor formalism of Sect. 152 as applied to the problem of constructing invariants of fields under the Euclidean group of transformations (rigid motions). It was noted in Sect. 152 that if $\boldsymbol{\Psi}$ was a world tensor satisfying the conditions

$$\Psi^{\Omega\Delta\cdots} t_\Delta = \Psi^{\Delta\Omega\cdots} t_\Delta = \cdots = 0, \tag{308.8}$$

then the non-vanishing components $\Psi^{rs\cdots}$ in a Euclidean frame transform as a 3-dimensional tensor under the group of Euclidean transformations. This special kind of world tensor was called a *space tensor*. Consider then the electromagnetic field $\boldsymbol{\varphi}$ and the world velocity vector $\boldsymbol{v}$ of a motion. In terms of these quantities we can define two associated space tensors

$$\mathfrak{E}^{\Omega} = g^{\Omega\Delta} \varphi_{\Delta\Theta} v^{\Theta}, \tag{308.9}$$

$$\mathfrak{B}^{\Omega\Delta} = g^{\Omega\Psi} g^{\Delta\Theta} \varphi_{\Psi\Theta}. \tag{308.10}$$

In a Euclidean frame we have (274.1), so that, in such a frame,

$$\mathfrak{E} = (\boldsymbol{E} + \boldsymbol{v}\times\boldsymbol{B}, 0), \qquad \mathfrak{B} = (\text{dual}\,\boldsymbol{B}, 0). \tag{308.11}$$

Consider next the polarization-magnetization world tensor $\boldsymbol{\pi}$ defined in (283.23). In terms of this tensor we defined the space tensors $\mathfrak{P}^{\Omega}$ and $\mathfrak{M}^{\Omega\Delta}$ called the *world polarization* and *magnetization densities* in (283.24) and (183.25). In a Euclidean frame we have

$$\mathfrak{P} = (\boldsymbol{P}, 0), \qquad \mathfrak{M} = (\text{dual}\,\boldsymbol{M}, 0). \tag{308.12}$$

Therefore, the world tensor equations

$$\mathfrak{P}^{\Omega} = \varepsilon_0 \sqrt{g}\, \chi\, \mathfrak{E}^{\Omega}, \qquad \mathfrak{M}^{\Omega\Delta} = 0, \tag{308.13}$$

reduce to (308.3) when the medium is at rest in a Euclidean frame. Moreover, since these equations involve only space tensors, they are *rotationally invariant*. Stated more explicitly, the proportionality of polarization and electromotive intensity and the vanishing of the magnetization are invariant conditions under the group of Euclidean transformations relating the class of reference frames moving rigidly with respect to one another. Now from (283.33), (283.24), (283.25), (279.13), and the aether relations (308.1) we have

$$\left.\begin{aligned} \Upsilon^{\Omega\Delta} &= \eta^{\Omega\Delta} - \pi^{\Omega\Delta}, \\ &= \sqrt{\frac{\varepsilon_0}{\mu_0}}\,(-\det\gamma)^{\frac{1}{2}} \Big[\overset{-1}{\gamma}{}^{\Omega\Psi} \overset{-1}{\gamma}{}^{\Delta\Theta} \varphi_{\Psi\Theta} - \frac{\chi}{c^2} g^{\Delta\Psi} v^{\Omega} v^{\Theta} \varphi_{\Psi\Theta} - \\ &\quad - \frac{\chi}{c^2} g^{\Omega\Theta} v^{\Delta} v^{\Psi} \varphi_{\Psi\Theta}\Big], \end{aligned}\right\} \tag{308.14}$$

or, in a frame for which we have the canonical forms (308.4),

$$\left.\begin{aligned} \mathfrak{D} &= \varepsilon_0 [\boldsymbol{E} + \chi(\boldsymbol{E} + \boldsymbol{v}\times\boldsymbol{B}], \\ \mathfrak{H} &= \frac{1}{\mu_0}\Big[\boldsymbol{B} + \frac{\chi}{c^2}\boldsymbol{v}\times(\boldsymbol{E} + \boldsymbol{v}\times\boldsymbol{B})\Big]. \end{aligned}\right\} \tag{308.15}$$

We have arrived at these constitutive equations for the potentials $\mathfrak{D}$ and $\mathfrak{H}$ by demanding that the relations between polarization and magnetization be invariant under the group of rigid (Euclidean) transformations. We are led to this idea of invariance by thinking of polarization and magnetization as quantities characteristic of the moving material medium and carried with it.

Thus, on the basis of classical thought, any relation between them and the electromagnetic field should be invariant under the group of rigid motions. This is an application of the principle of material indifference.

We wish now to contrast with the results (308.14) and (308.16) obtained by this classical procedure the corresponding results obtained by MINKOWSKI[1], whose reasoning was guided by the idea of Lorentz invariance of the constitutive relations in moving media.

β) The Lorentz-invariant constitutive equations for a moving Maxwellian dielectric. Suppose a Maxwellian dielectric is at rest in some Lorentz frame which, for our present purpose, we may imagine as coinciding with some Euclidean frame. Recall that the class of Lorentz frames are those for which the tensor $\overset{-1}{\gamma}$ has the canonical form $(308.4)_1$. Let us assume that, when the medium is at rest, it is characterized by the relations (308.5). Now suppose that such a medium is set into motion. Any selected particle of the medium undergoing the motion will be instantaneously at rest in some other Lorentz frame. MINKOWSKI reasoned that the relations (308.5) should hold in that Lorentz frame as a consequence of the correct constitutive equations for the moving medium. To put this idea into effect, let $L^{\Omega}{}_{\Delta}$ be the coefficients of a Lorentz transformation (cf. Sect. 282), so that by definition we have

$$L^{\Omega}{}_{\Delta} L^{\Psi}{}_{\Theta} \overset{-1}{\gamma}{}^{\Delta\Theta} = \overset{-1}{\gamma}{}^{\Omega\Psi}. \tag{308.16}$$

Let w^{Ω} be the *relativistic velocity vector* of the motion defined by

$$w^{\Omega} \equiv \frac{c\, v^{\Omega}}{\sqrt{-\gamma_{\Phi\Theta} v^{\Phi} v^{\Theta}}}, \qquad \gamma_{\Omega\Delta} w^{\Omega} w^{\Delta} = -c^2, \tag{308.17}$$

where, as before, $\boldsymbol{v}$ is the classical world velocity vector of the motion. In the reference frame in which the medium is moving we have

$$\boldsymbol{w} = \left(\frac{\boldsymbol{v}}{\sqrt{1-\frac{v^2}{c^2}}}, \frac{1}{\sqrt{1-\frac{v^2}{c^2}}} \right). \tag{308.18}$$

We can choose the coefficients $L^{\Omega}{}_{\Delta}$ of a Lorentz transformation so that

$$\overline{w}^{\Omega} = L^{\Omega}{}_{\Delta} w^{\Delta} \tag{308.19}$$

has the form

$$\overline{\boldsymbol{w}} = (0, 0, 0, 1) \tag{308.20}$$

at some selected event. If the medium is in uniform translatory motion, so that $\boldsymbol{v} = \text{const}$, then $\overline{\boldsymbol{w}}$ will have the form (308.20) throughout a space-time region, but, for general motions, we shall have (308.20) only at a single event. Let $\overline{\Upsilon}^{\Omega\Delta}$ and $\overline{\varphi}^{\Omega\Delta}$ denote the transformed components of $\boldsymbol{\Upsilon}$ and $\boldsymbol{\varphi}$. That is,

$$\overline{\Upsilon}^{\Omega\Delta} = L^{\Omega}{}_{\Psi} L^{\Delta}{}_{\Theta} \Upsilon^{\Psi\Theta}, \qquad \overline{\varphi}_{\Omega\Delta} = L^{\Psi}{}_{\Omega} L^{\Theta}{}_{\Delta} \varphi_{\Psi\Theta}. \tag{308.21}$$

Thus, in the new coordinate system, MINKOWSKI assumes that

$$\overline{\Upsilon}^{\Omega\Delta} = \sqrt{\frac{\varepsilon}{\mu_0}} (-\det \overline{\varkappa})^{\frac{1}{2}} \overline{\varkappa}^{\Omega\Psi} \overline{\varkappa}^{\Delta\Theta} \overline{\varphi}_{\Psi\Theta}, \tag{308.22}$$

where $\overline{\varkappa}$ has rest values consistent with (308.6). Applying the inverse Lorentz transformation $\overset{-1}{L}{}^{\Theta}{}_{\Delta} (\overset{-1}{L}{}^{\Omega}{}_{\Delta} L^{\Delta}{}_{\Psi} = \delta^{\Omega}_{\Psi})$ to (308.22) we see that the relation between

[1] [1910, 7].

the unbarred components of Υ and φ must be

$$\Upsilon^{\Omega\Delta} = \sqrt{\frac{\varepsilon}{\mu_0}}\,(-\det\varkappa)^{\frac{1}{2}}\,\varkappa^{\Omega\Psi}\varkappa^{\Delta\Theta}\varphi_{\Psi\Theta}, \tag{308.23}$$

where, for the moving medium,

$$\varkappa^{\Omega\Delta} = L^{\Omega}{}_{\Psi} L^{\Delta}{}_{\Theta}\,\bar{\varkappa}^{\Psi\Theta}. \tag{308.24}$$

To compute the values of the unbarred components $\varkappa^{\Omega\Delta}$ it is convenient to decompose $\bar{\varkappa}$ as follows:

$$\left.\begin{aligned} \bar{\varkappa}^{\Omega\Delta} &= \gamma^{\Omega\Delta} - \mu_0(\varepsilon-\varepsilon_0)\,\bar{w}^{\Omega}\bar{w}^{\Delta}, \\ &= \gamma^{\Omega\Delta} - \frac{\chi}{c^2}\,\bar{w}^{\Omega}\bar{w}^{\Delta}. \end{aligned}\right\} \tag{308.25}$$

It then follows easily from (308.16) and (308.19) that

$$\varkappa^{\Omega\Delta} = \gamma^{\Omega\Delta} - \frac{\chi}{c^2}\,w^{\Omega}w^{\Delta}. \tag{308.26}$$

Substituting this result into (308.23), we get the Minkowski relations in the form

$$\left.\begin{aligned} \Upsilon^{\Omega\Delta} &= \sqrt{\frac{\varepsilon_0}{\mu_0}}\,(-\det\gamma)^{\frac{1}{2}}\,\times \\ &\times\left[\overset{-1}{\gamma}{}^{\Omega\Psi}\overset{-1}{\gamma}{}^{\Delta\Theta}\varphi_{\Psi\Theta} - \frac{\chi}{c^2}\left(\overset{-1}{\gamma}{}^{\Omega\Psi}w^{\Delta}w^{\Theta}\varphi_{\Psi\Theta} - \overset{-1}{\gamma}{}^{\Delta\Psi}w^{\Omega}w^{\Theta}\varphi_{\Psi\Theta}\right)\right] \end{aligned}\right\} \tag{308.27}$$

or, in terms of $\mathfrak{D}$ and $\mathfrak{H}$,

$$\left.\begin{aligned} \mathfrak{D} &= \varepsilon_0\left[\boldsymbol{E} + \frac{\chi}{1-\frac{v^2}{c^2}}(\boldsymbol{E}+\boldsymbol{v}\times\boldsymbol{B}) - \frac{\chi}{c^2\left(1-\frac{v^2}{c^2}\right)}\,\boldsymbol{v}\,\boldsymbol{v}\cdot\boldsymbol{E}\right], \\ \mathfrak{H} &= \frac{1}{\mu_0}\left[\boldsymbol{B} + \frac{\chi}{c^2\left(1-\frac{v^2}{c^2}\right)}\,\boldsymbol{v}\times(\boldsymbol{E}+\boldsymbol{v}\times\boldsymbol{B})\right]. \end{aligned}\right\} \tag{308.28}$$

Eqs. (308.28) are to be compared with the classical expressions (308.15). We see that if one neglects all terms $O(v^2/c^2)$ in the Minkowski constitutive equations, they reduce to the classical ones. MINKOWSKI's method of deriving Lorentz-invariant electromagnetic constitutive relations was extended to the case of moving crystals by EINSTEIN and LAUB[1] and by BATEMAN[2].

γ) Moving surfaces of discontinuity, the velocity of light, and Fresnel's dragging formula. If we substitute the constitutive relations (308.15) into the electromagnetic boundary conditions (278.8), (278.9) and (283.44), (283.45) at a moving surface of discontinuity, we obtain the following system of equations:

$$\left.\begin{aligned} &\boldsymbol{n}\times[\boldsymbol{E}] - u_n[\boldsymbol{B}] = 0, \quad [\boldsymbol{B}]\cdot\boldsymbol{n} = 0, \quad [\boldsymbol{E}+\chi(\boldsymbol{E}+\boldsymbol{v}\times\boldsymbol{B})]\cdot\boldsymbol{n} = 0, \\ &\boldsymbol{n}\times\left[\boldsymbol{B} + \frac{\chi}{c^2}\boldsymbol{v}\times(\boldsymbol{E}+\boldsymbol{v}\times\boldsymbol{B})\right] + \frac{\chi}{c^2}[\boldsymbol{E}+\chi(\boldsymbol{E}+\boldsymbol{v}\times\boldsymbol{B})] = 0. \end{aligned}\right\} \tag{308.29}$$

The second and third of these conditions are satisfied as a consequence of the first and the fourth provided we assume $u_n \neq 0$. Let us suppose that the velocity $\boldsymbol{v}$ and the polarizability χ of the dielectric medium are continuous. In this case, the latter two vector equations constitute a system of six homogeneous linear

---

[1] [1908, *3*].
[2] [1922, *1*].

equations in the six quantities E_r and B^r. Therefore, the system admits non-zero solutions for the jumps $[\boldsymbol{E}]$ and $[\boldsymbol{B}]$ if and only if the determinant of the coefficients vanishes. To simplify matters, let us consider the case where the normal $\boldsymbol{n}$ to the surface of discontinuity is parallel to the velocity of the medium $\boldsymbol{v}$. In this special case, the vanishing of the determinant requires that the speed of propagation u_n satisfy the quadratic equation

$$K\left(\frac{u_n}{c}\right)^2 - \frac{2\chi v}{c}\left(\frac{u_n}{c}\right) + \frac{\chi v^2}{c^2} - 1 = 0. \tag{308.30}$$

The solution of this equation is

$$\frac{u_n}{c} = \frac{\chi \frac{v}{c} \pm \sqrt{K - \chi \frac{v^2}{c^2}}}{K}. \tag{308.31}$$

Now when the polarizability of the medium vanishes ($\chi = 0$, $K = 1$) the speed of propagation of the surface of discontinuity is $\pm c$. Thus the fundamental constant $c \equiv \sqrt{1/\varepsilon_0 \mu_0}$ is the speed of propagation of an electromagnetic discontinuity in vacuum or any medium in which the polarization and magnetization vanish and there is no free charge or current. We see that the motion of such a medium devoid of polarization has no effect on this result. Now set $n = \sqrt{K}$, where n is the *index of refraction*. We can then write the solution (308.31) in the form

$$\frac{u_n}{c} = \frac{1}{n}\left[\pm 1 + \frac{\chi v}{n c} + O(v^2/c^2)\right], \tag{308.32}$$

which, when the terms $O(v^2/c^2)$ are neglected, reduces to FRESNEL's classic formula[1] for the dragging of light by a moving polarizable medium.

If one substitutes MINKOWSKI's constitutive relations into the jump conditions, the equation analogous to (308.30) is

$$\frac{K}{1 - \frac{v^2}{c^2}}\left(\frac{u_n}{c}\right)^2 - \frac{2\chi}{c\left(1 - \frac{v^2}{c^2}\right)}\left(\frac{u_n}{c}\right) + \frac{\chi v^2}{c^2\left(1 - \frac{v^2}{c^2}\right)} - 1 = 0. \tag{308.33}$$

Again we see that neglecting the terms $O(v^2/c^2)$ leads to FRESNEL's result. However, the two equations for the determination of the speed of an electromagnetic discontinuity in a moving Maxwellian dielectric based on the classical and Minkowski constitutive equations of the moving medium differ by terms $O(v^2/c^2)$.

309. VOLTERRA's electromagnetic constitutive equations. We next mention VOLTERRA's generalization of MAXWELL's constitutive relations for $\mathfrak{D}$ and $\mathfrak{H}$ in stationary dielectric and magnets[2]. In most real materials, the magnetization is not a single-valued function of the magnetic flux $\boldsymbol{B}$, much less a linear function. The magnetization at any instant, however, can be considered as a functional of the history of the field $\boldsymbol{B}$ (cf. Sect. 305). If the "heredity" is linear, VOLTERRA writes

$$\boldsymbol{B}(\boldsymbol{x}, t) = \mu\,\mathfrak{H}(x, t) + \int_0^\infty \Phi(\tau)\,\mathfrak{H}(\boldsymbol{x}, t - \tau)\,d\tau, \tag{309.1}$$

where $\Phi(\tau)$ is the *coefficient of heredity*. When constitutive equations of this type are substituted into the differential field equations obtained from the conservation laws, one obtains a system of integro-differential equations which govern the evolution of the physical system rather than a system of partial differential equations as in the simpler theories.

[1] WHITTAKER [1951, *39*, p. 403].
[2] [1912, *7*], [1930, *9*, p. 195].

310. MIE's theory. The role of the potential fields η and $\boldsymbol{\alpha}$ in the classical theory is rather odd and asymmetrical. We see that the electromagnetic potential $\boldsymbol{\alpha}$ is but a subsidiary field which is generally introduced in order to facilitate the solution of problems. The whole system of electromagnetic equations is invariant under potential transformations of the electromagnetic potential (gauge transformations). On the other hand, the charge-current potential η plays a more fundamental role in the classical theory because of the Maxwell-Lorentz aether relations, which are not invariant under potential transformations of η. In MIE's theory[1], the potentials $\boldsymbol{\alpha}$ and η are made to enter the theory in a more symmetrical manner. In addition to the conservation laws of charge and magnetic flux, MIE assumes the existence of a "universal" function Λ such that

$$\eta^{\Omega\Delta} = \frac{\partial \Lambda}{\partial \varphi_{\Delta\Omega}}, \qquad \sigma^{\Omega} = \frac{\partial \Lambda}{\partial \alpha_{\Omega}}. \tag{310.1}$$

The function Λ is supposed to be a Lorentz-invariant function of the electromagnetic field $\boldsymbol{\varphi}$ and the electromagnetic potential $\boldsymbol{\alpha}$. MIE's theory is concerned with the fundamental question of the electromagnetic constitution of matter. VAN DANTZIG[2] proposed a theory somewhat similar to MIE's in which the potentials η and $\boldsymbol{\alpha}$ were assumed to be linear functionals of the fields $\boldsymbol{\varphi}$ and $\boldsymbol{\sigma}$. We cite these examples of constitutive relations to illustrate the variety of viewpoints which have been expressed as regards the appropriate constitutive relations to accompany the conservation laws of electromagnetic theory.

311. OHM's law for moving conductors. We consider the class of ideal materials such that, when they are at rest in a Euclidean frame, the current $\boldsymbol{J}$ is a linear isotropic function of the electric field $\boldsymbol{E}$:

$$\boldsymbol{J} = C\boldsymbol{E}, \qquad C = \text{const}. \tag{311.1}$$

This relation is called *Ohm's law*. We generalize OHM's law to the case of a moving medium in much the same way as the constitutive equations of a Maxwellian dielectric were generalized in Sect. 308. First we consider the classical or rotationally invariant generalization. In terms of the charge-current vector $\boldsymbol{\sigma}$ and the velocity vector $\boldsymbol{v}$ of a motion, we define two space tensors

$$\left.\begin{aligned} \mathfrak{Q} &\equiv \sigma^{\Omega} t_{\Omega}, \\ \mathfrak{J}^{\Omega} &= \sigma^{\Omega} - v^{\Omega}\sigma^{\Delta} t_{\Delta}. \end{aligned}\right\} \tag{311.2}$$

In a Euclidean frame we have

$$\mathfrak{Q} = Q, \qquad \boldsymbol{\mathfrak{J}} = (\mathfrak{J}, 0), \tag{311.3}$$

where Q is the charge density and $\mathfrak{J}$ is the conduction current. MAXWELL's generalization[3] of OHM's law to the case of a moving medium can then be stated in the form

$$\mathfrak{J}^{\Omega} = C\,\mathfrak{E}^{\Omega}. \tag{311.4}$$

Since $\mathfrak{J}$ and $\mathfrak{E}$ are space tensors, the 4-dimensional formalism insures that the constitutive equation (311.4) is rotationally invariant. In a Euclidean frame it assumes the form

$$\left.\begin{aligned} \mathfrak{J} &= C\,\mathfrak{E}, \\ \boldsymbol{J} - Q\boldsymbol{v} &= C(\boldsymbol{E} + \boldsymbol{v}\times\boldsymbol{B}), \end{aligned}\right\} \tag{311.5}$$

---

[1] [1912, *6*]. A summary of MIE's theory is given by WEYL [1921, *6*, § 26] [1950, *35*, § 28].
[2] [1934, *10*].
[3] [1873, *5*, § 609].

where $\mathfrak{E}$ is the electromotive intensity at a point moving with the medium. When the medium is at rest, (311.5) reduces to (311.1).

MINKOWSKI's generalization of OHM's law is obtained by requiring (311.1) to hold in the Lorentz frame in which the particle of the moving medium is instantaneously at rest. Thus we set

$$\left.\begin{aligned} \bar{\sigma}^{\Omega} &= L^{\Omega}{}_{\Delta}\,\sigma^{\Delta}, \\ \bar{\varphi}_{\Omega\Delta} &= L^{\Psi}{}_{\Omega}\,L^{\Theta}{}_{\Delta}\,\varphi_{\Psi\Theta}, \\ \bar{w}^{\Omega} &= (0, 0, 0, 1), \quad w^{\Omega} = \overset{-1}{L}{}^{\Omega}{}_{\Delta}\,\bar{w}^{\Delta}, \end{aligned}\right\} \tag{311.6}$$

so that the relation (311.1) in the moving Lorentz frame can be put in the form

$$\bar{\sigma}^{\Omega} + \frac{1}{c^2}\gamma_{\Psi\Theta}\,\bar{\sigma}^{\Psi}\,\bar{w}^{\Theta}\,\bar{w}^{\Omega} = C\,\gamma^{\Omega\Psi}\,\bar{\varphi}_{\Psi\Theta}\,\bar{w}^{\Theta}. \tag{311.7}$$

Applying the inverse Lorentz transformation, we then get

$$\sigma^{\Omega} + \frac{1}{c^2}\gamma_{\Psi\Theta}\,\sigma^{\Psi}\,w^{\Theta}\,w^{\Omega} = C\,\gamma^{\Omega\Psi}\,\varphi_{\Psi\Theta}\,w^{\Theta}. \tag{311.8}$$

The spatial component of this last equation is

$$\boldsymbol{J} + \frac{1}{c^2}\left(\frac{\boldsymbol{J}\cdot\boldsymbol{v} - c^2 Q}{1 - v^2/c^2}\right)\boldsymbol{v} = C\left(\frac{\boldsymbol{E} + \boldsymbol{v}\times\boldsymbol{B}}{1 - v^2/c^2}\right). \tag{311.9}$$

Taking the scalar product of (311.9) by $\boldsymbol{v}$, we obtain an equation which can be solved for $\boldsymbol{J}\cdot\boldsymbol{v}$:

$$\boldsymbol{J}\cdot\boldsymbol{v} = C\sqrt{1 - v^2/c^2}\,\boldsymbol{v}\cdot\boldsymbol{E} + v^2 Q. \tag{311.10}$$

We then eliminate $\boldsymbol{J}\cdot\boldsymbol{v}$ from (311.9) and obtain finally

$$\boldsymbol{J} - Q\,\boldsymbol{v} = \frac{C}{\sqrt{1 - v^2/c^2}}\left(\boldsymbol{E} + \boldsymbol{v}\times\boldsymbol{B} - \frac{1}{c^2}\boldsymbol{v}\,\boldsymbol{v}\cdot\boldsymbol{E}\right), \tag{311.11}$$

which is the relativistic form of OHM's law for moving media. Again we see that the relativistic or Lorentz-invariant constitutive equation (311.11) differs from the classical or Euclidean-invariant constitutive equation (311.5) only by terms $O(v^2/c^2)$.

VII. Electromechanical constitutive equations.

312. Elastic dielectrics. The phenomena of piezoelectricity, photoelasticity, and electrostriction in elastic solids are closely related. Because of their importance in engineering applications, the classical theories for these effects have become highly specialized disciplines[1]. Any such theory must be based on simultaneous application of the principles of mechanics and of electromagnetism. The laws of conservation of energy and momentum, agumented so as to include the effects of the electromagnetic field, have been formulated in Chap. F. The relevant equations of mechanics, aside from the boundary conditions, are conveniently summarized in the set of Eqs. (288.11) to (288.13) and (286.14). For dielectric media in the absence of free charge and current and of magnetization, the charge and current densities occurring in (288.11) and (288.12) are expressed in terms of the polarization $\boldsymbol{P}$ as in (283.22). If these expressions for the charge

[1] As sources of experimental and theoretical results and references to original and contemporary literature in this field we may cite the following works: VOIGT [1910, *10*], MASON [1950, *17*], CADY [1946, *1*], COKER and FILON [1931, *3*], STRATTON [1941, *8*].

and current are substituted into (288.11) and (288.12) we obtain the basic field equations of mechanics as applied to the case of dielectric media:

$$\left.\begin{aligned} \varrho \ddot{x}^r &= t^{rs}{}_{,s} - \frac{1}{\sqrt{g}} \operatorname{div} \boldsymbol{P}\, \mathfrak{E}^r + \frac{1}{\sqrt{g}} \left(\frac{d_c \boldsymbol{P}}{dt} \times \boldsymbol{B}\right)^r, \\ \varrho \dot{\varepsilon} &= t^r{}_s \dot{x}^s{}_{,r} + \frac{1}{\sqrt{g}} \frac{d_c \boldsymbol{P}}{dt} \cdot \mathfrak{E} + \operatorname{div} \boldsymbol{h}, \\ &\frac{\partial \varrho}{\partial t} + (\varrho \dot{x}^r)_{,r} = 0, \\ &t^{[rs]} = 0. \end{aligned}\right\} \tag{312.1}$$

We have written these equations as they appear in a general curvilinear inertial co-ordinate system. Under general time-independent transformations of the spatial co-ordinates x^k, P^r and B^r transform as vector densities of weight 1. The quantities ϱ, ε, t^{rs}, h^r and E_r transform as absolute tensors. A comma denotes, as usual, covariant differentiation based on the Christoffel symbols of the metric tensor g_{rs}. Eqs. (312.1) are supplemented by the conservation laws of charge and magnetic flux and by the aether relations of electromagnetic theory. The classical theory of piezoelectricity is based on the linearized version of this system of equations corresponding to infinitesimal deformations and weak fields and the linear piezoelectric constitutive equations of VOIGT, which can be expressed in the form

$$\left.\begin{aligned} t^{rs} &= c^{rsmn} e_{mn} + \frac{1}{\sqrt{g}} \gamma^{rs}{}_m P^m, \\ E_r &= \frac{1}{\sqrt{g}} \overset{-1}{\chi}_{rs} P^s + \gamma^{mn}{}_r e_{mn}, \end{aligned}\right\} \tag{312.2}$$

where $e_{rs} = u_{(r,s)}$ ($\boldsymbol{u}$ = displacement vector) is the classical measure of infinitesimal strain. Because of the relation $\boldsymbol{D} = \varepsilon_0 \boldsymbol{E} + \boldsymbol{P}$, the Voigt relations (312.2) can be written in various forms corresponding to different choices of independent variables.

A thermodynamic treatment of the Voigt relations can be based on the energy equation $(312.1)_2$ by making the assumptions necessary to yield the equation

$$\varrho\, \theta\, \dot{\eta} = \operatorname{div} \boldsymbol{h}, \tag{312.3}$$

where η is the entropy density and θ is the absolute temperature. Then by assuming that the internal energy ε, the stress $\boldsymbol{t}$, the electric field $\boldsymbol{E}$, and the temperature θ are functions of the infinitesimal strain measure $\boldsymbol{e}$, the polarization $\boldsymbol{P}$, and the entropy η, we obtain VOIGT'S relations in the linear approximation if the energy equation is assumed to be satisfied *identically* in the independent variables $\boldsymbol{e}$, $\boldsymbol{P}$, and η. The coefficients in the Voigt relations are then given by

$$c^{rsmn} = \varrho \frac{\partial \varepsilon}{\partial e_{rs}\, \partial e_{mn}}, \qquad \gamma^{mn}{}_s = \varrho \sqrt{g} \frac{\partial \varepsilon}{\partial e_{mn}\, \partial P^s}, \qquad \overset{-1}{\chi}_{rs} = \varrho\, g \frac{\partial \varepsilon}{\partial P^r\, \partial P^s}. \tag{312.4}$$

The classical linear theory of piezoelectricity outlined above has been generalized to the case of finite deformation and large field strengths by TOUPIN[1]. The non-linear theory stands in the same relation to the Voigt theory as the theory of finite elastic deformations stands in relation to linear elasticity theory. In the non-linear theory, the internal energy ε is assumed initially to be a general polynomial function of the displacement gradients $x^i{}_{;\lambda}$ of a deformation, the

[1] [1956, *20*].

polarization $\boldsymbol{P}$, and the entropy density η. One then shows that if the energy is invariant under the group of rigid motions, it must reduce to a function of the variables

$$C_{\mu\nu} = g_{rs}\, x^r{}_{;\mu}\, x^s{}_{;\nu}, \quad \Pi^\mu = \frac{1}{\sqrt{g}\,\varrho} X^\mu{}_{;r} P^r, \quad \eta. \tag{312.5}$$

Assuming that (312.3) holds for this ideal medium, that the electromotive intensity, the stress, and the absolute temperature are functions of the displacement gradients, polarization, and entropy, and that the energy equation is satisfied identically yields the constitutive relations of the general theory:

$$\left.\begin{aligned} t^{rs} &= 2\varrho \frac{\partial \varepsilon}{\partial C_{\mu\nu}} x^r{}_{;\mu}\, x^s{}_{;\nu} = t^{sr}, \\ \mathscr{E}_r &= \frac{\partial \varepsilon}{\partial \Pi^\mu} X^\mu{}_{;r}, \\ \theta &= \frac{\partial \varepsilon}{\partial \eta}. \end{aligned}\right\} \tag{312.6}$$

These relations generalize the Voigt relations (312.2) to the case of finite deformations, large field strengths, and moving media. They reduce to the Voigt relations in the linear approximation and for stationary media.

313. Magnetohydrodynamics. Electrodynamics of continuous media is currently enjoying a new birth of interest. This contemporary work is generally classified under the title, *magnetohydrodynamics.* The abundance of theoretical and experimental labor now directed toward the novel behavior of conducting fluids moving in a magnetic field should yield progress toward an understanding of the general theory of electrodynamics of continuous media. In the present theories of magnetohydrodynamics, the constitutive relations for the stress generally take the form

$$t^{rs} = -p\, g^{rs} + \lambda\, d^k{}_k\, g^{rs} + 2\nu\, d^{rs}, \tag{313.1}$$

as in the classical theory of viscous fluids. OHM's law for moving media (311.5) is generally assumed and a simple type of constitutive relation such as $\boldsymbol{M} = k\boldsymbol{B}$ is introduced to account for the effects of magnetization. The general equations of magnetohydrodynamics are obtained by substituting these rather special constitutive relations into (288.11) and augmenting this set of equations by the equations of conservation of charge and magnetic flux. The linear magnetohydrodynamic equations can then be obtained by casting away the non-linear terms in the general equations. Qualitative features of magnetohydrodynamic solutions are discussed by ELSASSER and ALFVÉN[1].

List of Works Cited.

(Italic numbers in parentheses following the reference indicate the sections where the work is mentioned.)

1678 *1.* HOOKE, R.: Lectures de Potentia Restitutiva, or of Spring Explaining the Power of Springing Bodies. London = R. T. GUNTHER: Early Science in Oxford **8**, 331–356 (1931). (*293*)

1687 *1.* NEWTON, I.: Philosophiae Naturalis Principia Mathematica. London. 3rd ed., ed. H. PEMBERTON, London 1726; trans. A. MOTTE, Sir Isaac Newton's Mathematical Principles of Natural Philosophy and his System of the World. London: 1729. There are many later editions, reprints, and translations. Our references are to the first edition. (*3, 77, 89, 196, 196A, 293, 294, 298*)

[1] [1956, *6*] [1950, *1*].

1736 *1.* EULER, L.: Mechanica sive Motus Scientia Analytice Exposita **1**, Petropoli = Opera omnia (2) **1**. (*4, 99, 155*)
1738 *1.* BERNOULLI, D.: Hydrodynamica sive de Viribus et Motibus Fluidorum Commentarii. Argentorati. (*66A, 77, 89*)
1743 *1.* BERNOULLI, J.: Hydraulica nunc primum detecta ac demonstrata directe ex fundamentis pure mechanicis, Opera **4**, 387–493. Virtually the same work appears under different titles in Comm. Acad. Sci. Petrop. **9** (1737), 3–49 (1744); **10** (1738), 207–260 (1747). (*200*)
2. D'ALEMBERT, J. L.: Traité de Dynamique ... Paris. (*196*)
1744 *1.* D'ALEMBERT, J. L.: Traité de l'Équilibre et du Mouvement des Fluides pour servir de Suite au Traité de Dynamique. Paris; 2nd ed., 1770. (*89*)
1745 *1.* CLAIRAUT, A.: Sur quelques principes qui donnent la solution d'un grand nombre de problèmes de dynamique. Mém. Acad. Sci. Paris (1742), 1–52. (*143, 197*)
2. EULER, L.: Neue Grundsätze der Artillerie, aus dem Englischen des Herrn **Benjamin Robins** übersetzt und mit vielen Anmerkungen versehen. Berlin = Opera omnia (2) **14**. (*66A, 70, 202, 248*)
1746 *1.* EULER, L.: De motu corporum in superficiebus mobilibus. Opusc. var. arg. **1**, 1–136 = Opera omnia (2) **6**, 75–174. (*168*)
1751 *1.* KOENIG, S.: De universali principio aequilibrii et motus, in vi viva reperto, deque nexu inter vim vivam et actionem, utriusque minimo, dissertatio. Nova acta erudit., 125–135, 162–176 = L. Euleri Opera omnia (2) **5**, 303–324. (*167*)
1752 *1.* D'ALEMBERT, J. L.: Essai d'une Nouvelle Théorie de la Résistance des Fluides. Paris. (*66A, 68, 70, 77, 86A, 88, 98, 108, 130, 140, 156, 202, 206*)
2. EULER, L.: Découverte d'un nouveau principe de mécanique. Mém. Acad. Sci. Berlin [**6**] (1750), 185–217 = Opera omnia (2) **5**, 81–108. (*1, 7, 168, 196*)
3. EULER, L.: Recherches sur l'effet d'une machine hydraulique proposée par Mr. **Segner** Professeur à Göttingue. Mém. Acad. Sci. Berlin [**6**] (1750), 311–354 = Opera omnia (2) **15**, 1–39. (*143, 197*)
1753 *1.* EULER, L.: Recherche sur une nouvelle manière d'élever de l'eau proposée par Mr. **De Mour**. Mém. Acad. Sci. Berlin [**7**] (1751), 305–330 = Opera omnia (2) **15**, 134–156. (*163*)
1755 *1.* SEGNER, A.: ... Specimen Theoriae Turbinum, Halae Typis Gebaverianis, 27 April 1755, 40 pp. (*168*)
1756 *1.* EULER, L.: Théorie plus complette des machines qui sont mises en mouvement par la réaction de l'eau. Mém. Acad. Sci. Berlin [**10**] (1754), 227–295 = Opera omnia (2) **15**, 157–218. (*163, 197*)
1757 *1.* EULER, L.: Principes généraux de l'état d'équilibre des fluides. Mém. Acad. Sci. Berlin [**11**] (1755), 217–273 = Opera omnia (2) **12**, 2–53. (*200, 207, 248*)
2. EULER, L.: Principes généraux du mouvement des fluides. Mém. Acad. Sci. Berlin [**11**] (1755), 274–315 = Opera omnia (2) **12**, 54–91. (*66A, 70, 76, 88, 89, 98, 108, 109, 110, 156, 205, 248, 297*)
3. EULER, L.: Continuation des recherches sur la théorie du mouvement des fluides. Mém. Acad. Sci. Berlin [**11**] (1755), 316–361 = Opera omnia (2) **12**, 92–132. (*89, 99, 109, 110, 125, 130, 137, 143, 161, 197, 248*)
1761 *1.* D'ALEMBERT, J. L.: Remarques sur les lois du mouvement des fluides. Opusc. **1**, 137–168. (*99, 108, 133, 140, 161*)
2. EULER, L.: Principia motus fluidorum (1752–1755). Novi Comm. Acad. Sci. Petrop. **6** (1756–1757), 271–311 = Opera omnia (2) **12**, 133–168. (*66A, 76, 77, 84, 86A, 88, 98, 108, 109, 130, 141*)
1762 *1.* EULER, L.: Lettre de M. **Euler** à M. **de La Grange**, Recherches sur la propagation des ébranlemens dans un milieu élastique. Misc. Taur. **2**^{2} (1760–1761), 1–10 = Opera omnia (2) **10**, 255–263 = Oeuvres de Lagrange **14**, 178–188. (*15, 17, 20, 66A, 156*)
2. LAGRANGE, J. L.: Nouvelles recherches sur la nature et la propagation du son. Misc. Taur. **2**^{2} (1760–1761), 11–172 = Oeuvres **1**, 151–316. (*20, 108*)
3. LAGRANGE, J. L.: Application de la méthode exposée dans le mémoire précédent à la solution de différens problèmes de dynamique. Misc. Taur. **2**^{2} (1760–1761), 196–298 = Oeuvres **1**, 365–468. (*86A, 140, 160, 233*)
1764 *1.* EULER, L.: De motu fluidorum a diverso caloris gradu oriundo. Novi Comm. Acad. Sci. Petrop. **11** (1765), 232–267 = Opera omnia (2) **12**, 244–271. (*248*)
1765 *1.* EULER, L.: Theoria Motus Corporum Solidorum seu Rigidorum ex Primis nostrae Cognitionis Principiis Stabilita et ad Omnis Motus, qui in hujusmodi Corpora Cadere Possunt, Accomodata. Rostock = Opera omnia (2) **3** and **4**, 3–293. (*165, 168, 196, 294*)

2. EULER, L.: Recherches sur la connaissance mécanique des corps. Mém. Acad. Sci. Berlin [**14**] (1758), 131—153. (*165, 168*)

3. EULER, L.: Du mouvement de rotation des corps solides autour d'un axe variable. Mém. Acad. Sci. Berlin [**14**] (1758), 154—193. (*143, 170*)

1766 *1.* EULER, L.: Supplément aux recherches sur la propagation du son. Mém. Acad. Sci. Berlin [**15**] (1759), 210—240 = Opera omnia (3) **1**, 452—483. (*17, 20, 66A, 156*)

1767 *1.* EULER, L.: Recherches sur le mouvement des rivières (1751). Mém. Acad. Sci. Berlin [**16**] (1760), 101—118 = Opera omnia (2) **12**, 212—288. (*66A, 125*)

1768 *1.* D'ALEMBERT, J. L.: Sur l'équation qui exprime la loi du mouvement des fluides. Opusc. math. **5**, No. 33, 95—131. (*161*)

2. D'ALEMBERT, J. L.: Suite des recherches sur le mouvement des fluides. Opusc. math. **5**, No. 34, 132—170. (*202*)

1769 *1.* EULER, L.: Sectio prima de statu aequilibrii fluidorum. Novi Comm. Acad. Sci. Petrop. **13** (1768), 305—416 = Opera omnia (2) **13**, 1—72. (*7, 248*)

1770 *1.* EULER, L.: Sectio secunda de principiis motus fluidorum. Novi Comm. Acad. Sci. Petrop. **14** (1769), 270—386 = Opera omnia (2) **13**, 73—153. (*15, 17, 20, 66A, 72, 76, 82A, 84, 88, 89, 109, 156, 160, 164, 210, 297*)

1771 *1.* EULER, L.: Sectio tertia de motu fluidorum lineari potissimum aquae. Novi Comm. Acad. Sci. Petrop. **15** (1770), 219—360 = Opera omnia (2) **13**, 154—261. (*248*)

2. EULER, L.: Genuina principia doctrinae de statu aequilibrii et motu corporum tam perfecte flexibilium quam elasticorum. Novi Comm. Acad. Sci. Petrop. **15** (1770), 381—413 = Opera omnia (2) **11**, 37—61. (*200, 214*)

1776 *1.* COULOMB, C. A.: Essai sur une application des règles de maximis et minimis à quelques problèmes de statique, relatifs à l'architecture. Mém. divers savants **7** (1773), 343—382. (*200, 204*)

2. EULER, L.: Formulae generales pro translatione quacunque corporum rigidorum. Novi Comm. Acad. Sci. Petrop. **20** (1775), 189—207. (*168, 196*)

3. EULER, L.: Nova methodus motum corporum rigidorum determinandi. Novi Comm. Acad. Sci. Petrop. **20** (1775), 208—238. (*170*)

4. EULER, L.: De gemina methodo tam aequilibrium quam motus corporum flexibilium determinandi et utriusque egregio consensu. Novi Comm. Acad. Sci. Petrop. **20** (1775), 286—303 = Opera omnia (2) **11**, 180—193. (*214*)

1783 *1.* LAGRANGE, J. L.: Mémoire sur la théorie du mouvement des fluides. Nouv. mém. Acad. Sci. Berlin (1781), 151—198 = Oeuvres **4**, 695—748. (*72, 74, 99, 134, 161, 163, 184, 206*)

1786 *1.* COUSIN, J. A. J.: Mémoire contenant quelques remarques sur la théorie mathématique du mouvement des fluides. Mém. Acad. Sci. Paris (1783), 665—692. (*72*)

1788 *1.* LAGRANGE, J. L.: Méchanique Analitique. Paris. Oeuvres **11**, **12**, are the 5th ed. Our references are to the first edititon. (*6, 15, 20, 72, 74, 99, 134, 196, 232*)

1806 *1.* EULER, L.: Die Gesetze des Gleichgewichts und der Bewegung flüssiger Körper. Leipzig (being a transl., with some changes and additions, by H. W. BRANDES, of [1769, *1*], [1770, *1*], [1771, *1*], and another paper). (*108*)

1821 *1.* NAVIER, C.-L.-M.-H.: Sur les lois des mouvements des fluides, en ayant égard à l'adhésion des molécules. Ann. chimie **19**, 244—260. (*298*)

1822 *1.* FOURIER, J.: Théorie Analytique de la Chaleur. Paris = Oeuvres **1**. (*241, 296*)

2. NAVIER, C.-L.-M.-H.: Sur les lois du mouvement des fluides, en ayant égard à l'adhésion de leurs molécules. Bull. Soc. Philomath. 75—79. (*298*)

1823 *1.* CAUCHY, A.-L.: Recherches sur l'équilibre et le mouvement intérieur des corps solides ou fluides, élastiques ou non élastiques. Bull. Soc. Philomath. 9—13 = Oeuvres (2) **2**, 300—304. (*7, 21, 26, 27, 82A, 200, 203, 298, 300, 301*)

2. CAUCHY, A.-L.: Mémoire sur une espèce particulière de mouvement des fluides. J. École Polytech. **12**, cahier 19, 204—214 = Oeuvres (2) **1**, 264—274. (*77, 99*)

3. NAVIER, C.-L.-M.-H.: Sur les lois de l'équilibre et du mouvement des corps solides élastiques (1821). Bull. Soc. Philomath., 177—181. (Abstract of [1827, *7*].) (*301*)

1824 *1.* CARNOT, S.: Reflexions sur la Puissance Motrice du Feu et sur les Machines Propres à Développer cette Puissance. Paris = Ann. École Norm. (2) **1**, 393—457 (1872). See [1878, *1*]. (*240*)

1825 *1.* NAVIER, C.-L.-M.-H.: Mémoire sur les lois du mouvement des fluides, en ayant égard à l'adhésion des molécules (1822). Bull. Soc. Philomath. 49—52. (Abstract of [1827, *6*].) (*298*)

2. PIOLA, G.: Sull' applicazione de' principj della meccanica analitica del **Lagrange** ai principali problemi. Milano: Regia Stamparia, VII + 252 pp. (*17, 82, 156*)

1827 *1.* CAUCHY, A.-L.: De la pression ou tension dans un corps solide. Ex. de math. **2**, 42—56 = Oeuvres (2) **7**, 60—78. (*21, 200, 203, 204, 205*)

2. CAUCHY, A.-L.: Sur la condensation et la dilatation des corps solides. Ex. de math. 2, 60–69 = Oeuvres (2) **7**, 82–83. (*21, 22, 26, 27, 30, 54, 57, 82A*)
3. CAUCHY, A.-L.: Sur les moments d'inertie. Ex. de math. **2**, 93–103 = Oeuvres (2) **7**, 124–136. (*168*)
4. CAUCHY, A.-L.: Sur les relations qui existent dans l'état d'équilibre d'un corps solide ou fluide, entre les pressions ou tensions et les forces accélératrices. Ex. de math. **2**, 108–111 = Oeuvres (2) **7**, 141–145. (*205*)
5. CAUCHY, A.-L.: Théorie de la propagation des ondes à la surface d'un fluide pesant d'une profondeur indéfinie (1815). Mém. divers savants (2) **1** (1816), 3–312 = Oeuvres (1) **1**, 5–318. (*20, 86A, 134, 206*)
6. NAVIER, C.-L.-M.-H.: Mémoire sur les lois du mouvement des fluides (1822). Mém. Acad. Sci. Inst. France (2) **6**, 389–440. (*289*)
7. NAVIER, C.-L.-M.-H.: Mémoire sur les lois de l'équilibre et du mouvement des corps solides élastiques (1821). Mém. Acad. Sci. Inst. France (2) **7**, 375–393. (*301*)

1828 1. CAUCHY, A.-L.: Sur les centres, les plans principaux et les axes principaux des surfaces du second degré. Ex. de math. **3**, 1–22 = Oeuvres (2) **8**, 9–35. (*28*)
2. CAUCHY, A.-L.: Sur les équations qui expriment les conditions d'équilibre, ou les lois du mouvement intérieur d'un corps solide, élastique, ou non élastique. Ex. de math. **3**, 160–187 = Oeuvres (2) **8**, 195–226. (*205, 298, 300, 301*)
3. CAUCHY, A.-L.: Sur quelques théorèmes relatifs à la condensation ou à la dilatation des corps. Ex. de math. **3**, 237–244 = Oeuvres (2) **8**, 278–287. (*27, 82A*)

1829 1. CAUCHY, A.-L.: Sur les pressions ou tensions supportées en un point donné d'un corps solide par trois plans perpendiculaires entre eux. Ex. de math. **4**, 30–40 = Oeuvres (2) **9**, 41–52. (*203*)
2. CAUCHY, A.-L.: Sur la relation qui existe entre les pressions ou tensions supportées par deux plans quelconques en un point donné d'un corps solide. Ex. de math. **4**, 41–46 = Oeuvres (2) **9**, 53–55. (*205*)
3. CAUCHY, A.-L.: Sur les corps solides ou fluides dans lesquels la condensation ou dilatation linéaire est la même en tous sens autour de chaque point. Ex. de math. **4**, 214–216 = Oeuvres (2) **9**, 254–258. (*142*)
4. CAUCHY, A.-L.: Sur l'équilibre et le mouvement intérieur des corps considérés comme des masses continues. Ex. de math. **4**, 293–319 = Oeuvres (2) **9**, 243–369. (*150, 293, 301, 304*)
5. POISSON, S.-D.: Mémoire sur l'équilibre et le mouvement des corps élastiques (1828). Mém. Acad. Sci. Inst. France (2) **8**, 357–570. (*205, 301*)

1830 1. CAUCHY, A.-L.: Sur les diverses méthodes à l'aide desquelles on peut établir les équations qui représentent les lois d'équilibre, ou le mouvement intérieur des corps solides ou fluides. Bull. sci. math. soc. prop. conn. **13**, 169–176. (*301*)

1831 1. FARADAY, M.: On a peculiar class of optical deceptions. J. Roy. Inst. Gt. Brit. **1**, 205–223 = Res. Chem. Phys. 291–309 (1859). (*70*)
2. POISSON, S.-D.: Mémoire sur les équations générales de l'équilibre et du mouvement des corps solides élastiques et des fluides (1829). J. École Polytech. **13**, cahier 20, 1–174. (*134, 184, 205, 298, 301*)

1832 1. DUHAMEL, J.-M.-C.: Mémoire sur les équations générales de la propagation de la chaleur dans les corps solides dont la conductibilité n'est pas la même dans tous les sens (1828). J. École Polytech. **13**, cahier 21, 356–399. (*296*)

1833 1. FOURIER, J.: Sur le mouvement de la chaleur dans les fluides. Mém. Acad. Sci. Inst. France (2) **12**, 507–530 = Oeuvres **2**, 595–614. (*241, 296*)
2. LAMÉ, G., et E. CLAPEYRON: Mémoire sur l'équilibre des corps solides homogènes. Mém. divers savants (2) **4**, 465–562. (*21*)
3. PIOLA, G.: La meccanica de' corpi naturalmente estesi trattata col calcolo delle variazioni. Opusc. mat. fis. di diversi autori. Milano: Giusti **1**, 201–236. (*29, 95, 210, 232*)
4. POISSON, S.-D.: Traité de Mécanique, 2nd ed. Paris. (*134, 184*)

1834 1. CLAPEYRON, E.: Mémoire sur la puissance motrice de la chaleur. J. École Polytech. **14**, cahier 23, 153–190. (*245, 252*)

1835 1. CORIOLIS, G.: Mémoire sur la manière d'établir les différents principes de la mécanique pour des systèmes de corps, en les considérant comme des assemblages de molécules. J. École Polytech. **15**, cahier 24, 93–132. (*86*)
2. CORIOLIS, G.: Mémoire sur les équations du mouvement relatif des systèmes de corps. J. École Polytech. **15**, cahier 24, 142–154. (*143*)

1836 1. PIOLA, G.: Nuova analisi per tutte le questioni della meccanica molecolare. Mem. Mat. Fis. Soc. Ital. Modena **21** (1835), 155–321. (*17, 26, 72, 82A, 95, 96, 101, 109, 156, 210*)

1837 *1.* Earnshaw, S.: On fluid motion, so far it is expressed by the equation of continuity. Trans. Cambridge Phil. Soc. **6** (1836—1838), 203—233. (*110*)
2. Jacobi, C. J. J.: Über die Reduktion der Integration der partiellen Differentialgleichungen erster Ordnung zwischen irgend einer Zahl Variabeln auf die Integration eines einzigen Systems gewöhnlicher Differentialgleichungen. J. reine angew. Math. **17**, 97—162 = Werke **4**, 57—127. (*219*)
3. Möbius, A. F.: Lehrbuch der Statik, 2 vols., xx + 356 + 314 pp. Leipzig: Göschen. (*219*)
4. Mohr, F.: Ansichten über die Natur der Wärme. Ann. der Pharm. **24**, 141—147. Transl., P. G. Tait, Views of the nature of heat. Phil. Mag. (5) **2** (1876), 110—114. (*240*)

1839 *1.* Green, G.: On the laws of reflection and refraction of light at the common surface of two non-crystallized media (1837). Trans. Cambridge Phil. Soc. **7** (1838—1842), 1—24 = Papers, 245—269. (*33A, 218, 232A, 262, 301, 303*)
2. Seguin, M.: De l'Influence des Chemins de Fer et de l'Art de les Tracer et de les Construire. Paris. (*240*)

1841 *1.* Cauchy, A.-L.: Mémoire sur les dilatations, les condensations et les rotations produits par un changement de forme dans un système de points matériels. Ex. d'an. phys. math. **2**, 302—330 = Oeuvres (2) **12**, 343—377. (*26, 27, 28, 33A, 36, 56, 57, 85, 86, 86A*)
2. Green, G.: On the propagation of light in crystallized media (1839). Trans. Cambridge Phil. Soc. **7** (1839—1842), 121—140 = Papers 293—311. (*26, 31, 262, 301, 303*)
3. Jacobi, C. J. J.: De determinantibus functionalibus. J. reine angew. Math. **22**, 319—359 = Werke **3**, 393—438. (*17*)
4. Lamé, G.: Mémoire sur les surfaces isostatiques dans les corps solides homogènes en équilibre d'élasticité. J. Math. pures appl. **6**, 37—60. (*205A, 209*)
5. Svanberg, A. F.: On fluides rörelse. K. Vetenskapsacad. handlingar (1839), 139—154. Transl., Sur le mouvement des fluides. J. reine angew. Math. **24** (1842), 153—163. (*99, 133*)

1842 *1.* Challis, J.: A general investigation of the differential equations applicable to the motion of fluids. Trans. Cambridge Phil. Soc. **7** (1839—1842), 371—396. (*136*)
2. Mayer, J. R.: Bemerkungen über die Kräfte der unbelebten Natur. Liebigs Ann. Chem. **42**, 233—240 = pp. 3—12 of J. R. Mayer, Die Mechanik der Wärme. Stuttgart 1867 = [1929, *8*, pp. 35—42]. Transl., G. C. Foster, Remarks on the forces of inorganic nature. Phil. Mag. (4) **24**, 371—377 (1862); transl., G. Sarton [1929, *8*, pp. 27—33]; transl., W. F. Magie, A Source Book in Physics pp. 197—203. New York and London 1934. (*240, 253*)
3. Power, J.: On the truth of the hydrodynamical theorem, that if $u\,dx + v\,dy + w\,dz$ be a complete differential with respect to x, y, z at any one instant, it is always so. Trans. Cambridge Phil. Soc. **7** (1839—1842), 455—464. (*134*)
4. Stokes, G. G.: On the steady motion of incompressible fluids. Trans. Cambridge Phil. Soc. **7** (1839—1842), 439—453 = Papers **1**, 1—16. (*125, 126, 133, 161, 162*)

1843 *1.* Cauchy, A.-L.: Note sur les pressions supportées, dans un corps solide ou fluide, par deux portions de surface très voisine, l'une extérieure, l'autre intérieure à ce même corps. C.R. Acad. Sci., Paris **16**, 151—156 = Oeuvres (1) **1**, 246—251. (*205*)
2. Joule, J. P.: On the caloric effects of magneto-electricity, and on the mechanical value of heat. Phil. Mag. (3) **23**, 263—276, 347—355, 435—443 = Papers **1**, 123—159. (Abstract in Rep. Brit. Assn. (1843), 33 = [1929, *8*, p. 43].) (*240*)
3. St. Venant, A.-J.-C. B. de: Mémoire sur le calcul de la résistance et de la flexion des pièces solides à simple ou à double courbure, en prenant simultanément en considération les divers efforts auxquels elles peuvent être soumises dans tous les sens. C. R. Acad. Sci., Paris **17**, 942—954, 1020—1031. (*60, 63A, 214*)
4. St. Venant, A.-J.-C. B. de: Note à joindre au mémoire sur la dynamique des fluides, présenté le 14 Avril 1834. C. R. Acad. Sci., Paris **17**, 1240—1243. (*298*)
5. Waterston, J. J.: Note on the physical constitution of gaseous fluids and a theory of heat, appendix to Thoughts on the Mental Functions. Edinburgh = Papers, 183—206. (*240*)

1844 *1.* Binet, J.: Mémoire sur l'intégration des équations de la courbe élastique à double courbure. C. R. Acad. Sci., Paris **18**, 1115—1119. (*63A*)
2. Jacobi, C. J. J.: Theoria novi multiplicatoris systemati aequationum differentialium vulgarium applicandi. J. reine angew. Math. **27**, 199—268 = Werke **4**, 317—394. (*17*)
3. St. Venant, A.-J.-C. B. de: Sur les pressions qui se développent à l'intérieur des corps solides lorsque les déplacements de leurs points, sans altérer l'élasticité, ne

peuvent cependant pas être considérés comme très petits. Bull. Soc. Philomath. **5**, 26–28. (*31, 55*)

4. Stokes, G. G.: On some cases of fluid motion. Trans. Cambridge Phil. Soc. **8** (1844–1849), 105–137 = Papers **1**, 17–68. (*114*)

1845 *1.* Joule, J. P.: On the changes of temperature produced by rarefaction and condensation of air (1844). Phil. Mag. (3) **26**, 369–383 = Papers **1**, 171–189. (*240*)

2. Joule, J. P.: On the existence of an equivalent relation between heat and the ordinary forms of mechanical power. Phil. Mag. (3) **27**, 205–207 = Papers **1**, 202–205. (*240*)

3. St. Venant, A.-J.-C. B. de: Note sur l'état d'équilibre d'une verge élastique à double courbure lorsque les déplacements éprouvés par ses points, par suite de l'action des forces qui la sollicitent, ne sont pas très petits. C. R. Acad. Sci., Paris **19**, 36–44, 181–187. (*63A*)

4. Stokes, G. G.: On the theories of the internal friction of fluids in motion, and of the equilibrium and motion of elastic solids. Trans. Cambridge Phil. Soc. **8** (1844–1849), 287–319 = Papers **1**, 75–129. (*72, 82A, 86, 90, 96, 134, 142, 298*)

1846 *1.* Stokes, G. G.: Report on recent researches in hydrodynamics. Rep. Brit. Assn. Pt. I, 1–20 = Papers **1**, 151–187. (*134*)

1847 *1.* Hopkins, W.: On the internal pressure to which rock masses may be subjected, and its possible influence in the production of the laminated structure. Trans. Cambridge Phil. Soc. **8** (1844–1849), 456–470. (*203, 204*)

2. Joule, J. P.: On matter, living forces, and heat. Manchester Courier, May 5 and 12 = Papers **1**, 265–276 = pp. 385–390 of E. C. Watson: **Joule**'s only general exposition of the principle of conservation of energy. Amer. J. Phys. **15**, 383–390 (1947). (*240*)

3. St. Venant, A.-J.-C. B. de: Mémoire sur l'équilibre des corps solides, dans les limites de leur élasticité, et sur les conditions de leur résistance, quand les déplacements ne sont pas très petits. C. R. Acad. Sci., Paris **24**, 260–263. (*31, 55*)

1848 *1.* MacCullagh, J.: An essay towards a dynamical theory of crystalline reflection and refraction (1839). Trans. Roy. Irish Acad. Sci. **21**, 17–50 = Works, 145–184. (*86A, 302*)

2. Piola, G.: Intorno alle equazioni fondamentali del movimento di corpi qualsivogliono, considerati secondo la naturale loro forma e costituzione (1845). Mem. Mat. Fis. Soc. Ital. Modena **24**^{1}, 1–186. (*17, 20, 26, 72, 82, 82A, 95, 96, 156, 204, 205, 210, 232*)

3. Stokes, G. G.: Notes on hydrodynamics (4), Demonstration of a fundamental theorem. Cambr. Dubl. Math. J. **3**, 209–219 = (with added notes) Papers **2**, 36–50. (*129, 134*)

4. Stokes, G. G.: On a difficulty in the theory of sound. Phil. Mag. **23**, 349–356. A drastically condensed version, with an added note, appears in Papers **2**, 51–55. (*7, 177, 185, 189, 205, 241, 297*)

5. Thomson, W. (Lord Kelvin): Notes on hydrodynamics (2), On the equation of the bounding surface. Cambr. Dubl. Math. J. **3**, 89–93 = Papers **1**, 83–87. (*74, 177, 184*)

1849 *1.* Haughton, S.: On the equilibrium and motion of solid and fluid bodies (1846). Trans. Roy. Irish Acad. **21**, 151–198. (*205, 232A*)

2. Schweins, F.: Fliehmomente, oder die Summe $\sum(xX+yY)$ bei Kräften in der Ebene, und $\sum(xX+yY+zZ)$ bei Kräften im Raume. J. reine angew. Math. **38**, 77–88. (*219*)

3. Thomson, W. (Lord Kelvin): Notes on hydrodynamics (5), On the vis-viva of a liquid in motion. Cambr. Dubl. Math. J. **4**, 90–94 = Papers **1**, 107–112. (*94, 113, 169*)

4. Thomson, W. (Lord Kelvin): An account of **Carnot's** theory of the motive power of heat; with numerical results deduced from **Regnault's** experiments on steam. Trans. Roy. Soc. Edinb. **16**, 541–574 = Ann. Chimie **35**, 248–255 (1852) = Papers **1**, 113–155. (*245*)

1850 *1.* Clausius, R.: Über die bewegende Kraft der Wärme und die Gesetze, welche sich daraus für die Wärmelehre selbst ableiten lassen. Ann. Physik (3) **19**, 368–398, 500–524 = Abh. **1**, 16–78. Transl., On the moving force of heat, and the laws regarding the nature of heat itself which are deducible therefrom. Phil. Mag. (4) **2** (1851), 1–21, 102–119; transl., W. F. Magie, On the motive power of heat, and on the laws which can be deduced from it for the theory of heat, pp. 65–107 of The Second Law of Thermodynamics. New York and London 1899. (*245, 248, 252, 253*)

2. KIRCHHOFF, G.: Über das Gleichgewicht und die Bewegung einer elastischen Scheibe. J. reine angew. Math. **40**, 51—88 = Ges. Abh. **237**—279. *(232A, 262)*

3. THOMSON, W. (Lord KELVIN): Remarks on the forces experienced by inductively magnetized ferromagnetic or diamagnetic non-crystalline substances. Phil. Mag. (3) **37**, 241—253 = Papers Electr. Magn., §§ 647—668. *(102)*

1851 1. RANKINE, W. J. M.: Laws of the elasticity of solid bodies. Cambr. Dubl. Math. J. **6**, 41—80, 178—181, 185—186 = Papers 67—101. *(25)*

2. STOKES, G. G.: On the effect of the internal friction of fluids on the motion of pendulums (1850). Trans. Cambridge Phil. Soc. 9^2, 8—106 = Papers **3**, 1—141. *(72, 170, 217, 298)*

3. STOKES, G. G.: On the conduction of heat in crystals. Cambr. Dubl. Math. J. **6**, 215—238 = Papers **3**, 203—227. *(241)*

1852 1. KIRCHHOFF, G.: Über die Gleichungen des Gleichgewichts eines elastischen Körpers bei nicht unendlich kleinen Verschiebungen seiner Theile. Sitzgsber. Akad. Wiss. Wien **9**, 762—773. (Not repr. in Abh.) *(19, 210, 217, 236A, 256A, 262, 303)*

2. LAMÉ, G.: Leçons sur la Théorie Mathématique de l'Élasticité. Also 2nd ed., Paris 1866. *(21)*

1853 1. RANKINE, W. J. M.: On the mechanical action of heat I—V (1850—1851). Trans. Roy. Soc. Edinb. **20**, 147—210. *(245)*

2. REECH, M. F.: Théorie générale des effets dynamiques de la chaleur. J. Math. pures appl. **18**, 357—568. *(245, 252)*

3. THOMSON, W. (Lord KELVIN): On the dynamical theory of heat I—V (1851). Trans. Roy. Soc. Edinb. **20**, 261—293, 475—483 = Phil. Mag. (4) **4**, (1852) 8—20, 105—117, 168—176, 424—434, **9** (1855), 523—531 = Papers **1**, 174—232. Pts. I—III = pp. 111—146 of The Second Law of Thermodynamics, ed. W. F. MAGIE. New York and London 1899. *(245)*

1854 1. CLAUSIUS, R.: Über eine veränderte Form des zweiten Hauptsatzes der mechanischen Wärmetheorie. Ann. Physik **93**, 481—506 = Abh. **1**, 126—154. *(245, 248, 249, 252, 258)*

2. SCHWEINS, F.: Theorie der Dreh- und Fliehmomente der parallelen Seitenkräfte, in welche Kräfte im Raume zerlegt werden können. J. reine angew. Math. **47**, 238—245. *(219)*

1855 1. FICK, A.: Über Diffusion. Ann. der Phys. **94**, 59—86. *(158, 295)*

2. HAUGHTON, S.: On a classification of elastic media, and the laws of plane waves propagated through them (1849). Trans. Roy. Irish Acad. Pt. I-Science **22**, 97—138. *(205, 232)*

3. MEISSEL, E.: Über einen speciellen Fall des Ausflusses von Wasser in einer verticalen Ebene. Ann. Physik (4) **5**, 276—283. *(161)*

4. THOMSON, W. (Lord KELVIN): On the thermo-elastic and thermo-magnetic properties of matter. Quart. J. Math. **1** (1855—1857) 55—77 = (with notes and additions) Phil. Mag. (5) **5** (1878), 4—27 = Pt. VII of On the dynamical theory of heat. Papers **1**, 291—316. *(218, 262)*

1856 1. RANKINE, W. J. M.: On axes of elasticity and crystalline forms. Phil. Trans. Roy. Soc. Lond. (146) **46**, 261—285 = Papers 119—149. *(200)*

2. THOMSON, W. (Lord KELVIN): Elements of a mathematical theory of elasticity. Phil. Trans. Roy. Soc. Lond. **146**, 481—498. (See [1877, *5*].) *(262)*

1857 1. CLEBSCH, A.: Über eine allgemeine Transformation der hydrodynamischen Gleichungen. J. reine angew. Math. **54**, 293—312. *(125, 136, 137, 162, 210)*

1858 1. HELMHOLTZ, H.: Über Integrale der hydrodynamischen Gleichungen, welche den Wirbelbewegungen entsprechen. J. reine angew. Math. **55**, 25—55 = Wiss. Abh. **1**, 101—134. Transl., P. G. TAIT, On integrals of the hydrodynamical equations, which express vortex-motion. Phil. Mag. (4) **33**, (1867) 485—512. *(8, 75, 88, 93, 106, 108, 113, 133, 175, 185, 186, 218)*

1859 1. CLEBSCH, A.: Über die Integration der hydrodynamischen Gleichungen. J. reine angew. Math. **56**, 1—10. *(136, 136A)*

2. KIRCHHOFF, G.: Über das Gleichgewicht und die Bewegung eines unendlich dünnen elastischen Stabes. J. reine angew. Math. **56**, 285—313 = Ges. Abh., 285—316. *(57, 63A, 214, 236A, 301)*

3. LAMÉ, G.: Leçons sur les Coordonnées Curvilignes et leurs divers Applications. Paris: Mallet-Bachelier. xxvii + 368 pp. *(205A, 209)*

1860 1. DIRICHLET, G. LEJEUNE: Untersuchungen über ein Problem der Hydrodynamik. Gött. Abh., math. Cl. **8** (1858—1859), 3—42 = J. reine angew. Math. **58**, 181—216 (1861) = Werke **2**, 263—301. *(66)*

2. MAXWELL, J. C.: Illustrations of the dynamical theory of gases. Phil. Mag. (4) **19**, 19—32 and **20**, 21—37 = Papers **1**, 377—409. *(295)*

3. NEUMANN, C.: Zur Theorie der Elasticität. J. reine angew. Math. **57**, 281–318. (*210*)

4. RIEMANN, B.: Ueber die Fortpflanzung ebener Luftwellen von endlicher Schwingungsweite. Gött. Abh., math. Cl. **8** (1858–1859), 43–65 = Werke, 145–164. (*185, 189, 205*)

1861 1. HANKEL, H.: Zur allgemeinen Theorie der Bewegung der Flüssigkeiten. Göttingen. (*66A, 88, 108, 128, 136*)

2. WARREN, J. W.: On invariant points, lines, and surfaces in space, and their physical significance. Amer. J. Math. **4**, 306–310. (*38*)

1862 1. CLAUSIUS, R.: Ueber die Anwendung des Satzes von der Aequivalenz der Verwandlungen auf die innere Arbeit. Vjschr. nat. Ges. Zürich **7**, 48–95 = Ann. Physik **116**, 73–112 = Abh. **1**, 242–279. (*245, 258*)

2. CLEBSCH, A.: Theorie der Elasticität fester Körper. Leipzig. (*63A, 214*)

3. EULER, L.: Fragmentum 97. Opera postuma **1**, 494–496 = Opera omnia (1) **29**, 437–440. (*33A*)

4. EULER, L.: Anleitung zur Natur-Lehre, worin die Gründe zu Erklärung aller in der Natur sich ereignenden Begebenheiten und Veränderungen festgesetzt werden (probably written between 1755 and 1759). Opera postuma **2**, 449–560 = Opera omnia (3) **1**, 16–178. (*76*)

5. WARREN, J. W.: On the internal pressure within an elastic solid. Quart. J. Math. **5**, 109–117. (*21*)

1863 1. AIRY, G. B.: On the strains in the interior of beams. Phil. Trans. Roy. Soc. Lond. **153**, 49–80. Abstract in Rep. Brit. Assn. 1862, 82–86 (1863). (*224*)

2. THOMSON, W. (Lord KELVIN): Dynamical problems regarding elastic spheroidal shells and spheroids of incompressible liquid. Phil. Trans. Roy. Soc. Lond. **153**, 583–616 = Papers **3**, 351–394. (*262, 301, 303*)

1864 1. JOULE, J. P.: Note on the history of the dynamical theory of heat. Phil. Mag. (4) **28**, 150–151. (*240*)

2. RANKINE, W. J. M.: On plane water-lines in two dimensions. Phil. Trans. Roy. Soc. Lond. **154**, 369–391 = Papers 495–521. (*71, 161*)

3. ST. VENANT, A.-J.-C. B. DE: Établissement élémentaire des formules et équations générales de la théorie de l'élasticité des corps solides. Appendix in: Résumé des Leçons données à l'École des Ponts et Chaussées sur l'Application de la Mécanique, première partie, première section, De la Résistance des Corps Solides, par C.-L.-M.-H. NAVIER, 3rd. ed. Paris. (*57*)

4. ST. VENANT, A.-J.-C. B. DE: Théorie de l'élasticité des solides, ou cinématique de leurs déformations. L'Institut **32**[1], 389–390. (*27*)

5. WARREN, J. W.: Note on a transformation of the general equation of wave propagation, due to internal force. Quart. J. Math. **6**, 137–139. (*209*)

1865 1. CLAUSIUS, R.: Ueber verschiedene für die Anwendung bequeme Formen der Hauptgleichungen der mechanischen Wärmetheorie. Vjschr. nat. Ges. Zürich **10**, 1–59 = Ann. der Phys. **125**. 353–400 = Abh. **2**, 1–44. (*245, 246, 258, 264*)

2. THOMSON, W. (Lord KELVIN): On the elasticity and viscosity of metals. Proc. Roy. Soc. Lond. **14**, 289–297. (See [1877, *5*].)

1866 1. DARBOUX, G.: Sur les surfaces orthogonales. Ann. École Norm. (1) **3**, 97–141. (*122*)

2. JACOBI, C. J. J.: Vorlesungen über Dynamik (1842–1843). Werke, Suppl. (*219*).

3. LIPSCHITZ, R.: Ueber einen algebraischen Typus der Bedingungen eines bewegten Massensystems. J. reine angew. Math. **66**, 363–374. (*219*)

4. VERDET, E.: Introduction, Oeuvres de **Fresnel 1**, IX–XCIX. (*301*)

1867 1. BERTRAND, J.: Théorème relatif au mouvement le plus général d'un fluide. C. R. Acad. Sci., Paris **66**, 1127–1230. (*78, 86, 86A, 89*)

2. MAXWELL, J. C.: On the dynamical theory of gases (1866). Phil. Trans. Roy. Soc. Lond. **157**, 49–88 = Phil. Mag. (4) **35** (1868), 129–145, 185–217 = Papers, **2**, 26–78. (*215, 243, 295, 305*)

3. THOMSON, W. (Lord KELVIN), and P. G. TAIT: Treatise on Natural Philosophy. Part I. Cambridge. (*8, 12, 21, 27, 35, 36, 38, 42, 45, 46, 167, 205, 262*)

1868 1. Anonymous: Les Mondes **17**, 620–623. (*89*)

2. BERTRAND, J.: Note relative à la théorie des fluides. Réponse à la communication de M. **Helmholtz**. C. R. Acad. Sci., Paris **67**, 267–269. (*86A*)

3. BERTRAND, J.: Observations nouvelles sur un mémoire de M. **Helmholtz**. C. R. Acad. Sci., Paris **67**, 469–472. (*86A*)

4. BERTRAND, J.: Réponse à la note de M. **Helmholtz**. C. R. Acad. Sci., Paris **67**, 773–775. (*86A*)

5. Fresnel, A.: Première mémoire sur la double réfraction (1821). Oeuvres **2**, 261—308. (*208, 209*)

6. Fresnel, A.: Supplément au mémoire sur la double réfraction (1822). Oeuvres **2**, 343—367. (*21, 208, 209*)

7. Fresnel, A.: Second supplément au mémoire sur la double réfraction (1822). Oeuvres **2**, 369—442. (*21, 203, 205, 208, 209*)

8. Helmholtz, H.: Sur le mouvement le plus général d'un fluide. Réponse à une communication précédente de M. J. **Bertrand**. C. R. Acad. Sci., Paris **67**, 221—225 = Wiss. Abh. **1**, 135—139. (*86A*)

9. Helmholtz, H.: Sur le mouvement des fluides. Deuxième réponse à M. J. **Bertrand**. C. R. Acad. Sci., Paris **67**, 754—757 = Wiss. Abh. **1**, 140—144. (*86A*)

10. Helmholtz, H.: Réponse à la note de M. J. **Bertrand**, du 19 Octobre. C. R. Acad. Sci., Paris **67**, 1034—1035 = Wiss. Abh. **1**, 145. (*86A*)

11. Kirchhoff, G.: Ueber den Einfluß der Wärmeleitung in einem Gase auf die Schallbewegung. Ann. Physik **134**, 177—193 = Ges. Abh. **1**, 540—556. (*241, 248*)

12. Maxwell, J. C.: On reciprocal diagrams in space, and their relation to **Airy's** function of stress. Proc. Lond. Math. Soc. (1) **2** (1865—1969), 58—60 = Papers **2**, 102—104. (*227*)

13. Riemann, B.: Über die Hypothesen, welche der Geometrie zu Grunde liegen (1854). Abh. Ges. Wiss. Göttingen **13**, (1866—1867) 133—150 = Werke, 2nd. ed. (1892), 272—287. (*34*)

14. Tait, P. G.: Sketch of Thermodynamics. London. (*240*)

15. Walter, A.: Anwendung der Methode **Hamiltons** auf die Grundgleichungen der mathematischen Theorie der Elasticität. Diss. Berlin. (We have not been able to see this work.) (*236*)

16. Warren, J.: Theorem with regard to the three axes of invariable direction in a strained elastic body. Quart. J. Math. **9**, 171—172. (*38*)

17. Weber, H.: Über eine Transformation der hydrodynamischen Gleichungen. J. reine angew. Math. **68**, 286—292. (*135*)

1869 *1.* Christoffel, E. B.: Über die Transformation der homogenen Differentialausdrücke 2ten Grades. J. reine angew. Math. **70**, 46—70 = Abh. **1**, 352—377 (*34*)

2. Christoffel, E. B.: Über ein die Transformation homogener Differentialausdrücke zweiten Grades betreffendes Theorem. J. reine angew. Math. **70**, 241—245 = Abh. **1**, 378—382 (*34*)

3. Lipschitz, R.: Untersuchungen in Betreff der ganzen homogenen Functionen von n Variablen. J. reine angew. Math. **70**, 71—102. (*34*)

4. Massieu, F.: Sur les fonctions caractéristiques des divers fluides. C. R. Acad. Sci., Paris **69**, 858—862. (*251*)

5. Massieu, F.: Addition au précédent mémoire sur les fonctions caractéristiques. C. R. Acad. Sci., Paris **69**, 1057—1061. (*251*)

6. St. Venant, A.-J.-C. B. de: Problème des mouvements que peuvent prendre les divers points d'une liquide, ou solide ductile, contenue dans un vase à parois verticales, pendant son écoulement par un orifice horizontal intérieur. C. R. Acad. Sci., Paris **68**, 221—237. (*89, 134*)

7. Thomson, W. (Lord Kelvin): On vortex motion. Trans. Roy. Soc. Edinb. **25**, 217—260 = Papers **4**, 13—66. (*79, 88, 105, 106, 108, 128, 129, 297*)

1870 *1.* Clausius, R.: Über einen auf die Wärme anwendbaren mechanischen Satz. Ann. Physik u. Chem. (5) **21** = **141**, 124—130. (*219*)

2. Lipschitz, R.: Fortgesetzte Untersuchung in Betreff der ganzen homogenen Funktionen von n Differentialen. J. reine angew. Math. **72**, 1—65. (*34*)

3. Lévy, M.: Mémoire sur les équations générales des mouvements intérieurs des corps solides ductiles au delà des limites où l'élasticité pourrait les ramener à leur premier état. C. R. Acad. Sci., Paris **70**, 1323—1325. (*300*)

4. Maxwell, J. C.: On reciprocal figures, frames, and diagrams of forces. Trans. Roy. Soc. Edinb. **26** (1869—1872), 1—40 = Papers **2**, 161—207. (*216, 224, 227*)

5. Maxwell, J. C.: On the displacement in a case of fluid motion. Proc. Lond. Math. Soc. **3** (1869—1871), 82—87 = Papers **2**, 208—214. (*71*)

6. Rankine, W. J. M.: On the thermodynamic theory of waves of finite longitudinal disturbance. Phil. Trans. Roy. Soc. Lond. **160**, 277—288 = Papers, 530—543. (*183, 189, 205, 241, 297*))

7. St. Venant, A.-J.-C. B. de: Mémoire sur l'établissement des équations différentielles des mouvements intérieurs opérés dans les corps solides ductiles au delà des limites où l'élasticité pourrait les ramener à leur premier état. C. R. Acad. Sci., Paris **70**, 473—480. (*300*)

8. Warren, J.: Note on a fundamental theorem in hydrodynamics. Quart. J. Math. **10**, 128–129. (*101*)

1871 *1.* Beltrami, E.: Sui principi fondamentali della idrodinamica. Mem. Acad. Sci. Bologna (3) **1**, 431–476; **2** (1872), 381–437; **3** (1873), 349–407; **5** (1874), 443–484 = Richerche sulla cinematica dei fluidi. Opere **2**, 202–379. (*78, 82A, 83, 84, 86, 101, 105, 129, 136*)

2. Boussinesq, J.: Étude nouvelle sur l'équilibre et le mouvement des corps solides élastiques dont certaines dimensions sont très petites par rapport à d'autres. Premier Mémoire. J. Math. pures appl. (2) **16**, 125–240. (*57*)

3. Cherbuliez, E.: Geschichtliche Mittheilungen aus dem Gebiete der mechanischen Wärmetheorie. Mittheil. Naturf. Ges. Bern **1870**, 291–324. (*240*)

4. Durrande, H.: Extrait d'une théorie du déplacement d'une figure qui se déforme. C. R. Acad. Sci., Paris **73**, 736–738. (*82A*)

5. Maxwell, J. C.: Theory of Heat. London. (*252*)

6. Stefan, J.: Über das Gleichgewicht und die Bewegung, insbesondere die Diffusion von Gasmengen. Sitzgsber. Akad. Wiss. Wien **63**², 63–124. (*158, 295*)

1872 *1.* Boussinesq, J.: Théorie des ondes liquides périodiques (1869). Mém. divers sav. **20**, 509–615. (*210*)

2. Rapport sur un mémoire de M. **Kleitz** intitulé: „Études sur les forces moléculaires dans les liquides en mouvement, et application à l'hydrodynamique" (St. Venant). C. R. Acad. Sci., Paris **74**, 426–438. (*82A, 208*)

3. Lipschitz, R.: Extrait d'une lettre. Bull. sci. math. astr. (1) **3**, 349–352. (*219*)

4. Villarceau, Y.: Sur un nouveau théorème de mécanique générale. C. R. Acad. Sci., Paris **75**, 232–240. (*219*)

1873 *1.* Bobylew, D.: Einige Betrachtungen über die Gleichungen der Hydrodynamik. Math. Ann. **6**, 72–84. (*211*)

2. Gibbs, J. W.: Graphical methods in the thermodynamics of fluids. Trans. Connecticut Acad. **2**, 309–342 = Works **1**, 1–32. (*245, 247, 248, 265*)

3. Gibbs, J. W.: A method of geometrical representation of the thermodynamic properties of substances by means of surfaces. Trans. Connecticut Acad. **2**, 382–404 = Works **1**, 33–54. (*245, 246, 247*)

4. Kleitz, C.: Études sur les forces moléculaires dans les liquides en mouvement et application à l'hydrodynamique. Paris. (*Cf.* [1872, *2*].) (*82A, 110, 204, 208, 293*)

5. Maxwell, J. C.: A Treatise on Electricity and Magnetism. 2 vols., Oxford. *Cf.* [1881, *4*]. (*175, 187, 200, 276, 311*)

6. Strutt, J. W. (Rayleigh): Some general theorems relating to vibrations. Proc. Lond. Math. Soc. **4** (1871–1873), 357–368 = Papers **1**, 170–181. (*241A*)

1874 *1.* Boltzmann, L.: Zur Theorie der elastischen Nachwirkung. Sitzungsber. Akad. Wiss. Wien **70**², 275–306 = Wiss. Abh. **1**, 616–639. (*305*)

2. Maxwell, J. C.: **Van der Waals** on the continuity of the gaseous and liquid states. Nature **10**, 477–480 = Papers **2**, 407–415. (*219*)

3. Meyer, O.-E.: Zur Theorie der inneren Reibung. J. reine angew. Math. **78**, 130 to 135. (*305*)

4. Meyer, O.-E.: Theorie der elastischen Nachwirkung. Ann. Physik (6) **1**, 108–119. (*305*)

5. Nanson, E. J.: Note on hydrodynamics. Mess. math. **3**, 120–121. (*75, 108*)

6. Umov, N.: Ableitung der Bewegungsgleichungen der Energie in continuirlichen Körpern. Z. Math. Phys. **19**, 418–431. (*217*)

1875 *1.* Gibbs, J. W.: On the equilibrium of heterogeneous substances. Trans. Connecticut Acad. **3** (1875–1878), 108–248, 343–524 = Works **1**, 55–353. (*14, 245, 246, 247, 251, 260, 262, 264, 265*)

2. Greenhill, A. G.: Hydromechanics. Encycl. Britt., 9th ed. (carried through 14th ed., 1926). (*99, 207*)

3. Lipschitz, R.: Determinazione della pressione nell' intorno d'un fluido incompressibile soggetto ad attrazioni interne ed esterne. Ann. mat. (2) **6** (1873–1875), 226–231. (*102, 141*)

4. Maxwell, J. C.: 4th. ed. of [1871, *5*]. (*245, 246*)

5. Meyer, O.-E.: Zusatz zu der Abhandlung zur Theorie der inneren Reibung. J. reine angew. Math. **80**, 315–316. (*305*)

1876 *1.* Cotterill, J.: On the distribution of energy in a mass of liquid in a state of steady motion. Phil. Mag. **1**, 108–111. (*125*)

2. Kirchhoff, G.: Vorlesungen über mathematische Physik: Mechanik. Leipzig. 2nd ed., 1877; 3rd ed., 1883. (*57, 63A, 113, 129, 196, 200, 205, 214, 232, 236, 301*)

3. MASSIEU, F.: Mémoire sur les fonctions caractéristiques de divers fluides et sur la théorie des vapeurs. Mém. divers savants **22**, No. 2, 92 pp. (*251*)

4. MAXWELL, J. C.: On the equilibrium of heterogeneous substances. Repts. South Kensington conf. spec. loan coll. sci. app. 144–150 = (with a note added by LARMOR) Phil. Mag. (6) **16**, 818–824 (1908). Condensed version, Proc. Cambridge Phil. Soc. **2**, 427–430 (1876) = Papers **2**, 498–500. (*260*)

5. RIEMANN, B.: Commentatio mathematica, qua respondere tentatur quaestioni ab Illma Academia Parisiensi propositae. Werke, 370–399 = Werke, 2nd ed. (1892), 391–404. (*34*)

6. TAIT, P. G.: Recent Advances in Physical Science. London. 1st and 2nd eds. (*240*)

7. ZHUKOVSKI, N. E.: Kinematics of fluid bodies [in Russian]. Mat. Sborn. **8**, 1–79, 163–238 = repr. separ., Moscow 1876 = Sobr. Soch. **2**, 7–144. (*8, 12, 70, 102, 108*)

1877 *1.* BOUSSINESQ, J.: Sur la construction géométrique des pressions que supportent les divers éléments plans se croisant en un même point d'un corps, et sur celle des déformations qui se produisent autour d'un tel point. J. Math. pures appl. (3) **3**, 147–152. (*23, 35, 208*)

2. CHRISTOFFEL, E. B.: Untersuchungen über die mit dem Fortbestehen linearer partieller Differentialgleichungen verträglichen Unstetigkeiten. Ann. Mat. (2) **8**, 81–113 = Abh. **2**, 51–80. (*173, 177, 180, 183, 189, 205, 297*)

3. CHRISTOFFEL, E. B.: Über die Fortpflanzung von Stößen durch elastische feste Körper. Ann. Mat. (2) **8**, 193–244 = Abh. **1**, 81–126. (*301*)

4. LAMB, H.: Note on a theorem in hydrodynamics. Mess. math. **7** (1877–1878), 41–42. (*76*)

5. THOMSON, W. (Lord KELVIN): Elasticity, Encycl. Britt., 9th ed. = (with revisions) Papers **3**, 1–112. The first part is a revision of [1865, *2*]; the second part, a revision of [1856, *2*]. (*33A*)

1878 *1.* Reprint of [1824, *1*], with the addition of MS notes of CARNOT (1824–1832) Paris. Transl., R. H. THURSTON, Reflections on the Motive Power of Heat. New York 1890; new ed., New York 1943. Transl., W. F. MAGIE, pp. 2–60 of The Second Law of Thermodynamics. New York and London 1899. The most important portions of the MS notes = [1929, *8*, pp. 33–34, 44]. (*240*)

2. CLIFFORD, W. K.: Elements of Dynamic, an Introduction to the Study of Motion and Rest in Solid and Fluid Bodies, I. Kinematic. London. (*86A*)

3. GILBERT, P.: Sur le problème de la composition des accelérations d'ordre quelconque. C. R. Acad. Sci., Paris **86**, 1390–1391. (*143*)

4. LAISANT, C.-A.: Note sur un théorème sur les mouvements relatifs. C. R. Acad. Sci., Paris **87**, 204–206. Cf. remarks on pp. 259 and 377. (*143*)

5. LAMB, H.: On the conditions for steady motion of a fluid. Proc. Lond. Math. Soc. **9**, (1877–1878), 91–92. (*122, 125, 133*)

6. LÉVY, M.: Sur la composition des accélérations d'ordre quelconque et sur un problème plus générale que celui de la composition des mouvements. C. R. Acad. Sci., Paris **86**, 1068–1071. (*143*)

7. MÜLLER, J.: Einleitung in die Hydrodynamik. Vjschr. naturf. Ges. Zürich **23**, 129–159, 242–265. (*107*)

8. NANSON, E. J.: Note on hydrodynamics. Mess. Math. **7** (for 1877–1878), 182–185. (*20*)

1879 *1.* FORSYTH, A. R.: On the motion of a viscous incompressible fluid. Mess. Math. **9** (1879–1880), 134–139. (*121*)

2. LAMB, H.: A Treatise on the Mathematical Theory of the Motion of Fluids. Cambridge. (See [1895, *2*], [1906, *4*] [1916, *3*], [1924, *9*], [1932, *8*].) (*117, 122, 125*)

3. OBERBECK, A.: Über die Wärmeleitung der Flüssigkeiten bei Berücksichtigung der Strömungen infolge von Temperaturdifferenzen. Ann. Physik (2) **7**, 271–292. (*241*)

4. SHILLER, N.: Certain applications of the mechanical theory of heat to the changes of state of elastic bodies [in Russian]. Zh. Rus. Fiz.-Chim. Obsch. **11**, Fiz. Otd. 55–77. (*246*)

1880 *1.* BOUSSINESQ, J.: Sur la manière dont les frottements entrent en jeu dans un fluide qui sort de l'état de repos, et sur leur effet pour empêcher l'existence d'une fonction des vitesses. C. R. Acad. Sci., Paris **90**, 736–739. (*134*)

2. BOUSSINESQ, J.: Quelques considérations à l'appui d'une note du 29 mars, sur l'impossibilité d'admettre, en général, une fonction des vitesses dans toute question d'hydraulique où les frottements ont un rôle notable. C. R. Acad. Sci., Paris **9**, 967–969. (*134*)

3. BRESSE, J.: Fonction des vitesses; extension des théorèmes de **Lagrange** au cas d'un fluide imparfait. C. R. Acad. Sci., Paris **90**, 501—504. (*134*)
4. BRESSE, J.: Réponse à une note de M. J. **Boussinesq**. C. R. Acad. Sci., Paris **90**, 857—858. (*134*)
5. CRAIG, T.: Motion of viscous fluids. J. Franklin Inst. **110**, 217—227. (*121, 125*)
6. CRAIG, T.: On steady motion in an incompressible viscous fluid. Phil. Mag. (5) **10**, 342—357. (*125*)
7. CRAIG, T.: On certain possible cases of steady motion in a viscous fluid. Amer. J. Math. **3**, 269—293. (*121*)
8. FITZGERALD, G. F.: On the electromagnetic theory of the reflection and refraction of light. Phil. Trans. Roy. Soc. Lond. **171**, 691—711. (*302*)
9. ROWLAND, H. A.: On the motion of a perfect incompressible fluid when no solid bodies are present. Amer. J. Math. **3**, 226—268. (*121*)
10. ST. VENANT, A.-J.-C. B. DE: Géométrie cinématique. — Sur celle des déformations des corps soit élastiques, soit plastiques, soit fluides. C. R. Acad. Sci., Paris **90**, 53—56, 209. (*8, 27*)

1881 1. BELTRAMI, E.: Sulle equazioni generali dell' elasticità. Ann. Mat. (2) **10** (1880—1882), 188—211 = Opere **3**, 383—407. (*232, 238*)
2. CRAIG, T.: Methods and Results, General Properties of the Equations of Steady Motion. U.S. Treasury Dept. (Coast and Geod. Survey) Document 71, Washington. (*122, 136*)
3. HILL, M. J. M.: Some properties of the equations of hydrodynamics. Quart. J. Math. **17**, 1—20, 168—174. (*66, 136 136A, 137*)
4. MAXWELL, J. C.: 2nd ed. of [1873, *5*]. (*175, 187, 276*)

1882 1. HELMHOLTZ, H.: Zur Geschichte der Entdeckung des Gesetzes von der Erhaltung der Kraft. Abh. **1**, 71—74. (*240*)
2. HELMHOLTZ, H.: Die Thermodynamik chemischer Vorgänge. Sitzgsber. Akad. Wiss. Berlin **1882**, 22—39 = Abh. **2**, 958—978. (*246, 249, 251, 263, 265*)
3. MOHR, O.: Über die Darstellung des Spannungszustandes und des Deformationszustandes eines Körperelementes und über die Anwendung derselben in der Festigkeitslehre. Civilingenieur **28**, 112—155. (*24*)
4. VOIGT, W.: Allgemeine Formeln für die Bestimmung der Elasticitätsconstanten von Krystallen durch die Beobachtung der Biegung und Drillung von Prismen. Ann. Physik (2) **16**, 273—321, 398—416. (*224*)

1884 1. GILBERT, P.: Abstract. Ann. Soc. Sci. Bruxelles **8** (1883—1884), A 53—A 56. (*143*)

1885 1. BOBYLEW, D.: Über die relative Bewegung eines Punktes in einem in continuirlicher Deformation begriffenen Medium. Z. Math. Phys. **30**, 336—344. (*144*)
2. HILL, M. J. M.: The differential equations of cylindrical annular vortices. Proc. Lond. Math. Soc. **16** (1884—1885), 171—183. (*161, 162*)
3. HUGONIOT, H.: Sur la propagation du mouvement dans un fluide indéfini (Première Partie). C. R. Acad. Sci., Paris **101**, 1118—1120. (*177, 180, 181, 183, 190, 297*)
4. HUGONIOT, H.: Sur la propagation du mouvement dans un fluide indéfini (Deuxième Partie). C. R. Acad. Sci., Paris **101**, 1229—1232. (*177, 181, 182, 183, 190, 246, 297*)
5. LAMB, H.: Proof of a hydrodynamical theorem. Mess. Math. **14**, 87—92. (*138*)
6. MORERA, G.: Sulle equazioni generali per l'equilibrio dei sistemi continui a tre dimensioni. Atti Accad. Sci. Torino **20** (1884—1885), 43—53. (*205A*) *209, 232*)
7. SOMOFF, P.: Über einen Satz von **Burmester**. Z. Math. Phys. **30**, 248—250. (*74*)
8. THOMSON, W. (Lord KELVIN): On the motion of a liquid within an ellipsoidal hollow. Proc. Roy. Soc. Edinb. **13**, 114, 370—378 = Papers **4**, 193—201. (*70*)

1886 1. BELTRAMI, E.: Sull' interpretazione meccanica delle formole di **Maxwell**. Mem. Accad. Sci. Bologna (4) **7**, 1—38 = Opere **4**, 190—223. (*57*)
2. DUHEM, P.: Le Potentiel Thermodynamique et ses Applications à la Mécanique Chimique et à l'étude des phénomènes électriques. Paris: Hermann. (*246, 252, 260*)
3. HUGONIOT, H.: Sur un théorème général relatif à la propagation du mouvement. C.R. Acad. Sci., Paris **102**, 858—860. (*194A, 301*)
4. TODHUNTER, I., and K. PEARSON: A History of the Theory of Elasticity and of the Strength of Materials from **Galilei** to the Present Time, **1**. Cambridge. (*21, 33A, 45, 204, 205A, 232*)

1887 1. HUGONIOT, H.: Mémoire sur la propagation du mouvement dans les corps et spécialement dans les gaz parfaits. J. École Polytech. **57**, 3—97; **58**, 1—125 (1889). The date of the memoir is 26 October 1885. (*180, 205, 241, 297*)
2. HUGONIOT, H.: Mémoire sur la propagation du mouvement dans un fluide indéfini. J. Math. pures appl. (4) **3**, 477—492; (4) **4**, 153—167. (*183, 190, 194A, 297*)

3. Planck, M.: Über das Princip der Vermehrung der Entropie. Dritte Abhandlung. Ann. Physik (2) **32**, 162–503. (*264*)

4. Reiff, R.: Zur Kinematik der Potentialbewegung. Math. naturw. Mitt. (Tübingen) $\mathbf{1}_3$ (1884–1886), 41–48. (*88*)

5. Voigt, W.: Theoretische Studien über die Elasticitätsverhältnisse der Krystalle. Abh. Ges. Wiss. Göttingen **34**, 100 pp. (*200, 217*)

1888 *1.* Basset, A. B.: A Treatise on Hydrodynamics, 2 vols. Cambridge. (*99, 125, 136, 136A, 137, 162*)

2. Brillouin, M.: Chaleur specifique pour une transformation quelconque et thermodynamique. J. phys. théor. appl. (2) **7**, 148–152. (*250*)

3. Brillouin, M.: Déformations permanentes et thermodynamique. J. phys. théor. appl. (2) **7**, 327–347; **8**, 169–179 (1889). Much of the first part appears in C. R. Acad. Sci., Paris **106**, 416–418, 482–485, 537–540, 589–592. (*250*)

4. Donati, L.: Sul lavoro di deformazione dei sistemi elastici. Mem. Accad. Sci. Bologna (4) **9**, 345–367. (*205*)

5. Gilbert, P.: Sur les composantes des accélérations d'ordre quelconque suivant trois directions rectangulaires variables. J. Math. pures appl. (4) **4**, 465–473. (*143*)

6. Love, A. E. H.: The small free vibrations and deformation of a thin elastic shell. Phil. Trans. Roy. Soc. Lond. A **179**, 491–546. (*212*)

7. Ricci, G.: Delle derivazioni covarianti e controvarianti e del loro uso nella analisi applicata. Studi Univ. Padova comm. ottavo centenar. Univ. Bologna **3**, No. 12, 23 pp. (*84*)

8. Riecke, E.: Beiträge zur Hydrodynamik. Nachr. Ges. Wiss. Göttingen, 347–357 = Ann. Physik (2) **36**, 322–334 (1889). (*71*)

1889 *1.* Beltrami, E.: Considerazione idrodinamiche. Rend. Ist. Lombardo (2) **22**, 121—130 = Opere **4**, 300–309. (*99*)

2. Beltrami, E.: Sur la théorie de la déformation infiniment petite d'un milieu. C. R. Acad. Sci., Paris **108**, 502–504 = Opere **4**, 344–347. (*57*)

3. Gilbert, P.: Recherches sur les accélérations en général. Ann. Soc. Sci. Bruxelles **13**, 261–315. (*143*)

4. Leonardo da Vinci: Les ... manuscrits F and I de la bibliothèque de l'institut publiés en fac-similés phototypiques ..., ed. Ravaisson-Mollien. Paris. (*70*)

5. Lippmann, M.: Cours de Thermodynamique. Paris. (*240*)

6. Morera, G.: Sui moti elicoidali dei fluidi. Rend. Lincei (4) **5**, 511–617. (*99, 112*)

7. Padova, E.: Sulle deformazioni infinitesimi. Rend. Lincei (4) $\mathbf{5}^1$, 174–178. (*34, 57*)

8. Poincaré, H.: Leçons sur la Théorie Mathématique de la Lumière, redigées par J. Blondin. Paris. (*233*)

9. Voigt, W.: Ueber adiabatische Elasticitätsconstanten. Ann. Physik **36**, 743–759. (*256A, 296*)

10. Voigt, W.: Über die innere Reibung der festen Körper, insbesondere der Krystalle. Abh. Ges. Wiss. Göttingen **36**, No. 1. (*305*)

1890 *1.* Basset, A. B.: On the extension and flexure of cylindrical and spherical thin elastic shells. Phil. Trans. Roy. Soc. Lond. A **181**, 433–480. (*213*)

2. Gilbert, Ph.: Sur quelques formules d'un usage général dans la physique mathématique. Ann. Soc. Sci. Bruxelles **14**, 1–18. (*99*)

3. Gosiewski, W.: O ciśnieniu kinetycznem w płynie nieściśliwym i jednorodnym. Pamiętnik Akad. Krakow, mat.-przyr. **17**, 128–134. (*121*)

4. Gosiewski, W.: O naturze ruchu wewnątrz elementu płynnego. Pamiętnik Akad. Krakow, mat.-przyr. **17**, 135–142. (*86, 88*)

5. Isé, E.: Sulla deformazione elastica di un corpo isotropo. Atti Accad. Pontaniana **20**, 241–260. (*42*)

6. Lamb, H.: On the deformation of an elastic shell. Proc. Lond. Math. Soc. **21**, 119–146. (*212, 213*)

7. Lévy, M.: L'hydrodynamique moderne et l'hypothèse des actions à distance. Rév. Gén. sci. pures appl. **1**, 721–728. (*86, 107, 108, 109, 110*)

8. Padova, E.: Il potenziale delle forze elastiche di mezzi isotropi. Atti Ist. Veneto sci. lett. arte **48** = (7) **1**, 445–451. (*61*)

1891 *1.* Brillouin, M.: Déformations homogènes finies. Énergie d'un corps isotrope. C. R. Acad. Sci., Paris **112**, 1500–1502. (*36*)

2. Duhem, P.: Sur les équations générales de la thermodynamique (1888). Ann. École Norm. (3) **8**, 231–266. (*246*)

3. Duhem, P.: Hydrodynamique, Élasticité, Acoustique, Paris: Hermann, 2 vols. (*194A, 241*)

4. NATANSON, L.: Thermodynamische Bemerkungen. Ann. Phys. **278** = (2) **42**, 178–185. *(251, 252)*

5. SAMPSON, R. A.: On Stokes' current function. Phil. Trans. Roy. Soc. Lond. A **182**, 449–518. *(162)*

1892 1. BASSET, A. B.: On the theory of elastic wires. Proc. Lond. Math. Soc. **23** (1891–1892), 105–127. *(214)*

2. BELTRAMI, E.: Osservazioni sulla nota precendente. Rend. Lincei (5) $\mathbf{1}_1$, 141–142 = Opere **4**, 510–512. *(227)*

3. DUHEM, P.: Commentaire aux principes de la thermodynamique, Première partie. J. math. pures appl. (4) **8**, 269–330. *(240)*

4. FINGER, J.: Über die gegenseitigen Beziehungen gewisser in der Mechanik mit Vortheil anwendbaren Flächen zweiter Ordnung nebst Anwendung auf Probleme der Astatik. Sitzgsber. Akad. Wiss. Wien (IIa) **101**, 1105–1142. *(14, 21, 22, 29, 30, 37)*

5. HELMHOLTZ, H.: Das Princip der kleinsten Wirkung in der Elektrodynamik. Sitzgsber. Akad. Wiss. Berlin **1892**, 459–475 = Ann. Phys. (2) **47**, 1–26 = Abh. **3**, 476–504. *(150)*

6. KILLING, W.: Über die Grundlagen der Geometrie. J. reine angew. Math. **109**, 121–186. *(84)*

7. LARMOR, J.: The equations of propagation of disturbances in gyrostatically loaded media, and of the circular polarization of light. Proc. Lond. Math. Soc. **23** (1891–1892), 127–135 = Papers **2**, 248–255. *(200)*

8. LOVE, A. E. H.: A Treatise on the Mathematical Theory of Elasticity, vol. 1. Cambridge. (See [1906, *5*] [1927, *6*].) *(31, 35, 36, 39, 44, 45, 46, 205)*

9. MORERA, G.: Soluzione generale delle equazioni indefinite dell' equilibrio di un corpo continuo. Rend. Lincei (5) $\mathbf{1}_1$, 137–141. *(227)*

10. MORERA, G.: Appendice alla Nota: Sulla soluzione più generale delle equazioni indefinite dell' equilibrio di un corpo continuo. Rend. Lincei (5) $\mathbf{1}_1$, 233–234. *(227)*

11. POINCARÉ, H.: Leçons sur la Théorie de l'Élasticité. Paris. *(55, 205, 210, 233)*

12. VOIGT, W.: Ueber innere Reibung fester Körper, insbesondere der Metalle. Ann. Phys. (2) **47**, 671–693. *(305)*

13. VOIGT, W.: Bestimmung der Constanten der Elasticität und Untersuchung der inneren Reibung für einige Metalle. Gött. Abh. **38**, No. 2. *(305)*

14. WIEN, W.: Über den Begriff der Localisirung der Energie. Ann. Phys. (2) **47**, 1–26 = Abh. **3**, 476–504. *(157, 217)*

1893 1. DUHEM, P.: Le potentiel thermodynamique et la pression hydrostatique. Ann. École Norm. (3) **10**, 187–230. *(60, 215, 241, 245, 247)*

2. DUHEM, P.: Dissolutions et mélanges. Trav. Mém. Fac. Lille **3**, Nos. 2 and 12. *(215, 255, 265, 295)*

3. DUHEM, P.: Commentaire aux principes de la thermodynamique, Deuxième partie. J. Math. pures appl. (4) **9**, 293–359. *(245)*

4. HEAVISIDE, O.: Electromagnetic Theory, vol. 1. London. *(302)*

5. LOVE, A. E. H.: A Treatise on the Mathematical Theory of Elasticity, vol. 2. Cambridge. (See [1906, *5*] [1927, *6*].) *(63A, 212, 213)*

6. POINCARÉ, H.: Leçons sur la Théorie des Tourbillons. Paris. *(106, 107, 109, 117, 122, 125, 134, 145)*

7. TOUCHE, P.: Transformation des équations générales du mouvement des fluides. Bull. Soc. Math. France **21**, 72–75. *(99)*

1894 1. CURIE, P.: Sur la symétrie dans les phénomènes physiques, symétrie d'un champ électrique et d'un champ magnétique. J. Physique (3) **3**, 393–415 = Oeuvres 118–141. *(258)*

2. DUHEM, P.: Commentaire aux principes de la thermodynamiques, Troisième partie. J. Math. pures appl. (4) **10**, 207–285. *(246, 249, 265)*

3. FABRI, C.: I moti vorticosi di ordine superiore al primo in relazione alle equazioni pel movimento dei fluidi viscosi. Mem. Accad. Sci. Bologna (5) **4**, 383–392. *(123)*

4. FINGER, J.: Über die allgemeinsten Beziehungen zwischen Deformationen und den zugehörigen Spannungen in aeolotropen und isotropen Substanzen. Sitzgsber. Akad. Wiss. Wien (IIa) **103**, 1073–1100. *(29, 30, 303)*

5. KIRCHHOFF, G.: Vorlesungen über die Theorie der Wärme. Leipzig: Teubner. x + 210 pp. *(241)*

6. METZLER, W. H.: Homogeneous strains. Ann. of Math. **8**, 148–156. *(42)*

7. NEUMANN, C.: Über die Bewegung der Wärme in compressiblen oder auch incompressiblen Flüssigkeiten. Ber. Verh. Ges. Wiss. Leipzig **46**, 1–24. *(241)*

8. SHILLER, N.: On the problem of the thermodynamic potential [in Russian]. Isv. imp. Obschestva Moskov. Univ. **7**, No. 1, 22–31. *(246)*

1895 *1.* BASSETT, A. B.: On the deformation of thin elastic wires. Amer. J. Math. **17**, 281–317. (*214*)
2. LAMB, H.: Hydrodynamics. Cambridge. (2nd ed. of [1879, *2*], see [1906, *4*], [1916, *3*], [1924, *9*], [1932, *8*].) (*94, 106, 108*)
3. OUMOFF, N.: Une expression générale du potentiel thermodynamique. Bull. Soc. Impér. Naturalistes de Moscou (2) **8** (1894), 138–145. (*246*)
4. PAINLEVÉ, P.: Leçons sur l'Intégration des Équations Différentielles de la mécanique et applications. 291 pp. Paris: Hermann. (Lithographed.) (*167, 170*)
5. SCHÜTZ, J. R.: Über die Herstellung von Wirbelbewegungen in idealen Flüssigkeiten durch conservative Kräfte. Ann. Physik (N. F.) **56**, 144–147. (*108*)
6. TOUCHE, P.: Équation d'une trajectoire fluide. Bull. Soc. Math. France **23**, 111–113. (*108*)
7. VOIGT, W.: Über Medien ohne innere Kräfte und eine durch sie gelieferte mechanische Deutung der Maxwell-Hertz'schen Gleichungen. Gött. Abh. **1894**, 72–79. (*200*)
8. VOIGT, W.: Kompendium der theoretischen Physik **1**. Leipzig. (*256A, 296*)
1896 *1.* COSSERAT, E. et F.: Sur la théorie de l'élasticité. Ann. Toulouse **10**, 1–116. (*12, 14, 26, 31, 46, 57, 59, 95, 210, 217*)
2. DUHEM, P.: Sur les déformations permanentes et l'hystérésis, Parts I–V. Mém. Couronn. Mém. Sav. Étrangers Acad. Belg. **54, 56** (1898); **62** (1902). (*250*)
3. MACH, E.: Die Prinzipien der Wärmelehre historisch-kritisch entwickelt. Leipzig: Johann Ambrosius Barth. (*240, 245, 253*)
1897 *1.* APPELL, P.: Sur les équations de l'hydrodynamique et la théorie des tourbillons. J. Math. pures appl. (5) **3**, 5–16. (*108, 129, 135*)
2. FERRARIS, G.: Teoria geometrica dei campi vettoriali come introduzione allo studio della elettricità, del magnetismo, ecc. Mem. Accad. Sci. Torino (2) **47**, 259–338. (*175*)
3. FINGER, J.: Über das innere Virial eines elastischen Körpers. Sitzgsber. Akad. Wiss. Wien (IIa) **106**, 722–738. (*219*)
4. FÖPPL, A.: Die Geometrie der Wirbelfelder. Leipzig. (*93, 118, 129*)
5. MEHMKE, R.: Zum Gesetz der elastischen Dehnungen. Z. Math. Phys. **42**, 327–338. (*25, 33A*)
6. MILLER, W. L.: On the second differential coefficients of **Gibbs'** function ζ. The vapour tensions, freezing and boiling points of ternary mixtures. J. Phys. Chem. **1**, 633–642. (*252*)
7. NEUMANN, C.: Die Anwendung des Hamilton'schen Princips in der Hydrodynamik und Aerodynamik. Ber. Verh. Ges. Wiss. Leipzig **94**, 611–615. (*150*)
8. TOUCHE, P.: Équations d'une trajectoire fluide dans le cas général. Bull. Soc. Math. France **27**, 5–8. (*108*)
9. TREVOR, J. E.: Inner thermodynamic equilibria. J. Phys. Chem. **1**, 205–220. (*251*)
1898 *1.* BJERKNES, V.: Über die Bildung von Circulationsbewegungen und Wirbeln in reibungslosen Flüssigkeiten. Vidensk. Skrift. No. 5. (*170*)
2. DUHEM, P.: On the general problem of chemical statics. J. Phys. Chem. **2**, 1–42, 91–115. (*265*)
1899 *1.* APPELL, P.: Lignes correspondentes dans la déformation d'un milieu; extension des théorèmes sur les tourbillons. J. Math. pures appl. (5) **5**, 137–153. Abstract in Bull. Soc. Math. France **26**, 135–136 (1898). (*132*)
2. BROCA, A.: Sur le principe de l'égalité de l'action et de la réaction. C. R. Acad. Sci., Paris **129**, 1016–1019. (*175*)
3. VOLTERRA, V.: Sul flusso di energia meccanica. Atti Accad. Sci. Torino **34** (1898—1899), 366–375. (*217*)
4. VOLTERRA, V.: Sul flusso di energia meccanica. Nuovo Cim. (4) **10**, 337–359. (*209*)
1900 *1.* BRILLOUIN, M.: Théorie moléculaire des gaz. Diffusion du mouvement et de l'énergie. Ann. chimie (7) **20**, 440–485. (*293*)
2. BROCA, A.: Sur les masses vectorielles de discontinuité. C. R. Acad. Sci., Paris **130**, 317–319. (*175*)
3. DUHEM, P.: Sur le théorème d'**Hugoniot** et quelques théorèmes analogues. C. R. Acad. Sci., Paris **131**, 1171–1173. (*180, 181, 187*)
4. HERTZ, H.: Electric Waves. London. (*276*)
5. HILBERT, D.: Mathematische Probleme. Nachr. Ges. Wiss. Göttingen 1900, 253–297. Repr. with addns., Arch. Math. Phys. **1**, 44–63, 213–217 (1901). Transl. with addns., Sur les problèmes futurs des mathématiques. C. R. 2me Congr. Int.

Math. Paris (1900), 58—114 (1902). Transl., Mathematical problems. Bull. Amer. Math. Soc. (2) **8**, 437—479 (1902). *(3)*

6. MICHELL, J. H.: On the direct determination of stress in an elastic solid, with applications to the theory of plates. Proc. Lond. Math. Soc. **31** (1899), 100—124. *(224)*

7. MICHELL, J. H.: The uniform torsion and flexure of incomplete tores, with application to helical springs. Proc. Lond. Math. Soc. **31**, (1899) 130—146. *(228)*

8. MOHR, O.: Welche Umstände bedingen die Elastizitätsgrenze und den Bruch eines Materials? Z. VDI **44**, 1524—1530. *(24)*

9. STRUTT, J. W. (Lord RAYLEIGH): On a theorem analogous to the virial theorem. Phil. Mag. (5) **50**, 210—213 = Papers **4**, 491—493. *(216)*

10. VIEILLE, P.: Étude sur le rôle des discontinuités dans les phénomènes de propagation. Mém. poudres salpêtres **10** (1899—1900), 177—260. *(297)*

11. ZORAWSKI, K.: Über die Erhaltung der Wirbelbewegung. C. R. Acad. Sci. Cracovie, 335—341. *(75, 80, 107, 153)*

12. ZORAWSKI, K.: Über gewisse Änderungsgeschwindigkeiten von Linienelementen bei der Bewegung eines continuirlichen materiellen Systems. C. R. Acad. Sci. Cracovie, 367—374. *(78, 153)*

1901 *1.* ABRAHAM, M.: Geometrische Grundbegriffe. Enz. math. Wiss. **4**[3], 3—47. *(163)*

2. APPELL, P.: Déformation spéciale d'un milieu continu; tourbillons de divers ordres. Bull. Soc. Math. France **29**, 16—17. *(20)*

3. DARBOUX, G.: Sur les déformations finies et sur les systèmes triples de surfaces orthogonales. Proc. Lond. Math. Soc. **32** (1900), 377—383. *(38)*

4. DUHEM, P.: Sur les équations de l'hydrodynamique. Ann. Toulouse (2) **3**, 253 to 279. *(99, 136)*

5. DUHEM, P.: Sur la condition supplémentaire en hydrodynamique. C. R. Acad. Sci., Paris **132**, 117—120. *(241)*

6. DUHEM, P.: Sur les théorèmes d'**Hugoniot**, les lemmes de M. **Hadamard** et la propagation des ondes dans les fluides visqueux. C. R. Acad. Sci., Paris **132**, 1163—1167. *(181, 298)*

7. DUHEM, P.: Recherches sur l'hydrodynamique. Ann. Toulouse (2) **3**, 315—377, 379—431; **4** (1902), 101—169; **5** (1903), 5—61, 197—255, 353—404 = repr. separ. Paris, 2 vols., 1903, 1904. *(113, 134, 181, 241, 246, 256, 258, 262, 297, 298)*

8. HADAMARD, J.: Sur la propagation des ondes. Bull. Soc. Math. France **29**, 50—60. *(134, 173, 176, 177, 181, 183, 187, 190)*

9. JOUGUET, E.: Sur la propagation des discontinuités dans les fluides. C. R. Acad. Sci., Paris **132**, 673—676. *(189, 205, 241, 258)*

10. MUIRHEAD, R. F.: Stress — its definition. Nature, Lond. **64**, 207. *(200)*

11. NATANSON, L.: Sur les lois de la viscosité. Bull. Int. Acad. Sci. Cracovie 95—111. *(305)*

12. NATANSON, L.: Sur la double réfraction accidentelle dans les liquides. Bull. Int. Acad. Sci. Cracovie 161—171. *(305)*

13. NATANSON, L.: Über die Gesetze der inneren Reibung. Z. physik. Chem. **38**, 690—704. *(305)*

14. RICCI, G., et T. LEVI-CIVITA: Méthodes de calcul différentiel absolu et leurs applications. Math. Ann. **54**, 125—137. *(205A)*

15. WEINGARTEN, G.: Sulle superficie di discontinuità nella teoria della elasticità dei corpi solidi. Rend. Accad. Lincei (5) **10**[1], 57—60. *(49, 175)*

16. WEINGARTEN, J.: Über die geometrischen Bedingungen, denen die Unstetigkeiten der Derivierten eines Systems dreier stetigen Funktionen des Ortes unterworfen sind, und ihre Bedeutung in der Theorie der Wirbelbewegung. Arch. Math. Phys. (3) **1**, 27—33. *(106, 175)*

17. ZORAWSKI, K.: Über gewisse Änderungsgeschwindigkeiten von Linienelementen bei der Bewegung eines continuierlichen materiellen Systems, 2. Mitth. Bull. Int. Acad. Sci. Cracovie 486—499. *(86)*

1902 *1.* BJERKNES, V.: Cirkulation relativ zu der Erde. Meteor. Z. **37**, 97—108. *(145)*

2. COMBEBIAC, G.: Sur les équations générales de l'élasticité. Bull. Soc. Math. France **30**, 108—110, 242—247. *(200, 217)*

3. COULON, J.: Sur l'Intégration des Équations aux Dérivées Partielles du Second Ordre per la Méthode des Caractéristiques. Paris: Hermann. 122 pp. *(175, 176, 180, 181)*

4. HELMHOLTZ, H.: Vorlesungen über theoretische Physik, **2**, Dynamik continuirlich verbreiteter Massen. Leipzig (lectures of 1894). *(4, 6)*

5. THOMSON, W. (Lord KELVIN): A new specifying method for stress and strain in an elastic solid. Proc. Roy. Soc. Edinb. **24** (1901—1903), 97—101 = Phil. Mag.

(6) **3**, 444–448 = Math. Phys. Papers **4**, 556–560. Prelim. note, Phil. Mag. (6) **3**, 95–97. (*33A*)

6. Natanson, L.: Sur la propagation d'un petit mouvement dans un fluide visqueux. Bull. Int. Acad. Sci. Cracovie 19–35. (*305*)

7. Natanson, L.: Sur la fonction dissipative d'un fluide visqueux. Bull. Int. Acad. Sci. Cracovie, 488–494. (*305*)

8. Natanson, L.: Sur la déformation d'un disque plastico-visqueux. Bull. Int. Acad. Sci. Cracovie, 494–512. (*305*)

9. Natanson, L.: Sur la conductibilité calorifique d'un gaz en mouvement. Bull. Int. Acad. Sci. Cracovie 137–146. (*305*)

10. Zermelo, E.: Hydrodynamische Untersuchungen über die Wirbelbewegungen in einer Kugelfläche. Z. Math. Phys. **47**, 201–237. (*162*)

1903 *1.* Appell, P.: Sur quelques fonctions de point dans le mouvement d'un fluide. J. Math. pures appl. (5) **9**, 5–19. An abstract is given in C. R. Acad. Sci., Paris **136**, 186–187 (1903). (*101, 102*)

2. Appell, P.: Sur quelques fonctions et vecteurs de point contenant uniquement les dérivées premières des composantes de la vitesse. Bull. Soc. Math. France **31**, 68–73. (*90, 101*)

3. Duhem, P.: Sur quelques formules de cinématique utiles dans la théorie générale de l'élasticité. C. R. Acad. Sci., Paris **136**, 139–141. (*82A*)

4. Duhem, P.: Sur la viscosité en un milieu vitreux. C. R. Acad. Sci., Paris **136**, 281–283. (*262, 305*)

5. Duhem, P.: Sur les équations du mouvement et la relation supplémentaire au sein d'un milieu vitreux. C. R. Acad. Sci., Paris **136**, 343–345. (*241, 262*)

6. Duhem, P.: Sur le mouvement des milieux vitreux, affectés de viscosité, et très peu déformés. C. R. Acad. Sci., Paris **136**, 592–595. (*305*)

7. Duhem, P.: Sur les ondes au sein d'un milieu vitreux, affecté de viscosité et très peu déformé. C. R. Acad. Sci., Paris **136**, 733–735. (*305*)

8. Duhem, P.: Des ondes du premier ordre par rapport à la vitesse au sein d'un milieu vitreux doué de viscosité, et affecté de mouvements finis. C. R. Acad. Sci., Paris **136**, 858–860. (*305*)

9. Duhem, P.: Des ondes du second ordre par rapport à la vitesse au sein des milieux vitreux, doués de viscosité, et affectés de mouvements finis. C. R. Acad. Sci., Paris **136**, 1032–1034. (*305*)

10. Duhem, P.: Sur une généralisation du théorème de **Reech**. Procès-Verbaux Soc. Sci. Bordeaux **1902–1903**, 65–73. (*252*)

11. Hadamard, J.: Leçons sur la Propagation des Ondes et les Équations de l'Hydrodynamique (lectures of 1898–1900). Paris. (*20, 65, 86, 88, 172, 173, 174, 175, 176, 177, 180, 181, 182, 183, 184, 185, 186, 187, 188, 189, 190, 191, 218, 241, 246, 297, 301*)

12. Natanson, L.: Über einige von Herrn **B. Weinstein** zu meiner Theorie der inneren Reibung gemachten Bemerkungen. Physik. Z. **4**, 541–543. (*305*)

13. Natanson, L.: Sur l'application des équations de **Lagrange** dans la théorie de la viscosité. Bull. Int. Acad. Sci. Cracovie, 268–283. (*305*)

14. Natanson, L.: Sur l'approximation de certaines équations de la théorie de la viscosité. Bull. Int. Acad. Sci. Cracovie, 283–311. (*305*)

15. Reynolds, O.: The Sub-Mechanics of the Universe. Papers 3. (*81, 158, 159, 215, 243*)

16. Stephansen, E.: Von der Bewegung eines Continuums mit einem Ruhepunkte. Arch. math. og naturv. **25**, No. 6, 29 pp. (*70*)

17. Zaremba, S.: Remarques sur les travaux de M. **Natanson** relatifs à la théorie de la viscosité. Bull. Int. Acad. Sci. Cracovie 85–93. (*305*)

18. Zaremba, S.: Sur une généralisation de la théorie classique de la viscosité. Bull. Int. Acad. Sci. Cracovie 380–403. (*256, 305*)

19. Zaremba, S.: Sur un problème d'hydrodynamique lié à un cas double réfraction accidentale dans les liquides et sur les considérations théoriques de M. **Natanson** relatives à ce phénomène. Bull. Int. Acad. Sci. Cracovie 403–423. (*305*)

20. Zaremba, S.: Sur une forme perfectionée de la théorie de la relaxation. Bull. Int. Acad. Sci. Cracovie 594–614. (*148, 293, 305*)

21. Zaremba, S.: Le principe des mouvements relatifs et les équations de la mécanique physique. Réponse à M. **Natanson**. Bull. Int. Acad. Sci. Cracovie 614–621. (*148, 150, 293*)

1904 *1.* Duhem, P.: Recherches sur l'élasticité. Ann. École Norm. (3) **21**, 99–139, 375–414; **22** (1905), 143–217; **23** (1906), 169–223 = repr. separ., Paris 1906. (*95, 200, 262, 301, 305*)

2. Duhem, P.: Sur la stabilité de l'équilibre en thermodynamique et les recherches de **J. W. Gibbs** au sujet de ce problème. Procès-Verbaux Soc. Sci. Bordeaux **1903–1904**, 112–121. (*245, 246, 265*)
3. Jouguet, E.: Remarques sur la propagation des percussions dans les gaz. C. R. Acad. Sci., Paris **138**, 1685–1688. (*258*)
4. Love, A. E. H.: Wave-motions with discontinuities at wave-fronts. Proc. Lond. Math. Soc. (2) **1**, 37–62. (*180*)
5. Manville, O.: Sur la déformation finie d'un milieu continu. Mém. soc. sci. Bordeaux (6) **2**, 83–162. According to Signorini [1943, *6*; Chap. I, ¶ 1], this paper is not exempt from errors. (*34*)
6. Saurel, P.: On the stability of the equilibrium of a homogeneous phase. J. Phys. Chem. **8**, 325–334. (*265*)
7. Saurel, P.: On the stability of equilibrium of bivariant systems. J. Phys. Chem. **8**, 436–439. (*265*)
8. Whittaker, E. T.: A Treatise on the Analytical Dynamics of Particles and Rigid Bodies; with an Introduction to the Problem of Three Bodies. Cambridge: Cambridge Univ. Press. xiii + 414 pp. (*99*)

1905 *1.* Hadamard, J.: Remarque sur la note de M. **Gyčzö Zemplén**. C. R. Acad. Sci., Paris **141**, 713. (*258*)
2. Jaumann, G.: Die Grundlagen der Bewegungslehre von einem modernen Standpunkte aus. Leipzig. (*8, 12, 66B, 79, 80, 81, 107*)
3. Klein, F., u. K. Wieghardt: Über Spannungsflächen und reziproke Diagramme, mit besonderer Berücksichtigung der Maxwell'schen Arbeiten. Arch. Math. Phys. (3) **8**, 1–10, 95–119. (*227*)
4. Marcolongo, R.: Le formule del **Saint-Venant** per le deformazioni finite. Rend. Circ. Nat. Palermo **19**, 151–155. (*34*)
5. Riquier, C.: Sur l'intégration d'un système d'équations aux dérivées partielles auquel conduit l'étude des déformations finies d'un milieu continu. Ann. École Norm. (3) **22**, 475–538. (*34*)
6. Volterra, V.: Sull'equilibrio dei corpi elastici più volte connessi. Rend. Accad. Lincei **14**[1], 193–202. (*49, 185*)
7. Zemplén, G.: Kriterien für die physikalische Bedeutung der unstetigen Lösungen der hydrodynamischen Bewegungsgleichungen. Math. Ann. **61**, 437–449. (*7, 9, 193, 205, 236A*)
8. Zemplén, G.: Sur l'impossibilité des ondes de choc négatives dans les gaz. C. R. Acad. Sci., Paris **141**, 710–712. (*258*)
9. Zemplén, G.: Besondere Ausführungen über unstetige Bewegungen in Flüssigkeiten. Enz. math. Wiss. **IV**[3], art. 19. (*172, 205, 236A, 241*)

1906 *1.* Cafiero, D.: La deformazione finita di un mezzo continuo. Atti Accad. Peloritana **21**, 179–228. (*12, 34, 42, 44, 45*)
2. Cesaro, E.: Sulle formole del **Volterra**, fondamentali nella teoria delle distorsioni elastiche. Rend. Accad. Napoli (3a) **12**, 311–321. (*57, 175*)
3. Cisotti, U.: Sul paradosso di **d'Alembert**. Atti Accad. Veneto **65**, 1291–1295. (*202*)
4. Lamb, H.: 3rd ed. of [1879, *2*]. (See [1895, *2*], [1916, *3*], [1924, *8*], [1932, *9*].) (*99, 136*)
5. Love, A. E. H.: 2nd ed. of [1892, *8*] and [1893, *5*]. (*12, 27, 41, 45, 195, 214, 218, 228, 301*)
6. Müller, C. H., u. A. Timpe: Die Grundgleichungen der mathematischen Elastizitätstheorie. Enz. math. Wiss. **IV**[4], art. 23, 1–54. (*231, 256A*)

1907 *1.* Cosserat, E. et F.: Sur la mécanique générale. C. R. Acad. Sci., Paris **145**, 1139–1142. (*60*)
2. Cosserat, E. et F.: Sur la statique de la ligne déformable. C. R. Acad. Sci., Paris **145**, 1409–1412. (*214*)
3. Hilbert, D.: Mechanik der Continua, lectures of 1906–1907; MS notes by A. R. Crathorne in Univ. Illinois Library. (*9, 158, 256*)
4. Hilbert, D.: ibid., more complete MS notes by W. Marshall in Purdue Univ. Library. (*9, 158, 245, 256*)
5. Lorentz, H. A.: Über die Entstehung turbulenter Flüssigkeitsbewegungen und über den Einfluß dieser Bewegung bei der Strömung durch Röhren. Abh. theor. Phys. 43–71. (*207*)
6. Neumann, E. R.: Über eine neue Reduktion bei hydrodynamischen Problemen. J. reine angew. Math. **132**, 189–215. (*224, 225*)
7. Volterra, V.: Sur l'équilibre des corps élastiques multiplement connexes. Ann. École Norm. (3) **24**, 401–517. (*57*)

8. Wilson, E. B.: Oblique reflections and unimodular strains. Trans. Amer. Math. Soc. **8**, 270–298. (*46*)

1908 *1.* Cosserat, E. et F.: Sur la statique de la surface déformable et la dynamique de la ligne déformable. C. R. Acad. Sci., Paris **146**, 68–71. (*63A*, *64*)

2. Cosserat, E. et F.: Sur la théorie des corps minces. C. R. Acad. Sci., Paris **146**, 169–172. (*60*)

3. Einstein, A., u. J. Laub: Über die elektromagnetischen Grundgleichungen für bewegte Körper. Ann. Physik, **26**, 532–540. (*308*)

4. Hamel, G.: Über die Grundlagen der Mechanik. Math. Ann. **66**, 350–397. (*2*, *231*)

5. Hargreaves, R.: On integral forms and their connection with physical equations. Cambr. Phil. Trans. **21**, 107. (*266*, *270*)

6. Planck, M.: Bemerkungen zum Prinzip der Aktion und Reaktion in der allgemeinen Dynamik. Verh. dtsch. phys. Ges. **10**, 728–732. (*289*)

1909 *1.* Abraham, M.: Zur Elektrodynamik bewegter Körper. Rend. Palermo **28**, 1–28. (*80*, *279*)

2. Brill, A.: Vorlesungen zur Einführung in die Mechanik raumerfüllender Massen. Leipzig u. Berlin. x + 236 pp. (*8*, *237*)

3. Carathéodory, C.: Untersuchungen über die Grundlagen der Thermodynamik. Math. Annalen **67**, 355–386. (*245*)

4. Cisotti, U.: Sul moto di un solido in un canale. Rend. Palermo **28**, 307–351. (*202*)

5. Cosserat, E. et F.: Théorie des Corps Déformables. Paris = pp. 953–1173 of O. D. Chwolson, Traité de Physique, transl. E. Davaux, 2nd ed., **2**. Paris 1909. (*12*, *60*, *61*, *63A*, *64*, *200*, *210*, *212*, *214*, *217*, *238*, *256*)

6. Gibbs, J. W., and E. B. Wilson: Vector Analysis. New Haven. (*66B*)

7. Ludwik, P.: Elemente der Technologischen Mechanik. Berlin. (*25*, *33*)

8. Mises, R. v.: Theorie der Wasserräder. Z. Math. Phys. **57**, 1–120. (*99*, *137*, *138*, *145*, *162*, *170*, *202*, *217*)

9. Usai, G.: Sulle deformazioni di 2^0 ordine di una superficie flessibile ed estendibile. Rend. Ist. Lombardo (2) **42**, 416–426. (*64*)

10. Volterra, V.: Sulle equazioni integro-differenziali della teoria dell' elasticità. Rend. Lincei (5) $\mathbf{18_1}$, 295–301 = Opere **3**, 288–293. (*305*)

11. Volterra, V.: Equazioni integro-differenziali della elasticità nel caso della isotropia. Rend. Lincei (5) $\mathbf{18_1}$, 577–586 = Opere **3**, 294–303. (*305*)

1910 *1.* Bateman, H.: The transformation of the electrodynamical equations. Proc. Lond. Math. Soc. (2) **8**, 223–264. (*266*, *270*, *282*)

2. Boggio, T.: Sul moto permanente di un solido in un fluido indefinito. Atti Accad. Veneto $\mathbf{69^2}$, 883–891. (*202*)

3. Callender, H. L.: art. Heat. Encycl. Britt., 11th ed. to present ed. (*240*, *245*)

4. Cisotti, U.: Sul moto permanente di un solido in un fluido indefinito. Atti Accad. Veneto $\mathbf{69^2}$, 427–445. (*202*)

5. Darboux, G.: Leçons sur les Systèmes Orthogonaux et les Coordonnées Curvilignes, 2nd ed. Paris. 567 pp. (*38*)

6. Kötter, F.: Über die Spannungen in einem ursprünglich geraden, durch Einzelkräfte in stark gekrümmter Gleichgewichtslage gehaltenen Stab. Sitzgsber. Preuss. Akad. Wiss., part 2, 895–922. (*303*)

7. Minkowski, H.: Die Grundgleichungen für die elektromagnetischen Vorgänge in bewegten Körpern. Math. Annalen **68**, 472–525. (*308*)

8. Strutt, J. W. (Lord Rayleigh): Aerial plane waves of finite amplitude. Proc. Roy. Soc. Lond. A **84**, 247–284 = Papers **5**, 573–610. (*258*)

9. Veblen, O., and J. W. Young: Projective Geometry. Boston. (*21*)

10. Voigt, W.: Lehrbuch der Kristallphysik. Leipzig (repr. 1928). (*256A*, *296*, *305*, *312*)

1911 *1.* Almansi, E.: Sulle deformazioni finite dei solidi elastici isotropi, I. Rend. Lincei (5A) $\mathbf{20^1}$, 705–714. (*31*, *303*)

2. Almansi, E.: Sulle deformazioni finite dei solidi elastici isotropi, III. Rend. Lincei (5A) $\mathbf{20^2}$, 287–296. (*30*)

3. Crudeli, V.: Sopra le deformazioni finite. Le equazioni del **De Saint-Venant.** Rend. Lincei (5A) $\mathbf{20^2}$, 306–308, 470. (*34*)

4. Duhem, P.: Traité d'Énergetique. Paris. 2 vols. (*201*, *241*, *247*, *262*, *265*)

5. Gwyther, R. F.: The conditions that the stresses in a heavy elastic body should be purely elastic stresses. Mem. Manchester Lit. Phil. Soc. **55**, No. 20 (12 pp.). (*227*)

6. Hamel, G.: Zum Turbulenzproblem. Nachr. Ges. Wiss. Göttingen, math.-phys. Kl., 261–270. (*116*)
7. Jaumann, G.: Geschlossenes System physikalischer und chemischer Differentialgesetze. Sitzgsber. Akad. Wiss. Wien (IIa) **120**, 385–530. (*6, 159, 215, 243, 245, 262, 295*)
8. Kármán, T. v.: Festigkeitsversuche unter allseitigem Druck. Z. VDI **55**, 1749–1757. (*24*)
9. Le Roux, J.: Sur les covariants fondamenteaux du second ordre dans la déformation finie d'un milieu continu. C. R. Acad. Sci., Paris **152**, 1002–1005. (*14, 20, 26*)
10. Le Roux, J.: Sur l'incurvation et la flexion dans les déformations finies. C. R. Acad. Sci., Paris **152**, 1655–1657. (*20*)
11. Le Roux, J.: Étude géométrique de la torsion et de la flexion dans la déformation infinitésimale d'un milieu continu. Ann. École Norm. (3) **28**, 523–579. (*20*)
12. Vessiot, E.: Sur les transformations infinitésimales et la cinématique des milieux continus. Bull. Sci. Math. (2) **35**[1], 233–244. (*107, 109*)
13. Zorawski, K.: Über stationäre Bewegungen kontinuierlicher Medien. Bull. Int. Acad. Sci. Cracovie A, 1–17. (*143, 146, 148, 153*)
14. Zorawski, K.: Invariantentheoretische Untersuchung gewisser Eigenschaften der Bewegungen kontinuierlicher Medien. Bull. Int. Acad. Sci. Cracovie A, 175–218. (*84, 90, 104, 143, 144, 145, 153*)

1912 1. Boussinesq, J.: Théorie géométrique, pour un corps non rigide, des déplacements bien continus, ainsi que les déformations et rotations de ses particules. J. Math. pures appl. (6) **8**, 211–227. Abstract, C. R. Acad. Sci., Paris **154**, 949–954. (*30, 57*)
2. De Donder, T.: Sur la cinématique des milieux continus. Bull. Acad. Belg., Cl. Sci., 243–251. (*134, 146*)
3. Gwyther, R. F.: The formal specification of the elements of stress in cartesian, and in cylindrical and spherical polar coordinates. Mem. Manchester Lit. Phil. Soc. **56**, No. 10 (13 pp.). (*227*)
4. Hamel, G.: Elementare Mechanik. Leipzig u. Berlin. (*30, 31*)
5. Karman, T. v.: Festigkeitsversuche unter allseitigen Druck. Forsch.-Arb. Ing.-Wes. **1912**, 37–68. (*24*)
6. Mie, G.: Grundlagen einer Theorie der Materie. Ann. Physik **37**, 511–534; **39**, 1–40; **40**, 1–66. (*276, 310*)
7. Volterra, V.: Sur les équations intégro-différentielles et leurs applications. Acta Math. **35**, 295–356. (*309*)
8. Zorawski, K.: Über gewisse Eigenschaften der Bewegungen kontinuierlicher Medien. Bull. Int. Acad. Sci. Cracovie A 269–292. (*104, 153*)

1913 1. Ferrari, M.: Flusso di energia e velocità di gruppo. Rend. Lincei (5) **22**, 761–766. (*157*)
2. Gwyther, R. F.: The specification of the elements of stress. Part II. A simplification of the specification given in Part I. Mem. Manchester Lit. Phil. Soc. **57**, No. 5 (4 pp.) (*227*)
3. Havelock, T. H.: The displacement of the particles in a case of fluid motion. Proc. Univ. Durham Phil. Soc. **4** (1911–1912), 62–74. (*71*)
4. Heun, K.: Ansätze und allgemeine Methoden der Systemmechanik. Enz. math. Wiss. **4**[2] (1904–1935), art. **11**. (*12, 66B, 159, 200, 212, 214, 236, 243*)
5. Le Roux, J.: Recherches sur la géométrie des déformations finies. Ann. École Norm. (3) **30**, 193–235. (*20*)
6. Mises, R. v.: Mechanik der festen Körper im plastisch-deformablen Zustand. Gött. Nachr. 582–592. (*300*)
7. Morton, W. B.: On the displacements of the particles and their paths in some cases of two-dimensional motion of a frictionless liquid. Proc. Roy. Soc. Lond. A **89**, 106–124. (*71*)

1914 1. Burgatti, P.: Sulle deformazioni finite dei corpi continui. Mem. Accad. Sci. Bologna (7) **1**, 237–244. (*34*)
2. Duhem, P.: Sur le paradoxe hydrodynamique de **d'Alembert**. C. R. Acad. Sci., Paris **159**, 592–595. (*202*)
3. Duhem, P.: Remarque sur le paradoxe hydrodynamique de **d'Alembert**. C. R. Acad. Sci., Paris **159**, 638–640. (*202*)
4. Hellinger, E.: Die allgemeinen Ansätze der Mechanik der Kontinua. Enz. math. Wiss. **4**[4], 602–694. (*7, 9, 195, 231, 232, 232A, 233, 236, 237, 238, 256*)
5. Hesselberg, Th., u. H. U. Sverdrup: Das Beschleunigungsfeld bei einfachen Luftbewegungen. Veröff. geophys. Inst. Univ. Leipzig (2), No. 5, 117–146. (*70*)

. HUNTINGTON, E. V.: The theorem of rotation in elementary mechanics. Amer. math. monthly **21**, 315–320. (*170*)

. MATTIOLI, G. D.: Sulle equazioni fondamentali della dinamica dei sistemi continui. Rend. Lincei (5) **23**², 328–334. (*157, 207, 217, 240*)

8. MOHR, O.: Abhandlungen aus dem Gebiete der technischen Mechanik, 2nd ed. Berlin. (*24*)

9. PICARD, E.: À propos du paradoxe hydrodynamique de **d'Alembert**. C. R. Acad. Sci., Paris **159**, 638. (*202*)

10. ROEVER, W. H.: The design and theory of a mechanism for illustrating certain systems of lines of force and stream lines. Z. Math. Phys. **62** (1913–1914), 376—384. Abstract in Bull. Amer. Soc. **18**, 435–536 (1912) and **19**, 220 (1913). (*70*)

11. SOMIGLIANA, C.: Sulla teoria delle distorsioni elastiche. Rend. Lincei (5) **23**₁, 463–472. (*175*)

1915 *1.* BÖKER, R.: Die Mechanik der bleibenden Formänderung in kristallinisch aufgebauten Körpern. Forsch.-Arb. Ing.-Wes. 1915, 1–51. (*24*)

2. EMDE, F.: Zur Vektorrechnung. Archiv Math. (3) **24**, 1–11. (*173*)

3. LORENTZ, H. A.: The Theory of Electrons. Leipzig. (*279*)

4. MORTON, W. B., and J. VINT: On the paths of the particles in some cases of motion of frictionless fluid in a rotating enclosure. Phil. Mag. (6) **30**, 284–287. (*71*)

1916 *1.* BIRKELAND, R.: Développements sur le mouvement d'un fluide parallèle à un plan fixe. C. R. Acad. Sci., Paris **163**, 200–202. (*161*)

2. FRIEDMANN, A.: Sur les tourbillons dans un liquide à température variable. C. R. Acad. Sci., Paris **163**, 219–222. (*110*)

3. LAMB, H.: 4th ed. of [1879, *2*]. (See [1895, *2*], [1906, *4*], [1924, *9*], [1932, *8*].) (*86A*)

4. PALATINI, A.: Sulle quadriche di deformazione per gli spazi S_3. Atti Ist. Veneto **76**, 125–148. (*84*)

5. SPIELREIN, J.: Lehrbuch der Vektorrechnung nach den Bedürfnissen in der Technischen Mechanik und Elektrizitätslehre. Stuttgart. (*66B, 79, 80, 81, 173*)

6. TAYLOR, G. I.: Motion of solids in fluids when the flow is not irrotational. Proc. Roy. Soc. Lond., Ser. A **93** (1916–1917), 99–113. (*143, 207*)

1917 *1.* ALMANSI, E.: L'ordinaria teoria dell' elasticità e la teoria delle deformazioni finite. Rend. Lincei (5) **26**², 3–8. (*55*)

2. APPELL, P.: Sur un théorème de **Joseph Bertrand** relatif à la cinématique des milieux continus. Bull. Sci. Math. (2) **41**, 23–28. (*101*)

3. APPELL, P.: Sur une extension des équations de la théorie des tourbillons et des équations de **Weber**. C. R. Acad. Sci., Paris **164**, 71–74. (*101, 136*)

4. CISOTTI, U.: Sulle azioni dinamiche di masse fluide continue. Rend. Lombardo (2) **50**, 502–515. (*170, 202*)

5. LOHR, E.: Entropieprinzip und geschlossenes Gleichungssystem. Denkschr. Akad. Wiss. Wien **93**, 339–421. (*6, 159, 243, 245, 262, 295*)

1918 *1.* CRUDELI, U.: Le formule del **Cauchy** e i fluidi viscosi. Rend. Lincei (5) **27**², 49–52. (*134*)

2. DIETZIUS, R.: Die Gestalt der Stromlinien in der Nähe der singulären Punkte. Beitr. Phys. fr. Atmosph. **8**, 29–52. (*70*)

3. JAUMANN, G.: Physik der kontinuierlichen Medien. Denkschr. Akad. Wiss. Wien **95**, 461–562. (*6, 245, 262, 295*)

1920 *1.* EINSTEIN, A.: Schallausbreitung in teilweise dissoziierten Gasen. Sitzgsber. Akad. Wiss. Berlin, math.-phys. Kl., 380–385. (*159A*)

2. PASCAL, M.: Circuitazione superficiale. Rend. Lincei (5) **29**², 353–356; **30**¹, 117—119, 249—251 (1921). (*105*)

3. PAULI, W.: Relativitätstheorie. Enz. math. Wiss. **5**, No. 2. (*284*)

1921 *1.* APPELL, P.: Traité de Mécanique Rationnelle, **3**, Équilibre et Mouvement des Milieux Continus, 3rd. ed. Paris. (*86, 101, 109, 122, 136, 138*)

2. BORN, M.: Kritische Betrachtungen zur traditionellen Darstellung der Thermodynamik. Physik. Z. **22**, 218–224, 249–254, 282–286. (*245*)

3. JAFFÉ, G.: Über den Transport von Vektorgrößen, mit Anwendung auf Wirbelbewegung in reibenden Flüssigkeiten. Physik. Z. **22**, 180–183. (*101, 131*)

4. MURNAGHAN, F. D.: The absolute significance of **Maxwell's** equations. Phys. Rev. (2) **17**, 73–88. (*266*)

5. PASCAL, M.: Circuitazione superficiale. Giorn. mat. **59**, 215–234. (*105*)

6. WEYL, H.: Raum-Zeit-Materie. 4th ed. Berlin: Springer. (*266, 310*)

1922 *1.* BATEMAN, H.: Equations for the description of electromagnetic phenomena. Bull. Nat. Res. Council **4**, 96–191. (*308*)

2. CARTAN, E.: Leçons sur les invariants intégraux. Paris: Hermann. x + 210 pp. (*129*)

3. KOTTLER, F.: Newton'sches Gesetz und Metrik. Sitzgsber. Akad. Wiss. Wien (IIa) **131**, 1—14. (*7*)
4. KOTTLER, F.: Maxwell'sche Gleichungen und Metrik. Sitzgsber. Akad. Wiss. Wien (IIa) **131**, 119—146. (*7, 266*)
5. LAMPE, E.: Review of [1914, *10*]. Jb. Fortschr. Math. 1914—1915, 963. (*70*)
6. MUNK, M.: Notes on aerodynamic forces, I. Rectilinear motion. Nat. Adv. Comm. Aero. T. N. 104. (*76*)

1923 1. CARTAN, E.: Sur les variétés à connexion affine et la théorie de la relativité généralisée. Ann. École Norm. (3) **40**, 325—412; **41**, 1—25 (1924); **42**, 17—88 (1925). (*7, 152. 211*)
2. CISOTTI, U.: Sull' energia cinetica di masse fluide continua: viriale degli sforzi. Rend. Lincei (5) **32**², 464—467. (*219*)
3. CISOTTI, U.: Considerazione sulla nota formula idrodinamica di **Daniele Bernoulli**. Boll. Un. Mat. Ital. **2**, 125—128. (*125*)
4. CISOTTI, U.: Sul carattere necessariamente vorticoso dei moti regolari permanenti di un fluido qualsiasi in ambienti limitati, eppure in quiete all' infinito. Boll. Un. Mat. Ital. **2**, 170—172. (*125*)
5. LEONARDO DA VINCI: Del moto e misura dell' acqua. Ed. CARUSI e FAVARO: Bologna. Parts of this work were printed in various publications from 1765 onward. (*77*)

1924 1. ARIANO, R.: Deformazioni finite dei sistemi continui. Rend. Palermo **48**, 97—120. (*12*)
2. CALDONAZZO, B.: Sulla geometria differenziale di superficie aventi interesse idrodinamico. Rend. Lincei (5A) **33**², 396—400. (*122*)
3. COLLINET, E.: Sur l'énergie interne d'un corps élastique. C. R. Acad. Sci., Paris **178**, 373—375. (*210*)
4. HEYMANS, P.: Note on a property of rectilinear lines of principal stress. J. Math. Phys. **3**, 182—185. (*223*)
5. INGRAM, W. H.: Note on the curl. J. Math. Phys. **3**, 186—187. (*86*)
6. JOUGUET, E.: Le potential interne des corps élastiques. C. R. Acad. Sci., Paris **178**, 840—842. (*210*)
7. JOUGUET, E., et M. ROY: Le paradoxe de **d'Alembert** dans le cas des fluides compressibles. C. R. Acad. Sci., Paris **178**, 1470—1472. *Cf.* remarks by CISOTTI, C. R. Acad. Sci., Paris **178**, 1792—1793 and by the authors, C. R. Acad. Sci., Paris **179**, 142—143. (*202*)
8. KELLOGG, O. D.: The theorem of the moment of momentum. Amer. Math. Monthly **31**, 429—432. (*170*)
9. LAMB, H.: 5th ed. of [1879, *2*]. (See [1895, *2*], [1906, *4*], [1916, *3*], [1932, *8*].) (*136*)
10. LOHR, E.: Das Entropieprinzip der Kontinuitätstheorie. Festschr. Techn. Hochschule Brünn 176—187. (*243, 245, 262*)
11. NOAILLON, P.: Réponse aux observations de M. **Pascal** sur la circulation superficielle. C. R. Acad. Sci., Paris **178**, 311—314. (*101, 105*)

1925 1. ARIANO, R.: Deformazioni finite di sistemi continui, I. Ann. Mat. (4) **2**, 217—261. (*12*)
2. BRILLOUIN, L.: Les lois de l'élasticité sous forme tensorielle valable pour des coordonnées quelconques. Ann. de Phys. (10) **3**, 251—298. (*12, 31, 210*)
3. CALDONAZZO, B.: Un' osservazione a proposito del teorema di **Bernoulli**. Boll. Un. Mat. Ital. **4**, 1—3. (*122*)
4. CARATHÉODORY, C.: Über die Bestimmung der Energie und der absoluten Temperatur mit Hilfe von reversiblen Prozessen. Ber. Akad. Wiss. Berlin, 39—47. (*245*)
5. EHRENFEST-AFANASSJEWA, T.: Zur Axiomatisierung des zweiten Hauptsatzes der Thermodynamik. Z. Physik **33**, 933—945; **34**, 638. (*245*)
6. FRIEDMANN, A.: Über Wirbelbewegung in einer kompressiblen Flüssigkeit. Z. angew. Math. Mech. **4**, 102—107. (*110*)
7. HENCKY, H.: Die Bewegungsgleichungen beim nichtstationären Fliessen elastischer Massen. Z. angew. Math. Mech. **5**, 144—146. (*66B, 150*)
8. LELLI, M.: Una forma più generale del principio di **Archimede**. Boll. Un. Mat. Ital. **4**, 63—64. (*202*)
9. LELLI, M.: Sui teoremi delle quantità di moto per i sistemi continui. Boll. Un. Mat. Ital. **4**, 204—206. (*202*)
10. LICHTENSTEIN, L.: Über einige Existenzprobleme der Hydrodynamik homogener, unzusammendrückbarer, reibungsloser Flüssigkeiten und die Helmholtzschen Wirbelsätze. Math. Z. **23**, 89—154. (*101*)

1926 *1.* Eisenhart, L. P.: Riemannian Geometry. Princeton. (*34*)
2. Finzi, B.: Constatation énergetique du paradoxe de d'Alembert dans les liquides visqueux. C. R. Acad. Sci., Paris **182**, 1077–1079. (*202*)
3. Kotchine, N. E.: Sur la théorie des ondes de choc dans un fluide. Rend. Circ. Mat. Palermo **50**, 305–344. (*7, 172, 175, 180, 185, 186, 189, 191, 193, 194, 205, 241, 297, 298*)
4. Landé, A.: Axiomatische Begründung der Thermodynamik durch Carathéodory. Handbuch Physik **9**, Chap. 4. (*245*)
1927 *1.* Bouligand, G.: Sur le signe de la pression dans un liquide pesant en mouvement irrotationnel. Verh. 2. Int. Congr. Techn. Mech. (1926) 460–461. (*121*)
2. Bouligand, G.: Un théorème relatif à la pression au sein d'un liquide parfait en mouvement irrotationnel. J. Math. pures appl. (9) 427–433. (*121*)
3. De Donder, T.: L'affinité. Mém. acad. Sci. Belg. in 8° (2) **9**, No. 7, 94 pp. Also issued separately, Paris 1927. 2nd ed., revised by P. van Rysselberghe. Paris: Gauthier-Villars 1936. xv + 142 pp. (*159A, 259*)
4. Eisenhart, L. P.: Non-Riemannian Geometry. New York: Amer. Math. Soc. (*152*)
5. Lichtenstein, L.: Über einige Existenzprobleme der Hydrodynamik II: Nichthomogene, unzusammenrückbare, reibungslose Flüssigkeiten. Math. Z. **26**, 196–323. (*101*)
6. Love, A. E. H.: 4th ed. of [1892, *8*] and [1893, *5*]. (*38, 256A, 301*)
7. Masotti, A.: Osservazioni sui moti di un fluido nei quali è stazionaria la distribuzione del vortice. Rend. Lincei (6) **6**, 224–228. (*111, 125*)
8. Veblen, O.: Invariants of quadratic differential forms. Cambr. tracts math. and math. phys. No. 24, Cambridge. (*34, 147*)
1928 *1.* Ariano, R.: Deformazioni finite di sistemi continui I–III. Ann. di mat. (4) **5**, 55–71; **6**, 265–282 (1929). (*12*)
2. Brillouin, L.: Les lois de l'élasticité en coordonnées quelconques. Proc. Int. Congr. Math. Toronto (1924) **2**, 73–97 (a preliminary version of [1925, *2*]). (*31, 210*)
3. Haldane, J. S.: Gases and Liquids. Edinburgh. (*245*)
4. Hencky, H.: Über die Form des Elastizitätsgesetzes bei ideal elastischen Stoffen. Z. techn. Phys. **9**, 214–223, 457. (*33*)
5. Levi-Civita, T.: Fondamenti di meccanica relativistica. Bologna: Zanichelli. vii + 185 pp. (*211*)
1929 *1.* De Donder, T.: L'affinité II. Bull. Sci. Acad. Belg. (5) **15**, 615–625, 900–912; **17**, 298–314, 507–515, 653–663, 780–787, 874–887, 1001–1007 (1931). Repr., with 2 suppl. papers, Paris 1931. (*259*)
2. Hencky, H.: Welche Umstände bedingen die Verfestigung bei der bildsamen Verformung von festen isotropen Körpern? Z. Physik **55**, 145–155. (*33*)
3. Hencky, H.: Das Superpositionsgesetz eines endlich deformierten relaxationsfähigen elastischen Kontinuums und seine Bedeutung für eine exakte Ableitung der Gleichungen für die zähe Flüssigkeit in der Eulerschen Form. Ann. Physik (5) **2**, 617–630. (*33, 95, 148*)
4. Lichtenstein, L.: Grundlagen der Hydromechanik. Berlin: Springer. (*88, 101, 105, 121, 172, 174, 182, 187, 189*)
5. Lichtenstein, L.: Bemerkung über einen Verzerrungssatz bei topologischer Abbildung in der Hydromechanik. Math. Z. **30**, 321–324. (*65*)
6. Polubarinova, P. Ya.: Singular points of stream lines of affine motion in space [in Russian]. Isv. Glavn. Geofiz. Observ. No. 1, 1–16. (*70*)
7. Pompeiu, D.: Sur la condition des vitesses dans un fluide incompressible. Bull. math. phys. École Polyt. Bucarest **1**, 42–43. (*160*)
8. Sarton, G.: The discovery of the law of conservation of energy. Isis **13** (1929—1930), 18–34. (*240*)
9. Southwell, R. V.: Mechanics. Encycl. Britt. 14th ed. In the 1944 printing, **15**, 156–168. (*3*)
10. Thirring, H.: Begriffssystem und Grundgesetze der Feldphysik. Handbuch Physik **4**, 81–177. (*293*)
1930 *1.* Finzi, B.: Sopra il tensore di deformazione di un velo. Rend. Ist. Lombardo (2) **63**, 975–982. (*34, 84*)
2. Kirchhoff, W.: Reduktion simultaner partieller Differentialgleichungen bei hydrodynamischen Problemen. J. reine angew. Math. **164**, 183–195. (*161, 225*)
3. Mises, R. v.: Über die bisherigen Ansätze in der klassischen Mechanik der Kontinua. Proc. 3rd Int. Congr. appl. Mech. Stockholm **2**, 1–9. (*7*)

4. SIGNORINI, A.: Sulle deformazioni finite dei sistemi continui. Rend. Lincei (6) **12**, 312–316. (*34*)
5. SIGNORINI, A.: Sulla meccanica dei sistemi continui. Rend. Lincei (6) **12**, 411–416. (*210*)
6. SIGNORINI, A.: Sulle deformazioni termoelastiche finite. Proc. 3rd Int. Congr. appl. Mech. Stockholm **2**, 80–89. (*33A, 43, 44, 210*)
7. TONOLO, A.: Forma intrinseca delle equazioni dell' equilibrio dei mezzi elastici. Rend. Lincei (6) **12**, 247–250, 347–351. (*205A*)
8. VILLAT, H.: Leçons sur la Théorie des Tourbillons. Paris. (*101*)
9. VOLTERRA, V.: Theory of Functionals and of Integral and Integro-differential Equations. London and Glasgow: Blackie. (*305, 309*)

1931 1. ANDRUETTO, G.: Sulle equazioni intrinseche dell' equilibrio elastico. Rend. Lincei (6) **13**, 489–494. (*205A*)
2. BATEMAN, H.: On dissipative systems and related variational principles. Phys. Rev. (2) **38**, 815–819. (*129*)
3. COKER, E. G., and L. N. G. FILON: Photoelasticity. London: Cambridge Univ. Press. (*31, 312*)
4. DE DONDER, T.: Théorie invariantive de l'élasticité à déformations finies. Bull. Sci. Acad. Belg. (5) **17**, 1152–1157. (*152, 245*)
5. DUPONT, Y.: Quelques contributions à la théorie invariantive de l'élasticité. Bull. Sci. Acad. Belg. (5) **17**, 441–459. (*19, 31, 104*)
6. DUPONT, Y.: Sur la théorie invariantive de l'élasticité à déformations finies. C. R. Acad. Sci., Paris **192**, 873–875. (*31*)
7. MIMURA, Y.: On the foundation of the second law of thermodynamics. J. Hiroshima Univ. **1** (1930–1931), 43–53. (*245*)
8. MIMURA, Y.: On the equations of motion in thermodynamics. J. Hiroshima Univ. **1** (1930–1931), 117–123. (*245*)
9. SBRANA, F.: Sulla validità del teorema di **Bernoulli** per un fluido reale. Boll. Un. Mat. Ital. **10**, 77–78. (*124*)
10. SLEBODZINSKI, W.: Sur les équations canoniques de **Hamilton**. Bull. Sci. Acad. Belg. (5) **17**, 864–870. (*150*)
11. WEISSENBERG, K.: Die Mechanik deformierbarer Körper. Abh. Akad. Wiss. Berlin No. 2. (*241*)

1932 1. ANDRUETTO, G.: Le formule di **Saint-Venant** per gli spazi curvi a tre dimensioni. Rend. Lincei (6) **15**, 214–218. (*84*)
2. ANDRUETTO, G.: Le formule di **Saint-Venant** per le varietà V_n a curvatura costante. Rend. Lincei (6) **15**, 792–797. (*84*)
3. BONVICINI, D.: Sulle deformazioni non infinitesimi. Rend. Lincei (6) **16**, 607–612. (*29, 86, 95, 104*)
4. CISOTTI, U.: Spostamenti rigidi finiti. Rend. Lincei (6) **16**, 381–386. (*55*)
5. DE DONDER, T.: L'affinité (III). Bull. Sci. Acad. Belg. (5) **18**, 578–595, 888–898, 1124–1137; **19**, 881–892, 1140–1152, 1364–1376 (1933); **20**, 268–281 (1934). Reprinted, with 8 supplementary papers, Paris 1934. (*259*)
6. DUPONT, Y.: Applications physico-chimiques de la thermodynamique des systèmes en mouvement. Bull. Sci. Acad. Belg. (5) **18**, 83–94 = reprint of [1932, *5*, suppl., 23–34]. (*259*)
7. IWATSUKI, T., and MIMURA: On the adiabatic process of the thermodynamical system in which the entropy cannot be defined. J. Hiroshima Univ. **2** (1931–1932), 127–138. (*245*)
8. LAMB, H.: 6th ed. of [1879, *2*]. (See [1895, *2*], [1906, *4*], [1916, *3*], [1924, *9*].) (*256*)
9. MASOTTI, A.: Sul teorema di **König**. Atti Pontif. Accad. Sci. Nuovi Lincei **85** (1931–1932), 37–42. (*167*)
10. ONSAGER, L., and R. M. FUOSS: Irreversible properties in electrolytes. Diffusion, conductance, and viscous flow in arbitrary mixtures of strong electrolytes. J. Physic. Chem. **36**, 2689–2978. (*295*)
11. SIGNORINI, A.: Alcune proprietà di media nella elastostatica ordinaria. Rend. Lincei (6) **15**, 151–156. (*220*)
12. SIGNORINI, A.: Sollecitazioni iperastatiche. Rend. Ist. Lombardo (2) **65**, 1–7. (*221*)
13. VAN DANTZIG, D.: Zur allgemeinen projektiven Differentialgeometrie, II. X_{n+1} mit eingliedriger Gruppe. Proc. Akad. Wet. Amsterd. **35**, 535–542. (*150*)
14. WUNDHEILER, A.: Kovariante Ableitung und die Cesarosche Unbeweglichkeitsbedingungen. Math. Z. **36**, 104–109. (*16, 150*)
15. WUNDHEILER, A. W.: Rheonome Geometrie. Absolute Mechanik. Proc. matem.-fiz. (Warsaw) **49**, 97–142. (*153*)

1933 *1.* Agostinelli, C.: Le condizioni di **Saint-Venant** per le deformazioni di una varietà riemanniana generica. Rend. Lincei (6) **18**, 529–533; (6) **19**, 22–26 (1934). (*84*)
2. Arrighi, G.: Una generalizzazione dell' equazione di continuità. Rend. Lincei (6) **18**, 302–307. (*157*)
3. Bjerknes, V., J. Bjerknes, H. Solberg u. T. Bergeron: Physikalische Hydrodynamik. Berlin: Springer. xviii + 797 pp. (*3, 185, 256A, 297*)
4. Fromm, H.: Stoffgesetze des isotropen Kontinuums, insbesondere bei zähplastischem Verhalten. Ing.-Arch. **4**, 432–466. (*148*)
5. Galli, A.: Sulle deformazioni pure infinitesime. Rend. Napoli (4) **3**, 55–58.
6. Hlavatý, V.: Über eine Art der Punktkonnexion. Math. Z. **38**, 135–145. (*16*)
7. Klotter, K.: Über die graphische Darstellung zugeordneter Spannungs- und Verzerrungszustand. Z. angew. Math. Mech. **13**, 433–434. (*24*)
8. Müller, W.: Über den Impulssatz der Hydrodynamik für bewegte Gefäßwände und die Berechnung der Reaktionskräfte der Flüssigkeit. Ann. Physik (5) **16**, 489–512. (*202*)
9. Odone, F.: Deformazioni finite e deformazioni infinitesime. Boll. Un. Mat. Ital. **11**, 238–241. (*57*)
10. Signorini, A.: Sopra alcune questioni di statica dei sistemi continui. Ann. Scuola Norm. Pisa (2) **2**, 231–257. (*216, 221, 222, 223*)
11. Tesar, V.: A graphical representation of two-dimensional stress distributions. J. Franklin Inst. **216**, 217–224. (*209*)
12. Valcovici, V.: Sur le mouvement d'un solide rigide. Bull. math. phys. École Polyt. Bucarest **3** (1932–1933), 171–179. (*160*)
13. Van Lehrberghe, G.: La conservation de l'énergie dans les milieux continus. Publ. Assoc. Ingen. Mons **47**, 601–618. (*241*)

1934 *1.* Brahtz, J.: Notes on the **Airy** stress function. Bull. Amer. Math. Soc. **40**, 427–430. (*224*)
2. Finzi, B.: Integrazione delle equazioni indefinite della meccanica dei sistemi continui. Rend. Lincei (6) **19**, 578–584, 620–623. (*84, 211, 224, 227, 229, 230*)
3. Palatini, A.: Sulle condizioni di **Saint-Venant** in una V_n qualsivoglia. Rend. Lincei (6) **19**, 466–469. (*84*)
4. Phillips, H. B.: Stress functions. J. Math. Phys. **13**, 421–425. (*224, 228*)
5. Pucher, A.: Über den Spannungszustand in gekrümmten Flächen. Beton und Eisen **33**, 298–304. (*229*)
6. Sbrana, F.: Moti fluidi dipendenti da due sole coordinate. Atti Pontif. Accad. Sci. Nuovi Lincei **87** (1933–1934), 256–262. (*162*)
7. Schouten, J., u. E. van Kampen: Beiträge zur Theorie der Deformation. Proc. mat.-fiz. **41**, 1–19. (*150*)
8. Schouten, J., u. J. Haantjes: Über die konforminvariante Gestalt der Maxwellschen Gleichungen und der elektromagnetischen Impulsenergiegleichungen. Physica, Haag **1**, 869–872. (*266*)
9. Tricomi, F.: Un'interpretazione intuitiva del rotore e della condizione d'irrotazionalità. Rend. Lincei (6) **19**, 399–401. (*87*)
10. Van Dantzig, D.: Electromagnetism, independent of metrical geometry, I. The foundations; II. Variational principles and a further generalization of the theory; III. Mass and Motion; IV. Momentum and energy; waves. Proc. Akad. Wet. Amsterd. **37**, 521–525, 526–531, 643–652, 825–836. (*166, 238, 270, 310*)
11. Van Dantzig, D.: The fundamental equations of electromagnetism, independent of metrical geometry. Proc. Cambridge Phil. Soc. **30**, 421–427. (*266*)

1935 *1.* Boggio, T.: Sull' integrazione delle equazioni idrodinamiche di **Helmholtz**. Rend. Lincei (6) **21**, 415–419. (*101*)
2. Bonvicini, D.: Sulla deformazione pura nel caso di spostamenti finiti e sulla relazione di essa colla tensione nei corpi anisotropi. Ann. di Mat. (4) **13**, 113–117. (*29, 95*)
3. Levi-Civita, T.: Movimenti di un sistema continuo che rispettano l'invariabilità sostanziale del baricentro. Atti Pontif. Accad. Sci. Nuovi Lincei **88** (1934–1935), 102–106. Abstract in Pontif. Accad. Sci. Nuncius No. 37, 4 (1935). (*165*)
4. Schouten, J. A., u. D. J. Struik: Einführung in die neueren Methoden der Differentialgeometrie I, 2nd ed. Groningen: Noordhoff. xii + 202 pp. (*150*)
5. Schwerdtfeger, A.: Über schwerpunktbehaltende Bewegungen materieller Systeme. Atti Pontif. Accad. Sci. Nuovi Lincei **88** (1934–1935), 302–307. (*165*)
6. Shaw, A. N.: The derivation of thermodynamical relations for a simple system. Phil. Trans. Roy. Soc. Lond. A **234**, 299–328. (*251*)
7. Sobrero, L.: Del significato meccanico della funzione di **Airy**. Ric. di Ingegneria **3**, 77–80 = Rend. Lincei (6) **21**, 264–269. (*224*)

8. SUDRIA, J.: L'action euclidienne de déformation et de mouvement. Mém. Sci. Phys. Paris. No. **29**, 56 pp. *(60)*

9. VAN MIEGHEN, J.: Thermodynamique des systèmes non-uniformes en vue des applications à la météorologie. Geofys. Publ. Norske Vid.-Akad. **10**, No. 14, 18 pp. *(243, 262)*

10. WEISSENBERG, K.: La mécanique des corps déformables. Arch. Sci. phys. nat. Genève (5) **17**, 44–106, 130–171. *(33, 42, 241)*

1936 1. BATEMAN, H.: Progressive waves of finite amplitude and some steady motions of an elastic fluid. Proc. Nat. Acad. Sci. U.S.A. **22**, 607–619. *(224)*

2. BERKER, A. R.: Sur quelques cas d'intégration des équations du mouvement d'un fluide visqueux incompressible. Paris and Lille. 161 pp. *(68)*

3. CROCCO, L.: Una nuova funzione di corrente per lo studio del moto rotazionale dei gas. Rend. Lincei (6A) **23**, 115–124; transl. Eine neue Stromfunktion für die Erforschung der Bewegung der Gase mit Rotation. Z. angew. Math. Mech. **17** (1937), 1–7. *(123, 133, 161, 162)*

4. HAMEL, G.: Ein allgemeiner Satz über den Druck bei der Bewegung volumbeständiger Flüssigkeiten. Mh. Math. Phys. **43**, 345–363. *(115, 121)*

5. KILCHEVSKI, N.: Foundation and generalization of **Hooke's** law on the basis of a law of economy of mass-energy and the problem of two non-linear measures in the theory of elasticity [in Ukrainian]. Zh. Inst. Mat. Akad. Nauk USRR. **1936–1937**, No. 2, 77–89. *(211)*

6. KOSTYUK, A.: Equation of continuity [in Russian]. Uchenie Zapiski Kazanskogo Univ. **96**^4–5^, 44–59. *(141)*

7. MUNK, M.: On the common space integrals of aerodynamics. J. Aeronaut. Sci. **3**, 243–247. *(76)*

8. PLATRIER, C.: Cinématique des milieux continus. Actual. sci. industr. No. 327, 34 pp. *(12)*

9. SYNGE, J. L.: Conditions satisfied by the velocity and the streamfunction in a viscous liquid moving in two dimensions between fixed parallel planes. Proc. Lond. Math. Soc. (2) **40**, 23–36. *(116)*

1937 1. EPSTEIN, S.: Textbook of Thermodynamics. New York. *(240, 265)*

2. HAMEL, G.: Potentialströmungen mit konstanter Geschwindigkeit. Sitzgsber. preuss. Akad. Wiss., phys.-math. Kl. 5–20. *(111)*

3. KÁRMÁN, T. v.: On the statistical theory of turbulence. Proc. Nat. Acad. Sci. U.S.A. **23**, 98–105. *(101)*

4. LAGALLY, M.: Die zweite Invariante des Verzerrungstensors. Z. angew. Math. Mech. **17**, 80–84. *(121)*

5. The Origin of CLERK MAXWELL's Electric Ideas, as Described in Familiar Letters to **William Thomson**. Ed. J. LARMOR. Cambridge. *(300)*

6. MERLIN, E.: Un théorème sur le mouvement des fluides dépourvus d'accélération. C. R. Acad. Sci., Paris **205**, 1128–1130. *(141)*

7. MURNAGHAN, F. D.: Finite deformations of an elastic solid. Amer. J. Math. **59**, 235–260. *(14, 30, 31, 95, 232)*

8. ODQVIST, F.: Équations de compatibilité pour un système de coordonnées triples orthogonaux quelconques. C. R. Acad. Sci., Paris **205**, 202–204. *(57)*

9. THEODORESCO, N.: Sur l'équilibre des milieux continus. Bull. math. phys. École Polyt. Bucarest **8** (1936–1937), 179–185. *(223)*

10. VAN DANTZIG, D.: Über das Verhältnis von Geometrie und Physik. C. r. Congr. Int. Math. Oslo (1936) **2**, 225–227. *(266)*

11. VAN DANTZIG, D.: Some possibilities of the future development of the notions of space and time. Erkenntnis **7**, 142–146. *(7, 266)*

12. ZAREMBA, S.: Sur une conception nouvelle des forces intérieures dans un fluide en mouvement. Mém. sci. math. No. 82. *(148, 293, 305)*

1938 1. BATEMAN, H.: The lift and drag functions for an elastic fluid in two-dimensional irrotational flow. Proc. Nat. Acad. Sci. U.S.A. **24**, 246–251. *(224)*

2. BRILLOUIN, L.: Les Tenseurs en Mécanique et en Élasticité. Paris. *(12, 31, 210, 246, 256A)*

3. DE DONDER, T.: Sur la vitesse réactionelle. Bull. Acad. Belg., Cl. Sci. (5) **24**, 15–18 *(159A, 245)*

4. ECKART, C.: The electrodynamics of material media. Phys. Rev. (2) **54**, 920–923. *(236A)*

5. KILCHEVSKI, N.: A new theory of the mechanics of continuous media [in Ukrainian]. Zbirnik Inst. Mat. Akad. Nauk URSR No. 1, 17–114. *(33A, 34, 152, 211, 212, 213)*

6. LAMPARIELLO, G.: Varietà sostanziali nel moto di un sistema continuo. Atti 1° Congr. Un. Mat. Ital. (1937), 391–393. (*75*)

7. MEISSNER, W.: Thermodynamische Behandlung stationärer Vorgänge in Mehrphasensystemen. Ann. Physik (5) **32**, 115–127. (*159, 243, 258*)

8. MERLIN, E.: Étude du mouvement d'un fluide parfait dépourvu d'accélération. Ann. École Norm. (3) **55**, 223–255. (*141*)

9. MILNE-THOMSON, L. M.: Theoretical Hydrodynamics. London: Macmillan. (*71, 202*)

10. PAILLOUX, H.: Mouvements fluides fournissant une suite de surfaces applicables C. R. Acad. Sci., Paris **207**, 319–321. (*83*)

11. RIABOUCHINSKI, D.: Comparaison de la méthode des variables ($\varphi, \psi_1, \psi_2, t$) à celles des variables d'**Euler** et de **Lagrange**. C. R. Acad. Sci., Paris **206**, 295–297. (*70*)

1939 *1.* BATEMAN, H.: The aerodynamics of reacting substances. Proc. Nat. Acad. Sci. U.S.A. **25**, 388–391. (*159*)

2. BIEZENO, C. B., u. R. GRAMMEL: Technische Dynamik. Berlin: Springer. (*33*)

3. BIOT, M. A.: Non-linear theory of elasticity and the linearized case for a body under initial stress. Phil. Mag. (7) **27**, 468–489. (*33A*)

4. BIOT, M. A.: Theory of elasticity with large displacements and rotations. Proc. 5th Int. Congr. appl. Mech., New York and London, 117–122. (*33A*)

5. BIOT, M. A.: Théorie de l'élasticité du second ordre avec application à la théorie du flambage. Ann. Soc. Sci. Bruxelles (1) **59**, 104–112. (*33A*)

6. CHAPMAN, S., and T. G. COWLING: The Mathematical Theory of Non-uniform Gases. Cambridge. (*215, 243, 295*)

7. EICHENWALD, A.: La funzione hamiltoniana nei mezzi continui. Rend. Sem. Mat. Fis. Milano **13**, 15–34. (*231*)

8. FRITSCHE, B.: Die Spannungsellipse. Ing.-Arch. **10**, 427–428. (*21*)

9. GUEST, J. J.: A graphical construction for stress. Phil. Mag. (7) **27**, 445–448. (*23*)

10. JARDETZKY, W.: Zur Frage der Axiomatik des zweiten Hauptsatzes der Thermodynamik. Bull. Acad. Serbe Sci. (A) **5**, 33–47. (*245*)

11. KAPPUS, R.: Zur Elasticitätstheorie endlicher Verschiebungen. Z. angew. Math. Mech. **19**, 271–285, 344–361. (*55*)

12. PUCHER, A.: Über die Spannungsfunktion beliebig gekrümmter dünner Schalen. Proc. 5th Int. Congr. appl. Mech. Cambridge (1938), 134–139. (*229*)

13. SIGNORINI, A.: Valori medi delle caratteristiche dello stress in stereodinamica. Rend. Lincei (6) **29**, 536–541. (*220*)

1940 *1.* BALLABH, R.: Superposable fluid motions. Proc. Benares Math. Soc. (2) **2**, 69–79. (*112, 114*)

2. BALLABH, R.: Self-superposable fluid motions of the type $\xi = \lambda u$, etc. Proc. Benares Math. Soc. (2) **2**, 85–89. (*114*)

3. BIOT, M. A.: Elastizitätstheorie zweiter Ordnung mit Anwendungen. Z. angew. Math. Mech. **20**, 89–99. (*33A*)

4. BRØNSTED, J. N.: The fundamental principles of energetics. Phil. Mag. (7) **29**, 449–470. (*245*)

5. BRØNSTED, J. N.: The derivation of the equilibrium conditions in physical chemistry on the basis of the work principle. J. Phys. Chem. **44**, 699–712. (*245*)

6. CISOTTI, U.: Elementi di media nella meccanica dei sistemi continui. Rend. Sem. Mat. Fis. Milano **14**, 128–138. (*216, 219*)

7. ECKART, C.: The thermodynamics of irreversible processes, I. The simple fluid. Phys. Rev. (2) **58**, 267–269. (*257*)

8. ECKART, C.: The thermodynamics of irreversible processes, II. Fluid mixtures. Phys. Rev. (2) **58**, 269–275, 924. (*159, 159A, 215, 243, 245, 255, 259, 261, 295*)

9. EISENHART, L. P.: An Introduction to Differential Geometry with the Use of the Tensor Calculus. Princeton: Univ. Press. x + 304 pp. (*64*)

10. FADLE, J.: Über eine zweckmäßige Darstellung des Mohrschen Spannungskreises zur unmittelbaren Auffindung sämtlicher Spannungs- Richtungspfeile. Ing.-Arch. **11**, 319–322. (*24*)

11. FINZI, B.: Principio variazionale nella meccanica dei continui. Rend. Accad. Ital. (7) **1**, 412–417. (*232A*)

12. GRIOLI, G.: Una proprietà di minimo nella cinematica delle deformazioni finite. Boll. Un. Mat. Ital. (2) **2**, 452–455. (*42*)

13. HELLUND, E. J.: Generalized theory of diffusion. Phys. Rev. (2) **57**, 319–333. (*295*)

14. LEVI-CIVITA, T.: Nozione adimensionale di vortice e sua applicazione alle onde trocoidale di **Gerstner**. Acta Pontif. Acad. Sci. **4**, 23–30. (*91*)

15. Locatelli, P.: Sulla congruenza delle deformazioni. Rend. Ist. Lombardo **73** = (3) **4** (1939—1940), 457—464. (*234, 235*)
16. Locatelli, P.: Sul principio di **Menabrea**. Boll. Un. Nat. Ital. (2) **2**, 342—347. (*234*)
17. Lohr, E.: Ein thermodynamischer Weg zum Planckschen Strahlungsgesetz. Z. Physik **116**, 454—468. (*262*)
18. Wise, J. A.: Circles of strain. J. Aeronaut. Sci. **7**, 438—440. (*24*)

1941 1. Deuker, E.-A.: Beitrag zur Theorie endlicher Verformungen und zur Stabilitätstheorie des elastischen Körpers. Deutsche Math. **5** (1940—1941), 546—562. (*65, 66, 95, 210*)
2. Meixner, J.: Zur Thermodynamik der Thermodiffusion. Ann. Physik (5) **39**, 333—356. (*159, 241, 243, 245, 255, 257, 295*)
3. Murnaghan, F. D.: The compressibility of solids under extreme pressures. Kármán Anniv. Vol., 121—136. (*33, 33A*)
4. Munk, M.: On some vortex theorems of hydrodynamics. J. Aeronaut. Sci. **5**, 90—96. (*118*)
5. Reissner, E.: A new derivation of the equations for the deformation of elastic shells. Amer. J. Math. **63**, 177—184. (*212*)
6. Serini, R.: Deduzione delle equazioni della dinamica dei sistemi continui senza far uso del principio di **d'Alembert**. Boll. Un. Mat. Ital. (2) **3**, 281—283. (*170*)
7. Signorini, A.: Sulle proprietà di media communi a tutti i sistemi continui. Rend. Accad. Ital. (7) **2**, 728—734. (*222*)
8. Stratton, J. A.: Electromagnetic Theory. New York. (*281, 283, 312*)
9. Synge, J. L., and W. Z. Chien: The intrinsic theory of elastic shells and plates. Kármán Anniv. Vol., 103—120. (*64, 212, 213*)

1942 1. Bergmann, P. G.: Introduction to the Theory of Relativity. New York. (*271, 282*)
2. Carstoiu, I.: Nouveaux points de vue sur quelques théorèmes fondamentaux de la mécanique des fluides. Bull. Politech. Bucarest **13**, 42—45. (*110*)
3. Cisotti, U.: Formule integrali relative alla meccanica dei sistemi continui. Atti 2⁰ Congr. Un. Mat. Ital. (1940), 404—414. (*216, 219*)
4. Ertel, H.: Ein neuer hydrodynamischer Erhaltungssatz. Naturwiss. **30**, 543—544. (*130*)
5. Ertel, H.: Ein neuer hydrodynamischer Wirbelsatz. Meteor. Z. **59**, 277—281. (*130*)
6. Ertel, H.: Über das Verhältnis des neuen hydrodynamischen Wirbelsatzes zum Zirkulationssatz von **V. Bjerknes**. Meteor. Z. **59**, 385—387. (*130*)
7. Ertel, H.: Über hydrodynamische Wirbelsätze. Physik. Z. **43**, 526—529. (*130*)
8. Hay, G. E.: The finite displacement of thin rods. Trans. Amer. Math. Soc. **51**, 65—102. (*63A, 214*)
9. Meixner, J.: Reversible Bewegungen von Flüssigkeiten und Gasen. Ann. Physik (5) **41**, 409—425. (*241, 259*)
10. Moisil, G. C.: Sur le passage des variables de **Lagrange** aux variables d'**Euler** en hydrodynamique. Bull. Math. Soc. Roumaine Sci. **44**, 55—58. (*66*)
11. Signorini, A.: Deformazioni elastiche finite: elasticità di 2⁰ grado. Atti 2⁰ Congr. Mat. Ital. Rome (1940), 56—71. (*34*)
12. Stewart, R. J.: The energy equation for a viscous compressible fluid. Proc. Nat. Acad. Sci. U.S.A. **28**, 161—164. (*241*)
13. Tedone, G.: Qualche applicazione di una proprietà di media dello stress. Boll. Un. Mat. Ital. (2) **4**, 93—99. (*220*)
14. Verschaffelt, J. E.: La thermomécanique de la diffusion des gaz. Bull. Sci. Acad. Roy. Belg. (5) **28**, 455—475. (*159, 243*)
15. Verschaffelt, J. E.: Sur la thermomécanique des fluides en mouvement. Bull. Sci. Acad. Roy. Belg. (5) **28**, 476—489. (*159, 243*)

1943 1. Boyer, C. B.: History of the measurement of heat. II. The conservation of energy. Sci. monthly **57**, 546—554. (*240*)
2. Meixner, J.: Zur Thermodynamik der irreveriblen Prozesse in Gasen mit chemisch reagierenden, dissoziierenden und anregbaren Komponenten. Ann. Physik (5) **43**, 244—270. (*243, 245, 255, 259, 295*)
3. Meixner, J.: Zur Thermodynamik der irreversiblen Prozesse. Z. physik. Chem., Abt. B **53**, 235—263. (*257, 259, 295*)
4. Novozhilov, V.: On a error in a hypothesis of the theory of shells. C. R. Acad. Sci. SSSR. (Doklady) (N.S.) **38**, 160—164. (*213*)
5. Novozhilov, V., and P. Finkelshtein: On an error in the hypothesis of **Kirchhoff** in the theory of shells [in Russian]. Prikl. Mat. Mekh. **7**, 333—340. (*213*)
6. Signorini, A.: Trasformazioni termoelastiche finite, Memoria 1ª. Ann. di Mat. (4) **22**, 33—143. (*12, 27, 30, 31, 34, 37, 38, 39, 41, 45, 47, 50, 59, 210*)

7. Tolotti, C.: Le equazioni lagrangiane della meccanica dei sistemi continui in coordinate generali. Rend. Acad. Napoli (4) **13**, 1–9. (*210*)
8. Tonolo, A.: Teoria tensoriale delle deformazioni finite dei corpi solidi. Rend. Sem. Mat. Padova **14**, 43–117. (*12, 29, 31, 34*)
1944 *1.* Carstoiu, I.: Sur la condition des accélérations dans un fluide incompressible. Math. Timişoara **20**, 172–173. (*160*)
2. Chien, W.-Z.: The intrinsic theory of thin shells and plates. Part I. General theory. Quart. appl. Math. **1**, 297–327. (*64*)
3. Cisotti, U.: Deformazione finite isotrope. Rend. Ist. Lombardo **77** = (3) **8**, 73—79. (*27, 31*)
4. Cisotti, U.: Influenza delle rotazione finite nelle deformazioni infinitesime di un solido elastico. Rend. Ist. Lombardo **77** = (3) **8**, 249–252. (*55, 57*)
5. Dupont, P.: Représentation triangulaire du tenseur des contraintes en élasticité. C. R. Acad. Sci., Paris **218**, 778–780. (*24*)
6. Jacob, C.: Sur une interprétation de l'équation de continuité hydrodynamique. Bull. Math. Soc. Roumaine Sci. **46**, 81–90. (*160*)
7. Lamm, O.: Theorie der Diffusion ternärer Lösungen. Arkiv. Kemi Miner. Geol. **18** A (1944/1945), No. 2, 10pp. (*295*)
8. Leaf, B.: The principles of thermodynamics. J. Chem. Phys. **12**, 89–98. (*245*)
9. Luneberg, R. K.: Mathematical Theory of Optics. Mimeographed notes of a series of lectures given at Brown University, Providence. (*278*)
10. Rivaud, J.: Remarques sur le problème de l'élasticité non linéaire. C. R. Acad. Sci., Paris **218**, 698–700. (*210*)
11. Seth, B. R.: Consistency equations of finite strain. Proc. Indian Acad. Sci. A **20**, 336–339. (*34*)
12. Vlasov, V. Z.: Equations of compatibility of deformation in curvilinear co-ordinates [in Russian]. Prikl. Mat. Mekh. **8**, 301–306. (*57*)
1945 *1.* Charreau, A.: Sur des représentations planes du tenseur des contraintes dans un milieu continu. C. R. Acad. Sci., Paris **220**, 642–643. (*24*)
2. Cowling, T. G.: The electrical conductivity of an ionized gas in a magnetic field, with applications to the solar atmosphere and the ionosphere. Proc. Roy. Soc. Lond., Ser. A **183**, 453–479. (*295*)
3. Kuzmin, R. O.: On **Maxwell's** formulae in the theory of elasticity. C. R. Acad. Sci. USSR. (Doklady) **49**, 326–328. (*227*)
4. Onsager, L.: Theories and problems of liquid diffusion. Ann. N.Y. Acad. Sci. **46**, 241–265. (*295*)
5. Reiner, M.: A mathematical theory of dilatancy. Amer. J. Math. **67**, 350–362. (*299*)
6. Truesdell, C.: The membrane theory of shells of revolution. Trans. Amer. Math. Soc. **58**, 96–166. (*213*)
1946 *1.* Cady, W. G.: Piezoelectricity. London: McGraw-Hill. (*312*)
2. Carstoiu, I.: Généralization des formules de **Helmholtz** et de **Cauchy** pour un fluide visqueux incompressible. C. R. Acad. Sci., Paris **223**, 1095–1096. (*101*)
3. Carstoiu, I.: Sur le vecteur tourbillon de l'accélération et les fonctions qui s'y rattachent. Bull. Cl. Sci. Acad. Roumaine **29**, 207–214. (*101, 109*)
4. Carstoiu, I.: Sur le mouvement tourbillonaire à $\Omega =$ const. d'un fluide parfait incompressible. Bull. Cl. Sci. Acad. Roumaine **28**, 589–592. (*138*)
5. Hayes, W. D.: Transformation groups of the thermodynamic variables. Quart. appl. Math. **4**, 227–232. (*251*)
6. Kampé de Feriet, J.: Sur la décroissance de l'énergie cinétique d'un fluide visqueux incompressible occupant un domaine plan borné. C. R. Acad. Sci., Paris **223**, 1096–1098. (*93, 116*)
7. Leaf, B.: Phenomenological theory of transport processes in fluids. Phys. Rev. (2) **70**, 748–758. (*215, 243, 255, 295*)
8. Torre, C.: Über den plastischen Körper von **Prandtl**. Zur Theorie der Mohrschen Grenzkurve. Öst. Ing.-Arch. **1**, 36–50. (*24*)
9. Valcovici, V.: Sur une interprétation cinématique du tourbillon et sur la rotation des directions principales de la déformation. Mathematica (Timişoara) **22**, 57–65. (*86*)
10. Weyl, H.: The Classical Groups, their Invariants and Representations. Princeton. (*268*)
1947 *1.* Caldonazzo, B.: Sui moti liberi di un mezzo continuo. Ann. di Mat. (4) **26**, 43—55. (*141*)
2. Carstoiu, I.: De la circulation dans un fluide visqueux incompressible. C. R. Acad. Sci., Paris **224**, 534–535. (*136*)

3. Carstoiu, I.: Recherches sur la théorie des tourbillons. MS thesis, Univ. Paris. (*123, 129*)
4. Dewulf, N.: Cercle de **Mohr** et coniques d'élasticité. Houille Blanche **2**, 481–487. (*24*)
5. Finck, J. L.: On metastable states of equilibrium in thermodynamics. J. Franklin Inst. **243**, 1–12. (*253*)
6. Hicks, B., P. Guenther and R. Wasserman: New formulation of the equations for compressible flow. Quart. appl. Math. **5**, 357–361. (*132*)
7. Jung, F.: Der Culmannsche und der Mohrsche Kreis. Öst. Ing.-Arch. **1**, 408–410. (*24*)
8. Kampé de Feriet, J.: On a property of the Laplacian of a function in a two dimensional bounded domain, when the first derivatives of the function vanish at the boundary. Math. Mag. **21**, 74–79. (*116*)
9. Michal, A. D.: Matrix and Tensor Calculus. New York: John Wiley & Sons; London: Chapman-Hall, Ltd. xiii + 132 pp. (*14, 19*)
10. Moufang, R.: Volumentreue Verzerrungen bei endlichen Formänderung. Z. angew. Math. Mech. **25–27**, 209–214. (*33*)
11. Pailloux, H.: Sur les équations du mouvement des fluides parfaits. C. R. Acad. Sci., Paris **225**, 1122–1124. (*211*)
12. Prigogine, I.: Étude Thermodynamique des Phénomènes Irréversibles. Paris and Liège: Dunod-Desoer. 143 pp. (*157, 158, 215, 243, 245, 255, 259, 261, 295, 306*)
13. Rivlin, R. S.: Hydrodynamics of non-Newtonian fluids. Nature **160**, 611–613. (*299*)
14. Sommerfeld, A.: Vorlesungen über theoretische Physik. II, Mechanik deformierbarer Medien, 2nd ed. Wiesbaden. (*302*)
15. Swainger, K.: Stress-strain compatibility in greatly deformed engineering metals. Phil. Mag. (7) **38**, 422–439. (*33A*)
16. Swainger, K. H.: Large strains and displacements in stress-strain problems. Nature, Lond. **160**, 399–400. (*33A*)
17. Truesdell, C. A., and R. N. Schwartz: The Newtonian mechanics of continua. U.S. Nav. Ord. Lab. Mem. 9223. (*293*)
18. Truesdell, C., and R. Prim: **Zorawski**'s kinematical theorems. U.S. Nav. Ord. Lab. Mem. 9354. (*75*)

1948 *1.* Ballabh, R.: On coincidence of vortex and stream lines in ideal liquids. Ganita **1**, 1–14. (*112*)
2. Bilimovitch, A.: Aires et volumes vélocidiques et hodographiques dans un mouvement du fluide. Acad. Serbe Sci. Publ. Inst. Math. **2**, 37–52. (*115, 160*)
3. Byushgens, S.: Geometry of steady flow of an ideal incompressible fluid [in Russian]. Isv. Akad. Nauk SSSR., Ser. Mat. **12**, 481–512. (*110*)
4. Carstoiu, I.: Sur certaines formules intégrales dans le mouvement d'un fluide. C. R. Acad. Sci., Paris **227**, 1337–1339. (*160*)
5. Castoldi, L.: Deduzione variazionale delle equazioni della dinamica dei continui deformabili. Nuovo Cim. (9) **5**, 140–149. (*210*)
6. Castoldi, L.: Superficie e linee di **Bernoulli** nel moto stazionario di un fluido reale. Atti Accad. Ligure **4** (1947), 21–25. (*124*)
7. Chien, W.-Z.: Derivation of the equations of equilibrium of an elastic shell from the general theory of elasticity. Sci. Rep. Tsing-Hua Univ. **A 5**, 240–251. (*213*)
8. Courant, R., and K. Friedrichs: Supersonic Flow and Shock Waves. New York: Interscience. xvi + 464 pp. (*241, 253, 265*)
9. Dorn, J. E., and A. J. Latter: Stress-strain relations for finite elastoplastic deformations. J. appl. Mech. **15**, 234–236. (*33A*)
10. Eckart, C.: The thermodynamics of irreversible processes IV: The theory of elasticity and anelasticity. Phys. Rev. (2) **73**, 373–382. (*34, 96*)
11. Finck, J. L.: Thermodynamics, Part I: The second law from the standpoint of the equation of state; Part II: Work, heat, and temperature concepts, and an examination of the temperature scale. J. Franklin Inst. **245**, 301–317, 365–378. (*253*)
12. Lohr, E.: Quantenstatistik und Kontinuumsphysik. Z. Naturfosch. **3**a, 625–636. (*262*)
13. Manarini, M.: Sulle equazioni della dinamica dei fluidi perfetti. Boll. Un. Mat. Ital. (3) **3**, 111–114. (*211*)
14. Manarini, M.: Sui paradossi di **d'Alembert** e di **Brillouin** nella dinamica dei fluidi. Rend. Lincei (8) **4**, 427–433. (*202*)
15. Mooney, M.: The thermodynamics of a strained elastomer. I. General analysis. J. appl. Phys. **19**, 434–444. (*33A*)

16. MOREAU, J.-J.: Sur deux théorèmes généraux de la dynamique d'un milieu incompressible illimité. C. R. Acad. Sci., Paris **226**, 1420–1422. (*117*)
17. NEMÉNYI, P., and R. PRIM: Some properties of rotational flow of a perfect gas. Proc. Nat. Acad. Sci. U.S.A. **34**, 119–124; **35**, 116 (1949). (*132*)
18. NOVOZHILOV, V. V.: Foundations of the Nonlinear Theory of Elasticity [in Russian]. (We have seen this work only in the English translation by F. BAGEMIHL, H. KROMM and W. SEIDEL, Rochester 1953.) (*12, 34, 36, 55, 56, 210*)
19. PACELLA, G. B.: Su una proprietà della meccanica dei corpi continui e una deduzione geometrica della legge di **Hooke**. Rend. Lincei (8) **5**, 31–37. (*209*)
20. PLATRIER, C.: Équations universelles de l'équilibre des milieux les plus généraux à tenseur symétrique. Bull. Acad. Belg., Cl. Sci. **34**, 274–277. (*210*)
21. PRIM, R.: Extension of **Crocco's** theorems to flows having a non-uniform stagnation enthalpy. Phys. Rev. (2) **73**, 186. (*133*)
22. PRIM, R. C.: On doubly-laminar flow fields having a constant velocity magnitude along each stream-line. U.S. Naval Ord. Lab. Mem. 9762. (*111*)
23. REINER, M.: Elasticity beyond the elastic limit. Amer. J. Math. **70**, 433–446. (*32, 33, 33A, 95, 303*)
24. RICHTER, H.: Das isotrope Elastizitätsgesetz. Z. angew. Math. Mech. **28**, 205–209. (*33, 33A*)
25. RICHTER, H.: Bemerkung zum Moufangschen Verzerrungsdeviator. Z. angew. Math. Mech. **28**, 126–127. (*33*)
26. RIVLIN, R. S.: The hydrodynamics of non-Newtonian fluids I. Proc. Roy. Soc. Lond. A **193**, 260–281. (*299*)
27. RIVLIN, R. S.: Large elastic deformations of isotropic materials IV. Further developments of the general theory. Phil. Trans. Roy. Soc. A **241**, 379–397. (*30*)
28. STRANG, J. A.: Superposable fluid motions. Comm. fac. sci. Ankara **1**, 1–32. (*114*)
29. SWAINGER, K. H.: Strain energy in greatly deformed elastic or inelastic anisotropic engineering metals. J. Franklin Inst. **245**, 501–516. (*33A*)
30. SWAINGER, K. H.: Large displacements with small strains in loaded structures. J. appl. Mech. **15**, 45–52. (*33A*)
31. TOLMAN, R. C., and P. C. FINE: On the irreversible production of entropy. Rev. Mod. Phys. **20**, 51–77. (*257*)
32. TRUESDELL, C.: On the differential equations of slip flow. Proc. Nat. Acad. Sci. U.S.A. **34**, 342–347. (*293*)
33. TRUESDELL, C.: A new definition of a fluid. U.S. Nav. Ord. Lab. Mem. 9487. (*293*)
34. TRUESDELL, C.: On the total vorticity of motion of a continuous medium. Phys. Rev. (2) **73**, 510–512. (*131*)
35. TRUESDELL, C.: On the transfer of energy in continuous media. Phys. Rev. (2) **73**, 513–515. (*241*)
36. TRUESDELL, C.: Généralisation de la formule de **Cauchy** et des théorèmes de **Helmholtz** au mouvement d'un milieu continu quelconque. C. R. Acad. Sci., Paris **227**, 757–759. (*101, 129*)
37. TRUESDELL, C.: Une formule pour le vecteur tourbillon d'un fluide visqueux élastique. C. R. Acad. Sci., Paris **227**, 821–823. (*101*)
38. WEBER, C.: Spannungsfunktionen des dreidimensionalen Kontinuums. Z. angew. Math. Mech. **28**, 193–197. (*227*)
39. ZELMANOV, A. N.: Application of convected co-ordinates in non-relativistic mechanics [in Russian]. Dokl. Akad. Nauk SSSR. **61**, 993–996. (*34, 66B, 95, 150, 210*)

1949 1. BERKER, R.: Sur certaines propriétés du rotationnel d'un champ vectoriel qui est nul sur la frontière de son domaine de définition. Bull. sci. math. (2) **73**, 163–176. Abstract, C. R. Acad. Sci., Paris **228**, 1630–1632. (*116, 119*)
2. BERKER, R.: Inégalité vérifée par l'énergie cinétique d'un fluide visqueux incompressible occupant un domaine spatial borné. Bull. tech. Univ. Istanbul **2**, 41–50. Summary, C. R. Acad. Sci., Paris **228**, 1327–1329. (*93*)
3. CORBEN, C.: The transformation theory of thermodynamics. Phys. Rev. (2) **76**, 166. (*251*)
4. CURTISS, C. F., and J. O. HIRSCHFELDER: Transport properties of multicomponent gas mixtures. J. Chem. Phys. **17**, 550–555. (*295*)
5. DONDER, T. DE, et F. H. VAN DEN DUNGEN: Sur les principes variationnels des milieux continus. Bull. Acad. Roy. Belg., Cl. Sci. (5) **35**, 841–846. (*236A*)
6. ERGUN, A. N.: Some cases of superposable fluid motions. Comm. fac. sci. Ankara **2**, 48–88. (*114*)
7. ERTEL, H., u. C.-G. ROSSBY: Ein neuer Erhaltungssatz der Hydrodynamik. Sitzgsber. Akad. Wiss. Berlin, math. nat. Kl. 1949, 1–11. (*130*)

8. Ertel, H., and C.-G. Rossby: A new conservation theorem of hydrodynamics. Geofisica pura appl. **14**, 189–193. (*130*)

9. Ertel, H., u. H. Köhler: Ein Theorem über die stationäre Wirbelbewegung kompressibler Flüssigkeiten. Z. angew. Math. Mech. **29**, 109–113. (*139, 163*)

10. Finck, J. L.: Thermodynamics: the meaning of the first law, and its relation to the behavior of material systems. Phys. Rev. (2) **67**, 166. (*253*)

11. Finzi, B., e M. Pastori: Calcolo Tensoriale e Applicazioni. Bologna: Zanichelli. vii + 427 pp. (*164, 226*)

12. Gleyzal, A.: A mathematical formulation of the general continuous deformation problem. Quart. appl. Math. **6**, 429–437. (*66B, 210*)

13. Hencky, H.: Mathematical principles of rheology. Research **2**, 437–443. (*33A, 90*)

14. Hicks, B. L.: On the characterization of fields of diabatic flow. Quart. appl. Math. **6**, 405–416. (*132*)

15. Kondo, K.: A proposal of a new theory concerning the yielding of materials based on Riemannian geometry. J. Japan Soc. appl. Mech. **2**, 123–128, 146–151. (*34*)

16. LaMer, V. K.: Some current misconceptions of **N. L. Sadi Carnot's** memoir and cycle. Science **109**, 598. (*240*)

17. Manarini, M.: Sui paradossi di **d'Alembert** e di **Brillouin** nella dinamica dei fluidi. Boll. Un. Mat. Ital. (3) **4**, 352–353. (*202*)

18. McVittie, G. C.: A systematic treatment of moving axes in hydrodynamics. Proc. Roy. Soc. Lond. A **196**, 285–300. (*144, 152, 154, 211, 241*)

19. Moreau, J.-J.: Sur l'interprétation tourbillonaire des surfaces de glissement. C. R. Acad. Sci., Paris **228**, 1923–1925. (*186*)

20. Moreau, J.-J.: Sur la dynamique d'un écoulement rotationnel. C. R. Acad. Sci., Paris **229**, 100–102. (*117*)

21. Neményi, P., and R. Prim: On the steady Beltrami flow of a perfect gas. Proc. 7th Int. Congr. Appl. Mech. (1948) **2**, 300–314. (*132*)

22. Partington, J. R.: An Advanced Treatise on Physical Chemistry, vol. 1. London, New York and Toronto. (*239, 240, 245*)

23. Peretti, G.: Significato del tensore arbitrario che interviene nell'integrale generale delle equazioni della statica dei continui. Atti Sem. Mat. Fis. Univ. Modena **3**, 77–82. (*227*)

24. Prigogine, I.: Le domaine de la validité de la thermodynamique des phénomènes irreversibles. Physica, Haag **15**, 272–284. (*259*)

25. Prim, R.: A note on the substitution principle for steady gas flow. J. appl. Phys. **20**, 448–450. (*132*)

26. Richter, H.: Verzerrungstensor, Verzerrungsdeviator, und Spannungstensor bei endlichen Formänderungen. Z. angew. Math. Mech. **29**, 65–75. (*32, 33*)

27. Signorini, A.: Trasformazioni termoelastiche finite, Memoria 2ª. Ann. Mat. (4) **30**, 1–72. (*256A*)

28. Storchi, E.: Integrazione delle equazioni indefinite della statica dei sistemi continui su una superficie di rotazione. Rend. Lincei (8) **7**, 227–231. (*229*)

29. Supino, G.: Sul moto irrotazionale dei liquidi viscosi, I. Rend. Lincei (8) **6**, 615–620. (*113*)

30. Swainger, K. H.: **Saint-Venant's** and **Filon's** finite strains: definitions non-linear in displacement gradients. Nature, Lond. **164**, 23–24. (*33A*)

31. Thomas, T. Y.: The fundamental hydrodynamical equations and shock conditions for gases. Math. Mag. **22**, 169–189. (*192*)

32. Tonolo, A.: Sopra una classe di deformazioni finite. Ann. Mat. (4) **29**, 99–114. (*27*)

33. Truesdell, C.: A new definition of a fluid, I. The Stokesian fluid. Proc. 7th Int. Congr. Appl. Mech. (1948) **2**, 351–364 = U.S. Naval Res. Lab. Rep. No. P-3457. (*293, 298, 299*)

34. Truesdell, C.: A new definition of a fluid, II. The Maxwellian fluid. U.S. Naval Res. Lab. Rep. No. P-3553 = [1351, *27*]. (*293, 307*)

35. Truesdell, C.: Deux formes de la transformation de **Green**. C. R. Acad. Sci., Paris **229**, 1199–1200. (*118*)

36. Weissenberg, K.: Geometry of rheological phenomena (1946–1947). The Principles of Rheological Measurement, pp. 36–65. London. (*32*)

37. Weissenberg, K.: Abnormal substances and abnormal phenomena of flow. Proc. Int. Congr. Rheology (1948), I–29 – I–46. (*42*)

38. Weyl, H.: Shock waves in arbitrary fluids. Comm. pure appl. Math. N.Y. Univ. **2**, 103–122. (*265*)

39. ZERNA, W.: Beitrag zur allgemeinen Schalenbiegetheorie. Ing.-Arch. **17**, 149–164. (*213*)

1950 1. ALFVÉN, H.: Cosmical Electrodynamics. London. (*313*)

2. BLOKH, V.: Stress functions in the theory of elasticity [in Russian]. Prikl. Mat. Mekh. **14**, 415–422. (*227, 228*)

3. CASTOLDI, L.: Linee sostanziali inestese nelle deformazioni finite dei continui materiali. Atti Accad. Ligure **6**, 165–169. (*26*)

4. CASTOLDI, L.: Linee sostanziali nel moto un continuo deformabile e moti con linee di flusso (e di corrente) „sostanzialmente permanente". Rend. Ist. Lombardo (3) **14**, 259–264. (*75*)

5. CROCCO, L.: On a kind of stress-function for the study of non-isentropic two-dimensional motion of gases. Proc. 7th Int. Congr. Appl. Mech. London (1948) **2**, 315–329. (*224*)

6. DELVAL, J.: Le principe de la moindre contrainte appliqué à la dynamique des fluides incompressibles. Bull. Acad. Roy. Belg., Cl. Sci. (5) **36**, 639–648. (*130*)

7. ERTEL, H.: Ein Theorem über asynchron-periodische Wirbelbewegungen kompressibler Flüssigkeiten. Misc. Acad. Berol. **1**, 62–68. (*135*)

8. GORDON, A. N.: A linear theory of finite strain. Nature, Lond. **166**, 657. (*33A*)

9. GREEN, A. E., and W. ZERNA: The equilibrium of thin elastic shells. Quart. J. Mech. appl. Math. **3**, 9–22. (*213*)

10. GREEN, A. E., and W. ZERNA: Theory of elasticity in general coordinates. Phil. Mag. (7) **41**, 313–336. (*12, 34, 66B*)

11. HÖLDER, E.: Über die Variationsprinzipe der Mechanik der Kontinua. Ber. Verh. Akad. Wiss. Leipzig **97**, No. 2, 13 pp. (*236A*)

12. IRVING, J., and J. KIRKWOOD: The statistical mechanical theory of transport processes. IV, The equations of hydrodynamics. J. Chem. Phys. **18**, 817–829. (*1*)

13. JOHNSON, M. H., and E. O. HULBERT: Diffusion in the ionosphere. Phys. Rev. (2) **79**, 802–807. (*295*)

14. KONDO, K.: On the dislocation, the group of holonomy and the theory of yielding. J. Japan Soc. appl. Mech. **3**, 107–110. (*34*)

15. KONDO, K.: On the fundamental equations of the theory of yielding. J. Japan Soc. appl. Mech. **3**, 184–188. (*34*)

16. KONDO, K.: The mathematical analyses of the yield point. J. Japan Soc. appl. Mech. **3**, 188–195; **4**, 4–8, 35–38 (1951). (*34*)

17. MASON, W. P.: Piezoelectric Crystals and their Application to Ultrasonics. Toronto, New York and London: D. van Nostrand. (*312*)

18. MOREAU, J.-J.: Relations générales directes entre les actions aerodynamiques et les éléments tourbillonaires. Actes Congr. Int. Méc. Poitiers **4**, 6 pp. (*117*)

19. MORINAGA, K., and T. NÔNO: On stress-functions in general coordinates. J. sci. Hiroshima Univ. A **14**, 181–194. (*227, 230*)

20. NADAI, A.: Theory of Flow and Fracture of Solids, **1**, 2nd ed. New York. (*24*)

21. Ein Kapitel aus der Vorlesung von F. NEUMANN über mechanische Wärmetheorie. Königsberg 1854/55. Ausgearbeitet von C. NEUMANN. Herausgeg. von E. R. NEUMANN. Abh. bayer. Akad. Wiss., math.-nat. Kl. (2), No. 59. (*245*)

22. OLDROYD, J. G.: On the formulation of rheological equations of state. Proc. Roy. Soc. Lond. A **200**, 523–541. (*66, 66B, 104, 150, 293*)

23. OLDROYD, J. G.: Finite strains in an anisotropic elastic continuum. Proc. Roy. Soc. Lond. A **202**, 345–358. (*33A, 34*)

24. PRIM, R., and C. TRUESDELL: A derivation of **Zorawski's** criterion for permanent vector-lines. Proc. Amer. Math. Soc. **1**, 32–34. (*75*)

25. SCHLÜTER, A.: Dynamik des Plasmas I. Z. Naturforsch. **5**a, 72–78. (*295*)

26. SEUGLING, W. R.: Equations of compatibility for finite deformation of a continuous medium. Amer. Math. monthly **57**, 679–681. (*34*)

27. STORCHI, E.: Sulle equazioni indefinite della statica delle membrane tese su generiche superficie. Rend. Lincei **8**, 116–120. (*229*)

28. STORCHI, E.: Integrazione delle equazioni indefinite della statica dei veli tesi su una generica superficie. Rend. Lincei **8**, 326–331. (*229*)

29. SWAINGER, K. H.: Non-coaxiality of principal normal stresses and the 'strain' ellipsoid in the classical theory on infinitesimal deformation. Nature, Lond. **165**, 159–160. (*33A*)

30. SWAINGER, K. H.: Reply to critism of **Gordon** [1950, *8*]. Nature, Lond. **166**, 657–659. (*33A*)

31. SYNGE, J. L.: Note on the kinematics of plane viscous motion. Quart. appl. Math. **8**, 107–108. (*116*)

32. TRUESDELL, C.: A new definition of a fluid, I. The Stokesian fluid. J. Math. pures appl. (9) **29**, 215–244. (*293, 298, 299*)

33. TRUESDELL, C.: **Bernoulli's** theorem for viscous compressible fluids. Phys. Rev. **77**, 535–536. (*124*)

34. TRUESDELL, C.: On the balance between deformation and rotation in the motion of a continuous medium. J. Washington Acad. Sci. **40**, 313–317. (*115, 160*)

35. WEYL, H.: Space-time-matter (English transl. of [1921, *6*], with a new preface). New York: Dover (*266, 310*)

36. ZERNA, W.: Allgemeine Grundgleichungen der Elasticitätstheorie. Ing.-Arch. **18**, 211–220. (*34*)

1951 1. BÉGHIN, H.: Sur la notation de travail dans la mécanique du continu. Ann. Inst. Fourier Grenoble **2** (1950), 173–184. (*217*)

2. BJØRGUM, O.: On Beltrami vector fields and flows, Part I. Univ. Bergen Arbok 1951, Naturv. rekke No. 1. (*112, 117*)

3. DEDECKER, P.: Sur le théorème de la circulation de **V. Bjerknes** et la théorie des invariants intégraux. Publ. Inst. R. Météorol. Belg. misc. No. **36**, 63 pp. (*129*)

4. DEDECKER, P.: Sur le théorème de la circulation de **V. Bjerknes**. Mém. Inst. R. Météorol. Belg. **48**, 4 pp. (*129*)

5. DIRAC, P. A. M.: A new classical theory of the electron, I. II, III. Proc. Roy. Soc. Lond. A **209**, 291–296; A **212**, 330–339; A **223**, 438–445. (*276*)

6. FINCK, J. L.: Thermodynamics from a Generalized Standpoint. New York. (*253*)

7. FILONENKO-BORODICH, M. M.: The problem of the equilibrium of an elastic parallelepiped subject to assigned loads on its boundaries [in Russian]. Prikl. Mat. Mekh. **15**, 137–148. (*227*)

8. FILONENKO-BORODICH, M. M.: Two problems on the equilibrium of an elastic parallelepiped [in Russian]. Prikl. Mat. Mekh. **15**, 563–574. (*227*)

9. GALIMOV, K. Z.: Invariant form of the conditions of compatibility in finite deformation [in Russian]. Dokl. Akad. Nauk SSSR. **77**, 577–580. (*34*)

10. GIESE, J. H.: Streamfunctions for three-dimensional flows. J. Math. Phys. **30**, 31–35. (*163*)

11. HAASE, R.: Zur Thermodynamik der irreversiblen Prozesse. Z. Naturforsch. **6**a, 420–437, 522–540. (*245*)

12. IRVING, J. H., and R. W. ZWANZIG: The statistical mechanical theory of transport processes. V. Quantum hydrodynamics. J. Chem. Phys. **19**, 1173–1180. (*1, 2*)

13. JARRE, G.: Sul moto relativo nei mezzi continui. Atti Accad. Torino **85**, 183–191. (*146*)

14. JOHNSON, M. H.: Diffusion as hydrodynamic motion. Phys. Rev. (2) **84**, 566–568. (*295*)

15. LODGE, A. S.: The compatibility conditions for large strains. Quart. J. Mech. appl. Math. **4**, 85–93. (*34*)

16. LODGE, A. S.: On the use of convected coordinate systems in the mechanics of continuous media. Proc. Cambridge Phil. Soc. **47**, 575–584. (*66, 66B, 293*)

17. MAZUR, P., et I. PRIGOGINE: Sur l'hydrodynamique des mélanges liquides de He^3 et He^4. Physica, Haag **17**, 680–693. (*215, 243, 259*)

18. MURNAGHAN, F. D.: Finite Deformation of an Elastic Solid. New York and London. (*12*)

19. POPOV, S. G.: Remark on the integrals of **Bernoulli** and **Lagrange (Cauchy)**. [in Russian]. Moskov. Gos. Univ. Zap. 152, Mekh. **3**, 43–46. (*125*)

20. PRAGER, W., and P. G. HODGE jr.: Theory of Perfectly Plastic Solids. New York. (*24*)

21. PRIGOGINE, I., et P. MAZUR: Sur deux formulations de l'hydrodynamique et le problème de l'hélium liquide II. Physica, Haag **17**, 661–679. (*215, 243, 258, 259*)

22. RIVLIN, R. S., and D. W. SAUNDERS: Large elastic deformations of isotropic materials, VII. Experiments on the deformation of rubber. Phil. Trans. Roy. Soc. Lond. A **243**, 251–288. (*40*)

23. SCHLÜTER, A.: Dynamik des Plasmas II. Z. Naturforsch. **6**a, 73–78. (*295*)

24. SMITH-WHITE, W. B.: The **Poisson-Kelvin** hypothesis and the theory of dielectrics. J. Roy. Soc. N.S. Wales **85**, 82–112. (*283*)

25. SYNGE, J. L.: Conditions satisfied by the expansion and vorticity of a viscous fluid in a fixed container. Quart. appl. Math. **9**, 319–322. (*116*)

26. TORRE, C.: Über die physikalische Bedeutung der Mohrschen Hüllkurve. Z. angew. Math. Mech. **31**, 275–277. (*24*)

27. TRUESDELL, C.: A new definition of a fluid. II. The Maxwellian fluid. J. Math. pures appl. **30**, 111–158. (*293, 304, 307*)

28. Truesdell, C.: Vorticity averages. Canad. J. Math. **3**, 69–86. *(117, 118, 131)*

29. Truesdell, C.: A form of **Green's** transformation. Amer. J. Math. **73**, 43–47. *(112, 117)*

30. Truesdell, C.: Caractérisation des champs vectoriels qui s'annulent sur une frontière fermée. C. R. Acad. Sci., Paris **232**, 1277–1279. *(116)*

31. Truesdell, C.: Analogue tri-dimensionnel au théorème de M. **Synge** sur les champs vectoriels plans qui s'annulent sur une frontière fermée. C. R. Acad. Sci., Paris **232**, 1396–1397. *(116)*

32. Truesdell, C.: Vereinheitlichung und Verallgemeinerung der Wirbelsätze ebener und rotationssymmetrischer Gasbewegungen. Z. angew. Math. Mech. **31**, 65–71. An abstract under the title "A new vorticity theorem" appears in Proc. Int. Congr. Math. (1950), **1**, 639–640 (1952). *(132, 133)*

33. Truesdell, C.: On **Ertel's** vorticity theorem. Z. angew. Math. Phys. **2**, 109–114. *(130)*

34. Truesdell, C.: Proof that **Ertel's** vorticity theorem holds in average for any medium suffering no tangential acceleration on the boundary. Geofis. pura appl. **19**, No. 3–4, 1–3. *(131)*

35. Truesdell, C.: On the equation of the bounding surface. Bull. Tech. Univ. Istanbul **3**, 71–77. *(182, 184)*

36. Truesdell, C.: On the velocity of sound in fluids. J. Aeronaut. Sci. **18**, 501. *(297)*

37. Van den Dungen, F.: Note on the **Hamel-Synge** theorem. Quart. appl. Math. **9**, 203–204. *(116)*

38. Verschaffelt, J. E.: La thermomécanique des phénomènes de transport. J.Phys Radium **12**, 93–98. *(159)*

39. Whittaker, E. T.: A History of the Theories of Aether and Electricity, vol. 1. New York. *(266, 276, 283, 289, 302, 308)*

1952 1. Arzhanikh, I. S.: Tensor stress functions of hydrodynamics [in Russian]. Dokl. Akad. Nauk SSSR. **83**, 195–198. *(211, 230)*

2. Coburn, N.: Intrinsic relations satisfied by the vorticity and velocity vectors in fluid flow theory. Michigan Math. J. **1**, 113–130; **2**, 41–44 (1953). *(99, 110)*

3. De Groot, S. R.: Thermodynamics of Irreversible Processes. Amsterdam: North-Holland Publ. Co. xvi + 242 pp. *(158, 243, 245, 255, 257, 259, 261, 295, 306)*

4. Ertel, H.: Über die physikalische Bedeutung von Funktionen, welche in der Clebsch-Transformation der hydrodynamischen Gleichungen auftreten. Sitzgsber. Akad. Wiss. Berlin, Kl. Math. allg. Naturw. **1952**, No. 3. *(135)*

5. Fényes, I.: Die Anwendung der mathematischen Prinzipien der Mechanik in der Thermodynamik. Z. Physik **132**, 140–145. *(251)*

6. Fényes, I.: Ergänzungen zur axiomatischen Begründung der Thermodynamik. Z. Physik **134**, 95–100. *(245)*

7. Grad, H.: Statistical mechanics, thermodynamics, and fluid dynamics of systems with an arbitrary number of integrals. Comm. pure appl. math. **5**, 455–494. *(157, 205)*

8. Grioli, G.: Relazioni quantitative per lo stato tensionale di un qualunque sistema continuo e per la deformazione di un corpo elastico in equilibrio. Ann. di Mat. (4) **33**, 239–246. *(221, 222)*

9. Grioli, G.: Integrazione del problema della statica delle piastre omogenee di spessore qualunque. Ann. Scuola Norm. Pisa (3) **6**, 31–49. *(222)*

10. Heinrich, G.: Der Energietransport in strömenden Medien. Z. angew. Math. Mech. **32**, 286–288. *(217)*

11. Hershey, A. V.: A review of the definitions of finite strain. Proc. 1st U.S. Nat. Congr. Appl. Mech. (1951), 473–478. *(33A)*

12. Kirkwood, J. G., and B. Crawford jr.: The macroscopic equations of transport J. Physic. Chem. **56**, 1048–1051. *(243, 285, 295)*

13. Moreau, J.-J.: Bilan dynamique d'un écoulement rotationnel. J. Math. pures appl. (9) **31**, 355–375; **32**, 1–78 (1953). *(117, 130, 186, 297)*

14. Nardini, R.: Sul valore medio dello stress per particolari sollecitazioni. Ann. Univ. Ferrara sez. VII (2) **1**, 89–91. *(220)*

15. Partington, J. R.: Advances in thermodynamics. Nature, Lond. **170**, 730–732. *(245)*

16. Prim, R. C.: Steady rotational flow of ideal gases (1949). J. Rational Mech. Anal. **1**, 425–497. *(111, 132, 133)*

17. Richter, H.: Zur Elasticitätstheorie endlicher Verformungen. Math. Nachr. **8**, 65–73. *(37)*

18. Storchi, E.: Le superficie eccezionali nella statica delle membrane. Rev. mat. Univ. Parma **3**, 339–360. *(229)*

19. Szebehely, V. G.: Generalization of the dimensionless frequency parameter in unsteady flows. David Taylor Model Basin Rep. 833. (*100*)

20. Tonolo, A.: Sopra un problema di **Darboux** della meccanica dei mezzi continui. Ann. Univ. Ferrara sez. VII (2) **1**, 103–109. (*38*)

21. Truesdell, C.: The mechanical foundations of elasticity and fluid dynamics. J. Rational Mech. Anal. **1**, 125–300. (*12, 14, 17, 29, 33A, 34, 44, 46, 50, 57, 95, 96, 210, 256A, 293, 298, 299, 303*)

22. Truesdell, C.: A program of physical research in classical mechanics. Z. angew. Math. Phys. **3**, 79–95. (*2, 293, 298, 299*)

23. Truesdell, C.: Vorticity and the Thermodynamic State in a Gas Flow. Mém. sci. math. No. 119, Paris. (*132, 133, 297*)

24. Whaples, G.: **Carathéodory's** temperature equations. J. Rational Mech. Anal. **1**, 301–307. (*245*)

1953 *1.* Bilimovitch, A.: Sur l'homogénisation des équations de nature vélocidique. Acad. Serbe Sci. Publ. Inst. Math. **5**, 29–34. (*115, 160*)

2. Bjørgum, O., and T. Godal: On Beltrami vector vields and flows, Part II. Univ. Bergen Årbok 1952, Naturv. rekke No. 13.

3. Carstoiu, I.: Sur la déformation d'une particule dans le mouvement d'un fluide. C. R. Acad. Sci., Paris **236**, 2209–2211. (*101*)

4. Castoldi, L.: Teoremi di Bernoulli per fluidi comprimibili viscosi. Atti Accad. Ligure **9** (1952), 215–222. (*125*)

5. Castoldi, L.: Sui moti di fluidi reali per cui si verifica una esatta linearizzazione della equazione dinamica. Atti Accad. Ligure **9** (1952), 222–227. (*122, 127*)

6. Corson, E. M.: Introduction to Tensors, Spinors and Relativistic Wave-Equations (Relation Structure). New York. (*282*)

7. Darwin, C.: Note on hydrodynamics. Proc. Cambridge Phil. Soc. **49**, 342–354. (*71*)

8. Defrise, P.: Analyse géométrique de la cinématique des milieux continus. Inst. R. Météorol. Belg. Publ. sér. B, No. 16, 63 pp. (*12, 152, 153*)

9. de Groot, S.: Hydrodynamics and thermodynamics. Proc. Symp. appl. Math. **4**, 87–99. (*245*)

10. Fortak, H.: Zur Bedeutung der in der Clebsch-Transformation der hydrodynamischen Gleichungen auftretenden Funktionen. Acta Hydrophys. **1**, 145–150. (*135*)

11. Geiringer, H.: Some recent results in the theory of an ideal plastic body. Adv. appl. Mech. **3**, 197–294. (*300*)

12. Grioli, G.: Proprietà di media ed equilibrio elastico. Atti 4to Congr. Un. Mat. Ital. (1951) **1**, 68–77. (*221, 222*)

13. Howard, L. N.: Constant speed flows. Thesis, MS in Princeton University Library. (*111*)

14. Kilchevski, N. A.: Stress, velocity, and density functions in static and dynamic problems in the mechanics of continuous media [in Russian]. Dokl. Akad. Nauk SSSR. **92**, 895–898. (*228, 230*)

15. Langhaar, H.: An invariant membrane stress function for shells. J. appl. Mech. **20**, 178–182. (*229*)

16. Langhaar, H.: The principle of complementary energy in nonlinear elasticity theory. J. Franklin Inst. **256**, 255–264. (*232A*)

17. Masuda, H.: A new proof of **Lagrange's** theorem in hydrodynamics. J. Phys. Soc. Japan **8**, 390–393. (*136*)

18. McVittie, G. C.: A method of solution of the equations of classical gas dynamics using **Einstein's** equations. Quart. appl. Math. **11**, 327–336. (*225, 230*)

19. Mişicu, M.: Echilibrul mediilor continue cu deformări mări. Acad. Repub. Pop. Romane. Stud. Cerc. Mec. Metalurgie **4**, 31–53. (*12*)

20. Nyborg, W. L.: Acoustic streaming equations: laws of rotational motion for fluid elements. J. Acoust. Soc. **25**, 938–944. (*170*)

21. Platrier, C.: Conditions d'intégrabilité du tenseur de déformation totale dans une transformation finie d'un milieu à trois dimensions. Ann. Ponts Chaussées **123**, 703–709. (*34*)

22. Platrier, C.: Conditions d'intégrabilité du tenseur de déformation totale dans une transformation finie d'un milieu à trois dimensions. Bull. Acad. Roy. Belg., Cl. Sci. (5) **39**, 490–494. (*34*)

23. Prager, W.: Three-dimensional plastic flow under uniform stress. Brown univ. tech. rep. No. 95, August = Rev. fac. sci. Univ. Istanbul **19**, 23–27 (1954). (*142*)

24. Pratelli, A.: Principi variazionali nella meccanica dei fluidi. Rend. Ist. Lombardo (3) **17** (86), 484–500. (*94, 163*)

25. PRATELLI, A.: Sulla stazionarietà di significativi integrali nella meccanica dei continui. Rend. Ist. Lombardo (3) **17** (86), 714–724. (*227, 235*)
26. REIK, H.: Zur Theorie irreversibler Vorgänge. Ann. Physik (6) **11**, 270–284, 407–419, 420–428; **13**, 73–96. (*245*)
27. REISSNER, E.: On a variational theorem for finite elastic deformations. J. Math. Phys. **32**, 129–135. (*232A*)
28. SCHAEFER, H.: Die Spannungsfunktionen des dreidimensionalen Kontinuums und des elastischen Körpers. Z. angew. Math. Mech. **33**, 356–362. (*226, 227, 229*)
29. STORCHI, E.: Sulle membrane aventi comportamento meccanico eccezionale. Rend. Ist. Lombardo (3) **17** (86), 462–483. (*229*)
30. SZEBEHELY, V. G.: A measure of unsteadiness of time-dependent flows. Proc. 3rd Midwest Conf. on Fluid Mech., Univ. Minn., 221–231. (*100*)
31. TRUESDELL, C.: Two measures of vorticity. J. Rational Mech. Anal. **2**, 173–217. (Partial abstract in Proc. Int. Congr. theor. appl. Mech. Istanbul 1952.) (*91, 100, 121*)
32. TRUESDELL, C.: Corrections and additions to "The mechanical foundations of elasticity and fluid dynamics". J. Rational Mech. Anal. **2**, 593–616. (*12, 14, 28, 37, 150, 304*)
33. TRUESDELL, C.: Generalization of a geometrical theorem of **Euler**. Comm. mat. Helv. **27**, 233–234. (*142*)
34. TRUESDELL, C.: La velocità massima nel moto di Gromeka-Beltrami. Rend. Lincei (8) **13** (1952), 378–379. (*121*)
35. WHITTAKER, E. T.: A History of the Theories of Aether and Elasticity, vol. II. London: Nelson. (*266*)
36. YOSHIMURA, Y.: On the natural shearing strain. Proc. 2nd Japan Mat. Congr. appl. Mech. (1952), 1–4. (*33A*)
37. YOSHIMURA, Y.: On the definition of stress in the finite deformation theory. J. Phys. Soc. Japan **8**, 669–673. (*210*)

1954 1. CARSTOIU, I.: Vorticity and deformation in fluid mechanics. J. Rational Mech. Anal. **3**, 691–712. (*101, 160*)
2. CASTOLDI, L.: Le „condizioni di congruenza" per deformazioni infinitesime non lineari. Atti Ist. Veneto, Cl. sci. mat. nat. **112**, 41–47. (*54, 57*)
3. CASTOLDI, L.: Sopra un classificazione dei comportamenti elastici dei mezzi deformabili. Atti Ist. Veneto, Cl. sci. mat. nat. **112**, 17–30. (*55*)
4. DEAN, W. R.: Note on the motion of an infinite cylinder in a rotating viscous liquid. Quart. J. Mech. appl. Math. **7**, 257–262. (*207*)
5. DUGAS, R.: La Mécanique au XVIIe Siècle. Ed. Griffon, Neuchatel. 620 pp. (*3*)
6. ERICKSEN, J. L., and R. S. RIVLIN: Large elastic deformations of homogeneous anisotropic materials. J. Rational Mech. Anal. **3**, 281–301. (*49, 62, 233*)
7. GREEN, A. E., and W. ZERNA: Theoretical Elasticity. Oxford: Clarendon Press. xiii + 442 pp. (*12, 66B*)
8. GÜNTHER, W.: Spannungsfunktionen und Verträglichkeitsbedingungen der Kontinuumsmechanik. Abh. Braunschweig. Wiss. Ges. **6**, 207–219. (*227, 229*)
9. HIRSCHFELDER, J. O., C. F. CURTISS, and R. B. BIRD: Molecular Theory of Gases and Liquids. New York: Wiley. xxvi + 1219 pp. (*215, 243, 259, 295*)
10. KONDO, K.: On the theory of the mechanical behavior of microscopically non-uniform materials. Research notes Res. Assn. appl. Geom. (Tokyo) (2), No. **4**, 37 pp. (*34*)
11. KRÖNER, E.: Die Spannungsfunktion der dreidimensionalen isotropen Elastizitätstheorie. Z. Physik **139**, 175–188. Correction, Z. Physik **143**, 374 (1955). (*227*)
11A. LAMM, O.: The formal theory of diffusion, and its relation to self-diffusion, sedimentation equilibrium, and viscosity. Acta Chem. Scand. **8**, 1120–1128. (*295*)
12. LANGHAAR, H., and M. STIPPES: Three-dimensional stress functions. J. Franklin Inst. **258**, 371–382. (*227*)
13. LODE, W.: Tensoren zur Berechnung großer Formänderungen. Kolloid-Z. **138**, 28–38. (*33A*)
14. MANACORDA, T.: Sopra un principio variazionale di **E. Reissner** per la statica dei mezzi continui. Boll. Un. Mat. Ital. (3) **9**, 154–159. (*232A*)
15. MAZZARELLA, F.: Determinazione delle componenti di secondo ordine della deformazione riferite ad un generico sistema di coordinate curvilinee. Rend. Accad. Napoli (4) **21**, 107–114. (*26*)
16. ORNSTEIN, W.: Stress functions of **Maxwell** and **Morera**. Quart. appl. Math. **12**, 198–201. (*227*)
17. OSBORN, H.: The existence of conservation laws. Appl. math. stat. lab. Stanford tech. rep. 27 (70 pp.). Also Ann. of Math. **69**, 105–118 (1959). (*157*)

18. PARKER, E. N.: Tensor virial equations. Phys. Rev. (2) **96**, 1686–1689. *(219)*
19. REIK, H.: Die Thermodynamik irreversibler Prozesse und ihre Anwendung auf Transportphänomene. Z. Physik **137**, 333–361, 463–493. *(245)*
20. REINER, M.: Second order effects in elasticity and hydrodynamics. Bull. Res. Council Israel **3**, 372–379. *(33A)*
21. SCHOUTEN, J. A.: Ricci-calculus, 2nd ed. Berlin: Springer. xx + 516 pp. *(34, 150, 152, 268)*
22. SIGNORINI, A.: Una espressiva applicazione delle proprietà di medio dello stress comuni a tutti i sistemi continui. Studies math. mech. pres. **R. v. Mises**. New York: Academic Press. 274–277. *(222)*
23. SWAINGER, K.: Analysis of Deformation. I. Mathematical Theory. London: xix + 285 pp. *(33A)*
24. TRUESDELL, C.: The Kinematics of Vorticity. Indiana Univ. sci. ser. No. 19. *(12, 66A, 72, 78, 101, 102, 108, 110, 112, 114, 118, 119, 120, 121, 124, 125, 129, 130, 132, 133, 134, 136, 137, 138, 145)*
25. TRUESDELL, C.: Rational fluid mechanics 1687–1765. **L. Euleri** Opera Omnia (2) **12**, IX–CXXV. *(66A, 120, 130, 200)*
26. WHITHAM, G. B.: A note on a paper by **G. C. McVittie**. Quart. appl. Math. **12**, 316–318. *(230)*

1955 1. ADKINS, J.: A note on the finite plane-strain equations for isotropic incompressible materials. Proc. Cambridge Phil. Soc. **51**, 363–367. *(47)*
2. BILBY, B. A., R. BULLOUGH and E. SMITH: Continuous distributions of dislocations: a new application of the methods of non-Riemannian geometry. Proc. Roy. Soc. Lond., Ser. A **231**, 263–273. *(34)*
3. BJØRGUM, O.: On the physical boundary conditions in fluid dynamics. Univ. Bergen Årbok, Naturv. rekke Nr. 4, 8 pp. *(105)*
4. COTTER, B., and R. S. RIVLIN: Tensors associated with time-dependent stress. Quart. appl. Math. **13**, 177–182. *(149, 150, 293, 305)*
5. ERICKSEN, J. L.: Note concerning the number of directions which, in a given motion, suffer no instantaneous rotation. J. Washington Acad. Sci. **45**, 65–66. *(78, 91)*
6. ERICKSEN, J. L.: Deformations possible in every compressible, isotropic, perfectly elastic material. J. Math. Phys. **34**, 126–128. *(34)*
7. ERTEL, H.: Kanonischer Algorithmus hydrodynamischer Wirbelgleichungen. Sitzgsber. Akad. Wiss. Berlin **1954**, No. 4, 11 pp. *(130)*
8. ERTEL, H.: Ein neues Wirbel-Theorem der Hydrodynamik. Sitzgsber. Akad. Wiss. Berlin, math. Kl. **1954**, No. 5, 12 pp. *(130)*
9. FINZI, L.: Sulle equazioni di **Pucher** nell' equilibrio delle strutture a guscio. Rend. Ist. Lombardo (3) **88**, 907–916. *(229)*
10. GRIOLI, G.: Limitazioni per lo stato tensionale di un qualunque sistema continuo. Ann. di Mat. (4) **39**, 255–266. *(222)*
11. GYORGYI, G.: Die Bewegung des Energiemittelpunktes und der Energie-Impuls-Tensor des elektromagnetischen Feldes in Dielektrika. Acta phys. Hung. **4**, 121–131. *(284)*
12. HEINRICH, G.: Der Energiestrom in elastischen Medien. Öst. Ing.-Arch. **9**, 148–156. *(217)*
13. HERIVEL, J.: The derivation of the equations of motion of an ideal fluid by **Hamilton's** principle. Proc. Cambridge Phil. Soc. **51**, 344–349. *(236A)*
14. HUNT, F. V.: Notes on the exact equations governing the propagation of sound in fluids. J. Acoust. Soc. Amer. **27**, 1019–1039. *(257)*
15. JAIN, M. K.: The motion of an infinite cylinder in rotating non-Newtonian liquid. Z. angew. Math. Mech. **35**, 379–381. *(207)*
16. MARGUERRE, K.: Ansätze zur Lösung der Grundgleichungen der Elastizitätstheorie. Z. angew. Math. Mech. **35**, 242–263. *(228)*
17. MCVITTIE, G. C.: Review of [1954, *24*]. Bull. Amer. Math. Soc. **61**, 356–358. *(99)*
18. NOLL, W.: On the continuity of the solid and fluid states. J. Rational Mech. Anal. **4**, 13–81. *(12, 37, 66B, 95, 207, 293, 303, 304, 305)*
19. NOLL, W.: Die Herleitung der Grundgleichungen der Thermomechanik der Kontinua aus der statischen Mechanik. J. Rational Mech. Anal. **4**, 627–646. *(1, 262)*
20. RICHTER, H.: Review of [1954, *23*]. Zbl. Math. **56**, 175–176. *(33A)*
21. RIVLIN, R. S., and J. L. ERICKSEN: Stress-deformation relations for isotropic materials. J. Rational Mech. Anal. **4**, 323–425. *(33, 104, 143, 144, 149, 293, 304, 305)*
22. SCHAEFER, H.: Die Spannungsfunktion einer Dyname. Abh. Braunschweig. Wiss. Ges. **7**, 107–112. *(227)*

23. SPEISER, A.: Einleitung. L. **Euleri** Opera Omnia (1) **28**, VII–XLIV. *(33A)*
24. THOMAS, T. Y.: On the structure of the stress-strain relations. Proc. Nat. Acad. Sci. U.S.A. **41**, 716–720. *(148, 293)*
25. THOMAS, T. Y.: Kinematically preferred co-ordinate systems. Proc. Nat. Acad. Sci. U.S.A. **41**, 762–770. *(148, 293)*
26. TRUESDELL, C.: I. The first three sections of **Euler's** treatise on fluid mechanics (1766); II. The theory of aerial sound (1687–1788); III. Rational fluid mechanics (1765–1788). L. **Euleri** Opera Omnia (2) **13**, VII–CXVIII. *(66A)*
27. TRUESDELL, C.: The simplest rate theory of pure elasticity. Comm. pure appl. Math. **8**, 123–132. *(142, 304)*
28. TRUESDELL, C.: Hypo-elasticity. J. Rational Mech. Anal. **4**, 83–133, 1019–1020. *(223, 293, 304)*
29. TRUESDELL, C.: Review of [1954, *23*]. Math. Rev. **16**, 307. *(33A)*

1956 1. BERKER, R.: Sur les équations de compatibilité relatives au mouvement d'un gaz. C. R. Acad. Sci., Paris **242**, 342–344. *(110)*
2. BRESSAN, A.: Sulla possibilità di stabilire limitazioni inferiori per le componenti intrinseche del tensore degli sforzi in coordinate generali. Rend. Sem. Mat. Padova **26**, 139–147. *(222)*
3. COLONNETTI, G.: L'équilibre des voiles minces hyperstatiques (Le cas des voiles de surface minimum). C. R. Acad. Sci., Paris **243**, 1087–1089, 1701–1704. *(229)*
4. DORN, W. S., and A. SCHILD: A converse to the virtual work theorem for deformable solids. Quart. appl. Math. **14**, 209–213. *(227, 234)*
5. DOYLE, T. C., and J. L. ERICKSEN: Nonlinear elasticity. Adv. appl. Mech. **4**, 53–115. *(12, 14, 18)*
6. ELSASSER, W. M.: Hydromagnetic dynamo theory. Rev. Mod. Phys. **28**, 135–163. *(313)*
7. FINZI, L.: Legame fra equilibrio e congruenza e suo significato fisics, Rend. Lincei (8) **20**, 205–211, 338–342. *(84, 224, 234)*
8. GEIS, T.: „Ähnliche" dreidimensionale Grenzschichten. J. Rational Mech. Anal. **5**, 643–686. *(163)*
9. GREEN, A. E.: Hypo-elasticity and plasticity. Proc. Roy. Soc. Lond., Ser. A **234**, 46–59. *(304)*
10. GREEN, A. E.: Hypo-elasticity and plasticity, II. J. Rational Mech. Anal. **5**, 725–734. *(304)*
11. HANIN, M., and M. REINER: On isotropic tensor-functions and the measure of deformation. Z. angew. Math. Phys. **7**, 377–393. *(33A)*
12. HASKIND, M. D.: Unsteady motion of a solid in an accelerated flow of an unlimited fluid [in Russian]. Prikl. Mat. Mekh. **20**, 120–123. *(202)*
13. IKENBERRY, E., and C. TRUESDELL: On the pressures and the flux of energy in a gas according to **Maxwell's** kinetic theory, I. J. Rational Mech. Anal. **5**, 1–54. *(307)*
14. KOPPE, E.: Methoden der nichtlinearen Elastizitätstheorie mit Anwendung auf die dünne Platte endlicher Durchbiegung. Z. angew. Math. Mech. **36**, 455–462. *(34)*
15. MARGULIES, G.: Remark on kinematically preferred co-ordinate systems. Proc. Nat. Acad. Sci. U.S.A. **42**, 152–153. *(148)*
16. OLDROYD, J. G., and R. H. THOMAS: The motion of a cylinder in a rotating liquid with general elastic and viscous properties. Quart. J. Mech. Appl. Math. **9**, 136–139. *(207)*
17. OSWATITSCH, K.: Über eine Verallgemeinerung des Potentials auf Strömungen mit Drehung. Öst. Ing.-Arch. **10**, 239–241. *(125)*
18. RAW, C. J. G., and W. YOURGRAU: "Acceleration" of chemical reactions. Nature, Lond. **178**, 809. *(159A)*
19. SPEISER, A.: Einleitung. L. **Euleri** Opera Omnia (1) **29**, VII–XLII. *(33A)*
20. TOUPIN, R. A.: The elastic dielectric. J. Rational Mech. Anal. **5**, 849–915. *(14, 19, 37, 312)*
21. TRUESDELL, C.: Zur Geschichte des Begriffes „Innerer Druck". Physik. Bl. **12**, 315–326. *(200)*
22. TRUESDELL, C.: Review of [1953, *20*]. Math. Rev. **17**, 97. *(170)*
23. TRUESDELL, C.: Experience, theory, and experiment. Proc. 6th Hydraulics Conf., bull. 36, Univ. Iowa Stud. Engin., 3–18. *(3)*

1957 1. BRAND, L.: The pi theorem of dimensional analysis. Arch. Rational Mech. Anal. **1** (1957/58), 34–45. *(293)*
2. BRDIČKA, M.: On the general form of the **Beltrami** equation and **Papkovich's** solution of the axially symmetrical problem of the classical theory of elasticity. Czech. J. Phys. **7**, 262–274. *(228)*

3. Duncan, W. R.: Analysis of a vector field and some applications to fluid motion. Aero. Quart. **8**, 207–214.
4. Filonenko-Borodich, M. M.: On the problem of **Lamé** for the parallelepiped in the general case of surface loads [in Russian]. Prikl. Mat. Mekh. **21**, 550–559. *(227)*
5. Graiff, F.: Sulle condizioni di congruenza per una membrana. Rend. Ist. Lombardo A **92**, 33–42. *(84)*
6. Green, A. E., and R. S. Rivlin: The mechanics of non-linear materials with memory, Part I. Arch. Rational Mech. Anal. **1** (1957/58), 1–21. *(305)*
7. Hayes, W.: The vorticity jump across a gasdynamic discontinuity. J. Fluid Mech. **2**, 595–600. *(179)*
8. Howard, L. N.: Divergence formulas involving vorticity. Arch. Rational Mech. Anal. **1** (1957/58), 113–123. *(119, 131)*
9. Milne-Thomson, L. M.: A general solution of the equations of hydrodynamics. J. Fluid Mech. **2**, 88. *(230)*
10. Nachbar, W., F. Williams and S. S. Penner: The conservation laws for independent coexistent continua and for multicomponent reacting gas mixtures. Lockheed aircraft corp. LMSD 2082, 15 March = Quart. appl. Math. **17**, 43–54 (1959). *(215, 243)*
11. Noll, W.: On the foundations of the mechanics of continuous media. Carnegie Inst. Tech. Rep. No. 17, Air Force Off. Sci. Res. *(3, 196, 196A, 211, 293, 305)*
12. Noll, W.: On the rotation of an incompressible continuous medium in plane motion. Quart. appl. Math. **15**, 317–319. *(161, 207)*
13. Storchi, E.: Una soluzione delle equazioni indefinite della meccanica dei continui negli spazi riemanniani. Rend. Ist. Lombardo **90** (1956), 369–378. *(229)*
14. Ting, T. W., and J. C. M. Li: Thermodynamics for elastic solids. General formulation. Phys. Rev. (2) **106**, 1165–1167. *(246)*
15. Thomas, T. Y.: Extended compatibility conditions for the study of surfaces of discontinuity in continuum mechanics. J. Math. Mech. **6**, 311–322, 907–908. *(175, 176, 177, 179, 180, 181)*
16. Truesdell, C.: Sulle basi della termomeccanica. Rend. Lincei (8) **22**, 33–38, 158–166. *(158, 159, 215, 243, 254, 255, 259)*
17. Truesdell, C.: General solution for the stresses in a curved membrane. Proc. Nat. Acad. Sci. U.S.A. **43**, 1070–1072. *(84, 229)*
18. Whitney, H.: Geometric Integration Theory. Princeton Univ. Press. *(269)*
19. Yih, C.-S.: Stream functions in three-dimensional flows. Houille Blanche, 445–450. *(164)*

1958 1. Ericksen, J. L., and C. Truesdell: Exact theory of stress and strain in rods and shells. Arch. Rational Mech. Anal. **1**, 295–323. *(60, 61, 63A, 64, 212, 214)*
1A. Ferrarese, G.: Sulla relazione simbolica della meccanica dei sistemi continui vincolati. Rend. Mat. e Appl. (5) **14**, 305–312. *(233)*
2. Godal, T.: On **Beltrami** vector fields and flows, Part III. Some considerations on the general case. Univ. Bergen Årbok 1957, Naturv. r. Nr. 12, 28 pp. *(108)*
3. Graiff, F.: Sulle condizioni di congruenza per deformazioni anche finite. Rend. Lincei (8) **24**, 415–422. *(84)*
4. Günther, W.: Zur Statik und Kinematik des Cosseratschen Kontinuums. Abh. Braunschweig. Wiss. Ges. **10**, 195–213. *(61, 200, 227)*
5. Kanwal, R. P.: Determination of the vorticity and the gradients of flow parameters behind a three-dimensional unsteady curved shock wave. Arch. Rational Mech. Anal. **1**, 225–232. *(175)*
6. Krzywoblocki, M. Z. v.: On the stream functions in nonsteady three-dimensional flow. J. Aeronaut. Sci. **25**, 67. *(160)*
7. Marshak, R. E.: Effect of radiation on shock wave behavior. Phys. of Fluids **1**, 24–29. *(241)*
8. Noll, W.: A mathematical theory of the mechanical behavior of continuous media. Arch. Rational Mech. Anal. **2** (1958/59), 197–226. *(3, 196, 200, 293, 305)*
9. Toupin, R. A.: World invariant kinematics. Arch. Rational Mech. Anal. **1** (1957/58), 181–211. *(152)*
10. Truesdell, C.: Geometric interpretation for the reciprocal deformation tensors. Quart. appl. Math. **15**, 434–435. *(29)*
11. Truesdell, C.: Intrinsic equations of spatial gas flow. Math. Res. Center U.S. Army, Univ. Wisconsin rep. No. 33, July. Cf. [1960, *5*]. *(99)*
12. Washizu, K.: A note on the conditions of compatibility. J. Math. Phys. **36**, 306–312. *(34)*

1959 1. Blankfield, J., and G. C. McVittie: Einstein's equations and classical hydrodynamics. Arch. Rational Mech. Anal. **2** (1958/59), 337–354. *(230)*

2. BLANKFIELD, J., and G. C. MCVITTIE: A method of solution of the equations of magnetohydrodynamics. Arch. Rational Mech. Anal. **2** (1958/59), 411–422. (*230*)
3. COLEMAN, B. D., and W. NOLL: On the thermostatics of continuous media. Arch. Rational Mech. Anal. **4** (1959/60), 97–128. (*262*)
4. DROBOT, S., and A. RYBARSKI: A variational principle of hydromechanics. Arch. Rational Mech. Anal. **2** (1958/59), 393–410. (*132*)
5. GRAIFF, F.: Soluzione generale delle equazioni indefinite di equilibrio per una membrana. Rend. Lincei (8) **26**, 189–196. (*229*)
6. GREEN, A. E., R. S. RIVLIN and A. J. M. SPENCER: The mechanics of non-linear materials with memory. Part II. Arch. Rational Mech. Anal. **3**, 82–90. (*305*)
7. GREEN, A. R.: The equilibrium of rods. Arch. Rational Mech. Anal. **3**, 417–421. (*214*)
8. KRÖNER, E., u. A. SEEGER: Nicht-lineare Elastizitätstheorie der Versetzungen und Eigenspannungen. Arch. Rational Mech. Anal. **3**, 97–119. (*20, 34*)
9. NOLL, W.: The foundations of classical mechanics in the light of recent advances in continuum mechanics. The Axiomatic Method, with special reference to geometry and physics (1957). Amsterdam: North Holland Co., 266–281. (*3, 196, 293*)
10. SCHAEFER, H.: Die Spannungsfunktionen des dreidimensionalen Kontinuums; statische Deutung und Randwerte. Ing.-Arch. **28**, 291–306. (*227*)
11. TRUESDELL, C.: Invariant and complete stress functions for general continua. Arch. Rational Mech. Anal. **4** (1959/60), 1–29. (*226*)
12. ZOLLER, K.: Die Wärmeleitgleichung bei Wärmespannungen. Ing.-Arch. **28**, 366–372. (*256A*)

1960 1A. COLEMAN, B., and C. TRUESDELL: On the reciprocal relations of **Onsager**. J. Chem. Phys. forthcoming. (*245*)
1. ERICKSEN, J. L.: Anisotropic fluids. Arch. Rational Mech. Anal. **4** (1959/60), 231–237. (*205, 298*)
2. HAYES, W. D.: Generalized **Kelvin's** minimum energy theorem. J. Fluid Mech., forthcoming. (*94*)
3. KRÖNER, E.: Allgemeine Kontinuumstheorie der Versetzungen und Eigenspannungen. Arch. Rational Mech. Anal. **4** (1959/60), 273–334. (*20, 61, 201, 266*)
4. TRUESDELL, C.: The rational mechanics of flexible or elastic bodies, 1638–1788. L. **Euleri** Opera Omnia (2) **11**, Part 2. (*33A, 66A, 200, 212*)
5. TRUESDELL, C.: Intrinsic equations of spatial gas flow. Z. angew. Math. Mech. **40**, 9–14. (*99*)
6. TRUESDELL, C.: Mechanical aspects of diffusion, forthcoming. (*295*)

Additional Bibliography K: Kinematics of special motions (geometrical theory).

1872 *K 1*. DURRANDE, H.: Propriétés générales du déplacement d'une figure de forme variable. C. R. Acad. Sci., Paris **74**, 1243–1247.
K 2. DURRANDE, H.: De l'accélération dans le déplacement d'un système de points qui reste homographique à lui-même. C. R. Acad. Sci., Paris **75**, 1177–1180.
1873 *K 1*. DURRANDE, H.: Essai sur le déplacement d'une figure de forme variable. Ann. Éc. Norm. (2) **2**, 81–121.
K 2. GROUARD: Sur le mouvement d'une figure, qui se déplace dans l'espace en restant semblable à elle-même. Bull. Soc. Philomath. (6) **9**, 47–49. Abstract: L'Institut (2) **1**, 163–164.
1874 *K 1*. BURMESTER, L.: Kinematisch-geometrische Untersuchungen der Bewegung ähnlich-veränderlicher ebener Systeme. Z. Math. Phys. **19**, 154–169.
K 2. BURMESTER, L.: Kinematisch-geometrische Untersuchungen der Bewegung affin-veränderlicher und collinear-veränderlicher ebener Systeme. Z. Math. Phys. **19**, 465–492.
K 3. DURRANDE, H.: Déplacement d'un système de points. Propriétés géométriques dépendant des paramètres différentiels du second ordre. C. R. Acad. Sci., Paris **78**, 1036–1040.
K 4. DURRANDE, H.: Étude de l'accélération dans le déplacement d'un système de forme variable. Ann. Éc. Norm. (2) **3**, 151–164.
1875 *K 1*. BURMESTER, L.: Kinematisch-geometrische Untersuchungen der Bewegung gesetzmäßig-veränderlicher Systeme. Z. Math. Phys. **20**, 381–422.
K 2. JORDAN, C.: Sur le mouvement des figures dans le plan et dans l'espace. Bull. Soc. Math. France **1**, 144–148.

K 3. LIGUINE, V.: Sur le lieu des points d'un système invariable mobile d'une manière générale dans l'espace, dont les accélérations du premier ordre sont constantes. Bull. Soc. Math. France **1**, 152—154.

1877 K 1. MÜLLER, R.: Über Selbsthüllcurven und Selbsthüllflächen in ähnlich-veränderlichen Systemen. Z. Math. Phys. **22**, 369—376.

1878 K 1. BURMESTER, L.: Kinematisch-geometrische Theorie der Bewegung der affin-veränderlichen, ähnlich veränderlichen und starren räumlichen oder ebenen Systeme. Z. Math. Phys. **23**, 108—131.

K 2. BURMESTER, L.: Über den Beschleunigungszustand ähnlich-veränderlicher und starrer ebener Systeme. Civilingenieur **24**.

1879 K 1. BURMESTER, L.: Über die Festlegung projectiv-veränderlicher ebener Systeme. Math. Ann. **14**, 472—497.

K 2. FOURET, G.: Sur le mouvement d'un corps qui se déplace et se déforme en restant homothétique à lui-même. C. R. Acad. Sci., Paris **88**, 227—230.

K 3. FROMENTI, C.: Movimento delle figure che se mantengono simili a se stesse. Giorn. math. **17**, 232—243.

K 4. GEISENHEIMER, L.: Untersuchung der Bewegung ähnlich-veränderlicher Systeme. Z. Math. Phys. **24**, 129—159.

K 5. GEISENHEIMER, L.: Die Bildung affiner Figuren durch ähnlich-veränderliche Systeme. Z. Math. Phys. **24**, 345—381.

1880 K 1. BURMESTER, L.: Über das bifocal-veränderliche System. Math. Ann. **16**, 89—111.

1881 K 1. SCHUMANN, A.: Beiträge zur Kinematik ähnlich-veränderlicher und affin-veränderlicher Gebilde. Z. Math. Phys. **26**, 157—179.

1883 K 1. MEHMKE, R.: Über die Geschwindigkeiten beliebiger Ordnung eines in seiner Ebene bewegten ähnlich-veränderlichen Systems. Civilingenieur (2) **29**, 487—508.

K 2. MEHMKE, R.: Über den geometrischen Ort der Punkte ohne Normalbeschleunigung in einer Phase eines starren oder affin-veränderlichen Systems. Civilingenieur (2) **29**, 581—582.

K 3. NICOLI, F.: Intorno ad un caso di movimento di una figura piana che si varia rimanendo simile a se stessa. Mem. Accad. Modena (2) **1**, 59—71.

K 4. NICOLI, F.: Intorno ad un caso di movimento di una figura piana che si conserva simile a se stessa. Mem. Accad. Modena (2) **1**, 171—178.

K 5. NICOLI, F.: Intorno a due casi di movimento di una figura solida che rimane simile a se stessa. Mem. Accad. Modena (2) **1**, 249—260.

1885 K 1. SOMOFF, P.: Über die Bewegung ähnlich-veränderlicher ebener Systeme. Z. Math. Phys. **30**, 193—209.

1888 K 1. BURMESTER, L.: Lehrbuch der Kinematik. Leipzig: A. Felix. xx + 942 pp. + Atlas. See §§ 329—351.

1889 K 1. SOMOV, P.: Some problems on the distribution of velocity in variable systems [in Russian]. Varshavskia Univ. Izv. **1889**, No. 4, 32 pp.

1890 K 1. MORLEY, F.: On the kinematics of a triangle of constant shape but varying size (with a note). Quart. J. Math. **24**, 359—369, 386.

K 2. SOMOW, P.: On the acceleration in collinearly variable systems [in Russian]. Proceedings of the 8th meeting of Russian Natural Scientists and Physicians, St. Petersburg, 1890. Math. and Astron. 41—44. (We have not been able to see this reference.)

1892 K 1. SHEBUEV, G. N.: Application of the theory of quaternions to the mechanics of similar and homogeneously variable systems [in Russian]. Izv. Fiz.-Mat. Obsh. Kazan (2) **3**, 111—160.

1894 K 1. MANNHEIM, A.: Principes et développements de géométrie cinématique. Paris: Gauthier-Villars. ix + 589 pp. See pp. 14—53, 457—475.

1897 K 1. DE SAUSSURE, R.: Calcul géométrique réglé. Amer. J. Math. **19**, 329—370.

1898 K 1. DE SAUSSURE, R.: Cinématique des fluides. Mouvement d'un fluide dans un plan. Arch. Sci. Phys. Nat. Geneve (4) **5**, 497—519.

1899 K 1. CAVALLI, E.: Le figure reciproche e la trasformazione quadratica nella cinematica. Atti Acc. Napoli (2) **9**, No. 12, 29 pp.

1901 K 1. DE SAUSSURE, R.: Sur le mouvement d'une droite qui possède trois degrés de liberté. C. R. Acad. Sci., Paris **133**, 1283—1285.

K 2. SEILIGER, D.: On a fundamental theorem in the statics of a variable system [in Russian]. Papers Univ. Kazan No. 718, 75—82. (We have not been able to see this reference.)

1902 K 1. BURMESTER, L.: Kinematisch-geometrische Theorie der Bewegung der affin-veränderlichen, ähnlich-veränderlichen und starren räumlichen oder ebenen Systeme, Teil 2. Z. Math. Phys. **47**, 128—156.

K 2. CARDINAAL, J.: Over de beweging van veranderlijke stelsels. Amst. Akad. Versl. **10**, 550–566, 687–691.

K 3. CARDINAAL, J.: Over de afbeelding van de beweging van veranderlijke stelsels. Amst. Akad. Versl. **11**, 466–471.

K 4. DE SAUSSURE, R.: Théorie géométrique du mouvement des corps. Arch. Sci. Phys. Nat. Genève (4) **13**, 425–461; **14**, 14–41, 209–231; **18**, 25–63 (1904); **21**, 36–55, 129–133 (1906). Also issued separately, in parts, Geneva 1902, 1904, 1906.

K 5. SCHOENFLIESS, A., u. M. GRÜBLER: Kinematik. Enzykl. Math. Wiss. **4**[1], 190–278.

K 6. SOMOV, P.: On hinged members with variable elements [in Russian]. Varshavskia Univ. Izv. **1902**, Part 8, No. 3, 45 pp.

1903 *K 1.* SOMOFF, P.: Über einige Gelenksysteme mit ähnlich-veränderlichen oder affin-veränderlichen Elementen. Z. Math. Phys. **49**, 25–61.

1907 *K 1.* KOENIGS, G.: Sur les déformations élastiques qui laissent invariables les longueurs d'une triple infinité de lignes droites. C. R. Acad. Sci., Paris **144**, 557–560.

1910 *K 1.* KRAUSE, M.: Zur Theorie der ebenen ähnlich veränderlichen Systeme. Jber. dtsch. Math.-Ver. **19**, 327–339.

K 2. MEHMKE, R.: Analytischer Beweis des Satzes von Herrn **Reinhold Müller** über die Erzeugung der Koppelkurve durch ähnlich-veränderliche Systeme. Z. Math. Phys. **58**, 257–259.

K 3. MÜLLER, R.: Erzeugung der Koppelkurve durch ähnlich-veränderliche Systeme. Z. Math. Phys. **58**, 247–251.

K 4. MÜLLER, R.: Über die Momentanbewegung eines ebenen ähnlich-veränderlichen Systems in seiner Ebene. Jber. dtsch. Math. Ver. **10**, 29–89.

K 5. SKUTSCH, R.: Über die von Herrn **Reinhold Müller** untersuchte besondere Bewegung eines ähnlich-veränderlichen Systems. Z. Math. Phys. **58**, 252–257.

K 6. STUDY, E.: Die Kinematik der Herren **de Saussure** und **Bricard**. Jber. dtsch. Math.-Ver. **19**, 255–263.

1911 *K 1.* KRAUSE, M.: Zur Theorie der affin veränderlichen ebenen Systeme. Sitzgsber. Akad. Wiss. Leipzig **63**, 271–288.

K 2. KRAUSE, M.: Über räumliche Bewegungen mit ebenen Bahnkurven. Sitzgsber. Akad. Wiss. Leipzig **63**, 515–533.

K 3. MEHMKE, R.: Beiträge zur Kinematik starrer und affin-veränderlicher Systeme, insonderheit über die Windung der Bahnen der Systempunkte. Z. Math. Phys. **59**, 90–94, 204–220, 440–442.

1912 *K 1.* HARTMANN, T.: Zur Theorie der Momentanbewegung eines ebenen ähnlich-veränderlichen Systems. Diss. Rostock, 144 pp.

1913 *K 1.* DE DONDER, T.: Sur divers modes de croissance des milieux continus. Bull. Acad. Sci. Belg. 614–621, 642–646.

K 2. HERRMANN, E.: Über die einförmige Bewegung des ebenen kreisverwandt-veränderlichen Systems. Diss. Tech. Hoch. Dresden, 93 pp.

1914 *K 1.* CARL, A.: Zur Theorie der ebenen ähnlich veränderlichen Systeme. Diss. Dresden, 125 pp.

K 2. WINKLER, R.: Über die Bewegung affin-veränderlicher ebener Systeme. Diss. Dresden, 73 pp.

1920 *K 1.* KRAUSE, M., Assisted by A. CARL: Analysis der ebenen Bewegung. Berlin. 216 pp.

1922 *K 1.* DELASSUS, E.: Stabilité de l'équilibre sur une liaison finie unilatérale. Bull. Sci. Math. (2) **46**, 283–304.

K 2. GAMBIER, B.: Mécanismes transformables ou déformables. Couples de surfaces qui s'en déduisent. J. Math. Pures Appl. (9) **1**, 19–76.

1932 *K 1.* ABRAMESCO, N.: Le mouvement d'une figure plane variable qui reste semblable à elle-même. Ann. Sci. Norm. Pisa (2) **1**, 155–164.

K 2. PASCAL, M.: Sul moto di un corpo deformabile che si mantiene simile a se stesso. I: Formola fondamentale e proprietà che se ne deducono. II: Centro istantaneo di velocità e conseguenze. Rend. Lincei (6) **15**, 871–874; **16**, 320–324.

1933 *K 1.* ABGHIRIADI, M.: Sur le mouvement d'une figure plane semblablement variable. Mathesis **47**, Suppl., 14 pp.

K 2. PASCAL, M.: Sul moto di una figura deformabile piana di area costante e che rimane affine a se stessa. Rend. Napoli (4) **3**, 71–77.

K 3. PASCAL, M.: Sul moto di una figura deformabile piana che si conserva affine a se stessa. Rend. Napoli (4) **3**, 78–82.

K 4. PASCAL, M.: Sul centro istantaneo di velocità nulla nel moto di una figura piana di area costante e a deformate affine. Rend. Napoli (4) **3**, 110–113.

K 5. PASCAL, M.: Sull'accelerazione nel moto di una figura piana di area costante e a deformate affine. Rend. Napoli (4) **3**, 123–126.

K 6. Pascal, M.: Sulla cinematica affine di una figura piana di area costante. Rend. Napoli (4) **3**, 142–144.
1934 *K 1.* Di Noi, S.: Considerazioni geometriche sul moto di un corpo deformabile che si mantiene simile a se stesso. Rend. Napoli (4) **3**, 176–181.
K 2. Pascal, M.: Sul moto di un corpo deformabile di volume costante e che rimane affine a se stesso. Atti Soc. ital. Progr. Sc. A **22**², 194–195.
1936 *K 1.* Abramesco, N.: Proprietăti geometrice ale mişcării unei figuri plane variabile care rămâne asemenea cu ea în săşi, când trei drepte ale figurii trei prin trei puncte fixe, sau când trei puncte descriu trei drepte fixe. Gaz. Mat. Bucarest **41**, 409–414.
K 2. Harmegnies, R.: Sur le mouvement d'une figure plane qui reste homographique à elle-même. C. R. Acad. Sci., Paris **202**, 1323–1324.

See also the following items from the "List of works cited": Durrande [1871, *4*], and the papers cited in Sects. 140–142.

Additional Bibliography N:
Non-relativistic kinematics and mechanics in generalized spaces.

1876 *N 1.* Beltrami, E.: Formules fondamentales de cinématique dans les espaces de courbure constante. Bull. Sci. Math. (1) **11**, 233–240.
1878 *N 1.* Lévy, M.: Sur la cinématique des figures continues sur les surfaces courbes, et, en général, dans les variétés planes ou courbes. C. R. Acad. Sci., Paris **86**, 812–818.
N 2. Lévy, M.: Sur les conditions que doit remplir un espace pour qu'on y puisse déplacer un système invariable, à partir de l'une quelconque de ses positions, dans une ou plusieurs directions. C. R. Acad. Si., Paris **86**, 875–878.
1881 *N 1.* Beltrami, E.: Sulle equazioni generali dell'elasticità. Ann. Mat. (2) **10**, 188–211 (1880–1882) = Opere **3**, 383–407.
1884 *N 1.* Beltrami, E.: Sull'uso delle coordinate curvilinee nelle teorie del potenziale e dell'elasticità. Mem. Accad. Sci. Bologna (4) **6**, 401–488 = Opere **4**, 136–179.
N 2. Heath, R. S.: On the dynamics of a rigid body in elliptic space. Phil. Trans. Roy. Soc. Lond. **175**, 281–324.
1885 *N 1.* Killing, W.: Die Mechanik in den nicht-Euklidischen Raumformen. J. reine angew. Math. **89**, 1–48.
N 2. Hill, M. J. M.: On some general equations which include the equations of hydrodynamics (1883). Trans. Cambridge Phil. Soc. **14**, 1–29.
1888 *N 1.* Cesaro, E.: Sur une récente communication de M. **Lévy**. C. R. Acad. Sci., Paris **107**, 520–522.
1889 *N 1.* Padova, E.: La teoria di **Maxwell** negli spazi curvi. Rend. Lincei (4) **5**¹, 875–880.
N 2. Somigliana, C.: Sopra la dilatazione cubica di un corpo elastico isotropo in uno spazio di curvatura costante. Ann. Mat. (2) **16**, 101–115.
1890 *N 1.* Padova, E.: Il potenziale delle forze elastiche di mezzi isotropi. Atti Ist. Veneto **48** = (7) **1**, 445–451.
1894 *N 1.* Cesaro, E.: Sulle equazioni dell'elasticità negli iperspazii. Rend. Lincei (5) **3**², 290–294.
1900 *N 1.* De Francesco, D.: Alcuni problemi di meccanica in uno spazio a tre dimensioni di curvatura costante, I and II. Atti Accad. Napoli (2) **10**, Nos. 4 (38 pp.) and 9 (33 pp.).
N 2. De Francesco, D.: Sul moto spontaneo di un corpo rigido in uno spazio di curvatura costante. Atti R. Accad. Sci. Torino **35**, 34–38, 231–243.
1901 *N 1.* Bohlin, K.: Sur l'extension d'une formule d'**Euler** et sur le calcul des moments d'inertie principaux d'un système de points materiels. C. R. Acad. Sci., Paris **133**, 530–532.
N 2. De Francesco, D.: Su alcuni problemi di meccanica, in uno spazio pseudosferico, analiticamente equivalenti a problemi nello spazio ordinario. Rend. Accad. Napoli (3ª) **7**, 28–38.
1902 *N 1.* De Francesco, D.: Alcune formole della meccanica dei fluidi in uno spazio a tre dimensioni di curvatura costante, I and II. Atti Accad. Napoli (2) **12**, Nos. 9 (18 pp.) and 10 (13 pp.).
N 2. Stäckel, P.: De ea mecanicae analyticae parte quae ad varietates complurium dimensionum spectat. Libellus **Ioannis Bolyai** . . . ad celebrandam memoriam . . . (Claudiopoli), 63–79.
1903 *N 1.* Stäckel, P.: Bericht über die Mechanik mehrfacher Mannigfaltigkeiten. Jber. dtsch. Math.-Ver. **12**, 469–481.

1907 *N 1.* RIQUIER, CH.: Sur les systèmes d'équations aux dérivées partielles auquels conduisent: 1° l'étude des déformations finies d'un milieu continu dans l'espace à n dimensions; 2° la détermination des systèmes de coordonnées curvilignes orthogonales à n variables. C. R. Acad. Sci., Paris **145**, 1137–1139.
1911 *N 1.* VESSIOT, E.: Sur la cinématique des milieux continus à n dimensions. C. R. Acad. Sci., Paris **152**, 1732–1735.
1912 *N 1.* DE DONDER, T.: Sur la cinématique des milieux continus. Bull. Acad. Roy. Belg. Cl. Sci., 243–251.
N 2. ZORAWSKI, K.: Über gewisse Pfaff'sche Systeme, welche bei Bewegungen kontinuierlicher Medien invariant bleiben. Bull. Int. Acad. Sci. Cracovie A **1912**, 436–461.
1913 *N 1.* DEL RE, A.: Sulle equazioni generali per la dinamica negli spazii ad n dimensioni ed a curvatura costante, Ann. Mat. (3) **22**, 63–70.
1926 *N 1.* SYNGE, J. L.: Applications of the absolute differential calculus to the theory of elasticity (1924). Proc. Lond. Math. Soc. (2) **24**, 103–108.
1930 *N 1.* TONOLO, A.: Une interprétation physique du tenseur de **Riemann** et des courbures principales d'une variété V_3. C. R. Acad. Sci., Paris **190**, 787–788.
N 2. TONOLO, A.: Equazioni intrinseche di equilibrio dell' elasticità negli spazî a curvatura costante. Rend. Sem. Mat. Padova **1**, 73–84.
1931 *N 1.* TONOLO, A.: Sistemi isostatici dei corpi elastici negli spazî a curvatura costante. Rend. Sem. Mat. Padova **2**, 152–163.
1933 *N 1.* CARTAN, E.: La cinématique newtonienne et la théorie des espaces réglées à connexion euclidienne. Ass. Franc. Avancem. Sci. 19–20.
N 2. CARTAN, E.: La cinématique newtonienne et les espaces à connexion euclidienne. Bull. Math. Soc. Roumaine Sci. **35**, 69–73.
N 3. FINZI, B.: Equazioni intrinseche della meccanica dei sistemi continui perfettamente od imperfettamente flessibili. Ann. Mat. (4) **11**, 215–245.
N 4. TEODORIA, L.: Sur la cinématique du corps solide dans l'espace euclidien à n dimensions. Bull. Math. Soc. Roumaine Sci. **35**, 243–247.
1934 *N 1.* PASTORI, M.: Sulle equazioni della meccanica dei mezzi isotropi non euclidei. Rend. Lincei (6) **19**, 566–572.
N 2. PASTORI, M.: Sulla dissipazione di energia nei fluidi viscosi. Rend. Ist. Lombardo (2) **67**, 823–848.
1935 *N 1.* WESTERGAARD, H. M.: General solution of the problem of elastostatics of an n-dimensional homogeneous isotropic solid in an n-dimensional space. Bull. Amer. Math. Soc. **41**, 695–699.
1936 *N 1.* LOTZE, A.: Die Grundgleichungen der Mechanik im elliptischen Raum. Jber. dtsch. Math.-Ver. **46**, 51–70.
1937 *N 1.* LAMPARIELLO, G.: Varietà sostanziali nel moto di un sistema continuo. Rend. Lincei (6a) **15**, 383–387.
1940 *N 1.* ORTVAY, R.: The physical implications of some new viewpoints in mathematics [in Hungarian]. Mat. Fiz. Lapok **47**, 111–138.
N 2. SKOLEM, T.: A little study on transfinite mechanics [in Norwegian]. Norsk Mat. Tidsskr. **22**, 5–9.
1942 *N 1.* BLASCHKE, W.: Nicht-Euklidische Geometrie und Mechanik. Hamburg. Math. Einzelschr. **34**.
1944 *N 1.* BLASCHKE, W.: Nicht-Euklidische Mechanik. Sitzgsber. Akad. Wiss. Heidelberg **1943**, No. 2, 10 pp.
1951 *N 1.* SANTALÒ, L. A.: On permanent vector-varieties in n dimensions. Portugaliae Math. **10**, 125–127.
N 2. SYNGE, J. L.: On permanent vector-lines in n dimensions. Proc. Amer. Math. Soc. **2**, 370–372.

See also the following items from the "List of Works Cited": CLEBSCH [1857, *1*; §§ 1–4], PADOVA [1889, *7*], ZORAWSKI [1900, *12*], [1901, *17*], ZERMELO [1902, *10*], ZORAWSKI [1911, *13* and *14*], DE DONDER [1912, *2*], ZORAWSKI [1912, *8*].

Additional Bibliography P: Principles of Mechanics.

Any partial bibliography of work on the concepts and axioms of mechanics from the origins through the time of LAGRANGE would be misleading. No adequate critical history has ever been written. The remarks on this subject given in treatises or general histories of physics are often mendacious and usually so incomplete and inaccurate as to be totally misinformative. Large extracts from some of the sources, along with helpful comments, may be found in:

JOUGUET, É.: Lectures de mécanique, 2 vols. Paris: Gauthier-Villars 1908, 1909. x + 210 + 284 pp.
DUHEM, P.: Le Système du Monde, Histoire des doctrines cosmologiques de Platon à Copernic. Paris: Hermann. 10 Vols., 1913–1959.
DUGAS, R.: Histoire de la mécanique. Neuchatel: Éd. du Griffon 1950. 649 pp.
DUGAS, R.: La mécanique au XVIIe-siècle. Neuchatel: Éd. du Griffon 1954. 620 pp.
CLAGETT, M.: The Science of Mechanics in the Middle Ages. Madison: Univ. Wisconsin Press 1959. 711 pp.

1883 *P 1*. MACH, E.: Die Mechanik in ihrer Entwicklung, historisch-kritisch dargestellt. Leipzig: Brockhaus. There are many subsequent editions and translations of this unreliable work.
1894 *P 1*. HERTZ, H.: Die Prinzipien der Mechanik in neuem Zusammenhange dargestellt. Leipzig: Johann Ambrosius Barth. xxix + 312 pp. Trans. D. E. JONES and J. T. WALLEY, The principles of mechanics. London: MacMillan & Co. 1899. xxviii + 276 pp.
1897 *P 1*. BOLTZMANN, L.: Vorlesungen über die Principe der Mechanik I. Leipzig. x + 241 pp. See §§ 1–12.
1905 *P 1*. PAINLEVÉ, P.: Les axiomes de la mécanique et le principe de causalité. Bull. Soc. franç. Philos. **5**, 27–50. See [1922, *P 1*].
1906 *P 1*. FARKAS, J.: Beiträge zu den Grundlagen der analytischen Mechanik. J. reine angew. Math. **131**, 165–201.
1909 *P 1*. PAINLEVÉ, P.: Les axiomes de la mécanique classique, chapter in La méthode dans les sciences. Paris: Alcan. See [1922, *P 1*].
1911 *P 1*. MARCOLONGO, R.: Theoretische Mechanik, 2 vol., Deutsch von H. E. TIMERDING. Leipzig: B. G. Teubner 1911 and 1912.
1922 *P 1*. PAINLEVÉ, P.: Les axiomes de la mécanique. Examen critique. Note sur la propagation de la lumière. Paris: Gauthier-Villars. xvii + 112 pp. (Reprint of [1905, *P 2*], [1909, *P 1*], and other material.)
1923 *P 1*. DIJKSTERHUIS, E. J.: De axioma's der mechanica. Christiaan Huygens **3** (1923—1924), 87–101.
1927 *P 1*. HAMEL, G.: Die Axiome der Mechanik. Handbuch Physik **5**, 1–42.
1928 *P 1*. DIJKSTERHUIS, E. J.: De historische behandelingswijze van de axiomata der mechanica van **Newton**. Euclides **4**, 245–255.
1934 *P 1*. ZAREMBA, S.: Sur la notion de force en mécanique. Bull. Soc. Math. France **62**, 110–119.
1936 *P 1*. LINDSAY, R., B., and H. MARGENAU: Foundations of Physics. New York: J. Wiley & Sons, Inc. xiii + 537 pp.
P 2. PLATRIER, C.: Les Axiomes de la Mécanique Newtonienne. Act. Sci. Indust. No. 427. Paris: Hermann.
1937 *P 1*. BARBILIAN, D.: Eine Axiomatisierung der klassischen Mechanik. C. R. Acad. Sci. Roumaine **2**, 9–16.
P 2. PENDSE, C. G.: A note on the definition and determination of mass in Newtonian mechanics. Phil. Mag. (7) **24**, 1012–1022.
1938 *P 1*. HERMES, H.: Eine Axiomatisierung der allgemeinen Mechanik. Leipzig: Hirzel 48 pp.
P 2. ROSSER, B.: Review of the foregoing. J. Symb. Logic **3**, 119–120.
1939 *P 1*. KRATZER, A.: Betrachtungen zu den Grundlagen der Mechanik. Semester-Ber. Math. Sem. Münster **14**, 1–15.
P 2. PENDSE, C. G.: A further note on the definition and determination of mass in Newtonian mechanics. Phil. Mag. (7) **27**, 51–61.
P 3. NARLIKAR, V. V.: The concept and determination of mass in Newtonian mechanics. Phil. Mag. (7) **27**, 33–36.
1940 *P 1*. PENDSE, C. G.: On mass and force in Newtonian mechanics. Phil. Mag. **29**, 477–484.
P 2. ZAREMBA, S.: Réflexions sur les fondaments de la mécanique rationnelle. Enseignement Math. **38**, 59–69.
1943 *P 1*. BRELOT, M.: Sur les principes mathématique de la mécanique classique. Ann. Univ. Grenoble (sect. sci.-méd.) **18**, 24 pp.
1944 *P 1*. BRELOT, M.: Sur quelques points de mécanique rationnelle. Ann. Univ. Grenoble (sect. sci.-méd.) **20**, 37 pp.
1947 *P 1*. SIMON, H. A.: Axioms of Newtonian mechanics. Phil. Mag. (7) **36**, 888–905.
1949 *P 1*. HAMEL, G.: Theoretische Mechanik. Berlin-Göttingen-Heidelberg: Springer. xvi + 796 pp. See Kap. 1.
1950 *P 1*. BANACH, S.: Mechanika w zakresie szkól akademickich. Krakow, Monogr. Mat. t. 8–9, 3rd ed., ix + 555 pp. Trans. E. J. SCOTT, Mechanics, Warsaw 1951. iv + 546 pp. See Chap. III, §§ 1–4.

1953 *P 1*. MCKINSEY, J. C. C., A. C. SUGAR and P. SUPPES: Axiomatic foundations of classical particle mechanics. J. Rational Mech. Anal. **2**, 253–272.
P 2. MCKINSEY, J. C. C., and P. SUPPES: Transformations of systems of classical particle mechanics. J. Rational Mech. Anal. **2**, 273–289.
1954 *P 1*. BRELOT, M.: Les Principes Mathématiques de la Mécanique Classique. Grenoble and Paris.
P 2. PLATRIER, C.: Mécanique Rationelle **1**. Paris: Dunod. See pp. 77–180.
See also the following items from the "List of works Cited": KIRCHHOFF [1876, *2*; Vorl. 1, §§ 1–2, 4, 7], WHITTAKER [1904, *8*; Ch. II], JAUMANN [1905, *2*], HAMEL [1908, *4*], NOLL [1957, *11*], [1958, *8*], [1959, *9*].

Additional Bibliography R: Relativistic Continuum Theories

1904 *R 1*. ABRAHAM, M.: Zur Theorie der Strahlung und des Strahlungsdruckes. Ann. d. Phys. **14**, 236–287.
1909 *R 1*. ABRAHAM, M.: Zur Elektrodynamik bewegter Körper. Rend. Pal. **28**, 1–28.
R 2. BORN, M.: Die Theorie des starren Elektrons in der Kinematik des Relativitätsprinzips. Ann. d. Phys. **30**, 1–56.
R 3. HERGLOTZ, G.: Über den vom Standpunkt des Relativitätsprinzips aus als „Starr" zu bezeichnenden Körper. Ann. d. Phys. **31**, 393–415.
R 4. NOETHER, F.: Zur Kinematik des starren Körpers in der Relativtheorie. Ann. d. Phys. **31**, 919–944.
1911 *R 1*. BORN, M.: Elastizitästheorie und Relativitätstheorie. Phys. Z. **12**, 569–575.
R 2. HERGLOTZ, G.: Über die Mechanik des deformierbaren Körpers vom Standpunkte der Relativitätstheorie. Ann. Phys. (4) **36**, 493–533.
R 3. IGNATOWSKY, W. v.: Zur Elastizitätstheorie vom Standpunkte des Relativitätsprinzips. Phys. Z. **12**, 164–169.
R 4. IGNATOWSKY, W. v.: Zur Hydrodynamik vom Standpunkte des Relativitätsprinzips. Phys. Z. **12**, 441–442.
R 5. JÜTTNER, F.: Die Dynamik eines bewegten Gases in der Relativitätstheorie. Ann. Phys. (4) **35**, 145–161.
R 6. LAUE, M. v.: Zur Dynamik der Relativitätstheorie. Ann. Phys. (4) **35**, 524–542.
R 7. NORDSTRÖM, G.: Zur Relativitätsmechanik deformierbarer Körper. Phys. Z. **12**, 854–857.
1912 *R 1*. LAMLA, E.: Über die Hydrodynamik des Relativitätsprinzips. Ann. Phys. (4) **37**, 772–796.
R 2. LÉMERAY, E. M.: Élasticité longitudinale et élasticité transversale. Soc. Franç. Phys. No. 29, 2.
1914 *R 1*. MATTIOLI, G. D.: La dinamica di relatività dei mezzi continui dedotta della dinamica classica colla modificazione di un solo principio. Rend. Lincei (5) $\mathbf{23}_2$, 427–432.
1915 *R 1*. DANIELL, P. J.: Rotation of elastic bodies and the principle of relativity. Phil. Mag. (6) **30**, 754–761.
R 2. EINSTEIN, A.: Der Energiesatz in der allgemeinen Relativitätstheorie. Sitzgsber. preuß. Akad. Wiss. **1915**, 778–786.
R 3. MATTIOLI, G. D.: La dinamica di relatività dei mezzi continui dedotta della dinamica classica colla modificazione di un solo principio. Nuovo Cim. (6) **9**, 263–270.
1916 *R 1*. EINSTEIN, A.: Die Grundlage der allgemeinen Relativitätstheorie. Ann. d. Phys. **49**, 769–822.
1917 *R 1*. NORDSTRÖM, G.: Die Mechanik der Continua in der Gravitationstheorie von **Einstein**. Handel. Nederl. Natuuren Geneesk. Congr. **16**, 176–178.
1918 *R 1*. EINSTEIN, A.: Über Gravitationswellen. Sitzgsber. preuss. Akad. Wiss. **1918**, 154–167.
R 2. THIRRING, H.: Über die Wirkung rotierender ferner Massen in den Einsteinschen Gravitationstheorie. Phys. Z. **19**, 33–39.
1922 *R 1*. BACH, R.: Neue Lösungen der Einsteinschen Gravitationsgleichungen. A. Das Feld in der Umgebung eines langsam rotierenden kugelähnlichen Körpers von beliebiger Masse in 1. und 2. Annäherung. Math. Z. **13**, 119–133.
1923 *R 1*. CARTAN, É.: Sur les variétés à connexion affine et la théorie de la relativité généralisée. Ann. Éc. Norm. (3) **40**, 325–412; **41**, 1–25 (1924); **42**, 17–88 (1925).
R 2. GORDON, W.: Zur Lichtfortpflanzung nach der Relativitätstheorie. Ann. d. Phys. **72**, 421–456.

R 3. Marcolongo, R.: Relatività, 2a edizione riveduta ed ampliata. Messina: G. Principato. ix + 235 pp., esp. p. 117.
1924 R 1. Eisenhart, L. P.: Space-time continua of perfect fluids in general relativity. Trans. Amer. Math. Soc. **26**, 205–220.
1926 R 1. Meksyn, D.: On equilibrium and motion of a continuous medium in four-dimensional space. Phil. Mag. (7) **2**, 994–1006.
1928 R 1. Ghosh, J.: On the gravitational field of an ideal fluid. Bull. Calcutta Math. Soc. **19**, 67–82.
R 2. Levi-Civita, T.: Fondamenti di meccanica relativistica. Bologna: Zanichelli. vii + 185 pp., esp. p. 70.
1931 R 1. Akeley, E. S.: The rotating fluid in the relativity theory. Phil. Mag. (7) **11**, 330–344.
R 2. Akeley, E. S.: The rotating fluid in the relativity theory. Phys. Rev. (2) **37**, 109–110.
1932 R 1. De Donder, Th., et Y. Dupont: Théorie relativiste de l'élasticité et de l'électromagnétostriction. Bull. Sci. Acad. Belg. (5) **18**, 680–691, 782–790, 899–910; **19**, 370–378 (1933).
R 2. Finzi, B.: Meccanica relativistica ereditaria. Atti Soc. Ital. Progr. Sc. **20**, 8–11.
1933 R 1. Tolman, R. C.: Thermodynamics and relativity. Science (2) **77**, 291–298, 313—317.
R 2. Tolman, R. C., and H. P. Robertson: On the interpretation of heat in relativistic thermodynamics. Phys. Rev. **43**, 564–568.
1934 R 1. Synge, J. L.: The energy-tensor of a continuous medium. Trans. Roy. Soc. Canada (3) **28**, 127–171.
R 2. Tolman, R. C.: Relativity, thermodynamics and cosmology. Oxford: Clarendon Press. xv + 502 pp.
1937 R 1. Synge, J. L.: Relativistic hydrodynamics. Proc. Lond. Math. Soc. (2) **43**, 376–416.
1939 R 1. Tamm, J.: **Cerenkov** radiation and symmetry of the stress tensor. J. Phys. USSR. **1**, 439.
R 2. Van Dantzig, D.: Stress tensor and particle density in special relativity. Nature, Lond. **143**, 855.
R 3. Van Dantzig, D.: On the phenomenological thermodynamics of moving matter. Physica, Haag **6**, 673–704.
R 4. Van Dantzig, D.: On relativistic thermodynamics. Proc. Kon. Nederl. Akad. Wetensch. **42**, 601–607 = Indagat. Math. **1**, 212–218.
R 5. Van Dantzig, D.: On relativistic gas theory. Proc. Kon. Nederl. Akad. Wetensch. **42**, 608–625 = Indagat. Math. **1**, 219–236.
1940 R 1. Eckart, C.: The thermodynamics of irreversible processes, III. Relativistic theory of the simple fluid. Phys. Rev. (2) **58**, 919–924.
R 2. Haantjes, J.: Die Gleichberechtigung gleichförmig beschleunigter Beobachter für die electromagnetischen Erscheinungen. Proc. Kon. Nederl. Akad. Wetensch. **43**, 1288–1299.
R 3. Lichnerowicz, A.: Sur un théorème d'hydrodynamique relativiste. C. R. Acad. Sci., Paris **211**, 117–119.
R 4. Rosenfeld, L.: Sur le tenseur à impulsion-énergie. Acad. Roy. Belg., Cl. Sci. Mém. Coll. **18**, fasc. 6, 30 pp.
R 5. Van Dantzig, D.: On the thermodynamics of perfectly perfect fluids. Kon. Nederl. Akad. Wetensch. Proc. **43**, 387–401, 609–618.
1941 R 1. Bergmann, P. G.: On relativistic thermodynamics. Phys. Rev. (2) **59**, 928.
R 2. Lichnerowicz, A.: Sur des théorèmes d'unicité relatifs aux équations gravitationnelles du cas intérieur. Bull. Sci. Math. (2) **65**, 54–72.
R 3. Lichnerowicz, A.: Sur l'invariant intégral de l'hydrodynamique relativiste. Ann. Éc. Norm. (3) **58**, 285–304.
1942 R 1. Berenda, C. W.: The problem of the rotating disk. Phys. Rev. **62**, 280–290.
1944 R 1. Costa de Beauregard, O.: La Relativité Restreinte et la Première Mécanique Broglienne. Paris: Gauthier-Villars. 71 pp.
R 2. Costa de Beauregard, O.: Sur les équations fondamentales, classiques, puis relativistes, de la dynamique des milieux continus. J. Math. Pures Appl. **23**, 211–217.
R 3. Einstein, A., and V. Bargmann: Bivector fields, I and II. Ann. Math. **45**, 1–14, 15–23.
1945 R 1. Costa de Beauregard, O.: Définition covariante de la force. C. R. Acad. Sci., Paris **221**, 743–747.
R 2. Mautner, F., and E. Schrödinger: Infinitesimal affine connection with twofold Einstein-Bargmann symmetry. Proc. Roy. Irish Acad. **50**, 223–231.

R 3. SCHRÖDINGER, E.: On distant affine connection. Proc. Roy. Irish Acad. **50**, 143–154.

1946 *R 1.* COSTA DE BEAUREGARD, O.: Sur la conservation de la masse propre. Sur la notion de fluide parfait. C. R. Acad. Sci., Paris **222**, 271–273.

R 2. COSTA DE BEAUREGARD, O.: Equations générales de l'hydrodynamique des fluides parfaits. C. R. Acad. Sci., Paris **222**, 369–371.

R 3. COSTA DE BEAUREGARD, O.: Sur la théorie des forces élastiques. C. R. Acad. Sci., Paris **222**, 477–479.

R 4. COSTA DE BEAUREGARD, O.: Sur la thermodynamique des fluides. C. R. Acad. Sci., Paris **222**, 590–592.

R 5. COSTA DE BEAUREGARD, O.: Retour sur la dynamique et la thermodynamique des milieux continus. C. R. Acad. Sci., Paris **222**, 1472–1474.

R 6. COSTA DE BEAUREGARD, O.: Dynamique relativiste des milieux continus. La variation de la masse propre en fonction du travail des forces superficielles. J. Math. Pures Appl. **25**, 187–207.

1947 *R 1.* DE WET, J. S.: Symmetric energy-momentum tensors in relativistic field theories. Proc. Cambridge Phil. Soc. **43**, 511–520.

R 2. HILL, E. L.: On the kinematics of uniformly accelerated motions and classical electromagnetic theory. Phys. Rev. **72**, 143–149.

R 3. ROSEN, N.: Notes on rotation and rigid bodies in relativity theory. Phys. Rev. **71**, 54.

R 4. WEYSSENHOFF, J., and A. ROABE: Relativistic dynamics of spin fluids and spin particles. Acta phys. Polonica **9**, 7–18.

1948 *R 1.* TAUB, A. H.: Relativistic Rankine-Hugoniot equations. Phys. Rev. (2) **74**, 328–334.

1949 *R 1.* CLARK, G. L.: The mechanics of continuous matter in relativity theory. Proc. Roy. Soc. Edinburgh, Sect. A **62**, 434–441.

R 2. COSTA DE BEAUREGARD, O.: La Théorie de la Relativité Restreinte. Paris: Masson & Cie. vi + 173 pp.

R 3. FOKKER, A. D.: On the space-time geometry of a moving rigid body. Rev. Mod. Phys. **21**, 406–408.

R 4. PAPAPETROU, A.: Non-symmetric stress-energy-momentum tensor and spin density. Phil. Mag. **40**, 937–946.

1950 *R 1.* BLOCH, C.: Variational principle and conservation equations in non-local field theory. Kgl. danske Vidensk. Selsk., mat.-fys. Medd. **26**, 30 pp.

R 2. CLARK, G. L.: The external gravitational and electromagnetic fields of rotating bodies. Proc. Roy. Soc. Lond., Ser. A **201**, 488–509.

R 3. DE HOFFMAN, F., and E. TELLER: Magneto-hydrodynamic shocks. Phys. Rev. (2) **80**, 692–702.

R 4. ROBERTSON, H. P.: The geometries of the thermal and gravitational fields. Amer. Math. Monthly **51**, 232–245.

1951 *R 1.* HILL, E. L.: The deformation of moving coordinate systems in relativistic theories. Phys. Rev. **84**, 1165–1168.

R 2. LEAF, B.: The continuum in special relativity theory. Phys. Rev. **84**, 345–350.

1952 *R 1.* GETHING, P. J. D.: Rotation and magnetism in the world-models of kinematic relativity. Monthly Notices Roy. Astronom. Soc. London **112**, 578–582.

R 2. MÖLLER, C.: The Theory of Relativity. Oxford.

R 3. TAKENO, H.: On relativistic theory of rotating disk. Progr. Theor. Phys. Japan **7**, 367–376.

R 4. UENO, Y., and H. TAKENO: On equivalent observers. Prog. Theor. Phys. Japan **8**, 291–301.

R 5. WHITROW, G. J.: The Fitzgerald-Lorentz contraction phenomenon and theories of the relativity of Galilean frames. Sci. Proc. Roy. Dublin Soc. **26**, 37–44.

1953 *R 1.* GÜRSEY, F.: Gravitation and cosmic expansion in conformal space-time. Proc. Cambridge Phil. Soc. **49**, 285–291.

R 2. KIRKWOOD, R. L.: The physical basis of gravitation. Phys. Rev. **92**, 1557–1562.

R 3. MATRAI, T.: A relativistic treatment of rigid motions. Nature, Lond. **172**, 858–859.

R 4. STEPHENSON, G., and C. W. KILMISTER: A unified theory of gravitation and electromagnetism. Nuovo Cim. **10**, 230–235.

R 5. TAYLOR, N. W.: The relativistic electromagnetic equations in a material medium. Austral. J. Phys. **6**, 1–9.

R 6. UENO, Y: On the wave theory of light in general relativity. Progr. Theor. Phys. Japan **10**, 442–450.

1954 *R 1.* GOMES, R. L.: The motion of rigid body in restricted relativity. Gaz. Mat. Lisboa **15**, No. 58, 9–11.

R 2. KILMISTER, C. W., and G. STEPHENSON: An axiomatic criticism of unified field theories, I and II. Nuovo Cim. **11**, Suppl. 91–105. 118–140.

R 3. KIRKWOOD, R. L.: Gravitational field equations. Phys. Rev. **95**, 1051–1056.

R 4. POUNDER, J. R.: On relativistic rigid surfaces of revolution. Comm. Dublin Inst. Adv. Study, Ser. A, No. 11, 53 pp.

R 5. SALZMAN, G., and A. H. TAUB: Born-type rigid motion in relativity. Phys. Rev. (6) **95**, 1659–1669.

R 6. SCHRÖDINGER, E.: On distant affine connection. Proc. Roy. Irish Acad. **50**, 143–154.

R 7. SYNGE, J. L.:Relativistically rigid surfaces, Studies in mathematics and mechanics presented to **Richard von Mises**, pp. 217–226. New York: Academic Press.

R 8. TAUB, A. H.: General relativistic variational principle for perfect fluids. Phys. Rev. **94**, 1468–1470.

1955 *R 1.* BOLAZS, N. L.: The propagation of light rays in moving media. J. Opt. Soc. Amer. **45**, 63–64.

R 2. FINKELSTEIN, D.: Internal structure of spinning particles. Phys. Rev. **100**, 924–931.

R 3. KREMPASKY, J.: The strain tensor in space-time as a result of motion. Mat. Fyz. Casopis Slovensk. Akad. **5**, 124–131.

R 4. LICHNEROWICZ, A.: Théories Relativistes de la Gravitation et de l'Electromagnetisme. Paris: Masson & Cie.

R 5. MARX, G., u. G. GYÖRGYI: Über den Energie-Impuls-Tensor des elektromagnetischen Feldes in Dielektrika. Ann. Phys. (6) **16**, 241–256.

1956 *R 1.* GOMES, R. L.: Kinematics of rigid bodies in relativity. Gaz. Fis. Lisboa **3**, 99–107.

R 2. SCHILD, A.: On gravitational theories of **Whitehead's** type. Proc. Roy. Soc. Lond., Ser. A **235**, 202–209.

R 3. SCHNUTZER, E.: Zur relativistischen Elektrodynamik in beliebigen Medien. Ann. Phys. (6) **18**, 171–180.

R 4. SYNGE, J. L.: Relativity: The Special Theory. Amsterdam: North Holland Publishing Co.; New York: Interscience Publ.

R 5. TAUB, A. H.: Isentropic hydrodynamics in plane symmetric space-times. Phys. Rev. (2) **103**, 454–467.

R 6. WINOGRADSKI, J.: Sur le tenseur impulsion-energie métrique et le théorème de **Noether**. Cahiers de Phys. **67**, 1–5.

1957 *R 1.* ARNOWITT, R. L.: Phenomenological approach to a unified field theory. Phys. Rev. **105**, 735–742.

R 2. MARX, G.: Innere Arbeit in der relativistischen Dynamik. Acta phys. Acad. Sci. Hungar. **6**, 353–379.

R 3. SYNGE, J. L.: The Relativistic Gas. Amsterdam.

R 4. TAUB, A. H.: Singular hypersurfaces in general relativity. Illinois J. Math. **1**, 370–388.

1959 *R 1.* SYNGE, J. L.: A theory of elasticity in general relativity. Math. Z. **12**, 82–87.

R 2. TAUB, A. H.: On circulation in relativistic hydrodynamics. Arch. Rational Mech. Anal. **3**, 312–324.

Appendix.

Tensor Fields.

By

J. L. Ericksen.

With 2 Figures.

I. Preliminaries.

1. The nature of this appendix. This appendix is a heterogeneous collection of results on tensor fields. The reader is assumed familiar with the elements of vector and tensor analysis and of matrix algebra. The author is indebted to Drs. W. Noll, R. A. Toupin, C. Truesdell and M. K. Zoller for bringing to his attention many references and helping to correct many errors. Such defects as remain are of course the responsibility of the author.

a) Notation.

2. Tensor notation. We employ the methods and notation of the tensor calculus and, to a lesser extent, matrix algebra and Gibbs' vector analysis. Until further notice, write $x^1, x^2, \ldots, x^n$ for the co-ordinates of points in an n-dimensional space. For abbreviation, we let $\boldsymbol{x}$ stand for this set of co-ordinates. Until further notice, italic lower case letters other than x will be used for the kernel indices and tensor indices of tensors. Also, for a given set of tensor components $a^{k\ldots m}{}_{p\ldots q}$ we write $\boldsymbol{a}$; for "the components $a^{k\ldots m}{}_{p\ldots q}$ of the tensor $\boldsymbol{a}$" or "the tensor having the components $a^{k\ldots m}{}_{p\ldots q}$", we write "the tensor $a^{k\ldots m}{}_{p\ldots q}$", or, more often, "the tensor $\boldsymbol{a}$". We employ the *summation convention* for diagonally repeated tensorial indices and indices of Christoffel symbols. The notation $a^{\dots}_{\dots;k}$ denotes the *covariant derivative*, except when otherwise noted.

A square bracket enclosing m running indices indicates that the tensor is completely antisymmetrized with respect to these indices. This is done by permuting the indices in all possible ways, attaching a positive (negative) sign to the tensors corresponding to even (odd) permutations, adding these tensors and dividing by $m!$. Parentheses inclosing m running indices indicate that the tensor is completely symmetrized with respect to the enclosed indices. This is done in the same way except that the positive sign is attached to all permutations.

For example

$$a^k_{[m\,p]} \equiv \frac{1}{2!}\left(a^k_{m\,p} - a^k_{p\,m}\right), \qquad a^k_{(m\,p)} \equiv \frac{1}{2!}\left(a^k_{m\,p} + a^k_{p\,m}\right). \tag{2.1}$$

We use the short notation

$$|\boldsymbol{x}/\boldsymbol{X}| \equiv \frac{\partial(x^1, x^2, \ldots, x^n)}{\partial(X^1, X^2, \ldots, X^n)} \tag{2.2}$$

when the x^k are given in terms of parameters X^K by a mapping $\boldsymbol{x} = \boldsymbol{x}(\boldsymbol{X})$, whether this be regarded as a co-ordinate transformation or as a point transformation.

Unless greater or lesser generality is asserted explicitly or is evident from the context, the space considered in this treatise is Euclidean 3-space with real co-ordinates. The components of associated tensors are regarded as different sets of components of the same tensor; the same kernel letter is used for all these components, and the context will make clear whether $\boldsymbol{a}$ stands for a particular set of components or is intended as a general symbol for all components. In particular, we shall often write $\mathbf{1}$ for the metric tensor g_{km}, its mixed components δ^k_m, and its inverse g^{km}.

Setting $g \equiv \det g_{km}$ and assuming $g > 0$, we use the notations

$$e_{k_1 k_2 \dots k_n} \equiv g^{\frac{1}{2}} \varepsilon_{k_1 k_2 \dots k_n}, \qquad e^{k_1 k_2 \dots k_n} \equiv g^{-\frac{1}{2}} \varepsilon^{k_1 k_2 \dots k_n}, \tag{2.3}$$

where n is the dimension of the space and the ε's are the usual permutation symbols, with $\varepsilon_{1\dots n} = \varepsilon^{1\dots n} = 1$ in all co-ordinate systems. The $\boldsymbol{e}$'s are examples of *axial tensors*, which transform as absolute tensors under co-ordinate transformations with positive Jacobian, and under those only.

In Euclidean space, a rectangular Cartesian co-ordinate system is available. If we write $\boldsymbol{z}$ or $z^1, z^2, \dots, z^n$ for the co-ordinates of points, *this in itself indicates that we are using rectangular Cartesian co-ordinates*, and at the same time we follow the notation and the summation convention of *Cartesian tensors*, writing all tensor indices as subscripts and summing on repeated indices.

Moreover, for the *position vector* from a fixed point to $\boldsymbol{x}$, we write p^k or p_k or $\boldsymbol{p}$. If the fixed point has rectangular Cartesian co-ordinates ${}_0\boldsymbol{z}$, then $\boldsymbol{p}$ is the vector whose covariant and contravariant components reduce to $\boldsymbol{z} - {}_0\boldsymbol{z}$ in rectangular Cartesian systems. If ${}_0\boldsymbol{z} = 0$, then in rectangular Cartesian systems and only in such systems we have $\boldsymbol{x} = \boldsymbol{p} = \boldsymbol{z}$. Formulae in which $\boldsymbol{p}$ occur should be invariant when ${}_0\boldsymbol{z}$ is replaced by a different fixed point ${}_0\boldsymbol{z}^*$; in most cases, verification of this invariance is left to the reader. In general co-ordinates, $\boldsymbol{p}$ and $\boldsymbol{x}$ are entirely different quantities. For the vector element of arc we write $d\boldsymbol{x}$ in general co-ordinates, $d\boldsymbol{z}$ in formulae whose validity is limited to rectangular Cartesian co-ordinates. Curvilinear components of $\boldsymbol{p}$ at $\boldsymbol{x}$ generally differ from those at the fixed point. Except where otherwise noted, we use the former.

3. Matrix and vector notations. For second order tensors, we sometimes use matrix or dyadic notation[1]. For example, we write $\boldsymbol{c} = \boldsymbol{a} \cdot \boldsymbol{b}$ for the equation $c^k_m = a^k_p b^p_m$; $\boldsymbol{a}^{-1}$ for the tensor, if it exists, which is related to $\boldsymbol{a}$ by $\boldsymbol{a} \cdot \boldsymbol{a}^{-1} = \mathbf{1}$, the components of $\boldsymbol{a}^{-1}$ being a_m^{-1k}; while $\boldsymbol{a}'$ denotes the transpose of $\boldsymbol{a}$, given by $a'_{km} = a_{mk}$.

More generally, the single dot operation is defined by the equations $\boldsymbol{a} \cdot \boldsymbol{b} \equiv \boldsymbol{t}$ and $\boldsymbol{b} \cdot \boldsymbol{a} \equiv \boldsymbol{u}$, where the tensors $\boldsymbol{t}$ and $\boldsymbol{u}$ are given by

$$t^{m\dots n} \equiv a_k b^{km\dots n}, \qquad u^{m\dots n} = a_k b^{m\dots nk}. \tag{3.1}$$

In three-dimensional spaces, also the single cross operation of Gibbs[2] may be defined. If $\boldsymbol{a}$ is any vector, $\boldsymbol{b}$ any tensor, then $\boldsymbol{a} \times \boldsymbol{b} \equiv \boldsymbol{c}$ and $\boldsymbol{b} \times \boldsymbol{a} \equiv \boldsymbol{d}$, where the tensors $\boldsymbol{c}$ and $\boldsymbol{d}$ are given by

$$c_k^{m\dots n} \equiv e_{krs} a^r b^{sm\dots n}, \qquad d_w^{k\dots m} \equiv e_{wrs} b^{k\dots mr} a^s. \tag{3.2}$$

The *cross* or *vector* $\boldsymbol{t}_\times$ of a second order tensor $\boldsymbol{t}$ is the axial vector given by

$$t^k_\times = e^{krs} t_{rs} = e^{krs} t_{[rs]}. \tag{3.3}$$

Thus apart from algebraic sign, the contravariant components of $\boldsymbol{t}_\times$ are the components of $2\sqrt{g}\, t_{[rs]}$, not of $\sqrt{g}\, t_{[rs]}$, and in particular $(\operatorname{grad} \boldsymbol{b})_\times = \operatorname{curl} \boldsymbol{b}$. Since a

[1] Alternatively, our direct notation may be interpreted in terms of linear transformations. Cf. e.g. Halmos [1942, *1*, Chap. II], and, for applications to kinematics and mechanics, Noll [1955, *1*].

[2] [1881, *1*], [1884, *1*], [1909, *1*].

skew-symmetric tensor of second order in a three-dimensional space has at most three independent components, in any equation where it appears it may be replaced by its cross. When both $\boldsymbol{t}$ and $\boldsymbol{u}$ are skew-symmetric,

$$\boldsymbol{t}_\times \cdot \boldsymbol{u}_\times = 2t^{km} u_{km}, \qquad t_\times^2 = 2t^{km} t_{km}. \tag{3.4}$$

We introduce the symbol Ψ, which may be read "trident", to stand for a scalar, vector, or tensor of any order, so long as the formulae in which it appears have a sense. By grad Ψ, div Ψ, and curl Ψ we mean the tensors $\boldsymbol{a}$, $\boldsymbol{b}$, and $\boldsymbol{c}$ given by

$$a_{km\ldots n} \equiv \Psi_{m\ldots n,k}, \qquad b^{k\ldots m} \equiv \Psi^{rk\ldots m}{}_{,k} \quad \text{or} \quad c^k_{m\ldots n} \equiv e^{ksr}\, \Psi_{rm\ldots n,s}, \tag{3.5}$$

where the last applies only in the three-dimensional case. These definitions are so framed that the identities div curl $\Psi = 0$ curl grad $\Psi = 0$, familiar when Ψ is a vector, hold for a tensor of any order.

For positive integral K, the *K-th power* $\boldsymbol{b}^{(K)}$ of a vector $\boldsymbol{b}$ is the tensor $b^{m_1}\ldots b^{m_k}$. The definition is extended to all integral K and put in inductive form as follows:

$$\boldsymbol{b}^{(-1)} \equiv 0, \qquad \boldsymbol{b}^{(0)} \equiv 1, \qquad \boldsymbol{b}^{(K+1)} \equiv \boldsymbol{b}^{(K)}\boldsymbol{b}. \tag{3.6}$$

We denote by $\{b^{(K)}\Psi\}$ the expression[1]

$$\{\boldsymbol{b}^{(K)}\Psi\} \equiv \boldsymbol{b}^{(K)}\Psi + \boldsymbol{b}^{(K-1)}\Psi\boldsymbol{b} + \cdots + \Psi\boldsymbol{b}^{(K)}. \tag{3.7}$$

4. Projections. If $\boldsymbol{n}$ is a unit vector, we define the *normal projection* Ψ_n and the *tangential projection* Ψ_t of Ψ onto the direction of $\boldsymbol{n}$ by

$$\Psi_n \equiv \boldsymbol{n}\cdot\Psi, \qquad \Psi_t \equiv \boldsymbol{n}\times\Psi \tag{4.1}$$

the latter formula being restricted to three-dimensional space. When $\Psi = \boldsymbol{c}$, a vector, we write c_n rather than $\boldsymbol{c}_n$ for the normal projection.

Again in 3-space, consider the special co-ordinate transformations which leave x^1 fixed: $x^{*1} = x^1$, $x^{*\alpha} = x^{*\alpha}(x^2, x^3)$, and write ${}_1g \equiv \det g_{\alpha\beta}$, where α, β have the range 2, 3. Then, considering only those co-ordinate systems such that ${}_1g > 0$, we see that the quantities ${}_1\boldsymbol{t}$ defined by

$$ {}_1t^\alpha_\beta \equiv \sqrt{{}_1g}\, \varepsilon_{\beta\gamma}\, t^{\alpha\gamma}, \tag{4.2}$$

where $\boldsymbol{t}$ is a three-dimensional tensor, transform as components of a tensor[2] on the surface $x^1 =$ const. The two-dimensional tensor ${}_1\boldsymbol{t}$ may be called the *skew projection* of $\boldsymbol{t}$ onto the x^1-direction. Taking the trace of Eq. (4.2) yields a relation[3] between the components of $\boldsymbol{t}_\times$ and the traces of the three skew projections ${}_k\boldsymbol{t}$:

$$t^k_\times = (g/{}_kg)^{\frac{1}{2}}\, {}_k t^\alpha_\alpha. \tag{4.3}$$

b) Use of complex co-ordinates.

5. Use of tensor methods in complex co-ordinates. The formalism of the tensor calculus applies to complex as well as real co-ordinate transformations; for problems in the Euclidean plane, complex co-ordinates are sometimes convenient. Suppose, for example, that we introduce complex co-ordinates $x^1 = z_1 + iz_2$, $x^2 = \overline{x}^1$, where the z_k are rectangular Cartesian and the superposed bar denotes the complex conjugate. The usual formulae of tensor transformation

---

[1] TRUESDELL [1949, *3*], [1951, *2*], [1951, *3*], [1954, *2*, Chap. I].

[2] This result and Eq. (4.2) are due in principle to CAUCHY [1841, *1*, Th. VIII]. Cf. TRUESDELL [1954, *6*, § 2].

[3] When both $t_\times$ and ${}_kt$ are referred to physical components (Sect. 13) in an orthogonal co-ordinate system, the factor $(g/{}_1g)^{\frac{1}{2}}$ disappears from Eq. (4.3).

yield the covariant components $t_{km}(\boldsymbol{x})$ in terms of the Cartesian components t_{km}:

$$\left.\begin{aligned} 4t_{11}(\boldsymbol{x}) &= 4\bar{t}_{22}(\boldsymbol{x}) = t_{11} - t_{22} - i(t_{12} + t_{21}),\\ 4t_{12}(\boldsymbol{x}) &= 4\bar{t}_{21}(\boldsymbol{x}) = t_{11} + t_{22} + i(t_{12} - t_{21}).\end{aligned}\right\} \tag{5.1}$$

For the special case[1] when $\boldsymbol{t}$ is the metric tensor, this yields $g_{11}(\boldsymbol{x}) = g_{22}(\boldsymbol{x}) = 0$, $g_{12}(\boldsymbol{x}) = g_{21}(\boldsymbol{x}) = \frac{1}{2}$; hence $g^{11}(\boldsymbol{x}) = g^{22}(\boldsymbol{x}) = 0$, $g^{12}(\boldsymbol{x}) = g^{21}(\boldsymbol{x}) = 2$. All results which do not depend essentially on conditions of reality carry over to complex co-ordinates. As is clear from the above example, the inequalities $g_{11} > 0$, $g_{11}g_{22} - g_{12}^2 > 0$ which, in real co-ordinates, express the fact that the g_{km} are coefficients of a real, positive definite quadratic form, do not. Since we occasionally use such conditions, our treatment is restricted to real co-ordinates except where otherwise noted.

6. Complex representation of plane second order Cartesian tensors. Returning to the example (5.1), we note that a rotation of the co-ordinates z^k induces on x^k the transformation $x^{*1} = \overline{x^{*2}} = x^1 e^{i\varphi}$, where φ is the angle of rotation, and that the tensor $t_{km}(\boldsymbol{x})$ transforms according to the simple laws

$$t_{12}(\boldsymbol{x}^*) = \overline{t_{21}(\boldsymbol{x}^*)} = t_{12}(\boldsymbol{x}) = \overline{t_{21}(\boldsymbol{x})}, \; t_{11}(\boldsymbol{x}^*) = \overline{t_{22}(x^*)} = t_{11}(\boldsymbol{x})\, e^{-2i\varphi}. \tag{6.1}$$

In linear elasticity, $\boldsymbol{t}$, interpreted as the stress tensor, satifies the equilibrium equations and compatibility conditions for generalized plane stress if and only if[2]

$$t_{12}(\boldsymbol{x}) = \bar{t}_{21}(\boldsymbol{x}) = t^k_k = f + \bar{f}, \quad t_{11}(\boldsymbol{x}) = g - x^2 f', \tag{6.2}$$

where f and g are arbitrary functions of the complex variable x^1. Similar results hold for plane strain. In this context, $t_{11}(\boldsymbol{x})$ is sometimes called the conjugate stress deviator to indicate that it is unaltered if $\boldsymbol{t}$ is replaced by $\boldsymbol{t} + K\boldsymbol{1}$, where K is any scalar and $\boldsymbol{1}$ is the two-dimensional unit matrix. Comparison of these results with their Cartesian analogues shows the simplicity that can result from using complex co-ordinates. In both applications all three Cartesian components of $\boldsymbol{t}$ may be calculated easily from the two complex components $t_{12}(\boldsymbol{x})$ and $t_{11}(\boldsymbol{x})$, the former of which may be regarded as a scalar. It follows easily from Eq. (5.1) that $t_{12}(\boldsymbol{x}) = 0$ if and only if t is symmetric and traceless, and that in the symmetric case $t_{11}(\boldsymbol{x}) = 0$ if and only if $\boldsymbol{t} = K\boldsymbol{1}$, where K is a scalar.

When $\boldsymbol{t}$ is symmetric, by Eq. $(5.1)_2$ we see that $t_{12}(\boldsymbol{x})$ is always real. As will be shown in Sect. 48, there exists a real rotation that renders $\boldsymbol{t}$ diagonal. Letting $x^* = z_1^* + i z_2^*$, from Eq. (5.1) we see that $t_{11}(\boldsymbol{x}^*)$ is real. By Eq. (6.1) we get $t_{11}(\boldsymbol{x}) = t_{11}(\boldsymbol{x}^*)\, e^{2i\varphi}$, and hence Eq. (5.1) yields[3]

$$\tan 2\varphi = \frac{2t_{12}}{t_{11} - t_{22}}. \tag{6.3}$$

II. Dimensions and physical components.

a) Dimensions of a tensor and its components.

7. Preliminary definitions. Among the transformations to which quantities occurring in physics are subject are those of dimensional units[4]. In constructing a system of measurement, in principle one must introduce a unit of measurement for each type of physical quantity to be considered. It is customary to lay down a finite set of units U_α $(\alpha = 1, \ldots, n)$ as *fundamental units*, and to require that

[1] For other examples, cf. GREEN and ZERNA [1954, *1*].

[2] MUSKHELISHVILI [1953, *1*, Chap. 5] gives a rigorous derivation of these results, which he attributes to KOLOSOV [1909, *4*].

[3] MAXWELL [1870, *3*, pp. 194–195].

[4] Dimensional analysis remains a controversial and somewhat obscure subject. We do not attempt a complete presentation here.

every unit be expressible in terms of them in the symbolic form $U_1^{K_1} \dots U_n^{K_n}$, where the K's are real exponents. A fundamental unit may be regarded as a measure of a definite type of entity, such as length or time, the "size" of the unit being assigned arbitrarily. To compensate for this arbitrariness, we regard any other system of units, obtainable from the given set by changing the size of the fundamental units independently, as equivalent or indistinguishable. More precisely, we require physical equations to be invariant under the transformations

$$U_a^* = U_a^{-1} U_a, \tag{7.1}$$

hereafter called *dimensional transformations*, the U_a being arbitrarily chosen real numbers. A function $Q(U)$ of the units $U_1, \dots, U_n$ is called a *measure number having the dimension* $[U_1^{K_1} \dots U_n^{K_n}]$ provided that it transforms under Eq. (7.1) according to the law $Q(U^*) = Q(U)\, U_1^{K_1} \dots U_n^{K_n}$. The appropriate units for Q are $U_1^{K_1} \dots U_n^{K_n}$, in accordance with the usual notion that the product of a measure number by its units is unaffected by dimensional transformations. We write $[1]$ for $[U_1^0 \dots U_n^0]$.

For a given group G of co-ordinate transformations, a set of functions $t_{p \dots q}^{k \dots m}(\boldsymbol{x}, U)$ are said to be *the components, relative to the co-ordinates* x^k *and units* U_a, *of a tensor of weight* w *having the dimension* $[U_1^{K_1} \dots U_n^{K_n}]$ provided that, under Eq. (7.1) and an arbitrary co-ordinate transformation $x^{*k} = x^{*k}(\boldsymbol{x})$ of G, these functions transform according to the law

$$t_{p \dots q}^{k \dots m}(\boldsymbol{x}^*, U^*) = |\boldsymbol{x}/\boldsymbol{x}^*|^{\mathrm{w}}\, t_{u \dots v}^{r \dots s}(\boldsymbol{x}, U) \frac{\partial x^u}{\partial x^{*p}} \cdots \frac{\partial x^v}{\partial x^{*q}} \frac{\partial x^{*k}}{\partial x^r} \cdots \frac{\partial x^{*m}}{\partial x^s} U_1^{K_1} \dots U_n^{K_n}. \tag{7.2}$$

With a fixed choice of co-ordinates, each component of $\boldsymbol{t}$ is clearly a measure number having the dimension $[U_1^{K_1} \dots U_n^{K_n}]$. Statements to the effect that physical equations are "in their simplest form" generally mean only that they reduce to equations relating such tensors.

8. Physical dimensions of tensors. In classical mechanics, it is customary to take G to be the orthogonal group[1], to require one of the fundamental units to be a unit L of length, and to assume that the co-ordinate differentials dz_k transform as a Cartesian vector having the dimension $[L]$, in agreement with the notion that $ds^2 = dz_k\, dz_k$ is an absolute scalar having the dimension $[L^2]$. Using physical laws and other devices, one determines the appropriate Cartesian tensor law to represent the entities to be considered. The dimensions so assigned in rectangular Cartesian systems we call *physical dimensions* to distinguish them from other dimensions to be introduced below. From this point of view the Cartesian metric tensor δ_{km} can consistently be regarded as a *dimensionless* tensor, by which is meant a tensor having the physical dimension $[1]$.

In passing to orthogonal curvilinear co-ordinates, one assigns a dimension to each variable introduced as a co-ordinate[2]. The common procedure is to use LAMÉ's theory of orthogonal curvilinear co-ordinates[3], or some other mathematically equivalent procedure, to obtain curvilinear components of Cartesian tensors. Viewed as scalars, these components have a common dimension, which is the physical dimension of the Cartesian tensor which they represent. Under co-ordinate transformations they do not transform as simply as do tensor com-

[1] For simplicity, we restrict ourselves to time independent co-ordinate transformations.

[2] In principle, this dimension may be arbitrary. In practice, each co-ordinate is a homogeneous function of the rectangular Cartesian co-ordinates, so that it is natural to assign it the dimension $[L^K]$, where K is the degree of the homogeneous function.

[3] The method is given in [1840, *1*], [1859, *1*, Leçons 1–2]; a typical application, in [1841, *2*, §§ VII–VIII], [1859, *1*, §§ 144–147].

ponents. Since RICCI and LEVI-CEVITA[1] set down the foundations of tensor calculus, there has been an increasing tendency to use tensor components in place of LAMÉ's, primarily because tensor components transform more simply under co-ordinate transformations. In treating Cartesian tensors having dimensions, one fixes the units and transforms the components according to the appropriate tensor law. Having done so, one can assign a dimension to each component of the tensor consistent with the tensor law, the dimension assigned to the Cartesian tensor, and the dimensions assigned to the co-ordinates. In general, these dimensions will be different for different components and from those of the Cartesian tensor. At least for an absolute tensor $\boldsymbol{t}$, the dimension of the scalar which we denote by the Russian letter Д,

$$Д_t \equiv \sqrt{t_{k\ldots m}\, t^{k\ldots m}}, \tag{8.1}$$

will be unaltered by co-ordinate transformations, and it is customary to regard this dimension as the physical dimension of $\boldsymbol{t}$.

Apparently MCCONNELL[2] was the first to give a satisfactory explanation of the relation between the "components" of mathematical physics and tensor components for tensors of arbitrary order. His method might be explained as follows. Given a tensor field referred to an orthogonal curvilinear co-ordinate system, let its *physical components*[3] at a given point be defined as its corresponding tensor components, at that point, in a rectangular Cartesian system whose axes are parallel to the co-ordinate curves at the point. In mathematical physics, the term "components" usually means these physical components, which henceforth we denote by $t\langle k\ldots m\rangle$. A simple method of calculating them will be given in Sect. 10.

Using tensor methods, one can easily obtain components of tensors in oblique co-ordinate systems. Various methods have been used to motivate definitions of physical components of tensors in such systems, so arranged that the dimension of each physical component of $\boldsymbol{t}$ is its physical dimension. Some of the different formulae, all of which are equivalent to MCCONNELL's when the co-ordinate system is orthogonal, will be discussed below.

9. Absolute dimensions of tensors. From the viewpoint of invariant theory, the kernel of the problem is that tensor components transform according to simple laws under co-ordinate, but not dimensional transformations, while, for the somewhat ambiguously defined physical components, the situation is reversed. If we desire components which transform simply under both types of transformations, as is much to be wished for dimensional analysis of tensor equations and for other questions of invariance, we must either in effect stay with Cartesian co-ordinate systems or else reshape our notions concerning dimensions of tensors. To begin the latter, it is natural to attempt to represent entities by tensors which retain a fixed dimension under arbitrary co-ordinate transformations. SCHOUTEN[4], and DORGELO and SCHOUTEN[5] have shown how this can be done in a logically sound manner. The essential idea is to regard co-ordinates as mere labels, devoid of metrical significance. One could argue, with some force, that science has been hampered by the needless habit of taking co-ordinates as me-

---

[1] [1901, *1*].

[2] [1931, *1*, Appendix, § 2]. THIRRING [1925, *3*, § 2] had explained the connection between physical and tensorial components of the stress tensor.

[3] The name "physical components" had been used earlier, e.g. by HENCKY [1925, *4*], who noted that normal physical components and normal mixed components of second order tensors coincide.

[4] [1951, *1*, Chap. 5].

[5] These authors [1946, *1*] remark that their discussion is extracted from what appears to be an early, unpublished version of [1951, *1*].

tric[1]. According to the usual intuitive interpretations, a dimensional transformation does not rearrange the points of physical space. In assigning dimensions to co-ordinates, one generally has in mind that the co-ordinates themselves are to change in a definite way when the units are transformed. Technically, any such change is a co-ordinate transformation[2]. If we regard co-ordinates as mere labels, we should not distinguish these transformations from any others or, more precisely, we should always transform the co-ordinate differentials according to the law

$$dx^{*k} = \frac{\partial x^{*k}}{\partial x^m} dx^m, \tag{9.1}$$

regardless of what dimensional transformations take place. In other words, $d\boldsymbol{x}^*$ is regarded as a contravariant vector having the dimension $[1]$. SCHOUTEN, and DORGELO and SCHOUTEN call the dimensions assigned to tensors by this scheme *absolute dimensions*. In setting up a correspondence between absolute and physical dimensions, one is guided by the notion that, for absolute scalars, they coincide. For example, the element of arc $ds^2 = g_{km}\, dx^k\, dx^m$ is regarded as an absolute scalar having the absolute and physical dimension $[L^2]$. Using Eq. (9.1), we have

$$\left.\begin{aligned} ds^2(L^*) &= g_{km}(\boldsymbol{x}^*, L^*)\, dx^{*k}\, dx^{*m}, \\ &= L^2\, ds^2(L) = L^2 g_{rs}(\boldsymbol{x}, L)\, dx^r dx^s = L^2 g_{rs}(\boldsymbol{x}, L) \frac{\partial x^r}{\partial x^{*k}} \frac{\partial x^s}{\partial x^{*m}}\, dx^{*k} dx^{*m} \end{aligned}\right\} \tag{9.2}$$

for arbitrary dx^k from which we conclude that

$$g_{km}(\boldsymbol{x}^*, L^*) = g_{rs}(\boldsymbol{x}, L) \frac{\partial x^r}{\partial x^{*k}} \frac{\partial x^s}{\partial x^{*m}} L^2, \tag{9.3}$$

assuming, as usual, that $g_{km} = g_{mk}$. In other words, the metric tensor is regarded as a covariant tensor having the absolute dimensions $[L^2]$. According to Eq. (9.3), g_{km} is not fixed uniquely until we specify the co-ordinate system x^k and the unit L.

To put it another way, in a given co-ordinate system, g_{km} is determined only to within a constant factor, L being assumed to be independent of position. Were we to allow L to vary with position, the multiplication by L^2 in Eq. (9.3) would be formally identical to the "gauge transformation" introduced by WEYL[3] in his unified theory of relativity, the intuitive motivation being similar also.

Mathematically, one can define a rectangular Cartesian co-ordinate system z^k based on the unit L of length to be one such that $g_{km}(\boldsymbol{z}, L) = \delta_{km}$. From Eq. (9.3), any two such systems are related by a transformation $z^{*k} = z^{*k}(\boldsymbol{z})$, $L^* = L^{-1} L$ such that

$$\delta_{km} = \delta_{rs} \frac{\partial z^r}{\partial z^{*k}} \frac{\partial z^s}{\partial z^{*m}} L^2. \tag{9.4}$$

If one regards a rectangular Cartesian co-ordinate system based on the unit of length as obtained by the usual intuitive construction wherein one measures off equal lengths in mutually orthogonal directions, one can check that Eq. (9.4) is the transformation law which should relate the metric tensors δ_{km} of any two such systems. Intuitively, if we have two such systems, z^{*k} and z^k, with

[1] In explaining why so much time elapsed between the appearance of the special and general theories of relativity, EINSTEIN [1949, *1*, p. 66] wrote: „Der hauptsächliche Grund liegt darin, daß man sich nicht so leicht von der Auffassung befreit, daß den Koordinaten eine unmittelbare metrische Bedeutung zukommen müsse."

[2] A clear description of what constitutes a co-ordinate system and a co-ordinate transformation is given by THOMAS [1944, *1*].

[3] [1918, *1*].

common co-ordinate axes, based on the units L^* and L, respectively, the numbers z^{*k} and z^k assigned as co-ordinates to the same geometric point will be related by the transformation $z^{*k} = L\, z^k$, which, in the present scheme, is to be regarded as a co-ordinate transformation, so that we transform co-ordinates and units simultaneously. With this interpretation, we have

$$\delta_{rs} \frac{\partial z^r}{\partial z^{*k}} \frac{\partial z^s}{\partial z^{*m}} L^2 = \delta_{rs} (L^{-1} \delta^r_k)(L^{-1} \delta^s_m) L^2 = \delta_{km}, \tag{9.5}$$

in agreement with Eq. (9.4). Cases where the co-ordinate axes of one system are rotated and translated relative to those of the other can be checked similarly.

To secure consistency with the notion that δ^k_m is a mixed tensor unaffected by dimensional transformations, we must regard the inverse g^{km} of the metric tensor as a contravariant tensor having the absolute dimension $[L^{-2}]$. It is clear that associated tensors, though we generally think of them as being different components of the same tensor, cannot have the same absolute dimensions. Thus there is a certain artificiality inherent in the concept of absolute dimension, but it is objectionable only if one attributes intrinsic significance to dimensions[1]. Otherwise it is no more disturbing than the fact that a tensor is represented by different sets of (associated) components. Some writers[2] have argued that convention plays such an important role in determining the dimensions we assign to quantities that, on logical grounds, there is little reason to impute intrinsic significance to them and that attempts to do so have only led to fruitless controversies. SCHOUTEN[3] and DORGELO and SCHOUTEN[4] seek to make plausible that absolute dimensions indicate the type of measurements required to determine the components to which they correspond, and point out that frequently one set of components is, in some sense, more natural than others. For example, dx^k is most naturally considered as a contravariant vector, though, in the usual terminology, $g_{km}\, dx^m$ is the same vector.

Given the absolute dimension of a tensor $\boldsymbol{t}$, we can calculate its physical dimensions taking these to be the absolute or, equivalently, physical dimensions of the scalar $\Pi_{\boldsymbol{t}}$, given by Eq. (8.1). For example, the physical dimension of the vector dx^k is that of $(dx_k\, dx^k)^{\frac{1}{2}} = ds$, i.e. $[L]$. It follows that the physical dimension of $\boldsymbol{t}$ can be obtained by multiplying the absolute dimensions of any one of the associated tensors representing it by L^K, where K is the number of contravariant indices less the number of covariant indices. We call a tensor *dimensionless* if and only if its physical dimension is $[1]$. The metric tensor is thus dimensionless, for $\Pi_1 = (g_{km}\, g^{km})^{\frac{1}{2}} = n^{\frac{1}{2}}$ in n dimensions.

10. Anholonomic components. In an n-dimensional space, consider n linearly independent contravariant vectors $e^k_{\mathfrak{a}}$ and the reciprocal set $e^{\mathfrak{a}}_k$ satisfying

$$e^{\mathfrak{a}}_k e^k_{\mathfrak{b}} = \delta^{\mathfrak{a}}_{\mathfrak{b}}, \qquad e^{\mathfrak{a}}_m e^k_{\mathfrak{a}} = \delta^k_m, \tag{10.1}$$

where we employ the summation convention for diagonally repeated German indices. For a given tensor $\boldsymbol{t}$, one can form the scalars

$$t^{\mathfrak{a}\ldots\mathfrak{b}}{}_{\mathfrak{c}\ldots\mathfrak{d}} = t^{k\ldots m}{}_{r\ldots s}\, e^{\mathfrak{a}}_k \ldots e^{\mathfrak{b}}_m e^r_{\mathfrak{c}} \ldots e^s_{\mathfrak{d}}, \tag{10.2}$$

equal in number to the tensor components. These are called *anholonomic components* of $\boldsymbol{t}$, more properly, anholonomic components of $\boldsymbol{t}$ relative to the

[1] ESNAULT-PELTERIE [1945, *1*] is perhaps the most ardent proponent of the notion that one should.

[2] Cf. LANGHAAR [1950, *3*] for references and discussion.

[3] [1951, *1*, Chap. 5].

[4] [1946, *1*].

vectors[1] $\boldsymbol{e}_{\mathfrak{a}}$. Inverting Eq. (10.2), we obtain

$$t^{k\dots m}{}_{r\dots s} = t^{\mathfrak{a}\dots\mathfrak{b}}{}_{\mathfrak{c}\dots\mathfrak{d}}\, e^k_{\mathfrak{a}} \dots e^m_{\mathfrak{b}}\, e^{\mathfrak{c}}_r \dots e^{\mathfrak{d}}_s . \tag{10.3}$$

For any given set of $\boldsymbol{e}$'s, associated tensors will, in general, have different anholonomic components unless $e^{\mathfrak{a}}_k = e^k_{\mathfrak{a}}$, which is the case if and only if the $\boldsymbol{e}_{\mathfrak{a}}$ are mutually orthogonal unit vectors. If the vectors $e^k_{\mathfrak{a}}$ have the absolute dimension $[L^{-1}]$, i.e. are dimensionless, the dimension of each of the scalars $t^{\mathfrak{a}\dots\mathfrak{b}}{}_{\mathfrak{c}\dots\mathfrak{d}}$ will be the physical dimension of $\boldsymbol{t}$.

If the $\boldsymbol{e}_{\mathfrak{a}}$ are chosen as unit vectors *tangent to the co-ordinate lines of a fixed orthogonal curvilinear co-ordinate system*, the scalars Eq. (10.2) become numerically equal to McCONNELL's physical components $t\langle k \dots m\rangle$ in that system. In three-dimensional spaces, the matrices of the two sets of $\boldsymbol{e}$'s become in this case

$$\|e^k_{\mathfrak{a}}\| = \begin{Vmatrix} \frac{1}{\sqrt{g_{11}}} & 0 & 0 \\ 0 & \frac{1}{\sqrt{g_{22}}} & 0 \\ 0 & 0 & \frac{1}{\sqrt{g_{33}}} \end{Vmatrix}, \qquad \|e^{\mathfrak{a}}_k\| = \begin{Vmatrix} \sqrt{g_{11}} & 0 & 0 \\ 0 & \sqrt{g_{22}} & 0 \\ 0 & 0 & \sqrt{g_{33}} \end{Vmatrix}, \tag{10.4}$$

so that

$$\left.\begin{aligned} t\langle k \dots m\, r \dots s\rangle &= t^{\mathfrak{a}\dots\mathfrak{b}}{}_{\mathfrak{c}\dots\mathfrak{d}}\, \delta^k_{\mathfrak{a}} \dots \delta^m_{\mathfrak{b}}\, \delta^{\mathfrak{c}}_r \dots \delta^{\mathfrak{d}}_s , \\ &= \left(\frac{g_{kk}\cdots g_{mm}}{g_{rr}\cdots g_{ss}}\right)^{\frac{1}{2}} t^{k\dots m}{}_{r\dots s} \quad \text{(unsummed)}, \\ &= (g_{kk}\cdots g_{ss})^{\frac{1}{2}}\, t^{k\dots s} = (g_{kk}\cdots g_{ss})^{-\frac{1}{2}}\, t_{k\dots s} \quad \text{(unsummed)}. \end{aligned}\right\} \tag{10.5}$$

b) Physical components.

11. Vectors as directed line segments. Since McCONNELL's definition of physical components in orthogonal co-ordinates is entirely satisfactory, whatever definition is used for an oblique system should be consistent with this. Since the only reason for introducing physical components is to obtain quantities simpler to interpret physically than are tensor components, the definition should rest on simple intuitive notions. We frequently think of a vector $\boldsymbol{v}$ as a directed line segment[2], regarding its "components" $v\langle k\rangle$ as directed line segments which are tangent to the co-ordinate lines and to be added according to the parallelogram rule to form $\boldsymbol{v}$ itself. For v^2 to be the squared length of a diagonal of a parallelepiped whose edges are tangent to the co-ordinate curves at $\boldsymbol{x}$ and are of lengths $v\langle k\rangle$, the physical components $v\langle k\rangle$ must be given by

$$v\langle k\rangle = \sqrt{g_{kk}}\, v^k = \sqrt{g_{kk}}\, g^{km} v_m \quad (k \text{ unsummed}). \tag{11.1}$$

Hence $v^2 = \sum_{k,m} c_{km}\, v\langle k\rangle\, v\langle m\rangle$, where c_{km} is the cosine of the angle between the x^k and x^m co-ordinate curves: $c_{km} = g_{km}/\sqrt{g_{kk} g_{mm}}$ (unsummed). The physical components (11.1) all have the same dimension, the dimension of $\boldsymbol{v}$, and in an orthogonal system they reduce to McCONNELL's components.

RICCI and LEVI-CIVITA[3] introduced these and three other sets of components, $v_k/\sqrt{g_{kk}}$, $v^k/\sqrt{g^{kk}}$ and $v_k\sqrt{g^{kk}}$ which share the properties just stated but have different geometrical interpretations.

---

[1] The reference system $\mathbf{e}_{\mathfrak{a}}$ and $\mathbf{e}^{\mathfrak{a}}$ is sometimes called an *anholonomic co-ordinate system*, provided the vectors $\mathbf{e}^{\mathfrak{a}}$ are not all gradient vectors. See SCHOUTEN [1954, *4*, Chap. II] for further discussion and references to relevant literature.

[2] We follow TRUESDELL [1953, *2*].

[3] [1901, *1*, Chap. I, § 4].

12. Physical components of vectors as anholonomic components. When co-ordinates are assigned dimensions, under change of units they suffer transformations of the form

$$x^{*k} = \lambda^k x^k, \tag{12.1}$$

where the λ's are positive constants. From the point of view of absolute dimensions, under the transformation (12.1) the components of the metric tensor transform according to Eq. (12.3), so that $\sqrt{g_{kk}}(\boldsymbol{x}^*) = \sqrt{g_{kk}}(\lambda^k)^{-1} L$ (unsummed). Thus the quantities e_k^a, given by

$$e_k^a = \sqrt{g_{kk}}\, \delta_k^a \tag{12.2}$$

in the co-ordinate systems related by Eq. (12.1), transform under Eq. (12.1) as covariant vectors having the absolute dimension $[L]$. This transformation law, together with the specification (12.2) of the $\boldsymbol{e}^a$ in these preferred co-ordinate systems, gives a consistent definition of these vectors in all co-ordinate systems. By Eq. (10.2), the anholonomic components of a contravariant vector $\boldsymbol{v}$ relative to these are given by

$$v^a = v^k e_k^a = v^k \sqrt{g_{kk}}\, \delta_k^a = v\langle k\rangle\, \delta_k^a, \tag{12.3}$$

the latter two equations holding only in the frames where the $\boldsymbol{e}^a$ are given by Eq. (12.2). Thus the physical components $v\langle k\rangle$, given by Eq. (11.1) in any of the co-ordinate systems related by Eq. (12.1), can be regarded as anholonomic components relative to the vectors $\boldsymbol{e}^a$. In the preferred systems, the components of the set $\boldsymbol{e}_a$ reciprocal to $\boldsymbol{e}^a$ are given by $e_a^k = (g_{kk})^{-\frac{1}{2}}\, \delta_a^k$. Hence $\boldsymbol{v}$, considered as a covariant vector, has anholonomic components

$$v_a = v_k e_a^k = v_k (g_{kk})^{-\frac{1}{2}}\, \delta_a^k = g_{mk} v^k (g_{mm})^{-\frac{1}{2}}\, \delta_a^m = \sum_{k,m} c_{km}\, v\langle k\rangle\, \delta_a^m, \tag{12.4}$$

where the c_{km} are the direction cosines defined in Sect. 11. Hence v_a and v^a can be regarded as covariant and contravariant components of $\boldsymbol{v}$ in a rectilinear but not necessarily rectangular co-ordinate system whose axes are tangent to the x^k co-ordinate curves at a given point, c_{km} being the metric tensor in this system.

13. Physical components of tensors. Several authors[1] have defined physical components of second order tensors. The definitions differ except in the case of orthogonal co-ordinates, when they are all equivalent to McConnell's. All authors make some attempt to avoid introducing physical components of the metric tensor, Truesdell, noting that in classical physics tensors are always defined in terms of vectors and scalars, in these definitions replaces the tensorial components of the vectors by their expressions in terms of physical components; the requirement that the physical components of the tensor being defined shall be related to the physical components of the given vectors in just the same way as the respective tensorial components are related then yields unique definitions for the physical components of the particular tensor being defined, as had been observed by other authors in special cases.

Even if one picks uniquely defined components, such as $v\langle k\rangle$, for vectors, one does not obtain a unique definition for physical components of all second order tensors. Truesdell uses equations defining the stress tensor in terms of vectors to motivate definitions which he later[2] shows correspond to the anholonomic components t_b^a and t_a^b relative to the vectors $\boldsymbol{e}^a$ discussed in Sect. 12. For tensors

---

[1] Synge and Schild [1949, *1*, § 5.1], Ollendorff [1950, *1*, III 5], Green and Zerna [1950, *2*, Eq. (5.16)], Truesdell [1953, *2*].

[2] [1954, *1*].

defined by scalar equations such as the strain tensor and the stretching tensor, one can motivate use of $t_{\mathfrak{a}\mathfrak{b}}$ for physical components. Considering the inverse of such a tensor when it exists, one can motivate use of $t^{\mathfrak{a}\mathfrak{b}}$. In orthogonal co-ordinates, these four sets coincide and correspond to McCONNELL's components. In general co-ordinates, the four sets $t^{\mathfrak{a}}_{\mathfrak{b}}$, $t^{\mathfrak{b}}_{\mathfrak{a}}$, $t_{\mathfrak{a}\mathfrak{b}}$, and $t^{\mathfrak{a}\mathfrak{b}}$ may be regarded as associated tensor components of $\boldsymbol{t}$ in a rectilinear but not necessarily rectangular co-ordinate system, based on the unit of length and having axes tangent at the given point to the co-ordinate curves of the given curvilinear system.

Plainly TRUESDELL's method is included as a special case in the method of rectilinear co-ordinates, just described. The latter has the advantage of being direct rather than inductive; the former, that out of the several sets of physical components provided by the latter it selects one which is suggested by the physical situation giving rise to the tensor. Both schemes are easily extended in the flexible method of anholonomic components.

Possibly a method sufficiently general as to cover every proposal which might appeal on some or another intuitive grounds could be constructed. In this connection we mention only that the method of anholonomic components, which appears quite general, does not include GREEN and ZERNA's definition. It seems, however, that in a given situation only one set of physical components is wanted, and appropriateness rather than generality should be the aim.

We shall use physical components only in orthogonal co-ordinate systems, when they are given by Eq. (10.5).

14. Covariant differentiation. Our object is a formula giving the anholonomic components of the covariant derivative of a tensor directly in terms of the derivatives of the anholonomic components. Set $\partial_{\mathfrak{a}} = e^k_{\mathfrak{a}} \dfrac{\partial}{\partial x^k}$. Then

$$v_{\mathfrak{a},\mathfrak{b}} \equiv e^k_{\mathfrak{a}} e^m_{\mathfrak{b}} v_{k,m} = e^k_{\mathfrak{a}} e^m_{\mathfrak{b}} \left(\frac{\partial v_k}{\partial x^m} - \left\{ {r \atop k\,m} \right\} v_r \right) = \partial_{\mathfrak{b}} (e^k_{\mathfrak{a}} v_k) - v_r \left(\partial_{\mathfrak{b}} e^r_{\mathfrak{a}} + \left\{ {r \atop k\,m} \right\} e^k_{\mathfrak{a}} e^m_{\mathfrak{b}} \right). \tag{14.1}$$

Using Eq. (10.2), we may rewrite this as

$$v_{\mathfrak{a},\mathfrak{b}} \equiv \partial_{\mathfrak{b}} v_{\mathfrak{a}} - v_{\mathfrak{c}} \Gamma^{\mathfrak{c}}_{\mathfrak{b}\mathfrak{a}}, \tag{14.2}$$

where

$$\Gamma^{\mathfrak{c}}_{\mathfrak{b}\mathfrak{a}} \equiv e^{\mathfrak{c}}_r \partial_{\mathfrak{b}} e^r_{\mathfrak{a}} + e^{\mathfrak{c}}_r \left\{ {r \atop k\,m} \right\} e^k_{\mathfrak{a}} e^m_{\mathfrak{b}} = - e^r_{\mathfrak{a}} \partial_{\mathfrak{b}} e^{\mathfrak{c}}_r + e^{\mathfrak{c}}_r \left\{ {r \atop k\,m} \right\} e^k_{\mathfrak{a}} e^m_{\mathfrak{b}}. \tag{14.3}$$

Similarly, for a tensor $\boldsymbol{t}$ of any order we have

$$\left. \begin{aligned} t^{\mathfrak{a}\ldots\mathfrak{b}}{}_{\mathfrak{c}\ldots\mathfrak{d},\mathfrak{e}} = \partial_{\mathfrak{e}} t^{\mathfrak{a}\ldots\mathfrak{b}}{}_{\mathfrak{c}\ldots\mathfrak{d}} + t^{\mathfrak{f}\ldots\mathfrak{b}}{}_{\mathfrak{c}\ldots\mathfrak{d}} \Gamma^{\mathfrak{a}}_{\mathfrak{e}\mathfrak{f}} + \cdots + t^{\mathfrak{a}\ldots\mathfrak{f}}{}_{\mathfrak{c}\ldots\mathfrak{d}} \Gamma^{\mathfrak{b}}_{\mathfrak{e}\mathfrak{f}} - \\ - t^{\mathfrak{a}\ldots\mathfrak{b}}{}_{\mathfrak{f}\ldots\mathfrak{d}} \Gamma^{\mathfrak{f}}_{\mathfrak{e}\mathfrak{c}} - \cdots - t^{\mathfrak{a}\ldots\mathfrak{b}}{}_{\mathfrak{c}\ldots\mathfrak{f}} \Gamma^{\mathfrak{f}}_{\mathfrak{e}\mathfrak{d}}, \end{aligned} \right\} \tag{14.4}$$

formally identical with the rule for covariant differentiation based on an affine connection $\Gamma^{\mathfrak{c}}_{\mathfrak{a}\mathfrak{b}}$.

It is important to note that in general $\Gamma^{\mathfrak{c}}_{\mathfrak{a}\mathfrak{b}} \neq \Gamma^{\mathfrak{c}}_{\mathfrak{b}\mathfrak{a}}$; the antisymmetric part, $\Gamma^{\mathfrak{c}}_{[\mathfrak{a}\mathfrak{b}]} = e^{\mathfrak{c}}_r \partial_{[\mathfrak{a}} e^r_{\mathfrak{b}]} = - e^k_{[\mathfrak{b}} e^m_{\mathfrak{a}]} \partial e^{\mathfrak{c}}_k / \partial x^m$, sometimes called "the object of anholonomity"[1], vanishes if and only if all the $\boldsymbol{e}^{\mathfrak{a}}$ be gradient vectors.

Turning now to McCONNELL's physical components, we use the special choice of $\boldsymbol{e}$'s given by Eq. (10.4). Writing

$$\left. \begin{aligned} \Gamma\langle k m n \rangle &\equiv \Gamma^{\mathfrak{c}}_{\mathfrak{b}\mathfrak{a}} \delta^{\mathfrak{b}}_k \delta^{\mathfrak{a}}_m \delta^n_{\mathfrak{c}}, \\ \frac{\partial}{\partial s_k} &\equiv \delta^{\mathfrak{a}}_k \partial_{\mathfrak{a}} = \frac{1}{\sqrt{g_{k\,k}}} \frac{\partial}{\partial x^k}, \end{aligned} \right\} \tag{14.5}$$

---

[1] Cf. SCHOUTEN [1954, *4*, Chap. II], [1951, *1*, Chap. IV].

the latter being the directional derivative along the x^k co-ordinate curve. By Eq. (25.3) we find that

$$\left.\begin{aligned} \Gamma\langle kmn\rangle &= 0 \quad \text{unless } m\neq n \quad \text{and} \quad k=m \quad \text{or} \quad n,\\ \Gamma\langle kkm\rangle &= -\frac{\partial}{\partial s_m}\log\sqrt{g_{kk}}, \quad k\neq m,\\ \Gamma\langle kmk\rangle &= \frac{\partial}{\partial s_m}\log\sqrt{g_{kk}}, \quad k\neq m. \end{aligned}\right\} \tag{14.6}$$

To calculate physical components of derivatives, one has only to substitute these formulae into

$$\left.\begin{aligned} t\langle k\ldots m,p\rangle = \frac{\partial}{\partial s_p}t\langle k\ldots m\rangle + \Gamma\langle pqk\rangle t\langle q\ldots m\rangle + \cdots +\\ + \Gamma\langle pqm\rangle t\langle k\ldots q\rangle, \end{aligned}\right\} \tag{14.7}$$

which is the form that Eq. (14.4) now assumes[1].

III. Double tensor fields.

a) Definition and examples.

15. Definition of double tensors. Let $\boldsymbol{x}$ and $\boldsymbol{X}$ be points in two spaces of dimensions n and N, respectively, over which differentiable co-ordinate transformations $\boldsymbol{x}^*=\boldsymbol{x}^*(\boldsymbol{x})$ and $\boldsymbol{X}^*=\boldsymbol{X}^*(\boldsymbol{X})$ are defined. The quantities $T^{K\ldots M}_{P\ldots Q}{}^{k\ldots m}_{p\ldots q}$ constitute a *double tensor* if they obey the transformation law for a tensor of type $T^{K\ldots M}_{P\ldots Q}$ when the $\boldsymbol{X}$ co-ordinates are transformed, for a tensor of type $t^{k\ldots m}_{p\ldots q}$ when the $\boldsymbol{x}$ co-ordinates are transformed. Precisely, the quantities $T^{*K_1\ldots K_M}_{Q_1\ldots Q_L}{}^{k_1\ldots k_m}_{q_1\ldots q_l}(\boldsymbol{X}^*,\boldsymbol{x}^*)$ and $T^{R_1\ldots R_M}_{S_1\ldots S_L}{}^{r_1\ldots r_m}_{s_1\ldots s_l}(\boldsymbol{X},\boldsymbol{x})$ are said to be the components[2] in the $\boldsymbol{X}^*$, $\boldsymbol{x}^*$ and $\boldsymbol{X}$, $\boldsymbol{x}$ co-ordinate systems, respectively, of a double tensor field of weight W, w, of the contravariant order M, m, and of covariant order L, l if[3]

$$\left.\begin{aligned} T^{*K_1\ldots K_M}_{Q_1\ldots Q_L}{}^{k_1\ldots k_m}_{q_1\ldots q_l} = |\boldsymbol{X}/\boldsymbol{X}^*|^{\mathrm{W}}\,|\boldsymbol{x}/\boldsymbol{x}^*|^{\mathrm{w}}\,\frac{\partial X^{*K_1}}{\partial X^{R_1}}\cdots\frac{\partial X^{*K_M}}{\partial X^{R_M}}\times\\ \times\frac{\partial X^{S_1}}{\partial X^{*Q_1}}\cdots\frac{\partial X^{S_L}}{\partial X^{*Q_L}}\,\frac{\partial x^{*k_1}}{\partial x^{r_1}}\cdots\frac{\partial x^{*k_m}}{\partial x^{r_m}}\,\frac{\partial x^{s_1}}{\partial x^{*q_1}}\cdots\frac{\partial x^{s_l}}{x^{*q_l}}\,T^{R_1\ldots R_M}_{S_1\ldots S_L}{}^{r_1\ldots r_m}_{s_1\ldots s_l}. \end{aligned}\right\} \tag{15.1}$$

As a special case, it follows that the components of a double tensor of the type $T^{K\ldots M}_{P\ldots Q}$ transform as *scalars* under $\boldsymbol{x}$-transformations: In other words, ordinary tensor fields are included as a special kind of double fields. On the other hand, it is legitimate to regard double tensors as ordinary tensors under a restricted group of transformations in the space whose points are the ordered pairs $(\boldsymbol{x},\boldsymbol{X})$, but this view does not usually promote the intended applications.

Generalization to the case of any finite number of spaces is immediate but will not be needed in this appendix.

---

[1] This rule was given in a more explicit form by TRUESDELL [1953, *1*], who recorded also fully explicit expressions for the gradients of a vector and of a second order tensor.

[2] There is possible ambiguity in specialization to particular components. E.g., is $T^1{}_2$ the result of putting $k=1$, $K=2$ in $T^k{}_K$, or the result of putting $k=2$, $K=1$ in $T^K{}_k$? In the few cases where such ambiguity might arise, we avoid it by using peculiar letters for co-ordinates. E.g., if the system at $\boldsymbol{X}$ is rectangular Cartesian (X, Y, Z), while that at $\boldsymbol{x}$ is cylindrical polar (r, ϑ, z), the two possibilities just mentioned will be denoted by $T^X{}_\vartheta$ and $T^r{}_Y$, respectively.

[3] In principle, this concept derives from CLEBSCH [1872, *4*, §§ 3–5]; cf. TUCKER [1931, *5*, § 1] and, from a different standpoint, EINSTEIN and BARGMANN [1944, *4*, § 1].

We shall use double tensor fields in three different interpretations:

α) $N<n$ and the space of $\boldsymbol{X}$ is a lesser-dimensional subspace of the space of $\boldsymbol{x}$, or, if we like, a *surface*[1] within it; the subspace is defined by a mapping

$$\boldsymbol{x} = \boldsymbol{x}(\boldsymbol{X}), \tag{15.2}$$

whereby a point with co-ordinates $\boldsymbol{X}$ in the subspace is regarded as the *same point* as that whose co-ordinates in the embedding space are $\boldsymbol{x}(\boldsymbol{X})$.

β) $N=n$, the subspace defined by Eq. (15.2) is of the same dimension as the embedding space and may coincide with it; the mapping (15.2) is to be interpreted as a *deformation* of the space.

γ) $N=n$, and no mapping (15.2) is presumed; $\boldsymbol{x}$ and $\boldsymbol{X}$ are regarded as co-ordinates of *two points* in the same space[2].

These cases, which do not exhaust the possible interpretations of double fields, we shall explore in reverse order.

16. Euclidean shifters. If in a Euclidean space we choose to employ Cartesian co-ordinates, the sum of the components of two tensors of the same order evaluated at different points of space, or the integral of a tensor field over a curve, surface, or region, is a Cartesian tensor of the same order. Under general co-ordinate transformations, such sums or integrals do not transform as tensors. In Euclidean space it is possible to define tensorially invariant operations which reduce to these sums or integrals in Cartesian co-ordinate systems. For such purposes, the operation of finite parallel displacement, or *shifting*, is fundamental. Let V^K denote the components in a rectangular Cartesian co-ordinate system of a vector $\boldsymbol{V}$ defined at the point $\boldsymbol{Z}$. At any other point $\boldsymbol{z}$, there exists a vector $\boldsymbol{v}$, whose components v^k in this frame, in order, have the same numerical values as those of $\boldsymbol{V}$. That is, trivially,

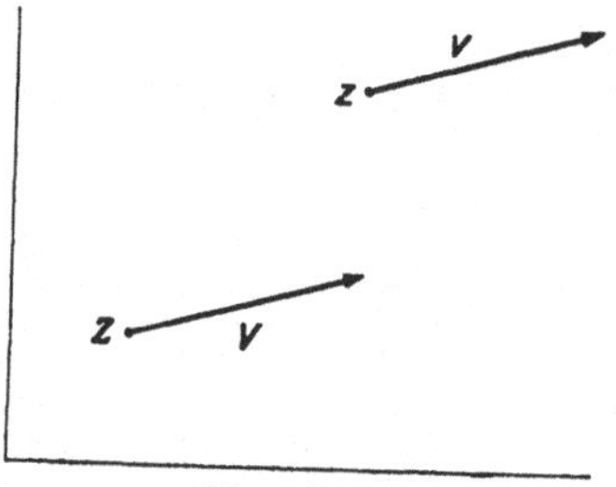

Fig. 1. Shifting.

$$v^k = \delta^k_K V^K, \qquad V^K = \delta^K_k v^k. \tag{16.1}$$

In Euclidean space, we are accustomed to regarding $\boldsymbol{v}$ and $\boldsymbol{V}$ as *the same* vector; we say $\boldsymbol{v}$ is obtained from $\boldsymbol{V}$ by (finite) parallel displacement from $\boldsymbol{Z}$ to $\boldsymbol{z}$ (Fig. 1), or for short, we *shift* the vector from $\boldsymbol{Z}$ to $\boldsymbol{z}$.

To put this concept in invariant[3] form, we introduce arbitrary[4] transformations $x^k = x^k(\boldsymbol{z})$, $X^K = X^K(\boldsymbol{Z})$, regard $\boldsymbol{v}$ and $\boldsymbol{V}$ as contravariant vectors, obtaining from (16.1)

$$v^m(\boldsymbol{x}) \frac{\partial z^k}{\partial x^m} = \delta^k_K V^M(\boldsymbol{X}) \frac{\partial Z^K}{\partial X^M}, \qquad V^M(\boldsymbol{X}) \frac{\partial Z^K}{\partial X^M} = \delta^K_k v^m(x) \frac{\partial z^k}{\partial x^m} \tag{16.2}$$

which can be rewritten as

$$v^m(\boldsymbol{x}) = g^m{}_M(\boldsymbol{x}, \boldsymbol{X}) V^M(\boldsymbol{X}), \qquad V^M(\boldsymbol{X}) = g^M{}_m(\boldsymbol{X}, \boldsymbol{x}) v^m(\boldsymbol{x}), \tag{16.3}$$

[1] The terms "variety" and "hypersurface" are often used.

[2] Michal [1927, *1* § 4].

[3] Einstein [1944, *5*, § 1]. We follow Doyle and Ericksen [1955, *1*] and Toupin [1956, *1*]. The geometrically inclined reader will see that the shifters may be regarded as operators for converting ordinary components to anholonomic components (Sect. 12), but the viewpoint of anholonomic components does not promote the intended interpretation.

[4] One can regard x^k and X^K as co-ordinates of different points in the same co-ordinate system, or, as is sometimes convenient, as co-ordinates of different points in independently selected co-ordinate systems.

where

$$g^m{}_M \equiv \delta^k_K \frac{\partial x^m}{\partial z^k} \frac{\partial Z^K}{\partial X^M}, \qquad g^M{}_m \equiv \delta^K_k \frac{\partial X^M}{\partial Z^K} \frac{\partial z^k}{\partial x^m}. \tag{16.4}$$

It follows easily that

$$g^k_M\, g^M{}_m = \delta^k_m, \qquad g^k{}_M\, g^K{}_k = \delta^K_M. \tag{16.5}$$

In the sense indicated by Eq. (16.3), $g^m{}_M$ and $g^M{}_m$ shift contravariant vectors from $\boldsymbol{X}$ to $\boldsymbol{x}$ and $\boldsymbol{x}$ to $\boldsymbol{X}$, respectively, so we call them *shifters*. If we proceed similarly but regard $\boldsymbol{v}$ and $\boldsymbol{V}$ as covariant vectors, we obtain

$$v_m(\boldsymbol{x}) = g_m{}^M(\boldsymbol{x}, \boldsymbol{X})\, V_m(\boldsymbol{X}), \qquad V_M(\boldsymbol{X}) = g_M{}^m(\boldsymbol{X}, \boldsymbol{x})\, v_m(\boldsymbol{x}), \tag{16.6}$$

where these inverse shifters are given by

$$g_m{}^M \equiv \delta^K_k \frac{\partial X^M}{\partial Z^K} \frac{\partial z^k}{\partial x^m}, \qquad g_M{}^m \equiv \delta^k_K \frac{\partial x^m}{\partial z^k} \frac{\partial Z^K}{\partial X^M}. \tag{16.7}$$

From Eqs. (16.4) and (16.7) we see that

$$g_k{}^K = g^K{}_k, \qquad g_K{}^k = g^k{}_K. \tag{16.8}$$

Using the rules of association $v_m(\boldsymbol{x}) = g_{mk}(\boldsymbol{x})\, v^k(\boldsymbol{x})$, $V_M(\boldsymbol{X}) = g_{MK}(\boldsymbol{X})\, V^K(\boldsymbol{X})$, or by direct calculation using the defining equations of the metric tensor, one can verify that

$$g_m{}^M = g_{mk}\, g^k_K\, g^{KM}, \qquad g_M{}^m = g_{MK}\, g^K{}_k\, g^{km}, \tag{16.9}$$

as suggested by the notation. From Eqs. (16.8) and (16.9), we see that all these shifters can be obtained from any one by raising or lowering indices with the appropriate metric tensor; in particular, the inverse shifters $g^k{}_K$ and $g^K{}_k$ are so related. It is thus unambiguous to use the same kernel index for all shifters and to write g^k_K in place of $g^k{}_K$ or $g_K{}^k$, and g^K_k in place of $g^K{}_k$ or $g_k{}^K$, which we henceforth do.

Under further co-ordinate transformations, g^k_K transforms as a contravariant vector under transformations of the form $\boldsymbol{x}^* = \boldsymbol{x}^*(\boldsymbol{x})$ and as a covariant vector under transformations of the form $\boldsymbol{X}^* = \boldsymbol{X}^*(\boldsymbol{X})$ while g^K_k transforms as in an inverse manner, as follows immediately from Eq. (16.4).

From Eq. (16.4) we have, for any triple of points $\boldsymbol{x}^*$, $\boldsymbol{x}$ and $\boldsymbol{X}$,

$$\left.\begin{aligned} g^k_K(\boldsymbol{x}^*, \boldsymbol{X}) &= \delta^m_M \frac{\partial x^{*k}}{\partial z^{*m}} \frac{\partial Z^M}{\partial X^K} = \delta^m_s \frac{\partial x^{*k}}{\partial z^{*m}} \frac{\partial z^s}{\partial x^r} \frac{\partial x^r}{\partial z^u} \delta^u_U \frac{\partial Z^U}{\partial X^K}, \\ &= g^k_r(\boldsymbol{x}^*, \boldsymbol{x})\, g^r_K(\boldsymbol{x}, \boldsymbol{X}), \\ g^K_k(\boldsymbol{X}, \boldsymbol{x}^*) &= g^r_k(\boldsymbol{x}, \boldsymbol{x}^*)\, g^K_r(\boldsymbol{X}, \boldsymbol{x}). \end{aligned}\right\} \tag{16.10}$$

These formulae show that the rule of composition for shifting is the ordinary matrix rule.

To shift from $\boldsymbol{X}$ to $\boldsymbol{x}$ any absolute tensor $T^{K \ldots M}_{U \ldots V}$ defined at $\boldsymbol{X}$, we transvect all indices with shifters, obtaining for the shifted components

$$T^{K \ldots M}_{U \ldots V}\, g^k_K \ldots g^m_M\, g^U_u \ldots g^V_v. \tag{16.11}$$

From Eq. (16.9) we obtain $g_{KM}\, g^K_k\, g^M_m = g_{km}$, so the shifted components of g_{KM} are g_{km}; hence length and angle are preserved by shifting, i.e. $g_{KM}\, V^K_1\, V^M_2 = g_{KM} \times g^K_k\, g^M_m\, v^k_1\, v^m_2 = g_{km}\, v^k_1\, v^m_2$ for any pair of vectors.

The process of transvecting a single index with a shifter we shall describe as *converting* that index.

Had we transformed $\boldsymbol{V}$ and $\boldsymbol{v}$ as relative vectors, we should have been led to a somewhat different rule for shifting. For a relative tensor $\boldsymbol{T}$ of weight w, reasoning similar to that used above leads one to regard

$$|\det g_r^R|^{\mathrm{w}}\, T^{K\ldots M}_{U\ldots V}\, g_K^k \cdots g_M^m\, g_u^U \cdots g_v^V \tag{16.12}$$

as the shifted components of $\boldsymbol{T}$.

As a check on Eqs. (16.11) and (16.12), we note from [Eq. (16.4)] that $g_k^K = \partial X^K/\partial x^k$ and $g_K^k = \partial x^k/\partial X^K$ when the two points coincide. Shifting then reduces to a co-ordinate transformation, and we should obtain the components of the same tensor in the transformed co-ordinate system, as indeed follows from Eqs. (16.11) and (16.12).

There are infinitely many relative tensors, of different weights, which correspond to a given Cartesian tensor. It is common to eliminate this ambiguity by using only absolute and axial tensors (Sect. 2). The shifted components of an axial tensor are given by

$$\varepsilon T^{K\ldots M}_{U\ldots V}\, g_K^k \cdots g_M^m\, g_u^U \cdots g_v^V, \tag{16.13}$$

where $\varepsilon = \operatorname{sgn}\det g_K^k$.

In Eqs. (16.4), leave the rectangular Cartesian system fixed but transform the general co-ordinate systems at $\boldsymbol{x}$ and $\boldsymbol{X}$ to arbitrary new systems $\boldsymbol{x}^* = \boldsymbol{x}^*(\boldsymbol{x})$ and $\boldsymbol{X}^* = \boldsymbol{X}^*(\boldsymbol{X})$. From Eq. $(16.4)_1$ it follows that

$$\left.\begin{aligned} g_M^{*m} &= \delta_K^k \frac{\partial x^{*m}}{\partial z^k}\frac{\partial Z^K}{\partial X^{*M}} = \delta_K^k \frac{\partial x^{*m}}{\partial x^p}\frac{\partial x^p}{\partial z^k}\frac{\partial Z^K}{\partial X^P}\frac{\partial X^P}{\partial X^{*M}},\\ &= g_P^p \frac{\partial x^{*m}}{\partial x^p}\frac{\partial X^P}{\partial X^{*M}}. \end{aligned}\right\} \tag{16.14}$$

Hence g_M^m is an absolute double tensor of covariant order 1, 0 and contravariant order 0, 1 consistently with the positions and numbers of its indices.

17. Addition and integration in Euclidean space. To obtain a tensorially invariant sum of tensors of the same type, defined at different points of a Euclidean space, one can shift all these to an arbitrarily selected point $\boldsymbol{x}$, then add. This sum is a tensor of the same type, defined at $\boldsymbol{x}$. For example, the vector equation $\boldsymbol{u} = \boldsymbol{V} + \boldsymbol{W}$, where $\boldsymbol{V}$ and $\boldsymbol{W}$ are vectors defined at $\boldsymbol{X}$ and $\boldsymbol{X}^*$, respectively, may be read

$$\left.\begin{aligned} u^k(\boldsymbol{x}) &= v^k(\boldsymbol{x}) + w^k(\boldsymbol{x}),\\ &\equiv g_K^k(\boldsymbol{x},\boldsymbol{X})\, V^K(\boldsymbol{X}) + g_K^k(\boldsymbol{x},\boldsymbol{X}^*)\, W^K(\boldsymbol{X}^*). \end{aligned}\right\} \tag{17.1}$$

As follows immediately from Eq. (16.10), adding at $\boldsymbol{x}$ and then shifting to $\boldsymbol{x}^*$ yields the same sum as is obtained by shifting directly to $\boldsymbol{x}^*$, then adding.

For example, if p^k and P^K are position vectors of the points $\boldsymbol{x}$ and $\boldsymbol{X}$ from the same fixed point ${}_0\boldsymbol{z}$, as defined in Sect. 2, then the vector from $\boldsymbol{X}$ to $\boldsymbol{x}$ has the co-ordinates $p^k - g_K^k P^K$ and $g_k^K p^k - P^K$ in the co-ordinates at $\boldsymbol{x}$ and $\boldsymbol{X}$, respectively.

Similarly, one can form a tensorially invariant integral of a tensor field by shifting the field to an arbitrary fixed point $\boldsymbol{x}$, then integrating the shifted components, so obtaining a tensor defined at $\boldsymbol{x}$, as indicated by the example below.

If x^k and X^K in a three-dimensional space are referred to the same cylindrical polar system given by $z^1 = r\cos\vartheta$, $z^2 = r\sin\vartheta$, $z^3 = z$, $Z^1 = R\cos\Theta$, $Z^2 = R\sin\Theta$, $Z^3 = Z$, by Eq. (16.4) we obtain

$$\|g_K^k\| = \left\|\begin{matrix} \cos(\vartheta-\Theta) & R\sin(\vartheta-\Theta) & 0\\ (R/r)\sin(\Theta-\vartheta) & (R/r)\cos(\Theta-\vartheta) & 0\\ 0 & 0 & 1 \end{matrix}\right\|. \tag{17.2}$$

Suppose that having the contravariant components A^R, A^Θ, A^Z of a vector $\boldsymbol{A}$ as functions of R, Θ, and Z, we are to calculate the vector integral $\mathfrak{A} \equiv \int_v \boldsymbol{A}\, dV$. Using Eq. (16.3), we then shift $\boldsymbol{A}$ to the *fixed* point (r, ϑ, z), obtaining components given by

$$\left.\begin{aligned} a^r &= A^R \cos(\vartheta - \Theta) + A^\Theta R \sin(\vartheta - \Theta), \\ r a^\vartheta &= A^R \sin(\Theta - \vartheta) + A^\Theta R \cos(\Theta - \vartheta), \\ a^z &= A^Z. \end{aligned}\right\} \tag{17.3}$$

In this co-ordinate system, $dV = R\, dR\, d\Theta\, dZ$. The vector $\mathfrak{A}$, thought of as defined at (r, ϑ, z), may be calculated from

$$\left.\begin{aligned} \mathfrak{A}^r &= \int_v a^r(r, \vartheta, z, R, \Theta, Z)\, R\, dR\, d\Theta\, dZ, \\ \mathfrak{A}^\vartheta &= \int_v a^\vartheta(r, \vartheta, z, R, \Theta, Z)\, R\, dR\, d\Theta\, dZ, \\ \mathfrak{A}^z &= \int_v a^z(r, \vartheta, z, R, \Theta, Z)\, R\, dR\, d\Theta\, dZ. \end{aligned}\right\} \tag{17.4}$$

Here (r, ϑ, z), being a fixed point, may be chosen so as to make the integrations as simple as possible.

More generally, to integrate an absolute tensor field $\boldsymbol{T}(\boldsymbol{X})$ over a curve, surface, or region, we introduce the element of arc, surface or volume in the X^K co-ordinate system, shift all components to a fixed point $\boldsymbol{x}$ as indicated by Eq. (16.11), then integrate these components. That these components are components of an absolute tensor defined at $\boldsymbol{x}$ follows at once from the transformation laws of the shifters and the fact that $\boldsymbol{x}$ is a fixed point. It follows from Eq. (16.10) that shifting from $\boldsymbol{x}$ to $\boldsymbol{x}^*$ after integrating yields the same components as would be obtained by shifting from $\boldsymbol{X}$ to $\boldsymbol{x}^*$ and then integrating, since the quantities $g^k_r(\boldsymbol{x}, \boldsymbol{x}^*)$ and $g^r_k(\boldsymbol{x}, \boldsymbol{x}^*)$ are independent of the variables of integration. Relative tensors can be treated similarly.

To illustrate these remarks we consider the integral $\mathfrak{A} = \int_v \boldsymbol{A}\, dV$, which, in component form, may be written

$$\left.\begin{aligned} \mathfrak{A}^k(\boldsymbol{x}^*) &= \int_v g^k_K(\boldsymbol{x}^*, \boldsymbol{X})\, A^K(\boldsymbol{X})\, dV(\boldsymbol{X}) = \int_v g^k_r(\boldsymbol{x}^*, \boldsymbol{x})\, g^r_K(\boldsymbol{x}, \boldsymbol{X})\, A^K(\boldsymbol{X})\, dV(\boldsymbol{X}) \\ &= g^k_r(\boldsymbol{x}^*, \boldsymbol{x}) \int_v g^r_K(\boldsymbol{x}, \boldsymbol{X})\, A^K(\boldsymbol{X})\, dV(\boldsymbol{X}), \\ &= g^k_r(\boldsymbol{x}^*, \boldsymbol{x})\, \mathfrak{A}^r(\boldsymbol{x}). \end{aligned}\right\} \tag{17.5}$$

If we take x^{*k} and x^m to be co-ordinates of the same point in different co-ordinate systems, then $g^k_r(\boldsymbol{x}^*, \boldsymbol{x}) = \partial x^{*k}/\partial x^r$, and Eq. (16.3) shows that A^r transforms as a contravariant vector. If we hold the x^{*k} co-ordinate system fixed but transform the co-ordinates X^K, the functions $g^k_K A^K$, and hence the quantities A^k, transform as absolute scalars.

18. Partial covariant differentiation of a double tensor field. Returning to the general concept of double field as presented in Sect. 15, suppose now that both the space of $\boldsymbol{x}$ and the space of $\boldsymbol{X}$ be affine spaces, with respective connections Γ^k_{pq} and Γ^K_{PQ}. We employ the notation $T^{\dots}_{\dots;M}$ to denote the covariant derivative of $T^{\dots}_{\dots}(\boldsymbol{x}, \boldsymbol{X})$ with respect to $\boldsymbol{X}$ *when $\boldsymbol{x}$ is kept constant.* Thus for example,

$$T^k{}_{K,M} \equiv \frac{\partial T^k{}_K}{\partial X^M} - \Gamma^P_{K\,M}\, T^k{}_P, \quad T^k{}_{K,m} \equiv \frac{\partial T^k{}_K}{\partial x^m} + \Gamma^k_{mp}\, T^p{}_K, \tag{18.1}$$

where in $\partial T^k{}_K/\partial X^M$ we hold constant not only X^P for $P \neq M$ *but also all the* x^m and analogously for $\partial T^k{}_K/\partial x^m$. It is easy to verify that $T^{\dots}_{\dots;M}$ and $T^{\dots}_{\dots;m}$ are double fields of the tensor character suggested by the positions and numbers of their indices, and that the covariant derivatives of a double field obey the usual formal rules of covariant differentiation.

For a tensor of the special type $T^{k\dots m}_{p\dots q}(\boldsymbol{x})$, we have

$$T^{k\dots m}_{p\dots q,K} = \frac{\partial T^{k\dots m}_{p\dots q}}{\partial X^K} = 0. \tag{18.2}$$

but no such result holds for a tensor of the type $T^{k\dots m}_{p\dots q}(\boldsymbol{x}, \boldsymbol{X})$.

Now consider the case when the space of $\boldsymbol{x}$ is metric, and the connection Γ^k_{pq} is based on the metric tensor $g_{rs}(\boldsymbol{x})$. Then

$$g_{rs,q} = 0, \quad g_{rs,K} = 0, \tag{18.3}$$

where the former result follows as in ordinary tensor analysis, and the latter is a consequence of Eq. (18.2).

Finally, consider a Euclidean space with shifter g^k_K as defined in Sect. 16. If we refer both $\boldsymbol{x}$ and $\boldsymbol{X}$ to the same rectangular Cartesian system, we have $g^k_K = \delta^k_K$, and, hence, in this system,

$$g^k_{K,M} = 0, \quad g^k_{K,m} = 0; \tag{18.4}$$

since these equations are of double tensor form, they are valid in all co-ordinate system[1].

These results show that in a Euclidean space the tensors g_{km}, g^k_K, g^K_k, and g_{KM}, considered as double tensor functions of pairs of points, behave as constants under partial covariant differentiation. Therefore,

$$T^{\cdots}_{\cdots k,K} = g^M_k T^{\cdots}_{\cdots M,K}, \tag{18.5}$$

so that *partial covariant differentiation commutes with shifting*. However, in general

$$T^{\cdots}_{\cdots k,m} g^m_K \neq T^{\cdots}_{\cdots k,K}. \tag{18.6}$$

Hence follows the ***caution***; *a differentiated index may not be converted.*

b) The total covariant derivative.

19. Double tensors associated with a surface. When there is a mapping (15.2) defining a surface, certain double tensor functions of points $\boldsymbol{x}$, $\boldsymbol{X}$ in the n-dimensional embedding space and the N-dimensional surface may be constructed directly from the mapping. Setting

$$x^k_{;K} \equiv x^k_{,K} \equiv \frac{\partial x^k}{\partial X^K}, \tag{19.1}$$

we find that under the double transformation $\boldsymbol{x}^* = \boldsymbol{x}^*(\boldsymbol{x})$, $\boldsymbol{X}^* = \boldsymbol{X}^*(\boldsymbol{X})$, we have

$$x^{*k}_{;K} \equiv \frac{\partial x^{*k}}{\partial X^{*K}} = \frac{\partial x^{*k}}{\partial x^m} x^m_{;M} \frac{\partial X^M}{\partial X^{*K}}. \tag{19.2}$$

Comparison with Eq. (15.1) shows that $x^k_{;K}$ *is an absolute double field of contravariant order* 0, 1 *and covariant order* 1, 0.

For the interpretation in surface theory, where $N < n$, it is convenient to write V^Γ in place of X^K. The quantity $x^k_{;\Gamma}$ is a *tangent vector* to the V^Γ co-ordinate curve on the surface, since a differential along that curve is given by $dx^k = x^k_{;\Gamma}\, dV^\Gamma$ (unsummed). At each point of the surface, N linearly independent tangent vectors are given by $x^k_{;\Gamma}$, $\Gamma = 1, 2, \ldots, N$.

A covariant tensor $\boldsymbol{n}$ is said to be *normal* to the N-dimensional surface if

$$n_{k_1 k_2 \ldots k_q} x^{k_1}_{;\Gamma_1} x^{k_2}_{;\Gamma_2} \ldots x^{k_q}_{;\Gamma_q} = 0. \tag{19.3}$$

A normal tensor of order $n - N$ is given by

$$\varepsilon_{\Gamma_1 \Gamma_2 \ldots \Gamma_N} n_{k_1 k_2 \ldots k_{n-N}} = \varepsilon_{k_1 k_2 \ldots k_{n-N} k_{n-N+1} \ldots k_n} x^{k_{n-N+1}}_{;\Gamma_1} x^{k_{n-N+2}}_{;\Gamma_2} \ldots x^{k_n}_{;\Gamma_N}. \tag{19.4}$$

Any $\boldsymbol{n}$ of order $n - N$ is a scalar multiple if the $\boldsymbol{n}$ just defined. For example, in three-dimensional space we may consider surfaces of dimensions 1 and 2, the

[1] TOUPIN [1956, 2, § 3].

former being called curves. In these two cases we have $x^k = x^k(V)$ and $x^k = x^k(V^1, V^2)$, respectively, and Eq. (19.4) yields

$$n_{km} = \varepsilon_{kmp} \frac{dx^p}{dV}, \qquad \varepsilon_{\Gamma\Delta} n_k = \varepsilon_{kmp} x^m_{;\Gamma} x^p_{;\Delta}. \tag{19.5}$$

The former equation gives a covariant, skew-symmetric normal tensor, and all vectors normal to the curve $x^k = x^k(V)$ may be obtained by taking the scalar product of this tensor n_{km} by other vectors; i.e., any normal vector is of the form $\boldsymbol{c} \times d\boldsymbol{x}/dV$ for some $\boldsymbol{c}$. The second of Eqs. (19.5) gives a covariant normal vector to the surface $x^k = x^k(V^1, V^2)$. The tensors $\boldsymbol{n}$ defined by Eqs. (19.4) and (19.5) are double tensors of weight 1, -1, not absolute tensors. In the case of a normal vector, Eq. (19.3) reduces to

$$n_k x^k_{;\Gamma} = 0; \quad \text{hence} \quad (\partial_\Delta n_k) x^k_{;\Gamma} + n_k \partial_\Delta x^k_{;\Gamma} = 0, \tag{19.6}$$

where $\partial_\Delta \equiv \partial/\partial V^\Delta$.

In a metric space with metric tensor $\boldsymbol{g}$ an induced surface metric $\boldsymbol{a}$ is given by

$$a_{\Gamma\Delta} = g_{km} x^k_{;\Gamma} x^m_{;\Delta}. \tag{19.7}$$

In a space with positive definite metric, it is customary to replace Eq. (19.4) by an analogous definition in which the absolute tensors $\boldsymbol{e}$, defined by Eq. $(2.3)_1$, replace the permutation symbols. The $\boldsymbol{n}$ so defined is a normal tensor according to Eq. (19.3) but is an absolute rather than relative tensor.

20. Definition of the total covariant derivative based on a mapping. The partial covariant derivatives $T^{\cdots}_{\cdots,k}$ and $T^{\cdots}_{\cdots,K}$ of a double field $\boldsymbol{T}(\boldsymbol{x}, \boldsymbol{X})$ were defined in Sect. 18. Supposing given a mapping $\boldsymbol{x} = \boldsymbol{x}(\boldsymbol{X})$, we now desire to define a *total covariant derivative* having the following properties:

1. $T^{\cdots}_{\cdots;R}$ is a double tensor field.

2. When T is of the special form $T^{K\ldots M}_{P\ldots Q}(\boldsymbol{X})$ or $T^{k\ldots m}_{p\ldots q}(\boldsymbol{x})$, then $T^{\cdots}_{\cdots;R}$ reduces to $T^{\cdots}_{\cdots,R}$ or $T^{\cdots}_{\cdots,r} x^r_{;R}$, respectively.

3. If we first apply the operation "$;R$" to $\boldsymbol{T}(\boldsymbol{x}, \boldsymbol{X})$, then replace $\boldsymbol{x}$ by $\boldsymbol{x}(\boldsymbol{X})$, we obtain the same result as if we apply the operation "$;R$" to $\boldsymbol{T}(\boldsymbol{x}(\boldsymbol{X}), (\boldsymbol{X})$.

When $\boldsymbol{T}$ reduces to a double scalar function of two variables $T = T(x, X)$, the partial covariant derivatives reduce to $\partial T/\partial x$ and $\partial T/\partial X$, while the desired total derivative is just the ordinary derivative:

$$\frac{dT}{dX} = \frac{d}{dX} T(x(X), X) = \frac{\partial x}{\partial X}\frac{\partial T}{\partial x} + \frac{\partial T}{\partial X}. \tag{20.1}$$

The generalization to arbitrary double tensor fields is given by[1]

$$T^{\cdots}_{\cdots;K} \equiv T^{\cdots}_{\cdots,K} + T^{\cdots}_{\cdots,k} x^k_{;K}. \tag{20.2}$$

The total derivative obeys the usual formal rules of covariant differentiation. In the metric case, for example, we have

$$g_{KM;P} = 0, \qquad g_{km;R} = 0. \tag{20.3}$$

In a Euclidean space, by Eqs. (18.4) we have also

$$g^k_{K;M} = 0, \qquad g^K_{k;M} = 0. \tag{20.4}$$

[1] This derivative was introduced by Bortolotti [1927, *4*] and van der Waerden [1927, *5*, § 3] as a special case of the multiple affine differentiation of R. Lagrange [1926, *3*, Chap. II, § 2]. Still more general derivatives were defined by Schouten and van Kampen [1930, *4*, § 4] and Tucker [1931, *6*, §§ 4, 7]; our notation follows the latter.

According to our rule of association, $T^{\dots K}_{\dots} = g^K_k T^{\dots k}_{\dots}$, and this notation is supported by the identity

$$T^{\dots K}_{\dots;M} = (g^K_k T^{\dots k}_{\dots})_{;M} = g^K_k T^{\dots k}_{\dots;M}. \tag{20.5}$$

(However, $T^{\dots}_{\dots;K} \neq g^k_K T^{\dots}_{\dots;k}$ in general.) We write $T^{\dots}_{\dots;KM} = (T^{\dots}_{\dots;K})_{;M}$. Also

$$T^{\dots;K}_{\dots} \equiv g^{KM} T^{\dots}_{\dots;M} = (g^{KM} T^{\dots}_{\dots})_{;M}. \tag{20.6}$$

All the theory presented in this section is valid when $N \leqq n$ and thus may be used in all three contexts listed at the end of Sect. 15. In the next section we consider some special properties of the case when $N=2$ and $n=3$, while in the following section we present an important special result valid when $n=N$.

21. Some formulae from the ordinary theory of surfaces. We list here invariant forms of those results concerning the theory of two-dimensional surfaces embedded in Euclidean three-dimensional space to which we shall later refer. Proofs, though usually only in rectangular Cartesian spatial co-ordinates, are given in works on differential geometry[1].

We employ the notations introduced in Sect. 19. The metrics $\boldsymbol{a}$ and $\boldsymbol{g}$ are related by Eq. (19.7);

$$g_{km;\Gamma} = 0, \quad a_{\Gamma\Delta;\Lambda} = 0; \tag{21.1}$$

and the metric $\boldsymbol{a}$ is used to raise and lower surface indices. Following the suggestion made at the end of Sect. 19, we discard the definition (19.5) and replace it by

$$e_{\Gamma\Delta}\, n_k = e_{kmp}\, x^m_{;\Gamma}\, x^p_{;\Delta}, \tag{21.2}$$

where $e_{\Gamma\Delta} \equiv a^{\frac{1}{2}}\, \varepsilon_{\Gamma\Delta}$, $e_{kmp} \equiv g^{\frac{1}{2}}\, \varepsilon_{kmp}$ in accord with Eq. (2.3). It follows that

$$n_k n^k = 1, \quad n_k\, x^k_{;\Gamma} = 0, \quad n_k\, x^k_{;\Gamma\Delta} + n_{k;\Delta}\, x^k_{;\Gamma} = 0, \tag{21.3}$$

the first of these being an assertion that $\boldsymbol{n}$ is a *unit normal* to the surface, while the second and third are consequences of Eqs. (19.6). Also

$$a^{\Gamma\Delta}\, x^k_{;\Gamma}\, x^m_{;\Delta} = g^{km} - n^k n^m, \quad e^k_{pm}\, x^m_{;\Gamma}\, n^p = a^{\Delta\Lambda}\, e_{\Gamma\Delta}\, x^k_{;\Lambda}. \tag{21.4}$$

The *second fundamental form* $\boldsymbol{b}$ is defined by

$$b_{\Gamma\Delta} \equiv n_k\, x^k_{;\Gamma\Delta} = -g_{km}\, x^k_{;\Gamma}\, n^m_{;\Delta} \tag{21.5}$$

and satisfies the equations of GAUSS and WEINGARTEN:

$$x^k_{;\Gamma\Delta} = b_{\Gamma\Delta}\, n^k, \quad n^k_{;\Gamma} = -b^\Delta_\Gamma\, x^k_{;\Delta}. \tag{21.6}$$

Also

$$\left.\begin{aligned} n^k_{;\Gamma\Delta} &= -b^\Lambda_{\Gamma;\Delta}\, x^k_{;\Lambda} - b^\Lambda_\Gamma\, b_{\Lambda\Delta}\, n^k, \\ n_k\, n^k_{;\Gamma\Delta} &= -b^\Lambda_\Gamma\, b_{\Lambda\Delta}, \quad g_{km}\, n^k_{;\Gamma\Delta}\, x^m_{;\Lambda} = -b_{\Gamma\Lambda;\Delta}. \end{aligned}\right\} \tag{21.7}$$

Conditions of integrability to be satisfied by tensors $\boldsymbol{a}$ and $\boldsymbol{b}$ as necessary and sufficient that there exist a surface $\boldsymbol{x}=\boldsymbol{x}(\boldsymbol{V})$ such that $\boldsymbol{a}$ and $\boldsymbol{b}$ are its first and second quadratic forms are the equations of MAINARDI-CODAZZI and GAUSS:

$$b_{\Gamma\Delta;\Lambda} = b_{\Gamma\Lambda;\Delta}, \quad R_{\Gamma\Delta\Lambda\Sigma} = b_{\Gamma\Lambda}\, b_{\Delta\Sigma} - b_{\Delta\Lambda}\, b_{\Gamma\Sigma}, \tag{21.8}$$

where $\boldsymbol{R}$ is the Riemann tensor based upon $\boldsymbol{a}$.

The normal curvature $K_{(\boldsymbol{t})}$ in the direction of a tangent vector $\boldsymbol{t}$ of unit length is given by

$$K_{(\boldsymbol{t})} = b_{\Gamma\Delta}\, t^\Gamma t^\Delta = b^\Gamma_\Delta\, t_\Gamma\, t^\Delta. \tag{21.9}$$

[1] MCCONNELL [1931, *1*, Chap. XV, § 5] gives the results in double tensor form but in the derivations always supposes $\boldsymbol{x}$ to be eliminated, although this is not necessary.

The *principal curvatures* K_1 and K_2 are the curvatures in the principal directions of $\boldsymbol{b}$. Writing $a \equiv \det a_{\Gamma\Delta}$, $b \equiv \det b_{\Gamma\Delta}$, for the total curvature K and mean curvature $\overline{K}$ of the surface we have the formulae

$$\left.\begin{aligned} K &= \frac{R_{1212}}{a} = \frac{1}{4} e^{\Gamma\Delta} e^{\Lambda\Sigma} R_{\Gamma\Delta\Lambda\Sigma} = \frac{b}{a} = \mathrm{II}_{\boldsymbol{b}} = K_1 K_2, \\ \overline{K} &= a^{\Gamma\Delta} b_{\Gamma\Delta} = b^{\Gamma}_{\Gamma} = \mathrm{I}_{\boldsymbol{b}} = K_1 + K_2. \end{aligned}\right\} \tag{21.10}$$

Also

$$\overline{\mathrm{II}}_{\boldsymbol{b}} = b^{\Gamma\Delta} b_{\Gamma\Delta} = \overline{K}^2 - 2K. \tag{21.11}$$

22. Properties of the total covariant derivative based on a homeomorphism. Returning to the total covariant derivative as defined in Sect. 20, consider the case when $n = N$ and the transformation $\boldsymbol{x} = \boldsymbol{x}(\boldsymbol{X})$ is a homeomorphism, so that there is a single-valued differentiable inverse $\boldsymbol{X} = \boldsymbol{X}(\boldsymbol{x})$. As mentioned in Sect. 15, this case may be visualized as a deformation of the space of $\boldsymbol{X}$ into the space of $\boldsymbol{x}$, or conversely.

We may now use the inverse mapping $\boldsymbol{X} = \boldsymbol{X}(\boldsymbol{x})$ in exactly the same way as we used the mapping $\boldsymbol{x} = \boldsymbol{x}(\boldsymbol{X})$ in Sect. 20. In particular, writing

$$X^K_{;k} \equiv X^K_{,k} \equiv \frac{\partial X^K}{\partial x^k}, \tag{22.1}$$

for a double field $\boldsymbol{T}(\boldsymbol{x}, \boldsymbol{X})$ we may set

$$T^{\cdots}_{\cdots;k} \equiv T^{\cdots}_{\cdots,k} + T^{\cdots}_{\cdots,K} X^K_{;k}. \tag{22.2}$$

Since

$$x^k_{;K} X^K_{;m} = \delta^k_m, \qquad X^K_{;k} x^k_{;M} = \delta^K_M, \tag{22.3}$$

from Eqs. (20.2) and (22.2) we have the *chain rule*[1]:

$$T^{\cdots}_{\cdots;K} = T^{\cdots}_{\cdots;k} x^k_{;K}, \qquad T^{\cdots}_{\cdots;k} = T^{\cdots}_{\cdots;K} X^K_{;k}. \tag{22.4}$$

IV. Integrals of tensor fields.

a) Preliminaries.

23. Conventions for integrals. While the operation of shifting explained and illustrated in Sects. 16 and 17 permits integration of tensors in curvilinear co-ordinate systems in Euclidean space, it is laborious. For the purpose of this treatise it suffices when integrating tensors of order greater than 0 to consider *rectangular Cartesian co-ordinates only*. The reader is to recall this *convention* without further reminder. There is no advantage in imposing such a restriction on integrals of scalars, for which we continue to employ general curvilinear co-ordinates.

Suppose ψ is a continuous tensor field of order greater than 0, and suppose

$$\int_{\mathfrak{v}} \psi \, dv = 0 \tag{23.1}$$

for all volumes $\mathfrak{v}$ in a region $\mathfrak{r}$. By the above convention, Eq. (23.1) is understood as written in rectangular Cartesian co-ordinates. From it follows

$$\psi = 0 \quad \text{in } \mathfrak{r} \tag{23.2}$$

with ψ again written in rectangular Cartesian co-ordinates. However, since ψ is a tensor, Eq. (23.2) holds in all co-ordinate systems. As a part of the above

[1] Bortolotti [1927, *4*].

convention we add the understanding that henceforth from an equation of the type (23.1) we shall *directly* infer Eq. (23.2) in *general co-ordinates*, without further mention.

24. Smoothness and order. Except in the few places where the degree of smoothness of regions, surfaces, and curves is a matter of especial interest, we tacitly assume whatever smoothness is necessary. For GREEN's transformation (Sect. 26), KELVIN's transformation (Sect. 28) and most other purposes, it suffices that bounded regions, surfaces, and curves be regular in the sense of KELLOGG[1] and that unbounded sets be such that their intersection with every sphere is regular.

We sometimes wish to state conditions concerning the vanishing of integrals at infinity. Consider an unbounded region $\mathfrak{v}$ of space, and let P be any interior point. Let $\mathfrak{s}_p$ be that portion of the surface of the sphere of radius p with center at P which lies in the interior of $\mathfrak{v}$. If Ψ be such that

$$\lim_{p\to\infty}\int_{\mathfrak{s}_p} d\boldsymbol{a}\cdot\Psi p^m = 0, \qquad \lim_{p\to\infty}\int_{\mathfrak{s}_p} d\boldsymbol{a}\times\Psi p^m = 0, \qquad \lim_{p\to\infty}\int_{\mathfrak{s}_p} d\boldsymbol{a}\,\Psi p^m = 0, \tag{24.1}$$

we write[2], respectively,

$$\Psi_n = \bar{o}(p^{-m-2}), \qquad \Psi_t = \bar{o}(p^{-m-2}), \qquad \Psi = \bar{o}(p^{-m-2}). \tag{24.2}$$

For (24.1) to hold it is clearly sufficient that each component of Ψ_n, Ψ_t, and Ψ, respectively, be $o(p^{-m-2})$ as $p\to\infty$, which motivates the notation. The symbol $\bar{o}$ may be read "lower mean order than".

We shall frequently wish to state boundary conditions of the following type:

1. On each finite boundary of $\mathfrak{v}$, $\Psi=0$.
2. In any portion of $\mathfrak{v}$ extending to ∞,

$$\Psi = \bar{o}(p^{-m-2}).$$

As a part of the ***convention*** of Sect. 23, we agree that henceforth

$$\Psi = 0 \quad \text{or} \quad \bar{o}(p^{-m-2}) \quad \text{on} \quad \mathfrak{s} \tag{24.3}$$

shall serve as an abbreviation for the statements 1 and 2. Analogous conventions are adopted for statements in terms of Ψ_n or Ψ_t.

b) Circulation, flux, total, and moments.

25. Definitions. Let Ψ be any integrable tensor field. The line integral

$$\int_{\mathfrak{c}} d\boldsymbol{z}\cdot\Psi = \int_{\mathfrak{c}} dz_k\,\Psi_{k\ldots m} \tag{25.1}$$

is called the *flow* of Ψ along the curve $\mathfrak{c}$. When $\mathfrak{c}$ is closed, it is called a *circuit*; the integral (25.1) is then called the *circulation of* Ψ *around* $\mathfrak{c}$ and is written

$$\oint_{\mathfrak{c}} d\boldsymbol{z}\cdot\Psi. \tag{25.2}$$

Two curves are ***reconcilable***[3] in a given point set provided one can be continuously deformed into the other while remaining in the set. A circuit which can be

[1] [1929, *1*, Chap. IV].

[2] The notations were introduced by TRUESDELL [1954, *2*, § 4].

[3] These concepts and names derive from KELVIN [1869, *1*, § 58 and § 60(a)]. Topologists use the word "homotopic" in place of "reconcilable", whereas the latter word is generally employed in mechanics. KELVIN stated that he applied RIEMANN's [1857, *1*] theory of multiple connectivity as presented by HELMHOLTZ [1858, *1*], but the concept of reconcilable curves does not appear in these papers.

continuously shrunk to a point while remaining in a given set is said to be *reducible* in the set.

The surface integral

$$\int_{\mathscr{s}} \boldsymbol{da} \cdot \Psi \tag{25.3}$$

is called the *flux of* Ψ *across* the surface $\mathscr{s}$. When $\mathscr{s}$ is closed and $\boldsymbol{da}$ is directed outward, the integral (25.3) is called[1] the *flux of* Ψ *out of* $\mathscr{s}$ and is written

$$\oint_{\mathscr{s}} \boldsymbol{da} \cdot \Psi. \tag{25.4}$$

A closed surface which can be continuously shrunk to a point while remaining in a given region is said to be *reducible* in it.

With the notation (3.7), the volume integral

$$\int_{\mathscr{v}} \{\boldsymbol{p}^{(K)}\, \Psi\}\, dv \tag{25.5}$$

is called the K-th *moment of* Ψ with respect to the origin. The zeroth moment,

$$\int_{\mathscr{v}} \Psi\, dv \tag{25.6}$$

is called[2] the *total* Ψ *in* $\mathscr{v}$.

c) The transformations of GREEN and KELVIN.

26. GREEN's transformation. We shall employ *Green's transformation*[3] in the forms

$$\int_{\mathscr{v}} \operatorname{grad} \Psi\, dv = \oint_{\mathscr{s}} \boldsymbol{da}\, \Psi, \quad \int_{\mathscr{v}} \operatorname{div} \Psi\, dv = \oint_{\mathscr{s}} \boldsymbol{da} \cdot \Psi, \quad \int_{\mathscr{v}} \operatorname{curl} \Psi\, dv = \int_{\mathscr{s}} \boldsymbol{da} \times \Psi, \tag{26.1}$$

where it is assumed that Ψ is single valued and continuous throughout the closure of the finite region $\mathscr{v}$, bounded by $\mathscr{s}$, that grad Ψ is continuous throughout a finite number of subregions of $\mathscr{v}$ whose sum is $\mathscr{v}$, and that the volume integrals are convergent. In the terminology of Sect. 25, Eq. $(26.1)_2$ states that the total divergence of a quantity in a region is equal to the flux of the quantity out of the boundary of the region.

Another form of GREEN's transformation is[4]

$$\left.\begin{aligned} &\int_{\mathscr{v}} [\boldsymbol{b}\{\boldsymbol{c}\,\boldsymbol{p}^{(K-1)}\} + \boldsymbol{c}\{\boldsymbol{b}\,\boldsymbol{p}^{(K-1)}\} - \boldsymbol{b}\cdot\boldsymbol{c} \operatorname{grad} \boldsymbol{p}^{(K)} + \\ &\qquad + (\boldsymbol{b} \operatorname{div} \boldsymbol{c} + \boldsymbol{c} \operatorname{div} \boldsymbol{b} + \operatorname{curl} \boldsymbol{c} \times \boldsymbol{b} + \operatorname{curl} \boldsymbol{b} \times \boldsymbol{c})\, \boldsymbol{p}^{(K)}]\, dv \\ &\qquad = \oint_{\mathscr{s}} [\boldsymbol{da}\cdot(\boldsymbol{b}\,\boldsymbol{c} + \boldsymbol{c}\,\boldsymbol{b})\,\boldsymbol{p}^{(K)} - \boldsymbol{da}\,\boldsymbol{b}\cdot\boldsymbol{c}\,\boldsymbol{p}^{(K)}], \end{aligned}\right\} \tag{26.2}$$

---

[1] This name derives from MAXWELL [1873, *2*, §§ 12–13], who called Eq. (25.3) "the surface integral of the flux".

[2] These names derive from TRUESDELL [1954, *2*, § 6]. FÖPPL [1897, *1*, § 4] introduced the quantity (25.6) for a vector, calling it the "Feldsumme".

[3] A result equivalent to those listed here was first given by GREEN [1828, *2*, pp. 23–26], special cases having been stated earlier by LAGRANGE [1762, *1*, § 45] and GAUSS [1813, *1*, §§ 3–5], OSTROGRADSKY [1831, *2*] stated an equivalent result as a formula for differentiating triple integrals. The name "GREEN's transformation" was suggested by LOVE [1892, *2*, § 14]; other common names are "GAUSS's theorem", "the divergence theorem", and "OSTROGRADSKY's theorem". In the early literature, use of this transformation is often called "integration by parts".

[4] TRUESDELL [1949, *4*], [1951, *3*], [1954, *1*, § 7]. In the formulae given in these papers, $\{\boldsymbol{p}^{(n-1)}\, 1\}$ should be replaced by grad $\boldsymbol{p}^{(n)}$. The case $K=0$ is due to BURGATTI [1931, *3*].

where $\boldsymbol{b}$ and $\boldsymbol{c}$ are arbitrary continuously differentiable vector fields, $\boldsymbol{p}$ is the position vector, and K is any integer.

27. POINCARÉ's generalization. In n dimensions, the tensor element of area of an oriented m-dimensional surface $\mathfrak{s}_m$, given parametrically by $x^k = x^k(\boldsymbol{u})$, is

$$d s_{(m)}^{r \ldots s} = m! \frac{\partial x^{[r}}{\partial u^1} \cdots \frac{\partial x^{s]}}{\partial u^m} d u^1 \ldots d u^m, \tag{27.1}$$

which transforms as an absolute contravariant tensor of order m under co-ordinate transformations, as an absolute scalar under parameter transformations with positive Jacobian. Let $\mathfrak{s}_m$ be bounded by the surface $\mathfrak{s}_{m-1}$, given parametrically by $x^k = x^k(\boldsymbol{v})$. Let $\boldsymbol{v}$ be a vector defined on $\mathfrak{s}_{m-1}$, directed outward relative to $\mathfrak{s}_m$. Arrange the parametrizations so that the vectors v^k, $\partial x^k/\partial v^1, \ldots, \partial x^k/v^{m-1}$ and $\partial x^k/\partial u^1, \ldots, \partial x^k/\partial u^m$, in these orders, have the same orientation relative to $\mathfrak{s}_m$. For any continuously differentiable covariant tensor field $\boldsymbol{t}$ of order m, we then have[1]

$$\int_{\mathfrak{s}_m} \frac{\partial}{\partial x^r} t_{u \ldots s} \, d s_{(m)}^{r u \ldots s} = \int_{\mathfrak{s}_m} \frac{\partial}{\partial x^{[r}} t_{u \ldots s]} \, d s_{(m)}^{r u \ldots s} = \oint_{\mathfrak{s}_{m-1}} t_{u \ldots s} \, d s_{(m-1)}^{u \ldots s}. \tag{27.2}$$

It then follows from the transformation properties of the integrands that each integral transforms as an absolute scalar. The underlying space need not be Euclidean or have any other geometrical structure. In metric spaces, or affine spaces with a symmetric connection, where covariant differentiation is defined, we have $\frac{\partial}{\partial x^{[u}} t_{r \ldots s]} = t_{[r \ldots s, u]}$, which makes the invariance of the integrals more obvious. Various alternative forms of Eq. (27.2) are available in the literature[2]. In three-dimensional Euclidean space, GREEN's transformation may be deduced from Eq. (27.2) with $m=3$, KELVIN's transformation (Sect. 28) from Eq. (27.2) with $m=2$.

28. KELVIN's transformation. KELVIN's transformation, commonly[3] called "STOKES' theorem", is

$$\int_{\mathfrak{s}} d\boldsymbol{a} \cdot \operatorname{curl} \Psi = \oint_{c} d\boldsymbol{z} \cdot \Psi, \tag{28.1}$$

subject to the usual right-handed screw convention connecting the sense of $d\boldsymbol{z}$ with that of $d\boldsymbol{a}$. Ψ is assumed continuously differentiable in a region whose boundary contains[4] $\mathfrak{s}$. It is important to note that the field Ψ need not be defined on both sides of $\mathfrak{s}$. In the terminology of Sect. 25, Eq. (28.1) states that the circulation of a quantity around a closed circuit equals the flux of its curl across any surface bounded by the circuit.

It is possible to give forms of Eq. (28.1) which contain $d\boldsymbol{z}\,\Psi$ and $d\boldsymbol{z} \times \Psi$ on the right-hand side, in the style of Eq. (28.1). Since these are awkward in dyadic notation, we write the former in Cartesian tensor notation as follows;

$$\int_{\mathfrak{s}} e_{m p r} \, d a_r \, \Psi_{p \ldots n, k} = \oint_{c} d z_k \, \Psi_{m \ldots n}. \tag{28.2}$$

Contracting this equation on k and m yields Eq. (28.1), while replacing km by $[km]$ yields the second alternative mentioned above.

[1] This result is due in essence to POINCARÉ [1887, *1*, § 2], [1895, *1*, § 7].

[2] See e.g. SCHOUTEN [1951, *1*, Chap. IV].

[3] We follow TRUESDELL [1954, *2*, § 8] in departing from usage. See the appendix to this section.

[4] See LICHTENSTEIN [1929, *2*, Kap. 2, § 3] for references to works proving the result under weaker assumptions.

Appendix. History of Kelvin's transformation[1]. The plane form of Eq. (28.1) was discovered by AMPÈRE [1826, *E1*]. The three-dimensional form was discovered by KELVIN, who communicated it to STOKES in a letter dated July 2, 1850 (LARMOR's annotation to the 1905 reprint of [1854, *1*]). Independent discovery, as well as priority in publication, is due to HANKEL [1861, *E1*, § 7] (*cf.* also ROCH [1863, *E1*, § 4]). KELVIN's proofs were published later [1867, *1*, § 190(j)] [1869, *E1*, § 60(q)], and in the second of these publications he claimed priority. The only connection of STOKES with the matter was to set proof of the result as an examination question [1854, *T2*]. Later KELVIN [1879 ed. of [1867, *1*, § 190(j)]] mentioned this fact, which had been noted earlier by MAXWELL [1873, *3*, § 24], after whose custom the common name has been adopted. It was KELVIN who first realized the significance of Eq. (28.1), showing how to use it for the proofs of important kinematical theorems. If additional names beyond KELVIN's are to be attached to Eq. (28.1), they should be those of AMPÈRE and HANKEL.

V. Vector fields.

a) Vector lines, sheets, and tubes.

29. Definitions. A *vector line* of a given vector field $\boldsymbol{c}$ is a curve everywhere tangent to $\boldsymbol{c}$. Parametric equations of a vector line are the solutions $x^k = x^k(u)$ of the differential system

$$\left.\begin{aligned} &\frac{d\boldsymbol{x}}{du}\times\boldsymbol{c}=0,\\ \text{or}\qquad &\frac{dx^{[k}}{du}c^{m]}=0. \end{aligned}\right\} \tag{29.1}$$

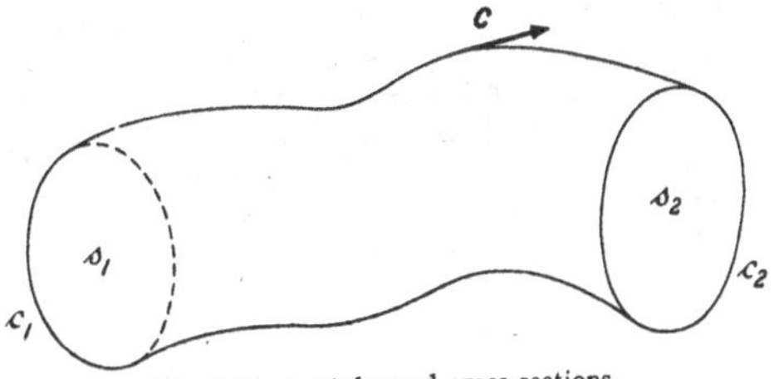

Fig. 2. Vector tube and cross-sections.

If $\boldsymbol{c}$ is continuous in a closed region, there exists at least one vector line through each interior point of the region; if $\boldsymbol{c}$ also satisfies a Lipschitz condition[2], there is exactly one through each point where $c \neq 0$.

A surface everywhere tangent to $\boldsymbol{c}$ is a *vector sheet* of $\boldsymbol{c}$. Such surfaces $f(\boldsymbol{x}) =$ const are non-constant solutions of

$$\boldsymbol{c}\cdot\operatorname{grad} f = 0, \quad \text{or} \quad c^k\frac{\partial f}{\partial x^k} = 0. \tag{29.2}$$

Equivalently, a parametric representation $\boldsymbol{x} = \boldsymbol{x}(u, v)$ of a vector sheet satisfies

$$\boldsymbol{c}\cdot\frac{\partial\boldsymbol{x}}{\partial u}\times\frac{\partial\boldsymbol{x}}{\partial v} = 0, \qquad \frac{\partial\boldsymbol{x}}{\partial u}\times\frac{\partial\boldsymbol{x}}{\partial v} \neq 0. \tag{29.3}$$

A vector line can always be represented locally as the intersection of two vector sheets. Consider a field $\boldsymbol{c}$ possessing a unique vector line through each point of a region r, and let c be a curve which is not a vector line of $\boldsymbol{c}$; then, locally, the surface swept out by the vector lines intersecting c define a unique vector sheet of $\boldsymbol{c}$ associated with c. When c is a circuit, this vector sheet is called a *vector tube* of $\boldsymbol{c}$.

Let s be a vector tube of $\boldsymbol{c}$, bounded by two simple circuits c_1 and c_2, each embracing the tube once. Let s_1 and s_2 be two surfaces whose complete boundaries are c_1 and c_2 respectively (Fig. 2); when the unit normals to s_1 and s_2 are given senses so as to subtend an acute angle with $\boldsymbol{c}$, s_1 and s_2 are *cross-sections* of the tube. Assuming $\boldsymbol{c}$ is integrable on the cross-sections s_1 and s_2, we call its flux through them the *strengths* of the vector tubes at these cross-sections. Letting $s_3 = s + s_1 + s_2$, we have

$$\oint_{s_3} d\boldsymbol{a}\cdot\boldsymbol{c} = \int_{s_1} d\boldsymbol{a}\cdot\boldsymbol{c} + \int_{s_2} d\boldsymbol{a}\cdot\boldsymbol{c}, \tag{29.4}$$

[1] Cf. TRUESDELL [1954, *2*, § 8].
[2] See e.g. KAMKE [1930, *1*, §§ 15, 16] for proof of uniqueness under weaker conditions.

since $d\boldsymbol{a} \cdot \boldsymbol{c} = 0$ on $\mathfrak{s}$. In Eq. (29.4), the normals to $\mathfrak{s}_1$ and $\mathfrak{s}_2$ are both taken outward. If $\boldsymbol{c}$ is directed inward relative to $\mathfrak{s}$ on $\mathfrak{s}_1$ and we take the unit normal on $\mathfrak{s}_1$ to be directed inward, we have

$$\oint_{\mathfrak{s}_3} d\boldsymbol{a} \cdot \boldsymbol{c} = \int_{\mathfrak{s}_2} d\boldsymbol{a} \cdot \boldsymbol{c} - \int_{\mathfrak{s}_1} d\boldsymbol{a} \cdot \boldsymbol{c}. \tag{29.5}$$

Hence the flux of $\boldsymbol{c}$ *out of a surface consisting of a vector tube of* $\boldsymbol{c}$ *and two cross-sections equals the difference between the strengths of the tube at these cross-sections.*

Let $\mathfrak{v}$ be the region bounded by $\mathfrak{s}_3$. Assuming sufficient smoothness for GREEN's transformation (26.1) to hold, from (29.5) we have

$$\int_{\mathfrak{v}} \operatorname{div} \boldsymbol{c}\, dv = \int_{\mathfrak{s}_2} d\boldsymbol{a} \cdot \boldsymbol{c} - \int_{\mathfrak{s}_1} d\boldsymbol{a} \cdot \boldsymbol{c}. \tag{29.6}$$

Hence the total divergence of the field $\boldsymbol{c}$ *in a region bounded by a vector tube of* $\boldsymbol{c}$ *and two cross-sections equals the difference between the strengths of the tube at these cross-sections.*

The above definitions may be extended to n-dimensional spaces which need not possess a metric tensor. If $\boldsymbol{c}$ is any relative or absolute contravariant vector, the vector lines of $\boldsymbol{c}$ are the curves obtained as solutions of Eq. $(29.1)_2$. As before, in a closed region where $\boldsymbol{c}$ satisfies a Lipschitz condition there exists a unique vector line through each interior point at which $\boldsymbol{c} \neq 0$. The vector lines intersecting an $m-1$-dimensional surface $\mathfrak{c}$, when $1 < m < n$, then sweep out an m-dimensional vector sheet $\mathfrak{s}$ of $\mathfrak{c}$, this being defined as an m-dimensional surface given parametrically by $x^k = x^k(u^1, \ldots, u^m)$, where

$$\frac{\partial x^{[k}}{\partial u^1} \cdots \frac{\partial x^r}{\partial u^m} c^{s]} = 0, \qquad \frac{\partial x^{[k}}{\partial u^1} \cdots \frac{\partial x^{r]}}{\partial u^m} \neq 0, \tag{29.7}$$

or, equivalently, as the set of points satisfying $f_\alpha(\boldsymbol{x}) = \text{const}$, $\alpha = 1, \ldots, n-m$, where the f's are functionally independent solutions of Eq. $(29.2)_2$. If $\mathfrak{c}$ be closed, $\mathfrak{s}$ is called a vector tube of $\mathfrak{c}$. For an n-dimensional vector tube of a relative vector of weight one, generalizations of Eqs. (29.4) and (29.5) are immediate, and a generalization of Eq. (29.6) is obtained by setting $t_{r \ldots s} = c^k \varepsilon_{kr \ldots s}$ in (27.2).

30. Invariants of vector lines. Consider a region in which the magnitude c of a vector field $\boldsymbol{c}$ is non-zero. Let $\boldsymbol{t} = \boldsymbol{c}/c$ denote the unit tangent to the vector lines of $\boldsymbol{c}$, and put

$$A \equiv \boldsymbol{t} \cdot \operatorname{curl} \boldsymbol{t}, \qquad D \equiv \operatorname{div} \boldsymbol{t}. \tag{30.1}$$

Then

$$A = \boldsymbol{t} \cdot \operatorname{curl}(\boldsymbol{c}/c) = (\boldsymbol{t}/c) \cdot \operatorname{curl} \boldsymbol{c} = c^{-2} \boldsymbol{c} \cdot \operatorname{curl} \boldsymbol{c}. \tag{30.2}$$

Now

$$\operatorname{grad} \boldsymbol{c} = \operatorname{grad}(c\, \boldsymbol{t}) = (\operatorname{grad} c)\, \boldsymbol{t} + c \operatorname{grad} \boldsymbol{t}, \tag{30.3}$$

so that

$$\operatorname{div} \boldsymbol{c} = \boldsymbol{t} \cdot \operatorname{grad} c + c \operatorname{div} \boldsymbol{t} = \frac{dc}{dt} + c\, D, \tag{30.4}$$

$$\operatorname{curl} \boldsymbol{c} = \operatorname{grad} c \times \boldsymbol{t} + c \operatorname{curl} \boldsymbol{t}, \tag{30.5}$$

where d/dt denotes differentiation with respect to arc length along the vector line[1].

For any unit vector $\boldsymbol{e}$ and any vector $\boldsymbol{v}$, we have the identity

$$\boldsymbol{v} = \boldsymbol{e}\boldsymbol{e} \cdot \boldsymbol{v} - \boldsymbol{e} \times (\boldsymbol{e} \times \boldsymbol{v}), \tag{30.6}$$

so that

$$\operatorname{curl} \boldsymbol{c} = \boldsymbol{t}\boldsymbol{t} \cdot \operatorname{curl} \boldsymbol{c} - \boldsymbol{t} \times (\boldsymbol{t} \times \operatorname{curl} \boldsymbol{c}). \tag{30.7}$$

---

[1] Here and elsewhere, we denote by d/de the directional derivative in the direction of a unit vector $\boldsymbol{e}$.

Using Eqs. (30.5) and (30.6) with $\boldsymbol{e}=\boldsymbol{t}$, $\boldsymbol{v}=\operatorname{grad} c$, we get

$$\left.\begin{aligned}\boldsymbol{t}\times\operatorname{curl}\boldsymbol{c} &= \boldsymbol{t}\times(\operatorname{grad} c\times\boldsymbol{t}) + c\,\boldsymbol{t}\times\operatorname{curl}\boldsymbol{t},\\ &= \operatorname{grad} c - \boldsymbol{t}\,\boldsymbol{t}\cdot\operatorname{grad} c + c\,\boldsymbol{t}\times\operatorname{curl}\boldsymbol{t}.\end{aligned}\right\} \tag{30.8}$$

Introducing the unit principal normal $\boldsymbol{n}$ and binormal $\boldsymbol{b}$ of the vector lines, we have

$$\operatorname{grad} = \boldsymbol{t}\,\boldsymbol{t}\cdot\operatorname{grad} + \boldsymbol{n}\,\boldsymbol{n}\cdot\operatorname{grad} + \boldsymbol{b}\,\boldsymbol{b}\cdot\operatorname{grad}, \tag{30.9}$$

which enables us to rewrite Eq. (30.8) as

$$\left.\begin{aligned}\boldsymbol{t}\times\operatorname{curl}\boldsymbol{c} &= \boldsymbol{n}\,\boldsymbol{n}\cdot\operatorname{grad} c + \boldsymbol{b}\,\boldsymbol{b}\cdot\operatorname{grad} c + \boldsymbol{c}\,\boldsymbol{t}\times\operatorname{curl}\boldsymbol{t},\\ &= \boldsymbol{n}\frac{dc}{dn} + \boldsymbol{b}\frac{dc}{db} + c\,\boldsymbol{t}\times\operatorname{curl}\boldsymbol{t},\\ &= \boldsymbol{n}\frac{dc}{dt} + \boldsymbol{b}\frac{dc}{db} - c\,\boldsymbol{t}\cdot\operatorname{grad}\boldsymbol{t},\\ &= \boldsymbol{n}\left(\frac{dc}{dn} - c\,\varkappa\right) + \boldsymbol{b}\frac{dc}{db},\end{aligned}\right\} \tag{30.10}$$

where $\varkappa$ is the curvature of the vector line. Hence

$$-\boldsymbol{t}\times(\boldsymbol{t}\times\operatorname{curl}\boldsymbol{c}) = \frac{dc}{db}\boldsymbol{n} + \left(c\,\varkappa - \frac{dc}{dn}\right)\boldsymbol{b}. \tag{30.11}$$

From Eqs. (30.2), (30.7), and (30.11) we then obtain an *intrinsic* relation for[1] curl $\boldsymbol{c}$:

$$\operatorname{curl}\boldsymbol{c} = c\,A\,\boldsymbol{t} + \frac{dc}{db}\boldsymbol{n} + \left(c\,\varkappa - \frac{dc}{dn}\right)\boldsymbol{b}. \tag{30.12}$$

The special case $c=1$ yields

$$\operatorname{curl}\boldsymbol{t} = A\,\boldsymbol{t} + \varkappa\,\boldsymbol{b}. \tag{30.13}$$

Following LEVI-CIVITA[2], we call the invariant A the *abnormality* of the field, this name being suggested by the following geometric interpretation, attributed to BERTRAND by LECORNU[3].

Consider any regular surface with unit normal $\boldsymbol{N}$ such that $\boldsymbol{N}=\boldsymbol{t}$ at some interior point P. On this surface, let $\mathfrak{s}$ be any region containing P and bounded by a circuit c, reducible on the surface, and let $\mathfrak{S}$ be the area of $\mathfrak{s}$. Then by KELVIN's transformation (28.1)

$$\mathfrak{S}^{-1}\oint_c d\boldsymbol{x}\cdot\boldsymbol{t} = \mathfrak{S}^{-1}\int_{\mathfrak{s}} \boldsymbol{N}\cdot\operatorname{curl}\boldsymbol{t}\,da. \tag{30.14}$$

Reducing c to the point P, so that $\mathfrak{S}\to 0$, $\boldsymbol{N}\to\boldsymbol{t}$, by using Eq. (30.2) we get

$$A|_P = \lim_{\mathfrak{S}\to 0}\mathfrak{S}^{-1}\oint_c d\boldsymbol{x}\cdot\boldsymbol{t} \tag{30.15}$$

If there exists a congruence of surfaces with $\boldsymbol{t}$ as normal, we may take $\mathfrak{s}$ to be a region on one of these, obtaining from Eq. (30.15) $A=0$. Thus A may be regarded as a measure of the departure of $\boldsymbol{c}$ from the property of having a normal congruence of surfaces.

b) Special classes of fields.

31. Solenoidal fields I: Integral properties. An integrable vector field $\boldsymbol{c}$ is called *solenoidal* provided its flux out of every reducible closed surface $\mathfrak{s}$ in its

[1] MASOTTI [1927, *2*]. Cf. BJØRGUM [1951, *4*, § 2.3], COBURN [1952, *1*, § 2]. GILBERT [1890, *2*] (cf. also KLEITZ [1873, *3*, §§ 2, 8, 40]) had given formulae for the gradient and divergence of a tensor in terms of the radii of curvature of the members of a triply orthogonal system.

[2] [1900, *1*]. The names "torsion of the curve system" and "torsion of neighboring vector lines" have also been proposed; cf. BJØRGUM [1951, *4*, § 2.4].

[3] [1919, *1*].

region of definition is zero. From Eq. $(26.1)_2$ follows ***Kelvin's*[1] *characterization:*** *A continuously differentiable field* $\mathbf{c}$ *is solenoidal if and only if its divergence vanishes*

$$\operatorname{div} \mathbf{c} = 0. \tag{31.1}$$

There is also the ***characterization of Helmholtz*[2]**: *A field* $\mathbf{c}$, *continuously differentiable*[3] *in a closed region, is solenoidal if and only if the strength of every vector tube is the same at all cross sections.* The necessity of this condition follows immediately from Eq. (29.6). If the strength of every vector tube is the same at all cross sections, by Eq. (29.5) it follows easily that Eq. (31.1) holds, hence that $\mathbf{c}$ is solenoidal.

In Sects. 31 and 32 we assume $\mathbf{c}$ is a continuously differentiable solenoidal field. Then by Eq. (13.7) we calculate

$$\operatorname{div}(\mathbf{c}\,\mathbf{p}^{(K+1)}) = \mathbf{c}\cdot\operatorname{grad}\mathbf{p}^{(K+1)} = \{\mathbf{p}^{(K)}\,\mathbf{c}\}; \tag{31.2}$$

hence by Eq. $(26.1)_2$ we obtain[4] for the moments (25.5) of $\mathbf{c}$

$$\mathfrak{c}_{(K)} \equiv \int_{v} \{\mathbf{p}^{(K)}\,\mathbf{c}\}\,dv = \oint_{s} d\mathbf{a}\cdot\mathbf{c}\,\mathbf{p}^{(K+1)} = \oint_{s} da\, c_n\,\mathbf{p}^{(K+1)}. \tag{31.3}$$

Hence the moments of a solenoidal field are determined by the boundary values of its normal projection c_n. *If* $c_n = 0$ *on* s, *then all moments of* $\mathbf{c}$ *over the volume* v *bounded by* s *vanish.*

More generally[5],

$$\operatorname{div}(\mathbf{c}\,\Psi) = \mathbf{c}\cdot\operatorname{grad}\Psi, \tag{31.4}$$

whence follows

$$\int_{v} \mathbf{c}\cdot\operatorname{grad}\Psi\,dv = \oint_{s} d\mathbf{a}\cdot\mathbf{c}\,\Psi = \oint_{s} da\,c_n\,\Psi. \tag{31.5}$$

Hence, *if* $\mathbf{c}$ *and* Ψ *are continuously differentiable and* $\mathbf{c}$ *is solenoidal, then the total* $\mathbf{c}\cdot\operatorname{grad}\Psi$ *in a region* v *is determined by the boundary values of* $c_n\,\Psi$. *If* $c_n = 0$ *on* s, *then the total* $\mathbf{c}\cdot\operatorname{grad}\Psi$ *in* v *is zero*[6].

---

[1] [1851, *1*, § 74].

[2] [1858, *1*, § 2]. From the fact that the vorticity field is solenoidal, Helmholtz [1858, *1*, § 2] concluded that „Es folgt hieraus auch, daß ein Wirbelfaden nirgends innerhalb der Flüssigkeit aufhören dürfe, sondern entweder ringförmig innerhalb der Flüssigkeit in sich zurücklaufen, oder bis an die Grenzen der Flüssigkeit reichen müsse. Denn wenn ein Wirbelfaden innerhalb der Flüssigkeit irgendwo endete, würde sich eine geschlossene Fläche construiren lassen, für welche das Integral $\int q \cos\vartheta\, d\omega$ nicht den Werth Null hätte". Truesdell [1954, *2*, § 10] has noted that this statement is misleading and incomplete. It is not trivial to reformulate it as a theorem susceptible of rigorous proof. One point glossed over in Helmholtz's statement is the fact that a vector tube of a field continuously differentiable in a compact region need not approach the boundary, be closed, or end in the interior. For example, in the plane, it may be bounded by curves having as a common limit set a limit cycle. One can show that this particular situation cannot occur if the field be solenoidal. Cf. Sect. 32, footnote 1.

[3] Truesdell [1954, *2*, § 10] remarks that it suffices to assume $\mathbf{c}$ continuous and piecewise continuously differentiable, but that the result need not hold for piecewise continuous fields.

[4] Truesdell [1949, *3*], [1954, *2*, § 10]. The case $K = 0$ was discussed by Föppl [1897, *1*, § 4].

[5] Truesdell [1951, *2*, § 9] has remarked that the more general result is implied by an example due to Kelvin [1849, *1*, § 7].

[6] Using Kelvin's transformation (28.1), one can show that this result is equivalent to the lemma 1 of Weyl [1940, *2*]. A special case is given by Berker [1949, *5*, Chap. III]. We follow Truesdell [1954, *2*, § 10].

Setting $\boldsymbol{b}=\boldsymbol{c}$ in Eq. (26.2) and employing Eq. (31.1), we obtain

$$\left.\begin{aligned}&\oint_{\mathfrak{s}} [d\boldsymbol{a}\cdot\boldsymbol{c}\,\boldsymbol{c}\,\boldsymbol{p}^{(K)} - \tfrac{1}{2}c^2\,d\boldsymbol{a}\,\boldsymbol{p}^{(K)}]\\ &\qquad = \int_{\mathfrak{v}} [\boldsymbol{c}\{\boldsymbol{c}\,\boldsymbol{p}^{(K-1)}\} - \tfrac{1}{2}c^2\,\mathrm{grad}\,\boldsymbol{p}^{(K)} + \mathrm{curl}\,\boldsymbol{c}\times\boldsymbol{c}\,\boldsymbol{p}^{(K)}]\,dv.\end{aligned}\right\} \tag{31.6}$$

The case when $K=0$ is POINCARÉ'S[1] identity

$$\oint_{\mathfrak{s}} [d\boldsymbol{a}\cdot\boldsymbol{c}\,\boldsymbol{c} - \tfrac{1}{2}c^2\,d\boldsymbol{a}] = \int_{\mathfrak{v}} \mathrm{curl}\,\boldsymbol{c}\times\boldsymbol{c}\,dv. \tag{31.7}$$

Recalling the convention of Sect. 24, we formulate conditions sufficient for the vanishing of the integral over $\mathfrak{s}$ and conclude that *if a field* $\boldsymbol{c}$ *which is continuously differentiable and solenoidal in a region* $\mathfrak{v}$ *satisfies the boundary conditions*

$$\left.\begin{aligned}c_n &= 0 \quad \text{or} \quad \boldsymbol{c}\,c_n = \bar{o}(p^{-2}) \quad \text{on } \mathfrak{s},\\ c^2 &= \text{const} \quad \text{or} \quad c^2 = \bar{o}(p^{-2}) \quad \text{on } \mathfrak{s},\end{aligned}\right\} \tag{31.8}$$

then

$$\int_{\mathfrak{v}} \mathrm{curl}\,\boldsymbol{c}\times\boldsymbol{c}\,dv = 0. \tag{31.9}$$

Sufficient for Eq. (31.8) to hold are the simpler conditions

$$c=0 \quad \text{or} \quad c=\bar{o}(p^{-1}) \quad \text{on } \mathfrak{s}. \tag{31.10}$$

For $K=1$, Eq. (31.6) becomes

$$\oint_{\mathfrak{s}} [d\boldsymbol{a}\cdot\boldsymbol{c}\,\boldsymbol{c}\,\boldsymbol{p} - \tfrac{1}{2}c^2\,d\boldsymbol{a}\,\boldsymbol{p}] = \int_{\mathfrak{v}} [\boldsymbol{c}\,\boldsymbol{c} - \tfrac{1}{2}c^2\,\boldsymbol{1} + \mathrm{curl}\,\boldsymbol{c}\times\boldsymbol{c}\,\boldsymbol{p}]\,dv. \tag{31.11}$$

Contracting Eq. (31.11) yields an identity of LAMB and J. J. THOMSON[2]

$$\tfrac{1}{2}\int_{\mathfrak{v}} c^2\,dv = \oint_{\mathfrak{s}} d\boldsymbol{a}\cdot(\tfrac{1}{2}c^2\boldsymbol{p} - \boldsymbol{c}\,\boldsymbol{c}\cdot\boldsymbol{p}) + \int_{\mathfrak{v}} \boldsymbol{p}\cdot\mathrm{curl}\,\boldsymbol{c}\times\boldsymbol{c}\,dv. \tag{31.12}$$

from which it follows that *if a field* $\boldsymbol{c}$ *which is continuously differentiable and solenoidal in a region* $\mathfrak{v}$ *satisfies the following conditions:*

1. $c=0$ *or* $c^2=\bar{o}(p^{-3})$, $\quad \boldsymbol{p}\cdot\boldsymbol{c}\,c_n=\bar{o}(p^{-2})$ on $\mathfrak{s}$, (31.13)

2. *Throughout* $\mathfrak{v}$, $\mathrm{curl}\,\boldsymbol{c}\times\boldsymbol{c}=0$; (31.14)

then $\boldsymbol{c}=0$ *throughout* $\mathfrak{v}$.

Sufficient for (31.13) to hold are the simpler conditions

$$c=0 \quad \text{or} \quad c=\bar{o}(p^{-\frac{3}{2}}) \quad \text{on } \mathfrak{s}. \tag{31.15}$$

The alternating part of Eq. (31.11) yields[3]

$$\oint_{\mathfrak{s}} [d\boldsymbol{a}\cdot\boldsymbol{c}\,\boldsymbol{p}\times\boldsymbol{c} + \tfrac{1}{2}c^2\,d\boldsymbol{a}\times\boldsymbol{p}] = \int_{\mathfrak{v}} \boldsymbol{p}\times(\mathrm{curl}\,\boldsymbol{c}\times\boldsymbol{c})\,dv, \tag{31.16}$$

from which we conclude that *if a field* $\boldsymbol{c}$ *which is continuously differentiable and solenoidal in a region* $\mathfrak{v}$ *satisfies the boundary conditions*

$$c_n=0 \quad \text{and} \quad c=\text{const} \quad \text{or} \quad \boldsymbol{p}\times\boldsymbol{c}\,c_n=\bar{o}(p^{-2}), \tag{31.17}$$

then

$$\int_{\mathfrak{v}} \boldsymbol{p}\times(\mathrm{curl}\,\boldsymbol{c}\times\boldsymbol{c})\,dv = 0. \tag{31.18}$$

Sufficient for Eq. (31.17) to hold are the simpler conditions (31.15).

---

[1] [1893, *1*, § 5]. See also TRUESDELL [1954, *2*, §§ 10, 64].

[2] [1879, *1*, § 136], [1932, *6*, § 153]; [1883, *1*, § 6]. Cf. also TRUESDELL [1954, *2*, § 10] for the results given in italics here.

[3] POINCARÉ [1893, *1*, § 115]. Cf. TRUESDELL [1954, *1*, § 10].

In Eq. (26.2), put $K=1$, $\boldsymbol{b}=\boldsymbol{p}$; then

$$\left.\begin{aligned}&\int_{v}[\boldsymbol{p}\boldsymbol{c}+\boldsymbol{c}\boldsymbol{p}-\boldsymbol{p}\cdot\boldsymbol{c}\,\mathbf{1}+(3\boldsymbol{c}+\operatorname{curl}\boldsymbol{c}\times\boldsymbol{p})\,\boldsymbol{p}]\,dv\\&\qquad=\oint_{s}[d\boldsymbol{a}\cdot(\boldsymbol{p}\boldsymbol{c}+\boldsymbol{c}\boldsymbol{p})\,\boldsymbol{p}-d\boldsymbol{a}\,(\boldsymbol{p}\cdot\boldsymbol{c})\,\boldsymbol{p}].\end{aligned}\right\}\tag{31.19}$$

Contracting this relation yields

$$2\int_{v}\boldsymbol{p}\cdot\boldsymbol{c}\,dv=\oint_{s}d\boldsymbol{a}\cdot\boldsymbol{c}\,p^{2},\tag{31.20}$$

while the alternating part is equivalent to[1]

$$\left.\begin{aligned}3\int_{v}\boldsymbol{p}\times\boldsymbol{c}\,dv&=\int_{v}\boldsymbol{p}\times(\boldsymbol{p}\times\operatorname{curl}\boldsymbol{c})\,dv-\\&\quad-\oint_{s}[(d\boldsymbol{a}\times\boldsymbol{c})\times\boldsymbol{p}]\times\boldsymbol{p}.\end{aligned}\right\}\tag{31.21}$$

J. J. THOMSON[2] derived two similar integral formulae which, while not restricted to solenoidal fields, are of interest mainly in the solenoidal case. First, in Eq. (26.2) put $K=0$, $\boldsymbol{b}=\boldsymbol{p}$, obtaining

$$\left.\begin{aligned}&\int_{v}(\operatorname{curl}\boldsymbol{c}\times\boldsymbol{p}+3\boldsymbol{c}+\boldsymbol{p}\operatorname{div}\boldsymbol{c})\,dv\\&\qquad=\oint_{s}[d\boldsymbol{a}\cdot(\boldsymbol{c}\boldsymbol{p}+\boldsymbol{p}\boldsymbol{c})-d\boldsymbol{a}\,(\boldsymbol{p}\cdot\boldsymbol{c})],\\&\qquad=\oint_{s}[(d\boldsymbol{a}\times\boldsymbol{c})\times\boldsymbol{p}-d\boldsymbol{a}\cdot\boldsymbol{c}\boldsymbol{p}].\end{aligned}\right\}\tag{31.22}$$

Since $\operatorname{div}(\boldsymbol{c}\boldsymbol{p})=\boldsymbol{p}\operatorname{div}\boldsymbol{c}+\boldsymbol{c}$, application of Eq. $(26.1)_2$ yields

$$2\int_{v}\boldsymbol{c}\,dv=\int_{v}\boldsymbol{p}\times\operatorname{curl}\boldsymbol{c}\,dv+\oint_{s}(d\boldsymbol{a}\times\boldsymbol{c})\times\boldsymbol{p}.\tag{31.23}$$

Second, if we integrate the identity

$$\operatorname{curl}(p^{2}\boldsymbol{c})=p^{2}\operatorname{curl}\boldsymbol{c}+2\boldsymbol{p}\times\boldsymbol{c}\tag{31.24}$$

over v and then apply Eq. $(26.1)_3$, we obtain

$$2\int_{v}\boldsymbol{p}\times\boldsymbol{c}\,dv=-\int_{v}p^{2}\operatorname{curl}\boldsymbol{c}\,dv+\oint_{s}d\boldsymbol{a}\times p^{2}\boldsymbol{c}.\tag{31.25}$$

For the various integral theorems we have given here, the region need not be simply-connected.

32. Solenoidal fields II: Differential properties. *If* $\boldsymbol{c}$ *is any continuously differentiable field, there exists an infinite number of scalar functions* m *such that* $m\boldsymbol{c}$ *is solenoidal.* To see this, one need only note that

$$\operatorname{div}(m\,\boldsymbol{c})=0,\tag{32.1}$$

being a first order differential equation for m, with continuous coefficients, admits an infinite number of solutions. Since $\boldsymbol{c}$ and $m\boldsymbol{c}$ have the same vector lines, *the vector lines of any continuously differentiable field* $\boldsymbol{c}$ *are also the vector lines of an infinite number of solenoidal fields*[3]. This result may mislead in seeming to imply that vector lines of solenoidal fields have no special properties. Distinctions can arise, for example, because of the fact that solutions m of Eq. (32.1)

[1] MUNK [1941, *1*, p. 95].

[2] [1883, *1*, §§ 4–5]. The second of these is usually presented in a non-invariant form; cf. e.g. LAMB [1932, *6*, § 195], where it is asserted that an error in J. J. THOMSON's formula was corrected by WELSH.

[3] APPELL [1897, *2*, § 5].

can possess singularities at points where $c=0$, even if $\boldsymbol{c}$ be analytic. For example, in the plane, one can show that an isolated point where $c=0$ is never a spiral point for the vector lines of a solenoidal field $\boldsymbol{c}$, whereas it can be if $\boldsymbol{c}$ is not solenoidal[1].

Any continuously differentiable solenoidal vector field $\boldsymbol{c}$ may be represented locally in EULER's form[2]

$$\boldsymbol{c} = \operatorname{grad} F \times \operatorname{grad} G \tag{32.2}$$

where F and G are scalar functions which satisfy

$$\boldsymbol{c} \cdot \operatorname{grad} F = 0, \qquad \boldsymbol{c} \cdot \operatorname{grad} G = 0. \tag{32.3}$$

The surfaces $F=\text{const}$ and $G=\text{const}$ are thus vector sheets of $\boldsymbol{c}$. One of the functions, say F, may be taken as an arbitrary non-constant solution of Eq. (32.3); the other may be taken as the function given by

$$G = -\int \frac{\boldsymbol{e} \times \boldsymbol{c} \cdot d\boldsymbol{x}}{\boldsymbol{e} \cdot \operatorname{grad} F}, \tag{32.4}$$

where $\boldsymbol{e}$ is an arbitrary continuous vector field such that $\boldsymbol{e} \cdot \operatorname{grad} F \neq 0$, and where the path of integration lies on one of the surfaces $F=\text{const}$.

Combining the results of the two foregoing paragraphs, we see that any continuously differentiable field may be represented locally in the form

$$\boldsymbol{c} = H \operatorname{grad} F \times \operatorname{grad} G, \tag{32.5}$$

the surfaces $F=\text{const}$ and $G=\text{const}$ being vector sheets of $\boldsymbol{c}$. In Eq. (32.5), F and G may be chosen as arbitrary functionally independent solutions of Eq. (32.3), and the function H is then determined uniquely. If $\boldsymbol{c}$ is solenoidal, Eq. (32.5) yields

$$0 = \operatorname{div} \boldsymbol{c} = \frac{\partial(H, F, G)}{\partial(x^1, x^2, x^3)}; \tag{32.6}$$

therefore $H = H(F, G)$. From Eqs. (32.5) and (32.6) follows ***Euler's theorem on solenoidal fields***[3]: *If F and G are any independent functions such that the surfaces $F=\text{const}$ and $G=\text{const}$ are vector sheets of a continuously differentiable solenoidal field $\boldsymbol{c}$, then $\boldsymbol{c}$ can be represented locally in the form*

$$\boldsymbol{c} = H(F, G) \operatorname{grad} F \times \operatorname{grad} G. \tag{32.7}$$

From Eq. (32.2) and the identity

$$\operatorname{curl}(F \operatorname{grad} G + \operatorname{grad} H) = \operatorname{grad} F \times \operatorname{grad} G, \tag{32.8}$$

it follows that *any continuously differentiable solenoidal field $\boldsymbol{c}$ may be represented locally as the curl of a vector $\boldsymbol{v}$,*

$$\boldsymbol{c} = \operatorname{curl} \boldsymbol{v}, \tag{32.9}$$

the field $\boldsymbol{v}$ being indeterminate to within an additive gradient. Conversely, if Eq. (32.9) holds, $\operatorname{div} \boldsymbol{c} = \operatorname{div} \operatorname{curl} \boldsymbol{v} = 0$[4].

---

[1] In connection with HELMHOLTZ's characterization (Sect. 31, footnote 2), many expositions assert that the vector lines of a solenoidal field cannot end in the interior. KELLOGG [1929, *1*, Chap. II, § 6] has remarked that this is false.

[2] [1770, *2*, §§ 26, 49], [1806, *1*, § 142]. Derivations are given in works on vector analysis, e.g. BRAND [1947, *2*, § 104, Chap. 3].

[3] [1757, *1*, §§ 47–49].

[4] More generally, any analytic field $\boldsymbol{v}$ yields the infinite sequence of solenoidal fields $\operatorname{curl}^K \boldsymbol{v}$ $(K=1, 2, \ldots)$. For $K \geqq 3$ we have $\operatorname{curl}^K \boldsymbol{v} = -\nabla^2 \operatorname{curl}^{K-2} \boldsymbol{v}$, whence follow volume integrals yielding $\operatorname{curl}^{K-2} \boldsymbol{v}$ in terms of $\operatorname{curl}^K \boldsymbol{v}$. These facts were noted by ROWLAND [1880, *1*] and FABRI [1892, *3*] who attempted a kinematical interpretation of $\operatorname{curl}^K \boldsymbol{v}$. For the case $K=2$, cf. BOGGIO-LERA [1887, *3*].

At points where $c \neq 0$, Eq. (30.4) shows that $\operatorname{div} \mathbf{c} = 0$ if and only if

$$\frac{d \log c}{dt} = -D, \tag{32.10}$$

which may be integrated along the vector lines of $\mathbf{c}$ to give

$$c = c_0 \exp[-\int D\, dt], \tag{32.11}$$

this characterization being due to BJØRGUM[1]. It follows immediately that a non-zero solenoidal field is determined by its vector lines and by its magnitude at one point on each. By successive differentiation of (30.4), we conclude that if c and D are analytic functions of arc length on a given vector line, c vanishes at one point if and only if it vanishes all along the vector line through the point.

33. Lamellar and complex-lamellar fields. Following BJØRGUM[2], we call any vector field proportional to a ... field a complex - ... field, wherein any name may be inserted for ... From results given in Sect. 32, an arbitrary continuously differentiable field may thus be called a complex-solenoidal field.

A field $\mathbf{c}$ is *lamellar*[3] in a region provided

$$\oint_c d\mathbf{x} \cdot \mathbf{c} = \oint_c dx^k c_k = 0 \tag{33.1}$$

for any reducible circuit c in the region[4]. It follows that *a field* $\mathbf{c}$, *continuous in a region* v, *is lamellar if and only if there exists a scalar* P, *called the potential of* $\mathbf{c}$, *such that*

$$\mathbf{c} = -\operatorname{grad} P. \tag{33.2}$$

The potential is single-valued if v *is simply connected, not so in general if* v *is multiply-connected.* A lamellar field is thus everywhere normal to the *equipotential surfaces* $P = \text{const}$. A complex-lamellar field, being by its definition representable in the form

$$\mathbf{c} = Q \operatorname{grad} P, \tag{33.3}$$

where Q is a scalar, also is normal to a family of surfaces. From KELVIN's transformation (28.1) it follows that *a continuously differentiable field* $\mathbf{c}$ *is lamellar if and only if*

$$\operatorname{curl} \mathbf{c} = 0, \quad \text{or} \quad \partial c_{[k} / \partial x^{m]} = 0. \tag{33.4}$$

Similarly, *a continuously differentiable field* $\mathbf{c}$ *is complex-lamellar if and only if*

$$\mathbf{c} \cdot \operatorname{curl} \mathbf{c} = 0, \quad \text{or} \quad c_{[k}\, \partial c_r / \partial x^{s]} = 0. \tag{33.5}$$

These results were established by KELVIN[5], a result equivalent to Eq. (33.5) having been obtained much earlier by EULER[6]. From $(33.5)_2$ it follows that $\mathbf{c}$ is complex-lamellar if there exists a co-ordinate system in which

$$c_1 = f(x^1, x^2), \quad c_2 = g(x^1, x^2), \quad c_3 = 0. \tag{33.6}$$

A field $\mathbf{c}$ is *plane* if Eq. (33.6) holds in some rectangular Cartesian co-ordinate system, *rotationally symmetric if* Eq. (33.6) holds in cylindrical co-ordinates, x^3

[1] [1951, *4*, § 3.1].

[2] [1951, *4*]. The reader will not confuse this usage of "complex" with "complex-valued".

[3] The names *lamellar* and *complex-lamellar* derive from KELVIN [1850, *1*], [1851, *1*, §§ 68–69, 75].

[4] WEYL [1940, *2*] generalizes the definitions of lamellar and solenoidal field given here.

[5] [1851, *1*, § 75]. VELTMANN also discussed these fields [1870, *1*, pp. 453–456]. For a modern proof that Eq. (33.5) is sufficient as well as necessary, see e.g. BRAND [1947, *2*, § 105]. The representation $\mathbf{c} = Q \operatorname{grad} P + \operatorname{grad} F(Q, P)$ is discussed by CASTOLDI [1955, *5*].

[6] [1770, *1*, § 1].

being the angle variable. It is essential that the covariant components of $\boldsymbol{c}$ be used in Eqs. $(33.4)_2$ and $(33.5)_2$.

For any vector field $\boldsymbol{c}$, any curve $\boldsymbol{x} = \boldsymbol{x}(u)$ satisfying the equation

$$\boldsymbol{c} \cdot \frac{d\boldsymbol{x}}{du} = c_k \frac{dx^k}{du} = 0 \tag{33.7}$$

is orthogonal to the vector lines of $\boldsymbol{c}$ which it intersects. A theorem of CARATHÉODORY[1] asserts that *in a closed and bounded region where $\boldsymbol{c}$ is non-zero and satisfies a Lipschitz condition, $\boldsymbol{c}$ is complex-lamellar if and only if in every neighborhood of an arbitrary interior point $\boldsymbol{x}$ there exist a point $\boldsymbol{x}^*$ such that no curve satisfying* Eq. (33.7) *joins $\boldsymbol{x}$ to $\boldsymbol{x}^*$.* If $\boldsymbol{c}$ is complex-lamellar, it follows from Eqs. (33.3) and (33.7) that $\boldsymbol{x}$ and $\boldsymbol{x}^*$ may be joined by a solution of Eq. (33.7) if and only if they lie on the same surface $P = \text{const}$, from which the necessity of this condition follows. We omit the more complicated proof of sufficiency.

With the exception of the definitions of plane and rotationally symmetric fields, all these results are easily extended to n-dimensional spaces which need not possess a metric tensor, $\boldsymbol{c}$ being taken to be a covariant field[2]. Eqs. (33.1) to (33.3), $(33.4)_2$, $(33.5)_2$ and (33.7) require no alteration, while Eq. (33.6) is to be replaced by

$$c_1 = f(x^1, x^2), \quad c_2 = g(x^1, x^2), \quad c_3 = \cdots = c_n = 0. \tag{33.8}$$

It is clear from Eq. (33.3) that the magnitude of a complex-lamellar field is in no way restricted by its vector lines. Indeed, it is immediate that a non-vanishing field is complex-lamellar if and only if its unit tangent is complex-lamellar. However, the condition that a field be lamellar is essentially different in that it connects the magnitude of the field with the geometric properties of its vector lines.

In the three-dimensional case, it is easy to render the above remarks explicit. In the first place, either directly from Eq. $(30.1)_1$ or by substitution in Eq. (33.5), or from Eq. (30.15), we conclude that *a non-zero continuously differentiable field is complex-lamellar if and only if its abnormality A is zero.* For a lamellar field, we have from Eq. (30.12) the much stronger necessary and sufficient conditions of BJØRGUM[3]: (1) $A = 0$ when $c \neq 0$, (2) c is constant along the vector lines of the binormal field b, and (3)

$$c = c_0 \exp \int \varkappa \, dn, \tag{33.9}$$

the integration being performed on the vector lines of the principal normal $\boldsymbol{n}$, which lie on the equipotential surfaces, as do those of $\boldsymbol{b}$. It follows that, in a sufficiently small region, a lamellar field is determined by its vector lines and by its magnitude at a single point on each equipotential surface.

When a field is complex-lamellar, its unit tangent $\boldsymbol{t}$ is the unit normal of the surfaces $P = \text{const}$. Hence the mean curvature $\overline{K}$ of those surfaces is given by[4]

$$\overline{K} = -\operatorname{div} \boldsymbol{t} = -D = \frac{d \log c}{dt} - \frac{1}{c} \operatorname{div} \boldsymbol{c}. \tag{33.10}$$

Therefore, if we are given a set of vector lines having a normal congruence, we may obtain all non-vanishing complex-lamellar fields having these vector

---

[1] [1909, *5*].

[2] In situations to which CARATHÉODORY's sufficiency condition has been applied, it is artificial and unnecessary to introduce a metric tensor. Cf. CARATHÉODORY [1909, *5*], ERICKSEN [1956, *1*].

[3] [1951, *4*, § 3.4].

[4] Cf. e.g. BRAND [1947, *2*, § 131]. The result is due to CHALLIS [1842, *1*], whose derivation was criticized and corrected by TARDY [1850, *1*]. The geometry of the surfaces $P =$ const is studied by CALDONAZZO [1924, *1*], [1925, *1*] and PASTORI [1927, *2*].

lines by prescribing the scalar field div $\boldsymbol{c}$ everywhere and the value of c at one point on each vector line. Also, *for a non-vanishing complex-lamellar field, any two of the following three conditions imply the third*[1]:

1. *The normal surfaces are minimal surfaces.*
2. *The field is solenoidal.*
3. *The magnitude of the field is constant on each vector line.*

34. Screw fields. From Eq. (33.5), a field $\boldsymbol{c}$ is complex-lamellar if and only if it be normal to its curl. The opposite extreme is furnished by a *screw field*[2], defined as being parallel to its curl:

$$\boldsymbol{c}\times \operatorname{curl}\boldsymbol{c} = 0, \qquad \operatorname{curl}\boldsymbol{c} \neq 0; \tag{34.1}$$

equivalently,

$$c_{[k,m]}\,c^k = 0, \qquad c_{[k,m]} \neq 0. \tag{34.2}$$

In either of these equations covariant derivatives may be replaced by partial derivatives. It follows from Eqs. (30.12) and $(34.1)_1$ that $\boldsymbol{c}$ is a screw field if and only if

$$\operatorname{curl}\boldsymbol{c} = A\,\boldsymbol{c} \neq 0, \tag{34.3}$$

so *the abnormality of a screw field* $\boldsymbol{c}$ *is the factor of proportionality between the curl and the field.* Since $\boldsymbol{c}$ and curl $\boldsymbol{c}$ have the same vector lines when Eq. (34.3) holds, and since the abnormality depends only on the vector lines, we may replace $\boldsymbol{c}$ by curl $\boldsymbol{c}$ in Eq. (30.2), obtaining[3]

$$A = c^{-2}\,\boldsymbol{c}\cdot\operatorname{curl}\boldsymbol{c} = \frac{\operatorname{curl}\boldsymbol{c}\cdot\operatorname{curl}\operatorname{curl}\boldsymbol{c}}{\operatorname{curl}\boldsymbol{c}\cdot\operatorname{curl}\boldsymbol{c}}. \tag{34.4}$$

If m is any scalar function such that $m\boldsymbol{c}$ is solenoidal (cf. Sect. 32),

$$0 = \operatorname{div} m\,\boldsymbol{c} = \operatorname{div}\left(\frac{m}{A}\operatorname{curl}\boldsymbol{c}\right) = \operatorname{grad}\frac{m}{A}\cdot\operatorname{curl}\boldsymbol{c} = A\,\boldsymbol{c}\cdot\operatorname{grad}\frac{m}{A}, \tag{34.5}$$

so the surfaces $m/A = \text{const}$ are vector sheets of $\boldsymbol{c}$ and of curl $\boldsymbol{c}$. Conversely, if, for some scalar function $f(\boldsymbol{x})$, the surfaces $f = \text{const}$ are vector sheets of a screw field $\boldsymbol{c}$, we may set $m = fA$, where A is the abnormality; reading Eq. (34.5) backwards, we obtain div $m\boldsymbol{c} = 0$. These results constitute the following theorem[4]: *For a twice continuously differentiable screw field* $\boldsymbol{c}$, *a necessary and sufficient*

[1] A special case is due to CALDONAZZO [1924, *2*, § 6]. We generalize an argument of PRIM [1948, *1*, § 3], [1952, *1*, Chap. V]. Cf. also CASTOLDI [1947, *5*], BYUSHGENIS [1948, *2*, § 2.1].

[2] Much of the literature follows CISOTTI [1923, *3*] in calling these fields "Beltrami fields", after the researches of BELTRAMI [1889, *1*], but in fact nearly all of BELTRAMI'S results were included in the prior and more extensive work of GROMEKA [1881, *2*]. BELTRAMI himself called them "helicoidal". We revert to the more descriptive term *screw field* introduced by CRAIG [1880, *2*, p. 225], [1880, *3*, p. 276], [1881, *5*, pp. 5–6], the first person to remark them. Earlier STOKES [1842, *2*, p. 3] had concluded that Eq. $(34.1)_1$ implies $\boldsymbol{c} = 0$, but later he realized his error (footnote 1, p. 3, 1880 reprint of [1842, *2*]).

[3] LECORNU [1919, *1*]. If we set $B \equiv |\operatorname{curl}\boldsymbol{c}|/c$, then $\boldsymbol{c}$ is a screw field if and only if $B = A$, the proof being immediate from Eq. (34.3); necessity was proved by APPELL [1921, *1*, § 763], sufficiency by CARSTOIU [1946, *4*]. A differential system for a screw field is given as Eq. (35.4) below. Another follows by taking the curl of Eq. (34.3) and then eliminating A by Eq. (34.4) (BALLABH [1948, *3*, § 4]), but a simpler one may be obtained by putting Eq. (34.4) into Eq. (34.3) directly. Other differential equations are derived by BJØRGUM [1951, *4*, Sect. 5], [1954, *7*].

[4] TRUESDELL [1954, *2*, § 12]. This generalizes theorems of NEMÉNYI and PRIM [1949, *6*, Th. 1], BELTRAMI [1889, *1*] (see also MORERA [1889, *2*]), and GROMEKA [1881, *2*, Gl. 2, § 9], the references being arranged in order of decreasing generality.

condition that the surfaces

$$\frac{A}{m} = \frac{(\operatorname{curl} \mathbf{c})_1}{m c_1} = \frac{(\operatorname{curl} \mathbf{c})_2}{m c_2} = \frac{(\operatorname{curl} \mathbf{c})_3}{m c_3} = \text{const} \tag{34.6}$$

be vector sheets is that $m\mathbf{c}$ *be solenoidal.* It is true *a fortiori that* m/A *is constant on each vector line of a screw field* $\mathbf{c}$ *if* $m\mathbf{c}$ *is solenoidal*[1]. Setting $m=1$, we obtain the corollaries: *The surfaces of constant abnormality of a screw field* $\mathbf{c}$ *are vector sheets if and only if* $\operatorname{div} \mathbf{c} = 0$; *the abnormality of a solenoidal screw field is constant along each vector line.* Thus from the vector lines alone, without knowledge of the magnitude of the field, we can determine whether or not a screw field is solenoidal. More generally, by putting $m=1$ in Eq. (34.5) we may derive

$$c^{-1} \operatorname{div} \mathbf{c} = -\frac{d \log |A|}{dt}. \tag{34.7}$$

A relation between a screw field and its vector lines may be read off from Eq. (30.12). This relation results from that given in Sect. 33 for lamellar fields if we replace "$A=0$" by "$A \neq 0$"[2]. Indeed, from Eqs. (30.4) and (34.3) follows

$$c A = c_0 A_0 \exp\left(-\int D\, dt\right). \tag{34.8}$$

Bjørgum[3] has shown that for a given screw field $\mathbf{c}$, it is always possible to choose a co-ordinate system such that the x^3-lines are the vector lines, $g_{13}=0$, $g_{23}=x^1 g_{33}$, and $c g_{33}=1$. Conversely, in a co-ordinate system such that $g_{13}=0$ and $g_{23}=x^1 g_{33}$, if a vector field $\mathbf{c}$ be tangent to the x^3-lines and of magnitude c so adjusted that $c g_{33}=1$, then $\mathbf{c}$ is a screw field.

From Eqs. (34.3) and (34.4) we conclude that

$$0 < \operatorname{curl} \mathbf{c} \cdot \operatorname{curl} \mathbf{c} = \mathbf{c} \cdot \operatorname{curl} \operatorname{curl} \mathbf{c}, \tag{34.9}$$

so a screw field and the curl of its curl always intersect at an acute angle. If this angle is zero, $\operatorname{curl} \mathbf{c}$ is also a screw field. It follows from taking the curl of Eq. (34.3) that *the curl of a screw field* $\mathbf{c}$ *is again a screw field if and only if the abnormality of* $\mathbf{c}$ *is uniform. Then* $\mathbf{c}$ *is solenoidal, and successive curls of* $\mathbf{c}$ *are screw fields having the same abnormality*[4].

In the case when A is uniform, that $\mathbf{c}$ is solenoidal follows alternatively as a corollary of the first corollary following Eq. (34.6). Hence taking the curl of Eq. (34.3) yields

$$\nabla^2 \mathbf{c} + A \mathbf{c} = 0. \tag{34.10}$$

This equation was derived by Gromeka[5], who based upon it a theory of determining a screw field of constant abnormality from appropriate boundary conditions. The same problem has been taken up by Bjørgum and Godal[6]; besides constructing many interesting examples, they have shown that such a field $\mathbf{c}$ can be represented in the form

$$\mathbf{c} = A \operatorname{grad} H \times \mathbf{e} + \mathbf{e} A^2 H + \mathbf{e} \cdot \operatorname{grad} \operatorname{grad} H, \tag{34.11}$$

where $\mathbf{e}$ is a fixed unit vector and $\nabla^2 H + A^2 H = 0$.

For the many more known properties of screw fields, the reader is referred to the treatise of Bjørgum[7].

---

[1] Truesdell [1954, *6*, § 52[7]] attributes to van Tuyl the remark that this theorem is an immediate consequence of the fact that any two solenoidal vector fields with common vector lines are proportional along them.

[2] Bjørgum [1951, *4*, § 3.3].

[3] [1957, *4*, § 5].

[4] Neményi and Prim [1949, *6*, Th. 3].

[5] [1881, *2*, Gl. 2, § 9]. Cf. also Stekloff [1908, *2*, §§ 39–52], Trkal [1919, *3*], Ballabh [1940, *5*, §§ 5–7].

[6] [1953, *3*], [1958, *2*]. Cf. also Bjørgum [1951, *4*, § 6].

[7] [1951, *7*]. Cf. also Truesdell [1954, *2*, §§ 12, 52].

In n-dimensional metric spaces, screw fields may be defined by Eq. (34.2). There is no immediate extension to spaces which are not metric since associated components do not exist. One natural generalization of Eq. (34.2) is obtained by requiring a covariant vector $\boldsymbol{b}$ and a contravariant vector $\boldsymbol{c}$ to satisfy

$$\partial b_{[k}/\partial x^{m]}\, c^k = 0, \qquad \partial b_{[k}/\partial x^{m]} \neq 0. \tag{34.12}$$

Such pairs of vectors occur in studies on relativity[1].

c) Potentials.

35. MONGE's potentials. If $\boldsymbol{c}$ be a twice continuously differentiable field, the field curl $\boldsymbol{c}$, being solenoidal, has a representation of the form (32.2), namely,

$$\operatorname{curl}\boldsymbol{c} = \operatorname{grad} F \times \operatorname{grad} G, \tag{35.1}$$

from which follows curl $(\boldsymbol{c} - F \operatorname{grad} G) = 0$. Thus $\boldsymbol{c} - F \operatorname{grad} G$ is lamellar, so there exists a scalar H such that

$$\boldsymbol{c} = \operatorname{grad} H + F \operatorname{grad} G. \tag{35.2}$$

In general, this representation is valid only locally[2]. The three scalars F, G, and H, called ***Monge potentials***[3] of $\boldsymbol{c}$, are not uniquely determined, but in most applications there is no need to specify one particular set rather than another.

From Eqs. (35.1) and (35.2) follows

$$\boldsymbol{c} \cdot \operatorname{curl}\boldsymbol{c} = \frac{\partial(H, F, G)}{\partial(x^1, x^2, x^3)}, \tag{35.3}$$

whence, by Eq. (33.5), $\boldsymbol{c}$ is complex-lamellar if and only if its three Monge potentials are functionally dependent. Directly from (35.1) we see that $\boldsymbol{c}$ is lamellar if and only if the two potentials F and G are functionally dependent.

MORERA[4] obtained differential equations to be satisfied by the Monge potentials of a screw field. In geometrical terms, these equations assert that for $\boldsymbol{c}$ to be a screw field the surfaces $F = \text{const}$ and $G = \text{const}$, which by Eq. (35.1) are always vector sheets of curl $\boldsymbol{c}$, must simultaneously be vector sheets of $\boldsymbol{c}$. Formally,

$$\left.\begin{aligned} \boldsymbol{c} \cdot \operatorname{grad} F &= (\operatorname{grad} H + F \operatorname{grad} G) \cdot \operatorname{grad} F = 0, \\ \boldsymbol{c} \cdot \operatorname{grad} G &= (\operatorname{grad} H + F \operatorname{grad} G) \cdot \operatorname{grad} G = 0. \end{aligned}\right\} \tag{35.4}$$

Conversely, if Eq. (35.4) hold and if F and G be functionally independent, then both $\boldsymbol{c}$ and curl $\boldsymbol{c}$ are perpendicular to grad F and to grad G; hence they are parallel, so $\boldsymbol{c}$ is a screw field.

36. STOKES' potentials. In a finite region $\mathscr{v}$, an arbitrary vector field $\boldsymbol{c}$ has a representation of the form[5]

$$\boldsymbol{c} = -\operatorname{grad} S + \operatorname{curl}\boldsymbol{v} \tag{36.1}$$

and hence is the sum of a lamellar and a solenoidal field. Functions S and $\boldsymbol{v}$ satisfying Eq. (36.1) are called, respectively, a *scalar potential* and a *vector potential* of $\boldsymbol{c}$; together, they are called the ***Stokes' potentials***. An infinite number of potentials correspond to a given field $\boldsymbol{c}$. Let $\boldsymbol{c}$ be piecewise differentiable in a finite region $\mathscr{v}$, bounded by $\mathscr{s}$; within $\mathscr{v}$, let $\boldsymbol{c}$ be continuous except upon a surface $\mathscr{s}'$, on each side of which it has finite limits; then a pair of potentials for $\boldsymbol{c}$

[1] E.g. VAN DANTZIG [1934, *1*, p. 646].

[2] Cf. HADAMARD [1903, *2*, p. 80].

[3] This result was implied, but not stated explicitly, by MONGE [1787, *1*, §§ XVI–XVIII, XX] and PFAFF [1818, *1*, § 4]. The above derivation is due to HANKEL [1861, *1*, § 11].

[4] [1889, *2*]. See also BJØRGUM [1951, *4*, § 5.1].

[5] STOKES [1851, *2*, Part I, Sect. 1, §§ 3–8]. Proofs are given in works on vector analysis, e.g. PHILLIPS [1933, *1*, § 83]. Four stronger decomposition theorems are given by WEYL [1940, *2*]. Cf. also BLUMENTHAL [1905, *1*].

is given by

$$\left.\begin{aligned} S &= \frac{1}{4\pi}\int_{v} \frac{\operatorname{div} \mathbf{c}}{d}\, dv + \frac{1}{4\pi}\int_{s'} \frac{d\mathbf{a}\cdot[\mathbf{c}]}{d} - \frac{1}{4\pi}\oint_{s} \frac{d\mathbf{a}\cdot\mathbf{c}}{d},\\ \mathbf{v} &= \frac{1}{4\pi}\int_{v} \frac{\operatorname{curl} \mathbf{c}}{d}\, dv + \frac{1}{4\pi}\int_{s'} \frac{d\mathbf{a}\times[\mathbf{c}]}{d} - \frac{1}{4\pi}\oint_{s} \frac{d\mathbf{a}\times\mathbf{c}}{d} \end{aligned}\right\} \tag{36.2}$$

where d is the distance from the point of integration to the point where S and $\mathbf{v}$ are being calculated, and where the bold-face bracket denotes the *jump* of $\mathbf{c}$ across s',

$$[\mathbf{c}] \equiv \mathbf{c}^+ - \mathbf{c}^-, \tag{36.3}$$

$\mathbf{c}^+$ and $\mathbf{c}^-$ being the limiting values of $\mathbf{c}$ on the two sides of s' and the sense of $d\mathbf{a}$ being fixed appropriately in terms of the choice of signs + and − for the two sides of s'. In the case of a region extending to infinity in all directions, if $\mathbf{c} = \bar{o}(p^{-2})$ the formulae (36.2) still hold, providing the integrals over s be omitted. For suitably selected potentials, e.g., those given by Eqs. (36.2), the representation (36.1) is valid globally[1] and $\mathbf{v}$ is solenoidal. We always assume the potentials are so selected.

Since curl $\mathbf{v}$ is solenoidal, we may replace it in Eq. (36.1) by an expression of the form (32.2), so obtaining a local representation

$$\mathbf{c} = -\operatorname{grad} S + \operatorname{grad} F \times \operatorname{grad} G. \tag{36.4}$$

VI. Tensors of order two.

a) Proper numbers and vectors.

37. Definitions and conditions of reality. A *proper number* a of a second order tensor $\boldsymbol{a}$ is a root of the equation

$$\det(a^k{}_m - a\,\delta^k_m) = 0. \tag{37.1}$$

Since in n dimensions Eq. (37.1) is a polynomial equation of degree n in a, there are always n proper numbers, which need not be real or distinct. The *left and right proper vectors of* $\boldsymbol{a}$ *corresponding to the proper number* a are, respectively, the non-zero vectors $\boldsymbol{m}$ and $\boldsymbol{q}$, in general complex, such that

$$m_k(a^k{}_r - a\,\delta^k_r) = 0, \qquad (a^r{}_k - a\,\delta^r_k)\,q^k = 0. \tag{37.2}$$

The directions of the vectors are the *principal directions* of $\boldsymbol{a}$.

A sufficient condition that all proper numbers of $\boldsymbol{a}$ *be real is that*[2]

$$\boldsymbol{a} = \boldsymbol{b}^{-1}\cdot\boldsymbol{c}, \text{ where } \left\{\begin{array}{l} \boldsymbol{b} \text{ is symmetric and positive definite}^3, \\ \boldsymbol{c} \text{ is symmetric.} \end{array}\right\} \tag{37.3}$$

Inded, if $\boldsymbol{a}$ be so expressible, its proper numbers are given by

$$(\det b_{km})\,[\det(a^m{}_r - a\,\delta^m_r) = \det(b_{km}\,a^m{}_r - a\,b_{kr}) = \det(c_{kr} - a\,b_{kr}) = 0; \tag{37.4}$$

[1] Hadamard [1903, 2, p. 80].

[2] An alternative statement is: For some choice of the metric tensor (viz. $g_{km} = b_{km}$), $\boldsymbol{a}$ is symmetric. In general, such a metric is not Euclidean.

[3] Throughout this work we write "the tensor $\boldsymbol{b}$ is positive definite" in place of "the components b_{km} are coefficients of a positive definite quadratic form".

the assertion then follows because each root a of an equation of this type is real[1]. Furthermore, if the roots of Eq. (37.4) are $a_1, \ldots, a_n$, there exist real linearly independent vectors $\boldsymbol{q}_1, \ldots, \boldsymbol{q}_n$ such that

$$c_{km}\, q^m_{\mathfrak{d}} = a_{\mathfrak{d}}\, b_{km}\, q^m_{\mathfrak{d}}, \qquad b_{km}\, q^k_{\mathfrak{d}}\, q^m_{\mathfrak{e}} = \delta_{\mathfrak{d}\mathfrak{e}} \tag{37.5}$$

whence follows

$$a^m{}_k\, q^k_{\mathfrak{d}} = \overset{-1}{b}{}^{mr}\, c_{rk}\, q^k_{\mathfrak{d}} = a_{\mathfrak{d}}\, q^m_{\mathfrak{d}}, \tag{37.6}$$

so $q_{\mathfrak{d}}$ is a right proper vector of $\boldsymbol{a}$ corresponding to the proper number $a_{\mathfrak{d}}$. Defining the vector $\boldsymbol{m}_{\mathfrak{d}}$ by $m_{\mathfrak{d}k} \equiv b_{ks}\, q^s_{\mathfrak{d}}$, from Eq. (37.5) we get

$$a^k{}_r\, m_{\mathfrak{d}k} = \overset{-1}{b}{}^{ks}\, c_{sr}\, b_{ku}\, q^u_{\mathfrak{d}} = a_d\, b_{rs}\, q^s_{\mathfrak{d}} = a_{\mathfrak{d}}\, m_{\mathfrak{d}r}, \tag{37.7}$$

so $\boldsymbol{m}_{\mathfrak{d}}$ is a left proper vector of $\boldsymbol{a}$ corresponding to the proper number $a_{\mathfrak{d}}$.

Were the above condition also necessary, then all proper numbers' being real would ensure the existence of n linearly independent right proper vectors. A counter-example is shown by the matrix

$$\begin{Vmatrix} 0 & 0 & 0 \\ 1 & 0 & 0 \\ 0 & 0 & 1 \end{Vmatrix}, \tag{37.8}$$

which has the real proper numbers 0, 0, 1, but only two linearly independent right proper vectors[2], e.g. (0, 1, 0) and (0, 0, 1). A necessary condition follows: *If the proper numbers $a_1, \ldots, a_n$ of $\boldsymbol{a}$ are all real, and if there exist corresponding right proper vectors $\boldsymbol{q}_1, \ldots, \boldsymbol{q}_n$ forming a linearly independent set*[3], *then* Eq. (37.3) *holds*. To show this, we first note that without loss of generality the $\boldsymbol{q}_{\mathfrak{d}}$ may be assumed real. Since they are linearly independent, there exists a unique real reciprocal set $\boldsymbol{m}_{\mathfrak{e}}$ such that

$$m_{\mathfrak{e}k}\, q^k_{\mathfrak{d}} = \delta_{\mathfrak{e}\mathfrak{d}}, \qquad \sum_{\mathfrak{d}=1}^{n} m_{\mathfrak{d}k}\, q^r_{\mathfrak{d}} = \delta^r_k. \tag{37.9}$$

If we set

$$b_{kr} \equiv \sum_{\mathfrak{d}=1}^{n} m_{\mathfrak{d}k}\, m_{\mathfrak{d}r} \tag{37.10}$$

then $\boldsymbol{b}$ is a symmetric positive definite tensor. Moreover,

$$m_{\mathfrak{d}k} = b_{kr}\, q^r_{\mathfrak{d}}, \qquad q^k_{\mathfrak{d}} = \overset{-1}{b}{}^{kr}\, m_{\mathfrak{d}r}. \tag{37.11}$$

Since the $\boldsymbol{q}_d$ are right proper vectors of $\boldsymbol{a}$, from Eqs. $(37.2)_2$, $(37.9)_2$, and $(37.11)_2$ it follows that

$$\left.\begin{aligned} a^p_k &= a^p{}_r\, \delta^r_k = a^p_r \sum_{\mathfrak{d}=1}^{n} m_{\mathfrak{d}k}\, q^r_{\mathfrak{d}} = \sum_{\mathfrak{d}=1}^{n} a_{\mathfrak{d}}\, m_{\mathfrak{d}k}\, q^p_{\mathfrak{d}} \\ &= \sum_{\mathfrak{d}=1}^{n} m_{\mathfrak{d}k}\, a_{\mathfrak{d}}\, \overset{-1}{b}{}^{pr}\, m_{\mathfrak{d}r} = \overset{-1}{b}{}^{pr}\, c_{rk}, \end{aligned}\right\} \tag{37.12}$$

---

[1] See e.g. COURANT and HILBERT [1931, *2*, p. 32]. The proof that the roots of Eq. $(31.4)_3$ are real is implicit in the work of CAUCHY [1828, *1*], [1829, *1*], who treated explicitly the case $b_{kr} = \delta_{kr}$, thereby showing that the proper numbers of a symmetric tensor are always real. HERMITE [1855, *1*] extended CAUCHY's result to complex matrices, showing that the proper numbers of a Hermitian matrix are always real.

[2] TRUESDELL [1954, *2*, § 22] attributes this example to WHAPLES and gives an example of a matrix having all its proper numbers real and all its proper vector proportional to a single vector. Those familiar with the theory of elementary divisors will see easily that the number of linearly independent proper vectors equals the number of elementary divisors.

[3] Sufficient for this is that the proper numbers all be distinct.

where we have set

$$c_{rk} \equiv \sum_{\mathfrak{d}=1}^{n} a_{\mathfrak{d}}\, m_{\mathfrak{d}\, r}\, m_{\mathfrak{d}\, k}. \tag{37.13}$$

Since $c_{rk} = c_{kr}$, Eq. (37.12) is the desired result.

The tensors $\boldsymbol{b}$ and $\boldsymbol{c}$ are not uniquely determined by $\boldsymbol{a}$. Given any admissible set of proper vectors $\boldsymbol{q}_{\mathfrak{d}}$, we can obtain an infinite number of different admissible sets by multiplying each vector $\boldsymbol{q}_{\mathfrak{d}}$ by an arbitrary non-zero scalar factor. As is easily seen from Eqs. (37.8), (37.10), and (37.13), different sets generally will determine different tensors $\boldsymbol{b}$ and $\boldsymbol{c}$. It follows from Eqs. (37.9) and (37.11) that $\boldsymbol{b}$ as defined by Eqs. (37.10) satisfies (37.5), whatever be the choice of the $\boldsymbol{q}_{\mathfrak{d}}$.

From Eqs. (37.11) and (37.7) we conclude that the reciprocal set of vectors $\boldsymbol{m}_{\mathfrak{e}}$, defined by the condition (37.9), are in fact left proper vectors corresponding to the proper numbers $a_{\mathfrak{e}}$. The intermediate formula $(37.12)_3$ is particularly important in that *it expresses the tensor $\boldsymbol{a}$ uniquely in terms of its proper numbers, supposed real, and a set of its independent right proper vectors, supposed n in number, and left proper vectors so chosen as to form a reciprocal set.* The invariance of this representation under possible different choices of the $\boldsymbol{q}_{\mathfrak{e}}$ is immediate from Eq. (37.9).

It can occur that the vectors $\boldsymbol{q}_{\mathfrak{e}}$ are mutually orthogonal. If they are normalized so as to be unit vectors, Eq. (37.10) implies that $b_{km} = g_{km}$, whence we conclude that $\boldsymbol{a} = \boldsymbol{a}'$. That is, *if all proper numbers of $\boldsymbol{a}$ are real, and if $\boldsymbol{a}$ has n linearly independent mutually orthogonal proper vectors, then $\boldsymbol{a}$ is symmetric*; this theorem is due in principle to KELVIN and TAIT[1]. The example (37.8), interpreted as being referred to rectangular Cartesian co-ordinates, shows that the reality of all proper numbers and orthogonality of a maximal linearly independent set of right proper vectors is insufficient to imply $\boldsymbol{a} = \boldsymbol{a}'$. There must exist n orthogonal proper vectors, which is not the case for the example.

Aside from the result of KELVIN and TAIT, there has been no need to introduce the metric tensor if $\boldsymbol{a}$ is taken as a mixed tensor, $\boldsymbol{b}$ and $\boldsymbol{c}$ as covariant tensors.

For any field $\boldsymbol{a}(\boldsymbol{x})$ such that Eq. (37.3) holds, it is possible to select a co-ordinate system such that at a given point P, the right proper vectors have the components $\delta^k_{\mathfrak{a}}$. Indeed, let the $\boldsymbol{q}_{\mathfrak{d}}(\boldsymbol{x})$ constitute any linearly independent set of right proper vectors, in any co-ordinate system; then we need only select $\boldsymbol{x}^* = \boldsymbol{x}^*(\boldsymbol{x})$ such that at P we have

$$q^k_{\mathfrak{d}}(\boldsymbol{x}^*) = \frac{\partial x^{*k}}{\partial x^m}\, q^m_{\mathfrak{d}}(\boldsymbol{x}) = \delta^k_{\mathfrak{d}}. \tag{37.14}$$

From Eq. $(37.9)_2$ it follows that $m_{\mathfrak{e}k}(\boldsymbol{x}^*) = \delta_{\mathfrak{e}k}$ at P. Hence

$$\frac{\partial x^r}{\partial x^{*k}}\, m_{\mathfrak{e}r}(\boldsymbol{x}) = \delta_{\mathfrak{e}k} \quad \text{at} \quad P. \tag{37.15}$$

From Eqs. $(37.12)_2$, (37.14), and (37.15), at P we derive

$$\left.\begin{aligned} a^k{}_r(\boldsymbol{x}^*) &= \frac{\partial x^{*k}}{\partial x^u}\,\frac{\partial x^s}{\partial x^{*r}}\, a^u{}_s(\boldsymbol{x}) = \sum_{\mathfrak{d}=1}^{n} \frac{\partial x^{*k}}{\partial x^u}\,\frac{\partial x^s}{\partial x^{*r}}\, m_{\mathfrak{d}s}\, q^u_{\mathfrak{d}}\, a^{\mathfrak{s}}_{\mathfrak{d}}, \\ &= \sum_{\mathfrak{d}=1}^{n} a_{\mathfrak{d}}\, \delta^k_{\mathfrak{d}}\, \delta_{\mathfrak{d}r}. \end{aligned}\right\} \tag{37.16}$$

This equation asserts that in the co-ordinate system $\boldsymbol{x}^*$, the matrix $\|a^k{}_r\|$ assumes diagonal form with entries $a_{\mathfrak{d}}$. Conversely, if there is a real, non-singular

---

[1] [1867, *1*, § 183].

co-ordinate transformation which diagonalizes $\|a^k{}_m\|$ at a point, then $\boldsymbol{a}=\boldsymbol{b}^{-1}\cdot\boldsymbol{c}$, where $\boldsymbol{b}$ and $\boldsymbol{c}$ are the tensors having at P the components δ_{km} and $\sum_{\mathfrak{d}=1}^{n} a_{\mathfrak{d}}\,\delta_{\mathfrak{d}k}\,\delta_{\mathfrak{d}m}$, respectively, in the system in which $\|a^k{}_m\|$ is diagonal. Such a co-ordinate system, hereafter called a *principal co-ordinate system*, need not be orthogonal, which means that $\|a^{km}\|$ and $\|a_{km}\|$ need not be diagonalized by the transformation. Combining these results, we have the following characterization: *A necessary and sufficient condition that $\boldsymbol{a}=\boldsymbol{b}^{-1}\cdot\boldsymbol{c}$, where $\boldsymbol{b}$ and $\boldsymbol{c}$ are symmetric and $\boldsymbol{b}$ is positive definite, is that at an arbitrarily selected point P, there exist a real non-singular co-ordinate transformation which diagonalizes the matrix* $\|a^k{}_m\|$. In the language of matrices, this result may be restated as follows: *A necessary and sufficient condition that $\boldsymbol{a}=\boldsymbol{b}^{-1}\cdot\boldsymbol{c}$, where $\boldsymbol{b}$ and $\boldsymbol{c}$ are symmetric and $\boldsymbol{b}$ is positive definite, is that $\boldsymbol{a}$ be similar to a diagonal matrix* modulo *the real matrices*, i.e. that $\boldsymbol{a}=\boldsymbol{f}\cdot\boldsymbol{d}\cdot\boldsymbol{f}^{-1}$, where $\boldsymbol{f}$ is real and non-singular, $\boldsymbol{d}$ real and diagonal. A restatement of the theorems of CAUCHY and KELVIN and TAIT is: *$\boldsymbol{f}$ may be taken as orthogonal*, i.e. $\boldsymbol{f}'=\boldsymbol{f}^{-1}$, *if and only if* $\boldsymbol{a}=\boldsymbol{a}'$.

In a principal co-ordinate system we have

$$a_k{}^m{}_{,p}=\frac{\partial a_k{}^m}{\partial x^p}+\left\{\begin{matrix} m\\ k\,p\end{matrix}\right\}(a_k{}^k-a_m{}^m)\quad\text{(unsummed)};\tag{37.17}$$

when $k=m$, the second term vanishes. In general, this formula holds only at the point where the principal co-ordinate system is defined, and only for the diagonal mixed components do covariant derivatives reduce to ordinary partial derivatives. Even if $\|a^{km}\|$ and $\|a_{km}\|$ also be diagonal matrices, we have $a^{kk}{}_{,p}\neq\partial a^{kk}/\partial x^p$ and $a_{kk,p}\neq\partial a_{kk}/\partial x^p$, in general. When the principal co-ordinate system is real and orthogonal, Eq. (37.17) may be read in terms of physical components:

$$a\langle km,p\rangle=\frac{\partial}{\partial s_p}a\langle km\rangle+\frac{\partial\log\sqrt{g_{mm}}}{\partial s_x}\left(a\langle kk\rangle-a\langle mm\rangle\right),\tag{37.18}$$

where we have used Eq. (25.6).

For the case of most interest here, namely $n=3$, we may make use of the obvious but useful fact that the number of real proper vectors is fixed by the sign of the discriminant of the left member of Eq. (37.1). We may use also the fact that the proper numbers of $\boldsymbol{a}$ are all real if and only if the same be true of $\boldsymbol{a}+m\,\boldsymbol{1}$, where m is any scalar, as follows immediately from Eq. (37.1). If $\boldsymbol{a}$ admits the decomposition (37.3), we have $\boldsymbol{a}+m\boldsymbol{1}=\boldsymbol{b}^{-1}\cdot(\boldsymbol{c}+m\boldsymbol{b})$ as a corresponding decomposition for $\boldsymbol{a}+m\boldsymbol{1}$.

FROBENIUS[1] showed that $\boldsymbol{a}$ has a proper number which is real, positive, simple, and greater in absolute value than any other proper number provided, for some choice of co-ordinates, the numbers $a^k{}_m$ be all positive.

BANG[2] noted that in three dimensions the proper numbers of $\boldsymbol{a}$ are all real if, in some co-ordinate system, $a^1{}_2\,a^2{}_3\,a^3{}_1=a^2{}_1\,a^3{}_2\,a^1{}_3$ and $a^1{}_2\,a^2{}_3/a^1{}_3$, $a^2{}_3\,a^3{}_1/a^2{}_1$, and $a^3{}_1\,a^1{}_2/a^3{}_2$ are of the same sign.

38. Principal and related invariants. The K-th *principal invariant* $\mathrm{I}_{\boldsymbol{a}}^{(K)}$ of a second order tensor $\boldsymbol{a}$ is the K-th elementary symmetric function of the proper numbers of $\boldsymbol{a}$,

$$\mathrm{I}_{\boldsymbol{a}}^{(K)}\equiv\sum_{\mathfrak{b}_1<\cdots<\mathfrak{b}_K}a_{\mathfrak{b}_1}a_{\mathfrak{b}_2}\ldots a_{\mathfrak{b}_K}\tag{38.1}$$

so that

$$(a_{\mathfrak{b}})^n-\mathrm{I}_{\boldsymbol{a}}^{(1)}(a_{\mathfrak{b}})^{n-1}+\cdots+(-1)^n\,\mathrm{I}_{\boldsymbol{a}}^{(n)}=0.\tag{38.2}$$

---

[1] [1908, *1*].

[2] [1893, *3*]. MUIR [1896, *2*] extended BANG's result to n dimensions.

Therefore the principal invariants determine the proper numbers uniquely up to order. When the proper numbers are real, unique determination is effected by the order convention $a_1 \geqq a_2 \geqq \cdots \geqq a_n$, so that, *for values of the* $\mathrm{I}_{\boldsymbol{a}}^{(K)}$ *such that all roots of* Eq. (38.2) *are real, any single-valued function of proper numbers equals a single-valued function of principal invariants.* Comparison of Eq. (38.2) with Eq. (37.1) shows that $\mathrm{I}_{\boldsymbol{a}}^{(K)}$ is the sum of the principal K-rowed minors of the matrix $\|a^k{}_m\|$:

$$\mathrm{I}_{\boldsymbol{a}}^{(K)} = \frac{1}{K!}\,\delta^{r_1\ldots r_K}_{m_1\ldots m_K}\, a^{m_1}{}_{r_1}\ldots a^{m_K}{}_{r_K}. \tag{38.3}$$

For a real tensor $\boldsymbol{a}$ *such that there is a real non-singular co-ordinate transformation which diagonalizes* $\|a^k{}_m\|$, *any single-valued absolute scalar invariant* $f(\boldsymbol{a})$ *under arbitrary co-ordinate transformations is expressible as a single-valued function of principal invariants*[1]. For proof, we evaluate f in a principal co-ordinate system; by Eq. (37.16), f is a function of the $a_{\mathfrak{b}}$, which are real; the assertion follows by the italicized statement in the paragraph preceding. Q.E.D.

This result does not extend to an arbitrary second order tensor $\boldsymbol{a}$, it being necessary to adjoin to the principal invariants WEYR's[2] *characteristics* of $\boldsymbol{a}$ to obtain a complete set. If $\boldsymbol{a}$ can be diagonalized, the WEYR characteristics of $\boldsymbol{a}$ are uniquely determined by the principal invariants. It is true that if f is a scalar polynomial in the $a^k{}_m$, then it is always expressible as a polynomial in the principal invariants[3]. The WEYR characteristics are not such polynomials; in fact they are not even continuous functions of $\boldsymbol{a}$ at $\boldsymbol{a}=0$. Of more interest in mechanics are corresponding results for scalar invariants of both a_{km} and g_{km} under arbitrary co-ordinate transformations, or of a_{km} under orthogonal transformations.

The K-th *moment* $\bar{\mathrm{I}}_{\boldsymbol{a}}^{(K)}$ of $\boldsymbol{a}$ is the sum of K-th powers of the proper numbers of $\boldsymbol{a}$:

$$\bar{\mathrm{I}}_{\boldsymbol{a}}^{(K)} \equiv \sum_{\mathfrak{b}=1}^{n} (a_{\mathfrak{b}})^K = a^{m_1}{}_{m_2}\, a^{m_2}{}_{m_3} \ldots a^{m_K}{}_{m_1}. \tag{38.4}$$

If we set

$$\mathit{M}_{\boldsymbol{a}}^{(K,L)} \equiv (-1)^{K+L} L! \sum \frac{(\mathrm{I}_{\boldsymbol{a}}^{(1)})^{M_1}\,(\mathrm{I}_{\boldsymbol{a}}^{(2)})^{M_2}\ldots(\mathrm{I}_{\boldsymbol{a}}^{(n)})^{M_n}}{M_1!\,M_2!\ldots M_n!} \tag{38.5}$$

where the sum runs over all sets of n non-negative integers $M_{\mathfrak{b}}$ such that $\sum_{\mathfrak{b}=1}^{n} M_{\mathfrak{b}} = L$, $\sum_{\mathfrak{b}=1}^{n} \mathfrak{b}\, M_{\mathfrak{b}} = K$, then an expression for the moments $\bar{\mathrm{I}}_{\boldsymbol{a}}^{(K)}$ in terms of the principal invariants $\mathrm{I}_{\boldsymbol{a}}^{(K)}$ is[4]

$$\bar{\mathrm{I}}_{\boldsymbol{a}}^{(K)} = K \sum_{L=1}^{K} \frac{1}{L}\, \mathit{M}_{\boldsymbol{a}}^{(K,L)}. \tag{38.6}$$

Except where otherwise noted, we henceforth assume $n=3$ and write[5] $\mathrm{I}_{\boldsymbol{a}}$, $\mathrm{II}_{\boldsymbol{a}}$, $\mathrm{III}_{\boldsymbol{a}}$, $\bar{\mathrm{I}}_{\boldsymbol{a}}$, $\overline{\mathrm{II}}_{\boldsymbol{a}}$, $\overline{\mathrm{III}}_{\boldsymbol{a}}$ in place of $\mathrm{I}_{\boldsymbol{a}}^{(1)}$, $\mathrm{II}_{\boldsymbol{a}}^{(2)}$, $\mathrm{I}_{\boldsymbol{a}}^{(3)}$, $\bar{\mathrm{I}}_{\boldsymbol{a}}^{(1)}$, $\bar{\mathrm{I}}_{\boldsymbol{a}}^{(2)}$, $\mathrm{I}_{\boldsymbol{a}}^{(3)}$, respectively. In this notation, a is a proper number of $\boldsymbol{a}$ if and only if

$$a^3 - \mathrm{I}_{\boldsymbol{a}}\, a^2 + \mathrm{II}_{\boldsymbol{a}}\, a - \mathrm{III}_{\boldsymbol{a}} = 0; \tag{38.7}$$

in particular, $a=1$ is a proper number if and only if

$$\mathrm{I}_{\boldsymbol{a}} - \mathrm{II}_{\boldsymbol{a}} + \mathrm{III}_{\boldsymbol{a}} = 1. \tag{38.8}$$

---

[1] For the symmetric case, RANKINE [1856, 2, § 3] refers to this result as a discovery of CAYLEY.

[2] [1885, 1] and [1890, 1]. See also MACDUFFEE [1933, 2, § 40].

[3] For $n=3$ this follows immediately from a result due to WEITZENBÖCK [1923, 1, pp. 65–66]. See also TRUESDELL [1952, 2, p. 132], [1953, 4, p. 594].

[4] BURNSIDE and PANTON [1901, 5, § 159] attribute this result to WARING.

[5] In a space whose co-ordinates are $\mathrm{I}_{\boldsymbol{a}}$, $\mathrm{II}_{\boldsymbol{a}}$, $\mathrm{III}_{\boldsymbol{a}}$, BORDONI [1955, 4] discusses the surface where the discriminant of (38.7) is constant.

From Eq. (38.6) we have

$$\left.\begin{aligned} \overline{\mathrm{II}}_{\boldsymbol a} &= \mathrm{I}_{\boldsymbol a}^2 - 2\,\mathrm{II}_{\boldsymbol a}, & \overline{\mathrm{III}}_{\boldsymbol a} &= \mathrm{I}_{\boldsymbol a}^3 - 3\,\mathrm{I}_{\boldsymbol a}\,\mathrm{II}_{\boldsymbol a} + 3\,\mathrm{III}_{\boldsymbol a}, \\ 2\,\mathrm{II}_{\boldsymbol a} &= \mathrm{I}_{\boldsymbol a}^2 - \overline{\mathrm{II}}_{\boldsymbol a}, & \mathrm{III}_{\boldsymbol a} &= \tfrac{1}{6}\,\mathrm{I}_{\boldsymbol a}^3 - \tfrac{1}{2}\,\mathrm{I}_{\boldsymbol a}\,\overline{\mathrm{II}}_{\boldsymbol a} + \tfrac{1}{3}\,\overline{\mathrm{III}}_{\boldsymbol a}. \end{aligned}\right\} \tag{38.9}$$

The inverse of $\boldsymbol a$, which exists when $\mathrm{III}_{\boldsymbol a} \neq 0$, is given explicitly by

$$\mathrm{III}_{\boldsymbol a}\,\boldsymbol a^{-1} = \boldsymbol a^2 - \mathrm{I}_{\boldsymbol a}\,\boldsymbol a + \mathrm{II}_{\boldsymbol a}\,\mathbf 1. \tag{38.10}$$

The *octahedral invariant*[1] $\amalg_{\boldsymbol a}$ is defined by

$$\left.\begin{aligned} \amalg_{\boldsymbol a} &\equiv \sum_{\mathfrak b<\mathfrak c} [\tfrac{1}{2}(a_{\mathfrak b} - a_{\mathfrak c})]^2 = \tfrac{1}{2}(\overline{\mathrm{II}}_{\boldsymbol a} - \mathrm{II}_{\boldsymbol a}), \\ &= \tfrac{1}{2}(\mathrm{I}_{\boldsymbol a}{}^2 - 3\,\mathrm{II}_{\boldsymbol a}) = \tfrac{1}{4}(3\,\overline{\mathrm{II}}_{\boldsymbol a} - \mathrm{I}_{\boldsymbol a}{}^2); \end{aligned}\right\} \tag{38.11}$$

it satisfies the identity $\amalg_{\boldsymbol a + m\mathbf 1} = \amalg_{\boldsymbol a}$, for an arbitrary scalar m. The *deviator*[2] ${}_0\boldsymbol a$ of $\boldsymbol a$ is its traceless part:

$${}_0a^k{}_m \equiv a^k{}_m - \tfrac{1}{3}\,I_{\boldsymbol a}\,\delta^k_m. \tag{38.12}$$

Hence $\boldsymbol a = {}_0\boldsymbol a$ if and only if $\mathrm{I}_{\boldsymbol a} = 0$; moreover, ${}_0\boldsymbol a = 0$ if and only if $\boldsymbol a$ is a scalar multiple of $\mathbf 1$, or, as we shall sometimes say, if $\boldsymbol a$ is *spherical*. For $\boldsymbol a$ to be spherical it is necessary and sufficient that it be symmetric and that all its proper numbers be equal. We note also that

$$\overline{\mathrm{II}}_{{}_0\boldsymbol a} = \overline{\mathrm{II}}_{\boldsymbol a} - \tfrac{1}{3}\,\mathrm{I}_{\boldsymbol a}^2. \tag{38.13}$$

From Eqs. (38.12) and (38.11) follows

$$\amalg_{\boldsymbol a} = \amalg_{{}_0\boldsymbol a} = \tfrac{3}{4}\,\overline{\mathrm{II}}_{{}_0\boldsymbol a}. \tag{38.14}$$

When the proper numbers of $\boldsymbol a$ are real, we have $\amalg_{\boldsymbol a} \geqq 0$ and $\overline{\mathrm{II}}_{\boldsymbol a} \geqq 0$ with equality holding in the former case if and only if all proper numbers be equal; in the latter, if and only if all be zero. The quantity $(\overline{\mathrm{II}}_{\boldsymbol a})^{\frac{1}{2}}$ is the *intensity* of $\boldsymbol a$. For symmetric[3] $\boldsymbol a$, *the intensity vanishes if and only if $\boldsymbol a$ vanishes*, whence it follows from Eq. (38.13) that $\amalg_{\boldsymbol a}$ *vanishes if and only if $\boldsymbol a$ be spherical.*

From Eq. (3.3) we find that the squared magnitude of the cross of $\boldsymbol a$ is given by[4]

$$\left.\begin{aligned} \tfrac{1}{2}\,a_{\times}^2 &= \overline{\mathrm{II}}_{{}_s\boldsymbol a} - \overline{\mathrm{II}}_{\boldsymbol a} = -\,\overline{\mathrm{II}}_{{}_s\boldsymbol a} + \mathrm{I}_{\boldsymbol a}^2 - 2\,\mathrm{II}_{{}_s\boldsymbol a}, \\ &= 2\,(\mathrm{II}_{\boldsymbol a} - \mathrm{II}_{{}_s\boldsymbol a}), \end{aligned}\right\} \tag{38.15}$$

where ${}_s\boldsymbol a \equiv \tfrac{1}{2}(\boldsymbol a + \boldsymbol a')$.

We have[5]

$$\left.\begin{aligned} &\frac{\partial \mathrm{I}_{\boldsymbol a}}{\partial a^k{}_m} = \delta^m_k, \quad \frac{\partial \mathrm{II}_{\boldsymbol a}}{\partial a^k{}_m} = \mathrm{I}_{\boldsymbol a}\,\delta^m_k - a^m{}_k, \quad \frac{\partial \overline{\mathrm{II}}_{\boldsymbol a}}{\partial a^k{}_m} = 2\,a^m{}_k, \\ &\frac{\partial \mathrm{III}_{\boldsymbol a}}{\partial a^k{}_m} = a^m{}_r\,a^r{}_k - \mathrm{I}_{\boldsymbol a}\,a^m{}_k + \mathrm{II}_{\boldsymbol a}\,\delta^m_k = \mathrm{III}_{\boldsymbol a}\,\overset{-1}{a}{}^m{}_k, \quad \frac{\partial \overline{\mathrm{III}}_{\boldsymbol a}}{\partial a^k{}_m} = 3\,a^m{}_r\,a^r{}_k, \end{aligned}\right\} \tag{38.16}$$

---

[1] The invariance and significance of $\amalg_{\boldsymbol a}$ were known to MAXWELL in 1856 [1937, *10*, pp. 32–38]; it was introduced by v. MISES [1913, *2*, § 1]. The name "octahedral invariant" is usually attached to $(2/\sqrt{3})\,\amalg_{\boldsymbol a}^{\frac{1}{2}}$ on the basis of a geometrical interpretation given by NADAI and LODE [1933, *3*, § II], [1937, *2*, pp. 206–207].

[2] KLEITZ [1873, *3*, § 23] was the first to make an explicit study of the deviator.

[3] More generally, for any $\boldsymbol a$ that can be diagonalized by real transformations.

[4] LIPSCHITZ [1875, *1*]; HAMEL [1936, *1*, § 1].

[5] For the symmetric case, these results are given by MURNAGHAN [1937, *3*, § 3], SIGNORINI [1943, *1*, § 17], and REINER [1945, *3*, § 4].

where Eq. $(38.16)_5$, which follows from Eq. $(38.16)_4$ by Eq. (38.10), holds only when $\boldsymbol{a}^{-1}$ exists. Also[1], when $\boldsymbol{a}^{-1}$ exists, we have $d\boldsymbol{a} = -\boldsymbol{a} \cdot d\boldsymbol{a}^{-1} \cdot \boldsymbol{a}$ and hence

$$\frac{\partial}{\partial \overset{-1}{a^k}_m} = -a^r{}_k a^m{}_s \frac{\partial}{\partial a^r{}_s}. \tag{38.17}$$

39. Inequalities[2]. CAUCHY[3] was first to establish bounds for the proper numbers of matrices. Let the proper numbers of a real symmetric three-dimensional matrix be ordered so that $a_1 \geqq a_2 \geqq a_3$. Then, if $\boldsymbol{a}$ be referred to a rectangular Cartesian co-ordinate system, CAUCHY's results are that a_1 is never less than, while a_3 is never greater than any one of the six quantities

$$\tfrac{1}{2}\left[a_{kk} + a_{mm} \pm \sqrt{(a_{kk} - a_{mm})^2 + 4a_{km}^2}\right] \qquad (k \neq m) \tag{39.1}$$

and that, for each k and m, a_2 lies between the two numbers given by the expressions (39.1). Of course, one may interpret the components in (39.1) as physical components of $\boldsymbol{a}$ in an arbitrary orthogonal co-ordinate system.

Let $\boldsymbol{a}$ be any second order tensor, real or complex, in n dimensions, set $\boldsymbol{b} \equiv \frac{1}{2}(\boldsymbol{a} + \bar{\boldsymbol{a}}')$, $\boldsymbol{c} \equiv \frac{1}{2}(\boldsymbol{a} - \bar{\boldsymbol{a}}')$, where the bar denotes the complex conjugate; select any proper number a of $\boldsymbol{a}$, and write it in the form $a = p + iq$, where p and q are real. Refer $\boldsymbol{a}$ to any rectangular Cartesian co-ordinate system and let A, B, and C be the maxima of the absolute values of the components of $\boldsymbol{a}$, $\boldsymbol{b}$, and $\boldsymbol{c}$, respectively. We then have HIRSCH's inequalities[4]

$$|a| \leqq nA, \qquad |p| \leqq nB, \qquad |q| \leqq nC. \tag{39.2}$$

One corollary is HERMITE's theorem[5]: *A sufficient condition that the proper numbers of $\boldsymbol{a}$ all be real is that $\boldsymbol{a}$ be Hermitian,* i.e. *that* $\boldsymbol{c} = 0$. Another is WEIERSTRASS' theorem[6]: *A sufficient condition that the proper numbers of $\boldsymbol{a}$ all be pure imaginary is that $\boldsymbol{a}$ be skew-Hermitian,* i.e., that $\boldsymbol{b} = 0$. HIRSCH[7] showed also that when $\boldsymbol{b}$ is real, $q \leqq [\frac{1}{2}(n-1)]^{\frac{1}{2}} C$, and that, if the proper numbers of $\boldsymbol{b}$, which in this case HERMITE's theorem shows to be real, be ordered so that $b_1 \geqq \cdots \geqq b_n$, then $b_1 \geqq p \geqq b_n$. When the proper numbers of the Hermitian tensor $\boldsymbol{d} = \boldsymbol{a} \cdot \bar{\boldsymbol{a}}'$ are similarly ordered, BROWNE's inequality[8] asserts that any proper number a of $\boldsymbol{a}$ satisfies $d_1 \geqq a\bar{a} \geqq d_n$.

We again restrict attention to the case when $n = 3$ and $\boldsymbol{a}$ is real. From Eq. (38.15) follows

$$\overline{\mathrm{II}}_{\boldsymbol{a}} \leqq \overline{\mathrm{II}}_{s\boldsymbol{a}}, \qquad \mathrm{II}_{s\boldsymbol{a}} \leqq \mathrm{II}_{\boldsymbol{a}}, \tag{39.3}$$

the equality holding if and only if $\boldsymbol{a}$ be symmetric. Further inequalities follow from the observation that since complex proper numbers appear in conjugate pairs, all symmetric functions of them are real, so that an inequality may be inferred from every identity in which the square of such a function occurs. From (38.15) we thus obtain

$$\overline{\mathrm{II}}_{\boldsymbol{a}} \leqq \overline{\mathrm{II}}_{s\boldsymbol{a}}, \tag{39.4}$$

---

[1] SIGNORINI [1943, *1*, § 17].

[2] Further discussion of results in this section is given by MACDUFFEE [1933, *2*, § 18].

[3] [1828, *1*], [1830, *1*, Th. 1]; in [1829, *1*, Th. 1] he extended the result to n dimensions.

[4] [1901, *2*].

[5] [1855, *1*]. For real $\boldsymbol{a}$, this theorem was proved earlier by CAUCHY [1828, *1*], [1829, *1*].

[6] [1879, *2*]. For real $\boldsymbol{a}$, the theorem was proved earlier by CLEBSCH [1863, *2*].

[7] [1901, *2*]. For real $\boldsymbol{a}$, the results are due to BENDIXSON [1901, *3*].

[8] [1928, *1*, Th. V].

with equality holding if and only if $\boldsymbol{a}$ be symmetric. From Eqs. (38.9)$_1$ and (38.11) it follows that

$$\overline{\mathrm{II}}_{\boldsymbol{a}} \geqq -2\,\mathrm{II}_{\boldsymbol{a}}, \qquad 3\,\overline{\mathrm{II}}_{\boldsymbol{a}} \geqq 4\,\amalg_{\boldsymbol{a}} \geqq -6\,\mathrm{II}_{\boldsymbol{a}}; \tag{39.5}$$

From Eq. (38.13) we get

$$\overline{\mathrm{II}}_{{}_0\boldsymbol{a}} \leqq \overline{\mathrm{II}}_{\boldsymbol{a}}. \tag{39.6}$$

In all cases of Ineqs. (39.5) and (39.6), equality holds if and only if $\mathrm{I}_{\boldsymbol{a}}=0$.

When we add the condition that all proper numbers of $\boldsymbol{a}$ be real, as is the case when $\boldsymbol{a}$ is symmetric, from the fact that then $\overline{\mathrm{II}}_{\boldsymbol{a}} \geqq 0$ with equality holding if and only if $\boldsymbol{a}=0$, a further sequence of inequalities may be inferred. From Eqs. (38.9) and (38.13) we thus get

$$3\,\overline{\mathrm{II}}_{\boldsymbol{a}} \geqq \mathrm{I}_{\boldsymbol{a}}^2 \geqq 2\,\mathrm{II}_{\boldsymbol{a}}, \tag{39.7}$$

where equality holds on the left if and only if $\boldsymbol{a}$ is spherical; on the right, if and only if $\boldsymbol{a}=0$. From Eq. (38.11) follows

$$\amalg_{\boldsymbol{a}} \geqq -\tfrac{3}{2}\,\mathrm{II}_{\boldsymbol{a}}, \tag{39.8}$$

where equality holding if and only if $\boldsymbol{a}=0$. Also from Eq. (38.11), since $\amalg_{\boldsymbol{a}} \geqq 0$, we derive

$$\mathrm{I}_{\boldsymbol{a}}^2 \geqq 3\,\mathrm{II}_{\boldsymbol{a}}, \tag{39.9}$$

with equality holding if and only if $\boldsymbol{a}$ is spherical.

When all proper numbers are non-negative, we have obviously[1]

$$\mathrm{I}_{\boldsymbol{a}} \geqq 0, \qquad \mathrm{II}_{\boldsymbol{a}} \geqq 0, \qquad \mathrm{I}_{\boldsymbol{a}}^3 \geqq 27\,\mathrm{III}_{\boldsymbol{a}} \geqq 0. \tag{39.10}$$

For a real symmetric tensor referred to rectangular Cartesian co-ordinates, from Eq. (38.4)$_2$ we see that

$$(a_{km})^2 \leqq \overline{\mathrm{II}}_{\boldsymbol{a}}. \tag{39.11}$$

There is also HADAMARD's inequality[2]

$$\mathrm{III}_{\boldsymbol{a}}^2 \leqq \prod_{k=1}^{3} \sum_{m=1}^{3} (a_{km})^2. \tag{39.12}$$

If $|\mathrm{I}_{\boldsymbol{a}}| \leqq K$ and $|\mathrm{II}_{\boldsymbol{a}}| \leqq K$, Eq. (38.9)$_1$ and Ineq. (39.11) imply that for a symmetric tensor we have

$$(a_{km})^2 \leqq K^2 + 2K. \tag{39.13}$$

In Ineqs. (39.11) to (39.13), a_{km} can be interpreted as a physical component of $\boldsymbol{a}$ in any orthogonal co-ordinate system. In any co-ordinate system, we have

$$|\overline{\mathrm{II}}_{\boldsymbol{a}}| = |a^k{}_m\, a^m{}_k| \leqq 9M^2, \tag{39.14}$$

where M is the maximum of the absolute values of the mixed components of $\boldsymbol{a}$.

Results of WEDDERBURN[3] imply for any two symmetric tensors $\boldsymbol{a}$ and $\boldsymbol{b}$ the inequalities

$$\overline{\mathrm{II}}_{\boldsymbol{a}+\boldsymbol{b}}^{\frac{1}{2}} \leqq \overline{\mathrm{II}}_{\boldsymbol{a}}^{\frac{1}{2}} + \overline{\mathrm{II}}_{\boldsymbol{b}}^{\frac{1}{2}}, \qquad \overline{\mathrm{II}}_{\boldsymbol{a}\boldsymbol{b}} \leqq \overline{\mathrm{II}}_{\boldsymbol{a}}^{\frac{1}{2}}\,\overline{\mathrm{II}}_{\boldsymbol{b}}^{\frac{1}{2}}. \tag{39.15}$$

---

[1] The third inequality, for the symmetric case, was given in more complicated form by SIGNORINI [1949, 7, Chap. II, § 5].

[2] [1893, 2], SCHUR [1909, 2] gives an extension of this inequality as well as several others. Cf. also MUIR [1930, 2, Chap. I (a)].

[3] [1925, E2].

b) Powers and matrix polynomials.

40. Integral powers. In this Part, the dimension of the underlying space is arbitrary. When K is an integer, the K-th *power* of a tensor $\boldsymbol{a}$ is defined inductively by[1]

$$\boldsymbol{a}^K \equiv \boldsymbol{a}^{K-1} \cdot \boldsymbol{a} = \boldsymbol{a} \cdot \boldsymbol{a}^{K-1}, \qquad \boldsymbol{a}^0 \equiv \mathbf{1} \tag{40.1}$$

where K is restricted to be positive when $\boldsymbol{a}$ has no inverse. It follows from Eq. (37.2) that

$$\boldsymbol{m} \cdot (\boldsymbol{a}^K - a^K \mathbf{1}) = (\boldsymbol{a}^K - a^K \mathbf{1}) \cdot \boldsymbol{q} = 0; \tag{40.2}$$

hence *the* K-th *power of any proper number of* $\boldsymbol{a}$ *is a proper number of* $\boldsymbol{a}^K$, *and every right (left) proper vector of* $\boldsymbol{a}$ *is a right (left) proper vector of* $\boldsymbol{a}^K$. To obtain a sharper result, we note the following identities, valid for any scalar m and any tensor $\boldsymbol{a}$:

$$\boldsymbol{a}^K - m\,\mathbf{1} = (\boldsymbol{a} - m_1 \mathbf{1}) \ldots (\boldsymbol{a} - m_K \mathbf{1}) \quad \text{when} \quad K > 0, \tag{40.3}$$

$$\boldsymbol{a}^K - m\,\mathbf{1} = (\boldsymbol{a}^{-1} - m_1 \mathbf{1}) \ldots (\boldsymbol{a}^{-1} - m_K \mathbf{1}) \quad \text{when} \quad K < 0, \tag{40.4}$$

where $m_1, \ldots, m_K$ are the $|K|$ complex $|K|$-th roots of m. Taking the determinant of both sides of Eqs. (40.3) and (40.4) shows that m *is a proper number of* $\boldsymbol{a}^K$ *if and only if at least one* $|K|$-*th root of* m *is a proper number*[2] *of* $\boldsymbol{a}$ *when* $K>0$, *or of* $\boldsymbol{a}^{-1}$ *when* $K<0$. The proper numbers of $\boldsymbol{a}^{-1}$ are the reciprocals[3] of those of $\boldsymbol{a}$.

It can occur for some K that there exist proper vectors of $\boldsymbol{a}^K$ which are not proper vectors of $\boldsymbol{a}$. E.g., if $\|a^K{}_m\| = \left\|\begin{matrix}1 & 0\\ 0 & -1\end{matrix}\right\|$, then $\boldsymbol{a}^2 = \mathbf{1}$, so that an arbitrary vector is a proper vector of $\boldsymbol{a}^2$, while the only right proper vectors of $\boldsymbol{a}$ are $(b, 0)$ and $(0, c)$.

By the Hamilton-Cayley theorem, we may replace $(a_{\mathfrak{b}})^K$ by $\boldsymbol{a}^K$ in Eq. (38.2), obtaining

$$\boldsymbol{a}^n - \mathrm{I}_{\boldsymbol{a}}^{(1)}\, \boldsymbol{a}^{n-1} + \cdots + (-1)^n\, \mathrm{I}_{\boldsymbol{a}}^{(n)}\, \mathbf{1} = 0. \tag{40.5}$$

This formula makes it possible to write any power of $\boldsymbol{a}$ as *a linear combination of* $\mathbf{1}, \boldsymbol{a}, \ldots, \boldsymbol{a}^{n-1}$, *with scalar coefficients that are polynomials in the principal invariants if the power is positive, rational functions if it is negative.* For positive K, RANUM[4] gave the formula

$$\boldsymbol{a}^{n+K} = \sum_{L=1}^{N} \text{И}_{\boldsymbol{a}}^{(K,L)}\, \boldsymbol{a}^{\{L\}}, \tag{40.6}$$

where $\text{И}_{\boldsymbol{a}}^{(K,L)}$ is defined by Eq. (38.5), and where

$$\boldsymbol{a}^{\{L\}} \equiv \sum_{Q=0}^{n-L} (-1)^{n-Q+1}\, \mathrm{I}_{\boldsymbol{a}}^{(n-Q)}\, \boldsymbol{a}^{Q+L-1}. \tag{40.7}$$

41. Real powers of positive tensors. We call a tensor $a^k{}_m$ *positive* when its proper numbers are real and positive and it possesses n linearly independent proper vectors. In particular, a positive definite (symmetric) tensor is positive. For a positive tensor we have the representation $(37.12)_3$:

$$a^r{}_s = \sum_{\mathfrak{b}=1}^{n} a_{\mathfrak{b}}\, q^r_{\mathfrak{b}}\, m_{\mathfrak{b}s}, \tag{41.1}$$

---

[1] STICKELBURGER [1881, *4*] was first to define general powers of matrices.

[2] BORCHARDT [1846, *1*], [1847, *1*], when $K>0$.

[3] This was known to SPOTTISWOODE [1856, *1*]; the result then follows for all $K \neq 0$.

[4] [1911, *1*].

where $a_{\mathfrak{b}} > 0$ and where the $\boldsymbol{q}_{\mathfrak{b}}$ and $\boldsymbol{m}_{\mathfrak{b}}$ are reciprocal sets of right and left proper vectors. For any real K, the K-th *power* of a positive tensor $\boldsymbol{a}$ is defined to be the positive tensor $\boldsymbol{a}^K$ whose components $\overset{K}{a}{}^r_s$ are given by

$$\overset{K}{a}{}^r_s = \sum_{\mathfrak{b}=1}^{n} (a_{\mathfrak{b}})^K q^r_{\mathfrak{b}} m_{\mathfrak{b}s}, \tag{41.2}$$

wherein $(a_{\mathfrak{b}})^K$ is the positive real K-th power of $a_{\mathfrak{b}}$. That is, $\boldsymbol{a}^K$ is the unique tensor having the same proper vectors as $\boldsymbol{a}$ and having as its proper numbers the positive K-th powers of the proper numbers of $\boldsymbol{a}$. This definition is equivalent to Eq. (40.1) when both are applicable. The usual laws of exponents apply to $\boldsymbol{a}^K$ as defined by Eq. (41.2). For example, by Eq. $(37.9)_1$ we obtain

$$\sum_{\mathfrak{b}=1}^{n} (a_{\mathfrak{b}})^{K+L} q^r_{\mathfrak{b}} m_{\mathfrak{b}s} = \sum_{\mathfrak{b},\mathfrak{c}=1}^{n} (a_{\mathfrak{b}})^K q^r_{\mathfrak{b}} m_{\mathfrak{b}u} (a_{\mathfrak{c}})^L q^u_{\mathfrak{c}} m_{\mathfrak{c}s}, \tag{41.3}$$

so that $\boldsymbol{a}^{K+L} = \boldsymbol{a}^K \cdot \boldsymbol{a}^L$.

For a *non-negative* tensor, i.e., a tensor having n linearly independent proper vectors and real proper numbers which are positive or zero[1], Eq. (41.2) serves to define a unique K-th power when $K \geqq 0$.

If we accept the possibility that $\boldsymbol{a}^K$ may be complex and multivalued, by using different determinations of the K-th powers of the several proper numbers in Eq. (41.2) we may define various K-th powers of any matrix having a representation of the form (41.1). Such a definition is unsatisfactory in two respects. First, once one admits the possibility that $\boldsymbol{a}^K$ be multivalued, it seems preferable that $\boldsymbol{a}^{1/M}$ should represent any solution $\boldsymbol{x}$ of

$$\boldsymbol{x}^M = \boldsymbol{a} \tag{41.4}$$

when M is an integer, whereas the definition just mentioned excludes some solutions. Second, there is little or no motivation for excluding tensors not representable in the form (41.1). A more satisfactory definition of $\boldsymbol{a}^K$ is easily obtained by regarding it as a complex multivalued function of $\boldsymbol{a}$ in the sense of CIPOLLA[2]. We omit the details, as the definitions already given are adequate for this appendix. Several writers, beginning with CAYLEY[3], have studied the solution of Eq. (41.4).

42. Matrix functions. A matrix $\boldsymbol{b}$ given by

$$\boldsymbol{b} = \boldsymbol{b}(\boldsymbol{a}) = c_0 \boldsymbol{1} + \cdots + c_K \boldsymbol{a}^K, \tag{42.1}$$

where the $c_{\mathfrak{G}}$ are scalar constants, is a *matrix polynomial* in the variable matrix $\boldsymbol{a}$. Matrix polynomials may be set into one-to-one correspondence with the polynomials in a scalar variable x which are defined by the same set of constants, and the algebra of matrix polynomials is isomorphic to the algebra of polynomials in a scalar indeterminate.

Similarly, an infinite set of constants $c_{\mathfrak{G}}$ defines a formal *matrix power series* in $\boldsymbol{a}$, corresponding to the formal scalar power series $b(x)$ determined by the same coefficients. *The matrix power series thus obtained converges if and only if every proper number of $\boldsymbol{a}$ lies inside or on the circle of convergence of the scalar series*

---

[1] For real tensors, this is a slight generalization of the usual definition of non-negative tensors or linear transformations. See e.g. HALMOS [1942, *1*, § 56].

[2] [1932, *2*]. See also MACDUFFEE [1933, *2*, § 50].

[3] [1858, *2*], [1872, *1*]. WEITZENBÖCK [1932, *3*] gave a method for determining all solutions. AUTONNE [1902, *1*], [1903, *1*] showed that in the complex field, if $\boldsymbol{a}$ is non-negative and Hermitian, then there is a unique solution which is non-negative and Hermitian. A similar analysis shows that if $\boldsymbol{a}$ is non-negative according to the definition given above, then there is a unique solution which is non-negative. For further references, see MACDUFFEE [1933, *2*, §§ 48, 50] and WEDDERBURN [1934, *2*, p. 171].

$b(x)$ and, for every proper number a of multiplicity m, the power series for the formal $m-1^{\text{st}}$ derivative $b^{(m-1)}(a)$ converges[1].

Each proper number of a matrix polynomial $\boldsymbol{b}(\boldsymbol{a})$ is a function of a single proper number of $\boldsymbol{a}$, as follows from a more general and more explicit theorem of FROBENIUS: *Let $r(x, \ldots, y) = p(x, \ldots, y)/q(x, \ldots, y)$ be a rational function of the scalar indeterminates $x, \ldots, y$, p and q being polynomials; let $\boldsymbol{a}, \ldots, \boldsymbol{b}$ be commutative matrices such that the matrix $\boldsymbol{q}(\boldsymbol{a}, \ldots, \boldsymbol{b})$ is non-singular; then the proper numbers $a_{\mathfrak{b}}, \ldots, b_{\mathfrak{b}}$ of $\boldsymbol{a}, \ldots, \boldsymbol{b}$ can be ordered so that $r(a_1, \ldots, b_1), \ldots, r(a_n, \ldots, b_n)$ are the proper numbers of $\boldsymbol{r} = \boldsymbol{p}(\boldsymbol{a}, \ldots, \boldsymbol{b}) \cdot \boldsymbol{q}^{-1}(\boldsymbol{a}, \ldots, \boldsymbol{b}) = \boldsymbol{q}^{-1} \cdot \boldsymbol{p}$, the ordering being the same for all rational functions*[2]. Loosely related to this is SYLVESTER's assertion that, whether or not $\boldsymbol{a}$ and $\boldsymbol{b}$ commute, *the proper numbers* of $\boldsymbol{a} \cdot \boldsymbol{b}$ and $\boldsymbol{b} \cdot \boldsymbol{a}$ coincide[3].

RANUM's equation (40.6) serves to reduce any matrix polynomial or power series to the form

$$\boldsymbol{b}(\boldsymbol{a}) = d_0 \mathbf{1} + \cdots + d_{n-1} \boldsymbol{a}^{n-1}, \tag{42.2}$$

where the coefficients $d_{\mathfrak{G}}$ are, respectively, polynomials or power series[4] in the principal invariants of $\boldsymbol{a}$. There are many functions representable in this form that are neither matrix polynomials nor matrix power series: for example, $\boldsymbol{b}(\boldsymbol{a}) = I_{\boldsymbol{a}}^{(1)} \mathbf{1}$. Any function of the form (42.2) satisfies

$$\boldsymbol{b}(\boldsymbol{f} \cdot \boldsymbol{a} \cdot \boldsymbol{f}^{-1}) = \boldsymbol{f} \cdot \boldsymbol{b}(\boldsymbol{a}) \cdot \boldsymbol{f}^{-1}, \tag{42.3}$$

where $\boldsymbol{f}$ is an arbitrary non-singular matrix. Conversely, if each component of $\boldsymbol{b}$ be a polynomial in the components of $\boldsymbol{a}$ and if $\boldsymbol{b}(\boldsymbol{a})$ satisfy Eq. (42.3) for arbitrary non-singular $\boldsymbol{f}$, then $\boldsymbol{b}$ is representable[5] in the form (42.2) with coefficients $d_{\mathfrak{G}}$ which are polynomials in the principal invariants of $\boldsymbol{a}$. DIRAC[6] proposed Eq. (42.3), with $\boldsymbol{a}$, $\boldsymbol{b}$, and $\boldsymbol{f}$ interpreted as elements of any algebra[7], as a part of the definition of $\boldsymbol{b}$'s being a function of $\boldsymbol{a}$. He noted that Eq. (42.3) implies that $\boldsymbol{f}$ commutes with $\boldsymbol{b}$ whenever $\boldsymbol{f}$ commutes with $\boldsymbol{a}$. TURNBULL and AITKEN[8] showed that if $\boldsymbol{a}$ and $\boldsymbol{b}$ are complex $n \times n$ matrices and if $\boldsymbol{b}$ commutes with every complex matrix that commutes with $\boldsymbol{a}$, then $\boldsymbol{b}$ is expressible as a linear combination of $\mathbf{1}, \boldsymbol{a}, \ldots, \boldsymbol{a}^{n-1}$. An analogous result for matrices over the real field is readily established. Consequently Eq. (42.3) implies that $\boldsymbol{b}$ is a linear combination of $\mathbf{1}, \boldsymbol{a}, \ldots, \boldsymbol{a}^{n-1}$ with coefficients depending on $\boldsymbol{a}$ in such a way as to be scalars under arbitrary symmetry transformations. Under reasonably general conditions, these scalar coefficients are expressible as functions of the principal invariants of $\boldsymbol{a}$, as follows from the theorems given at the beginning of Sect. 38.

---

[1] HENSEL [1926, *1*]. WEYR [1887, *2*] had previously treated the case where no proper number of $\boldsymbol{a}$ lies on the circle of convergence. PHILLIPS [1919, *2*] gave sufficient conditions for the convergence of a matrix power series in any finite set of commuting matrices.

[2] According to MACDUFFEE [1933, *2*, p. 23], BROMWICH [1901, *4*] noted that this theorem may fail to hold when $r(x, \ldots, y)$ is not rational. We find no explicit statement to this effect in BROMWICH's paper, though it may follow from results given in his § 3.

[3] SYLVESTER [1883, *2*] stated this without proof. MACDUFFEE [1933, *2*, p. 23] gives an elegant proof.

[4] Expressions of the type (42.2) are sometimes called "polynomials in $\boldsymbol{a}$" even when the $d_{\mathfrak{G}}$ are not polynomials in $\boldsymbol{a}$. See e.g. MACDUFFEE [1933, *2*, Th. 15.3]

[5] For the case $n = 3$, this follows immediately from a theorem of WEITZENBÖCK [1923, *1*, pp. 65—66]. The result for arbitrary n can be established similarly.

[6] [1926, *2*].

[7] The elements of some, but not all algebras are representable by square matrices of finite order.

[8] [1932, *5*, p. 150].

Various writers[1] have considered the problem of inverting a matrix polynomial to obtain $\boldsymbol{a}$ in terms of $\boldsymbol{b}$. The inverse $\boldsymbol{a}(\boldsymbol{b})$, when it exists[2], is in general complex and multivalued. It can be shown that for arbitrary non-singular $\boldsymbol{f}$ we have

$$\boldsymbol{f}\cdot\boldsymbol{a}(\boldsymbol{b})\cdot\boldsymbol{f}^{-1}=\boldsymbol{a}(\boldsymbol{f}\cdot\boldsymbol{b}\cdot\boldsymbol{f}^{-1}), \tag{42.4}$$

this being interpreted in the sense that for given $\boldsymbol{f}$ each value of $\boldsymbol{a}(\boldsymbol{b})$ is equal to some value of $\boldsymbol{f}^{-1}\cdot[\boldsymbol{a}(\boldsymbol{f}\cdot\boldsymbol{b}\cdot\boldsymbol{f}^{-1})]\cdot\boldsymbol{f}$.

Contrary to what might be expected from the results of WEITZENBÖCK and of TURNBULL and AITKEN concerning Eq. (42.3), in general there are values of $\boldsymbol{a}(\boldsymbol{b})$ which are not expressible as linear combinations of $1, \ldots, \boldsymbol{b}^{n-1}$; values, that is which cannot be regarded as matrix polynomials or power series. Recognizing this, algebraists have attempted to devise more general definitions of matrix functions[3] to include such possibilities, still insisting upon a correspondence between matrix functions and functions of a single scalar variable. According to these definitions, a matrix function $\boldsymbol{x}(\boldsymbol{a})$ always satisfies Eq. (42.3), this being interpreted as was Eq. (42.4) in cases where $\boldsymbol{b}(\boldsymbol{a})$ is multivalued. Such definitions are not easily generalized to the case of matrices depending on several matrices.

The reader who has even slight familiarity with recent development in continuum mechanics will see that here we have come up against a central problem in modern theories of materials. We now formulate ***functional definitions*** which seem particularly well suited for application in classical mechanics, if perhaps not so well for other fields:

1. If each component of the matrix $\boldsymbol{b}$ be a function of the components of the matrices $\boldsymbol{a}, \ldots, \boldsymbol{c}$, we say that $\boldsymbol{b}$ is a *matrix function* of $\boldsymbol{a}, \ldots, \boldsymbol{c}$.

2. A matrix function $\boldsymbol{b}(\boldsymbol{a}, \ldots, \boldsymbol{c})$ such that

$$\boldsymbol{b}(\boldsymbol{o}\cdot\boldsymbol{a}\cdot\boldsymbol{o}^{-1}, \ldots, \boldsymbol{o}\cdot\boldsymbol{c}\cdot\boldsymbol{o}^{-1})=\boldsymbol{o}\cdot\boldsymbol{b}(\boldsymbol{a}, \ldots, \boldsymbol{c})\cdot\boldsymbol{o}^{-1} \tag{42.5}$$

for all orthogonal matrices $\boldsymbol{o}$, i.e., for all $\boldsymbol{o}$ such that

$$\boldsymbol{o}^{-1}=\boldsymbol{o}', \tag{42.6}$$

is an *isotropic*[4] *matrix function* of $\boldsymbol{a}, \ldots, \boldsymbol{c}$.

3. If each component of the matrix function $\boldsymbol{b}(\boldsymbol{a}, \ldots, \boldsymbol{c})$ be a polynomial in the components of $\boldsymbol{a}, \ldots, \boldsymbol{c}$, then $\boldsymbol{b}(\boldsymbol{a}, \ldots, \boldsymbol{c})$ is a *polynomial matrix function* of $\boldsymbol{a}, \ldots, \boldsymbol{c}$.

4. An *isotropic polynomial matrix function* is one satisfying both 2 and 3.

c) Decompositions.

43. Invariant decompositions. The group of transformations of rectangular Cartesian co-ordinate systems decomposes the linear vector space of second order tensors into three linear subspaces, the tensors comprising these spaces being, respectively, *the spherical tensors, the symmetric traceless tensors, and the skew-symmetric tensors*. Any orthogonal transformation maps each subspace onto itself; the only tensor common to two of them is the zero tensor; none possesses a proper linear subspace mapped into itself by every orthogonal transformation. In modern terminology, these three are the irreducible invariant subspaces of

[1] Cf. MACDUFFEE [1933, *2*, §§ 47, 48] for references. RUTHERFORD [1932, *4*] gives certain rather explicit solutions for a fairly general class of equations. MACDUFFEE [1933, *2*, p. 94] indicates how all solutions can be obtained in the case when $\boldsymbol{b}$ is spherical.

[2] WEITZENBÖCK [1932, *3*, p. 161] gives a simple example of a case where no inverse exists for the equation $\boldsymbol{b}=\boldsymbol{a}^2$.

[3] Cf. MACDUFFEE [1933, *2*, § 50] for references.

[4] The terminology agrees with that used by RIVLIN and ERICKSEN [1954, *5*, § 21].

the space of tensors or order two with respect to the orthogonal group[1]. Explicitly, we have the decomposition

$$a_{km} = n^{-1} \mathrm{I}_{\boldsymbol{a}}^{(1)} g_{km} + [a_{(km)} - n^{-1} \mathrm{I}_{\boldsymbol{a}}^{(1)} g_{km}] + a_{[km]}, \tag{43.1}$$

which is, in the sense indicated above, maximal.

An arbitrary matrix $\boldsymbol{a}$ can be written in infinitely many ways as the product of two symmetric matrices, one of which is non-singular[2]. This can be formulated in the following two ways:

$$a^k{}_m = b^{kr} c_{rm}, \quad b^{[km]} = c_{[km]} = 0, \quad \det b^{km} \neq 0, \tag{43.2}$$

$$a^k{}_m = d^{kr} e_{rm}, \quad d^{[km]} = e_{[km]} = 0, \quad \det e_{km} \neq 0. \tag{43.3}$$

As is clear from the results at the beginning of Sect. 37, $\boldsymbol{b}$ cannot always be chosen to be positive definite; the same applies to $\boldsymbol{e}$.

Any non-singular matrix $\boldsymbol{a}$ may be written in the forms

$$\boldsymbol{a} = \boldsymbol{s} \cdot \boldsymbol{o} = \boldsymbol{o} \cdot \boldsymbol{s}^*, \tag{43.4}$$

where $\boldsymbol{o}$ is orthogonal, and $\boldsymbol{s}$ and $\boldsymbol{s}^*$ are symmetric and positive definite[3]; *$\boldsymbol{o}$, $\boldsymbol{s}$, and $\boldsymbol{s}^*$ are uniquely determined.* We now give a proof of this ***polar decomposition theorem***[4]. Since $\boldsymbol{a} \cdot \boldsymbol{a}'$ is a positive definite[5] symmetric tensor, by the results of Sect. 41, it has a unique positive square root $\boldsymbol{s}$; since $\boldsymbol{a}$ is non-singular, so is $\boldsymbol{s}$. Therefore we may set

$$\boldsymbol{s} \equiv (\boldsymbol{a} \cdot \boldsymbol{a}')^{\frac{1}{2}}, \quad \boldsymbol{o} \equiv \boldsymbol{s}^{-1} \cdot \boldsymbol{a}. \tag{43.5}$$

Now $\boldsymbol{s}^{-1} \cdot \boldsymbol{a} \cdot \boldsymbol{a}' \cdot \boldsymbol{s}^{-1} = \boldsymbol{1}$, or $(\boldsymbol{s}^{-1} \cdot \boldsymbol{a}) \cdot (\boldsymbol{s}^{-1} \cdot \boldsymbol{a})' = \boldsymbol{1}$; therefore $\boldsymbol{o}$ is orthogonal. From Eq. $(43.5)_2$ follows Eq. $(43.4)_1$. In the same way we obtain a decomposition $\boldsymbol{a} = \boldsymbol{o}^* \cdot \boldsymbol{s}^*$, with

$$\boldsymbol{s}^* \equiv (\boldsymbol{a}' \cdot \boldsymbol{a})^{\frac{1}{2}}, \quad \boldsymbol{o}^* \equiv \boldsymbol{a} \cdot \boldsymbol{s}^{-1}. \tag{43.6}$$

---

[1] The corresponding decomposition of the space of tensors of any given order is derived and discussed in some detail by WEYL [1946, *3*, Chap. V B], references to relevant literature being given on pp. 310–311.

[2] FROBENIUS [1910, *1*]. VOSS [1878, *1*, p. 343] previously established this for non-singular $\boldsymbol{a}$. HILTON [1914, *1*] characterized the matrices which can be written as the product of two skew-symmetric matrices or as the product of a symmetric and a skew-symmetric matrix.

[3] We need to interpret this theorem also in terms of linear transformations; equivalently, any second order tensor with non-vanishing determinant may be expressed as the product of a unique orthogonal tensor by a unique positive definite symmetric tensor. An orthogonal matrix was defined by Eq. (42.6). An *orthogonal tensor* $o^k{}_m$ is defined by the property that under the transformation $\bar{v}^k = o^k{}_p v^p$, $\bar{w}^m = o^m{}_p w^p$, for arbitrary vectors $\boldsymbol{v}$ and $\boldsymbol{w}$, the inner product $g_{km} v^k w^m$ is invariant. That is,

$$g_{km} \bar{v}^k \bar{w}^m = g_{km} o^k{}_p o^m{}_q v^p w^q = g_{km} v^k w^m. \tag{A}$$

In order for this relation to hold for arbitrary $\boldsymbol{v}$ and $\boldsymbol{w}$, $\boldsymbol{o}$ must satisfy

$$g_{km} o^k{}_p o^k{}_q = g_{pq}; \tag{B}$$

alternatively, $o_k{}^m o^k{}_p = \delta^m_p$.

In a rectangular Cartesian system, the matrix $o^k{}_m$ is an orthogonal matrix. In general co-ordinates, (B) asserts that $(\boldsymbol{g} \cdot \boldsymbol{o})' = (\boldsymbol{o} \cdot \boldsymbol{g}^{-1})^{-1}$, where $\boldsymbol{g} = \|g_{km}\|$.

[4] An equivalent algebraic statement and all the essential ideas for an algebraic proof were given by FINGER [1892, *4*, Eq. (25)]; the first algebraic proof, by E. and F. COSSERAT [1896, *3*, § 6] (*cf.* also BURGATTI [1914, *3*], SIGNORINI [1930, *3*]). The ideas of FINGER and the COSSERATS were put into matrix notation and extended to complex matrices by AUTONNE [1902, *T9*, Lemma II], his form of the proof being that given above. If $\boldsymbol{a}$ is singular, decompositions $\boldsymbol{a} = \boldsymbol{s} \cdot \boldsymbol{o} = \boldsymbol{o}^* \cdot \boldsymbol{s}^*$ exist but are not unique; *cf.* HALMOS [1942, *1*, § 67].

[5] If we set $\boldsymbol{w} \equiv \boldsymbol{v} \cdot \boldsymbol{a}$, then, since $\boldsymbol{a}$ is non-singular, $\boldsymbol{w} = 0$ if and only if $\boldsymbol{v} = 0$. Since $\boldsymbol{v} \cdot (\boldsymbol{a} \cdot \boldsymbol{a}') \cdot \boldsymbol{v} = \boldsymbol{w} \cdot \boldsymbol{w}$, it follows that $\boldsymbol{a} \cdot \boldsymbol{a}'$ is positive definite.

We shall show presently that both decompositions of $\boldsymbol{a}$ are unique. Since $\boldsymbol{a} = \boldsymbol{o} \cdot (\boldsymbol{o}^{-1} \cdot \boldsymbol{s} \cdot \boldsymbol{o})$, it is a consequence of this uniqueness that

$$\boldsymbol{s}^* = \boldsymbol{o}^{-1} \cdot \boldsymbol{s} \cdot \boldsymbol{o}, \qquad \boldsymbol{o}^* = \boldsymbol{o}. \tag{43.7}$$

To prove uniqueness[1], we note that $\boldsymbol{s} \cdot \boldsymbol{o} = \boldsymbol{s}^{\backprime} \cdot \boldsymbol{o}^{\backprime}$ implies $\boldsymbol{o} = \boldsymbol{s}^{-1} \cdot \boldsymbol{s}^{\backprime} \cdot \boldsymbol{o}^{\backprime}$ and $\boldsymbol{o}' = \boldsymbol{o}^{-1} = (\boldsymbol{o}^{\backprime})^{-1} \cdot \boldsymbol{s}^{\backprime} \cdot \boldsymbol{s}^{-1}$, or $\boldsymbol{o} = \boldsymbol{s} \cdot (\boldsymbol{s}^{\backprime})^{-1} \cdot \boldsymbol{o}^{\backprime}$. Hence $(\boldsymbol{s}^{-1} \cdot \boldsymbol{s}^{\backprime} - \boldsymbol{s} \cdot (\boldsymbol{s}^{\backprime})^{-1}) \cdot \boldsymbol{o}^{\backprime} = 0$, whence follows $\boldsymbol{s}^{-1} \cdot \boldsymbol{s}^{\backprime} = \boldsymbol{s} \cdot (\boldsymbol{s}^{\backprime})^{-1}$, or $\boldsymbol{s}^2 = (\boldsymbol{s}^{\backprime})^2$; therefore, since $\boldsymbol{s}$ and $\boldsymbol{s}^{\backprime}$ are positive definite, we derive $\boldsymbol{s} = \boldsymbol{s}^{\backprime}$, and hence $\boldsymbol{o} = \boldsymbol{o}^{\backprime}$. Q.E.D. By this uniqueness, it follows also that $\boldsymbol{s}$ and $\boldsymbol{o}$ commute if and only if $\boldsymbol{s} = \boldsymbol{s}^*$; that is, if and only if $\boldsymbol{a} \cdot \boldsymbol{a}' = \boldsymbol{a}' \cdot \boldsymbol{a}$, in which case $\boldsymbol{a}$ is called *normal*[2]. The class of normal matrices includes all symmetric, skew-symmetric, or orthogonal matrices.

If $\boldsymbol{a}$ is non-singular, there exists a non-singular matrix $\boldsymbol{b}$ such that[3]

$$\boldsymbol{a} = \boldsymbol{b}^2 \tag{43.8}$$

and, if $\boldsymbol{a}$ is also symmetric, a matrix $\boldsymbol{c}$ such that[4]

$$\boldsymbol{a} = \boldsymbol{c}' \cdot \boldsymbol{c}. \tag{43.9}$$

In Eqs. (43.8) and (43.9), $\boldsymbol{b}$ and $\boldsymbol{c}$ in general are complex and are not uniquely determined.

44. Certain canonical forms. For any second order tensor $\boldsymbol{a}$, there exists a co-ordinate transformation, which may be chosen to be unitary, reducing $\boldsymbol{a}$ at a given point to *superdiagonal form*[5]:

$$\|a^k{}_m\| = \begin{Vmatrix} a^1{}_1 & a^1{}_2 \dots a^1{}_n \\ 0 & a^2{}_2 \dots \;\cdot \\ \vdots & \qquad\;\; \vdots \\ 0 & 0 \;\;\dots a^n{}_n \end{Vmatrix}. \tag{44.1}$$

When $\boldsymbol{a}$ is real and all its proper numbers are real, this transformation is real and may be chosen to be orthogonal. When there are complex proper numbers, the transformation is necessarily complex.

In three dimensions a real tensor field has either three real proper numbers or one real, two complex conjugate. In the latter case, there is a real transformation, which may be taken to be orthogonal, reducing $\boldsymbol{a}$ to the form[6]

$$\|a^k{}_m\| = \begin{Vmatrix} a & b & c \\ -b & d & e \\ 0 & 0 & f \end{Vmatrix}. \tag{44.2}$$

When Eq. (44.1) holds in rectangular Cartesian co-ordinates, $\boldsymbol{a}$ is normal (Sect. 43) if and only if the matrix (44.1) is in fact diagonal. Hence a tensor all of whose proper numbers are real is normal if and only if it is symmetric. A tensor of the form (44.2) is normal if and only if $c = e = 0$, $a = d$.

It is known[7] that a real symmetric traceless matrix $\boldsymbol{a}$ can be transformed by orthogonal transformations to a system in which all diagonal components of $\boldsymbol{a}$

---

[1] MURNAGHAN and WINTNER [1931, *5*].
[2] MURNAGHAN and WINTNER [1931, *4*], [1931, *5*].
[3] Cf. MACDUFFEE [1933, *2*, §§ 35, 48].
[4] This follows immediately from the familiar result, used by LAGRANGE [1759, *1*], that a quadratic form can be reduced to a sum of squares by linear transformations.
[5] SCHUR [1909, *2*].
[6] MURNAGHAN and WINTNER [1931, *4*]. There is a generalization to n dimensions.
[7] LOVE [1906, *1*, § 16] stated this without proof. WHAPLES has shown us a proof and an extension of the theorem to matrices defined over essentially arbitrary fields.

vanish. Now consider the matrix (43.1); in every co-ordinate system the diagonal components of **1** are all equal, those of the skew-symmetric part $\frac{1}{2}(\boldsymbol{a}-\boldsymbol{a}')$ vanish, and the remaining tensor in the matrix (43.1) is symmetric and traceless, whence it follows by the preceding theorem that *in a suitably chosen rectangular Cartesian co-ordinate system, all diagonal components of* $\boldsymbol{a}$ *are equal.* When $n=3$, we have

$$\boldsymbol{a}=\begin{Vmatrix} a & b & c \\ d & a & e \\ f & g & a \end{Vmatrix}. \tag{44.3}$$

When the proper numbers of $\boldsymbol{a}$ are all distinct, $\|a^k{}_m\|$ can be transformed by real, but not necessarily orthogonal transformations to the form[1]

$$\|a^k{}_m\|=\begin{Vmatrix} c_1 & c_2 & \dots & c_n \\ 1 & 0 & \dots & 0 \\ \vdots & & & \\ 0 & 0 & \dots 1 & 0 \end{Vmatrix}, \tag{44.4}$$

where $c_K=(-1)^{K+1}\,I_{\boldsymbol{a}}^{(K)}$. When $\boldsymbol{a}$ has multiple proper numbers it can be reduced to a direct sum of matrices of the type (44.4). Denote by $\boldsymbol{b}$ the matrix on the right of Eq. (44.4). Then $\boldsymbol{c}^{-1}\cdot\boldsymbol{b}\cdot\boldsymbol{c}$ is diagonal[2], where

$$\boldsymbol{c}=\begin{Vmatrix} (a_1)^{n-1} & (a_2)^{n-1} & \dots & (a_n)^{n-1} \\ (a_1)^{n-2} & (a_2)^{n-2} & \dots & (a_n)^{n-2} \\ \vdots & & & \\ 1 & 1 & \dots & 1 \end{Vmatrix}. \tag{44.5}$$

This provides another proof of the result, noted in Sect. 37, that $\|a^k{}_m\|$ can be reduced to diagonal form by real transformations if its proper numbers be real and distinct.

Any matrix can be reduced by a complex transformation to a direct sum of matrices of the form[3]

$$\begin{Vmatrix} a & 1 & 0 & \dots & 0 \\ 0 & a & 1 & \dots & 0 \\ \vdots & & & & \\ & \dots & & a & 1 \\ 0 & 0 & \dots & 0 & a \end{Vmatrix}. \tag{44.6}$$

If all the proper numbers are real, the transformation may be taken as real also. NOLL has informed us that any real matrix can be transformed by a real transformation to a direct sum of matrices of the two forms (44.6) and

$$\begin{Vmatrix} 0 & 1 & 0 & \dots & & & 0 \\ a & b & 1 & 0 & \dots & & 0 \\ 0 & 0 & 1 & 0 & \dots & & 0 \\ 0 & a & b & 1 & 0 & \dots & 0 \\ 0 & 0 & 0 & 1 & 0 & \dots & 0 \\ \vdots & & & & & & \vdots \\ 0 & & & & a & b & 1 \\ 0 & & & & 0 & 0 & 1 \\ 0 & \dots & & & \dots & a & b \end{Vmatrix}. \tag{44.7}$$

When all proper numbers are distinct, the forms (44.6) and (44.7) reduce to $\|a\|$ and $\begin{Vmatrix} 0 & 1 \\ a & b \end{Vmatrix}$, respectively. This furnishes still another proof that a matrix with real and distinct proper numbers can be diagonalized by a real transformation.

---

[1] Cf. MACDUFFEE [1933, *2*, § 39] for references and discussion.

[2] SCHUR [1909, *3*].

[3] JORDAN [1870, *4*, p. 114].

d) Normal and shear components.

45. Definitions. Except where otherwise noted, the second order tensors considered in the remainder of this chapter are assumed to be symmetric, there then being no distinction between right and left proper vectors. The scalar $a_{km} v^k v^m$, where $\boldsymbol{v}$ is any unit vector, is called the *normal component of* $\boldsymbol{a}$ *for the direction* $\boldsymbol{v}$. The scalar $a_{km} u^k v^m$, where $\boldsymbol{u}$ and $\boldsymbol{v}$ are unit vectors, is the *shear component of* $\boldsymbol{a}$ *for the directions* $\boldsymbol{u}$ and $\boldsymbol{v}$, the normal components being included as a special case. When $\boldsymbol{u}$ and $\boldsymbol{v}$ are perpendicular unit vectors, we call $a_{km} u^k v^m$ the corresponding *orthogonal shear component.*

In orthogonal co-ordinates, the physical component $a\langle kk\rangle$ is the normal component of $\boldsymbol{a}$ for the direction of the tangent to the k-th co-ordinate curve, while $a\langle km\rangle$, $k \neq m$, is the shear component for the directions of the tangents to the k-th and m-th co-ordinate curves. In passages where a fixed orthogonal co-ordinate system has been laid down, the phrases "normal components" and "shear components", with no further qualification, mean the components $a\langle kk\rangle$ and $a\langle km\rangle$, $k \neq m$, respectively.

Because of the existence of the canonical form exemplified by Eq. (44.3), *it is always possible to refer* $\boldsymbol{a}$ *to an orthogonal co-ordinate system in which the normal components of* $\boldsymbol{a}$ *at a given point are all equal.* They can all be made to vanish if and only if $\boldsymbol{a}$ is traceless. In order for the normal components to be zero in all orthogonal co-ordinate system, it is necessary and sufficient that $\boldsymbol{a}=0$. Since by assumption $\boldsymbol{a}$ is symmetric, there exist orthogonal co-ordinate systems in which $\|a\langle km\rangle\|$ is diagonal at a given point; that is, *it is always possible to refer* $\boldsymbol{a}$ *to an orthogonal co-ordinate system* in which at a given point all the shear components are zero. These co-ordinate systems are principal co-ordinate systems as defined in Sect. 37. In order for the shear components to vanish in all orthogonal co-ordinate systems, it is necessary and sufficient that $\boldsymbol{a}$ be spherical; for a spherical tensor, the shear component for the directions $\boldsymbol{u}$ and $\boldsymbol{v}$ is zero if and only if $\boldsymbol{u}$ and $\boldsymbol{v}$ be orthogonal.

The proper numbers of $\boldsymbol{a}$, which are real, we order as follows:

$$a_1 \geqq \cdots \geqq a_n. \tag{45.1}$$

46. Extremal properties. Setting

$$C \equiv a_{km} u^k v^m, \tag{46.1}$$

for a fixed tensor $\boldsymbol{a}$, we consider the problem of finding the extremes of the scalar C when the vectors $\boldsymbol{u}$ and $\boldsymbol{v}$ are varied subject to certain constraints.

Problem 1. Let the constraints be $\boldsymbol{u}=\boldsymbol{v}$, $u_k u^k = 1$, so that C is the *normal component* of $\boldsymbol{a}$ for the direction $\boldsymbol{u}$. Referring $\boldsymbol{a}$ to a rectangular Cartesian principal co-ordinate system, we then have

$$C = a_1 (u_1)^2 + \cdots + a_n (u_n)^2, \quad (u_1)^2 + \cdots + (u_n)^2 = 1, \tag{46.2}$$

and from Eq. (45.1) it follows that

$$a_1 = a_1\{(u_1)^2 + \cdots + (u_n)^2\} \geqq a_1 (u_1)^2 + \cdots + a_n (u_n)^2 \geqq a_n\{(u_1)^2 + \cdots + (u_n)^2\} = a_n, \tag{46.3}$$

so that $a_1 \geqq C \geqq a_n$. To see when equality can hold on the left, we set $a_1 = C$ in Eq. $(46.2)_1$ and use Eq. $(46.2)_2$, so obtaining

$$\left.\begin{aligned} 0 &= a_1\{(u_1)^2 - 1\} + a_2 (u_2)^2 + \cdots + a_n (u_n)^2 \\ &= (a_2 - a_1)(u_2)^2 + \cdots + (a_n - a_1)(u_n)^2, \end{aligned}\right\} \tag{46.4}$$

whence it follows by Eq. (45.1) that $u_k=0$ when $k>1$ unless $a_k=a_1$. Therefore, letting K be the largest integer for which $a_K=a_1$, so that $a_{\mathfrak{l}}=a_1$ when $\mathfrak{l}\leq K$, we conclude that $u_{K+1}=\cdots=u_n=0$, whence it follows by direct calculation that $a^k_m u^m=a_1 u^k$. Thus *the normal component of* $\boldsymbol{a}$ *is greatest for the directions which are principal directions corresponding to the greatest proper number, and the value of the greatest normal component is the greatest proper number.* Similary, C takes on its minimum value a_n if and only if $\boldsymbol{u}$ be a proper vector of $\boldsymbol{a}$ corresponding to the proper number a_n. The remaining proper numbers of $\boldsymbol{a}$ are extremal values of C; if $\mathfrak{b}>1$, $a_{\mathfrak{b}}$ is the maximum value of C when $\boldsymbol{u}$ ranges over the vectors which satisfy the constraints and are perpendicular to $\mathfrak{b}-1$ mutually orthogonal proper vectors of $\boldsymbol{a}$ corresponding to the proper numbers[1] $a_1,\ldots,a_{\mathfrak{b}-1}$. In general C takes on these intermediate values for infinitely many vectors $\boldsymbol{u}$ which do not satisfy the orthogonality conditions and are not proper vectors of $\boldsymbol{a}$.

Problem 2. Let the constraints be $u_k u^k=v_k v^k=1$, so that C is the *shear component* for the directions $\boldsymbol{u}$ and $\boldsymbol{v}$. Extremal conditions for C are

$$\frac{\partial C}{\partial u^k}=a_{km}v^m=a\,u_k,\qquad \frac{\partial C}{\partial v^k}=a_{km}u^m=b\,v_k, \tag{46.5}$$

where a and b are multipliers. Since $u^k\,\partial C/\partial u^k=v^k\,\partial C/\partial v^k=C$, we conclude that $a=b$; hence

$$a_{km}(u^m+v^m)=a\,(u_k+v_k),\qquad a_{km}(u^m-v^m)=-\,a\,(u_k-v_k). \tag{46.6}$$

If $\boldsymbol{a}$ has two proper numbers $\pm a$ that are numerically equal but opposite in sign, Eq. (46.6) can be satisfied by choosing $\boldsymbol{u}$ and $\boldsymbol{v}$ to be non-parallel vectors such that $\boldsymbol{u}+\boldsymbol{v}$ and $\boldsymbol{u}-\boldsymbol{v}$ are proper vectors corresponding to a and $-a$. This includes the limiting case where zero is a multiple proper number of $\boldsymbol{a}$. Otherwise, we must have $\boldsymbol{u}\pm\boldsymbol{v}=0$; the case $\boldsymbol{u}=\boldsymbol{v}$ is that considered in Problem 1, while the case $\boldsymbol{u}=-\boldsymbol{v}$ is similar. We have shown that *extrema of shear components are always to be found among those for the coincident directions, which are the normal components, and among those for opposite directions; for additional extrema of the shear components to exist, it is necessary and sufficient that* $\boldsymbol{a}$ *have a pair of proper numbers* $\pm a$.

Problem 3. In Problem 2, we add the further constraint $\boldsymbol{u}\cdot\boldsymbol{v}=0$, thus confining attention to *orthogonal shear components*. Clearly, the largest (smallest) value of C attainable with this additional constraint is never greater (less) than that attainable with the constraints of Problem 2. Extremal conditions are

$$\frac{\partial C}{\partial u^k}=a_{km}v^m=a\,u_k+b\,v_k,\qquad \frac{\partial C}{\partial v^k}=a_{km}u^m=b\,u_k+c\,v_k, \tag{46.7}$$

where a, b, and c are multipliers. It follows that $a=c$ and the corresponding extremal value of C is a. Therefore

$$a_{km}(u^m+v^m)=(a+b)\,(u_k+v_k),\qquad a_{km}(u^m-v^m)=(b-a)\,(u_k-v_k). \tag{46.8}$$

The constraints require that the vectors $\boldsymbol{u}\pm\boldsymbol{v}$ be non-null, whence it follows that these are proper vectors corresponding to the proper numbers $(b\pm a)$ of $\boldsymbol{a}$. This result is the ***theorem of Coulomb and Hopkins***[2]: *The maximum and minimum*

---

[1] See, e.g., Courant and Hilbert [1931, *2*, pp. 20–23]. Cauchy [1829, *1*], [1830, *1*, Chap. II] noted that the proper numbers of $\boldsymbol{a}$ are the (real) extremal values of $a_{km}\times u^k u^m/u_r u^r$.

[2] The theorem was derived in the plane case by Coulomb [1776, *1*, § VIII]; in the general case, by Hopkins [1847, *2*, §§ 4, 5].

orthogonal shear components of $\boldsymbol{a}$ *are given by*

$$\max C = \tfrac{1}{2}(a_1 - a_n), \quad \min C = \tfrac{1}{2}(a_n - a_1) \tag{46.9}$$

respectively, these being taken on if and only if $\sqrt{2}\boldsymbol{u} = \boldsymbol{m}_1 + \boldsymbol{m}_n$, $\sqrt{2}\boldsymbol{v} = \boldsymbol{m}_1 - \boldsymbol{m}_n$, *where* $\boldsymbol{m}_1$ *and* $\boldsymbol{m}_n$ *are orthogonal unit proper vectors of* $\boldsymbol{a}$ *corresponding to the greatest and least proper numbers* a_1 *and* a_n. When $a_1 = a_n$, $\boldsymbol{a}$ is spherical, so that all orthogonal shear components vanish, as follows also from (46.9). The remaining extremal values of C, as given by (46.8), are $\frac{1}{2}(a_{\mathfrak{d}} - a_{\mathfrak{e}})$, $\mathfrak{d} \neq \mathfrak{e}$. Each of these is a minimax unless $a_{\mathfrak{d}} - a_{\mathfrak{e}} = \pm(a_1 - a_n)$. The octahedral invariant $\text{Ш}_{\boldsymbol{a}}$, defined by Eq. (46.11), is thus *the sum of the squares of the extremal orthogonal shear components of* $\boldsymbol{a}$.

From Eq. (46.9) we observe that

$$\tfrac{1}{2}(a_n - a_1) \leq a_{km} u^k v^m (u_r u^r v_s v^s)^{-\frac{1}{2}} \leq \tfrac{1}{2}(a_1 - a_n) \tag{46.10}$$

for all non-null orthogonal vectors $\boldsymbol{u}$ and $\boldsymbol{v}$.

In a rectangular Cartesian co-ordinate system such that at a given point $n-2$ axes are principal axes of $\boldsymbol{a}$ and the remaining two are directions satisfying Eq. (46.7), $\boldsymbol{a}$ assumes the form

$$\|a_{km}\| = \begin{Vmatrix} \boldsymbol{d} & 0 & 0 \\ 0 & b & a \\ 0 & a & b \end{Vmatrix} \tag{46.11}$$

where $\boldsymbol{d}$ is a diagonal $(n-2)\times(n-2)$ matrix[1].

Problem 4. We now take $\boldsymbol{v}$ as a *fixed* unit vector and vary $\boldsymbol{u}$ subject to the constraints $u_k u^k = 1$, $u_k v^k = 0$. Thus we seek the greatest shear components among all directions perpendicular to a given direction. As conditions that C be extremal, we obtain

$$a_{km} v^m = a\,u_k + b\,v_k. \tag{46.12}$$

Therefore $b = a_{km} v^k v^m$; that is, b is the normal component of $\boldsymbol{a}$ for the direction $\boldsymbol{v}$, and a, which satisfies

$$a^2 = a^2 u_k u^k = (a_{km} v^m - b\,v_k)(a^k_r v^r - b\,v^k), \tag{46.13}$$

is the extremal value of C. When $\boldsymbol{v}$ is a proper vector of $\boldsymbol{a}$, then $C = 0$ for all $\boldsymbol{u}$ satisfying the constraints. This is no more than the statement that the shear component for a principal direction and any direction orthogonal to it is zero. Otherwise Eq. (46.13) determine a unique value for a^2, and therefore Eq. (46.12) determines $\boldsymbol{u}$ up to sign. It can be shown that[2]

$$a^2 = \sum_{\mathfrak{d}<\mathfrak{e}} (a_{\mathfrak{d}} - a_{\mathfrak{e}})^2 \cos^2\vartheta_{\mathfrak{d}} \cos^2\vartheta_{\mathfrak{e}}, \tag{46.14}$$

where $\vartheta_{\mathfrak{d}}$ is the angle between $\boldsymbol{v}$ and the proper vector of $\boldsymbol{a}$ corresponding to $a_{\mathfrak{d}}$. It is easily seen from the results of the preceding paragraph or from Eq. (46.14) that $a^2 \leq [\frac{1}{2}(a_1 - a_n)]^2$, the equality holding if and only if $\boldsymbol{v}$ lies in the plane of and bisects the angle between a pair of orthogonal proper vectors corresponding to a_1 and a_n. A direction for which a^2 is maximum is called a *direction of maximum shear*[3]. When $a_1 \neq a_2$ and $a_{n-1} \neq a$, there are precisely two such directions[4], and these are orthogonal. Otherwise, there are infinitely many.

---

[1] SHARP [1882, *1*, p. 229] showed that $\boldsymbol{a}$ could always be reduced to this form when $n = 3$.

[2] This and other results of this paragraph are due in essence to KLEITZ [1872, *5*], [1873, *3*, § 12].

[3] When $n = 3$, BOUSSINESQ [1877, *1*] noted that a^2, calculated for any unit vector $\boldsymbol{v}$ lying in the plane of the principal directions corresponding to a_1 and a_3, is never less than the value of a^2 for any unit vector $\boldsymbol{v}^*$ such that $a_{km} v^{*k} v^{*m} = a_{km} v^k v^m$.

[4] Here $\boldsymbol{v}$ and $-\boldsymbol{v}$ are counted as having the same direction.

Problem 5. We impose the constraints $a_{km} u^k u^m = a_{km} v^k v^m = 1$, $u_k v^k = 0$, and we require $\boldsymbol{a}$ to be positive definite. Proceeding as before, we obtain

$$a_{km} v^m = a\, a_{km} u^m + b\, v_k, \qquad a_{km} u^m = a\, a_{km} v^m + b\, u_k; \tag{46.15}$$

hence

$$a_{km}(u^m + v^m)(1-a) = b(u_k + v_k), \qquad a_{km}(u^m - v^m)(1+a) = b(u_k - v_k). \tag{46.16}$$

The constraints imply that $\boldsymbol{u} \pm \boldsymbol{v} \neq 0$. Since $\boldsymbol{a}$ is positive definite, $a_{km}(u^m \pm v^m) \neq 0$, whence it follows from Eq. (46.15) that $a \neq \pm 1$. Therefore the directions giving C extreme values are the same as those which are extremal for the orthogonal shears, but now $b/(1-a)$ and $b/(1+a)$ are proper numbers of $\boldsymbol{a}$. A simple calculation then yields the extremal values

$$C = a = (a_{\mathfrak{d}} - a_{\mathfrak{e}})/(a_{\mathfrak{d}} + a_{\mathfrak{c}}), \qquad \mathfrak{d} \neq \mathfrak{e}, \tag{46.17}$$

taken on when $\boldsymbol{u}$ and $\boldsymbol{v}$ lie in the plane of and bisect the angle between orthogonal proper vectors corresponding to $a_{\mathfrak{d}}$ and $a_{\mathfrak{e}}$. Since $\boldsymbol{a}$ is positive definite, its proper numbers are all positive, so Eq. (45.1) implies that

$$0 \leq a_1(a_{\mathfrak{e}} - a_n) + a_n(a_1 - a_{\mathfrak{d}}) = a_1 a_{\mathfrak{e}} - a_n a_{\mathfrak{d}}, \tag{46.18}$$

whence follows

$$(a_1 - a_n)(a_{\mathfrak{d}} + a_{\mathfrak{e}}) \geqq (a_{\mathfrak{d}} - a_{\mathfrak{e}})(a_1 + a_n). \tag{46.19}$$

Comparing Eqs. (46.17) and (46.19) shows that the maximum value of C is $(a_1 - a_n)/(a_1 + a_n)$, the minimum value $(a_n - a_1)/(a_1 + a_n)$. One or the other of these values is taken on when the directions of $\boldsymbol{u}$ and $\boldsymbol{v}$ are both directions of maximum shear. These results may be rephrased as follows: We have

$$(a_n - a_1)/(a_1 + a_n) \leq a_{km} u^k v^m (a_{rs} u^r u^s a_{tu} v^t v^u)^{-\frac{1}{2}} \leq (a_1 - a_n)/(a_1 + a_n) \tag{46.20}$$

for all non-null orthogonal vectors $\boldsymbol{u}$ and $\boldsymbol{v}$; when the directions of $\boldsymbol{u}$ and $\boldsymbol{v}$ are directions of maximum shear, either the right or left equality holds in Eq. (46.20).

e) Tensor lines and sheets.

47. Definitions. There are direction fields, such as the principal directions and the directions of maximum shear, which are intrinsically related to a given second order tensor field. For any such direction field, one can determine vector lines or sheets as indicated in Sect. 29. Vector lines of the principal directions and of the directions of maximum shear are called *principal* and *shear trajectories*, respectively. These are special kinds of *tensor lines*, defined as curves $\boldsymbol{x} = \boldsymbol{x}(u)$ satisfying differential equations of the form $\boldsymbol{F}(\boldsymbol{a}, d\boldsymbol{x}/du) = 0$, where $\boldsymbol{F}$ is a tensor invariant of the indicated arguments[1]. For example, for principal trajectories, we may take $\boldsymbol{F}$ to be the invariant given by

$$F^{kr} = \frac{dx^m}{du} a^{[k}{}_m \frac{dx^{r]}}{du}, \tag{47.1}$$

so that the vanishing of $\boldsymbol{F}$ is necessary and sufficient that $a^k{}_m\, dx^m/du$ and dx^k/du be parallel, i.e. that $d\boldsymbol{x}/du$ be a proper vector of $\boldsymbol{a}$.

Similarly, a *tensor sheet* is a surface $f(\boldsymbol{x}) = \text{const}$ satisfying differential equations of the form $\boldsymbol{G}(\boldsymbol{a}, \operatorname{grad} f) = 0$, where $\boldsymbol{G}$ is a tensor invariant of the indicated arguments. As a special case, the *isostatic surfaces*, defined as surfaces whose

[1] It is understood that $\boldsymbol{F}$ may depend on the metric tensor $\boldsymbol{1}$ and may be multi-valued.

normals are proper vectors of $\boldsymbol{a}$, satisfy e.g.

$$G_{mr} \equiv f_{,k}\, a^k{}_{[m}\, f_{,r]} = 0. \tag{47.2}$$

A necessary and sufficient condition that there exist a congruence of isostatic surfaces is that some proper vector be lamellar, all proportional proper vectors thus being complex-lamellar [1]. The normal trajectories of these surfaces are principal trajectories. When isostatic surfaces $f = \text{const}$ corresponding to the proper number a exist, they are the solutions of the equations

$$(a^k{}_m - a\,\delta^k_m)\, f_{,k} = 0. \tag{47.3}$$

When a tensor $\boldsymbol{a}$ enjoys n independent families of isostatic surfaces, these may be selected as co-ordinate surfaces of an *isostatic co-ordinate system*, in general not orthogonal. When $\boldsymbol{a}$ is referred to isostatic co-ordinates, the components $a^k{}_m$ vanish everywhere when $k \neq m$. Therefore the expression (37.17) for the covariant derivative of $\boldsymbol{a}$ reduces to a particularly simple form [2]:

$$\left.\begin{aligned} a_{k,p}^{k} &= \frac{\partial a_k{}^k}{\partial x^p} \quad \text{(unsummed)}, \\ a_{k,p}^{m} &= \left\{ {m \atop k\,p} \right\} (a_k{}^k - a_m{}^m) \quad (k \neq m, \text{ unsummed}), \end{aligned}\right\} \tag{47.4}$$

and Eq. (37.18) may be simplified analogously. Conversely, if $a^k{}_m = 0$ when $k \neq m$, the co-ordinate system is isostatic, and all proper vectors of $\boldsymbol{a}$ are complex-lamellar.

Surfaces of maximum shear are surfaces whose normals are directions of maximum shear. In order that there exist a congruence of such surfaces, it is necessary and sufficient that at least one of the directions of maximum shear be complex-lamellar. The normal trajectories of these surfaces are shear trajectories.

48. Isostatic surfaces of symmetric tensors. Consider first a symmetric field $\boldsymbol{a}$ with n distinct proper numbers. To obtain necessary and sufficient conditions that its proper vectors all be complex-lamellar, we may use the criterion of RICCI and LEVI-CIVITA [3], which asserts that n mutually orthogonal unit vector fields $\boldsymbol{m}_{\mathfrak{a}}$ are all complex-lamellar if and only if

$$m^k_{\mathfrak{a}}\, m^r_{\mathfrak{b}}\, m_{\mathfrak{c}\,k,r} = 0, \qquad \mathfrak{a}, \mathfrak{b}, \mathfrak{c} \neq . \tag{48.1}$$

If we choose the $\boldsymbol{m}_{\mathfrak{a}}$ to be unit proper vectors of a symmetric tensor $\boldsymbol{a}$ with distinct proper numbers, Eq. (48.1) provides necessary and sufficient conditions that all proper vectors of $\boldsymbol{a}$ be complex-lamellar. EISENHART [4] noted that Eq. (48.1) can be replaced by

$$a_{kr,s}\, m^k_{\mathfrak{a}}\, m^r_{\mathfrak{b}}\, m^s_{\mathfrak{c}} = 0, \qquad a, \mathscr{b}, c \neq, \tag{48.2}$$

still assuming that $\boldsymbol{a}$ has distinct proper numbers. Analysis of WEINGARTEN [5] shows that, in three dimensions, Eq. (48.2) can be replaced by the three condi-

[1] LAMÉ [1841, *2*, Introd.], [1851, *3*, § 90], [1859, *1*, § 148] introduced isostatic surfaces and presumed that they always exist. BOUSSINESQ [1872, *2*] was first to note that their existence is exceptional.

[2] It was these formulae, due in principle to LAMÉ [1841, *2*, § XI], [1859, *1*, § 149], that motivated the introduction of isostatic surfaces.

[3] [1901, *1*, p. 151].

[4] [1913, *2*, Eq. (29)]. EISENHART claims also to derive necessary and sufficient conditions for the case where two proper numbers of $\boldsymbol{a}$ coincide in a region. His analysis is in error, his Eqs. (36) and (37) being sufficient but not necessary conditions that his Eqs. (34) and (35) hold.

[5] [1881, *3*]. FRIEDMANN [1912, *1*] characterized the subset of tensors satisfying WEINGARTEN's conditions which satisfy also $a^{km}{}_{,m} = 0$ and are such that each proper number is constant over the isostatic surface corresponding to it.

tions

$$a_{kr,s}\, a_p^k\, \varepsilon^{rsp} = a_{kr,s}\, b_p^k\, \varepsilon^{rsp} = b_{kr,s}\, b_p^k\, \varepsilon^{rsp} = 0 \tag{48.3}$$

where $2b_q^k \equiv \varepsilon^{krs}\, \varepsilon_{qtp}\, a_r^t\, a_s^p$. Equivalent conditions are[1]

$$a_{kr,s}\, a_p^k\, \varepsilon^{rsp} = a_{kr,s}\, a_p^{2k}\, \varepsilon^{rsp} = a_{(kr,s)}\, a_p^k\, a_q^{2r}\, \varepsilon^{spq} = 0. \tag{48.4}$$

When a three-dimensional tensor has two but not three coincident proper numbers throughout a region, it can be shown that WEINGARTEN'S and DOYLE'S conditions are necessary and sufficient that a field of proper vectors corresponding to the simple proper number be lamellar. If and only if the corresponding isostatic surfaces can be embedded in a triply orthogonal family of surfaces, which is not always true[2], there will exist three mutually orthogonal complex-lamellar proper vectors of $\boldsymbol{a}$. The same conditions are necessary and sufficient for the existence of an orthogonal co-ordinate system in which the shear components of $\boldsymbol{a}$ vanish identically. If all three proper numbers of $\boldsymbol{a}$ coincide in a region, $\boldsymbol{a}$ is diagonal in any orthogonal co-ordinate system, so that triply orthogonal families of isostatic surfaces always exist trivially in this case.

Consider a vector $\boldsymbol{v}$ of the form

$$\boldsymbol{v} = A^{\mathfrak{a}}\, \boldsymbol{m}_{\mathfrak{a}}, \qquad A^{\mathfrak{a}} = \text{const} \tag{48.5}$$

where the $\boldsymbol{m}_{\mathfrak{a}}$ are the independent unit proper vectors of $\boldsymbol{a}$. Such a vector is complex-lamellar if and only if[3]

$$\sum_{\mathfrak{b} \neq \mathfrak{c}} (a_{\mathfrak{c}} - a_{\mathfrak{b}})^{-1}\, \varepsilon^{krs}\, a_{uv,s}\, m_{\mathfrak{c}}^u\, m_{\mathfrak{b}}^v\, v_k\, A^{\mathfrak{c}}\, m_{\mathfrak{b}r} = 0, \tag{48.6}$$

it being assumed that the proper numbers of $\boldsymbol{a}$ are all distinct. Necessary and sufficient conditions for the existence of a congruence of isostatic surfaces corresponding to the proper number a result from setting $A^{\mathfrak{a}} = \delta_{\mathfrak{c}}^{\mathfrak{a}}$ in Eqs. (48.5) and (48.6). No one has obtained equivalent conditions for the existence of isostatic surfaces corresponding to a single proper number which are expressible as the vanishing of polynomials in a_{km} and $a_{km,r}$. It seems probable that this is impossible.

49. Isostatic surfaces of asymmetric tensors. We restrict attension to tensors whose proper numbers are real and distinct, and we avoid introducing a metric tensor. From Eqs. (37.2), (37.12)$_3$, and (37.9) we then have

$$(a^k{}_r - a_{\mathfrak{b}}\, \delta_r^k)\, m_{\mathfrak{b}k} = 0, \qquad (a^r{}_k - a_{\mathfrak{b}}\, \delta_k^r)\, q_{\mathfrak{b}}^k = 0, \tag{49.1}$$

$$a_k^s = \sum_{\mathfrak{b}=1}^{n} a_{\mathfrak{b}}\, q_{\mathfrak{b}}^s\, m_{\mathfrak{b}k}, \qquad m_{\mathfrak{a}k}\, q_{\mathfrak{b}}^k = \delta_{\mathfrak{a}\mathfrak{b}}, \tag{49.2}$$

with $a_{\mathfrak{b}} \neq a_{\mathfrak{c}}$ when $\mathfrak{b} \neq \mathfrak{c}$. We seek conditions that $\boldsymbol{m}_{\mathfrak{b}}$ be complex-lamellar. Differentiating Eq. (49.1)$_1$, we obtain

$$(a_s^k - a_{\mathfrak{b}}\, \delta_s^k)\, \partial m_{\mathfrak{b}k}/\partial x^r = -(\partial a_s^k/\partial x^r - \partial a_{\mathfrak{b}}/\partial x^r\, \delta_s^k)\, m_{\mathfrak{b}k}. \tag{49.3}$$

Since the vectors $\boldsymbol{m}_{\mathfrak{c}}$ are linearly independent, there exist quantities $A_{\mathfrak{b}r}^{\mathfrak{c}}$ such that

$$\partial m_{\mathfrak{b}k}/\partial x^r = m_{\mathfrak{c}k}\, A^{\mathfrak{c}}{}_{\mathfrak{b}r}. \tag{49.4}$$

---

[1] DOYLE [1958, *1*].

[2] CAYLEY [1872, *3*], [1873, *1*], using a method suggested by LÉVY [1870, *2*], first worked out necessary and sufficient conditions that it be possible to embed one, two, or three given congruences of surfaces in a triply orthogonal system. Cf. FORSYTH [1920, *1*, Chap. XI].

[3] THOMAS [1944, *2*, Eq. (11)].

Making this substitution in Eq. (49.3) and using Eq. $(49.1)_1$, we obtain

$$\left.\begin{aligned}(a^k{}_s - a_{\mathfrak{b}}\delta^k_s)\, m_{\mathfrak{c}k} A^{\mathfrak{c}}{}_{\mathfrak{b}r} &= (a_{\mathfrak{c}} - a_{\mathfrak{b}})\, m_{\mathfrak{c}s} A^{\mathfrak{c}}{}_{\mathfrak{b}r},\\ &= -(\partial a^k{}_s/\partial x^r - \partial a_{\mathfrak{b}}/\partial x^r \delta^k_s)\, m_{\mathfrak{b}k}.\end{aligned}\right\} \tag{49.5}$$

We multiply by $q^s_{\mathfrak{d}}$, sum on s and use Eq. $(49.2)_2$; when $\mathfrak{d} \neq \mathfrak{b}$, we get

$$A^{\mathfrak{d}}{}_{\mathfrak{b}r} = (a_{\mathfrak{b}} - a_{\mathfrak{d}})^{-1}\, \partial a^k{}_s/\partial x^r\, q^s_{\mathfrak{d}}\, m_{\mathfrak{b}k}, \tag{49.6}$$

so that Eq. (49.4) can be written as

$$\partial m_{\mathfrak{b}k}/\partial x^r = m_{\mathfrak{b}k} A^{\mathfrak{b}}{}_{\mathfrak{b}r} + \sum_{\mathfrak{c}\neq\mathfrak{b}} m_{\mathfrak{c}k} (a_{\mathfrak{b}} - a_{\mathfrak{c}})^{-1}\, \partial a^p_s/\partial x^r\, m_{\mathfrak{b}p}\, q^s_{\mathfrak{c}} \ (\mathfrak{b} \text{ unsummed}). \tag{49.7}$$

By Eq. $(33.5)_2$, for $\boldsymbol{m}_{\mathfrak{b}}$ to be complex-lamellar it is necessary and sufficient that

$$0 = m_{\mathfrak{b}[s}\, \partial m_{\mathfrak{b}k}/\partial x^{r]} = M_{\mathfrak{b}[skr]}, \tag{49.8}$$

where, from Eq. (49.7),

$$M_{\mathfrak{b}skr} = \sum_{\mathfrak{c}\neq\mathfrak{b}} m_{\mathfrak{b}s} m_{\mathfrak{c}k} (a_{\mathfrak{b}} - a_{\mathfrak{c}})^{-1}\, \partial a^p{}_t/\partial x^r\, m_{\mathfrak{b}p}\, q^t_{\mathfrak{c}}, \tag{49.9}$$

the term involving $A^{\mathfrak{b}}{}_{\mathfrak{b}r}$ contributing nothing to $M_{\mathfrak{b}[skr]}$.

50. Surfaces of maximum shear. As was noted in Sect. 46, for a symmetric tensor $\boldsymbol{a}$ with distinct proper numbers there are two directions of maximum shear, given by $\sqrt{2}\,\boldsymbol{u} = \boldsymbol{m}_1 + \boldsymbol{m}_n$, $\sqrt{2}\,\boldsymbol{v} = \boldsymbol{m}_1 - \boldsymbol{m}_n$, where $\boldsymbol{m}_1$ and $\boldsymbol{m}_n$ are unit proper vectors of $\boldsymbol{a}$ corresponding respectively to the largest and smallest proper numbers. We assume $n = 3$. From Eqs. (48.5) and (48.6), by setting $A^1 = A^2 = 2^{-\frac{1}{2}}$, $A^3 = 0$ and $A^1 = -A^2 = 2^{-\frac{1}{2}}$, $A^3 = 0$ we obtain necessary and sufficient conditions that $\boldsymbol{u}$ and $\boldsymbol{v}$, respectively, be complex-lamellar. These conditions reduce to

$$\left.\begin{aligned}a_{kr,s}\{&(a_1 - a_2)^{-1} m^k_1 m^r_2 (m^s_3 - m^s_1) - 2(a_1 - a_3)^{-1} m^k_1 m^r_3 m^s_2 +\\ &+ (a_2 - a_3)^{-1} m^k_2 m^r_3 (m^s_1 - m^s_3)\} = 0,\\ a_{kr,s}\{&(a_1 - a_2)^{-1} m^k_1 m^r_2 (m^s_3 + m^s_1) - 2(a_1 - a_3)^{-1} m^k_1 m^r_3 m^s_2 +\\ &+ (a_2 - a_3)^{-1} m^k_2 m^r_3 (m^s_1 + m^s_3)\} = 0.\end{aligned}\right\} \tag{50.1}$$

Adding and subtracting these formulae yields the slightly simpler equivalent conditions[1]

$$\left.\begin{aligned}a_{kr,s}\{&(a_1 - a_2)^{-1} m^k_1 m^r_2 m^s_3 - 2(a_1 - a_3)^{-1} m^k_1 m^r_3 m^s_2 +\\ &+ (a_2 - a_3)^{-1} m^k_2 m^r_3 m^s_1\} = 0,\\ a_{kr,s}\{&(a_1 - a_2)^{-1} m^k_1 m^r_2 m^s_1 + (a_2 - a_3)^{-1} m^k_3 m^r_2 m^s_3\} = 0.\end{aligned}\right\} \tag{50.2}$$

Either Eqs. (50.1) or Eqs. (50.2) are necessary and sufficient for the existence of two congruences of surfaces of maximum shear, necessarily orthogonal. If there exist a congruence of surfaces normal to both, i.e. if there exists a congruence of isostatic surfaces corresponding to the proper number a_2, the triply orthogonal family can be introduced as co-ordinate surfaces. In such a system, the matrix of physical components of $\boldsymbol{a}$ takes the form (46.10).

References.

1757 *1.* EULER, L.: Continuation des recherches sur la théorie du mouvement des fluides. Hist. Acad. Berlin **1755**, 316–361.

1759 *1.* LAGRANGE, J.L.: Recherches sur la méthode de maximums et minimums. Misc. Taur. **1**, 18–33 = Oeuvres **1**, 3–20.

[1] THOMAS [1944, *2*, p. 172] gives these in principal co-ordinates.

1762 *1.* Lagrange, J.L.: Nouvelles recherches sur la nature et la propagation du son. Misc. Taur. **2**², 11—172 (1760—1761) = Oeuvres **1**, 151—316.

1770 *1.* Euler, L.: Institutionum Calculi Integralis Volumen Tertium. Petropolis = Opera (1) **13**.

2. Euler, L.: Sectio secunda de principiis motus fluidorum. Novi Comm. Petrop. **14**, 270—386 (1769).

1771 *1.* Euler, L.: Sectio tertia de motu fluidorum lineari potissimum aquae. Novi Comm. Petrop. **15**, 219—360 (1770).

1772 *1.* Euler, L.: Sectio quarta de motu aerio in tubis. Novi Comm. Petrop. **16**, 281—425 (1770).

1776 *1.* Coulomb, C.A.: Essai sur une application des règles de maximis et minimis à quelques problèmes de statique, relatifs à l'architecture. Mém. Math. Phys. Acad. Sci. divers savans **7**, 343—382 (1773).

1787 *1.* Monge, G.: Supplément, où l'on fait voir que les équations aux différences ordinaires, pour lesquelles les conditions d'intégrabilité ne sont pas satisfaites, sont susceptibles d'une véritable intégration, et que c'est de cette intégration que dépend celle des équations aux différences partielles élevées. Mém. Acad. Sci. Paris **1784**, 502—576.

1806 *1.* Euler, L.: Die Gesetze des Gleichgewichts und der Bewegung flüssiger Körper. Leipzig, Translation by H.W. Brandes of (1770, *1*), (1771, *1*), (1772, *1*).

1813 *1.* Gauss, C.F.: Theoriae attractionis corporum sphaerodicorum ellipticorum homogeneorum methodus nova tractata. Comm. Soc. Sci. Götting. **2** = Werke **5**, 3—22.

1818 *1.* Pfaff, J.F.: Methodus generalis, aequationes differentiarum partialium, nec non aequationes differentiales vulgares, utrasque primi ordinis, inter quascunque variabiles, complete integrandi (1815). Abh. Akad. Wiss. Berlin **1814—1815**, 76—136; trans. G. Kowalewski: Allgemeine Methode, partielle Differentialgleichungen zu integrieren. Ostwald's Klass. **129**, Leipzig, 1902.

1826 *1.* Ampère, A.M.: Mémoire sur la théorie mathématique des phénomènes électrodynamiques uniquement déduite de l'expérience. Paris.

1828 *1.* Cauchy, A.L.: Sur les centres, les planes principaux et les axes principaux des surfaces du second degré. Ex de math. **3** = Oeuvres (2) **8**, 9—35.

2. Green, G.: An essay on the application of mathematical analysis to the theories of electricity and magnetism. Nottingham = J. reine angew. Math. **39**, 73—89 (1850); **44**, 356—374 (1852); **47**, 161—221 (1854) = Papers 3—115.

1829 *1.* Cauchy, A.L.: Sur l'équation à l'aide de laquelle on détermine les inégalités séculaires des mouvement des planètes. Ex. de math. **4**, 140—160 = Oeuvres (2) **9**, 174—195.

1830 *1.* Cauchy, A.L.: Mémoire sur l'équation qui a pour racines les moments d'inertie principaux d'un corps solide et sur diverses équations du même genre. Mém. Acad. Sci. Paris **9**, 111—113 = Oeuvres (1) **2**, 79—81.

1831 *1.* Ostrogradsky: Note sur une intégrale qui se rencontre dans le calcul de l'attraction des sphéroides. Mém. Acad. Sci. St. Pétersb., Ser. Math. (6), **1**, 39—53.

1833 *1.* Lamé, G., and E. Clapeyron: Mémoire sur l'équilibre des corps solides homogènes. Mém. divers savants (2) **4**, 465—562.

1840 *1.* Lamé, G.: Mémoire sur les coordonnées curvilignes. J. Math. pures appl. **5**, 313 to 347.

1841 *1.* Cauchy, A.L.: Mémoire sur les dilatations, les condensations et les rotations produits par un changement de forme dans un système de points matériels. Ex. Am. phys. math. **2** = Oeuvres (2) **12**, 343—377.

2. Lamé, G.: Mémoire sur les surfaces isostatiques dans les corps solides homogènes en équilibre d'élasticité. J. Math. pures appl. **6**, 37—60.

1842 *1.* Challis, J.: A general investigation of the differential equations applicable to the motion of fluids. Trans. Cambridge Phil. Soc. **7**, 371—393.

2. Stokes, G.G.: On the steady motion of incompressible fluids. Trans. Cambridge Phil. Soc. **8**, 287—319 (1844—1849) = Papers **1**, 1—16.

1846 *1.* Borchardt, C.W.: Neue Eigenschaft der Gleichung, mit deren Hülfe man die säcularen Störungen der Planeten bestimmt. J. reine angew. Math. **30**, 38—46.

1847 *1.* Borchardt, C.W.: Développements sur l'équation à l'aide de laquelle on détermine les inégalities séculaires du mouvement des planètes. J. Math. pures appl. (1) **12**, 50—67.

2. HOPKINS, W.: On the internal pressure to which rock masses may be subjected, and its possible influence in the production of laminated structure. Trans. Cambridge Phil. Soc. **8**, 456–470 (1844–1849).

1849 1. THOMSON, W.: (Lord KELVIN), Notes on Hydrodynamics (5), On the vis-viva of a liquid in motion. Cambridge Dubl. Math. J. **4**, 90–94 = Papers **1**, 107–112.

1850 1. TARDY, P.: Some observations on a new equation in hydrodynamics. Phil. Mag. **36**, 171–178.

1851 1. THOMSON, W.: (Lord KELVIN), A mathematical theory of magnetism. Phil. Trans. Roy. Soc. Lond. **141**, 243–285 = Papers Electr. Magn. §§ 432–523.

2. STOKES, G.G.: On the dynamical theory of diffraction (1849). Trans. Cambridge Phil. Soc. **9**[1], 1–62 = Papers **2**, 243–328.

3. LAMÉ, G.: Leçons sur la théorie mathématique de l'élasticité des corps solides. Paris.

1854 1. STOKES, G.G.: SMITH's Prize Examination Papers (February 1954). Papers **5**, 320–322.

1855 1. HERMITE, C.: Remarque sur un théorème de M. CAUCHY. C. R. Acad. Sci., Paris **41**, 181–183.

1856 1. SPOTTISWOODE, W.: Elementary theorems relating to determinants. J. reine angew. Math. **51**, 209–271, 328–381.

2. RANKINE, W.J.M.: On axes of elasticity and crystalline forms. Phil. Trans. Roy. Soc. Lond. **146**, 261–285.

1857 1. RIEMANN, B.: Lehrsätze aus der Analysis Situs für die Theorie der Integrale von zweigliedrigen vollständigen Differentialen. J. reine angew. Math. **54**, 105–109 = Werke, 2nd. ed. 91–96.

1858 1. HELMHOLTZ, H.: Über Integrale der hydrodynamischen Gleichungen, welche den Wirbelbewegungen entsprechen. J. reine angew. Math. **55**, 25–55 = Wiss. Abh. **1**, 101–134; trans. P.G. TAIT: On integrals of the hydrodynamical equations, which express vortex-like motion. Phil. Mag. (4) **33**, 485–512 (1867).

2. CAYLEY, A.: A memoir on the automorphic transformation of a bipartite quadric function. Phil. Trans. Roy. Soc. Lond. **148**, 39–46 = Works **2**, 497–505.

1859 1. LAMÉ, G.: Leçons sur les coordonnées curvilignes. Paris.

1861 1. HANKEL, H.: Zur allgemeinen Theorie der Bewegung der Flüssigkeiten. Göttingen.

1862 1. WARREN, J.W.: On the internal pressure within an elastic solid. Quart. J. Pure Appl. Math. **5**, 109–117.

1863 1. ROCK, G.: Anwendung der Potentialausdrücke auf die Theorie der molekularphysikalischen Fernewirkungen und der Bewegung der Electricität in Leitern. J. reine angew. Math. **61**, 283–308.

2. CLEBSCH, A.: Über eine Classe von Gleichungen, welche nur reelle Wurzeln besitzen. J. reine angew. Math. **62**, 232–245.

1867 1. THOMSON, W. (Lord KELVIN), and P.G. TAIT: Treatise on Natural Philosophy, Part I. Cambridge.

1869 1. THOMSON, W. (Lord KELVIN): On vortex motion. Trans. roy. Soc. Edinb. **25**, 217 to 260 = Papers **4**, 13–66.

1870 1. VELTMANN, W.: Die Helmholtz'sche Theorie der Flüssigkeitswirbel. Z. Math. Phys. **15**, 450–463.

2. LÉVY, M.: Mémoire sur les coordonnées curvilignes orthogonales. J. Ecole Polytech. **26**, 157–200.

3. MAXWELL, J.C.: On reciprocal figures, frames, and diagrams of forces. Trans. Roy. Soc. Edinb. **26**, 1–40 (1869–1872) = Papers **2**, 161–207.

4. JORDAN, C.: Traité des substitutions et des équations algébriques. Paris.

1872 1. CAYLEY, A.: On the extraction of the square root of a matrix of the third order. Proc. Roy. Soc. Edinb. **6**, 92–93 = Papers **1**, 74–75.

2. BOUSSINESQ, J.: Lois géométriques de la distribution des pressions, dans un solide homogène et ductile soumis à des déformations planes. C. R. Acad. Sci., Paris **74**, 242–246.

3. CAYLEY, A.: Sur la condition pour qu'une famille de surfaces données puisse faire partie d'un système orthogonal. C. R. Acad. Sci., Paris **75**, 177–185, 246–250, 324–330, 381–385 and 1801–1803 = Papers **8**, 269–291.

4. CLEBSCH, A.: Ueber eine Fundamentalaufgabe der Invariantentheorie. Abh. Ges. Wiss. Göttingen, math.-phys. Kl. **17**, 1–62.

5. St. Venant, A.-J.-C. de: Rapport sur un mémoire de M. Kleitz intitulé „Etudes sur les forces moléculaires dans les liquides en mouvement et application à l'hydrodynamique". C. R. Acad. Sci., Paris **74**, 426–438.

1873 1. Cayley, A.: On curvature and orthogonal surfaces. Phil. Trans. Roy. Soc. Lond. **163**, 229–251.

2. Maxwell, J. C.: A Treatise on Electricity and Magnetism. 2 vols. Oxford.

3. Kleitz, M.: Études sur les forces moléculaires dans le liquides en mouvement et application a l'hydrodynamique. Paris.

1875 1. Lipschitz, R.: Determinazione della pressione nell'intorno d'un fluido incompressibile suggetto ad attrazioni interne ed esterne. Ann. di Mat. (2) **6**, 226–231 (1873–1875).

1877 1. Boussinesq, J.: Sur la construction géométrique des pressions qui supportent les divers éléments plans se croisant en un même point d'un corps, et sur celle des déformations qui se produisent autor d'un tel point. J. Math. pures appl. (3) **3**, 147–152.

1878 1. Voss, A.: Zur Theorie der orthogonalen Substitutionen. Math. Ann. **13**, 320–374.

1879 1. Lamb, H.: A Treatise on the Mathematical Theory of the Motion of Fluids. Cambridge.

1880 1. Rowland, H. A.: On the motion of a perfect incompressible fluid when no solid bodies are present. Amer. J. Math. **3**, 226–268.

2. Craig, T.: Motion of viscous fluids. J. Franklin Inst. **110**, 217–227.

3. Craig, T.: On steady motion in an incompressible viscous fluid. Phil. Mag. (5) **10**, 342–357.

1881 1. Gibbs, J. W.: Elements of Vector Analysis, Part I. New Haven = Works **2**, 17–36.

2. Gromeka, I.: Certain Cases of Motion of an Incompressible Fluid [In Russian], Kazan Sobr. Sotch. 76–148.

3. Weingarten, J.: Zur Theorie der isotatischen Flächen. J. reine angew. Math. **90**, 18–33.

4. Stickelburger, L.: Zur Theorie der linearen Differentialgleichungen. Leipzig.

5. Craig, T.: Methods and Results, General Properties of the Equations of Steady Motion. U.S. Treasury Dep't. (Coast and Geod. Surveys) Document **71**, Washington.

1882 1. Sharp, W. J. C.: On the invariants of a certain orthogonal transformation, with special reference to the theory of the strains and stresses of an elastic solid. Proc. Lond. Math. Soc. **13**, 216–239.

2. Mohr, O.: Über die Darstellung des Spannungszustandes und des Deformationszustandes eines Körperelementes und über die Anwendung derselben in der Festigkeitslehre. Civilingenieur **28**, 112–155.

1883 1. Thomson, J. J.: A Treatise on the Motion of Vortex Rings. London.

2. Sylvester, J. J.: On the equation to the secular inequalities in the planetary theory. Phil. Mag. (5) **16**, 267–269.

1884 1. Gibbs, J. W.: Elements of Vector Analysis, Part II. New Haven = Works **2**, 50–90.

1885 1. Weyr, E.: Répartition des matrices en espèces et formation de toutes les espèces. C. R. Acad. Sci., Paris **100**, 966–969.

1886 1. Todhunter, L., and K. Pearson: A History of the Theory of Elasticity, Vol. **1**. Cambridge.

1887 1. Poincaré, H.: Sur les résidus des intégrales doubles. Acta math. **6**, 321–380.

2. Weyr, E.: Note sur la théorie des quantités complexes formées avec *n* unités principales. Bull. Sci. math. (2) **11**, 205–215.

3. Boggio-Lera, E.: Sulla cinematica dei mezzi continui. Nuovo Cim. (3) **22**, 63–69, 143–149, 231–240; **23**, 41–55, 158–162 (1888) = Ann. R. Scuola Norm. Pisa **4**, 53–99.

1889 1. Beltrami, E.: Considerazioni idrodinamiche. Rend. Ist. Lombardo (2) **22**, 121–130 = Opere **4**, 300–309.

2. Morera, Z. G.: Sui moti elicoidali dei fluidi. Rend. Lincei (4) **5**, 611–617.

1890 1. Weyr, E.: Zur Theorie der bilinearen Formen. Mh. Math. Phys. **1**, 163–236.

2. Gilbert, P. H.: Sur quelques formules d'un usage général dans la physique mathématique. Ann. Soc. Sci., Bruxelles **14**, 1–18.

1892 1. Finger, J.: Über die gegenseitigen Beziehungen von gewissen in der Mechanik mit Vortheil anwendbaren Flächen zweiter Ordnung nebst Anwendungen auf Probleme der Astatik. Wien. Sitzgsber. (IIa) **101**, 1105–1142.

2. LOVE, A. E. H.: A Treatise on the Mathematical Theory of Elasticity. Cambridge.
3. FABRI, C.: Sulla teoria dei moti vorticosi nei fluidi incompressibili. Nuovo Cim. (3), **31**, 135–145, 221–227 = Ann. R. Scuola Pisa **7**, No. 4, 35 pp. (1895).
4. FINGER, J.: Über die gegenseitigen Beziehungen von gewissen in der Mechanik mit Vortheil anwendbaren Flächen zweiter Ordnung nebst Anwendungen auf Probleme der Astatik. Wien. Sitzgsber. (IIa) **101**, 1105–1142.

1893 1. POINCARÉ, H.: Leçons sur la théorie des tourbillons. Paris.
2. HADAMARD, J.: Résolution d'une question relative aux déterminants. Bull. Sci. math. (2), **17**, 240–246.
3. BANG, A. S.: Om en Trediegradsligning. Nyt. Tidss. Math. **4**, B, 57–60; **9** B, 94–96.

1895 1. POINCARÉ, H.: Analysis situs. J. École Polytech. (2) **1**, 1–123.

1896 1. FROBENIUS, G.: Über vertauschbare Matrizen. Sitzgsber. preuß. Akad. Wiss. **1**, 601–614.
2. MUIR, T.: On LAGRANGE's determinantal equation. Phil. Mag. **43**, 220–226.
3. COSSERAT, E. et F.: Sur la théorie de l'élasticité. Ann. Toulouse **10**, I 1–I 116.

1897 1. FÖPPL, A.: Die Geometrie der Wirbelfelder. Leipzig.
2. APPELL, P.: Sur les équations de l'hydrodynamique et la théorie des tourbillons. J. Math. pures appl. (5) **3**, 5–16.

1900 1. LEVI-CIVITA, T.: Complementi al teorema de Malus-Dupin. Rend. Lincei (5a) **9**^{1}, 185–189, 237–245.
2. MOHR, O.: Welche Umstände bedingen die Elastizitätsgrenze und den Bruch eines Materials? Z. Ver. Dtsch. Ing. **44**, 1524–1530.

1901 1. RICCI, G., et T. LEVI-CIVITA: Méthodes de calcul différentiel absolu et leurs applications. Math. Ann. **54**, 125–201.
2. HIRSCH, A.: Sur les racines d'une équation fondamentale. Acta math. **25**, 367–370.
3. BENDIXON, I.: Sur les racines d'une équation fondamentale. Acta math. **25**, 359 to 365.
4. BROMWICH, T. J. I. A.: Theorems on matrices and bilinear forms. Proc. Cambridge Phil. Soc. **11**, 75–91.
5. BURNSIDE, W. S., and A. W. PANTON: The Theory of Equations with an Introduction to the Theory of Binary Algebraic Forms. Dublin.

1902 1. AUTONNE, L.: Sur l'hermetien. Rend. Circ. Mat. Palermo **16**, 104–128.
2. AUTONNE, L.: Sur les groupes linéaires, réels, et orthogonaux. Bull. Soc. Math. France **30**, 121–134.

1903 1. AUTONNE, L.: Sur l'hypohermétien. Bull. Soc. Math. France **31**, 140–155.
2. HADAMARD, J.: Leçons sur la propagation des ondes et les équations de l'hydrodynamique. Paris.

1905 1. BLUMENTHAL, O.: Über die Zerlegung unendlicher Vektorfelder. Math. Annalen **61**, 235–250.

1908 1. FROBENIUS, G.: Über Matrizen aus positiven Elementen. Sitzgsber. preuß. Akad. Wiss. **1**, 471–476.
2. STEKLOFF, W.: Sur la théorie des tourbillons. Ann. Fac. Sci. Toulouse (2) **10**, 271–334.

1909 1. GIBBS, J. W., and E. B. WILSON: Vector Analysis. New Haven.
2. SCHUR, L.: Über die characteristischen Wurzeln einer linearen Substitution. Math. Annalen **66**, 488–510.
3. SCHUR, I.: Beiträge zur Theorie der Gruppen linearer homogener Substitutionen. Trans. Amer. Math. Soc. **10**, 159–175.
4. KOLOSOV, V. V.: On an application of complex function theory to a plane problem of the mathematical theory of elasticity. Yuriev. (We have been unable to see this reference.)
5. CARATHÉODORY, C.: Untersuchungen über die Grundlagen der Thermodynamik. Math. Annalen **67**, 355–386.

1910 1. FROBENIUS, G.: Über die mit einer Matrix vertauschbaren Matrizen. Sitzgsber. preuß. Akad. Wiss. **2**, 3–15.
2. VEBLEN, O., and J. W. YOUNG: Projective Geometry. Boston.

1911 1. RANUM, A.: The general term of a recurring series. Bull. Amer. Math. Soc. **17**, 457–461.
2. KÁRMÁN, TH. v.: Festigkeitsversuche unter allseitigem Druck. Z. Ver. Dtsch. Ing. **55**, 1749–1757.

1912 1. FRIEDMAN, A.: Sur la recherche des surfaces isodynamiques. C. R. Acad. Sci., Paris **154**, 864—865.
2. KÁRMÁN, TH. v.: Festigkeitsversuche unter allseitigem Druck. Forsch.-Arb. Ing.-Wes. **1912**, 37—68.
1914 1. HILTON, H.: Homogeneous Linear Substitutions. Oxford.
2. MOHR, O.: Abhandlungen aus dem Gebiete der Technischen Mechanik, 2nd ed. Berlin.
3. BURGATTI, P.: Sulle deformazioni finite dei corpi continui. Mem. Accad. Sci. Bologna (7) **1**, 237—244.
1915 1. BOKER, R.: Die Mechanik der bleibenden Formänderung in kristallinisch aufgebauten Körpern. Forsch.-Arb. Ing.-Wes. **1915**, 1—15.
1918 1. WEYL, H.: Gravitation und Electrizität. Sitzgsber. preuß. Akad. Wiss. **1918**, 465—480.
1919 1. LECORNU, L.: Sur les tourbillons d'une veine fluide. C. R. Acad. Sci., Paris **168**, 923—926.
2. PHILLIPS, H.B.: Functions of matrices. Amer. J. Math. **41**, 266—278.
3. TRKAL, V.: Poznámka k hydrodynamice vazkych tekutin. Časopis Pěst. Mat. Fys. **48**, 302—320.
1920 1. FORSYTH, A.R.: Lectures on the Differential Geometry of Curves and Surfaces. Cambridge.
1921 1. APPELL, P.: Traité de Mécanique Rationelle, Vol. 3, Équilibre et Mouvement des Milieux Continus, 3rd. ed. Paris.
1923 1. WEITZENBÖCK, R.: Invariantentheorie. Groningen.
2. EISENHART, L.P.: Orthogonal systems of hypersurfaces in a general Riemann space. Trans. Amer. Math. Soc. **25**, 259—280.
3. CISOTTI, U.: Considerazione sulla nota formula idrodinamica di Daniele Bernoulli. Bull. Un. Mat. Ital. **2**, 125—128.
1924 1. COLDONAZZO, B.: Sulla geometria differenziale di superficie aventi interesse idrodinamico. Rend. Lincei (5^A) **33**^{2}, 396—400.
1925 1. COLDONAZZO, B.: Un' osservazione a proposito del teorema di Bernoulli. Boll. Un. Mat. Ital. **4**, 1—3.
2. WEDDERBURN, J.H.M.: The absolute value of the product of two matrices. Bull. Amer. Math. Soc. **31**, 304—308.
3. THIRRING, H.: Zur tensoranalytischen Darstellung der Elastizitätstheorie. Phys. Z. **26**, 518—522.
4. HENCKY, H.: Die Bewegungsgleichungen beim nichtstationären Fließen plastischer Massen. Z. angew. Math. Mech. **5**, 144—146.
1926 1. HENSEL, K.: Über Potenzreihen von Matrizen. Sitzgsber. preuß. Akad. Wiss. **1**, 601—614.
2. DIRAC, P.A.M.: On quantum algebra. Proc. Cambridge Phil. Soc. **33**, 412—418.
3. LAGRANGE, R.: Calcul differentiel absolu. Mém. Sci. Math. **19**, Paris.
1927 1. MICHAL, A.D.: Functionals of r-dimensional manifolds admitting continuous groups of point transformations. Trans. Amer. Math. Soc. **29**, 612—646.
2. PASTORI, M.: Sulle superficie ortogonali a una congruenza normale di curve. Rend. Ist. Lombardo (2), **60**, 111—119.
3. MASOTTI, A.: Decomposizione intrinseca del vortice e sue applicazioni. Rend. Ist. Lombardo (2) **60**, 869—874.
4. BORTOLOTTI, E.: Spazi subordinati: equazioni di Gauss e Codazzi. Boll. Un. Mat. Ital. **6**, 134—137.
5. WAERDEN, B.L. VAN DER: Differentialkovarianten von n-dimensionalen Mannigfaltigkeiten in Riemannschen m-dimensionalen Räumen. Abh. Math. Sem. Hamburg Univ. **5**, 154—160.
1928 1. BROWNE, E.T.: The characteristic equation of a matrix. Bull. Amer. Math. Soc. **34**, 363—368.
1929 1. KELLOGG, O.D.: Foundations of Potential Theory. New York.
1930 1. KAMKE, E.: Differentialgleichungen reeller Funktionen. Leipzig.
2. MUIR, T.: Contributions to the History of the Theory of Determinants, 1900—1920. London.
3. SIGNORINI, A.: Sulle deformazioni finite dei sistemi continui. Rend. Lince (6) **12**, 312—316.
4. SCHOUTEN, J.A., and E.R. VAN KAMPEN: Zur Einbettungs- und Krümmungstheorie nichtholonomer Gebilde. Math. Annalen **103**, 752—783.

1931 *1.* MCCONNELL, A. J.: Applications of the Absolute Differential Calculus. Glasgow.
2. COURANT, R., and D. HILBERT: Methoden der Mathematischen Physik, Bd. **1**. Berlin.
3. BURGATTI, P.: Intorno a una formula generale di trasformazione di un integrale di spazio in uno di superficie e alle sue varie deduzioni. Boll. Un. Mat. Ital. **10**, 1–5.
4. MURNAGHAN, F. D., and A. WINTNER: A canonical form for real matrices under orthogonal transformations. Proc. Nat. Acad. Sci. Wash. **17**, 417–420.
5. WINTNER, A., and F. D. MURNAGHAN: On a polar representation of nonsingular square matrices. Proc. Nat. Acad. Sci. Wash. **17**, 676–678.
6. TUCKER, A. W.: On generalised covariant differentiation. Ann. of Math. (2) **32**, 451–460.

1932 *1.* PLANCK, M.: General mechanics, art. 28 of Introduction to Theoretical Physics, Vol. 1. London.
2. CIPOLLA, M.: Sulle matrici espressioni analitiche di un'altra. Rend. Circ. Mat. Palermo **56**, 144–154.
3. WEITZENBÖCK, R.: Über die Matrixgleichung $X^2 = A$. Proc. Kon. Ned. Acad. Wet. **35**, 157–161.
4. RUTHERFORD, D. E.: On the canonical form of a rational function of a matrix. Proc. Edinb. Math. Soc. (2) **3**, 135–143.
5. TURNBULL, H. W., and A. C. AITKEN: Canonical Matrices. London.
6. LAMB, H.: Hydrodynamics, 6th. ed. Cambridge.

1933 *1.* KLOTTER, K.: Über die graphische Darstellung zugeordneter Spannungs- und Verzerrungszustände. Z. angew. Math. Mech. **13**, 433–434.
2. MCDUFFEE, C. C.: The Theory of Matrices. Berlin.
3. NÁDAI, A.: Theories of strength. Trans. Amer. Soc. Mech. Engrs. **55**, APM-55-15, 111–129.

1934 *1.* DANTZIG, D. VAN: Electromagnetism, independent of metrical geometry 3. Mass and motion. Proc. Kon. Ned. Akad. Wet. **34**, 643–652.
2. WEDDERBURN, J. H. M.: Lectures on Matrices. Amer. Math. Soc. Coll. Publ. **17**.

1936 *1.* HAMEL, G.: Ein allgemeiner Satz über den Druck bei der Bewegung volumbeständiger Flüssigkeiten. Mh. Math. Phys. **43**, 345–363.

1937 *1.* The Origin of CLERK MAXWELL's electric ideas, as described in familiar letters to WILLIAM THOMSON, ed. by J. LARMOR. Cambridge.
2. NÁDAI, A.: Plastic behavior of metals in the strain-hardening range, Part I. J. Appl. Phys. **8**, 205–213.
3. MURNAGHAN, F. D.: Finite deformation of an elastic solid. Amer. J. Math. **59**, 235–260.

1939 *1.* GUEST, J. J.: A graphical construction for stress. Phil. Mag. (7), **27**, 445–448.
2. FRITSCHE, B.: Die Spannungsellipse. Ing.-Arch. **10**, 427–429.

1940 *1.* EISENHART, L. P.: Introduction to Differential Geometry. Princeton.
2. WEYL, H.: The method of orthogonal projection in potential theory. Duke Math. J. **7**, 411–444.
3. FADLE, J.: Über eine zweckmäßige Darstellung des Mohrschen Spannungskreises zur unmittelbaren Auffindung sämtlicher Spannungs-Richtungspfeile. Ing.-Arch. **11**, 319–322.
4. WISE, J. A.: Circles of strain. J. Aeronaut. Sci. **7**, 438–440.
5. BALLABH, R.: Superposable fluid motions. Proc. Benares Math. Soc. (2) **2**, 69–79.

1941 *1.* MUNK, M.: On some vortex theorems of hydrodynamics. J. Aeronaut. Sci. **5**, 90–96.

1942 *1.* HALMOS, P. R.: Finite Dimensional Vector Spaces. Princeton.

1943 *1.* SIGNORINI, A.: Trasformazioni termoelastiche finite. Memoria 1ª. Ann. di Mat. (4) **22**, 33–143.

1944 *1.* THOMAS, T. Y.: The Concept of Invariance in Mathematics. Berkeley.
2. THOMAS, T. Y.: Surfaces of maximum shearing stress. J. Math. Phys. **23**, 167–172.
3. DUPONT, P.: Représentation triangulaire du tenseur des constraintes en élasticité. C. R. Acad. Sci., Paris **218**, 778–780.
4. EINSTEIN, A., and V. BARGMANN: Bivector fields. Ann. of Math. (2) **45**, 1–14.
5. EINSTEIN, A.: Bivector fields II. Ann. of Math. (2) **45**, 15–23.

1945 *1.* ESNAULT-PELTERIE, R.: L'Analyse Dimensionelle. Lausanne.
2. CHARREAU, A.: Sur des représentations planes du tenseur des constraintes dans un milieu continu. C. R. Acad. Sci., Paris **220**, 642–643.

3. REINER, M.: A mathematical theory of dilatancy. Amer. J. Math. **67**, 350–362.

1946 1. DORGELO, H.B., and J.A. SCHOUTEN: On unities and dimensions. Proc. Kon. Ned. Akad. Wet. **48**, 124–131.
2. WEYL, H.: The Classical Groups, Their Invariants and Representations. Princeton.
3. TORRE, C.: Über den plastischen Körper von PRANDTL. Zur Theorie der Mohrschen Grenzkurve. Öst. Ing.-Arch. **1**, 36–50.
4. CARSTOIU, I.: Sur le vecteur tourbillon de l'accélération et les fonctions qui s'y rattachent. Bull. Cl. Sci. Acad. Roumaine **29**, 207–214.

1947 1. MICHAL, A.D.: Matrix and Tensor Calculus. New York.
2. BRAND, L.: Vector and Tensor Analysis. New York.
3. JUNG, F.: Der Culmannsche und der Mohrsche Kreis. Öst. Ing.-Arch. **1**, 408–410.
4. DEWULF, N.: Cercle de Mohr et coniques d'élasticité. Houille bl. **2**, 481–487.
5. CASTOLDI, L.: Sopra una proprietà dei moti permanenti di fluidi incomprimibili in cui le linee di corrente formano una congruenza normale di linee isotache. Rend. Accad. Lincei (8) **3**, 333–337.

1948 1. PRIM, R.C.: On doubly-laminar flow fields having a constant velocity magnitude along each stream-line. U.S. Naval Ordnance Lab. Mem. **9762**.
2. BYUSHGENIS, S.: Geometry of steady flow of an ideal incompressible fluid [in Russian]. Isv. Akad. Nauk, SSSR., Ser. Mat. **12**, 481–512.
3. BALLABH, R.: On coincidence of vortex and streamlines in ideal liquids. Ganita **1**, 1–4.

1949 1. EINSTEIN, A.: Autobiographisches. From ALBERT EINSTEIN; Philosopher Scientist, New York.
2. SYNGE, J.L., and A. SCHILD: Tensor Calculus. Toronto.
3. TRUESDELL, C.: Deux formes de la transformation de Green. C. R. Acad. Sci., Paris **229**, 1199–1200.
4. TRUESDELL, C.: Généralisation de la formule de Cauchy et des théorèmes de Helmholtz au mouvement d'un milieu continu quelconque. C. R. Acad. Sci., Paris **227**, 757–759.
5. BERKER, R.: Sur certaines propriétés du rotationnel d'un champ vectoriel qui est nul sur la frontière de son domaine de définition. Bull. Sci. Math. (2) **73**, 163 to 176. Abstract C. R. Acad. Sci., Paris **228**, 1630–1632.
6. NEMÉNYI, P., and R. PRIM: Some properties of rotational flow of a perfect gas. Proc. Nat. Acad. Sci. Wash. **34**, 119–124. Erratum, Proc. Nat. Acad. Sci. Wash. **35**, 116.
7. SIGNORINI, A.: Trasformazioni termoelastiche finite. Memoria 2ª. Ann. di Mat. (4) **30**, 1–72.

1950 1. OLLENDORF, F.: Die Welt der Vektoren. Vienna.
2. GREEN, A.E., and W. ZERNA: Theory of Elasticity in General Coordinates. Phil. Mag. (7) **41**, 313–336.
3. LANGHAAR, H.L.: Dimensional Analysis and the Theory of Models. New York.
4. NÁDAI, A.: Theory of Flow and Fracture of Solids, Vol. 1, 2nd ed. New York.

1951 1. SCHOUTEN, J.A.: Tensor Analysis for Physicists. Oxford.
2. TRUESDELL, C.: Vorticity averages. Canad. J. Math. **3**, 69–86.
3. TRUESDELL, C.: A form of GREEN's transformation. Amer. J. Math. **73**, 43–47.
4. BJØRGUM, O.: On Beltrami vector fields and flows, Part I. Univ. Bergen Årbok 1951, Naturv. rekke No. 1.
5. PRAGER, W., and P.G. HODGE jr.: Theory of Perfectly Plastic Solids. New York.
6. TORRE, C.: Über die physikalische Bedeutung der Mohrschen Hüllkurve. Z. angew. Math. Mech. **31**, 275–277.

1952 1. COBURN, N.: Intrinsic relations satisfied by the vorticity and velocity vectors in fluid flow theory. Michigan Math. J. **1**, 113–130.
2. TRUESDELL, C.: The mechanical foundations of elasticity and fluid dynamics. J. Rational Mech. Anal. **1**, 125–300.
3. PRIM, R.C.: Steady rotational flow of ideal gases (1949). J. Rational Mech. Anal. **1**, 425–497.

1953 1. MUSKHELISHVILI, N.I.: Some Basic Problems of the Mathematical Theory of Elasticity, trans. by J.R.M. RADOK, Groningen.
2. TRUESDELL, C.: The physical components of vectors and tensors. Z. angew. Math. Mech. **33**, 345–356.
3. BJØRGUM, O., and T. GODAL: On Beltrami vector fields and flows, Part II. Univ. Bergen Årbok 1952, Naturv. rekke No. 13.
4. TRUESDELL, C.: Corrections and additions to "The mechanical foundations of elasticity and fluid dynamics". J. Rational Mech. Anal. **2**, 593–616.

1954 *1.* GREEN, A.E., and W. ZERNA: Theoretical Elasticity. Oxford.
2. TRUESDELL, C.: The Kinematics of Vorticity. Bloomington.
3. TRUESDELL, C.: Remarks on the paper "The physical components of vectors and tensors". Z. angew. Math. Mech. **34**, 69–70.
4. SCHOUTEN, J.A.: Ricci-Calculus. Berlin.
5. RIVLIN, R.S., and J.L. ERICKSEN: Stress-deformation relations for isotropic materials. J. Rational Mech. Anal. **4**, 323–425.
6. TRUESDELL, C.: The Kinematics of Vorticity. Bloomington.
7. BJØRGUM, O.: On the analytic representation of Beltrami vector fields ($\nabla \times \boldsymbol{v} = \Omega \boldsymbol{v}$). Proc. Internat. Congr. Math. Amsterdam **2**, 323–324.

1955 *1.* DOYLE, T.C., and J.L. ERICKSEN: Nonlinear elasticity. Adv. Appl. Mech. **4**, 53–115.
2. TOUPIN, R.A.: The elastic dielectric. J. Rational Mech. Anal. **5**, 849–915.
3. NOLL, W.: On the continuity of the solid and fluid states. J. Rational Mech. Anal. **4**, 3–81.
4. BORDONI, P.G.: Limitazioni per gli invarianti di deformazione. Rend. Mat. e Appl. (5) **14**, 269–279.
5. CASTOLDI, L.: Rappresentazioni equivalenti di moti stazionari di fluidi incomprimibili con congruenza normale di linee di corrente. In particolare: moti quasi-Euleriani. Rend. Sem. Fac. Sci. Univ. Cagliari **25**, 32–43.

1956 *1.* ERICKSEN, J.L.: Stress deformation relations for solids. Canad. J. Phys. **34**, 226–227.

1958 *1.* DOYLE, T.C.: Higher order invariants of stress or deformation tensors. J. Math. Phys. **36**, 297–305.
2. GODAL, T.: On Beltrami vector fields and flows, Part III. Some considerations on the general case. Univ. Bergen Årbok 1957, Naturv. rekke Nr. 12, 28 pp.

Sachverzeichnis.

(Deutsch-Englisch.)

Bei gleicher Schreibweise in beiden Sprachen sind die Stichwörter nur einmal aufgeführt.

Subject Index.

(English-German.)

Where English and German spelling of a word is identical the German version is omitted.